The Elements

Name	Symbol	Atomic Number	Relative Atomic Mass	Name	Symbol	Atomic Number	Relative Atomic Mass
Actinium	Ac	89	227.028	Mercury	Hg		
Aluminum	Al	13	26.9815	Molybdenum	Mo		
Americium	Am	95	(243)	Neodymium	Nd	60	
Antimony	Sb	51	121.75	Neon	Ne	10	20.1797
Argon	Ar	18	39.948	Neptunium	Np	93	237.048
Arsenic	As	33	74.9216	Nickel	Ni	28	58.69
Astatine	At	85	(210)	Nielsbohrium	Ns	107	(262)
Barium	Ba	56	137.327	Niobium	Nb	41	92.9064
Berkelium	Bk	97	(247)	Nitrogen	N	7	14.0067
Beryllium	Be	4	9.01218	Nobelium	No	102	(259)
Bismuth	Bi	83	208.980	Osmium	Os	76	190.23
Boron	B	5	10.811	Oxygen	O	8	15.9994
Bromine	Br	35	79.904	Palladium	Pd	46	106.42
Cadmium	Cd	48	112.411	Phosphorus	P	15	30.9738
Calcium	Ca	20	40.078	Platinum	Pt	78	195.08
Californium	Cf	98	(251)	Plutonium	Pu	94	(244)
Carbon	C	6	12.011	Polonium	Po	84	(209)
Cerium	Ce	58	140.115	Potassium	K	19	39.0983
Cesium	Cs	55	132.905	Praseodymium	Pr	59	140.908
Chlorine	Cl	17	35.4527	Promethium	Pm	61	(145)
Chromium	Cr	24	51.9961	Protactinium	Pa	91	231.036
Cobalt	Co	27	58.9332	Radium	Ra	88	226.025
Copper	Cu	29	63.546	Radon	Rn	86	(222)
Curium	Cm	96	(247)	Rhenium	Re	75	186.207
Dysprosium	Dy	66	162.50	Rhodium	Rh	45	102.906
Einsteinium	Es	99	(252)	Rubidium	Rb	37	85.4678
Erbium	Er	68	167.26	Ruthenium	Ru	44	101.07
Europium	Eu	63	151.965	Rutherfordium	Rf	104	(261)
Fermium	Fm	100	(257)	Samarium	Sm	62	150.36
Fluorine	F	9	18.9984	Scandium	Sc	21	44.9559
Francium	Fr	87	(223)	Seaborgium	Sg	106	(263)
Gadolinium	Gd	64	157.25	Selenium	Se	34	78.96
Gallium	Ga	31	69.723	Silicon	Si	14	28.0855
Germanium	Ge	32	72.61	Silver	Ag	47	107.868
Gold	Au	79	196.967	Sodium	Na	11	22.9898
Hafnium	Hf	72	178.49	Strontium	Sr	38	87.62
Hahnium	Ha	105	(262)	Sulfur	S	16	32.066
Hassium	Hs	108	(265)	Tantalum	Ta	73	180.948
Helium	He	2	4.00260	Technetium	Tc	43	(98)
Holmium	Ho	67	164.930	Tellurium	Te	52	127.60
Hydrogen	H	1	1.00794	Terbium	Tb	65	158.925
Indium	In	49	114.818	Thallium	T1	81	204.383
Iodine	I	53	126.904	Thorium	Th	90	232.038
Iridium	Ir	77	192.22	Thulium	Tm	69	168.934
Iron	Fe	26	55.847	Tin	Sn	50	118.710
Krypton	Kr	36	83.80	Titanium	Ti	22	47.88
Lanthanum	La	57	138.906	Tungsten	W	74	183.85
Lawrencium	Lr	103	(260)	Uranium	U	92	238.029
Lead	Pb	82	207.2	Vanadium	V	23	50.9415
Lithium	Li	3	6.941	Xenon	Xe	54	131.29
Lutetium	Lu	71	174.967	Ytterbium	Yb	70	173.04
Magnesium	Mg	12	24.3050	Yttrium	Y	39	88.9059
Manganese	Mn	25	54.9381	Zinc	Zn	30	65.39
Meitnerium	Mt	109	(266)	Zirconium	Zr	40	91.224
Mendelevium	Md	101	(258)				

Atomic masses in this table are relative to carbon-12 and limited to six significant figures, although some atomic masses are known more precisely. For certain radioactive elements the numbers listed (in parentheses) are the mass numbers of the most stable isotopes.

General Chemistry

Principles and Modern Applications

General Chemistry

Principles and Modern Applications

SEVENTH EDITION

*R*alph H. Petrucci
California State University, San Bernardino

*W*illiam S. Harwood
University of Maryland, College Park

PRENTICE HALL *Upper Saddle River, New Jersey 07458*

Library of Congress Cataloging-in-Publication Data

Petrucci, Ralph H.
  General chemistry : principles and modern applications / Ralph H.
  Petrucci, William S. Harwood. -- 7th ed.
    p.   cm.
  Includes index.
  ISBN 0–13–533498-5
  1. Chemistry.   I. Harwood, William S.   II. Title.
  QD31.2.P48 1997
  540--dc20   96–43719
                                                   CIP

Acquisitions Editor: Ben Roberts
Development Editors: Alan MacDonell, Karen Karlin
Editor in Chief: Paul F. Corey
Editorial Director: Tim Bozik
Production Editor: York Production Services
Art Director/Cover Designer: Joseph Sengotta
Executive Managing Editor: Kathleen Schiaparelli
Creative Director: Paula Maylahn
Art Manager: Gus Vibal
Vice President of Production and Manufacturing: David W. Riccardi
Editor in Chief, Development: Ray Mullaney
Interior Designer: Lisa Jones
Associate Editor: Mary Hornby
Marketing Manager: Linda Taft MacKinnon
Manufacturing Manager: Trudy Pisciotti
Manufacturing Buyer: Ben Smith
Photo Administrator: Melinda Reo
Photo Editor: Lorinda Morris-Nantz
Photo Researcher: Cindy-Lee Overton
Editorial Assistant: Ashley Scattergood
Marketing Assistant: Elizabeth A. Kell
Cover photograph and frontispiece: Richard Megna, Fundamental
Photographs, NYC. Sodium, an alkali metal, reacts violently with
water to form hydrogen gas and an aqueous solution of sodium
hydroxide. The hydrogen gas spontaneously ignites.

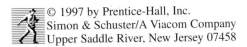

Printed in the United States of America
10 9 8 7 6 5 4 3

ISBN 0-13-533498-5

Prentice-Hall International (UK) Limited, London
Prentice-Hall of Australia Pty. Limited, Sydney
Prentice-Hall Canada Inc., Toronto
Prentice-Hall Hispanoamericana, S.A., Mexico
Prentice-Hall of India Private Limited, New Delhi
Prentice-Hall of Japan, Inc., Tokyo
Simon & Schuster Asia Pte. Ltd., Singapore
Editora Prentice-Hall do Brasil, Ltda., Rio de Janeiro

Brief Contents

Contents

7   Thermochemistry   202

8   The Atmospheric Gases and Hydrogen   240

9   Electrons in Atoms   270

10   The Periodic Table and Some Atomic Properties   314

11   Chemical Bonding I: Basic Concepts   344

12   Chemical Bonding II: Additional Aspects   386

16 Principles of Chemical Equilibrium 544

17 Acids and Bases 580

18 Additional Aspects of Acid–Base Equilibria 620

19 Solubility and Complex Ion Equilibria 654

20   Spontaneous Change: Entropy and Free Energy   688

21   Electrochemistry   724

22   Main Group Elements I: Metals   764

23   Main Group Elements II: Nonmetals   794

24   The Transition Elements   830

25   Complex Ions and Coordination Compounds   864

Appendixes

A Mathematical Operations A1

B Some Basic Physical Concepts A11

C SI Units A15

D Data Tables A17

E Glossary A31

F Answers to Practice Examples and Selected Exercises A47

About the Authors

Ralph H. Petrucci

Ralph Petrucci received his B.S. in Chemistry from Union College and his Ph.D. from the University of Wisconsin–Madison. Following several years of teaching, research, consulting, and directing NSF Institutes for Secondary School Science Teachers at Case Western Reserve University, Professor Petrucci joined the planning staff of the new California State University campus at San Bernardino in 1964. There, in addition to his faculty appointment, he has served as Chairman of the Natural Sciences Division and Dean of Academic Planning. Professor Petrucci is the author of several books, including *General Chemistry* with John W. Hill.

William S. Harwood

Bill Harwood received his Ph.D. in Inorganic Chemistry from Purdue University in 1986. He is currently Assistant Dean for Undergraduate Studies at the University of Maryland at College Park. Previously, Dr. Harwood directed the undergraduate program in the Department of Chemistry and Biochemistry. In his current role, Dr. Harwood continues to teach General Chemistry and conduct research in chemical education. He has received several awards for teaching excellence and is locally popular for his chemistry demonstrations. Dr. Harwood is also active in the American Chemical Society and the Division of Chemical Education and is a consultant to AAAS Project 2061. In Maryland, he is involved in the science reform efforts at both the precollege and college levels. His research focuses on how best to use technology to improve learning in chemistry.

Preface

We are aware that most general chemistry students have career interests not in chemistry but in biology, medicine, engineering, and environmental and agricultural sciences as well as in many other fields. We also know that general chemistry will be the only college chemistry course for many students and their only opportunity to learn some practical applications of chemistry. We have designed this text for these "typical" students.

Students of this text probably have already studied some chemistry, but those with no prior background will find that the early chapters develop fundamental concepts from the most elementary ideas. Students who do plan to become professional chemists will also find opportunities in the text to pursue their own special interests.

The typical student needs help in identifying and applying principles and in visualizing their physical significance. The pedagogical features of this text are designed to provide this help. At the same time, we hope the text serves to sharpen student skills in problem solving and critical thinking. Throughout, we have tried to strike the proper balances between principles and applications, qualitative and quantitative discussions, and rigor and simplification.

Organizational Changes

In the previous edition, we made some major organizational changes. We retain the overall organization in this new edition but have made a number of smaller organizational changes. Several subsections have been moved from their sixth-edition locations to new chapters in this edition. We think these changes will improve the flow of information to the student.

For example, we have moved the introduction to the periodic table from Chapters 3 and 10 to Chapter 2. The periodic table is an extraordinarily useful tool, and presenting it early allows us to use the periodic table in new ways throughout the early chapters of the text. We have also moved the discussion of electron configurations and the periodic table from Chapter 10 to Chapter 9. The section on energy changes in the formation of ionic crystals has been moved from Chapter 11 to Chapter 13, where we discuss crystal structure more fully. The discussion of formal charge in Chapter 11 is more focused than in the previous edition. The discussion of bonding in hypercoordinated species (Chaper 12) carefully reflects the current consensus among theorists, though in a limited way that is appropriate for a first-year chemistry course. The brief discussion of the acidity of aquated metal ions has been moved from Chapter 17 to Chapter 25, where it becomes part of a larger discussion of the acid–base reactions of complex ions. As with previous editions, we have emphasized real-world chemistry in the final descriptive chapters, and we have tried to make this material easy to bring forward into earlier parts of the text. Moreover, many topics in these chapters can be covered selectively, without requiring the study of entire chapters.

Throughout the text we also provide real-world examples to enhance the discussion. For instance, we include examples relevant to the biological sciences, engineering, and the environmental sciences.

Special Features

The text contains a number of pedagogical features:

Important Expressions. The most significant equations, concepts, and rules are highlighted with a yellow panel so that students can readily find them.

Summary/Key Terms/Glossary. Each chapter concludes with a comprehensive verbal *Summary* of important concepts and factual information. The Summary is followed by a list of *Key Terms*—terms that appear in boldface type in the text and are defined again in the *Glossary* (Appendix E). Students can use Key Terms lists and the Glossary to help them master the terminology of general chemistry.

Are You Wondering . . . . To help clarify matters that often puzzle students, questions are posed and answered under this heading. These questions are cast in the form in which students typically ask them. Some are designed to help students avoid common pitfalls; others provide analogies or alternative explanations of a concept. In response to reviewer and student comments and suggestions, several of these have been changed in this edition, and some new ones have been added. A few such questions have been incorporated directly into the text.

In-Text Illustrative Examples. In each chapter, most concepts—especially those that students will be expected to apply in homework assignments and examinations—are illustrated with worked-out examples. In many cases a line drawing or photograph accompanies an example to help students visualize what is going on in the problem.

Practice Examples. These were first introduced in the previous edition as part of the in-text illustrative examples. They were designed to give students immediate practice in applying the principle(s) illustrated in the example. In this new edition, we have expanded the *Practice Examples:* We now offer two for every illustrative example. The first Practice Example provides the immediate practice in a problem very similar to the illustrative example. The second Practice Example generally takes the student one step further than the illustrative example. This combination helps students to integrate and extend their knowledge and problem-solving skills. Answers to all Practice Examples are given in Appendix F. Complete solutions are given in the Student Study Guide.

Summarizing Examples. Each chapter concludes with a multi-part example, generally of a practical nature. *Summarizing Examples* link various important problem types introduced in the chapter—with each other and, in many cases, with problem types from earlier chapters. As in the previous edition, the solution is only outlined, and intermediate results and a final answer are given. Students are expected to work out the details of the solution. Complete solutions are given in the Student Study Guide.

End-of-Chapter Exercises. Each chapter ends with four categories of exercises. *Review Questions* require straightforward application of principles introduced in the chapter, each generally involving a single concept. *Exercises* are grouped by subject matter and are of a broader nature than the Review Questions. In this edition, the Exercises have also been paired, so that there are now two problems covering the same idea. The *Advanced Exercises* are not grouped by type. As expected, some of these are more difficult than those in the previous sections, but some are advanced only in the sense that they introduce new ideas or pursue certain ideas further than is done in the text. The fourth category, *Feature Problems,* is discussed in the next paragraph. Answers to all Review Questions, odd-numbered Exercises, selected Advanced Exercises, and Feature Problems are given in Appendix F.

Feature Problems. New to this edition are end-of-chapter *Feature Problems,* which are of special interest. Some of these problems retrace aspects of the history of chemistry; a few deal with classic experiments; others require students to interpret data or graphs; some present new material; some suggest alternative techniques for problem solving; and a few summarize main points of the chapter in a comprehensive manner. The Feature Problems are a resource that can be used in several ways: for example, as discussion points in class, as assigned homework for individuals, or for collaborative group work. The Feature Problems are cited in margin notes in the text itself.

Focus On. We believe that relevant applications should be an integral part of the text, that asides should be limited to margin notes and occasional Are You Wondering . . . features, and that interesting but less vital issues should follow the main text of a chapter. With this view in mind, we have concluded the text of each chapter with a short essay on a practical topic appropriate to the chapter content. These essays, which may be considered optional reading, focus on ideas introduced in the chapter.

Supplements

For the Instructor

MATTER: The Visual Presentation Resource CD-ROM **[IBM (0-13-596909-3); Macintosh (0-13-700006-5)]** includes animated concept sequences, video laboratories, current and informative ABC News Video segments, still illustrations, and version 3.0 of chemistry Presentation Manager.

The ***Instructor's Resource Manual* (0-13-271370-5) by Robert K. Wismer, Millersville University,** integrates ancillary material, offers Notes for the Instructor and key concepts, lists Chapter Objectives, and provides worked-out solutions to all Advanced Exercises.

The ***Laboratory Demonstration Videos* (0-13-129158-0) by David Humphreys, McMaster University,** is a two-cassette series with approximately 100 laboratory demonstrations of experiments that may be either too expensive or dangerous for the classroom.

The ***Chemical Concepts & Techniques Video* (0-13-338666-X)** is a 100-minute videotape divided into 10 segments; each highlights important concepts through a variety of visual media, including electron microscopy and computer animation.

The ***ABC NEWS/Prentice Hall Video Library* (0-13-578428-X)** contains 5-20 minute segments from award-winning shows such as *20/20, World News Tonight,* and *The American Agenda.* This innovative resource shows chemical principles at work and teaches students to analyze media messages critically based on their scientific knowledge.

The *Chemical Animation Video Series* **(0-13-719022-0)** contains over 50 animated concept sequences authored by Theodore L. Brown. This series shows chemistry in action. They provide students with a view of matter and its transformation beyond what can be illustrated on the printed page.

The *Transparencies* **(0-13-578394-1)** include approximately 250 full-color images from this text.

The *Prentice Hall Custom Test* **[Macintosh (0-13-271412-4); Windows (0-13-271420-5); Test Item File (Hard Copy Manual) (0-13-271404-3)]** consists of more than 1900 multiple-choice questions. PH Custom Test allows you to create and tailor exams to your own needs.

The *Solutions Manual* **(0-13-271347-0)** and *Selected Solutions* **(0-13-271354-3) by Robert K. Wismer, Millersville University,** contain the solutions to the end-of-chapter exercises. With instructor permission, these manuals may be made available to students.

For the Laboratory

Experiments in General Chemistry **(0-13-271362-4)** is a comprehensive laboratory manual containing 37 experiments that parallel the text, including a final group of six experiments on qualitative cation analysis. There is an accompanying instructor's manual **(0-13-271338-8).**

For the Student

The *Prentice Hall Interactive General Chemistry CD-ROM* **(0-13-598012-7) by Steven D. Gammon, et al.,** is an interactive tutorial that focuses on a number of core chemical concepts covered in General Chemistry. The student can interact with and visualize chemical concepts in ways that are not possible through traditional learning programs.

The award-winning *LOGAL Chemistry Explorer Runtime Simulation Software Version 2.0* **[Windows (0-13-51986-0); Mac (0-13-350646-0)]** allows students to explore the molecular world in over 75 interactive experiments by altering parameters and exploring "what if" scenarios.

The *Student Study Guide* **(0-13-281990-2)** is organized around a set of learning objectives for each chapter and features brief discussions of these objectives, drill problems, self quizzes, and sample tests.

The New York Times/Prentice Hall Themes of the Times Supplement, in newspaper format, brings together a collection of recent chemistry articles from the pages of *The New York Times.* This free supplement, available in quantity, encourages students to make connections between the classroom and the world around them.

Acknowledgments

Many people have given of their time, creativity, and support during the preparation of this textbook. Numerous colleagues from California State University, San Bernardino, have offered helpful suggestions through the several editions of this text. From the University of Maryland, D. Webster Keogh and Allessandra Quadrelli acted as sounding boards and gave excellent suggestions for a number of the Feature Problems. Robert K. Wismer of Millersville University has again provided many of the supplementary materials and has done a careful and complete examination of the manuscript. We cannot thank him enough for his tireless efforts.

We are grateful to many people at Prentice Hall for their help and encouragement, starting with our editor, Ben Roberts. Al MacDonell and Karen Karlin, our development editors, have done a fabulous job of keeping us on track, handling the myriad details, maintaining a sense of humor, and asking dozens of questions after the manner of the best students. We would also like to thank Kirsten Kauffman, our production editor at York Production Services, for her careful and caring work in leading this book through production. It has been a real joy to work with all of them. Cindy-Lee Overton and her crew of photo researchers have sought out and found many exciting images in a brief amount of time. Her grace under pressure and good humor are much appreciated. We applaud Richard Megna, who shot several special photographs, for his willingness to try almost anything and for his artfulness in capturing the drama of chemistry.

We are especially grateful for the patience and encouragement of our wives, Ruth Petrucci and Diana Harwood. They have been willing to give up precious family time so that we could produce this textbook. Without their love and support, this book would not have been possible.

We also appreciate the significant contributions of those who commented on the sixth edition or reviewed manuscript chapters of this edition. Their comments and suggestions have been very helpful.

J. Atherton
Memorial University of Newfoundland

Ronald M. Backus
American River College

Richard Bretz
The University of Michigan—Dearborn

Albert W. Burgstahler
The University of Kansas

Donald Campbell
University of Wisconsin—Eau Claire

Robert Crabtree
Yale University

Roberta Day
University of Massachusetts at Amherst

Bob Desiderato
University of North Texas

Daryl J. Doyle
GMI Engineering & Management Institute

John Forsberg
Saint Louis University

Frank Garland
The University of Michigan—Dearborn

Carter Gilmer
The University of Michigan—Dearborn

Peter L. Gold
The Pennsylvania State University

Stan Granda
University of Nevada, Las Vegas

Alton Hassell
Baylor University

Sherman Henzel
Monroe Community College

Andrew J. Holder
University of Missouri—Kansas City

Charles Keilin
Laney College

Donald Kleinfelter
University of Tennessee

Richard W. Kopp
East Tennessee State University

Edwin H. Lane
Williams Jewell College

Stacey Lowery-Bretz
The University of Michigan—Dearborn

John Maguire
Southern Methodist University

Maureen Scharberg
San Jose State University

Patricia A. Metz
Texas Tech University

Anil K. Sharma
Mississippi Valley State University

Quichee Mir
Yakima Valley College

Julie Stewart
El Camino College

Donald Newlin
Montgomery College

Tamar Y. Susskind
Oakland Community College

Robert Porod
Rock Valley College

Duane Swank
Pacific Lutheran University

Bernard L. Powell
University of Texas at San Antonio

Carl A. von Frankenberg
University of Delaware

Bruce Prall
Marian College

Garth W. Welch
Weber State University

Paul Reinbold
Southern Nazarene University

Ronald Wikholm
University of Connecticut

George E. Shankle
Angelo State University

Warren Yeakel
Henry Ford Community College

R.H.P.
W.S.H.

WARNING: Many of the compounds described or pictured in this text are hazardous, as are many of the chemical reactions. The reader should not attempt any experiment pictured or implied in the text. Experiments should be performed only in authorized laboratory settings and under adequate supervision.

A Student's Guide to Using This Text

*T*o some students, chemistry is an almost magical art: There are both mystery and excitement to chemistry. No one knows all the answers, and many aspects of our science are not well understood. Chemists are continuously seeking new materials, new methods, and, most important, new ideas that enhance our understanding of basic scientific principles and may be applied to improve our lives.

In studying chemistry, you will discover some of the ideas and concepts chemists use to understand and direct chemical changes. We urge you not to settle for memorizing facts and mathematical equations. Demand of yourself, this text, and your instructors explanations of the *why* and *how* of chemistry. Seek the concepts behind the facts and equations. Skill in algebra and the assimilation of facts are important, but imagination is the key to mastering chemistry. At the heart of most chemistry problems is perceiving a connection between an observation in the "real," macroscopic world and an imagined change in the microscopic world—the world of atoms, ions, and molecules. Once you perceive this connection, finding the solution to a problem should become much simpler.

This book contains a number of special features designed to help you gain an understanding of the concepts and methods used by chemists. We illustrate several of them on these next few pages.

CHAPTER OPENERS

Each chapter opens with a striking full-page photograph that illustrates a topic discussed in the chapter. On the facing page in the margin, a descriptive caption ties the photo in to the chapter concepts. In each chapter, an outline of chapter sections leads you into introductory text.

Gases

6

6-1   Properties of Gases: Gas Pressure
6-2   The Simple Gas Laws
6-3   The Ideal Gas Equation
6-4   Applications of the Ideal Gas Equation
6-5   Gases in Chemical Reactions
6-6   Mixtures of Gases
6-7   Kinetic-Molecular Theory of Gases
6-8   Gas Properties Relating to the Kinetic-Molecular Theory
6-9   Nonideal (Real) Gases
Focus On   The Chemistry of Air-Bag Systems

A bicycle tire must not be overinflated because it might rupture. An aerosol can should not be discarded in an incinerator. Carbon dioxide gas vaporizing from a block of dry ice sinks to the floor. A balloon filled with helium rises in air, and so does a balloon filled with hot air. One should never search for a gas leak with an open flame. These and many other observations concerning gases can be explained by concepts considered in this chapter.

For example, the behavior of the bicycle tire and the aerosol can are based on relationships among the variables pressure, temperature, volume, and amount of gas. An understanding of the lifting power of lighter-than-air balloons comes in large part from a knowledge of gas densities and their dependence on molar mass, temperature, and pressure. Predicting how far and how fast gas molecules migrate through air requires knowing something about the phenomenon of diffusion.

To do quantitative calculations about the behavior of gases, we will use some simple gas laws and a more general expression called the ideal gas equation. To explain these laws, we will use a theory known as the kinetic-molecular theory of gases. Ideas presented in this chapter reappear in new contexts in several later chapters.

Hot-air balloons have intrigued people from the time the simple gas laws fundamental to their operation came to be understood over 200 years ago.

163

Standard Conditions of Temperature and Pressure

Because gas properties depend on temperature and pressure, at times we find it useful to choose standard conditions of temperature and pressure. This is especially true when we compare gases to one another. The standard temperature for gases is taken to be 0 °C = 273.15 K and standard pressure, as 1 atm = 760 mmHg, exactly. Standard conditions of temperature and pressure are usually abbreviated as **STP**.

TWO PRACTICE EXAMPLES

Each worked Example is followed by *two* Practice Examples. Practice Example A allows you to work a problem similar to the worked Example that precedes it. Practice Example B takes you one step further or calls on a previously learned concept. These sets of Practice Examples should help you integrate and extend your knowledge and problem-solving skills and prepare you for the end-of-chapter Exercises.

▶ Feature Problem C on page 201 tests your qualitative understanding of the gas laws at both the microscopic and molecular levels.

▶ If the volume of $N_2(g)$ at STP had been somewhat *greater* than 1.20 L, we could not have answered the question with qualitative reasoning alone. Where calculations are required, we can do these most readily with the ideal gas equation (Section 6-3).

EXAMPLE 6-5

Testing Your Qualitative Understanding of the Gas Laws. Which will have the greater volume when the gases are compared at STP: (a) 1.20 L $N_2(g)$ at 25 °C and 748 mmHg or (b) 1.25 L $O_2(g)$ at STP?

Solution

First, we can assume that the simple gas laws apply to both gases. The fact that one gas is N_2 and the other O_2 does not enter into our assessment. Next, we see that the gas in (b) is already at STP and its volume is 1.25 L. The crux of the matter is whether the volume of gas increases or decreases when the conditions in (a) are changed to STP. Lowering the temperature from 25 to 0 °C causes the volume to decrease (Charles's law). Raising the pressure from 748 to 760 mmHg also causes the volume to decrease (Boyle's law). At STP the gas in (a) will occupy a volume smaller than its initial 1.20 L. Thus the greater volume at STP is the 1.25 L $O_2(g)$.

Practice Example A: Suppose that the initial condition in case (a) in Example 6-5 is 0.60 L $N_2(g)$ at 150 K and 0.98 atm. Which gas, the N_2 or the O_2, will have the greater volume at STP?

Practice Example B: Suppose that the initial condition in case (a) in Example 6-5 is 1.20 L $N_2(g)$ at 4.2 °C and 758 mmHg, whereas that in (b) is 1.25 L $O_2(g)$ at −2.5 °C and 1.02 atm. Which gas will have the greater volume at STP?

FEATURE PROBLEMS

Some key ideas in the chapter are called out in margin notes (as above) and featured in problems of special interest. These Feature Problems may involve the interpretation of data or graphs, retrace aspects of the history of chemistry, present new material or alternative techniques of problem solving, or address situations with a practical application. These problems vary in difficulty and may be discussed as in-class exercises, worked on collaboratively in groups, or done individually.

Feature Problems

A. The figure shows how the emission of pollutants is related to the air/fuel ratio in an internal combustion engine.
1. Use information given in this chapter to provide a general interpretation of the figure.
2. Establish by calculation that the stoichiometric ratio of air to fuel (on a mass basis) is about 14.5 : 1. To do so, assume that (a) octane, C_8H_{18}, is a typical gasoline molecule; (b) the maximum quantity of heat is evolved when $C_8H_{18}(l)$ and $O_2(g)$ are in their stoichiometric proportions and $CO_2(g)$ and $H_2O(l)$ are the products; (c) air used in the combustion is the naturally occurring mixture of $O_2(g)$, $N_2(g)$, and $Ar(g)$.

B. In some research that required the careful measurement of gas densities, John Rayleigh, a physicist, found that the density of $O_2(g)$ had the same value whether the gas was obtained from air or derived from one of its compounds. The situation with $N_2(g)$ was different, however. The density of $N_2(g)$ had the same value when the $N_2(g)$ was derived from any of various compounds, but a *different* value if the $N_2(g)$ was extracted from air. In 1894, Rayleigh enlisted the aid of William Ramsay, a chemist, to solve this apparent mystery; in the course of their work they discovered the noble gases.
1. Why do you suppose that the $N_2(g)$ extracted from liquid air did *not* have the same density as $N_2(g)$ obtained from its compounds?
2. Which gas do you suppose had the greater density, $N_2(g)$ extracted from air or $N_2(g)$ prepared from nitrogen compounds? Explain.
3. The way in which Ramsay proved that nitrogen gas extracted from air was itself a mixture of gases involved allowing this nitrogen to react with magnesium metal to form magnesium nitride. Explain the significance of this experiment.
4. Use data on the composition of air from Table 8.1 to calculate the *percent difference* in the densities at STP of

Rayleigh's $N_2(g)$ extracted from air and $N_2(g)$ derived from nitrogen compounds.

C. Various thermochemical cycles are being explored as possible sources of $H_2(g)$. The object is to find a series of reactions that can be conducted at moderate temperatures (about 500 °C) and has as its net result the decomposition of water into H_2 and O_2. Show that the following series of reactions meets this requirement. (*Hint:* Balance the equations, multiply by the appropriate coefficients, and combine them into a net equation.)

$$FeCl_2 + H_2O \longrightarrow Fe_3O_4 + HCl + H_2$$
$$Fe_3O_4 + HCl + Cl_2 \longrightarrow FeCl_3 + H_2O + O_2$$
$$FeCl_3 \longrightarrow FeCl_2 + Cl_2$$

Exercises

The Periodic Law

19. Use data from Figure 10-1 and Equation (10.1) to estimate the density expected for the undiscovered element 114. Assume a mass number of 298.

21. The following densities, in grams per cubic centimeter, are for the listed elements in their standard states at 298 K. Show that density is a periodic property of these elements: Al, 2.699; Ar, 0.0018; As, 5.778; Br, 3.100; Ca, 1.550; Cl, 0.0032; Ga, 5.904; Ge, 5.323; Kr, 0.0037; Mg, 1.738; P, 1.823; K, 0.856; Se, 4.285; Si, 2.336; Na, 0.968; S, 2.069.

20. Suppose that lanthanum ($Z = 57$) were a newly discovered element having a density of 6.145 g/cm^3. Estimate its molar mass. (*Hint:* Use Figure 10-1.)

22. The following melting points are in degrees Celsius. Show that melting point is a periodic property of these elements: Al, 660; Ar, -189; Be, 1278; B, 2300; C, 3350; Cl, -101; F, -220; Li, 179; Mg, 651; Ne, -249; N, -210; O, -218; P, 590; Si, 1410; Na, 98; S, 119.

The Periodic Table

23. Mendeleev's periodic table did not preclude the possibility of a new group of elements that would fit within the existing table, as was the case with the noble gases. Moseley's work did preclude this possibility. Explain this difference.

25. Assuming that the seventh period is 32 members long, what should be the atomic number of the noble gas following radon (Rn)? of the alkali metal following francium (Fr)? What would you expect their approximate atomic masses to be?

24. Explain why the several periods in the periodic table do not all have the same number of members.

26. Concerning the incomplete seventh period of the periodic table, what should be the atomic number of the element **(a)** for which the filling of the $6d$ subshell is completed; **(b)** that should most closely resemble bismuth; **(c)** that should be a noble gas?

---

the masses of other atoms can be established by mass spectrometry. From the masses of the different isotopes of an element and their percent natural abundances, the weighted-average atomic mass (weight) of an element can be determined.

of atoms ... ned its molar mass. Molar ... and the Avogadro constant are used in a variety of calculations involving the mass, amount (in moles), or number of atoms in a sample of an element.

Summarizing Example

A stainless steel ball bearing has a radius of 6.35 mm and a density of 7.75 g/cm^3. Iron is the principal element in steel. Carbon is a key minor element. The ball bearing consists of 0.25% carbon, by mass.

Given that the percent natural abundance of ^{13}C is 1.108%, how many ^{13}C atoms are present in the ball bearing?

1. *Determine the volume of the ball bearing, in cubic centimeters.* The formula for the volume (V) of a sphere is $V = \frac{4}{3}\pi r^3$. The radius (r) must be expressed in centimeters. *Result:* 1.07 cm^3.

2. *Determine the mass of carbon present.* (a) Use the density of steel and the volume of the ball bearing to calculate its

mass. (b) Use the percent carbon in the steel to convert from mass of steel to mass of carbon. *Result:* 0.021 g C.

3. *Determine the total number of carbon atoms present.* (a) Use the molar mass of carbon to convert the mass of carbon to the amount of carbon expressed in moles. (b) Use the Avogadro constant to establish the number of carbon atoms. *Result:* 1.1×10^{21} C atoms.

4. *Determine the number of ^{13}C atoms.* The percent natural abundance leads to the conversion factor: 1.108 ^{13}C atoms/100 C atoms. Multiply the result of step **3** by this factor to obtain the number of ^{13}C atoms. *Answer:* 1.2×10^{19} ^{13}C atoms.

Key Terms

actinides (2-6)	group (2-6)	neutron (2-3)
alpha (α) particle (2-2)	ion (2-4)	neutron number (2-3)
atomic mass (weight) (2-5)	isotope (2-4)	nonmetals (2-6)
atomic mass unit (u) (2-3)	lanthanides (2-6)	nuclide (2-4)
atomic number (proton number), Z (2-3)	law of conservation of mass (2-1)	percent natural abundance (2-4)
	law of constant composition (definite proportions) (2-1)	period (2-6)
Avogadro constant, N_A (2-7)		periodic table (2-6)
beta (β) particle (2-2)	main group elements (2-6)	proton (2-3)
cathode ray (2-2)	mass number, A (2-3)	radioactivity (2-2)
chemical symbol (2-4)	mass spectrometer (2-4)	transition elements (transition metals) (2-6)
electron (2-2)	metals (2-6)	
family (2-6)	molar mass, $\mathcal{M}$ (2-7)	
gamma (γ) ray (2-2)	mole (2-7)	

ARE YOU WONDERING . . .

New ideas may prompt you to raise important questions as you read the text. These boxes highlight some questions that students frequently ask and provide responses at that point in the text where such questions usually arise.

---

48 CHAPTER 2 *Atoms and the Atomic Theory*

Are You Wondering . . .

Why some atomic masses (for example, F = 18.9984032 u) are stated so much more precisely than others?

There is *one* naturally occurring type of fluorine atom: fluorine-19. Determining the atomic mass of fluorine means establishing the mass of this type of atom as precisely as possible. On the other hand, krypton, with an atomic mass 83.80 u, has *six* naturally occurring isotopes. Because the percent distribution of the isotopes of krypton may differ very slightly from one sample to another, we can't state the weighted-average atomic mass of krypton with high precision.

2-6 Introduction to the Periodic Table

Scientists spend a lot of time organizing information into useful patterns. Before they can organize information, however, they must possess it, and it must be correct. Botanists had enough information about plants to organize their field in the eighteenth century. Because of uncertainties in atomic masses and because several elements remained undiscovered, chemists were not able to organize the elements until a century later.

We can distinguish one element from all others through its particular set of observable physical properties. For example, sodium has a low density of 0.971 g/cm^3 and a low melting point of 97.81 °C. No other element has this same combination of density and melting point. Potassium, though, also has a low density (0.862 g/cm^3) and low melting point (63.65 °C), much like sodium. Sodium and potassium further resemble each other in that both are good conductors of heat and electricity. Moreover, both react vigorously with water to liberate hydrogen gas. Gold, on the other hand, has a density (19.32 g/cm^3) and melting point (1064 °C) that are very much larger than those of sodium or potas...

Focus On   **Fats, Carbohydrates, and Energy Storage**

Aerobic exercise is an excellent way to burn fat and build muscle.

Aerobics, tennis, weight lifting, and jogging are popular forms of exercise, which is important to maintaining a healthy body. Where do we get the energy to do these things? Surprisingly, mostly from fat, the body's chief energy storage system.

During exercise, fat molecules react with water (hydrolyze) to form a group of compounds called fatty acids. Through a complex series of reactions, these fatty acids are converted to carbon dioxide and water. Energy released in these reactions is used to power muscles. A

typical human fatty acid is palmitic acid, $CH_3(CH_2)_{14}COOH$.

The direct combustion of $CH_3(CH_2)_{14}COOH$ in a bomb calorimeter yields the same products as its metabolism in the body, together with a large quantity of heat.

$$CH_3(CH_2)_{14}COOH + 23 \ O_2(g) \longrightarrow$$
$$16 \ CO_2(g) + 16 \ H_2O(l)$$
$$\Delta H = -9977 \ kJ$$

A hydrocarbon with a similar carbon and hydrogen content, $C_{16}H_{34}$, yields a similar quantity of heat on complete combustion, $-10,700$ kJ. The fat stored in our bodies (our fuel) is comparable to the jet fuel in an airplane. In both cases the fuel adds weight, but can be burned to produce the large amount of energy needed for movement and life support.

Energy from simple carbohydrates is released more rapidly than from fats, and this is why we consume sugars (fruit juice, candy bars, . . .) for a quick burst of energy. The combustion of 1 mol $C_{12}H_{22}O_{11}$ (sucrose or cane sugar), however, yields much less energy

(−5640 kJ/mol) than does a fatty acid of similar carbon and hydrogen content. In the enthalpy diagram of Figure 7-15, we can think of the sucrose molecule as in a condition somewhere between a hydrocarbon and the ultimate oxidation products: CO_2 and H_2O. A fatty acid, on the other hand, is in a condition closer to that of the hydrocarbon.

A common energy unit for expressing the energy values of food is the food Calorie (Cal), which is actually a kilocalorie.

$$1 \ Cal = 1 \ kcal = 4.184 \ kJ$$

As energy sources, fats yield about 9 Cal/g (38 kJ/g), whereas carbohydrates and proteins both yield about 4 Cal/g (17 kJ/g).

FIGURE 7-15
Energy values of a fatty acid, a carbohydrate, and a hydrocarbon.
Each of these three comparable 12-carbon compounds produces 12 mol $CO_2(g)$ and some $H_2O(l)$ as combustion products. The hydrocarbon, $C_{12}H_{24}$, releases the most energy on combustion. The fatty acid lauric acid, $CH_3(CH_2)_{10}COOH$, yields nearly as much, but the carbohydrate sucrose, $C_{12}H_{22}O_{11}$, yields considerably less.

Consider a 65-kg (143-lb) person of average height and build. Typically this person would have about 11 kg (24 lb) of stored fats—a fuel reserve of about 4.2×10^5 kJ. On the other hand, if the body stored only carbohydrates, the stored reservoir would have to be about 25 kg. The "65-kg" person would have to weigh about 80 kg (176 lb). A more dramatic picture might be that of a bird carrying its energy reserves as carbohydrates instead of fat. The excess "baggage" might make it impossible for the bird to fly.

It is important to understand some other matters about fat. First, the quantity of stored fat that an individual typically carries is greater than what is needed to meet energy

requirements. Excess body fat leads to an increased risk of heart disease, diabetes, and other medical problems. Second, one does not have to consume fats in order to store body fats. All food types are broken down into small molecules during digestion, and these small molecules are reassembled into the more complex structures that the body needs. In the body, fat molecules can be synthesized from the small molecules produced from carbohydrates, for example. Current medical findings indicate that the most healthful diet provides the bulk of food calories through carbohydrates, with the recommended percent of caloric intake from fats being less than 30%.

FOCUS ON

Practical and real-world examples of chemistry are integrated throughout the text. And at the end of each chapter is a short essay on a practical topic that relates the chapter subject to real-world situations. This feature gives you an opportunity to read about a current topic in more depth.

General Chemistry

Principles and Modern Applications

Matter—Its Properties and Measurement

1

The restoration of art objects, such as this stained glass window, is based in large measure on the physical and chemical properties of materials.

*I*n recent times the general public has become increasingly aware of chemistry because of issues like acid rain and the threat to the ozone layer reported in the popular press. Popular accounts, however, usually do not provide much depth of understanding of basic principles. And it is these principles that are needed when applying chemical knowledge to real-world problems. Mastering principles requires a more systematic approach to the subject of chemistry.

In this chapter we will consider some of the most elementary terms that chemists use and the general methods they employ in making measurements and expressing their results. You may already be familiar with some of this material from your earlier studies. In the following chapters we will focus on ideas more specific to chemistry and some practical applications of these ideas. When we reach Chapter 8, we will have studied enough fundamentals to permit an in-depth look at several applications involving the atmospheric gases. Following this, we will return to a mix of theory and application throughout the middle of the text and concentrate again on applications in the last few chapters.

Dr. Susan Solomon, a chemist and Head Project Scientist of the National Ozone Expedition to Antarctica in 1986–87, is one of the world's leading experts in the interdisciplinary study of stratospheric ozone depletion.

As a biochemist, Percy Julian (1899–1975) developed a number of pharmaceuticals based on soybean extracts. An early accomplishment was a treatment for glaucoma. Eventually, Dr. Julian founded his own company where treatments for arthritis and other conditions were developed.

1-1   The Scope of Chemistry

Everything is made up of chemicals, and much of what we do with things involves chemical reactions. We use chemical reactions to cook foods, and, after we eat, our bodies carry out complex chemical reactions to extract nutrients from these foods. The gasoline that fuels our automobiles is a mixture of dozens of different chemicals. The burning of this mixture provides the energy that propels the automobile. Unfortunately, some of the substances produced in the combustion of gasoline are involved in the formation of smog. Paradoxically, although many of the environmental problems that beset modern society are of a "chemical" origin, the methods of controlling and correcting these problems are also largely of a chemical nature. In many ways, then, we are all practicing chemists—chemistry touches everyone.

Chemistry is sometimes called the "central science" because it relates to so many areas of human endeavor and curiosity. Chemists who develop new materials to improve electronic devices—such as solar cells, transistors, and fiber optic cables—work at the interfaces of chemistry with physics and engineering. Those who develop new pharmaceuticals for use against cancer or AIDS work at the interfaces of chemistry with pharmacology and medicine.

Many chemists work in more traditional fields of chemistry. Biochemists are interested in chemical processes that occur in living organisms. Physical chemists work with fundamental principles of physics and chemistry in an attempt to answer the basic questions that apply to all of chemistry: Why do some substances react with one another but others do not? How fast will a particular chemical reaction occur? How much useful energy can be extracted from a chemical reaction? Analytical chemists study ways to separate and identify chemical substances. Many of the techniques developed by analytical chemists are used extensively by environmental scientists. Organic chemists focus their attention on substances that contain carbon and hydrogen in combination with a few other elements. The vast majority of substances are organic chemicals. For example, living cells consist of water and organic chemicals with a small amount of various salts. Inorganic chemists focus on most of the elements other than carbon, though the fields of organic and inorganic chemistry overlap in some ways.

Although chemistry is a "mature" science, the landscape of chemistry is dotted with unanswered questions and challenges. Modern technology calls for new materials with unusual properties, and chemists must devise methods of producing these materials. Modern medicine needs drugs targeted to perform specific tasks in the human body, and chemists must design strategies to synthesize these drugs from relatively simple starting materials. Society requires improved methods of pollution control, substitutes for scarce materials, nonhazardous means of disposing of toxic wastes, and more efficient ways to extract energy from fuels. Chemists work in all these areas.

Progress in science comes about in the way scientists do their work—asking the right questions, designing the right experiments to supply the right answers, and formulating plausible explanations of their findings. Let's look further into the scientific method.

1-2   The Scientific Method

Science differs from other fields of study in the *method* that scientists use to acquire knowledge, and the special significance of this knowledge. Scientific knowledge can be used to *predict* future events. The principles of rocket propulsion, established by the scientific method, are so well understood that we can predict precisely how long it will take for a rocket to reach the moon. On the other hand, no one can predict with certainty who will be President of the United States in the year 2008. Politics, so strongly linked to human behavior, economic conditions, international affairs, and unforeseen events, cannot be studied by the scientific method.

The ancient Greeks developed some powerful methods of acquiring knowledge, particularly in mathematics. The Greek approach was to start with certain basic assumptions or premises. Then, by the method known as *deduction,* certain conclusions must logically follow. For example, if $a = b$ and if $b = c$, then $a = c$. Deduction alone is not enough for obtaining scientific knowledge, however. The Greek philosopher Aristotle *assumed* four fundamental substances: air, earth, water, and fire. All other materials, he believed, were formed by combinations of these four elements. Chemists of several centuries ago (more commonly referred to as alchemists) tried, in vain, to apply the four-element idea to turn lead into gold. They failed for many reasons, one being that the four-element assumption is false.

The scientific method originated in the seventeenth century with such people as Galileo, Francis Bacon, and Isaac Newton. The key to the method is to make no initial assumptions, but rather to make careful observations of natural phenomena. When enough observations have been made that a pattern begins to emerge, one then formulates a generalization or **natural law** describing the phenomenon. This process—observations leading to a general statement or natural law—is called *induction.* For example, early in the 16th century the Polish astronomer Nicolas Copernicus (1473–1543), through a careful study of astronomical observations, concluded that Earth revolves around the sun in a circular orbit. We can think of this statement as a generalization or natural law.

To test a natural law a scientist designs a controlled situation, an *experiment,* to see if conclusions deduced from the natural law agree with experimental results. We judge the success of a natural law by its ability to summarize observations and predict new phenomena. Copernicus's work was a great success because he was able to predict future positions of the planets more accurately than his contemporaries. We should not think of a natural law as an *absolute* truth, however. Future experiments may require us to modify the law or to discard it altogether. Copernicus's ideas were refined a half-century later by Johannes Kepler, who showed that planets travel in elliptical orbits, not circular orbits.

A **hypothesis** is a tentative explanation of a natural law. If a hypothesis survives testing by experiments, it is often referred to as a theory. We can use this term in a broader sense, though. A **theory** is a model or way of looking at nature that can be used to explain and to make further predictions about natural phenomena. When differing or conflicting theories are proposed, the one that is most successful in its predictions is generally chosen. Also, the theory that involves the smallest number of assumptions—the simplest theory—is preferred. Over time, as new evidence accumulates, most scientific theories undergo modification, and some are discarded.

The **scientific method** is the combination of observations, experimentation, and the formulation of laws, hypotheses, and theories. But it is wrong to suppose that merely following a set of procedures, rather like a cookbook, will guarantee scientific success. Occasionally scientists develop a pattern of thinking about their field, known as a *paradigm,* that is at first successful but then becomes less so. A new paradigm may be needed. For example, in psychiatry it was long believed that all mental illness was a product of the mind, not the body. A new paradigm in psychiatry recognizes that some mental illness is caused by

▶ Feature Problem A on page 30 requires you to formulate a plausible hypothesis (or hypotheses) to explain some observations made in weighing pennies.

Louis Pasteur (1822–1895). This great practitioner of the scientific method was the developer of the germ theory of disease, the sterilization of milk by "pasteurization," and vaccination against rabies. He has been called the greatest physician of all time by some. He was, in fact, not a physician at all but a chemist—by training and by profession.

chemical imbalances in the body. And, finally, many discoveries (X-rays, radioactivity, and penicillin, to name a few) have been made by accident. Such chance discoveries are referred to as serendipity. In 1839, the American inventor Charles Goodyear was searching for a treatment of natural rubber that would make it less brittle when cold and less tacky when warm. In the course of this work, he accidentally spilled a rubber–sulfur mixture on a hot stove and found that the resulting product had exactly the properties he was seeking. So scientists (and inventors) always need to be alert to unexpected observations. Perhaps no one has been more aware of this than Louis Pasteur, who wrote "Chance favors the prepared mind."

1-3 Properties of Matter

Dictionary definitions of chemistry usually include the terms *matter, composition,* and *properties,* as in the statement that "chemistry is the science that deals with the composition and properties of various forms of matter." In this and the next section we will consider some basic ideas relating to these three terms in hopes of gaining a better understanding of what chemistry is all about.

Matter is anything that occupies space, displays a property known as *mass,* and possesses inertia. Every human being is an object of matter. We all occupy space, and we describe our mass through a related property, our weight. (Mass and weight are described in more detail in Section 1-5. Inertia is described in Appendix B.) All the objects that we see around us are objects of matter. The gases of the atmosphere, even though they are invisible, are examples of matter; they occupy space and have mass. Sunlight is *not* matter; rather, it is a form of energy. We can wait until later chapters, however, to introduce the concept of energy.

Composition refers to the parts or components of a sample of matter and their relative proportions. Ordinary water is made up of two simpler substances—hydrogen and oxygen—present in certain fixed proportions. A chemist would say that the composition of water is 11.19% hydrogen and 88.81% oxygen by mass. Hydrogen peroxide, a substance used in bleaches and antiseptics, is also made up of hydrogen and oxygen, but it has a different composition. Hydrogen peroxide is 5.93% hydrogen and 94.07% oxygen by mass.

Properties are those qualities or attributes that we can use to distinguish one sample of matter from others. In some cases we can establish properties visually. Thus, we can distinguish between the reddish brown solid, copper, and the yellow solid, sulfur, by the property of *color* (Figure 1-1). Properties of matter are generally grouped into two broad categories: physical and chemical.

FIGURE 1-1
Physical properties compared.
Copper (left) can be obtained as pellets, hammered into a thin foil, or drawn into a wire. A lump of sulfur crumbles into a fine powder when hammered (right).

Physical Properties and Physical Changes

A **physical property** is one that a sample of matter displays without changing its composition. Copper can be hammered into a thin sheet or foil (see Figure 1-1). Solids that have this ability are said to be *malleable*. Sulfur is not malleable. If we strike a chunk of sulfur with a hammer, it crumbles into a powder. Sulfur is *brittle*. Other physical properties of copper that sulfur does not share are the ability to be drawn into a fine wire and the ability to conduct heat and electricity.

Sometimes a sample of matter changes in its physical appearance; it undergoes a physical change. In a **physical change** some of the physical properties of a sample of matter may change, but its composition remains *unchanged*. When liquid water freezes into solid water (ice), it certainly looks different and in many ways it is different. Yet, the water remains 11.19% hydrogen and 88.81% oxygen by mass.

Chemical Properties and Chemical Change

In a **chemical change,** or **chemical reaction,** one or more samples of matter are converted to new samples with *different* compositions. The key to identifying chemical change, then, comes in observing a *change* in composition. The burning of paper involves a chemical change. Paper is a complex material, but its principal components are carbon, hydrogen, and oxygen. The chief products of the combustion are two gases, one consisting of carbon and oxygen (carbon dioxide) and the other of hydrogen and oxygen (water, as steam). The ability of paper to burn is an example of a chemical property. A **chemical property** is the ability (or inability) of a sample of matter to undergo a change in composition under stated conditions.

Zinc reacts with hydrochloric acid solution to produce hydrogen gas and a water solution of zinc chloride (Figure 1-2). The ability of zinc to react with hydrochloric acid is one of zinc's distinctive chemical properties. The *inability* of gold to react with hydrochloric acid is one of gold's chemical properties (see Figure 1-2). Sodium reacts not only with hydrochloric acid but also with water. In some of their physical properties, zinc, gold, and sodium are similar. For example, each is malleable and a good conductor of heat and electricity. In most of their chemical properties, though, zinc, gold, and sodium are quite different. Knowing these differences helps us to understand why zinc, which does not react with water, is used in roofing nails, for roof flashings, and in rain gutters, and sodium is not. Also, we can appreciate why gold, because of its chemical inertness, is prized for jewelry and coins: it does not tarnish or rust. In our study of chemistry we hope to learn not only how differences in properties determine the ways in which we use materials, but also what the underlying causes of these differences are.

FIGURE 1-2
Reaction with hydrochloric acid: a chemical property of zinc and gold compared.
The zinc-plated (galvanized) nail reacts with hydrochloric acid, producing bubbles of hydrogen gas. The gold bracelet is unaffected by hydrochloric acid. In this photograph the zinc plating has been completely consumed, exposing the underlying iron nail. The reaction of iron with hydrochloric acid imparts some color to the acid solution.

1-4 Classification of Matter

As we will describe more fully in later chapters, matter is built up from very tiny units called **atoms.** Presently we know of 112 different types of atoms, and *all* matter is made up of just these 112 types! A chemical **element** is a substance made up of only a *single* type of atom. The 112 known elements range from such common substances as carbon, iron, and silver to uncommon ones, such as lutetium and thulium. We can obtain about 90 of the elements from natural sources. The remainder do not occur naturally and we can only create them artificially. On the inside front cover you will find a complete listing of the ele-

▶ The type of atom is established by a feature called its atomic number, introduced on page 42. In recent years, research in nuclear physics (considered in Section 26–4) has led to several new elements. Elements number 110 and 111 were discovered in 1994, and element 112 in 1996.

ments and also a special tabular arrangement known as the periodic table. The periodic table is the chemist's directory of the elements. We will describe it in Chapter 2 and use it throughout most of the text.

Chemical **compounds** are substances in which atoms of *different* elements are combined with one another. Scientists have identified millions of different chemical compounds. In some cases we can isolate a molecule of a compound. A **molecule** is the smallest entity having the same proportions of the constituent atoms as does the compound as a whole. A molecule of water consists of three atoms: two hydrogen atoms joined to a single oxygen atom. By contrast, a molecule of the blood protein gamma globulin is made up of 19,996 atoms altogether, but of just four types: carbon, hydrogen, oxygen, and nitrogen.

The composition and properties of an element or compound are uniform throughout a given sample and from one sample to another. Elements and compounds are called **substances.** (In the chemical sense, the term substance should be used only for elements and compounds.)

Solutions, or **homogeneous mixtures,** are uniform in composition and properties throughout a given sample, but the composition and properties may vary from one sample to another. A given solution of sucrose (cane sugar) in water is uniformly "sweet" throughout the solution, but the sweetness of another sucrose solution might be quite different. Ordinary air is a homogeneous mixture of several gases, principally the *elements* nitrogen and oxygen. Seawater is a solution of the *compounds* water, sodium chloride (salt), and a host of others.

In **heterogeneous mixtures**—sand and water, for example—the components separate into distinct regions. Thus, the composition and physical properties vary from one part of the mixture to another. Salad dressing, a slab of concrete, and the leaf of a plant are all heterogeneous. Usually heterogeneous mixtures can easily be distinguished from homogeneous ones, but at times this may be more difficult.

A scheme classifying matter into elements and compounds and homogeneous and heterogeneous mixtures is summarized in Figure 1-3.

Separating Mixtures

A mixture can be separated into its components by appropriate *physical* changes. Consider again a heterogeneous mixture of sand in water. When we pour this mixture into a funnel lined with porous filter paper, liquid water passes through and sand is retained on the paper. This process of separating a solid from a liquid in which it is suspended is called *filtration.* You will probably use this procedure in the laboratory. On the other hand, we cannot separate seawater, a ho-

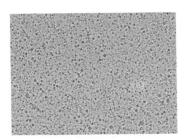

Is it homogeneous or heterogeneous? When viewed through a microscope, homogenized milk is seen to consist of globules of fat dispersed in a watery medium. Homogenized milk is a *heterogeneous* mixture.

The components of a water color mixture are separated by a modern technique known as paper chromatography.

FIGURE 1-3
A classification scheme for matter.
Every sample of matter is either a substance or a mixture. If a substance, it is either an element or a compound. If a mixture, it is either homogeneous or heterogeneous.

mogeneous mixture (solution), by filtration. All components pass through the paper. We can, however, use a different physical change to achieve a partial separation. We can boil the solution. Pure liquid water is obtained from the vapor given off by the boiling solution. Salt and other seawater solids remain behind.

Decomposing Compounds

A chemical compound retains its identity during physical changes, but it can be *decomposed* into its constituent elements by chemical changes. Figure 1-4 shows a water solution of a copper–sulfur–oxygen compound known as copper sulfate. By adding solid zinc to this solution, the copper sulfate is partially decomposed to yield solid copper. Zinc takes the place of copper in solution, and the water solution becomes one of zinc sulfate. Additional chemical changes are required to recover the sulfur and oxygen that were part of the original copper sulfate, but this is a more difficult matter.

States of Matter

Another classification scheme is based on the three *states of matter.* In a **solid,** atoms or molecules are in close contact, sometimes in a highly organized arrangement called a *crystal.* A solid occupies a definite volume and has a definite shape. In a **liquid,** the atoms or molecules are usually separated by greater distances than in a solid. Movement of these atoms or molecules gives a liquid its most distinctive property—the ability to flow, covering the bottom and assuming the shape of its container. In a **gas,** distances between atoms or molecules are much greater than in a liquid. A gas always expands to fill its container. Depending on conditions, a substance may exist in only one state of matter, or it may be in two or three states. Thus, as the ice in a small pond begins to melt in the spring, water is in two states: solid and liquid (actually, three states if we also consider water vapor in the air above the pond).

Figure 1-5 contrasts the three states of matter at two levels. The *macroscopic* level refers to how we perceive matter with our eyes, through the outward appearance of objects. The *microscopic* level describes matter as chemists conceive of it—in terms of atoms and molecules and their behavior. In this text we

(a)   **(b)**

FIGURE 1-4
Decomposing a chemical compound: a chemical change.
(a) A blue solution of copper sulfate in water and a sample of solid zinc.
(b) Zinc is added to the copper sulfate solution and a chemical reaction occurs. The products are a colorless water solution of zinc sulfate and a reddish brown deposit of solid copper.

An important application of this type of reaction is in the extraction of gold from natural sources. Gold is displaced from a water solution of one of its compounds by zinc.

(a) **(b)** **(c)**

FIGURE 1-5
Macroscopic and microscopic views of matter
Pictured at the bottom are the three states of matter as we perceive them *macroscopically,* exemplified by solid calcium (a), liquid water (b), and gaseous helium (c). The circular insets show how chemists conceive of these states of matter *microscopically.* In the solid element calcium the structural units are atoms. In the liquid compound water, the units are molecules, each molecule consisting of one oxygen and two hydrogen atoms. The gaseous element helium is made up of widely separated helium atoms.

will describe many macroscopic, observable properties of matter, but to explain these properties we will often shift our view to the atomic or molecular level—the microscopic level.

1-5   Measurement of Matter: SI (Metric) Units

▶Nonnumerical information is *qualitative,* such as the color blue.

Chemistry is a *quantitative* science. This means that in many cases we can measure a property of a substance and compare it with a standard having a known value of the property. We express the measurement as the product of a *number* and a *unit.* The unit indicates the standard against which the measured quantity is being compared. When we say that the length of the playing field in football is 100 yd, we mean that the field is 100 times longer than a standard of length called the yard (yd). In this section we will introduce some basic units of measurement that are important to chemists.

▶The meter was originally defined as 1/10,000,000 of the distance from the Equator to the North Pole.

The scientific system of measurement is called the *Système Internationale d'Unités* (International System of Units) and abbreviated SI. It is a modern version of the metric system, a system based on the unit of length called a *meter* (m). The seven base quantities in SI are listed in Table 1.1.

SI is a *decimal* system. Quantities differing from the base unit by powers of ten are noted by the use of prefixes written before the base unit. For example, the

TABLE 1.1 SI Base Quantities		
Physical Quantity	**Unit**	**Abbreviation**
length	meter[a]	m
mass	kilogram	kg
time	second	s
temperature	kelvin	K
amount of substance[b]	mole	mol
electric current[c]	ampere	A
luminous intensity[d]	candela	cd

[a]The official spelling of this unit is "metre," but we will use the more common American spelling.

[b]The mole is introduced in Section 2-7.

[c]Electric current is described in Appendix B and in Chapter 21.

[d]Luminous intensity is not discussed in this text.

prefix *kilo* means *one thousand times* (10^3) the base unit; it is abbreviated as k. Thus 1 *kilo*meter = *1000* meters, or 1 km = 1000 m. Some of the more commonly used SI prefixes are listed in Table 1.2.

Most measurements in chemistry are made in SI units. Sometimes we must convert between units within SI, as when converting kilometers to meters. At other times we must convert measurements expressed in non-SI units into SI units or from SI units into non-SI units. In all of these cases we can use a *conversion factor* or a series of conversion factors in a scheme called a conversion pathway. In the following sections we discuss various physical quantities encountered in chemistry and the units used in their measurement.

▶ Problem-solving with conversion factors is a skill usually mastered at the precollegiate level. Conversion factors are an essential feature of this text, and you should carefully study the discussion in Appendix A if you are unfamiliar with their use.

Mass

Mass describes the quantity of matter in an object. In SI, the standard of mass itself carries the prefix *kilo;* it is one *kilogram* (kg). The kilogram is a fairly large unit for most applications in chemistry. More commonly we use the unit *gram* (g) (about the mass of three aspirin tablets).

TABLE 1.2 Some Common SI Prefixes		
Multiple	**Prefix**	**Abbreviation**
10^{12}	tera	T
10^9	giga	G
10^6	mega	M
10^3	kilo	k
10^{-1}	deci	d
10^{-2}	centi	c
10^{-3}	milli	m
10^{-6}	micro	μ[a]
10^{-9}	nano	n
10^{-12}	pico	p

[a]The Greek letter μ (pronounced "mew").

Weight is the force of gravity on an object. It is directly proportional to mass, as shown in the mathematical equations

$$W \propto m \quad \text{and} \quad W = g \cdot m \tag{1.1}$$

▶ The symbol ∝ means "proportional to." It can be replaced by an equality sign and a proportionality constant. A more general statement than (1.1) is that force is the product of mass and acceleration, $F = m \cdot a$. When the acceleration is that due to gravity (g), the force is called weight. See Appendix B.

An object of matter has a fixed mass (m), regardless of where or how the measurement is made. Its weight (W), on the other hand, may vary because the acceleration due to gravity (g) varies slightly from one point on Earth to another. Thus, an object that weighs 100.0 kg in St. Petersburg, Russia, weighs only 99.6 kg in Panama (about 0.4% less). The same object would weigh only about 17 kg on the moon. Conversely, the mass of the object is the same in all three locations. The terms *weight* and *mass* are often used interchangeably, but only *mass* is a measure of the quantity of matter. A common laboratory device for measuring mass is called a balance. (In everyday language a balance is often called, incorrectly, a scale.)

The principle used in a balance is that of counteracting the force of gravity on an "unknown" mass with a force of equal magnitude that can be precisely measured. In older types of balances this counterbalancing is achieved through the force of gravity on objects of precisely known mass called "weights." In the type of balance most commonly seen in laboratories today—the electronic balance—the counter-balancing force is a magnetic force produced by passing an electric current through an electromagnet. First, an initial balance condition is achieved when no object is present on the balance pan. When the object to be weighed is placed on the pan, the initial balance condition is upset. To restore the balance condition, additional electric current must be passed through the electromagnet. The magnitude of this additional current is proportional to the mass of the object being weighed and is translated into a mass reading that is displayed on the balance.

An electronic balance.

Time

In daily use we measure time in seconds, minutes, hours, and years, depending on whether we are dealing with short intervals (such as the time for a 100-m race) or long ones (such as the time before the next appearance of Halley's comet, in 2062). We use all these units in scientific work, although in SI the standard of time is the *second* (s).

Temperature

To establish a temperature scale we arbitrarily set certain fixed points and temperature increments called degrees. Two commonly used fixed points are the temperature at which ice melts and the temperature at which water boils, both at standard atmospheric pressure.*

On the *Fahrenheit* temperature scale the melting point of ice is 32 °F, the boiling point of water is 212 °F, and the interval between is divided into 180 equal parts, called Fahrenheit degrees. On the *Celsius* (centigrade) scale the melting point of ice is 0 °C, the boiling point of water is 100 °C, and the interval between is divided into 100 equal parts, called Celsius degrees. Thus, a range of 100 degrees of Celsius temperature is equal to a range of 180 degrees of Fahrenheit temperature, or 10 degrees of C to 18 degrees of F, or 5 degrees of C to 9 degrees of F. Figure 1-6 compares the Fahrenheit and Celsius temperature scales.

*Standard atmospheric pressure is defined in Section 6-1. The effect of pressure on melting and boiling points is described in Chapter 13.

FIGURE 1-6
A comparison of temperature scales.
(a) The melting point of ice. (b) The boiling point of water. (c) Comparison of Fahrenheit (°F) and Celsius (°C) temperature scales. Shown below are equations for converting temperatures between scales: Kelvin from Celsius; and Celsius from Fahrenheit.

$$T \text{ (K)} = t \, (°C) + 273.15$$
$$t \, (°C) = \tfrac{5}{9}[t \, (°F) - 32]$$

The SI temperature scale, called the Kelvin scale, assigns a value of zero to the lowest conceivable temperature. This zero—0 K—comes at -273.15 °C. The Kelvin scale is an absolute temperature scale; there are no negative Kelvin temperatures. The degree interval on the Kelvin scale, called a *kelvin,* is the same size as the Celsius degree.

▶ We reintroduce and use Kelvin temperature in Chapter 6.

In the laboratory we mostly measure Celsius temperature. Often these temperatures must be converted to the Kelvin scale (in describing the behavior of gases, for example). At other times, particularly in many engineering applications, temperatures must be converted between Celsius and Fahrenheit scales. Temperature conversions can be made in a straightforward way by using the algebraic equations given in the caption to Figure 1-6.

EXAMPLE 1-1

Converting Between Fahrenheit and Celsius Temperatures. A recipe calls for roasting a cut of meat at 350 °F. What is this temperature on the Celsius scale?

Solution

We are given a Fahrenheit temperature and seek a Celsius temperature. We need the algebraic equation from Figure 1-6 that expresses t (°C) as a function of t (°F).

$$t \, (°C) = \frac{5}{9}[t \, (°F) - 32]$$

$$t \, (°C) = \frac{5}{9}[350 - 32] = 177 \ °C$$

[Note that if a problem requires converting Celsius to Fahrenheit temperature we need to use the equation in the alternate form, t (°F) $= \tfrac{9}{5}t$ (°C) $+ 32$.]

Practice Example A: The predicted high temperature for New Delhi, India on a given day is 41 °C. Is this temperature higher or lower than the predicted daytime high of 103 °F in Phoenix, Arizona for the same day?

▶ Answers to Practice Examples are given in Appendix F.

Practice Example B: A particular automobile engine coolant has antifreeze protection to a temperature of -22 °C. Will this coolant offer protection to temperatures as low as -15 °F?

Derived Units

The seven units listed in Table 1-1 are the SI units for the fundamental quantities of length, mass, time, and so on. Many measured properties are expressed as combinations of certain of these fundamental, or base, quantities. We refer to the units of such properties as *derived units*. For example, velocity is a distance divided by the time required to travel that distance. The unit of velocity is that of length divided by time, such as m/s or m s^{-1}.

An important measurement that chemists express through derived units is *volume*. Volume has the unit (length)^3, and the SI standard unit of volume is the *cubic meter* (m^3). A more commonly used volume unit is the *cubic centimeter* (cm^3), and still another is the *liter* (L). One liter is defined as a volume of 1000 cm^3, which means that one *milliliter* (1 mL) is equal to 1 cm^3. The liter is also equal to one *cubic decimeter* (1 dm^3). Several volume units are depicted in Figure 1-7.

Non-SI Units

Although in the United States we are growing more accustomed to expressing distances in kilometers and volumes in liters, most units used in everyday life are still non-SI. Masses are given in pounds, room dimensions in feet, and so on. We will not routinely use these non-SI units, but we will occasionally introduce them in examples and end-of-chapter exercises. When they are used, any necessary relationships between non-SI and SI units will be given or can be found on the inside back cover. Figure 1-8 may help you to develop a frame of reference between some non-SI and SI units.

FIGURE 1-7
Some metric volume units compared.
The largest volume, shown in part, is the SI standard—1 cubic meter (m^3). A cube with a length of 10 cm (1 dm) on edge (in blue) has a volume of 1000 cm^3 (1 dm^3) and is called 1 liter (1 L). The smallest cube is 1 cm on edge (red) and has a volume of 1 cm^3 = 1 mL.

FIGURE 1-8
Some familiar non-SI and SI units compared.
The green ribbon is 1 cm wide and is wrapped around a stick that is 1 m long. The yellow ribbon is 1 inch (in.) wide and is wrapped around a stick 1 yard (yd) long. The meter stick is about 10% longer than the yard stick. 1 in. = 2.54 cm, *exactly.*

Of the two identical beakers, the left one contains 1 kg of candy and the right one, 1 pound (lb). 1 lb = 0.4536 kg = 453.6 g.

The two identical volumetric flasks hold 1 L when filled to the mark. The flask on the left and the carton behind it each contain 1 quart (qt) of milk. The flask on the right and the bottle behind it each contain 1 L of orange juice. 1 qt = 0.9464 L.

1-6 Density and Percent Composition: Their Use in Problem Solving

Throughout this text we will encounter new concepts about the structure and behavior of matter. One means of firming up our understanding of some of these concepts is to work problems relating ideas that we already know to those we are trying to understand. In this section we will introduce two quantities frequently required in problem solving: density and percent composition.

Density

Here is an old riddle: "What weighs more, a ton of bricks or a ton of feathers?" If you answer that they weigh the same, you demonstrate a clear understanding of the meaning of mass—a measure of quantity of matter. Anyone who answers that the bricks weigh more than the feathers has confused the concepts of mass and density. Matter in a brick is more concentrated than in a feather—that is, the matter in a brick is confined to a smaller volume. Bricks are more dense than feathers. **Density** is the ratio of mass to volume.

$$\text{density } (d) = \frac{\text{mass } (m)}{\text{volume } (V)} \tag{1.2}$$

The SI base units of mass and volume are kilograms and cubic meters, respectively, but chemists generally express mass in grams and volume in cubic centimeters or milliliters. The most commonly encountered density unit, then, is grams per cubic centimeter (g/cm^3), or the identical grams per milliliter (g/mL).

The mass of 1.000 L of water at 4 °C is 1.000 kg. The density of water at 4 °C is 1000 g/1000 mL = 1.000 g/mL. At 20 °C, the density of water is 0.9982 g/mL. *Density is a function of temperature* because volume varies with temperature whereas mass remains constant. One reason why global warming is a concern is because if the average temperature of seawater increases, the water becomes less dense. Because the mass of water does not change, its volume must increase and sea level rises—all before any ice melts at the polar caps.

In addition to temperature, the state of matter affects the density of a substance. In general, solids are denser than liquids and both are denser than gases. There are significant overlaps, however. The following are some observations about the numerical values of densities that should prove useful in problem-solving situations.

- Solid densities: from about 0.2 g/cm^3 to 20 g/cm^3.
- Liquid densities: from about 0.5 g/mL to 3–4 g/mL.
- Gas densities: mostly in the range of a few grams per *liter.*

In general, densities of liquids are known more precisely than those of solids (which may have imperfections in their microscopic structures). Also, densities of elements and compounds are known more precisely than densities of materials with variable compositions (such as wood or rubber).

An important consequence of the differing densities of solids and liquids is that liquids and solids of lower density will float on a liquid of higher density (so long as the liquids and solids do not form solutions with each other).

Density in Conversion Pathways

If we measure the mass of an object and its volume, simple division gives us its density. Once we know the density of an object, we can use density as a conversion factor to determine the object's mass or volume. For example, a cube of osmium 1.000 cm on edge weighs 22.48 g. The density of osmium (the densest of the elements) is 22.48 g/cm^3. What would be the mass of another cube of osmium that is 1.00 in. on edge (1 in. = 2.54 cm)? To solve this problem we can map out a *conversion pathway:* length (cm) → volume (cm^3) → mass (g). First, we determine the volume of the cube as $V = (2.54 \text{ cm})^3 = 16.4 \text{ cm}^3$. Then we use the density of osmium as a conversion factor from volume to mass.

$$\text{mass} = 16.4 \text{ cm}^3 \times \frac{22.48 \text{ g}}{\text{cm}^3} = 369 \text{ g}$$

We have previously noted that the density of water at 20 °C is 0.9982 g/mL. Suppose we wish to know the volume of 1.000 kg of water at 20 °C. We should proceed by (a) identifying the known information: 1.000 kg of water and $d = 0.9982$ g/mL (at 20 °C); (b) noting precisely what we are trying to determine—a volume in milliliters (? mL water); and (c) looking for the relevant conversion factors. We are helped in finding these conversion factors by outlining the *conversion pathway:*

kg water → g water → mL water

We need the factor 1000 g/kg to convert from kilograms to grams. Density provides the factor to convert from mass to volume. But in this instance we need to use density in the *inverted* form. That is,

▶ Some thought-provoking questions that can be answered with the concept of density and a minimum of calculations are illustrated in Feature Problems B, C, and D on pages 30 and 31.

$$? \text{ mL water} = 1.000 \text{ kg} \times \frac{1000 \text{ g}}{1 \text{ kg}} \times \frac{1 \text{ mL water}}{0.9982 \text{ g}} = 1002 \text{ mL water}$$

Examples 1-2 and 1-3 further illustrate that numerical calculations involving density are generally of two types: determining density from mass and volume measurements and using density as a conversion factor to relate mass and volume.

EXAMPLE 1-2

FIGURE 1-9
Measuring the volume of a regularly shaped object—Example 1-2 visualized.
The volume of the rectangular solid is the product of its length, width, and height: $V = l \times w \times h$.

Calculating the Density of an Object from Its Mass and Volume. The block of wood pictured in Figure 1-9 has a mass of 2.52 kg. What is the density of the wood, expressed in grams per cubic centimeter?

Solution

We can use the geometric formula in Figure 1-9 to calculate the volume of the rectangular block. We need the volume in cubic centimeters, however, so each dimension of the block must be expressed in centimeters. The length, 1.08 m, must be changed to centimeters.

$$l = 1.08 \, \text{m} \times \frac{100 \text{ cm}}{1 \, \text{m}} = 108 \text{ cm}$$

Now we can calculate the volume in cubic centimeters.

$$V = 108 \text{ cm} \times 5.1 \text{ cm} \times 6.2 \text{ cm} = 3400 \text{ cm}^3$$

The mass of the block must be expressed in grams.

$$m = 2.52 \, \text{kg} \times \frac{1000 \text{ g}}{1 \, \text{kg}} = 2520 \text{ g}$$

The density of the wood is

$$d = \frac{m}{V} = \frac{2520 \text{ g}}{3400 \text{ cm}^3} = 0.74 \text{ g/cm}^3$$

▶ If the mass were mistakenly taken as 2.52 g, the result would be $d = 7.4 \times 10^{-4}$ g/cm^3. If the length were mistakenly taken as 1.08 cm, the result would be $d = 74$ g/cm^3. Both values are well outside the range of densities expected for a solid.

Practice Example A: To determine the density of trichloroethylene, a liquid used to degrease electronic components, a flask is first weighed empty (108.6 g). Then it is filled with 125 mL of the trichloroethylene to give a total mass of 291.4 g. What is the density of trichloroethylene in grams per milliliter?

Practice Example B: A graduated cylinder is filled with 33.8 mL of water. A stone with a mass of 28.4 g is placed in the graduated cylinder. The new volume measured on the cylinder is 44.1 mL. What is the density of the stone?

EXAMPLE 1-3

Calculating the Mass of a Liquid from Its Volume and Density. What is the mass of a 275-mL sample of ethanol (ethyl alcohol) at 20 °C? The density of ethanol at 20 °C is 0.789 g/mL.

Solution

The product of the volume of liquid and density as a conversion factor gives us our result.

$$? \text{ g ethanol} = 275 \, \text{mL ethanol} \times \frac{0.789 \text{ g ethanol}}{1 \, \text{mL ethanol}} = 217 \text{ g ethanol}$$

Practice Example A: What is the mass of 125 mL of a sugar solution (sucrose in water) that has a density of 1.081 g/mL at 20 °C?

Practice Example B: What is the volume, in liters, occupied by 50.0 kg ethanol at 20 °C? The density of ethanol at 20 °C is 0.789 g/mL.

▶ An alternative is to solve the density equation, $d = m/V$, for mass: $m = d \cdot V = 0.789$ g ethanol/mL $\times 275$ mL $= 217$ g ethanol.

Percent as a Conversion Factor

In Section 1-3 we described composition as an identifying characteristic of a sample of matter, and a common way of referring to composition is through percentages. The Latin word *centum* means 100. **Percent** (*percentum*) is the number of parts of a constituent in 100 parts of the whole. To say that a seawater sample contains 3.5% sodium chloride, by mass, means that in every 100 g of the seawater there is 3.5 g of sodium chloride present. We make the statement in terms of grams because we are talking about percent *by mass*. We can express this percent by writing the following ratios

$$\frac{3.5 \text{ g sodium chloride}}{100 \text{ g seawater}} \quad \text{and} \quad \frac{100 \text{ g seawater}}{3.5 \text{ g sodium chloride}} \tag{1.3}$$

In Example 1-4, we will use this type of ratio as a conversion factor.

EXAMPLE 1-4

Using Percent as a Conversion Factor. A 75-g sample of sodium chloride (salt) is to be produced by evaporating to dryness a quantity of seawater containing 3.5% sodium chloride by mass. How many *liters* of seawater must be taken for this purpose? Assume a density of 1.03 g/mL for seawater.

Solution

To convert from "grams of sodium chloride" to "grams of seawater," we need the conversion factor with grams of seawater in the numerator and grams of sodium chloride in the denominator. In addition we need to make the conversions g seawater → mL seawater → L seawater.

$$? \text{ L seawater} = 75 \text{ g sodium chloride} \times \frac{100 \text{ g seawater}}{3.5 \text{ g sodium chloride}}$$

$$\times \frac{1 \text{ mL seawater}}{1.03 \text{ g seawater}} \times \frac{1 \text{ L seawater}}{1000 \text{ mL seawater}}$$

$$= 2.1 \text{ L seawater}$$

Practice Example A: How many kilograms of ethanol are present in 25 L of a "gasohol" solution that is 90% gasoline-10% ethanol by mass? The density of the "gasohol" is 0.71 g/mL.

Practice Example B: Common rubbing alcohol is a solution of 70.0% isopropyl alcohol in water, by mass. If a 25.0-mL sample of rubbing alcohol contains 15.0 g of isopropyl alcohol, what is the density of the rubbing alcohol?

1-7 Uncertainties in Scientific Measurements

All measurements are subject to error. To some extent, measuring instruments have built-in, or inherent, errors, called **systematic errors.** (For example, a kitchen scale might consistently yield results that are 25 g too high, or a thermometer a reading that is 2° too low.) Limitations in an experimenter's skill or ability to read a scientific instrument also lead to errors and give results that may be either too high or too low. Such errors are called **random errors.**

Are You Wondering . . .

In doing a problem with percentages, when to multiply and when to divide?

A common way of dealing with a percentage is to convert it to decimal form (3.5% becomes 0.035) and then to multiply or divide by this decimal. Students sometimes can't decide which to do. Expressing percentage as a conversion factor and using it to produce a cancellation of units gets around this difficulty. Also, remember that

- The quantity of a *component* must always be *less* than the quantity of the whole mixture. (*Multiply by percentage.*)
- The quantity of a *mixture* must always be *greater* than the quantity of any of its components. (*Divide by percentage.*)

If in Example 1-4 we had not been careful about the cancellation of units and had multiplied by percentage (3.5/100) instead of dividing by it (100/3.5), we would have gotten the numerical answer 2.5×10^{-3}. This would be a 2.5-mL sample of seawater, weighing about 2.5 g. Clearly, a sample of seawater that would *contain* 75 g sodium chloride must have a mass *greater than* 75 g.

Precision refers to the degree of reproducibility of a measured quantity—that is, the closeness of agreement when the same quantity is measured several times. The precision of a series of measurements is *high* (or good) if each of a series of measurements deviates by only a small amount from the average. Conversely, if there is wide deviation among the measurements, the precision is *poor* (or low). **Accuracy** refers to how close a measured value is to the accepted, or "real," value. High-precision measurements are not always accurate—a large systematic error could be present. Still, it is more likely that measurements of high precision rather than low precision are accurate.

To illustrate some of these ideas, consider the mass of an object measured on two different balances. One is a relatively crude balance called a platform balance, and the other is a sophisticated analytical balance.

Glassware of varying precision. Beakers and flasks provide the *least* precise means of measuring liquid volumes. Graduated cylinders are much more precise measuring tools. In the foreground of the photo are two pipets. Pipets provide one of the most precise methods of measuring liquid volume. Just behind the pipets is another very precise measuring tool, the buret.

	Platform Balance	Analytical Balance
three measurements	10.4, 10.2, 10.3 g	10.3107, 10.3108, 10.3106 g
their average	10.3 g	10.3107 g
reproducibility	±0.1 g	±0.0001 g
precision	low or poor	high or good

1-8   Significant Figures

For the measurements on the platform balance just considered—that is, 10.4, 10.2, and 10.3 g—the result we would report is the average of the three: 10.3 g. The way a scientist would interpret our result is that the first two digits—

10—are known with certainty, and the last digit—3—because it was estimated, is uncertain by a unit or two. The measurement 10.3 g is said to have *three* **significant figures.** If we express this mass in kilograms rather than grams, 10.3 g = 0.0103 kg, the measurement is still expressed to *three* significant figures, even though more than three digits are shown. For the measurements on the analytical balance, the corresponding reported value would be 10.3107 g—a value with *six* significant figures. The number of significant figures in a measured quantity, then, gives an indication of the capabilities of the measuring device and the precision of the measurements.

One of our basic needs is to be able to determine the number of significant figures in an expressed quantity. The rules for doing this, with which you should become thoroughly familiar, are presented in Figure 1-10. Even with the ideas outlined in Figure 1-10, we remain uncertain over how many significant figures to assign to the measurement 7500 m. Do we mean 7500 m, measured to the nearest meter? 10 meters? or 100 meters? If all the zeros are significant—if the value has *four* significant figures—we can write 7500. m. That is, by writing a decimal point that is not otherwise needed, we show that all zeros preceding the decimal point are significant. This technique does not help if only one of the zeros, or if neither zero, is significant. The best approach here is to use exponential notation. (Review Appendix A if necessary.) The coefficient establishes the number of significant figures and the power of ten locates the decimal point.

2 significant figures	*3 significant figures*	*4 significant figures*
7.5×10^3 m	7.50×10^3 m	7.500×10^3 m

Significant Figures in Numerical Calculations

An important requirement in calculations is that precision can be neither gained nor lost in calculations involving measured quantities. There are several exact methods of determining how precisely to express the result of a calculation, but it is usually sufficient just to observe some simple rules involving significant figures:

> The result of multiplication or division may contain only as many significant figures as the *least* precisely known quantity in the calculation.

▶ A more exact rule on multiplication/division is that the result should have about the same percent error as the least precisely known quantity. Usually the significant figure rule conforms to this requirement; occasionally it does not (see Exercise 67).

In the following chain multiplication to determine the volume of a rectangular block of wood, we should round off the result to *three* significant figures. Figure 1-11 may help you to understand this.

$$14.79 \text{ cm} \times 12.11 \text{ cm} \times 5.05 \text{ cm} = 904 \text{ cm}^3$$
(4 sig. fig.)   (4 sig. fig.)   (3 sig. fig.)   (3 sig. fig.)

FIGURE 1-10
Determining the number of significant figures in a quantity.
The quantity shown here, 0.004004500, has *seven* significant figures. All nonzero digits are significant, as are the indicated zeros.

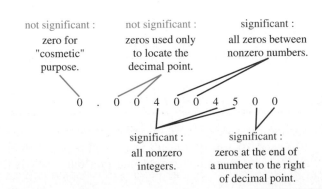

not significant :
zero for
"cosmetic"
purpose.

not significant :
zeros used only
to locate the
decimal point.

significant :
all zeros between
nonzero numbers.

0 . 0 0 4 0 0 4 5 0 0

significant :
all nonzero
integers.

significant :
zeros at the end of
a number to the right
of decimal point.

FIGURE 1-11 Significant figure rule in multiplication.
In forming the product 14.79 cm × 12.11 cm × 5.05 cm, the least precisely known quantity is 5.05 cm. Shown on the calculators are the products of 14.79 and 12.11 with 5.04, 5.05, and 5.06, respectively. In the three results only the first two digits, "90. . .," are identical. Variations begin in the third digit. We are certainly not justified in carrying digits beyond the third. We express the volume as 904 cm³. Usually, instead of a detailed analysis of the type done here, we can use a simpler idea: *The result of a multiplication may contain only as many significant figures as does the least precisely known quantity.*

In adding and subtracting numbers the applicable rule is that

The result of addition or subtraction must be expressed with the same number of decimal places as the quantity carrying the *smallest* number of decimal places.

Consider the following sum of masses:

$$
\begin{array}{r}
15.02 \ \text{g} \\
9{,}986.0 \ \ \text{g} \\
\underline{3.518 \ \text{g}} \\
10{,}004.538 \ \text{g}
\end{array}
$$

The sum has the same uncertainty, ±0.1 g, as does the term carrying the smallest number of decimal places, 9,986.0 g. Note that this calculation is *not* limited by significant figures. In fact the sum has more significant figures (6) than do any of the terms in the addition. Nor is the calculation limited by the least precisely known term (9,986.0 g is known with greater precision than either 15.02 g or 3.518 g).

There are two situations when a quantity appearing in a calculation may be *exact,* that is, when it is not subject to errors in measurement. This may occur

- By definition (such as 3 ft = 1 yd, or 1 in. = 2.54 cm)

or as a result of

- Counting (such as *six* faces on a cube, or *two* hydrogen atoms in a water molecule).

Exact numbers can be considered to have an unlimited number of significant figures.

▶Later in the text we will need to apply ideas about significant figures to logarithms. This matter is discussed in Appendix A.

"Rounding Off" Numerical Results

To three significant figures, we should express 15.453 as 15.5, and 14,775 as 1.48×10^4. If our need is to drop just one digit, that is, to "round off" a number, the simplest rule to follow is to increase the final digit by one unit if the digit dropped is 5, 6, 7, 8, or 9, and to leave the final digit unchanged if the digit dropped is 0, 1, 2, 3, or 4. To three significant figures, 15.55 rounds off to 15.6 and 15.54 rounds off to 15.5.

EXAMPLE 1-5

▶ As added practice in working with significant figures, review the calculations in Section 1-6. You will note that they conform to the significant figure rules presented here.

Applying Significant Figure Rules: Multiplication/Division. Express the result of the following calculation with the correct number of significant figures.

$$\frac{0.225 \times 0.0035}{2.16 \times 10^{-2}} = ?$$

Solution

By inspecting the three quantities, we see that the least precisely known is 0.0035, with *two* significant figures. Our result must also contain only *two* significant figures. The fact that two of the quantities are given in decimal form and one in exponential form is immaterial. Simply enter the numbers, as written, into an electronic calculator and read off the answer: 0.0364583. Express the result to *two* significant figures: 0.036. Alternatively, write the answer in exponential form: 3.6×10^{-2}.

Practice Example A: Perform the following calculation and express the result with the appropriate number of significant figures.

$$\frac{62.356}{0.000456 \times 6.422 \times 10^3} = ?$$

Practice Example B: Perform the following calculation and express the result with the appropriate number of significant figures.

$$\frac{8.21 \times 10^4 \times 1.3 \times 10^{-3}}{0.00236 \times 4.071 \times 10^{-2}} = ?$$

EXAMPLE 1-6

Applying Significant Figure Rules: Addition/Subtraction. Express the result of the following calculation with the correct number of significant figures.

$$(2.06 \times 10^2) + (1.32 \times 10^4) - (1.26 \times 10^3) = ?$$

Solution

If this calculation is performed with an electronic calculator, the quantities can be entered just as they are written, and the answer expressed to the correct number of significant figures. To help establish the correct number of significant figures, we must first realize that the largest value will determine the last significant digit. In this case, 1.32×10^4, or 13,200, shows the hundreds place is the last significant digit. Adding these values on a calculator yields

$$(2.06 \times 10^2) + (1.32 \times 10^4) - (1.26 \times 10^3) = 1.2146 \times 10^4$$

To put the last significant digit in the hundreds place, this value is rounded off to 1.21×10^4. In this case, the answer has three significant digits. It is possible, when

Are You Wondering . . .

At what point(s) to round off numerical results during a calculation?

What's nice about using electronic calculators is that we don't have to write down intermediate results. On occasion, however, when we take a stepwise approach to an in-text example, we will write down intermediate results. We may round off these intermediate results as well as the final result. In general, though, you should store all intermediate results in your electronic calculator without regard for significant figures. Round off to the correct number of significant figures only in the final answer.

adding or subtracting, to have a result with more or fewer significant digits than any of the numbers that are in the calculation.

Practice Example A: Express the result of the following calculation with the appropriate number of significant figures.

$$0.236 + 128.55 - 102.1 = ?$$

Practice Example B: Perform the following calculation and express the result with the appropriate number of significant figures.

$$\frac{(1.302 \times 10^3) + 952.7}{(1.57 \times 10^2) - 12.22} = ?$$

Summary

Matter is classified as either a substance—element or compound—or a mixture—homogeneous or heterogeneous—and in a certain state—solid, liquid, or gas. Chemistry deals with the study of properties of matter and how these may be changed by physical or chemical means. Like other branches of science, chemistry makes use of the scientific method, a series of activities involving observations and experimentation that culminate in natural laws and theories to explain and predict natural phenomena.

The system of measurement used in science is called SI. Four of the fundamental, or base, quantities in SI are length, mass, time, and temperature. Other quantities described in the chapter have derived units, such as volume, expressed through

the unit (length)^3. When measuring a property of matter, the precision of the measurement needs to be shown. This is done through the proper use of significant figures. Moreover, calculations must be performed so that the calculated result is stated no more precisely than warranted by the measured quantities.

The method of problem solving employed in this chapter, called the conversion factor method, is described in detail in Appendix A. Many of the conversion factors introduced are based on relationships among SI units or between SI and non-SI units. Also shown to be conversion factors are the physical property, density, and the composition unit, percent.

Summarizing Example

Methanol (methyl alcohol, or wood alcohol) is a potential automotive fuel, either pure or mixed with gasoline. Some fleet vehicles, such as municipal buses, have been modified to burn methanol-containing fuels. These fuels are also used by some race cars.

An automobile, modified to use a mixture of 85.0% methanol and 15.0% gasoline, by mass, gets 25.5 mi/gal. The fuel has a density of 0.775 g/mL. How many kilograms of methanol does the auto consume in a trip of 808 km?

Focus On

The Scientific Method at Work: Polywater

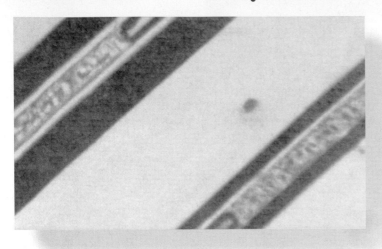

The melting behavior of polywater–water mixtures. Two polywater–water mixtures at −10 °C. The sample at the left has just begun to melt. When the temperature reaches −6 °C, both samples will have completely melted. Pure water, by contrast, does not melt until the temperature reaches 0 °C.

sample of water was confined in a quartz capillary tube (a very small diameter tube of a special high-purity glass), some liquid water condensed in another part of the tube. The recondensed water had some peculiar properties, including a high boiling point (over 150 °C), a low freezing point (less than −30 °C), and a density greater than that of ordinary water.

At first, this discovery was largely overlooked by other scientists, probably because it was published in specialized Russian-language journals. Only after the Russian scientists presented an interpretation of their results at an international meeting did other scientists begin to take note. The Russian results were repeated by many laboratories and thus met the first criterion of a valid scientific result: It must be reproducible. There was no dispute, then, about

About 30 years ago scientists were heatedly debating the reported discovery of a new form of water called "polywater." The initial experiments were done in Russia by scientists who were highly regarded as careful experimenters. These scientists reported that if a

---

1. *Convert the trip length from kilometers to miles.* This can be a single-step conversion if a conversion factor between kilometers and miles is available. Alternatively, a lengthier, but just as acceptable, pathway of conversions can be used, such as km → m → in. → ft → mi. *Result:* 502 mi.

2. *Determine the volume of fuel consumed, in gallons.* The fuel mileage can be expressed as the conversion factor 1 gal/25.5 mi. Then take the product of this factor and the distance from step **1**. *Result:* 19.7 gal.

3. *Convert the volume of fuel to mass of fuel.* Convert 19.7 gal to an equivalent volume in milliliters (for example, gal → qt → L → mL). Then multiply by the density of the fuel. *Result:* 5.78×10^4 g.

4. *Determine the mass of methanol in the fuel.* Multiply the result of step **3** by the percent composition of the fuel, that is, by the conversion factor 85.0 g methanol/100.0 g fuel. *Answer:* 49.1 kg methanol.

Key Terms (See Glossary for definitions of these terms)

accuracy (1-7)
atom (1-4)
chemical change (reaction) (1-3)
chemical property (1-3)
composition (1-3)
compound (1-4)
density (1-6)
element (1-4)
gas (1-4)
heterogeneous mixture (1-4)

homogeneous mixture (solution) (1-4)
hypothesis (1-2)
liquid (1-4)
mass (1-5)
matter (1-3)
molecule (1-4)
natural law (1-2)
percent (1-6)
physical change (1-3)

physical property (1-3)
precision (1-7)
property (1-3)
random error (1-7)
scientific method (1-2)
significant figures (1-8)
solid (1-4)
substance (1-4)
systematic error (1-7)
theory (1-2)

the facts—but their interpretation sparked active debate.

The properties of the recondensed water could be explained in one of two ways, and each explanation was consistent with the observations. One explanation was that the water contained impurities that had dissolved during the preparation of the water sample. The other was that the water was of a form in which simple water molecules (each consisting of two hydrogen atoms and one oxygen atom) aggegrate into large clusters. Such "super" molecules are known as *polymers.*

In one critical experiment some American researchers subjected recondensed water to very careful examination by a method known as spectroscopy. The method is designed to reveal the structure of a substance at the molecular level. The spectroscopic data suggested that the recondensed water was a pure substance, not a solution. Further, the data suggested a polymer-like structure for the water. Hence the name "polywater" was coined. The publication of these results in 1969 created a sensation, both among scientists and in the popular press.

Those who firmly believed in polywater employed the reasoning of Sherlock Holmes, a fictional but ardent follower of the scientific method: "When you have eliminated the impossible, whatever remains, however improbable, must be the truth." The trouble was that the impossible had not been eliminated. Within a year or so, new reports pointed to dissolved impurities in "polywater." At about the same time, new spectroscopic data indicated that the results interpreted as due to "polywater" could not be due to any form of water. They were now attributed to impurities in the water. Finally, in 1973, the Russian scientists who had made the original observations reported that they could not produce "polywater" in capillaries that had been freed of all impurities. And so ended this brief but fascinating chapter in the history of science.

The "polywater" story may seem to reveal a failure of science, but actually it represents a great success of the scientific method. A novel idea was thoroughly examined by many scientists by a variety of techniques. Hypotheses were proposed, experiments were conducted, and the experimental results were discussed openly. After all of this, the scientific community reached a consensus: "polywater" is an impure form of water. A crucial aspect of the scientific method was also revealed: the self-correcting nature of science. Science may get off course at times, but eventually it always gets back on the right track.

Review Questions

1. In your own words define or explain the following terms or symbols: **(a)** m^3; **(b)** % by mass; **(c)** °C; **(d)** density; **(e)** element.

2. Briefly describe each of the following ideas: **(a)** SI base units; **(b)** significant figures; **(c)** natural law.

3. Explain the important distinctions between each pair of terms: **(a)** mass and weight; **(b)** intensive and extensive property; **(c)** substance and mixture; **(d)** precision and accuracy; **(e)** hypothesis and theory.

4. Perform the following conversions.
(a) 1.55 kg = _____ g
(b) 642 g = _____ kg
(c) 2896 mm = _____ cm
(d) 0.086 cm = _____ mm

5. Perform the following conversions.
(a) 0.127 L = _____ mL
(b) 15.8 mL = _____ L
(c) 981 cm^3 = _____ L
(d) 2.65 m^3 = _____ cm^3

6. Perform the following conversions from non-SI to SI units. (Use information from Figure 1-8, as necessary.)
(a) 68.4 in. = _____ cm
(b) 94 ft = _____ m
(c) 1.42 lb = _____ g
(d) 248 lb = _____ kg
(e) 1.85 gal = _____ L
(f) 3.72 qt = _____ mL

7. Determine the number of
(a) square meters (m^2) in 1 square kilometer (km^2);
(b) square centimeters (cm^2) in 1 square meter (m^2);
(c) square meters (m^2) in 1 square mile (mi^2) (1 mi = 5280 ft, 1 ft = 12 in., 1 in. = 2.54 cm).

8. *Without doing a calculation,* explain which of the following represents the higher temperature: 204 °F or 102 °C.

9. *Without doing a detailed calculation,* explain which of the following represents the greatest *mass:* 80.0 g ethanol (d = 0.79 g/mL), 100.0 mL benzene (d = 0.87 g/mL), or 90.0 mL carbon disulfide (d = 1.26 g/mL).

10. A 2.18-L sample of butyric acid, a substance present in rancid butter, has a mass of 2088 g. What is the density of butyric acid in grams per milliliter?

11. A 385-mL sample of liquid mercury has a mass of 5.23 kg. What is the density of liquid mercury in grams per milliliter?

12. Ethylene glycol, an antifreeze, has a density of 1.11 g/mL at 20 °C.
(a) What is the mass, in grams, of 452 mL ethylene glycol?
(b) What is the mass, in kilograms, of 18.6 L ethylene glycol?
(c) What is the volume, in milliliters, occupied by 65.0 g ethylene glycol?
(d) What is the volume, in liters, occupied by 23.9 kg ethylene glycol?

13. A solution consisting of 8.50% acetone–91.5% water, by mass, has a density of 0.9867 g/mL. What mass of acetone, in kilograms, is present in 7.50 L of the antifreeze?

14. A brand of vinegar is found to contain 5.4% acetic acid, by mass. What mass of acetic acid, in grams, is contained per pound of this vinegar? (1 lb = 453.6 g.)

15. What mass, in grams, of a solution containing 12.62% by mass of sucrose (cane sugar) is needed for an application that requires 1.00 kg sucrose?

16. A fertilizer contains 21% nitrogen by mass. What mass of this fertilizer, in kilograms, is required for an application requiring 775 g of nitrogen?

17. Express each number in exponential notation (see Appendix A). **(a)** 8950.; **(b)** 10,700.; **(c)** 0.0240; **(d)** 0.0047; **(e)** 938.3; **(f)** 275,482

18. Express each number in common decimal form (see Appendix A). **(a)** 3.21×10^{-2}; **(b)** 5.08×10^{-4}; **(c)** 121.9×10^{-5}; **(d)** 16.2×10^{-2}

19. How many significant figures are shown in each of the following? If indeterminate, give the possible range of significant figures. **(a)** 450; **(b)** 98.6; **(c)** 0.0033; **(d)** 902.10; **(e)** 0.02173; **(f)** 7000; **(g)** 7.02; **(h)** 67,000,000

20. Express each of the following to *four* significant figures. **(a)** 3984.6; **(b)** 422.04; **(c)** 186,000; **(d)** 33,900; **(e)** 6.321×10^4; **(f)** 5.0472×10^{-4}

21. Perform the following calculations, expressing each answer in exponential form and with the appropriate number of significant figures.
(a) $0.406 \times 0.0023 =$
(b) $0.1357 \times 16.80 \times 0.096 =$
(c) $0.458 + 0.12 - 0.037 =$
(d) $32.18 + 0.055 - 1.652 =$

22. Perform the following calculations, expressing each number and the answer in exponential form and with the appropriate number of significant figures.
(a) $\dfrac{320 \times 24.9}{0.080} =$
(b) $\dfrac{432.7 \times 6.5 \times 0.002300}{62 \times 0.103} =$
(c) $\dfrac{32.44 + 4.9 - 0.304}{82.94} =$
(d) $\dfrac{8.002 + 0.3040}{13.4 - 0.066 + 1.02} =$

23. Calculate the mass of a block of iron ($d = 7.86$ g/cm^3) with dimensions 52.8 cm × 6.74 cm × 3.73 cm.

24. Calculate the mass of a cylinder of stainless steel ($d = 7.75$ g/cm^3) with height 18.35 cm and radius 1.88 cm. (*Hint:* The volume of a cylinder is given by the formula $V = \pi r^2 h$.)

Exercises

Scientific Method

25. Is it possible to predict how many experiments are required to verify a natural law? Explain.

27. An important premise of science is that there exists an underlying order to nature. Einstein described this belief in the words "God is subtle but He is not malicious." Explain more fully what he meant by this remark.

26. What are the principal reasons one theory might be adopted over a conflicting one?

28. Explain why the common saying, "The exception proves the rule," is incompatible with the scientific method.

Properties and Classification of Matter

29. State whether each property is physical or chemical.
(a) An iron nail is attracted to a magnet.
(b) Charcoal lighter fluid is ignited with a match.
(c) A bronze statue develops a green coating (patina) over time.
(d) A block of wood floats on water.

30. State whether each property is physical or chemical.
(a) A piece of sliced apple turns brown.
(b) A slab of marble feels cool to the touch.
(c) A sapphire is blue.
(d) A clay pot is fired in a kiln.

31. Indicate whether each sample of matter listed is a substance or a mixture, and, if a mixture, whether homogeneous or heterogeneous.
(a) a sliver of wood
(b) ink
(c) deionized water
(d) hot chocolate
33. What type of change—physical or chemical—is necessary to bring about the following separations? (*Hint:* Refer to a listing of the elements.)
(a) the chocolate chips from a cookie
(b) iron from iron oxide (rust)
(c) sulfur from sulfuric acid (battery acid)
(d) pure water from seawater

32. Indicate whether each sample of matter listed is a substance or a mixture, and, if a mixture, whether homogeneous or heterogeneous.
(a) A breath of fresh air
(b) A brass door knob
(c) garlic salt
(d) ice
34. Suggest physical changes by which the following mixtures can be separated.
(a) iron filings and wood chips
(b) sugar and sand
(c) water and gasoline
(d) gold flakes and water

Exponential Arithmetic (see Appendix A)

35. Express each value in exponential form. Where appropriate include units in your answer.
(a) speed of sound (sea level): 34,000 centimeters per second
(b) equatorial radius of Earth: six thousand three hundred seventy eight kilometers
(c) $\dfrac{(2.2 \times 10^3) + (4.7 \times 10^2)}{5.8 \times 10^{-3}} =$

36. Express each value in exponential form. Where appropriate include units in your answer.
(a) solar radiation received by Earth: 173 thousand trillion watts
(b) average human cell diameter: ten millionths of a meter
(c) $\dfrac{5.07 \times 10^4 \times (1.8 \times 10^{-3})^2}{0.065 + (3.3 \times 10^{-2})} =$

Significant Figures

37. Indicate whether each of the following is an exact number or a measured quantity subject to uncertainty.
(a) the number of soda cans in a case
(b) the volume of milk in a gallon jug
(c) the distance between Earth and the sun
39. Perform the following calculations and retain the appropriate number of significant figures in each result.
(a) $38.4 \times 10^{-3} \times 6.36 \times 10^5 =$
(b) $\dfrac{1.45 \times 10^2 \times 8.76 \times 10^{-4}}{(9.2 \times 10^{-3})^2} =$
(c) $24.6 + 18.35 - 2.98 =$
(d) $(1.646 \times 10^3) - (2.18 \times 10^2) + (1.36 \times 10^4 \times 5.17 \times 10^{-2}) =$

38. Indicate whether each of the following is an exact number or a measured quantity subject to uncertainty.
(a) the number of pages in this text
(b) the number of days in the month of January
(c) the area of a city lot
40. Express the result of each of the following calculations in exponential form and with the appropriate number of significant figures.
(a) $4.65 \times 10^4 \times 2.95 \times 10^{-2} \times 6.663 \times 10^{-3} \times 8.2 =$
(b) $\dfrac{1912 \times (0.0077 \times 10^4) \times 3.12 \times 10^{-3}}{(4.18 \times 10^{-4})^3} =$
(c) $(3.46 \times 10^3) \times 0.087 \times 15.26 \times 1.0023 =$
(d) $\dfrac{(4.505 \times 10^{-2})^2 \times 1.080 \times 1545.9}{0.03203 \times 10^3} =$

41. A press release describing the 1986 nonstop, round-the-world trip by the aircraft *Voyager* included the following data.

> flight distance: 25,012 mi
> flight time: 9 days, 3 minutes, 44 seconds
> fuel capacity: nearly 9000 lb
> fuel remaining at end of flight: 14 gal

To the maximum number of significant figures permitted, calculate
(a) the average speed of the aircraft, in miles per hour;
(b) the fuel consumption, in miles per pound of fuel. (Assume a density of 0.70 g/mL for the fuel.)

42. Use the concept of significant figures to criticize the way in which the following information was presented. "The estimated proved reserve of natural gas as of January 1, 1982, was 2,911,346 trillion cubic feet."

Units of Measurement

43. Which is the greater mass, 2172 μg or 0.00515 mg? Explain.

45. The non-SI unit, the hand (used in horsemanship), is 4 inches. What is the height, in meters, of a horse that stands 15 hands high? (1 in. = 2.54 cm.)

47. A sprinter runs the 100-yd dash in 9.3 s. At this same rate, **(a)** How long would it take the sprinter to run 100.0 m? (1 yd = 36 in., 1 in. = 2.54 cm.) **(b)** What is the sprinter's speed in meters per second? **(c)** How long would it take the sprinter to run a distance of 1.45 km?

49. In SI units land area is measured by the *hectare,* defined as 1 hm^2. 1 hectometer (hm) = 100 m. How many acres correspond to 1 hectare? (1 mi^2 = 640 acres, 1 mi = 5280 ft, 1 ft = 12 in.)

51. A typical pressure for optimal performance of automobile tires is 32 lb/in^2. What is this pressure in units of grams per square centimeter and kilograms per square meter?

44. Which is the greater mass, 3257 mg or 0.00475 kg? Explain.

46. The unit *furlong* is used in horseracing. The units *chain* and *link* are used in surveying. There are exactly 8 furlongs in 1 mi, 10 chains in 1 furlong, and 100 links in 1 chain. To three significant figures, what is the length of 1 link in inches?

48. A non-SI unit of mass used in pharmaceutical work is the grain (gr). (15 gr = 1.0 g) An aspirin tablet contains 5.0 gr of aspirin. A 155-lb arthritic individual takes two aspirin tablets per day. (1 lb = 453.6 g) **(a)** What is the quantity of aspirin in two tablets, expressed in milligrams? **(b)** What is the dosage rate of aspirin, expressed in milligrams of aspirin per kilogram of body mass? **(c)** At the given rate of consumption of aspirin tablets, how many days would it take to consume 1.0 lb of aspirin?

50. A tungsten steel alloy used in permanent magnets has a density of 0.292 lb/in.^{3}, what is its density in grams per cubic centimeter?

52. An airplane flying at the speed of sound has a speed of Mach 1; Mach 1.5 is 1.5 times the speed of sound; and so on. If the speed of sound in air is 1130 ft/s, what is the speed, in kilometers per hour, of an airplane flying at Mach 1.38?

Temperature Scales

53. A table of climatic data lists the highest and lowest temperatures on record for San Bernardino, California, as 118 and 17 °F, respectively. What are these temperatures on the Celsius scale?

55. A home economics class is given an assignment in candy making that requires a sugar mixture to be brought to a "soft ball" stage (234–240 °F). A student borrows a thermometer having a range from −10 to 110 °C from the chemistry laboratory to do this assignment. Will this thermometer serve the purpose? Explain.

54. We wish to mark off a thermometer in both Celsius and Fahrenheit temperatures. On the Celsius scale, the lowest temperature mark is at −15 °C, and the highest temperature mark is at 60 °C. What Fahrenheit temperatures should be shown for these same two marks?

56. The absolute zero of temperature is found at −273.15 °C. Should it be possible to achieve a temperature of −465 °F? Explain.

Density

57. To determine the density of acetone, a 55.0-gal drum is weighed twice. The 55.0-gal drum weighs 75.0 lb when empty. When it is filled with acetone, the drum weighs 437.5 lb. What is the density of acetone, expressed in grams per milliliter? (1 gal = 3.785 L, 1 lb = 453.6 g)

59. The following densities are given at 20 °C: water, 0.998 g/cm^3; iron, 7.86 g/cm^3; aluminum 2.70 g/cm^3. Arrange the following items in terms of *increasing* mass. (1) a rectangular bar of iron, 81.5 cm × 2.1 cm × 1.6 cm (2) a sheet of aluminum foil, 12.12 m × 3.62 m × 0.003 cm (3) 4.051 L of water

58. To determine the volume of an irregularly shaped glass vessel, the vessel is weighed empty (121.3 g) and when filled with carbon tetrachloride (283.2 g). What is the volume capacity of the vessel, in milliliters, given that the density of carbon tetrachloride is 1.59 g/mL?

60. The density of aluminum is 2.70 g/cm^3. A square piece of aluminum foil, 9.0 in. on a side, is found to weigh 2.568 g. What is the thickness of this foil, in millimeters? (1 in. = 2.54 cm.)

61. To determine the approximate mass of a small spherical shot of copper, the following experiment is performed. 125 pieces of the shot are counted out and added to 8.4 mL of water in a graduated cylinder; the total volume becomes 8.9 mL. The density of copper is 8.92 g/cm³. Determine the approximate mass of a single piece of shot, assuming that all the pieces are of the same dimensions. (*Hint:* What is the volume of the shot?)

62. The angle iron pictured here is made of steel with a density of 7.78 g/cm³. What is the mass, in grams, of this object? (*Hint:* What is its volume?)

Percent Composition

63. In a class of 76 students the results of a particular examination were 9 A, 21 B, 36 C, 8 D, 2 F. What was the percent distribution of grades, that is, % A, % B, and so on?

64. A class of 84 students had a final grade distribution of 18% A, 25% B, 32% C, 13% D, 12% F. How many students received each grade?

65. A water solution that is 28.0% sucrose by mass has a density of 1.118 g/mL. What mass of sucrose, in grams, is contained in 2.75 L of this solution?

66. A water solution containing 12.0% sodium hydroxide by mass has a density of 1.131 g/mL. What volume of this solution, in liters, must be used in an application requiring 3.50 kg of sodium hydroxide?

Advanced Exercises

67. According to the rules on significant figures, the product 99.9 × 1.008 should be expressed to three significant figures—101. Yet, in this case it would be more appropriate to express the result to *four* significant figures—100.7. Explain why. (*Hint:* Recall the marginal note on page 20.)

68. A solution used to chlorinate a home swimming pool contains 7% chlorine by mass. An ideal chlorine level for the pool is one part per million (1 ppm). (Think of 1 ppm as being 1 g chlorine per million grams of water.) If you assume densities of 1.10 g/mL for the chlorine solution and 1.00 g/mL for the swimming pool water, what volume of the chlorine solution, in liters, is required to produce a chlorine level of 1 ppm in a 18,000-gallon swimming pool?

69. A standard kilogram mass is to be cut from a cylindrical bar of steel with a diameter of 1.50 in. The density of the steel is 7.70 g/cm³. How many inches long must the section be? (1 in. = 2.54 cm.)

70. The volume of seawater on Earth is about 330,000,000 mi³. If seawater is 3.5% sodium chloride by mass and has a density of 1.03 g/mL, what is the approximate mass of sodium chloride, in tons, dissolved in the seawater on Earth? (1 ton = 2000 lb, 1 lb = 453.6 g.)

71. The diameter of metal wire is often referred to by its American wire gauge number. A 16-gauge wire has a diameter of 0.05082 in. What length of wire, in meters, is there in a 1.00-lb spool of 16-gauge copper wire? The density of copper is 8.92 g/cm³.

72. Magnesium metal can be extracted from seawater by the Dow process (described in Section 22-2). Magnesium occurs in seawater to the extent of 1.4 g magnesium per kilogram of seawater. The annual production of magnesium in the United States is about 10⁵ tons. If all this magnesium were extracted from seawater, what volume of seawater, in cubic meters, would have to be processed? (1 ton = 2000 lb, 1 lb = 453.6 g. Assume a density of 1.025 g/mL for seawater.)

73. A typical rate of deposit of dust ("dustfall") from unpolluted air is 10 tons per square mile per month. **(a)** What is this dustfall, expressed in milligrams per square meter per hour? **(b)** If the dust has an average density of 2 g/cm³, how long would it take to accumulate a layer of dust 1 mm thick?

74. The volume of irrigation water is usually expressed in acre-feet. One acre-foot is a volume of water sufficient to cover 1 acre of land to a depth of 1 ft (640 acres = 1 mi²). The principal lake in the California Water Project is Lake Oroville, whose water storage capacity is listed as 3.54 × 10⁶ acre-feet. Express the volume of Lake Oroville in **(a)** cubic feet; **(b)** cubic meters; **(c)** gallons. (1 mi = 5280 ft; 1 ft = 12 in., 1 in. = 2.54 cm, 3.785 L = 1 gal.)

75. A Fahrenheit and a Celsius thermometer are immersed in the same medium, whose temperature is to be measured. At what Celsius temperature will the numerical reading on the Fahrenheit thermometer be
(a) the same as that on the Celsius thermometer?

(b) twice that on the Celsius thermometer?

(c) one-eighth that on the Celsius thermometer?

(d) 300° more than that on the Celsius thermometer?

76. The accompanying illustration shows a 100.0-mL graduated cylinder half-filled with 8.0 g of diatomaceous earth, a material consisting mostly of silica and used as a filtering medium in swimming pools. How many milliliters of water are required to fill the cylinder to the 100.0-mL mark? The diatomaceous earth is insoluble in water and has a density of 2.2 g/cm³.

77. The simple device pictured here, a pycnometer, is used for precise density determinations. From the data presented, together with the fact that the density of water at 20 °C is 0.99821 g/mL, determine the density of methanol, in grams per milliliter.

empty:	filled with water	filled with methanol
25.601 g	at 20 °C: 35.552 g	at 20 °C: 33.490 g

78. A pycnometer (see Exercise 77) weighs 25.60 g empty and 35.55 g when filled with water at 20 °C. The density of water at 20 °C is 0.9982 g/mL. When 10.20 g lead is placed in the pycnometer and the pycnometer again filled with water at 20 °C, the total mass is 44.83 g. What is the density of the lead, in grams per cubic centimeter?

Feature Problems

A. In an attempt to determine any possible relationship between the year in which a U.S. penny was minted and its current mass, students weighed an assortment of pennies and obtained the following data.

1968	1973	1977	1980	1982	1983	1985
3.11	3.14	3.13	3.12	3.12	2.51	2.54
3.08	3.06	3.10	3.11	2.53	2.49	2.53
3.09 g	3.07 g	3.06 g	3.08 g	2.54 g	2.47 g	2.53 g

What valid conclusion(s) might they have drawn about the relationship between the masses of the pennies within a given year and from year to year?

B. The Greek mathematician Archimedes is credited with discovering an important principle in the 3rd century B.C. that is useful in density determinations. The story told is that King Hiero of Syracuse (in Sicily) asked Archimedes to verify that an ornate crown made for him by a goldsmith consisted of pure gold and not a gold-silver alloy. He had to do this, of course, without damaging the crown in any way. Tell how Archimedes did this, or if you don't know the rest of the story, rediscover Archimedes' principle and tell how it can be used to settle the question.

C. The Galileo thermometer shown in the photograph is based on the dependence of density on temperature. The liquid in the outer cylinder and the liquid in the partially filled floating glass balls are the same, except that a colored dye has been added to the liquid in the balls. Can you explain how the Galileo thermometer works?

D. The canoe gliding gracefully along the water in the photograph is made of concrete. Concrete has a density of about 2.4 g/cm³. Explain why the canoe does not sink.

Atoms and the Atomic Theory

2

*O*ur current knowledge of the structure of atoms comes mostly from the field of physics. Beginning over 200 years ago, however, discoveries made by chemists led to the first theory of atomic structure. After a brief survey of these early chemical discoveries, we will describe physical evidence leading to the modern picture of a *nuclear* atom: protons and neutrons combined into a nucleus with electrons found outside this nucleus.

We will also introduce the Periodic Table as the primary means of organizing elements into groups with similar properties. Finally, we will introduce the Avogadro constant and the concept of the mole, which are the principal tools for counting atoms and molecules and measuring amounts of substances. We will use these tools throughout the text.

Individual carbon atoms in graphite. The hypothesis that all matter is made up of atoms has existed for over two thousand years. It is only within the last 30 years, however, that techniques have been developed that can make individual atoms visible.

2-1   Early Chemical Discoveries and the Atomic Theory

Chemistry is a rather mature science. Important chemicals, such as sulfuric acid (oil of vitriol), nitric acid (aqua fortis), and sodium sulfate (Glauber's salt) were all well known and used several hundred years ago. Before the end of the eighteenth century, the principal gases of the atmosphere—nitrogen and oxygen—had been isolated, and natural laws had been proposed describing the physical behavior of gases. Yet, chemistry cannot be said to have entered the modern age until the process of combustion was explained. In this section we explore the direct link between the explanation of combustion and Dalton's atomic theory.

Law of Conservation of Mass

The process of combustion—burning—is so familiar it is hard to realize what a difficult riddle this posed for early scientists. Some of the difficult-to-explain observations are described in Figure 2-1.

▶ Lavoisier's data for this experiment are presented in Feature Problem A, page 63.

In 1774, Antoine Lavoisier (1743–1794) performed an experiment in which he heated a sealed glass vessel containing a sample of tin and some air. He found that the mass before heating (glass vessel + tin + air) and after heating (glass vessel + "tin calx" + remaining air) were the same. Through further experiments he showed that the product of the reaction, "tin calx" (we now call it tin oxide), consisted of the original tin together with a portion of the air. Experiments like this proved to Lavoisier that oxygen from air is essential to combustion and also led him to formulate the **law of conservation of mass:** *The total mass of substances present after a chemical reaction is the same as the total mass of substances before the reaction.* Stated another way, this law says that matter can neither be created nor destroyed in a chemical reaction.

FIGURE 2-1

Two combustion reactions.
The apparent product of the combustion of the match—the ash—weighs *less* than the match. The product of the combustion of the magnesium ribbon (the "smoke") weighs *more* than the ribbon. Actually, in each case the total mass remains *unchanged*. To understand this requires knowing that oxygen gas enters into both combustions, and that water and carbon dioxide are also products of the combustion of the match.

EXAMPLE 2-1

Applying the Law of Conservation of Mass. A 0.455-g sample of magnesium is allowed to burn in 2.315 g oxygen gas. The sole product is magnesium oxide. After the reaction the mass of unreacted oxygen is 2.015 g. What mass of magnesium oxide was produced?

Solution

To answer this question you need to identify the substances present before and after the reaction. The total mass is unchanged.

mass before reaction = 0.455 g magnesium + 2.315 g oxygen = 2.770 g

mass after reaction = ? g magnesium oxide + 2.015 g oxygen = 2.770 g

? g magnesium oxide = 2.770 g − 2.015 g = 0.755 g

Practice Example A: A 0.382-g sample of magnesium reacts with 2.652 g nitrogen gas. The sole product is magnesium nitride. After the reaction the mass of unreacted nitrogen is 2.505 g. What mass of magnesium nitride was produced?

Practice Example B: A 7.12-g sample of magnesium is heated with 1.80 g of bromine. All the bromine is used up, and 2.07 g of magnesium bromide is the only product. What mass of magnesium remains *unreacted?*

Law of Constant Composition

In 1799, Joseph Proust (1754–1826) reported, "One hundred pounds of copper, dissolved in sulfuric or nitric acids and precipitated by the carbonates of soda or potash, invariably gives 180 pounds of green carbonate."* This and similar observations became the basis of the **law of constant composition,** or the **law of definite proportions:** *All samples of a compound have the same composition—the same proportions by mass of the constituent elements.*

To see how the law of constant composition works, consider the compound water. Water is made up of two atoms of hydrogen (H) for every atom of oxygen (O), a fact that can be represented symbolically by a *chemical formula,* the familiar H_2O. The two samples described below have the same proportions of the two elements, expressed as percentages by mass. To determine the percent by mass of hydrogen, for example, simply divide the mass of hydrogen by the sample mass and multiply by 100%. For each sample you will obtain the same result: 11.19% H.

The mineral *malachite* is basic copper carbonate and has the same composition (the very same percent copper, for example) as did the basic copper carbonate prepared by Proust in 1799.

Sample A	Composition	Sample B
10.000 g		27.000 g
1.119 g H	% H = 11.19	3.021 g H
8.881 g O	% O = 88.81	23.979 g O

EXAMPLE 2-2

Using the Law of Constant Composition. A 0.100-g sample of magnesium, when combined with oxygen, yields 0.166 g of magnesium oxide. A second magnesium sample with a mass of 0.144 g is also combined with oxygen. What mass of magnesium oxide is produced from this second sample?

Solution

We can use data from the first experiment to establish the proportion of magnesium in magnesium oxide.

$$\frac{0.100 \text{ g magnesium}}{0.166 \text{ g magnesium oxide}}$$

This proportion must be the same in *all* samples. Note that for the second sample we must invert the factor before we use it, because we must convert from mass of magnesium to mass of magnesium oxide.

$$0.144 \text{ g magnesium} \times \frac{0.166 \text{ g magnesium oxide}}{0.100 \text{ g magnesium}} = 0.239 \text{ g magnesium oxide}$$

▶ The mass of magnesium oxide must be greater than the mass of magnesium. The conversion factor must have a value greater than 1.

Practice Example A: Use information provided in the example to determine the mass of magnesium contained in 0.500 g of magnesium oxide.

Practice Example B: We wish to make exactly 2.00 g of magnesium oxide. What masses of magnesium and oxygen must we combine to do this?

*The substance Proust worked with is actually a more complex substance called *basic* copper carbonate. Proust's results were valid because, like all compounds, basic copper carbonate has a definite composition.

DALTON COLLECTING MARSH FIRE GAS

John Dalton (1766–1844)—developer of the atomic theory. Dalton has not been considered a particularly good experimenter, perhaps because of his color blindness (a condition now called daltonism). However, he did skillfully use the data of others in formulating his atomic theory.

Dalton's Atomic Theory

From 1803 to 1808, John Dalton, an English schoolteacher, used the two fundamental laws of chemical combination that we have just described as the basis of an atomic theory. His theory involved three assumptions.

1. Each chemical element is composed of minute, indestructible particles called atoms. Atoms can neither be created nor destroyed during a chemical change.

2. All atoms of an element are alike in mass (weight) and other properties, but the atoms of one element are different from those of all other elements.

3. In each of their compounds, different elements combine in a simple numerical ratio: for example, one atom of A to one of B (AB), or one atom of A to two of B (AB$_2$), or. . . .

If atoms of an element are indestructible (assumption 1), then the *same* atoms must be present after a chemical reaction as before. The total mass remains unchanged. Dalton's theory explains the law of conservation of mass. If all atoms of an element are alike in mass (assumption 2), and if atoms unite in *fixed* numerical ratios (assumption 3), the percent composition of a compound must have a unique value, regardless of the origin of the sample analyzed. Dalton's theory also explains the law of constant composition.

The characteristic masses of the atoms of an element became known as atomic weights, and throughout the nineteenth century chemists worked at establishing reliable values of relative atomic weights. Mostly, however, chemists directed their attention to discovering new elements, synthesizing new compounds, developing techniques for analyzing materials, and, in general, building up a vast body of chemical facts. Efforts to unravel the structure of the atom became the focus of physicists, as we see in the next several sections.

2-2   Electrons and Other Discoveries in Atomic Physics

Fortunately, we can acquire a qualitative understanding of atomic structure without having to retrace all the discoveries that preceded atomic physics. We do, however, need a few key ideas about the interrelated phenomena of electricity and magnetism, which we briefly discuss here with the help of Figures 2-2 and 2-3, and consider in more detail in Appendix B.

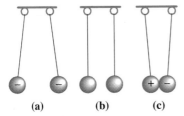

(a)   **(b)**   **(c)**

FIGURE 2-2
Forces between electrically charged objects.
The objects in (a) both possess a property of negative electric charge. Objects carrying a like charge repel each other. The objects in (b) lack the property of electric charge and exert no forces on each other. The objects in (c) carry opposite charges—one positive and one negative—and attract each other.

In the center of Figure 2-2 two small objects are shown suspended by thin threads from a support. The objects, because they have mass, hang straight down due to the force of gravity, which draws them as close to the center of Earth as possible. Certain objects, in addition to mass, may display a property called electric charge, which can be either of two types: *positive* or *negative*. Two charged objects exert a force on each other. Objects with like charges (either both positive or both negative) repel each other, as represented in Figure 2-2a. As shown in Figure 2-2c, objects with unlike charges (one positive and one negative) attract each other. The force (F) of attraction or repulsion is *directly* proportional to the product of the magnitudes or quantities of the charges (Q_1 and Q_2) and *inversely* proportional to the square of the distance between them (r^2). In mathematical terms

$$F \propto \frac{Q_1 \cdot Q_2}{r^2}$$

A positive force is a force of repulsion, and a negative force is one of attraction.

As we learn in this section, all objects of matter are made up of charged particles. An electrically *neutral* object has equal numbers of positively and negatively charged particles and carries no net charge. If the number of positive charges exceeds the number of negative charges, an object has a net *positive* charge. If negative charges exceed positive charges in number, an object has a net *negative* charge.

Figure 2-3 shows how charged particles behave when they move through the field of a magnet. They are deflected from their straight-line path into a curved path in a plane perpendicular to the field. Think of the field or region of influence of the magnet as represented by a series of invisible lines ("lines of force") running from the north pole to the south pole of the magnet.

The Discovery of Electrons

CRT, the abbreviation for cathode ray tube, has become a familiar acronym. The CRT is the heart of the familiar computer monitor or TV set. The first cathode ray tube was made by Michael Faraday (1791–1867) about 150 years ago. In passing electricity through evacuated glass tubes, Faraday discovered **cathode rays,** a type of radiation emitted by the negative terminal or *cathode* that crossed the evacuated tube to the positive terminal or *anode*. Later scientists found that cathode rays travel in straight lines and have properties that are independent of the cathode material (that is, whether it is iron, platinum, etc.). As suggested by Figure 2-4, cathode rays are invisible, and they can be detected only by the light emitted by materials that they strike. (*Fluorescence* is the term used to describe

FIGURE 2-3
Effect of a magnetic field on charged particles.
When charged particles travel through a magnetic field in such a way that their path is perpendicular to the field, they are deflected by the field. Negatively charged particles are deflected in one direction, and positively charged particles are deflected in the opposite direction. Several of the phenomena described in this section depend on this behavior.

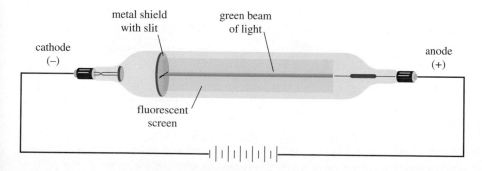

metal shield with slit

green beam of light

cathode (−)

anode (+)

fluorescent screen

High voltage source

FIGURE 2-4
A cathode-ray tube.
The high-voltage source of electricity creates a negative charge on the electrode at the left (cathode) and a positive charge on the electrode at the right (anode). Cathode rays pass from the cathode, through the slit in the metal shield, and to the anode. The rays are visible only through the green fluorescence they produce along the zinc sulfide-coated screen. They are invisible in other parts of the tube.

FIGURE 2-5
Deflection of cathode rays in a magnetic field.
The beam of cathode rays is deflected as it travels from left to right in the field of the magnet. The deflection corresponds to that expected of negatively charged particles.

the emission of light by a material when it is struck by energetic radiation.) Another significant observation about cathode rays is that they are deflected by electric and magnetic fields in the manner expected for *negatively charged* particles (see Figure 2-5).

In 1897, by the method outlined in Figure 2-6, J. J. Thomson (1856–1940) established the ratio of mass (*m*) to electric charge (*e*) for cathode rays, that is, *m/e*. Also, Thomson concluded that cathode rays are negatively charged *fundamental* particles of matter found in *all* atoms. Cathode rays subsequently became known as **electrons,** a term first proposed by George Stoney in 1874.

Robert Millikan (1868–1953) determined the electronic charge *e* through a series of "oil drop" experiments (1906–1914), described in Figure 2-7. The currently accepted value of the electronic charge *e* (to five significant figures) is -1.6022×10^{-19} C. By combining this value with an accurate value of the mass-to-charge ratio for an electron, we get the mass of an electron: 9.1094×10^{-28} g.

Once the electron was seen to be a fundamental particle of matter found in all atoms, atomic physicists began to speculate on how these particles were in-

▶ Some of Millikan's original data are presented in Feature Problem B, page 63.

▶ The coulomb (C) is the SI unit of electric charge (see also Appendix B).

FIGURE 2-6
Determining the mass-to-charge ratio, *m/e*, for cathode rays.
The cathode-ray beam strikes the end screen, undeflected, if the forces exerted on it by the electric and magnetic fields are counterbalanced. By knowing the strengths of the electric and magnetic fields, together with other data, a value of *m/e* can be obtained. Precise measurements yield a value of -5.6857×10^{-9} gram per coulomb. (Because cathode rays carry a negative charge, the sign of the mass-to-charge ratio is also negative.)
C, cathode; A, anode (perforated to allow the passage of a narrow beam of cathode rays); E, electrically charged condenser plates; M, magnet; F, fluorescent screen.

FIGURE 2-7
Millikan's oil drop experiment.
Ions (charged atoms or molecules) are produced by energetic radiation such as X-rays (X). Some of these ions become attached to oil droplets, giving them a net charge.

The fall of a droplet in the electric field between the condenser plates is speeded up or slowed down, depending on the magnitude and sign of the charge on the droplet. By analyzing data from a large number of droplets, Millikan concluded that the magnitude of the charge, q, on a droplet is an *integral* multiple of the electronic charge, e. That is, $q = ne$ (where $n = 1, 2, 3, \ldots$).

corporated into atoms. The commonly accepted model was that proposed by J. J. Thomson. Thomson thought that the positive charge needed to counterbalance the negative charges of electrons in a neutral atom was in the form of a nebulous cloud. Electrons, he suggested, floated in this diffuse cloud of positive charge (rather like a lump of Jello with electron "fruit" embedded in it). This model became known as the "plum pudding" model because of its similarity to a popular English dessert. The "plum pudding" model is illustrated in Figure 2-8 for a neutral atom and for atomic species, called *ions,* which carry a net charge.

X-rays and Radioactivity

Cathode ray research had many important spin-offs. In particular, two natural phenomena of immense theoretical and practical significance were discovered in the course of other investigations.

In 1895, Wilhelm Roentgen (1845–1923) noticed that when cathode ray tubes were operating, certain materials *outside* the tubes would glow or fluoresce. He showed that this fluorescence was caused by radiation emitted by the cathode ray tubes. Because of the unknown nature of this radiation, Roentgen coined the term *X-ray.* We now recognize the X-ray as a form of high-energy electromagnetic radiation, which is discussed in Chapter 9.

Antoine Henri Becquerel (1852–1908) associated X-rays with fluorescence and wondered if naturally fluorescent materials produce X-rays. To test this idea he wrapped a photographic plate with black paper, placed a coin on the paper, covered the coin with a uranium-containing fluorescent material, and exposed the entire assembly to sunlight. When he developed the film, a clear image of the coin could be seen. The fluorescent material had emitted radiation (presumably X-rays) that penetrated the paper and exposed the film. On one occasion, because the sky was overcast, Becquerel placed the experimental assembly inside a desk drawer for a few days while waiting for the weather to clear. On resuming the experiment, Becquerel decided to replace the original photographic film, expecting that it may have become slightly exposed. He did develop the original film, however, and found that instead of the expected feeble image he got a very sharp one. The film had become strongly exposed. The material had emitted radiation continuously, even when it was not fluorescing. Becquerel had discovered **radioactivity.**

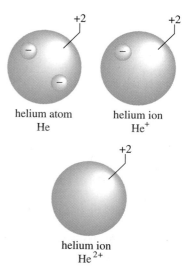

FIGURE 2-8
The "plum pudding" atomic model.
According to this model, a hydrogen atom consists of a "cloud" of positive charge (+1) and one electron (−1). Helium would have a +2 cloud and two electrons (−2). If a helium atom loses one electron, the atom becomes charged and is called an *ion.* This ion, referred to as He+, has a net charge of +1. If the helium atom loses both electrons, the He^{2+} ion forms. Ions are discussed further on page 44.

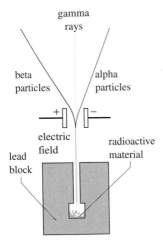

FIGURE 2-9
Three types of radiation from radioactive materials.
The radioactive material is enclosed in a lead block. All the radiation except that passing through the narrow opening is absorbed by the lead. When this radiation is passed through an electric field, it splits into three beams. One beam is undeflected—these are gamma (γ) rays. A second beam is attracted to the negatively charged plate. These are the positively charged alpha (α) particles. The third beam, of negatively charged beta (β) particles, is deflected toward the positive plate.

▶ Perhaps because he found sitting in the dark and counting spots of light on a zinc sulfide screen tedious, Geiger was motivated to develop an automatic radiation detector. The result was the well-known Geiger counter.

Ernest Rutherford (1871–1937) identified two types of radiation from radioactive materials, alpha (α) and beta (β). **Alpha particles** carry two fundamental units of positive charge and have the same mass as helium atoms. Alpha particles are identical to He^{2+} ions (see Figure 2-8). **Beta particles** are negatively charged particles produced by changes occurring within the nuclei of radioactive atoms and have the same properties as electrons. A third form of radiation that is not affected by an electric field was discovered by Paul Villard (1900). This radiation, called **gamma rays** (γ), is not made up of particles; it is electromagnetic radiation of extremely high penetrating power. These three forms of radioactivity are illustrated in Figure 2-9.

By the early 1900s additional radioactive elements were discovered, principally by Marie and Pierre Curie. Rutherford and Frederick Soddy made another profound finding: The chemical properties of a radioactive element *change* as it undergoes radioactive decay. This observation suggests that radioactivity involves fundamental changes at the *subatomic* level—that in radioactive decay one element is changed into another, a process known as *transmutation*.

2-3 The Nuclear Atom

In 1909, Rutherford, with his assistant Hans Geiger, began a line of research using alpha particles as probes to study the inner structure of atoms. Based on Thomson's "plum pudding" model (recall Figure 2-8), Rutherford expected that a beam of alpha particles would pass through thin sections of matter largely undeflected. He believed, however, that some alpha particles would be slightly scattered or deflected as they encountered electrons. By studying these scattering patterns, he hoped to deduce something about the distribution of electrons in atoms.

The apparatus used for these studies is pictured in Figure 2-10. Alpha particles were detected by the flashes of light they produced when they struck a zinc sulfide screen mounted on the end of a telescope. When Geiger and Ernst Marsden, a student, bombarded very thin foils of gold with α-particles, here is what they observed:

- The majority of α-particles penetrated the foil undeflected.
- Some α-particles experienced slight deflections.
- A few (about one in every 20,000) suffered rather serious deflections as they penetrated the foil.
- A similar number did not pass through the foil at all but "bounced back" in the direction from which they had come.

FIGURE 2-10
The scattering of α-particles by metal foil.
The telescope travels in a circular track around an evacuated chamber containing the metal foil. Most α-particles pass through the metal foil undeflected, but some are deflected through large angles.

The large-angle scattering greatly puzzled Rutherford. As he commented some years later, this observation was "about as credible as if you had fired a 15-inch shell at a piece of tissue paper and it came back and hit you." By 1911, though, Rutherford had an explanation. He based his explanation on a model of the atom known as the *nuclear* atom and having these features:

1. Most of the mass and all of the positive charge of an atom are centered in a very small region called the *nucleus. The atom is mostly empty space.*

2. The magnitude of the positive charge is different for different atoms and is approximately one half the atomic weight of the element.

3. There exist as many electrons outside the nucleus as there are units of positive charge on the nucleus. The atom as a whole is electrically neutral.

Rutherford's initial expectation and his explanation of the α-particle experiments are described in Figure 2-11.

Protons and Neutrons

Rutherford's nuclear atom suggested the existence of positively charged fundamental particles of matter in the nuclei of atoms. Rutherford himself discovered these particles, called **protons,** in 1919, in studies involving the scattering of α-particles by nitrogen atoms in air. The protons were freed as a result of collisions between α-particles and the nuclei of nitrogen atoms. At about this same

FIGURE 2-11 **Explaining the results of α-particle scattering experiments.**
(a) Rutherford's expectation. Small, positively charged α-particles should pass through the nebulous, positively charged cloud of the Thomson atomic model largely undeflected. Some would be slightly deflected by passing near electrons (present to neutralize the positive charge of the cloud). **(b)** Rutherford's explanation. With an atomic model based on a small, dense, positively charged nucleus and extranuclear electrons, one would expect the four different types of paths actually observed:
1. undeflected straight-line paths exhibited by most of the α-particles.
2. slight deflections of α-particles passing close to electrons.
3. severe deflections of α-particles passing close to a nucleus.
4. reflections from the foil of α-particles approaching a nucleus "head-on."

(a)   (b)

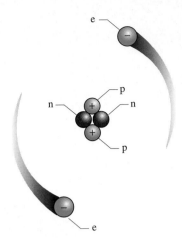

FIGURE 2-12
The nuclear atom—illustrated by the helium atom.
In this drawing electrons are shown much closer to the nucleus than is the case. The actual situation is more like this: If the entire atom were represented by a room, 5 m × 5 m × 5 m, the nucleus would only occupy about as much space as the period at the end of this sentence.

time, Rutherford predicted the existence in the nucleus of electrically neutral fundamental particles. In 1932, James Chadwick showed that a newly discovered penetrating radiation consisted of beams of *neutral* particles. These particles, called **neutrons,** originated from the nuclei of atoms. Thus, it has been only for about the past 65 years that we have had the atomic model suggested by Figure 2-12.

Fundamental Particles: A Summary

In this and the preceding section we described three fundamental particles of matter: electrons, protons, and neutrons. Table 2.1 presents the charges and masses of these three fundamental particles in two ways.

An electron carries an atomic unit of negative charge. A proton carries an atomic unit of positive charge. The **atomic mass unit** (described more fully on page 45) is defined as exactly 1/12 of the mass of the atom known as carbon-12 ("carbon twelve"). An atomic mass unit is referred to by the abbreviation, amu, and the unit, u. As we see from Table 2.1, the proton and neutron masses are just slightly greater than 1 u. By comparison the mass of an electron is seen to be only about 1/2000th the mass of the proton or neutron.

The number of protons in an atom is called the **atomic number,** or the **proton number, Z.** In an atom, which must be electrically neutral, the number of electrons is also equal to Z. The total number of protons and neutrons in an atom is called the **mass number, A.** The number of neutrons, the **neutron number,** is $A - Z$.

2-4 Chemical Elements

Now that we have acquired some fundamental ideas about atomic structure, we can more thoroughly discuss the concept of chemical elements that we introduced in the opening pages of this text.

All the atoms of an element have the same atomic number, Z. That is, atoms with the same number of protons are atoms of the same element. At present the elements with atomic numbers ranging from $Z = 1$ to 112 are known. Each element has a name and a distinctive symbol. For most elements the **chemical symbol** is an abbreviation of the English name and consists of one or two letters. The first (but never the second) letter of the symbol is capitalized; for example: carbon, C; oxygen, O; nitrogen, N; sulfur, S; neon, Ne; and silicon, Si. Some elements known since ancient times have symbols based on their Latin names, such as Fe for iron (*ferrum*) and Pb for lead (*plumbum*).

TABLE 2.1	Properties of Three Fundamental Particles			
	Electric Charge		**Mass**	
	SI (C)	**Atomic**	**SI (g)**	**Atomic (u)[a]**
proton	$+1.602 \times 10^{-19}$	$+1$	1.673×10^{-24}	1.0073
neutron	0	0	1.675×10^{-24}	1.0087
electron	-1.602×10^{-19}	-1	9.109×10^{-28}	0.0005486

[a] u is the SI unit for atomic mass unit (abbreviated as amu).

The element sodium has the symbol Na, based on the Latin *natrium* for sodium carbonate. Potassium has the symbol K, based on the Latin *kalium* for potassium carbonate. The symbol for tungsten, W, is based on the German *wolfram*.

Elements beyond uranium ($Z = 92$) do not occur naturally and must be synthesized in particle accelerators (described in Chapter 26). Elements of the very highest atomic numbers have only been produced on a limited number of occasions, a few atoms at a time. Inevitably, controversies have arisen about which research team discovered a new element and, in fact, whether a discovery was made at all. Although some international discussion continues regarding the names of the elements beyond $Z = 103$, the American Chemical Society has officially adopted names and symbols for these elements, and these are the ones we use in the inside front cover.

Isotopes

To represent the composition of any particular atom we need to specify the number of protons (p), neutrons (n), and electrons (e) in the atom. We can do this with the symbolism

$$\text{number p + number n} \searrow \quad {}^{A}_{Z}E \longleftarrow \text{symbol of element} \qquad (2.1)$$
$$\text{number p} \nearrow$$

This symbolism indicates that the atom is of the element E. It has an atomic number Z and a mass number A. For example, an atom of aluminum represented as ${}^{27}_{13}Al$ has 13 protons and 14 neutrons in its nucleus and 13 electrons outside the nucleus. (Recall that an atom has the same number of electrons as protons.)

Contrary to what Dalton thought, we now know that atoms of an element do not necessarily all have the same mass. In 1912, J. J. Thomson measured the mass-to-charge ratios of positive ions formed in neon gas. He found that about 91% of the atoms had one mass and that the remaining atoms were about 10% heavier. All neon atoms have ten protons in their nuclei, and most have ten neutrons as well. A very few neon atoms, however, have 11 neutrons and some have 12. We can represent these three different types of neon atoms as

$$ {}^{20}_{10}Ne \qquad {}^{21}_{10}Ne \qquad {}^{22}_{10}Ne $$

Two or more atoms having the *same* atomic number (Z) but *different* mass numbers (A) are called **isotopes.** Of all Ne atoms on Earth, 90.48% are ${}^{20}_{10}Ne$. The percentages of ${}^{21}_{10}Ne$ and ${}^{22}_{10}Ne$ are 0.27% and 9.26%, respectively. These percentages—90.48%, 0.27%, 9.26%—are called the **percent natural abundances** of the three neon isotopes. Sometimes the mass numbers of isotopes are incorporated into the names of elements, such as neon-20 (read this as "neon twenty"). Some elements, as they exist in nature, consist of just a single type of atom and therefore do not have naturally occurring isotopes.* Aluminum, for example, consists only of aluminum-27 atoms.

▶ Because neon is the only element with $Z = 10$, the symbols ${}^{20}Ne$, ${}^{21}Ne$, and ${}^{22}Ne$ convey the same meaning as ${}^{20}_{10}Ne$, ${}^{21}_{10}Ne$, and ${}^{22}_{10}Ne$.

▶ Percent natural abundances of isotopes are given on a *number* basis, not a mass basis. Thus, 905 out of every 1000 neon atoms are neon-20 atoms.

*****Nuclide** is the general term used to describe an atom with a particular atomic number and mass number. Although there are several elements with only one naturally occurring nuclide, it is possible to produce additional nuclides of these elements—isotopes—by artificial means (Section 26-3). The artificial isotopes are radioactive, however. In all, the number of synthetic isotopes exceeds the number of naturally occurring ones several fold.

Ions

When an atom either loses or gains electrons, the species formed is called an **ion** and carries a net charge. Because an electron is negatively charged, adding one or more electrons to an electrically neutral atom produces a negatively charged ion. Removing one or more electrons results in a positively charged ion. The number of protons never changes when an atom becomes an ion. For example, $^{20}Ne^+$ and $^{22}Ne^{2+}$ are ions. The first one has ten protons, ten neutrons, and *nine* electrons. The second one also has ten protons, but 12 neutrons and *eight* electrons. The charge on an ion is equal to the number of protons *minus* the number of electrons. That is

$$\text{number p + number n} \searrow\atop\nearrow\ \ \ ^A_Z E^{\pm ?} \longleftarrow \text{number p} - \text{number e} \qquad (2.2)$$
$$\text{number p}$$

Another example is the $^{16}O^{2-}$ ion. In this ion there are eight protons (atomic number 8), eight neutrons (mass number − atomic number), and *ten* electrons ($8 - 10 = -2$).

EXAMPLE 2-3

Relating the Numbers of Protons, Neutrons, and Electrons in Atoms and Ions. **(a)** Indicate the number of protons, neutrons, and electrons in $^{35}_{17}Cl$. **(b)** Write an appropriate symbol for the species consisting of 29 protons, 34 neutrons, and 27 electrons.

Solution

In using the symbolism $^A_Z E$ pay particular attention to whether the species is a neutral atom or an ion. If it is a neutral atom, number e = number p = Z (the atomic number). If the species is an ion, determine whether the number of electrons is smaller (positive ion) or larger (negative ion) than the number of protons. Whether the species is an atom or ion, the number of neutrons is equal to $A - Z$.

a. $^{35}_{17}Cl$: $Z = 17$, $A = 35$, and no charge indicated: a neutral atom.

number p = 17   number e = 17   number n = $A - Z$ = 35 − 17 = 18

b. The element with $Z = 29$ is copper (Symbol: Cu, see inside front cover). The mass number A = number p + number n = 29 + 34 = 63. Because the species has only 27 electrons, it must be an ion with a net charge = number p − number e = 29 − 27 = +2. It should be represented as $^{63}_{29}Cu^{2+}$.

Practice Example A: Write an appropriate symbol for the species with 47 protons, 61 neutrons, and 47 electrons.

Practice Example B: Determine the numbers of protons, neutrons, and electrons in an ion of sulfur-35 that carries the charge 2−.

Isotopic Masses

▶ Ordinarily we expect like-charged objects (such as protons) to repel each other. The forces holding protons and neutrons together in the nucleus are very much stronger than ordinary electrical forces (Section 26-6).

We cannot determine the mass of an individual atom just by adding up the masses of its fundamental particles. When an atomic nucleus is created from protons and neutrons, a small quantity of the original mass is converted to energy. This is the nuclear binding energy that holds the protons and neutrons together, and we cannot predict exactly how much this will be. Determining the masses of individual atoms, then, is something that must be done *by experiment*, in the following way:

velocity selector
plates

photographic plate

198   200   202

196   199   201   204

mass spectrum of
mercury vapor

FIGURE 2-13
A mass spectrometer.
A gaseous sample is ionized by bombardment with electrons in the lower part of the apparatus (not shown). The positive ions thus formed are subjected to an electrical force by the electrically charged velocity selector plates and a magnetic force by a magnetic field perpendicular to the page. Only ions with a particular velocity pass through and are deflected into circular paths by the magnetic field. Ions with different masses strike the detector (here a photographic plate) in different regions. The more ions of a given type, the greater the response of the detector (intensity of line on the photographic plate).

We *arbitrarily* choose one atom and assign it a certain mass. By international agreement this standard is an atom of the isotope carbon-12, which is assigned a mass of *exactly* 12 atomic mass units, that is, *12 u.* Next, we determine the masses of other atoms relative to the value of 12 u for carbon-12. To do this we use a **mass spectrometer.** In this device a beam of gaseous ions separates into components of differing mass as it passes through electric and magnetic fields. The separated ions are focused on a measuring instrument, which records their presence and amounts. Figure 2-13 illustrates mass spectrometry. A typical mass spectrum is pictured in Figure 2-14.

Although mass numbers are whole numbers, the *actual* masses of individual atoms (in atomic mass units, u) are never *exact* whole numbers, except for carbon-12. However, they are very close in value to the corresponding mass numbers. This means, for example, that we should expect the mass of oxygen-16 to be very nearly 16 u. In Example 2-4 we see that it is.

▶ This definition also establishes that one atomic mass unit (1 u) is *exactly* $\frac{1}{12}$ the mass of a carbon-12 atom.

▶ For a discussion of an earlier atomic mass scale based on oxygen, see Feature Problem C, page 63.

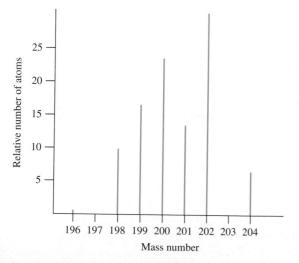

FIGURE 2-14
Mass spectrum for mercury.
The response of the ion detector in Figure 2-13 (intensity of lines on photographic plate) has been converted to a scale of relative numbers of atoms. The percent natural abundances of the mercury isotopes are ^{196}Hg, 0.146%; ^{198}Hg, 10.02%; ^{199}Hg, 16.84%; ^{200}Hg, 23.13%; ^{201}Hg, 13.22%; ^{202}Hg, 29.80%; ^{204}Hg, 6.85%.

EXAMPLE 2-4

Establishing Isotopic Masses by Mass Spectrometry. With mass spectral data the ratio of the mass of ^{16}O to ^{12}C is found to be 1.33291. What is the mass of an ^{16}O atom?

Solution

The ratio of the masses is $^{16}O/^{12}C = 1.33291$. The mass of the ^{16}O atom is 1.33291 times the mass of ^{12}C.

$$\text{mass of } ^{16}O = 1.33291 \times 12 \text{ u} = 15.9949 \text{ u}$$

Practice Example A: By mass spectrometry an atom of ^{16}O is found to have 1.06632 times the mass of ^{15}N. What is the mass of an ^{15}N atom, expressed in atomic mass units?

Practice Example B: What is the ratio of masses for $^{202}Hg/^{12}C$, if the isotopic mass for ^{202}Hg is 201.970617 u?

2-5 Atomic Masses

▶ Carbon-14, used for radiocarbon dating, is formed in the upper atmosphere. The amount of carbon-14 on Earth is too small to affect the atomic mass of carbon.

In a table of atomic masses the value listed for carbon is 12.011, yet the atomic mass standard is *exactly* 12. Why the difference? The atomic mass standard is based on a sample of carbon containing *only* atoms of carbon-12, whereas *naturally occurring* carbon contains some carbon-13 atoms as well. The existence of these two isotopes causes the *observed* atomic mass to be greater than 12. The **atomic mass (weight)**[*] of an element is the average of the isotopic masses, *weighted* according to the naturally occurring abundances of the isotopes of the element. In a "weighted" average we must assign greater importance—give greater weight—to the quantity that occurs more frequently. Since carbon-12 atoms are much more abundant than carbon-13, the weighted average must lie much closer to 12 than to 13. This is the result that we get by applying the following general equation, where the right hand side of the equation includes one term for each naturally occurring isotope.

$$
\begin{array}{l}
\text{at. mass} \\
\text{of an} \\
\text{element}
\end{array}
=
\left(
\begin{array}{l}
\text{fractional} \quad\quad \text{mass of} \\
\text{abundance of} \times \text{isotope} \\
\text{isotope (1)} \quad\quad (1)
\end{array}
\right)
+
\left(
\begin{array}{l}
\text{fractional} \quad\quad \text{mass of} \\
\text{abundance of} \times \text{isotope} \\
\text{isotope (2)} \quad\quad (2)
\end{array}
\right)
+ \cdots
$$

(2.3)

The mass spectrum of carbon shows that 98.892% of carbon atoms are carbon-12 with a mass of 12 u, *exactly,* and 1.108% are carbon-13 with a mass of 13.00335 u. In the set-up below we calculate the contribution of each isotope to the weighted average separately and then add these contributions together, as required by Equation (2.3).

$$
\begin{array}{l}
\text{contribution to} \\
\text{at. mass by } ^{12}C
\end{array}
=
\begin{array}{l}
\text{fraction of carbon} \\
\text{atoms that are } ^{12}C
\end{array}
\times \text{mass } ^{12}C \text{ atom}
$$

$$= 0.98892 \quad\quad \times 12 \text{ u} \quad\quad = 11.867 \text{ u}$$

---

[*]Since Dalton's time atomic masses have been called "atomic weights." They still are by most chemists. Yet, what we are describing here is mass, not weight. Old habits die hard.

$$\begin{aligned}\text{contribution to}\\ \text{at. mass by } ^{13}\text{C}\end{aligned} = \begin{aligned}\text{fraction of carbon}\\ \text{atoms that are } ^{13}\text{C}\end{aligned} \times \text{mass } ^{13}\text{C atom}$$

$$= 0.01108 \qquad \times 13.00335 \text{ u} \quad = 0.1441 \text{ u}$$

$$\begin{aligned}\text{at. mass of naturally}\\ \text{occurring carbon}\end{aligned} = (\text{contribution by } ^{12}\text{C}) + (\text{contribution by } ^{13}\text{C})$$

$$= \qquad 11.867 \text{ u} \qquad + 0.1441 \text{ u}$$
$$= 12.011 \text{ u}$$

▶For an analogy comparing a weighted-average atomic mass to the grade-point average of an academic record, see Feature Problem D, page 63.

To determine the atomic mass of an element having three naturally occurring isotopes, such as potassium, we would have to combine three contributions to the weighted average, and so on.

The percent natural abundances of the elements remain remarkably constant from one sample of matter to another. For example, the proportions of ^{12}C and ^{13}C atoms are the same whether we analyze a sample of pure carbon (diamond), carbon dioxide gas, or a mineral form of calcium carbonate (calcite). We can treat all natural carbon-containing materials as if there were a single *hypothetical* type of carbon atom with a mass of 12.011 u. This means that once weighted-average atomic masses have been determined and tabulated,* we can simply use these values in calculations requiring atomic masses.

In addition to being able to do calculations, it is also important to have a *qualitative* understanding of the relationship between isotopic masses, percent natural abundances, and weighted-average atomic masses, as illustrated in Example 2-5. The Practice Examples accompanying Example 2-5 provide additional applications of Equation (2.3).

EXAMPLE 2-5

Understanding the Meaning of a Weighted-Average Atomic Mass. The two naturally occurring isotopes of lithium, lithium-6 and lithium-7, have masses of 6.01513 and 7.01601 u, respectively. Which of these two occurs in greater abundance?

Solution

From a table of atomic masses (inside the front cover) we see that of lithium is 6.941 u. Because this value—a weighted-average atomic mass—is much closer to 7.01601 than to 6.01513, lithium-7 must be the more abundant isotope.

Practice Example A: The two naturally occurring isotopes of boron, boron-10 and boron-11, have masses of 10.012937 u and 11.009305 u, respectively. Which of these two occurs in greater abundance?

Practice Example B: Use the weighted-average atomic mass of 6.941 u in Equation (2.3) to determine the percent natural abundances of lithium-6 and lithium-7. [*Hint:* Let x be the *fractional* abundance of one isotope and $1 - x$, the other. Solve the equation for x, and convert fractional to percent abundances (multiply by 100%).]

*Atomic mass (atomic weight) values in tables are often written without units, especially if they are referred to as *relative* atomic masses. This simply means that the values listed are in relation to *exactly* 12 (rather than 12 u) for carbon-12. We will use the atomic mass unit (u) when referring to atomic masses (atomic weights). Most chemists do.

Are You Wondering . . .

Why some atomic masses (for example, F = 18.9984032 u) are stated so much more precisely than others?

There is *one* naturally occurring type of fluorine atom: fluorine-19. Determining the atomic mass of fluorine means establishing the mass of this type of atom as precisely as possible. On the other hand, krypton, with an atomic mass 83.80 u, has *six* naturally occurring isotopes. Because the percent distribution of the isotopes of krypton may differ very slightly from one sample to another, we can't state the weighted-average atomic mass of krypton with high precision.

2-6   Introduction to the Periodic Table

Scientists spend a lot of time organizing information into useful patterns. Before they can organize information, however, they must possess it, and it must be correct. Botanists had enough information about plants to organize their field in the eighteenth century. Because of uncertainties in atomic masses and because several elements remained undiscovered, chemists were not able to organize the elements until a century later.

We can distinguish one element from all others through its particular set of observable physical properties. For example, sodium has a low density of 0.971 g/cm³ and a low melting point of 97.81 °C. No other element has this same combination of density and melting point. Potassium, though, also has a low density (0.862 g/cm³) and low melting point (63.65 °C), much like sodium. Sodium and potassium further resemble each other in that both are good conductors of heat and electricity. Moreover, both react vigorously with water to liberate hydrogen gas. Gold, on the other hand, has a density (19.32 g/cm³) and melting point (1064 °C) that are very much higher than those of sodium or potassium, and it does not react with water or even with ordinary acids. Gold does resemble sodium and potassium in its ability to conduct heat and electricity, however. Chlorine is very different still from sodium, potassium, and gold. It is a gas under ordinary conditions, which means that the melting point of solid chlorine is far below room temperature (−101 °C). Also, chlorine is a nonconductor of heat and electricity.

Even from these very limited data, we get an inkling of a useful classification scheme of the elements. If the scheme is to group together elements with similar properties, then sodium and potassium should appear in the same group. And, if the classification scheme is in some way to distinguish between elements that are good conductors of heat and electricity and those that are not, chlorine should be set apart from sodium, potassium, and gold. The classification system we seek is the one shown in Figure 2-15 (and inside the front cover) and known as the **periodic table** of the elements. In a later chapter we will describe how the periodic table was formulated, and we will also learn its theoretical basis. For the present we will consider only a few features of the table.

Periodic Table of the Elements

FIGURE 2-15
Periodic Table of the Elements.
Atomic masses are relative to carbon-12. For certain radioactive elements the numbers listed (in parentheses) are the mass numbers of the most stable isotopes. Metals are shown in orange, nonmetals in blue, and metalloids in green. The noble gases (also non-metals) are shown in purple.

Features of the Periodic Table

Elements in the periodic table are listed according to increasing atomic number, starting at the upper left and arranged in a series of horizontal rows. This arrangement places similar elements in vertical **groups,** or **families**. For example, sodium and potassium are found together in a group labeled 1A (called the alkali metals). We should expect other members of the group, such as cesium and rubidium, to have similar properties to sodium and potassium. Chlorine is found at the other end of the table in a group labeled 7A. Some of the groups are given distinctive names, mostly related to an important property of the elements in the group. For example, the Group 7A elements are called the halogens, a term derived from Greek and meaning "salt former."

Each element is listed by placing its symbol in the middle of a box in the table. The atomic number (Z) of the element is shown above the symbol and the weighted-average atomic mass of the element is shown below its symbol. Some periodic tables provide other information, such as density and melting point, but the atomic number and atomic mass are generally sufficient for our needs. Elements with atomic masses in parentheses, such as plutonium, Pu (244), are produced synthetically, and the number shown is the mass number of the most stable isotope.

It is customary also to divide the elements into two broad categories known as **metals** and **nonmetals.** In Figure 2-15 colored backgrounds are used to distinguish the metals (orange) from the nonmetals (blue and purple). Among the features of metals are that they are solids at room temperature (except mercury, a liquid). Metals generally are malleable (capable of being flattened into thin sheets) and ductile (capable of being drawn into fine wires). They are good conductors of heat and electricity and have a lustrous or shiny appearance. Nonmetals generally have the "opposite" properties of metals, such as being poor conductors of heat and electricity. Several of the nonmetals, such as nitrogen, oxygen, and chlorine, are gases at room temperature. Some, like silicon and sulfur, are brittle solids. One, bromine, is a liquid.

Two other highlighted categories in Figure 2-15 are a special group of nonmetals known as the *noble gases* (purple), and a small group of elements, often called *metalloids* (green), that have some metallic and some nonmetallic properties. We will apply the metal/nonmetal designations often in this text.

The *horizontal* rows of the table are called **periods.** (The periods are numbered at the extreme left in the periodic table inside the front cover.) The first period of the table consists of just two elements, hydrogen and helium. This is followed by two periods of eight elements each, lithium through neon and sodium through argon. The fourth and fifth periods contain 18 elements each, ranging from potassium through krypton and from rubidium through xenon. The sixth period is a long one of 32 members. To fit this period to a table that is held to a maximum width of 18 members, we extract 14 members of the period and place them at the bottom of the table. This series of 14 elements follows lanthanum ($Z = 57$) and these elements are called the **lanthanides.** The seventh and final period is incomplete (some members are yet to be discovered), but it is known to be a long one. A 14-member series is also extracted from the seventh period and placed at the bottom of the table. Because the elements in this series follow actinium ($Z = 89$), they are called the **actinides.**

Currently, three different systems are used to designate the vertical groups in the periodic table, and we will describe the relationship between them in Chapter 10. Two systems are shown in the periodic table inside the front cover, but in this text we will use Arabic numerals (1 through 8) with the letters A and B. Thus we refer to Groups 1A, 2A, . . . , and 1B, 2B, . . . , and so on. The designation "A" refers to *main-group* elements, whereas "B" indicates *transition* elements, terms we will explore shortly.

Useful Relationships from the Periodic Table

The periodic table helps chemists describe and predict the properties of chemical compounds and the outcomes of chemical reactions. Throughout this text we will use it as an aid in understanding chemical concepts. One application of the table worth mentioning here is how it can be used to predict likely charges on simple ions.

Main-group elements are those in "A Groups," from Group 1A through 8A. When they form ions, the main-group metal atoms generally lose the same number of electrons as the group number. Thus, Na atoms (Group 1A) tend to lose one electron to become Na^+, and Ca atoms (Group 2A) tend to lose two electrons to become Ca^{2+}.

When they form ions, nonmetal atoms tend to gain electrons. The number of electrons a nonmetal atom gains is normally 8 minus the group number. Thus, an O atom (Group 6A) gains $8 - 6 = 2$ electrons to become O^{2-}, and a Cl atom (Group 7A) gains $8 - 7 = 1$ electron to become Cl^-. The "8 minus group number" rule suggests that an atom of Ne (Group 8A) gains no electrons: $8 - 8 = 0$. The very limited tendency of the noble gas atoms to form ions is one of several characteristics of this family of elements that we will consider from time to time.

The Group-B elements are the **transition elements,** and because all of them are metals, they are also called the **transition metals.** Like the main-group metals, the transition metals form positive ions, but the number of electrons lost is generally not related to the group-B number. Moreover, most transition metals are able to form two or more ions of differing charge. We will discuss transition metal ions in more detail in several later chapters.

EXAMPLE 2-6

Describing Relationships Based on the Periodic Table. Refer to the periodic table on the inside front cover and indicate
a. The element that is in Group 4A and the fourth period.
b. Two elements with properties similar to those of molybdenum (Mo).
c. The ion most likely formed from a strontium atom.

Solution

a. The elements in the fourth period range from K ($Z = 19$) to Kr ($Z = 36$). Those in Group 4A are C, Si, Ge, Sn, and Pb. The only element that is common to both of these groupings is Ge ($Z = 32$).
b. Molybdenum is in Group 6B. Two other members of this group that should resemble it are chromium (Cr) and tungsten (W).
c. Strontium (Sr) is in Group 2A. It should form the ion Sr^{2+}.

Practice Example A: Write a symbol for the ion most likely formed by an atom of Li, S, Ra, F, I, and Al.

Practice Example B: Classify each of the following elements as being a main-group or transition element. Also, identify which are metals, metalloids, and nonmetals: Na, Re, S, I, Kr, Mg, U, Si, B, Al, As, H.

2-7   The Avogadro Constant and the Concept of the Mole

Our most frequent use of the periodic table will undoubtedly be for its listing of atomic masses. As we will learn in subsequent chapters, atomic masses are essential to determining the compositions of chemical compounds and the quantities of substances that are produced in chemical reactions. In this section, though, we encounter another concept that is just as fundamental as atomic mass.

Starting with Dalton, chemists have recognized the importance of relative numbers, as in the statement that *two* hydrogen atoms and *one* oxygen atom combine to form *one* molecule of water. Yet we cannot count atoms in the usual sense. We must resort to some other measurement, usually mass. This means that we need a relationship between the *measured* mass of an element and some *known* but *uncountable* number of atoms. Consider a practical example of mass substituting for a desired number of items: If you want to nail down new floor boards on the deck of a mountain cabin, you need a certain number of nails. You do not attempt to count out the number of nails you need however, you buy them by the pound.

The SI quantity that describes an amount of substance by relating it to a number of particles of that substance is called the *mole* (abbreviated *mol*). A **mole** is an amount of substance that contains the same number of elementary entities as there are carbon-12 atoms in *exactly* 12 g of carbon-12. The "number of elementary entities (atoms, molecules, . . .)" referred to in this definition is the **Avogadro constant,** N_A (also called Avogadro's number).

$$N_A = 6.022137 \times 10^{23} \text{ mol}^{-1}$$

▶ The most precise value for N_A is obtained by X-ray diffraction studies of crystals, as we will see in Chapter 13.

Often we will round off the value of N_A to 6.022×10^{23} mol^{-1}, or even to 6.02×10^{23} mol^{-1}. The unit "mol^{-1}" signifies that the entities being counted are those present in one mole.

If a substance contains atoms of a single isotope, we may write

$$1 \text{ mol } ^{12}\text{C} = 6.02214 \times 10^{23} \text{ }^{12}\text{C atoms} = 12.0000 \text{ g}$$

▶ We established the atomic mass of ^{16}O relative to that of ^{12}C in Example 2-4.

$$1 \text{ mol } ^{16}\text{O} = 6.02214 \times 10^{23} \text{ }^{16}\text{O atoms} = 15.9949 \text{ g} \quad \text{and so on.}$$

Most elements are composed of mixtures of two or more isotopes. The atoms to be "counted out" to yield one mole are not all of the same mass. They must be taken in their naturally occurring proportions. Thus, in 1 mol of carbon most of the atoms are carbon-12 but some are carbon-13. In 1 mol of oxygen most of the atoms are oxygen-16 but some are oxygen-17 and some, oxygen-18. As a result,

▶ We calculated the weighted-average atomic mass of carbon on page 47.

$$1 \text{ mol of } carbon = 6.02214 \times 10^{23} \text{ C atoms} = 12.011 \text{ g}$$

$$1 \text{ mol of } oxygen = 6.02214 \times 10^{23} \text{ O atoms} = 15.9994 \text{ g} \quad \text{and so on.}$$

▶ Note that molar mass has the unit g/mol.

We can easily establish the mass of one mole of atoms, called the **molar mass,** $\mathcal{M}$, from a table of atomic masses—for example, 6.941 g Li/mol Li. Figure 2-16 illustrates the meaning of one mole of atoms of an element, and Figure 2-17 pictures one mole each of four common elements.

FIGURE 2-16
An attempt to picture one mole of atoms.
One mole comprises an enormously large number of atoms.
(a) There is only one type of fluorine atom, ^{19}F (shown in red).
(b) In chlorine, 75.77% of the atoms are ^{35}Cl (red) and the remainder are ^{37}Cl (blue).
(c) Magnesium has one principal isotope, ^{24}Mg (red), and two minor ones, ^{25}Mg (gray) and ^{26}Mg (blue).
(d) Lead has four naturally occurring isotopes: 1.4% ^{204}Pb (yellow), 24.1% ^{206}Pb (blue), 22.1% ^{207}Pb (gray), and 52.4% ^{208}Pb (red).

(a) 6.02214×10^{23} F atoms
$= 18.9984$ g

(b) 6.02214×10^{23} Cl atoms
$= 35.4527$ g

(c) 6.02214×10^{23} Mg atoms
$= 24.3050$ g

(d) 6.02214×10^{23} Pb atoms
$= 207.2$ g

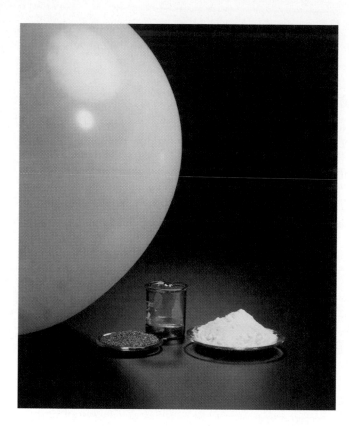

FIGURE 2-17
One mole of an element.
The watch glasses contain one mole of copper atoms (left) and one mole of sulfur atoms (right). The beaker contains one mole of mercury atoms as liquid mercury, and the balloon contains one mole of helium atoms in the gaseous state.

Thinking About the Avogadro Constant

Avogadro's constant (6.02214×10^{23}) is an enormously large number and practically inconceivable in terms of ordinary experience. Suppose that the piles of spheres in Figure 2-16 were garden peas instead of atoms. If the typical pea had a volume of about 0.1 cm^3, the required pile to form "one mole of peas" would cover the United States to a depth of about 6 km (4 mi). Or, imagine that a mole of objects, say grains of wheat, could be counted at the rate of 100 per minute. A given individual might be able to count out about 4 billion objects in a lifetime. Even so, if all the people currently on Earth were to spend their lives counting grains of wheat they could not reach Avogadro's constant. In fact, if all the people who ever lived on Earth had spent their lifetimes counting grains of wheat, the total would still be far less than Avogadro's constant. (And one mole of grains of wheat is far more wheat than has been produced in human history.) Now consider a much more efficient counting device, a modern supercomputer; it is capable of counting about 1 billion per second. The task of counting out Avogadro's constant would still take about 20 million years!

   Avogadro's constant is clearly not a useful number for counting ordinary objects. On the other hand, when this inconceivably large number is used to count inconceivably small objects, such as atoms and molecules, the result is quantities of materials that are easily within our grasp.

2-8   Using the Mole Concept in Calculations

Throughout the text, the mole concept will provide us with conversion factors for problem-solving situations. As we encounter each new situation we will explore how the mole concept applies. For now, we will deal with the relationship between numbers of atoms and the mole. Consider the statement: 1 mol S = 6.022×10^{23} S atoms = 32.07 g S. This allows us to write the conversion factors

$$\frac{1 \text{ mol S}}{6.022 \times 10^{23} \text{ S atoms}} \quad \text{and} \quad \frac{32.07 \text{ g S}}{1 \text{ mol S}}$$

We use these two factors in Example 2-8. Example 2-7 is perhaps the simplest possible application of the mole concept: relating the number of atoms in a sample to the number of moles of atoms.

In calculations requiring the Avogadro constant, students often ask when to multiply and when to divide by N_A. One answer is always to use the constant in a way that gives you the proper cancellation of units. Another answer is to think in terms of the expected result. In calculating a number of atoms, we expect the answer to be a *very large* number, and certainly *never* smaller than 1. The number of *moles* of atoms, on the other hand, is generally a number of more modest size. It will often be smaller than 1.

EXAMPLE 2-7

Relating Number of Moles and Total Number of Atoms. A sample of iron metal is described as being 2.35 mol Fe. How many iron atoms are present in this sample?

Solution

The conversion factor we need is based on the fact that 1 mol Fe = 6.022×10^{23} Fe atoms.

▶ From this point on we will not routinely show the cancellation of units. However, you should always assure yourself that the proper cancellation does occur.

$$? \text{ Fe atoms} = 2.35 \text{ mol Fe} \times \frac{6.022 \times 10^{23} \text{ Fe atoms}}{1 \text{ mol Fe}}$$

$$= 1.42 \times 10^{24} \text{ Fe atoms}$$

Practice Example A: How many atoms of gold are present in 5.07×10^{-3} mol Au?

Practice Example B: How many lead-206 atoms are present in 8.27×10^{-3} mol Pb? [*Hint:* What proportion of all Pb atoms are ^{206}Pb (see Figure 2-16)?]

EXAMPLE 2-8

Relating Number of Atoms of an Element to an Amount in Moles and a Mass in Grams. **(a)** How many moles of sulfur are present in a sample containing 7.65×10^{22} S atoms? **(b)** What is the mass of this sample?

Solution

a. As in Example 2-7, we need the Avogadro constant as a conversion factor, but here the factor is *inverted* from the form used there.

$$? \text{ mol S} = 7.65 \times 10^{22} \text{ S atoms} \times \frac{1 \text{ mol S}}{6.022 \times 10^{23} \text{ S atoms}}$$

$$= 0.127 \text{ mol S}$$

b. We can begin with the result of part **(a)**—0.127 mol S—and use molar mass as a conversion factor.

Are You Wondering . . .

How many significant figures to carry in atomic masses or the Avogadro constant in calculations?

As a general rule, to ensure the maximum precision allowable, carry at least one more significant figure in well-known physical constants than in other measured quantities. For example, in calculating the mass of 0.127 mol of sulfur, the answers 0.127 mol S $\times$ 32.066 g S/mol S = 4.07 g S or 0.127 mol S $\times$ 32.07 g S/mol S = 4.07 g S are more accurate responses than 0.127 mol S $\times$ 32.1 g S/mol S = 4.08 g S.

$$? \text{ g S} = 0.127 \text{ mol S} \times \frac{32.07 \text{ g S}}{1 \text{ mol S}} = 4.07 \text{ g S}$$

Or, we can simply combine the set-ups for parts **(a)** and **(b)**.

$$? \text{ g S} = 7.65 \times 10^{22} \text{ S atoms} \times \frac{1 \text{ mole S}}{6.022 \times 10^{23} \text{ S atoms}} \times \frac{32.07 \text{ g S}}{1 \text{ mol S}}$$

$$= 4.07 \text{ g S}$$

Figure 2-18 shows how we can obtain this quantity of sulfur by weighing.

Practice Example A: What is the mass of 2.35×10^{24} atoms of Cu?

Practice Example B: How many He atoms are present in a 22.6-g sample of He gas? [*Hint:* This calculation is the "inverse" of part **(b)**.]

FIGURE 2-18
Measurement of 7.65×10^{22} S atoms (0.127 mol S)—Example 2-8 illustrated.
The balance is set to zero (tared) when just the weighing boat is present. The sample of sulfur weighs 4.07 g.

Example 2-9 is perhaps the most representative way in which we use the mole concept—as part of a larger problem that requires other, unrelated, conversion factors as well. One approach is to outline a series of conversions—chart a path—to get from the given to the desired information. The path in Example 2-9, from mg K $\rightarrow$ g K $\rightarrow$ mol K $\rightarrow$ K atoms $\rightarrow$ ^{40}K atoms, is represented schematically in Figure 2-19. The first step requires a conversion factor between SI mass units that you should know. The second step requires molar mass as a conversion factor. This number you can obtain from a table of atomic masses. In the third step you need Avogadro's constant, a quantity that you should remember. In the fourth step the necessary conversion factor, the fractional abundance of ^{40}K, is related to the percent natural abundance of ^{40}K, which must be given in the statement of the problem.

▶ For an interesting practical application involving a series of conversions, see Feature Problem E, page 63.

EXAMPLE 2-9

Combining Several Factors in a Calculation—Molar Mass, the Avogadro Constant, Percent Abundances. Potassium-40 is one of the few naturally occurring radioactive isotopes of elements of low atomic number. Its percent natural abundance among K isotopes is 0.012%. How many ^{40}K atoms do you ingest by drinking one cup of whole milk containing 371 mg K?

▶ The radioactivity associated with milk is part of the natural background radiation to which we are all exposed and is not considered a matter of concern.

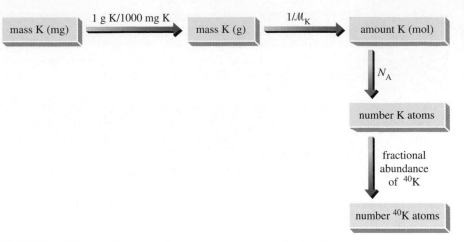

FIGURE 2-19 **Outlining a calculation—Example 2-9 visualized.**

Solution

Following the path charted in Figure 2-19, convert the mass of K to number of moles of K. For this, use molar mass in the *inverse* manner to Example 2-8.

$$? \text{ mol K} = 371 \text{ mg K} \times \frac{1 \text{ g K}}{1000 \text{ mg K}} \times \frac{1 \text{ mol K}}{39.10 \text{ g K}} = 9.49 \times 10^{-3} \text{ mol K}$$

Then convert the number of moles of K to the number of K atoms.

$$? \text{ K atoms} = 9.49 \times 10^{-3} \text{ mol K} \times \frac{6.022 \times 10^{23} \text{ K atoms}}{1 \text{ mol K}}$$

$$= 5.71 \times 10^{21} \text{ K atoms}$$

Finally, use the percent natural abundance of ^{40}K to formulate a factor to convert from number of K atoms to number of ^{40}K atoms.

$$? \,^{40}\text{K atoms} = 5.71 \times 10^{21} \text{ K atoms} \times \frac{0.012 \,^{40}\text{K atoms}}{100 \text{ K atoms}}$$

$$= 6.9 \times 10^{17} \,^{40}\text{K atoms}$$

Outlining an approach to a multistep problem, as in Figure 2-19, may also make it easier to see how to do a problem through a single set-up. Thus, we can write

$$? \,^{40}\text{K atoms} = 371 \text{ mg K} \times \frac{1 \text{ g K}}{1000 \text{ mg K}} \times \frac{1 \text{ mol K}}{39.10 \text{ g K}}$$

$$\times \frac{6.022 \times 10^{23} \text{ K atoms}}{1 \text{ mol K}} \times \frac{0.012 \,^{40}\text{K atoms}}{100 \text{ K atoms}}$$

$$= 6.9 \times 10^{17} \,^{40}\text{K atoms}$$

Practice Example A: How many Pb atoms are present in a small piece of lead with a volume of 0.105 cm³? The density of Pb = 11.34 g/cm³.

Practice Example B: Rhenium-187 is a radioactive isotope that can be used to determine the age of meteorites. A 0.100-mg sample of Re contains 2.02×10^{17} atoms of ^{187}Re. What is the percent abundance of rhenium-187 in the sample?

Summary

Modern chemistry began with eighteenth-century discoveries that led to the formulation of the law of conservation of mass and the law of constant composition, followed by Dalton's atomic theory. Important nineteenth-century developments leading to an understanding of the structure of the atom occurred mostly in the field of physics, however.

Cathode-ray research led to the discovery of the electron—a fundamental particle of all matter and the basic unit of negative electric charge. The discoveries of X-rays and radioactivity were a consequence of cathode-ray studies. The discovery of isotopes and the development of modern mass spectrometry also have their origins in cathode-ray research. Studies on the scattering of α-particles by thin metal foils led to the concept of the nuclear atom. A more complete description of the atomic nucleus was made possible by the later discovery of the proton and neutron.

By assigning a mass of *exactly* 12 u to a carbon-12 atom, the masses of other atoms can be established by mass spectrometry. From the masses of the different isotopes of an element and their percent natural abundances, the weighted-average atomic mass (weight) of an element can be determined.

It is these weighted averages that are listed in tables of atomic masses (weights).

The periodic table is a listing of the elements organized into horizontal rows (periods) and vertical columns (groups) and according to atomic number. Each group in the table consists of elements with similar physical and chemical properties. The periodic table can also be subdivided into broad categories of elements: metals, nonmetals, metalloids, and noble gases. The periodic table has many uses that are examined throughout the text. Emphasis in this chapter is on using the table as a listing of atomic masses and as an aid in writing symbols for simple ions.

The Avogadro constant, $N_A = 6.02214 \times 10^{23}$ mol^{-1}, represents the number of carbon-12 atoms in *exactly* 12 g of carbon-12. More generally, it is the number of elementary entities present in one mole of a substance. The mass of one mole of atoms of an element is called its molar mass. Molar mass and the Avogadro constant are used in a variety of calculations involving the mass, amount (in moles), or number of atoms in a sample of an element.

Summarizing Example

A stainless steel ball bearing has a radius of 6.35 mm and a density of 7.75 g/cm^3. Iron is the principal element in steel. Carbon is a key minor element. The ball bearing consists of 0.25% carbon, by mass.

Given that the percent natural abundance of ^{13}C is 1.108%, how many ^{13}C atoms are present in the ball bearing?

1. *Determine the volume of the ball bearing, in cubic centimeters.* The formula for the volume (V) of a sphere is $V = \frac{4}{3}\pi r^3$. The radius (r) must be expressed in centimeters. *Result:* 1.07 cm^3.
2. *Determine the mass of carbon present.* (a) Use the density of steel and the volume of the ball bearing to calculate its

mass. (b) Use the percent carbon in the steel to convert from mass of steel to mass of carbon. *Result:* 0.021 g C.
3. *Determine the total number of carbon atoms present.* (a) Use the molar mass of carbon to convert the mass of carbon to the amount of carbon expressed in moles. (b) Use the Avogadro constant to establish the number of carbon atoms. *Result:* 1.1 × 10^{21} C atoms.
4. *Determine the number of ^{13}C atoms.* The percent natural abundance leads to the conversion factor: 1.108 ^{13}C atoms/100 C atoms. Multiply the result of step **3** by this factor to obtain the number of ^{13}C atoms. *Answer:* 1.2 × 10^{19} ^{13}C atoms.

Key Terms

actinides (2-6)
alpha (α) particle (2-2)
atomic mass (weight) (2-5)
atomic mass unit (u) (2-3)
atomic number (proton number), Z
  (2-3)
Avogadro constant, N_A (2-7)
beta (β) particle (2-2)
cathode ray (2-2)
chemical symbol (2-4)
electron (2-2)
family (2-6)
gamma (γ) ray (2-2)

group (2-6)
ion (2-4)
isotope (2-4)
lanthanides (2-6)
law of conservation of mass (2-1)
law of constant composition (definite
  proportions) (2-1)
main group elements (2-6)
mass number, A (2-3)
mass spectrometer (2-4)
metals (2-6)
molar mass, M (2-7)
mole (2-7)

neutron (2-3)
neutron number (2-3)
nonmetals (2-6)
nuclide (2-4)
percent natural abundance (2-4)
period (2-6)
periodic table (2-6)
proton (2-3)
radioactivity (2-2)
transition elements (transition
  metals) (2-6)

Focus On

Occurrence and Abundances of the Elements

Iron is one of the most abundant elements in Earth's crust. Iron compounds give soils a characteristic red color, such as in the Navajo Sandstone Paria Canyon Wilderness Area of Arizona.

What is the most abundant element? This seemingly simple question does not have a simple answer. If we consider the entire universe, hydrogen accounts for about 90% of all the atoms and 75% of the mass, and helium accounts for most of the rest. If we consider only the elements present on Earth, iron is probably the most abundant element. However, most of this iron is believed to be in Earth's core. The currently accessible elements are those present in Earth's atmosphere, oceans, and solid continental crust. The relative abundances of some elements and common materials containing these elements are listed in Table 2.2.

Not all the known elements exist in Earth's crust. Trace amounts of neptunium ($Z = 93$) and plutonium ($Z = 94$) are found in uranium minerals, but for practical purposes elements with atomic numbers higher than 92 can be produced only artificially by nuclear processes. Moreover, most of the elements do not occur *free* in nature—that is, as the uncombined element. Only about 20% of them do. The remaining elements occur only in chemical combination with other elements.

We cannot assume that the ease and cost of obtaining a pure element from its natural sources are necessarily related to the relative abundance of the element. Aluminum is the most abundant of the metals in Earth's crust, but it cannot be produced as cheaply as iron. This is partly because concentrated deposits of iron-containing compounds— iron *ores*—are more common than are ores of aluminum. Some elements whose percent abundances are quite low are nevertheless widely used because their ores are fairly common. This is the case with copper, for example, whose abundance in Earth's crust is only 0.005%. On the other hand, some el-

Review Questions

1. In your own words define or explain the following terms or symbols: (a) $_Z^A E$; (b) β-particle; (c) isotope; (d) ^{16}O; (e) molar mass.

2. Briefly describe each of the following ideas: (a) law of conservation of mass; (b) Rutherford's nuclear atom; (c) weighted-average atomic mass; (d) radioactivity.

3. Explain the important distinctions between each pair of terms: (a) cathode rays and X-rays; (b) protons and neutrons; (c) nuclear charge and ionic charge; (d) period and group of the periodic table; (e) metal and nonmetal; (f) main group and transition element; (g) Avogadro's constant and the mole.

4. A 0.406-g sample of magnesium reacts with oxygen, producing 0.674 g magnesium oxide as the only product. What mass of oxygen was consumed in the reaction?

5. A 1.205-g sample of potassium reacts with 6.815 g chlorine to produce potassium chloride as the only product. After the reaction 3.300 g chlorine remains unreacted. What mass of potassium chloride was formed?

6. When a sample of pure sulfur is burned, there is no measurable solid residue. How can this observation conform to the law of conservation of mass?

7. If sodium and chlorine atoms combine in the proportions 1:1 in sodium chloride, what must be the percent by mass of sodium in the compound? (*Hint:* Use the tabulated atomic masses of Na and Cl.)

8. In Example 2-2 we established that the proportion of magnesium to magnesium oxide was 0.100 g magnesium/0.166 g magnesium oxide.
(a) What is the proportion of oxygen to magnesium oxide?
(b) What is the proportion of oxygen to magnesium (i.e., g O/g Mg)?
(c) What is the percent by mass of magnesium in magnesium oxide?

TABLE 2.2　Abundances of the Elements in Earth's Crust[a]

Element	Abundance, Mass (%)	Principal Materials Containing the Element
oxygen	49.3	water; silica; silicates; metal oxides; the atmosphere
silicon	25.8	silica (sand, quartz, agate, flint); silicates (feldspar, clay, mica)
aluminum	7.6	silicates (clay, feldspar, mica); oxide (bauxite)
iron	4.7	oxide (hematite, magnetite)
calcium	3.4	carbonate (limestone, marble, chalk); sulfate (gypsum); fluoride (fluorite); silicates (feldspar, zeolites)
sodium	2.7	chloride (rock salt, ocean waters); silicates (feldspar, zeolites)
potassium	2.4	chloride; silicates (feldspar, mica)
magnesium	1.9	carbonate; chloride (seawater); sulfate (Epsom salts)
hydrogen	0.7	oxide (water); natural gas and petroleum; organic matter
titanium	0.4	oxide
chlorine	0.2	common salt (rock salt, ocean waters)
phosphorus	0.1	phosphate rock; organic matter
all others[b]	0.8	

[a] Earth's crust is taken to consist of the solid crust, terrestrial waters, and the atmosphere.

[b] This figure includes C, N, and S—all essential to life—and less abundant, although commercially important, elements, such as B, Be, Cr, Cu, F, I, Pb, Sn, and Zn.

ements have a fairly high abundance but no characteristic ores of their own. They are not easily obtainable, as in the case of rubidium, the sixteenth most abundant element.

In later chapters we will describe the ways in which elements are obtained from natural sources: oxygen, nitrogen, and argon from the atmosphere; hydrogen from natural gas; magnesium and bromine from seawater; sulfur from underwater deposits; sodium and chlorine from ordinary rock salt; and several metals from their ores.

9. Samples of pure carbon weighing 3.62, 5.91, and 7.07 g were burned in an excess of air. The masses of carbon dioxide obtained (the sole product in each case) were 13.26, 21.66, and 25.91 g, respectively.
(a) Do these data establish that carbon dioxide has a fixed composition?
(b) What is the composition of carbon dioxide, expressed in %C and %O, by mass?
10. When 4.15 g magnesium and 82.6 g bromine react, we find that (1) all the magnesium is used up, (2) some bromine remains unreacted, and (3) magnesium bromide is the only product. With this information alone, is it possible to deduce the mass of magnesium bromide produced? Explain.
11. Complete the following table. What minimum amount of information is required to characterize completely an atom or ion? (Note that not all rows can be completed.)

Name	Symbol	Number protons	Number electrons	Number neutrons	Mass number
sodium	$^{23}_{11}Na$	11	11	12	23
silicon	___	___	___	14	___
___	___	37	___	___	85
___	^{40}K	___	___	___	___
___	___	___	33	42	___
___	$^{20}Ne^{2+}$	___	___	___	___
___	___	___	___	___	80
___	___	___	___	126	___

12. Arrange the following species in order of increasing **(a)** number of electrons; **(b)** number of neutrons; **(c)** mass.

$^{112}_{50}Sn$　$^{40}_{18}Ar$　$^{122}_{52}Te$　$^{59}_{29}Cu$　$^{120}_{48}Cd$　$^{58}_{27}Co$　$^{39}_{19}K$

13. All of these radioactive isotopes have applications in medicine. Write their symbols in the form $^A_Z E$. **(a)** cobalt-60; **(b)** phosphorus-32; **(c)** iodine-131; **(d)** sulfur-35.

14. For the nuclide $^{138}_{56}Ba$, express the percentage, *by number,* of the fundamental particles in the nucleus that are neutrons.

15. There are *two* principal isotopes of iridium (atomic mass = 192.22 u). One of these, ^{191}Ir, has a mass of 190.9609 u. Which of these must be the second isotope: ^{190}Ir, ^{192}Ir, ^{193}Ir? Explain.

16. An isotope with mass number 38 has two more neutrons than protons. This is an isotope of what element?

17. The following data on isotopic masses are given in a handbook. What is the ratio of each of these masses to that of $^{12}_6C$? **(a)** $^{35}_{17}Cl$, 34.96885 u; **(b)** $^{26}_{12}Mg$, 25.98259 u; **(c)** $^{222}_{86}Rn$, 222.0175 u.

18. The following ratios of masses were obtained with a mass spectrometer: $^{19}_9F/^{12}_6C = 1.5832$; $^{35}_{17}Cl/^{19}_9F = 1.8406$; $^{81}_{35}Br/^{35}_{17}Cl = 2.3140$. Determine the mass of a $^{81}_{35}Br$ atom in atomic mass units. (*Hint:* What is the mass of a ^{12}C atom?)

19. In naturally occurring argon, 99.600% of the atoms are $^{40}_{18}Ar$ with mass 39.9624 u; 0.337%, $^{36}_{18}Ar$ with mass 35.96755 u; and 0.063%, $^{38}_{18}Ar$ with mass 37.96272 u. Calculate the atomic mass of naturally occurring argon.

20. Refer to the periodic table inside the front cover and identify

(a) an element that is in both Group 3A and the fifth period;

(b) an element similar to, and one unlike, sulfur;

(c) the alkali metal in the sixth period;

(d) the halogen element in the fifth period;

(e) an element with atomic number greater than 50 that has similar properties to the element with atomic number 18;

(f) the group number of the element E that forms an ion E^{3-}.

(g) an element M that you would expect to form the ion M^{2+}.

21. Assuming that the seventh period of the periodic table is 32 members long, what should be the atomic number of the noble gas following radon (Rn)? of the alkali metal following francium (Fr)?

22. What is the total number of atoms in each of the following samples? **(a)** 12.7 mol Ca; **(b)** 0.00361 mol Ne; **(c)** 1.8×10^{-12} mol Pu.

23. Calculate the quantities indicated.

(a) the number of moles represented by 2.18×10^{26} Fe atoms

(b) the mass, in grams, of 7.71 mol Kr

(c) the mass, in mg, of a sample containing 6.15×10^{19} Au atoms

(d) the number of atoms in 112 cm^3 of Fe ($d = 7.86$ g/cm^3)

24. *Without doing detailed calculations,* indicate which of the following quantities contains the greatest number of atoms: 6.02×10^{23} Ni atoms, 25.0 g nitrogen, 52.0 g Cr, 10.0 cm^3 Fe ($d = 7.86$ g/cm^3). Explain your reasoning.

25. How many ^{204}Pb atoms are present in a piece of lead weighing 215 mg? The percent natural abundance of ^{204}Pb is 1.4%.

26. A particular lead–cadmium alloy is 8.0% cadmium by mass. What mass of this alloy, in grams, must you weigh out to obtain a sample containing 6.50×10^{23} Cd atoms?

Exercises

Law of Conservation of Mass

27. When an iron object rusts its mass increases. When a match burns its mass decreases. Do these observations violate the law of conservation of mass? Explain.

28. When a strip of magnesium metal is burned in air (recall Figure 2-1), it produces a white powder that weighs more than the original metal. When a strip of magnesium is burned in a photoflash bulb, the bulb weighs the same before and after it is flashed. Explain the difference in these observations.

29. When a solid mixture consisting of 10.500 g of calcium hydroxide and 11.125 g of ammonium chloride is strongly heated, gaseous products are evolved and 14.336 g of a solid residue remains. The gases are passed into 62.316 g of water, and the mass of the resulting solution is 69.605 g. Within the limits of experimental error, show that these data conform to the law of conservation of mass.

30. Within the limits of experimental error, show that the law of conservation of mass was obeyed in the following experiment: 10.00 g calcium carbonate (found in limestone) was dissolved in 100.0 mL hydrochloric acid ($d = 1.148$ g/mL). The products were 120.40 g solution (a mixture of hydrochloric acid and calcium chloride) and 2.22 L carbon dioxide gas ($d = 1.9769$ g/L).

Law of Constant Composition

31. In one experiment 2.18 g sodium was allowed to react with 16.12 g chlorine. All the sodium was used up, and 5.54 g sodium chloride (salt) was produced. In a second experiment 2.10 g chlorine was allowed to react with 10.00 g sodium. All the chlorine was used up, and 3.46 g sodium chloride was produced. Show that these results are consistent with the law of constant composition.

32. When 3.06 g hydrogen was allowed to react with an excess of oxygen, 27.35 g water was obtained. In a second experiment a sample of water was decomposed by electrolysis, resulting in 1.45 g hydrogen and 11.51 g oxygen. Are these results consistent with the law of constant composition? Demonstrate why or why not.

33. In one experiment the burning of 0.312 g sulfur produced 0.623 g sulfur dioxide as the sole product of the reaction. In a second experiment 0.842 g sulfur dioxide was obtained. What mass of sulfur must have been burned in the second experiment?

34. In one experiment the reaction of 1.00 g mercury and an excess of sulfur yielded 1.16 g of a sulfide of mercury as the sole product. In a second experiment the same sulfide was produced in the reaction of 1.50 g mercury and 1.00 g sulfur. **(a)** What mass of the sulfide of mercury was produced in the second experiment? **(b)** What mass of which element (mercury or sulfur) remained *unreacted?*

Fundamental Particles

35. Cite the evidence that most convincingly established that electrons are fundamental particles of all matter.

36. Why could not the same methods that had been used to characterize electrons be used to isolate and detect neutrons?

Fundamental Charges and Mass-to-Charge Ratios

37. These observations were made for a series of 5 oil drops in an experiment similar to Millikan's (see Figure 2-7). Drop 1 carried a charge of 1.28×10^{-18} C; drops 2 and 3 each carried $\frac{1}{2}$ the charge of drop 1; drop 4 carried $\frac{1}{8}$ the charge of drop 1; drop 5 had a charge four times that of drop 1. Are these data consistent with the value of the electronic charge given in the text? Could Millikan have inferred the charge on the electron from this particular series of data? Explain.

38. In an experiment similar to that described in Exercise 37, drop 1 carried a charge of 6.41×10^{-19} C; drop 2 had $\frac{1}{2}$ the charge of drop 1; drop 3 had twice the charge of drop 1; drop 4 had a charge of 1.44×10^{-18} C; and drop 5 had $\frac{1}{3}$ the charge of drop 4. Are these data consistent with the value of the electronic charge given in the text? Could Millikan have inferred the charge on the electron from this particular series of data? Explain.

39. Use data from Table 2.1 to verify the following statements. **(a)** The mass of electrons is about 1/2000 that of H atoms. **(b)** The mass-to-charge ratio, m/e, for positive ions is considerably larger than for the electron.

40. Determine the approximate value of the mass-to-charge ratio, m/e, in coulombs per gram, for the ions $^{127}_{53}I^-$ and $^{32}_{16}S^{2-}$. Why are these values only approximate?

Atomic Number, Mass Number, and Isotopes

41. For the atom ^{108}Pd, with mass 107.90389 u, determine **(a)** the numbers of protons, neutrons, and electrons in the atom; **(b)** the ratio of the mass of this atom to that of an atom of $^{12}_{6}C$.

42. For the ion $^{228}Ra^{2+}$, with a mass of 228.030 u, determine **(a)** the numbers of protons, neutrons, and electrons in the atom; **(b)** the ratio of the mass of this atom to that of an atom of ^{16}O (refer to Example 2-4).

43. An isotope of silver has a mass that is 6.68374 times that of oxygen-16. What is the mass (in u) of this isotope? (*Hint:* Refer to Example 2-4.)

44. The ratio of the masses of the two naturally occurring isotopes of indium is 1.0177:1. The heavier of the two isotopes has 7.1838 times the mass of ^{16}O. What are the masses (in u) of the two isotopes? (Refer to Example 2-4.)

45. Given the following species: $^{24}Mg^{2+}$, ^{47}Cr, $^{60}Co^{3+}$, $^{35}Cl^-$, $^{124}Sn^{2+}$, ^{226}Th, ^{90}Sr. Which of these species **(a)** has equal numbers of neutrons and protons? **(b)** has protons contributing more than 50% of the mass? **(c)** has 50% more neutrons than protons?

46. Given the same species as listed in Exercise 45, which **(a)** has equal numbers of neutrons and electrons? **(b)** has protons, neutrons, and electrons in the ratio 9:11:8? **(c)** has a number of neutrons equal to the number of protons plus one half the number of electrons?

Atomic Mass Units, Atomic Masses

47. The mass of a carbon-12 atom is taken to be *exactly* 12 u. Are there likely to be any other atoms with an *exact* integral (whole number) mass, expressed in u? Explain.

48. Which statement is probably true concerning the masses of *individual* copper atoms: that *all*, *some*, or *none* have a mass of 63.546 u? Explain.

49. There are *three* naturally occurring isotopes of magnesium. Their masses and percent natural abundances are 23.985042 u, 78.99%; 24.985837 u, 10.00%; and 25.982593 u, 11.01%. Calculate the weighted-average atomic mass of magnesium.

50. There are *four* naturally occurring isotopes of chromium. Their masses and percent natural abundances are 49.9461 u, 4.35%; 51.9405 u, 83.79%; 52.9407 u, 9.50%; and 53.9389 u, 2.36%. Calculate the weighted-average atomic mass of chromium.

51. The *two* naturally occurring isotopes of silver have the abundances: ^{107}Ag, 51.84%; ^{109}Ag, 48.16%. The mass of ^{107}Ag is 106.905092 u. What is the mass of ^{109}Ag? (*Hint:* What is the weighted-average atomic mass of Ag?)

53. The three isotopes of naturally occurring potassium are ^{39}K, 38.963707 u; ^{40}K, 39.963999 u; and ^{41}K. The percent natural abundances of ^{39}K and ^{41}K are 93.2581% and 6.7302%, respectively. Determine the isotopic mass of ^{41}K.

52. Bromine has *two* naturally occurring isotopes. One of them, bromine-79, has a mass of 78.918336 u and a natural abundance of 50.69%. What must be the mass and percent natural abundance of the other isotope, bromine-81?

54. Boron has a weighted-average atomic mass of 10.811 u. What are the percent natural abundances of its two isotopes, ^{10}B and ^{11}B? These isotopes have masses of 10.012937 u and 11.009305 u, respectively.

Mass Spectrometry

55. A mass spectrum of germanium displayed peaks at mass numbers 70, 72, 73, 74, and 76, with relative heights of 20.5, 27.4, 7.8, 36.5, and 7.8, respectively.
(a) In the manner of Figure 2-14, sketch this mass spectrum
(b) Estimate the weighted-average atomic mass of germanium and state why this result is only approximately correct.

56. Hydrogen and chlorine atoms form simple diatomic molecules with H and Cl in a 1:1 ratio, that is, HCl. The natural abundances of the chlorine isotopes are 75.53% ^{35}Cl and 24.47% ^{37}Cl. The natural abundances of ^{2}H and ^{3}H are 0.015% and less than 0.001%, respectively.
(a) How many different HCl molecules are possible and what are their mass numbers (i.e., the sum of the mass numbers of the H and Cl atoms)?
(b) Which is the most abundant of the possible HCl molecules? Which is the second most abundant?

The Avogadro Constant and the Mole

57. Determine
(a) the number of moles of Rb in a 167.0-g sample of rubidium metal;
(b) the number of Fe atoms in 363.2 kg iron;
(c) the mass of a one-trillion (1.0×10^{12}) atom sample of metallic silver.

59. How many Ag atoms are present in a piece of sterling silver jewelry weighing 38.7 g? Sterling silver contains 92.5% Ag, by mass.

61. Medical experts generally believe a level of 30 μg Pb per deciliter of blood poses a significant health risk (1 dL = 0.1 L). Express this level **(a)** in the unit mol Pb/L blood; **(b)** as the number of Pb atoms per milliliter blood.

58. Determine
(a) the number of Ar atoms in a 5.25-mg sample of argon;
(b) the molar mass, $\mathcal{M}$, of an element if the mass of a 2.80×10^{22}-atom sample of the element is 4.24 g;
(c) the mass of a sample of aluminum that contains the same number of atoms as 35.55 g of zinc.

60. What is the total number of atoms present in a 75.0-cm^3 sample of plumber's solder, an alloy containing 67% Pb and 33% Sn by mass and having a density of 9.4 g/cm^3?

62. During a severe air pollution episode the concentration of lead in air was observed to be 3.01 μg Pb/m^3. How many Pb atoms would be present in a 0.500-L sample of this air (the approximate lung capacity of a human adult)?

Advanced Exercises

63. William Prout (1815) proposed that all other atoms are built up of hydrogen atoms, suggesting that all elements should have integral atomic masses based on an atomic mass of 1 for hydrogen. This hypothesis appeared discredited by the discovery of atomic masses, such as 24.3 u for magnesium and 35.5 u for chlorine. In terms of modern knowledge, explain why Prout's hypothesis is actually quite reasonable.

64. Fluorine has a single atomic species, ^{19}F. Determine the atomic mass of ^{19}F by summing the masses of its protons, neutrons, and electrons, and compare with the value listed on the inside front cover. Explain why the agreement is poor.

65. Use 1×10^{-13} cm as the approximate diameter of the spherical nucleus of the hydrogen-1 atom, together with data from Table 2.1, to estimate the density of matter in a proton.

66. Use fundamental definitions and statements from the text to establish the fact that 6.022×10^{23} u = 1.000 g.

67. In each case, identify the element in question if

(a) the mass number of an atom is 234 and the atom has 60.0% more neutrons than protons.
(b) an ion with a 2+ charge has 10.0% more protons than electrons.
(c) an ion with a mass number of 110 and a 2+ charge has 25.0% more neutrons than electrons.

68. Identify the isotope E if its nucleus contains one more neutron than protons and the mass number is nine times larger than the charge on the ion E^{3+}.

69. Determine the only possible 2+ ion for which the following two conditions are both satisfied.
• the net ionic charge is *one tenth* the nuclear charge, and
• the number of neutrons is *four* more than the number of electrons.

70. Determine the only possible isotope (E) for which the following conditions are met.
• The mass number of E is 2.50 times its atomic number.

- The atomic number of E is equal to the mass number of another isotope (Y). In turn, the isotope Y has a neutron number that is 1.33 times the atomic number of Y and equal to the neutron number of selenium-82.

71. Suppose we redefined the atomic mass scale by arbitrarily assigning to the naturally occurring *mixture* of chlorine isotopes an atomic mass of 35.00000 u.
(a) What would be the atomic masses of helium, sodium, and iodine on this new atomic mass scale?
(b) Why do these three elements have nearly integral (whole number) atomic masses based on carbon-12 but not based on naturally occurring chlorine?

72. The two naturally occurring isotopes of nitrogen have masses of 14.0031 and 15.0001 u, respectively. Determine the percentage of ^{15}N atoms in naturally occurring nitrogen.

73. The masses of the naturally occurring mercury isotopes are ^{196}Hg, 195.9658 u; ^{198}Hg, 197.9668 u; ^{199}Hg, 198.9683 u; ^{200}Hg, 199.9683 u; ^{201}Hg, 200.9703 u; ^{202}Hg, 201.9706 u; and ^{204}Hg, 203.9735 u. Use these data, together with data from Figure 2-14, to calculate the atomic mass of mercury.

74. Silicon has one major isotope, ^{28}Si (27.97693 u, 92.21% natural abundance), and two minor ones, ^{29}Si (28.97649 u) and ^{30}Si (29.97376 u). What are the percent natural abundances of the two minor isotopes?

75. Deuterium, ^{2}H (2.0140 u), can be used to replace ordinary ^{1}H atoms in chemical studies. The percent natural abundance of ^{2}H is 0.015%. What mass of hydrogen would have to be processed to extract 1.00 g of pure ^{2}H atoms?

76. An alloy that melts at about the boiling point of water has Bi, Pb, and Sn atoms in the ratio 10:6:5, respectively. What mass of alloy contains a total of one mole of atoms?

77. A particular silver solder (used in the electronics industry to join electrical components) is to have the *atom* ratio 5.00 Ag/4.00 Cu/1.00 Zn. What masses of the three metals must be melted together to prepare 1.00 kg of the solder?

78. A low-melting Sn–Pb–Cd alloy called *eutectic alloy* is analyzed. The *mole* ratio of tin to lead is found to be 2.73 : 1.00, and the *mass* ratio of lead to cadmium is found to be 1.78 : 1.00. What is the mass percent composition of this alloy?

Feature Problems

A. The data Lavoisier obtained in the experiment described on page 34 are as follows.

Before heating: glass vessel + tin + air =
                              13 onces, 2 gros, 2.50 grains
After heating: glass vessel + tin calx + remaining air =
                              13 onces, 2 gros, 5.62 grains

How closely did Lavoisier's results conform to the law of conservation of mass? 1 livre = 16 onces; 1 once = 8 gros; 1 gros = 72 grains. In modern terms, 1 livre = 30.59 g.

B. Some of Millikan's oil drop data are shown here. The measured quantities were not actual charges on oil drops but were proportional to these charges. Show that these data are consistent with the idea of a fundamental electronic charge.

Observation	Measured quantity	Observation	Measured quantity
1	19.66	8	53.91
2	24.60	9	59.12
3	29.62	10	63.68
4	34.47	11	68.65
5	39.38	12	78.34
6	44.42	13	83.22
7	49.41		

C. Prior to 1961, physicists used as the standard for atomic weights the isotope ^{16}O, to which they assigned a value of exactly 16. At the same time chemists assigned a value of exactly 16 to the naturally occurring mixture of the isotopes ^{16}O, ^{17}O, and ^{18}O. Would you expect atomic masses listed in a 50-year old text to be the same, generally higher, or generally lower than in this text? Explain.

D. A student's grade point average is analogous to a weighted-average atomic mass. The simple average of a grade of "B" (3.0) and one of "D" (1.0) is a "C" (2.0). The grade point average, however, requires that the grades be weighted according to the numbers of units assigned to the courses being averaged. Set up an equation for grade point average similar to Equation (2.3) on page 46, and compute the grade point average for a student earning 5 units of A (4.0), 3 units of B (3.0), and 4 units of C (2.0).

E. The German chemist Fritz Haber proposed paying off the reparations imposed against Germany after World War I by extracting gold from seawater. Given that (a) the amount of the reparations was $28.8 billion dollars, (b) the value of gold at the time was about $21.25 per troy ounce (12 troy ounces = 1 lb), and (c) gold occurs in seawater to the extent of 4.67 $\times 10^{17}$ atoms per ton of seawater (1 ton = 2000 lb), how many cubic kilometers of seawater would have had to be processed to obtain the required amount of gold? Assume that the density of seawater is 1.03 g/cm^3. (Haber's scheme proved not to be commercially feasible, and the reparations were never fully repaid.)

Chemical Compounds

3

Water, ammonia, carbon monoxide, and carbon dioxide are familiar substances to almost everyone. They are all rather simple chemical compounds. Only slightly less familiar are sucrose (cane sugar), acetylsalicylic acid (aspirin), and ascorbic acid (vitamin C). They too are chemical compounds. In fact, the entire study of chemistry is mostly about chemical compounds, and in this chapter we will consider a few basic ideas about compounds.

Chemical compounds are composed of two or more different elements. We have already seen how the periodic table is used to classify the elements. Now we can use these classifications to establish broad categories for classifying compounds. Just as we represent elements by symbols, we represent compounds by chemical formulas. In much of the chapter we will learn how to use the many kinds of information incorporated into chemical formulas. Finally, we will present some general rules that will serve most of our needs in relating chemical names and formulas.

Blue crystals of copper(II) sulfate pentahydrate, $CuSO_4 \cdot 5H_2O$. Chemical compounds, their formulas, and names are topics discussed in this chapter.

3-1 Types of Chemical Compounds and Their Formulas

In Chapter 2 we learned to represent elements by their chemical symbols. To represent a compound we use a combination of symbols known as a chemical formula. Whatever other information it conveys, at a minimum the **chemical formula** of a compound indicates

- the elements present, and
- the relative numbers of atoms of each element in the compound.

In the following formula for water the elements are denoted by their symbols. The relative numbers of atoms are indicated by *subscript* numbers. (Where no subscript is written, the number 1 is understood.)

the two elements present

$$H_2O$$

lack of subscript means one atom of O per formula unit

two H atoms per formula unit

A **formula unit** is the *smallest* collection of atoms on which the formula of a compound can be based. A **molecule** of a compound is a group of bonded atoms that actually exists and can be identified as a distinct entity. The formula of water is based on a group of two hydrogen atoms and one oxygen atom (a formula unit). Also, this group of three atoms can be isolated and identified as a distinct entity (a molecule). For water, the formula unit and the molecule are one and the same. We can always identify a formula unit of a compound, but some compounds do not consist of discrete molecules, as we see next.

Ionic Compounds

Combinations of metals and nonmetals usually result in an *ionic compound*. An **ionic compound** is a compound made up of positive and negative ions joined together by electrostatic forces of attraction (recall the attraction of oppositely charged objects pictured in Figure 2-2). The atoms of metallic elements tend to lose one or more electrons when they combine with nonmetal atoms, and the nonmetal atoms tend to gain one or more electrons. As a result of this electron transfer, the metal atom becomes a positive ion (called a **cation**) and the nonmetal atom becomes a negative ion (called an **anion**). We can usually deduce the magnitude of the charge on a main group cation or anion from the group to which an element belongs (recall Section 2-6), and thus the periodic table can help us to write the formulas of ionic compounds.

In the formation of sodium chloride—ordinary table salt—each sodium atom gives up one electron to become a sodium ion, Na^+, and each chlorine atom gains one electron to become a chloride ion, Cl^-. This fact conforms to the relationship between periodic table locations of the elements and the charges on their simple ions (see page 50). For the formula unit of sodium chloride to be electrically neutral, we require one Na^+ ion for each Cl^- ion ($+1 - 1 = 0$). Thus, the formula is NaCl.

As suggested by Figure 3-1, each Na^+ ion in sodium chloride is surrounded by *six* Cl^- ions (and vice versa), and we cannot say that any one of these six Cl^- ions belongs exclusively to a given Na^+ ion. Yet, the ratio of Cl^- to Na^+ ions in sodium chloride *is* 1:1, and so we arbitrarily select a combination of one Na^+ ion and one Cl^- ion as a formula unit. Because this formula unit is buried

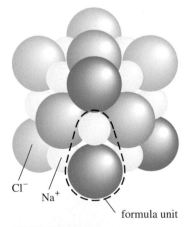

Cl^- Na^+

formula unit

FIGURE 3-1
Portion of an ionic crystal and a formula unit of NaCl.
Solid sodium chloride consists of enormous numbers of Na^+ and Cl^+ ions in a network called a crystal. The combination of one Na^+ and one Cl^+ ion is the smallest collection of ions from which we can deduce the formula NaCl. It is a formula unit.

in a vast array of ions, it does not exist as a distinct entity. It is *inappropriate* to call it a molecule of solid sodium chloride.

The situation with magnesium chloride is similar. In magnesium chloride, found as a trace impurity in table salt, magnesium atoms lose two electrons to become magnesium ions, Mg^{2+} (Mg is in Group 2A). To obtain an electrically neutral formula unit, we need two Cl^- ions, each with a charge of $1-$, for every Mg^{2+} ion with its $2+$ charge. The formula of magnesium chloride is $MgCl_2$.

The ions Na^+, Mg^{2+}, and Cl^- are *monatomic,* meaning that each consist of a single ionized atom. By contrast, a *polyatomic* ion is made up of two or more atoms. In the nitrate ion, NO_3^-, the subscripts signify that *three* O atoms and *one* N atom are joined into the single ion NO_3^-. Magnesium nitrate is an ionic compound made up of magnesium and nitrate ions. An electronically neutral formula unit of this compound must consist of *one* Mg^{2+} ion and *two* NO_3^- ions. We can denote this by writing a formula in which we enclose "NO_3" in parentheses, followed by the subscript "2": $Mg(NO_3)_2$. We will return to a discussion of polyatomic ions in Section 3-5, where we will emphasize the relationship between names and formulas.

Molecular Compounds

A formula unit consisting of four chlorine atoms bonded to a carbon atom *does* exist as a distinct entity. It is both a formula unit and a *molecule* of carbon tetrachloride, CCl_4 (see Figure 3-2). We refer to substances such as H_2O and CCl_4 as molecular compounds. A **molecular compound** is made up of discrete molecules consisting, typically, of a small number of different nonmetal atoms held together by forces known as *covalent bonds*. We will study covalent bonds in some detail in Chapters 11 and 12.

Formulas for ionic compounds simply tell us the combining ratio of the ions in the compound. For molecular compounds, on the other hand, we can write formulas that tell us more than just the combining ratio of the elements. To illustrate, consider the carbon–oxygen–hydrogen compound acetic acid, the acid constituent that gives vinegar its sour taste. Figure 3-3 shows several representations of this compound, described further on page 68.

FIGURE 3-2
A molecule of CCl_4.

FIGURE 3-3
Several representations of the compound acetic acid.
In the molecular model the black spheres are carbon; red, oxygen; and white, hydrogen. To show that one H atom in the molecule is fundamentally different from the other three, the formula of acetic acid is often written as $HC_2H_3O_2$ (see Section 5-3). To show that this H atom is bonded to an O atom, the formulas CH_3COOH and CH_3CO_2H are also used. For a few chemical compounds you can expect to encounter different versions of chemical formulas in different sources.

Molecular model:
("ball and stick")

Empirical formula: CH_2O

Molecular formula: $C_2H_4O_2$

Structural formula:

$$\begin{array}{c} \quad\ \ H\ \ \ O \\ \quad\ \ |\ \ \ \ \| \\ H-C-C-O-H \\ \quad\ \ | \\ \quad\ \ H \end{array}$$

The **empirical formula** of a compound is the simplest formula we can write; it tells us the types of atoms present and their relative numbers. The subscripts in an empirical formula are reduced to their simplest whole-number ratio. The empirical formula of a compound represented as P_4O_{10}, for example, is P_2O_5. Generally, the empirical formula does not tell us a great deal about a compound. Acetic acid, formaldehyde (used to make certain plastics and resins), and glucose (blood sugar) all have the empirical formula CH_2O.

The **molecular formula** is based on an actual molecule of the compound. In some cases the empirical and molecular formulas are identical, CH_2O in the case of formaldehyde, for example. In other cases the molecular formula is a "multiple" of the empirical formula. A molecule of acetic acid, for example, consists of eight atoms—two C atoms, four H atoms, and two O atoms. This is twice the number of atoms in the formula unit (CH_2O). The molecular formula of acetic acid is $C_2H_4O_2$. That of glucose, on the other hand, is $C_6H_{12}O_6$.

A **structural formula** shows the order in which atoms are bonded together in a molecule and by what types of bonds. Thus, the structural formula of acetic acid tells us that three of the four H atoms are bonded to one of the C atoms, the remaining H atom is bonded to an O atom. Both of the O atoms are bonded to one of the C atoms, and the two C atoms are bonded to each other. The covalent bonds in the structural formula are represented by lines or dashes (—). One of the bonds is represented by a double dash (=) and is called a *double* covalent bond. We will learn some of the differences between single and double bonds later in the text. For now, just think of a double bond as being a stronger or "tighter" bond than a single bond.

Finally, molecules occupy space; they have a three-dimensional shape. When we write empirical or molecular formulas we do not convey any information about the spatial arrangements of atoms. With a structural formula we can, at times, do some of this, but usually the only satisfactory way to represent the three-dimensional structures of molecules is by constructing models. In a "ball-and-stick" model we represent the centers of the bonded atoms by small spheres and the bonds between atoms by rods. Such models help us to visualize distances between the centers of atoms (bond lengths) and the geometrical shapes of molecules. In later discussions we will see just how profoundly the properties of a substance can be affected by the geometrical shape of its molecules. The fact that the three atoms in the water molecule, H_2O, are arranged in a triangular rather than linear (straight-line) fashion is in many ways essential to the existence of life on Earth.

▶ The triangular, or bent, shape of a water molecule is suggested by the structural formula

3-2 The Mole Concept and Chemical Compounds

Once we have written the chemical formula of a compound, it is an easy matter to describe its formula mass. **Formula mass** is the mass of a formula unit relative to a mass of exactly 12 u for carbon-12. We can base the formulas for molecular compounds on actual molecules (that is, on molecular formulas) and thus speak of the molecular mass. **Molecular mass** is the mass of a molecule relative to a mass of exactly 12 u for carbon-12.

Because atomic masses are also relative to carbon-12, we can get weighted-average formula and molecular masses just by adding up weighted-average

▶ The terms *formula weight* and *molecular weight* are often used in place of formula mass and molecular mass.

atomic masses (those on the inside front cover). Thus, for the molecular compound water, H_2O, we get

$$\text{molecular mass } H_2O = (2 \times \text{atomic mass H}) + (\text{atomic mass O})$$

$$= (2 \times 1.008 \text{ u}) + 15.999 \text{ u}$$

$$= 18.015 \text{ u}$$

For the ionic compound magnesium chloride, $MgCl_2$,

$$\text{formula mass } MgCl_2 = \text{atomic mass Mg} + (2 \times \text{atomic mass Cl})$$

$$= 24.305 \text{ u} + (2 \times 35.453 \text{ u})$$

$$= 95.211 \text{ u}$$

and for the ionic compound magnesium nitrate, $Mg(NO_3)_2$,

$$\text{formula mass } Mg(NO_3)_2 = \text{atomic mass Mg} +$$

$$2 \times [\text{atomic mass N} + (3 \times \text{atomic mass O})]$$

$$= 24.305 \text{ u} + 2 \times [14.007 \text{ u} + (3 \times 15.999 \text{ u})]$$

$$= 148.313 \text{ u}$$

▶ The terms *formula mass* and *molecular mass* have essentially the same meaning, although when referring to ionic compounds, such as NaCl and $MgCl_2$, *formula mass* is the term that should be used.

Mole of a Compound

Recall that in Chapter 2 we defined a mole as an amount of substance having the same number of elementary entities as there are carbon-12 atoms in exactly 12 g of pure carbon-12. This definition carefully avoids saying that the entities to be counted are always atoms. As a result we can apply the concept of a mole to any quantity that we can represent by a symbol or formula—atoms, ions, formula units, molecules, . . . Specifically, a *mole of compound* is an amount of compound containing Avogadro's number (6.02214×10^{23}) of formula units or molecules. The *molar mass* is the mass of one mole of compound—one mole of molecules of a molecular compound and one mole of formula units of an ionic compound.

Because the weighted-average molecular mass of H_2O is 18.015 u compared with a mass of exactly 12 u for a carbon-12 atom, when we scale up a sample size by Avogadro's number we get a mass of 18.015 g H_2O compared with exactly 12 g for carbon-12. The molar mass of H_2O is 18.015 g H_2O/mol H_2O. If we know the formula of a compound, we can equate the following terms, as illustrated for H_2O, $MgCl_2$, and $Mg(NO_3)_2$.

▶ The terms *molecular* mass and *molar* mass sound similar but they are quite different. Molecular mass is the weighted-average mass of one molecule (expressed in atomic mass units, u). Molar mass is the mass of Avogadro's number of molecules (expressed in grams per mole, g/mol).

$$1 \text{ mol } H_2O = 18.015 \text{ g } H_2O = 6.02214 \times 10^{23} \text{ H}_2\text{O molecules}$$

$$1 \text{ mol } MgCl_2 = 95.211 \text{ g } MgCl_2 = 6.02214 \times 10^{23} \text{ MgCl}_2 \text{ formula units}$$

$$1 \text{ mol } Mg(NO_3)_2 = 148.313 \text{ g } Mg(NO_3)_2 = 6.02214 \times 10^{23} \text{ Mg(NO}_3)_2 \text{ formula units}$$

Expressions such as these provide several different types of conversion factors that can be applied in a variety of problem-solving situations. The strategy that works best for a particular problem will depend, in part, on how one most readily visualizes the necessary conversions. For instance, in Example 3-1 we convert from mass to an amount in moles and then to a number of formula units. Generally, the central focus of a problem is the conversion of a mass in grams to an amount in moles (or vice versa). Often, this conversion must be preceded or followed by other conversions involving volumes, densities, percentages, and so on. As we saw in Chapter 2, one helpful tool in problem solving is to establish a "conversion pathway" (recall Figure 2-19).

▶ Feature Problem A (p. 95) presents a method for establishing the numerical value of the Avogadro constant, N_A.

EXAMPLE 3-1

Relating Molar Mass, the Avogadro Constant, and Formula Units of an Ionic Compound. An analytical balance can detect a mass of 0.1 mg. What is the total number of ions present in this minimally detectable quantity of $MgCl_2$?

Solution

After making the mass conversion, mg $\rightarrow$ g, we use the molar mass to convert from mass to number of moles of $MgCl_2$. Then, with the Avogadro constant as a conversion factor, we convert from moles to number of formula units. Our final factor uses the fact that there are *three* ions (*one* Mg^{2+} and *two* Cl^-) per formula unit (f.u.) of $MgCl_2$.

As we noted previously, it is often helpful to map out a conversion pathway that starts with the information given and proceeds through a series of conversion factors to the information sought. In Example 2-9 we did this through a diagram (Figure 2-19). Usually we can map the pathway more simply. For example, here we begin with milligrams of $MgCl_2$ and make the series of conversions: mg $\rightarrow$ g $\rightarrow$ mol $\rightarrow$ f.u. $\rightarrow$ number of ions.

$$? \text{ ions} = 0.1 \text{ mg MgCl}_2 \times \frac{1 \text{ g MgCl}_2}{1000 \text{ mg MgCl}_2} \times \frac{1 \text{ mol MgCl}_2}{95 \text{ g MgCl}_2}$$

$$\times \frac{6.0 \times 10^{23} \text{ f.u. MgCl}_2}{1 \text{ mol MgCl}_2} \times \frac{3 \text{ ions}}{1 \text{ f.u. MgCl}_2}$$

$$= 2 \times 10^{18} \text{ ions}$$

Practice Example A: Zinc oxide, ZnO, is used in sunscreen preparations. What is the total number of ions present in a 1.0-g sample of ZnO?

Practice Example B: How many grams of $MgCl_2$ should you measure out if you seek a sample with a total of 5.0×10^{23} Cl^- ions? (*Hint:* How many moles of $MgCl_2$ do you need?)

In Example 3-2, we use additional factors—for example, density—before using factors based on the concept of a mole of compound. Overall, the conversion pathway is $\mu L \rightarrow L \rightarrow mL \rightarrow g \rightarrow mol \rightarrow$ number of molecules. Again, notice that we start with a value given in the problem and use conversion factors to make a path to the desired solution.

EXAMPLE 3-2

Combining Several Factors in a Calculation Involving Molar Mass. The volatile liquid ethyl mercaptan, C_2H_6S, is one of the most odoriferous substances known. It is used in natural gas to make gas leaks detectable. How many C_2H_6S molecules are contained in a 1.0-μL sample? ($d = 0.84$ g/mL.)

▶ Given that a normal drop of liquid is about 0.05 mL, a volume of 1.0 μL is only *0.02 of a drop.*

Solution

First, we convert from *micro*liters to liters, and then to milliliters. At this point we bring in density as a conversion factor. The remaining conversions are from the mass of the substance to the number of moles and, finally, to the number of molecules.

$$? \text{ molecules } C_2H_6S = 1.0 \text{ } \mu L \times \frac{1 \times 10^{-6} \text{ L}}{1 \text{ } \mu L} \times \frac{1000 \text{ mL}}{1 \text{ L}} \times \frac{0.84 \text{ g } C_2H_6S}{1 \text{ mL}}$$

$$\times \frac{1 \text{ mol } C_2H_6S}{62.1 \text{ g } C_2H_6S} \times \frac{6.02 \times 10^{23} \text{ molecules } C_2H_6S}{1 \text{ mol } C_2H_6S}$$

$$= 8.1 \times 10^{18} \text{ molecules } C_2H_6S$$

Practice Example A: Gold has a density of 19.32 g/cm^3. A piece of gold leaf is 2.50 cm on each side and 0.100 mm thick. How many atoms of gold are in this piece of gold leaf?

Practice Example B: If 1.0 μL of liquid ethyl mercaptan, C_2H_6S ($d = 0.84$ g/mL), is allowed to evaporate and distribute itself throughout a 1500-m^3 chemistry lecture hall, will the vapor be detectable in the room? The limit of detectability is 9×10^{-4} $\mu mol/m^3$. (*Hint:* 1 μmol = 1×10^{-6} mol. How many micromoles C_2H_6S are there in 1.0 μL?)

Mole of an Element—A Second Look

In Chapter 2 we took one mole of an element to be 6.02214×10^{23} *atoms* of the element. This is the only definition possible for elements such as iron, magnesium, sodium, and copper, in which enormous numbers of individual spherical atoms are clustered together, much like marbles in a can. But in some elements, atoms of the same kind are joined together to form molecules. Bulk samples of these elements are composed of collections of molecules. A molecule of S_8 is represented in Figure 3-4. The molecular formulas of elements that you should become familiar with are

H_2 O_2 N_2 F_2 Cl_2 Br_2 I_2 P_4 S_8

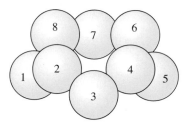

FIGURE 3-4
Sulfur atoms and a sulfur molecule.
Sulfur atoms are joined into puckered rings with eight members, S_8. In a sample of solid sulfur there are eight times as many atoms as there are molecules. Thus, 1 mol S_8 consists of 8 mol S atoms.

For these elements we can speak of an *atomic* mass or a *molecular* mass. Also, we can express molar mass in two ways. For hydrogen, for example, the atomic mass is 1.008 u and the molecular mass is 2.016 u; the molar mass can be expressed as 1.008 g H/mol H or 2.016 g H_2/mol H_2.

Are You Wondering . . .

What to use for the molar mass when dealing with "a mole of hydrogen?"

The phrase "a mole of hydrogen" is ambiguous. One must always say either a mole of hydrogen *atoms* or a mole of hydrogen *molecules*. Better still is to write 1 mol H or 1 mol H_2. If this is done, then you'll want to use 1.008 g H/mol H when dealing with H atoms and 2.016 g H_2/mol H_2 when working with H_2 molecules. This is very much like the distinction between one dozen socks and one dozen pairs of socks. A dozen socks means 12 individual socks (like 12 individual H atoms). A dozen *pairs* of socks means 12 *pairs* of socks—24 individual socks (like 24 H atoms joined into 12 H_2 molecules).

3-3   Composition of Chemical Compounds

A chemical formula conveys considerable quantitative information about a compound and its constituent elements. We have already learned how to determine the molar mass of a compound, and in this section we consider some other types of calculations based on the chemical formula.

The colorless, volatile liquid halothane has been used as a fire extinguisher and also as an inhalation anesthetic. Its empirical and molecular formulas are both $C_2HBrClF_3$; its molecular mass is 197.38 u and its molar mass is 197.38 g/mol. We can obtain this molar mass by using atomic masses from the periodic table.

$$\mathcal{M}_{C_2HBrClF_3} = 2\mathcal{M}_C + \mathcal{M}_H + \mathcal{M}_{Br} + \mathcal{M}_{Cl} + 3\mathcal{M}_F$$

$$= (2 \times 12.01)\ g + 1.01\ g + 79.90\ g + 35.45\ g + (3 \times 19.00)\ g$$

$$= 197.38\ g/mol$$

▶ The structural formula of halothane is

$$
\begin{array}{ccc}
 & H & F \\
 & | & | \\
Br- & C-C & -F \\
 & | & | \\
 & Cl & F \\
\end{array}
$$

The molecular formula of $C_2HBrClF_3$ tells us that *per mole* of halothane there are two moles of C atoms, one mole each of H, Br, and Cl atoms, and three moles of F atoms. This factual statement can be turned into conversion factors to answer questions such as, "How many C atoms are present per mole of halothane?" In this case the factor needed is 2 mol C/mol $C_2HBrClF_3$. That is,

$$? C\ atoms = 1.000\ mol\ C_2HBrClF_3 \times \frac{2\ mol\ C}{1\ mol\ C_2HBrClF_3} \times \frac{6.022 \times 10^{23}\ C\ atoms}{1\ mol\ C}$$

$$= 1.204 \times 10^{24}\ C\ atoms$$

In Example 3-3 we use a different conversion factor derived from the same formula, that is, from $C_2HBrClF_3$. This factor is shown in blue in the set-up. In addition we use other familiar factors to make our conversion pathway: mL → g → mol $C_2HBrClF_3$ → mol F.

EXAMPLE 3-3

Using Relationships Derived from a Chemical Formula. How many moles of F atoms are in a 75.0-mL sample of halothane (d = 1.871 g/mL)?

Solution

First, convert the volume of the sample to mass; this requires density as a conversion factor. The next conversion is from mass of halothane to its amount in moles; this requires the inverse of the molar mass as a conversion factor. Then comes the final conversion factor, this one based on the formula of halothane.

$$? mol\ F = 75.0\ mL\ C_2HBrClF_3 \times \frac{1.871\ g\ C_2HBrClF_3}{1\ mL\ C_2HBrClF_3}$$

$$\times \frac{1\ mol\ C_2HBrClF_3}{197.4\ g\ C_2HBrClF_3} \times \frac{3\ mol\ F}{1\ mol\ C_2HBrClF_3}$$

$$= 2.13\ mol\ F$$

▶ Feature Problem B (p. 95) considers the use of mole ratios to confirm a nineteenth-century law of chemical combination known as the law of multiple proportions.

Practice Example A: How many grams of C are contained in 75.0 mL of halothane (d = 1.871 g/mL)?

Practice Example B: How many milliliters of halothane, $C_2HBrClF_3$, must be measured out to contain 100.0 g Br? (*Hint:* How many moles and how many grams of halothane are required?)

Calculating Percent Composition
from a Chemical Formula

When a chemist believes that he or she has synthesized a new compound, a sample is generally sent to an analytical laboratory where its percent composition is determined. This experimentally determined percent composition is then compared with the percent composition calculated from the formula of the expected compound. In this way one can see if the compound obtained was indeed the one expected. Let us outline how the calculation is done.

Establish the molar mass of the compound, keeping track of the contribution of each element to the molar mass. For each element, formulate the ratio of its mass contribution to the mass of the compound as a whole. Multiply this ratio by 100% to obtain the mass percent of the element. This approach is illustrated in Example 3-4.

EXAMPLE 3-4

Calculating the Mass Percent Composition of a Compound. What is the mass percent composition of halothane, $C_2HBrClF_3$?

Solution

First, determine the molar mass of $C_2HBrClF_3$. This was shown on page 72 to be 197.38 g/mol.

Then, for one mole of compound, formulate mass ratios and percents.

$$\% \ C = \frac{(2 \times 12.01) \ g}{197.38 \ g} \times 100\% = 12.17\%$$

$$\% \ H = \frac{1.01 \ g}{197.38 \ g} \times 100\% = 0.51\%$$

$$\% \ Br = \frac{79.90 \ g}{197.38 \ g} \times 100\% = 40.48\%$$

$$\% \ Cl = \frac{35.45 \ g}{197.38 \ g} \times 100\% = 17.96\%$$

$$\% \ F = \frac{(3 \times 19.00) \ g}{197.38 \ g} \times 100\% = 28.88\%$$

Practice Example A: Calculate the mass percent composition of acetic acid, the compound featured in Figure 3-3.

Practice Example B: Adenosine triphosphate (ATP) is the main energy storage molecule in cells. Its chemical formula is $C_{10}H_{11}N_5P_3O_{13}$. What is its mass percent composition?

The percentages of the elements in a compound should add up to 100.00%, and we can use this fact in one of two ways.

1. Check the accuracy of the computations by ensuring that the percentages do total 100.00%. As applied to the results of Example 3-4,

$$12.17\% + 0.51\% + 40.48\% + 17.96\% + 28.88\% = 100.00\%$$

▶ Feature Problem C (p. 95) illustrates an application in which the composition of a compound may be expressed in terms other than the percentages of its elements.

2. Determine the percentages of all the elements but one. Obtain that one by difference (subtraction). In Example 3-4,

$$\% \text{ H} = 100.00\% - \% \text{ C} - \% \text{ Br} - \% \text{ Cl} - \% \text{ F}$$

$$= 100.00\% - 12.17\% - 40.48\% - 17.96\% - 28.88\%$$

$$= 0.51\%$$

Establishing Formulas from the Experimentally Determined Percent Composition of Compounds

At times a chemist isolates a chemical compound—say, from an exotic tropical plant—and has no idea what it is. A report from an analytical laboratory on the percent composition of the compound provides data that can be used to determine its formula.

Percent composition establishes the relative proportions of the elements in a compound on a *mass* basis. A chemical formula requires these proportions to be on a *mole* basis, that is, in terms of *numbers* of atoms. Consider the following five-step approach to determining a formula from the experimentally determined percent composition of a compound. Let's apply it to the compound 2-deoxyribose, a sugar that is a basic constituent of DNA (deoxyribonucleic acid). The mass percent composition of 2-deoxyribose is 44.77% C, 7.52% H, and 47.71% O.

1. Although we could choose any sample size, if we take one of *exactly* 100 g, the masses of the elements are numerically equal to their percentages, that is, 44.77 g C, 7.52 g H, and 47.71 g O.

2. Convert the masses of the elements in the 100.0-g sample to amounts in moles.

$$? \text{ mol C} = 44.77 \text{ g C} \times \frac{1 \text{ mol C}}{12.011 \text{ g C}} = 3.727 \text{ mol C}$$

$$? \text{ mol H} = 7.52 \text{ g H} \times \frac{1 \text{ mol H}}{1.008 \text{ g H}} = 7.46 \text{ mol H}$$

$$? \text{ mol O} = 47.71 \text{ g O} \times \frac{1 \text{ mol O}}{15.999 \text{ g O}} = 2.982 \text{ mol O}$$

3. Write a tentative formula based on the numbers of moles just determined.

$$C_{3.727}H_{7.46}O_{2.982}$$

4. Attempt to convert the subscripts in the tentative formula to small whole numbers. This requires dividing each of the subscripts by the smallest one (2.982).

$$C_{\frac{3.727}{2.982}}H_{\frac{7.46}{2.982}}O_{\frac{2.982}{2.982}} = C_{1.25}H_{2.50}O$$

5. If the subscripts at this point differ only slightly from whole numbers, round them off to whole numbers to obtain the final formula. If one or more subscripts is not a whole number, multiply all subscripts by a small whole number that will make all subscripts integral. Here, we must multiply by 4.

$$C_{(4 \times 1.25)}H_{(4 \times 2.50)}O_{(4 \times 1)} = C_5H_{10}O_4$$

The formula that we got by the method just outlined, $C_5H_{10}O_4$, is the simplest possible formula—the *empirical formula*. The actual *molecular formula* may be equal to it or to some multiple of the empirical formula, such as $C_{10}H_{20}O_8$, $C_{15}H_{30}O_{12}$, $C_{20}H_{40}O_{16}$, and so on. To find the multiplying factor, we must compare the formula mass based on the empirical formula with the true molecular mass of the compound. We can establish the molecular mass from a separate experiment (by methods introduced in Chapters 6 and 14). The experimentally determined molecular mass of 2-deoxyribose is 134 u. The formula mass based on the empirical formula, $C_5H_{10}O_4$, is 134.1 u. The measured molecular mass is the same as the empirical formula mass. The molecular formula is also $C_5H_{10}O_4$.

We apply the five-step approach outlined on page 74 in Example 3-5, where we will find that the empirical formula and the molecular formula are not the same.

EXAMPLE 3-5

Determining the Empirical and Molecular Formulas of a Compound from Its Mass Percent Composition. Dibutyl succinate is an insect repellent used against household ants and roaches. Its composition is 62.58% C, 9.63% H, and 27.79% O. Its experimentally determined molecular mass is 230 u. What are the empirical and molecular formulas of dibutyl succinate?

Solution

The first five steps are the same as those illustrated previously.

Step 1. Determine the mass of each element in a 100.0-g sample.

$$62.58 \text{ g C} \qquad 9.63 \text{ g H} \qquad 27.79 \text{ g O}$$

Step 2. Convert each of these masses to an amount in moles.

$$? \text{ mol C} = 62.58 \text{ g C} \times \frac{1 \text{ mol C}}{12.011 \text{ g C}} = 5.210 \text{ mol C}$$

$$? \text{ mol H} = 9.63 \text{ g H} \times \frac{1 \text{ mol H}}{1.008 \text{ g H}} = 9.55 \text{ mol H}$$

$$? \text{ mol O} = 27.79 \text{ g O} \times \frac{1 \text{ mol O}}{15.999 \text{ g O}} = 1.737 \text{ mol O}$$

Step 3. Write a tentative formula based on these numbers of moles.

$$C_{5.21}H_{9.55}O_{1.74}$$

Step 4. Divide each of the subscripts of the tentative formula by the smallest (1.74).

$$C_{\frac{5.21}{1.74}}H_{\frac{9.55}{1.74}}O_{\frac{1.74}{1.74}} = C_{2.99}H_{5.49}O$$

Step 5. Round off any subscripts from step 4 that differ only slightly from whole numbers

$$C_3H_{5.49}O$$

and multiply all the subscripts by a small whole number chosen to make them integral. Here, we multiply by 2: $2 \times 5.49 = 10.98 \approx 11$.

$$C_{2\times3}H_{2\times5.49}O_{2\times1} = C_6H_{10.98}O_2$$

Empirical formula: $C_6H_{11}O_2$

The empirical formula mass is $[(6 \times 12.0) + (11 \times 1.0) + (2 \times 16.0)] = 115$ u. Because the experimentally determined molecular mass (230 u) is twice the empirical formula mass, we conclude

Molecular formula: $C_{12}H_{22}O_4$

Practice Example A: Diacetoneglucose has a molecular mass of 260 u and the mass percent composition: 55.37% C, 7.75% H, and 36.88% O. What are the empirical and molecular formulas of this substance?

Practice Example B: Sorbitol, used as a sweetener in some "sugar-free" foods, has a molecular mass of 182 u and a mass percent composition: 39.56% C, 7.74% H, and 52.70% O. What are the empirical and molecular formulas of sorbitol?

Are You Wondering . . .

How much rounding off to do to get integral subscripts in an empirical formula, and what factors to use to convert fractional to whole numbers?

The rounding off that is justified depends on how precisely the elemental analysis is done. Also, the effect of rounding off excessively is more likely to produce errors in formulas of high molecular mass. As a result, there is no ironclad rule on the matter. For the examples in this text, if you carry all the significant figures allowable in a calculation, you can generally round off a subscript that is within one or two hundredths of a whole number (for example, 3.98 rounds off to 4). If the deviation is more than this, you may need to adjust subscripts to integral values by multiplying by the appropriate constant. In choosing this constant, you'll find it helpful to recognize decimal equivalents of some common fractions: $0.50 = \frac{1}{2}$; $0.333 = \frac{1}{3}$; $0.25 = \frac{1}{4}$; $0.20 = \frac{1}{5}$; and so on. For example, for the subscript 1.25 we use the factor 4. That is, $1.25 = \frac{5}{4}$, and $\frac{5}{4} \times 4 = 5$.

Combustion Analysis

Figure 3-5 illustrates an experimental method for establishing an empirical formula for compounds that are easily burned, such as compounds containing carbon and hydrogen with oxygen, nitrogen, and a few other elements. In **combustion analysis** a weighed sample of a compound is burned in a stream of oxygen gas. The water vapor and carbon dioxide gas produced in the combustion are absorbed by appropriate substances. The increases in mass of these absorbers correspond to the masses of water and carbon dioxide. We can think of the matter in this way.

$$C_xH_yO_z + O_2 \longrightarrow CO_2 + H_2O$$

All the C atoms in the sample appear as C atoms in CO_2. All the H atoms appear in the H_2O. Furthermore, the only source of these atoms is the sample being analyzed. Oxygen atoms in the CO_2 and H_2O may come from both the sam-

FIGURE 3-5
Apparatus for combustion analysis.
Oxygen gas (A) passes through the combustion tube containing the sample being analyzed (B). This portion of the apparatus is enclosed in a high-temperature furnace (C).
Products of the combustion are absorbed as they leave the furnace—water vapor by magnesium perchlorate (D) and carbon dioxide gas by sodium hydroxide (E), producing sodium
carbonate. The differences in mass of the absorbers D and E, after and before the combustion, yield the masses of H_2O and CO_2 produced in the combustion reaction.

ple and the oxygen gas consumed in the combustion. We have to determine the
quantity of oxygen in the sample indirectly. These ideas are applied in Example 3-6.

EXAMPLE 3-6

Determining an Empirical Formula from Combustion Analysis Data. Vitamin C is
essential for the prevention of scurvy (and large doses may be effective in preventing colds). Combustion of a 0.2000-g sample of this carbon–hydrogen–oxygen compound yields 0.2998 g CO_2 and 0.0819 g H_2O. What is the empirical formula of vitamin C?

Solution

We can base our calculation directly on the 0.2000-g sample. All we need to do is
determine the number of moles of C, H, and O present. Once we have done this, we
can write the empirical formula by the method presented earlier. There is a complication, however. We can easily get the numbers of moles of C and H from the quantities of CO_2 and H_2O produced. To get the number of moles of O in the sample,
however, we must know the mass of O present. And we can only get the mass of O
by difference. What all this means is that we are forced to determine the masses of
C and H, as well as their amounts in moles.

$$? \text{ mol C} = 0.2998 \text{ g CO}_2 \times \frac{1 \text{ mol CO}_2}{44.010 \text{ g CO}_2} \times \frac{1 \text{ mol C}}{1 \text{ mol CO}_2}$$

$$= 0.006812 \text{ mol C}$$

$$? \text{ g C} = 0.006812 \text{ mol C} \times \frac{12.011 \text{ g C}}{1 \text{ mol C}}$$

$$= 0.08182 \text{ g C}$$

$$? \text{ mol H} = 0.0819 \text{ g H}_2\text{O} \times \frac{1 \text{ mol H}_2\text{O}}{18.02 \text{ g H}_2\text{O}} \times \frac{2 \text{ mol H}}{1 \text{ mol H}_2\text{O}}$$

$$= 0.00909 \text{ mol H}$$

$$? \text{ g H} = 0.00909 \text{ mol H} \times \frac{1.008 \text{ g H}}{1 \text{ mol H}}$$

$$= 0.00916 \text{ g H}$$

Now we can get the mass of O in the vitamin C sample by difference,

$$? \text{ g O} = 0.2000 \text{ g sample} - 0.08182 \text{ g C} - 0.00916 \text{ g H} = 0.1090 \text{ g O}$$

and then the number of moles of O.

$$? \text{ mol O} = 0.1090 \text{ g O} \times \frac{1 \text{ mol O}}{15.999 \text{ g O}} = 0.006813 \text{ mol O}$$

As trial subscripts for the empirical formula we can write

$$C_{0.006812}H_{0.00909}O_{0.006813}$$

Next, divide each subscript by the smallest—0.006812—to obtain

$$CH_{1.33}O$$

Finally, multiply each subscript by 3 (because $3 \times 1.33 = 3.99 \approx 4.00$).

Empirical formula of vitamin C: $C_3H_4O_3$

Practice Example A: Isobutyl propionate is the substance that provides the flavor for rum extract. Combustion of a 1.152-g sample of this carbon–hydrogen–oxygen compound yields 2.726 g CO_2 and 1.116 g H_2O. What is the empirical formula of isobutyl propionate?

Practice Example B: Complete combustion of a 1.505-g sample of thiophene, a carbon–hydrogen–sulfur compound, yields 3.149 g CO_2, 0.645 g H_2O, and 1.146 g SO_2. What is the empirical formula of thiophene? (*Hint:* All the sulfur appears as SO_2.)

▶An alternative approach to establishing an empirical formula from combustion analysis data is to calculate the mass percent composition of a compound and then to proceed as in Example 3-5.

We have just seen how combustion reactions can be used to analyze chemical substances, but not all samples can be easily burned. Fortunately, several other types of reactions can be used for chemical analyses. Also, modern methods in chemistry rely much more on physical measurements with instruments than on chemical reactions. We will cite some of these methods later in the text.

3-4   Oxidation States: A Useful Tool in Describing Chemical Compounds

Most basic concepts in chemistry deal with properties or phenomena that actually can be measured. In a few instances, though, a concept is devised to achieve specific purposes, more for convenience than because of any fundamental significance. This is the case with **oxidation state** (oxidation number),* which is related to the number of electrons that an atom loses, gains, or otherwise appears to use in joining with other atoms in compounds.

Consider NaCl. In this compound an Na atom, a metal, loses one electron to a Cl atom, a nonmetal. The compound consists of the ions Na^+ and Cl^-, as pictured in Figure 3-1. Na^+ is in the oxidation state $+1$ and Cl^-, -1.

*Because oxidation state refers to a number, the term *oxidation number* is often used synonymously. We will use the two terms interchangeably.

In $MgCl_2$, an Mg atom loses two electrons to become Mg^{2+}, and each Cl atom gains one electron to become Cl^-. As in NaCl, the oxidation state of Cl is -1, but that of Mg is $+2$. If we take the *total* of the oxidation states of all the atoms (ions) in a formula unit of $MgCl_2$, we get $+2 - 1 - 1 = 0$.

In the molecule Cl_2, the two Cl atoms are identical and should have the *same* oxidation state. But if their total is to be zero, each oxidation state must itself be 0. Thus, the oxidation state of an element can vary, depending on the compounds in which it occurs. In the molecule H_2O, we *arbitrarily* assign H the oxidation state $+1$. Then, because the total of the oxidation states of the atoms must be zero, the oxidation state of oxygen must be -2.

From these examples you can see that we need some conventions or rules for assigning oxidation states. The following seven rules are sufficient to deal with most cases in this text, with this important qualification: *Whenever two rules appear to contradict each other (which they often will), follow the rule that appears higher in the list.* Some examples are given for each rule, and the rules are applied together in Example 3-7.

1. *The oxidation state (O.S.) of an individual atom in a free element (uncombined with other elements) is 0.*
    [*Examples:* The O.S. of an isolated Cl atom is 0; the two Cl atoms in the molecule Cl_2 both have an O.S. of 0.]

2. *The total of the oxidation states of all the atoms in*
    (a) *a neutral species—an isolated atom, a molecule, a formula unit—is 0.*
    [*Examples:* The sum of the O.S. of all the atoms in CH_3OH and of all the ions in $MgCl_2$ is 0.]
    (b) *an ion is equal to the charge on the ion.*
    [*Examples:* The O.S. of Fe in Fe^{3+} is $+3$. The sum of the O.S. of all the atoms in MnO_4^- is -1.]

3. *In their compounds the Group 1A metals have an O.S. of $+1$ and the Group 2A metals have an O.S. of $+2$.*
    [*Examples:* The O.S. of K is $+1$ in KCl and in K_2CO_3; the O.S. of Mg is $+2$ in $MgBr_2$ and in $Mg(NO_3)_2$.]

4. *In its compounds, the O.S. of fluorine is -1.*
    [*Examples:* The O.S. of F is -1 in HF, ClF_3, and SF_6.]

5. *In its compounds, hydrogen has an O.S. of $+1$.*
    [*Examples:* The O.S. of H is $+1$ in HI, H_2S, NH_3, and CH_4.]

6. *In its compounds, oxygen has an O.S. of -2.*
    [*Examples:* The O.S. of O is -2 in H_2O, CO_2, and $KMnO_4$.]

7. *In their binary (two-element) compounds with metals, the Group 7A elements have an O.S. of -1; the Group 6A elements, -2; and the Group 5A elements, -3.*
    [*Examples:* The O.S. of Br is -1 in $MgBr_2$; the O.S. of S is -2 in Li_2S; and the O.S. of N is -3 in Li_3N.]

▶ The principal exceptions to rule 5 occur when H is bonded to metals, as in LiH, NaH, and CaH_2; exceptions to rule 6 occur in compounds where O atoms are bonded to one another, as in H_2O_2, and KO_2.

EXAMPLE 3-7

Assigning Oxidation States. What is the oxidation state of the underlined element in each of the following? **(a)** $\underline{P}_4$; **(b)** $\underline{Al}_2O_3$; **(c)** $\underline{Mn}O_4^-$; **(d)** Na$\underline{H}$; **(e)** $H_2\underline{O}_2$; **(f)** $\underline{Fe}_3O_4$.

Solution

(a) P_4: This is the formula of a molecule of elemental phosphorus. For an atom of a free element the O.S. = 0 (rule 1). The O.S. of P in P_4 is 0.

(b) Al_2O_3: The total of the oxidation states of all the atoms in this formula unit is 0 (rule 2). The O.S. of oxygen is -2 (rule 6). The total for three O atoms is -6. The total for two Al atoms is $+6$. The O.S. of Al is $+3$.

(c) MnO_4-: This is the formula for permanganate *ion*. The total of the oxidation states of all the atoms in the ion is -1 (rule 2). The total for the four O atoms is -8. The O.S. of Mn is $+7$.

(d) NaH: This is a formula unit of the *ionic* compound sodium hydride. Rule 3 states that Na should have O.S. $+1$. Rule 5 indicates that H should also have O.S. $+1$. If both atoms had O.S. $+1$, the total for the formula unit would be $+2$. This violates rule 2. *Rules 2 and 3 take precedence over rule 5.* Na has O.S. $+1$; the total for the formula unit is 0; and the O.S. of H is -1.

(e) H_2O_2: This is hydrogen peroxide. Rule 5, stating that H has O.S. $+1$, takes precedence over rule 6 (which says that oxygen has O.S. -2). The sum of the oxidation states of the two H atoms is $+2$ and that of the two O atoms must be -2. The O.S. of O is -1.

(f) Fe_3O_4: The total of the oxidation states of four O atoms is -8 (-2 for each atom). For three Fe atoms the total must be $+8$. The O.S. per Fe atom is $\frac{8}{3}$ or $+2\frac{2}{3}$.

Practice Example A: What is the oxidation state of the underlined element in each of the following: $\underline{S}_8$; $\underline{Cr}_2O_7{}^{2-}$; $\underline{Cl}_2O$; $K\underline{O}_2$?

Practice Example B: What is the oxidation state of the underlined element in each of the following: $\underline{S}_2O_3{}^{2-}$; $\underline{Hg}_2Cl_2$; $K\underline{Mn}O_4$; $H_2\underline{C}O$?

In part (f) of Example 3-7, we got the somewhat surprising answer of $+2\frac{2}{3}$ for the oxidation state of the iron atoms in Fe_3O_4. Prior to that we saw only integral values for oxidation state. How does this fractional value come about? Generally, it comes from the assumption that all the atoms of an element have the same oxidation state in a given compound. Usually they do, but not always. Fe_3O_4, for example, is probably better represented as $FeO \cdot Fe_2O_3$, that is, through a combination of two simpler formula units. In FeO the Fe atom is in O.S. $+2$. In Fe_2O_3 *two* Fe atoms are in O.S. $+3$. When we *average* the oxidation states over all three Fe atoms, we get a nonintegral value: $(2 + 3 + 3)/3 = \frac{8}{3} = 2\frac{2}{3}$.

Also, we may at times need to "fragment" a formula into its constituent parts before assigning oxidation states. The ionic compound NH_4NO_3, for instance, consists of the ions $NH_4{}^+$ and $NO_3{}^-$. The oxidation state of N in $NH_4{}^+$ is -3, and in $NO_3{}^-$, $+5$, and we do *not* want to average them. It is far more useful to know the oxidation states of the individual N atoms than it is to deal with an average oxidation state of $+1$ for the two N atoms.

Our first use of oxidation states comes in the naming of chemical compounds in the next section.

3-5 Naming Inorganic Compounds

Throughout this chapter we have referred to compounds mostly by their formulas, but we do need to give them names. When we know the name of a compound, we can look up its properties in a handbook, locate a chemical on a storeroom shelf, or discuss an experiment with a colleague. Later in the text we will see cases in which different compounds have the same formula. In these instances we will find it essential to distinguish among compounds by name. We cannot give two substances the same name, yet we do want some similarities in the names of similar substances (see Figure 3-6). If all compounds were referred to by a common or trivial name, such as *water* (H_2O) or *ammonia* (NH_3), we would have to learn millions of unrelated names—an impossibility.

What we need is a *systematic* method of assigning names—a system of **nomenclature.** Several systems are used, and we will introduce each at an appropriate point in the text. Compounds formed by carbon and hydrogen or carbon and hydrogen together with oxygen, nitrogen, and a few other elements are organic compounds. They are generally considered in a special branch of chemistry with its own set of nomenclature rules—organic chemistry. Compounds that do not fit this description are inorganic compounds. For now we will consider only the naming of inorganic compounds.

Binary Compounds of Metals and Nonmetals

Binary compounds are those formed between *two* elements. If one of the elements is a metal and the other a nonmetal, the binary compound is usually made up of ions, that is, a binary ionic compound. To name a binary compound of a metal and a nonmetal

- write the *unmodified* name of the metal, followed by
- the name of the nonmetal, modified to end in "ide."

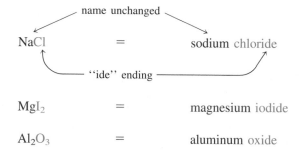

MgI_2 = magnesium iodide

Al_2O_3 = aluminum oxide

Ionic compounds, though made up of positive and negative ions, must be *electrically neutral.* The net, or total, charge of the ions in a formula unit must be *zero.* This means *one* Na^+ to *one* Cl^- (in NaCl); *one* Mg^{2+} to *two* I^- (in MgI_2); *two* Al^{3+} to *three* O^{2-} in Al_2O_3; and so on. Table 3.1 lists the names and symbols of simple ions formed by metals and nonmetals. You will find this list useful when writing names and formulas of binary compounds of metals and nonmetals. Notice that the main group metals and nonmetals form ions related to their group number, as discussed in Section 2-6. The transition metals (the

FIGURE 3-6
Two oxides of lead.
These two compounds contain the same elements—lead and oxygen—but in different proportions. Their names and formulas must convey this fact: lead(IV) oxide = PbO_2 (red-brown); lead(II) oxide = PbO (yellow).

		TABLE 3.1 Some Simple Ions		

Name	Symbol	Name	Symbol
		Positive ions (cations)	
lithium ion	Li^+	chromium(II) ion	Cr^{2+}
sodium ion	Na^+	chromium(III) ion	Cr^{3+}
potassium ion	K^+	iron(II) ion	Fe^{2+}
rubidium ion	Rb^+	iron(III) ion	Fe^{3+}
cesium ion	Cs^+	cobalt(II) ion	Co^{2+}
magnesium ion	Mg^{2+}	cobalt(III) ion	Co^{3+}
calcium ion	Ca^{2+}	copper(I) ion	Cu^+
strontium ion	Sr^{2+}	copper(II) ion	Cu^{2+}
barium ion	Ba^{2+}	mercury(I) ion	Hg_2^{2+}
aluminum ion	Al^{3+}	mercury(II) ion	Hg^{2+}
zinc ion	Zn^{2+}	tin(II) ion	Sn^{2+}
silver ion	Ag^+	lead(II) ion	Pb^{2+}
		Negative ions (anions)	
hydride ion	H^-	oxide ion	O^{2-}
fluoride ion	F^-	sulfide ion	S^{2-}
chloride ion	Cl^-	nitride ion	N^{3-}
bromide ion	Br^-		
iodide ion	I^-		

Group-B elements), however, may form several ions; it is important to distinguish between these in naming their compounds.

The metal iron, for example, forms *two* common ions, Fe^{2+} and Fe^{3+}. To distinguish between them we call the first *iron(II)* ion and the second *iron(III)* ion. The Roman numeral immediately following the name of the metal indicates its oxidation state or simply the charge on the ion. To name $FeCl_2$ and $FeCl_3$, we call the former *iron(II)* chloride and the latter *iron(III)* chloride. This system of nomenclature based on Roman numerals for oxidation states is called the *Stock system.* It was proposed by Alfred Stock shortly after World War I and began to receive widespread acceptance after World War II.

An earlier system of nomenclature that is still used to some extent is the "ous/ic" system. With it we can distinguish between two binary compounds containing the same two elements but in different proportions, such as Cu_2O and CuO. In Cu_2O, the oxidation state of copper is $+1$, and in CuO it is $+2$. Cu_2O is assigned the name cupr*ous* oxide and CuO, cupr*ic* oxide. Similarly, $FeCl_2$ is ferr*ous* chloride and $FeCl_3$ is ferr*ic* chloride. The idea is to use the "ous" ending for the lower oxidation state of the metal and "ic" for the higher oxidation state. The "ous/ic" system has several inadequacies; for example, there is no way to apply it in naming these four oxides of vanadium: VO, V_2O_3, VO_2, and V_2O_5. We will therefore not use it in this text.

EXAMPLE 3-8

Writing Formulas When Names of Compounds Are Given. Write formulas for the compounds barium oxide, calcium fluoride, and iron(III) sulfide.

Solution

In each case identify the cations and their charges, based on the group number or Stock number: Ba^{2+}, Ca^{2+}, and Fe^{3+}; then the anions and their charges: O^{2-}, F^-, and S^{2-}. Combine the cations and anions in the relative numbers required to produce electrically *neutral* formula units.

$$\text{barium oxide:} \qquad \textit{one } Ba^{2+} \text{ and } \textit{one } O^{2-} = BaO$$

$$\text{calcium fluoride:} \qquad \textit{one } Ca^{2+} \text{ and } \textit{two } F^{-} = CaF_2$$

$$\text{iron(III) sulfide:} \qquad \textit{two } Fe^{3+} \text{ and } \textit{three } S^{2-} = Fe_2S_3$$

Note that in the first case an electrically neutral formula unit results from the combination of the charges $2+$ and $2-$; in the second case, $2+$ and $2 \times (1-)$; and in the third case, $2 \times (3+)$ and $3 \times (2-)$.

Practice Example A: Write formulas for the compounds lithium oxide, tin(II) fluoride, lithium nitride.

Practice Example B: Write formulas for the compounds aluminum sulfide, magnesium nitride, and vanadium(III) oxide.

EXAMPLE 3-9

Naming Compounds When Their Formulas Are Given. Write acceptable names for the compounds Na_2S, AlF_3, Cu_2O.

Solution

This task is generally easier than that of Example 3-8 because all you need to do is name the ions present. However, you must recognize that copper forms *two* different ions and that the cation in Cu_2O is Cu^{+}, copper(I).

$$Na_2S: \quad \text{sodium sulfide}$$

$$AlF_3: \quad \text{aluminum fluoride}$$

$$Cu_2O: \quad \text{copper(I) oxide}$$

Practice Example A: Write acceptable names for the compounds CsI, CaF_2, FeO, $CrCl_3$.

Practice Example B: Write acceptable names for the compounds CaH_2, CuCl, Ag_2S, Hg_2Cl_2.

Are You Wondering . . .

Why we don't use names such as sodium(I) chloride for NaCl and magnesium(II) chloride for $MgCl_2$?

Each proposed name clearly indicates the compound in question, but as a general rule chemists always write the *simplest name possible.* The metals of periodic table Group 1A (including Na) and Group 2A (including Mg) have *only one ionic form,* one oxidation state. Roman numerals designating these oxidation states are superfluous. Later in the text you'll acquire a better understanding of whether an element can exist in several oxidation states. For now, use the information in Table 3-1 as a guide.

Binary Compounds of Two Nonmetals

If the two elements in a binary compound are both nonmetals instead of a metal and a nonmetal, the compound is a molecular compound. The method of naming these compounds is similar to what we have already learned, however. For example,

$$HCl = \text{hydrogen chloride}$$

In both the formula and the name, we write the element with the positive oxidation state first: HCl and *not* ClH.

Some pairs of nonmetals form more than a single binary molecular compound, and we need to distinguish among them. Generally, we do not use the Stock system but one in which relative numbers of atoms are shown through *prefixes*.

mono = 1   di = 2   tri = 3   tetra = 4   penta = 5   hexa = 6

Thus, for the two principal oxides of sulfur we write

$$SO_2 = \text{sulfur dioxide}$$

$$SO_3 = \text{sulfur trioxide}$$

and for the following boron–bromine compound

$$B_2Br_4 = \text{diboron tetrabromide}$$

Additional examples are given in Table 3.2. Note that in these examples the prefix *mono* is treated in a special way. We do not use it for the first named element. Thus, NO is called nitrogen monoxide, *not mono*nitrogen *mon*oxide. Finally, several substances have common or trivial names that are so well established that we almost never use their systematic names. For example,

$$H_2O = \text{water (dihydrogen monoxide)}$$

$$NH_3 = \text{ammonia } (H_3N = \text{trihydrogen mononitride})$$

TABLE 3.2   Naming Binary Molecular Compounds

Formula	Name[a]
BCl_3	boron trichloride
CCl_4	carbon tetrachloride
CO	carbon monoxide
CO_2	carbon dioxide
NO	nitrogen monoxide
NO_2	nitrogen dioxide
N_2O	dinitrogen monoxide
N_2O_3	dinitrogen trioxide
N_2O_4	dinitrogen tetroxide
N_2O_5	dinitrogen pentoxide
PCl_3	phosphorus trichloride
PCl_5	phosphorus pentachloride
SF_6	sulfur hexafluoride

[a]When the prefix ends in "a" or "o" and the element name begins with "a" or "o," we drop the final vowel of the prefix for ease of pronunciation. For example, carbon *monox*ide, not carbon *mono*oxide, and dinitrogen *tetrox*ide, not dinitrogen *tetra*oxide. However, PI_3 is phosphorus *triio*dide, not phosphorus *tri*odide.

Are You Wondering . . .

Why isn't MgCl₂ called magnesium dichloride and FeCl₃, iron trichloride?

These names do give a clear indication of the compounds we mean, but they do *not* conform to the conventions that chemists use. Because the magnesium ion can only be Mg^{2+}, the name *magnesium chloride* can only mean $MgCl_2$. Similarly, *iron(III) chloride* can only refer to a compound of Fe^{3+} and Cl^-, that is, $FeCl_3$. On the other hand, to name the several oxides of nitrogen we do need prefixes (mono, di, tri, . . .).

Binary Acids

Chemists have a particular way of defining substances known as acids, which we consider later in the text. For now, so that we can recognize these substances and name them when we see their formulas, let us say that an **acid** is a substance that produces hydrogen ions (H^+)* when dissolved in water. **Binary acids,** then, are certain compounds of H with other nonmetal atoms. HCl, when dissolved in water, ionizes or breaks down into hydrogen ions (H^+) and chloride ions (Cl^-); it is an acid. NH_3 in water is *not* an acid. It shows practically no tendency to produce H^+ under any conditions.

Even though we use names like hydrogen chloride for pure binary molecular compounds, we sometimes want to emphasize that their aqueous solutions are acids. In these cases, we use the prefix "hydro," followed by the other nonmetal name modified to an "ic" ending. The most important binary acids are listed below.

$$HF(aq) = \textit{hydro}\text{fluor}\textit{ic}\text{ acid}$$
$$HCl(aq) = \textit{hydro}\text{chlor}\textit{ic}\text{ acid}$$
$$HBr(aq) = \textit{hydro}\text{brom}\textit{ic}\text{ acid}$$
$$HI(aq) = \textit{hydro}\text{iod}\textit{ic}\text{ acid}$$
$$H_2S(aq) = \textit{hydro}\text{sulfur}\textit{ic}\text{ acid}$$

▶ NH_3 belongs to a complementary category of substances called bases. As described in Chapter 5, bases yield hydroxide ion (OH^-) in aqueous solutions.

▶ The symbol (aq) signifies a substance in aqueous (water) solution.

Polyatomic Ions

The ions listed in Table 3.1 (with the exception of Hg_2^{2+}) are monatomic ions. Each consists of a single atom. In **polyatomic ions** two or more atoms are joined together by covalent bonds. These ions are commonly encountered, especially among the nonmetals. A number of polyatomic ions and compounds containing them are listed in Table 3.3. From this table you can see that

1. Polyatomic anions are more common than polyatomic cations. The most familiar polyatomic cation is the ammonium ion, NH_4^+.
2. Very few polyatomic anions carry the "ide" ending in their names. Of those listed, only OH^- (hydroxide ion) and CN^- (cyanide ion) do. The common endings are *ite* and *ate,* and some names carry the prefixes *hypo* or *per.*

*The species produced in aqueous solution is actually more complex than the simple ion H^+. The H^+ combines with an H_2O molecule to produce an ion known as the hydronium ion, H_3O^+. Chemists often use the simple symbol H^+ for H_3O^+, and that is what we will do until we discuss this matter more fully in Chapter 17.

▶ Learn the information in this table as quickly as you can. You will find a lot of use for it as we proceed through the text.

TABLE 3.3	Some Common Polyatomic Ions	
Name	**Formula**	**Typical Compound**
Cation		
ammonium ion	NH_4^+	NH_4Cl
Anions		
acetate ion	$C_2H_3O_2^-$	$NaC_2H_3O_2$
carbonate ion	CO_3^{2-}	Na_2CO_3
hydrogen carbonate ion[a] (or bicarbonate ion)	HCO_3^-	$NaHCO_3$
hypochlorite ion	ClO^-	$NaClO$
chlorite ion	ClO_2^-	$NaClO_2$
chlorate ion	ClO_3^-	$NaClO_3$
perchlorate ion	ClO_4^-	$NaClO_4$
chromate ion	CrO_4^{2-}	Na_2CrO_4
dichromate ion	$Cr_2O_7^{2-}$	$Na_2Cr_2O_7$
cyanide ion	CN^-	$NaCN$
hydroxide ion	OH^-	$NaOH$
nitrite ion	NO_2^-	$NaNO_2$
nitrate ion	NO_3^-	$NaNO_3$
oxalate ion	$C_2O_4^{2-}$	$Na_2C_2O_4$
permanganate ion	MnO_4^-	$NaMnO_4$
phosphate ion	PO_4^{3-}	Na_3PO_4
hydrogen phosphate ion[a]	HPO_4^{2-}	Na_2HPO_4
dihydrogen phosphate ion[a]	$H_2PO_4^-$	NaH_2PO_4
sulfite ion	SO_3^{2-}	Na_2SO_3
hydrogen sulfite ion[a] (or bisulfite ion)	HSO_3^-	$NaHSO_3$
sulfate ion	SO_4^{2-}	Na_2SO_4
hydrogen sulfate ion[a] (or bisulfate ion)	HSO_4^-	$NaHSO_4$
thiosulfate ion	$S_2O_3^{2-}$	$Na_2S_2O_3$

[a]These anion names are sometimes written as a single word—for example, hydrogencarbonate, hydrogenphosphate, etc.

3. An element common to many polyatomic anions is *oxygen*. The oxygen is usually combined with another nonmetal. Such anions are called **oxoanions.**

4. Certain nonmetals (such as Cl, N, P, and S) form a series of oxoanions containing different numbers of oxygen atoms. Their names are related to the oxidation state of the nonmetal atom to which the O atoms are bonded, ranging from "hypo" (lowest) to "per" (highest) according to the scheme

Increasing oxidation state of nonmetal →

hypo_____ite   _____ite   _____ate   per_____ate

Increasing number of oxygen atoms →

5. All the common oxoanions of Cl, Br, and I carry a charge of 1−.

6. Some series of oxoanions also contain varying numbers of H atoms and are named accordingly. For example, HPO_4^{2-} is the *hydrogen phosphate* ion and $H_2PO_4^-$, the *dihydrogen phosphate* ion.

TABLE 3.4 Nomenclature of Some Oxoacids and Their Salts

Oxidation State	Formula of Acid[a]	Name of Acid[b]	Formula of Salt	Name of Salt[b]
Cl: +1	HClO	*hypo*chlor*ous* acid	NaClO	sodium *hypo*chlor*ite*
Cl: +3	HClO$_2$	chlor*ous* acid	NaClO$_2$	sodium chlor*ite*
Cl: +5	HClO$_3$	chlor*ic* acid	NaClO$_3$	sodium chlor*ate*
Cl: +7	HClO$_4$	*per*chlor*ic* acid	NaClO$_4$	sodium *per*chlor*ate*
N: +3	HNO$_2$	nitr*ous* acid	NaNO$_2$	sodium nitr*ite*
N: +5	HNO$_3$	nitr*ic* acid	NaNO$_3$	sodium nitr*ate*
S: +4	H$_2$SO$_3$	sulfur*ous* acid	Na$_2$SO$_3$	sodium sulf*ite*
S: +6	H$_2$SO$_4$	sulfur*ic* acid	Na$_2$SO$_4$	sodium sulf*ate*

[a]In all these acids H atoms are bonded to O atoms, not the central nonmetal atom. Often formulas are written to reflect this fact, such as HOCl instead of HClO and HOClO instead of HClO$_2$.

[b]In general the *ic* and *ate* names are assigned to compounds in which the central nonmetal atom has an oxidation state equal to the periodic group number. Halogen compounds are exceptional in that the *ic* and *ate* names are assigned to compounds in which the halogen has an oxidation state of +5 (even though the group number is 7, that is, 7A).

7. The prefix "thio" signifies that a sulfur atom has been substituted for an oxygen atom. (The sulfate ion has *one* S and *four* O atoms; thiosulfate ion has *two* S and *three* O atoms.)

Oxoacids

The majority of acids are **ternary compounds.** They contain *three* different elements—hydrogen, *oxygen,* and another nonmetal—and are called **oxoacids.** Or, think of oxoacids as combinations of hydrogen ions (H$^+$) and oxoanions. The scheme for naming oxoacids is similar to that outlined for oxoanions, except that the ending "ous" is used instead of "ite" and "ic" instead of "ate." Several oxoacids are listed in Table 3.4. Also listed are the names and formulas of the compounds in which the hydrogen of the oxoacid has been replaced by a metal such as sodium. Although we will consider more precise definitions later, these compounds are called **salts.** Acids are molecular compounds, and salts are ionic compounds.

EXAMPLE 3-10

Applying Various Rules in Naming Compounds. Name the compounds (a) CuCl$_2$; (b) ClO$_2$; (c) HIO$_4$; (d) Ca(H$_2$PO$_4$)$_2$.

Solution

(a) In this compound, the oxidation state of Cu is +2. Because Cu can also exist in the oxidation state +1, we must clearly distinguish between the two possible chlorides. CuCl$_2$ is copper(II) chloride.

(b) Both Cl and O are nonmetals. ClO$_2$ is a binary molecular compound called chlorine dioxide.

(c) The oxidation state of I is +7. By analogy to the chlorine-containing oxoacids in Table 3.4, we should name this compound periodic acid (pronounced "purr eye oh dic" acid).

▶ Do not make the common mistake of calling ClO$_2$ chlorite. ClO$_2^-$ is the chlorite *ion.* ClO$_2$ is the *compound* chlorine dioxide. There can be no compound consisting of just a single type of ion. There is no substance called chloride or chlorite or chlorate, and so on.

(d) The polyatomic anion $H_2PO_4^-$ is dihydrogen phosphate ion. Two of these ions are present for every Ca^{2+} ion in the compound calcium dihydrogen phosphate.

Practice Example A: Name the compounds SF_6, HNO_2, $Ca(HCO_3)_2$, $FeSO_4$.

Practice Example B: Name the compounds NH_4NO_3, PCl_3, $HBrO$, $AgClO_4$, $Fe_2(SO_4)_3$.

EXAMPLE 3-11

Applying Various Rules in Writing Formulas. Write the formula of the compound (**a**) tetranitrogen tetrasulfide; (**b**) ammonium chromate; (**c**) bromic acid; (**d**) calcium hypochlorite.

Solution

(**a**) Molecules of this compound consist of *four* N atoms and *four* S atoms. The formula is N_4S_4.

(**b**) Two ammonium ions (NH_4^+) must be present for every chromate ion (CrO_4^{2-}). Place parentheses around NH_4^+, followed by the subscript 2. The formula is $(NH_4)_2CrO_4$. (This formula is read as "N-H-4, taken twice, C-R-O-4.")

(**c**) The "ic" acid for the oxoacids of the halogens (Group 7A) has the halogen in oxidation state +5. Bromic acid is $HBrO_3$ (analogous to $HClO_3$ in Table 3.4).

(**d**) Here there are *one* Ca^{2+} and *two* ClO^- ions in a formula unit. This leads to the formula $Ca(ClO)_2$.

▶ Note the importance of the parentheses in (d). If we left them out, we would have $CaClO_2$, an *incorrect* formula for both calcium hypochlorite and calcium chlorite.

Practice Example A: Write formulas for the following compounds: boron trifluoride, potassium dichromate, sulfuric acid, calcium chloride.

Practice Example B: Write formulas for the following compounds: aluminum nitrate, tetraphosphorous decoxide, chromium(III) hydroxide, iodic acid.

Some Compounds of Greater Complexity

The copper compound that Joseph Proust used to establish the law of constant composition (page 35) is referred to in different ways. If you look up Proust's compound in a handbook of minerals, you will find it listed as *malachite,* with the formula $Cu_2(OH)_2CO_3$. In a handbook that specializes in pharmaceutical applications, this same compound is listed as *basic cupric carbonate,* with the formula $CH_2Cu_2O_5$. In a chemistry handbook it is listed as *copper(II) carbonate dihydroxide,* with the formula $CuCO_3 \cdot Cu(OH)_2$. All that is important for you to understand at this point is that regardless of the formula you use, you should obtain the same molar mass (221.12 g/mol), the same mass percent copper (57.48% Cu), the same H-to-O mole ratio (2 mol H/5 mol O), and so on. In short, you should be able to interpret a formula, no matter how complex its appearance.

▶ In general, we use dots (•) to show that a formula is a composite of two or more simpler formulas.

Some complex substances you are certain to encounter are known as hydrates. In a **hydrate** each formula unit of the compound has associated with it a certain number of water molecules. This does not mean that the compounds are "wet," however. The water molecules are incorporated in the solid structure of the compound. The formula shown below signifies *six* H_2O molecules per formula unit of $CoCl_2$.

$$CoCl_2 \cdot 6H_2O$$

FIGURE 3-7
Effect of moisture on CoCl₂.
The piece of filter paper was soaked in a water solution of cobalt(II) chloride and then allowed to dry. When kept in dry air, the paper is blue in color (anhydrous CoCl₂). In humid air, the paper changes to a pink color (CoCl₂·6H₂O).

Using the prefix for six—*hexa*—we call this compound cobalt(II) chloride *hexa*-hydrate. Its formula mass is that of $CoCl_2$ *plus* that associated with six H_2O: 129.8 u + (6 × 18.02 u) = 237.9 u. We can speak of the mass percent water in a hydrate. For $CoCl_2·6H_2O$ this is

$$\% \; H_2O = \frac{(6 \times 18.02) \; g \; H_2O}{237.9 \; g \; CoCl_2·6H_2O} \times 100\% = 45.45\%$$

The water present in compounds as water of hydration can generally be removed, in part or totally, usually by heating. When the water is totally removed, the resulting compound is said to be *anhydrous* ("without water"). Anhydrous compounds can be used as water absorbers, as in the use of anhydrous magnesium perchlorate in combustion analysis (recall Figure 3-5). $CoCl_2$ gains and loses water quite readily and indicates this through a color change. Anhydrous $CoCl_2$ is blue, whereas the hexahydrate is pink. This fact can be used to make a simple moisture detector (see Figure 3-7).

Summary

Chemical formulas, once established, provide a wealth of information about chemical compounds, serving as the basis for determining

- formula (molecular) masses;
- molar masses;
- percent compositions;
- various relationships among the elements present.

Chemical formulas can be related to experimentally determined percent compositions of compounds. For organic compounds this usually involves combustion analysis. The formulas determined from experimental data are empirical formulas—the simplest formulas that can be written. Molecular formulas can be related to empirical formulas in those cases in which experimentally determined molecular masses are available.

Compounds are designated by name as well as by formula. Two important ideas are brought forward from Chapter 2 to help relate names and formulas (nomenclature): the periodic table of the elements and the classification of elements as metals and nonmetals. A new idea, the concept of oxidation state, is introduced in this chapter. Together, these three ideas are used in naming and writing formulas of binary ionic and molecular compounds, polyatomic ions, oxoacids and their salts, and hydrates.

Summarizing Example

Copper(II) sulfate is the most widely used copper compound, finding applications in electroplating, in pesticide mixtures, as a wood preservative, and as an algicide (for killing algae in lakes). Whether in the laboratory or in chemical plants, copper(II) sulfate is usually obtained as a hydrate, commonly called "blue vitriol" (see photograph on page 64).

Focus On

Polymers—Macromolecular Substances

In 1934, Wallace Carothers and his associates at E. I. du Pont de Nemours & Company succeeded in producing the first synthetic fiber—*nylon*. Chemistry students can now carry out a variation of this polymerization in the general chemistry laboratory.

The molecules studied in this chapter all have molecular masses in the range from 2 u (for H_2) to about 200 u. Some molecules, however, have molecular masses up to several million atomic mass units. These are macromolecules, or polymers. *Polymers* are made up of simple molecules with low molecular masses joined together into extremely large molecules. Polymers with molecular masses below about 20,000 u are called low polymers and those above 20,000 u, high polymers.

One familiar polymer is *polyethylene*. As its name implies, its basic unit, called a *monomer*, is the ethylene molecule, which has the structural formula

$$H_2C=CH_2$$

In the polymerization reaction, the double bonds in the molecules "open up." Instead of each C atom forming a double bond to one other C atom, it now forms single bonds to two other C atoms, as represented in the structure

$$\cdots-\overset{\displaystyle H}{\underset{\displaystyle H}{\overset{|}{\underset{|}{C}}}}-\overset{\displaystyle H}{\underset{\displaystyle H}{\overset{|}{\underset{|}{C}}}}-\overset{\displaystyle H}{\underset{\displaystyle H}{\overset{|}{\underset{|}{C}}}}-\overset{\displaystyle H}{\underset{\displaystyle H}{\overset{|}{\underset{|}{C}}}}-\overset{\displaystyle H}{\underset{\displaystyle H}{\overset{|}{\underset{|}{C}}}}-\overset{\displaystyle H}{\underset{\displaystyle H}{\overset{|}{\underset{|}{C}}}}-\cdots$$

The final *macromolecule* has hundreds of the C_2H_4 units.

Another polymer in which the monomer units are joined end to end is latex, or natural rubber.

rubber

Early rubber products were of limited use because they were sticky in hot weather and stiff in

---

A 2.574-g sample of hydrated copper(II) sulfate, $CuSO_4 \cdot xH_2O$, was heated to drive off the water of hydration. After the anhydrous $CuSO_4$ had cooled to room temperature, its mass was found to be 1.647 g. What is the formula of the hydrate (that is, what is the value of x in the formula $CuSO_4 \cdot xH_2O$)?

1. *Determine the mass of water in the original sample.* This is simply the difference in mass between the original and heated samples. *Result:* 0.927 g H_2O.

2. *Determine the amounts of H_2O and of $CuSO_4$, in moles, in the original sample.* You need to use the molar masses of H_2O and $CuSO_4$ in conversion factors. *Result:* 0.0514 mol H_2O, 0.01032 mol $CuSO_4$.

3. *Determine the ratio mol H_2O/mol $CuSO_4$ and thus the formula. Answer:* $CuSO_4 \cdot 5H_2O$.

Key Terms

acid (3-5)	**formula unit** (3-1)	**nomenclature** (3-5)
anion (3-1)	**formula mass** (3-2)	**oxidation state** (3-4)
binary acid (3-5)	**hydrate** (3-5)	**oxoacid** (3-5)
binary compound (3-5)	**ionic compound** (3-1)	**oxoanion** (3-5)
cation (3-1)	**molecular compound** (3-1)	**polyatomic ion** (3-5)
chemical formula (3-1)	**molecular formula** (3-1)	**structural formula** (3-1)
combustion analysis (3-3)	**molecular mass** (3-2)	**ternary compound** (3-5)
empirical formula (3-1)	**molecule** (3-1)	

cold weather. In 1839, Charles Goodyear accidentally discovered that by heating a sulfur–rubber mixture a product could be made that was stronger, more elastic, and more resistant to heat and cold than natural rubber. This process is now called vulcanization (after Vulcan, the Roman god of fire). The purpose of vulcanization is to form *crosslinks* between long polymer chains. An example of a crosslink through two sulfur atoms is shown below.

Polymers are familiar products in the modern world. Nylon, one of the first polymers developed, is like an artificial silk and is used in making clothing, ropes, and sails. The fluorine-containing polymer Teflon (polytetrafluoroethylene) is used in nonstick frying and baking pans. Polyvinyl chloride (PVC) is used in food wrap, hoses, pipes, and floor tile. In all, the polymer industry is a very large one. It has been estimated, for example, that about one half of all chemists work with polymers.

In addition to latex, which comes from the rubber tree *(Hevea brasiliensis)*, many *natural* polymers occur. Cellulose, the basic structural material of plants, is a polymer with common glucose $(C_6H_{12}O_6)$ as its monomer. Proteins are high polymers with amino acids as their monomers. DNA, sometimes called the "thread of life," is also a macromolecular substance. We will learn more about these natural polymers later in the text.

Review Questions

1. In your own words define or explain the following terms or symbols: **(a)** formula unit; **(b)** S_8; **(c)** ionic compound; **(d)** oxoacid; **(e)** hydrate.

2. Briefly describe each of the following ideas or methods: **(a)** molecule of an element; **(b)** structural formula; **(c)** oxidation state; **(d)** carbon–hydrogen–oxygen determination by combustion analysis.

3. Explain the important distinctions between each pair of terms: **(a)** chemical symbol and chemical formula; **(b)** empirical and molecular formula; **(c)** systematic and trivial or common name; **(d)** binary and ternary acid.

4. Explain clearly the meaning of each of the following terms as it applies to the element oxygen: **(a)** atomic mass; **(b)** molecular mass; **(c)** molar mass.

5. Calculate the total number of **(a)** atoms in one molecule of nitroglycerine, $C_3H_5(NO_3)_3$; **(b)** atoms in 0.00102 mol C_2H_6; **(c)** F atoms in 12.15 mol $C_2HBrClF_3$.

6. Determine the *mass*, in grams, of **(a)** 7.34 mol N_2O_4; **(b)** 3.16×10^{24} O_2 molecules; **(c)** 18.6 mol $CuSO_4 \cdot 5H_2O$; **(d)** 4.18×10^{24} molecules of $C_2H_4(OH)_2$.

7. Determine the number of moles of Br_2 in a sample consisting of **(a)** 8.08×10^{22} Br_2 molecules; **(b)** 2.17×10^{24} Br atoms; **(c)** 11.3 kg bromine; **(d)** 2.65 L liquid bromine ($d = 3.10$ g/mL).

8. The amino acid methionine is essential in human diets. It has the molecular formula $C_5H_{11}NO_2S$. Determine **(a)** its molecular mass; **(b)** the number of moles of H atoms per mole of methionine; **(c)** the number of grams of C per mole of methionine; **(d)** the number of C atoms in 9.07 mol methionine.

9. Determine the mass percent H in the hydrocarbon decane, $C_{10}H_{22}$.

10. Determine the mass percent O in the mineral malachite, $Cu_2(OH)_2CO_3$.

11. Determine the mass percent H_2O in the hydrate $Cr(NO_3)_3 \cdot 9H_2O$.

12. Determine the percent, by mass, of the indicated element in each of the following.

(a) Pb in tetraethyl lead, $Pb(C_2H_5)_4$, once extensively used as an antiknock additive in gasoline
(b) Fe in Prussian blue, $Fe_4[Fe(CN)_6]_3$, a pigment used in paints and printing inks

13. *Without doing detailed calculations,* explain which of the following has the greatest mass percent of sulfur: SO_2, S_2Cl_2, Na_2S, or $Na_2S_2O_3$.

14. An oxide of chromium used as a green pigment in paints contains 68.42% Cr and 31.58% O. What is the empirical formula of the oxide?

15. Diethylene glycol, used to de-ice aircraft, is a carbon–hydrogen–oxygen compound with 45.27% C and 9.50% H, by mass. What is its empirical formula?

16. The food flavor enhancer monosodium glutamate (MSG) has the composition 13.6% Na, 35.5% C, 4.8% H, 8.3% N, 37.8% O, by mass. What is the empirical formula of MSG?

17. A compound of C, H, and O, known as terephthalic acid, is used in the manufacture of Dacron (polyethylene terephthalate). Its molecular mass is 166.1 u, and by combustion analysis it is found to have 57.83% C and 3.64% H, by mass. What is the molecular formula of terephthalic acid?

18. Ibuprofen is a carbon–hydrogen–oxygen compound used in painkillers. When a 2.174-g sample is burned completely, it yields 6.029 g CO_2 and 1.709 g H_2O.
(a) What is the percent composition, by mass, of ibuprofen?
(b) What is the empirical formula of ibuprofen?

19. An oxide of chromium used in chrome plating has a formula mass of 100.0 u and contains *four* atoms per formula unit. Establish the formula of this compound, *with a minimum of calculation.*

20. Supply the missing information (name or formula) for each *ion.*
(a) lead(II), _____
(b) _____, Co^{3+}
(c) _____, Ba^{2+}
(d) chromium(II), _____
(e) periodate, _____
(f) _____, ClO_2^-
(g) gold(III), _____
(h) _____, HSO_3^-
(i) hydrogen carbonate, _____
(j) cyanide, _____

21. Name the binary compounds **(a)** KBr; **(b)** $SrCl_2$; **(c)** ClF_3; **(d)** N_2O_4; **(e)** PCl_5.

22. Name the ternary compounds **(a)** KCN; **(b)** HClO; **(c)** $(NH_4)_2SO_4$; **(d)** KIO_3.

23. Indicate the oxidation state of the underlined element in **(a)** $\underline{Zn}$; **(b)** Ba$\underline{S}$; **(c)** $\underline{N}O_2$; **(d)** H$\underline{N}O_2$; **(e)** $\underline{V}O^{2+}$; **(f)** $H_2\underline{P}O_4^-$.

24. Write correct formulas for the compounds **(a)** magnesium bromide; **(b)** barium oxide; **(c)** mercury(II) acetate; **(d)** iron(III) oxalate; **(e)** strontium perchlorate; **(f)** potassium hydrogen sulfate; **(g)** nitrogen trichloride; **(h)** bromine pentafluoride.

25. Supply the name of each of the following acids: **(a)** $HClO_2$; **(b)** H_2SO_3; **(c)** H_2Se; **(d)** HNO_2.

26. Supply the formula of each of the following acids: **(a)** hydroiodic acid; **(b)** nitric acid; **(c)** phosphoric acid; **(d)** sulfuric acid.

Exercises

The Avogadro Constant and the Mole

27. *Without doing detailed calculations,* determine which of the following has the greatest number of S atoms: 0.12 mol of solid sulfur, S_8; 0.50 mol of gaseous S_2O; 65 g of gaseous SO_2; 75 mL of liquid thiophene, C_4H_4S ($d = 1.064$ g/mL).

29. Determine the *mass,* in grams, of
(a) 6.25×10^{-2} mol P_4;
(b) 4.03×10^{24} molecules of stearic acid, $C_{18}H_{36}O_2$;
(c) a quantity of the amino acid lysine, $C_6H_{14}N_2O_2$, containing 1.15 mol N atoms;

31. In rhombic sulfur, S atoms are joined into S_8 molecules (see Figure 3-4). If the density of rhombic sulfur is 2.07 g/cm^3, determine for a crystal of volume 0.568 mm^3: **(a)** the number of moles of S_8 present; **(b)** the total number of S atoms.

28. *Without doing detailed calculations,* explain which of the following has the greatest number of N atoms: 50.0 g N_2O; 17.0 g NH_3; 150 mL of liquid pyridine, C_5H_5N ($d = 0.983$ g/mL); 1.0 mol N_2.

30. Determine the number of *moles* of
(a) N_2O_4 in a 82.5-g sample;
(b) N atoms in 106 g of $Mg(NO_3)_2$;
(c) N atoms in a sample of $C_7H_5(NO_2)_3$ that has the same number of O atoms as 56.5 g $C_6H_{12}O_6$.

32. The hemoglobin content of blood is about 15.5 g/100 mL blood. The molar mass of hemoglobin is about 64,500 g/mol, and there are four iron (Fe) atoms in a hemoglobin molecule. Approximately how many Fe atoms are present in the 6 L of blood in a typical adult?

Chemical Formulas

33. Explain which of the following statements is (are) correct concerning glucose (blood sugar), $C_6H_{12}O_6$.
(1) The percentages, by mass, of C and O are the same as in CO.
(2) The ratio of C to H to O atoms is the same as in dihydroxyacetone, $(CH_2OH)_2CO$.
(3) The proportions, by mass, of C and O are equal.
(4) The highest percentage, by mass, is that of H.

34. Explain which of the following statement(s) is (are) correct for sorbic acid, $C_6H_8O_2$, a mold and yeast inhibitor.
(1) It has a C:H:O mass ratio of 3:4:1.
(2) It has the same mass percent composition as the aquatic herbicide, acrolein, C_3H_4O.
(3) It has the same empirical formula as aspidinol, $C_{12}H_{16}O_4$, a drug used to kill parasitic worms.
(4) It has four times as many H atoms as O atoms, but four times the mass of O as of H.

35. For the compound halothane, $C_2HBrClF_3$, determine
(a) the total number of atoms in one formula unit;
(b) the ratio, by number, of F atoms to C atoms;
(c) the ratio, by mass, of Br to F;
(d) the element present in the greatest mass percent;
(e) the mass of compound required to contain 1.00 g F.

36. For the compound $Ge[S(CH_2)_4CH_3]_4$, determine
(a) the total number of atoms in one formula unit;
(b) the ratio, by number, of C atoms to H atoms;
(c) the ratio, by mass, of Ge to S;
(d) the number of g S in 1 mol of the compound;
(e) the number of C atoms in 33.10 g of the compound.

Percent Composition of Compounds

37. Determine the mass percent of each of the elements in the fatty acid, stearic acid, $C_{18}H_{36}O_2$.
39. All of the following minerals are semiprecious or precious stones. Determine the mass percent of the indicated element in each one. **(a)** Zr in zircon, $ZrSiO_4$; **(b)** Be in beryl (emerald), $Be_3Al_2Si_6O_{18}$; **(c)** Fe in almandine (garnet), $Fe_3Al_2Si_3O_{12}$; **(d)** S in lazurite (lapis lazuli), $Na_4SSi_3Al_3O_{12}$.
41. *Without doing detailed calculations,* arrange the following in order of increasing % Cr, by mass, and explain your reasoning: CrO, Cr_2O_3, CrO_2, CrO_3.

38. Determine the mass percent of each of the elements in the antimalarial drug quinine, $C_{20}H_{24}N_2O_2$.
40. Determine the mass percent of the indicated elements in the following forms of penicillin. **(a)** K in penicillin G potassium, $C_{16}H_{17}KN_2O_4S$; **(b)** N in penicillin N, $C_{14}H_{21}N_3O_6S$; **(c)** S in penicillin S potassium, $C_{14}H_{18}ClKN_2O_4S_2$; **(d)** C in penicillin G calcium, $Ca(C_{16}H_{17}N_2O_4S)_2$.
42. *Without doing detailed calculations,* arrange the following in order of increasing % P, by mass, and explain your reasoning: $Ca(H_2PO_4)_2$, $(NH_4)_2HPO_4$, H_3PO_4, Na_3PO_4.

Chemical Formulas from Percent Composition

43. A compound of carbon and hydrogen consists of 93.71% C and 6.29% H, by mass. The molecular mass of the compound is found to be 128 u. What is its molecular formula?

45. Determine the empirical formula of
(a) the rodenticide (rat killer) warfarin, which consists of 74.01% C, 5.23% H, and 20.76% O, by mass;
(b) the chemical warfare agent, mustard gas, which consists of 30.20% C, 5.07% H, 44.58% Cl, and 20.16% S, by mass.

47. Indigo, the dye for blue jeans, has a percent composition, by mass, of 73.27% C, 3.84% H, 10.68% N, and the remainder as oxygen. Its molecular mass is 262.3 u. What is the molecular formula of indigo?
49. The compound XF_3 consists of 65% F, by mass. What is the atomic mass of X?

44. Selenium, an element used in the manufacture of photoelectric cells and solar energy devices, forms two oxides. One has 28.8% O, by mass, and the other, 37.8% O. What are the formulas of these oxides? Propose acceptable names for them.
46. Determine the empirical formula of
(a) benzo[*a*]pyrene, a suspected carcinogen found in cigarette smoke, consisting of 95.21% C and 4.79% H, by mass.
(b) hexachlorophene, used in germicidal soaps, which consists of 38.37% C, 1.49% H, 52.28% Cl, and 7.86% O by mass.

48. Beta-carotene provides the orange color in carrots and is important for good vision in humans. Its percent composition, by mass, is 89.49% C and 10.51% H. The molecular mass of β-carotene is 536.9 u. What is its molecular formula?
50. The element X forms the chloride XCl_4 containing 75.0% Cl, by mass. What is the element X?

Combustion Analysis

51. A 0.2612-g sample of a hydrocarbon produces 0.8661 g CO_2 and 0.2216 g H_2O in combustion analysis. Its molecular mass is found to be 106 u. For this hydrocarbon, determine **(a)** its mass percent composition; **(b)** its empirical formula; **(c)** its molecular formula.

53. Dimethylhydrazine is a carbon–hydrogen–nitrogen compound used in rocket fuels. When burned completely, a 0.505-g sample yields 0.741 g CO_2 and 0.605 g H_2O. The nitrogen content of a 0.486-g sample is converted to 0.226 g N_2. What is the empirical formula of dimethylhydrazine?
55. *Without doing detailed calculations,* explain which of the following compounds produces the *greatest* mass of CO_2 when 1.00 mol of the compound undergoes complete combustion: CH_4, C_2H_5OH, $C_{10}H_8$, C_6H_5OH.

52. *Para*-cresol (*p*-cresol) is used as a disinfectant and in the manufacture of herbicides. A 0.3654-g sample of this carbon–hydrogen–oxygen compound yields 1.0420 g CO_2 and 0.2437 g H_2O in combustion analysis. Its molecular mass is 108.1 u. For *p*-cresol, determine **(a)** its mass percent composition; **(b)** its empirical formula; **(c)** its molecular formula.
54. The organic solvent thiophene is a carbon–hydrogen–sulfur compound that yields CO_2, H_2O, and SO_2 on complete combustion. When subjected to combustion analysis, a 1.086-g sample of thiophene produces 2.272 g CO_2, 0.465 g H_2O, and 0.827 g SO_2. What is the empirical formula of thiophene?
56. *Without doing detailed calculations,* explain which of the following compounds produces the *greatest* mass of H_2O when 1.00 g of the compound undergoes complete combustion: CH_4, C_2H_5OH, $C_{10}H_8$, C_6H_5OH.

Oxidation States

57. Indicate the oxidation state of the underlined element in (**a**) $\underline{C}H_4$; (**b**) $\underline{S}F_4$; (**c**) $Na_2\underline{O}_2$; (**d**) $\underline{C}_2H_3O_2^-$; (**e**) $\underline{Fe}O_4^{2-}$.

59. Chromium forms three principal oxides. Write appropriate formulas for these compounds if the oxidation states of Cr in them are +3, +4, and +6, respectively.

58. Indicate the oxidation state of S in each of the following: (**a**) SO_3^{2-}; (**b**) $S_2O_3^{2-}$; (**c**) $S_2O_8^{2-}$; (**d**) HSO_4^-; (**e**) $S_4O_6^{2-}$.

60. Nitrogen forms five oxides. Write appropriate formulas for these compounds if the oxidation states of N in them are +1, +2, +3, +4, and +5, respectively.

Nomenclature

61. Name the compounds (**a**) SrO; (**b**) ZnS; (**c**) K_2CrO_4; (**d**) Cs_2SO_4; (**e**) Cr_2O_3; (**f**) $Fe_2(SO_4)_3$; (**g**) $Mg(HCO_3)_2$; (**h**) $(NH_4)_2HPO_4$; (**i**) $Ca(HSO_3)_2$; (**j**) $Cu(OH)_2$; (**k**) HNO_3; (**l**) $KClO_4$; (**m**) $HBrO_3$; (**n**) H_3PO_3.

63. Assign suitable names to the compounds (**a**) CS_2; (**b**) SiF_4; (**c**) ClF_5; (**d**) N_2O_5; (**e**) SF_6; (**f**) I_2Cl_6.

65. Write correct formulas for the compounds (**a**) aluminum sulfate; (**b**) ammonium dichromate; (**c**) silicon tetrafluoride; (**d**) iron(III) oxide; (**e**) tricarbon disulfide; (**f**) cobalt(II) nitrate; (**g**) strontium nitrite; (**h**) hydrobromic acid; (**i**) iodic acid; (**j**) phosphorus dichloride trifluoride.

67. Write a formula for (**a**) the chloride of titanium having Ti in the O.S. +4; (**b**) the sulfate of iron having Fe in the O.S. +3; (**c**) an oxide of chlorine with Cl in the O.S. +7; (**d**) an oxoanion of sulfur in which the apparent O.S. of S is +7 and the ionic charge is 2−.

62. Name the compounds (**a**) $Ba(NO_3)_2$; (**b**) HNO_2; (**c**) CrO_2; (**d**) KIO_3; (**e**) $LiCN$; (**f**) KIO; (**g**) $Fe(OH)_2$; (**h**) $Ca(H_2PO_4)_2$; (**i**) H_3PO_4; (**j**) $NaHSO_4$; (**k**) $Na_2Cr_2O_7$; (**l**) $NH_4C_2H_3O_2$; (**m**) MgC_2O_4; (**n**) $Na_2C_2O_4$.

64. Assign suitable names to the compounds (**a**) ICl; (**b**) ClF_3; (**c**) SF_4; (**d**) BrF_5; (**e**) N_2O_4; (**f**) S_4N_4.

66. Write correct formulas for the compounds (**a**) magnesium perchlorate; (**b**) lead(II) acetate; (**c**) tin(IV) oxide; (**d**) hydroiodic acid; (**e**) chlorous acid; (**f**) sodium hydrogen sulfite; (**g**) calcium dihydrogen phosphate; (**h**) aluminum phosphate; (**i**) dinitrogen tetroxide; (**j**) disulfur dichloride.

68. Write a formula for (**a**) an oxide of nitrogen with N in the O.S. +5; (**b**) an oxoacid of nitrogen with N in the O.S. +3; (**c**) an oxide of carbon in which the apparent O.S. of C is $+\frac{4}{3}$; (**d**) a sulfur-containing oxoanion in which the apparent O.S. of S is +2.5 and the ionic charge 2−.

Hydrates

69. *Without performing detailed calculations,* indicate which of the following hydrates has the greatest % H_2O, by mass: $CuSO_4 \cdot 5H_2O$, $Cr_2(SO_4)_3 \cdot 18H_2O$, $MgCl_2 \cdot 6H_2O$, and $LiC_2H_3O_2 \cdot 2H_2O$.

71. Anhydrous $CuSO_4$ can be used to dry liquids in which it is insoluble. The $CuSO_4$ is converted to $CuSO_4 \cdot 5H_2O$, which can be filtered off from the liquid. What minimum mass of anhydrous $CuSO_4$ would be required to remove 8.5 g H_2O that had become mixed with a tankful of gasoline?

70. *Without performing detailed calculations,* determine the hydrate of Na_2SO_3 that contains almost exactly 50% H_2O, by mass.

72. Anhydrous sodium sulfate, Na_2SO_4, absorbs water vapor and is converted to the *deca*hydrate, $Na_2SO_4 \cdot 10H_2O$. How much would the mass of 3.50 g of anhydrous Na_2SO_4 increase if converted completely to the decahydrate?

Advanced Exercises

73. The mineral spodumene has the empirical formula $LiAlSi_2O_6$. Given that the percentage of lithium-6 atoms in naturally occurring lithium is 7.40%, how many lithium-6 atoms are present in a 518-g sample of spodumene?

74. Two compounds of Cl and X are found to have molecular masses and % Cl, by mass, as follows: 137 u, 77.5% Cl; 208 u, 85.1% Cl. What is the element X? What is the formula of each compound?

75. Chlorophyll contains Mg to the extent of 2.72% by mass. Assuming one Mg atom per chlorophyll molecule, what is the molecular mass of chlorophyll?

76. A certain hydrate is found to have the composition 20.3% Cu, 8.95% Si, 36.3% F, and 34.5% H_2O, by mass. What is the empirical formula of this hydrate?

77. A sample of $MgSO_4 \cdot xH_2O$ weighing 8.129 g is heated until all the water of hydration is driven off. The resulting anhydrous compound, $MgSO_4$, weighs 3.967 g. What is the formula of the hydrate?

78. A particular type of brass contains Cu, Sn, Pb, and Zn. A 1.1713-g sample is treated in such a way as to convert the Sn to 0.245 g SnO_2, the Pb to 0.115 g $PbSO_4$, and the Zn to 0.246 g $Zn_2P_2O_7$. What is the mass percent of each element in the sample?

79. A 1.562-g sample of the hydrocarbon C_7H_{16} is burned in an excess of oxygen. What masses of CO_2 and H_2O should be obtained?

80. Refer to the compound ethyl mercaptan described in Example 3-2. Assuming that the complete combustion of this compound produces CO_2, H_2O, and SO_2, what masses of each of these three products would be produced in the complete combustion of 1.50 mL of ethyl mercaptan?

81. What mass of the food preservative butylated hydroxyanisole (BHA), $C_{11}H_{16}O_2$, should be burned to produce 0.5000 g CO_2 in combustion analysis?

82. In a saturated hydrocarbon, C_nH_{2n+2}, the C atoms are bonded to one another and the H atoms to C atoms, all by single bonds. Furthermore, every C atom forms *four* bonds. For example, for $n = 3$, the hydrocarbon is propane, C_3H_8.

$$H-\underset{\underset{H}{|}}{\overset{\overset{H}{|}}{C}}-\underset{\underset{H}{|}}{\overset{\overset{H}{|}}{C}}-\underset{\underset{H}{|}}{\overset{\overset{H}{|}}{C}}-H$$

Can there be a saturated hydrocarbon that yields a greater mass of H_2O than of CO_2 on combustion? Explain.

83. A hydrocarbon mixture consists of 60.0% by mass of C_3H_8 and 40.0% of C_xH_y. When 10.0 g of this mixture is burned, it yields 29.0 g CO_2 and 18.8 g H_2O as the only products. What is the formula of the unknown hydrocarbon?

84. A 0.732-g mixture of methane, CH_4, and ethane, C_2H_6, is burned, yielding 2.064 g CO_2. What is the percent composition of this mixture **(a)** by mass; **(b)** on a mole basis?

85. A sample of the compound MSO_4 weighing 0.1131 g reacts with barium chloride and yields 0.2193 g $BaSO_4$. What must be the atomic mass of the metal M? (*Hint:* All the SO_4^{2-} from the MSO_4 appears in the $BaSO_4$.)

86. The metal M forms the sulfate $M_2(SO_4)_3$. A 0.738-g sample of this sulfate is converted to 1.511 g $BaSO_4$. What is the atomic mass of M? (*Hint:* Refer to Exercise 85.)

87. A 0.622-g sample of a metal oxide with the formula M_2O_3 is converted to the sulfide, MS, yielding 0.685 g. What is the atomic mass of the metal M?

88. $MgCl_2$ often occurs as an impurity in table salt (NaCl) and is responsible for "caking" of the salt. A 0.5200-g sample of table salt is found to contain 61.10% Cl, by mass. What is the % $MgCl_2$ in the sample? Why is the precision of this calculation so poor?

89. When 2.750 g of the oxide of lead Pb_3O_4 is strongly heated, it decomposes and produces 0.0640 g of oxygen gas and 2.686 g of a second oxide of lead. What is the empirical formula of this second oxide?

90. A 1.013-g sample of $ZnSO_4 \cdot xH_2O$ is dissolved in water and the sulfate ion precipitated as $BaSO_4$. The mass of pure, dry $BaSO_4$ obtained is 0.8223 g. What is the formula of the zinc sulfate hydrate?

Feature Problems

A. To deposit exactly one mole of Ag from an aqueous solution containing Ag^+ requires a quantity of electricity known as one faraday ($\mathcal{F}$). The electrodeposition requires that each Ag^+ ion gain one electron to become an Ag atom. Use appropriate physical constants listed on the inside back cover to obtain a precise value of the Avogadro constant, N_A.

B. While developing his atomic theory, Dalton proposed a law of chemical combination known as the *law of multiple proportions:* When a pair of elements form more than one compound, the masses of one element that combine with a fixed mass of the second element are in the ratio of small whole numbers. For example, one oxide of carbon (CO) has a mass ratio of oxygen to carbon of 16 g O to 12 g C and another oxide (CO_2), 32 g O to 12 g C; the masses of oxygen in the two compounds are in the ratio 16 g O to 32 g O = 1 to 2 (small whole numbers). Dalton assigned the formulas N_2O, NO, and NO_2 to the three oxides of nitrogen known in his time. Show that these formulas are consistent with the law of multiple proportions. Would you expect to find an oxide of nitrogen containing 26.40% N? Explain.

C. All-purpose fertilizers contain the essential elements nitrogen, phosphorous, and potassium. A typical fertilizer carries numbers on its label, such as "5-10-5." These numbers represent the % N, % P_2O_5, and % K_2O, respectively. The N is contained in the form of a nitrogen compound such as $(NH_4)_2SO_4$, NH_4NO_3, or $CO(NH_2)_2$ (urea). The P is generally present as a phosphate and the K as KCl. The expressions "% P_2O_5" and "% K_2O" were devised in the nineteenth century, before the nature of chemical compounds was fully understood. To convert from % P_2O_5 to % P and from % K_2O to % K, we must use the factors 2 mol P/mol P_2O_5 and 2 mol K/mol K_2O, together with molar masses.

1. What is the percent composition of the "5-10-5" fertilizer in % N, % P, and % K?

2. What is the % P_2O_5 in the following compounds, both common fertilizers: **(a)** $Ca(H_2PO_4)_2$; **(b)** $(NH_4)_2HPO_4$?

Chemical Reactions

4

We are all aware that iron rusts and natural gas (for cooking or heating) burns. These processes are chemical reactions. Chemical reactions are the central concern, not just of this chapter, but of the entire science of chemistry.

In this chapter we will first learn to represent chemical reactions by chemical equations, and then we will use chemical equations to establish quantitative (numerical) relationships among the reactants and products of a reaction, a topic known as reaction stoichiometry. Because many chemical reactions occur in solution, we will also introduce a method of describing the composition of a solution, called solution molarity. Throughout the chapter we will discuss new aspects of problem solving and more uses for the mole concept.

In this vigorous reaction between iron(III) oxide and powdered aluminum metal, called the thermite reaction, the products are liquid iron and aluminum oxide. The thermite reaction is used in on-site welding of large iron objects.

FIGURE 4-1
Precipitation of silver chromate.
When aqueous solutions of silver nitrate and potassium chromate are mixed, the disappearance of the distinctive yellow color of chromate ion and the appearance of red-brown solid silver chromate provide physical evidence of a reaction.

▶Sometimes products react, partially or completely, to re-form the original reactants. Such reactions are called *reversible* reactions and are designated by a double arrow (⇌). In this chapter we assume that any reverse reactions are negligible and that reactions go only in the forward direction.

4-1 Chemical Reactions and the Chemical Equation

A **chemical reaction** is a process in which one set of substances called **reactants** is converted to a new set of substances called **products.** In other words, a chemical reaction is the *process* by which a chemical change occurs. In many cases, though, nothing happens when substances are mixed; they retain their original compositions and properties. We need evidence before we can say that a reaction has occurred. Some of the types of physical evidence to look for are

- a color change (Figure 4-1);
- formation of a solid (precipitate) within a clear solution (Figure 4-1);
- evolution of a gas (Figure 4-2);
- evolution or absorption of heat (Figure 4-2).

At times, none of these signs of a chemical reaction appears. Then, we need *chemical* evidence to decide whether a reaction has occurred. This requires a detailed chemical analysis of the reaction mixture to discover whether any new substances are present.

Just as we use symbols for elements and formulas for compounds, we have a symbolic, or shorthand, way of representing a chemical reaction—the **chemical equation.** In a chemical equation, formulas of the reactants are written on the *left* side of the equation and formulas of the products are written on the *right*. The two sides of the equation are joined by an arrow (→) or an equal sign (=). We say that the reactants *yield* the products. Consider the reaction of colorless nitrogen monoxide and oxygen gases to form red-brown nitrogen dioxide gas, a reaction that occurs in the manufacture of nitric acid.

$$\text{nitrogen monoxide} + \text{oxygen} \longrightarrow \text{nitrogen dioxide}$$

1. Substitute chemical *formulas* for names. The result is a *formula expression.*

$$NO + O_2 \longrightarrow NO_2$$

2. *Balance* the formula expression to obtain a *chemical equation.**

$$2\,NO + O_2 \longrightarrow 2\,NO_2$$

Because atoms are neither created nor destroyed in a chemical reaction, an equation must be balanced. In a **balanced equation** the total number of atoms of each element is the same on both sides. In step 1 there are *three* O atoms on the left side (*one* in the molecule NO and *two* in the molecule O_2). On the right side there are only *two* O atoms (in the molecule NO_2). In the balancing step (2), we place the coefficient 2 in front of the formulas NO and NO_2. This means that *two* molecules of NO are consumed and *two* molecules of NO_2 are produced for every molecule of O_2 consumed. In the balanced equation there are *two* N atoms and *four* O atoms on each side. The coefficients required to balance a chemical equation are called **stoichiometric coefficients.** As we will shortly learn, these coefficients are essential in calculating the amounts of re-

---

*An equation—whether mathematical or chemical—must have the left and right sides equal. We should not call a formula expression an equation until it is balanced. And the term chemical equation automatically signifies that this balance exists. Although unnecessary, the terms *unbalanced* and *balanced* are both commonly used when referring to a chemical equation.

actants used or products formed in a chemical reaction. Such calculations, called **stoichiometry** (stoy-key-om-eh-tree), have the balanced chemical equation as their central focus.

In balancing a chemical equation keep in mind that the equation can be balanced *only* by adjusting the *coefficients* of formulas, as necessary. In particular,

⃠ *Never introduce extraneous formulas.*

$$\cancel{NO + O_2 \longrightarrow NO_2 + O}$$

Although the above equation is "balanced," it is *incorrect*. No atomic oxygen (O) is produced in this reaction.

⃠ *Never change formulas for the purpose of balancing an equation.*

$$\cancel{NO + O_2 \longrightarrow NO_3}$$

Again, the equation is "balanced," but it is *incorrect*. The sole product of the reaction is NO_2. No NO_3 is formed.

The method of equation balancing we have been describing is called *balancing by inspection*. Balancing by inspection means to adjust stoichiometric coefficients by "trial and error" until a balanced condition is found. Although we can generally balance the elements in any order, equation balancing need not be a "hit or miss" affair. Here, for example, are two useful strategies for balancing equations.

- If an element occurs in only one compound on each side of the equation, try balancing this element *first*.
- When one of the reactants or products exists as the *free* element, balance this element *last*.

At this point, you should be able (a) to write formulas for the reactants and products of a reaction and (b) to balance the equation. A third task that we will undertake in later chapters is to *predict* the products that should form when certain reactants are brought together. Even now, however, based on what we learned in the previous chapter, we can predict the products of a *combustion reaction*. The combustion of hydrocarbons and of carbon–hydrogen–oxygen compounds in a plentiful supply of oxygen gas produces carbon dioxide gas and liquid water as the *only* products. If the compound contains sulfur as well, sulfur dioxide is also a product. These ideas and an equation-balancing strategy are illustrated in Examples 4-1 and 4-2.

FIGURE 4-2
Decomposition of ammonium dichromate.
When solid ammonium dichromate is ignited, the evolution of a gas (nitrogen) and of heat provide physical evidence of a reaction.

EXAMPLE 4-1

Writing and Balancing an Equation: The Combustion of a Hydrocarbon. Propane gas, C_3H_8, is easily liquefied, stored, and transported for use as a fuel. It is commonly encountered in backyard gas-fired barbecue grills. Write a balanced chemical equation to represent its complete combustion.

Solution

The combustion of a hydrocarbon (carbon–hydrogen compound) in an excess of oxygen yields carbon dioxide and water as the sole products.

$$\text{propane} + \text{oxygen} \longrightarrow \text{carbon dioxide} + \text{water}$$

Formula expression: $C_3H_8 + O_2 \longrightarrow CO_2 + H_2O$

▶ For many students, substituting correct formulas for names is the most difficult part of writing and balancing equations. The nomenclature rules from Chapter 3 can be helpful here.

According to the equation-balancing suggestions noted earlier, we could start with either C or H. Each occurs in just one reactant and one product. We should, however, balance oxygen last because it occurs in both products (on the right) and as the free element as a reactant (on the left). As we set each coefficient, we keep it fixed while we work on the next one, and so on, until we achieve a final balance.

Balance C:	$C_3H_8 + O_2 \longrightarrow 3\,CO_2 + H_2O$
Balance H:	$C_3H_8 + O_2 \longrightarrow 3\,CO_2 + 4\,H_2O$
Balance O:	$C_3H_8 + 5\,O_2 \longrightarrow 3\,CO_2 + 4\,H_2O$
Balanced equation:	$C_3H_8 + 5\,O_2 \longrightarrow 3\,CO_2 + 4\,H_2O$

Self-Check. To be certain that an equation is balanced, run a final check. *Left side:* 3 C atoms and 8 H atoms (in one molecule of C_3H_8); 10 O atoms (in five molecules of O_2). *Right side:* 3 C atoms (in three molecules of CO_2); 8 H atoms (in four molecules of H_2O); 10 O atoms (six from the three CO_2 molecules and four from the four H_2O molecules).

Practice Example A: Write a balanced equation for the complete combustion of ethylene, C_2H_4, the starting material for producing polyethylene plastics.

Practice Example B: Write a balanced equation for the complete combustion of thiophene, C_4H_4S, a compound used in the manufacture of pharmaceuticals.

EXAMPLE 4-2

Writing and Balancing an Equation: The Combustion of a Carbon–Hydrogen–Oxygen Compound. Liquid triethylene glycol, $C_6H_{14}O_4$, is used as a solvent and plasticizer for vinyl and polyurethane plastics. Write a balanced chemical equation for its complete combustion.

Solution

Carbon–hydrogen–oxygen compounds, like hydrocarbons, yield carbon dioxide and water when burned in oxygen gas.

Formula expression:	$C_6H_{14}O_4 + O_2 \longrightarrow CO_2 + H_2O$
Balance C:	$C_6H_{14}O_4 + O_2 \longrightarrow 6\,CO_2 + H_2O$
Balance H:	$C_6H_{14}O_4 + O_2 \longrightarrow 6\,CO_2 + 7\,H_2O$

At this point, the right side of the expression has 19 O atoms (12 in six CO_2 molecules and 7 in seven H_2O molecules). To get 19 O atoms on the left, we start with 4 in a molecule of $C_6H_{14}O_4$ and need 15 more. This requires a *fractional* coefficient of $\frac{15}{2}$ for O_2.

Balance O: $\quad C_6H_{14}O_4 + \frac{15}{2}\,O_2 \longrightarrow 6\,CO_2 + 7\,H_2O \quad$ (balanced)

Final Adjustment of Coefficients. Although fractional coefficients are quite acceptable, general practice is to remove them by multiplying all coefficients by the same whole number—in this case "2," the denominator of the fractional coefficient.

$$2\,C_6H_{14}O_4 + 15\,O_2 \longrightarrow 12\,CO_2 + 14\,H_2O \quad \text{(balanced)}$$

Self-Check

Left:	$(2 \times 6) = 12\ \text{C};\ (2 \times 14) = 28\ \text{H};\ [(2 \times 4) + (15 \times 2)] = 38\ \text{O}$
Right:	$(12 \times 1) = 12\ \text{C};\ (14 \times 2) = 28\ \text{H};\ [(12 \times 2) + (14 \times 1)] = 38\ \text{O}$

Practice Example A: Glycerol, $C_3H_8O_3$, is used in the food industry (called glycerin) and is also an important biological molecule. Write a balanced equation for its combustion.

Practice Example B: Write a balanced equation for the combustion of thiosalicylic acid, $C_7H_6O_2S$, used in the manufacture of indigo dyes.

States of Matter

Propane (Example 4-1) is a gas, but triethylene glycol (Example 4-2) is a liquid. Sometimes we need to represent this kind of information in a chemical equation. We can show the state of matter or physical form of reactants and products through parenthetical symbols. Four common symbols and their meanings are

(g) = gas   (l) = liquid   (s) = solid   (aq) = aqueous (water) solution

Thus, for the combustion of triethylene glycol we can write

$$2\ C_6H_{14}O_4(l) + 15\ O_2(g) \longrightarrow 12\ CO_2(g) + 14\ H_2O(l)$$

▶ At the high combustion temperature, water is present as $H_2O(g)$. But when the reaction products are returned to the initial temperature, the water condenses to a liquid, $H_2O(l)$.

Reaction Conditions

The equation for a chemical reaction alone does not provide enough information to plan how to carry out the reaction in the laboratory or a chemical plant. An important aspect of modern chemical research involves working out the conditions for a reaction. We often write the reaction conditions above or below the arrow in an equation. For example, the Greek capital letter delta, Δ, means that a high temperature is required. That is, the reaction mixture must be heated, as in the *decomposition* of silver oxide.

$$2\ Ag_2O(s) \overset{\Delta}{\longrightarrow} 4\ Ag(s) + O_2(g)$$

▶ In a *decomposition* reaction a substance is broken down into simpler substances (for instance, into its elements).

An even more explicit statement of reaction conditions is shown below for the BASF (Badische Anilin- & Soda-Fabrik) process for the *synthesis* of methanol from CO and H_2. This reaction occurs at 350 °C, under a total gas pressure that is 340 times greater than the normal pressure of the atmosphere, and on the surface of a mixture of ZnO and Cr_2O_3 acting as a catalyst. (As we will learn later on, a *catalyst* is a substance that enters into a reaction in such a way as to make the reaction go faster without itself being consumed in the reaction.)

▶ In a *synthesis* reaction a new compound is formed from the reaction of two or more simpler substances, usually called the reactants or starting materials.

$$CO(g) + 2\ H_2(g) \xrightarrow[\substack{340\ \text{atm} \\ ZnO,\ Cr_2O_3}]{350\ °C} CH_3OH(g)$$

It is important to be able to calculate how much of a particular product will be produced when certain quantities of the reactants are consumed. In the next section, we will see how to use chemical equations to set up conversion factors that we can use for these and related calculations.

4-2   The Chemical Equation and Stoichiometry

In Greek, the word *stoicheion* means element. The term *stoichiometry* means, literally, to measure the elements. Its more practical meaning, though, includes all the quantitative relationships involving atomic and formula masses, chemical formulas, and the chemical equation. We considered the quantitative meaning of chemical formulas in Chapter 3, and now we will explore the quantitative aspects of chemical equations.

The coefficients in the chemical equation

$$2 H_2(g) + O_2(g) \longrightarrow 2 H_2O(l) \qquad (4.1)$$

mean that

$$2 \text{ molecules } H_2 + 1 \text{ molecule } O_2 \longrightarrow 2 \text{ molecules } H_2O$$

or that

$$2x \text{ molecules } H_2 + x \text{ molecules } O_2 \longrightarrow 2x \text{ molecules } H_2O.$$

Suppose that we let $x = 6.02214 \times 10^{23}$ (the Avogadro constant). Then x molecules represents *1 mole*. Thus the chemical equation also means that

$$2 \text{ mol } H_2 + 1 \text{ mol } O_2 \longrightarrow 2 \text{ mol } H_2O$$

The coefficients in the chemical equation allow us to make statements, such as

- *Two* moles of H_2O are *produced* for every *two* moles of H_2 *consumed.*
- *Two* moles of H_2O are *produced* for every *one* mole of O_2 *consumed.*
- *Two* moles of H_2 are *consumed* for every *one* mole of O_2 *consumed.*

Moreover, we can turn such statements into conversion factors, called stoichiometric factors. A **stoichiometric factor** relates the amounts of any two substances involved in a chemical reaction, on a *mole* basis. In the examples that follow stoichiometric factors are printed in blue.

EXAMPLE 4-3

Relating the Numbers of Moles of a Product and a Reactant. How many moles of H_2O are produced in reaction (4.1) by burning 2.72 mol H_2 in an excess of O_2?

Solution

The statement "an excess of O_2" means that there is more than enough O_2 available to permit the complete conversion of 2.72 mol H_2 to H_2O. The conversion factor we need from Equation (4.1) is based on the fact that 2 mol H_2O is produced for every 2 mol H_2. Think of this as 2 mol H_2O = 2 mol H_2.

$$? \text{ mol } H_2O = 2.72 \text{ mol } H_2 \times \frac{2 \text{ mol } H_2O}{2 \text{ mol } H_2} = 2.72 \text{ mol } H_2O$$

Practice Example A: How many moles of O_2 are produced from the decomposition of 1.76 moles of potassium chlorate: $2 KClO_3(s) \rightarrow 2 KCl(s) + 3 O_2(g)$?

Practice Example B: How many moles of Ag are produced in the decomposition of 1.00 kg of silver(I) oxide: $2 Ag_2O(s) \rightarrow 4 Ag(s) + O_2(g)$? (*Hint:* How many moles of Ag_2O are in 1.00 kg?)

Most reaction stoichiometry calculations are more complex than Example 4-3, or at least they may seem so. In some ways, finding the correct set-up is like finding a path through a maze. But there is a key for getting out of the maze—a stoichiometric factor from the balanced equation. Because a stoichiometric factor is always written on a mole-to-mole basis, the part of the set-up *preceding* this factor must involve conversions from other quantities to moles of a substance. The part of the set-up *following* the stoichiometric factor involves conversions from moles of a substance to other quantities. So we set up a conversion pathway to get us to the center of this maze, where we use a stoichiometric factor to convert moles of one substance into moles of another. Then we use another conversion pathway to get us out of the maze—to the answer in the desired units.

Example 4-4 and Figure 4-3 apply this strategy to calculating the mass of a product we expect from a given mass of reactant. This is a practical type of calculation because we commonly measure quantities of substances by their masses.

EXAMPLE 4-4

Relating the Mass of a Reactant and a Product. What mass of H_2O is formed in the reaction of 4.16 g H_2 with an excess of O_2?

Solution

The general strategy for reaction stoichiometry problems, outlined earlier and illustrated in Figure 4-3, suggests these three steps.

1. Convert the quantity of H_2 from grams to moles. [Use molar mass of H_2.]
2. From the number of moles of H_2, calculate the number of moles of H_2O formed. [Use stoichiometric factor from Equation (4.1).]
3. Convert the quantity of H_2O from moles to grams. [Use molar mass of H_2O.]

FIGURE 4-3
Strategy for a stoichiometric calculation.
The object of the calculation is to relate the masses of two substances: one a reactant and the other a product, or both reactants, or both products (tan panels). In each step, a quantity in a horizontal panel is multiplied by a factor in a vertical panel. The key conversion in the overall calculation is from mol A to mol B (pink panels), through the use of a stoichiometric factor (green panel). The stoichiometric factor is the ratio of the stoichiometric coefficients from the balanced equation. Molar masses are also required, in the manner indicated by the gray panels. In Example 4-4, H_2 is the substance A and H_2O is the substance B. In Example 4-5, H_2 is again the substance A and O_2 is the substance B.

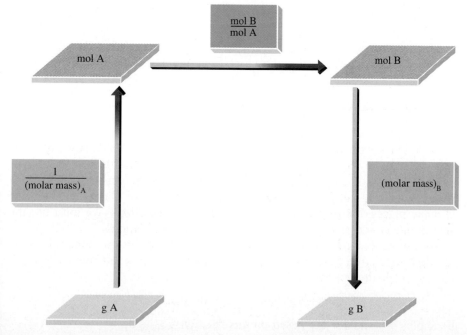

Although we think in terms of these steps, we can just as easily write a single set-up. The individual steps are noted in the following through the small numbers above the arrows.

$$(g\ H_2 \xrightarrow{\ 1\ } mol\ H_2 \xrightarrow{\ 2\ } mol\ H_2O \xrightarrow{\ 3\ } g\ H_2O)$$

$$?\ g\ H_2O = 4.16\ g\ H_2 \times \frac{1\ mol\ H_2}{2.016\ g\ H_2} \times \frac{2\ mol\ H_2O}{2\ mol\ H_2} \times \frac{18.02\ g\ H_2O}{1\ mol\ H_2O}$$

$$= 37.2\ g\ H_2O$$

Practice Example A: How many grams of magnesium nitride are produced by the reaction of 3.82 g Mg with an excess of N_2? $3\ Mg + N_2 \rightarrow Mg_3N_2$

Practice Example B: How many grams of $H_2(g)$ are required to produce 1.00 kg methanol, CH_3OH, by the reaction: $CO + 2\ H_2 \rightarrow CH_3OH$? (*Hint:* Note that here we seek information about a reactant rather than a product.)

EXAMPLE 4-5

Relating the Masses of Two Reactants to Each Other. What mass of O_2 is consumed in the complete combustion of 6.86 g H_2 in reaction (4.1)?

Solution

As in Example 4-4, we need molar masses and the appropriate stoichiometric factor from the balanced equation (4.1) to use as conversion factors. We use them in the following three-stage set-up.

$$(g\ H_2 \longrightarrow mol\ H_2 \longrightarrow mol\ O_2 \longrightarrow g\ O_2)$$

$$?\ g\ O_2 = 6.86\ g\ H_2 \times \frac{1\ mol\ H_2}{2.016\ g\ H_2} \times \frac{1\ mol\ O_2}{2\ mol\ H_2} \times \frac{32.00\ g\ O_2}{1\ mol\ O_2}$$

$$= 54.4\ g\ O_2$$

Practice Example A: How many grams of H_2 are consumed per gram of O_2 in reaction (4.1)?

Practice Example B: How many grams of O_2 are consumed per gram of octane, C_8H_{18}, in the combustion of octane? (*Hint*: What is the balanced chemical equation?)

What lends great variety to stoichiometric calculations is that many other conversions may be required before and after the mol A $\rightarrow$ mol B step at the center of the stoichiometric maze. These other conversions may use such factors as volume, density, and percent composition. As a result, we may enter the maze in many places, but to get out—to complete the calculation—we must use the appropriate stoichiometric factor from the chemical equation.

To sample this greater range of possibilities, let us shift our attention to the reaction pictured in Figure 4-4, a simple laboratory method of preparing small volumes of hydrogen gas.

$$2\ Al(s) + 6\ HCl(aq) \longrightarrow 2\ AlCl_3(aq) + 3\ H_2(g) \tag{4.2}$$

Examples 4-6 and 4-7 are based on this reaction.

FIGURE 4-4
The reaction 2 Al(s) + 6 HCl(aq) →
2 AlCl$_3$(aq) + 3 H$_2$(g)
HCl(aq) is introduced to the flask on the left. The reaction occurs within the flask. The liberated H$_2$(g) is conducted to a gas-collection apparatus, where it displaces water. Hydrogen is only very slightly soluble in water.

EXAMPLE 4-6

Additional Conversion Factors in a Stoichiometric Calculation: Volume, Density, and Percent Composition. An alloy used in aircraft structures consists of 93.7% Al and 6.3% Cu by mass. The alloy has a density of 2.85 g/cm^3. A 0.691-cm^3 piece of the alloy reacts with an excess of HCl(aq). If we assume that *all* the Al, but *none* of the Cu, reacts with HCl(aq), what is the mass of H$_2$ obtained?

Solution

One approach to this calculation is to follow the strategy diagrammed in Figure 4-5 (page 106), followed by the strategy we outlined in Figure 4-3. Although such diagrams are helpful, we can generally make a simpler outline by using arrows to represent conversions. Each numbered arrow refers to a conversion factor that changes the unit on the left to the one on the right. That is,

$$\text{cm}^3 \text{ alloy} \xrightarrow{1} \text{g alloy} \xrightarrow{2} \text{g Al} \xrightarrow{3} \text{mol Al} \xrightarrow{4} \text{mol H}_2 \xrightarrow{5} \text{g H}_2$$

We can do the calculation in five distinct steps, reaching our final answer in step 5. Alternatively, we can do the calculation through a single set-up in which five conversions are written in sequence. This approach, which we use below, avoids the need to write down intermediate results.

$$(\text{cm}^3 \text{ alloy} \xrightarrow{\ 1\ } \text{g alloy} \xrightarrow{\ 2\ } \text{g Al} \xrightarrow{\ 3\ } \text{mol Al}$$

$$?\ \text{g H}_2 = 0.691\ \text{cm}^3 \text{ alloy} \times \frac{2.85\ \text{g alloy}}{1\ \text{cm}^3 \text{ alloy}} \times \frac{93.7\ \text{g Al}}{100.0\ \text{g alloy}} \times \frac{1\ \text{mol Al}}{26.98\ \text{g Al}}$$

$$\xrightarrow{\ 4\ } \text{mol H}_2 \xrightarrow{\ 5\ } \text{g H}_2)$$

$$\times \frac{3\ \text{mol H}_2}{2\ \text{mol Al}} \times \frac{2.016\ \text{g H}_2}{1\ \text{mol H}_2} = 0.207\ \text{g H}_2$$

FIGURE 4-5
Strategy for the calculation in Example 4-6.
The steps shown here are for the conversion of the volume of Al-Cu alloy to the mass of aluminum present. The remainder of the calculation follows the strategy outlined in Figure 4-3. The mass of aluminum corresponds to the panel marked "g A" in Figure 4-3, and the mass of hydrogen, to the panel marked "g B." The stoichiometric factor is 3 mol H_2/2 mol Al. As in Figure 4-3, the quantities in the horizontal panels are multiplied by the factors in the vertical panels.

Practice Example A: What volume of the aluminum–copper alloy described in Example 4-6 must be dissolved in an excess of HCl(aq) to produce 1.00 g H_2? (*Hint:* You may think of this as the "inverse" of Example 4-6.)

Practice Example B: A fresh sample of the aluminum-copper alloy described in Example 4-6 yielded 1.31 g H_2. How many grams of *copper* were present in the sample?

EXAMPLE 4-7

Additional Conversion Factors in a Stoichiometric Calculation: Volume, Density, and Percent Composition of a Solution. A hydrochloric acid solution consists of 28.0% HCl, by mass, and has a density of 1.14 g/mL. What volume of this solution is required to react with 1.87 g Al in reaction (4.2)?

Solution

A challenge here is to determine where to begin. Although the problem refers to 28.0% HCl and a density of 1.14 g/mL, the appropriate starting point is with the "given" information—1.87 g Al. The goal of our calculation is a solution volume—mL HCl solution.

$$\text{g Al} \xrightarrow{1} \text{mol Al} \xrightarrow{2} \text{mol HCl} \xrightarrow{3} \text{g HCl} \xrightarrow{4} \text{g HCl solution} \xrightarrow{5} \text{mL HCl solution}$$

The conversion factors in the calculation involve (1) the molar mass of Al, (2) stoichiometric coefficients from Equation (4.2), (3) the molar mass of HCl, (4) the percent composition of the HCl solution, and (5) the density of the HCl solution. The strategy is outlined below.

$$(\text{g Al} \xrightarrow{1} \text{mol Al} \xrightarrow{2} \text{mol HCl} \xrightarrow{3} \text{g HCl}$$

$$? \text{ mL HCl soln} = 1.87 \text{ g Al} \times \frac{1 \text{ mol Al}}{26.98 \text{ g Al}} \times \frac{6 \text{ mol HCl}}{2 \text{ mol Al}} \times \frac{36.46 \text{ g HCl}}{1 \text{ mol HCl}}$$

$$\xrightarrow{4} \text{ g HCl soln } \xrightarrow{5} \text{mL HCl soln)}$$

$$\times \frac{100.0 \text{ g HCl soln}}{28.0 \text{ g HCl}} \times \frac{1 \text{ mL HCl soln}}{1.14 \text{ g HCl soln}}$$

$$= 23.8 \text{ mL HCl soln}$$

Practice Example A: How many milligrams of H_2 are produced when 1 drop (0.05 mL) of the hydrochloric acid solution described in Example 4-7 reacts with an excess of aluminum in reaction (4.2)?

Practice Example B: A particular vinegar contains 4.0% $HC_2H_3O_2$, by mass. It reacts with $NaHCO_3$ (baking soda) to produce carbon dioxide.

$$HC_2H_3O_2(aq) + NaHCO_3(s) \longrightarrow NaC_2H_3O_2(aq) + H_2O(l) + CO_2(g)$$

How many grams of CO_2 are produced by the reaction of 5.00 mL of this vinegar with an excess of $NaHCO_3$? The density of the vinegar is 1.01 g/mL.

4-3   Chemical Reactions in Solution

Most chemical reactions in the general chemistry laboratory are carried out in solutions. This is partly because mixing the reactants in solution helps to achieve the close contact between atoms, ions, or molecules necessary for a reaction to occur. To describe the stoichiometry of reactions in solutions, we can use the same ideas as for other reactions. In fact, we did this in Example 4-7. We also need, however, a few new ideas that apply specifically to solution stoichiometry.

One component of a solution, called the **solvent,** determines whether the solution exists as a solid, liquid, or gas. In this discussion we will limit ourselves to solutions in which liquid water is the solvent commonly referred to as *aqueous* solutions. The other components of a solution, called **solutes,** are said to be dissolved in the solvent. By writing NaCl(aq), for example, we are describing a solution in which liquid water is the solvent and NaCl, the solute. What the expression "NaCl(aq)" does not tell us, however, is the composition of the solution—the relative proportions of NaCl and H_2O. For this purpose we prefer something other than the percent by mass composition that we have used previously.

Molarity

The concentration or **molarity** of a solution is defined as

$$\text{molarity (M)} = \frac{\text{amount of solute (in moles)}}{\text{volume of solution (in liters)}} \qquad (4.3)$$

If 0.455 mol urea, $CO(NH_2)_2$, is dissolved in enough water to make 1.000 L of solution, the solution concentration, or molarity, is

$$\frac{0.455 \text{ mol } CO(NH_2)_2}{1.000 \text{ L soln}} = 0.455 \text{ M } CO(NH_2)_2$$

▶ Because the volume of a solution is rarely equal to that of the solvent used, molarity must be based on the volume of *solution.*

The unit of molarity, mol (solute)/L (solution), is often replaced by the symbol M and called "molar." We refer to this solution as "0.455 molar urea."

Of course, we must relate an amount in moles to other quantities that we can measure directly. Also, we need not always work with exactly 1 liter of solution. Sometimes a much smaller volume is sufficient; sometimes we need larger volumes. In Example 4-8, we relate the mass of a liquid solute to its volume by using density as a conversion factor. Then, we use molar mass to convert from the mass of solute to an amount of solute in moles. Also, we convert the solution volume from milliliters (mL) to liters (L).

EXAMPLE 4-8

Calculating a Molarity from Measured Quantities. A solution is prepared by dissolving 25.0 mL ethanol, C_2H_5OH ($d = 0.789$ g/mL), in enough water to produce 250.0 mL solution. What is the molarity of ethanol in the solution?

Solution

To determine the molarity, we must first calculate how many moles of ethanol are in the solution. To calculate the number of moles of ethanol in a 25.0-mL sample, we need conversion factors based on density and molar mass.

$$? \text{ mol } C_2H_5OH = 25.0 \text{ mL } C_2H_5OH \times \frac{0.789 \text{ g } C_2H_5OH}{1 \text{ mL } C_2H_5OH} \times \frac{1 \text{ mol } C_2H_5OH}{46.07 \text{ g } C_2H_5OH}$$

$$= 0.428 \text{ mol } C_2H_5OH$$

Now, we note that 250.0 mL = 0.2500 L, and then we use the definition of molarity in expression (4.3).

$$\text{molarity} = \frac{0.428 \text{ mol } C_2H_5OH}{0.2500 \text{ L soln}} = 1.71 \text{ M } C_2H_5OH$$

Practice Example A: A 22.3-g sample of acetone, $(CH_3)_2CO$, is dissolved in enough water to produce 1.25 L of solution. What is the molarity of acetone in this solution?

Practice Example B: 15.0 mL of acetic acid, $HC_2H_3O_2$ ($d = 1.048$ g/mL), is dissolved in enough water to produce 500.0 mL of solution. What is the molarity of acetic acid in the solution?

Figure 4-6 illustrates a method commonly used to prepare a solution. We weigh out a solid sample and dissolve it in sufficient water to produce a solution of *known* volume—250.0 mL in Figure 4-6. To fill a beaker to a 250-mL mark is not nearly precise enough as a volume measurement. As discussed in Section 1-7, there would be a systematic error because the beaker is not calibrated with sufficient precision. (The error in volume could be 10–20 mL or more.) Likewise, adding 250 mL water to a beaker from a graduated cylinder is also not sufficiently precise. Although the graduated cylinder is calibrated more precisely than the beaker, the error might still be 1–2 mL or more. On the other hand, the volumetric flask pictured in Figure 4-6 contains 250.0 mL, with less than 0.1 mL of error when filled to the calibration mark.

In Example 4-9, we calculate the mass of solute needed to produce the solution in Figure 4-6. The key to this calculation is to formulate a conversion factor based on molarity. Thus, when we write 0.250 M K_2CrO_4, we are in effect saying that 0.250 mol K_2CrO_4 = 1 L soln. The conversion pathway required in Example 4-9 is L soln → mol solute → g solute.

(a)

(b)

(c)

FIGURE 4-6 **Preparation of 0.250 M K₂CrO₄—Example 4-9 illustrated.**
The solution *cannot* be prepared just by adding 12.1 g K_2CrO_4(s) to 250.0 mL water. Instead, (a) the weighed quantity of K_2CrO_4(s) is first added to a clean dry volumetric flask;
(b) the K_2CrO_4(s) is dissolved in less than 250 mL of water; and (c) the flask is filled to the
250.0-mL calibration mark by the careful addition (dropwise) of the remaining water.

EXAMPLE 4-9

Calculating the Mass of Solute in a Solution of Known Molarity. We want to prepare
exactly 0.2500 L (250.0 mL) of an 0.250 M K_2CrO_4 solution in water. What mass
of K_2CrO_4 should we use (see Figure 4-6)?

Solution

As just stated, we can use solution molarity (printed in blue below) to convert between volume of solution and amount of solute. The other conversion factor we need
is the molar mass of K_2CrO_4.

$$(\text{L soln} \longrightarrow \text{mol } K_2CrO_4 \longrightarrow \text{g } K_2CrO_4)$$

$$? \text{ g } K_2CrO_4 = 0.2500 \text{ L soln} \times \frac{0.250 \text{ mol } K_2CrO_4}{1 \text{ L soln}} \times \frac{194.2 \text{ g } K_2CrO_4}{1 \text{ mol } K_2CrO_4}$$

$$= 12.1 \text{ g } K_2CrO_4$$

Practice Example A: An aqueous solution saturated with $NaNO_3$ at 25 °C is
10.8 M $NaNO_3$. How many grams of $NaNO_3$ are present in 125 mL of this solution
at 25 °C?

Practice Example B: How many grams of $Na_2SO_4 \cdot 10H_2O$ are needed to prepare 355 mL of 0.445 M Na_2SO_4?

▶ A *concentrated* solution has a relatively large amount of dissolved solute; a *dilute* solution has a small amount.

▶ Feature Problem A on page 127 describes dilution from a biological perspective.

Solution Dilution

A common sight in chemistry storerooms and laboratories is rows of bottles containing solutions for use in chemical reactions. It is not practical, however, to store solutions of every possible concentration. Instead, we generally store fairly concentrated solutions, so-called *stock* solutions. It is a rather simple matter to prepare more dilute solutions by adding water to stock solutions. The principle involved, suggested by Figure 4-7, is that *all the solute taken from the initial, more concentrated solution appears in the final diluted solution.*

This statement and the definition of molarity are all that you need to work out dilution problems. You may, however, prefer a method based on expression (4.3). Suppose we refer to a solution concentration (molarity) as *M,* a solution volume, in liters, as *V,* and the amount of solute, in moles, as *n*. Then, expression (4.3) becomes, $M = n/V$, which we can rearrange to

$$n = M \times V$$

When a solution is diluted, the amount of solute *remains constant* between the initial (i) solution taken and the final (f) solution produced. That is,

$$M_i \times V_i = n_i = n_f = M_f \times V_f$$

or

$$M_i \times V_i = M_f \times V_f \qquad (4.4)$$

Figure 4-8 illustrates the laboratory procedure for preparing a solution by dilution. Example 4-10 explains the calculation that would have to be done.

EXAMPLE 4-10

Preparing a Solution by Dilution. A particular analytical chemistry procedure requires 0.0100 M K_2CrO_4. What volume of 0.250 M K_2CrO_4 must we dilute with water to prepare 0.250 L of 0.0100 M K_2CrO_4?

Solution

To solve this problem, imagine reversing the arrow in Figure 4-7, that is, start with the final solution and determine something about the initial solution. First, let's calculate the amount of solute that must be present in the *final* solution.

$$? \text{ mol } K_2CrO_4 = 0.250 \text{ L soln} \times \frac{0.0100 \text{ mol } K_2CrO_4}{1 \text{ L soln}} = 0.00250 \text{ mol } K_2CrO_4$$

FIGURE 4-7
Visualizing the dilution of a solution.
The final solution is prepared by extracting 1/8th of the initial solution—1 cm³—and diluting it with water to a volume of 8 cm³. The number of dots in the 8 cm³ of final solution, representing the number of solute particles, is the same as in the 1 cm³ of initial solution.

├─── 1 cm ───┤

├─── 2 cm ───┤
initial solution

├─── 2 cm ───┤
final solution

(a)   (b)   (c)

FIGURE 4-8   **Preparing a solution by dilution—Example 4-10 illustrated.**
(a) A pipet is used to withdraw a 10.0-mL sample of 0.250 M K_2CrO_4(aq). (b) The pipetful of 0.250 M K_2CrO_4 is discharged into a 250.0-mL volumetric flask. (c) Following this, water is added to bring the level of the solution to the calibration mark on the neck of the flask. At this point the solution is 0.0100 M K_2CrO_4.

Because all the solute in the final, diluted solution comes from the initial, more concentrated solution, we must answer the question: What volume of 0.250 M K_2CrO_4 contains 0.00250 mol K_2CrO_4?

$$? \text{ L soln} = 0.00250 \text{ mol } K_2CrO_4 \times \frac{1 \text{ L soln}}{0.250 \text{ mol } K_2CrO_4} = 0.0100 \text{ L soln}$$

Alternatively, we can use Equation (4.4). We know the volume of solution that we wish to prepare (V_f = 250.0 mL) and the concentrations of the final (0.0100 M) and initial (0.250 M) solutions. We must solve for the initial volume, V_i. Note that, although in deriving Equation (4.4) we expressed volumes in liters, in applying it we can use any volume unit, as long as we use the same unit for both V_i and V_f (milliliters in the present case). The term needed to convert volumes to the unit, L, would appear on each side of the equation and cancel out.

$$V_i = V_f \times \frac{M_f}{M_i} = 250.0 \text{ mL} \times \frac{0.0100 \text{ M}}{0.250 \text{ M}} = 10.0 \text{ mL}$$

Practice Example A: A 15.00-mL sample of 0.450 M K_2CrO_4 is diluted to 100.00 mL. What is the concentration of the new solution?

Practice Example B: When left in an open beaker for a period of time, the volume of 275 mL of 0.105 M NaCl is found to decrease to 237 mL, due to the evaporation of water. What is the new concentration of the solution?

Stoichiometry of Reactions in Solution

When we measure the volume of a solution of known molarity, we obtain a fixed number of moles of solute. Therefore, in stoichiometric calculations, which require us to work in the unit mol, we can use molarities and volumes instead. Example 4-11 illustrates this idea. Viewed from the standpoint of the stoichiometric maze introduced in the preceding section (Figures 4-3 and 4-5), the key to the maze in Example 4-11 is the same as in previous examples—the appropriate stoichiometric factor. What differs from previous examples is that we need to use molarity as a conversion factor from solution volume to number of moles of reactant.

EXAMPLE 4-11

Relating the Mass of a Product to the Volume and Molarity of a Reactant Solution. A 25.00-mL pipetful of 0.250 M K_2CrO_4(aq) is added to an excess of $AgNO_3$(aq). What mass of Ag_2CrO_4(s) will precipitate from the solution?

$$K_2CrO_4(aq) + 2\ AgNO_3(aq) \longrightarrow Ag_2CrO_4(s) + 2\ KNO_3(aq)$$

Solution

▶ In general, the first step in a stoichiometric calculation is to write a balanced equation for the reaction. Either this must be given, or you must be able to supply your own.

The steps involved in this calculation are (1) determine the amount of K_2CrO_4 that reacts; (2) use a stoichiometric factor from the chemical equation to determine the amount (in moles) of Ag_2CrO_4(s) produced; (3) convert the amount of Ag_2CrO_4(s) to its mass in grams. This conversion pathway is: mL K_2CrO_4 → mol K_2CrO_4 → mol Ag_2CrO_4 → g Ag_2CrO_4. We can combine these steps into a single set-up.

$$? \text{ g } Ag_2CrO_4 = 25.00 \text{ mL} \times \frac{1 \text{ L}}{1000 \text{ mL}} \times \frac{0.250 \text{ mol } K_2CrO_4}{1 \text{ L}}$$

$$\times \frac{1 \text{ mol } Ag_2CrO_4}{1 \text{ mol } K_2CrO_4} \times \frac{331.7 \text{ g } Ag_2CrO_4}{1 \text{ mol } Ag_2CrO_4}$$

$$= 2.07 \text{ g } Ag_2CrO_4(s)$$

Practice Example A: How many milliliters of 0.250 M K_2CrO_4 must be added to excess $AgNO_3$(aq) to produce 1.50 g Ag_2CrO_4? (*Hint:* Think of this as the "inverse" of Example 4-11.)

Practice Example B: How many milliliters of 0.150 M $AgNO_3$(aq) must be added to an excess of K_2CrO_4 (aq) to produce exactly 1.00 g Ag_2CrO_4?

We will consider a number of additional examples of stoichiometric calculations involving solutions in Chapter 5.

4-4 Determining the Limiting Reactant

▶ Feature Problem B on page 127 deals with an experimental method of determining stoichiometric proportions.

▶ By an excess of a reactant we mean that more of the reactant is present than is consumed in the reaction. Some is left over.

▶ Chemicals are often referred to as *reagents,* and the limiting reactant in a reaction is sometimes called a *limiting reagent.*

When all the reactants are completely and simultaneously consumed in a chemical reaction we say that the reactants are in **stoichiometric proportions**—in the mole ratios dictated by the coefficients in the balanced equation. We sometimes require this condition, for example, in certain chemical analyses. At other times, such as in a precipitation reaction, we convert one of the reactants completely into products by using an *excess* of all the other reactants. The reactant that is completely consumed—the **limiting reactant**—determines the quantities of products that form. In the reaction illustrated in Figure 4-1 (page 98), K_2CrO_4 is the limiting reactant and $AgNO_3$ is present in excess. Up to this point we have stated the reactant(s) in excess, and, by implication, the limiting reactant. In some cases, however, the limiting reactant will not be indicated explicitly, and you must first do a calculation to identify it. Consider the following analogy.

title page — 87 copies

instructions — 83 copies

data sheets — 168 copies

graph paper — 328 copies

an assembled "hand out" experiment

4 sheets of graph paper

2 data sheets

1 sheet of instructions

1 title page

FIGURE 4-9   An analogy to determining the limiting reactant in a chemical reaction—assembling a handout experiment.
From the numbers of copies available of each of the parts, we conclude that only 82 complete handouts are possible and that the graph paper is the limiting reactant.

Your chemistry instructor asks you to assemble a "handout" experiment for use in the laboratory and wants to know how many copies will be available. As shown in Figure 4-9, a handout is assembled in the following way.

title page + instruction sheet + 2 data sheets + 4 sheets of graph paper $\longrightarrow$ handout

Based on the number of title pages available, you would say that there can be no more than 87 complete handouts. From the available instruction sheets you would lower your assessment to 83. Based on the data sheets, you would say $168/2 = 84$. Finally, based on the graph paper, you would conclude $328/4 = 82$ copies. You would report to your instructor that the available number of handouts is 82—the *smallest* of the four possible answers.

This problem-solving strategy is somewhat different from what we have used before in the text. Instead of a single step-by-step progression from the given information to the final result, we do two (or more) calculations in parallel and from the two (or more) results we choose the correct one. Example 4-12 applies this line of reasoning to a chemical reaction.

EXAMPLE 4-12

Determining the Limiting Reactant in a Reaction. Phosphorus trichloride, PCl_3, is a commercially important compound used in the manufacture of pesticides, gasoline additives, and a number of other products. It is made by the direct combination of phosphorus and chlorine.

$$P_4(s) + 6\,Cl_2(g) \longrightarrow 4\,PCl_3(l)$$

What mass of $PCl_3(l)$ forms in the reaction of 125 g P_4 with 323 g Cl_2?

Solution

The method suggested by Figure 4-9 is to do *two* calculations. In each, determine the mass of PCl_3 that would be produced from one of the reactants, assuming that the other is in excess. The correct answer is the *smaller* of the two results.

Assuming that P_4 is the limiting reactant:

$$? \text{ g } PCl_3 = 125 \text{ g } P_4 \times \frac{1 \text{ mol } P_4}{123.9 \text{ g } P_4} \times \frac{4 \text{ mol } PCl_3}{1 \text{ mol } P_4} \times \frac{137.3 \text{ g } PCl_3}{1 \text{ mol } PCl_3}$$

$$= 554 \text{ g } PCl_3$$

Assuming that Cl_2 is the limiting reactant:

$$? \text{ g PCl}_3 = 323 \text{ g Cl}_2 \times \frac{1 \text{ mol Cl}_2}{70.91 \text{ g Cl}_2} \times \frac{4 \text{ mol PCl}_3}{6 \text{ mol Cl}_2} \times \frac{137.3 \text{ g PCl}_3}{1 \text{ mol PCl}_3}$$

$$= 417 \text{ g PCl}_3$$

The smaller value is 417 g PCl_3. When this mass of PCl_3 has been formed, the Cl_2 is completely consumed and the reaction stops. *Cl_2 is the limiting reactant.*

There are other approaches to limiting reactant problems. For example, we can calculate mole ratios of reactants and compare these ratios to stoichiometric factors from the balanced equations to determine limiting reactants. However, with the aid of modern calculators the parallel-calculation method illustrated here is quick and straightforward.

Practice Example A: If 215 g P_4 is allowed to react with 725 g Cl_2 in Example 4-12, how many grams of PCl_3 are formed?

Practice Example B: If 1.00 kg each of PCl_3, Cl_2, and P_4O_{10} are allowed to react, how many kilograms of $POCl_3$ will be formed?

$$6 \text{ PCl}_3(l) + 6 \text{ Cl}_2(g) + \text{P}_4\text{O}_{10}(s) \longrightarrow 10 \text{ POCl}_3(l)$$

▶ Feature Problem C on page 127 presents an analogy to reaction stoichiometry and the concept of a limiting reactant from the world of commerce.

Sometimes in a limiting reactant problem our interest is in determining how much of an excess reactant remains, as well as how much product is formed. This additional calculation is illustrated in Example 4-13.

EXAMPLE 4-13

Determining the Quantity of Excess Reactant(s) Remaining After a Reaction. What mass of P_4 remains in excess following the reaction in Example 4-12?

Solution

The key to this problem is to calculate the mass of P_4 that is consumed, and we can base this calculation either on the mass of Cl_2 consumed

$$? \text{ g P}_4 = 323 \text{ g Cl}_2 \times \frac{1 \text{ mol Cl}_2}{70.91 \text{ g Cl}_2} \times \frac{1 \text{ mol P}_4}{6 \text{ mol Cl}_2} \times \frac{123.9 \text{ g P}_4}{1 \text{ mol P}_4} = 94.1 \text{ g P}_4$$

or on the mass of PCl_3 produced.

$$? \text{ g P}_4 = 417 \text{ g PCl}_3 \times \frac{1 \text{ mol PCl}_3}{137.3 \text{ g PCl}_3} \times \frac{1 \text{ mol P}_4}{4 \text{ mol PCl}_3} \times \frac{123.9 \text{ g P}_4}{1 \text{ mol P}_4} = 94.1 \text{ g P}_4$$

The mass of P_4 remaining after the reaction is simply the difference between what was originally present and what was consumed, that is,

$$125 \text{ g P}_4 \text{ initially} - 94.1 \text{ g P}_4 \text{ consumed} = 31 \text{ g P}_4 \text{ remaining}$$

◀ We lose one significant figure because in subtraction we can retain only the smallest number of decimal places. Here, 125 is significant only in the "ones" place.

Practice Example A: In Practice Example 4-12A, which reactant is in excess and what mass of that reactant remains after the reaction to produce PCl_3?

Practice Example B: If 12.2 g H_2 and 154 g O_2 are allowed to react, which gas and what mass of that gas remains after the reaction?

$$2 \text{ H}_2(g) + \text{O}_2(g) \longrightarrow 2 \text{ H}_2\text{O}(l)$$

4-5   Other Practical Matters

There are a few added matters of concern regarding reaction stoichiometry in both the laboratory and the manufacturing plant. For one, the calculated outcome of a reaction may not be what is actually observed. Specifically, the amount of product may be, unavoidably, less than expected. Also, the route to producing a desired chemical may require several reactions carried out in sequence. And in some cases two or more reactions may occur simultaneously. These are the matters we describe in this section.

Theoretical Yield, Actual Yield, and Percent Yield

The **theoretical yield** of a reaction is the *calculated* quantity of product that one expects from given quantities of reactants. The quantity of product that is *actually* produced is called the **actual yield.** The **percent yield** is defined as

$$\text{percent yield} = \frac{\text{actual yield}}{\text{theoretical yield}} \times 100\% \qquad (4.5)$$

In many reactions the actual yield almost exactly equals the theoretical yield, and the reactions are said to be *quantitative.* We can use such reactions in performing quantitative chemical analyses, for example. On the other hand, in some reactions the actual yield is *less than* the theoretical yield, and the percent yield is *less than* 100%.

The yield may be less than 100% for many reasons. The product of a reaction rarely appears in a pure form, and in the necessary purification steps some product may be lost through handling. This reduces the yield. In many cases the reactants may participate in reactions other than the one of central interest. These are called **side reactions** and the unintended products are called **by-products.** To the extent that side reactions occur, the yield of the main product is reduced. Finally, if a reverse reaction occurs, some of the expected product may react to re-form the reactants, and again the yield is less than expected.

At times, the apparent yield is greater than 100%. Because one cannot get something for nothing, this situation usually indicates an error in technique. Some products are formed as a precipitate from a solvent. The product may be wet with solvent, thus giving a larger-than-expected mass for the wet product. More thorough drying of the product would give a more accurate yield determination. Another possibility is that the product is contaminated with an excess reactant or a by-product. This makes the mass of product appear larger than expected. In any case, a product must be purified before its yield is determined.

In Example 4-14, we establish the theoretical, actual, and percent yields of an important industrial process.

EXAMPLE 4-14

Determining Theoretical, Actual, and Percent Yields. Billions of pounds of urea, $CO(NH_2)_2$, are produced annually for use as a fertilizer. The reaction used is

$$2\,NH_3 + CO_2 \longrightarrow CO(NH_2)_2 + H_2O$$

The typical starting reaction mixture has a $3:1$ mole ratio of NH_3 to CO_2. If 47.7 g urea forms *per mole* of CO_2 that reacts, what is the (a) theoretical yield; (b) actual yield; and (c) percent yield in this reaction?

Solution

a. The stoichiometric proportions are 2 mol NH_3 : 1 mol CO_2. Because the mole ratio of NH_3 to CO_2 used is 3:1, NH_3 is in excess and CO_2 is the limiting reactant. Because the quantity of urea is given per mole of CO_2, we should base the calculation on 1.00 mol CO_2.

$$\text{theoretical yield} = 1.00 \text{ mol } CO_2 \times \frac{1 \text{ mol } CO(NH_2)_2}{1 \text{ mol } CO_2} \times \frac{60.1 \text{ g } CO(NH_2)_2}{1 \text{ mol } CO(NH_2)_2}$$

$$= 60.1 \text{ g } CO(NH_2)_2$$

b.
$$\text{actual yield} = 47.7 \text{ g } CO(NH_2)_2$$

c.
$$\% \text{ yield} = \frac{47.7 \text{ g } CO(NH_2)_2}{60.1 \text{ g } CO(NH_2)_2} \times 100\% = 79.4\%$$

Practice Example A: Formaldehyde, CH_2O, can be made from methanol by the following reaction, using a copper catalyst:

$$CH_3OH(g) \longrightarrow CH_2O(g) + H_2(g)$$

If 25.7 g $CH_2O(g)$ is produced per mole of methanol that reacts, what are **(a)** the theoretical yield; **(b)** the actual yield; and **(c)** the percent yield in this reaction?

Practice Example B: What is the percent yield if the reaction of 25.0 g P_4 with 91.5 g Cl_2 produces 104 g PCl_3? $P_4 + 6 Cl_2 \rightarrow 4 PCl_3$ (*Hint:* Which is the limiting reactant?)

If we know that a yield will be less than 100%, we need to make an adjustment in the amounts of reactants that we use to produce the desired amount of product. That is, we cannot simply use the theoretical amounts of reactants; we must use more. This point is illustrated in Example 4-15.

EXAMPLE 4-15

Adjusting the Quantities of Reactants in Accordance with the Percent Yield of a Reaction. When heated with sulfuric or phosphoric acid, cyclohexanol, $C_6H_{11}OH$, is converted to cyclohexene, C_6H_{10}.

$$C_6H_{11}OH(l) \longrightarrow C_6H_{10}(l) + H_2O(l) \tag{4.6}$$

If the percent yield of this reaction is 83%, what mass of cyclohexanol must we use to obtain 25 g cyclohexene?

Solution

First, let us answer this question: If the 25 g of C_6H_{10} represents only 83% of the theoretical yield, what is the theoretical yield? Then we can calculate the quantity of $C_6H_{11}OH$ required to produce this theoretical yield of C_6H_{10}.

▶ Think of this first step as an algebra problem. If 83% of a number is equal to 25, what is the number? $0.83x = 25$; $x = 25/0.83 = 30$.

Step 1. Rearrange Equation (4.5) and calculate the theoretical yield.

$$\text{theoretical yield} = \frac{\text{actual yield} \times 100\%}{\text{percent yield}} = \frac{25 \text{ g} \times 100\%}{83\%} = 30 \text{ g}$$

Step 2. Calculate the quantity of $C_6H_{11}OH$ needed to produce 30 g C_6H_{10}.

$$? \text{ g } C_6H_{11}OH = 30 \text{ g } C_6H_{10} \times \frac{1 \text{ mol } C_6H_{10}}{82.1 \text{ g } C_6H_{10}} \times \frac{1 \text{ mol } C_6H_{11}OH}{1 \text{ mol } C_6H_{10}}$$

$$\times \frac{100.2 \text{ g } C_6H_{11}OH}{1 \text{ mol } C_6H_{11}OH}$$

$$= 37 \text{ g } C_6H_{11}OH$$

Practice Example A: If the percent yield for the formation of urea in Example 4-14 were 87.5%, what mass of CO_2, together with an excess of NH_3, would have to be used to obtain 50.0 g $CO(NH_2)_2$?

Practice Example B: Calculate the mass of cyclohexanol ($C_6H_{11}OH$) needed to produce 45.0 g cyclohexene (C_6H_{10}) by reaction (4.6) if the reaction has a 86.2% yield and the cyclohexanol is 92.3% pure. (*Hint:* As an extra step you must convert from a quantity of *pure* $C_6H_{11}OH$ to a quantity of the *impure* starting material.)

Consecutive Reactions, Simultaneous Reactions, and Net Reactions

Where possible, whether in the laboratory or the manufacturing plant, we prefer processes that yield a product through a single reaction. Often such processes give a higher yield because there is no need to remove products from one reaction mixture for further processing in subsequent reactions. But in many cases a multistep reaction is unavoidable. Reactions that are carried out one after another in sequence to yield a final product are called **consecutive reactions.** In **simultaneous reactions,** two or more substances react independently of each other in separate reactions occurring at the same time.

Example 4-16 presents an industrial process that is carried out in two reactions occurring consecutively. The key to the calculation is to use a stoichiometric factor for each reaction. Practice Example 4-16B deals with two reactions that occur simultaneously to produce a common product, hydrogen gas.

EXAMPLE 4-16

Calculating the Quantity of a Substance Produced by Reactions Occurring Consecutively. Titanium dioxide, TiO_2, is the most widely used white pigment for paints, having displaced most lead-based pigments, which are environmental hazards. Before it can be used, however, naturally occurring TiO_2 must be freed of colored impurities. One process for doing this converts impure $TiO_2(s)$ to $TiCl_4(g)$, which is then converted back to pure $TiO_2(s)$. What mass of carbon is consumed in producing 1.00 kg of pure $TiO_2(s)$ in this process?

$$2 \text{ TiO}_2 \text{ (impure)} + 3 \text{ C(s)} + 4 \text{ Cl}_2(g) \longrightarrow 2 \text{ TiCl}_4(g) + \text{CO}_2(g) + 2 \text{ CO(g)}$$

$$\text{TiCl}_4(g) + \text{O}_2(g) \longrightarrow \text{TiO}_2(s) + 2 \text{ Cl}_2(g)$$

Solution

This calculation requires us to begin with the product, TiO_2, and work backwards to one of the reactants, C. First, we convert (a) from moles of TiO_2 to moles of $TiCl_4$ in the *second* reaction and (b) from moles of $TiCl_4$ to moles of C in the *first* reaction. Together with other conversions between grams and moles of substances, the overall conversion pathway is

$$? \text{ g C} = 1.00 \text{ kg TiO}_2 \times \frac{1000 \text{ g TiO}_2}{1 \text{ kg TiO}_2} \times \frac{1 \text{ mol TiO}_2}{79.88 \text{ g TiO}_2} \times \frac{1 \text{ mol TiCl}_4}{1 \text{ mol TiO}_2}$$

$$\times \frac{3 \text{ mol C}}{2 \text{ mol TiCl}_4} \times \frac{12.01 \text{ g C}}{1 \text{ mol C}} = 226 \text{ g C}$$

Practice Example A: Nitric acid, HNO_3, is produced from ammonia and oxygen by the following consecutive reactions

$$4 \text{ NH}_3(g) + 5 \text{ O}_2(g) \longrightarrow 4 \text{ NO}(g) + 6 \text{ H}_2\text{O}(g)$$

$$2 \text{ NO}(g) + \text{O}_2(g) \longrightarrow 2 \text{ NO}_2(g)$$

$$3 \text{ NO}_2(g) + \text{H}_2\text{O}(l) \longrightarrow 2 \text{ HNO}_3(aq) + \text{NO}(g)$$

How many grams of nitric acid can be obtained from 1.00 kg $NH_3(g)$, if $NO(g)$ in the third reaction is *not* recycled?

Practice Example B: Magnalium alloys are widely used in aircraft construction. One magnalium alloy contains 70.0% Al and 30.0% Mg, by mass. How many grams of $H_2(g)$ are produced in the reaction of a 0.710-g sample of this alloy with excess HCl(aq)?

$$2 \text{ Al}(s) + 6 \text{ HCl}(aq) \longrightarrow 2 \text{ AlCl}_3(aq) + 3 \text{ H}_2(g)$$

$$\text{Mg}(s) + 2 \text{ HCl}(aq) \longrightarrow \text{MgCl}_2(aq) + \text{H}_2(g)$$

[*Hint:* Determine the mass of $H_2(g)$ produced in each reaction and combine the results.]

Usually, we can combine a series of chemical equations to obtain a single equation to represent the overall reaction. The overall reaction is called the **net reaction,** and the equation for this net reaction is the **net equation.** At times we can use the net equation for problem solving instead of working with the individual equations. This strategy does not work, however, if the substance of interest is not a starting material or final product but appears only in one of the intermediate reactions. Any substance that is produced in one step and consumed in another step of a multistep process is called an **intermediate.** In Example 4-16, $TiCl_4(g)$ is an intermediate.

To write a net equation for Example 4-16, multiply the coefficients in the second equation by the factor 2, add it to the first equation, and cancel any substances that appear on both sides of the net equation.

$$2 \text{ TiO}_2 \text{ (impure)} + 3 \text{ C}(s) + 4 \text{ Cl}_2(g) \longrightarrow 2 \text{ TiCl}_4(g) + \text{CO}_2(g) + 2 \text{ CO}(g)$$

$$\underline{2 \text{ TiCl}_4(g) + 2 \text{ O}_2(g) \longrightarrow 2 \text{ TiO}_2(s) + 4 \text{ Cl}_2(g)}$$

Net equation: $3 \text{ C}(s) + 2 \text{ O}_2(g) \longrightarrow \text{CO}_2(g) + 2 \text{ CO}(g)$

This is an interesting result. It suggests that (a) we should obtain as much TiO_2 in the second reaction as we started with in the first; (b) the $Cl_2(g)$ produced in the second reaction can be recycled back into the first reaction; and (c) the only substances actually consumed in the net reaction are $C(s)$ and $O_2(g)$. You can also see that we could not have used the net equation in the calculation of Example 4-16; $TiO_2(s)$ does not appear in the net equation.

Summary

Chemical reactions are represented by chemical equations, and the equations must be balanced to show that the total number of atoms of each element remains unchanged in the reaction. Physical states of the reactants and products and reaction conditions can also be indicated. Calculations based on the chemical equation use conversion factors derived from the equation (stoichiometric factors). These calculations often require the use of molar masses, densities, and percent compositions as well.

The molarity of a solution is the amount of solute (in moles) per liter of solution. Molarity can be treated as a conversion factor between solution volume and amount of solute. Molarity as a conversion factor may be applied to individual solutions, in cases where solutions are mixed or diluted by adding more solvent, and to reactions occurring in solution.

Additional features sometimes arise in stoichiometric calculations: The single reactant that determines the amount of product formed—the limiting reactant—may have to be identified. Some reactions yield exactly the quantity of product calculated. They have 100% yield. Others have an actual yield less than the theoretical value. Their percent yield is less than 100%. In some cases, a final product may be produced in a sequence of reactions. Sometimes it is possible to replace a series of equations for consecutive reactions by a single *net*

Summarizing Example

Sodium nitrite is used in the production of dyes for coloring fabrics, as a preservative in meat processing (to prevent botulism), as a bleach for fibers, and in photography. It can be prepared by passing nitrogen monoxide and oxygen gases into an aqueous solution of sodium carbonate. Carbon dioxide gas is another product of the reaction.

In one reaction with a 95.0% yield, 225 mL of 1.50 M $Na_2CO_3(aq)$, 22.1 g NO, and a large excess of O_2 are allowed to react. What mass of $NaNO_2$ is obtained?

1. *Write a chemical equation for the reaction.* This requires substituting formulas for names of substances and balancing the formula expression. *Result:*

$$2\,Na_2CO_3(aq) + 4\,NO(g) + O_2(g) \longrightarrow 4\,NaNO_2(aq) + 2\,CO_2(g)$$

2. *Determine the limiting reactant and the theoretical yield.* Calculate the theoretical yield of $NaNO_2$, first assuming that Na_2CO_3, and then, NO, is the limiting reactant (oxygen is in excess). To do this, you will need to determine the number of moles of Na_2CO_3 in 225 mL 1.50 M Na_2CO_3 and the number of moles of NO in 22.1 g NO. You will also need to use the stoichiometric factors 4 mol $NaNO_2$/2 mol Na_2CO_3 and 4 mol $NaNO_2$/4 mol NO, and the molar mass of NO. *Result:* Na_2CO_3 is the limiting reactant, and the theoretical yield is 46.6 g $NaNO_2$.

3. *Determine the actual yield.* Use expression (4.5), which relates percent yield (95.0%), theoretical yield (46.6 g $NaNO_2$), and the actual yield. *Answer:* 44.3 g $NaNO_2$.

Key Terms

actual yield (4-5)
balanced equation (4-1)
by-product (4-5)
chemical equation (4-1)
chemical reaction (4-1)
consecutive reactions (4-5)
intermediate (4-5)
limiting reactant (4-4)

molarity (4-3)
net equation (4-5)
net reaction (4-5)
percent yield (4-5)
product (4-1)
reactant (4-1)
side reaction (4-5)
simultaneous reactions (4-5)

solute (4-3)
solvent (4-3)
stoichiometric coefficient (4-1)
stoichiometric factor (4-2)
stoichiometric proportions (4-4)
stoichiometry (4-1)
theoretical yield (4-5)

Focus On Industrial Chemistry

Chemicals are manufactured on a large scale and sold for use in everything from food to pharmaceuticals. Every day, tons of chemicals are made to be used directly or to serve as reactants in the production of other materials (Table 4.1). It is important that these chemicals be made in high yield and be of high purity, but this is not always easy. Industrial chemists and engineers must consider many factors when making decisions about building and operating a chemical plant.

Let us examine some of these factors using as an example the production of hydrazine, N_2H_4, by a manufacturing method known as the Raschig process. Hydrazine is used as a rocket fuel, as a starting material to make pesticides, and in water treatment to remove dissolved oxygen from water in hot-water heating systems.

Stepwise Reactions, Intermediates, and the Net Chemical Equation

An industrial process is usually carried out in steps. The Raschig process involves three steps.

1. $Cl_2(g)$ and $NaOH(aq)$ react to produce sodium hypochlorite, $NaOCl(aq)$.

$$2\,NaOH + Cl_2 \longrightarrow NaOCl + NaCl + H_2O$$

2. $NH_3(aq)$ reacts with $NaOCl(aq)$ to form chloramine, $NH_2Cl(aq)$.

$$NaOCl + NH_3 \longrightarrow NH_2Cl + NaOH$$

3. Additional $NH_3(l)$ reacts with $NH_2Cl(aq)$ to form hydrazine, $N_2H_4(aq)$.

$$NH_3 + NH_2Cl + NaOH \longrightarrow N_2H_4 + NaCl + H_2O$$

$NaOCl(aq)$ and $NH_2Cl(aq)$ are *intermediates* in the process. Their presence is crucial to the overall process, but they are consumed immediately after their formation. Hydrazine is the *end product* of the process, and the overall process can be represented through the *net chemical equation*

$$2\,NaOH(aq) + Cl_2(g) + 2\,NH_3(aq) \longrightarrow N_2H_4(aq) + 2\,NaCl(aq) + 2\,H_2O \quad (4.7)$$

Notice that the intermediates do not show up as part of the net chemical equation for this reaction.

By-products and Side Reactions

Equation (4.7) shows two products in addition to N_2H_4: $NaCl$ and H_2O. Substances formed along with the desired end product are called *by-products*. The by-products themselves may be commercially valuable, a feature that improves the economic success of a process.

By-products may also form as a result of *side reactions* that compete with the main reaction. For example, one intermediate, chloramine, NH_2Cl, reacts with hydrazine to produce NH_4Cl and N_2 as by-products.

$$2\,NH_2Cl(aq) + N_2H_4(aq) \longrightarrow 2\,NH_4Cl(aq) + N_2(g) \quad (4.8)$$

Side reactions lower the yield of the desired product, so it is important for industrial chemists and chemical engineers to find conditions that minimize them.

Reaction Conditions

Even if the conditions of a chemical reaction are shown in a chemical equation (above the arrow), the reasons for these conditions are probably not. Temperature, pressure, and the presence of catalysts that help speed up the reaction are all important conditions that need to be controlled in order to get the best yield of the end product.

Modifications are always being made to industrial reactions. For

Review Questions

1. In your own words define or explain the following terms or symbols: (a) $\xrightarrow{\Delta}$; (b) (aq); (c) stoichiometric coefficient; (d) net equation.

2. Briefly describe each of the following ideas or methods: (a) balancing a chemical equation; (b) preparing a solution by dilution; (c) determining the limiting reactant in a reaction.

TABLE 4.1 The Top Ten Chemicals Produced in the United States (1995)

Rank	Billions of Pounds	Chemical	Formula	Principal Uses and End Products
1	95.36	sulfuric acid	H_2SO_4	fertilizers (70%); metallurgy (10%); petroleum refining (5%); manufacture of chemicals (5%)
2	68.04	nitrogen	N_2	inert atmospheres for manufacture of chemicals (25%); electronics manufacturing (20%); petroleum recovery (20%); metals processing (5%); freezing foods (5%)
3	53.48	oxygen	O_2	steel-making and other metallurgical processes (55%); chemical manufacture (20%)
4	46.97	ethylene	C_2H_4	manufacture of plastics (70%), antifreeze (10%), fibers (5%), and solvents (5%)
5	41.23	calcium oxide (lime)	CaO	metallurgy (40%); pollution control and wastewater treatment (15%); manufacture of chemicals (10%); municipal water treatment (10%)
6	35.60	ammonia	NH_3	fertilizers (80%); manufacture of plastics and fibers (10%) and explosives (5%)
7	26.19	phosphoric acid	H_3PO_4	fertilizers (95%)
8	26.19	sodium hydroxide	$NaOH$	manufacture of chemicals (50%), pulp and paper (20%), and soaps and detergents (5%); oil refining (5%)
9	25.69	propylene	C_3H_6	plastics (60%); fibers (15%); solvents (10%)
10	25.09	chlorine	Cl_2	manufacture of organic chemicals and plastics (65%), pulp and paper (15%)

example, a modern modification of the reaction shown in Equation (4.7), has the reaction being carried out in acetone, $(CH_3)_2CO$, rather than water. This change of solvent changes the mechanism of the reaction, that is, the path by which the reactant molecules become the product molecules. As a result the side reaction, Equation (4.8), is largely eliminated and the yield of hydrazine increases from 60–80% to nearly 100%. Currently most chemical plants in the United States use the modified process but some still use the original process. This illustrates two points:

1. Industrial methods of manufacturing chemicals undergo constant change, and

2. The changeover to a new process takes place over time, as new plants are built or old ones converted.

Purification

Purifying the product is an essential last step of any process. Rarely is the end product sufficiently pure for its intended uses. Hydrazine produced in reaction (4.7) is in a dilute aqueous solution, together with NH_3, $NaCl$, NH_4Cl, and traces of $NaOH$. Ammonia and water are removed by evaporating them from the solution (distillation). Substances such as $NaCl$, NH_4Cl, and $NaOH$ deposit as solids as water is evaporated from solution (crystallization). The solids are removed by filtration. Ultimately, a product that is more than 98% N_2H_4 is obtained. The NH_3 recovered in the purification process is reused in step 2 of the process. This illustrates another important principle of industrial chemistry. *Materials are recycled whenever possible.*

3. Explain the important distinctions between each pair of terms: (a) chemical formula and chemical equation; (b) decomposition and synthesis reaction; (c) solute and solvent;

(d) actual yield and percent yield.
4. Balance the following equations by inspection.
(a) $Na_2SO_4(s) + C(s) \rightarrow Na_2S(s) + CO(g)$

(b) $HCl(g) + O_2(g) \rightarrow H_2O(l) + Cl_2(g)$
(c) $PCl_5(s) + H_2O(l) \rightarrow H_3PO_4(aq) + HCl(aq)$
(d) $PbO(s) + NH_3(g) \rightarrow Pb(s) + N_2(g) + H_2O(l)$
(e) $Mg_3N_2(s) + H_2O(l) \rightarrow Mg(OH)_2(s) + NH_3(g)$

5. Write balanced equations for the following reactions.
(a) magnesium + oxygen → magnesium oxide
(b) nitrogen monoxide + oxygen → nitrogen dioxide
(c) ethane (C_2H_6) + oxygen → carbon dioxide + water
(d) aqueous silver sulfate + aqueous barium iodide → solid barium sulfate + solid silver iodide

6. Write a balanced chemical equation to represent
(a) the complete combustion of C_7H_{16};
(b) the complete combustion of C_4H_9OH;
(c) the reaction (neutralization) of hydroiodic acid with aqueous sodium carbonate to produce sodium iodide, water and carbon dioxide gas;
(d) the precipitation of iron(III) hydroxide by the mixing of aqueous solutions of iron(III) chloride and sodium hydroxide.

7. On an examination students were asked to write a *balanced* equation for the decomposition of potassium chlorate to yield potassium chloride and oxygen gas. Among the incorrect responses were the following three. Tell what is wrong with each expression.
(a) $KClO_3(s) \rightarrow KCl(s) + 3\ O(g)$
(b) $KClO_3(s) \rightarrow KCl(s) + O_2(g) + O(g)$
(c) $KClO_3(s) \rightarrow KClO(s) + O_2(g)$

8. Which of the following are correct statements for the reaction: $2\ H_2S + SO_2 \rightarrow 3\ S + 2\ H_2O$. Explain your reasoning. (1) 3 mol S is produced per mole of H_2S; (2) 3 g S is produced for every gram of SO_2 consumed; (3) 1 mol H_2O is produced per mole of H_2S consumed; (4) two thirds of the S produced comes from H_2S; (5) the total number of moles of products equals the number of moles of reactants consumed.

9. Iron metal reacts with chlorine gas as follows.
$$2\ Fe(s) + 3\ Cl_2(g) \longrightarrow 2\ FeCl_3(s)$$
How many moles of $FeCl_3$ are obtained when 7.26 mol Cl_2 reacts with excess Fe?

10. How many grams of O_2 are produced in the decomposition of 43.4 g $KClO_3$?
$$2\ KClO_3(s) \xrightarrow{\Delta} 2\ KCl(s) + 3\ O_2(g)$$

11. If 0.337 mol PCl_3 is produced by the reaction
$$6\ Cl_2 + P_4 \longrightarrow 4\ PCl_3$$
how many grams each of Cl_2 and P_4 are consumed?

12. The following reaction of potassium superoxide, KO_2, is used in life-support systems to replace $CO_2(g)$ in expired air by $O_2(g)$.
$$4\ KO_2(s) + 2\ CO_2(g) \longrightarrow 2\ K_2CO_3(s) + 3\ O_2(g)$$
(a) How many moles of $O_2(g)$ are produced by the reaction of 156 g $CO_2(g)$ with excess $KO_2(s)$?
(b) How many grams of $KO_2(s)$ are consumed per 100.0 g $CO_2(g)$ removed from expired air?
(c) How many O_2 molecules are produced per milligram of KO_2 consumed?

13. What are the molarities of the following solutes when dissolved in water?
(a) 2.92 mol CH_3OH in 7.16 L of solution
(b) 7.69 mmol C_2H_5OH in 50.00 mL of solution

(c) 25.2 g $CO(NH_2)_2$ in 275 mL of solution
(d) 18.5 mL of $C_3H_5(OH)_3$ ($d = 1.26$ g/mL) in 375 mL of solution

14. How much
(a) NaI, in moles, is needed to prepare 2.55×10^3 L of 0.125 M NaI?
(b) Na_2CO_3, in grams, is needed to produce 475 mL of 0.398 M Na_2CO_3?
(c) $CaCl_2$, in mg, is present *per milliliter* of 0.148 M $CaCl_2(aq)$?

15. *Without doing detailed calculations*, explain which of the following represents a 1.00 M KCl solution: 1.00 L of solution containing 100 g KCl; 500 mL of solution containing 74.6 g KCl; a solution containing 7.46 mg KCl/mL; 5.00 L of solution containing 373 g KCl.

16. What volume of 0.650 M $AgNO_3$, in milliliters, must be diluted with water to prepare 250.0 mL of 0.423 M $AgNO_3$?

17. Water is evaporated from 135 mL of 0.188 M K_2SO_4 solution until the solution volume becomes 105 mL. What is the molarity of K_2SO_4 in the solution that results?

18. Excess $NaHCO_3(s)$ is added to 415 mL of 0.275 M $Cu(NO_3)_2(aq)$. How many grams of $CuCO_3(s)$ will precipitate from solution?
$$Cu(NO_3)_2(aq) + 2\ NaHCO_3(s) \longrightarrow$$
$$CuCO_3(s) + 2\ NaNO_3(aq) + H_2O(l) + CO_2(g)$$

19. How many milliliters of 2.35 M $HCl(aq)$ are required to dissolve a 1.75-g piece of calcium carbonate (marble)?
$$CaCO_3(s) + 2\ HCl(aq) \longrightarrow CaCl_2(aq) + H_2O + CO_2(g)$$

20. A 0.78-mol sample of calcium hydride (CaH_2) and 1.52 mol of water are allowed to react. How many moles of H_2 will be produced? (*Hint:* Which is the limiting reactant?)
$$CaH_2(s) + 2\ H_2O(l) \longrightarrow Ca(OH)_2(s) + 2\ H_2(g)$$

21. A 0.696-mol sample of Cu is added to 136 mL of 6.0 M $HNO_3(aq)$. Assuming the following reaction is the only one that occurs, will the Cu react completely? (*Hint:* Which is the limiting reactant?)
$$3\ Cu(s) + 8\ HNO_3(aq) \longrightarrow$$
$$3\ Cu(NO_3)_2(aq) + 4\ H_2O + 2\ NO(g)$$

22. In the reaction of 1.80 mol CCl_4 with an excess of HF, 1.55 mol CCl_2F_2 is obtained. What are the **(a)** theoretical, **(b)** actual, and **(c)** percent yields of this reaction?
$$CCl_4 + 2\ HF \longrightarrow CCl_2F_2 + 2\ HCl$$

23. In the following reaction, 100.0 g $C_6H_{11}OH$ yielded 64.0 g C_6H_{10}.
$$C_6H_{11}OH \longrightarrow C_6H_{10} + H_2O$$
(a) What is the theoretical yield of the reaction?
(b) What is the percent yield?
(c) What mass of $C_6H_{11}OH$ should have been used to produce 100.0 g C_6H_{10}, if the percent yield is that determined in part **(b)**?

24. Chalkboard chalk is made from calcium carbonate and calcium sulfate, with minor impurities such as SiO_2. Only the $CaCO_3$ reacts with dilute $HCl(aq)$. What is the mass percent $CaCO_3$ in a piece of chalk if a 3.28-g sample yields 0.981 g $CO_2(g)$? (*Hint:* What mass of $CaCO_3$ produces 0.981 g CO_2?)
$$CaCO_3(s) + 2\ HCl(aq) \longrightarrow CaCl_2(aq) + H_2O(l) + CO_2(g)$$

Exercises

Writing and Balancing Chemical Equations

25. Balance the following equations by inspection.
(a) $Cr_2O_3(s) + Al(s) \xrightarrow{\Delta} Al_2O_3(s) + Cr(l)$
(b) $CaC_2(s) + H_2O(l) \rightarrow Ca(OH)_2(s) + C_2H_2(g)$
(c) $H_2(g) + Fe_2O_3(s) \xrightarrow{\Delta} Fe(l) + H_2O(g)$
(d) $NCl_3(g) + H_2O(l) \rightarrow NH_3(g) + HOCl(aq)$

27. Write balanced equations to represent the complete combustion of **(a)** butane, C_4H_{10}; **(b)** isopropyl alcohol, C_3H_7OH; **(c)** lactic acid, $HC_3H_5O_3$.

29. Write balanced equations to represent the following reactions.
(a) the decomposition, by heating, of solid ammonium nitrate to produce dinitrogen monoxide gas (laughing gas) and water vapor
(b) the reaction of aqueous sodium carbonate with hydrochloric acid to produce water, carbon dioxide gas, and aqueous sodium chloride
(c) The reaction of methane (CH_4), ammonia, and oxygen gases to form gaseous hydrogen cyanide and water vapor.

26. Balance the following equations by inspection.
(a) $(NH_4)_2Cr_2O_7(s) \xrightarrow{\Delta} Cr_2O_3(s) + N_2(g) + H_2O(g)$
(b) $NO_2(g) + H_2O(l) \rightarrow HNO_3(aq) + NO(g)$
(c) $H_2S(g) + SO_2(g) \rightarrow S(g) + H_2O(g)$
(d) $SO_2Cl_2 + HI \rightarrow H_2S + H_2O + HCl + I_2$

28. Write balanced equations to represent the complete combustion of **(a)** propylene, C_3H_6; **(b)** glycerol, $CH_2OHCHOHCH_2OH$; **(c)** thiobenzoic acid, C_6H_5COSH.

30. Write balanced equations to represent the following reactions.
(a) the reaction of sulfur dioxide gas with oxygen gas to produce sulfur trioxide gas (one reaction involved in the industrial preparation of sulfuric acid)
(b) the dissolving of limestone (calcium carbonate) in water containing dissolved carbon dioxide to produce calcium hydrogen carbonate (a reaction producing temporary hardness in groundwater)
(c) The reaction of ammonia and nitrogen monoxide to form nitrogen and water vapor.

Stoichiometry of Chemical Reactions

31. A laboratory method of preparing $O_2(g)$ involves the decomposition of $KClO_3(s)$.
$$2\ KClO_3(s) \xrightarrow{\Delta} 2\ KCl(s) + 3\ O_2(g)$$
(a) How many moles $O_2(g)$ can be produced by the decomposition of 32.8 g $KClO_3$?
(b) How many grams of $KClO_3$ must be decomposed to produce 50.0 g O_2?
(c) How many grams of KCl are formed, together with 28.3 g O_2, in the decomposition of $KClO_3$?

33. How many grams of Ag_2CO_3 must have been decomposed if 75.1 g Ag was obtained in the reaction
$$Ag_2CO_3(s) \xrightarrow{\Delta} Ag(s) + CO_2(g) + O_2(g) \text{ (not balanced)}$$

35. Iron ore is impure Fe_2O_3. When Fe_2O_3 is heated with an excess of carbon (coke), iron metal is produced. From a sample of ore weighing 938 kg, 523 kg of pure iron is obtained. What is the % Fe_2O_3, by mass, in the ore sample?
(*Hint:* Write a balanced equation for the reaction. How much Fe_2O_3 is required to produce the 523 kg iron?)

37. A piece of aluminum foil measuring 10.25 cm × 5.50 cm × 0.601 mm is dissolved in excess $HCl(aq)$. What mass of $H_2(g)$ is produced? Use Equation (4.2) and 2.70 g/cm^3 as the density of Al.

39. *Without performing detailed calculations,* determine which of the following produces the maximum quantity of $O_2(g)$ *per gram of reactant.*
(1) $2\ NH_4NO_3(s) \xrightarrow{\Delta} 2\ N_2(g) + 4\ H_2O(l) + O_2(g)$
(2) $2\ Ag_2O(s) \xrightarrow{\Delta} 4\ Ag(s) + O_2(g)$
(3) $2\ HgO(s) \xrightarrow{\Delta} 2\ Hg(l) + O_2(g)$
(4) $2\ Pb(NO_3)_2(s) \xrightarrow{\Delta} 2\ PbO(s) + 4\ NO_2(g) + O_2(g)$

32. A commercial method of manufacturing hydrogen involves the reaction of iron and steam.
$$3\ Fe(s) + 4\ H_2O(g) \xrightarrow{\Delta} Fe_3O_4(s) + 4\ H_2(g)$$
(a) How many moles of H_2 can be produced from 42.7 g Fe and an excess of $H_2O(g)$ [steam]?
(b) How many grams of H_2O are consumed in the conversion of 63.5 g Fe to Fe_3O_4?
(c) If 7.36 mol $H_2(g)$ is produced, how many grams of Fe_3O_4 must also be produced?

34. How many kilograms of HNO_3 are consumed to produce 125 kg $Ca(H_2PO_4)_2$ in the reaction
$Ca_3(PO_4)_2 + HNO_3 \longrightarrow$
$$Ca(H_2PO_4)_2 + Ca(NO_3)_2 \quad \text{(not balanced)}$$

36. Silver oxide decomposes at temperatures in excess of 300 °C, yielding metallic silver and oxygen gas. A 3.13-g sample of *impure* silver oxide yields 0.187 g $O_2(g)$. If one assumes that $Ag_2O(s)$ is the only source of $O_2(g)$, what is the % Ag_2O, by mass, in the sample? (*Hint:* Write a balanced equation for the reaction that occurs.)

38. An excess of aluminum foil is allowed to react with 225 mL of an aqueous solution of HCl that has a density of 1.088 g/mL and contains 18.0% HCl by mass. What mass of $H_2(g)$ is produced? [*Hint:* Use equation (4.2).]

40. *Without performing detailed calculations,* determine which of the following metals yields the greatest amount of H_2 *per gram* of metal reacting with $HCl(aq)$: Na, Mg, Al, or Zn? [*Hint:* Write equations similar to (4.2).]

41. How many grams of HCl are consumed in the reaction of 425 g of a mixture containing 35.2% $MgCO_3$ and 64.8% $Mg(OH)_2$, by mass?

$$Mg(OH)_2 + 2\ HCl \longrightarrow MgCl_2 + 2\ H_2O$$
$$MgCO_3 + 2\ HCl \longrightarrow MgCl_2 + H_2O + CO_2(g)$$

43. Dichlorodifluoromethane, once widely used as a refrigerant, can be prepared by the following reactions.

$$CH_4 + 4\ Cl_2 \longrightarrow CCl_4 + 4\ HCl$$
$$CCl_4 + 2\ HF \longrightarrow CCl_2F_2 + 2\ HCl$$

How many moles of Cl_2 must be consumed in the first reaction to produce 2.25 kg CCl_2F_2 in the second? Assume that all the CCl_4 produced in the first reaction is consumed in the second.

42. How many grams of CO_2 are produced in the complete combustion of 406 g of a bottled gas that consists of 72.7% propane (C_3H_8) and 27.3% butane (C_4H_{10}), by mass? (*Hint:* Write balanced equations for the combustion of the two gases.)

44. $CO_2(g)$ produced in the combustion of a sample of ethane is absorbed in $Ba(OH)_2(aq)$, producing 0.506 g $BaCO_3(s)$. How many grams of ethane (C_2H_6) must have been burned?

$$2\ C_2H_6(g) + 7\ O_2(g) \longrightarrow 4\ CO_2(g) + 6\ H_2O(l)$$
$$CO_2(g) + Ba(OH)_2(aq) \longrightarrow BaCO_3(s)$$

Molarity

45. What are the molarities of the following solutes?
(a) sucrose ($C_{12}H_{22}O_{11}$), if 150.0 g is dissolved per 250.0 mL of water solution
(b) urea, $CO(NH_2)_2$, if 98.3 mg of the 97.9% pure solid is dissolved in 5.00 mL of aqueous solution
(c) methanol, CH_3OH ($d = 0.792$ g/mL), if 125.0 mL is dissolved in enough water to make 15.0 L of solution

47. How much
(a) glucose, $C_6H_{12}O_6$, in grams, must be dissolved in water to produce 75.0 mL of 0.350 M $C_6H_{12}O_6$?
(b) methanol, CH_3OH ($d = 0.792$ g/mL), in milliliters, must be dissolved in water to produce 2.25 L of 0.485 M CH_3OH?

49. Which has the higher concentration of sucrose: a 46% sucrose solution by mass with a density of 1.21 g/mL, or 1.50 M $C_{12}H_{22}O_{11}$? Explain your reasoning.

51. A 10.00-mL sample of 2.05 M KNO_3 is diluted to a volume of 250.0 mL. What is the concentration of the diluted solution?

53. Given a 0.250 M K_2CrO_4 stock solution, describe how you would prepare a solution that is 0.0125 M K_2CrO_4. That is, what combination(s) of pipet and volumetric flask would you use? Typical sizes of volumetric flasks found in a general chemistry laboratory are 100.0, 250.0, 500.0, and 1000.0 mL, and typical sizes of volumetric pipets are 1.00, 5.00, 10.00, 25.00, and 50.00 mL.

46. What are the molarities of the following solutes?
(a) aspartic acid ($H_2C_4H_5NO_4$), if 0.405 g is dissolved in enough water to make 100.0 mL of solution
(b) acetone, C_3H_6O ($d = 0.790$ g/mL), if 35.0 mL is dissolved in enough water to make 425 mL of solution
(c) diethyl ether, $(C_2H_5)_2O$, if 8.8 mg is dissolved in enough water to make 3.00 L of solution

48. How much
(a) ethanol, C_2H_5OH ($d = 0.789$ g/mL), in liters, must be dissolved in water to produce 200.0 L of 1.65 M C_2H_5OH?
(b) concentrated hydrochloric acid solution (36.0% HCl by mass; $d = 1.18$ g/mL), in milliliters, is required to produce 12.0 L of 0.234 M HCl?

50. Which has the greater molarity of ethanol: a white wine ($d = 0.95$ g/mL) with 11% C_2H_5OH by mass, or the solution described in Example 4-8? Explain your reasoning.

52. A 25.0-mL sample of HCl(aq) is diluted to a volume of 500.0 mL. If the concentration of the diluted solution is found to be 0.085 M HCl, what was the concentration of the original solution?

54. Given a two-liter supply of 0.496 M KCl, describe how you would use this solution to prepare 250.0 mL of 0.175 M KCl. Give sufficient details that another student could follow your instructions.

Chemical Reactions in Solution

55. For the reaction

$$2\ AgNO_3(aq) + Na_2S(aq) \longrightarrow Ag_2S(s) + 2\ NaNO_3(aq)$$

(a) How many grams of $Na_2S(s)$ are required to react completely with 27.8 mL of 0.163 M $AgNO_3$?
(b) How many grams of $Ag_2S(s)$ are obtained from the reaction in part (a)?

57. Refer to Example 4-7 and Equation (4.2). For the conditions stated in Example 4-7, determine **(a)** the number of moles of $AlCl_3$ and **(b)** the *molarity* of the $AlCl_3(aq)$, if the solution volume is simply the 23.8 mL calculated in the example.

56. For the reaction

$$Ca(OH)_2(s) + 2\ HCl(aq) \longrightarrow CaCl_2(aq) + 2\ H_2O(l)$$

(a) How many grams of $Ca(OH)_2$ are required to react completely with 415 mL of 0.477 M HCl?
(b) How many kilograms of $Ca(OH)_2$ are required to react with 324 L of an HCl solution that is 24.28% HCl, by mass, and has a density of 1.12 g/mL?

58. Refer to the Summarizing Example. If 138 g Na_2CO_3 in 1.42 L of aqueous solution is treated with an excess of NO(g) and $O_2(g)$, what is the molarity of the $NaNO_2(aq)$ solution that results? (Assume that the reaction goes to completion.)

59. How many milliliters of 0.650 M K_2CrO_4 are needed to precipitate all the silver ion in 415 mL of 0.186 M $AgNO_3$ as $Ag_2CrO_4(s)$?

$$2 \, AgNO_3(aq) + K_2CrO_4(aq) \longrightarrow Ag_2CrO_4(s) + 2 \, KNO_3(aq)$$

61. How many grams of sodium must react with 155 mL H_2O to produce a solution that is 0.175 M NaOH? (Assume a final solution volume of 155 mL.)

$$2 \, Na(s) + 2 \, H_2O(l) \longrightarrow 2 \, NaOH(aq) + H_2(g)$$

63. A 0.3126-g sample of oxalic acid, $H_2C_2O_4$, requires 26.21 mL of a particular concentration of NaOH(aq) to complete the following reaction. What is the molarity of the NaOH(aq)?

$$H_2C_2O_4(s) + 2 \, NaOH(aq) \longrightarrow Na_2C_2O_4(aq) + 2 \, H_2O(l)$$

60. How many grams of Ag_2CrO_4 will precipitate if excess $K_2CrO_4(aq)$ is added to the 415 mL of 0.186 M $AgNO_3$ in Exercise 59?

62. A method of adjusting the concentration of HCl(aq) is to allow the solution to react with a small quantity of Mg.

$$Mg(s) + 2 \, HCl(aq) \longrightarrow MgCl_2(aq) + H_2(g)$$

How many milligrams of Mg must be added to 250.0 mL of 1.023 M HCl to reduce the solution concentration to exactly 1.000 M HCl?

64. A 25.00-mL sample of HCl(aq) was added to a 0.1000-g sample of $CaCO_3$. All the $CaCO_3$ reacted, leaving some excess HCl(aq).

$$CaCO_3(s) + 2 \, HCl(aq) \longrightarrow CaCl_2(aq) + H_2O(l) + CO_2(g)$$

The excess HCl(aq) required 43.82 mL of 0.01185 M $Ba(OH)_2(aq)$ to complete the following reaction.

$$2 \, HCl(aq) + Ba(OH)_2(aq) \longrightarrow BaCl_2(aq) + 2 \, H_2O(l)$$

What was the molarity of the original HCl(aq)?

Determining the Limiting Reactant

65. How many grams of NO(g) can be produced in the reaction of 1.00 mol $NH_3(g)$ and 1.00 mol $O_2(g)$?

$$4 \, NH_3(g) + 5 \, O_2(g) \xrightarrow{\Delta} 4 \, NO(g) + 6 \, H_2O(l)$$

67. A side reaction in the manufacture of rayon from wood pulp is

$$3 \, CS_2 + 6 \, NaOH \longrightarrow 2 \, Na_2CS_3 + Na_2CO_3 + 3 \, H_2O$$

How many grams of Na_2CS_3 are produced in the reaction of 92.5 mL of liquid CS_2 ($d = 1.26$ g/mL) and 2.78 mol NaOH?

69. Ammonia can be generated by heating together the solids NH_4Cl and $Ca(OH)_2$; $CaCl_2$ and H_2O are also formed. If a mixture containing 33.0 g each of NH_4Cl and $Ca(OH)_2$ is heated, how many grams of NH_3 will form? Which reactant remains in excess, and in what mass?
(*Hint:* Write a balanced equation for the reaction.)

Theoretical, Actual, and Percent Yields

71. The reaction of 15.0 g C_4H_9OH, 22.4 g NaBr, and 32.7 g H_2SO_4 yields 17.1 g C_4H_9Br in the reaction

$$C_4H_9OH + NaBr + H_2SO_4 \longrightarrow C_4H_9Br + NaHSO_4 + H_2O$$

What are **(a)** the theoretical yield, **(b)** the actual yield, and **(c)** the percent yield of this reaction?

73. How many grams of commercial acetic acid (97% $C_2H_4O_2$ by mass) must be allowed to react with an excess of PCl_3 to produce 75 g of acetyl chloride (C_2H_3OCl), if the reaction has a 78.2% yield?

$$C_2H_4O_2 + PCl_3 \longrightarrow C_2H_3OCl + H_3PO_3 \quad \text{(not balanced)}$$

66. How many grams of $H_2(g)$ are produced by the reaction of 1.84 g Al with 75.0 mL of 2.95 M HCl? [*Hint:* Use Equation (4.2).]

68. Lithopone is a brilliant white pigment used in water-based interior paints. It is a mixture of $BaSO_4$ and ZnS produced by the reaction

$$\underset{\text{lithopone}}{BaS(aq) + ZnSO_4(aq) \longrightarrow ZnS(s) + BaSO_4(s)}$$

How many grams of lithopone are produced in the reaction of 315 mL of 0.275 M $ZnSO_4$ and 285 mL of 0.315 M BaS?

70. Chlorine can be generated by heating together calcium hypochlorite and hydrochloric acid. Calcium chloride and water are also formed. If 50.0 g $Ca(OCl)_2$ and 275 mL of 6.00 M HCl are allowed to react, how many grams of chlorine gas will form? Which reactant, $Ca(OCl)_2$ or HCl, remains in excess, and in what mass? (*Hint:* Write a balanced equation for the reaction.)

72. Azobenzene, an intermediate in the manufacture of dyes, can be prepared from nitrobenzene by reaction with triethylene glycol in the presence of Zn and KOH. In one reaction 0.10 L of nitrobenzene ($d = 1.20$ g/mL) and 0.30 L of triethylene glycol ($d = 1.12$ g/mL) yields 55 g azobenzene. What are **(a)** the theoretical yield, **(b)** the actual yield, and **(c)** the percent yield of this reaction?

$$\underset{\text{nitrobenzene} \quad\quad \text{triethylene glycol}}{2 \, C_6H_5NO_2 + 4 \, C_6H_{14}O_4} \xrightarrow[\text{KOH}]{\text{Zn}}$$

$$\underset{\text{azobenzene}}{(C_6H_5N)_2 + 4 \, C_6H_{12}O_4 + 4 \, H_2O}$$

74. Suppose that each of the following reactions has a 92% yield.

(1) $CH_4 + Cl_2 \rightarrow CH_3Cl + HCl$
(2) $CH_3Cl + Cl_2 \rightarrow CH_2Cl_2 + HCl$

Starting with 112 g CH_4 in reaction (1) and an excess of $Cl_2(g)$, how many grams of CH_2Cl_2 are formed in reaction (2)?

75. An essentially 100% yield is necessary for a chemical reaction used to *analyze* a compound, but it is almost never expected for a reaction that is used to *synthesize* a compound. Explain this difference.

76. Suppose we carry out the precipitation of $Ag_2CrO_4(s)$ described in Example 4-11. If we obtain 2.058 g of precipitate, we might conclude that it is nearly pure $Ag_2CrO_4(s)$, but if we obtain 2.112 g, we can be quite sure that the precipitate is not pure. Explain this difference.

Advanced Exercises

77. Write chemical equations to represent the following reactions.
(a) Limestone rock (calcium carbonate) is heated (calcined) and decomposes to calcium oxide and carbon dioxide gas.
(b) Zinc sulfide ore is heated in air (roasted) and is converted to zinc oxide and sulfur dioxide gas. (Note that oxygen gas in the air is also a reactant.)
(c) Propane gas (C_3H_8) reacts with gaseous water to produce a mixture of carbon monoxide and hydrogen gases (called *synthesis gas* and used in the chemical industry to produce a variety of other chemicals).

78. Write chemical equations to represent the following reactions.
(a) Sulfur dioxide gas is passed into an aqueous solution containing sodium sulfide and sodium carbonate. The reaction products are carbon dioxide gas and a water solution of sodium thiosulfate.
(b) Copper metal reacts with gaseous oxygen, carbon dioxide, and water to form green basic copper carbonate, $Cu_2(OH)_2CO_3$ (a reaction responsible for the formation of the green patina, or coating, often seen on outdoor bronze statues).
(c) Calcium dihydrogen phosphate reacts with sodium hydrogen carbonate (bicarbonate), producing (tri)calcium phosphate, (di)sodium hydrogen phosphate, carbon dioxide, and water (the principal reaction occurring in the action of ordinary baking powder in the baking of cakes, bread, and biscuits).

79. $H_2(g)$ is passed over $Fe_2O_3(s)$ at 400 °C. Water vapor is formed, together with a black residue—a compound consisting of 72.3% Fe and 27.7% O. Write a balanced equation for this reaction.

80. A sulfide of iron, containing 36.5% S, by mass, is heated in $O_2(g)$, and the products are sulfur dioxide and an oxide of iron containing 27.6% O, by mass. Write a balanced chemical equation for this reaction. (*Hint:* The formulas of the sulfide and oxide are not necessarily those that you would predict from Table 3-1.)

81. What is the molarity of NaCl(aq) if a solution has 1.52 ppm Na? [Assume that NaCl is the only source of Na and that the solution density is 1.00 g/mL. Parts per million (ppm) can be taken to mean g Na per million grams of solution.]

82. How many milligrams $Ca(NO_3)_2$ must be present in 50.0 L of a solution containing 2.35 ppm Ca? (*Hint:* See also Exercise 81.)

83. A drop (0.05 mL) of 12.0 M HCl is spread over a sheet of thin aluminum foil. Assume that all the acid reacts with and thus dissolves through the foil. What will be the area, in cm², of the cylindrical hole produced? (density of Al = 2.70 g/cm³; thickness of the foil = 0.10 mm.)
$$2\ Al(s) + 6\ HCl(aq) \longrightarrow 2\ AlCl_3(aq) + 3\ H_2(g)$$

84. A small piece of zinc is dissolved in 50.00 mL of 1.035 M HCl.
$$Zn(s) + 2\ HCl(aq) \longrightarrow ZnCl_2(aq) + H_2(g)$$
At the conclusion of the reaction, the concentration of the 50.00-mL sample is redetermined and found to be 0.812 M HCl. What must have been the mass of the piece of zinc that dissolved?

85. How many milliliters of 0.715 M NH_4NO_3 solution must be diluted with water to produce 1.00 L of a solution with a concentration of 2.37 mg N per mL?

86. A seawater sample has a density of 1.03 g/mL and 2.8% NaCl, by mass. A saturated solution of NaCl in water is 5.45 M NaCl. How many liters of water would have to be evaporated from 1.00×10^6 L of the seawater before NaCl would begin to crystallize? (A saturated solution contains the maximum amount of dissolved solute possible.)

87. A 99.8-mL sample of a solution that is 12.0% KI by mass ($d = 1.093$ g/mL) is added to 96.7 mL of another solution that is 14.0% $Pb(NO_3)_2$ by mass ($d = 1.134$ g/mL). How many grams of PbI_2 should form?
$$Pb(NO_3)_2(aq) + 2\ KI(aq) \longrightarrow PbI_2(s) + 2\ KNO_3(aq)$$

88. $CaCO_3(s)$ reacts with HCl(aq) to form H_2O, $CaCl_2(aq)$ and $CO_2(g)$. If a 45.0-g sample of $CaCO_3(s)$ is added to 1.25 L of HCl(aq) that is 25.7% HCl by mass ($d = 1.13$ g/mL), what will be the molarity of HCl in the solution after the reaction is completed? (Assume that the solution volume remains constant.)

89. A 2.05-g sample of an iron–aluminum alloy (ferroaluminum) is dissolved in excess HCl(aq) to produce 0.105 g $H_2(g)$. What is the percent composition, by mass, of the ferroaluminum?
$$Fe(s) + 2\ HCl(aq) \longrightarrow FeCl_2(aq) + H_2(g)$$
$$2\ Al(s) + 6\ HCl(aq) \longrightarrow 2\ AlCl_3(aq) + 3\ H_2(g)$$

90. A 0.155-g sample of an Al–Mg alloy reacts with an excess of HCl(aq) to produce 0.0163 g H_2. What is the percent Mg in the alloy?
[*Hint:* Write equations similar to (4.2).]

91. An organic liquid is either methyl alcohol (CH_3OH), ethyl alcohol (C_2H_5OH), or a mixture of the two. A 0.220-g sample of the liquid is burned in an excess of $O_2(g)$ and yields 0.352 g $CO_2(g)$. Is the liquid a pure alcohol or a mixture of the two?

92. The manufacture of ethyl alcohol, C_2H_5OH, yields diethyl ether, $(C_2H_5)_2O$, as a by-product. The complete combustion of a 1.005-g sample of the product of this process yields

1.963 g CO_2. What must be the mass percents of C_2H_5OH and of $(C_2H_5)_2O$ in this sample?

93. Under appropriate conditions, copper sulfate, potassium chromate, and water react to form a product containing Cu^{2+}, CrO_4^{2-}, and OH^- ions. Analysis of the compound yields 48.7% Cu^{2+}, 35.6% CrO_4^{2-}, and 15.7% OH^-.
(a) Determine the empirical formula of the compound.
(b) Write a plausible equation for the reaction.

94. Write a chemical equation to represent the complete combustion of malonic acid, a compound with 34.62% C, 3.88% H, and 61.50% O, by mass.

95. A mixture of 4.800 g H_2 and 36.40 g O_2 reacts.

(a) Write a balanced equation for this reaction.
(b) What substances are present after the reaction?
(c) Show that the total mass of substances present before and after the reaction is the same.

96. Silver nitrate is a very expensive chemical. For a particular experiment, a student needs 100.0 mL of 0.0750 M $AgNO_3$(aq). She has available 60 mL of 0.0500 M $AgNO_3$. She decides to (1) pipet exactly 50.00 mL of the solution that she does have into a 100.0-mL flask, (2) add an appropriate mass of $AgNO_3$, and (3) dilute the resulting solution to exactly 100.0 mL. What mass of $AgNO_3$ must she use?

Feature Problems

A. It is often difficult to determine the concentration of a species in solution, particularly if it is a biological species that takes part in complex reaction pathways. One way to do this is through a dilution experiment with labeled molecules. Instead of molecules, however, we will use fish.

A fisher wants to know the number of fish in a particular pond. The fisher puts an indelible mark on 100 fish and adds them to the pond. After waiting for the fish to spread throughout the pond, the fisher starts fishing. Eventually, the fisher catches 18 fish. Of these, five are marked. What is the total number of fish in the pond?

B. Lead nitrate and potassium iodide react in aqueous solution to form a yellow precipitate of lead iodide. In one series of experiments the masses of the two reactants were varied, but the *total* mass of the two was held constant at 5.000 g. The lead iodide formed was filtered from solution, washed, and dried. Data for a series of reactions are presented below, together with a blank graph.

Experiment	Mass Of Lead Nitrate, g	Mass of Lead Iodide, g
1	0.500	0.692
2	1.000	1.388
3	1.500	2.093
4	3.000	2.778
5	4.000	1.391

1. Complete the graph; that is, plot the data and draw the appropriate curve(s) connecting the data points. What is the maximum mass of precipitate that can be obtained?
2. Explain why the maximum mass of precipitate is obtained when the reactants are in their stoichiometric proportions. What are these stoichiometric proportions expressed (a) as a mass ratio, and (b) as a mole ratio?
3. Show how the stoichiometric proportions determined in part 2 are related to the balanced equation for the reaction.

C. Some situations in the business world require balancing supplies (reactants) and products. For example, a business firm makes necklaces using beads of three colors: red, blue, and green. Each necklace contains 70 beads, shown in the sketch. The firm has on hand 8 boxes of red beads, 11 boxes of blue beads, and 10 boxes of green beads. Each box contains 10.0 kg of beads, and the average masses of the beads are red: 1.98 g; blue: 3.05 g; green: 1.82 g.

1. Write a "balanced equation" for forming a necklace from individual beads.
2. How many necklaces can be made with the supplies on hand?
3. Explain the limitations on the precision of your answer.

Introduction to Reactions in Aqueous Solutions

5

*P*recipitation, the formation of a solid when solutions are mixed, is probably the most common evidence of a chemical reaction that general chemistry students see. A practical application of precipitation is in determining the presence of certain ions in solution. If, for example, we do not know whether a bottle contains distilled water or a barium chloride solution, we can easily find out by adding a few drops of silver nitrate solution to a small sample of the liquid. If a white solid (AgCl) forms, the sample is a barium chloride solution; if nothing happens, it is water. Precipitation reactions are the first type we study in this chapter.

$Mg(OH)_2(s)$, though insoluble in water, is soluble in hydrochloric acid, HCl(aq), as a result of an acid–base reaction. This is the reaction by which milk of magnesia neutralizes excess stomach acid. Magnesium hydroxide is a base, and a survey of acids, bases, and acid–base reactions is another topic in this chapter.

The third category of reactions presented in this chapter, oxidation–reduction reactions, can be found in all aspects of life, from reactions in organisms, to processes for manufacturing chemicals, to such practical matters as bleaching fabrics, purifying water, and destroying toxic chemicals.

Reaction between $PCl_3(l)$ and $H_2O(l)$ produces $H_3PO_3(aq)$, phosphorous acid, and HCl(aq). Acids and acid–base reactions are among the topics considered in this chapter.

129

5-1 The Nature of Aqueous Solutions

Reactions in aqueous (water) solution are important because (1) water is inexpensive and able to dissolve a vast number of substances, (2) in aqueous solution many substances are dissociated into ions, which can participate in chemical reactions, and (3) aqueous solutions are found everywhere, from seawater to living systems.

We have learned that ions are individual atoms or groupings of atoms that acquire a net electric charge by gaining or losing electrons. Thus, Mg^{2+} is a positively charged ion (a cation) formed when a Mg atom loses two electrons. Cl^- is a negatively charged ion (an anion) formed when a Cl atom gains one electron. When a group of three O atoms covalently bonded to one N atom acquires an additional electron, it becomes the polyatomic nitrate anion, NO_3^-.

With the apparatus pictured in Figure 5-1 we can detect the presence of ions in an aqueous solution by measuring how well the solution conducts electricity. The ability to conduct electricity requires that ions be present in the solution, and these ions must come from the solute. Water itself is a nonconductor. Depending on which of the following three observations we make, we label the solute a nonelectrolyte, a strong electrolyte, or a weak electrolyte.

- *The lamp fails to light up.* Conclusion: There are no ions present (or, if there are some present, their concentration is extremely low). A **nonelectrolyte** is a substance that is not ionized and does not conduct electric current.

- *The lamp lights up brightly.* Conclusion: The concentration of ions in solution is *high*. A **strong electrolyte** is a substance that is completely ionized in aqueous solution, and the solution is a good electrical conductor.

FIGURE 5-1
Electrical conductivity of aqueous solutions.
For electric current to flow, charged particles must pass through the solution between the graphite electrodes. The only particles able to do this are positively and negatively charged *ions*. We judge the ability of the solution to conduct electricity by the brilliance of the incandescent lamp. (a) The solute is a nonelectrolyte; (b) the solute is a strong electrolyte; (c) the solute is a weak electrolyte.

(a)   (b)   (c)

- *The lamp lights up only dimly.* Conclusion: The concentration of ions in solution is *low*. A **weak electrolyte** is *partially* ionized in aqueous solution, and the solution is only a fair conductor of electricity.

The following generalization will help us decide whether a particular solute in an aqueous solution is most likely to be a nonelectrolyte, a strong electrolyte, or a weak electrolyte.

- Essentially all soluble ionic compounds and only a relatively few molecular compounds are *strong electrolytes.*
- Most molecular compounds are either *nonelectrolytes* or *weak electrolytes.*

Now let us consider how best to represent these three types of substances in chemical equations. For water-soluble $MgCl_2$, an ionic compound, we may write

$$MgCl_2(s) \xrightarrow{H_2O} Mg^{2+}(aq) + 2\ Cl^-(aq)$$

By this equation we mean that in the presence of water—that is, in an aqueous solution—formula units of $MgCl_2$ are dissociated into the separate ions. The best representation of $MgCl_2(aq)$, then, is $Mg^{2+}(aq) + 2\ Cl^-(aq)$.*

For a molecular compound that is a *strong electrolyte,* we similarly can write

$$HCl(g) \xrightarrow{H_2O} H^+(aq) + Cl^-(aq)$$

which means that the best representation of hydrochloric acid, $HCl(aq)$, is $H^+(aq) + Cl^-(aq)$.

For a *weak electrolyte* the situation is best described as a reaction that does not go to completion. Of all the molecules placed in solution, only a fraction are ionized. An equation written with a double arrow, such as

$$HC_2H_3O_2(aq) \rightleftharpoons H^+(aq) + C_2H_3O_2{}^-(aq)$$

indicates that a process is *reversible*. While some $HC_2H_3O_2$ (acetic acid) molecules ionize, some H^+ and $C_2H_3O_2{}^-$ ions in solution recombine to form new $HC_2H_3O_2$ molecules. Nevertheless, in any given acetic acid solution, the relative proportions of ionized and nonionized (molecular) acid remain fixed. The predominant species is the molecule $HC_2H_3O_2$, and the solution is best represented by $HC_2H_3O_2(aq)$, *not* $H^+(aq) + C_2H_3O_2{}^-(aq)$. Think of the matter this way: If we could tag a particular $C_2H_3O_2$ group and watch it over time, we would sometimes see it as the ion $C_2H_3O_2{}^-$, but most of the time it would be in a molecule, $HC_2H_3O_2$.

For a *nonelectrolyte,* we write its molecular formula in chemical equations. Thus, for a solution of ethanol (ethyl alcohol) in water we would write $C_2H_5OH(aq)$.

With this new information about the nature of aqueous solutions we can introduce a useful notation for solution concentrations. In a solution that is 0.010 M NaCl, we assume that the NaCl is completely dissociated into ions and say that the solution is 0.010 M Na^+ and 0.010 M Cl^-. Better still, let us introduce

---

*To say that a strong electrolyte is completely dissociated into individual ions in aqueous solution is a good approximation, but somewhat of an oversimplification. Some of the cations and anions in solution may become associated into units called *ion pairs*. Generally, though, at the solution concentrations we will be using, assuming complete dissociation will not seriously affect our results.

a special symbol for the concentration of a species in solution—the bracket symbol [ ]. The statement $[Na^+] = 0.010$ M means that the concentration of the species within the brackets—that is, Na^+—is 0.010 mol/L. Thus,

in 0.010 M NaCl: $[Na^+] = 0.010$ M $[Cl^-] = 0.010$ M $[NaCl] = 0$

Although we do not usually write expressions like $[NaCl] = 0$, we do so here to emphasize that there is essentially no undissociated NaCl in the solution.

Figure 5-2 is a visual summary of some ideas presented in this section. It is a *microscopic* interpretation of what we observe at the *macroscopic* level in Figure 5-1. Example 5-1 shows how to calculate the concentrations of ions in a strong electrolyte solution.

EXAMPLE 5-1

Calculating Ion Concentrations in a Solution of a Strong Electrolyte. What are the aluminum and sulfate ion concentrations in 0.0165 M $Al_2(SO_4)_3$(aq)?

Solution

First, identify the solute as a *strong electrolyte* and write an equation to represent its dissociation. The solute is an ionic compound consisting of the ions Al^{3+} and SO_4^{2-}.

$$Al_2(SO_4)_3(s) \xrightarrow{H_2O} 2\,Al^{3+}(aq) + 3\,SO_4^{2-}(aq)$$

FIGURE 5-2   Nonelectrolytes, strong electrolytes, and weak electrolytes—a summary.

This is a "microscopic" view of Figure 5-1. (a) In the nonelectrolyte C_2H_5OH, essentially no ionization occurs. No electric current flows, and the lamp remains unlit. (b) $MgCl_2$, a strong electrolyte, is dissociated into ions. Two Cl^- ions and one Mg^{2+} ion are formed per formula unit of $MgCl_2$. Electricity passes through this solution in the form of Mg^{2+} ions attracted to the negative electrode and Cl^- ions attracted to the positive electrode. The lamp lights up brilliantly. (c) Only a small fraction (about 5%) of the weak electrolyte molecules, $HC_2H_3O_2$, ionize into H^+ and $C_2H_3O_2^-$ ions, and only these ions are capable of conducting electricity through the solution. A weak electrolyte conducts electricity only weakly.

$\bigcirc$ = C_2H_5OH

$\bullet$ = Mg^{2+}
$\bullet$ = Cl^-

$\circ$ = H^+
$\bigcirc$ = $C_2H_3O_2^-$
$\bullet$ = $HC_2H_3O_2$

(a) C_2H_5OH a non-
electrolyte

(b) $MgCl_2$ a strong
electrolyte

(c) $HC_2H_3O_2$ a weak
electrolyte

Now, with stoichiometric factors from this equation we can devise a conversion pathway to relate $[Al^{3+}]$ and $[SO_4^{2-}]$ to the given molarity of $Al_2(SO_4)_3$. The stoichiometric factors shown in blue in the following equations are derived from the fact that 1 mol $Al_2(SO_4)_3$ consists of 2 mol Al^{3+} and 3 mol SO_4^{2-}.

$$[Al^{3+}] = \frac{0.0165 \text{ mol } Al_2(SO_4)_3}{1 \text{ L}} \times \frac{2 \text{ mol } Al^{3+}}{1 \text{ mol } Al_2(SO_4)_3} = \frac{0.0330 \text{ mol } Al^{3+}}{1 \text{ L}}$$

$$= 0.0330 \text{ M}$$

$$[SO_4^{2-}] = \frac{0.0165 \text{ mol } Al_2(SO_4)_3}{1 \text{ L}} \times \frac{3 \text{ mol } SO_4^{2-}}{1 \text{ mol } Al_2(SO_4)_3} = \frac{0.0495 \text{ mol } SO_4^{2-}}{1 \text{ L}}$$

$$= 0.0495 \text{ M}$$

Practice Example A: Na_3PO_4 is commonly used as a cleaning agent under the trade name TSP, trisodium phosphate. What are the sodium and phosphate ion concentrations in a 0.358 M solution of TSP?

Practice Example B: The chief ions in seawater are Na^+, Mg^{2+}, and Cl^-. Seawater is approximately 0.438 M NaCl and 0.0512 M $MgCl_2$. What is the molarity of Cl^-—that is, the *total* $[Cl^-]$—in seawater?

Are You Wondering . . .

How to tell whether a solute is a strong electrolyte?

To do calculations similar to those in Example 5-1, you need to be able to identify strong electrolytes. Common compounds of the *metals* are ionic. When you look at a chemical formula and see that a metal is present, you can generally assume that, if the ionic compound is water-soluble, it will be a *strong electrolyte*. In this assessment treat the ammonium ion, NH_4^+, as if it were a Group 1A ion similar to K^+. So few common molecular compounds are strong electrolytes that you should be able to remember them when we mention them later (Section 5-3).

5-2   Precipitation Reactions

Some metal salts, such as NaCl, are quite soluble in water. Others, such as AgCl, are not very soluble at all. In fact, so little AgCl will dissolve in water that we generally consider this compound to be *insoluble*. Precipitation reactions occur when certain cations and anions combine to produce an insoluble ionic solid called a **precipitate.** One laboratory use of precipitation reactions is in identifying the ions present in a solution, an aspect of chemical analysis we consider in a later chapter. In industry, precipitation reactions are used in the manufacture of numerous chemicals. In the extraction of magnesium metal from seawater, for instance, the first step is to precipitate Mg^{2+} as $Mg(OH)_2(s)$. Our objective in this section is to represent precipitation reactions by chemical equations and to apply some simple rules for predicting precipitation reactions.

A qualitative test for Mg^{2+} ion in hard water involves the addition of a concentrated solution of NaOH(aq). The formation of a precipitate of $Mg(OH)_2(s)$ confirms the presence of Mg^{2+}.

Net Ionic Equations

The reaction of silver nitrate and sodium iodide in a water solution yields sodium nitrate in solution and a yellow precipitate of silver iodide (Figure 5-3). Applying the principles of equation writing from Chapter 4, we can write

$$AgNO_3(aq) + NaI(aq) \longrightarrow AgI(s) + NaNO_3(aq) \tag{5.1}$$

You might note a contradiction, however, between Equation (5.1) and something we learned earlier in this chapter. In their water solutions, the soluble ionic compounds $AgNO_3$, NaI, and $NaNO_3$—all *strong* electrolytes—should be represented by their separate ions.

$$Ag^+(aq) + NO_3^-(aq) + Na^+(aq) + I^-(aq) \longrightarrow$$
$$AgI(s) + Na^+(aq) + NO_3^-(aq) \tag{5.2}$$

▶ The term *molecular* equation is sometimes used for equations like (5.1). We will not use this term because it implies that all substances exist as molecules in solution, which, of course, they do not.

We might say that Equation (5.1) is the "whole formula" form of the equation, whereas Equation (5.2) is the "ionic" form. Notice also this interesting feature of Equation (5.2): $Na^+(aq)$ and $NO_3^-(aq)$ appear on both sides of the equation. These ions are not reactants; they go through the reaction unchanged. We might call them "spectator" ions. If we eliminate the spectator ions, all that remains is the net ionic Equation (5.3).

$$Ag^+(aq) + I^-(aq) \longrightarrow AgI(s) \tag{5.3}$$

▶ Although the insoluble solid consists of ions, we do not represent ionic charges in the whole formula. That is, we write AgI(s), not Ag^+I^-(s).

A **net ionic equation** is an equation that includes only the actual participants in a reaction, with each participant denoted by the symbol or formula that best represents it. Symbols are written for individual ions [such as $Ag^+(aq)$], and whole formulas are written for insoluble solids [such as AgI(s)]. Because net ionic equations include electrically charged species—ions—a net ionic equation must be balanced *both* for the numbers of atoms of all types and for electric charge. The same net electric charge must appear on both sides of the equation. Throughout the remainder of this chapter, we will represent most chemical reactions by net ionic equations.

Predicting Precipitation Reactions

Suppose we are asked whether precipitation occurs when the following aqueous solutions are mixed.

$$AgNO_3(aq) + KBr(aq) \longrightarrow ? \tag{5.4}$$

A good way to begin is to rewrite expression (5.4) in the ionic form.

$$Ag^+(aq) + NO_3^-(aq) + K^+(aq) + Br^-(aq) \longrightarrow ? \tag{5.5}$$

There are only *two* possibilities. Either some cation–anion combination leads to an insoluble solid—a precipitate—or no such combination is possible and there is no reaction at all.

To *predict* what will happen, that is, without going into the laboratory to do experiments, we need some information about which sorts of ionic compounds are water-soluble and which are water-insoluble. We expect the insoluble ones to form when the appropriate ions are mixed in solution. The most concise form for this information is a set of *solubility rules*. On page 135, we set out some of the simpler rules.

FIGURE 5-3
A precipitate of silver iodide.
When a water solution of $AgNO_3$ is added to one of NaI, insoluble yellow AgI(s) precipitates from solution.

Compounds that are mostly *soluble*.

▶ Some solubility rules.

- Those of the alkali metals (Group 1A) and the ammonium ion (NH_4^+).
- Nitrates and acetates.
- Chlorides, bromides, and iodides.
  (Those of Pb^{2+}, Ag^+, and Hg_2^{2+} are *insoluble*.)
- Sulfates.
  (Those of Sr^{2+}, Ba^{2+}, Pb^{2+}, and Hg_2^{2+} are *insoluble*.)

Compounds that are mostly *insoluble*.

- Hydroxides and sulfides.
  (Those of the Group 1A metals and NH_4^+ are *soluble*. Sulfides of the Group 2A metals are soluble. The hydroxides of Ca^{2+}, Sr^{2+}, and Ba^{2+} are slightly soluble.)
- Carbonates and phosphates.
  (Those of the Group 1A metals are *soluble*.)

▶ In principle, all ionic compounds dissolve in water to some extent, though this may be very slight. For practical purposes, we consider a compound to be insoluble if the maximum amount we can dissolve is less than about 0.01 mole per liter.

According to these rules, AgBr(s) is insoluble in water and should precipitate. Written as an ionic equation, expression (5.5) becomes

$$Ag^+(aq) + NO_3^-(aq) + K^+(aq) + Br^-(aq) \longrightarrow AgBr(s) + K^+(aq) + NO_3^-(aq)$$

For the *net ionic equation,* we have

$$Ag^+(aq) + Br^-(aq) \longrightarrow AgBr(s) \qquad (5.6)$$

The three predictions concerning precipitation reactions made in Example 5-2 are verified in Figure 5-4.

EXAMPLE 5-2

Using Solubility Rules to Predict Precipitation Reactions. Predict whether a reaction will occur in each of the following cases. If so, write a net ionic equation for the reaction.
(a) $NaOH(aq) + MgCl_2(aq) \longrightarrow$?
(b) $BaS(aq) + CuSO_4(aq) \longrightarrow$?
(c) $(NH_4)_2SO_4(aq) + ZnCl_2(aq) \longrightarrow$?

Solution

(a) Because all common Na compounds are water-soluble, Na^+ remains in solution. The combination of Mg^{2+} and OH^- produces *insoluble* $Mg(OH)_2$. With this information we can write

$$2\ Na^+(aq) + 2\ OH^-(aq) + Mg^{2+}(aq) + 2\ Cl^-(aq) \longrightarrow$$
$$Mg(OH)_2(s) + 2\ Na^+(aq) + 2\ Cl^-(aq)$$

and with the elimination of spectator ions we obtain

$$2\ OH^-(aq) + Mg^{2+}(aq) \longrightarrow Mg(OH)_2(s)$$

▶ As you gain experience, you should be able to go directly to a net ionic equation, without writing an ionic equation that includes spectator ions.

FIGURE 5-4
Verifying the predictions made in Example 5-2.
(a) When NaOH(aq) is added to MgCl₂(aq), a white precipitate of Mg(OH)₂(s) forms.
(b) When colorless BaS(aq) is added to blue CuSO₄(aq), a dark precipitate forms. The precipitate is a mixture of white BaSO₄(s) and black CuS(s). (A slight excess of CuSO₄ remains in solution.)
(c) No reaction occurs when colorless (NH₄)₂SO₄(aq) is added to colorless ZnCl₂(aq).

(b) From the solubility rules we conclude that the insoluble combinations formed when Ba^{2+}, S^{2-}, Cu^{2+}, and SO_4^{2-} are found together in solution are $BaSO_4(s)$ and $CuS(s)$. The net ionic equation is

$$Ba^{2+}(aq) + S^{2-}(aq) + Cu^{2+}(aq) + SO_4^{2-}(aq) \longrightarrow BaSO_4(s) + CuS(s)$$

(c) A careful review of the solubility rules shows that all the possible ion combinations lead to water-soluble compounds.

$$2\, NH_4^+(aq) + SO_4^{2-}(aq) + Zn^{2+}(aq) + 2\, Cl^-(aq) \longrightarrow \text{no reaction}$$

Practice Example A: Indicate whether a precipitate forms by completing each of the following as a net ionic equation. If no reaction occurs, so state.
(a) $AlCl_3(aq) + KOH(aq) \longrightarrow$?
(b) $K_2SO_4(aq) + FeBr_3(aq) \longrightarrow$?
(c) $CaI_2(aq) + Pb(NO_3)_2(aq) \longrightarrow$?

Practice Example B: Indicate whether a precipitate forms by completing each of the following as a net ionic equation. If no reaction occurs, so state.
(a) $Na_3PO_4(aq) + AlCl_3(aq) \longrightarrow$?
(b) $Al_2(SO_4)_3(aq) + BaCl_2(aq) \longrightarrow$?
(c) $(NH_4)_2CO_3(aq) + Pb(NO_3)_2(aq) \longrightarrow$?

5-3   Acid–Base Reactions

Ideas about acids and bases (or alkalis) date back to ancient times. The word *acid* is derived from the Latin *acidus* (sour). *Alkali* (base) comes from the Arabic *al-qali*, referring to the ashes of certain plants from which alkaline substances can be extracted. The acid–base concept is a major theme in the history of chemistry. The view we describe in this section, proposed by Svante Arrhenius in 1884, is still used in many applications. Later in the text we will explore more modern theories.

FIGURE 5-5
An acid, a base, and an acid–base indicator.
The acidic nature of apple juice is shown by the color change in blue litmus to *red*. The basic nature of soap is indicated by the change in color of red litmus to *blue*.

Acids

From a practical standpoint, we can identify acids through their sour taste, their ability to react with a variety of metals and carbonate minerals, and their effect on the colors of substances called *acid–base indicators* (Figure 5-5). From a chemist's point of view, however, we define an **acid** as a substance that provides hydrogen ions (H^+)* in aqueous solution.

When hydrogen chloride dissolves in water, complete ionization into H^+ and Cl^- occurs (HCl is a strong electrolyte).

$$HCl(g) \xrightarrow{\text{H}_2\text{O}} H^+(aq) + Cl^-(aq)$$

A water solution of nitric acid, also a strong electrolyte, can be represented as

$$H^+(aq) + NO_3^-(aq)$$

Strong acids are molecular compounds that are completely ionized in aqueous solutions—for example, HCl and HNO_3. There are actually so few common strong acids that they make only a short list (see Table 5.1).

As we described on page 131, the ionization of acetic acid does not go to completion; it is *reversible*. Acetic acid is a *weak electrolyte*.

$$\underset{\text{acetic acid}}{HC_2H_3O_2(aq)} \rightleftharpoons H^+(aq) + C_2H_3O_2^-(aq)$$

Acids that are incompletely ionized in aqueous solution are called **weak acids.** Like acetic acid, the vast majority of acids are weak acids.

We can use the same method to calculate ion concentrations in a strong acid as we did in Example 5-1 for other strong electrolytes. Because the ionization of a weak acid does not go to completion, however, it is a more difficult matter to calculate ion concentrations in a weak acid solution. We will not take up these calculations at this time.

---

*The simple hydrogen ion, H^+, does not exist in aqueous solutions. Its actual form is as *hydronium* ion, H_3O^+, in which an H^+ ion is attached to a H_2O molecule. The hydronium ion is discussed in Chapter 17. For the present we will follow the practice, common among chemists, of using H^+ as a shorthand notation for H_3O^+.

TABLE 5.1 Common Strong Acids and Strong Bases	
Acids	**Bases**
HCl	LiOH
HBr	NaOH
HI	KOH
HClO₄	RbOH
HNO₃	CsOH
H₂SO₄[a]	Ca(OH)₂
	Sr(OH)₂
	Ba(OH)₂

[a]H_2SO_4 ionizes in two distinct steps. It is a strong acid only in its first ionization step (see Section 17-6).

Bases

From a practical standpoint, we can identify bases through their bitter taste, slippery feel, and effect on the colors of acid–base indicators (see Figure 5-5). A better definition is that a **base** is a substance capable of producing hydroxide ions (OH^-) in aqueous solution. Consider a soluble ionic hydroxide such as NaOH. In the solid state this compound consists of Na^+ and OH^- ions. When the solid dissolves in water, the ions become completely dissociated from each other.

$$NaOH(s) \xrightarrow{H_2O} Na^+(aq) + OH^-(aq)$$

A substance that dissociates completely in water solution and yields OH^- ions is a **strong base.** As is true of strong acids, the number of common strong bases is small (see Table 5.1). They are primarily the hydroxides of Group 1A and some Group 2A metals.

Certain substances produce OH^- ions by *reacting* with water, not just by dissolving in it. Such substances are also bases, as is ammonia, for example.

$$NH_3(aq) + H_2O \rightleftharpoons NH_4^+(aq) + OH^-(aq) \tag{5.7}$$

NH_3 is a weak electrolyte; its reaction with water does not go to completion. A base that is incompletely ionized in aqueous solution is a **weak base.** Most basic substances are weak bases.

▶ The formula of $NH_3(aq)$ is sometimes written as NH_4OH (ammonium hydroxide), and its ionization represented as
$NH_4OH(aq) \rightleftharpoons$
$NH_4^+(aq) + OH^-(aq)$
There is no hard evidence for the existence of NH_4OH, however.

Neutralization

Perhaps the most significant property of acids and bases is the ability of each to cancel or neutralize the properties of the other. In a **neutralization** reaction an acid and a base react to form water and an aqueous solution of an ionic compound called a **salt.** Thus, in *ionic* form

$$\underbrace{H^+(aq) + Cl^-(aq)}_{(acid)} + \underbrace{Na^+(aq) + OH^-(aq)}_{(base)} \longrightarrow \underbrace{Na^+(aq) + Cl^-(aq)}_{(salt)} + \underbrace{H_2O(l)}_{(water)}$$

By eliminating the spectator ions we discover the essential nature of a neutralization reaction: H^+ ions from an acid and OH^- ions from a base combine to form water.

$$H^+(aq) + OH^-(aq) \longrightarrow H_2O(l)$$

In a neutralization involving the weak base $NH_3(aq)$, we can think of H^+ from an acid combining directly with NH_3 molecules to form NH_4^+. The neutralization can be represented by an ionic equation, such as

$$\underbrace{H^+(aq) + Cl^-(aq)}_{(acid)} + \underbrace{NH_3(aq)}_{(base)} \longrightarrow \underbrace{NH_4^+(aq) + Cl^-(aq)}_{(salt)}$$

or by a net ionic equation.

$$H^+(aq) + NH_3(aq) \longrightarrow NH_4^+(aq)$$

Recognizing Acids and Bases

Acids contain *ionizable* hydrogen atoms. We generally identify ionizable hydrogen atoms by the way we write the formula of an acid. For now we'll do this by writing ionizable H atoms first. Thus, $HC_2H_3O_2$ has *one* H atom that is ionizable (and *three* that are not). It *is* an acid. CH_4 has four H atoms, but they are *not* ionizable. CH_4 is *not* an acid, nor is it a base.

We expect a substance to be a base if its formula indicates a combination of OH^- ions with metal ions (for example, NaOH). To identify a weak base, we usually need a chemical equation for the ionization reaction, as in Equation (5.7). The main weak base we will work with for the present is NH_3. Metal salts of carbonate (CO_3^{2-}) and hydrogen carbonate (HCO_3^-) are also common weak bases. Sodium hydrogen carbonate, $NaHCO_3$, for instance, is often used in the laboratory to neutralize acid spills. Note that ethanol, C_2H_5OH, is *not* a base. The OH group is not present as OH^-, either in pure ethanol or in its aqueous solutions.

More Acid–Base Reactions

$Mg(OH)_2$ is a base because it contains OH^-, but this compound is quite insoluble in water. Its finely divided solid particles form a suspension in water that is the familiar milk of magnesia, used as an antacid. In this suspension, $Mg(OH)_2(s)$ does dissolve very slightly, producing some OH^- in solution. If an acid is added, H^+ from the acid combines with this OH^- to form water—neutralization occurs. More $Mg(OH)_2(s)$ dissolves to produce more OH^- in solution, which is neutralized by more H^+, and so on. In this way, the neutralization reaction results in the dissolving of otherwise insoluble $Mg(OH)_2(s)$. The net equation for the reaction of $Mg(OH)_2(s)$ with a *strong* acid is

$$Mg(OH)_2(s) + 2\ H^+(aq) \longrightarrow Mg^{2+}(aq) + 2\ H_2O(l) \qquad (5.8)$$

$Mg(OH)_2(s)$ also reacts with a weak acid such as acetic acid. In the net ionic equation, we write acetic acid in its *molecular* form. But remember that some H^+ and $C_2H_3O_2^-$ ions are always present in an acetic acid solution. The H^+ ions react with OH^- ions, as in reaction (5.8), followed by further ionization of $HC_2H_3O_2$, more neutralization, and so on. If enough acetic acid is present, the $Mg(OH)_2$ will dissolve completely.

$$Mg(OH)_2(s) + 2\ HC_2H_3O_2(aq) \longrightarrow$$
$$Mg^{2+}(aq) + 2\ C_2H_3O_2^-(aq) + 2\ H_2O(l) \quad (5.9)$$

Calcium carbonate (present in limestone and marble) is another water-insoluble solid that is soluble in strong and weak acids. Here the solid produces a low concentration of CO_3^{2-} ions, which combine with H^+ to form the weak acid H_2CO_3. This causes more of the solid to dissolve, and so on. Carbonic acid, H_2CO_3, is a very unstable substance that decomposes into H_2O and $CO_2(g)$. Thus, a gas is given off when $CaCO_3(s)$ reacts with an acid and dissolves. Calcium carbonate, like magnesium hydroxide, is used as an antacid.

$$CaCO_3(s) + 2\ H^+(aq) \longrightarrow Ca^{2+}(aq) + H_2O(l) + CO_2(g) \qquad (5.10)$$

To treat reaction (5.10) as an acid–base reaction, we must think of CO_3^{2-} as a base. The definition that we have been using recognizes only OH^- as a base, but when we reconsider acids and bases (Chapter 17) our expanded definitions will identify CO_3^{2-} and many other anions as bases. Table 5.2 lists several com-

This marble statue has been eroded by acid rain. Marble is a material having $CaCO_3$ as its primary component. Acids react with and dissolve the marble through the reaction described in Equation (5.10).

TABLE 5.2	Some Common Gas-Forming Reactions
Ion	**Reaction**
HSO_3^-	$HSO_3^- + H^+ \longrightarrow SO_2(g) + H_2O$
SO_3^{2-}	$SO_3^{2-} + 2\ H^+ \longrightarrow SO_2(g) + H_2O$
HCO_3^-	$HCO_3^- + H^+ \longrightarrow CO_2(g) + H_2O$
CO_3^{2-}	$CO_3^{2-} + 2\ H^+ \longrightarrow CO_2(g) + H_2O$
S^{2-}	$S^{2-} + 2\ H^+ \longrightarrow H_2S(g)$
NH_4^+	$NH_4^+ + OH^- \longrightarrow NH_3(g) + H_2O$

mon anions and one cation that participate in acid–base reactions to produce gases.

EXAMPLE 5-3

Writing Equation(s) for Acid–Base Reactions. Write a net ionic equation to represent the reaction of **(a)** aqueous strontium hydroxide with nitric acid; **(b)** solid aluminum hydroxide with hydrochloric acid.

Solution

In each case we begin by writing the reactants in the "whole-formula" form. Then we substitute ionic forms, where appropriate. Finally we complete the equation as a net ionic equation.

(a) *"Whole-formula" form:* $HNO_3(aq) + Sr(OH)_2(aq) \longrightarrow ?$

   Ionic form:

$$2\,H^+(aq) + 2\,NO_3^-(aq) + Sr^{2+}(aq) + 2\,OH^-(aq) \longrightarrow$$
$$Sr^{2+}(aq) + 2\,NO_3^-(aq) + 2\,H_2O(l)$$

   Net ionic equation: Remove the spectator ions (Sr^{2+} and NO_3^-).

$$2\,H^+(aq) + 2\,OH^-(aq) \longrightarrow 2\,H_2O(l)$$

   or, more simply,

$$H^+(aq) + OH^-(aq) \longrightarrow H_2O(l)$$

(b) *"Whole-formula" form:* $Al(OH)_3(s) + HCl(aq) \longrightarrow ?$

   Ionic form:

$$Al(OH)_3(s) + 3\,H^+(aq) + 3\,Cl^-(aq) \longrightarrow Al^{3+}(aq) + 3\,Cl^-(aq) + 3\,H_2O(l)$$

   Net ionic equation: Remove the spectator ion (Cl^-).

$$Al(OH)_3(s) + 3\,H^+(aq) \longrightarrow Al^{3+}(aq) + 3\,H_2O(l)$$

Practice Example A: Write a net ionic equation to represent the reaction of aqueous ammonia with propionic acid, $HC_3H_5O_2$. What is the formula and name of the salt that results from this neutralization?

Practice Example B: Calcium carbonate is a major constituent of the hard water deposits found in tea kettles and automatic coffee makers. Vinegar, which is essentially a dilute aqueous solution of acetic acid, is commonly used to remove such deposits. Write a net ionic equation for the reaction that occurs. [*Hint:* Recall Equations (5.9) and (5.10).]

5-4   Oxidation–Reduction: Some General Principles

Iron ores are minerals with a high iron content. One of them, *hematite,* Fe_2O_3, is chemically very similar to ordinary iron rust. In simplified fashion, the reaction in which iron metal is produced from hematite in a blast furnace is described as

$$Fe_2O_3(s) + 3\,CO(g) \xrightarrow{\Delta} 2\,Fe(l) + 3\,CO_2(g) \tag{5.11}$$

The tarnishing of silver is an oxidation-reduction reaction. The silver metal reacts with sulfur-containing compounds in the air to produce silver sulfide (Ag_2S)—tarnish.

In this reaction we can think of the $CO(g)$ as taking O atoms away from Fe_2O_3 to produce $CO_2(g)$ and the free element iron. A commonly used term to describe a reaction in which a substance gains O atoms is "oxidation," and one in which a substance loses O atoms, "reduction." In reaction (5.11) $CO(g)$ is oxidized and $Fe_2O_3(s)$ is reduced. An oxidation and a reduction must always occur together, and a reaction in which they do is called an **oxidation–reduction** (or **redox**) **reaction.**

Definitions of oxidation and reduction based solely on the transfer of O atoms are too restrictive. By using broader definitions, we can, for example, describe many reactions in aqueous solution as oxidation–reduction reactions, even when the reactions do not involve oxygen. We can look at oxidation–reduction reactions in two other ways, which we describe next.

▶ Because it is easier to say, the term *redox* is often used instead of oxidation–reduction.

Oxidation State Changes

Suppose we rewrite Equation (5.11) and indicate the oxidation states (O.S.) of the elements on both sides of the equation by using the rules listed on page 79.

$$\overset{+3\ -2}{Fe_2O_3} + 3\ \overset{+2\ -2}{CO} \longrightarrow 2\ \overset{0}{Fe} + 3\ \overset{+4\ -2}{CO_2}$$

The O.S. of oxygen is -2 everywhere it appears in this equation. That of iron (shown in red) changes. It *decreases* from $+3$ in Fe_2O_3 to 0 in the free element, Fe. The O.S. of carbon (shown in blue) also changes. It *increases* from $+2$ in CO to $+4$ in CO_2. In terms of oxidation state changes, in an *oxidation* process the O.S. of some element *increases,* and in *reduction* the O.S. of an element *decreases.*

EXAMPLE 5-4

Identifying Oxidation–Reduction Reactions. Indicate whether each of the following is an oxidation–reduction reaction.
(a) $MnO_2(s) + 4\ H^+(aq) + 2\ Cl^-(aq) \longrightarrow Mn^{2+}(aq) + 2\ H_2O + Cl_2(g)$
(b) $H_2PO_4^-(aq) + OH^-(aq) \longrightarrow HPO_4^{2-}(aq) + H_2O$

Solution

In each case indicate the oxidation states of the elements on both sides of the equation and look for changes.
(a) The O.S. of Mn *decreases* from $+4$ in MnO_2 to $+2$ in Mn^{2+}. MnO_2 is reduced to Mn^{2+}. The O.S. of O remains at -2 throughout the reaction, and that of H,

▶ Even though we assess oxidation state changes by element, oxidation and reduction involve an entire formula unit or ion. Thus, we say that MnO_2 is reduced, not just Mn; and Cl^- is oxidized, not Cl.

at $+1$. The O.S. of Cl *increases* from -1 in Cl^- to 0 in Cl_2. Cl^- is oxidized to Cl_2. The reaction is an oxidation–reduction reaction.

(b) The O.S. of H is $+1$ on both sides of the equation. Oxygen remains in the O.S. -2 throughout. The O.S. of phosphorus is $+5$ in $H_2PO_4^-$ and also $+5$ in HPO_4^{2-}. There are no changes in O.S. This is *not* an oxidation–reduction reaction. (It is, in fact, an acid–base reaction.)

Practice Example A: Identify whether each of the following is an oxidation–reduction reaction.
(a) $(NH_4)_2SO_4(aq) + Ba(NO_3)_2(aq) \longrightarrow BaSO_4(s) + 2\ NH_4NO_3(aq)$
(b) $2\ Pb(NO_3)_2(s) \longrightarrow 2\ PbO(s) + 4\ NO_2(g) + O_2(g)$

Practice Example B: Identify the species that is oxidized and the one that is reduced in the reaction

$$5\ VO^{2+}(aq) + MnO_4^-(aq) + H_2O \longrightarrow 5\ VO_2^+(aq) + Mn^{2+}(aq) + 2\ H^+(aq)$$

Oxidation and Reduction Half-reactions

The reaction illustrated in Figure 5-6 is an oxidation–reduction reaction.

$$Zn(s) + Cu^{2+}(aq) \longrightarrow Zn^{2+}(aq) + Cu(s)$$

We can show this by evaluating changes in oxidation state. But there is another way to establish that it is an oxidation–reduction reaction that we will find especially useful. Think of the reaction as involving two **half-reactions** occurring at the same time. The overall, or *net, reaction* is the sum of the two half-reactions. We can represent the half-reactions by *half-equations* and the net reaction by a *net equation.*

Oxidation:	$Zn(s) \longrightarrow Zn^{2+}(aq) + 2\ e^-$	(5.12)
Reduction:	$Cu^{2+}(aq) + 2\ e^- \longrightarrow Cu(s)$	(5.13)
Net equation:	$Zn(s) + Cu^{2+}(aq) \longrightarrow Zn^{2+}(aq) + Cu(s)$	(5.14)

In half-equation (5.12) the oxidation state of zinc *increases* from 0 to $+2$, corresponding to a *loss* of two electrons by each zinc atom. $Zn(s)$ is *oxidized,*

FIGURE 5-6

An oxidation–reduction reaction.
(a) A blue solution of copper(II) sulfate in water and solid zinc. (b) Zinc is added to the copper(II) sulfate solution and an oxidation–reduction reaction occurs. The products are a colorless aqueous solution of zinc(II) sulfate and a reddish-brown deposit of solid copper.

(a)   （b)

and we label the half-equation representing this as *oxidation.* In half-equation (5.13) the oxidation state of copper *decreases* from $+2$ to 0, corresponding to the *gain* of two electrons by each Cu^{2+} ion. $Cu^{2+}(aq)$ is *reduced,* and we label the half-equation representing this as *reduction.* To summarize,

- **Oxidation** is a process in which the O.S. of some element *increases,* and in which electrons appear on the *right* side of a half-equation.
- **Reduction** is a process in which the O.S. of some element *decreases,* and in which electrons appear on the *left* side of a half-equation.
- Oxidation and reduction half-reactions must always occur together, and the total number of electrons associated with the oxidation must equal the total number associated with the reduction.

EXAMPLE 5-5

Expressing an Oxidation–Reduction Reaction Through Half-Equations and a Net Equation. Write a net ionic equation to represent the reaction of iron with hydrochloric acid solution to produce $H_2(g)$ and $Fe^{2+}(aq)$ (Figure 5-7).

Solution

Iron displaces H^+ ions as $H_2(g)$. The iron appears in solution as $Fe^{2+}(aq)$. Iron is oxidized and $H^+(aq)$ is reduced.

$$
\begin{array}{ll}
\textit{Oxidation:} & Fe(s) \longrightarrow Fe^{2+}(aq) + 2\ e^- \\
\textit{Reduction:} & \underline{2\ H^+(aq) + 2\ e^- \longrightarrow H_2(g)} \\
\textit{Net equation:} & Fe(s) + 2\ H^+(aq) \longrightarrow Fe^{2+}(aq) + H_2(g)
\end{array}
$$

Practice Example A: Represent the reaction of aluminum with hydrochloric acid (reaction 4.2) by oxidation and reduction half-equations and a net equation.

Practice Example B: Represent the reaction of chlorine gas with aqueous sodium bromide to produce liquid bromine and aqueous sodium chloride by oxidation and reduction half-equations and a net equation.

(a)   (b)   (c)

FIGURE 5-7
Displacement of $H^+(aq)$ by iron metal—Example 5-5 illustrated.
(a) An iron nail is wrapped in a piece of copper screen. (b) The nail and screen are placed in HCl(aq). Hydrogen gas is evolved as the nail reacts. (c) The nail reacts completely and produces $Fe^{2+}(aq)$, but the copper does not react.

Figure 5-7 and Example 5-5 suggest some fundamental questions about oxidation–reduction. For example,

- Why does Fe react with HCl(aq), displacing $H_2(g)$, whereas Cu does not?
- Why does Fe form Fe^{2+} and not Fe^{3+} in this reaction?

Even now you can probably see that the answers to these questions lie in the relative abilities of Fe and Cu atoms to give up electrons—to become oxidized. Fe gives up electrons more easily than does Cu; also Fe is more readily oxidized to Fe^{2+} than it is to Fe^{3+}. We can give more complete answers after we develop specific criteria describing electron loss and gain. For now the information in Table 5.3 should be helpful. The table lists some common metals that do react with acids to displace $H_2(g)$, and a few that do not. As noted in the table, most of the Group 1A and 2A metals are so reactive that they will react with cold water to produce $H_2(g)$ and a solution of the metal hydroxide.

5-5   Balancing Oxidation–Reduction Equations

The same principles of equation-balancing apply to oxidation–reduction (redox) equations as to other equations—balance for numbers of atoms and balance for electric charge. Often, though, we find it a little more difficult to apply these principles in redox equations. In fact, we can balance only a small proportion of redox equations by simple inspection. We need a systematic approach. Several methods are available, but we emphasize the one described here.

The Half-reaction (Ion-electron) Method
In this method of balancing a redox equation, we

- Write and balance separate half-equations for oxidation and reduction.
- Adjust coefficients in the two half-equations so that the same number of electrons appears in each half-equation.
- Add together the two half-equations (cancelling out electrons) to obtain the balanced net equation.

We apply this method in a stepwise fashion in Example 5-6.

TABLE 5.3 Behavior of Some Common Metals with Non-oxidizing Acids[a]	
React to Produce $H_2(g)$	**Do Not React**
alkali metals (Group 1A)[b]	Cu, Ag, Au, Hg
alkaline earth metals (Group 2A)[b]	
Al, Zn, Fe, Sn, Pb	

[a]A non-oxidizing acid (for example, HCl, HBr, HI) is one in which the only possible reduction half-reaction is the reduction of H^+ to H_2. Additional possibilities for metal–acid reactions are considered in Chapter 21.

[b]With the exception of Be and Mg, all Group 1A and Group 2A metals also react with cold water to produce $H_2(g)$. (The metal hydroxide is the other product.)

EXAMPLE 5-6

Balancing the Equation for a Redox Reaction in Acidic Solution. The reaction described by expression (5.15) is used to determine the sulfite ion concentration present in wastewater from a papermaking plant. Write the balanced equation for this reaction in acidic solution.

$$SO_3^{2-}(aq) + MnO_4^-(aq) \longrightarrow SO_4^{2-}(aq) + Mn^{2+}(aq) \qquad (5.15)$$

Solution

Step 1. *Write "skeleton" half-equations based on the species undergoing oxidation and reduction.* The O.S. of sulfur increases from $+4$ in SO_3^{2-} to $+6$ in SO_4^{2-}. The O.S. of Mn decreases from $+7$ in MnO_4^- to $+2$ in Mn^{2+}. The "skeleton" half-equations are

$$SO_3^{2-}(aq) \longrightarrow SO_4^{2-}(aq)$$

$$MnO_4^-(aq) \longrightarrow Mn^{2+}(aq)$$

Step 2. *Balance each half-equation "atomically," in this order.*

- atoms other than H and O
- O atoms by adding H_2O with the appropriate coefficient
- H atoms by adding H^+ with the appropriate coefficient

The "other" atoms (S and Mn) are already balanced in the skeleton half-equations. To balance O atoms, we add one H_2O molecule to the left side of the first half-equation and four to the right side of the second.

$$SO_3^{2-}(aq) + H_2O \longrightarrow SO_4^{2-}(aq)$$

$$MnO_4^-(aq) \longrightarrow Mn^{2+}(aq) + 4\,H_2O$$

To balance H atoms, we add two H^+ ions to the right side of the first half-equation and eight to the left side of the second.

$$SO_3^{2-}(aq) + H_2O \longrightarrow SO_4^{2-}(aq) + 2\,H^+(aq)$$

$$MnO_4^-(aq) + 8\,H^+(aq) \longrightarrow Mn^{2+}(aq) + 4\,H_2O$$

Step 3. *Balance each half-equation "electrically."* Add the number of electrons necessary to get the same electric charge on both sides of each half-equation. By doing this you will see that the half-equation in which electrons appear on the *right* side is the *oxidation* half-equation. The other half-equation, with electrons on the *left* side, is the *reduction* half-equation.

Oxidation: $SO_3^{2-}(aq) + H_2O \longrightarrow SO_4^{2-}(aq) + 2\,H^+(aq) + 2\,e^-$
(net charge on each side, -2)

Reduction: $MnO_4^-(aq) + 8\,H^+(aq) + 5\,e^- \longrightarrow Mn^{2+}(aq) + 4\,H_2O$
(net charge on each side, $+2$)

Step 4. *Obtain the net redox equation by combining the half-equations.* Multiply through the oxidation half-equation by *5* and through the reduction half-equation by *2*. This results in 10 e^- on *each* side of the net equation. These terms cancel out. *Electrons must not appear in the final net equation.*

$$5\,SO_3^{2-}(aq) + 5\,H_2O \longrightarrow 5\,SO_4^{2-}(aq) + 10\,H^+(aq) + \cancel{10\,e^-}$$
$$2\,MnO_4^-(aq) + 16\,H^+(aq) + \cancel{10\,e^-} \longrightarrow 2\,Mn^{2+}(aq) + 8\,H_2O$$
$$\overline{5\,SO_3^{2-}(aq) + 2\,MnO_4^-(aq) + 5\,H_2O + 16\,H^+(aq) \longrightarrow}$$
$$5\,SO_4^{2-}(aq) + 2\,Mn^{2+}(aq) + 8\,H_2O + 10\,H^+(aq)$$

▶ Try balancing expression (5.15) by inspection just for number of atoms. You will see that the electric charge does not balance.

▶ In most cases you should be able to identify the key species in skeleton half-equations without having to evaluate oxidation states.

Step 5. *Simplify.* The net equation should not contain the same species on both sides. Subtract *five* H_2O from each side of the equation in step 4. This leaves *three* H_2O on the right. Also subtract *ten* H^+ from each side, leaving *six* on the left.

$$5\,SO_3^{2-}(aq) + 2\,MnO_4^-(aq) + 6\,H^+(aq) \longrightarrow$$
$$5\,SO_4^{2-}(aq) + 2\,Mn^{2+}(aq) + 3\,H_2O$$

Step 6. *Verify.* Check the final net equation to ensure that it is balanced both "atomically" and "electrically." For example, show that in the balanced equation from Step 5, the net charge on each side of the equation is $6-$.

Practice Example A: Balance the equation for this reaction in acidic solution.

$$Fe^{2+}(aq) + MnO_4^-(aq) \longrightarrow Fe^{3+}(aq) + Mn^{2+}(aq)$$

Practice Example B: Balance the equation for this reaction in acidic solution.

$$UO^{2+}(aq) + Cr_2O_7^{2-}(aq) \longrightarrow UO_2^{2+}(aq) + Cr^{3+}(aq)$$

Balancing Redox Equations in Basic Solution

To balance equations for redox reactions in *basic* solution we generally have to add a step or two to the procedure used in Example 5-6. The problem is this: In basic solution, OH^-, not H^+, must appear in the final balanced equation. Because *both* OH^- and H_2O contain H and O atoms, at times it is hard to decide on which side of the half-equations to put each one. One simple approach is to treat the reaction *as if* it occurs in an acidic solution and balance it as in Example 5-6. Then, add to each side of the net redox equation a number of OH^- ions equal to the number of H^+ ions. Where H^+ and OH^- appear on the same side of the equation, combine them to produce H_2O molecules. Simplify, in the event that H_2O now appears on both sides of the equation. This method is illustrated in Example 5-7.

EXAMPLE 5-7

Balancing the Equation for a Redox Reaction in Basic Solution. Balance the equation for the reaction in which permanganate ion oxidizes cyanide ion to cyanate ion in basic solution and is itself reduced to $MnO_2(s)$.

$$MnO_4^-(aq) + CN^-(aq) \longrightarrow MnO_2(s) + OCN^-(aq) \qquad (5.16)$$

Solution

Initially, we treat the half-reactions and the net reaction *as if* they occur in an acidic solution, and finally we adjust the net equation to a basic solution.

Step 1. *Write two skeleton half-equations and balance them for Mn, C, and N atoms.*

$$MnO_4^-(aq) \longrightarrow MnO_2(s)$$

$$CN^-(aq) \longrightarrow OCN^-(aq)$$

Step 2. *Balance the half-equations for O and H atoms. Add H_2O and/or H^+ as required.*

$$MnO_4^- + 4\,H^+(aq) \longrightarrow MnO_2(s) + 2\,H_2O$$

$$CN^-(aq) + H_2O \longrightarrow OCN^-(aq) + 2\,H^+(aq)$$

Step 3. *Balance the half-equations for electric charge by adding the appropriate numbers of electrons.*

$$\text{Reduction:} \quad MnO_4^-(aq) + 4\ H^+(aq) + 3\ e^- \longrightarrow MnO_2(s) + 2\ H_2O$$

$$\text{Oxidation:} \quad CN^-(aq) + H_2O \longrightarrow OCN^-(aq) + 2\ H^+(aq) + 2\ e^-$$

Step 4. *Combine the half-equations to obtain a net redox equation.* Multiply the reduction half-equation by 2 and the oxidation half-equation by 3. Make the appropriate cancellations of $H^+(aq)$ and H_2O.

$$2\ MnO_4^-(aq) + 8\ H^+(aq) + \cancel{6\ e^-} \longrightarrow 2\ MnO_2(s) + 4\ H_2O$$
$$\underline{3\ CN^-(aq) + 3\ H_2O \longrightarrow 3\ OCN^-(aq) + 6\ H^+(aq) + \cancel{6\ e^-}}$$

$$\text{Net:} \quad 2\ MnO_4^-(aq) + 3\ CN^-(aq) + 2\ H^+(aq) \longrightarrow$$
$$2\ MnO_2(s) + 3\ OCN^-(aq) + H_2O$$

Step 5. *Change from an acidic to a basic medium by adding OH^- to both sides of the net equation; combine H^+ and OH^- to form H_2O and simplify.*

$$2\ MnO_4^-(aq) + 3\ CN^-(aq) + 2\ H^+(aq) + 2\ OH^-(aq) \longrightarrow$$
$$2\ MnO_2(s) + 3\ OCN^-(aq) + H_2O + 2\ OH^-(aq)$$

$$2\ MnO_4^-(aq) + 3\ CN^-(aq) + \cancel{2}\ H_2O \longrightarrow$$
$$2\ MnO_2(s) + 3\ OCN^-(aq) + \cancel{H_2O} + 2\ OH^-(aq)$$

The final net balanced redox equation for reaction (5.16) is

$$2\ MnO_4^-(aq) + 3\ CN^-(aq) + H_2O \longrightarrow$$
$$2\ MnO_2(s) + 3\ OCN^-(aq) + 2\ OH^-(aq)$$

Practice Example A: Balance the equation for this reaction in basic solution.

$$S(s) + OCl^-(aq) \longrightarrow SO_3^{2-}(aq) + Cl^-(aq)$$

Practice Example B: Balance the equation for this reaction in basic solution.

$$MnO_4^-(aq) + SO_3^{2-}(aq) \longrightarrow MnO_2(s) + SO_4^{2-}(aq)$$

Disproportionation Reactions

In some oxidation–reduction reactions, called **disproportionation reactions,** the same substance is both oxidized *and* reduced. Some of these reactions have practical significance. The decomposition of hydrogen peroxide, H_2O_2, produces $O_2(g)$. It is the $O_2(g)$ that has a germicidal effect when a dilute aqueous solution of hydrogen peroxide (usually 3%) is used as an antiseptic.

$$2\ H_2O_2(aq) \longrightarrow 2\ H_2O + O_2(g) \qquad (5.17)$$

In reaction (5.17) the oxidation state of oxygen changes from -1 in H_2O_2 to -2 in H_2O (a reduction) and to 0 in $O_2(g)$ (an oxidation). H_2O_2 is both oxidized and reduced.

Solutions of sodium thiosulfate ($Na_2S_2O_3$) are often used in the laboratory in redox reactions. The disproportionation of $S_2O_3^{2-}$ produces sulfur as one of its products, which accounts for the fact that old stock solutions of $Na_2S_2O_3$ sometimes have a light yellow deposit in them.

$$S_2O_3^{2-}(aq) + 2\ H^+ \longrightarrow S(s) + SO_2(g) + H_2O \qquad (5.18)$$

The oxidation states of S are $+2$ in $S_2O_3^{2-}$, 0 in $S(s)$, and $+4$ in $SO_2(g)$.

The same substance appears on the left side in each half-equation for a disproportionation reaction. The balanced half-equations and net equation for reaction (15.18) are

Oxidation: $\quad S_2O_3{}^{2-}(aq) + H_2O \longrightarrow 2\ SO_2(g) + 2\ H^+(aq) + 4\ e^-$

Reduction: $\quad \underline{S_2O_3{}^{2-}(aq) + 6\ H^+(aq) + 4\ e^- \longrightarrow 2\ S(s) + 3\ H_2O}$

$\qquad\qquad\quad 2\ S_2O_3{}^{2-}(aq) + 4\ H^+(aq) \longrightarrow 2\ S(s) + 2\ SO_2(g) + 2\ H_2O$

Net: $\qquad\quad S_2O_3{}^{2-}(aq) + 2\ H^+(aq) \longrightarrow S(s) + SO_2(g) + H_2O$

Are You Wondering . . .

How to balance equations for redox reactions in the gaseous state?

Generally you will be able to do this by inspection, just as you would balance the equation for the combustion of a hydrocarbon (a redox reaction). Occasionally, you may encounter an equation that seems difficult to balance by inspection. In these cases, you can sometimes treat the reaction as if it occurs in an acidic aqueous solution, with H^+ and any extraneous H_2O cancelling out of the net equation (see Exercise 46).

5-6 Oxidizing and Reducing Agents

Chemists frequently use the terms *oxidizing agent* and *reducing agent* to describe certain of the reactants in redox reactions, as in statements like "fluorine gas is a powerful oxidizing agent," or "calcium metal is a good reducing agent." Let us briefly consider the meaning of these terms.

In a redox reaction, the substance that makes it possible for some other substance to be *oxidized* is called the **oxidizing agent,** or **oxidant.** In doing so, the oxidizing agent is itself reduced. Similarly, the substance that causes some other substance to be *reduced* is called the **reducing agent,** or **reductant.** In the reaction the reducing agent is itself oxidized. Or, stated in other ways,

An oxidizing agent (oxidant):

- contains an element whose oxidation state *decreases* in a redox reaction;
- "gains" electrons (electrons are found on the left side of its half-equation).

A reducing agent (reductant):

- contains an element whose oxidation state *increases* in a redox reaction;
- "loses" electrons (electrons are found on the right side of its half-equation).

These statements are summarized in Figure 5-8. In general, a substance with an element in one of its highest possible oxidation states is an oxidizing agent. If the element is in one of its lowest possible oxidation states, the substance is a reducing agent. Certain substances in which the oxidation state of an element is between its highest and lowest possible values may act as an oxidizing agent in some instances and a reducing agent in others.

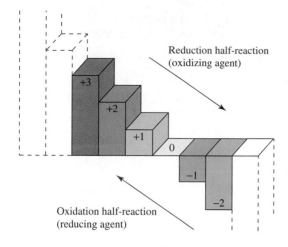

FIGURE 5-8
Identifying oxidizing and reducing agents.
In a reduction half-reaction, the oxidation state of some element decreases and the substance containing that element is an *oxidizing* agent. In an oxidation half-reaction, the oxidation state of some element increases and the substance containing that element is a *reducing* agent.

EXAMPLE 5-8

Identifying Oxidizing and Reducing Agents. Hydrogen peroxide, H_2O_2, is a versatile chemical. Its uses include bleaching wood pulp and fabrics and substituting for chlorine in water purification. One reason for its versatility is that it can be either an oxidizing or a reducing agent. For the following reactions, identify whether hydrogen peroxide is an oxidizing or a reducing agent.

(a) $H_2O_2(aq) + 2 Fe^{2+}(aq) + 2 H^+(aq) \longrightarrow 2 H_2O + 2 Fe^{3+}(aq)$
(b) $5 H_2O_2(aq) + 2 MnO_4^-(aq) + 6 H^+(aq) \longrightarrow 8 H_2O + 2 Mn^{2+}(aq) + 5 O_2(g)$

Solution

(a) Fe^{2+} is oxidized to Fe^{3+} and H_2O_2 makes this possible; H_2O_2 is an oxidizing agent. Looking at the matter another way, we see that the oxidation state (O.S.) of oxygen in H_2O_2 is -1. In H_2O its O.S. is -2. Hydrogen peroxide is *reduced* and thereby acts as an oxidizing agent.

(b) MnO_4^- is reduced to Mn^{2+} and H_2O_2 makes this possible; H_2O_2 is a reducing agent. Or, the O.S. of oxygen *increases* from -1 in H_2O_2 to 0 in O_2. Hydrogen peroxide is *oxidized* and thereby acts as a reducing agent.

▶ When H_2O_2 acts as an oxidizing agent, it is reduced to H_2O in acidic solution or to OH^- in basic solution. When it acts as a reducing agent, it is oxidized to $O_2(g)$.

Practice Example A: Is $H_2(g)$ an oxidizing or reducing agent in the following reaction? Explain.

$$2 NO_2(g) + 7 H_2(g) \longrightarrow 2 NH_3(g) + 4 H_2O(g)$$

Practice Example B: Identify the oxidizing agent and the reducing agent in the following reaction.

$$4 Au(s) + 8 CN^-(aq) + O_2(g) + 2 H_2O \longrightarrow 4 [Au(CN)_2]^-(aq) + 4 OH^-(aq)$$

Permanganate ion, MnO_4^-, is a versatile oxidizing agent with many uses in the chemical laboratory. In the next section we describe its use in the quantitative analysis of iron—that is, the determination of the exact (quantitative) amount of iron in an iron-containing material. *Ozone,* $O_3(g)$, a triatomic form of oxygen, is an oxidizing agent used in water purification, as in the oxidation of the organic compound phenol, C_6H_5OH.

$$C_6H_5OH(aq) + 14 O_3(g) \longrightarrow 6 CO_2(g) + 3 H_2O + 14 O_2(g)$$

Thiosulfate ion, $S_2O_3{}^{2-}$, is an important reducing agent. One of its industrial uses is as an "antichlor," to destroy residual chlorine from the bleaching of fibers.

$$S_2O_3{}^{2-}(aq) + 4\ Cl_2(aq) + 5\ H_2O \longrightarrow 2\ HSO_4{}^-(aq) + 8\ H^+(aq) + 8\ Cl^-(aq)$$

Oxidizing and reducing agents also play important roles in biological systems: in photosynthesis (to store the sun's energy), in metabolism (oxidizing glucose), and in the transport of oxygen.

5-7 Stoichiometry of Reactions in Aqueous Solutions: Titrations

If our objective is to get the maximum yield of a product, we generally choose one of the reactants (usually the most expensive one) as the limiting reactant, and use excess amounts of the other reactants in a reaction. This is the case in most precipitation reactions. In some instances, as in determining the concentration of a solution, we may not be interested in the products of a reaction but only in the relationship between one reactant and another. Then, we have to carry out the reaction in such a way that *neither* reactant is in excess. A method that has long been used for doing this is known as *titration.*

We place a solution of one reactant in a small beaker or flask. We place another reactant, also in solution, in a *buret,* a long graduated tube equipped with a stopcock valve. We can slowly add the second solution to the first by manipulating the stopcock. **Titration** is a reaction carried out by the carefully controlled addition of one solution to another. The trick is to stop the titration at the point where both reactants have been consumed simultaneously, a condition called the **equivalence point** of the titration. In a titration we need some means of signaling when the equivalence point is reached. In modern chemical laboratories this is commonly done with an appropriate measuring instrument. Still widely used, though, is a technique in which a very small quantity of a substance added to the reaction mixture changes color at or very near the equivalence point. Such substances are called **indicators.** Figure 5-9 illustrates the neutralization of an acid by a base by the titration technique. Calculations that use

▶ The key to a successful acid–base titration is in selecting the right indicator. We learn how to do this when we consider theoretical aspects of titration (Chapter 18).

FIGURE 5-9
An acid–base titration—Example 5-10 illustrated.
(a) A 5.00-mL sample of vinegar, a small quantity of water, and a few drops of phenolphthalein indicator are added to a flask.
(b) 0.1000 M NaOH from a previously filled buret is slowly added.
(c) As long as the acid is in excess, the solution in the flask remains colorless. When the acid has been neutralized, an additional drop of NaOH(aq) causes the solution to become slightly basic. The phenolphthalein indicator turns a light pink. The first lasting appearance of the pink color is taken to be the equivalence point of the titration.

(a)

(b)

(c)

titration data are much the same as those introduced in Chapter 4, as illustrated in Example 5-9.

EXAMPLE 5-9

Using Titration Data to Establish the Concentrations of Acids and Bases. Vinegar is a dilute aqueous solution of acetic acid produced by the bacterial fermentation of apple cider, wine, or other carbohydrate material. The legal minimum acetic acid content of vinegar is 4% by mass. A 5.00-mL sample of a particular vinegar was titrated with 38.08 mL of 0.1000 M NaOH(aq). Does this sample exceed the minimum limit? (Vinegar has a density of about 1.01 g/mL.)

Solution

With ideas from Section 5-3 we can write a net ionic equation for the neutralization reaction.

$$HC_2H_3O_2(aq) + OH^-(aq) \longrightarrow C_2H_3O_2^-(aq) + H_2O(l)$$

Then, using familiar conversion factors, we proceed in this stepwise fashion: mL NaOH → L NaOH → mol NaOH → mol OH$^-$ → mol HC$_2$H$_3$O$_2$ → concn HC$_2$H$_3$O$_2$.

$$? \text{ mol } HC_2H_3O_2 = 38.08 \text{ mL} \times \frac{1 \text{ L}}{1000 \text{ mL}} \times \frac{0.1000 \text{ mol NaOH}}{1 \text{ L}}$$
$$\times \frac{1 \text{ mol OH}^-}{1 \text{ mol NaOH}} \times \frac{1 \text{ mol } HC_2H_3O_2}{1 \text{ mol OH}^-}$$
$$= 3.808 \times 10^{-3} \text{ mol } HC_2H_3O_2$$

$$\text{concn } HC_2H_3O_2 = \frac{3.808 \times 10^{-3} \text{ mol } HC_2H_3O_2}{5.00 \times 10^{-3} \text{ L}} = 0.762 \text{ M } HC_2H_3O_2$$

Now, we can use the definition of molarity, the molar mass of HC$_2$H$_3$O$_2$, and the density of the vinegar to determine the mass percent of HC$_2$H$_3$O$_2$.

$$\% \ HC_2H_3O_2 = \frac{0.762 \text{ mol } HC_2H_3O_2}{1000 \text{ mL vinegar}} \times \frac{60.05 \text{ g } HC_2H_3O_2}{1 \text{ mol } HC_2H_3O_2}$$
$$\times \frac{1 \text{ mL vinegar}}{1.01 \text{ g vinegar}} \times 100\%$$
$$= 4.53\% \ HC_2H_3O_2$$

The vinegar sample exceeds the legal minimum limit, but only a bit. There is also a standard for the maximum amount of acetic acid allowed in vinegar. A vinegar producer might use this titration technique to ensure that the vinegar stays between these limits. The competitors' products might also be monitored in this way.

▶ Feature Problem B on page 161 presents a novel use of an acid–base titration in the chemical research laboratory.

Practice Example A: A particular solution of NaOH is supposed to be approximately 0.100 M. To determine the molarity of the NaOH, a solid acid, KHC$_8$H$_4$O$_4$ (KHP), is used in the titration of this NaOH solution. A 0.5000-g sample of KHP is dissolved in water and titrated with 24.03 mL of the NaOH solution. What is the actual molarity of the NaOH?

$$HC_8H_4O_4^-(aq) + OH^-(aq) \longrightarrow C_8H_4O_4^{2-}(aq) + H_2O(l)$$

Practice Example B: A 0.235-g sample of a solid that is 92.5% KOH, 7.5% Ba(OH)$_2$ by mass requires 45.6 mL of an HCl(aq) solution for its titration. What is the molarity of the HCl(aq)? (*Hint:* Write a net ionic equation, and note that there are two sources of OH$^-$.)

(a) (b) (c)

FIGURE 5-10   Standardizing a solution of an oxidizing agent through a redox titration—Example 5-11 illustrated.
(a) The solution contains a known amount of Fe^{2+} and the buret is filled with the intensely colored $KMnO_4(aq)$ to be standardized. (b) As it is added to the strongly acidic solution of $Fe^{2+}(aq)$, the $KMnO_4(aq)$ is immediately decolorized as a result of reaction (5.19). (c) When all the Fe^{2+} has been oxidized to Fe^{3+}, additional $KMnO_4(aq)$ has nothing left to oxidize and the solution turns a distinctive pink. Even a fraction of a drop of the $KMnO_4(aq)$ beyond the equivalence point is sufficient to cause this pink coloration.

Suppose we need a $KMnO_4(aq)$ solution of exactly known molarity, close to 0.020 M. We cannot prepare this solution by weighing out the required amount of $KMnO_4(s)$ and dissolving it in water. The solid is *not pure,* and its actual purity (that is, % $KMnO_4$) is *not known.* On the other hand, we can obtain iron wire in essentially pure form, and allow the wire to react with an acid to yield $Fe^{2+}(aq)$. $Fe^{2+}(aq)$ is *oxidized* to $Fe^{3+}(aq)$ by $KMnO_4(aq)$ in an acidic solution. By determining the volume of $KMnO_4(aq)$ required to oxidize a known quantity of $Fe^{2+}(aq)$, we can calculate the exact molarity of the $KMnO_4(aq)$. Example 5-10 and Figure 5-10 illustrate this procedure, called **standardization of a solution.**

EXAMPLE 5-10

Standardizing a Solution for Use in Redox Titrations. A piece of iron wire weighing 0.1568 g is converted to $Fe^{2+}(aq)$ and requires 26.24 mL of a $KMnO_4(aq)$ solution for its titration. What is the molarity of the $KMnO_4(aq)$?

$$5\ Fe^{2+}(aq) + MnO_4^-(aq) + 8\ H^+(aq) \longrightarrow 5\ Fe^{3+}(aq) + Mn^{2+}(aq) + 4\ H_2O$$
(5.19)

Solution

First, determine the amount of $KMnO_4$ consumed in the titration.

$$? \text{ mol KMnO}_4 = 0.1568 \text{ g Fe} \times \frac{1 \text{ mol Fe}}{55.847 \text{ g Fe}} \times \frac{1 \text{ mol Fe}^{2+}}{1 \text{ mol Fe}}$$

$$\times \frac{1 \text{ mol MnO}_4^-}{5 \text{ mol Fe}^{2+}} \times \frac{1 \text{ mol KMnO}_4}{1 \text{ mol MnO}_4^-}$$

$$= 5.615 \times 10^{-4} \text{ mol KMnO}_4$$

The volume of solution containing the 5.615×10^{-4} mol KMnO$_4$ is 26.24 mL = 0.02624 L, which means that

$$\text{concn KMnO}_4(aq) = \frac{5.615 \times 10^{-4} \text{ mol KMnO}_4}{0.02624 \text{ L}} = 0.02140 \text{ M KMnO}_4(aq)$$

Practice Example A: Another substance that may be used to standardize KMnO$_4$(aq) is sodium oxalate, Na$_2$C$_2$O$_4$. If 0.2482 g Na$_2$C$_2$O$_4$ is dissolved in water and titrated with 23.68 mL KMnO$_4$, what is the molarity of the KMnO$_4$(aq)?

$$2 \text{ MnO}_4^-(aq) + 5 \text{ C}_2\text{O}_4^{2-}(aq) + 16 \text{ H}^+(aq) \longrightarrow$$
$$2 \text{ Mn}^{2+}(aq) + 8 \text{ H}_2\text{O} + 10 \text{ CO}_2(g)$$

Practice Example B: A 0.376-g sample of an iron ore is dissolved in acid, the iron reduced to Fe^{2+}(aq), and then titrated with 41.25 mL of 0.02140 M KMnO$_4$. Determine the % Fe by mass in the iron ore. [*Hint:* Use Equation (5.19).]

▶Feature Problem C on page 161 considers a chemical analysis that requires two oxidation–reduction reactions in succession.

Summary

Substances in aqueous solution are nonelectrolytes, weak electrolytes, or strong electrolytes, depending on the extent to which they produce ions. Strong electrolytes are completely dissociated into ions, and the concentration of a solution can be expressed in terms of these ions.

Some reactions in solution involve the combination of ions to yield water–insoluble solids—precipitates. Precipitation reactions can be predicted through a small number of solubility rules.

Still other reactions involve the combination of H$^+$ and OH$^-$ ions to form H$_2$O (HOH). The source of H$^+$ is called an *acid* and the source of OH$^-$, a *base*. The reaction is an acid–base, or *neutralization*, reaction. By extending the definition of acids and bases to certain ions other than H$^+$ and OH$^-$, some reactions in which gases are evolved can also be treated as acid–base reactions.

In an *oxidation–reduction (redox)* reaction certain atoms undergo an increase in oxidation state, a process called oxidation. Others undergo a decrease in oxidation state—reduction. An especially useful representation of redox reactions is through separate *half-equations* for oxidation and reduction and a *net* equation obtained by combining the two half-equations. This approach also underlies a technique for balancing redox equations.

A common technique for carrying out a reaction in solution is known as *titration*. Titration data can be used to establish the molarities of solutions or to provide other information about the compositions of samples being analyzed.

Summarizing Example

Sodium dithionite, Na$_2$S$_2$O$_4$, is an important reducing agent. One interesting use is the reduction of chromate ion to insoluble chromium(III) hydroxide by dithionite ion, S$_2$O$_4^{2-}$, in basic solution. Sulfite ion is another product. The chromate ion may be present in wastewater from a chrome-plating plant, for example.

What mass of Na$_2$S$_2$O$_4$ is consumed in a reaction with 100.0 L of wastewater having [CrO$_4^{2-}$] = 0.0148 M?

1. *Write an ionic expression representing the reaction.* The participants in the reaction are named in the opening paragraph. Use information from Chapter 3 to substitute symbols

and formulas for names. *Result:* CrO$_4^{2-}$ + S$_2$O$_4^{2-}$ + OH$^-$ → Cr(OH)$_3$(s) + SO$_3^{2-}$

2. *Balance the redox equation written in part 1.* The skeleton half-equations involve CrO$_4^{2-}$ → Cr(OH)$_3$ and S$_2$O$_4^{2-}$ → SO$_3^{2-}$. First, balance the equation *as if* the reaction occurred in acidic solution. Then eliminate H$^+$ by adding the appropriate number of OH$^-$ ions. *Result:* 3 S$_2$O$_4^{2-}$ + 2 CrO$_4^{2-}$ + 2 H$_2$O + 2 OH$^-$ → 6 SO$_3^{2-}$ + 2 Cr(OH)$_3$(s)

3. *Complete the calculation of mass of Na$_2$S$_2$O$_4$.* This requires using the solution volume, chromate ion molarity, the stoichiometric factor, 3 mol S$_2$O$_4^{2-}$/2 mol CrO$_4^{2-}$, and the molar mass of Na$_2$S$_2$O$_4$.   *Answer:* 387 g Na$_2$S$_2$O$_4$.

Focus On

Water Treatment

Aeration and chemicals are used to treat effluent wastewater from papermaking.

Water is not just water. Water often contains impurities that make it unsuitable for some purposes. For instance, a chemist would not use ordinary tap water to prepare an aqueous solution of silver nitrate. The solution would acquire a milky cast from the reaction of $Ag^+(aq)$ with traces of $Cl^-(aq)$ to form $AgCl(s)$. Yet this same tap water is probably perfectly safe to drink.

The way in which water is purified depends on how it is to be used or on how it has been used. Water for a certain industrial application may require a different prior treatment than does water for domestic use. Also, water that has been used in a chemical plant may require a different treatment after use than does domestic sewage. Still, these varying treatments of water share a common feature. They include reactions in aqueous solutions of the general types described in this chapter. A few specific examples follow.

Removal of Iron from Drinking Water

Certain wells deliver water with up to 25 mg of iron per liter, but the federal recommended limit in drinking water in the U.S. is 0.3 mg/L. Methods of removing excess iron generally involve (a) making the chlorinated water slightly basic with slaked lime, $Ca(OH)_2$; (b) oxidizing Fe^{2+} to Fe^{3+} with hypochlorite ion, OCl^-; and (c) precipitating $Fe(OH)_3(s)$ from the basic solution. The reactions can be represented by the equations

Key Terms

acid (5-3)
base (5-3)
disproportionation reaction (5-5)
equivalence point (5-7)
half-reaction (5-4)
half-reaction method (5-5)
indicator (5-7)
net ionic equation (5-2)
neutralization (5-3)

nonelectrolyte (5-1)
oxidation (5-4)
oxidation–reduction (redox) reaction (5-4)
oxidizing agent (oxidant) (5-6)
precipitate (5-2)
reducing agent (reductant) (5-6)
reduction (5-4)
salt (5-3)

standardization of a solution (5-7)
strong acid (5-3)
strong base (5-3)
strong electrolyte (5-1)
titration (5-7)
weak acid (5-3)
weak base (5-3)
weak electrolyte (5-1)

Review Questions

1. In your own words define or explain the following terms or symbols: (a) $\rightleftharpoons$; (b) [ ]; (c) "spectator" ion; (d) a weak acid.
2. Briefly describe each of the following ideas or methods: (a) half-reaction method of balancing redox equations; (b)

disproportionation reaction; (c) titration; (d) standardization of a solution.
3. Explain the important distinctions between each pair of terms: (a) strong electrolyte and strong acid; (b) oxidizing and

$$Cl_2(g) + 2\ OH^-(aq) \longrightarrow$$
$$Cl^-(aq) + OCl^-(aq) + H_2O$$

$$2\ Fe^{2+}(aq) + OCl^-(aq) + H_2O \longrightarrow$$
$$2\ Fe^{3+}(aq) + Cl^-(aq) + 2\ OH^-(aq)$$

$$Fe^{3+}(aq) + 3\ OH^-(aq) \longrightarrow$$
$$Fe(OH)_3(s)$$

While all of this is occurring, the OCl^- is also serving its primary function: destroying pathogenic microorganisms in the water.

Removal of Oxygen from Boiler Water

High-temperature boilers are used to convert water to steam in electric power plants. Dissolved oxygen in the water is highly objectionable because it promotes corrosion of the steel from which many of the boiler parts are made. Because $O_2(g)$ is a good oxidizing agent, a reducing agent such as hydrazine, N_2H_4, is needed to remove it.

$$O_2(aq) + N_2H_4(aq) \longrightarrow$$
$$2\ H_2O + N_2(g)$$

Removal of Phosphates from Domestic Sewage

Eutrophication is the term used to describe a series of events leading to the rapid growth of algae, the killing of fish, and other deleterious effects in freshwater bodies. These events are caused by an excess of nutrients, principally phosphates. The treatment of domestic sewage includes the removal of phosphates. A particularly simple method is to precipitate the phosphates with slaked lime, $Ca(OH)_2$. The phosphates may be present in a number of different forms, including hydrogen phosphate ion, HPO_4^{2-}.

$$5\ Ca^{2+}(aq) + 3\ HPO_4^{2-}(aq) +$$
$$4\ OH^-(aq) \longrightarrow$$
$$Ca_5OH(PO_4)_3(s) + 3\ H_2O$$

The precipitate has the composition of the mineral hydroxyapatite.

Destruction of Cyanide Ion in Industrial Operations

Cyanide compounds are used in metal-cleaning operations, in electroplating, and in extracting gold from gold-bearing rocks in mining operations. The cyanide ion must be destroyed in waste solutions from these operations. This can be done through an oxidation–reduction reaction, such as

$$2\ CN^-(aq) + 5\ OCl^-(aq) +$$
$$2\ OH^-(aq) \longrightarrow N_2(g) +$$
$$2\ CO_3^{2-}(aq) + 5\ Cl^-(aq) + H_2O$$

The poisonous $CN^-(aq)$ is converted to innocuous $N_2(g)$ and $CO_3^{2-}(aq)$.

Later in the text we will consider additional examples of water treatment that have important domestic and industrial uses, such as reverse osmosis and water softening. Also, we must not overlook the biological processes that play an important role in water treatment, particularly in the treatment of wastewater.

---

reducing agent; **(c)** precipitation and neutralization reactions; **(d)** half-reaction and net reaction.

4. From the following solutions, select the **(a)** best and **(b)** poorest electrical conductor and explain the reason for your choices: 0.10 M NH_3, 0.10 M NaCl, 0.10 M $HC_2H_3O_2$ (acetic acid), 0.10 M C_2H_5OH (ethanol).

5. Identify each of the following substances as a strong acid, weak acid, strong base, weak base, or salt: **(a)** Na_2SO_4; **(b)** $Ba(OH)_2$; **(c)** $Ba(NO_3)_2$; **(d)** H_3PO_4; **(e)** HBr; **(f)** HNO_2; **(g)** NH_3; **(h)** NH_4I; **(i)** KOH.

6. *Without doing detailed calculations,* indicate which of the following solutions has the greatest $[SO_4^{2-}]$: 0.075 M H_2SO_4, 0.22 M $MgSO_4$, 0.15 M Na_2SO_4, 0.080 M $Al_2(SO_4)_3$, 0.20 M $CuSO_4$.

7. Determine the concentration of the ion indicated in each of the following solutions: **(a)** $[K^+]$ in 0.238 M KNO_3; **(b)** $[NO_3^-]$ in 0.167 M $Ca(NO_3)_2$; **(c)** $[Al^{3+}]$ in 0.083 M $Al_2(SO_4)_3$; **(d)** $[Na^+]$ in 0.209 M Na_3PO_4.

8. Which of the following contains the greatest amount of chloride ion? 200.0 mL of 0.35 M NaCl, 500.0 mL of 0.065 M $MgCl_2$, 1.00 L of 0.068 M HCl.

9. A solution is prepared by dissolving 0.132 g $Ba(OH)_2 \cdot 8H_2O$ in 275 mL of water solution. What is $[OH^-]$ in this solution?

10. A solution is 0.126 M KCl and 0.148 M $MgCl_2$. What are $[K^+]$, $[Mg^{2+}]$, and $[Cl^-]$ in this solution?

11. How many milligrams of MgI_2 must be added to 250.0 mL of 0.0876 M KI to produce a solution with $[I^-] = 0.1000$ M?

12. Which of the following compounds is (are) *insoluble* in water? Explain. $CuCl_2$, NaI, $BaSO_4$, $Zn(NO_3)_2$, $Pb(C_2H_3O_2)_2$, $Al(OH)_3$.

13. Which of the following react(s) with HCl(aq) to produce a gas? Explain. Na_2SO_4, $KHSO_3$, $Zn(OH)_2$, Ca, $CaCl_2$.

14. Complete each of the following as a *net ionic equation,* indicating whether a precipitate forms. If no reaction occurs, so state.

(a) $Na^+ + Br^- + Pb^{2+} + 2 NO_3^- \rightarrow$
(b) $Mg^{2+} + 2 Cl^- + Cu^{2+} + SO_4^{2-} \rightarrow$
(c) $Fe^{3+} + 3 NO_3^- + Na^+ + OH^- \rightarrow$
(d) $Ca^{2+} + 2 I^- + 2 Na^+ + CO_3^{2-} \rightarrow$
(e) $Ba^{2+} + S^{2-} + 2 Na^+ + SO_4^{2-} \rightarrow$
(f) $2 K^+ + S^{2-} + Ca^{2+} + 2 Cl^- \rightarrow$

15. Complete each of the following as a *net ionic equation.* If no reaction occurs, so state.
(a) $Ba^{2+} + 2 OH^- + HC_2H_3O_2 \rightarrow$
(b) $H^+ + Cl^- + HC_3H_5O_2 \rightarrow$
(c) $FeS(s) + H^+ + I^- \rightarrow$
(d) $K^+ + HCO_3^- + H^+ + NO_3^- \rightarrow$
(e) $Mg(s) + H^+ \rightarrow$
(f) $Cu(s) + H^+ \rightarrow$

16. What volume of 0.0962 M NaOH is required exactly to neutralize 10.00 mL 0.128 M HCl? [*Hint:* The net equation is $H^+(aq) + OH^-(aq) \rightarrow H_2O(l)$.]

17. The exact neutralization of 10.00 mL of 0.1012 M $H_2SO_4(aq)$ requires 23.31 mL of NaOH(aq). What must be the molarity of the NaOH(aq)?

$$H_2SO_4(aq) + 2 NaOH(aq) \longrightarrow Na_2SO_4(aq) + 2 H_2O(l)$$

18. A 23.58-mL sample of 0.1278 M KOH is added to 25.13 mL of 0.1264 M HCl. Is the resulting mixture acidic, basic, or just exactly neutralized? Explain. (*Hint:* Write a net ionic equation for the reaction. Is there a limiting reactant?)

19. Assign oxidation states to the elements involved in the following reactions. Indicate which are redox reactions and which are not.
(a) $MgCO_3(s) + 2 H^+(aq) \rightarrow Mg^{2+}(aq) + H_2O + CO_2(g)$
(b) $Cl_2(aq) + 2 Br^-(aq) \rightarrow 2 Cl^-(aq) + Br_2(aq)$
(c) $Ag(s) + 2 H^+(aq) + NO_3^-(aq) \rightarrow$
$$Ag^+(aq) + H_2O + NO_2(g)$$
(d) $2 Ag^+(aq) + CrO_4^{2-}(aq) \rightarrow Ag_2CrO_4(s)$

20. Assign oxidation states to the elements in the following redox reactions. Indicate which are the oxidizing and which are the reducing agents.
(a) $2 NO(g) + 5 H_2(g) \rightarrow 2 NH_3(g) + 2 H_2O(g)$
(b) $3 Cu(s) + 8 H^+(aq) + 2 NO_3^-(aq) \rightarrow$
$$3 Cu^{2+}(aq) + 4 H_2O + 2 NO(g)$$
(c) $3 Cl_2(g) + 6 OH^-(aq) \rightarrow$
$$5 Cl^-(aq) + ClO_3^-(aq) + 3 H_2O$$

21. Complete and balance the following half-equations, and indicate whether oxidation or reduction is involved.
(a) $SO_3^{2-} \rightarrow S_2O_3^{2-}$
(b) $HNO_3 \rightarrow N_2O(g)$ (acidic solution)
(c) $I^- \rightarrow IO_3^-$ (acidic solution)
(d) $Al(s) \rightarrow Al(OH)_4^-$ (basic solution)

22. Balance these equations for redox reactions in acidic solution.
(a) $Zn(s) + NO_3^- \rightarrow Zn^{2+} + NO(g)$
(b) $Zn(s) + NO_3^- \rightarrow Zn^{2+} + NH_4^+$
(c) $Fe^{2+} + Cr_2O_7^{2-} \rightarrow Fe^{3+} + Cr^{3+}$
(d) $H_2O_2 + MnO_4^- \rightarrow Mn^{2+} + O_2(g)$

23. Balance these equations for redox reactions in basic solution.
(a) $MnO_2(s) + ClO_3^- \rightarrow MnO_4^- + Cl^-$
(b) $Fe(OH)_3(s) + OCl^- \rightarrow FeO_4^{2-} + Cl^-$
(c) $ClO_2 \rightarrow ClO_3^- + Cl^-$

24. A 25.12-mL sample of a KMnO$_4$(aq) solution is required to titrate 0.2879 g sodium oxalate, $Na_2C_2O_4$, in a redox reaction occuring in acidic solution.

$$C_2O_4^{2-}(aq) + MnO_4^-(aq) \longrightarrow Mn^{2+}(aq) + CO_2(g)$$

What is the molarity of the KMnO$_4$(aq)? (*Hint:* You must first balance the redox equation.)

Exercises

Strong Electrolytes, Weak Electrolytes, and Nonelectrolytes

25. Using information from this chapter, indicate whether each of the following substances in aqueous solution is a nonelectrolyte, weak electrolyte, or strong electrolyte. Explain. **(a)** HC_6H_5O; **(b)** Li_2SO_4; **(c)** MgI_2; **(d)** $(CH_3CH_2)_2O$; **(e)** $Sr(OH)_2$.

26. NH$_3$(aq) conducts electric current only weakly. The same is true for $HC_2H_3O_2$(aq). When these solutions are mixed, however, the resulting solution conducts electric current very well. How do you explain this?

Ion Concentrations

27. These data are given for several cations in solution. Express them as molarities. **(a)** 35.0 mg Ca^{2+}/L; **(b)** 25.6 mg K^+/100 mL; **(c)** 0.168 mg Zn^{2+}/mL.

28. What molarity of NaF(aq) corresponds to a fluoride ion content of 0.9 mg F^-/L, the Federal Government's recommended limit for fluoride ion in drinking water?

29. Which of the following aqueous solutions has the highest concentration of Na$^+$? **(a)** 0.208 M Na_2SO_4; **(b)** a solution containing 1.05 g NaCl/100 mL; **(c)** a solution having 14.7 mg Na^+/mL.

30. Which of the following aqueous solutions has the greatest [H$^+$]: 0.011 M $HC_2H_3O_2$, 0.010 M HCl, 0.010 M H_2SO_4, 1.00 M NH$_3$. Explain.

31. If one assumes the volumes are additive, what is [Cl$^-$] in a solution obtained by mixing 225 mL of 0.625 M KCl and 615 mL of 0.385 M MgCl$_2$?

32. If one assumes the volumes are additive, what is [NO$_3^-$] in a solution obtained by mixing 275 mL of 0.283 M KNO$_3$, 328 mL of 0.421 M Mg(NO$_3$)$_2$, and 784 mL of H$_2$O?

Predicting Precipitation Reactions

33. Predict whether a reaction is likely to occur in each of the following cases. If so, write a net ionic equation.
(a) $HI(aq) + Zn(NO_3)_2(aq) \rightarrow$
(b) $CuSO_4(aq) + Na_2CO_3(aq) \rightarrow$
(c) $Cu(NO_3)_2(aq) + Na_3PO_4(aq) \rightarrow$

35. What reagent solution might you use to separate the cations in the following mixtures, that is, with one ion appearing in solution and the other in a precipitate? (*Hint:* Refer to the solubility rules on page 135 and consider water also to be a reagent.)
(a) $BaCl_2(s)$ and $CaCl_2(s)$
(b) $MgCO_3(s)$ and $Na_2CO_3(s)$
(c) $AgNO_3(s)$ and $Cu(NO_3)_2(s)$

37. You are provided with $NaOH(aq)$, $K_2SO_4(aq)$, $Mg(NO_3)_2(aq)$, $BaCl_2(aq)$, $NaCl(aq)$, $Sr(NO_3)_2(aq)$, $Ag_2SO_4(s)$, and $BaSO_4(s)$. Write net ionic equations to show how you would use one or more of these reagents to obtain **(a)** $SrSO_4(s)$; **(b)** $Mg(OH)_2(s)$; **(c)** $KCl(aq)$.

34. Predict whether a reaction is likely to occur in each of the following cases. If so, write a net ionic equation.
(a) $AgNO_3(aq) + CuCl_2(aq) \rightarrow$
(b) $Na_2S(aq) + FeCl_2(aq) \rightarrow$
(c) $Na_2CO_3(aq) + AgNO_3(aq) \rightarrow$

36. What reagent solution might you use to separate the cations in each of the following mixtures? (*Hint:* Refer to Exercise 35.)
(a) $PbSO_4(s)$ and $Cu(NO_3)_2(s)$
(b) $Mg(OH)_2(s)$ and $BaSO_4(s)$
(c) $PbCO_3(s)$ and $CaCO_3(s)$

38. Write net ionic equations to show how you would use one or more of the reagents in Exercise 37 to obtain **(a)** $BaSO_4(s)$; **(b)** $AgCl(s)$; **(c)** $KNO_3(aq)$.

Acid–Base Reactions

39. Every antacid contains one or more ingredients capable of reacting with excess stomach acid (HCl). The essential neutralization products are CO_2 and/or H_2O. Write net ionic equations to represent the neutralizing action of the following popular antacids.
(a) Alka-Seltzer (sodium bicarbonate)
(b) Tums (calcium carbonate)
(c) Milk of magnesia (magnesium hydroxide)
(d) Maalox (magnesium hydroxide; aluminum hydroxide)
(e) Rolaids $[NaAl(OH)_2CO_3]$

41. In this chapter we described an acid as a substance capable of producing H^+ and a salt as the ionic compound formed in the neutralization of an acid by a base. Write ionic equations to show that sodium hydrogen sulfate has both the characteristics of a salt and of an acid (it is sometimes called an *acid salt*).

40. Suppose you are given the following separate solids and solvents: solids, Na_2CrO_4, $BaCO_3$, $Al(OH)_3$, and $ZnSO_4$; solvents, $H_2O(l)$, $HCl(aq)$, and $H_2SO_4(aq)$. Your task is to prepare four solutions, each containing one of the cations (that is, one with Na^+, one with Ba^{2+}, and so on). What solvent would you use to prepare each solution? Explain.

42. The neutralization between an acid and a base is a common method of preparing useful salts. Give net ionic equations showing how the following salts could be prepared in this way: $(NH_4)_2HPO_4$, NH_4NO_3, and $(NH_4)_2SO_4$.

Oxidation–Reduction (Redox) Equations

43. Explain why the following reactions cannot occur as written. (*Hint:* Is it possible to balance the equations?)
(a) $Fe^{3+}(aq) + MnO_4^-(aq) + H^+(aq) \rightarrow$
$$Mn^{2+}(aq) + Fe^{2+}(aq) + H_2O$$
(b) $H_2O_2(aq) + Cl_2(aq) \rightarrow ClO^-(aq) + O_2(g) + H^+(aq)$

45. These reactions do not occur in aqueous solutions. Balance their equations by inspection.
(a) $(NH_4)_2Cr_2O_7(s) \rightarrow Cr_2O_3(s) + N_2(g) + H_2O(g)$
(b) $Pb(NO_3)_2(s) \rightarrow PbO(s) + NO_2(g) + O_2(g)$

44. A newspaper account of an accidental spill of hydrochloric acid in an area where sodium hydroxide solution was also stored spoke of the potential hazardous release of chlorine gas if the two solutions should come into contact. Was this an accurate accounting of the hazard involved? Explain.

46. These reactions do not occur in aqueous solutions. First, balance their equations by inspection, and then try applying the half-reaction method, as suggested in the Are You Wondering feature on page 148.
(a) $NH_3(g) + O_2(g) \rightarrow NO(g) + H_2O(g)$
(b) $NO(g) + H_2(g) \rightarrow NH_3(g) + H_2O(g)$

47. Balance the following equations for redox reactions occurring in *acidic* solution.
(a) $MnO_4^- + I^- \rightarrow Mn^{2+} + I_2(s)$
(b) $BrO_3^- + N_2H_4 \rightarrow Br^- + N_2$
(c) $VO_4^{3-} + Fe^{2+} \rightarrow VO^{2+} + Fe^{3+}$
(d) $UO^{2+} + NO_3^- \rightarrow UO_2^{2+} + NO(g)$

48. Balance the following equations for redox reactions occurring in *acidic* solution.
(a) $P_4(s) + NO_3^- \rightarrow H_2PO_4^- + NO(g)$
(b) $S_2O_3^{2-} + MnO_4^- \rightarrow SO_4^{2-} + Mn^{2+}$
(c) $HS^- + HSO_3^- \rightarrow S_2O_3^{2-}$
(d) $Fe^{3+} + NH_3OH^+ \rightarrow Fe^{2+} + N_2O(g)$

49. Balance the following equations for redox reactions occurring in *basic* solution.
(a) $CN^- + MnO_4^- \rightarrow MnO_2(s) + CNO^-$
(b) $[Fe(CN)_6]^{3-} + N_2H_4 \rightarrow [Fe(CN)_6]^{4-} + N_2(g)$
(c) $Fe(OH)_2(s) + O_2(g) \rightarrow Fe(OH)_3(s)$
(d) $C_2H_5OH(aq) + MnO_4^- \rightarrow C_2H_3O_2^- + MnO_2(s)$

51. Balance the following redox equations.
In acidic solution:
(a) $S_2O_3^{2-} + Cl_2(g) \rightarrow SO_4^{2-} + Cl^-$
(b) $Sn^{2+} + Cr_2O_7^{2-} \rightarrow Sn^{4+} + Cr^{3+}$
In basic solution:
(c) $S_8(s) \rightarrow S^{2-} + S_2O_3^{2-}$
(d) $As_2S_3 + H_2O_2 \rightarrow AsO_4^{3-} + SO_4^{2-}$

50. Balance the following equations for *disproportionation* reactions occurring in *basic* solution, except where noted.
(a) $Cl_2(g) \rightarrow Cl^- + ClO_3^-$
(b) $S_2O_4^{2-} \rightarrow S_2O_3^{2-} + HSO_3^-$ (acidic solution)
(c) $MnO_4^{2-} \rightarrow MnO_2(s) + MnO_4^-$
(d) $P_4(s) \rightarrow H_2PO_2^- + PH_3(g)$

52. Write a balanced equation for each of the following redox reactions.
(a) the oxidation of nitrite ion to nitrate ion by permanganate ion, MnO_4^-, in acidic solution (MnO_4^- ion is reduced to Mn^{2+}).
(b) the reaction of H_2S, a reducing agent, with SO_2, an oxidizing agent. (Both reactants are converted to elemental sulfur, and the other product is water.)
(c) the reaction of sodium metal with hydroiodic acid.
(d) the reduction of vanadyl ion (VO^{2+}) to vanadic ion (V^{3+}) in acidic solution with zinc metal as the reducing agent.

Oxidizing and Reducing Agents

53. What are the oxidizing and reducing agents in the following redox reactions?
(a) $5\ SO_3^{2-} + 2\ MnO_4^- + 6\ H^+ \rightarrow$
$$5\ SO_4^{2-} + 2\ Mn^{2+} + 3\ H_2O$$
(b) $2\ NO_2(g) + 7\ H_2(g) \rightarrow 2\ NH_3(g) + 4\ H_2O(g)$
(c) $2\ [Fe(CN)_6]^{4-} + H_2O_2 + 2\ H^+ \rightarrow$
$$2\ [Fe(CN)_6]^{3-} + 2\ H_2O$$

54. Thiosulfate ion, $S_2O_3^{2-}$, is a reducing agent that can be oxidized to different products, depending on the strength of the oxidizing agent and other conditions. By adding H^+, H_2O, and/or OH^- as necessary, write redox equations to show the oxidation of $S_2O_3^{2-}$ to
(a) $S_4O_6^{2-}$ by I_2; iodide ion is another product.
(b) HSO_4^- by Cl_2; chloride ion is another product.
(c) SO_4^{2-} by OCl^- in basic solution; chloride ion is another product.

Neutralization and Acid–Base Titrations

55. A NaOH(aq) solution cannot be made up to an exact concentration simply by weighing out the required mass of NaOH. (The NaOH is not pure, and water vapor condenses on the solid as it is being weighed.) The solution must be standardized by titration. For this purpose a 25.00-mL sample of an NaOH(aq) solution requires 28.34 mL of 0.1085 M HCl. What is the molarity of the NaOH(aq)?
$$HCl(aq) + NaOH(aq) \longrightarrow NaCl(aq) + H_2O(l)$$

57. How many milliliters of 0.2155 M KOH(aq) are required to titrate 25.00 mL of 0.3057 M $HC_3H_5O_2$(aq) (propionic acid)? (*Hint:* What is the net reaction?)

59. We want to determine the acetylsalicyclic acid content of a series of aspirin tablets by titration with NaOH(aq).

$$HC_9H_7O_4(aq) + OH^-(aq) \longrightarrow C_9H_7O_4^-(aq) + H_2O(l)$$

Each of the tablets is expected to contain about 0.32 g of $HC_9H_7O_4$. What molarity NaOH should we use for titration volumes of about 23 mL? (This procedure ensures good precision and allows the titration of two samples with the contents of a 50-mL buret.)

56. Household ammonia, used as a window cleaner and for other cleaning purposes, is NH₃(aq). A 28.72-mL sample of 1.021 M HCl(aq) is required to neutralize the NH₃ present in a 5.00-mL sample. The net ionic equation for the neutralization is

$$NH_3(aq) + H^+(aq) \longrightarrow NH_4^+(aq)$$

What is the molarity of NH₃ in the sample?

58. How many milliliters of 0.0884 M Ba(OH)₂(aq) are required to titrate 50.00 mL of 0.0526 M HNO₃(aq)? (*Hint:* What is the net reaction?)

60. For use in titrations we wish to prepare 20 L of HCl(aq) of a concentration known to *four* significant figures. We use a two-step procedure. First, we prepare a solution having a concentration of about 0.10 M HCl. Then, we titrate a sample of this dilute HCl(aq) with an NaOH(aq) solution of known concentration.
(a) How many milliliters of concentrated HCl(aq) ($d = 1.19$ g/mL; 38% HCl, by mass) must we dilute to 20.0 L with water to prepare 0.10 M HCl?
(b) A 25.00-mL sample of the approximately 0.10 M HCl prepared in part (a) requires 20.93 mL of 0.1186 M NaOH for its titration. What is the molarity of the HCl(aq)?
(c) Why do you suppose a titration is necessary? That is, why could we not prepare a standard solution of 0.1000 M HCl simply by an appropriate dilution of the concentrated HCl(aq)?

61. 25.00 mL of 0.132 M HNO_3 and 10.00 mL of 0.318 M KOH are mixed. Is the resulting solution acidic, basic, or exactly neutralized? (*Hint:* Is there a limiting reactant?)

63. Refer to Example 5-9. Suppose the analysis of all vinegar samples uses 5.00 mL of the vinegar and 0.1000 M NaOH for the titration. What volume of the 0.1000 M NaOH would represent the legal minimum 4.0%, by mass, acetic acid content of the vinegar? That is, calculate the volume of 0.1000 M NaOH such that if a titration requires more than this volume, the legal minimum limit is met (less than this volume and the limit is not met).

Stoichiometry of Oxidation–Reduction Reactions

65. A $KMnO_4(aq)$ solution is to be standardized by titration against $As_2O_3(s)$. A 0.1078-g sample of As_2O_3 requires 22.15 mL of the $KMnO_4(aq)$ for its titration. What is the molarity of the $KMnO_4(aq)$?

$$5\ As_2O_3 + 4\ MnO_4^- + 9\ H_2O + 12\ H^+ \longrightarrow$$
$$10\ H_3AsO_4 + 4\ Mn^{2+}$$

67. An iron ore sample weighing 0.9132 g is dissolved in HCl(aq) and the iron is obtained as $Fe^{2+}(aq)$. This solution is then titrated with 28.72 mL of 0.05051 M $K_2Cr_2O_7(aq)$. What is the % Fe, by mass, in the ore sample?

$$6\ Fe^{2+} + 14\ H^+ + Cr_2O_7^{2-} \longrightarrow 6\ Fe^{3+} + 2\ Cr^{3+} + 7\ H_2O$$

69. The titration of 50.0 mL of a saturated solution of sodium oxalate, $Na_2C_2O_4$, requires 25.8 mL of 0.02140 M $KMnO_4$ in acidic solution. What mass of $Na_2C_2O_4$, in grams, would be present in 1.00 L of this saturated solution? (*Hint:* First you must complete and balance the equation for the titration reaction.)

$$C_2O_4^{2-} + MnO_4^- \longrightarrow Mn^{2+} + CO_2(g)$$

62. A 7.55-g sample of $Na_2CO_3(s)$ is added to 125 mL of a vinegar that is 0.762 M $HC_2H_3O_2$. Will the resulting solution still be acidic? Explain. (*Hint:* What is the reaction that occurs?)

64. The electrolyte in a lead storage battery must have a concentration between 4.8 and 5.3 M H_2SO_4 if the battery is to be most effective. A 5.00-mL sample of a battery acid requires 49.74 mL of 0.935 M NaOH for its complete reaction (neutralization). Does the concentration of the battery acid fall within the desired range? (*Hint:* The H_2SO_4 produces two H^+ ions per formula unit. Write a net ionic equation for the neutralization reaction.)

66. Refer to Example 5-6. Assume that the only reducing agent present in a particular wastewater is SO_3^{2-}. If a 25.00-mL sample of this wastewater requires 31.46 mL of 0.02237 M $KMnO_4$ for its titration, what is the molarity of SO_3^{2-} in the wastewater?

68. $Mn^{2+}(aq)$ can be determined by titration with $MnO_4^-(aq)$ in basic solution.

$$Mn^{2+} + MnO_4^- \longrightarrow MnO_2(s) \quad \text{(not balanced)}$$

A 25.00-mL sample of $Mn^{2+}(aq)$ requires 37.21 mL of 0.04162 M $KMnO_4(aq)$ for its titration. What is $[Mn^{2+}]$ in the sample?

70. Refer to the Summarizing Example. In the treatment of 1.00×10^2 L of a wastewater solution that is 0.0126 M in CrO_4^{2-},
(a) how many grams of $Cr(OH)_3(s)$ would precipitate?
(b) how many grams of $Na_2S_2O_4$ would be consumed?

Advanced Exercises

71. Following are some laboratory methods occasionally used for the preparation of small quantities of chemicals. Write a balanced equation for each.
(a) Preparation of $H_2S(g)$: HCl(aq) is heated with FeS(s).
(b) Preparation of $Cl_2(g)$: HCl(aq) is heated with $MnO_2(s)$; $MnCl_2(aq)$ and H_2O are other products.
(c) Preparation of N_2: Br_2 and NH_3 react in aqueous solution; NH_4Br is another product.
(d) Preparation of chlorous acid: an aqueous suspension of solid barium chlorite is treated with dilute $H_2SO_4(aq)$. (*Hint:* What is the other probable product in addition to chlorous acid?)

72. When concentrated $CaCl_2(aq)$ is added to $Na_2HPO_4(aq)$, a white precipitate forms that is 38.7% Ca, by mass. Write a net ionic equation to represent the probable reaction that occurs. (*Hint:* What is the probable calcium-containing product?)

73. You have available a solution that is 0.0250 M $Ba(OH)_2$ and the following pieces of equipment: 1.00, 5.00, 10.00, 25.00,

and 50.00-mL pipets and 100.0, 250.0, 500.0, and 1000.0-mL volumetric flasks. Describe how you would use this equipment to produce a solution that has $[OH^-] = 0.0100$ M.

74. Sodium hydroxide used to make standard NaOH(aq) solutions for acid–base titrations is invariably contaminated with some sodium carbonate.
(a) Explain why, except in the most precise work, the presence of this sodium carbonate generally does not seriously affect the results obtained, for example, when NaOH(aq) is used to titrate HCl(aq).
(b) On the other hand, show that if Na_2CO_3 comprises more than 1–2% of the solute in NaOH(aq), the titration results are affected.

75. A 110.520-g sample of mineral water is analyzed for its magnesium content. The Mg^{2+} in the sample is first precipitated as $MgNH_4PO_4$, and this precipitate is then converted to $Mg_2P_2O_7$, which is found to weigh 0.0549 g. Express the quantity of magnesium in the sample in parts per million (that is, in grams of Mg per million grams of H_2O).

76. What volume of 0.248 M $CaCl_2$ must be added to 335 mL of 0.186 M KCl to produce a solution with a concentration of 0.250 M Cl^-?

77. Balance the following redox equations in acidic solution.
(a) $IBr + BrO_3^- + H^+ \rightarrow IO_3^- + Br^- + H_2O$
(b) $C_2H_5NO_3 + Sn \rightarrow NH_2OH + C_2H_5OH + Sn^{2+}$
(c) $As_2S_3 + NO_3^- \rightarrow H_3AsO_4 + S(s) + NO(g)$
(d) $H_5IO_6 + I_2 \rightarrow IO_3^- + H^+ + H_2O$
(e) $S_2F_2(g) + H_2O(l) \rightarrow S_8(s) + H_2S_4O_6(aq) + HF(aq)$

78. Balance the following redox equations in basic solution.
(a) $Fe_2S_3(s) + H_2O + O_2(g) \rightarrow Fe(OH)_3(s) + S(s)$
(b) $O_2^-(aq) + H_2O \rightarrow OH^-(aq) + O_2(g)$
(c) $CrI_3(s) + H_2O_2 \rightarrow CrO_4^{2-} + IO_4^-$
(d) $Ag(s) + CN^- + O_2(g) + OH^- \rightarrow [Ag(CN)_2]^- + H_2O$
(e) $B_2Cl_4 + OH^- \rightarrow BO_2^- + Cl^- + H_2O + H_2(g)$
(f) $C_2H_5OH(aq) + MnO_4^-(aq) \rightarrow C_2H_3O_2^-(aq) + MnO_2(s)$

79. A method of producing phosphine, PH_3, from elemental phosphorus, P_4, involves heating the P_4 with H_2O. An additional product is phosphoric acid, H_3PO_4. Write a balanced equation for this oxidation–reduction reaction.

80. When iron pyrites, waste products in the mining of coal, are exposed to the environment, their sulfur content is oxidized to sulfuric acid. This creates the environmental problem known as *acid mine drainage*. Following are two of the main reactions involved. Balance these redox equations.
(a) $FeS_2(s) + O_2 + H_2O \rightarrow Fe^{2+} + SO_4^{2-} + H^+$
(b) $FeS_2(s) + Fe^{3+} + H_2O \rightarrow Fe^{2+} + SO_4^{2-} + H^+$

81. A sample of battery acid is to be analyzed for its sulfuric acid content. A 1.00-mL sample weighs 1.239 g. This 1.00-mL sample is diluted to 250.0 mL, and 10.00 mL of this diluted acid requires 32.44 mL of 0.00498 M $Ba(OH)_2$ for its titration. What is the mass percent of H_2SO_4 in the battery acid? (Assume that complete ionization and neutralization of the H_2SO_4 occurs.)

82. A piece of marble (assume it to be pure $CaCO_3$) reacts with 2.00 L of 2.52 M HCl. After dissolution of the marble, a 10.00-mL sample of the remaining HCl(aq) is withdrawn, added to some water, and titrated with 24.87 mL of 0.9987 M NaOH. What must have been the mass of the piece of marble? Comment on the precision of this method, that is, how many significant figures are justified in the result? (*Hint:* Write net equations for the two reactions involved.)

83. The reaction below can be used as a laboratory method of preparing small quantities of $Cl_2(g)$. If a 62.6-g sample that is 98.5% $K_2Cr_2O_7$, by mass, is allowed to react with 325 mL of HCl(aq) with a density of 1.15 g/mL and 30.1% HCl, by mass, how many grams of $Cl_2(g)$ are produced?
$$Cr_2O_7^{2-} + H^+ + Cl^- \longrightarrow Cr^{3+} + H_2O + Cl_2(g)$$
(not balanced)

84. Refer to Example 5-10. Suppose that the $KMnO_4(aq)$ described in this example were standardized by reaction with As_2O_3 instead of iron wire. If a 0.1304-g sample that is 99.96% As_2O_3, by mass, had been used in the titration, how many milliliters of the $KMnO_4(aq)$ would have been required?
$$As_2O_3 + MnO_4^- + H^+ + H_2O \longrightarrow H_3AsO_4 + Mn^{2+}$$
(not balanced)

85. A new method for water treatment under development uses chlorine dioxide rather than chlorine itself. One method of producing ClO_2 involves passing $Cl_2(g)$ into a concentrated solution of sodium chlorite; NaCl(aq) is the other product. If the reaction has a 97% yield, what mass of ClO_2 is produced per gallon of 2.0 M $NaClO_2$(aq) treated in this way? (*Hint:* Write a net equation for the redox reaction. Also, 1 gal = 3.785 L.)

Feature Problems

A. Write net ionic equations for the series of reactions shown in the photos. [*Hint:* Identify *all* the reactants and products for each of these three reactions before attempting to write a balanced chemical equation.]

1. Sodium metal reacts with water to produce hydrogen gas.

2. An excess of $FeCl_3$(aq) is added to the solution produced from **1.**

3. The precipitate from **2** is collected, and an excess of HCl(aq) is added to it.

B. Sodium cyclopentadienide (NaCp), NaC_5H_5, is a common reducing agent in the chemical laboratory, but there is a problem in using it: NaCp is contaminated with tetrahydrofuran (THF), C_4H_8O, a solvent used in its preparation. The THF is present as $NaCp\cdot(THF)_x$, and it is generally necessary to know exactly how much of this $NaCp\cdot(THF)_x$ is present. This is ac-

complished by allowing a small amount of the $NaCp\cdot(THF)_x$ to react with water,

$$NaCp\cdot(C_4H_8O)_x + H_2O \longrightarrow NaOH(aq) + Cp\text{—}H + x\,C_4H_8O$$

followed by titration of the NaOH(aq) with a standard acid.

From the sample data tabulated below, determine the value of x in the formula $NaCp\cdot(THF)_x$.

	Trial 1	Trial 2
mass of $NaCp\cdot(THF)_x$	0.242 g	0.199 g
volume of 0.1001 M HCl required to titrate NaOH(aq)	14.92 mL	11.99 mL

C. Manganese is derived from pyrolusite ore, an impure manganese dioxide. To analyze a pyrolusite ore for its MnO_2 content this procedure is used: A 0.533-g sample is treated with 1.651 g oxalic acid ($HC_2O_4\cdot2H_2O$) in an acidic medium. Following this reaction, the *excess* oxalic acid is titrated with 0.1000 M $KMnO_4$, 30.06 mL being required. What is the % MnO_2, by mass, in the ore? The *unbalanced* equations are

$$H_2C_2O_4 + MnO_2 + H^+ \longrightarrow Mn^{2+} + H_2O + CO_2$$
$$H_2C_2O_4 + MnO_4^- + H^+ \longrightarrow Mn^{2+} + H_2O + CO_2$$

Gases

6

Hot-air balloons have intrigued people from the time the simple gas laws fundamental to their operation came to be understood over 200 years ago.

A bicycle tire must not be overinflated because it might rupture. An aerosol can should not be discarded in an incinerator. Carbon dioxide gas vaporizing from a block of dry ice sinks to the floor. A balloon filled with helium rises in air, and so does a balloon filled with hot air. One should never search for a gas leak with an open flame. These and many other observations concerning gases can be explained by concepts considered in this chapter.

For example, the behavior of the bicycle tire and the aerosol can are based on relationships among the variables pressure, temperature, volume, and amount of gas. An understanding of the lifting power of lighter-than-air balloons comes in large part from a knowledge of gas densities and their dependence on molar mass, temperature, and pressure. Predicting how far and how fast gas molecules migrate through air requires knowing something about the phenomenon of diffusion.

To do quantitative calculations about the behavior of gases, we will use some simple gas laws and a more general expression called the ideal gas equation. To explain these laws, we will use a theory known as the kinetic-molecular theory of gases. Ideas presented in this chapter reappear in new contexts in several later chapters.

6-1 Properties of Gases: Gas Pressure

Several characteristics of gases are familiar to just about everyone. Gases expand to fill and assume the shapes of their containers. They diffuse into one another and mix in all proportions. They are invisible—that is, there are no visible particles of a gas, although some gases are colored, such as chlorine, bromine, and iodine (Figure 6-1). Some gases are combustible, such as hydrogen and methane, whereas others are chemically unreactive, such as helium and neon.

Four properties determine the physical behavior of a gas: the amount of gas and its volume, temperature, and pressure. If we know any three of these, we can usually calculate the value of the remaining one. To do this we can use a mathematical equation called an equation of state, as we will see later in the chapter. To some extent we have already discussed the properties of amount, volume, and temperature, but we need to consider the idea of pressure.

The Concept of Pressure

A balloon expands when it is inflated with air, but what keeps the balloon in its distended shape? A plausible hypothesis is that molecules of a gas are in constant motion, frequently colliding with one another and with the walls of their container. In their collisions the gas molecules exert a *force* on the container walls. This force keeps the balloon distended. It is not easy, however, to measure the total force exerted by a gas. Instead of focusing on this total force, we consider instead the gas pressure. **Pressure** is defined as a force per unit area, that is, a force divided by the area over which the force is distributed. Figure 6-2 illustrates the idea of pressure exerted by a solid.

$$P = \frac{F}{A} \tag{6.1}$$

▶ Refer to Appendix B for a review of fundamental physical quantities.

In SI units, force should be expressed in newtons (N) and area in square meters (m^2). The corresponding *force per unit area*—the pressure—is in the unit N/m^2, also called a *pascal* (Pa). Thus, one pascal is a pressure of $1 \ N/m^2$. A pascal is a rather small unit of pressure, and the unit *kilopascal* (kPa) is more commonly used. The pressure unit pascal honors Blaise Pascal (1623–1662), who studied pressure and its transmission through liquids—the basis of modern hydraulics. We will not apply these definitions just now because we will soon introduce some more commonly used pressure units.

FIGURE 6-1

The gaseous states of three halogens (Group 7A).
The greenish-yellow gas is $Cl_2(g)$; the brownish-red gas is $Br_2(g)$ above a small pool of liquid bromine; the violet gas is $I_2(g)$ in contact with grayish-black solid iodine. Most other common gases, such as H_2, O_2, N_2, CO, and CO_2, are colorless.

Liquid Pressure

Because it is not easy to measure the total force exerted by gas molecules, it is also difficult to apply Equation (6.1) to gases. The pressure of a gas is usually measured *indirectly,* by comparing it with a liquid pressure. Figure 6-3 illustrates the concept of liquid pressure and suggests that the pressure of a liquid depends only on the height of the liquid column and the density of the liquid. To confirm this statement, consider a liquid with density d, contained in a cylinder with cross-sectional area A, filled to a height h.

Now, recall these facts. (1) Weight is a force, and weight and mass are proportional: $W = g \cdot m$. (2) The mass of a liquid is the product of its volume and density: $m = V \cdot d$. (3) The volume of a cylinder is the product of its height and cross-sectional area: $V = h \cdot A$. We use these ideas to derive Equation (6.2).

$$P = \frac{F}{A} = \frac{W}{A} = \frac{g \cdot m}{A} = \frac{g \cdot V \cdot d}{A} = \frac{g \cdot h \cdot A \cdot d}{A}$$

$$= g \cdot h \cdot d \tag{6.2}$$

Thus, because g is a constant, liquid pressure is directly proportional to the liquid density and the height of the liquid column.

Barometric Pressure

In 1643, Evangelista Torricelli constructed the device pictured in Figure 6-4 to measure the pressure exerted by the atmosphere. This device is called a **barometer.**

If we stand a tube with *both ends open* upright in a container of mercury (see Figure 6-4a), the mercury levels inside and outside the tube are the same. To create the situation in Figure 6-4b, we seal a long glass tube at one end, fill the tube with Hg(l), cover the open end, and invert the tube into a container of Hg(l). Then we reopen the end under the mercury. The mercury level in the tube falls to a certain height and stays there. Something keeps the mercury at a greater height inside the tube than outside. Some tried to explain this phenomenon in terms of forces *within* the tube, but Torricelli understood that these forces exist *outside* the tube.

In the open-end tube (Figure 6-4a) the atmosphere exerts the same pressure on the surface of the mercury, both inside and outside the tube. The liquid levels are equal. Inside the closed-end tube (Figure 6-4b) there is no air above the mercury (only a trace of mercury vapor). The atmosphere exerts a force on the surface of the mercury in the outside container that is transmitted through the liquid, holding up a column of mercury. This column exerts a downward pressure that depends on its height and the density of Hg(l). For a particular height the pressure at the bottom of the mercury column and that of the atmosphere are equal. Thus, the column is maintained.

The height of mercury in a barometer, called the **barometric pressure,** varies with atmospheric conditions and with altitude. The **standard atmosphere (atm)** is defined as the pressure exerted by a mercury column of *exactly* 760 mm in height, when the density of mercury = 13.5951 g/cm³ (0 °C) and the acceleration due to gravity $g = 9.80665$ m s⁻², exactly. This statement relates two useful units of pressure, the standard atmosphere (atm) and the millimeter of mercury (mmHg).

$$1 \text{ atm} = 760 \text{ mmHg} \tag{6.3}$$

FIGURE 6-2
Illustrating the pressure exerted by a solid.
The two cylinders have the same mass and exert the same force on the supporting surface ($F = g \cdot m$). The tall, thin one has a smaller area of contact, however, and exerts a greater pressure ($P = F/A$).

FIGURE 6-3
The concept of liquid pressure.
All the interconnected vessels fill to the same height. As a result the liquid pressures are the same, despite the different shapes and volumes of the containers.

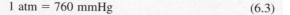

FIGURE 6-4

Measurement of atmospheric pressure with a mercury barometer.

Arrows represent the pressure exerted by the atmosphere.
(a) The liquid mercury levels are equal inside and outside the open-end tube.
(b) A column of mercury 760 mm high is maintained in the closed-end tube. The pressure at the bottom of the column of liquid mercury is equal to that exerted by the atmosphere. The region above the mercury column is devoid of air and contains only a trace of mercury vapor.

air pressure

(a)

760 mm

air pressure

(b)

The pressure unit *torr* honors Torricelli. It is defined as exactly 1/760 of a standard atmosphere. Or, 1 atm = 760 torr. Thus, we can use the pressure units torr and millimeters of mercury interchangeably.

Mercury is a relatively rare, expensive, and poisonous liquid. Why use it rather than water in a barometer? As we will see in Example 6-1, we use mercury because of the extreme height required for a water barometer.

EXAMPLE 6-1

Comparing Liquid Pressures. What is the height of a column of water that exerts the same pressure as a column of mercury 76.0 cm (760 mm) high?

Solution

We can use Equation (6.2) to describe the pressure of the mercury column of known height and the pressure of the water column of unknown height. Then we can set the two pressures equal to each other.

$$\text{pressure of Hg column} = g \cdot h_{Hg} \cdot d_{Hg} = g \times 76.0 \text{ cm} \times 13.6 \text{ g/cm}^3$$

$$\text{pressure of H}_2\text{O column} = g \cdot h_{H_2O} \cdot d_{H_2O} = g \times h_{H_2O} \times 1.00 \text{ g/cm}^3$$

$$\cancel{g} \times h_{H_2O} \times 1.00 \text{ g/cm}^3 = \cancel{g} \times 76.0 \text{ cm} \times 13.6 \text{ g/cm}^3$$

$$h_{H_2O} = 76.0 \text{ cm} \times \frac{13.6 \text{ g/cm}^3}{1.00 \text{ g/cm}^3} = 1.03 \times 10^3 \text{ cm} = 10.3 \text{ m}$$

Thus, we see that the column height is inversely proportional to the liquid density: for a given liquid pressure, the lower the liquid density the greater the height of the liquid column.

▶ Whereas atmospheric pressure can be measured with a mercury barometer less than 1 meter high, a water barometer would have to be as tall as a three-story building.

Practice Example A: A barometer is filled with perchloroethylene ($d = 1.62$ g/cm³). The liquid height is found to be 6.38 m. What is the barometric (atmospheric) pressure expressed in millimeters of mercury?

Practice Example B: A barometer is filled with ethanol. The liquid height is found to be 13.04 m when the atmospheric pressure is 757 mmHg. What is the density of ethanol?

When you use a drinking straw, you reduce the air pressure above the liquid inside the straw by inhaling. Atmospheric pressure outside the straw then pushes

FIGURE 6-5   Measurement of gas pressure with an open-end manometer.
The possible relationships between barometric pressure and a gas pressure under measurement are pictured here and described in Example 6-2.

the liquid into your mouth. An old-fashioned hand suction pump for pumping water (once common in rural areas) works on the same principle. The result of Example 6-1 indicates, however, that even if all the air could be removed from inside a pipe, atmospheric pressure outside the pipe could not raise water to a height of more than about 10 m. A suction pump works only for shallow wells. To pump water from a deep well, a mechanical pump is required. The force of the mechanical pump pushing the water upward is greater than the force of the atmosphere pushing the water downward.

Manometers

Although a mercury barometer is indispensable for measuring the pressure of the atmosphere, rarely can we use it alone to measure other gas pressures. The difficulty is in placing a barometer inside the container of gas whose pressure we wish to measure. We can, however, *compare* the gas pressure and barometric pressure with a **manometer.** Figure 6-5 illustrates the principle of an open-end manometer. As long as the gas pressure being measured and the prevailing atmospheric (barometric) pressure are equal, the heights of the mercury columns in the two arms of the manometer are equal. A difference in height of the two arms signifies a difference between the gas pressure and barometric pressure.

EXAMPLE 6-2

Using a Manometer to Measure Gas Pressure. What is the gas pressure P_{gas}, if the conditions in Figure 6-5c are these: the manometer is filled with liquid mercury ($d = 13.6$ g/cm^3); barometric pressure is 748.2 mmHg; and the difference in mercury levels is 8.6 mmHg?

Solution

Whether the situation corresponds to Figure 6-5b or c, the pressure we seek is $P_{gas} = P_{bar} + \Delta P$. The crucial difference between the two situations is that in Figure 6-5b the measured gas pressure is *greater than* barometric pressure; ΔP is positive. In Figure 6-5c P_{gas} is *less than* barometric pressure; ΔP is negative. In the question we have been asked, ΔP is *negative*. (Because all pressures are expressed in millimeters of mercury, the density of mercury does not enter into the calculation.)

$$P_{gas} = P_{bar} + \Delta P = 748.2 \text{ mmHg} - 8.6 \text{ mmHg} = 739.6 \text{ mmHg}$$

Practice Example A: Suppose that the mercury level in Example 6-2 is 7.8 mm higher in the arm open to the atmosphere than in the closed arm. What would be the value of P_{gas}?

Practice Example B: Suppose P_{bar} and P_{gas} are those described in Example 6-2, but the manometer is filled with liquid glycerol ($d = 1.26$ g/cm^3) instead of mercury. What would be the difference in the two levels of the liquid? (*Hint:* You need to compare two liquid pressures, as in Example 6-1. To do this you need the densities of both liquids.)

Units of Pressure: A Summary

Table 6.1 lists several different units used to express pressure. In this text we will use primarily the first three units (shown in red). When we use the units millimeters of mercury and torr we substitute the height of a liquid column (mercury) for an actual pressure. Similarly, the fourth and fifth units are based on a mass (rather than a force) per unit area. The unit pounds per square inch (psi) is used most commonly in engineering work. The next three units (shown in blue) are the ones preferred in the SI system. The unit bar differs from the kilopascal by a factor of 100. The unit millibar is commonly used by meteorologists.

6-2 The Simple Gas Laws

In this section we consider relationships involving pressure, volume, temperature, and amount of a gas. Specifically, we will see how one variable depends on another, as the remaining two are held fixed. Collectively, we refer to these relationships as the simple gas laws. You can use these laws in problem solving, but you will probably prefer the equation we develop in the next section—the ideal gas equation. You may find that the greatest use of the simple gas laws is in firming up your qualitative understanding of the behavior of gases.

Boyle's Law

▶Boyle's apparatus and some of his original data are the subject of Feature Problem A on page 201.

In 1662, working with air, Robert Boyle discovered the first of the simple gas laws, now known as **Boyle's law.**

> For a fixed amount of gas at a constant temperature, gas volume is inversely proportional to gas pressure.

TABLE 6.1 Some Common Pressure Units	
atmosphere (atm)	
millimeter of mercury (mmHg)	1 atm = 760 mmHg
torr (torr)	= 760 torr
pounds per square inch (lb/in.^{2})	= 14.696 lb/in.^{2}
kilogram per square centimeter (kg/cm^2)	= 1.0333 kg/cm^2
newton per square meter (N/m^2)	= 101,325 N/m^2
pascal (Pa)	= 101,325 Pa
kilopascal (kPa)	= 101.325 kPa
bar (bar)	= 1.01325 bar
millibar (mb)	= 1013.25 mb

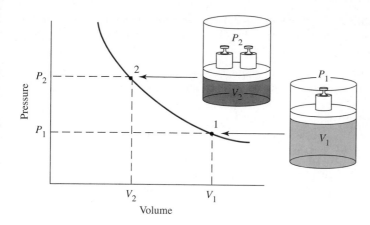

FIGURE 6-6
Relationship between gas volume and pressure—Boyle's law.
When the temperature and amount of gas are held constant, gas volume is inversely proportional to the pressure: a doubling of the pressure causes the volume to decrease to one-half its original value.

Consider the gas in Figure 6-6. It is confined in a cylinder closed off by a freely moving "weightless" piston. The pressure of the gas depends on the total weight placed on top of the piston. This weight (a force), divided by the area of the piston, yields the gas pressure. If the weight on the piston is doubled, the pressure *doubles* and the gas volume decreases to *one half* of its original value. If the pressure of the gas is tripled, the volume decreases to *one third*. On the other hand, if the pressure is reduced to *one half,* the volume *doubles,* and so on. Mathematically, we can express this inverse relationship between pressure and volume as

$$P \propto \frac{1}{V} \quad \text{or} \quad PV = a \text{ (a constant)} \tag{6.4}$$

When we replace the proportionality sign ($\propto$) by an equal sign and a proportionality constant, we see that the product of the pressure and volume of a fixed amount of gas at a given temperature is a constant (a). The value of a depends on the amount of gas and the temperature. The graph in Figure 6-6 is that of $PV = a$. It is called a rectangular hyperbola.

▶ Feature Problem B on page 201 describes a different method of plotting Boyle's law data.

EXAMPLE 6-3

Relating Gas Volume and Pressure—Boyle's Law. The volume of a large, irregularly shaped, closed tank is determined as follows. The tank is first evacuated, and then it is connected to a 50.0-L cylinder of compressed nitrogen gas. The gas pressure in the cylinder, originally at 21.5 atm, falls to 1.55 atm after it is connected to the evacuated tank. What is the volume of the tank? (Figure 6-7 on page 170.)

Solution

One way to solve this problem is by writing Equation (6.4) twice: for the initial condition (1) and for the final condition (2).

$$P_1V_1 = a = P_2V_2$$

Now solve for the final volume, V_2.

$$V_2 = V_1 \times \frac{P_1}{P_2} = 50.0 \text{ L} \times \frac{21.5 \text{ atm}}{1.55 \text{ atm}} = 694 \text{ L}$$

Of this volume, 50.0 L is that of the cylinder. The volume of the tank is 694 L − 50.0 L = 644 L.

▶ Reasoning qualitatively from Boyle's law, if the pressure of the gas decreases we expect the volume to increase; we expect V_2 to be larger than V_1.

Initial condition

21.5 atm

50.0 L

Final condition

1.55 atm

tank

50.0 L

V = ?

FIGURE 6-7
An application of Boyle's law—Example 6-3 illustrated.
The final volume is the volume of the cylinder plus the volume of the tank. The amount of gas and its temperature remain constant when the cylinder is connected to the tank, but the pressure drops from 21.5 atm (the initial pressure) to 1.55 atm.

Practice Example A: A cylinder like that pictured in Figure 6-6 contains a gas at 5.25 atm pressure. When the gas is allowed to expand to a final volume of 12.5 L, the pressure drops to 1.85 atm. What was the original volume of the gas?

Practice Example B: 1.50 L of gas at 2.25 atm pressure expands to a final volume of 8.10 L. What is the final gas pressure in millimeters of mercury?

Charles's Law

The relationship between gas volume and temperature was discovered by the French physicist Jacques Charles in 1787 and, independently, by Josef Louis Gay-Lussac, who published it in 1802.

Figure 6-8 pictures a fixed amount of gas confined in a cylinder. The pressure is held constant while the temperature is varied. The volume of gas increases as the temperature is raised or decreases as the temperature is lowered. The relationship is linear (a straight line).

One point in common to the lines in Figure 6-8 is their intersection with the temperature axis. Although they differ at every other temperature, the gas volumes all reach a value of zero at the same temperature. The temperature at which

FIGURE 6-8
Gas volume as a function of temperature.
Volume is plotted against temperature on two different scales—Celsius (blue) and Kelvin (red). Gases *A*, *B*, and *C*, are shown at three different initial conditions, all at 1 atm. Gas *C*, for example, starts at 344 K (71 °C) and 60.0 cm³. As the temperature is lowered, the volume decreases as predicted by Charles's law. Thus, at 172 K (−101 °C), half that of the initial 344 K, the volume has become one-half of the original 60.0 cm³, that is, 30.0 cm³. Although the relationship between volume and temperature is linear for both the Celsius and Kelvin temperature scales, the volume is *directly proportional* to temperature only on the Kelvin scale.

the volume of a *hypothetical* gas becomes zero is the absolute zero of temperature: −273.15 °C.*

Now suppose that in Figure 6-8 we (a) construct a new temperature scale (in red) that has −273.15 °C as its zero and call this the **absolute** or **Kelvin** scale, (b) define the temperature interval of one kelvin as equal to one degree Celsius, and (c) place the volume axis (in red) at 0 K. As a result we can see that the straight-line graphs of volume versus temperature pass through the origin ($T = 0$, $V = 0$) of the new graph, and that this simple relationship exists between Kelvin and Celsius temperature.

$$T \text{ (K)} = t \text{ (°C)} + 273.15 \qquad (6.5)$$

Now we can state **Charles's law.**

> The volume of a fixed amount of gas at constant pressure is directly proportional to the Kelvin (absolute) temperature. $\qquad$ (6.6)

In mathematical terms

$$V \propto T \qquad \text{or} \qquad V = bT \qquad \text{(where } b \text{ is a constant)} \qquad (6.7)$$

The value of the constant b depends on the amount of gas and the pressure. It does *not* depend on what gas is involved.

From either expression (6.6) or (6.7) we see that doubling the Kelvin temperature of a gas causes its volume to double. Reducing the Kelvin temperature by one-half (say, from 300 to 150 K) causes the volume to decrease to one-half, and so on. It is not difficult to see that the temperature used in Equation (6.7) *must* be on the Kelvin scale. This is further verified by the graphs of the three gas conditions in Figure 6-8. We certainly do not expect that raising the temperature of a gas from 1 to 2 °C or from 1 to 2 °F (in each case an apparent "doubling") would cause the gas volume to double.

Charles experimented with the first hydrogen-filled balloons, much like the one shown here, though smaller. Most of the features of modern ballooning were invented by Charles, such as the suspended basket and the valve to release excess gas.

▶ Charles's ideas about the effect of temperature on the volume of a gas were probably influenced by his passion for hot-air balloons, a popular craze of the late eighteenth century.

EXAMPLE 6-4

Relating a Gas Volume and Temperature—Charles's Law. A balloon is inflated to a volume of 2.50 L in a warm living room (24 °C). Then it is taken outside on a very cold winter's day (−25 °C). Assume that the quantity of air in the balloon and its pressure both remain constant. What will be the volume of the balloon when it is outdoors?

Solution

As in Example 6-3, write Equation (6.7) twice: for the initial (1) and the final (2) conditions. In doing this, though, first rearrange the equation and solve for b. That is, $b = V/T$.

$$\frac{V_1}{T_1} = b = \frac{V_2}{T_2}$$

---

*All gases condense to liquids or solids before the temperature approaches absolute zero. Also, when we speak of the volume of a gas we mean the free volume among the gas molecules, not the volume of the molecules themselves. Thus, the gas we refer to here is *hypothetical*. It is a gas whose molecules have mass but no volume and that does not condense to a liquid or solid.

Now, solve the above equation for V_2, but do not forget to change temperatures to the Kelvin scale—that is, $T_1 = 24\ °C + 273 = 297\ K$, and $T_2 = -25\ °C + 273 = 248\ K$.

$$V_2 = V_1 \times \frac{T_2}{T_1} = 2.50\ L \times \frac{248\ K}{297\ K} = 2.09\ L$$

► Reasoning qualitatively from Charles's law, if the temperature of the gas is lowered we expect its volume to decrease; we expect V_2 to be less than V_1.

Practice Example A: A gas at 25 °C and 0.987 atm pressure is heated under a piston in a cylinder. The volume expands from 0.250 L to 1.65 L. What is the new temperature of the gas, assuming the pressure is constant?

Practice Example B: If an aerosol can contains a gas at 1.82 atm pressure at 22 °C, what will be the gas pressure in the can if it is dumped into an incinerator at 935 °C? (*Hint:* Because the volume cannot increase as the temperature is raised, the pressure increases instead. For a fixed amount of gas in a constant volume, pressure is directly proportional to Kelvin temperature.)

Standard Conditions of Temperature and Pressure

Because gas properties depend on temperature and pressure, at times we find it useful to choose standard conditions of temperature and pressure. This is especially true when we compare gases to one another. The standard temperature for gases is taken to be $0\ °C = 273.15\ K$ and standard pressure, as 1 atm = 760 mmHg, exactly. Standard conditions of temperature and pressure are usually abbreviated as **STP**.

► Feature Problem C on page 201 tests your qualitative understanding of the gas laws at both the microscopic and molecular levels.

► If the volume of $N_2(g)$ at STP had been somewhat *greater* than 1.20 L, we could not have answered the question with qualitative reasoning alone. Where calculations are required, we can do these most readily with the ideal gas equation (Section 6-3).

EXAMPLE 6-5

Testing Your Qualitative Understanding of the Gas Laws. Which will have the greater volume when the gases are compared at STP: (a) 1.20 L $N_2(g)$ at 25 °C and 748 mmHg or (b) 1.25 L $O_2(g)$ at STP?

Solution

First, we can assume that the simple gas laws apply to both gases. The fact that one gas is N_2 and the other O_2 does not enter into our assessment. Next, we see that the gas in (b) is already at STP and its volume is 1.25 L. The crux of the matter is whether the volume of gas increases or decreases when the conditions in (a) are changed to STP. Lowering the temperature from 25 to 0 °C causes the volume to decrease (Charles's law). Raising the pressure from 748 to 760 mmHg also causes the volume to decrease (Boyle's law). At STP the gas in (a) will occupy a volume smaller than its initial 1.20 L. Thus the greater volume at STP is the 1.25 L $O_2(g)$.

Practice Example A: Suppose that the initial condition in case (a) in Example 6-5 is 0.60 L $N_2(g)$ at 150 K and 0.98 atm. Which gas, the N_2 or the O_2, will have the greater volume at STP?

Practice Example B: Suppose that the initial condition in case (a) in Example 6-5 is 1.20 L $N_2(g)$ at 4.2 °C and 758 mmHg, whereas that in (b) is 1.25 L $O_2(g)$ at -2.5 °C and 1.02 atm. Which gas will have the greater volume at STP?

Avogadro's Law

In 1808, Gay-Lussac reported that gases react by volumes in the ratio of small whole numbers. One proposed explanation was that equal volumes of gases at the same temperature and pressure contain equal numbers of atoms. Dalton did not agree with this proposition, however. He viewed the reaction of hydrogen and oxygen to be $H(g) + O(g) \rightarrow HO(g)$, for which the combining volumes should have been in the ratio $1:1:1$, not the $2:1:2$ ratio that was observed.

In 1811, Amedeo Avogadro resolved this dilemma by proposing not only the "equal volumes–equal numbers" hypothesis, but that molecules of a gas may break up into half-molecules when they react. Using modern terminology we would say that O_2 molecules split into atoms and then combine with molecules of H_2 to form H_2O molecules. In this way, only one-half the volume of oxygen is needed as of hydrogen. Avogadro's reasoning is outlined in Figure 6-9.

Avogadro's "equal volumes–equal numbers" hypothesis can be stated in either of two ways.

1. Equal volumes of different gases compared at the same temperature and pressure contain equal numbers of molecules.

2. Equal numbers of molecules of different gases compared at the same temperature and pressure occupy equal volumes.

Another relationship, known as **Avogadro's law,** is that

At a fixed temperature and pressure, the volume of a gas is directly proportional to the amount of gas.   (6.8)

If the number of moles of gas (n) is doubled, the volume doubles, and so on. A mathematical statement of this fact is

$$V \propto n \quad \text{and} \quad V = c \cdot n$$

At STP the number of molecules contained in 22.4 L of a gas is 6.02×10^{23}, that is, *1 mol.*

$$1 \text{ mol gas} = 22.4 \text{ L gas} \quad \text{(at STP)} \quad (6.9)$$

Figure 6-10 should help you to visualize 22.4 L of a gas.

▶ Avogadro's hypothesis and statements derived from it apply *only to gases*. There is *no* similar relationship for liquids or solids.

Amedeo Avogadro (1776–1856)—a scientist ahead of his time. Avogadro's hypothesis and its ramifications were not understood by his contemporaries but were effectively communicated by Stanislao Cannizzaro (1826–1910) about fifty years later.

2 vol. H_2 1 vol. O_2 2 vol. H_2O

FIGURE 6-9
Formation of water—actual observation and Avogadro's hypothesis.
In the reaction $2 H_2(g) + O_2(g) \longrightarrow 2 H_2O(g)$ only one-half as many O_2 molecules are required as of H_2. If equal volumes of gases contain equal numbers of molecules, this means the volume of $O_2(g)$ is one-half that of $H_2(g)$. The combining ratio by volume is $2:1:2$.

FIGURE 6-10
Molar volume of a gas visualized.
The wooden cube has the same volume as a mole of gas at STP: 22.4 L. By contrast, the volume of the basketball is 7.5 L; the soccer ball, 6.0 L; and the football, 4.4 L.

EXAMPLE 6-6

Using the Molar Volume of a Gas at STP as a Conversion Factor. What is the mass of 1.00 L of cyclopropane gas, C_3H_6 (used as an anesthetic), when measured at STP?

Solution

Expression (6.9) allows us to convert directly from volume at STP to number of moles of gas. The conversion from moles of gas to mass of gas requires the molar mass.

$$? \text{ g } C_3H_6 = 1.00 \text{ L} \times \frac{1 \text{ mol } C_3H_6}{22.4 \text{ L } C_3H_6} \times \frac{42.08 \text{ g } C_3H_6}{1 \text{ mol } C_3H_6} = 1.88 \text{ g } C_3H_6$$

Practice Example A: Propane, C_3H_8, is stored as a liquid under pressure for use in backyard barbecue units. The liquid vaporizes to a gas when the pressure is released. A small tank of propane is opened and releases 30.0 L of gas at STP. What mass of gas was released?

Practice Example B: A 128-g piece of solid carbon dioxide ("dry ice") sublimes (evaporates without first melting) into $CO_2(g)$. What is the volume of this gas at STP, in liters?

▶ If a gas is *not* at STP, relating the amount of gas and its volume is best done with the ideal gas equation (Section 6-3).

6-3 The Ideal Gas Equation

At this time we want to replace the simple gas laws with a single equation that includes all four gas variables: volume, pressure, temperature, and amount of gas. From the three simple gas laws it seems reasonable that the volume of a gas should be *directly* proportional to the amount of gas, *directly* proportional to the Kelvin temperature, and *inversely* proportional to pressure. That is,

$$V \propto \frac{nT}{P} \quad \text{and} \quad V = \frac{RnT}{P}$$

▶ The ideal gas equation.

$$PV = nRT \tag{6.10}$$

Equation (6.10) is called the **ideal gas equation,** and any gas that obeys this equation is said to be an **ideal gas** (or perfect gas). Under suitable conditions enough real gases do obey the equation so as to make it a very useful one.

Before we can apply Equation (6.10) to specific situations, we need a numerical value for R, called the **ideal gas constant.** One of the simplest ways to obtain this value is to substitute into Equation (6.10) the molar volume of an ideal gas at STP, 22.4141 L.

$$R = \frac{PV}{nT} = \frac{1 \text{ atm} \times 22.4141 \text{ L}}{1 \text{ mol} \times 273.15 \text{ K}} = 0.082058 \frac{\text{L atm}}{\text{mol K}} \tag{6.11}$$

This value of R is the one that we will mostly use in this chapter (usually rounded off to 0.08206 or 0.0821). The units are "liter atmospheres per mole per kelvin" and they can be written as in (6.11) or as L atm mol^{-1} K^{-1}. Some applications require different units, and two different expressions of R are listed on the inside back cover.

In Examples 6-7 and 6-8, to help organize the data, we first list the known quantities and the unknown. As you become more familiar with the ideal gas equation, you may be able to bypass this step and substitute directly into the equation. A useful check of the calculated result is to make certain the units cancel properly.

EXAMPLE 6-7

Calculating a Gas Volume with the Ideal Gas Equation. What is the volume occupied by 13.7 g $Cl_2(g)$ at 45 °C and 745 mmHg?

Solution

$$P = 745 \text{ mmHg} \times \frac{1 \text{ atm}}{760 \text{ mmHg}} = \frac{745}{760} \text{ atm} = 0.980 \text{ atm}$$

$$V = ?$$

$$n = 13.7 \text{ g } Cl_2 \times \frac{1 \text{ mol } Cl_2}{70.91 \text{ g } Cl_2} = 0.193 \text{ mol } Cl_2$$

$$R = 0.08206 \text{ L atm mol}^{-1} \text{ K}^{-1}$$

$$T = 45 \text{ °C} + 273 = 318 \text{ K}$$

Divide both sides of the ideal gas equation by P to solve for V.

$$\frac{PV}{P} = \frac{nRT}{P} \quad \text{and} \quad V = \frac{nRT}{P}$$

$$V = \frac{0.193 \text{ mol} \times 0.08206 \text{ L atm mol}^{-1} \text{ K}^{-1} \times 318 \text{ K}}{0.980 \text{ atm}} = 5.14 \text{ L}$$

▶ When checking the cancellation of units, note that a unit such as mol^{-1} is the same as 1/mol. Thus, mol $\times$ mol^{-1} = 1. Also, K $\times$ K^{-1} = 1.

Practice Example A: What is the volume occupied by 20.2 g $NH_3(g)$ at −25 °C and 752 mmHg?

Practice Example B: At what temperature will a 13.7-g Cl_2 sample exert a pressure of 745 mmHg when confined in a 7.50-L container? (*Hint:* You can use some information from Example 6-7, but now you must solve the ideal gas equation for the Kelvin temperature.)

EXAMPLE 6-8

Calculating an Amount of Gas with the Ideal Gas Equation. How many molecules of N_2 remain in an ultrahigh vacuum system of 128-mL volume when the pressure is reduced to 5×10^{-10} mmHg at 25 °C?

Solution

The chief need here is to determine the number of moles of gas. For this we can use the ideal gas equation. Then we can convert from number of moles to number of molecules with the Avogadro constant.

$$P = 5 \times 10^{-10} \text{ mmHg} \times \frac{1 \text{ atm}}{760 \text{ mmHg}} = 7 \times 10^{-13} \text{ atm}$$

$$V = 128 \text{ mL} \times \frac{1 \text{ L}}{1000 \text{ mL}} = 0.128 \text{ L}$$

$$n = ?$$

$$R = 0.0821 \text{ L atm mol}^{-1} \text{ K}^{-1}$$

$$T = 25 \text{ °C} + 273 = 298 \text{ K}$$

$$n = \frac{PV}{RT} = \frac{7 \times 10^{-13} \text{ atm} \times 0.128 \text{ L}}{0.0821 \text{ L atm mol}^{-1} \text{ K}^{-1} \times 298 \text{ K}} = 4 \times 10^{-15} \text{ mol}$$

$$? \text{ molecules N}_2 = 4 \times 10^{-15} \text{ mol N}_2 \times \frac{6.02 \times 10^{23} \text{ molecules N}_2}{1 \text{ mol N}_2}$$

$$= 2 \times 10^9 \text{ molecules N}_2$$

Practice Example A: How many moles of He(g) are in a 5.00-L storage tank filled with helium at 10.5 atm pressure and at 30.0 °C?

Practice Example B: What is the pressure exerted by 1.00×10^{20} N_2 molecules when they are confined in a 3.05-L volume at 175 °C?

In Examples 6-7 and 6-8 we applied the ideal gas equation to a *single* set of conditions (P, V, n, and T). Sometimes a gas is described under *two* different sets of conditions. Here we have to apply the ideal gas equation *twice,* to an initial and a final condition. That is,

(1) Initial condition   **(2) Final condition**

$$P_1V_1 = n_1RT_1 \qquad\qquad P_2V_2 = n_2RT_2$$

$$\frac{P_1V_1}{n_1T_1} = R \qquad\qquad \frac{P_2V_2}{n_2T_2} = R$$

The above expressions are equal to each other because each is equal to R.

$$\frac{P_1V_1}{n_1T_1} = \frac{P_2V_2}{n_2T_2} \tag{6.12}$$

Expression (6.12) is called the **general gas equation.** We often apply it in cases in which one or two of the gas properties are held constant, and we can simplify the equation by eliminating these constants. This is what we do in Example 6-9, which establishes the simple relationship between gas pressure and temperature known as *Amonton's law.*

(a) Ice bath

EXAMPLE 6-9

Applying the General Gas Equation. Pictured in Figure 6-11 is a 1.00-L flask of $O_2(g)$, first at STP and then at 100 °C. What is the pressure at 100 °C?

Solution

Identify the terms in the general gas equation that *remain constant.* Cancel out these terms and solve the remaining equation. In this case the amount of $O_2(g)$ is constant ($n_1 = n_2$) and the volume is constant ($V_1 = V_2$).

$$\frac{P_1 V_1}{n_1 T_1} = \frac{P_2 V_2}{n_2 T_2} \quad \text{and} \quad \frac{P_1}{T_1} = \frac{P_2}{T_2} \quad \text{and} \quad P_2 = P_1 \times \frac{T_2}{T_1}$$

Because P_1 is standard pressure = 1.00 atm,

$$P_2 = 1.00 \text{ atm} \times \frac{(100 + 273) \text{ K}}{273 \text{ K}} = 1.00 \text{ atm} \times \frac{373 \text{ K}}{273 \text{ K}} = 1.37 \text{ atm}$$

We know intuitively that heating a gas in a closed container will cause its pressure to rise. The result here is consistent with our expectation.

Practice Example A: A 1.00-mL sample of $N_2(g)$ at 36.2 °C and 2.14 atm is heated to 37.8 °C and the pressure changed to 1.02 atm. What volume does the gas occupy at this temperature and pressure?

Practice Example B: Suppose that in Figure 6-11 we wish the pressure to remain at 1.00 atm when the $O_2(g)$ is heated to 100 °C. What mass of $O_2(g)$ must we release from the flask? (*Hint:* How many moles of O_2 are present in the 1.00-L flask at STP? Recall Example 6-6.)

(b) Boiling water

FIGURE 6-11
Pressure of a gas as a function of temperature—Example 6-9 visualized.
The amount of gas and volume are held constant.
(a) 1.00 L $O_2(g)$ at STP; (b) 1.00 L $O_2(g)$ at 100 °C.

Are You Wondering . . .

When to use the ideal gas equation and when to use the general gas equation?

In many instances you can use either, and your choice will probably depend on the method that seems clearer. In general, however, use the *ideal gas equation* if you are dealing with *a single set of conditions* and three of the four gas variables (P, V, n, T) are given. You may find the *general gas equation* easier to use when you deal with *two* sets of conditions (initial and final).

6-4   Applications of the Ideal Gas Equation

Although the ideal gas equation can always be used as it was presented in Equation (6.10), you may find it helpful to recast the equation into slightly different forms for some applications. We will consider two such applications in this section: determination of molar masses and gas densities.

Molar Mass Determination

If we know the volume occupied by a gas at a fixed temperature and pressure, we can solve the ideal gas equation for the amount of gas expressed in moles, n. Because the number of moles of gas, n, is equal to the mass of gas, m, divided by the molar mass, $\mathcal{M}$, if we know the mass of gas we can solve the expression $n = m/\mathcal{M}$ for the unknown molar mass, $\mathcal{M}$. An alternative is to make the substitution $n = m/\mathcal{M}$ directly into the ideal gas equation.

$$PV = \frac{mRT}{\mathcal{M}} \qquad (6.13)$$

We apply Equation (6.13) to a molar mass determination in Example 6-10, but note that the equation can also be used when the molar mass of a gas is known and the mass of a particular sample of the gas is sought.

EXAMPLE 6-10

Determining a Molar Mass with the Ideal Gas Equation. Propylene is an important commercial chemical (about no. 9 among manufactured chemicals). It is used in the synthesis of other organic chemicals and in plastics production (polypropylene). A glass vessel weighs 40.1305 g when clean, dry, and evacuated; 138.2410 g when filled with water at 25.0 °C (density of water = 0.9970 g/mL); and 40.2959 g when filled with propylene gas at 740.3 mmHg and 24.0 °C. What is the molar mass of propylene?

Solution

Our first task is to determine the volume of the glass vessel and hence the volume of the gas.

$$\text{mass of water to fill vessel} = 138.2410 \text{ g} - 40.1305 \text{ g} = 98.1105 \text{ g}$$

$$\text{volume of water (volume of vessel)} = 98.1105 \text{ g H}_2\text{O} \times \frac{1 \text{ mL H}_2\text{O}}{0.9970 \text{ g H}_2\text{O}}$$

$$= 98.41 \text{ mL} = 0.09841 \text{ L}$$

Now we can list values of the other variables, starting with the mass of gas, which we get by difference.

$$\text{mass of gas} = 40.2959 \text{ g} - 40.1305 \text{ g} = 0.1654 \text{ g}$$

$$\text{temperature} = 24.0 \text{ °C} + 273.15 = 297.2 \text{ K}$$

$$\text{pressure} = 740.3 \text{ mmHg} \times \frac{1 \text{ atm}}{760.0 \text{ mmHg}} = 0.9741 \text{ atm}$$

Finally, we substitute into a rearranged form of Equation (6.13).

$$\mathcal{M} = \frac{mRT}{PV} = \frac{0.1654 \text{ g} \times 0.08206 \text{ L atm mol}^{-1} \text{ K}^{-1} \times 297.2 \text{ K}}{0.9741 \text{ atm} \times 0.09841 \text{ L}}$$

$$= 42.08 \text{ g/mol}$$

Practice Example A: The same glass vessel used in Example 6-10 is filled with an unknown gas at 772 mmHg and 22.4 °C. The vessel filled with gas weighs 40.4868 g. What is the molar mass of the gas?

Practice Example B: A 1.27-g sample of an oxide of nitrogen, believed to be either NO or N_2O, occupies a volume of 1.07 L at 25 °C and 737 mmHg. Which oxide is it?

Consider now that we wish to determine the formula of an unknown hydrocarbon. With combustion analysis we can establish the mass percent composition of the hydrocarbon, and from this we can determine its empirical formula. The method of Example 6-10 gives us a molar mass (g/mol), which is numerically equal to the molecular mass (u). This is all the information we need to establish the true molecular formula of the hydrocarbon (see Exercise 83).

Gas Densities

To determine the density of a gas, we can start with the density equation, $d = m/V$. Then we can express the mass of gas as the product of the number of moles of gas and the molar mass: $m = n \times \mathcal{M}$. This leads to

$$d = \frac{m}{V} = \frac{n \times \mathcal{M}}{V} = \frac{n}{V} \times \mathcal{M}$$

Now, with the ideal gas equation, we can replace n/V by its equivalent, P/RT, to obtain

$$d = \frac{m}{V} = \frac{\mathcal{M}P}{RT} \tag{6.14}$$

The density of a gas at STP can easily be calculated by dividing its molar mass by the molar volume (22.4 L/mol). For $O_2(g)$ at STP, for example, the density is 32.0 g/22.4 L = 1.43 g/L. Under other conditions of temperature and pressure, we can use Equation (6.14).

▶ Gas densities are much smaller than the densities of solids and liquids. They are usually expressed in grams per liter instead of grams per milliliter.

EXAMPLE 6-11

Using the Ideal Gas Equation to Calculate a Gas Density. What is the density of oxygen gas (O_2) at 298 K and 0.987 atm?

Solution

We can get the terms for the right side of Equation (6.14) easily. Density (m/V) is the left side of the equation.

$$d = \frac{m}{V} = \frac{\mathcal{M}P}{RT} = \frac{32.00 \text{ g mol}^{-1} \times 0.987 \text{ atm}}{0.08206 \text{ L atm mol}^{-1} \text{ K}^{-1} \times 298 \text{ K}} = 1.29 \text{ g/L}$$

Practice Example A: What is the density of helium gas at 298 K and 0.987 atm? Based on your answer, explain why we can say that helium is "lighter than air."

Practice Example B: At what temperature will the density of $O_2(g)$ be 1.00 g/L, if the pressure is kept at 745 mmHg? [*Hint:* You must again use Equation (6.14), but solve for a different term than in Example 6-11.]

Gas densities differ from solid and liquid densities in two important ways.

1. Gas densities depend strongly on pressure and temperature, increasing as the gas pressure increases and decreasing as the temperature increases. Densities of liquids and solids also depend somewhat on temperature, but they depend far less on pressure.

2. The density of a gas is directly proportional to its molar mass. No simple relationship exists between density and molar mass for liquids and solids.

FIGURE 6-12
The lifting power of a helium-filled balloon.
The helium-filled balloon exerts a lifting force on the 100.00-g weight, so that the balloon and weight together weigh only 99.75 g.

An important application of gas densities is in establishing conditions for lighter-than-air balloons. A gas-filled balloon will rise in the atmosphere only if the density of the gas is less than that of air. Because gas densities are directly proportional to molar masses, the lower the molar mass of the gas the greater is its lifting power. The lowest molar mass is that of hydrogen, but hydrogen is inflammable and forms explosive mixtures with air. The explosion of the dirigible *Hindenburg* in 1937 spelled the end to transoceanic travel by hydrogen-filled airships. Now, airships (such as the Goodyear blimps) use helium, which has a molar mass only twice that of hydrogen (Figure 6-12). Hydrogen is still used for weather and other observational balloons.

Another alternative is to fill a balloon with *hot* air. Equation (6.14) indicates that the density of a gas is *inversely* proportional to temperature. Hot air is less dense than cold air. However, because the density of air decreases rapidly with altitude, there is a limit to how high a hot-air balloon or any gas-filled balloon can rise.

6-5 Gases in Chemical Reactions

Reactions involving gases as reactants or products (or both) are no strangers to us. Now, though, we have a new tool to apply to reaction stoichiometry calculations: the ideal gas equation. Specifically, we can now handle information about gases in terms of volumes, temperatures, and pressures, as well as by mass and amount in moles.

In many cases the best approach is to (a) use stoichiometric factors to relate the amount of a gas to amounts of other reactants or products and (b) use the ideal gas equation to relate the amount of gas to volume, temperature, and pressure. In Example 6-12 we determine the number of moles of $N_2(g)$ from the stoichiometry of a reaction. Then we use the ideal gas equation to determine a gas volume.

EXAMPLE 6-12

Using the Ideal Gas Equation in Reaction Stoichiometry Calculations. The decomposition of sodium azide, NaN_3, at high temperatures produces $N_2(g)$. Together with the necessary devices to initiate the reaction and trap the sodium metal formed, this reaction is used in air-bag safety systems (see Focus On, page 194). What volume of $N_2(g)$, measured at 735 mmHg and 26 °C, is produced when 70.0 g NaN_3 is decomposed?

$$2 \, NaN_3(s) \xrightarrow{\Delta} 2 \, Na(l) + 3 \, N_2(g)$$

Solution

$$? \text{ mol } N_2 = 70.0 \text{ g } NaN_3 \times \frac{1 \text{ mol } NaN_3}{65.01 \text{ g } NaN_3} \times \frac{3 \text{ mol } N_2}{2 \text{ mol } NaN_3} = 1.62 \text{ mol } N_2$$

$$P = 735 \text{ mmHg} \times \frac{1 \text{ atm}}{760 \text{ mmHg}} = 0.967 \text{ atm}$$

$$V = ?$$

$$n = 1.62 \text{ mol}$$

$$R = 0.08206 \text{ L atm mol}^{-1} \text{ K}^{-1}$$

$$T = 26 \text{ °C} + 273 = 299 \text{ K}$$

$$V = \frac{nRT}{P} = \frac{1.62 \text{ mol} \times 0.08206 \text{ L atm mol}^{-1} \text{ K}^{-1} \times 299 \text{ K}}{0.967 \text{ atm}} = 41.1 \text{ L}$$

Practice Example A: How many grams of NaN_3 are needed to produce 20.0 L of $N_2(g)$ at 30.0 °C and 776 mmHg? (*Hint:* This calculation is essentially the "inverse" of Example 6-12.)

Practice Example B: What mass of Na(l), in grams, is produced, *per liter* of $N_2(g)$ at 25 °C and 751 mmHg, in the decomposition of sodium azide?

Law of Combining Volumes

At times, if the reactants and products involved in a stoichiometric calculation are gases, we can use a particularly simple approach. Consider this reaction.

$$2\ NO(g)\ +\ O_2(g)\ \longrightarrow\ 2\ NO_2(g)$$

$$2\ mol\ NO(g) + 1\ mol\ O_2(g) \longrightarrow 2\ mol\ NO_2(g)$$

Suppose the gases are compared at the same T and P. Under these conditions one mole of gas occupies a particular volume, call it V liters; 2 mol gas, $2V$ liters; and so on.

$$2V\ L\ NO(g) + V\ L\ O_2(g) \longrightarrow 2V\ L\ NO_2(g)$$

Now divide each coefficient by V.

$$2\ L\ NO(g) + 1\ L\ O_2(g) \longrightarrow 2\ L\ NO_2(g)$$

When we describe the chemical reaction in this way, we can write conversion factors based on the statements

$$2\ L\ NO_2(g) = 2\ L\ NO(g) \qquad 2\ L\ NO_2(g) = 1\ L\ O_2(g) \qquad 2\ L\ NO(g) = 1\ L\ O_2(g)$$

What we have just done is to develop, in modern terms, Gay-Lussac's law of combining volumes, the idea presented on page 172 that gases react by volumes in the ratio of small whole numbers. We apply this law in Example 6-13.

EXAMPLE 6-13

Applying the Law of Combining Volumes. Zinc blende, ZnS, is the most important zinc ore. Roasting (strong heating) of ZnS is the first step in the commercial production of zinc.

$$2\ ZnS(s) + 3\ O_2(g) \xrightarrow{\Delta} 2\ ZnO(s) + 2\ SO_2(g)$$

What volume of $SO_2(g)$ forms *per liter* of $O_2(g)$ consumed? Both gases are measured at 25 °C and 745 mmHg.

Solution

The reactant and product being compared are both gases *and both are at the same temperature and pressure,* so we can derive a ratio of combining volumes from the balanced equation (in blue) and use it as follows.

$$?\ L\ SO_2(g) = 1.00\ L\ O_2(g) \times \frac{2\ L\ SO_2(g)}{3\ L\ O_2(g)} = 0.667\ L\ SO_2(g)$$

Practice Example A: The first step in making nitric acid is to convert ammonia to nitrogen monoxide. This is done under conditions of high temperature and in the presence of a platinum catalyst. What volume of $O_2(g)$ is consumed per liter of NO(g) formed?

$$4\ NH_3(g) + 5\ O_2(g) \xrightarrow[850\ °C]{Pt} 4\ NO(g) + 6\ H_2O(g)$$

Practice Example B: If all gases are measured at the same temperature and pressure, what volume of $NH_3(g)$ is produced when 225 L $H_2(g)$ is consumed in the reaction $N_2(g) + H_2(g) \rightarrow NH_3(g)$ (not balanced)?

Are You Wondering . . .

How to proceed if the gases involved in a stoichiometric calculation are *not* at the same temperature and pressure?

If the gases are not at *identical* temperatures and pressures, you can't use the law of combining volumes. In such cases the best approach is to convert information about gases to a mole basis. Then use a mole ratio from the chemical equation, as in Example 6-12. Note also that, when the law of combining volumes does apply, you don't need to know the temperature and pressure values. Thus, in Example 6-13 we didn't use the conditions "25 °C and 745 mmHg."

6-6 Mixtures of Gases

The simple gas laws, such as Boyle's and Charles's laws, were based on the behavior of air—a mixture of gases. So, the simple gas laws and the ideal gas equation apply to a *mixture* of nonreactive gases as well as to individual gases. Where we are able to do so, the simplest approach to working with gaseous mixtures is just to use for the value of n the *total* amount, in moles, of the gaseous mixture (n_{tot}).

EXAMPLE 6-14

Applying the Ideal Gas Equation to a Mixture of Gases. What is the pressure exerted by a mixture of 1.0 g H_2 and 5.00 g He when the mixture is confined to a volume of 5.0 L at 20 °C?

Solution

$$n_{tot} = \left(1.0 \text{ g } H_2 \times \frac{1 \text{ mol } H_2}{2.02 \text{ g } H_2}\right) + \left(5.00 \text{ g He} \times \frac{1 \text{ mol He}}{4.003 \text{ g He}}\right)$$

$$= 0.50 \text{ mol } H_2 + 1.25 \text{ mol He} = 1.75 \text{ mol gas}$$

$$P = \frac{n_{tot}RT}{V} \tag{6.15}$$

$$= \frac{1.75 \text{ mol} \times 0.0821 \text{ L atm mol}^{-1} \text{ K}^{-1} \times 293 \text{ K}}{5.0 \text{ L}} = 8.4 \text{ atm}$$

Practice Example A: If we add 12.5 g Ne to the mixture of gases described in Example 6-14 and then raise the temperature to 55 °C, what will be the total gas pressure? (*Hint:* What is the new number of moles of gas? What effect does raising the temperature have on the pressure of a gas at constant volume?)

Practice Example B: 2.0 L of $O_2(g)$ and 8.0 L of $N_2(g)$, each at STP, are mixed together. The nonreactive gaseous mixture is compressed to occupy 2.0 L at 298 K. What is the pressure exerted by this mixture?

John Dalton made an important contribution to the study of gaseous mixtures. He proposed that in a mixture each gas expands to fill the container and exerts the same pressure, called its **partial pressure,** that it would if it were alone in the container. **Dalton's law of partial pressures** states that the total pressure of a mixture of gases is the sum of the partial pressures of the components of the mixture (Figure 6-13). For a mixture of gases, A, B, . . . ,

$$P_{tot} = P_A + P_B + \cdots \tag{6.16}$$

In a gaseous mixture of n_A moles of A, n_B moles of B, and so on, the volume each gas would individually occupy at a pressure equal to P_{tot} is

$$V_A = n_A RT/P_{tot}; \quad V_B = n_B RT/P_{tot}; \text{ and so on.}$$

The total volume of the gaseous mixture is $V_{tot} = V_A + V_B + \cdots$; and the commonly used expression *percent by volume* is

$$\text{Volume \% A} = \frac{V_A}{V_{tot}} \times 100; \quad \text{Volume \% B} = \frac{V_B}{V_{tot}} \times 100; \text{ and so on.}$$

We can derive a particularly useful expression by taking the ratios

$$\frac{P_A}{P_{tot}} = \frac{n_A(RT/V_{tot})}{n_{tot}(RT/V_{tot})} = \frac{n_A}{n_{tot}} \quad \text{and} \quad \frac{V_A}{V_{tot}} = \frac{n_A(RT/P_{tot})}{n_{tot}(RT/P_{tot})} = \frac{n_A}{n_{tot}}$$

which means that

$$\frac{n_A}{n_{tot}} = \frac{P_A}{P_{tot}} = \frac{V_A}{V_{tot}} \tag{6.17}$$

FIGURE 6-13
Dalton's law of partial pressures illustrated.
The pressure of each gas is proportional to the number of moles of gas. The total pressure is the sum of the partial pressures of the individual gases.

The term n_A/n_{tot} is given a special name, the mole fraction of A. The **mole fraction** of a component in a mixture is the fraction of all the molecules in the mixture contributed by that component. The sum of all the mole fractions in a mixture is 1.

As illustrated in Example 6-15, you will often find it possible to think about mixtures of gases in more than one way.

EXAMPLE 6-15

Calculating the Partial Pressures in a Gaseous Mixture. What are the partial pressures of H_2 and He in the gaseous mixture described in Example 6-14?

Solution

One approach involves a direct application of Dalton's law. That is, calculate the pressure each gas would exert if it were alone in the container.

$$P_{H_2} = \frac{n_{H_2} \cdot RT}{V} = \frac{0.50 \text{ mol} \times 0.0821 \text{ L atm mol}^{-1} \text{ K}^{-1} \times 293 \text{ K}}{5.0 \text{ L}} = 2.4 \text{ atm}$$

$$P_{He} = \frac{n_{He} \cdot RT}{V} = \frac{1.25 \text{ mol} \times 0.0821 \text{ L atm mol}^{-1} \text{ K}^{-1} \times 293 \text{ K}}{5.0 \text{ L}} = 6.0 \text{ atm}$$

Expression (6.17) gives us a simpler way to answer the question because we already know the number of moles of each gas and the total pressure from Example 6-14 ($P_{tot} = 8.4$ atm).

$$P_{H_2} = \frac{n_{H_2}}{n_{tot}} \times P_{tot} = \frac{0.50}{1.75} \times 8.4 \text{ atm} = 2.4 \text{ atm}$$

$$P_{He} = \frac{n_{He}}{n_{tot}} \times P_{tot} = \frac{1.25}{1.75} \times 8.4 \text{ atm} = 6.0 \text{ atm}$$

▶ As you might expect, the sum of these partial pressures is the total pressure—8.4 atm (see Example 6-14).

Practice Example A: A mixture of 0.197 mol $CO_2(g)$ and 0.00278 mol $H_2O(g)$ is held at 30.0 °C and 2.50 atm. What is the partial pressure of each gas?

Practice Example B: The percent composition of air by volume is 78.08% N_2, 20.95% O_2, 0.93% Ar, and 0.036% CO_2. What are the partial pressures of these four gases in a sample of air at a barometric pressure of 748 mmHg? [*Hint:* Use expression (6.17) and work with pressure and volume ratios.]

▶ Some early experimenters used mercury in the pneumatic trough, so as to be able to collect gases soluble in water.

The device pictured in Figure 6-14, a *pneumatic trough,* played a crucial role in isolating gases in the early days of chemistry. The method only works, of course, for gases that are insoluble in the liquid being displaced. But many important gases are. For example, H_2, O_2, and N_2 are all insoluble in water.

A gas collected in a pneumatic trough filled with water is said to be *collected over water* and is "wet." It is a mixture of the desired gas and water vapor. The gas being collected expands to fill the container and exerts its partial pressure, P_{gas}. Water vapor, formed by the evaporation of liquid water, also fills the container and exerts a partial pressure, P_{H_2O}. The pressure of the water vapor depends only on the temperature of the water. Water vapor pressure data are readily available in tabulated form (see Table 13.2).

According to Dalton's law, the *total* pressure of the "wet" gas is the sum of the two partial pressures. The total pressure can be made equal to the prevailing pressure of the atmosphere (barometric pressure).

$$P_{tot} = P_{bar} = P_{gas} + P_{H_2O} \quad \text{or} \quad P_{gas} = P_{bar} - P_{H_2O}$$

Once P_{gas} has been established, it can be used in stoichiometric calculations as illustrated in Example 6-16.

FIGURE 6-14
Collecting a gas over water.
The bottle is filled with water and its open end is held below the water level in the container. Gas from a gas-generating apparatus is directed into the bottle. As gas accumulates in the bottle, water is displaced from the bottle into the container. To make the total gas pressure in the bottle equal to barometric pressure, the position of the bottle must be adjusted so that the water levels inside and outside the bottle are the same.

EXAMPLE 6-16

Collecting a Gas over a Liquid (Water). In the following reaction 81.2 mL of $O_2(g)$ is collected over water at 23 °C and barometric pressure 751 mmHg. What must have been the mass of $Ag_2O(s)$ decomposed? (The vapor pressure of water at 23 °C is 21.1 mmHg.)

$$2 Ag_2O(s) \longrightarrow 4 Ag(s) + O_2(g)$$

Solution

First we need to calculate the number of moles of $O_2(g)$ produced. We can do this with the ideal gas equation. The key to this calculation is to see that the gas collected is "wet," that is, a *mixture* of $O_2(g)$ and water vapor.

$$P_{O_2} = P_{bar} - P_{H_2O} = 751 \text{ mmHg} - 21.1 \text{ mmHg} = 730 \text{ mmHg}$$

$$= 730 \text{ mmHg} \times \frac{1 \text{ atm}}{760 \text{ mmHg}} = 0.961 \text{ atm}$$

$$V = 81.2 \text{ mL} = 0.0812 \text{ L}$$

$$n = ?$$

$$R = 0.08206 \text{ L atm mol}^{-1} \text{ K}^{-1}$$

$$T = 23 \text{ °C} + 273 = 296 \text{ K}$$

$$n = \frac{PV}{RT} = \frac{0.961 \text{ atm} \times 0.0812 \text{ L}}{0.08206 \text{ L atm mol}^{-1} \text{ K}^{-1} \times 296 \text{ K}} = 0.00321 \text{ mol}$$

From the chemical equation we obtain a factor to convert from moles of O_2 to moles of Ag_2O. The molar mass of Ag_2O provides the final factor.

$$? \text{ g Ag}_2\text{O} = 0.00321 \text{ mol O}_2 \times \frac{2 \text{ mol Ag}_2\text{O}}{1 \text{ mol O}_2} \times \frac{231.7 \text{ g Ag}_2\text{O}}{1 \text{ mol Ag}_2\text{O}}$$

$$= 1.49 \text{ g Ag}_2\text{O}$$

Practice Example A: The reaction of aluminum with hydrochloric acid produces hydrogen gas.

$$2 Al(s) + 6 HCl(aq) \longrightarrow 2 AlCl_3(aq) + 3 H_2(g)$$

If 35.5 mL of $H_2(g)$ is collected over water at 26 °C and a barometric pressure of 755 mmHg, how many moles of HCl must have been consumed? (The vapor pressure of water at 26 °C is 25.2 mmHg.)

Practice Example B: An 8.07-g sample that is 88.3% Ag_2O by mass is decomposed and the $O_2(g)$ is collected over water at 25 °C and 749.2 mmHg barometric pressure: $2 Ag_2O(s) \rightarrow 4 Ag(s) + O_2(g)$. The vapor pressure of water at 25 °C is 23.8 mmHg. What is the volume of gas collected?

6-7 Kinetic-Molecular Theory of Gases

Let us apply some terminology that we introduced in Section 1-2 on the scientific method: The simple gas laws and the ideal gas equation are used to predict gas behavior. They are *natural laws*. To *explain* the gas laws, we need a *theory*. One theory developed during the middle nineteenth century is called the **kinetic-**

FIGURE 6-15
Visualizing molecular motion.
Molecules of a gas (•) are in constant motion and undergo collisions with one another and with the container wall.

molecular theory of gases. It is based on the *model* illustrated in Figure 6-15 and outlined as follows.

1. A gas is composed of a very large number of extremely small particles (molecules or, in some cases, atoms) in *constant, random, straight-line* motion.

2. Molecules of a gas are separated by great distances. The gas is mostly empty space. (The molecules are treated as if they have mass but no volume, so-called "point masses.")

3. Molecules collide with one another and with the walls of their container. These collisions occur very rapidly, however, and most of the time molecules are not colliding.

4. There are assumed to be no forces between molecules except very briefly during collisions. That is, each molecule acts independently of all the others and is unaffected by their presence.

5. Individual molecules may gain or lose energy as a result of collisions. In a collection of molecules at constant temperature, however, *the total energy remains constant.*

Because pressure is a force per unit area, the key to deriving a kinetic-molecular theory equation for pressure is in assessing the force of molecular collisions. This depends on several factors.

One factor is the *frequency* of molecular collisions—the number of collisions per second. The higher this frequency, the greater the total force of these collisions. Collision frequency increases with the number of molecules per unit volume and with molecular speeds.

A second factor is the amount of *translational kinetic energy* that molecules have. Translational kinetic energy is energy possessed by objects moving through space. Like speeding bullets, gas molecules have this energy of motion. The translational kinetic energy of a molecule is represented as e_k and has the value $e_k = \frac{1}{2}mu^2$, where m is the mass of the molecule and u is its speed. The faster the molecules move, the greater their translational kinetic energies and the greater the force of their collisions.

Still another factor is that molecules move in all directions and at different speeds. When all factors are properly considered, the result is the following expression for the pressure of a gas, the basic equation of the kinetic-molecular theory.

$$P = \frac{1}{3}\frac{N}{V}m\overline{u^2} \tag{6.18}$$

▶ The bar over a quantity means that the quantity can have a range of values and that the *average* value is intended.

In Equation (6.18) N is the number of molecules present in the volume V. The molecules each have the mass m, and the *average* of the *squares* of their speeds is $\overline{u^2}$.* Equation (6.18) leads to some interesting results, as we see next.

Distribution of Molecular Speeds

The averaging of molecular speeds referred to in establishing Equation (6.18) is required because molecules of a gas do not all have the same speed. There is a distribution of speeds, as shown in Figure 6-16.

*Consider three automobiles with speeds of 40, 50, and 60 mi/h. The average of their *speeds* is 50 mi/h, but the average of the *squares* of their speeds is not 2500 mi²/h², but

$$\frac{[(40)^2 + (50)^2 + (60)^2]}{3} = 2567 \text{ mi}^2/\text{h}^2$$

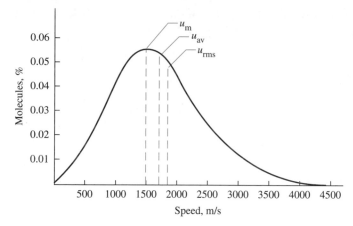

FIGURE 6-16
Distribution of molecular speeds—hydrogen gas at 0 °C.
The percentages of molecules with a certain speed are plotted as a function of the speed. Three different speeds are noted on the graph and discussed in the text.

Three characteristic speeds are identified in Figure 6-16. More molecules have the speed u_m, the most probable, or modal, speed, than any other single speed. The average speed, u_{av} (or $\bar{u}$), is the simple average. The **root-mean-square speed**, u_{rms}, is the *square root* of the *average* of the *squares* of the speeds of all the molecules in a sample.*

We can derive a useful equation for u_{rms} by combining Equation (6.18) with the ideal gas equation. Consider 1 mol of an ideal gas. The number of molecules present is $N = N_A$ (Avogadro's number), and the ideal gas equation becomes $PV = RT$ (that is, $n = 1$ in the equation $PV = nRT$). First, replace N by N_A and multiply both sides of Equation (6.18) by V. This leads to

▶ If the postulates of the kinetic-molecular theory hold, a gas is automatically an ideal gas.

$$PV = \tfrac{1}{3}N_A m \overline{u^2} \qquad (6.19)$$

Next, replace PV by RT and multiply both sides of the equation by 3.

$$3RT = N_A m \overline{u^2}$$

Now, note that the product $N_A m$ represents the mass of 1 mol of molecules, the molar mass, $\mathcal{M}$.

$$3RT = \mathcal{M}\overline{u^2}$$

Finally, solve for $\overline{u^2}$ and then $\sqrt{\overline{u^2}}$, which is u_{rms}.

$$u_{rms} = \sqrt{\frac{3RT}{\mathcal{M}}} \qquad (6.20)$$

To use Equation (6.20) in calculating a root-mean-square speed, we must express the gas constant as

$$R = 8.3145 \text{ J mol}^{-1} \text{ K}^{-1}$$

Then, to get units to cancel properly, we have to express the unit joule (J) in terms of mass, length, and time. Because kinetic energy is given by $e_k = \tfrac{1}{2}mu^2$, the joule must have the units of (mass) × (velocity)^2 = (kg)(m/s)^2, leading to

▶ The joule (J) and other energy units are discussed in Appendix B.

$$R = 8.3145 \text{ kg m}^2 \text{ s}^{-2} \text{ mol}^{-1} \text{ K}^{-1}$$

---

*To continue the analogy of the three automobiles from the previous footnote, $\bar{u} = u_{av} = 50$ mi/h, but $u_{rms} = \sqrt{2567 \text{ mi}^2/\text{h}^2\text{s}} = 50.7$ mi/h. A root-mean-square speed is weighted toward molecules with higher speeds.

Finally, again to get units to cancel properly, we must express molar mass in *kilograms* per mole. All of these ideas are illustrated in Example 6-17.

EXAMPLE 6-17

Calculating a Root-Mean-Square Speed. Which is the greater speed, that of a bullet from a high-powered M-16 rifle (2180 mi/h) or the root-mean-square speed of H_2 molecules at 25 °C?

Solution

Determine u_{rms} of H_2 with Equation (6.20). In doing so, recall the points about R and $\mathcal{M}$ noted above.

$$u_{rms} = \sqrt{\frac{3 \times 8.3145 \text{ kg m}^2 \text{ s}^{-2} \text{ mol}^{-1} \text{ K}^{-1} \times 298 \text{ K}}{2.016 \times 10^{-3} \text{ kg mol}^{-1}}}$$

$$= \sqrt{3.69 \times 10^6 \text{ m}^2/\text{s}^2} = 1.92 \times 10^3 \text{ m/s}$$

The remainder of the problem requires us either to convert 1.92×10^3 m/s to a speed in miles per hour, or 2180 mi/h to meters per second. Then we can compare the two speeds. When we do this, we find that 1.92×10^3 m/s corresponds to 4.29×10^3 mi/h. The root-mean-square speed of H_2 molecules at 25 °C is greater than the speed of the high-powered rifle bullet.

Practice Example A: Which has the greater root-mean-square speed at 25 °C, $NH_3(g)$ or $HCl(g)$? Calculate u_{rms} for the one with the greater speed.

Practice Example B: At what temperature are u_{rms} of H_2 and the speed of the M-16 rifle bullet given in Example 6-17 the same? [*Hint:* Convert the speed of the bullet to meters per second. Solve Equation (6.20) for a temperature. Will this temperature be higher or lower than 25 °C?]

The Meaning of Temperature

We can gain an important insight into the meaning of temperature by starting with Equation (6.19), the basic equation of the kinetic-molecular theory written for 1 mol of gas. We modify it slightly by replacing the fraction $\frac{1}{3}$ by the equivalent product $\frac{2}{3} \times \frac{1}{2}$.

$$PV = \tfrac{1}{3}N_A m\overline{u^2} = \tfrac{2}{3}N_A(\tfrac{1}{2}m\overline{u^2})$$

The terms in this equation that are gathered into $(\tfrac{1}{2}m\overline{u^2})$ represent the average translational kinetic energy, $\overline{e_k}$, of a collection of molecules. When we make this substitution and replace PV by RT, we obtain

$$RT = \tfrac{2}{3}N_A\overline{e_k}$$

Then, we can solve this equation for $\overline{e_k}$.

$$\overline{e_k} = \frac{3}{2}\frac{R}{N_A}(T) \tag{6.21}$$

Because R and N_A are constants, Equation (6.21) simply states that $\overline{e_k} = con\text{-}stant \times T$. This leads to an interesting new idea about temperature:

The Kelvin temperature (T) of a gas is directly proportional to the average translational kinetic energy ($\overline{e_k}$) of its molecules.

Also, we have a new conception of what changes in temperature mean—changes in the intensity of translational molecular motion. When heat flows from one body to another, molecules in the hotter body (higher temperature) give up some of their kinetic energy through collisions with molecules in the colder body (lower temperature). The flow of heat continues until the average translational kinetic energies of the molecules become equal, that is, until the temperatures become equalized. Finally, we have a new way of looking at the absolute zero of temperature: *It is the temperature at which translational molecular motion should cease.*

▶ The absolute zero is unattainable, although current attempts have resulted in temperatures as low as just a few nanokelvins.

6-8 Gas Properties Relating to the Kinetic-Molecular Theory

A molecular speed of 1500 m/s corresponds to about 1 mi/s, or 3600 mi/h. It might seem that a given gas molecule can travel very long distances over a very short time, but this is not quite the case. Every gas molecule undergoes frequent collisions with other gas molecules and as a result keeps changing direction. Gas molecules follow a tortuous zigzag path, which slows them down in getting from one point to another in a gas. Still, the net rate at which gas molecules move in a particular direction does depend on their average speeds.

Diffusion is the migration or intermingling of molecules of different substances as a result of random molecular motion. Figure 6-17 pictures a phenomenon commonly seen in a chemistry laboratory, the diffusion of $NH_3(g)$ and $HCl(g)$ through air and the subsequent reaction of the gases to form the white solid $NH_4Cl(s)$. A related phenomenon, **effusion,** is the escape of gas molecules from their container through a tiny orifice or pin hole. The effusion of a hypothetical mixture of two gases is suggested by Figure 6-18. The rates at which diffusion and effusion occur are directly proportional to molecular speeds. That is, molecules with high speeds diffuse and effuse faster than molecules with low speeds. Let us consider the effusion of two different gases at the same temperature and pressure. We can first compare effusion rates with root-mean-square speeds and then substitute expression (6.20) for these speeds.

$$\frac{\text{rate of effusion of A}}{\text{rate of effusion of B}} = \frac{(u_{rms})_A}{(u_{rms})_B} = \sqrt{\frac{3RT/\mathcal{M}_A}{3RT/\mathcal{M}_B}} = \sqrt{\frac{\mathcal{M}_B}{\mathcal{M}_A}} \qquad (6.22)$$

FIGURE 6-17
Diffusion of $NH_3(g)$ and $HCl(g)$.
$NH_3(g)$ escapes from the aqueous ammonia solution (labeled ammonium hydroxide in this photograph) and $HCl(g)$ from the hydrochloric acid solution. The gases diffuse toward each other through the air. Where the two gases meet, they react to produce a white "smoke" of solid ammonium chloride: $NH_3(g) + HCl(g) \longrightarrow NH_4Cl(s)$.

FIGURE 6-18
Effusion through an orifice.
Average speeds of the two types of molecules are suggested by the lengths of the arrows. The faster molecules (shown in red) effuse faster.

The result shown in Equation (6.22) is a kinetic-theory statement of a nineteenth-century law called **Graham's law.***

> The rates of effusion (or diffusion) of two different gases are inversely proportional to the square roots of their molar masses.

When compared at the same temperature, two different gases have the same value of $\overline{e_k} = \frac{1}{2}mu^2$. This means that molecules with a smaller mass (m) have a higher speed (u_{rms}). When we use Equation (6.22), we find that a lighter gas (for example, H_2) effuses *faster* than a heavier gas. The gas that effuses *fastest* takes the *shortest* time to do so. Also the gas that effuses *fastest* travels *farthest* in a given period of time. We need to use variations of Equation (6.22) to describe these alternative ways of looking at effusion (and diffusion). An effective method for doing this is to note that in every case a ratio of effusion rates, times, distances, . . . is equal to the *square root* of a ratio of molar masses. That is,

ratio of: (1) molecular speeds

(2) rates of effusion

(3) effusion times $= \sqrt{\text{ratio of two molar masses}}$ (6.23)

(4) distance traveled by molecules

(5) amount of gas effused

In using Equation (6.23), first reason *qualitatively* whether the ratio of properties should be *greater or less than one*. Then set up the ratio of molar masses accordingly. Examples 6-18 and 6-19 illustrate this line of reasoning.

EXAMPLE 6-18

Comparing Amounts of Gases Effusing Through an Orifice. 2.2×10^{-4} mol $N_2(g)$ effuses through a tiny hole in 105 s. How much $H_2(g)$ would effuse through the same orifice in 105 s?

Solution

First, let us reason *qualitatively.* H_2 molecules have less mass than N_2 molecules. They should have the greater speed when the gases are compared at the same temperature. Because $H_2(g)$ effuses faster, more H_2 than N_2 molecules should effuse in a given length of time. In the following set-up we need a ratio of molar masses *greater than 1.*

$$\frac{? \text{ mol } H_2}{2.2 \times 10^{-4} \text{ mol } N_2} = \sqrt{\frac{\mathcal{M}_{N_2}}{\mathcal{M}_{H_2}}} = \sqrt{\frac{28.014}{2.016}} = 3.728$$

$$? \text{ mol } H_2 = 3.728 \times 2.2 \times 10^{-4} = 8.2 \times 10^{-4} \text{ mol } H_2$$

---

*Graham's law applies only if certain conditions are met. For effusion, the gas pressure must be very low, so that molecules escape individually, not as a jet of gas. The orifice must be tiny, so that no collisions occur as molecules pass through. The situation for gaseous diffusion is even more restrictive because molecules of a diffusing gas undergo collisions with each other and with the gas into which they are diffusing. Some move in an opposite direction to the net flow. Nevertheless we still get useful qualitative answers by relating rates of effusion and diffusion to molar masses.

Practice Example A: In Example 6-18, how much $O_2(g)$ would effuse through the same orifice in 105 s?

Practice Example B: In Example 6-18, how long would it take for 2.2×10^{-4} mol H_2 to effuse through the same orifice as the 2.2×10^{-4} mol N_2? (*Hint:* Here you must work with a ratio of effusion times.)

EXAMPLE 6-19

Relating Effusion Times and Molar Masses. A sample of Kr(g) escapes through a tiny hole in 87.3 s, and an unknown gas requires 42.9 s under identical conditions. What is the molar mass of the unknown gas?

Solution

Again, start with *qualitative* reasoning. Because the unknown gas effuses faster, it must have a *smaller* molar mass than Kr. The ratio of molar masses in the following set-up must be *smaller than 1*. $\mathcal{M}_{unk}$ goes in the *numerator.*

$$\frac{\text{effusion time for unknown}}{\text{effusion time for Kr}} = \frac{42.9 \text{ s}}{87.3 \text{ s}} = \sqrt{\frac{\mathcal{M}_{unk}}{\mathcal{M}_{Kr}}} = 0.491$$

$$\mathcal{M}_{unk} = (0.491)^2 \times \mathcal{M}_{Kr} = (0.491)^2 \times 83.80 = 20.2 \text{ g/mol}$$

Practice Example A: Under the same conditions as in Example 6-19, another unknown gas requires 131.3 s to escape. What is the molar mass of this unknown gas?

Practice Example B: Given all the same conditions as in Example 6-19, how long would it take for a sample of ethane gas, C_2H_6, to effuse?

Applications of Diffusion

The diffusion of gases into one another has many practical applications. Natural gas and liquefied petroleum gas (LPG) are odorless, and for commercial use a small quantity of a gaseous organic sulfur compound, methyl mercaptan, CH_3SH, is added to them. The mercaptan has an odor that can be detected in parts per billion (ppb) or less. When a leak occurs, which can present a serious explosion hazard, we rely on the diffusion of this odorous compound for a warning.

During World War II the Manhattan Project (the secret government-run plan for developing the atomic bomb) used a method involving gaseous diffusion in order to separate the desired isotope ^{235}U from the predominant ^{238}U. The method is now also used to obtain ^{235}U-enriched nuclear fuels for nuclear power plants. One of the few compounds of uranium that can be obtained as a gas at moderate temperatures is uranium hexafluoride, UF_6. When high-pressure $UF_6(g)$ is forced through a porous barrier, molecules containing the isotope ^{235}U pass through the barrier slightly faster than those containing ^{238}U. As a result the $UF_6(g)$ contains a slightly higher ratio of ^{235}U to ^{238}U than it did previously. The gas has become "enriched" in ^{235}U. By continuing this slight enrichment through several thousand stages, a product with a high proportion of ^{235}U is finally obtained.

FIGURE 6-19

The behavior of real gases—compressibility factor as a function of pressure at 0 °C.
Values of the compressibility factor less than 1 signify that intermolecular forces of attraction are largely responsible for deviations from ideal gas behavior. Values greater than 1 are found when the volume of the gas molecules themselves is a significant fraction of the total gas volume.

▶Feature Problem D on page 201 describes how at times data for a real gas can be extrapolated to conditions for an ideal gas.

The mineral form in which uranium is most commonly found is as the solid oxide U_3O_8. Through a complex series of reactions, the U_3O_8 is purified and then converted first to a nitrate by treatment with nitric acid and then to the oxide $UO_2(s)$. $UO_2(s)$, when allowed to react with $HF(g)$, is converted first to $UF_4(s)$ and then to $UF_6(g)$. After enrichment by the process of gaseous diffusion, $UF_6(g)$ enriched in the ^{235}U isotope is reduced either to $UO_2(s)$ or to uranium metal. The natural abundance of ^{235}U in the original U_3O_8 is 0.72%; in enriched fuels for nuclear power reactors it is about 4%; and for nuclear weapons it is about 90%.

6-9 Nonideal (Real) Gases

In introducing the ideal gas equation we remarked that real gases obey the ideal gas law under suitable conditions. We should comment briefly on what those "suitable" conditions are, and on what to do when the conditions are not "suitable." A useful measure of how much a gas deviates from ideal gas behavior is found in its compressibility factor. The *compressibility factor* of a gas is the ratio PV/nRT. From the ideal gas equation ($PV = nRT$) we see that for an ideal gas $PV/nRT = 1$. For a real gas, the closeness of the experimentally determined ratio, PV/nRT, to 1 is a measure of how ideally the gas behaves. In Figure 6-19 the compressibility factor is plotted as a function of pressure for three different gases. Our principal conclusion from this figure is that all gases behave ideally at sufficiently low pressures, say, below 1 atm, but that deviations set in at increased pressures. At very high pressures the compressibility factor is always greater than 1.

Here is how we might explain nonideal gas behavior: Boyle's law predicts that at very high pressures a gas volume becomes extremely small and approaches zero. This cannot be, however, because the molecules themselves occupy space and are practically incompressible. The PV product is larger than predicted for an ideal gas and the compressibility factor is greater than 1. We must also allow for the fact that intermolecular forces exist in gases. Figure 6-20 suggests that because of attractive forces among the molecules, the force of the collisions of gas molecules with the container walls is less than expected for an ideal gas. Intermolecular forces of attraction account for compressibility factors less than 1. These forces become increasingly important at *low temperatures,* where translational molecular motion slows down. To summarize:

- Gases tend to behave *ideally* at *high temperatures and low pressures.*
- Gases tend to behave *nonideally* at *low temperatures and high pressures.*

The van der Waals Equation

A number of equations can be used for real gases, equations that apply over a wider range of temperatures and pressures than the ideal gas equation. Such equations are not as general as the ideal gas equation. They contain terms that have specific but different values for different gases. Such equations must correct for the volume associated with the molecules themselves and for intermolecular forces of attraction. In the **van der Waals equation**

$$\left(P + \frac{n^2a}{V^2}\right)(V - nb) = nRT \qquad (6.24)$$

V is the volume of *n* moles of gas. The term n^2a/V^2 is related to intermolecular forces of attraction. It is added to *P* because the measured pressure is lower than expected (recall Figure 6-20). The value *b*, called the *excluded* volume per mole, is related to the volume of the gas molecules, and *nb* is subtracted from the measured volume to represent the *free* volume within the gas: *V* − *nb*. Both *a* and *b* have specific values for particular gases, values that vary somewhat with temperature and pressure. In Example 6-20 we calculate the pressure of a real gas using the van der Waals equation. Solving the equation for either *n* or *V* is more difficult, however (see Exercise 97).

FIGURE 6-20
Intermolecular forces of attraction.
Attractive forces of the red molecules for the green molecule cause the green molecule to exert less force when it collides with the wall than if these attractions did not exist.

EXAMPLE 6-20

Using the van der Waals Equation to Calculate the Pressure of a Gas. Use the van der Waals equation to calculate the pressure exerted by 1.00 mol $Cl_2(g)$ when it is confined to a volume of 2.00 L at 273 K. The value of $a = 6.49$ L^2 atm mol^{-2}, and that of $b = 0.0562$ L mol^{-1}.

Solution

Solve equation (6.24) for *P*.

$$P = \frac{nRT}{V - nb} - \frac{n^2a}{V^2}$$

Then substitute the following values into the equation.

$n = 1.00$ mol; $V = 2.00$ L; $T = 273$ K; $R = 0.08206$ L atm mol^{-1} K^{-1}

$$n^2a = (1.00)^2 \text{ mol}^2 \times 6.49\frac{\text{L}^2 \text{ atm}}{\text{mol}^2} = 6.49 \text{ L}^2 \text{ atm}$$

$$nb = 1.00 \text{ mol} \times 0.0562 \text{ L mol}^{-1} = 0.0562 \text{ L}$$

$$P = \frac{1.00 \text{ mol} \times 0.08206 \text{ L atm mol}^{-1} \text{ K}^{-1} \times 273 \text{ K}}{(2.00 - 0.0562) \text{ L}} - \frac{6.49 \text{ L}^2 \text{ atm}}{(2.00)^2 \text{ L}^2}$$

$$P = 11.5 \text{ atm} - 1.62 \text{ atm} = 9.9 \text{ atm}$$

The pressure calculated with the ideal gas equation is 11.2 atm. By including only the *b* term in the van der Waals equation we get a value of 11.5 atm. Including the *a* term reduces the calculated pressure by 1.62 atm. Under the conditions of this problem, intermolecular forces of attraction are the main cause of the departure from ideal behavior.

▶ Although the deviation from ideality here is rather large, in problem-solving situations you can generally assume that the ideal gas equation will give satisfactory results.

Practice Example A: Substitute $CO_2(g)$ for $Cl_2(g)$ in Example 6-20. The values of *a* and *b* are $a = 3.59$ L^2 atm mol^{-2} and $b = 0.0427$ L mol^{-1}. Which gas, CO_2 or Cl_2, shows the greater departure from ideal gas behavior? (*Hint:* For which gas do you find the greater difference in calculated pressures, using first the ideal gas equation and then the van der Waals equation?)

Practice Example B: Substitute $CO(g)$ for $Cl_2(g)$ in Example 6-20. The values of *a* and *b* are $a = 1.49$ L^2 atm mol^{-2} and $b = 0.0399$ L mol^{-1}. Including CO_2 from Practice Example 6-20A, which of the three gases—Cl_2, CO_2, or CO—shows the greatest departure from ideal gas behavior?

Focus On

The Chemistry of Air-Bag Systems

Air bags inflating during a crash test.

Air bags in automobiles have saved thousands of lives. The idea behind them is simple: When a crash occurs, a plastic bag rapidly inflates with a gas, protecting the driver from hitting the dashboard or steering column. The development of a workable air-bag system required the combined efforts of chemists and engineers.

The air-bag system has many special requirements. The air bag must not inflate accidentally. The gas used must be nontoxic in case of leakage after inflation. The gas must be "cool," so as not to produce burn injuries. The gas must be produced very rapidly, ideally inflating the bag within 20–60 ms. Finally, the gas-producing chemicals must be easy to handle and stable for long periods.

Nitrogen is the best choice for a nontoxic gas. After all, nitrogen

Summary

A gas is described through its pressure, temperature, volume, and amount. Gas pressure is most readily measured by comparing it with the pressure exerted by a liquid column, usually mercury. Atmospheric pressure is measured with a mercury barometer, and other gas pressures with a manometer. Pressure can be expressed in a variety of units.

The most common simple gas laws are Boyle's law, relating gas pressure and volume; Charles's law, relating gas volume and temperature; and Avogadro's law, relating volume and amount of gas. Some important ideas that originate from the simple gas laws are the Kelvin temperature scale, the standard condition of temperature and pressure (STP), and the molar volume of a gas at STP—22.4 L/mol.

The simple gas laws can be combined into the ideal gas equation: $PV = nRT$. This equation can be solved for any one of the variables when the others are known. It can also be applied in determining molar masses and gas densities. Still other uses of the ideal gas equation are in describing (a) the gaseous reactants or products of a chemical reaction and (b) mixtures of gases. Collecting gases over water is a common procedure involving a mixture of gases—the particular gas being isolated and water vapor.

The kinetic-molecular theory establishes one relationship involving the root-mean-square speed of molecules, temperature, and molar mass of a gas and another between average molecular translational kinetic energy and Kelvin temperature. Also, the diffusion and effusion of gases can be related to their molar masses through the kinetic-molecular theory.

Real gases generally behave ideally only at high temperatures and low pressures. Other equations of state, such as the van der Waals equation, often work where the ideal gas equation fails.

Summarizing Example

Some fleet vehicles have been modified to burn natural gas, which is mostly methane, $CH_4(g)$. The combustion can be controlled to produce CO_2 and H_2O, with a minimum of pollutants (CO and oxides of nitrogen). The ideal air–fuel ratio uses methane and oxygen (from the air) in *stoichiometric* proportions (meaning that neither is in excess).

makes up about 78% of air by volume. A good source of nitrogen is the decomposition of alkali metal azides, such as sodium azide, NaN_3.

$$2\ NaN_3(s) \xrightarrow{\Delta} 2\ Na(l) + 3\ N_2(g)$$

The air-bag system is activated by sensors that detect the initial crash and electrically initiate the explosion of a small charge. This in turn starts the rapid burning of a pellet containing sodium azide, which releases a large volume of $N_2(g)$ to fill the bag (recall Example 6-12).

By 1980, the engineering challenges of the air-bag safety system had been resolved, but not the chemical challenges. The problems posed by the use of sodium azide were that it does not make a good pellet, its reaction does not go to completion rapidly, and one of the reaction products—sodium metal—reacts violently with water.

To solve these problems, investigators tried adding other compounds to the sodium azide. To make a good pellet-forming mixture, a lubricant, typically molybdenum disulfide (MoS_2), was added to the sodium azide. This mixture did not burn well, however. Sulfur, a familiar constituent of gunpowder, was then added to produce smooth-burning pellets. The nitrogen gas produced was "cool," and the sodium metal was converted mainly to the sulfate, though the solid residue was finely powdered and difficult to contain. Some air-bag systems on the market today use the MoS_2–S–NaN_3 pellet, but the latest ones use a pellet that is still more complex.

In earlier experimental work, a mixture of sodium azide and iron(III) oxide had proved satisfactory for trapping the sodium metal

by converting it to an easy-to-handle solid residue. This mixture did not burn well, however. Researchers then tried the obvious solution: Mix together all the compounds that give the gas-forming pellets their most desirable properties, sodium azide, iron(III) oxide, molybdenum disulfide, and sulfur. In scientific research, the "obvious" often turns up some unexpected results. In this case, however, the hoped-for final result was achieved: a rapidly burning pellet producing cool, odor-free nitrogen gas and a nonreactive solid residue that is easily trapped. At this point the widespread use of air-bag collision systems in automobiles became a reality.

What volume of air, measured at 22 °C and 745 mmHg, is required for the complete combustion of 1.00 L of compressed $CH_4(g)$ at 22 °C and 3.55 atm? [Air contains 20.95% $O_2(g)$, by volume.]

1. *Write a chemical equation for the complete combustion of methane. Result:*

$$CH_4(g) + 2\ O_2(g) \longrightarrow CO_2(g) + 2\ H_2O(g)$$

2. *Determine the number of moles of $O_2(g)$ consumed in the combustion.* Solve the ideal gas equation for n_{CH_4}. Then, use the stoichiometric factor 2 mol O_2/1 mol CH_4.

Result: 0.293 mol $O_2(g)$.

3. *Determine the volume of $O_2(g)$ required.* Substitute n_{O_2} from part **2** into the ideal gas equation.

Result: 7.24 L $O_2(g)$.

4. *Determine the volume of air required.* The volume percent $O_2(g)$ in air leads to the conversion factor 100 L air/20.95 L $O_2(g)$.

Answer: 34.6 L air.

Key Terms

Avogadro's law (hypothesis) (6-2)	general gas equation (6-3)	mole fraction (6-6)
barometer (6-1)	Graham's law (6-8)	partial pressure (6-6)
barometric pressure (6-1)	ideal (perfect) gas (6-3)	pressure (6-1)
Boyle's law (6-2)	ideal gas constant, R (6-3)	root-mean-square speed (6-7)
Charles's law (6-2)	ideal gas equation (6-3)	standard atmosphere (atm) (6-1)
Dalton's law of partial pressures (6-6)	Kelvin (absolute) temperature (6-2)	standard conditions of temperature and pressure (STP) (6-2)
diffusion (6-8)	kinetic-molecular theory of gases (6-7)	van der Waals equation (6-9)
effusion (6-8)	manometer (6-1)	

Review Questions

1. In your own words define or explain the following terms or symbols: **(a)** atm; **(b)** STP; **(c)** R; **(d)** partial pressure; **(e)** u_{rms}.

2. Briefly describe each of the following ideas, phenomena, or methods: **(a)** absolute zero of temperature; **(b)** collection of a gas over water; **(c)** effusion of a gas; **(d)** law of combining volumes.

3. Explain the important distinctions between each pair of terms: **(a)** barometer and manometer; **(b)** Celsius and Kelvin temperature; **(c)** ideal gas equation and general gas equation; **(d)** ideal gas and real gas.

4. Convert each pressure to an equivalent pressure in standard atmospheres: **(a)** 736 mmHg; **(b)** 58.2 cm Hg; **(c)** 892 torr; **(d)** 48 psi.

5. Calculate the height of a mercury column required to produce a pressure **(a)** of 0.984 atm; **(b)** of 928 torr; **(c)** equal to that of a column of water 142 ft high.

6. What is P_{gas} for the manometer readings (millimeters of mercury) in the figure when barometric pressure is 744 mmHg?

7. A sample of $O_2(g)$ has a volume of 26.7 L at 762 mmHg. What is the new volume if, with the temperature and amount of gas held constant, the pressure is **(a)** lowered to 385 mmHg; **(b)** increased to 3.68 atm?

8. A 886-mL sample of $Ne(g)$ is at 752 mmHg and 26 °C. What will be the new volume if, with the pressure and amount of gas held constant, the temperature is **(a)** increased to 98 °C; **(b)** lowered to −20 °C?

9. We want to increase the volume of a fixed amount of gas from 57.3 to 165 mL while holding the pressure constant. To what temperature must we heat this gas if the initial temperature is 22 °C?

10. What is the volume at STP of a 49.6-g sample of acetylene gas, C_2H_2 (burned with oxygen to generate high temperatures for welding)?

11. What volume of $Cl_2(g)$ would you measure at STP if you needed a 250.0-g sample of $Cl_2(g)$?

12. *Without doing detailed calculations,* determine which of the following gases has the greatest density at STP: Cl_2, SO_3, N_2O, PF_3. Explain.

13. What is the volume, in milliliters, occupied by 89.2 g $CO_2(g)$ at 37 °C and 737 mmHg?

14. A 40.0-L cylinder contains 285 g $SO_2(g)$ at 27 °C. What is the pressure, in atmospheres, exerted by this gas?

15. A 0.418-g sample of gas has a volume of 115 mL at 66.3 °C and 743 mmHg. What is the molar mass of this gas?

16. What is the density, in grams per liter, of $CO_2(g)$ at 32.7 °C and 758 mmHg?

17. How many liters of $H_2(g)$ at STP are produced per gram of $Al(s)$ consumed in the following reaction?

$$2\ Al(s) + 6\ HCl(aq) \longrightarrow 2\ AlCl_3(aq) + 3\ H_2(g)$$

18. A method of removing $CO_2(g)$ from a spacecraft is to allow the CO_2 to react with LiOH. How many liters of $CO_2(g)$ at 25.9 °C and 751 mmHg can be removed per kilogram of LiOH consumed?

$$2\ LiOH(s) + CO_2(g) \longrightarrow Li_2CO_3(s) + H_2O(l)$$

19. What is the volume, in liters, occupied by a mixture of 15.2 g $Ne(g)$ and 34.8 g $Ar(g)$ at 7.15 atm pressure and 26.7 °C?

20. A balloon filled with $H_2(g)$ at STP has a volume of 2.24 L. To the balloon is added 0.10 mol $He(g)$. Then the temperature is raised to 100 °C while the pressure and amount of gas are held constant. What is the final gas volume?

21. An 89.3-mL sample of "wet" $O_2(g)$ is collected over water at 21.3 °C at a barometric pressure of 756 mmHg (vapor pressure of water at 21.3 °C = 19 mmHg). **(a)** What is the *partial* pressure of $O_2(g)$, in millimeters of mercury, in the sample collected? **(b)** What is the volume percent O_2 in the gas collected? **(c)** How many grams of O_2 are present in the sample?

22. Which of the following give(s) a true statement when comparing 0.50 mol $H_2(g)$ and 1.0 mol $He(g)$ at STP? The two gases have equal (1) average molecular kinetic energies; (2) molecular speeds; (3) volumes; (4) effusion rates.

23. A sample of $Cl_2(g)$ effuses through a tiny hole in 28.6 s. How long would it take for a sample of $NO(g)$ to effuse under the same conditions?

24. Under which of these conditions is Cl_2 most likely to behave like an ideal gas? Explain. (1) 100 °C and 10.0 atm; (2) 0 °C and 0.50 atm; (3) 200 °C and 0.50 atm; (4) −100 °C and 10.0 atm.

Exercises

Pressure and Its Measurement

25. Calculate the height of a column of liquid benzene (d = 0.879 g/cm³), in meters, required to exert a pressure of 0.970 atm.

26. Calculate the height of a column of liquid glycerol (d = 1.26 g/cm³), in meters, required to exert the same pressure as 3.02 m of $CCl_4(l)$ (d = 1.59 g/cm³).

27. The mercury level in the open arm of an open-end manometer is 276 mm above a reference point. In the arm connected to a container of gas, the level is 49 mm above the same reference point. If barometric pressure is 749 mmHg, what is the pressure of the gas in the container? (*Hint:* Refer to the drawing on page 196.)

28. A gas is collected over water when the barometric pressure is 756.2 mmHg; but the water level inside the container of gas is 4.5 cm higher than outside. What is the total pressure of the gas inside the container, in millimeters of mercury? (*Hint:* Refer to the drawing on page 184.)

The Simple Gas Laws

29. A 35.8-L cylinder of Ar(g) is connected to an evacuated 1875-L tank. If the temperature is held constant and the final pressure is 721 mmHg, what must have been the original gas pressure in the cylinder, in atmospheres?

30. A sample of $N_2(g)$ occupies a volume of 42.0 mL under the existing barometric pressure. Increasing the pressure by 85 mmHg reduces the volume to 37.7 mL. What is the prevailing barometric pressure, in millimeters of mercury?

31. A fixed amount of gas, maintained in a constant volume of 275 mL, exerts a pressure of 798 mmHg at 23.4 °C. At what temperature, in degrees Celsius, will the pressure of the gas become exactly 1.00 atm?

32. A 12.5-g sample of gas is added to an evacuated, constant-volume container at 21 °C. The *pressure of the gas is to remain constant* as the temperature is raised to 210 °C. How many grams of gas must be released?

33. A 27.6-mL sample of $PH_3(g)$ (used in the manufacture of flame-retardant chemicals) is obtained at STP.
(a) What is the mass of this gas, in milligrams?
(b) How many molecules of PH_3 are present?

34. A 5.0×10^{17}-atom sample of radon gas is obtained.
(a) What is the mass of this sample, in micrograms?
(b) What is the volume of this sample at STP, in microliters?

35. You purchase a bag of potato chips at an ocean beach to take on a picnic in the mountains. At the picnic you notice that the bag has become inflated, almost to the point of bursting. Use your knowledge of gas behavior to explain this phenomenon.

36. Scuba divers know that they must not ascend quickly from deep underwater because of a condition known as the "bends," discussed in Chapter 14. Another concern is that they must constantly exhale during their ascent to prevent damage to the lungs and blood vessels. Describe what would happen to the lungs of a diver who inhaled compressed air at a depth of 30 meters and held her breath while rising to the surface.

Ideal Gas Equation

37. A 12.8-L cylinder contains 35.8 g O_2 at 46 °C. What is the pressure of this gas, in atmospheres?

38. $Kr(g)$ in a 18.5-L cylinder exerts a pressure of 11.2 atm at 28.2 °C. How many grams of gas are present?

39. A 72.8-L constant-volume cylinder containing 1.85 mol He is heated until the pressure reaches 3.50 atm. What is the final temperature, in degrees Celsius?

40. A sample of gas has a volume of 4.25 L at 25.6 °C and 748 mmHg. What will be the volume of this gas at 26.8 °C and 742 mmHg?

41. A 34.0-L cylinder contains 305 g $O_2(g)$ at 22 °C. How many grams of $O_2(g)$ must be released to reduce the pressure in the cylinder to 1.15 atm?

42. A 10.0-g sample of a gas has a volume of 5.25 L at 25 °C and 762 mmHg. If to this *constant* 5.25-L volume is added 2.5 g of the same gas and the temperature raised to 62 °C, what is the new gas pressure?

Determining Molar Mass

43. Refer to Example 6-10. By combustion analysis, the mass percent composition of propylene is found to be 85.63% C and 14.37% H. What is the *molecular* formula of propylene?

44. A 2.650-g sample of a gaseous compound occupies 428 mL at 24.3 °C and 742 mmHg. The compound consists of 15.5% C, 23.0% Cl, and 61.5% F, by mass. What is its molecular formula?

45. A gaseous hydrocarbon weighing 0.231 g occupies a volume of 102 mL at 23 °C and 749 mmHg. What is the molar mass of this compound? What conclusion can you draw about its molecular formula? (*Hint:* Each C atom forms four bonds to other atoms, and each H atom is bonded to a C atom.)

46. A 132.10-mL glass vessel weighs 56.1035 g when evacuated and 56.2445 g when filled with the gaseous hydrocarbon acetylene at 749.3 mmHg and 20.02 °C. What is the molar mass of acetylene? What conclusion can you draw about its molecular formula?

Gas Densities

47. A particular application calls for $N_2(g)$ with a density of 1.80 g/L at 32 °C. What must be the pressure of the $N_2(g)$ in millimeters of mercury?

48. Monochloroethylene is used to make polyvinylchloride (PVC). It has a density of 2.56 g/L at 22.8 °C and 756 mmHg. What is the molar mass of monochloroethylene?

49. In order for a gas-filled balloon to rise in air, the density of the gas in the balloon must be less than that of air.
(a) Consider air to have a molar mass of 28.96 g/mol air and determine the density of air at 25 °C and 1 atm, in g/L.
(b) Show, by calculation, that a balloon filled with carbon dioxide at 25 °C and 1 atm would not be expected to rise in air at 25 °C.

51. The density of phosphorus vapor at 310 °C and 775 mmHg is 2.64 g/L. What is the molecular formula of the phosphorus under these conditions?

Gases in Chemical Reactions

53. What volume of $O_2(g)$ is consumed in the combustion of 75.6 L $C_3H_8(g)$ if both gases are measured at STP? (*Hint:* What is the combustion reaction?)

55. A particular coal sample contains 3.28% S, by mass. When the coal is burned, the sulfur is converted to $SO_2(g)$. What volume of $SO_2(g)$, measured at 23 °C and 738 mmHg, is produced by burning 2.7×10^6 lb of this coal?

57. Hydrogen peroxide, H_2O_2, is used to disinfect contact lenses.

$$2\ H_2O_2(aq) \longrightarrow 2\ H_2O(l) + O_2(g)$$

Calculate the volume, in milliliters, of $O_2(g)$ at 22 °C and 752 mmHg that could be liberated from 10.0 mL of a water solution containing 3.00% H_2O_2, by mass. The density of the aqueous solution of H_2O_2 is 1.01 g/mL.

Mixtures of Gases

59. A gas cylinder of 53.7 L volume contains $N_2(g)$ at a pressure of 28.2 atm and 26 °C. How many grams of Ne(g) must we add to this same cylinder to raise the total pressure to 75.0 atm?

61. A mixture of 4.0 g $H_2(g)$ and 10.0 g He(g) in a 4.3-L flask is maintained at 0 °C.
(a) What is the total pressure in the container?
(b) What is the partial pressure of each gas?

63. Air that is exhaled (expired) differs from normal air. A typical analysis of expired air at 37 °C and 1.00 atm, expressed as percent *by volume*, is 74.2% N_2, 15.2% O_2, 3.8% CO_2, 5.9% H_2O, and 0.9% Ar. The composition of ordinary air is given in Practice Example 6-15B.
(a) Is the density of expired air greater or less than that of ordinary air at the same temperature and pressure? Explain.
(b) What is the ratio of the partial pressure of $CO_2(g)$ in expired air to that in ordinary air?

Collecting Gases over Liquids

65. A 1.65-g sample of Al reacts with excess HCl and the liberated H_2 is collected over water at 25 °C at a barometric pressure of 744 mmHg. What *total* volume of gas, in liters, is collected? (Vapor pressure of water at 25 °C = 23.8 mmHg.)

$$2\ Al(s) + 6\ HCl(aq) \longrightarrow 2\ AlCl_3(aq) + 3\ H_2(g)$$

50. One way to reduce the density of a gas is to heat it. Refer to Exercise 49, and determine the minimum temperature to which $CO_2(g)$ would have to be heated before a balloon filled with the gas at 1 atm and 25 °C would rise in air? (Ignore the mass of the balloon itself.)

52. A particular gaseous hydrocarbon that is 82.7% C and 17.3% H, by mass, has a density of 2.33 g/L at 23 °C and 746 mmHg. What is the molecular formula of this hydrocarbon?

54. Calculate the volume of $H_2(g)$, measured at 26 °C and 751 mmHg, required to react with 28.5 L CO(g), measured at 0 °C and 760 mmHg, in the following reaction.

$$3\ CO(g) + 7\ H_2(g) \longrightarrow C_3H_8(g) + 3\ H_2O(l)$$

56. A 3.57-g sample of a KCl–KClO₃ mixture is decomposed by heating and produces 119 mL $O_2(g)$, measured at 22.4 °C and 738 mmHg? What is the mass percent of $KClO_3$ in the mixture? (*Hint:* The KCl is unchanged.)

$$2\ KClO_3(s) \longrightarrow 2\ KCl(s) + 3\ O_2(g)$$

58. The Haber process is the principal method for fixing nitrogen (converting N_2 to nitrogen compounds).

$$N_2(g) + 3\ H_2(g) \longrightarrow 2\ NH_3(g)$$

Assume complete conversion of the reactant gases to $NH_3(g)$ and that the gases behave ideally.
(a) What volume of $NH_3(g)$ can be produced from 313 L of $H_2(g)$ if the gases are measured at 515 °C and 525 atm?
(b) What volume of $NH_3(g)$, measured at 25 °C and 727 mmHg, can be produced from 313 L $H_2(g)$, measured at 515 °C and 525 atm?

60. A 2.35-L container of $H_2(g)$ at 762 mmHg and 24 °C is connected to a 3.17-L container of He(g) at 728 mmHg and 24 °C. After mixing, what is the *total* gas pressure, in millimeters of mercury, with the temperature remaining at 24 °C?

62. A 2.00-L container is filled with Ar(g) at 752 mmHg and 35 °C. A 0.728-g sample of C_6H_6 vapor is then added.
(a) What is the total pressure in the container?
(b) What is the partial pressure of Ar and of C_6H_6?

64. Producer gas is a type of fuel gas made by passing air or steam through a bed of hot coal or coke. A typical producer gas has the following composition in percent by volume: 8.0% CO_2, 23.2% CO, 17.7% H_2, 1.1% CH_4, and 50.0% N_2.
(a) What is the density of this gas at 23 °C and 763 mmHg, in grams per liter?
(b) What is the partial pressure of CO in this mixture at STP?

66. A 367-mL sample of Ar(g) at 25 °C and at a barometric pressure of 748 mmHg is passed through water at 25 °C. What is the volume of gas when saturated with water vapor and again measured at 25 °C and 748 mmHg barometric pressure? (Vapor pressure of water at 25 °C = 23.8 mmHg.)

67. A sample of $O_2(g)$ is collected over water at 24 °C. The volume of gas is 1.16 L. In a subsequent experiment it is determined that the mass of O_2 present is 1.46 g. What must have been barometric pressure at the time the gas was collected? (Vapor pressure of water at 24 °C = 22.4 mmHg.)

68. A 1.072-g sample of He(g) is found to occupy a volume of 8.446 L when collected over hexane at 25.0 °C and 738.6 mmHg barometric pressure. Use these data to determine the vapor pressure of hexane at 25 °C.

Kinetic-Molecular Theory

69. Calculate u_{rms}, in meters per second, for $Cl_2(g)$ molecules at 30 °C.

70. The root-mean-square speed, u_{rms}, of H_2 molecules at 273 K is 1.84×10^3 m/s. At what temperature is u_{rms} for H_2 twice this value?

71. Refer to Example 6-17. What must be the molecular mass of a gas if its molecules are to have a root-mean-square speed at 25 °C equal to that of the M-16 rifle bullet?

72. Refer to Example 6-17. Noble gases (Group 8A) are monatomic gases (existing as atoms, not molecules). Cite one noble gas whose u_{rms} at 25 °C is greater than the rifle bullet's and one whose u_{rms} is smaller.

73. At what temperature will u_{rms} for Ne(g) be the same as u_{rms} for He at 300 K?

74. Determine u_m, $\bar{u}$, and u_{rms} for a group of ten automobiles clocked by radar at speeds of 38, 44, 45, 48, 50, 55, 55, 57, 58, and 60 mi/h, respectively.

Diffusion and Effusion of Gases

75. If 0.00484 mol $N_2O(g)$ effuses through an orifice in a certain period of time, how much $NO_2(g)$ would effuse in the same time under the same conditions?

76. A sample of $N_2(g)$ effuses through a tiny hole in 38 s. What must be the molar mass of a gas that requires 64 s to effuse under identical conditions?

77. What are the ratios of the diffusion rates for the following pairs of gases? **(a)** N_2 and O_2; **(b)** H_2O and D_2O (D = deuterium, i.e., $_1^2H$); **(c)** $^{14}CO_2$ and $^{12}CO_2$; **(d)** $^{235}UF_6$ and $^{238}UF_6$.

78. A common laboratory demonstration of Graham's law of diffusion is pictured here (also, recall Figure 6-17). The diffusing species are $NH_3(g)$ and HCl(g). Where these gases meet, they react to form a white cloud of $NH_4Cl(s)$. If initially the $NH_3(aq)$ and HCl(aq) are 1 m apart, at what point between them would you expect $NH_4Cl(s)$ to form?

Nonideal Gases

79. Use the van der Waals equation to calculate the pressure exerted by 1.50 mol of $SO_2(g)$ when it is confined to a volume of 5.00 L at 298 K. The value of $a = 6.71$ L^2 atm mol^{-2}, and that of $b = 0.0564$ L mol^{-1}. Compare this pressure to what you would calculate assuming sulfur dioxide was an ideal gas.

80. Refer to Example 6-20. Recalculate the pressure of $Cl_2(g)$, using both the ideal gas equation and van der Waals equation for the data given, at the following temperatures: **(a)** 100 °C; **(b)** 200 °C; **(c)** 400 °C. From the results, confirm the statement that a gas tends to be more ideal at high temperatures than at low temperatures.

Advanced Exercises

81. Explain why it is necessary to include the density of Hg(l) and the value of the acceleration due to gravity, g, in a precise definition of a standard atmosphere of pressure (page 165).

82. Start with the conditions at points A, B, and C in Figure 6-8. Use Charles's law to calculate the volume of each gas at 0, −100, −200, −250, and −270 °C; and show that indeed the volume of each gas becomes zero at −273.15 °C.

83. Complete combustion of 1.110 g of a gaseous hydrocarbon yields 3.613 g CO_2 and 1.109 g H_2O. A 0.288-g sample

of the hydrocarbon occupies a volume of 131 mL at 24.8 °C and 753 mmHg. What is the molecular formula of the hydrocarbon?

84. A balloon is inflated with 2.15 ft^3 of He(g) at STP and released. What is the gas pressure in the balloon when it has expanded to a volume of 155 L? Assume a temperature of −20 °C at this altitude.

85. In the drawings, the diver toy floats at the top of the liquid (left), but when the bottle is squeezed, the diver descends (right). Explain the phenomena illustrated by these drawings.

86. In the photographs below, we see two balloons connected by a stopcock. In the top, one balloon is inflated to a greater volume than the other. In the bottom, we see what happens when the stopcock is opened. Explain this observation, which seems contrary to what we would expect intuitively.

87. A 3.05-g sample of $NH_4NO_3(s)$ is introduced into an evacuated 2.18-L flask and then heated to 250 °C. What is the *total* gas pressure, in atmospheres, in the flask at 250 °C when the NH_4NO_3 has completely decomposed?

$$NH_4NO_3(s) \longrightarrow N_2O(g) + 2\,H_2O(g)$$

88. Show that the statements in the text of **(a)** Boyle's law, **(b)** Charles's law, and **(c)** Avogadro's hypothesis are consistent with Equations (6.19) and (6.21).

89. In the reaction of $CO_2(g)$ and sodium peroxide (Na_2O_2), sodium carbonate and oxygen gas are formed.

$$2\,Na_2O_2(s) + 2\,CO_2(g) \longrightarrow 2\,Na_2CO_3(s) + O_2(g)$$

This reaction is used in submarines and space vehicles to remove expired $CO_2(g)$ and to generate some of the $O_2(g)$ required for breathing. Assume the following. Volume of gases exchanged in the lungs: 4.0 L/min; CO_2 content of expired air: 3.8% CO_2, by volume. If the $CO_2(g)$ and $O_2(g)$ in the above reaction are measured at the same temperature and pressure, **(a)** how many milliliters of $O_2(g)$ are produced per minute; **(b)** at what rate is the $Na_2O_2(s)$ consumed, in grams per hour, if you assume that the gases are at 25 °C and 735 mmHg pressure?

90. What is the partial pressure of $Cl_2(g)$, in millimeters of mercury, in a gaseous mixture at STP that consists of 46.5% N_2, 12.7% Ne, and 40.8% Cl_2, *by mass?*

91. A gaseous mixture of He and O_2 has a density of 0.518 g/L at 25 °C and 721 mmHg. What is the % He, by mass, in the mixture?

92. When working with a mixture of gases it is sometimes convenient to use an *apparent molar mass* (a weighted-average molar mass). Think in terms of replacing the mixture by a *hypothetical* single gas. What is the apparent molar mass of air, given that air is 78.08% N_2, 20.95% O_2, 0.93% Ar, and 0.036% CO_2, by volume? (*Hint:* This situation is similar to establishing the atomic mass of an element as a weighted average of its isotopic masses.)

93. Gas cylinder A has a volume of 48.2 L and contains $N_2(g)$ at 8.35 atm at 25 °C. Gas cylinder B, of unknown volume, contains He(g) at 9.50 atm at 25 °C. When the two cylinders are connected and the gases mixed, the pressure in each cylinder becomes 8.71 atm. What is the volume of cylinder B?

94. A *mixture* of $H_2(g)$ and $O_2(g)$ is prepared by electrolyzing 1.32 g water, and the mixture of gases is collected over water at 30 °C when the barometric pressure is 748 mmHg. The volume of "wet" gas obtained is 2.90 L. What must be the vapor pressure of water at 30 °C?

$$2\,H_2O(l) \xrightarrow{\text{electrolysis}} 2\,H_2(g) + O_2(g)$$

95. A 0.168-L sample of $O_2(g)$ is collected over water at 26 °C and a barometric pressure of 737 mmHg. In the gas that is collected, what is the percent water vapor **(a)** by volume; **(b)** by number of molecules; **(c)** by mass? (Vapor pressure of water at 26 °C = 25.2 mmHg.)

96. **(a)** Use SI units and the ideal gas equation to express the gas constant R in the units kPa dm^3 mol^{-1} K^{-1}.
(b) Use the value found in **(a)**, together with information from Appendix B, to obtain R in the units J mol^{-1} K^{-1}.
(c) Calculate the pressure, in kilopascals, exerted by 1242 g CO(g) confined to a tank of 1.55-m^3 volume at 285 K.

97. If the van der Waals equation is solved for volume, a cubic equation is obtained.
(a) Derive the following equation by rearranging Equation (6.24).

$$V^3 - n\left(\frac{RT + bP}{P}\right)V^2 + \left(\frac{n^2 a}{P}\right)V - \frac{n^3 ab}{P} = 0$$

(b) What is the volume, in liters, occupied by 185 g $CO_2(g)$ at a pressure of 12.5 atm and 286 K? For $CO_2(g)$, $a = 3.59$ L^2 atm mol^{-2} and $b = 0.0427$ L mol^{-1}.

Feature Problems

A. Shown here is a diagram of Boyle's original apparatus. At the start of the experiment, the length of the air column (A) on the left was 30.5 cm and the heights of mercury in each arm of the tube were equal. When mercury was added to the right arm of the tube, a difference in mercury levels (B) was produced and the entrapped air on the left was compressed to a shorter length (smaller volume). For example, in the illustration A = 27.9 cm and B = 7.1 cm. Boyle's values of A and B, in centimeters, are listed as follows:

A:	30.5	27.9	25.4	22.9	20.3
B:	0.0	7.1	15.7	25.7	38.3
A:	17.8	15.2	12.7	10.2	7.6
B:	53.8	75.4	105.6	147.6	224.6

Barometric pressure at the time of the experiment was 739.8 mmHg. Assuming that the length of the air column (A) is proportional to the volume of air, show that these data do in fact conform reasonably well to Boyle's law.

B. An alternative to Figure 6-6 is to plot P against $1/V$. The resulting graph is a straight line passing through the origin. Use Boyle's data from Feature Problem A to draw such a straight-line graph. What factors would affect the *slope* of this straight line? Explain.

C. The "initial" sketch illustrates, both at the macroscopic and molecular levels, an initial condition: 1 mole of a gas at STP. The "final" sketch suggests a final condition. Supply the sketch with as much detail as possible to represent the final condition after each of the following changes.

1. The pressure is changed to 250 mmHg while standard temperature is maintained.
2. The temperature is changed to 140 K while standard pressure is maintained.
3. The pressure is changed to 0.5 atm while the temperature is changed to 550 K.
4. An additional 0.5 mol of gas is introduced into the cylinder, the temperature is changed to 135 °C, and the pressure is changed to 2.25 atm.

Initial   Final

D. The equation $d/P = \mathcal{M}/RT$, which can be derived from Equation (6.14), suggests that the ratio of the density (d) to pressure (P) of a gas at constant temperature should be a constant. The following gas density data were obtained for $O_2(g)$ at various pressures at 273.15 K.

P, mmHg:	760.00	570.00	380.00	190.00
d, g/L:	1.428962	1.071485	0.714154	0.356985

1. Calculate values of d/P, and with a graph or by other means determine the best value of the term d/P for $O_2(g)$ at 273.15 K. [*Hint:* The best value is that associated with a perfect (ideal) gas.]

2. Use the value of d/P from part **1** to calculate a precise value of the atomic weight of oxygen and compare this value with that listed on the inside front cover.

Thermochemistry

7

Natural gas consists mostly of methane, CH_4. As we learned in Chapter 4, the complete combustion of a hydrocarbon such as methane yields carbon dioxide and water as products. More important, however, is another "product" of this reaction that we have not previously mentioned: heat. We can use this heat to produce hot water in a water heater, to heat a house, or to cook food.

Thermochemistry is the branch of chemistry concerned with heat effects accompanying chemical reactions, and much of this chapter deals with determining quantities of heat—by measurement and by calculation. Some of these calculations will allow us to establish, indirectly, a quantity of heat that would be difficult or impossible to measure directly. This type of calculation, which relies on compilations of tabulated data, will come up again in later chapters. Finally, with concepts introduced in this chapter we can answer a host of practical questions, such as why natural gas is a better fuel than coal and why the energy value of fats is greater than that of carbohydrates.

A stream of boiling water transfers heat to the ice cube, causing it to melt. The transfer of heat between substances is an important aspect of thermochemistry.

7-1   Getting Started: Some Terminology

In this section we will introduce and briefly define some basic terms; most of these terms are described in more detail in later sections. Your understanding of them should increase as you proceed through the chapter.

The part of the universe we choose to study is called a **system.** The parts of the universe with which the system interacts are called the **surroundings.** Figure 7-1 identifies three types of systems: *open, closed,* and *isolated.* Interactions refer to the transfer of energy or matter between a system and its surroundings, and it is these interactions that we generally focus on. Energy transfers can occur as heat (q) or in several other forms, known collectively as work (w). Energy transfers occurring as heat or work affect the total amount of energy contained within a system, its internal energy (E). The components of internal energy of special interest to us are **thermal energy**—energy associated with random molecular motion—and **chemical energy**—energy associated with chemical bonds and intermolecular forces.

In the remainder of this section we will say more, in a general way, about energy. This is followed by sections on work and heat. We will consider internal energy again in Section 7-5.

Energy.   Like many other scientific terms, *energy* is derived from Greek. It means "work within." **Energy** is the capacity to do work. **Work** is done when a force acts through a distance. Moving objects do work when they slow down or are stopped. Thus, when one billiard ball strikes another and sets it in motion, work is done. The energy of a moving object is called **kinetic energy,** the word *kinetic* meaning "motion" in Greek. The kinetic energy (K.E.) of an object is related to its mass (m) and velocity (u) through the equation

▶ A derivation of this equation is given in Appendix B.

$$\text{K.E.} = \tfrac{1}{2}mu^2 \tag{7.1}$$

The units in Equation (7.1) are kg m^2 s^{-2} = joule (J).

The bouncing ball in Figure 7-2 suggests something about the nature of energy and work. First, to lift the ball to the starting position we have to apply a force through a distance (to overcome the force of gravity). The work we do is "stored" in the ball as energy. This stored energy has the potential to do work

FIGURE 7-1

Systems and their surroundings.

(a) **Open system.** Both energy and matter can be transferred between a system and its surroundings. The beaker of hot coffee loses heat to the surroundings; it cools. Matter is also transferred in the form of water vapor.

(b) **Closed system.** Energy can be transferred between the system and its surroundings, but not matter. The stoppered glass flask of hot coffee loses heat to its surroundings; it cools. No water vapor escapes.

(c) **Isolated system.** Neither energy nor matter is exchanged between the system and its surroundings. Hot coffee in a vacuum flask approximates an isolated system. Energy is slowly transferred, however. In time the coffee in the flask cools to room temperature.

(a)   (b)   (c)

FIGURE 7-2
Potential energy (P.E.) and kinetic energy (K.E.).
The energy of the bouncing tennis ball changes continuously from potential to kinetic energy, back to potential energy, and so on. The maximum in potential energy is at the top of each bounce, and the maximum in kinetic energy occurs at the moment of impact. The sum of P.E. and K.E. decreases with each bounce as the thermal energies of the ball and the surroundings increase. The ball soon comes to rest.

when it is released and is therefore called potential energy. **Potential energy** is energy due to condition, position, or composition; it is an energy associated with forces of attraction or repulsion between objects.

When we release the ball, it is pulled toward Earth's center by the force of gravity—it falls. Potential energy is converted to kinetic energy during this fall. The kinetic energy reaches its maximum just as the ball strikes the surface. On its rebound, the kinetic energy of the ball decreases (the ball slows down) and its potential energy increases (the ball rises). If the collision of the ball with the surface were perfectly *elastic,* like collisions between molecules in the kinetic-molecular theory, the sum of the potential and kinetic energies of the ball would remain constant. The ball would reach the same maximum height on each rebound, and the ball would bounce forever. But we know that this doesn't happen. The bouncing ball soon comes to rest.

All the energy originally invested in the ball as potential energy (by raising it to its initial position) eventually appears as additional kinetic energy of the atoms and molecules that make up the ball, the surface, and the surrounding air—*thermal energy.* In discussing energy changes, we need to make a distinction between those produced by the action of forces through distances—*work*—and those involving the transfer of thermal energy—*heat.*

7-2   Work

The most common type of work associated with chemical reactions is **pressure–volume work.** This is work involved in the expansion or compression of gases. In Figure 7-3, a quantity of gas is confined in a cylinder by a freely mov-

FIGURE 7-3
Pressure–volume work.
The expanding gas does work as it lifts an object of mass m through the distance h. The increase in volume of gas (ΔV) is the product $A \times h$.

ing piston. The gas is confined to its initial volume by two objects, each with a mass *m*. When one of the objects is removed, the remaining one is raised through the distance *h*. Now let us see how pressure and volume enter into determining the quantity of work done by the expanding gas.

The "pressure" part of pressure–volume work is simply the pressure exerted by the remaining object on the piston: $P = F/A$ (Equation 6.1). But we can also turn this expression around and write $F = P \times A$. Now, let us use the fact that work is done when a force acts through a distance. In Figure 7-3 the force (F) exerted by the expanding gas acts through the distance (h).

$$\text{Work } (w) = \text{force } (F) \times \text{distance } (h) = P \times A \times h$$

▶ The Greek letter *delta,* Δ, indicates a *change* in some quantity.

The product $A \times h$ is the *change* in volume, ΔV, that results from the expansion. It is the "volume" part of "pressure–volume" work. The expression for pressure–volume work, then, is

▶ On page 214 we introduce a sign convention that governs the use of Equation 7.2 in the first law of thermodynamics.

$$w = P\Delta V \tag{7.2}$$

If pressure is stated in atmospheres and volume in liters, the unit of work is the liter-atmosphere, L atm. The SI unit of work is the joule, J. The conversion factor between these two units of work can be obtained from the gas constant, R.

▶ To obtain this conversion factor, simply equate two values of the gas constant R: 0.082058 L atm mol^{-1} K^{-1} = 8.3145 J mol^{-1} K^{-1}.

$$\frac{8.3145 \text{ J mol}^{-1} \text{ K}^{-1}}{0.082058 \text{ L atm mol}^{-1} \text{ K}^{-1}} = 101.32 \text{ J/L atm}$$

EXAMPLE 7-1

Calculating Pressure–Volume Work. What is the quantity of work, in joules, done by the gas in Figure 7-3 if it expands against a constant pressure of 0.980 atm and the change in volume (ΔV) is 25.0 L?

Solution

We begin by substituting values for P and ΔV, in the units given, into Equation (7.2).

$$w = P\Delta V = 0.980 \text{ atm} \times 25.0 \text{ L} = 24.5 \text{ L atm}$$

Then we apply the conversion factor between L atm and J.

$$w = 24.5 \text{ L atm} \times \frac{101.3 \text{ J}}{1 \text{ L atm}} = 2.48 \times 10^3 \text{ J} = 2.48 \text{ kJ}$$

Practice Example A: What is the work done on a gas, in joules, when the gas is compressed from an initial volume of 35.0 L to a final volume of 23.5 L under a constant pressure of 0.987 atm?

Practice Example B: What is the work, in joules, done by 0.225 mol $N_2(g)$ when it expands at a constant temperature of 23 °C from an initial pressure of 2.15 atm to a final pressure of 746 mmHg? (*Hint:* Use the final pressure as that against which expansion occurs. What are the initial and final volumes?)

Pressure–volume work is the type of work performed by explosives and by the gases formed in the combustion of gasoline in an automobile engine. We will say more about pressure–volume work in chemical reactions in Section 7-6.

7-3   Heat

Heat is energy transferred between a system and its surroundings as a result of a temperature difference. Energy, as heat, passes from a warmer body (with a higher temperature) to a colder body (with a lower temperature). At the molecular level, molecules of the warmer body, through collisions, lose kinetic energy to those of the colder body. Thermal energy is transferred—heat "flows"—until the average molecular kinetic energies of the two bodies become the same, until the temperatures become equal.* Heat, like work, describes energy in transit between a system and its surroundings.

Although we commonly use expressions like "heat is lost," "heat is gained," "heat flows," and "the system loses heat to the surroundings," you should not take these statements to mean that a system contains heat. It does not. The energy contained in a system is internal energy. Heat is simply a form in which a quantity of energy may be *transferred* across a boundary between a system and its surroundings.

It is reasonable to expect that the quantity of heat energy, q, required to change the temperature of a substance depends on

- how much the temperature is to be changed
- the quantity of substance
- the nature of the substance (type of atoms or molecules).

Historically, the quantity of heat required to change the temperature of one *gram* of water by one degree Celsius has been called the **calorie (cal).** The calorie is a small unit of energy and the unit *kilocalorie* (kcal) has also been widely used. The SI unit for heat is simply the basic SI energy unit, the joule (J).

$$1 \text{ cal} = 4.184 \text{ J} \tag{7.3}$$

Although we will use the joule almost exclusively in this text, you should also be familiar with the calorie. It is widely encountered in older scientific literature and still used to some extent.

The quantity of heat required to change the temperature of a system by one degree is called the **heat capacity** of the system. If the system is a mole of substance, we can use the term *molar* heat capacity. If it is one gram of substance, we call it the *specific* heat capacity or more commonly **specific heat** (sp ht).† The specific heat of *water* depends somewhat on temperature, but, over the range from 0 to 100 °C, its value is about

$$\frac{1.00 \text{ cal}}{\text{g °C}} = 1.00 \text{ cal g}^{-1} \text{ °C}^{-1} = \frac{4.18 \text{ J}}{\text{g °C}} = 4.18 \text{ J g}^{-1} \text{ °C}^{-1} \tag{7.4}$$

James Joule (1818–1889)—an amateur scientist. Joule's primary occupation was to run a brewery, but he also conducted scientific research in a home laboratory. His precise measurements of quantities of heat formed the basis of the law of conservation of energy.

---

*Although the ultimate result of heat flow is to equalize the temperature of two bodies, the temperature of a body may remain constant for a time as heat enters or leaves it. This is what happens, for example, when a block of ice (0 °C) absorbs heat from the surrounding air. Its temperature does not change until all the ice has melted. A process occurring at a constant temperature is said to be *isothermal.*

†The original meaning of specific heat was that of a *ratio:* the quantity of heat required to change the temperature of a mass of substance divided by the quantity of heat required to produce the same temperature change in the same mass of water. This would make specific heat dimensionless. The meaning given here is more commonly used.

In Example 7-2, our objective is to calculate a quantity of heat, based on how much of a substance we have, the specific heat of that substance, and its temperature change. In some later examples in the chapter, we will routinely do calculations of the type in Example 7-2 as part of a larger problem.

► Feature Problem A on page 239 deals with a common way to heat water—in a microwave oven.

EXAMPLE 7-2

Calculating a Quantity of Heat. How much heat is required to raise the temperature of 7.35 g of water from 21.0 to 98.0 °C? (Assume the specific heat of water is 4.18 J g^{-1} °C^{-1} throughout this temperature range.)

Solution

The heat needed to change the temperature of 7.35 g of water is 7.35 times as great as that needed to change the temperature of 1 g of water. And the heat required to change the temperature by 77.0 °C (from 21.0 to 98.0 °C) is 77 times as great as that needed to raise the temperature by 1 °C. We must multiply the specific heat of water by 7.35 and by 77.0.

$$? \text{ J} = 7.35 \text{ g water} \times \frac{4.18 \text{ J}}{\text{g water °C}} \times (98.0 - 21.0) \text{ °C} = 2.37 \times 10^3 \text{ J}$$

Practice Example A: How much heat, in kilojoules (kJ), is required to raise the temperature of 8.00 ounces of ice water (237 g) from 4.0 °C to 37.0 °C (body temperature)?

Practice Example B: How much heat, in kilojoules (kJ), is required to raise the temperature of 2.50 kg Hg(l) from −20.0 to −6.0 °C? Assume a density of 13.6 g/mL and a molar heat capacity of 28.0 J mol^{-1} °C^{-1} for Hg(l).

The line of reasoning we used in Example 7-2 can be summarized in the following basic equation relating a mass of substance, a temperature change, and a quantity of heat.

$$\text{quantity of heat} = \underbrace{\text{mass of substance} \times \text{specific heat}}_{\text{heat capacity}} \times \text{temp. change} \qquad (7.5)$$

$$q = m \times \text{specific heat} \times \Delta T \qquad (7.6)$$

In Equation (7.6), the temperature change is expressed as $\Delta T = T_f - T_i$, where T_f is the final temperature and T_i is the initial temperature.

► The symbol > means "greater than" and < means "less than."

When the temperature of a system increases ($T_f > T_i$), ΔT is *positive*. A *positive q* signifies that heat is *absorbed* or *gained* by the system. When the temperature of a system decreases ($T_f < T_i$), ΔT is *negative*. A *negative q* signifies that heat is *evolved* or *lost* to the surroundings.

Another idea that enters into calculations of quantities of heat is the **law of conservation of energy:** In interactions between a system and its surroundings, the total energy remains *constant*—energy is neither created nor destroyed. Applied to the exchange of heat, this means that

$$q_{\text{system}} + q_{\text{surroundings}} = 0 \qquad (7.7)$$

Thus, heat *lost* by a system is *gained* by its surroundings, and vice versa.

$$q_{\text{system}} = -q_{\text{surroundings}} \qquad (7.8)$$

FIGURE 7-4
Determining the specific heat of lead—Example 7-3 illustrated.
(a) A 150.0-g sample of lead is heated to the temperature of boiling water (100.0 °C).
(b) A 50.0-g sample of water is added to a thermally insulated beaker, and its temperature is found to be 22.0 °C.
(c) The hot lead is dumped into the cold water, and the final lead–water mixture has a temperature of 28.8 °C.

Experimental Determination of Specific Heats

Let us consider how the law of conservation of energy is used in the experiment outlined in Figure 7-4. The object is to determine the specific heat of the metal, lead. The transfer of energy, as heat, from the lead to the cooler water causes the temperature of the lead to decrease and that of the water to increase, until the point is reached where both the lead and water are at the same temperature. Either the lead or the water can be considered the system. If we consider lead to be the system, we can write $q_{lead} = q_{system}$ and $q_{water} = q_{surroundings}$. Then, applying Equation (7.8), we have

$$q_{lead} = -q_{water} \tag{7.9}$$

If we consider water to be the system and lead the surroundings, Equation (7.8) leads to $q_{water} = -q_{lead}$, which is equivalent to Equation (7.9). We complete the calculation in Example 7-3.

EXAMPLE 7-3

Determining a Specific Heat from Experimental Data. Use data presented in Figure 7-4 to calculate the specific heat of lead.

Solution

First, let us use Equation (7.6) to calculate q_{water}.

$$q_{water} = 50.0 \text{ g water} \times \frac{4.18 \text{ J}}{\text{g water °C}} \times (28.8 - 22.0) \text{ °C} = 1.4 \times 10^3 \text{ J}$$

From Equation (7.9) we can write

$$q_{lead} = -q_{water} = -1.4 \times 10^3 \text{ J}$$

TABLE 7.1 Specific Heats of Several Solid Elements (in J g^{-1} °C^{-1})	
metals	
lead	0.128
copper	0.385
iron	0.449
aluminum	0.903
magnesium	1.024
nonmetals	
selenium	0.321
sulfur	0.706
phosphorus	0.777
metalloids	
tellurium	0.202
arsenic	0.329

Now, from Equation (7.6) again, we obtain

$$q_{lead} = 150.0 \text{ g lead} \times \text{specific heat of lead} \times (28.8 - 100.0) \text{ °C}$$

$$= -1.4 \times 10^3 \text{ J}$$

$$\text{specific heat of lead} = \frac{-1.4 \times 10^3 \text{ J}}{150.0 \text{ g lead} \times (28.8 - 100.0) \text{ °C}}$$

$$= \frac{-1.4 \times 10^3 \text{ J}}{150.0 \text{ g lead} \times -71.2 \text{ °C}} = 0.13 \text{ J g}^{-1} \text{ °C}^{-1}$$

Practice Example A: When 1.00 kg lead (specific heat = 0.13 J g^{-1} °C^{-1}) at 100.0 °C is added to a quantity of water at 28.5 °C, the final temperature of the lead–water mixture is 35.2 °C. What is the mass of water present? [*Hint:* If you know any four of the five quantities—*q, m,* specific heat, T_f, T_i—you can solve Equation (7.6) for the remaining one.]

Practice Example B: A 100.0-g copper sample (specific heat = 0.385 J g^{-1} °C^{-1}) at 100.0 °C is added to 50.0 g water at 26.5 °C. What is the final temperature of the copper-water mixture?

Significance of Specific-Heat Values

Table 7.1 lists the specific heats of several solid elements. The relatively high specific heat of aluminum compared with other metals helps to account for its use in "miracle thaw" products designed to thaw frozen foods rapidly. The aluminum cools only slowly as it transfers heat to the frozen food, which thaws more quickly than when simply exposed to air.

Because of their greater complexity at the molecular level, compounds generally have more ways of storing internal energy than do the elements; they tend to have higher heat capacities. Water, for example, has a specific heat that is more than 30 times greater than that of lead. We need a much larger quantity of heat to change the temperature of a sample of water than of an equal mass of a metal. To return to the food-thawing example, we expect frozen food placed in room-temperature water to thaw much faster than if it is just exposed to room-temperature air.

An environmental consequence of the high specific heat of water is found in the effect of large lakes on local climates. Because a lake takes much longer to heat up in summer and to cool in winter than do other types of terrain, lakeside communities tend to be cooler in summer and warmer in winter than communities more distant from the lake.

▶ Feature Problem B on page 239 describes how specific heats of the elements were used in an early attempt to establish atomic weights.

An ice cube thawing on a block of aluminum alloy ("miracle thaw").

7-4 Heats of Reaction and Calorimetry

In the preceding section we used measured quantities of heat to determine the specific heat of a substance. Our primary interest in heat measurements, though, is in evaluating a property called the heat of reaction. If we think of a chemical reaction as a process in which some chemical bonds are broken and others are formed, then, in general, we expect the chemical energy of a system to change as a result of a reaction. Furthermore, we might expect some of this energy change to appear as heat. A **heat of reaction**, q_{rxn}, is the quantity of heat exchanged between a system and its surroundings when a chemical reaction occurs within the system, at *constant temperature*. One of the most common reactions studied is the combustion reaction. This is such a common reaction that we of-

▶ By a reaction at constant temperature we mean that, after the reaction, the products return to the initial temperature of the reactants by exchanging heat with the surroundings.

ten refer to the *heat of combustion* when describing the heat energy released by a combustion reaction. Heats of reaction are experimentally determined in a **calorimeter,** a device for measuring quantities of heat. We will consider two types of calorimeters in this section, and we will treat both of them as *isolated* systems.

If a reaction occurs in an *isolated* system, that is, one that exchanges no matter or energy with its surroundings, the reaction produces a change in the thermal energy of the system—the temperature either increases or decreases. Now, imagine that the previously isolated system is allowed to interact with its surroundings. The heat of reaction is the quantity of heat exchanged between the system and its surroundings as the system is restored to its initial temperature. In actual practice, we do not physically restore the system to its initial temperature. We calculate the quantity of heat that *would be* exchanged in this restoration. We do this by having a probe (thermometer) within the system to record the temperature change produced by the reaction, as we see in Examples 7-4 and 7-5. Before turning to these examples, however, let us introduce two terms regarding heats of reaction that are widely used by chemists.

An **exothermic** reaction is one that produces a temperature increase in an isolated system or, in a nonisolated system, gives off heat to the surroundings. For an exothermic reaction, the heat of reaction is a negative quantity ($q_{rxn} < 0$). In an **endothermic** reaction the corresponding situations are a temperature decrease in an isolated system, or a gain of heat from the surroundings by a nonisolated system. In this case, the heat of reaction is a positive quantity ($q_{rxn} > 0$). Figure 7-5 pictures one exothermic and one endothermic reaction.

Bomb Calorimetry

The type of calorimeter shown in Figure 7-6, called a **bomb calorimeter,** is ideally suited for measuring the heat evolved in a combustion reaction. The *system* is everything within the double-walled outer jacket of the calorimeter. This includes the bomb and its contents, the water in which the bomb is immersed, the thermometer, the stirrer, and so on. The system is *isolated* from its surroundings. When the combustion reaction occurs, chemical energy is converted to thermal energy, and the temperature of the system rises. The heat of reaction, as we described earlier, is the quantity of heat that the system would have to *lose* to its surroundings to be restored to its initial temperature. This quantity

(a)

(b)

FIGURE 7-5
Exothermic and endothermic reactions.
(a) **An exothermic reaction.**
Slaked lime [$Ca(OH)_2$] is produced by the action of water on quicklime (CaO). The reactants are mixed at room temperature, but the temperature of the mixture rises to 40.5 °C.
$$CaO(s) + H_2O(l) \rightarrow Ca(OH)_2(s)$$
(b) **An endothermic reaction.**
$Ba(OH)_2 \cdot 8\ H_2O(s)$ and $NH_4Cl(s)$ are mixed at room temperature, and the temperature falls to 5.8 °C in the reaction:
$$Ba(OH)_2 \cdot 8\ H_2O(s) + 2\ NH_4Cl(s) \rightarrow$$
$$BaCl_2 \cdot 2\ H_2O(s) + 2\ NH_3(aq)$$
$$+ 8\ H_2O(l)$$

thermometer
wire for ignition
stirrer
water
reactants
steel "bomb"

FIGURE 7-6
A bomb calorimeter assembly.
An iron wire is imbedded in the sample in the lower half of the bomb. The bomb is assembled and filled with $O_2(g)$ at high pressure. The assembled bomb is immersed in water in the calorimeter, and the initial temperature is measured. A short pulse of electric current heats the sample to the point that it ignites. The final temperature of the calorimeter assembly is determined after the combustion.

of heat, in turn, is just the *negative* of the thermal energy gained by the calorimeter and its contents ($q_{calorim.}$).

$$q_{rxn} = -q_{calorim.} \quad (\text{where } q_{calorim.} = q_{bomb} + q_{water} + \dots) \quad (7.10)$$

If we assemble the calorimeter in exactly the same way each time we use it—that is, use the same bomb, the same quantity of water, and so on—we can define a *heat capacity of the calorimeter*. This is the quantity of heat required to raise the temperature of the calorimeter assembly by one degree Celsius. When we multiply this heat capacity by the observed temperature change we get $q_{calorim.}$.

$$q_{calorim.} = \text{heat capacity of calorim.} \times \Delta T \quad (7.11)$$

And, from $q_{calorim.}$ we then establish q_{rxn}, as in Example 7-4.

EXAMPLE 7-4

Using Bomb Calorimetry Data to Determine a Heat of Reaction. The combustion of 1.010 g sucrose, $C_{12}H_{22}O_{11}$, in a bomb calorimeter causes the temperature to rise from 24.92 to 28.33 °C. The heat capacity of the calorimeter assembly is 4.90 kJ/°C. **(a)** What is the heat of combustion of sucrose, expressed in kilojoules per mole of $C_{12}H_{22}O_{11}$? **(b)** Verify the claim of sugar producers that one teaspoon of sugar (about 4.8 g) "contains only 19 Calories."

Solution

(a) First we can calculate $q_{calorim.}$ with Equation (7.11).

$$q_{calorim.} = 4.90 \text{ kJ/°C} \times (28.33 - 24.92) \text{ °C} = 4.90 \times 3.41 = 16.7 \text{ kJ}$$

Now, using Equation (7.10) we get

$$q_{rxn} = -q_{calorim.} = -16.7 \text{ kJ}$$

This is the heat of combustion of the 1.010-g sample.

Per gram $C_{12}H_{22}O_{11}$:

$$q_{rxn} = \frac{-16.7 \text{ kJ}}{1.010 \text{ g } C_{12}H_{22}O_{11}} = -16.5 \text{ kJ/g } C_{12}H_{22}O_{11}$$

Per mole $C_{12}H_{22}O_{11}$:

$$q_{rxn} = \frac{-16.5 \text{ kJ}}{\text{g } C_{12}H_{22}O_{11}} \times \frac{342.3 \text{ g } C_{12}H_{22}O_{11}}{1 \text{ mol } C_{12}H_{22}O_{11}}$$

$$= -5.65 \times 10^3 \text{ kJ/mol } C_{12}H_{22}O_{11}$$

▶ The negative sign of the heat of reaction signifies that the reaction is *exothermic*.

(b) We can use the heat of combustion per gram of sucrose determined in part **(a)**, together with a factor to convert from kilojoules to kilocalories. (Because 1 cal = 4.184 J, 1 kcal = 4.184 kJ.)

$$? \text{ kcal} = 4.8 \text{ g } C_{12}H_{22}O_{11} \times \frac{-16.5 \text{ kJ}}{\text{g } C_{12}H_{22}O_{11}} \times \frac{1 \text{ kcal}}{4.184 \text{ kJ}} = -19 \text{ kcal}$$

1 food Calorie (1 Calorie) is actually 1000 cal, or 1 kcal. Therefore, 19 kcal = 19 Calories. The claim is justified.

Practice Example A: Vanillin is a natural constituent of vanilla. It is also manufactured for use in artificial vanilla flavoring. The combustion of 1.013 g of vanillin, $C_8H_8O_3$, in the same bomb calorimeter as in Example 7-4, causes the temperature to rise from 24.89 °C to 30.09 °C. What is the heat of combustion of vanillin, expressed in kilojoules per mole of $C_8H_8O_3$?

Practice Example B: The combustion of a 1.176-g sample of benzoic acid ($HC_7H_5O_2$) causes a temperature *increase* of 4.96 °C in a bomb calorimeter assembly. What is the heat capacity of the assembly? The heat of combustion of benzoic acid is −26.42 kJ/g. (*Hint*: What is the quantity of heat liberated in the combustion? What effect does this heat have on the temperature of the assembly?)

▶ The heat capacity of a bomb calorimeter must be determined by experiment.

The "Coffee-Cup" Calorimeter

In the general chemistry laboratory you are much more likely to run into the simple calorimeter pictured in Figure 7-7 than a bomb calorimeter. We mix the reactants (generally in aqueous solution) in a Styrofoam cup and measure the temperature change. Styrofoam is a good heat insulator, so there is very little heat transfer between the cup and the surrounding air. We treat the system—the cup and its contents—as an *isolated* system.

As with the bomb calorimeter, we define the heat of reaction as the quantity of heat that would be exchanged with the surroundings in restoring the calorimeter to its initial temperature. But, again, we do not physically restore the calorimeter to its initial conditions. We simply take the heat of reaction to be the *negative* of the quantity of heat producing the temperature change in the calorimeter. That is, we use Equation (7.10): $q_{rxn} = -q_{calorim.}$.

In Example 7-5 we make certain assumptions to simplify the calculation, but for more precise measurements these assumptions would not be made (see Exercise 47).

FIGURE 7-7
A Styrofoam "coffee-cup" calorimeter.
The reaction mixture is in the inner cup. The outer cup provides additional thermal insulation from the surrounding air. The cup is closed off with a cork stopper through which a thermometer and stirrer are inserted and immersed into the reaction mixture.

EXAMPLE 7-5

Determining a Heat of Reaction from Calorimetric Data. In the neutralization of a strong acid with a strong base the essential reaction is the combination of $H^+(aq)$ and $OH^-(aq)$ to form water (recall page 138).

$$H^+(aq) + OH^-(aq) \longrightarrow H_2O(l)$$

Two solutions, 100.0 mL of 1.00 M HCl(aq) and 100.0 mL of 1.00 M NaOH(aq), both initially at 21.1 °C, are added to a Styrofoam cup calorimeter and allowed to react. The temperature rises to 27.8 °C. Determine the heat of reaction for the neutralization reaction, expressed per mole of H_2O formed. Is the reaction endothermic or exothermic?

In addition to assuming that the calorimeter is an isolated system, assume that all there is in the system to absorb heat is 200.0 mL of water. This assumption ignores the fact that 0.10 mol of each of NaCl and H_2O are formed in the reaction, that the density of the resulting NaCl(aq) is not quite 1.00 g/mL, and that its specific heat is not quite 4.18 J g^{-1} °C^{-1}. Also, ignore the small heat capacity of the Styrofoam cup itself.

Solution

Because the reaction is a neutralization reaction, let us call the heat of reaction $q_{neutr.}$. Now, according to Equation (7.10), $q_{neutr.} = -q_{calorim.}$, and if we make the assumptions described above,

$$q_{calorim.} = 200.0 \text{ mL} \times \frac{1.00 \text{ g}}{\text{mL}} \times \frac{4.18 \text{ J}}{\text{g °C}} \times (27.8 - 21.1) \text{ °C} = 5.6 \times 10^3 \text{ J}$$

$$q_{neutr.} = -q_{calorim.} = -5.6 \times 10^3 \text{ J} = -5.6 \text{ kJ}$$

In 100.0 mL of 1.00 M HCl the amount of H^+ is

$$? \text{ mol } H^+ = 0.1000 \text{ L} \times \frac{1.00 \text{ mol HCl}}{1 \text{ L}} \times \frac{1 \text{ mol } H^+}{1 \text{ mol HCl}} = 0.100 \text{ mol } H^+$$

▶ Feature Problem C on page 239 examines the way in which the heat of a neutralization reaction depends on the proportions of acid and base used.

Similarly, in 100.0 mL of 1.00 M NaOH there is 0.100 mol OH^-. Thus, the H^+ and the OH^- combine to form 0.100 mol H_2O. (The two are in *stoichiometric* proportions; neither is in excess.)

Per mole of H_2O produced:

$$q_{\text{neutr.}} = \frac{-5.6 \text{ kJ}}{0.100 \text{ mol } H_2O} = -56 \text{ kJ/mol } H_2O$$

Because $q_{\text{neutr.}}$ is a *negative* quantity, the neutralization reaction is *exothermic*.

Practice Example A: Two solutions, 100.0-mL of 1.00 M $AgNO_3(aq)$ and 100.0 mL of 1.00 M $NaCl(aq)$, both initially at 22.4 °C, are added to a Styrofoam cup calorimeter and allowed to react. The temperature rises to 30.2 °C. Determine q_{rxn} per mole of $AgCl(s)$ in the reaction.

$$Ag^+(aq) + Cl^-(aq) \longrightarrow AgCl(s)$$

Practice Problem B: Two solutions, 100.0-mL of 1.020 M HCl and 50.0 mL of 1.988 M NaOH, both initially at 24.52 °C, are mixed in a Styrofoam cup calorimeter. What will be the final temperature of the mixture? Make the same assumptions and use the heat of neutralization established in Example 7-5. (*Hint:* The reactants are not in stoichiometric proportions. Which is the limiting reactant?)

7-5 The First Law of Thermodynamics

We have considered the nature of work and heat, and have described methods of determining heats of reaction. Now we want to bring the concept of internal energy, briefly introduced in Section 7-1, into the discussion of heats of reaction. Before doing so, let us first say something in this section about the first law of thermodynamics.

Heat and work are the means by which a system *exchanges* energy with its surroundings. Heat and work only exist during a *change*. A system *does not* contain energy in the form of heat or work. The energy contained within a system is the **internal energy, E.** This is the total energy (kinetic and potential) associated with chemical bonds, intermolecular attractions, kinetic energy of molecules, and so on. The relationship involving heat (q), work (w), and *changes* in internal energy (ΔE) is dictated by the law of conservation of energy, also known as **the first law of thermodynamics.**

▶ The symbol U is also commonly used for internal energy.

$$\Delta E = q + w \qquad (7.12)$$

In using equation (7.12) keep these important points in mind.

- Any energy *entering* the system carries a *positive* sign. Thus, if heat is *absorbed* by the system, $q > 0$. If work is done *on* the system, $w > 0$.
- Any energy *leaving* the system carries a *negative* sign. Thus, if heat is *given off* by the system, $q < 0$. If work is done *by* the system, $w < 0$.

• In general, the internal energy of a system changes as a result of energy entering or leaving the system as heat and/or work. If, on balance more energy enters the system than leaves, ΔE is *positive*. If more energy leaves than enters, ΔE is *negative*.

These ideas are summarized in Figure 7-8 and illustrated in Example 7-6.

FIGURE 7-8
Sign conventions used in thermodynamics.
Arrows represent the direction of flow of heat ($\rightarrow$) and work ($\rightarrow$). The + signs refer to energy entering the system from the surroundings. The − signs signify energy leaving the system and going into the surroundings.

EXAMPLE 7-6

Relating ΔE, q, and w Through the First Law of Thermodynamics. A gas, while expanding (recall Figure 7-3), absorbs 25 J of heat and does 243 J of work. What is ΔE for the gas?

Solution

The key to problems of this type lies in assigning the correct signs to the quantities of heat and work. Because heat is absorbed by (enters) the system, q is *positive*. Because work done *by* the system represents energy *leaving* the system, w is *negative*. You may find it useful to represent the values of q and w, with their correct signs, within parentheses. Then, complete the algebra.

$$\Delta E = q + w = (+25 \text{ J}) + (-243 \text{ J}) = 25 \text{ J} - 243 \text{ J} = -218 \text{ J}$$

Practice Example A: In compressing a gas, 355 J of work is done on the system. At the same time, 185 J of heat escapes from the system. What is ΔE for the system?

Practice Example B: If the internal energy of a system *decreases* by 125 J at the same time it *absorbs* 54 J of heat, does the system do work or have work done on it? How much?

Functions of State. To describe a system completely, we need to indicate its temperature and pressure and the kinds and amounts of substances it contains. With this information we say that we have specified the *state* of the system. Any property having a unique value when the state of the system is defined is called a **function of state** or **state function.** Internal energy E, even though we cannot actually measure it, is a function of state. Heat (q) and work (w) are *not* functions of state.

Perhaps the mountain-climbing analogy in Figure 7-9 will help you understand this matter of states and functions of state. The initial state, *state 1*, is the base of the mountain and the final state, *state 2*, is the summit. The elevation at any point on the mountain is analogous to internal energy E. The *difference* in elevation between the base and summit of the mountain (known as the gain in elevation) is analogous to ΔE. Heat (q) and work (w) are analogous to the path chosen to climb the mountain. Path (a) is short but steep; path (b) is longer and more gradual. You can probably see that the time it takes to climb the mountain depends on the path chosen, but the total elevation gain is *fixed*. Furthermore, after a trip to the top of the mountain and back down, the total elevation gain is *zero*. If a system passes from *state 1* to *state 2* and back to *state 1* (the initial condition), all functions of state, including internal energy, return to their initial values.

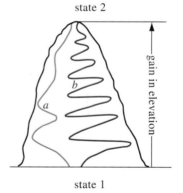

FIGURE 7-9
An analogy to a function of state.
The gain in elevation in climbing from the base to the summit of a mountain is independent of the path chosen. This elevation gain is analogous to ΔE in a system.

7-6 Heats of Reaction: ΔE and ΔH

We can think of the reactants in a chemical reaction as representing one state of a system, state 1, with an internal energy E_1. The products represent a different state, state 2, with an internal energy E_2.

$$\text{reactants} \longrightarrow \text{products}$$
$$\text{(state 1)} \qquad \text{(state 2)}$$
$$E_1 \qquad\qquad E_2$$

Accompanying the reaction is a change in internal energy: $\Delta E = E_2 - E_1$. According to the first law of thermodynamics we can also represent this change as $\Delta E = q + w$. In this expression, q is the quantity we have previously introduced as the heat of reaction, that is, q_{rxn}.

Let us think again about a combustion reaction carried out in a bomb calorimeter (recall Figure 7-6). Because the original reactants and the products are confined within the bomb, we can say the reaction occurs at *constant volume*. And, because the volume remains constant, no work is done, $w = 0$. If we represent the heat of reaction as $q_{rxn} = q_V$ and apply the first law of thermodynamics, we have

$$\Delta E = q_V + 0 = q_V = q_{rxn} \tag{7.13}$$

The heat of reaction that we measure in a bomb calorimeter is equal to ΔE, the change in internal energy for a reaction.

But we do not ordinarily carry out chemical reactions in bomb calorimeters. The metabolism of sucrose occurs under the conditions present in the human body. The combustion of methane (natural gas) in a water heater occurs in an open flame. This question then arises: "How does the heat of a reaction measured in a bomb calorimeter compare with the heat of reaction if the reaction is carried out in some other way?" The usual "other" way is in beakers, flasks, and other containers open to the atmosphere and under the *constant pressure* of the atmosphere. The neutralization reaction of Example 7-5 is typical of this more common method of conducting chemical reactions.

At a fixed temperature, the heat of combustion of sucrose turns out to be the same, whether at constant volume (q_V) or constant pressure (q_P). Only heat is transferred between the reaction mixture and the surroundings; no pressure–volume work is done. In many reactions in the open atmosphere, however, a small amount of pressure–volume work is done as the system expands or contracts. In these cases the measured heat of reaction, q_P, is slightly different from ΔE. Because of this we find it useful to define a new property that is closely related to internal energy but has the distinction of being equal to q_P.

This property is called **enthalpy change,** denoted by the symbol ΔH. Let us begin by defining **enthalpy,** H, as the sum of the internal energy and the pressure–volume product of a system: $H = E + PV$. The *change* in enthalpy for a process carried out at a constant temperature and pressure and with work limited to pressure–volume work is given by the expression: $\Delta H = \Delta E + P \, \Delta V$. Now, let us substitute into this expression (a) the first law of thermodynamics: $\Delta E = q + w$, (b) the heat of reaction at constant pressure: q_P, and (c) the work done *by* the system: $w = -P \, \Delta V$.

$$\Delta H = \Delta E + P \, \Delta V = (q_P + w) + P \, \Delta V = q_P - \cancel{P \, \Delta V} + \cancel{P \, \Delta V} = q_P$$

What this all boils down to is that at a constant temperature and pressure, and with work limited to pressure–volume work

$$\Delta H = q_P \tag{7.14}$$

Are You Wondering . . .

Why introduce the term ΔH anyway? Why not work just with ΔE, q, and w?

It's mainly a matter of convenience. Think of an analogous situation from daily life—buying gasoline at a filling station. The gasoline price posted on the pump is actually the sum of a base price and various federal, state, and perhaps local taxes. This breakdown is important to the accountants who must determine how much tax is to be paid to which agencies. To the consumer, however, it's easier to be given just the total cost per gallon. After all, this determines what he or she must pay. In thermochemistry our chief interest is generally in heats of reaction, not pressure–volume work. And because most reactions are carried out under atmospheric pressure, it's helpful to have a function of state, enthalpy, H, whose change is exactly equal to something we can measure: q_P.

Representing ΔH in a Chemical Equation

Because internal energy (E) and the pressure–volume product (PV) of a system are both functions of state, so too is their sum, enthalpy (H). We cannot determine *absolute* values of enthalpy, H, but fortunately, we do not need absolute values. Because enthalpy is a function of state, *changes* in enthalpy, ΔH, have unique values. For chemical reactions, this means that the *difference* in enthalpy between the products and reactants—the *change* in enthalpy of a reaction, ΔH—has a unique value. Because the volume of a system is almost entirely determined by the volume of gas, and because 12 mol $CO_2(g)$ occupies the same volume as 12 mol $O_2(g)$, there is no change in volume in the combustion of sucrose: $q_P = q_V$. Thus, we can represent the result of Example 7–4 as

$$C_{12}H_{22}O_{11}(s) + 12\ O_2(g) \longrightarrow 12\ CO_2(g) + 11\ H_2O(l)$$
$$\Delta H = -5.65 \times 10^3 \text{ kJ} \quad (7.15)$$

That is, 1 mol $C_{12}H_{22}O_{11}(s)$ reacts with 12 mol $O_2(g)$ to produce 12 mol $CO_2(g)$, 11 mol $H_2O(l)$, and 5.65×10^3 kJ of evolved heat.*

For the formation of 1 mol $H_2O(l)$ in the strong acid–strong base neutralization of Example 7-5,

$$H^+(aq) + OH^-(aq) \longrightarrow H_2O(l) \qquad \Delta H = -56 \text{ kJ} \qquad (7.16)$$

In most reactions, such as (7.16), the heat of reaction we measure is ΔH. In some reactions, notably combustion reactions, we measure ΔE (that is, q_V). In reaction (7.15), $\Delta E = \Delta H$, but this is not always the case. Where it is not, we can obtain a value of ΔH from ΔE by adding the term $P\Delta V$. The $P\Delta V$ term is usually quite small, however, so that even in these cases ΔH and ΔE are nearly equal. In this text we will treat all heats of reactions as ΔH values.

▶ We calculated a typical pressure–volume work in Example 7-1. It was 2.48 kJ. In Example 7-4 we determined a typical heat of reaction. It was -5.65×10^3 kJ. Can you see that $P\Delta V$ work is generally negligible compared with a heat of reaction?

---

*Strictly speaking, the unit for ΔH should be kilojoules per mole, meaning per mole of reaction. "One mole of reaction" relates to the amounts of reactants and products in the equation as written. Thus, reaction (7.15) involves 1 mol $C_{12}H_{22}O_{11}(s)$, 12 mol $O_2(g)$, 12 mol $CO_2(g)$, 11 mol $H_2O(l)$, and -5.65×10^3 kJ of enthalpy change *per mol reaction*. The /mol part of the unit of ΔH is often dropped, but there are times we need to carry it to achieve the proper cancellation of units. We will find this to be the case in Chapters 20 and 21.

Example 7-7 shows how enthalpy changes can provide conversion factors for problem solving.

EXAMPLE 7-7

Stoichiometric Calculations Involving Quantities of Heat. What quantity of heat is associated with the complete combustion of 1.00 kg of sucrose, $C_{12}H_{22}O_{11}$?

Solution

First, express the quantity of sucrose in moles.

$$? \text{ mol} = 1.00 \text{ kg C}_{12}H_{22}O_{11} \times \frac{1000 \text{ g C}_{12}H_{22}O_{11}}{1 \text{ kg C}_{12}H_{22}O_{11}} \times \frac{1 \text{ mol C}_{12}H_{22}O_{11}}{342.3 \text{ g C}_{12}H_{22}O_{11}}$$

$$= 2.92 \text{ mol C}_{12}H_{22}O_{11}$$

Now, formulate a conversion factor (shown in blue) based on the information in Equation (7.15)—that is, -5.65×10^3 kJ of heat is associated with the combustion of 1 mol $C_{12}H_{22}O_{11}$.

$$? \text{ kJ} = 2.92 \text{ mol C}_{12}H_{22}O_{11} \times \frac{-5.65 \times 10^3 \text{ kJ}}{1 \text{ mol C}_{12}H_{22}O_{11}} = -1.65 \times 10^4 \text{ kJ}$$

The negative sign denotes that heat is *given off* in the combustion.

Practice Example A: What mass of sucrose must be burned to produce 1.00×10^3 kJ of heat? [*Hint:* Again use the enthalpy change of Equation (7.15) as a conversion factor. Here, however, you must invert it.]

Practice Example B: A 25.0-mL sample of 0.1045 M HCl(aq) was neutralized by NaOH(aq). Use Equation (7.16) to determine the heat evolved in this neutralization.

Enthalpy Diagrams

The negative sign of ΔH in Equation (7.15) means that the enthalpy of the products is lower than that of the reactants. This *decrease* in enthalpy appears as heat evolved to the surroundings. The reaction is exothermic. In the reaction

$$N_2(g) + O_2(g) \longrightarrow 2 \text{ NO}(g) \qquad \Delta H = +180.50 \text{ kJ} \qquad (7.17)$$

the products have a *higher* enthalpy than the reactants; ΔH is positive. To produce this increase in enthalpy, heat is absorbed from the surroundings. The reaction is endothermic. An **enthalpy diagram** is a diagrammatic representation of enthalpy changes in a process. Figure 7-10 shows how we can represent exothermic and endothermic reactions through such diagrams.

7-7 Indirect Determination of ΔH: Hess's Law

One of the reasons the enthalpy concept is so useful is that we can calculate large numbers of heats of reaction from a small number of measurements. The following features of enthalpy change (ΔH) make this possible.

1. ΔH is an Extensive Property. Consider the enthalpy change in the formation of NO(g).

$$N_2(g) + O_2(g) \longrightarrow 2 \text{ NO}(g) \qquad \Delta H = 180.50 \text{ kJ}$$

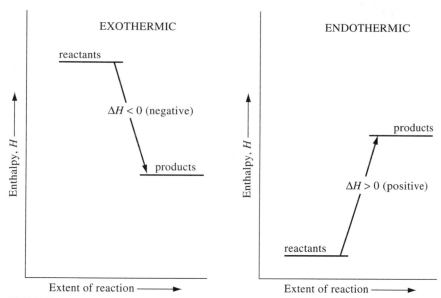

FIGURE 7-10
Enthalpy diagrams.
Because we cannot measure an absolute value of H, we cannot assign numerical values on the enthalpy axis. What we can say, though, is that enthalpy increases in the direction of the arrow along the enthalpy axis. In an *exothermic* reaction the products have a lower enthalpy than the reactants. Enthalpy decreases in the reaction; $\Delta H < 0$. In an *endothermic* reaction the products have a higher enthalpy than the reactants. Enthalpy increases in the reaction; $\Delta H > 0$.

To express the enthalpy change in terms of *one mole* of NO(g), we divide all coefficients *and the* ΔH *value* by *two.*

$$\tfrac{1}{2} N_2(g) + \tfrac{1}{2} O_2(g) \longrightarrow NO(g) \qquad \Delta H = \tfrac{1}{2} \times 180.50 = 90.25 \text{ kJ}$$

Enthalpy change is directly proportional to the amounts of substances in a system.

2. *ΔH Changes Sign When a Process is Reversed.* As in the mountain-climbing analogy in Figure 7-9, if we reverse a process, the change in a function of state reverses sign. Thus, ΔH for the *decomposition* of one mole of NO(g) is $-\Delta H$ for the *formation* of one mole of NO(g).

$$NO(g) \longrightarrow \tfrac{1}{2} N_2(g) + \tfrac{1}{2} O_2(g) \qquad \Delta H = -90.25 \text{ kJ}$$

3. *Hess's Law of Constant Heat Summation.* To describe the enthalpy change for the formation of $NO_2(g)$ from $N_2(g)$ and $O_2(g)$,

$$\tfrac{1}{2} N_2(g) + O_2(g) \longrightarrow NO_2(g) \qquad \Delta H = ?$$

we can think of the reaction as proceeding in two steps: First we form NO(g) from $N_2(g)$ and $O_2(g)$, and then $NO_2(g)$ from NO(g) and $O_2(g)$. When we add the equations for these two steps, together with their individual and distinctive ΔH values, we get the net equation and ΔH value that we are seeking.

$$\tfrac{1}{2} N_2(g) + \tfrac{1}{2} O_2(g) \longrightarrow \cancel{NO(g)} \qquad \Delta H = +90.25 \text{ kJ}$$
$$\cancel{NO(g)} + \tfrac{1}{2} O_2(g) \longrightarrow NO_2(g) \qquad \Delta H = -57.07 \text{ kJ}$$
$$\overline{\tfrac{1}{2} N_2(g) + O_2(g) \longrightarrow NO_2(g) \qquad \Delta H = +33.18 \text{ kJ}}$$

▶ An *extensive* property, such as mass, volume, or enthalpy change, depends on the quantity of matter under study. By contrast, an *intensive* property, such as density, melting point, or boiling point, is independent of the quantity of matter included in the observation.

▶ Although we have avoided fractional coefficients previously, we need them here. We require the coefficient of NO(g) to be 1.

▶ This is like determining the total elevation gain in a hike by summing the elevation gain (or loss) in stages. For example, $+1000 \text{ m} - 200 \text{ m} + 400 \text{ m} \dots$

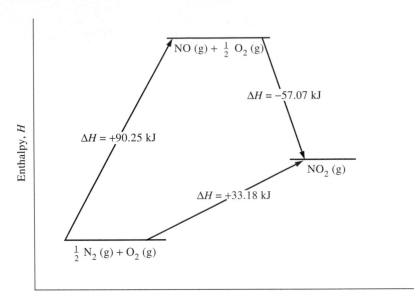

FIGURE 7-11
An enthalpy diagram illustrating Hess's law.
Whether the reaction occurs through a single step (red arrow) or in two steps (black arrows), the enthalpy change is $\Delta H = +33.18$ kJ for the net reaction: $\frac{1}{2}$ N$_2$(g) + O$_2$(g) → NO$_2$(g).

Note that in summing the two equations we canceled out NO(g), a species that would have appeared on both sides of the net equation. Figure 7-11 illustrates what we just did through an enthalpy diagram. **Hess's law** states the principle we used.

▶ Feature Problem D on page 239 describes a simple experiment to confirm Hess's law.

If a process occurs in stages or steps (even if only hypothetically), the enthalpy change for the overall (net) process is the sum of the enthalpy changes for the individual steps.

Suppose we want the enthalpy change for the reaction

$$3 \text{ C(graphite)} + 4 \text{ H}_2(g) \longrightarrow \text{C}_3\text{H}_8(g) \qquad \Delta H = ? \qquad (7.18)$$

How should we proceed? If we try to get graphite and hydrogen to react, a slight reaction will occur, but it will not go to completion. Furthermore, the product will not be limited to propane (C$_3$H$_8$). Several other hydrocarbons will form as well. The fact is we cannot measure ΔH for reaction (7.18) directly. Instead, we must resort to an *indirect calculation* from ΔH values that can be established by experiment. Here is where Hess's law is of greatest value. It permits us to calculate ΔH values that we cannot measure directly. In Example 7-8, we use the following heats of combustion to calculate ΔH for reaction (7.18).

$$\Delta H_{\text{combustion}}: \quad \text{C}_3\text{H}_8(g) \quad = -2219.1 \text{ kJ/mol C}_3\text{H}_8(g)$$
$$\text{C(graphite)} = -393.5 \text{ kJ/mol C(graphite)}$$
$$\text{H}_2(g) \quad = -285.8 \text{ kJ/mol H}_2(g)$$

EXAMPLE 7-8

Applying Hess's Law. Use the heat of combustion data listed above to determine ΔH for reaction (7.18): 3 C(graphite) + 4 H$_2$(g) → C$_3$H$_8$(g) $\Delta H = ?$

Solution

To determine an enthalpy change with Hess's law, we need to combine the appropriate chemical equations. A good starting point is to write chemical equations for the given combustion reactions, based on *one mole* of the indicated reactant. Recall from Section 3-4 that the products of the combustion of carbon–hydrogen–oxygen compounds are $CO_2(g)$ and $H_2O(l)$.

(a) $C_3H_8(g) + 5\,O_2(g) \longrightarrow 3\,CO_2(g) + 4\,H_2O(l)$ $\Delta H = -2219.1$ kJ
(b) $C(graphite) + O_2(g) \longrightarrow CO_2(g)$ $\Delta H = -393.5$ kJ
(c) $H_2(g) + \frac{1}{2}\,O_2(g) \longrightarrow H_2O(l)$ $\Delta H = -285.8$ kJ

Because our objective in reaction (7.18) is to *produce* $C_3H_8(g)$, the next step is to find a reaction in which $C_3H_8(g)$ is formed—the *reverse* of reaction (a).

(a): $3\,CO_2(g) + 4\,H_2O(l) \longrightarrow C_3H_8(g) + 5\,O_2(g)$ $\Delta H = -(-2219.1)$ kJ

$$= +2219.1 \text{ kJ}$$

Now, we turn our attention to the reactants, C(graphite) and $H_2(g)$. To get the proper number of moles of each we must multiply equation (b) by 3 and equation (c) by 4.

3 × (b): $3\,C(graphite) + 3\,O_2(g) \longrightarrow 3\,CO_2(g)$ $\Delta H = 3 \times (-393.5$ kJ)

$$= -1181 \text{ kJ}$$

4 × (c): $4\,H_2(g) + 2\,O_2(g) \longrightarrow 4\,H_2O(l)$ $\Delta H = 4 \times (-285.8$ kJ)

$$= -1143 \text{ kJ}$$

Here is the net change we have described: 3 mol C(graphite) and 4 mol $H_2(g)$ have been consumed and 1 mol $C_3H_8(g)$ has been produced. This is exactly what is required in Equation (7.18). We can now combine the three modified equations.

−(a):   3 ~~CO_2~~(g) + 4 ~~H_2O~~(l) $\longrightarrow C_3H_8(g) + $ ~~5 O_2~~(g) $\Delta H = +2219.1$ kJ

3 × (b): $3\,C(graphite) + $ ~~3 O_2~~(g) $\longrightarrow$ 3 ~~CO_2~~(g) $\Delta H = -1181$ kJ

4 × (c):   4 H_2(g) + ~~2 O_2~~(g) $\longrightarrow$ 4 ~~H_2O~~(l) $\Delta H = -1143$ kJ

$\overline{}$

    $3\,C(graphite) + 4\,H_2(g) \longrightarrow C_3H_8(g)$ $\Delta H = -105$ kJ

Practice Example A: The heat of combustion of propene, $C_3H_6(g)$, is -2058 kJ/mol $C_3H_6(g)$. Use this value and other data from this example to determine ΔH for the hydrogenation of propene to propane.

$$C_3H_6(g) + H_2(g) \longrightarrow C_3H_8(g)\qquad \Delta H = ?$$

Practice Example B: From the data in Practice Example A and the following reaction

$$C_3H_6(g) + H_2O(l) \longrightarrow C_3H_7OH(l)\qquad \Delta H = -36.8 \text{ kJ}$$

determine the enthalpy of combustion of one mole of 1-propanol, $C_3H_7OH(l)$.

7-8   Standard Enthalpies of Formation

As we noted in the caption to Figure 7-10, we cannot determine *absolute* values of enthalpy, H. Because enthalpy is a function of state, however, *changes* in enthalpy, ΔH, have unique values and we can deal just with these changes. Nevertheless, as with many other properties, it is useful to have a starting point, a "zero" value.

Consider again the mountain-climbing analogy of Figure 7-9. What is the elevation of the mountaintop? Do we mean by this the vertical distance between the mountaintop and the center of Earth? Between the mountaintop and the deepest trench in the ocean? No, by agreement we mean the vertical distance between the mountaintop and mean sea level. We *arbitrarily* assign to mean sea level an elevation of zero and to all other points on Earth, an elevation relative to this zero. The elevation of Mt. Everest is $+8848$ m; that of Badwater, Death Valley, California is -86 m. We do something similar with enthalpies. In effect, we relate our zero to the enthalpies of certain forms of the elements and determine the enthalpies of other substances relative to this zero. Before we can do this, however, we first need to describe *standard* states of substances.

The **standard state** of a solid or liquid substance is the pure element or compound at 1 atm pressure* and at the temperature of interest. For a gaseous substance the standard state is the (hypothetical) pure gas behaving as an ideal gas at 1 atm pressure and the temperature of interest. The **standard enthalpy of reaction** is the enthalpy *change* for a reaction in which the reactants and products are in their standard states; it is denoted with a superscript degree symbol, $\Delta H°$.

▶ We use the expression "enthalpy" of formation, although what we are describing are actually enthalpy *changes*.

The **standard enthalpy of formation** ($\Delta H_f°$) of a substance is the enthalpy *change* that occurs in the formation of 1 mole of the substance in the standard state from the reference forms of the elements in their standard states. The reference forms of the elements are the most stable forms of the elements at 1 atm pressure and the given temperature. The superscript degree symbol denotes that the enthalpy change is a standard enthalpy change, and the subscript "f" signifies that the reaction is one in which a substance is formed from its elements. Because the formation of the most stable form of an element from itself is no change at all,

the standard enthalpy of formation of a pure element in its most stable form is 0.

Listed below are the most stable forms of several elements at 298 K, the temperature at which thermochemical data are commonly tabulated.

$$Na(s) \quad H_2(g) \quad N_2(g) \quad O_2(g) \quad C(graphite) \quad Br_2(l)$$

TABLE 7.2 Some Standard Enthalpies of Formation at 298 K			
Substance	$\Delta H_{f, 298}°$, **kJ/mol**[a]	**Substance**	$\Delta H_{f, 298}°$, **kJ/mol**[a]
$CO(g)$	-110.5	$HBr(g)$	-36.40
$CO_2(g)$	-393.5	$HI(g)$	26.48
$CH_4(g)$	-74.81	$H_2O(g)$	-241.8
$C_2H_2(g)$	226.7	$H_2O(l)$	-285.8
$C_2H_4(g)$	52.26	$H_2S(g)$	-20.63
$C_2H_6(g)$	-84.68	$NH_3(g)$	-46.11
$C_3H_8(g)$	-104.7	$NO(g)$	90.25
$C_4H_{10}(g)$	-125.6	$N_2O(g)$	82.05
$CH_3OH(l)$	-238.7	$NO_2(g)$	33.18
$C_2H_5OH(l)$	-277.7	$N_2O_4(g)$	9.16
$HF(g)$	-271.1	$SO_2(g)$	-296.8
$HCl(g)$	-92.31	$SO_3(g)$	-395.7

[a]Values are for reactions in which one mole of substance is formed. Most of the data have been rounded off to four significant figures.

*The International Union of Pure and Applied Chemistry (IUPAC) recommends that the standard-state pressure be changed from 1 atm (101,325 Pa) to 1 bar (1×10^5 Pa). The effects of this change are minor, and we will continue to use 1 atm as the standard-state pressure in this text.

FIGURE 7-12
Some standard enthalpies of formation at 298 K.
Enthalpies of formation of elements are shown in the central plane, with $\Delta H_f^\circ = 0$. Substances whose enthalpies of formation are positive are represented above this plane. Those with negative enthalpies of formation are below the plane.

The situation with carbon is an interesting one. In addition to graphite, carbon also exists in the form of diamond. Because an enthalpy difference can be measured between them, these two forms of carbon cannot both be assigned an enthalpy of formation of zero. Of the two forms, graphite is assigned an enthalpy of formation of zero. Similarly, only $Br_2(l)$ can be assigned an enthalpy of formation of zero at 298 K and 1 atm. If we attempt to obtain $Br_2(g)$ at 298 K and 1 atm pressure, the gas condenses to the liquid. This is what we mean by the liquid being more stable than the gas at 298 K and 1 atm.

Enthalpies of formation of some common substances are presented in Table 7.2. A more extensive listing can be found in Appendix D. Figure 7-12 is a three-dimensional representation of a few ΔH_f° values. We will use standard enthalpies of formation in a variety of calculations. Often, the first thing we must do is write the chemical equation to which a ΔH_f° value applies, as in Example 7-9.

EXAMPLE 7-9

Relating a Standard Enthalpy of Formation to a Chemical Equation. The enthalpy of formation of formaldehyde, HCHO(g), at 298 K is $\Delta H_f^\circ = -108.6$ kJ/mol. Write the chemical equation to which this value applies.

Solution

The equation must be written for the formation of *one mole* of *gaseous* HCHO. The most stable forms of the elements at 298 K and 1 atm are *gaseous* H_2 and O_2 and solid carbon in the form of *graphite*. Note that we need one fractional coefficient in this equation. Also, see Figure 7-13.

$$H_2(g) + \tfrac{1}{2} O_2(g) + C(\text{graphite}) \longrightarrow HCHO(g) \qquad \Delta H_f^\circ = -108.6 \text{ kJ}$$

Practice Example A: The enthalpy of formation for the amino acid leucine, $C_6H_{13}O_2N(s)$, is -637.3 kJ/mol. Write the chemical equation to which this value applies.

Practice Example B: How is ΔH for the following reaction related to the enthalpy of formation of $NH_3(g)$ listed in Table 7.2? What is the value of ΔH?

$$2 \, NH_3(g) \longrightarrow N_2(g) + 3 \, H_2(g) \qquad \Delta H = ?$$

Are You Wondering . . .

What is the significance of the sign of a ΔH_f° value?

A compound having a positive value of ΔH_f° is formed from its elements by an endothermic reaction. If the reaction is reversed, the compound decomposes into its elements in an exothermic reaction. We say that the compound is "unstable" with respect to its elements. This does not mean that the compound cannot be made, but it does suggest a tendency for the compound to enter into chemical reactions yielding products with lower enthalpies of formation.

For now, we will consider enthalpy change to be a fair indicator of the likelihood of a chemical reaction occurring—exothermic reactions generally being more likely to occur unassisted than endothermic ones. We'll present much better criteria later in the text.

Standard Enthalpies of Reaction

We have previously stated that if all the reactants and products are in their standard states the enthalpy change is called the *standard* enthalpy of reaction, which we can denote either as ΔH° or ΔH_{rxn}°. The primary use of standard enthalpies of formation is in calculating standard enthalpies of reaction.

FIGURE 7-13
Standard enthalpy of formation of formaldehyde, HCHO(g).
The formation of HCHO(g) from its elements in their standard states is an *exothermic* reaction. The heat evolved per mole of HCHO(g) formed is the standard enthalpy (heat) of formation.

Let us apply Hess's law and other ideas from Section 7-7 to calculate the standard enthalpy of combustion of ethane, $C_2H_6(g)$, a component of natural gas.

$$C_2H_6(g) + \tfrac{7}{2} O_2(g) \longrightarrow 2 CO_2(g) + 3 H_2O(l) \qquad \Delta H° = ? \qquad (7.19)$$

We can add three equations to obtain Equation (7.19).

$$\text{(a)}\ C_2H_6(g) \longrightarrow 2\ C(graphite) + 3\ H_2(g)\quad \Delta H° = -\Delta H_f°[C_2H_6(g)]$$

$$\text{(b)}\ 2\ C(graphite) + 2\ O_2(g) \longrightarrow 2\ CO_2(g) \qquad\qquad \Delta H° = 2 \times \Delta H_f°[CO_2(g)]$$

$$\underline{\text{(c)}\ 3\ H_2(g) + \tfrac{3}{2} O_2(g) \longrightarrow 3\ H_2O(l) \qquad\qquad \Delta H° = 3 \times \Delta H_f°[H_2O(l)]}$$

$$C_2H_6(g) + \tfrac{7}{2} O_2(g) \longrightarrow 2\ CO_2(g) + 3\ H_2O(l) \qquad \Delta H° = ?$$

Equation (a) is the *reverse* of the equation representing the formation of one mole of $C_2H_6(g)$ from its elements. This means that $\Delta H°$ for reaction (a) is the *negative* of $\Delta H_f°[C_2H_6(g)]$. Equations (b) and (c) represent the formation of *two* moles of $CO_2(g)$ and *three* moles of $H_2O(l)$, respectively. Thus, we can express the value of $\Delta H°$ for the combustion reaction as

$$\Delta H° = \{2 \times \Delta H_f°[CO_2(g)] + 3 \times \Delta H_f°[H_2O(l)]\} - \Delta H_f°[C_2H_6(g)] \qquad (7.20)$$

Equation (7.20) is a specific application of this more general relationship for the standard enthalpy of a reaction.

$$\Delta H° = \Sigma\, v_p \Delta H_f°(\text{products}) - \Sigma\, v_r \Delta H_f°(\text{reactants}) \qquad (7.21)$$

The symbol Σ (Greek, sigma) means "the sum of." The terms that are added together are the products of the standard enthalpies of formation ($\Delta H_f°$) and their stoichiometric coefficients, v. One sum is required for the reaction products and another for the initial reactants. The enthalpy change of the reaction is the sum of terms for the products *minus* the sum of terms for the reactants. Equation (7.21) is applied in Example 7-10.

EXAMPLE 7-10

Calculating $\Delta H°$ from Tabulated Values of $\Delta H_f°$. Complete the calculation of $\Delta H°$ for reaction (7.19).

Solution

The reaction is

$$C_2H_6(g) + \tfrac{7}{2} O_2(g) \longrightarrow 2\ CO_2(g) + 3\ H_2O(l)$$

The relationship we need is (7.21). The data we substitute into the relationship are from Table 7.2.

$$\Delta H° = \{2 \times \Delta H_f°[CO_2(g)] + 3 \times \Delta H_f°[H_2O(l)]\} - \Delta H_f°[C_2H_6(g)] - \tfrac{7}{2} \Delta H_f°[O_2(g)]$$

$$= 2 \times (-393.5\ kJ) + 3 \times (-285.8\ kJ) - (-84.7\ kJ) - \tfrac{7}{2} \times (0)$$

$$= -787.0\ kJ - 857.4\ kJ + 84.7\ kJ = -1559.7\ kJ$$

Practice Example A: Calculate the heat of combustion *per mole* of a gaseous fuel that contains $C_3H_8(g)$ and $C_4H_{10}(g)$ in the mole fractions 0.62 and 0.38, respectively. [*Hint:* Write equations for the combustion reactions, and recall the definition of mole fraction (page 183).]

Practice Example B: Use $\Delta H°$ for the following reaction, together with data from Table 7.2, to determine the enthalpy of formation of benzene, $\Delta H_f°[C_6H_6(l)]$.

$$2 \, C_6H_6(l) + 15 \, O_2(g) \longrightarrow 12 \, CO_2(g) + 6 \, H_2O(l) \qquad \Delta H° = -6.535 \times 10^3 \text{ kJ}$$

[*Hint:* Use Equation (7.21), but with a *known* $\Delta H°$ and one *unknown* value of $\Delta H_f°$. Note that the $\Delta H°$ is based on two moles of $C_6H_6(l)$.]

Ionic Reactions in Solutions

Many chemical reactions in aqueous solution are best thought of as reactions between ions and best represented by net ionic equations. Consider the neutralization of a strong acid by a strong base. Using a somewhat more accurate enthalpy of neutralization than we obtained in Example 7-5, we can write

$$H^+(aq) + OH^-(aq) \longrightarrow H_2O(l) \qquad \Delta H° = -55.8 \text{ kJ} \qquad (7.22)$$

We should also be able to *calculate* this enthalpy of neutralization by using enthalpy of formation data in expression (7.21). But this requires us to have enthalpy of formation data for individual ions. And there is a slight problem in getting these. We cannot create ions of a single type in a chemical reaction. We always produce cations and anions simultaneously, as in the reaction of sodium and chlorine to produce Na^+ and Cl^- in NaCl. We must choose a particular ion to which we assign an enthalpy of formation of *zero* in its aqueous solutions. We then compare the enthalpies of formation of other ions to this reference ion. The ion we choose for our zero is $H^+(aq)$. Now let us see how we can use expression (7.21) and data from Equation (7.22) to determine the enthalpy of formation of $OH^-(aq)$.

▶ The standard enthalpy of formation of $H^+(aq)$ is arbitrarily assigned a value of *zero*.

$$\Delta H° = -55.8 \text{ kJ} = \Delta H_f°[H_2O(l)] - \{\Delta H_f°[H^+(aq)] + \Delta H_f°[OH^-(aq)]\}$$

$$-55.8 \text{ kJ} = -285.8 \text{ kJ} - 0 - \Delta H_f°[OH^-(aq)]$$

$$\Delta H_f°[OH^-(aq)] = -285.8 \text{ kJ} + 55.8 \text{ kJ} = -230.0 \text{ kJ}$$

Table 7.3 lists data for several common ions in aqueous solution. Enthalpies of formation in solution depend on the solute concentration. These data are repre-

TABLE 7.3   Some Standard Enthalpies of Formation of Ions in Aqueous Solution

Ion	$\Delta H_{f,298}°$, kJ/mol	Ion	$\Delta H_{f,298}°$, kJ/mol
H^+	0	OH^-	−230.0
Li^+	−278.5	Cl^-	−167.2
Na^+	−240.1	Br^-	−121.6
K^+	−252.4	I^-	−55.19
NH_4^+	−132.5	NO_3^-	−205.0
Ag^+	+105.6	CO_3^{2-}	−677.1
Mg^{2+}	−466.9	S^{2-}	+33.05
Ca^{2+}	−542.8	SO_4^{2-}	−909.3
Ba^{2+}	−537.6	$S_2O_3^{2-}$	−648.5
Cu^{2+}	+64.77	PO_4^{3-}	−1277
Al^{3+}	−531		

sentative for *dilute* aqueous solutions (about 1 M), the type of solution that we normally deal with. Some of these data are used in Example 7-11.

EXAMPLE 7-11

Calculating the Enthalpy Change in an Ionic Reaction. Given that $\Delta H_f^\circ[BaSO_4(s)] = -1473$ kJ/mol, what is the standard enthalpy change for the precipitation of barium sulfate?

Solution

The net ionic equation for the precipitation reaction is

$$Ba^{2+}(aq) + SO_4^{2-}(aq) \longrightarrow BaSO_4(s) \qquad \Delta H^\circ = ?$$

The enthalpy of formation of $BaSO_4(s)$ is given and those of $Ba^{2+}(aq)$ and $SO_4^{2-}(aq)$ are found in Table 7.3.

$$\Delta H^\circ = \Delta H_f^\circ[BaSO_4(s)] - \Delta H_f^\circ[Ba^{2+}(aq)] - \Delta H_f^\circ[SO_4^{2-}(aq)]$$

$$= -1473 \text{ kJ} - (-537.6 \text{ kJ}) - (-909.3 \text{ kJ})$$

$$= -26 \text{ kJ}$$

Practice Example A: Given that $\Delta H_f^\circ[AgI(s)] = -61.84$ kJ/mol, what is the standard enthalpy change for the precipitation of silver iodide?

Practice Example B: Given that $\Delta H_f^\circ[Mg(OH)_2(s)] = -924.5$ kJ/mol, what is the enthalpy change for the reaction of magnesium chloride and sodium hydroxide? (*Hint:* What precipitate forms? What is the net equation? How do the enthalpies of formation of the spectator ions enter in?)

7-9   Fuels as Sources of Energy

One of the most important uses of thermochemical measurements and calculations is in assessing materials as energy sources. For the most part these materials, called fuels, liberate heat through the process of combustion. We will briefly survey some common fuels, emphasizing matters that a thermochemical background helps us to understand.

Fossil Fuels

The bulk of current energy needs are met by petroleum, natural gas, and coal—so-called fossil fuels. These fuels are derived from plant and animal life of millions of years ago. The original source of the energy locked into these fuels is solar energy. In the process of *photosynthesis,* CO_2 and H_2O, in the presence of chlorophyll (a catalyst) and sunlight, are converted to *carbohydrates.* These are compounds with formulas $C_m(H_2O)_n$, where m and n are integers. For example, in the sugar glucose $m = n = 6$, that is, $C_6(H_2O)_6 = C_6H_{12}O_6$. Its formation through photosynthesis is an *endothermic* process, represented as

$$6\ CO_2 + 6\ H_2O \xrightarrow[\text{sunlight}]{\text{chlorophyll}} C_6H_{12}O_6 + 6\ O_2 \qquad \Delta H = +2.8 \times 10^3 \text{ kJ} \quad (7.23)$$

When reaction (7.23) is reversed, as in the combustion of glucose, heat is evolved. The combustion reaction is *exothermic.*

▶ Although the formula $C_m(H_2O)_n$ suggests a "hydrate" of carbon, in carbohydrates there are no H_2O units as there are in hydrates such as $CuSO_4 \cdot 5H_2O$. H and O atoms are simply found in the same numerical ratio as in H_2O.

The complex carbohydrate cellulose, with molecular masses ranging up to 500,000 u, is the principal structural material of plants. When plant life decomposes, in the presence of bacteria and out of contact with air, O and H atoms are removed and the approximate carbon content of the residue increases in the progression

peat $\longrightarrow$ lignite (32% C) $\longrightarrow$ subbituminous coal (40% C) $\longrightarrow$
bituminous coal (60% C) $\longrightarrow$ anthracite coal (80% C)

For this process to proceed all the way to anthracite coal may take about 300 million years. Coal, then, is a combustible organic rock consisting of carbon, hydrogen, and oxygen, together with small quantities of nitrogen, sulfur, and mineral matter (ash). (One proposed formula for a "molecule" of bituminous coal is $C_{153}H_{115}N_3O_{13}S_2$.)

Petroleum and natural gas have formed in a somewhat different way. Plants and animals living in ancient seas were deposited on the ocean floor, decomposed by bacteria, and covered with sand and mud. Over time, the sand and mud were converted to sandstone. The high pressures and temperatures resulting from this overlying sandstone rock formation transformed the original organic matter into petroleum and natural gas. The ages of these deposits range from about 250 to 500 million years.

A typical natural gas consists of about 85% methane (CH_4), 10% ethane (C_2H_6), 3% propane (C_3H_8), and small quantities of other combustible and noncombustible gases. A typical petroleum consists of several hundred different hydrocarbons, which range in complexity from C_1 molecules (i.e., CH_4) to C_{40} or higher (such as $C_{40}H_{82}$).

One way to compare different fuels is through their heats of combustion: in general, *the higher the heat of combustion the better the fuel*. Table 7.4 lists approximate heats of combustion for the fossil fuels. These data show that *biomass* (living matter or materials derived from it—wood, alcohols, municipal waste) is a viable fuel, but that fossil fuels yield more energy per unit mass.

Problems Posed by Fossil-Fuel Use

There are two fundamental problems with the use of fossil fuels. First, the formation of new fossil fuels, if it is occurring at all, cannot possibly match the rate at which existing resources are being depleted. Fossil fuels are essentially *nonrenewable* energy sources. As shown in Figure 7-14, they will be used up in a relatively short time (at least compared with the span of human history).

The second problem is their environmental effect. Sulfur impurities in fuels produce oxides of sulfur. The high temperatures associated with combustion cause the reaction of N_2 and O_2 in air to form oxides of nitrogen. Oxides of sulfur and nitrogen are implicated in air pollution and are important contributors

TABLE 7.4
Approximate Heats of Combustion of Some Fuels

Fuel	Heat of Combustion kJ/g
municipal waste	−12.7
cellulose	−17.5
pine wood	−21.2
methanol	−22.7
peat	−20.8
bituminous coal	−28.3
isooctane (a component of gasoline)	−47.8
natural gas	−49.5

FIGURE 7-14
Estimated world production of fossil fuels.
The exact shape of each curve and the predicted time when the fuel will be used up depend on the estimate used for the total quantity of recoverable fuel and the rate of production. These curves are based on no increases in current rates of consumption, which is probably an unrealistic assumption. Even a growth rate of 1% per year would cause the rate of consumption to double in 70 years and shorten the time until the remaining supplies are exhausted.

to the environmental problem known as acid rain. Another inevitable product of the combustion of fossil fuels is carbon dioxide, one of the "greenhouse" gases that may lead to global warming and changes in Earth's climate.

In the United States, coal reserves far exceed those of petroleum and natural gas. Despite this relative abundance, however, the use of coal has not increased significantly in recent years. In addition to the environmental effects cited above, the expense and hazards involved in the deep mining of coal are considerable. Surface mining, which is less hazardous and expensive than deep mining, is also more damaging to the environment. One promising possibility for using coal reserves is to convert coal to gaseous or liquid fuels, either in surface installations or while the coal is still underground.

▶ Each of these environmental issues is discussed more fully elsewhere in the text.

Gasification of Coal

Before cheap natural gas became available in the 1940s, gas produced from coal (variously called producer gas, town gas, or city gas) was widely used in the United States. This gas was manufactured by passing steam and air through heated coal and involved reactions such as

$$C(s) + H_2O(g) \longrightarrow CO(g) + H_2(g) \qquad \Delta H° = +131.3 \text{ kJ} \qquad (7.24)$$

$$CO(g) + H_2O(g) \longrightarrow CO_2(g) + H_2(g) \qquad \Delta H° = -41.2 \text{ kJ} \qquad (7.25)$$

$$2 \, C(s) + O_2(g) \longrightarrow 2 \, CO(g) \qquad \Delta H° = -221.0 \text{ kJ} \qquad (7.26)$$

$$C(s) + 2 \, H_2(g) \longrightarrow CH_4(g) \qquad \Delta H° = -74.8 \text{ kJ} \qquad (7.27)$$

The principal gasification reaction (7.24) is highly endothermic. The heat requirements for this reaction are met by the carefully controlled partial burning of coal (reaction 7.26).

A typical producer gas consists of about 23% CO, 18% H_2, 8% CO_2, and 1% CH_4, by volume. It also contains about 50% N_2 because air is used in its production. Because the N_2 and CO_2 are noncombustible, producer gas has only about 10–15% of the heat value of natural gas. Modern gasification processes include the following features:

1. Use $O_2(g)$ instead of air, thereby eliminating $N_2(g)$ in the product.

2. Provide for the removal of noncombustible $CO_2(g)$ and of sulfur impurities. For example,

$$CaO(s) + CO_2(g) \longrightarrow CaCO_3(s)$$

$$2H_2S(g) + SO_2(g) \longrightarrow 3 \, S(s) + 2 \, H_2O(g)$$

3. Include a step (called *methanation*) to convert CO and H_2, in the presence of a catalyst, to CH_4.

$$CO(g) + 3 \, H_2(g) \longrightarrow CH_4(g) + H_2O(l)$$

The product is called *substitute natural gas* (SNG), a gaseous mixture with composition and heat value similar to that of natural gas.

Liquefaction of Coal

The first step in obtaining liquid fuels from coal generally involves gasification of coal, as in reaction (7.24). This step is followed by catalytic reactions in which liquid hydrocarbons are formed.

$$n \, CO + (2n + 1) \, H_2 \longrightarrow C_nH_{2n+2} + n \, H_2O$$

Focus On

Fats, Carbohydrates, and Energy Storage

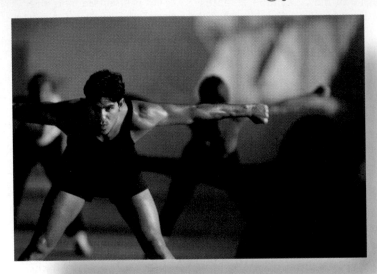

Aerobic exercise is an excellent way to burn fat and build muscle.

typical human fatty acid is palmitic acid, $CH_3(CH_2)_{14}COOH$.

The direct combustion of $CH_3(CH_2)_{14}COOH$ in a bomb calorimeter yields the same products as its metabolism in the body, together with a large quantity of heat.

$$CH_3(CH_2)_{14}COOH + 23\ O_2(g) \longrightarrow$$
$$16\ CO_2(g) + 16\ H_2O(l)$$
$$\Delta H = -9977\ kJ$$

A hydrocarbon with a similar carbon and hydrogen content, $C_{16}H_{34}$, yields a similar quantity of heat on complete combustion, $-10{,}700$ kJ. The fat stored in our bodies (our fuel) is comparable to the jet fuel in an airplane. In both cases the fuel adds weight, but can be burned to produce the large amount of energy needed for movement and life support.

Aerobics, tennis, weight lifting, and jogging are popular forms of exercise, which is important to maintaining a healthy body. Where do we get the energy to do these things? Surprisingly, mostly from fat, the body's chief energy storage system.

During exercise, fat molecules react with water (hydrolyze) to form a group of compounds called fatty acids. Through a complex series of reactions, these fatty acids are converted to carbon dioxide and water. Energy released in these reactions is used to power muscles. A

Energy from simple carbohydrates is released more rapidly than from fats, and this is why we consume sugars (fruit juice, candy bars, . . .) for a quick burst of energy. The combustion of 1 mol $C_{12}H_{22}O_{11}$ (sucrose or cane sugar), however, yields much less energy

In still another process liquid methanol is formed.

$$CO(g) + 2\ H_2(g) \longrightarrow CH_3OH(l) \tag{7.28}$$

In 1942, some 32 million gallons of aviation fuel were made from coal in Germany. In South Africa, the Sasol process for coal liquefaction is a major source of gasoline and a variety of other petroleum products and chemicals.

▶ Isooctane, $(CH_3)_3CCH_2CH(CH_3)_2$, is an excellent, smooth-burning automotive fuel; it is given an octane number of 100. Heptane, $CH_3(CH_2)_5CH_3$, burns very unevenly and causes engine "knock"; it is given an octane number of 0. Gasolines are rated against mixtures of these two hydrocarbons: A 92-octane premium gasoline is equivalent to a mixture of 92% isooctane and 8% heptane.

Methanol

Methanol, CH_3OH, can be obtained from coal by reaction (7.28). It can also be produced by thermal decomposition (pyrolysis) of wood, manure, sewage, or municipal waste. The heat of combustion of methanol is only about one half that of a typical gasoline on a mass basis, but methanol has a high octane number—106—compared with 100 for the gasoline hydrocarbon isooctane and about 92 for premium gasoline. Methanol has been tested and used as a fuel in internal combustion engines and found to burn more cleanly than gasoline. Methanol can also be used for space heating, electric power generation, fuel cells, and as a reactant to make a variety of other organic compounds.

Each of these three comparable 12-carbon compounds produces 12 mol $CO_2(g)$ and some $H_2O(l)$ as combustion products. The hydrocarbon, $C_{12}H_{24}$, releases the most energy on combustion. The fatty acid lauric acid, $CH_3(CH_2)_{10}COOH$, yields nearly as much, but the carbohydrate sucrose, $C_{12}H_{22}O_{11}$, yields considerably less.

(-5640 kJ/mol) than does a fatty acid of similar carbon and hydrogen content. In the enthalpy diagram of Figure 7-15, we can think of the sucrose molecule as in a condition somewhere between a hydrocarbon and the ultimate oxidation products: CO_2 and H_2O. A fatty acid, on the other hand, is in a condition closer to that of the hydrocarbon.

A common energy unit for expressing the energy values of food is the food Calorie (Cal), which is actually a kilocalorie.

$$1 \text{ Cal} = 1 \text{ kcal} = 4.184 \text{ kJ}$$

As energy sources, fats yield about 9 Cal/g (38 kJ/g), whereas carbohydrates and proteins both yield about 4 Cal/g (17 kJ/g).

Consider a 65-kg (143-lb) person of average height and build. Typically this person would have about 11 kg (24 lb) of stored fats—a fuel reserve of about 4.2×10^5 kJ. On the other hand, if the body stored only carbohydrates, the stored reservoir would have to be about 25 kg. The "65-kg" person would have to weigh about 80 kg (176 lb). A more dramatic picture might be that of a bird carrying its energy reserves as carbohydrates instead of fat. The excess "baggage" might make it impossible for the bird to fly.

It is important to understand some other matters about fat. First, the quantity of stored fat that an individual typically carries is greater than what is needed to meet energy requirements. Excess body fat leads to an increased risk of heart disease, diabetes, and other medical problems. Second, one does not have to consume fats in order to store body fats. All food types are broken down into small molecules during digestion, and these small molecules are reassembled into the more complex structures that the body needs. In the body, fat molecules can be synthesized from the small molecules produced from carbohydrates, for example. Current medical findings indicate that the most healthful diet provides the bulk of food calories through carbohydrates, with the recommended percent of caloric intake from fats being less than 30%.

Ethanol

Ethanol, C_2H_5OH, is produced mostly from ethylene, C_2H_4, which in turn is derived from petroleum. Current interest centers on the production of ethanol by the fermentation of organic matter, a process known throughout recorded history. Ethanol production by fermentation is probably in its most advanced development in Brazil, where sugar cane and cassava (manioc) are the plant matter ("biomass") used. In the United States ethanol is used chiefly as 90% gasoline–10% ethanol mixture called *gasohol*. Ethanol is also used to raise the octane number of gasoline.

Hydrogen

Another fuel with great potential is hydrogen. Its most attractive features are that

- on a per gram basis, its heat of combustion is more than twice that of methane and about three times that of gasoline, and
- the product of its combustion is H_2O, not CO and CO_2 as with gasoline.

Currently, the bulk of hydrogen used commercially is made from petroleum and natural gas. (Alternative methods of producing hydrogen, and the prospects of developing an economy based on hydrogen are discussed in the next chapter.)

Alternative Energy Sources

Combustion reactions are only one means of extracting useful energy from materials. For example, reactions that yield the same products as combustion can be carried out in electrochemical cells, called *fuel cells*. The energy is released as electricity rather than as heat (Section 21-5). Solar energy can be used directly, without recourse to photosynthesis. Nuclear processes can be used in place of chemical reactions (Chapter 26).

Alternative energy sources, including those intended for automobiles, are expected to increase in importance dramatically in the 21st century.

Summary

Thermochemistry is concerned with heat and work accompanying chemical reactions. Reactions in which heat is *evolved* by the system to the surroundings are called *exothermic*. Those in which heat is *absorbed* by the system are called *endothermic*. Heat effects are most commonly measured through temperature changes in a calorimeter. The most common type of work in chemical reactions is that associated with the expansion and compression of gases, pressure–volume work.

Combustion reactions can be carried out in a bomb calorimeter. Because these reactions occur at constant volume, the measured heats of reaction are $q_V = \Delta E$. Most reactions are carried out in containers open to the atmosphere, and for these reactions we use the enthalpy change (ΔH). For a reaction conducted at constant pressure and with work limited to pressure–volume work, the heat of reaction is $q_P = \Delta H$.

Thermochemical calculations are of two main types.

1. Calculating heat effects from calorimetric data—heat capacities, specific heats, masses, and temperature changes.

2. Manipulating known ΔH values to obtain an unknown value. (In some cases these manipulations require the use of Hess's law of constant heat summation. In others, they are based on the concepts of standard states and standard enthalpies of formation of pure substances and of ions in aqueous solution.)

One of the chief applications of thermochemistry is to the study of the combustion of fuels as energy sources. Currently, the principal fuels are the fossil fuels, but potential alternative fuels are also mentioned in this chapter and discussed in more depth later in the text.

Summarizing Example

The partial burning of coal in the presence of O_2 and H_2O produces a mixture of CO(g) and H_2(g) called *synthesis gas*. This gas can be used to synthesize organic compounds or it can be burned as a fuel. A typical synthesis gas consists of 55.0% CO(g), 33.0% H_2(g), and 12.0% noncombustible gases (mostly CO_2), *by volume*.

To what temperature can 1.00 kg water at 25.0 °C be heated with the heat liberated by the combustion of 1.00 L (STP) of this typical synthesis gas?

1. *Determine the number of moles of CO(g) and H_2(g) in a 1.00-L sample.* Use the molar volume at STP to determine the total number of moles of gas. Then use expression (6.17) to convert from volume ratios to mole ratios of the components. *Result: $n_{CO} = 0.0245$ mol; $n_{H_2} = 0.0147$ mol.*

2. *Use tabulated data to determine the enthalpy of combustion of the 1.00 L of synthesis gas.* Write an equation for the combustion of H_2(g). H_2O(l) is the product. Determine $\Delta H_{comb.}$ from enthalpy of formation data. Use this ΔH as a conversion factor (as in Example 7-7). Do the same for CO(g), with CO_2(g) as the product of combustion. *Result: -4.20 kJ (for H_2); -6.93 kJ (for CO); total: -11.13 kJ.*

3. *Apply Equation (7.6).* In the expression

q_{water} = mass water × specific heat of water × ΔT

solve for ΔT and add this to the initial temperature, 25.0 °C. Recall that $q_{water} = -q_{comb.}$; $q_{comb.} = -11.13$ kJ. Answer: 27.7 °C.

Key Terms

bomb calorimeter (7-4)
calorie (cal) (7-3)
calorimeter (7-4)
chemical energy (7-1)
closed system (7-1)
endothermic (7-4)
energy (7-1)
enthalpy (H) (7-6)
enthalpy change (ΔH) (7-6)
enthalpy diagram (7-6)
exothermic (7-4)

first law of thermodynamics (7-5)
function of state (state function)
 (7-5)
heat (7-3)
heat capacity (7-3)
heat of reaction (7-4)
Hess's law (7-7)
internal energy (E) (7-5)
isolated system (7-1)
kinetic energy (7-1)
law of conservation of energy (7-3)

open system (7-1)
potential energy (7-1)
pressure–volume work (7-2)
specific heat (7-3)
standard enthalpy of formation (7-8)
standard enthalpy of reaction (7-8)
standard state (7-8)
surroundings (7-1)
system (7-1)
thermal energy (7-1)
work (7-1, 7-2)

Review Questions

1. In your own words define or explain the following terms or symbols: **(a)** ΔH; **(b)** $P\Delta V$; **(c)** ΔH_f°; **(d)** standard state; **(e)** fossil fuel.

2. Briefly describe each of the following ideas or methods: **(a)** law of conservation of energy; **(b)** bomb calorimetry; **(c)** function of state; **(d)** enthalpy diagram; **(e)** Hess's law.

3. Explain the important distinctions between each pair of terms: **(a)** system and surroundings; **(b)** heat and work; **(c)** specific heat and heat capacity; **(d)** endothermic and exothermic.

4. Calculate the quantity of heat
(a) in *kilocalories,* required to raise the temperature of 9.25 L of water from 22.0 to 29.4 °C;
(b) in *kilojoules,* associated with a 33.5 °C *decrease* in temperature in a 5.85-kg aluminum bar (specific heat of aluminum = 0.903 J g^{-1} °C^{-1}).

5. Calculate the final temperature that results when
(a) a 12.6-g sample of water at 22.9 °C *absorbs* 875 J of heat;
(b) a 1.59-kg sample of platinum at 78.2 °C *gives off* 1.05 kcal of heat (specific heat of Pt = 0.032 cal g^{-1} °C^{-1}).

6. From the information given, determine
(a) the specific heat of toluene (C_7H_8), given that 186 J of heat is required to raise the temperature of a 15.0-g sample from 22.3 to 29.6 °C;
(b) the final temperature, when 2.75 kcal of heat is removed from a 2.25-kg sample of water initially at 23.1 °C.

7. A mixture of 118 g of copper and 197 g of water is heated from 22.7 to 79.2 °C. How much heat, in kilojoules, is absorbed by the copper–water mixture? (Specific heats: water, 4.18 J g^{-1} °C^{-1}; copper, 0.385 J g^{-1} °C^{-1}.)

8. A 1.22-kg piece of iron at 126.5 °C is dropped into 981 g water at 22.1 °C. The temperature rises to 34.4 °C. Determine the specific heat of iron, in J g^{-1} °C^{-1}.

9. *Without doing detailed calculations,* decide which of the following is a plausible final temperature when 75.0 mL of water at 80.0 °C is added to 100.0 mL of water at 20.0 °C: (1) 40 °C; (2) 46 °C; (3) 50 °C; (4) 28 °C. Explain your reasoning.

10. What are the internal energy changes, ΔE, if a system
(a) absorbs 67 J of heat and does 67 J of work?

(b) absorbs 356 J of heat and does 592 J of work?
(c) loses 38 J of heat and has 171 J of work done on it.
(d) absorbs no heat and does 416 J of work?

11. Upon complete combustion, the following quantities of heat are evolved by the substances indicated. Express each heat of combustion in kilojoules per mole of substance.
(a) 0.584 g of propane, $C_3H_8(g)$, yields 29.4 kJ.
(b) 0.136 g of camphor, $C_{10}H_{16}O$, yields 1.26 kcal.
(c) 2.35 mL of acetone, $(CH_3)_2CO(l)$ ($d = 0.791$ g/mL), yields 58.3 kJ.

12. A sample gives off 5228 cal when burned in a bomb calorimeter. The temperature of the calorimeter assembly increases by 4.39 °C. Calculate the heat capacity of the calorimeter, in kilojoules per degree Celsius.

13. The following substances undergo complete combustion in a bomb calorimeter. The calorimeter assembly has a heat capacity of 5.136 kJ/°C. In each case, what is the final temperature if the initial water temperature is 22.43 °C?
(a) 0.3268 g caffeine, $C_8H_{10}O_2N_4$; heat of combustion = -1014.2 kcal/mol caffeine.
(b) 1.35 ml of methyl ethyl ketone, $C_4H_8O(l)$ ($d = 0.805$ g/mL); heat of combustion = -2444 kJ/mol methyl ethyl ketone.

14. A bomb calorimetry experiment is performed with xylose, $C_5H_{10}O_5(s)$, as the combustible substance. The data obtained are

mass of xylose burned: 1.183 g
heat capacity of calorimeter: 4.728 kJ/°C
initial calorimeter temperature: 23.29 °C
final calorimeter temperature: 27.19 °C

(a) What is the heat of combustion of xylose, in kilojoules per mole?
(b) Write the chemical equation for the complete combustion of xylose and represent the value of ΔH in this equation. (Assume for this reaction that $\Delta E \approx \Delta H$.)

15. A coffee cup calorimeter contains 100.0 mL of 0.300 M HCl at 20.3 °C. When 1.82 g Zn(s) is added, the temperature rises to 30.5 °C. What is the heat of reaction per mol Zn? Make the same assumptions as in Example 7-5, and also that there is no heat lost to the $H_2(g)$ that escapes.

$$Zn(s) + 2\,H^+(aq) \longrightarrow Zn^{2+}(aq) + H_2(g)$$

16. A 0.75-g sample of KCl is added to 35.0 g H_2O in a Styrofoam cup and stirred until it dissolves. The temperature of the solution drops from 24.8 to 23.6 °C.
(a) Is the process endothermic or exothermic?
(b) What is the heat of solution of KCl expressed in kilojoules per mole of KCl?

17. The heat of solution of potassium acetate, $KC_2H_3O_2(s)$, in water is -15.3 kJ/mol $KC_2H_3O_2$. If 0.136 mol $KC_2H_3O_2$ is dissolved in 525 mL water that is initially at 25.1 °C, what will be the final solution temperature?

18. Write the balanced chemical equations that have the following as their standard enthalpy changes.
(a) $\Delta H_f^\circ = +82.05$ kJ/mol $N_2O(g)$
(b) $\Delta H_f^\circ = -394.1$ kJ/mol $SO_2Cl_2(l)$
(c) $\Delta H_{combustion}^\circ = -1527$ kJ/mol $CH_3CH_2COOH(l)$

19. The complete combustion of butane, $C_4H_{10}(g)$, is represented by the equation

$$C_4H_{10}(g) + \tfrac{13}{2} O_2(g) \longrightarrow 4 CO_2(g) + 5 H_2O(l)$$
$$\Delta H = -2877 \text{ kJ}$$

How much heat, in kilojoules, is evolved in the complete combustion of **(a)** 1.325 g $C_4H_{10}(g)$; **(b)** 28.4 L $C_4H_{10}(g)$ at STP; **(c)** 12.6 L $C_4H_{10}(g)$ at 23.6 °C and 738 mmHg?

20. The enthalpy of formation of $NH_3(g)$ is -46.11 kJ/mol NH_3. What is ΔH for the following reaction?

$$\tfrac{2}{3} NH_3(g) \longrightarrow \tfrac{1}{3} N_2(g) + H_2(g) \qquad \Delta H = ?$$

21. Use Hess's law to determine ΔH for the reaction $CO(g) + \tfrac{1}{2} O_2(g) \rightarrow CO_2(g)$, given that
(1) $C(graphite) + \tfrac{1}{2} O_2(g) \rightarrow CO(g) \quad \Delta H = -110.54$ kJ
(2) $C(graphite) + O_2(g) \rightarrow CO_2(g) \quad \Delta H = -393.51$ kJ

22. Use Hess's law to determine ΔH for the reaction $C_3H_4(g) + 2 H_2(g) \rightarrow C_3H_8(g)$, given that
(1) $H_2(g) + \tfrac{1}{2} O_2(g) \rightarrow H_2O(l) \qquad \Delta H = -285.8$ kJ
(2) $C_3H_4(g) + 4 O_2(g) \rightarrow 3 CO_2(g) + 2 H_2O(l)$
$$\Delta H = -1937 \text{ kJ}$$
(3) $C_3H_8(g) + 5 O_2(g) \rightarrow 3 CO_2(g) + 4 H_2O(l)$
$$\Delta H = -2219.1 \text{ kJ}$$

23. Given the following information:
$$\tfrac{1}{2} N_2(g) + \tfrac{3}{2} H_2(g) \longrightarrow NH_3(g) \qquad \Delta H_1$$
$$NH_3(g) + \tfrac{5}{4} O_2(g) \longrightarrow NO(g) + \tfrac{3}{2} H_2O(l) \qquad \Delta H_2$$
$$H_2(g) + \tfrac{1}{2} O_2(g) \longrightarrow H_2O(l) \qquad \Delta H_3$$
Determine ΔH for the following reaction, expressed in terms of ΔH_1, ΔH_2, and ΔH_3.
$$N_2(g) + O_2(g) \longrightarrow 2 NO(g) \qquad \Delta H = ?$$

24. Use enthalpies of formation from Table 7.2 in Equation (7.21) to determine the enthalpy changes in the following reactions.
(a) $C_3H_8(g) + H_2(g) \rightarrow C_2H_6(g) + CH_4(g)$
(b) $2 H_2S(g) + 3 O_2(g) \rightarrow 2 SO_2(g) + 2 H_2O(l)$

25. Use enthalpies of formation from Tables 7.2 and 7.3 and Equation (7.21) to determine the enthalpy change of the following reaction.
$$NH_4^+(aq) + OH^-(aq) \longrightarrow H_2O(l) + NH_3(g)$$

26. Use the information given here, data from Appendix D, and Equation (7.21) to calculate the enthalpy of formation, *per mole*, of $ZnS(s)$.
$$2 ZnS(s) + 3 O_2(g) \longrightarrow 2 ZnO(s) + 2 SO_2(g)$$
$$\Delta H^\circ = -878.2 \text{ kJ}$$

Exercises

Work

27. Calculate the quantity of work, in joules, associated with a 3.5-L expansion of a gas (ΔV) against a pressure of 748 mmHg in the units **(a)** liter atmospheres (L atm); **(b)** joules (J); **(c)** calories (cal). [*Hint:* You can use the different units of the gas constant R (from the inside back cover) to establish conversion factors among energy units.]

28. Calculate the quantity of work, in joules, associated with the compression of a gas from 5.62 L to 3.37 L by a constant pressure of 1.23 atm.

Heat Capacity (Specific Heat)

29. Refer to Example 7-3. The experiment is repeated with several different metals substituting for the lead. The masses of metal and water and the initial temperatures of the metal and water are the same as in Figure 7-4. The final temperatures are **(a)** Zn, 38.9 °C; **(b)** Pt, 28.8 °C; **(c)** Al, 52.7 °C. What is the specific heat of each metal, expressed in J g^{-1} °C^{-1}?

30. A 75.0-g piece of Ag metal is heated to 80.0 °C and dropped into 50.0 g of water at 23.2 °C. The final temperature of the Ag–H_2O mixture is 27.6 °C. What is the specific heat of silver?

31. A 465-g chunk of iron is removed from an oven and plunged into 375 g water in an insulated container. The temperature of the water increases from 26 to 87 °C. If the specific heat of iron is 0.449 J g^{-1} °C^{-1}, what must have been the original oven temperature?

32. A piece of stainless steel (specific heat = 0.50 J g^{-1} °C^{-1}) is transferred from an oven (183 °C) to 125 mL of water at 23.2 °C, into which it is immersed. The water temperature rises to 51.5 °C. What is the mass of the steel? How precise is this method of mass determination? Explain.

33. A 1.00-kg sample of magnesium at 40.0 °C is added to 1.00 L of water maintained at 20.0 °C in an insulated container. What will be the final temperature of the Mg–H_2O mixture? (Specific heat of Mg = 1.024 J g^{-1} °C^{-1}.)

35. A 74.8-g sample of copper at 143.2 °C is added to an insulated vessel containing 165 mL of glycerol, $C_3H_8O_3$(l) (d = 1.26 g/mL), at 24.8 °C. The final temperature is 31.1 °C. The specific heat of copper is 0.385 J g^{-1} °C^{-1}. What is the heat capacity of glycerol in J mol^{-1} °C^{-1}?

34. Brass has a density of 8.40 g/cm^3 and a specific heat of 0.385 J g^{-1} °C^{-1}. A 15.2-cm^3 piece of brass at an initial temperature of 163 °C is dropped into an insulated container with 150.0 g water initially at 22.4 °C. What will be the final temperature of the brass–water mixture?

36. What volume of 18.5 °C water must be added, together with a 1.23-kg piece of iron at 68.5 °C, so that the temperature of the water in the insulated container shown in the figure remains constant at 25.6 °C?

Heats of Reaction

37. How much heat, in kilojoules, is associated with the production of 283 kg of slaked lime, Ca(OH)$_2$?

$$CaO(s) + H_2O(l) \longrightarrow Ca(OH)_2(s) \qquad \Delta H = -65.2 \text{ kJ}$$

39. The combustion of methane gas, the principal constituent of natural gas, is represented by the equation

$$CH_4(g) + 2 O_2(g) \longrightarrow CO_2(g) + 2 H_2O(l)$$
$$\Delta H = -890.3 \text{ kJ}$$

(a) What mass of methane, in kilograms, must be burned to liberate 2.80 × 10^7 kJ of heat?
(b) What quantity of heat, in kilojoules, is liberated in the complete combustion of 1.65 × 10^4 L of CH_4(g), measured at 18.6 °C and 768 mmHg?
(c) If the quantity of heat calculated in part **(b)** could be transferred with 100% efficiency to water, what volume of water, in liters, could be heated from 8.8 to 60.0 °C as a result?

41. The combustion of hydrogen–oxygen mixtures is used to produce very high temperatures (ca. 2500 °C) needed for certain types of welding operations. Consider the reaction to be

$$H_2(g) + \tfrac{1}{2} O_2(g) \longrightarrow H_2O(g) \qquad \Delta H = -241.8 \text{ kJ}$$

What is the quantity of heat, in kilojoules, evolved when a 180-g mixture containing *equal* parts of H_2 and O_2, *by mass*, is burned?

38. The enthalpy change in the combustion of the hydrocarbon octane is $\Delta H = -5.48 \times 10^3$ kJ/mol C_8H_{18}(l). How much heat, in kilojoules, is liberated *per gallon* of octane burned? (Density of octane = 0.703 g/mL; 1 gal = 3.785 L.)
40. Refer to the Summarizing Example (parts **2** and **3**). What volume of the synthesis gas, measured at STP and burned in an open flame (constant-pressure process), is required to heat 40.0 gal of water from 15.2 to 65.0 °C? (1 gal = 3.785 L.)

42. Thermite mixtures are used for certain types of welding. The thermite reaction is highly exothermic.

$$Fe_2O_3(s) + 2 Al(s) \longrightarrow Al_2O_3(s) + 2 Fe(s)$$
$$\Delta H = -852 \text{ kJ}$$

1.00 mol Fe_2O_3 and 2.00 mol Al are mixed at room temperature (25 °C) and a reaction is initiated. The liberated heat is retained within the products, whose combined specific heats over a broad temperature range is about 0.8 J g^{-1} °C^{-1}. The melting point of iron is 1530 °C. Show that the quantity of heat liberated is more than sufficient to raise the temperature of the products to the melting point of iron.

43. A 0.205-g pellet of potassium hydroxide, KOH, is added to 55.9 g water in a Styrofoam coffee cup. The water temperature rises from 23.5 to 24.4 °C. [Assume that the specific heat of dilute KOH(aq) is the same as that of water.]
(a) What is the approximate heat of solution of KOH, expressed as kilojoules per mole of KOH?
(b) How could the precision of this measurement be improved *without* modifying the apparatus?

45. You are planning a lecture demonstration to illustrate an *endothermic* process. You desire to lower the temperature of 1400 mL water in an insulated container from 25 to 10 °C. Approximately what mass of NH₄Cl(s) should you dissolve in the water to achieve this result? The heat of solution of NH₄Cl is +14.7 kJ/mol NH₄Cl.

47. Refer to Example 7-5. The product of the neutralization is 0.500 M NaCl. For this solution, assume a density of 1.02 g/mL and a specific heat of $4.02 \text{ J g}^{-1} \text{ °C}^{-1}$. Also, assume a heat capacity of the Styrofoam cup of 10 J/°C, and recalculate the heat of neutralization.

44. The heat of solution of KI(s) in water is +20.3 kJ/mol KI. If a quantity of KI is added to sufficient water at 23.5 °C in a Styrofoam cup to produce 150.0 mL of 2.50 M KI, what will be the final temperature? (Assume a density of 1.30 g/mL and a specific heat of $2.7 \text{ J g}^{-1} \text{ °C}^{-1}$ for 2.50 M KI.)

46. Care must be taken in preparing solutions of solutes that liberate heat on dissolving. The heat of solution of NaOH is −44.5 kJ/mol NaOH. To what maximum temperature may a sample of water, originally at 21 °C, be raised in the preparation of 500 mL of 7.0 M NaOH? Assume the solution has a density of 1.08 g/mL and specific heat of $4.00 \text{ J g}^{-1} \text{ °C}^{-1}$.

48. The heat of neutralization of HCl(aq) by NaOH(aq) is −55.84 kJ/mol H₂O produced. If 50.00 mL of 1.05 M NaOH is added to 25.00 mL of 1.86 M HCl, with both solutions originally at 24.72 °C, what will be the final solution temperature? (Assume that no heat is lost to the surrounding air and that the solution produced in the neutralization reaction has a density of 1.02 g/mL and a specific heat of $3.98 \text{ J g}^{-1} \text{ °C}^{-1}$.)

Bomb Calorimetry

49. A 1.620-g sample of naphthalene, $C_{10}H_8(s)$, is completely burned in a bomb calorimeter assembly and a temperature increase of 8.44 °C is noted. If the heat of combustion of naphthalene is −5156.1 kJ/mol $C_{10}H_8$, what is the heat capacity of the bomb calorimeter?

50. Salicylic acid, $C_7H_6O_3$, has been suggested as a calorimetric standard. Its heat of combustion is -3.023×10^3 kJ/mol $C_7H_6O_3$. From the following data determine the heat capacity of a bomb calorimeter assembly (i.e., of the bomb, water, stirrer, thermometer, wires, . . .).

mass of salicylic acid burned:	1.201 g
initial calorimeter temperature:	23.68 °C
final calorimeter temperature:	29.82 °C

51. Refer to Example 7-4. Based on the heat of combustion of sucrose established in the example, what should be the temperature change (ΔT) produced by the combustion of 1.227 g $C_{12}H_{22}O_{11}$ in a bomb calorimeter assembly with a heat capacity of 3.87 kJ/°C?

52. A 1.397-g sample of thymol, $C_{10}H_{14}O(s)$ (a preservative and a mold and mildew preventative), is burned in a bomb calorimeter assembly. The temperature increase is 11.23 °C and the heat capacity of the bomb calorimeter is 4.68 kJ/°C. What is the heat of combustion of thymol, expressed in kilojoules per mole of $C_{10}H_{14}O$?

Hess's Law

53. For the reaction $C_2H_4(g) + Cl_2(g) \rightarrow C_2H_4Cl_2(l)$ determine ΔH, given that

$$4 \text{ HCl}(g) + O_2(g) \longrightarrow 2 \text{ Cl}_2(g) + 2 \text{ H}_2\text{O}(l)$$
$$\Delta H = -202.4 \text{ kJ}$$
$$2 \text{ HCl}(g) + C_2H_4(g) + \tfrac{1}{2} O_2(g) \longrightarrow C_2H_4Cl_2(l) + H_2O(l)$$
$$\Delta H = -318.7 \text{ kJ}$$

55. Substitute natural gas (SNG) is a gaseous mixture containing CH₄(g) that can be used as a fuel. One reaction for the production of SNG is

$$4 \text{ CO}(g) + 8 \text{ H}_2(g) \longrightarrow 3 \text{ CH}_4(g) + CO_2(g) + 2 \text{ H}_2\text{O}(l)$$
$$\Delta H = ?$$

Use appropriate data from the following listing to determine ΔH for this SNG reaction.

$$\text{C(graphite)} + \tfrac{1}{2} O_2(g) \longrightarrow CO(g) \qquad \Delta H = -110.5 \text{ kJ}$$
$$CO(g) + \tfrac{1}{2} O_2(g) \longrightarrow CO_2(g) \qquad \Delta H = -283.0 \text{ kJ}$$
$$H_2(g) + \tfrac{1}{2} O_2(g) \longrightarrow H_2O(l) \qquad \Delta H = -285.8 \text{ kJ}$$
$$\text{C(graphite)} + 2 \text{ H}_2(g) \longrightarrow CH_4(g) \qquad \Delta H = -74.81 \text{ kJ}$$
$$CH_4(g) + 2 \text{ O}_2(g) \longrightarrow CO_2(g) + 2 \text{ H}_2\text{O}(l) \quad \Delta H = -890.3 \text{ kJ}$$

54. Determine ΔH for the reaction

$$N_2H_4(l) + 2 \text{ H}_2O_2(l) \longrightarrow N_2(g) + 4 \text{ H}_2\text{O}(l)$$

from these data.

$$N_2H_4(l) + O_2(g) \longrightarrow N_2(g) + 2 \text{ H}_2\text{O}(l) \qquad \Delta H = -622.2 \text{ kJ}$$
$$H_2(g) + \tfrac{1}{2} O_2(g) \longrightarrow H_2O(l) \qquad \Delta H = -285.8 \text{ kJ}$$
$$H_2(g) + O_2(g) \longrightarrow H_2O_2(l) \qquad \Delta H = -187.8 \text{ kJ}$$

56. CCl₄, an important commercial solvent, is prepared by the reaction of Cl₂(g) with a carbon compound. Determine ΔH for the reaction

$$CS_2(l) + 3 \text{ Cl}_2(g) \longrightarrow CCl_4(l) + S_2Cl_2(l)$$

Use appropriate data from the following listing.

$$CS_2(l) + 3 \text{ O}_2(g) \longrightarrow CO_2(g) + 2 \text{ SO}_2(g) \quad \Delta H = -1077 \text{ kJ}$$
$$2 \text{ S}(s) + Cl_2(g) \longrightarrow S_2Cl_2(l) \qquad \Delta H = -58.2 \text{ kJ}$$
$$\text{C}(s) + 2 \text{ Cl}_2(g) \longrightarrow CCl_4(l) \qquad \Delta H = -135.4 \text{ kJ}$$
$$\text{S}(s) + O_2(g) \longrightarrow SO_2(g) \qquad \Delta H = -296.8 \text{ kJ}$$
$$SO_2(g) + Cl_2(g) \longrightarrow SO_2Cl_2(l) \qquad \Delta H = +97.3 \text{ kJ}$$
$$\text{C}(s) + O_2(g) \longrightarrow CO_2(g) \qquad \Delta H = -393.5 \text{ kJ}$$
$$CCl_4(l) + O_2(g) \longrightarrow COCl_2(g) + Cl_2O(g) \quad \Delta H = -5.2 \text{ kJ}$$

57. Use Hess's law and the following data

$$CH_4(g) + 2 O_2(g) \longrightarrow CO_2(g) + 2 H_2O(g) \qquad \Delta H = -802 \text{ kJ}$$
$$CH_4(g) + CO_2(g) \longrightarrow 2 CO(g) + 2 H_2(g) \qquad \Delta H = +247 \text{ kJ}$$
$$CH_4(g) + H_2O(g) \longrightarrow CO(g) + 3 H_2(g) \qquad \Delta H = +206 \text{ kJ}$$

to determine ΔH for the reaction

$$CH_4(g) + \tfrac{1}{2} O_2(g) \longrightarrow CO(g) + 2 H_2(g),$$

an important commercial source of hydrogen gas.

58. The heats of combustion (ΔH) per mole of 1,3-butadiene [$C_4H_6(g)$], butane [$C_4H_{10}(g)$], and $H_2(g)$ are -2540.2, -2877.6, and -285.8 kJ, respectively. Use these data to calculate the heat of hydrogenation of 1,3-butadiene to butane.

$$C_4H_6(g) + 2 H_2(g) \longrightarrow C_4H_{10}(g) \qquad \Delta H = ?$$

[*Hint:* Write equations for the combustion reactions. In each combustion the products are $CO_2(g)$ and $H_2O(l)$.]

Standard Enthalpies of Formation

59. Why is it that the standard enthalpies of formation of some compounds have positive values whereas others have negative values? Are there likely to be many compounds that have a standard enthalpy of formation of zero? Explain.

60. Only one of the following statements is correct. Choose the correct one and tell what is wrong with the others. The standard enthalpy of formation of $CO_2(g)$ is (a) 0; (b) the standard enthalpy of combustion of C(graphite); (c) the sum of the standard enthalpies of formation of CO(g) and $O_2(g)$; (d) the standard enthalpy of combustion of CO(g).

61. Use standard enthalpies of formation from Table 7.2 to determine the enthalpy change at 25 °C for the reaction

$$2 Cl_2(g) + 2 H_2O(l) \longrightarrow 4 HCl(g) + O_2(g) \qquad \Delta H° = ?$$

63. Use data from Table 7.2 to determine the heat of combustion of $C_2H_5OH(l)$, if reactants and products are maintained at 25 °C and 1 atm.

62. Use data from Appendix D to calculate the enthalpy change for the following reaction at 25 °C.

$$Fe_2O_3(s) + 3 CO(g) \longrightarrow 2 Fe(s) + 3 CO_2(g) \qquad \Delta H° = ?$$

64. Use data from Table 7.2, together with the fact that $\Delta H° = -3509$ kJ for the complete combustion of one mole of pentane, $C_5H_{12}(l)$, to calculate $\Delta H°$ for the synthesis of 1 mol $C_5H_{12}(l)$ from CO(g) and $H_2(g)$.

$$5 CO(g) + 11 H_2(g) \longrightarrow C_5H_{12}(l) + 5 H_2O(l) \qquad \Delta H° = ?$$

(*Hint:* You can get $\Delta H_f°[C_5H_{12}(l)]$ from the combustion equation.)

65. Use data from Table 7.2 and $\Delta H°$ for the following reaction to determine the enthalpy of formation of $CCl_4(g)$ at 25 °C and 1 atm.

$$CH_4(g) + 4 Cl_2(g) \longrightarrow CCl_4(g) + 4 HCl(g)$$
$$\Delta H° = -397.3 \text{ kJ}$$

67. Use data from Table 7.3 and Appendix D to determine the standard enthalpy change in the reaction

$$Al^{3+}(aq) + 3 OH^-(aq) \longrightarrow Al(OH)_3(s) \qquad \Delta H° = ?$$

69. The decomposition of limestone, $CaCO_3(s)$, into quicklime, CaO(s), and $CO_2(g)$ is carried out in a gas-fired kiln. Use data from Appendix D to determine how much heat is required to decompose 1.35×10^3 kg $CaCO_3(s)$. (Assume that heats of reaction are the same as at 25 °C and 1 atm.)

66. Use data from Table 7.2 and $\Delta H°$ for the following reaction to determine the enthalpy of formation of hexane, $C_6H_{14}(l)$, at 25 °C and 1 atm.

$$2 C_6H_{14}(l) + 19 O_2(g) \longrightarrow 12 CO_2(g) + 14 H_2O(l)$$
$$\Delta H° = -8326 \text{ kJ}$$

68. Use data from Table 7.3 and Appendix D to determine the enthalpy change in the reaction

$$Mg(OH)_2(s) + 2 NH_4^+(aq) \longrightarrow$$
$$Mg^{2+}(aq) + 2 H_2O(l) + 2 NH_3(g) \qquad \Delta H° = ?$$

70. Use data from Table 7.2 to calculate the volume of butane, $C_4H_{10}(g)$, measured at 24.6 °C and 756 mmHg, that must be burned to liberate 5.00×10^4 kJ of heat.

Advanced Exercises

71. A British thermal unit (Btu) is defined as the quantity of heat required to change the temperature of 1 lb of water by 1 °F. Assume the specific heat of water to be independent of temperature. How much heat is required to raise the temperature of the water in a 40-gal water heater from 48 to 145 °F (a) in Btu; (b) in kcal; (c) in kJ? (1 L = 1.06 qt; 4 qt = 1 gal; 1 kg = 2.205 lb.)

72. A 7.26-kg shot (as used in the sporting event the shot put) is dropped from the top of a building 168 m high. What is the maximum temperature increase that could occur in the shot? Assume a specific heat of 0.47 J g^{-1} °C^{-1} for the shot. Why

would the actual measured temperature increase likely be less than the calculated value? [*Hint:* Use Equations (B.1) and (B.2) from Appendix B.]

73. *The internal energy of a fixed quantity of an ideal gas depends only on its temperature.* A sample of an ideal gas is allowed to expand at a constant temperature (isothermal expansion). (a) Does the gas do work? (b) Does the gas exchange heat with its surroundings? (c) What happens to the temperature of the gas? (d) What is ΔE for the gas?

74. In an *adiabatic* process a system is thermally insulated from its surroundings. There is no exchange of heat ($q = 0$).

For the adiabatic expansion of an ideal gas **(a)** Does the gas do work? **(b)** Does the internal energy of the gas increase, decrease, or remain constant? **(c)** What happens to the temperature of the gas? (*Hint:* Also refer to Exercise 73.)

75. An alternative approach to bomb calorimetry is to establish the heat capacity of the calorimeter, *exclusive* of the water it contains. The heat absorbed by the water and by the rest of the calorimeter must be calculated separately and then added together.

A bomb calorimeter assembly containing 983.5 g water is calibrated by the combustion of 1.354 g anthracene. The temperature of the calorimeter rises from 24.87 to 35.63 °C. When 1.053 g citric acid is burned in the same assembly, but containing 968.6 g water, the temperature increases from 25.01 to 27.19 °C. The heat of combustion of anthracene, $C_{14}H_{10}(s)$, is -7067 kJ/mol $C_{14}H_{10}$. What is the heat of combustion of citric acid, $C_6H_8O_7$, expressed in kilojoules per mole?

76. The method of Exercise 75 is used in some bomb calorimetry experiments. A 1.148-g sample of benzoic acid is burned in an excess of $O_2(g)$ in a bomb immersed in 1181 g of water. The temperature of the water rises from 24.96 to 30.25 °C. The heat of combustion of benzoic acid is -26.42 kJ/g. In a second experiment, a 0.895-g powdered coal sample is burned in the same calorimeter assembly. The temperature of 1162 g of water rises from 24.98 to 29.81 °C.
(a) What is the heat capacity of the bomb calorimeter assembly (exclusive of the water)?
(b) What is the heat of combustion of the coal, expressed in kilojoules per gram?
(c) How many metric tons (1 metric ton = 1000 kg) of this coal would have to be burned to release 2.15×10^9 kJ of heat?

77. A handbook lists two different values for the heat of combustion of hydrogen: 33.88 kcal/g H_2 if $H_2O(l)$ is formed, and 28.67 kcal/g H_2 if $H_2O(g)$ (steam) is formed. Explain why these two values are different, and indicate what property this difference represents. Devise a means of verifying your conclusions with data from Table 7.2.

78. Use data from Appendix D, Hess's law, or other means to complete the diagram given here. That is, determine the missing values of $\Delta H°$.

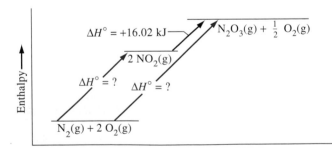

79. A particular natural gas consists, in mole percents, of 83.0% CH_4, 11.2% C_2H_6, and 5.8% C_3H_8. The heats of combustion (ΔH) of these gases are -890.3 kJ/mol CH_4, -1559.7 kJ/mol C_2H_6, and -2219.1 kJ/mol C_3H_8. A 385-L sample of

this gas, measured at 22.6 °C and 739 mmHg, is burned at constant pressure in an excess of oxygen gas. How much heat, in kilojoules, is evolved in the combustion reaction?

80. A net reaction for a coal gasification process is
$$2\ C(s) + 2\ H_2O(g) \longrightarrow CH_4(g) + CO_2(g)$$
Show that this net equation can be established by an appropriate combination of equations from Section 7-9.

81. Which of the following gases has the greater fuel value, on a per liter (STP) basis? That is, which has the greater heat of combustion? (*Hint:* The only combustible gases are CH_4, C_3H_8, CO, and H_2. Also, use data from Table 7.2.)
(a) coal gas: 49.7% H_2, 29.9% CH_4, 8.2% N_2, 6.9% CO, 3.1% C_3H_8, 1.7% CO_2, and 0.5% O_2, by volume.
(b) sewage gas: 66.0% CH_4, 30.0% CO_2, and 4.0% N_2, by volume.

82. A calorimeter that measures an exothermic heat of reaction by the quantity of ice that can be melted is called an ice calorimeter. Now consider that 0.100 L of methane gas, $CH_4(g)$, at 25.0 °C and 744 mmHg is burned completely at constant pressure in an excess of air. The heat liberated is captured and used to melt 10.7 g ice at 0 °C. (The heat required to melt ice, called the heat of fusion, is 333.5 J/g.)
(a) Write an equation for the combustion reaction.
(b) What is ΔH for this reaction?

83. For the reaction
$$C_2H_4(g) + 3\ O_2(g) \longrightarrow 2\ CO_2(g) + 2\ H_2O(l)$$
$$\Delta H° = -1410.9\ \text{kJ}$$
if the H_2O were obtained as a gas rather than a liquid, **(a)** would the heat of reaction be greater (more negative) or smaller (less negative) than that indicated in the equation? **(b)** Explain your answer. **(c)** Now calculate the value of ΔH in this case.

84. Some of the butane, $C_4H_{10}(g)$, in a 200.0-L cylinder at 26.0 °C is withdrawn and burned at a constant pressure in an excess of air. As a result the pressure of the gas in the cylinder falls from 2.35 atm to 1.10 atm. The liberated heat is used to raise the temperature of 132.5 L of water from 26.0 to 62.2 °C. Assume that the combustion products are $CO_2(g)$ and $H_2O(l)$ exclusively and determine the efficiency of the water heater. (That is, what percent of the heat of combustion was absorbed by the water?)

85. The metabolism of glucose, $C_6H_{12}O_6$, yields $CO_2(g)$ and $H_2O(l)$ as products. Heat released in the process is converted to useful work with about 70% efficiency. Calculate the mass of glucose metabolized by a 58.0-kg person in climbing a mountain with an elevation gain of 1450 m. Assume that the work performed in the climb is about four times that required simply to lift 58.0 kg by 1450 m. $\Delta H_f°$ of $C_6H_{12}O_6(s)$ is -1273.3 kJ/mol.

86. An alkane hydrocarbon has the formula C_nH_{2n+2}. Whatever the value of the subscript for carbon, n, the subscript for hydrogen is "twice n plus 2." The enthalpies of formation of the alkanes decrease (become more negative) as the numbers of C atoms increase. Starting with butane, $C_4H_{10}(g)$, for each additional CH_2 group in the formula the enthalpy of formation, $\Delta H_f°$, changes by about -21 kJ/mol. Use this fact, and data from Table 7.2, to estimate the heat of combustion of heptane, $C_7H_{16}(l)$.

87. Upon complete combustion, a 1.00-L sample (at STP) of a natural gas gives off 43.6 kJ of heat. If the gas is a mixture of $CH_4(g)$ and $C_2H_6(g)$, what is its percent composition, *by volume?* [*Hint:* What are the heats of combustion of $CH_4(g)$ and $C_2H_6(g)$?]

Feature Problems

A. Several factors are involved in determining the cooking times required for foods in a microwave oven. One of these factors is specific heat. Determine the approximate time required to warm 250 mL of chicken broth from 4 °C (a typical refrigerator temperature) to 50 °C in a 700-W microwave oven. Assume that the density of chicken broth is about 1 g/mL and that its specific heat is approximately 4.2 J g^{-1} °C^{-1}. A power of 1 W = 1 J s^{-1}.

B. Based on specific heat measurements, Pierre Dulong and Alexis Petit proposed in 1818 that the specific heat of an element is inversely related to its atomic weight (atomic mass). Thus, by measuring the specific heat of a new element, its atomic weight could be readily established.

1. Use data from Table 7.1 and inside the front cover to plot a *straight-line* graph relating atomic mass and specific heat. Write the equation for this straight line.

2. Use the measured specific heat of 0.23 J g^{-1} °C^{-1} and the equation derived in part 1 to obtain an approximate value of the atomic mass of cadmium, an element discovered in 1817.

3. To raise the temperature of 75.0 g of a particular metal by 15 °C requires 450 J of heat. What might this metal be?

C. We can use the heat liberated by a neutralization reaction as a means of establishing the stoichiometry of the reaction. The data in the table below are for the reaction of 1.00 M NaOH with 1.00 M citric acid, $C_6H_8O_7$, in a total solution volume of 60.0 mL.

mL 1.00 M NaOH used	mL 1.00 M citric acid used	ΔT, °C
20.0	40.0	4.7
30.0	30.0	6.3
40.0	20.0	8.2
50.0	10.0	6.7
55.0	5.0	2.7

1. Plot ΔT versus mL 1.00 M NaOH and identify the exact stoichiometric proportions of NaOH and citric acid at the equivalence point of the neutralization reaction. (*Hint:* Recall Feature Problem B of Chapter 4.)

2. Why is the temperature change in the neutralization greatest when the reactants are in their exact stoichiometric proportions? That is, why doesn't this maximum temperature change occur when one of the reactants is in excess?

3. Rewrite the formula of citric acid to reflect more precisely its acidic properties. Then write a balanced "whole formula" ionic equation for the neutralization reaction.

D. In a student experiment to confirm Hess's law, the reaction

$$NH_3(\text{conc aq}) + HCl(aq) \longrightarrow NH_4Cl(aq)$$

was carried out in two different ways. First, 8.00 mL of concentrated $NH_3(aq)$ was added to 100.0 mL of 1.00 M HCl in a calorimeter. [The $NH_3(aq)$ was slightly in excess.] The reactants were initially at 23.8 °C and the final temperature after neutralization was 35.8 °C.

In a second experiment, outlined in part in the sketch, air was bubbled through 100.0 mL of concentrated $NH_3(aq)$, sweeping out $NH_3(g)$. The $NH_3(g)$ was neutralized in 100.0 mL of 1.00 M HCl. The temperature of the concentrated $NH_3(aq)$ fell from 19.3 to 13.2 °C. At the same time, the temperature of the 1.00 M HCl rose from 23.8 to 42.9 °C as it was neutralized by $NH_3(g)$.

Assume that all solutions have densities of 1.00 g/mL and specific heats of 4.18 J g^{-1} °C^{-1}.

1. Write the two equations and ΔH values for the processes occurring in the second experiment. Show that the sum of these two equations is the same as the equation for the reaction in the first experiment.

2. Show that, within the limits of experimental error, ΔH for the net reaction is the same in the two experiments, thereby confirming Hess's law.

1.00 M HCl   Concentrated $NH_3(aq)$

The Atmospheric Gases and Hydrogen

8

*T*he first seven chapters have presented a number of fundamental ideas or principles—the "nuts and bolts" of chemistry. Although many of these ideas are interesting in themselves, the main reason for studying principles is to provide a basis for learning about chemical substances. The application of principles to describing and explaining chemical behavior is often called *descriptive chemistry.*

Our aim in this chapter is to convey the essence of descriptive chemistry by studying the properties and uses of a small number of elements and compounds—hydrogen and some of the elements and compounds associated with the atmosphere. Among the topics we will consider are several important environmental issues, such as smog formation, ozone destruction, and global warming. The principles we will apply are primarily those dealing with stoichiometry, gases, and thermochemistry. We will also point out how our appreciation of certain topics will increase as we learn more principles later in the text. So, we can think of this chapter as serving the dual purpose of applying some principles and previewing others.

The aurora borealis, or "northern lights" occur in the part of the atmosphere known as the ionosphere. Charged particles from the solar wind excite atoms, ions and molecules. These excited species emit light in a variety of colors that, from the ground, appear as a shimmering curtain in the air.

8-1 The Atmosphere

When astronauts look at Earth from space, they see it surrounded by a thin blue shell—the atmosphere. Commonly, we use the word *air* to describe the substances that make up the atmosphere. Air is a mixture of nitrogen and oxygen gases, with smaller quantities of argon, carbon dioxide, and the other substances listed in Table 8.1. This blanket of air surrounding Earth protects us from harmful radiation and is a major source of a number of chemicals necessary for life to exist.

Structure of the Atmosphere

About 90% of the mass of the atmosphere is in the first 12 km above Earth's surface. This familiar layer of air is called the **troposphere.** It is the region in which weather phenomena occur. Temperatures in the troposphere fall continuously with increasing altitude to a minimum of about 220 K (Figure 8-1).

The region from about 12 to 55 km above Earth's surface, the **stratosphere,** is less familiar to us, although supersonic aircraft fly in its lower reaches. The temperature–altitude profile of the stratosphere is much different from that of the troposphere. At altitudes from about 12 to 25 km, the temperature is fairly constant at about 220 K, and it then rises to about 280 K at 50 km. In part this is due to exothermic reactions that occur between the atoms in the thinning atmosphere and ultraviolet radiation from the sun. These reactions produce the *ozone layer* and will be discussed in more detail later.

Beyond the stratosphere the atmosphere is very "thin," with densities rapidly falling to the range of micrograms and nanograms per liter. In the region from 55 to 80 km, the *mesophere,* the temperature falls continuously to about 180 K. Beyond this, in a region known as the *thermosphere* or *ionosphere,* temperatures rise to about 1500 K. Here, the atmosphere consists of positive and nega-

TABLE 8.1 Composition of Dry Air (Near Sea Level)	
Component	**Volume Percent**[a]
nitrogen (N_2)	78.084
oxygen (O_2)	20.946
argon (Ar)	0.934
carbon dioxide (CO_2)	0.036
neon (Ne)	0.001818
helium (He)	0.000524
methane (CH_4)	0.0002
krypton (Kr)	0.000114
hydrogen (H_2)	0.00005
dinitrogen monoxide (N_2O)	0.00005
xenon (Xe)	0.000009
ozone (O_3)	
sulfur dioxide (SO_2)	
nitrogen dioxide (NO_2)	trace
ammonia (NH_3)	
carbon monoxide (CO)	
iodine (I_2)	

[a]Recall the meaning of percent by volume of a gaseous mixture introduced in Section 6-6.

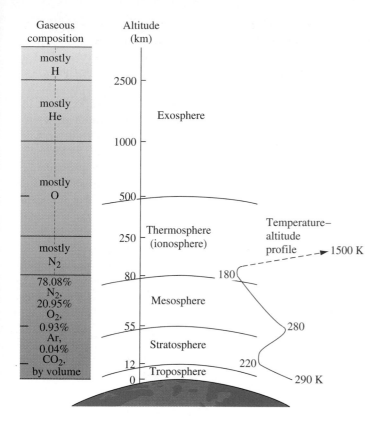

FIGURE 8-1
Regions of the atmosphere: composition and temperature–altitude profile.
The panel at the left describes the composition of the atmosphere at different altitudes. The temperature–altitude profile at the right gives approximate temperatures for the first 80 km and indicates that the temperature rises to 1500 K in the thermosphere.

The values given here are only approximate. For example, the height of the troposphere varies from about 8 km at the poles to 16 km at the equator. Also, temperatures in the thermosphere vary greatly between day and night.

tive ions, free electrons, neutral atoms, and molecules. The dissociation of molecules into atoms and the ionization of atoms into positive ions and free electrons require the absorption of energy. The source of that energy is electromagnetic radiation from the sun. An interesting natural phenomenon associated with the ionosphere is the *aurora borealis,* or "northern lights," pictured in Figure 8-2.

Normally, at 1500 K an iron object glows a bright red, but a cold iron object brought to the thermosphere would not glow. From the kinetic theory of gases (Section 6-7), we know that at 1500 K gaseous atoms and molecules travel at high speeds. And these energetic atoms and molecules transfer energy as heat when they collide with a cold object. Because the concentration of gaseous particles in the thermosphere is so low, however, molecular collisions are infrequent and very little heat is transferred. An object does not "heat up" the way it would in the lower atmosphere. Thus, as meteors fall toward Earth they do not begin to glow in the high-temperature thermosphere. They do so only at lower elevations where collisions with molecules of air cause surface atoms of the meteor to vaporize, ionize, and emit light.

Water Vapor in the Atmosphere

Table 8.1 lists the components of dry air, but air is not ordinarily dry. It contains water vapor, $H_2O(g)$, in quantities ranging from traces to as much as 4% by volume. This water vapor plays an essential role in the hydrologic (water) cycle: Ocean water evaporates; air masses move from the oceans to the land; water vapor in the atmosphere, when cooled, forms clouds; and the clouds produce rain. Water falling on land returns to the oceans through systems of lakes, rivers, and groundwater.

FIGURE 8-2
The aurora borealis.
The aurora borealis, or "northern lights," observed at high northern latitudes are light emissions from atoms, ions, and molecules in the ionosphere.

The amount of water vapor (n_{H_2O}) in a sample of air and the *partial pressure* of the water vapor (P_{H_2O}) are proportional, as we see from the ideal gas equation: $P = nRT/V$. The maximum possible partial pressure of water vapor at a given temperature is a quantity that we have previously introduced: the *vapor pressure* of water (see page 184). For example, at 25 °C, the vapor pressure of water is 23.8 mmHg. This is the highest partial pressure of water vapor that can be maintained at 25 °C. If a sample of air at 25 °C has a water vapor pressure higher than 23.8 mmHg, we expect some of the vapor to condense to liquid water. A common method of describing the water vapor content of air is through its relative humidity. **Relative humidity** is the ratio of the partial pressure of water vapor to the vapor pressure of water at the same temperature, expressed on a percent basis. Thus, if the partial pressure of water vapor in air at 25 °C is 12.2 mmHg, the relative humidity of the air is 51.3%.

$$\text{Relative humidity} = \frac{\text{partial pressure of water vapor}}{\text{vapor pressure of water}} \times 100\%$$

$$= \frac{12.2 \text{ mmHg}}{23.8 \text{ mmHg}} \times 100\% = 51.3\%$$

Because the vapor pressure of water increases with temperature, air having $P_{H_2O} = 12.2$ mmHg would have a relative humidity of only 38.4% at 30 °C, where the vapor pressure of water is 31.8 mmHg. Conversely, at 10 °C, where the vapor pressure of water is only 9.2 mmHg, some water vapor in the air sample would condense as liquid. This fluctuation of the relative humidity with temperature accounts for the formation of dew that is sometimes observed in the early morning hours.

The Atmosphere as a Source of Chemicals

We do not normally think of water as a commercial chemical, but it is one, in addition to being essential to life. And, as we indicated previously, fresh water on Earth comes mostly from seawater, by way of the atmosphere. Nitrogen and oxygen are among the most widely used manufactured chemicals, as we will see in the next two sections. They, as well as argon and other noble gases, are obtained by the fractional distillation of liquid air. This process involves only physical changes and is described in Figure 8-3.

FIGURE 8-3

The fractional distillation of liquid air—a simplified representation.

Clean air is fed into a compressor and cooled by refrigeration. The cold air expands through a nozzle and is cooled still further—enough to cause it to liquefy. The liquid air is filtered to remove solid CO_2 and hydrocarbons and then distilled. Liquid air enters the top of the column where nitrogen, the most volatile component (lowest boiling point), passes off as a gas. In the middle of the column gaseous argon is removed. Liquid oxygen, the least volatile component, collects at the bottom. The normal boiling points of nitrogen, argon, and oxygen are 77.4, 87.5, and 90.2 K, respectively.

EXAMPLE 8-1

Using the Atmosphere as a Source of Chemicals. What volume of air must be processed to produce 5.00 L of $N_2(g)$?

Solution

From Table 8.1 we find that air contains 78.084% N_2 by volume. To find the volume of air needed we must use this value as a conversion factor.

$$\text{volume of air} = 5.00 \text{ L } N_2 \times \frac{100.00 \text{ L air}}{78.084 \text{ L } N_2} = 6.40 \text{ L}$$

Practice Example A: What volume of air must be processed to produce 5.00 L of $Ar(g)$?

Practice Example B: What volume of air, measured at 1 atm and 298 K, must be processed to produce 5.00 g of $CO_2(s)$?

8-2   Nitrogen

Nitrogen occurs mainly in the atmosphere. Its abundance in Earth's solid crust is only 0.002% by mass. The only important nitrogen-containing minerals are KNO_3 (niter, or saltpeter) and $NaNO_3$ (sodaniter, or Chile saltpeter), found in a few desert regions. Other natural sources of nitrogen-containing compounds are plant and animal protein and fossilized remains of ancient plant life, such as coal.

Until about 100 years ago sources of pure nitrogen and its compounds were quite limited. This all changed with the invention of a process for the liquefaction of air in 1895 (see Figure 8-3) and a process for converting nitrogen to ammonia in 1908. A host of nitrogen compounds can be made from ammonia. Nitrogen gas has many important uses of its own in addition to being a precursor of manufactured nitrogen compounds. Some of these uses are listed in Table 8.2.

The substance from which all nitrogen compounds are ultimately derived, $N_2(g)$, is unusually stable. One explanation of the limited reactivity of the N_2 molecule is based on its electronic structure, a topic that we will consider in Chapter 11. There we will learn that the bond between the two N atoms in N_2 is a *triple* covalent bond, an unusually strong bond that is difficult to break. In thermochemical terms we say that the enthalpy change associated with breaking the bonds in one mole of N_2 molecules is very high—the dissociation reaction is highly endothermic.

$$N{\equiv}N(g) \longrightarrow 2 \text{ N}(g) \qquad \Delta H° = +945.4 \text{ kJ}$$

Also, the enthalpies of formation of many nitrogen compounds are positive, which means that their formation reactions are *endothermic*. For $NO(g)$,

$$\tfrac{1}{2} N_2(g) + \tfrac{1}{2} O_2(g) \longrightarrow NO(g) \qquad \Delta H° = 90.25 \text{ kJ}$$

As we noted in Chapter 7 (page 224), our general expectation is that highly endothermic reactions, such as the formation of $NO(g)$ from its elements, do not occur to any significant extent at normal temperatures. Just imagine the situa-

TABLE 8.2   Production and Uses of Nitrogen Gas

Production

1995: 34.02×10^6 tons
Ranking, by mass, among manufactured chemicals in the U.S.: No. 2

Uses

provide a blanketing (inert) atmosphere for the production of chemicals and electronic components
pressurized gas for enhanced oil recovery
metals treatment
refrigerant (e.g., fast freezing of foods)

Liquid nitrogen freezes the delicate petals of a flower. The frozen petals are so brittle, they easily shatter on the table top.

tion if $\Delta H_f^\circ[\text{NO(g)}] = -90.25$ kJ/mol instead of $+90.25$ kJ/mol. The reaction of N_2(g) and O_2(g) to form NO(g) would likely proceed to a far greater extent. With an atmosphere depleted in O_2(g) and rich in noxious NO(g), life as we know it would not be possible on Earth.

In the remainder of this section, we consider several important nitrogen compounds and their uses.

Ammonia and Related Compounds

Ammonia, NH_3, is one of the most useful chemicals we know. Each year it ranks about sixth, by mass, among manufactured chemicals in the United States. Ammonia's importance stems from the ease with which it can be converted into a wide variety of other nitrogen-containing chemicals.

It was not until the present century that Fritz Haber (1908) worked out the conditions for the ammonia synthesis reaction and tested these in the laboratory. Converting Haber's method into a manufacturing process was one of the most challenging engineering problems of its time. This work was led by Carl Bosch at Badische Anilin- & Soda-Fabrik (BASF) in Germany. By 1913, a plant producing 30,000 kg NH_3 per day was in operation. A modern ammonia plant has about 50 times this capacity.

The essential difficulty in the ammonia synthesis reaction is that under most conditions it does not go to completion. It is reversible (recall page 131).

$$N_2(g) + 3\,H_2(g) \rightleftharpoons 2\,NH_3(g) \tag{8.1}$$

A high yield of ammonia requires (a) a high temperature (400 °C), (b) a catalyst to speed up the reaction, and (c) a high pressure (about 200 atm). The key to achieving essentially 100% yield is continuous removal of NH_3 and recycling of the unreacted N_2(g) and H_2(g). The NH_3 is removed by liquefaction. The Haber–Bosch process is outlined in Figure 8-4. A critical aspect of the process is having a source of H_2(g). Mostly this is made from natural gas (see page 258).

Ammonia is the starting point for making most other nitrogen compounds, but it has some direct uses of its own. Its most important use is as a fertilizer. The highest concentration in which nitrogen fertilizer can be applied to fields is as pure liquid NH_3, known as "anhydrous ammonia." NH_3(aq) is also applied in a variety of household cleaning products, such as commercial glass cleaners. In these products the ammonia acts as an inexpensive base to produce OH^-(aq).

Fritz Haber (1868–1934). Haber's perfection of the ammonia synthesis reaction, which made the manufacture of inexpensive explosives possible, was of critical importance to Germany in World War I. After the war, Haber again applied his chemical knowledge for his country's benefit by attempting, unsuccessfully, to extract gold from seawater for use in paying war reparations. Despite his past services, this Jewish scientist was driven from his academic post by the Nazi regime in 1933.

FIGURE 8-4

Ammonia synthesis reaction—the Haber–Bosch process.
The gaseous N_2–H_2 mixture is introduced into a reactor at high temperature and pressure in the presence of a catalyst. The gaseous N_2–H_2–NH_3 mixture leaves the reactor and is cooled as it passes through a condenser. Liquefied NH_3 is removed, and the remaining N_2–H_2 mixture is compressed and returned to the reactor. Essentially 100% yield is obtained.

The $OH^-(aq)$ reacts with grease and oil molecules to convert them into compounds that are more soluble in water and that do not hold dirt well. In addition, the aqueous ammonia solution dries quickly, leaving few streaks on glass.

Because ammonia is a base, a simple approach to producing certain nitrogen compounds is to neutralize ammonia with an appropriate acid. The acid–base reaction that forms ammonium sulfate, an important solid fertilizer, is

$$2\,NH_3(aq) + H_2SO_4(aq) \longrightarrow (NH_4)_2SO_4(aq) \qquad (8.2)$$

Ammonium chloride, made by the reaction of $NH_3(aq)$ and $HCl(aq)$, is used in the manufacture of dry cell batteries, in cleaning metals, and as an agent to help solder flow smoothly when soldering metals. Ammonium nitrate, made by the reaction of $NH_3(aq)$ and $HNO_3(aq)$, is used both as a fertilizer and as an explosive. The explosive power of ammonium nitrate was not fully appreciated until a shipload of this material exploded in Texas City, Texas, in 1947, killing many people. More recently, mixtures of ammonium nitrate and fuel oil were used as explosives in the terrorist attacks on the World Trade Center in New York City in 1993 and the federal building in Oklahoma City in 1995. The reaction of $NH_3(aq)$ and $H_3PO_4(aq)$ yields ammonium phosphates [such as $NH_4H_2PO_4$ and $(NH_4)_2HPO_4$]. These compounds are good fertilizers because they supply two vital plant nutrients, N and P; they are also used as fire retardants.

Urea is often manufactured at ammonia synthesis plants, by the simple reaction

$$2\,NH_3 + CO_2 \longrightarrow \underset{\text{urea}}{CO(NH_2)_2} + H_2O \qquad (8.3)$$

Urea contains 46% nitrogen, by mass. It is an excellent fertilizer, either as a pure solid, as a solid mixed with ammonium salts, or in a very concentrated aqueous solution mixed with NH_4NO_3 or NH_3 (or both). Urea is also used as a feed supplement for cattle and in the production of polymers and pesticides.

Currently, urea ranks seventeenth, by mass, among the top chemicals produced in the United States; ammonium nitrate ranks fifteenth and ammonium sulfate, thirty-first (based on 1995 data).

Nitrogen Oxides

Nitrogen forms a series of oxides in which the oxidation state of N can have every value ranging from +1 to +5. Table 8.3 lists their formulas and names them according to the scheme introduced in Chapter 3. The oxides of nitrogen are not as familiar as several other nitrogen compounds, but we do encounter

TABLE 8.3 Nomenclature of Some Common Oxides of Nitrogen

O.S. of N	Formula	Name
+1	N_2O	dinitrogen monoxide
+2	NO	nitrogen monoxide
+3	N_2O_3	dinitrogen trioxide
+4	NO_2	nitrogen dioxide
+4	N_2O_4	dinitrogen tetroxide
+5	N_2O_5	dinitrogen pentoxide

them in a number of situations. N_2O has anesthetic properties and finds some use in dentistry ("laughing gas"). NO_2 is employed in the manufacture of nitric acid. N_2O_4 is used extensively as an oxidizer in rocket fuels. NO is the most important oxide of nitrogen from a biological standpoint. In humans, it plays a role in maintaining blood pressure, aids in the immune response in killing foreign organisms, and is essential to the establishment of long-term memory. In 1996, scientists discovered that hemoglobin carries NO as well as O_2. The NO causes the walls of blood vessels to thin, making it easier for oxygen to be transported to the surrounding tissues.

Dinitrogen monoxide (nitrous oxide), $N_2O(g)$, can be produced in the laboratory by an interesting disproportionation reaction, the decomposition of $NH_4NO_3(s)$.

$$NH_4NO_3(s) \xrightarrow{200\text{--}260\ ^\circ C} N_2O(g) + 2\ H_2O(g)$$

The N atom in NH_4^+ is in the oxidation state (O.S.) -3; in NO_3^- the O.S. of N is $+5$. In N_2O both N atoms are in the O.S. $+1$. The loss of O.S. in one N atom is just matched by the gain of O.S. in the other, which makes the redox equation very easy to balance.

Nitrogen monoxide (nitric oxide), $NO(g)$, is produced commercially by the oxidation of $NH_3(g)$ in the presence of a catalyst (the Ostwald process). This oxidation is the first step in converting NH_3 to a number of other nitrogen compounds.

$$4\ NH_3(g) + 5\ O_2(g) \xrightarrow[850\ ^\circ C]{Pt} 4\ NO(g) + 6\ H_2O(g) \tag{8.4}$$

Another source of NO, usually unwanted, is in high-temperature combustion processes, such as those that occur in automobile engines and in electric power plants. At the same time that the fuel combines with oxygen from air to produce a high temperature, $N_2(g)$ and $O_2(g)$ in the hot air combine to a limited extent to form $NO(g)$.

$$N_2(g) + O_2(g) \xrightarrow{\Delta} 2\ NO(g) \tag{8.5}$$

We often see brown nitrogen dioxide, $NO_2(g)$, in reactions involving nitric acid. An example is the reaction of $Cu(s)$ with warm concentrated $HNO_3(aq)$.

$$Cu(s) + 4\ H^+(aq) + 2\ NO_3^-(aq) \longrightarrow Cu^{2+}(aq) + 2\ H_2O + 2\ NO_2(g) \tag{8.6}$$

Of particular interest to atmospheric chemists is the key role of $NO_2(g)$ in the formation of photochemical smog (page 250).

Nitric Acid and Nitrates

An **acid anhydride** is a nonmetallic oxide that reacts with water to produce a ternary acid as the sole product. The acid anhydride of nitric acid is N_2O_5.

$$N_2O_5(s) + H_2O(l) \longrightarrow 2\ HNO_3(aq)$$

The commercial synthesis of nitric acid does not use N_2O_5, however. It involves the following three reactions, the first of which—the Ostwald process—we described before. $NO(g)$ from reaction (8.8) is recycled into reaction (8.7).

$$4\ NH_3(g) + 5\ O_2(g) \xrightarrow[850\ ^\circ C]{Pt} 4\ NO(g) + 6\ H_2O(g) \tag{8.4}$$

$$2\ NO(g) + O_2(g) \longrightarrow 2\ NO_2(g) \tag{8.7}$$

$$3\ NO_2(g) + H_2O(l) \longrightarrow 2\ HNO_3(aq) + NO(g) \tag{8.8}$$

A copper penny reacting with nitric acid. The reaction is that given by Equation (8.6). The blue-green color of the solution is due to $Cu^{2+}(aq)$, and the reddish brown color is that of nitrogen dioxide, $NO_2(g)$.

Nitric acid is used in the preparation of various dyes; drugs; fertilizers (ammonium nitrate); and explosives, such as nitroglycerin, nitrocellulose, and trinitrotoluene (TNT). Nitric acid is also used in metallurgy and in reprocessing spent nuclear fuels. Nitric acid is about fourteenth, by mass, among the top chemicals produced in the United States.

Nitrates can be made by neutralizing nitric acid with appropriate bases. Nitric acid is also a good oxidizing agent. For example, copper reacts with dilute $HNO_3(aq)$ primarily producing NO or, with concentrated $HNO_3(aq)$, NO_2 (reaction 8.6). With a more active metal such as Zn, the reduction product has N in one of its lower oxidation states, as in NH_4^+ in some cases.

The preceding paragraph refers to several chemical reactions without explicitly supplying equations for them. If you read the descriptions of these reactions carefully and compare them with other reactions for which equations are given, however, you should be able to supply your own equations. Example 8-2 illustrates how to do this.

Flash paper is used by magicians for dramatic effect. It can be made by treating paper with nitric and sulfuric acids. This process converts the cellulose fibers into nitrocellulose, which burns cleanly and rapidly.

EXAMPLE 8-2

Writing Chemical Equations. Write an equation to represent the synthesis of a calcium nitrate solution by the reaction of aqueous nitric acid and solid calcium hydroxide.

Solution

As we learned in Chapter 4, when a chemical reaction is described through the names of its reactants and products, we can substitute formulas for names, complete the formula expression, and then balance the equation.

$$Ca(OH)_2(s) + 2\ HNO_3(aq) \longrightarrow Ca(NO_3)_2(aq) + 2\ H_2O(l)$$

Practice Example A: Lead(IV) oxide reacts with nitric acid to produce water, oxygen gas, and an aqueous solution of lead(II) nitrate. Write a chemical equation for this reaction.

Practice Example B: Write an equation to represent the reaction of zinc with dilute nitric acid solution to form aqueous zinc(II) nitrate, ammonium nitrate, and water. [*Hint:* This is an oxidation–reduction reaction similar to reaction (8.6). Balance it by the half-reaction method of Section 5-5.]

An Environmental Issue Involving Oxides of Nitrogen: Smog

About 100 years ago a new word entered the English language—**smog.** It referred to a condition, common in London, in which a combination of *smoke* and *fog* obscured visibility and produced health hazards (including death). These conditions are often associated with heavy industry. This type of smog is now called *industrial* smog.

The form of air pollution more commonly thought of as smog results from the action of sunlight on the products of combustion. Chemical reactions brought about by light are called *photochemical* reactions, and the smog formed by such reactions is *photochemical* smog.

Photochemical smog results from high-temperature combustion processes, such as those that occur in automobile engines. Because the combustion of gasoline takes place in air, rather than in pure oxygen, NO(g) produced by reaction (8.5) is inevitably found in the exhaust from automobiles. Other products found

▶ Air quality in London has been greatly improved by control measures, such as elimination of coal as a household fuel, introduced after a severe smog episode in 1952.

FIGURE 8-5
Smog in Mexico City.
At times, the topographical features, climatic conditions, traffic congestion, and heavy industrial pollution combine to create severe smog conditions in Mexico City.

in the exhaust are hydrocarbons (unburned gasoline) and partially oxidized hydrocarbons. These, then, are the starting materials—the precursors of photochemical smog.

Many substances have been identified in smoggy air, including NO, NO_2, ozone (a form of the element oxygen with the formula O_3, discussed further in Section 8-3), and a variety of organic compounds derived from gasoline hydrocarbons. Ozone is very reactive and is largely responsible for the breathing difficulties that some people experience during smog episodes. Another obnoxious substance found in smog is an organic compound known as peroxyacetyl nitrate (PAN). PAN is a powerful lacrimator—that is, it causes tear formation in the eyes. Photochemical smog components cause heavy crop damages (to oranges, for example) and the deterioration of rubber goods. And, of course, the best known symbol of photochemical smog is the hazy brown air that results in reduced visibility (Figure 8-5).

The challenge to chemists is to determine the mechanism(s) by which photochemical smog is formed. Knowing the mechanism(s) might lead scientists and civic leaders to ideas about how to reduce or eliminate the formation of this smog. Chemists who have been studying photochemical smog formation over the past several decades have determined that the precursors listed earlier are converted to the observable smog components through the action of sunlight. Because the chemical reactions involved are very complex and still not totally understood, we will give only a very brief, simplified scheme showing how photochemical smog is formed.

The precursor in smog formation is NO(g), produced by reaction (8.5) occurring in automobile engines.

$$N_2(g) + O_2(g) \xrightarrow{\Delta} 2 \, NO(g)$$

NO(g) is then converted to NO_2(g), which absorbs ultraviolet radiation from sunlight and decomposes.

$$NO_2 + sunlight \longrightarrow NO + O \tag{8.9}$$

This is followed by a reaction forming ozone, O_3.

$$O + O_2 \longrightarrow O_3 \tag{8.10}$$

A large buildup of ozone in photochemical smog, then, requires a plentiful source of NO_2. At one time this source was believed to be reaction (8.7).

$$2 \, NO + O_2 \longrightarrow 2 \, NO_2 \tag{8.7}$$

▶ We will discuss the rate of reaction (8.7) in Chapter 15.

It is now well established, however, that reaction (8.7) occurs much too slowly to yield the required levels of NO_2 in photochemical smog. NO is rapidly converted to NO_2 when it reacts with O_3,

$$O_3 + NO \longrightarrow O_2 + NO_2 \tag{8.11}$$

but although this reaction accounts for the formation of NO_2, it leads to the *destruction* of ozone. Thus, photochemical smog formation cannot occur just through the reaction sequence: (8.5), (8.11), (8.9), and (8.10). The ozone would be consumed as quickly as it formed and there would be no ozone buildup at all.

We now know that organic compounds, particularly unburned hydrocarbons in automotive exhaust, provide a pathway for the conversion of NO to NO_2. The reaction sequence that follows involves some highly reactive molecular fragments known as *free radicals* and represented by formulas written with a bold dot. RH represents a hydrocarbon molecule, and R· is a fragment of a hydrocarbon molecule, a free radical. Oxygen atoms, fragments of the O_2 molecule, are also represented as free radicals, as are hydroxyl groups, fragments of the H_2O molecule.

$$RH + O\cdot \longrightarrow R\cdot + \cdot OH$$

$$RH + \cdot OH \longrightarrow R\cdot + H_2O$$

$$R\cdot + O_2 \longrightarrow RO_2\cdot$$

$$RO_2\cdot + NO \longrightarrow RO\cdot + NO_2$$

The final step in this sequence accounts for the rapid conversion of NO to NO_2 that seems essential to smog formation.

The role of NO_2 in the formation of the smog component PAN is suggested by the equation

$$\underset{CH_3\overset{\displaystyle O}{\overset{\|}{C}}-O-O\,\cdot\ +\ NO_2}{} \longrightarrow \underset{\underset{PAN}{CH_3\overset{\displaystyle O}{\overset{\|}{C}}-O-ONO_2}}{}$$

The details of smog formation have been worked out in part through the use of smog chambers. By varying experimental conditions in these chambers, scientists have been able to create polluted atmospheres very similar to smog. For example, they have found that if hydrocarbons are omitted from the starting materials in the smog chamber no ozone is formed. The reaction scheme just proposed is consistent with this observation. Figure 8-6 gives a typical result from a smog chamber.

To control smog, automobiles are now provided with a *catalytic converter.* CO and hydrocarbons are oxidized to CO_2 and H_2O in the presence of an *oxidation* catalyst (such as platinum or palladium metal). NO must be *reduced* to N_2, and this requires a *reduction* catalyst. A dual-catalyst system uses both types of catalysts. Alternatively, the air–fuel ratio of the engine is set to produce some CO and unburned hydrocarbons. These then act as reducing agents to reduce NO to N_2.

$$2\ CO(g) + 2\ NO(g) \longrightarrow 2\ CO_2(g) + N_2(g)$$

Next, the exhaust gases are passed through an oxidation catalyst to oxidize the remaining hydrocarbons and CO to CO_2 and H_2O. Future control measures may include the use of alternative fuels, such as methanol or hydrogen, and the development of electric-powered automobiles.

8-3   Oxygen

Although nitrogen is the most abundant element in the atmosphere, it occurs to a very limited extent in Earth's crust. By contrast, oxygen, also a major component of the atmosphere, is found far more extensively in compounds in Earth's crust. In fact, oxygen is the most abundant of the elements, constituting 45.5% by mass of Earth's solid crust.

Oxygen is central to a study of chemistry. It forms compounds with all the elements except the Group 8A elements of low atomic numbers (He, Ne, Ar). We constantly find ourselves considering properties of oxygen and its compounds as we encounter new chemical principles. Most of the discussion of oxy-

FIGURE 8-6
Smog component profile.
Data from a smog chamber show how the concentrations of smog components change with time. For example, the concentrations of hydrocarbon (**RH**) and nitrogen monoxide (**NO**) fall continuously, whereas that of nitrogen dioxide (NO_2) rises to a maximum and then drops off. The concentrations of ozone (O_3) and peroxyacetyl nitrate (**PAN**) build up more slowly. Any reaction scheme proposed to explain smog formation must be consistent with observations such as these. Under actual smog conditions, the pattern of concentration changes shown here repeats itself on a daily basis.

▶ The importance of the air–fuel ratio in automobile engines is the subject of Feature Problem A on page 269.

▶ Most elements in Earth's crust, because they are in contact with a highly oxygenated atmosphere, occur combined with oxygen.

gen and its compounds, then, comes in later chapters. Because of its many uses and its role in atmospheric chemistry, however, it merits discussion here also.

Preparation and Uses

Even though oxygen occurs and is used mostly in combined form, oxygen gas is itself an important commercial chemical. Some of its more important uses are listed in Table 8.4. It is obtained mostly from air (page 244). With its ready commercial availability, O_2 is not commonly prepared in the laboratory.

In submarines, spacecraft, and in emergency breathing apparatus, however, it is necessary to generate small quantities of oxygen from solids. The reaction of potassium *superoxide* with CO_2 works well for this purpose; it removes CO_2 while O_2 is being formed.

▶Potassium superoxide contains the superoxide ion, O_2^-, formed when an O_2 molecule gains an electron (see page 771).

$$4\ KO_2(s) + 2\ CO_2(g) \longrightarrow 2\ K_2CO_3(s) + 3\ O_2(g)$$

Oxygen and hydrogen gases can be obtained simultaneously through the electrolysis of water. **Electrolysis** is the decomposition of a substance through the passage of electric current. This method is only cost-effective for producing small quantities of these gases, though, unless electric power is unusually cheap. (Many other materials, because of their higher value, can be economically produced using electrolysis.)

$$2\ H_2O(l) \xrightarrow[H_2SO_4(aq)]{\text{electrolysis}} 2\ H_2(g) + O_2(g)$$

The relationship between electricity and chemical change, referred to as *electrochemistry,* is so important that we devote an entire later chapter to a discussion of this topic. Figure 8-7 illustrates the electrolysis of water and provides some background information that will meet our present needs.

EXAMPLE 8-3

Redox Chemistry. From the description given in the caption to Figure 8-7, write equations for the half-reactions occurring at each electrode and then a net equation for the electrolysis of water.

FIGURE 8-7
The electrolysis of water.
The passage of electric current through a liquid involves the migration of ions. To make water an electrical conductor, an electrolyte such as H_2SO_4 must be added. The electrolysis occurs in $H_2SO_4(aq)$. H^+ ions in this acidic solution are attracted to the negative electrode (cathode). Here they gain electrons to form H atoms, and H atoms join to form molecules of $H_2(g)$. *Reduction* occurs at the cathode. SO_4^{2-} ions are attracted to the positive electrode (anode), but they undergo no change. Instead, a reaction occurs in which H_2O molecules decompose to replace the H^+ ions lost at the cathode and release $O_2(g)$. *Oxidation* occurs at the anode. The net reaction is:

$$2\ H_2O(l) \longrightarrow 2\ H_2(g) + O_2(g).$$

Solution

The description of the electrolysis indicates the type of half-reaction occurring at each electrode and names the reactants and products. This information and the general approach to writing half-equations presented in Section 5-5 lead us to write

Reduction (at cathode): $\quad 2 H^+(aq) + 2 e^- \longrightarrow H_2(g)$

Oxidation (at anode): $\qquad\qquad 2 H_2O(l) \longrightarrow 4 H^+(aq) + O_2(g) + 4 e^-$

To obtain the net equation, multiply the reduction half-equation by two, add it to the oxidation half-equation, and simplify.

$$\cancel{4 H^+(aq)} + 2 H_2O(l) + \cancel{4 e^-} \longrightarrow 2 H_2(g) + \cancel{4 H^+(aq)} + O_2(g) + \cancel{4 e^-}$$

Net equation: $\qquad 2 H_2O(l) \longrightarrow 2 H_2(g) + O_2(g)$

Practice Example A: The electrolysis of dilute NaOH(aq) yields $H_2(g)$ at the cathode (reduction electrode) and $O_2(g)$ at the anode (oxidation electrode). Write plausible half-equations for this electrolysis. (*Hint:* The net equation is the same as in Example 8-3.)

Practice Example B: Small amounts of oxygen for use in spacecraft can be obtained from lithium peroxide, Li_2O_2. Write a balanced chemical equation for the reaction of lithium peroxide with carbon dioxide to form lithium carbonate and oxygen as products.

TABLE 8.4 Production and Uses of Oxygen Gas
Production
1995: 26.74×10^6 tons ranking, by mass, among manufactured chemicals in the U.S.: No. 3
Uses
manufacture of iron and steel manufacture and fabrication of other metals (cutting and welding) chemicals manufacture and other oxidation processes water treatment oxidizer of rocket fuels medicinal uses petroleum refining

Ozone: An Allotrope of Oxygen

Although the name oxygen conjures up the formula O_2, there are actually two different oxygen molecules. Familiar oxygen is *dioxygen*, O_2; the other is *trioxygen*—ozone, O_3. The term used to describe the existence of two or more forms of an element that differ in their bonding and molecular structure is **allotropy.** O_2 and O_3 are *allotropes* of oxygen.

▶ Bonding in O_2 and O_3 will be described in Chapter 11.

Normally, the quantity of O_3 in the atmosphere is quite limited at low altitudes, about 0.04 part per million (ppm). As we saw on page 250, however, its level increases (perhaps severalfold) in smog situations. Ozone levels exceeding 0.12 ppm are considered unhealthful.

The reaction producing $O_3(g)$ directly from $O_2(g)$ is highly endothermic and occurs only rarely in the lower atmosphere.

$$3 O_2(g) \longrightarrow 2 O_3(g) \qquad \Delta H° = +285 \text{ kJ}$$

This reaction does occur in high-energy environments such as electrical storms. If you have ever smelled a pungent odor around heavy-duty electrical equipment or xerographic office copiers, it was probably O_3. The chief method of producing ozone in the laboratory, in fact, is to pass an electric discharge (high-energy electrons) through $O_2(g)$. Because ozone is unstable and decomposes back to $O_2(g)$, it is always generated at the point where it is to be used.

Ozone is an excellent oxidizing agent. Its oxidizing ability is surpassed by few other substances (two are F_2 and OF_2). Its most important use is as a substitute for chlorine in purifying drinking water. Its advantages are that it does not impart a taste to the water and it does not form the potentially carcinogenic chlorination products that chlorine can. Its main disadvantage is that O_3 is unstable and quickly disappears from the water after it is treated. Thus, the water is not as well protected against bacterial contamination *after* leaving the waterworks as is water treated with chlorine.

An Environmental Issue Involving Ozone: The Ozone Layer

In the stratosphere, at altitudes between 25 and 35 km, the concentration of O_3 (expressed in molecules per cubic centimeter) is several times higher than at ground level. When expressed in proportion to the other gases present, the O_3 content of this part of the stratosphere—as much as 8 ppm—is considerably greater than at ground level (0.04 ppm). This belt of the stratosphere is called the **ozone layer.**

Stratospheric ozone plays a vital role in protecting life on Earth. First, ozone absorbs certain ultraviolet (UV) radiation that at Earth's surface causes skin cancer and eye damage in humans and is harmful to other biological organisms as well. Second, in absorbing ultraviolet radiation, O_3 molecules dissociate, evolving heat, and thus help to maintain a heat balance in the atmosphere.

The chemical reactions that produce $O_3(g)$ in the upper atmosphere are

$$O_2 + \text{UV radiation} \longrightarrow O + O \tag{8.12}$$

$$O_2 + O + M \longrightarrow O_3 + M \tag{8.13}$$

Equation (8.12) describes how an O_2 molecule absorbs ultraviolet radiation and dissociates. Atomic and molecular oxygen then react to form ozone (8.13). The "third body," M [for example, $N_2(g)$], carries off excess energy; otherwise the O_3 formed would be too energetic and simply decompose.

Equation (8.14) illustrates the first of the two important functions of ozone: absorbing ultraviolet radiation. The second function, that of releasing heat into the atmosphere, is illustrated by Equation (8.15).

► The UV radiations absorbed by O_2 in reaction (8.12) and by O_3 in reaction (8.14) differ in a quality known as wavelength, as we will describe in the next chapter.

$$O_3 + \text{UV radiation} \longrightarrow O_2 + O \tag{8.14}$$

$$O_3 + O \longrightarrow 2\,O_2 \qquad \Delta H = -389.8 \text{ kJ} \tag{8.15}$$

Reaction (8.15) is an ozone-destroying reaction that occurs naturally. There are other natural ozone-destroying processes that have (8.15) as their net reaction, such as

$$NO + O_3 \longrightarrow NO_2 + O_2$$

$$\underline{NO_2 + O \longrightarrow NO + O_2}$$

$$\textit{Net reaction:} \qquad O_3 + O \longrightarrow 2\,O_2$$

What is interesting about this pair of reactions is that NO consumed in the first reaction is replenished in the second. A little NO goes a long way. Atmospheric NO is produced mainly from N_2O released by soil bacteria. Thus, natural events account for the continuous formation and destruction of stratospheric ozone and the maintenance of a "steady-state" concentration of about 8 ppm.

It has recently become apparent that certain gases produced by human activities are contributing to ozone depletion and threatening the integrity of the ozone layer. For example, NO produced by combustion in supersonic jets operating in the stratosphere might contribute to ozone destruction. Most worrisome, though, are gaseous chlorofluorocarbons (CFCs). These molecules have a very long lifetime in the atmosphere and eventually rise to the stratosphere in low concentrations. In the stratosphere they can absorb ultraviolet radiation and dissociate. For example,

► Dichlorodifluoromethane, CCl_2F_2, once widely used in refrigeration and air-conditioning systems, is a typical chlorofluorocarbon. Its desirability was based, in part, on the fact that it is nontoxic and was thought to be completely inert.

$$CCl_2F_2 + \text{UV radiation} \longrightarrow CClF_2 + Cl$$

FIGURE 8-8
Ozone levels over Antarctica, recorded on October 2, 1994.
The gray, pink, and purple areas at the center constitute the ozone hole, with the concentration of ozone being lowest in the gray region. Higher concentrations of ozone are found in the yellow, green, and brown regions.

Cl atoms from this reaction can then set up an ozone-destroying cycle.

$$Cl + O_3 \longrightarrow ClO + O_2$$
$$\frac{ClO + O \longrightarrow Cl + O_2}{\textit{Net reaction:} \quad O_3 + O \longrightarrow 2\,O_2} \tag{8.15}$$

Currently, the most convincing evidence that depletion of stratospheric ozone is indeed occurring comes from studies in Antarctica. With the coming of spring (October) a large depletion of O_3 occurs over a period of several weeks, before more normal levels return (Figure 8-8). The chemical reactions postulated to account for this are much more complex than the simplified scheme outlined here. In addition, meteorological conditions play an important role. Evidence is mounting that ozone destruction also occurs in Arctic regions and perhaps worldwide. The most significant control measure taken to date is an international agreement among industrialized nations to reduce the production and use of chlorofluorocarbons by up to 50% by the year 2000 or sooner. Some chemical companies have already ceased chlorofluorocarbon production.

8-4   The Noble Gases

In 1785, Henry Cavendish, the discoverer of hydrogen, passed electric discharges through air to form oxides of nitrogen. (A similar process occurs during lightning storms.) He then dissolved these oxides in water to form nitric acid. Even by using excess oxygen, Cavendish was unable to get all the air to react. He suggested that air contained an unreactive gas making up "not more than 1/120 of the whole." John Rayleigh and William Ramsay isolated this gas one century later (1894) and named it argon. The name argon is derived from the Greek *argos* "lazy one," meaning inert. Its *inability* to form chemical compounds with

Mixtures of the noble gases are used to make a variety of colors of "neon" lights.

▶ The observation that led to the discovery of the noble gases was an apparent anomaly in the density of $N_2(g)$, as described in Feature Problem B on page 269.

any of the other elements—its chemical inertness—was found to be argon's most notable feature. Because argon resembled no other known element, Ramsay placed it in a separate group of the periodic table and reasoned that there should be other members of this group.

Ramsay then began a systematic search for other "inert" gases. In 1895, he extracted helium from a uranium mineral. A few years later, by very carefully distilling liquid argon, he was able to extract three additional inert gases: neon, krypton, and xenon. The final member of the group of inert gases, a radioactive element called radon, was discovered in 1900. In 1962, compounds of Xe were first prepared, and so the inert gases proved not to be completely inert after all. Since that time this group of gases has been called the *noble gases*. They are found in Group 8A of the periodic table on the inside front cover.

Occurrence

Air contains 0.000524% He, 0.001818% Ne, and 0.934% Ar, by volume. The proportion of Kr is about 1 ppm by volume, and that of Xe, 0.05 ppm. The atmosphere is the only source of all these gases except helium. The alternative for helium is certain natural gas wells in the western United States that produce natural gas containing up to 8% He by volume. It is cost-effective to extract helium from natural gas even down to levels of about 0.3%. Underground helium accumulates as a result of alpha particle emission by radioactive elements in Earth's crust. Whereas the abundance of He on Earth is very limited, it is second only to hydrogen in the universe as a whole.

Are You Wondering . . .

Why argon is so much more abundant in the atmosphere than all the other noble gases?

Most of the noble gases have escaped from the atmosphere since Earth was formed, but Ar is an exception. The concentration of Ar remains quite high because it is constantly being formed by the radioactive decay of potassium-40, a reasonably abundant naturally occurring radioactive isotope. Helium is also constantly produced through α-particle emissions by radioactive decay processes. But because the molar mass of He is 10 times less than that of Ar, it escapes from the atmosphere into outer space at a higher rate.

Properties and Uses

The lighter noble gases are commercially important, in part because they are chemically inert. Helium has several unique physical properties. Best known of these is the fact that it exists as a liquid at temperatures approaching 0 K. All other substances freeze to solids at temperatures well above 0 K. (The melting point of solid H_2, for example, is 14 K.)

Because of their inertness, both He and Ar are used to blanket materials that need to be protected from nitrogen and oxygen in the air, as in certain types of welding, in metallurgical processes, and in the preparation of ultrapure Si and Ge, and other semiconductor materials. Helium mixed with oxygen is used as a breathing mixture for deep-sea diving and in certain medical applications. The efficiency and life of electric light bulbs are increased when they are filled with an argon–nitrogen mixture. Electric discharge through neon-filled glass or plas-

▶ Using helium instead of nitrogen in deep-sea diving breathing mixtures avoids nitrogen narcosis ("rapture of the deep"). It also helps avoid a condition called the bends (page 481) because it is more smoothly and rapidly expelled by the body than is nitrogen.

tic tubes produces a distinctive red light ("neon light"). Krypton and xenon are used in lasers and in flash lamps in photography.

Large quantities of liquid helium are used to maintain low temperatures (cryogenics). Metals essentially lose their electrical resistivity at liquid He temperatures and become *superconductors*. Powerful magnets can be made by immersing the coils of electromagnets in liquid helium. Such magnets are used in particle accelerators and in nuclear fusion research. More familiar uses of large electromagnets cooled by liquid helium are nuclear magnetic resonance (NMR) instruments in research laboratories and magnetic resonance imaging (MRI) devices in hospitals. Helium is also used to fill lighter-than-air airships (blimps).

Compounds of xenon are of special interest to research chemists because they can be fairly easily made and are useful in studies of chemical bonding. We will consider some xenon compounds in Chapter 23.

An Environmental Issue Involving Radon

All atoms with an atomic number greater than 83 are radioactive. The nuclei of these atoms are unstable and emit α, β, or γ radiation, eventually breaking down to more stable elements with lower atomic numbers. Radon-222, a colorless, odorless gas is produced by the loss of α-particles by radium-226, which in turn results from the radioactive decay, through several steps, of uranium-238.

▶ We first mentioned the emission of α, β, and γ rays on page 40.

In December 1984, a worker at a nuclear power plant in New Jersey registered high readings on a radiation detector during a routine safety check. But the radiation to which he had been exposed came not from within the plant but from his own home. This incident led to the recognition that a number of individuals may be exposed to high levels of radioactivity from radon. The possible harmful effects of this exposure, primarily an increased risk of lung cancer, are unclear and a topic of continuing research and debate.

In some instances the source of radon is in wastes from uranium mining or phosphate production. In most cases it is emitted by the radioactive decay of ^{238}U present in small amounts in rocks and soils. Because radon is a gas, it readily passes through air passages in the body and is breathed in and out. The product formed when a ^{222}Rn atom gives up an α-particle is the isotope polonium-218, which also emits α-particles. Unlike radon, polonium is a solid. Health hazards posed by radon seem to be from radioactive decay products such as ^{218}Po becoming attached to dust particles in the air and then being breathed into the lungs.

Fortunately, indoor radon can be rather easily detected by its radioactivity. The chief method of reducing radon levels is through improved ventilation and by venting subsoil radon to keep it from concentrating within a building. In the future, minimizing indoor radon should become a conscious part of building construction.

8-5   Oxides of Carbon

The chief oxides of carbon are carbon monoxide, CO, and carbon dioxide, CO_2. There are about 360 ppm of CO_2 in air (0.036% by volume). CO occurs to a much lesser extent. Although they are only minor constituents of air, these two oxides are important in many ways, as we will see in this section.

Combustion of Carbon Compounds

Carbon dioxide is the only oxide of carbon formed when carbon or carbon-containing compounds are burned in an *excess* of air (providing an abundance of O_2). This condition exists when a "fuel-lean" mixture is burned in an automobile engine. Thus, for the combustion of the gasoline component octane,

$$C_8H_{18}(l) + \tfrac{25}{2} O_2(g) \longrightarrow 8 CO_2(g) + 9 H_2O(l) \qquad (8.16)$$

If the combustion occurs in a *limited* quantity of air, carbon monoxide is also produced. This condition prevails when a "fuel-rich" mixture is burned in an automobile engine. One possibility for the incomplete combustion of octane is

$$C_8H_{18}(l) + 12 O_2(g) \longrightarrow 7 CO_2(g) + CO(g) + 9 H_2O(l) \qquad (8.17)$$

▶ The significance of the terms "fuel rich" and "fuel lean" is explored more fully in Feature Problem A on page 269.

Air pollution by CO comes chiefly from the incomplete combustion of fossil fuels in automobile engines. CO is an inhalation poison because CO molecules bond to Fe atoms in hemoglobin in blood and displace the O_2 molecules that the hemoglobin normally carries (Figure 8-9).

Not only does incomplete combustion of gasoline contribute to air pollution, but it represents a loss of efficiency. A given quantity of gasoline evolves less heat if $CO(g)$ is formed as a combustion product rather than $CO_2(g)$ (see Exercise 46).

Preparation and Uses

Although carbon dioxide can be obtained directly from the atmosphere as a by-product in the liquefaction of air (see Figure 8-3), this is not an important source. Some of the principal commercial sources of CO_2 are summarized in Table 8.5.

The major use of carbon dioxide (about 50%) is as a refrigerant, in the form of dry ice for freezing, preserving, and transporting food. Carbonated beverages account for about 20% of CO_2 consumption. Other important uses are in oil recovery in oil fields and in fire-extinguishing systems. Of course, the major use is not by humans but by plants. Atmospheric CO_2 is the source of all the carbon-containing compounds (organic compounds) in plant life (see Focus On: The Carbon Cycle, page 264).

A modern method of making carbon monoxide is the steam reforming of natural gas, represented as

$$CH_4(g) + H_2O(g) \longrightarrow CO(g) + 3 H_2(g)$$

The term *steam* refers to gaseous water. *Reforming* refers to the restructuring of a carbon compound, such as from CH_4 to CO. The reforming of natural gas (mostly CH_4) is the principal source of $H_2(g)$ for use in the synthesis of NH_3 (page 246).

FIGURE 8-9
CO bound to hemoglobin.
Carbon monoxide binds to the iron atoms in hemoglobin more strongly than does oxygen. Thus, toxic amounts of carbon monoxide can cause death by oxygen deprivation. The portion of the hemoglobin molecule shown here is called a *heme* group.

TABLE 8.5   Some Industrial Methods of Preparing CO_2

Method	Chemical Reaction
recovery from exhaust stack gases in the combustion of carbonaceous fuels, such as the combustion of coke	$C(s) + O_2(g) \longrightarrow CO_2(g)$
recovery in ammonia plants from steam-reforming reactions used to produce hydrogen	$CH_4(g) + 2\ H_2O(g) \longrightarrow$ $CO_2(g) + 4\ H_2(g)$
decomposition (calcination) of limestone at about 900 °C	$CaCO_3(s) \longrightarrow CaO(s) + CO_2(g)$
fermentation by-product in the production of ethanol	$C_6H_{12}O_6(aq) \longrightarrow$ a sugar $2\ C_2H_5OH(aq) + 2\ CO_2(g)$

There are three main uses of carbon monoxide: One is in synthesizing other compounds. Often this is done by producing a mixture of CO and H_2 by reforming methane or some other hydrocarbon. This mixture, which is known as **synthesis gas,** is then converted to a new organic chemical product, such as in the synthesis of methanol.

$$CO(g) + 2\ H_2(g) \longrightarrow CH_3OH(l)$$

Another use of CO is as a reducing agent. In a blast furnace, for instance, **coke,** a rather pure form of carbon produced by heating coal in the absence of air, is converted to CO. The CO then reduces iron oxide to iron.

$$Fe_2O_3(s) + 3\ CO(g) \longrightarrow 2\ Fe(l) + 3\ CO_2(g)$$

A third use of CO is as a fuel, usually mixed with CH_4, H_2, and other combustible gases (recall Section 7-9).

Carbon Dioxide and Carbonates

$CO_2(g)$ dissolves in water to produce a solution generally referred to as carbonic acid, $H_2CO_3(aq)$. H_2CO_3 is a weak acid that ionizes in two steps. When the acid is neutralized in the first step, a salt, variously called a hydrogen carbonate, an acid carbonate, or a bicarbonate, is obtained. If the neutralization is carried through the second step by neutralizing the acid carbonate, a carbonate is obtained.

▶ Carbonic acid, H_2CO_3, does not exist as a stable molecule, but $CO_2(aq)$ is often referred to as $H_2CO_3(aq)$ to emphasize its acidic properties. This matter is discussed further in Chapter 17.

$$H_2CO_3(aq) + Na^+ + OH^- \longrightarrow \underbrace{Na^+ + HCO_3^-}_{\text{sodium hydrogen carbonate}} + H_2O(l)$$

$$Na^+ + HCO_3^- + Na^+ + OH^- \longrightarrow 2\ \underbrace{Na^+ + CO_3^{2-}}_{\text{sodium carbonate}} + H_2O(l)$$

The previous reactions can be reversed by adding an acid to a carbonate. The carbonic acid breaks down into CO_2 and H_2O, and this is a simple way of preparing $CO_2(g)$ in the laboratory.

$$Na_2CO_3(aq) + 2\ H^+(aq) \longrightarrow 2\ Na^+(aq) + H_2O(l) + CO_2(g)$$

▶ This type of reaction accounts for the damage done by acid rain to marble ($CaCO_3$) statues.

The Group 1A metal carbonates are water-soluble and occur in natural salt water solutions called brines. Group 2A and other metal carbonates are water-insoluble and many of these are naturally occurring minerals. Calcite ($CaCO_3$) is one such mineral. Dolomite ($CaCO_3 \cdot MgCO_3$) is another. Carbonate minerals are discussed later in the text, as are other aspects of the chemistry of carbonic

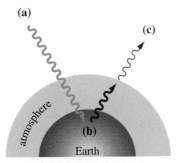

FIGURE 8-10
The "greenhouse" effect.
(a) Sunlight is received by Earth. Some incoming radiation is reflected back into space by the atmosphere, and some is absorbed, such as certain ultraviolet light by stratospheric ozone. Much of the radiation, however, reaches Earth's surface.
(b) Earth's surface emits infrared radiation.
(c) Infrared radiation leaving Earth's atmosphere is less intense than that emitted by Earth's surface. The infrared radiation absorbed by CO_2 and other "greenhouse" gases warms the atmosphere.

acid and carbonates. Aspects of carbonic acid/carbonate chemistry range from the maintenance of the proper acidity (pH) of blood to the formation of limestone caves and the formation of hard water.

An Environmental Issue Involving Carbon Dioxide: Global Warming

We do not think of CO_2 as an air pollutant because it is essentially nontoxic. Its ultimate effect on the environment, however, could be very significant. A buildup of $CO_2(g)$ in the atmosphere may disturb the energy balance on Earth.

Earth's atmosphere is largely transparent to visible and ultraviolet radiation from the sun. This radiation is absorbed at Earth's surface and warms it. But some of this absorbed energy is reradiated as infrared radiation. Certain atmospheric gases, primarily CO_2 and water vapor, absorb some of this infrared radiation. Energy thus retained in the atmosphere produces a warming effect. This process, outlined in Figure 8-10, is often compared to the retention of thermal energy in a greenhouse and called the "greenhouse effect."* The natural greenhouse effect is crucial to maintaining the proper temperature for life on Earth. Without it, Earth would be permanently covered with ice.

From 1880 to 1980 the CO_2 content of the atmosphere increased from 275 to 339 ppm. The current content is about 360 ppm (Figure 8-11). These increases come from the burning of *all* carbon-containing fuels (wood, coal, natural gas, gasoline, . . .) and from the deforestation of tropical regions (trees and other plants, through photosynthesis, consume CO_2 from the atmosphere). The expected effect of a CO_2 buildup is an increase in Earth's average temperature, a **global warming.** Some estimates are that a doubling of the CO_2 content in air from its preindustrial values may cause a global temperature increase of from 1.5 to 4.5 °C. At the present rate of buildup the effects of this doubling could be experienced as early as the middle of the next century.

The probable effects of a CO_2 buildup in the atmosphere are quite uncertain. Predicting these effects is done largely through computer models, and it is very

---

*Glass, like CO_2, is transparent to visible and some ultraviolet light but absorbs infrared radiation. The glass in a greenhouse, though, acts primarily to prevent the bulk flow of warm air out of the greenhouse.

FIGURE 8-11
Increasing carbon dioxide content of the atmosphere.
Shown here is the atmospheric content of CO_2 (expressed in parts per million) measured at Mauna Loa, Hawaii. The peaks come about in April and the valleys in October and reflect different levels of photosynthetic activity at different times of the year (photosynthesis consumes CO_2). The increase in CO_2 content is clearly seen, an 11% increase over a period of about 30 years.

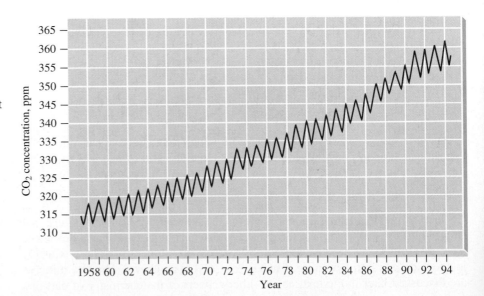

difficult to know all the factors that should be included in these models and their relative importance. For example, global warming could lead to the increased evaporation of water and increased cloud formation. In turn, an increased cloud cover could reduce the amount of solar radiation reaching Earth's surface and produce a *cooling* effect. Two of the most cited probable effects of global warming are

- local temperature changes (Washington, D.C. might experience 12 days a year of 100 °F temperatures instead of the current average of 1);
- rise in mean sea level (thermal expansion of seawater and increased melting of polar ice caps could raise sea level by several meters, with profound effects on coastal cities).

Although much current thinking about global warming is speculative, some evidence supports its likelihood. For example, analyses of tiny air bubbles trapped in the Antarctic ice cap show a strong correlation between the atmospheric CO_2 content and temperature for the past 160,000 years—low temperatures during periods of low CO_2 levels and higher temperatures with higher levels.

CO_2 is not the only greenhouse gas. Several gases are even stronger infrared absorbers—specifically, methane (CH_4), ozone (O_3), nitrous oxide (N_2O), and chlorofluorocarbons (CFCs). Furthermore, atmospheric concentrations of some of these gases have been growing at a faster rate than that of CO_2. No single strategy beyond curtailing the use of fossil fuels has emerged for countering a possible global warming. Like several other major environmental issues, many facets of the global warming problem are poorly understood, and research and debate are likely to occur simultaneously for a long time to come.

8-6 Hydrogen

Think how important hydrogen is in the study of chemistry. John Dalton based atomic masses on a value of 1 for the H atom. Humphry Davy (1810) proposed that hydrogen is the key element in acids. In the next chapter we will see that theoretical studies of the H atom provide us with our modern view of atomic structure. In Chapter 12 we will find that the H_2 molecule is the usual starting point for modern theories of molecular structure. For all of its theoretical significance, though, hydrogen is also of great practical importance, as we will emphasize in this section.

An ice core from the ice sheet in Antarctica is cut into sections in a refrigerated *clean room*. The ice core will be analyzed to determine the amount and type of trapped gases and trace elements it contains. These data will provide information regarding past changes in climate and current trends in the pollution of the atmosphere.

Occurrence and Preparation

Hydrogen is a very minor component of the atmosphere, about 0.5 ppm until one gets to very high altitudes. Above 2500 km the atmosphere is mostly atomic hydrogen at extremely low pressures. In the universe as a whole, hydrogen accounts for about 90% of the atoms and 75% of the mass. And even here on Earth, *hydrogen occurs in more compounds than any other element.*

The free element can easily be produced, but only from a few of its compounds. Our first choice might be H_2O—the most abundant hydrogen compound. To extract hydrogen from water means reducing the oxidation state of H from +1 in H_2O to 0 in H_2. To do this we need an appropriate *reducing* agent, such as carbon (coal or coke), carbon monoxide, or a hydrocarbon—particularly methane (natural gas). The first pair of reactions that follow are called the **water gas** reactions; they represent a way of making combustible gases—CO and H_2— from steam.

Water gas reactions: $C(s) + H_2O(g) \longrightarrow CO(g) + H_2(g)$

$CO(g) + H_2O(g) \longrightarrow CO_2(g) + H_2(g)$

Reforming of methane: $CH_4(g) + H_2O(g) \longrightarrow CO(g) + 3\ H_2(g)$

Another source of $H_2(g)$ is as a by-product in petroleum refining.

Often we use methods in the chemical laboratory that are not commercially feasible. Electrolysis of water (page 252) is one useful laboratory method; another involves the reaction of active metals (recall the list in Table 5.3) in acidic solutions, such as

$$Zn(s) + 2\ H^+(aq) \longrightarrow Zn^{2+}(aq) + H_2(g)$$

Compounds of Hydrogen

Hydrogen forms binary compounds, **hydrides,** with most of the other elements. Binary hydrides are usually grouped into the three broad categories: covalent, ionic, and metallic. *Covalent* hydrides are those formed between hydrogen and nonmetals. Some of these hydrides are simple molecules that can be formed by the direct union of hydrogen and the second element, such as

$$H_2(g) + Cl_2(g) \longrightarrow 2\ HCl(g)$$

$$3\ H_2(g) + N_2(g) \rightleftharpoons 2\ NH_3(g)$$

Ionic hydrides form between hydrogen and the most active metals, particularly those of Groups 1A and 2A. In these compounds hydrogen exists as the hydride *ion,* H^-.

$$2\ M(s) + H_2(g) \longrightarrow 2\ MH(s) \qquad M(s) + H_2(g) \longrightarrow MH_2(s)$$
$$\text{(M is any Group 1A metal)} \qquad\qquad \text{(M is Ca, Sr, or Ba)}$$

Ionic hydrides react vigorously with water to produce $H_2(g)$. CaH_2, a gray solid, has been used as a "portable" source of $H_2(g)$ for filling weather observation balloons.

$$CaH_2(s) + 2\ H_2O(l) \longrightarrow Ca(OH)_2(s) + 2\ H_2(g)$$

Metallic hydrides are commonly formed with the transition elements—the B-group elements in the middle of the periodic table. A distinctive feature of these hydrides is that in many cases they are *nonstoichiometric*—the ratio of H atoms to metal atoms is variable, not fixed. This is because H atoms can enter the voids or holes among the metal atoms in a crystalline lattice and fill some but not others.

Uses of Hydrogen

Hydrogen is not listed among the top chemicals produced because only a small percentage is ever sold to customers. Most hydrogen is produced and used on the spot. In these terms, its most important use (about 42%) is in the manufacture of NH_3 (reaction 8.1). Its next most important use (about 38%) is in petroleum refining, where H_2 is produced in some operations and consumed in others, such as in the production of the high-octane gasoline component, isooctane, from diisobutylene.

diisobutylene   isooctane

In similar reactions, called **hydrogenation reactions,** hydrogen atoms, in the presence of a catalyst, can be added to double or triple bonds in other molecules. This type of reaction, for example, will convert liquid oleic acid, $C_{17}H_{33}COOH$, to solid stearic acid, $C_{17}H_{35}COOH$. Similar reactions serve as the basis for converting oils that contain carbon–carbon double bonds, such as vegetable oils, into solid or semisolid fats, such as shortening.

Another important chemical manufacturing process that uses hydrogen is the synthesis of methyl alcohol (methanol), an alternative fuel.

$$CO(g) + 2\ H_2(g) \xrightarrow{\text{catalyst}} CH_3OH(g)$$

Hydrogen gas is an excellent reducing agent and in some cases is used to produce metals from their oxide ores. For example, at 850 °C

$$WO_3(s) + 3\ H_2(g) \longrightarrow W(s) + 3\ H_2O(g)$$

The uses of hydrogen described here, together with several others, are listed in Table 8.6.

Liquid vegetable oils contain long molecules with some carbon-to-carbon double bonds. When some of these double bonds in the molecules are hydrogenated to give carbon-to-carbon single bonds, the result is conversion of the liquid to a solid "partially hydrogenated vegetable oil."

An Environmental Issue Involving Hydrogen:
A Hydrogen Economy

As we contemplate the eventual decline in the world's supplies of fossil fuels, hydrogen emerges as an attractive means of storing, transporting, and using energy. For example, when an automobile engine burns hydrogen rather than gasoline, its exhaust is essentially pollution-free. The range of supersonic aircraft could be increased if they used liquid hydrogen as a fuel. A hypersonic airplane (the space plane) might also become possible.

The basic problems are in finding a cheap source and an effective means of storing hydrogen. One possibility is to use hydrogen made by the electrolysis of seawater. This possibility requires an abundant energy source, however; perhaps nuclear fusion energy, if it can be developed.

Another alternative is the thermal decomposition of water. The problem here is that even at 2000 °C water is only about 1% decomposed. What is needed is a thermochemical cycle, a series of reactions that have as their net reaction: $2\ H_2O(l) \rightarrow 2\ H_2(g) + O_2(g)$. Ideally, no single reaction in the cycle would require a very high temperature. Still another alternative being studied involves the use of solar energy to decompose water—photodecomposition. One method

▶ One thermochemical cycle for producing hydrogen from water is considered in Feature Problem C on page 269.

TABLE 8.6 Some Uses of Hydrogen

synthesis of:
  ammonia, NH_3
  hydrogen chloride, HCl
  methanol, CH_3OH
hydrogenation reactions in:
  petroleum refining
  converting oils to fats
reduction of metal oxides, such as those of iron,
  cobalt, nickel, copper, tungsten, molybdenum
metal cutting and welding with atomic and oxyhydrogen torches
rocket fuel, usually $H_2(l)$ in combination with $O_2(l)$
fuel cells for generating electricity, in combination
  with $O_2(g)$

Focus On

The Carbon Cycle

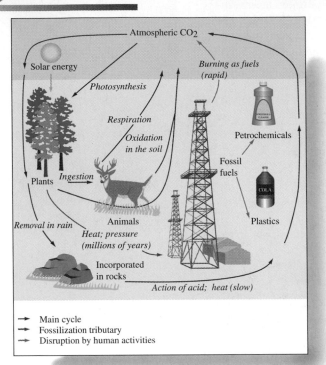

- → Main cycle
- → Fossilization tributary
- → Disruption by human activities

*Labels in diagram:*
Atmospheric CO_2
Solar energy
Photosynthesis
Respiration
Oxidation in the soil
Burning as fuels (rapid)
Petrochemicals
Fossil fuels
Plants
Ingestion
Animals
COLA
Plastics
Removal in rain
Heat; pressure (millions of years)
Incorporated in rocks
Action of acid; heat (slow)

Unlike energy, which flows to Earth constantly as sunlight, the pool of elements essential to life is fixed. After these elements have served their purposes in living matter, they are recycled by natural processes. Perhaps the most familiar of these so-called nutrient cycles is the nitrogen cycle, through which elemental nitrogen from the atmosphere is converted to plant and animal protein and subsequently returned to the air as N_2 (see page 572). But other natural cycles are of interest as well. Here we briefly consider the *carbon cycle,* a portion of which—the major exchanges that occur between the atmosphere and the land masses on Earth—is represented at the left.

The only source of carbon available to plants for making organic

of using hydrogen that is already available combines H_2 and O_2 to form H_2O in an electrochemical device called a *fuel cell.* This device, used in space vehicles for example, converts chemical energy to electricity rather than heat. The subsequent conversion of electrical energy to mechanical energy (work) can be carried out much more efficiently than can the conversion of heat to mechanical energy. Fuel cells are described more fully in Chapter 21.

Storage of gaseous hydrogen is difficult because of the bulk of the gas. When liquefied, hydrogen occupies a much smaller volume, but because of its very low boiling point ($-253\ °C$) the $H_2(l)$ must be stored at very low temperatures. Also, hydrogen must be maintained out of contact with oxygen or air with which it forms explosive mixtures. One approach may be to dissolve $H_2(g)$ in a metal or metal alloy, such as an iron–titanium alloy. The gas can be released by mild heating. In an automobile this storage system would replace the gasoline tank. The heat required to release hydrogen from the metal would come from the engine exhaust.

If the problems described here can be solved, not only can hydrogen be used to supplant gasoline as a fuel for transportation, but it can also replace natural gas for space heating. Because H_2 is a good reducing agent, it can replace carbon (as coal or coke) in metallurgical processes. And, of course, it would be abundantly available for reaction with N_2 to produce NH_3 for the manufacture of fertilizers. The combination of all these potential uses of hydrogen could lead to a fundamental change in our way of life and give rise to what is called a **hydrogen economy.**

compounds, through a process known as *photosynthesis,* is atmospheric CO_2. The process is extremely complicated and has been known for only a few decades (Melvin Calvin, Nobel Prize: 1961). It involves up to 100 sequential steps for the conversion of 6 mol CO_2 to 1 mol $C_6H_{12}O_6$ (glucose). The net change is

$$6\ CO_2 + 6\ H_2O \xrightarrow[\text{sunlight}]{\text{chlorophyll}}$$
$$C_6H_{12}O_6 + 6\ O_2$$
$$\Delta H = +2.8 \times 10^3\ kJ$$

The net reaction is highly endothermic. The required energy comes from sunlight. Chlorophyll, a green pigment in plants, also plays a crucial role.

Here are some of the ideas illustrated in the diagram. When animals consume plants, carbon atoms pass on to the animals. Some carbon is returned to the atmosphere as CO_2 when the animals breathe (respiration) and expel gas (methane). Additional CO_2 returns to the atmosphere as plants and animals die and decay. Some carbon is locked up, as decaying organic matter is converted to coal, petroleum, and natural gas—the fossilization tributary. Not represented is the fact that CO_2 also cycles through the oceans of the world. Phytoplankton convert CO_2 to organic compounds by photosynthesis, and other organisms feed on the plankton. Huge quantities of carbon have accumulated in the form of carbonate rocks (mostly $CaCO_3$). These come from the shells of decayed molluscs in ancient seas.

Human activities now play a far more significant role in the carbon cycle than in preindustrial times. Slash-and-burn agricultural methods—particularly in rain forests—are returning CO_2 to the atmosphere more rapidly than naturally occurring fires do. And the combustion of fossil fuels is replacing stored carbon by carbon dioxide to an even greater extent. We have already seen the possible consequences of this distortion of the carbon cycle: an increased level of atmospheric CO_2 and a possible future global warming (page 260). Human disruption of the natural carbon cycle has become a widely debated issue.

Summary

The atmosphere consists of several regions that differ in their temperature–altitude profiles and the gases of which they are composed. In the portions of the atmosphere closest to Earth's surface, the major components are nitrogen, oxygen, and argon. These gases are obtained by the fractional distillation of liquid air. Important minor constituents of air are carbon dioxide, carbon monoxide, and the noble gases other than argon. All of these gases, as well as hydrogen, are the subjects of this chapter.

Some of the topics considered are the methods of obtaining nitrogen compounds, particularly NH_3 and HNO_3, the uses of these compounds, and the role of nitrogen oxides in smog formation. The discussion of oxygen centers on methods of preparing the allotropes O_2 and O_3 and their uses. The environmental threat resulting from ozone destruction in the stratosphere is also explored. The major point of interest concerning the noble gases is some of the special uses made possible by their physical properties. The greatest interest concerning radon gas is in the potential environmental hazards associated with its radioactivity.

The focus of the discussion of CO and CO_2 is their formation in combustion processes, their industrial preparation and uses, and the chemistry of carbonates. The environmental issue associated with CO is air pollution, and with CO_2, global warming. The discussion of hydrogen emphasizes its preparation—both commercially and in the laboratory—its uses, and some of its binary compounds—hydrides. Hydrogen is so useful that a hydrogen-based economy replacing a petroleum-based one is a future possibility.

Summarizing Example

The oxidation of $NH_3(g)$ to $NO(g)$ in the Ostwald process (reaction 8.4) must be very carefully controlled in terms of temperature, pressure, and contact time with the catalyst. This is because the oxidation of $NH_3(g)$ can yield any one of the products $N_2(g)$, $N_2O(g)$, $NO(g)$, and $NO_2(g)$, depending on conditions. Show that oxidation of $NH_3(g)$ to $N_2(g)$ is the most exothermic (has the most negative $\Delta H°$) of the four possible reactions.

1. *Write equations for each reaction.* Each of these equations has $NH_3(g)$ and $O_2(g)$ as reactants. As products, write the nitrogen-containing species and $H_2O(g)$. Because a comparison is to be made among the four reactions, it is best to write each equation in terms of *one* mole of $NH_3(g)$. *Result:*

$$NH_3(g) + \tfrac{3}{4} O_2(g) \longrightarrow \tfrac{1}{2} N_2(g) + \tfrac{3}{2} H_2O(g)$$

$$NH_3(g) + O_2(g) \longrightarrow \tfrac{1}{2} N_2O(g) + \tfrac{3}{2} H_2O(g)$$

$$NH_3(g) + \tfrac{5}{4} O_2(g) \longrightarrow NO(g) + \tfrac{3}{2} H_2O(g)$$

$$NH_3(g) + \tfrac{7}{4} O_2(g) \longrightarrow NO_2(g) + \tfrac{3}{2} H_2O(g)$$

2. *Use standard enthalpies of formation from Appendix D and expression (7.21) to obtain enthalpies of reaction.* You can do four separate calculations using the expression $\Delta H^\circ = \Sigma \Delta H_f^\circ(\text{products}) - \Sigma \Delta H_f^\circ(\text{reactants})$, and choose the most negative result. There is a simpler way, though. In each application of (7.21) the terms based on $\tfrac{3}{2} H_2O(g)$ and $NH_3(g)$ will be the same. The most negative ΔH° will be for the reaction in which ΔH_f° of the nitrogen-containing species is *smallest*. Refer to Appendix D and you will see that all oxides of nitrogen have *positive* values of ΔH_f°. The smallest ΔH_f°—zero—is for the element $N_2(g)$. *Answer:*   The most exothermic reaction is $NH_3(g) + \tfrac{3}{4} O_2(g) \rightarrow \tfrac{1}{2} N_2(g) + \tfrac{3}{2} H_2O(g)$.

Key Terms

acid anhydride (8-2)	**hydride** (8-6)	**smog** (8-2)
allotropy (8-3)	**hydrogen economy** (8-6)	**stratosphere** (8-1)
coke (8-5)	**hydrogenation reactions** (8-6)	**synthesis gas** (8-5)
electrolysis (8-3)	**ozone layer** (8-3)	**troposphere** (8-1)
global warming (8-5)	**relative humidity** (8-1)	**water gas** (8-6)

Review Questions

1. In your own words define the following terms: **(a)** allotropy; **(b)** electrolysis; **(c)** chlorofluorocarbon; **(d)** nonstoichiometric compound.

2. Briefly describe each of the following ideas, methods, or phenomena: **(a)** noble gas; **(b)** water gas reactions; **(c)** hydrogenation reaction; **(d)** global warming through the "greenhouse effect."

3. Explain the important distinctions between each pair of terms: **(a)** troposphere and stratosphere; **(b)** allotrope and isotope; **(c)** "fuel-lean" and "fuel-rich"; **(d)** metallic and ionic hydride.

4. Supply a name or formula for **(a)** O_3; **(b)** dinitrogen monoxide; **(c)** potassium superoxide; **(d)** CaH_2; **(e)** Mg_3N_2; **(f)** potassium carbonate; **(g)** ammonium dihydrogen phosphate.

5. Describe these commercial materials by chemical formulas: **(a)** ammonia; **(b)** coke; **(c)** urea; **(d)** limestone; **(e)** synthesis gas.

6. Write the formula of a compound mentioned in this chapter that has the indicated element in the designated oxidation state: **(a)** N, + 4; **(b)** O, $-\tfrac{1}{2}$; **(c)** N, +1; **(d)** H, −1; **(e)** C, +2.

7. Give a practical laboratory method that you might use to produce small quantities of **(a)** O_2; **(b)** N_2O; **(c)** H_2; **(d)** CO_2.

8. What volume of air must be processed to produce 5.00 L of He?

9. Complete and balance these equations for the reactions of substances with water.
(a) $LiH(s) + H_2O \rightarrow$
(b) $C(s) + H_2O \xrightarrow{\Delta}$
(c) $NO_2(g) + H_2O \rightarrow$

10. Complete and balance these equations for the reactions of substances with acids.
(a) $Mg(s) + HCl(aq) \rightarrow$
(b) $NH_3(g) + HNO_3(aq) \rightarrow$
(c) $MgCO_3(s) + HCl(aq) \rightarrow$
(d) $NaHCO_3(s) + HC_2H_3O_2(aq) \rightarrow$

11. Write an equation to represent the complete neutralization of an aqueous sulfuric acid solution by an aqueous ammonia solution.

12. Identify the oxidizing and reducing agents in reactions (8.4), (8.8), and (8.16).

13. Hydrogen gas reduces an oxide of iron with Fe in the oxidation state +3 to pure iron metal. Write a plausible equation for this reaction.

14. Write an equation similar to (8.6) in which copper reacts with nitric acid to produce $NO(g)$.

15. The following compounds give off a gas when heated: $KClO_3$, $CaCO_3$, NH_4NO_3. Identify the gaseous product from each, and write an equation to show its production.

16. In 1968, before pollution controls were introduced, over 75 billion gal of gasoline was used in the United States as a motor fuel. Assume an emission of oxides of nitrogen of 5 g per vehicle mile and an average mileage of 15 mi/gal of gasoline. How many kilograms of nitrogen oxides were released to the atmosphere in the United States in 1968?

17. The noble gases have many characteristics in common, such as their inertness. Why, then, is helium found in certain natural gas deposits whereas none of the other noble gases are?

18. *Without doing detailed calculations,* explain in which of the following materials you would expect to find the greatest

mass percent of hydrogen: seawater, the atmosphere, natural gas (CH_4), ammonia.

19. How many grams of $CaH_2(s)$ are required to generate sufficient $H_2(g)$ to fill a 235-L weather observation balloon at 722 mmHg and 19.7 °C?

$$CaH_2(s) + 2\ H_2O \longrightarrow Ca(OH)_2(aq) + 2\ H_2(g)$$

20. Use enthalpy of formation data from Appendix D to determine the enthalpy change in the reaction referred to in the Summarizing Example.

$$NH_3(g) + \tfrac{3}{4}\ O_2(g) \longrightarrow \tfrac{1}{2}\ N_2(g) + \tfrac{3}{2}\ H_2O(g) \qquad \Delta H° = ?$$

Exercises

The Atmosphere

21. Show that the composition of air on a mole percent basis is the same as given in Table 8.1 on a volume percent basis.

22. Explain why the composition of air expressed on a mass percent basis is not the same as on a volume percent basis.

Nitrogen

23. Write balanced equations for the following important commercial reactions involving nitrogen and its compounds.
(a) The principal artificial method of fixing atmospheric N_2.
(b) Oxidation of ammonia to NO.
(c) Preparation of nitric acid from NO.

24. When heated, each of the following substances decomposes to the products indicated. Write balanced equations for these reactions.
(a) $NH_4NO_3(s)$ to $N_2(g)$, $O_2(g)$, and $H_2O(g)$.
(b) $NaNO_3(s)$ to sodium nitrite and oxygen gas.
(c) $Pb(NO_3)_2(s)$ to lead(II) oxide, nitrogen dioxide, and oxygen.

25. Pure liquid HNO_3 decomposes, even at low temperatures, to produce N_2O_4, O_2, and water. Write a balanced equation for this reaction.

26. Sodium nitrite can be made by passing oxygen and nitrogen monoxide gases into an aqueous solution of sodium carbonate. Write a balanced equation for this reaction.

27. A chemistry magazine listed the 1995 demand for $N_2(g)$ as 9.39×10^{11} ft^3 at STP. What is this quantity of N_2 expressed in kilograms? (1 in. = 2.54 cm; 1 ft = 12 in.)

28. Concentrated $HNO_3(aq)$ used in laboratories is usually 15 M HNO_3 and has a density of 1.41 g/mL. What is the percent by mass of HNO_3 in this concentrated acid?

Oxygen

29. Each of the following compounds decomposes to produce $O_2(g)$ when heated. Write plausible equations for the reactions that occur: **(a)** $HgO(s)$; **(b)** $KClO_4(s)$.

30. $O_3(g)$ is a powerful oxidizing agent. Write equations to represent oxidation of **(a)** I^- to I_2 in acidic solution; **(b)** sulfur in the presence of moisture to sulfuric acid; **(c)** $[Fe(CN)_6]^{4-}$ to $[Fe(CN)_6]^{3-}$ in basic solution. In each case $O_3(g)$ is reduced to $O_2(g)$.

31. *Without performing detailed calculations,* determine which of the following compounds has the greatest percent oxygen by mass: dinitrogen tetroxide, aluminum oxide, tetraphosphorus hexoxide, or carbon dioxide.

32. *Without performing detailed calculations,* determine which decomposition yields the most $O_2(g)$ **(a)** per mole and **(b)** per gram of substance.
(1) ammonium nitrate $\longrightarrow$ nitrogen + oxygen + water
(2) hydrogen peroxide $\longrightarrow$ oxygen + water
(3) potassium chlorate $\longrightarrow$ potassium chloride + oxygen

33. The natural abundance of O_3 in unpolluted air at ground level is about 0.04 parts per million (ppm) by volume. What is the approximate partial pressure of O_3 under these conditions, expressed in millimeters of mercury?

34. A typical concentration of O_3 in the ozone layer is 5×10^{12} O_3 molecules/cm^3. What is the partial pressure of O_3, expressed in millimeters of mercury? Assume a temperature of 220 K.

35. Explain why the volumes of $H_2(g)$ and $O_2(g)$ in the electrolysis of water pictured in Figure 8-7 are not the same.

36. In the electrolysis of a sample of water in an apparatus similar to that of Figure 8-7, 22.83 mL of $O_2(g)$ was collected at 25.0 °C at an oxygen partial pressure of 736.7 mmHg. Determine the mass of water that was decomposed.

The Noble Gases

37. A 55-L cylinder contains Ar at 145 atm and 26 °C. What minimum volume of air at STP must have been liquefied and distilled to produce this Ar? Air contains 0.934% Ar, by volume.

38. Some sources of natural gas contains 8% He by volume. How many liters of such a natural gas must be processed at STP to produce 5.00 g of He?

39. A breathing mixture is prepared in which He is substituted for N_2. The gas is 79% He and 21% O_2, by volume. What is the density of this mixture, in grams per liter, at 25 °C and 1 atm?

Carbon

41. Write equations for the reactions that you would expect when
(a) $C_6H_{14}(l)$ is burned in an excess of air;
(b) $CO(g)$ is heated with $PbO(s)$;
(c) $CO_2(g)$ is bubbled into $KOH(aq)$;
(d) $MgCO_3(s)$ is added to $HCl(aq)$.

43. A laboratory method for preparing CO_2 is the reaction of an acid with a carbonate. Would you expect any CO to form in this reaction? Explain.

45. Determine the quantity of heat evolved in the *complete* combustion of 1.00 gal of octane, $C_8H_{18}(l)$, in reaction (8.16). Use $\Delta H_f^\circ[C_8H_{18}(l)] = -250.0$ kJ/mol and data for other substances from Appendix D. The density of $C_8H_{18}(l)$ is 0.703 g/mL; 1 gal = 3.785 L.

Hydrogen

47. Write chemical equations for the following reactions.
(a) The displacement of $H_2(g)$ from $HCl(aq)$ by $Al(s)$.
(b) The reforming of propane gas (C_3H_8) with steam.
(c) The reduction of $MnO_2(s)$ to $Mn(s)$ with $H_2(g)$.
49. $CaH_2(s)$ reacts with water to produce $Ca(OH)_2$ and $H_2(g)$. $Ca(s)$ reacts with water to produce the same products. $Na(s)$ reacts with water to form $NaOH$ and $H_2(g)$. *Without doing detailed calculations,* determine **(a)** which of these reactions produces the most H_2 per liter of water used, and **(b)** which solid—CaH_2, Ca, or Na—produces the most $H_2(g)$ *per gram* of the solid.

40. Concerning the breathing mixture in Exercise 39, at what pressure would the He–O_2 mixture have the same density as does air at 25 °C and 1 atm? (*Hint:* What are the apparent molar masses of air and of the He–O_2 mixture?)

42. Refer to Section 8-5 and write equations for the following reactions.
(a) The action of vinegar (aqueous acetic acid) on baking soda (sodium hydrogen carbonate).
(b) Carbon monoxide as a reducing agent in the reduction of zinc oxide to zinc metal.
(c) The production from synthesis gas of CH_2OHCH_2OH (ethylene glycol), used as an antifreeze.
44. Chlorofluorocarbons are implicated in two environmental issues: ozone destruction and global warming. Carbon dioxide is implicated only in the problem of global warming. Explain this difference.
46. Use Hess's law to show that if 1.00 mol $C_8H_{18}(l)$ burns according to Equation (8.17) rather than (8.16) the heat of combustion is less, by an amount equal to ΔH° for the reaction $CO(g) + \frac{1}{2} O_2(g) \rightarrow CO_2(g)$. What is this quantity of heat? What percent of the maximum possible heat of combustion [ΔH° for reaction (8.16)] does this represent? Use data from Appendix D, together with the value $\Delta H_f^\circ[C_8H_{18}(l)] = -250.0$ kJ/mol.

48. Write equations to show how to prepare $H_2(g)$ from each of the following substances: **(a)** H_2O; **(b)** $HI(aq)$; **(c)** $Mg(s)$; **(d)** $CO(g)$. Use other common laboratory reactants as necessary, that is, water, acids or bases, metals, and so on.
50. What volume of $H_2(g)$ at 25 °C and 752 mmHg is required to hydrogenate oleic acid, $C_{17}H_{33}COOH(l)$, to produce one mole of stearic acid, $C_{17}H_{35}COOH(s)$. Assume the reaction proceeds with a 95% yield.

Advanced Exercises

51. Explain why, expressed as a *mass percent*, the % N_2 in air is less than the 78.084% listed in Table 8.1 whereas the % O_2 is greater than the 20.946%.
52. Recompute the data in Table 8.1 to express the percentages of atmospheric gases on a mass basis.
53. A 0.25-mL sample of $H_2O(l)$ at 20 °C ($d = 0.998$ g/mL) is allowed to vaporize into a closed 18.5-L container of dry air at 20 °C. What is the relative humidity of this sample of air? The vapor pressure of water at 20 °C = 17.5 mmHg.
54. A 0.1052-g sample of $H_2O(l)$ in a 8.050-L sample of dry air at 30.1 °C evaporates completely. To what temperature must the air be cooled to give a relative humidity of 80.0%? Vapor pressures of water: 20 °C, 17.54 mmHg; 19 °C, 16.48 mmHg; 18 °C, 15.48 mmHg; 17 °C, 14.53 mmHg; 16 °C, 13.63 mmHg; 15 °C, 12.79 mmHg.

55. Suppose that no attempt is made to separate the $H_2(g)$ and $O_2(g)$ produced by the electrolysis of water. What volume of an H_2/O_2 mixture, saturated with $H_2O(g)$ and collected at 23 °C and 755 mmHg, would be produced by electrolyzing 17.3 g water? Assume that the water vapor pressure of the dilute electrolyte solution is 20.5 mmHg.
56. One measure for reducing $CO_2(g)$ emissions is to use fuels that produce a large quantity of heat per mole of $CO_2(g)$ produced. Viewed in this way, which is the better fuel, gasoline [assume it to be octane, $C_8H_{18}(l)$] or methane, $CH_4(g)$? The enthalpy of formation of $C_8H_{18}(l)$ is −250.0 kJ/mol. Obtain other enthalpies of formation from Appendix D.
57. A few drops of a deep purple acidic solution of $KMnO_4$ impart a pink color when added to $NaNO_3(aq)$ but are decolorized when added to $NaNO_2(aq)$. Explain this difference in behavior.

58. Carbon suboxide, a compound consisting of 52.96 %C and 47.04 %O by mass, is the acid anhydride of malonic acid, an acid with two ionizable H atoms per molecule and consisting of 34.62 %C, 61.50 %O, and 3.87 %H by mass. Write an equation for the reaction of carbon suboxide with water. (*Hint:* Recall the definition of an acid anhydride on page 248.)

59. For every volume of $O_2(g)$ inhaled a person expires 0.82 volume of $CO_2(g)$. Oxygen-generating systems for enclosed spaces, such as spacecraft, should therefore have the capacity to consume 0.82 L of $CO_2(g)$ for every liter of $O_2(g)$ produced. Does the reaction of potassium superoxide and carbon dioxide, producing potassium carbonate and oxygen, meet this requirement? Explain.

60. Zn can reduce NO_3^- to $NH_3(g)$ in basic solution. (The following equation is *not* balanced.)

$$NO_3^- + Zn(s) + OH^- + H_2O \longrightarrow [Zn(OH)_4]^{2-} + NH_3(g)$$

The NH_3 can be neutralized with an excess of HCl(aq). Then, the unreacted HCl can be titrated with NaOH. In this way a quantitative determination of NO_3^- can be achieved. A 25.00-mL sample of nitrate solution was treated with zinc in basic solution. The $NH_3(g)$ was passed into 50.00 mL of 0.1500 M

HCl. The excess HCl required 32.10 mL of 0.1000 M NaOH for its titration. What was the $[NO_3^-]$ in the original sample?

61. The net reaction in producing nitric acid from ammonia is $NH_3 + 2 O_2 \rightarrow HNO_3 + H_2O$. Show that Equations (8.4), (8.7), and (8.8) on page 248 can be combined to yield this net equation.

62. Refer to the smog formation discussion on page 250 to explain why the time at which O_3 levels reach a maximum in the smog component profile in Figure 8-6 corresponds to the time at which NO levels reach a minimum.

63. Oxygen atoms are an important constituent of the thermosphere, a layer of the atmosphere with temperatures up to 1500 K. Calculate the average translational kinetic energy of O atoms at 1500 K.

64. One reaction for the production of adipic acid, used in the synthesis of nylon, involves the oxidation of cyclohexanone by a nitric acid solution. Balance the equation representing this reaction.

$$H^+ + NO_3^- + C_6H_{11}O \xrightarrow{\text{catalyst}} H_2C_6H_8O_4 + N_2O(g)$$
$$\text{adipic acid}$$

Feature Problems

A. The figure shows how the emission of pollutants is related to the air/fuel ratio in an internal combustion engine.

1. Use information given in this chapter to provide a general interpretation of the figure.

2. Establish by calculation that the stoichiometric ratio of air to fuel (on a mass basis) is about 14.5 : 1. To do so, assume that (a) octane, C_8H_{18}, is a typical gasoline molecule; (b) the maximum quantity of heat is evolved when $C_8H_{18}(l)$ and $O_2(g)$ are in their stoichiometric proportions and $CO_2(g)$ and $H_2O(l)$ are the products; (c) air used in the combustion is the naturally occurring mixture of $O_2(g)$, $N_2(g)$, and $Ar(g)$.

B. In some research that required the careful measurement of gas densities, John Rayleigh, a physicist, found that the density of $O_2(g)$ had the same value whether the gas was obtained from air or derived from one of its compounds. The situation with $N_2(g)$ was different, however. The density of $N_2(g)$ had the same value when the $N_2(g)$ was derived from any of various compounds, but a *different* value if the $N_2(g)$ was extracted from air. In 1894, Rayleigh enlisted the aid of William Ramsay, a chemist, to solve this apparent mystery; in the course of their work they discovered the noble gases.

1. Why do you suppose that the $N_2(g)$ extracted from liquid air did *not* have the same density as $N_2(g)$ obtained from its compounds?

2. Which gas do you suppose had the greater density, $N_2(g)$ extracted from air or $N_2(g)$ prepared from nitrogen compounds? Explain.

3. The way in which Ramsay proved that nitrogen gas extracted from air was itself a mixture of gases involved allowing this nitrogen to react with magnesium metal to form magnesium nitride. Explain the significance of this experiment.

4. Use data on the composition of air from Table 8.1 to calculate the *percent difference* in the densities at STP of

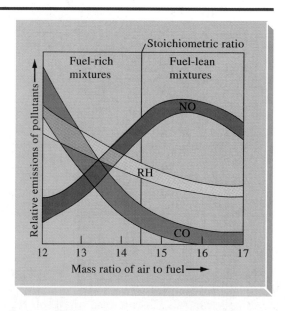

Rayleigh's $N_2(g)$ extracted from air and $N_2(g)$ derived from nitrogen compounds.

C. Various thermochemical cycles are being explored as possible sources of $H_2(g)$. The object is to find a series of reactions that can be conducted at moderate temperatures (about 500 °C) and has as its net result the decomposition of water into H_2 and O_2. Show that the following series of reactions meets this requirement. (*Hint:* Balance the equations, multiply by the appropriate coefficients, and combine them into a net equation.)

$$FeCl_2 + H_2O \longrightarrow Fe_3O_4 + HCl + H_2$$
$$Fe_3O_4 + HCl + Cl_2 \longrightarrow FeCl_3 + H_2O + O_2$$
$$FeCl_3 \longrightarrow FeCl_2 + Cl_2$$

Electrons in Atoms

9

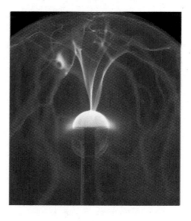

Streams of electrons excite noble gas atoms to produce a colorful display.

*S*ome observers of the scientific scene at the close of the nineteenth century believed the time was near for closing the books on the field of physics. With the accumulated knowledge of the previous two or three centuries, all that remained was to apply this body of physics—classical physics—to fields such as chemistry and biology.

Only a few fundamental problems remained, such as explaining certain details of light emission and a phenomenon known as the photoelectric effect. But the solution of these problems, rather than marking an end, spelled the beginning of a new golden age of physics. These problems were solved through a bold new proposal—the quantum theory—a scientific breakthrough of epic proportions. In this chapter we will see that to explain phenomena at the atomic and molecular level classical physics is inadequate—only the quantum theory will do.

The aspect of quantum mechanics that we will emphasize is how electrons are described through features known as quantum numbers and electron orbitals. The model of atomic structure we develop here will help us to explain many of the topics discussed in the next several chapters: periodic trends in the physical and chemical properties of the elements, chemical bonding, and intermolecular forces.

9-1   Electromagnetic Radiation

Our main topic in this chapter is the electronic structures of atoms. We can learn about electrons in atoms by studying the interactions of electromagnetic radiation and matter. We will present background information about electromagnetic radiation in this section and consider connections between electromagnetic radiation and atomic structure in the sections that follow.

Electromagnetic radiation is a form of energy transmission through a vacuum (empty space) or a medium (such as glass) in which electric and magnetic fields are propagated as waves. A **wave** is a disturbance that transmits energy through a medium. Anyone who has sat in a small boat on a large body of water has experienced wave motion. The wave moves across the surface of the water and the disturbance alternately lifts the boat and allows it to drop. Although water waves may be more familiar, let us use a simpler example of wave motion—a traveling wave in a string—to illustrate some important ideas and essential terminology about waves.

Imagine tying one end of a long string to a door knob and holding the other end in your hand (Figure 9-1). Imagine further that you have colored one small segment of the string with red ink. As you move your hand up and down, you set up a wave motion in the string. The wave moves to the right, but the colored segment simply moves up and down. In relation to the center line (the broken line in Figure 9-1), the wave consists of *crests,* where the string is at a maximum, or high point, above the center line, and *troughs,* low points or minima, below the center line. The maximum height of the wave above the center line, or the maximum depth below, is called the *amplitude.* The distance between the tops of two successive crests (or the bottoms of two troughs) is called the **wavelength,** designated by the Greek letter lambda, **λ.**

Wavelength is one important characteristic of a wave. Another feature, **frequency,** designated by the Greek letter nu, **ν,** is the number of crests or troughs that pass through a given point per unit of time. Frequency has the unit, time^{-1}, usually s^{-1} (per second), meaning the number of events or cycles per second. The product of the length of a wave (λ) and the frequency (ν) shows how far the wave front travels in a unit of time. This is the *velocity* of the wave. Thus, if the wavelength in Figure 9-1 were 0.5 m and the frequency, 3 s^{-1} (meaning three complete up-and-down hand motions per second), the velocity of the wave would be 0.5 m $\times$ 3 s^{-1} = 1.5 m/s.

Electromagnetic radiation is represented in Figure 9-2. As the figure shows, the radiation component associated with the magnetic field lies in a plane perpendicular to that of the electric field component. An electric field is the region around an electrically charged particle. We can detect the presence of an electric field by measuring the force on an electrically charged object when it is brought into the field. A magnetic field is found in the region surrounding a magnet. According to a theory proposed by James Clerk Maxwell (1831–1879) in 1865, electromagnetic radiation—a propagation of electric and magnetic fields—is produced by an accelerating electrically charged particle (a charged particle whose velocity changes). Radio waves, for example, are a form of electromagnetic radiation produced by causing oscillations (fluctuations) of the electric current in a specially designed electrical circuit. With visible light, another form of electromagnetic radiation, the accelerating charged particles are the electrons in atoms or molecules.

▶Water waves, sound waves, and seismic waves (which produce earthquakes) are unlike electromagnetic radiation. They require a material medium for their transmission.

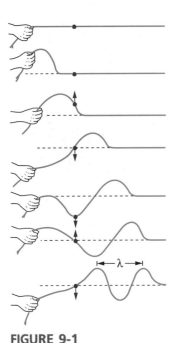

FIGURE 9-1
The simplest wave motion—traveling wave in a string.
As a result of the up-and-down hand motion (top to bottom), waves pass along the long string from left to right. This one-dimensional moving wave is called a traveling wave. The wavelength of the wave, λ—the distance between two successive crests—is identified.

FIGURE 9-2
Electromagnetic waves.
This sketch of two different electromagnetic waves shows the propagation of mutually perpendicular oscillating electric and magnetic fields. For a given wave, the wavelengths, frequencies, and amplitudes of the electric and magnetic field components are identical. Note that the wave with the *longer* wavelength (a) has a *lower* frequency, and the wave with the *shorter* wavelength (b) has a *higher* frequency.

Frequency, Wavelength, and Velocity of Electromagnetic Radiation

The SI unit for frequency, s^{-1}, is the **hertz (Hz),** and the basic SI wavelength unit is the meter (m). Because many types of electromagnetic radiation have very short wavelengths, however, smaller units, including those listed here, are also used. The angstrom, named for the Swedish physicist Anders Ångström (1814–1874), is not an SI unit.

1 centimeter (cm) = 1×10^{-2} m
1 micrometer (μm) = 1×10^{-6} m
1 nanometer (nm) = 1×10^{-9} m = 1×10^{-7} cm = 10 Å
1 angstrom (Å) = 1×10^{-10} m = 1×10^{-8} cm

A distinctive feature of electromagnetic radiation is its *constant* velocity of 2.997925×10^8 m s^{-1} in a vacuum, often referred to as the *speed of light.* The speed of light is represented by the symbol c, and the relationship between this speed and the frequency and wavelength of electromagnetic radiation is

▶ The speed of light is commonly rounded off to 3.00×10^8 m s^{-1}.

$$c = \nu \cdot \lambda \qquad (9.1)$$

Figure 9-3 indicates the wide range of possible wavelengths and frequencies for some common types of electromagnetic radiation, and illustrates this important fact: The wavelength of electromagnetic radiation is shorter for high frequencies and longer for low frequencies. Example 9-1 illustrates an application of Equation (9.1).

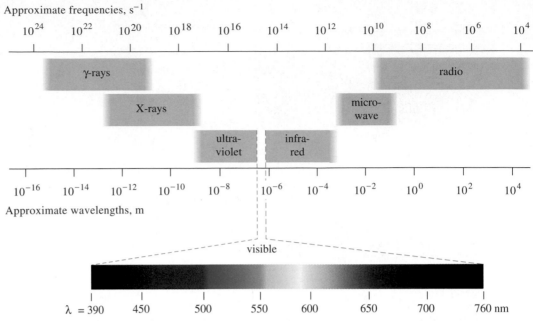

FIGURE 9-3   The electromagnetic spectrum.
The visible region, which extends from violet at the shortest wavelength to red at the longest wavelength, is only a small portion of the entire spectrum. The approximate wavelength and frequency ranges of some other forms of electromagnetic radiation are also indicated.

EXAMPLE 9-1

Relating the Frequency and Wavelength of Electromagnetic Radiation. Most of the light from a sodium vapor lamp has a wavelength of 589 nm. What is the frequency of this radiation?

Solution

We know the speed of light in the unit meters per second (m/s). If we convert the wavelength of the light emitted by the sodium vapor lamp from nanometers to meters, we can use Equation (9.1).

$$c = 2.998 \times 10^8 \text{ m/s}$$

$$\lambda = 589 \text{ nm} \times \frac{1 \times 10^{-9} \text{ m}}{1 \text{ nm}} = 5.89 \times 10^{-7} \text{ m}$$

$$\nu = ?$$

Rearrange Equation (9.1) to the form $\nu = c/\lambda$, and solve for ν.

$$\nu = \frac{c}{\lambda} = \frac{2.998 \times 10^8 \text{ m s}^{-1}}{5.89 \times 10^{-7} \text{ m}} = 5.09 \times 10^{14} \text{ s}^{-1} = 5.09 \times 10^{14} \text{ Hz}$$

Practice Example A   The light from red LEDs (light-emitting diodes) is commonly seen in many electronic devices. A typical LED produces 690-nm light. What is the frequency of this light?

Practice Example B   An FM radio station broadcasts on a frequency of 91.5 megahertz (MHz). What is the wavelength of these radio waves, in meters?

The Visible Spectrum

In a medium like glass, the speed of light is lower than in a vacuum. As a consequence, light is refracted, or bent, when it passes from one medium to another (Figure 9-4). Also, although they have the same speed in a vacuum, electromagnetic waves of differing wavelength have slightly differing speeds in air or other media. "White" light consists of a large number of light waves with different wavelengths. When a beam of white light is passed through a transparent medium, the wavelength components are refracted differently. The light is dispersed into a band, or spectrum, of colors. In Figure 9-5 white light passes through a narrow opening, or slit, and is dispersed by a glass prism. The colored components are then recorded on a photographic film. The device incorporating these features is called a spectrograph.

Example 9-2 tests your familiarity with the colors found in the visible spectrum of white light.

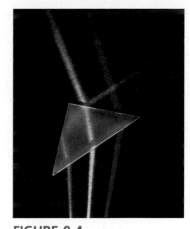

FIGURE 9-4
Refraction of light.
Light is refracted (bent) as it passes from air into the glass prism, and again as it emerges from the prism into air. This photograph shows that red light is refracted the least and blue light the most. The blue light strikes the prism at such an angle that the beam undergoes an internal reflection before it emerges from the prism.

EXAMPLE 9-2

Relating the Frequencies, Wavelengths, and Colors of Components of White Light. Which of these common street lamps emits light of the higher frequency: mercury vapor or sodium vapor? (see Figure 9-6.)

Solution

To answer this question you need to

1. Understand, from Equation (9.1), that wavelength and frequency are inversely proportional: the *longer* the wavelength, the *lower* the frequency.
2. Have a sense, from Figure 9-3, of the relationship between colors and wavelengths. The order of *decreasing* wavelength of the common colors is: red > orange > yellow > green > blue > indigo > violet.

light source

slit   prism

red

violet

photographic film

FIGURE 9-5
The spectrum of "white" light.
Red light is refracted the least and violet light the most when "white" light is passed through a glass prism. The other colors of the visible spectrum are found between red and violet.

FIGURE 9-6
Sodium and mercury vapor lamps.
A sodium vapor lamp (foreground) produces light of a distinctly yellow cast, whereas the light from mercury vapor lamps (background) is predominantly blue. In each lamp, a beam of electrons passing through the gaseous element excites atoms. The energetically excited atoms give off light as they revert to their normal states.

Because of its *shorter* wavelength, the predominantly bluish light of the mercury vapor lamp is of *higher* frequency than the yellow light of the sodium vapor lamp.

Practice Example A   Which device uses electromagnetic radiation of higher frequency: a microwave oven or a dentist's X-ray machine?

Practice Example B   Arrange the following four light sources according to *increasing* wavelength: the lights you see from a traffic signal when it indicates "stop," "go," and "caution," and the light from the mercury vapor lamp of Figure 9-6.

9-2 Atomic Spectra

In Figure 9-5 the light source can be sunlight or the heated filament of an ordinary electric light bulb. Each wavelength component of the white light yields an image of the slit in the form of a line. There are so many of these lines that they blend together into an unbroken band of color from red to violet: the spectrum of white light is *continuous*. On the other hand, the spectra produced by certain gaseous substances consist of only a limited number of colored lines with dark spaces between them. These *discontinuous* spectra are called **atomic,** or **line spectra,** and Figure 9-7 suggests how such a spectrum can be produced.

Each element has its own distinctive line spectrum—a kind of atomic fingerprint. Robert Bunsen (1811–1899) and Gustav Kirchhoff (1824–1887) developed the first spectroscope and used it to identify elements. In 1860, they discovered a new element and named it cesium (L. *caesius*, sky blue) because of the distinctive blue lines in its spectrum. They discovered rubidium in 1861 in a similar way (L. *rubidius*, deepest red). Still another element characterized by its unique spectrum is helium (Gr. *helios*, the sun). Its spectrum was observed during the solar eclipse of 1868, but helium was not isolated on Earth for another 27 years.

Among the most extensively studied atomic spectra has been that of hydrogen. Light from a hydrogen lamp appears to the eye as a reddish purple color (Figure 9-8). The principal wavelength component of this light is red light of wavelength 656.3 nm. Three other lines appear in the visible spectrum of atomic

▶ The notion of an emission spectrum as an "atomic fingerprint" is explored in Feature Problem A on page 312.

▶ Bunsen designed a special gas burner for his spectroscopic studies. This burner, the common laboratory Bunsen burner, produces very little background radiation to interfere with the spectral observations.

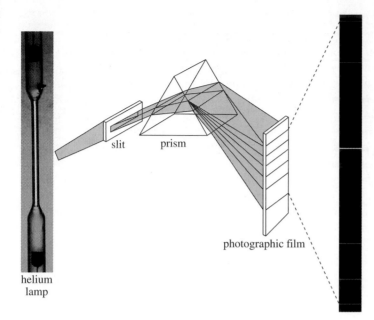

FIGURE 9-7
Production of an atomic, or line, spectrum.
The light source is a lamp containing helium gas at a low pressure. When an electric discharge is passed through the lamp, helium atoms absorb energy, which they reemit as light. The visible spectrum of helium consists of six lines bright enough to be seen with the unaided eye. The apparatus pictured here, in which the spectral lines are photographed, is called a *spectrograph*. If observations are made by sight, the apparatus is called a *spectroscope*.

FIGURE 9-8
Light emission from a hydrogen lamp.
Light is emitted by excited gaseous hydrogen atoms produced when an electric discharge is passed through low-pressure hydrogen gas. Hydrogen molecules do not emit visible light.

hydrogen, however: a greenish blue line at 486.1 nm, a violet line at 434.0 nm, and another violet line at 410.1 nm. This atomic spectrum of hydrogen is shown in Figure 9-9. In 1885, Johann Balmer, apparently through trial and error, deduced a formula for the wavelengths of these spectral lines. Balmer's equation, written in a form proposed by Johannes Rydberg, is

$$\nu = 3.2881 \times 10^{15} \text{ s}^{-1}\left(\frac{1}{2^2} - \frac{1}{n^2}\right) \qquad (9.2)$$

In this equation *n* must be an *integer* (whole number) *greater than 2,* and ν is the frequency of a spectral line. If $n = 3$ is substituted into the equation, the frequency of the red line is obtained. If $n = 4$, the frequency of the greenish blue line is obtained, and so on.

The fact that atomic spectra consist of only limited numbers of well-defined wavelength lines provides a great opportunity to learn about the structures of atoms. For example, it suggests that only a limited number of energy values are available to excited gaseous atoms. Classical (nineteenth-century) physics, however, was not able to provide an explanation of atomic spectra. The key to this puzzle lay in a great breakthrough of modern science—the quantum theory.

▶ A graphical method for understanding Balmer's and Rydberg's work is explored in Feature Problem B on page 312.

410.1 nm   434.0 nm   486.1 nm   656.3 nm

FIGURE 9-9
The Balmer series for hydrogen atoms—a line spectrum.
The four lines shown are the only ones visible to the unaided eye. Additional, closely spaced lines lie in the ultraviolet region.

9-3 Quantum Theory

▶Objects emit radiation at all temperatures, not just at high temperatures. For example, infrared radiation emitted by objects makes them visible at night through night-vision goggles.

We are aware that hot objects emit light of different colors, from the dull red of an electric-stove heating element to the bright white of a light-bulb filament. As with atomic spectra, classical physics could not provide a complete explanation of light emission by heated solids, known as *blackbody radiation.* In 1900, Max Planck (1858–1947), to explain certain aspects of blackbody radiation, made a revolutionary proposal: Energy, like matter, is discontinuous. Classical physics places no limitations on the amount of energy a system may possess, but quantum theory limits this energy to a discrete set of specific values. The *difference* between two of the allowed energies of a system also has specific value, called a **quantum** of energy. This means that when the energy increases from one allowed value to the next, it increases by a tiny jump, or quantum. We can think of a quantum of energy in relation to the total energy of a system as being like a single atom in relation to an entire sample of matter.

Planck postulated that the energy of a quantum of electromagnetic radiation is proportional to the frequency of the radiation—the higher the frequency, the greater the energy. Planck's equation is

$$E = h\nu \tag{9.3}$$

The proportionality constant, h, called **Planck's constant,** has a value of 6.626 $\times 10^{-34}$ J s photon^{-1}.

Only after the quantum hypothesis was successfully applied to phenomena other than blackbody radiation did it acquire status as a great new scientific theory. The first of these successes came in 1905 with Albert Einstein's quantum explanation of the photoelectric effect.

▶Comparing the energy restrictions of the quantum theory and the lack of such restrictions in classical theory is like comparing a long stairway and a ramp leading into a building. In climbing the stairs (the "quantum" case), a person can only rest on a step (a discrete energy state), not between steps. On the ramp, one can rest anywhere that one chooses (the "classical" case).

Are You Wondering . . .

How to relate the quantum theory to everyday experiences?

We don't need quantum theory to explain the behavior of everyday objects, whether they be golf balls in flight, automobiles on a freeway, or the Space Shuttle in orbit. These macroscopic objects have so much energy that the gain or loss of a few quanta is undetectable. The energy of macroscopic objects appears to be continuous, and this is why we don't seem to experience quantum effects in our daily lives. On the other hand, some common experiences are analogous to the quantum world. Think about a coin-operated vending machine that requires exact change and accepts only nickels, dimes, and quarters. You know that items in this machine can only be priced at five-cent intervals. Prices of $0.45, $0.50, $0.55, . . . are all possible, but not a price of $0.57. The "quantum" in this case is a nickel.

The Photoelectric Effect

Figure 9-10 describes the photoelectric effect, discovered by H. Hertz in 1888 and studied extensively by P. Lenard. In the **photoelectric effect** a beam of electrons ("electric") is produced by shining light ("photo") on certain metal sur-

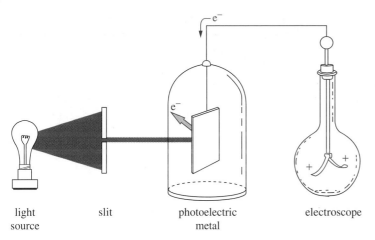

FIGURE 9-10
The photoelectric effect.
A light beam strikes the metal surface and knocks out electrons. The photoelectric metal, by losing electrons, acquires a positive charge. This positively charged metal draws electrons away from the electroscope, causing the metal-foil leaves to acquire a positive charge. Having like charges, the leaves repel each other.

light source   slit   photoelectric metal   electroscope

faces. Although the *number* of ejected electrons depends on the intensity, or brightness, of the light used, their *energies* do not. The electron energies depend on the *frequency* (color) of the light. The freed electrons have greater energies when a photoelectric material is struck by a feeble blue light (low intensity, high frequency) than by a bright red one (high intensity, low frequency). The photoelectric effect was another unsolvable puzzle for classical physics, which viewed the energy of light as depending on its intensity but not on its frequency.

In 1905, Einstein proposed that electromagnetic radiation has particle-like qualities and that "particles" of light, called **photons,** have a characteristic energy, given by Planck's equation, $E = h\nu$. We can think of the energy of light as being concentrated into photons whose energies depend on the frequency of the light. In the photoelectric effect these photons transfer energy during collisions with electrons. Even though a photon is not a particle of matter, as is an electron, a photon can still collide with an electron and transfer its energy. In each collision a photon gives up its entire energy—a quantum of energy—to the electron. A higher frequency of light implies more energetic photons, more energy transferred to electrons, and greater kinetic energies for the ejected electrons. The particle-like nature of light is suggested by Figure 9-11.

The photoelectric effect has many practical uses. In an "electric eye" antitheft system, a laser light is beamed across a space that is to be protected. The beam strikes a photomultiplier tube, a device based on the photoelectric effect. A photoelectric material ejects electrons, which become part of an electric circuit that prevents an alarm from sounding as long as electric current flows. When an object, say a hand, breaks the laser beam, the current is momentarily interrupted and the alarm goes off.

Photons of Light and Chemical Reactions

In Chapter 8 we encountered chemical reactions that are induced by light, *photochemical* reactions. In these reactions photons are "reactants" and we can indicate them in chemical equations by the symbol $h\nu$. Thus, the reactions by which ozone is produced from oxygen in the atmosphere can be represented as

$$O_2 + h\nu \longrightarrow O + O$$

$$O_2 + O + M \longrightarrow O_3 + M$$

The radiation required in the first reaction is ultraviolet radiation with wavelength less than 242.4 nm. O atoms from the first reaction then combine with

A thief in a Hollywood movie (Bruce Willis) uses an aerosol to see the laser beams that are part of a security system. As described in the text, such a system makes use of the photoelectric effect.

▶ To escape from a photoelectric surface, an electron must do so with the energy from a single photon collision. The electron cannot accumulate the energy from several hits by photons.

A light beam having this appearance

actually consists of "particles" called photons.

FIGURE 9-11
Photons of light visualized.

O_2 to form O_3. In the second reaction, a "third body," M, such as $N_2(g)$, is needed to carry away excess energy to prevent immediate dissociation of O_3 molecules.

Photochemical reactions involving ozone are the subject of Example 9-3. There we see that the product of Planck's constant (h) and frequency (ν) yields the energy of a single photon of electromagnetic radiation in the unit joule. Invariably, this energy is only a tiny fraction of a joule. Often it is useful to deal with the much larger energy of a mole of photons (6.02214×10^{23} photons).

EXAMPLE 9-3

Using Planck's Equation to Calculate the Energy of Photons of Light. For radiation of wavelength 242.4 nm, the longest wavelength that will bring about the photodissociation of O_2, what is the energy of (a) one photon and (b) a mole of photons of this light?

Solution

a. To use Planck's equation we need the frequency of the radiation. This we can get from Equation (9.1), after first expressing the wavelength in meters.

$$\nu = \frac{c}{\lambda} = \frac{2.998 \times 10^8 \text{ m s}^{-1}}{242.4 \times 10^{-9} \text{ m}} = 1.237 \times 10^{15} \text{ s}^{-1}$$

Planck's equation is written for *one* photon of light. We emphasize this by including the unit, photon^{-1}, in the value of h.

$$E = h\nu = 6.626 \times 10^{-34} \frac{\text{J s}}{\text{photon}} \times 1.237 \times 10^{15} \text{ s}^{-1}$$

$$= 8.196 \times 10^{-19} \text{ J/photon}$$

b. Once we have the energy per photon, we can multiply it by the Avogadro constant to convert to a per mole basis.

$$E = 8.196 \times 10^{-19} \text{ J/photon} \times 6.022 \times 10^{23} \text{ photons/mol}$$

$$= 4.936 \times 10^5 \text{ J/mol}$$

(This quantity of energy is sufficient to raise the temperature of 10.0 L of water by 11.8 °C.)

Practice Example A: The protective action of ozone in the atmosphere comes through ozone's absorption of ultraviolet radiation in the 230–290-nm wavelength range. What is the energy, in kilojoules per mole, associated with radiation in this wavelength range?

Practice Example B: Chlorophyll absorbs light at energies of 3.056×10^{-19} J/photon and 4.414×10^{-19} J/photon. To what color and frequency do these absorptions correspond?

9-4 The Bohr Atom

The Rutherford model of a nuclear atom (Section 2-3) does not indicate how electrons are arranged outside the nucleus of an atom. According to classical physics, stationary negatively charged electrons would be pulled into the positively charged nucleus. This suggests that the electrons in an atom must be in motion, like the orbiting motion of the planets around the sun. However, again according to classical physics, orbiting electrons should be constantly accelerating and should radiate energy. By losing energy, the electrons would be drawn ever closer to the nucleus and soon spiral into it. This unstable situation is unsatisfactory as an atomic model; it is pictured in Figure 9-12.

In 1913, Niels Bohr (1885–1962) resolved this dilemma by using Planck's quantum hypothesis. In an interesting blend of classical and quantum theory, Bohr postulated that for a hydrogen atom

1. The electron moves in circular orbits about the nucleus, with the motion described by classical physics.

2. The electron has only a *fixed set of allowed orbits,* called *stationary states.* The allowed orbits are those in which certain properties of the electron have unique values. Even though classical theory would predict otherwise, *as long as an electron remains in a given orbit its energy is constant and no energy is emitted.*

3. An electron can pass only from one allowed orbit to another. In such transitions, fixed discrete quantities of energy (quanta) are involved, in accordance with Planck's equation, $E = h\nu$.

The atomic model of hydrogen based on these ideas is pictured in Figure 9-13. The allowed states for the electron are numbered, $n = 1$, $n = 2$, $n = 3$, and so on. These *integral* numbers are called **quantum numbers.** *

The Bohr theory predicts the radii of the allowed orbits in a hydrogen atom.

$$r_n = n^2 a_0, \quad \text{where } n = 1, 2, 3, \dots \text{ and } a_0 = 0.53 \text{ Å (53 pm)} \quad (9.4)$$

The theory also allows us to calculate the electron velocities in these orbits, and, most important, the energy. When the electron is free of the nucleus, by convention, it is said to be at a *zero* of energy. When a free electron is attracted to the nucleus and confined to the orbit n, energy is *emitted* in the form of a photon. The electron energy becomes negative, with its value lowered to

$$E_n = \frac{-R_H}{n^2} \quad (9.5)$$

R_H is a numerical constant with a value of 2.179×10^{-18} J.

FIGURE 9-12
An unsatisfactory atomic model.
In this model, as energy is lost through light emission, the electron is drawn ever closer to the nucleus and eventually spirals into it. The collapse of the atom would occur very quickly.

Niels Bohr (1885–1962). In addition to his work on the hydrogen atom, Bohr headed the Institute of Theoretical Physics in Copenhagen, which became a mecca for theoretical physicists in the 1920s and 1930s.

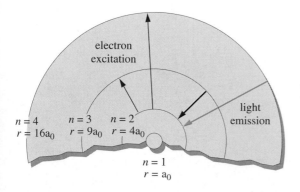

electron excitation

$n = 4$ $n = 3$ $n = 2$
$r = 16a_0$ $r = 9a_0$ $r = 4a_0$

light emission

$n = 1$
$r = a_0$

FIGURE 9-13
Bohr model of the hydrogen atom.
A portion of the hydrogen atom is pictured. The nucleus is at the center, and the electron is found in one of the discrete orbits, $n = 1$, 2, ... Excitation of the atom raises the electron to higher numbered orbits, as shown through black arrows. Light is emitted when the electron falls to a lower numbered orbit. Two transitions are shown that produce lines in the Balmer series of the hydrogen spectrum, in the approximate colors of the spectral lines.

*The particular property of the electron having only certain allowed values, leading to only a discrete set of allowed orbits, is called the angular momentum. Its possible values are $nh/2\pi$, where n must be an integer. Thus $n = 1$ for the first orbit; $n = 2$ for the second orbit; and so on.

With expression (9.5) we can calculate the energies of the allowed energy states, or *energy levels,* of the hydrogen atom. These levels can be represented schematically as in Figure 9-14. This representation is called an **energy-level diagram.** Example 9-4 emphasizes why only a discrete set of energy levels appears in the diagram.

EXAMPLE 9-4

Understanding the Meaning of Quantization of Energy. Is it likely that there is an energy level for the hydrogen atom, $E_n = -1.00 \times 10^{-20}$ J?

Solution

We really do not need to do a detailed calculation to conclude that the answer is very likely *no.* We cannot expect some *random* value to correspond to one of the *unique* set of allowed energy levels. Let us rearrange Equation (9.5), solve for n^2, and then for n.

$$n^2 = -\frac{R_H}{E_n}$$

$$= \frac{-2.179 \times 10^{-18} \text{ J}}{-1.00 \times 10^{-20} \text{ J}} = 2.179 \times 10^2 = 217.9$$

$$n = \sqrt{217.9} = 14.76$$

Because the value of n is not an integer, this is not an allowed energy level for the hydrogen atom.

Practice Example A: Is there an energy level for the hydrogen atom, $E_n = -2.69 \times 10^{-20}$ J?

Practice Example B: Is it likely that one of the electron orbits in the Bohr atom has a radius of 1.00 nm? [*Hint:* Recall expression (9.4).]

FIGURE 9-14
Energy-level diagram for the hydrogen atom.
If the electron acquires 2.179×10^{-18} J of energy, it moves to the orbit $n = \infty$; ionization of the H atom occurs (black arrow). Energy emitted when the electron falls from higher numbered orbits to the orbit $n = 1$ is in the form of ultraviolet light, which produces a spectral series called the Lyman series (gray lines). Electron transitions to the orbit $n = 2$ yield lines in the Balmer series (recall Figure 9-9); three of the lines are shown here (in color). Transitions to $n = 3$ yield spectral lines in the infrared.

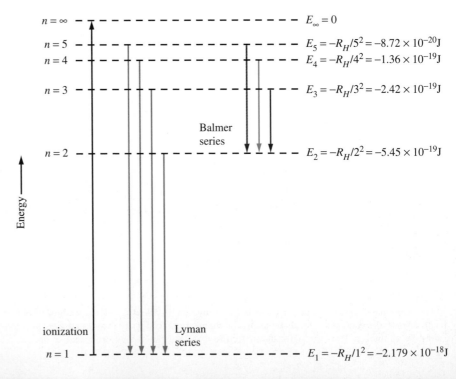

Normally the electron in a hydrogen atom is found in the orbit closest to the nucleus ($n = 1$). This is the lowest allowed energy, or the **ground state.** When the electron gains a quantum of energy it moves to a higher level ($n = 2, 3, \ldots$) and the atom is in an **excited state.** When the electron drops from a higher to a lower numbered orbit, a unique quantity of energy is emitted—the difference in energy between the two levels. We can use Equation (9.5) to derive an expression for the *difference* in energy between two levels, where n_f is the final level and n_i is the initial one.

▶ This phenomenon is often described as being like a person going up stair steps (excitation) and down stair steps (emission). The person must land on a step—the in-between levels are not available.

$$\Delta E = E_f - E_i = \frac{-R_H}{n_f^2} - \frac{-R_H}{n_i^2} = R_H\left(\frac{1}{n_i^2} - \frac{1}{n_f^2}\right) =$$

$$2.179 \times 10^{-18} \text{ J}\left(\frac{1}{n_i^2} - \frac{1}{n_f^2}\right) \quad (9.6)$$

▶ Notice the resemblance of this equation to the Balmer equation (9.2).

In Example 9-5 we use Equation (9.6) as a basis for calculating the lines in the hydrogen emission spectrum. Because the differences between energy levels are limited in number, so too are the energies of the emitted photons. Therefore only certain wavelengths (or frequencies) are observed for the spectral lines.

EXAMPLE 9-5

Calculating the Wavelength of a Line in the Hydrogen Spectrum. Determine the wavelength of the line in the Balmer series of hydrogen corresponding to the transition from $n = 5$ to $n = 2$.

Solution

The specific data for Equation (9.6) are $n_i = 5$ and $n_f = 2$.

$$\Delta E = 2.179 \times 10^{-18} \text{ J}\left(\frac{1}{5^2} - \frac{1}{2^2}\right)$$

$$= 2.179 \times 10^{-18} \times (0.04000 - 0.25000)$$

$$= -4.576 \times 10^{-19} \text{ J}$$

The negative sign of ΔE signifies that energy is *emitted,* just as we expect it to be. For the remainder of the calculation, though, we are interested only in the *magnitude* of ΔE—its numerical value without regard for its sign. The magnitude of ΔE, the energy difference between two energy levels, is equal to the energy of the photon emitted. The photon frequency can be found from the Planck equation: $\Delta E = E_{photon} = h\nu$, which we rearrange to the form

$$\nu = \frac{E_{photon}}{h} = \frac{4.576 \times 10^{-19} \text{ J photon}^{-1}}{6.626 \times 10^{-34} \text{ J s photon}^{-1}} = 6.906 \times 10^{14} \text{ s}^{-1}$$

Finally, to calculate the wavelength of this line we use Equation (9.1).

$$\lambda = \frac{c}{\nu} = \frac{2.998 \times 10^8 \text{ m s}^{-1}}{6.906 \times 10^{14} \text{ s}^{-1}} = 4.341 \times 10^{-7} \text{ m} = 434.1 \text{ nm}$$

Note the good agreement between this result and data in Figure 9-9.

Practice Example A: Determine the wavelength of light *absorbed* in an electron transition from $n = 2$ to $n = 4$ in a hydrogen atom.

Practice Example B: What is the *longest* wavelength of the lines in the Lyman series of the hydrogen spectrum? (*Hint:* Refer to Figure 9-14. Which transition produces the longest wavelength line in the Lyman series?)

The great value of the Bohr theory was in providing a simple model for interpreting the atomic spectrum of hydrogen. The model also works for hydrogen-like species, such as the ions He^+ and Li^{2+}, which have only one electron. For these species the nuclear charge (atomic number) appears in the energy-level expression. That is,

$$E_n = \frac{-Z^2 R_H}{n^2} \qquad (9.7)$$

Even with several modifications, the Bohr model does not do a good job of predicting atomic spectra of multielectron atoms. For this, a new quantum theory is needed, based on ideas presented in the next section.

9-5 Two Ideas Leading to a New Quantum Mechanics

In the previous section we examined some successes of the Bohr theory and pointed out its inability to deal with multielectron atoms. A decade or so after Bohr's work on hydrogen, two landmark ideas stimulated a new approach to quantum mechanics. We consider those two ideas in this section and the new quantum mechanics—wave mechanics—in the next.

Wave–Particle Duality

To explain the photoelectric effect Einstein suggested that light has particle-like properties, embodied in photons. Other phenomena, however, such as the dispersion of light into a spectrum by a prism, are best understood in terms of the wave theory of light. Light, then, appears to have a *dual* nature.

In 1924, Louis de Broglie, considering the nature of light and matter, offered a startling proposition: *Small particles of matter may at times display wave-like properties.* De Broglie described matter waves in mathematical terms, but his proposal was verified in 1927 by experiments in which electron beams were diffracted, or dispersed, by crystals in much the same way as X-rays. With the discovery of the wave-like properties of electrons, the feasibility of an electron microscope was established. The electron microscope has revolutionized science. For example, biological macromolecules are routinely studied with modern electron microscopes.

According to de Broglie's hypothesis, the wavelength associated with a particle is related to the particle momentum, p, and Planck's constant, h. Momentum is the product of mass, m, and velocity, v.

$$\lambda = \frac{h}{p} = \frac{h}{mv} \qquad (9.8)$$

In Equation (9.8) wavelength is in meters, mass is in kilograms, and velocity is in meters per second. Planck's constant must also be expressed in units of mass, length, and time. This requires replacing the unit, joule, by the equivalent units $kg\ m^2\ s^{-2}$.

A transmission electron microscope uses an electron beam to make an image of a sample. Because the wavelength of an electron beam is much shorter than the wavelength of visible light, the image of the sample reveals details that are not even visible with an optical microscope.

EXAMPLE 9-6

Calculating the Wavelength Associated with a Beam of Particles. What is the wavelength associated with electrons traveling at one-tenth the speed of light?

Solution

The electron mass, expressed in kilograms, is 9.109×10^{-31} kg (recall Table 2.1). The electron velocity is $v = 0.100 \times c = 0.100 \times 3.00 \times 10^8$ m s^{-1} $= 3.00 \times 10^7$ m s^{-1}. Planck's constant $h = 6.626 \times 10^{-34}$ J s $= 6.626 \times 10^{-34}$ kg m^2 s^{-2} s $= 6.626 \times 10^{-34}$ kg m^2 s^{-1}. Substituting these data into Equation (9.8), we obtain

$$\lambda = \frac{6.626 \times 10^{-34} \text{ kg m}^2 \text{ s}^{-1}}{(9.109 \times 10^{-31} \text{ kg})(3.00 \times 10^7 \text{ m s}^{-1})}$$

$$= 2.42 \times 10^{-11} \text{ m} = 24.2 \text{ pm}$$

Practice Example A: Assuming Superman has a mass of 91 kg, what is the wavelength associated with him if he is traveling at one-fifth the speed of light?

Practice Example B: To what velocity (speed) must a beam of protons be accelerated to display a de Broglie wavelength of 10.0 pm? Obtain the proton mass from Table 2.1.

The wavelength calculated in Example 9-6, 24.2 pm, is about one-half the radius of the first Bohr orbit of a hydrogen atom. It is only when wavelengths are comparable to atomic or nuclear dimensions that wave–particle duality is important. The concept has little meaning when applied to large (macroscopic) objects, such as baseballs and automobiles, because their wavelengths are too small to measure. For these macroscopic objects the laws of classical physics are quite adequate.

The Uncertainty Principle

The laws of classical physics permit us to make precise predictions. For example, we can *calculate* the exact point at which a rocket will land after it is fired. The more precisely we measure the variables that affect the rocket's trajectory (path), the more accurate our calculation (prediction) will be. In effect, there is no limit to the accuracy we can achieve. In classical physics nothing is left to chance—physical behavior can be predicted with certainty.

During the 1920s Niels Bohr and Werner Heisenberg considered hypothetical experiments to establish just how precisely the behavior of subatomic particles can be determined. The two variables that must be measured are the position of the particle (x) and its momentum (p). [Recall that momentum is the product of mass (m) and velocity (v).] The conclusion they reached is that there must *always* be uncertainties in measurement such that the product of the uncertainty in position, Δx, and the uncertainty in momentum, Δp, is

$$\Delta x \, \Delta p \geq \frac{h}{4\pi} \tag{9.9}$$

The significance of this expression, called the **Heisenberg uncertainty principle,** is that we cannot measure position and momentum with great precision simultaneously. If we design an experiment to locate the position of a particle with great precision, we cannot measure its *momentum* precisely, and vice versa. In simpler terms, if we know precisely where a particle is, we cannot also know precisely where it has come from or where it is going. If we know precisely how a particle is moving, we cannot also know precisely where it is. In the subatomic world, things must always be "fuzzy." Why should this be so?

Suppose we wish to study an electron in a hydrogen atom by looking at it with a microscope. What kind of microscope should this be? The smallest par-

▶ The uncertainty principle is not easy for most people to accept. Einstein spent a good deal of time from the middle 1920s until his death in 1955 attempting, unsuccessfully, to disprove it.

FIGURE 9-15
The uncertainty principle.
A photon of "light" strikes an electron and is reflected (left). In the collision the photon transfers momentum to the electron. The reflected photon is seen through the microscope, but the electron is out of focus (right). Its exact position cannot be determined.

ticle we can see with a microscope has about the same dimensions as the wavelength of the light used to illuminate it. Let us try working out the energy of electromagnetic radiation needed to see an electron and what effect that might have on the electron we are trying to observe.

The first Bohr orbit in a hydrogen atom is 53 pm, and the diameter of the atom is about 100 pm (10^{-10} m). Electrons are much smaller than the atoms in which they are found. Suppose we assume a diameter of about 10^{-14} m for an electron. Light of this wavelength would have a frequency of 3×10^{22} s^{-1} ($\nu = c/\lambda$) and an energy per photon of 2×10^{-11} J ($E = h\nu$). But from Figure 9-14 we can see that this energy is far, far in excess of the 2.179×10^{-18} J required to ionize the electron in a hydrogen atom, that is, to strip it completely away from the atomic nucleus. So a photon of this energy would not help us to see the electron; instead, it would end up knocking the electron out of the atom! Just in trying to look at the electron, we would drastically change the situation from an electron bound in a hydrogen atom to an ionized electron free of the atom. This situation is pictured in Figure 9-15. On the other hand, in trying to see and make measurements on a golf ball rolling across a green, we can use visible light. Light with a frequency of about 5×10^{14} s^{-1} has an energy per photon of 3×10^{-19} J. This energy, even if multiplied millions of times over, is much smaller than the kinetic energy of the ball. The light would not interfere with the measurement at all. We do not need to apply the Heisenberg uncertainty principle to study macroscopic objects.

9-6 Wave Mechanics

The ideas presented in Section 9-5 yield important clues about electrons in atoms: Heisenberg's uncertainty principle sets limits on how precisely we can determine an electron's position and momentum. De Broglie's relationship suggests that electrons should display wave-like properties and that they can be thought of as *matter waves.*

Erwin Schrödinger, in 1927, met with great success in treating electrons as matter waves. But how are we to think of an electron in an atom by this wave picture? Does the electron no longer exist as a particle? Is it literally smeared out into a wave? Heisenberg's uncertainty principle says that we do not know and *cannot* know. Maybe we can gain some insights, however, by returning to a matter introduced early in the chapter—wave motion in a string (see Figure 9-1).

Standing Waves

In the *traveling wave* in Figure 9-1 every portion of a very long string goes through an identical up-and-down motion. The wave transmits energy along the entire length of the string. For an electron confined to an atom as a matter wave, however, the situation is more like wave motion in a *short* string with fixed ends, a type of wave called a **standing wave.**

Think of the vibrations in a plucked guitar string, suggested by Figure 9-16. Segments of the string experience up-and-down displacements with time and oscillate or vibrate between the limits set by the blue curves. Of special interest is the fact that the magnitudes of the oscillations differ from point to point along the wave, including certain points, called *nodes,* that undergo no displacement at all.

(a) $l = \dfrac{\lambda}{2}$

(b) $l = 2\dfrac{\lambda}{2}$

(c) $l = 3\dfrac{\lambda}{2}$

FIGURE 9-16
Standing waves in a string.
The string can be set into motion by plucking it. The blue boundaries outline the range of displacements at each point for each standing wave. The relationships between the wavelength, string length, and the number of nodes—points that are not displaced—are given by Equation (9.10). The nodes are marked by bold dots.

We might say that the permitted wavelengths of standing waves are quantized. They are related to the length of the string (*l*), which must be equal to a whole number (*n*) times one-half the wavelength (λ/2).

$$l = n(\lambda/2) \qquad \text{where } n = 1, 2, 3, \ldots$$

$$\text{and the total number of nodes} = n + 1 \quad (9.10)$$

The plucked guitar string represents a *one*-dimensional standing wave. As an example of a *two*-dimensional standing wave we might consider a tapped drumskin. The electron as a matter wave is still more complex. It is a *three*-dimensional standing wave. Despite this complexity, the simplified representation of a matter wave in Figure 9-17 should help to establish the idea that only certain wave patterns are acceptable.

Wave Functions

Just as we can describe the allowed wave patterns for a vibrating string through mathematical equations (such as 9.10), the allowed wave patterns for electrons can be described through similar but more complex equations. Writing these equations is well beyond the scope of our study, so let us just say that the acceptable solutions of these wave equations are called **wave functions,** denoted by the Greek letter psi, ψ. The mathematical procedure producing acceptable wave functions requires the use of three integral parameters called **quantum numbers,** a situation similar to the requirement of integers in Equation (9.10). When specific values are assigned to these three quantum numbers, the resulting wave function is called an **orbital.**

An orbital is a mathematical function, *not* a physical entity. Specifically, it is *not* an orbit or precise trajectory or path that an electron travels in an atom. The closest we can come to giving it physical meaning is to think of it as a mathematical description of a region in an atom where an electron has some likelihood of being found, if we think of the electron as a particle; or as a region of high electron charge density, if we think of the electron as a matter wave. From the "electron-as-particle" standpoint, then, we have a special interest in the *probability* that the electron is at some particular point; from the "electron-as-wave" standpoint, our interest is in *electron charge density*. In a classical wave (such as visible light) the amplitude of the wave corresponds to ψ, and the intensity of the wave to ψ^2. The intensity relates to the number of photons present in a region, the photon density. For an electron wave, then, ψ^2 relates to electron charge density. Electron probability is proportional to electron charge density, and both these quantities are associated with ψ^2.

We will describe the three quantum numbers and four types of orbitals in some detail in the next section and include the symbolism used to represent orbitals. For the present, let's just consider the electron probabilities and charge densities associated with two orbitals. These are shown in three different ways in Figure 9-18 for the orbital known as 1*s*. Figure 9-18a shows ψ^2 as a function of distance from the nucleus along a line through the nucleus. The pattern of dots in Figure 9-18b represents the distribution of electron probability and charge densities in a plane with the nucleus at its center. Figure 9-18c shows a spherical envelope within which there is a certain probability (say 90%) of finding an electron. This is the way a 1*s* orbital is usually depicted.

Because a very low electron charge density exists even at large distances from the nucleus, we cannot represent a 1*s* orbital encompassing all the electron charge. The sphere drawn in Figure 9-18c contains about 90% of the electron charge. There is a 90% probability of finding an electron within this sphere. To

(a)

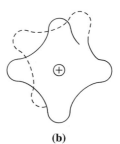

(b)

FIGURE 9-17
The electron as a matter wave.
These patterns are two-dimensional cross sections of a much more complicated three-dimensional wave. The wave pattern in (a), a *standing wave,* is an acceptable representation. It has an integral number of wavelengths (four) about the nucleus; successive waves reinforce one another. The pattern in (b) is unacceptable. The number of wavelengths is nonintegral (about 4.5), and successive waves tend to cancel each other, that is, the crest in one part of the wave overlaps a trough in another part of the wave, and there is no resultant wave at all.

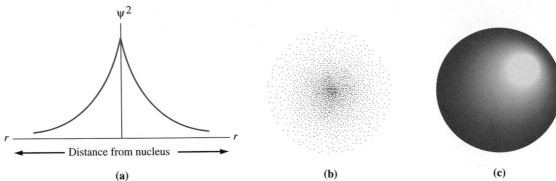

FIGURE 9-18 **Three representations of the electron probability and charge density for a 1s orbital.**
(a) The highest electron probability or charge density (ψ^2) exists at points near the nucleus (see *Are You Wondering*).
(b) The pattern of dots represents the electron probability or charge density in a plane with the nucleus at its center. The closer the spacing between dots, the higher is the probability of finding an electron and the greater is the electron charge density.
(c) A spherical envelope containing 90% of the electron charge density or a 90% probability of finding an electron.

FIGURE 9-19
Dartboard analogy to a 1s orbital.
Imagine that a single dart (electron) is thrown at a dartboard 1500 times. The board contains 90% of all the holes; it is analogous to the 1s orbital. Where is a thrown dart most likely to hit?

The number of holes per unit area is greatest in the "50" region—that is, the "50" region has the greatest probability density. The most likely score is "30," however. Even though the density of holes is not as great, the total area, and hence the total number of hits, is greater in the "30" ring than in the "50" ring. The probability of scoring "30" is greater than that of scoring "50."

We consider the dartboard analogy again in Feature Problem C on page 312.

Are You Wondering . . .

What it means for the electron charge density and probability for a 1s orbital to be highest at the nucleus? Does this mean that the electron is most likely to be at the nucleus itself?

The probability of finding an electron at a given *point* for a 1s orbital is highest at the nucleus: ψ^2 is largest there. A more meaningful way to describe electron probability, though, is to add together the probabilities at equivalent points. For example, the electron probability at 100 pm is the sum of the probabilities at *all* points 100 pm from the nucleus. These points lie within a thin *spherical* shell with a radius of 100 pm. When we deal with probabilities this way, we find the greatest probability for the electron in a 1s orbital to be in a spherical shell of radius 53 pm (0.53 Å). This is the same radius as that of the first Bohr orbit. The boundary surface within which there is a 90% probability of finding an electron (see Figure 9-18c) is a much larger sphere with a radius of about 141 pm.

Figure 9-19 presents an analogy to the distinction between electron probability at a point and in a spherically symmetric region of equivalent points. The appearance of a well-used dartboard is one with a number of hits in the bull's-eye region, yet in the hands of the average player the most probable dart hit will not be a bull's-eye. The board itself contains 90% of the hits, analogous to the sphere depicted in Figure 9-18c. The most probable area hit, however, is the "30" *ring*, analogous to a spherical *shell* of 53 pm radius within the larger sphere representing the 1s orbital.

encompass 90% of the electron probability for an orbital of the 2s type, we need a larger sphere than for 1s. Another feature of the 2s orbital, shown in Figure 9-20, is an interior spherical shell in which the electron probability is *zero*. This spherical region is a node, equivalent to the nodes in a standing wave in a string. We continue the discussion of orbitals in the next section after learning about the three quantum numbers required by wave mechanics.

9-7 Quantum Numbers and Electron Orbitals

In the preceding section we stated that by specifying *three* quantum numbers in a wave function (ψ) we obtain an orbital. Here we explore the combinations of quantum numbers that produce different orbitals. First, though, we need to learn more about the nature of these three quantum numbers.

Assigning Quantum Numbers

The following relationships involving the three quantum numbers arise from the mathematical process used to obtain wave functions. In this mathematical process, the values of the quantum numbers are fixed *in the order listed*.

The first number to be fixed is the *principal* quantum number, n, which may have only a *positive, nonzero integral* value.

$$n = 1, 2, 3, 4, \ldots \tag{9.11}$$

Second is the *orbital angular momentum* quantum number, ℓ, which may be *zero* or a *positive* integer, but not larger than $n - 1$ (where n is the principal quantum number).

$$\ell = 0, 1, 2, 3, \ldots, n - 1 \tag{9.12}$$

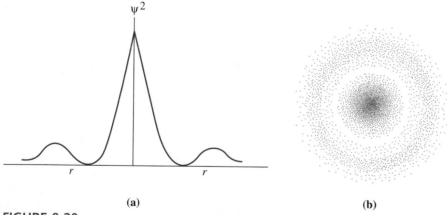

(a)   (b)

FIGURE 9-20
Electron probability and charge density for a 2s orbital.
(a) The electron probability and charge density are highest near the nucleus ($r = 0$) and drop to zero at a particular distance from the nucleus. The probability and charge density then rise to a second maximum at a somewhat greater distance and again gradually decrease.
(b) There are two regions with a high density of dots. These correspond to the two regions of high value of ψ^2 in (a), one close to the nucleus and the other farther away. As in the dartboard analogy of Figure 9-19, the most probable location of the electron is in a spherical shell found in the outer region of dots. To encompass 90% of the electron probability and charge density, the spherical envelope representing a 2s orbital is larger than for the 1s orbital.

Third is the *magnetic* quantum number, m_ℓ. It may be a negative or positive integer, including zero, and ranging from $-\ell$ to $+\ell$ (where ℓ is the orbital angular momentum quantum number).

$$m_\ell = -\ell, -\ell + 1, -\ell + 2, \ldots, 0, 1, 2, \ldots, +\ell \qquad (9.13)$$

► Chemists often refer to the three quantum numbers simply as the "n," "ℓ," and "m_ℓ" quantum numbers rather than designating them by name.

EXAMPLE 9-7

Applying Relationships Among Quantum Numbers. Can an orbital have the quantum numbers $n = 2$, $\ell = 2$, and $m_\ell = 2$?

Solution

No. The ℓ quantum number cannot be greater than $n - 1$. Thus, if $n = 2$, ℓ can be only 0 or 1. And if ℓ can be only 0 or 1, m_ℓ cannot be 2; m_ℓ must be 0 if $\ell = 0$ and may be -1, 0, or $+1$ if $\ell = 1$.

Practice Example A: Can an orbital have the quantum numbers $n = 3$, $\ell = 0$, and $m_\ell = 0$?

Practice Example B: For an orbital with $n = 3$ and $m_\ell = -1$, what is (are) the possible value(s) of ℓ?

Principal Shells and Subshells

All orbitals with the same value of n are in the same **principal electronic shell** or **principal level,** and all orbitals with the same n and ℓ values are in the same **subshell** or **sublevel.**

Principal electronic shells are numbered according to the value of n. The *first* principal shell consists of orbitals with $n = 1$; the *second* principal shell, of orbitals with $n = 2$; and so on. The value of n relates to the energy and most probable distance of an electron from the nucleus. The higher the value of n, the greater the electron energy and the farther, on average, the electron is from the nucleus.

The number of subshells found in a principal electronic shell is equal to the number of allowed values of the orbital angular momentum quantum number, ℓ. In the first principal shell, with $n = 1$, the only allowed value of ℓ is 0, and there is a single subshell. The second principal shell ($n = 2$), with the allowed ℓ values of 0 and 1, consists of two subshells; the third principal shell ($n = 3$) has three subshells ($\ell = 0$, 1, and 2); and so on. Or, to put the matter in another way, because there are n possible values of the ℓ quantum number, that is, 0, 1, 2, $\ldots$ $(n - 1)$, the number of subshells in a principal shell is equal to the principal quantum number. Thus, there is one subshell in the principal shell with $n = 1$, two subshells in the principal shell with $n = 2$, and so on. The name given to a subshell, regardless of the principal shell in which it is found, depends on the value of the ℓ quantum number. The first four subshells are

s subshell	p subshell	d subshell	f subshell
$\ell = 0$	$\ell = 1$	$\ell = 2$	$\ell = 3$

The number of orbitals in a subshell is equal to the number of allowed values of m_ℓ for the particular value of ℓ. Recall that the allowed values of m_ℓ are 0, ± 1, ± 2, $\ldots \pm \ell$, and you can see that the number of orbitals in a subshell is

$2\ell + 1$. The names of the orbitals are the same as the names of the subshells in which they appear.

s orbitals	*p* orbitals	*d* orbitals	*f* orbitals
$\ell = 0$	$\ell = 1$	$\ell = 2$	$\ell = 3$
$m_\ell = 0$	$m_\ell = 0, \pm 1$	$m_\ell = 0, \pm 1, \pm 2$	$m_\ell = 0, \pm 1, \pm 2, \pm 3$
one *s* orbital	three *p* orbitals	five *d* orbitals	seven *f* orbitals
in an *s* subshell	in a *p* subshell	in a *d* subshell	in an *f* subshell

To designate the particular principal shell in which a given subshell or orbital is found, we use a combination of a number and a letter. The symbol 1*s* designates the *s* subshell—an *s* orbital—in the first principal shell; 2*s*, the *s* subshell—an *s* orbital—of the second principal shell. The symbol 2*p* is used to designate both the *p* subshell of the second principal shell and any of the three *p* orbitals in that subshell.

We showed the electron probabilities and charge densities corresponding to the 1*s* and 2*s* orbitals in Figures 9-18 and 9-20. All *s* orbitals are spherically symmetric, meaning that the electron probability is the same at all points equidistant from the nucleus. The boundary surface of the orbital (recall Figure 9-18c) is a sphere with the nucleus at its center. Figure 9-21 gives three views of electron probabilities and charge densities in a 2*p* orbital. The chief difference between *s* and *p* orbitals is that the electron probability or charge density in a *p* orbital is *not* spherically symmetric. The greatest probability of finding an electron in a 2*p* orbital is within the dumbbell-shaped region of Figure 9-21c. The two lobes of this region are separated by a plane in which the electron probability or charge density drops to zero. This is called a *nodal plane*. The set of three 2*p* orbitals and their nodal planes are shown in Figure 9-22.

The geometrical shapes representing the *d* orbitals, which are more complex than those for *s* and *p* orbitals, are shown in Figure 9-23. We will not attempt to depict *f* orbitals in this text.

FIGURE 9-21   Three representations of electron probability and charge density for a 2*p* orbital.
(a) The value of ψ^2 is zero at the nucleus, rises to a maximum on either side, and then falls off with distance (*r*) along a line through the nucleus (i.e., along the *x*, *y*, or *z* axis).
(b) The dots represent electron probability and charge density in a *plane* passing through the nucleus, for example, the *xz* plane.
(c) Electron probabilities and charge densities represented in three dimensions. The greatest probability of finding an electron is within the two lobes of the dumbbell-shaped region. Note that this region is *not* spherically symmetric. Note also that the probability drops to zero in the shaded plane—the nodal plane (the *yz* plane).

(a)   (b)   (c)

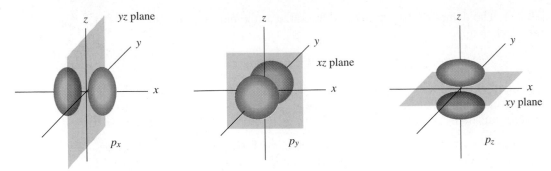

FIGURE 9-22 **Representations of the three 2*p* orbitals.**
The *p* orbitals are usually represented as directed along the perpendicular *x*, *y*, and *z* axes, and the symbols p_x, p_y, and p_z are often used. The p_z orbital has $m_\ell = 0$. The situation with p_x and p_y is more complex: Each of these orbitals has contributions from both $m_\ell = 1$ and $m_\ell = -1$. Our main concern is just to recognize that *p* orbitals occur in sets of three and can be represented in the orientation shown here. In higher numbered shells *p* orbitals have a somewhat different appearance, but we will use these general shapes for all *p* orbitals.

Some of the points discussed in the preceding paragraphs are illustrated in Example 9-8.

EXAMPLE 9-8

Relating Orbital Designations and Quantum Numbers. Write an orbital designation corresponding to the quantum numbers $n = 4$, $\ell = 2$, $m_\ell = 0$.

Solution

The magnetic quantum number, m_l, is not reflected in the orbital designation. The type of orbital is determined by the ℓ quantum number. Because $\ell = 2$, the orbital is of the *d* type. Because $n = 4$, the designation is $4d$.

FIGURE 9-23 **Representations of the five *d* orbitals.**
The designations *xy*, *xz*, *yz*, and so on are related to the values of the quantum number m_ℓ, but this is a detail that we will not pursue in the text. The number of nodal surfaces for an orbital is equal to the ℓ quantum number. For *d* orbitals there are two such surfaces. The nodal planes for the d_{xy} orbital are shown here. (The nodal surfaces for the d_{z^2} orbital are actually cone-shaped.)

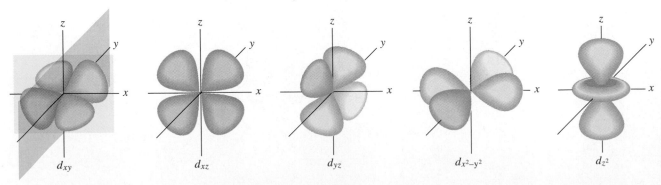

Practice Example A: Write an orbital designation corresponding to the quantum numbers $n = 3$, $\ell = 1$, and $m_\ell = 1$.

Practice Example B: Specify three quantum numbers corresponding to the orbital designation $5s$.

Are You Wondering . . .

How an electron gets from one lobe to the other in the representation of a *p* orbital, given that there is zero probability of an electron being in the nodal plane?

The problem here is in thinking of the electron as a classical moving particle. The electron acts as a matter wave found in many places at the same time. The electron charge density or electron probability is equal in the two lobes. The nodal plane is just a region in which the wave function has a value of zero. This situation is analogous to standing waves in a string (see Figure 9-16). Think of the entire vibrating string as equivalent to an electron. The points on the string that are not displaced, the nodal points, are equivalent to the nodal planes in the *p* orbitals.

9-8 Electron Spin: A Fourth Quantum Number

Wave mechanics provides three quantum numbers with which we can develop a description of electron orbitals. But we need a fourth quantum number as well. In 1925, George Uhlenbeck and Samuel Goudsmit proposed that some unexplained features of the hydrogen spectrum could be understood by assuming that an electron acts as if it spins, much as Earth spins on its axis. As suggested by Figure 9-24, there are two possibilities for **electron spin.** The electron spin quantum number, m_s, may have a value of $+\frac{1}{2}$ (also denoted by the arrow ↑) or $-\frac{1}{2}$ (denoted by the arrow ↓); the value of m_s does not depend on any of the other three quantum numbers.

But what is the evidence that such a phenomenon as electron spin exists? An experiment by Otto Stern and Walter Gerlach in 1920, though designed for another purpose, seems to yield this proof (Figure 9-25). Silver was vaporized in a furnace and a beam of silver atoms was passed through a nonuniform magnetic field. The beam split in two. Here is a simplified explanation.

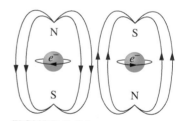

FIGURE 9-24
Electron spin visualized.
Two possibilities for electron spin are shown with their associated magnetic fields. Two electrons with opposing spins have opposing magnetic fields that cancel, leaving no net magnetic field for the pair.

oven   slit   magnet   detector

Ag atoms

FIGURE 9-25
The Stern–Gerlach experiment.
Ag atoms vaporized in the oven are collimated into a beam by the slit, and the beam is passed through a nonuniform magnetic field. The beam splits in two. (The beam of atoms would not experience a force if the magnetic field were uniform. The field strength must be stronger in certain directions than in others.)

1. An electron, because of its spin, generates a magnetic field.

2. A pair of electrons with opposing spins has no net magnetic field.

3. In a silver atom, 23 electrons have a spin of one type and 24 of the opposite type. The direction of the net magnetic field produced depends only on the spin of the unpaired electron.

4. In a beam of a large number of silver atoms there is an equal chance that the unpaired electron will have a spin of $+\frac{1}{2}$ or $-\frac{1}{2}$. The magnetic field induced by the silver atoms interacts with the nonuniform field and the beam of silver atoms splits into two beams.

9-9 Multielectron Atoms

Schrödinger developed his wave equation for the hydrogen atom—an atom containing just one electron. For multielectron atoms a new factor arises: mutual repulsions between electrons. Because exact electron positions are not known, electron repulsions can only be approximated, and the solutions of the wave equation are also approximate. The approach taken is to consider the electrons, one by one, in the environment established by the nucleus and other electrons. When this is done, the electron orbitals obtained are of the same types as those obtained for the hydrogen atom; they are called *hydrogen-like* orbitals.

Through Equation (9.5) and Figure 9-14 we saw that the electron energies in the Bohr hydrogen atom depend only on the Bohr orbit. That is, energy is a function only of the distance of the electron from the nucleus. Wave mechanics yields a similar result for the hydrogen atom: all orbitals with the same principal quantum number n have the same energy. Orbitals with the same energy are said to be *degenerate*. In a hydrogen atom the orbitals $2s$ and $2p$ are degenerate, as are $3s$, $3p$, and $3d$.

In multielectron atoms the attractive force of the nucleus for a given electron increases as the nuclear charge increases. As a result, we find that orbital energies become lower (more negative) with increasing atomic number of the atom. Also, orbital energies in multielectron atoms depend on the type of orbital; the orbitals are not degenerate.

Think about the attractive force of the atomic nucleus for one particular electron some distance from the nucleus. Electrons in orbitals closer to the nucleus reduce the effectiveness of the nucleus in attracting this particular electron. They *screen*, or *shield*, the electron from the full effects of the nucleus. In effect they reduce the nuclear charge to a *net* charge called an *effective nuclear charge*, Z_{eff}.

The effectiveness of the shielding by inner electrons depends on the type of orbital in which the affected electron is found. An electron in an s orbital, for example, spends more time close to the nucleus than does an electron in a p orbital of the same electronic shell (compare the probability distributions for $2s$ and $2p$ orbitals in Figures 9-20 and 9-21). The electron in the s orbital is not as well screened as is the one in a p orbital. The s electron experiences a higher Z_{eff}, is held more tightly, and is at a lower energy than is the p electron. The s orbital is at a lower energy than a p orbital of the same principal shell. In turn, a p orbital is at a lower energy than a d orbital of the same principal shell. The energy level of a principal shell is split into separate levels for its subshells. There is no further splitting of energies within a subshell, however. All three p orbitals of a principal shell have the same energy; all five d orbitals have the same energy; and so on.

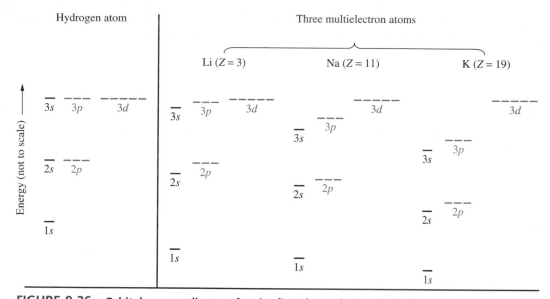

FIGURE 9-26   Orbital energy diagram for the first three electronic shells.
Energy levels are shown for a hydrogen atom (left) and three typical multielectron atoms (right). Each multielectron atom has its own energy level diagram. Note that for the hydrogen atom orbital energies within a principal shell—for example, 3s, 3p, 3d—are alike (degenerate), but in a multielectron atom they become rather widely separated. Another feature of the diagram described in the text is the steady decrease in all orbital energies with increasing atomic number.

The ideas presented in this section are illustrated in Figure 9-26, which represents orbital energies for the first three principal shells of hydrogen and some typical multielectron atoms.

9-10   Electron Configurations

Even though we have stressed that orbitals are mathematical functions and not themselves physical regions in space, it is customary to refer to an electron described by a particular orbital as being "in the orbital." The **electron configuration** of an atom is a designation of how electrons are distributed among various orbitals in principal shells and subshells. In later chapters we will find that many of the physical and chemical properties of elements can be correlated with electron configurations. In this section we will see how the results of wave mechanics, expressed as a set of rules, can help us to write probable electron configurations for the elements.

Rules for Assigning Electrons to Orbitals
1. Electrons occupy orbitals in a way that minimizes the energy of the atom.
Figure 9-26, an energy level diagram for the first three electronic shells, suggests the order in which electrons occupy orbitals in these shells, first the 1s, then 2s, 2p, and so on. The fact that a principal energy level splits into different sublevels has an important consequence, however: At higher quantum levels the energies of certain sublevels are so close that the order of increasing orbital energy and the order in which electrons occupy orbitals may not quite match. Thus, even though the 3d subshell is at a slightly lower energy than 4s, the 4s fills first. The order of filling of orbitals has been established by *experi-*

ment, principally through spectroscopy and magnetic studies, and it is this order that we must follow in assigning electron configurations to the elements. Except for a few elements, the order in which orbitals fill is

$$1s, 2s, 2p, 3s, 3p, 4s, 3d, 4p, 5s, 4d, 5p, 6s, 4f, 5d, 6p, 7s, 5f, 6d, 7p \quad (9.14)$$

Some students find the diagram pictured in Figure 9-27 a useful way to remember this order. In Section 9-11 we will discuss a superior method of establishing the order of filling of orbitals based on the periodic table.

2. No two electrons in an atom may have all four quantum numbers alike— the Pauli exclusion principle. In 1926, Wolfgang Pauli explained complex features of emission spectra associated with atoms in magnetic fields by proposing that no two electrons in an atom can have all four quantum numbers alike. The first three quantum numbers, n, ℓ, and m_ℓ determine a specific orbital. Two electrons may have these three quantum numbers alike; but if they do, they must have different values of m_s, the spin quantum number. Another way to state this result is that *only two electrons may occupy the same orbital and these electrons must have opposing spins.*

Because of this limit of two electrons per orbital, the capacity of a subshell for electrons can be obtained by *doubling* the number of orbitals in the subshell. Thus, the *s* subshell consists of *one* orbital with a capacity of *two* electrons; the *p* subshell consists of *three* orbitals with a total capacity of *six* electrons; and so on.

3. When orbitals of identical energy are available, electrons initially occupy these orbitals singly. As a result of this rule, known as **Hund's rule,** an atom tends to have as many unpaired electrons as possible. This behavior can be rationalized by saying that electrons, because they all carry the same electric charge, try to get as far apart as possible. They do this by seeking out empty orbitals of similar energy in preference to pairing up with other electrons in half-filled orbitals.

Representing Electron Configurations

Before we assign electron configurations to atoms of the different elements, we need to introduce methods of representing these configurations. The electron configuration of an atom of carbon is shown in three different ways:

spdf notation (condensed):   C $1s^2 2s^2 2p^2$

spdf notation (expanded):   C $1s^2 2s^2 2p_x^1 2p_y^1$

orbital diagram: C   [see figure]

In each of these methods we assign *six* electrons because the atomic number of carbon is 6. Two of these electrons are in the $1s$ subshell, two in the $2s$, and two in the $2p$. The condensed *spdf notation* only denotes the total number of electrons in each subshell; it does not show how electrons are distributed among orbitals of equal energy. In the expanded *spdf* notation Hund's rule is reflected in the assignment of electrons to the $2p$ subshell—two $2p$ orbitals, each singly occupied. (One $2p$ orbital remains empty.) The **orbital diagram** breaks down each subshell into individual orbitals (drawn as boxes). This notation is similar to an energy level diagram, except that the direction of increasing energy is from left to right instead of vertically.

▶ This order of filling corresponds roughly to the order of increasing orbital energy, but the overriding principle governing the order of filling of orbitals is that the energy of the atom as a whole be kept at a minimum.

FIGURE 9-27
The order of filling of electronic subshells.
Beginning with the top line, follow the arrows and the order obtained is the same as in expression (9.14).

▶ When listed in tables, as in Appendix D, electron configurations are usually written in the condensed *spdf* notation.

Electrons in orbitals are shown as arrows. An arrow pointing up corresponds to one type of spin $(+\frac{1}{2})$ and an arrow pointing down to the other $(-\frac{1}{2})$. Electrons in the same orbital with opposing (opposite) spins are said to be *paired* ($\uparrow\downarrow$). The electrons in the 1*s* and 2*s* orbitals of the carbon atom are paired. Electrons in different singly occupied orbitals of the same subshell have the same or *parallel* spins (arrows pointing in the same direction). This fact is conveyed in the orbital diagram for carbon, where we write [$\uparrow$][$\uparrow$][ ] rather than [$\uparrow$][$\downarrow$][ ] for the 2*p* subshell. Both experiment and theory confirm that an electron configuration in which electrons in singly occupied orbitals have parallel spins is a better representation of the lowest energy state of an atom than any other electron configuration that we can write.

The most stable or the most energetically favorable configurations for isolated atoms, those discussed here, are called ground state electron configurations. Later in the text we will briefly mention some electron configurations that are not the most stable. Atoms with such configurations are said to be in an excited state.

The Aufbau Process

To write electron configurations we will use the **Aufbau process.** *Aufbau* is a German word that means "building up," and what we do is assign electron configurations to the elements in order of increasing atomic number. To proceed from one atom to the next we add a proton and some neutrons to the nucleus and then describe the orbital into which the added electron goes.

Z = 1, H.   The lowest energy state for the electron is the 1*s* orbital. The electron configuration is $1s^1$.

Z = 2, He.   A second electron goes into the 1*s* orbital, and the two electrons have opposing spins, $1s^2$.

Z = 3, Li.   The third electron cannot be accommodated in the 1*s* orbital (Pauli exclusion principle). It goes into the lowest energy orbital available, 2*s*. The electron configuration is $1s^2 2s^1$.

Z = 4, Be.   The configuration is $1s^2 2s^2$.

Z = 5, B.   Now the 2*p* subshell begins to fill: $1s^2 2s^2 2p^1$.

Z = 6, C.   A second electron goes into the 2*p* subshell, but into one of the remaining empty *p* orbitals (Hund's rule), and with a spin parallel to the first 2*p* electron.

Z = 7–10, N through Ne.   In this series of four elements, the filling of the 2*p* subshell is completed. The number of unpaired electrons reaches a maximum (3) with nitrogen and then decreases to zero with neon.

$Z = 11-18$, Na through Ar. This series of eight elements closely parallels what we have already seen for the eight elements from Li through Ne, except that electrons go into $3s$ and $3p$ orbitals. Each element has the $1s$, $2s$ and $2p$ subshells filled. Because the configuration $1s^2 2s^2 2p^6$ is that of neon, we will call this the neon "core," represent it as [Ne], and concentrate on the electrons beyond the core. Electrons that are added to the electronic shell of highest principal quantum number (the outermost or valence shell) are called **valence electrons.** The electron configuration of Na is written below with the neon core, and for the other third period elements only the valence shell electron configuration is shown.

Na	Mg	Al	Si	P	S	Cl	Ar
$[Ne]3s^1$	$3s^2$	$3s^2 3p^1$	$3s^2 3p^2$	$3s^2 3p^3$	$3s^2 3p^4$	$3s^2 3p^5$	$3s^2 3p^6$

$Z = 19$ and 20, K and Ca. After argon, instead of $3d$ the next subshell to fill is $4s$. Using the symbolism [Ar] to represent the core electrons for the configuration $1s^2 2s^2 2p^6 3s^2 3p^6$, we get the configurations shown below for K and Ca.

$$\text{K:} \quad [Ar]4s^1 \quad \text{and} \quad \text{Ca:} \quad [Ar]4s^2$$

$Z = 21-30$, Sc through Zn. This next series of elements is characterized by electrons filling the d orbitals of the third shell. The d subshell has a total capacity of ten electrons—ten elements are involved. For scandium we have two possible ways to write the configuration.

$$\text{(a) Sc:} \quad [Ar]3d^1 4s^2 \quad \text{or} \quad \text{(b) Sc:} \quad [Ar]4s^2 3d^1$$

▶ Although method (b) conforms better to the order in which orbitals fill, method (a) better represents the order in which electrons are lost on ionization, as we will see in the next chapter.

Both methods are commonly used. Method (a) groups together all the subshells of a principal shell and places last subshells of the highest principal quantum level. Method (b) lists orbitals in the apparent order in which they fill. In this text we will use method (a).

The electron configurations of this series of ten elements are listed below in both the orbital diagram and the *spdf* notation.

The *d* orbitals fill in a fairly regular fashion in this series but there are two exceptions: chromium (Cr) and copper (Cu). These exceptions are usually explained in terms of a special stability for configurations in which a 3*d* subshell is half-filled with electrons, as with Cr ($3d^5$), or completely filled, as with Cu ($3d^{10}$).

$Z = 31–36$, Ga through Kr. In this series of six elements the 4*p* subshell is filled, ending with krypton.

$$\text{Kr: } [\text{Ar}]3d^{10}4s^24p^6$$

$Z = 37–54$, Rb to Xe. In this series of 18 elements the subshells fill in the order 5*s*, 4*d*, and 5*p*, ending with the configuration of xenon.

$$\text{Xe: } [\text{Kr}]4d^{10}5s^25p^6$$

$Z = 55–86$, Cs to Rn. In this series of 32 elements, with a few exceptions, the subshells fill in the order 6*s*, 4*f*, 5*d*, 6*p*. The configuration of radon is

$$\text{Rn: } [\text{Xe}]4f^{14}5d^{10}6s^26p^6$$

$Z = 87–?$, Fr to ?. Francium starts a series of elements in which the subshells that fill are 7*s*, 5*f*, 6*d*, and presumably 7*p*, though elements in which the 7*p* subshell might be occupied are not yet known.

Appendix D gives a complete listing of probable electron configurations.

▶ Feature Problem D on page 313 summarizes a number of ideas about electron configurations.

9-11 Electron Configurations and the Periodic Table

We have just described the Aufbau process of making probable assignments of electrons to the orbitals in atoms. Although electron configurations may seem rather abstract, they actually lead us to a better understanding of the periodic table. Beginning about 1920 Niels Bohr began to promote the connection between the periodic table and quantum theory. The chief link, he pointed out, is in electron configurations. *Elements in the same group of the table have similar electron configurations.*

To construct Table 9.1 on page 300 we have taken three groups of elements from the periodic table and written their electron configurations. The similarity in electron configuration within each group is readily apparent. If we label the shell of highest principal quantum number—the outermost, or valence, shell—as *n*, then

- The Group 1A atoms (alkali metals) have a *single* outer-shell (valence) electron in an *s* orbital, that is, ns^1.

- The Group 7A atoms (halogens) have *seven* outer-shell (valence) electrons, in the configuration ns^2np^5.

- The Group 8A atoms (noble gases)—with the exception of helium, which has only two electrons—have outermost shells with *eight* electrons, in the configuration ns^2np^6.

Although it is not correct in all details, Figure 9-28 on page 301 relates the Aufbau process to the periodic table by dividing the table into the following four blocks of elements, according to the subshells being filled.

TABLE 9.1 Electron Configurations of Some Groups of Elements		
Group	**Element**	**Configuration**
1A	H	$1s^1$
	Li	$[He]2s^1$
	Na	$[Ne]3s^1$
	K	$[Ar]4s^1$
	Rb	$[Kr]5s^1$
	Cs	$[Xe]6s^1$
	Fr	$[Rn]7s^1$
7A	F	$[He]2s^22p^5$
	Cl	$[Ne]3s^23p^5$
	Br	$[Ar]3d^{10}4s^24p^5$
	I	$[Kr]4d^{10}5s^25p^5$
	At	$[Xe]4f^{14}5d^{10}6s^26p^5$
8A	He	$1s^2$
	Ne	$[He]2s^22p^6$
	Ar	$[Ne]3s^23p^6$
	Kr	$[Ar]3d^{10}4s^24p^6$
	Xe	$[Kr]4d^{10}5s^25p^6$
	Rn	$[Xe]4f^{14}5d^{10}6s^26p^6$

- ***s-block.*** The *s* orbital of highest principal quantum number (*n*) fills. The *s*-block consists of Groups 1A and 2A.

- ***p-block.*** The *p* orbitals of highest quantum number (*n*) fill. The *p*-block consists of Groups 3A, 4A, 5A, 6A, 7A, and 8A.

- ***d-block.*** The *d* orbitals of the electronic shell *n* − 1 (the next to outermost) fill. The *d*-block includes Groups 3B, 4B, 5B, 6B, 7B, 8B, 1B, and 2B.

- ***f-block.*** The *f* orbitals of the electronic shell *n* − 2 fill. The *f*-block elements are the lanthanides and the actinides.

Dividing the periodic table into blocks can simplify the task of assigning electron configurations. For the main group elements (A groups), the group number is also the number of electrons in the *s* and *p* subshells of the outermost or valence shell. For example, aluminum is in Period 3 and Group 3A. Its valence shell electron configuration is $3s^23p^1$. The remaining ten electrons in Al (Z = 13) are in a neon core configuration:

$$Al: [Ne]3s^23p^1$$

Gallium is also in Group 3A, but in Period 4. Its valence shell electron configuration is $4s^24p^1$. If we follow the order implied by Figure 9-28, the following subshells must fill consecutively before the first electron enters the 4*p* subshell: 1*s*, 2*s*, 2*p*, 3*s*, 3*p*, 4*s*, and 3*d*. The electron configuration of gallium is

$$Ga: 1s^22s^22p^63s^23p^63d^{10}4s^24p^1$$

Alternatively, we can start with the electron configuration of the noble gas that closes the third period, argon, and add to it the subshells that fill in the fourth period: 4*s*, 3*d*, and 4*p*. The 3*d* subshell must fill with 10 electrons before the 4*p* subshell begins to fill, and, as a consequence, the electron configuration of gallium must be

Main group elements

FIGURE 9-28 Electron configurations and the periodic table.
To use this figure as a guide to the Aufbau process, locate the position of an element in the table. Subshells listed ahead of this position are filled. For example, germanium ($Z = 32$) is located in Group 4A of the blue $4p$ row. The filled subshells are $1s^2$, $2s^2$, $2p^6$, $3s^2$, $3p^6$, $4s^2$, and $3d^{10}$. At $Z = 32$, a second electron has entered the $4p$ subshell. The electron configuration of Ge is $[Ar]3d^{10}4s^24p^2$.

Exceptions to the orderly filling of subshells suggested here are found among a few of the d-block and some of the f-block elements.

$$\text{Ga: } [Ar]3d^{10}4s^24p^1$$

Thallium is in Group 3A and Period 6. Its valence shell electron configuration is $6s^26p^1$. To write its electron configuration in the expanded form we could again refer to Figure 9-28 and fill all the subshells up to $6p^1$. More commonly, though, we would indicate the electron configuration of xenon ($Z = 54$) as a core and add the subshells that fill in the sixth period: $6s$, $4f$, $5d$, and $6p$.

$$\text{Tl: } [Xe]4f^{14}5d^{10}6s^26p^1$$

For the transition elements (B groups), the group numerals do not correspond to the number of valence electrons, except for Groups 1B and 2B. Instead, most transition metal atoms have 2 electrons in the s subshell of the valence shell, but some have only one. In general, if we know the number of valence electrons in an atom of a transition metal, we can readily work out its electron configuration. Thus, vanadium ($Z = 23$), which has two valence electrons in the $4s$ subshell and core electrons in the configuration of the noble gas argon, must have three $3d$ electrons $(2 + 18 + 3 = 23)$.

$$\text{V: } [\text{Ar}]3d^34s^2$$

Chromium $(Z = 24)$, on the other hand, which has only one valence electron in the $4s$ subshell and core electrons in the argon configuration, must have five $3d$ electrons $(1 + 18 + 5 = 24)$.

$$\text{Cr: } [\text{Ar}]3d^54s^1$$

Copper $(Z = 29)$ also has only one valence electron, but we do not have to be told this. We can deduce this fact from copper's location in the periodic table—Group 1B. In addition to its argon core and one valence electron, the copper atom must have ten $3d$ electrons $(1 + 18 + 10 = 29)$.

$$\text{Cu: } [\text{Ar}]3d^{10}4s^1$$

Examples 9-9 through 9-11 provide several more illustrations of the assignment of electron configurations.

EXAMPLE 9-9

Using spdf *Notation for an Electron Configuration.*
(a) Identify the element having the electron configuration

$$1s^22s^22p^63s^23p^5$$

(b) Write out the electron configuration of arsenic.

Solution

a. All electrons must be accounted for in an electron configuration. Add up the superscript numerals $(2 + 2 + 6 + 2 + 5)$ to obtain the atomic number 17. The element with this atomic number is chlorine.
b. Arsenic $(Z = 33)$ is in Period 4 and Group 5A. Its valence shell electron configuration is $4s^24p^3$. The noble gas that closes the third period is Ar $(Z = 18)$, and the subshells that fill in the fourth period are $4s$, $3d$, and $4p$, in that order. Note that we account for 33 electrons in the configuration

$$\text{As: } [\text{Ar}]3d^{10}4s^24p^3$$

Practice Example A: Identify the element having the electron configuration $1s^22s^22p^63s^23p^63d^24s^2$.

Practice Example B: Use *spdf* notation to show the electron configuration of iodine. How many electrons does the I atom have in its $3d$ subshell? How many unpaired electrons are there in an I atom?

EXAMPLE 9-10

Representing Electron Configurations. Write the **(a)** electron configuration of mercury, and **(b)** an orbital diagram for the electron configuration of tin.

Solution

a. Mercury, in Period 6 and Group 2B, is the transition element at the end of the third transition series, in which the $5d$ subshell fills $(5d^{10})$. Because of its position in Group 2B, we know that mercury's valence shell electron configuration is $6s^2$. The noble gas that closes Period 5 is xenon, and the lanthanide series intervenes between xenon and mercury. When we put all these facts together, we conclude that the electron configuration of mercury is

$$[Xe]4f^{14}5d^{10}6s^2$$

b. Tin is in Period 5 and Group 4A. Its valence shell electron configuration is $5s^25p^2$. The noble gas that closes the fourth period is Kr ($Z = 36$), and the subshells that fill in the fifth period are $5s$, $4d$, and $5p$. Note that all subshells are filled in the orbital diagram except for $5p$, where two of the $5p$ orbitals are singly occupied by electrons with parallel spins and one $5p$ orbital remains empty.

Sn: [Kr] $4d$ $5s$ $5p$

Practice Example A: Represent the electron configuration of iron with an orbital diagram.

Practice Example B: Represent the electron configuration of bismuth with an orbital diagram.

EXAMPLE 9-11

Relating Electron Configurations to the Periodic Table. Indicate the number of **(a)** valence electrons in an atom of bromine; **(b)** $5p$ electrons in an atom of tellurium; **(c)** unpaired electrons in an atom of indium; **(d)** $3d$ and $4d$ electrons in a silver atom.

Solution

Determine the atomic number and location in the periodic table of each element. Then establish the significance of the location.

a. Bromine ($Z = 35$) is in Group 7A. There are *seven* outer shell, or valence, electrons in all atoms in this group.
b. Tellurium ($Z = 52$) is in Period 5 and Group 6A. All 6A atoms have six outer shell electrons, two of them s and the other four, p. The valence shell electron configuration of tellurium is $5s^25p^4$; the tellurium atom has *four* $5p$ electrons.
c. Indium ($Z = 49$) is in Period 5 and Group 3A. The electron configuration of its inner shells is $[Kr]4d^{10}$. All the electrons in this inner shell configuration are paired. The valence shell electron configuration is $5s^25p^1$. The two $5s$ electrons are paired and the $5p$ electron is unpaired. The In atom has *one* unpaired electron.
d. Ag ($Z = 47$) is in Period 5 and Group 1B. For the transition elements in Group 1B, the number of valence shell electrons *does* equal the group number. The valence shell electron configuration is $5s^1$, and the noble gas core is that of Kr ($Z = 36$): $1s^22s^22p^63s^23p^63d^{10}4s^24p^6$. The electron configuration of silver is

$$Ag: [Kr]4d^{10}5s^1$$

There are *ten* $3d$ electrons and *ten* $4d$ electrons in a silver atom.

Practice Example A: For an atom of Sn, indicate the number of **(a)** electronic shells that are either filled or partially filled; **(b)** $3p$ electrons; **(c)** $5d$ electrons; **(d)** unpaired electrons.

Practice Example B: Indicate the number of **(a)** $3d$ electrons in Y atoms; **(b)** $4p$ electrons in Ge atoms; and **(c)** unpaired electrons in Au atoms.

Focus On   He–Ne Lasers

Helium–neon laser device for use as a bar-code scanner.

Laser devices seem to be in use everywhere—from reading bar codes in supermarket checkout stands to maintaining quality control in factory assembly lines. Lasers are used in compact disc players, laboratory instruments, and, increasingly, as a tool in surgery. The word *laser* is an acronym for *l*ight *a*mplification by *s*timulated *e*mission of *r*adiation.

Consider a neon sign. An electric discharge produces high-energy electrons that strike Ne atoms and excite certain Ne electrons to a higher energy state. When an excited electron in a neon atom returns to a lower energy orbital, a photon of light is emitted. Although we know how long Ne atoms remain in an excited state *on average,* we do not know precisely when an atom will emit a photon, nor do we

know in what direction the photon will travel. Light emission in a neon sign is a spontaneous, or random, event. *Spontaneous* light emission is pictured in Figure 9-29a.

Like a neon sign, a helium–neon laser also produces red light (633 nm), but the similarity ends here. The He–Ne laser works on this principle: If a 633-nm photon interacts with an excited Ne atom *before* the atom spontaneously emits a photon, the atom will be induced, or *stimulated*, to emit its photon at the precise time of the interaction. Moreover, this second photon will be *coherent*, or *in phase*, with the first; that is, the crests and troughs of the two waves will match exactly. *Stimulated* light emission is pictured in Figure 9-29b.

An He–Ne laser consists of a tube containing a helium–neon

mixture at a pressure of about 1 mmHg. One end of the tube is a totally reflecting mirror, and the other end, a mirror that allows about 1% of the light to leave. An electric discharge is used to produce excited He atoms. These atoms transfer their excitation energy to Ne atoms through collisions. The Ne atoms enter a *metastable* state, an excited state that is stable for a relatively long time before the spontaneous emission of a photon occurs. Most of the Ne atoms must remain "pumped up" to this metastable state during the operation of the laser.

Following the first spontaneous release of a photon, other excited atoms release photons of the same frequency and in phase with the stimulating photon. This is rather like lining up a string of dominoes and knocking over the first one. This is the beginning of the laser beam. As shown in Figure 9-30, the beam is *amplified* as it bounces back and forth between the two mirrors. The output is the portion of the laser beam that escapes through the partially reflecting mirror. Because the photons of light travel in phase and in the same direction, laser light can be produced at a much higher intensity than would normally be achieved with spontaneous light emission.

Shown to the right are the transitions that produce (1) exited He atoms, (2) Ne atoms in the metastable state, and (3) laser light. After the laser light emission, (4) Ne atoms return to the ground state by way of two additional transitions. Ground state Ne atoms are then pumped back up to the metastable state by collisions with excited He atoms. (The symbols * and # represent electronically excited species.)

(1) $He \xrightarrow{e^-} He^*$
$\quad 1s^2 \qquad 1s^1 2s^1$

(2) $He^* + Ne \rightarrow He + Ne^*$ (metastable)
$\qquad\quad [He]2s^2 2p^6 \quad [He]2s^2 2p^5 5s^1$

(3) $Ne^* \rightarrow Ne^{\#} \quad + \quad h\nu$
$\qquad [He]2s^2 2p^5 3p^1 \quad (633\text{-nm})$

(4) $Ne^{\#} \rightarrow \rightarrow Ne$ (return to step 2)

(a) Spontaneous emission

(b) Stimulated emission

FIGURE 9-29
Spontaneous and stimulated light emission by Ne atoms.
(a) The waves a, b, c, and d, although having the same frequency and wavelength, are emitted in different directions. Waves a and b are out of phase. The crests of one wave line up with the troughs of the other. The two waves cancel and they transmit no energy at all.
(b) Photon (1) interacts with a Ne atom in a metastable energy state and stimulates it to emit photon (2). Photon (2) stimulates another Ne atom to emit photon (3), and so on. The waves are coherent, or in phase. The crests and troughs of the waves match perfectly.

totally reflecting mirror

partially reflecting mirror

laser beam

FIGURE 9-30
Operation of a He–Ne laser.
Pictured here is a hypothetical laser tube arbitrarily divided into three sections representing the same portion of the tube at different times. At the top left a 633-nm photon (→) stimulates light emission from an excited Ne atom. Now there are two photons, and one of them in turn stimulates emission of a third photon. The photons are reflected by the mirror on the right, and their transit of the tube in the opposite direction is shown in the center. Three photons are amplified to five. At the bottom, the five photons are amplified to seven, and so on. A small portion of the photons is drawn off as the laser beam.

Summary

Understanding electromagnetic radiation is important to learning about atomic structure. The dispersion of "white" light produces a continuous spectrum—a rainbow. Light produced by excited gaseous atoms yields a line spectrum when dispersed—a series of colored lines. The simplest line spectrum is that of hydrogen atoms, which can be described through the Balmer equation.

To explain phenomena at the atomic and molecular level requires thinking about energy in tiny discrete units—quanta. Einstein used the quantum theory to explain the photoelectric effect, and Bohr applied it to an atomic model that explains the observed spectrum of hydrogen.

Wave mechanics is a more sophisticated form of quantum mechanics than that used by Bohr. Two ideas contributing to wave mechanics are de Broglie's concept of wave–particle duality (matter waves) and Heisenberg's uncertainty principle. Schrödinger used these ideas to provide a new model of the hydrogen atom.

An essential aspect of the Schrödinger atom is to view an electron as a cloud of negative electric charge with a particular geometrical shape. Another view is that the electron behaves like a particle and has a high probability of being found in a three-dimensional region called an *orbital*. The specific orbital types are *s*, *p*, *d*, and *f*. The key parameters that distinguish among orbitals are the three quantum numbers, n, ℓ, and m_ℓ. A fourth parameter required to describe an electron in an atom is the spin quantum number, m_s.

The wave-mechanical model of the hydrogen atom can be modified to apply to multielectron atoms, and, through a set of three rules, electrons can be assigned to the orbitals in principal shells and subshells. These assignments—*electron configurations*—are made in Section 9-10 by a method known as the Aufbau process. Finally, the connection between electron configurations and the periodic table is explored.

Summarizing Example

Microwave ovens have become increasingly popular in kitchens around the world. They are also useful in the chemical laboratory, particularly in drying samples for chemical analysis. A typical microwave oven uses microwave radiation with a wavelength of 12.2 cm.

Are there any electronic transitions in the hydrogen atom that could *conceivably* produce microwave radiation of wavelength 12.2 cm?

1. *Calculate the frequency of the microwave radiation.* Microwaves are a form of electromagnetic radiation and thus travel at the speed of light, 2.998×10^8 m s^{-1}. Convert the wavelength to meters, and then use the equation $\nu = c/\lambda$. Result: 2.46×10^9 Hz.

2. *Calculate the energy associated with one photon of the microwave radiation.* This is a direct application of Planck's equation, $E = h\nu$. Result: 1.63×10^{-24} J.

3. *Determine if there are any electronic transitions in the hydrogen atom with an energy per photon of 1.63×10^{-24} J.* Look at Figure 9-14, the energy level diagram for the Bohr hydrogen atom. Energy differences between the low-lying levels are of the order 10^{-19} to 10^{-20} J. This is orders of magnitude (10^4–10^5 times) greater than the energy per photon of 1.63×10^{-24} J from part **2**. Note, however, that the energy differences become progressively smaller for higher numbered orbits. As n approaches ∞ the energy differences approach zero. *Answer:* Some transitions between high-numbered orbits do correspond to microwave radiation (see also Exercise 80).

Key Terms

atomic (line) spectra (9-2)
Aufbau process (9-10)
d-block (9-11)
electromagnetic radiation (9-1)
electron configuration (9-10)
electron spin (9-8)
energy level diagram (9-4)
excited state (9-4)
f-block (9-11)
frequency, ν (9-1)
ground state (9-4)

Heisenberg uncertainty principle (9-5)
hertz, Hz (9-1)
Hund's rule (9-10)
orbital (9-6)
orbital diagram (9-10)
Pauli exclusion principle (9-10)
p-block (9-11)
photoelectric effect (9-3)
photon (9-3)
Planck's constant, h (9-3)

principal electronic shell (level) (9-7)
quantum (9-3)
quantum numbers (9-4, 9-6, 9-7)
s-block (9-11)
spdf notation (9-10)
standing wave (9-6)
subshell (sublevel) (9-7)
valence electrons (9-10)
wave (9-1)
wave function, ψ (9-6)
wavelength, λ (9-1)

Review Questions

1. In your own words define the following terms or symbols: **(a)** λ; **(b)** ν; **(c)** h; **(d)** ψ; **(e)** principal quantum number, n.

2. Briefly describe each of the following ideas or phenomena: **(a)** atomic (line) spectrum; **(b)** photoelectric effect; **(c)** matter wave; **(d)** Heisenberg uncertainty principle; **(e)** electron spin; **(f)** Pauli exclusion principle.

3. Explain the important distinctions between each pair of terms: **(a)** frequency and wavelength; **(b)** ultraviolet and infrared light; **(c)** continuous and discontinuous spectrum; **(d)** traveling and standing wave; **(e)** quantum number and orbital; **(f)** *spdf* notation and orbital diagram; **(g)** *s*-block and *p*-block; **(h)** main group and transition element.

4. Restate each of the following wavelengths in the unit indicated.
(a) 1625 Å = _____ nm
(b) 3880 Å = _____ μm
(c) 7.27×10^{-3} m = _____ nm
(d) 546 nm = _____ m
(e) 1.12 cm = _____ nm
(f) 2.6×10^4 Å = _____ cm

5. What are the wavelengths, in meters, associated with radiation of the following frequencies? In what portion of the electromagnetic spectrum is each radiation found? **(a)** 6.8×10^{12} s^{-1}; **(b)** 9.8×10^{15} s^{-1}; **(c)** 2.54×10^7 Hz; **(d)** 1.07×10^8 Hz.

6. *Without doing detailed calculations,* determine which of the following wavelengths represents light of the *highest* frequency: **(a)** 5.9×10^{-4} cm; **(b)** 1.13 mm; **(c)** 860 Å; **(d)** 6.92 μm.

7. A hypothetical electromagnetic wave is pictured here. What is the wavelength of this radiation?

1.17 nm

8. For the electromagnetic wave described in Exercise 7, what are **(a)** the frequency, in hertz, and **(b)** the energy, in joules per photon?

9. For electromagnetic radiation transmitted through a vacuum, state whether each of the following properties is directly proportional to, inversely proportional to, or independent of the frequency: **(a)** velocity; **(b)** wavelength; **(c)** energy per mole. Explain.

10. Use the Balmer equation (9.2) to determine
(a) the frequency, in s^{-1}, of the radiation corresponding to $n = 5$;
(b) the wavelength, in nanometers, of the line in the Balmer series corresponding to $n = 7$;
(c) the value of n corresponding to the Balmer series line at 380 nm.

11. How would the Balmer equation (9.2) have to be modified to predict lines in the infrared spectrum of hydrogen? [*Hint:* Compare Equations (9.2) and (9.6).]

12. Use Planck's Equation (9.3) to determine
(a) the energy, in joules per photon, of radiation of frequency, 8.62×10^{15} s^{-1};
(b) the energy, in kilojoules per mole, of radiation of frequency, 1.53×10^{14} s^{-1}.

13. Use Planck's equation (9.3) to determine
(a) the frequency, in hertz, of radiation having an energy of 4.18×10^{-21} J/photon;
(b) the wavelength, in nanometers, of radiation with 215 kJ/mol of energy.

14. *Without doing detailed calculations,* indicate which of the following electromagnetic radiations has the *greatest* energy per photon and which has the *least*. (a) 662 nm; (b) 2.1×10^{-5} cm; (c) 3.58 μm; (d) 4.1×10^{-6} m.

15. For the Bohr hydrogen atom determine
(a) the radius of the orbit $n = 4$;
(b) whether there is an orbit having a radius of 4.00 Å;
(c) the energy level corresponding to $n = 8$;
(d) whether there is an energy level at -5.00×10^{-18} J.

16. What is ΔE for the transition of an electron from $n = 5$ to $n = 3$ in a Bohr hydrogen atom? What is the frequency of the spectral line produced?

17. *Without doing detailed calculations,* indicate which of the following electron transitions in the hydrogen atom results in the *emission* of light of the *longest* wavelength. (a) $n = 4$ to $n = 3$; (b) $n = 1$ to $n = 2$; (c) $n = 1$ to $n = 6$; (d) $n = 3$ to $n = 2$.

18. If traveling at equal speeds, which of the following matter waves has the *longest* wavelength? Explain. (a) electron; (b) proton; (c) neutron; (d) alpha particle (He^{2+}).

19. Give a possible value for the missing quantum number(s) in each of the following sets. **(a)** $n = 3$, $\ell = 1$, $m_\ell = ?$; **(b)** $n = 4$, $\ell = ?$, $m_\ell = -1$; **(c)** $n = ?$, $\ell = 1$, $m_\ell = +1$.

20. Write the values corresponding to n and ℓ for each of the following orbital designations. **(a)** $4s$; **(b)** $3p$; **(c)** $5f$; **(d)** $3d$.

21. Which of the following sets of quantum numbers is (are) *not* allowed? Why not?
(a) $n = 3$, $\ell = 2$, $m_\ell = -1$
(b) $n = 2$, $\ell = 3$, $m_\ell = -1$
(c) $n = 3$, $\ell = 0$, $m_\ell = +1$
(d) $n = 6$, $\ell = 2$, $m_\ell = -1$
(e) $n = 4$, $\ell = 4$, $m_\ell = +4$
(f) $n = 4$, $\ell = 3$, $m_\ell = -1$

22. How many orbitals can there be of each of the following types? Explain. **(a)** $2s$; **(b)** $3f$; **(c)** $4p$; **(d)** $5d$; **(e)** $5f$; **(f)** $6p$.

23. Use *spdf* notation to write out the complete electron configuration of **(a)** bromine; **(b)** sulfur; **(c)** antimony; **(d)** silicon. Do not use symbols such as [Ne] or [Ar] for noble-gas cores.

24. Use an orbital diagram to show the electron configuration of **(a)** calcium; **(b)** scandium; **(c)** gallium; **(d)** nickel; **(e)** bismuth. You may use symbols like [Ne] or [Ar] for noble-gas cores.

25. Indicate the number of unpaired electrons in an atom of (a) magnesium; (b) thallium; (c) tellurium; (d) aluminum.
26. Refer to the periodic table on the inside front cover and identify (a) a main group metal; (b) a main group nonmetal; (c) a noble gas; (d) a *d*-block element; (e) an inner transition element.
27. Use the periodic table as a guide to write electron configurations for (a) In; (b) Cd; (c) Sb; (d) Au. Compare your results with the electron configurations shown in Appendix D.

28. Use Figure 9-28 as a guide to indicate the number of (a) 4*s* electrons in K; (b) 5*p* electrons in I; (c) 3*d* electrons in Zn; (d) 2*p* electrons in S; (e) 4*f* electrons in Pb; (f) 3*d* electrons in Ni.

Exercises

Electromagnetic Radiation

29. The magnesium spectrum has a line at 266.8 nm. Which of these statements is (are) correct concerning this radiation? Explain.
(a) It has a higher frequency than radiation with wavelength 402 nm.
(b) It is visible to the eye.
(c) It has a greater speed in a vacuum than does red light of wavelength 652 nm.
(d) Its wavelength is longer than that of X-rays.
31. How long does it take light from the sun, 93 million miles away, to reach Earth?

30. The most intense line in the cerium spectrum is at 418.7 nm.
(a) Determine the frequency of the radiation producing this line.
(b) In what part of the electromagnetic spectrum does this line occur?
(c) Is it visible to the eye? If so, what color is it? If not, is this line at higher or lower energy than visible light?

32. In astronomy, distances are measured in *light-years,* the distance that light travels in one year. What is the distance of one light-year expressed in kilometers?

Atomic Spectra

33. Calculate the wavelengths, in nanometers, of the first four lines of the Balmer series of the hydrogen spectrum, starting with the *longest* wavelength component.
35. To what value of *n* in Equation (9.2) does the line in the Balmer series at 389 nm correspond?

34. A line is detected in the hydrogen spectrum at 1880 nm. Is this line in the Balmer series? Explain.

36. The Lyman series of the hydrogen spectrum can be represented by the equation

$$\nu = 3.2881 \times 10^{15} \text{ s}^{-1}\left(\frac{1}{1^2} - \frac{1}{n^2}\right)$$

(where *n* = 2, 3, . . .)

(a) Calculate the maximum and minimum wavelength lines, in nanometers, in this series.
(b) What value of *n* corresponds to a spectral line at 95.0 nm?
(c) Is there a line at 108.5 nm? Explain.

Quantum Theory

37. A certain radiation has a wavelength of 474 nm. What is the energy, in joules, of (a) one photon; (b) a mole of photons of this radiation?
39. In what region of the electromagnetic spectrum would you expect to find radiation having an energy per photon 100 times that associated with 988-nm radiation? (*Hint:* Refer to Figure 9-3.)

38. What is the wavelength, in nanometers, of light with an energy content of 1799 kJ/mol? In what portion of the electromagnetic spectrum is this light?
40. High-pressure sodium vapor lamps are used in street lighting. The two brightest lines in the sodium spectrum are at 589.00 and 589.59 nm. What is the *difference* in energy per photon of the radiations corresponding to these two lines?

The Photoelectric Effect

41. The lowest frequency light that will produce the photo-electric effect is called the *threshold frequency*.
(a) The threshold frequency for indium is 9.96×10^{14} s^{-1}. What is the energy, in joules, of a photon of this radiation? **(b)** Will indium display the photoelectric effect with ultraviolet light? infrared light? Explain.

42. Sir James Jeans described the photoelectric effect in this way: "It not only prohibits killing two birds with one stone, but also the killing of one bird with two stones." Comment on the appropriateness of this analogy in reference to the marginal note on page 279.

The Bohr Atom

43. Use the description of the Bohr atom given in the text to determine **(a)** the radius, in nanometers, of the sixth Bohr orbit for hydrogen; **(b)** the energy, in joules, of the electron when it is in this orbit.

45. What are the **(a)** frequency, in s^{-1}, and **(b)** wavelength, in nanometers, of the light emitted when the electron in a hydrogen atom drops from the energy level $n = 7$ to $n = 4$? **(c)** In what portion of the electromagnetic spectrum is this light?

47. What electron transition in a hydrogen atom, starting from the orbit $n = 7$, will produce infrared light of wavelength 2170 nm? (*Hint:* In what orbit must the electron end up?)

44. Calculate the increase in **(a)** distance from the nucleus and **(b)** energy when an electron is excited from the first to the third Bohr orbit.

46. *Without doing detailed calculations,* indicate which of the following electron transitions requires the greatest amount of energy to be *absorbed* by a hydrogen atom: from **(a)** $n = 1$ to $n = 2$; **(b)** $n = 2$ to $n = 4$; **(c)** $n = 3$ to $n = 9$; **(d)** $n = 10$ to $n = 1$.

48. What electron transition in a hydrogen atom, ending in the orbit $n = 5$, will produce light of wavelength 3740 nm? (*Hint:* From what orbit must the electron fall?)

Wave–Particle Duality

49. Which must possess a greater velocity to produce matter waves of the same wavelength (such as 1 nm), protons or electrons? Explain your reasoning.

51. Calculate the de Broglie wavelength, in nanometers, associated with a 145-g baseball traveling at a speed of 168 km/h. How does this wavelength compare with typical nuclear or atomic dimensions?

50. What must be the velocity, in meters per second, of a beam of electrons if they are to display a de Broglie wavelength of 1 μm?

52. What is the wavelength, in nanometers, associated with a 1000-kg automobile traveling at a speed of 25 m/s, that is, considering the automobile to be a "matter" wave? Comment on the feasibility of an experimental measurement of this wavelength.

The Heisenberg Uncertainty Principle

53. Describe the ways in which the Bohr model of the hydrogen atom appears to violate the Heisenberg uncertainty principle.

54. Although Einstein made some early contributions to quantum theory, he was never able to accept the Heisenberg uncertainty principle. He stated, "God does not play dice with the Universe." What do you suppose Einstein meant by this remark? In reply to Einstein's remark, Niels Bohr is supposed to have said, "Albert, stop telling God what to do." What do you suppose Bohr meant by this remark?

Wave Mechanics

55. A standing wave in a string 42 cm long has a *total* of six nodes (including those at the ends). What is the wavelength, in centimeters, of this standing wave? (*Hint:* Make a sketch of the standing wave.)
57. Describe some of the differences between orbits of the Bohr atom and orbitals of the wave-mechanical atom. Are there any similarities?

56. What is the length of a string that has a standing wave with four nodes (including those at the ends) and $\lambda = 17$ cm.

58. The greatest probability of finding the electron in a small-volume element of the $1s$ orbital of the hydrogen atom is at the nucleus. Yet the most probable distance of the electron from the nucleus is 53 pm. How can you reconcile these two statements?

Quantum Numbers and Electron Orbitals

59. Select the correct answer and explain your reasoning. An electron having $n = 3$ and $m_\ell = 0$ (a) must have $m_s = +\frac{1}{2}$; (b) must have $\ell = 1$; (c) may have $\ell = 0, 1,$ or 2; (d) must have $\ell = 2$.

60. Write an acceptable value for each of the missing quantum numbers.
(a) $n = 3$, $\ell = ?$, $m_\ell = 2$, $m_s = +\frac{1}{2}$
(b) $n = ?$, $\ell = 2$, $m_\ell = -1$, $m_s = -\frac{1}{2}$
(c) $n = 4$, $\ell = 2$, $m_\ell = 0$, $m_s = ?$
(d) $n = ?$, $\ell = 0$, $m_\ell = ?$, $m_s = ?$

61. What type of orbital (i.e., $3s$, $4p$, . . .) is designated by the quantum numbers
(a) $n = 5$, $\ell = 1$, $m_\ell = 0$;
(b) $n = 4$, $\ell = 2$, $m_\ell = -2$;
(c) $n = 2$, $\ell = 0$, $m_\ell = 0$?

62. Which of the following statements is (are) correct for an electron with $n = 4$ and $m_\ell = -2$? Explain.
(a) The electron is in the fourth principal shell.
(b) The electron may be in a d orbital.
(c) The electron may be in a p orbital.
(d) The electron must have $m_s = +\frac{1}{2}$.

63. Concerning the concept of subshells and orbitals,
(a) How many subshells are found in the $n = 4$ level?
(b) What are the names of the subshells in the $n = 3$ level?
(c) How many orbitals have the values $n = 4$ and $\ell = 3$?
(d) How many orbitals have the values $n = 4$, $\ell = 3$, and $m = -2$?
(e) What is the total number of orbitals in the $n = 4$ level?

64. Concerning the electrons in the shells, subshells, and orbitals of an atom, how many can have
(a) $n = 3$, $\ell = 2$, $m_\ell = 0$, and $s = +\frac{1}{2}$?
(b) $n = 3$, $\ell = 2$, and $m_\ell = 0$?
(c) $n = 3$ and $\ell = 2$?
(d) $n = 3$?
(e) $n = 3$, $\ell = 2$, and $s = +\frac{1}{2}$?

Electron Configurations

65. Which of the following is the correct orbital diagram for the ground state electron configuration of phosphorus? Explain what is wrong with each of the others.

66. Which of the following is the correct orbital diagram for the ground state electron configuration of molybdenum? Explain what is wrong with each of the others.

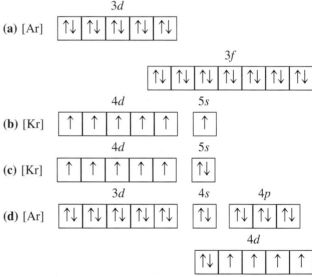

67. Use the basic rules for electron configurations to indicate the number of (a) unpaired electrons in an atom of P; (b) $3d$ electrons in an atom of Br; (c) $4p$ electrons in an atom of Ge; (d) $6s$ electrons in an atom of Ba; (e) $4f$ electrons in an atom of Au.

68. Use orbital diagrams to show the distribution of electrons among the orbitals in (a) the $4p$ subshell of Br; (b) the $3d$ subshell of Co^{2+}, given that the two electrons lost are $4s$; (c) the $5d$ subshell of Pb.

69. On the basis of the periodic table and rules for electron configurations, indicate the number of (a) $2p$ electrons in N; (b) $4s$ electrons in Rb; (c) $4d$ electrons in As; (d) $4f$ electrons in Au; (e) unpaired electrons in Pb; (f) elements in Group 4A of the periodic table; (g) elements in the sixth period of the periodic table.

70. Based on the relationship between electron configurations and the periodic table, give the number of (a) outer shell electrons in an atom of Sb; (b) electrons in the fourth principal electronic shell of Pt; (c) elements whose atoms have six outer shell electrons; (d) unpaired electrons in an atom of Te; (e) transition elements in the sixth period.

71. The element that the unknown and undiscovered element 114 should most closely resemble is Pb.
(a) Write the electron configuration of Pb.
(b) Propose a plausible electron configuration for element 114.

72. Without referring to any tables or listings in the text, mark an appropriate location in the blank periodic table provided for each of the following: **(a)** the fifth period noble gas; **(b)** a sixth period element whose atoms have three unpaired p electrons; **(c)** a d-block element having one $4s$ electron; **(d)** a p-block element that is a metal.

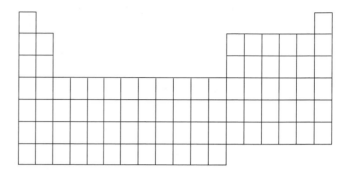

Advanced Exercises

73. Electromagnetic radiation can be transmitted through a vacuum or empty space. Can heat or sound be similarly transferred? Explain.

74. The *work function* is the energy that must be supplied to cause the release of an electron from a photoelectric material. The corresponding photon frequency is the threshold frequency. The higher the energy of the incident light, the more kinetic energy the electrons have in moving away from the surface. The work function for mercury is equivalent to 435 kJ/mol photons.
(a) Can the photoelectric effect be obtained with mercury by using visible light? Explain.
(b) What is the kinetic energy, in joules, of the ejected electrons when light of 215 nm strikes a mercury surface?
(c) What is the velocity, in meters per second, of the ejected electrons in **(b)**?

75. Derive the Balmer equation (9.2) from Equation (9.6).

76. The Pfund series of the hydrogen spectrum has as its *longest* wavelength component a line at 7400 nm. Describe the electron transitions that produce this series. That is, give a Bohr quantum number that is "common" to this series.

77. Between which two orbits of the Bohr hydrogen atom must an electron fall to produce light of wavelength 1876 nm?

78. Use Equation (9.7) and other appropriate relationships to determine the wavelength of the line in the emission spectrum of the ion He^+ resulting from the electron transition from $n = 5$ to $n = 2$.

79. Draw an energy level diagram similar to Figure 9-14 that represents all the lines that might arise in the emission spectrum of hydrogen atoms that drop from the excited state $n = 5$, ultimately to the ground state $n = 1$. (*Hint:* In principle, how many lines might be observed?)

80. Refer to the Summarizing Example. Assume that the microwave radiation could be produced by an electronic transi-

tion from the Bohr orbit $(n + 1)$ to the orbit n. Show that the value of n is 138.

81. Infrared lamps are used in cafeterias to keep food warm. How many photons per second are produced by an infrared lamp that consumes energy at the rate of 95 W (watts) (95 J/s) and is 14% efficient in converting this energy to infrared radiation? Assume that the radiation has a wavelength of 1525 nm.

82. A proton is accelerated to one-tenth the velocity of light, and this velocity can be measured with a precision of ±1%. What is the uncertainty in the position of this proton?

83. Show that the uncertainty principle is not significant when applied to large objects such as automobiles. Assume that m is precisely known; assign a reasonable value to either Δx or Δv and estimate a value of the other.

84. What must be the velocity of electrons if their associated wavelength is to equal the radius of the first Bohr orbit of the hydrogen atom?

85. What must be the velocity of electrons if their associated wavelength is to equal the *shortest* wavelength line in the Lyman series? (*Hint:* Refer to Figure 9-14.)

86. In a plucked guitar string, the frequency of the *longest* wavelength standing wave is called the *fundamental frequency*. The standing wave with one interior node is called the first *overtone*, and so on. What is the wavelength of the second overtone of a 24-in. guitar string?

87. If all other rules governing electron configurations were valid, what would be the electron configuration of cesium if **(a)** there were *three* possibilities for electron spin? **(b)** the quantum number ℓ could have the value n?

88. Ozone, O_3, absorbs ultraviolet radiation and dissociates into O_2 molecules and O atoms: $O_3 + h\nu \rightarrow O_2 + O$. A 1.00-L sample of air at 22 °C and 748 mmHg contains 0.25 ppm of O_3. How much energy, in joules, must be absorbed if all the O_3 molecules in the sample of air are to dissociate? Assume that each photon absorbed causes one O_3 molecule to dissociate, and that the wavelength of the radiation is 254 nm.

89. Radio signals from *Voyager 1* in the 1970s were broadcast at a frequency of 8.4 GHz (gigahertz). On Earth this radiation was received by an antenna able to detect signals as weak as 4×10^{-21} W (1 watt = 1 J/s). How many photons per second does this detection limit represent?

90. Certain metal compounds impart colors to flames: e.g., sodium compounds, yellow; lithium, red; barium, green. "Flame tests" can be used to detect these elements.

(a) At a flame temperature of 800 °C can collisions between gaseous atoms with average kinetic energies supply the energies required for the emission of visible light?

(b) If not, how do you account for the excitation energy?

91. The angular momentum of an electron in the Bohr hydrogen atom is mvr, where m is the mass of the electron, v, its velocity, and r, the radius of the Bohr orbit. The angular momentum can have only the values $nh/2\pi$, where n is an integer (the number of the Bohr orbit). Show that the *circumferences* of the various Bohr orbits are integral multiples of the de Broglie wavelengths of the electron treated as a matter wave.

Feature Problems

A. We have noted that an emission spectrum is a kind of "atomic fingerprint." The various steels are alloys of iron and carbon, usually containing one or more other metals. Based on the principal lines of their atomic spectra, which of the metals in the table below are likely to be present in a steel sample whose hypothetical emission spectrum is pictured? Is it likely that still other metals are present in the sample? Explain.

Principal spectral lines of some Period 4

transition metals, nm

V	306.64	309.31	318.40	318.54	327.11	437.92	438.47	439.00
Cr	357.87	359.35	360.53	361.56	425.44	427.48	428.97	520.45
Mn	257.61	259.37	279.48	279.83	403.08	403.31	403.45	
Fe	344.06	358.12	372.00	373.49	385.99			
Ni	341.48	344.63	345.85	346.17	349.30	351.51	352.45	361.94

Hypothetical emission spectrum.
In a real spectrum the photographic images of the spectral lines would differ in depth and thickness depending on the strengths of the emissions producing them. Some of the spectral lines would not be seen because of their faintness.

B. Balmer seems to have deduced his formula for the visible spectrum of hydrogen just by manipulating numbers. A more common scientific procedure is to graph experimental data and then find a mathematical equation to describe the graph. Show that Equation (9.2) describes a straight line. Indicate which variables must be plotted and determine the numerical values of the slope and intercept of this line. Use data from Figure 9-9 to confirm that the four lines in the visible spectrum of hydrogen fall on the straight line graph.

C. The graph shown here represents the relative probability of finding the 1s electron within a very thin spherical shell as a function of the distance, r, of the surface of the shell from the nucleus of a hydrogen atom. (*Hint:* Think of the very thin shell in relation to the entire atom as somewhat like the skin of an apple relative to the whole apple.)

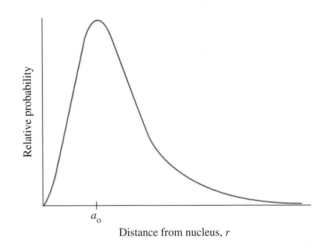

Distance from nucleus, r

1. Figure 9-18a is also an electron probability plot for the 1s electron of hydrogen. Explain why that plot differs from the one here.

2. The graph shown here is related to the dartboard analogy of Figure 9-19. Plot a graph of the dartboard analogy using the scoring summary given below. That is, plot the number of hits as a function of the scoring ring: 50, 40, ...

Summary of scoring (1500 darts)	
200 darts score	"50"
300	"40"
400	"30"
250	"20"
200	"10"
150	off the board

3. Compare the graph drawn in part 2 to the one shown above, explaining the similarities and differences between the two.

D. The accompanying diagrams can help us summarize some ideas about the electron configurations of the elements. The diagrams contain some features marked with a question mark (?) and others by leaders (straight line pointers). Replace each (?) by the symbol of the appropriate element or noble gas core. Use terms from the following list to attach a label to each leader. You may use a given label only once, and you will find that some labels are inappropriate.

(a) an unpaired electron   (b) a filled subshell
(c) a valence shell   (d) a valence electron
(e) an *f* subshell   (f) a principal shell
(g) a half-filled subshell (h) a *d* orbital
(i) a *p* subshell   (j) a *d* subshell
(k) an electron pair   (l) an *s* subshell

1. (?): $1s^2 2s^2 2p^6 3s^2 3p^6 3d^5 4s^2$

2. Nb: [?]

3. (?):   [?]

4*d*

4. (?):   [?]

5*f*

The Periodic Table and Some Atomic Properties

10

Potassium, an alkali metal, reacts with water to liberate hydrogen gas, which bursts into flame. Phenolphthalein indicator in the water turns magenta, signaling the formation of hydroxide ions, another product of the reaction. The vigor of this reaction will not seem surprising, once we have explored the significance of potassium's location in Group 1A of the periodic table.

*B*y the middle of the nineteenth century, chemists had discovered a large number of elements, determined their relative atomic masses, and measured a host of their properties. Chemists had assembled what amounted to the "white pages" of a chemistry directory, but what they needed was a set of "yellow pages"—an arrangement that would group similar elements together. This tabulation would help chemists focus on similarities and differences among the known elements and predict properties of elements still undiscovered. In this chapter we will continue to study the first successful tabulation—the periodic table of the elements. Chemists value the periodic table as a means of organizing their field, and they would continue to use it even if they had never figured out why it works. But the underlying rationale of the periodic table was discovered about 50 years after the table was proposed.

The basis of the periodic table is in the electron configurations of the elements, a topic we studied in Chapter 9. In this chapter, we will use the table as a backdrop for a discussion of some properties of the elements—atomic radii, ionization energies, electron affinities. We will use these atomic properties in the discussion of chemical bonding in the following two chapters. And the periodic table itself will be our indispensable guide throughout much of the remainder of the text.

10-1 Classifying the Elements: The Periodic Law and the Periodic Table

In 1869, Dmitri Mendeleev and Lothar Meyer, independently, proposed the **periodic law:**

When the elements are arranged in order of increasing atomic mass, certain sets of properties recur periodically.

Meyer's first example of a periodic property was one called atomic volume—the atomic mass of an element divided by the density of its solid form. We now just call this the molar volume.

$$\text{atomic (molar) volume (cm}^3\text{/mol)} = \text{molar mass (g/mol)} \times \frac{1}{d} \text{ (cm}^3\text{/g)} \quad (10.1)$$

Meyer presented his results graphically, and Figure 10-1 is an updated version of his graph. Notice how high atomic volumes recur periodically for the alkali metals Li, Na, K, Rb, and Cs. Meyer examined other physical properties of the elements and their compounds and found that many of these also recur periodically.

Mendeleev's Periodic Table

We have already described the periodic table as a tabular arrangement of the elements that groups similar elements together. Mendeleev succeeded where others had failed for two reasons: He left blank spaces in his table for undiscovered elements, and he corrected some atomic mass values. The blanks in his

FIGURE 10-1

An illustration of the periodic law—variation of atomic volume with atomic number.

This adaptation of Meyer's 1870 graph plots atomic volumes against atomic numbers. Of course, a number of elements, such as the noble gases, were undiscovered in Meyer's time. The graph shows peaks at the alkali metals (Li, Na, K, . . .). Nonmetals fall on the ascending portions of the curve and metals at the peaks, on the descending portions, and in the valleys.

Reihen	Gruppe I. — R^2O	Gruppe II. — RO	Gruppe III. — R^2O^3	Gruppe IV. RH^4 RO^2	Gruppe V. RH^3 R^2O^5	Gruppe VI. RH^2 RO^3	Gruppe VII. RH R^2O^7	Gruppe VIII. RO^4
1	H = 1							
2	Li = 7	Be = 9,4	B = 11	C = 12	N = 14	O = 16	F = 19	
3	Na = 23	Mg = 24	Al = 27,3	Si = 28	P = 31	S = 32	Cl = 35,5	
4	K = 39	Ca = 40	– = 44	Ti = 48	V = 51	Cr = 52	Mn = 55	Fe = 56, Co = 59, Ni = 59, Cu = 63.
5	(Cu = 63)	Zn = 65	– = 68	– = 72	As = 75	Se = 78	Br = 80	
6	Rb = 85	Sr = 87	?Yt = 88	Zr = 90	Nb = 94	Mo = 96	– = 100	Ru = 104, Rh = 104, Pd = 106, Ag = 108
7	(Ag = 108)	Cd = 112	In = 113	Sn = 118	Sb = 122	Te = 125	J = 127	
8	Cs = 133	Ba = 137	?Di = 138	?Ce = 140	–	–	–	
9	(–)		–					
10	–	–	?Er = 178	?La = 180	Ta = 182	W = 184	–	Os = 195, Ir = 197, Pt = 198, Au = 199
11	(Au = 199)	Hg = 200	Tl = 204	Pb = 207	Bi = 208			
12	–	–	–	Th = 231	–	U = 240		

Dmitri Mendeleev (1834–1907). Mendeleev's discovery of the periodic table came as a result of attempting to systematize properties of the elements for presentation in a chemistry textbook. His highly influential book went through eight editions in his lifetime and five more after his death.

In his periodic table, Mendeleev arranged the elements into eight groups (Gruppe) and twelve rows (Reihen). The formulas R^2O, RO, . . . , are those of the element oxides (such as Li_2O, MgO, . . .); the formulas RH^4, RH^3, . . . , are those of the element hydrides (such as CH_4, NH_3, . . .).

table came at atomic masses 44, 68, 72, and 100, for the elements we now know as scandium, gallium, germanium, and technetium. Two of the atomic mass values he corrected were those of indium and uranium.

In Mendeleev's table, similar elements fall in vertical groups, and their properties change gradually from top to bottom in the group. As an example, we have seen that the alkali metals (Mendeleev's Group I) have high molar volumes (see Figure 10-1). They also have low melting points, which decrease in the order

$$\text{Li (174 °C)} > \text{Na (97.8 °C)} > \text{K (63.7 °C)} > \text{Rb (38.9 °C)} > \text{Cs (28.5 °C)}$$

In their compounds, the alkali metals exhibit the oxidation state $+1$, forming ionic compounds, such as $NaCl$, KBr, CsI, Li_2O, and so on.

▶ We consider additional properties of the alkali metals in Section 10-8.

Discovery of New Elements

Two of the elements predicted by Mendeleev were discovered shortly after the appearance of his 1871 periodic table (gallium, 1875; scandium, 1879). Table 10.1 illustrates how closely Mendeleev's predictions for eka-silicon agree with the observed properties of the element germanium, discovered in 1886. Often, new ideas in science take hold slowly, but the success of Mendeleev's predictions stimulated chemists to adopt his table fairly quickly.

A New Group for the Periodic Table

One group of elements that Mendeleev did not anticipate was the noble gases. He left no blanks for them. As we noted in Chapter 8, William Ramsay, their discoverer, proposed placing them in a separate group of the table. Because ar-

TABLE 10.1   Properties of Germanium: Predicted and Observed

Property	Predicted: Eka-Silicon (1871)	Observed: Germanium (1886)
atomic mass	72	72.6
density, g/cm^3	5.5	5.47
color	dirty gray	grayish white
density of oxide, g/cm^3	EsO_2: 4.7	GeO_2: 4.703
boiling point of chloride	$EsCl_4$: below 100 °C	$GeCl_4$: 86 °C
density of chloride, g/cm^3	$EsCl_4$: 1.9	$GeCl_4$: 1.887

▶ The term "eka" is derived from Sanskrit and means "first." That is, eka-silicon means, literally, first comes silicon (and then comes the unknown element).

gon, the first member of the group discovered, had an atomic mass greater than that of chlorine and comparable to that of potassium, Ramsay placed the new group, which he called Group 0, between the halogen elements (Group VII) and the alkali metals (Group I).

Atomic Number as the Basis for the Periodic Law

Mendeleev had to place certain elements out of order to get them into the proper groups of his periodic table. He assumed this was because of errors in atomic masses. With improved methods of determining atomic masses and with the discovery of argon (Group 0, at. mass 39.9), which was placed ahead of potassium (Group I, at. mass 39.1), it became clear that a few elements might always remain "out of order." At the time, these out-of-order placements were justified by chemical evidence. Elements were placed in the groups that their chemical behavior dictated. There was no theoretical explanation for this reordering. Matters changed in 1913 as a result of some research by H. G. J. Moseley on the X-ray spectra of the elements.

As we learned in Chapter 2, X rays are a high-frequency form of electromagnetic radiation. They are produced when a cathode-ray (electron) beam strikes the anode of a cathode-ray tube. The anode is called the target. Moseley was familiar with Bohr's atomic model, which explains X-ray emission in terms of transitions in which electrons drop into orbits close to the atomic nucleus (see Figure 9-13). Moseley reasoned that because the energies of electron orbits depend on the nuclear charge, the frequencies of emitted X rays should depend on the nuclear charges of atoms in the target. Using techniques newly developed by the father–son team W. Henry Bragg and W. Lawrence Bragg, Moseley obtained photographic images of X-ray spectra and assigned frequencies to the spectral lines. His spectra for the elements from Ca to Zn are reproduced in Figure 10-2.

Moseley was able to correlate X-ray frequencies to numbers equal to the nuclear charges and corresponding to the positions of elements in Mendeleev's periodic table. For example, aluminum, the thirteenth element in the table, was assigned an *atomic number* of 13. Moseley's equation is $\nu = A(Z - b)^2$, where ν is the X-ray frequency, Z is the atomic number, and A and b are constants. Moseley used this relationship to predict three new elements ($Z = 43$, 61, and 75), which were discovered in 1937, 1945, and 1925, respectively. Also, he proved that in the portion of the periodic table with which he worked (from $Z = 13$ to $Z = 79$) there could be no new elements. All available atomic numbers had been assigned. From the standpoint of Moseley's work, then, we should restate the periodic law.

Henry G. J. Moseley (1887–1915). Moseley was one of a group of brilliant scientists whose careers were launched under Ernest Rutherford. He was tragically killed at Gallipoli in Turkey in World War I.

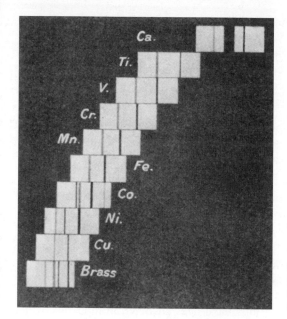

FIGURE 10-2
Moseley's X-ray spectra of several elements.
In this photograph from Moseley's 1913 paper, you can see two lines for each element, beginning with Ca at the top. With each successive element the lines are displaced to the left, the direction of increasing X-ray frequency in these experiments. Where more than two lines appear, the sample had one or more other elements as an impurity. Notice, for example, that one line in the Co spectrum matches a line in the Fe spectrum, and another matches a line in the Ni spectrum. Brass, which is an alloy of copper and zinc, shows two lines for Cu and two for Zn.

Similar properties recur periodically when elements are arranged according to increasing atomic number.

10-2 Description of a Modern Periodic Table: The Long Form

Mendeleev's periodic table is in the "short" form, consisting of eight groups. Most modern periodic tables arrange the elements in 18 groups, called the "long" form (see the inside front cover). We gave a description of the periodic table in Section 2-6. Let us at this time briefly review that description.

The vertical groups bring together elements with similar properties. The horizontal periods of the table are arranged in order of increasing atomic number from left to right. The groups are numbered at the top and the periods at the extreme left in the periodic table on the inside front cover. The first two groups—the *s*-block—and the last six groups—the *p*-block—together constitute the *main group* elements. Because they come between the *s*-block and the *p*-block, the *d*-block elements are known as the *transition* elements. The *f*-block elements, sometimes called the *inner transition* elements, would extend the table to a width of 32 members if incorporated in the main body of the table. The table would generally be too wide to fit on a printed page, and so the *f*-block elements are extracted from the table and placed at the bottom. The 14 elements following lanthanum ($Z = 57$) are called the *lanthanides* and the 14 following actinium ($Z = 89$) are called the *actinides*.

There are several controversies about the details of the periodic table that are still unresolved despite the table's existence for more than a century. For instance, there are three different systems currently used to designate the vertical groups in the periodic table. The format of the periodic table we are using is called the "American" table. An alternative format, called the "European" table, also uses the numerals 1 through 8 but handles the letters A and B differently: All groups to the left of Fe, Ru, and Os are labeled A and the groups to the right of Ni, Pd, and Pt are labeled B. As a result, the distinction between main group and transition elements is lost (that is, elements of each type are found among both the "A" and "B" groups).

To resolve this "A/B" controversy, the International Union of Pure and Applied Chemistry (IUPAC) has recommended a new format. In the IUPAC table, the groups are numbered consecutively from left to right—from 1 to 18. The letters A and B are not used. One advantage of this system is that the groups headed by Fe, Co, and Ni have separate designations (8, 9, and 10, respectively) instead of being lumped together as Group 8B. The IUPAC system does pose a problem for *p*-block elements. Here, instead of representing the number of outer shell electrons, the group numerals are "10 + no. of outer shell electrons." Thus, oxygen, with six outer shell electrons ($2s^22p^4$), is in Group 16.

At this time chemists have not agreed on any single form of the table, and you are likely to encounter all three types of group designations. Both the "American" and the IUPAC systems are shown on the periodic table on the inside front cover.

Another area of contention has been in the naming of elements with atomic number greater than 92 (the *transuranium elements*). Traditionally, the group that discovers a new element names the element; however, for a number of the newer elements several separate groups have been involved in their discovery, and it has been difficult for these groups to reach an agreement. For many years some of these elements have had no name other than their atomic number. The groups with conflicting claims have now reached agreement on the names of elements 104 to 109. IUPAC, however, has chosen not to accept some of these names and has proposed other names instead. For example, IUPAC objects to naming an element for a living scientist and has rejected the name seaborgium (Sg) for element 106. Glenn Seaborg, who won the Nobel Prize in 1951 for discovering several transuranium elements, is also credited with proposing the current placement of the lanthanides and actinides in the periodic table. The American Chemical Society has accepted the names originally submitted to IUPAC, and these are the names and symbols shown inside the front cover. IUPAC is still discussing the issue of naming some of the newer elements, but no final decision had been reached as this text went to press.

10-3   Metals and Nonmetals and Their Ions

In Section 2-6, to help us write names and formulas, we established two categories of elements: *metal* and *nonmetal*. At that time we described metals and nonmetals in terms of physical properties: Most metals are good conductors of heat and electricity, are malleable and ductile, and have moderate to high melting points. In general, nonmetals are nonconductors of heat and electricity and are nonmalleable (brittle) solids, though a number of nonmetals are gases at room temperature.

Through the color scheme of the periodic table on the inside front cover, we see that the majority of the elements are metals (orange) and that nonmetals (blue) are confined to the right side of the table. The noble gases (purple) are treated as a special group of nonmetals. Metals and nonmetals are often separated by a stair-step diagonal line, and several elements along this line are often called metalloids (green). **Metalloids** are elements that look like metals and in some ways behave like metals, but that also have some nonmetallic properties.

Originally, the positions of the elements in the periodic table were related to their properties. In Chapter 9 we saw that the positions are also related to the electron configurations of the elements. We might now wonder if there is a relationship between the properties of elements and their electron configurations. Generally speaking there is. The properties of an element seem to be determined largely by the electron configuration of the *outermost,* or valence, electronic shell. Adjacent members of a series of main group elements in the same period (such as P, S, and Cl) have significantly different properties because they differ in their valence electron configurations. Within a transition series, differences in electron configurations are mostly in inner shells, and so a transition element has some similarities to neighboring transition elements in the same period. We particularly find many similar properties for adjacent members of the same period within the *f*-block. In fact, the strong similarities among the lanthanide elements presented a particular challenge to the nineteenth-century chemists who tried to separate and identify them.

Let us now begin our exploration of the crucial role that electron configurations play in establishing the chemistry of both metals and nonmetals. We will start by examining the noble gases.

Noble Gases

Atoms of the noble gases have the maximum number of electrons permitted in the valence shell of an atom, two in helium ($1s^2$) and eight in the other noble gas atoms (ns^2np^6). These electron configurations are very difficult to alter and seem to confer a high degree of chemical inertness to the noble gases. It is interesting to note, then, that the *s*-block metals, together with Al in Group 3A, tend to lose enough electrons to acquire the electron configurations of the noble gases. On the other hand, nonmetals tend to gain enough electrons to achieve the same configurations.

Main Group Metal Ions

The electron configurations of the atoms of Groups 1A and 2A—the most active metals—differ from those of the noble gas of the preceding period by only one and two electrons in the *s* orbital of a new electronic shell. If a K atom is stripped of its outer shell electron, it becomes the *positive ion* K^+, with the electron configuration [Ar]. A Ca atom acquires the [Ar] configuration following the removal of two electrons.

$$K \ ([Ar]4s^1) \longrightarrow K^+ \ ([Ar]) + e^-$$

$$Ca \ ([Ar]4s^2) \longrightarrow Ca^{2+} \ ([Ar]) + 2 \ e^-$$

Although metal atoms do not lose electrons spontaneously, the energy required to bring about ionization is often provided by other processes occurring at the same time (such as an attraction between positive and negative ions). Aluminum is the only *p*-block metal that forms an ion with a noble gas electron configuration—Al^{3+}. The electron configurations of the other *p*-block metal ions are summarized in Table 10.2.

Main Group Nonmetal Ions

The atoms of Groups 7A and 6A—the most active nonmetals—have one and two electrons fewer than the noble gas at the end of the period. Groups 7A and 6A atoms can acquire the electron configurations of noble gas atoms by *gaining* the appropriate numbers of electrons.

$$Cl \ ([Ne]3s^23p^5) + e^- \longrightarrow Cl^- \ ([Ar])$$

$$S \ ([Ne]3s^23p^4) + 2 \ e^- \longrightarrow S^{2-} \ ([Ar])$$

TABLE 10.2 Electron Configurations of Some Main Group Metal Ions			
"Noble Gas"		"Pseudo Noble Gas"[a]	"18 + 2"[b]
Li^+ Be^{2+}		Ga^{3+}	In^+
Na^+ Mg^{2+}		Tl^{3+}	Tl^+
K^+ Ca^{2+}			Sn^{2+}
Rb^+ Sr^{2+}			Pb^{2+}
Cs^+ Ba^{2+}			Sb^{3+}
Fr^+ Ra^{2+}			Bi^{3+}
Al^{3+}			

[a]In the configuration labeled "pseudo noble gas" all electrons of the outermost shell have been lost. The next-to-outermost electron shell of the atom becomes the outermost shell of the ion and contains 18 electrons, for example, Ga^{3+}: $[Ne]3s^2 3p^6 3d^{10}$.

[b]In the configuration labeled "18 + 2" all outer shell electrons except the two *s* electrons are lost, producing an ion with 18 electrons in the next-to-outermost shell and 2 electrons in the outermost, for example, Sn^{2+}: $[Ar]3d^{10}4s^2 4p^6 4d^{10}5s^2$.

In most cases a nonmetal atom will gain a single electron spontaneously, but energy is required to force it to accept additional electrons. Often other processes occurring simultaneously supply the necessary energy (such as an attraction between positive and negative ions). Nonmetal ions with a charge of 3− are rare, but some metal nitrides are often described as containing the nitride ion, N^{3-}.

Transition Metal Ions

We have seen in the Aufbau process that we fill the *ns* subshell before placing electrons in the $(n-1)d$ subshell (page 298), but we have also noted that the energy levels of these two subshells are nearly the same (page 295). Thus, it is not surprising that when transition metal atoms ionize, the *ns* subshell is emptied. For example, the electron configuration of Ti is $[Ar]3d^2 4s^2$, and that of Ti^{2+} is $[Ar]3d^2$. Moreover, we may expect that in some cases one or more $(n-1)d$ electrons might be lost together with the *ns* electrons. This happens in the formation of Ti^{4+}, which has the electron configuration [Ar].

A few transition metal atoms acquire noble gas electron configurations when forming cations, as does Ti in Ti^{4+} and Sc in Sc^{3+}. Most transition metal atoms *do not*. An iron atom does not acquire a noble gas electron configuration when it loses its $4s^2$ electrons to form the ion Fe^{2+},

$$Fe\ ([Ar]3d^6 4s^2) \longrightarrow Fe^{2+}\ ([Ar]3d^6) + 2\ e^-$$

nor does it with the loss of an additional $3d$ electron to form the ion Fe^{3+}.

$$Fe\ ([Ar]3d^6 4s^2) \longrightarrow Fe^{3+}\ ([Ar]3d^5) + 3\ e^-$$

The $3d$ subshell in Fe^{3+} is half-filled, a fact that helps to account for the observed ease of oxidation of iron(II) to iron(III) compounds. Electron configurations with half-filled or filled *d* or *f* subshells have a special stability, and we find that a number of transition metal ions have such configurations.

Are You Wondering . . .

Why hydrogen, a nonmetal, is included with the Group 1A metals in the periodic table?

Although all the other elements have a definite place in the periodic table, hydrogen doesn't. Its uniqueness stems from the fact that its atoms have only one electron, in the configuration $1s^1$. This is what causes us to put hydrogen in Group 1A, even though we classify it as a nonmetal.* Because, like the halogens, hydrogen is one electron short of having a noble gas electron configuration ($1s^2$), in some periodic tables it is placed in Group 7A. Hydrogen, however, doesn't resemble the halogens very much. For example, F_2 and Cl_2 are excellent oxidizing agents, but H_2 isn't. Still another alternative that one sometimes sees is hydrogen placed by itself at the top of the periodic table and near the center.

10-4 The Sizes of Atoms and Ions

In earlier chapters we discovered the importance of atomic masses in matters relating to stoichiometry. To understand certain physical and chemical properties, we need to know something about atomic sizes. In this section we describe atomic radius, the first of a group of *atomic* properties that we will examine in this chapter.

Atomic Radius

Unfortunately, atomic radius is hard to define. The probability of finding an electron decreases with increasing distance from the nucleus, but nowhere does the probability fall to zero. There is no precise outer boundary to an atom. We might describe an *effective* atomic radius as, say, the distance from the nucleus within which 90% of all the electron charge density is found. But, in fact, all that we can *measure* is the distance between the nuclei of adjacent atoms (internuclear distance). Even though this distance varies, depending on whether atoms are chemically bonded or merely in contact without forming a bond, we define atomic radius in terms of internuclear distance.

Because we are primarily interested in bonded atoms, we will emphasize an atomic radius based on the distance between the nuclei of two atoms joined by a chemical bond. The **covalent radius** is one-half the distance between the nuclei of two identical atoms joined by a single covalent bond. The **ionic radius** is based on the distance between the nuclei of ions joined by an ionic bond. Because the ions are not identical in size, this distance must be properly apportioned between the cation and anion. For metals, we define a **metallic radius** as one-half the distance between the nuclei of two atoms in contact in the crystalline solid metal.

The angstrom unit, Å, has long been used for atomic dimensions (1 Å = 10^{-10} m). Angstrom, however, is not a recognized SI unit. The SI units are the nanometer (nm) and picometer (pm).

$$1 \text{ nm} = 1000 \text{ pm} = 1 \times 10^{-9} \text{ m} \qquad (10.2)$$

▶ Despite the SI convention, the angstrom unit, Å, is still widely used by X-ray crystallographers and others who work with atomic and molecular dimensions.

*Hydrogen appears to become metallic when subjected to pressures of about 2 million atm, but these are hardly ordinary laboratory conditions.

Covalent radius:

—157 pm

Na / Na

Metallic radius:

—186 pm

Na / Na

Ionic radius:

—99 pm

Na^+ / Cl^-

FIGURE 10-3
Covalent, metallic, and ionic radii compared.
Atomic radii are represented by colored arrows. The covalent radius is based on the diatomic molecule $Na_2(g)$, found only in *gaseous* sodium. The metallic radius is based on adjacent atoms in solid sodium, $Na(s)$. To obtain an ionic radius, the internuclear distance must be apportioned between a cation and an anion. This requires a series of comparisons. For example, starting with a radius of 140 pm for O^{2-}, the radius of Mg^{2+} can be obtained from the internuclear distance in MgO, the radius of Cl^- from the internuclear distance in $MgCl_2$, and the radius of Na^+ from the internuclear distance in NaCl.

▶ A way to think about the imperfect screening by core electrons is that outer shell electrons, especially *s* electrons, have some probability of being near the nucleus. They penetrate the inner core and experience a greater attractive force than otherwise expected.

Figure 10-3 illustrates the definitions of covalent, ionic, and metallic radii by comparing these three radii for sodium. Figure 10-4 is a plot of atomic radius against atomic number for a large number of elements. In such a plot it is customary to use metallic radii for metals and covalent radii for nonmetals; this is what we have done. Figure 10-4 suggests certain trends in atomic radii, for example, large radii for Group 1A, decreasing across the periods to smaller radii for Group 7A. We consider next the three most important trends among atomic radii in relation to the periodic table.

1. Variation of Atomic Radii Within a Group of the Periodic Table. The higher the principal quantum number of an electronic shell, the farther from the nucleus will significant electron charge density still exist. We should expect that the more electronic shells in an atom, the larger the atom. This idea works for the group members of lower atomic numbers, where the increase in radius from one period to the next is large (as from Li to Na to K in Group 1A). At higher atomic numbers the increase in radius is smaller (as from K to Rb to Cs in Group 1A). In these elements of higher atomic number, outer shell electrons are held somewhat more tightly than otherwise expected because inner shell electrons in *d* and *f* subshells are less effective than *s* and *p* electrons in screening outer shell electrons from the nucleus (recall the discussion in Section 9-9). Nevertheless, in general,

> The more electronic shells in an atom, the larger the atom. Atomic radius increases from top to bottom through a group of elements.

2. Variation of Atomic Radii Within a Period of the Periodic Table. From Figure 10-4 we see that in general atomic radii decrease from left to right across a period. A careful look at Figure 10-4 reveals that this decrease in atomic radii does not apply to the transition elements. Let us look first at the general trend of decreasing radii and then at what is special about the transition elements.

Consider the hypothetical process of building up each atom in the third period from the atom preceding it, beginning with sodium. In this process the number of inner shell, or core, electrons is fixed at ten, in the configuration $1s^2 2s^2 2p^6$. As a first approximation, let us *assume* that the core electrons completely cancel an equivalent charge on the nucleus. In this way the core electrons *shield,* or *screen,* the outer shell electrons from the full attractive force of the nucleus. Let us also assume that the outer shell electrons do not screen one another. And, let us define an **effective nuclear charge, Z_{eff},** as the true nuclear charge minus the charge that is screened out by electrons.

In sodium, with atomic number 11, the ten core electrons would screen out 10 units of nuclear charge, leaving an effective nuclear charge of $11 - 10 = +1$. In magnesium, with atomic number 12, Z_{eff} would be +2 (that is, $12 - 10$). In aluminum Z_{eff} would be +3, and so on across the period. As Z_{eff} increases, the core and outer shell electrons are attracted more strongly to the nucleus. This results in an overall contraction of the size of the atom.

Actually, neither of the assumptions is correct. In sodium, the ten core electrons cancel only about 9 units of nuclear charge. Z_{eff} is more nearly +2 than +1. Also, outer shell electrons do screen one another somewhat. Each outer shell electron is about one-third effective in screening the other outer shell electrons. Thus, the Z_{eff} that each of the two outer shell electrons in magnesium ex-

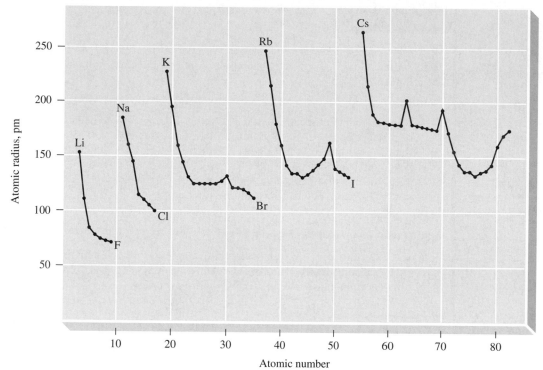

FIGURE 10-4 **Atomic radii.**
The values plotted are metallic radii for metals and covalent radii for nonmetals. Data for the noble gases are not included because of the difficulty of measuring covalent radii for these elements (only Kr and Xe compounds are known). (The explanations usually given for the several small peaks in the middle of some periods are beyond the scope of this discussion.)

periences is about $12 - 9 - \frac{1}{3} = +2\frac{2}{3}$ (Figure 10-5). In any case, Z_{eff} increases left to right through a period. Whether we deal with Z_{eff} or just the nuclear charge Z, the result is that

The atomic radius decreases from left to right through a period of elements.

3. Variation of Atomic Radii Within a Transition Series. With the transition elements the situation is a little different. In Figure 10-4, it is apparent that the atomic radii of transition elements tend to be about the same across a period with a few unusual peaks. It is beyond the scope of this text to explain the exceptions; however, the general trend is not difficult to understand. In a series of transition elements, additional electrons go into an *inner* electron shell, where they participate in shielding outer shell electrons from the nucleus. At the same time the number of electrons in the *outer* shell tends to remain constant. Thus, the outer shell electrons experience a roughly comparable force of attraction to the nucleus throughout a transition series. Consider Fe, Co, and Ni. Fe has 26 protons in the nucleus and 24 inner shell electrons. In Co ($Z = 27$) there are 25 inner shell electrons, and in Ni ($Z = 28$) there are 26. In each case the two outer shell electrons are under the influence of about the same net charge (about $+2$). Thus, atomic radii do not change very much for this series of three elements.

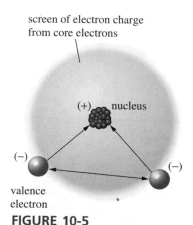

FIGURE 10-5
The shielding effect and effective nuclear charge, Z_{eff}.
Two valence electrons (blue) are attracted to the nucleus of a Mg atom. The atom's $+12$ nuclear charge is screened by the 10 core electrons (gray), but not perfectly. The valence electrons also screen each other somewhat. The result is an *effective* nuclear charge, Z_{eff}, closer to $+3$ than to $+2$.

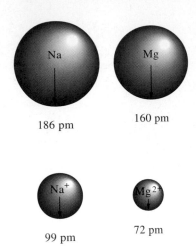

FIGURE 10-6
A comparison of atomic and ionic sizes.
Metallic radii are shown for Na and Mg and ionic radii for Na$^+$ and Mg^{2+}.

186 pm   160 pm

99 pm   72 pm

EXAMPLE 10-1

Applying Generalizations Relating Atomic Sizes to Positions in the Periodic Table. Refer only to the periodic table on the inside front cover and determine which is the largest atom: Sc, Ba, or Se.

Solution

Sc and Se are both in the fourth period and we should expect Sc to be larger than Se because it is much closer to the beginning of the period (to the left). Ba is in the sixth period and so has more electronic shells than either Sc or Se. Furthermore, it lies even closer to the left side of the table (Group 2A) than does Sc (Group 3B). We can say with confidence that the Ba atom should be largest of the three. [The actual atomic radii are Se (117 pm), Sc (161 pm), and Ba (217 pm).]

Practice Example A: Use the periodic table on the inside front cover to predict which is the smallest atom: As, I, or S.

Practice Example B: Which of the following atoms do you think is closest in size to the Na atom: Br, Ca, K, or Al? Explain your reasoning, and do not use any tabulated data from the chapter in reaching your conclusion.

Ionic Radius

When a metal atom loses one or more electrons to form a positive ion, there is an excess of nuclear charge over the number of electrons in the resulting cation. The nucleus draws the electrons in closer, and as a consequence

Cations are *smaller* than the atoms from which they are formed.

Figure 10-6 compares four species: the atoms Na and Mg and the ions Na$^+$ and Mg^{2+}. As we should expect, the Mg atom is smaller than the Na atom, and the cations are smaller than the corresponding atoms. Na$^+$ and Mg^{2+} are **isoelectronic**—they have equal numbers of electrons (10) in identical configurations, $1s^2 2s^2 2p^6$. Mg^{2+} is smaller than Na$^+$ because its nuclear charge is larger (+12, compared with +11 for Na).

For isoelectronic cations, the more positive the charge, the smaller the ionic radius.

When a nonmetal atom gains one or more electrons to form a negative ion (anion), the nuclear charge remains constant, but Z_{eff} is reduced because of the additional electron(s). The electrons are not held as tightly. Repulsions among the electrons increase. The electrons spread out more, and the size of the atom increases, as suggested in Figure 10-7.

Anions are *larger* than the atoms from which they are formed. For isoelectronic anions, the more negative the charge, the larger the ionic radius.

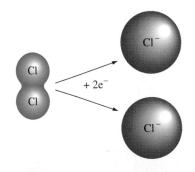

covalent radius   ionic radius

99 pm   181 pm

FIGURE 10-7
Covalent and anionic radii compared.
The two Cl atoms in a Cl$_2$ molecule gain one electron each to form two Cl$^-$ ions.

EXAMPLE 10-2

Comparing the Sizes of Cations, Anions, and Neutral Atoms. Refer only to the periodic table on the inside front cover and arrange the following species in order of increasing size: Ar, K$^+$, Cl$^-$, S^{2-}, Ca^{2+}.

Solution

The key lies in recognizing that the five species are *isoelectronic,* having the electron configuration of Ar, $1s^2 2s^2 2p^6 3s^2 3p^6$. For isoelectronic cations, the higher the charge, the smaller the ion. This means that Ca^{2+} is smaller than K^+. Because K^+ has a higher nuclear charge than Ar ($Z = 19$ compared with $Z = 18$), it is smaller than Ar. And Ar is smaller than Cl^-, again because of a higher nuclear charge ($Z = 18$ compared with $Z = 17$). For isoelectronic anions, the higher the charge, the larger the ion. S^{2-} is larger than Cl^-. The order of increasing size is

$$Ca^{2+} < K^+ < Ar < Cl^- < S^{2-}$$

Practice Example A: Refer only to the periodic table on the inside front cover, and arrange the following species in order of increasing size: Ti^{2+}, V^{3+}, Ca^{2+}, Br^-, and Sr^{2+}.

Practice Example B: Refer only to the periodic table on the inside front cover, and determine which species is in the *middle* position when the following five are ranked according to size: the atoms N, Cs, and As and the ions Mg^{2+} and Br^-.

▶ We can summarize the generalizations about isoelectronic atoms and ions into a single statement: among *isoelectronic* species, the greater the atomic number, the smaller the size.

Knowledge of atomic and ionic radii can be used to vary certain physical properties. One example concerns the strengthening of glass. Normal window glass contains Na^+ and Ca^{2+} ions. The glass is brittle and shatters easily when struck a hard blow. One way to strengthen the glass is to replace the Na^+ ions on the surface with K^+ ions. The K^+ ions are larger and fill up the surface sites, leaving less opportunity for cracking than is the case with the smaller Na^+ ions. The result is a shatter-resistant glass.

Another example is the striking result when Cr^{3+} ions replace about 1% of the Al^{3+} ions in aluminum oxide, Al_2O_3. This substitution is possible because Cr^{3+} ions are only slightly larger (by 9 pm) than Al^{3+} ions. Pure aluminum oxide is colorless, but with this small amount of chromium(III) ion, it is a beautiful red color and is the gem known as a ruby. Rubies and other gemstones can be made artificially and are used as jewelry and in devices such as lasers. The color of the ruby is further discussed in Chapter 25.

Figure 10-8, arranged in the format of the periodic table, shows relative sizes of typical atoms and ions. It summarizes the generalizations described in this section.

10-5   Ionization Energy

In discussing metals we talked about metal atoms losing electrons and thereby altering their electron configurations. But atoms do not eject electrons spontaneously. Electrons are attracted to the positive charge on the nucleus of an atom and energy is needed to overcome that attraction. The more easily its electrons are lost, the more metallic we consider an atom to be. The **ionization energy,** **I,** is the quantity of energy a *gaseous* atom must absorb so that an electron is stripped from the atom. The electron lost is the one most loosely held.

Ionization energies are measured through experiments in which gaseous atoms at low pressures are bombarded with beams of electrons (cathode rays). Here are two typical values.

$$Mg(g) \longrightarrow Mg^+(g) + e^- \qquad I_1 = 738 \text{ kJ/mol}$$

$$Mg^+(g) \longrightarrow Mg^{2+}(g) + e^- \qquad I_2 = 1451 \text{ kJ/mol}$$

FIGURE 10-8 A comparison of some atomic and ionic radii.
The values given, in picometers (pm), are metallic radii for metals, single covalent radii for nonmetals, and ionic radii for the ions indicated.

Element	Radius	Ion	Ionic radius
Li	152	Li⁺	59
Be	111	Be²⁺	27
B	88		
C	77		
N	75	N³⁻	171
O	73	O²⁻	140
F	71	F⁻	133
Na	186	Na⁺	99
Mg	160	Mg²⁺	72
Al	143	Al³⁺	53
Si	117		
P	110	P³⁻	212
S	104	S²⁻	184
Cl	99	Cl⁻	181
K	227	K⁺	138
Ca	197	Ca²⁺	100
Sc	161	Sc³⁺	75
Ti	145	Ti²⁺	86
V	132	V²⁺ / V³⁺	79 / 64
Cr	125	Cr²⁺ / Cr³⁺	82 / 62
Mn	124	Mn²⁺	83
Fe	124	Fe²⁺ / Fe³⁺	77 / 65
Co	125	Co²⁺ / Co³⁺	75 / 61
Ni	125	Ni²⁺	70
Cu	128	Cu⁺ / Cu²⁺	96 / 73
Zn	133	Zn²⁺	75
Ga	122	Ga³⁺	62
Ge	122		
As	121		
Se	117	Se²⁻	198
Br	114	Br⁻	196
Rb	248	Rb⁺	149
Sr	215	Sr²⁺	113
Ag	144	Ag⁺	115
Cd	149	Cd²⁺	95
In	163	In³⁺	79
Sn	141	Sn²⁺	93
Sb	140	Sb³⁺	76
Te	137	Te²⁻	221
I	133	I⁻	220

The symbol I_1 stands for the *first* ionization energy—the energy required to strip one electron from a *neutral* gaseous atom.* I_2 is the *second* ionization energy—the energy to strip an electron from a gaseous ion with a charge of 1+. Further ionization energies are I_3, I_4, ... Invariably we find that each succeeding ionization energy is larger than the preceding ones. In the case of magnesium, for example, in the second ionization the electron, once freed, has to move away from an ion with a charge of 2+ (Mg^{2+}). More energy must be invested than for a freed electron to move away from an ion with a charge of 1+ (Mg^+). This is a direct consequence of Coulomb's law, which states that the force of attraction between oppositely charged particles is directly proportional to the magnitudes of the charges.

*Ionization energies are sometimes expressed in the unit electronvolt (eV). One electronvolt is the energy acquired by an electron as it falls through an electric potential difference of 1 volt. It is a very small energy unit, especially suited to describing processes involving individual atoms. When ionization is based on a *mole* of atoms, kJ/mol is the preferred unit. 1 eV/atom = 96.49 kJ/mol. Sometimes the term *ionization potential* is used instead of ionization energy. Further, the quantities I_1, I_2, . . . , may be replaced by enthalpy changes, ΔH_1, ΔH_2, ...

FIGURE 10-9 **First ionization energies as a function of atomic number.**
Because their electron configurations are so stable, more energy is required to ionize noble gas atoms than to ionize atoms of the elements immediately preceding or following them. The maxima on the graph come at the atomic numbers of the noble gases. The alkali metals are the most easily ionized of all groups. The minima in the graph come at their atomic numbers.

First ionization energies (I_1) for most of the elements are plotted in Figure 10-9. To explain how the values in this figure vary with atomic number, we need to consider several factors. First, in general, the farther an electron is from the nucleus, the more easily it can be extracted.

Ionization energies *decrease* as atomic radii increase.

▶ This generalization works well for main group elements but less so for transition metal elements, where there are several exceptions.

By this measure, atoms lose electrons more easily (become *more* metallic) as we move from top to bottom in a group of the periodic table. The decreases in ionization energy and the parallel increases in atomic radii are outlined in Table 10.3 for Group 1A.

Table 10.4 lists ionization energies for the third-period elements. With minor exceptions the trend in moving across a period (follow the colored stripe) is that atomic radii decrease, ionization energies increase, and the elements become *less* metallic, or *more* nonmetallic, in character. Table 10.4 lists stepwise ionization energies ($I_1, I_2, \ldots$). Note particularly the large breaks that occur along the zigzag diagonal line. Consider magnesium as an example. To remove a *third* electron, as measured by I_3, requires breaking into the especially stable noble gas subshell electron configuration $2s^2 2p^6$. I_3 is *much* larger than I_2—so much

TABLE 10.3 Atomic Radii and First Ionization Energies of the Alkali Metal (Group 1A) Elements

	Atomic Radius, pm	Ionization Energy (I_1), kJ/mol
Li	152	520.2
Na	186	495.8
K	227	418.8
Rb	248	403.0
Cs	265	375.7

larger that Mg^{3+} cannot be produced in ordinary chemical processes. Similarly, we do not expect to find the ions Na^{2+} or Al^{4+}.

Now let us turn to the obvious exceptions to the regular trend in I_1 values for the third-period elements and ask: Why is I_1 of Al smaller than that of Mg and I_1 of S smaller than that of P?

We expect I_1 of Al to be *larger* than for Mg. The reversal occurs because of the particular electrons lost. Mg loses a $3s$ electron, Al, a $3p$ electron. *More* energy is required to strip an electron from the lower energy $3s$ orbital in Mg ($[\uparrow\downarrow]_{3s}[\ ][\ ][\ ]_{3p}$) than from the $3p$ orbital in Al ($[\uparrow\downarrow]_{3s}[\uparrow][\ ][\ ]_{3p}$). (Recall the orbital energy diagrams in Figure 9-26.) I_1 for S is slightly lower than for P for a different reason. In this case, all the $3p$ orbitals are at the same energy level. However, we can think of repulsion between the pair of electrons in the filled $3p$ orbital of an S atom ($[\uparrow\downarrow][\uparrow][\uparrow]_{3p}$) as making it easier to remove one of those electrons than an unpaired electron from a half-filled $3p$ orbital of a P atom ($[\uparrow][\uparrow][\uparrow]_{3p}$).

EXAMPLE 10-3

Relating Ionization Energies and Atomic Radii. Refer to the periodic table on the inside front cover, and arrange the following in order of *increasing* first ionization energy, I_1: As, Sn, Br, Sr.

Solution

If we arrange these four atoms according to *decreasing* radius, we will automatically have arranged them according to *increasing* ionization energy. The largest atoms are to the left and the bottom of the periodic table. Of the four atoms, the one that best

TABLE 10.4 Ionization Energies of the Third-Period Elements (in kJ/mol)

	Na	Mg	Al	Si	P	S	Cl	Ar
I_1	495.8	737.7	577.6	786.5	1012	999.6	1251.1	1520.5
I_2	4562	1451	1817	1577	1903	2251	2297	2666
I_3		7733	2745	3232	2912	3361	3822	3931
I_4			11580	4356	4957	4564	5158	5771
I_5				16090	6274	7013	6542	7238
I_6					21270	8496	9362	8781
I_7						27110	11020	12000

fits the large-atom category is *Sr*. The smallest atoms are to the right and toward the top of the periodic table. Although none of the four atoms is particularly close to the top of the table, *Br* is the best fit for this category. This fixes the two extremes: Sr with the lowest ionization energy and Br with the highest. In comparing As and Sn, we see that Sn is *below* and *to the left* of As in the periodic table. It should have a lower ionization energy than As. The order of *increasing* first ionization energies is Sr < Sn < As < Br.

Practice Example A: Refer to the periodic table on the inside front cover, and arrange the following in order of increasing first ionization energy, I_1: Cl, K, Mg, S.

Practice Example B: Refer to the periodic table on the inside front cover, and determine which element is in the *middle* position when the following five elements are arranged according to first ionization energy, I_1: Rb, As, Sb, Br, Sr.

10-6   Electron Affinity

Ionization energy concerns the *loss* of electrons. **Electron affinity, *EA*,** is a measure of the energy change that occurs when a *gaseous* atom *gains* an electron. For example,

$$F(g) + e^- \longrightarrow F^-(g) \qquad EA = -328 \text{ kJ/mol}$$

When an F atom gains an electron, energy is given off. The process is *exothermic,* and, following the thermochemical conventions we established in Chapter 7, the electron affinity is a *negative* quantity.*

Why should a neutral fluorine atom so readily gain an electron? A plausible explanation is that when a free electron approaches an F atom from an "infinite" distance away, the electron "sees" a center of positive charge—the atomic nucleus—to which it is attracted. This attraction is offset to some extent by the repulsive effect of other electrons in the atom. But, so long as the attractive force on the additional electron exceeds the repulsive force, the electron is gained and energy is given off. In becoming F^-, a fluorine atom acquires the very stable electron configuration of the noble gas neon (Ne). That is,

$$F\ (1s^2 2s^2 2p^5) + e^- \longrightarrow F^-\ (1s^2 2s^2 2p^6)$$

Even metal atoms can form negative ions *in the gaseous state,* where they will only rarely contact one another. This is the situation for gaseous Li atoms, for example, where the added electron enters the half-filled 2*s* orbital.

$$\text{Li}(g) + e^- \longrightarrow \text{Li}^-(g) \qquad EA = -59.6 \text{ kJ/mol}$$
$$(1s^2 2s^1) \qquad\qquad (1s^2 2s^2)$$

For some atoms, there is no tendency to gain an electron. This is the case with the noble gases, where the added electron would have to enter the empty *s* orbital of the next electronic shell; the Groups 2A and 2B elements, where the electron would have to enter the *p* subshell of the valence shell; and a few other elements, such as Mn, where the electron would have to enter either the *p* subshell of the valence shell or a half-filled 3*d* subshell.

---

*We have defined electron affinity to reflect the tendency for a neutral atom to gain an electron. An alternative definition refers to the energy change in the process: $X^-(g) \longrightarrow X(g) + e^-$, that is, reflecting the tendency of an anion to lose an electron. This alternative definition leads to the opposite signs for *EA* values from those written in this text. You should be prepared to see electron affinities expressed in both ways in the chemical literature.

FIGURE 10-10

Electron affinities of representative elements.
Values are in kilojoules per mole for the process $X(g) + e^- \rightarrow X^-(g)$.

1A							8A
H −72.8							He −−
	2A	3A	4A	5A	6A	7A	
Li −59.6	Be −−	B −26.7	C −153.9	N −7	O −141.0	F −328.0	Ne −−
Na −52.9	Mg −−	Al −42.5	Si −133.6	P −72	S −200.4	Cl −349.0	Ar −−
K −48.4	Ca −−	Ga −28.9	Ge −119.0	As −78	Se −195.0	Br −324.6	Kr −−
Rb −46.9	Sr −−	In −28.9	Sn −107.3	Sb −103.2	Te −190.2	I −295.2	Xe −−
Cs −45.5	Ba −−	Tl −19.2	Pb −35.1	Bi −91.2	Po −186	At −270	Rn −−

In considering the gain of a *second* electron by a nonmetal atom we encounter *positive* electron affinities. Here the electron to be added is approaching not a neutral atom but a *negative ion*. A strong repulsion is felt, and the energy of the system increases. Thus, for an element like oxygen the first electron affinity is negative and the second is positive.

$$O(g) + e^- \longrightarrow O^-(g) \qquad EA_1 = -141.0 \text{ kJ/mol}$$

$$O^-(g) + e^- \longrightarrow O^{2-}(g) \qquad EA_2 = +744 \text{ kJ/mol}$$

The high positive value of EA_2 makes the formation of *gaseous* O^{2-} seem very unlikely. The ion O^{2-} can exist, however, in ionic compounds such as MgO(s), where its formation is accompanied by other energetically favorable processes.

Some representative electron affinities are listed in Figure 10-10. Generalizations about electron affinities are more difficult to make than about ionization energies. We can see that the smaller atoms to the right of the periodic table (for example, Group 7A) tend to have large negative electron affinities.* Electron affinities tend to become less negative in progressing toward the bottom of a group, but the second period group members (for example, N, O, and F) do not conform to this trend. It is likely that for these small atoms an incoming electron encounters strong repulsions from other electrons in the atom and is thereby not as tightly bound as we might otherwise expect.

10-7   Magnetic Properties

Another property of atoms and ions, their behavior in a magnetic field, is also helpful in establishing electron configurations. A spinning electron is an electric charge in motion. It induces a magnetic field (recall the discussion on page

---

*It is somewhat awkward to speak of "larger" and "smaller" with the term *electron affinity*. A strong tendency to gain an electron, which implies a high "affinity" for an electron, as with F and Cl, is reflected through a *low* value of EA—a large *negative* value.

(a) No magnetic field

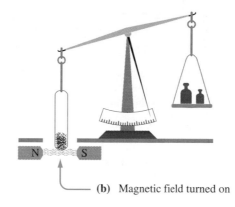
(b) Magnetic field turned on

FIGURE 10-11

Paramagnetism illustrated.
(a) A sample is weighed in the absence of a magnetic field. (b) When the field is turned on, the balanced condition is upset. The sample gains weight because it is now subjected to two attractive forces: the force of gravity *and* the force of interaction of the external magnetic field and the unpaired electrons.

294). In a **diamagnetic** atom or ion all electrons are paired, and these individual magnetic effects cancel out. A diamagnetic species is weakly repelled by a magnetic field. A **paramagnetic** atom or ion has *unpaired* electrons, and the individual magnetic effects do not cancel out. The unpaired electrons induce a magnetic field that causes the atom or ion to be attracted into an external magnetic field. The more unpaired electrons present, the stronger this attraction.

A straightforward way to measure magnetic properties is to weigh a substance "in" and "out" of a magnetic field, as illustrated in Figure 10-11. The mass of the substance is the same in either case. If the substance is diamagnetic, however, it is slightly repelled by a magnetic field and *weighs* less in the field. On the other hand, if the substance is paramagnetic, it *weighs* more.

Manganese has a paramagnetism corresponding to five unpaired electrons, which is consistent with the electron configuration

		3d					4s
Mn: [Ar]	↑	↑	↑	↑	↑		↑↓

When a manganese atom loses *two* electrons it becomes the ion Mn^{2+}. Mn^{2+} is paramagnetic, and the strength of its paramagnetism corresponds to *five* unpaired electrons.

		3d					4s
Mn^{2+}: [Ar]	↑	↑	↑	↑	↑		

When a third electron is lost to produce Mn^{3+}, we find that the ion has a paramagnetism corresponding to *four* unpaired electrons. The third electron lost is one of the unpaired 3d electrons.

		3d				4s
Mn^{3+}: [Ar]	↑	↑	↑	↑		

▶ The attraction of iron objects into a magnetic field is far stronger than can be accounted for by unpaired electrons alone. This special property of iron and a few other metals and alloys, called *ferromagnetism,* is discussed in Chapter 24.

EXAMPLE 10-4

Determining the Magnetic Properties of an Atom or Ion. Which of the following would you expect to be diamagnetic and which paramagnetic?
a. the Na atom   **b.** the Mg atom   **c.** the Cl^- ion   **d.** the Ag atom

Solution

a. Paramagnetic. The Na atom has a single 3s electron outside the Ne core. This electron is unpaired.

b. Diamagnetic. The Mg atom has *two* 3*s* electrons outside the Ne core. They must be paired, as are all the other electrons.

c. Diamagnetic. Cl^- is isoelectronic with Ar, and Ar has all electrons paired ($1s^2 2s^2 2p^6 3s^2 3p^6$).

d. Paramagnetic. We do not need to work out the exact electron configuration of Ag. Because the atom has 47 electrons—an odd number—at least one of the electrons must be unpaired (recall the Stern–Gerlach experiment, page 293).

Practice Example A: Identify which of the following are paramagnetic and which are diamagnetic: Zn, Cl, K^+, O^{2-}, and Al.

Practice Example B: Which has the greater number of unpaired electrons, Cr^{2+} or Cr^{3+}? Explain.

10-8   Periodic Properties of the Elements

As we indicated at the beginning of the chapter, the periodic law and periodic table can be used in making predictions about the properties of elements and compounds. In this concluding section of the chapter, we will make comparisons and predictions dealing with three types of properties: atomic, physical, and chemical.

Atomic Properties

Earlier in the chapter we examined some atomic properties (atomic radius, ionization energy, electron affinity) and described how they vary among the groups and periods of the periodic table. We summarize these trends in relation to the periodic table in Figure 10-12. Trends are generally easy to apply within a group: the atomic radius of Sr is greater than that of Mg; both elements are in Group 2A. Usually, there is no difficulty in applying trends within a period either: the first ionization energy of P is greater than that of Mg; both elements are in the third period. When the elements we compare are not within the same group or period, we can encounter difficulties. We have no problem seeing that the atomic radius of Sr is greater than that of P. Sr is farther down in its group of the periodic table and much farther to the left in its period than is P. Each of these directions is that of increasing atomic radius. On the other hand, we cannot eas-

FIGURE 10-12
Atomic properties and the periodic table—a summary.
Atomic radius refers to metallic radius for metals and covalent radius for nonmetals. Ionization energies refer to first ionization energy. For electron affinity, the direction is that in which values become *more negative*. Metallic character relates generally to the ability to lose electrons, and nonmetallic character, to gain electrons.

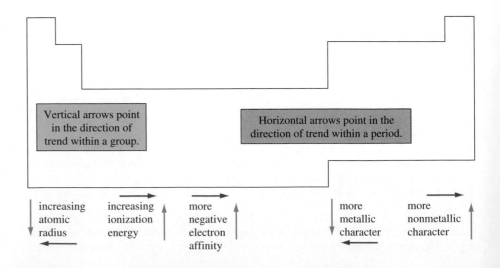

Vertical arrows point in the direction of trend within a group.

Horizontal arrows point in the direction of trend within a period.

increasing atomic radius   increasing ionization energy   more negative electron affinity   more metallic character   more nonmetallic character ily predict whether Mg or I has the larger atomic radius. The position of Mg to the left in its period suggests that Mg should have the larger radius, but the position of I toward the bottom of its group argues for I. Despite this limitation, you should find Figure 10-12 helpful in most cases.

Physical Properties

We will attempt to explain and predict several physical properties. Out of this discussion should come the realization that in some cases predictions are fairly reliable, but in others they are not.

Variation of Physical Properties Within a Group.   Table 10.5 lists some properties of three of the halogens (Group 7A). The table has two blank spaces for bromine. We fill in these blanks in Example 10-5 by making an assumption that works often enough to make it useful: *The value of a property often changes uniformly from top to bottom in a group of elements in the periodic table.*

First, let us make some predictions about fluorine, the halogen not listed in Table 10.5. Its closest neighbor in Group 7A is chlorine, which has a boiling point of 239 K ($-34\ °C$); chlorine is a *gas* at room temperature (about 298 K). The other halogens are *liquid* bromine and *solid* iodine (Figure 10-13). We should expect flourine to have a lower melting point and lower boiling point than chlorine, and to also be a *gas* at room temperature. (Observed values for F_2: mp = 53 K; bp = 85 K.)

EXAMPLE 10-5

Using the Periodic Table to Estimate Physical Properties. Use data from Table 10.5 to estimate the boiling point of bromine.

Solution

The atomic number of bromine (35) is intermediate to the atomic numbers of chlorine (17) and iodine (53). Its atomic mass (79.90 u) is also about intermediate to those of chlorine and iodine. (The average of the atomic masses of Cl and I is 81.18 u.) It is reasonable to expect that the boiling point of liquid bromine might also be intermediate to the boiling points of chlorine and iodine.

$$\text{bp } Br_2 = \frac{239\ K + 458\ K}{2} = 349\ K$$

The observed boiling point is 332 K.

Practice Example A: Estimate the melting point of bromine.

Practice Example B: Estimate the boiling point of astatine, At.

TABLE 10.5   Some Properties of Three Halogen (Group 7A) Elements

	Atomic Number	Atomic Mass, u	Molecular Form	Melting Point, K	Boiling Point, K
Cl	17	35.45	Cl_2	172	239
Br	35	79.90	Br_2	?	?
I	53	126.90	I_2	387	458

FIGURE 10-13
Three halogen elements.
Chlorine is a yellow-green gas. Bromine is a dark red liquid. Iodine is a grayish black solid.

TABLE 10.6 Melting Points of Two Series of Compounds		
	Molecular Mass, u	**Melting Point, °C**
CF_4	88.0	−183.7
CCl_4	153.8	−22.9
CBr_4	331.6	90.1
CI_4	519.6	171
HF	20.0	−83.6
HCl	36.5	−114.2
HBr	80.9	−86.8
HI	127.9	−50.8

The generalization that a property varies uniformly within a group of the periodic table can work for compounds as well as for elements. Table 10.6 lists the melting points of two sets of compounds, binary carbon-halogen compounds and the *hydrogen halides,* HX (where X = F, Cl, Br, or I). We see that the melting points increase fairly uniformly with increasing molecular mass for the carbon-halogen compounds. This trend between melting point (and boiling point) and molecular mass can be explained in terms of intermolecular forces, as we shall see in Chapter 13. Based on the melting points of HCl, HBr, and HI, we should expect the melting point of HF to be about −145 °C, but the observed value is −83.6 °C. Some factor other than molecular mass must be the determining factor here. In Chapter 13 we will find that in HF there is a special intermolecular force of attraction that is missing or unimportant in the other compounds in Table 10.6. In conclusion, then, where the same factor underlies a property of all the members of a group, we can generally make reasonably good predictions of group trends. Where different factors are at play we cannot.

Variation of Physical Properties Across a Period. A few properties vary regularly across a period. The ability to conduct heat and the ability to conduct electricity are two that do. Thus, among the third-period elements, the metals Na, Mg, and Al have good thermal and electrical conductivities. The metalloid Si is only a fair conductor of heat and electricity. The nonmetals P, S, Cl, and Ar have poor thermal and electrical conductivities.

In some cases, though, the trend in a property reverses direction in the period (similar to our experience above with the trend in melting points of the hydrogen halides reversing direction within a group). Consider, for example, the melting points of the third-period elements, shown in a bar graph in Figure 10-14. Melting involves destruction of the orderly arrangement of atoms or molecules found in a crystalline solid. The amount of thermal energy needed for melting to occur, and hence the melting point temperature, depends on the strength of the attractive forces between atoms or molecules in the solid. For the metals Na, Mg, and Al these forces are *metallic bonds,* which, roughly speaking, become stronger as the number of electrons available to participate in the bonding increases. Sodium, therefore, has the lowest melting point (371 K) of the third-period metals. With silicon the forces between atoms are strong *covalent bonds* extending throughout the crystalline solid. Silicon has the highest melting point (1683 K) of the third-period elements. Phosphorus, sulfur, and chlorine exist as discrete molecules (P_4, S_8, and Cl_2). The bonds between atoms within molecules are strong, but *intermolecular forces,* the forces between mol-

▶ Metallic bonds are described in Section 12-7; covalent bonding in substances like silicon is discussed in Section 13-7; and the topic of intermolecular forces is examined throughout Chapter 13.

FIGURE 10-14
Melting points of the third-period elements.
The trend outlined in this bar graph is discussed in the text.

ecules, become progressively weaker. The melting points decrease. Argon atoms do not form molecules, and the forces between Ar atoms in solid argon are especially weak. Argon's melting point is lowest for the entire period (84 K). The property of hardness also depends on forces between atoms and molecules in a solid. So the hardness of the solid third-period elements vary in much the same way as do melting points. Silicon has the greatest hardness.

Chemical Properties

As we did in our brief survey of the periodic nature of physical properties, in examining chemical properties we will consider some variations within groups of elements and one variation across a period.

Reducing Abilities of Group 1A and 2A Metals. We learned in Chapter 5 that a reducing agent makes possible a reduction half-reaction. The reducing agent itself, by losing electrons, is oxidized. In the following reactions M, a Group 1A or 2A metal, is the reducing agent and H_2O is the substance that is reduced.

$$2\ M(s) + 2\ H_2O(l) \longrightarrow 2\ M^+(aq) + 2\ OH^-(aq) + H_2(g)$$

(M = Group 1A metal)

$$M(s) + 2\ H_2O(l) \longrightarrow M^{2+}(aq) + 2\ OH^-(aq) + H_2(g)$$

(M = Ca, Sr, Ba, or Ra)

At a first guess, we might think that the lower the energy requirement for extracting electrons—the lower the ionization energy—the better the metal is as a reducing agent and the more vigorous its reaction with water. Potassium, for instance, has a lower ionization energy ($I_1 = 419$ kJ/mol) than does the next member of the fourth period, calcium ($I_1 = 590; I_2 = 1145$ kJ/mol). Our expectation is that potassium should react more vigorously with water than calcium would. This indeed is the case (Figure 10-15). Mg and Be do not react with cold water as do the other alkaline earth metals. We might explain this in terms of the higher ionization energies for those two metals (Mg: $I_1 = 738, I_2 = 1451$ kJ/mol; Be: $I_1 = 900, I_2 = 1757$ kJ/mol).

Attributing these Group 1A and 2A metal reactivities just to ionization energies is an oversimplification, however. As long as the differences in ionization

(b)

FIGURE 10-15
The reactions of potassium and of calcium with water: a comparison.
(a) Potassium, a Group 1A metal, reacts so rapidly that the hydrogen evolved bursts into flame. Notice that the metal is less dense than water. (b) Calcium, a Group 2A metal, reacts more slowly than does potassium. Also, calcium is denser than water. The pink color of the acid–base indicator phenolphthalein signals the buildup of OH^- ions.

FIGURE 10-16
Reaction of sodium metal and chlorine gas.
The contents of the flask glow at the high temperatures reached in this exothermic reaction between the active metal Na(s) and active nonmetal $Cl_2(g)$. The product is the ionic solid NaCl(s).

energies are very large, we may get by with considering only this factor. Where differences in ionization energies are smaller, though, other factors must be considered when making comparisons.

Oxidizing Abilities of the Halogen Elements (Group 7A). An oxidizing agent gains the electrons that are lost in an oxidation half-reaction. The oxidizing agent, by gaining electrons, is itself reduced. Electron affinity is the atomic property introduced in this chapter that is related to the gain of electrons. We might expect an atom with a strong tendency to gain electrons (a large *negative* electron affinity) to take electrons away from atoms with low ionization energies—metals. In these terms, it is understandable that active metals form ionic compounds with active nonmetals. If M is a Group 1A metal and X a Group 7A nonmetal (halogen), this exchange of an electron leads to the formation of M^+ and X^- ions. In some cases the reaction is especially vigorous (Figure 10-16).

$$2\,M + X_2 \longrightarrow 2\,MX \quad [\text{e.g., } 2\,Na(s) + Cl_2(g) \longrightarrow 2\,NaCl(s)]$$

Another interesting oxidation–reduction reaction involving the halogens is a *displacement* reaction. Two halogens, one in molecular form and the other in ionic form, exchange places, as in this reaction (Figure 10-17).

$$Cl_2(g) + 2\,I^-(aq) \longrightarrow I_2(aq) + 2\,Cl^-(aq)$$

Think of this reaction as a competition between Cl atoms (actually Cl_2 molecules) and I atoms for the extra electron that the I atoms originally have (as I^-). The Cl atoms win out; they have a more negative electron affinity. With this line of reasoning can you see why no reaction occurs for this combination?

$$Br_2(l) + Cl^-(aq) \longrightarrow \text{no reaction}$$

Predictions such as those in the preceding paragraph work well for the halogens Cl_2, Br_2, and I_2, but not for F_2. We cannot account for the *observed* fact that F_2 is the strongest oxidizing agent among all chemical substances just by considering electron affinities.

Acid–Base Nature of Element Oxides. Some metal oxides, such as Li_2O, react with water to produce basic metal hydroxides. These metal oxides are called *basic* oxides (base anhydrides).

$$\underset{\substack{\text{a basic} \\ \text{oxide}}}{Li_2O(s)} + H_2O \longrightarrow \underset{\text{lithium hydroxide}}{2\,Li^+(aq) + 2\,OH^-(aq)}$$

Some nonmetal oxides, like CO_2, produce acidic solutions in water. These nonmetal oxides are *acidic* oxides (acid anhydrides).

$$\underset{\substack{\text{an acidic} \\ \text{oxide}}}{CO_2(g)} + H_2O \longrightarrow \underset{\text{carbonic acid}}{H_2CO_3(aq)}$$

Now let us examine the acid–base properties of the oxides of the third-period elements. We should expect the metal oxides at the left of the period to be basic and the nonmetal oxides at the right to be acidic. But where and how does the changeover occur? Na_2O and MgO yield basic solutions in water. Cl_2O, SO_2, and P_4O_{10} produce acidic solutions. SiO_2 (quartz) does not dissolve in water. However, it does dissolve slightly in strongly basic solutions to produce silicates (similar to the carbonates formed by CO_2 in basic solutions). For this reason we consider SiO_2 to be an acidic oxide.

     (a)   **(b)**

FIGURE 10-17
Displacement of I^-(aq) by Cl_2(g).
(a) $Cl_2(g)$ is bubbled through a colorless, dilute solution containing iodide ion, I^-(aq). (b) I_2 produced in the aqueous solution is extracted into CCl_4(l), in which it is much more soluble (purple layer).

Aluminum, a good conductor of heat and electricity, is clearly metallic in its physical properties. Al_2O_3, however, can act as either an acidic *or* a basic oxide. Oxides with this ability are called *amphoteric* (from the Greek word *amphos*, meaning "both"). Al_2O_3 is insoluble in water, but exhibits its amphoterism by reacting with both acidic and basic solutions. This "in-between" nature of aluminum oxide represents the changeover from basic to acidic oxides that we referred to earlier.

▶ We will study the classification of element oxides more fully in Chapter 17.

Summary

The *experimental* basis of the periodic table of the elements is the periodic law: Certain properties recur periodically when the elements are arranged by increasing atomic number. The *theoretical* basis is that the properties of an element are related to the electron configuration of its atoms, and elements in the same group of the periodic table have similar electron configurations.

The metallic and nonmetallic characters of atoms can be related to a set of *atomic* properties. In general, large atomic radii and low ionization energies are associated with metals; small atomic radii, high ionization energies, and large negative electron affinities are associated with nonmetals. The magnetic properties of an atom or ion stem from the presence or absence of unpaired electrons and are useful in establishing electron configurations. Several examples are given in the chapter of predictions and comparisons of properties based on the periodic table.

Summarizing Example

Francium ($Z = 87$) is an extremely rare, radioactive element formed when actinium ($Z = 89$) undergoes alpha particle emission. Francium occurs in natural uranium minerals, but estimates are that there is no more than 15 g of francium in the top 1 km of Earth's crust. Few of francium's properties have been measured, but some can be inferred from its position in the periodic table.

Estimate the melting point, density, and atomic (metallic) radius of francium.

1. *Locate Fr (Z = 87) in the periodic table.* Once you have done this you will know which group of elements it resembles. *Result:* Group 1A.

2. *Determine the melting point of Fr.* From the trend in melting points of the alkali metals (Group 1A) listed on page 317, estimate a melting point for Fr. *Answer:* 24° C.

3. *Determine the atomic volume and density of Fr.* Estimate the atomic volume from Figure 10-1. Approximate the molar mass of Fr from the values given for neighboring elements ($Z = 86$ and $Z = 88$). Then use the relationship between atomic (molar) volume, molar mass, and density (Equation 10.1) to obtain the density. *Answer:* 2.8 g/cm^3.

4. *Determine the atomic radius of Fr.* Use Figure 10-4 to make this estimate. *Answer:* 280 pm.

Key Terms

covalent radius (10-4)
diamagnetic (10-7)
effective nuclear charge, Z_{eff} (10-4)
electron affinity, *EA* (10-6)

ionic radius (10-4)
ionization energy, *I* (10-5)
isoelectronic (10-4)
metallic radius (10-4)

metalloid (10-3)
paramagnetic (10-7)
periodic law (10-1)

Review Questions

1. In your own words define the following terms: **(a)** isoelectronic; **(b)** valence shell electrons; **(c)** metal; **(d)** nonmetal; **(e)** metalloid.
2. Briefly describe each of the following ideas or phenomena: **(a)** the periodic law; **(b)** ionization energy; **(c)** electron affinity; **(d)** paramagnetism.
3. Explain the important distinctions between each pair of terms: **(a)** actinide and lanthanide element; **(b)** covalent and metallic radius; **(c)** atomic number and effective nuclear charge; **(d)** ionization energy and electron affinity; **(e)** paramagnetic and diamagnetic.
4. Find three pairs of elements that are out of order in the periodic table in terms of their atomic masses. Why is it necessary to invert their order in the table?
5. For the atom $^{119}_{50}Sn$ indicate the number of **(a)** protons in the nucleus; **(b)** neutrons in the nucleus; **(c)** 4d electrons; **(d)** 3s electrons; **(e)** 5p electrons; **(f)** electrons in the valence shell.

Mercury is a liquid at room temperature. You are probably aware of this fact, but to chemists it is surprising. Our expectation from the periodic law is that mercury should be a solid. Zinc and cadmium, the elements above mercury in Group 2B, are both solids. Their melting points are 419.6 and 320.9 °C, respectively. Based on these values and applying the predictive methods of Section 10-8, we should expect mercury to be a soft, solid metal with a melting point of about 200 °C.

Suppose we base our prediction on mercury's sixth-period neighbors. To the immediate left of mercury we find the metals Pt ($Z = 78$, mp = 1172 °C) and Au ($Z = 79$,

6. Among the following ions, several pairs are *isoelectronic*. Identify these pairs. Fe^{2+}, Sc^{3+}, Ca^{2+}, F^-, Co^{2+}, Co^{3+}, Sr^{2+}, Cu^+, Zn^{2+}, Al^{3+}.

7. Is it possible for two different atoms to be isoelectronic? two different cations? two different anions? a cation and an anion? Explain.

8. Refer to the periodic table on the inside front cover, and indicate **(a)** the most nonmetallic element; **(b)** the transition metal with lowest atomic number; **(c)** a metalloid whose atomic number is exactly midway between those of two noble gas elements.

9. For each of the following pairs, indicate the atom that has the *larger* size: **(a)** Te or Br; **(b)** K or Ca; **(c)** Ca or Cs; **(d)** N or O; **(e)** O or P; **(f)** Al or Au.

10. Indicate the *smallest* and the *largest* species (atom or ion) in the following group: the Al atom, the F atom, the As atom, the Cs^+ ion, the I^- ion, the N atom.

11. Use principles established in this chapter to arrange the following atoms in order of *increasing* value of the first ionization energy: Sr, Cs, S, F, As.

12. Give the symbol of the element **(a)** in Group 4A that has the smallest atoms; **(b)** in Period 5 that has the largest atoms; **(c)** in Group 7A that has the lowest first ionization energy.

13. How much energy, in joules, must be absorbed to convert to Na^+ all the atoms present in 1.00 mg of *gaseous* Na? The first ionization energy of Na is 495.8 kJ/mol.

14. Refer only to the periodic table on the inside front cover, and indicate which of the atoms Bi, S, Ba, As, and Ca **(a)** is most metallic; **(b)** is most nonmetallic; **(c)** has the intermediate value when the five are arranged in order of increasing first ionization energy.

15. Arrange the following elements in order of *decreasing* metallic character: Sc, Fe, Rb, Br, O, Ca, F, Te.

16. Which of the following species would you expect to be diamagnetic and which paramagnetic? **(a)** K^+; **(b)** Cr^{3+}; **(c)** Zn^{2+}; **(d)** Cd; **(e)** Co^{3+}; **(f)** Sn^{2+}; **(g)** Br.

17. Match each of the lettered items in the column on the left with the most appropriate numbered item(s) in the column on the right. Some of the numbered items may be used more than once and some, not at all.

(a) Tl	1. an alkaline earth metal
(b) $Z = 70$	2. element in Period 5 and Group 5A
(c) Ni	3. largest atomic radius of all the elements
(d) $[Ar]4s^2$	4. an element in Period 4 and Group 6A
(e) a metalloid	5. $3d^8$
(f) a nonmetal	6. one p electron in the shell of highest n
	7. lowest ionization energy of all the elements
	8. an f-block element

18. The boiling points of the noble gases increase down the group as follows: He, 4.2; Ne, 27.1; Ar, 87.3; Kr, 119.7; and Xe, 165 K. What would you predict to be the boiling point of radon?

mp = 1064 °C). To the immediate right are two more metals: Tl (Z = 81, mp = 303 °C) and Pb (Z = 82, mp = 328 °C). From these data we might expect a melting point for Hg (Z = 80) of well over 300 °C. Instead, we observe a melting point of −38.86 °C.

The reason why mercury is a liquid at room temperature is something of a mystery and not completely understood. In this limited discussion we can only indicate the most likely explanation, and this comes from an unlikely source—Einstein's theory of relativity. According to this theory the mass of a particle *increases* if the particle travels at speeds approaching the speed of light. The value listed in Table 2-1 for the mass of an electron is its *rest mass*. We are justified in using the rest mass as long as an electron travels at low to moderate speeds. In atoms of high atomic number, though, because of the high nuclear charge, electrons that come close to the nucleus are accelerated to high speeds, and their masses increase.

In the wave-mechanical model the relativity effect is especially pronounced for s electrons because these are the electrons that have a rather high probability of being found near the nucleus. With increased speed and, therefore, mass of an electron, there is a corresponding decrease in size of the electron orbital and a lowering of the orbital energy. In Hg, because of the lowering of the energy of the 6s orbital and the stability of the $5s^2 5p^6 5d^{10} 6s^2$ electron configuration, the bonds between Hg atoms are too weak to maintain a solid structure at room temperature. Mercury becomes almost as stable as a noble gas—sort of a noble liquid.

The relativistic shrinking of s orbitals affects all of the heavy metal atoms, but it is at a maximum with mercury. As new heavy metals are characterized, such as the recently discovered element with Z = 112 or the yet-to-be-discovered element with Z = 114, chemists may find other significant deviations from the periodic law.

Exercises

The Periodic Law

19. Use data from Figure 10-1 and Equation (10.1) to estimate the density expected for the undiscovered element 114. Assume a mass number of 298.

20. Suppose that lanthanum (Z = 57) were a newly discovered element having a density of 6.145 g/cm³. Estimate its molar mass. (*Hint:* Use Figure 10-1.)

21. The following densities, in grams per cubic centimeter, are for the listed elements in their standard states at 298 K. Show that density is a periodic property of these elements: Al, 2.699; Ar, 0.0018; As, 5.778; Br, 3.100; Ca, 1.550; Cl, 0.0032; Ga, 5.904; Ge, 5.323; Kr, 0.0037; Mg, 1.738; P, 1.823; K, 0.856; Se, 4.285; Si, 2.336; Na, 0.968; S, 2.069.

22. The following melting points are in degrees Celsius. Show that melting point is a periodic property of these elements: Al, 660; Ar, −189; Be, 1278; B, 2300; C, 3350; Cl, −101; F, −220; Li, 179; Mg, 651; Ne, −249; N, −210; O, −218; P, 590; Si, 1410; Na, 98; S, 119.

The Periodic Table

23. Mendeleev's periodic table did not preclude the possibility of a new group of elements that would fit within the existing table, as was the case with the noble gases. Moseley's work did preclude this possibility. Explain this difference.

25. Assuming that the seventh period is 32 members long, what should be the atomic number of the noble gas following radon (Rn)? of the alkali metal following francium (Fr)? What would you expect their approximate atomic masses to be?

24. Explain why the several periods in the periodic table do not all have the same number of members.

26. Concerning the incomplete seventh period of the periodic table, what should be the atomic number of the element **(a)** for which the filling of the 6d subshell is completed; **(b)** that should most closely resemble bismuth; **(c)** that should be a noble gas?

Atomic Radii

27. Explain why the radii of atoms do not simply increase uniformly with increasing atomic number.

29. Which is **(a)** the smallest atom in Group 3A; **(b)** the smallest of the following atoms: Te, In, Sr, Po, Sb? Why?

31. All the isoelectronic species illustrated in the text had the electron configurations of noble gases. Can two ions be isoelectronic *without* having noble-gas electron configurations? Explain.

33. Arrange the following in expected order of *increasing* radius: Br, Li^+, Se, I^-. Explain your answer.

28. The masses of individual atoms can be determined with great precision, yet there is considerable uncertainty about the exact size of an atom. Explain why this is the case.

30. How would you expect the sizes of the hydrogen ion, H^+, and the hydride ion, H^-, to compare with that of the H atom and the He atom? Explain.

32. The following species are *isoelectronic* with the noble gas krypton. Arrange them in order of *increasing* radius and comment on the principles involved in doing so: Rb^+, Y^{3+}, Br^-, Kr, Sr^{2+}, Se^{2-}.

34. Explain why we cannot use the generalizations presented in Figure 10-12 to answer the question, "Which is larger, an Al atom or an I atom?"

Ionization Energies; Electron Affinities

35. Are there any atoms for which the second ionization energy (I_2) is smaller than the first (I_1)? Explain.

37. How much energy, in kilojoules, is required to remove all the third-shell electrons in a mole of gaseous silicon atoms? (Hint: see Table 10.4.)

39. The production of gaseous bromide ions from bromine molecules can be considered a two-step process in which the first step is

$$Br_2(g) \longrightarrow 2\ Br(g) \qquad \Delta H = +193\ kJ$$

Is the formation of $Br^-(g)$ from $Br_2(g)$ an endothermic or exothermic process? (*Hint:* What is *EA* for Br?)

41. The Na^+ ion and the Ne atom are isoelectronic. The ease of loss of an electron by a gaseous Ne atom, I_1, has a value of 2081 kJ/mol. The ease of loss of an electron from a gaseous Na^+ ion, I_2, has a value of 4562 kJ/mol. Why are these values not the same?

36. Some electron affinities are negative quantities and some are zero or positive. Why is this not also the case with ionization energies?

38. What is the maximum number of Cs^+ ions that can be produced per joule of energy absorbed by a sample of gaseous Cs atoms? (Hint: see Table 10.3.)

40. Use ionization energies and electron affinities listed in the text to determine whether the following reaction is endothermic or exothermic.

$$Mg(g) + 2\ F\ (g) \longrightarrow Mg^{2+}(g) + 2\ F^-(g)$$

42. From the data given in Figure 10-10 the formation of a gaseous *anion* Li^- appears energetically favorable. That is, energy is given off when gaseous Li atoms accept electrons. Comment on the likelihood of forming a stable compound containing the Li^- ion, such as Li^+Li^- or Na^+Li^-.

Magnetic Properties

43. Unpaired electrons are found in only one of the following species. Indicate which is that one and explain why: F^-, Ca^{2+}, Fe^{2+}, S^{2-}.

45. Must all atoms with an odd atomic number be paramagnetic? Must all atoms with an even atomic number be diamagnetic? Explain.

44. Write electron configurations consistent with the following data on number of unpaired electrons: Ni^{2+}, two; Cu^{2+}, one; Cr^{3+}, three.

46. Neither Fe^{2+} nor Fe^{3+} has $4s$ electrons in its electron configuration. How many unpaired electrons would you expect to find in each of these ions? Explain.

Predictions Based on the Periodic Table

47. Use ideas presented in this chapter to indicate
(a) three metals that you would expect to exhibit the photoelectric effect with visible light, and three that you would not;
(b) the noble gas element that should have the highest density when in the liquid state;
(c) the approximate I_1 of fermium ($Z = 100$);
(d) the approximate density of solid radium ($Z = 88$).

49. Gallium is a commercially important element (used to make gallium arsenide for the semiconductor industry). Gallium was unknown in Mendeleev's time, and he predicted properties of this element. Predict the following for gallium: **(a)** its density; **(b)** the formula and percent composition of its oxide. [*Hint:* Use Figure 10-1, Equation (10.1), and Mendeleev's periodic table (page 317).]

48. The heat of atomization for an element is the amount of energy required to convert an appropriate amount of an element in its standard state to a mole of atoms in the gaseous state. The heats of atomization, in kilojoules per mole, for three of the Group 4A elements are: carbon, 717; silicon, 452; and tin, 302 kJ/mol, respectively. Predict the heat of atomization of germanium.

50. Estimate the missing boiling point in the following series of compounds.
(a) CH_4, -164 °C; SiH_4, -112 °C; GeH_4, -90 °C; SnH_4, ? °C.
(b) H_2O, ? °C ; H_2S, -61 °C; H_2Se, -41 °C; H_2Te, -2 °C.

Does your estimate in (b) agree with the known value?

51. For the following groups of elements select the one that has the property noted.
(a) The largest atom: Mg, Mn, Mo, Ba, Bi, Br.
(b) The lowest first ionization energy: B, Sr, Al, Br, Mg, Pb.
(c) The most negative electron affinity: As, B, Cl, K, Mg, S.
(d) The largest number of unpaired electrons: F, N, S^{2-}, Mg^{2+}, Sc^{3+}, Ti^{3+}.

52. Match each of the lettered items on the left with an appropriate numbered item on the right. All the numbered items should be used, and some more than once.

(a) $Z = 32$ 1. two unpaired p electrons
(b) $Z = 8$ 2. diamagnetic
(c) $Z = 53$ 3. more negative electron affinity than elements on either side of it in the same period
(d) $Z = 38$
(e) $Z = 48$
(f) $Z = 20$ 4. first ionization energy lower than that of Ca but greater than that of Cs

Advanced Exercises

53. Discuss the likelihood that element 168, should it ever be synthesized in sufficient quantity, would be a "noble liquid" at 298 K and 1 atm. Some data that may be of use are tabulated. Might element 168 be a "noble solid" at 298 K and 1 atm? Use *spdf* notation to show the electron configuration you would expect for element 168.

Element	Atomic Mass, u	mp, K	bp, K
Argon	39.948	83.95	87.45
Helium	4.0026	—	4.25
Krypton	83.80	116.5	120.9
Neon	20.179	24.48	27.3
Radon	222	202	211.4
Xenon	131.29	161.3	166.1

54. Sketch a periodic table that would include *all* the elements in the main body of the table. How many "numbers" wide would the table be?

55. In Mendeleev's time indium oxide, which is 82.5% In, by mass, was thought to be InO. If this were the case, in which group of Mendeleev's table (page 317) should indium be placed? (*Hint:* What would be the atomic mass of indium?)

56. Instead of accepting the atomic mass of indium implied by the data in Exercise 55, Mendeleev proposed that the formula of indium oxide is In_2O_3. Show that this assumption places indium in the proper group of Mendeleev's periodic table on page 317.

57. Listed below are two atomic properties of the element germanium. Refer only to the periodic table on the inside front cover and indicate probable values for each of the following elements, expressed as greater than, about equal to, or less than the value for Ge.

	Atomic radius	First ionization energy
Ge	122 pm	762 kJ/mol
Al	?	?
In	?	?
Se	?	?

58. In estimating the boiling point and freezing point of bromine in Example 10-5, could we have used Celsius or Fahrenheit instead of Kelvin temperature? Explain.

59. In the formula X_2, if the two X atoms are the same halogen, the substance is a halogen element (e.g., Cl_2, Br_2). If the two X atoms are different (such as Cl and Br), we describe an *interhalogen* compound. For BrCl and ICl, predict whether each is a solid, liquid, or gas at room temperature (25 °C). (*Hint:* Refer to Example 10-5 and Table 10-5.)

60. Assume that atoms are hard spheres, and use the metallic radius of 186 pm for Na to estimate the volume of one Na atom and of one mole of Na atoms. How does your result compare with the atomic volume found in Figure 10-1? Why is there so much disagreement between the two values?

61. When sodium chloride is strongly heated in a flame, the flame takes on the yellow color associated with the emission spectrum of sodium atoms. The reaction that occurs in the *gaseous* state is

$$Na^+(g) + Cl^-(g) \longrightarrow Na(g) + Cl(g).$$

Show that this reaction is *exothermic*.

62. Use Equation (9.7) to determine the lowest energy level for the He^+ ion. Use this information to calculate the *second* ionization energy (I_2) for the helium atom. Compare your result with the tabulated value of 5251 kJ/mol.

63. Refer only to the periodic table on the inside front cover, and arrange the following ionization energies in order of *increasing* value: I_1 for F; I_2 for Ba; I_3 for Sc; I_2 for Na; I_3 for Mg. Explain the basis of any uncertainties.

64. Refer to the footnote on page 328. Then, use values of basic physical constants and other data from the appendices to show that 1 eV/atom = 96.49 kJ/mol.

Chemical Bonding I: Basic Concepts

11

*C*onsider all that we already know about chemical compounds. We can determine their compositions and write their formulas. We can represent the reactions of compounds by chemical equations and perform stoichiometric and thermochemical calculations based on these equations. And we can do all this without really having to consider the ultimate structure of matter—the structure of atoms and molecules. Yet the shape of a molecule—that is, the arrangement of its atoms in space—often defines the chemistry of that molecule. If water had a different shape its properties would be significantly different from those familiar to us, and life as we know it would not be possible.

In this chapter we will describe the interactions between atoms called chemical bonds. Most of our discussion will center on the simplest method of representing chemical bonding, known as the Lewis theory. We will also explore, however, another relatively simple theory, one for predicting probable molecular shapes. And, throughout the chapter, we will try to relate these theories to what is known about molecular structures from experimental measurements. In Chapter 12 we will examine the subject of chemical bonding in more depth, and in Chapter 13 we will describe forces between molecules—intermolecular forces—and explore further the relationship between molecular shape and the properties of substances.

Computer-generated, three-dimensional molecular models of methanol (bottom right), CH_3OH, and ethanol (top left), CH_3CH_2OH. In this chapter we study ideas that enable us to predict the geometric shapes of molecules.

11-1 Lewis Theory: An Overview

In the period 1916–1919, two Americans, G. N. Lewis and Irving Langmuir, and a German, Walther Kossel, advanced an important proposal about chemical bonding: Something unique in the electron configurations of noble gas atoms accounts for their inertness, and atoms of other elements combine with one another to acquire electron configurations like the noble gas atoms. The theory that grew out of this model has been most closely associated with G. N. Lewis and is called the **Lewis theory.** Some fundamental ideas in Lewis's theory are

▶ Since 1962 a number of compounds of Xe and Kr have been synthesized. This does not alter the usefulness of ideas about noble gas electron configurations, however.

1. Electrons, especially those of the outermost (valence) electronic shell, play a fundamental role in chemical bonding.

2. In some cases electrons are *transferred* from one atom to another. Positive and negative ions are formed and attract each other through electrostatic forces called **ionic bonds.**

3. In other cases one or more pairs of electrons are *shared* between atoms; this sharing of electrons is called a **covalent bond.**

4. Electrons are transferred, or shared, in such a way that each atom acquires an especially stable electron configuration. Usually this is a noble gas configuration, one with eight outer shell electrons, or an **octet.**

▶ The term *covalent* was introduced by Irving Langmuir.

Lewis Symbols and Lewis Structures

Lewis developed a special set of symbols for his theory. A **Lewis symbol** consists of a chemical symbol to represent the nucleus and *core* (inner shell) electrons of an atom, together with dots placed around the symbol to represent the *valence* (outer shell) electrons. Thus, the Lewis symbol for silicon, which has the electron configuration $[Ne]3s^2 3p^2$ is

$$\cdot \overset{\cdot}{Si} \cdot$$

Electron spin had not yet been proposed when Lewis framed his theory, and so he did not show that two of the valence electrons ($3s^2$) are paired and two ($3p^2$), unpaired. We will write Lewis symbols in the way Lewis did. We will place single dots on the sides of the symbol, up to a maximum of four. Then we will pair up dots until we reach an octet. Lewis symbols are commonly written for main group elements but much less often for transition elements. Lewis symbols for several main group elements are written in Example 11-1.

Gilbert Newton Lewis (1875–1946). Lewis's contribution to the study of chemical bonding is evident throughout this text. Equally important, however, was his pioneering introduction of thermodynamics into chemistry.

EXAMPLE 11-1

Writing Lewis Symbols. Write Lewis symbols for the following elements: **(a)** N, P, As, Sb, Bi; **(b)** Al, I, Se, Ar.

Solution

a. These are the elements of Group 5A. Their atoms all have *five* valence electrons ($ns^2 np^3$). The Lewis symbols all have *five* dots.

$$\cdot \overset{\cdot \cdot}{N} \cdot \qquad \cdot \overset{\cdot \cdot}{P} \cdot \qquad \cdot \overset{\cdot \cdot}{As} \cdot \qquad \cdot \overset{\cdot \cdot}{Sb} \cdot \qquad \cdot \overset{\cdot \cdot}{Bi} \cdot$$

b. Al is in Group 3A; I, in Group 7A; Se, in Group 6A; Ar, in Group 8A.

$$\cdot \overset{\cdot}{Al} \cdot \qquad : \overset{\cdot \cdot}{I} \cdot \qquad : \overset{\cdot \cdot}{Se} \cdot \qquad : \overset{\cdot \cdot}{Ar} :$$

Note that for main group elements the number of valence electrons, and hence the number of dots appearing in a Lewis symbol, is equal to the periodic table group number.

Practice Example A: Write Lewis symbols for Mg, Ge, K, and Ne.

Practice Example B: Write Lewis symbols for: Sn, Br^-, Tl^+, S^{2-}. (*Hint:* How are the electron configurations of the ions related to those of the atoms from which they are formed?)

A **Lewis structure** is a combination of Lewis symbols that represents either the transfer or the sharing of electrons in a chemical bond.

Ionic bonding (transfer of electrons):

$$Na\times + \ \cdot \overset{..}{\underset{..}{Cl}}: \longrightarrow [Na]^+[\overset{..}{\underset{..}{\times Cl}}:]^- \qquad (11.1)$$

Lewis symbols   Lewis structure

Covalent bonding (sharing of electrons):

$$H\times + \ \cdot \overset{..}{\underset{..}{Cl}}: \longrightarrow H\times \overset{..}{\underset{..}{Cl}}: \qquad (11.2)$$

Lewis symbols   Lewis structure

In these two examples the electrons from one atom are designated as · and from the other atom, as ×. It is impossible to distinguish between electrons, however. Henceforth we will use only dots · to represent electrons.

Lewis's work dealt mostly with covalent bonding, which we will emphasize throughout this chapter. Lewis's ideas also apply to ionic bonding, though, and we briefly describe this application next.

Lewis Structures for Ionic Compounds

In Section 3-2 we learned that the formula unit of an ionic compound is the simplest electrically neutral collection of cations and anions from which we can establish the chemical formula of the compound. The Lewis structure of sodium chloride (structure 11.1) represents its formula unit. In the Lewis structure of an ionic compound of main group elements (1) the Lewis symbol of the metal ion has no dots if all the valence electrons are lost, and (2) the ionic charges of both ions are shown. These ideas are further illustrated through Example 11-2.

EXAMPLE 11-2

Writing Lewis Structures of Ionic Compounds. Write Lewis structures for the following compounds: **(a)** BaO; **(b)** $MgCl_2$; **(c)** aluminum oxide.

Solution

a. Write the Lewis symbol and determine how many electrons each atom must gain or lose to acquire a noble gas electron configuration. Ba loses two electrons and O gains two.

$$Ba \cdot \!\!\! \curvearrowright \!\!\! + \overset{\cdot}{\underset{.}{O}}: \longrightarrow [Ba]^{2+}[:\overset{..}{\underset{..}{O}}:]^{2-}$$

Lewis structure

b. A Cl atom can accept only *one* electron because it already has seven valence electrons. One more will give it a complete octet. On the other hand, a Mg atom must lose *two* electrons to have the electron configuration of the noble gas, neon. So two Cl atoms are required for each Mg atom.

$$\text{Mg} + \overset{..}{\underset{..}{\text{Cl}}} \longrightarrow [:\overset{..}{\underset{..}{\text{Cl}}}:]^- [\text{Mg}]^{2+} [:\overset{..}{\underset{..}{\text{Cl}}}:]^-$$

Lewis structure

c. We do not need to be given the formula of aluminum oxide. It follows directly from the Lewis structure that we write. The combination of one Al atom, which loses three electrons, and one O atom, which gains two, leaves an excess of one electron lost. To match the numbers of electrons lost and gained, we need a formula unit based on *two* Al atoms and *three* O atoms.

$$2[\text{Al}]^{3+} 3[:\overset{..}{\underset{..}{\text{O}}}:]^{2-}$$

Lewis structure

Practice Example A: Write plausible Lewis structures for **(a)** Na_2S and **(b)** Mg_3N_2.

Practice Example B: Write plausible Lewis structures for **(a)** calcium iodide; **(b)** barium sulfide; **(c)** lithium oxide.

The compounds described in Example 11-2 are *binary* ionic compounds consisting of monatomic cations and monatomic anions. Commonly encountered *ternary* ionic compounds consist of monatomic and *poly*atomic ions, and bonding between atoms in the polyatomic ions is covalent. We will consider some ternary ionic compounds later in the chapter.

With the exception of ion pairs such as (Na^+Cl^-) that may be found in the *gaseous* state, formula units of ionic compounds do not exist as separate entities. Instead, each cation surrounds itself with anions, and each anion with cations. These very large numbers of ions are arranged in an orderly network called an *ionic crystal* (Figure 11-1). We will describe ionic crystal structures and energy changes accompanying the formation of ionic crystals in Chapter 13.

11-2   Covalent Bonding: An Introduction

A chlorine atom shows a tendency to gain an electron, as indicated by its electron affinity (-349 kJ/mol). From which atom, sodium or hydrogen, can the electron most readily be extracted? Neither atom gives up an electron freely, but the energy required to extract an electron from Na ($I_1 = 496$ kJ/mol) is much smaller than that for H ($I_1 = 1312$ kJ/mol). In Chapter 10 we learned that the lower its ionization energy, the more metallic an element is; *sodium is much more metallic than hydrogen* (recall Figure 10-12). In fact, hydrogen is considered to be a *nonmetal*. A hydrogen atom does not *give up* an electron to another nonmetal atom. Bonding between a hydrogen atom and a chlorine atom involves the *sharing* of electrons. This sharing leads to a *covalent bond*.

FIGURE 11-1
Portion of an ionic crystal.
This structure of alternating Na^+ and Cl^- ions extends in all directions and involves countless numbers of ions.

To emphasize the sharing of electrons let us think of the Lewis structure of HCl in this manner.

$$(H \vdots Cl \vdots)$$

The broken circles represent the outermost electronic shells of the bonded atoms. The number of dots lying on or within each circle represents the effective number of electrons in each valence shell. The H atom has two dots, as in the electron configuration of He. The Cl atom has eight dots, corresponding to the outer shell configuration of Ar. Note that we counted the two electrons between H and Cl (:) *twice*. These two electrons are shared by the H and Cl atoms. This shared pair of electrons constitutes the covalent bond. Notice that the following Lewis structures, as well as the one for HCl, conform to the **octet rule,** a requirement of eight valence shell electrons for each atom (except H).

$$H : \overset{..}{\underset{..}{O}} : H \qquad H : \overset{..}{N} : H \qquad : \overset{..}{\underset{..}{Cl}} : \overset{..}{\underset{..}{O}} : \overset{..}{\underset{..}{Cl}} :$$
$$H$$

| water | ammonia | dichlorine oxide |

Lewis theory helps us to understand why elemental hydrogen and chlorine exist as *diatomic* molecules, H_2 and Cl_2. In each case a pair of electrons is shared between the two atoms. The sharing of a *single* pair of electrons between bonded atoms produces a **single covalent bond.** To underscore the importance of electron *pairs* in the Lewis theory, we introduce the term **bond pair** for a pair of electrons in a covalent bond and **lone pair** for electron pairs that are not involved in bonding. Also it is customary to replace some electron pairs by dash signs (—), especially for bond pairs. These features are shown in the following Lewis structures.

▶ The Lewis structures written for H_2O and Cl_2O suggest that these molecules have a linear shape. They do not. Lewis theory, by itself, does not address the question of molecular shape. (See Section 11-7.)

$$H\cdot \; + \; \cdot H \longrightarrow H : H \quad or \quad H-H \qquad (11.3)$$

bond pair

$$: \overset{..}{\underset{..}{Cl}}\cdot \; + \; \cdot \overset{..}{\underset{..}{Cl}} : \longrightarrow : \overset{..}{\underset{..}{Cl}} : \overset{..}{\underset{..}{Cl}} : \quad or \quad : \overset{..}{\underset{..}{Cl}}-\overset{..}{\underset{..}{Cl}} : \Longleftarrow \text{lone pairs} \qquad (11.4)$$

bond pair

Multiple Covalent Bonds

If we apply the ideas about Lewis structures we have just learned to the molecule N_2, we have a problem.

$$: \overset{.}{N}\cdot \; + \cdot \overset{.}{N} : \longrightarrow : \overset{.}{N} : \overset{.}{N} : \quad (incorrect)$$

Each N atom appears to have only *six* outer shell electrons, not the expected eight. We can correct the situation by bringing the four unpaired electrons into the region between the N atoms and using them for additional bond pairs. In all, we now show the sharing of *three* pairs of electrons between the N atoms. The bond between the N atoms in N_2 is a **triple covalent bond** ($\equiv$).

$$: N : N : \longrightarrow : N \vdots\vdots\vdots N : \quad or \quad : N\equiv N : \qquad (11.5)$$

The triple covalent bond in N_2 is a very strong bond that is difficult to break in a chemical reaction. The unusual strength of this bond makes $N_2(g)$ quite inert. As a result, $N_2(g)$ coexists with $O_2(g)$ in the atmosphere and forms oxides of nitrogen only in trace amounts at high temperatures. The lack of reactivity of N_2 toward O_2 is an essential condition for life on Earth. The inertness of $N_2(g)$ also makes it difficult to synthesize nitrogen compounds artificially.

Another molecule whose Lewis structure features a multiple bond is O_2. Here the bond involves *two* pairs of electrons and is a **double covalent bond.**

$$:\overset{..}{O}\cdot \; + \; \cdot\overset{..}{O}: \; \longrightarrow \; :\overset{..}{O}:\overset{..}{O}: \; \longrightarrow \; \overset{..}{O} \; \overset{..}{O} \; \longrightarrow \; :\overset{..}{O}{=}\overset{..}{O}: \; ? \qquad (11.6)$$

We have put a question mark on structure (11.6) to suggest that there is something wrong with it. The structure satisfies all the requirements of the Lewis theory but fails to account for the fact that oxygen is *paramagnetic* (Figure 11-2). The O_2 molecule has *unpaired* electrons. We will examine the reasons for oxygen's paramagnetism in Chapter 12, but the point to note here is that merely being able to write a plausible Lewis structure does *not* prove that it is the correct electronic structure.

Bond Order and Bond Lengths

The case of O_2 underscores the importance of verifying a Lewis structure through *experimental* evidence. The evidence that calls into question structure (11.6) involves magnetic properties. Another measured property that sheds light on the electronic structures of molecules is *bond length*. The measurements involve interpreting interactions of electromagnetic radiation (for example, X-rays) or matter waves (electrons or neutrons) with the substance under study.

The term **bond order** describes whether a covalent bond is *single* (bond order = 1), *double* (bond order = 2), or *triple* (bond order = 3). Think of electrons as the "glue" that binds atoms together in covalent bonds. The higher the bond order—that is, the more electrons present (the more glue)—the more tightly the atoms are held together.

Bond length is the distance between the centers of two atoms joined by a covalent bond. As we might expect, a double bond between atoms is shorter than a single bond, and a triple bond is shorter still. You can see this relationship clearly in Table 11.1 by comparing the three different bond lengths for the nitrogen-to-nitrogen bond. For example, the measured nitrogen-to-nitrogen bond length in N_2 is 109.8 pm—a triple bond—whereas the nitrogen-to-nitrogen bond length in hydrazine, H_2N—NH_2, is 147 pm—a single bond.

FIGURE 11-2
Paramagnetism of oxygen.
Liquid oxygen is attracted into the magnetic field of a large magnet.

TABLE 11.1   Some Average Bond Lengths[a]

Bond	Bond Length, pm	Bond	Bond Length, pm	Bond	Bond Length, pm
H—H	74.14	C—C	154	N—N	145
H—C	110	C=C	134	N=N	123
H—N	100	C≡C	120	N≡N	109.8
H—O	97	C—N	147	N—O	136
H—S	132	C=N	128	N=O	120
H—F	91.7	C≡N	116	O—O	145
H—Cl	127.4	C—O	143	O=O	121
H—Br	141.4	C=O	120	F—F	143
H—I	160.9	C—Cl	178	Cl—Cl	199
				Br—Br	228
				I—I	266

[a]Most values (C—H, N—H, C—H, . . .) are averaged over a number of species containing the indicated bond and may vary by a few picometers. Where a diatomic molecule exists, the value given is the actual bond length in that molecule (H_2, N_2, HF, . . .) and is known more precisely.

Perhaps you can now also better understand the meaning of covalent radius that we introduced in Section 10-5. The single covalent radius is one-half the distance between the centers of identical atoms joined by a *single* covalent bond. Thus, the single covalent radius of chlorine in Figure 10-8 (99 pm) is one-half the bond length given in Table 11.1, that is, $\frac{1}{2} \times$ 199 pm. Furthermore, as a *rough* generalization, *the length of the covalent bond between two atoms can be approximated as the sum of the covalent radii of the two atoms.*

Some of these ideas about bond length are applied in Example 11-3.

EXAMPLE 11-3

Estimating Bond Lengths. Provide the best estimate you can of these bond lengths: **(a)** the nitrogen-to-hydrogen bonds in NH_3; **(b)** the bromine-to-chlorine bond in BrCl.

Solution

a. The Lewis structure of ammonia (page 349) shows the nitrogen-to-hydrogen bonds as single bonds. The value listed in Table 11.1 for the N—H bond is 100 pm, so this is the value we would predict. (The measured N—H bond length in NH_3 is 101.7 pm.)

b. We do not find a bromine-to-chlorine bond length in Table 11.1. We need to calculate an approximate bond length with the relationship between bond length and covalent radii. BrCl contains a Br—Cl *single* bond [just imagine substituting one Br atom for one Cl atom in structure (11.4).] The length of the Br—Cl bond is one-half the Cl—Cl bond length *plus* one-half the Br—Br bond length: $(\frac{1}{2} \times 199$ pm$) + (\frac{1}{2} \times 228$ pm$) = 214$ pm. (The measured bond length is 213.8 pm.)

Practice Example A: Estimate the bond lengths of the carbon-to-hydrogen bonds and the carbon-to-bromine bond in CH_3Br.

Practice Example B: In a CO_2 molecule each O atom is bonded to the C atom with a bond length of 116.3 pm. Write a plausible Lewis structure for CO_2 that is consistent with this experimental evidence.

Later in this chapter, and throughout the text, information on bond order and bond lengths will help us in writing Lewis structures.

11-3 Polar Covalent Bonds

We could keep our discussion of chemical bonds simple if all bonds were clearly of just two types: ionic bonds involving a *complete transfer* of electrons and covalent bonds involving an *equal sharing* of electron pairs. Such is not the case, however, and many chemical bonds have features of both types.

A covalent bond in which electrons are not shared equally between two atoms is called a **polar covalent bond.** In such a bond, electrons are displaced toward the more nonmetallic element. In Figure 11-3a, note the even distribution of electron charge density between the two H atoms in H_2 and between the two Cl atoms in Cl_2. In both molecules each atom has the same electron affinity, and electrons are not displaced toward either atom. The centers of positive and negative charge coincide; each is found at a point midway between the atomic nuclei. Both the H—H and Cl—Cl bonds are *nonpolar.* In HCl, on the other hand, Cl attracts electrons more strongly than does H. The electron charge density is greater near the Cl atom than near the H atom. The center of negative charge lies closer to the Cl nucleus than does the center of positive charge. We say that there is a separation of charge in the H—Cl bond and that the bond is *polar.* We can represent the polar bond in HCl by a Lewis structure in which the bond pair of electrons lies closer to the Cl than to the H.

$$\delta^+H : \ddot{C}\ddot{l} : {}^{\delta-}$$

The $\delta+$ signifies that the center of positive charge is displaced toward the H nucleus, and $\delta-$ signifies that the center of negative charge is displaced toward the Cl nucleus. We say that there is a separation of charge, with a *partial* positive charge on H and a *partial* negative charge on Cl.

Electronegativity

We concluded that the H—Cl bond should be polar because the Cl atom has a greater affinity for electrons than does the H atom. Electron affinity is an atomic property, however, and more meaningful predictions about bond polarities are

---

FIGURE 11-3

Nonpolar and polar covalent bonds.
(a) In the diatomic *nonpolar* molecules H_2 and Cl_2, the center of positive charge lies midway between the atomic nuclei along a line between them. The center of negative charge comes at the same point. There is no charge separation. (b) In the H—Cl bond the center of positive charge lies much closer to the Cl nucleus than to the H nucleus, but this is because the Cl nucleus has 17 units of positive charge compared with just 1 unit for H. The center of negative charge lies still closer to the Cl nucleus. This is because of the stronger attraction of Cl than of H for the electron pair in the H—Cl bond. In a *polar covalent* bond the centers of positive and negative charge are separated.

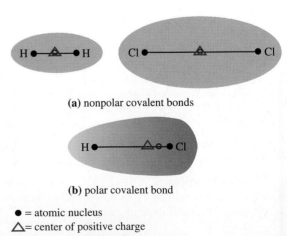

(a) nonpolar covalent bonds

(b) polar covalent bond

● = atomic nucleus
△ = center of positive charge
○ = center of negative charge

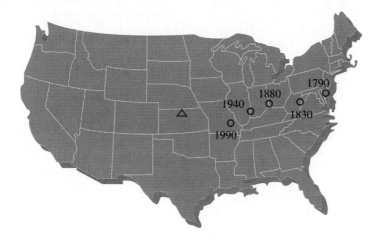

FIGURE 11-4
An analogy to a polar covalent bond.
The geographical center of the contiguous 48 states of the United States remains fixed ($\triangle$), but the population center (O) is moving to the south and west. The separation between these two centers is analogous to the separation of the centers of positive and negative charge in a polar covalent bond. As the distance between the centers becomes smaller, the bond becomes less polar.

those based on a molecular property, one that relates to the ability of atoms to lose or gain electrons when they are part of a molecule rather than isolated from other atoms. **Electronegativity (EN)** describes an atom's ability to compete for electrons with other atoms to which it is bonded. As such, electronegativity is related to ionization energy (I) and electron affinity (EA). A widely used electronegativity scale, with values given in Figure 11-5, is one devised by Linus Pauling (1901–1994). Pauling's EN values range from about 0.7 to 4.0. In general, the lower its EN, the more metallic an element is; and the higher the EN, the more nonmetallic it is. From Figure 11-5 we also see that electronegativity decreases from top to bottom in a group and increases from left to right in a period of the periodic table.

▶ The basis of Pauling's electronegativity scale is explored in Feature Problem A on page 385.

Are You Wondering . . .

If there is an analogy you can use in thinking about polar covalent bonds?

Consider the map in Figure 11-4 that locates the geographic center of the contiguous 48 states of the United States. Here is a way to show the significance of this center: Imagine cutting out the map and then trying to balance the cut-out on the point of a pin. The geographic center is the point on the map where you would have to put the pin to achieve this balance. Also shown on the map is the center of population density in the United States at various times in its history. To locate the population center, imagine putting a pencil dot for each inhabitant at his or her geographic location on a weightless cut-out of the map; then search for the new balance point.

The analogy to a polar bond is to think of the geographic center as the center of positive charge and the population center as the center of negative charge. Two hundred years ago there was a great separation between these two centers, analogous to a highly polar bond. With time, however, the distance between these two centers has shortened, analogous to a less polar bond. If the two centers were eventually to coincide, this would be analogous to a nonpolar bond.

1A																	
H 2.1																	

below 1.0   2.0 - 2.4

1.0 - 1.4   2.5 - 2.9

1.5 - 1.9   3.0 - 4.0

	2A											3A	4A	5A	6A	7A
Li 1.0	Be 1.5											B 2.0	C 2.5	N 3.0	O 3.5	F 4.0
Na 0.9	Mg 1.2	3B	4B	5B	6B	7B	⌐—— 8B ——⌐			1B	2B	Al 1.5	Si 1.8	P 2.1	S 2.5	Cl 3.0
K 0.8	Ca 1.0	Sc 1.3	Ti 1.5	V 1.6	Cr 1.6	Mn 1.5	Fe 1.8	Co 1.8	Ni 1.8	Cu 1.9	Zn 1.6	Ga 1.6	Ge 1.8	As 2.0	Se 2.4	Br 2.8
Rb 0.8	Sr 1.0	Y 1.2	Zr 1.4	Nb 1.6	Mo 1.8	Tc 1.9	Ru 2.2	Rh 2.2	Pd 2.2	Ag 1.9	Cd 1.7	In 1.7	Sn 1.8	Sb 1.9	Te 2.1	I 2.5
Cs 0.8	Ba 0.9	La* 1.1	Hf 1.3	Ta 1.5	W 2.4	Re 1.9	Os 2.2	Ir 2.2	Pt 2.2	Au 2.4	Hg 1.9	Tl 1.8	Pb 1.8	Bi 1.9	Po 2.0	At 2.2
Fr 0.7	Ra 0.9	Ac† 1.1														

*Lanthanides: 1.1 - 1.3
†Actinides: 1.3 - 1.5

FIGURE 11-5   Electronegativities of the elements.
As a general rule, electronegativities *decrease* from top to bottom in a group and *increase* from left to right in a period of elements. The values are from L. Pauling, *The Nature of the Chemical Bond,* 3rd edition, Cornell University, Ithaca, NY, 1960, p. 93. They may be somewhat different from values based on other scales.

▶ The absolute value means the numerical value without regard for its sign. Thus, for $\Delta EN = 3.0 - 3.5$, we write 0.5 rather than -0.5.

With electronegativity values we are able to gain an insight into the amount of polar character in a covalent bond. We do this by describing the **electronegativity difference, ΔEN,** as the absolute value of the difference in EN values of the bonded atoms. If ΔEN for two atoms is very small, the bond between them is "essentially covalent." If ΔEN is large, a bond is "essentially ionic." For intermediate values of ΔEN the bond is described as polar covalent. A useful rough relationship between ΔEN and percent ionic character of a bond is presented in Figure 11-6.

Large EN differences are found between the more metallic and the more nonmetallic elements. Combinations of these elements are expected to produce bonds that are "essentially ionic." Small EN differences are expected for two nonmetal atoms, and the bond between them should be "essentially covalent." Thus, even without a compilation of EN values at hand, you should be able to

FIGURE 11-6
Percent ionic character of a chemical bond as a function of electronegativity difference.

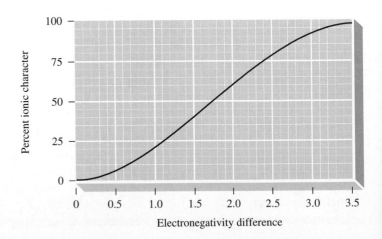

predict the essential character of a bond between two atoms. Simply assess the metallic/nonmetallic characters of the bonded elements from the periodic table (recall Figure 10-12).

EXAMPLE 11-4

Assessing Electronegativity Differences and the Polarity of Bonds.
a. Which bond is more polar, H—Cl or H—O?
b. What is the percent ionic character of each of these bonds?

Solution

a. Look up EN values of H, Cl, and O in Figure 11-5 and compute ΔEN. $EN_H =$ 2.1; $EN_{Cl} = 3.0$; $EN_O = 3.5$. For the H—Cl bond, $\Delta EN = 3.0 - 2.1 = 0.9$. For the H—O bond, $\Delta EN = 3.5 - 2.1 = 1.4$. Because its ΔEN is somewhat greater, we expect the H—O bond to be the more polar bond.
b. Determine the percent ionic character from Figure 11-6.

$$H\text{—}Cl \text{ bond:} \quad \Delta EN = 0.9; \quad \approx 20\% \text{ ionic}$$

$$H\text{—}O \text{ bond:} \quad \Delta EN = 1.4; \quad \approx 35\% \text{ ionic}$$

Practice Example A: Which of the following bonds is the most polar, that is, has the greatest ionic character: H—Br, N—H, N—O, P—Cl?

Practice Example B: Which is the most polar bond: C—S, C—P, P—O, or O—F?

11-4   Writing Lewis Structures

With the background acquired in the preceding three sections, we are now in a position to write a variety of Lewis structures. Let us begin with a reminder of some basic ideas.

- *All* the valence electrons of the atoms in a Lewis structure must appear in the structure.
- *Usually,* all the electrons in a Lewis structure are paired.
- *Usually,* each atom acquires an outer shell octet of electrons. Hydrogen, however, is limited to two outer shell electrons.
- *Sometimes* multiple covalent bonds (double or triple bonds) are needed. Multiple covalent bonds are formed most readily by C, N, O, P, and S atoms.

Another requirement is to start with the correct **skeleton structure.** This is an arrangement of the atoms in the order in which they are bonded together. For example, a water molecule has the skeleton structure H—O—H, *not* O—H—H. In a skeleton structure it is generally useful to distinguish between central atom(s) and terminal atoms. A **central atom** is bonded to two or more other atoms. A **terminal atom** is bonded to just one other atom.

H atoms are always terminal atoms, and C atoms are almost always central atoms.

At this point, we could start writing Lewis structures by a trial-and-error method. That is, we could just move electron dots around until we satisfy all the requirements for a plausible Lewis structure. Most students feel more comfortable, however, when they have a specific *strategy* to follow in writing a Lewis structure. This is the strategy we will use.

► A strategy for writing Lewis structures.

1. Determine the total number of valence electrons in the structure.
2. Identify the central atom(s). This is most often the atom with the lowest electronegativity. Hydrogen, however, is never a central atom.
3. Write the skeleton structure, and join the atoms in this structure by *single* covalent bonds.
4. For each single bond thus formed, subtract *two* from the total number of valence electrons.
5. With the valence electrons remaining, first complete the octets of the terminal atoms. Then, to the extent possible, complete the octets of the central atom(s).
6. At this point, if the central atom(s) lacks an octet, form multiple covalent bonds by converting lone pair electrons from terminal atoms into bond pairs.

EXAMPLE 11-5

Applying the General Strategy for Writing Lewis Structures. Write the Lewis structure of HCN, the poisonous gas hydrogen cyanide.

Solution

Step 1. Obtain the number of valence electrons for each atom in the structure from its position in the periodic table. These numbers are *one* for H (Group 1A), *four* for C (Group 4A), and *five* for N (Group 5A). The total number of valence electrons is 10.

Step 2. We have three reasons to assume that C is the central atom in the skeleton structure: (a) it has a lower electronegativity than N, (b) H is always a terminal atom, and (c) C is almost always a central atom.

Step 3. Now, join the atoms in the skeleton structure by single covalent bonds.

$$H\text{—}C\text{—}N$$

Step 4. The bonds in the skeleton structure account for 4 of the 10 valence electrons. This leaves six electrons.

Step 5. Complete the octet of the N atom. (H can only accommodate two electrons in its valence shell.)

► Note that in this strategy we do not need to keep track of which valence electrons come from which atoms.

$$H\text{—}C\text{—}\ddot{\underset{..}{N}}: \quad \text{(incorrect)}$$

Step 6. We have used up all the available electrons, but the central C atom has only *four* electrons around it, not an octet. We can correct this situation by moving two lone pairs of electrons of the N atom into the region between the C and N atom to form a *triple covalent bond*.

$$H\text{—}C\overset{\curvearrowleft}{\underset{\curvearrowright}{\text{—}}}\ddot{N}: \quad \longrightarrow \quad H\text{—}C\equiv N:$$

Practice Example A: Draw plausible Lewis structures for **(a)** NI_3 and **(b)** CS_2.

Practice Example B: Draw a plausible Lewis structure for dinitrogen difluoride, a compound discovered in 1952.

Formal Charge

We had no difficulty in determining that C is the central atom in the Lewis structure of HCN in Example 11-5. Can we similarly determine which is the central atom in the Lewis structure of nitrosyl chloride? Most reference sources list the formula of nitrosyl chloride as NOCl, but this doesn't seem right. The skeleton structure NOCl places O (EN = 3.5), the *most* electronegative of the three atoms, as a central atom. Deciding between N and Cl as the central atom also presents a problem: the two atoms have essentially the same electronegativity (EN = 3.0). The best way to establish the skeleton structure and the most plausible Lewis structure of nitrosyl chloride is through the concept of *formal charge*.

Formal charges are apparent charges on certain atoms in a Lewis structure that arise when the atoms have not made equal contributions of electrons to the covalent bond joining them. The **formal charge** on an atom in a Lewis structure is *the number of valence electrons in the free neutral atom minus the number of electrons assigned to that atom in the Lewis structure.* We assign electrons to the atoms in a Lewis structure in this way:

▶ The chief structures having O atoms as central atoms are those having a peroxo linkage (—O—O—) or a hydroxy group (—O—H). Otherwise, expect O atoms to be terminal atoms.

- Count all *lone-pair* electrons as belonging entirely to the atom in which they are found.

- Divide all *bond-pair* electrons *equally* between the bonded atoms.

If we write a Lewis structure based on the (incorrect) skeleton structure NOCl, one of the possibilities is

$$: \ddot{N}\!\!=\!\!\ddot{O}\!\!-\!\!\ddot{C}l :$$

When we assign electrons (e^-) in the manner just described, the formal charges (FC) we obtain are

$$
\begin{array}{llll}
N: & \textit{five} \text{ valence } e^- \text{ in N} & - & (\textit{four} \text{ lone-pair } e^-) & - & \tfrac{1}{2}(\textit{four} \text{ bond-pair } e^-) \\
\text{FC} = & 5 & - & 4 & - & 2 & = -1
\end{array}
$$

$$
\begin{array}{llll}
O: & \text{six valence } e^- \text{ in O} & - & (\textit{two} \text{ lone-pair } e^-) & - & \tfrac{1}{2}(\textit{six} \text{ bond-pair } e^-) \\
\text{FC} = & 6 & - & 2 & - & 3 & = +1
\end{array}
$$

$$
\begin{array}{llll}
Cl: & \textit{seven} \text{ valence } e^- \text{ in Cl} & - & (\textit{six} \text{ lone-pair } e^-) & - & \tfrac{1}{2}(\textit{two} \text{ bond-pair } e^-) \\
\text{FC} = & 7 & - & 6 & - & 1 & = 0
\end{array}
$$

To distinguish between formal charges and actual charges, we will use small encircled numbers to represent formal charges, as in the Lewis structure

$$: \overset{\ominus_1}{\ddot{N}}\!\!=\!\!\overset{\oplus_1}{\ddot{O}}\!\!-\!\!\ddot{C}l :$$ (11.7)

The assessment of formal charges that we have just made amounts to using the equation

$$\text{FC} = \text{no. valence } e^- \text{ in free atom} - \text{no. lone-pair } e^- - \tfrac{1}{2}(\text{no. bond-pair } e^-)$$ (11.8)

As a general rule, the preferred Lewis structure of a molecule is one with no formal charges, but failing this, we should note that

- Where formal charges are required, these should be as small as possible.
- Negative formal charges usually appear on the most electronegative atoms and positive formal charges on the least electronegative atoms.

▶ We will see some exceptions to the idea that formal charges should be kept to a minimum in Section 11-6.

- The sum of the formal charges of the atoms in a Lewis structure must equal *zero* for a neutral molecule and must equal the ionic charge for a polyatomic ion.

Notice that structure (11.7) does not conform to the second of these points; that is, the positive formal charge is on the more electronegative O atom and the negative formal charge on the less electronegative N atom. In Example 11-6 we present a strategy for determining the most plausible Lewis structure of nitrosyl chloride.

EXAMPLE 11-6

Using the Formal Charge Concept in Writing Lewis Structures. Write the most plausible Lewis structure of nitrosyl chloride, NOCl, one of the oxidizing agents present in *aqua regia,* a mixture of concentrated nitric and hydrochloric acids capable of dissolving gold.

Solution

We have already established that the skeleton structure is not NOCl, and now we must determine whether it is OClN or ONCl. Let's do this by applying the strategy outlined on page 356 and then assessing the resulting Lewis structures for formal charges.

Establish the number of dots that must appear in the Lewis structure. This is the total number of valence electrons in NOCl.

$$\underset{5}{\underset{From\ N}{}} + \underset{6}{\underset{From\ O}{}} + \underset{7}{\underset{From\ Cl}{}} = 18$$

Join each of the terminal atoms to the central atom by a *single* covalent bond. Then place the remaining electrons as lone pairs, first around the terminal atoms and then around the central atom. To complete the octet of the central atom, move a pair of electrons from a lone-pair position to form an additional bond to the central atom. At this point it becomes apparent that there are *two* possibilities based on structure (a) and *two* based on structure (b). These are structures (a$_1$), (a$_2$), (b$_1$), and (b$_2$), respectively.

(a)		(b)
O—Cl—N	1. Assign four electrons.	O—N—Cl
:Ö—Cl—N̈:	2. Assign twelve more electrons.	:Ö—N—Cl̈:
:Ö—Cl—N̈:	3. Assign the last two electrons.	:Ö—N—Cl̈:
	4. Complete the octet on the central atom.	

(a$_1$)	(a$_2$)	(b$_1$)	(b$_2$)
:Ö=Cl—N̈:	:Ö—Cl=N̈:	:Ö=N—Cl̈:	:Ö—N=Cl̈:

Evaluate formal charges using equation (11.8). In structure (a$_1$), for the N atom, $FC = 5 - 6 - \frac{1}{2}(2) = -2$; for the O atom, $FC = 6 - 4 - \frac{1}{2}(4) = 0$; for the Cl atom, $FC = 7 - 2 - \frac{1}{2}(6) = +2$. Proceed in a similar manner for the other three structures. Summarize the formal charges for the four structures.

	(a$_1$)	(a$_2$)	(b$_1$)	(b$_2$)
N:	−2	−1	0	0
O:	0	−1	0	−1
Cl:	+2	+2	0	+1

Select the best Lewis structure in terms of formal charge rules. First, note that all four structures obey the requirement that formal charges add up to zero. In structure (a_1) the formal charges are large (+2 on Cl and −2 on N) and the negative formal charge is not on the most electronegative atom. Structure (a_2) has formal charges on all atoms, one of them large (+2 on Cl). Structure (b_1) is the ideal we seek—no formal charges. In structure (b_2), we again have formal charges. The best Lewis structure of nitrosyl chloride is

$$: \ddot{O} = \ddot{N} - \ddot{\underset{..}{Cl}} :$$

▶ Based on structure (b_1), a better way to write the formula of nitrosyl chloride is ONCl.

Practice Example A: Use the formal charge concept to determine which of the following structures is a more plausible one for nitric acid, HNO_3.

(a)   (b)

Practice Example B: Write two Lewis structures for cyanamide, NH_2CN, an important chemical of the fertilizer and plastics industries. Use the formal charge concept to choose the more plausible structure.

Polyatomic Ions

Polyatomic ions consist of two or more atoms, and the forces holding atoms together *within* such ions are covalent bonds. Consider the hydroxide ion, OH^-. Its Lewis structure must involve *eight* electrons: six from O, one from H, and one additional electron to account for the ionic charge of 1−.

$$[: \ddot{O} : H]^-$$

The *ionic* compound sodium hydroxide consists of simple Na^+ cations and *polyatomic* anions, OH^-.

Figure 11-7 shows the formation of the ionic compound ammonium chloride by the reaction of HCl and NH_3. It also illustrates bonding in the polyatomic *cation* NH_4^+, the ammonium ion. To determine the number of valence electrons for the Lewis structure of NH_4^+ we count *five* from N, *one* each from *four* H atoms, *minus one* electron to account for the ionic charge 1+. This gives a total of eight electrons. Another way to look at the situation in Figure 11-7 is that in the formation of NH_4^+, the fourth hydrogen atom that attaches itself to the N atom of NH_3 leaves its electron behind with the Cl atom (converting it to Cl^-). In the bond between the N atom and this fourth H atom, the N atom supplies *both* electrons (its lone pair) for the covalent bond. A covalent bond in which one atom contributes both electrons is called a **coordinate covalent bond.** Once such a bond is formed, we cannot tell it from a regular covalent bond. In other words, in NH_4^+ all four N—H bonds are the same (for example, all have the same bond length). Note also that the formal charge on N is +1, and that the sum of the formal charges for all the atoms is equal to the ionic charge.

FIGURE 11-7
Formation of the ammonium ion, NH_4^+.
The H atom of HCl leaves its electron with the Cl atom and, as H^+, attaches itself to the NH_3 molecule through the lone pair electrons on the N atom. The ions NH_4^+ and Cl^- are formed.

▶ The formation of coordinate covalent bonds leads to formal charges in a Lewis structure.

EXAMPLE 11-7

Writing a Lewis Structure for a Polyatomic Ion. Data from spacecraft have established that the atmosphere of Mars is composed mainly of carbon dioxide. Through photochemical reactions in this atmosphere, small quantities of the ion CO_2H^+ are formed. Write a plausible Lewis structure for this ion. Identify the atom that carries the positive formal charge.

Solution

In the usual fashion, we should begin by assessing the total number of valence electrons that must appear in the Lewis structure.

From C   From O   From H   To establish charge of 1+

$$4 \; + (2 \times 6) + \; 1 \; - \; 1 \; = 16$$

Carbon, with the lowest electronegativity, is the central atom. The skeleton structure involves a central C atom, two O atoms bonded to it, and an H atom bonded to one of the O atoms. When 16 electrons are distributed among these atoms in the usual way, we get

$$[:\overset{..}{O}-C-\overset{..}{\underset{..}{O}}-H]^+$$

In this structure the central C atom is lacking an octet; it needs two more electron pairs. These can be acquired by shifting a lone pair of electrons from each O atom into the bond with the C atom to obtain

$$[:\overset{..}{O}-C-\overset{..}{\underset{..}{O}}-H]^+ \longrightarrow [:\overset{..}{O}=C=\overset{..}{O}-H]^+$$

To determine which atom carries the positive formal charge we can apply the rules for formal charges. The hydrogen atom has one bond and no formal charge. The left-hand oxygen atom also has no formal charge: $FC = 6 - 4 - \frac{1}{2}(4) = 0$.

The carbon atom has no formal charge: $FC = 4 - 0 - \frac{1}{2}(8) = 0$.

This leaves the oxygen atom to which carbon and hydrogen atoms are bonded. We now expect it to carry the formal charge: $FC = 6 - 2 - \frac{1}{2}(6) = +1$.

Practice Example A: Write a Lewis structure for the BF_4^- ion. Identify the atom that carries the negative formal charge.

Practice Example B: Write a Lewis structure for sodium peroxide, Na_2O_2. (*Hint:* The compound contains both ionic and covalent bonds.)

11-5   Resonance

The ideas presented in the previous section allow us to write many Lewis structures, but some structures still present problems. We describe these problems in the next two sections.

As we learned in Chapter 8, although oxygen commonly occurs as *diatomic* molecules O_2, it can also exist as *triatomic* molecules of *ozone,* O_3. Ozone is found naturally in the stratosphere and is also produced in the lower atmosphere as a constituent of smog.

When we apply the usual rules for Lewis structures for ozone we come up with these *two* possibilities.

$$:\overset{..}{O}=\overset{..}{O}-\overset{..}{\underset{..}{O}}: \qquad :\overset{..}{\underset{..}{O}}-\overset{..}{O}=\overset{..}{O}:$$

But there is something wrong with both structures. Each suggests that one oxygen-to-oxygen bond is single and the other is double. Yet, experimental evidence indicates that the two oxygen-to-oxygen bonds are the same; each has a length of 127.8 pm. This bond length is shorter than the O—O single-bond length of 147.5 pm in hydrogen peroxide, H—Ö—Ö—H, but it is longer than the double-bond length of 120.74 pm in diatomic oxygen, :O=O:. The bonds in ozone are intermediate between a single and a double bond. The difficulty is resolved if we say that the *true* Lewis structure of O_3 is *neither* of the previously proposed structures, but a composite, or *hybrid*, of the two, a fact that we can represent as

$$:Ö=Ö—Ö: \longleftrightarrow :Ö—Ö=Ö: \qquad (11.9)$$

The situation in which two or more plausible Lewis structures can be written but the "correct" structure cannot be written at all is called **resonance.** The true structure is a *resonance hybrid* of plausible contributing structures. Acceptable contributing structures to a resonance hybrid must all have the same skeleton structure; they can differ only in how electrons are distributed within the structure. In expression (11.9) the two contributing structures are joined by a double-headed arrow. The arrow does *not* mean that the molecule has one structure part of the time and the other structure the rest of the time. It has the *same* structure *all* of the time. By averaging the single bond in one structure with the double bond in the other, we might say that the oxygen-to-oxygen bonds in ozone have a bond order of about $\frac{3}{2}$, or 1.5.

▶ Think of a mule as a resonance hybrid of a horse and a donkey. The mule is a distinctive animal with features of both a horse and a donkey, but it is not half horse–half donkey. Neither is it a horse half of the time and a donkey the rest.

EXAMPLE 11-8

Representing the Lewis Structure of a Resonance Hybrid. Write the Lewis structure of the nitrate ion NO_3^-. Nitrogen is the central atom, and all three nitrogen-to-oxygen bonds in the ion are identical.

Solution

Our structure must show 24 valence electrons: *five* from N, *six* each from *three* O atoms, and *one* extra electron to produce the ionic charge of $1-$. Our first attempt at distributing 24 valence electrons leads to a structure in which the central N atom does not have an octet.

$$\left[\begin{array}{c} :\ddot{O}: \\ | \\ :\ddot{O}—N—\ddot{O}: \end{array} \right]^-$$

We need to shift a lone pair of electrons from one of the O atoms into one of the nitrogen-to-oxygen bonds.

$$\left[\begin{array}{c} :\ddot{O} \\ \| \\ :\ddot{O}—N—\ddot{O}: \end{array} \right]^-$$

This structure is plausible, but it does not conform to the observation that the nitrogen-to-oxygen bonds are identical. No single structure will do this. We need the resonance hybrid of these contributing structures.

$$\left[\begin{array}{c} :\ddot{O} \\ \| \\ :\ddot{O}—N—\ddot{O}: \end{array} \right]^- \longleftrightarrow \left[\begin{array}{c} :\ddot{O}: \\ | \\ :\ddot{O}=N—\ddot{O}: \end{array} \right]^- \longleftrightarrow \left[\begin{array}{c} :\ddot{O}: \\ | \\ :\ddot{O}—N=\ddot{O}: \end{array} \right]^- \qquad (11.10)$$

Practice Example A: Draw Lewis structures to represent resonance in the acetate ion, $CH_3CO_2^-$.

Practice Example B: Sodium azide, NaN_3, is the nitrogen gas-forming substance used in automobile air-bag systems. It is an ionic compound containing the azide ion, N_3^-. In this ion, the two nitrogen-to-nitrogen bond lengths are 116 pm. Describe the resonance hybrid Lewis structure of this ion. (*Hint:* Use bond length data from Table 11.1 and show that there are *three* contributing structures.)

Are You Wondering . . .

Why we don't write structures like the following for nitrate ion?

$$\left[:\overset{..}{O}=\overset{..}{N}-\overset{..}{\underset{..}{O}}-\overset{..}{\underset{..}{O}}: \right]^- \quad (incorrect)$$

1. This structure is not equivalent to those in expression (11.10); it has a different skeleton structure.

2. Even though this structure conforms to our Lewis structure rules, it doesn't conform to the *experimental* evidence. It has only *two* nitrogen-to-oxygen bonds, and an oxygen-to-oxygen bond (peroxo bond) that is not found in NO_3^-.

3. Although some molecules have several central atoms, usually simple structures are compact. Where possible, we write Lewis structures with a single central atom.

11-6 Exceptions to the Octet Rule

The octet rule has been a mainstay in writing Lewis structures, and it will continue to be one. Yet at times we must depart from the octet rule, as we will see in this section.

Odd-Electron Species

The molecule NO has 11 valence electrons, an *odd* number. If the number of valence electrons in a Lewis structure is *odd*, there must be an unpaired electron somewhere in the structure. Lewis theory deals with electron pairs and does not tell us where to put the unpaired electron; it could be on either the N or the O atom. To obtain a structure free of formal charges, however, we will put the unpaired electron on the N atom.

$$\cdot \overset{..}{N}=\overset{..}{O}:$$

The presence of unpaired electrons causes odd-electron species to be *paramagnetic*. NO is paramagnetic. Molecules with an *even* number of electrons are expected to have all electrons paired and to be *diamagnetic*. This is usually the case, yet O_2, with 12 valence electrons, is *paramagnetic*. Lewis theory does not provide a good electronic structure for O_2, but the molecular orbital theory that we will consider in the next chapter is much more successful.

▶ Experimental evidence for the paramagnetism of O_2 is shown in Figure 11-2 on page 350.

The number of stable odd-electron molecules is quite limited. More common are highly reactive molecular fragments with one or more unpaired electrons, called **free radicals,** or simply *radicals*. The formulas of free radicals are usually written with a "dot" to emphasize the presence of an unpaired electron, such as in the *methyl* radical, $\cdot CH_3$, and the *hydroxyl* radical, $\cdot OH$. The Lewis structures of these two free radicals are

$$
\begin{array}{c}
\text{H} \\
| \\
\text{H---C---H} \qquad\qquad \cdot\ddot{\text{O}}\text{---H} \\
| \\
\text{H}
\end{array}
$$

Both free radicals are commonly encountered as transitory species in flames. In addition, $\cdot OH$ is formed in the atmosphere in trace amounts as a result of photochemical reactions. Many important atmospheric reactions involve free radicals as reactants, such as in the oxidation of CO to CO_2.

$$\cdot OH + CO \longrightarrow CO_2 + \cdot H$$

Incomplete Octets

To write the Lewis structure of boron trifluoride, BF_3, our initial attempt leads to a structure in which the B atom has only six electrons in its valence shell—an incomplete octet. We have learned to correct this situation by shifting a lone pair of electrons from a terminal atom into a bond with the central atom, and here this means a boron-to-fluorine double bond. Also, we have learned that when equivalent structures are possible we should represent the Lewis structure as a resonance hybrid: In BF_3 there are three structures with a boron-to-fluorine double bond. In view of its molecular properties and chemical behavior, the best representation of BF_3 appears to be a resonance hybrid based on *four* contributing structures, including one with an incomplete octet.

$$
:\!\ddot{\text{F}}\text{---B---}\ddot{\text{F}}\!: \longleftrightarrow :\!\ddot{\text{F}}\text{---}\overset{\ominus}{\text{B}}\!=\!\overset{\oplus}{\text{F}}\!: \longleftrightarrow :\!\overset{\oplus}{\text{F}}\!=\!\overset{\ominus}{\text{B}}\text{---}\ddot{\text{F}}\!: \longleftrightarrow :\!\ddot{\text{F}}\text{---}\overset{\ominus}{\text{B}}\!=\!\ddot{\text{F}}\!: \qquad (11.11)
$$

The contributing structures with boron-to-fluorine double bonds have formal charges of -1 on B and $+1$ on F, contrary to the rules we have been using. The Lewis structure with an incomplete octet has no formal charges and might therefore appear to be the most favorable. On the other hand, the B—F bond length in BF_3 (130 pm) is less than expected for a single bond, underscoring the importance of the double-bonded structures to the resonance hybrid. Whichever BF_3 structure we choose to emphasize, an important characteristic of BF_3 is its strong tendency to form a coordinate covalent bond with a species capable of donating an electron pair to the B atom. This can be seen in the formation of the BF_4^- ion.

▶ The primary industrial uses of BF_3 are not to produce chemicals containing the elements boron or fluorine. Rather, it is used because of properties stemming from its electronic structure. In most cases the BF_3 is recovered and recycled.

$$
:\!\ddot{\text{F}}\!:^- + \text{B}\!\!-\!\!\ddot{\text{F}}\!: \longrightarrow \left[:\!\ddot{\text{F}}\text{---}\overset{\displaystyle :\ddot{\text{F}}:}{\underset{\displaystyle :\ddot{\text{F}}:}{\text{B}}}\text{---}\ddot{\text{F}}\!: \right]^-
$$

In BF_4^- the bonds are single bonds and the bond length is 145 pm.

The number of species with incomplete octets is limited to some beryllium, boron, and aluminum compounds. Perhaps the best examples are the boron hydrides. Bonding in the boron hydrides will be discussed in Chapter 23.

"Expanded" Octets

We have consistently tried to write Lewis structures in which all atoms except H have a complete octet, that is, in which each atom has 8 valence electrons. There are a few Lewis structures that break this rule by having 10 or even 12 valence electrons around the central atom, creating what is called an **expanded octet.** Describing bonding in these structures is an area of current active interest among chemists.

Molecules with expanded octets typically involve nonmetal atoms of the third period and beyond that are bonded to highly electronegative atoms. For example, phosphorus forms two chlorides, PCl_3 and PCl_5. We can write a Lewis structure for PCl_3 with the octet rule. In PCl_5, with five Cl atoms bonded directly to the central P atom, the outer shell of the P atom appears to have *ten* electrons. We might say that the octet has been "expanded" to 10 electrons. In the SF_6 molecule the octet appears to expand to 12.

octet   expanded octet   expanded octet

The problem with expanded octet structures, of course, is where do the "extra" electrons go? This expansion has been rationalized by assuming that after the $3s$ and $3p$ subshells fill to capacity in the central atom (eight electrons), extra electrons go into the empty $3d$ subshell. The thought is that the energy difference between the $3p$ and $3d$ levels is not very large, and so the octet expansion scheme seems reasonable.

Recently, however, some chemists have used sophisticated computer modeling programs to suggest that the $3d$ subshell is not significantly involved in bonding in molecules such as PCl_5 and SF_6, but a discussion of the evidence in support of this assertion is beyond the scope of this text. In any case, our need is just to recognize that to write useful Lewis structures for certain species we may at times have to employ an expanded octet on the central atom.

The concept of expanded octets has also been used at times in cases where they appear to give a better Lewis structure than strict adherence to the octet rule. Shown below are two Lewis structures for the sulfate ion, one based on the octet rule and the other on an expanded octet.

normal octet   expanded octet

The argument for including the expanded-octet structure is that it reduces formal charges. Also, the experimentally determined sulfur-to-oxygen bond lengths seem to be shorter than single bonds. The expanded octet structure is suggestive of this partial double bond character, whereas the octet structure is not. A newer proposal, however, argues that the reduction in bond length can be attributed to the partial ionic character of the sulfur-to-oxygen bonds, and this partial ionic character is emphasized in the octet structure with its positive and negative formal charges.

Although unresolved questions about expanded octets may seem unsettling to you, the point to keep in mind is that the unmodified octet rule works perfectly well for most of our uses of Lewis structures. It will continue to be our mainstay for the remainder of the chapter.

11-7 The Shapes of Molecules

When we write the Lewis structure for water

$$H:\overset{..}{\underset{..}{O}}:H$$

we get the impression that the atoms are arranged in a straight line. The experimentally determined shape of the molecule is *not* linear, however. The molecule is *bent*, as shown in Figure 11-8. Does it really matter that the H_2O molecule is bent rather than linear?

Of course, the answer is yes. The bent shape of water molecules helps to account for the fact that water is a liquid rather than a gas at room temperature. It also accounts for the ability of liquid water to dissolve so many different substances. We will see a number of examples of the importance of molecular shapes later in the text.

What we seek in this section is a simple model that allows us to predict the approximate shape of a molecule. Unfortunately, Lewis theory does not tell us anything about the shapes of molecules. It cannot serve as our model, but it is an excellent place to begin. The next step is to use an idea based on repulsions between valence-shell electron pairs. We will discuss this idea after defining a few terms.

Some Terminology

By molecular shape we mean the geometric figure we get by joining the nuclei of bonded atoms by straight lines. Figure 11-8 depicts the *triatomic* (three-atom) water molecule through a ball-and-stick model. The balls represent the three atoms in the molecule and the straight lines (sticks) the bonds between atoms. In reality the atoms in the molecule are in close contact, but for clarity we show only the centers of the atoms. To have a complete description of the shape of a molecule, we need to know two quantities,

- **bond lengths,** the distances between the nuclei of bonded atoms, and
- **bond angles,** the angles between adjacent lines representing bonds.

Of these two quantities, we will concentrate on bond angles.

A *diatomic* molecule has only one bond and no bond angle. Because the geometric shape determined by two points is a straight line, *all diatomic molecules are linear.* A *triatomic* molecule has two bonds and one bond angle. If the bond angle is 180°, the three atoms lie on a straight line and the molecule is *linear.* For any other bond angle the triatomic molecule is said to be *angular, bent,* or *V-shaped.* Some *polyatomic* molecules with more than three atoms have planar or even linear shapes. More commonly, however, the centers of the atoms in these molecules define a three-dimensional geometric figure.

Valence-Shell Electron-Pair Repulsion (VSEPR) Theory

The shape of a molecule is established by experiment or by a quantum mechanical calculation confirmed by experiment. There is a way of thinking about molecules, however, that lets us predict their probable shapes. These predictions

FIGURE 11-8
Geometric shape of a molecule.
The triatomic H_2O molecule is represented. To establish the shape of this molecule we need to determine the distances between the nuclei of the bonded atoms and the angle between adjacent bonds. In H_2O the bond lengths $d_1 = d_2 = 95.8$ pm and the bond angle $\alpha = 104.45°$.

FIGURE 11-9
Balloon analogy to valence-shell electron-pair repulsion.
When two elongated balloons are twisted together, they separate into four lobes. To minimize interferences, the lobes spread out into a tetrahedral pattern. (A regular tetrahedron has four faces, each an equilateral triangle.) The lobes are analogous to valence-shell electron pairs.

are usually in good agreement with experimental results. The **valence-shell electron-pair repulsion theory** (written **VSEPR** and pronounced "vesper") focuses on *pairs* of electrons in the *valence* electronic shell. *Electron pairs repel each other, whether they are in chemical bonds (bond pairs) or unshared (lone pairs). Electron pairs assume orientations about an atom to minimize repulsions.* This, in turn, results in particular geometric shapes for molecules.

Before we examine individual molecules, let us focus not just on *pairs* of electrons, but rather on *electron groups.* A group of electrons can be a pair, either a lone pair or a bond pair, but it can also be a single unpaired electron on an atom with an incomplete octet, such as in NO. A group can also be a double or triple bond between two atoms. So that in C≡O, there is only one electron group located between the two atoms.

Consider the methane molecule, CH_4, in which the central C atom has acquired the electron configuration of Ne by forming covalent bonds with four H atoms.

$$\begin{array}{c} H \\ \cdot\cdot \\ H : C : H \\ \cdot\cdot \\ H \end{array}$$

What orientation will the four electron groups (bond pairs) assume? The balloon analogy of Figure 11-9 suggests that electron group repulsions will force the groups as far apart as possible—to the corners of a tetrahedron having the C atom at its center. The VSEPR method predicts, correctly, that CH_4 is a *tetrahedral* molecule.

In NH_3 and H_2O the central atom is also surrounded by four groups of electrons,

$$\begin{array}{cc} H & H \\ \cdot\cdot & \cdot\cdot \\ H : N : H \qquad \text{and} \qquad : O : H \\ \cdot\cdot & \cdot\cdot \end{array}$$

but these molecules do not have a tetrahedral shape. Here is the situation: VSEPR theory predicts the distribution of electron groups, and in these molecules electron groups are arranged tetrahedrally about the central atom. The shape of a molecule, however, is determined by the location of the atomic nuclei. To avoid confusion, we will call the geometric distribution of electron groups the **electron group geometry** and the geometric arrangement of the atomic nuclei—the actual determinant of the molecular shape—the **molecular geometry.**

In the NH_3 molecule only *three* of the electron groups are bond pairs; the fourth is a *lone pair.* By joining the N nucleus to the H nuclei by straight lines, we outline a *pyramid* (called a trigonal pyramid). This pyramid has the N atom at the apex and the three H atoms at the base. This is not the same as a tetrahedron, which would have the N atom at the center. We say that the electron group geometry is tetrahedral and the molecular geometry is trigonal-pyramidal.

In the H_2O molecule two of the four electron groups are *bond* pairs and two are *lone* pairs. The molecular shape is obtained by joining the two H nuclei to the O nucleus with straight lines. For H_2O, the electron group geometry is tetrahedral and the molecular geometry is V-shaped, or bent. The geometric shapes of CH_4, NH_3, and H_2O are shown in Figure 11-10, together with "space-filling" molecular models. Unlike a ball-and-stick model, a "space-filling" model uses balls proportioned in size to actual atomic sizes, and the balls are in contact.

In the VSEPR notation used in Figure 11-10, A is the central atom, X is a terminal atom or group of atoms bonded to the central atom, and E is a lone

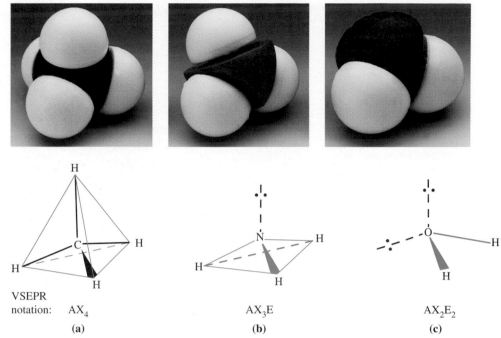

VSEPR
notation: AX_4 AX_3E AX_2E_2

(a) (b) (c)

FIGURE 11-10 **Molecular shapes based on tetrahedral electron group geometry—CH_4, NH_3, and H_2O.**
Molecular shapes are established by the blue lines. Lone pair electrons are shown as red dots along broken lines originating at the central atom. (a) All electron groups around the central atom are bond pairs. The blue lines that outline the molecule are different from the black lines representing the carbon-to-hydrogen bonds. (b) The lone pair of electrons is directed to the "missing" corner of the tetrahedron. The nitrogen-to-hydrogen bonds form three of the edges of a trigonal pyramid. (c) The H_2O molecule is a bent molecule outlined by the two oxygen-to-hydrogen bonds.

pair of electrons. Thus, the symbol AX_2E_2 signifies that *two* atoms or groups (X) are bonded to the central atom (A). The central atom also has *two* lone pairs of electrons (E). H_2O is an example of a molecule of the AX_2E_2 type.

For tetrahedral electron group geometry we expect bond angles of 109.5°, known as the *tetrahedral* bond angle. In the CH_4 molecule the measured bond angles are, in fact, 109.5°. The bond angles in NH_3 and H_2O are slightly smaller: 107° for the H—N—H bond angle and 104.5° for the H—O—H bond angle. We can explain these less-than-tetrahedral bond angles by assuming that the charge cloud of the lone pair electrons spreads out. This forces the bond pair electrons closer together and reduces the bond angles.

▶ VSEPR theory works best for second-period elements. The predicted bond angle of 109.5° for H_2O is close to the measured angle of 104.5°. For H_2S, however, the predicted value of 109.5° is not in good agreement with the observed 92°.

Possibilities for Electron Group Distributions

In general, we will encounter situations in which central atoms have two, three, four, five, or six electron groups distributed around them. The electron group geometries for these cases are

- two electron groups: linear
- three electron groups: trigonal-planar
- four electron groups: tetrahedral
- five electron groups: trigonal-bipyramidal
- six electron groups: octahedral

FIGURE 11-11
Several electron group geometries illustrated.
The electron group geometries pictured are trigonal-planar (orange), tetrahedral (blue-green), trigonal-bipyramidal (pink), and octahedral (yellow). The atoms at the ends of the balloons are not shown and are not important in this model. The relationship between electron group geometry and molecular geometry is summarized in Table 11.2.

Figure 11-11 extends the "balloon" analogy to these cases. The cases for five or six electron groups would be typified by expanded octet molecules like PCl_5 or SF_6.

The molecular geometry is the same as the electron group geometry *only* when all electron groups are bond pairs. These are for the VSEPR notation AX_n (that is, AX_2, AX_3, AX_4, . . .). In Table 11.2 the AX_n cases are illustrated by photographs of ball-and-stick models. If one or more electron groups are lone pairs, the molecular geometry is *different* from the electron group geometry, although still derived from it. To understand all the cases in Table 11.2 we need two more ideas.

- *The closer together two groups of electrons are forced, the stronger the repulsion between them.* The repulsion between two electron groups is much stronger at an angle of 90° than at 120° or 180°.

- *Lone pair electrons spread out more than do bond pair electrons.* As a result the repulsion of one lone pair of electrons for another lone pair is greater than, say, between two bond pairs. We can set up this order of repulsive forces, from strongest to weakest.

  lone pair–lone pair > lone pair–bond pair > bond pair–bond pair

Consider SF_4 (with the VSEPR notation AX_4E). There are *two* possibilities for the lone pair electrons. In Table 11.2 we put the lone pair into the central plane of the bipyramid. As a result *two* of the lone pair-bond pair interactions are 90°. If we had put the lone pair of electrons at the top of the bipyramid, above the central plane, *three* lone pair-bond pair interactions would have been at 90°, a less favorable arrangement.

Applying VSEPR Theory

Let us use the following four-step strategy for predicting the shapes of molecules.

1. Draw a plausible Lewis structure of the species (molecule or polyatomic ion).

2. Determine the number of electron groups around the central atom and identify them as being either *bond* pairs or *lone* pairs.

3. Establish the electron group geometry around the central atom—linear, trigonal-planar, tetrahedral, trigonal-bipyramidal, or octahedral.

TABLE 11.2   Molecular Geometry as a Function of Electron Group Geometry

Number of Electron Groups	Electron Group Geometry	Number of Lone Pairs	VSEPR Notation	Molecular Geometry	Ideal Bond Angles	Example	
2	linear	0	AX_2	X—A—X (linear)	180°	$BeCl_2$	(BeCl₂)
3	trigonal-planar	0	AX_3	(trigonal-planar)	120°	BF_3	(BF₃)
	trigonal-planar	1	AX_2E	(angular)	120°	SO_2[a]	
4	tetrahedral	0	AX_4	(tetrahedral)	109.5°	CH_4	(CH₄)
	tetrahedral	1	AX_3E	(trigonal-pyramidal)	109.5°	NH_3	
	tetrahedral	2	AX_2E_2	(angular)	109.5°	OH_2	
5	trigonal-bipyramidal	0	AX_5	(trigonal-bipyramidal)	90°, 120°	PCl_5	(PCl₅)

(continues)

TABLE 11.2   (Continued)

Number of Electron Groups	Electron Group Geometry	Number of Lone Pairs	VSEPR Notation	Molecular Geometry	Ideal Bond Angles	Example
	trigonal-bipyramidal	1	AX_4E^b	(seesaw)	90°, 120°	SF_4
	trigonal-bipyramidal	2	AX_3E_2	(T-shaped)	90°	ClF_3
	trigonal-bipyramidal	3	AX_2E_3	(linear)	180°	XeF_2
6	octahedral	0	AX_6	(octahedral)	90°	SF_6
	octahedral	1	AX_5E	(square-pyramidal)	90°	BrF_5
	octahedral	2	AX_4E_2	(square-planar)	90°	XeF_4

(SF_6)

[a]For a discussion of the structure of SO_2, see page 371.

[b]For a discussion of the placement of the lone pair electrons in this structure, see page 368.

4. Determine the molecular geometry from the positions around the central atom occupied by the other atomic nuclei—that is, from data in Table 11.2.

EXAMPLE 11-9

Using VSEPR Theory to Predict a Geometric Shape. Predict the molecular geometry of the polyatomic anion ICl_4^-.

Solution

Apply the four steps outlined above.

Step 1. Write the Lewis structure. The number of valence electrons is

From I	*From Cl*	*To establish ionic charge of −1*	
$(1 \times 7) +$	$(4 \times 7) +$	1	$= 36$

To join 4 Cl atoms to the central I atom and to provide octets for all the atoms we need 32 electrons. In order to account for all 36 valence electrons we need to place an *additional* four electrons around the I atom as lone pairs. That is, we are forced to go to an expanded octet for the I atom. It does not matter, though, how we show the two lone pairs of electrons around the I atom in the Lewis structure.

$$\left[\begin{array}{c} \ddot{\underset{..}{Cl}} \quad \ddot{\underset{..}{Cl}} \\ \ddot{I} \\ \ddot{\underset{..}{Cl}} \quad \ddot{\underset{..}{Cl}} \end{array} \right]^-$$

Step 2. There are six electron groups around the I atom, four *bond* pairs and two *lone* pairs.

Step 3. The electron group geometry, that is, the orientation of six electron groups, is *octahedral*.

Step 4. The ICl_4^- anion is of the type AX_4E_2, which according to Table 11.2 leads to a molecular geometry that is square planar.

Figure 11-12 suggests two possibilities for distributing bond pairs and lone pairs in ICl_4^-. The square planar structure is correct because the lone pair–lone pair interaction is kept at 180°. In the incorrect structure this interaction is at 90°, which results in a strong repulsion.

Practice Example A: Predict the molecular geometry of nitrogen trichloride.

Practice Example B: Predict the molecular geometry of phosphoryl chloride, $POCl_3$, an important chemical in the manufacture of gasoline additives, hydraulic fluids, and fire retardants.

▶ Despite the reservations about the expanded octet concept discussed on page 364, expanded octets combined with VSEPR theory do a good job of describing molecular shapes of species with five and six electron group geometries.

(incorrect)

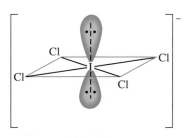

(correct)

FIGURE 11-12
Two predictions of the structure of ICl_4^-—Example 11-9 illustrated.
The observed structure is the square planar structure.

Structures with Multiple Covalent Bonds

In a multiple covalent bond, all electrons in the bond are confined to the region between the bonded atoms and together constitute one group of electrons. Let us test this idea by predicting the molecular geometry of sulfur dioxide. S is the central atom, and the total number of valence electrons is 18: (3×6). The Lewis structure is the resonance hybrid of two contributing structures.

$$\ddot{\underset{..}{O}} = \overset{\textstyle \ddot{S}}{\underset{\textstyle \ddot{\underset{..}{O}}}{\diagup}} \longleftrightarrow \overset{\textstyle \ddot{S}}{\underset{\textstyle \ddot{\underset{..}{O}}}{\diagdown}} \ddot{\underset{..}{O}}$$

It is immaterial which structure we use. In either case, we count the electrons in the double covalent bond as one group. This bond and the sulfur-to-oxygen single bond account for *two* electron groups. The third electron group around the central S atom is a *lone* pair. The electron group geometry around the central S atom is that of *three* electron groups—*trigonal-planar.* Of the three electron groups, two are bonding groups and one is a lone pair. This is the case of AX_2E (see Table 11.2). The molecular shape is *angular* or *bent,* with an expected bond angle of 120°. (The measured bond angle in SO_2 is 119°.)

EXAMPLE 11-10

Using VSEPR Theory to Predict the Shape of a Molecule with a Multiple Covalent Bond. Predict the molecular geometry of formaldehyde, H_2CO, used to make a number of polymers such as melamine resins.

Solution

The Lewis structure has a total of 12 valence electrons and C as the central atom. If all the bonds to the carbon atom were single bonds, C would lack an octet. This deficiency is corrected by moving a lone pair of electrons from the O atom into the carbon-to-oxygen bond, making it a double bond.

We count *three* electron groups around the C atom, two groups in the carbon-to-hydrogen single bonds and the third group in the carbon-to-oxygen double bond. The electron group geometry for three electron groups is *trigonal planar.* Because all the electron groups are involved in bonding, the VSEPR notation for this molecule is AX_3. The molecular geometry is also trigonal planar.

Practice Example A: Predict the shape of the COS molecule.

Practice Example B: Nitrous oxide, N_2O, is the familiar "laughing gas" used as an anesthetic in density. Predict the shape of the N_2O molecule.

Molecules with More Than One Central Atom

Although many of the structures of interest to us have only a single central atom, we can apply VSEPR theory to molecules or polyatomic anions with more than one central atom. What we have to do is to work out the geometric distribution of terminal atoms around *each* central atom and then combine the results into a single description of the molecular shape. We use this idea in Example 11-11.

EXAMPLE 11-11

Applying VSEPR Theory to a Molecule with More Than One Central Atom. Methyl isocyanate, CH_3NCO, is used in the manufacture of insecticides, such as carbaryl (Sevin). In the CH_3NCO molecule, the three H atoms and the O atom are terminal atoms and the two C and one N atom are central atoms. Draw a sketch of this molecule.

Solution

To apply the VSEPR method let us begin with a plausible Lewis structure. The number of valence electrons in the structure is

From C   From N   From O   From H

$(2 \times 4) + (1 \times 5) + (1 \times 6) + (3 \times 1) = 22$

In drawing the skeleton structure and assigning valence electrons, we first obtain a structure with incomplete octets. By shifting the indicated electrons, we can give each atom an octet.

$$\underset{\underset{H}{|}}{\overset{\overset{H}{|}}{H-C-N-C-O:}} \longrightarrow \underset{\underset{H}{|}}{\overset{\overset{H}{|}}{H-C-N=C=O:}}$$

   The C atom on the left has four electron groups around it—all bond pairs. The shape of this end of the molecule is *tetrahedral*. The C atom to the right, by forming two double bonds, is treated as having *two* groups of electrons around it. This distribution is *linear*. For the N atom *three* groups of electrons are distributed in a *trigonal-planar* manner. The C—N—C bond angle should be about 120°.

Practice Example A: Draw a sketch of the methanol molecule, CH_3OH. Indicate the bond angles in this molecule.

Practice Example B: Glycine, an amino acid, has the formula H_2NCH_2COOH. Sketch the glycine molecule and indicate the various bond angles.

Molecular Shapes and Dipole Moments

Let us recall some facts that we learned about polar covalent bonds in Section 11-3. In the HCl molecule, the Cl atom is more electronegative than the H atom. Electrons are displaced toward the Cl atom. The HCl molecule is a **polar molecule.** In the following representation, we use a cross-base arrow ($\longleftrightarrow$) that points to the atom that attracts electrons more strongly.

$$H \longleftrightarrow Cl$$

   The extent of the charge displacement in a polar covalent bond is given by the **dipole moment, μ.** The dipole moment is the product of a partial charge (δ) and distance (d).

$$\mu = \delta d \qquad (11.12)$$

If the product, δd, has a value of 3.34×10^{-30} coulomb · meter (C · m), the dipole moment, μ, has a value called 1 *debye, D.*

   The polarity of the H—Cl bond, as we demonstrated on page 352, involves a shift of the electron charge density toward the Cl atom and a separation of the centers of positive and negative charge. Suppose, instead, we think of the equiv-

(a) CCl$_4$: a nonpolar molecule

(b) CHCl$_3$: a polar molecule

FIGURE 11-13
Molecular shapes and dipole moments.
(a) The resultant of two of the C—Cl bond moments is shown as a red arrow, and that of the other two, as a blue arrow. The red and blue arrows point in opposite directions and cancel. The CCl$_4$ molecule is nonpolar. (b) The individual bond moments do combine to yield a resultant dipole moment (red arrow) of 1.01 D.

▶ In Feature Problem B on page 385 we relate the measured dipole moment of water to the H—O—H bond angle.

alent transfer of a *fraction* of the charge of an electron from the H atom to the Cl atom through the entire internuclear distance. Let us determine the magnitude of this partial charge, δ. To do this we need the measured dipole moment, 1.03 D; the H—Cl bond length, 127.4 pm; and Equation (11.12) rearranged to

$$\delta = \frac{\mu}{d} = \frac{1.03 \text{ D} \times 3.34 \times 10^{-30} \text{ C} \cdot \text{m/D}}{127.4 \times 10^{-12} \text{ m}} = 2.70 \times 10^{-20} \text{ C}$$

This charge is about 17% of the charge on an electron (1.60×10^{-19} C) and suggests that HCl is about 17% ionic. This assessment of the percent ionic character of the H—Cl bond agrees well with the 20% we made based on electronegativity differences (recall Example 11-4).

CO$_2$. The molecule CO$_2$ is *nonpolar.* To understand this observation, we need to distinguish between the displacement of electron charge density in a particular bond and in the molecule as a whole. The electronegativity difference between C and O causes a displacement of electron charge density toward the O atom in each carbon-to-oxygen bond and gives rise to a *bond* moment. Because the two bond moments are equal in magnitude and point in opposite directions, however, they cancel each other and lead to a *resultant* dipole moment of zero for the molecule.

$$O \xleftarrow{\;+\;} C \xrightarrow{\;+\;} O \qquad \mu = 0$$

The fact that CO$_2$ is nonpolar is experimental proof that CO$_2$ is a linear molecule. Of course, we might also predict that CO$_2$ is a linear molecule with the VSEPR theory, based on the Lewis structure

$$: \overset{..}{O} = C = \overset{..}{O} :$$

H$_2$O. The molecule H$_2$O is *polar.* It has bond moments because of the electronegativity difference between H and O, and the bond moments combine to produce a resultant dipole moment of 1.84 D. The molecule cannot be linear, for this would lead to a cancellation of bond moments, just as with CO$_2$. We have predicted with the VSEPR theory that the H$_2$O molecule is bent, and the observation that it is a polar molecule simply confirms the prediction.

CCl$_4$. The molecule CCl$_4$ is *nonpolar.* Based on the electronegativity difference between Cl and C, we expect a bond moment for the C—Cl bond. The fact that the resultant dipole moment is *zero* means that the bond moments must be oriented in such a way that they cancel. The tetrahedral molecular geometry of CCl$_4$ provides the symmetrical distribution of bond moments that leads to this cancellation, as shown in Figure 11-13. Can you see that the molecule will be polar if we replace one of the Cl atoms by an atom with a different electronegativity, say H? In the molecule CHCl$_3$ there is a resultant dipole moment (see Figure 11-13).

EXAMPLE 11-12

Determining the Relationship Between Geometric Shapes and the Resultant Dipole Moments of Molecules. Which of these molecules would you expect to be polar? Cl$_2$, ICl, BF$_3$, NO, SO$_2$.

Solution

Polar: ICl, NO, SO$_2$.   ICl and NO are diatomic molecules with an electronegativity difference between the bonded atoms. SO$_2$ is a bent molecule with an electronegativity difference between the S and O atoms.

Nonpolar: Cl$_2$ and BF$_3$.   Cl$_2$ is a diatomic molecule of identical atoms; hence no electronegativity difference. For BF$_3$ refer to Table 11.2. BF$_3$ is a symmetrical planar molecule (120° bond angles). The B—F bond moments cancel each other.

Practice Example A: Only one of the following molecules is polar. Which is it, and why? SF$_6$, H$_2$O$_2$, C$_2$H$_4$.

Practice Example B: Only one of the following molecules is *nonpolar.* Which is it, and why? Cl$_3$CCH$_3$, PCl$_5$, CH$_2$Cl$_2$, NH$_3$.

11-8   Bond Energies

We have stressed that Lewis structures must be in harmony with *measured* molecular properties. The property that we have mostly used for this purpose is bond length. Another property that can be used to assess a proposed Lewis structure is bond energy. Bond energy and length are related in the sense that the more multiple character in a bond, the shorter the bond and the greater is the bond energy.

Energy is *released* when isolated atoms join to form a covalent bond, and energy must be *absorbed* to break apart covalently bonded atoms. **Bond dissociation energy, *D*,** is the quantity of energy required to break one mole of covalent bonds in a *gaseous* species. The SI units are kilojoules per mole of bonds (kJ/mol).

In the manner of Chapter 7, we can think of bond dissociation energy as an enthalpy change or a heat of reaction. For example,

Bond breakage: $H_2(g) \longrightarrow 2\ H(g)$ $\Delta H = D(H\!-\!H) = +435.93$ kJ/mol

Bond formation: $2\ H(g) \longrightarrow H_2(g)$ $\Delta H = -D(H\!-\!H) = -435.93$ kJ/mol

It is not hard to picture the meaning of bond energy for a *diatomic* molecule because there is only one bond in the molecule. It is also not difficult to see that the bond dissociation energy of a diatomic molecule can be expressed rather precisely, as is that of H$_2$(g). With a polyatomic molecule such as H$_2$O the situation is different (Figure 11-14). The energy needed to dissociate one mole of H atoms by breaking one O—H bond per H$_2$O molecule

$$H\!-\!OH(g) \longrightarrow H(g) + OH(g)   \Delta H = D(H\!-\!OH) = +498.7 \text{ kJ/mol}$$

is different from the energy required to dissociate one mole of H atoms by breaking the bonds in OH(g).

$$O\!-\!H(g) \longrightarrow H(g) + O(g)   \Delta H = D(O\!-\!H) = +428.0 \text{ kJ/mol}$$

The two O—H bonds in H$_2$O are identical; therefore, they should have identical energies. This energy, which we can call the O—H bond energy in H$_2$O, is the average of the two values listed above: 463.4 kJ/mol. The O—H bond energy in other molecules containing the OH group will be somewhat different from that in H—O—H. For example, in methanol, CH$_3$OH, the O—H bond dissociation energy, which we can represent as $D(H\!-\!OCH_3)$, is 436.8 kJ/mol. The

FIGURE 11-14
Some bond energies compared.
The same quantity of energy, 435.93 kJ/mol, is required to break all H—H bonds. In H$_2$O, more energy is required to break the first bond (498.7 kJ/mol) than to break the second (428.0 kJ/mol). The second bond broken is that in the OH radical. The O—H bond energy in H$_2$O is the average of the two values: 463.4 kJ/mol.

435.93 kJ/mol

H ⟞ H

498.7 kJ/mol

H ⟞ O — H

428.0 kJ/mol

O ⟞ H

	Bond Energy, kJ/mol	**Bond**	**Bond Energy, kJ/mol**	**Bond**	**Bond Energy, kJ/mol**
TABLE 11.3	**Some Average Bond Energies**[a]				
Bond					
H—H	436	C—C	347	N—N	163
H—C	414	C=C	611	N=N	418
H—N	389	C≡C	837	N≡N	946
H—O	464	C—N	305	N—O	222
H—S	368	C=N	615	N=O	590
H—F	565	C≡N	891	O—O	142
H—Cl	431	C—O	360	O=O	498
H—Br	364	C=O	736	F—F	159
H—I	297	C—Cl	339	Cl—Cl	243
				Br—Br	193
				I—I	151

[a]Although all data are listed with about the same precision (three significant figures), some values are actually known more precisely. Specifically, the values for the diatomic molecules: H_2, HF, HCl, HBr, HI, N_2 (N≡N), O_2 (O=O), F_2, Cl_2, Br_2, and I_2 are actually bond dissociation energies, rather than average bond energies.

usual method of tabulating bond energies (Table 11.3) is as *averages*. An **average bond energy** is the average of bond dissociation energies for a number of different species containing the particular bond. Understandably, average bond energies cannot be stated as precisely as specific bond dissociation energies.

As you can see from Table 11.3, double bonds have higher bond energies than do single bonds between the same atoms, but they are *not* twice as large. Triple bonds are stronger still, but their bond energies are *not* three times as large as single bonds between the same atoms. This observation about bond order and bond energy will seem quite reasonable when we describe multiple bonds more fully in the next chapter.

Bond energies also have some interesting uses in thermochemistry. For a reaction involving *gases,* visualize the process

gaseous reactants ⟶ gaseous atoms ⟶ gaseous products

> ▶ We stress *gases* in our discussion of bond energies because tabulated bond energies are for isolated molecules, not molecules in the close contact found in liquids and solids.

In this hypothetical process, we first break all the bonds in reactant molecules and form gaseous atoms. For this step the enthalpy change is ΔH(bond breakage) = ΣBE (reactants), where the symbol BE stands for bond energy. Next, we allow the gaseous atoms to recombine into product molecules. In this step bonds are formed and ΔH(bond formation) = $-\Sigma$BE(products). The enthalpy change of the reaction, then, is

$$\Delta H_{rxn} = \Delta H(\text{bond breakage}) + \Delta H(\text{bond formation})$$

$$= \sum \text{BE(reactants)} - \sum \text{BE(products)} \qquad (11.13)$$

When *average* bond energies are used in expression (11.13), a number of terms often cancel out, because some of the same types of bonds appear in the products as in the reactants. We can base the calculation of ΔH_{rxn} just on the *net* number and types of bonds broken and formed, as illustrated in Example 11.13.

414 kJ/mol   243 kJ/mol   −339 kJ/mol   −431 kJ/mol

$$
\begin{array}{c}
\quad\;\; H \\
\quad\;\; | \\
H{-}C{-}H \\
\quad\;\; | \\
\quad\;\; H
\end{array}
\;+\;
Cl{-}Cl
\;\longrightarrow\;
\begin{array}{c}
\quad\;\; H \\
\quad\;\; | \\
H{-}C{-}Cl \\
\quad\;\; | \\
\quad\;\; H
\end{array}
\;+\;
H{-}Cl
$$

FIGURE 11-15

Net bond breakage and formation in a chemical reaction—Example 11-13 illustrated.

Bonds that are broken are shown in red, and bonds that are formed, in blue. Bonds that remain unchanged are in black. The net change is that *one* C—H and *one* Cl—Cl bond break and *one* C—Cl and *one* H—Cl bond form.

EXAMPLE 11-13

Calculating an Enthalpy of Reaction from Bond Energies. The reaction of methane (CH_4) and chlorine produces a mixture of products called chloromethanes. One of these is monochloromethane, CH_3Cl, used in the preparation of silicones. Calculate ΔH for the reaction

$$CH_4(g) + Cl_2(g) \longrightarrow CH_3Cl(g) + HCl(g)$$

Solution

To assess which bonds are broken and formed, it helps to draw structural formulas (or Lewis structures), as in Figure 11-15. To apply expression (11.13) literally, we would break *four* C—H bonds and *one* Cl—Cl bond and form *three* C—H bonds, *one* Cl—Cl bond, and *one* H—Cl bond. The *net* change, however, is the breaking of *one* C—H bond and *one* Cl—Cl bond, followed by the formation of *one* C—Cl bond and *one* H—Cl bond.

ΔH for net bond breakage:	1 mol C—H bonds = +414 kJ
	1 mol Cl—Cl bonds = +243 kJ
	sum: +657 kJ
ΔH for net bond formation:	1 mol C—Cl bonds = −339 kJ
	1 mol H—Cl bonds = −431 kJ
	sum: −770 kJ
Enthalpy of reaction:	$\Delta H_{rxn} = 657 − 770 = −113$ kJ

Practice Example A: Use bond energies to estimate the enthalpy change for the reaction

$$2\,H_2(g) + O_2(g) \longrightarrow 2\,H_2O(g)$$

Practice Example B: Use bond energies to estimate the enthalpy of formation of $NH_3(g)$. (*Hint:* What is the reaction having the required enthalpy change?)

There is no advantage to using bond energies over enthalpy-of-formation data. Enthalpies of formation are generally known rather precisely, whereas bond energies are only average values. But when enthalpy-of-formation data are lacking bond energies can prove particularly useful.

Another way to use bond energies is in predicting whether a reaction will be *endothermic* or *exothermic.* In general, if

$$
\underset{\text{(reactants)}}{\text{weak bonds}} \longrightarrow \underset{\text{(products)}}{\text{strong bonds}} \qquad \Delta H < 0 \quad \textit{(exothermic)}
$$

and

$$\underset{\text{(reactants)}}{\text{strong bonds}} \longrightarrow \underset{\text{(products)}}{\text{weak bonds}} \qquad \Delta H > 0 \quad \textit{(endothermic)}$$

Example 11-14 applies this idea to a reaction involving highly reactive, unstable species for which enthalpies of formation are not normally listed.

EXAMPLE 11-14

Using Bond Energies to Predict Exothermic and Endothermic Reactions. One of the steps in the formation of monochloromethane (Example 11-13) is the reaction of a gaseous chlorine *atom* (a chlorine radical) with a molecule of methane. The products are an unstable methyl radical, $\cdot CH_3$, and $HCl(g)$. Is this reaction endothermic or exothermic?

$$CH_4(g) + \cdot Cl(g) \longrightarrow \cdot CH_3(g) + HCl(g)$$

Solution

For every molecule of CH_4 that reacts, *one* C—H bond breaks, requiring 414 kJ per mole of bonds; and *one* H—Cl bond forms, releasing 431 kJ per mole of bonds. Because more energy is released in forming new bonds than is absorbed in breaking old ones, we predict that the reaction is exothermic.

Practice Example A: One naturally occurring reaction involved in the sequence of reactions leading to the destruction of ozone (page 254) is

$$NO_2(g) + O(g) \longrightarrow NO(g) + O_2(g)$$

Is this reaction endothermic or exothermic?

Practice Example B: Predict whether the following reaction should be exothermic or endothermic: $H_2O(g) + Cl_2(g) \longrightarrow \frac{1}{2} O_2(g) + 2 HCl(g)$. (*Hint:* Write Lewis structures and use data from Table 11.3.)

Summary

A Lewis symbol represents the valence electrons of an atom through dots placed around the chemical symbol. A Lewis structure is a combination of Lewis symbols used to represent chemical bonding. Normally, all the electrons in a Lewis structure are paired, and each atom in the structure acquires eight electrons in its valence shell (octet rule).

Most bonds have both partial ionic and partial covalent character. If the centers of positive and negative charge in a bond become separated because one member of the bond attracts electrons more strongly than the other, the bond is said to be *polar*. Whether a bond is polar can be predicted through the concept of *electronegativity*. The greater the electronegativity difference between two atoms, the more polar is the bond, and the more ionic the character of the bond.

To draw the Lewis structure of a covalent molecule one needs to know the skeleton structure—that is, which is the central atom and what atoms are bonded to it. Typically, the atom with the lowest electronegativity is a central atom. At times the concept of formal charge is useful in selecting a skeleton structure and assessing the plausibility of a Lewis structure.

Often more than one plausible Lewis structure can be written for a species. In these cases the true structure is a *resonance hybrid* of two or more contributing structures. At times a molecule may have unpaired electrons (such as NO), and in some compounds of nonmetals of the third period and beyond, the valence shell of the central atom must be expanded to 10 or 12 electrons in order to write a Lewis structure.

A powerful method for predicting molecular shapes is the valence-shell electron-pair repulsion (VSEPR) theory. The shape of a molecule (or polyatomic ion) depends on the geometric distribution of valence-shell electron groups and whether these groups are bonding electrons or lone pairs. An important use of information about the shapes of molecules is in establishing whether bond moments in a molecule combine to produce a resultant dipole moment. Molecules with a resultant dipole moment are *polar*. Those with no resultant dipole moment are *nonpolar*.

Molecular properties, such as bond length and bond energy, are used to assess whether a particular Lewis structure is plausible. For example, they can be used to establish whether a covalent bond has a multiple-bond character. Also, bond energies can be used to estimate enthalpy changes for reactions involving gases.

Summarizing Example

Nitryl fluoride is a reactive gas useful in rocket propellants. Its composition is 21.55% N, 49.23% O, and 29.23% F, by mass. The density of the gas at 20 °C and 1 atm is 2.7 g/L.

Describe the nitryl fluoride molecule as completely as you can, that is, determine the formula, Lewis structure, molecular shape, polarity.

1. *Determine the empirical formula.* Start with the percent composition of the compound and proceed as in Example 3-5 (page 75). *Result:* NO_2F.
2. *Determine the true molecular formula.* Use the gas density and ideal gas law to determine the molar mass, as in Example 6-10 (page 178). The empirical and molecular formulas prove to be identical. *Answer:* NO_2F.
3. *Write a Lewis structure.* N is the central atom (lowest EN). The other atoms are terminal atoms. One nitrogen-to-oxygen bond is a double bond; the other bonds are single. *Answer:* Two contributing structures to a resonance hybrid.
4. *Apply the VSEPR theory.* Use the method of Examples 11-9 to 11-11. *Answer:* FNO_2 is a trigonal-planar molecule with 120° bond angles.
5. *Assess the polarity of FNO_2.* The factors that enter in are the symmetrical shape of the molecule and the differing electronegativities of F and O. *Answer:* The molecule has a small resultant dipole moment.

Key Terms

average bond energy (11-8)	**electronegativity difference (ΔEN)** (11-3)	**octet** (11-1)
bond angle (11-7)		**octet rule** (11-2)
bond dissociation energy, *D* (11-8)	**electron group geometry** (11-7)	**polar covalent bond** (11-3)
bond length (11-2, 11-7)	**expanded octet** (11-6)	**polar molecule** (11-7)
bond order (11-2)	**formal charge** (11-4)	**resonance** (11-5)
bond pair (11-2)	**free radical** (11-6)	**single covalent bond** (11-2)
central atom (11-4)	**ionic bond** (11-1)	**skeleton structure** (11-4)
coordinate covalent bond (11-4)	**Lewis structure** (11-1)	**terminal atom** (11-4)
covalent bond (11-1, 11-2)	**Lewis symbol** (11-1)	**triple covalent bond** (11-2)
dipole moment, μ (11-7)	**Lewis theory** (11-1)	**valence-shell electron-pair repulsion (VSEPR) theory** (11-7)
double covalent bond (11-2)	**lone pair** (11-2)	
electronegativity (EN) (11-3)	**molecular geometry** (11-7)	

Review Questions

1. In your own words define the following terms: **(a)** valence electrons; **(b)** electronegativity; **(c)** bond dissociation energy; **(d)** double covalent bond; **(e)** coordinate covalent bond.
2. Briefly describe each of the following ideas: **(a)** formal charge; **(b)** resonance; **(c)** expanded octet; **(d)** bond energy.
3. Explain the important distinctions between each pair of terms: **(a)** ionic and covalent bond; **(b)** lone pair and bond pair electrons; **(c)** molecular geometry and electron group geometry; **(d)** bond moment and resultant dipole moment; **(e)** polar molecule and nonpolar molecule.
4. Write Lewis symbols for the following atoms and ions: **(a)** H^-; **(b)** Kr; **(c)** Sn^{2+}; **(d)** K^+; **(e)** Br^-; **(f)** Ge; **(g)** N; **(h)** Ca; **(i)** Se^{2-}; **(j)** Sc^{3+}.
5. Write Lewis structures for the following ionic compounds: **(a)** calcium chloride; **(b)** barium sulfide; **(c)** lithium oxide; **(d)** sodium fluoride; **(e)** magnesium nitride.

6. Write plausible Lewis structures for the following molecules, which contain only single covalent bonds: **(a)** ICl; **(b)** Br_2; **(c)** OF_2; **(d)** NI_3; **(e)** H_2Se.
7. Each of the following molecules contains at least one multiple (double or triple) covalent bond. Give a plausible Lewis structure for each. **(a)** CS_2; **(b)** $(CH_3)_2CO$; **(c)** Cl_2CO; **(d)** FNO.
8. Describe what is wrong with each of the following Lewis structures.

(a) H—H—$\overset{\cdots}{\underset{}{N}}$—$\overset{\cdots}{\underset{\cdots}{O}}$—H

(b) $:\overset{\cdots}{\underset{\cdots}{O}}$—$\overset{\cdots}{\underset{\cdots}{Cl}}$—$\overset{\cdots}{\underset{\cdots}{O}}:$

(c) $[\cdot\,\overset{}{C}=\overset{}{N}:]^-$

(d) Ca—$\overset{\cdots}{\underset{\cdots}{O}}$

Focus On

Fullerenes: "Buckyballs"

Geodesic domes at the Mitchell Park Horticultural Conservatory.

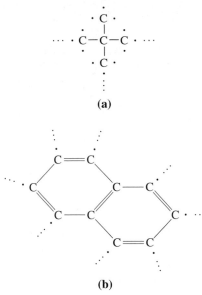

(a)

(b)

FIGURE 11-16

Attempts at writing a Lewis structure for C$_{60}$.
(a) In this structure each C atom attempts to form four single bonds. In a molecule with 60 atoms, those at the ends would be left with unpaired electrons—"dangling" bonds.
(b) By using alternating single and double bonds, the carbon atoms can be arranged in a sheet, but this still leaves "dangling" bonds on many of the C atoms.

The geodesic dome is an architectural form pioneered by R. Buckminster Fuller. The surface of the dome is a pattern of regular hexagons (⬡) and pentagons (⬠). Recently, carbon molecules with this "geodesic" shape have been identified. Their discovery promises to extend our knowledge of chemical bonding. These molecules, at first called buckminsterfullerenes, are now known as fullerenes, or "buckyballs" for short.

In 1985, at Rice University in Houston, Texas, a team of scientists used a special apparatus to mimic conditions found near giant red stars to see if small carbon molecules might be produced.* With a mass spectrometer they did detect a number of different small molecules, but the strongest peak in the mass spectrum came at a molecular mass of 720 u, corresponding to the molecular formula C$_{60}$. Propos-

*H. W. Kroto, J. R. Heath, S. C. O'Brien, R. F. Curl, and R. E. Smalley, *Nature,* 318:162 (1985).

9. Each of the following ionic compounds consists of a combination of monatomic and polyatomic ions. Represent these compounds with Lewis structures. **(a)** Ca(OH)$_2$; **(b)** NH$_4$Br; **(c)** Ca(OCl)$_2$.

10. Assign formal charges to each of the atoms in the following structures.
(a) [H—C≡C:]$^-$

(b)

(c) [CH$_3$—CH—CH$_3$]$^+$

(d) :I—I:

(e)

(f)

11. Which of the following species would you expect to be diamagnetic and which paramagnetic? **(a)** OH$^-$; **(b)** OH; **(c)** NO$_3$; **(d)** SO$_3$; **(e)** SO$_3^{2-}$; **(f)** HO$_2$.

12. Draw plausible Lewis structures for the following species, using the notion of expanded octets where necessary. **(a)** Cl$_2$O; **(b)** PF$_3$; **(c)** CO$_3^{2-}$; **(d)** BrF$_5$.

13. One each of the following is linear, angular (bent), planar, tetrahedral, and octahedral. Indicate the correct shape of **(a)** H$_2$S; **(b)** N$_2$O$_4$; **(c)** HCN; **(d)** SbCl$_6^-$; **(e)** BF$_4^-$.

ing a geometric structure for this molecule was a challenge.

As shown in Figure 11-16a, a Lewis structure based on the Lewis symbol ·C· leads to difficulties: incomplete octets and "dangling" bonds at the edges of the structure. A structure employing double bonds, such as shown in Figure 11-16b, presents similar difficulties. Species with "dangling" bonds should be highly reactive, but C_{60} appears to be remarkably stable. The "dangling" bond problem was solved by wrapping the structure back onto itself in the form of a sphere or ball. But this requires straight bonds between atoms to generate a stable curved structure. Imagine trying to create such a structure in a ball-and-stick fashion. One must not put too much strain on the sticks or they will break. The model that works is the geodesic dome, a combination of geometric figures that, taken as a whole, approximates the shape of a sphere. The most stable "sphere" of 60 vertices, with each vertex representing a carbon atom, consists of 12 pentagonal and 20 hexagonal faces. (This same combination of pentagons and hexa-gons is seen on the cover of a soccer ball.)

The next task lay in confirming this hypothetical structure by experiment. This was done by X-ray crystallography, which analyzes the interaction of X-rays with a solid (a method discussed in Section 13-8). To apply this method requires crystals of the solid. Relatively large quantities of C_{60} can be prepared by heating a graphite rod in an atmosphere of helium. The C_{60} is soluble in the solvent toluene, and crystalline C_{60} can be recovered from the toluene solution. Results of the X-ray analysis of C_{60} yielded the structure shown in Figure 11-17, confirming the proposed geometric shape.

Subsequent to the discovery of C_{60}, other fullerenes have been found, including C_{70}, C_{74}, and C_{82}. Buckyballs can form compounds, mainly by attaching atoms or groups of atoms to their surfaces, such as in the compound $C_{60}F_{36}$. Another discovery has been that fullerenes form compounds with metals. The compound designated La@C_{60}, for example, consists of molecules with an La atom caged inside a C_{60} fullerene. And the dis-

FIGURE 11-17
Computer-generated molecular model of fullerene, C_{60}.

coveries keep coming.

Hundreds of scientific papers have been written in the past few years on these remarkable molecules, testifying to the rapidity with which developments often occur in science.

---

14. Predict the shapes of the following sulfur-containing species: **(a)** SO_2; **(b)** SO_3^{2-}; **(c)** SO_4^{2-}.

15. Predict the geometric shapes of **(a)** CO; **(b)** $SiCl_4$; **(c)** PH_3; **(d)** ICl_3; **(e)** $SbCl_5$; **(f)** SO_2; **(g)** AlF_6^{3-}.

16. Use data from Tables 11.1 and 11.3 to determine for each bond in the following structure **(a)** the bond length and **(b)** the bond energy.

$$H-\overset{\overset{\displaystyle O}{\|}}{C}-\overset{\overset{\displaystyle H}{|}}{\underset{\underset{\displaystyle H}{|}}{C}}-Cl$$

17. *Without* referring to tables in the text, indicate which of the following bonds you would expect to have the greatest bond length and give your reasons. (a) O_2; (b) N_2; (c) Br_2; (d) BrCl.

18. A reaction occurring in the upper atmosphere involved in the formation of ozone is $O_2 \rightarrow 2\ O$. *Without* referring to Table 11.3, indicate whether this reaction is endothermic or exothermic. Explain.

19. Use data from Table 11.3, but *without performing detailed calculations,* determine whether each of the following reactions is exothermic or endothermic.
    (a) $CH_4(g) + I(g) \rightarrow \cdot CH_3(g) + HI(g)$
    (b) $H_2(g) + I_2(g) \rightarrow 2\ HI(g)$

20. Given the bond dissociation energies: nitrogen-to-oxygen bond in NO, 631 kJ/mol; H—H in H_2, 436 kJ/mol; N—H in NH_3, 389 kJ/mol; O—H in H_2O, 463 kJ/mol; calculate ΔH for the reaction

$$2\ NO(g) + 5\ H_2(g) \longrightarrow 2\ NH_3(g) + 2\ H_2O(g)$$

21. Without referring to tables or figures in the text other than the periodic table, indicate which of the following atoms, Bi, S, Ba, As, or Mg, has the intermediate value when the five are arranged in order of increasing electronegativity.

22. Use your knowledge of electronegativities, but *do not* refer to tables or figures in the text, to arrange the following bonds in terms of *increasing* ionic character: C—H; F—H; Na—Cl; Br—H; K—F.

23. Which of the following molecules would you expect to have a resultant dipole moment (μ)? Explain. (a) F_2; (b) NO_2; (c) BF_3; (d) HBr; (e) H_2CCl_2; (f) SiF_4; (g) OCS.

24. The following statements are not made as carefully as they might be. Criticize each one.
(a) Lewis structures with formal charges are incorrect.
(b) Triatomic molecules have a planar shape.
(c) Molecules in which there is an electronegativity difference between the bonded atoms are polar.

Exercises

Lewis Theory

25. Give several examples for which the following statement proves to be incorrect. "All atoms in a Lewis structure have an octet of electrons in their valence shells."

26. Which of the following have Lewis structures that *do not* obey the octet rule: NH_3, BF_3, SF_6, SO_3, NH_4^+, SO_4^{2-}, NO_2? Explain.

27. By means of Lewis structures represent bonding between the following pairs of elements. Your structures should show whether the bonding is essentially ionic or covalent. (a) Cs and Br; (b) H and Sb; (c) B and Cl; (d) Cs and Cl; (e) Li and O; (f) Cl and I.

28. Suggest reasons why the following do not exist as stable molecules. (a) H_3 (b) HHe; (c) He_2; (d) H_3O.

29. Only one of the following Lewis structures is correct. Select that one and indicate the errors in the others.
(a) cyanate ion, $[:\ddot{O}—C\!\!=\!\!\ddot{N}:]^-$
(b) carbide ion, $[C\!\!\equiv\!\!C:]^{2-}$
(c) hypochlorite ion, $[:\ddot{C}l—\ddot{O}:]^-$
(d) nitrogen(II) oxide, $:N\!\!=\!\!\ddot{O}:$

30. Indicate what is wrong with each of the following Lewis structures. Replace each with a more acceptable structure.
(a) Mg $:\ddot{O}:$
(b) $[:\ddot{C}l]^+[:\ddot{O}:]^{2-}[\ddot{C}l:]^+$
(c) $[:\ddot{O}—N\!\!=\!\!\ddot{O}:]^+$
(d) $[:\ddot{S}—C\!\!=\!\!\ddot{N}:]^-$

Ionic Bonding

31. Derive the correct formulas for the following ionic compounds by writing Lewis structures. (a) lithium sulfide; (b) sodium fluoride; (c) calcium iodide; (d) scandium chloride.

32. Under appropriate conditions both hydrogen and nitrogen can form monatomic anions. What are the Lewis symbols for these ions? What are the Lewis structures of the compounds (a) lithium hydride; (b) calcium hydride; (c) magnesium nitride?

Formal Charge

33. Both oxidation state and formal charge involve conventions for assigning valence electrons to bonded atoms in compounds, but clearly they are not the same. Describe several ways in which these concepts differ.

34. Although the notion that a Lewis structure in which formal charges are zero or held to a minimum seems to apply in most instances, describe several significant situations in which this appears not to be the case.

35. What is the formal charge of the indicated atom in each of the following structures?
(a) the central O atom in O_3 (page 361)
(b) B in BF_4^- (page 363)
(c) N in NO_3^- (page 361)
(d) P in PCl_5 (page 364)
(e) I in ICl_4^- (page 371)

36. Assign formal charges to the atoms in the following species, and then select the more likely skeleton structure.
(a) H_2NOH or H_2ONH;
(b) SCS or CSS;
(c) NFO or FNO;
(d) $SOCl_2$ or $OSCl_2$ or OCl_2S.

Lewis Structures

37. Write acceptable Lewis structures for the following molecules: (a) H_2NOH; (b) $HOClO_2$; (c) HONO; (d) O_2SCl_2.

38. Two molecules that have the same formulas but different structures are said to be isomers. (In isomers the same atoms are present but linked together in different ways.) Draw acceptable Lewis structures for *two* isomers of S_2F_2. [*Hint:* What is(are) the central atom(s)?]

39. The following polyatomic anions involve covalent bonds between O atoms and the central nonmetal atom. Propose an acceptable Lewis structure for each. (a) SO_3^{2-}; (b) NO_2^-; (c) CO_3^{2-}; (d) HO_2^-.

40. Represent the following ionic compounds by Lewis structures: (a) barium hydroxide; (b) sodium nitrite; (c) magnesium iodate; (d) aluminum sulfate.

41. Draw a plausible Lewis structure for crotonaldehyde, $CH_3CHCHCHO$, a substance used in tear gas and insecticides.

42. Draw a plausible Lewis structure for C_3O_2, a substance known as carbon suboxide.

Bond Lengths

43. A relationship between bond lengths and single covalent radii of atoms is given on page 351. Use this relationship together with appropriate data from Table 11.1 to estimate these single-bond lengths. (a) I—Cl; (b) O—Cl; (c) C—F; (d) C—Br.

44. In which of the following molecules would you expect the oxygen-to-oxygen bond to be the *shortest*: (a) H_2O_2, (b) O_2, (c) O_3? Explain. (*Hint:* Draw Lewis structures.)

45. Refer to the Summarizing Example. Use data from the chapter to estimate the length of the N—F bond in FNO_2. (*Hint:* What is the nature of the N—F bond? single? double?)

46. Write a Lewis structure of the hydroxylamine molecule, H_2NOH. Then, with data from Table 11.1, determine all the bond lengths.

Polar Covalent Bonds

47. What is the percent ionic character of each of the following bonds? (a) S—H; (b) O—Cl; (c) Al—O; (d) As—O.

48. Plot the data of Figure 11-5 as a function of atomic number. Does the property of electronegativity conform to the periodic law? Do you think it should?

Resonance

49. In Example 11-7 the concept of resonance was illustrated for the nitrate ion. Resonance is also involved in the nitrite ion. Show this through appropriate Lewis structures.

51. Dinitrogen oxide (nitrous oxide, or "laughing gas") is sometimes used as an anesthetic. Here are some data about the N_2O molecule: N—N bond length = 113 pm; N—O bond length = 119 pm. Use these data and other information from the chapter to comment on the plausibility of each of the following Lewis structures. Are they all valid? Which ones do you think contribute most to the resonance hybrid?

$$:N\equiv N-\ddot{O}: \qquad :\ddot{N}=N=\ddot{O}:$$
$$\quad (1) \qquad\qquad\qquad (2)$$
$$:\ddot{N}-N\equiv O: \qquad :\ddot{N}=O=\ddot{N}:$$
$$\quad (3) \qquad\qquad\qquad (4)$$

50. Which of the following species requires a resonance hybrid for its Lewis structure? Explain. (a) CO_2; (b) OCl^-; (c) CO_3^{2-}; (d) OH^-.

52. The Lewis structure of nitric acid, $HONO_2$, is a resonance hybrid. How important do you think the contribution of the following structure is to the resonance hybrid? Explain.

$$H-\ddot{O}=N\genfrac{}{}{0pt}{}{\overset{\displaystyle :\ddot{O}:}{\diagup}}{\underset{\displaystyle :\ddot{O}:}{\diagdown}}$$

Odd-Electron Species

53. Write plausible Lewis structures for the following odd-electron species: (a) CH_3; (b) ClO_2; (c) NO_3.

54. Write a plausible Lewis structure for NO_2 and indicate whether the molecule is diamagnetic or paramagnetic. Two NO_2 molecules can join together (*dimerize*) to form N_2O_4. Write a plausible Lewis structure for N_2O_4 and comment on the magnetic properties of the molecule.

Expanded Octets

55. In which of the following species is it *necessary* to employ an expanded octet to represent the Lewis structure: PO_4^{3-}, PI_3, ICl_3, $OSCl_2$, SF_4, ClO_4^-? Explain your choices.

56. Describe the carbon-to-sulfur bond in H_2CSF_4. That is, is it most likely a single, double, or triple bond?

Molecular Shapes

57. Use VSEPR theory to predict the geometric shapes of the following molecules and ions. (a) N_2; (b) HCN; (c) NH_4^+; (d) NO_3^-; (e) NSF.

59. One of the following ions has a *trigonal-planar* shape. Which ion is it? Explain. SO_3^{2-}; PO_4^{3-}; PF_6^-; CO_3^{2-}.

58. Use VSEPR theory to predict the geometric shapes of the following molecules and ions. (a) PCl_3; (b) SO_4^{2-}; (c) $SOCl_2$; (d) SO_3; (e) BrF_4^+.

60. Two of the following have the same shape. Which are these two? What is their shape? What are the shapes of the other two? NI_3, HCN, SO_3^{2-}, NO_3^-.

61. Each of the following molecules contains one or more multiple covalent bonds. Draw plausible Lewis structures to represent this fact, and predict the shape of each molecule. (a) CO_2; (b) Cl_2CO; (c) $ClNO_2$.

63. Use the VSEPR theory to predict the shapes of the anions (a) ClO_4^-; (b) $S_2O_3^{2-}$ (i.e., SSO_3^{2-}); (c) PF_6^-; (d) I_3^-.

65. The molecular shape of BF_3 (Table 11.2) is planar. If a fluoride ion is attached to the B atom of BF_3 through a coordinate covalent bond, the ion BF_4^- results. What is the shape of this ion?

Polar Molecules

67. Predict the shapes of the following molecules, and then predict which you would expect to have resultant dipole moments. (a) SO_2; (b) NH_3; (c) H_2S; (d) C_2H_4; (e) SF_6; (f) CH_2Cl_2.

69. The molecule H_2O_2 has a resultant dipole moment, $\mu = 2.2$ D. The bonding is H—O—O—H. Can the molecule be linear? Explain.

Bond Energies

71. Use data from Table 11.3 to estimate the enthalpy change (ΔH) for the following reaction. (*Hint:* Draw Lewis structures of the reactants and products.)
$$C_2H_6(g) + Cl_2(g) \longrightarrow C_2H_5Cl(g) + HCl(g) \qquad \Delta H = ?$$

73. Estimate the standard enthalpies of formation at 25 °C and 1 atm of (a) $OH(g)$; (b) $N_2H_4(g)$. Write Lewis structures and use data from Table 11.3, as necessary.

75. Use data from Table 11.3 to estimate the enthalpy of combustion of ethanol, CH_3CH_2OH.
$$CH_3CH_2OH(g) + 3\,O_2(g) \longrightarrow 2\,CO_2(g) + 3\,H_2O(g)$$
$$\Delta H = ?$$

62. Draw a sketch of the probable geometric shape of a molecule of (a) N_2O_4 (O_2NNO_2); (b) C_2N_2 (NCCN); (c) C_2H_6 (H_3CCH_3); (d) C_2H_6O (H_3COCH_3).

64. Use the VSEPR theory to predict the shape of (a) the molecule OSF_2; (b) the molecule O_2SF_2; (c) the ion SF_5^-; (d) the ion ClO_4^-; (e) the ion ClO_3^-.

66. The contributing structures (11.9) to the resonance hybrid for ozone suggest that the molecule is linear. Do you think this is the true shape of the O_3 molecule? Explain.

68. Which of the following molecules would you expect to be polar? Give reasons for your conclusions. (a) HCN; (b) SO_3; (c) CS_2; (d) OCS; (e) $SOCl_2$; (f) SiF_4; (g) POF_3.

70. Refer to the Summarizing Example. A compound related to nitryl fluoride is nitrosyl fluoride, FNO. For this molecule indicate (a) a plausible Lewis structure and (b) the geometric shape. (c) Explain why the measured resultant dipole moment for FNO is larger than the value for FNO_2.

72. One of the chemical reactions that occurs in the formation of photochemical smog is $O_3 + NO \rightarrow NO_2 + O_2$. Estimate the enthalpy change of this reaction by using data from Table 11.3 and the Lewis structures for each species.

74. Use ΔH for the reaction in Example 11-13 and other data from Appendix D to estimate $\Delta H_f^\circ[CH_3Cl(g)]$.

76. Equations (1) and (2) can be combined to yield the equation for the formation of $CH_4(g)$ from its elements.

(1)	$C(s)$	$\longrightarrow C(g)$	$\Delta H = 717$ kJ
(2)	$C(g) + 2\,H_2(g)$	$\longrightarrow CH_4(g)$	$\Delta H = ?$
Net:	$C(s) + 2\,H_2(g)$	$\longrightarrow CH_4(g)$	$\Delta H_f^\circ = -75$ kJ/mol

Use the preceding data and a bond energy of 436 kJ/mol for H_2 to estimate the C—H bond energy. Compare your result with the value listed in Table 11.3. [*Hint:* What is the value of ΔH for reaction (2)?]

Advanced Exercises

77. A compound consists of 47.5% S and 52.5% Cl by mass. Write a Lewis structure based on the empirical formula of this compound and comment on its deficiencies. Write a more plausible structure with the *same* ratio of S to Cl.

78. A 0.325-g sample of a gaseous hydrocarbon occupies a volume of 193 mL at 749 mmHg and 26.1 °C. Determine the molecular mass and draw a plausible Lewis structure for this hydrocarbon.

79. A 1.24-g sample of a hydrocarbon, when completely burned in an excess of $O_2(g)$, yields 4.04 g CO_2 and 1.24 g H_2O. Draw a plausible Lewis structure for the hydrocarbon molecule. (*Hint:* There is more than one possible arrangement of the C and H atoms.)

80. Assess formal charges for the Lewis structure of CO_2H^+

in Example 11-7. Is it possible to draw a better Lewis structure? Explain.

81. Draw Lewis structures for two different molecules with the formula C_3H_4. Is either of these molecules linear? Explain.

82. Explain what is wrong with the following statement: Because the ions ICl_2^+ and ICl_2^- differ by only two electrons, and because electrons are so small compared to the size of the entire polyatomic anions, we should expect these two ions to have the same geometric shape. Give a clear statement comparing the molecular geometries of the two ions.

83. Use the bond dissociation energies of $N_2(g)$ and $O_2(g)$ in Table 11.3, together with data from Appendix D, to estimate the bond dissociation energy of NO(g).

84. Hydrogen azide, HN_3, is a liquid that explodes violently when subjected to shock. In the HN_3 molecule one nitrogen-to-nitrogen bond length is 113 pm and the other, 124 pm. The H—N—N bond angle is 112°. Draw Lewis structures and a sketch of the molecule consistent with these facts.

85. Carbon suboxide has the formula C_3O_2. The carbon-to-carbon bond lengths are 130 pm and carbon-to-oxygen, 120 pm. Propose a plausible Lewis structure to account for these bond lengths, and predict the shape of the molecule.

86. In certain polar solvents PCl_5 undergoes an ionization reaction in which a Cl^- ion leaves one PCl_5 molecule and attaches itself to another. The products of the ionization are PCl_4^+ and PCl_6^-. Draw a sketch to show the changes in geometric shapes that occur in this ionization (i.e., what are the shapes of PCl_5, PCl_4^+, and PCl_6^-?).

$$2\ PCl_5 \rightleftharpoons PCl_4^+ + PCl_6^-$$

87. Use bond energies from Table 11.3 to estimate the enthalpy change (ΔH) for the following reaction.
$$C_2H_2(g) + H_2(g) \longrightarrow C_2H_4(g) \qquad \Delta H = ?$$

88. Estimate the enthalpy of formation of HCN using bond energies from Table 11.3 and the reaction scheme outlined as follows.

(1)	$C(s)$	$\longrightarrow$	$C(g)$	$\Delta H = 717$ kJ
(2)	$C(g) + \frac{1}{2} N_2(g) + \frac{1}{2} H_2(g) \longrightarrow$		$HCN(g)$	$\Delta H = ?$
Net:	$C(s) + \frac{1}{2} N_2(g) + \frac{1}{2} H_2(g) \longrightarrow$		$HCN(g)$	$\Delta H_f^\circ = ?$

89. The enthalpy of formation of $H_2O_2(g)$ is -136 kJ/mol. Use this value, with other appropriate data from the text, to estimate the oxygen-to-oxygen single-bond energy. Compare your result with the value listed in Table 11.3.

90. Ozone, O_3, has a resultant dipole moment of 0.53 D. How do you account for this observation, given that all the atoms in this molecule are O atoms?

Feature Problems

A. Pauling's reasoning in establishing his original electronegativity scale went something like this: The bond energy associated with the bond A—B, if we *assume* that the bond is nonpolar, is the average of the bond energies of A—A and B—B. The *difference* between the calculated and measured bond energies of the bond A—B is attributable to the partial ionic character of the bond and is called the *ionic resonance energy* (IRE). If the IRE is expressed in kilojoules per mole, the relationship between IRE and the electronegativity difference is $(\Delta EN)^2 = IRE/96$. To test this basis for an electronegativity scale,

1. Use data from Table 11.3 to determine IRE for the H—Cl bond.

2. Determine ΔEN for the H—Cl bond.

3. Establish the approximate percent ionic character in the H—Cl bond by using the result of part 2 and Figure 11-6. Compare this result with that obtained in Example 11-4.

B. On page 374 the bond angle in the H_2O molecule is given as 104° and the resultant dipole moment is $\mu = 1.84$ D.

1. By an appropriate geometric calculation, determine the value of the H—O bond moment in H_2O.

2. Use the same method as in part 1 to estimate the bond angle in H_2S, given that the H—S bond moment is 0.67 D and that the resultant dipole moment is $\mu = 0.93$ D.

Chemical Bonding II: Additional Aspects

12

*U*seful as it has been in our discussion of chemical bonding, Lewis theory does have its shortcomings. It does not help us explain why metals conduct electricity or how a semiconductor works, for example. Although we will continue to use Lewis theory for most purposes, we need more sophisticated approaches in some cases.

In one such method we will work with the familiar *s*, *p*, and *d* atomic orbitals, or with mixed-orbital types called *hybrid* orbitals. In a second approach we will create a set of orbitals that belong to a molecule as a whole. Then we will assign electrons to these *molecular* orbitals.

Our purpose in this chapter is not to try to master theories of covalent bonding in all their details. We want simply to discover how these theories provide models that yield deeper insights into the nature of chemical bonding than do Lewis structures alone.

A computer graphics representation of one of the molecular orbitals of the benzene molecule showing the buildup of electron charge density between the nuclei. Another molecular orbital representation of benzene is presented on page 408.

12-1   What a Bonding Theory Should Do

Imagine bringing together two atoms that are initially very far apart. Three types of interactions occur: Electrons are attracted to the two nuclei, electrons repel one another, and the two nuclei repel each other. We can plot potential energy—the net energy of interaction of the atoms—as a function of the distance between the atomic nuclei. In this plot, negative energies correspond to a net attractive force between the atoms; positive energies, to a net repulsion.

Figure 12-1 shows the energy of interaction of two H atoms. This starts at zero when the atoms are very far apart. At very small internuclear distances, repulsive forces exceed attractive forces and the potential energy is positive. At intermediate distances, attractive forces predominate and the potential energy is negative. In fact, at one particular internuclear distance (74 pm) the potential energy reaches its lowest value (-436 kJ/mol). This is the condition in which the two H atoms combine into an H_2 molecule through a covalent bond. The nuclei continuously move back and forth, that is, the molecule vibrates, but the average internuclear distance remains constant. This internuclear distance corresponds to the *bond length*, and the potential energy, to the negative of the *bond dissociation energy*. What a theory of covalent bonding should do is help us understand why a given molecule has its particular set of observed properties—bond dissociation energies, bond lengths, bond angles, and so on.

It is important to realize that there are several approaches to understanding bonding. Often the approach used depends on the situation. Different methods have different strengths and weaknesses. The strength of the Lewis theory is in the ease with which we can apply it; we can write a Lewis structure rather quickly. With VSEPR theory we can propose molecular shapes that are gener-

FIGURE 12-1

Energy of interaction of two hydrogen atoms.
This graph shows
• a zero of energy when two H atoms are separated by great distances;
• a drop in potential energy (net attraction) as the two atoms approach each other;
• a minimum in potential energy (-436 kJ/mol) at a particular internuclear distance (74 pm) corresponding to the stable molecule H_2;
• an increase in potential energy as the atoms approach more closely.

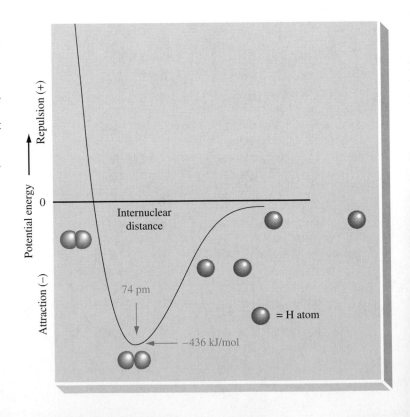

ally in good agreement with experiment. These methods do not yield quantitative information about bond energies and bond lengths, however. Also, Lewis theory has problems with odd-electron species and situations in which it is not possible to represent a molecule through a single electronic structure (resonance).

12-2   Introduction to the Valence-Bond Method

As the two H atoms of Figure 12-1 approach each other, their $1s$ orbitals *overlap*. We have previously described the covalent bond as a strong force of attraction between two atoms due to their sharing a pair of electrons. Now we can say that associated with the covalent bond is a region of high electron charge density resulting from the overlap of atomic orbitals between the bonded atoms. A description of covalent bond formation in terms of atomic orbital overlap is called the **valence-bond method.** The valence-bond method gives a *localized* electron model of bonding: Core electrons and lone pair valence electrons retain the same orbital locations as in the separated atoms, and the charge density of the bonding electrons is concentrated in the region of orbital overlap.

Figure 12-2 shows the imagined overlap of atomic orbitals involved in the formation of hydrogen-to-sulfur bonds in hydrogen sulfide. Note especially that maximum overlap between the $1s$ orbital of an H atom and a $3p$ orbital of an S atom occurs along a line joining the centers of the H and S atoms. The two half-filled $3p$ orbitals of sulfur involved in orbital overlap in H_2S are perpendicular to each other, and the valence-bond method suggests an H—S—H bond angle of 90°. This is in good agreement with the observed angle of 92°.

Because the energies of atomic orbitals differ from one type to another (recall Figure 9-26), the valence-bond method implies different bond energies for different bonds. A quantitative application of the valence-bond method would show that the $1s$—$1s$ overlap in H_2 produces a greater bond energy than the $1s$—$3p$ overlap in an H—S bond in H_2S. The Lewis structures H—H and H—$\ddot{\text{S}}$—H provide no information about bond energies.

▶ What we are calling "overlap" is actually an interpenetration of two orbitals.

FIGURE 12-2   Bonding in H_2S represented by atomic orbital overlap.
Orbitals with a single electron are gray. Those with an electron pair are in color. For S, only $3p$ orbitals are shown. The $1s$ orbitals of the two H atoms overlap with the $3p_x$ and $3p_z$ orbitals of the S atom.

bonding orbitals of P atom

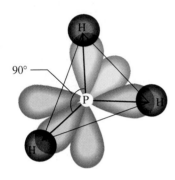

covalent bonds formed

FIGURE 12-3
Bonding and structure of the PH$_3$ molecule—Example 12-1 illustrated.
Orbitals with a single electron are gray. Those with electron pairs are in color. Only bonding orbitals are shown. The 1s orbitals of three H atoms overlap with the three 3p orbitals of the P atom.

EXAMPLE 12-1

Using the Valence-Bond Method to Describe a Molecular Structure. Describe the phosphine molecule, PH$_3$, by the valence-bond method.

Solution

Here is a four-step approach to applying the valence-bond method.

Step 1. Draw valence-shell orbital diagrams for the separate atoms.

Step 2. Sketch the orbitals of the central atom (P) that are involved in the overlap. These are the half-filled 3p orbitals (Figure 12-3).

Step 3. Complete the structure by bringing together the bonded atoms and representing the orbital overlap.

Step 4. Describe the structure. PH$_3$ is a *trigonal-pyramidal* molecule. The three H atoms lie in the same plane. The P atom is situated at the top of the pyramid above the plane of the H atoms, and the three H—P—H bond angles are 90°.

(The experimentally measured H—P—H bond angles are 93–94°.)

Practice Example A: Use the valence-bond method to describe bonding and the expected molecular geometry in nitrogen triiodide, NI$_3$.

Practice Example B: Describe the molecular geometry of NH$_3$, first using the VSEPR method and then using the valence-bond method. How do your answers differ? Which method seems to be more appropriate in this case? Explain.

Are You Wondering . . .

Which method to use in describing the geometric shape of a molecule, the VSEPR method (Section 11-7) or the valence-bond method?

There is no "correct" method for describing molecular structures. The only correct information is the *experimental* evidence from which the structure is established. Once this experimental evidence is in hand, you may find it easier to *rationalize* this evidence by one method or another. For H$_2$S, the valence-bond method, which suggests a bond angle of 90°, seems to do a better job of explaining the observed 92° bond angle than does the VSEPR theory. For the Lewis structure, H—$\ddot{\text{S}}$—H, VSEPR theory predicts a tetrahedral electron group geometry, which in turn suggests a tetrahedral bond angle—that is, 109.5°. However, by modifying this initial VSEPR prediction to accommodate lone pair/lone pair and lone pair/bond pair repulsions (see page 368), the predicted bond angle is less than 109.5°.

VSEPR theory gives reasonably good results in the majority of cases. Unless you have specific information to suggest otherwise, describing a molecular shape with the VSEPR theory is still a good bet.

12-3 Hybridization of Atomic Orbitals

If we try to extend the unmodified valence-bond method of Section 12-2 to a greater number of molecules, we are quickly disappointed. In most cases our descriptions of molecular geometry do not conform to observed measurements. For example, based on the *ground state* electron configuration of the valence shell of carbon

Ground state   C

2s	2p	
↑↓	↑	↑	

and employing only half-filled orbitals, we expect the existence of a molecule with the formula CH_2 and a bond angle of 90°. CH_2 does not exist as a stable molecule, however.

The simplest stable hydrocarbon is methane, CH_4, consistent with our expectation from the octet rule of the Lewis theory. To obtain this molecular formula by the valence-bond method, we need an orbital diagram for carbon in which there are *four* unpaired electrons, so that orbital overlap leads to four C—H bonds. To get such a diagram, imagine that one of the 2s electrons in a ground state C atom absorbs energy and is promoted to the empty 2p orbital. The resulting electron configuration is that of an *excited state,* which we can represent as

Excited state   C

2s	2p		
↑	↑	↑	↑

The electron configuration of this excited state suggests a molecule with three mutually perpendicular C—H bonds based on the 2p orbitals of the C atom (90° bond angles). The fourth bond would be directed to whatever position in the molecule could accommodate the fourth H atom. This description, however, does not agree with the experimentally determined H—C—H bond angles, all four of which are found to be 109.5°, the same as predicted by VSEPR theory (Figure 12-4). A bonding scheme based on the excited state electron configuration does a poor job of explaining the bond angles in CH_4.

The problem is not with the theory but with the way we have defined the situation. We have been describing *bonded* atoms as if they have the same kinds of orbitals (that is, *s*, *p*, . . .) as *isolated nonbonded* atoms. This assumption worked rather well for H_2S, and PH_3, but why should we expect these unmodified pure atomic orbitals to work equally well in all cases?

One way that we can respond to this situation is by modifying the atomic orbitals of the bonded atoms. Recall that atomic orbitals are mathematical expressions of the electron waves in an atom. We need to combine the wave equations of the 2s and three 2p orbitals of the carbon atom to produce a new set of four identical orbitals. These new orbitals, which are directed in a tetrahedral fashion, have energies that are intermediate between those of the 2s and 2p orbitals. This mathematical process of replacing pure atomic orbitals by reformulated atomic orbitals for bonded atoms is called **hybridization,** and the new orbitals are called **hybrid orbitals.** Figure 12-5 pictures the hybridization of an *s* and three *p* orbitals into a new set of four sp^3 **hybrid orbitals.**

In a hybridization scheme *the number of hybrid orbitals equals the total number of atomic orbitals that are combined.* The symbols identify the numbers and kinds of atomic orbitals involved. Thus, sp^3 signifies that *one s* and *three p* or-

FIGURE 12-4
Ball-and-stick model of methane, CH_4.
The molecule has a tetrahedral structure, and the H—C—H bond angles are 109.5°.

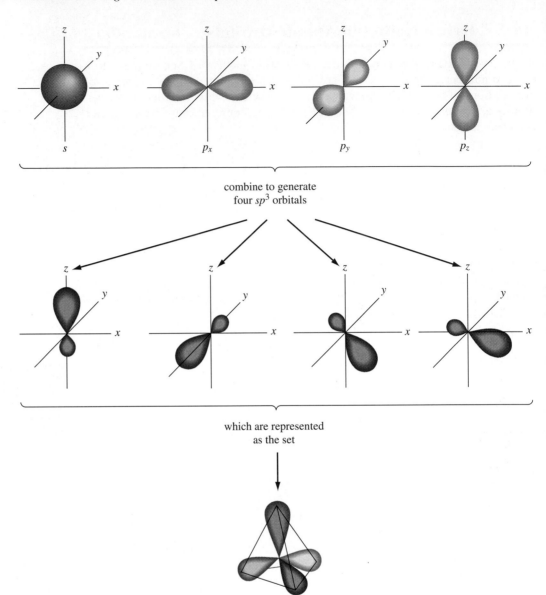

FIGURE 12-5   The *sp*³ hybridization scheme.

bitals are combined. A useful representation of *sp*³ hybridization of the valence-shell orbitals of carbon is

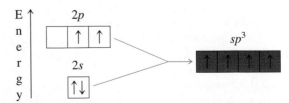

Figure 12-6 pictures *sp*³ hybrid orbitals and bond formation in methane.

The objective of a hybridization scheme is an after-the-fact rationalization of the *experimentally observed* shape of a molecule. Hybridization is not an actual physical phenomenon. We cannot observe electron charge distributions changing from those of pure orbitals to those of hybrid orbitals. Moreover, for some covalent bonds no single hybridization scheme works well. Nevertheless this hybridization idea works very well for carbon-containing molecules and is therefore used a great deal in organic chemistry.

Bonding in H_2O and NH_3

Applied to H_2O and NH_3, VSEPR theory describes a tetrahedral electron group geometry for *four* electron groups. This, in turn, requires an sp^3 hybridization scheme for the central atoms in H_2O and NH_3. This scheme suggests angles of 109.5° for the H—O—H bond in water and the H—N—H bonds in NH_3. These angles are in reasonably good agreement with the experimentally observed bond angles of 104.5° in water and 107° in NH_3. We can describe bonding in NH_3, for example, in terms of the following valence-shell orbital diagram for nitrogen.

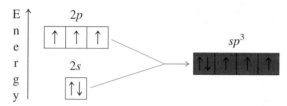

Because one of the sp^3 orbitals is occupied by a lone pair of electrons, only the three half-filled sp^3 orbitals are involved in bond formation. This suggests the trigonal-pyramidal molecular geometry depicted in Figure 12-7, just as does VSEPR theory.

Even though the sp^3 hybridization scheme seems to work quite well for H_2O and NH_3, there is both theoretical and experimental (spectroscopic) evidence that favors a description based on *unhybridized p* orbitals of the central atoms. The H—O—H and H—N—H bond angle expected for $1s$ and $2p$ atomic orbital overlaps is 90°, which does not conform to the observed bond angles. One possible explanation is that, because O—H and N—H bonds have considerable ionic character, repulsions between the positive partial charges associated with the H atoms force the H—O—H and H—N—H bonds to "open up" to values greater than 90°. The issue of how best to describe the bonding orbitals in H_2O and NH_3 is still unsettled and underscores the occasional difficulty of finding a single theory that is consistent with *all* the available evidence.

sp^2 Hybrid Orbitals

Carbon's Group 3A neighbor, boron, has *four* orbitals but only *three* electrons in its valence shell. For most boron compounds the appropriate hybridization scheme combines the $2s$ and *two* $2p$ orbitals into *three* sp^2 **hybrid orbitals** and leaves one p orbital unhybridized. Valence-shell orbital diagrams for this hybridization scheme for boron are shown here, and the scheme is further outlined in Figure 12-8.

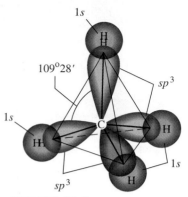

FIGURE 12-6

Bonding and structure of CH_4.
The four carbon orbitals are sp^3 hybrid orbitals (violet). Those of the hydrogen atoms (red) are $1s$. The structure is tetrahedral, with H—C—H bond angles of 109.5° (more precisely, 109°28′).

▶ Notice in this orbital diagram how hybrid orbitals can accommodate lone pair electrons as well as bonding electrons.

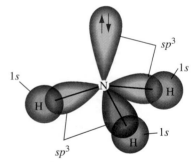

FIGURE 12-7
sp^3 hybrid orbitals and bonding in NH_3.
An sp^3 hybridization scheme yields a molecular geometry in close agreement with what is observed experimentally. The portion of the figure *exclusive* of the orbital occupied by a lone pair of electrons is a trigonal pyramid.

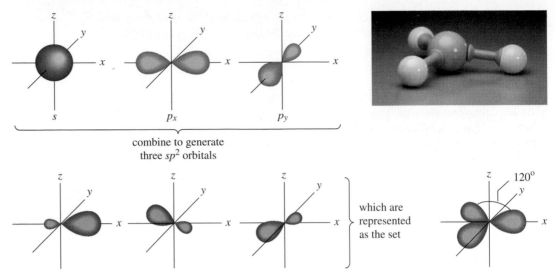

FIGURE 12-8 The *sp²* hybridization scheme.

The sp^2 hybridization scheme corresponds to trigonal-planar electron group geometry and 120° bond angles, as in BF_3.

sp Hybrid Orbitals

Boron's Group 2A neighbor, beryllium, has *four* orbitals and only *two* electrons in the valence shell. In the hybridization scheme that best describes certain *gaseous* beryllium compounds, the $2s$ and *one* $2p$ orbital of Be are hybridized into *two* **sp hybrid orbitals** and the remaining two $2p$ orbitals are left unhybridized. Valence-shell orbital diagrams of beryllium in this hybridization scheme are shown here, and the scheme is further outlined in Figure 12-9.

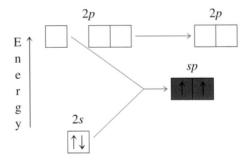

The sp hybridization scheme corresponds to a linear electron group geometry and a 180° bond angle, as in $BeCl_2(g)$.

Hybrid Orbitals and the Valence-Shell Electron-Pair Repulsion Theory

It should come as no surprise that molecular geometries predicted by hybridization schemes in the valence bond method are the same as those predicted by VSEPR theory. To be of value, each approach has to yield the results generally observed by experiment. A useful way to go about choosing the likely hybridization scheme for a central atom in a structure in the valence bond method is to

• Write a plausible Lewis structure for the species of interest.

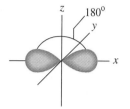

combine to generate
two *sp* orbitals

which are
represented
as the set

FIGURE 12-9 **The *sp* hybridization scheme.**

- Use VSEPR theory to predict the probable electron group geometry of the central atom.
- Select the hybridization scheme corresponding to the electron group geometry.

As suggested by Table 12.1, the hybridization scheme adopted for a central atom should be the one producing the same number of hybrid orbitals as there are valence shell electron groups, and in the same geometric orientation. Thus, an sp^3 hybridization scheme for the central atom predicts that four hybrid orbitals are distributed in a tetrahedral fashion. This results in molecular structures that are tetrahedral, trigonal-pyramidal, or angular, depending on how many hybrid orbitals are involved in orbital overlap and how many contain lone pair electrons, corresponding to the VSEPR notations AX_4, AX_3E, and AX_2E_2, respectively.

To describe hybridization schemes to correspond to the 5 and 6 electron group geometries of VSEPR theory, we need to go beyond the *s* and *p* subshells of the valence shell, and traditionally this has meant including *d*-orbital contributions. We can achieve the *five* half-filled orbitals of phosphorus to account for the five P—Cl bonds in PCl_5 and its trigonal-bipyramidal molecular geometry through

▶ Because one of the main purposes of hybridizing orbitals is to describe molecular geometry (for example, bond angles), we generally apply hybridization schemes only to central atoms, not to terminal atoms.

TABLE 12.1 Some Hybrid Orbitals and Their Geometric Orientations		
Hybrid Orbitals	**Geometric Orientation**	**Example**
sp	linear	$BeCl_2$
sp^2	trigonal-planar	BF_3
sp^3	tetrahedral	CH_4
sp^3d	trigonal-bipyramidal	PCl_5
sp^3d^2	octahedral	SF_6

the hybridization of the *s,* three *p,* and one *d* orbital of the valence shell into *five sp³d* **hybrid orbitals.**

$$sp^3d$$

↑	↑	↑	↑	↑

We can achieve the *six* half-filled orbitals of sulfur to account for the six S—F bonds in SF_6 and its octahedral molecular geometry through the hybridization of the *s,* three *p,* and two *d* orbitals of the valence shell into *six sp³d²* **hybrid orbitals.**

$$sp^3d^2$$

↑	↑	↑	↑	↑	↑

A structure for which we might use the sp^3d^2 hybridization scheme in which some of the orbitals are occupied by lone pairs of electrons is XeF_4. For this square-planar structure (VSEPR notation: AX_4E_2) we would write the following Lewis structure and orbital diagram.

```
        :F:
         |  .. 
  ..     | ..   ..
 :F—Xe—F:
  ..     |  ..
         |
        :F:
         ..
```

$$sp^3d^2$$

↑↓	↑↓	↑	↑	↑	↑

The sp^3d and sp^3d^2 hybrid orbitals and two of the molecular geometries they predict are featured in Figure 12-10.

We have previously stated that hybridization is not a real phenomenon but an after-the-fact rationalization of an experimentally determined result (see page 392). Perhaps there is no better illustration of this point than the issue of the sp^3d and sp^3d^2 hybrid orbitals. In discussing the concept of the expanded octet in Chapter 11 we noted that octet expansion would seem to require *d* electrons in bonding schemes, but recent theoretical considerations cast serious doubt on *d* electron participation. The same doubt, of course, extends to the use of *d* orbitals in hybridization schemes.

Despite the difficulty posed by hybridization schemes involving *d* orbitals, the *sp*, sp^2, and sp^3 hybridization schemes are well established and very commonly encountered, particularly among the second-period elements. Thus *s* and *p* orbital hybridization schemes are especially important in organic compounds, whose principal elements are C, O, and N, in addition to H. We will consider some important applications to organic chemistry in the next section.

12-4 Multiple Covalent Bonds

In this section we will find that two different types of orbital overlap occur when multiple bonds are described by the valence-bond method. In our discussion we will use as specific examples the carbon-to-carbon double bond in ethylene, C_2H_4, and the carbon-to-carbon triple bond in acetylene, C_2H_2.

Bonding in C_2H_4

Ethylene has a carbon-to-carbon double bond in its Lewis structure.

```
      H   H
      |   |
  H—C=C—H
```

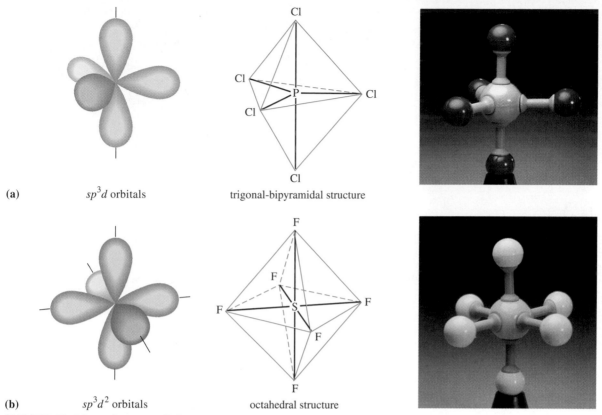

(a) sp^3d orbitals   trigonal-bipyramidal structure

(b) sp^3d^2 orbitals   octahedral structure

FIGURE 12-10 *sp^3d and sp^3d^2 hybrid orbitals.*

Ethylene is a *planar* molecule with 120° H—C—H and H—C—C bond angles. VSEPR theory treats each C atom as being surrounded by *three* electron groups in a trigonal-planar arrangement. VSEPR theory does *not* dictate that the two —CH₂ groups be coplanar, but as we shall see, valence-bond theory does.

The hybridization scheme that produces a set of hybrid orbitals with a trigonal-planar orientation is sp^2. The valence-shell orbital diagrams of carbon for this scheme are

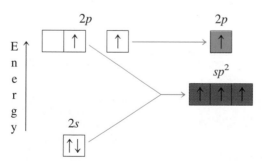

The $sp^2 + p$ orbital set is pictured in Figure 12-11. One of the bonds between the carbon atoms results from the overlap of sp^2 hybrid orbitals from each atom. This overlap occurs along the line joining the nuclei of the two atoms. Orbitals that overlap in this "end-to-end" fashion produce a **sigma bond,** designated σ **bond.** Figure 12-11 shows that a second bond between the C atoms results from the overlap of the unhybridized p orbitals. In this bond there is a region of high electron charge density above and below the plane of the carbon and hydrogen the set of orbitals $sp^2 + p$ sigma (σ) bonds

FIGURE 12-11
Sigma (σ) and pi (π) bonding in C_2H_4.
The reddish purple orbitals are sp^2 hybrid orbitals and the blue orbitals, $2p$. The sp^2 hybrid orbitals overlap along the line joining the bonded atoms—a σ bond. The $2p$ orbitals overlap in a side-to-side fashion and form a π bond.

overlap of p orbitals leading to pi (π) bond

atoms. The bond produced by this "side-to-side" overlap of two parallel orbitals is called a **pi bond,** designated π **bond.**

Bonding in ethylene is illustrated by the ball-and-stick model in Figure 12-12. This model helps to illustrate that.

- The shape of a molecule is determined only by the orbitals forming σ bonds (the σ bond framework).

- Rotation about the double bond is severely restricted. In the ball-and-stick model we could easily twist or rotate the terminal H atoms about the σ bonds that join them to a C atom. To twist one —CH_2 group out of the plane of the other, however, would reduce the amount of overlap of the p orbitals and weaken the π bond. The double bond is rigid, and the C_2H_4 molecule is planar.

Additionally, regarding carbon-to-carbon multiple bonds, the σ bond involves more extensive overlap than does the π bond. As a result, a carbon-to-carbon double bond ($\sigma + \pi$) is stronger than a single bond (σ), but not twice as strong (from Table 11.3, C—C, 347 kJ/mol; C=C, 611 kJ/mol; C≡C, 837 kJ/mol).

Bonding in C_2H_2

Bonding in acetylene, C_2H_2, is similar to that in C_2H_4, but with these differences: The Lewis structure of C_2H_2 features a triple covalent bond, H—C≡C—H. The molecule is *linear*, as found by experiment and expected from VSEPR theory. A hybridization scheme to produce hybrid orbitals in a linear orientation is sp. The valence-shell orbital diagrams representing sp hybridization are

FIGURE 12-12
Ball-and-stick model of ethylene, C_2H_4.
The H—C—H and C—C—H bond angles are 120°. The model also distinguishes between the σ bond between the C atoms (the straight plastic tube) and the π bond extending above and below the plane of the molecule. The picture of the π bond suggested by the white plastic "arches" is somewhat distorted, but the model does convey the idea that the π bond places a high electron charge density above and below the plane of the molecule.

Formation of σ bonds

Formation of π bonds

Space-filling model

FIGURE 12-13   Sigma (σ) and pi (π) bonding in C_2H_2.
The σ bond framework joins the atoms H—C—C—H through 1s orbitals of the H atoms and sp orbitals of the C atoms. There are two π bonds. Each π bond consists of two parallel cigar-shaped segments. The four segments shown actually merge into a hollow cylindrical shell with the carbon-to-carbon σ bond as its axis.

In the triple bond in C_2H_2, one of the carbon-to-carbon bonds is a σ bond and *two* are π bonds, as suggested in Figure 12-13.

▶ Always think of a triple bond as consisting of one σ and two π bonds.

EXAMPLE 12-2

Proposing Hybridization Schemes Involving σ and π Bonds. Formaldehyde gas, H_2CO, is used in the manufacture of plastics; in water solution, it is the familiar biological preservative called formalin. Describe the molecular geometry and a bonding scheme for the H_2CO molecule.

Solution

1. *Write the Lewis structure.* C is the central atom and H and O are terminal atoms. The total number of valence electrons is 12. Note that this structure requires a carbon-to-oxygen double bond.

$$\begin{array}{c} H \\ | \\ H{-}C{=}\ddot{O}: \end{array}$$

2. *Determine the electron group geometry of the central C atom.* The σ bond framework is based on three electron groups around the central C atom. VSEPR theory, based on the distribution of three electron groups, suggests a trigonal-planar molecule with 120° bond angles.

3. *Identify the orbitals that conform to the electron group geometry.* A trigonal-planar orientation of orbitals is associated with sp^2 hybrid orbitals.

4. *Sketch the orbitals of the central atom that are involved in orbital overlap.* The C atom is hybridized to produce the orbital set $sp^2 + p$, as in C_2H_4. Two of the sp^2 hybrid orbitals are used to form σ bonds with the H atoms. The remaining sp^2 hybrid orbital is used to form a σ bond with oxygen. The unhybridized p orbital of the C atom is used to form a π bond with O.

5. *Sketch the bonding orbitals of the terminal atoms.* The H atoms have only 1s orbitals available for bonding. For oxygen we can use a half-filled 2p orbital for end-to-end overlap in the σ bond to carbon and a half-filled 2p orbital to participate in the "sidewise" overlap leading to a π bond. That is, the valence-shell orbital diagram we can use for oxygen is

2s   2p

The bonding and structure of the formaldehyde molecule are suggested by the three-dimensional sketch in Figure 12-14.

▶ Now we see the VSEPR theory rationale of treating a multiple bond as a single group of electrons. The number of electron groups around the central atom is equal to the number of σ bonds in the σ-bond framework.

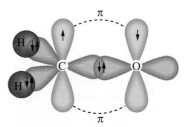

FIGURE 12-14
Bonding and structure of the H_2CO molecule—Example 12-2 illustrated.
The orbital set $sp^2 + p$ is used for the C atom, 1s orbitals for H, and two half-filled 2p orbitals for O. For simplicity, only bonding orbitals of the valence shells are shown.

Practice Example A: Describe a plausible bonding scheme and the molecular geometry of dimethyl ether, CH_3OCH_3.

Practice Example B: Acetic acid, the acidic component of vinegar, has the formula CH_3COOH. Describe the molecular geometry and a bonding scheme for this molecule.

π: $C(2p) - O(2p)$

σ: $H(1s) - C(sp^2)$ 120° C = O

120°

σ: $C(sp^2) - O(2p)$

FIGURE 12-15
Bonding in H_2CO—a schematic representation.

Drawing three-dimensional sketches to show orbital overlaps, as in Figure 12-14, is not easy. A simpler, two-dimensional representation of the bonding scheme for formaldehyde is shown in Figure 12-15. Bonds between atoms are drawn as straight lines. They are labeled σ or π, and the orbitals that overlap are indicated.

We have stressed a Lewis structure as the first step in describing a bonding scheme. Sometimes the starting point is a description of the species *obtained by experiment*. Example 12-3 illustrates such a case.

EXAMPLE 12-3

Using Experimental Data to Assist in Selecting a Hybridization and Bonding Scheme. Formic acid, HCOOH, is an irritating substance released by ants when they sting (Latin, *formica,* ant). A structural formula with bond angles is given here. Propose a hybridization and bonding scheme consistent with this structure.

124°
O
118° 108°
C
H O
H

Solution

The 118° H—C—O bond angle on the left is very nearly the 120° angle for a trigonal-planar distribution of three groups of electrons. This requires an sp^2 hybridization scheme for the C atom. The 124° O—C—O bond angle is also close to the 120° expected for sp^2 hybridization. The C—O—H bond angle of 108° is close to the tetrahedral angle—109.5°. The O atom on the right employs an sp^3 hybridization scheme. The four σ and one π bonds and the orbital overlaps producing them are indicated in Figure 12-16.

Practice Example A: Acetonitrile is an industrial solvent. Propose a hybridization and bonding scheme consistent with its structure:

H
|
H—C—C≡N
|
H

Practice Example B: A reference source on molecular structures lists the following data for dinitrogen monoxide (nitrous oxide), N_2O: Bond lengths: N—N = 113 pm; N—O = 119 pm; bond angle: 180°. Show that the Lewis structure of N_2O is a resonance hybrid of two contributing structures, and describe a plausible hybridization and bonding scheme for each.

σ: $C(sp^2) - O(2p)$

π: $C(2p) - O(2p)$

σ: $C(sp^2) - H(1s)$

O

C H

H O

σ: $O(sp^3) - H(1s)$

σ: $C(sp^2) - O(sp^3)$

FIGURE 12-16
Bonding and structure of HCOOH—Example 12-3 illustrated.

12-5 Molecular Orbital Theory

Lewis structures, VSEPR theory, and the valence-bond method make a potent combination for describing covalent bonding and molecular structures. They are satisfactory for most of our purposes. Sometimes, however, chemists need a greater understanding of molecular structures and properties than these methods provide. None of these methods, for instance, provides an explanation of the electronic spectra of molecules, why oxygen is paramagnetic, or why H_2^+ is a stable species. To address these questions, we need a different method of describing chemical bonding.

This method, called **molecular orbital theory,** starts with a simple picture of molecules, but it quickly becomes complex in its details. We can provide only an overview here. The theory assigns the electrons in a molecule to a series of orbitals that belong to the molecule as a whole. These are called molecular orbitals. Like atomic orbitals, molecular orbitals are *mathematical* functions, but we can relate them to the probability of finding electrons in certain regions of a molecule. Also like an atomic orbital, a molecular orbital can accommodate just two electrons, and the electrons must have opposing spins.

Figure 12-17 suggests the combination of two $1s$ orbitals of H atoms into two molecular orbitals in an H_2 molecule. One of the molecular orbitals, designated σ_{1s}, is at a *lower* energy than the $1s$ atomic orbitals. It is a **bonding molecular orbital** because it places a high electron charge density between the two nuclei. A high electron charge density between atomic nuclei reduces repulsions between the positively charged nuclei and promotes a strong bond.

The other molecular orbital, designated σ_{1s}^*, is at a *higher* energy than the $1s$ atomic orbitals. It places a very low electron charge density between the two nuclei and is an **antibonding molecular orbital.** With a low electron

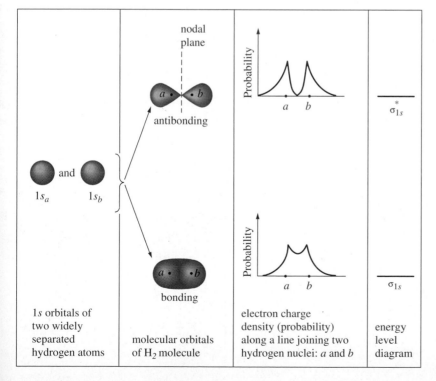

1s orbitals of two widely separated hydrogen atoms	molecular orbitals of H_2 molecule	electron charge density (probability) along a line joining two hydrogen nuclei: *a* and *b*	energy level diagram

FIGURE 12-17
The interaction of two hydrogen atoms according to the molecular orbital theory. The energy of the σ_{1s} bonding molecular orbital is lower, and that of the σ_{1s}^* antibonding molecular orbital is higher, than the energies of the $1s$ atomic orbitals. Electron charge density in a bonding molecular orbital is high in the internuclear region. In an antibonding orbital it is high in parts of the molecule away from the internuclear region.

charge density between atomic nuclei, the nuclei are not screened from each other, strong repulsions occur, and the bond is weakened (hence the term "antibonding").

Basic Ideas Concerning Molecular Orbitals

Here are some useful ideas about molecular orbitals and how electrons are assigned to them.

1. The number of molecular orbitals (MO) formed is equal to the number of atomic orbitals combined.
2. Of the two MOs formed when two atomic orbitals are combined, one is a *bonding* MO at a *lower* energy than the original atomic orbitals. The other is an *antibonding* MO at a *higher* energy.
3. Electrons normally enter the lowest energy MOs available to them.
4. The maximum number of electrons in a given MO is two (Pauli exclusion principle).
5. Electrons enter MOs of identical energies *singly* before they pair up (Hund's rule).

If a molecular species is to be stable, it must have more electrons in bonding than in antibonding orbitals. For example, if the excess of bonding over antibonding electrons is *two*, this corresponds to a *single* covalent bond in Lewis theory. In molecular orbital theory we say that the bond order is 1. **Bond order** is one-half the difference between the number (no.) of bonding and antibonding electrons (e^-), that is,

$$\text{bond order} = \frac{\text{no. } e^- \text{ in bonding MOs} - \text{no. } e^- \text{ in antibonding MOs}}{2} \quad (12.1)$$

Diatomic Molecules of the First-Period Elements

Let's use the ideas just outlined to describe some molecular species of the first-period elements, H and He. Figure 12-18 helps us to do this.

H_2^+.   This species has a single electron. It enters the σ_{1s} orbital, a bonding molecular orbital. With Equation (12.1) we see that the bond order is

---

FIGURE 12-18

Molecular-orbital diagrams for the diatomic molecules and ions of the first-period elements.

The 1s energy levels of the isolated atoms are shown to the left and right of each diagram. The line segments in the middle represent the molecular orbital energy levels—lower than the 1s levels for σ_{1s} and higher for σ_{1s}^*.

$(1 - 0)/2 = \frac{1}{2}$. This is equivalent to a one-electron, or *"half,"* bond, a bond type that is not easily described by the Lewis theory.

H$_2$.   This molecule has two electrons, both in the σ_{1s} orbital. The bond order is $(2 - 0)/2 = 1$. With Lewis theory and the valence-bond method we describe the bond in H$_2$ as single covalent.

He$_2^+$.   This ion has three electrons. Two electrons are in the σ_{1s} orbital and one is in the σ_{1s}^* orbital. This species exists as a stable ion with a bond order of $(2 - 1)/2 = \frac{1}{2}$.

He$_2$.   Two electrons are in the σ_{1s} orbital and two are in the σ_{1s}^*. The bond order is $(2 - 2)/2 = 0$. No bond is produced. He$_2$ is not a stable species.

EXAMPLE 12-4

Relating Bond Energy and Bond Order. The bond energy of H$_2$ is 436 kJ/mol. Estimate the bond energies of H$_2^+$ and He$_2^+$.

Solution

The bond order in H$_2$ is 1, equivalent to a single bond. In H$_2^+$ and in He$_2^+$ the bond order is $\frac{1}{2}$. We should expect the bonds in these two species to be only about one-half as strong as in H$_2$—about 220 kJ/mol. (Actual values: H$_2^+$, 255 kJ/mol; He$_2^+$, 251 kJ/mol.)

Practice Example A: The bond energy of Li$_2$ is 106 kJ/mol. Estimate the bond energy of Li$_2^+$.

Practice Example B: Do you think the ion H$_2^-$ is stable? Explain.

Molecular Orbitals of the Second-Period Elements

For diatomic molecules and ions of H and He we had to combine only $1s$ orbitals. In the second period the situation is more interesting because we must work with both $2s$ and $2p$ orbitals. This results in *eight* molecular orbitals. Let's see how this comes about.

The molecular orbitals formed by combining $2s$ orbitals are similar to those from $1s$ orbitals, except they are at a higher energy. The situation for combining $2p$ atomic orbitals, however, is different. Two possible ways for $2p$ atomic orbitals to combine into molecular orbitals are shown in Figure 12-19: end-to-end and side-to-side. The best overlap for p orbitals is along a straight line (that is, end-to-end). This combination produces σ-type molecular orbitals: σ_{2p} and σ_{2p}^*.

Only one pair of p orbitals can combine in an end-to-end fashion. The other two pairs must combine in a parallel or sidewise fashion to produce π-type molecular orbitals: π_{2p} and π_{2p}^*. There are *four* π-type molecular orbitals (two bonding and two antibonding) because there are *two* pairs of $2p$ atomic orbitals arranged in a parallel fashion. In Figure 12-19 we again see that bonding molecular orbitals place a high electron charge density between the atomic nuclei. In antibonding molecular orbitals there are nodal planes between the nuclei, where the electron charge density falls to zero.

The energy level diagram for the molecular orbitals formed from atomic orbitals of the second principal electronic shell is related to the atomic orbital energy levels. For example, molecular orbitals formed from $2s$ orbitals are at a

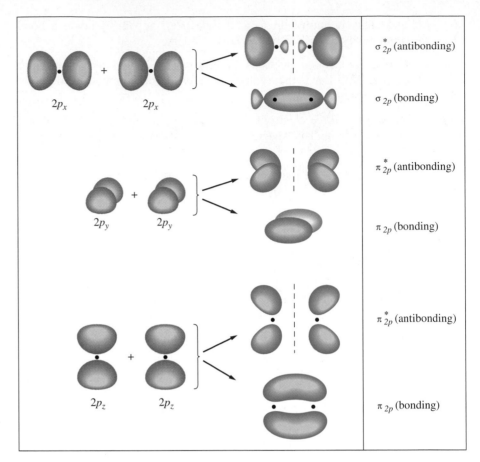

FIGURE 12-19
Molecular orbitals formed from 2*p* atomic orbitals.
These diagrams suggest the electron charge distributions for the several orbitals. They are not exact in all details. Nodal planes for antibonding orbitals are represented by broken lines.

lower energy than those formed from 2*p* orbitals, the same relationship as between the 2*s* and 2*p* atomic orbitals. Another expectation is that σ-type bonding orbitals should have lower energies than π-type because end-to-end overlap of 2*p* orbitals should be more extensive than sidewise overlap and result in a lower energy. This expectation does appear to be the case for the molecular orbitals in O_2 and F_2. For other diatomic molecules of the second-period elements (for example, C_2 and N_2) the situation is reversed, however. That is, the π_{2p} orbitals are at a lower energy than σ_{2p}.*

Here is how we assign electrons to the molecular orbitals: We start with the σ_{1s} and σ_{1s}^* orbitals filled. Then we add electrons, in order of increasing energy, to the available molecular orbitals of the second principal shell. Figure 12-20 shows the electron assignments for the actual and hypothetical homonuclear diatomic molecules (X_2) of the second-period elements. Some molecular properties are also listed in the figure.

▶ A *homonuclear* diatomic molecule is one in which both atoms are of the same kind. In a *hetero*nuclear diatomic molecule the two atoms are different.

A Special Look at O_2

Molecular orbital theory helps us understand some of the previously unexplained features of the O_2 molecule. Each O atom brings six valence electrons to the

---

*The situation with O_2 and F_2 is that expected when the energy difference between 2*s* and 2*p* orbitals is large. Where the difference is smaller, 2*s*-2*p* orbital interactions affect the way in which atomic orbitals combine, reversing the molecular orbital energy levels—that is, with π_{2p} below σ_{2p}.

diatomic molecule, O_2. In Figure 12-20, we see that when we assign 12 valence electrons to molecular orbitals the molecule has *two unpaired electrons.* This explains the *paramagnetism* of O_2. The number of valence electrons in bonding orbitals is *eight;* the number in antibonding orbitals is *four;* and the bond order is 2. This corresponds to a double covalent bond in O_2.

A convenient way to summarize these facts about the O_2 molecule is through a molecular orbital diagram. Just as we might arrange the valence-shell atomic orbitals of an O atom, we arrange the second-shell *molecular* orbitals of the O_2 molecule in their order of increasing energy.

$$\begin{array}{ccccccc} & \sigma_{2s} & \sigma_{2s}^* & \sigma_{2p} & \pi_{2p} & \pi_{2p}^* & \sigma_{2p}^* \\ O_2 & \boxed{\uparrow\downarrow} & \boxed{\uparrow\downarrow} & \boxed{\uparrow\downarrow} & \boxed{\uparrow\downarrow}\,\boxed{\uparrow\downarrow} & \boxed{\uparrow}\,\boxed{\uparrow} & \boxed{} \end{array} \qquad (12.2)$$

FIGURE 12-20
Molecular orbital diagrams for actual and hypothetical diatomic molecules of the second-period elements.
In all cases the σ_{1s} and σ_{1s}^* molecular orbitals are filled but not shown.

	Li_2	Be_2	B_2	C_2	N_2
Bond order	1	0	1	2	3
Magnetism	dia-magnetic	—	para-magnetic	dia-magnetic	dia-magnetic

	O_2	F_2	Ne_2
Bond order	2	1	0
Magnetism	para-magnetic	dia-magnetic	—

EXAMPLE 12-5

Writing a Molecular Orbital Diagram and Determining Bond Order. Represent bonding in O_2^+ with a molecular orbital diagram, and determine the bond order in this ion.

Solution

The ion O_2^+ has 11 valence electrons. Assign these to the available molecular orbitals in accordance with the ideas stated on page 402. [Or simply remove one of the electrons from a π_{2p}^* orbital in expression (12.2).] In the following diagram we find an excess of *five* bonding electrons over antibonding ones. The bond order is 2.5.

$$O_2^+ \quad \overset{\sigma_{2s}}{\boxed{\uparrow\downarrow}} \quad \overset{\sigma_{2s}^*}{\boxed{\uparrow\downarrow}} \quad \overset{\sigma_{2p}}{\boxed{\uparrow\downarrow}} \quad \overset{\pi_{2p}}{\boxed{\uparrow\downarrow}\boxed{\uparrow\downarrow}} \quad \overset{\pi_{2p}^*}{\boxed{\uparrow}\boxed{}} \quad \overset{\sigma_{2p}^*}{\boxed{}}$$

Practice Example A: Refer to Figure 12-20. Write a molecular orbital diagram and determine the bond order of **(a)** N_2^+; **(b)** Ne_2^+; **(c)** C_2^{2-}.

Practice Example B: The bond lengths for O_2^+, O_2, O_2^-, and O_2^{2-} are 112 pm, 121 pm, 128 pm, and 149 pm, respectively. Are these bond lengths consistent with the bond order determined from the molecular orbital diagram? Explain.

12-6 Delocalized Electrons: Bonding in the Benzene Molecule

▶ The term "aromatic" relates to the fragrant aromas associated with some (but by no means all) of these compounds.

In Section 12-4 we discussed localized π bonds, such as those in ethylene, C_2H_4. Some molecules have a network of π bonds, such as benzene, C_6H_6, and substances related to it—aromatic compounds. In this section we will consider bonding in benzene with the bonding theories we have studied so far, and the conclusions we reach will help us understand other cases of bonding as well.

Bonding in Benzene

In 1865, Friedrich Kekulé advanced the first good proposal for the structure of benzene. He suggested that the C_6H_6 molecule consists of a flat, hexagonal ring of six carbon atoms joined by alternating single and double covalent bonds. Each C atom is joined to two other C atoms and to one H atom. To explain the fact that the carbon-to-carbon bonds are all alike, Kekulé suggested that the single and double bonds continually oscillate from one position to another. Today, we say that the two possible Kekulé structures are actually contributing structures to a resonance hybrid. This view is suggested by Figure 12-21.

We can gain a more thorough understanding of bonding in the benzene molecule by combining the valence-bond and molecular orbital methods. We can construct a σ bond framework for the observed *planar* structure with 120° bond angles by using sp^2 hybridization at each carbon atom. End-to-end overlap of the sp^2 orbitals produces σ bonds. Sidewise overlap of $2p$ orbitals yields three π bonds. Figure 12-22 gives a valence-bond theory interpretation of bonding in C_6H_6.

We do not need to think in terms of an oscillation between two structures (Kekulé) or of a resonance hybrid for benzene. The π bonds in Figure 12-22 are not localized between specific carbon atoms but are spread out around the

Dame Kathleen Londsdale first determined the X-ray crystal structure of benzene. Her experiment demonstrated that the benzene molecule is flat, as predicted by theorists.

(a)   **(b)**   **(c)**

FIGURE 12-21
Resonance in the benzene molecule and the Kekulé structures.
(a) Lewis structure for C_6H_6, showing alternate carbon-to-carbon single and double bonds.
(b) Two equivalent Kekulé structures for benzene. A carbon atom is at each corner of the
hexagonal structure, and a hydrogen atom is bonded to each carbon atom. (The symbols
for carbon and hydrogen, as well as the C—H bonds, are customarily omitted in these
structures.) (c) A space-filling model.

six-membered ring. To represent this *delocalized* π bonding, the symbol for benzene is often written as a hexagon with a circle inscribed within it (see Figure
12-22).

We can best understand delocalized π bonds through molecular orbital theory. Six $2p$ atomic orbitals of the C atoms combine to form six molecular orbitals of the π type. Three of these π-type molecular orbitals are bonding and
three are antibonding. The three bonding orbitals fill with six electrons (one $2p$
electron from each C atom), and the three antibonding orbitals remain empty.
The bond order associated with the six electrons in π-bonding molecular orbitals is $(6 - 0)/2 = 3$. The three bonds are distributed among the six C atoms,
which amounts to 3/6, or a half-bond, between each pair of C atoms. Add to

(a) σ bond framework

FIGURE 12-22
**Bonding in benzene, C_6H_6, by the valence-
bond method.**
(a) Carbon atoms use sp^2 and p orbitals. Each
carbon atom forms three σ bonds, two with
neighboring C atoms in the hexagonal ring
and a third with an H atom. (b) The overlap in
sidewise fashion of $2p$ orbitals produces three
π bonds. Thus there are three double bonds
($\sigma + \pi$) between carbon atoms in the ring.
(c) Because the three π bonds are delocalized
around the benzene ring, the molecule is often
represented through a hexagon with an inscribed circle.

(b) π bonding   **(c)** Symbolic representation

FIGURE 12-23
π molecular-orbital diagram for C_6H_6.
Of the six π molecular orbitals, three are bonding orbitals and each is filled with an electron pair. The three antibonding molecular orbitals at a higher energy remain empty.

FIGURE 12-24
Molecular orbital representation of π bonding in benzene.
A computer-generated model of the benzene molecule. The planar σ bond framework is clearly visible. The π orbitals above and below the C and H plane are highlighted.

this the σ bonds in the σ bond framework and we have a bond order of 1.5 for each carbon-to-carbon bond. This is exactly what we also get by averaging the two Kekulé structures of Figure 12-21. The π molecular orbital diagram for benzene is shown in Figure 12-23.

The three bonding π molecular orbitals in C_6H_6 describe the distribution of π electron charge in the molecule. We can think of this in terms of two doughnut-shaped regions: one above and one below the plane of the C and H atoms. Because they are spread out among all six C atoms instead of being concentrated between pairs of C atoms, these molecular orbitals are called **delocalized molecular orbitals.** Figure 12-24 pictures these delocalized molecular orbitals in the benzene molecule.

Other Structures with Delocalized Molecular Orbitals

By using delocalized bonding schemes we can avoid writing two or more contributing structures to a resonance hybrid, as is so often required in the Lewis theory. Consider the ozone molecule, O_3, that we used to introduce the concept of resonance in Section 11-5. In place of the resonance hybrid based on these contributing structures,

we can write the single structure shown in Figure 12-25. These are the ideas that lead to Figure 12-25.

1. With VSEPR theory we predict a trigonal-planar electron group geometry (the measured bond angle is 117°). The hybridization scheme chosen for the central O atom is sp^2, and, although we normally do not need to invoke hybridization for terminal atoms, this case is simplified if we assume sp^2 hybridization for the terminal O atoms as well. Thus, each O atom uses the orbital set $sp^2 + p$.

2. Of the 18 valence electrons in O_3, assign 14 to the sp^2 hybrid orbitals of the σ bond framework. Four of these are bonding electrons (red) and ten are lone-pair electrons (blue).

3. The three unhybridized $2p$ orbitals combine to form *three* molecular orbitals of the π type. One of these orbitals is a bonding molecular orbital and another is antibonding. The third is a type we have not described before—a *nonbonding* molecular orbital. A nonbonding molecular orbital has the same energy as the atomic orbitals from which it is formed and neither adds to nor detracts from bond formation.

4. The remaining four valence electrons are assigned to the π molecular orbitals. Two go into the bonding orbital and two into the nonbonding orbital. The antibonding orbital remains empty.

5. The bond order associated with the π molecular orbitals is $(2 - 0)/2 = 1$. This π bond is distributed between the two O—O bonds and amounts to one-half of a π bond for each.

The points listed here lead to a total bond order of 1.5 for the O—O bonds in O_3. This is equivalent to averaging the two Lewis structures. The O—O bond length suggested by this treatment was described in Section 11-5.

12-7   Bonding in Metals

In nonmetal atoms the valence shells generally have more electrons than they do orbitals. To illustrate, an F atom has *four* valence-shell orbitals ($2s$, $2p_x$, $2p_y$, $2p_z$) and *seven* valence-shell electrons. Whether fluorine exists as a solid, liquid, or gas, F atoms join in pairs to form F_2 molecules. One pair of electrons is shared in the F—F bond, and the other electron pairs are lone pairs, as seen in the Lewis structure $:\ddot{F}—\ddot{F}:$. By contrast, the metal atom Li has the same four valence-shell orbitals as F but only *one* valence-shell electron ($2s^1$). This may account for the formation of the *gaseous* molecule Li:Li, but in the solid metal each Li atom is bonded, somehow, to *eight* neighbors. The challenge to a bonding theory for metals is to explain how so much bonding can occur with so few electrons. Also, the theory should account for certain properties that metals display to a far greater extent than nonmetals, such as a lustrous appearance, an ability to conduct electricity, and an ease of deformation (metals are easily flattened into sheets and drawn into wires).

The Electron Sea Model

An oversimplified theory, which can explain some of the properties of metals just cited, pictures a solid metal as a network of positive ions immersed in a "sea of electrons." In lithium, for instance, the ions are Li^+ and one electron per atom is contributed to the sea of electrons. Electrons in the sea are *free* (not attached to any particular ion) and they are mobile. Thus, if electrons from an external source enter a metal wire at one end, free electrons pass through the wire and leave the other end at the same rate. In this way electrical conductivity is explained.

Free electrons (those in the electron sea) are not limited in their ability to absorb photons of visible light as are electrons bound to an atom. Thus metals absorb visible light; they are opaque. Electrons at the surface of a metal are able to reradiate, at the same frequency, light that strikes the surface, which explains the lustrous appearance of metals. The ease of deformation of metals can be explained in this way: If one layer of metal ions is forced across another, perhaps by hammering, no bonds are broken, the internal structure of the metal remains essentially unchanged, and the sea of electrons rapidly adjusts to the new situation (Figure 12-26).

Band Theory

The electron sea model is a simple qualitative description of the metallic state, but for most purposes the theory of metallic bonding used is a form of molecular orbital theory called **band theory.**

Recall the formation of molecular orbitals and the bonding between two Li atoms (see Figure 12-20). Each Li atom contributes one $2s$ orbital to the production of two molecular orbitals: σ_{2s} and σ_{2s}^*. The electrons originally described as the $2s^1$ electrons of the Li atoms enter and half-fill these molecular orbitals. That is, they fill the σ_{2s} orbital and leave the σ_{2s}^* empty. If we extend this combination of Li atoms to a third Li atom, three molecular orbitals are formed, which contain a total of three electrons. Again the set of molecular orbitals is half-filled. We can extend this process to an enormously large number (N) of atoms—the total number of atoms in a crystal of Li. Here is the result we get: A set of N molecular orbitals with an extremely small energy separation be-

(a) σ bond framework

(b) Delocalized π molecular orbital

FIGURE 12-25
Structure of the ozone molecule O_3.
(a) The σ bond framework and the assignment of bond pair (:) and lone pair (:) electrons to sp² hybrid orbitals are discussed in items 1 and 2 on page 408.
(b) The π molecular orbitals and assignments of electrons to them are discussed in items 3 and 4.

▶A dull metal surface usually signifies that the surface is coated with a compound of the metal (for example, an oxide, sulfide, or carbonate).

FIGURE 12-26

The electron sea model of metals.

A network of positive ions is immersed in a "sea of electrons," derived from the valence shells of the metal atoms and belonging to the crystal as a whole. One particular ion (red), its nearest neighboring ions (brown), and nearby electrons in the electron sea (blue) are emphasized.

At the bottom of the figure, a force is applied (from left to right). The highlighted cation is unaffected; its immediate environment is unchanged. The electron sea model explains the ease of deformation of metals.

FIGURE 12-27

Formation of an energy band in lithium metal.

As more and more Li atoms are added to the growing "molecule," Li_2, Li_3, . . . , additional energy levels are added and the spacing between levels becomes increasingly smaller. In an entire crystal of N atoms, the energy levels merge into a band of N closely spaced levels. The lowest $N/2$ levels are filled with electrons and the upper $N/2$ levels are empty.

tween each pair of successive levels. This collection of very closely spaced molecular orbital energy levels is called an *energy band* (Figure 12-27).

In the band just described there are N electrons (a $2s$ electron from each Li atom) occupying, in pairs, $N/2$ molecular orbitals of lowest energy. These are the electrons responsible for bonding the Li atoms together. They are valence electrons, and the band in which they are found is called a *valence band*. Because the energy differences between the occupied and unoccupied levels in the valence band are so small, however, electrons can be easily excited from the highest filled levels to the unfilled levels that lie immediately above them in energy. This excitation, which has the effect of producing mobile electrons, can be accomplished by applying a small electric potential difference across the crystal. This is how the band theory explains the ability of metals to conduct electricity. The essential feature for electrical conductivity, then, is *an energy band that is only partly filled with electrons.* Such an energy band is called a *conduction band.* In lithium the $2s$ band is both a valence band and a conduction band.

Let's extend our discussion to N atoms of beryllium, which has the electron configuration $1s^2 2s^2$. We expect the band formed from $2s$ atomic orbitals to be filled—N molecular orbitals and $2N$ electrons. But how can we reconcile this with the fact that Be is a good electric conductor? At the same time that $2s$ orbitals are being combined into a $2s$ band, $2p$ orbitals combine to form an *empty* $2p$ band. The lowest levels of the $2p$ band are at a *lower* energy than the highest levels of the $2s$ band. The bands overlap. As a consequence, empty molecular orbitals are available to the valence electrons in beryllium.

In an *electrical insulator* like diamond or silica (SiO_2), not only is the valence band filled, but there is a large energy gap between the valence band and the conduction band. Very few electrons are able to make the transition between the two. In a **semiconductor,** such as silicon or germanium, the filled valence band and empty conduction band are separated only by a small energy gap. Electrons in the valence band may acquire enough thermal energy to jump to a level in the conduction band. The greater the thermal energy, the more electrons can make the transition. In this way band theory explains the observation that the electrical conductivity of semiconductors increases with temperature. Figure 12-28 compares metals, semiconductors, and insulators in terms of band theory.

(a) Metal   **(b)** Metal   **(c)** Semiconductor   **(d)** Insulator

FIGURE 12-28   Metals, semiconductors, and insulators as viewed by band theory.
(a) In some metals the valence band (blue) is only partially filled (e.g., the half-filled $3s$ band in Na). The valence band also serves as a conduction band (outlined in black). (b) In other metals the valence band is full, but a conduction band overlaps it (e.g., the empty $3p$ band of Mg overlaps the full $3s$ valence band). (c) In a semiconductor the valence band is full, and the conduction band is empty. The energy gap (ΔE) between the two is small enough, however, that some electrons make the transition between the two just by acquiring thermal energy. (d) In an insulator the valence band is filled with electrons, and a large energy gap (ΔE) separates the valence band from the conduction band. Few electrons can make the transition between bands, and the insulator does not conduct electricity.

Summary

Valence-bond theory considers a covalent bond in terms of the overlap of atomic orbitals of the bonded atoms. Some molecules can be described through the overlap of simple orbitals, but often *hybrid* orbitals are needed. The hybridization scheme chosen is the one that produces an orientation of hybrid orbitals to match the electron group geometry predicted by the VSEPR theory.

End-to-end overlap of orbitals produces σ bonds. Sidewise overlap of two p orbitals produces a π bond. Single covalent bonds are σ bonds. A double bond consists of one σ and one π bond. A triple bond consists of one σ and two π bonds. The geometric shape of a species is determined by its σ bond framework, not its π bonds.

In molecular orbital theory, electrons are assigned to molecular orbitals. The numbers and kinds of molecular orbitals are related to the atomic orbitals used to generate them. Electron charge density between atoms is high in bonding molecular orbitals and very low in antibonding orbitals. Bond order is one-half the difference between the numbers of electrons in bonding and in antibonding molecular orbitals. Molecular orbital energy-level diagrams and an Aufbau process can be used to describe the electronic structure of a molecule, similar to what was done for atomic electron configurations in Chapter 9.

Bonding in the benzene molecule, C_6H_6, is based on the concept of delocalized molecular orbitals. These are regions of high electron charge density that extend over several atoms in a molecule. Delocalized molecular orbitals also provide an alternative to the concept of resonance in other molecules and ions.

Finally, molecular orbital theory, in the form called band theory, can be applied to metals, semiconductors, and insulators.

Summarizing Example

Hydrogen azide, HN_3, and its salts (metal azides) are unstable substances used in detonators for high explosives. Sodium azide, NaN_3, is used in air-bag safety systems in automobiles (see page 194). A reference source lists the following data for HN_3. (The subscripts a, b, and c distinguish the three N atoms from one another.) Bond lengths: $N_a-N_b = 124$ pm; $N_b-N_c = 113$ pm. Bond angles: $H-N_a-N_b = 112.7°$; $N_a-N_b-N_c = 180°$.

Write two contributing structures to the resonance hybrid for HN_3, and describe a plausible hybridization scheme for each structure.

1. *Compare bond lengths with values from Table 11.1.* The bond lengths establish whether there is multiple-bond char-acter to the N_a-N_b and N_b-N_c bonds. *Result:* N_a-N_b is essentially a double bond, and N_b-N_c has some triple bond character.

2. *Draw plausible Lewis structures.* The H and N_c atoms are terminal atoms. N_a and N_b are central atoms. 16 valence electrons are used. *Result:* Structure (I), N_a-N_b and N_b-N_c are both double bonds; structure (II), N_a-N_b is a single bond; and N_b-N_c is a triple bond.

3. Determine hybridization schemes for the central atoms. Use VSEPR theory to predict the electron group geometry around N_a and N_b. Indicate the hybridization schemes to match the bond angle data. *Answer:* Structure (I), N_a is sp^2 and N_b is sp; structure (II), N_a is sp^3 and N_b is sp.

Focus On   Semiconductors

Solar cells in an array provide electric power for a telephone.

Much of modern electronics depends on the use of semiconductor materials. Light-emitting diodes (LEDs), transistors, and solar cells are among the familiar electronic components using semiconductors. Semiconductors such as cadmium yellow (CdS) and vermillion (HgS) are brilliantly colored and artists use them in paints.

What determines the electronic properties of a semiconductor is the energy gap (band gap) between the valence band and the conduction band (recall Figure 12-28). In some materials, such as CdS, this gap is of a fixed size. These materials are called *intrinsic* semiconductors. When white light interacts with the semiconductor, electrons are excited (promoted) to the conduction band. CdS absorbs violet and some blue light, but other frequencies contain less energy than that needed to excite an electron above the energy gap. These frequencies are reflected, and the color we see is yellow. Some semiconductors, such as GaAs and PbS, have a small enough band gap that all frequencies of visible light are absorbed. There is no reflected visible light, and the materials have a black color.

In many semiconductors, called *extrinsic* semiconductors, the size of the band gap is controlled by carefully adding impurities—a process called *doping*. Let's consider what doping does to one of the most common semiconductors, silicon.

When silicon is doped with phosphorus, the energy level of the P atoms lies just below the *conduction* band of the silicon, as shown in Figure 12-29. Each P atom uses *four* of its five valence electrons to form bonds to four neighboring Si atoms, and thermal energy alone is enough to cause the "extra" valence electron to be promoted to the conduction band, leaving behind an immobile positive P^+ ion. The P atoms are called *donor* atoms, and electrical conductivity in this type of semiconductor involves primarily the movement through the conduction band of electrons coming from the donor atoms. This type of semiconductor is called an *n-type*, where *n* refers to negative—the type of electric charge carried by electrons.

When silicon is doped with aluminum, the energy level of the Al atoms, called *acceptor* atoms, lies just above the *valence* band of the silicon (see Figure 12-29). Because an Al atom has only *three* valence electrons, it forms regular electron pair bonds with three neighboring Si atoms but only a one-electron bond with a fourth Si atom. An electron is easily promoted from the valence band to an Al atom in the acceptor level, however, forming an immobile negative Al^- ion. When this occurs, a *positive hole* is created in the valence band. Because electrical conductivity in this type of semiconductor consists primarily in the migration of positive holes, it is called a *p-type* semiconductor.

Figure 12-30 suggests how semiconductors are used in photovoltaic (solar) cells. A thin layer of a *p-type* semiconductor is in contact with an *n-type* semiconductor in a region called the *junction*. Normally, migration of electrons and positive holes across the junction is very limited because such a migra-

Key Terms

antibonding molecular orbital (12-5)
band theory (12-7)
bond order (12-5)
bonding molecular orbital (12-5)
delocalized molecular orbital (12-6)
hybrid orbital (12-3)

hybridization (12-3)
molecular orbital theory (12-5)
pi (π) bond (12-4)
semiconductor (12-7)
sigma (σ) bond (12-4)
sp hybrid orbital (12-3)

sp^2 hybrid orbital (12-3)
sp^3 hybrid orbital (12-3)
sp^3d hybrid orbital (12-3)
sp^3d^2 hybrid orbital (12-3)
valence-bond method (12-2)

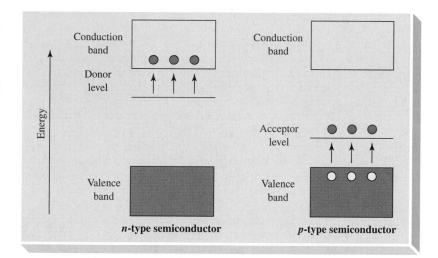

FIGURE 12-29
p- and n-type semiconductors.
In a semiconductor with donor atoms (for example, P in Si), the donor level lies just beneath the conduction band. Electrons (●) are easily promoted into the conduction band. The semiconductor is of the *n*-type. In a semiconductor with acceptor atoms (for example, Al in Si), the acceptor level lies just above the valence band. Electrons (●) are easily promoted to the acceptor level, leaving positive holes (○) in the valence band. The semiconductor is of the *p*-type.

positive holes, can easily cross the junction into the *n*-type semiconductor. This sets up a flow of electrons, an electric current. Electrons can be carried by wires through an external load (lights, electric motors, . . .) and eventually returned to the *p*-type semiconductor, where they fill positive holes. Further light absorption creates more conduction electrons and positive holes, and the process continues as long as light shines on the solar cell.

tion would lead to a separation of charge: Positive holes crossing the junction from the *p*-type semiconductor would have to move away from immobile Al^- ions, and electrons crossing from the *n*-type semiconductor would have to move away from immobile P^+ ions.

Now imagine that the *p*-type semiconductor is struck by a beam of light. Electrons in the valence band can absorb energy and be promoted to the conduction band, creating positive holes in the valence band. Conduction electrons, unlike

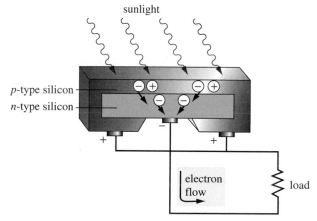

FIGURE 12-30
A photovoltaic (solar) cell using silicon-based semiconductors.

Review Questions

1. In your own words define the following terms or symbols: **(a)** sp^2 hybrid orbital; **(b)** π_{2p} molecular orbital; **(c)** bond order; **(d)** π bond.
2. Briefly describe each of the following ideas: **(a)** hybridization of atomic orbitals; **(b)** σ-bond framework; **(c)** Kekulé structures of benzene, C_6H_6; **(d)** band theory of metallic bonding.

3. Explain the important distinctions between the terms in each of the following pairs: **(a)** σ and π bonds; **(b)** localized and delocalized electrons; **(c)** valence-bond method and VSEPR theory; **(d)** metal and semiconductor.
4. The measured bond angle in H_2Se is 91°. Which orbitals of the H and Se atoms are most likely involved in orbital overlap in H—Se bonds? Explain.

5. In which of the following would you expect to find sp^2 hybridization of the central atom? Explain. CO_3^{2-}, SO_2, CCl_4, CO, NO_2^-.

6. Only one of the following statements is correct regarding bonding in carbon–hydrogen–oxygen compounds. Identify the correct statement and indicate the errors in the other statements. (a) All oxygen-to-hydrogen bonds are π bonds; (b) all carbon-to-hydrogen bonds are σ bonds; (c) all carbon-to-carbon bonds consist of a σ and a π bond; (d) all carbon-to-carbon bonds are π bonds.

7. In the manner of Example 12-1, describe the probable structure and bonding in (a) HI; (b) BrCl; (c) H_2Se; (d) OCl_2.

8. Explain why the molecular structure of BF_3 cannot adequately be described through overlaps involving pure s and p orbitals.

9. In the manner of Figure 12-14, indicate the structures of the following molecules in terms of the overlap of simple atomic orbitals and hybrid orbitals: (a) CH_2Cl_2; (b) OCN^-; (c) BF_3.

10. Match each of the following species with one of these hybridization schemes: sp, sp^2, sp^3, sp^3d, sp^3d^2. (a) PF_6^-; (b) COS; (c) $SiCl_4$; (d) NO_3^-; (e) AsF_5. (*Hint:* What is the electron group geometry of the central atom in each case?)

11. Propose a hybridization scheme to account for bonds formed by the central carbon atom in each of the following molecules: (a) hydrogen cyanide, HCN; (b) methyl alcohol, CH_3OH; (c) acetone, $(CH_3)_2CO$; (d) carbamic acid,

12. Indicate which of the following molecules and ions are linear, which are planar, and which are neither. Then propose

hybridization schemes for the central atoms. (a) $Cl_2C=CCl_2$; (b) $N\equiv C-C\equiv N$; (c) $F_3C-C\equiv N$; (d) $[S-C\equiv N]^-$.

13. For the following molecules write Lewis structures and then label each σ and π bond. (a) HCN; (b) C_2N_2; (c) $CH_3CHCHCCl_3$; (d) HONO.

14. Represent bonding in the carbon dioxide molecule, CO_2, by (a) a Lewis structure and (b) the valence-bond method, identifying σ and π bonds, the necessary hybridization scheme, and orbital overlap.

15. What is the total number of (a) σ bonds and (b) π bonds in the molecule CH_3NCO? (*Hint:* Draw a plausible Lewis structure.)

16. For the following pairs of molecular orbitals, indicate the one you expect to have the lower energy and state the reason for your choice. (a) σ_{1s} or σ_{1s}^*; (b) σ_{2s} or σ_{2p}; (c) σ_{1s}^* or σ_{2s}; (d) σ_{2p} or π_{2p}^*.

17. Which of the diatomic molecules do you think has the greater bond energy, Li_2 or C_2? Explain.

18. For each of the species C_2^+; O_2^-; and F_2^+:
(a) Write the molecular orbital diagram (as in Example 12-5).
(b) Determine the bond order and state whether you expect the species to be stable or unstable.
(c) Determine if the species is diamagnetic or paramagnetic, and if paramagnetic, indicate the number of unpaired electrons.

19. Describe the electron sea model for metallic bonding. How does it differ from the band theory of metals?

20. In what type of material is the energy gap between the valence band and the conduction band greatest: metal, semiconductor, or insulator? Explain.

Exercises

Valence-Bond Method

21. Indicate several ways in which the valence-bond method is superior to Lewis structures in describing covalent bonds.

22. State why it is necessary to hybridize atomic orbitals when applying the valence-bond method. That is, why are there so few molecules that can be described through the overlap of pure atomic orbitals only?

23. Describe the molecular geometry of H_2O suggested by each of the following methods: (a) Lewis theory; (b) valence-bond method using simple atomic orbitals; (c) VSEPR theory; (d) valence-bond method using hybridized atomic orbitals.

24. Describe the molecular geometry of CCl_4 suggested by each of the following methods: (a) Lewis theory; (b) valence-bond method using simple atomic orbitals; (c) VSEPR theory; (d) valence-bond method using hybridized atomic orbitals.

25. For each of the following species identify the central atom(s) and propose a hybridization scheme for the atom(s). (a) CO_2; (b) $HONO_2$; (c) ClO_3^-; (d) BF_4^-.

26. Propose a plausible Lewis structure, geometric structure, and hybridization scheme for the NSF molecule.

27. Describe a hybridization scheme for the central Cl atom in the molecule ClF_3 that is consistent with the geometric shape pictured in Table 11.2. Which orbitals of the Cl atom are involved in overlaps and which are occupied by lone pair electrons?

28. Describe a hybridization scheme for the central S atom in the molecule SF_4 that is consistent with the geometric shape pictured in Table 11.2. Which orbitals of the S atom are involved in overlaps and which are occupied by lone pair electrons?

29. Use the method of Figure 12-15 to represent bonding in each of the following molecules: (a) CCl_4; (b) ONCl; (c) HONO; (d) $COCl_2$.

30. Use the method of Figure 12-15 to represent bonding in each of the following ions: (a) NO_2^-; (b) I_3^-; (c) $C_2O_4^{2-}$; (d) HCO_3^-.

31. Malic acid is a common organic acid found in unripe apples and other fruit. With the help of VSEPR theory, sketch a three-dimensional structure of the molecule. Indicate the type of orbital overlaps that are consistent with this structure.

$$
\underset{\underset{\text{H}}{\underset{|}{\text{C}}}}{\overset{\text{O}}{\overset{||}{\text{C}}}}
$$

H—O—C—C—C—C—O—H

(with O, H, H, O on top and H, O—H below the chain)

32. Glycine is a common amino acid found in many proteins. It has the formula NH_2CH_2COOH. With the help of VSEPR theory, sketch a three-dimensional structure of the molecule. Indicate plausible orbital overlaps in the bonding scheme.

33. Propose a bonding scheme that is consistent with the structure for propynal. (*Hint:* Consult Table 11.1 to assess the multiple-bond character in some of the bonds.)

$$
\begin{array}{c}
\text{H}\underset{106\text{ pm}}{-}\text{C}\underset{120.4\text{ pm}}{-}\text{C}\underset{146\text{ pm}}{-}\text{C}\overset{123°}{\underset{120°}{<}}\overset{\text{O} \;\; 121\text{ pm}}{\underset{\text{H} \;\; 108\text{ pm}}{}}
\end{array}
$$

34. The structure of the molecule allene, CH_2CCH_2 is shown here. Propose hybridization schemes for the C atoms in this molecule.

Molecular Orbital Theory

35. Explain the essential difference in the ways that the valence-bond method and molecular orbital theory describe a covalent bond.

36. Describe the bond order of diatomic carbon, C_2, with Lewis theory and molecular orbital theory, and explain why the results are different.

37. $N_2(g)$ has an exceptionally high bond energy. Would you expect either N_2^- or N_2^{2-} to be a stable diatomic species in the gaseous state? Explain.

38. The paramagnetism of gaseous B_2 has been established. Explain how this observation confirms that the π_{2p} orbitals are at a lower energy than the σ_{2p} orbital for B_2.

39. In our discussion of bonding we have not encountered a bond order higher than triple. Use the energy level diagrams of Figure 12-20 to show why this is to be expected.

40. Is it correct to say that when a diatomic molecule loses an electron the bond energy always decreases, that is, that the bond is always weakened? Explain.

41. Assume that the energy level diagrams of Figure 12-20 are applicable and write plausible molecular orbital diagrams for the following *heteronuclear* diatomic species: (a) NO; (b) NO^+; (c) CO; (d) CN; (e) CN^-; (f) CN^+; (g) BN.

42. We have used the term "isoelectronic" to refer to atoms with identical electron configurations. In molecular orbital theory this term can be applied to molecular species as well. Which of the species in Exercise 41 are isoelectronic?

43. Assume that the energy level diagrams of Figure 12-20 apply to the diatomic ions NO^+ and N_2^+.
(a) Predict the bond order of each.
(b) Which of these ions is paramagnetic? diamagnetic?
(c) Which of these ions do you think has the greater bond length? Explain.

44. Assume that the energy level diagrams of Figure 12-20 apply to the diatomic ions CO^+ and CN^-.
(a) Predict the bond order of each.
(b) Which of these ions is paramagnetic? diamagnetic?
(c) Which of these ions do you think has the greater bond length? Explain.

Delocalized Molecular Orbitals

45. Explain why the concept of delocalized molecular orbitals is essential to an understanding of bonding in the benzene molecule, C_6H_6.

46. Explain how it is possible to avoid the concept of resonance by using molecular orbital theory.

47. In which of the following molecules would you expect to find delocalized molecular orbitals? Explain. (a) C_2H_4; (b) SO_2; (c) H_2CO.

48. In which of the following ions would you expect to find delocalized molecular orbitals? Explain. (a) HCO_2^-; (b) CO_3^{2-}; (c) CH_3^+.

Metallic Bonding

49. Which of the following factors are especially important in determining whether a substance has metallic properties? Explain. (a) atomic number; (b) atomic mass; (c) number of valence electrons; (d) number of vacant atomic orbitals; (e) total number of electronic shells in the atom.

50. Based on the ground state electron configurations of the atoms, how would you expect the melting points and hardnesses of sodium, iron, and zinc to compare? Explain.

51. How many energy levels are present in the $3s$ conduction band of a single crystal of sodium weighing 26.8 mg? How many electrons are present in this band?

52. Magnesium is an excellent electrical conductor even though it has a full $3s$ subshell with the electron configuration: $[Ne]3s^2$. Use band theory to explain why magnesium conducts electricity.

Advanced Exercises

53. The Lewis structure of N_2 indicates that the nitrogen-to-nitrogen bond is a triple covalent bond. Other evidence suggests that the σ bond in this molecule involves the overlap of sp hybrid orbitals.
(a) Draw orbital diagrams for the N atoms to describe bonding in N_2.
(b) Can this bonding be described by either sp^2 or sp^3 hybridization of the N atoms? Can bonding in N_2 be described in terms of unhybridized orbitals? Explain.
54. Show that both the valence-bond method and molecular orbital theory provide an explanation for the existence of the covalent molecule Na_2 in the gaseous state. Would you predict Na_2 by the Lewis theory?
55. Assume that the energy level diagrams of Figure 12-20 apply, and describe bonding in CO and in CO^+. Indicate in which of these species you would expect the greater **(a)** bond energy; **(b)** bond length.
56. Lewis theory is satisfactory to explain bonding in the ionic compound K_2O. It does not readily explain formation of the ionic compounds potassium superoxide, KO_2, and potassium peroxide, K_2O_2, however.
(a) Show that molecular orbital theory can provide this explanation.
(b) Write Lewis structures consistent with this explanation.
57. The compound potassium sesquoxide has the empirical formula K_2O_3. Show that this compound can be described by an appropriate combination of potassium, peroxide, and superoxide ions. Write a Lewis structure for a formula unit of the compound. (*Hint:* Refer also to Exercise 56.)
58. Draw a Lewis structure for the urea molecule $CO(NH_2)_2$, and predict its geometric shape with the VSEPR theory. Then revise your assessment of this molecule given the fact that all the atoms lie in the same plane and all the bond angles are 120°. Propose a hybridization and bonding scheme consistent with these experimental observations.
59. Methyl nitrate, CH_3NO_3, is used as a rocket propellant. The skeleton structure of the molecule is CH_3ONO_2. The N and three O atoms all lie in the same plane, but the CH_3 group is not in the same plane as the NO_3 group. The bond angle C—O—N is 105° and the bond angle O—N—O is 125°. One nitrogen-to-oxygen bond length is 136 pm and the other two are 126 pm.
(a) Draw a sketch of the molecule, showing its geometric shape.
(b) Label all the bonds in the molecule as σ or π and indicate the probable orbital overlaps involved.
(c) Explain why all three nitrogen-to-oxygen bond lengths are not the same.
60. Fluorine nitrate, $FONO_2$, is an oxidizing agent used as a rocket propellant. A reference source lists the following data for FO_aNO_2. (The subscript "a" shows that this O atom is different from the other two.)
Bond lengths: N—O = 129 pm; N—O_a = 139 pm; O_a—F = 142 pm
Bond angles: O—N—O = 125°; F—O_a—N = 105°
NO_aF plane is perpendicular to the O_2NO_a plane.
Use these data to construct a Lewis structure(s), a three-dimensional sketch of the molecule, and a plausible bonding scheme showing hybridization and orbital overlaps.
61. Draw a Lewis structure(s) for the nitrite ion, NO_2^-. Then propose a bonding scheme to describe the σ and π bonding in this ion. What conclusion can you reach about the number and types of π molecular orbitals in this ion? Explain.
62. Think of the reaction shown here as involving the transfer of a fluoride ion from ClF_3 to AsF_5 to form the ions ClF_2^+ and AsF_6^-. As a result, the hybridization scheme of each central atom must change. For each reactant molecule and product ion indicate **(a)** its geometric structure and **(b)** the hybridization scheme for its central atom.
$$ClF_3 + AsF_5 \longrightarrow (ClF_2^+)(AsF_6^-)$$
63. In the gaseous state, HNO_3 molecules have two nitrogen-to-oxygen bond distances of 121 pm and one of 140 pm. Draw a plausible Lewis structure(s) to represent this fact, and propose a bonding scheme in the manner of Figure 12-15.
64. He_2 does not exist as a stable molecule, but there is evidence that such a molecule can be formed between

electronically excited He atoms. Write a molecular orbital diagram to account for this.

65. The molecule formamide, $HCONH_2$, has the following approximate bond angles: H—C—O, 123°; H—C—N, 113°; N—C—O, 124°; C—N—H, 119°; H—N—H, 119°. The C—N bond length is 138 pm. Two Lewis structures can be written for this molecule, with the true structure being a resonance hybrid of the two. Propose a hybridization and bonding scheme for each structure.

66. Pyridine, C_5H_5N, is used in the synthesis of vitamins and drugs. The molecule can be thought of in terms of replacing one CH unit in benzene with an N atom. Draw orbital diagrams to show the orbitals of the C and N atoms involved in the σ and π bonding in pyridine. How many bonding and antibonding π-type molecular orbitals are present? How many delocalized electrons are present?

67. One of the characteristics of antibonding molecular orbitals is the presence of a nodal plane. Which of the *bonding* molecular orbitals considered in this chapter have nodal planes? Explain how a molecular orbital can have a nodal plane and still be a bonding molecular orbital.

68. The ion F_2Cl^- is linear, but the ion F_2Cl^+ is bent. Describe hybridization schemes for the central Cl atom consistent with this difference in structure.

69. Shown below are ball-and-stick models. Describe hybridization and orbital-overlap schemes consistent with these structures.

(a) S_2O **(b)** BrF_3 **(c)** CH_3OH

Liquids, Solids, and Intermolecular Forces

13

Two of the many natural phenomena described in this chapter—the universal tendency of liquids to form spherical drops and the inability of drops of water to spread across a waxed surface—are seen in this photograph.

*I*n our study of gases, we intentionally sought conditions in which forces between molecules—intermolecular forces—are negligible. This approach allowed us to describe gases with the ideal gas equation and to explain their behavior with the kinetic-molecular theory of gases.

To describe the other states of matter—liquids and solids—we turn the situation around: We purposely seek situations in which intermolecular forces are significant and consider some interesting properties of liquids and solids related to the strengths of these forces. Learning about these properties is a worthwhile goal in itself, for they have some important applications. By studying intermolecular forces, however, we also can understand *why* some liquids and solids have the properties they do, and we can even *predict* some properties. We will not formulate any general equations of state for liquids and solids; our treatment will be less mathematical and more qualitative than was the case for gases.

13-1 Intermolecular Forces and Some Properties of Liquids

In our study of gases we noted that at high pressures and low temperatures intermolecular forces cause gas behavior to depart from ideality. When these forces are sufficiently strong, a gas condenses to a liquid. That is, the intermolecular forces keep the molecules in close enough proximity that the molecules are confined to a definite volume, as expected for the liquid state.

Intermolecular forces are important in establishing the surface tension and viscosity of a liquid, two properties described in this section. Other properties of liquids closely associated with intermolecular forces will be described in Section 13-2.

▶ Solids, liquids, and gases were compared at the macroscopic and microscopic levels in Figure 1-5.

Surface Tension

The observation of a needle floating on water, as pictured in Figure 13-1, is puzzling. Steel is much denser than water and should not float. Something must overcome the force of gravity on the needle, allowing it to remain suspended on the surface of the water. The floating needle is made possible by a special quality associated with the surface of a liquid.

Figure 13-2 suggests an important difference in the forces between molecules within the bulk of a liquid and at the surface: Interior molecules have more neighbors to which they are attracted by intermolecular forces than do surface molecules. This increased attraction by its neighboring molecules places an interior molecule in a lower energy state than a surface molecule. As a consequence, as many molecules as possible try to get into the bulk of a liquid, and as few as possible remain at the surface.

The tendency for liquids to maintain a minimum surface area is most clearly seen in the formation of spherical drops in a rain shower. A sphere has a smaller ratio of surface area to volume than does any other geometric figure. To increase the surface area of a liquid requires that molecules be moved from the interior to the surface of a liquid, and this requires that work be done. **Surface tension** is the energy or work required to increase the surface area of a liquid. Surface tension is often represented by the Greek letter gamma (γ) and has the units of

FIGURE 13-1
An effect of surface tension illustrated.
Despite being denser than water, the needle is supported on the surface of the water. The property of surface tension accounts for this unexpected behavior.

FIGURE 13-2
Intermolecular forces in a liquid.
Molecules at the surface are attracted only by other surface molecules and by molecules below the surface. Molecules in the interior experience forces from neighboring molecules in all directions.

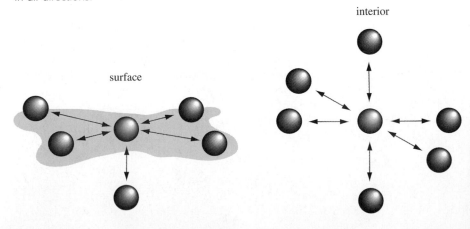

interior

surface

energy per unit area, typically joules per square meter (J/m^2). As the temperature—and hence the intensity of molecular motion—increases, intermolecular forces become less effective. Less work is required to extend the surface of a liquid, meaning that surface tension *decreases* with *increased* temperature.

The steel needle of Figure 13-1 remains suspended on the surface of the water because the work that would be done by the force of gravity acting on the needle is not as great as the energy required to spread the surface of the water over the top of the needle.

When a drop of liquid spreads into a film across a surface, we say that the liquid *wets* the surface. Whether a drop of liquid wets a surface or retains its spherical shape and stands on the surface depends on the strengths of two types of intermolecular forces. **Cohesive forces** are the intermolecular forces between *like* molecules and **adhesive forces,** between *unlike* molecules. If cohesive forces are strong compared with adhesive forces, a drop maintains its shape. If adhesive forces are strong enough, on the other hand, the energy requirement for spreading the drop into a film is met through the work done by the collapsing drop.

Water wets many surfaces, such as glass and certain fabrics. This characteristic is essential to its use as a cleaning agent. If glass is coated with a film of oil or grease, water no longer wets the surface, and water droplets stand on the glass, as shown in Figure 13-3. When we clean glassware in the laboratory, we have done a good job if water forms a uniform thin film on the glass. When we wax a car, we have done a good job if water uniformly beads up all along the surface.

Adding a detergent to water has two effects: The detergent solution dissolves grease to expose a clean surface, and the detergent lowers the surface tension of water. Lowering the surface tension means lowering the energy required to spread drops into a film. Substances that reduce the surface tension of water and allow it to spread more easily are known as *wetting agents*. They are used in applications ranging from dishwashing to industrial processes.

Figure 13-4 illustrates another familiar observation. If the liquid in the glass tube is water, the water is drawn slightly up the walls of the tube by adhesive forces between water and glass. The interface between the water and the air above it, called a *meniscus,* is *concave* or curved upward. With liquid mercury the meniscus is *convex* or curved downward. Cohesive forces in mercury, consisting of metallic bonds between Hg atoms, are strong; mercury does not wet

▶ The surface tension of water at 20 °C, for example, is 7.28×10^{-2} J/m^2, and that of mercury is 47.2×10^{-2} J/m^2.

FIGURE 13-4
Meniscus formation.
Water wets glass (left). The meniscus is concave—the bottom of the meniscus is below the level of the water/glass contact line. Mercury does not wet glass. The meniscus is convex—the top of the meniscus is above the mercury/glass contact line.

FIGURE 13-3
Wetting of a surface.
Water spreads into a thin film on a clean glass surface (left). If the glass is coated with oil or grease, the adhesive forces between water and the oil are not strong enough to spread the water. Drops of water stand on the surface (right).

FIGURE 13-5
Capillary action.
A thin film of water spreads up the inside walls of the capillary because of strong adhesive forces between water and glass (water wets glass). The pressure below the meniscus falls slightly. Atmospheric pressure then pushes a column of water up the tube to eliminate the pressure difference. The *smaller* the diameter of the capillary, the *higher* the liquid rises. Because its magnitude is also directly proportional to surface tension, capillary rise provides a simple experimental method of determining surface tension, described in Feature Problem A on page 465.

glass. The effect of meniscus formation is greatly magnified in tubes of small diameter, called *capillary* tubes. In the *capillary action* shown in Figure 13-5, the water level inside the capillary is noticeably higher than outside. The soaking action of a sponge depends on the rise of water into capillaries of a fibrous material such as cellulose. The penetration of water into soils also depends in part on capillary action. On the other hand, mercury, with its strong cohesive forces and weaker adhesive forces, does not show a capillary rise. Rather, mercury in a glass capillary tube will have a *lower* level than the mercury outside the capillary.

Viscosity

Another property at least partly related to intermolecular forces is **viscosity**—a liquid's resistance to flow. The stronger the intermolecular forces of attraction, the greater the viscosity. When a liquid flows, one portion of the liquid moves with respect to neighboring portions. Cohesive forces within the liquid create an "internal friction," which reduces the rate of flow. The effect is weak in liquids of low viscosity, such as ethyl alcohol and water. They flow easily. Liquids like honey and heavy motor oil flow much more sluggishly. We say that they are *viscous*. One method of measuring viscosity is to time the fall of a steel ball through a certain depth of liquid. The greater the viscosity of the liquid, the longer it takes for the ball to fall. Because intermolecular forces of attraction can be offset by higher molecular kinetic energies, viscosity generally *decreases* with increased temperature.

13-2 Vaporization of Liquids: Vapor Pressure

The relationship between intermolecular forces of attraction and molecular kinetic energies helps us understand how surface tension and viscosity vary with temperature. This relationship is also helpful in explaining other common phenomena.

As we learned in our study of the kinetic-molecular theory (Section 6-7), the speeds and kinetic energies of molecules vary over a wide range at any given temperature, and we generally work with *average* values. Some molecules, moreover, have sufficiently high kinetic energies that they are able to overcome intermolecular forces and escape from a liquid. The passage of molecules from the surface of a liquid into the gaseous, or vapor, state is called **vaporization.** The term *evaporation* is also used. As you might expect, the tendency for a liquid to evaporate *increases* with *increased* temperature and *decreased* strength of intermolecular forces.

Enthalpy of Vaporization

Because the molecules lost through evaporation are more energetic than average, the average kinetic energy of the remaining molecules decreases. The temperature of the evaporating liquid *falls*. This accounts for the cooling sensation you feel when a volatile liquid such as ethyl alcohol evaporates from your skin.

Suppose we allow a liquid to vaporize but wish to keep its temperature *constant*. How can we do this? We must replace the excess kinetic energy carried away by the vaporizing molecules by adding heat to the liquid. The *enthalpy of vaporization* is the quantity of heat that must be absorbed if a certain quantity of liquid is vaporized at a constant temperature. Stated in another way,

$$\Delta H_{\text{vaporization}} = H_{\text{vapor}} - H_{\text{liquid}}$$

Although we cannot measure the absolute enthalpies H_{vapor} and H_{liquid}, we know these are functions of state; they have unique values. Their difference, $\Delta H_{vaporization}$, or more simply, ΔH_{vap}, is also a unique and a *measurable* quantity (see Example 13-1). Because vaporization is an *endothermic* process, ΔH_{vap} is always positive. In this text enthalpies of vaporization are expressed in terms of *one mole* of liquid vaporized (Table 13.1).

EXAMPLE 13-1

Determining an Enthalpy of Vaporization. To vaporize 1.75 g of acetone, $(CH_3)_2CO$, at 298 K, 767 J of heat is required. What is the enthalpy of vaporization, ΔH_{vap}, of acetone, in kilojoules per mole?

Solution

Determine the number of moles of acetone in 1.75 g. Convert 767 J to kilojoules and divide this quantity of heat by the number of moles of acetone.

$$\Delta H_{vap} = \frac{767 \text{ J} \times \dfrac{1 \text{ kJ}}{1000 \text{ J}}}{1.75 \text{ g } (CH_3)_2CO \times \dfrac{1 \text{ mol } (CH_3)_2CO}{58.08 \text{ g } (CH_3)_2CO}}$$

$$= 25.5 \text{ kJ/mol } (CH_3)_2CO$$

Practice Example A: How much heat is required to vaporize a 2.35-g sample of diethyl ether at 298 K? (*Hint:* Use data from Table 13.1.)

Practice Example B: How much heat is required to raise the temperature of 215 g of liquid CH_3OH from 20.0 to 30.0° C, followed by vaporization of the CH_3OH at 30.0 °C? The specific heat of CH_3OH is 2.53 J g^{-1} $°C^{-1}$. Assume that the value of ΔH_{vap} listed in Table 13.1 holds at 30.0 °C. (*Hint:* You must calculate two quantities of heat and add them together. One calculation is similar to that of Example 7-2.)

The conversion of a gas or vapor to a liquid is called **condensation.** From a thermochemical standpoint, condensation is the reverse of vaporization.

$$\Delta H_{condensation} = H_{liquid} - H_{vapor} = -\Delta H_{vap}$$

Because it is opposite in sign but equal in magnitude to ΔH_{vap}, $\Delta H_{condensation}$ is always negative. Condensation is an *exothermic* process. This explains why burns produced by a given mass of steam (vaporized water) are very much more

TABLE 13.1 Some Enthalpies of Vaporization at 298 K[a]	
Liquid	ΔH_{vap}, **kJ/mol**
diethyl ether, $(C_2H_5)_2O$	29.1
methyl alcohol, CH_3OH	38.0
ethyl alcohol, CH_3CH_2OH	42.6
water, H_2O	44.0

[a]ΔH_{vap} values are somewhat temperature-dependent (see Exercise 85).

FIGURE 13-6
Establishing liquid–vapor equilibrium.
(a) A liquid is allowed to evaporate into a closed container. Initially only vaporization occurs. (b) Condensation begins. The rate at which molecules evaporate is greater than the rate at which they condense, and the number of molecules in the vapor state continues to increase. (c) The rate of condensation is equal to the rate of vaporization. The number of vapor molecules remains constant over time, as does the pressure exerted by this vapor.

○ molecules in vapor state
◑→ molecules undergoing vaporization
◑→ molecules undergoing condensation

(a)　　　　**(b)**　　　　**(c)**

severe than burns produced by the same mass of hot water. Hot water burns by releasing heat as it cools. Steam releases a large quantity of heat when it condenses to liquid water followed by the further release of heat as the hot water cools.

Vapor Pressure

We know that water left in an open beaker will completely evaporate. A different condition results if vaporization occurs into a *closed* vapor volume. As shown in Figure 13-6, in a container with both liquid and vapor present, vaporization and condensation occur simultaneously. If sufficient liquid is present, eventually a condition is reached in which no *additional* vapor is formed. This condition is one of *dynamic equilibrium.* Dynamic equilibrium always implies that two opposing processes are occurring simultaneously and at equal rates. As a result there is no net change with time once equilibrium has been established.

The pressure exerted by a vapor in dynamic equilibrium with its liquid is called the **vapor pressure.** Liquids with high vapor pressures at room temperature are said to be *volatile,* and those with very low vapor pressures are *nonvolatile.* Whether a liquid is volatile or not is determined primarily by the strengths of intermolecular forces—the *weaker* these forces, the more volatile the liquid (the higher its vapor pressure). Diethyl ether and acetone are volatile liquids; at 25 °C their vapor pressures are 534 and 231 mmHg, respectively. Water at ordinary temperatures is a moderately volatile liquid; at 25 °C its vapor pressure is 23.8 mmHg. Mercury is essentially a nonvolatile liquid; at 25 °C its vapor pressure is 0.0018 mmHg.

▶ Gasoline is a mixture of volatile hydrocarbons. Gasoline vapor is readily detected by its odor.

As an excellent first approximation, the vapor pressure of a liquid depends only on the particular liquid and its temperature. As long as some of each is present at equilibrium, the vapor pressure does not depend on the amount of either the liquid or the vapor. We can determine approximate values of vapor pressures with a simple mercury barometer by the method outlined in Figure 13-7. A graph

FIGURE 13-7
Measurement of vapor pressure.
(a) A mercury barometer. (b) A small volume of liquid is introduced to the top of the mercury column. The pressure of the vapor in equilibrium with the liquid depresses the mercury level. (c) The vapor pressure is greater than in (b): vapor pressure increases with increased temperature.

P_{bar}　　P_{vapor}　—vapor　—liquid　P_{vapor}

(a)　　　　**(b)**　　　　**(c)**

TABLE 13.2 Vapor Pressure of Water at Various Temperatures

Temperature, °C	Pressure, mmHg	Temperature, °C	Pressure, mmHg	Temperature, °C	Pressure, mmHg
0.0	4.6	29.0	30.0	93.0	588.6
10.0	9.2	30.0	31.8	94.0	610.9
20.0	17.5	40.0	55.3	95.0	633.9
21.0	18.7	50.0	92.5	96.0	657.6
22.0	19.8	60.0	149.4	97.0	682.1
23.0	21.1	70.0	233.7	98.0	707.3
24.0	22.4	80.0	355.1	99.0	733.2
25.0	23.8	90.0	525.8	100.0	760.0
26.0	25.2	91.0	546.0	110.0	1074.6
27.0	26.7	92.0	567.0	120.0	1489.1
28.0	28.3				

of vapor pressure as a function of temperature is known as a **vapor pressure curve.** Vapor pressure curves always have the appearance of those in Figure 13-8: *vapor pressure increases with temperature.* Vapor pressures of water at different temperatures are presented in Table 13.2.

Boiling and the Boiling Point

When a liquid is heated in a container *open to the atmosphere,* there is a particular temperature at which vaporization occurs throughout the liquid rather than simply at the surface. Vapor bubbles form within the bulk of the liquid, rise to the surface, and escape. The pressure exerted by escaping molecules equals that exerted by molecules of the atmosphere, and **boiling** is said to occur. Dur-

▶ When a pan of water is put on the stove to boil, usually small bubbles are observed as the water begins to warm. These are bubbles of dissolved air being expelled. Once the water boils, however, all the dissolved air is expelled and the bubbles consist only of water vapor.

FIGURE 13-8
Vapor pressure curves of several liquids.
(a) diethyl ether, $C_4H_{10}O$; (b) benzene, C_6H_6; (c) water, H_2O; (d) toluene, C_7H_8; (e) aniline, C_6H_7N. The normal boiling points are the temperatures of intersection of the dashed line at $P = 760$ mmHg with the vapor pressure curves.

ing boiling, energy absorbed as heat is used only to convert molecules of liquid to vapor. The temperature remains constant until all the liquid has boiled away, as is dramatically illustrated in Figure 13-9. The temperature at which the vapor pressure of a liquid is equal to standard atmospheric pressure (1 atm = 760 mmHg) is the **normal boiling point.** In other words, the normal boiling point is the boiling point of a liquid at 1 atm pressure. The normal boiling points of several liquids can be determined from the intersection of the dashed line in Figure 13-8 with the vapor pressure curves for the liquids.

Figure 13-8 also helps us to see that the boiling point of a liquid varies significantly with barometric pressure. Shift the dashed line shown at $P = 760$ mmHg to higher or lower pressures, and the new points of intersection with the vapor pressure curves come at different temperatures. Barometric pressures below 1 atm are commonly encountered at high altitudes. At an altitude of 1609 m (that of Denver, Colorado) barometric pressure is about 630 mmHg. The boiling point of water at this pressure is 95 °C (203 °F). To cook foods under these conditions of lower boiling temperatures, longer cooking times are needed. A "3-minute" boiled egg takes longer than 3 min to cook. We can counteract the effect of high altitudes by using a pressure cooker. In a pressure cooker the cooking water is maintained under higher than atmospheric pressure and its boiling temperature increases, for example, to about 120 °C at 2 atm pressure.

▶ A more extreme case is that on the summit of Mt. Everest, where a climber would barely be able to heat a cup of tea to 70 °C.

▶ Feature Problem B, page 465, presents another interesting phenomenon dealing with vapor pressure and boiling.

The Critical Point

In describing boiling we made an important qualification: boiling occurs "in a container open to the atmosphere." If a liquid is heated in a *sealed* container, boiling does not occur. Instead, the temperature and vapor pressure rise continuously. Pressures many times atmospheric pressure may be attained. If we seal just the right quantity of liquid in a glass tube and heat it, as in Figure 13-10, we observe that

- The density of the liquid decreases; that of the vapor increases; and eventually the two densities become equal.
- The surface tension of the liquid approaches zero. The meniscus between the liquid and vapor becomes less distinct and eventually disappears.

FIGURE 13-9

Boiling water in a paper cup.
An empty paper cup heated over a Bunsen burner quickly bursts into flame. If a paper cup is filled with water, it can be heated for an extended time as the water boils. This is possible for three reasons: (1) Because of the high heat capacity of water, heat from the burner goes primarily into heating the water, not the cup. (2) As the water boils, large quantities of heat (ΔH_{vap}) are required to convert the liquid to its vapor. (3) The temperature of the cup does not rise above the boiling point of water as long as liquid water remains.

The boiling point of 99.9 instead of 100.0 °C suggests that the prevailing barometric pressure was slightly below 1 atm.

The **critical point** is the point at which these conditions are reached, and the liquid and vapor become indistinguishable. The temperature at the critical point is the critical temperature, T_c, and the pressure is the critical pressure, P_c. The critical point is the highest point on a vapor pressure curve and represents the highest temperature at which the liquid can exist. Several critical temperatures and pressures are listed in Table 13.3.

A gas can be liquefied only at temperatures *below* its critical temperature, T_c. If room temperature is *below* T_c, this liquefaction can be accomplished just by applying sufficient pressure. If room temperature is *above* T_c, however, added pressure *and* a lowering of temperature to a value below T_c are required.

Using Vapor Pressure Data

One use of vapor pressure data that we encountered in Section 6-6 concerns collecting gases over liquids, particularly water. Another use is in predicting whether a substance exists solely as a vapor (gas) or as a liquid and vapor in equilibrium.

The sketch in Figure 13-11 suggests that a substance exists exclusively as a gas or vapor if it is represented by a point that lies *below* the vapor pressure curve. Points *above* the curve represent a liquid under pressure and the absence of vapor. Thus, we can predict that it is not possible to maintain water exclusively as vapor at 760 mmHg at 75 °C—the point (75 °C, 760 mmHg) lies above the vapor pressure curve. Under these conditions water exists exclusively as liquid, but *only* if the liquid completely fills its container. If the amount of liquid is insufficient to fill the container, as is often the case, some of the liquid vaporizes; a liquid–vapor equilibrium is established; and the point representing the system drops to the vapor pressure curve.

In Example 13-2 we make another prediction, this time with quantitative data.

▶ Although the term "gas" can be used exclusively, sometimes the term "vapor" is used for the gaseous state at temperatures *below* T_c and gas at temperatures *above* T_c.

about 10 °C   about 1 °C   critical
below T_c below T_c temp T_c

FIGURE 13-10
Attainment of the critical point.
The meniscus separating a liquid from its vapor disappears at the critical point. The liquid and vapor become indistinguishable.

TABLE 13.3 Some Critical Temperatures, T_c, and Critical Pressures, P_c		
Substance	T_c, K	P_c, atm
"Permanent" gases[a]		
H_2	33.3	12.8
N_2	126.2	33.5
O_2	154.8	50.1
CH_4	191.1	45.8
"Nonpermanent" gases[b]		
CO_2	304.2	72.9
HCl	324.6	82.1
NH_3	405.7	112.5
SO_2	431.0	77.7
H_2O	647.3	218.3

[a]"Permanent" gases cannot be liquefied at 25 °C (298 K).
[b]"Nonpermanent" gases can be liquefied at 25 °C.

FIGURE 13-11
A prediction based on the vapor pressure curve of water.
At 75 °C and 760 mmHg, the system is one of liquid water (red point) only if the amount of liquid is sufficient to fill the container. Otherwise, a liquid–vapor equilibrium is established and the pressure is that of a point (black) on the vapor pressure curve. Water can be maintained exclusively as a vapor at 760 mmHg only above 100 °C.

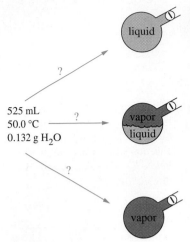

FIGURE 13-12
Predicting states of matter—Example 13-2 illustrated.
For the conditions given on the left, which of the final conditions pictured on the right will result? The possibility that the sample might exist as liquid only can be ruled out because 0.132 g $H_2O(l)$ is not nearly enough to fill a 525-mL flask.

EXAMPLE 13-2

Making Quantitative Predictions with Vapor Pressure Data. As a result of a chemical reaction, 0.132 g H_2O is produced and maintained at a temperature of 50.0 °C in a closed flask of 525-mL volume. Will the water be present as vapor only or as liquid and vapor in equilibrium (Figure 13-12)?

Solution

First, use the ideal gas equation to calculate the pressure that would exist if the water were present as vapor only.

$$P = \frac{nRT}{V}$$

$$= \frac{0.132 \text{ g } H_2O \times \dfrac{1 \text{ mol } H_2O}{18.02 \text{ g } H_2O} \times 0.08206 \text{ L atm mol}^{-1} \text{ K}^{-1} \times 323.2 \text{ K}}{0.525 \text{ L}}$$

$$= 0.370 \text{ atm} \times \frac{760 \text{ mmHg}}{1 \text{ atm}} = 281 \text{ mmHg}$$

Now, obtain the vapor pressure of water at 50.0 °C from Table 13.2. Because the calculated pressure (281 mmHg) exceeds the vapor pressure (92.5 mmHg), some of the vapor must condense to liquid. The water is present as liquid and vapor in equilibrium at 92.5 mmHg.

Practice Example A: If the reaction described in this example resulted in H_2O produced and maintained at 80.0 °C, would the water be present as vapor only or as liquid and vapor in equilibrium? Explain.

Practice Example B: For the situation described in this example, what mass of water is present as liquid and what mass as vapor?

An Equation for Expressing Vapor Pressure Data

If you seek vapor pressure data on a liquid in a handbook or data tables, you are unlikely to find graphs like Figure 13-8. Also, with the exception of a few liquids, such as water and mercury, you are unlikely to find data tables like Table 13.2. What you will find instead are mathematical equations relating vapor pressures and temperatures. One such equation can summarize in one line data that might otherwise take a full page. A particularly common form of vapor pressure equation is the one shown here, which expresses the natural logarithm (ln) of vapor pressure as a function of the reciprocal of the Kelvin temperature ($1/T$). The relationship is that of a straight line, and the straight-line plots for the liquids featured in Figure 13-8 are drawn in Figure 13-13.

$$\ln P = -A \left(\frac{1}{T} \right) + B \tag{13.1}$$

Equation of straight line: $\quad y = m \times x + b$

To use Equation (13.1), we need to have values for the two constants, A and B. The constant A is related to the enthalpy of vaporization of the liquid: $A = \Delta H_{vap}/R$, where ΔH_{vap} is expressed in the unit J/mol and the value used for R is 8.3145 J mol^{-1} K^{-1}. It is customary to eliminate B by rewriting (13.1) in a form called the Clausius-Clapeyron equation. We apply Equation (13.2) in Example 13-3.

▶ Refer to Appendix A-4 to see how the constant B is eliminated to give Equation (13.2).

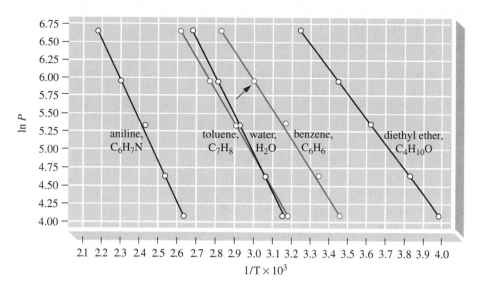

FIGURE 13-13
Vapor pressure data plotted as ln *P* vs. 1/*T*.
Pressures are in millimeters of mercury and temperatures are in kelvins. Data from Figure 13-8 have been recalculated and replotted as in the following example: For benzene at 60 °C, the vapor pressure is 400 mmHg; ln P = ln 400 = 5.99. T = 60 + 273 = 333 K; $1/T$ = 1/333 = 0.00300 = 3.00×10^{-3}; $1/T \times 10^3$ = $3.00 \times 10^{-3} \times 10^3$ = 3.00. The point corresponding to (3.00, 5.99) is marked by the arrow ($\rightarrow$).

$$\ln \frac{P_2}{P_1} = \frac{\Delta H_{vap}}{R}\left(\frac{1}{T_1} - \frac{1}{T_2}\right) \qquad (13.2)$$

EXAMPLE 13-3

Applying the Clausius–Clapeyron Equation. Calculate the vapor pressure of water at 35.0 °C with data from Tables 13.1 and 13.2.

Solution

Let P_1 be the unknown vapor pressure at temperature T_1 = 35.0 °C = 308.2 K. From Table 13.2, choose for P_2 and T_2 known data at a temperature close to 35 °C. At T_2 = 40.0 °C = 313.2 K, P_2 = 55.3 mmHg. We need a value for ΔH_{vap}. To simplify the problem, we will assume ΔH_{vap} is independent of temperature. From Table 13.1 we obtain the value ΔH_{vap} = 44.0 kJ/mol, which we convert to the unit J/mol. Substitute these values into Equation (13.2) to obtain

$$\ln \frac{55.3 \text{ mmHg}}{P_1} = \frac{44.0 \times 10^3 \text{ J mol}^{-1}}{8.3145 \text{ J mol}^{-1} \text{ K}^{-1}}\left(\frac{1}{308.2} - \frac{1}{313.2}\right)\text{K}^{-1}$$

$$= 5.29 \times 10^3 (0.003245 - 0.003193) = 0.28$$

Next, determine that $e^{0.28}$ = 1.32 (see Appendix A). Thus,

$$\frac{55.3 \text{ mmHg}}{P_1} = e^{0.28} = 1.32$$

$$P_1 = 55.3 \text{ mmHg}/1.32 = 41.9 \text{ mmHg}$$

(The experimentally determined vapor pressure of water at 35.0 °C is 42.175 mmHg.)

Practice Example A: A handbook lists the vapor pressure of methyl alcohol as 100 mmHg at 21.2 °C. What is its vapor pressure at 25.0 °C?

▶ Agreement between the calculated and experimentally determined values would be better if we were to use a value of ΔH_{vap} at 35 °C rather than at 25 °C.

Practice Example B: A handbook lists the normal boiling point of isooctane, a gasoline component, as 99.2 °C and its enthalpy of vaporization (ΔH_{vap}) as 35.76 kJ/mol C_8H_{18}. Calculate the vapor pressure of isooctane at 25 °C. (*Hint:* What is the vapor pressure at the normal boiling point?)

13-3   Some Properties of Solids

We mentioned some properties of solids (for example, malleability, ductility) at the beginning of this text, and we will continue to consider additional properties. For now, we will comment on some properties that allow us to think of solids in relation to the other states of matter—liquids and gases.

Melting, Melting Point, and Heat of Fusion

As a crystalline solid is heated, its atoms, ions, or molecules vibrate more vigorously. Eventually a temperature is reached at which these vibrations disrupt the ordered crystalline structure; atoms, ions, or molecules can slip past one another; the solid loses its definite shape and is converted to a liquid. This process is called **melting** (or fusion), and the temperature at which it occurs is the **melting point.** The reverse process, the conversion of a liquid to a solid, is called **freezing** (or solidification), and the temperature at which it occurs is the **freezing point.** The melting point of a solid and the freezing point of its liquid are identical. At this temperature, solid and liquid coexist in equilibrium.

If we add heat uniformly to a solid–liquid mixture at equilibrium, the temperature *remains constant* while the solid melts. Only when all the solid has melted does the temperature begin to rise. Conversely, if we remove heat uniformly from a solid–liquid mixture at equilibrium, the liquid freezes *at a constant temperature.* The quantity of heat required to melt a solid is the *enthalpy of fusion,* ΔH_{fus}. Some typical enthalpies of fusion, expressed in kilojoules per mole, are listed in Table 13.4. Perhaps the most familiar example of a melting (and freezing) point is that of water, 0 °C. This is the temperature at which liquid and solid water, in contact with air and under standard atmospheric pressure, are in equilibrium. The enthalpy of fusion of water is 6.01 kJ/mol, which we can express as

$$H_2O(s) \longrightarrow H_2O(l) \qquad \Delta H_{fus} = +6.01 \text{ kJ/mol} \qquad (13.3)$$

TABLE 13.4   Some Enthalpies of Fusion

Substance	Melting Point, °C	ΔH_{fus}, kJ/mol
mercury, Hg	−38.9	2.30
sodium, Na	97.8	2.60
methyl alcohol, CH_3OH	−97.7	3.21
ethyl alcohol, CH_3CH_2OH	−114	5.01
water, H_2O	0.0	6.01
benzoic acid, C_6H_5COOH	122.4	18.08
naphthalene, $C_{10}H_8$	80.2	18.98

Here is an easy way to determine the freezing point of a liquid. Allow the liquid to cool and measure the liquid temperature as it decreases with time. When freezing begins, the temperature *remains constant* until all the liquid has frozen. Then the temperature is again free to fall as the solid cools. If we plot temperatures against time we get a graph known as a *cooling curve.* Figure 13-14 is a cooling curve for water. We can also run this process backwards, that is, by starting with the solid and adding heat. Now the temperature remains constant while melting occurs. This temperature–time plot is called a *heating curve.*

Often an experimentally determined cooling curve does not look quite like the solid-line plot in Figure 13-14. The temperature may drop *below* the freezing point without any solid appearing. This condition is known as *supercooling.* For a crystalline solid to start forming from a liquid at the freezing point, the liquid must contain some small particles (for example, suspended dust particles) on which crystals can form. If a liquid contains a very limited number of particles on which crystals can grow, it may supercool for a time before freezing. When a supercooled liquid does begin to freeze, however, the temperature rises back to the normal freezing point while freezing is completed. We can always recognize supercooling through a slight dip in a cooling curve just before the straight-line portion.

Sublimation

Like liquids, solids can also give off vapors, although because of the stronger intermolecular forces present, solids are generally not as volatile as liquids at a given temperature. The direct passage of molecules from the solid to the vapor state is called **sublimation.** The reverse process, the passage of molecules from the vapor to the solid state, is called *deposition.* When sublimation and deposition occur at equal rates, a dynamic equilibrium exists between a solid and its vapor. The vapor exerts a characteristic pressure called the *sublimation pressure.* A plot of sublimation pressure as a function of temperature is called a *sublimation curve.* The *enthalpy of sublimation* is the quantity of heat needed to convert a solid to vapor. At the melting point, sublimation (solid → vapor) is equivalent to melting (solid → liquid) followed by vaporization (liquid → vapor). This suggests the following relationship between ΔH_{fus}, ΔH_{vap}, and ΔH_{sub} at the melting point.

$$\Delta H_{sub} = \Delta H_{fus} + \Delta H_{vap} \qquad (13.4)$$

The value of ΔH_{sub} obtained with Equation (13.4) can replace the enthalpy of vaporization in the Clausius–Clapeyron equation [Equation (13.2)], enabling us to calculate sublimation pressures as a function of temperature.

Two familiar solids with significant sublimation pressures are ice and dry ice (solid carbon dioxide). If you live in a cold climate you are aware that snow may disappear from the ground even though the temperature may fail to rise above 0 °C. Under these conditions the snow does not melt; it sublimes. The sublimation pressure of ice at 0 °C is 4.58 mmHg. That is, the solid ice has a vapor pressure of 4.58 mmHg at 0 °C. If the air is not already saturated with water vapor, the ice will sublime. The sublimation and deposition of iodine are pictured in Figure 13-15.

13-4 Phase Diagrams

Imagine constructing a pressure–temperature graph in which each point on the graph represents a condition under which a substance might be found. At low temperatures and high pressures, such as the red points in Figure 13-16, we ex-

FIGURE 13-14
Cooling curve for water.
The broken-line portion represents the condition of supercooling that occasionally occurs. (l) = liquid; (s) = solid.

▶ The relationship in Equation (13.4) plays a key role in Feature Problem C on page 465.

FIGURE 13-15
Sublimation of iodine.
Even at temperatures well below its melting point of 114 °C, solid iodine exhibits an appreciable sublimation pressure. Here, purple iodine vapor is produced at about 70 °C. Deposition of the vapor to solid iodine occurs on the colder walls of the flask.

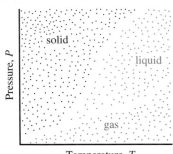

FIGURE 13-16
Temperatures, pressures, and states of matter.
The outline of a phase diagram is suggested by the distribution of dots. (See also Figures 13-17 and 13-18.)

pect the atoms, ions, or molecules of a substance to be in a close orderly arrangement—a solid. At high temperatures and low pressures—the green points in Figure 13-16—we expect the gaseous state, and at intermediate temperatures and pressures we expect a liquid.

Figure 13-16 suggests a **phase diagram,** a graphical representation of the conditions of temperature and pressure at which solids, liquids, and gases (vapors) exist, either as single phases or states of matter or as two or more phases in equilibrium with one another. The different regions of the diagram correspond to single phases, or states, of matter. Straight or curved lines where single-phase regions adjoin represent two phases in equilibrium.

Iodine

A typical phase diagram, and one of the simplest, is that of iodine shown in Figure 13-17. The curve *OC* is the vapor pressure curve of liquid iodine, and *C* is the critical point. *OB* is the sublimation curve of solid iodine. The effect of pressure on the melting point of iodine is represented by the line *OD*; it is called the *fusion curve*. The point *O* has a special significance. It gives the *unique* temperature and pressure at which the *three* phases solid, liquid, and vapor coexist at equilibrium and is called a **triple point.** For iodine the triple point is at 114 °C and 91 mmHg. The normal melting point (114 °C) and boiling point (184 °C) are the temperatures at which a line at $P = 1$ atm intersects the fusion and vapor pressure curves, respectively.

Carbon Dioxide

The case of carbon dioxide, shown in Figure 13-18, differs from that of iodine in one important respect—the pressure at the triple point *O* is greater than 1 atm. A line at $P = 1$ atm intersects the *sublimation curve,* not the vapor pressure curve. If solid CO_2 is heated in an open container, it sublimes away at a *constant* temperature of -78.5 °C. *It does not melt* (and so is called "dry ice"). Because it maintains a low temperature and does not produce a liquid by melting, dry ice is widely used in freezing and preserving foods.

FIGURE 13-17
Phase diagram for iodine.
Note that the melting-point and triple-point temperatures for iodine are essentially the same. Generally, large pressure increases are required to produce even small changes in solid–liquid equilibrium temperatures. The pressure and temperature axes on a phase diagram are generally not drawn to scale so that the significant features of the diagram can be more readily emphasized.

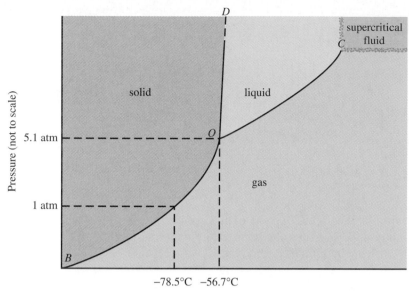

FIGURE 13-18
Phase diagram for carbon dioxide.

We can obtain liquid CO_2 at pressures above 5.1 atm and encounter it most frequently in CO_2 fire extinguishers. All three states of matter are involved in the action of these fire extinguishers. When the liquid CO_2 is released, most of it quickly vaporizes. The heat required for this vaporization is extracted from the remaining $CO_2(l)$, which has its temperature lowered to the point that it freezes and falls as a $CO_2(s)$ "snow." In turn the $CO_2(s)$ quickly sublimes to $CO_2(g)$. All of this helps to quench a fire by (a) displacing the air around the fire with a "blanket" of $CO_2(g)$ and (b) cooling down the area somewhat.

Supercritical Fluids

It is difficult to know what to call the phase that exists at temperatures and pressures above the critical point, for the liquid and gaseous phases become identical and indistinguishable at the critical point. For example, this phase has the high density of a liquid, but the low viscosity of a gas. One term that is finding increased use is *supercritical fluid* (SCF). We have marked "gray" areas in Figures 13-17 and 13-18 and called them the supercritical fluid region.

Although we do not ordinarily think of liquids or solids as being soluble in gases, volatile ones are. The mole fraction solubility is simply the ratio of the vapor pressure (or the sublimation pressure) to the total gas pressure. And liquids and solids become much more soluble in a gas above its critical pressure and temperature. This is mostly because the density of the SCF is high and approaches that of a liquid. Molecules in supercritical fluids, being in much closer proximity than in ordinary gases, can exert strong attractive forces on the molecules of a liquid or solid solute. SCFs display solvent properties similar to ordinary liquid solvents. To vary the pressure of an SCF means to vary its density and also its solvent properties. Thus, a given SCF, such as carbon dioxide, can be made to behave like many different solvents.

Until recently, the principal method of decaffeinating coffee was to extract the caffeine with a solvent such as methylene chloride (CH_2Cl_2). This solvent is objectionable because it is hazardous in the workplace and difficult to remove completely from the coffee. Now, supercritical fluid CO_2 is used. In one process green coffee beans are brought in contact with CO_2 at about 90 °C and

Decaffeinated coffee.
"Naturally" decaffeinated coffee is made through a process that uses supercritical fluid CO_2 as a solvent to dissolve the caffeine in green coffee beans. Afterwards the beans are roasted and sold to consumers.

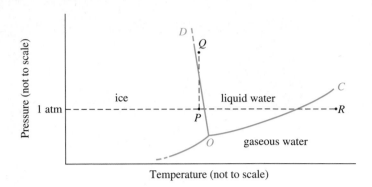

FIGURE 13-19
Phase diagram for water.
Point *O*, the triple point, is at +0.0098 °C and 4.58 mmHg. The critical point, *C*, is at 374.1 °C and 218.2 atm. The negative slope of the fusion curve *OD* is exaggerated in this diagram. The significance of the broken straight lines is described in Example 13-4.

160–220 atm. The caffeine content of the coffee is reduced from its normal 1–3% to about 0.02%. When the temperature and pressure of the CO_2 are reduced, the caffeine precipitates. Then the CO_2 is recycled.

Water

The phase diagram for water (Figure 13-19) presents an interesting new feature: The fusion curve *OD* has a *negative* slope, that is, it slopes toward the pressure axis. The melting point of ice *decreases* with increased pressure, which is rather unusual behavior for a solid (bismuth and antimony are two other examples of solids that behave in this way). We do not readily observe this behavior in natural phenomena, however. The example most commonly given is that of ice skating, where the pressure of the skate blade presumably melts the ice. The skater skims along on a thin lubricating film of water. This explanation may be valid for certain types of skates (figure skates) on ice at temperatures only slightly below the normal melting point.* The fact that ice is less dense than liquid water—ice floats on liquid water—is the cause of the negative slope of the fusion curve *OD*.

▶ An increase in pressure to 125 atm lowers the freezing point of water only by about 1 °C.

Are You Wondering . . .

Where the fusion curves in phase diagrams terminate?

The broken portions of the fusion curves (curve OD) in Figures 13-17, 13-18, and 13-19 signify that the curves extend indefinitely. In the phase diagrams for I_2 and CO_2, at sufficiently high pressures the fusion curve reaches temperatures beyond the critical temperature. It is reasonable to expect that under extreme pressures substances are compressed to solids, even if they can only exist as gases at lower pressures. The fusion curve in the phase diagram for H_2O starts off with a *negative* slope. At pressures above about 2000 atm, however, ordinary ice is transformed into forms of ice whose fusion curves have *positive* slopes. So even solid water exists at very high temperatures if the pressures are extreme.

---

* For outdoor skating at much lower temperatures, it is more likely that frictional heating of the ice produces the liquid film. A complete phase diagram for water helps to establish that this is probably the case. The fusion curve of ordinary ice, with its negative slope, terminates at −22.0 °C, at which point a different phase of ice appears. Below this temperature no liquid can be produced, no matter how high the pressure.

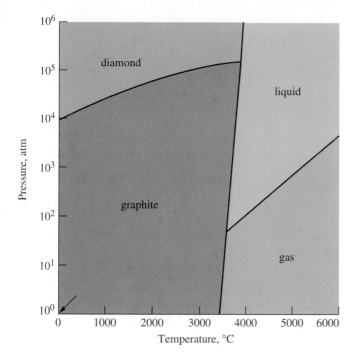

FIGURE 13-20
Phase diagram for carbon.
This is a simplified phase diagram for carbon. To include the vast range of pressures required (from 1 to 1,000,000 atm), pressure is plotted on a logarithmic scale. The arrow marks the point 1 atm, 25 °C.

Phase versus State of Matter

What's the difference between a phase and a state of matter? These terms tend to be used synonomously, but there is a small distinction between them. As we have already noted, there are just *three* states of matter: solid, liquid, and gas. A *phase* is any sample of matter with definite composition and uniform properties that is distinguishable from other phases with which it is in contact. Thus, we can describe liquid water in equilibrium with its vapor as a two-phase mixture. The liquid is one phase and the gas, or vapor, is the other. In this case the phases, liquid and gas, are the same as the states of matter present, liquid and gas.

In describing equilibrium between solid graphite and diamond, we refer to a two-phase mixture, even though both phases exist within a single state of matter—solid (Figure 13-20). For *mixtures* of substances it is possible to have several phases in the liquid state as well as in the solid state. Because different phases in the same state of matter can be represented in the same pressure–temperature diagram, it is appropriate to call the diagram a *phase* diagram.

Example 13-4 illustrates how we can use a phase diagram to explain the phase transitions that a substance can undergo.

▶ The crossing of a two-phase curve in a phase diagram is called a *transition*. Specific names are
    melting (s → l)
    freezing (l → s)
    vaporization (l → v)
    condensation (v → l)
    sublimation (s → v)
    deposition (v → s)
    transition (s_1 → s_2)

EXAMPLE 13-4

Interpreting a Phase Diagram. A sample of ice is maintained at 1 atm and at a temperature represented by point *P* in Figure 13-19. Describe what happens when **(a)** the temperature is raised, at constant pressure, to point *R*, and **(b)** the pressure is raised, at constant temperature, to point *Q*. The experimental apparatus and the starting condition, point *P*, are suggested by Figure 13-21.

▶ Interpretation of a more complex phase diagram, that of sulfur, is the subject of Feature Problem D on page 465.

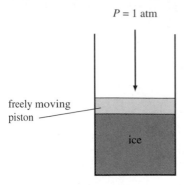

FIGURE 13-21
Observing phase changes—Example 13-4 illustrated.
This condition prevails at point *P* in Figure 13-19. The sample can be heated under the constant pressure of the atmosphere [Example 13-4, part (a)]. Or, the pressure can be increased by adding weights to the piston [Example 13-4, part (b)].

P = 1 atm

freely moving piston

ice

Solution

a. When the temperature reaches a point on the fusion curve *OD* (0 °C), ice begins to melt. The temperature remains constant as ice is converted to liquid. When melting is complete, the temperature again increases. No vapor appears in the cylinder until the temperature reaches 100 °C, at which point the vapor pressure is 1 atm. When all the liquid has vaporized, the temperature is again free to rise to a final value of *R*.

b. Because solids are not very compressible, very little change occurs until the pressure reaches the point of intersection of the constant-temperature line *PQ* with the fusion curve *OD*. Here melting begins. A significant *decrease* in volume occurs (about 10%) as ice is converted to liquid water. After melting, additional pressure produces very little change because liquids are not very compressible.

Practice Example A: Figure 13-20 gives a simplified phase diagram for carbon. Describe what happens when a sample of graphite is **(a)** heated to 1000 °C and then put under increasing pressure until a pressure of 10^5 atm is reached, and **(b)** heated to approximately 3800 K followed by an increase in pressure to a value of 10^6 atm.

Practice Example B: Sketch the *heating* curve obtained when a sample of ice is slowly heated from point *P* to point *R* of Figure 13-19. [*Hint:* How does the general shape of a *heating* curve compare with that of a *cooling* curve (Figure 13-14)?]

13-5 Van der Waals Forces

Because helium forms no stable chemical bonds, we might expect helium to remain a gas, right down to 0 K. Although helium does remain gaseous to very low temperatures, it does condense to a liquid at 4 K and freeze to a solid (at 25 atm pressure) at 1 K. These data suggest that intermolecular forces, even though very weak, must exist among He atoms. If the temperature is sufficiently low, these forces overcome thermal agitation and cause helium to condense. In this section we will examine the types of intermolecular forces known collectively as **van der Waals forces.** These forces are the ones accounted for in the van der Waals equation for nonideal gases (see Section 6-9).

Instantaneous and Induced Dipoles

In describing electronic structures we speak of electron charge density or the *probability* that an electron is in a certain region at a given time. One probability is that at some particular instant—purely by chance—electrons are concentrated in one region of an atom or molecule. This displacement of electrons causes a normally nonpolar species to become momentarily polar. An *instantaneous dipole* is formed. That is, the molecule has an instantaneous dipole moment (see page 373 for discussion of dipole moment). After this, electrons in a neighboring atom or molecule may be displaced to also produce a dipole. This is a process of induction (Figure 13-22), and the newly formed dipole is called an *induced dipole.*

Taken together, these two events lead to an intermolecular force of attraction (Figure 13-23). We can call this an instantaneous dipole/induced dipole attraction, but the names more commonly used are **dispersion force** and **London force.** (Fritz London offered a theoretical explanation of these forces in 1928.)

The ease with which a dipole can be induced in an atom or molecule is called **polarizability.** Polarizability increases with increased numbers of electrons, and the number of electrons increases with increased molecular mass. Also, in large

FIGURE 13-22
The phenomenon of induction.
The attraction of a balloon to a surface is a commonplace example of induction. The balloon is charged by rubbing, and the charged balloon induces an opposite charge on the surface. (See also Appendix B).

FIGURE 13-23
Instantaneous and induced dipoles.
(a) *Normal condition.* A nonpolar molecule has a symmetrical charge distribution. (b) *Instantaneous condition.* A displacement of the electronic charge produces an instantaneous dipole with a charge separation represented as δ+ and δ−. (c) *Induced dipole.* The instantaneous dipole on the left induces a charge separation in the molecule on the right. The result is a dipole–dipole attraction.

molecules some electrons, being farther from atomic nuclei, are less firmly held. These electrons are more easily displaced, and the polarizability of the molecule increases. Because dispersion forces become stronger as polarizability increases, molecules attract each other more strongly, with the result that melting points and boiling points of covalent substances generally increase with increasing molecular mass. For instance, helium, with a molecular (atomic) mass of 4 u, has a boiling point of 4 K, whereas radon (atomic mass, 222 u) has a boiling point of 211 K. The melting points and boiling points of the halogens increase in a similar way in the series F_2, Cl_2, Br_2, I_2 (recall Table 10.5).

The strength of dispersion forces also depends on *molecular shape*. Electrons in elongated molecules are more easily displaced than are those in small, compact, symmetrical molecules; the elongated molecules are more polarizable. Two substances with identical numbers and kinds of atoms but different molecular shapes (*isomers*) may have different properties. This idea is illustrated in Figure 13-24.

Dipole–Dipole Interactions

In a *polar* substance, the molecules have permanent dipole moments. The result is that the molecules try to line up with the positive end of one dipole directed toward the negative ends of neighboring dipoles (Figure 13-25). This additional partial ordering of molecules can cause a substance to persist as a solid or liq-

(a) Neopentane
b.p. = 9.5 °C

(b) Pentane
b.p. = 36.1 °C

FIGURE 13-24
Molecular shapes and polarizability.
The elongated pentane molecule is more easily polarized than the compact neopentane molecule. Intermolecular forces are stronger in pentane than in neopentane. Pentane boils at a higher temperature than neopentane.

FIGURE 13-25
Dipole–dipole interactions.
Dipoles tend to arrange themselves with the positive end of one dipole pointed toward the negative end of a neighboring dipole. Ordinarily, thermal motion upsets this orderly array. Nevertheless, this tendency for dipoles to align themselves can affect physical properties, such as the melting points of solids and the boiling points of liquids.

uid at temperatures higher than otherwise expected. Consider N_2, O_2, and NO. There are no electronegativity differences in N_2 and O_2, and both substances are *nonpolar*. In NO, on the other hand, there is an electronegativity difference, and the molecule has a slight dipole moment. Considering only dispersion forces, we would expect the boiling point of NO(l) to be intermediate to those of N_2(l) and O_2(l), but in the following comparison we see that it is not. NO(l) has the highest boiling point of the three.

N_2	NO	O_2
$\mu = 0$ (nonpolar)	$\mu = 0.153$ D (polar)	$\mu = 0$ (nonpolar)
mol. mass 28 u	mol. mass 30 u	mol. mass 32 u
b.p. 77.34 K	b.p. 121.39 K	b.p. 90.19 K

EXAMPLE 13-5

Comparing Physical Properties of Polar and Nonpolar Substances. Which would you expect to have the *higher* boiling point, the hydrocarbon fuel butane, C_4H_{10}, or the organic solvent acetone, $(CH_3)_2CO$?

Solution

Let's start with a comparison of molecular masses. Because both substances have a molecular mass of 58 u, we have to base our prediction on some other factor.

Next, let's see whether either molecule is polar. The electronegativity difference between C and H is so small that we generally expect hydrocarbons to be *nonpolar.* The acetone molecule contains an O atom, and we do expect a strong carbon-to-oxygen bond moment. At times we need to sketch the structure of a molecule to see whether symmetrical features cause bond moments to cancel. Because only one bond has a large bond moment, however, drawing the structure is not necessary with the acetone molecule. Acetone has a resultant dipole moment and is a *polar* molecule. Given two substances with the same molecular mass, one polar and one nonpolar, we expect the *polar* substance—acetone—to have the *higher* boiling point. (The measured boiling points are butane, -0.5 °C; acetone, 56.2 °C.)

Practice Example A: Which of the following substances would you expect to have the *highest* boiling point: C_3H_8, CO_2, CH_3CN? Explain.

Practice Example B: Arrange the following in the expected order of increasing boiling point: C_8H_{18}, $CH_3CH_2CH_2CH_3$, $(CH_3)_3CH$, C_6H_5CHO, SO_3 (octane, butane, isobutane, benzaldehyde, and sulfur trioxide, respectively).

Summary of Van der Waals Forces

When assessing the importance of van der Waals forces, consider the following:

- *Dispersion (London) forces* involve displacements of all the electrons in molecules. The strengths of these forces increase with increased molecular mass and also depend on molecular shapes. Dispersion forces exist between all molecules.

- *Forces associated with permanent dipoles* involve displacements of electron pairs in bonds rather than in molecules as a whole. They are found only in substances with resultant dipole moments (polar molecules). Their existence *adds* to the effect of dispersion forces also present.

- *When comparing substances of roughly comparable molecular masses,* dipole forces can produce significant differences in such properties as melting point, boiling point, and enthalpy of vaporization.

- *When comparing substances of widely different molecular masses,* dispersion forces are usually more significant than dipole forces.

Let's see how these statements relate to the data in Table 13.5, which includes a rough breakdown of van der Waals forces into dispersion forces and those due to dipoles. HCl and F_2 have comparable molecular masses, but because HCl is polar it has a significantly larger ΔH_{vap} and a higher boiling point than does F_2. Within the series HCl, HBr, and HI, molecular mass increases sharply, and ΔH_{vap} and boiling points increase in the order HCl < HBr < HI. The more polar nature of HCl and HBr relative to HI is not sufficient to reverse the trends produced by the increasing molecular masses—dispersion forces are the predominant intermolecular forces.

EXAMPLE 13-6

Relating Intermolecular Forces and Physical Properties. Arrange the following substances in the order in which you would expect their boiling points to increase: CCl_4, Cl_2, ClNO, N_2.

Solution

Three of the substances are nonpolar. For these, the strengths of dispersion forces, and hence the boiling points, should increase with increasing molecular mass, that is, $N_2 < Cl_2 < CCl_4$. ClNO has a molecular mass (65.5 u) comparable to that of Cl_2 (70.9 u), but the ClNO molecule is polar (bond angle ≈ 120°). This suggests stronger intermolecular forces and a higher boiling point for ClNO than for Cl_2. We should not expect the boiling point of ClNO to be higher than that of CCl_4, however, because of the large difference in their molecular masses (65.5 u compared with 154 u). The expected order is $N_2 < Cl_2 < ClNO < CCl_4$. (The observed boiling points are 77.3, 239.1, 266.7, and 349.9 K, respectively.)

Practice Example A: Arrange the following in the expected order of increasing boiling point: Ne, He, Cl_2, $(CH_3)_2CO$, O_2, O_3.

Practice Example B: Following are some values of ΔH_{vap} for several liquids at their normal boiling points: H_2, 0.92 kJ/mol; CH_4, 8.16 kJ/mol; C_6H_6, 31.0 kJ/mol; CH_3NO_2, 34.0 kJ/mol. Explain the differences among these values.

TABLE 13.5   Intermolecular Forces and Properties of Selected Substances

| | Molecular Mass, u | Dipole Moment, D | Van der Waals Forces | | ΔH_{vap}, kJ/mol | Boiling Point, K |
			% Dispersion	% Dipole		
F_2	38.00	0	100	0	6.86	85.01
HCl	36.46	1.08	81.4	18.6	16.15	188.11
HBr	80.92	0.82	94.5	5.5	17.61	206.43
HI	127.91	0.44	99.5	0.5	19.77	237.80

13-6 Hydrogen Bonding

Life is made possible by chemical reactions involving complex structures, such as DNA and proteins. Certain bonds in these structures must be easily broken and re-formed. Only one type of bonding involves bond energies of the right magnitude to allow this—hydrogen bonding.

To see what is unique about hydrogen bonding, let's first study Figure 13-26, in which the boiling points of a series of similar compounds are plotted as a function of molecular mass. The hydrogen compounds (hydrides) of the Group 4A elements display normal behavior; that is, the boiling point increases regularly with increasing molecular mass, from that of CH_4 to that of SnH_4. But note the three striking exceptions in Groups 5A, 6A, and 7A: NH_3, H_2O, and HF, respectively. Their boiling points are as high or higher than that of any other hydride in their group—not lowest, as we would expect. The cause of this exceptional behavior is hydrogen bonding. This type of bonding is illustrated for hydrogen fluoride in Figure 13-27, which emphasizes these points:

1. The alignment of HF dipoles places an H atom between two F atoms. Because of the very small size of the H atom, the dipoles come close together and produce strong *dipole–dipole* attractions.

2. Although an H atom is covalently bonded to one F atom, it is also weakly bonded to the F atom of a nearby HF molecule. This occurs through a lone pair of electrons on the F atom. In hydrogen bonding an H atom acts as a bridge between two nonmetal atoms.

3. The bond angle between two nonmetal atoms bridged by an H atom (that is, the bond angle, F—H···F) is about 180°.

FIGURE 13-27
Hydrogen bonding in gaseous hydrogen fluoride.
In gaseous hydrogen fluoride many of the HF molecules are associated into cyclic $(HF)_6$ structures of the type pictured here. Each H atom is bonded to one F atom by a single covalent bond (—) and to another F atom through a hydrogen bond (····).

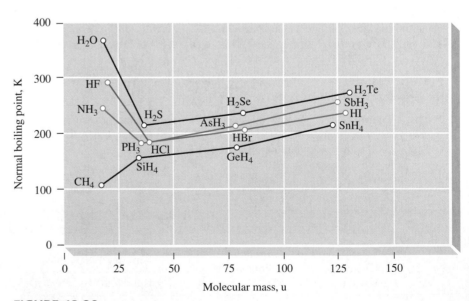

FIGURE 13-26
Comparison of boiling points of some hydrides of the elements of Groups 4A, 5A, 6A, and 7A.
The values for NH_3, H_2O, and HF are unusually high compared with those of other members of their groups.

A **hydrogen bond** is formed when an H atom bonded to one highly electronegative atom is simultaneously attracted to a highly electronegative atom of a neighboring molecule. Only F, O, and N easily meet the electronegativity requirement for hydrogen bonding. Hydrogen bonds are rather strong intermolecular forces, with energies of the order of 15 to 40 kJ/mol. (Recall that a typical covalent bond energy is over 150 kJ/mol. See Table 11.3 for further comparisons.)

In hydrogen bond formation, the highly electronegative atom to which an H atom is covalently bonded pulls the shared electron pair away from the H nucleus, a proton. This leaves the proton unshielded, and it is attracted to a lone pair of electrons on a highly electronegative atom of a neighboring molecule. Hydrogen bonds are possible only with certain hydrogen-containing compounds, because all atoms other than H have inner shell electrons to shield their nuclei from attraction by lone-pair electrons of nearby atoms.

Ordinary water is certainly the most common substance in which hydrogen bonding occurs. Figure 13-28 shows how one water molecule is held to four neighbors in a tetrahedral arrangement by hydrogen bonds. In ice, hydrogen bonds hold the water molecules in a rigid but rather open structure. As ice melts, only a fraction of the hydrogen bonds are broken. One indication of this is the relatively low heat of fusion of ice (6.01 kJ/mol). It is much less than we would expect if all the hydrogen bonds were to break during melting.

The open structure of ice shown in Figure 13-28 gives ice a low density. When ice melts, some of the hydrogen bonds are broken. This allows the water molecules to be more compactly arranged, accounting for the increase in density when ice melts. That is, the number of H_2O molecules per unit volume is greater in the liquid than in the solid.

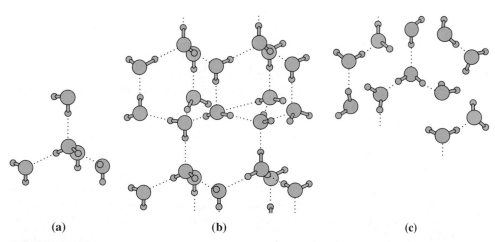

(a)   (b)   (c)

FIGURE 13-28
Hydrogen bonding in water.
(a) Each water molecule is linked to four others through hydrogen bonds. The arrangement is tetrahedral. Each H atom is situated along a line joining two O atoms, but closer to one O atom (100 pm) than to the other (180 pm). (b) The crystal structure of ice. H atoms lie between pairs of O atoms, again closer to one O atom than to the other. Molecules behind the plane of the page are shaded light blue. O atoms are arranged in bent hexagonal rings arranged in layers. This characteristic pattern is revealed in the hexagonal shapes of snowflakes. (c) In the liquid, water molecules have hydrogen bonds to only some of their neighbors. This allows the water molecules to pack more densely in the liquid than in the solid.

▶Feature Problem E on page 465 considers the properties expected for water if it were a "normal" liquid.

FIGURE 13-29
Solid and liquid densities compared.
The sight of ice cubes floating on liquid water (left) is a familiar one. Ice is less dense than liquid water. The more common situation, however, is that of paraffin wax (right). Solid paraffin is denser than the liquid and sinks to the bottom of the beaker.

Sequoia trees.
The mystery as to how these trees can bring water to leaves that are hundreds of feet up may be explained by hydrogen bonding.

FIGURE 13-30
An acetic acid dimer.
Hydrogen bonding permits acetic acid molecules to exist in stable pairs (dimers).

As liquid water is heated above the melting point, hydrogen bonds continue to break. The molecules become even more closely packed, and the density of the liquid water continues to increase. Liquid water attains its maximum density at 3.98 °C. Above this temperature the water behaves in a "normal" fashion: its density decreases as temperature increases. This behavior explains why a freshwater lake freezes from the top down. When the water temperature falls below 4 °C, the denser water sinks to the bottom of the lake and the colder surface water freezes. The ice at the top of the lake then tends to insulate the water below the ice from further heat loss. This allows fish to survive the winter in a lake that has been frozen over. Without hydrogen bonding, all lakes would freeze from the bottom up; and fish, small bottom-feeding animals, and aquatic plants would not survive the winter. The density relationship between liquid water and ice is compared with the more common liquid/solid density relationship in Figure 13-29.

Hydrogen bonding can explain other natural phenomena. For example, hydrogen bonding in water seems to be the principal reason why trees can grow higher than 30 feet. Redwood trees can reach heights of over 100 m (340 feet). How does water get to the tallest leaves in such a tree? We know from our discussion of pressure in Chapter 6 that atmospheric pressure will raise a column of water about 10 m—far too short for these trees. In Chapter 14 we will learn about osmotic pressure, which can transport solutions to considerable heights, but hydrogen bonding seems to be what allows these beautiful trees to reach such heights. Thin columns of water (in xylem) extend to the top of the tree. In the column, water molecules are hydrogen-bonded to each other, producing a high cohesion between the molecules. You can imagine each molecule as a link in a chain that goes from root tip to leaf tip. As water in the leaves evaporates a new molecule moves into the space just vacated. This movement pulls the next molecule in the chain further up the column. Biologists have done a variety of experiments to prove this *cohesion–tension* theory, and it is the best theory so far for explaining the dramatic heights to which some trees can grow.

Hydrogen bonding can also be seen in some unexpected properties of other compounds. In acetic acid, CH_3COOH, hydrogen bonding leads to the formation of *dimers* (double molecules) in both the liquid and the vapor states. Not all the hydrogen bonds between molecules are broken when acetic acid vaporizes, and as a result the heat of vaporization is abnormally *low*. Figure 13-30 shows a dimer of acetic acid. Hydrogen bonding can also help us to understand certain trends in viscosity. Alcohols are carbon-based compounds containing the O—H group. The H atoms in alcohols can form hydrogen bonds with O atoms in neighboring molecules. The more O—H groups found in a molecule, the more possibilities there are for hydrogen bonding. This leads to a greater resistance to flow and a greater viscosity. Thus ethanol, CH_3CH_2OH, is a free-flowing liquid with a viscosity comparable to that of water. Glycerol, $CH_2OHCHOHCH_2OH$, on the other hand, is a viscous liquid that flows sluggishly.

Throughout this section we have been describing hydrogen bonding between two molecules—*inter*molecular hydrogen bonding. Another possibility is for a hydrogen atom to bridge two nonmetal atoms within the *same* molecule. This is called *intra*molecular hydrogen bonding (see Exercise 58).

Finally, although most cases of hydrogen bonding involve hydrogen-containing compounds of N, O, and F, weak hydrogen bonding may occur between an H atom of one molecule and the nonmetal atoms Cl or S of a neighboring molecule.

13-7 Chemical Bonds as Intermolecular Forces

In Chapters 11 and 12 we discussed chemical bonds, that is, forces between atoms—*intra*molecular forces. In this chapter we have been emphasizing *inter*molecular forces—forces between molecules. In some substances *intra*molecular and *inter*molecular forces are one and the same. We will consider some examples in this section.

Network Covalent Solids

In most covalent substances, intermolecular forces are quite weak compared with the bonds between atoms within molecules. This is why covalent substances of low molecular mass (those with weak dispersion forces) are generally gaseous at room temperature. Others, usually of somewhat higher molecular masses (with stronger dispersion forces), are liquids. Still others are solids with moderately low melting points.

In a few substances, known as *network covalent* solids, covalent bonds extend throughout a crystalline solid. In these cases the entire crystal is held together by strong forces. Two of the best examples are two of the allotropic forms in which pure carbon occurs—diamond and graphite.

Diamond.   Figure 13-31 shows one way carbon atoms can bond one to another in a very extensive array or crystal. The two-dimensional Lewis structure (Figure 13-31a) is useful only in showing that this bonding scheme involves ever-increasing numbers of C atoms leading to a giant molecule. It does not give us any insight into the three-dimensional structure of the molecule. For this we need the portion of the crystal shown in Figure 13-31b. Each atom is bonded to four others. Atoms 1, 2, and 3 lie in a plane with atom 4 above the plane. Atoms 1, 2, 3, and 5 define a tetrahedron with atom 4 at its center. The type of hybridization that corresponds to four bonds directed from a central atom to the corners of a tetrahedron is sp^3. When viewed from a particular direction, a nonplanar hexagonal arrangement of carbon atoms (gray) is also seen.

If silicon atoms are substituted for one-half the carbon atoms, the resulting structure is that of silicon carbide (carborundum). Both diamond and silicon carbide are extremely hard, and this accounts for their extensive use as abrasives. In fact, diamond is the hardest substance known. To scratch or break diamond or silicon carbide crystals, covalent bonds must be broken. These two materials are also nonconductors of electricity and do not melt or sublime until very high temperatures are reached. SiC sublimes at 2700 °C, and diamond melts above 3500 °C.

▶ A third allotropic form of carbon called fullerene, C_{60}, was discovered in 1985. It was discussed in the Focus On on page 380.

The world's largest uncut diamond (860 carats). The yellow color results from nitrogen atoms replacing some carbon atoms in the crystal structure.

▶ Another silicon-containing network covalent solid is ordinary silica—silicon dioxide, SiO_2.

(a)   **(b)**

FIGURE 13-31
The diamond structure.
(a) A portion of the Lewis structure. (b) Crystal structure. Each carbon atom is bonded to four others in a tetrahedral fashion. The segment of the entire crystal shown here is called a unit cell.

FIGURE 13-32
The graphite structure.

Graphite conducts electricity.
Delocalized electrons in graphite allow the conduction of electricity. In this photo pencil "lead," a mixture of graphite and clay, is used as electrodes to complete the circuit. The beaker contains a solution of ions that carry the current between the pencil electrodes.

Graphite.   Carbon atoms can bond together in a different way to produce a solid with properties very different from those of diamond. This bonding involves the orbital set $sp^2 + p$. The three sp^2 orbitals are directed in a plane at angles of 120°. The p orbital is perpendicular to the plane, directed above and below it. These orbitals are the same ones used for carbon atoms in benzene, C_6H_6 (described in Figure 12-22).

This type of bonding produces the crystal structure shown in Figure 13-32. Each carbon atom forms strong covalent bonds with three neighboring carbon atoms in the same plane, which gives rise to layers of carbon atoms in a hexagonal arrangement. The p electrons of the carbon atoms are *delocalized* (recall the discussion in Section 12-6). Bonding within layers is strong, but between layers it is much weaker. We can see this through bond distances. The C—C bond distance within a layer is 142 pm (compared with 139 pm in benzene); between layers it is 335 pm.

Its unique crystal structure gives graphite some distinctive properties. Because bonding between layers is weak, the layers can glide over one another rather easily. As a result, graphite is a good lubricant, either in dry form or suspended in oil. If we apply a mild pressure to a piece of graphite, layers of the graphite flake off; this is what happens when we use a graphite pencil. Also, because the p electrons are *delocalized,* they migrate through the planes of carbon atoms when an electric field is applied; graphite conducts electricity. An important use of graphite is as electrodes in batteries and in industrial electrolysis processes. Diamond is not an electrical conductor because all of its valence electrons are localized or permanently fixed into single covalent bonds.

Interionic Forces

When predicting properties of an ionic solid, we often face this question: How difficult is it to break up an ionic crystal and separate its ions? This question is addressed by the *lattice energy* of a crystal. **Lattice energy** is the energy given off when separated *gaseous* ions, positive and negative, come together to form *1 mol* of a *solid* ionic compound. Lattice energies can be useful in making predictions about the melting points and water solubilities of ionic compounds. We will examine how to calculate lattice energies in Section 13-9. At times, however, we only need to make *qualitative* comparisons of interionic forces, and the following generalization works quite well.

> The attractive force between a pair of oppositely charged ions increases with increased charge on the ions and with decreased ionic sizes.

This idea is illustrated through Figure 13-33.

For most ionic compounds, lattice energies are great enough that ions do not readily detach themselves from the crystal and pass into the gaseous state. Ionic solids do not sublime at ordinary temperatures. We can melt ionic solids by supplying enough thermal energy to disrupt the crystalline lattice. In general, the higher the lattice energy, the higher the melting point.

The energy required to break up an ionic crystal when it dissolves results from the interaction of ions in the crystal with molecules of the solvent (for example, water molecules). The extent to which an ionic solid dissolves in a solvent, however, depends only in part on the lattice energy of the ionic solid. As a rough rule, though, the lower the lattice energy, the greater the quantity of an ionic solid that can be dissolved in a given quantity of solvent.

Relative attractive force:

Radius: $Na^+ = 99$ pm
$Cl^- = 181$ pm

Radius sum = distance between center of ions: 280 pm

$Mg^{2+} = 72$ pm
$O^{2-} = 140$ pm

212 pm

FIGURE 13-33
Interionic forces of attraction.
Because of the higher charges on the ions and the closer proximity of their centers, the interionic attractive force between Mg^{2+} and O^{2-} is about seven times as great as between Na^+ and Cl^-.

EXAMPLE 13-7

Predicting Physical Properties of Ionic Compounds. Which has the higher melting point, KI or CaO?

Solution

Ca^{2+} and O^{2-} are more highly charged than K^+ and I^-. Also, Ca^{2+} is smaller than K^+, and O^{2-} is smaller than I^-. We certainly expect the interionic forces in crystalline CaO to be much larger than in KI. CaO should have the higher melting point. (The observed melting points are 677 °C for KI and 2590 °C for CaO.)

Practice Example A: Cite one ionic compound that you would expect to have a lower melting point than KI and one with a higher melting point than CaO.

Practice Example B: Which would you expect to have the higher solubility in water, NaI or $MgCl_2$? Explain.

13-8 Crystal Structures

Crystals—whether as ice, rock salt, quartz, or gemstones—have aroused interest from earliest times. Yet, only in relatively recent times have we come to a fundamental understanding of the crystalline state. This understanding started with the invention of the optical microscope and was greatly expanded following the discovery of X-rays. The key idea, now supported by countless experiments, is that the regularity we observe in crystals at the macroscopic level is due to an underlying regular pattern in the arrangement of atoms, ions, or molecules.

Crystal Lattices

You can probably think of a number of situations in which you have had to deal with repeating patterns in one or two dimensions. These might include projects like sewing a decorative border on a piece of material, wallpapering a room, or creating a design with floor tiles.

To describe the structures of crystals, however, we have to work with *three*-dimensional patterns. We do this through three sets of parallel planes called a *lattice*. We have chosen a special case for Figure 13-34: The planes are equidistant and mutually perpendicular (intersecting at 90° angles). This is called a

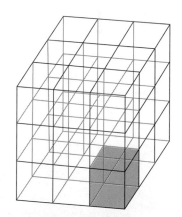

FIGURE 13-34
The cubic space lattice.
A typical parallelepiped formed by the intersection of mutually perpendicular planes is shaded in green. It is a cube. An endless lattice can be generated by simple displacements of the green cube in the three perpendicular directions (that is, left and right, up and down, and forward and backward).

cubic lattice. We can use it to describe some crystals. For others, the appropriate lattice may involve planes that are not equidistant or that intersect at angles other than 90°. In all there are seven possibilities for crystal lattices, but we will emphasize only the cubic lattice.

Lattice planes intersect to produce three-dimensional figures having six faces arranged in three sets of parallel planes. These figures are called *parallelepipeds*. In Figure 13-34 these parallelepipeds are cubes. A parallelepiped that can be used to generate the entire lattice by simple straight-line displacements is called a **unit cell.** Where possible, we arrange the three-dimensional space lattice so that structural units of the crystal (atoms, ions, or molecules) are situated at lattice points. If a unit cell has structural particles only at its corners, it is called a *primitive* unit cell or simple cubic cell, the simplest unit cell that we can consider. But sometimes we choose a unit cell that has more structural particles. In the **body-centered cubic (bcc)** structure, a structural particle of the crystal is found at the center of the cube as well as at each corner. In a **face-centered cubic (fcc)** structure there is a structural particle at the center of each face as well as at each corner. These unit cells are shown in Figure 13-35.

Closest Packed Structures

Unlike boxes, which can be stacked to fill all space, when spheres are stacked together there must always be some unfilled space. In some sphere arrangements, however, the spheres come into as close contact as possible and the holes or voids are kept to a minimum. These are known as closest packed structures and are the basis of a number of crystal structures.

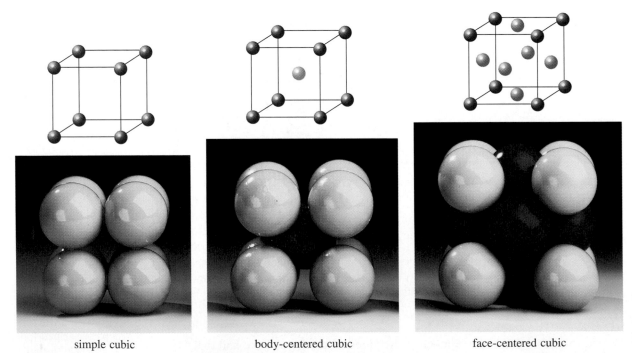

simple cubic   body-centered cubic   face-centered cubic

FIGURE 13-35
Unit cells in the cubic crystal system.
In the line-and-ball drawings in the top row only the centers of spheres (atoms) are shown at their respective positions in the unit cells. The space-filling models in the bottom row show contacts between spheres (atoms). In the simple cubic cell, spheres come into contact along each edge. In the body-centered cubic (bcc) cell, contact of the spheres is along the cube diagonal. In the face-centered cubic (fcc) cell, contact is along the diagonal of each face.

To analyze the closest packed structures in Figure 13-36, let us imagine one layer of spheres, layer A (red), in which each sphere is in contact with six others arranged in a hexagonal fashion around it. Among the spheres we see open spaces or voids. Once the first sphere is placed in the next layer, layer B (yellow), the entire pattern for that layer is fixed. Again, there are holes or voids in layer B, but the holes are of two different types. *Tetrahedral* holes fall directly over spheres in layer A and have this shape ◁. *Octahedral* holes fall directly over holes in layer A and have this shape ✿. Now, there are two possibilities for C, the third layer. In one arrangement, called **hexagonal closest packed (hcp),** all the tetrahedral holes are covered. Layer C is identical to layer A, and the structure begins to repeat itself. In the other arrangement, called **cubic closest packed,** all the octahedral holes are covered. The spheres in layer C (blue) are out of line with those in layer A. Only when the fourth layer is added does the structure begin to repeat itself.

Study Figure 13-37 on page 448 and you will see that the cubic closest packed structure has a face-centered cubic unit cell. The unit cell of the hexagonal closest packed structure is shown in Figure 13-38 on page 448. In both the hcp and the fcc structures voids account for only 25.96% of the total volume. Another arrangement in which the packing of spheres is not quite so close has a body-centered cubic unit cell. In this structure voids account for 31.98% of the total volume. The best examples of crystal structures based on the closest packing of spheres are found among the metals. Some examples are listed in Table 13.6.

▶ A method of calculating the percent voids in crystal structures is outlined in Exercise 96.

Coordination Number and Number of Atoms per Unit Cell

In close-packed structures of atoms, each atom is in contact with several others. For example, can you see in Figure 13-35 that the center atom in the bcc

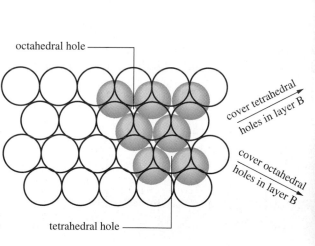

octahedral hole

cover tetrahedral holes in layer B

cover octahedral holes in layer B

tetrahedral hole

A
B
A

hexagonal closest packed

A
C
B
A

cubic closest packed

FIGURE 13-36
Closest packed structures.
Spheres in layer A are red. Those in layer B are yellow, and in layer C, blue.

FIGURE 13-37

A face-centered cubic unit cell for the cubic closest packing of spheres.
The 14 spheres on the left are extracted from a larger array of spheres in a cubic closest packed structure. The two middle layers each have six atoms; the top and bottom layers, one. Rotation of the group of 14 spheres reveals the fcc unit cell (right).

unit cell is in contact with each corner atom? We call the number of atoms with which a given atom is in contact the *coordination number.* For the bcc structure this is 8. For the fcc and hcp structures the coordination number is 12. The easiest way to see this is from the layering of spheres described in Figure 13-36. Each sphere is in contact with *six* others in the same layer, *three* in the layer above, and *three* in the layer below.

Although we use nine atoms to draw the bcc unit cell, we are wrong if we conclude that the unit cell consists of nine atoms. As shown in Figure 13-39, only the center atom belongs *entirely* to the bcc unit cell. The corner atoms are shared among eight adjoining unit cells. Only one eighth of each corner atom should be thought of as belonging to a given unit cell. Thus, the eight corner atoms collectively contribute the equivalent of *one* atom to the unit cell. The total number of atoms in a bcc unit cell, then, is *two* [that is, $1 + (8 \times \frac{1}{8})$]. For the hcp unit cell of Figure 13-38 we also get *two* atoms per unit cell if we use the correct counting procedure. The corner atoms account for $\frac{1}{8} \times 8 = 1$ atom and the central atom belongs entirely to the unit cell. In the fcc unit cell the corner atoms account for $\frac{1}{8} \times 8 = 1$ atom, and those in the center of the faces for $\frac{1}{2} \times 6 = 3$ atoms. The fcc unit cell contains *four* atoms.

X-Ray Diffraction

We can see macroscopic objects using visible light and our eyes. To "see" how atoms, ions, or molecules are arranged in a crystal, we need light of much shorter

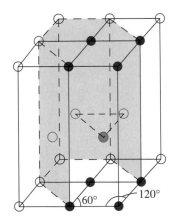

FIGURE 13-38

The hexagonal closest packed (hcp) crystal structure.
A unit cell is highlighted in heavy black. The atoms that are part of that cell are in solid color. Note that the unit cell is a parallelepiped, but not a cube. Three adjoining unit cells are also shown. The highlighted unit cell and broken-line regions together show the layering (ABA) described in Figure 13-36.

TABLE 13.6 Some Features of Close-Packed Structures in Metals			
	Coordination Number	**Number of Atoms per Unit Cell**	**Examples**
hexagonal closest packed (hcp)	12	2	Cd, Mg, Ti, Zn
face-centered cubic (fcc)	12	4	Al, Cu, Pb, Ag
body-centered cubic (bcc)	8	2	Fe, K, Na, W

wavelength. When a beam of X-rays encounters atoms, X-rays interact with electrons in the atoms and the original beam is reradiated, diffracted, or scattered in all directions. The pattern of this scattered radiation is related to the distribution of electronic charge in the atoms. Our eyes and brains cannot perceive X-rays, however. We must make the scattered X-rays produce a visible pattern, as on a photographic film. Then we have to infer the microscopic structure of the substance from this visible pattern. How successful we are in making inferences depends on the amount of the scattered radiation we recover, that is, on how much "information" we gather. The power of the X-ray diffraction method has been greatly increased by the use of high-speed computers to process vast amounts of X-ray data.

Figure 13-40 suggests a method of scattering X-rays from a crystal. X-ray data can be explained by a geometric analysis proposed by W. H. Bragg and W. L. Bragg in 1912 and illustrated in Figure 13-41 on page 450. Here are pictured two rays in a monochromatic (single-wavelength) X-ray beam, labeled *a* and *b*. Wave *a* is reflected by one plane of atoms or ions in a crystal and wave *b* from the next plane below. Wave *b* travels a greater distance than wave *a*. The additional distance is $2d \sin \theta$. The intensity of the scattered radiation will be greatest if waves *a* and *b* reinforce each other, that is, if their crests and troughs line up. And, to satisfy this requirement, the additional distance traveled by wave *b* must be an integral multiple of the wavelength of the X-rays.

$$n\lambda = 2d \sin \theta \tag{13.4}$$

From the measured angle θ that yields the maximum intensity for the scattered X-rays, together with the X-ray wavelength (λ), one can calculate the spacing (d) between atomic planes. With different orientations of the crystal one can determine atomic spacings and electron densities for different directions through the crystal, in short, the crystal structure.

Once a crystal structure is known, certain other properties can be determined by calculation. In Example 13-8 we calculate a metallic radius, and in Example 13-9 we estimate the density of a crystalline solid. For both of these calculations we need to sketch, or in some way visualize, a unit cell of the crystal. In particular, we need to see which atoms are in direct contact.

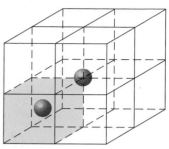

FIGURE 13-39
Apportioning atoms among bcc unit cells.
Eight unit cells are outlined. Attention is directed to the blue unit cell. For clarity, only the centers of two atoms are pictured. The atom in the center of the cell belongs entirely to that cell. The corner atom is seen to be shared by all eight unit cells.

▶ The X-ray diffraction method was originated by Max von Laue (Nobel Prize, 1914), but carried further by the Braggs. William Lawrence Bragg was only 25 years old when he and his father, William Henry Bragg, won the Nobel Prize in 1915.

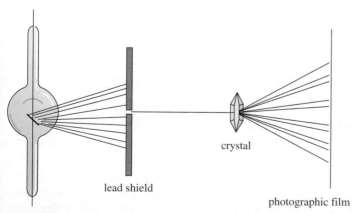

crystal

lead shield

photographic film

X-ray tube

FIGURE 13-40
Diffraction of X-rays by a crystal.

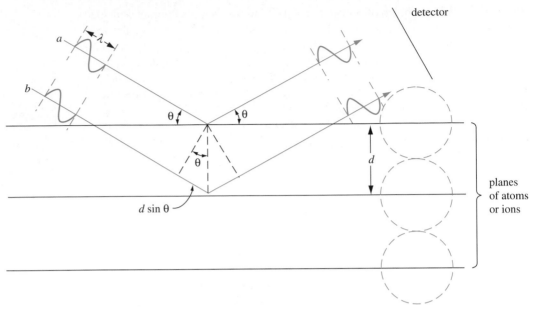

FIGURE 13-41
Determination of crystal structure by X-ray diffraction.
The two triangles outlined by dotted lines are identical. The hypotenuse of each triangle is equal to the interatomic distance, *d*. The side opposite the angle θ thus has a length of *d* sin θ. Wave *b* travels farther than wave *a* by the distance 2*d* sin θ.

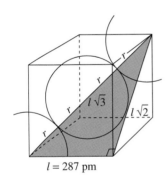

l = 287 pm

l = 287 pm

FIGURE 13-42
Determination of the atomic radius of iron—Example 13-8 illustrated.
The right triangle must conform to the Pythagorean formula $a^2 + b^2 = c^2$. That is, with *l* an edge of the cube, $(l)^2 + (l\sqrt{2})^2 = (l\sqrt{3})^2$, or $l^2 + 2(l^2) = 3(l^2)$.

EXAMPLE 13-8

Using X-ray Data to Determine an Atomic Radius. At room temperature, iron crystallizes in a bcc structure. By X-ray diffraction, the edge of the cubic cell corresponding to Figure 13-42 is found to be 287 pm. What is the radius of an iron atom?

Solution

Nine atoms are associated with a bcc unit cell. One atom is located at each of the eight corners of the cube and one at the center. The three atoms along a cube diagonal are in contact. The length of the cube diagonal (the distance from the farthest upper-right corner to the nearest lower-left corner) is four times the atomic radius. But also shown in Figure 13-42 is that the diagonal of a cube is equal to $\sqrt{3} \times l$. The length of an edge, *l*, is what is given.

$$4r = l\sqrt{3} \qquad r = \frac{\sqrt{3} \times 287 \text{ pm}}{4} = \frac{1.732 \times 287 \text{ pm}}{4} = 124 \text{ pm}$$

Practice Example A: Potassium crystallizes in the bcc structure. What is the length of the unit cell in this structure? Use the metallic radius of potassium given in Figure 10-8.

Practice Example B: Aluminum crystallizes in an fcc structure. Given that the atomic radius of Al is 143.1 pm, what is the volume of a unit cell? (*Hint:* Draw a sketch of the unit cell. What dimension of the unit cell can you relate to the atomic radius?)

EXAMPLE 13-9

Relating Density to Crystal Structure Data. Use data from Example 13-8, together with the molar mass of Fe and the Avogadro constant, to calculate the density of iron.

Solution

In Example 13-8 we saw that the length of a unit cell is $l = 287$ pm $= 287 \times 10^{-12}$ m $= 2.87 \times 10^{-8}$ cm. The volume of the unit cell is $V = l^3 = (2.87 \times 10^{-8})^3$ cm^3 $= 2.36 \times 10^{-23}$ cm^3.

From Table 13.6 we find that there are *two* Fe atoms per bcc unit cell. We need the mass of these two atoms, and the key to getting this is a conversion factor based on the fact that 1 mol Fe $= 6.022 \times 10^{23}$ Fe atoms $= 55.85$ g Fe.

$$\text{mass} = 2 \text{ Fe atoms} \times \frac{55.85 \text{ g Fe}}{6.022 \times 10^{23} \text{ Fe atoms}} = 1.855 \times 10^{-22} \text{ g Fe}$$

Density is the ratio of mass to volume.

$$\text{density of Fe} = \frac{m}{V} = \frac{1.855 \times 10^{-22} \text{ g Fe}}{2.36 \times 10^{-23} \text{ cm}^3} = 7.86 \text{ g Fe/cm}^3$$

Practice Example A: Use the result of Practice Example 13-8A, together with the molar mass of K and the Avogadro constant, to calculate the density of potassium.

Practice Example B: Use the result of Practice Example 13-8B, together with the molar mass of Al and its density (2.6984 g/cm^3), to evaluate the Avogadro constant, N_A. (*Hint:* From the volume of a unit cell and the density of Al you can determine the mass of a unit cell. Knowing the number of Al atoms in the fcc unit cell, you can determine the mass per Al atom. How many atoms does it take to yield a total mass equal to the molar mass of Al?)

FIGURE 13-43
The sodium chloride unit cell.
For clarity, only the centers of the ions are shown. Oppositely charged ions are actually in contact. We can think of this structure as an fcc lattice of Cl$^-$ ions, with Na$^+$ ions filling the octahedral holes.

Ionic Crystal Structures

If we try to apply the packing-of-spheres model to an ionic crystal we run into two complications: (1) Some of the ions are positively charged and some are negatively charged, and (2) the cations and anions are of different sizes. What we can expect, however, is that oppositely charged ions will come into close proximity. Generally we think of them as being in contact. Like-charged ions, because of mutual repulsions, are not in direct contact. We can think of some ionic crystals in this way: a fairly closely packed arrangement of ions of one type with holes or voids filled by ions of the opposite charge. The relative sizes of cations and anions are important in establishing a particular packing arrangement. In defining a unit cell of an ionic crystal we must choose a unit cell that

- by translation in three dimensions generates the entire crystal;
- indicates the coordination numbers of the ions;
- is consistent with the formula of the compound.

Unit cells of crystalline NaCl and CsCl are pictured in Figures 13-43 and 13-44.

To establish the coordination number in an ionic crystal, count the number of nearest neighbor ions of opposite charge to any given ion in the crystal. Note the Na$^+$ ion in the center of the unit cell of Figure 13-43. It is surrounded by

FIGURE 13-44
The cesium chloride unit cell.
The Cs$^+$ ion is in the center of the cube with Cl$^-$ ions at the corners. In reality each Cl$^-$ is in contact with the Cs$^+$ ion. An alternative unit cell has Cl$^-$ at the center and Ca$^+$ at the corners.

six Cl^- ions. The coordination numbers of both Na^+ and Cl^- are *six*. By contrast, the coordination numbers of Cs^+ and Cl^- in Figure 13-44 are *eight*.

We must apportion the 27 ions in Figure 13-43 among the unit cell and its neighboring unit cells in the following way: Each Cl^- ion in a corner position is shared by *eight* unit cells, and each Cl^- in the center of a face is shared by *two* unit cells. This leads to a total number of Cl^- ions in the unit cell of $(8 \times \frac{1}{8}) + (6 \times \frac{1}{2}) = 1 + 3 = 4$. There are 12 Na^+ ions along the edges of the unit cell, and each edge is shared by *four* unit cells. The Na^+ ion in the very center of the unit cell belongs entirely to that cell. Thus, the total number of Na^+ ions in a unit cell is $(12 \times \frac{1}{4}) + (1 \times 1) = 3 + 1 = 4$. The unit cell has the equivalent of 4 Na^+ and 4 Cl^- ions. The ratio of Na^+ to Cl^- is 4:4 = 1:1, corresponding to the formula NaCl.

Are You Wondering . . .

Why we don't select one of the smaller cubes in Figure 13-43 as the unit cell of NaCl?

There are eight smaller cubes in the unit cell, each of which has an Na^+ ion at four of the corners and a Cl^- ion at the other four. Such a cube is consistent with the formula NaCl, but will not generate the entire lattice by simple displacements. Move any one of these cubes to the adjoining cube in any direction and an Na^+ ion takes the position where one should find a Cl^- ion.

EXAMPLE 13-10

Relating Ionic Radii and the Dimensions of a Unit Cell of an Ionic Crystal. The ionic radii of Na^+ and Cl^- in NaCl are 99 and 181 pm, respectively. What is the length of the unit cell of NaCl?

Solution

Again the key to solving this problem lies in understanding geometric relationships in the unit cell. Along each edge of the unit cell (see Figure 13-43) two Cl^- ions are in contact with one Na^+. The edge length is equal to the radius of one Cl^- plus the diameter of Na^+, plus the radius of another Cl^-. That is,

$$\text{length} = (r_{Cl^-}) + (r_{Na^+}) + (r_{Na^+}) + (r_{Cl^-})$$

$$= 2(r_{Na^+}) + 2(r_{Cl^-})$$

$$= [(2 \times 99) + (2 \times 181)] = 560 \text{ pm}$$

Practice Example A: The ionic radius of Cs^+ is 167 pm. Use Figure 13-44 and information in Examples 13-8 and 13-10 to determine the length of the unit cell of CsCl.

Practice Example B: Use the length of the unit cell of NaCl obtained in Example 13-10, together with the molar mass of NaCl and the Avogadro constant, to estimate the density of NaCl. (*Hint:* What is the volume of a unit cell? How many formula units of NaCl are contained in the unit cell? What is the mass of this number of formula units?)

Ionic compounds of the type $M^{2+}X^{2-}$ (for example, MgO, BaS, CaO) may form crystals of the NaCl type. For substances with the formulas MX_2 or M_2X, the crystal structures are more complex. Because the cations and anions occur in unequal numbers, the crystals have *two* coordination numbers, one for the cation and another for the anion. Two typical structures of a more complex type are shown in Figure 13-45.

In CaF_2 (the fluorite structure) there are twice as many fluoride ions as calcium ions. The coordination number of Ca^{2+} is *eight,* and that of F^- is *four.* This is easiest to see by looking at the Ca^{2+} ion in the middle of a face. There are four F^- ions within the unit cell that are nearest neighbors. In addition, the four F^- ions in the next unit cell (the one that shares the face centered Ca^{2+} ion) are also nearest neighbors. This gives a coordination number of eight for the Ca^{2+}. The F^- ions each have one corner Ca^{2+} ion and three face-centered Ca^{2+} ions as nearest neighbors, giving a coordination number of four. In TiO_2 (the rutile structure) Ti^{4+} has a coordination number of *six* and O^{2-}, *three.* In this structure two of the O^{2-} ions are within the interior of the cell, two are in the top face, and two in the bottom face of the cell. Ti^{4+} ions are at the corners and the center of the cell.

Types of Crystalline Solids: A Summary

In this section we have emphasized crystal structures composed of metal atoms and those composed of ions. But the structural particles of crystalline solids can be atoms, ions, or molecules. To picture a crystal structure having molecules as its structural units, consider solid methane, CH_4. The crystal structure is fcc, which means that the unit cell has a CH_4 molecule at each corner and at the center of each face. Each CH_4 molecule occupies a volume equivalent to a sphere with a radius of 228 pm.

The intermolecular forces operating among the structural units of a crystal may be metallic bonds, interionic attractions, van der Waals forces, hydrogen bonds, or covalent bonds. Table 13.7 on page 454 lists the basic types of crystalline solids, the intermolecular forces within them, some of their characteristic properties, and examples of each.

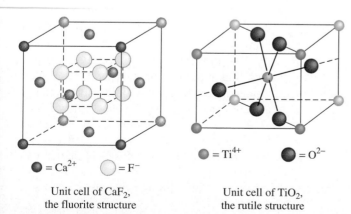

= Ca^{2+} = F^-

Unit cell of CaF_2,
the fluorite structure

= Ti^{4+} = O^{2-}

Unit cell of TiO_2,
the rutile structure

FIGURE 13-45
Some unit cells of greater complexity.

TABLE 13.7 Characteristics of Crystalline Solids

Type	Structural Particles	Intermolecular Forces	Typical Properties	Examples
Metallic	cations and delocalized electrons	metallic bonds	hardness varies from soft to very hard; melting point varies from low to very high; lustrous; ductile; malleable; very good conductors of heat and electricity	Na; Mg; Al; Fe; Sn; Cu; Ag; W
Ionic	cations and anions	electrostatic attractions	hard; moderate to very high melting points; nonconductors as solids, but good electric conductors as liquids; many are soluble in polar solvents like water.	NaCl; MgO; NaNO$_3$
Network covalent	atoms	covalent bonds	most are very hard and either sublime or melt at very high temperatures; most are nonconductors of electricity	C (diamond); C (graphite); SiC; AlN; SiO$_2$
Molecular *nonpolar*	atoms or nonpolar molecules	dispersion forces	soft; extremely low to moderate melting points (depending on molar mass); sublime in some cases; soluble in some nonpolar solvents	He; Ar; H$_2$; CO$_2$; CCl$_4$; CH$_4$; I$_2$
polar	polar molecules	dispersion forces and dipole–dipole attractions	low to moderate melting points; soluble in some polar and some nonpolar solvents	(CH$_3$)$_2$O; CHCl$_3$; HCl
hydrogen-bonded	molecules with H bonded to N, O, or F	hydrogen bonds	low to moderate melting points; soluble in some hydrogen-bonded solvents and some polar solvents	

13-9 Energy Changes in the Formation of Ionic Crystals

The concept of lattice energy, which we introduced qualitatively in Section 13-7, is most useful when stated in *quantitative* terms. It is difficult, however, to calculate a lattice energy directly. The problem is that oppositely charged ions attract one another and like-charged ions repel one another, and these interactions must be considered at the same time. More commonly, lattice energy is determined *indirectly* through an application of Hess's law known as the Born–Fajans–Haber cycle, named after its originators Max Born, Kasimir Fajans, and Fritz Haber. The crux of the method is to design a sequence of steps in which enthalpy changes are known for all the steps but one—the step in which a crystal lattice is formed from gaseous ions.

Figure 13-46 illustrates the method for NaCl through five steps:

(1) sublime one mole of solid Na;

(2) dissociate 0.5 mol Cl$_2$(g) into one mole of Cl(g);

(3) ionize one mole of Na(g) to $Na^+(g)$;

(4) convert one mole of Cl(g) to $Cl^-(g)$;

(5) allow the $Na^+(g)$ and $Cl^-(g)$ to form one mole of NaCl(s);

The net change in these five steps is the same as the reaction in which NaCl(s) is formed from its elements in their standard states—that is, $\Delta H_{net} = \Delta H_f^\circ[NaCl(s)]$. From Appendix D, we see that $\Delta H_f^\circ[NaCl(s)] = -411$ kJ/mol, and so in the following set-up, the lattice energy of NaCl is the only unknown.

1. $Na(s) \longrightarrow Na(g)$ $\Delta H_1 = \Delta H_{sublimation} = +107$ kJ

2. $\frac{1}{2} Cl_2(g) \longrightarrow Cl(g)$ $\Delta H_2 =$
$\frac{1}{2}$ Cl—Cl bond energy $= +122$ kJ

3. $Na(g) \longrightarrow Na^+(g) + e^-$ $\Delta H_3 = $ 1st ioniz. energy $= +496$ kJ

4. $Cl(g) + e^- \longrightarrow Cl^-(g)$ $\Delta H_4 = $ EA of Cl $= -349$ kJ

5. $Na^+(g) + Cl^-(g) \longrightarrow NaCl(s)$ $\Delta H_5 = $ lattice energy of NaCl $= ?$

net: $Na(s) + \frac{1}{2}Cl_2(g) \longrightarrow NaCl(s)$

$$\Delta H_{net} = -411 \text{ kJ} = \Delta H_1 + \Delta H_2 + \Delta H_3 + \Delta H_4 + \Delta H_5$$

$$-411 \text{ kJ} = 107 \text{ kJ} + 122 \text{ kJ} + 496 \text{ kJ} - 349 \text{ kJ} + \Delta H_5$$

$$\Delta H_5 = \text{lattice energy} = (-411 - 107 - 122 - 496 + 349) \text{ kJ} = -787 \text{ kJ}$$

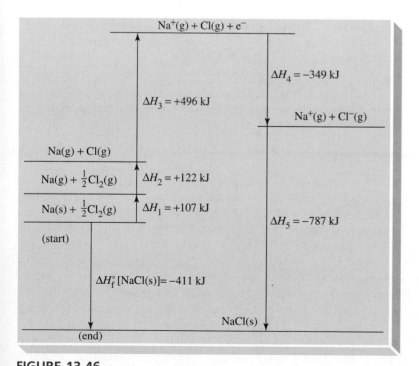

FIGURE 13-46
Enthalpy diagram for the formation of an ionic crystal.
Shown here is a five-step sequence for the formation of one mole of NaCl(s) from its elements in their standard states. The sum of the five enthalpy changes gives $\Delta H_f^\circ[NaCl(s)]$. The equivalent one-step reaction for the formation of NaCl(s) directly from Na(s) and $Cl_2(g)$ is shown in color. (ΔH values are not to scale.)

One way to use the concept of lattice energy is in making predictions about the possibility of synthesizing ionic compounds. In Example 13-11 we predict the likelihood of obtaining the compound MgCl(s) by evaluating its enthalpy of formation.

EXAMPLE 13-11

Relating Enthalpy of Formation, Lattice, Energy, and Other Energy Quantities. With the following data, calculate ΔH_f° per mol MgCl(s). (1) Enthalpy of sublimation of 1 mol Mg(s): $+146$ kJ; (2) enthalpy of dissociation of $\frac{1}{2}$ mol $Cl_2(g)$: $+122$ kJ; (3) first ionization energy of 1 mol Mg(g): $+738$ kJ; (4) electron affinity of 1 mol Cl(g): -349 kJ; (5) lattice energy of 1 mol MgCl(s): -676 kJ.

Solution

In this case the lattice energy (ΔH_5) is known and the unknown is ΔH_{net}, which is the enthalpy of formation of MgCl(s).

$$Mg(s) \longrightarrow Mg(g) \qquad\qquad \Delta H_1 = +146 \text{ kJ}$$

$$\tfrac{1}{2} Cl_2(g) \longrightarrow Cl(g) \qquad\qquad \Delta H_2 = +122 \text{ kJ}$$

$$Mg(g) \longrightarrow Mg^+(g) + e^- \qquad \Delta H_3 = +738 \text{ kJ}$$

$$Cl(g) + e^- \longrightarrow Cl^-(g) \qquad\qquad \Delta H_4 = -349 \text{ kJ}$$

$$Mg^+(g) + Cl^-(g) \longrightarrow MgCl(s) \qquad \Delta H_5 = -676 \text{ kJ}$$

$$\textit{net}: Mg(s) + \tfrac{1}{2} Cl_2(g) \longrightarrow MgCl(s)$$

$$\Delta H_{net} = \Delta H_f^\circ[MgCl(s)] = \Delta H_1 + \Delta H_2 + \Delta H_3 + \Delta H_4 + \Delta H_5$$

$$= 146 \text{ kJ} + 122 \text{ kJ} + 738 \text{ kJ} - 349 \text{ kJ} - 676 \text{ kJ} = -19 \text{ kJ}$$

Practice Example A: The enthalpy of sublimation of cesium is 78.2 kJ/mol, and $\Delta H_f^\circ[CsCl(s)] = -442.8$ kJ/mol. Use these values, together with other data from the text, to calculate the lattice energy of CsCl(s).

Practice Example B: Given the following data, together with data included in Example 13-11, calculate ΔH_f° per mol $CaCl_2$(s): Enthalpy of sublimation of Ca(s), $+178.2$ kJ/mol; first ionization energy of Ca(g), $+590$ kJ/mol; second ionization energy of Ca(g), $+1145$ kJ/mol; lattice energy of $CaCl_2$(s), -2255 kJ/mol.

Example 13-11 suggests that we can obtain MgCl(s) as a stable compound—it has a slightly negative enthalpy of formation. Why have we been writing $MgCl_2$ all this time instead of MgCl? You might think that because MgCl has a Mg-to-Cl ratio of 1:1 and $MgCl_2$ has a ratio of 1:2, MgCl should form if Mg(s) reacts with a limited amount of $Cl_2(g)$. But this is not the case. No matter how limited the amount of $Cl_2(g)$ available, the only compound that forms is $MgCl_2$. To understand this, repeat the calculation of Example 13-11 for the formation of $MgCl_2$(s), and you will obtain an enthalpy of formation that is very much more negative than that for MgCl(s) (see Exercise 77). Even though the energy requirement to produce Mg^{2+} is larger than to produce Mg^+, the lattice energy is very much greater for $MgCl_2$(s) than for MgCl(s). This is because the *doubly* charged Mg^{2+} ions exert a much stronger force on Cl^- ions than do *singly* charged Mg^+ ions. The reaction between Mg and Cl atoms does not stop at MgCl, but continues on to the more stable $MgCl_2$.

Is $NaCl_2$ a stable compound? Here the answer is no. The additional lattice energy associated with $NaCl_2$ over NaCl is not nearly enough to compensate for the very high second ionization energy of sodium (see Exercise 98).

Summary

Two liquid properties related to intermolecular forces are surface tension and viscosity. Familiar phenomena, such as drop shape, meniscus formation, and capillary action depend on surface tension. Vapor pressure—the pressure exerted by the vapor in equilibrium with a liquid—is a measure of the volatility of a liquid. It too is related to the strength of intermolecular forces. Properties of a solid affected by intermolecular forces are its sublimation (vapor) pressure and its melting point.

A phase diagram is a graphical plot of conditions under which solids, liquids, and gases (vapors) exist, as single phases or in equilibrium with one another. Especially significant points on a phase diagram are the triple point, melting point, boiling point, and critical point. The critical point is the condition of temperature and pressure at which a liquid and its vapor become indistinguishable.

The most common intermolecular forces of attraction are those between instantaneous and induced dipoles (dispersion forces). In polar substances there are also dipole–dipole forces. In some substances containing H and certain highly electronegative elements, hydrogen bonding is most significant. In network covalent solids, chemical bonds extend throughout a crystalline structure, as they do also in ionic crystals. For these substances the chemical bonds are themselves intermolecular forces.

Crystal structures can be described in terms of the close packing of spheres. With ionic crystals we must also note that the ions are not all of the same size or charge. An important idea for crystals of all types is the unit cell. Its dimensions can be used in calculating atomic radii and densities. Lattice energies of ionic crystals can be related to certain atomic and thermodynamic properties.

Summarizing Example

Enthalpies of fusion, vaporization, and sublimation can be combined with other thermochemical properties, such as specific heats, to make predictions about the final result when different phases of matter are brought together. This is an extension of calculations of the type introduced in Chapter 7.

To an insulated container with 100.0 g water at 20.0 °C are added 175 g steam at 100.0 °C and 1.65 kg ice at 0.0 °C. What mass of ice remains unmelted after equilibrium is established? (For water: $\Delta H_{fus} = 6.01$ kJ/mol, $\Delta H_{vap} = 40.7$ kJ/mol at 100 °C.)

1. *Determine the heat loss by 100.0 g water in cooling from 20.0 °C to 0.0 °C.* Because not all of the ice melts, the final temperature of the mixture must be 0.0 °C. Use the mass of liquid H_2O, its specific heat, and its temperature change. *Result:* $q = -8.36$ kJ.

2. *Determine the heat loss when 175 g steam condenses to liquid water at 100.0 °C and then cools to 0.0 °C.* This calculation can be done in two steps: Determine the heat loss on condensation and then the further loss as the liquid cools. *Result:* $q = (-395 - 73.2)$ kJ $= -468$ kJ.

3. *Determine the amount of ice that melts.* The heat available for melting ice is $(8.36 + 468)$ kJ $= 476$ kJ. *Result:* 79.2 mol H_2O melts.

4. *Establish the quantity of unmelted ice.* Convert the quantity of ice from part 3 to the mass of ice melted. Obtain the answer by difference. *Answer:* 0.22 kg ice.

Key Terms

adhesive force (13-1)
body-centered cubic (bcc) (13-8)
boiling (13-2)
cohesive force (13-1)
condensation (13-2)
critical point (13-2)
cubic closest packed (13-8)
dispersion (London) forces (13-5)
face-centered cubic (fcc) (13-8)

freezing (13-3)
freezing point (13-3)
hexagonal closest packed (13-8)
hydrogen bond (13-6)
lattice energy (13-7, 13-9)
melting (13-3)
melting point (13-3)
normal boiling point (13-2)
phase diagram (13-4)

polarizability (13-5)
sublimation (13-3)
surface tension (13-1)
triple point (13-4)
unit cell (13-8)
van der Waals forces (13-5)
vaporization (evaporation) (13-2)
vapor pressure (13-2)
vapor pressure curve (13-2)
viscosity (13-1)

Review Questions

1. In your own words define or explain the following terms or symbols: **(a)** ΔH_{vap}; **(b)** T_c; **(c)** instantaneous dipole; **(d)** coordination number; **(e)** unit cell.
2. Briefly describe each of the following phenomena or methods: **(a)** surface tension; **(b)** viscosity; **(c)** sublimation; **(d)** su-

percooling; **(e)** determining the freezing point of a liquid from a cooling curve.
3. Explain the important distinctions between each pair of terms: **(a)** adhesive and cohesive forces; **(b)** vaporization and condensation; **(c)** triple point and critical point; **(d)** face-

Focus On

Liquid Crystals

The flashlight provides enough radiant energy to raise slightly the temperature of this liquid crystalline material. As their temperature changes, the liquid crystals undergo a change in color. This effect can be used to make liquid crystal thermometers.

The acronym LCD needs little introduction. Most people know that this stands for liquid crystal display and that such displays are found in many places—calculators, watches and clocks, thermometers, ... After a study of this chapter, though, "liquid crystal" sounds like a contradiction. Liquids are one state of matter, and crystals are something we associate with another state of matter—solids. Liquid crystals, discovered as laboratory curiosities about 100 years ago, are forms of matter that are intermediate between the liquid and solid state. Liquid crystals have the fluid properties of liquids and the optical properties of solids.

Liquid crystals are observed most commonly in organic compounds that have cylindrically shaped (rod-like) molecules with masses of 200 to 500 u and lengths four to eight times their diameters. Potentially this represents about 0.5% of all organic compounds.

In the **nematic** (meaning thread-like) form of the liquid crystalline state, the rod-like molecules are arranged in a parallel fashion. They are free to move in all directions but they can rotate only on their long axes. (Imagine the ways in which you might move a particular pencil in a box of loosely packed pencils.) In the **smectic** (meaning grease-like) form, rod-like molecules are arranged in layers, with the long axes of the molecules perpendicular to the planes of the layers. The molecular motions possible here are a translation within but not between layers and a rotation about the long axis. The **cholesteric** form is related to the smectic form, but the orientation in each layer is different from that in the layer above and below. These three forms of liquid crystals are illustrated in Figure 13-47.

In a cholesteric structure each particular orientation is repeated over a span of several layers. The distance between planes with molecules in the same orientation is a distinctive feature of a cholesteric liquid crystal. Certain properties of light reflected by a liquid crystal depend on this characteristic distance. For example, because this distance is very temperature sensitive, the reflected light changes

---

centered and body-centered cubic unit cell; **(e)** tetrahedral and octahedral hole.
4. Three types of intermolecular forces considered in this chapter were instantaneous dipole/induced dipole, dipole–dipole, and hydrogen bonds. Describe the essential nature of each of these forces.
5. Based on the discussion of phase diagrams, must every substance have **(a)** a normal melting point; **(b)** a normal boiling point; **(c)** a critical point? Explain. (*Hint:* What does "normal" mean in these cases?)
6. Which of the following quantities (expressed in kilojoules per mole) would you expect to be largest for a substance: **(a)** heat capacity of the liquid; **(b)** enthalpy of fusion; **(c)** enthalpy of vaporization; **(d)** enthalpy of sublimation? Explain.
7. Which of the following factors affect the vapor pressure of a liquid? Explain.
(a) intermolecular forces in the liquid;

(b) volume of liquid in the liquid–vapor equilibrium;
(c) volume of vapor in the liquid–vapor equilibrium;
(d) the size of the container holding the liquid–vapor equilibrium mixture;
(e) temperature of the liquid.
8. A television commercial claims that a product makes water "wetter." Can there be any basis to this claim? Explain.
9. A double boiler is used when a careful control of temperature is required in cooking. Water is boiled in an outside container to produce steam and the steam condenses on the outside walls of an inner container in which cooking occurs. (A related laboratory device is called a steam bath.)
(a) How is heat energy conveyed to the food to be cooked?
(b) What is the maximum temperature that can be reached in the inside container?
10. At its normal boiling point the enthalpy of vaporization of chloroform, $CHCl_3$, is 247 J/g.

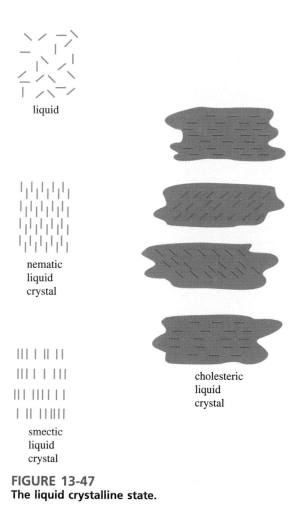

liquid

nematic
liquid
crystal

smectic
liquid
crystal

cholesteric
liquid
crystal

FIGURE 13-47
The liquid crystalline state.

color with changing temperature. This phenomenon is the basis of liquid crystal temperature-sensing devices, which can detect temperature changes as small as 0.01 °C.

The orientation of molecules in a thin film of nematic liquid crystals is easily altered by pressure and by an electric field. The altered orientation affects the optical properties of the film, such as causing the film to become opaque. Suppose electrodes are arranged in certain patterns (say, in the shape of numbers). When an electric field is imposed through these electrodes onto a thin film of liquid crystals, the patterns of the electrodes become visible. This principle is used in liquid crystal display devices.

Liquid crystals occur widely in living matter. Cell membranes and certain tissues have structures that can be described as liquid crystalline. Hardening of the arteries is caused by the deposition of liquid crystalline compounds of cholesterol. Liquid crystalline properties have also been identified in various synthetic polymers, such as DuPont Kevlar fiber.

(a) How many grams of $CHCl_3$ can be vaporized with 6.62 kJ of heat?

(b) What is ΔH_{vap} of $CHCl_3$, expressed in *kilojoules per mole*?

(c) How much heat, in kilojoules, is *evolved* when 19.6 g $CHCl_3(g)$ condenses?

11. From Figure 13-8, estimate (a) the vapor pressure of C_6H_7N at 100 °C; (b) the normal boiling point of C_7H_8.

12. Use data in Figure 13-13 to estimate (a) the normal boiling point of aniline; (b) the vapor pressure of diethyl ether at 25 °C.

13. Equilibrium is established between $Br_2(l)$ and $Br_2(g)$ at 25.0 °C. A 250.0-mL sample of the vapor weighs 0.486 g. What is the vapor pressure of bromine at 25.0 °C, in millimeters of mercury.

14. Cyclohexanol has a vapor pressure of 10.0 mmHg at 56.0 °C and 100.0 mmHg at 103.7 °C. Calculate its enthalpy of vaporization, ΔH_{vap}.

15. The vapor pressure of methyl alcohol is 40.0 mmHg at

5.0 °C. Estimate its normal boiling point. $\Delta H_{vap} = 38.0$ kJ/mol.

16. How much heat, in kilojoules, is required to melt a cube of ice that measures 5.08 cm on an edge? The density of ice is 0.92 g/cm^3 and the enthalpy of fusion is 6.01 kJ/mol.

17. At its normal melting point, ΔH_{fus} of Cu is 13.05 kJ/mol.

(a) How much heat, in kilojoules, is *evolved* when a 3.78-kg sample of molten Cu freezes?

(b) How much heat, in kilojoules, must be absorbed to melt a bar of copper that is 75 cm × 15 cm × 12 cm? (Assume $d = 8.92$ g/cm^3 for Cu.)

18. A 80.0-g piece of dry ice, $CO_2(s)$, is placed in a 0.500-L container and the container is sealed. If this container is held at 25 °C, what state(s) of matter must be present? (*Hint:* Refer to Table 13.3 and Figure 13-18.)

19. In each of the following pairs, which would you expect to have the higher boiling point? (a) C_7H_{16} or $C_{10}H_{22}$; (b) C_3H_8 or $(CH_3)_2O$; (c) CH_3CH_2SH or CH_3CH_2OH.

20. One of the substances is out of order in the following list based on *increasing* boiling point. Identify it and put it in its proper place: N_2, O_3, F_2, Ar, Cl_2. Explain your reasoning.

21. Arrange the following substances in the expected order of increasing melting point: KI, Ne, K_2SO_4, C_3H_8, CH_3CH_2OH, MgO, $CH_2OHCHOHCH_2OH$.

22. Hydrazine, $H_2N—NH_2$, has a normal boiling point of 113.5 °C. Do you think hydrogen bonding is an important intermolecular force in hydrazine? Explain.

23. Place each of the following substances in the appropriate category in Table 13.7 and state your reason for each placement. **(a)** Si; **(b)** $SiCl_4$; **(c)** $CaCl_2$; **(d)** Ag; **(e)** SO_2.

24. Only one of the following statements is true about the unit cell of an ionic crystal. Which is the correct statement, and what is wrong with the others? The unit cell **(a)** is the same as the formula unit; **(b)** is any portion of the ionic crystal with a cubic shape; **(c)** shares some of its ions with other unit cells; **(d)** always contains the same number of cations and anions.

25. In the manner illustrated in the text for NaCl, show that the formula of CsCl is consistent with the unit cell in Figure 13-44.

26. The fcc unit cell is a cube with atoms at each of the corners and in the center of each face, as shown here. Copper has the fcc crystal structure. Assume an atomic radius of 128 pm for a Cu atom.
(a) What is the length of the unit cell of Cu?
(b) What is the volume of the unit cell?
(c) How many atoms belong to the unit cell?
(d) What is the mass of a unit cell?
(e) Calculate the density of copper.

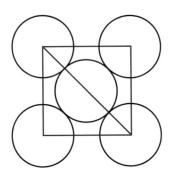

Exercises

Surface Tension; Viscosity

27. Silicone oils are used in water repellents for treating tents, hiking boots, and similar items. Explain how they function.

28. Surface tension, viscosity, and vapor pressure are all related in some way to intermolecular forces. Why do surface tension and viscosity decrease with temperature whereas vapor pressure increases with temperature?

Vaporization

29. When a liquid evaporated from an open container, the liquid temperature was observed to remain roughly constant. When the same liquid evaporated from a thermally insulated container (a vacuum bottle, or Dewar flask), its temperature was observed to drop. How would you account for this difference?

30. Estimate the heat absorbed when 1.00 g of Instant Car Kooler™ vaporizes. Comment on the effectiveness of this spray in cooling the interior of a car. Assume the spray is 10% $C_2H_5OH(aq)$ by mass, the temperature is 55 °C, the heat capacity of air is 29 J mol^{-1} K^{-1}, and use the ΔH_{vap} data from Table 13.1.

31. The enthalpy of vaporization of benzene, $C_6H_6(l)$, is 33.9 kJ/mol at 298 K. How many liters of $C_6H_6(g)$, measured at 298 K and 95.1 mmHg, are formed when 1.54 kJ of heat is absorbed by $C_6H_6(l)$ at a constant temperature of 298 K?

32. A vapor volume of 1.17 L forms when a sample of liquid acetonitrile, CH_3CN, absorbs 1.00 kJ of heat at its normal boiling point (81.6 °C and 1 atm). What is ΔH_{vap}, in kilojoules per mole of CH_3CN?

33. How many liters of $CH_4(g)$, measured at 23.4 °C and 768 mmHg, must be burned to provide the heat needed to vaporize 3.78 L of water at 100 °C? $\Delta H_{combustion} = -8.90 \times 10^2$ kJ/mol CH_4. For $H_2O(l)$ at 100 °C, $d = 0.958$ g/cm³ and $\Delta H_{vap} = 40.7$ kJ/mol.

34. A 50.0-g piece of iron at 152 °C is dropped into 20.0 g $H_2O(l)$ in a thermally insulated container at 89 °C. How much water would you expect to vaporize? The specific heats of iron and water are 0.45 and 4.21 J g^{-1} $°C^{-1}$, respectively, and $\Delta H_{vap} = 40.7$ kJ/mol H_2O. (*Hint:* What is the final temperature?)

Vapor Pressure and Boiling Point

35. Pressure cookers achieve a high cooking temperature to speed the cooking process by heating a small amount of water under a constant pressure. If the pressure is set at 2 atm, what is the boiling point of the water? Use information from Table 13.2.

36. Use data from Table 13.2 to estimate **(a)** the boiling point of water in Santa Fe, New Mexico, if the prevailing atmospheric pressure is 640 mmHg; **(b)** the prevailing atmospheric pressure at Lake Arrowhead, California, if the observed boiling point of water is 94 °C.

37. A 25.0-L volume of He(g) at 30.0 °C is passed through 6.220 g of liquid aniline ($C_6H_5NH_2$) at 30.0 °C. The liquid remaining after the experiment weighs 6.108 g. Assume that the He(g) becomes saturated with aniline vapor and that the total gas volume and temperature remain constant. What is the vapor pressure of aniline at 30.0 °C?

38. A 7.53-L sample of N_2(g) at 742 mmHg and 45.0 °C is bubbled through CCl_4(l) at 45.0 °C. Assuming the gas becomes saturated with CCl_4(g), what is the volume of the resulting gaseous mixture, if the total pressure remains at 742 mmHg and the temperature remains at 45 °C? The vapor pressure of CCl_4 at 45 °C is 261 mmHg.

39. Some vapor pressure data for Freon-12, CCl_2F_2, once a common refrigerant, are -12.2 °C, 2.0 atm; 16.1 °C, 5.0 atm; 42.4 °C, 10.0 atm; 74.0 °C, 20.0 atm. Also, bp $= -29.8$ °C, $T_c = 111.5$ °C, $P_c = 39.6$ atm. Use these data to plot the vapor pressure curve of Freon-12. What approximate pressure would be required in the compressor of a refrigeration system to convert Freon-12 vapor to liquid at 25 °C?

40. A 10.0-g sample of liquid water is sealed in a 1515-mL flask and allowed to come to equilibrium with its vapor at 27 °C. What is the mass of H_2O(g) present when equilibrium is established? (*Hint:* Use vapor pressure data from Table 13.2.)

The Clausius–Clapeyron Equation

41. The normal boiling point of acetone, an important laboratory and industrial solvent, is 56.2 °C and its ΔH_{vap} is 25.5 kJ/mol. At what temperature does acetone have a vapor pressure of 375 mmHg?

42. Hydrazine, N_2H_4, has a normal boiling point of 113.5 °C and a critical point at 380 °C and 145.4 atm. Estimate its vapor pressure at 100.0 °C. (*Hint:* First determine ΔH_{vap}.)

Critical Point

43. Which substances listed in Table 13.3 can exist as liquids at room temperature (about 20 °C)? Explain.

44. Can SO_2 be maintained as a liquid under a pressure of 100 atm at 0 °C? Can liquid methane be obtained under the same conditions? (Refer to Table 13.3.)

States of Matter and Phase Diagrams

45. Describe what happens to the following samples in a device like that pictured in Figure 13-21. Be as specific as you can about the temperatures and pressures at which changes occur.
(a) A sample of H_2O is heated from -20 to 200 °C at a constant pressure of 600 mmHg.
(b) The pressure on a sample of I_2 is increased from 90 mmHg to 100 atm at a constant temperature of 114.5 °C.
(c) A sample of CO_2 at 35 °C is cooled to -100 °C at a constant pressure of 50 atm. (*Hint:* Refer also to Table 13.3.)

46. In the portion of the phase diagram for phosphorus shown here
(a) Indicate the phases present in the regions labeled (?).
(b) A sample of solid red phosphorus cannot be melted by heating in a container open to the atmosphere. Explain why this is so.
(c) Trace the phase changes that occur when the pressure on a sample is reduced from point A to B, at constant temperature.

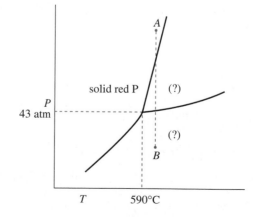

47. A 0.240-g sample of H_2O(l) is sealed into an evacuated 3.20-L flask. What is the pressure of the vapor in the flask if the temperature is **(a)** 30.0 °C; **(b)** 50.0 °C; **(c)** 70.0 °C? (*Hint:* Use data from Table 13.2. Does any of the water remain as liquid or does it vaporize completely?)

48. A 2.50-g sample of H_2O(l) is sealed in a 5.00-L flask at 120.0 °C.
(a) Show that the sample exists completely as vapor.
(b) Estimate the temperature to which the flask must be cooled before liquid water condenses.

49. Is it likely that any of the following will occur naturally at or near Earth's surface anywhere on Earth? Explain. **(a)** $CO_2(s)$; **(b)** $CH_4(l)$; **(c)** $SO_2(g)$; **(d)** $I_2(l)$; **(e)** $O_2(l)$. (*Hint:* Use the appropriate phase diagrams and data from Table 13.3.)

51. Refer to the Summarizing Example. What *additional* mass of steam should be introduced into the insulated container to just melt all of the ice?

53. You decide to cool a can of soda pop quickly in the freezer compartment of a refrigerator. When you take out the can, the soda pop is still liquid; but when you open the can, the soda pop immediately freezes. Explain why this happens.

Van der Waals Forces

55. For each of the following substances describe the importance of dispersion (London) forces, dipole–dipole interactions, and hydrogen bonding. **(a)** HCl; **(b)** Br_2; **(c)** ICl; **(d)** HF; **(e)** CH_4.

Hydrogen Bonding

57. One of the following substances is a liquid at room temperature, whereas the others are gaseous. Which do you think is the liquid? Explain. CH_3OH; C_3H_8; N_2; N_2O.

50. Trace the phase changes that occur as a sample of $H_2O(g)$, originally at 1.00 mmHg and $-0.10\ °C$, is compressed at constant temperature until the pressure reaches 100 atm. (*Hint:* See Figure 13-19.)

52. A 54-cm^3 ice cube at at $-25.0\ °C$ is added to a thermally insulated container with 400.0 mL $H_2O(l)$ at 32.0 °C. What will be the final temperature and what state(s) of matter will be present? Specific heats: $H_2O(s)$, 2.01 J $g^{-1}\ °C^{-1}$; $H_2O(l)$, 4.18 J $g^{-1}\ °C^{-1}$. Densities: $H_2O(s)$, 0.917 g/cm^3; $H_2O(l)$, 0.998 g/cm^3. Also, ΔH_{fus} of ice = 6.01 kJ/mol.

54. Why is the triple point of water (ice–liquid–vapor) a better fixed point for establishing a thermometric scale than either the melting point of ice or the boiling point of water?

56. When another atom or group of atoms is substituted for one of the hydrogen atoms in benzene, C_6H_6, the boiling point changes. Explain the order of the following boiling points: C_6H_6, 80 °C; C_6H_5Cl, 132 °C; C_6H_5Br, 156 °C; C_6H_5OH, 182 °C.

58. In some cases a hydrogen bond can form *within* a single molecule. This is called an *intra*molecular hydrogen bond. Do you think this type of bonding is an important factor in **(a)** C_2H_6; **(b)** CH_3CH_2OH; **(c)** CH_3COOH; **(d)** *ortho*-phthalic acid?

ortho-phthalic acid

Network Covalent Solids

59. Based on data presented in the text, would you expect diamond or graphite to have the greater density? Explain.

61. Silicon carbide, SiC, crystallizes in a form similar to diamond, whereas boron nitride, BN, crystallizes in a form similar to graphite.
(a) Sketch the SiC structure as in Figure 13-31.
(b) Propose a bonding scheme for BN.

Ionic Bonding and Properties

63. The melting points of NaF, NaCl, NaBr, and NaI are 988, 801, 755, and 651 °C, respectively. Are these data consistent with ideas developed in Section 13-7? Explain.

60. Diamond is often used as a cutting medium in glass cutters. What property of diamond makes this possible? Could graphite function as well?

62. Buckminsterfullerene (see page 380) is an allotrope of carbon. It is not a network covalent solid. What makes it different from diamond and graphite? It has been shown recently that carbon can form chains in which every other carbon atom is bonded to the next carbon atom by a triple bond. Is this newest allotrope of carbon a network covalent solid? Explain.

64. Use Coulomb's law (see Appendix B) to verify the conclusion concerning the relative strengths of the attractive forces in the ion pairs Na^+Cl^- and $Mg^{2+}O^{2-}$ presented in Figure 13-33.

Crystal Structures

65. Explain why there are *two* arrangements for the closest packing of spheres rather than a single one.

66. Argon, copper, sodium chloride, and carbon dioxide all crystallize in the fcc structure. How can this be when their physical properties are so different?

67. For the two-dimensional lattice shown here

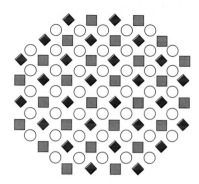

(a) Identify a unit cell.
(b) How many of each of the following elements are in the unit cell: ■, ■, and ○?
(c) Indicate some simpler units than the unit cell, but explain why they cannot function as a unit cell.

69. Tungsten has a body-centered cubic crystal structure. Using a metallic radius of 139 pm for the W atom, calculate the density of tungsten.

Ionic Crystal Structures

71. Show that the unit cells for CaF_2 and TiO_2 in Figure 13-45 are consistent with their formulas.

73. The crystal structure of magnesium oxide, MgO, is of the NaCl type (Figure 13-43). Use this fact, together with ionic radii from Figure 10-8, to establish
(a) the coordination numbers of Mg^{2+} and O^{2-};
(b) the number of formula units in the unit cell;
(c) the length and volume of a unit cell;
(d) the density of MgO.

Lattice Energy

75. *Without doing calculations,* indicate how you would expect the lattice energies of LiCl(s), KCl(s), RbCl(s), and CsCl(s) to compare with the value of -787 kJ/mol determined for NaCl(s) on page 455. (*Hint:* Assume that the enthalpies of sublimation of the alkali metals are comparable in value. What atomic properties from Chapter 10 should you compare?)
77. Refer to Example 13-11. Together with data given there, use the data below to calculate ΔH_f° for 1 mol $MgCl_2$(s). Explain why you would expect $MgCl_2$ to be much more stable a compound than MgCl. Second ionization energy of Mg, I_2 = 1451 kJ/mol; Lattice energy of $MgCl_2$(s) = -2526 kJ/mol $MgCl_2$. (*Hint:* Recall that I_2 is the energy requirement for the process $Mg^+ \rightarrow Mg^{2+} + e^-$.)

68. As we saw in Section 13-8, the stacking of spheres always leaves open space or voids. Consider the corresponding situation in two dimensions: Squares can be arranged to cover all the area but circles cannot. For the arrangement of circles pictured here, what percentage of the area remains uncovered?

70. Magnesium crystallizes in the hcp arrangement shown in Figure 13-38. The dimensions of the unit cell are height, 520 pm; length on an edge, 320 pm. Calculate the density of Mg(s) and compare with the measured value of 1.738 g/cm³.

72. Using methods similar to Examples 13-9 and 13-10, calculate the density of CsCl. (*Hint:* What is the crystal structure of CsCl? Use 169 pm as the radius of Cs^+.)
74. Potassium chloride has the same crystal structure as NaCl. Careful measurement of the internuclear distance between K^+ and Cl^- ions gave a value of 314.54 pm. The density of KCl is 1.9893 g/cm³. Use these data to evaluate the Avogadro constant, N_A. (*Hint:* Recall Practice Example 13-9B.)

76. Determine the lattice energy of KF(s) from the following data: ΔH_f°[KF(s)] = -567.3 kJ/mol; enthalpy of sublimation of K(s), 89.24 kJ/mol; enthalpy of dissociation of F_2(g), 159 kJ/mol F_2; I_1 for K(g), 418.9 kJ/mol; EA for F(g), -328 kJ/mol.

78. In ionic compounds with certain metals, hydrogen exists as hydride ion, H^-. Determine the electron affinity of hydrogen [i.e., ΔH for the process: $H(g) + e^- \rightarrow H^-(g)$]. To do so, use (1) data from Section 13-9; (2) the bond energy of H_2(g) from Table 11.3; (3) -812 kJ/mol NaH for the lattice energy of NaH(s); and (4) -57 kJ/mol NaH for the enthalpy of formation of NaH(s).

Advanced Exercises

79. Explain why vaporization occurs only at the surface of a liquid until the boiling point temperature is reached. That is, why does not vapor form throughout the liquid at all temperatures?

80. When a wax candle is burned, the fuel consists of *gaseous* hydrocarbons appearing at the end of the candle wick. Describe the phase changes and processes by which the solid wax is ultimately consumed.

81. A dramatic lecture demonstration consists of continuously evacuating water vapor from an open container of liquid water in a bell jar. If the vacuum pump is sufficiently powerful, the water can be made to freeze. Describe the principles involved in this demonstration.

82. The normal boiling point of isooctane (a gasoline component with a high octane rating) is 99.2 °C and its ΔH_{vap} is 35.76 kJ/mol. Because isooctane and water have nearly identical boiling points, will they have nearly equal vapor pressures at room temperature? If not, which would you expect to be more volatile? Explain.

83. A supplier of cylinder gases warns customers to determine how much gas remains in a cylinder by weighing the cylinder and comparing this mass to the original mass of the full cylinder. In particular, the customer is told not to try to estimate the mass of gas available from the measured gas pressure. Explain the basis of this warning. (*Hint:* There are cases where a measurement of the gas pressure *can* be used as a measure of the remaining available gas. What are they?)

84. Use data from Appendix D to determine the quantity of heat needed to vaporize 15.0 mL of liquid mercury at 25 °C. The density of Hg(l) = 13.6 g/mL.

85. To vaporize 1.000 g water at 20 °C requires 2447 J of heat. At 100 °C, 10.00 kJ of heat will convert 4.430 g $H_2O(l)$ to $H_2O(g)$. Does ΔH_{vap} of water increase or decrease with temperature? Does this observation conform to what you would expect in terms of intermolecular forces? Explain.

86. What is the difference in the boiling point temperature of water at normal atmospheric pressure (1 atm) and at a pressure of 1 bar?

87. One handbook lists the sublimation pressure of *solid* benzene as a function of *Kelvin* temperature, T, as log P (mmHg) = 9.846 − 2309/T. Another handbook lists the vapor pressure of *liquid* benzene as a function of *Celsius* temperature, t, as log P (mmHg) = 6.90565 − 1211.033/(220.790 + t). Use these equations to estimate the normal melting point of benzene, and compare your result with the listed value of 5.5 °C.

88. By the method used to graph Figure 13-13, plot ln P vs. 1/T for liquid yellow phosphorus, and estimate **(a)** its normal boiling point and **(b)** its enthalpy of vaporization, ΔH_{vap}, in kJ/mol. Vapor-pressure data: 76.6 °C, 1 mmHg; 128.0 °C, 10; 166.7 °C, 40; 197.3 °C, 100; 251.0 °C, 400.

89. Assume that a skater has a mass of 80 kg and that his skates make contact with 2.5 cm² of ice.
(a) Calculate the pressure in atm exerted by the skates on the ice. (*Hint:* Review the discussion of pressure in Section 6-1.)
(b) If the melting point of ice decreases by 1.0 °C for every 125 atm of pressure, what would be the melting point of the ice under the skates?

90. Estimate the boiling point of water in Leadville, Colorado, elevation 3170 m. To do this, use the barometric formula relating pressure and altitude: $P = P_0 \times 10^{-Mgh/2.303RT}$ (where P = pressure in atm; P_0 = 1 atm; g = acceleration due to gravity; M = molar mass of air = 0.02896 kg/mol; R = 8.3145 J mol⁻¹ K⁻¹; and T is the Kelvin temperature). Assume the air temperature is 10.0 °C and that ΔH_{vap} = 41 kJ/mol H_2O.

91. A cylinder containing 151 lb Cl_2 has an inside diameter of 10 in. and a height of 45 in. The gas pressure is 100 psi (1

atm = 14.7 psi) at 20 °C. Cl_2 melts at −103 °C, boils at −35 °C, and has its critical point at 144 °C and 76 atm. In what state(s) of matter does the Cl_2 exist in the cylinder?

92. In acetic acid vapor some molecules exist as monomers and some as dimers (see Figure 13-30). If the density of the vapor at 350 K and 1 atm is 3.23 g/L, what percent of the molecules must exist as dimers? Would you expect this percent to increase or decrease with temperature?

93. A 685-mL sample of Hg(l) at 20 °C is added to a large quantity of liquid N_2 kept at its boiling point in a thermally insulated container. What mass of $N_2(l)$ is vaporized as the Hg is brought to the temperature of the liquid N_2? For the specific heat of Hg(l) from 20 to −39 °C use 0.138 J g⁻¹ °C⁻¹, and for Hg(s) from −39 to −196 °C, 0.126 J g⁻¹ °C⁻¹. The density of Hg(l) is 13.6 g/mL, its melting point is −39 °C, and its enthalpy of fusion is 2.30 kJ/mol. The boiling point of $N_2(l)$ is −196 °C, and its ΔH_{vap} is 5.58 kJ/mol.

94. The crystal structure of zinc sulfide (ZnS), is pictured here. The length of the unit cell is 6.00×10^2 pm. For this structure, determine
(a) the coordination numbers of Zn^{2+} and S^{2-};
(b) the number of formula units in the unit cell;
(c) the density of ZnS.

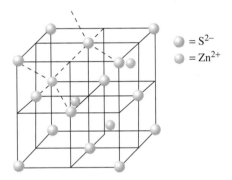

$\bigcirc = S^{2-}$
$\bigcirc = Zn^{2+}$

95. Refer to Figures 13-41 and 13-43. Suppose that the two planes of ions pictured in Figure 13-41 correspond to the top and middle planes of ions in the NaCl unit cell in Figure 13–43. If the X-rays used have a wavelength of 154.1 pm, at what angle θ would the diffracted beam have its greatest intensity? [*Hint:* Use $n = 1$ in Equation (13.4).]

96. Use the analysis of a bcc structure in Example 13-8 and the fcc structure in Exercise 26 to prove that the % voids in their respective packing-of-sphere arrangements are 31.98% for bcc and 25.96% for fcc. (*Hint:* What is the volume of the unit cell? How many atoms are present in the cell? What is their volume? Use the general case of the packing of spheres without reference to a particular metal.)

97. The enthalpy of formation of NaI(s) is −288 kJ/mol. Use this value, together with other data in the text, to calculate the lattice energy of NaI(s). [*Hint:* An extra step is required for the sublimation of iodine: $I_2(s) \rightarrow I_2(g)$. Use data from Appendix D.]

98. Show that the formation of $NaCl_2(s)$ is very unfavorable, that is, $\Delta H_f^\circ[NaCl_2(s)]$ is a large *positive* quantity. To do this, use data from Section 13-9 and assume that the lattice energy for $NaCl_2$ would be about the same as that of $MgCl_2$, -2.5×10^3 kJ/mol.

99. We cannot measure the second electron affinity of oxygen directly.

$$O^-(g) + e^- \longrightarrow O^{2-}(g) \qquad EA_2 = ?$$

The O^{2-} ion can exist in the solid state, however, where the high energy requirement for its formation is offset by the large lattice energies of ionic oxides.

(a) Show that EA_2 can be calculated from the enthalpy of formation and lattice energy of $MgO(s)$, enthalpy of sublimation of $Mg(s)$, ionization energies of Mg, bond energy of O_2, and EA_1 for $O(g)$.
(b) The lattice energy of MgO is -3925 kJ/mol. Combine this with other values in the text to estimate EA_2 for oxygen.

Feature Problems

A. In a capillary rise experiment, the height (h) to which a liquid rises depends on the density (d) and surface tension (γ) of the liquid and the radius of the capillary (r). The equation relating these quantities and the acceleration due to gravity (g) is $h = 2\gamma/dgr$. The sketch provides data obtained with ethanol. What is the surface tension of ethanol?

$$\rightarrow| \; |\leftarrow r = 0.50 \text{ mm}$$

$$h = 1.1 \text{ cm}$$

$$d = 0.789 \text{ g/mL}$$

B. One popular demonstration in chemistry labs is performed by boiling a small quantity of water in a metal can (such as a used soda can), picking up the can with tongs and quickly submerging it upside-down in cold water. The can collapses with a loud and satisfying pop. Give an explanation of this crushing of the can. (Note: If you try this demonstration, do not heat the can over an open flame.)

C. Use appropriate values of properties of hydrazine, N_2H_4, listed in the table to answer the following questions:

Property	Value
Freezing point	2.0 °C
Boiling point	113.5 °C
Critical temperature	380 °C
Critical pressure	145.4 atm
Enthalpy of fusion	12.66 kJ/mol
Specific heat of liquid	3.1 J mol^{-1} °C^{-1}
Density of liquid at 25 °C	1.0045 g/mL
Vapor pressure at 25 °C	14.4 mmHg

1. Sketch a phase diagram for hydrazine, and label significant points on the diagram.
2. Determine the ΔH_{vap} of hydrazine.
3. Determine the ΔH_{sub} of hydrazine at its melting point.
4. A sample of hydrazine is maintained in a closed container in an ice-water bath. What is the pressure of hydrazine vapor

in this container? (*Hint:* What is the temperature? In what state(s) of matter does hydrazine exist at this temperature?)

D. The sketch is a portion of the phase diagram for the element sulfur. The transition between solid orthorhombic (S_α) and solid monoclinic (S_β) sulfur, in the presence of sulfur vapor, is at 95.3 °C. The triple point involving monoclinic sulfur, liquid sulfur, and sulfur vapor is at 119 °C.

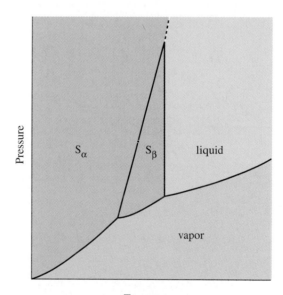

Temperature

1. How would you modify the phase diagram to represent the melting of orthorhombic sulfur that is sometimes observed at 113 °C? (*Hint:* What would the phase diagram look like if the monoclinic sulfur did not form?)
2. Account for the observation that if a sample of rhombic sulfur is melted at 113 °C and then heated, the liquid sulfur freezes at 119 °C upon cooling.

E. In the text (see page 442) we indicated that the hydrogen bonding in water makes life possible. This is a dramatic statement that bears investigating. If water were a typical liquid—that is, one without hydrogen bonding between molecules—what would you expect to find for its **(a)** boiling point; **(b)** freezing point; **(c)** temperature of maximum density of the liquid; **(d)** relative densities of the solid and liquid states? What likely impact would these altered physical properties have on life on Earth as we know it?

Solutions and Their Physical Properties

14

The dissolving of a cube of sugar (sucrose) is seen here as swirls of a higher density sucrose solution falling through the lower density water.

*R*esidents of cold climate regions know they must add antifreeze to the water in the cooling system of an automobile in the winter. The antifreeze–water mixture has a much lower freezing point than does pure water. In this chapter we will learn why.

To restore body fluids to a dehydrated individual by intravenous injection, pure water cannot be used. A solution with just the right value of a physical property known as osmotic pressure is necessary, and this requires a solution of a particular concentration. Again, in this chapter we will learn why.

Altogether, we will explore several solution properties whose values depend on solution concentration. Our emphasis will be on describing solution phenomena and their applications and explaining these phenomena in terms of what is happening at the molecular level.

TABLE 14.1 Some Common Solutions	
Solution	**Components**
Gaseous solutions	
air	$N_2 + O_2$ + several others
natural gas	$CH_4 + C_2H_6$ + several others
Liquid solutions	
seawater	$H_2O + NaCl$ + many others
vinegar	$H_2O + HC_2H_3O_2$ (acetic acid)
soda pop	$H_2O + CO_2 + C_{12}H_{22}O_{11}$ (sucrose) + several others
Solid solutions	
yellow brass	$Cu + Zn$
palladium/hydrogen	$Pd + H_2$

14-1 Types of Solutions: Some Terminology

In Chapters 1 and 4 we learned that a solution is a *homogeneous mixture.* It is *homogeneous* because its composition and properties are uniform, and it is a *mixture* because it contains two or more substances in proportions that can be varied. The **solvent** is the component that is present in the greatest quantity or that determines the state of matter in which a solution exists. A **solute** is a solution component said to be dissolved in the solvent. A *concentrated* solution has a relatively large quantity of dissolved solute(s), and a *dilute* solution has only a small quantity. Consider solutions containing sucrose (cane sugar) as one of their solutes in the solvent water: Pancake syrup is a concentrated solution, whereas a sweetened cup of coffee is much more dilute.

Although liquid solutions are most common, solutions can exist in the gaseous and solid states as well. For instance, the U.S. five-cent nickel is a solid solution of 75% Cu and 25% Ni. Solid solutions with a metal as the solvent are also called **alloys.*** Table 14.1 lists a few common solutions.

14-2 Solution Concentration

In Chapters 4 and 5 we learned that to describe a solution fully we must know its *concentration,* a measure of the quantity of solute present in a given quantity of solvent (or solution). The unit we stressed there was molarity. In this section we describe several methods of expressing concentration, each of which serves a different purpose.

Mass Percent, Volume Percent, and Mass/Volume Percent

If we dissolve 5.00 g NaCl in 95.0 g H_2O, we get 100.0 g of a solution that is 5.00% NaCl, by *mass.* Mass percent is widely used in industrial chemical applications. Thus, we might read that the action of 78% H_2SO_4(aq) on phosphate rock [$3Ca_3(PO_4)_2 \cdot CaF_2$] produces 46% H_3PO_4(aq).

Because liquid volumes are so easily measured, some solutions are prepared on a *volume* percent basis. For example, a handbook lists a freezing point of

---

*The term *alloy* can also apply to certain heterogeneous mixtures, such as the common two-phase solid mixture of lead and tin known as *solder,* or to intermetallic compounds, such as the silver–tin compound Ag_3Sn that is mixed with mercury in *dental amalgam.*

$-15.6\,^{\circ}$C for a methyl alcohol–water antifreeze solution that is 25.0% CH_3OH, by volume. This solution is prepared by dissolving 25.0 mL $CH_3OH(l)$ for every 100.0 mL of aqueous solution.

Another possibility is to express the *mass* of solute and *volume* of solution. An aqueous solution with 0.9 g NaCl in 100.0 mL of solution is said to be 0.9% NaCl (*mass/vol*). This unit is extensively used in medicine and pharmacy.

Mole Fraction and Mole Percent

To relate certain physical properties (such as vapor pressure) to solution concentration we need a concentration unit in which all solution components are expressed on a mole basis. This we can do with the mole fraction. The **mole fraction** of component i, designated χ_i, is the fraction of all the molecules in a solution that are of type i. The mole fraction of component j is χ_j, and so on. The mole fraction of a solution component is defined as

$$\chi_i = \frac{\text{amount of component } i \text{ (in moles)}}{\text{total amount of all solution components (in moles)}}$$

and

the sum of the mole fractions of all the solution components is 1.

$$\chi_i + \chi_j + \chi_k + \ldots = 1$$

The **mole percent** of a solution component is the percent of all the molecules in solution that are of a given type. Mole percents are mole fractions multiplied by 100%.

Molarity

In Chapters 4 and 5 we introduced molarity to provide a conversion factor relating the amount of solute and the volume of solution. We used it in various stoichiometric calculations. As we learned at that time,

$$\text{molarity (M)} = \frac{\text{amount of solute (in moles)}}{\text{volume of solution (in liters)}}$$

Molality

Suppose we prepare a solution at 20 °C by using a volumetric flask calibrated at 20 °C. Then suppose we warm this solution to 25 °C. As the temperature increases from 20 to 25 °C, the amount of solute remains *constant,* but the solution volume *increases* slightly (by about 0.1%). The number of moles of solute per liter—the molarity—*decreases* slightly (by about 0.1%).

The temperature dependence of molarity can be a problem in experiments demanding a high precision. If the solution is used at a different temperature from that at which it was prepared, its molarity is not exactly the one written on the label. A concentration unit that is *independent* of temperature and also proportional to mole fraction in dilute solutions is **molality,** the amount of solute (in moles) per kilogram of *solvent* (not of solution). A solution in which 1.00 mol of urea, $CO(NH_2)_2$, is dissolved in 1.00 kg of water is described as a 1.00 molal solution and designated as 1.00 m $CO(NH_2)_2$. Molality is defined as

$$\text{molality } (m) = \frac{\text{amount of solute (in moles)}}{\text{mass of solvent (in kilograms)}}$$

—10.00 mL C₂H₅OH

←—100.0 mL—→

water

ethanol–water
solution:
$d = 0.982$ g/mL

FIGURE 14-1
**Preparation of an
ethanol–water solution—
Example 14-1 illustrated.**
A 10.00-mL sample of C_2H_5OH is
added to some water in the volumetric flask. The solution is mixed
and more water added to bring
the total volume to 100.0 mL.

Illustrative Examples

In Example 14-1 the concentration of a solution is expressed in several different ways. The calculation in Example 14-2 is perhaps more typical: A concentration is converted from one unit (molarity) to another (mole fraction).

EXAMPLE 14-1

Expressing a Solution Concentration in Various Units. An ethanol–water solution
is prepared by dissolving 10.00 mL of ethanol, C_2H_5OH ($d = 0.789$ g/mL), in a
sufficient volume of water to produce 100.0 mL of a solution with a density of
0.982 g/mL (Figure 14-1). What is the concentration of ethanol in this solution, expressed as **(a)** volume percent; **(b)** mass percent; **(c)** mass/volume percent; **(d)** mole
fraction; **(e)** mole percent; **(f)** molarity; **(g)** molality?

Solution

a. Volume percent ethanol

$$\text{volume percent ethanol} = \frac{10.00 \text{ mL ethanol}}{100.0 \text{ mL soln}} \times 100\% = 10.00\%$$

b. Mass percent ethanol

$$\text{mass ethanol} = 10.00 \text{ mL ethanol} \times \frac{0.789 \text{ g ethanol}}{1 \text{ mL ethanol}}$$

$$= 7.89 \text{ g ethanol}$$

$$\text{mass soln} = 100.0 \text{ mL soln} \times \frac{0.982 \text{ g soln}}{1 \text{ mL soln}} = 98.2 \text{ g soln}$$

$$\text{mass percent ethanol} = \frac{7.89 \text{ g ethanol}}{98.2 \text{ g soln}} \times 100\% = 8.03\%$$

c. Mass/volume percent ethanol

$$\text{mass/volume percent ethanol} = \frac{7.89 \text{ g ethanol}}{100.0 \text{ mL soln}} \times 100\% = 7.89\%$$

d. Mole fraction of ethanol
   Convert the mass of ethanol from part **(b)** to an amount in moles.

$$? \text{ mol } C_2H_5OH = 7.89 \text{ g } C_2H_5OH \times \frac{1 \text{ mol } C_2H_5OH}{46.07 \text{ g } C_2H_5OH}$$

$$= 0.171 \text{ mol } C_2H_5OH$$

Determine the mass of water present in 100.0 mL of solution.

$$98.2 \text{ g soln} - 7.89 \text{ g ethanol} = 90.3 \text{ g water}$$

Convert the mass of water to the number of moles present.

$$? \text{ mol } H_2O = 90.3 \text{ g } H_2O \times \frac{1 \text{ mol } H_2O}{18.02 \text{ g } H_2O} = 5.01 \text{ mol } H_2O$$

$$\chi_{C_2H_5OH} = \frac{0.171 \text{ mol } C_2H_5OH}{0.171 \text{ mol } C_2H_5OH + 5.01 \text{ mol } H_2O} = \frac{0.171}{5.18} = 0.0330$$

e. Mole percent ethanol

$$\text{mole percent } C_2H_5OH = \chi_{C_2H_5OH} \times 100\% = 0.0330 \times 100\% = 3.30\%$$

f. Molarity of ethanol

Divide the number of moles of ethanol from part **(d)** by the solution volume, 100.0 mL = 0.1000 L.

$$\text{molarity} = \frac{0.171 \text{ mol } C_2H_5OH}{0.1000 \text{ L soln}} = 1.71 \text{ M } C_2H_5OH$$

g. Molality of ethanol

First, convert the mass of water present in 100.0 mL of solution [from part **(d)**] to the unit kg.

$$? \text{ kg } H_2O = 90.3 \text{ g } H_2O \times \frac{1 \text{ kg } H_2O}{1000 \text{ g } H_2O} = 0.0903 \text{ kg } H_2O$$

Use this result and the number of moles of C_2H_5OH from part **(d)** to establish the molality.

$$\text{molality} = \frac{0.171 \text{ mol } C_2H_5OH}{0.0903 \text{ kg } H_2O} = 1.89 \text{ } m \text{ } C_2H_5OH$$

Practice Example A: A solution that is 20.0% ethanol, by volume, is found to have a density of 0.977 g/mL. Use this fact, together with data from Example 14-1, to determine the mass percent ethanol in this solution.

Practice Example B: A 11.3-mL sample of CH_3OH ($d = 0.793$ g/mL) is dissolved in enough water to produce 75.0 mL of a solution with a density of 0.980 g/mL. What is the solution concentration expressed as **(a)** mole fraction H_2O; **(b)** molarity of CH_3OH; **(c)** molality of CH_3OH?

▶ The "proof" system of describing the composition of ethanol–water mixtures is the subject of Feature Problem A on page 502.

EXAMPLE 14-2

Converting Molarity to Mole Fraction. Laboratory ammonia is 14.8 M NH_3(aq) with a density of 0.8980 g/mL. What is χ_{NH_3} in this solution?

Solution

No volume of solution is stated, which suggests that our calculation can be based on any fixed volume of our choice. A convenient volume to work with is one liter. We need to determine the number of moles of NH_3 and of H_2O in one liter of the solution.

$$\text{moles of } NH_3 = 1.00 \text{ L} \times \frac{14.8 \text{ mol } NH_3}{1 \text{ L}} = 14.8 \text{ mol } NH_3$$

To get the number of moles of H_2O we can proceed as follows.

$$\text{mass of soln} = 1000.0 \text{ mL soln} \times \frac{0.8980 \text{ g soln}}{1 \text{ mL soln}} = 898.0 \text{ g soln}$$

$$\text{mass of } NH_3 = 14.8 \text{ mol } NH_3 \times \frac{17.03 \text{ g } NH_3}{1 \text{ mol } NH_3} = 252 \text{ g } NH_3$$

$$\text{mass of } H_2O = 898.0 \text{ g soln} - 252 \text{ g } NH_3 = 646 \text{ g } H_2O$$

$$\text{moles of } H_2O = 646 \text{ g } H_2O \times \frac{1 \text{ mol } H_2O}{18.02 \text{ g } H_2O} = 35.8 \text{ mol } H_2O$$

$$\chi_{NH_3} = \frac{14.8 \text{ mol } NH_3}{14.8 \text{ mol } NH_3 + 35.8 \text{ mol } H_2O} = 0.292$$

Practice Example A: A 16.00% by mass aqueous solution of glycerol, $C_3H_5(OH)_3$, has a density of 1.037 g/mL. What is the mole fraction of $C_3H_5(OH)_3$ in this solution?

Practice Example B: A 10.00% by mass aqueous solution of sucrose, $C_{12}H_{22}O_{11}$, has a density of 1.040 g/mL. What is (a) the molarity; (b) the molality; and (c) the mole fraction of $C_{12}H_{22}O_{11}$ in this solution?

14-3 Intermolecular Forces and the Solution Process

If even a little water enters the fuel tank of an automobile, the driver will soon notice that the engine misfires. This problem would not occur if water were soluble in gasoline, but why is it that water does not form solutions with gasoline? To understand a process we can often analyze the process's energy requirements; this approach can help us to explain why some substances mix to form solutions and others do not. In this section we focus on the behavior of molecules in solution, specifically on intermolecular forces.

Enthalpy of Solution

In the formation of some solutions, heat is given off to the surroundings; in many other cases heat is absorbed. An enthalpy of solution, ΔH_{soln}, can be rather easily measured—for example, in the "coffee cup" calorimeter of Figure 7-7—but why should some solution processes be exothermic whereas others are endothermic?

Let's think in terms of a three-step approach to ΔH_{soln}. First, solvent molecules must be separated from one another to make room for the solute molecules. This will take some energy to overcome the forces of attraction between solvent molecules. As a result, this step should be an endothermic one: $\Delta H > 0$. Second, the solute molecules must also be separated from one another. This step, too, will take energy and should be endothermic. Finally, we can imagine that we allow the separated solvent and solute molecules to be attracted to one another. These attractions will bring the molecules closer together and energy should be released. This is an exothermic step: $\Delta H < 0$. The enthalpy of solution is the sum of the three enthalpy changes just described, and depending on their relative values, ΔH_{soln} is either positive (endothermic) or negative (exothermic). This three-step process is summarized through Equation (14.1) and in Figure 14-2.

(a)	pure solvent	$\longrightarrow$ separated solvent molecules	$\Delta H_a > 0$
(b)	pure solute	$\longrightarrow$ separated solute molecules	$\Delta H_b > 0$
(c)	separated solvent and solute molecules	$\longrightarrow$ solution	$\Delta H_c < 0$
Net:	pure solvent + pure solute $\longrightarrow$ solution		

$$\Delta H_{soln} = \Delta H_a + \Delta H_b + \Delta H_c \qquad (14.1)$$

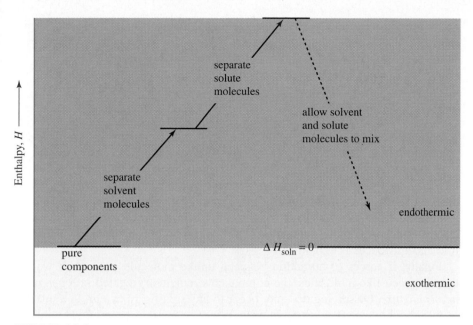

FIGURE 14-2
Enthalpy diagram for solution formation.
Depending on whether the broken arrow ends above, below, or on the line, the solution process is endothermic, exothermic, or has $\Delta H_{soln} = 0$, respectively.

Intermolecular Forces in Mixtures

We see from Equation (14.1) that the magnitude (and sign) of ΔH_{soln} depends on the values of the three terms ΔH_a, ΔH_b, and ΔH_c. These, in turn, depend on the strengths of the *three* kinds of intermolecular forces of attraction represented in Figure 14-3. Four possibilities for the relative strengths of these intermolecular forces are described in the discussion that follows.

If all intermolecular forces of attraction are of about equal strength, a random intermingling of molecules occurs. A homogeneous mixture or solution results. Because properties of solutions of this type can generally be predicted from the properties of the pure components, they are called **ideal solutions.** There is no net enthalpy change in forming an ideal solution from its components, and $\Delta H_{soln} = 0$. This means that ΔH_c in Equation (14.1) is equal in magnitude and opposite in sign to the sum of ΔH_a and ΔH_b. Many mixtures of liquid hydrocarbons fit this description (Figure 14-4).

If forces of attraction between unlike molecules *exceed* those between like molecules, a solution also forms. The properties of such solutions generally cannot be predicted, however; the solutions are *nonideal.* Interactions between solute

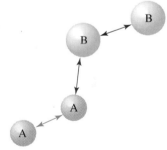

FIGURE 14-3
Representing intermolecular forces in a solution.
The intermolecular forces of attraction represented here are between: (1) solvent molecules A ($\leftrightarrow$), (2) solute molecules B ($\leftrightarrow$), and (3) solvent A and solute B ($\leftrightarrow$).

(a)   **(b)**

FIGURE 14-4
Two components of an ideal solution.
Think of the —CH_3 group in toluene (b) as a small "bump" on the planar benzene ring (a). Substances with similar molecular structures have similar intermolecular forces of attraction.

FIGURE 14-5
Intermolecular force between unlike molecules leading to a nonideal solution.
Hydrogen bonding between $CHCl_3$ (chloroform) and $(CH_3)_2CO$ (acetone) molecules produces forces of attraction between unlike molecules that exceed those between like molecules.

▶ An old adage with the same general meaning is "oil and water don't mix."

and solvent molecules (ΔH_c) release more heat than the heat absorbed to separate the solvent and solute molecules ($\Delta H_a + \Delta H_b$). The solution process is *exothermic* ($\Delta H_{soln} < 0$). Solutions of acetone and chloroform fit this type. As suggested by Figure 14-5, weak *hydrogen bonding* occurs between the two kinds of molecules, but the conditions for hydrogen bonding are not met in either of the pure liquids alone.*

If forces of attraction between solute and solvent molecules are somewhat *weaker* than between molecules of the same kind, complete mixing may still occur, but the solution formed is *nonideal*. The solution has a higher enthalpy than the pure components, and the solution process is *endothermic*. This type of behavior is observed in mixtures of carbon disulfide (CS_2), a *nonpolar* liquid, and acetone, a *polar* liquid. In these mixtures the acetone molecules show a preference for other acetone molecules as neighbors, to which they are attracted by dipole–dipole interactions. How a solution process can be endothermic and still occur is explained on page 476.

Finally, if forces of attraction between unlike molecules are much weaker than between like molecules, the components remain segregated in a *heterogeneous mixture*. Dissolving does not occur to any significant extent. In a mixture of water and octane (a constituent of gasoline) strong hydrogen bonds hold water molecules together in clusters. The nonpolar octane molecules are not able to exert a strong attractive force on the water molecules, and the two liquids do not mix. Thus, we now have an answer to the question of why water does not dissolve in gasoline, posed at the beginning of this section.

As a somewhat oversimplified summary of the four cases described in the preceding paragraphs, we can say that "like dissolves like." That is, substances with similar molecular structures are likely to exhibit similar intermolecular forces of attraction and to be soluble in one another. Substances with dissimilar structures are likely not to form solutions. Of course, in many cases the structures may be similar in part and dissimilar in part. Then it is a matter of trying to establish which part is most important, a matter we explore in Example 14-3.

EXAMPLE 14-3

Using Intermolecular Forces to Predict Solution Formation. Predict whether you would expect a solution to form in each of the following mixtures. If so, tell whether the solution is likely to be ideal. **(a)** ethyl alcohol, CH_3CH_2OH, and water (HOH); **(b)** the hydrocarbons hexane, $CH_3(CH_2)_4CH_3$, and octane, $CH_3(CH_2)_6CH_3$; **(c)** octyl alcohol, $CH_3(CH_2)_6CH_2OH$, and water (HOH).

Solution

a. If we think of water as H—OH, ethyl alcohol is similar to water. (Just substitute the group CH_3CH_2— for one of the H atoms in water.) Both molecules meet the requirements of hydrogen bonding as an important intermolecular force. The strengths of the hydrogen bonds between like and between unlike molecules are likely to differ, however. We should expect ethyl alcohol and water to form *nonideal* solutions.

b. In hexane the carbon chain is six atoms long and in octane, eight. Both substances are virtually nonpolar, and intermolecular attractive forces (of the dispersion type)

---

*In most cases H atoms bonded to C atoms cannot participate in hydrogen bonding. In a molecule like $CHCl_3$, however, the three Cl atoms have a strong electron-withdrawing effect on electrons in the C—H bond ($\mu = 1.01$ D). The H atom is then attracted to a lone pair of electrons on the O atom of $(CH_3)_2CO$.

in the solution and in the pure liquids should be quite similar. We expect a solution to form and for it to be nearly *ideal*.

c. At first sight, this case may seem similar to **(a)**, with the substitution of a hydrocarbon group for an H atom in H—OH. Here, however, the carbon chain is *eight* members long. This long carbon chain is much more important than the terminal —OH group in establishing the physical properties of octyl alcohol. Viewed from this perspective, octyl alcohol and water are quite *dissimilar*. We do not expect a solution to form.

Practice Example A: Which of the following organic compounds do you think is most readily soluble in water? Explain.

(a) toluene, $\langle\bigcirc\rangle$—CH$_3$

(b) oxalic acid, $H-O-\overset{\overset{\displaystyle O}{\|}}{C}-\overset{\overset{\displaystyle O}{\|}}{C}-O-H$

(c) benzaldehyde, $\langle\bigcirc\rangle$—$\overset{\displaystyle C}{\underset{\displaystyle O}{\|}}$—H

Practice Example B: In which solvent is I$_2$ likely to be more soluble, water or CCl$_4$? Explain.

▶ Ideal or nearly ideal solutions are not too common. They require the solvent and solute to be quite similar in structure.

▶ Butyl alcohol, CH$_3$CH$_2$CH$_2$CH$_2$OH, has a limited solubility in water (9 grams per 100 grams of water). The aqueous solubilities of alcohols fall off fairly rapidly as the hydrocarbon chain length increases beyond four.

Formation of Ionic Solutions

To assess the energy requirements for the formation of aqueous solutions of ionic compounds, we turn to the process pictured in Figure 14-6. Water dipoles are shown clustered around ions at the surface of a crystal. The negative ends of water dipoles are pointed toward the positive ions and the positive ends of water dipoles toward negative ions. If these ion–dipole forces of attraction are strong enough to overcome the interionic forces of attraction in the crystal, dis-

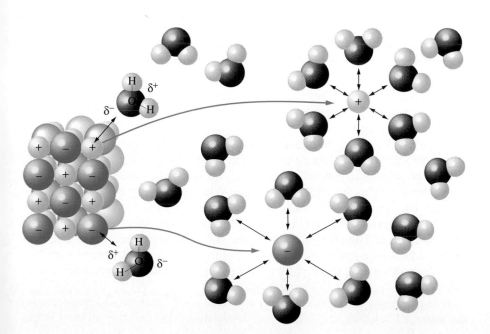

FIGURE 14-6
An ionic crystal dissolving in water.
Clustering of water dipoles around the surface of the ionic crystal and the formation of hydrated ions in solution are the key factors in the dissolving process.

solving will occur. Moreover, these ion–dipole forces also persist in the solution. An ion surrounded by a cluster of water molecules is said to be *hydrated*. Energy is *released* when ions become hydrated. The greater the hydration energy compared with the energy needed to separate ions from the ionic crystal, the more likely that the ionic solid will dissolve in water.

We can again use a *hypothetical* three-step process to describe the dissolving of an ionic solid. The energy requirement to dissociate a mole of an ionic solid into separated gaseous ions, an endothermic process, is the *negative* of the lattice energy. Energy is released in the next two steps—hydration of the gaseous cations and anions. The enthalpy of solution is the sum of these three ΔH values, described below for NaCl.

▶ We discussed lattice energy in Section 13-9.

$$NaCl(s) \longrightarrow Na^+(g) + Cl^-(g) \qquad \Delta H_1 = (-\text{lattice energy of NaCl}) > 0$$

$$Na^+(g) \xrightarrow{H_2O} Na^+(aq) \qquad \Delta H_2 = (\text{hydration energy of Na}^+) < 0$$

$$\underline{Cl^-(g) \xrightarrow{H_2O} Cl^-(aq) \qquad \Delta H_3 = (\text{hydration energy of Cl}^-) < 0}$$

$$NaCl(s) \xrightarrow{H_2O} Na^+(aq) + Cl^-(aq) \qquad \Delta H_{soln} = \Delta H_1 + \Delta H_2 + \Delta H_3 \approx +5 \text{ kJ/mol}$$

The dissolving of sodium chloride in water is *endothermic,* as is also the case for the vast majority (about 95%) of soluble ionic compounds. Why does NaCl dissolve in water if the process is endothermic? One might think that an endothermic process will not occur because of the increase in enthalpy. Since we observe that NaCl does dissolve in water, there must be a factor that we overlooked or were unaware of. In fact, we need to consider *two* factors to determine if a process will occur spontaneously. Enthalpy change is only one of them. The other factor, which we will introduce in Section 20-2 and call *entropy,* concerns the natural tendency for groups of atoms, ions, or molecules to become as disordered or mixed up as possible. The greater disorder found in NaCl(aq) compared with pure NaCl(s) and $H_2O(l)$ offsets the $+5$ kJ/mol increase in enthalpy in the solution process. In summary, if the hypothetical three-step process is *exothermic* we expect dissolving to occur; but we also expect a solution to form for an endothermic solution process, as long as ΔH_{soln} is not too large.

14-4   Solution Formation and Equilibrium

In the previous section we described what happens at the molecular (or microscopic) level when solutions form. In this section we will describe solution formation in terms of phenomena that we can actually observe. Let's call this a "macroscopic" view.

Figure 14-7 suggests what happens when a solid solute and liquid solvent are mixed. At first only dissolving occurs, but soon the reverse process of crystallization becomes increasingly important, and some dissolved atoms, ions, or molecules return to the undissolved state. When dissolving and crystallization occur at the same rate, the solution is in a state of dynamic equilibrium. The quantity of dissolved solute remains constant with time, and the solution is said to be **saturated.** The concentration of the saturated solution is called the **solubility** of the solute in the given solvent. Solubility varies with temperature, and a solubility–temperature graph is called a *solubility curve.* Some typical solubility curves are shown in Figure 14-8.

▶ In some solutions the solute and solvent are miscible (dissolve) in all ratios. The solution never becomes saturated. Ethyl alcohol–water solutions are an example.

If in preparing a solution we start with less solute than would be present in the saturated solution, the solute completely dissolves, and the solution is **un-**

(a)   (b)   (c)

FIGURE 14-7
Formation of a saturated solution.
The lengths of the arrows represent the rate of dissolving (↑) and the rate of crystallization (↓). (a) When solute is first placed in the solvent only dissolving occurs. (b) After a time the rate of crystallization becomes significant. (c) The solution becomes saturated when the rates of dissolving and crystallization become equal.

saturated. On the other hand, suppose we prepare a saturated solution at one temperature and then change the temperature to a value at which the solubility is lower (generally this means a lower temperature). Usually the excess solute crystallizes from solution, but occasionally all the solute may remain in solution. Because in these cases the quantity of solute is greater than in a normal saturated solution, the solution is said to be **supersaturated.** A supersaturated solution is unstable, and if a few crystals of solute are added to serve as particles on which crystallization can occur, the excess solute crystallizes. Figure 14-8 shows how unsaturated and supersaturated solutions can be represented with a solubility curve.

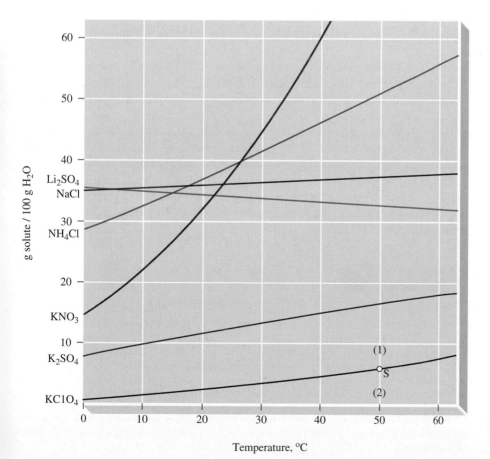

FIGURE 14-8
Water solubility of several salts as a function of temperature.
Solubilities can be expressed in many ways: as molarities, as mass percent, or, as in this figure, grams of solute per 100 g H_2O. For each solubility curve (as shown here for $KClO_4$) points on the curve (S) represent saturated solutions. Regions above the curve (1) correspond to supersaturated solutions and below the curve (2), to unsaturated solutions.

FIGURE 14-9
Recrystallization of KNO$_3$.
Crystals of KNO$_3$ separated from
an aqueous solution of KNO$_3$
and CuSO$_4$ (an impurity). The
pale blue color of the solution is
produced by Cu^{2+}, which remains
in solution.

Solubility as a Function of Temperature

As a general observation, the solubilities of ionic substances (about 95% of
them) *increase* with increasing temperature. Exceptions to this generalization
tend to be found among compounds containing the anions SO$_3^{2-}$, SO$_4^{2-}$,
AsO$_4^{3-}$, and PO$_4^{3-}$.

In Chapter 16 we will learn to predict how an equilibrium condition changes
with such variables as temperature and pressure by using an idea known as *Le
Châtelier's principle.* One statement of the principle is that heat added to a system at equilibrium stimulates the heat-absorbing, or endothermic, reaction. This
suggests that when $\Delta H_{soln} > 0$, raising the temperature stimulates dissolving and
increases the solubility of the solute. Conversely, if $\Delta H_{soln} < 0$ (exothermic),
the solubility *decreases* with increasing temperature. In this case crystallization,
being endothermic, is favored rather than dissolving.

We must be careful in applying the relationship we just described. The particular value of ΔH_{soln} that establishes whether solubility increases or decreases
with increased temperature is that associated with dissolving a small quantity
of solute in a solution that is already saturated or very nearly so. In some cases
this heat effect is altogether different from what we observe by adding a solute
to the pure solvent. For example, when we dissolve NaOH in water we observe
a sharp increase in temperature—an *exothermic* process. This fact suggests that
the solubility of NaOH in water should decrease as the temperature is raised.
What we observe, though, is that the solubility of NaOH in water *increases* with
increased temperature. This is because when a small quantity of NaOH is added
to a solution that is already nearly saturated, heat is *absorbed,* not evolved.*

Fractional Crystallization

Compounds synthesized in chemical reactions are generally impure, but the fact
that the solubilities of most solids increase with increased temperature provides
the basis for one simple method of purification. Usually, the impure solid consists of a high proportion of the desired compound and lesser proportions of the
impurities. Suppose both the compound and its impurities are soluble in a particular solvent and that we prepare a concentrated solution at a high temperature. Then we let the concentrated solution cool. At lower temperatures the solution becomes saturated in the desired compound. The excess compound
crystallizes from solution. The impurities remain in solution because the temperature is still too high for these to crystallize.† This method of purifying a
solid, called **fractional crystallization,** or **recrystallization,** is pictured in Figure 14-9. Example 14-4 illustrates how solubility curves can be used to predict
the outcome of a fractional crystallization.

▶Fractional crystallization
works best when (a) the quantities of impurities are small and
(b) the solubility curve rises
steeply with temperature.

EXAMPLE 14-4

Applying Solubility Data in Fractional Crystallization. A solution is prepared by dissolving 95 g NH$_4$Cl in 200.0 g H$_2$O at 60 °C. **(a)** What mass of NH$_4$Cl will recrystallize when the solution is cooled to 20 °C? **(b)** How might we improve the yield of
NH$_4$Cl?

---

*The solid in equilibrium with saturated NaOH(aq) over a range of temperatures around 25 °C
is NaOH·H$_2$O(s). It is actually the temperature dependence of the solubility of this hydrate that
we have been discussing.
†This is the usual behavior, but at times one or more impurities may form a solid solution with
the compound being recrystallized. In these cases simple recrystallization does not work as a method
of purification.

Solution

a. From Figure 14-8 we see that the solubility of NH_4Cl at 20 °C is 37 g NH_4Cl/100 g H_2O. The quantity of NH_4Cl in the saturated solution at 20 °C is

$$200.0 \text{ g } H_2O \times \frac{37 \text{ g } NH_4Cl}{100 \text{ g } H_2O} = 74 \text{ g } NH_4Cl$$

The mass of NH_4Cl recrystallized is $95 - 74 = 21$ g NH_4Cl.

b. The yield of NH_4Cl in **(a)** is rather poor—21 g out of 95 g, or 22%. We can do better: (1) The solution at 60 °C, although concentrated, is not saturated. A saturated solution at 60 °C has 55 g NH_4Cl/100 g H_2O. Thus, the 95 g NH_4Cl requires less than 200.0 g H_2O to make a saturated solution. At 20 °C, a smaller quantity of saturated solution would contain less NH_4Cl than in **(a)**, and the yield of recrystallized NH_4Cl would be greater. (2) Instead of cooling the solution only to 20 °C, we might cool it to 0 °C. Here the solubility of NH_4Cl is less than at 20 °C, and more solid would recrystallize. (3) Still another possibility is to start with a solution at a higher temperature than 60 °C, say closer to 100 °C. The mass of water needed for the saturated solution would be less than at 60 °C. Note that options (1) and (3) both require changing the conditions by using a different amount of water from that originally specified.

Practice Example A: Calculate the quantity of NH_4Cl that would be obtained if suggestions (1) and (2) in part **(b)** were followed. (*Hint:* Use data from Figure 14-8. What mass of water is needed to produce a saturated solution containing 95 g NH_4Cl at 60 °C?)

Practice Example B: Use Figure 14-8 to examine the solubility curves for the three potassium salts: $KClO_4$, K_2SO_4, and KNO_3. If saturated solutions of each of these salts at 40 °C are cooled to 20 °C, rank the salts in order of highest percent yield for the recrystallization.

14-5   Solubilities of Gases

Why does a freshly opened can of cola fizz, and why does the cola go flat after a time? To answer questions like these requires an understanding of the solubilities of gases. As we see in this section, the effect of temperature is generally different from what we find with solid solutes, and the pressure of a gas strongly affects its solubility.

Effect of Temperature

We cannot make an all-inclusive generalization about the effect of temperature on the solubilities of gases in solvents. It is certainly true, though, that the solubilities of most gases in water *decrease* with an increase in temperature, for example, those of $N_2(g)$ and $O_2(g)$—the major components of air—and of air itself (Figure 14-10). This fact helps to explain why many types of fish can survive only in cold water. There is not enough dissolved air (oxygen) in warm water to sustain them.

For solutions of gases in organic solvents, the situation is often the reverse of that just described; that is, gases may become more soluble at higher temperatures. The solubility behavior of the noble gases in water is more complex. The solubility of each gas decreases with an increase in temperature, reaching a minimum at a certain temperature; then the solubility trend reverses direction, with the gas becoming more soluble with an increase in temperature. For example, for helium at 1 atm pressure, this minimum in water solubility comes at 35 °C.

FIGURE 14-10
Effect of temperature on the solubilities of gases.
Dissolved air is released as water is heated, even at temperatures well below the boiling point.

The unopened bottle of soda water is under a high pressure of $CO_2(g)$. When a similar bottle is opened, the pressure quickly drops and some of the $CO_2(g)$ is released from solution (bubbles).

Effect of Pressure

Pressure affects the solubility of a gas in a liquid much more than does temperature. The English chemist William Henry (1775–1836) found that *the solubility of a gas increases as the gas pressure is increased.* A mathematical statement of **Henry's law** is

$$C = k \cdot P_{gas} \qquad (14.2)$$

In this equation C represents the solubility of a gas in a particular solvent, P_{gas} is the partial pressure of the gas above the solution, and k is a proportionality constant. We do not need to evaluate the proportionality constant k if we can treat a problem as a comparison of gas solubilities at two different pressures.

For example, the aqueous solubility of $N_2(g)$ at 0 °C and 1.00 atm is 23.54 mL N_2 per liter. Suppose we wish to increase the solubility of the $N_2(g)$ to a value of 100.0 mL N_2 per liter. Equation (14.2) suggests that to do so we must increase the pressure of $N_2(g)$ above the solution. That is,

$$k = \frac{C}{P_{gas}} = \frac{23.54 \text{ mL } N_2/L}{1.00 \text{ atm}} = \frac{100.0 \text{ mL } N_2/L}{P_{N_2}}$$

The required gas pressure, $P_{N_2} = (100/23.54) \times 1.00$ atm $= 4.25$ atm.

At times we are required to change the units used to express a gas solubility at the same time that the pressure is changed. This variation is illustrated in Example 14-5.

To avoid the painful and dangerous condition of the "bends," divers must not surface too quickly from great depths (see page 481).

EXAMPLE 14-5

Using Henry's Law. At 0 °C and an O_2 pressure of 1.00 atm, the aqueous solubility of $O_2(g)$ is 48.9 mL O_2 per liter. What is the molarity of O_2 in a saturated water solution when the O_2 is under its normal partial pressure in air, 0.2095 atm?

Solution

Think of this as a two-part problem. (1) Determine the molarity of the saturated O_2 solution at 0 °C and 1 atm. (2) Use Henry's law in the manner outlined above.

Determine the molarity of O_2 at 0 °C when $P_{O_2} = 1$ atm.

$$\text{molarity} = \frac{0.0489 \text{ L } O_2(\text{STP}) \times \dfrac{1 \text{ mol } O_2}{22.4 \text{ L } O_2(\text{STP})}}{1 \text{ L soln}} = 2.18 \times 10^{-3} \text{ M } O_2$$

Apply Henry's law.

$$k = \frac{C}{P_{gas}} = \frac{2.18 \times 10^{-3} \text{ M } O_2}{1.00 \text{ atm}} = \frac{C}{0.2095 \text{ atm}}$$

$$C = (0.2095/1.00) \times 2.18 \times 10^{-3} \text{ M } O_2 = 4.57 \times 10^{-4} \text{ M } O_2$$

Practice Example A: Use data from Example 14-5 to determine the partial pressure of O_2 above a 8.23×10^{-4} M O_2 saturated aqueous solution at 0 °C?

Practice Example B: A handbook lists the solubility of carbon monoxide in water at 0 °C and 1 atm pressure as 0.0354 mL CO per milliliter of H_2O. What should be the pressure of CO(g) above the solution to obtain 0.0100 M CO?

Here is how we can rationalize Henry's law: In a saturated solution the rate of evaporation of gas molecules from solution and the rate of condensation of gas molecules into the solution are equal. These rates both depend on the number of molecules per unit volume. As the number of molecules per unit volume increases in the gaseous state (through an increase in the gas pressure), the number of molecules per unit volume must also increase in the solution (through an increase in concentration). Figure 14-11 illustrates this rationalization.

We see a practical application of Henry's law in soft drinks. The dissolved gas is carbon dioxide, and the higher the gas pressure maintained above the soda pop, the more CO_2 that dissolves. When we open a can of pop, we release some gas. As the gas pressure above the solution drops, dissolved CO_2 is expelled, usually fast enough to cause fizzing. In sparkling wines the dissolved CO_2 is also under pressure, but the CO_2 is produced by a fermentation process within the bottle, rather than being added artificially as in soda pop.

Deep-sea diving provides us with still another example. Divers must carry a supply of air to breathe while underwater. If they are to stay submerged for any period of time, divers must breathe compressed air. High-pressure air, however, is much more soluble in blood and other body fluids than is air at normal pressures. When a diver returns to the surface, excess dissolved $N_2(g)$ is released as tiny bubbles and can cause severe pain in the limbs and joints, probably by interfering with the nervous system. This dangerous condition, known as "the bends," can be avoided if the diver ascends very slowly or spends time in a decompression chamber. Another effective method is to substitute a helium–oxygen mixture for compressed air. Helium is less soluble in blood than is nitrogen.

Henry's law (Equation 14.2) fails for gases at high pressures; it also fails if the gas ionizes in water or reacts with water. For example, at 20 °C and with $P_{HCl} = 1$ atm, a saturated solution of HCl(aq) is about 20 M. But to prepare 10 M HCl we do not need to maintain $P_{HCl} = 0.5$ atm above the solution, nor is $P_{HCl} = 0.05$ atm above 1 M HCl. We cannot even detect HCl(g) above 1 M HCl by its odor. The reason we cannot is that HCl ionizes in aqueous solutions, and in dilute solutions there are almost no molecules of HCl.

$$HCl(g) \xrightarrow{H_2O} H^+(aq) + Cl^-(aq)$$

We expect Henry's law to apply only to equilibrium between molecules of a gas and the same *molecules* in solution.

FIGURE 14-11
Effect of pressure on the solubility of a gas.
The concentration of dissolved gas (suggested by the depth of color) is proportional to the pressure of the gas above the solution (suggested by the density of the dots).

14-6   Vapor Pressures of Solutions

Separating compounds from one another is a task that chemists commonly face. If the compounds are volatile liquids, this separation often can be achieved by *distillation*. To understand how distillation works, we need to know something about the vapor pressures of solutions. Also, an understanding of the vapor pressures of solutions provides us with a springboard for dealing with other important solution properties, such as boiling points, freezing points, and osmotic pressures.

To simplify the discussion we will consider only solutions that contain two components, a solvent A and a solute B. In the 1880s the French chemist F. M. Raoult found that a dissolved solute *lowers* the vapor pressure of the solvent. **Raoult's law** states that the partial pressure exerted by solvent vapor above an ideal solution, P_A, is the product of the mole fraction of solvent in the solution, χ_A, and the vapor pressure of the pure solvent at the given temperature, P_A°.

$$P_A = \chi_A P_A^\circ \qquad (14.3)$$

To see how Equation (14.3) relates to Raoult's observation that a dissolved solute lowers the vapor pressure of the solvent, note that because $\chi_A + \chi_B = 1.00$, χ_A must be less than 1.00, and P_A must be smaller than P_A°.

▶ In an *ideal* solution Raoult's law applies to all solution components, not just the solvent.

To explain Raoult's law requires the notion of entropy or disorder that we briefly mentioned on page 476 in explaining endothermic solution processes. Rather than attempt the explanation now, however, we will wait until Section 20-4, after we have said more about entropy.

EXAMPLE 14-6

Predicting Vapor Pressures of Ideal Solutions. The vapor pressures of pure benzene and toluene at 25 °C are 95.1 and 28.4 mmHg, respectively. A solution is prepared in which the mole fractions of benzene and toluene are both 0.500. What are the partial pressures of the benzene and toluene above this solution? What is the total vapor pressure?

Solution

We saw through Figure 14-4 that benzene–toluene solutions should be ideal. We expect Raoult's law to apply to both solution components.

$$P_{benz} = \chi_{benz} P_{benz}^\circ = 0.500 \times 95.1 \text{ mmHg} = 47.6 \text{ mmHg}$$

$$P_{tol} = \chi_{tol} P_{tol}^\circ = 0.500 \times 28.4 \text{ mmHg} = 14.2 \text{ mmHg}$$

$$P_{total} = P_{benz} + P_{tol} = 47.6 \text{ mmHg} + 14.2 \text{ mmHg} = 61.8 \text{ mmHg}$$

▶ Because the mole fractions of the two components are both 0.500, the total vapor pressure is just the average of the vapor pressures of the two components. This is a very special case, however.

Practice Example A: The vapor pressure of pure hexane and pentane at 25 °C are 149.1 mmHg and 508.5 mmHg, respectively. If a hexane–pentane solution has a mole fraction of hexane of 0.750, what are the vapor pressures of hexane and pentane above the solution? What is the total vapor pressure?

Practice Example B: Calculate the vapor pressures of benzene, C_6H_6, and toluene, C_7H_8, and the total pressure at 25 °C above a solution with equal *masses* of the two liquids. (*Hint:* You may choose any amount of solution. How are mole fractions related to the mass? Use the vapor pressure data given in Example 14-6.)

EXAMPLE 14-7

Calculating the Composition of Vapor in Equilibrium with a Liquid Solution. What is the composition of the vapor in equilibrium with the benzene–toluene solution of Example 14-6?

Solution

The ratio of each partial pressure to the total pressure is the mole fraction of that component in the vapor. (This is another application of Equation 6.17.) The mole-fraction composition of the vapor is

$$\chi_{\text{benz}} = \frac{P_{\text{benz}}}{P_{\text{total}}} = \frac{47.6 \text{ mmHg}}{61.8 \text{ mmHg}} = 0.770$$

$$\chi_{\text{tol}} = \frac{P_{\text{tol}}}{P_{\text{total}}} = \frac{14.2 \text{ mmHg}}{61.8 \text{ mmHg}} = 0.230$$

Practice Example A: What is the composition of the vapor in equilibrium with the hexane–pentane solution described in Practice Example 14-6A?

Practice Example B: What is the composition of the vapor in equilibrium with the benzene–toluene solution described in Practice Example 14-6B?

Liquid–Vapor Equilibrium: Ideal Solutions

The results of Examples 14-6 and 14-7, together with similar data for other benzene–toluene solutions, are plotted in Figure 14-12. This figure consists of four lines—three straight and one curved—spanning the entire concentration range.

The red line shows how the vapor pressure of benzene varies with the solution composition. Because benzene in benzene–toluene solutions obeys Raoult's law, the red line has the equation $P_{\text{benz}} = \chi_{\text{benz}}P^{\circ}_{\text{benz}}$. The blue line shows how the vapor pressure of toluene varies with solution composition and indicates that toluene also obeys Raoult's law. The broken black line shows how the *total* vapor pressure varies with the solution composition. Can you see that each pressure on this black line is the sum of the pressures on the two straight lines that

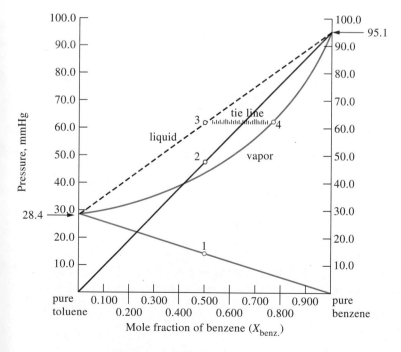

FIGURE 14-12
Liquid–vapor equilibrium for benzene–toluene mixtures at 25 °C.

——————— vapor pressure of benzene
——————— vapor pressure of toluene
- - - - - - - total vapor pressure (and liquid composition)
——————— vapor composition

FIGURE 14-13
"Moonshine" whiskey obtained by distillation.
Primitive stills like this one have been used for generations. The liquid in the distillation pot over the fire contains a fermented mash (for example, from corn). Vapor from the mash, enriched in ethanol, passes into the copper coil. Cooling water from a nearby stream passes over the coil, causing an aqueous ethanol solution (whiskey) to condense from the vapor and collect in a jug. The overall process is a simple distillation. (Fractional distillation requires repeated vaporization and condensation steps.)

lie below it? Point 3 represents the total vapor pressure (point 1 + point 2) of a benzene–toluene solution in which $\chi_{benz} = 0.500$ (recall Example 14-6).

As we calculated in Example 14-7, the vapor in equilibrium with a solution in which $\chi_{benz} = 0.500$ is richer still in benzene. The vapor has $\chi_{benz} = 0.770$ (point 4). The line joining points 3 and 4 is called a *tie line*. Imagine establishing a series of tie lines throughout the composition range. The vapor ends of these tie lines can be joined to form the green curve in Figure 14-12. From the relative placement of the liquid and vapor curves we see that for ideal solutions of two components, *the vapor phase is richer in the more volatile component than is the liquid phase.*

Fractional Distillation

The concluding statement of the preceding paragraph is the basis of an important laboratory and industrial method of separating the volatile components of a solution. The tie line drawn in Figure 14-12 shows that from a solution with $\chi_{benz} = 0.500$ we obtain a vapor with $\chi_{benz} = 0.770$. Imagine that we remove some of this vapor and condense it to a liquid, say by cooling it. This liquid will have $\chi_{benz} = 0.770$. If we now allow this solution to vaporize, the vapor will again be richer in benzene, the more volatile component, than is the solution. In fact, in this new vapor $\chi_{benz} = 0.92$. We could repeat the condensation–vaporization process several times and obtain vapor with an ever-increasing χ_{benz}. Eventually we would obtain virtually pure benzene in the vapor.

The hypothetical process just described was at a constant vaporization temperature of 25 °C. In actual practice one works with boiling solutions spread throughout a long column, called a *fractionating column*. Condensation–vaporization temperatures in fractionating columns vary over a wide range. The lowest temperatures are at the top of the column and the highest are at the bottom, closest to the heat source. In **fractional distillation** the most volatile component is obtained at the top of the column and the least volatile component collects at the bottom.

In Chapter 8 we described how N_2, O_2, and Ar in liquid air are separated by fractional distillation (Figure 8-3). Probably the most important industrial application is in the fractional distillation of petroleum. One of the oldest uses of distillation (simple distillation) is to make ethanol (Figure 14-13).

Liquid–Vapor Equilibrium: Nonideal Solutions

For nonideal solutions the liquid–vapor equilibrium diagram cannot be constructed in the simple manner illustrated in Figure 14-12. For example, vapor pressures in acetone–chloroform solutions are *lower* than we would predict for ideal solutions. These solutions exhibit *negative* deviations from ideality. In acetone–carbon disulfide solutions, on the other hand, vapor pressures are *higher* than predicted. These solutions exhibit *positive* deviations from ideality.

In Figure 14-5 we saw that forces of attraction between unlike molecules are greater than between like molecules in acetone–chloroform mixtures. It is not unreasonable to expect the components in such solutions to show a reduced tendency to vaporize and to have lower-than-predicted vapor pressures. With acetone–carbon disulfide solutions the situation is the reverse: forces of attraction between unlike molecules are weaker than between like molecules. This leads to greater tendencies for vaporization and higher vapor pressures than predicted by Raoult's law.

In nonideal solutions, if the departures from ideal solution behavior are sufficiently great, certain solutions may vaporize to produce a vapor that has the *same* composition as the liquid. These solutions, called **azeotropes,** boil at a constant temperature, and because the liquid and vapor have the same composition, the components cannot be separated by fractional distillation. A good example is that consisting of 95.6% ethanol (C_2H_5OH) and 4.4% water, by mass, and having a boiling point of 78.2 °C. Dilute ethanol–water solutions can be distilled to produce the azeotrope, but the remaining water cannot be removed by ordinary distillation. As a result, most ethanol used in the laboratory or in industry is only 95.6% C_2H_5OH. To obtain absolute, or 100%, C_2H_5OH requires special measures.

▶ Another common azeotrope is described in Feature Problem B on page 502.

14-7   Osmotic Pressure

To this point we have emphasized solutions containing a volatile solvent and volatile solute. Another common type of solution is one with a volatile solvent, such as water, but a *nonvolatile* solute(s), such as glucose, sucrose, or urea. Raoult's law still applies to the solvent in such solutions—the vapor pressure of the solvent is lowered.

Figure 14-14a pictures two aqueous solutions of a nonvolatile solute within the same enclosure. They are labeled A and B. The curved arrow indicates that water vaporizes from A and condenses into B. What is the driving force behind this? It must be that the vapor pressure of H_2O above A is greater than that above B. Solution A is more dilute; it has a higher mole fraction of H_2O. How long will this transfer of water continue? Solution A becomes more concentrated as it loses water and solution B becomes more dilute as it gains water. When the mole fraction of H_2O is the same in each solution, the transfer of H_2O stops.

A related phenomenon occurs when $CaCl_2 \cdot 6H_2O(s)$ is exposed to air (Figure 14-14b). Water vapor from the air condenses on the solid, and the solid begins to dissolve, a phenomenon known as *deliquescence.* In order for a solid to deliquesce, the partial pressure of water vapor in the air must be greater than the vapor pressure of water above a saturated aqueous solution of the solid. This requirement is often met for certain solids under conditions of an appropriate relative humidity. The deliquescence of $CaCl_2 \cdot 6H_2O$ occurs when the relative humidity exceeds 32%. (Relative humidity was described in Section 8-1.)

Like the case just described, Figure 14-15 also pictures the flow of solvent molecules. Here, however, the flow is not through the vapor phase. An aqueous sucrose (sugar) solution in a long glass tube is separated from pure water by a

▶ The relationship between deliquescence and relative humidity is explored further in Feature Problem C on page 502.

(a)

(b)

FIGURE 14-14
Observing the direction of flow of water vapor.
(a) Water passes, as vapor, from the more dilute solution (higher mole fraction of H_2O) to the more concentrated solution.
(b) Water vapor in air condenses onto solid calcium chloride hexahydrate, $CaCl_2 \cdot 6H_2O$. The liquid water dissolves some of the solid. The eventual result could be an unsaturated solution.

▶ Semipermeable membranes are materials, such as pig's bladder, parchment, or cellophane, containing submicroscopic holes. The holes permit the passage of solvent molecules but not those of the solute.

semipermeable membrane (permeable to water only). Water molecules can pass through the membrane in either direction, and they do. But because the concentration of water molecules is *greater* in the pure water than in the solution, there is a net flow *from* the pure water *into* the solution. This net flow, called **osmosis,** causes the solution to rise up the tube. The more concentrated the sucrose solution, the higher the solution level rises.

Applying a pressure to the sucrose solution slows down the net flow of water across the membrane into the solution. With a sufficiently high pressure the net influx of water can be stopped altogether. The necessary pressure to stop osmotic flow is called the **osmotic pressure** of the solution. For a 20% sucrose solution, this pressure is about 15 atm.

The magnitude of osmotic pressure depends only on the *number* of solute particles per unit volume of solution. It does not depend on the identity of the solute. Properties of this sort, whose values depend only on the concentration of solute particles in solution and *not* on what the solute is, are called **colligative properties.** The following equation works quite well for calculating osmotic pressures of *dilute* solutions of nonelectrolytes. The osmotic pressure is represented by the symbol π, R is the gas constant (0.08206 L atm mol^{-1} K^{-1}); and T is the Kelvin temperature. The term n represents the amount of *solute* (in moles) and V is the volume (in liters) of *solution.* Notice that it is similar to the equation for the ideal gas law. In this case, however, it is convenient to rearrange terms to yield Equation (14.4). The ratio, n/V, then, is the *molarity* of the solution, represented by the symbol M.

▶ The adjustment required to apply Equation (14.4) to electrolyte solutions is discussed in Section 14-9.

$$\pi V = nRT$$

$$\pi = \frac{n}{V}RT = \mathrm{M} \cdot RT \qquad (14.4)$$

FIGURE 14-15
An illustration of osmosis.
Water molecules pass through pores in the membrane and create a pressure within the funnel that causes the solution to rise and overflow. After a time, the solution inside the funnel becomes more dilute and the pure water outside becomes a sucrose solution. Liquid flow stops when the compositions of the solutions separated by the membrane have become nearly equal. Can you see the similarity between this phenomenon and that pictured in Figure 14-14a? The only difference is that here water remains in the liquid phase throughout.

water — aqueous solution of sucrose

EXAMPLE 14-8

Calculating Osmotic Pressure. What is the osmotic pressure at 25 °C of an aqueous solution that is 0.0010 M C$_{12}$H$_{22}$O$_{11}$ (sucrose)?

Solution

We just need to substitute the data into Equation (14.4).

$$\pi = \frac{0.0010 \text{ mol} \times 0.08206 \text{ L atm mol}^{-1} \text{ K}^{-1} \times 298 \text{ K}}{1 \text{ L}}$$

$$= 0.024 \text{ atm (18 mmHg)}$$

Practice Example A: What is the osmotic pressure at 25 °C of an aqueous solution that contains 1.50 g C$_{12}$H$_{22}$O$_{11}$ in 125 mL of solution?

Practice Example B: What mass of urea [CO(NH$_2$)$_2$] would you dissolve in 225 mL of solution to obtain an osmotic pressure of 0.015 atm at 25 °C? (*Hint:* What molarity solution has this osmotic pressure?)

The pressure difference of 18 mmHg we calculated in Example 14-8 is easy to measure. (It corresponds to a solution height of about 25 cm.) This means that we can easily use the measurement of osmotic pressure for determining molar masses when we are dealing with very dilute solutions or solutes with high molar masses (or both). Example 14-9 shows how osmotic pressure measurements can be used to determine molar mass.

EXAMPLE 14-9

Establishing a Molar Mass from a Measurement of Osmotic Pressure. A 50.00-mL sample of an aqueous solution is prepared that contains 1.08 g of a blood plasma protein, human serum albumin. The solution has an osmotic pressure of 5.85 mmHg at 298 K. What is the molar mass of the albumin?

Solution

First we need to express the osmotic pressure in atmospheres.

$$\pi = 5.85 \text{ mmHg} \times \frac{1 \text{ atm}}{760 \text{ mmHg}} = 7.70 \times 10^{-3} \text{ atm}$$

Now, we can modify Equation (14.4) slightly [that is, with the number of moles of solute (n) represented by the mass of solute (m) divided by the molar mass ($\mathcal{M}$)], and solve it for $\mathcal{M}$.

$$\pi = \frac{(m/\mathcal{M})\,RT}{V} \qquad \text{and} \qquad \mathcal{M} = \frac{mRT}{\pi V}$$

$$\mathcal{M} = \frac{1.08 \text{ g} \times 0.08206 \text{ L atm mol}^{-1}\text{ K}^{-1} \times 298 \text{ K}}{7.70 \times 10^{-3} \text{ atm} \times 0.0500 \text{ L}} = 6.86 \times 10^4 \text{ g/mol}$$

Practice Example A: Creatinine is a by-product of nitrogen metabolism and can be used to provide an indication of renal function. A 4.04-g sample of creatinine is dissolved in enough water to make 100.0 mL of solution. The osmotic pressure of the solution is 8.73 mmHg at 298 K. What is the molar mass of creatinine?

Practice Example B: What should be the osmotic pressure at 37.0 °C if a 2.12-g sample of human serum albumin is present in 75.00 mL of aqueous solution? Use the molar mass determined in Example 14-9.

A normal red blood cell (right) and two red blood cells from a hypertonic solution (left).

FIGURE 14-16
Desalination of saltwater by reverse osmosis.
The membrane is permeable to water but not to ions. The normal flow of water is from side A to side B. If we exert a pressure on side B that exceeds the osmotic pressure of the saltwater, a net flow of water occurs in the *reverse* direction—from the saltwater to the pure water. The lengths of the arrows suggest the magnitudes of the flow of water molecules in each direction.

Practical Applications

Some of the best examples of osmosis are those associated with living organisms. For instance, consider red blood cells. If we place red blood cells in pure water, the cells expand and eventually burst as a result of water entering through osmosis. The osmotic pressure associated with the fluid inside the cell is equivalent to that of 0.92% (mass/vol) NaCl(aq). Thus, if we place a cell in a sodium chloride (saline) solution of this concentration, there will be no net flow of water through the cell walls and the cell will remain stable. A solution with the same osmotic pressure as body fluids is said to be *isotonic*. If we place cells in a solution with concentration greater than 0.92% NaCl, water flows out of the cells and the cells shrink. The solution is *hypertonic*. If the NaCl concentration is less than 0.92%, a *hypotonic* solution, water flows into the cells. Fluids that are intravenously injected into patients to combat dehydration or to supply nutrients must be adjusted so that they are isotonic with blood. The osmotic pressure of the fluids must be the same as that of 0.92% (mass/vol) NaCl.

One recent application of osmosis goes to the very definition of osmotic pressure. Suppose in the device shown in Figure 14-16 we apply a pressure to the right side (side B) that is *less than* the osmotic pressure of the saltwater. The net flow of water molecules through the membrane will be from side A to side B. This is the process of *osmosis*. If we apply a pressure *greater than* the osmotic pressure to side B, we can cause a net flow of water in the *reverse* direction, from the saltwater into the pure water. This is the condition known as

▶ Solutions used in the medical treatment known as oral rehydration therapy are isotonic, as we establish in Feature Problem D on page 503.

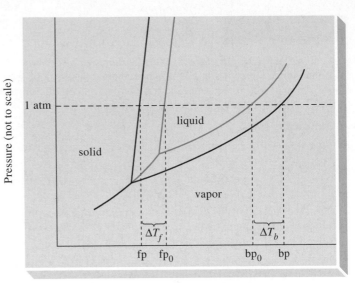

FIGURE 14-17
Vapor pressure lowering by a nonvolatile solute.
The normal freezing point and normal boiling point of the pure solvent are fp_0 and bp_0, respectively. The corresponding points for the solution are fp and bp. The freezing point depression, ΔT_f, and the boiling point elevation, ΔT_b, are indicated. Because the solute is assumed to be insoluble in the solid solvent, the sublimation curve of the solvent is unaffected by the presence of solute in the liquid solution phase. That is, the curve is the same for the two phase diagrams.

reverse osmosis. Reverse osmosis can be used in the *desalination* of seawater, to supply drinking water for emergency situations or as an actual source of municipal water. Another application of reverse osmosis is the removal of dissolved materials from industrial or municipal wastewater before it is discharged into the environment.

14-8 Freezing Point Depression and Boiling Point Elevation of Nonelectrolyte Solutions

In Section 14-6 we examined the lowering of the vapor pressure of a solvent produced by a dissolved solute. Vapor pressure lowering itself is not so frequently measured as are certain properties directly related to it. To aid in this discussion, let's refer to Figure 14-17. The blue curves represent the vapor pressure, fusion, and sublimation curves in the phase diagram for a pure solvent. The red curves represent the vapor pressure and fusion curves of the solvent in a solution. The sublimation curve for the solid solvent that freezes from the solution is shown in purple. Two assumptions are implicit in Figure 14-17. One is that the solute is nonvolatile, and the other is that the solid that freezes from a solution is pure solvent. For many mixtures these requirements are easily met.*

The vapor pressure curve of the solution (red) intersects the sublimation curve at a lower temperature than is the case for the pure solvent. The solid–liquid fusion curve, because it originates at the intersection of the sublimation and vapor pressure curves, is also displaced to lower temperatures. Now recall how we establish normal melting points and boiling points in a phase diagram. They are the temperatures at which a line at $P = 1$ atm intersects the fusion and vapor

---

*Actually, the equation for freezing point depression (14.5) applies even if the solute is volatile.

pressure curves, respectively. We have highlighted four points of intersection in Figure 14-17—the freezing points and the boiling points of the pure solvent and of the solvent in a solution. The freezing point of the solvent in solution is *depressed*, and the boiling point is *elevated*.

The extent to which the freezing point is lowered or the boiling point raised is proportional to the mole fraction of solute (just as is vapor pressure lowering). In *dilute* solutions the solute mole fraction is proportional to its molality, and so we can write

$$\Delta T_f = -K_f \times m \tag{14.5}$$

$$\Delta T_b = K_b \times m \tag{14.6}$$

In these equations ΔT_f and ΔT_b are the freezing point depression and boiling point elevation, respectively; m is the solute molality; and K_f and K_b are proportionality constants. The value of K_f depends on the melting point, enthalpy of fusion, and molar mass of the solvent. The value of K_b depends on the boiling point, enthalpy of vaporization, and molar mass of the solvent. The units of K_f and K_b are $°C\ m^{-1}$, and you can think of their values as representing the freezing point depression and boiling point elevation for a 1 *m* solution. In practice, though, Equations (14.5) and (14.6) often fail for solutions as concentrated as 1 *m*. Table 14.2 lists some typical values of K_f and K_b.

Historically, chemists have used the group of colligative properties—vapor pressure lowering, freezing point depression, boiling point elevation, and osmotic pressure—for molecular mass determinations. In Example 14-9, we showed how this could be accomplished with osmotic pressure. Example 14-10 shows how freezing point depression can be used to determine a molar mass and, with other information, a molecular formula. To help you understand how this is done, we present a three-step procedure in the form of answers to three separate questions. In other cases, you should be prepared to work out your own stepwise procedure.

TABLE 14.2 Freezing Point Depression and Boiling Point Elevation Constants

Solvent	K_f	K_b
acetic acid	3.90	3.07
benzene	5.12	2.53
nitrobenzene	8.1	5.24
phenol	7.27	3.56
water	1.86	0.512

Values correspond to freezing point depressions and boiling point elevations, in degrees Celsius, due to 1 mol of solute particles dissolved in 1 kg of solvent. Units: $°C$ kg solvent (mol solute)$^{-1}$ or $°C\ m^{-1}$.

EXAMPLE 14-10

Establishing a Molecular Formula with Freezing Point Data. Nicotine, extracted from tobacco leaves, is a liquid completely miscible with water at temperatures below 60 °C. **(a)** What is the *molality* of nicotine in an aqueous solution that starts to freeze at −0.450 °C? **(b)** If this solution is obtained by dissolving 1.921 g of nicotine in 48.92 g H_2O, what must be the molar mass of nicotine? **(c)** Combustion analysis shows nicotine to consist of 74.03% C, 8.70% H, and 17.27% N, by mass. What is the molecular formula of nicotine?

Solution

a. We can establish the molality of the nicotine by using Equation (14.5) with the value of K_f for water listed in Table 14.2. Note that $T_f = -0.450$ °C, and that $\Delta T_f = -0.450$ °C − 0.000 °C = −0.450 °C.

$$\text{molality} = \frac{\Delta T_f}{-K_f} = \frac{-0.450 \text{ °C}}{-1.86 \text{ °C } m^{-1}} = 0.242 \ m$$

b. Here we can use the definition of molality, but with a known molality (0.242 *m*) and an unknown molar mass of solute (*M*). The number of moles of solute is simply 1.921 g/*M*.

$$\text{molality} = \frac{1.921 \text{ g}/M}{0.04892 \text{ kg water}} = \frac{0.242 \text{ mol}}{\text{kg water}}$$

$$M = \frac{1.921 \text{ g}}{(0.04892 \times 0.242) \text{ mol}} = 162 \text{ g/mol}$$

c. To establish the empirical formula of nicotine we need to use the method of Example 3-5. (This calculation is left as an exercise for you to do.) The result you should obtain is C_5H_7N. The formula mass based on this empirical formula is 81 u. The molecular mass obtained from the molar mass in part **b** is exactly twice this value—162 u. The molecular formula is twice C_5H_7N, or $C_{10}H_{14}N_2$.

Practice Example A: Vitamin B_2, riboflavin, is soluble in water. If 0.833 g of riboflavin is dissolved in 18.1 g H_2O, the resulting solution has a freezing point of −0.227 °C. **(a)** What is the molality of the solution? **(b)** What is the molar mass of riboflavin? **(c)** What is the molecular formula of riboflavin if combustion analysis shows it to consist of 54.25% C, 5.36% H, 25.51% O, and 14.89% N?

Practice Example B: An aqueous solution that is 0.205 *m* urea $[CO(NH_2)_2]$ is found to boil at 100.025 °C. Is the prevailing barometric pressure above or below 760.0 mmHg? [*Hint:* At what temperature would you predict that the solution would begin to boil if atmospheric pressure were 760.0 mmHg? Use Equation (14.6) and a value of K_b from Table 14.2.]

▶ The adjustment required to apply these equations to electrolyte solutions is discussed in Section 14-9.

Molar mass determination by freezing point depression or boiling point elevation has its limitations. First, Equations (14.5) and (14.6) apply only to dilute solutions of nonelectrolytes, usually much less than 1 *m*. This requires the use of special thermometers so that temperatures can be measured very precisely, say to ±0.001 °C. Because boiling points depend on barometric pressure, precise measurements require that pressure be held constant. As a consequence, boiling point elevation is not much used. The precision of the freezing point depression method can be improved by using a solvent with a larger K_f value than that of water. For example, for cyclohexane $K_f = 20.0$ °C m^{-1} and for camphor $K_f = 37.7$ °C m^{-1}.

Practical Applications

The typical automobile antifreeze is ethylene glycol, $C_2H_4(OH)_2$. It is a good idea to leave the ethylene glycol–water mixture in the cooling system at all times to provide all-weather protection. In summer the ethylene glycol helps by raising the boiling point of water and preventing cooling system boilover.

Citrus growers faced with an impending freeze know they must take preventive measures only if the temperature drops below 0 °C by several degrees. The juice in the fruit has enough dissolved solutes to lower the freezing point by a degree or two. The growers also know they must protect lemons sooner than oranges because oranges have a higher concentration of dissolved solutes (sugars) than do lemons.

Salts such as NaCl can be used to prepare a *slush bath,* a mixture used to cool or freeze something. One example is the mixture of ice and NaCl(s) used to freeze ice cream in a home ice cream maker. The mixture has a much lower melting point than does pure ice. Because the slush bath is at a temperature well below 0 °C, it is easy to freeze the sugar and milk mixture that makes up the ice cream. NaCl is also useful for de-icing roads. It is effective in melting ice at temperatures as low as -21 °C (-6 °F). This is the lowest freezing point of a NaCl(aq) solution.

Water sprayed on citrus fruit releases its heat of fusion as it freezes into a layer of ice that acts as a thermal insulator. For a time, the temperature remains at 0 °C. The juice of the fruit, having a freezing point below 0 °C, is protected from freezing.

14-9 Solutions of Electrolytes

Our discussion of the electrical conductivities of solutions in Section 5-1 retraced some of the work done by the Swedish chemist Svante Arrhenius for his doctoral dissertation (1883). Prevailing opinion at the time was that ions form only with the passage of electric current. Arrhenius, however, reached the conclusion that in some cases ions exist in a solid substance and become dissociated from each other when the solid dissolves in water. Such is the case with NaCl, for example. In other cases, as with HCl, ions do not exist in the substance but are formed when it dissolves in water. In any event, electricity is not required to produce ions.

Although Arrhenius developed his theory of electrolytic dissociation to explain the electrical conductivities of solutions, he was able to apply it more widely. One of his first successes came in explaining certain anomalous values of colligative properties described by the Dutch chemist Jacobus van't Hoff (1852–1911).

Svante Arrhenius (1859–1927). At the time Arrhenius was awarded the Nobel Prize in chemistry (1903), his results were described thus: "Chemists would not recognize them as chemistry; nor physicists as physics. They have in fact built a bridge between the two." The field of physical chemistry had its origins in Arrhenius's work.

"Anomalous" Colligative Properties

Certain solutes produce a greater effect on colligative properties than expected. For example, consider a 0.0100 *m* aqueous solution. The predicted freezing point depression of this solution is

$$\Delta T_f = -K_f \times m = -1.86 \text{ °C } m^{-1} \times 0.0100 \text{ } m = -0.0186 \text{ °C}$$

We expect the solution to have a freezing point of -0.0186 °C. If the 0.0100 *m* solution is 0.0100 *m* urea, the measured freezing point is just about -0.0186 °C. If the solution is 0.0100 *m* NaCl, however, the measured freezing point is about -0.0361 °C.

Van't Hoff defined the factor *i* as the ratio of the measured value of a colligative property to the expected value if the solute were a nonelectrolyte. For 0.0100 *m* NaCl,

$$i = \frac{\text{measured } \Delta T_f}{\text{expected } \Delta T_f} = \frac{-0.0361 \text{ °C}}{-1.86 \text{ °C } m^{-1} \times 0.0100 \text{ } m} = 1.94$$

▶ Why the experimentally determined *i* for 0.0100 *m* NaCl is 1.94 instead of 2 is addressed later in this section.

Arrhenius's theory of electrolytic dissociation allows us to explain different values of the van't Hoff factor *i* for different solutes. For solutes such as urea, glycerol, and sucrose (all nonelectrolytes), *i* = 1. For a strong electrolyte such as NaCl, which produces *two* moles of ions in solution per mole of solute dissolved, we should expect the effect on freezing point depression to be twice as great as for a nonelectrolyte. We should expect *i* = 2. Similarly, for $MgCl_2$ our expectation would be that *i* = 3. For the weak acid $HC_2H_3O_2$ (acetic acid), which is only slightly ionized in aqueous solution, we expect *i* to be slightly larger than 1 but not nearly equal to 2.

This discussion suggests that Equations (14.4), (14.5), and (14.6) should all be rewritten in the form

$$\pi = i \times M \times RT$$

$$\Delta T_f = -i \times K_f \times m$$

$$\Delta T_b = i \times K_b \times m$$

If these equations are used for nonelectrolytes, simply substitute *i* = 1. For strong electrolytes, predict a value of *i*, as suggested in Example 14-11.

EXAMPLE 14-11

Predicting Colligative Properties for Electrolyte Solutions. Predict the freezing point of aqueous 0.00145 *m* $MgCl_2$.

Solution

First determine the value of *i* for $MgCl_2$. We can do this by writing an equation to represent the dissociation of $MgCl_2(s)$.

$$MgCl_2(s) \xrightarrow{H_2O} Mg^{2+}(aq) + 2\,Cl^-(aq)$$

Because *three* moles of ions are obtained per mole of formula units dissolved, we expect the value *i* = 3.

Now use the expression

$$\Delta T_f = -i \times K_f \times m$$
$$= -3 \times 1.86\ °C\ m^{-1} \times 0.00145\ m$$
$$= -0.0081\ °C$$

▶ Because the value of *i* for $MgCl_2$ is not exactly 3, we are not justified in carrying more than one or two significant figures in our answer.

The predicted freezing point is 0.0081 °C below the normal freezing point of water, that is, a freezing point of −0.0081 °C.

Practice Example A: What is the expected osmotic pressure of a 0.0530 M $MgCl_2$ solution at 25 °C?

Practice Example B: You wish to prepare an aqueous solution that has a freezing point of −0.100 °C. How many milliliters of 12.0 M HCl would you use to prepare 250.0 mL of such a solution? (*Hint:* What must be the molality of the solution? Note that in a *dilute* aqueous solution molality and molarity are essentially numerically equal.)

Interionic Attractions

Despite its initial successes, deficiencies in Arrhenius's theory were also readily apparent. The electrical conductivities of concentrated solutions of strong electrolytes are not as great as expected, and values of the van't Hoff factor *i* depend on the solution concentrations, as shown in Table 14.3. If strong elec-

TABLE 14.3   Variation of the van't Hoff Factor, *i*, with Solution Molality

Solute	Molality, *m*					
	1.0	0.10	0.010	0.0010	. . .	Inf. dil.
NaCl	1.81	1.87	1.94	1.97	. . .	2
MgSO$_4$	1.09	1.21	1.53	1.82	. . .	2
Pb(NO$_3$)$_2$	1.31	2.13	2.63	2.89	. . .	3

The limiting values: $i = 2$, 2, and 3 are reached when the solution is infinitely dilute. Note that a solute whose ions are singly charged (for example, NaCl) approaches its limiting value more quickly than does a solute whose ions carry higher charges. Interionic attractions are greater in solutes with more highly charged ions.

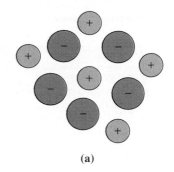

(a)

trolytes exist completely in ionic form in aqueous solutions, we would expect $i = 2$ for NaCl, $i = 3$ for MgCl$_2$, and so on, regardless of the solution concentration.

These difficulties can be resolved with a theory of electrolyte solutions proposed by Peter Debye and Erich Hückel in 1923. This theory continues to view strong electrolytes as existing only in ionic form in aqueous solution, but the ions in solution *do not behave independently of one another.* Instead, each positive ion is surrounded by a cluster of predominantly negative ions, and each negative ion by a cluster in which positive ions predominate. In short, each ion is enveloped by an *ionic atmosphere* with a net charge opposite to its own, as pictured in Figure 14-18.

In an electric field the mobility of each ion is reduced because of the attraction or drag exerted by its ionic atmosphere. Similarly, the magnitudes of colligative properties are reduced. This explains why, for example, the value of i for 0.010 m NaCl is 1.94 rather than 2.00. What we can say then is that each type of ion in an aqueous solution has two "concentrations." One is called the *stoichiometric concentration* and is based on the amount of solute dissolved. The other is an "effective" concentration, called the *activity*, which takes into account interionic attractions. Accurate stoichiometric calculations of the type presented in Chapters 4 and 5 can be made with stoichiometric concentrations. No other calculations involving solution properties are 100% accurate, however, if stoichiometric concentrations are used. Activities are needed instead. The activity of a solution is related to its stoichiometric concentration through a factor called an *activity coefficient.*

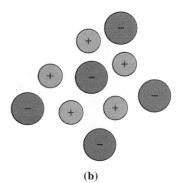

(b)

FIGURE 14-18
Interionic attractions in aqueous solution.
(a) A positive ion in aqueous solution is surrounded by a shell of negative ions. (b) A negative ion attracts positive ions to its immediate surroundings.

Are You Wondering . . .

How much you need to know about activities?

First, it is worth noting that in sufficiently dilute solutions, activity coefficients approximate 1: Activities and stoichiometric concentrations are the same. In doing calculations we'll only work with stoichiometric concentrations because a quantitative treatment of activity coefficients is beyond the scope of this text. Mainly, you should be aware of the distinction between activity and stoichiometric concentration because it helps us to explain a number of chemical phenomena, as we shall see again in a few other instances in the text.

Focus On   Colloidal Mixtures

The flashlight beam is not visible as it passes through a true solution (left), but it is readily seen as it passes through the colloidal dispersion of Fe_2O_3 (right).

Everybody knows that in a mixture of sand and water the sand (silica, SiO_2) quickly settles to the bottom. Yet, mixtures can be prepared in which up to 40% by mass of silica remains dispersed in water for many years. The silica is not present as dissolved ions or molecules in these mixtures. Rather, such mixtures consist of submicroscopic particles of SiO_2 suspended in water. The mixtures are said to be colloidal.

To be classified *colloidal,* a material must have one or more of its dimensions (length, width, or thickness) in the approximate range of 1–1000 nm. If all the dimensions are smaller than 1 nm, the particles are of molecular size. If all the dimensions exceed 1000 nm, the particles are of ordinary, or macroscopic, size (even if they are only visible under a microscope). The particles in colloidal silica have a spherical shape. Some colloidal particles are rod-shaped, and some, like gamma globulin in human blood plasma, have a disc-like shape. Thin films, like an oil slick on water, are colloidal. And some colloids, such as cellulose fibers, are randomly coiled filaments.

One method of determining whether a mixture is a true solution or colloidal is illustrated in the photograph. When light passes through a true solution, an observer viewing from a direction perpendicular to the light beam sees no light. In a colloidal dispersion light is scattered in many directions and is readily seen. This effect, first studied by John Tyndall in 1869, is known as the *Tyndall effect.* A common example is the scattering of light by dust particles in a flashlight beam.

What keeps the SiO_2 particles suspended in colloidal silica? The

Summary

To describe the composition of a solution one must indicate how much solute and solvent (or solution) are present. Percentage composition units have practical importance, but the units with more scientific value are molarity, molality, and mole fraction. At times it is necessary to convert among these various units.

To predict whether two substances will mix to form a solution, one must compare intermolecular forces involving like and unlike molecules. This approach also allows us to distinguish between ideal and nonideal solutions. Generally, a solvent has a limited ability to dissolve a solute to produce a saturated solution. Solute solubilities as a function of temperature

FIGURE 14-19
Coagulation of colloidal iron oxide.
On the left is red colloidal hydrous Fe_2O_3, obtained by adding $FeCl_3(aq)$ to boiling water. When a few drops of $Al_2(SO_4)_3(aq)$ are added, the suspended particles rapidly coagulate into a precipitate of $Fe_2O_3(s)$ (right).

most important factor is that the surfaces of the particles *adsorb*, or attach to themselves, ions from the solution, and they preferentially adsorb one type of ion over others. In the case of SiO_2 the preferred ions that are adsorbed are OH^-, and as a result, the particles acquire a net negative charge. Having like charges, the particles repel one an-

other. These mutual repulsions overcome the force of gravity, and the particles remain suspended indefinitely.

Although electric charge can be important in stabilizing a colloid, a high concentration of ions can also bring about the *coagulation*, or precipitation, of a colloid (Figure 14-19). The ions responsible for the

coagulation are those carrying a charge opposite to that on the colloidal particles themselves. *Dialysis,* a process similar to osmosis, can be used to remove excess ions from a colloidal mixture. Molecules of solvent and molecules or ions of solute pass through a semipermeable membrane, but the much larger colloidal particles do not. In some cases the process is more effective when carried out in an electric field. In *electrodialysis,* ions are attracted out of a colloidal mixture to an electrode carrying the opposite charge. A human kidney dialyzes blood, a colloidal mixture, to remove excess electrolytes produced in metabolism. In certain diseases the kidneys lose this ability, but a dialysis machine, external to the body, can function for the kidneys.

As so aptly put by Wilder Bancroft, an American pioneer in the field of colloid chemistry, ". . . colloid chemistry is essential to anyone who really wishes to understand . . . oils, greases, soaps, . . . glue, starch, adhesives, . . . paints, varnishes, lacquers, . . . cream, butter, cheese, . . . cooking, washing, dyeing, . . . colloid chemistry is the chemistry of life."

are represented by solubility curves. These curves can be used when planning the fractional crystallization of a solute. The solubilities of gases depend on pressure as well as temperature, and many familiar phenomena are related to gas solubilities. A simple mathematical relationship, called Henry's law, can be used to relate the concentration of a gas in solution to its pressure above the solution.

The physical properties studied in this chapter—vapor pressure lowering, osmotic pressure, freezing point depres-

sion, and boiling point elevation—have many practical applications. The values of these colligative properties depend only on the concentration of solute particles and not on the identity of the solute. In dealing with the physical properties of electrolyte solutions, however, one must work with the total concentration of ions in solution. Simple equations can be used to calculate the several colligative properties. For electrolyte solutions these equations are modified through the van't Hoff factor i.

Summarizing Example

Cinnamaldehyde is the chief constituent of cinnamon oil, obtained from the twigs and leaves of cinnamon trees. Cinnamon oil is used in the manufacture of food flavorings, perfumes, and cosmetics.

Cinnamaldehyde consists of 81.79% C, 6.10% H, and 12.11% O. A 0.100% by mass solution of cinnamaldehyde in water has an osmotic pressure of 141 mmHg at 298 K, and $d = 1.00$ g/mL. What is the molecular formula of cinnamaldehyde?

1. *Determine the empirical formula.* Use the percent composition data in the method of Example 3-5. *Result:* C_9H_8O.
2. *Determine the molarity of the cinnamaldehyde solution.*

First, convert the osmotic pressure from millimeters of mercury to atmospheres. Then solve Equation (14.4) for molarity. *Result:* 7.59×10^{-3} M.
3. *Express 0.100% by mass as molarity.* Work with 1-kg of solution which, because $d = 1.00$ g/mL, is also 1 L. The mass of cinnamaldehyde in this solution is 1.00 g. The number of moles of cinnamaldehyde is $1.00/\mathcal{M}$. *Result:* molarity = $(1.00/\mathcal{M})$ M.
4. *Use the results of parts 2 and 3 to solve for $\mathcal{M}$. Result:* $\mathcal{M} = 132$ g/mol.
5. *Establish the molecular formula of cinnamaldehyde.* Determine the formula mass for C_9H_8O. Compare it with the result of step **4**. They are the same. *Answer:* C_9H_8O.

Key Terms

alloy (14-1)
azeotrope (14-6)
colligative properties (14-7)
fractional crystallization (recrystallization) (14-4)
fractional distillation (14-6)
Henry's law (14-5)

ideal solution (14-3)
molality (m) (14-2)
mole fraction (14-2)
mole percent (14-2)
osmosis (14-7)
osmotic pressure (14-7)
Raoult's law (14-6)

reverse osmosis (14-7)
saturated solution (14-4)
solubility (14-4)
solute (14-1)
solvent (14-1)
supersaturated solution (14-4)
unsaturated solution (14-4)

Review Questions

1. In your own words define or explain the following terms or symbols: **(a)** χ_B; **(b)** P_A°; **(c)** K_f; **(d)** i; **(e)** activity.
2. Briefly describe each of the following ideas or phenomena: **(a)** Henry's law; **(b)** freezing point depression; **(c)** recrystallization; **(d)** hydrated ion; **(e)** deliquescence.
3. Explain the important distinctions between each pair of terms. **(a)** molality and molarity; **(b)** ideal and nonideal solution; **(c)** unsaturated and supersaturated solution; **(d)** fractional crystallization and fractional distillation; **(e)** osmosis and reverse osmosis.
4. A saturated aqueous solution of NaBr at 20 °C contains 116 g NaBr/100 g H_2O. Express this composition in the more conventional percent by mass, that is, as grams of NaBr per 100 grams of solution.
5. An aqueous solution with density 0.988 g/mL at 20 °C is prepared by dissolving 12.8 mL $CH_3CH_2CH_2OH$ ($d = 0.803$ g/mL) in enough water to produce 75.0 mL of solution. What is the percent $CH_3CH_2CH_2OH$ expressed as **(a)** percent by volume; **(b)** percent by mass; **(c)** percent (mass/vol)?
6. A certain brine has 3.87% NaCl by mass. A 75.0-mL sample weighs 76.9 g. How many liters of this solution should be evaporated to dryness to obtain 725 kg NaCl?

7. You are asked to prepare 125.0 mL of 0.0321 M $AgNO_3$. How many grams would you need of a sample known to be 99.81% $AgNO_3$ by mass?
8. What is the molality of *para*-dichlorobenzene in a solution prepared by dissolving 2.65 g $C_6H_4Cl_2$ in 50.0 mL of benzene ($d = 0.879$ g/mL)?
9. An aqueous solution is 6.00% methanol, CH_3OH, by mass, with $d = 0.988$ g/mL. What is the molarity of CH_3OH in this solution?
10. A solution is prepared by mixing 1.28 mol C_7H_{16}, 2.92 mol C_8H_{18}, and 2.64 mol C_9H_{20}. What are the **(a)** mole fraction and **(b)** mole percent of each component of the solution?
11. A solution ($d = 1.159$ g/mL) is 62.0% glycerol, $C_3H_8O_3$, and 38.0% H_2O, by mass. Determine **(a)** the molarity of $C_3H_8O_3$ (with H_2O as the solvent); **(b)** the molarity of H_2O (with $C_3H_8O_3$ as the solvent); **(c)** the molality of H_2O in $C_3H_8O_3$; **(d)** the mole fraction of $C_3H_8O_3$; **(e)** the mole percent H_2O.
12. A solution prepared by dissolving 1.12 mol NH_4Cl in 150.0 g H_2O is brought to a temperature of 30 °C. Use Figure 14-8 to determine whether the solution is unsaturated or whether excess solute will crystallize.

13. Which of the following mixtures in a single liquid phase is most likely to be an ideal solution? Explain. (a) NaCl–H_2O; (b) $C_2H_5OH(l)$–$C_6H_6(l)$;   (c) $C_7H_{16}(l)$–$H_2O(l)$;   (d) $C_7H_{16}(l)$–$C_8H_{18}(l)$.

14. Which of the following do you expect to be most water-soluble and why? $C_{10}H_8(s)$, $NH_2OH(s)$, $C_6H_6(l)$, $CaCO_3(s)$.

15. Which one of the following do you expect to be moderately soluble both in water and benzene [$C_6H_6(l)$], and why? (a) butyl alcohol, C_4H_9OH; (b) naphthalene, $C_{10}H_8$; (c) hexane, C_6H_{14}; (d) NaCl(s).

16. Under an $O_2(g)$ pressure of 1.00 atm, 28.31 mL of $O_2(g)$ at 25 °C dissolves in 1.00 L H_2O at 25 °C. What will be the molarity of O_2 in the saturated solution at 25 °C when the O_2 pressure is 3.86 atm? (Assume that the solution volume remains at 1.00 L.)

17. What are the partial and total vapor pressures of a solution obtained by mixing 35.8 g benzene, C_6H_6, and 56.7 g toluene, C_7H_8, at 25 °C? Vapor pressures at 25 °C: C_6H_6 = 95.1 mmHg; C_7H_8 = 28.4 mmHg.

18. Determine the composition of the vapor above the benzene–toluene solution described in Exercise 17.

19. *Without doing detailed calculations,* determine which of the following aqueous solutions probably has the *lowest* freezing point. **(a)** 0.010 *m* $MgSO_4$; **(b)** 0.011 *m* NaCl; **(c)** 0.050 *m* C_2H_5OH (ethanol); **(d)** 0.010 *m* MgI_2.

20. Use your knowledge of strong, weak, and nonelectrolytes to arrange the following 0.0010 *m* aqueous solutions in the probable order of *decreasing* freezing point: C_2H_5OH, NaCl, $MgBr_2$, $HC_2H_3O_2$, and $Al_2(SO_4)_3$.

21. NaCl(aq) isotonic with blood is 0.92% NaCl (mass/vol). For this solution, what is **(a)** [Na^+]; **(b)** the total molarity of ions; **(c)** the osmotic pressure at 37 °C; **(d)** the approximate freezing point? (Assume that the solution has a density of 1.005 g/mL.)

22. A 0.72-g sample of polyvinyl chloride (PVC) is dissolved in 250.0 mL of a suitable solvent at 25 °C. The solution has an osmotic pressure of 1.67 mmHg. What is the molar mass of the PVC?

23. 1.10 g of an unknown compound reduces the freezing point of 75.22 g benzene from 5.53 to 4.92 °C. What is the molar mass of the compound?

24. The freezing point of an 0.010 *m* aqueous solution of a nonvolatile solute is −0.072 °C. What would you expect the normal boiling point of this same solution to be?

Exercises

Homogeneous and Heterogeneous Mixtures

25. Substances that dissolve in water do not generally dissolve in benzene. Some substances are moderately soluble in both solvents, however. One of the following is such a substance. Which do you think it is and why?
(a) *para*-dichlorobenzene (a moth repellent)

Cl—⟨◯⟩—Cl

(b) salicyl alcohol (a local anesthetic)

CH_2OH
OH

(c) diphenyl (a heat transfer agent)

⟨◯⟩—⟨◯⟩

(d) hydroxyacetic acid (used in textile dyeing)

H O
| ‖
HO—C—C—OH
|
H

26. Some vitamins are water-soluble and some, fat-soluble. (Fats are substances whose molecules have long hydrocarbon chains.) The structural formulas of two vitamins are shown here. One is water-soluble and one is fat-soluble. Identify each and explain your reasoning.

vitamin C

vitamin E

27. Two of the substances listed here are highly soluble in water; two are only slightly soluble in water; and two are insoluble in water. Indicate the situation you expect for each one.
(a) iodoform, CHI_3
(b) benzoic acid,

(c) formic acid, $H-\overset{O}{\overset{\|}{C}}-OH$
(d) butyl alcohol, $CH_3CH_2CH_2CH_2OH$
(e) chlorobenzene,

(f) propylene glycol, $CH_3CHOHCH_2OH$

29. In light of the factors outlined on page 476, which of the following ionic fluorides would you expect to be most water-soluble on a moles per liter basis: MgF_2, NaF, KF, CaF_2? Explain your reasoning.

28. Benzoic acid, C_6H_5COOH, is much more soluble in $NaOH(aq)$ than it is in pure water. Can you suggest a reason for this? The structural formula for benzoic acid is given in Exercise 27(b).

30. Explain the observation that all metal nitrates are water-soluble, whereas many metal sulfides are not water-soluble. Among metal sulfides, which would you expect to be most soluble?

Percent Concentration

31. According to Example 14-1 the mass percent ethanol in a particular aqueous solution is less than the volume percent in the same solution. Explain why this is also true for *all* aqueous solutions of ethanol. Would it be true of all ethanol solutions, regardless of the other component? Explain.

33. A certain vinegar is 6.02% acetic acid ($HC_2H_3O_2$), by mass. How many grams of $HC_2H_3O_2$ are contained in a 355-mL bottle of vinegar. Assume a density of 1.01 g/mL.

32. Is either mass percent or volume percent independent of temperature? Explain your answer.

34. 6.00 M sulfuric acid, $H_2SO_4(aq)$, has a density of 1.338 g/mL. What is the percent by mass of sulfuric acid in this solution?

Molarity

35. How many milliliters of the ethanol–water solution described in Example 14-1 should be diluted with water to produce 825 mL of 0.235 M C_2H_5OH?

36. A 30.00%-by-mass solution of nitric acid, HNO_3, in water has a density of 1.18 g/cm^3 at 20 °C. What is the molarity of HNO_3 in this solution?

Molality

37. How many grams of iodine, I_2, must be dissolved in 725 mL of carbon disulfide, CS_2 ($d = 1.261$ g/mL), to produce a 0.236 m solution?

39. An aqueous solution is 34.0% H_3PO_4 by mass and has a density of 1.209 g/mL. What are the molarity and molality of this solution?

38. How many grams of water would you add to 1.00 kg of 1.38 m $CH_3OH(aq)$ to reduce the molality to 1.00 m CH_3OH?

40. A 10.00%-by-mass solution of ethanol, C_2H_5OH, in water has a density of 0.9831 g/mL at 15 °C and 0.9804 g/mL at 25 °C. Calculate the molality of the ethanol–water solution at these two temperatures. Does the molality differ at the two temperatures (that is, at 15 and 25 °C)? Would you expect the molarities to differ? Explain.

Mole Fraction, Mole Percent

41. Calculate the mole fraction of solute in the following aqueous solutions: (a) 21.7% C_2H_5OH, by mass; (b) 0.684 m $CO(NH_2)_2$ (urea).

43. Refer to Example 14-1. How many grams of C_2H_5OH must be added to 100.0 mL of the solution described in part (d) to increase the mole fraction of C_2H_5OH to 0.0525?

42. Calculate the mole fraction of the solute in the following aqueous solutions: (a) 0.112 M $C_6H_{12}O_6$ ($d = 1.006$ g/mL); (b) 3.20% ethanol, by volume ($d = 0.993$ g/mL; pure C_2H_5OH, $d = 0.789$ g/mL).

44. How many milliliters of glycerol, $C_3H_8O_3$ ($d = 1.26$ g/mL), must be added per kilogram of water to produce a solution with 4.85 mol % $C_3H_8O_3$?

Solubility Equilibrium

45. Refer to Figure 14-8 and determine the molality of NH_4Cl in a saturated aqueous solution at 40 °C.

47. A solution of 20.0 g $KClO_4$ in 500.0 g of water is brought to a temperature of 40 °C.
(a) Refer to Figure 14-8 and determine whether the solution is unsaturated or supersaturated at 40 °C.
(b) Approximately what mass of $KClO_4$, in grams, must be added to saturate the solution (if originally unsaturated) or what mass of $KClO_4$ can be crystallized (if originally supersaturated)?

46. Refer to Figure 14-8 and estimate the temperature at which a saturated aqueous solution of $KClO_4$ is 0.200 *m*.

48. One way to recrystallize a solute from a solution is to change the temperature. Another way is to evaporate solvent from the solution. A 335-g sample of a saturated solution of $KNO_3(s)$ in water is prepared at 25.0 °C. If 55 g H_2O is evaporated from the solution, at the same time the temperature is reduced from 25.0 to 0.0 °C, what mass of $KNO_3(s)$ will recrystallize? (Refer to Figure 14-8.)

Solubility of Gases

49. Most natural gases consist of about 90% methane, CH_4. Assume that the solubility of natural gas at 20 °C and 1 atm gas pressure is about the same as that of CH_4, 0.02 g/kg water. If a sample of natural gas under a pressure of 20 atm is kept in contact with 1.00×10^3 kg of water, what mass of natural gas will dissolve?

51. The aqueous solubility at 20 °C of Ar at 1.00 atm is equivalent to 33.7 mL Ar(g), measured at STP, per liter of water. What is the molarity of Ar in water at 20 °C that is saturated with air at 1.00 atm? Air contains 0.934% Ar, by volume. Assume that the volume of water does not change when it becomes saturated with air.

50. At 1.00 atm, the solubility of O_2 in water is 2.18×10^{-3} M at 0 °C and 1.26×10^{-3} M at 25 °C. What volume of $O_2(g)$, measured at 25 °C and 1.00 atm, is expelled when 515 mL of water saturated with O_2 is heated from 0 to 25 °C?

52. The aqueous solubility at 20 °C of CO_2 at 1.00 atm is equivalent to 87.8 mL $CO_2(g)$, measured at STP, per 100 mL of water. What should be the molarity of CO_2 in water at 20 °C that is saturated with air at 1.00 atm? The volume percent of CO_2 in air is 0.0360%. Assume that the volume of the water does not change when it becomes saturated with air.

Raoult's Law and Liquid–Vapor Equilibrium

53. Calculate the vapor pressure at 25 °C of a solution containing 165 g of the *nonvolatile* solute, glucose, $C_6H_{12}O_6$, in 685 g H_2O. The vapor pressure of water at 25 °C is 23.8 mmHg.

55. Styrene, used in the manufacture of polystyrene plastics, is made from ethylbenzene by the extraction of hydrogen atoms. The product obtained contains about 38% styrene (C_8H_8) and 62% ethylbenzene (C_8H_{10}), by mass. The mixture is separated by fractional distillation at 90 °C. Determine the composition of the vapor in equilibrium with this 38%–62% mixture at 90 °C given the vapor pressures of the two components: ethylbenzene, 182 mmHg; styrene, 134 mmHg.

54. Calculate the vapor pressure at 20 °C of a saturated solution of the *nonvolatile* solute, urea, $CO(NH_2)_2$, in methanol, CH_3OH. The solubility is 17 g urea/100 mL methanol. The density of methanol is 0.792 g/mL and its vapor pressure at 20 °C is 95.7 mmHg.

56. Calculate $\chi_{C_6H_6}$ in a benzene–toluene liquid solution that is in equilibrium at 25 °C with a vapor phase that contains 62.0 mol % C_6H_6. (Use data from Exercise 17.)

Osmotic Pressure

57. When the stems of cut flowers are held in concentrated NaCl(aq), the flowers wilt. In a similar solution a fresh cucumber shrivels up (becomes pickled). Explain the basis of these phenomena.

59. In what volume of water must 1 mol of a nonelectrolyte be dissolved if the solution is to have an osmotic pressure of 1 atm at 273 K? Which of the gas laws does this result resemble?

61. An 0.50-g sample of polyisobutylene (a polymer used in synthetic rubber) in 100.0 mL of benzene solution has an osmotic pressure at 25 °C that supports a 5.1-mm column of solution ($d = 0.88$ g/mL). What is the molar mass of the polyisobutylene? (For Hg, $d = 13.6$ g/mL.)

58. Some fish live in saltwater environments and some in freshwater, but in either environment they need water to survive. Saltwater fish drink water, but freshwater fish do not. Explain this difference between the two types of fish. (*Hint:* How do you suppose osmotic pressure enters in?)

60. The molecular mass of hemoglobin is 6.86×10^4 u. What mass of hemoglobin must be present per 100.0 mL of a solution to exert an osmotic pressure of 7.25 mmHg at 25 °C?

62. Use the concentration of an isotonic saline solution, 0.92% NaCl (mass/vol), to determine the osmotic pressure of blood at body temperature, 37.0 °C. (*Hint:* Assume that NaCl is completely dissociated in aqueous solutions.)

63. What approximate pressure is required in the reverse osmosis depicted in Figure 14-16 if the saltwater contains 2.5% NaCl, by mass. (*Hint:* Assume that NaCl is completely dissociated in aqueous solutions. Also, assume a temperature of 25 °C.)

64. The two solutions pictured here are separated by a semipermeable membrane that permits only the passage of water molecules. In what direction will a net flow of water occur; that is, from left to right or right to left?

14.0 g $C_3H_8O_3$ in 55.2 mL solution

17.2 g $C_{12}H_{22}O_{11}$ in 62.5 mL solution

Freezing Point Depression and Boiling Point Elevation

65. Adding 1.00 g of benzene, C_6H_6, to 80.00 g cyclohexane, C_6H_{12}, lowers the freezing point of the cyclohexane from 6.5 to 3.3 °C.
(a) What is the value of K_f for cyclohexane?
(b) Which is the better solvent for molar mass determinations by freezing point depression, benzene or cyclohexane? Explain.

66. The boiling point of water at 749.2 mmHg is 99.60 °C. What mass percent sucrose ($C_{12}H_{22}O_{11}$) should be present in an aqueous sucrose solution to raise the boiling point to 100.00 °C at this pressure?

67. A compound is 42.9% C, 2.4% H, 16.7% N, and 38.1% O, by mass. Addition of 6.45 g of this compound to 50.0 mL benzene, C_6H_6 ($d = 0.879$ g/mL), lowers the freezing point from 5.53 to 1.37 °C. What is the molecular formula of this compound?

68. Nicotinamide is a water-soluble vitamin important in metabolism. A deficiency in this vitamin results in the debilitating condition known as pellagra. Nicotinamide is 59.0% C, 5.0% H, 22.9% N, 13.1% O, by mass. Addition of 3.88 g of nicotinamide to 30.0 mL nitrobenzene, $C_6H_5NO_2$ ($d = 1.204$ g/mL), lowers the freezing point from 5.7 to −1.4 °C. What is the molecular formula of this compound?

69. Thiophene (fp, −38.3; bp, 84.4 °C) is a sulfur-containing hydrocarbon sometimes used as a solvent in place of benzene. Combustion of a 2.348-g sample of thiophene produces 4.913 g CO_2, 1.005 g H_2O, and 1.788 g SO_2. When a 0.867-g sample of thiophene is dissolved in 44.56 g of benzene (C_6H_6), the freezing point is lowered by 1.183 °C. What is the molecular formula of thiophene?

70. Coniferin is a glycoside (a derivative of a sugar) found in conifers such as fir trees. When a 1.205-g sample of coniferin is subjected to combustion analysis (recall Section 3-3), the products are 0.698 g H_2O and 2.479 g CO_2. A 2.216-g sample is dissolved in 48.68 g H_2O, and the normal boiling point of this solution is found to be 100.068 °C. What is the molecular formula of coniferin?

Strong Electrolytes, Weak Electrolytes, and Nonelectrolytes

71. Predict the approximate freezing points of 0.10 *m* solutions of the following solutes dissolved in water: **(a)** $CO(NH_2)_2$ (urea); **(b)** NH_4NO_3; **(c)** HCl; **(d)** $CaCl_2$; **(e)** $MgSO_4$; **(f)** C_2H_5OH (ethanol); **(g)** $HC_2H_3O_2$ (acetic acid).

72. Calculate the van't Hoff factors of the following weak electrolyte solutions.
(a) 0.050 *m* $HCHO_2$, which begins to freeze at −0.0986 °C;
(b) 0.100 M HNO_2, which has a hydrogen ion (and nitrite ion) concentration of 6.91×10^{-3} M.

73. $NH_3(aq)$ conducts electric current only weakly. The same is true for acetic acid, $HC_2H_3O_2(aq)$. When these solutions are mixed, however, the resulting solution conducts electric current very well. Propose an explanation.

74. An isotonic solution is described as 0.92% NaCl (mass/vol). Would this also be the required concentrations for isotonic solutions of other salts, such as KCl, $MgCl_2$, or $MgSO_4$? Explain

Advanced Exercises

75. A typical root beer contains 0.13% of a 75% H_3PO_4 solution by mass. How many milligrams of phosphorus are contained in a 12-oz can of this root beer? Assume a solution density of 1.00 g/mL; also, 1 oz = 29.6 mL.

76. Consider two NaCl(aq) solutions. *Solution 1* is a saturated solution with undissolved NaCl(s) present. *Solution 2* is unsaturated.

(a) Above which solution is the vapor pressure of water, P_{H_2O}, greater? Explain.

(b) Above one of these solutions the vapor pressure of water, P_{H_2O}, remains *constant*, even as water evaporates from solution. Which solution is this? Explain.

(c) Which of these solutions has the *higher* boiling point? Explain.

77. Cooks often add some salt to water before boiling it. Some people say this helps the cooking process by raising the boiling point of the water. Others say not enough salt is usually added to make any noticeable difference. Approximately how many grams of NaCl must be added to a liter of water at 1 atm pressure to raise the boiling point by 2 °C? Is this a typical amount of salt that you might add to cooking water?

78. Aqueous solutions of acetone (used in nail polish remover), $(CH_3)_2CO$, are prepared in which the concentration of acetone is **(a)** 0.10%, by mass; **(b)** 0.10 M; **(c)** 0.10 m; and **(d)** $\chi_{acetone} = 0.10$. Which of these solutions has the highest vapor pressure of water? the lowest freezing point?

79. The formation of NaCl(aq) is an *endothermic* process ($\Delta H_{soln} \approx 5$ kJ/mol). By contrast, for LiCl(aq) $\Delta H_{soln} = -35$ kJ/mol, an *exothermic* process. What factor(s) do you suppose accounts for this difference in solution behavior?

80. Water and phenol are only partially miscible at temperatures below 66.8 °C. In a mixture prepared at 29.6 °C from 50.0 g water and 50.0 g phenol, 32.8 g of a phase consisting of 92.50% water and 7.50% phenol by mass is obtained. This is a saturated solution of phenol in water. What is the mass percent of water in the second phase—a saturated solution of water in phenol?

81. An aqueous solution has 109.2 g KOH/L solution. The solution density is 1.09 g/mL. Your task is to use 100.0 mL of this solution to prepare 0.250 *m* KOH. What mass of which component, KOH or H_2O, would you add to the 100.0-mL of solution?

82. A solid mixture consists of 85.0% KNO_3 and 15.0% K_2SO_4, by mass. A 60.0-g sample of this solid is added to 130.0 g of water at 60 °C. With reference to Figure 14-8

(a) Will all the solid dissolve at 60 °C?

(b) If the resulting solution is cooled to 0 °C, what mass of KNO_3 should crystallize?

(c) Will K_2SO_4 also crystallize at 0 °C?

83. Henry's law can be stated this way: The mass of a gas dissolved by a given quantity of solvent at a fixed temperature is directly proportional to the pressure of the gas. Show how this statement is related to Equation (14.2).

84. Still another statement of Henry's law is this: A given quantity of liquid at a fixed temperature dissolves the same volume of gas at all pressures. What is the connection between this statement and the one given in Exercise 83? Under what conditions is this second statement not valid?

85. A benzene–toluene solution with $\chi_{benz} = 0.300$ has a normal boiling point of 98.6 °C. The vapor pressure of pure toluene at 98.6 °C is 533 mmHg. What must be the vapor pressure of pure benzene at 98.6 °C? (Assume ideal solution behavior.)

86. Suppose you have available 2.50 L of a solution ($d = 0.9767$ g/mL) that is 13.8% ethanol (C_2H_5OH), by mass. From

this solution you would like to make the *maximum* quantity of ethanol–water antifreeze solution that will offer protection to -2.0 °C. Would you add more ethanol or more water to the solution? What mass of liquid would you add?

87. An important test for the purity of an organic compound is to measure its melting point. Usually, if the compound is not pure it begins to melt at a *lower* temperature than the pure compound.

(a) Why is this the case, rather than the melting point being higher in some cases and lower in others?

(b) Are there any conditions under which the melting point of the impure compound may be *higher* than that of the pure compound? Explain.

88. Stearic acid ($C_{18}H_{36}O_2$) and palmitic acid ($C_{16}H_{32}O_2$) are common fatty acids. Commercial grades of stearic acid usually contain palmitic acid as well. A 1.115-g sample of a commercial-grade stearic acid is dissolved in 50.00 mL benzene ($d = 0.879$ g/mL). The freezing point of the solution is found to be 5.072 °C. The freezing point of pure benzene is 5.533 °C and K_f for benzene is 5.12 °C m^{-1}. What is the mass percent of palmitic acid in the stearic acid sample?

89. Nitrobenzene, $C_6H_5NO_2$, and benzene, C_6H_6, are completely miscible in each other. Other properties of the two liquids are *nitrobenzene*: fp = 5.7 °C, $K_f = 8.1$ °C m^{-1}; *benzene*: fp = 5.5 °C, $K_f = 5.12$ °C m^{-1}. With these two liquids it is possible to prepare *two* different solutions having a freezing point of 0.0 °C. What are the compositions of these two solutions, expressed as mass percent nitrobenzene?

90. Refer to Figure 14-14a. Initially, solution A contains 0.515 g urea, $CO(NH_2)_2$, dissolved in 92.5 g H_2O, and solution B, 2.50 g sucrose, $C_{12}H_{22}O_{11}$, dissolved in 85.0 g H_2O. What are the compositions of the two solutions when equilibrium is reached, that is, when the two have the same vapor pressure?

91. In Figure 14-15, why does the transfer of water stop when the two solutions are of *nearly* equal concentrations rather than of *exactly* equal concentrations?

92. At 20 °C liquid benzene has a density of 0.879 g/mL and liquid toluene, 0.867 g/mL. Assume that benzene–toluene solutions are ideal. Let the volume percent benzene be V and show that $d = 1/100[0.879\ V + 0.867(100 - V)]$.

93. Demonstrate that

(a) for a *dilute aqueous* solution the numerical value of the molality is essentially equal to that of the molarity;

(b) in a *dilute* solution the solute mole fraction is proportional to the molality;

(c) In a *dilute aqueous* solution the solute mole fraction is proportional to the molarity.

94. At 25 °C and under an $O_2(g)$ pressure of 1 atm, the solubility of $O_2(g)$ in water is 28.31 mL/1.00 L H_2O. Under an $N_2(g)$ pressure of 1 atm, the solubility of $N_2(g)$ at 25 °C is 14.34 mL/1.00 L H_2O. The composition of the atmosphere is 78.08% N_2 and 20.95% O_2, by volume. What is the composition of air dissolved in water at 25 °C, expressed as volume percents of N_2 and O_2?

95. A saturated solution is prepared at 70 °C containing 32.0 g $CuSO_4$ per 100.0 g solution. A 335-g sample of this solution is then cooled to 0 °C and $CuSO_4 \cdot 5H_2O$ crystallizes out. If the concentration of a saturated solution at 0 °C is 12.5 g $CuSO_4$/100 g soln, what mass of $CuSO_4 \cdot 5H_2O$ would be obtained? (*Hint:* Note that the solution composition is stated in terms of $CuSO_4$ and that the solid that crystallizes is the hydrate $CuSO_4 \cdot 5H_2O$.)

Feature Problems

A. The term "proof," still used to describe the ethanol content of alcoholic beverages, originated in seventeenth-century England. A sample of whiskey was poured on gunpowder and set afire. If the gunpowder ignited after the whiskey had burned off, this "proved" that the whiskey had not been watered down. The minimum ethanol content for a positive test was about 50%, by volume. The 50% ethanol solution became known as "100 proof." Thus, the label pictured here shows the 80-proof whiskey to be 40% CH_3CH_2OH by volume. Listed in the table are some data for several aqueous solutions of ethanol. *With a minimum amount of calculation,* determine which of the solutions are more than "100 proof."

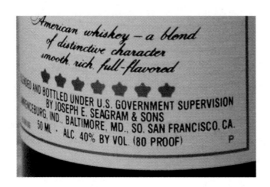

Molarity of Ethanol	Density of Solution
4.00 M	0.970 g/mL
5.00	0.963
6.00	0.955
7.00	0.947
8.00	0.936
9.00	0.926
10.00	0.913

B. The phase diagram on the right is for mixtures of HCl and H_2O at a pressure of 1 atm. The red curve represents the normal boiling points of solutions of HCl(aq) of various mole fractions. The blue curve represents the compositions of the vapors in equilibrium with boiling solutions.

1. As a solution containing $\chi_{HCl} = 0.50$ begins to boil, will the vapor have a mole fraction of HCl equal to, less than, or greater than 0.50? Explain.

2. As a solution containing $\chi_{HCl} = 0.05$ begins to boil, will the vapor have a mole fraction of HCl equal to, less than, or greater than 0.05? Explain.

3. Unlike in the boiling of a pure liquid, the composition of HCl(aq) *changes* as the solution boils in an open container. Explain why this is so?

4. One particular solution (the azeotrope) is an exception to the observation stated in part **3**; that is, its composition remains *unchanged* during boiling. What is the approximate composition and boiling point of this solution?

5. A 5.00-mL sample of the azeotrope (*d* = 1.099 g/mL) requires 30.32 mL of 1.006 M NaOH for its titration in an acid-base reaction. Use these data to determine a more precise value of the composition of the azeotrope, expressed as the mole fraction of HCl.

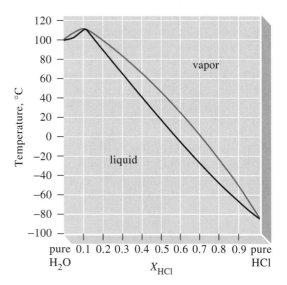

C. The laboratory device pictured on page 503 is called a *desiccator.* It can be used to maintain a constant relative humidity within an enclosure. The materials(s) used to control the relative humidity are placed in the bottom compartment, and the substance being subjected to a controlled relative humidity is placed in a container on the platform.

1. If the material in the bottom compartment is a saturated solution of NaCl(aq) in contact with NaCl(s), what should be the approximate relative humidity in the container at a temperature of 20 °C? Obtain the solubility of NaCl from Figure 14-8 and vapor pressure data for water from Table 13.2; use the definition of Raoult's law from page 482 and that of relative humidity from page 244; and assume that the NaCl is completely dissociated into its ions.

2. If the material placed in the beaker is dry $CaCl_2 \cdot 6H_2O(s)$, will the solid deliquesce? Explain. (*Hint:* Recall the discussion on page 485.)

3. To maintain $CaCl_2 \cdot 6H_2O(s)$ in the dry state in a desiccator, should the substance in the saturated solution in the bottom compartment be one with a high or a low water solubility? Explain.

D. Every year, oral rehydration therapy (ORT)—the feeding of an electrolyte solution—saves the lives of countless children worldwide who become severely dehydrated as a result of diarrhea. One requirement of the solution used is that it be *isotonic* with human blood.

1. One definition of an isotonic solution given in the text is that it have the same osmotic pressure as 0.92% NaCl(aq) (mass/volume). Another definition is that the solution have a freezing point of -0.52 °C. Show that these two definitions are in reasonably close agreement, given that we are using solution concentrations rather than activities.

2. Use the freezing-point definition from part **1** to show that an ORT solution containing 3.5 g NaCl, 1.5 g KCl, 2.9 g $Na_3C_6H_5O_7$ (sodium citrate), and 20.0 g $C_6H_{12}O_6$ (glucose) per liter meets the requirement of being isotonic. (*Hint:* Which of the solutes are nonelectrolytes and which are strong electrolytes?)

Chemical Kinetics

15

*I*n burning a rocket fuel we seek a rapid release of gaseous products and energy to give the rocket maximum thrust. We store milk in a refrigerator to slow down the chemical reactions that cause it to spoil. The strategies we are forming to reduce the rate of deterioration of the ozone layer seek to deprive the ozone-consuming reaction cycle of key intermediates that come from chlorofluorocarbons (CFCs). These examples all point to the need to know how quickly chemical reactions occur.

In addition to these practical applications, one of the main reasons for wanting to know how fast a reaction occurs is for the insight this information provides into the mechanism of the reaction—the step-by-step molecular pathway leading from reactants to products. *Chemical kinetics,* the topic of this chapter, is the study of rates of chemical reactions, rate laws of chemical reactions, and reaction mechanisms.

We will begin the chapter by describing what we mean by the term "rate of a reaction" and presenting some ideas that deal with measuring rates of reaction. We will follow this by introducing mathematical equations, called rate laws, that relate the rates of reactions to the concentrations of the reactants. Finally, with this information as background, we will turn to one of our central purposes: relating rate laws to plausible reaction mechanisms.

The decomposition of hydrogen peroxide, H_2O_2, to H_2O and O_2 is a highly exothermic reaction that is strongly catalyzed by platinum metal. Both the rate of this decomposition and the nature of catalysis are explored in this chapter.

15-1 The Rate of a Chemical Reaction

Rate, or speed, refers to something that happens in a unit of time. A car traveling at 60 mph, for example, covers a distance of 60 miles in one hour. For chemical reactions, the **rate of reaction** describes how fast the concentration of a reactant or product changes with time.

To illustrate, let's consider the decomposition of an aqueous solution of hydrogen peroxide, H_2O_2.

$$2 H_2O_2(aq) \longrightarrow 2 H_2O(l) + O_2(g) \tag{15.1}$$

▶ H_2O_2, in a 3% aqueous solution, is a common antiseptic (Figure 15-17 on page 533).

Suppose we start with 1.000 M H_2O_2 and discover that 10.0 s later the solution concentration is 0.983 M H_2O_2. In a period of time, $\Delta t = 10.0$ s, the change in concentration $\Delta[H_2O_2] = 0.983$ M $- 1.000$ M $= -0.017$ M. The average rate of change of $[H_2O_2]$ during this time interval is the change in H_2O_2 concentration divided by the length of time.

▶ Recall that the symbol [ ] means "the molarity of." Also, Δ means "the change in," that is, the final value minus the initial value.

$$\text{rate of change of } [H_2O_2] = \frac{\Delta[H_2O_2]}{\Delta t} = \frac{-0.017 \text{ M}}{10.0 \text{ s}}$$

$$= -1.7 \times 10^{-3} \text{ M s}^{-1}$$

To avoid using a negative quantity for the rate of reaction of a substance consumed in a reaction, let's take its rate of reaction to be the *negative* of the rate of change of its concentration.

$$\text{rate of reaction of } H_2O_2 = -(\text{rate of change of } [H_2O_2])$$

$$= -(-1.7 \times 10^{-3} \text{ M s}^{-1}) = 1.7 \times 10^{-3} \text{ M s}^{-1}$$

Note that the units of a rate of reaction are mol L^{-1} (time)$^{-1}$, such as mole per liter per second, which we can write as mol L^{-1} s^{-1} or M s^{-1}.

If two or more reactants are involved, we have a bit of a problem describing the rate of a chemical reaction, as in the following example.

$$4 NH_3(g) + 5 O_2(g) \longrightarrow 4 NO(g) + 6 H_2O(g) \tag{15.2}$$

We can describe the rate of reaction (15.2) in terms of either reactant, but these rates are not the same. *Five* moles of O_2 are consumed for every *four* moles of NH_3; the rate of reaction of O_2 is somewhat greater than that of NH_3. We can express this fact as

$$\text{rate of reaction of } O_2 = \frac{-\Delta[O_2]}{\Delta t} = -\frac{5}{4}\frac{\Delta[NH_3]}{\Delta t} = \frac{5}{4}(\text{rate of reaction of } NH_3)$$

Rates of reaction are more often expressed in terms of the reactants than of the products, but should we want to know the rate of formation of a product in reaction (15.2), we can still use the stoichiometric coefficients to write, for example,

$$\text{rate of formation of NO} = \text{rate of reaction of } NH_3$$

$$\text{rate of formation of } H_2O = \frac{6}{4}(\text{rate of reaction of } NH_3)$$

In this text we will usually indicate the particular reactant or product whose rate of change in concentration is the basis of the rate of reaction.*

---

*An IUPAC recommendation (not followed in this text) is to define a *general* rate of reaction, which for reaction (15.2) takes the form:

$$\textit{rate of reaction} = -\frac{1}{4}(\text{rate of change of } [NH_3]) = -\frac{1}{5}(\text{rate of change of } [O_2])$$

$$= \frac{1}{4}(\text{rate of change of } [NO]) = \frac{1}{6}(\text{rate of change of } [H_2O])$$

EXAMPLE 15-1

Expressing the Rate of a Reaction. Suppose that at some point in the reaction

$$A + 3 B \longrightarrow 2 C + 2 D$$

[A] = 0.9986 M, and that 13.20 min later [A] = 0.9746 M. What is the average rate of reaction of A during this time period, expressed in M s^{-1}?

Solution

The rate of change of [A] is the *change* in molarity, Δ[A], divided by the time interval over which this change occurs, Δt: Δ[A] = 0.9746 M − 0.9986 M = −0.0240 M; Δt = 13.20 min.

$$\text{rate of reaction of A} = -(\text{rate of change of [A]})$$

$$= \frac{-\Delta[A]}{\Delta t} = \frac{-(-0.0240 \text{ M})}{13.20 \text{ min}}$$

$$= 1.82 \times 10^{-3} \text{ M min}^{-1}$$

To express the rate of reaction in moles per liter per second means converting from min^{-1} to s^{-1}. We can do this with the conversion factor 1 min/60 s.

$$\text{rate of reaction of A} = 1.82 \times 10^{-3} \text{ M min}^{-1} \times \frac{1 \text{ min}}{60 \text{ s}}$$

$$= 3.03 \times 10^{-5} \text{ M s}^{-1}$$

Alternatively, we could have converted 13.20 min to 792 s and used Δt = 792 s in evaluating the rate of reaction of A.

Practice Example A: At some point in the reaction 2 A + B → C + D, [A] = 0.3629 M. At a time 8.25 min later [A] = 0.3187 M. What is the average rate of reaction of A during this time interval, expressed in M s^{-1}?

Practice Example B: In the reaction 2 A → 3 B, [A] drops from 0.5684 M to 0.5522 M in 2.50 min. What is the average rate of formation of B during this time interval, expressed in M s^{-1}? (*Hint:* How is the rate of formation of B related to the rate of reaction of A?)

15-2   Measuring Reaction Rates

To determine a rate of reaction we need to measure changes in concentration and in time. A change in time can be measured with a stopwatch or other timing device, but how do we measure concentration changes during a chemical reaction? Also, why have we used the term *average* in referring to a rate of reaction? These are two of the questions we will answer in this section.

▶ You will encounter all of the ideas introduced in Section 15-2, and some from elsewhere in the chapter, in Feature Problem A on page 542.

Following a Chemical Reaction

In reaction (15.1), the decomposition of H_2O_2, O_2(g) escapes from the reaction mixture and the reaction goes to completion.*

$$2 H_2O_2(aq) \longrightarrow 2 H_2O(l) + O_2(g)$$

*Although reaction (15.1) goes to completion, it does so very slowly. Generally, a catalyst is used to speed up the reaction. We describe the function of a catalyst in this reaction in Section 15-11.

FIGURE 15-1
Experimental set-up for determining the rate of decomposition of H$_2$O$_2$.
Oxygen gas given off by the reaction mixture is trapped, and its volume is measured in the gas buret. The amount of H$_2$O$_2$ consumed and the remaining concentration of H$_2$O$_2$ can be calculated from the measured volume of O$_2$(g).

TABLE 15.1
Decomposition of H$_2$O$_2$

Time, s	[H$_2$O$_2$], M
0	2.32
200	2.01
400	1.72
600	1.49
1200	0.98
1800	0.62
3000	0.25

▶One's instantaneous speed can result in a speeding ticket even if the average speed never exceeds the speed limit.

We can follow the progress of the reaction by focusing either on the formation of O$_2$(g) or on the disappearance of H$_2$O$_2$. That is, we can

- Measure the volumes of O$_2$(g) (Figure 15-1) and relate these volumes to decreases in concentration of H$_2$O$_2$.
- Remove small samples of the reaction mixture from time to time and analyze these samples for their H$_2$O$_2$ content. We can do this, for example, by titration with KMnO$_4$ in acidic solution. The net ionic equation for this oxidation–reduction reaction is

$$2\ MnO_4^- + 5\ H_2O_2 + 6\ H^+ \longrightarrow 2\ Mn^{2+} + 8\ H_2O + 5\ O_2(g) \qquad (15.3)$$

Table 15.1 lists typical data for the decomposition of H$_2$O$_2$, and Figure 15-2 displays comparable data graphically.

Rate of Reaction Expressed as $-\Delta[H_2O_2]/\Delta t$

We have extracted some data from Figure 15-2 and listed them in Table 15.2 (in blue). Column III lists the molarities of H$_2$O$_2$ at the times shown in column I. Column II lists the *arbitrary* time interval we have chosen between data points—400 s. Column IV reports the concentration changes that occur for each 400-s interval. The rates of reaction of H$_2$O$_2$ are shown in column V. Can you see that the reaction rate is not constant? The lower the remaining concentration of H$_2$O$_2$, the more slowly the reaction proceeds.

Rate of Reaction Expressed as the Slope of a Tangent Line

When we express the rate of reaction as $-\Delta[H_2O_2]/\Delta t$, we simply get an *average* value for the time interval Δt. For example, in the interval from 1200 to 1600 s the rate averages 6.3×10^{-4} M s^{-1} (fourth entry in column V of Table 15.2). We can think of this as the reaction rate at about the middle of the interval—1400 s. We could just as well have chosen $\Delta t = 200$ s in Table 15.2. In this case, to obtain the rate of reaction at $t = 1400$ s, we would use concentration data at $t = 1300$ s and $t = 1500$ s. The rate of reaction would be slightly less than 6.3×10^{-4} M s^{-1}. The smaller the interval we choose, the closer we will come to the actual rate at $t = 1400$ s. As Δt approaches 0 s, the rate of reaction approaches the *negative of the slope of the tangent line* to the graph of Figure 15-2. The rate of reaction determined from the slope of a tangent line to a concentration–time graph is the **instantaneous rate of reaction** at the point in the reaction at which the tangent line is drawn. We show two tangent lines in Figure 15-2, one at $t = 0$ and one at $t = 1400$ s. Our best estimate of the rate of reaction of H$_2$O$_2$ at 1400 s, then, is from the slope of the blue tangent line in Figure 15-2: $-(-6.1 \times 10^{-4}$ M s$^{-1}) = 6.1 \times 10^{-4}$ M s^{-1}.

As an analogy to average and instantaneous rates, think of taking a 108-mi highway trip in 2.00 h. The *average* speed or rate is 54.0 mi/h. The *instantaneous* speed is the speedometer reading at any instant.

Initial Rate of Reaction

Sometimes we simply want to find the rate of reaction when the reactants are first brought together—the **initial rate of reaction.** One way to get this is from the tangent line to the concentration–time graph at $t = 0$. An alternative way is

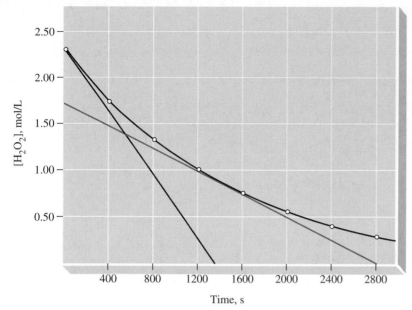

FIGURE 15-2
Graphical representation of kinetic data for the reaction:
2 H$_2$O$_2$(aq) → 2 H$_2$O(l) + O$_2$(g).
This is the usual form in which concentration–time data are plotted. Reaction rates are determined from the slopes of the tangent lines. The blue line has a slope of -1.70 M/2800 s $= -6.1 \times 10^{-4}$ M s^{-1}. The slope of the black line and its relation to the initial rate of reaction are described in Example 15-2.

to measure the concentration of the chosen reactant as soon as possible after mixing, in this way obtaining Δ[reactant] for a very short time interval (Δt) at essentially $t = 0$. These two approaches give the same result if we limit ourselves to the time interval in which the tangent line and the concentration–time graph practically coincide. In Figure 15-2 this occurs for about the first 200 s.

TABLE 15.2	Decomposition of H$_2$O$_2$—Derived Rate Data			
I	**II**	**III**	**IV**	**V**
				Reaction Rate =
Time, s	**Δt, s**	**[H$_2$O$_2$], M**	**Δ[H$_2$O$_2$], M**	**$-\Delta[\text{H}_2\text{O}_2]/\Delta t$, M s^{-1}**
0		2.32		
	400		-0.60	15.0×10^{-4}
400		1.72		
	400		-0.42	10.5×10^{-4}
800		1.30		
	400		-0.32	8.0×10^{-4}
1200		0.98		
	400		-0.25	6.3×10^{-4}
1600		0.73		
	400		-0.19	4.8×10^{-4}
2000		0.54		
	400		-0.15	3.8×10^{-4}
2400		0.39		
	400		-0.11	2.8×10^{-4}
2800		0.28		

EXAMPLE 15-2

Determining and Using an Initial Rate of Reaction. From the data in Table 15.1 and Figure 15-2 for the decomposition of H_2O_2, **(a)** determine the initial rate of decomposition of H_2O_2, and **(b)** $[H_2O_2]_t$ at $t = 100$ s.

Solution

a. To determine the initial rate of decomposition from the slope of the tangent line, we use the intersections of the black tangent line with the axes: $t = 0$, $[H_2O_2] = 2.32$ M; $t = 1360$ s, $[H_2O_2] = 0$.

$$\text{initial rate} = -(\text{slope of tangent line}) = \frac{-(0 - 2.32)\ M}{(1360 - 0)\ s}$$

$$= 1.71 \times 10^{-3}\ M\ s^{-1}$$

An alternative method is to use data from Table 15.1: $[H_2O_2] = 2.32$ M at $t = 0$ and $[H_2O_2] = 2.01$ M at $t = 200$ s.

$$\text{initial rate} = \frac{-\Delta[H_2O_2]}{\Delta t} = \frac{-(2.01 - 2.32)\ M}{200\ s}$$

$$= 1.6 \times 10^{-3}\ M\ s^{-1}$$

The agreement between the two methods is fairly good, although it might be better if the time interval were less than 200 s. Of the two results, that based on the tangent line is presumably more precise, because it is expressed with more significant figures. On the other hand, the reliability of the tangent line depends on how carefully the tangent line is constructed.

▶ Another reason for favoring a graphical method is that it tends to minimize the effect of errors that may be found in individual data points.

b. We assume that the rate determined in **(a)** remains essentially constant for at least 100 s. Because

$$\text{rate of reaction of } H_2O_2 = \frac{-\Delta[H_2O_2]}{\Delta t}$$

then

$$1.71 \times 10^{-3}\ M\ s^{-1} = \frac{-\Delta[H_2O_2]}{100\ s}$$

$$-(1.71 \times 10^{-3}\ M\ s^{-1})(100\ s) = \Delta[H_2O_2] = [H_2O_2]_t - [H_2O_2]_0$$

$$-1.71 \times 10^{-1}\ M = [H_2O_2]_t - 2.32\ M$$

$$[H_2O_2]_t = 2.32\ M - 0.17\ M = 2.15\ M$$

Practice Example A: For reaction (15.1), determine **(a)** the *instantaneous* rate of reaction at 2400 s and **(b)** $[H_2O_2]$ at 2450 s. (*Hint:* Draw a tangent line to Figure 15-2 at 2400 s. Assume that the rate at 2400 s holds constant for the next 50 s.)

Practice Example B: Use data *only* from Table 15.2 to determine $[H_2O_2]$ at $t = 100$ s. Compare this value to the one calculated in Example 15-2b. Explain the reason for the difference.

15-3   The Effect of Concentrations on Rates of Reactions: The Rate Law

One of the goals in a chemical kinetics study is to derive an equation we can use to predict how a rate of reaction depends on the concentrations of reactants.

Are You Wondering . . .

How long a time interval Δt you're allowed to use in calculating a reaction rate with the expression $-\Delta[\text{reactant}]/\Delta t$?

There's no set time interval that applies to all reactions. Instead, use this rule of thumb: The method generally works well if only a small percentage of the reactants are consumed, say less than 5%. In the 200-s interval in Example 15-2(a), $[H_2O_2]$ decreased from 2.32 to 2.01 M. About 13% of the H_2O_2 was consumed. The 200-s value of Δt was a little too large.

Such an experimentally determined equation is called a **rate law,** or **rate equation.**

Consider the hypothetical reaction

$$a\text{A} + b\text{B} + \cdots \longrightarrow g\text{G} + h\text{H} + \cdots \qquad (15.4)$$

where $a, b, \ldots$ stand for coefficients in the balanced equation. We can often express the rate of such a reaction as*

$$\text{rate of reaction} = k[\text{A}]^m[\text{B}]^n \cdots \qquad (15.5)$$

The terms [A], [B], . . . represent reactant molarities. The exponents, $m, n, \ldots$ are generally small, positive whole numbers, although in some cases they may be zero, fractional, and/or negative. They are *not* generally related to the stoichiometric coefficients $a, b, \ldots$. That is, often $m \neq a$, $n \neq b$, and so on.

The term *order* is used in two ways in describing a rate of reaction: (1) If $m = 1$, we say that the reaction is *first order in A.* If $n = 2$, the reaction is *second order in B,* and so on. (2) The *overall* **order of reaction** is the sum of all the exponents: $m + n + \cdots$. The proportionality constant k relates the rate of a reaction to reactant concentrations and is called the **rate constant** of the reaction. Its value depends on the specific reaction, the presence of a catalyst (if any), and the temperature. *The larger the value of k, the faster a reaction goes.* The units of k depend on the order of the reaction (that is, on the values of the exponents $m, n, \ldots$).

With the rate law for a reaction, we can

- calculate rates of reaction for known concentrations of reactants.
- derive an equation that expresses a reactant concentration as a function of time.

But how do we establish the rate law? We need to use *experimental* data of the type described in Section 15-2. The method we describe next works especially well.

---

*We assume that reaction (15.4) goes to completion. If it is reversible, the rate equation is more complex than (15.5). Even for reversible reactions, though, Equation (15.5) applies to the *initial rate* of reaction because in the early stages of the reaction there are not enough products for a significant reverse reaction to occur.

▶ The method of initial rates is used in Feature Problem B, on page 543, together with additional relationships from later sections of this chapter.

Method of Initial Rates

As its name implies, this method requires us to work with *initial* rates of reaction. As an example, let's look at a specific reaction: that between mercury(II) chloride and oxalate ion.

$$2\ HgCl_2(aq) + C_2O_4{}^{2-}(aq) \longrightarrow 2\ Cl^-(aq) + 2\ CO_2(g) + Hg_2Cl_2(s) \qquad (15.6)$$

The tentative rate law that we can write for this reaction is

$$\text{rate of reaction of } C_2O_4{}^{2-} = k[HgCl_2]^m[C_2O_4{}^{2-}]^n \qquad (15.7)$$

We can follow the reaction by measuring the quantity of $Hg_2Cl_2(s)$ formed as a function of time. In Example 15-3, we will use some of the representative data in Table 15.3 to illustrate the method of initial rates.

TABLE 15.3   Kinetic Data for the Reaction:
$2\ HgCl_2 + C_2O_4{}^{2-} \longrightarrow 2\ Cl^- + 2\ CO_2 + Hg_2Cl_2$

Experiment	[HgCl₂], M	[C₂O₄²⁻], M	Initial rate, M min⁻¹
1	$[HgCl_2]_1 = 0.105$	$[C_2O_4{}^{2-}]_1 = 0.15$	1.8×10^{-5}
2	$[HgCl_2]_2 = 0.105$	$[C_2O_4{}^{2-}]_2 = 0.30$	7.1×10^{-5}
3	$[HgCl_2]_3 = 0.052$	$[C_2O_4{}^{2-}]_3 = 0.30$	3.5×10^{-5}

EXAMPLE 15-3

Establishing the Order of a Reaction by the Method of Initial Rates. Use data from Table 15.3 to establish the order of reaction (15.6) with respect to $HgCl_2$ and $C_2O_4{}^{2-}$ and also the overall order of the reaction.

Solution

We need to determine the values of m and n in Equation (15.7). In comparing Experiment 2 with Experiment 3, note that $[HgCl_2]$ is essentially doubled (0.105 M $\approx$ 2 × 0.052 M) while $[C_2O_4{}^{2-}]$ is held constant (at 0.30 M). Note also that $R_2 = 2 \times R_3$ (7.1 × 10⁻⁵ ≈ 2 × 3.5 × 10⁻⁵). Rather than use the actual concentrations and rates in the following rate equation, let's work with their symbolic equivalents.

$$R_2 = k \times [HgCl_2]_2^m \times [C_2O_4{}^{2-}]_2^n = k \times (2 \times [HgCl_2]_3)^m \times [C_2O_4{}^{2-}]_3^n$$

$$R_3 = k \times [HgCl_2]_3^m \times [C_2O_4{}^{2-}]_3^n$$

$$\frac{R_2}{R_3} = \frac{2 \times R_3}{R_3} = 2 = \frac{k \times 2^m \times [HgCl_2]_3^m \times [C_2O_4{}^{2-}]_3^n}{k \times [HgCl_2]_3^m \times [C_2O_4{}^{2-}]_3^n} = 2^m$$

In order that $2^m = 2$, $m = 1$.

To determine the value of n we can form the ratio R_2/R_1. Now, $[C_2O_4{}^{2-}]$ is doubled and $[HgCl_2]$ is held constant. This time let's use actual concentrations instead of symbolic equivalents. Also, we now have the value $m = 1$.

$$R_2 = k \times [HgCl_2]_2^1 \times [C_2O_4{}^{2-}]_2^n = k \times (0.105)^1 \times (2 \times 0.15)^n$$

$$R_1 = k \times [HgCl_2]_1^1 \times [C_2O_4{}^{2-}]_1^n = k \times (0.105)^1 \times (0.15)^n$$

$$\frac{R_2}{R_1} = \frac{7.1 \times 10^{-5}}{1.8 \times 10^{-5}} \approx 4 = \frac{k \times (0.105)^1 \times 2^n \times (0.15)^n}{k \times (0.105)^1 \times (0.15)^n} = 2^n$$

In order that $2^n = 4$, $n = 2$.

In summary, the reaction is *first* order in $HgCl_2$ ($m = 1$), *second* order in $C_2O_4^{2-}$ ($n = 2$), and *third* order overall ($m + n = 1 + 2 = 3$).

Practice Example A: The decomposition of N_2O_5 is given by the following equation:

$$2\ N_2O_5 \longrightarrow 4\ NO_2 + O_2$$

At an initial $[N_2O_5] = 3.15$ M, the initial rate of reaction $= 5.45 \times 10^{-5}$ M s^{-1}, and when $[N_2O_5] = 0.78$ M, the initial rate of reaction $= 1.35 \times 10^{-5}$ M s^{-1}. Determine the order of this decomposition reaction.

Practice Example B: Consider a hypothetical Experiment 4 in Table 15.3 in which the initial conditions are $[HgCl_2]_4 = 0.025$ M and $[C_2O_4^{2-}]_4 = 0.045$ M. Predict the initial rate of reaction. (*Hint:* Work with a ratio of two initial rates but now with known values of m and n.)

In Example 15-3 we made an important observation: if a reaction is *first order* in one of the reactants, doubling the initial concentration of that reactant causes the initial rate of reaction to double. In general, the effect on the initial rate caused by a *doubling* of the concentration of a reactant is related to the reaction order in that reactant in the following way: For a zero-order reaction the initial rate is *unaffected* ($2^0 = 1$); for a first-order reaction the initial rate *doubles* ($2^1 = 2$). With a second-order reaction the initial rate increases *fourfold* ($2^2 = 4$), and with a third-order reaction, *eightfold* ($2^3 = 8$).

Once we have the exponents in a rate equation, we can determine the value of the rate constant k. To do this, all we need is the rate of reaction corresponding to known initial concentrations of reactants, as illustrated in Example 15-4.

EXAMPLE 15-4

Using the Rate Equation. Use the results of Example 15-3 and data from Table 15.3 to establish the value of k in the rate equation (15.7).

Solution

We can use data from any one of the three experiments of Table 15.3, together with the values $m = 1$ and $n = 2$. First, we solve Equation (15.7) for k.

$$k = \frac{R_1}{[HgCl_2][C_2O_4^{2-}]^2} = \frac{1.8 \times 10^{-5}\ \text{M min}^{-1}}{0.105\ \text{M} \times (0.15)^2\ \text{M}^2}$$

$$= 7.6 \times 10^{-3}\ \text{M}^{-2}\ \text{min}^{-1}$$

▶ Notice that the units of k, $\text{M}^{-2}\ \text{min}^{-1}$, lead to the correct units for the rate of the reaction: rate of reaction $=$ $k \times [HgCl_2][C_2O_4^{2-}]^2 =$ $\text{M}^{-2}\ \text{min}^{-1} \times \text{M} \times \text{M}^2 =$ M min^{-1}

Practice Example A: A reaction has the rate law: rate $= k[A]^2[B]$. When $[A] = 1.12$ M and $[B] = 0.87$ M, the rate of reaction $= 4.78 \times 10^{-2}$ M s^{-1}. What is the value of the rate constant, k?

Practice Example B: What is the rate of reaction (15.6) at the point where $[HgCl_2] = 0.050$ M and $[C_2O_4^{2-}] = 0.025$ M? [*Hint:* Use the rate equation (15.7) with its known values of m, n, and k.]

Are You Wondering . . .

Just what is the difference between the rate of a reaction and the rate *constant* of a reaction?

Many students have difficulty with this distinction. Remember, a *rate of reaction* tells how the amount of a reactant or product changes with time and is usually expressed as moles per liter per time. A rate of reaction can be established through an expression of the type $-\Delta[A]/\Delta t$, from a tangent to a concentration–time graph, and by calculation from a rate equation. In most cases the rate of a reaction strongly depends on reactant concentrations.

The *rate constant* of a reaction (k) relates the rate of a reaction to reactant concentrations. Generally, it is not itself a rate of reaction, but we can use it to calculate rates of reaction. Once the value of k at a given temperature has been established, *this value stays fixed*. Whereas the units of rates of reaction do not depend on the order of a reaction, those of k do, as we indicate in the next three sections.

15-4 Zero-Order Reactions

▶ Perhaps the most important examples of zero-order reactions are found in the action of enzymes. Enzyme-catalyzed reactions are discussed in Section 15-11.

A **zero-order** reaction has a rate equation in which the sum of the exponents, $m + n + \cdots$ is equal to 0. As an example, let's take a reaction in which a single reactant A decomposes to products

$$A \longrightarrow \text{products}$$

If the reaction is zero order, the rate equation is

$$\text{rate of reaction} = k[A]^0 = k = \text{constant} \qquad (15.8)$$

Other features of this zero-order reaction are

- The concentration–time graph is a *straight line* with a *negative* slope (Figure 15-3).
- The rate of reaction, which is equal to k and remains *constant* throughout the reaction, is the *negative* of the slope of this line.
- The units of k are the *same* as the units of the rate of a reaction: mol L^{-1} (time)$^{-1}$, for example, mol L^{-1} s^{-1}, or M s^{-1}.

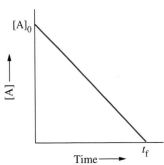

FIGURE 15-3

A zero-order reaction: A → products.

The initial concentration of the reactant A is [A]$_0$, that is, [A] = [A]$_0$ at $t = 0$. [A] decreases at a constant rate until the reaction stops. This occurs at the time, t_f, where [A] = 0. The slope of the line is $(0 - [A]_0)/(t_f - 0) = -[A]_0/t_f$. The rate constant is the *negative* of the slope: $k = -\text{slope} = [A]_0/t_f$.

Equation (15.8) is the *rate law* for a zero-order reaction. Another useful equation, called an **integrated rate equation,** expresses the *concentration of a reactant as a function of time*. We can establish this equation rather easily from the graph in Figure 15-3. Let's start with the general equation for a straight line

$$y = mx + b$$

and substitute $y = [A]_t$ (the concentration of A at some time t); $x = t$ (time); $b = [A]_0$ (the initial concentration of A at time, $t = 0$); and $m = -k$ (m, the slope of the straight line, is obtained as indicated in Figure 15-3).

$$[A]_t = -kt + [A]_0 \qquad (15.9)$$

15-5   First-Order Reactions

A **first-order** reaction has a rate equation in which the sum of the exponents, $m + n + \cdots$ is equal to 1. A particularly common type of first-order reaction, and the only type we will consider, is that in which a single reactant decomposes into products. Reaction (15.1), the decomposition of H_2O_2 that we described in Section 15-2,

$$2\, H_2O_2(aq) \longrightarrow 2\, H_2O(l) + O_2(g)$$

is a first-order reaction. The rate of decomposition depends on the concentration of H_2O_2 raised to the *first* power, that is,

$$\text{rate of reaction of } H_2O_2 = k[H_2O_2] \tag{15.10}$$

It is easy to establish that reaction (15.1) is first order by the *method of initial rates* (recall Practice Example 15-3A). But there are other ways of recognizing a first-order reaction, and for this purpose we will use an equation other than (15.10).

An Integrated Rate Equation for a First-Order Reaction

Let's begin our discussion of first-order reactions as we did zero-order reactions in Section 15-4, by examining a hypothetical reaction

$$A \longrightarrow \text{products}$$

for which the rate law is

$$\text{rate of reaction} = -(\text{rate of change of } [A]) = k[A] \tag{15.11}$$

We can obtain the *integrated* rate equation for this first-order reaction by applying the calculus technique known as *integration* to Equation (15.11). The result of this derivation is*

$$\ln\frac{[A]_t}{[A]_0} = -kt \quad \text{or} \quad \ln[A]_t = -kt + \ln[A]_0 \tag{15.12}$$

$[A]_t$ is the concentration of A at time t, $[A]_0$ is its concentration at $t = 0$, and k is the rate constant. Because the logarithms of numbers have no units (are dimensionless), the product $-k \times t$ must also be without units. This means that the unit of k in a first-order reaction is $(\text{time})^{-1}$, such as s^{-1} or $\min^{-1}$. Equation (15.12) is that of a straight line.

▶ We can also see that the unit of k must be $(\text{time})^{-1}$ from the rate law (15.11): rate of reaction = $k[A] = (\text{time})^{-1} \times \text{mol L}^{-1} = \text{mol L}^{-1} (\text{time})^{-1} = M (\text{time})^{-1}$.

$$\ln[A]_t = (-k)t + \ln[A]_0$$

Equation of straight line $\underbrace{y}\ \ =\ \underbrace{m \cdot x}\ +\ \underbrace{b}$

An easy test for a first-order reaction is to plot the logarithm of a reactant concentration versus time and see if the graph is linear. The data from Table

---

*The steps in the derivation are
(1) Replace the rate of change of [A] in Equation (15.11) by $d[A]/dt$.
(2) Rearrange (15.11) to the form, $d[A]/[A] = -k\,dt$.
(3) Integrate between the limits $[A]_0$ at time $t = 0$ and $[A]_t$ at time t.

$$\int_{[A]_0}^{[A]_t} \frac{d[A]}{[A]} = -k \int_0^t dt, \qquad \text{yielding } \ln\frac{[A]_t}{[A]_0} = -kt$$

FIGURE 15-4
Test for a first-order reaction: decomposition of H_2O_2(aq).
Here we plot ln $[H_2O_2]$ vs. t. The data are based on Table 15.1 and are listed below. The slope of the line is used in the text.

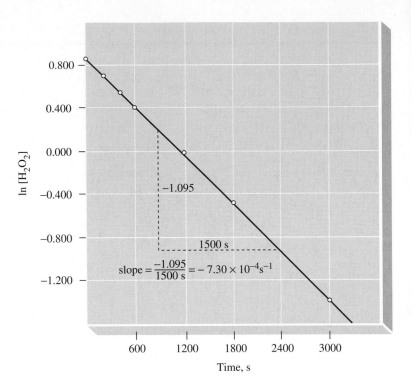

t, s	$[H_2O_2]$, M	ln $[H_2O_2]$
0	2.32	0.842
200	2.01	0.698
400	1.72	0.542
600	1.49	0.399
1200	0.98	−0.020
1800	0.62	−0.48
3000	0.25	−1.39

15.1 are plotted in Figure 15-4 and the rate constant k is derived from the slope of the line: $k = -\text{slope} = -(-7.30 \times 10^{-4} \text{ s}^{-1}) = 7.30 \times 10^{-4} \text{ s}^{-1}$. An alternative, nongraphical approach, illustrated in Practice Example 15-5B, is to substitute data points into Equation (15.12) and solve for k.

EXAMPLE 15-5

Using the Integrated Rate Equation for a First-Order Reaction. H_2O_2(aq), initially at a concentration of 2.32 M, is allowed to decompose. What will $[H_2O_2]$ be at $t = 1200$ s? Use $k = 7.30 \times 10^{-4} \text{ s}^{-1}$ for this first-order decomposition.

Solution

We have values for three of the four terms in Equation (15.12)

$$k = 7.30 \times 10^{-4} \text{ s}^{-1} \qquad t = 1200 \text{ s}$$

$$[H_2O_2]_0 = 2.32 \text{ M} \qquad [H_2O_2]_t = ?$$

which we substitute into the expression

$$\ln [H_2O_2]_t = -kt + \ln[H_2O_2]_0$$

$$= -(7.30 \times 10^{-4} \text{ s}^{-1} \times 1200 \text{ s}) + \ln 2.32$$

$$= \qquad -0.876 \qquad + 0.842 = -0.034$$

$$[H_2O_2]_t = e^{-0.034} = 0.97 \text{ M}$$

▶ To find the number whose natural logarithm is −0.034, raise e to the −0.034 power.

This calculated value (0.97 M) agrees well with the experimentally determined value of 0.98 M.

Practice Example A: The reaction 2 A → 2 B + C is first order. If the initial [A] = 2.80 M and $k = 3.02 \times 10^{-3} \text{ s}^{-1}$, what is the value of [A] after 325 s?

Practice Example B: Use data tabulated in Figure 15-4, together with Equation (15.12), to show that the decomposition of H_2O_2 is a first-order reaction. (*Hint:* Use $[H_2O_2]_0 = 2.32$ M. For some later time—for example, 600 s—use the corresponding $[H_2O_2]_t$—in this case, 1.49 M. Solve for k. Repeat this calculation using other values of t and $[H_2O_2]_t$.)

Although until now we have used only concentrations in kinetics equations, we can at times work directly with the masses of reactants. Another possibility is to work with a fraction of reactant consumed, as with the concept of half-life.

The **half-life** of a reaction is the time required for one-half of a reactant to be consumed. It is the time during which the amount of reactant or its concentration decreases to one-half of its initial value. That is, at $t = t_{1/2}$, $[A]_t = \frac{1}{2}[A]_0$. At this time, Equation (15.12) takes the form

$$\ln \frac{[A]_t}{[A]_0} = \ln \frac{\frac{1}{2}[A]_0}{[A]_0} = \ln \frac{1}{2} = -\ln 2 = -k \times t_{1/2}$$

$$t_{1/2} = \frac{\ln 2}{k} = \frac{0.693}{k} \tag{15.13}$$

For the decomposition of $H_2O_2(aq)$, we conclude that the half-life is

$$t_{1/2} = \frac{0.693}{7.30 \times 10^{-4}\ \text{s}^{-1}} = 9.49 \times 10^2\ \text{s} = 949\ \text{s}$$

Equation (15.13) indicates that *the half-life is constant for a first-order reaction.* Thus, regardless of the value of $[A]_0$ at the time we begin to follow a reaction, at $t = t_{1/2}$, $[A] = \frac{1}{2}[A]_0$. After *two* half-lives, that is, at $t = 2 \times t_{1/2}$, $[A] = \frac{1}{2} \times \frac{1}{2}[A]_0 = \frac{1}{4}[A]_0$. At $t = 3 \times t_{1/2}$, $[A] = \frac{1}{8}[A]_0$, and so on. The constancy of half-life can be used as a test for a first-order reaction. Try it with the simple concentration–time graph of Figure 15-2. That is, starting with $[H_2O_2] = 2.32$ M at $t = 0$, at what time is $[H_2O_2] \approx 1.16$ M, ≈ 0.58 M, ≈ 0.29 M? Starting with $[H_2O_2] = 1.50$ M at $t = 600$ s, at what time is $[H_2O_2] = 0.75$ M?

As illustrated in Example 15-6, a first-order reaction can also be described in terms of the percent of a reactant consumed or remaining.

EXAMPLE 15-6

Expressing Fraction (or Percent) of Reactant Consumed in a First-Order Reaction. Use a value of $k = 7.30 \times 10^{-4}\ \text{s}^{-1}$ for the first-order decomposition of $H_2O_2(aq)$ to determine the percent H_2O_2 that has decomposed in the first 500.0 s after the reaction begins.

Solution

The ratio $[H_2O_2]_t/[H_2O_2]_0$ represents the fractional part of the initial amount of H_2O_2 that remains *unreacted* at time t. Our problem is to evaluate this ratio at $t = 500.0$ s.

$$\ln \frac{[H_2O_2]_t}{[H_2O_2]_0} = -kt = -7.30 \times 10^{-4}\ \text{s}^{-1} \times 500.0\ \text{s} = -0.365$$

$$\frac{[H_2O_2]_t}{[H_2O_2]_0} = e^{-0.365} = 0.694 \qquad \text{and} \qquad [H_2O_2]_t = 0.694[H_2O_2]_0$$

The fractional part of the H_2O_2 remaining is 0.694, or 69.4%. The percent of H_2O_2 that has decomposed is 100.0% − 69.4% = 30.6%.

Practice Example A: Consider the first-order reaction A → products, with $k = 2.95 \times 10^{-3} \text{ s}^{-1}$. What percent of A remains after 150 s?

Practice Example B: At what time after the start of the reaction is a sample of $H_2O_2(aq)$ two-thirds decomposed? $k = 7.30 \times 10^{-4} \text{ s}^{-1}$. (*Hint:* What is the fraction of H_2O_2 remaining? Does the actual value of $[H_2O_2]_0$ make any difference?)

Reactions Involving Gases

For gaseous reactions, rates are often measured in terms of gas pressures. For the hypothetical reaction, A(g) → products, the initial partial pressure, $(P_A)_0$, and the partial pressure at some time t, $(P_A)_t$, are related through the expression

$$\ln \frac{(P_A)_t}{(P_A)_0} = -kt \tag{15.14}$$

To see how this equation is derived, start with the ideal gas equation written for reactant A: $P_A V = n_A RT$. Note that the ratio n_A/V is the same as [A]. So, $[A]_0 = (P_A)_0/RT$ and $[A]_t = (P_A)_t/RT$. Substitute these terms into Equation (15.12), and note that the RT terms cancel in the numerator and denominator and leave the simple ratio $(P_A)_t/(P_A)_0$.

Di-*t*-butyl peroxide (DTBP) is used as a catalyst in the manufacture of polymers. In the gaseous state DTBP decomposes into acetone and ethane by a first-order reaction.

$$\underset{\text{DTBP}}{C_8H_{18}O_2(g)} \longrightarrow \underset{\text{acetone}}{2\ C_3H_6O(g)} + \underset{\text{ethane}}{C_2H_6(g)} \tag{15.15}$$

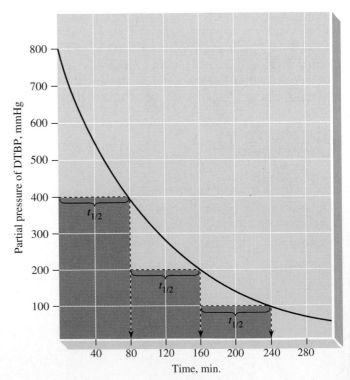

FIGURE 15-5
Decomposition of di-*t*-butyl peroxide (DTBP) at 147 °C.
The decomposition reaction is described through Equation (15.15). In this graph of the partial pressure of DTBP as a function of time, three successive half-life periods of 80 min each are indicated. This constancy of the half-life is proof that the reaction is first order.

Partial pressures of DTBP are plotted as a function of time in Figure 15-5, and the half-life of the reaction is indicated.

EXAMPLE 15-7

Applying First-Order Kinetics to a Reaction Involving Gases. Reaction (15.15) is started with pure DTBP at 147 °C and 800.0 mmHg pressure in a flask of constant volume. **(a)** What is the value of the rate constant k? **(b)** At what time will the partial pressure of DTBP be 50.0 mmHg?

Solution

a. From Figure 15-5 we see that $t_{1/2} = 8.0 \times 10^1$ min. For a first-order reaction, $t_{1/2} = 0.693/k$, or

$$k = 0.693/t_{1/2} = 0.693/8.0 \times 10^1 \text{ min} = 8.7 \times 10^{-3} \text{ min}^{-1}$$

b. A DTBP partial pressure of 50.0 mmHg is $\frac{1}{16}$ of the starting pressure of 800.0 mmHg, that is, $P_{DTBP} = (\frac{1}{2})^4 \times 800.0 = 50.0$ mmHg. The reaction must go through *four* half-lives; $t = 4 \times t_{1/2} = 4 \times 8.0 \times 10^1 = 3.2 \times 10^2$ min.

Practice Example A: Use data from Table 15.4 to determine the partial pressure of ethylene oxide after 30.0 h in a reaction vessel at 415 °C if the initial partial pressure of $(CH_2)_2O(g)$ is 782 mmHg.

Practice Example B: Start with DTBP at a pressure of 800.0 mmHg at 147 °C. What will be the pressure of DTBP at $t = 125$ min, if $t_{1/2} = 8.0 \times 10^1$ min? [*Hint:* Because 125 min is not an exact multiple of the half-life, you must use Equation (15.14). But can you see that the answer is between 200 and 400 mmHg?]

Examples of First-Order Reactions

One of the most familiar examples of a first-order process is radioactive decay. For example, the isotope iodine-131, used in treating thyroid disorders, has a half-life of 8.04 days. Whatever number of iodine-131 atoms we have in a sample at this moment, we will have half that number in 8.04 days; one-quarter of that number in 8.04 + 8.04 = 16.08 days; and so on. The rate constant for the

▶ We will explore radioactive decay in some detail in Chapter 26.

TABLE 15.4   Some Typical First-Order Processes

Process	Half-Life, $t_{1/2}$	Rate Constant k, s^{-1}
radioactive decay of $^{238}_{92}U$	4.51×10^9 years	4.87×10^{-18}
radioactive decay of $^{14}_{6}C$	5.73×10^3 years	3.83×10^{-12}
radioactive decay of $^{32}_{15}P$	14.3 days	5.61×10^{-7}
$C_{12}H_{22}O_{11}(aq) + H_2O(l) \xrightarrow{15 \text{ °C}} C_6H_{12}O_6(aq) + C_6H_{12}O_6(aq)$ sucrose $\qquad\qquad$ glucose $\qquad$ fructose	8.4 h	2.3×10^{-5}
$(CH_2)_2O(g) \xrightarrow{415 \text{ °C}} CH_4(g) + CO(g)$ ethylene oxide	56.3 min	2.05×10^{-4}
$2 N_2O_5 \xrightarrow[45 \text{ °C}]{\text{in } CCl_4} 2 N_2O_4 + O_2(g)$	18.6 min	6.21×10^{-4}
$HC_2H_3O_2(aq) \longrightarrow H^+(aq) + C_2H_3O_2^-(aq)$	8.9×10^{-7} s	7.8×10^5

decay is $k = 0.693/t_{1/2}$, and in Equation (15.12) we can use numbers of atoms, that is, N_t for $[A]_t$ and N_0 for $[A]_0$. Table 15.4 lists several examples of first-order processes. Note the great range of values of $t_{1/2}$ and k. The processes range from very slow to ultrafast.

15-6 Second-Order Reactions

A **second-order** reaction has a rate equation in which the sum of the exponents, $m + n + \cdots$ is equal to 2. As with the discussion of zero- and first-order reactions, we will limit our discussion to reactions involving the decomposition of a single reactant

$$A \longrightarrow \text{Products}$$

that follow the rate law

$$\text{rate of reaction} = -(\text{rate of change of } [A]) = k[A]^2 \qquad (15.16)$$

And, again, our primary interest will be in the *integrated* rate equation derived from the rate law by the methods of *integral* calculus.* The equation obtained is that of a straight-line graph.

$$\frac{1}{[A]_t} = kt + \frac{1}{[A]_0} \qquad (15.17)$$

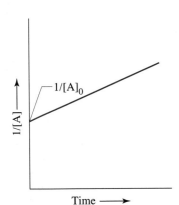

FIGURE 15-6
A straight-line plot for the second-order reaction A → products.
The *reciprocal* of the concentration, 1/[A], is plotted against time. As the reaction proceeds, [A] decreases and 1/[A] increases in a linear fashion. The slope of the line is the rate constant k.

Figure 15-6 is a plot of $1/[A]_t$ against time. The slope of the line is k, and the intercept is $1/[A]_0$. From the graph we can see that the units of k must be the *reciprocal* of concentration divided by time: $M^{-1}/(\text{time})$ or $M^{-1}(\text{time})^{-1}$—for example, $M^{-1}\,s^{-1}$ or $M^{-1}\,\text{min}^{-1}$. We can reach this same conclusion by determining the units of k that produce the required units for the rate of a reaction, that is, moles per liter per time. From Equation (15.16): units of rate of reaction = units of $k \times (\text{molarity})^2 = M^{-1}\,(\text{time})^{-1} \times M^2 = M\,(\text{time})^{-1}$.

For the half-life of the second-order reaction A → products, we can substitute $t = t_{1/2}$ and $[A]_t = \frac{1}{2}[A]_0$ into Equation (15.17).

$$\frac{1}{[A]_0/2} = kt_{1/2} + \frac{1}{[A]_0}; \qquad \frac{2}{[A]_0} = kt_{1/2} + \frac{1}{[A]_0}; \qquad \text{and}$$

$$t_{1/2} = \frac{1}{k[A]_0} \qquad (15.18)$$

From Equation (15.18) we see that the half-life depends on both the rate constant *and* the initial concentration $[A]_0$. *The half-life is not constant.* Its value depends on the concentration of reactant at the start of each half-life interval. Because the starting concentration is always one half that of the previous half-life, each successive half-life is twice as long as the one before it.

*The steps in this derivation are
(1) Replace the rate of change of [A] in Equation (15.16) by $d[A]/dt$.
(2) Rearrange (15.16) to the form, $d[A]/[A]^2 = -kdt$.
(3) Integrate between the limits $[A]_0$ at time $t = 0$ and $[A]_t$ at time t.

$$\int_{[A]_0}^{[A]_t} \frac{d[A]}{[A]^2} = -k\int_0^t dt, \quad \text{yielding} \quad -\frac{1}{[A]_t} + \frac{1}{[A]_0} = -kt \quad \text{or} \quad \frac{1}{[A]_t} = kt + \frac{1}{[A]_0}$$

TABLE 15.5 Kinetic Data for Example 15-8			
Time, min	[A], M	ln [A]	1/[A]
0	1.00	0.00	1.00
5	0.63	−0.46	1.6
10	0.46	−0.78	2.2
15	0.36	−1.02	2.8
25	0.25	−1.39	4.0

EXAMPLE 15-8

Graphing Data to Determine the Order of a Reaction. The data listed in Table 15.5 were obtained for the decomposition reaction A → products. **(a)** Establish the order of the reaction. **(b)** What is the rate constant, k? **(c)** What is the half-life, $t_{1/2}$, if $[A]_0 = 1.00$ M?

Solution

a. Plot the following three graphs.

1. [A] versus time. (If a straight line, reaction is zero order.)
2. ln [A] versus time. (If a straight line, reaction is first order.)
3. 1/[A] versus time. (If a straight line, reaction is second order.)

These graphs are plotted in Figure 15-7. The reaction is second order.

b. The slope of graph 3 in Figure 15-7 is

$$k = \frac{(4.00 - 1.00) \text{ L/mol}}{25 \text{ min}} = 0.12 \text{ M}^{-1} \text{ min}^{-1}$$

c. According to Equation (15.18)

$$t_{1/2} = \frac{1}{k[A]_0} = \frac{1}{0.12 \text{ M}^{-1} \text{ min}^{-1} \times 1.00 \text{ M}} = 8.3 \text{ min}$$

Practice Example A: In the decomposition reaction B → products, the following data are obtained. $t = 0$ s, $[B] = 0.88$ M; 25 s, 0.74 M; 50 s, 0.62 M; 75 s, 0.52 M; 100 s, 0.44 M; 150 s, 0.31 M; 200 s, 0.22 M; 250 s, 0.16 M. What are the order of this reaction and its rate constant k? (*Hint:* You should be able to answer this question without graphing the data.)

Practice Example B: The following data are obtained for the reaction A → products: $t = 0$ min, $[A] = 0.250$ M; $t = 4.22$ min, $[A] = 0.210$ M; $t = 6.60$ min, $[A] = 0.188$ M; $t = 10.61$ min, $[A] = 0.150$ M; $t = 14.48$ min, $[A] = 0.114$ M; $t = 18.00$ min, $[A] = 0.083$ M. What are the order of this reaction and its rate constant, k?

Pseudo-First-Order Reactions

At times, it's possible to simplify the kinetic study of complex reactions by getting them to behave like reactions of a lower order. Then their rate laws become easier to work with. Consider the hydrolysis of ethyl acetate, which is second order overall.

$$\underset{\text{ethyl acetate}}{CH_3COOC_2H_5} + H_2O \longrightarrow \underset{\text{acetic acid}}{CH_3COOH} + \underset{\text{ethanol}}{C_2H_5OH}$$

(1)

(2)

(3)

FIGURE 15-7
Testing for the order of a reaction—Example 15-8 illustrated.
The straight-line plot is obtained for 1/[A] vs. *t*, graph (3). The reaction is second order.

Suppose we follow the hydrolysis of 1 L of aqueous 0.01 M ethyl acetate to completion. $[CH_3COOC_2H_5]$ decreases from 0.01 M to essentially zero. This means that 0.01 mol $CH_3COOC_2H_5$ is consumed, and along with it, 0.01 mol H_2O. But now consider what happens to the molarity of the H_2O. Initially the solution contains about 1000 g H_2O, or about 55.5 mol H_2O. When the reaction is completed, there is still 55.5 mol H_2O (that is, $55.5 - 0.01 \approx 55.5$). The molarity of the water remains essentially constant throughout the reaction—55.5 M. The rate of reaction does not appear to depend on $[H_2O]$. So, the reaction appears to be *zero* order in H_2O, *first* order in $CH_3COOC_2H_5$, and *first* order overall. A second-order reaction that is made to behave like a first-order reaction by holding one reactant concentration constant is called a *pseudo*-first-order reaction. We can treat the reaction with the methods of first-order reaction kinetics. Other reactions of higher order can be made to behave like reactions of lower order under certain conditions. Thus, a third-order reaction might be converted to pseudo-second order, or even to pseudo-first order.

15-7   Reaction Kinetics: A Summary

Let's pause briefly to review what we have learned about rates of reaction, rate constants, and reaction orders. Although a problem often can be solved in several different ways, these approaches are generally most direct.

1. To calculate a rate of reaction when the rate law is known, use this expression: rate of reaction $= k[A]^m[B]^n \cdots$.

2. To determine a rate of reaction when the rate law is not given, use

   - the slope of an appropriate tangent line to the graph of $[A]$ versus t.
   - the expression $-\Delta[A]/\Delta t$, with a short time interval Δt.

3. To determine the reaction order, depending on the data given,

   - use the method of initial rates;
   - find the graph of rate data that yields a straight line;
   - test for the constancy of the half-life (good only for first-order);
   - substitute data into integrated rate equations to find the equation that gives a constant k.

4. To relate reactant concentrations and times, use the appropriate integrated rate equation, after first determining k.

15-8   Theoretical Models for Chemical Kinetics

We can describe the practical aspects of reaction kinetics without ever considering the behavior of individual molecules, but to acquire more insight we must turn to the molecular level. For example, it is easy to show by experiment that the decomposition of H_2O_2 is a first-order reaction, but we may wonder *why* it is first order. In the remainder of the chapter we will look mostly at theoretical aspects of chemical kinetics that help us to answer such questions.

Are You Wondering . . .

If you can determine the order of a reaction just from a single graph of [A] vs. time?

You can for reactions of the type A → products and orders limited to zero, first, and second. If the [A] versus t graph is a straight line, the reaction is *zero order*. If the graph is not linear, apply the half-life test as in Figure 15-5. If $t_{1/2}$ is constant, the reaction is *first order*. If the half-life is not constant, the reaction is *second order*. (For a second-order reaction you'll find that the half-life *doubles* for every successive half-life period.)

Collision Theory

Although we will not attempt to do so, with the kinetic-molecular theory (Section 6-7) it is possible to calculate the number of molecular collisions per unit time—the **collision frequency.** In a typical reaction involving gases, the collision frequency is of the order of 10^{30} collisions per second. If each collision yielded product molecules, the rate of reaction would be about 10^6 M s^{-1}, an extremely rapid rate. The typical gas-phase reaction would go essentially to completion in a fraction of a second. Gas-phase reactions generally proceed at a much slower rate, perhaps on the order of 10^{-4} M s^{-1}. This must mean that, generally, *only a fraction of the collisions among gaseous molecules lead to chemical reaction.*

This is a reasonable conclusion; we would not expect every collision to result in a reaction. For a reaction to occur following a collision between molecules, there must be a redistribution of energy that puts enough energy into certain key bonds to break them. We would not expect two slow-moving molecules to bring enough kinetic energy into their collision to permit bond breakage. We would expect two fast-moving molecules to do so, however, or perhaps one extremely fast molecule colliding with a slow-moving one. The **activation energy** of a reaction is the minimum energy above the average kinetic energy that molecules must bring to their collisions for a chemical reaction to occur.

We can find a common analogy to molecular collisions in automobile collisions. Two slow-moving cars on an icy roadway may be involved in a "fender bender" with minor damage, but collisions involving speeding automobiles, even those in which a car crashes into a stationary object, are much more damaging.

The kinetic-molecular theory can be used to establish the fraction of all the molecules in a mixture that possess certain kinetic energies. The results of this calculation are depicted in Figure 15-8. On this graph a hypothetical energy is noted, and the fraction of all molecules having energies in excess of this value is identified. Let's assume that these are the molecules whose molecular collisions are most likely to lead to chemical reaction. The rate of a reaction, then, depends on the product of the collision frequency *and* the fraction of these "activated" molecules—in other words, on how often molecules with sufficient kinetic energy to react are likely to collide with each other. Because the fraction of high-energy molecules is generally so small, the rate of reaction is usually much smaller than the collision frequency. Moreover, the *higher* the activation energy of a reaction, the *smaller* the fraction of energetic collisions and the slower the reaction.

Another factor that affects the rate of a reaction is the orientation of molecules at the time of their collision. Suppose that during collisions in the reaction

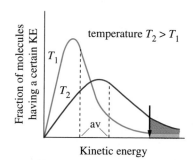

FIGURE 15-8
Distribution of molecular kinetic energies.
At both temperatures, the fraction of all molecules having kinetic energies in excess of the value marked by the heavy black arrow is small. (Note the shaded areas on the right.) At the higher temperature T_2 (red), however, this fraction is considerably larger than at the lower temperature T_1 (blue).

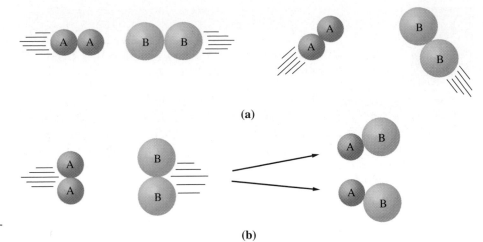

FIGURE 15-9
Molecular collisions and chemical reactions.
(a) Two unfavorable collisions for chemical reaction to produce AB.
(b) A favorable collision for chemical reaction to produce AB.

$$A_2(g) + B_2(g) \longrightarrow 2\,AB(g)$$

the bonds A—A and B—B break and the bond A—B forms. As a result, reactant molecules A_2 and B_2 are converted to the product molecules AB. As suggested by Figure 15-9, however, a particular orientation of the molecules may be required if a collision is to be effective in producing a chemical reaction. The number of unfavorable collisions often exceeds the number of favorable ones.

To continue the automobile collision analogy, the collision of two slow-moving automobiles, bumper to bumper, may produce no damage at all, whereas if the bumper of one car strikes the side of the other, the damage may be considerable.

Transition State Theory

In a theory proposed by Henry Eyring (1901–1981) and others, special emphasis is placed on a hypothetical species believed to exist in an intermediate state that lies between the reactants and the products. We call the transitory intermediate state the **transition state,** and the hypothetical species, the **activated complex.** The activated complex, formed through collisions, either dissociates back into the original reactants or forms product molecules. We can represent an activated complex for a hypothetical reaction $A_2 + B_2 \rightarrow 2\,AB$ in this way.

$$
\begin{array}{ccc}
\begin{array}{c} A \\ | \\ A \end{array} + \begin{array}{c} B \\ | \\ B \end{array}
& \rightleftharpoons &
\begin{array}{c} A\cdots B \\ \vdots\quad\vdots \\ A\cdots B \end{array}
\longrightarrow
\begin{array}{c} A{-}B \\ \\ A{-}B \end{array}
\end{array}
$$

reactants   activated complex   products

Figure 15-10 suggests a graphical way of looking at activation energy, the reaction profile. In a **reaction profile,** energies are plotted on the vertical axis against a quantity called "progress of reaction" on the horizontal axis. Think of the progress of reaction as representing the extent of the reaction. That is, the reaction starts with reactants on the left, progresses through a transition state, and ends with products on the right.

The difference in energies between the reactants and products is ΔH for the reaction. The formation of HI(g) is a slightly exothermic reaction. The difference in energy between the activated complex and the reactants, 171 kJ, is the *activation energy* of the reaction. Thus, a large energy barrier separates the reactants from the products, and only especially energetic molecules can pass over

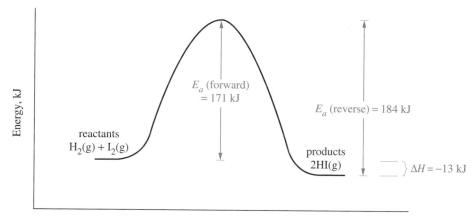

FIGURE 15-10
A reaction profile for the reaction $H_2(g) + I_2(g) \rightarrow 2\,HI(g)$.
This simplified reaction profile traces energy changes during the course of a reaction, from reactants through the transition state to products. The enthalpy change of the forward reaction is seen to conform to the expression: $\Delta H = E_a(\text{forward}) - E_a(\text{reverse})$.

this barrier. Figure 15-11 suggests an analogy to activation energy and the reaction profile.

Figure 15-10 describes both the forward reaction and its reverse—the decomposition of HI(g) into $H_2(g)$ and $I_2(g)$. The activation energy for the reverse reaction is 184 kJ; this reverse reaction is slightly endothermic. Figure 15-10 also illustrates two useful ideas: (1) The enthalpy change of a reaction is equal to the difference in activation energies of the forward and reverse reactions and (2) for an *endothermic* reaction the activation energy must be equal to or greater than the enthalpy of reaction (and usually it is greater).

Attempts at purely theoretical predictions of rate constants have not been very successful. The principal value of reaction rate theories is to help us explain *experimentally observed* reaction rate data. For example, in the next section we will see how the concept of activation energy enters into a discussion of the effect of temperature on reaction rates.

FIGURE 15-11
An analogy to reaction profile and activation energy.
A hike (red path) is taken from the valley on the left (reactants) over the ridge to the valley on the right (products). The ridge above the starting point corresponds to the transition state. It is probably the height of this ridge (activation energy) more than anything else that determines how many people are willing to take the hike, regardless of the fact that it is all "downhill" on the other side.

15-9   The Effect of Temperature on Reaction Rates

From practical experience we expect chemical reactions to go faster at higher temperatures. To speed up the biochemical reactions involved in cooking we raise the temperature. And to slow down other reactions we lower the temperature, as in refrigerating milk to prevent it from souring.

In 1889, Svante Arrhenius demonstrated that the rate constants of many chemical reactions vary with temperature in accordance with the expression

$$k = Ae^{-E_a/RT}$$

By taking the natural logarithm of both sides of this equation, we obtain the following expression.

$$\ln k = \frac{-E_a}{RT} + \ln A \qquad (15.19)$$

▶ The form of this equation means that a rate constant *increases* as the temperature *increases* and as the activation energy *decreases*.

A graph of ln k versus $1/T$ is that of a straight line, and we can use Equation (15.19) for a graphical determination of the activation energy of a reaction, as in Figure 15-12. Or, we can derive a useful variation of the equation by writing it for two sets of k and temperature values and eliminating the constant ln A. The result, also called the Arrhenius equation, is

▶ This is the same technique used for the Clausius–Clapeyron equation on page 429 and illustrated in Appendix A-4.

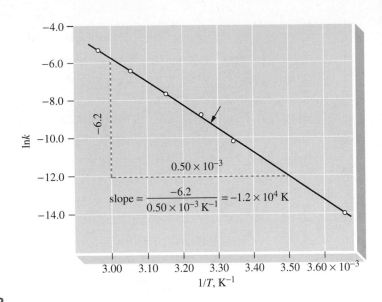

FIGURE 15-12
**Temperature dependence of the rate constant k for the reaction
$2 N_2O_5$ (in CCl_4) → $2 N_2O_4$ (in CCl_4) + $O_2(g)$.**
Data are plotted as follows, for the representative point in black.

$t = 25\ °C = 298\ K$

$1/T = 1/298 = 0.00336 = 3.36 \times 10^{-3}\ K^{-1}$

$k = 3.46 \times 10^{-5}\ s^{-1};\ \ln k = \ln 3.46 \times 10^{-5} = -10.272$

To evaluate E_a,

slope of line $= -E_a/R = -1.2 \times 10^4\ K$

$E_a = 8.3145\ J\ mol^{-1}\ K^{-1} \times 1.2 \times 10^4\ K$

$= 1.0 \times 10^5\ J/mol = 1.0 \times 10^2\ kJ/mol$

(A more precise plot yields a value of $E_a = 106$ kJ/mol. The arrow points to data referred to in Example 15-9.)

$$\ln \frac{k_2}{k_1} = \frac{E_a}{R}\left(\frac{1}{T_1} - \frac{1}{T_2}\right) \qquad (15.20)$$

In Equation (15.20) T_2 and T_1 are two kelvin temperatures; k_2 and k_1 are the rate constants at these temperatures; E_a is the activation energy in joules per mole. R is the gas constant expressed as $8.3145\ J\ mol^{-1}\ K^{-1}$.

EXAMPLE 15-9

Applying the Arrhenius Equation. Use data from Figure 15-12 to determine the temperature at which $t_{1/2}$ for the first-order decomposition of N_2O_5 in CCl_4 is 2.00 h.

Solution

First we need to find the rate constant k corresponding to a 2.00-h half-life. For a *first-order reaction,*

$$k = \frac{\ln 2}{t_{1/2}} = \frac{0.693}{2.00\ h} = \frac{0.693}{7200\ s} = 9.63 \times 10^{-5}\ s^{-1}$$

Now, we can proceed in one of two ways.

Graphical method. We want the temperature at which $k = 9.63 \times 10^{-5}$ and $\ln k = \ln 9.63 \times 10^{-5} = -9.248$. We have marked this point on Figure 15-12 with an arrow. Corresponding to $\ln k = -9.248$, $1/T = 3.28 \times 10^{-3} \text{ K}^{-1}$.

$$T = 1/(3.28 \times 10^{-3}) \text{ K} = 305 \text{ K} = 32 \text{ °C}$$

With Equation (15.20). Take T_2 to be the temperature at which $k = k_2 = 9.63 \times 10^{-5} \text{ s}^{-1}$. T_1 is some other temperature at which a value of k is known. Suppose we take $T_1 = 298 \text{ K}$ and $k_1 = 3.46 \times 10^{-5} \text{ s}^{-1}$, a point referred to in the caption of Figure 15-12. The activation energy is 106 kJ/mol = 1.06×10^5 J/mol (the more precise value given in Figure 15-12). Now we can solve Equation (15.20) for T_2. (For simplicity we have omitted units below, but the temperature is obtained in kelvins.)

$$\ln \frac{k_2}{k_1} = \frac{E_a}{R}\left(\frac{1}{T_1} - \frac{1}{T_2}\right)$$

$$\ln\frac{9.63 \times 10^{-5}}{3.46 \times 10^{-5}} = \frac{1.06 \times 10^5}{8.3145}\left(\frac{1}{298} - \frac{1}{T_2}\right)$$

$$1.024 = 1.27 \times 10^4\left(0.00336 - \frac{1}{T_2}\right) = 42.7 - (1.27 \times 10^4)/T_2$$

$$(1.27 \times 10^4)/T_2 = 42.7 - 1.024 = 41.7$$

$$T_2 = (1.27 \times 10^4)/41.7 = 305 \text{ K}$$

Practice Example A: What is the half-life of the first-order decomposition of N_2O_5 at 75.0 °C? [*Hint:* Use data from Example 15-9 in Equation (15.20). How is $t_{1/2}$ related to k?]

Practice Example B: At what temperature will it take 1.50 h for two-thirds of a sample of N_2O_5 in CCl_4 to decompose in Example 15-9?

Although Arrhenius established Equation (15.20) before the collision theory of chemical reactions had been developed (by fitting the data with his equation), his equation is consistent with the collision theory presented in the preceding section. There we discussed the importance of (1) the frequency of molecular collisions, (2) the fraction of collisions sufficiently energetic to produce a reaction, and (3) the need for favorable orientations during collisions. Let's represent the collision frequency by the symbol Z_0. From kinetic-molecular theory, the fraction of sufficiently energetic collisions proves to be $e^{-E_a/RT}$. The probability of favorable orientations of colliding molecules is p. In collision theory, the rate constant of a reaction can be expressed as the product of these three terms. If we replace the product $Z_0 \times p$ by the term A, collision theory yields a result identical to Arrhenius's experimentally determined equation.

$$k = Z_0 \cdot p \cdot e^{-E_a/RT} = Ae^{-E_a/RT}$$

The rate of chirping of tree crickets and the flashing of fireflies both roughly double for a 10 °C temperature rise. This corresponds to an activation energy of about 50 kJ/mol and suggests that the physiological processes governing these phenomena involve chemical reactions.

15-10 Reaction Mechanisms

In the discussion of photochemical smog in Section 8-2, we indicated a key role played by $NO_2(g)$, but it is unlikely that very much of this gas is formed in the atmosphere by the direct reaction

$$2 \text{ NO(g)} + O_2(g) \longrightarrow 2 \text{ NO}_2(g) \tag{15.21}$$

For this reaction to occur in a single step in the manner suggested by Equation (15.21), *three* molecules would have to collide simultaneously, or very nearly so. A three-molecule collision is an unlikely event. The reaction appears to follow a different mechanism or pathway. One of the main purposes in determining rate laws of chemical reactions is to relate them to probable reaction mechanisms.

A **reaction mechanism** is a detailed description of a chemical reaction presented as a series of one-step changes called elementary processes. An **elementary process** is any molecular event that significantly alters a molecule's energy or geometry or produces a new molecule(s). Two requirements of a plausible reaction mechanism are that it must

- be consistent with the stoichiometry of the overall (net) reaction and
- account for the *experimentally determined* rate law.

In this section we will first explore the nature of elementary processes and then apply these processes to two simple types of reaction mechanisms.

Elementary Processes

To propose plausible reaction mechanisms we need to use the following important characteristics of elementary processes.

1. The exponents of the concentration terms in the rate equation for an *elementary* process are the *same* as the stoichiometric coefficients in the balanced equation for the process. (Note that this is unlike the case of the *overall* rate law, for which the exponents are *not* necessarily related to the stoichiometric coefficients in the net equation.)

2. Elementary processes are either **unimolecular**—a process in which a single molecule dissociates—or **bimolecular**—a process involving the collision of two molecules. A *termolecular* process, which would involve the simultaneous collision of three molecules, is unlikely as an elementary process.

3. Elementary processes are reversible, and some may reach a condition of equilibrium in which the rates of the forward and reverse processes are equal.

4. Certain species are produced in one elementary process and consumed in another. In a proposed reaction mechanism, such intermediates must not appear in either the net chemical equation or the overall rate equation.

5. One elementary process may occur much more slowly than all the others, and in some cases may determine the rate of the overall reaction. Such a process is called the **rate-determining step** (Figure 15-13).

FIGURE 15-13
The San Ysidro/Tijuana border station: an analogy to a rate-determining step.
Note the tie-up of traffic on the Mexican side of the border (top) and the relatively few cars on the United States side (bottom). This station is a "bottleneck" and hence the rate-determining part of the trip by car from Tijuana, Mexico, to San Diego, CA, two cities located only about a dozen miles apart.

A Mechanism with a Slow Step Followed by a Fast Step

The reaction between gaseous iodine monochloride and gaseous hydrogen produces iodine and hydrogen chloride as gaseous products.

$$H_2(g) + 2 \, ICl(g) \longrightarrow I_2(g) + 2 \, HCl(g)$$

The experimentally determined rate law for this reaction is

$$\text{rate of reaction} = k[H_2][ICl]$$

Let's *postulate* the following two steps that, when added together, yield the *net* reaction.

$$\text{(1) Slow:} \quad H_2 + ICl \longrightarrow HI + HCl$$
$$\text{(2) Fast:} \quad HI + ICl \longrightarrow I_2 + HCl$$

$$\text{Net:} \quad H_2 + 2\, ICl \longrightarrow I_2 + 2\, HCl$$

The rate law expressions we can write for the elementary steps are

$$\text{rate(1)} = k_1[H_2][ICl] \quad \text{and} \quad \text{rate(2)} = k_2[HI][ICl]$$

Now, let's further postulate that step (1) occurs *slowly* but step (2) occurs *rapidly.* This suggests that HI is consumed in the second step just as fast as it is formed in the first. The first step is the rate-determining step, and the rate of the overall reaction is governed just by the rate at which HI is formed in this first step, that is, by rate (1). This explains why the observed rate law for the net reaction is: rate of reaction $= k[H_2][ICl]$. Figure 15-14 gives an analogy to this reaction mechanism.

A Mechanism with a Fast, Reversible First Step, Followed by a Slow Step

The rate law for the reaction of NO(g) and O_2(g)

$$2\, NO(g) + O_2(g) \longrightarrow 2\, NO_2(g) \tag{15.21}$$

is found to be

$$\text{rate of reaction} = k[NO]^2[O_2] \tag{15.22}$$

Even though it is consistent with this rate law, we have already noted that the one-step *termolecular* mechanism suggested by Equation (15.21) is *highly improbable* for this reaction. Let's explore instead the following mechanism.

$$\text{Fast:} \quad 2\, NO(g) \underset{k_2}{\overset{k_1}{\rightleftharpoons}} N_2O_2(g) \tag{15.23}$$

$$\text{Slow:} \quad N_2O_2(g) + O_2(g) \xrightarrow{k_3} 2\, NO_2(g) \tag{15.24}$$

$$\text{Net:} \quad 2\, NO(g) + O_2(g) \longrightarrow 2\, NO_2(g) \tag{15.21}$$

In this mechanism there is a rapid equilibrium as the first step, and N_2O_2 is consumed in the second slower step. The rate equation for the slow, or *rate-determining,* step (15.24) is

$$\text{rate of reaction} = k_3[N_2O_2][O_2] \tag{15.25}$$

Because N_2O_2 is an *intermediate,* however, we must eliminate it from the rate equation. We can do this by assuming that the fast, reversible step (15.23) reaches a *steady-state condition,* in which N_2O_2 is produced and consumed at equal rates and from which it is withdrawn only very slowly through the rate-determining step. That is, $[N_2O_2]$ remains constant throughout most of the reaction. We can use the steady-state assumption to express $[N_2O_2]$ in terms of [NO].

$$\frac{\Delta[N_2O_2]}{\Delta t} = \text{rate of formation of } N_2O_2 + \text{rate of disappearance of } N_2O_2 = 0$$

$$\text{rate of formation of } N_2O_2 = -(\text{rate of disappearance of } N_2O_2)$$

$$k_1[NO]^2 = k_2[N_2O_2]$$

and
$$[N_2O_2] = \frac{k_1}{k_2}[NO]^2$$

FIGURE 15-14
Analogy to a reaction mechanism with a slow step followed by a fast step.
The two funnels differ only in the diameters of their stems. Liquid flows through the bottom funnel just as quickly as it arrives from the top funnel. There is no buildup of liquid in the bottom funnel.

Now, if we substitute this value of $[N_2O_2]$ into the rate equation (15.25) and replace k_1k_3/k_2 by k, we obtain for the net reaction

$$\text{rate of reaction} = \frac{k_1k_3}{k_2}[NO]^2[O_2] = k[NO]^2[O_2] \qquad (15.22)$$

We have just shown that the proposed mechanism is consistent with (a) the reaction stoichiometry and (b) the experimentally determined rate law. Whether this mechanism is the actual reaction path, we cannot say, however. All that we can say is that the mechanism is *plausible*.

EXAMPLE 15-10

Testing a Reaction Mechanism. An alternative mechanism of the reaction $2\,NO(g) + O_2(g) \rightarrow 2\,NO_2(g)$ follows. Show that this mechanism is consistent with the rate equation (15.22).

$$\textit{Fast:} \qquad NO(g) + O_2(g) \underset{k_2}{\overset{k_1}{\rightleftharpoons}} NO_3(g)$$

$$\textit{Slow:} \qquad \underline{NO_3(g) + NO(g) \xrightarrow{k_3} 2\,NO_2(g)}$$

$$\textit{Net:} \qquad 2\,NO(g) + O_2(g) \longrightarrow 2\,NO_2(g)$$

Solution

The rate equation for the rate-determining step is

$$\text{rate of reaction} = k_3[NO_3][NO]$$

To eliminate $[NO_3]$ we use the steady-state condition.

$$\frac{\Delta[NO_3]}{\Delta t} = \text{rate of formation of } NO_3 + \text{rate of disappearance of } NO_3 = 0$$

$$\text{rate of formation of } NO_3 = -(\text{rate of disappearance of } NO_3)$$

$$k_1[NO][O_2] = k_2[NO_3]$$

and

$$[NO_3] = \frac{k_1}{k_2}[NO][O_2]$$

Finally, we substitute this value of $[NO_3]$ into the rate equation for the rate-determining step.

$$\text{rate of reaction} = \frac{k_1k_3}{k_2}[NO]^2[O_2] = k[NO]^2[O_2] \qquad (15.22)$$

Practice Example A: In a proposed two-step mechanism for the reaction $CO + NO_2 \rightarrow CO_2 + NO$, the second, fast step is $NO_3 + CO \rightarrow NO_2 + CO_2$. What must be the *first* step? What would you expect the rate law of the reaction to be? Explain.

Practice Example B: Show that the proposed mechanism for the reaction $2\,NO_2(g) + F_2(g) \rightarrow 2\,NO_2F(g)$ is plausible. The rate of reaction is rate $= k[NO_2][F_2]$.

$$\textit{Fast:} \qquad NO_2(g) + F_2(g) \rightleftharpoons NO_2F_2(g)$$

$$\textit{Slow:} \qquad NO_2F_2(g) \longrightarrow NO_2F(g) + F(g)$$

$$\textit{Fast:} \qquad F(g) + NO_2(g) \longrightarrow NO_2F(g)$$

15-11   Catalysis

A reaction can generally be made to go faster by raising the temperature. Another way to speed up a reaction is to use a catalyst. A **catalyst** provides an alternative reaction pathway of *lower* activation energy. It participates in a chemical reaction without undergoing permanent change. As a result, the formula of a catalyst does not appear in the net chemical equation. (Its formula is generally placed over the reaction arrow.)

The success of a chemical process often hinges on finding the right catalyst, as in the manufacture of nitric acid. By conducting the oxidation of $NH_3(g)$ very quickly (less than 1 ms) in the presence of a Pt–Rh catalyst, $NO(g)$ can be obtained as a product instead of $N_2(g)$. The formation of $HNO_3(aq)$ from $NO(g)$ then follows easily (see page 248).

In this section the two basic types of catalysis—homogeneous and heterogeneous—are described first. This is followed by discussions of the catalyzed decomposition of $H_2O_2(aq)$ and the biological catalysts called enzymes.

Homogeneous Catalysis

Figure 15-15 shows reaction profiles for the decomposition of formic acid (HCOOH). In the uncatalyzed reaction an H atom must be transferred from one part of the formic acid molecule to another, shown by the arrow. Then a C—O bond breaks. Because the energy requirement for this atom transfer is high, the activation energy is high and the reaction is slow.

In the acid-catalyzed decomposition of formic acid a hydrogen ion from solution attaches itself to the O atom that is singly bonded to the C atom to form $(HCOOH_2)^+$. The C—O bond breaks, and an H atom attached to a *carbon* atom in the intermediate species $(HCO)^+$ is released to the solution as H^+.

$$\text{H}-\overset{\overset{\text{O}}{\|}}{\text{C}}-\text{O}-\text{H} + \text{H}^+ \longrightarrow \left(\text{H}-\overset{\overset{\text{O}}{\|}}{\text{C}}-\overset{\overset{\text{H}}{|}}{\text{O}}-\text{H} \right)^+ \longrightarrow \left(\text{H}-\underset{\underset{\text{H}^+ + \text{C}\equiv\text{O}}{\downarrow}}{\overset{\overset{\text{O}}{\|}}{\text{C}}} \right)^+ + \text{H}_2\text{O}$$

This catalyzed reaction pathway does not require an H atom to be transferred within the formic acid molecule. It has a lower activation energy than does the uncatalyzed reaction and proceeds at a faster rate. Because the reactants and

FIGURE 15-15
An example of homogeneous catalysis.
The activation energy is lowered in the presence of H^+, a catalyst for the decomposition of HCOOH.

Energy / Progress of reaction

(a) Uncatalyzed Reaction

Energy / Progress of reaction

(b) Catalyzed Reaction

▶ Continuing the analogy of Figure 15-11, a catalyst is like a guide who leads hikers to an easier (less steep) trail.

products of this reaction are all present throughout the solution or *homogeneous* mixture, this type of catalysis is called *homogeneous* catalysis.

Heterogeneous Catalysis

Many reactions can be catalyzed by allowing them to occur on an appropriate solid surface. Essential reaction intermediates are found on the surface. This type of catalysis is called *heterogeneous* catalysis because the catalyst is present in a different phase of matter than are the reactants and products. The precise mechanism of heterogeneous catalysis is imperfectly understood, but the availability of *d* electrons and *d* orbitals in surface atoms may play a role. Catalytic activity is associated with many transition elements and their compounds.

A key feature of heterogeneous catalysis is that reactants from a gaseous or solution phase are *adsorbed*, or attached, to the surface of the catalyst. Not all surface atoms are equally effective for catalysis; those that are, are called **active sites.** Basically, heterogeneous catalysis involves (1) *adsorption* of reactants; (2) diffusion of reactants along the surface; (3) reaction at an active site to form adsorbed product; (4) *desorption* of the product.

In Chapter 8 we described the *oxidation* of CO to CO_2 and *reduction* of NO to N_2 in automotive exhaust gases as a smog-control measure (page 251). Figure 15-16 shows how this reaction is thought to occur on the surface of rhodium metal in a catalytic converter.

The Catalyzed Decomposition of Hydrogen Peroxide

As we have previously noted (footnote on page 507), the decomposition of $H_2O_2(aq)$ is a slow reaction and generally must be catalyzed. Iodide ion is a good catalyst that seems to function by means of the following two-step mechanism.

$$Slow: \quad H_2O_2 + I^- \longrightarrow OI^- + H_2O$$

$$Fast: \quad H_2O_2 + OI^- \longrightarrow H_2O + I^- + O_2(g)$$

$$Net: \quad 2\,H_2O_2 \longrightarrow 2\,H_2O + O_2(g)$$

As required for a catalyzed reaction, the formula of the catalyst does not appear in the net equation. Neither does the intermediate species OI^-. The rate of reaction of H_2O_2 is determined by the rate of the slow first step.

FIGURE 15-16
Heterogeneous catalysis in the reaction $2\,CO + 2\,NO \xrightarrow{Rh} 2\,CO_2 + N_2$.
(a) Molecules of CO and NO are adsorbed on the rhodium surface. (b) The adsorbed NO molecules dissociate into adsorbed N and O atoms. (c) Adsorbed CO molecules and O atoms combine to form CO_2 molecules, which desorb into the gaseous state. Two N atoms combine and are desorbed as an N_2 molecule.

FIGURE 15-17
Decomposition of hydrogen peroxide solution.
A platinum metal catalyst is contained in a small ceramic disk. Hydrogen peroxide solution decomposed by this catalyst yields $O_2(g)$ for disinfecting soft contact lenses (contained in the white plastic holder). Bubbles of oxygen gas also help to dislodge dirt and debris from the lenses.

$$\text{rate of reaction of } H_2O_2 = k[H_2O_2][I^-] \qquad (15.26)$$

Because I^- is constantly regenerated, its concentration is constant throughout a given reaction. If we replace the product of the constant terms $k[I^-]$ by a new constant k', we can rewrite the rate law as

$$\text{rate of reaction of } H_2O_2 = k'[H_2O_2] \qquad (15.27)$$

Equation (15.26) indicates that the rate of decomposition of $H_2O_2(aq)$ is affected by the initial concentration of I^-. For each initial concentration of I^- we get a different rate constant, k', in Equation (15.27).

We have just described the *homogeneous* catalysis of the decomposition of hydrogen peroxide. We can also use *heterogeneous* catalysis for this decomposition, as indicated by the photograph in Figure 15-17.

Enzymes as Catalysts

Unlike platinum, which catalyzes a wide variety of reactions, the catalytic action of high molar mass proteins known as **enzymes** is very specific. For example, in the digestion of milk, lactose, a more complex sugar, breaks down into two simpler ones, glucose and galactose. This occurs in the presence of the enzyme *lactase*.

$$\underset{\text{"milk sugar"}}{\text{lactose}} \xrightarrow{\text{lactase}} \text{glucose} + \text{galactose}$$

Biochemists describe enzyme activity with the "lock-and-key" model (Figure 15-18 on page 534). The reacting substance, the **substrate** (S), attaches itself to the enzyme (E) at a particular point called an active site to form the complex (ES). The complex decomposes to form products (P) and regenerate the enzyme.

$$E + S \rightleftharpoons ES \longrightarrow E + P$$

Most enzyme reactions proceed fastest at about 37 °C (body temperature). If the temperature is raised much higher than this, the structure of the enzyme changes, the active sites become distorted, and the enzyme activity is lost.

A computer graphics representation of the enzyme phosphoglycerate kinase (carbon backbone shown as blue ribbon). A molecule of ATP, the substrate, is shown in green.

▶ Many people lose the ability to produce lactase when they become adults. In them, lactose passes through the small intestine into the colon, where it ferments and may cause severe gastric disturbances.

FIGURE 15-18
Lock-and-key model of enzyme action.
(a) The substrate attaches itself to an active site on an enzyme molecule. (b) Reaction occurs. (c) Product molecules detach themselves from the site, freeing the enzyme molecule to attach another molecule of substrate.

The substrate and enzyme must have complementary structures to produce a complex, hence the term *lock-and-key*.

FIGURE 15-19
Effect of substrate concentration on the rate of an enzyme reaction.

Determining the rates of enzyme reactions is an important part of enzyme studies. Figure 15-19, a plot of reaction rate against substrate concentration, illustrates what we generally observe. Along the rising portion of the graph, the rate of reaction is proportional to the substrate concentration, [S]. The reaction is first order: rate of reaction = $k[S]$. At high substrate concentrations, the rate is independent of [S]. The reaction follows a zero-order rate equation: rate of reaction = k.

This situation is much like what one observes at a bank of telephones in an airport terminal. The telephones are like active sites and the callers are like substrate molecules. When the terminal is relatively empty (low [S]) the rate at which calls are made depends on the number of people present. When the terminal is crowded, with callers waiting in line, calls are made at essentially a constant rate, which depends on the number of telephones, not the number of patrons.

Summary

The rate of a reaction is related to the rate of change of concentration of a reactant or product. Rates can be calculated through an expression of the type $-\Delta[A]/\Delta t$ or determined from the slope of a tangent line to a concentration–time graph. The relationship between the rate of a reaction and the concentrations of the reactants is called a rate law and has the form: rate of reaction = $k[A]^m[B]^n$ . . . .

The order of a reaction refers to the exponents $m, n, \ldots$ in the rate equation. In this chapter main consideration is given to zero-, first-, and second-order reactions, mostly of the type A → products. One method of determining the order of a reaction is by measuring the initial rates of reaction for different initial concentrations of reactants. Another method requires plotting an appropriate function of concentration against time. And a third method is to do a series of calculations of the rate constant k by using the appropriate integrated rate equation.

The theoretical basis of chemical kinetics is that reactions involve molecular collisions, and that only collisions in which the molecules have sufficient energy and a proper orientation result in reaction. The rates of chemical reactions can be increased by raising the temperature or by using a catalyst. A catalyst changes the reaction pathway to one of lower activation energy. For reactions in living systems these catalysts are called enzymes.

To propose a mechanism for a chemical reaction, a series of elementary processes is postulated. Rate equations are written for these elementary processes and combined into a rate law for the net reaction. To be *plausible*, the reaction mechanism must be consistent with the stoichiometry of the net reaction and its experimentally determined rate law.

Summarizing Example

Peroxyacetyl nitrate (PAN) is an air pollutant produced in photochemical smog by the reaction of hydrocarbons, oxides of nitrogen, and sunlight. PAN is unstable and dissociates into peroxyacetyl radicals and $NO_2(g)$. Its presence in polluted air is like a reservoir for NO_2 storage.

$$CH_3\overset{\displaystyle O}{\overset{\displaystyle \|}{C}}OONO_2 \longrightarrow CH_3\overset{\displaystyle O}{\overset{\displaystyle \|}{C}}OO + NO_2$$

PAN   peroxyacetyl
radical

The first-order decomposition of PAN has a half-life of 30.0 min at 25 °C and an activation energy of 1.2×10^2 kJ/mol. At what temperature will a sample of air containing 5.0×10^{14} PAN molecules per liter decompose at the rate of 1.0×10^{12} PAN molecules per liter per minute?

1. *Determine k at the unknown temperature.* Because the reaction is first order, the rate equation is rate = k[PAN]. You can use the rate of reaction and concentration of PAN in molecules per liter rather than the more customary moles per liter. Solve for k. *Result:* $k = 0.0020$ min^{-1}.
2. *Determine k at 25 °C.* Use the expression $k = 0.693/t_{1/2}$. *Result:* 0.0231 min^{-1}.
3. *Determine the unknown temperature.* You now have two values of k, the value of E_a, one known temperature, and one unknown temperature. Solve the Arrhenius equation (15.20) for the unknown temperature. *Answer:* 283 K.

Key Terms

activated complex (15-8)	**half-life** (15-5)	**reaction mechanism** (15-10)
activation energy (15-8)	**initial rate of reaction** (15-2)	**reaction profile** (15-8)
active sites (15-11)	**instantaneous rate of reaction** (15-2)	**second-order reaction** (15-6)
bimolecular process (15-10)	**integrated rate equation** (15-4)	**substrate** (15-11)
catalyst (15-11)	**order of reaction** (15-3)	**transition state** (15-8)
collision frequency (15-8)	**rate constant, k** (15-3)	**unimolecular process** (15-10)
elementary process (15-10)	**rate-determining step** (15-10)	**zero-order reaction** (15-4)
enzyme (15-11)	**rate law (rate equation)** (15-3)	
first-order reaction (15-5)	**rate of reaction** (15-1)	

Review Questions

1. In your own words define or explain the following terms or symbols: **(a)** $[A]_0$; **(b)** k; **(c)** $t_{1/2}$; **(d)** zero-order reaction; **(e)** catalyst.
2. Briefly describe each of the following ideas, phenomena, or methods: **(a)** the method of initial rates; **(b)** activated complex; **(c)** reaction mechanism; **(d)** heterogeneous catalysis; **(e)** rate-determining step.
3. Explain the important distinctions between each pair of terms: **(a)** first-order and second-order reactions; **(b)** rate equation and *integrated* rate equation; **(c)** activation energy and enthalpy of reaction; **(d)** elementary process and net reaction; **(e)** enzyme and substrate.
4. In the reaction A → products, the initial concentration of A is 0.2643 M and 35 min later, 0.1832 M. What is the initial rate of reaction of A, expressed in **(a)** M min^{-1}; **(b)** M s^{-1}?
5. In the reaction 2 A + B → C + 3 D, reactant A is found to react at the rate of 6.2×10^{-4} M s^{-1}. **(a)** What is the rate of reaction of B? **(b)** What is the rate of formation of D?
6. From Figure 15-2 estimate the rate of reaction of H_2O_2 at **(a)** $t = 800$ s; **(b)** the time at which $[H_2O_2] = 0.50$ M.
7. A *first-order* reaction A → products has a half-life of 75 s. Which of the following statements is correct about the reaction, and what is wrong with the other three? **(a)** The reac-

tion goes to completion in 150 s; **(b)** the quantity of A remaining after 150 s is half of what remains after 75 s; **(c)** the same quantity of A is consumed for every 75 s of the reaction; **(d)** one-quarter of the original quantity of A is consumed in the first 37.5 s of the reaction.
8. The reaction A + B → C + D is *second order* in A and *zero order* in B. The value of k is 0.0103 M^{-1} min^{-1}. What is the rate of this reaction when [A] = 0.116 M and [B] = 3.83 M?
9. A *first-order* reaction A → products has a half-life of 19.8 min. What is the rate of the reaction when [A] = 0.632 M? (*Hint:* What is k?)
10. A reaction is 50% complete in 30.0 min. How long after the start of the reaction will it be 75% complete, if the reaction is **(a)** first order; **(b)** zero order?
11. Substance A decomposes by a *first-order* reaction. Starting initially with [A] = 2.00 M, after 126 min [A] = 0.250 M. For this reaction what is **(a)** $t_{1/2}$; **(b)** k?
12. The reaction A → products is *first order* in A.
(a) If 1.60 g A is allowed to decompose for 38 min, the mass of A remaining undecomposed is found to be 0.40 g. What is the half-life, $t_{1/2}$, of this reaction?
(b) Start with 1.60 g A. What is the mass of A remaining undecomposed after 1.00 h?

Focus On · Combustion and Explosions

The main thrusters that provide the energy to put the space shuttle into orbit are powered by the controlled combustion of hydrogen and oxygen.

At times, only a fine line separates combustion reactions and explosions. The typical explosion is a self-sustaining combustion reaction that proceeds at an ever-increasing rate and cannot be stopped. A thorough understanding of the mechanisms of combustion and explosion reactions can help us to achieve the one and avoid the other.

A dramatic contrast can be seen in the reaction

$$2\ H_2 + O_2 \longrightarrow 2\ H_2O$$

When it is carried out in a controlled fashion, this reaction provides the energy and thrust needed to propel the Space Shuttle into orbit.

One type of explosion, called a *thermal* explosion, occurs because the exothermic heat of a reaction cannot be conducted away from the reaction mixture fast enough. The temperature of the reaction mixture rises, and with the increase in temperature the rate constant increases unchecked. Another cause of explosive reactions is found in the *branched-chain* mechanism, outlined below in simplified form for the hydrogen–oxygen reaction.

Initiation:
(1) $H_2 + O_2 \longrightarrow HO_2\cdot\ + H\cdot$

Propagation:
(2) $HO_2\cdot\ + H_2 \longrightarrow HO\cdot + H_2O$
(3) $HO\cdot + H_2 \longrightarrow H\cdot + H_2O$

Branching:
(4) $H\cdot\ + O_2 \longrightarrow O\cdot\ + HO\cdot$
(5) $O\cdot\ + H_2 \longrightarrow HO\cdot\ + H\cdot$

Termination:
(6) $O\cdot\ + O\cdot\ + M \longrightarrow O_2 + M^*$

etc.

---

13. In the *first-order* reaction $A \rightarrow$ products, $[A] = 0.816$ M initially and 0.632 M after 16.0 min.
(a) What is the value of the *rate constant, k*?
(b) What is the *half-life* of this reaction?
(c) At what time will $[A] = 0.235$ M?
(d) What will $[A]$ be after 2.5 h?

14. The initial rate of the reaction $2\ A + 2\ B \rightarrow C + D$ is determined for different initial conditions, with the results listed in the following table.

Expt	[A], M	[B], M	Initial rate, M s^{-1}
1	0.185	0.133	3.35×10^{-4}
2	0.185	0.266	1.35×10^{-3}
3	0.370	0.133	6.75×10^{-4}
4	0.370	0.266	2.70×10^{-3}

(a) What is the order of reaction with respect to A and to B?
(b) What is the overall reaction order?
(c) What is the value of the rate constant, k?

15. A kinetic study of the reaction $A \rightarrow$ products yields the data: $t = 0$ s, $[A] = 2.00$ M; 500 s, 1.00 M; 1500 s, 0.50 M; 3500 s, 0.25 M. *Without performing detailed calculations*, determine the order of this reaction, and indicate your method of reasoning.

16. The reaction $A + B \rightarrow C + D$ has $\Delta H = +25$ kJ. Which of the following is true concerning the activation energy of the reaction? Explain. $E_a =$ (a) -25 kJ; (b) $+25$ kJ; (c) less than $+25$ kJ; (d) more than $+25$ kJ.

17. The rate constant for the reaction $H_2(g) + I_2(g) \rightarrow 2\ HI(g)$ has been determined at the following temperatures: 599 K, $k = 5.4 \times 10^{-4}$ M^{-1} s^{-1}; 683 K, $k = 2.8 \times 10^{-2}$ M^{-1} s^{-1}.
(a) Calculate the activation energy for the reaction.
(b) At what temperature will the rate constant have the value $k = 5.0 \times 10^{-3}$ M^{-1} s^{-1}?

Three different sets of data of $[A]$ versus time are given in the following table for the reaction $A \rightarrow$ products. (*Hint:* There are several ways of arriving at answers for each of the following six questions.)

Data for Exercises 18–23

I		II		III	
Time, s	[A], M	Time, s	[A], M	Time, s	[A], M
0	1.00	0	1.00	0	1.00
25	0.78	25	0.75	25	0.80
50	0.61	50	0.50	50	0.67
75	0.47	75	0.25	75	0.57
100	0.37	100	0.00	100	0.50
150	0.22			150	0.40
200	0.14			200	0.33
250	0.08			250	0.29

As we have learned previously (page 250), a *free radical* is a highly reactive molecular fragment with one or more unpaired electrons. Free radicals are produced in reactions such as step (1) above. In steps (2) and (3), for each free radical consumed a molecule of the product H_2O is formed, together with another free radical. These steps continue the "chain" of reaction and are called *propagation* steps.

In steps (4) and (5), a free radical and a molecule combine to form *two* new free radicals—the chain branches. As more and more branches are produced in the "chain" of reaction, more and more free radicals are produced, and the reaction quickly becomes explosive.

One method of preventing a combustion reaction from becoming explosive is to keep the concentration of one of the reactants very low (*lean*). The rate of reaction is slowed down because of this low concentration of the "lean" reactant. Another method is to introduce a species M into the reaction mixture that can remove energy from free radicals and cause them to combine, as in the *termination* step (6). Energy from the excited species M* is dissipated without the further formation of free radicals. Also, a species M can be chosen that reacts with free radicals; this again reduces their number. Still another way to avoid explosions is to keep the gaseous mixture at a *low pressure*. Then, many of the free radicals formed are able to migrate to the walls of the container without undergoing further reaction. At the walls, the free radicals lose energy and combine in reactions similar to step (6). At higher pressures, however, the free radi-

cals are more likely to collide with molecules of the reactants before they reach the container walls. Here again, an explosion rather than a steady combustion is the likely result.

18. Which of these sets of data corresponds to a **(a)** zero-order, **(b)** first-order, **(c)** second-order reaction?

19. What is the value of the rate constant k of the zero-order reaction?

20. What is the approximate half-life of the first-order reaction?

21. What is the approximate initial rate of the second-order reaction?

22. What is the approximate rate of reaction at $t = 75$ s for the **(a)** zero-order, **(b)** first-order, **(c)** second-order reaction?

23. What is the approximate concentration of A remaining after 110 s in the **(a)** zero-order, **(b)** first-order, **(c)** second-order reaction?

24. For the reaction $A + 2\,B \rightarrow C + D$, the rate law is rate of reaction $= k[A][B]$.

(a) Show that the following mechanism is consistent with the stoichiometry of the net reaction and with the rate law.

$$A + B \longrightarrow I \qquad \text{(slow)}$$
$$I + B \longrightarrow C + D \qquad \text{(fast)}$$

(b) Show that the following mechanism is consistent with the stoichiometry of the net reaction, but *not* with the rate law.

$$2\,B \underset{k_2}{\overset{k_1}{\rightleftarrows}} B_2 \qquad \text{(fast)}$$

$$A + B_2 \overset{k_3}{\longrightarrow} C + D\,\text{(slow)}$$

Exercises

Rates of Reactions

25. In the reaction $A \rightarrow$ products, [A] is found to be 0.485 M at $t = 71.5$ s and 0.474 M at $t = 82.4$ s. What is the average rate of the reaction during this time interval?

26. In the reaction $A \rightarrow$ products, at $t = 0$, $[A] = 0.1565$ M. After 1.00 min, $[A] = 0.1498$ M, and after 2.00 min, $[A] = 0.1433$ M.

(a) Calculate the average rate of the reaction during the first minute and during the second minute.

(b) Why are these two rates not equal?

27. In the reaction A → products, 4.40 min after the reaction is started [A] = 0.588 M. The rate of reaction at this point is found to be rate = $-\Delta[A]/\Delta t = 2.2 \times 10^{-2}$ M min^{-1}. Assume that this rate remains constant for a short period of time.
(a) What is [A] 5.00 min after the reaction is started?
(b) At what time after the reaction is started will [A] = 0.565 M?

29. The reaction A + 2 B → 2 C has a rate of 1.76×10^{-5} mol A L^{-1} s^{-1} at the time when [A] = 0.3580 M.
(a) What is the rate of formation of C?
(b) What will [A] be 1.00 min later?
(c) Assume the rate remains at 1.76×10^{-5} M s^{-1}. How long would it take for [A] to change from 0.3580 to 0.3500 M?

31. In the reaction A(g) → 2 B(g) + C(g), the *total* pressure increases while the *partial* pressure of A(g) decreases. If the initial pressure of A(g) in a vessel of constant volume is 1000 mmHg,
(a) What will be the total pressure when the reaction has gone to completion?
(b) What will be the total gas pressure when the partial pressure of A(g) has fallen to 800 mmHg?

28. Refer to Experiment 2 of Table 15.3 and to reaction (15.6) and rate law (15.7). Exactly 1.00 h after the reaction is started, what are (a) [HgCl$_2$] and (b) [C$_2$O$_4^{2-}$] in the mixture?

30. In reaction (15.1), if the rate of decomposition of H$_2$O$_2$(aq) is 5.7×10^{-4} M s^{-1}, what is the rate of production of O$_2$(g) from 1.00 L of the H$_2$O$_2$(aq), expressed as (a) mol O$_2$ s^{-1}; (b) mol O$_2$ min^{-1}; (c) mL O$_2$(STP) min^{-1}?

32. At 65 °C the half-life for the first-order decomposition of N$_2$O$_5$(g) is 2.38 min.
$$2 \text{ N}_2\text{O}_5(g) \longrightarrow 4 \text{ NO}_2(g) + \text{O}_2(g)$$
If 1.00 g of N$_2$O$_5$ is introduced into an evacuated 15-L flask at 65 °C,
(a) What is the initial partial pressure, in mmHg, of N$_2$O$_5$(g)?
(b) What is the partial pressure, in mmHg, of N$_2$O$_5$(g) after 2.38 min?
(c) What is the total gas pressure, in mmHg, after 2.38 min?

Method of Initial Rates

33. For the reaction A + B → C + D the following initial rates of reaction were found. What is the rate law for this reaction?

Expt	[A], M	[B], M	Initial rate, M min^{-1}
1	0.50	1.50	4.2×10^{-3}
2	1.50	1.50	1.3×10^{-2}
3	3.00	3.00	5.2×10^{-2}

34. The following data are obtained for the initial rates of reaction in the reaction A + 2 B + C → 2 D + E.

Expt	[A], M	[B], M	[C], M	Initial rate
1	1.40	1.40	1.00	R_1
2	0.70	1.40	1.00	$R_2 = \frac{1}{2} \times R_1$
3	0.70	0.70	1.00	$R_3 = \frac{1}{4} \times R_2$
4	1.40	1.40	0.50	$R_4 = 16 \times R_3$
5	0.70	0.70	0.50	$R_5 = ?$

(a) What are the reaction orders with respect to A, B, and C?
(b) What is the value of R_5 in terms of R_1?

First-Order Reactions

35. One of the following statements is true and the other is false regarding the *first-order* reaction 2 A → B + C. Identify the true statement and the false one, and explain your reasoning.
(a) The rate of the reaction decreases as more and more of B and C form.
(b) The time required for one-half of substance A to react is directly proportional to the quantity of A present initially.

37. The first-order reaction A → products has $t_{1/2} = 180$ s.
(a) What percent of a sample of A remains *unreacted* 900 s after a reaction has been started?
(b) What is the rate of reaction when [A] = 0.50 M?

39. In the first-order reaction A → products, it is found that 99% of the original amount of reactant A decomposes in 137 min. What is the half-life, $t_{1/2}$, of this decomposition reaction?

36. One of the following statements is true and the other is false regarding the *first-order* reaction 2 A → B + C. Identify the true statement and the false one, and explain your reasoning.
(a) A graph of [A] versus time is a straight line.
(b) The rate of formation of C is one-half the rate of reaction of A.

38. The reaction A → products is first order in A. Initially, [A] = 0.800 M; and after 54 min, [A] = 0.100 M.
(a) At what time is [A] = 0.025 M?
(b) What is the rate of reaction when [A] = 0.025 M?

40. The half-life of the radioactive isotope phosphorus-32 is 14.3 days. How long does it take for a sample of phosphorus-32 to lose 99% of its radioactivity?

41. Acetoacetic acid, CH_3COCH_2COOH, a reagent used in organic synthesis, decomposes in acidic solution, producing acetone and $CO_2(g)$.

$$CH_3COCH_2COOH(aq) \longrightarrow CH_3COCH_3(aq) + CO_2(g)$$

This first-order decomposition has a half-life of 144 min.
(a) How long will it take for a sample of acetoacetic acid to be 65% decomposed?
(b) How many liters of $CO_2(g)$, measured at STP, are produced as a 10.0-g sample of CH_3COCH_2COOH decomposes for 575 min? (Neglect the aqueous solubility of $CO_2(g)$.)

43. For the reaction A $\longrightarrow$ products, the following data give [A] as a function of time: $t = 0$ s, [A] = 0.600 M; 100 s, 0.497 M; 200 s, 0.413 M; 300 s, 0.344 M; 400 s, 0.285 M; 600 s, 0.198 M; 1000 s, 0.094 M.
(a) Show that the reaction is first order.
(b) What is the value of the rate constant, k?
(c) What is [A] at $t = 750$ s?

42. The following first-order reaction occurs in $CCl_4(l)$ at 45 °C: 2 $N_2O_5 \rightarrow 2$ $N_2O_4 + O_2(g)$. The rate constant is $k = 6.2 \times 10^{-4}$ s^{-1}. An 80.0-g sample of N_2O_5 in $CCl_4(l)$ is allowed to decompose at 45 °C.
(a) How long does it take for the quantity of N_2O_5 to be reduced to 2.5 g?
(b) How many liters of O_2, measured at 745 mmHg and 45 °C, are produced up to this point?

44. The decomposition of dimethyl ether at 504 °C is

$$(CH_3)_2O(g) \longrightarrow CH_4(g) + H_2(g) + CO(g)$$

The following data are partial pressures of dimethyl ether (DME) as a function of time: $t = 0$ s, $P_{DME} = 312$ mmHg; 390 s, 264 mmHg; 777 s, 224 mmHg; 1195 s, 187 mmHg; 3155 s, 78.5 mmHg.
(a) Show that the reaction is first order.
(b) What is the value of the rate constant, k?
(c) What is the total gas pressure at 390 s?
(d) What is the total gas pressure when the reaction has gone to completion?
(e) What is the total gas pressure at $t = 1000$ s?

Second-Order Reactions

45. For the disproportionation of *p*-toluenesulfinic acid,

$$3 \text{ ArSO}_2\text{H} \longrightarrow \text{ArSO}_2\text{SAr} + \text{ArSO}_3\text{H} + \text{H}_2\text{O}$$

[where Ar = *p*-$CH_3C_6H_4$—], the following data were obtained: $t = 0$ min, [ArSO$_2$H] = 0.100 M; 15 min, 0.0863 M; 30 min, 0.0752 M; 45 min, 0.0640 M; 60 min, 0.0568 M; 120 min, 0.0387 M; 180 min, 0.0297 M; 300 min, 0.0196 M.
(a) Show that this reaction is second order.
(b) What is the value of the rate constant, k?
(c) At what time would [ArSO$_2$H] = 0.0500 M?
(d) At what time would [ArSO$_2$H] = 0.0250 M?
(e) At what time would [ArSO$_2$H] = 0.0350 M?

46. The following data were obtained for the dimerization of 1,3-butadiene, 2 $C_4H_6(g) \rightarrow C_8H_{12}(g)$, at 600 K: $t = 0$ min, [C_4H_6] = 0.0169 M; 12.18 min, 0.0144 M; 24.55 min, 0.0124 M; 42.50 min, 0.0103 M; 68.05 min, 0.00845 M.
(a) Show that this reaction is second order.
(b) What is the value of the rate constant, k?
(c) At what time would [C_4H_6] = 0.00423 M?
(d) At what time would [C_4H_6] = 0.0050 M?

Establishing the Order of a Reaction

47. For the reaction A $\longrightarrow$ products, the following data are obtained.

First experiment	
[A] = 1.512 M	$t = 0$ min
[A] = 1.490 M	$t = 1.0$ min
[A] = 1.469 M	$t = 2.0$ min

Second experiment	
[A] = 3.024 M	$t = 0$ min
[A] = 2.935 M	$t = 1.0$ min
[A] = 2.852 M	$t = 2.0$ min

(a) Determine the *initial rate* of reaction (that is, $-\Delta[A]/\Delta t$) in each of the two experiments.
(b) Determine the order of the reaction.

48. For the reaction A $\rightarrow 2$ B + C, the following data are obtained for [A] as a function of time: $t = 0$ min, [A] = 0.80 M; 8 min, 0.60 M; 24 min, 0.35 M; 40 min, 0.20 M.
(a) By suitable means establish the order of the reaction.
(b) What is the value of the rate constant, k?
(c) Calculate the rate of formation of B at $t = 30$ min.

49. The following data were obtained for the decomposition of N_2O_5 in $CCl_4(l)$ at 45 °C: $t = 0$ s, $[N_2O_5] = 1.46$ M; 423 s, 1.09 M; 753 s, 0.89 M; 1116 s, 0.72 M; 1582 s, 0.54 M; 1986 s, 0.43 M; 2343 s, 0.35 M. Determine a value of k for the reaction $2\ N_2O_5 \rightarrow 2\ N_2O_4 + O_2(g)$.

51. In three different experiments the following results were obtained for the reaction $A \rightarrow$ products: $[A]_0 = 1.00$ M, $t_{1/2} = 50$ min; $[A]_0 = 2.00$ M, $t_{1/2} = 25$ min; $[A]_0 = 0.50$ M, $t_{1/2} = 100$ min. Write the rate equation for this reaction and indicate the value of k.

50. For the reaction $A \rightarrow$ products, the following data were obtained: $t = 0$ s, $[A] = 0.715$ M; 22 s, 0.605 M; 74 s, 0.345 M; 132 s, 0.055 M. **(a)** What is the order of this reaction? **(b)** What is the half-life of the reaction?

52. Ammonia decomposes on the surface of a hot tungsten wire. Following are the half-lives that were obtained at 1100 °C for different initial concentrations of NH_3: $[NH_3]_0 = 0.0031$ M, $t_{1/2} = 7.6$ min; 0.0015 M, 3.7 min; 0.00068 M, 1.7 min. For this decomposition reaction, what is **(a)** the order of the reaction; **(b)** the rate constant k?

Collision Theory; Activation Energy

53. Explain why
(a) A reaction rate cannot be calculated from the collision frequency alone.
(b) The rate of a chemical reaction may increase dramatically with temperature, whereas the collision frequency increases much more slowly.
(c) The addition of a catalyst to a reaction mixture can have such a pronounced effect on the rate of a reaction, even if the temperature is held constant.

55. For the reversible reaction $A + B \rightleftharpoons C + D$, the enthalpy change of the forward reaction is $+21$ kJ/mol. The activation energy of the forward reaction is 84 kJ/mol.
(a) What is the activation energy of the reverse reaction?
(b) In the manner of Figure 15-10 sketch the reaction profile of this reaction.

54. If even a tiny spark is introduced into a mixture of $H_2(g)$ and $O_2(g)$, a highly exothermic explosive reaction occurs. Without the spark the mixture remains unreacted indefinitely.
(a) Explain this difference in behavior.
(b) Why is the nature of the reaction independent of the size of the spark?

56. By an appropriate sketch indicate why there is some relationship between the enthalpy change and the activation energy for an endothermic reaction but not for an exothermic reaction.

Effect of Temperature on Rates of Reaction

57. A treatise on atmospheric chemistry lists the following rate constants for the decomposition of PAN described in the Summarizing Example on page 535: 0 °C, $k = 5.6 \times 10^{-6}$ s^{-1}; 10 °C, $k = 3.2 \times 10^{-5}$ s^{-1}; 20 °C, $k = 1.6 \times 10^{-4}$ s^{-1}; 30 °C, $k = 7.6 \times 10^{-4}$ s^{-1}.
(a) Construct a graph of $\ln k$ versus $1/T$.
(b) What is the activation energy, E_a, of the reaction?
(c) Calculate the half-life of the decomposition reaction at 40 °C?

59. The first-order reaction $A \rightarrow$ products has a half-life, $t_{1/2}$, of 46.2 min at 25 °C and 2.6 min at 102 °C.
(a) Calculate the activation energy of this reaction.
(b) At what temperature would the half-life be 10.0 min?

58. The reaction $C_2H_5I + OH^- \rightarrow C_2H_5OH + I^-$ was studied in an ethanol (C_2H_5OH) solution, and the following rate constants were obtained: 15.83 °C, $k = 5.03 \times 10^{-5}$; 32.02 °C, 3.68×10^{-4}; 59.75 °C, 6.71×10^{-3}; 90.61 °C, 0.119 M^{-1} s^{-1}.
(a) Determine E_a for this reaction by a graphical method.
(b) Determine E_a by the use of Equation (15.20).
(c) Calculate the value of the rate constant k at 100.0 °C.

60. For the first-order reaction
$$2\ N_2O_5(g) \longrightarrow 4\ NO_2(g) + O_2(g)$$
$t_{1/2} = 22.5$ h at 20 °C and 1.5 h at 40 °C.
(a) Calculate the activation energy of this reaction.
(b) If the Arrhenius constant, $A = 2.05 \times 10^{13}$ s^{-1}, determine the value of k at 30 °C. (*Hint:* see page 525.)

61. A commonly stated rule of thumb is that reaction rates double for a temperature increase of about 10 °C. (This rule is very often wrong.)
(a) What must be the approximate activation energy for this statement to be true for reactions at about room temperature?
(b) Would you expect this rule of thumb to apply to the formation of HI from its gaseous elements at room temperature? Explain. (*Hint:* Refer to Figure 15-10.)

62. Concerning the rule of thumb stated in Exercise 61, estimate how much faster cooking will occur in a pressure cooker in which the vapor pressure of water is 2.00 atm than in water under normal boiling conditions. (*Hint:* Refer to Table 13.2.)

Catalysis

63. The following statements about catalysis are not stated as carefully as they might be. What slight modifications would you make in them?
(a) A catalyst is a substance that speeds up a chemical reaction but does not take part in the reaction.
(b) The function of a catalyst is to lower the activation energy for a chemical reaction.
65. What is the principal difference between the catalytic activity of platinum metal and of an enzyme?

64. The following substrate concentration, [S], versus time data are obtained in an enzyme reaction: $t = 0$ min, $[S] = 1.00$ M; 20 min, 0.90 M; 60 min, 0.70 M; 100 min, 0.50 M; 160 min, 0.20 M. What is the order of this reaction with respect to S in the concentration range studied?

66. Certain gas-phase reactions on a heterogeneous catalyst are first order at low gas pressures and zero order at high pressures. Can you suggest a reason for this?

Reaction Mechanisms

67. We have used the terms *order of a reaction* and *molecularity of an elementary process* (that is, unimolecular, bimolecular). What is the relationship, if any, between these two terms?

69. The reaction $2 NO + 2 H_2 \rightarrow N_2 + 2 H_2O$ is second order in [NO] and first order in [H_2]. A *three-step* mechanism has been proposed. The first, fast step is the elementary process given as Equation (15.23) on page 529. The third step, also fast, is $N_2O + H_2 \rightarrow N_2 + H_2O$. Propose an entire three-step mechanism and show that it conforms to the experimentally determined reaction order.

68. According to collision theory chemical reactions occur through molecular collisions. A unimolecular elementary process in a reaction mechanism involves dissociation of a *single* molecule. How can these two ideas be compatible? Explain.

70. The observed rate law for the reaction $2 O_3(g) \rightarrow 3 O_2(g)$ is

$$\text{rate} = k\frac{[O_3]^2}{[O_2]}$$

Propose a two-step mechanism for this reaction that consists of a fast, reversible first step followed by a slow second step.

Advanced Exercises

71. [A]$_t$ as a function of time for the reaction A → products is plotted in the graph shown here. Use data from this graph to determine **(a)** the order of the reaction; **(b)** the rate constant, k; **(c)** the rate of the reaction at $t = 3.5$ min, using the results of parts (a) and (b); **(d)** the rate of the reaction at $t = 5.0$ min, from the slope of the tangent line; **(e)** the initial rate of the reaction.

Time, min.

72. Exactly 300 s after decomposition of $H_2O_2(aq)$ begins (reaction 15.1), a 5.00-mL sample is removed and immediately titrated with 37.1 mL of 0.1000 M $KMnO_4$. What is [H_2O_2] at this 300-s point in the reaction? [*Hint:* The H_2O_2–$KMnO_4$ reaction is described by Equation (15.3).]
73. Use the method of Exercise 72 to determine the volume of 0.1000 M $KMnO_4$ required to titrate 5.00-mL samples of $H_2O_2(aq)$ for each of the entries in Table 15.1. Plot these volumes of $KMnO_4(aq)$ as a function of time, and show that from this graph you can get the same rate of reaction at 1400 s as that obtained in Figure 15-2.
74. The initial rate of decomposition of H_2O_2 (reaction 15.1) is found to be 1.7×10^{-3} M s^{-1}. Assume that this rate holds for 2 minutes. Start with 175 mL of 1.55 M $H_2O_2(aq)$ at $t = 0$. How many milliliters of $O_2(g)$, measured at 24 °C and 757 mmHg, are released from solution in the first minute of the reaction?
75. We have seen that the unit of k depends on the overall order of a reaction. Derive a general expression for the units of k for a reaction of any overall order, based on the order of the reaction (o) and the units of concentration (M) and time (s).
76. At room temperature (20 °C) milk turns sour in about 64 hours. In a refrigerator at 3 °C milk can be stored *three times as long* before it sours. **(a)** Estimate the activation energy of the reaction that causes the souring of milk. **(b)** How long should it take milk to sour at 40 °C? [*Hint:* How is the rate constant, k, related to the souring time?]

77. Hydroxide ion is involved in the mechanism of the following reaction but is not consumed in the net reaction.

$$OCl^- + I^- \xrightarrow{OH^-} OI^- + Cl^-$$

(a) From the data given, determine the order of the reaction with respect to OCl^-, I^-, and OH^-.
(b) What is the overall reaction order?
(c) Write the rate equation and determine the value of the rate constant, k.

$[OCl^-]$, M	$[I^-]$, M	$[OH^-]$, M	Rate formation OI^-, $M\ s^{-1}$
0.0040	0.0020	1.00	4.8×10^{-4}
0.0020	0.0040	1.00	5.0×10^{-4}
0.0020	0.0020	1.00	2.4×10^{-4}
0.0020	0.0020	0.50	4.6×10^{-4}
0.0020	0.0020	0.25	9.4×10^{-4}

78. The half-life for the first-order decomposition of nitramide, $NH_2NO_2(aq) \rightarrow N_2O(g) + H_2O(l)$, is 123 min at 15 °C. If 165 mL of a 0.105 M NH_2NO_2 solution is allowed to decompose, how long must the reaction proceed to produce 50.0 mL of "wet" $N_2O(g)$ measured at 15 °C and a barometric pressure of 756 mmHg? (The vapor pressure of water at 15 °C is 12.8 mmHg.)

79. The decomposition of ethylene oxide at 690 K is followed by measuring the *total* gas pressure as a function of time. The data obtained are $t = 10$ min, $P_{tot} = 139.14$ mmHg; 20 min, 151.67 mmHg; 40 min, 172.65 mmHg; 60 min, 189.15 mmHg; 100 min, 212.34 mmHg; 200 min, 238.66 mmHg; ∞, 249.88 mmHg. What is the order of the reaction $(CH_2)_2O(g) \rightarrow CH_4(g) + CO(g)$? [*Hint:* How is the total gas pressure related to the partial pressure of $(CH_2)_2O(g)$? What is the initial partial pressure of $(CH_2)_2O(g)$?]

80. Refer to Example 15-7. For the decomposition of di-*t*-butyl peroxide (DTBP), determine the time at which the *total* gas pressure is 2100 mmHg. (*Hint:* See Exercise 79.)

81. The following data are for the reaction $2\ A + B \rightarrow$ products. Establish the order of this reaction with respect to A and to B.

Expt 1, [B] = 1.00 M		Expt 2, [B] = 0.50 M	
Time, min	[A], M	Time, min	[A], M
0	1.000×10^{-3}	0	1.000×10^{-3}
1	0.951×10^{-3}	1	0.975×10^{-3}
5	0.779×10^{-3}	5	0.883×10^{-3}
10	0.607×10^{-3}	10	0.779×10^{-3}
20	0.368×10^{-3}	20	0.607×10^{-3}

82. Show that the following mechanism is consistent with the rate law established for the iodide–hypochlorite reaction in Exercise 77.

(fast) $\quad OCl^- + H_2O \underset{k_2}{\overset{k_1}{\rightleftharpoons}} HOCl + OH^-$

(slow) $\quad I^- + HOCl \xrightarrow{k_3} HOI + Cl^-$

(fast) $\quad HOI + OH^- \underset{k_5}{\overset{k_4}{\rightleftharpoons}} H_2O + OI^-$

83. In the hydrogenation of a compound containing a carbon-to-carbon triple bond, two products are possible, as in the reaction

$$CH_3-C\equiv C-CH_3 + H_2 \longrightarrow$$

The amount of each product can be controlled by using an appropriate catalyst. The Lindlar catalyst is a *heterogeneous* catalyst that produces only one of these products. Which product, I or II, do you think is produced and why? Draw a sketch of how the reaction might occur.

Feature Problems

A. Benzenediazonium chloride decomposes by a first-order reaction in water, yielding $N_2(g)$ as a product.

$$C_6H_5N_2Cl \longrightarrow C_6H_5Cl + N_2(g)$$

The reaction can be followed by measuring the volume of $N_2(g)$ as a function of time. The data in the table were obtained for the decomposition of an 0.071 M solution at 50 °C, where $t = \infty$ corresponds to the completed reaction.

To convert the volume of $N_2(g)$ to the concentration of $C_6H_5N_2Cl$, consider that after 3 min the volume of $N_2(g)$

Time, min	$N_2(g)$, mL	Time, min	$N_2(g)$, mL
0	0	18	41.3
3	10.8	21	44.3
6	19.3	24	46.5
9	26.3	27	48.4
12	32.4	30	50.4
15	37.3	∞	58.3

evolved is 10.8 mL of a total 58.3 mL. The fraction of the $N_2(g)$ evolved is 10.5 mL/58.3 mL = 0.185. The fraction of $C_6H_5N_2Cl$ remaining unreacted is $1.000 - 0.185 = 0.815$. $[C_6H_5N_2Cl]$ at 3 min = 0.815×0.071 M = 0.058 M.

1. Convert the information given here into a table with one column for time and the other for $[C_6H_5N_2Cl]$.

2. Construct a table similar to Table 15.2, in which the time interval is $\Delta t = 3$ min.

3. Plot graphs similar to Figure 15-2 showing both the formation of $N_2(g)$ and the disappearance of $C_6H_5N_2Cl$ as a function of time.

4. From the graph of part 3 determine the rate of reaction at $t = 21$ min and compare your result with the reported value of 1.1×10^{-3} M min^{-1}.

5. Determine the initial rate of reaction.

6. Write the rate law for the first-order decomposition of $C_6H_5N_2Cl$ and estimate a value of k based on the rate determined in parts 4 and 5.

7. Determine $t_{1/2}$ for the reaction, by estimation from the graph of the rate data and by calculation.

8. At what time would the decomposition of the sample be three-fourths complete?

9. Plot ln $[C_6H_5N_2Cl]$ versus time and show that the reaction is indeed first order.

10. Determine k from the slope of the graph of part 9.

B. Our object is to study the kinetics of the reaction between peroxodisulfate and iodide ions.

(a) $S_2O_8^{2-}(aq) + 3\,I^-(aq) \longrightarrow 2\,SO_4^{2-}(aq) + I_3^-(aq)$

The I_3^- formed in reaction (a) is actually a complex of iodine, I_2, and iodide ion, I^-. Thiosulfate ion, $S_2O_3^{2-}$, also present in the reaction mixture, reacts with I_3^- just as fast as it is formed.

(b) $2\,S_2O_3^{2-}(aq) + I_3^-(aq) \longrightarrow S_4O_6^{2-} + 3\,I^-(aq)$

When all of the thiosulfate ion present initially has been consumed by reaction (b), a third reaction occurs between $I_3^-(aq)$ and starch, which is also present in the reaction mixture.

(c) $I_3^-(aq) + \text{starch} \longrightarrow \text{blue complex}$

The rate of reaction (a) is inversely related to the time required for the blue color of the starch-iodine complex to appear. That is, the faster reaction (a) proceeds, the more quickly the thiosulfate ion is consumed in reaction (b), and the sooner the blue color appears in reaction (c). One of the photographs shows the initial colorless solution and an electronic timer set at $t = 0$; the other photograph shows the very first appearance of the blue complex (after 49.89 s). Tables I and II list some actual student data obtained in this study.

Table I

Reaction conditions at 24 °C: 25.0 mL of the $(NH_4)_2S_2O_8(aq)$ listed, 25.0 mL of the KI(aq) listed, 10.0 mL of 0.010 M $Na_2S_2O_3(aq)$, and 5.0 mL starch solution are mixed. The time is that for the first appearance of the starch-iodine complex.

Experiment	Initial concentrations, M		Time, s
	$(NH_4)_2S_2O_8$	KI	
1	0.20	0.20	21
2	0.10	0.20	42
3	0.050	0.20	81
4	0.20	0.10	42
5	0.20	0.050	79

Table II

Reaction conditions: Those listed in Table I for Experiment 4, but at the temperatures listed.

Experiment	Temperature, °C	Time, s
6	3	189
7	13	88
8	24	42
9	33	21

1. Use the data in Table I to establish the order of reaction (a) with respect to $S_2O_8^{2-}$ and to I^-. What is the overall reaction order? (*Hint:* You can work simply with the times required for the blue complex to appear, without determining actual rates of reaction.)

2. Calculate the initial rate of reaction of $S_2O_8^{2-}$ in Experiment 1, expressed in M s^{-1}. [*Hint:* Here you must take into account the dilution that occurs when the various solutions are mixed, as well as the reaction stoichiometry indicated by Equations (a), (b), and (c).]

3. Calculate the value of the rate constant, k, based on Experiments 1 and 2. (*Hint:* Recall Example 15-4.)

4. Calculate the rate constant, k, for the four different temperatures in Table II.

5. Determine the activation energy, E_a, of the peroxodisulfate-iodide ion reaction.

6. The following mechanism has been proposed for reaction (a). The first step is slow and the others are fast.

$$I^- + S_2O_8^{2-} \longrightarrow IS_2O_8^{3-}$$
$$IS_2O_8^{3-} \longrightarrow 2\,SO_4^{2-} + I^+$$
$$I^+ + I^- \longrightarrow I_2$$
$$I_2 + I^- \longrightarrow I_3^-$$

Show that this mechanism is consistent with both the stoichiometry and the rate law of reaction (a). Explain why it is reasonable to expect the first step in the mechanism to be slower than the others.

Principles of Chemical Equilibrium

16

*U*p to this point in our discussion of chemical reactions, we have stressed reactions that go to completion and the concepts of stoichiometry that allow us to calculate the outcomes of such reactions. We have made occasional references to situations involving both a forward and a reverse reaction—reversible reactions—but in this chapter we will look at them in a detailed and systematic way.

Our emphasis in this chapter will be on the condition of equilibrium that is reached when forward and reverse reactions proceed at the same rate. Our main tool in dealing with equilibrium will be the equilibrium constant, a quantity we introduce early in the chapter. We will first present some key relationships involving equilibrium constants; then we will make qualitative predictions about the condition of equilibrium; and finally we will do various equilibrium calculations.

As we will discover in this chapter and on numerous occasions in the remainder of the text, the condition of equilibrium plays a role in a number of natural phenomena and affects the methods used to produce many important industrial chemicals.

During electrical storms $N_2(g)$ and $O_2(g)$ combine to produce small quantities of $NO(g)$ in a reversible chemical reaction. The condition of equilibrium in such reactions is the subject of this chapter.

(a) **(b)**

FIGURE 16-1
Dynamic equilibrium in a physical process.
(a) A yellow-brown saturated solution of I_2 in water (top layer) is brought into contact with colorless $CCl_4(l)$ (bottom layer). (b) I_2 molecules distribute themselves between the H_2O and CCl_4. When equilibrium is reached, $[I_2]$ in the CCl_4 (violet, bottom layer) is about 85 times greater than in the water (colorless, top layer).

16-1 The Condition of Dynamic Equilibrium

Let's begin by describing three simple *physical* phenomena. These will help us get a feeling for what happens in a system at **equilibrium**—two opposing processes take place at equal rates.

1. When a liquid vaporizes within a closed container, after a time vapor molecules condense to the liquid state at the same rate at which liquid molecules vaporize. Even though molecules continue to pass back and forth between liquid and vapor (a *dynamic* process), the pressure exerted by the vapor remains constant with time. *The vapor pressure of a liquid is a property associated with an equilibrium condition.*

2. When a solute dissolves in a solvent, a time is reached when the rate of dissolving is just matched by the rate at which dissolved solute crystallizes—that is, the solution is saturated. Even though solute particles continue to pass back and forth between the saturated solution and the undissolved solute, the concentration of dissolved solute remains constant. *The solubility of a solute is a property associated with an equilibrium condition.*

3. When an aqueous solution of iodine, I_2, is shaken with pure carbon tetrachloride, $CCl_4(l)$, I_2 molecules move into the CCl_4 layer. As the concentration of I_2 builds up in the CCl_4, the rate of return of I_2 to the water layer becomes significant. When I_2 molecules pass between the two liquids at equal rates—a condition of dynamic equilibrium—the concentration of I_2 in each layer remains constant. At this point $[I_2]$ in the CCl_4 is about 85 times greater than in the H_2O (Figure 16-1). The ratio of concentrations of a solute in two immiscible solvents is called the *distribution coefficient. The distribution coefficient of a solute between two immiscible solvents is a property associated with an equilibrium condition.*

 ## Are You Wondering . . .

How we know that an equilibrium is DYNAMIC—that forward and reverse reactions continue even after equilibrium is reached?

Suppose we have an equilibrium mixture of AgI(s) and its saturated aqueous solution.

$$AgI(s) \rightleftharpoons AgI(satd\ aq)$$

Now let's add to this mixture some AgI(s) containing a trace of radioactive iodine-131 as iodide ion. If both the forward and reverse processes stopped at equilibrium, radioactivity would be confined to a few "hot" spots in the undissolved solid. What we find, though, is that radioactivity shows up in the saturated solution and, over time, distributes itself throughout the undissolved solid as well. The only way this can happen is if dissolving of the solid solute and its crystallization from the saturated solution continue indefinitely. The equilibrium condition is *dynamic.*

The properties described in the three situations on page 546—vapor pressure, solubility, distribution coefficient—are all examples of a general quantity known as an *equilibrium constant,* the subject of the next section.

16-2   The Equilibrium Constant Expression

Methanol (methyl alcohol) is synthesized from a carbon monoxide–hydrogen mixture called synthesis gas. This reaction is likely to become increasingly important as methanol and its mixtures with gasoline find greater use as motor fuels. Methanol has a high octane rating, and its combustion produces much less air pollution than does gasoline.

Methanol synthesis is a *reversible* reaction, which means that at the same time $CH_3OH(g)$ is being formed

$$CO(g) + 2 H_2(g) \longrightarrow CH_3OH(g) \qquad (16.1)$$

it decomposes in the reverse reaction

$$CH_3OH(g) \longrightarrow CO(g) + 2 H_2(g) \qquad (16.2)$$

Initially, only the forward reaction (16.1) occurs, but just as soon as some CH_3OH forms, the reverse reaction (16.2) occurs as well. As time passes, the forward reaction slows down because of the decreasing concentrations of CO and H_2. The reverse reaction speeds up as more CH_3OH accumulates. Eventually the forward and reverse reactions proceed at equal rates, and the reaction mixture reaches a condition of dynamic equilibrium, which we can represent with a double arrow $\rightleftharpoons$.

$$CO(g) + 2 H_2(g) \rightleftharpoons CH_3OH(g) \qquad (16.3)$$

One consequence of the equilibrium condition is that the amounts of the reactants and products remain constant with time. These equilibrium amounts, however, depend on the particular amounts of reactants and products that were present initially. To illustrate, we have listed data for three hypothetical experiments in Table 16.1. All three experiments are conducted in a 10.0-L flask at 483 K. In the first experiment only CO and H_2 are present initially; in the second, only CH_3OH; and in the third, CO, H_2, and CH_3OH. The data from Table 16.1 are plotted in Figure 16-2, and from these graphs we see that

- in no case is any reacting species completely consumed;
- the equilibrium amounts of reactants and products in these three cases appear to have nothing in common.

Although it is not obvious from a cursory inspection of the data, a particular ratio involving equilibrium *concentrations* of product and reactants has a *constant* value, independent of how the equilibrium is reached. This ratio, which is central to the study of chemical equilibrium, can be derived theoretically, but it can also be established by trial and error. Three reasonable attempts at formulating the desired ratio for reaction (16.3) are outlined in Table 16.2 on page 549, and the ratio that works is identified.

For the methanol synthesis reaction, the ratio of equilibrium concentrations shown in the following equation has a constant value of 14.5 at 483 K.

$$K_c = \frac{[CH_3OH]}{[CO][H_2]^2} = 14.5 \qquad (16.4)$$

TABLE 16.1 Three Approaches to Equilibrium in the Reaction[a]
$CO(g) + 2 H_2(g) \rightleftharpoons CH_3OH(g)$

	CO(g)	H₂(g)	CH₃OH(g)
Experiment 1			
initial amounts, mol	1.000	1.000	0.000
equilibrium amounts, mol	0.911	0.822	0.0892
equilibrium concentrations, mol/L	0.0911	0.0822	0.00892
Experiment 2			
initial amounts, mol	0.000	0.000	1.000
equilibrium amounts, mol	0.753	1.506	0.247
equilibrium concentrations, mol/L	0.0753	0.151	0.0247
Experiment 3			
initial amounts, mol	1.000	1.000	1.000
equilibrium amounts, mol	1.380	1.760	0.620
equilibrium concentrations, mol/L	0.138	0.176	0.0620

The concentrations printed in blue are used in the calculations in Table 16.2.

[a]Reaction carried out in a 10.0-L flask at 483 K.

FIGURE 16-2
Three approaches to equilibrium in the reaction
$CO(g) + 2 H_2(g) \rightleftharpoons CH_3OH(g).$
The initial and equilibrium amounts for each of these three cases are listed in Table 16.1.
t_e = time for equilibrium to be reached.

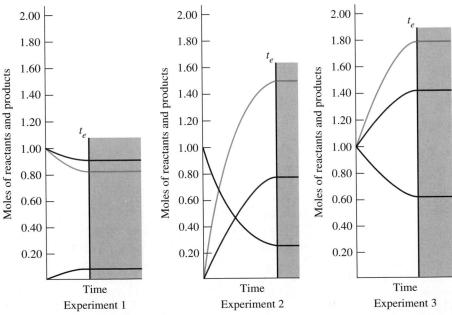

t_e = time for equilibrium to be reached
— mol CO
— mol H₂
— mol CH₃OH

TABLE 16.2   Three Attempts to Find a Constant Ratio of Equilibrium Concentrations in the Reaction $CO + 2\,H_2 \rightleftharpoons CH_3OH$

Expt	Trial 1: $\dfrac{[CH_3OH]}{[CO][H_2]}$	Trial 2: $\dfrac{[CH_3OH]}{[CO](2 \times [H_2])}$	Trial 3: $\dfrac{[CH_3OH]}{[CO][H_2]^2}$
1	$\dfrac{0.00892}{0.0911 \times 0.0822} = 1.19$	$\dfrac{0.00892}{0.0911(2 \times 0.0822)} = 0.596$	$\dfrac{0.00892}{0.0911(0.0822)^2} = 14.5$
2	$\dfrac{0.0247}{0.0753 \times 0.151} = 2.17$	$\dfrac{0.0247}{0.0753(2 \times 0.151)} = 1.09$	$\dfrac{0.0247}{0.0753(0.151)^2} = 14.4$
3	$\dfrac{0.0620}{0.138 \times 0.176} = 2.55$	$\dfrac{0.0620}{0.138(2 \times 0.176)} = 1.28$	$\dfrac{0.0620}{0.138(0.176)^2} = 14.5$

Equilibrium concentration data are from Table 16.1. In Trial 1, the equilibrium concentration of CH_3OH is placed in the numerator and the product of the equilibrium concentrations, $[CO] \times [H_2]$, in the denominator. In Trial 2, each concentration is multiplied by its stoichiometric coefficient. In Trial 3, each concentration is raised to a power equal to its stoichiometric coefficient. Trial 3 has essentially the same value for each experiment. This value is the equilibrium constant K_c.

This ratio is called the **equilibrium constant expression,** and its numerical value is the **equilibrium constant,** denoted by the symbol K_c. The subscript "c" indicates that concentrations (expressed as molarities) are used.

What makes the equilibrium constant expression useful is that it allows us to calculate equilibrium concentrations of reactants and products. We will concentrate on this application in Section 16-7, but as a preview, consider Example 16-1.

▶ One of the classic experiments providing a basis for the concept of the equilibrium constant is described in Feature Problem A on page 579.

EXAMPLE 16-1

Relating Equilibrium Concentrations of Reactants and Products. These equilibrium concentrations are measured in reaction (16.3) at 483 K: $[CO] = 1.03$ M and $[CH_3OH] = 1.56$ M. What is the equilibrium concentration of H_2?

Solution

Substitute the known *equilibrium* concentrations into the equilibrium constant expression (16.4). Then solve for $[H_2]$.

▶ Any concentrations substituted into an equilibrium constant expression, or obtained from it, must be *equilibrium* concentrations.

$$K_c = \frac{[CH_3OH]}{[CO][H_2]^2} = \frac{1.56}{1.03 \times [H_2]^2} = 14.5$$

$$[H_2]^2 = \frac{1.56}{1.03 \times 14.5} = 0.104; \qquad [H_2] = \sqrt{0.104} = 0.322 \text{ M}$$

Practice Example A: In another experiment, equal concentrations of CH_3OH and CO are found at equilibrium in reaction (16.3). What must be the equilibrium concentration of H_2? (*Hint:* Do the actual values of $[CH_3OH]$ and $[CO]$ matter?)

Practice Example B: At a certain temperature $K_c = 1.8 \times 10^4$ for the reaction $N_2(g) + 3\,H_2(g) \rightleftharpoons 2\,NH_3(g)$. If the equilibrium concentrations of N_2 and NH_3 are 0.015 M and 2.00 M, respectively, what is the equilibrium concentration of H_2?

A General Expression for K_c

The equilibrium constant expression (16.4) for the methanol synthesis reaction is just a specific example of a more general case. For the hypothetical, generalized reaction

$$aA + bB + \cdots \rightleftharpoons gG + hH + \cdots$$

The equilibrium constant expression has the form

$$K_c = \frac{[G]^g[H]^h \cdots}{[A]^a[B]^b \cdots} \qquad (16.5)$$

The *numerator* of an equilibrium constant expression is the product of the concentrations of the species on the right side of the equation ([G], [H], . . .), and each concentration is raised to a power given by the stoichiometric coefficient (*g, h,* . . .). The *denominator* is the product of the concentrations of the species on the left side of the equation ([A], [B], . . .), and again each concentration term is raised to a power given by the stoichiometric coefficient (*a, b,* . . .).

The numerical value of an equilibrium constant, K_c, depends on the particular reaction and on the temperature. We will explore the significance of these numerical values in Section 16-4.

The Thermodynamic Equilibrium Constant, K_{eq}: A Preview

As we will learn in Chapter 20, the most fundamental description of equilibrium is provided by thermodynamics, through the *thermodynamic equilibrium constant, K_{eq}.* The terms used in the thermodynamic equilibrium constant are *dimensionless* quantities known as *activities* (recall Section 14-9). Activities are defined in such a way that the numerical values of molarities for solutes in aqueous solutions can generally be substituted into K_c expressions. In the next section, when we describe equilibrium constant expressions in terms of gas pressures, we will substitute numerical values of pressures in atmospheres (atm) for the activities of gases. And to pure solids and pure liquids we will assign activities of 1. The main point to understand now is why we do not write a unit to accompany the numerical value of an equilibrium constant: We are anticipating a later switch to a form of the equilibrium constant based on activities, and activities are dimensionless quantities.

16-3 Relationships Involving Equilibrium Constants

Before assessing an equilibrium situation, it may be necessary to make some preliminary calculations or decisions to get the appropriate equilibrium constant expression. This section presents some useful ideas in working with equilibrium constants.

Relationship of K_c to the Balanced Chemical Equation

We must always make certain that the expression for K_c matches the corresponding balanced equation. In doing so,

- When we *reverse* an equation, we *invert* the value of K_c.

- When we *multiply* the coefficients in a balanced equation by a common factor (2, 3, . . .), we *raise* the equilibrium constant to the corresponding power (2, 3, . . .).
- When we *divide* the coefficients in a balanced equation by a common factor (2, 3, . . .), we *take* the corresponding *root* of the equilibrium constant (square root, cube root, . . .).

Suppose that in discussing the synthesis of CH_3OH from CO and H_2 we had written the reverse of Equation (16.3)—that is,

$$CH_3OH(g) \rightleftharpoons CO(g) + 2\,H_2(g) \qquad K_c' = ?$$

Now, according to the generalized equilibrium constant expression (16.5), we should write

$$K_c' = \frac{[CO][H_2]^2}{[CH_3OH]} = \frac{1}{\dfrac{[CH_3OH]}{[CO][H_2]^2}} = \frac{1}{K_c} = \frac{1}{14.5} = 0.0690$$

In the preceding expression, the terms printed in blue are the equilibrium constant expression and K_c value originally written as expression (16.4). We see that $K_c' = 1/K_c$.

Suppose for a certain application we want an equation based on synthesizing *two* moles of $CH_3OH(g)$.

$$2\,CO(g) + 4\,H_2(g) \rightleftharpoons 2\,CH_3OH(g) \qquad K_c'' = ?$$

Here, $K_c'' = (K_c)^2$. That is,

$$K_c'' = \frac{[CH_3OH]^2}{[CO]^2[H_2]^4} = \left(\frac{[CH_3OH]}{[CO][H_2]^2}\right)^2 = (K_c)^2 = (14.5)^2 = 2.10 \times 10^2$$

EXAMPLE 16-2

Relating K_c to the Balanced Chemical Equation. The following K_c value is given at 298 K for the synthesis of $NH_3(g)$ from its elements.

$$N_2(g) + 3\,H_2(g) \rightleftharpoons 2\,NH_3(g) \qquad K_c = 3.6 \times 10^8$$

What is the value of K_c at 298 K for the reaction

$$NH_3 \rightleftharpoons \tfrac{1}{2}N_2(g) + \tfrac{3}{2}H_2(g) \qquad K_c = ?$$

Solution

First, let's reverse the given equation. This will put $NH_3(g)$ on the left side of the equation where we need it. The equilibrium constant K_c' becomes $1/(3.6 \times 10^8)$.

$$2\,NH_3(g) \rightleftharpoons N_2(g) + 3\,H_2(g) \qquad K_c' = 1/(3.6 \times 10^8) = 2.8 \times 10^{-9}$$

Then, to base the equation on 1 mol $NH_3(g)$, let's *divide* all coefficients by "2." This requires us to take the *square root* of K_c'.

$$NH_3(g) \rightleftharpoons \tfrac{1}{2}N_2(g) + \tfrac{3}{2}H_2(g) \qquad K_c = \sqrt{2.8 \times 10^{-9}} = 5.3 \times 10^{-5}$$

Practice Example A: Use data from Example 16-2 to determine the value of K_c at 298 K for the reaction

$$\tfrac{1}{3}N_2(g) + H_2(g) \rightleftharpoons \tfrac{2}{3}NH_3(g)$$

Practice Example B: For the reaction $NO(g) + \frac{1}{2} O_2(g) \rightleftharpoons NO_2(g)$ at 184 °C, $K_c = 7.5 \times 10^2$. What is the value of K_c at 184 °C for the reaction $2 NO_2(g) \rightleftharpoons 2 NO(g) + O_2(g)$?

Combining Equilibrium Constant Expressions

In Section 7-7, through Hess's law, we showed how to combine a series of equations into a single net equation. To obtain the enthalpy change of the net reaction we added together the enthalpy changes of the individual reactions. We can use a similar procedure when working with equilibrium constants, but with this important difference:

> When individual equations are combined (that is, added), their equilibrium constants are *multiplied* to obtain the equilibrium constant for the net reaction.

Suppose we seek the equilibrium constant for the reaction

$$N_2O(g) + \tfrac{1}{2} O_2(g) \rightleftharpoons 2 NO(g) \qquad K_c = ? \qquad (16.6)$$

and know the K_c values of these two equilibria:

$$N_2(g) + \tfrac{1}{2} O_2(g) \rightleftharpoons N_2O(g) \qquad K_c = 2.7 \times 10^{-18} \qquad (16.7)$$

$$N_2(g) + O_2(g) \rightleftharpoons 2 NO(g) \qquad K_c = 4.7 \times 10^{-31} \qquad (16.8)$$

We can get Equation (16.6) by *reversing* Equation (16.7) and adding it to (16.8). When we do this, we must also take the *reciprocal* of the K_c value of Equation (16.7).

(a) $\qquad N_2O(g) \rightleftharpoons N_2(g) + \tfrac{1}{2} O_2(g) \qquad K_c(a) = 1/(2.7 \times 10^{-18})$

$$= 3.7 \times 10^{17}$$

(b) $\qquad \underline{N_2(g) + O_2(g) \rightleftharpoons 2 NO(g) \qquad\qquad K_c(b) = 4.7 \times 10^{-31}}$

Net: $\quad N_2O(g) + \tfrac{1}{2} O_2(g) \rightleftharpoons 2 NO(g) \qquad K_c(\text{net}) = ? \qquad (16.6)$

According to the general expression (16.5),

$$K_c(\text{net}) = \frac{[NO]^2}{[N_2O][O_2]^{1/2}} = \underbrace{\frac{[N_2][O_2]^{1/2}}{[N_2O]}}_{K_c(a)} \times \underbrace{\frac{[NO_2]^2}{[N_2][O_2]}}_{K_c(b)} = K_c(a) \times K_c(b)$$

$$= 3.7 \times 10^{17} \times 4.7 \times 10^{-31} = 1.7 \times 10^{-13}$$

Equilibria Involving Gases: The Equilibrium Constant, K_p

Mixtures of gases are as much solutions as are mixtures with a liquid solvent. Thus, concentrations in a gaseous mixture can be expressed on a mole per liter basis, and this is what we do when we write and use K_c expressions. Yet, in our preview of the thermodynamic equilibrium constant expression on page 550, we indicated a need to describe gases through their partial pressures in atmospheres. We do this when we write an equilibrium constant expression, K_p, based on the *partial pressures* of gaseous reactants and products.

To derive this alternative form of an equilibrium constant, let's consider the formation of $SO_3(g)$, a key step in the manufacture of sulfuric acid.

$$2 \, SO_2(g) + O_2(g) \rightleftharpoons 2 \, SO_3(g) \tag{16.9}$$

$$K_c = \frac{[SO_3]^2}{[SO_2]^2[O_2]} = 2.8 \times 10^2 \text{ (at 1000 K)}$$

Now, let's use the ideal gas law, $PV = nRT$, to write

$$[SO_3] = \frac{n_{SO_3}}{V} = \frac{P_{SO_3}}{RT} \qquad [SO_2] = \frac{n_{SO_2}}{V} = \frac{P_{SO_2}}{RT} \qquad [O_2] = \frac{n_{O_2}}{V} = \frac{P_{O_2}}{RT}$$

Next, substitute these terms for the concentration terms in K_c.

$$K_c = \frac{(P_{SO_3}/RT)^2}{(P_{SO_2}/RT)^2(P_{O_2}/RT)} = \frac{(P_{SO_3})^2}{(P_{SO_2})^2(P_{O_2})} \times RT \tag{16.10}$$

The ratio of equilibrium partial pressures, expressed in atmospheres and shown in blue, is the equilibrium constant, K_p. The relationship between K_p and K_c for reaction (16.9) is

$$K_c = K_p \times RT \qquad \text{or} \qquad K_p = \frac{K_c}{RT} = K_c(RT)^{-1} \tag{16.11}$$

If we carried out a similar derivation for the general reaction

$$aA(g) + bB(g) + \cdots \rightleftharpoons gG(g) + hH(g) + \cdots$$

this is the result we would get:

$$K_p = K_c(RT)^{\Delta n_{gas}} \tag{16.12}$$

where Δn_{gas} is the difference in the stoichiometric coefficients of *gaseous* products and reactants; that is, $\Delta n_{gas} = (g + h + \cdots) - (a + b + \cdots)$. In reaction (16.9), $\Delta n_{gas} = 2 - (2 + 1) = -1$, just as we found in Equation (16.11).

Although we have decided not to use units for equilibrium constants, we do need to use the correct units for terms *within* equilibrium constant expressions. These are *molarity* for K_c expressions and pressures in *atmospheres* for K_p expressions. These choices then require that we use a value of $R = 0.08206$ L atm mol^{-1} K^{-1} in Equation (16.12).

▶ If $\Delta n_{gas} = 0$, $K_p = K_c(RT)^0$. Because any number raised to the zero power equals 1, for this case $K_p = K_c$.

EXAMPLE 16-3

Relating K_c and K_p. Complete the calculation of K_p for reaction (16.9) from the data given.

Solution

$$K_p = K_c(RT)^{-1} = 2.8 \times 10^2(0.08206 \times 1000)^{-1} = \frac{2.8 \times 10^2}{0.08206 \times 1000}$$

$$= 3.4$$

Practice Example A: For the reaction $2 \, NH_3(g) \rightleftharpoons N_2(g) + 3 \, H_2(g)$ at 298 K, $K_c = 2.8 \times 10^{-9}$. What is the value of K_p for this reaction?

Practice Example B: At 1065 °C, for the reaction $2 \, H_2S(g) \rightleftharpoons 2 \, H_2(g) + S_2(g)$, $K_p = 1.2 \times 10^{-2}$. What is the value of K_c for this reaction at 1065 °C?

Equilibria Involving Pure Liquids and Solids

Up to now, all our examples in this chapter have involved gaseous reactions. In subsequent chapters we will emphasize equilibria in aqueous solutions. Gas-phase reactions and those in solution are *homogeneous* reactions: they occur within a single phase. Let's extend our coverage now to include reactions involving one or more condensed phases—solids and liquids—in contact with a gas or solution phase. These are called *heterogeneous* reactions. One of the most important ideas about heterogeneous reactions is that

> *Equilibrium constant expressions* do not *contain concentration terms for solid or liquid phases of a single component (pure solids and liquids).*

We can think about this statement in either of two ways: (1) An equilibrium constant expression includes terms *only* for reactants and products whose concentrations and/or partial pressures can *change* during a chemical reaction. The concentration of the single component within a pure solid or liquid phase *cannot change*. (2) When we shift to the thermodynamic equilibrium constant based on activities (see again page 550), we will replace concentrations by activities. Because the activities of pure liquid and solids are set equal to 1, the effect on the numerical value of the thermodynamic equilibrium constant is the same as not including terms for pure solids and liquids at all.

The water-gas reaction (see Section 8-6), used to make combustible gases from coal, has reacting species in both gaseous and *solid* phases.

$$C(s) + H_2O(g) \rightleftharpoons CO(g) + H_2(g)$$

Although solid carbon must be present for the reaction to occur, the equilibrium constant expression contains terms only for the species in the homogeneous gas phase: H_2O, CO, and H_2.

$$K_c = \frac{[CO][H_2]}{[H_2O]}$$

The decomposition of calcium carbonate (limestone) is also a heterogeneous reaction. The equilibrium constant expression contains just a single term.

$$CaCO_3(s) \rightleftharpoons CaO(s) + CO_2(g) \qquad K_c = [CO_2] \qquad (16.13)$$

We can write K_p for reaction (16.13) by using Equation (16.12), with $\Delta n_{gas} = 1$.

$$K_p = P_{CO_2} \qquad \text{and} \qquad K_p = K_c(RT) \qquad (16.14)$$

Equation (16.14) indicates that the equilibrium pressure of $CO_2(g)$ in contact with $CaCO_3(s)$ and $CaO(s)$ is a *constant*, equal to K_p. Its value is *independent* of the quantities of $CaCO_3$ and CaO (as long as both solids are present).

One of our examples in Section 16-1 was liquid–vapor equilibrium. This is a *physical* equilibrium because no chemical reactions are involved. Consider the liquid–vapor equilibrium for water.

$$H_2O(l) \rightleftharpoons H_2O(g)$$

$$K_c = [H_2O(g)] \qquad K_p = P_{H_2O} \qquad K_p = K_c(RT)$$

So, equilibrium vapor pressures such as P_{H_2O} are just values of K_p. As we have seen before, these values do *not* depend on the quantities of liquid or vapor at equilibrium, as long as some of each is present.

EXAMPLE 16-4

Writing Equilibrium Constant Expressions for Reactions Involving Pure Solids or Liquids. At equilibrium in the following reaction at 60 °C, the partial pressures of the gases are found to be $P_{HI} = 3.65 \times 10^{-3}$ atm and $P_{H_2S} = 9.96 \times 10^{-1}$ atm. What is the value of K_p for the reaction?

$$H_2S(g) + I_2(s) \rightleftharpoons 2 \, HI(g) + S(s) \qquad K_p = ?$$

Solution

Recall that terms for pure solids do *not* appear in an equilibrium constant expression. The value of K_p is

$$K_p = \frac{(P_{HI})^2}{(P_{H_2S})} = \frac{(3.65 \times 10^{-3})^2}{9.96 \times 10^{-1}} = 1.34 \times 10^{-5}$$

Practice Example A: Teeth are made principally from the mineral hydroxyapatite, $Ca_5(PO_4)_3OH$, which can be dissolved in acidic solution such as that produced by bacteria in the mouth. The reaction is: $Ca_5(PO_4)_3OH(s) + 4 \, H^+(aq) \rightleftharpoons 5 \, Ca^{2+}(aq) + 3 \, HPO_4{}^{2-}(aq) + H_2O(l)$. Write the equilibrium constant expression K_c for this reaction.

Practice Example B: The steam–iron process is used to generate $H_2(g)$, mostly for use in hydrogenating oils. Iron metal and steam [$H_2O(g)$] react to produce $Fe_3O_4(s)$ and $H_2(g)$. Write expressions for K_c and K_p for this reversible reaction. How are the values of K_c and K_p related to each other? Explain.

16-4 Significance of the Magnitude of an Equilibrium Constant

In principle every chemical reaction has an equilibrium constant, but often they are not used. Why is this so? Table 16.3 lists equilibrium constants for several reactions mentioned in this chapter or previously in the text. The first of these reactions is the synthesis of H_2O from its elements. We have always assumed that this reaction *goes to completion,* that is, that the reverse reaction is negligible and the net reaction proceeds only in the forward direction. If a reaction goes to completion, one (or more) of the reactants is used up. A term in the *denominator* of the equilibrium constant expression approaches *zero* and makes

TABLE 16.3 Equilibrium Constants of Some Common Reactions

Reaction	Equilibrium Constant, K_p
$2 \, H_2(g) + O_2(g) \rightleftharpoons 2 \, H_2O(l)$	1.4×10^{83} at 298 K
$CaCO_3(s) \rightleftharpoons CaO(s) + CO_2(g)$	1.9×10^{-23} at 298 K
	1.0 at about 1200 K
$2 \, SO_2(g) + O_2(g) \rightleftharpoons 2 \, SO_3(g)$	3.4 at 1000 K
$C(s) + H_2O(g) \rightleftharpoons CO(g) + H_2(g)$	1.6×10^{-21} at 298 K
	10.0 at about 1100 K

the value of the equilibrium constant very large. A very large numerical value of K_c or K_p signifies that the forward reaction, as written, *goes to completion* or very nearly so. Because the value of K_p for the water synthesis reaction is 1.4×10^{83}, we are entirely justified in saying that the reaction goes to completion at 298 K.

On the other hand, if the equilibrium constant is so large, why is a mixture of hydrogen and oxygen gases stable at room temperature? The value of the equilibrium constant relates to thermodynamic stability: $H_2O(l)$ is much more thermodynamically stable than is a mixture of $H_2(g)$ and $O_2(g)$; it lies at a lower energy state. As we learned in Chapter 15, however, the rate of a chemical reaction is strongly governed by the activation energy, E_a. Because E_a is very high for the synthesis of $H_2O(l)$ from $H_2(g)$ and $O_2(g)$, the rate of reaction is inconsequential at 298 K. To get the reaction to occur at a measurable rate, we either must raise the temperature or use a catalyst. A chemist would say that the synthesis of $H_2O(l)$ at 298 K is a *kinetically* controlled reaction (as opposed to *thermodynamically* controlled).

From Table 16.3 we see that for the decomposition of $CaCO_3(s)$ (limestone), K_p is very small at 298 K (only 1.9×10^{-23}). To account for a very small numerical value of an equilibrium constant, the *numerator* must be very small (approaching zero). A very small numerical value of K_c or K_p signifies that the forward reaction, as written, *does not occur to any significant extent.* Although limestone does not decompose at ordinary temperatures, the partial pressure of $CO_2(g)$ in equilibrium with $CaCO_3(s)$ and $CaO(s)$ increases with temperature. It becomes 1 atm at about 1200 K. An important application of this decomposition reaction is in the commercial production of quicklime (CaO) discussed on page 776.

The conversion of $SO_2(g)$ and $O_2(g)$ to $SO_3(g)$ at 1000 K has an equilibrium constant that is neither very large nor very small (see Table 16.3). Both the forward and reverse reactions are important, and we expect significant amounts of both reactants and products to be present at equilibrium. A similar situation exists for the reaction of C(s) and $H_2O(g)$ at 1100 K, but not at 298 K where the forward reaction does not occur to any significant extent ($K_p = 1.6 \times 10^{-21}$). In conclusion, then,

A reaction is most likely to reach a state of equilibrium in which significant quantities of both reactants and products are present if the numerical value of K_c or K_p is *neither very large nor very small,* roughly in the range from about 10^{-10} to 10^{10}.

Ideas from this section should help you, at times, to decide whether you need to do a further equilibrium calculation to obtain amounts of reactants and products in a chemical reaction.

16-5 The Reaction Quotient, Q: Predicting the Direction of a Net Reaction

Let's return briefly to the set of three experiments that we discussed in Section 16-2 regarding the reaction

$$CO(g) + 2\,H_2(g) \rightleftharpoons CH_3OH(g) \qquad K_c = 14.5$$

In Experiment 1 we start with just the reactants CO and H_2. An overall, or net, change has to occur in which some CH_3OH forms. Only in this way can an equilibrium condition be reached in which all reacting species are present. We say that a net reaction occurs in the *forward* direction (*to the right*).

In Experiment 2 we start with just the product, CH_3OH. Here a net change has to occur in which some CH_3OH decomposes back to CO and H_2. We say that a net reaction occurs in the *reverse* direction (*to the left*).

In Experiment 3 all the reacting species are present initially—CO, H_2, and CH_3OH. Here, it is not so obvious in what direction a net change occurs to establish equilibrium.

The ability to predict the direction of net change in establishing equilibrium is important to us for two reasons:

- At times we do not need detailed equilibrium calculations. We may only need a qualitative description of the changes that occur in establishing equilibrium from a given set of initial conditions.

- In some equilibrium calculations it is helpful to determine the direction of net change as a first step.

For any set of *initial* concentrations in a reaction mixture we can set up a ratio of concentrations having the same form as the equilibrium constant expression. This ratio is called the **reaction quotient** and designated Q_c. For a hypothetical generalized reaction, the reaction quotient is

$$Q_c = \frac{[G]^g [H]^h \cdots}{[A]^a [B]^b \cdots} \qquad (16.15)$$

If $Q_c = K_c$, a reaction is at equilibrium, but our primary interest in the relationship between Q_c and K_c is for a reaction mixture that is *not* at equilibrium. To see what this relationship is, let's turn again to the experiments in Table 16.1.

In Expt 1 the *initial* concentrations of CO and H_2 are 1.000 mol/10.0 L = 0.100 M. Initially there is *no* CH_3OH. The value of Q_c is

$$Q_c = \frac{[CH_3OH]_{init}}{[CO]_{init}[H_2]_{init}^2} = \frac{0}{(0.100)(0.100)^2} = 0 \qquad (16.16)$$

We know that a net reaction occurs *to the right* so as to produce some CH_3OH. As it does, the numerator in expression (16.16) increases, the denominator decreases, the value of Q_c *increases,* and eventually $Q_c = K_c$.

If $Q_c < K_c$, a net reaction proceeds from left to right (the forward reaction).

In Expt 2 the *initial* concentration of CH_3OH is 1.000 mol/10.0 L = 0.100 M. Initially there is *no* CO or H_2. The value of Q_c is

$$Q_c = \frac{[CH_3OH]_{init}}{[CO]_{init}[H_2]_{init}^2} = \frac{0.100}{0 \times 0} = \infty \qquad (16.17)$$

We know that a net reaction occurs *to the left* so as to produce some CO and H_2. As it does, the numerator in expression (16.17) decreases, the denominator increases, the value of Q_c *decreases,* and eventually $Q_c = K_c$.

If $Q_c > K_c$, a net reaction proceeds from right to left (the reverse direction).

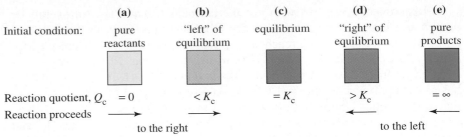

(a)	(b)	(c)	(d)	(e)
Initial condition:	pure reactants	"left" of equilibrium	equilibrium	"right" of equilibrium	pure products

Reaction quotient, Q_c $= 0$ $< K_c$ $= K_c$ $> K_c$ $= \infty$

Reaction proceeds $\longrightarrow$ $\longrightarrow$ $\longleftarrow$ $\longleftarrow$

to the right   to the left

FIGURE 16-3
Predicting the direction of net change in a reversible reaction.
Five possibilities for the relationship of initial and equilibrium conditions are shown. From Table 16.1 and Figure 16-2, Experiment 1 corresponds to initial condition (a); Experiment 2 to condition (e); and Experiment 3 to (d). The situation in Example 16-5 also corresponds to condition (d).

Now let us turn to a case for which we really need a criterion for the direction of net change. In Expt 3 the *initial* concentrations of all three species are 1.000 mol/10.0 L = 0.100 M. The value of Q_c is

$$Q_c = \frac{[CH_3OH]_{init}}{[CO]_{init}[H_2]^2_{init}} = \frac{0.100}{(0.100)(0.100)^2} = 100$$

Because $Q_c > K_c$ (100 compared with 14.5), a net reaction occurs in the *reverse direction*. (Note how you can verify this conclusion from Figure 16-2. The amounts of CO and H_2 at equilibrium are greater than they were initially, and the amount of CH_3OH is less.)

The criteria for predicting the direction of a net chemical change in a reversible reaction are summarized in Figure 16-3 and applied in Example 16-5.

EXAMPLE 16-5

Predicting the Direction of a Net Chemical Reaction in Establishing Equilibrium. To increase the yield of $H_2(g)$ in the reaction of $C(s)$ and $H_2O(g)$, an additional reaction called the "water-gas shift reaction" is generally used. For the water-gas shift reaction,

$$CO(g) + H_2O(g) \rightleftharpoons CO_2(g) + H_2(g)$$

$K_c = 1.00$ at about 1100 K. The following amounts of substances are brought together and allowed to react at 1100 K: 1.00 mol CO, 1.00 mol H_2O, 2.00 mol CO_2, and 2.00 mol H_2. Compared with their initial amounts, which of the substances will be present in a greater amount and which, in a lesser amount, when equilibrium is established?

Solution

Our task is to determine the direction of net change. This means evaluating Q_c. To substitute concentrations into the expression for Q_c, we can assume an arbitrary volume V. Its value is immaterial because the volume cancels out in this case.

▶ Volume terms cancel in a reaction quotient or equilibrium constant expression whenever the sum of the exponents in the numerator equals that in the denominator.

$$Q_c = \frac{[CO_2][H_2]}{[CO][H_2O]} = \frac{(2.00/V)(2.00/V)}{(1.00/V)(1.00/V)} = 4.00$$

Because $Q_c > K_c$ (that is, $4.00 > 1.00$), a net reaction occurs to the *left*. When equilibrium is established, the amounts of CO and H_2O will be *greater* than initially and the amounts of CO_2 and H_2 will be *less* than initially.

Practice Example A: For the reaction $PCl_5(g) \rightleftharpoons PCl_3(g) + Cl_2(g)$, $K_c = 0.0454$ at 261 °C. If a vessel is filled with these gases such that the initial concentrations are $[PCl_3] = 0.50$ M, $[Cl_2] = 0.20$ M, and $[PCl_5] = 4.50$ M, in which direction will the net reaction proceed?

Practice Example B: In Example 16-5, equal masses of CO, H_2O, CO_2, and H_2 are mixed at a temperature of 1100 K. When equilibrium is established, which substance(s) will be present in greater amounts than initially and which in lesser amounts? (*Hint:* You do not need to know the actual masses or the reaction volume.)

16-6 Altering Equilibrium Conditions: Le Châtelier's Principle

At times we want only to make qualitative statements about a reversible reaction: the direction of a net reaction, whether the amount of a substance will increase or decrease when equilibrium is reached, and so on. Also, we may not have the data needed for a quantitative calculation. In these cases we can use a statement attributed to the French chemist Henri Le Châtelier (1884). **Le Châtelier's principle** is hard to state unambiguously, but its essential meaning is that

> When an equilibrium system is subjected to a change in temperature, pressure, or concentration of a reacting species, the system responds by attaining a new equilibrium that *partially* offsets the impact of the change.

As we shall see in the examples that follow, it is generally not difficult to predict the outcome of changing one or more variables in a system at equilibrium.

▶ Le Châtelier's principle and its use to maximize the yield of an industrially important reaction are explored in Feature Problem B on page 579.

Effect of Changing the Amounts of Reacting Species on Equilibrium

Let's return to the reaction

$$2 SO_2(g) + O_2(g) \rightleftharpoons 2 SO_3(g) \qquad K_c = 2.8 \times 10^2 \text{ at } 1000 \text{ K} \qquad (16.9)$$

Suppose we start with certain equilibrium amounts of SO_2, O_2, and SO_3, as suggested by Figure 16-4a. Now let's create a disturbance in the equilibrium mixture by forcing into the 10.0-L flask an additional 1.00 mol SO_3 (Figure 16-4b). How will the amounts of the reacting species change to reestablish equilibrium?

According to Le Châtelier's principle, if the system is to partially offset an action that increases the equilibrium concentration of one of the reacting species, it must do so by favoring the reaction in which that species is consumed. This is the *reverse* reaction—conversion of some of the added SO_3 to SO_2 and O_2. In the new equilibrium there will be greater amounts of all the substances than in the original equilibrium, although the additional amount of SO_3 will be less than the 1.00 mol that was added (Figure 16-4c).

Another way to look at the matter is to evaluate the reaction quotient immediately after adding the SO_3.

Original equilibrium

$$Q_c = \frac{[SO_3]^2}{[SO_2]^2[O_2]} = K_c$$

Following disturbance

$$Q_c = \frac{[SO_3]^2}{[SO_2]^2[O_2]} > K_c$$

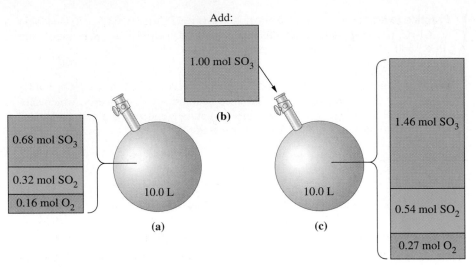

FIGURE 16-4
Changing equilibrium conditions by changing the amount of a reactant
$2 SO_2(g) + O_2(g) \rightleftharpoons 2 SO_3(g)$, $K_c = 2.8 \times 10^2$ at 1000 K.
(a) The original equilibrium condition. (b) Disturbance caused by adding 1.00 mol SO_3. (c)
The new equilibrium condition (as calculated in Exercise 72).
    The amount of SO_3 in the new equilibrium mixture (c), 1.46 mol, is greater than the
original 0.68 mol (a), but it is not as great as immediately after the addition of 1.00 mol
SO_3 in (b), 1.68 mol. The effect of adding SO_3 to an equilibrium mixture is *partially* offset
when equilibrium is restored.

Adding any quantity of SO_3 to a constant-volume equilibrium mixture makes
Q_c larger than K_c. A net reaction occurs in the direction that reduces [SO_3], that
is, to the left, or in the *reverse* direction. (Notice that reaction in the reverse direction increases [SO_2] and [O_2], further decreasing the value of Q_c.)

EXAMPLE 16-6

*Applying Le Châtelier's Principle: Effect of Adding More of a Reactant to an Equi-
librium Mixture.* Predict the effect of adding more $H_2(g)$ to a constant-volume equilibrium mixture of N_2, H_2, and NH_3.

$$N_2(g) + 3 H_2(g) \rightleftharpoons 2 NH_3(g)$$

Solution

The action of increasing [H_2] stimulates the forward reaction and a shift of the equilibrium condition "to the right." However, only a portion of the added H_2 is consumed
in this reaction. When equilibrium is reestablished, there will be more H_2 than present originally. The amount of NH_3 will also be greater, but the amount of N_2 will be
smaller. Some of the original N_2 must be consumed in converting some of the added
H_2 to NH_3.

Practice Example A: Given the reaction $2 CO(g) + O_2(g) \rightleftharpoons 2 CO_2(g)$ at equilibrium and constant volume, what is the effect of adding $O_2(g)$?

Practice Example B: Calcination of limestone (decomposition by heating),
$CaCO_3(s) \rightleftharpoons CaO(s) + CO_2(g)$, is the commercial source of quicklime, $CaO(s)$. What
is the effect on this equilibrium mixture caused by *adding* some **(a)** $CaO(s)$; **(b)**
$CO_2(g)$?

Effect of Changes in Pressure or Volume on Equilibrium

There are three ways we can change the pressure of a constant-temperature equilibrium mixture.

1. *Add or remove a gaseous reactant or product from an equilibrium mixture.* The effect of these actions on the equilibrium condition is simply that due to adding or removing a reaction component, as described previously.

2. *Add an inert gas to the constant-volume reaction mixture.* This has the effect of increasing the *total* pressure, but the partial pressures of the reacting species are all unchanged. As a result, an inert gas added in this way has no effect on the equilibrium condition.

3. *Change the pressure by changing the volume of the system.* The pressure can be increased by decreasing the volume of the system, or decreased by increasing the system volume. The effect of a pressure change can thus be thought of as one of a volume change.

Let's explore the third situation first. Consider, again, the formation of $SO_3(g)$ from $SO_2(g)$ and $O_2(g)$.

$$2\ SO_2(g) + O_2(g) \rightleftharpoons 2\ SO_3(g) \qquad K_c = 2.8 \times 10^2 \text{ at 1000 K} \qquad (16.9)$$

The equilibrium mixture in Figure 16-5a has its volume reduced to one-tenth of its original value by increasing the external pressure. To see how the equilibrium amounts of the gases change, let's first rearrange the equilibrium constant expression to the form

$$K_c = \frac{[SO_3]^2}{[SO_2]^2[O_2]} = \frac{(n_{SO_3}/V)^2}{(n_{SO_2}/V)^2(n_{O_2}/V)} = \frac{(n_{SO_3})^2}{(n_{SO_2})^2(n_{O_2})} \times V = 2.8 \times 10^2 \qquad (16.18)$$

From Equation (16.18) we see that if V is *reduced* by a factor of 10, the ratio

$$\frac{(n_{SO_3})^2}{(n_{SO_2})^2(n_{O_2})}$$

FIGURE 16-5
Effect of pressure change on equilibrium in the reaction
$2\ SO_2(g) + O_2(g) \rightleftharpoons 2\ SO_3(g)$.
An increase in external pressure causes a decrease in the reaction volume and a shift in equilibrium "to the right." (See Exercise 72 for a calculation of the new equilibrium amounts.)

0.68 mol SO_3

0.32 mol SO_2

0.16 mol O_2

10.0 L

1.00 L

0.83 mol SO_3

0.17 mol SO_2

0.085 mol O_2

(a)

(b)

must *increase* by a factor of 10. In this way, the value of K_c is restored, which it must be when equilibrium is re-established. There is only one way in which the ratio of moles of gases will increase in value: the number of moles of SO_3 must increase and the numbers of moles of SO_2 and O_2 must decrease. The equilibrium shifts in the direction producing more SO_3—"to the right."

Notice that in reaction (16.9) 3 moles of *gas* on the left produce 2 moles of *gas* on the right. When compared at the same T and P, two moles of $SO_3(g)$ occupies a smaller volume than does a mixture of two moles of $SO_2(g)$ and one mole of $O_2(g)$. Given this fact and the observation from Equation (16.18) that a decrease in volume favors the production of additional SO_3, we can formulate a statement that is especially easy to apply.

> *When the volume of an equilibrium mixture of gases is* reduced, *a net reaction occurs in the direction producing fewer moles of gases. When the volume is* increased, *a net reaction occurs in the direction producing more moles of gases.*

Figure 16-5, as we have seen, suggests a way of *decreasing* the volume of gaseous mixture at equilibrium—by increasing the external pressure. One way to *increase* the volume is to lower the external pressure. Another way is to transfer the equilibrium mixture from its original container to one of larger volume. A third method is to add an inert gas at *constant pressure;* the volume of the mixture must increase to make room for the added gas. The effect on the equilibrium, however, is the same for all three methods: Equilibrium shifts in the direction of the reaction producing the greater number of moles of gas.

▶ An inert gas has no effect on an equilibrium condition if the gas is added to a system maintained at constant volume, but it can have an effect if added at constant pressure.

Equilibria between condensed phases are not much affected by changes in external pressure because solids and liquids are not easily compressible. Also, we cannot assess whether the forward or reverse reaction is favored by these changes just by examining the chemical equation, as we can with gases.

EXAMPLE 16-7

Applying Le Châtelier's Principle: The Effect of Changing Volume. An equilibrium mixture of $N_2(g)$, $H_2(g)$, and $NH_3(g)$ is transferred from a 1.50-L flask to a 5.00-L flask. In which direction does a net reaction occur to restore equilibrium?

$$N_2(g) + 3 H_2(g) \rightleftharpoons 2 NH_3(g)$$

Solution

When the gaseous mixture is transferred to the larger flask, the partial pressure of each gas and the total pressure drop. Whether we think in terms of a *decrease* in pressure or an *increase* in volume, we reach the same conclusion. Equilibrium shifts in such a way as to produce a larger number of moles of gas. Some of the NH_3 originally present decomposes back to N_2 and H_2. A net reaction occurs to the left in restoring equilibrium.

Practice Example A: The reaction $N_2O_4(g) \rightleftharpoons 2 NO_2(g)$ is at equilibrium in a 3.00-L cylinder. What would be the effect on the concentrations of $N_2O_4(g)$ and $NO_2(g)$ if the pressure were doubled (cylinder volume decreased to 1.50 L)?

Practice Example B: How is the equilibrium amount of $H_2(g)$ produced in the "water-gas shift reaction" affected by changing the total gas pressure or the system volume? Explain.

$$CO(g) + H_2O(g) \rightleftharpoons CO_2(g) + H_2(g)$$

Effect of Temperature on Equilibrium

We can think of changing the temperature of an equilibrium mixture in terms of adding heat (raising the temperature) or removing heat (lowering the temperature). According to Le Châtelier's principle, adding heat favors the reaction in which heat is absorbed (*endothermic* reaction). Removing heat favors the reaction in which heat is evolved (*exothermic* reaction). Stated in terms of changing temperature,

> Raising the temperature *of an equilibrium mixture shifts the equilibrium condition in the direction of the* endothermic *reaction.* Lowering the temperature *causes a shift in the direction of the* exothermic *reaction.*

The principal effect of temperature on equilibrium is in changing the value of the equilibrium constant. In Chapter 20 we will learn how to *calculate* equilibrium constants as a function of temperature. For the present, however, we will limit ourselves to qualitative predictions.

EXAMPLE 16-8

Applying Le Châtelier's Principle: Effect of Temperature on Equilibrium. Consider the reaction

$$2 \, SO_2(g) + O_2(g) \rightleftharpoons 2 \, SO_3(g) \qquad \Delta H° = -197.8 \text{ kJ}$$

Will the amount of $SO_3(g)$ formed from given amounts of $SO_2(g)$ and $O_2(g)$ be greater at high or low temperatures?

Solution

Raising the temperature favors the endothermic reaction, the reverse reaction. To favor the forward (exothermic) reaction requires that we lower the temperature. Therefore, an equilibrium mixture would have a higher concentration of SO_3 at lower temperatures. The conversion of SO_2 to SO_3 is more favored at *low* temperatures.

Practice Example A: The reaction $N_2O_4(g) \rightleftharpoons 2 \, NO_2(g)$ has $\Delta H° = +57.2$ kJ. Will the amount of $NO_2(g)$ formed from $N_2O_4(g)$ be greater at high or low temperatures? (See Figure 16-6 on page 565.)

Practice Example B: The enthalpy of formation of NH_3 is $\Delta H_f°[NH_3(g)] = -46.11$ kJ/mol NH_3. Will the concentration of NH_3 in an equilibrium mixture with its elements be greater at -100 or at 300 °C? Explain. (*Hint:* What is the relevant reaction, and what is its enthalpy change?)

Effect of a Catalyst on Equilibrium

A catalyst in a reaction mixture speeds up *both* the forward and reverse reactions. Equilibrium is achieved more rapidly, but the equilibrium amounts are *unchanged* by the catalyst. Consider again reaction (16.9)

$$2 \, SO_2(g) + O_2(g) \rightleftharpoons 2 \, SO_3(g) \qquad K_c = 2.8 \times 10^2 \text{ at } 1000 \text{ K}$$

For a given set of reaction conditions the equilibrium amounts of SO_2, O_2, and SO_3 have fixed values. This is true whether the reaction is carried out by a slow homogeneous reaction, catalyzed in the gas phase, or conducted as a heterogeneous reaction on the surface of a catalyst. Stated another way, the presence of a catalyst does not change the numerical value of the equilibrium constant.

We now have two thoughts about a catalyst to reconcile: one from the preceding chapter and one from this discussion.

- The function of a catalyst is to change the mechanism of a reaction to one having a lower activation energy.
- A catalyst has no effect on the condition of equilibrium in a reversible reaction.

Taken together these two statements must mean that an equilibrium condition is *independent* of the reaction mechanism. Thus, even though we have described equilibrium in terms of opposing reactions occurring at equal rates, we do not have to concern ourselves with the kinetics of chemical reactions to work with the equilibrium concept. This observation is still another indication of the fact that the equilibrium constant is a thermodynamic quantity, as we plan to describe more fully in Chapter 20.

16-7 Equilibrium Calculations: Some Illustrative Examples

▶Feature Problem C on page 579 explores and extends the ideas in this section.

We are now ready to tackle the problem of describing, in quantitative terms, the condition of equilibrium in a reversible reaction. Part of the approach we use may seem unfamiliar at first—it has an algebraic look to it. But as you adjust to this "new look," do not lose sight of the fact that we continue to use some familiar and important ideas—molar masses, molarities, and stoichiometric factors from the balanced equation, for example.

The five numerical examples that follow apply the general equilibrium principles described earlier in the chapter. The first four involve gases, and the fifth deals with equilibrium in an aqueous solution. The study of equilibria in aqueous solutions is the principal topic of the next three chapters. Each example includes a brief section labeled "comments" and printed on a colored background. Think of the collection of these "comments" as the basic methodology of equilibrium calculations. You may want to refer back to these comments from time to time in later chapters.

Example 16-9 is relatively straightforward. It demonstrates how to determine the equilibrium constant of a reaction when the equilibrium concentrations of the reactants and products are known.

EXAMPLE 16-9

Determining a Value of K_c from the Equilibrium Quantities of Substances. Dinitrogen tetroxide, $N_2O_4(l)$, is an important component of rocket fuels—for example, as an oxidizer of liquid hydrazine in the Titan rocket. At 25 °C, N_2O_4 exists as a colorless gas, which partially dissociates into NO_2, a red-brown gas. The color of an equilibrium mixture of these two gases depends on their relative proportions, which in turn depends on the temperature (Figure 16-6).

Equilibrium is established in the reaction $N_2O_4(g) \rightleftharpoons 2\ NO_2(g)$ at 25 °C. The quantities of the two gases present in a 3.00-L vessel are 7.64 g N_2O_4 and 1.56 g NO_2. What is the value of K_c for this reaction?

(a) **(b)**

FIGURE 16-6
The equilibrium $N_2O_4(g) \rightleftharpoons 2 NO_2(g)$.
(a) At dry ice temperatures N_2O_4 exists as a solid. The gas in equilibrium with the solid is mostly colorless N_2O_4 with only a trace of brown NO_2. (b) When warmed to room temperature and above, the N_2O_4 melts and vaporizes. The proportion of $NO_2(g)$ at equilibrium increases over that at low temperatures, and the equilibrium mixture of $N_2O_4(g)$ and $NO_2(g)$ has a red-brown color.

Solution

For both N_2O_4 and NO_2 we need the conversions g → mol → mol/L.

$$[N_2O_4] = \frac{7.64 \text{ g } N_2O_4 \times \dfrac{1 \text{ mol } N_2O_4}{92.01 \text{ g } N_2O_4}}{3.00 \text{ L}} = 0.0277 \text{ M}$$

$$[NO_2] = \frac{1.56 \text{ g } NO_2 \times \dfrac{1 \text{ mol } NO_2}{46.01 \text{ g } NO_2}}{3.00 \text{ L}} = 0.0113 \text{ M}$$

$$K_c = \frac{[NO_2]^2}{[N_2O_4]} = \frac{(0.0113)^2}{0.0277} = 4.61 \times 10^{-3}$$

Practice Example A: Equilibrium is established in a 3.00-L flask at 1405 K for the reaction $2 H_2S(g) \rightleftharpoons 2 H_2(g) + S_2(g)$. At equilibrium there is 0.11 mol $S_2(g)$, 0.22 mol $H_2(g)$, and 2.78 mol $H_2S(g)$. What is the value of K_c for this reaction?

Practice Example B: Equilibrium is established at 25 °C in the reaction $N_2O_4(g) \rightleftharpoons 2 NO_2(g)$, $K_c = 4.61 \times 10^{-3}$. If $[NO_2] = 0.0236$ M in a 2.26-L flask, how many grams of N_2O_4 are also present? (*Hint:* What must $[N_2O_4]$ be at equilibrium?)

Comments
1. The quantities required in an equilibrium constant expression, K_c, are equilibrium *concentrations* in *moles per liter,* not simply equilibrium amounts in moles or masses in grams. You will find it helpful to organize all the equilibrium data and carefully label each item.

Example 16-10 is somewhat more involved. We are still interested in determining the equilibrium constant for a reaction, but we do not have the same sort of information as in Example 16-9. We are given the *initial* concentrations of all the reactants and products, but the equilibrium concentration of only one substance. In this case, we need a little algebra and some careful "bookkeeping." We will introduce a tabular system for keeping track of the changing concentrations of the reactants and products that some call an **"ICE" table.** It notes the **i**nitial, **c**hange, and **e**quilibrium concentrations of all the species. It is a very helpful device that we will use throughout the next three chapters.

EXAMPLE 16-10

Determining a Value of K_p from Initial and Equilibrium Amounts of Substances: Relating K_c and K_p. Equilibrium involving $SO_2(g)$, $O_2(g)$, and $SO_3(g)$ is important in sulfuric acid production. When a 0.0200-mol sample of SO_3 is introduced into an evacuated 1.52-L vessel at 900 K, 0.0142 mol SO_3 is found to be present at equilibrium. What is the value of K_p for the dissociation of $SO_3(g)$ at 900 K?

$$2\ SO_3(g) \rightleftharpoons 2\ SO_2(g) + O_2(g) \qquad K_p = ?$$

Solution

Let's first determine K_c and then convert to K_p by using Equation (16.12). In the following table of data the key item is the *change* in amount of SO_3: In progressing from 0.0200 mol SO_3 to 0.0142 mol SO_3, 0.0058 mol SO_3 is dissociated. We use the *negative* sign ("−0.0058 mol") to indicate that this amount of SO_3 is *consumed* in establishing equilibrium. In the row labeled "change" we must relate the changes in amounts of SO_2 and O_2 to the change in amount of SO_3. For this we use the stoichiometric coefficients from the balanced equation: 2, 2, and 1. That is, *two* moles of SO_2 and *one* mole of O_2 are produced for every *two* moles of SO_3 that dissociate.

the reaction:	$2\ SO_3(g)$	$\rightleftharpoons$	$2\ SO_2(g)$	+	$O_2(g)$
initial amounts:	0.0200 mol		0.00 mol		0.00 mol
change:	−0.0058 mol		+0.0058 mol		+0.0029 mol
equil amounts:	0.0142 mol		0.0058 mol		0.0029 mol
equil concn:	$[SO_3]$ = 0.0142 mol/1.52 L		$[SO_2]$ = 0.0058 mol/1.52 L		$[O_2]$ = 0.0029 mol/1.52 L
	= 9.34×10^{-3} M		= 3.8×10^{-3} M		= 1.9×10^{-3} M

$$K_c = \frac{[SO_2]^2[O_2]}{[SO_3]^2} = \frac{(3.8 \times 10^{-3})^2(1.9 \times 10^{-3})}{(9.34 \times 10^{-3})^2} = 3.1 \times 10^{-4}$$

$$K_p = K_c(RT)^{\Delta n_{gas}} = 3.1 \times 10^{-4}(0.0821 \times 900)^{(2+1)-2}$$

$$= 3.1 \times 10^{-4}(0.0821 \times 900)^1 = 2.3 \times 10^{-2}$$

Practice Example A: A 5.00-L evacuated flask is filled with 1.86 mol NOBr. After equilibrium is established at 25 °C, it is found that there is 0.082 mol of Br_2 present. Determine K_c and K_p for the reaction: $2\ NOBr(g) \rightleftharpoons 2\ NO(g) + Br_2(g)$.

Practice Example B: 0.100 mol SO_2 and 0.100 mol O_2 are introduced into an evacuated 1.52-L flask at 900 K. When equilibrium is reached, the amount of SO_3 found is 0.0916 mol. Use these data to determine K_p for the reaction $2\ SO_3(g) \rightleftharpoons 2\ SO_2(g) + O_2(g)$. (*Hint:* Equilibrium is approached from the opposite direction to that in Example 16-10.)

Comments

2. The chemical equation for a reversible reaction serves *both* to establish the form of the equilibrium constant expression *and* to provide the conversion factors (stoichiometric factors) to relate the equilibrium quantity of one species to those of the others.

3. For equilibria involving gases you can use either K_c or K_p. In general, if the data given involve amounts of substances and volumes, it is easier to work with K_c. If data are given as partial pressures, then work with K_p. Whether working with K_c or K_p or the relationship between them, you must always base these expressions on the chemical equation that is given, not on what you may have used in other situations.

Example 16-11 demonstrates that we can often determine several pieces of useful information about an equilibrium system, given only the equilibrium constant and the reaction equation.

EXAMPLE 16-11

Determining Equilibrium Partial and Total Pressures from a Value of K_p. Ammonium hydrogen sulfide, $NH_4HS(s)$, used as a photographic developer, is unstable and dissociates at room temperature.

$$NH_4HS(s) \rightleftharpoons NH_3(g) + H_2S(g) \qquad K_p(atm) = 0.108 \text{ at } 25 \text{ °C}$$

A sample of $NH_4HS(s)$ is introduced into an evacuated flask at 25 °C. What is the total gas pressure at equilibrium?

Solution

K_p for this reaction is just the product of the equilibrium partial pressures of $NH_3(g)$ and $H_2S(g)$, each stated in atmospheres. (There is no term for NH_4HS because it is a solid.) Moreover, because these gases are produced in equimolar amounts, $P_{NH_3} = P_{H_2S}$. Let us first find P_{NH_3} and then P_{tot}.

$$K_p = (P_{NH_3})(P_{H_2S}) = (P_{NH_3})(P_{NH_3}) = (P_{NH_3})^2 = 0.108$$

$$P_{NH_3} = \sqrt{0.108} = 0.329 \text{ atm} \qquad P_{H_2S} = P_{NH_3} = 0.329 \text{ atm}$$

$$P_{tot} = P_{NH_3} + P_{H_2S} = 0.329 \text{ atm} + 0.329 \text{ atm} = 0.658 \text{ atm}$$

Practice Example A: Sodium hydrogen carbonate (baking soda) decomposes at elevated temperatures and is one of the sources of $CO_2(g)$ when this compound is used in baking.

$$2 \text{ NaHCO}_3(s) \rightleftharpoons Na_2CO_3(s) + H_2O(g) + CO_2(g) \qquad K_p(atm) = 0.231 \text{ at } 100 \text{ °C}$$

What is the partial pressure of $CO_2(g)$ when this equilibrium is established starting with $NaHCO_3(s)$?

Practice Example B: If enough additional $NH_3(g)$ is added to the flask in Example 16-11 to raise its partial pressure to 0.500 atm at equilibrium, what will be the *total* gas pressure when equilibrium is reestablished? (*Hint:* Do the partial pressures of NH_3 and H_2S remain equal?)

Comments

4. When using K_p expressions, look for relationships among partial pressures of the reactants. If you need to relate the total pressure to the partial pressures of the reactants, you should be able to do this with some equations presented in Chapter 6 (for example, Equations 6.15, 6.16, and 6.17).

Example 16-12 brings back the ICE format that we introduced in Example 16-10, but with a twist. This time we know the value of the equilibrium constant but do not have any information about the equilibrium concentrations of the reactants or products. That means that we do not know how much the initial values will change. We show this by using an "x" in that part of the table. The setup will be quite algebraic; in fact, we must use the quadratic formula to obtain a solution.

(a)   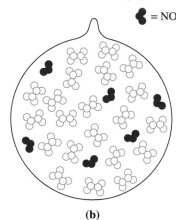 $= N_2O_4$

$= NO_2$

FIGURE 16-7
Equilibrium in the reaction
$N_2O_4(g) \rightleftharpoons 2\,NO_2(g)$
at 25 °C—Example 16-12 illustrated.
Each "molecule" illustrated represents 0.001 mol.
(a) Initially, the bulb contains 0.024 mol N_2O_4, represented by 24 "molecules." (b) At equilibrium,
some molecules of N_2O_4 have dissociated to NO_2.
The 21 "molecules" of N_2O_4 and 6 of NO_2 correspond to 0.021 mol N_2O_4 and 0.006 mol NO_2 at
equilibrium.

(b)

EXAMPLE 16-12

Calculating Equilibrium Concentrations from Initial Conditions. A 0.0240-mol sample of $N_2O_4(g)$ is allowed to come to equilibrium with $NO_2(g)$ in a 0.372-L flask at
25 °C. Calculate the amount of N_2O_4 present at equilibrium (Figure 16-7).

$$N_2O_4(g) \rightleftharpoons 2\,NO_2(g) \qquad K_c = 4.61 \times 10^{-3} \text{ at } 25\text{ °C}$$

Solution

We need to determine the amount of N_2O_4 that dissociates to establish equilibrium.
For the first time we introduce an algebraic unknown, x. Suppose we let $x =$ the number of moles of N_2O_4 that dissociate. In the following table we enter the value $-x$
into the row labeled "change." The amount of NO_2 produced is $+2x$ because the stoichiometric coefficient of NO_2 is *2* and that of N_2O_4 is *1*.

the reaction:	$N_2O_4(g)$	$\rightleftharpoons$	$2\,NO_2(g)$
initial amounts:	0.0240 mol		0.00 mol
change:	$-x$ mol		$+2x$ mol
equil amounts:	$(0.0240 - x)$ mol		$2x$ mol
equil concn:	$[N_2O_4] = (0.0240 - x \text{ mol})/0.372 \text{ L}$		$[NO_2] = 2x \text{ mol}/0.372 \text{ L}$

$$K_c = \frac{[NO_2]^2}{[N_2O_4]} = \frac{\left(\dfrac{2x}{0.372}\right)^2}{\left(\dfrac{0.0240 - x}{0.372}\right)} = \frac{4x^2}{0.372(0.0240 - x)} = 4.61 \times 10^{-3}$$

$$4x^2 = 4.12 \times 10^{-5} - (1.71 \times 10^{-3})x$$

$$x^2 + (4.28 \times 10^{-4})x - 1.03 \times 10^{-5} = 0$$

$$x = \frac{-4.28 \times 10^{-4} \pm \sqrt{(4.28 \times 10^{-4})^2 + 4 \times 1.03 \times 10^{-5}}}{2}$$

$$= \frac{-4.28 \times 10^{-4} \pm \sqrt{(1.83 \times 10^{-7}) + (4.12 \times 10^{-5})}}{2}$$

$$= \frac{-4.28 \times 10^{-4} \pm \sqrt{4.14 \times 10^{-5}}}{2}$$

$$= \frac{-4.28 \times 10^{-4} \pm 6.43 \times 10^{-3}}{2}$$

$$= \frac{-4.28 \times 10^{-4} + 6.43 \times 10^{-3}}{2} = \frac{6.00 \times 10^{-3}}{2}$$

$$= 3.00 \times 10^{-3} \text{ mol } N_2O_4$$

▶ For a discussion of quadratic equations see Appendix A.

▶ The symbol $\pm$ signifies that there are two possible roots. In this problem x must be a *positive* quantity smaller than 0.0240.

The amount of N_2O_4 at equilibrium is $(0.0240 - x) = (0.0240 - 0.0030)$
$$= 0.0210 \text{ mol } N_2O_4$$

Practice Example A: If 0.150 mol $H_2(g)$ and 0.200 mol $I_2(g)$ are introduced into a 15.0-L flask at 445 °C and allowed to come to equilibrium, how many moles of $HI(g)$ will be present?

$$H_2(g) + I_2(g) \rightleftharpoons 2 HI(g) \qquad K_c = 50.2 \text{ at } 445 °C$$

Practice Example B: Suppose the equilibrium mixture of Example 16-12 is transferred to a 10.0-L flask. **(a)** Will the equilibrium amount of N_2O_4 increase or decrease? Explain. **(b)** Calculate the number of moles of N_2O_4 in the new equilibrium. (*Hint:* Start with the initial conditions of Example 16-12 but use a 10.0-L volume.)

Comments
5. When you need to introduce an algebraic unknown, x, into an equilibrium calculation,

- introduce x into the ICE set-up in the row labeled "change";
- decide which change to label as x, the amount of a reactant consumed or of a product formed. Usually, we base this on the species that has the smallest stoichiometric coefficient in the balanced chemical equation;
- use stoichiometric factors to relate the other changes to x (that is, $2x$, $3x$, . . .);
- consider that equilibrium amounts = initial amounts + "change." (If you have assigned the correct signs to the changes, then equilibrium amounts will also be correct.)

▶ You will frequently encounter quadratic equations in equilibrium calculations, which you can solve with the quadratic formula. You may occasionally run into some higher degree equations, however, and Feature Problem C on page 579 outlines a straightforward method of dealing with these.

Our final example is similar to the previous one, but with this slight complication: initially we don't know whether a net reaction occurs to the right or to the left to establish equilibrium. We can find out, though, by using the reaction quotient, Q_c. Also, because the reactants and products are in solution, we can work only with concentrations and K_c (not partial pressures and K_p, as we usually can with gaseous equilibria).

EXAMPLE 16-13

Using the Reaction Quotient, Q_c, in an Equilibrium Calculation. Solid silver is added to a solution with these initial concentrations: $[Ag^+] = 0.200$ M, $[Fe^{2+}] = 0.100$ M, and $[Fe^{3+}] = 0.300$ M. The following reversible reaction occurs.

$$Ag^+(aq) + Fe^{2+}(aq) \rightleftharpoons Ag(s) + Fe^{3+}(aq) \qquad K_c = 2.98$$

What are the ion concentrations when equilibrium is established?

Solution

▶ If *all* of the reacting species are present initially, use Q_c to determine the direction of net change. If one or more of the reactants or products is not present initially, a net reaction must occur to produce some of that substance(s).

Because all reactants and products are present initially, we can use the reaction quotient Q_c to determine the direction in which a net reaction occurs.

$$Q_c = \frac{[Fe^{3+}]}{[Ag^+][Fe^{2+}]} = \frac{0.300}{(0.200)(0.100)} = 15.0$$

Because Q_c (15.0) is larger than K_c (2.98), a net reaction must proceed *to the left.* Let's define x as the change in molarity of Fe^{3+}. Because a net reaction occurs *to the left,* we designate the changes for the species on the left side of the equation as positive and on the right side as negative. The relevant data are tabulated as follows.

the reaction:	$Ag^+(aq)$	+	$Fe^{2+}(aq)$	$\rightleftharpoons Ag(s) + Fe^{3+}(aq)$
initial concn:	0.200 M		0.100 M	0.300 M
changes:	$+x$ M		$+x$ M	$-x$ M
equil concn:	$(0.200 + x)$ M		$(0.100 + x)$ M	$(0.300 - x)$ M

$$K_c = \frac{[Fe^{3+}]}{[Ag^+][Fe^{2+}]} = \frac{(0.300 - x)}{(0.200 + x)(0.100 + x)} = 2.98$$

This equation, which is solved in Appendix A, is a quadratic equation for which the acceptable root is $x = 0.11$. To obtain the equilibrium concentrations, we substitute this value of x into the terms shown in the table of data.

$$[Ag^+]_{equil} = 0.200 + 0.11 = 0.31 \text{ M}$$

$$[Fe^{2+}]_{equil} = 0.100 + 0.11 = 0.21 \text{ M}$$

$$[Fe^{3+}]_{equil} = 0.300 - 0.11 = 0.19 \text{ M}$$

Checking the Results

If we have done the calculation correctly, we should obtain a value very close to that given for K_c when we substitute the *calculated* equilibrium concentrations into the reaction quotient, Q_c. We do.

$$Q_c = \frac{[Fe^{3+}]}{[Ag^+][Fe^{2+}]} = \frac{0.19}{(0.31)(0.21)} = 2.9 \qquad (K_c = 2.98)$$

Practice Example A: Excess Ag(s) is added to 1.20-M $Fe^{3+}(aq)$. Given that

$$Ag^+(aq) + Fe^{2+}(aq) \rightleftharpoons Ag(s) + Fe^{3+}(aq) \qquad K_c = 2.98$$

what are the equilibrium concentrations of the species in solution?

Practice Example B: A solution is prepared with $[V^{3+}] = [Cr^{2+}] = 0.0100$ M and $[V^{2+}] = [Cr^{3+}] = 0.150$ M. The following reaction occurs.

$$V^{3+}(aq) + Cr^{2+}(aq) \rightleftharpoons V^{2+}(aq) + Cr^{3+}(aq) \qquad K_c = 7.2 \times 10^2$$

What are the ion concentrations when equilibrium is established? (*Hint:* The algebra can be greatly simplified by extracting the square root of both sides of the equation at the appropriate point.)

Comments

6. It is sometimes helpful to compare the reaction quotient, Q_c (or Q_p), to the equilibrium constant, K_c (or K_p), to determine the direction of the net reaction.

7. In many equilibrium calculations—often those in aqueous solutions—you can work with molarities directly, without having to work with moles of reactants and solution volumes.

8. Where possible, check your calculation, for instance by substituting *calculated* equilibrium concentrations into the reaction quotient, Q_c (or Q_p), to see if its numerical value is close to that of K_c (or K_p).

Summary

A *reversible* reaction reaches a point at which the rates of the forward and reverse reactions are equal. This condition of dynamic equilibrium is described through an equilibrium constant expression. In one case—K_c—the equilibrium condition is expressed through concentrations. In another—K_p—partial pressures of gases are used. The *form* of an equilibrium constant expression is established from the balanced chemical equation. Its *numerical value* is determined by experiment. Other ideas in the chapter deal with the effect of the following situations on K_c or K_p:

• the presence of pure solids or liquids;

• reversing a chemical equation;

• multiplying a chemical equation by a factor;

• combining several equations into a net chemical equation.

Le Châtelier's principle is used to make *qualitative* predictions of the effects of different variables on an equilibrium condition. This principle describes how an equilibrium condition is modified, or "shifts," whenever an equilibrium is disturbed by adding or removing reactants or by changing the reaction volume, external pressure, or temperature. Catalysts, by speeding up equally both the forward and reverse reactions, have no effect on an equilibrium condition.

Another *qualitative* prediction concerns the direction of net change leading to equilibrium. This prediction can be made with a ratio of *initial* concentrations, formed in the same manner as K_c or K_p and called the reaction quotient, Q_c or Q_p.

For *quantitative* equilibrium calculations, a few basic principles and algebraic techniques illustrated in the concluding section of the chapter are required.

Summarizing Example

In the manufacture of ammonia, the chief source of hydrogen gas is this reaction for the reforming of methane at high temperatures.

$$CH_4(g) + 2\ H_2O(g) \rightleftharpoons CO_2(g) + 4\ H_2(g) \quad (16.19)$$

The following data are also given.

$$(1)\ CO(g) + H_2O(g) \rightleftharpoons CO_2(g) + H_2(g)$$

$$\Delta H° = -40\ kJ;\ K_c = 1.4\ \text{at } 1000\ K$$

$$(2)\ CO(g) + 3\ H_2(g) \rightleftharpoons H_2O(g) + CH_4(g)$$

$$\Delta H° = -230\ kJ;\ K_c = 190\ \text{at } 1000\ K$$

A mixture containing 1.00 mol each of $CH_4(g)$ and $H_2O(g)$ is brought to equilibrium in reaction (16.19) in a 10.0-L flask at 1000 K. The equilibrium mixture is found to contain 12.4 g of *unreacted* $CH_4(g)$. Calculate the mass of $H_2(g)$ also present at equilibrium. Describe the effect on the equilibrium

amount of H_2 produced by each of these actions: **(a)** add a catalyst; **(b)** add CH_4; **(c)** add CO_2; **(d)** add He(g) at constant volume; **(e)** remove H_2O; **(f)** raise the temperature to 1200 K; **(g)** transfer the mixture to a 15.0-L flask.

1. *Calculate the mass of H_2 at equilibrium.* Because both the initial and equilibrium quantities of one of the reactants (CH_4) are known, *all* the equilibrium quantities can be related through stoichiometry alone. An equilibrium calculation is not required. *Answer:* 1.83 g H_2.

2. *Determine $\Delta H°$ for reaction (16.19).* This quantity is needed for part 3(f) and can be obtained by combining equation (1) and the reverse of equation (2). Note that, again, a value of K_c is not needed. *Result:* $\Delta H° = +190\ kJ$

3. *Make qualitative predictions using Le Châtelier's principle.* These questions relate to the discussion in Section 16-6. *Answer:* **(a)** no change; **(b)** increase; **(c)** decrease; **(d)** no change; **(e)** decrease; **(f)** increase; **(g)** increase.

Focus On   The Nitrogen Cycle

The Nitrogen Cycle

Nitrogen is an essential element required for the synthesis of proteins by plants and animals. The atmosphere contains a vast pool of elemental nitrogen, but plants and animals require nitrogen-containing compounds rather than $N_2(g)$. The conversion of elemental atmospheric nitrogen to nitrogen compounds is called *nitrogen fixation.* The simplified nitrogen cycle depicted here describes how elemental nitrogen is fixed, becomes part of plants, animals, and soil, and then returns again to the atmosphere. The numbers in parenthesis refer to the various parts of the cycle described below.

(1) A few leguminous plants, such as beans, peas, and alfalfa, have bacteria residing on their roots that convert nitrogen to compounds used to make plant protein. The decay of these plants produces a

Key Terms

equilibrium (16-1)
equilibrium constant (16-2)
equilibrium constant expression (16-2)

ICE table (16-7)
K_c (16-2)
K_p (16-3)

Le Châtelier's principle (16-6)
reaction quotient, Q_c (16-5)

Review Questions

1. In your own words define or explain the following terms or symbols: **(a)** K_p; **(b)** Q_c; **(c)** Δn_{gas}.

2. Briefly describe each of the following ideas or phenomena: **(a)** dynamic equilibrium; **(b)** direction of a net reaction; **(c)** Le Châtelier's principle; **(d)** effect of a catalyst on equilibrium.

3. Explain the important distinctions between each pair of terms: **(a)** reaction that goes to completion and reversible reaction; **(b)** K_c and K_p; **(c)** reaction quotient (Q) and equilibrium constant expression (K); **(d)** homogeneous and heterogeneous reaction.

4. Write equilibrium constant expressions, K_c, for the reactions

(a) $2 NO(g) + O_2(g) \rightleftharpoons 2 NO_2(g)$
(b) $Zn(s) + 2 Ag^+(aq) \rightleftharpoons Zn^{2+}(aq) + 2 Ag(s)$
(c) $Mg(OH)_2(s) + CO_3^{2-}(aq) \rightleftharpoons MgCO_3(s) + 2 OH^-(aq)$

5. Write equilibrium constant expressions, K_p, for the reactions

(a) $CS_2(g) + 4 H_2(g) \rightleftharpoons CH_4(g) + 2 H_2S(g)$
(b) $Ag_2O(s) \rightleftharpoons 2 Ag(s) + \frac{1}{2} O_2(g)$
(c) $2 NaHCO_3(s) \rightleftharpoons Na_2CO_3(s) + CO_2(g) + H_2O(g)$

6. Write an equilibrium constant expression, K_c, for the formation of 1 mol of each of the following *gaseous* compounds from its *gaseous* elements: **(a)** N_2O; **(b)** HBr; **(c)** NH_3; **(d)** ClF_3; **(e)** NOF_2.

reservoir in the soil of nitrogen compounds such as nitrates and ammonia. (2) These nitrogen compounds in the soil are used by other plants to synthesize proteins. (3) Animals derive nitrogen by feeding on plants or their fruits. (4) Soil bacteria feed on dead plant and animal matter, returning the nitrogen to the soil as nitrates and ammonia. (5) Other soil bacteria produce N_2O and N_2, which are returned directly to the atmosphere, completing the cycle.

Reversible chemical reactions play an important role in the nitrogen cycle. One example is the reaction of $N_2(g)$ and $O_2(g)$ to form $NO(g)$.

$$N_2(g) + O_2(g) \rightleftharpoons 2\,NO(g)$$

$$K_p = 4.7 \times 10^{-31} \text{ at 298 K}$$

$$= 1.3 \times 10^{-4} \text{ at 1800 K}$$

The reaction does not occur to any measurable extent at 298 K, but at 1800 K the situation is a little different. An equilibrium mixture of $N_2(g)$ and $O_2(g)$ at 1800 K contains about 1 or 2% $NO(g)$. There are two common ways in which $NO(g)$ is introduced into the nitrogen cycle through this high-temperature reaction: One occurs *naturally* in lightning flashes during thunderstorms (6), and the other occurs artificially in high-temperature combustion processes, as in internal combustion engines (7). Further reactions account for the formation of $HNO_3(aq)$ in rainstorms and the introduction of nitrates into the soil.

$$2\,NO(g) + O_2(g) \longrightarrow 2\,NO_2(g)$$

$$3\,NO_2(g) + H_2O(l) \longrightarrow$$

$$2\,HNO_3(aq) + NO(g)$$

Still another human activity that intervenes in the nitrogen cycle is the synthesis of ammonia.

$$N_2(g) + 3\,H_2(g) \rightleftharpoons 2\,NH_3(g)$$

From $NH_3(g)$ a host of nitrogen-containing compounds are made for use in the manufacture of explosives, pharmaceuticals, and plastics. And most important from the standpoint of the nitrogen cycle is (8), the direct introduction of $NH_3(l)$, ammonium compounds, and nitrate compounds into the soil as fertilizers. So much nitrogen is now being fixed artificially that fixed nitrogen is accumulating in the environment somewhat faster than it is being returned to the atmosphere. This causes environmental problems such as the buildup of nitrates in groundwater.

7. From these values of K_c

$$CO(g) + H_2O(g) \rightleftharpoons CO_2(g) + H_2(g)$$

$$K_c = 23.2 \text{ at 600 K}$$

$$SO_2(g) + \tfrac{1}{2} O_2(g) \rightleftharpoons SO_3(g)$$

$$K_c = 56 \text{ at 900 K}$$

$$2\,H_2S(g) \rightleftharpoons 2\,H_2(g) + S_2(g)$$

$$K_c = 2.3 \times 10^{-4} \text{ at 1405 K}$$

determine values of K_c for the following reactions at the corresponding temperatures.
(a) $CO_2(g) + H_2(g) \rightleftharpoons CO(g) + H_2O(g)$
(b) $2\,SO_2(g) + O_2(g) \rightleftharpoons 2\,SO_3(g)$
(c) $H_2(g) + \tfrac{1}{2} S_2(g) \rightleftharpoons H_2S(g)$

8. Determine K_c for the reaction

$$\tfrac{1}{2} N_2(g) + \tfrac{1}{2} O_2(g) + \tfrac{1}{2} Br_2(g) \rightleftharpoons NOBr(g)$$

from the following information (at 298 K).

$$2\,NO(g) \rightleftharpoons N_2(g) + O_2(g) \qquad K_c = 2.1 \times 10^{30}$$

$$NO(g) + \tfrac{1}{2} Br_2(g) \rightleftharpoons NOBr(g) \qquad K_c = 1.4$$

9. In the reversible reaction $H_2(g) + I_2(g) \rightleftharpoons 2\,HI(g)$, an initial mixture contains 2 mol H_2 and 1 mol I_2. Which of the following is the amount of HI expected at equilibrium? Explain.
(a) 1 mol; (b) 2 mol; (c) more than 2 but less than 4 mol; (d) less than 2 mol.

10. Equilibrium is established in the reversible reaction $2\,A + B \rightleftharpoons 2\,C$. The equilibrium concentrations are [A] = 0.55 M, [B] = 0.33 M, [C] = 0.43 M. What is the value of K_c for this reaction?

11. A 0.0040-mol sample of $S_2(g)$ is allowed to dissociate in an 0.500-L flask at 1000 K. When equilibrium is established, 2.0×10^{-11} mol $S(g)$ is present. What is the value of K_c for the reaction $S_2(g) \rightleftharpoons 2\,S(g)$?

12. An equilibrium mixture of SO_2, SO_3, and O_2 gases is maintained in a 2.05-L flask at a temperature at which $K_c = 35.5$ for the reaction

$$2\,SO_2(g) + O_2(g) \rightleftharpoons 2\,SO_3(g)$$

(a) If the numbers of moles of SO_2 and SO_3 in the flask are equal, how many moles of O_2 are present?
(b) If the number of moles of SO_3 in the flask is twice the number of moles of SO_2, how many moles of O_2 are present?

13. Determine the numerical values of K_p for reactions (a), (b), and (c) in Review Question 7.

14. For the reaction $2\,NO_2(g) \rightleftharpoons 2\,NO(g) + O_2(g)$, $K_c = 1.8 \times 10^{-6}$ at 184 °C. What is the value at 184 °C of K_p for this reaction?

$$NO(g) + \tfrac{1}{2} O_2(g) \rightleftharpoons NO_2(g)$$

15. An equilibrium mixture at 1000 K contains 0.276 mol H_2, 0.276 mol CO_2, 0.224 mol CO, and 0.224 mol H_2O.

$$CO_2(g) + H_2(g) \rightleftharpoons CO(g) + H_2O(g)$$

(a) Show that for this reaction K_c is independent of the reaction volume, V.
(b) Determine the value of K_c and of K_p.
16. The two common chlorides of phosphorus, PCl_3 and PCl_5, both important in the production of other phosphorus compounds, coexist in equilibrium through the reaction

$$PCl_3(g) + Cl_2(g) \rightleftharpoons PCl_5(g)$$

At 250 °C, an equilibrium mixture in a 2.50-L flask contains 0.105 g PCl_5, 0.220 g PCl_3, and 2.12 g Cl_2. What are the values of **(a)** K_c and **(b)** K_p for this reaction at 250 °C?
17. When 1.00 mol $I_2(g)$ is introduced into an evacuated 1.00-L flask at 1200 °C, 5% of the I_2 molecules dissociate into iodine atoms. For the reaction $I_2(g) \rightleftharpoons 2 I(g)$ at 1200 °C, what are the values of **(a)** K_c and **(b)** K_p?
18. If 0.455 mol SO_2, 0.183 mol O_2, and 0.568 mol SO_3 are introduced simultaneously into a 1.90-L vessel at 1000 K,
(a) Is this mixture at equilibrium?
(b) If not, in which direction must a net reaction occur?

$$2 SO_2(g) + O_2(g) \rightleftharpoons 2 SO_3(g)$$

$$K_c = 2.8 \times 10^2 \text{ at 1000 K}$$

19. What effect does increasing the volume of the system have on the equilibrium condition in each of the following reactions?
(a) $C(s) + H_2O(g) \rightleftharpoons CO(g) + H_2(g)$
(b) $Ca(OH)_2(s) + CO_2(g) \rightleftharpoons CaCO_3(s) + H_2O(g)$
(c) $4 NH_3(g) + 5 O_2(g) \rightleftharpoons 4 NO(g) + 6 H_2O(g)$
20. For which of the following reactions would you expect the extent of the forward reaction to increase with increasing temperatures? Explain.

(a) $NO(g) \rightleftharpoons \frac{1}{2} N_2(g) + \frac{1}{2} O_2(g)$ $\Delta H° = -90.2$ kJ
(b) $SO_3(g) \rightleftharpoons SO_2(g) + \frac{1}{2} O_2(g)$ $\Delta H° = +98.9$ kJ
(c) $N_2H_4(g) \rightleftharpoons N_2(g) + 2 H_2(g)$ $\Delta H° = -95.4$ kJ
(d) $COCl_2(g) \rightleftharpoons CO(g) + Cl_2(g)$ $\Delta H° = +108.3$ kJ
21. The Deacon process for producing chlorine gas from hydrogen chloride is used in situations where by-product HCl is available from other chemical processes.

$$4 HCl(g) + O_2(g) \rightleftharpoons 2 H_2O(g) + 2 Cl_2(g)$$

$$\Delta H° = -114 \text{ kJ}$$

A mixture of HCl, O_2, H_2O, and Cl_2 is brought to equilibrium at 400 °C. What is the effect on the equilibrium amount of $Cl_2(g)$ if
(a) Additional $O_2(g)$ is added to the mixture at constant volume?
(b) HCl(g) is removed from the mixture at constant volume?
(c) The mixture is transferred to a vessel of twice the volume?
(d) A catalyst is added to the reaction mixture?
(e) The temperature is raised to 500 °C?
22. A mixture consisting of 0.150 mol H_2 and 0.150 mol I_2 is brought to equilibrium at 445 °C in a 3.25-L flask. What are the equilibrium amounts of H_2, I_2, and HI?

$$H_2(g) + I_2(g) \rightleftharpoons 2 HI(g) \qquad K_c = 50.2 \text{ at 445 °C}$$

23. A 0.150-mol sample of HI is brought to equilibrium at 445 °C in a 3.25-L flask. How many moles of I_2 will be present?

$$H_2(g) + I_2(g) \rightleftharpoons 2 HI(g) \qquad K_c = 50.2 \text{ at 445 °C}$$

24. A sample of $NH_4HS(s)$ is introduced into a 2.58-L flask containing 0.100 mol NH_3. What will be the total gas pressure when equilibrium is established at 25 °C?

$$NH_4HS(s) \rightleftharpoons NH_3(g) + H_2S(g)$$

$$K_p(\text{atm}) = 0.108 \text{ at 25 °C}$$

Exercises

Writing Equilibrium Constant Expressions

25. Based on these descriptions, write a balanced equation and the corresponding K_c expression for each reversible reaction.
(a) Carbonyl fluoride, $COF_2(g)$, decomposes into gaseous carbon dioxide and gaseous carbon tetrafluoride.
(b) Copper metal displaces silver(I) ion from aqueous solution, producing silver metal and an aqueous solution of copper(II) ion.
(c) Peroxodisulfate ion, $S_2O_8^{2-}$, oxidizes iron(II) ion to iron(III) ion in aqueous solution, and is itself reduced to sulfate ion.

26. Based on these descriptions, write a balanced equation and the corresponding K_p expression for each reversible reaction.
(a) Oxygen gas oxidizes gaseous ammonia to gaseous nitrogen and water vapor.
(b) Hydrogen gas reduces gaseous nitrogen dioxide to gaseous ammonia and water vapor.
(c) Nitrogen gas reacts with the solids sodium carbonate and carbon to produce solid sodium cyanide and carbon monoxide gas.

27. Determine values of K_c from the K_p values given.
(a) $SO_2Cl_2(g) \rightleftharpoons SO_2(g) + Cl_2(g)$
$$K_p = 2.9 \times 10^{-2} \text{ at } 303 \text{ K}$$
(b) $2 NO(g) + O_2(g) \rightleftharpoons 2 NO_2(g)$
$$K_p = 1.48 \times 10^4 \text{ at } 184 \text{ °C}$$
(c) $Sb_2S_3(s) + 3 H_2(g) \rightleftharpoons 2 Sb(s) + 3 H_2S(g)$
$$K_p = 0.429 \text{ at } 713 \text{ K}$$

29. The vapor pressure of water at 25 °C is 23.8 mmHg. Write K_p for the vaporization of water, with pressures in atmospheres. What is the value of K_c for the vaporization process?

31. Given the equilibrium constant values

$N_2(g) + \frac{1}{2} O_2(g) \rightleftharpoons N_2O(g)$	$K_c = 2.7 \times 10^{-18}$
$N_2O_4(g) \rightleftharpoons 2 NO_2(g)$	$K_c = 4.6 \times 10^{-3}$
$\frac{1}{2} N_2(g) + O_2(g) \rightleftharpoons NO_2(g)$	$K_c = 4.1 \times 10^{-9}$

Determine a value of K_c for the reaction
$$2 N_2O(g) + 3 O_2(g) \rightleftharpoons 2 N_2O_4(g)$$

28. Determine the values of K_p from the K_c values given.
(a) $N_2O_4(g) \rightleftharpoons 2 NO_2(g)$ $K_c = 4.61 \times 10^{-3}$ at 25 °C
(b) $2 CH_4(g) \rightleftharpoons C_2H_2(g) + 3 H_2(g)$
$$K_c = 0.154 \text{ at } 2000 \text{ K}$$
(c) $2 H_2S(g) + CH_4(g) \rightleftharpoons 4 H_2(g) + CS_2(g)$
$$K_c = 5.27 \times 10^{-8} \text{ at } 973 \text{ K}$$

30. If $K_c = 5.12 \times 10^{-3}$ for the equilibrium established between liquid benzene and its vapor at 25 °C, what is the vapor pressure of C_6H_6 at 25 °C, expressed in millimeters of mercury?

32. Use the following data at 1200 K to estimate a value of K_p for the reaction $2 H_2(g) + O_2(g) \rightleftharpoons 2 H_2O(g)$, $K_p = ?$

$C(\text{graphite}) + CO_2(g) \rightleftharpoons 2 CO(g)$	$K_c = 0.64$
$CO_2(g) + H_2(g) \rightleftharpoons CO(g) + H_2O(g)$	$K_c = 1.4$
$C(\text{graphite}) + \frac{1}{2} O_2(g) \rightleftharpoons CO(g)$	$K_c = 1 \times 10^8$

Experimental Determination of Equilibrium Constants

33. 1.00×10^{-3} mol PCl_5 is introduced into a 250.0-mL flask, and equilibrium is established at 284 °C: $PCl_5(g) \rightleftharpoons PCl_3(g) + Cl_2(g)$. The quantity of $Cl_2(g)$ present at equilibrium is found to be 9.65×10^{-4} mol. What is the value of K_c for the dissociation reaction at 284 °C?

34. A mixture of 1.00 g H_2 and 1.06 g H_2S in a 0.500-L flask comes to equilibrium at 1670 K: $2 H_2(g) + S_2(g) \rightleftharpoons 2 H_2S(g)$. The equilibrium amount of $S_2(g)$ found is 8.00×10^{-6} mol. Determine the value of K_p at 1670 K.

Equilibrium Relationships

35. Equilibrium is established at 1000 K, where $K_c = 281$ for the reaction $2 SO_2(g) + O_2(g) \rightleftharpoons 2 SO_3(g)$. The equilibrium amount of $O_2(g)$ in a 0.185-L flask is 0.00247 mol. What is the ratio of $[SO_2]$ to $[SO_3]$ in this equilibrium mixture?

37. In the Ostwald process for oxidizing ammonia, a variety of products is possible—N_2, N_2O, NO, and NO_2—depending on the conditions. One possibility is
$$NH_3(g) + \frac{5}{4} O_2(g) \rightleftharpoons NO(g) + \frac{3}{2} H_2O(g)$$
$$K_p(\text{atm}) = 2.11 \times 10^{19} \text{ at } 700 \text{ K}$$
For the decomposition of NO_2 at 700 K,
$$NO_2(g) \rightleftharpoons NO(g) + \frac{1}{2} O_2(g) \qquad K_p(\text{atm}) = 0.524$$
(a) Write a chemical equation for the oxidation of $NH_3(g)$ to $NO_2(g)$.
(b) Determine K_p for the chemical equation you have written.

36. For the dissociation of $I_2(g)$ at about 1200 °C, $I_2(g) \rightleftharpoons 2 I(g)$, $K_c = 1.1 \times 10^{-2}$. What volume flask should we use if we want 0.37 mol I to be present for every 1.00 mol I_2 at equilibrium?

38. At 2000 K, $K_c = 0.154$ for the reaction $2 CH_4(g) \rightleftharpoons C_2H_2(g) + 3 H_2(g)$. If a 1.00-L equilibrium mixture at 2000 K contains 0.10 mol each of $CH_4(g)$ and $H_2(g)$,
(a) What is the mole fraction of $C_2H_2(g)$ present?
(b) Is the conversion of $CH_4(g)$ to $C_2H_2(g)$ favored at high or low pressures?
(c) If the equilibrium mixture at 2000 K is transferred from a 1.00-L flask to a 2.00-L flask, will the number of moles of $C_2H_2(g)$ increase, decrease, or remain unchanged?

Direction and Extent of Chemical Change

39. Can a mixture of 2.2 mol O_2, 3.6 mol SO_2, and 1.8 mol SO_3 be maintained indefinitely in a 7.2-L flask at a temperature at which $K_c = 100$ in this reaction? Explain.
$$2 SO_2(g) + O_2(g) \rightleftharpoons 2 SO_3(g)$$

41. Starting with 0.280 mol $SbCl_3$ and 0.160 mol Cl_2, how many moles of $SbCl_5$, $SbCl_3$, and Cl_2 are present when equilibrium is established at 248 °C in a 2.50-L flask?
$$SbCl_5(g) \rightleftharpoons SbCl_3(g) + Cl_2(g)$$
$$K_c = 2.5 \times 10^{-2} \text{ at } 248 \text{ °C}$$

40. Is a mixture of 0.0205 mol $NO_2(g)$ and 0.750 mol $N_2O_4(g)$ in a 5.25-L flask at 25 °C at equilibrium? If not, in which direction will the reaction proceed, toward products or reactants?
$$N_2O_4(g) \rightleftharpoons 2 NO_2(g) \qquad K_c = 4.61 \times 10^{-3} \text{ at } 25 \text{ °C}$$

42. Starting with 0.3500 mol $CO(g)$ and 0.05500 mol $COCl_2(g)$ in a 3.050-L flask at 668 K, what will be the number of moles of $Cl_2(g)$ when equilibrium is established?
$$CO(g) + Cl_2(g) \rightleftharpoons COCl_2(g) \qquad K_c = 1.2 \times 10^3 \text{ at } 668 \text{ K}$$

43. 1.00 g *each* of CO, H_2O, and H_2 are sealed in a 1.41-L vessel and brought to equilibrium at 600 K. How many grams of CO_2 will be present in the equilibrium mixture?

$$CO(g) + H_2O(g) \rightleftharpoons CO_2(g) + H_2(g) \qquad K_c = 23.2$$

45. For the following reaction, $K_c = 2.00$ at 1000 °C.

$$2\ COF_2(g) \rightleftharpoons CO_2(g) + CF_4(g)$$

If a 5.00-L mixture contains 0.145 mol COF_2, 0.262 mol CO_2, and 0.074 mol CF_4 at a temperature of 1000 °C,
(a) Will the mixture be at equilibrium?
(b) If the gases are not at equilibrium, in what direction will a net reaction occur?
(c) How many moles of each gas is present at equilibrium?

47. A gaseous mixture containing 0.125 mol *each* of $H_2(g)$ and $I_2(g)$ is introduced into a 6.14-L flask at 445 °C and equilibrium is established. What is the mole percent HI in the equilibrium mixture?

$$H_2(g) + I_2(g) \rightleftharpoons 2\ HI(g) \qquad K_c = 50.2$$

44. Equilibrium is established in a 2.50-L flask at 250 °C for the reaction

$$PCl_5(g) \rightleftharpoons PCl_3(g) + Cl_2(g) \qquad K_c = 3.8 \times 10^{-2}$$

How many moles of PCl_5, PCl_3, and Cl_2 are present at equilibrium, if
(a) 0.550 mol each of PCl_5 and PCl_3 are initially introduced into the flask?
(b) 0.610 mol PCl_5 alone is introduced into the flask?

46. In the reaction

$$C_2H_2OH + CH_3COOH \rightleftharpoons CH_3COOC_2H_5 + H_2O$$
$$K_c = 4.0$$

A reaction is allowed to occur in a mixture of 17.2 g C_2H_5OH, 23.8 g CH_3COOH, 48.6 g $CH_3COOC_2H_5$, and 71.2 g H_2O.
(a) In what direction will a net reaction occur?
(b) How many grams of each substance will be present at equilibrium?

48. The N_2O_4–NO_2 equilibrium mixture in the flask on the left in the figure is allowed to expand into the evacuated flask on the right. What is the composition of the gaseous mixture when equilibrium is reestablished in the system consisting of the two flasks? (*Hint:* In what direction does a net reaction occur?)

$$N_2O_4(g) \rightleftharpoons 2\ NO_2(g) \qquad K_c = 4.61 \times 10^{-3} \text{ at } 25 \text{ °C}$$

0.971 mol N_2O_4
0.0580 mol NO_2

0.750 L
25 °C

2.25 L
25 °C

49. Formamide, used in the manufacture of pharmaceuticals, dyes, and agricultural chemicals, decomposes at high temperatures.

$$HCONH_2(g) \rightleftharpoons NH_3(g) + CO(g) \qquad K_c = 4.84 \text{ at } 400 \text{ K}$$

If 0.186 mol $HCONH_2(g)$ dissociates in a 2.16-L flask at 400 K, what will be the *total* pressure at equilibrium?

51. An aqueous solution that is 1.00 M in $AgNO_3$ and 1.00 M in $Fe(NO_3)_2$ is allowed to come to equilibrium. What are the equilibrium values of $[Ag^+]$, $[Fe^{2+}]$, and $[Fe^{3+}]$?

$$Ag^+(aq) + Fe^{2+}(aq) \rightleftharpoons Fe^{3+}(aq) + Ag(s) \qquad K_c = 2.98$$

50. A mixture of 1.00 mol $NaHCO_3(s)$ and 1.00 mol $Na_2CO_3(s)$ is introduced into a 2.50-L flask in which the partial pressure of $CO_2(g)$ is 2.10 atm and that of $H_2O(g)$ is 715 mmHg. When equilibrium is established at 100 °C, will the partial pressures of $CO_2(g)$ and $H_2O(g)$ be greater or less than their initial partial pressures? Explain.

$$2\ NaHCO_3(s) \rightleftharpoons Na_2CO_3(s) + CO_2(g) + H_2O(g)$$
$$K_p(atm) = 0.23 \text{ at } 100 \text{ °C}$$

52. Lead metal is added to 0.100 M $Cr^{3+}(aq)$. What is $[Pb^{2+}]$ when equilibrium is established in the reaction? (*Hint:* To simplify the calculation, let $[Pb^{2+}] = x$ and assume that x is very much smaller than 0.100.)

$$Pb(s) + 2\ Cr^{3+}(aq) \rightleftharpoons Pb^{2+}(aq) + 2\ Cr^{2+}(aq)$$
$$K_c = 3.2 \times 10^{-10}$$

Partial Pressure Equilibrium Constant, K_p

53. The following reaction is used in some self-contained breathing devices as a source of $O_2(g)$.

$$4 KO_2(s) + 2 CO_2(g) \rightleftharpoons 2 K_2CO_3(s) + 3 O_2(g)$$
$$K_p = 28.5 \text{ at } 25 \text{ °C}$$

Suppose that a sample of $CO_2(g)$ is added to an evacuated flask containing $KO_2(s)$ and equilibrium is established. If the equilibrium partial pressure of $CO_2(g)$ is found to be 0.0721 atm, what are the equilibrium partial pressure of $O_2(g)$ and the total gas pressure?

55. 1.00 mol *each* of CO and Cl_2 are introduced into an evacuated 1.75-L flask and the following equilibrium is established at 668 K.

$$CO(g) + Cl_2(g) \rightleftharpoons COCl_2(g) \qquad K_p = 22.5$$

For this equilibrium, calculate **(a)** the partial pressure of $CO_2Cl_2(g)$; **(b)** the total gas pressure.

54. Concerning the reaction in Exercise 53, if $KO_2(s)$ and $K_2CO_3(s)$ are maintained in contact with air at 1.00 atm and 25 °C, in which direction will a net reaction occur to establish equilibrium, to the left or to the right? Explain. [*Hint:* Recall Equation (6.17) and the composition of air (Table 8.1.)]

56. Refer to Example 16-4. $H_2S(g)$ at 747.6 mmHg pressure and a 1.85-g sample of $I_2(s)$ are introduced into a 725-mL flask at 60 °C. What will be the *total pressure* in the flask at equilibrium?

$$H_2S(g) + I_2(s) \rightleftharpoons 2 HI(g) + S(s)$$
$$K_p(\text{atm}) = 1.34 \times 10^{-5} \text{ at } 60 \text{ °C}$$

Le Châtelier's Principle

57. Continuous removal of one of the products of a chemical reaction has the effect of causing the reaction to go to completion. Explain this fact in terms of Le Châtelier's principle.

59. Explain how each of the following factors affects the amount of H_2 present in an equilibrium mixture in the reaction

$$3 Fe(s) + 4 H_2O(g) \rightleftharpoons Fe_3O_4(s) + 4 H_2(g)$$
$$\Delta H° = -150 \text{ kJ}$$

(a) Raise the temperature of the mixture; **(b)** introduce more $H_2O(g)$; **(c)** double the volume of the container holding the mixture; **(d)** add an appropriate catalyst.

61. The reaction $N_2(g) + O_2(g) \rightleftharpoons 2 NO(g)$, $\Delta H° = +181$ kJ, occurs in high-temperature combustion processes carried out in air. Oxides of nitrogen produced from the nitrogen and oxygen in air are intimately involved in the production of photochemical smog. What effect does increasing the temperature have on **(a)** the equilibrium production of $NO(g)$; **(b)** the rate of this reaction?

63. In both the ammonia synthesis reaction $[N_2(g) + 3 H_2(g) \rightleftharpoons 2 NH_3(g)]$ and the conversion of $SO_2(g)$ to $SO_3(g)$ $[2 SO_2(g) + O_2(g) \rightleftharpoons 2 SO_3(g)]$, the mole fraction at equilibrium of the desired product (NH_3, SO_3) is greater at lower temperatures. Yet, in the commercial production of these substances relatively high temperatures are used. Explain why this is so.

58. We can represent the freezing of $H_2O(l)$ at 0 °C as H_2O (l, $d = 1.00$ g/cm³) $\rightleftharpoons H_2O(s, d = 0.92$ g/cm³). Explain why increasing the pressure on ice causes it to melt. Is this the behavior you expect for solids in general? Explain.

60. In the gas phase, iodine reacts with cyclopentene (C_5H_8) by a free radical mechanism to produce cyclopentadiene (C_5H_6) and hydrogen iodide. Explain how each of the following factors affects the amount of $HI(g)$ present in the equilibrium mixture in the reaction

$$I_2(g) + C_5H_8(g) \rightleftharpoons C_5H_6(g) + 2 HI(g) \qquad \Delta H° = 92.5 \text{ kJ}$$

(a) Raise the temperature of the mixture; **(b)** introduce more $C_5H_6(g)$; **(c)** double the volume of the container holding the mixture; **(d)** add an appropriate catalyst; **(e)** add an inert gas such as He to a constant volume reaction mixture.

62. Use data from Appendix D to determine if the forward reaction is favored by high or low temperatures.
(a) $PCl_3(g) + Cl_2(g) \rightleftharpoons PCl_5(g)$
(b) $SO_2(g) + 2 H_2S(g) \rightleftharpoons 2 H_2O(g) + 3 S(s)$
(c) $2 N_2(g) + 3 O_2(g) + 4 HCl(g) \rightleftharpoons$
$$4 NOCl(g) + 2 H_2O(g)$$

64. For the reaction

$$A(s) \rightleftharpoons B(s) + 2 C(g) + \tfrac{1}{2} D(g) \qquad \Delta H° = 0$$

(a) Will K_p increase, decrease, or remain constant with temperature? Explain.
(b) If a *constant-volume* mixture at equilibrium at 298 K is heated to 400 K and equilibrium re-established, will the number of moles of $D(g)$ increase, decrease, or remain constant? Explain.

Advanced Exercises

(In some of these exercises you may find the approximation method introduced in Feature Problem C on page 579 to be useful.)

65. Explain why the percent of molecules that dissociate into atoms in reactions of the type $I_2(g) \rightleftharpoons 2\ I(g)$ *always* increases with an increase in temperature.

66. A 1.100-L flask contains $CO_2(g)$ at 1.00 atm pressure at 25 °C in contact with 100.0 mL of a saturated aqueous solution in which $[CO_2(aq)] = 3.29 \times 10^{-2}$ M.
(a) What is the value of K_c at 25 °C for the equilibrium:
$$CO_2(g) \rightleftharpoons CO_2(aq) \qquad K_c = ?$$
(b) If 0.01000 mol of radioactive $^{14}CO_2$ is added to the flask, how many moles of the $^{14}CO_2$ will be found in the gas phase and in the aqueous solution when equilibrium is re-established? (*Hint:* The radioactive $^{14}CO_2$ distributes itself between the two phases in exactly the same manner as the non-radioactive $^{12}CO_2$.)

67. Refer to Example 16-13. Suppose that 0.100 L of the *equilibrium mixture* is diluted to 0.250 L with water. What will be the new concentrations when equilibrium is reestablished?

68. In the equilibrium described in Example 16-12, the percent dissociation of N_2O_4 can be expressed as
$$\frac{3.00 \times 10^{-3}\ \text{mol}\ N_2O_4\ \text{dissociated}}{0.0240\ \text{mol}\ N_2O_4\ \text{initially}} \times 100\% = 12.5\%$$
What must be the *total* pressure of the gaseous mixture if $N_2O_4(g)$ is to be 10.0% dissociated at 298 K?
$$N_2O_4 \rightleftharpoons 2\ NO_2(g) \qquad K_p\ (\text{atm}) = 0.113\ \text{at}\ 298\ K$$

69. Starting with $SO_3(g)$ at 1.00 atm, what will be the *total* pressure when equilibrium is reached in the following reaction at 700 K?
$$2\ SO_3(g) \rightleftharpoons 2\ SO_2(g) + O_2(g) \qquad K_p = 1.6 \times 10^{-5}$$

70. Refer to Exercise 54. What will be the partial pressures of $CO_2(g)$ and $O_2(g)$ when equilibrium is established between air at 1.00 atm and 25 °C and the solids $KO_2(s)$ and $K_2CO_3(s)$?

71. A sample of air with a mole ratio of N_2 to O_2 of 79:21 is heated to 2500 K. When equilibrium is established in a closed container, with air initially at 1.00 atm, the mole percent of NO is found to be 1.8%. Calculate K_p for the reaction
$$N_2(g) + O_2(g) \rightleftharpoons 2\ NO(g) \qquad K_p\ \text{at}\ 2500\ K = ?$$
(*Hint:* The result is independent of both volume and total pressure.)

72. Derive, by calculation, the equilibrium amounts of SO_2, O_2, and SO_3 listed in **(a)** Figure 16-4(c); **(b)** Figure 16-5(b).

73. One of the key reactions in the gasification of coal is the methanation reaction, in which methane is produced from synthesis gas—a mixture of CO and H_2.
$$CO(g) + 3\ H_2(g) \rightleftharpoons CH_4(g) + H_2O(g)$$
$$\Delta H = -230\ \text{kJ};\ K_c = 190\ \text{at}\ 1000\ K$$
(a) Is the equilibrium conversion of synthesis gas to methane favored at higher or lower temperatures? Higher or lower pressures?
(b) Assume you have 4.00 mol of synthesis gas with a 3:1 mol ratio of $H_2(g)$ to $CO(g)$ in a 15.0-L flask. What will be the mole fraction of $CH_4(g)$ at equilibrium at 1000 K?

74. A sample of pure $PCl_5(g)$ is introduced into an evacuated flask and allowed to dissociate.
$$PCl_5(g) \rightleftharpoons PCl_3(g) + Cl_2(g)$$
If the fraction of PCl_5 molecules that dissociate is denoted by α, and if the total gas pressure is P, show that
$$K_p = \frac{\alpha^2 P}{1 - \alpha^2}$$

75. Nitrogen dioxide obtained as a cylinder gas is always a mixture of $NO_2(g)$ and $N_2O_4(g)$. A 5.00-g sample obtained from such a cylinder is sealed in a 0.500-L flask at 298 K. What is the mole fraction of NO_2 in this mixture?
$$N_2O_4(g) \rightleftharpoons 2\ NO_2(g) \qquad K_c = 4.61 \times 10^{-3}$$

76. What is the apparent molar mass of the gaseous mixture that results when $COCl_2(g)$ is allowed to dissociate at 395 °C and a total pressure of 3.00 atm?
$$COCl_2(g) \rightleftharpoons CO(g) + Cl_2(g)$$
$$K_p = 4.44 \times 10^{-2}\ \text{at}\ 395\ °C$$
Think of the apparent molar mass as the molar mass of a hypothetical single gas that is equivalent to the gaseous mixture.

77. Show that in terms of mole fractions of gases and *total* gas pressure the equilibrium constant expression for
$$N_2(g) + 3\ H_2(g) \rightleftharpoons 2\ NH_3(g)$$
is
$$K_p = \frac{(\chi_{NH_3})^2}{(\chi_{N_2})(\chi_{H_2})^3} \times \frac{1}{(P_{tot})^2}$$

78. For the synthesis of ammonia at 500 K, $N_2(g) + 3\ H_2(g) \rightleftharpoons 2\ NH_3(g)$, $K_p = 9.06 \times 10^{-2}$. Assume that N_2 and H_2 are mixed in the mole ratio 1:3 and that the total pressure is maintained at 1.00 atm. What is the mole percent NH_3 at equilibrium? (*Hint:* Use the equation from Exercise 77.)

79. A mixture of $H_2S(g)$ and $CH_4(g)$ in the mole ratio 2:1 was brought to equilibrium at 700 °C and a total pressure of 1 atm. The *equilibrium* mixture was analyzed for the amount of H_2S; 9.54×10^{-3} mol H_2S was found. The CS_2 present at equilibrium was converted successively to H_2SO_4 and then to $BaSO_4$: 1.42×10^{-3} mol $BaSO_4$ was obtained. Use these data to determine K_p at 700 °C for the reaction
$$2\ H_2S(g) + CH_4(g) \rightleftharpoons CS_2(g) + 4\ H_2(g)$$
$$K_p\ \text{at}\ 700\ °C = ?$$

80. A solution is prepared having these initial concentrations: $[Fe^{3+}] = [Hg_2^{2+}] = 0.5000$ M; $[Fe^{2+}] = [Hg^{2+}] = 0.03000$ M. The following reaction occurs among the ions at 25 °C.
$$2\ Fe^{3+}(aq) + Hg_2^{2+}(aq) \rightleftharpoons 2\ Fe^{2+}(aq) + 2\ Hg^{2+}(aq)$$
$$K_c = 9.14 \times 10^{-6}$$
What will be the ion concentrations when equilibrium is established?

81. Refer to the Summarizing Example. A gaseous mixture is prepared containing 0.100 mol each of $CH_4(g)$, $H_2O(g)$, $CO_2(g)$, and $H_2(g)$ in a 5.00-L flask. Then the mixture is allowed to come to equilibrium at 1000 K in reaction (16.19). What will be the equilibrium amount, in moles, of each gas?

82. Concerning the reaction in Exercise 38 and the situation descibed in part (c) of that exercise, will the mole fraction of $C_2H_2(g)$ increase, decrease, or remain unchanged when equilibrium is re-established? Explain.

Feature Problems

A. A classic experiment in equilibrium studies dating from 1862 involved the reaction in solution of ethanol (C_2H_5OH) and acetic acid (CH_3COOH) to produce ethyl acetate and water.

$$C_2H_5OH + CH_3COOH \rightleftharpoons CH_3COOC_2H_5 + H_2O$$

The reaction can be followed by analyzing the equilibrium mixture for its acetic acid content.

$$2\ CH_3COOH(aq) + Ba(OH)_2(aq) \longrightarrow$$
$$Ba(CH_3COO)_2(aq) + 2\ H_2O(l)$$

In one experiment a mixture of 1.000 mol acetic acid and 0.5000 mol ethanol is brought to equilibrium. A sample containing exactly one-hundredth of the equilibrium mixture requires 28.85 mL 0.1000 M $Ba(OH)_2$ for its titration. Calculate the equilibrium constant, K_c, for the ethanol–acetic acid reaction based on this experiment.

B. Refer to the discussion of the Haber–Bosch process for the synthesis of ammonia on page 246 and in Figure 8-4. Explain why the stated conditions for the process were chosen, noting particularly which conditions conform to ideas presented in this chapter and which do not.

$$N_2(g) + 3\ H_2(g) \rightleftharpoons 2\ NH_3(aq) \qquad \Delta H^\circ = -92.2\ kJ$$

boiling points are: N_2, 77.4 K; H_2, 20.4 K; NH_3, 239.8 K

C. Recall the formation of methanol from synthesis gas, the reversible reaction with which we began our discussion of the equilibrium constant expression on page 547.

$$CO(g) + 2\ H_2(g) \rightleftharpoons CH_3OH(g) \qquad K_c = 14.5\ \text{at}\ 483\ K$$

A particular synthesis gas consisting of 35.0 mole percent CO(g) and 65.0 mole percent $H_2(g)$ at a total pressure of 100.0 atm at 483 K is allowed to come to equilibrium. Your task is to determine the partial pressure of $CH_3OH(g)$ in the equilibrium mixture. Consider the following stepwise approach.
1. Determine K_p at 483 K.
2. Determine the initial partial pressures of CO(g) and $H_2(g)$.
3. In the ICE tabular format, let P represent the partial pressure of $CH_3OH(g)$ at equilibrium, and express the other partial pressures in terms of their initial values and P.
4. Substitute equilibrium partial pressures into the K_p expression. At this point you should have the following equation:

$$\frac{P}{(35 - P)(65 - 2P)^2} = 9.23 \times 10^{-3}$$

This is a third degree (cubic) equation and is more difficult to solve than a quadratic equation. Do not attempt to rearrange or simplify the equation.
5. Can you see that P cannot possibly be as large as 32.5? [This would make the term $(65 - 2P)$ equal to zero.] "Guess" at a possible value of P, say 30.0 atm. Substitute this value into the left side of the expression above and you will have formulated a Q_p. You will find that $Q_p > K_p$. As a second "guess," try a smaller value of P, say 20.0 atm. Now you will find that $Q_p < K_p$. The answer you are seeking is 20.0 atm $< P <$ 30.0 atm. Make a series of approximations until you find the value of P that makes $Q_p \approx K_p$. (The method may sound tedious, but you will find that with an electronic calculator you can make several successive approximations in a matter of minutes.)

Acids and Bases

17

*T*he general public is regularly exposed to ideas about acids and bases. The environmental problem of "acid rain" is a popular topic in newspapers and magazines, and television commercials mention "pH" in relation to a variety of products, such as deodorants, shampoos, and antacids.

Chemists have been classifying substances as acids and bases for a long time. Lavoisier thought that the common element in all acids was oxygen, a fact conveyed by its name. Oxygen means "acid former" in Greek. In 1810, Humphry Davy showed that hydrogen is the element that acids have in common. In 1884, Arrhenius developed the theory of acids and bases that we introduced in Chapter 5. There we emphasized the stoichiometry of acid–base reactions.

Some of the topics we will study in this chapter are modern acid–base theories, factors affecting the strengths of acids and bases, the pH scale, and the calculation of ion concentrations in solutions of weak acids and bases. At the end of the chapter, we will bring some of these ideas together in a discussion of acid rain.

Fruit juices contain a variety of acids. Orange juice, for example, is an excellent source of citric acid, $H_3C_6H_5O_7$. Citric acid is a polyprotic acid, a type of acid discussed in Section 17-6.

17-1   The Arrhenius Theory: A Brief Review

Some aspects of the behavior of acids and bases can be explained adequately with the theory developed by Svante Arrhenius as part of his studies of electrolytic dissociation (Section 14-9). Arrhenius proposed that in an aqueous solution a strong electrolyte exists only in the form of ions, whereas a weak electrolyte exists partly as ions and partly as molecules. When the acid HCl dissolves in water, the HCl molecules ionize completely, yielding hydrogen ions, H^+, as one of the products.

$$HCl(g) \xrightarrow{H_2O} H^+(aq) + Cl^-(aq)$$

When the base NaOH dissolves in water, the Na^+ and OH^- ions present in the solid become dissociated from one another through the action of H_2O molecules (recall Figure 14-6).

$$NaOH(s) \xrightarrow{H_2O} Na^+(aq) + OH^-(aq)$$

We can represent the neutralization reaction of HCl and NaOH with the ionic equation

$$\underset{\text{an acid}}{H^+(aq) + Cl^-(aq)} + \underset{\text{a base}}{Na^+(aq) + OH^-(aq)} \longrightarrow \underset{\text{a salt}}{Na^+(aq) + Cl^-(aq)} + \underset{\text{water}}{H_2O(l)}$$

or, perhaps better still, with the net ionic equation

$$\underset{\text{an acid}}{H^+(aq)} + \underset{\text{a base}}{OH^-(aq)} \longrightarrow \underset{\text{water}}{H_2O(l)} \qquad (17.1)$$

Equation (17.1) illustrates an essential idea of the Arrhenius theory: *A neutralization reaction involves the combination of hydrogen ions and hydroxide ions to form water.*

Despite its early successes and continued usefulness, the Arrhenius theory does have limitations. One of the most glaring is in its treatment of the weak base ammonia, NH_3. The Arrhenius theory suggests that all bases *contain* OH^-. Where is the OH^- in NH_3? To get around this difficulty, chemists began to think of aqueous solutions of NH_3 as containing the compound ammonium hydroxide, NH_4OH, which as a weak base is partially ionized into NH_4^+ and OH^- ions.

$$NH_3(g) + H_2O(l) \longrightarrow NH_4OH(aq)$$

$$NH_4OH(aq) \rightleftharpoons NH_4^+(aq) + OH^-(aq)$$

The problem with this formulation is that there is no compelling evidence for the existence of NH_4OH in aqueous solutions. We should always question a hypothesis or theory that postulates the existence of hypothetical substances. As we shall see in the next section, the essential failure of the Arrhenius theory is in not recognizing the key role of the *solvent* in the ionization of a solute.

A holdover of the Arrhenius theory. Although there is no compelling evidence for the existence of NH_4OH molecules in $NH_3(aq)$, solutions are commonly labeled this way.

▶Recall that hydrogen atoms consist of one proton and one electron. Thus, the hydrogen ion, H^+, is simply a proton. So, a "proton donor" is a donor of H^+ ions.

17-2   Brønsted–Lowry Theory of Acids and Bases

In 1923, J. N. Brønsted in Denmark and T. M. Lowry in Great Britain independently proposed a new acid–base theory. In their theory an acid is a **proton donor** and a base is a **proton acceptor.** To describe the behavior of ammonia as a base, which we found difficult to do with the Arrhenius theory, we can write

$$NH_3 + H_2O \longrightarrow NH_4^+ + OH^- \qquad (17.2)$$
$$\quad \text{base} \quad\;\; \text{acid}$$

In reaction (17.2) H_2O acts as an *acid*. It denotes a proton, H^+, which is gained by NH_3, a *base*. As a result of this transfer the polyatomic ions NH_4^+ and OH^- are formed—the same ions produced by the ionization of the hypothetical NH_4OH of the Arrhenius theory. Because $NH_3(aq)$ is a *weak* base, we should also consider the *reverse* of (17.2). In the reverse reaction NH_4^+ is an *acid* and OH^- is a *base*.

$$NH_4^+ + OH^- \longrightarrow NH_3 + H_2O \qquad (17.3)$$
$$\;\text{acid} \qquad \text{base}$$

The conventional way to represent a reversible reaction is to use the double arrow notation. Also we can place "acid" and "base" labels under all four species in the reaction.

$$NH_3 \;+\; H_2O \;\rightleftharpoons\; NH_4^+ +\; OH^- \qquad (17.4)$$
$$\text{base(1)} \quad \text{acid(2)} \qquad \text{acid(1)} \quad \text{base(2)}$$

We label the NH_3/NH_4^+ combination as "(1)" and H_2O/OH^- as "(2)." Each combination is called a conjugate pair. An NH_3 molecule acts as a base by accepting a proton, and an NH_4^+ ion is the **conjugate acid** of NH_3. Similarly, in reaction (17.4) H_2O is an acid and OH^- is its **conjugate base.** Figure 17-1 illustrates the proton transfer involved in the forward and reverse reactions of (17.4).

On the basis of what we learned in Chapter 16, we might write as the equilibrium constant expression for reaction (17.4)

$$K_c = \frac{[NH_4^+][OH^-]}{[NH_3][H_2O]}$$

In aqueous ammonia, however, H_2O molecules are present in such overwhelming numbers compared with NH_3 molecules and NH_4^+ and OH^- ions that water, the solvent, is essentially a pure liquid with an activity of one (recall page 554). For this reason, the equilibrium constant expression we want does not have the term $[H_2O]$.

$$K_b = \frac{[NH_4^+][OH^-]}{[NH_3]} = 1.8 \times 10^{-5}$$

The equilibrium constant K_b is called the **base ionization constant.**

For the ionization of acetic acid we can write

$$HC_2H_3O_2 +\; H_2O \;\rightleftharpoons\; C_2H_3O_2^- +\; H_3O^+$$
$$\text{acid(1)} \qquad \text{base(2)} \qquad \text{base(1)} \qquad \text{acid(2)}$$

▶ In the designations (1) and (2), which conjugate pair we call (1) and which we call (2) does not matter, nor does the order in which the acid and base are written on each side of the equation.

FIGURE 17-1
Brønsted–Lowry acid–base reaction.
The arrows in this figure represent the proton transfer in reaction (17.4). The red arrows represent the forward reaction and the blue arrows, the reverse reaction. Because NH_4^+ is a stronger acid than H_2O and OH^- is a stronger base than NH_3, the reverse reaction proceeds to a greater extent than the forward reaction. Hence, NH_3 is only slightly ionized.

$$NH_3 \;+\; H_2O \;\rightleftharpoons\; NH_4^+ \;+\; OH^-$$
$$\text{base(1)} \qquad \text{acid(2)} \qquad \text{acid(1)} \qquad \text{base(2)}$$

▶ The Lewis structure of hydronium ion is

$$\left[\begin{array}{c} \overset{\displaystyle \cdot\cdot}{H-O-H} \\ | \\ H \end{array} \right]^{+}$$

The hydronium ion is described in more detail in Section 17-3.

Here, acetate ion, $C_2H_3O_2^-$, is the conjugate base of the acid $HC_2H_3O_2$. And this time H_2O acts as a base. Its conjugate acid is called *hydronium* ion, H_3O^+. Following the approach we used for $NH_3(aq)$, for the ionization of acetic acid we can write

$$K_a = \frac{[C_2H_3O_2^-][H_3O^+]}{[HC_2H_3O_2]} = 1.8 \times 10^{-5}$$

The equilibrium constant K_a is called the **acid ionization constant.** (The fact that K_a of acetic acid and K_b of ammonia have the same value is just a coincidence.)

We can represent the ionization of HCl in the same way that we did for acetic acid. In this case, however, because K_a is so large (about 10^6) we can treat the ionization of HCl as a reaction that goes to completion. We denote this by writing the ionization equation with a single arrow.

$$HCl + H_2O \longrightarrow Cl^- + H_3O^+ \tag{17.5}$$

In Example 17-1 we identify acids and bases in some typical acid–base reactions. In working through this example, notice the following additional features: (1) Any species that is an acid by the Arrhenius theory remains an acid in the Brønsted–Lowry theory; the same is true of bases. (2) Certain species, even though they do not contain the OH group, produce OH^- in aqueous solution—for example, OCl^-. As such they are Brønsted–Lowry bases. (3) The Brønsted–Lowry theory accounts for substances that can act either as an acid or a base; they are said to be *amphiprotic*. The Arrhenius theory does not account for amphiprotic behavior.

EXAMPLE 17-1

Identifying Brønsted–Lowry Acids and Bases and Their Conjugates. For each of the following identify the acids and bases in both the forward and reverse reactions, in the manner shown in Equation (17.4).

a. $HClO_2 + H_2O \rightleftharpoons ClO_2^- + H_3O^+$
b. $OCl^- + H_2O \rightleftharpoons HOCl + OH^-$
c. $NH_3 + H_2PO_4^- \rightleftharpoons NH_4^+ + HPO_4^{2-}$
d. $HCl + H_2PO_4^- \rightleftharpoons Cl^- + H_3PO_4$

Solution

Consider $HClO_2$ in reaction (a). It loses a proton, H^+, to become ClO_2^-. Therefore, $HClO_2$ is an acid, and ClO_2^- is its conjugate base. Now consider H_2O. It accepts the proton and becomes H_3O^+. H_2O is a base, and H_3O^+ is its conjugate acid. In reaction (b), OCl^- is a base and gains a proton from water. OH^- produced in this reaction is the conjugate base of H_2O. Taken together, reactions (a) and (b) show that H_2O is amphiprotic. Reactions (c) and (d) illustrate the same for $H_2PO_4^-$.

a. $\underset{\text{acid(1)}}{HClO_2} + \underset{\text{base (2)}}{H_2O} \rightleftharpoons \underset{\text{base(1)}}{ClO_2^-} + \underset{\text{acid(2)}}{H_3O^+}$

b. $\underset{\text{base(1)}}{OCl^-} + \underset{\text{acid(2)}}{H_2O} \rightleftharpoons \underset{\text{acid(1)}}{HOCl} + \underset{\text{base(2)}}{OH^-}$

c. $\underset{\text{base(1)}}{NH_3} + \underset{\text{acid(2)}}{H_2PO_4^-} \rightleftharpoons \underset{\text{acid(1)}}{NH_4^+} + \underset{\text{base(2)}}{HPO_4^{2-}}$

d. $\underset{\text{acid(1)}}{HCl} + \underset{\text{base(2)}}{H_2PO_4^-} \rightleftharpoons \underset{\text{base(1)}}{Cl^-} + \underset{\text{acid(2)}}{H_3PO_4}$

Practice Example A: For each of the following identify the acids and bases in both the forward and reverse reactions.

a. $HF + H_2O \rightleftharpoons F^- + H_3O^+$

b. $HSO_4^- + NH_3 \rightleftharpoons SO_4^{2-} + NH_4^+$

c. $C_2H_3O_2^- + HCl \rightleftharpoons HC_2H_3O_2 + Cl^-$

Practice Example B: Of the following, one is acidic, one is basic, and one is amphiprotic in their reactions with water: HNO_2, PO_4^{3-}, HCO_3^-. Write equations to represent these facts. (*Hint: Four* equations are needed.)

The ionization of HCl in aqueous solution (reaction 17.5) goes to completion because HCl, a strong acid, has a strong tendency to donate a proton to H_2O. At the same time, Cl^- ion, the conjugate base of HCl, has a very limited tendency to accept a proton from H_3O^+; it is a weak base. This observation suggests the generalization that

In an acid–base reaction, the favored direction of the reaction is from the stronger to the weaker member of a conjugate acid–base pair.

With this generalization we can predict that the neutralization of HCl by OH^- should go to completion.

$$HCl + OH^- \longrightarrow Cl^- + H_2O$$

$$\underset{\substack{\text{acid(1)} \\ \text{strong}}}{HCl} + \underset{\substack{\text{base(2)} \\ \text{strong}}}{OH^-} \longrightarrow \underset{\substack{\text{base(1)} \\ \text{weak}}}{Cl^-} + \underset{\substack{\text{acid(2)} \\ \text{weak}}}{H_2O}$$

And we would predict that the following reaction should occur almost exclusively in the reverse direction.

$$\underset{\substack{\text{acid(1)} \\ \text{weak}}}{H_2O} + \underset{\substack{\text{base(2)} \\ \text{weak}}}{I^-} \longleftarrow \underset{\substack{\text{base(1)} \\ \text{strong}}}{OH^-} + \underset{\substack{\text{acid(2)} \\ \text{strong}}}{HI}$$

To be able to apply the generalization more broadly, though, we need a tabulation of acid and base strengths, such as that in Table 17.1. The strongest acids are at the top of the column on the left, and the strongest bases are at the bottom of the column on the right.

Both HCl and $HClO_4$ are strong acids. This is because H_2O is a sufficiently strong base to accept protons from either acid in a reaction that goes to completion. Water is said to have a *leveling effect* on the two acids. How can we say, then, as we do in Table 17.1, that $HClO_4$ is a stronger acid than HCl?

To determine whether $HClO_4$ or HCl is the stronger acid, we need to use a solvent that is a weaker base than water, a solvent that will accept protons from the stronger of the two acids more readily than from the weaker one. In the solvent $(C_2H_5)_2O$, diethyl ether, $HClO_4$ is completely ionized but HCl is only partially ionized. Thus, $HClO_4$ is a stronger acid than is HCl.

$$HClO_4 + C_2H_5 - \overset{..}{\underset{..}{O}} - C_2H_5 \longrightarrow ClO_4^- + [C_2H_5 - \overset{\overset{\displaystyle H}{|}}{\underset{..}{O}} - C_2H_5]^+$$

$$HCl + C_2H_5 - \overset{..}{\underset{..}{O}} - C_2H_5 \rightleftharpoons Cl^- + [C_2H_5 - \overset{\overset{\displaystyle H}{|}}{\underset{..}{O}} - C_2H_5]^+$$

TABLE 17.1   Relative Strengths of Some Common Brønsted–Lowry Acids and Bases

	Acid		Conjugate Base	
	perchloric acid	$HClO_4$	perchlorate ion	ClO_4^-
	hydroiodic acid	HI	iodide ion	I^-
	hydrobromic acid	HBr	bromide ion	Br^-
	hydrochloric acid	HCl	chloride ion	Cl^-
	sulfuric acid	H_2SO_4	hydrogen sulfate ion	HSO_4^-
	nitric acid	HNO_3	nitrate ion	NO_3^-
	hydronium ion[a]	H_3O^+	water[a]	H_2O
	hydrogen sulfate ion	HSO_4^-	sulfate ion	SO_4^{2-}
	nitrous acid	HNO_2	nitrite ion	NO_2^-
	acetic acid	$HC_2H_3O_2$	acetate ion	$C_2H_3O_2^-$
	carbonic acid	H_2CO_3	hydrogen carbonate ion	HCO_3^-
	ammonium ion	NH_4^+	ammonia	NH_3
	hydrogen carbonate ion	HCO_3^-	carbonate ion	CO_3^{2-}
	water	H_2O	hydroxide ion	OH^-
	methanol	CH_3OH	methoxide ion	CH_3O^-
	ammonia	NH_3	amide ion	NH_2^-

increasing acid strength (left, pointing up)   *increasing base strength* (right, pointing down)

[a]The hydronium ion/water combination refers to the ease with which a proton is passed from one water molecule to another; that is, $H_3O^+ + H_2O \rightleftharpoons H_2O + H_3O^+$.

17-3   The Self-ionization of Water and the pH Scale

Even when it is pure, water contains a very low concentration of ions that can be detected in precise electrical conductivity measurements. We can think of some water molecules as donating protons and others as accepting the protons. In the **self-ionization** of water, for each H_2O molecule that acts as an acid another acts as a base, and hydronium (H_3O^+) and hydroxide (OH^-) ions are formed. The reaction is reversible, and in the reverse reaction H_3O^+ donates a proton to OH^-. In fact, the reverse reaction is far more significant than the forward reaction. *Equilibrium is displaced far to the left.* Acid(1) and base(2) are *much* stronger than are acid(2) and base(1).

$$
\underset{\text{base(1)}}{:\overset{..}{\underset{..}{O}}:H} + \underset{\text{acid(2)}}{:\overset{..}{\underset{..}{O}}:H} \rightleftharpoons \underset{\text{acid(1)}}{\left(:\overset{H}{\underset{H}{\overset{..}{O}}}:H\right)^+} + \underset{\text{base(2)}}{\left(:\overset{..}{\underset{..}{O}}:H\right)^-} \tag{17.6}
$$

Again following the approach we used in writing equilibrium constants for the ionization of the weak base NH_3 and the weak acid $HC_2H_3O_2$, for the self-ionization of water

$$H_2O + H_2O \rightleftharpoons H_3O^+ + OH^-$$

we can write

$$K_c = [H_3O^+][OH^-]$$

You can also see from Equation (17.6) that $[H_3O^+]$ and $[OH^-]$ are equal in pure water. There are several experimental methods of determining these concentrations. All lead to this result.

At 25 °C in pure water: $[H_3O^+] = [OH^-] = 1.0 \times 10^{-7}$ M

The equilibrium constant for the self-ionization of water is called the **ion product of water** and is symbolized as K_w. At 25 °C,

$$K_w = [H_3O^+][OH^-] = 1.0 \times 10^{-14} \tag{17.7}$$

The significance of Equation (17.7) is that it applies to *all* aqueous solutions, not just to pure water, as we shall see shortly.

▶ Like other equilibrium constants, K_w varies with temperature. At 60 °C, $K_w = 9.6 \times 10^{-14}$; at 100 °C, $K_w = 5.5 \times 10^{-13}$.

The Nature of the Hydronium Ion

Because the H^+ ion is very small, the positive charge of the ion is concentrated in a small region; the ion has a high positive charge density. We should expect H^+ ions (protons) to seek out centers of negative charge with which to form bonds. This happens when an H^+ ion attaches itself to a lone pair of electrons in an O atom in H_2O, as suggested in reaction (17.6).

For many years the **hydronium ion,** H_3O^+, was thought to be one of a series of *hydrated* protons, $[H(H_2O)_n]^+$, the one in which $n = 1$, that is, $[H(H_2O)]^+ = H_3O^+$. Recently, however, chemists have found experimental evidence for the existence of H_3O^+ in the gaseous state, in solids, and in aqueous solution. For example, what was once thought to be a monohydrate of perchloric acid, $HClO_4 \cdot H_2O$, has been shown through X-ray studies to be $H_3O^+ClO_4^-$. This salt, which we might call hydronium perchlorate, is structurally quite similar to ammonium perchlorate, $NH_4^+ClO_4^-$. The current view of the hydronium ion in aqueous solution is that of a central H_3O^+ and a cluster of surrounding water molecules, perhaps something like that pictured in Figure 17-2. In this text we will represent the hydronium ion in solution as H_3O^+ or $H_3O^+(aq)$.

FIGURE 17-2
A hydrated hydronium ion.
This species, $H_9O_4^+$, consists of a central H_3O^+ ion hydrogen-bonded to three H_2O molecules. It has been identified in the solid hydrate $HBr \cdot 4H_2O$.

pH and pOH

With the value of their product fixed at 1.0×10^{-14}, $[H_3O^+]$ and $[OH^-]$ are generally small quantities—typically less than 1 M and often very much less. Exponential notation is useful in these situations, for example, $[H_3O^+] = 2.2 \times 10^{-13}$ M. But we now want to consider an even more convenient way to describe hydronium and hydroxide ion concentrations.

In 1909, the Danish biochemist Søren Sørensen proposed the term **pH** to refer to the "potential of hydrogen ion." He defined pH as the *negative of the logarithm of* $[H^+]$. Restated in terms of $[H_3O^+]$*

$$pH = -\log[H_3O^+] \tag{17.8}$$

▶ The symbol H^+, a remnant of the Arrhenius theory, is still often used as a shorthand representation of H_3O^+.

▶ Note that this definition uses logarithms to the base 10 (log), not natural logarithms (ln).

Thus, in a solution that is 0.0025 M HCl,

$[H_3O^+] = 2.5 \times 10^{-3}$ M   and $pH = -\log(2.5 \times 10^{-3}) = 2.60$

To determine $[H_3O^+]$ corresponding to a pH value, we do an inverse calculation. In a solution with a pH = 4.50,

$\log[H_3O^+] = -4.50$ and $[H_3O^+] = 10^{-4.50} = 3.2 \times 10^{-5}$

▶ The determination of logarithms and inverse logarithms (antilogarithms) is discussed in Appendix A. Significant figure rules for logarithms are also presented there.

*Strictly speaking we should use the *activity* of H_3O^+, $a_{H_3O^+}$, a dimensionless quantity. But we will not use activities here, just as we did not use them in Chapter 16. We will substitute the numerical value of the molarity of H_3O^+ for its activity and recognize that some pH calculations may be only approximations.

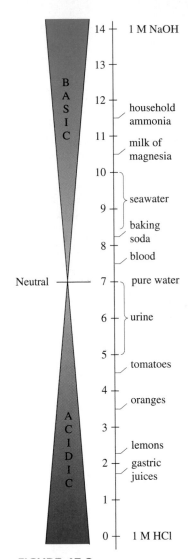

FIGURE 17-3
The pH scale and pH values of some common materials.
The scale shown here ranges from pH 0 to pH 14. Slightly negative pH values are also possible, perhaps to about −1 (corresponding to $[H_3O^+] \approx 10$ M). Also possible are pH values up to about 15 (corresponding to $[OH^-] \approx 10$ M). For practical purposes, however, the pH scale is only useful in about the range $2 < pH < 12$. This is because in concentrated acids and bases the molarities of H_3O^+ and OH^- may differ significantly from their true activities.

We can also define the quantity **pOH**.

$$pOH = -\log [OH^-] \quad (17.9)$$

And we can derive still another useful expression by taking the *negative logarithm* of the K_w expression (written for 25 °C) and introducing the symbol pK_w.

$$K_w = [H_3O^+][OH^-] = 1.0 \times 10^{-14}$$
$$-\log K_w = -(\log [H_3O^+][OH^-]) = -\log(1.0 \times 10^{-14})$$
$$pK_w = -(\log [H_3O^+] + \log [OH^-]) = -(-14.00)$$
$$= -\log [H_3O^+] - \log [OH^-] = 14.00$$

$$pK_w = pH + pOH = 14.00 \quad (17.10)$$

An aqueous solution with $[H_3O^+] = [OH^-]$ is said to be *neutral*. In pure water at 25 °C, $[H_3O^+] = [OH^-] = 1.0 \times 10^{-7}$ M and pH = 7.00. Thus, at 25 °C all aqueous solutions with pH = 7.00 are *neutral*. If the pH is *less than* 7.00, a solution is *acidic;* if the pH is *greater than* 7.00, a solution is *basic* or alkaline.

The pH values of a number of materials are depicted in Figure 17-3. These values and the many examples in this chapter and the next should help to familiarize you with the pH concept. Later, we will consider two methods for measuring pH: with acid–base indicators (Section 18-3) and through electrical measurements (Section 21-4).

EXAMPLE 17-2

Relating $[H_3O^+]$, $[OH^-]$, pH, and pOH. In a laboratory experiment, students measured the pH of samples of rainwater and household ammonia. Determine **(a)** $[H_3O^+]$ in the rainwater if its pH measured 4.35; **(b)** $[OH^-]$ in the ammonia if its pH measured 11.28.

Solution

a. By definition, pH = $-\log [H_3O^+]$, or

$$\log [H_3O^+] = -pH = -4.35$$
$$[H_3O^+] = 10^{-4.35} = 4.5 \times 10^{-5} \text{ M}$$

b. First, determine pOH with Equation (17.10).

$$pOH = 14.00 - pH = 14.00 - 11.28 = 2.72$$

Now, use the definition pOH = $-\log [OH^-]$.

$$\log [OH^-] = -pOH = -2.72$$
$$[OH^-] = 10^{-2.72} = 1.9 \times 10^{-3} \text{ M}$$

Practice Example A: Students found a sample of yogurt to have a pH of 2.85. What are $[H^+]$ and $[OH^-]$ in the yogurt?

Practice Example B: The pH of a solution of HCl in water is found to be 2.50. What volume of water would you add to 1.00 L of this solution to raise the pH to 3.10? (*Hint:* Determine $[H_3O^+]$ corresponding to each pH value and complete the calculation as a dilution problem.)

17-4   Strong Acids and Strong Bases

As we saw through Equation (17.5), the ionization of HCl in dilute aqueous solutions

$$HCl + H_2O \longrightarrow Cl^- + H_3O^+$$

goes essentially to completion.* By contrast, we learned in the preceding section that the self-ionization of water occurs only to a very slight extent (reaction 17.6). As a result we conclude that in calculating $[H_3O^+]$ in an aqueous solution of a strong acid, the strong acid is the only significant source of H_3O^+. The contribution due to the self-ionization of water can be ignored *unless the solution is extremely dilute.*

▶ Calculating $[H_3O^+]$ in an extremely dilute solution of a strong acid, say when $[H_3O^+]$ is less than about 10^{-6} M, is considerably more difficult than the calculation in Example 17-3. The self-ionization of water cannot be ignored in these cases.

EXAMPLE 17-3

Calculating Ion Concentrations in an Aqueous Solution of a Strong Acid. Calculate $[H_3O^+]$, $[Cl^-]$, and $[OH^-]$ in 0.015 M HCl(aq).

Solution

We can assume that HCl is completely ionized and is the sole source of H_3O^+ in solution. Therefore,

$$[H_3O^+] = 0.015 \text{ M}$$

Furthermore, because one Cl^- ion is produced for every H_3O^+ ion,

$$[Cl^-] = [H_3O^+] = 0.015 \text{ M}$$

To calculate $[OH^-]$ we muse use the following facts:

1. All the OH^- is derived from the self-ionization of water (17.6).
2. $[OH^-]$ and $[H_3O^+]$ must have values consistent with K_w for water.

$$K_w = [H_3O^+][OH^-] = 1.0 \times 10^{-14}$$

$$(0.015)[OH^-] = 1.0 \times 10^{-14}$$

$$[OH^-] = \frac{1.0 \times 10^{-14}}{1.5 \times 10^{-2}} = 0.67 \times 10^{-12} = 6.7 \times 10^{-13} \text{ M}$$

Practice Example A: A 0.0025 M solution of HI(aq) has $[H_3O^+] = 0.0025$ M. Calculate the $[I^-]$, $[OH^-]$, and pH of the solution.

Practice Example B: If 535 mL of *gaseous* HCl, at 26.5 °C and 747 mmHg, is dissolved in enough water to prepare 625 mL of solution, what is the pH of this solution? [*Hint:* What is the number of moles of HCl(g)?]

The common strong bases are ionic hydroxides. When these bases dissolve in water, H_2O molecules completely dissociate the cations and anions (OH^-) of the base from each other. The self-ionization of water, because it occurs to so very limited an extent, is an inconsequential source of OH^-. This means that in

*In very concentrated aqueous solutions, HCl does not exist exclusively as the separated ions H_3O^+ and Cl^-. One indication of this is that we can smell HCl in the vapor above such solutions.

TABLE 17.2
The Common Strong Acids and Strong Bases

Acids	Bases
HCl	LiOH
HBr	NaOH
HI	KOH
$HClO_4$	RbOH
HNO_3	CsOH
H_2SO_4[a]	$Mg(OH)_2$
	$Ca(OH)_2$
	$Sr(OH)_2$
	$Ba(OH)_2$

[a]H_2SO_4 ionizes in two distinct steps. It is a strong acid only in its first ionization (see page 601).

calculating $[OH^-]$ in an aqueous solution of a strong base, the strong base is the only significant source of OH^-, *unless the solution is extremely dilute.*

As you can see through Table 17.2, the number of common strong acids and strong bases is quite small; they should be committed to memory.

EXAMPLE 17-4

Calculating the pH of an Aqueous Solution of a Strong Base. Calcium hydroxide (slaked lime) is the cheapest strong base available and is generally used for industrial operations in which a high concentration of OH^- is not required. $Ca(OH)_2(s)$ is soluble in water only to the extent of 0.16 g $Ca(OH)_2$/100.0 mL solution at 25 °C. What is the pH of saturated $Ca(OH)_2(aq)$ at 25 °C?

Solution

First, we need to express the solubility of $Ca(OH)_2$ on a molarity basis.

$$\text{molarity} = \frac{0.16 \text{ g } Ca(OH)_2 \times \dfrac{1 \text{ mol } Ca(OH)_2}{74.1 \text{ g } Ca(OH)_2}}{0.1000 \text{ L}} = 0.022 \text{ M } Ca(OH)_2$$

Next, we can relate the molarity of OH^- to the molarity of $Ca(OH)_2$.

$$[OH^-] = \frac{0.022 \text{ mol } Ca(OH)_2}{1 \text{ L}} \times \frac{2 \text{ mol } OH^-}{1 \text{ mol } Ca(OH)_2} = 0.044 \text{ M } OH^-$$

Now we can calculate the pOH and the pH.

$$pOH = -\log [OH^-] = -\log 0.044 = 1.36$$
$$pH = 14.00 - pOH = 14.00 - 1.36 = 12.64$$

Practice Example A: Milk of magnesia is a saturated solution of $Mg(OH)_2$. The solubility is 9.63 mg $Mg(OH)_2$/100.0 mL solution at 20 °C. What is the pH of saturated $Mg(OH)_2$ at 20 °C?

Practice Example B: Calculate the pH of an aqueous solution that is 3.00% KOH, by mass, and has a density of 1.0242 g/mL. (*Hint:* You must combine ideas from Example 14-1 with those illustrated here.)

▶ A common error is to assume that you have calculated the pH when in fact you have found the pOH. The pH cannot be 1.36. The solution must be *basic*; it must have pH > 7. Use qualitative reasoning to avoid simple errors.

17-5 Weak Acids and Weak Bases

Figure 17-4 illustrates two ways of showing that ionization has occurred in an acid solution: one is by the color of an acid–base indicator, and the other, from the response of a pH meter. The pink color of the solution in Figure 17-4a tells us the pH of 0.10 M HCl is *less than 1.2.* The pH meter registers a value of 1.0, just what we expect for a *strong* acid solution with $[H_3O^+] = 0.10$ M. In Figure 17-4b, the yellow color indicates that in 0.10 M $HC_2H_3O_2$ (acetic acid) the pH is *2.8 or greater.* The pH meter registers 2.8.

So we see that two acids can have identical molarities but *different* pH values. The acid molarity simply states what was put into the solution, but $[H_3O^+]$ and pH depend on what *happens* in the solution. In both solutions some self-ionization of water occurs, but this is negligible. Ionization of HCl, a strong acid, can be assumed to go to completion, as previously indicated in Equation

▶ Most laboratory pH meters can be read to the nearest 0.01 unit. Some pH meters for research work can be read to 0.001 unit, but unless unusual precautions are taken, the reading of the meter may not correspond to the true pH. The meter used in Figure 17-4 registers pH only to the nearest 0.1 unit.

(a)   **(b)**

FIGURE 17-4
Strong and weak acids compared.
Thymol blue indicator, which is present in both solutions, has a color that depends on the pH of the solution.

pH	<	1.2	<	pH	<	2.8	<	pH
red				orange				yellow

The principle of the pH meter is discussed in Section 21-4. (a) 0.10 M HCl has pH ≈ 1. (b) The pH of 0.10 M $HC_2H_3O_2$ is about 2.8.

(17.5). As we have previously noted, ionization of $HC_2H_3O_2$, a weak acid, is a reversible reaction that reaches a condition of equilibrium.*

$$HC_2H_3O_2 + H_2O \rightleftharpoons H_3O^+ + C_2H_3O_2^- \qquad (17.11)$$

The equilibrium constant expression for reaction (17.11) is

$$K_a = \frac{[H_3O^+][C_2H_3O_2^-]}{[HC_2H_3O_2]} = 1.8 \times 10^{-5} \qquad (17.12)$$

Just as pH is a convenient shorthand designation related to $[H_3O^+]$, pK is related to an equilibrium constant. That is, p$K = -\log K$. Thus for acetic acid,

$$pK_a = -\log K_a = -\log(1.8 \times 10^{-5}) = -(-4.74) = 4.74$$

As with other equilibrium constants, the *larger* the value of K_a (or K_b for a base), the farther the equilibrium condition lies in the direction of the forward reaction. And the more extensive the ionization, the greater the concentrations of the ions produced.

Ionization constants must be determined *by experiment*. A few values for weak acids and weak bases are listed in Table 17.3, and a more extensive listing is given in Appendix D.

Identifying a Weak Acid

We know that the key element in an acid is hydrogen, but just because a substance contains H atoms does not make it an acid. One or more H atoms must be *ionizable*. Let's explore this idea further with acetic acid, whose formula is written in five different ways below.

Empirical formula:	CH_2O
Molecular formula:	$C_2H_4O_2$
"Acid" formula:	$HC_2H_3O_2$
Condensed structural formula:	CH_3COOH
Lewis structural formula:	(structure)

Lewis structure:

$$\begin{array}{c} \ \ \ \ \text{H} \ \ \ \ \overset{..}{\text{O}}: \\ \ \ \ \ | \ \ \ \ \ || \\ \text{H}-\text{C}-\text{C}-\overset{..}{\underset{..}{\text{O}}}-\text{H} \\ \ \ \ \ | \\ \ \ \ \ \text{H} \end{array}$$

Acetic acid, CH_3COOH.

▶ Feature Problem A on page 618 considers the type of experimental data required to establish formulas such as these.

---

*We have been writing ionization equations in the form: acid(1) + base(2) ⇌ base(1) + acid(2). Here we have written acid(1) + base(2) ⇌ acid(2) + base(1) to highlight H_3O^+, the species that is usually the subject of a calculation.

TABLE 17.3 Ionization Constants of Some Weak Acids and Weak Bases in Water at 25 °C

	Ionization Equilibrium	Ionization Constant K	pK
Acid		$K_a =$	p$K_a =$
acetic acid	$HC_2H_3O_2 + H_2O \rightleftharpoons H_3O^+ + C_2H_3O_2^-$	1.8×10^{-5}	4.74
benzoic acid	$HC_7H_5O_2 + H_2O \rightleftharpoons H_3O^+ + C_7H_5O_2^-$	6.3×10^{-5}	4.20
chloroacetic acid	$HC_2H_2ClO_2 + H_2O \rightleftharpoons H_3O^+ + C_2H_2ClO_2^-$	1.4×10^{-3}	2.85
chlorous acid	$HClO_2 + H_2O \rightleftharpoons H_3O^+ + ClO_2^-$	1.1×10^{-2}	1.96
formic acid	$HCHO_2 + H_2O \rightleftharpoons H_3O^+ + CHO_2^-$	1.8×10^{-4}	3.74
hydrocyanic acid	$HCN + H_2O \rightleftharpoons H_3O^+ + CN^-$	6.2×10^{-10}	9.21
hydrofluoric acid	$HF + H_2O \rightleftharpoons H_3O^+ + F^-$	6.6×10^{-4}	3.18
hypochlorous acid	$HOCl + H_2O \rightleftharpoons H_3O^+ + OCl^-$	2.9×10^{-8}	7.54
nitrous acid	$HNO_2 + H_2O \rightleftharpoons H_3O^+ + NO_2^-$	7.2×10^{-4}	3.14
phenol	$HOC_6H_5 + H_2O \rightleftharpoons H_3O^+ + C_6H_5O^-$	1.0×10^{-10}	10.00
Base		$K_b =$	p$K_b =$
ammonia	$NH_3 + H_2O \rightleftharpoons NH_4^+ + OH^-$	1.8×10^{-5}	4.74
aniline	$C_6H_5NH_2 + H_2O \rightleftharpoons C_6H_5NH_3^+ + OH^-$	7.4×10^{-10}	9.13
ethylamine	$C_2H_5NH_2 + H_2O \rightleftharpoons C_2H_5NH_3^+ + OH^-$	4.3×10^{-4}	3.37
hydroxylamine	$HONH_2 + H_2O \rightleftharpoons HONH_3^+ + OH^-$	9.1×10^{-9}	8.04
methylamine	$CH_3NH_2 + H_2O \rightleftharpoons CH_3NH_3^+ + OH^-$	4.2×10^{-4}	3.38
pyridine	$C_5H_5N + H_2O \rightleftharpoons C_5H_5NH^+ + OH^-$	1.5×10^{-9}	8.82

There is no clue in either the empirical or the molecular formula that acetic acid is an acid. By titrating the acid with a base, we can show that there is only one ionizable H atom in the acetic acid molecule. We can reflect this fact by writing the "acid" formula, $HC_2H_3O_2$, which shows that one H atom is different from the other three, and that it is this H atom that ionizes when acetic acid is added to water. In the condensed structural formula, we are more specific about how the one H atom differs from the other three. It is bonded to an O atom; the others are bonded to a carbon atom. The complete Lewis structural formula is even more specific in indicating the bonding arrangements. We will use the simple formula $HC_2H_3O_2$ in doing numerical calculations and one of the structural formulas when discussing phenomena affected by the molecular structure, such as in a discussion of acid strength in Section 17-8.

A large number of weak acids have the same structural feature as acetic acid—that is, a –COOH group as part of the molecule. This group is called a *carboxyl group* and is a common feature of many organic acids, including such biologically important acids as lactic acid and all the amino acids. We will use a number of carboxylic acids as examples in this and later chapters.

In general, to distinguish between a weak acid and a strong acid, you need only recall that the half dozen strong acids listed in Table 17.2 are the most common strong acids. Unless informed to the contrary, you may assume that any acid not listed in Table 17.2 is a weak acid.

Identifying a Weak Base

At first glance, weak bases seem more difficult to identify than weak acids—there is no distinctive element written first in the formula. On the other hand, if you study the bases in Table 17.3 you will see that all but one of them (pyridine) can be viewed as an ammonia molecule in which some other group

Methylamine, CH_3NH_2.

(—C_6H_5, —C_2H_5, —OH, —CH_3) has been substituted for one of the H atoms. The substitution of a methyl group, —CH_3, for an H atom is suggested in the following structural formulas,

We can represent the ionization of methylamine in this way:

and the ionization constant expression as

$$K_b = \frac{[CH_3NH_3^+][OH^-]}{[CH_3NH_2]} = 4.2 \times 10^{-4} \qquad (17.13)$$

Of course, not all weak bases contain N, but so many of them do that the similarity to NH_3 outlined here is well worth remembering. This large group of weak bases derived from ammonia are known as *amines*.

Illustrative Examples

For some students, solution equilibrium calculations are among the most challenging in general chemistry. At times the difficulty is in sorting out what is relevant to a given problem. The number of different types of calculations seems very large, although in fact it is quite limited. The key to solving solution equilibrium problems is in being able to picture in your mind's eye what is going on. Here are some questions to ask yourself.

- Which are the principal species in solution?
- What are the chemical reactions that produce them?
- Are there some reactions (for example, the self-ionization of water) that can be ignored?
- Are there assumptions that can be made to simplify the equilibrium calculations?
- What is a reasonable answer to the problem? For instance, should the final solution be acidic (pH < 7) or basic (pH > 7)?

In short, first think through a problem *qualitatively*. At times you may not even have to do a calculation. Another point is to organize the relevant data in a clear, logical manner. This action alone can get you on the right track. We recommend writing the balanced chemical equation and using it as the basis for an ICE table as discussed in Chapter 16. In this way many problems that may at first appear new to you will take on a familiar pattern. Look for other helpful hints as you proceed through this and the following two chapters.

EXAMPLE 17-5

Determining a Value of K_a from the pH of a Solution of a Weak Acid. Butyric acid, $HC_4H_7O_2$, is used to make compounds employed in artificial flavorings and syrups. A 0.250 M aqueous solution of $HC_4H_7O_2$ is found to have a pH of 2.72. Determine K_a for butyric acid.

$$HC_4H_7O_2 + H_2O \rightleftharpoons H_3O^+ + C_4H_7O_2^- \qquad K_a = ?$$

Solution

Because K_a for $HC_4H_7O_2$ is likely to be much larger than K_w, we can assume that self-ionization of water is unimportant and that ionization of the butyric acid is the only source of H_3O^+. Let's treat the situation as if $HC_4H_7O_2$ first dissolves in molecular form and then the molecules ionize until equilibrium is reached. At first we will represent the concentrations of H_3O^+ and $C_4H_7O_2^-$ at equilibrium as x M.

	$HC_4H_7O_2$	$+$	H_2O	$\rightleftharpoons$	H_3O^+	$+$	$C_4H_7O_2^-$
initial concn:	0.250 M				—		—
changes:	$-x$ M				$+x$ M		$+x$ M
equil concn:	$(0.250 - x)$ M				x M		x M

But x is *not* unknown. It is the $[H_3O^+]$ in solution, and we can determine this from the pH.

$$\log [H_3O^+] = -pH = -2.72$$
$$[H_3O^+] = 10^{-2.72} = 1.9 \times 10^{-3} = x$$

Now we can solve the following expression for K_a, substituting in the known value for x.

$$K_a = \frac{[H_3O^+][C_4H_7O_2^-]}{[HC_4H_7O_2]} = \frac{x \cdot x}{0.250 - x}$$
$$= \frac{(1.9 \times 10^{-3})(1.9 \times 10^{-3})}{0.250 - (1.9 \times 10^{-3})} = 1.5 \times 10^{-5}$$

Notice that our original assumption was correct—K_a is much larger than K_w.

Practice Example A: Hypochlorous acid, HOCl, is used in water treatment and as a disinfectant in swimming pools. A 0.150 M solution of HOCl has a pH of 4.18. Determine K_a for hypochlorous acid.

Practice Example B: The much-abused drug cocaine is an alkaloid. Alkaloids are noted for their bitter taste, an indication of their basic properties. Cocaine, $C_{17}H_{21}O_4N$, is soluble in water to the extent of 0.17 g/100 mL solution, and a saturated solution has a pH = 10.08. What is the value of K_b for cocaine?

$$C_{17}H_{21}O_4N + H_2O \rightleftharpoons C_{17}H_{21}O_4NH^+ + OH^- \qquad K_b = ?$$

(*Hint:* First convert the solubility to a molarity basis. Also, use the relationship pH + pOH = 14.00.)

▶For a weak base, $[OH^-]$ and the molarity of a cation appear in the numerator of the K_b expression (similar to $[H_3O^+]$ and the molarity of an anion in the numerator of the K_a expression of a weak acid).

Example 17-6 presents a common problem involving weak acids and weak bases: calculating the pH of a solution of known molarity. The calculation invariably involves a quadratic equation, but very often we can make a simplifying assumption leading to a shortcut that saves both time and effort.

EXAMPLE 17-6

Calculating the pH of a Weak Acid Solution. Show by calculation that the pH of 0.100 M $HC_2H_3O_2$ should be about the value shown on the pH meter in Figure 17-4, that is, pH ≈ 2.8.

Solution

Here we know that K_a is much larger than K_w. Let's again treat the situation as if $HC_2H_3O_2$ first dissolves in molecular form and then ionizes until equilibrium is reached. In other words, we set up the ICE table. In this case, the quantity x is an actual unknown that we must obtain by an algebraic solution.

$$HC_2H_3O_2 \quad + \quad H_2O \rightleftharpoons H_3O^+ \quad + \quad C_2H_3O_2^-$$

initial concn:	0.100 M	—	—
changes:	−x M	+x M	+x M
equil concn:	(0.100 − x) M	x M	x M

$$K_a = \frac{[H_3O^+][C_2H_3O_2^-]}{[HC_2H_3O_2]} = \frac{x \cdot x}{0.100 - x} = 1.8 \times 10^{-5}$$

We could solve this equation using the quadratic formula, but let's instead make a simplifying assumption that is often valid. Assume that x is very small compared with 0.100. That is, assume that $(0.100 - x) \approx 0.100$.

$$x^2 = 0.100 \times 1.8 \times 10^{-5} = 1.8 \times 10^{-6}$$

$$x = [H_3O^+] = \sqrt{1.8 \times 10^{-6}} = 1.3 \times 10^{-3} \text{ M}$$

Now, we must check our assumption: $0.100 - 0.0013 = 0.099 \approx 0.100$. Our assumption is good to about 1 part per 100 (1%) and is valid for a calculation involving two or three significant figures.

$$pH = -\log [H_3O^+] = -\log(1.3 \times 10^{-3}) = -(-2.89) = 2.89$$

Practice Example A: Substituting halogen atoms for hydrogen atoms bound to carbon increases the strength of carboxylic acids. Show that the pH of 0.100 M $HC_2H_2FO_2$, fluoroacetic acid, is lower than that calculated in Example 17-6 for 0.100 M $HC_2H_3O_2$.

$$HC_2H_2FO_2 + H_2O \rightleftharpoons H_3O^+ + C_2H_2FO_2^- \qquad K_a = 2.6 \times 10^{-3}$$

Practice Example B: Acetylsalicylic acid, $HC_9H_7O_4$, is the active component in aspirin. It causes the stomach upset that some people get when taking aspirin. Two extra-strength aspirin tablets, each containing 500 mg of acetylsalicylic acid, are dissolved in 325 mL of water. What is the pH of this solution?

$$HC_9H_7O_4 + H_2O \rightleftharpoons H_3O^+ + C_9H_7O_4^- \qquad K_a = 3.3 \times 10^{-4}$$

EXAMPLE 17-7

Dealing with the Failure of a Simplifying Assumption. What is the pH of a solution that is 0.00250 M $HNO_2(aq)$? (Obtain the K_a value from Table 17.3.)

Solution

We can begin by describing the ionization equilibrium in the customary format.

$$HNO_2 \quad + \quad H_2O \rightleftharpoons H_3O^+ \quad + \quad NO_2^-$$

initial concn:	0.00250 M		—	—
changes:	$-x$ M		$+x$ M	$+x$ M
equil concn:	$(0.00250 - x)$ M		x M	x M

$$K_a = \frac{[H_3O^+][NO_2^-]}{[HNO_2]} = \frac{x \cdot x}{0.00250 - x} = 7.2 \times 10^{-4}$$

Now let's assume that x is very much less than 0.00250 and that $0.00250 - x \approx 0.00250$.

$$\frac{x^2}{0.00250} = 7.2 \times 10^{-4} \qquad x^2 = 1.8 \times 10^{-6}$$

$$[H_3O^+] = x = 1.3 \times 10^{-3} \text{ M}$$

The value of x is about half as large as 0.00250—too large to ignore.

$$\frac{0.0013}{0.00250} \times 100\% = 52\%$$

An alternative to the quadratic formula is the method of successive approximations described in Feature Problem B on page 619.

Our assumption failed, so we must use an exact method. This means using the *quadratic* formula.

$$\frac{x^2}{0.00250 - x} = 7.2 \times 10^{-4}$$

$$x^2 + 7.2 \times 10^{-4} x - 1.8 \times 10^{-6} = 0$$

$$x = \frac{-7.2 \times 10^{-4} \pm \sqrt{(7.2 \times 10^{-4})^2 + 4 \times 1.8 \times 10^{-6}}}{2}$$

$$x = [H_3O^+] = \frac{-7.2 \times 10^{-4} \pm 2.8 \times 10^{-3}}{2} = 1.0 \times 10^{-3} \text{ M}$$

$$pH = -\log [H_3O^+] = -\log(1.0 \times 10^{-3}) = 3.00$$

Practice Example A: What is the pH of 0.015 M $HC_2H_2FO_2$(aq)?

$$HC_2H_2FO_2 + H_2O \rightleftharpoons H_3O^+ + C_2H_2FO_2^- \qquad K_a = 2.6 \times 10^{-3}$$

Practice Example B: Piperidine is a base found in small amounts in black pepper. What is the pH of 315 mL of a water solution containing 114 mg piperidine?

$$C_5H_{11}N + H_2O \rightleftharpoons C_5H_{11}NH^+ + OH^- \qquad K_b = 1.6 \times 10^{-3}$$

Percent Ionization

We can describe how extensive the ionization of a weak acid or weak base is by determining the degree of ionization or percent ionization.

For the ionization $HA + H_2O \rightleftharpoons H_3O^+ + A^-$, the degree of ionization is the fraction of the acid molecules that ionize. Thus, if in 1.00 M HA, ionization produces $[H_3O^+] = [A^-] = 0.05$ M, the degree of ionization = 0.05 M/1.00 M = 0.05. **Percent ionization** gives the proportion of ionized molecules on a percentage basis.

$$\text{percent ionization} = \frac{\text{molarity of } H_3O^+ \text{ derived from HA}}{\text{initial molarity of HA}} \times 100\% \qquad (17.14)$$

A weak acid with a degree of ionization of 0.05 has a percent ionization of 5%.

Are You Wondering . . .

If there is a way of knowing when a simplifying assumption will work?

The usual simplifying assumption is that of treating a weak acid or weak base as if it remains essentially nonionized (so that $M - x \approx M$). In general, this assumption will work if the molarity of the weak acid, M_A, or the weak base, M_B, exceeds the value of K_a or K_b by a factor of at least 100. That is,

$$\frac{M_A \text{ (or } M_B)}{K_a \text{ (or } K_b)} > 100$$

In any case, it is important to test the validity of any assumption that you make. If the assumption is good to within a few percent (say, less than 5%), then it is generally valid. In Example 17-6 the simplifying assumption was good to about 1%, but in Example 17-7 it was off by 52%.

Figure 17-5 compares percent ionization and solution molarity for a weak acid and a strong acid. Example 17-8 shows by calculation that the percent ionization of a weak acid or a weak base *increases* as the solution becomes *more dilute,* a fact that we can also demonstrate by a simple analysis of the ionization reaction

$$HA + H_2O \rightleftharpoons H_3O^+ + A^-$$

At equilibrium, n_{HA} moles of the acid HA, $n_{H_3O^+}$ moles of H_3O^+, and n_{A^-} moles of A^- are present in a volume of V liters; the K_a expression is

$$K_a = \frac{(n_{H_3O^+}/V)(n_{A^-}/V)}{n_{HA}/V} = \frac{(n_{H_3O^+})(n_{A^-})}{n_{HA}} \times \frac{1}{V}$$

When we dilute the solution, V increases, $1/V$ decreases, and the ratio: $(n_{H_3O^+})(n_{A^-})/n_{HA}$ must increase to maintain the constant value of K_a. This requires an increase in $n_{H_3O^+}$ and n_{A^-} and a decrease in n_{HA}, signifying an increase in the percent ionization.

FIGURE 17-5
Percent ionization of an acid as a function of concentration.
Over the concentration range shown, HCl, a strong acid, is 100% ionized. The percent ionization of $HC_2H_3O_2$ increases from about 4% in 0.010 M $HC_2H_3O_2$ to essentially 100% when the solution is extremely dilute. Note that in extremely dilute solutions the two acids would have virtually the same pH.

EXAMPLE 17-8

Determining Percent Ionization as a Function of Weak Acid Concentration. What is the percent ionization of acetic acid in 1.0 M, 0.10 M, and 0.010 M $HC_2H_3O_2$?

Solution

Let us use the ICE format to describe 1.0 M $HC_2H_3O_2$.

$$HC_2H_3O_2 + H_2O \rightleftharpoons H_3O^+ + C_2H_3O_2^-$$

initial concn:	1.0 M	—	—
changes:	$-x$ M	$+x$ M	$+x$ M
equil concn:	$(1.0 - x)$ M	x M	x M

We need to calculate $x = [H_3O^+] = [C_2H_3O_2^-]$. In doing so, let's make the usual assumption: $1.0 - x \approx 1.0$.

$$K_a = \frac{[H_3O^+][C_2H_3O_2^-]}{[HC_2H_3O_2]} = \frac{x \cdot x}{1.0 - x} = \frac{x^2}{1.0} = 1.8 \times 10^{-5}$$

$$x = [H_3O^+] = [C_2H_3O_2^-] = \sqrt{1.8 \times 10^{-5}} = 4.2 \times 10^{-3} \text{ M}$$

The percent ionization of 1.0 M $HC_2H_3O_2$ is

$$\% \text{ ionization} = \frac{[H_3O^+]}{[HC_2H_3O_2]} \times 100\% = \frac{4.2 \times 10^{-3} \text{ M}}{1.0 \text{ M}} \times 100\% = 0.42\%$$

The calculations for 0.10 M $HC_2H_3O_2$ and 0.010 M $HC_2H_3O_2$ are very similar and yield the results

0.10 M $HC_2H_3O_2$ is 1.3% ionized; 0.010 M $HC_2H_3O_2$ is 4.2% ionized.

Practice Example A: What is the percent ionization of hydrofluoric acid in 0.20 M HF and in 0.020 M HF?

Practice Example B: An 0.0284 M aqueous solution of lactic acid, a substance that accumulates in the blood and muscles during physical activity, is found to be 6.7% ionized. Determine a value of K_a for lactic acid.

$$HC_3H_5O_3 + H_2O \rightleftharpoons H_3O^+ + C_3H_5O_3^- \qquad K_a = ?$$

(*Hint:* Begin by determining $[H_3O^+]$ from the solution molarity and percent ionization.)

17-6 Polyprotic Acids

All the acids listed in Table 17.3 are weak *monoprotic* acids, meaning that their molecules have only one ionizable H atom, even though several contain more than one H atom. But some acids have *more than one* ionizable H atom per molecule. These are **polyprotic acids.** Table 17.4 lists ionization constants for several of them. We will use phosphoric acid, H_3PO_4, as the focus of our discussion.

Phosphoric Acid

Phosphoric acid, H_3PO_4, ranks second only to sulfuric acid among the important commercial acids. Annual production in the United States typically exceeds 10 million tons. Its principal use is in the manufacture of phosphate fertilizers.

Phosphoric acid, H_3PO_4.

TABLE 17.4 Ionization Constants of Some Common Polyprotic Acids

Acid	Ionization Equilibria	Ionization Constants, K	pK
carbonic[a]	$H_2CO_3 + H_2O \rightleftharpoons H_3O^+ + HCO_3^-$	$K_{a_1} = 4.4 \times 10^{-7}$	$pK_{a_1} = 6.36$
	$HCO_3^- + H_2O \rightleftharpoons H_3O^+ + CO_3^{2-}$	$K_{a_2} = 4.7 \times 10^{-11}$	$pK_{a_2} = 10.33$
hydrosulfuric[b]	$H_2S + H_2O \rightleftharpoons H_3O^+ + HS^-$	$K_{a_1} = 1.0 \times 10^{-7}$	$pK_{a_1} = 6.96$
	$HS^- + H_2O \rightleftharpoons H_3O^+ + S^{2-}$	$K_{a_2} = 1 \times 10^{-19}$	$pK_{a_2} = 19.0$
phosphoric	$H_3PO_4 + H_2O \rightleftharpoons H_3O^+ + H_2PO_4^-$	$K_{a_1} = 7.1 \times 10^{-3}$	$pK_{a_1} = 2.15$
	$H_2PO_4^- + H_2O \rightleftharpoons H_3O^+ + HPO_4^{2-}$	$K_{a_2} = 6.3 \times 10^{-8}$	$pK_{a_2} = 7.20$
	$HPO_4^{2-} + H_2O \rightleftharpoons H_3O^+ + PO_4^{3-}$	$K_{a_3} = 4.2 \times 10^{-13}$	$pK_{a_3} = 12.38$
phosphorous	$H_3PO_3 + H_2O \rightleftharpoons H_3O^+ + H_2PO_3^-$	$K_{a_1} = 3.7 \times 10^{-2}$	$pK_{a_1} = 1.43$
	$H_2PO_3^- + H_2O \rightleftharpoons H_3O^+ + HPO_3^{2-}$	$K_{a_2} = 2.1 \times 10^{-7}$	$pK_{a_2} = 6.68$
sulfurous[c]	$H_2SO_3 + H_2O \rightleftharpoons H_3O^+ + HSO_3^-$	$K_{a_1} = 1.3 \times 10^{-2}$	$pK_{a_1} = 1.89$
	$HSO_3^- + H_2O \rightleftharpoons H_3O^+ + SO_3^{2-}$	$K_{a_2} = 6.2 \times 10^{-8}$	$pK_{a_2} = 7.21$
sulfuric[d]	$H_2SO_4 + H_2O \rightleftharpoons H_3O^+ + HSO_4^-$	$K_{a_1} = $ very large	$pK_{a_1} < 0$
	$HSO_4^- + H_2O \rightleftharpoons H_3O^+ + SO_4^{2-}$	$K_{a_2} = 1.1 \times 10^{-2}$	$pK_{a_2} = 1.96$

[a]H_2CO_3 cannot be isolated. It is in equilibrium with H_2O and dissolved CO_2. The value given for K_{a_1} is actually for the reaction

$$CO_2(aq) + 2 H_2O \rightleftharpoons H_3O^+ + HCO_3^-$$

Generally, aqueous solutions of CO_2 are treated *as if* the $CO_2(aq)$ were first converted to H_2CO_3, followed by ionization of the H_2CO_3.
[b]The value for K_{a_2} of H_2S most commonly found in the older literature is about 1×10^{-14}, but current evidence suggests that the best value is considerably smaller.
[c]H_2SO_3 is a hypothetical, nonisolable species. The value listed for K_{a_1} is actually for the reaction

$$SO_2(aq) + 2 H_2O \rightleftharpoons H_3O^+ + HSO_3^-$$

[d]H_2SO_4 is completely ionized in the first step.

In addition, various sodium, potassium, and calcium phosphates are used in the food industry.

The H_3PO_4 molecule has *three* ionizable H atoms; it is a *triprotic* acid. It ionizes in three steps, and for each step we can write an ionization equation and an acid ionization constant expression with a distinctive value of K_a. The three steps are

(1) $H_3PO_4 + H_2O \rightleftharpoons H_3O^+ + H_2PO_4^-$

$$K_{a_1} = \frac{[H_3O^+][H_2PO_4^-]}{[H_3PO_4]} = 7.1 \times 10^{-3}$$

(2) $H_2PO_4^- + H_2O \rightleftharpoons H_3O^+ + HPO_4^{2-}$

$$K_{a_2} = \frac{[H_3O^+][HPO_4^{2-}]}{[H_2PO_4^-]} = 6.3 \times 10^{-8}$$

(3) $HPO_4^{2-} + H_2O \rightleftharpoons H_3O^+ + PO_4^{3-}$

$$K_{a_3} = \frac{[H_3O^+][PO_4^{3-}]}{[HPO_4^{2-}]} = 4.2 \times 10^{-13}$$

There is a ready explanation for the relative magnitudes of the ionization constants, that is, that $K_{a_1} > K_{a_2} > K_{a_3}$. When ionization occurs in step (1), a proton (H^+) moves away from an ion with a $1-$ charge ($H_2PO_4^-$). In step (2) the proton moves away from an ion with a $2-$ charge (HPO_4^{2-}), a more difficult separation. As a result, the ionization constant in the second step is smaller than in the first. Ionization is more difficult still in the third step.

The key ideas about the ionization of phosphoric acid illustrated in Example 17-9 are that

1. K_{a_1} is so much larger than K_{a_2} and K_{a_3} that essentially all the H_3O^+ is produced in the first ionization step alone.

2. So little of the $H_2PO_4^-$ formed in the first ionization step ionizes any further that we can assume $[H_2PO_4^-] = [H_3O^+]$ in the solution.

3. $[HPO_4^{2-}] \approx K_{a_2}$, regardless of the molarity of the acid.*

EXAMPLE 17-9

Calculating Ion Concentrations in a Polyprotic Acid Solution. For a 3.0 M H_3PO_4 solution, calculate **(a)** $[H_3O^+]$; **(b)** $[H_2PO_4^-]$; **(c)** $[HPO_4^{2-}]$; and **(d)** $[PO_4^{3-}]$.

Solution

a. Because K_{a_1} is so much larger than K_{a_2}, let's assume that all the H_3O^+ is formed in the *first* ionization step. This is equivalent to thinking of H_3PO_4 as if it were a monoprotic acid, ionizing only in the first step.

	H_3PO_4	$+$	H_2O	$\rightleftharpoons$	H_3O^+	$+$	$H_2PO_4^-$
initial concn:	3.0 M				—		—
changes:	$-x$ M				$+x$ M		$+x$ M
after 1st ionization:	$(3.0 - x)$ M				x M		x M

▶ In the assumption $3.0 - x \approx 3.0$, $x = 0.14$, which is 4.7% of 3.0. This is about the maximum error that can be tolerated for an acceptable assumption.

Following the usual assumption that x is much smaller than 3.0 and that $3.0 - x \approx 3.0$, we obtain

$$K_{a_1} = \frac{[H_3O^+][H_2PO_4^-]}{[H_3PO_4]} = \frac{x \cdot x}{(3.0 - x)} = \frac{x^2}{3.0} = 7.1 \times 10^{-3}$$

$$x^2 = 0.021 \qquad x = [H_3O^+] = 0.14 \text{ M}$$

b. From part (a), $x = [H_2PO_4^-] = [H_3O^+] = 0.14$ M

c. To determine $[H_3O^+]$ and $[H_2PO_4^-]$, we assumed that the second ionization is insignificant. Here we do have to consider the second ionization, no matter how slight; otherwise we would have no source of the ion HPO_4^{2-}. We can represent the second ionization, as shown in the following table. *Note especially how the results of the first ionization enter in.* We start with a solution in which $[H_2PO_4^-] = [H_3O^+] = 0.14$ M.

	$H_2PO_4^-$	$+$	H_2O	$\rightleftharpoons$	H_3O^+	$+$	HPO_4^{2-}
from 1st ionization:	0.14 M				0.14 M		—
changes:	$-y$ M				$+y$ M		$+y$ M
after 2nd ionization:	$(0.14 - y)$ M				$(0.14 + y)$ M		y M

If we assume that y is much smaller than 0.14, then $(0.14 + y) \approx (0.14 - y) \approx 0.14$. This leads to

$$K_{a_2} = \frac{[H_3O^+][HPO_4^{2-}]}{[H_2PO_4^-]} = \frac{(0.14 + y)(y)}{(0.14 - y)} = \frac{(0.14)(y)}{(0.14)} = 6.3 \times 10^{-8}$$

$$y = [HPO_4^{2-}] = 6.3 \times 10^{-8} \text{ M}$$

▶ The fact that the concentration of anion produced in the second ionization step of a polyprotic acid is equal to K_{a_2} can be used to get this result directly.

Note that the assumption is valid.

---

*If we assume that $[H_2PO_4^-] = [H_3O^+]$, the second ionization expression reduces to

$$\frac{[H_3O^+][HPO_4^{2-}]}{[H_2PO_4^-]} = K_{a_2}$$

d. The PO_4^{3-} ion is formed only in the third ionization step. When we write this acid ionization constant expression, we see that we have already calculated the ion concentrations other than $[PO_4^{3-}]$. We can simply solve the K_{a_3} expression for $[PO_4^{3-}]$.

$$K_{a_3} = \frac{[H_3O^+][PO_4^{3-}]}{[HPO_4^{2-}]} = \frac{0.14 \times [PO_4^{3-}]}{6.3 \times 10^{-8}} = 4.2 \times 10^{-13}$$

$$[PO_4^{3-}] = \frac{4.2 \times 10^{-13} \times 6.3 \times 10^{-8}}{0.14} = 1.9 \times 10^{-19} \text{ M}$$

Practice Example A: Malonic acid, $HOOCCH_2COOH$, is a diprotic acid used in the manufacture of barbiturates.

$$HOOCCH_2COOH + H_2O \rightleftharpoons H_3O^+ + HOOCCH_2COO^- \qquad K_{a_1} = 1.4 \times 10^{-3}$$

$$HOOCCH_2COO^- + H_2O \rightleftharpoons H_3O^+ + {}^-OOCCH_2COO^- \qquad K_{a_2} = 2.0 \times 10^{-6}$$

Calculate $[H_3O^+]$, $[HOOCCH_2COO^-]$, and $[{}^-OOCCH_2COO^-]$ in a 1.00 M solution of malonic acid.

Practice Example B: Oxalic acid, found in the leaves of rhubarb and other plants, is a diprotic acid.

$$H_2C_2O_4 + H_2O \rightleftharpoons H_3O^+ + HC_2O_4^- \qquad K_{a_1} = \,?$$

$$HC_2O_4^- + H_2O \rightleftharpoons H_3O^+ + C_2O_4^{2-} \qquad K_{a_2} = \,?$$

An aqueous solution that is 1.05 M $H_2C_2O_4$ has pH = 0.67. The free oxalate ion concentration in this solution is $[C_2O_4^{2-}] = 5.3 \times 10^{-5}$ M. Determine the values of K_{a_1} and K_{a_2} for oxalic acid.

A Somewhat Different Case: H_2SO_4

Sulfuric acid differs from most polyprotic acids in this important respect: It is a *strong* acid in its first ionization and a *weak* acid in the second. Ionization is complete in the first step, which means that in most $H_2SO_4(aq)$ solutions $[H_2SO_4] \approx 0$. Thus, if a solution is 0.50 M H_2SO_4, we can treat it as if it were 0.50 M H_3O^+ and 0.50 M HSO_4^- initially. Then we can determine the extent to which ionization of HSO_4^- produces additional H_3O^+ and SO_4^{2-}.

Sulfuric acid, H_2SO_4.

EXAMPLE 17-10

Calculating Ion Concentrations in Sulfuric Acid Solutions: Strong Acid Ionization Followed by Weak Acid Ionization. Calculate $[H_3O^+]$, $[HSO_4^-]$, and $[SO_4^{2-}]$ in 0.50 M H_2SO_4.

Solution

Let us use a similar approach to that of Example 17-9.

	H_2SO_4	+ H_2O	$\longrightarrow$	H_3O^+	+	HSO_4^-
initial concn:	0.50 M			—		—
changes:	−0.50 M			+0.50 M		+0.50 M
after 1st ionization:	≈ 0			0.50 M		0.50 M

	HSO_4^-	+ H_2O	$\rightleftharpoons$	H_3O^+	+	SO_4^{2-}
from 1st ionization:	0.50 M			0.50 M		—
changes:	−x M			+x M		+x M
after 2nd ionization:	(0.50 − x) M			(0.50 + x) M		x M

We need to deal only with the ionization constant expression for K_{a_2}. If we assume that x is much smaller than 0.50, then $(0.50 + x) \approx (0.50 - x) \approx 0.50$ and

$$K_{a_2} = \frac{[H_3O^+][SO_4^{2-}]}{[HSO_4^-]} = \frac{(0.50 + x) \cdot x}{(0.50 - x)} = \frac{0.50 \cdot x}{0.50} = 1.1 \times 10^{-2}$$

Our results, then, are

$$[H_3O^+] = 0.50 + x = 0.51 \text{ M}; \qquad [HSO_4^-] = 0.50 - x = 0.49 \text{ M}$$

$$[SO_4^{2-}] = x = K_{a_2} = 0.011 \text{ M}$$

Practice Example A: Calculate $[H_3O^+]$, $[HSO_4^-]$, and $[SO_4^{2-}]$ in 0.20 M H_2SO_4.

Practice Example B: Calculate $[H_3O^+]$, $[HSO_4^-]$, and $[SO_4^{2-}]$ in 0.020 M H_2SO_4. (*Hint:* Is the assumption that $[HSO_4^-] = [H_3O^+]$ valid?)

17-7 Ions as Acids and Bases

In our discussion to this point we have emphasized the behavior of *neutral molecules* as acids (for example, HCl, $HC_2H_3O_2$, H_3PO_4) or as bases (for example, NH_3, CH_3NH_2). We have also seen, however, that ions can act as acids or bases. For instance, in the second and subsequent ionization steps of a polyprotic acid an anion acts as an acid.

$$H_2PO_4^- + H_2O \rightleftharpoons H_3O^+ + HPO_4^{2-} \qquad K = K_{a_2} = 6.3 \times 10^{-8}$$

Now let's think about how each of the following can be described as an acid–base reaction.

$$\underset{\text{acid(1)}}{NH_4^+} + \underset{\text{base(2)}}{H_2O} \rightleftharpoons \underset{\text{base(1)}}{NH_3} + \underset{\text{acid(2)}}{H_3O^+} \qquad (17.15)$$

$$\underset{\text{base(1)}}{C_2H_3O_2^-} + \underset{\text{acid(2)}}{H_2O} \rightleftharpoons \underset{\text{acid(1)}}{HC_2H_3O_2} + \underset{\text{base(2)}}{OH^-} \qquad (17.16)$$

In reaction (17.15) NH_4^+ is an *acid,* donating a proton to water, a *base.* We describe equilibrium in this reaction through the *acid ionization constant* of the ammonium ion, NH_4^+.

$$K_a = \frac{[NH_3][H_3O^+]}{[NH_4^+]} = ? \qquad (17.17)$$

Two of the concentration terms in expression (17.17)—$[NH_3]$ and $[NH_4^+]$—are the same as in the K_b expression for NH_3, the conjugate base of NH_4^+. It seems that K_a for NH_4^+ and K_b for NH_3 should bear some relationship to each other, and they do. The easiest way to see this is to multiply both the numerator and denominator of (17.17) by $[OH^-]$. The product $[H_3O^+] \times [OH^-]$ is the ion product of water, K_w, shown in red. The other terms, shown in blue, represent the *inverse* of K_b for NH_3. The value obtained, 5.6×10^{-10}, is the missing value of K_a in expression (17.17).

$$K_a = \frac{[NH_3][H_3O^+][OH^-]}{[NH_4^+][OH^-]} = \frac{K_w}{K_b} = \frac{1.0 \times 10^{-14}}{1.8 \times 10^{-5}} = 5.6 \times 10^{-10}$$

The result we just obtained is an important consequence of the Brønsted–Lowry theory:

The product of the ionization constants of an acid and its conjugate base equals the ion product of water.

$$K_a \text{ (acid)} \times K_b \text{ (its conjugate base)} = K_w$$
$$K_b \text{ (base)} \times K_a \text{ (its conjugate acid)} = K_w$$

(17.18)

In reaction (17.16) $C_2H_3O_2{}^-$ acts as a *base* by accepting a proton from water, an *acid*. Here, equilibrium is described through the *base ionization constant* of the acetate ion, $C_2H_3O_2{}^-$, and with expression (17.18) we can evaluate K_b.

$$K_b = \frac{[HC_2H_3O_2][OH^-]}{[C_2H_3O_2{}^-]} = \frac{K_w}{K_a(HC_2H_3O_2)} = \frac{1.0 \times 10^{-14}}{1.8 \times 10^{-5}} = 5.6 \times 10^{-10}$$

In many tabulations of ionization constants only K_a values are listed, whether for neutral molecules or for ions. We can obtain the values of their conjugates with expression (17.18).

Hydrolysis

We have learned that in pure water at 25 °C, $[H_3O^+] = [OH^-] = 1.0 \times 10^{-7}$ M and pH = 7.00. *Pure water is pH neutral.* When NaCl dissolves in water at 25 °C, complete dissociation into Na^+ and Cl^- ions occurs, and the pH of the solution remains at 7.00. We can represent this fact with the equation

$$Na^+ + Cl^- + H_2O \longrightarrow \text{no reaction}$$

As shown in Figure 17-6, when NH_4Cl is added to water, the pH falls below 7. This means that in the solution $[H_3O^+] > [OH^-]$. A reaction producing H_3O^+ must occur.

$$Cl^- + H_2O \longrightarrow \text{no reaction}$$
$$NH_4{}^+ + H_2O \rightleftharpoons NH_3 + H_3O^+$$

The reaction between $NH_4{}^+$ and H_2O is fundamentally no different from other acid–base reactions. A reaction between an ion and water, however, is often

(a)   (b)   (c)

FIGURE 17-6
Ions as acids and bases.
Each of these 1 M solutions contains bromothymol blue indicator, which has the following colors.

pH < 7	pH = 7	pH > 7
yellow	green	blue

(a) $NH_4Cl(aq)$ is acidic. (b) $NaCl(aq)$ is neutral.
(c) $NaC_2H_3O_2(aq)$ is basic.

called a **hydrolysis** reaction. We say that ammonium ion *hydrolyzes* (and chloride ion does not).

When sodium acetate is dissolved in water, the pH rises above 7 (see Figure 17-6). This means that in the solution $[OH^-] > [H_3O^+]$. Here, acetate ion hydrolyzes.

$$Na^+ + H_2O \longrightarrow \text{no reaction}$$

$$C_2H_3O_2^- + H_2O \rightleftharpoons HC_2H_3O_2 + OH^-$$

We are now in a position to make both *qualitative* predictions and *quantitative* calculations concerning the pH values of aqueous solutions of salts. Whichever of these tasks is called for, note that hydrolysis takes place only if a chemical reaction producing a weak acid or weak base occurs. The following generalizations are useful.

- Salts of strong bases and strong acids (for example, NaCl) *do not hydrolyze.* The solution pH = 7.
- Salts of strong bases and *weak* acids (for example, $NaC_2H_3O_2$) *hydrolyze:* pH > 7. (The *anion* acts as a *base.*)
- Salts of *weak* bases and strong acids (for example, NH_4Cl) *hydrolyze:* pH < 7. (The *cation* acts as an *acid.*)
- Salts of *weak* bases and *weak* acids (for example, $NH_4C_2H_3O_2$) *hydrolyze.* (The cations are acids, and the anions are bases. Whether the solution is acidic or basic, however, depends on the relative values of K_a and K_b for the ions.)

EXAMPLE 17-11

Making Qualitative Predictions About Hydrolysis Reactions. Predict whether you would expect each of the following solutions to be acidic, basic, or pH neutral: **(a)** NaOCl(aq); **(b)** KCl(aq); **(c)** NH_4NO_3(aq).

Solution

a. The ions present are Na^+, which does not hydrolyze, and OCl^-, which does. OCl^- is the conjugate base of HOCl and forms a basic solution.

$$OCl^- + H_2O \rightleftharpoons HOCl + OH^-$$

b. Neither K^+ nor Cl^- hydrolyzes. KCl(aq) is neutral—pH = 7.
c. NH_4^+ hydrolyzes, but NO_3^- does not (HNO_3 is a strong acid).

$$NH_4^+ + H_2O \rightleftharpoons NH_3 + H_3O^+$$

Because of the hydrolysis reaction, $[H_3O^+] > [OH^-]$, and the solution is acidic.

Practice Example A: Predict whether you would expect each of the following 1.0 M solutions to be acidic, basic, or pH neutral: **(a)** $CH_3NH_3^+NO_3^-$(aq); **(b)** NaI(aq); **(c)** $NaNO_2$(aq).

Practice Example B: A water solution containing $H_2PO_4^-$ has a pH of about 4.7. Write equations for *two* reactions of $H_2PO_4^-$ with water and explain which reaction occurs to the greater extent. (*Hint:* One reaction is an acid ionization and the other is a hydrolysis. What are their K values?)

EXAMPLE 17-12

Evaluating Ionization Constants for Hydrolysis Reactions. Sodium nitrite, $NaNO_2$, and sodium benzoate, $NaC_7H_5O_2$, are both used as food preservatives. If solutions of these two salts of equal molarities are compared, which will have the *higher* pH?

Solution

Each of these substances is the salt of a strong base (NaOH) and a *weak acid*. The anions should ionize as *bases,* making their solutions somewhat basic.

$$NO_2^- + H_2O \rightleftharpoons HNO_2 + OH^- \qquad K_b(NO_2^-) = ?$$

$$C_7H_5O_2^- + H_2O \rightleftharpoons HC_7H_5O_2 + OH^- \qquad K_b(C_7H_5O_2^-) = ?$$

Our task is to determine the K_b values, neither of which is listed in a table in the chapter. We do have listings of K_a for the conjugate acids, however (Table 17.3). We can use expression (17.18) to write

$$K_b \text{ of } NO_2^- = \frac{K_w}{K_a(HNO_2)} = \frac{1.0 \times 10^{-14}}{7.2 \times 10^{-4}} = 1.4 \times 10^{-11}$$

$$K_b \text{ of } C_7H_5O_2^- = \frac{K_w}{K_a(HC_7H_5O_2)} = \frac{1.0 \times 10^{-14}}{6.3 \times 10^{-5}} = 1.6 \times 10^{-10}$$

We see that the K_b of $C_7H_5O_2^-$ is about 10 times larger than the K_b of NO_2^-. We expect the benzoate ion to hydrolyze to a somewhat greater extent than the nitrite ion to give a solution with a higher $[OH^-]$. A sodium benzoate solution should be more basic and have a higher pH than a sodium nitrite solution of the same concentration.

Practice Example A: The organic bases cocaine ($pK_b = 8.41$) and codeine ($pK_b = 7.95$) react with hydrochloric acid to form salts (similar to the formation of NH_4Cl by the reaction of NH_3 and HCl). If equimolar solutions of the following salts are compared, which solution would have the higher pH: cocaine hydrochloride, $C_{17}H_{21}O_4NH^+Cl^-$, or codeine hydrochloride, $C_{18}H_{21}ClO_3NH^+Cl^-$? (*Hint:* What are the K_a values of the ions that hydrolyze?)

Practice Example B: Predict whether the solution $NH_4CN(aq)$ is acidic, basic, or neutral. [*Hint:* Which ion(s) hydrolyze? What are the relevant values of K_a and K_b?]

▶ To predict the pH of $NH_4CN(aq)$ is a matter of some difficulty, but the calculation is facilitated by the alternative approach to equilibrium calculations presented in Feature Problem C on page 619.

EXAMPLE 17-13

Calculating the pH of a Solution in Which Hydrolysis Occurs. Sodium cyanide is an extremely poisonous substance, but it has very useful applications in gold and silver metallurgy and in the electroplating of metals. Aqueous solutions of cyanides are especially hazardous if they become acidified, because of the release of toxic hydrogen cyanide gas, HCN(g). Are NaCN(aq) solutions normally acidic, basic, or pH neutral? What is the pH of 0.50 M NaCN(aq)?

Solution

Na^+ does not hydrolyze but, as represented below, CN^- does to produce a basic solution. In the tabulation of the concentrations of the species involved in the hydrolysis reaction, we let $[OH^-] = x$.

▶ Solutions containing cyanide ion must be handled with extreme caution. They should only be handled in a fume hood by an operator wearing protective clothing.

▶ $Na^+(aq)$ is pH neutral. The aqueous solutions of some metal ions, however, are weakly acidic. These acidic metal ion solutions are discussed in Chapter 25.

$$\text{CN}^- \quad + \quad \text{H}_2\text{O} \rightleftharpoons \text{HCN} \quad + \quad \text{OH}^-$$

initial concn:	0.50 M		—	—
changes:	$-x$ M		$+x$ M	$+x$ M
equil concn:	$(0.50 - x)$ M		x M	x M

To obtain a value of K_b we need to use expression (17.18).

$$K_b = \frac{K_w}{K_a(\text{HCN})} = \frac{1.0 \times 10^{-14}}{6.2 \times 10^{-10}} = 1.6 \times 10^{-5}$$

Now we can return to the tabulated data.

$$K_b = \frac{[\text{HCN}][\text{OH}^-]}{[\text{CN}^-]} = \frac{x \cdot x}{0.50 - x} = \frac{x^2}{0.50 - x} = 1.6 \times 10^{-5}$$

▶The symbol $\ll$ means "much less than."

Next let's make a familiar assumption: $x \ll 0.50$ and $0.50 - x \approx 0.50$.

$$x^2 = 0.50 \times 1.6 \times 10^{-5} = 0.80 \times 10^{-5} = 8.0 \times 10^{-6}$$

$$x = [\text{OH}^-] = (8.0 \times 10^{-6})^{1/2} = 2.8 \times 10^{-3}$$

▶The value of x: 2.8×10^{-3}, is less than 1% of 0.500 M, and our assumption is valid.

$$\text{pOH} = -\log [\text{OH}^-] = -\log(2.8 \times 10^{-3}) = 2.55$$

$$\text{pH} = 14.00 - \text{pOH} = 14.00 - 2.55 = 11.45$$

Practice Example A: Sodium fluoride, NaF, is found in some toothpaste formulations as an anticavity agent. What is the pH of 0.10 M NaF(aq)?

Practice Example B: The pH of an aqueous solution of NaCN is found to be 10.38. What is $[\text{CN}^-]$ in this solution? (*Hint:* What are $[\text{OH}^-]$ and [HCN]?)

17-8 Molecular Structure and Acid–Base Behavior

We have now dealt with a number of aspects of acid–base chemistry, both qualitatively and quantitatively. Yet some very fundamental questions have not yet been answered, such as these: Why is HCl a strong acid, whereas HF is a weak acid? Why is acetic acid (CH_3COOH) a stronger acid than ethanol (CH_3CH_2OH), but a weaker acid than chloroacetic acid ($ClCH_2COOH$)?

These questions all involve relative acid strengths, and in this section we will examine the relationship between molecular structure and the strengths of acids and bases.

Strengths of Binary Acids

Because acid behavior requires the loss of a proton, we expect acid strength to be related to bond strength. It is an oversimplification, however, to try to relate the acidities of the HX binary acids to the mere breaking of the H—X bond. For one thing, bond energies are based on the dissociation of gaseous species, and here we are dealing with the ionization of species in solution. Nevertheless, it does seem reasonable that the stronger the H—X bond, the *weaker* the acid should be, and strong bonds are characterized by short bond lengths and high bond dissociation energies. For the binary acids of Group 7A, bond lengths *decrease* and bond dissociation energies *increase* in the following order

	HI		**HBr**		**HCl**		**HF**
bond length:	160.9	>	141.4	>	127.4	>	91.7 pm
bond dissociation energy:	297	<	368	<	431	<	569 kJ/mol

and acid strengths *decrease* in the order

$$\begin{array}{ccccccc} \textbf{HI} & & \textbf{HBr} & & \textbf{HCl} & & \textbf{HF} \\ K_a = 10^9 & > & 10^8 & > & 1.3 \times 10^6 & > & 6.6 \times 10^{-4} \end{array}$$

$$\underbrace{}_{\text{strong}} \qquad \underbrace{\phantom{6.6 \times 10^{-4}}}_{\text{weak}}$$

That HF is a weaker acid than the other hydrogen halides is expected, but that it should be so much weaker has always seemed an anomaly. Explanations of this behavior center on the tendency for hydrogen bonding in HF (recall Figure 13-27). For example, in HF(aq) ion pairs are held together by strong hydrogen bonds, which keeps the concentration of free H_3O^+ from being as large as otherwise expected.

$$HF + H_2O \longrightarrow (^-F \cdots H_3O^+) \rightleftharpoons H_3O^+ + F^-$$
$$\text{ion pair}$$

When comparing the strengths of binary acids within a *period,* bond polarity proves to be the predominant factor. The greater the electronegativity difference (ΔEN) in the bond H—X, the more polar is the bond. We might expect the loss of H^+ to a base to occur more readily from a polar bond, where partial ionic charges already exist, than from a nonpolar bond.

$$\begin{array}{ccccccc} & \textbf{CH}_4 & & \textbf{NH}_3 & & \textbf{H}_2\textbf{O} & & \textbf{HF} \\ \Delta EN: & 0.4 & < & 0.9 & < & 1.4 & < & 1.9 \end{array}$$

CH_4 and NH_3 do not have acidic properties in water. The acidity of H_2O is very limited ($K_w = 1.0 \times 10^{-14}$), but HF is an acid of moderate strength ($K_a = 6.6 \times 10^{-4}$).

Acidic, Basic, and Amphoteric Oxides

In Section 10-8 we characterized oxides as being acidic, basic, or amphoteric. Let us explore this scheme a little more fully in terms of the hypothetical oxide, E_xO_y, formed by the element E. Let us suppose that this oxide reacts with water to form a hydroxo compound—that is, a compound with an E—O—H linkage. There may also be other atoms or groups of atoms bonded to E, through the bonds represented by --- in the following equation.

$$E_xO_y + H_2O \longrightarrow \text{ } {\Large \diagdown}\!E\text{—O—H}$$

If the E atom has a high electronegativity, it attracts electrons away from the O—H bond in the E—O—H linkage. This weakens the O—H bond and causes the hydroxo compound to ionize as an *acid.* In the presence of a strong base such as OH^-, the products formed are H_2O and an oxoanion of the element E.

$$\text{ } {\Large \diagdown}\!E\text{—O—H} + OH^- \longrightarrow \left({\Large \diagdown}\!E\text{—O} \right)^- + H_2O$$

This is the expected behavior when E is a *nonmetal,* such as Cl, S, N, or P. In such a case, E_xO_y is an **acid anhydride,** and the hydroxo compound is an **oxoacid.**

If the E atom has a low electronegativity, bond breakage in the E—O—H linkage is in the E—O bond. The hydroxo compound ionizes to produce OH^-. The reaction of the hydroxo compound with an acid produces H_2O and a cation of the element E.

$$\overset{\diagdown}{\underset{\diagup}{E}}\text{—O—H} + H_3O^+ \longrightarrow \left(\overset{\diagdown}{\underset{\diagup}{E}}\right)^+ + 2\ H_2O$$

This is the type of behavior expected when E is a *metal.* E_xO_y is then a **base anhydride.** If the metal E has an especially low electronegativity (as do Na and K, for example) there is an actual transfer of an electron from E to the OH group in the hydroxo compound. That is, the compound is an ionic hydroxide—clearly a base (for example, NaOH or KOH).

In some cases the hydroxo compound may act either as an acid or a base. Here we say that the oxide is **amphoteric.** Amphoterism is associated with elements having electronegativities in an "intermediate" range (about 1.6–2.0).

The classification scheme we have been describing for element oxides is summarized in Table 17.5.

Strengths of Oxoacids

To describe the relative strengths of oxoacids we need to focus on the attraction toward the central atom of electrons from the O—H bond. The following factors promote this electron withdrawal from O—H bonds: (1) a high electronegativity (EN) of the central atom and (2) a large number of terminal O atoms in the acid molecule.

Neither HOCl nor HOI has a terminal O atom. The major difference between the two acids is one of electronegativity—Cl is more electronegative than I. As expected, HOCl is more acidic than HOI.

$$\text{H—O—Cl:} \qquad\qquad \text{H—O—I:}$$
$$EN_{Cl} = 3.0 \qquad\qquad EN_I = 2.5$$
$$K_a = 2.9 \times 10^{-8} \qquad K_a = 2.3 \times 10^{-11}$$

To compare the acid strengths of H_2SO_4 and H_2SO_3, we have to look beyond the central atom; it is S in each acid.

TABLE 17.5 Classification of Some Oxides

Acidic		Basic	Amphoteric	
Main group elements				
Cl_2O	P_4O_{10}	Na_2O	BeO	PbO
SO_2	CO_2	K_2O	Al_2O_3	Sb_2O_3
SO_3	SiO_2	MgO	SnO	
N_2O_5	B_2O_3	CaO		
Transition elements				
MoO_3		Sc_2O_3	ZnO	
WO_3		TiO_2	Cr_2O_3	
Mn_2O_7		ZrO_2		

$$
\begin{array}{c}
\ddot{\ddot{O}} \\
| \\
H-\ddot{\ddot{O}}-S-\ddot{\ddot{O}}-H \\
| \\
\ddot{\ddot{O}} \\
K_{a_1} \approx 10^3
\end{array}
\qquad
\begin{array}{c}
\ddot{\ddot{O}} \\
| \\
H-\ddot{\ddot{O}}-S-\ddot{\ddot{O}}-H \\
\\
K_{a_1} = 1.3 \times 10^{-2}
\end{array}
$$

We might expect, though, that a highly electronegative terminal O atom tends to withdraw electrons from the O—H bonds, weakening the bonds and increasing the acidity of the molecule. Because H_2SO_4 has *two* terminal O atoms whereas H_2SO_3 has only *one,* we should expect the electron-withdrawing effect to be greater in H_2SO_4. As a result, H_2SO_4 should be a stronger acid than H_2SO_3.

Strengths of Organic Acids

We conclude our discussion of the relationship between molecular structure and acid strength with a brief consideration of some organic compounds. Consider first the case of ethanol and acetic acid. Both have an O—H group bonded to a carbon atom, but acetic acid is a much stronger acid than ethanol.

$$
\begin{array}{c}
\quad H \quad \ddot{O} \\
\quad | \quad \| \\
H-C-C-\ddot{O}-H \\
\quad | \\
\quad H \\
\text{acetic acid} \\
K_a = 1.8 \times 10^{-5}
\end{array}
\qquad
\begin{array}{c}
\quad H \quad H \\
\quad | \quad | \\
H-C-C-\ddot{O}-H \\
\quad | \quad | \\
\quad H \quad H \\
\text{ethanol} \\
K_a = 1.3 \times 10^{-16}
\end{array}
$$

To rationalize the large difference in the acidities of these two compounds, we can say that the highly electronegative terminal O atom in acetic acid withdraws electrons from the O—H bond. The bond is weakened, and a proton (H^+) is lost more readily. Acetic acid is more acidic than ethanol. A more satisfactory explanation focuses on the anions formed in the ionization.

$$
\begin{array}{c}
H \quad \ddot{O} \\
| \quad / \\
H-C-C \\
| \quad \backslash \\
H \quad \ddot{O}^-
\end{array}
\longleftrightarrow
\begin{array}{c}
H \quad \ddot{O}^- \\
| \quad / \\
H-C-C \\
| \quad \backslash \\
H \quad \ddot{O}
\end{array}
\qquad
\begin{array}{c}
H \quad H \\
| \quad | \\
H-C-C-\ddot{O}^- \\
| \quad | \\
H \quad H
\end{array}
$$

acetate ion   ethoxide ion

We can write two plausible structures of the acetate ion, and these structures suggest that each carbon-to-oxygen bond is a "$\frac{3}{2}$" bond and each O atom carries "$\frac{1}{2}$" unit of negative charge. In short, the excess unit of negative charge in CH_3COO^- is spread out. This reduces the ability of either O atom to attach a proton and makes acetate ion only a weak Brønsted–Lowry base. In ethoxide ion, on the other hand, the unit of negative charge is localized on the single O atom. Ethoxide ion is a much stronger base than acetate ion. The stronger the conjugate base, the weaker the corresponding acid.

Other atoms or groups of atoms may also affect the acid strength of an organic molecule. If we replace one of the H atoms bonded to carbon in acetic acid with a Cl atom, the result is chloroacetic acid.

$$
\begin{array}{c}
\quad Cl \quad O \\
\quad | \quad \| \\
H-C-C-O-H \\
\quad | \\
\quad H \\
\text{chloroacetic acid} \\
K_a = 1.4 \times 10^{-3}
\end{array}
$$

▶ The group >C=O is called the *carbonyl* group,

$$
\begin{array}{c}
O \\
\| \\
-C-O-H
\end{array}
$$

and $-\overset{O}{\underset{\|}{C}}-O-H$ is the *carboxyl* group. Both of these groups are discussed in Chapter 27.

The highly electronegative Cl atom helps to draw electrons away from the O—H bond. The O—H bond is weakened, the proton is lost more readily, and the acid is a stronger acid than acetic acid. This effect falls off rapidly as the substituent atom or group is moved farther away from the O—H bond in an organic acid.

Example 17-14 illustrates some of the factors affecting acid strength discussed in this section.

EXAMPLE 17-14

Identifying Factors That Affect the Strengths of Acids. Explain which member of each of the following pairs is the stronger acid.

a. (I)   H—O—P—O—H   or   (II) $\ddot{\text{O}}$—Cl—O—H

(with phosphoric acid structure showing central P bonded to terminal O above and O—H below; and chloric acid structure)

b. (I) $\ddot{\text{Cl}}$—C—C—C—O—H   or   (II) H—C—C—C—O—H

(3-chloropropanoic acid and 2-chloropropanoic acid structures)

Solution

a. Phosphoric acid, H_3PO_4, has four O atoms to the three in $HClO_3$, but it is the number of *terminal* O atoms that we must consider, not just the total number of O atoms in the molecule. $HClO_3$ has *two* terminal O atoms and H_3PO_4 has *one*. Also, the Cl atom (EN = 3.0) is considerably more electronegative than the P atom (EN = 2.1). These facts point to chloric acid (II) being the stronger of the two acids. ($K_a \approx 5 \times 10^2$ for $HClO_3$ and $K_{a_1} = 7.1 \times 10^{-3}$ for H_3PO_4.)

b. The Cl atom withdraws electrons more strongly when it is directly adjacent to the carboxyl group. Compound (II) (2-chloropropanoic acid, $K_a = 1.4 \times 10^{-3}$) is a stronger acid than compound (I) (3-chloropropanoic acid, $K_a = 1.0 \times 10^{-4}$).

Practice Example A: Explain which is the stronger acid, HNO_3 or $HClO_4$; CH_2FCOOH or $CH_2BrCOOH$. (*Hint:* Draw plausible Lewis structures.)

Practice Example B: Explain which is the stronger acid, H_3PO_4 or H_2SO_3; CCl_3CH_2COOH or CCl_2FCH_2COOH. (*Hint:* Draw plausible Lewis structures.)

17-9   Lewis Acids and Bases

In the previous section we presented ideas about the molecular structures of acids and bases. An acid–base theory closely related to bonding and structure was proposed by G. N. Lewis in 1923. Lewis acid–base theory is not limited to reactions involving H^+ and OH^-, and it extends acid–base concepts to reactions in gases and in solids.

A **Lewis acid** is a species (an atom, ion, or molecule) that is an electron pair *acceptor,* and a **Lewis base** is a species that is an electron pair *donor.* A reaction between a Lewis acid and a Lewis base results in the formation of a covalent bond between them. In general, to identify Lewis *acids* we should look for species with vacant orbitals that can accommodate electron pairs, and for Lewis *bases,* species having lone pair electrons available for sharing.

By these definitions OH^-, a Brønsted–Lowry base, is also a Lewis base because of the presence of lone pair electrons on the O atom. So too is NH_3 a Lewis base. HCl, on the other hand, is not a Lewis acid: it is not an electron pair acceptor. We can think of HCl as producing H^+, however, and H^+ is a Lewis acid. H^+ seeks out an electron pair with which to form a coordinate covalent bond.

We might expect substances with an incomplete valence shell to be Lewis acids. When a coordinate covalent bond is formed with a Lewis base, the octet is completed. A good example of octet completion is the reaction of BF_3 and NH_3.

$$\text{:}\overset{..}{F}\text{:} \quad H \qquad \qquad \text{:}\overset{..}{F}\text{:}\; H$$
$$\text{:}F—B \quad + \quad \text{:}N—H \longrightarrow \text{:}F—B—N—H$$
$$\text{:}\underset{..}{F}\text{:} \quad H \qquad \qquad \text{:}\underset{..}{F}\text{:}\; H$$

One of the most important applications of the Lewis acid–base theory is in describing the formation of *complex ions.* Complex ions are polyatomic ions that contain a central metal ion to which other ions or small molecules are attached. For instance, Zn^{2+} combines with NH_3 to form the complex ion $[Zn(NH_3)_4]^{2+}$. We have indicated that NH_3 is a Lewis base; it has a lone pair of electrons available for forming a coordinate covalent bond. The central Zn^{2+} ion accepts the electrons; it is a Lewis acid. We will discuss the application of Lewis acid–base theory to complex ions in Chapter 25.

The reaction of lime (CaO) with sulfur dioxide is an important reaction for reducing SO_2 emissions from coal-fired power plants (see Figure 17-7). This reaction between a solid and a gas underscores that Lewis acid–base reactions can occur in all states of matter. The small curved arrow suggests that a rearrangement of an electron pair occurs in the Lewis structure in reaction (17.19).

$$Ca^{2+}\text{:}\overset{..}{O}\text{:}^{2-} \; + \; S\overset{O\text{:}}{\underset{\text{:}O\text{:}}{}} \longrightarrow Ca^{2+}\left[\text{:}\overset{\text{:}O\text{:}}{O—S}\text{:}\underset{\text{:}O\text{:}}{}\right]^{2-} \qquad (17.19)$$

EXAMPLE 17-15

Identifying Lewis Acids and Bases. According to the Lewis theory, each of the following is an acid–base reaction. Which species is the acid and which is the base?
a. $BF_3 + F^- \rightarrow BF_4^-$
b. $OH^-(aq) + CO_2(aq) \rightarrow HCO_3^-(aq)$

Solution

a. In BF_3 the B atom has a vacant orbital and an incomplete octet. The fluoride ion has an outer shell octet of electrons. BF_3 is the electron pair acceptor—the acid. F^- is the electron pair donor—the base.
b. We have already identified OH^- as a Lewis base, so we might suspect that it is the base and that $CO_2(aq)$ is the Lewis acid. We can see that this is the case with the following Lewis structures. As in reaction (17.19), a rearrangement of an electron pair at one of the double bonds is also required, as indicated by the arrow.

$$H—\overset{..}{O}\text{:}^- \; + \; \overset{O\text{:}}{\underset{O\text{:}}{C}} \longrightarrow \left[H—\overset{..}{O}—\overset{\text{:}O\text{:}}{\underset{O\text{:}}{C}}\right]^-$$

FIGURE 17-7
Fluidized-bed combustion.
As shown in this schematic diagram, powdered coal, limestone, and air are introduced into a combustion chamber, where water circulating through coils is converted to steam. Combustion is carried out at a relatively low temperature (760–870 °C), which minimizes the production of $NO(g)$ from $N_2(g)$ and $O_2(g)$ in the air. At the same time, $SO_2(g)$ from sulfur in the coal is trapped by reacting with $CaO(s)$ to form $CaSO_3(s)$ in reaction (17.19). Other components of the electric-power-generating system are also shown in the diagram.

Practice Example A: Identify the Lewis acids and bases in the reactions
(a) $BF_3 + NH_3 \rightarrow F_3BNH_3$
(b) $Cr^{3+} + 6\ H_2O \rightarrow [Cr(H_2O)_6]^{3+}$

Practice Example B: Identify the Lewis acids and bases in the reactions
(a) $Al(OH)_3 + OH^- \rightarrow [Al(OH)_4]^-$
(b) $SnCl_4 + 2\ Cl^- \rightarrow [SnCl_6]^{2-}$
(*Hint:* Draw plausible Lewis structures for the reactants and products.)

Summary

The Brønsted–Lowry theory describes an acid as a proton donor and a base as a proton acceptor. In an acid–base reaction a proton (H^+) is transferred from an acid to a base. In general, acid–base reactions are reversible, but equilibrium is displaced in the direction from the *stronger* acids and bases to their *weaker* conjugates.

In pure water and in aqueous solutions, self-ionization occurs to a very slight extent, producing H_3O^+ and OH^-, as described by the equilibrium constant K_w. The designations pH and pOH are often used to describe the concentrations of H_3O^+ and OH^- in aqueous solutions. Strong acids ionize completely in aqueous solutions to produce H_3O^+, and strong bases dissociate completely to produce OH^-. With weak acids and weak bases ionization is reversible and must be described through the ionization constants, K_a and K_b. In polyprotic acids, different constants K_{a_1}, K_{a_2}, . . . apply to each ionization step. In reactions between ions and water—hydrolysis reactions—the ions react as weak acids or weak bases. Calculations involving ionization equilibria are in many ways similar to those introduced in Chapter 16, although some additional considerations are necessary for polyprotic acids.

Molecular composition and structure are the keys to determining whether a substance is acidic, basic, or amphoteric.

In addition, molecular structure affects whether an acid or base is strong or weak. In assessing acid strength, for example, we must consider factors that affect the strength of the bond that must be broken to release H^+.

The Lewis acid–base theory views an acid as an electron pair acceptor and a base as an electron pair donor. Its greatest use is in situations that cannot be described through proton transfers, for example, in reactions involving gases and solids.

Summarizing Example

Sometimes the questions that prove to be most challenging do not require extensive calculations, but only a good insight into the concepts involved. The following question requires the application of many of the ideas discussed in this chapter.

Several aqueous solutions are prepared. *Without consulting any tables in the text,* arrange these solutions in order of *increasing* pH: 1.0 M NaBr, 0.05 M $HC_2H_3O_2$ (CH_3COOH), 0.05 M NH_3, 0.02 M $KC_2H_3O_2$, 0.05 M $Ba(OH)_2$, 0.05 M H_2SO_4, 0.10 M HI, 0.06 M NaOH, 0.05 M NH_4Cl, 0.05 M $HC_2H_2ClO_2$ ($CH_2ClCOOH$).

1. *Establish the ordering that is sought.* Low pH corresponds to strongly acidic solutions. Arranging the solutions by increasing pH means starting with the most acidic and proceeding to the most basic.
2. *Indicate whether each solution is acidic, basic, or neutral and why.* For example, 0.05 M $HC_2H_3O_2$ is a weak acid;

0.02 M $KC_2H_3O_2$ is basic because of hydrolysis of $C_2H_3O_2^-$; and so on. Result: *acidic*: 0.05 M $HC_2H_3O_2$, 0.05 M H_2SO_4, 0.10 M HI, 0.05 M NH_4Cl, 0.05 M $HC_2H_2ClO_2$; *basic*: 0.05 M NH_3, 0.02 M $KC_2H_3O_2$, 0.05 M $Ba(OH)_2$, 0.06 M NaOH; *neutral*: 1.0 M NaBr.
3. *Sort out the solutions in each category. Which is most acidic, next most acidic, and so on?* Thus, in 0.10 M HI, $[H_3O^+] = 0.10$ M. In 0.05 M H_2SO_4, 0.05 M $< [H_3O^+] <$ 0.10 M, because H_2SO_4 is a strong acid in the first ionization and weak in the second. $[H_3O^+]$ in 0.05 M $HC_2H_2ClO_2$ is greater than in 0.05 M $HC_2H_3O_2$ because of the electron-withdrawing effect of the Cl atom on the O—H bond. And so on.
4. *Combine the three sets of results from Step 3 into a final order of increasing pH. Answer:* 0.10 M HI $<$ 0.05 M H_2SO_4 $<$ 0.05 M $HC_2H_2ClO_2$ $<$ 0.05 M $HC_2H_3O_2$ $<$ 0.05 M NH_4Cl $<$ 1.0 M NaBr $<$ 0.05 M $KC_2H_3O_2$ $<$ 0.05 M NH_3 $<$ 0.06 M NaOH $<$ 0.05 M $Ba(OH)_2$.

Key Terms

acid anhydride (17-8)	**conjugate base** (17-2)	**pH** (17-3)
acid ionization constant, K_a (17-2, 17-5)	**hydrolysis** (17-7)	**pK** (17-5)
	hydronium ion, H_3O^+ (17-3)	**pOH** (17-3)
amphoterism (17-8)	**ion product of water, K_w** (17-3)	**polyprotic acid** (17-6)
base anhydride (17-8)	**Lewis acid** (17-9)	**proton acceptor** (17-2)
base ionization constant, K_b (17-2, 17-5)	**Lewis base** (17-9)	**proton donor** (17-2)
	oxoacid (17-8)	**self-ionization** (17-3)
conjugate acid (17-2)	**percent ionization** (17-5)	

Review Questions

1. In your own words define or explain the following terms or symbols: **(a)** K_w; **(b)** pH; **(c)** pK_a; **(d)** hydrolysis; **(e)** Lewis acid.
2. Briefly describe each of the following ideas or phenomena: **(a)** conjugate base; **(b)** percent ionization; **(c)** acid anhydride; **(d)** amphoterism.
3. Explain the important distinctions between each pair of terms: **(a)** Brønsted–Lowry acid and base; **(b)** $[H_3O^+]$ and

pH; **(c)** K_a for NH_4^+ and K_b for NH_3; **(d)** leveling effect and electron-withdrawing effect.
4. According to the Brønsted–Lowry theory, which of the following would you expect to be acids and which, bases? **(a)** HNO_2; **(b)** OCl^-; **(c)** NH_2^-; **(d)** NH_4^+; **(e)** $CH_3NH_3^+$.
5. Write the formula of the conjugate base in the reaction of each acid with water: **(a)** HIO_3; **(b)** C_6H_5COOH; **(c)** HPO_4^{2-}; **(d)** $C_2H_5NH_3^+$.

Focus On   Acid Rain

The photo on the left shows a marble statute of George Washington as it appeared in 1935. On the right, we see the same statue 59 years later. The effects of acid rain are clearly evident.

consider both the natural sources of acidity in rainwater and how human activities have contributed to acid rain.

$CO_2(g)$ is an acidic oxide. Although H_2CO_3 cannot be isolated, it is customary to consider that CO_2 reacts with water to form carbonic acid. The acidity of rainwater arises mostly through the first ionization step of this acid.

$$CO_2(g) + H_2O \longrightarrow H_2CO_3$$

$$H_2CO_3 + H_2O \rightleftharpoons H_3O^+ + HCO_3^-$$

$$K_{a_1} = 4.4 \times 10^{-7}$$

Rainwater in equilibrium with atmospheric $CO_2(g)$ at its normal partial pressure of 0.0036 atm has a pH of about 5.6.

$NO(g)$ is formed in lightning storms through the high-temperature reaction of $N_2(g)$ and $O_2(g)$ in air. This is followed by the air ox-

Pure water has pH = 7, but not rainwater—it is acidic. The natural pH of rainwater should be about 5.6. In recent years, however, the pH of rainfall in northeastern

United States has averaged about 4.5. Some rainfall has been found to be still more acidic, and in one episode on the West coast in 1982 the pH of fog was 1.7. Let us briefly

---

6. Calculate $[H_3O^+]$ and $[OH^-]$ for each solution: **(a)** 0.00165 M HNO_3; **(b)** 0.0087 M KOH; **(c)** 0.00213 M $Sr(OH)_2$; **(d)** 5.8×10^{-4} M HI.

7. What is the pH of each of the following solutions? **(a)** 0.0045 M HCl; **(b)** 6.14×10^{-4} M HNO_3; **(c)** 0.00683 M NaOH; **(d)** 4.8×10^{-3} M $Ba(OH)_2$.

8. *Without doing detailed calculations,* which value would you expect for $[H_3O^+]$ in 0.10 M H_2SO_4: 0.10 M, 0.05 M, 0.11 M, 0.20 M? Explain.

9. What is the pH of the solution obtained by mixing 24.80 mL of 0.248 M HNO_3 and 15.40 mL of 0.394 M KOH? (*Hint:* What reaction occurs? Is the final solution acidic, basic, or neutral?)

10. Only one of the following expressions is correct concerning 0.10 M $CH_3NH_2(aq)$: (1) $[H_3O^+] = 0.10$ M; (2) $[OH^-] = 0.10$ M; (3) pH < 7; (4) pH < 13. Choose the correct one and explain your choice.

11. A 625-mL sample of an aqueous solution containing 0.275 mol propionic acid, $HC_3H_5O_2$, has $[H_3O^+] = 0.00239$ M. What is the value of K_a for propionic acid?

$$HC_3H_5O_2 + H_2O \rightleftharpoons H_3O^+ + C_3H_5O_2^- \qquad K_a = ?$$

12. For the ionization of phenylacetic acid,

$$HC_8H_7O_2 + H_2O \rightleftharpoons H_3O^+ + C_8H_7O_2^- \qquad K_a = 4.9 \times 10^{-5}$$

(a) What is $[C_8H_7O_2^-]$ in 0.186 M $HC_8H_7O_2$?
(b) What is the pH of 0.121 M $HC_8H_7O_2$?

13. What mass of benzoic acid, $HC_7H_5O_2$, would you dissolve in 350.0 mL of water to produce a solution with a pH = 2.85?

$$HC_7H_5O_2 + H_2O \rightleftharpoons H_3O^+ + C_7H_5O_2^- \qquad K_a = 6.3 \times 10^{-5}$$

14. What must be the molarity of an aqueous solution of trimethylamine, $(CH_3)_3N$, if it is to have a pH = 11.12?

$$(CH_3)_3N + H_2O \rightleftharpoons (CH_3)_3NH^+ + OH^- \qquad K_b = 6.3 \times 10^{-5}$$

15. For 0.045 M H_2CO_3, a weak diprotic acid, calculate **(a)** $[H_3O^+]$, **(b)** $[HCO_3^-]$, and **(c)** $[CO_3^{2-}]$. Use data from Table 17.4 as necessary.

16. Complete the following equations in those instances in which a reaction (hydrolysis) will occur. If no reaction occurs, so state.

(a) $NH_4^+(aq) + NO_3^-(aq) + H_2O \rightarrow$
(b) $Na^+(aq) + NO_2^-(aq) + H_2O \rightarrow$
(c) $K^+(aq) + C_7H_5O_2^-(aq) + H_2O \rightarrow$
(d) $K^+(aq) + Cl^-(aq) + Na^+(aq) + I^-(aq) + H_2O \rightarrow$
(e) $C_6H_5NH_3^+(aq) + Cl^-(aq) + H_2O \rightarrow$

17. Calculate the pH of an aqueous solution that is 1.68 M NH_4Cl.

$$NH_4^+ + H_2O \rightleftharpoons H_3O^+ + NH_3 \qquad K_a = 5.6 \times 10^{-10}$$

idation of NO(g) to NO_2(g). Nitric acid forms by the reaction

$$3 NO_2(g) + H_2O(l) \longrightarrow 2 HNO_3(aq) + NO(g)$$

To some extent, then, naturally produced nitric acid contributes to the acidity of rain. More important, though, is HNO_3(aq) derived from artificial sources of NO(g), primarily high-temperature combustion processes in air, such as occur in automobile engines and power plants. In southern California, for example, where the oxides of nitrogen are seriously implicated in the formation of photochemical smog, HNO_3 is a major contributor to acid rain. On average, nitric acid accounts for about one-quarter of the acidity of acid rain.

SO_2(g) is normally only a trace constituent of the atmosphere, being produced mostly by biological decay and volcanic activity. Artifi-

cial sources of SO_2(g) are substantial, however, and create higher than normal concentrations in many areas. The chief source of SO_2 is the combustion of sulfur-containing coal (up to 8% S) in electric power plants. Another source is in the extraction of certain metals from their sulfide ores. In the atmosphere, SO_2(g) can be oxidized by any of several routes to SO_3(g), and SO_3 is the acid anhydride of sulfuric acid. Sulfuric acid accounts for well over one-half of the acidity of acid rain.

Some effects of acid rain are clearly visible in urban areas, particularly in the degradation of statues, monuments, and buildings made of sandstone, limestone, and marble. In India, acidic air pollutants from nearby industrial plants threaten to destroy the famous 350-year old marble mausoleum, the Taj Mahal. The most serious ecological

effects, though, are on freshwater lakes and streams and on forests. Fish and other aquatic life may be killed in acidic lakes. To maintain healthy forests the soil must furnish certain nutrients at the same time that it immobilizes other species, such as Al^{3+}, that interfere with the nutritional cycle of trees. Acidic rainwater often leaches important nutrients from the soil and at the same time frees up or mobilizes undesirable soil species.

Measures to reduce SO_2(g) and NO(g) emissions include the use of low-sulfur fuels, control of combustion temperatures to reduce NO(g) emissions, and trapping offending flue gases by various means. One promising pollution control method, fluidized bed combustion, was described in Figure 17-7 on page 612.

18. From data in Table 17.3, determine a value of **(a)** K_a for $C_5H_5NH^+$; **(b)** K_b for CHO_2^-; **(c)** K_b for $C_6H_5O^-$.
19. Determine the pH of 2.05 M $NaC_2H_2ClO_2$ (sodium chloroacetate). (*Hint:* Write an equation for the hydrolysis reaction. Use data from Table 17.3 to obtain a value of K_b for the chloroacetate ion.)
20. Explain why trichloroacetic acid, CCl_3COOH, is a stronger acid than acetic acid, CH_3COOH.
21. Which is the stronger acid of each of the following pairs of acids? Explain your reasoning. **(a)** HBr or HI; **(b)** HOClO or HOI; **(c)** $I_3CCH_2CH_2COOH$ or $CH_3CH_2CCl_2COOH$.

22. Indicate whether each of the following is a Lewis acid or base. **(a)** OH^-; **(b)** $(C_2H_5)_3B$; **(c)** CH_3NH_2. (*Hint:* You may find it helpful to draw Lewis structures.)
23. Each of the following is a Lewis acid–base reaction. Which reactant is the acid and which is the base? Explain.
(a) $SO_3 + H_2O \rightarrow H_2SO_4$
(b) $Zn(OH)_2(s) + 2 OH^-(aq) \rightarrow [Zn(OH)_4]^{2-}(aq)$
24. Indicate which of the following 0.10 M aqueous solutions has the highest pH and which has the lowest. Explain the basis of your reasoning. **(a)** NH_4Cl, **(b)** NH_3, **(c)** $NaC_2H_3O_2$, **(d)** KCl.

Exercises

Brønsted–Lowry Theory of Acids and Bases

25. For each of the following identify the acids and bases involved in both the forward and reverse directions.
(a) $HOBr + H_2O \rightleftharpoons OBr^- + H_3O^+$
(b) $HSO_4^- + H_2O \rightleftharpoons SO_4^{2-} + H_3O^+$
(c) $HS^- + H_2O \rightleftharpoons H_2S + OH^-$
(d) $C_6H_5NH_3^+ + OH^- \rightleftharpoons C_6H_5NH_2 + H_2O$

26. Which of the following species are *amphiprotic* in water solution? For each one that is, write two representative equations, one showing it acting as an acid and the other as a base: OH^-, NH_4^+, H_2O, HS^-, NO_2^-, HCO_3^-, HBr.

27. With which of the following bases will the ionization of acetic acid, $HC_2H_3O_2$, proceed furthest toward completion (to the right): (1) H_2O; (2) NH_3; (3) Cl^-; (4) NO_3^-?

29. With the aid of Table 17.1, predict the direction (forward or reverse) favored in each of the following acid–base reactions.
(a) $NH_4^+ + OH^- \rightleftharpoons H_2O + NH_3$
(b) $HSO_4^- + NO_3^- \rightleftharpoons HNO_3 + SO_4^{2-}$
(c) $CH_3OH + C_2H_3O_2^- \rightleftharpoons HC_2H_3O_2 + CH_3O^-$

28. In a similar manner to expression (17.6) represent the self-ionization of the following liquid solvents: (a) NH_3; (b) HF; (c) CH_3OH; (d) $HC_2H_3O_2$; (e) H_2SO_4.

30. With the aid of Table 17.1, predict the direction (forward or reverse) favored in each of the following acid–base reactions.
(a) $HC_2H_3O_2 + CO_3^{2-} \rightleftharpoons HCO_3^- + C_2H_3O_2^-$
(b) $HNO_2 + ClO_4^- \rightleftharpoons HClO_4 + NO_2^-$
(c) $H_2CO_3 + CO_3^{2-} \rightleftharpoons HCO_3^- + HCO_3^-$

Strong Acids, Strong Bases, and pH

31. Calculate $[H_3O^+]$ and pH in saturated $Ba(OH)_2(aq)$, which contains 3.9 g $Ba(OH)_2 \cdot 8H_2O$ per 100 mL of solution.

33. What is $[H_3O^+]$ in a solution obtained by dissolving 205 mL $HCl(g)$, measured at 23 °C and 751 mmHg, in 4.25 L of water solution?

35. How many milliliters of concentrated $HCl(aq)$ (36.0% HCl by mass, $d = 1.18$ g/mL) are required to produce 12.5 L of a solution with pH = 2.10?

37. What volume of 6.15 M $HCl(aq)$ is required to exactly neutralize 1.25 L of 0.265 M $NH_3(aq)$?

$$NH_3(aq) + H_3O^+(aq) \longrightarrow NH_4^+(aq) + H_2O$$

39. 50.00 mL of 0.0155 M $HI(aq)$ is mixed with 75.00 mL of 0.0106 M $KOH(aq)$. What is the pH of the final solution? (*Hint:* Is the final solution acidic, basic, or pH neutral?)

32. A saturated aqueous solution of $Ca(OH)_2$, a slightly soluble strong base, has a pH of 12.35. What is the solubility of $Ca(OH)_2$, expressed in milligrams per 100 mL of solution?

34. What is the pH of the solution obtained when 125 mL of 0.606 M NaOH is diluted to 15.0 L with water?

36. How many milliliters of a 15.0%, by mass, solution of $KOH(aq)$ ($d = 1.14$ g/mL) are required to produce 25.0 L of a solution with pH = 11.55?

38. A 28.2-L volume of $HCl(g)$, measured at 742 mmHg and 25.0 °C, is dissolved in water. What volume of $NH_3(g)$, measured at 762 mmHg and 21.0 °C, must be absorbed by the same solution to neutralize the HCl? (*Hint:* Does the volume of solution make any difference?)

40. 25.00 mL of a $HNO_3(aq)$ solution with a pH of 2.12 is mixed with 25.00 mL of a $KOH(aq)$ solution with a pH of 12.65. What is the pH of the final solution? (*Hint:* Is the final solution acidic, basic, or pH neutral?)

Weak Acids, Weak Bases, and pH

(Use data from Table 17.3 as necessary.)

41. What are the $[H_3O^+]$ and pH of 0.143 M HNO_2?

43. Fluoroacetic acid occurs in gifblaar, one of the most poisonous of all plants. A 0.318 M solution of the acid is found to have a pH = 1.56. Calculate K_a of fluoroacetic acid.
$$CH_2FCOOH(aq) + H_2O \rightleftharpoons$$
$$H_3O^+(aq) + CH_2FCOO^-(aq) \qquad K_a = ?$$

45. What are $[H_3O^+]$, $[OH^-]$, pH, and pOH of 0.55 M $HCHO_2$?

47. The solubility of 1-naphthylamine, $C_{10}H_7NH_2$, a substance used in the manufacture of dyes, is given in a handbook as 1 gram per 590 g H_2O. What is the approximate pH of a saturated aqueous solution of 1-naphthylamine?
$$C_{10}H_7NH_2 + H_2O \rightleftharpoons C_{10}H_7NH_3^+ + OH^- \qquad pK_a = 3.92$$

49. A particular vinegar is found to contain 5.7% acetic acid, $HC_2H_3O_2$, by mass. What mass of this vinegar should be diluted with water to produce 0.750 L of a solution with pH = 4.52?

42. What are the $[H_3O^+]$ and pH of 0.085 M $C_2H_5NH_2$?

44. Caproic acid, $HC_6H_{11}O_2$, found in small amounts in coconut and palm oils, is used in making artificial flavors. A saturated aqueous solution of the acid contains 11 g/L and has pH = 2.94. Calculate K_a for the acid.
$$HC_6H_{11}O_2 + H_2O \rightleftharpoons H_3O^+ + C_6H_{11}O_2^- \qquad K_a = ?$$

46. What are $[H_3O^+]$, $[OH^-]$, pH, and pOH of 0.386 M CH_3NH_2?

48. A saturated aqueous solution of *o*-nitrophenol, $HOC_6H_4NO_2$, has pH = 4.53. What is the solubility of *o*-nitrophenol in water, in grams per liter?
$$HOC_6H_4NO_2 + H_2O \rightleftharpoons H_3O^+ + {}^-OC_6H_4NO_2$$
$$pK_a = 7.23$$

50. A particular household ammonia solution ($d = 0.97$ g/mL) is 6.8% NH_3 by mass. How many milliliters of this solution should be diluted with water to produce 625 mL of a solution with pH = 11.55?

Percent Ionization

51. What is the (a) degree of ionization and (b) percent ionization of propionic acid in a solution that is 0.45 M $HC_3H_5O_2$?
$$HC_3H_5O_2 + H_2O \rightleftharpoons H_3O^+ + C_3H_5O_2^- \qquad pK_a = 4.89$$

52. What is the (a) degree of ionization and (b) percent ionization of trichloroacetic acid in a 0.035 M $HC_2Cl_3O_2$ solution?
$$HC_2Cl_3O_2 + H_2O \rightleftharpoons H_3O^+ + C_2Cl_3O_2^- \qquad pK_a = 0.52$$

53. What must be the molarity of an aqueous solution of NH_3 if it is to be 4.2% ionized?

54. Continuing the dilutions described in Example 17-8, should we expect to find that the percent ionization is 13% in 0.0010 M $HC_2H_3O_2$ and 42% in 0.00010 M $HC_2H_3O_2$? Explain.

Polyprotic Acids

(Use data from Table 17.4 as necessary.)

55. Explain why $[PO_4^{3-}]$ in 1.00 M H_3PO_4 is *not* simply $\frac{1}{3}[H_3O^+]$ but much, much less than $\frac{1}{3}[H_3O^+]$.

56. Cola drinks have a phosphoric acid content that is described as "from 0.057 to 0.084% of 75% phosphoric acid, by mass." Estimate the pH range of cola drinks corresponding to this range of H_3PO_4 content.

57. Determine $[H_3O^+]$, $[HS^-]$, and $[S^{2-}]$ for the following $H_2S(aq)$ solutions: **(a)** 0.075 M H_2S; **(b)** 0.0050 M H_2S; **(c)** 1.0×10^{-5} M H_2S.

58. Calculate $[H_3O^+]$, $[HSO_4^-]$, and $[SO_4^{2-}]$ in **(a)** 0.75 M H_2SO_4; **(b)** 0.075 M H_2SO_4; **(c)** 7.5×10^{-4} M H_2SO_4. (*Hint:* Check any assumptions that you make.)

59. Adipic acid, $HOOC(CH_2)_4COOH$, is among the top 50 manufactured chemicals in the U.S. (nearly 1 million tons annually). Its chief use is in the manufacture of nylon. It is a *diprotic* acid having $K_{a_1} = 3.9 \times 10^{-5}$ and $K_{a_2} = 3.9 \times 10^{-6}$. A saturated solution of adipic acid is about 0.10 M $HOOC(CH_2)_4COOH$. Calculate the concentration of each ionic species found in this solution. (*Hint:* Demonstrate that the usual simplifying assumptions work.)

60. The antimalarial drug quinine, $C_{20}H_{24}O_2N_2$, is a *diprotic base* with a water solubility of 1.00 g/1900 mL of solution. **(a)** Write equations for the ionization equilibria corresponding to $pK_{b_1} = 6.0$ and $pK_{b_2} = 9.8$. **(b)** What is the pH of saturated aqueous quinine?

Ions as Acids and Bases (Hydrolysis)

61. Predict whether a solution of each of the following salts is acidic, basic, or pH neutral: **(a)** KCl; **(b)** KF; **(c)** $NaNO_3$; **(d)** $Ca(OCl)_2$; **(e)** NH_4NO_2.

62. Arrange the following 0.010 M solutions in order of *increasing* pH: $NH_3(aq)$, $HNO_3(aq)$, $NaNO_2(aq)$, $HC_2H_3O_2(aq)$, $NaOH(aq)$, $NH_4C_2H_3O_2(aq)$, $NH_4ClO_4(aq)$.

63. What is the pH of an aqueous solution that is 0.089 M NaOCl?

64. What is the pH of an aqueous solution that is 0.123 M NH_4Cl?

65. Sorbic acid, $HC_6H_7O_2$ ($pK_a = 4.77$), is widely used in the food industry as a preservative. For example, its potassium salt (potassium sorbate) is added to cheese to inhibit the formation of mold. What is the pH of 0.37 M $KC_6H_7O_2(aq)$?

66. Pyridine, C_5H_5N ($pK_b = 8.82$), forms a salt, pyridinium chloride, as a result of a reaction with HCl. Write an ionic equation to represent the hydrolysis of the pyridinium ion, and calculate the pH of 0.0482 M $C_5H_5NH^+Cl^-(aq)$?

67. For each of the following ions write two equations: one showing its ionization as an acid and the other as a base: **(a)** HSO_3^-; **(b)** HS^-; **(c)** HPO_4^-. Then use data from Table 17.4 to predict whether each ion makes the solution acidic or basic.

68. It is desired to produce an aqueous solution of pH = 8.65 by dissolving one of the following salts in water. Which salt would you use and at what molarity? (a) NH_4Cl; (b) $KHSO_4$; (c) KNO_2; (d) $NaNO_3$.

Molecular Structure and Acid Strength

69. Predict which is the stronger acid **(a)** $HClO_2$ or $HClO_3$; **(b)** H_2CO_3 or HNO_2; **(c)** H_2SiO_3 or H_3PO_4. Explain.

70. Indicate which of the following is the *weakest* acid, and give reasons for your choice: HBr; $CH_2ClCOOH$; CH_3CH_2COOH; CH_2FCH_2COOH; CI_3COOH.

Lewis Theory of Acids and Bases

71. The three following reactions are acid–base reactions according to the Lewis theory. Draw Lewis structures and identify the Lewis acid and base in each reaction.
(a) $B(OH)_3 + OH^- \rightarrow [B(OH)_4]^-$
(b) $N_2H_4 + H_3O^+ \rightarrow N_2H_5^+ + H_2O$
(c) $(C_2H_5)_2O + BF_3 \rightarrow (C_2H_5)_2OBF_3$

72. $CO_2(g)$ can be removed from confined quarters (such as in spacecraft) by allowing it to react with an alkali metal hydroxide. Show that this is a Lewis acid–base reaction. For example,

$$CO_2(g) + LiOH(s) \longrightarrow LiHCO_3(s)$$

Advanced Exercises

73. The Brønsted–Lowry theory can be applied to acid–base reactions in *nonaqueous* solvents. In nonaqueous solvents the relative strengths of acids and bases can differ from what they are in aqueous solutions. Indicate whether each of the following would be an acid, a base, or amphiprotic in pure liquid acetic acid, $HC_2H_3O_2$, as a solvent. **(a)** $C_2H_3O_2^-$; **(b)** H_2O; **(c)** $HC_2H_3O_2$; **(d)** $HClO_4$. (*Hint:* Think of analogous situations in water. Also, refer to Table 17.1.)

74. A 10.0-mL sample of saturated $Ca(OH)_2(aq)$ is diluted to 250.0 mL in a volumetric flask. The solubility of $Ca(OH)_2$ is 0.17 g/100 mL soln. A 10.0-mL sample of the diluted $Ca(OH)_2(aq)$ is transferred to a beaker and some water is added. The resulting solution requires 25.1 mL of an HCl solution for its titration. What is the molarity of this HCl solution?

75. A handbook states that the solubility of methylamine, $CH_3NH_2(g)$, in water at 1 atm pressure and 25 °C is 959 volumes of the $CH_3NH_2(g)$ per volume of water.
(a) Estimate the maximum pH that can be attained by dissolving methylamine in water.
(b) What molarity $NaOH(aq)$ would be required to yield the same pH?

76. One handbook lists a value of 9.5 for pK_b of quinoline, C_9H_7N, a weak base used as a preservative for anatomical specimens and as an antimalarial. Another handbook lists the solubility of quinoline in water at 25 °C as 0.6 g/100 mL. Use this information to calculate the pH of a saturated solution of quinoline in water.

77. Show that when $[H_3O^+]$ is reduced to one-half of its original value, the pH of a solution increases by 0.30 unit, *regardless of the initial pH*. Is it also true that when any solution is diluted to one-half of its original concentration the pH increases by 0.30 unit? Explain.

78. Explain why $[H_3O^+]$ in a strong acid solution *doubles* as the total acid concentration doubles, whereas in a weak acid solution $[H_3O^+]$ increases only by about a factor of $\sqrt{2}$.

79. The pH of 1×10^{-2} M HCl is 2.0 and that of 1×10^{-4} M HCl is 4.0. Why is the pH of 1×10^{-8} M HCl *not* 8.0? Is the solution acidic? Explain.

80. To follow up on the question posed in Exercise 79, calculate the pH of a solution that is 1.5×10^{-7} M HCl. (*Hint:* Is it valid to assume that all H_3O^+ comes from the HCl?)

81. From the observation that 0.0500 M vinylacetic acid has a freezing point of -0.096 °C, determine K_a for this acid.
$$HC_4H_5O_2 + H_2O \rightleftharpoons H_3O^+ + C_4H_5O_2^- \qquad K_a = ?$$

82. You are asked to prepare a 100.0-mL sample of a solution with a pH of 5.50 by dissolving the appropriate amount of a solute in water with pH = 7.00. Which one of these solutes would you use and in what quantity? Explain your choice. (Note that there is more than one possibility.) **(a)** 15 M $NH_3(aq)$; **(b)** 12 M $HCl(aq)$; **(c)** $NH_4Cl(s)$; **(d)** glacial (pure) acetic acid, $HC_2H_3O_2$.

83. The solubility of $CO_2(g)$ in H_2O at 25 °C and under a $CO_2(g)$ pressure of 1 atm is 1.45 g CO_2/L. Air contains 0.036% CO_2, by volume. Use this information, together with data from Table 17.4, to show that rainwater saturated with CO_2 has a pH $\approx$ 5.6 (referred to in the Focus On feature on page 614 as the "normal" pH for rainwater). [*Hint:* Recall Henry's law. What is the partial pressure of $CO_2(g)$ in air?]

84. It is possible to write simple equations to relate pH, pK, and molarities (M) of various solutions. Three such equations are
Weak acid: $\quad pH = \frac{1}{2}pK_a - \frac{1}{2}\log M$
Weak base: $\quad pH = 14.00 - \frac{1}{2}pK_b + \frac{1}{2}\log M$
Salt of weak acid (pK_a) and strong base:
$$pH = 14.00 - \frac{1}{2}pK_w + \frac{1}{2}pK_a + \frac{1}{2}\log M$$
(a) Derive these three equations, and point out the assumptions involved in the derivations.
(b) Use these equations to determine the pH of 0.10 M $HC_2H_3O_2(aq)$, 0.10 M $NH_3(aq)$, and 0.10 M $NaC_2H_3O_2$. Verify that the equations give correct results by determining these pH values in the "usual" way.

85. A handbook lists the following formula for the percent ionization of a weak acid.
$$\% \text{ ionized} = \frac{100}{1 + 10^{(pK - pH)}}$$
(a) Derive this equation. What assumptions must you make in this derivation?
(b) Use the equation to determine the percent ionization of a formic acid solution, $HCOOH(aq)$, with a pH of 2.50.
(c) A 0.150 M solution of propionic acid has a pH of 2.85. What is K_a for propionic acid?
$$HC_3H_5O_2 + H_2O \rightleftharpoons H_3O^+ + C_3H_5O_2^- \qquad K_a = ?$$

86. The weak diprotic acid oxalic acid, $HOOCCOOH$, has $pK_{a_1} = 1.25$ and $pK_{a_2} = 3.81$. A related diprotic acid, suberic acid, $HOOC(CH_2)_8COOH$ has $pK_{a_1} = 4.21$ and $pK_{a_2} = 5.40$. Offer a plausible reason as to why the *difference* between pK_{a_1} and pK_{a_2} is so much greater for oxalic acid than for suberic acid.

87. What mass of acetic acid, $HC_2H_3O_2$, must be dissolved per liter of aqueous solution if the solution is to have the same freezing point as 0.150 M $HC_2H_2ClO_2$ (chloroacetic acid)?

88. What is the pH of a solution that is 0.68 M H_2SO_4 and 1.5 M $HCHO_2$ (formic acid)? (*Hint:* You must consider two ionization equilibria that occur simultaneously, but you can make some simplifying assumptions.)

Feature Problems

A. Maleic acid is a carbon-hydrogen-oxygen compound used in dyeing and finishing fabrics and as a preservative of oils and fats. In a combustion analysis, a 1.054-g sample of maleic acid yields 1.599 g CO_2 and 0.327 g H_2O. In a freezing-point

depression experiment, a 0.615-g sample of maleic acid dissolved in 25.10 g of glacial acetic acid, $CH_3COOH(l)$ (which has the freezing point depression constant $K_f = 3.90$ °C m^{-1} and in which maleic acid does not ionize), lowers the freezing point by 0.82 °C. In a titration experiment, a 0.4250-g sample of maleic acid is dissolved in water and requires 34.03 mL of 0.2152 M KOH for its complete neutralization. The pH of a 0.215-g sample of maleic acid dissolved in 50.00 mL of aqueous solution is found to be 1.80.

1. Determine the empirical and molecular formulas of maleic acid. [*Hint:* Which experiment(s) provide the necessary data?]
2. Use the results of part 1 and the titration data to rewrite the molecular formula to reflect the number of ionizable H atoms in the molecule.
3. Given that the ionizable H atom(s) is(are) associated with the carboxyl group(s), write the plausible condensed structural formula of maleic acid. (*Hint:* Recall the discussion on page 592 and the marginal note on page 609.)
4. Determine the ionization constant(s) of maleic acid. If the data supplied are insufficient, indicate what additional data would be needed.
5. Calculate the expected pH of a 0.0500 M aqueous solution of maleic acid? Indicate any assumptions required in this calculation.

B. In Example 17-7, rather than use the quadratic formula to solve the quadratic equation, we could have proceeded as follows: Substitute the value yielded by our failed assumption— $x = 0.0013$—into the *denominator* of the quadratic equation, and solve for a new value of x.

$$\frac{x^2}{0.00250 - 0.0013} = \frac{x^2}{0.0012} = 7.2 \times 10^{-4}$$

$$x = \sqrt{0.0012 \times 7.2 \times 10^{-4}} = 9.3 \times 10^{-4}.$$

When we repeat this procedure, we get

$$\frac{x^2}{0.00250 - 0.0009} = \frac{x^2}{0.0016} = 7.2 \times 10^{-4}$$

$$x = \sqrt{0.0016 \times 7.2 \times 10^{-4}} = 1.1 \times 10^{-3}.$$

A third attempt leads to

$$\frac{x^2}{0.00250 - 0.0011} = \frac{x^2}{0.0014} = 7.2 \times 10^{-4}$$

$$x = \sqrt{0.0014 \times 7.2 \times 10^{-4}} = 1.0 \times 10^{-3}.$$

This is the same value we got by using the quadratic formula in Example 17-7.

Use this method of successive approximations to determine the pH of 0.500 M $HClO_2$.
$$HClO_2 + H_2O \rightleftharpoons H_3O^+ + ClO_2^- \qquad K_a = 1.1 \times 10^{-2}$$

C. Here is an alternative method of solving equilibrium problems that is useful in more complex situations: (1) Identify the species present in any significant amounts (excluding H_2O molecules) and consider their concentrations as "unknowns." (2) Find the same number of equations as there are unknowns. (3) Solve the system of equations for the unknowns. The equa-

tions are of three types: (a) equilibrium constant expressions, (b) material balance equations, and (c) an electroneutrality condition. Consider the question: "What molarity $H_2SO_4(aq)$ solution will have a pH of 2.15?"

Possible Species
H_2SO_4, H_3O^+, HSO_4^-, SO_4^{2-}, $\cancel{OH^-}$.
We can eliminate H_2SO_4, because its ionization goes to completion in the first step. We can eliminate OH^-, because $[OH^-]$ is exceedingly small at pH = 2.15.

Unknowns
$\cancel{[H_3O^+]}$, $[HSO_4^-]$, $[SO_4^{2-}]$, and M [the molarity of the $H_2SO_4(aq)$].
We can easily eliminate $[H_3O^+]$: pH = $-\log [H_3O^+]$ = 2.15; $[H_3O^+] = 10^{-2.15} = 0.0071$. We are left with three unknowns.

Equations
(a) The K_a expression.
$$HSO_4^- + H_2O \rightleftharpoons H_3O^+ + SO_4^{2-}$$

$$\frac{[H_3O^+][SO_4^{2-}]}{[HSO_4^-]} = K_{a_2} = 1.1 \times 10^{-2} \qquad (a)$$

(b) Material balance. This equation accounts for the concentrations of the "sulfate-containing" species. The sum of concentrations is equal to the unknown solution molarity, M.
$$[HSO_4^-] + [SO_4^{2-}] = M \qquad (b)$$
(c) Electroneutrality condition. The solution must be electrically neutral. The positive charge must equal the sum of the negative charges. In this equation, $[SO_4^{2-}]$ is multiplied by *two* because each sulfate ion carries a charge of $2-$.
$$[H_3O^+] = [HSO_4^-] + 2 [SO_4^{2-}] = 0.0071 \qquad (c)$$

Solution
Solve equation (c) for $[HSO_4^-]$:
$[HSO_4^-] = 0.0071 - 2[SO_4^{2-}]$. Substitute this result, the known $[H_3O^+] = 0.0071$, and the unknown $[SO_4^{2-}]$ into equation (a).

$$\frac{0.0071 \times [SO_4^{2-}]}{0.0071 - 2[SO_4^{2-}]} = 1.1 \times 10^{-2}$$

Solve, to obtain the result $[SO_4^{2-}] = 0.0027$ M. Substitute this result into equation (c) to obtain $[HSO_4^-] = 0.0017$ M. Finally, according to equation (b), $[HSO_4^-] + [SO_4^{2-}] = 0.0044$ M. The required solution, then, is 0.0044 M $H_2SO_4(aq)$.

Apply the method outlined here to determine the pH of
1. A solution that is 0.315 M $HC_2H_3O_2$ and 0.250 M $HCHO_2$. (*Hint:* Make the usual simplifying assumptions, that is, $[C_2H_3O_2^-] \ll [HC_2H_3O_2]$ and $[CHO_2^-] \ll [HCHO_2]$.)
2. A solution that contains 1.55 g CH_3NH_2 and 12.5 g NH_3 in 375 mL. (*Hint:* Make the same assumptions as in part 1, that is, $[NH_4^+] \ll [NH_3]$ and $[CH_3NH_3^+] \ll [CH_3NH_2]$.)
3. 1.0 M $NH_4CN(aq)$. (*Hint:* You should identify six species and six equations. A simplifying assumption is that $[NH_4^+] = [CN^-]$.)

Additional Aspects of Acid–Base Equilibria

18

*I*n our study of acid rain (page 614) we learned that a very small amount of atmospheric $CO_2(g)$ dissolves in rainwater. Yet this is sufficient to lower the pH by nearly 2 units. And when acid-forming air pollutants, such as SO_2, SO_3, and NO_2, also dissolve in rainwater, it becomes even more acidic. A chemist would say that water has no "buffer capacity." Water is unable to resist a change in pH when acids or bases are dissolved in it.

One of the main topics of this chapter is buffer solutions—solutions that *are* able to resist a change in pH when acids or bases are added to them. We will consider how such solutions are prepared, how they maintain a nearly constant pH, and ways in which they are used. At the end of the chapter we will consider perhaps the most important buffer of all, that which maintains the constant pH of blood.

A second topic that we will explore is acid–base titrations. Here our aim is to calculate how pH changes during a titration. We can use this information to select an appropriate indicator for a titration and to determine, in general, which acid–base titrations work well and which do not. For the most part, we will find the calculations in this chapter to be extensions of those we did in Chapter 17.

HCl(aq) is slowly added to an aqueous solution containing the base NH_3 and the indicator methyl red. The indicator color changes from yellow to red as the pH changes from 6.2 to 4.4. The equivalence point of the neutralization is reached when the solution acquires an orange color. The selection of indicators for acid–base titrations is one of the topics considered in this chapter.

18-1 The Common Ion Effect in Acid–Base Equilibria

The questions we answered in Chapter 17 were mostly of the type, "What is the pH of 0.10 M $HC_2H_3O_2$, of 0.10 M NH_3, of 0.10 M H_3PO_4, of 0.10 M NH_4Cl?" In each of these cases we think of dissolving a *single* substance in aqueous solution and determining the concentrations of the species present at equilibrium. In most of this chapter we will study situations in which, in addition to a weak acid or weak base, a solution initially contains a second source of one of the ions produced in the ionization of the weak acid or weak base. We can say that this is an ion "common" to the weak acid or weak base, and the presence of this common ion can have some important consequences.

Solutions of Weak Acids and Strong Acids

Consider a solution that is at the same time 0.100 M $HC_2H_3O_2$ and 0.100 M HCl. We can write separate equations for the ionizations of the acids, one weak and the other strong.

$$HC_2H_3O_2 + H_2O \rightleftharpoons H_3O^+ + C_2H_3O_2^- \qquad K_a = 1.8 \times 10^{-5}$$
$$(0.100 - x)\text{ M} \qquad\qquad x \text{ M} \qquad x \text{ M}$$

$$HCl + H_2O \longrightarrow H_3O^+ + Cl^-$$
$$0.100 \text{ M} \qquad 0.100 \text{ M}$$

Of course, there can be only a single concentration of H_3O^+ in the solution, and this must be $[H_3O^+] = (0.100 + x)$ M. Because it is formed in both ionization processes, we say that H_3O^+ is a **common ion.** The weak acid/strong acid mixture described here is pictured in Figure 18-1. Although we might reasonably expect that the pH would be lower than 1.0, we can see from Figure 18-1 that this is not the case.

In Example 18-1 we calculate the concentrations of the species present in this mixture of a weak acid and a strong acid. Then we will comment on the significance of the result.

FIGURE 18-1
A weak acid/strong acid mixture.
The solution pictured is 0.100 M $HC_2H_3O_2$–0.100 M HCl. The reading on the pH meter (1.0) indicates that essentially all the H_3O^+ comes from the strong acid HCl. The red color of the solution is that of thymol blue indicator. Compare this figure with Figure 17-4, in which the separate acids 0.100 M $HC_2H_3O_2$ and 0.100 M HCl are shown.

EXAMPLE 18-1

Demonstrating the Common Ion Effect: A Solution of a Weak Acid and a Strong Acid.
(a) Determine $[H_3O^+]$ and $[C_2H_3O_2^-]$ in 0.100 M $HC_2H_3O_2$. (b) Then determine these same quantities in a solution that is 0.100 M in *both* $HC_2H_3O_2$ and HCl.

Solution

a. We did this calculation in Example 17-6 (see page 595). We found that in 0.100 M $HC_2H_3O_2$, $[H_3O^+] = [C_2H_3O_2^-] = 1.3 \times 10^{-3}$ M.

b. Instead of writing two separate ionization equations, as we did above, let's just write the ionization equation for $HC_2H_3O_2$ and enter information about the common ion, H_3O^+, in this format.

$$HC_2H_3O_2 \quad + \quad H_2O \rightleftharpoons \quad H_3O^+ \quad + \quad C_2H_3O_2^-$$

initial concn:			
weak acid:	0.100 M	—	—
strong acid:	—	0.100 M	—
changes:	$-x$ M	$+x$ M	$+x$ M
equil concn:	$(0.100 - x)$ M	$(0.100 + x)$ M	x M

As customary, we begin with the assumption that x is very small compared with 0.100. Thus, $0.100 - x \approx 0.100 + x \approx 0.100$.

$$K_a = \frac{[H_3O^+][C_2H_3O_2^-]}{[HC_2H_3O_2]} = \frac{(0.100 + x)(x)}{0.100 - x} = \frac{\cancel{0.100}\,(x)}{\cancel{0.100}} = 1.8 \times 10^{-5}$$

$$x = [C_2H_3O_2^-] = 1.8 \times 10^{-5}\ M \qquad 0.100 + x = [H_3O^+] = 0.100\ M$$

Practice Example A: Determine $[H_3O^+]$ and $[HF]$ in 0.500 M HF. Then determine these concentrations in a solution that is 0.100 M HCl and 0.500 M HF.

Practice Example B: How many drops of 12 M HCl would you add to 1.00 L of 0.100 M $HC_2H_3O_2$ to make $[C_2H_3O_2^-] = 1.0 \times 10^{-4}$ M? Assume 1 drop = 0.050 mL and that the volume of solution remains 1.00 L after dilution of the 12 M HCl. (*Hint:* What $[H_3O^+]$ must be maintained in the ionization equilibrium of $HC_2H_3O_2$? This $[H_3O^+]$ is established by the HCl.)

▶ By the criterion $M_a > 100K_a$ (see Are You Wondering on page 597), we expect the assumption to be valid. Actually, the assumption is even better than in Example 17-6 because the ionization of the weak acid is suppressed by the presence of the common ion, H_3O^+.

Now we see the consequence of adding a strong acid (HCl) to a weak acid ($HC_2H_3O_2$): The concentration of the anion $[C_2H_3O_2^-]$ is greatly reduced. In Example 18-1, between parts (a) and (b), $[C_2H_3O_2^-]$ is lowered from 1.3×10^{-3} M to 1.8×10^{-5} M—almost 100-fold. Another way to state this result is through Le Châtelier's principle. Increasing the concentration of one of the products of a reaction—the common ion—shifts the equilibrium condition in the *reverse* direction. The **common ion effect** is the suppression of the ionization of a weak electrolyte caused by the addition of an ion that is also a product of the ionization equilibrium of the weak electrolyte. The common ion effect of H_3O^+ on the ionization of acetic acid is suggested as follows.

$$HC_2H_3O_2 + H_2O \rightleftharpoons H_3O^+ + C_2H_3O_2^- \qquad K_a = 1.8 \times 10^{-5}$$

When a strong acid supplies the common ion H_3O^+, the equilibrium shifts.

The ionization of a weak base such as NH_3 is suppressed when a strong base such as NaOH is added. Here OH^- is the common ion, and its increased concentration shifts the equilibrium to the left.

$$NH_3 + H_2O \rightleftharpoons NH_4^+ + OH^- \qquad K_b = 1.8 \times 10^{-5}$$

When a strong base supplies the common ion OH^-, the equilibrium shifts.

Solutions of Weak Acids and Their Salts

The salt of a weak acid is a strong electrolyte—its ions become completely dissociated from one another in aqueous solution. One of the ions, the *anion*, is an ion common to the ionization equilibrium of the weak acid. The presence of this common ion suppresses the ionization of the weak acid. For example, we can represent the effect of acetate salts on the acetic acid equilibrium as

$$NaC_2H_3O_2(aq) \longrightarrow Na^+(aq) + C_2H_3O_2^-(aq)$$
$$HC_2H_3O_2 + H_2O \rightleftharpoons H_3O^+(aq) + C_2H_3O_2^-(aq) \qquad K_a = 1.8 \times 10^{-5}$$

When a salt supplies the common anion $C_2H_3O_2^-$, the equilibrium shifts.

The common ion effect of acetate ion on the ionization of acetic acid is pictured in Figure 18-2 and illustrated through Example 18-2. In solving "common ion"

FIGURE 18-2
A mixture of a weak acid and its salt.
Bromophenol blue indicator is present in both solutions. Its color dependence on pH is

pH < 3.0 <	pH < 4.6 <	pH
yellow	green	blue-violet

(a) 0.100 M $HC_2H_3O_2$ has a calculated pH of 2.89, but (b) if the solution is also 0.100 M in $NaC_2H_3O_2$ the calculated pH is 4.74. (The readability of the pH meters used here is 0.1 unit, and their accuracy is probably somewhat less than that.)

problems like Example 18-2, assume that ionization of the weak acid (or base) does not begin until both the weak acid (or base) and its salt have been placed in solution. Then consider that ionization occurs until equilibrium is reached.

EXAMPLE 18-2

Demonstrating the Common Ion Effect: A Solution of a Weak Acid and a Salt of the Weak Acid. Calculate $[H_3O^+]$ and $[C_2H_3O_2^-]$ in a solution that is 0.100 M in both $HC_2H_3O_2$ and $NaC_2H_3O_2$.

Solution

The set-up below is very similar to that in Example 18-1(b), except that here $NaC_2H_3O_2$ is the source of the common ion.

$$HC_2H_3O_2 \quad + \quad H_2O \rightleftharpoons H_3O^+ \quad + \quad C_2H_3O_2^-$$

initial concn:			
weak acid:	0.100 M	—	—
salt:	—	—	0.100 M
changes:	$-x$ M	$+x$ M	$+x$ M
equil concn:	$(0.100 - x)$ M	x M	$(0.100 + x)$ M

Because the salt suppresses the ionization of $HC_2H_3O_2$, we should expect $[H_3O^+] = x$ to be very small, and $0.100 - x \approx 0.100 + x \approx 0.100$. This proves to be a valid assumption.

$$K_a = \frac{[H_3O^+][C_2H_3O_2^-]}{[HC_2H_3O_2]} = \frac{(x)(0.100 + x)}{0.100 - x} = \frac{(x)\cancel{0.100}}{\cancel{0.100}} = 1.8 \times 10^{-5}$$

$$x = [H_3O^+] = 1.8 \times 10^{-5} \text{ M} \qquad 0.100 + x = [C_2H_3O_2^-] = 0.100 \text{ M}$$

As in Example 18-1, ionization of $HC_2H_3O_2$ is reduced about 100-fold. This time the large decrease in concentration is that of H_3O^+.

Practice Example A: Calculate $[H_3O^+]$ and $[CHO_2^-]$ in a solution that is 0.100 M $HCHO_2$ and 0.150 M $NaCHO_2$.

Practice Example B: What mass of $NaC_2H_3O_2$ should be added to 1.00 L of 0.100 M $HC_2H_3O_2$ to produce a solution with pH = 5.00? Assume that the volume remains 1.00 L. (*Hint:* What must $[C_2H_3O_2^-]$ be in this solution?)

(a) **(b)**

FIGURE 18-3
A mixture of a weak base and its salt.
Thymolphthalein indicator is colorless if pH < 10 and blue if pH > 10. (a) The pH of 0.100 M NH_3 is above 10 (calculated value: 11.11). (b) If the solution is also 0.100 M NH_4Cl, the pH drops below 10 (calculated value: 9.26). The ionization of NH_3 is suppressed in the presence of added NH_4^+. $[OH^-]$ decreases, $[H_3O^+]$ increases, and the pH is lowered.

Solutions of Weak Bases and Their Salts

The common ion effect of a salt of a weak base is similar to the weak acid/anion situation just described. The suppression of the ionization of NH_3 by the common *cation*, NH_4^+, is pictured in Figure 18-3 and represented as follows.

$$NH_4Cl(aq) \longrightarrow NH_4^+(aq) + Cl^-(aq)$$

$$NH_3 + H_2O \rightleftharpoons NH_4^+(aq) + OH^-(aq) \qquad K_b = 1.8 \times 10^{-5}$$

◀ *When a salt supplies the common cation NH_4^+, the equilibrium shifts.*

1.00 L of 0.001 M NaOH
with bromothymol blue

1.00 L of water at pH 7
with bromothymol blue

1.00 L of 0.001 M HCl
with bromothymol blue

pH

14 —
13 — B
12 — A
11 — S
10 — I
9 — C

8 —
Neutral 7 —
6 —
5 —
4 —
3 — A
2 — C
1 — I
0 — D
      C

40 mg NaOH
+
1.00 L H_2O
+
1.00 mL of 1 M HCl

FIGURE 18-4
Pure water has no buffering ability.
Bromothymol blue indicator is green at pH 7, yellow at pH < 7, and blue at pH > 7. Pure water has pH = 7.0. Addition of 0.001 mole H_3O^+ (1.00 mL of 1 M HCl) to 1.00 L water produces $[H_3O^+]$ = 0.001 M and pH = 3.0. Addition of 0.001 mol OH^- (40 mg · NaOH) to 1.00 L of water produces $[OH^-]$ = 0.001 M and pH = 11.0.

18-2 Buffer Solutions

Figure 18-4 illustrates a statement made in the chapter introduction: Pure water has no buffer capacity. There are some water solutions, however, called **buffer** (or **buffered**) **solutions,** whose pH values change only very slightly on the addition of small amounts of either an acid or a base.

What buffer solutions require are two components, one of which is able to neutralize acids, and the other able to neutralize bases. But, of course, the two components must not neutralize each other. This rules out mixtures of a strong acid and a strong base. Instead, common buffer solutions are described either as a mixture of

- *a weak acid and its conjugate base* or
- *a weak base and its conjugate acid.*

To show that such mixtures function as buffer solutions, let's consider a solution that has the equilibrium concentrations $[HC_2H_3O_2] = [C_2H_3O_2^-]$. As summarized in expression (18.1), in this solution $[H_3O^+] = K_a = 1.8 \times 10^{-5}$ M.

$$K_a = \frac{[H_3O^+][C_2H_3O_2^-]}{[HC_2H_3O_2]} = 1.8 \times 10^{-5}$$

$$[H_3O^+] = K_a \times \frac{[HC_2H_3O_2]}{[C_2H_3O_2^-]} = 1.8 \times 10^{-5} \text{ M} \qquad (18.1)$$

As a result, pH = $-\log [H_3O^+] = -\log K_a = -\log 1.8 \times 10^{-5} = 4.74$.

Now, imagine adding a *small* amount of a strong acid to this buffer solution. A reaction occurs in which a *small* amount of the base $C_2H_3O_2^-$ is converted to its conjugate acid $HC_2H_3O_2$.

$$C_2H_3O_2^- + H_3O^+ \longrightarrow HC_2H_3O_2 + H_2O$$

After the neutralization of the added H_3O^+, we find that in expression (18.1) $[HC_2H_3O_2]$ has increased *slightly* and $[C_2H_3O_2^-]$ has decreased *slightly*. The ratio $[HC_2H_3O_2]/[C_2H_3O_2^-]$ is only *slightly* greater than 1, and $[H_3O^+]$ has changed hardly at all. The buffer solution has resisted a change in pH following the addition of a small amount of acid; the pH remains close to the original 4.74.

Now, imagine adding a *small* amount of a strong base to the original buffer solution with $[HC_2H_3O_2] = [C_2H_3O_2^-]$. A reaction occurs in which a *small* amount of the weak acid $HC_2H_3O_2$ is converted to its conjugate base $C_2H_3O_2^-$.

$$HC_2H_3O_2 + OH^- \longrightarrow C_2H_3O_2^- + H_2O$$

Here we find that $[C_2H_3O_2^-]$ has increased *slightly* and $[HC_2H_3O_2]$ has decreased *slightly*. The ratio $[HC_2H_3O_2/[C_2H_3O_2^-]$ is only *slightly* smaller than 1, and again $[H_3O^+]$ has changed hardly at all. The buffer solution has resisted a change in pH following the addition of a small amount of base; again the pH remains close to the original 4.74.

Later in this section we will be more specific about what constitutes *small* additions of acid or base and *slight* changes in the concentrations of the buffer components and pH. Also we will discover that an acetic acid/sodium acetate buffer is only good for maintaining a nearly constant pH in a range of about 2 pH units centered on the pH = pK_a = 4.74. To prepare a buffer solution that maintains a nearly constant pH outside this range we must use different buffer components, as suggested in Example 18-3.

EXAMPLE 18-3

Predicting Whether a Solution is a Buffer Solution. Show that an NH_3–NH_4Cl solution is a buffer solution. In what pH range would you expect it to function?

Solution

To show that a solution has buffer properties, first identify a component in the solution that neutralizes acids and a component that neutralizes bases. In the present case these components are NH_3 and NH_4^+, respectively.

$$NH_3 + H_3O^+ \longrightarrow NH_4^+ + H_2O$$

$$NH_4^+ + OH^- \longrightarrow NH_3 + H_2O$$

In *all* aqueous solutions containing NH_3 and NH_4^+ we know that

$$NH_3 + H_2O \rightleftharpoons NH_4^+ + OH^- \qquad \text{and}$$

$$K_b = \frac{[NH_4^+][OH^-]}{[NH_3]} = 1.8 \times 10^{-5}$$

If a solution has approximately equal concentrations of NH_4^+ and NH_3, $[OH^-] \approx 1.8 \times 10^{-5}$ M; pOH ≈ 4.74; and pH ≈ 9.26. Ammonia/ammonium chloride solutions are *basic* buffer solutions that function in the pH range from about 8 to 10.

Practice Example A: Describe how a mixture of a strong acid (such as HCl) and the salt of a weak acid (such as $NaC_2H_3O_2$) can be a buffer solution. (*Hint:* What is the reaction that relates these two substances? Are certain proportions of the two substances needed in order to produce a buffer?)

Practice Example B: Describe how a mixture of NH_3 and HCl can result in a buffer solution. (*Hint:* What proportions of the two should you use?)

A common need is to calculate the pH of a buffer solution. At a minimum, such a calculation requires us to use the ionization constant expression for a weak acid or weak base. Aspects of solution stoichiometry may also be required. In Example 18-4, we first determine the stoichiometric concentrations of the buffer components. Then we perform an equilibrium calculation in the same fashion as in Examples 18-1 and 18-2.

EXAMPLE 18-4

Calculating the pH of a Buffer Solution. What is the pH of a buffer solution prepared by dissolving 25.5 g $NaC_2H_3O_2$ in a sufficient volume of 0.550 M $HC_2H_3O_2$ to make 500.0 mL of the buffer?

Solution

First we need to determine the molarity of $C_2H_3O_2^-$ corresponding to 25.5 g $NaC_2H_3O_2$ in 500.0 mL of solution.

$$\text{amount of } C_2H_3O_2^- = 25.5 \text{ g } NaC_2H_3O_2 \times \frac{1 \text{ mol } NaC_2H_3O_2}{82.04 \text{ g } NaC_2H_3O_2}$$

$$\times \frac{1 \text{ mol } C_2H_3O_2^-}{1 \text{ mol } NaC_2H_3O_2}$$

$$= 0.311 \text{ mol } C_2H_3O_2^-$$

$$[C_2H_3O_2^-] = \frac{0.311 \text{ mol } C_2H_3O_2^-}{0.500 \text{ L}} = 0.622 \text{ M } C_2H_3O_2^-$$

Equilibrium Calculation:

$$HC_2H_3O_2 \quad + \quad H_2O \rightleftharpoons H_3O^+ \quad + \quad C_2H_3O_2^-$$

initial concn:			
weak acid:	0.550 M	—	—
salt:	—	—	0.622 M
changes:	$-x$ M	$+x$ M	$+x$ M
equil concn:	$(0.550 - x)$ M	$+x$ M	$(0.622 + x)$ M

In our customary fashion, let's assume that x is very small so that $0.550 - x \approx 0.550$ and $0.622 + x \approx 0.622$. We will find this assumption to be valid.

$$K_a = \frac{[H_3O^+][C_2H_3O_2^-]}{[HC_2H_3O_2]} = \frac{(x)(0.622)}{0.550} = 1.8 \times 10^{-5}$$

$$x = [H_3O^+] = \frac{0.550}{0.622} \times 1.8 \times 10^{-5} = 1.6 \times 10^{-5}$$

$$pH = -\log [H_3O^+] = -\log(1.6 \times 10^{-5}) = 4.80$$

▶ Can you see that pH = 4.80 is a reasonable answer? We have seen that pH = pK_a = 4.74 when acetic acid and acetate ion are present in equal concentrations. Here, the concentration of the conjugate *base* (acetate ion) is greater than that of the acetic acid. The solution should be more basic (less acidic) than pH = 4.74.

Practice Example A: What is the pH of a buffer solution prepared by dissolving 23.1 g $NaCHO_2$ in a sufficient volume of 0.432 M $HCHO_2$ to make 500.0 mL of the buffer?

Practice Example B: A handbook states that to prepare a particular buffer solution mix 63.0 mL of 0.200 M $HC_2H_3O_2$ with 37.0 mL of 0.200 M $NaC_2H_3O_2$. What is the pH of this buffer? (*Hint:* What are the concentrations of the components in the resulting 100.0 mL of buffer solution?)

A point especially worth noting in Example 18-4 is that if a solution is to be an effective buffer, the assumptions $(M - x) \approx M$ and $(M + x) \approx M$ will always be valid. That is, the *equilibrium* concentrations of the buffer components will be very nearly the same as their *stoichiometric* concentrations. As a result, in Example 18-4 we could have gone directly from the stoichiometric concentrations of the buffer components to the expression

$$K_a = \frac{[H_3O^+][C_2H_3O_2^-]}{[HC_2H_3O_2]} = \frac{[H_3O^+](0.622)}{0.550} = 1.8 \times 10^{-5}$$

without setting up the ICE table. We can formalize this procedure through the special equation that we introduce next.

An Equation for Buffer Solutions: The Henderson–Hasselbalch Equation

Although we can continue to use the format illustrated in Example 18-4 for buffer calculations, it is often useful to describe a buffer solution through an equation known as the Henderson–Hasselbalch equation. This equation is commonly used by biochemists and molecular biologists, for example. To derive this variation of the ionization constant expression, let's consider a mixture of a hypothetical weak acid, HA (such as $HC_2H_3O_2$), and its salt, NaA (such as $NaC_2H_3O_2$). We start with the familiar expressions

$$HA + H_2O \rightleftharpoons H_3O^+ + A^-$$

$$K_a = \frac{[H_3O^+][A^-]}{[HA]}$$

and rearrange the right side of the K_a expression to obtain

$$K_a = [H_3O^+] \times \frac{[A^-]}{[HA]}$$

Next, we take the *negative logarithm* of each side of this equation.

$$-\log K_a = -\log [H_3O^+] - \log \frac{[A^-]}{[HA]}$$

Now, recalling that $pH = -\log [H_3O^+]$ and $pK_a = -\log K_a$,

$$pK_a = pH - \log \frac{[A^-]}{[HA]}$$

Then, we rearrange the equation to solve for pH.

$$pH = pK_a + \log \frac{[A^-]}{[HA]}$$

By recognizing that A^- is the conjugate base of the weak acid HA we can write the somewhat more general Equation (18.2).

$$pH = pK_a + \log \frac{[\text{conjugate base}]}{[\text{acid}]} \qquad (18.2)$$

To apply this equation to an acetic acid/sodium acetate buffer, we use pK_a for $HC_2H_3O_2$ and these concentrations: $[HC_2H_3O_2]$ for [acid] and $[C_2H_3O_2^-]$ for [conjugate base]. To apply it to an ammonia/ammonium chloride buffer, we use pK_a for NH_4^+ and these concentrations: $[NH_4^+]$ for [acid] and $[NH_3]$ for [conjugate base].

Equation (18.2) is useful only when we can substitute stoichiometric or initial concentrations for equilibrium concentrations, thus avoiding the need to set up an ICE table. This places important limitations on the equation's validity, however. Later, we will see that there are also conditions that must be met if a mixture is to be an effective buffer solution. Although the following rules may be overly restrictive in some cases, a reasonable approach to the twin concerns of effective buffer action and the validity of Equation (18.2) is to ensure that

1. the ratio [conjugate base]/[acid] is within the limits

$$0.10 < \frac{[\text{conjugate base}]}{[\text{acid}]} < 10 \qquad (18.3)$$

2. the molarity of each buffer component exceeds the value of K_a by a factor of at least 100.

▶ Another way to view the matter is that the Henderson–Hasselbalch equation works only for those cases in which we can make the simplifying assumption $M - x \approx M$. If a quadratic equation is required to solve the equilibrium constant expression, the Henderson-Hasselbalch equation will likely fail.

Preparing Buffer Solutions

Suppose we need a buffer solution with pH = 5.09. Equation (18.2) suggests two alternatives. One is to find a weak acid, HA, that has pK_a = 5.09 and prepare a solution with equal molarities of the acid and its salt.

$$pH = pK_a + \log \frac{[A^-]}{[HA]} = 5.09 + \log 1 = 5.09$$

Although this alternative is simple in concept, generally it is not practical. We are not likely to find a *readily available,* water-soluble weak acid with pK_a = 5.09. The second alternative is to use a cheap common weak acid like acetic acid, $HC_2H_3O_2$ (pK_a = 4.74), and establish an appropriate ratio of $[C_2H_3O_2^-]/[HC_2H_3O_2]$ to obtain a pH of 5.09.

EXAMPLE 18-5

Preparing a Buffer Solution of a Desired pH. What mass of $NaC_2H_3O_2$ must be dissolved in 0.300 L of 0.25 M $HC_2H_3O_2$ to produce a solution with pH = 5.09? (Assume that the solution volume remains constant at 0.300 L.)

Solution

Equilibrium among the buffer components is expressed through the equation

$$HC_2H_3O_2 + H_2O \rightleftharpoons H_3O^+ + C_2H_3O_2^- \qquad K_a = 1.8 \times 10^{-5}$$

and the ionization constant expression for acetic acid.

$$K_a = \frac{[H_3O^+][C_2H_3O_2^-]}{[HC_2H_3O_2]} = 1.8 \times 10^{-5}$$

Each of the three concentration terms appearing in a K_a expression should be an equilibrium concentration. The $[H_3O^+]$ corresponding to a pH of 5.09 is the equilibrium concentration. For $[HC_2H_3O_2]$ we will assume that the equilibrium concentration is equal to the stoichiometric or initial concentration. The value of $[C_2H_3O_2^-]$ that we calculate with the K_a expression is the equilibrium concentration, and we will assume that it is also the same as the initial concentration. The basis of these two assumptions is that neither the ionization of $HC_2H_3O_2$ to form $C_2H_3O_2^-$ nor the hydrolysis of $C_2H_3O_2^-$ to form $HC_2H_3O_2$ occurs to a sufficient extent to produce much of a difference between the initial and equilibrium concentrations of the buffer components. These assumptions work well if the conditions stated in expression (18.3) are met. The relevant concentration terms, then, are

$$[H_3O^+] = 10^{-pH} = 10^{-5.09} = 8.1 \times 10^{-6} \text{ M}$$

$$[HC_2H_3O_2] = 0.25 \text{ M}$$

$$[C_2H_3O_2^-] = ?$$

The required acetate ion concentration in the buffer solution is

$$[C_2H_3O_2^-] = K_a \times \frac{[HC_2H_3O_2]}{[H_3O^+]} = 1.8 \times 10^{-5} \times \frac{0.25}{8.1 \times 10^{-6}} = 0.56 \text{ M}$$

We complete the calculation of the mass of sodium acetate with some familiar ideas of solution stoichiometry.

$$\text{mass} = 0.300 \text{ L} \times \frac{0.56 \text{ mol } C_2H_3O_2^-}{1 \text{ L}} \times \frac{1 \text{ mol } NaC_2H_3O_2}{1 \text{ mol } C_2H_3O_2^-}$$

$$\times \frac{82.0 \text{ g } NaC_2H_3O_2}{1 \text{ mol } NaC_2H_3O_2} = 14 \text{ g } NaC_2H_3O_2$$

Practice Example A: How many grams of $(NH_4)_2SO_4$ must be dissolved in 0.500 L of 0.35 M NH_3 to produce a solution with pH = 9.00? (Assume that the solution volume remains at 0.500 L.)

Practice Example B: In Practice Example 18-3A we established that an appropriate mixture of a *strong* acid and the *salt* of a *weak* acid is a buffer solution. The formation of such a buffer solution is illustrated in Figure 18-5. Show that the solution in Figure 18-5(b) should have pH $\approx$ 5.1.

In Example 18-5 we achieved the desired ratio of $[C_2H_3O_2^-]/[HC_2H_3O_2]$ by adding 14 g of sodium acetate to the previously prepared 0.25 M $HC_2H_3O_2$ solution. This is a common method for obtaining a buffer solution. Other methods are sometimes useful. Sufficient NaOH(aq) could be added to $HC_2H_3O_2$(aq) to partially neutralize the acid, *producing* $C_2H_3O_2^-$ as a product. Or, as illustrated in Figure 18-5, enough $NaC_2H_3O_2$(s) can be added to HCl(aq) to convert all the HCl to $HC_2H_3O_2$ and leave some $C_2H_3O_2^-$ in excess.

Calculating pH Changes in Buffer Solutions

To calculate the change in pH produced by adding a small amount of a strong acid or base to a buffer solution, we must first use *stoichiometric* principles to establish how much of one buffer component is consumed and how much of the other is produced. Then we can use the new concentrations of weak acid (or weak base) and its salt to calculate the pH of the buffer solution. Essentially, we solve this problem in two steps. First, we allow the neutralization reaction

(a)

(b)

add 0.0060 mol H_3O^+

add 0.0060 mol OH^-

(c)

(d)

FIGURE 18-5
Formation and action of a buffer solution.
Thymol blue indicator.

pH < 1.2 < pH < 2.8 < pH

red   orange   yellow

(a) To 0.300 L of 0.250 M HCl (pH = 0.60) is added 33.05 g $NaC_2H_3O_2 \cdot 3H_2O$. (b) As a result of the reaction: $H_3O^+ + C_2H_3O_2^- \rightarrow HC_2H_3O_2 + H_2O$, the strong acid HCl is replaced by the weak acid $HC_2H_3O_2$, and excess $C_2H_3O_2^-$ remains. The calculated pH of the resulting buffer solution is 5.09. Pictured with the buffer are 5.00 mL of 1.2 M HCl (0.0060 mol H_3O^+) and 240 mg NaOH (0.0060 mol OH^-). (c) Adding 0.0060 mol H_3O^+ causes little change in the pH of the buffer. (d) Adding 0.0060 mol OH^- also has little effect on the pH.

to proceed *to completion*. Then we determine new stoichiometric concentrations. Second, we use the stoichiometric concentrations in the equilibrium constant expression to solve for $[H_3O^+]$ and then determine pH. This method is applied in Example 18-6, with the results illustrated in Figure 18-5.

EXAMPLE 18-6

Calculating pH Changes in a Buffer Solution. What is the effect on the pH of adding **(a)** 0.0060 mol HCl and **(b)** 0.0060 mol NaOH to 0.300 L of a buffer solution that is 0.250 M $HC_2H_3O_2$ and 0.560 M $NaC_2H_3O_2$? (This is the solution described in Figure 18-5.)

Solution

a. *Stoichiometric calculation:* All concentrations are converted to amounts in moles, and the neutralization is assumed to go to completion. Essentially, this is a limiting reactant calculation, but perhaps simpler than many of those in Chapter 4: In neutralizing the added H_3O^+, 0.0060 mol $C_2H_3O_2^-$ is converted to 0.0060 mol $HC_2H_3O_2$.

	$C_2H_3O_2^-$	+	H_3O^+	$\longrightarrow$	$HC_2H_3O_2$	+	H_2O
original buffer:	0.300 L × 0.560 M				0.300 L × 0.250 M		
	0.168 mol				0.0750 mol		
add:			0.0060 mol				
changes:	−0.0060 mol		−0.0060 mol		+0.0060 mol		
final buffer:							
amount	0.162 mol		(?)		0.0810 mol		
concn	0.162 mol/0.300 L		(?)		0.0810 mol/0.300 L		
	0.540 M				0.270 M		

Equilibrium Calculation: We can proceed as in Example 18-4 but substitute stoichiometric concentrations for the equilibrium concentrations in the K_a expression. Then we can calculate $[H_3O^+]$ and determine the pH.

$$K_a = \frac{[H_3O^+][C_2H_3O_2^-]}{[HC_2H_3O_2]} \quad \text{and} \quad [H_3O^+] = K_a \times \frac{[HC_2H_3O_2]}{[C_2H_3O_2^-]}$$

$$[H_3O^+] = 1.8 \times 10^{-5} \times \frac{0.270}{0.540} = 9.0 \times 10^{-6} \text{ M}$$

$$pH = -\log [H_3O^+] = -\log(9.0 \times 10^{-6}) = 5.05$$

▶ The equilibrium concentrations are $[HC_2H_3O_2] = 0.270 - x$ and $[C_2H_3O_2^-] = 0.540 + x$; the stoichiometric concentrations are 0.270 M and 0.540 M; and $x = [H_3O^+]$ is very small compared with these concentrations.

b. *Stoichiometric calculation:* In neutralizing the added OH^-, 0.0060 mol $HC_2H_3O_2$ is converted to 0.0060 mol $C_2H_3O_2^-$. The calculation of the new initial concentrations is shown on the last line of the following table.

	$HC_2H_3O_2$	+	OH^-	$\longrightarrow$	$C_2H_3O_2^-$	+	H_2O
original buffer:	0.300 L × 0.250 M				0.300 L × 0.560 M		
	0.0750 mol				0.168 mol		
add:			0.0060 mol				
changes:	−0.0060 mol		−0.0060 mol		+0.0060 mol		
final buffer:							
amount	0.0690 mol		(?)		0.174 mol		
concn	0.0690 mol/0.300 L		(?)		0.174 mol/0.300 L		
	0.230 M				0.580 M		

Equilibrium Calculation: This is the same type of calculation as in part (a), but here, as an alternative, let us use the Henderson–Hasselbalch Equation (18.2).

$$pH = pK_a + \log \frac{[C_2H_3O_2^-]}{[HC_2H_3O_2]}$$

$$= 4.74 + \log \frac{0.580}{0.230} = 4.74 + 0.40 = 5.14$$

Practice Example A: A 1.00-L volume of buffer is made with concentrations of 0.350 M $NaCHO_2$ (sodium formate) and 0.550 M $HCHO_2$ (formic acid). **(a)** What is the initial pH? **(b)** What is the pH after the addition of 0.0050 mol HCl(aq)? (Assume that the volume remains 1.00 L.) **(c)** What would be the pH after the addition of 0.0050 mol NaOH to the original buffer?

Practice Example B: How many, in milliliters of 6.0 M HNO_3 would you add to 300.0 mL of the buffer solution of Example 18-6 to change the pH from 5.09 to 5.03? (*Hint:* What must be the ratio $[C_2H_3O_2^-]/[HC_2H_3O_2]$ in the buffer *after* the added HNO_3 is neutralized?)

Perhaps you have already noticed a way to simplify the calculation in Example 18-6. Because the buffer components are always present in the same solution of volume V, we can substitute numbers of moles directly into Equation (18.2), without regard for the particular value of V. Thus, in Example 18-6(b),

$$\text{pH} = 4.74 + \log \frac{[C_2H_3O_2^-]}{[HC_2H_3O_2]} = 4.74 + \log \frac{0.174 \; \text{mol}/V}{0.0690 \; \text{mol}/V} = 4.74 + 0.40 = 5.14$$

This expression is also consistent with the observation that buffer solutions resist pH changes on dilution. Diluting a buffer solution means increasing its volume V by adding water. This action produces the same change in the numerator and the denominator of the ratio $[C_2H_3O_2^-]/[HC_2H_3O_2]$. The ratio itself remains unchanged, as does the pH.

Buffer Capacity and Buffer Range

It is not difficult to see that if we add more than 0.0750 mol OH^- to the buffer solution described in Example 18-6, the $HC_2H_3O_2$ will be completely converted to $C_2H_3O_2^-$, and an excess of OH^- will remain. The solution then becomes rather strongly basic.

Buffer capacity refers to the amount of acid or base that a buffer can neutralize before its pH changes appreciably. In general, the maximum buffer capacity exists when the concentrations of a weak acid and its conjugate base are kept *large* and *approximately equal to each other*. The **buffer range** is the pH range in which a buffer effectively neutralizes added acids and bases and maintains a fairly constant pH. As suggested by Equation (18.2),

▶ Consider as a "rule of thumb" that the amounts of the buffer components should be at least ten times as great as the amount of acid or base the buffer solution is meant to neutralize.

$$\text{pH} = pK_a + \log \frac{[\text{conjugate base}]}{[\text{acid}]}$$

when the ratio [conjugate base]/[acid] = 1, pH = pK_a. When the ratio falls to 0.10, the pH *decreases* by 1 pH unit from pK_a because log 0.10 = −1. If the ratio increases to a value of 10, the pH *increases* by 1 unit because log 10 = 1. For practical purposes this range of 2 pH units is the maximum range to which a buffer solution should be exposed. For acetic acid/sodium acetate buffers the effective range is about pH 3.7–5.7; for ammonia/ammonium chloride buffers, about pH 8.3–10.3.

▶ The buffer range for the ammonia/ammonium chloride solution is based on the equation
$$\text{pH} = 9.26 + \log\frac{[NH_3]}{[NH_4^+]}$$
where 9.26 is pK_a for NH_4^+. Note that $pK_a(NH_4^+) = 14.00 - pK_b(NH_3)$.

Applications of Buffer Solutions

An important example of a buffered system is that found in blood, which needs to be maintained at a pH of 7.4. We will consider the buffering of blood in the Focus On feature at the end of the chapter. But there are other important applications of buffers too.

Protein studies must often be performed in buffered media because the structures of protein molecules, including the magnitude and kind of electric charges they carry, depend on the pH (see Section 28-4). The typical enzyme is a protein capable of catalyzing a biochemical reaction, so enzyme activity is closely linked to protein structure and hence pH. Most enzymes in the body have their maximum activity between pH 6 and pH 8. Studying enzyme activity in the laboratory usually means working with media buffered in this pH range.

We will consider the importance of buffer solutions in solubility/precipitation processes in the next chapter.

18-3 Acid–Base Indicators

Several of the photographs in this and the preceding chapter have shown acid–base indicators in use. The indicator chosen depended on just how acidic or basic the solution was. In this section we will consider how an acid–base indicator works and how an appropriate indicator is selected for a pH measurement.

An **acid–base indicator** is a substance whose color depends on the pH of the solution to which it is added. The indicator exists in two forms: a weak acid, represented symbolically as HIn and having one color, and its conjugate base, represented as In^- and having a different color. When a small amount of indicator is added to a solution, the indicator does not affect the pH of the solution because so little of it is present. Instead, the ionization equilibrium of the indicator is itself affected by the prevailing $[H_3O^+]$ in solution. The indicator assumes either the acid (HIn) color or the base (In^-) color or a mixture of the two colors. The solution takes on the color that predominates in the indicator.

For the ionization equilibrium

$$HIn + H_2O \rightleftharpoons H_3O^+ + In^-$$
acid color   base color

▶ According to Le Châtelier's principle, increasing $[H_3O^+]$ displaces this reaction to the left (acid color). Decreasing $[H_3O^+]$ displaces the reaction to the right (base color).

we can write an equation similar to Equation (18.2), with pK_a of the indicator represented as pK_{HIn} and the substitutions $[HIn] = [acid]$ and $[In^-] = [conjugate base]$.

$$pH = pK_{HIn} + \log \frac{[In^-]}{[HIn]} \tag{18.4}$$

▶ The acid "color" of a few indicators is colorless.

In general, if 90% or more of an indicator is in the form HIn, the solution will take on the acid color. If 90% or more is in the form In^-, the solution takes on the base (or anion) color. If the concentrations of HIn and In^- are about equal, the indicator is in the process of changing from one form to the other and has an intermediate color. The complete change in color occurs over a range of about *2 pH units,* with pH = pK_{HIn} at about the middle of the range. The colors and pH ranges of several acid–base indicators are shown in Figure 18-6. A summary of these ideas is presented through the tabulated data following.

Acid color	Intermediate color	Base color
$[In^-]/[HIn] < 0.10$	$[In^-]/[HIn] \approx 1$	$[In^-]/[HIn] > 10$
$pH < pK_{HIn} + \log 0.10$	$pH \approx pK_{HIn} + \log 1$	$pH > pK_{HIn} + \log 10$
$pH < pK_{HIn} - 1$	$pH \approx pK_{HIn}$	$pH > pK_{HIn} + 1$

Example: *bromothymol blue,* $pK_{HIn} = 7.1$

pH < 6.1 (yellow)	pH ≈ 7.1 (green)	pH > 8.1 (blue)

An acid–base indicator is usually prepared as a solution (in water, ethanol, or some other solvent). In acid–base titrations a few drops of the indicator solution are added to the solution being titrated. In other applications, porous paper is impregnated with an indicator solution and dried. When this paper is moist-

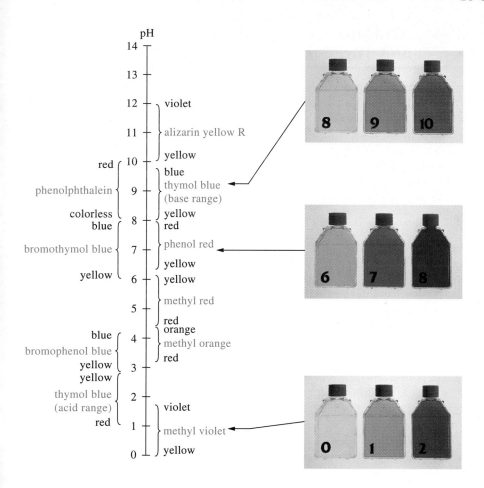

FIGURE 18-6
pH and color changes for some common acid–base indicators.
The indicators pictured and the pH values at which they change color are (a) thymol blue (pH 8–10); (b) phenol red (pH 6–8); (c) methyl violet (pH 0–2).

ened with the solution being tested, it acquires a color determined by the pH of the solution. This paper is usually called pH test paper.

Applications

Acid–base indicators find their greatest use when only an approximate pH determination is needed. For example, they are used in soil-testing kits to establish the pH of soils. Soils are usually acidic in regions of high rainfall and heavy vegetation and alkaline in more arid regions. The pH can vary considerably with local conditions, however. If a soil is found to be too acidic for a certain crop, its pH can be raised by adding slaked lime [$Ca(OH)_2$]. To reduce the pH of a soil, organic matter might be added.

In swimming pools, there is an optimal pH at which chlorinating agents are most effective, the growth of algae is avoided, and the corrosion of pool plumbing is minimized. This optimal pH is about 7.4, and phenol red is a common indicator used in testing swimming pool water. If chlorination is carried out with $Cl_2(g)$, the pool water becomes acidic because of the reaction of Cl_2 with H_2O: $Cl_2 + 2 H_2O \rightarrow H_3O^+ + Cl^- + HOCl$. In this case a basic substance such as sodium carbonate is used to raise the pH. Another widely used chlorinating agent is sodium hypochlorite, $NaOCl(aq)$, made by the reaction of $Cl_2(g)$ with *excess* $NaOH(aq)$: $Cl_2 + 2 OH^- \rightarrow Cl^- + OCl^- + H_2O$. The excess NaOH raises the pH of the pool water. The pH is adjusted by adding an acid such as HCl or H_2SO_4 to the pool.

18-4   Neutralization Reactions and Titration Curves

As we learned in our discussion of the stoichiometry of titration reactions (Section 5-7), the **equivalence point** of a neutralization reaction is the point in the reaction at which *both* acid and base have been consumed—that is, in which *neither* is in excess.

In a titration, one of the solutions to be neutralized—say, the acid—is placed in a flask or beaker together with a few drops of an acid–base indicator. Then the other solution (the base), contained in a buret, is added to the acid, first rapidly and then dropwise, up to the equivalence point (recall Figure 5-9). The equivalence point is located by noting the color change of the acid–base indicator. The point in a titration at which the indicator changes color is called the **endpoint** of the indicator. What is needed is to match the indicator endpoint with the equivalence point of the neutralization. We can do this by choosing an indicator whose color change occurs over a pH range that includes the pH of the equivalence point.

A graph of pH versus volume of titrant (the solution in the buret) is called a **titration curve.** Titration curves are most easily constructed by measuring the pH during a titration with a pH meter and plotting the data with a recorder. In this section we will emphasize the calculations necessary to establish the pH at various points in a titration. These calculations will serve as a review of aspects of acid–base equilibria considered in this and the preceding chapter.

▶ In a titration, the solution added from a buret is called the *titrant.*

▶ A different type of graph that is related to a titration curve is the subject of Feature Problem A on page 653.

The Millimole

In a typical titration the volume of solution delivered from a buret is less than 50 mL (more usually about 20–25 mL), and the molarity of the solution used for the titration is generally less than 1 M. The typical amount of OH^- (or H_3O^+) delivered from the buret during a titration is only a few thousandths of a mole, for example, 5.00×10^{-3} mol. In some calculations it is easier to work with millimoles instead of moles. The symbol **mmol** stands for a **millimole,** that is, one thousandth of a mole, or 10^{-3} mol.

We can derive an alternative definition of molarity by converting from moles to millimoles and from liters to milliliters.

$$M = \frac{mol}{L} = \frac{mol/1000}{L/1000} = \frac{mmol}{mL}$$

Thus, the expression from Chapter 4 that the amount of solute is the product of molarity and solution volume (page 110) can be based either on mol/L × L = mol or mmol/mL × mL = mmol.

Titration of a Strong Acid with a Strong Base

Suppose we place 25.00 mL of 0.100 M HCl (a *strong* acid) in a small flask or beaker and then add 0.100 M NaOH (a *strong* base) from a buret. We can calculate the pH of the accumulated solution at different points in the titration and plot these pH values against the volume of NaOH added. From this titration curve we can establish the pH at the equivalence point and identify an appropriate indicator for the titration. Some typical calculations are outlined in Example 18-7.

EXAMPLE 18-7

Calculating Points on a Titration Curve: Strong Acid Titrated with a Strong Base.
What is the pH at each of the following points in the titration of 25.00 mL of 0.100 M HCl with 0.100 M NaOH?
a. Before the addition of any NaOH (*initial pH*).
b. After the addition of 24.00 mL 0.100 M NaOH (*before equiv point*).
c. After the addition of 25.00 mL 0.100 M NaOH (*at equiv point*).
d. After the addition of 26.00 mL 0.100 M NaOH (*beyond equiv point*).

Solution

First, let's write the titration equation in the ionic and net ionic forms.

$$H_3O^+(aq) + Cl^-(aq) + Na^+(aq) + OH^-(aq) \longrightarrow$$

$$Na^+(aq) + Cl^-(aq) + 2\ H_2O(l)$$

$$H_3O^+(aq) + OH^-(aq) \longrightarrow 2\ H_2O(l)$$

a. Before any NaOH is added we are dealing with 0.100 M HCl. This solution has $[H_3O^+] = 0.100$ M and pH $= 1.00$.

b. The number of millimoles of H_3O^+ to be titrated is

$$25.00\ \text{mL} \times \frac{0.100\ \text{mmol}\ H_3O^+}{1\ \text{mL}} = 2.50\ \text{mmol}\ H_3O^+$$

The number of millimoles of OH^- present in 24.00 mL of 0.100 M NaOH is

$$24.00\ \text{mL} \times \frac{0.100\ \text{mmol}\ OH^-}{1\ \text{mL}} = 2.40\ \text{mmol}\ OH^-$$

Now we can represent the neutralization reaction in a familiar format.

	H_3O^+	+	OH^-	$\longrightarrow$	$2\ H_2O$
initially present:	2.50 mmol		—		
add:			2.40 mmol		
changes:	−2.40 mmol		−2.40 mmol		
after reaction:	0.10 mmol		≈0		

The remaining 0.10 mmol of H_3O^+ is present in 49.00 mL of solution (25.00 mL original acid + 24.00 mL added base).

$$[H_3O^+] = \frac{0.10\ \text{mmol}\ H_3O^+}{49.00\ \text{mL}} = 2.0 \times 10^{-3}\ \text{M}$$

$$\text{pH} = -\log\ [H_3O^+] = -\log(2.0 \times 10^{-3}) = 2.70$$

c. The equivalence point is the point at which the HCl is completely neutralized, and no excess NaOH is present. The solution at the equivalence point is simply NaCl(aq). And, as we learned in Section 17-7, because neither Na^+ nor Cl^- hydrolyzes in water, pH $= 7.00$.

d. To determine the pH of the solution beyond the equivalence point we can return to the format in **b**, except that now OH^- is in excess. The amount of OH^- added is 26.00 mL $\times$ 0.100 mmol/L $= 2.60$ mmol.

	H_3O^+	+	OH^-	$\longrightarrow$	$2\ H_2O$
initially present:	2.50 mmol		—		
add:			2.60 mmol		
changes:	−2.50 mmol		−2.50 mmol		
after reaction:	≈0		0.10 mmol		

The excess 0.10 mmol of NaOH is present in 51.00 mL of solution (25.00 mL original acid + 26.00 mL added base). The concentration of OH^- in this solution is

$$[OH^-] = \frac{0.10 \text{ mmol } OH^-}{51.00 \text{ mL}} = 2.0 \times 10^{-3} \text{ M}$$

$$pOH = -\log(2.0 \times 10^{-3}) = 2.70 \qquad pH = 14.00 - 2.70 = 11.30$$

Practice Example A: For the titration of 25.00 mL of 0.150 M HCl with 0.250 M NaOH calculate **(a)** the initial pH; **(b)** the pH when neutralization is 50.0% complete; **(c)** the pH when neutralization is 100.0% complete; and **(d)** the pH when 1.00 mL of NaOH is added beyond the equivalence point.

Practice Example B: For the titration of 50.00 mL of 0.00812 M $Ba(OH)_2$ with 0.0250 M HCl, calculate **(a)** the initial pH; **(b)** the pH when neutralization is 50.0% complete; **(c)** the pH when neutralization is 100.0% complete.

Figure 18-7 presents pH–volume data and the titration curve for the HCl–NaOH titration. From this figure we can establish these principal features of the titration curve for the titration of a *strong acid with a strong base.*

- The pH has a low value at the beginning of the titration.
- The pH changes slowly until just before the equivalence point.
- At the equivalence point the pH rises very sharply, perhaps by 6 units for an addition of only 0.10 mL (2 drops) of base.

▶A similar set of points can be made about the titration of a *strong base with a strong acid* by substituting pOH for pH in each statement. Titration curves involving pOH are described in Feature Problem B on page 653.

Titration Data	
mL NaOH(aq)	**pH**
0.00	1.00
10.00	1.37
20.00	1.95
22.00	2.19
24.00	2.70
25.00	7.00
26.00	11.30
28.00	11.75
30.00	11.96
40.00	12.36
50.00	12.52

FIGURE 18-7
Titration curve for the titration of a strong acid with a strong base—25.00 mL of 0.100 M HCl with 0.100 M NaOH.
The indicators whose color ranges fall along the steep portion of the titration curve are all suitable for this titration. Thymol blue changes color too soon and alizarin yellow-R, too late.

- Beyond the equivalence point the pH again rises only slowly.
- Any indicator whose color changes in the pH range from about 4 to 10 is suitable for this titration.

Titration of a Weak Acid with a Strong Base

Several important differences exist between the titration of a weak acid and a strong acid with a strong base, but one feature is *unchanged* when comparing the two: *For equal volumes of acid solutions of the same molarity, the volume of base required to titrate to the equivalence point is independent of whether the acid is strong or weak.* We can think of the neutralization of a weak acid such as $HC_2H_3O_2$ as involving the direct transfer of protons from $HC_2H_3O_2$ molecules to OH^- ions. In the neutralization of a strong acid the transfer is from H_3O^+ ions. In either case, the acid and base react in a 1:1 mole ratio.

$$HC_2H_3O_2 + OH^- \longrightarrow H_2O + C_2H_3O_2^-$$
$$H_3O^+ + OH^- \longrightarrow H_2O + H_2O$$

In Example 18-8 and in Figure 18-8 on page 641 we consider the titration of 25.00 mL of 0.100 M $HC_2H_3O_2$ with 0.100 M NaOH.

EXAMPLE 18-8

Calculating Points on a Titration Curve: Weak Acid Titrated with a Strong Base. What is the pH at each of the following points in the titration of 25.00 mL of 0.100 M $HC_2H_3O_2$ with 0.100 M NaOH?
a. Before the addition of any NaOH (*initial pH*).
b. After the addition of 10.00 mL 0.100 M NaOH (*before equiv point*).
c. After the addition of 12.50 mL 0.100 M NaOH (*half-neutralization*).
d. After the addition of 25.00 mL 0.100 M NaOH (*equiv point*).
e. After the addition of 26.00 mL 0.100 M NaOH (*beyond equiv point*).

Solution

a. The initial $[H_3O^+]$ is obtained by the calculation in Example 17-6 (page 595).
pH $= -\log(1.3 \times 10^{-3}) = 2.89$.
b. The number of millimoles of $HC_2H_3O_2$ to be neutralized is

$$25.00 \text{ mL} \times \frac{0.100 \text{ mmol } HC_2H_3O_2}{1 \text{ mL}} = 2.50 \text{ mmol } HC_2H_3O_2$$

At this point in the titration the number of millimoles of OH^- added is

$$10.00 \text{ mL} \times \frac{0.100 \text{ mmol } OH^-}{1 \text{ mL}} = 1.00 \text{ mmol } OH^-$$

The total solution volume = 25.00 mL original acid + 10.00 mL added base = 35.00 mL. We enter this information at appropriate points into the following setup.

Stoichiometric Calculation

	$HC_2H_3O_2$	+	OH^-	$\longrightarrow$	$C_2H_3O_2^-$	+	H_2O
initially present:	2.50 mmol		—		—		
add:			1.00 mmol				
changes:	−1.00 mmol		−1.00 mmol		+1.00 mmol		
after reaction:							
mmol	1.50 mmol				1.00 mmol		
concn	1.50 mmol/35.00 mL		≈0		1.00 mmol/35.00 mL		
	0.0429 M				0.0286 M		

▶ According to this analysis, can you see that the Henderson–Hasselbalch equation might be inaccurate in the very early stages of the titration and very near the equivalence point. In the first instance, $[C_2H_3O_2{}^-]/[HC_2H_3O_2] < 0.10$ and in the second, $[C_2H_3O_2{}^-]/[HC_2H_3O_2] > 10$.

Equilibrium Calculation: The most direct approach is to recognize that the acetic acid/sodium acetate solution is a buffer solution whose pH can be calculated with the Henderson–Hasselbalch equation. We are justified in using the equation because (a) the ratio: $[C_2H_3O_2{}^-]/[HC_2H_3O_2] = 0.0286/0.0429 = 0.667$ (satisfying the requirement on page 629 that it be between 0.10 and 10), and (b) $[C_2H_3O_2{}^-]$ and $[HC_2H_3O_2]$ exceed K_a (1.8×10^{-5}) by the factors 1.6×10^3 and 2.4×10^3, respectively (well above the factor of 100 cited on page 597).

$$pH = pK_a + \log \frac{[A^-]}{[HA]} = 4.74 + \log \frac{0.0286}{0.0429} = 4.74 - 0.18 = 4.56$$

Simpler still, would be to substitute the numbers of millimoles of $C_2H_3O_2{}^-$ and $HC_2H_3O_2$ directly into the Henderson–Hasselbalch equation, without converting to molarities. That is,

$$pH = pK_a + \log \frac{[A^-]}{[HA]} = 4.74 + \log \frac{1.00 \; \cancel{mmol}/V}{1.50 \; \cancel{mmol}/V} = 4.74 - 0.18 = 4.56$$

c. When we have added 12.50 mL of 0.100 M NaOH, we have added $12.50 \times 0.100 = 1.25$ mmol OH^-. As we see from the following set-up, this is enough base to neutralize exactly *one-half* of the acid.

	$HC_2H_3O_2$	+	OH^-	$\longrightarrow$	$C_2H_3O_2{}^-$	+	H_2O
initially present:	2.50 mmol		—		—		
add:			1.25 mmol				
changes:	−1.25 mmol		−1.25 mmol		+1.25 mmol		
after reaction:	1.25 mmol				1.25 mmol		

Again applying the Henderson–Hasselbalch equation,

$$pH = pK_a + \log \frac{[C_2H_3O_2{}^-]}{[HC_2H_3O_2]} = 4.74 + \log \frac{1.25 \; \cancel{mmol}/V}{1.25 \; \cancel{mmol}/V}$$

$$= 4.74 + \log 1 = 4.74$$

d. At the equivalence point, neutralization is complete and 2.50 mmol $NaC_2H_3O_2$ has been produced in 50.00 mL of solution, leading to 0.0500 M $NaC_2H_3O_2$. The question becomes "What is the pH of 0.0500 M $NaC_2H_3O_2$?" To answer this question we need to recognize that $C_2H_3O_2{}^-$ hydrolyzes (and Na^+ does not). The hydrolysis reaction and value of K_b are

$$C_2H_3O_2{}^- + H_2O \rightleftharpoons HC_2H_3O_2 + OH^-$$

$$K_b = \frac{K_w}{K_a} = \frac{1.0 \times 10^{-14}}{1.8 \times 10^{-5}} = 5.6 \times 10^{-10}$$

With a format similar to that used in the hydrolysis calculation of Example 17-13 (page 605), we obtain the following expression, where $x = [OH^-]$ and $x \ll 0.0500$.

$$K_b = \frac{[HC_2H_3O_2][OH^-]}{[C_2H_3O_2{}^-]} = \frac{x \cdot x}{0.0500 - x} = 5.6 \times 10^{-10}$$

$$x^2 = 2.8 \times 10^{-11} \qquad x = [OH^-] = 5.3 \times 10^{-6} \; M$$

$$pOH = -\log(5.3 \times 10^{-6}) = 5.28$$

$$pH = 14.00 - pOH = 14.00 - 5.28 = 8.72$$

e. The amount of OH^- added is $26.00 \; mL \times 0.100 \; mmol/mL = 2.60$ mmol. The volume of solution is 25.00 mL acid + 26.00 mL base = 51.00 mL. The 2.60 mmol OH^- neutralizes the 2.50 mmol of available acid, and 0.10 mmol OH^- remains in *excess*. Beyond the equivalence point, the pH of the solution is determined by the excess *strong* base.

$$[OH^-] = \frac{0.10 \text{ mmol OH}^-}{51.00 \text{ mL}} = 2.0 \times 10^{-3} \text{ M}$$

$$pOH = -\log(2.0 \times 10^{-3}) = 2.70 \qquad pH = 14.00 - 2.70 = 11.30$$

Practice Example A: A 20.00-mL sample of 0.150 M HF solution is titrated with 0.250 M NaOH. Calculate **(a)** the initial pH; and the pH when neutralization is **(b)** 25.0%, **(c)** 50.0%, and **(d)** 100.0% complete. (*Hint:* What is the initial amount of HF, and what amounts remain unneutralized at the points in question?)

Practice Example B: For the titration of 50.00 mL of 0.106 M NH_3 with 0.225 M HCl, calculate **(a)** the initial pH; and the pH when neutralization is **(b)** 25.0% complete; **(c)** 50.0% complete; **(d)** 100.0% complete.

Here are the principal features of the titration curve for a weak acid titrated with a strong base (see Figure 18-8).

1. The initial pH is higher (less acidic) than in the titration of a strong acid. (The weak acid is only partially ionized).

2. There is an initial increase in pH at the start of the titration. (The anion produced by the neutralization of the weak acid is a common ion that reduces the extent of ionization of the acid.)

3. Over a long section of the curve preceding the equivalence point the pH changes only gradually. (Solutions corresponding to this portion of the curve are buffer solutions.)

4. At the point of half-neutralization, pH = pK_a, because [HA] = [A$^-$].

FIGURE 18-8
Titration curve for the titration of a weak acid with a strong base—25.00 mL of 0.100 M $HC_2H_3O_2$ with 0.100 M NaOH.
Phenolphthalein is a suitable indicator for this titration, but methyl red is not. When exactly one-half of the acid is neutralized, [$HC_2H_3O_2$] = [$C_2H_3O_2^-$] and pH = pK_a = 4.74.

Titration Data	
mL NaOH(aq)	pH
0.00	2.89
5.00	4.14
10.00	4.57
12.50	4.74
15.00	4.92
20.00	5.35
24.00	6.12
25.00	8.72
26.00	11.30
30.00	11.96
40.00	12.36
50.00	12.52

▶ A similar set of points can be made about the titration of a *weak base with a strong acid* by substituting pOH for pH in each statement. Titration curves involving pOH are described in Feature Problem B on page 653.

5. The pH at the equivalence point is greater than 7. (The conjugate base of a weak acid hydrolyzes, producing OH^-.)

6. Beyond the equivalence point the titration curve is identical to that of a strong acid with a strong base. (In this portion of the titration the pH is established entirely by the concentration of unreacted OH^-.)

7. The steep portion of the titration curve at the equivalence point is over a relatively short pH range (from about pH 7 to 10).

8. The selection of indicators available for the titration is more limited than in a strong acid/strong base titration. (One cannot use an indicator whose color change occurs below pH 7.)

As illustrated in Example 18-8 and suggested by Figure 18-9, the necessary calculations for a weak acid/strong base titration curve are of four distinct types, depending on the portion of the titration curve being described. One type of titration that we generally cannot perform successfully is that of a weak acid with a weak base (or vice versa). The change in pH with volume of titrant is too gradual for us to locate the equivalence point with an acid–base indicator.

Titration of a Weak Polyprotic Acid

▶ This stepwise neutralization is observed only if successive ionization constants (K_{a_1}, K_{a_2}, ...) differ by a factor of 10^3 or more. Otherwise, the second neutralization step begins before the first step is completed, and so on.

The most striking evidence that a polyprotic acid ionizes in distinct steps comes through its titration curve. For a polyprotic acid, we expect to see a separate equivalence point for each acidic hydrogen. Thus, for H_3PO_4 we expect to see three equivalence points when it is titrated with NaOH(aq). In the neutralization of phosphoric acid by sodium hydroxide, essentially all the H_3PO_4 molecules are first converted to the salt, NaH_2PO_4. Then all the NaH_2PO_4 is converted to Na_2HPO_4; and finally the Na_2HPO_4 is converted to Na_3PO_4.

The titration of 10.0 mL of 0.100 M H_3PO_4 with 0.100 M NaOH is pictured in Figure 18-10. Notice that the first two equivalence points come at equal intervals on the volume axis, at 10.0 mL and at 20.0 mL. Although we expect a third equivalence point at 30.0 mL, it is not realized in this titration. The pH of the strongly hydrolyzed Na_3PO_4 solution at the third equivalence point—approaching 13—is higher than can be reached by adding 0.100 M NaOH to water. Na_3PO_4(aq) is nearly as basic as the NaOH(aq) used in the titration (see Section 18-5).

Let's focus on a few details of this titration. For each mole of H_3PO_4, 1 mol NaOH is required to reach the first equivalence point. At this first equivalence point, the solution is essentially NaH_2PO_4(aq). This is an *acidic* solution because $K_{a_2} > K_b$ for $H_2PO_4^-$: the reaction that produces H_3O^+ predominates over the one that produces OH^-.

FIGURE 18-9
Constructing the titration curve for a weak acid with a strong base.
The calculations needed to plot this graph, illustrated in Example 18-8, can be divided into *four* types.

- pH of a pure weak acid (initial pH).
- pH of a buffer solution of a weak acid and its salt (over a broad range before the equivalence point).
- pH of a salt solution undergoing hydrolysis (equiv point).
- pH of a solution of a strong base (over a broad range beyond the equiv point).

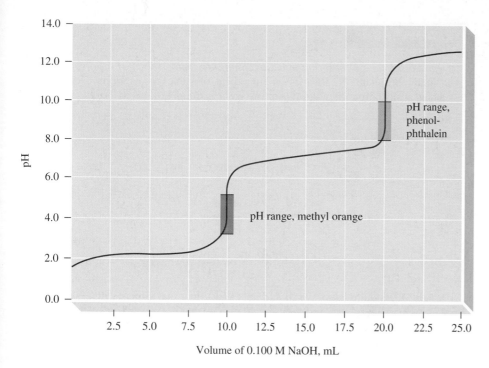

FIGURE 18-10
Titration of a weak polyprotic acid—10.0 mL of 0.100 M H_3PO_4 with 0.100 M NaOH.
A 10.0-mL volume of 0.100 M NaOH is required to reach the first equivalence point. The additional volume of 0.100 M NaOH required to reach the second equivalence point is also 10.0 mL.

$$H_2PO_4^- + H_2O \rightleftharpoons H_3O^+ + HPO_4^{2-} \qquad K_{a_2} = 6.3 \times 10^{-8}$$

$$H_2PO_4^- + H_2O \rightleftharpoons H_3PO_4 + OH^- \qquad K_b = 1.4 \times 10^{-12}$$

The pH at the equivalence point falls within the pH range over which the color of methyl orange indicator changes from red to orange.

An additional mole of NaOH is required to convert 1 mol $H_2PO_4^-$ to 1 mol HPO_4^{2-}. At this second equivalence point in the titration of H_3PO_4, the solution is *basic* because for HPO_4^{2-}, $K_b > K_{a_3}$.

$$HPO_4^{2-} + H_2O \rightleftharpoons H_2PO_4^- + OH^- \qquad K_b = 1.6 \times 10^{-7}$$

$$HPO_4^{2-} + H_2O \rightleftharpoons H_3O^+ + PO_4^{3-} \qquad K_{a_3} = 4.2 \times 10^{-13}$$

Phenolphthalein is an appropriate indicator for this equivalence point; its color change is from colorless to light pink.

▶ Other examples of titration curves with more than one equivalence point are presented in Feature Problem C on page 653.

18-5   Solutions of Salts of Polyprotic Acids

In discussing the neutralization of phosphoric acid by a strong base, we found that the first equivalence point should come in a somewhat acidic solution and the second in a mildly basic solution. We reasoned that the third equivalence point could be reached only in a strongly basic solution. The pH at this third equivalence point is not difficult to calculate. It corresponds to that of $Na_3PO_4(aq)$, and PO_4^{3-} can ionize (hydrolyze) only as a base.

$$PO_4^{3-} + H_2O \rightleftharpoons HPO_4^{2-} + OH^- \qquad K_b = K_w/K_{a_3}$$

$$= \frac{1.0 \times 10^{-14}}{4.2 \times 10^{-13}}$$

$$= 2.4 \times 10^{-2}$$

EXAMPLE 18-9

Determining the pH of a Solution Containing the Anion (A^{n-}) of a Polyprotic Acid. Sodium phosphate, Na_3PO_4, is an ingredient of some preparations used to clean painted walls before they are repainted. What is the pH of 1.0 M $Na_3PO_4(aq)$?

Solution

In the usual fashion we can write

$$PO_4^{3-} \; + \; H_2O \rightleftharpoons HPO_4^{2-} \; + \; OH^- \qquad K_b = 2.4 \times 10^{-2}$$

initial concn:	1.0 M	—	—
changes:	$-x$ M	$+x$ M	$+x$ M
equil concn:	$(1.0 - x)$ M	x M	x M

$$K_b = \frac{[HPO_4^{2-}][OH^-]}{[PO_4^{3-}]} = \frac{x \cdot x}{1.0 - x} = 2.4 \times 10^{-2}$$

Because K_b is quite large, we should not expect the usual simplifying assumption to work here. That is, x is *not* very much smaller than 1.0. Solution of the quadratic equation, $x^2 + 0.024x - 0.024 = 0$, yields $x = [OH^-] = 0.14$ M.

$$pOH = -\log [OH^-] = -\log 0.14 = +0.85$$

$$pH = 14.00 - 0.85 = 13.15$$

Practice Example A: Calculate the pH of an aqueous solution that is 1.0 M Na_2CO_3. (*Hint:* Use data from Table 17.4 to establish K_b for CO_3^{2-}.)

Practice Example B: Calculate the pH of an aqueous solution that is 0.500 M Na_2SO_3. (*Hint:* Use data from Table 17.4 to establish K_b for SO_3^{2-}.)

▶ The ratio $M_b/K_b = 1.0/0.024 = 42$. This is smaller than the minimum value of 100 that we have been using as the usual criterion.

▶ Sodium phosphate is often sold under the name *tri*sodium *phosphate* (TSP). One should always wear protective gloves when working with TSP. Fats and greases, including those found in human skin, are solubilized in strongly basic solutions.

It is more difficult to calculate the pH values of $NaH_2PO_4(aq)$ and $Na_2HPO_4(aq)$ than of $Na_3PO_4(aq)$. This is because with both $H_2PO_4^-$ and HPO_4^{2-} two equilibria must be considered *simultaneously:* ionization as an acid and ionization as a base (hydrolysis). For solutions that are reasonably concentrated (say 0.10 M or greater), the pH values prove to be *independent* of the solution concentration. Shown here are general expressions, printed in blue, and their application to $H_2PO_4^-(aq)$ and $HPO_4^{2-}(aq)$. (pK_a values are from Table 17.4.)

$$\text{for } H_2PO_4^-: \quad pH = \tfrac{1}{2}(pK_{a_1} + pK_{a_2}) = \tfrac{1}{2}(2.15 + 7.20) = 4.68 \qquad (18.5)$$

$$\text{for } HPO_4^{2-}: \quad pH = \tfrac{1}{2}(pK_{a_2} + pK_{a_3}) = \tfrac{1}{2}(7.20 + 12.38) = 9.79 \qquad (18.6)$$

18-6 Acid–Base Equilibrium Calculations: A Summary

In this and the preceding chapter we have considered a variety of acid–base equilibrium calculations. When you are faced with a new problem-solving situation, you may often find it helpful to relate the new problem to a type that you have encountered before. It is best not to rely exclusively on "labeling" a problem, however. Some problems may not fit a recognizable category. Instead, keep in mind some principles that apply regardless of the particular problem, as suggested by the following questions.

1. What are the species potentially present in solution, and how large are their concentrations likely to be?

   In a solution containing similar amounts of HCl and $HC_2H_3O_2$, the only significant *ionic* species are H_3O^+ and Cl^-. HCl is a completely ionized strong acid, and, in the presence of a strong acid, the weak acid $HC_2H_3O_2$ is only very slightly ionized because of the common ion effect. In a mixture containing similar amounts of two *weak* acids of similar strengths, such as $HC_2H_3O_2$ and HNO_2, each acid partially ionizes and all of these concentrations would be significant: $[HC_2H_3O_2]$, $[C_2H_3O_2^-]$, $[HNO_2]$, $[NO_2^-]$, and $[H_3O^+]$. In a solution containing phosphoric acid or a phosphate salt (or both), H_3PO_4, $H_2PO_4^-$, HPO_4^{2-}, PO_4^{3-}, OH^-, H_3O^+, and possibly other cations might be present. If the solution is simply $H_3PO_4(aq)$, however, the only species present in significant concentrations are those associated with the first ionization: H_3PO_4, H_3O^+, and $H_2PO_4^-$. On the other hand, if the solution is described as $Na_3PO_4(aq)$, the significant species are Na^+, PO_4^{3-}, and the ions associated with the hydrolysis of PO_4^{3-}—that is, HPO_4^{2-}, and OH^-.

2. Are there reactions possible among any of the solution components, and if so, what is their stoichiometry?

   If you are asked to calculate $[OH^-]$ in a solution made to be 0.10 M NaOH and 0.20 M NH_4Cl, before answering that $[OH^-] = 0.10$ M stop to consider whether a solution can be *simultaneously* 0.10 M in OH^- and 0.20 M in NH_4^+. It cannot; any solution containing both NH_4^+ and OH^- must also contain NH_3. The OH^- and NH_4^+ react in a 1:1 mole ratio until OH^- is almost totally consumed

$$NH_4^+ + OH^- \longrightarrow NH_3 + H_2O$$

   and you are now dealing with the *buffer* solution 0.10 M NH_3–0.10 M NH_4^+.

3. What are the equilibrium equations that apply to the particular situation? Which are the most significant?

   One equation that applies to all acids and bases in aqueous solutions is $K_w = [H_3O^+][OH^-] = 1.0 \times 10^{-14}$. In many calculations, however, this equation is not significant compared with others. One situation in which it *is* significant is in calculating $[OH^-]$ in an *acidic* solution or $[H_3O^+]$ in a *basic* solution. After all, an acid does not produce OH^- and a base does not produce H_3O^+. Another situation when K_w is likely to be significant is in a solution with a pH near 7.

   Often you will find that the ionization equilibrium with the largest K value is the most significant, but this will not always be the case. The amounts of the various species in solution are also an important consideration. When one drop of 1.00 M H_3PO_4 ($K_{a_1} = 7.1 \times 10^{-3}$) is added to 1.00 L of 0.100 M $HC_2H_3O_2$ ($K_a = 1.8 \times 10^{-5}$), the acetic acid ionization is most important in establishing the pH of the solution. The solution contains far more acetic acid than it does phosphoric acid.

Summary

The ionization of a weak acid, HA, is suppressed by the presence in solution of a common ion, either H_3O^+ (from a strong acid) or A^- (from a salt of the weak acid). The weak acid/salt combination is a *buffer* solution, as is also the combination weak base/salt. A buffer solution maintains a nearly constant pH in the presence of small added amounts of acids or bases.

The pH at which a buffer solution functions is determined by the pK_a value of the weak acid and the molarities of the weak acid and its conjugate base. The buffer exhibits its greatest capacity to neutralize added acids and bases when the concentrations of the buffer components are equal. The effective buffer range is about 1 pH unit on either side of the pK_a.

An acid–base indicator is a weak acid that has one color

Focus On   Buffers in Blood

Extreme hyperventilation is needed to compensate for the very low partial pressures of O_2 encountered by high-altitude mountain climbers. This leads to the condition called alkalosis. Climbers reaching the summit of Mt. Everest (8848 m = 29,028 ft) without supplemental oxygen have had their blood pH rise to 7.7 or 7.8.

One characteristic of blood that we rarely think about is its pH, yet maintaining the proper pH in blood and in intracellular fluids is crucial not only to human health but to life itself. This is primarily because the functioning of enzymes is sharply pH-dependent. The normal pH of blood is 7.4. Severe illness or death can result from sustained variations of just a few tenths of a pH unit.

Among the factors that can lead to *acidosis,* a condition in which the pH of blood decreases below normal, are heart failure, kidney failure, diabetes, persistent diarrhea, and a long-term high-protein diet. Prolonged, extensive exercise also can produce a temporary condition of acidosis. *Alkalosis,* a condition of an increased pH of blood, may occur as a result of severe vomiting, hyperventilation, or exposure to high altitudes (altitude sickness).

when present as the nonionized acid, HIn, and another color when present as the conjugate base, the anion In^-. The observed indicator color depends on the pH of the solution.

An acid–base titration curve is a graph of pH versus volume of titrating agent (titrant) added. At the equivalence point of the titration, the solution contains only the salt formed in the neutralization. Whether this solution is acidic, basic, or neutral depends on whether ions of the salt can ionize (hy-

drolyze) as acids or bases. Generally, a titration curve shows a sharp change in pH at the equivalence point. In the titration of a weak acid (or weak base) with a strong base (or strong acid), pH = pK_a at the point of half-neutralization. In the titration of a polyprotic acid, one steep portion of the curve is generally observed for each ionizable H atom. For a titration one chooses an acid–base indicator whose pK_{HIn} is close to the pH at the equivalence point.

Summarizing Example

An important use of an acid–base titration curve is to assist in selecting an indicator for the titration. For example, in Figure 18-7 several indicators are identified that can be used in the titration of HCl with NaOH.

The acid form (HIn) of the indicator bromocresol green has a yellow color, and its conjugate base form (In^-) is blue. The acid ionization constant $K_{HIn} = 2.1 \times 10^{-5}$. Is bromocresol green an appropriate indicator to use in the titration of 10.00 mL of 1.00 M NH_3 with 0.350 M HCl?

1. *Determine the color range for bromocresol green.* Determine pK_{HIn} for the indicator and recall how the color range is

established (see the table on page 634). *Result:* 3.7 < pH < 5.7.
2. *Determine the composition of the solution at the equivalence point.* The neutralization reaction is $NH_3(aq) + H_3O^+ \rightarrow NH_4^+(aq) + H_2O$. Determine the volume of 0.350 M HCl required for the titration, the amount of NH_4^+ present at the equivalence point, and then $[NH_4^+]$. *Result:* $[NH_4^+]$ = 0.259 M.
3. *Calculate the pH of a solution with $[NH_4^+]$ = 0.259 M.* The pH of the solution is governed by the acid ionization (hydrolysis) of NH_4^+.

$$NH_4^+ + H_2O \rightleftharpoons NH_3 + H_3O^+ \qquad K_a = K_w/K_b$$

Blood as a buffered solution

Human blood has a high buffer capacity. The addition of 0.01 mol HCl to 1 L of blood lowers the pH only from 7.4 to 7.2. The same amount of HCl added to a saline (NaCl) solution isotonic with blood lowers the pH from 7.0 to 2.0. The saline solution has no buffer capacity.

Several factors are involved in the control of blood pH. A particularly important one is the ratio of dissolved HCO_3^- (hydrogen carbonate ion) to H_2CO_3 (carbonic acid). Even though $CO_2(g)$ is only partially converted to H_2CO_3 when it dissolves in water, we generally treat the solution as if this conversion were complete. Moreover, although H_2CO_3 is a weak *diprotic* acid, we deal only with the first ionization step in the carbonic acid/hydrogen carbonate buffer system: H_2CO_3 is the weak acid and HCO_3^- is the conjugate base.

$$CO_2(g) + H_2O \longrightarrow H_2CO_3(aq)$$

$$H_2CO_3 + H_2O \rightleftharpoons H_3O^+ + HCO_3^-$$

$$K_{a_1} = 4.4 \times 10^{-7}$$

Carbon dioxide enters the bloodstream from tissues as the by-product of metabolic reactions. In the lungs $CO_2(g)$ is exchanged for $O_2(g)$, which is transported throughout the body by the blood.

Using Equation (18.2), a value of $pK_a = -\log(4.4 \times 10^{-7}) = 6.4$, and a normal blood pH of 7.4, we can write

$$pH = 7.4 = 6.4 + 1.0$$

$$= pK_{a_1} + \log \frac{[HCO_3^-]}{[H_2CO_3]}$$

$$= 6.4 + \log\left(\frac{10}{1}\right)$$

The large ratio of $[HCO_3^-]$ to $[H_2CO_3]$ (10:1) seems to place this buffer somewhat outside the range of its maximum buffer capacity. (Recall the discussion of buffer capacity on page 633.) The situation is rather complex but some of the factors involved are the following:

1. The need to neutralize excess acid (lactic acid produced by exercise) is generally greater than the need to neutralize excess base. The high proportion of HCO_3^- helps in this regard.

2. If additional H_2CO_3 is needed to neutralize excess alkalinity, $CO_2(g)$ in the lungs can be reabsorbed to build up the H_2CO_3 content of the blood.

3. Other components, such as the phosphate buffer system, $H_2PO_4^-/HPO_4^{2-}$, and some plasma proteins, contribute to maintaining blood pH at 7.4.

Use the method of Example 17-13. *Result:* pH = 4.92.
4. *Determine the suitability of the indicator.* Compare the pH at the equivalence point and the pH range for the indicator. *Answer:* Bromocresol green is a suitable indicator.

Key Terms

acid–base indicator (18-3)
buffer capacity (18-2)
buffer range (18-2)

buffer solution (18-2)
common ion effect (18-1)
endpoint (18-4)

equivalence point (18-4)
millimole (mmol) (18-4)
titration curve (18-4)

Review Questions

1. In your own words define or explain the following terms or symbols: (a) mmol; (b) HIn; (c) equivalence point of a titration; (d) a titration curve.
2. Briefly describe each of the following ideas, phenomena, or methods: (a) the common ion effect; (b) use of a buffer solution to maintain a constant pH; (c) determination of pK_a of a weak acid from a titration curve; (d) measurement of pH with an acid–base indicator.
3. Explain the important distinctions between each pair of terms: (a) buffer capacity and buffer range; (b) hydrolysis and neutralization; (c) first and second equivalence point in the titration of a weak diprotic acid; (d) equivalence point of a titration and endpoint of an indicator.
4. For a solution that is 0.275 M $HC_3H_5O_2$ (propionic acid, $K_a = 1.3 \times 10^{-5}$) and 0.0892 M HI, calculate (a) $[H_3O^+]$; (b) $[OH^-]$; (c) $[C_3H_5O_2^-]$; (d) $[I^-]$.
5. For a solution that is 0.164 M NH_3 and 0.102 M NH_4Cl, calculate (a) $[OH^-]$; (b) $[NH_4^+]$; (c) $[Cl^-]$; (d) $[H_3O^+]$.
6. Write equations to show how each of the following buffer solutions reacts with a small added amount of a strong acid or a strong base: (a) $HCHO_2$–$KCHO_2$; (b) $C_6H_5NH_2$–$C_6H_5NH_3^+Cl^-$; (c) KH_2PO_4–Na_2HPO_4.

7. Calculate the pH of a buffer that is (a) 0.012 M $HC_7H_5O_2$ ($K_a = 6.3 \times 10^{-5}$) and 0.033 M $NaC_7H_5O_2$; (b) 0.408 M NH_3 and 0.153 M NH_4Cl.

8. What concentration of formate ion, $[CHO_2^-]$, should be present in 0.366 M $HCHO_2$ to produce a buffer solution with pH = 4.06?

$$HCHO_2 + H_2O \rightleftharpoons H_3O^+ + CHO_2^- \qquad K_a = 1.8 \times 10^{-4}$$

9. What concentration of ammonia, $[NH_3]$, should be present in a solution with $[NH_4^+] = 0.732$ M to produce a buffer solution with pH = 9.12? $[K_b(NH_3) = 1.8 \times 10^{-5}]$

10. *Without performing detailed calculations,* determine which of the following will *raise* the pH of 1.00 L of 0.50 M HCl to the greatest extent: 0.40 mol NaOH; 0.50 mol $HC_2H_3O_2$; 0.60 mol $NaC_2H_3O_2$; 0.70 mol NaCl. Explain.

11. Lactic acid, $HC_3H_5O_3$, is found in sour milk. A solution containing 1.00 g $NaC_3H_5O_3$ in 100.0 mL of 0.0500 M $HC_3H_5O_3$ has pH = 4.11. What is K_a of lactic acid?

12. A $HCHO_2$–$NaCHO_2$ buffer solution is to be prepared $[K_a(HCHO_2) = 1.8 \times 10^{-4}]$.
(a) How many grams of $NaCHO_2$ must be dissolved in 0.250 L of 0.465 M $HCHO_2$ to produce a pH of 3.82?
(b) If to the 0.250 L of buffer solution in part (a) is added one small pellet of NaOH (0.20 g), what will be the new pH?

13. A handbook lists the following data.

Indicator	K_{HIn}	Color change Acid → Anion
bromophenol blue	1.4×10^{-4}	yellow → blue
bromocresol green	2.1×10^{-5}	yellow → blue
bromothymol blue	7.9×10^{-8}	yellow → blue
2,4-dinitrophenol	1.3×10^{-4}	colorless → yellow
chlorophenol red	1.0×10^{-6}	yellow → red
thymolphthalein	1.0×10^{-10}	colorless → blue

(a) Which of these indicators change color in acidic solution, which in basic solution, and which near the neutral point?
(b) What is the approximate pH of a solution if bromocresol green indicator assumes a green color; if chlorophenol red assumes an orange color?

14. With reference to the indicators listed in Exercise 13, what would be the color of each combination?
(a) 2,4-dinitrophenol in 0.100 M HCl(aq)
(b) chlorophenol red in 1.00 M NaCl(aq)
(c) thymolphthalein in 1.00 M NH_3(aq)
(d) bromocresol green in seawater (recall Figure 17-3)

15. What volume of 0.146 M KOH is needed for the *complete* neutralization of (a) 25.00 mL of 0.212 M HI; (b) 20.00 mL of 0.0942 M H_2SO_4.

16. Sketch the titration curves (pH vs. volume of titrant) that you would expect to obtain in the following titrations. Select a suitable indicator for each titration from Figure 18-6.
(a) NaOH(aq) is titrated with HNO_3(aq);
(b) NH_3(aq) is titrated with HCl(aq);
(c) $HC_2H_3O_2$(aq) is titrated with KOH(aq);
(d) NaH_2PO_4 is titrated with KOH(aq).

17. Calculate the pH at the points in the titration of 25.00 mL of 0.160 M HCl when (a) 10.00 mL and (b) 15.00 mL of 0.242 M KOH have been added.

18. Calculate the pH at the points in the titration of 20.00 mL of 0.275 M KOH when (a) 15.00 mL and (b) 20.00 mL of 0.350 M HCl have been added.

19. Calculate the pH at the points in the titration of 25.00 mL of 0.132 M HNO_2 when (a) 10.00 mL and (b) 20.00 mL of 0.116 M NaOH have been added. For HNO_2, $K_a = 7.2 \times 10^{-4}$.

$$HNO_2 + OH^- \longrightarrow H_2O + NO_2^-$$

20. Calculate the pH at the points in the titration of 20.00 mL of 0.318 M NH_3 when (a) 10.00 mL and (b) 15.00 mL of 0.475 M HCl have been added. For NH_3, $K_b = 1.8 \times 10^{-5}$.

$$NH_3(aq) + HCl(aq) \longrightarrow NH_4^+(aq) + Cl^-(aq)$$

21. A 25.00-mL sample of 0.0100 M $HC_7H_5O_2$ ($K_a = 6.3 \times 10^{-5}$) is titrated with 0.0100 M $Ba(OH)_2$. Calculate the pH (a) of the initial acid solution; (b) after the addition of 6.25 mL of 0.0100 M $Ba(OH)_2$; (c) at the equivalence point; (d) after the addition of a total of 15.00 mL of 0.0100 M $Ba(OH)_2$.

22. *Without performing detailed calculations,* determine which of the following 0.10 M aqueous solutions is the most acidic: Na_2S; $NaHSO_4$; $NaHCO_3$; Na_2HPO_4. Explain your choice.

Exercises

The Common Ion Effect

(Use data from Table 17.3, as necessary.)

23. Calculate the *change* in pH that results by adding (a) 0.100 mol $NaNO_2$ to 1.00 L of 0.100 M HNO_2(aq); (b) 0.100 mol $NaNO_3$ to 1.00 L of 0.100 M HNO_3(aq). Why are the changes not the same? Explain.

25. Calculate $[H_3O^+]$ in a solution that is (a) 0.035 M HCl and 0.075 M HOCl; (b) 0.100 M $NaNO_2$ and 0.0550 M HNO_2; (c) 0.0525 M HCl and 0.0768 M $NaC_2H_3O_2$. (*Hint:* What reaction occurs?)

24. In Example 17-8 we calculated the percent ionization of $HC_2H_3O_2$ in (a) 1.0 M; (b) 0.10 M; and (c) 0.010 M $HC_2H_3O_2$ solutions. Recalculate these percent ionizations if each solution is also made 0.10 M $NaC_2H_3O_2$. Explain why the results are different from those of Example 17-8.

26. Calculate $[OH^-]$ in a solution that is (a) 0.0062 M $Ba(OH)_2$ and 0.0105 M $BaCl_2$; (b) 0.315 M $(NH_4)_2SO_4$ and 0.486 M NH_3; (c) 0.196 M NaOH and 0.264 M NH_4Cl. (*Hint:* What reaction occurs?)

Buffer Solutions

(Use data from Tables 17.3 and 17.4, as necessary.)

27. Indicate which of the following aqueous solutions are buffer solutions and explain your reasoning. (*Hint:* Consider any reactions that may occur between solution components.)
(a) 0.100 M NaCl
(b) 0.100 M NaCl–0.100 M NH_4Cl
(c) 0.100 M CH_3NH_2–0.150 M $CH_3NH_3^+Cl^-$
(d) 0.100 M HCl–0.050 M $NaNO_2$
(e) 0.100 M HCl–0.200 M $NaC_2H_3O_2$
(f) 0.100 M $HC_2H_3O_2$–0.125 M $NaC_3H_5O_2$

29. What is the pH of a solution obtained by adding 1.15 mg of aniline hydrochloride ($C_6H_5NH_3^+Cl^-$) to 3.18 L of 0.105 M aniline ($C_6H_5NH_2$)? (*Hint:* Check any assumptions that you make.)

31. You wish to prepare a buffer solution with pH = 9.45.
(a) How many grams of $(NH_4)_2SO_4$ would you add to 425 mL of 0.258 M NH_3 to do this? Assume that the solution volume remains constant.
(b) Which buffer component, and how much (in grams), would you add to 0.100 L of the buffer in **(a)** to change its pH to 9.30? Assume that the solution volume remains constant.

33. If 0.55 mL of 12 M HCl is added to 0.100 L of the buffer solution in Exercise 31a, what will be the pH of the resulting solution?

35. You are asked to prepare a buffer solution with pH 3.50 and have available the following solutions, all 0.100 M: $HCHO_2$, $HC_2H_3O_2$, H_3PO_4, $NaCHO_2$, $NaC_2H_3O_2$, and NaH_2PO_4. Describe how you would prepare this buffer solution. (*Hint:* What volumes of which solutions would you use?)

37. Given 1.00 L of a solution that is 0.100 M $HC_3H_5O_2$ and 0.100 M $KC_3H_5O_2$,
(a) Over what pH range will this solution be an effective buffer?
(b) What is the buffer capacity of the solution? That is, how many millimoles of strong acid or strong base can be added to the solution before any significant change in pH occurs?

39. A solution of 75.0 mL volume is prepared containing 15.5 mmol $HCHO_2$ and 8.50 mmol $NaCHO_2$.
(a) What is the pH of this solution?
(b) If 0.25 mmol $Ba(OH)_2$ is added to the solution, what is the pH?
(c) If 1.05 mL of 12 M HCl is added to the original solution, what is the pH?

41. A handbook lists various procedures for preparing buffer solutions. To obtain a pH = 9.00, the handbook says to mix 36.00 mL of 0.200 M NH_3 with 64.00 mL of 0.200 M NH_4Cl.
(a) Show by calculation that the pH of this solution is 9.00.
(b) Would you expect the pH of this solution to remain at pH = 9.00 if the 100.00 mL of buffer solution were diluted to 1.00 L? To 1000 L? Explain.
(c) What will be the pH of the original 100.00 mL of buffer solution if 0.20 mL of 1.00 M HCl is added to it?
(d) What is the maximum volume of 1.00 M HCl that can be added to 100.00 mL of the original buffer solution so that the pH does not drop below 8.90?

28. The $H_2PO_4^-$/HPO_4^{2-} combination plays a role in maintaining the pH of blood.
(a) Write equations to show how a solution containing these ions functions as a buffer.
(b) Verify that this buffer is most effective at pH 7.2.
(c) Calculate the pH of a buffer solution in which $[H_2PO_4^-]$ = 0.050 M and $[HPO_4^{2-}]$ = 0.150 M. (*Hint:* Focus on the second step of the phosphoric acid ionization.)

30. What is the pH of a solution prepared by dissolving 8.50 g of aniline hydrochloride ($C_6H_5NH_3^+Cl^-$) in 750 mL of 0.215 M aniline ($C_6H_5NH_2$)? Would this solution be an effective buffer? Explain. (*Hint:* Check any assumptions you make.)

32. You prepare a buffer solution by dissolving 2.00 g each of benzoic acid, $HC_7H_5O_2$, and sodium benzoate, $NaC_7H_5O_2$, in 750.0 mL of water.
(a) What is the pH of this buffer? Assume that the solution volume is 750.0 mL.
(b) Which buffer component, and how much (in grams), would you add to the 750.0 mL of buffer solution to change its pH to 4.00?

34. If 0.35 mL of 15 M NH_3 is added to 0.750 L of the buffer solution in Exercise 32, what will be the pH of the resulting solution?

36. You are asked to reduce the pH of the 0.300 L of buffer solution in Example 18-5 from 5.09 to 5.00. How many milliliters and of which of these solutions would you use: 0.100 M NaCl, 0.150 M HCl, 0.100 M $NaC_2H_3O_2$, 0.125 M NaOH? Explain your reasoning.

38. Given 125 mL of a solution that is 0.0500 M CH_3NH_2 and 0.0500 M $CH_3NH_3^+Cl^-$,
(a) Over what pH range will this solution be an effective buffer?
(b) What is the buffer capacity of the solution? That is, how many millimoles of strong acid or strong base can be added to the solution before any significant change in pH occurs?

40. A solution of 0.500 L volume is prepared containing 1.68 g NH_3 and 4.05 g $(NH_4)_2SO_4$.
(a) What is the pH of this solution?
(b) If 0.88 g NaOH is added to the solution what is the pH?
(c) How many milliliters of 12 M HCl must be added to 0.500 L of the original solution to change its pH to 9.00?

42. An acetic acid/sodium acetate buffer can be prepared by the reaction

$$\underset{\text{(from }NaC_2H_3O_2\text{)}}{C_2H_3O_2^-} + \underset{\text{(from HCl)}}{H_3O^+} \longrightarrow HC_2H_3O_2 + H_2O$$

(a) If 12.0 g $NaC_2H_3O_2$ is added to 0.300 L of 0.200 M HCl, what is the pH of the resulting solution?
(b) If 1.00 g $Ba(OH)_2$ is added to the solution in part **(a)**, what is the new pH?
(c) What is the maximum mass of $Ba(OH)_2$ that can be neutralized by the buffer solution of part **(a)**?
(d) What is the pH of the solution in part **(a)** following the addition of 5.50 g $Ba(OH)_2$?

Acid–Base Indicators

(Use data from Tables 17.3 and 17.4, as necessary.)

43. In the use of acid–base indicators
(a) Why is it generally sufficient to use a *single* indicator in an acid–base titration but often necessary to use *several* indicators to establish the approximate pH of a solution?
(b) Why must the quantity of indicator used in a titration be kept as small as possible?

45. Phenol red indicator changes from yellow to red in the pH range from 6.6 to 8.0. *Without making detailed calculations,* state what color the indicator will assume in each of the following solutions. **(a)** 0.10 M KOH; **(b)** 0.10 M $HC_2H_3O_2$; **(c)** 0.10 M NH_4NO_3; **(d)** 0.10 M HBr; **(e)** 0.10 M NaCN; **(f)** 0.10 M $HC_2H_3O_2$–0.10 M $NaC_2H_3O_2$.

44. The indicator methyl red has a $pK_{HIn} = 4.95$. It changes from red to yellow over the pH range from 4.4 to 6.2.
(a) If the indicator is placed in a buffer solution of pH = 4.55, what percent of the indicator will be present in the acid form, HIn, and what percent will be present in the base or anion form, In^-?
(b) Which form of the indicator has the "stronger" (that is, more visible) color—the acid (red) form or base (yellow) form? Explain.

46. Thymol blue indicator has *two* pH ranges. It changes color from red to yellow in the pH range from 1.2 to 2.8, and from yellow to blue in the pH range from 8.0 to 9.6. What is the color of the indicator in the following situations?
(a) The indicator is placed in 350.0 mL of 0.205 M HCl.
(b) To the solution in part **(a)** is added 250.0 mL of 0.500 M $NaNO_2$.
(c) To the solution in part **(b)** is added 150.0 mL of 0.100 M NaOH.
(d) To the solution in part **(c)** is added 5.00 g $Ba(OH)_2$.

Neutralization Reactions

47. A 25.00-mL sample of $H_3PO_4(aq)$ requires 31.15 mL of 0.2420 M KOH for titration to the second equivalence point. What is the molarity of this $H_3PO_4(aq)$? (*Hint:* Write an equation for the neutralization reaction.)
49. Two aqueous solutions are mixed: 50.0 mL of 0.0150 M H_2SO_4 and 50.0 mL of 0.0385 M NaOH. What is the pH of the resulting solution?

48. A 20.00-mL sample of $H_3PO_4(aq)$ requires 18.67 mL of 0.1885 M NaOH for titration *between* the first and second equivalence points. What is the molarity of this $H_3PO_4(aq)$? (*Hint:* Write an equation for the neutralization reaction.)
50. Two solutions are mixed: 100.0 mL of HCl(aq) with pH 2.50 and 100.0 mL of NaOH(aq) with pH 11.00. What is the pH of the resulting solution?

Titration Curves

51. Explain why the volume of 0.100 M NaOH required to reach the equivalence point in the titration of 25.00 mL of 0.100 M HA is the same, regardless of whether HA is a strong or a weak acid, yet the pH at the equivalence point is not the same.
53. Sketch the following titration curves. Indicate the initial pH and the pH corresponding to the equivalence point. Indicate the volume of titrant required to reach the equivalence point, and select a suitable indicator from Figure 18-6.
(a) 25.0 mL of 0.100 M KOH with 0.200 M HI
(b) 10.0 mL of 1.00 M NH_3 with 0.250 M HCl

55. In the titration of 20.00 mL of 0.175 M NaOH, calculate the number of milliliters of 0.200 M HCl that must be added to reach a pH of **(a)** 12.55, **(b)** 10.80, **(c)** 4.25. (*Hint:* Solve an algebraic equation in which the number of milliliters is *x*. Which reactant is in excess at each pH?)

52. Explain whether you would expect the equivalence point of each of the following titrations to be below, above, or at pH 7. **(a)** $NaHCO_3(aq)$ is titrated with NaOH(aq); **(b)** HCl(aq) is titrated with $NH_3(aq)$; **(c)** KOH(aq) is titrated with HI(aq).

54. Determine the following characteristics of the titration curve for 20.0 mL of 0.275 M $NH_3(aq)$ titrated with 0.325 M HI(aq).
(a) The initial pH.
(b) The volume of 0.325 M HI(aq) at the equivalence point.
(c) The pH at the half-neutralization point.
(d). The pH at the equivalence point.

56. In the titration of 25.00 mL of 0.100 M $HC_2H_3O_2$, calculate the number of milliliters of 0.200 M NaOH that must be added to reach a pH of **(a)** 3.85, **(b)** 5.25, **(c)** 11.10. (*Hint:* Solve an algebraic equation in which the number of milliliters is *x*. Which reactant is in excess at each pH?)

57. Sketch a titration curve (pH versus milliliters of titrant) for each of the following three hypothetical weak acids when titrated with 0.100 M NaOH. Select suitable indicators for the titrations from Figure 18-6. (*Hint:* Select a few key points at which to estimate the pH of the solution.)
(a) 10.00 mL of 0.100 M HX; $K_a = 7.0 \times 10^{-3}$
(b) 10.00 mL of 0.100 M HY; $K_a = 3.0 \times 10^{-4}$
(c) 10.00 mL of 0.100 M HZ; $K_a = 2.0 \times 10^{-8}$

58. Sketch a titration curve (pH versus milliliters of titrant) for each of the following hypothetical weak bases when titrated with 0.100 M HCl. (Think of these bases as involving the substitution of organic groups, R, for one of the H atoms of NH_3.) Select suitable indicators for the titrations from Figure 18-6. (*Hint:* Select a few key points at which to estimate the pH of the solution.)
(a) 10.00 mL of 0.100 M RNH_2; $K_b = 1 \times 10^{-3}$
(b) 10.00 mL of 0.100 M $R'NH_2$; $K_b = 3 \times 10^{-6}$
(c) 10.00 mL of 0.100 M $R''NH_2$; $K_b = 7 \times 10^{-8}$

Salts of Polyprotic Acids

(Use data from Table 17.4 or Appendix D, as necessary.)

59. Is a solution that is 0.10 M $Na_2S(aq)$ likely to be acidic, basic, or pH neutral? Explain.

60. Is a solution of sodium dihydrogen citrate, NaH_2Cit, likely to be acidic, basic, or neutral? Explain. Citric acid, H_3Cit, is $H_3C_6H_5O_7$.

61. Sodium phosphate, Na_3PO_4, is made commercially by first neutralizing phosphoric acid with sodium carbonate to obtain Na_2HPO_4. The Na_2HPO_4 is further neutralized to Na_3PO_4 with NaOH.
(a) Write net ionic equations for the reactions described here.
(b) Na_2CO_3 is a much cheaper base than is NaOH. Why do you suppose that NaOH must be used as well as Na_2CO_3 to produce Na_3PO_4? (*Hint:* Compare the pH values of equimolar solutions of OH^-, CO_3^{2-}, and PO_4^{3-}.)

62. Sodium hydrogen carbonate (sodium bicarbonate) and sodium hydroxide can both be used to neutralize acid spills. What is the pH of 1.00 M $NaHCO_3(aq)$ and of 1.00 M NaOH(aq)? On a per liter basis, do these two solutions have an equal capacity to neutralize acids? Explain. On a per gram basis, do the two solids, $NaHCO_3(s)$ and NaOH(s), have an equal capacity to neutralize acids? Explain. Why do you suppose that $NaHCO_3$ is often preferred to NaOH in neutralizing acid spills?

General Acid–Base Equilibria

63. What stoichiometric concentration of the indicated substance is required to obtain an aqueous solution with the pH value shown? **(a)** $Ba(OH)_2$ for pH = 11.88; **(b)** $HC_2H_3O_2$ in 0.294 M $NaC_2H_3O_2$ for pH = 4.52.

64. What stoichiometric concentration of the indicated substance is required to obtain an aqueous solution with the pH value shown? **(a)** aniline, $C_6H_5NH_2$, for pH = 8.95; **(b)** NH_4Cl for pH = 5.12.

65. Using appropriate equilibrium constants, but *without doing detailed calculations,* determine whether a solution can be simultaneously
(a) 0.10 M NH_3 and 0.10 M NH_4Cl, with pH = 6.07;
(b) 0.10 M $NaC_2H_3O_2$ and 0.058 M HI;
(c) 0.10 M KNO_2 and 0.25 M KNO_3;
(d) 0.050 M $Ba(OH)_2$ and 0.65 M NH_4Cl;
(e) 0.018 M $HC_7H_5O_2$ and 0.018 M $NaC_7H_5O_2$, with pH = 4.20;
(f) 0.68 M KCl, 0.42 M KNO_3, 1.2 M NaCl, and 0.55 M $NaC_2H_3O_2$, with pH = 6.4.

66. This single equilibrium equation applies to different phenomena described in this or the preceding chapter.
$$HC_2H_3O_2 + H_2O \rightleftharpoons H_3O^+ + C_2H_3O_2^-$$
Indicate which of these phenomena—ionization of pure acid, common ion effect, buffer solution, hydrolysis—occurs if
(a) $[H_3O^+]$ and $[HC_2H_3O_2]$ are high; $[C_2H_3O_2^-]$ is very low.
(b) $[C_2H_3O_2^-]$ is high; $[HC_2H_3O_2]$ and $[H_3O^+]$ are very low.
(c) $[HC_2H_3O_2]$ is high; $[H_3O^+]$ and $[C_2H_3O_2^-]$ are low.
(d) $[HC_2H_3O_2]$ and $[C_2H_3O_2^-]$ are high; $[H_3O^+]$ is low.

Advanced Exercises

67. You are given 250.0 mL of 0.100 M $HC_3H_5O_2$ (propionic acid, $K_a = 1.35 \times 10^{-5}$). You wish to adjust its pH by adding an appropriate solution. What volume would you add of **(a)** 1.00 M HCl to lower the pH to 1.00; **(b)** 1.00 M $NaC_3H_5O_3$ to raise the pH to 4.00; **(c)** water to raise the pH by 0.15 unit?

68. Even though the carbonic acid/hydrogen carbonate buffer system is crucial to the maintenance of the pH of blood, it has no practical use as a laboratory buffer solution. Can you think of a reason(s) for this? (*Hint:* Refer to data in Exercise 83 of Chapter 17.)

69. Sodium hydrogen sulfate, $NaHSO_4$, is an acidic salt with a number of uses, such as metal pickling (removal of surface deposits). $NaHSO_4$ is made by the reaction of H_2SO_4 with NaCl. To determine the percent NaCl impurity in sodium hydrogen sulfate, a 1.016-g sample is titrated with NaOH(aq); 36.56 mL of 0.225 M NaOH is required.
(a) Write the net equation for the neutralization reaction.
(b) What is the % NaCl in the sample titrated? (*Hint:* What mass of $NaHSO_4$ is present? What happens to NaCl during the titration?)
(c) Select a suitable indicator(s) from Figure 18-6.

70. To measure the pH of a solution with a pH meter it is necessary to standardize the meter (compare it) with a buffer solution of known pH. One solution used for this purpose is 0.0500 M potassium hydrogen phthalate, $KHC_8H_4O_4$. What is the approximate pH of this solution?

$$H_2C_8H_4O_4 + H_2O \rightleftharpoons H_3O^+ + HC_8H_4O_4^-$$
$$K_{a_1} = 1.1 \times 10^{-3}$$
$$HC_8H_4O_4^- + H_2O \rightleftharpoons H_3O^+ + C_8H_4O_4^{2-}$$
$$K_{a_2} = 3.7 \times 10^{-6}$$

71. Thymol blue in its acid range is not a suitable indicator for the titration of HCl by NaOH. Suppose that a student uses thymol blue by mistake in the titration of Figure 18-7, and suppose that the indicator endpoint is taken to be pH = 2.0.
(a) Would there be a sharp color change [produced by adding a single drop of NaOH(aq)]?
(b) Approximately what percent of the HCl remains unneutralized at pH = 2.0?

72. Rather than calculate the pH for different volumes of titrant, an alternative procedure for establishing a titration curve is to calculate the volume of titrant required to reach certain pH values. Determine the volumes of 0.100 M NaOH required to reach the following pH values in the titration of 20.00 mL of 0.150 M HCl. Then plot the titration curve.
(a) pH = 2.00; **(b)** 3.50; **(c)** 5.00; **(d)** 10.50; **(e)** 12.00.

73. Use the method of Exercise 72 to determine the volume of titrant required to reach the indicated pH values in the following titrations.
(a) 25.00 mL of 0.250 M NaOH titrated with 0.300 M HCl; pH = 13.00, 12.00, 10.00, 4.00, 3.00.
(b) 50.00 mL of 0.0100 M benzoic acid ($HC_7H_5O_2$) titrated with 0.0500 M KOH: pH = 4.50, 5.50, 11.50. ($K_a = 6.3 \times 10^{-5}$)

74. A buffer solution can be prepared by starting with a weak acid, HA, and converting some of the weak acid to its salt, for example, NaA, by titration with a strong base. The *fraction* of the original acid that is converted to the salt is designated f.
(a) Derive an equation similar to Equation (18.2) but expressed in terms of f rather than concentrations.
(b) What is the pH at the point in the titration of phenol, HOC_6H_5, at which $f = 0.27$ (pK_a of phenol = 10.00)?

75. You are asked to prepare a KH_2PO_4–Na_2HPO_4 solution that has the same pH as human blood, 7.40.
(a) What should be the *ratio* of concentrations $[HPO_4^{2-}]/[H_2PO_4^-]$ in this solution?
(b) Suppose you had to prepare 1.00 L of the solution described in **(a)**, and suppose that this solution had to be isotonic with blood (with the same osmotic pressure as blood). What mass of KH_2PO_4 and of $Na_2HPO_4 \cdot 12H_2O$ would you use? (*Hint:* Refer to the definition of isotonic on page 487. Recall that a solution of NaCl with 9.2 g NaCl/L soln is isotonic with blood, and assume that NaCl is completely ionized in aqueous solution.)

76. You are asked to bring the pH of 0.500 L of 0.500 M NH_4Cl(aq) to a value of 7.00. How many drops (1 drop = 0.05 mL) of which of the following solutions would you use: 10.0 M HCl or 10.0 M NH_3?

77. Because an acid–base indicator is a weak acid, it can be titrated with a strong base. Suppose you titrate 25.00 mL of a 0.0100 M solution of the indicator *p*-nitrophenol, $HC_6H_4NO_3$, with 0.0200 M NaOH. The pK_a of *p*-nitrophenol is 7.15, and it changes from colorless to yellow in the pH range from 5.6 to 7.6.
(a) Sketch the titration curve for this titration.
(b) Show the pH range over which *p*-nitrophenol changes color.
(c) Explain why *p*-nitrophenol cannot serve as its own indicator in this titration.

78. The neutralization of NaOH by HCl is represented in Equation (1) and the neutralization of NH_3 by HCl in Equation (2).
(1) $OH^- + H_3O^+ \rightleftharpoons 2 H_2O$ $K = ?$
(2) $NH_3 + H_3O^+ \rightleftharpoons NH_4^+ + H_2O$ $K = ?$
(a) Determine the equilibrium constant K for each reaction.
(b) Explain why each neutralization reaction can be considered to go to completion.

79. The titration of a weak acid by a weak base is not a satisfactory procedure because the pH does not increase sharply at the equivalence point. Demonstrate this fact by sketching a titration curve for the neutralization of 10.00 mL of 0.100 M $HC_2H_3O_2$ with 0.100 M NH_3.

80. At times a salt of a weak base can be titrated by a strong base. Use appropriate data from the text to sketch a titration curve for the titration of 10.00 mL of 0.0500 M $C_6H_5NH_3^+Cl^-$ with 0.100 M NaOH.

81. Carbonic acid is a weak diprotic acid (H_2CO_3) with $K_{a_1} = 4.4 \times 10^{-7}$ and $K_{a_2} = 4.7 \times 10^{-11}$. The equivalence points for the titration come at approximately pH 4 and 9. Suitable indicators for use in titrating carbonic acid or carbonate solutions are methyl orange and phenolphthalein.
(a) Sketch the titration curve that would be obtained in titrating a sample of $NaHCO_3$(aq) with 1.00 M HCl.
(b) Sketch the titration curve for Na_2CO_3(aq) with 1.00 M HCl.
(c) What volume of 0.100 M HCl is required for the complete neutralization of 1.00 g $NaHCO_3$(s)?
(d) What volume of 0.100 M HCl is required for the complete neutralization of 1.00 g Na_2CO_3(s)?
(e) A sample of NaOH contains a small amount of Na_2CO_3. For titration to the phenolphthalein endpoint, 0.1000 g of this sample requires 23.98 mL of 0.1000 M HCl. An additional 0.78 mL is required to the methyl orange endpoint. What is the % Na_2CO_3, by mass, in the sample?

82. Consider a solution of NaH_2PO_4(aq) of molarity M and derive Equation (18.5) by showing that the pH is independent of M. Explain why the equation fails, however, for very dilute solutions, say with M < 0.01.

83. A solution is prepared that is 0.150 M $HC_2H_3O_2$ and 0.250 M $NaCHO_2$.
(a) Show that this is a buffer solution.
(b) Calculate the pH of this buffer solution.
(c) What is the final pH if to 1.00 L of this buffer solution is added 1.00 L of 0.100 M HCl?
(*Hint:* Refer to the alternative method of solving equilibrium problems given in Feature Problem C of Chapter 17, on page 619.)

Feature Problems

A. The graph, which is related to a titration curve, shows the fraction (f) of the stoichiometric amount of acetic acid that is present as nonionized $HC_2H_3O_2$ and as acetate ion, $C_2H_3O_2^-$, as a function of the pH of the solution containing these species.

1. Explain the significance of the point at which the two curves cross. What are the fractions and the pH at this point?

2. Sketch a comparable set of curves for carbonic acid, H_2CO_3. (*Hint:* How many "carbonate containing" species should appear in the graph? How many points of intersection should there be? at what pH values?)

3. Sketch a comparable set of curves for phosphoric acid, H_3PO_4. (*Hint:* How many "phosphate containing" species should appear in the graph? How many points of intersection should there be? at what pH values?)

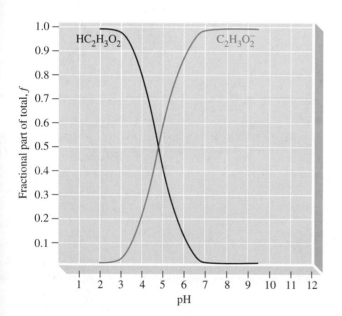

B. On two occasions in the discussion of titration curves we noted that we can graph pOH versus volume of a titrant, rather than pH versus volume of titrant (see marginal notes on pages 638 and 642).

1. For the titration of 25.00 mL of 0.100 M NaOH with 0.100 M HCl, calculate the pOH at a few representative points in the titration, sketch the titration curve of pOH versus volume of titrant, and show that it has exactly the same form as Figure 18-7. Then sketch the titration curve of pH versus volume of titrant for the same titration. [*Hint:* Can you see that the easiest way to do this is simply to use the calculated pOH values and to plot (14.00 − pOH) versus volume of titrant?]

2. Repeat part 1 for the titration of 25.00 mL 0.100 M NH_3 with 0.100 M HCl.

C. In some cases, the titration curve for a mixture of *two* acids will have the same appearance as for a single acid. In other cases it will not.

1. Sketch the titration curve (pH versus volume of titrant) for the titration with 0.200 M NaOH of 25.00 mL of a solution that is 0.100 M in HCl and 0.100 M in HNO_3. Does this curve differ in any way from what would be obtained in the titration of 25.00 mL of 0.200 M HCl with 0.200 M NaOH? Explain.

2. The titration curve shown was obtained when 10.00 mL of a solution containing *both* HCl and H_3PO_4 was titrated with 0.216 M NaOH. From this curve determine the stoichiometric molarities of both the HCl and the H_3PO_4. (*Hint:* In what portion of the curve is HCl neutralized? In what portion is H_3PO_4 neutralized to Na_2HPO_4? In what portion is NaH_2PO_4 neutralized to Na_2HPO_4?)

3. A 10.00-mL solution that is 0.0400 M H_3PO_4 and 0.0150 M NaH_2PO_4 is titrated with 0.0200 M NaOH. Sketch the titration curve obtained. (*Hint:* How many equivalence points are there? Are they equally spaced along the volume axis?)

Solubility and Complex Ion Equilibria

19

The White Cliffs of Dover, England, are made of chalk, a soft form of limestone, $CaCO_3$. These beautiful cliffs have persisted over a long time because of the low aqueous solubility of $CaCO_3$. Equilibria between sparingly soluble solutes and their ions in solution, expressed through the solubility product constant expression, K_{sp}, are discussed in this chapter.

*T*he dissolving and precipitation of limestone ($CaCO_3$) underlie a variety of natural phenomena, such as the formation of limestone caverns. Whether precipitation occurs from a solution containing Ca^{2+} and CO_3^{2-} ions depends on the concentrations of these ions. In turn, the CO_3^{2-} ion concentration depends on the pH of the solution. To develop a better understanding of the conditions under which $CaCO_3$ precipitates or dissolves, we need to consider equilibrium relationships between Ca^{2+} and CO_3^{2-}, and between CO_3^{2-}, H_3O^+, and HCO_3^-. This suggests a need to combine ideas about acid–base equilibria from Chapters 17 and 18 with ideas about the new types of equilibria to be introduced in this chapter.

Silver chloride is a familiar precipitate in the general chemistry laboratory, yet it does *not* precipitate from a solution having a moderate to high concentration of $NH_3(aq)$. This is because the silver ion and ammonia combine to form a species, called a complex ion, that remains in solution. Complex ion formation and equilibria involving complex ions are additional topics discussed in this chapter.

19-1 The Solubility Product Constant, K_{sp}

Gypsum, $CaSO_4 \cdot 2H_2O$, is an important calcium mineral. It is slightly soluble in water, and groundwater that comes into contact with gypsum often contains some dissolved calcium sulfate. This water cannot be used for certain applications, such as in evaporative cooling systems in power plants, because the calcium sulfate in the water might precipitate. The equilibrium between Ca^{2+}(aq) and SO_4^{2-}(aq) and undissolved $CaSO_4$(s) can be represented as

$$CaSO_4(s) \rightleftharpoons Ca^{2+}(aq) + SO_4^{2-}(aq)$$

We can write the equilibrium constant expression for this equilibrium as we learned to do in Chapter 16 (page 554)—that is, including concentration terms for ions in solution but not for the pure solid solute. Also, it is customary to represent the equilibrium constant by a special symbol: K_{sp}.

$$K_c = K_{sp} = [Ca^{2+}][SO_4^{2-}] = 9.1 \times 10^{-6} \quad \text{(at 25 °C)} \quad (19.1)$$

The **solubility product constant, K_{sp},** is the equilibrium constant for the equilibrium established between a solid solute and its ions in a saturated solution. Table 19.1 lists several K_{sp} values and the solubility equilibria to which they apply.

TABLE 19.1 Several Solubility Product Constants at 25 °C*

Solute	Solubility Equilibrium	K_{sp}
aluminum hydroxide	$Al(OH)_3(s) \rightleftharpoons Al^{3+}(aq) + 3\ OH^-(aq)$	1.3×10^{-33}
barium carbonate	$BaCO_3(s) \rightleftharpoons Ba^{2+}(aq) + CO_3^{2-}(aq)$	5.1×10^{-9}
barium sulfate	$BaSO_4(s) \rightleftharpoons Ba^{2+}(aq) + SO_4^{2-}(aq)$	1.1×10^{-10}
calcium carbonate	$CaCO_3(s) \rightleftharpoons Ca^{2+}(aq) + CO_3^{2-}(aq)$	2.8×10^{-9}
calcium fluoride	$CaF_2(s) \rightleftharpoons Ca^{2+}(aq) + 2\ F^-(aq)$	5.3×10^{-9}
calcium sulfate	$CaSO_4(s) \rightleftharpoons Ca^{2+}(aq) + SO_4^{2-}(aq)$	9.1×10^{-6}
chromium(III) hydroxide	$Cr(OH)_3(s) \rightleftharpoons Cr^{3+}(aq) + 3\ OH^-(aq)$	6.3×10^{-31}
iron(III) hydroxide	$Fe(OH)_3(s) \rightleftharpoons Fe^{3+}(aq) + 3\ OH^-(aq)$	4×10^{-38}
lead(II) chloride	$PbCl_2(s) \rightleftharpoons Pb^{2+}(aq) + 2\ Cl^-(aq)$	1.6×10^{-5}
lead(II) chromate	$PbCrO_4(s) \rightleftharpoons Pb^{2+}(aq) + CrO_4^{2-}(aq)$	2.8×10^{-13}
lead(II) iodide	$PbI_2(s) \rightleftharpoons Pb^{2+}(aq) + 2\ I^-(aq)$	7.1×10^{-9}
magnesium carbonate	$MgCO_3(s) \rightleftharpoons Mg^{2+}(aq) + CO_3^{2-}(aq)$	3.5×10^{-8}
magnesium fluoride	$MgF_2(s) \rightleftharpoons Mg^{2+}(aq) + 2\ F^-(aq)$	3.7×10^{-8}
magnesium hydroxide	$Mg(OH)_2(s) \rightleftharpoons Mg^{2+}(aq) + 2\ OH^-(aq)$	1.8×10^{-11}
magnesium phosphate	$Mg_3(PO_4)_2(s) \rightleftharpoons 3\ Mg^{2+}(aq) + 2\ PO_4^{3-}(aq)$	1×10^{-25}
mercury(I) chloride	$Hg_2Cl_2(s) \rightleftharpoons Hg_2^{2+}(aq) + 2\ Cl^-(aq)$	1.3×10^{-18}
silver bromide	$AgBr(s) \rightleftharpoons Ag^+(aq) + Br^-(aq)$	5.0×10^{-13}
silver chloride	$AgCl(s) \rightleftharpoons Ag^+(aq) + Cl^-(aq)$	1.8×10^{-10}
silver chromate	$Ag_2CrO_4(s) \rightleftharpoons 2\ Ag^+(aq) + CrO_4^{2-}(aq)$	1.1×10^{-12}
silver iodide	$AgI(s) \rightleftharpoons Ag^+(aq) + I^-(aq)$	8.5×10^{-17}
strontium carbonate	$SrCO_3(s) \rightleftharpoons Sr^{2+}(aq) + CO_3^{2-}(aq)$	1.1×10^{-10}
strontium sulfate	$SrSO_4(s) \rightleftharpoons Sr^{2+}(aq) + SO_4^{2-}(aq)$	3.2×10^{-7}

*A more extensive listing of K_{sp} values is given in Appendix D.

EXAMPLE 19-1

Writing Solubility Product Constant Expressions for Slightly Soluble Solutes. Write the solubility product constant expression for the solubility equilibrium of
a. calcium fluoride, CaF_2 (one of the products formed when a fluoride treatment is applied to teeth).
b. copper arsenate, $Cu_3(AsO_4)_2$ (used as an insecticide and fungicide).

Solution

The K_{sp} expression is formulated for the ionic species appearing in the equation for the solubility equilibrium. This equation is written for one mole of the slightly soluble solute. That is, the coefficient "1" is understood for the slightly soluble solute. The coefficients for the ions in solution are whatever is needed to balance the equation. The coefficients then establish the powers to which the ion concentrations are raised in the K_{sp} expression.
a. $CaF_2(s) \rightleftharpoons Ca^{2+}(aq) + 2\ F^-(aq)$ $K_{sp} = [Ca^{2+}][F^-]^2$
b. $Cu_3(AsO_4)_2(s) \rightleftharpoons 3\ Cu^{2+}(aq) + 2\ AsO_4^{3-}(aq)$ $K_{sp} = [Cu^{2+}]^3[AsO_4^{3-}]^2$

Practice Example A: Write the solubility product constant expression for **(a)** $MgCO_3$ (one of the components of dolomite, a form of limestone), and **(b)** Ag_3PO_4 (used in photographic emulsions).

Practice Example B: A handbook lists $K_{sp} = 1 \times 10^{-7}$ for calcium hydrogen phosphate, a substance used in dentifrices and as an animal feed supplement. Write **(a)** the equation for the solubility equilibrium and **(b)** the solubility product constant expression for this slightly soluble solute.

19-2   The Relationship Between Solubility and K_{sp}

The term *solubility* suggests that there should be a relationship between the solubility product constant, K_{sp}, of a solute and its *molar solubility*—its molarity in a saturated aqueous solution. As shown in Examples 19-2 and 19-3, there is a definite relationship between them. As we will point out in Section 19-4, calculations involving K_{sp} are somewhat more subject to error than those involving other equilibrium constants, but the results are suitable for most purposes. In Example 19-2, we start with an experimentally determined solubility and obtain a value of K_{sp}.

▶ Feature Problem A on page 687 describes a different experimental method of determining the K_{sp} of $CaSO_4$.

EXAMPLE 19-2

Calculating K_{sp} of a Slightly Soluble Solute from Its Solubility. A handbook lists the aqueous solubility of $CaSO_4$ at 25 °C as 0.20 g $CaSO_4$/100 mL. What is the K_{sp} of $CaSO_4$ at 25 °C?

$$CaSO_4(s) \rightleftharpoons Ca^{2+}(aq) + SO_4^{2-}(aq) \qquad K_{sp} = ?$$

Solution

We need to construct a conversion pathway that begins with finding $[Ca^{2+}]$ and $[SO_4^{2-}]$, which we can then substitute into the K_{sp} expression. That is,

g $CaSO_4$/100 mL $\longrightarrow$ mol $CaSO_4$/L $\longrightarrow$ $[Ca^{2+}]$ and $[SO_4^{2-}]$ $\longrightarrow$ K_{sp}

In the first step we obtain the molar solubility of $CaSO_4$. (Note that in this set-up we have replaced 100 mL by 0.100 L.)

$$\text{mol } CaSO_4/\text{L satd soln} = \frac{0.20 \text{ g } CaSO_4}{0.100 \text{ L soln}} \times \frac{1 \text{ mol } CaSO_4}{136 \text{ g } CaSO_4}$$

$$= 0.015 \text{ M } CaSO_4$$

The key factors in the next step are shown in color. They indicate that one mole of Ca^{2+} and one mole of SO_4^{2-} appear in solution for each mole of $CaSO_4$ that dissolves.

$$[Ca^{2+}] = \frac{0.015 \text{ mol } CaSO_4}{1 \text{ L}} \times \frac{1 \text{ mol } Ca^{2+}}{1 \text{ mol } CaSO_4} = 0.015 \text{ M}$$

$$[SO_4^{2-}] = \frac{0.015 \text{ mol } CaSO_4}{1 \text{ L}} \times \frac{1 \text{ mol } SO_4^{2-}}{1 \text{ mol } CaSO_4} = 0.015 \text{ M}$$

Finally, we substitute these ion concentrations into the solubility-product expression.

$$K_{sp} = [Ca^{2+}][SO_4^{2-}] = (0.015)(0.015) = 2.3 \times 10^{-4}$$

▶ The rather large discrepancy between this result and the K_{sp} value in expression (19.1) is explained on page 662.

Practice Example A: A handbook lists the aqueous solubility of AgOCN as 7 mg/100 mL at 20 °C. What is the K_{sp} of AgOCN at 20 °C?

Practice Example B: A handbook lists the aqueous solubility of lithium phosphate at 18 °C as 0.034 g Li_3PO_4/100 mL soln. What is the K_{sp} of Li_3PO_4 at 18 °C?

The "inverse" of Example 19-2 is the calculation of the solubility of a solute from its K_{sp} value. When we do this, as in Example 19-3, the result is always a molar solubility—a molarity. If we need the solubility in units other than moles per liter, additional conversions are required, as in Practice Example 19-3B.

EXAMPLE 19-3

Calculating the Solubility of a Slightly Soluble Solute from Its K_{sp} Value. Lead iodide, PbI_2, is a dense, golden yellow, "insoluble" solid used in bronzing and in ornamental work requiring a golden color (such as mosaic gold). Calculate the molar solubility of lead iodide in water at 25 °C, given that its $K_{sp} = 7.1 \times 10^{-9}$.

Solution

The solubility equilibrium equation

$$PbI_2(s) \rightleftharpoons Pb^{2+}(aq) + 2 \, I^-(aq)$$

shows that for each mole of PbI_2 that dissolves, *one* mole of Pb^{2+} and *two* moles of I^- appear in solution. If we let s represent the number of moles of PbI_2 found dissolved per liter of saturated solution, then in this solution

$$[Pb^{2+}] = s \quad \text{and} \quad [I^-] = 2s$$

These concentrations must also satisfy the K_{sp} expression.

$$K_{sp} = [Pb^{2+}][I^-]^2 = (s)(2s)^2 = 7.1 \times 10^{-9}$$

$$4s^3 = 7.1 \times 10^{-9}$$

$$s^3 = 1.8 \times 10^{-9}$$

$$s = (1.8 \times 10^{-9})^{1/3} = 1.2 \times 10^{-3}$$

$$= \text{molar solubility of } PbI_2 = 1.2 \times 10^{-3} \text{ M}$$

Practice Example A: The K_{sp} of Fe(OH)$_3$ at 25 °C is 4×10^{-38}. What is the molar solubility of Fe(OH)$_3$ in H$_2$O at 25 °C?

Practice Example B: BaSO$_4$(s), a good absorber of X-rays, is used in a "barium milkshake" to coat the gastrointestinal tract so that it can be X-rayed. Even though Ba^{2+}(aq) is poisonous, BaSO$_4$ can be used because of its very low aqueous solubility. How many milligrams of BaSO$_4$ are dissolved in a 225-mL sample of saturated BaSO$_4$(aq)? $K_{sp} = 1.1 \times 10^{-10}$.

Are You Wondering . . .

When comparing molar solubilities, is a solute with a larger value of K_{sp} always more soluble than one with a smaller value?

If the solutes being compared are of the same type (MX or MX$_2$ or M$_2$X . . .), their molar solubilities will be related in the same way as their K_{sp} values. That is, the solute with the largest K_{sp} value will have the greatest molar solubility. Thus, AgCl ($K_{sp} = 1.8 \times 10^{-10}$) is more soluble than AgBr ($K_{sp} = 5.0 \times 10^{-13}$). For these particular solutes the molar solubility is $s = \sqrt{K_{sp}}$.

If the solutes are *not* of the same type, you'll have to calculate, or at least estimate, each molar solubility and compare the results. Thus, even though its solubility product constant is smaller, Ag$_2$CrO$_4$ ($K_{sp} = 1.1 \times 10^{-12}$) is *more soluble* than is AgCl ($K_{sp} = 1.8 \times 10^{-10}$). For Ag$_2CrO_4$, the molar solubility is $s = (K_{sp}/4)^{1/3} = 6.5 \times 10^{-5}$ M, whereas for AgCl it is $s = \sqrt{K_{sp}} = 1.3 \times 10^{-5}$ M.

19-3   The Common Ion Effect in Solubility Equilibria

In Examples 19-2 and 19-3, the ions in the saturated solutions came from a *single* source, the pure solid solute. Suppose that to the saturated solution of PbI$_2$ in Example 19-3 we add some I$^-$—a *common ion*—from a source such as KI(aq). The situation would be much like the common ion effect that we encountered in Chapter 18 (page 623).

Le Châtelier's principle tells us that an equilibrium mixture responds to a forced increase in the concentration of one of its reactants by a shift in the direction in which that reactant is consumed. In the lead iodide solubility equilibrium,

$$PbI_2(s) \rightleftharpoons Pb^{2+}(aq) + 2\ I^-\ (aq)$$

if some of the common ion, I$^-$, is added, the reverse reaction (blue arrow) is favored, leading to a new equilibrium in which more PbI₂ precipitates | [Pb²⁺] is less than in the original equilibrium | [I⁻] is greater than in the original equilibrium.

The solubility of a slightly soluble ionic compound is lowered in the presence of a second solute that furnishes a common ion.

The common ion effect is illustrated in Figure 19-1, and it is applied quantitatively in Example 19-4.

EXAMPLE 19-4

Calculating the Solubility of a Slightly Soluble Solute in the Presence of a Common Ion. What is the molar solubility of PbI_2 in 0.10 M KI(aq)?

Solution

To solve this problem let's set up the familiar ICE table using s instead of x in representing changes in concentrations. Think of producing a saturated solution of PbI_2, but instead of using pure water as the solvent use 0.10 M KI(aq). Thus, we begin with $[I^-] = 0.10$ M. Now let s represent the amount of PbI_2, in moles, that dissolves to produce 1 L of saturated solution. The additional concentrations appearing in this solution are s mol Pb^{2+}/L and $2s$ mol I^-/L.

$$PbI_2(s) \rightleftharpoons Pb^{2+}(aq) + 2\,I^-(aq)$$

	Pb^{2+}	I^-
initial concn, M:		0.10
from PbI_2, M:	s	$2s$
equil concn, M:	s	$(0.10 + 2s)$

The usual K_{sp} relationship must be satisfied, that is,

$$K_{sp} = [Pb^{2+}][I^-]^2 = (s)(0.10 + 2s)^2 = 7.1 \times 10^{-9}$$

To simplify the solution to this equation, let's assume that s is much smaller than 0.10 M, so that $0.10 + 2s \approx 0.10$.

$$s(0.10)^2 = 7.1 \times 10^{-9}$$

$$s = \frac{7.1 \times 10^{-9}}{(0.10)^2} = 7.1 \times 10^{-7} \text{ M}$$

Our assumption is well justified: 7.1×10^{-7} is much smaller than 0.10, and

$$s = \text{molar solubility of } PbI_2 = 7.1 \times 10^{-7} \text{ M}$$

Practice Example A: What is the molar solubility of PbI_2 in 0.10 M $Pb(NO_3)_2$(aq)? (*Hint:* To which ion concentration should the solubility be related?)

Practice Example B: What is the molar solubility of $Fe(OH)_3$ in a buffered solution with pH = 8.20? (*Hint:* What is the $[OH^-]$ in this solution?)

The solubility of PbI_2 in the presence of 0.10 M I^- calculated in Example 19-4 is about 2000 times less than its value in pure water (Example 19-3). If you work out Practice Example 19-4A, you will see that the effect of added Pb^{2+} in reducing the solubility of PbI_2 is not as striking as that of I^-, but it is significant nevertheless.

(a)   **(b)**

FIGURE 19-1
The common ion effect in solubility equilibrium.
(a) A clear saturated solution of lead(II) iodide from which excess undissolved solute has been filtered off. (b) When a small volume of a concentrated solution of KI (containing the common ion, I^-) is added, a small quantity of $PbI_2(s)$ precipitates. A common ion reduces the solubility of a sparingly soluble solute.

A common student error in problems like Example 19-4 is to double the common ion concentration, for example, by writing that the initial $[I^-] = (2 \times 0.10)$ M. In *any* aqueous solution of PbI_2, the $[I^-]$ derived from PbI_2 must be *twice* the molarity of the PbI_2 that dissolves, that is, $2s$. But the $[I^-]$ that comes from a soluble strong electrolyte is determined only by the molarity of the strong electrolyte. Thus, $[I^-]$ in 0.10 M KI(aq) is 0.10 M. In 0.10 M KI(aq) that is also saturated with PbI_2, the *total* $[I^-] = (0.10 + 2s)$ M. In short, no relationship exists between the stoichiometry of the dissolving of PbI_2, which requires a factor of 2 in establishing $[I^-]$, and that of KI, which does not.

19-4   Limitations of the K_{sp} Concept

We have repeatedly used the term *slightly soluble* in describing the solutes for which we have written K_{sp} expressions. You may wonder if we can write K_{sp} expressions for moderately or highly soluble ionic compounds, such as NaCl, KNO_3, and NaOH. The answer is that we can, but the K_{sp} must be based on ion *activities* rather than concentrations. And in ionic solutions of moderate to high concentrations, activities and concentrations are *not* equal (recall Section 14-9). If we are not able to use molarities in place of activities, much of the simplicity of the solubility product concept is lost. Thus, K_{sp} values are usually limited to "slightly soluble" (essentially insoluble) solutes, and ion molarities are used in place of activities. In addition, there are several other limitations of the K_{sp} concept, which are discussed in the following sections.

The Diverse ("Uncommon") Ion Effect: The Salt Effect

We have explored the effect of common ions on a solubility equilibrium, but what effect do ions *different* from those involved in the equilibrium have on solute solubilities? The effect of "uncommon" ions is not as striking as the common ion effect. Moreover, uncommon ions tend to *increase* rather than decrease solubility. As the total ionic concentration of a solution increases, interionic attractions become more important. Activities (effective concentrations) become smaller than the stoichiometric or measured concentrations. For the ions involved in the solution process, this means that higher concentrations must appear in solution before equilibrium is established—*the solubility increases.* Figure 19-2 compares the effects of common ions and uncommon ions.

The uncommon, or diverse, ion effect is more commonly called the **salt effect.** Because of the salt effect, the numerical value of a K_{sp} based on molarities will vary depending on the ionic atmosphere. Tabulated values of K_{sp} usu-

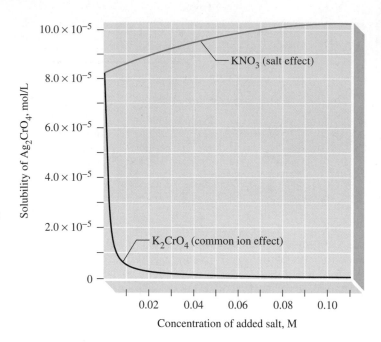

FIGURE 19-2
Comparison of the common ion effect and the salt effect on the molar solubility of Ag_2CrO_4.
The presence of CrO_4^{2-} ions, derived from $K_2CrO_4(aq)$, reduces the solubility of Ag_2CrO_4 by a factor of about 35 over the concentration range shown (from 0 to 0.10 M added salt). Over this same concentration range, the solubility of Ag_2CrO_4 is increased by the presence of the uncommon, or diverse, ions from KNO_3, but only by about 25%.

ally are based on activities rather than molarities, thus avoiding the problem of the salt effect.

Ion Pair Formation

In performing calculations involving K_{sp} values and solubilities, we have *assumed* that all the dissolved solute appears in solution as separated cations and anions, but this assumption is often *not* valid. For example, in a saturated solution of magnesium fluoride, in addition to Mg^{2+} and F^- ions there exist combinations of one Mg^{2+} and one F^- ion, MgF^+, called **ion pairs.** To the extent that ion pair formation occurs, the concentrations of the dissociated ions are reduced. This means that the amount of solute that must dissolve to maintain the required ion concentrations to satisfy the K_{sp} expression increases. *Solubility increases when ion pair formation occurs in solution.* Ion pair formation is increasingly likely when the cations or anions in solution carry multiple charges (for example, Mg^{2+}, SO_4^{2-}, . . .).

Assessing the Limitations of K_{sp}

Let's assess the importance of the effects discussed in this section—molarities substituting for activities, the salt effect, and ion pair formation. All of these apply to $CaSO_4$. Recall that we calculated K_{sp} for $CaSO_4$ in Example 19-2 based on its measured solubility. Our result was $K_{sp} = 2.3 \times 10^{-4}$. This value is about 25 times larger than the value listed in Table 19.1, which is $K_{sp} = 9.1 \times 10^{-6}$.

These conflicting results for $CaSO_4$ are understandable. The K_{sp} value listed in Table 19.1 is based on ion activities, and the K_{sp} value calculated from the experimentally determined solubility is based on ion concentrations. We will continue to substitute molarities for activities of ions, and the case of $CaSO_4$ simply suggests that some of our results, although of the appropriate general magnitude (that is, within a factor of 10 or 100), may not be highly accurate. These "order-of-magnitude" results, however, still allow us to make correct predictions and to apply the K_{sp} concept in some useful ways.

19-5 Criteria for Precipitation and Its Completeness

Silver iodide is a light-sensitive compound used in photographic film and in cloud seeding to produce rain. Its solubility equilibrium and K_{sp} are represented as

$$AgI(s) \rightleftharpoons Ag^+(aq) + I^-(aq)$$

$$K_{sp} = [Ag^+][I^-] = 8.5 \times 10^{-17}$$

Suppose we mix solutions of $AgNO_3(aq)$ and $KI(aq)$ so as to obtain a mixed solution that has $[Ag^+] = 0.010$ M and $[I^-] = 0.015$ M. Is this solution unsaturated, saturated, or supersaturated?

Recall the reaction quotient, Q, that we introduced in Chapter 16. It has the same form as an equilibrium constant expression but uses initial concentrations rather than equilibrium concentrations. Here, initially,

$$Q_{sp} = [Ag^+]_{init} \times [I^-]_{init} = (0.010)(0.015) = 1.5 \times 10^{-4} > K_{sp}$$

The fact that $Q_{sp} > K_{sp}$ indicates that the concentrations of Ag^+ and I^- are already higher than they would be in a saturated solution, and that a net reaction should occur *to the left.* The solution is *supersaturated.* As is generally the case with supersaturated solutions, excess AgI should precipitate from solution. If we had found $Q_{sp} < K_{sp}$, the solution would have been *unsaturated.* No precipitate would form from such a solution.

When applied to solubility equilibria, Q_{sp} is generally called the **ion product** because its form is that of the product of ion concentrations raised to appropriate powers. The criterion for determining whether ions in a solution will combine to form a precipitate requires comparing the ion product with K_{sp}.

- Precipitation *should occur* if $Q_{sp} > K_{sp}$.
- Precipitation *cannot occur* if $Q_{sp} < K_{sp}$.
- A solution is just saturated if $Q_{sp} = K_{sp}$.

This criterion is illustrated in Figure 19-3 and Example 19-5. The example emphasizes the important point that *any possible dilutions that may occur must be considered before the criterion for precipitation is applied.*

(a)

(b)

FIGURE 19-3
Applying the criterion for precipitation from solution—Example 19-5 illustrated.
(a) When three drops of 0.20 M KI are added to 100.0 mL of 0.010 M $Pb(NO_3)_2$, at first a precipitate forms because K_{sp} is exceeded in the immediate vicinity of the drops. (b) When the KI becomes uniformly mixed in the $Pb(NO_3)_2(aq)$, K_{sp} is no longer exceeded and the precipitate redissolves. The criterion for precipitation must be applied *after* dilution has occurred.

EXAMPLE 19-5

Applying the Criterion for Precipitation of a Slightly Soluble Solute. Three drops of 0.20 M KI are added to 100.0 mL of 0.010 M $Pb(NO_3)_2$. Will a precipitate of lead iodide form? (Assume 1 drop = 0.05 mL.)

$$PbI_2(s) \rightleftharpoons Pb^{2+}(aq) + 2\,I^-(aq) \qquad K_{sp} = 7.1 \times 10^{-9}$$

Solution

We need to compare the product $[Pb^{2+}][I^-]^2$ formulated for the initial concentrations with the K_{sp} for PbI_2. For $[Pb^{2+}]$ we can simply use 0.010 M. For $[I^-]$, however, we must consider the great reduction in concentration that occurs when the three drops of 0.20 M KI are diluted to 100.0 mL.

Dilution Calculation

$$\text{amount I}^- = 3 \text{ drops} \times \frac{0.05 \text{ mL}}{1 \text{ drop}} \times \frac{1 \text{ L}}{1000 \text{ mL}} \times \frac{0.20 \text{ mol KI}}{\text{L}}$$

$$\times \frac{1 \text{ mol I}^-}{1 \text{ mol KI}} = 3 \times 10^{-5} \text{ mol I}^-$$

$$[\text{I}^-] = \frac{3 \times 10^{-5} \text{ mol I}^-}{0.1000 \text{ L}} = 3 \times 10^{-4} \text{ M}$$

Applying Precipitation Criterion

$$Q_{sp} = [\text{Pb}^{2+}][\text{I}^-]^2 = (0.010)(3 \times 10^{-4})^2 = 9 \times 10^{-10}$$

Because a Q_{sp} of 9×10^{-10} is *smaller than* a K_{sp} of 7.1×10^{-9}, we conclude that $\text{PbI}_2(s)$ should *not* precipitate.

Practice Example A: Three drops of 0.20 M KI are added to 100.0 mL of a 0.010 M solution of AgNO_3. Will a precipitate of silver iodide form?

Practice Example B: We saw in Example 19-5 that a 3-drop volume of 0.20 M KI is insufficient to cause precipitation in 100.0 mL of 0.010 M $\text{Pb(NO}_3)_2$. How many drops would be required to produce the first precipitate? (*Hint:* What $[\text{I}^-]$ must be present if the solution is to be saturated in PbI_2, that is, with $Q_{sp} = K_{sp}$? How many drops of 0.20 M KI, diluted to 100.0 mL, are required to produce this $[\text{I}^-]$? To produce the first trace of precipitate, consider that you need this many drops plus one more.)

Precipitation of a solute is considered to be complete only if the amount remaining in solution is very small. A useful rule of thumb in most applications is that precipitation is complete if 99.9% or more of a particular ion has precipitated, leaving less than 0.1% of the ion in solution. In Example 19-6 we will calculate the concentration of Mg^{2+} remaining in a solution from which $\text{Mg(OH)}_2(s)$ has precipitated. We will compare this remaining $[\text{Mg}^{2+}]$ to the initial $[\text{Mg}^{2+}]$ to determine the completeness of the precipitation.

EXAMPLE 19-6

Assessing the Completeness of a Precipitation Reaction. The first step in a commercial process in which magnesium is obtained from seawater involves precipitating Mg^{2+} as $\text{Mg(OH)}_2(s)$. The magnesium ion concentration in seawater is about 0.059 M. If a seawater sample is treated so that its $[\text{OH}^-]$ is maintained at 2.0×10^{-3} M, **(a)** what will be $[\text{Mg}^{2+}]$ remaining in solution when precipitation stops ($K_{sp} = 1.8 \times 10^{-11}$)? **(b)** Can we say that precipitation of $\text{Mg(OH)}_2(s)$ is complete under these conditions?

▶ One method of maintaining a constant pH during a precipitation is to carry out the precipitation from a buffer solution.

Solution

a. There is no question that precipitation will occur because the ion product, $Q_{sp} = [\text{Mg}^{2+}][\text{OH}^-]^2 = (0.059)(2.0 \times 10^{-3})^2 = 2.4 \times 10^{-7}$, exceeds K_{sp}. Precipitation of $\text{Mg(OH)}_2(s)$ will continue as long as the ion product exceeds K_{sp} but will stop when it is equal to K_{sp}. At the point at which the ion product equals K_{sp}, whatever $[\text{Mg}^{2+}]$ is in solution remains in solution.

$$[\text{Mg}^{2+}][\text{OH}^-]^2 = [\text{Mg}^{2+}](2.0 \times 10^{-3})^2 = 1.8 \times 10^{-11} = K_{sp}$$

$$[\text{Mg}^{2+}]_{\text{remaining}} = \frac{1.8 \times 10^{-11}}{(2.0 \times 10^{-3})^2} = 4.5 \times 10^{-6} \text{ M}$$

b. $[Mg^{2+}]$ in seawater is reduced from 0.059 to 4.5×10^{-6} M as a result of the precipitation reaction. Expressed as a percentage.

$$\% \ [Mg^{2+}]_{remaining} = \frac{4.5 \times 10^{-6} \ M}{0.059 \ M} \times 100\% = 0.0076\%$$

Because less than 0.1% of the Mg^{2+} remains, we conclude that precipitation is essentially complete.

Practice Example A: A typical Ca^{2+} concentration in seawater is 0.010 M. Will the precipitation of $Ca(OH)_2$ be complete from a seawater sample in which $[OH^-]$ is maintained at 0.040 M?

Practice Example B: What $[OH^-]$ should be maintained in a solution if, after precipitation of Mg^{2+} as $Mg(OH)_2(s)$, the remaining Mg^{2+} is to be at a level of 1 $\mu g \ Mg^{2+}/L$?

Are You Wondering . . .

What conditions favor completeness of precipitation?

The key factors in determining whether the target ion is essentially completely removed from solution in a precipitation are the value of K_{sp}, the initial concentration of the target ion, and the concentration of the common ion. In general, completeness of precipitation is favored by

- *a very small value of K_{sp}.* (The concentration of the target ion remaining in solution will be very small.)
- *a high initial concentration of the target ion.* (The concentration of the target ion remaining in solution will be only a very small fraction of that initially present.)
- *a concentration of common ion much larger than that of the target ion.* (The common ion concentration will remain nearly constant during the precipitation.)

19-6 Fractional Precipitation

If a large excess of $AgNO_3(s)$ is added to a solution containing the ions CrO_4^{2-} and Br^-, a mixed precipitate of $Ag_2CrO_4(s)$ and $AgBr(s)$ is obtained. There is a way of adding $AgNO_3$, however, that will cause $AgBr(s)$ to precipitate but leave CrO_4^{2-} in solution.

Fractional precipitation is a technique in which two or more ions in solution, each capable of being precipitated by the same reagent, are *separated* by the proper use of that reagent: *One ion is precipitated while the other(s) remains in solution.* The primary condition for a successful fractional precipitation is that there be a significant difference in the solubilities of the substances being separated. (Usually this means a significant difference in their K_{sp} values.) The key to the technique is the slow addition of a concentrated solution of the precipitating reagent to the solution from which precipitation is to occur, as from a buret (Figure 19-4).

FIGURE 19-4
Fractional precipitation—Example 19-7 illustrated.
(a) $AgNO_3(aq)$ is slowly added to a solution that is
0.010 M in Br^- and 0.010 M in CrO_4^{2-}. (b) Essentially
all the Br^- has precipitated as pale yellow AgBr(s), with
$[Br^-]$ in solution $= 3.3 \times 10^{-8}$ M. Red-brown
$Ag_2CrO_4(s)$ is just about to precipitate.

In figure: $[CrO_4^{2-}] = [Br^-] = 0.010$ M

$[Ag^+] = 1.0 \times 10^{-5}$ M
$[CrO_4^{2-}] = 0.010$ M
$[Br^-] = 3.3 \times 10^{-8}$ M
AgBr(s)

(a) (b)

Example 19-7 considers the separation of $CrO_4^{2-}(aq)$ and $Br^-(aq)$ through
the use of $Ag^+(aq)$. The necessary data are the solubility equilibrium equations
and the K_{sp} values of Ag_2CrO_4 and AgBr.

$$Ag_2CrO_4(s) \rightleftharpoons 2\,Ag^+(aq) + CrO_4^{2-}(aq) \qquad K_{sp} = 1.1 \times 10^{-12}$$

$$AgBr(s) \rightleftharpoons Ag^+(aq) + Br^-(aq) \qquad K_{sp} = 5.0 \times 10^{-13}$$

EXAMPLE 19-7

Separating Ions by Fractional Precipitation. $AgNO_3(aq)$ is slowly added to a solution that has $[CrO_4^{2-}] = 0.010$ M and $[Br^-] = 0.010$ M.
(a) Show that AgBr(s) should precipitate before $Ag_2CrO_4(s)$.
(b) When $Ag_2CrO_4(s)$ begins to precipitate, what is $[Br^-]$ remaining in solution?
(c) Is complete separation of $Br^-(aq)$ and $CrO_4^{2-}(aq)$ by fractional precipitation feasible?

Solution

a. The required values of $[Ag^+]$ for precipitation to start are

AgBr ppt: $\qquad Q_{sp} = [Ag^+][Br^-] = [Ag^+](0.010) = 5.0 \times 10^{-13} = K_{sp}$

$\qquad [Ag^+] = 5.0 \times 10^{-11}$ M

Ag_2CrO_4 ppt: $\qquad Q_{sp} = [Ag^+]^2[CrO_4^{2-}] = [Ag^+]^2(0.010)$

$\qquad\qquad = 1.1 \times 10^{-12} = K_{sp}$

$[Ag^+]^2 = 1.1 \times 10^{-10} \qquad$ and $\qquad [Ag^+] = 1.0 \times 10^{-5}$ M

Because the $[Ag^+]$ required to start the precipitation of AgBr(s) is much less than
that for $Ag_2CrO_4(s)$, AgBr(s) precipitates first. As long as AgBr(s) is forming, the
silver ion concentration can only slowly approach the value required for the precipitation of $Ag_2CrO_4(s)$.
b. As more and more AgBr(s) precipitates, $[Br^-]$ gradually decreases; this permits
$[Ag^+]$ to increase. When $[Ag^+]$ reaches 1.0×10^{-5} M, precipitation of $Ag_2CrO_4(s)$
begins. To determine $[Br^-]$ at the point at which $[Ag^+] = 1.0 \times 10^{-5}$ M, we use
K_{sp} for AgBr and solve for $[Br^-]$.

$$K_{sp} = [Ag^+][Br^-] = (1.0 \times 10^{-5})[Br^-] = 5.0 \times 10^{-13}$$

$$[Br^-] = \frac{5.0 \times 10^{-13}}{1.0 \times 10^{-5}} = 5.0 \times 10^{-8}\ M$$

c. Before $Ag_2CrO_4(s)$ begins to precipitate, $[Br^-]$ will have been reduced from 1.0 $\times 10^{-2}$ M to 5.0×10^{-8} M. Essentially all of the Br^- will have precipitated from solution as $AgBr(s)$, whereas the CrO_4^{2-} remains in solution. Fractional precipitation is feasible for separating mixtures of Br^- and CrO_4^{2-}.

Practice Example A: $AgNO_3(aq)$ is slowly added to a solution with $[Cl^-] = 0.115$ M and $[Br^-] = 0.264$ M. What percent of the Br^- remains *unprecipitated* at the point at which $AgCl(s)$ begins to precipitate?

$$AgCl: K_{sp} = 1.8 \times 10^{-10} \qquad AgBr: K_{sp} = 5.0 \times 10^{-13}$$

Practice Example B: A solution has $[Ni^{2+}] = [Co^{2+}] = 0.10$ M. Use data from Appendix D to choose an appropriate precipitating agent, and show that it will effectively separate these two ions. What is the concentration of the first ion to precipitate when the second ion begins to precipitate?

How might the titration illustrated in Figure 19-4 and Example 19-7 be carried out? In other words, how can we stop the titration just as $Ag_2CrO_4(s)$ starts to precipitate. One possibility, suggested in Figure 19-4, is to look for a color change in the precipitate from pale yellow (AgBr) to red-brown (Ag_2CrO_4). A more effective method is to follow $[Ag^+]$ during the titration. $[Ag^+]$ increases very rapidly between the points at which AgBr has finished precipitating and Ag_2CrO_4 is about to begin. We will discuss an electrometric method of determining very low ion concentrations in Chapter 21.

▶ The basis of a related titration, known as the Mohr method, is the subject of Feature Problem B on page 687.

19-7   Solubility and pH

$Mg(OH)_2(s)$, when suspended in water, is known as milk of magnesia; it is a popular antacid. Dissociated hydroxide ions react with hydronium ions (excess stomach acid) to form water.

$$Mg(OH)_2(s) \rightleftharpoons Mg^{2+}(aq) + 2\ OH^-(aq) \qquad (19.2)$$

$$OH^-(aq) + H_3O^+(aq) \longrightarrow 2\ H_2O(l) \qquad (19.3)$$

According to Le Châtelier's principle we expect reaction (19.2) to be displaced to the right—$Mg(OH)_2$ dissolves to replace OH^- ions drawn off by the neutralization reaction (19.3). We can obtain the *net ionic* equation by doubling Equation (19.3) and adding it to Equation (19.2). At the same time we do this, we can apply the method of combining equilibrium constants that we learned in Chapter 16 (page 552). The result we obtain is

$$Mg(OH)_2(s) \rightleftharpoons Mg^{2+}(aq) + 2\ \cancel{OH^-(aq)} \qquad K_{sp} = 1.8 \times 10^{-11}$$

$$\underline{2\ \cancel{OH^-(aq)} + 2\ H_3O^+(aq) \rightleftharpoons 4\ H_2O(l) \qquad \qquad K' = 1/K_w^2 = 1.0 \times 10^{28}}$$

$$Mg(OH)_2(s) + 2\ H_3O^+ \rightleftharpoons Mg^{2+}(aq) + 4\ H_2O(l) \qquad (19.4)$$

$$K = K_{sp}/K_w^2 = 1.8 \times 10^{-11} \times 1.0 \times 10^{28} = 1.8 \times 10^{17}$$

The large value of K for reaction (19.4) indicates that the reaction goes essentially to completion and that $Mg(OH)_2$ is highly soluble in acidic solutions.

Other slightly soluble solutes having *basic anions* (such as $ZnCO_3$, MgF_2, and CaC_2O_4) also become more soluble in *acidic* solutions. For these solutes we can write net ionic equations for solubility equilibria and corresponding K values based on K_{sp} for the solutes and K_a for the conjugate acids of the anions.

▶ This matter is the subject of Exercises 83 and 84.

▶ Feature Problem C on page 687 describes the effect of various factors on the molar solubility of $Mg(OH)_2$.

Although $Mg(OH)_2$ is soluble in acidic solution, in moderately or strongly basic solutions it is not. In Example 19-8 we calculate $[OH^-]$ in a solution of the weak base NH_3 and then use the criterion for precipitation to see if $Mg(OH)_2(s)$ will precipitate. In Example 19-9, we determine how to adjust $[OH^-]$ to prevent precipitation of $Mg(OH)_2(s)$. This adjustment is made by adding NH_4^+ to the $NH_3(aq)$, thereby converting it to a *buffer* solution.

EXAMPLE 19-8

Determining Whether a Precipitate Will Form in a Solution in Which There Is Also an Ionization Equilibrium. Should $Mg(OH)_2(s)$ precipitate if a solution is 0.010 M $MgCl_2$ and also 0.10 M NH_3?

Solution

The key here is in understanding that $[OH^-]$ is established by the ionization of $NH_3(aq)$.

$$NH_3(aq) + H_2O \rightleftharpoons NH_4^+(aq) + OH^-(aq) \qquad K_b = 1.8 \times 10^{-5}$$

If we set this up in the usual way, the equilibrium values are $x = [NH_4^+] = [OH^-]$ and $[NH_3] = (0.10 - x) \approx 0.10$. Then we obtain

$$K_b = \frac{[NH_4^+][OH^-]}{[NH_3]} = \frac{x \cdot x}{0.10} = 1.8 \times 10^{-5}$$

$$x^2 = 1.8 \times 10^{-6} \qquad x = [OH^-] = 1.3 \times 10^{-3} \text{ M}$$

Now we can rephrase the original question: Should $Mg(OH)_2(s)$ precipitate from a solution in which $[Mg^{2+}] = 1.0 \times 10^{-2}$ M and $[OH^-] = 1.3 \times 10^{-3}$ M? We must compare the ion product, Q_{sp}, with K_{sp}.

$$Q_{sp} = [Mg^{2+}][OH^-]^2 = (1.0 \times 10^{-2})(1.3 \times 10^{-3})^2$$
$$= 1.7 \times 10^{-8} > K_{sp} = 1.8 \times 10^{-11}$$

Precipitation should occur.

Practice Example A: Should $Mg(OH)_2(s)$ precipitate from a solution that is 0.010 M $MgCl_2(aq)$ and also 0.10 M $NaC_2H_3O_2$? $K_{sp}[Mg(OH)_2] = 1.8 \times 10^{-11}$; $K_a(HC_2H_3O_2) = 1.8 \times 10^{-5}$. [*Hint:* OH^- comes from the ionization of $C_2H_3O_2^-$ as a base (hydrolysis). Recall that $K_b = K_w/K_a$.]

Practice Example B: Will a precipitate of $Fe(OH)_3$ form from a solution that is 0.013 M Fe^{3+} in a buffer solution that is 0.150 M $HC_2H_3O_2$ and 0.250 M $NaC_2H_3O_2$?

EXAMPLE 19-9

Controlling an Ion Concentration, Either to Cause Precipitation or to Prevent It. What $[NH_4^+]$ must be maintained to prevent precipitation of $Mg(OH)_2$ from a solution that is 0.010 M $MgCl_2$ and 0.10 M NH_3?

Solution

The maximum value of the ion product, Q_{sp}, before precipitation occurs is 1.8×10^{-11}, the value of K_{sp} for $Mg(OH)_2$. This allows us to determine the maximum concentration of OH^- that can be tolerated.

$$[Mg^{2+}][OH^-]^2 = (1.0 \times 10^{-2})[OH^-]^2 = 1.8 \times 10^{-11}$$

$$[OH^-]^2 = 1.8 \times 10^{-9}$$

$$[OH^-] = 4.2 \times 10^{-5} \text{ M}$$

Next let's determine what $[NH_4^+]$ must be present in 0.10 M NH_3 to maintain $[OH^-] = 4.2 \times 10^{-5}$ M.

$$NH_3(aq) + H_2O \rightleftharpoons NH_4^+(aq) + OH^-(aq) \qquad K_b = 1.8 \times 10^{-5}$$

$$K_b = \frac{[NH_4^+][OH^-]}{[NH_3]} = \frac{[NH_4^+](4.2 \times 10^{-5})}{0.10} = 1.8 \times 10^{-5}$$

$$[NH_4^+] = \frac{0.10 \times 1.8 \times 10^{-5}}{4.2 \times 10^{-5}} = 0.043 \text{ M}$$

In order to keep the $[OH^-]$ at 4.2×10^{-5} *or less,* and thus prevent the precipitation of $Mg(OH)_2(s)$, $[NH_4^+]$ should be maintained at 0.043 M, or *greater.*

Practice Example A: What minimum $[NH_4^+]$ must be present to prevent precipitation of $Mn(OH)_2(s)$ from a solution that is 0.0050 M $MnCl_2$ and 0.025 M NH_3? For $Mn(OH)_2$, $K_{sp} = 1.9 \times 10^{-13}$.

Practice Example B: An acetic acid/sodium acetate buffer solution is also 0.010 M $AlCl_3$. **(a)** At what minimum pH can $Al(OH)_3(s)$ be precipitated from this solution? **(b)** What ratio of concentrations, $[C_2H_3O_2^-]/[HC_2H_3O_2]$, should be maintained to prevent the precipitation of $Al(OH)_3(s)$? $K_{sp}[Al(OH)_3] = 1.3 \times 10^{-33}$; $K_a(HC_2H_3O_2) = 1.8 \times 10^{-5}$.

19-8   Complex Ions and Coordination Compounds: An Introduction

We have seen that some slightly soluble ionic compounds become more soluble in acidic solutions. Another circumstance that can lead to an increased solubility is complex ion formation. A **complex ion** is a polyatomic cation or anion composed of a central metal ion to which are bonded other groups (molecules or ions). Substances containing complex ions belong to a category of compounds called *coordination compounds.* Our present concern is with the solubilities of coordination compounds.

The following formulas describe a series of three coordination compounds represented as if they consisted of the simpler compounds cobalt(III) chloride and ammonia.

▶ Later in the text we will encounter some coordination compounds in which there are no ions, and some in which the central metal is present as a neutral atom rather than a cation.

$$CoCl_3 \cdot 6NH_3 \qquad CoCl_3 \cdot 5NH_3 \qquad CoCl_3 \cdot 4NH_3 \qquad (19.5)$$
$$\text{(a)} \qquad\qquad\quad \text{(b)} \qquad\qquad\quad \text{(c)}$$

To explain the existence of compounds such as these, in 1893 the Swiss chemist Alfred Werner proposed that certain metal atoms, primarily those of the transition metals, have *two types of valence.* One, the *primary* valence, is based on the number of electrons the atom loses in forming the metal ion. A *secondary,* or auxiliary, valence is responsible for the bonding of other groups, called **ligands,** to the central metal ion. Werner's theory describes the compounds listed in (19.5) with the following formulas.

$$[Co(NH_3)_6]^{3+} + 3\ Cl^- \qquad [CoCl(NH_3)_5]^{2+} + 2\ Cl^- \qquad [CoCl_2(NH_3)_4]^+ + Cl^-$$

or   or   or

$$[Co(NH_3)_6]Cl_3 \qquad\qquad [CoCl(NH_3)_5]Cl_2 \qquad\qquad [CoCl_2(NH_3)_4]Cl \qquad (19.6)$$

(a)   (b)   (c)

▶ A coordination compound

complex ion   anions

$$[Co(NH_3)_6]^{3+} \qquad 3\ Cl^-$$

central   coordination
ion   ligands   number

In each of the formulas in (19.6) we show that six ligands (blue) are bonded directly to the central Co^{3+} ion. It is this combination of a central metal ion and its ligands, enclosed in square brackets, that we call a complex ion. The region surrounding the central metal ion and containing the ligands is known as the **coordination sphere**. The **coordination number** of the central metal ion is the number of positions in the coordination sphere at which ligands can attach themselves. Some of these terms are further illustrated in the margin.

One of the early successes of Werner's theory was in explaining observations such as this: In the presence of an excess of $AgNO_3(aq)$, compound (a) yields *three moles* AgCl(s) per mole of compound; compound (b) yields only *two moles* AgCl(s); and compound (c) only *one mole*. It is not at all apparent that this should be the case with the formulas shown in (19.5). These results, however, are easily explained with Werner's formulations. In compound (a) all six ligands are NH_3 molecules. The three Cl^- ions (red) are free anions outside the coordination sphere. So, one mole of compound (a) yields *three* mol AgCl. In compound (b) five NH_3 molecules and one Cl^- ion are the ligands and *two* Cl^- ions are free anions. Compound (b) produces only *two* mol AgCl per mole of compound. Compound (c) has only *one* free Cl^- anion and produces only *one* mol AgCl per mole of compound.

Figure 19-5 illustrates two of the coordination compounds that we have been describing. In the next section we will consider equilibria involving complex ions and show how these equilibria affect the dissolution and precipitation of slightly soluble solutes. In Chapter 25 we will explore bonding, structures, properties, and uses of complex ions.

19-9 Equilibria Involving Complex Ions

When moderately concentrated $NH_3(aq)$ is added to AgCl(s), the solid dissolves. Whereas the *simple* compound AgCl(s) is insoluble in water, the *coordination* compound $Ag(NH_3)_2Cl(s)$ is soluble.

$$AgCl(s) + 2\ NH_3(aq) \longrightarrow [Ag(NH_3)_2]^+(aq) + Cl^-(aq) \qquad (19.7)$$

The key to this dissolving action is that Ag^+ from AgCl combines with NH_3 to

FIGURE 19-5
Two coordination compounds.
The compound shown on the left is $[Co(NH_3)_6]Cl_3$. The compound on the right is $[CoCl(NH_3)_5]Cl_2$.

form the complex ion $[Ag(NH_3)_2]^+$, which together with Cl^- represents the soluble coordination compound $[Ag(NH_3)_2]Cl$. It helps to think of reaction (19.7) as involving two simultaneous equilibria.

$$AgCl(s) \rightleftharpoons Ag^+(aq) + Cl^-(aq) \qquad (19.8)$$

$$Ag^+(aq) + 2\,NH_3(aq) \rightleftharpoons [Ag(NH_3)_2]^+(aq) \qquad (19.9)$$

Equilibrium in reaction (19.9) is shifted far to the right—$[Ag(NH_3)_2]^+$ is a stable complex ion. The equilibrium concentration of $Ag^+(aq)$ in (19.9) is kept so low that the ion product $[Ag^+][Cl^-]$ fails to exceed K_{sp}, and AgCl remains in solution. Let's first apply additional qualitative reasoning of this sort in Example 19-10. Then we can turn to some of the quantitative calculations that are also possible.

EXAMPLE 19-10

Predicting Reactions Involving Complex Ions. Predict what will happen if nitric acid is added to a solution of $[Ag(NH_3)_2]Cl$ in $NH_3(aq)$.

Solution

Nitric acid neutralizes free NH_3 in the solution. Because $HNO_3(aq)$ is a strong acid, we represent it as being completely ionized and write only the net ionic equation.

$$H_3O^+(aq) + NH_3(aq) \longrightarrow NH_4^+(aq) + H_2O$$

To replace free NH_3 lost in this neutralization, equilibrium in reaction (19.9) shifts to the *left*. This causes an increase in $[Ag^+]$. When $[Ag^+]$ increases to the point at which the ion product $[Ag^+][Cl^-]$ exceeds K_{sp}, $AgCl(s)$ precipitates (Figure 19-6).

Practice Example A: Copper(II) ion forms an insoluble hydroxide and also the complex ion $[Cu(NH_3)_4]^{2+}$. Write equations to represent the expected reaction when **(a)** $CuSO_4(aq)$ and $NaOH(aq)$ are mixed; **(b)** an *excess* of $NH_3(aq)$ is added to the product of (a); and **(c)** an excess of $HNO_3(aq)$ is added to the product of (b).

Practice Example B: Zinc(II) ion forms an insoluble hydroxide as well as the complex ions $[Zn(OH)_4]^{2-}$ and $[Zn(NH_3)_4]^{2+}$. Write four equations to represent the reactions of **(a)** $NH_3(aq)$ with $ZnSO_4(aq)$; followed by **(b)** enough $HNO_3(aq)$ to make the product of (a) acidic; **(c)** enough $NaOH(aq)$ to make the product of (b) slightly basic; **(d)** enough $NaOH(aq)$ to make the product of (c) strongly basic.

To describe the ionization of a weak acid, we use the ionization constant K_a. For a solubility equilibrium we use the solubility product constant K_{sp}. The equilibrium constant that is used to deal with a complex ion equilibrium is called the **formation constant, K_f**. The **formation constant, K_f**, of a complex ion is the equilibrium constant describing the formation of a complex ion from a central ion and its ligands. For reaction (19.9) this equilibrium constant expression is

$$K_f = \frac{[[Ag(NH_3)_2]^+]}{[Ag^+][NH_3]^2} = 1.6 \times 10^7$$

Table 19.2 lists some representative formation constants, K_f.

One feature that distinguishes K_f from most other equilibrium constants that we have been considering is that K_f values are usually large numbers. This fact can affect the way we approach certain calculations. We sometimes find that

FIGURE 19-6
Reprecipitating AgCl(s).
The reagent being added to the solution containing $[Ag(NH_3)_2]^+$ and Cl^- is $HNO_3(aq)$. H_3O^+ from the acid reacts with $NH_3(aq)$ to form $NH_4^+(aq)$. This upsets the equilibrium between $[Ag(NH_3)_2]^+$, Ag^+, and NH_3. The complex ion is destroyed, $[Ag^+]$ quickly rises to the point at which K_{sp} for AgCl is exceeded, and a precipitate forms.

TABLE 19.2 Formation Constants for Some Complex Ions[a]

Complex Ion	Equilibrium Reaction[b]	K_f
$[Co(NH_3)_6]^{3+}$	$Co^{3+} + 6\ NH_3 \rightleftharpoons [Co(NH_3)_6]^{3+}$	4.5×10^{33}
$[Cu(NH_3)_4]^{2+}$	$Cu^{2+} + 4\ NH_3 \rightleftharpoons [Cu(NH_3)_4]^{2+}$	1.1×10^{13}
$[Fe(CN)_6]^{4-}$	$Fe^{2+} + 6\ CN^- \rightleftharpoons [Fe(CN)_6]^{4-}$	1×10^{37}
$[Fe(CN)_6]^{3-}$	$Fe^{3+} + 6\ CN^- \rightleftharpoons [Fe(CN)_6]^{3-}$	1×10^{42}
$[PbCl_3]^-$	$Pb^{2+} + 3\ Cl^- \rightleftharpoons [PbCl_3]^-$	2.4×10^1
$[Ag(NH_3)_2]^+$	$Ag^+ + 2\ NH_3 \rightleftharpoons [Ag(NH_3)_2]^+$	1.6×10^7
$[Ag(CN)_2]^-$	$Ag^+ + 2\ CN^- \rightleftharpoons [Ag(CN)_2]^-$	5.6×10^{18}
$[Ag(S_2O_3)_2]^{3-}$	$Ag^+ + 2\ S_2O_3{}^{2-} \rightleftharpoons [Ag(S_2O_3)_2]^{3-}$	1.7×10^{13}
$[Zn(NH_3)_4]^{2+}$	$Zn^{2+} + 4\ NH_3 \rightleftharpoons [Zn(NH_3)_4]^{2+}$	4.1×10^8
$[Zn(CN)_4]^{2-}$	$Zn^{2+} + 4\ CN^- \rightleftharpoons [Zn(CN)_4]^{2-}$	1×10^{18}
$[Zn(OH)_4]^{2-}$	$Zn^{2+} + 4\ OH^- \rightleftharpoons [Zn(OH)_4]^{2-}$	4.6×10^{17}

[a]A more extensive tabulation is given in Appendix D.

[b]Tabulated here are *overall* formation reactions and the corresponding formation constants. In Section 25-7 we describe the formation of complex ions in a *stepwise* fashion and introduce formation constants for individual steps.

when working with large K values, it is convenient to solve a problem in two steps. First, assume that the forward reaction goes to completion; and second, assume that a small change occurs in the reverse direction to establish the equilibrium. This approach is illustrated in Example 19-11.

EXAMPLE 19-11

Determining Whether a Precipitate Will Form in a Solution Containing Complex Ions. A 0.10-mol sample of $AgNO_3$ is dissolved in 1.00 L of 1.00 M NH_3. If 0.010 mol NaCl is added to this solution, will AgCl(s) precipitate?

Solution

Let's begin by assuming that, because of the very large value of K_f for $[Ag(NH_3)_2]^+$, the following reaction initially goes to completion, with the results shown.

	$Ag^+(aq)$	$+$	$2\ NH_3(aq)$	$\longrightarrow$	$[Ag(NH_3)_2]^+(aq)$
initial concn, M:	0.10		1.00		
change, M:	−0.10		−0.20		+0.10
after reaction, M:	≈0		0.80		0.10

Of course, the concentration of uncomplexed silver ion, though very small, is not zero. To determine the value of $[Ag^+]$ let's start with $[[Ag(NH_3)_2]^+]$ and $[NH_3]$ in solution and establish $[Ag^+]$ at equilibrium.

	Ag^+	$+$	$2\ NH_3$	$\rightleftharpoons$	$[Ag(NH_3)_2]^+$
initial concn, M:			0.80		0.10
change, M:	$+x$		$+2x$		$-x$
equil concn, M:	x		$0.80 + 2x$		$0.10 - x$

When substituting into the following expression we make the assumption that $x \ll 0.10$, which we will find to be the case.

$$\frac{[[Ag(NH_3)_2]^+]}{[Ag^+][NH_3]^2} = \frac{0.10 - x}{x(0.80 + 2x)^2} \approx \frac{0.10}{x(0.80)^2} = 1.6 \times 10^7$$

$$x = [Ag^+] = \frac{0.10}{(1.6 \times 10^7)(0.80)^2} = 9.8 \times 10^{-9} \text{ M}$$

Finally, we must compare $Q_{sp} = [Ag^+][Cl^-]$ with K_{sp} for AgCl (that is, 1.8×10^{-10}). $[Ag^+]$ is the value of x we just calculated. Because the solution contains 0.010 mol NaCl/L, $[Cl^-] = 0.010 \text{ M} = 1.0 \times 10^{-2} \text{ M}$, and

$$Q_{sp} = (9.8 \times 10^{-9})(1.0 \times 10^{-2}) = 9.8 \times 10^{-11} < 1.8 \times 10^{-10}$$

AgCl will not precipitate.

Practice Example A: Will AgCl(s) precipitate from 1.50 L of a solution that is 0.100 M $AgNO_3$ and 0.225 M NH_3 if 1.00 mL of 3.50 M NaCl is added? (*Hint:* What are $[Ag^+]$ and $[Cl^-]$ immediately after the addition of the 1.00 mL of 3.50 M NaCl? Take into account the dilution of the NaCl(aq), but assume the total volume remains at 1.50 L.)

Practice Example B: A solution is prepared that is 0.100 M in $Pb(NO_3)_2$ and 0.250 M in the ethylenediaminetetraacetate anion, $EDTA^{4-}$. Pb^{2+} and $EDTA^{4-}$ form the complex ion $[PbEDTA]^{2-}$. If the solution is also made 0.10 M in I^-, will $PbI_2(s)$ precipitate? For PbI_2, $K_{sp} = 7.1 \times 10^{-9}$, and for $[PbEDTA]^{2-}$, $K_f = 2 \times 10^{18}$.

Just as some precipitation reactions can be controlled by using a buffer solution (Example 19-9), precipitation from a solution of complex ions can be controlled by fixing the concentration of ligands. This is illustrated for the precipitation of AgCl in Example 19-12.

EXAMPLE 19-12

Controlling a Ligand Concentration to Cause or Prevent Precipitation. What is the *minimum* concentration of NH_3 required to *prevent* AgCl(s) from precipitating from 1.00 L of a solution containing 0.10 mol $AgNO_3$ and 0.010 mol NaCl?

Solution

The $[Cl^-]$ that must be maintained in solution is 1.0×10^{-2} M. If no precipitation is to occur, $[Ag^+][Cl^-] \le K_{sp}$.

$$[Ag^+](1.0 \times 10^{-2}) \le K_{sp} = 1.8 \times 10^{-10} \qquad [Ag^+] \le 1.8 \times 10^{-8} \text{ M}$$

Thus, the maximum concentration of *uncomplexed* Ag^+ permitted in solution is 1.8×10^{-8} M. This means that essentially all the Ag^+ (0.10 mol/L) must be tied up (complexed) in the complex ion, $[Ag(NH_3)_2]^+$. We need to solve the following expression for $[NH_3]$.

$$K_f = \frac{[[Ag(NH_3)_2]^+]}{[Ag^+][NH_3]^2} = \frac{1.0 \times 10^{-1}}{1.8 \times 10^{-8} [NH_3]^2} = 1.6 \times 10^7$$

$$[NH_3]^2 = \frac{1.0 \times 10^{-1}}{1.8 \times 10^{-8} \times 1.6 \times 10^7} = 0.35 \qquad [NH_3] = 0.59 \text{ M}$$

The concentration just calculated is that of *free, uncomplexed* NH_3. Considering as well the 0.20 mol NH_3/L complexed in the 0.10 M $[Ag(NH_3)_2]^+$, the total concentration of NH_3(aq) required is

$$[NH_3]_{tot} = 0.59 \text{ M} + 0.20 \text{ M} = 0.79 \text{ M}$$

Practice Example A: What $[NH_3]_{total}$ is necessary to keep AgCl from precipitating from a solution that is 0.13 M $AgNO_3$ and 0.0075 M NaCl?

Practice Example B: What minimum concentration of thiosulfate ion, $S_2O_3^{2-}$, should be present in 0.10 M $AgNO_3(aq)$ so that AgCl(s) does not precipitate when the solution is also made 0.010 M in Cl^-? AgCl: $K_{sp} = 1.8 \times 10^{-10}$; $[Ag(S_2O_3)_2]^{3-}$: $K_f = 1.7 \times 10^{13}$. (*Hint:* What is the maximum $[Ag^+]$ that can be tolerated in the solution? What must the *total* concentration of $S_2O_3^{2-}$ be to keep the free $[Ag^+]$ below this level?)

On page 671 we described qualitatively how the solubility of AgCl increases in the presence of $NH_3(aq)$. In Example 19-13 we show how the actual solubility of AgCl in $NH_3(aq)$ can be calculated.

EXAMPLE 19-13

Determining the Solubility of a Solute When Complex Ions Are Formed. What is the molar solubility of AgCl in 0.100 M $NH_3(aq)$?

Solution

As we have already seen, Equation (19.7) describes the solubility equilibrium.

$$AgCl(s) + 2\,NH_3(aq) \rightleftharpoons [Ag(NH_3)_2]^+(aq) + Cl^-(aq) \qquad (19.7)$$

Let's base our calculation on the equilibrium constant K for reaction (19.7), and let's consider two ways of obtaining this value: By one method we obtain Equation (19.7) as the sum of Equations (19.8) and (19.9) on page 671, and we obtain its K value as the product of a K_{sp} and a K_f.

▶ This is similar to the method we used to obtain K for reaction (19.4) from Equations (19.2) and (19.3) on page 667.

$$AgCl(s) \rightleftharpoons Ag^+(aq) + Cl^-(aq) \qquad K_{sp} = 1.8 \times 10^{-10}$$

$$Ag^+(aq) + 2\,NH_3(aq) \rightleftharpoons [Ag(NH_3)_2^+](aq) \qquad K_f = 1.6 \times 10^7$$

$$AgCl(s) + 2\,NH_3(aq) \rightleftharpoons [Ag(NH_3)_2^+](aq) + Cl^-(aq)$$

$$K = K_{sp} \times K_f = 1.8 \times 10^{-10} \times 1.6 \times 10^7 = 2.9 \times 10^{-3}$$

▶ This is similar to the method we used to obtain K_a for ammonium ion from K_b for ammonia and the ion product of water, K_w, on page 602.

By a second method, we first write the equilibrium constant expression for reaction (19.7), and then multiply numerator and denominator by $[Ag^+]$.

$$K = \frac{[Ag(NH_3)_2^+][Cl^-]}{[NH_3]^2} = \frac{[Ag(NH_3)_2^+][Cl^-][Ag^+]}{[NH_3]^2[Ag^+]} = K_f \times K_{sp} = 2.9 \times 10^{-3}$$

The expression printed in red is K_f for $[Ag(NH_3)_2]^+$ and the one in blue is K_{sp} for AgCl. The K value for reaction (19.7) is the product of the two.

According to Equation (19.7), if s mol AgCl/L dissolves (the molar solubility), the expected concentrations of $[Ag(NH_3)_2]^+$ and Cl^- are also equal to s.

▶ The molar solubility s is actually the *total* concentration of silver in solution: $[Ag^+] + [[Ag(NH_3)_2]^+]$. When K_f is large and the concentration of ligands is sufficiently high, as is the case here, we can ignore $[Ag^+]$ in comparison with $[[Ag(NH_3)_2]^+]$.

	AgCl(s) +	2 NH$_3$(aq) $\rightleftharpoons$	[Ag(NH$_3$)$_2$]$^+$(aq) +	Cl$^-$(aq)
initial concn, M:		0.100		
change, M:		$-2s$	$+s$	$+s$
equil concn, M:		$(0.100 - 2s)$	s	s

$$K = \frac{[[Ag(NH_3)_2]^+][Cl^-]}{[NH_3]^2} = \frac{s \cdot s}{(0.100 - 2s)^2} = \left(\frac{s}{0.100 - 2s}\right)^2 = 2.9 \times 10^{-3}$$

We can solve this equation by taking the square root of both sides.

$$\frac{s}{0.100 - 2s} = \sqrt{2.9 \times 10^{-3}} = 5.4 \times 10^{-2}$$

$$s = 5.4 \times 10^{-3} - 0.11s$$

$$1.11s = 5.4 \times 10^{-3}$$

$$s = 4.9 \times 10^{-3}$$

The molar solubility of AgCl(s) in 0.100 M NH_3(aq) is 4.9×10^{-3} M.

Note: The usual simplifying assumption, that is, $(0.100 - 2s) \approx 0.100$, would not have worked too well in this calculation. The value of s obtained would have been 5.4×10^{-3} M, and $(0.100 - 2 \times 0.0054) \neq 0.100$. Fortunately, the quadratic equation was easily solved by the square root extraction. If this had not been so, we could have used either the quadratic formula or the method of successive approximations (Feature Problem B of Chapter 17, page 619).

Practice Example A: What is the molar solubility of $Fe(OH)_3$ in a solution containing 0.100 M $C_2O_4^{2-}$? $K_f[Fe(C_2O_4)_3^{3-}] = 2 \times 10^{20}$.

Practice Example B: Show, without doing detailed calculations, that the order of *decreasing* solubility in 0.100 M NH_3(aq) should be AgCl > AgBr > AgI. (*Hint:* Use K_{sp} values from Table 19.1 and portions of the solution of Example 19-13.)

19-10   Qualitative Analysis of Cations

In *qualitative* analysis we determine what substances are present in a mixture but *not* their quantities. If the analysis aims at identifying the cations present in a mixture, it is called **qualitative cation analysis.** Qualitative cation analysis is not as important a method as it once was because now most qualitative and quantitative analyses are done with instruments. Its current value is in the wealth of illustrations it provides of precipitation (and dissolution) equilibria, acid–base equilibria, and oxidation–reduction reactions. Also, in the general chemistry laboratory it offers the challenge of unraveling a mystery—solving a qualitative analysis "unknown."

In the scheme in Figure 19-7 on page 676, cations are divided into five groups depending on differing solubilities of their compounds. The first cations separated are those with insoluble *chlorides*—Pb^{2+}, Hg_2^{2+}, and Ag^+. The reagent used is HCl(aq). All other cations remain in solution because their chlorides are soluble. After removal of the chloride group precipitate, the solution is treated with H_2S in an acidic medium. Under these conditions a group of sulfides precipitates. Next, the solution containing the remaining cations is treated with H_2S in a basic medium to yield a mixture of insoluble hydroxides and sulfides. The fourth group that is precipitated consists of the carbonates of Mg^{2+}, Ca^{2+}, Sr^{2+}, and Ba^{2+}. At the end of this series of precipitations, the resulting solution contains only Na^+, K^+, and NH_4^+, all of whose common salts are water-soluble.

In this section, we will discuss the chemistry of the chloride and sulfide groups. The chemistry of metal carbonates is discussed in Chapter 22.

> ► We can think of Hg_2^{2+} as an ion formed by the covalent bonding of an Hg atom to an Hg^{2+} ion.

Cation Group 1: The Chloride Group

If a precipitate forms when a solution is treated with HCl(aq), one or more of these cations must be present: Pb^{2+}, Hg_2^{2+}, Ag^+. To establish the presence or absence of each of these three cations, the chloride group precipitate is filtered off and subjected to further testing.

FIGURE 19-7
Outline of a qualitative analysis scheme for cations.
Various aspects of this scheme are described in the text. A sample containing all 25 cations can be separated into five groups by the indicated reagents.

FIGURE 19-8
Chloride group precipitates.
(a) Group precipitate: $PbCl_2$ (white), Hg_2Cl_2 (white), and AgCl (white). (b) Test for Hg_2^{2+}: a mixture of Hg (black) and $HgNH_2Cl$ (white). (c) Test for Pb^{2+}: $PbCrO_4$ (yellow).

▶ Metallic mercury in bulk appears as a silvery liquid. Here, the particles of mercury are so small that they seem black in color. A black color is common for finely divided metals.

Of the three chlorides in the group precipitate, $PbCl_2(s)$ is the most soluble; its K_{sp} is much larger than those of AgCl and Hg_2Cl_2. When the precipitate is washed with hot water, a sufficient quantity of $PbCl_2$ dissolves to permit a test for Pb^{2+} in the solution. In this test a lead compound less soluble than $PbCl_2$, such as lead chromate, is precipitated (Figure 19-8).

$$Pb^{2+}(aq) + CrO_4^{2-}(aq) \longrightarrow PbCrO_4(s)$$

The portion of the chloride group precipitate that is insoluble in hot water is then treated with $NH_3(aq)$. Two things happen. One is that any AgCl(s) present dissolves and forms the complex ion $[Ag(NH_3)_2^+]$.

$$AgCl(s) + 2\ NH_3(aq) \longrightarrow [Ag(NH_3)_2]^+(aq) + Cl^-(aq) \qquad (19.7)$$

At the same time, any $Hg_2Cl_2(s)$ present undergoes an *oxidation–reduction* reaction. One of the products of the reaction is finely divided *black* mercury.

$$Hg_2Cl_2(s) + 2\ NH_3(aq) \longrightarrow \underbrace{Hg(l) + HgNH_2Cl(s)}_{\text{dark gray}} + NH_4^+(aq) + Cl^-(aq)$$

The appearance of a dark gray mixture of black mercury and white $HgNH_2Cl$ [mercury(II) amidochloride] is the qualitative analysis confirmation of mercury(I) ion (see Figure 19-8).

When the solution from reaction (19.7) is acidified with $HNO_3(aq)$, any silver ion present precipitates as $AgCl(s)$. This is the reaction we predicted in Example 19-10 and pictured in Figure 19-6.

Cation Groups 2 and 3:
Equilibria Involving Hydrogen Sulfide

Figure 19-7 suggests that aqueous hydrogen sulfide (hydrosulfuric acid) is the key reagent in the analysis of cation groups 2 and 3. In aqueous solution H_2S is a *weak diprotic acid*.

$$H_2S(aq) + H_2O \rightleftharpoons H_3O^+(aq) + HS^-(aq) \qquad K_{a_1} = 1.0 \times 10^{-7}$$

$$HS^-(aq) + H_2O \rightleftharpoons H_3O^+(aq) + S^{2-}(aq) \qquad K_{a_2} = 1 \times 10^{-19}$$

The extremely small value of K_{a_2} suggests that sulfide ion is a very strong base, as seen in this hydrolysis reaction and K_b value.

$$S^{2-} + H_2O \rightleftharpoons HS^- + OH^- \qquad K_b = K_w/K_{a_2}$$
$$= (1.0 \times 10^{-14})/(1 \times 10^{-19}) = 1 \times 10^5$$

The hydrolysis of S^{2-} should go nearly to completion, which means that very little S^{2-} can exist in an aqueous solution and that sulfide ion is probably not the precipitating agent for sulfides.

One way to handle the precipitation and dissolution of sulfide precipitates is to restrict our discussion to acidic solutions. Then we can write an equilibrium constant expression in which we eliminate concentration terms for HS^- and S^{2-}. This is a reasonable approach because most sulfide separations are carried out in acidic solution.

Consider (1) the solubility equilibrium equation for PbS written to reflect hydrolysis of S^{2-}, (2) an equation written for the reverse of the first ionization of H_2S, and (3) an equation that is the reverse of the self-ionization of water. We can combine these three equations into a net equation that shows the dissolving of $PbS(s)$ in an acidic solution. The equilibrium constant for this net equation is generally referred to as K_{spa}.*

(1) $\qquad PbS(s) + H_2O \rightleftharpoons Pb^{2+}(aq) + \cancel{HS^-(aq)} + \cancel{OH^-(aq)} \qquad K_{sp} = 3 \times 10^{-28}$

(2) $\quad H_3O^+(aq) + \cancel{HS^-(aq)} \rightleftharpoons H_2S(aq) + H_2O \qquad\qquad 1/K_{a_1} = 1/1.0 \times 10^{-7}$

(3) $\quad H_3O^+(aq) + \cancel{OH^-(aq)} \rightleftharpoons \cancel{H_2O} + H_2O \qquad\qquad 1/K_w = 1/1.0 \times 10^{-14}$

net: $\quad PbS(s) + 2\,H_3O^+(aq) \rightleftharpoons Pb^{2+}(aq) + H_2S(aq) + 2\,H_2O \qquad K_{spa} = ?$

$$K_{spa} = \frac{K_{sp}}{K_{a_1} \times K_w} = \frac{3 \times 10^{-28}}{(1.0 \times 10^{-7})(1.0 \times 10^{-14})} = 3 \times 10^{-7}$$

Example 19-14 illustrates the use of K_{spa} in the type of calculation needed to sort the sulfides into two different qualitative analysis groups. Pb^{2+} is in qualitative analysis cation group 2 and Fe^{2+} is in group 3. The conditions cited in the example are those generally used.

*See R. J. Myers, *J. Chem. Educ.* **63**, 687 (1986).

▶ $H_2S(g)$ has a familiar "rotten egg" odor, especially noticeable in volcanic areas and near sulfur hot springs. It can cause nausea and headaches at levels of 10 ppm in air, and it can produce paralysis or death at levels of 100 ppm. The gas is detectable through its odor at levels of 1 ppm, although exposure to the gas deadens the sense of smell.

▶ $PbCl_2$ is sufficiently soluble that enough Pb^{2+} ions from cation group 1 remain in solution to precipitate again as $PbS(s)$ in cation group 2.

▶ The value of K_{spa} of FeS can be derived from K_{sp} of FeS (6×10^{-19}) by the method outlined previously for PbS.

EXAMPLE 19-14

Separating Metal Ions by Selective Precipitation of Metal Sulfides. Show that PbS(s) should precipitate and FeS(s) should not precipitate from a solution that is 0.010 M in Pb^{2+}, 0.010 M in Fe^{2+}, saturated in H_2S (0.10 M H_2S), and maintained with $[H_3O^+] = 0.30$ M. For PbS, $K_{spa} = 3 \times 10^{-7}$ and for FeS, $K_{spa} = 6 \times 10^2$.

Solution

We must determine whether, for the stated conditions, equilibrium is displaced in the forward or reverse direction in reactions of the type

$$MS(s) + 2\, H_3O^+(aq) \rightleftharpoons M^{2+}(aq) + H_2S(aq) + 2\, H_2O \qquad (19.10)$$

where M represents either Pb or Fe. In each case we can compare the Q_{spa} expression to the appropriate K_{spa} value for reaction (19.10). If $Q_{spa} > K_{spa}$, a net reaction should occur to the *left*, which means that MS(s) should precipitate. If $Q_{spa} < K_{spa}$, a net reaction should occur to the *right*. This means that some of the metal sulfide could actually be dissolved in the solution and certainly signifies that no precipitation should occur.

$$Q_{spa} = \frac{[M^{2+}][H_2S]}{[H_3O^+]^2} = \frac{0.010 \times 0.10}{(0.30)^2} = 1.1 \times 10^{-2}$$

For PbS: Q_{spa} of $1.1 \times 10^{-2} > K_{spa}$ of 3×10^{-7}. We expect precipitation of PbS(s) to occur.

For FeS: Q_{spa} of $1.1 \times 10^{-2} < K_{spa}$ of 6×10^2. Precipitation of FeS(s) should not occur.

Practice Example A: Show that $Ag_2S(s)$ $(K_{spa} = 6 \times 10^{-30})$ should precipitate and FeS(s) $(K_{spa} = 6 \times 10^2)$ should not precipitate from a solution that is 0.010 M Ag^+ and 0.020 M Fe^{2+} and otherwise is under the same conditions as in Example 19-14.

Practice Example B: What is the minimum pH of a solution that is 0.015 M Fe^{2+} and saturated in H_2S (0.10 M) from which FeS(s) $(K_{spa} = 6 \times 10^2)$ can be precipitated? (*Hint:* What $[H_3O^+]$ is required for Q_{spa} to equal K_{spa}?)

Dissolving Metal Sulfides

In the qualitative analysis scheme it is necessary both to precipitate and to redissolve sulfides. Here, we look at several methods of dissolving metal sulfides. One way to increase the solubility of any sulfide is to allow it to react with an acid, as suggested by Equation (19.10). According to Le Châtelier's principle, the solubility increases as the solution is made more acidic—equilibrium is shifted to the right.

If our interest is in calculating the molar solubility of a particular metal sulfide in a solution with a known $[H_3O^+]$, we can set up the equilibrium constant expression for reaction (19.10) and solve for $[M^{2+}]$. In this case, of course, we must have a numerical value of K_{sp} or K_{spa} of the metal sulfide.

Another way to promote the dissolving of metal sulfides is to use an *oxidizing* acid such as $HNO_3(aq)$. In this case sulfide ion is oxidized to elemental sulfur and the free metal ion appears in solution, as in the dissolving of CuS(s).

$$3\, CuS(s) + 8\, H^+(aq) + 2\, NO_3{}^-(aq) \longrightarrow$$

$$3\, Cu^{2+}(aq) + 3\, S(s) + 2\, NO(g) + 4\, H_2O \qquad (19.11)$$

Figure 19-9 shows insoluble solid sulfur and $Cu^{2+}(aq)$ produced in reaction (19.11). To render the $Cu^{2+}(aq)$ more visible, it is converted to the deeply colored complex ion, $[Cu(NH_3)_4]^{2+}(aq)$, in a reaction in which NH_3 molecules replace H_2O molecules as ligands in the complex ion.

$$[Cu(H_2O)_4]^{2+}(aq) + 4\ NH_3(aq) \longrightarrow [Cu(NH_3)_4]^{2+}(aq) + 4\ H_2O$$
$$\text{pale blue} \qquad\qquad\qquad\qquad\qquad \text{deep blue}$$

A few metal sulfides dissolve in a basic solution with a high concentration of HS^-, just as acidic oxides dissolve in solutions with a high concentration of OH^-. This property is used to advantage in separating the eight sulfides of cation group 2, the hydrogen sulfide group, into two subgroups. The subgroup consisting of HgS, PbS, CuS, Bi_2S_3, and CdS remains undissolved after treatment with an alkaline solution with an excess of HS^-, but As_2S_3, Sb_2S_3, and SnS_2 dissolve.

(a)   **(b)**

FIGURE 19-9
The dissolving of CuS(s) and a test for $Cu^{2+}(aq)$.
(a) CuS(s) reacts with $HNO_3(aq)$ to produce an aqueous solution containing Cu^{2+} and a precipitate of elemental sulfur. (b) Addition of $NH_3(aq)$ to the $Cu^{2+}(aq)$ produces the deep blue complex ion $[Cu(NH_3)_4]^{2+}$.

Summary

Equilibrium between a slightly soluble ionic compound and its ions in solution is expressed through the *solubility product constant, K_{sp}*. Although K_{sp} and molar solubility are not equal, they are related to each other in a way that makes it possible to calculate one when the other is known. The solubility of a slightly soluble solute is greatly reduced in a solution containing an ion in common with the solubility equilibrium—a common *ion*.

A comparison of the *ion product* (reaction quotient) Q_{sp} with K_{sp} provides a criterion for precipitation: If $Q_{sp} > K_{sp}$, precipitation should occur; if $Q_{sp} < K_{sp}$, the solution remains unsaturated. A comparison of K_{sp} values is a factor in determining the feasibility of *fractional precipitation*. This is a process in which one type of ion is removed by precipitation while others remain in solution. At times, because of acid–base equilibria that occur simultaneously, the selective precipitation or dissolving of ionic solutes depends on the pH.

The formation of a complex ion from a central metal ion and ligands is an equilibrium process with an equilibrium constant called the *formation constant, K_f*. In general, if the formation constant of a complex ion is large, the concentration of uncomplexed metal ion in equilibrium with the complex ion is very small. Complex ion formation can render certain insoluble materials quite soluble in appropriate aqueous solutions, such as AgCl in $NH_3(aq)$.

Precipitation, acid–base, oxidation–reduction, and complex ion formation reactions are all used extensively in *qualitative analysis*. Such an analysis can provide a rapid means of determining the presence or absence of certain ions in an unknown material.

Focus On   Shells, Teeth, and Fossils

Chambered nautilus.

Although most of Earth's minerals have formed by cooling from molten material called magma, a few have been produced by precipitation processes. For example, warm seawater is nearly saturated in Ca^{2+} and HCO_3^- ions and can yield the mineral *calcite* ($CaCO_3$) by precipitation:

$$Ca^{2+}(aq) + 2\ HCO_3^-(aq) \longrightarrow$$
$$CaCO_3(s) + H_2O + CO_2(g)$$

Chemical precipitation, however, is not the principal source of calcite deposits in the ocean. Most calcite and related carbonate deposits originate from the biological precipitation of minerals to form the shells of certain organisms. These organisms can extract Ca^{2+} and HCO_3^- ions from seawater and concentrate them in solution in specialized cells. The organisms then secrete crystals of $CaCO_3$ to build shells within which they live.

Some organisms secrete harder materials than carbonates and use

Summarizing Example

Because of its limited solubility, $Ca(OH)_2(s)$ cannot be used to prepare aqueous solutions of high pH.

$$Ca(OH)_2(s) \rightleftharpoons Ca^{2+}(aq) + 2\ OH^-(aq)$$
$$K_{sp} = 5.5 \times 10^{-6}$$

On the other hand, when $Ca(OH)_2(s)$ reacts with a *soluble* carbonate such as $Na_2CO_3(aq)$, the solution produced has a much higher pH. Equilibrium is displaced to the right in reaction (19.13) because $CaCO_3$ is much less soluble than $Ca(OH)_2$.

$$Ca(OH)_2(s) + CO_3^{2-}(aq) \rightleftharpoons CaCO_3(s) + 2\ OH^-(aq)$$
$$(19.13)$$

Assume an initial $[CO_3^{2-}] = 1.0$ M in reaction (19.13), and show that the equilibrium pH should indeed be higher than in saturated $Ca(OH)_2(aq)$.

1. *Calculate the pH of saturated $Ca(OH)_2(aq)$.* Write the K_{sp} expression for $Ca(OH)_2$. Let s be the molar solubility, and $[OH^-] = 2s$. Obtain $[OH^-]$, pOH, and pH. *Result:* pH = 12.34.

2. *Determine the value of K_c for reaction (19.13).* Combine solubility equilibrium equations for $Ca(OH)_2$ and $CaCO_3$ to obtain (19.13) as the net equation. Combine K_{sp} values of $Ca(OH)_2$ and $CaCO_3$ in the appropriate way to obtain the net K_c. *Result:* $K_{sp}[Ca(OH)_2]/K_{sp}[CaCO_3] = [OH^-]^2/[CO_3^{2-}] = 2.0 \times 10^3$.

3. *Calculate the equilibrium pH in reaction (19.13).* Start with $[CO_3^{2-}] = 1.0$ M. At equilibrium, $[CO_3^{2-}] = 1.0 - x$ and $[OH^-] = 2x$. Solve a quadratic equation to obtain $[OH^-]$, and then calculate pOH and pH. *Answer:* pH = 14.3.

them to form bones and teeth. Human bones and tooth enamel, for example, are made principally of the mineral *hydroxyapatite*, $Ca_5(PO_4)_3OH$. Hydroxyapatite is somewhat soluble in acidic solutions, and this is a matter of concern in good dental health. Bacteria in the mouth produce acids from food trapped between teeth. These acids cause tooth enamel to dissolve through reactions such as

$$Ca_5(PO_4)_3OH(s) + 4\ H_3O^+(aq) \longrightarrow$$
$$5\ Ca^{2+}(aq) + 3\ HPO_4^{2-}(aq) + 5\ H_2O$$

Dissolution of tooth enamel, referred to as *demineralization*, results in the formation of cavities if it proceeds unchecked. Fortunately, removal of the bacteria and food particles from teeth by brushing and flossing is one effective strategy to combat tooth decay. Another strategy is to alter the composition of the tooth enamel to make it more resistant to attack by acids.

Fluorapatite, $Ca_5(PO_4)_3F$, is less soluble in acids than is hydroxyapatite because F^- is a weaker base than OH^-. Brushing the teeth with a fluoride-containing toothpaste sets up a competitive equilibrium that results in the conversion of some hydroxyapatite to fluorapatite, a process called *remineralization.*

$$Ca_5(PO_4)_3OH(s) + F^-(aq) \rightleftharpoons$$
$$Ca_5(PO_4)_3F(s) + OH^-(aq) \quad (19.12)$$

In addition to its lower solubility in acids, fluorapatite forms crystals that pack more densely than do crystals of hydroxyapatite, making the fluorapatite structure less porous and more resistant to penetration by acidic solutions.

Another source of fluoride ions for promoting the remineralization reaction (19.12) is drinking water. Some groundwater is naturally fluoridated, but in most communities that use fluoridated water the fluoridation is achieved by adding a soluble fluoride, enough to raise the F^- concentration in the water to about 1 ppm.

The same replacement of OH^- by F^- as described for fighting tooth decay occurs in the fossilization of bones and in natural deposits of hydroxyapatite. Fluoride ion in groundwater promotes reaction (19.12). As a result, phosphate rock, used in the manufacture of phosphorus and phosphorus-containing fertilizers, is a mixture of $Ca_5(PO_4)_3OH$ and $Ca_5(PO_4)_3F$. As a further consequence, industrial operations that use phosphate rock as a source of phosphorus invariably generate large quantities of CaF_2 as an unwanted by-product. Ancient bones recovered by archeologists are composed almost entirely of fluorapatite. The fossilized (remineralized) bones retain the form of the original bones, including nicks, cuts, and breaks, and thus preserve a record of the life of the original animal or person.

Key Terms

complex ion (19-8)
coordination number (19-8)
coordination sphere (19-8)
formation constant, K_f (19-9)

fractional precipitation (19-6)
ion pairs (19-4)
ion product, Q_{sp} (19-5)
ligands (19-8)

qualitative cation analysis (19-10)
salt effect (19-4)
solubility product constant, K_{sp} (19-1)

Review Questions

1. In your own words define the following terms or symbols: **(a)** K_{sp}; **(b)** K_f; **(c)** ligand; **(d)** coordination number.
2. Briefly describe each of the following ideas, methods, or phenomena: **(a)** common ion effect in solubility equilibrium; **(b)** fractional precipitation; **(c)** ion pair formation; **(d)** qualitative cation analysis.

3. Explain the important distinction between each pair of terms: **(a)** solubility and solubility product constant; **(b)** common ion effect and salt effect; **(c)** complex ion and coordination compound.
4. Write K_{sp} expressions for the following equilibria. For example, for the reaction $AgCl(s) \rightleftharpoons Ag^+(aq) + Cl^-(aq)$, $K_{sp} = [Ag^+][Cl^-]$.

(a) $Ag_2SO_4(s) \rightleftharpoons 2\ Ag^+(aq) + SO_4^{2-}(aq)$
(b) $Ra(IO_3)_2(s) \rightleftharpoons Ra^{2+}(aq) + 2\ IO_3^-(aq)$
(c) $Ni_3(PO_4)_2(s) \rightleftharpoons 3\ Ni^{2+}(aq) + 2\ PO_4^{3-}(aq)$
(d) $PuO_2CO_3(s) \rightleftharpoons PuO_2^{2+}(aq) + CO_3^{2-}(aq)$

5. Write solubility equilibrium equations that are described by the following K_{sp} expressions. For example, $K_{sp} = [Ag^+][Cl^-]$ represents $AgCl(s) \rightleftharpoons Ag^+(aq) + Cl^-(aq)$.
(a) $K_{sp} = [Fe^{3+}][OH^-]^3$
(b) $K_{sp} = [BiO^+][OH^-]$
(c) $K_{sp} = [Hg_2^{2+}][I^-]^2$
(d) $K_{sp} = [Pb^{2+}]^3[AsO_4^{3-}]^2$

6. The following K_{sp} values are found in a handbook. Write the solubility product expression to which each one applies. For example, $K_{sp}(AgCl) = [Ag^+][Cl^-] = 1.8 \times 10^{-10}$.
(a) $K_{sp}(CrF_3) = 6.6 \times 10^{-11}$
(b) $K_{sp}[Au_2(C_2O_4)_3] = 1 \times 10^{-10}$
(c) $K_{sp}[Cd_3(PO_4)_2] = 2.1 \times 10^{-33}$
(d) $K_{sp}(SrF_2) = 2.5 \times 10^{-9}$

7. Calculate the aqueous solubility, in moles per liter, of each of the following.
(a) $BaCrO_4$, $K_{sp} = 1.2 \times 10^{-10}$
(b) $PbBr_2$, $K_{sp} = 4.0 \times 10^{-5}$
(c) CeF_3, $K_{sp} = 8 \times 10^{-16}$
(d) $Mg_3(AsO_4)_2$, $K_{sp} = 2.1 \times 10^{-20}$

8. The following aqueous solubility data, expressed as molarities, are from a handbook. What are the K_{sp} values of these solutes? (a) $CsMnO_4$, 3.8×10^{-3} M; (b) $Pb(ClO_2)_2$, 2.8×10^{-3} M; (c) Li_3PO_4, 2.9×10^{-3} M

9. Pure water is saturated with slightly soluble PbI_2. Which of these is a correct statement concerning the lead ion concentration in the solution, and what is wrong with the others? (a) $[Pb^{2+}] = [I^-]$; (b) $[Pb^{2+}] = K_{sp}$ of PbI_2; (c) $[Pb^{2+}] = \sqrt{K_{sp}}$ of PbI_2; (d) $[Pb^{2+}] = 0.5[I^-]$.

10. Calculate the molar solubility of $Mg(OH)_2$ in (a) pure water; (b) 0.0862 M $MgCl_2$; (c) 0.0355 M KOH(aq). K_{sp} of $Mg(OH)_2 = 1.8 \times 10^{-11}$.

11. How would you expect the presence of each of the following solutes to affect the molar solubility of $CaCO_3$ in water: (a) Na_2CO_3; (b) HCl; (c) $NaHSO_4$? Explain.

12. Predict whether a precipitate is expected to form in a solution with the ion concentrations listed.
(a) $[Mg^{2+}] = 0.0037$ M, $[CO_3^{2-}] = 0.0068$ M
K_{sp} of $MgCO_3 = 3.5 \times 10^{-8}$
(b) $[Ag^+] = 0.018$ M, $[SO_4^{2-}] = 0.0062$ M
K_{sp} of $Ag_2SO_4 = 1.4 \times 10^{-5}$
(c) $[Cr^{3+}] = 0.038$ M, pH $= 3.20$
K_{sp} of $Cr(OH)_3 = 6.3 \times 10^{-31}$

13. A solution that is 0.103 M $CaCl_2$ is also made to be 0.750 M K_2SO_4. What percent of the Ca^{2+} remains *unprecipitated*? Would you say that the precipitation goes es-

sentially to completion? Assume that $[SO_4^{2-}]$ remains constant at 0.750 M. K_{sp} of $CaSO_4 = 9.1 \times 10^{-6}$.

14. KI(aq) is slowly added to a solution with $[Pb^{2+}] = [Ag^+] = 0.10$ M. $K_{sp}(PbI_2) = 7.1 \times 10^{-9}$; $K_{sp}(AgI) = 8.5 \times 10^{-17}$.
(a) Which precipitate should form first, PbI_2 or AgI?
(b) What $[I^-]$ is required for the *second* cation to begin to precipitate?
(c) What concentration of the first cation to precipitate remains in solution at the point at which the second cation begins to precipitate?
(d) Can $Pb^{2+}(aq)$ and $Ag^+(aq)$ be effectively separated by fractional precipitation?

15. In which of the following solutions would you expect $Mg(OH)_2(s)$ to be most soluble: NaOH(aq), Na_2CO_3(aq), $NaHSO_4$(aq)? Explain your choice.

16. Complete and balance the following equations. If no reaction occurs, so state.
(a) $Ag^+(aq) + NO_3^-(aq) + Na^+(aq) + Br^-(aq) \rightarrow$
(b) $Cu^{2+}(aq) + NO_3^-(aq) + H_3O^+(aq) + Cl^-(aq) \rightarrow$
(c) $Fe^{2+}(aq) + H_2S(aq, \text{in } 0.3 \text{ M HCl}) \rightarrow$
(d) $Cu(OH)_2(s) + NH_3(aq) \rightarrow$
(e) $Fe^{3+}(aq) + NH_3(aq) + H_2O \rightarrow$
(f) $Ag_2SO_4(s) + NH_3(aq) \rightarrow$
(g) $CaSO_3(s) + H_3O^+(aq) \rightarrow$

17. $Cu(OH)_2(s)$ is insoluble in water but reacts to dissolve in each of the following: HCl(aq), NH_3(aq), HNO_3(aq). Write net ionic equations for these reactions.

18. Cu^{2+} and Ag^+ are both present in the same aqueous solution. Which of these reagents would work best in separating these ions, precipitating one and leaving the other in solution: $(NH_4)_2CO_3$(aq), HCl(aq), NaOH(aq)? Explain your choice.

19. Mg^{2+} and Cu^{2+} are both present in the same aqueous solution. Which of these reagents would work best in separating these ions, precipitating one and leaving the other in solution: NaOH(aq), HCl(aq), NH_3(aq)? Explain your choice.

20. Will $Al(OH)_3(s)$ precipitate from a buffer solution that is 0.45 M $HC_2H_3O_2$ and 0.35 M $NaC_2H_3O_2$ and also 0.275 M in Al^{3+}(aq)? K_{sp} of $Al(OH)_3 = 1.3 \times 10^{-33}$; K_a of $HC_2H_3O_2 = 1.8 \times 10^{-5}$.

21. Will AgI(s) precipitate from a solution with $[[Ag(CN)_2]^-] = 0.012$ M, $[CN^-] = 1.05$ M, and $[I^-] = 2.0$ M? K_{sp} of AgI $= 8.5 \times 10^{-17}$; K_f of $[Ag(CN)_2]^- = 5.6 \times 10^{18}$. (*Hint:* What is $[Ag^+]$ in this solution? Is $[Ag^+]$ large enough that $Q_{sp} = [Ag^+][I^-] > K_{sp}$?)

22. A solution is 1.6 M in $[Ag(NH_3)_2]^+$ and 1.25 M in *free* NH_3. What is the maximum $[Cl^-]$ that can be maintained in the solution without forming a precipitate of AgCl(s)? K_{sp} of AgCl $= 1.8 \times 10^{-10}$; K_f of $[Ag(NH_3)_2]^+ = 1.6 \times 10^7$. (*Hint:* What is $[Ag^+]$ in this solution?)

Exercises

(Use data from Chapters 17 and 19 and Appendix D, as needed.)

K_{sp} and Solubility

23. Arrange the following solutes in order of *increasing* molar solubility in water: $AgCN$, $AgIO_3$, AgI, $AgNO_2$, Ag_2SO_4. Explain your reasoning.

25. Fluoridated drinking water contains about 1 part per million (ppm) of F^-. Is CaF_2 sufficiently soluble in water to be used as the source of fluoride ion for the fluoridation of drinking water? Explain. (*Hint:* Think of 1 ppm as signifying 1 g F^- per 10^6 g solution.)

27. A solution is saturated with magnesium palmitate $[Mg(C_{16}H_{31}O_2)_2$, a component of bathtub ring] at 50 °C. How many milligrams of magnesium palmitate will precipitate from 965 mL of this solution when it is cooled to 25 °C? For magnesium palmitate, $K_{sp} = 4.8 \times 10^{-12}$ at 50 °C and 3.3×10^{-12} at 25 °C.

29. A 25.00-mL sample of a clear *saturated* solution of PbI_2 requires 13.3 mL of a certain $AgNO_3(aq)$ for its titration. What is the molarity of this $AgNO_3(aq)$?

$$I^-(\text{satd } PbI_2) + Ag^+(\text{from } AgNO_3) \longrightarrow AgI(s)$$

31. To precipitate as $Ag_2S(s)$ all the Ag^+ present in 338 mL of a saturated solution of $AgBrO_3$ requires 30.4 mL of $H_2S(g)$ measured at 23 °C and 748 mmHg. What is K_{sp} for $AgBrO_3$?

$$2 Ag^+(aq) + H_2S(g) \longrightarrow Ag_2S(s) + 2 H^+(aq)$$

The Common Ion Effect

33. Describe the effects of the salts KI and $AgNO_3$ on the solubility of AgI in water.

35. A 0.150 M Na_2SO_4 solution also saturated with Ag_2SO_4 has $[Ag^+] = 9.7 \times 10^{-3}$ M. What is the value of K_{sp} for Ag_2SO_4 obtained with these data?

37. What $[Pb^{2+}]$ should be maintained in $Pb(NO_3)_2(aq)$ to produce a solubility of 1.5×10^{-4} mol PbI_2/L when $PbI_2(s)$ is added?

39. It is desired to lower the solubility of Ag_2CrO_4 to 5.0×10^{-8} mol Ag_2CrO_4/L. Do you think this can be done by using CrO_4^{2-} as the common ion? by using Ag^+? Explain.

41. A particular water sample that is saturated in CaF_2 is found to have a Ca^{2+} content of 115 ppm (that is, 115 g Ca^{2+} per 10^6 g of water sample). What is the F^- ion content of the water in parts per million?

24. Which of the following saturated aqueous solutions would you expect to have the highest $[Mg^{2+}]$: **(a)** $MgCO_3$; **(b)** MgF_2; **(c)** $Mg_3(PO_4)_2$? Explain.

26. In the qualitative analysis scheme Bi^{3+} is detected by the appearance of a white precipitate of bismuthyl hydroxide, BiOOH(s).

$$BiOOH(s) \rightleftharpoons BiO^+(aq) + OH^-(aq) \qquad K_{sp} = 4 \times 10^{-10}$$

Calculate the pH of a saturated aqueous solution of BiOOH.

28. A 725-mL sample of a saturated aqueous solution of calcium oxalate, CaC_2O_4, at 95 °C is cooled to 13 °C. How many milligrams of calcium oxalate will precipitate? For calcium oxalate, $K_{sp} = 1.2 \times 10^{-8}$ at 95 °C and 2.7×10^{-9} at 13 °C.

30. A 250-mL sample of saturated $CaC_2O_4(aq)$ requires 4.8 mL of 0.00134 M $KMnO_4(aq)$ for its titration. What is the value of K_{sp} for CaC_2O_4 obtained with these data? The titration reaction is

$$5 C_2O_4^{2-} + 2 MnO_4^- + 16 H^+ \longrightarrow$$
$$2 Mn^{2+} + 8 H_2O + 10 CO_2(g)$$

32. Excess $Ca(OH)_2(s)$ is shaken with water to produce a saturated solution. A 50.00-mL sample of the clear, saturated solution is withdrawn and requires 10.7 mL of 0.1032 M HCl for its titration. What is K_{sp} for $Ca(OH)_2$?

34. Describe the effect of the salt KNO_3 on the solubility of AgI in water, and explain why it is different from the effects noted in Exercise 33.

36. If 100.0 mL of 0.0025 M $Na_2SO_4(aq)$ is also saturated with $CaSO_4$ how many grams of $CaSO_4$ would be present in the solution? (*Hint:* Does the usual simplifying assumption hold?)

38. What $[I^-]$ should be maintained in KI(aq) to produce a solubility of 1.5×10^{-5} mol PbI_2/L when $PbI_2(s)$ is added?

40. A handbook lists the K_{sp} values 1.1×10^{-10} for $BaSO_4$ and 5.1×10^{-9} for $BaCO_3$. When 0.50 M $Na_2CO_3(aq)$ is added to saturated $BaSO_4(aq)$, a precipitate of $BaCO_3(s)$ forms. How do you account for this fact, given that $BaCO_3$ has a larger K_{sp} than does $BaSO_4$?

42. Assume that to be visible to the unaided eye a precipitate must weigh more than 1 mg. If you add 1.0 mL of 1.0 M NaCl(aq) to 100.0 mL of a clear saturated aqueous AgCl solution, will you be able to see AgCl(s) precipitated as a result of the common ion effect? Explain.

Criterion for Precipitation from Solution

43. Should precipitation of $MgF_2(s)$ occur if a 22.5-mg sample of $MgCl_2 \cdot 6H_2O$ is added to 325 mL of 0.035 M KF?

45. What is the minimum pH at which precipitation of $Cd(OH)_2(s)$ will occur from a solution that is 0.0055 M in $Cd^{2+}(aq)$?

47. Should precipitation occur in the following cases?
(a) 0.10 mg NaCl is added to 1.0 L of 0.10 M $AgNO_3(aq)$.
(b) One drop (0.05 mL) of 0.10 M KBr is added to 250 mL of a saturated solution of AgCl.
(c) One drop (0.05 mL) of 0.0150 M NaOH(aq) is added to 3.0 L of a solution with 2.0 mg Mg^{2+} per liter.

49. Determine if 1.50 g $H_2C_2O_4$ (oxalic acid: $K_{a_1} = 5.2 \times 10^{-2}$, $K_{a_2} = 5.4 \times 10^{-5}$) can be dissolved in 0.200 L of 0.150 M $CaCl_2$ without the formation of $CaC_2O_4(s)$ ($K_{sp} = 1.3 \times 10^{-9}$). (*Hint:* What is $[C_2O_4{}^{2-}]$ in this solution?)

44. Should precipitation of $PbCl_2(s)$ occur when 155 mL of 0.016 M KCl(aq) is added to 245 mL of 0.175 M $Pb(NO_3)_2(aq)$?

46. What is the minimum pH at which precipitation of $Cr(OH)_3(s)$ will occur from a solution that is 0.086 M in $Cr^{3+}(aq)$?

48. The electrolysis of $MgCl_2(aq)$ can be represented as
$$Mg^{2+}(aq) + 2\ Cl^-(aq) + 2\ H_2O \longrightarrow$$
$$Mg^{2+}(aq) + 2\ OH^-(aq) + H_2(g) + Cl_2(g)$$
The electrolysis of a 315-mL sample of 0.185 M $MgCl_2$ is continued until 0.652 L $H_2(g)$ at 22 °C and 752 mmHg has been collected. Will $Mg(OH)_2(s)$ precipitate when electrolysis is carried to this point? (*Hint:* Notice that $[Mg^{2+}]$ remains constant throughout the electrolysis, but $[OH^-]$ *increases*.)

50. 100.0 mL of a clear saturated solution of Ag_2SO_4 is added to 250.0 mL of a clear saturated solution of $PbCrO_4$. Will any precipitate form? (*Hint:* Take into account the dilutions that occur. What are the possible precipitates?)

Completeness of Precipitation

51. When 200.0 mL of 0.350 M $K_2CrO_4(aq)$ is added to 200.0 mL of 0.0100 M $AgNO_3(aq)$, what percent of the Ag^+ is left *unprecipitated*? (*Hint:* Assume that $[CrO_4{}^{2-}]$ remains constant during the precipitation.)

52. If a constant $[Cl^-] = 0.100$ M is maintained in a solution in which the initial $[Pb^{2+}] = 0.065$ M, what percent of the Pb^{2+} will remain in solution after $PbCl_2(s)$ precipitates? What $[Cl^-]$ should be maintained to ensure that only 1.0% of the Pb^{2+} remains unprecipitated?

Fractional Precipitation

53. Explain why in the fractional precipitation described in Example 19-7 it was not necessary to specify the concentration of the $AgNO_3(aq)$ that was used.

55. A solution is 0.010 M in both $CrO_4{}^{2-}$ and $SO_4{}^{2-}$. To this solution is slowly added 0.50 M $Pb(NO_3)_2(aq)$.
(a) What should be the first anion to precipitate from solution?
(b) What is $[Pb^{2+}]$ at the point at which the second anion begins to precipitate?
(c) Are the two anions effectively separated by this fractional precipitation?

54. Which one of the following solutions can be used to separate the cations in an aqueous solution in which $[Ba^{2+}] = [Ca^{2+}] = 0.050$ M: 0.10 M NaCl(aq), 0.05 M $Na_2SO_4(aq)$, 0.001 M NaOH(aq), and 0.50 M $Na_2CO_3(aq)$. Explain why.

56. To a solution that is 0.250 M NaCl and also 0.0022 M KBr is slowly added $AgNO_3(aq)$.
(a) Which anion should precipitate first, Cl^- or Br^-?
(b) What is $[Ag^+]$ at the point at which the second anion begins to precipitate?
(c) Can the Cl^- and Br^- be separated effectively by this fractional precipitation?

Solubility and pH

57. Which of the following solids is (are) likely to be more soluble in an acidic solution than in pure water: KCl, $MgCO_3$, FeS, $Ca(OH)_2$, C_6H_5COOH? Explain.

59. The solubility of $Mg(OH)_2$ in a particular buffer solution is found to be 0.65 g/L. What must be the pH of the buffer solution? (*Hint:* What is $[OH^-]$ in the buffer?)

58. Which of the following solids is (are) likely to be more soluble in a basic solution than in pure water: $BaSO_4$, $H_2C_2O_4$, $Fe(OH)_3$, $NaNO_3$, MnS? Explain.

60. To 0.350 L of 0.150 M NH_3 is added 0.150 L of 0.100 M $MgCl_2$. How many grams of $(NH_4)_2SO_4$ should be present to prevent precipitation of $Mg(OH)_2(s)$?

61. For the equilibrium

$$Al(OH)_3(s) \rightleftharpoons Al^{3+}(aq) + 3\ OH^-(aq)$$

$$K_{sp} = 1.3 \times 10^{-33}$$

(a) What is the *minimum pH* at which $Al(OH)_3(s)$ will precipitate from a solution that is 0.075 M in Al^{3+}?
(b) A solution has $[Al^{3+}] = 0.075$ M and $[HC_2H_3O_2] = 1.00$ M. What is the maximum quantity of $NaC_2H_3O_2$ that can be added to 250.0 mL of this solution before precipitation of $Al(OH)_3(s)$ begins? (*Hint:* The acetic acid/sodium acetate mixture is a buffer solution.)

62. Should the following precipitates form under the given conditions?
(a) $PbI_2(s)$ from a solution that is 1.05×10^{-3} M HI, 1.05×10^{-3} M NaI, and 1.1×10^{-3} M $Pb(NO_3)_2$.
(b) $Mg(OH)_2(s)$, from 2.50 L of 0.0150 M $Mg(NO_3)_2$ to which is added 1 drop (0.05 mL) of 6.00 M NH_3.
(c) $Al(OH)_3(s)$, from a solution that is 0.010 M in Al^{3+}, 0.010 M $HC_2H_3O_2$, and 0.010 M $NaC_2H_3O_2$.

Complex Ion Equilibria

63. $PbCl_2(s)$ is found to be considerably more soluble in HCl(aq) than in pure water, but its solubility in $HNO_3(aq)$ is not much different from what it is in water. Explain this difference in behavior.

65. Calculate K_f of $[Cu(CN)_4]^{3-}$, if it is found that in a solution that is 0.0500 M in $[Cu(CN)_4]^{3-}$ and 0.80 M in free CN^-, the concentration of Cu^+ is 6.1×10^{-32} M.

$$Cu^+(aq) + 4\ CN^-(aq) \rightleftharpoons [Cu(CN)_4]^{3-}(aq) \qquad K_f = ?$$

67. Can the following ion concentrations be maintained in the same solution *without* a precipitate forming: $[[Ag(S_2O_3)_2]^{3-}] = 0.048$ M, $[S_2O_3^{2-}] = 0.76$ M, and $[I^-] = 2.0$ M? (*Hint:* What is $[Ag^+]$ in this solution?)

69. A 0.10-mol sample of $AgNO_3(s)$ is dissolved in 1.00 L of 1.00 M NH_3. How many grams of KI can be dissolved in this same solution *without* a precipitate of AgI(s) forming?

64. Which of the following would be most effective and which would be least effective in reducing the concentration of $[Zn(NH_3)_4]^{2+}$ in a solution containing this complex ion: HCl, NH_3, NH_4Cl? Explain your choices.

66. Calculate $[Cu^{2+}]$ in a 0.10 M $CuSO_4(aq)$ solution that is also 6.0 M in free NH_3.

$$Cu^{2+}(aq) + 4\ NH_3(aq) \rightleftharpoons [Cu(NH_3)_4]^{2+}(aq)$$

$$K_f = 1.1 \times 10^{13}$$

68. A solution is 0.10 M in *free* NH_3, 0.10 M NH_4Cl, and 0.015 M $[Cu(NH_3)_4]^{2+}$. Should $Cu(OH)_2(s)$ precipitate from this solution? K_{sp} of $Cu(OH)_2$ is 2.2×10^{-20}.

70. A solution is prepared that has $[NH_3] = 1.00$ M and $[Cl^-] = 0.100$ M. How many grams of $AgNO_3$ can be dissolved in 1.00 L of this same solution *without* a precipitate of AgCl(s) forming?

Precipitation and Solubilities of Metal Sulfides

71. Can Fe^{2+} and Mn^{2+} be separated by precipitating FeS(s) and not MnS(s)? Assume $[Fe^{2+}] = [Mn^{2+}] = [H_2S] = 0.10$ M. Choose a $[H_3O^+]$ that ensures maximum precipitation of FeS(s) but not MnS(s). Will the separation be complete? K_{spa}: FeS, 6×10^2; MnS, 3×10^7.

73. A buffer solution is 0.25 M $HC_2H_3O_2$–0.15 M $NaC_2H_3O_2$, saturated in H_2S (0.10 M), and has $[Mn^{2+}] = 0.15$ M.
(a) Show that MnS will *not* precipitate from this solution (K_{spa} for MnS = 3×10^7).
(b) Which buffer component would you increase in concentration and to what minimum value, to ensure that precipitation of MnS(s) begins? Assume that the concentration of the other buffer component is held constant. [*Hint:* Recall equation (19.10).]

72. A solution is 0.05 M in Cu^{2+}, in Hg^{2+}, and in Mn^{2+}. Which sulfides will precipitate if the solution is made to be 0.10 M $H_2S(aq)$ and 0.010 M HCl(aq)? K_{spa}: CuS, 6×10^{-16}; HgS, 2×10^{-32}; MnS, 3×10^7.

74. The following expressions pertain to the precipitation or dissolving of metal sulfides. Use information about the qualitative analysis scheme to predict whether a reaction proceeds to a significant extent in the forward direction and what the products are in each case.
(a) $Cu^{2+}(aq) + H_2S(satd\ aq) \rightarrow$
(b) $Mg^{2+}(aq) + H_2S(satd\ aq) \xrightarrow{\ 0.3\ M\ HCl\ }$
(c) $PbS(s) + HCl\ (0.3\ M) \rightarrow$
(d) $ZnS(s) + HNO_3(aq) \rightarrow$

Qualitative Analysis

75. Suppose you did a cation group 1 analysis and treated the chloride precipitate with $NH_3(aq)$, without first treating it with hot water. What might you observe and what valid conclusions could you reach about cations present, cations absent, and cations in doubt?

76. Show that in qualitative analysis group 1, if you obtain 1.00 mL of saturated $PbCl_2(aq)$ at 25 °C, there should be sufficient Pb^{2+} present to produce a precipitate of $PbCrO_4(s)$. Assume that you use 1 drop (0.05 mL) of 1.0 M K_2CrO_4 for the test.

77. Addition of HCl(aq) to a solution containing several different cations produces a white precipitate. The filtrate is removed and treated with H_2S(aq) in 0.3 M HCl. No precipitate forms. Which of the following conclusions is(are) valid? Explain.
(a) Ag^+ or Hg_2^{2+} (or both) probably present;
(b) Mg^{2+} probably not present;
(c) Pb^{2+} probably not present;
(d) Fe^{2+} probably not present.

78. Write net ionic equations for the following qualitative analysis procedures.
(a) Precipitation of $PbCl_2$(s) from a solution containing Pb^{2+}.
(b) Dissolving of $Zn(OH)_2$(s) in a solution of NaOH(aq).
(c) Dissolving of $Fe(OH)_3$(s) in HCl(aq).
(d) Precipitation of CuS(s) from an acidic solution of Cu^{2+} and H_2S.

Advanced Exercises

(Use data from Chapters 17 and 19 and Appendix D, as needed.)

79. A particular water sample has 131 ppm of $CaSO_4$ (131 g $CaSO_4$ per 10^6 g water). If this water is boiled in a tea kettle, approximately what fraction of the water must be evaporated before $CaSO_4$(s) begins to precipitate? Assume that the solubility of $CaSO_4$(s) does not change much in the temperature range 0 to 100 °C.

80. A handbook lists the solubility of $CaHPO_4$ as 0.32 g $CaHPO_4 \cdot 2H_2O$/L and lists a K_{sp} value of
$$CaHPO_4(s) \rightleftharpoons Ca^{2+}(aq) + HPO_4^{2-}(aq)$$
$$K_{sp} = 1 \times 10^{-7}$$
(a) Are these data consistent? (That is, are the molar solubilities the same when derived in two different ways?)
(b) How do you account for the "discrepancy"? (*Hint:* Recall the nature of phosphate species in solution.)

81. A 50.0-mL sample of 0.152 M Na_2SO_4(aq) is added to 50.0 mL of 0.0125 M $Ca(NO_3)_2$(aq). What percent of the Ca^{2+} remains *unprecipitated*?

82. What percent of the Ba^{2+} in solution is precipitated as $BaCO_3$(s) if *equal volumes* of 0.0020 M Na_2CO_3(aq) and 0.0010 M $BaCl_2$(aq) are mixed?

83. Determine the molar solubility of lead azide, $Pb(N_3)_2$, in a buffer solution with pH = 3.00, given that
$$Pb(N_3)_2(s) \rightleftharpoons Pb^{2+}(aq) + 2 N_3^-(aq)$$
$$K_{sp} = 2.5 \times 10^{-9}$$
$$HN_3(aq) + H_2O \rightleftharpoons H_3O^+(aq) + N_3^-(aq)$$
$$K_a = 1.9 \times 10^{-5}$$

84. Calculate the molar solubility of $Mg(OH)_2$ in 1.00 M NH_4Cl(aq). (*Hint:* Consider the reaction to be $Mg(OH)_2$(s) + 2 NH_4^+(aq) $\rightleftharpoons Mg^{2+}$(aq) + 2 NH_3(aq) + 2 H_2O. What is K for this reaction?)

85. The chief compound in marble is $CaCO_3$. Marble has been widely used for statues and ornamental work on buildings. Marble is readily attacked by acids. Determine the solubility of marble (that is, $[Ca^{2+}]$ in a saturated solution) in **(a)** normal rainwater of pH = 5.6; **(b)** "acid" rainwater of pH 4.20. Assume that the net reaction that occurs is $CaCO_3$(s) + H_3O^+(aq) $\rightleftharpoons Ca^{2+}$(aq) + HCO_3^-(aq) + H_2O. (*Hint:* What is K for this reaction?)

86. What is the solubility of MnS, in grams per liter, in a buffer solution that is 0.100 M $HC_2H_3O_2$–0.500 M $NaC_2H_3O_2$? K_{spa} of MnS = 3×10^7. (*Hint:* Assume that $[H_3O^+]$ remains constant in the dissolution reaction.)

87. Write net ionic equations to represent the following observations.
(a) When concentrated $CaCl_2$(aq) is added to Na_2HPO_4(aq), a white precipitate is formed that is 38.7% Ca, by mass.
(b) When a piece of dry ice $[CO_2(s)]$ is placed into a clear dilute solution of "lime water" $[Ca(OH)_2(aq)]$, bubbles of gas are evolved. At first a white precipitate forms, but then it redissolves.

88. For the titration that is the subject of Example 19-7, verify the assertion on page 667 that $[Ag^+]$ increases very rapidly between the point when AgBr has finished precipitating and Ag_2CrO_4 is about to begin.

89. The solubility of AgCN(s) in 0.200 M NH_3(aq) is 8.8 × 10^{-6} mol/L. Calculate the value of K_{sp} for AgCN.

90. The solubility of $CdCO_3$(s) in 1.00 M KI(aq) is 1.2 × 10^{-3} mol/L. Given that K_{sp} of $CdCO_3$ is 5.2 × 10^{-12}, what is K_f for $[CdI_4]^{2-}$?

91. Use K_{sp} for $PbCl_2$ and K_f for $[PbCl_3]^-$ to determine the molar solubility of $PbCl_2$ in 0.10 M HCl(aq). (*Hint:* What is the total concentration of lead species in solution?)

92. A mixture of $PbSO_4$(s) and PbS_2O_3(s) is shaken with pure water until a saturated solution is formed. Both solids remain in excess. What is $[Pb^{2+}]$ in the saturated solution? (*Hint:* Both of the following equilibrium expressions are required, as well as a third equation.)
$$PbSO_4(s) \rightleftharpoons Pb^{2+}(aq) + SO_4^{2-}(aq)$$
$$K_{sp} = 1.6 \times 10^{-8}$$
$$PbS_2O_3(s) \rightleftharpoons Pb^{2+}(aq) + S_2O_3^{2-}(aq)$$
$$K_{sp} = 4.0 \times 10^{-7}$$

93. Use the method of Exercise 92 to determine $[Pb^{2+}]$ in a saturated solution in contact with a mixture of $PbCl_2$(s) and $PbBr_2$(s).

94. A 2.50-g sample of Ag_2SO_4(s) is added to a beaker containing 0.150 L of 0.025 M $BaCl_2$.
(a) Write an equation for any reaction that occurs.
(b) Describe the final contents of the beaker—that is, the masses of any precipitates present and the concentrations of the ions in solution.

Feature Problems

A. In an experiment to measure K_{sp} of $CaSO_4$, [D. Masterman, *J. Chem. Educ.* **64**, 409 (1987)], a saturated solution of $CaSO_4(aq)$ is poured into the ion exchange column pictured (and described in Section 22-3). As the solution passes through the column, Ca^{2+} is retained by the ion exchange medium and H_3O^+ is released: *two* H_3O^+ ions appear in the effluent solution for every Ca^{2+} ion. As the drawing suggests, a 25.00-mL sample is added to the column, and the effluent is collected and diluted to 100.0 mL in a volumetric flask. A 10.00-mL portion of the diluted solution requires 8.25 mL of 0.0105 M NaOH for its titration. Use these data to obtain a value of K_{sp} for $CaSO_4$.

B. In the Mohr titration $Cl^-(aq)$ is titrated with $AgNO_3(aq)$ in solutions that are at about pH = 7. Thus, it is suitable for determining the chloride ion content of drinking water. The indicator used in the titration is $K_2CrO_4(aq)$. A red-brown precipitate of $Ag_2CrO_4(s)$ forms after all the Cl^- has been precipitated. The titration reaction is $Ag^+(aq) + Cl^-(aq) \rightarrow AgCl(s)$. At the equivalence point of the titration, the titration mixture consists of $AgCl(s)$ and a solution having neither Ag^+ nor Cl^- in excess. Also, no $Ag_2CrO_4(s)$ is present but should form immediately after the equivalence point.
1. How many mL of 0.01000 M $AgNO_3(aq)$ would be required to titrate 100.0 mL of a municipal water sample having 29.5 mg Cl^-/L?
2. What is $[Ag^+]$ at the equivalence point of the Mohr titration?

3. What should $[CrO_4^{2-}]$ in the titration mixture be to meet the requirement of no precipitation of $Ag_2CrO_4(s)$ until immediately after the equivalence point?
4. Describe the effect on the results of the titration you would expect if $[CrO_4^{2-}]$ were (a) greater than that calculated in part 3 and (b) less than that calculated?
5. Do you think the Mohr titration would work if the reactants were exchanged, that is, with $Cl^-(aq)$ as the titrant and $Ag^+(aq)$ in the sample being analyzed? Explain.

C. The accompanying drawing suggests a series of manipulations starting with saturated $Mg(OH)_2(aq)$. Calculate $[Mg^{2+}](aq)$ at each of the numbered stages.
1. 0.500 L of saturated $Mg(OH)_2(aq)$ is in contact with $Mg(OH)_2(s)$.
2. 0.500 L of H_2O is added to the 0.500 L of solution in (1), and the solution is vigorously stirred. Undissolved $Mg(OH)_2(s)$ remains.
3. 100.0 mL of the clear solution in (2) is removed and added to 0.500 L of 0.100 M $HCl(aq)$.
4. 25.00 mL of the clear solution in (2) is removed and added to 250.0 mL of 0.065 M $MgCl_2(aq)$.
5. 50.00 mL of the clear solution in (2) is removed and added to 150.0 mL of 0.150 M $KOH(aq)$.

Spontaneous Change: Entropy and Free Energy

The natural tendency toward increased disorder serves as a criterion for spontaneous change. A process that produces increased order is *nonspontaneous,* as in the use of paving blocks in the construction of a roadway.

*I*n both Chapters 8 and 16 we noted that the reaction of nitrogen and oxygen gases, which does not occur appreciably in the forward direction at room temperature, produces significant equilibrium amounts of NO(g) at *high* temperatures.

$$N_2(g) + O_2(g) \rightleftharpoons 2\,NO(g)$$

Another reaction involving oxides of nitrogen is the conversion of NO(g) to NO_2(g).

$$2\,NO(g) + O_2(g) \rightleftharpoons 2\,NO_2(g)$$

This reaction, unlike the first, yields its greatest equilibrium amounts of NO_2(g) at *low* temperatures.

What is there about these two reactions that causes the forward reaction of the one to be favored at high temperatures and that of the other to be favored at low temperatures? Our primary objective in this chapter is to develop concepts to help us answer questions like this. This chapter, taken together with ideas from Chapter 7, shows the great power of thermodynamics to provide explanations of many chemical phenomena.

20-1 Spontaneity: The Meaning of Spontaneous Change

Most of us have played with spring-wound toys, whether a toy automobile, a top, or a music box. In every case, once the wound-up toy is released it keeps running until the stored energy in the spring has been released; then the toy stops. The toy never rewinds itself. Human intervention is necessary (winding by hand). The running down of a wound-up spring is an example of a *spontaneous* process. The rewinding of the spring is a *nonspontaneous* process. Now let's explore the scientific meaning of these two items.

▶ Spontaneous . . . "arising from internal forces or causes; independent of external agencies; self-acting." *The Random House Dictionary of the English Language,* 2nd ed., 1987.

A **spontaneous process** is a process that occurs in a system left to itself; once started, no action from outside the system (external action) is necessary to make the process continue. On the other hand, a **nonspontaneous process** will not occur *unless* some external action is continuously applied. Consider the rusting of an iron pipe exposed to the atmosphere. Although the process may occur quite slowly, it does so continuously. As a result, the amount of iron decreases and the amount of rust increases, until a final state of equilibrium is reached in which essentially all the iron has been converted to iron(III) oxide. We say that the reaction

$$4 \text{ Fe(s)} + 3 \text{ O}_2(\text{g}) \longrightarrow 2 \text{ Fe}_2\text{O}_3(\text{s})$$

is *spontaneous.* Now consider the reverse situation: extracting pure iron from iron(III) oxide. We should not say that the process is impossible, but it is certainly *nonspontaneous.* In fact, this nonspontaneous reverse process is used to manufacture iron from iron ore.

We will consider specific quantitative criteria for spontaneous change later in the chapter, but even now we can identify some spontaneous processes intuitively, as in Example 20-1.

EXAMPLE 20-1

Identifying Spontaneous and Nonspontaneous Processes. Use your general knowledge to indicate whether each of the following processes is spontaneous or nonspontaneous.
a. The reaction of NaOH(aq) and HCl(aq).
b. The electrolysis of liquid water.
c. The melting of an ice cube.

Solution

a. This is a neutralization reaction. The net ionic equation is

$$\text{H}_3\text{O}^+(\text{aq}) + \text{OH}^-(\text{aq}) \longrightarrow 2 \text{ H}_2\text{O}(\text{l})$$

There is little tendency for the reverse reaction (self-ionization of water) to occur. The neutralization reaction is spontaneous.
b. We have previously learned that we need to use an electric current (external action) to decompose liquid water into its elements.

$$2 \text{ H}_2\text{O}(\text{l}) \xrightarrow{\text{electrolysis}} 2 \text{ H}_2(\text{g}) + \text{O}_2(\text{g})$$

Left to itself, liquid water does not decompose. Therefore, the electrolysis of liquid water is a nonspontaneous process.

c. Here is an interesting case. From experience we know that the melting of an ice cube is spontaneous if the temperature is *above* the melting point (0 °C) and nonspontaneous if the temperature is *below* the melting point. We also know that the ice and liquid water are at equilibrium if the temperature is 0 °C. Certain processes are at equilibrium at one temperature, spontaneous at some temperatures, and nonspontaneous at other temperatures.

Practice Example A: Indicate whether each of the following processes is spontaneous or nonspontaneous: **(a)** the "souring" of cream; **(b)** obtaining gold nuggets by "panning"; **(c)** formation of a green patina (surface coating) on an outdoor bronze statue.

Practice Example B: Indicate whether each of the following processes is spontaneous or nonspontaneous: **(a)** the combustion of gasoline; **(b)** socks pairing in the laundry dryer; **(c)** riding on a playground swing; **(d)** sugar dissolving in hot coffee.

From our discussion of spontaneity to this point, we can conclude that

- If a process is spontaneous, the reverse process is *nonspontaneous.*

- Both spontaneous and nonspontaneous processes are *possible,* but only spontaneous processes will occur *without intervention.* Nonspontaneous processes require the system to be acted on by an outside force.

We would like to do more, however. We want to be able to *predict* whether the forward or the reverse direction is the direction of spontaneous change in a process. We cannot simply base predictions on common knowledge as we did in Example 20-1. We need a *criterion* for spontaneous change. To begin, let's look to mechanical systems for a clue. A ball rolls downhill, and water flows to a lower level. A common feature of these processes is that *potential energy decreases.*

For chemical systems the property analogous to the potential energy of a mechanical system is the internal energy (E) or the closely related property of enthalpy (H). In the 1870s, the chemists P. Berthelot and J. Thomsen proposed that the direction of spontaneous change is that in which the enthalpy of a system decreases. An enthalpy decrease means that heat is given off by the system to the surroundings. Bertholet and Thomsen concluded that *exothermic* reactions should be spontaneous. In fact, exothermic processes generally are spontaneous. On the other hand, some *endothermic* reactions are also spontaneous. Thus, we cannot predict whether a process is spontaneous from its enthalpy change alone. Here are three examples of *spontaneous, endothermic* processes.

- The melting of ice at room temperature.
- The evaporation of liquid diethyl ether from an open beaker.
- The dissolving of ammonium nitrate in water.

In looking beyond enthalpy change (ΔH) as a criterion for spontaneous change, let's try to find some other thermodynamic function with the properties suggested by Figure 20-1.

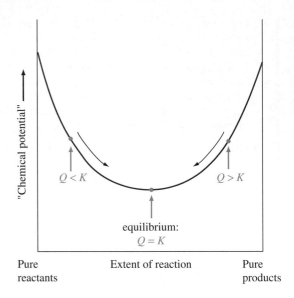

FIGURE 20-1
Search for a criterion for spontaneous chemical change.
A property called the "chemical potential" attains a minimum value in the course of the reaction. At this point the reaction is at equilibrium, with the reaction quotient Q equal to the equilibrium constant K. For a condition on either side of the equilibrium point, spontaneous reaction occurs toward the equilibrium point in the direction of the arrow. In the hypothetical reaction shown here, the equilibrium point is about midway between the reactants and the products, but usually this point lies closer to one side (reactants) or the other (products).

20-2   Spontaneity and Disorder: The Concept of Entropy

To continue our search for a criterion for spontaneous change, let's turn again to a mechanical analogy. Suppose we have ten identical balls, each colored half red and half white, and release them at the top of a ramp. Which of the three arrangements of the balls pictured in Figure 20-2 should we expect to see? We would *not* expect arrangement (a) because this is not a condition in which the potential energy has reached a minimum. Arrangements (b) and (c) both have this minimum in potential energy, but we would confidently predict that arrangement (c) is the expected result. A scrambled, or *disordered,* arrangement, as in (c), is far more likely to occur than the very orderly arrangement in (b). This example suggests that *two* factors underlie spontaneous change: Does the system gain or lose energy, and what happens to the degree of order or disorder in the system?

Let's consider another situation in which we see these two factors at work. In Figure 20-3a one ideal gas, labeled A, fills a glass bulb at a pressure of 1.00 atm. A second ideal gas, B, also at a pressure of 1.00 atm, fills a second bulb identical to the first. The two bulbs are joined by a stopcock valve. Assume that the two gases do not react and imagine what happens when the valve between

FIGURE 20-2
The importance of randomness or disorder.
If ten red-and-white balls are released at the top of the ramp, the arrangement in (c), in which all the balls are at the bottom of the ramp with a random mix of red and white faces up, is what we expect to find. The arrangement in (a), in which some of the balls remain on the ramp, is not possible. The arrangement in (b), in which all the balls have their red faces up, is extremely unlikely.

(a)   (b)   (c)

the two bulbs is opened. We expect the condition pictured in Figure 20-3b: Each gas expands into the bulb containing the other. The mixing continues until the partial pressure of each gas becomes 0.50 atm in each bulb.

One of the characteristics of an ideal gas is that the internal energy (E) and enthalpy (H) depend only on temperature and not on the gas pressure or volume. Because there are no intermolecular forces, when ideal gases mix at constant temperature, $\Delta E = \Delta H = 0$. Thus, enthalpy change is *not* the driving force behind the spontaneous mixing of gases in Figure 20-3. The driving force is simply the tendency of the molecules of the two gases to achieve the maximum state of mixing, or *disorder,* possible. **Entropy** is the thermodynamic property related to the degree of disorder in a system, and we designate it by the symbol S.

The greater the degree of randomness, or disorder, in a system, the greater its entropy.

Like internal energy and enthalpy, entropy is a function of state (see page 215). It has a unique value for a system whose temperature, pressure, and composition are specified. The **entropy change, ΔS,** is the difference in entropy between two states, and it also has a unique value.

To represent the mixing of gases symbolically, we can write

$$A(g) + B(g) \longrightarrow \text{mixture of A(g) and B(g)}$$

$$\Delta S = S_{\text{mix of gases}} - [S_{A(g)} + S_{B(g)}] > 0$$

In the mixing of gases, entropy *increases* and ΔS is a *positive* quantity, that is, $\Delta S > 0$. Now let's see what we can say about entropy changes in the three spontaneous, endothermic processes listed at the end of Section 20-1 on page 691.

As suggested by Figure 20-4, (1) in the melting of ice a crystalline solid is replaced by a less structured liquid. Disorder increases during melting, and so does entropy. (2) Molecules in the gaseous state, because of the large free volume in which they can move, have a much higher entropy than in the liquid state. The process of evaporation is accompanied by an increase in entropy. (3) In the dissolving of ammonium nitrate in water, a crystalline solid and a pure liquid are replaced by a mixture of ions and water molecules in the liquid (solution) state. Although some ordering of water molecules occurs around ions in solution (referred to as hydration of the ions), this ordering tendency is generally not as great as the disorder produced by destroying the original crystalline solid. We expect both disorder and entropy to increase in the dissolving process. In each of the three cases we have described, the increase in entropy ($\Delta S > 0$) outweighs the fact that heat must be absorbed ($\Delta H > 0$), and the process is spontaneous.

In summary, we generally expect entropy *increases* when

- pure liquids or liquid solutions are formed from solids;
- gases are formed, from either solids or liquids;
- the number of molecules of gas increases as a result of a chemical reaction;
- the temperature of a substance increases. (Increased temperature means increased molecular motion, whether it be vibrational motion of atoms or ions in a solid or translational motion of molecules in a liquid or gas. This increased molecular motion represents increased disorder.)

We apply these generalizations in Example 20-2.

(a) Before mixing

(b) After mixing

• gas A   • gas B

FIGURE 20-3
The mixing of ideal gases.
The total volume of the system and the total gas pressure remain fixed. The net change is that (a) before mixing, each gas is confined to one half of the total volume (a single bulb) at a pressure of 1.00 atm, and (b) after mixing, each gas has expanded into the total volume (both bulbs) and exerts a partial pressure of 0.50 atm.

▶ Even though the entropy of a substance increases with increased temperature, a temperature increase itself is not necessarily spontaneous. A temperature increase will be spontaneous only if the system is in contact with warmer surroundings, or if chemical energy is converted to thermal energy in an exothermic reaction occurring within an isolated system.

(a) Melting: $S_{liquid} > S_{solid}$

(b) Vaporization: $S_{vapor} > S_{liquid}$

FIGURE 20-4
Entropy and disorder—three processes in which entropy increases.
Each of the processes pictured here—the melting of a solid, the evaporation of a liquid, and the dissolving of a solute—results in a more disordered system and an increase in entropy. For the third situation, the generalization works best for nonelectrolyte solutions, in which ion-dipole forces do not exist.

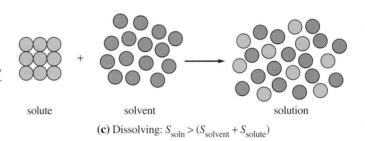

solute   solvent   solution

(c) Dissolving: $S_{soln} > (S_{solvent} + S_{solute})$

EXAMPLE 20-2

Making Qualitative Predictions of Entropy Changes in Chemical Processes. Predict whether each of the following processes involves an increase or decrease in entropy, or whether the outcome is uncertain.

a. The decomposition of ammonium nitrate (a fertilizer and a high explosive):
$2 NH_4NO_3(s) \rightarrow 2 N_2(g) + 4 H_2O(g) + O_2(g)$

b. The conversion of SO_2 to SO_3 (a key step in the manufacture of sulfuric acid):
$2 SO_2(g) + O_2(g) \rightarrow 2 SO_3(g)$

c. The extraction of sucrose from cane sugar juice: $C_{12}H_{22}O_{11}(aq) \rightarrow C_{12}H_{22}O_{11}(s)$

d. The "water gas shift" reaction (involved in the gasification of coal):
$CO(g) + H_2O(g) \rightarrow CO_2(g) + H_2(g)$

Solution

a. Here a solid yields a large quantity of gas. Entropy increases.

b. Three moles of gaseous reactants produce 2 moles of gaseous products. The *loss* of one mole of gas suggests that the S and O atoms are in a state of greater organization in SO_3 than in SO_2 and O_2. Entropy decreases.

c. The sucrose molecules achieve a high degree of order when they leave the solution and arrange themselves into a crystalline state. Entropy decreases.

d. The entropies of the four gases are likely to be different because of differences in their molecular structures. Because the number of moles of gases is the same on both sides of the equation, however, the entropy change is likely to be small, assuming a constant temperature. Based on the generalizations listed (page 693), whether entropy increases or decreases is uncertain.

Practice Example A: Predict whether entropy increases or decreases in the following reactions. **(a)** The Claus process for removing H_2S from natural gas: $2 H_2S(g) + SO_2(g) \rightarrow 3 S(s) + 2 H_2O(g)$; **(b)** the decomposition of mercury(II) oxide: $2 HgO(s) \rightarrow 2 Hg(l) + O_2(g)$.

Practice Example B: Predict whether entropy increases or decreases, or whether the outcome is uncertain in the following reactions. **(a)** $Zn(s) + Ag_2O(s) \rightarrow ZnO(s) + 2\ Ag(s)$; **(b)** The chlor-alkali process, $2\ Cl^-(aq) + 2\ H_2O(l) \xrightarrow{electrolysis} 2\ OH^-(aq) + H_2(g) + Cl_2(g)$.

Although we will often apply the entropy concept in a qualitative way, we should say a little about how entropy changes are measured and expressed quantitatively. Entropy is based on two *measurable* quantities that affect the degree of disorder in a system: heat and temperature. For example, when a solid at its melting point absorbs a quantity of heat, it is converted to the more disordered liquid state (entropy increases). And the ability of a given quantity of heat to produce disorder is much greater if the heat is absorbed by a highly ordered system (low temperature) than by a system that is already highly disordered (high temperature). Thus, an entropy change should be *directly* related to the quantity of heat absorbed (q) and *inversely* related to the temperature (T), as indicated by this equation.

$$\Delta S = \frac{q_{rev}}{T} \tag{20.1}$$

Equation (20.1) appears simple, but it is not. If S is to be a function of state, ΔS for a system must be *independent* of the path by which heat is lost or gained. On the other hand, because the value of q ordinarily depends on the path chosen, Equation (20.1) holds only for a carefully defined path. The path must be of a type called *reversible*, for which $q = q_{rev}$. A **reversible process** is one that can be made to reverse its direction by making just an infinitesimal change in some system property (Figure 20-5), but we do not need to pursue the question of reversibility at this time. It is important to note, however, that because q has the unit J and $1/T$ has the unit K^{-1}, the unit of entropy change, ΔS, is J/K, or $J\ K^{-1}$.

To vacuum

Sand

Gas

To vacuum

Sand

Gas

FIGURE 20-5
Pressure–volume work conducted nearly reversibly.
A reversible process can be made to reverse direction through just an infinitesimal change. The sand pictured here is being removed one grain at a time and the gas slowly expands. If at any time one grain of sand is added to the piston instead of one being removed, the piston will reverse direction and the gas will be compressed. This process is not strictly reversible because the grains of sand have more than an infinitesimal mass. Note that the similar expansion of a gas shown in Figure 7-3 on page 205 is *not* reversible. There the confining pressure is reduced in just one rather than an infinite number of stages.

20-3 Criteria for Spontaneous Change: The Second Law of Thermodynamics

If we attempt to use an increase in entropy of a system as the sole criterion for spontaneous change, we immediately encounter difficulties. For example, how do we explain the spontaneous freezing of water at $-10\ °C$? Because crystalline ice is a more ordered state than liquid water, the freezing of water is a process for which entropy *decreases.* The answer to this puzzle lies in the fact that we must always simultaneously consider *two* entropy changes—the entropy change of the system itself and the entropy change of the surroundings—in order to determine the total of the two, called the entropy change of the "universe."

$$\Delta S_{total} = \Delta S_{universe} = \Delta S_{system} + \Delta S_{surroundings} \tag{20.2}$$

Although it is beyond the scope of this discussion to verify the following expression, it provides the basic criterion for spontaneous change. It is one of the many forms used in stating the **second law of thermodynamics.**

$$\Delta S_{univ} = \Delta S_{sys} + \Delta S_{surr} > 0 \tag{20.3}$$

All spontaneous processes produce an increase in the entropy of the universe.

According to expression (20.3), if a process produces *positive* entropy changes in *both* the system *and* its surroundings, the process is surely *spontaneous.* And if *both* these entropy changes are *negative,* the process is just as surely *nonspontaneous.* The freezing of water produces a *negative* entropy change in the *system,* but in the *surroundings,* which absorb heat, the entropy change is *positive.* As long as the temperature is below 0 °C, the entropy of the surroundings increases more than the entropy of the system decreases. Because the *total* entropy change is *positive,* the freezing of water is indeed *spontaneous.*

Free Energy and Free Energy Change

We could use expression (20.3) as our basic criterion for spontaneity (spontaneous change), but we would find it very difficult to apply. To evaluate a total entropy change (ΔS_{univ}), we always have to evaluate ΔS for the surroundings. At best, this process is tedious, and in many cases it is not even possible because we cannot figure out all the interactions between a system and its surroundings. Surely it would be preferable to have a criterion that we could apply *to the system itself,* without having to worry about changes that occur in the surroundings.

To develop this new criterion, let's explore a hypothetical process conducted at constant temperature and pressure and with work limited to pressure–volume work. This process is accompanied by a heat effect, q_p, which, as we saw in Section 7-6, is equal to ΔH for the system (ΔH_{sys}). The heat effect experienced by the surroundings is the *negative* of that for the system: $q_{surr} = -q_p = -\Delta H_{sys}$. Furthermore, if the hypothetical surroundings are large enough, the path by which heat enters or leaves the surroundings can be made *reversible.* That is, the quantity of heat can be made to produce only an infinitesimal change in the temperature of the surroundings. In this case, according to Equation (20.1), the entropy change in the *surroundings* is $\Delta S_{surr} = -\Delta H_{sys}/T$.* Now we substitute this value of ΔS_{surr} into Equation (20.2).

$$\Delta S_{univ} = \Delta S_{sys} - \frac{\Delta H_{sys}}{T}$$

*We cannot similarly substitute $\Delta H_{sys}/T$ for ΔS_{sys}. A process that occurs spontaneously is generally far removed from an equilibrium condition and is therefore *irreversible.* We cannot substitute q for an irreversible process into Equation (20.1).

We next multiply by T to obtain

$$T\,\Delta S_{univ} = T\,\Delta S_{sys} - \Delta H_{sys} = -(\Delta H_{sys} - T\,\Delta S_{sys})$$

and then multiply by -1 (change signs).

$$-T\,\Delta S_{univ} = \Delta H_{sys} - T\,\Delta S_{sys} \qquad (20.4)$$

This is the significance of Equation (20.4): The equation has on its right side terms involving *only the system.* On the left side appears the term ΔS_{univ}, which embodies the criterion for spontaneous change: for a spontaneous process, $\Delta S_{univ} > 0$.

Equation (20.4) is generally cast in a somewhat different form, which re-quires that we introduce a new thermodynamic function called the Gibbs **free energy, G.** The Gibbs free energy for a system is defined by the equation

$$G = H - TS$$

and the **free energy change, ΔG,** for a process at constant T is

$$\Delta G = \Delta H - T\,\Delta S \qquad (20.5)$$

Note that in Equation (20.5) *all* the terms apply to measurements *on the system.* All reference to the surroundings has been eliminated. Also, when we compare Equations (20.4) and (20.5) we see that

$$\Delta G = -T\,\Delta S_{univ}$$

Now, by noting that when ΔS_{univ} is *positive* ΔG is *negative,* we have our final criterion for spontaneous change, based on properties just of the system itself.

For a process occurring at constant T and P, if

- $\Delta G < 0$ *(negative)*, the process is *spontaneous.*
- $\Delta G > 0$ *(positive)*, the process is *nonspontaneous.*
- $\Delta G = 0$ *(zero)*, the process is *at equilibrium.*

We can see that free energy is indeed an energy term by evaluating the units in Equation (20.5). ΔH has the unit joules (J), and the product, $T\,\Delta S$, has the units $K \times J/K = J$. ΔG is the difference in two quantities with units of energy.

J. Willard Gibbs (1839–1903)— the United States' great "un-known" scientist. Gibbs, a Yale University professor of mathemat-ical physics, lived most of his ca-reer without recognition, partly because of the abstractness of his work and partly because his im-portant publications were in little-read journals. Yet, today, Gibbs's ideas serve as the basis of most of chemical thermodynamics.

▶ Sometimes the term "$T\,\Delta S$" is referred to as "organizational energy" because ΔS is related to the degree of order (or disorder) in a system.

Applying the Free Energy Criterion for Spontaneous Change

Later, we will look at quantitative applications of Equation (20.5), but for now let's use the equation to make some *qualitative* predictions. Altogether there are *four* possibilities for ΔG based on the signs of ΔH and ΔS. These possibilities are outlined in Table 20.1 and illustrated in Example 20-3.

If ΔH is *negative* and ΔS is *positive,* the expression $\Delta G = \Delta H - T\,\Delta S$ is *neg-ative* at all temperatures. The process is spontaneous at all temperatures. This corresponds to the situation we noted previously in which ΔS_{sys} and ΔS_{surr} are both positive and ΔS_{univ} is also positive.

Unquestionably, if a process is accompanied by an *increase* in enthalpy (heat is absorbed) and a *decrease* in entropy (increased order), ΔG is *positive* at all temperatures and the process is *nonspontaneous.* This corresponds to a situa-tion in which both ΔS_{sys} and ΔS_{surr} are negative and ΔS_{univ} is also negative.

TABLE 20.1   Criterion for Spontaneous Change: $\Delta G = \Delta H - T\Delta S$

Case	ΔH	ΔS	ΔG	Result	Example
1	–	+	–	spontaneous at all temp	$2\ N_2O(g) \longrightarrow 2\ N_2(g) + O_2(g)$
2	–	–	$\begin{cases} - \\ + \end{cases}$	spontaneous at low temp / nonspontaneous at high temp	$H_2O(l) \longrightarrow H_2O(s)$
3	+	+	$\begin{cases} + \\ - \end{cases}$	nonspontaneous at low temp / spontaneous at high temp	$2\ NH_3(g) \longrightarrow N_2(g) + 3\ H_2(g)$
4	+	–	+	nonspontaneous at all temp	$3\ O_2(g) \longrightarrow 2\ O_3(g)$

▶ For cases 2 and 3 there is a particular temperature at which a process switches from being spontaneous to being nonspontaneous. In Section 20-6 we will show how to determine such a temperature.

The questionable cases are those in which the entropy and enthalpy factors work in opposition, that is, with ΔH and ΔS both *negative* or *both positive*. Here, whether a reaction is spontaneous or not, that is, whether ΔG is negative or positive, depends on temperature. In general, if a reaction has negative values for both ΔH and ΔS, it is spontaneous at *lower* temperatures, whereas if ΔH and ΔS are both positive, it is spontaneous at *higher* temperatures.

EXAMPLE 20-3

Using Enthalpy and Entropy Changes to Predict the Direction of Spontaneous Change. Under what temperature conditions would you expect the following reactions to occur spontaneously?
a. $2\ NH_4NO_3(s) \rightarrow 2\ N_2(g) + 4\ H_2O(g) + O_2(g)$ $\Delta H° = -236.0$ kJ
b. $I_2(g) \rightarrow 2\ I(g)$

Solution

▶ It is the high activation energy of the decomposition reaction that accounts for the fact that $NH_4NO_3(s)$ even exists at ordinary temperatures.

a. The reaction is exothermic, and in Example 20-2 we concluded that $\Delta S > 0$, because of the large quantities of gases produced. With $\Delta H < 0$ and $\Delta S > 0$ we conclude that this reaction should be spontaneous at all temperatures (case 1 of Table 20.1).

b. Because one mole of gaseous reactant produces two moles of gaseous product, we expect entropy to increase. But what is the sign of ΔH? We could calculate ΔH from enthalpy of formation data, but there is no need to. In the reaction, covalent bonds in $I_2(g)$ are broken and no new bonds are formed. Because energy is absorbed to break bonds, ΔH must be positive. With $\Delta H > 0$ and $\Delta S > 0$, this is case 3 in Table 20.1. ΔH is larger than $T\Delta S$ at low temperatures, and the reaction is nonspontaneous. At high temperatures the $T\Delta S$ term becomes larger than ΔH, ΔG becomes negative, and the reaction is spontaneous.

Practice Example A: Which of the four cases in Table 20.1 would you expect to apply to the following reactions? **(a)** $N_2(g) + 3\ H_2(g) \rightarrow 2\ NH_3(g)$, $\Delta H° = -92.22$ kJ; **(b)** $2\ C(graphite) + 2\ H_2(g) \rightarrow C_2H_4(g)$, $\Delta H° = 52.26$ kJ.

Practice Example B: Under what temperature conditions would you expect the following reactions to occur spontaneously? **(a)** $CaCO_3(s) \rightarrow CaO(s) + CO_2(g)$, $\Delta H° = 178$ kJ; **(b)** $2\ ZnS(s) + 3\ O_2(g) \rightarrow 2\ ZnO(s) + 2\ SO_2(g)$, $\Delta H° = -878.2$ kJ.

▶ A related observation is that only a small fraction of the mass of the universe is in molecular form.

Example 20-3(b) helps us to understand why there is an upper temperature limit for the stabilities of chemical compounds. No matter how positive the value of ΔH for dissociation of a molecule into its atoms, the term $T\Delta S$ will eventually exceed ΔH in magnitude as the temperature increases. Known temperatures range from near absolute zero to the interior temperatures of stars (about 3 ×

10^7 K). Molecules exist only at limited temperatures (up to about 1×10^4 K or about 0.03% of this total temperature range).

The method of Example 20-3 is adequate for making predictions about the sign of ΔG, but we will also want to use Equation (20.5) to calculate numerical values. To do this we must be able to evaluate entropy changes, ΔS, which is the topic of the next section.

Are You Wondering . . .

Just what the term "free energy" signifies?

For an exothermic reaction at constant T and P, ΔH represents a release of energy to the surroundings. We might think of the energy absorbed by the surroundings, $-\Delta H$, as being available to do work. If the entropy of the system *decreases,* however, (case 2 in Table 20.1), the quantity of energy associated with producing the more highly ordered system, $-T\Delta S$, is unavailable for use in the surroundings. Thus, the quantity of energy *freely* available for use in the surroundings, the negative of the *free* energy change, $-\Delta G$, is less than $-\Delta H$. On the other hand, if an exothermic reaction is accompanied by an *increase* in entropy (case 1 in Table 20.1), an additional amount of energy is made available for use in the surroundings. Now the available energy or *free* energy change, $-\Delta G$, is greater than $-\Delta H$. In Chapter 21 we will see how the free energy change of a reaction can be converted to electrical work. In any case, do not think of "free energy" as being "free" energy. There are always costs involved in tapping an energy source.

20-4 Evaluating Entropy and Entropy Changes

As with internal energy and enthalpy, we are concerned mainly with *changes* in entropy. On the other hand, *absolute* entropy values *can* be determined, something that cannot be done for internal energy and enthalpy. In this section we will work with both entropy changes and absolute entropies. We will begin by describing entropy changes in phase transitions.

Phase Transitions

Starting with Equation (20.5) and applying the criterion for equilibrium (that is, $\Delta G = 0$), we see that

$$\Delta G = \Delta H - T\,\Delta S = 0$$

$$\Delta S = \frac{\Delta H}{T}$$

at equilibrium

In writing ΔH and ΔS, we often use descriptive subscripts, such as "fus" (fusion) and "vap" (vaporization). If the transitions involve standard-state conditions (1 atm pressure), we also use the ° sign. Thus for the melting (fusion) of ice at its normal melting point

$$H_2O(s, 1 \text{ atm}) \rightleftharpoons H_2O(l, 1 \text{ atm}) \qquad \Delta H^\circ_{fus} = 6.02 \text{ kJ at } 273.15 \text{ K}$$

▶ Recall that the normal melting or boiling point is determined at 1 atm pressure.

the standard entropy change is

$$\Delta S^{\circ}_{\text{fus}} = \frac{\Delta H^{\circ}_{\text{fus}}}{T_{\text{mp}}} = \frac{6.02 \text{ kJ/mol}}{273.15 \text{ K}} = 2.20 \times 10^{-2} \text{ kJ mol}^{-1} \text{ K}^{-1}$$

$$= 22.0 \text{ J mol}^{-1} \text{ K}^{-1}$$

Entropy changes depend on the quantities of substances involved and are usually expressed on a "per mole" basis.

EXAMPLE 20-4

Determining the Entropy Change for a Phase Transition. What is the standard molar entropy of vaporization of water at 373 K given that the standard molar enthalpy of vaporization is 40.7 kJ/mol?

Solution

Although we do not specifically need a chemical equation, writing one helps us to see the process for which we seek the value of $\Delta S^{\circ}_{\text{vap}}$.

$$\text{H}_2\text{O(l, 1 atm)} \rightleftharpoons \text{H}_2\text{O(g, 1 atm)} \qquad \Delta H^{\circ}_{\text{vap}} = 40.7 \text{ kJ/mol H}_2\text{O}$$

$$\Delta S^{\circ}_{\text{vap}} = ?$$

$$\Delta S^{\circ}_{\text{vap}} = \frac{\Delta H^{\circ}_{\text{vap}}}{T_{\text{bp}}} = \frac{40.7 \text{ kJ/mol}}{373 \text{ K}} = 0.109 \text{ kJ mol}^{-1} \text{ K}^{-1}$$

$$= 109 \text{ J mol}^{-1} \text{ K}^{-1}$$

Practice Example A: What is the standard molar entropy of vaporization, $\Delta S^{\circ}_{\text{vap}}$, for CCl_2F_2, a chlorofluorocarbon that once was heavily used in refrigeration systems. The normal boiling point is -29.79 °C and $\Delta H^{\circ}_{\text{vap}} = 20.2 \text{ kJ mol}^{-1}$.

Practice Example B: The entropy change for the transition from solid rhombic sulfur to solid monoclinic sulfur at 95.5 °C is $\Delta S^{\circ}_{\text{tr}} = 1.09 \text{ J mol}^{-1} \text{ K}^{-1}$. What is the standard molar enthalpy change, $\Delta H^{\circ}_{\text{tr}}$, for this transition?

A useful generalization, known as **Trouton's rule,** states that for many liquids at their normal boiling points the standard molar *entropy of vaporization* has a value of about 87 J mol^{-1} K^{-1}.

$$\Delta S^{\circ}_{\text{vap}} = \frac{\Delta H^{\circ}_{\text{vap}}}{T_{\text{bp}}} \approx 87 \text{ J mol}^{-1} \text{ K}^{-1} \tag{20.6}$$

For example, the values for hexane (C_6H_{14}) and benzene (C_6H_6) are about 84 and 87 J mol^{-1} K^{-1}, respectively. If the degree of disorder produced in transferring one mole of molecules from liquid to vapor at one atmosphere pressure is roughly comparable for different liquids, then we should expect similar values of $\Delta S^{\circ}_{\text{vap}}$. Instances in which Trouton's rule fails are equally understandable. In water and ethanol, for example, hydrogen bonding among molecules produces a greater degree of order than would otherwise be expected in the liquid state. The degree of disorder produced in the vaporization process is greater than normal, and $\Delta S^{\circ}_{\text{vap}} > 87$ J mol^{-1} K^{-1}.

The entropy concept helps to explain Raoult's law (Section 14-6). Recall that for an ideal solution $\Delta H_{\text{soln}} = 0$ and intermolecular forces of attraction are the same as in the pure liquid solvent (page 473). We expect the molar ΔH_{vap} to be the same, whether vaporization of solvent occurs from an ideal solution or from

▶ We found the value of $\Delta S^{\circ}_{\text{vap}}$ for water at 373 K to be 109 J mol^{-1} K^{-1} in Example 20-4.

the pure solvent at the same temperature. So too should ΔS_{vap} be the same, for $\Delta S_{vap} = \Delta H_{vap}/T$. When one mole of solvent is transferred from liquid to vapor at the equilibrium vapor pressure $P°$, entropy increases by the amount ΔS_{vap}. The ideal solution, because it is more disordered than the pure solvent, is in a higher entropy state than is the pure solvent. Increasing the entropy by ΔS_{vap} through vaporization of solvent from the ideal solution requires the resulting vapor to be in a higher state of disorder than it is at the pressure $P°$. As a result, the vapor spreads out into a larger volume and exerts a pressure, P, that is less than $P°$. This corresponds to Raoult's law: $P_A = \chi_A P_A°$ (see Figure 20-6).

The Third Law of Thermodynamics

To establish *absolute* values of entropies, we look for a condition in which a substance is perfectly ordered and take this as the zero of entropy. Then entropy changes are evaluated as the substance is brought to other conditions of temperature and pressure. Add together these entropy changes, and a numerical value of the entropy is obtained. The principle that permits what we have just outlined is the **third law of thermodynamics,** which can be stated as follows.

The entropy of a pure perfect crystal at 0 K is zero.

Figure 20-7 illustrates the method outlined in the preceding paragraph for determining absolute entropy as a function of temperature. Where phase transitions occur, $\Delta S_{tr}° = \Delta H_{tr}°/T_{tr}$. Over temperature ranges in which there are no transitions, $\Delta S°$ values are obtained from measurements of specific heats as a function of temperature.

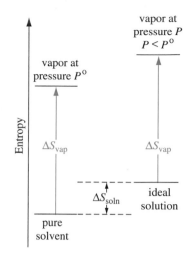

FIGURE 20-6
An entropy-based rationale of Raoult's law.
As described in the text, if ΔS_{vap} has the same value for vaporization from the pure solvent and from an ideal solution, the equilibrium vapor pressure is lower above the solution: $P < P°$.

FIGURE 20-7
Molar entropy as a function of temperature.
The standard molar entropy of a hypothetical substance is represented. By the third law of thermodynamics, an entropy of *zero* is expected at 0 K. Experimental methods cannot be carried to this temperature, however. An extrapolation is required (broken-line portion of the graph). A transition from solid(I) to solid(II) occurs at T_{tr}. Melting occurs at the normal melting point, T_{mp}, and vaporization at the normal boiling point, T_{bp}.

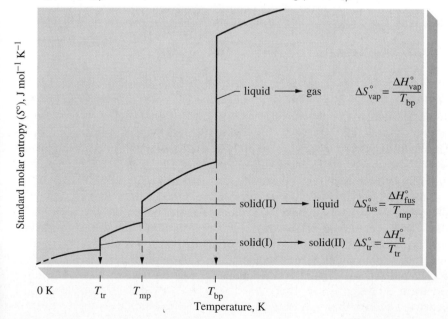

Standard molar entropies of a number of substances at 25 °C are tabulated in Appendix D. To use these values to *calculate* the entropy change of a reaction we use an equation with a familiar form (recall Equation 7.21).

$$\Delta S° = \left[\sum v_p S°(\text{products}) - \sum v_r S°(\text{reactants}) \right] \tag{20.7}$$

The symbol $\sum$ means "the sum of," and the terms added are the products of the standard molar entropies and their stoichiometric coefficients, v. Example 20-5 shows how to use this equation.

EXAMPLE 20-5

Calculating Entropy Changes from Standard Molar Entropies. Use data from Appendix D to calculate the standard molar entropy change for the conversion of nitrogen monoxide to nitrogen dioxide (a step in the manufacture of nitric acid).

$$2\,NO(g) + O_2(g) \longrightarrow 2\,NO_2(g) \qquad \Delta S°_{298\,K} = ?$$

Solution

Equation (20.7) takes the form shown here.

$$\Delta S° = 2S°_{NO_2(g)} - 2S°_{NO(g)} - S°_{O_2(g)}$$

$$= (2 \times 240.0) - (2 \times 210.7) - 205.0 = -146.4 \text{ J K}^{-1}$$

As a useful check on this calculation we can apply some *qualitative* reasoning: Because three moles of gaseous reactants produce only two moles of gaseous products, we should expect the entropy to *decrease*; $\Delta S°$ should be *negative*.

Practice Example A: Use data from Appendix D to calculate the standard molar entropy change for the synthesis of ammonia from its elements.

$$N_2(g) + 3\,H_2(g) \longrightarrow 2\,NH_3(g) \qquad \Delta S°_{298\,K} = ?$$

Practice Example B: N_2O_3 is an unstable oxide that decomposes at 25 °C in the reaction $N_2O_3(g) \rightarrow NO(g) + NO_2(g)$; $\Delta S° = 138.5$ J K^{-1}. What is the standard molar entropy of $N_2O_3(g)$ at 25 °C? [*Hint:* Use Equation (20.7) and other data from Example 20-5.]

▶The cancellation of units (for example for NO_2) is of the form 2 mol $\times$ 240.0 J mol^{-1} K^{-1} = 480.0 J K^{-1}.

In Example 20-5 we used the standard molar entropies of $NO_2(g)$ and $NO(g)$. We might wonder why the value for $NO_2(g)$, 240.0 J mol^{-1} K^{-1}, is greater than that of $NO(g)$: 210.7 J mol^{-1} K^{-1}. Entropy increases when a substance absorbs heat (recall that $\Delta S = q_{rev}/T$). Some of this heat goes simply into raising the average translational kinetic energies of molecules. But there are other ways for heat energy to be used. One possibility, pictured in Figure 20-8, is that the vibrational energies of molecules can be increased. In the *diatomic* molecule $NO(g)$ there is only one type of vibration possible, but in the *triatomic* molecule $NO_2(g)$ there are *three*. Because there are more possible ways of distributing energy among NO_2 molecules than among NO molecules, $NO_2(g)$ has a higher molar entropy than does $NO(g)$ at the same temperature. We should add the following to our generalizations about entropy on page 693.

FIGURE 20-8
Vibrational energy and entropy.
The movement of atoms is suggested by the arrows. The NO molecule (a) has only one type of vibrational motion, whereas the NO_2 molecule (b) has three. This difference helps account for the fact that the molar entropy of $NO_2(g)$ is greater than that of $NO(g)$.

- In general, the more complex their molecules (that is, the greater the number of atoms present), the greater the molar entropies of substances.

Now that we have a way of obtaining entropy changes from tabulated data, we can turn once more to the topic of free energy changes.

20-5   Standard Free Energy Change, $\Delta G°$

Because free energy is related to enthalpy ($G = H - TS$), we cannot establish absolute values of G, any more than we can for H. We must work with free energy *changes,* ΔG. In the next section we will find a special use for the **standard free energy change, $\Delta G°$,** corresponding to reactants and products in their standard states. The standard state conventions are the same as those we introduced for $\Delta H°$ in Chapter 7.

The **standard free energy of formation, $\Delta G_f°$,** is the free energy *change* for a reaction in which a substance in its standard state is formed from its elements in their most stable forms in their standard states. And, as was the case when establishing enthalpies of formation in Section 7-8, this definition leads to values of *zero* for the free energies of formation of the elements in their most stable forms at one atmosphere pressure. Other free energies of formation are related to this condition of zero and are generally tabulated per mole of substance (see Appendix D).

Some additional relationships involving free energy changes are similar to those presented for enthalpy in Section 7-7: (1) ΔG changes sign when a process is reversed; and (2) ΔG for a net or overall process can be obtained by summing the ΔG values for the individual steps. The expressions written below are both useful in calculating $\Delta G°$ values, depending on the data available. The first expression can be used at any temperature for which $\Delta H°$ and $\Delta S°$ values are known. The second expression can only be used at temperatures at which $\Delta G_f°$ values are known, and the only temperature at which tabulated data are commonly given is 298.15 K. The first expression is applied in Example 20-6 and Practice Example 20-6a, and the second expression in Practice Example 20-6b.

$$\Delta G° = \Delta H° - T \Delta S°$$
$$\Delta G° = \left[\sum v_p \, \Delta G_f°(\text{products}) - \sum v_r \, \Delta G_f°(\text{reactants}) \right]$$

EXAMPLE 20-6

Calculating $\Delta G°$ for a Reaction. Determine $\Delta G°$ at 298.15 K for the reaction

$$2\,NO(g) + O_2(g) \longrightarrow 2\,NO_2(g) \quad \text{(at 298.15 K)} \qquad \Delta H° = -114.1 \text{ kJ}$$
$$\Delta S° = -146.4 \text{ J K}^{-1}$$

Solution

Because we have values of $\Delta H°$ and $\Delta S°$, the most direct method of calculating $\Delta G°$ is to use the expression $\Delta G° = \Delta H° - T\,\Delta S°$. In doing so we must first convert all the data to a common energy unit (for example, kJ).

$$\Delta G° = -114.1 \text{ kJ} - (298.15 \text{ K} \times -0.1464 \text{ kJ K}^{-1})$$
$$= -114.1 \text{ kJ} + 43.65 \text{ kJ}$$
$$= -70.5 \text{ kJ}$$

Practice Example A: Determine $\Delta G°$ at 298.15 K for the reaction $4\,Fe(s) + 3\,O_2(g) \rightarrow 2\,Fe_2O_3(s)$. $\Delta H° = -1648$ kJ and $\Delta S° = -549.3$ J K^{-1}.

Practice Example B: Determine $\Delta G°$ for the reaction in Example 20-6 by using data from Appendix D. Compare the two results. (*Hint:* The expression you need is $\Delta G° = 2\,\Delta G_f°[NO_2(g)] - 2\,\Delta G_f°[NO(g)] - \Delta G_f°[O_2(g)]$.)

20-6   Free Energy Change and Equilibrium

We have seen that $\Delta G < 0$ for a spontaneous process and $\Delta G > 0$ for a nonspontaneous process. If $\Delta G = 0$, the forward and reverse processes show an equal tendency to occur, and the system is at *equilibrium*. At this point, even an infinitesimal change in one of the system variables (such as temperature or pressure) will cause a net change to occur. But if a system at equilibrium is left undisturbed there is no net change with time.

Let's consider the hypothetical process outlined in Figure 20-9. If we start at the left side of the figure, we see that ΔH exceeds $T\,\Delta S$, ΔG is positive, and the process is *nonspontaneous*. The magnitude of ΔG decreases with increasing tem-

FIGURE 20-9

Free energy change as a function of temperature.
The value of ΔG is given by the distance between the two lines; that is, $\Delta G = (\Delta H - T\,\Delta S)$. In this illustration, both ΔH and ΔS have positive values at all temperatures. If ΔH is greater than the $T\,\Delta S$ value, $\Delta G > 0$ and the reaction is *nonspontaneous*. If ΔH is less than $T\,\Delta S$, $\Delta G < 0$ and the reaction is *spontaneous*. Equilibrium occurs at the temperature at which the two lines intersect: $\Delta G = 0$. An assumption made here is that ΔH and ΔS are essentially independent of temperature.

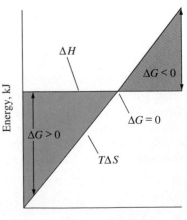

perature. At the right side of the figure, $T \Delta S$ exceeds ΔH, ΔG is negative, and the process is *spontaneous*. At the temperature at which the two lines intersect, $\Delta G = 0$, and the system is at equilibrium.

For the vaporization of water, *with both liquid and vapor in their standard states* (which means that $\Delta G = \Delta G°$), the intersection of the two lines in Figure 20-9 is at $T = 373.15$ K (100.00 °C). That is, for a reaction at 1 atm,

$$H_2O(l) \rightleftharpoons H_2O(g) \qquad \Delta G° = 0 \text{ at } 373.15 \text{ K}$$

At 25 °C, the $\Delta H°$ line lies above the $T \Delta S°$ line in Figure 20-9. This means that $\Delta G° > 0$.

$$H_2O(l) \longrightarrow H_2O(g) \qquad \Delta G° = +8.590 \text{ kJ at } 298.15 \text{ K}$$

The positive value of $\Delta G°$ does not mean that vaporization of water will not occur. From common experience we know that water spontaneously evaporates at room temperature. What the positive value means is that liquid water will not spontaneously produce $H_2O(g)$ at *1 atm pressure* at 25 °C. Instead, we find that $H_2O(g)$ is produced with a vapor pressure that is less than 1 atm pressure. The equilibrium vapor pressure of water at 25 °C is 23.76 mmHg $= 0.03126$ atm, that is,

$$H_2O(l, 0.03126 \text{ atm}) \rightleftharpoons H_2O(g, 0.03126 \text{ atm}) \qquad \Delta G = 0$$

▶ The vaporization of water at 298.15 K is the subject of further study in Feature Problem A on page 722.

▶ The liquid–vapor equilibrium represented here is out of contact with the atmosphere. In the presence of the atmosphere, the pressure on the liquid would be barometric pressure, whereas that of the vapor would remain essentially unchanged at 0.03126 atm.

Relationship of $\Delta G°$ to the Equilibrium Constant K_{eq}

If you think about the situation just described for the vaporization of water, there is not much value in describing equilibrium in a process through its $\Delta G°$ value. At only one temperature are the reactants in their standard states in equilibrium with products in their standard states; that is, at only one temperature does $\Delta G° = 0$. We want to be able to describe equilibrium for a variety of conditions, typically *nonstandard* conditions. For this we need to work with ΔG, not $\Delta G°$.

But here is an interesting prospect. We can relate ΔG for a reaction under *any* set of conditions to its value for standard conditions, that is, to $\Delta G°$. The key term in relating the two is the *reaction quotient, Q*, formulated for the actual, nonstandard conditions. A derivation of this relationship is beyond the scope of this text, but we can make use of the result, which is

$$\Delta G = \Delta G° + RT \ln Q \tag{20.8}$$

▶ In Equation (20.8) R is the familiar ideal gas constant, expressed as $R = 8.3145$ J K^{-1} mol^{-1}.

In Chapter 16 we learned that if a system is at equilibrium, $Q_c = K_c$, or $Q_p = K_p$ or, in general, $Q = K_{eq}$. And in this chapter we have seen that at equilibrium $\Delta G = 0$. These two facts allow us to write that, *at equilibrium,*

$$\Delta G = \Delta G° + RT \ln K_{eq} = 0$$

which means that

$$\Delta G° = -RT \ln K_{eq} \tag{20.9}$$

If we have a value of $\Delta G°$ at a given temperature, we can *calculate* an equilibrium constant K_{eq} with Equation (20.9). This means that the tabulation of thermodynamic data in Appendix D can serve as a direct source of countless equilibrium constant values at 298.15 K.

We need to say a few words about the units required in Equation (20.9). Because logarithms can only be taken of dimensionless numbers, K_{eq} has no units, nor does ln K_{eq}. The right side of Equation (20.9) has the units of "RT": J mol^{-1} K^{-1} × K = J mol^{-1}. $\Delta G°$ on the left side of the equation must have the same unit: J mol^{-1}. The "mol^{-1}" part of this unit means "per mole of reaction." One mole of reaction is simply the reaction based on the stoichiometric coefficients chosen for the balanced equation. Often the "mol^{-1}" portion of this unit is dropped. We have dropped it in most instances in the past, and we will continue to do so except in those cases in which we need mol^{-1} for the proper cancellation of units, as in Equation (20.9).

Criterion for Spontaneous Change: Our Search Concluded

In Figure 20-1 we plotted a hypothetical function that would reach a minimum at a point in a reaction at which equilibrium is reached. Figure 20-10 is a similar plot and demonstrates that free energy is the function we have been seeking. To construct this plot in all its details is beyond the scope of this text, but we can rationalize its general shape, based on two ideas: (a) Every chemical reaction consists of both a forward and a reverse reaction, even if one of these occurs only to a very slight extent. (b) The direction of spontaneous change in both the forward and reverse reaction is that in which free energy decreases ($\Delta G < 0$). As a consequence, free energy reaches a minimum at some point between the left and right sides of the graph. This is the equilibrium point.

Now consider that the vertical distance between the two endpoints of the graph represents $\Delta G°$ of the reaction. If, as in Figure 20-10, $\Delta G°$ of a reaction is small, either positive or negative, the equilibrium condition is one in which significant amounts of both reactants and products will be found. If $\Delta G°$ is a *large, positive* quantity, the equilibrium point lies far to the left—that is, very close to the reactants side, and we can say that the reaction occurs hardly at all. If $\Delta G°$ is a *large, negative* quantity, the equilibrium point lies far to the right—that is, very close to the products side, and we can say that the reaction goes essentially to completion. The caption of Figure 20-10 gives a simple analogy that is consistent with our remarks about the location of the equilibrium point. Table 20.2 summarizes the conclusions of this discussion, gives approximate magnitudes to the terms *small* and *large*, and relates $\Delta G°$ values to values of K_{eq}.

FIGURE 20-10
Free energy change, equilibrium, and the direction of spontaneous change.
Free energy is plotted against the extent of reaction for a hypothetical reaction. $\Delta G°$ is the difference between the standard molar free energies of formation of products and reactants. The equilibrium point lies somewhere between pure reactants and pure products. Mixture A, with $Q < K_{eq}$, and mixture B, with $Q > K_{eq}$, both undergo spontaneous change in the direction of lower free energy, eventually reaching the equilibrium point.

As an analogy, consider a length of rope attached to two parallel, upright poles (analogous to the sides of the graph). The rope, somewhat longer than the distance between the two poles, sags under the force of gravity, and the shape of the rope is analogous to the curve in the graph. The position of the minimum point on the rope (analogous to the equilibrium point on the curve) depends on the difference in vertical height between the two points of attachment (analogous to $\Delta G°$).

TABLE 20.2 Significance of the Magnitude of $\Delta G°$ (at 298 K)		
$\Delta G°$	K_{eq}	**Significance**
+200 kJ/mol	9.1×10^{-36}	No reaction
+100	3.0×10^{-18}	
+50	1.7×10^{-9}	
+10	1.8×10^{-2}	
+1.0	6.7×10^{-1}	Equilibrium calculation is necessary
0	1.0	
−1.0	1.5	
−10	5.6×10^{1}	
−50	5.8×10^{8}	
−100	3.3×10^{17}	Reaction goes to completion
−200	1.1×10^{35}	

The Thermodynamic Equilibrium Constant

To use Equation (20.9) we need to write equilibrium constant expressions in a particular way. One reason for this, as we have noted previously, is that an equilibrium constant must be a *dimensionless* number, so that we may take its logarithm. We need to use the **thermodynamic equilibrium constant, K_{eq},** an equilibrium constant based on activities. Having said this, though, we would still like to write equilibrium constant expressions in more familiar terms—molarities and gas pressures. We can do this by using the following conventions.

- *For pure solids and liquids:* The activity, $a = 1$.
- *For gases:* With ideal gas behavior assumed, the activity is replaced by the gas pressure in atmospheres.
- *For solutes in aqueous solution:* With ideal solution behavior (for example, no interionic attractions) assumed, the activity is replaced by the molarity.

▶ We introduced activities in Section 14-9 as "effective concentrations" designed to give the best agreement between the compositions and physical properties of solutions.

Thermodynamic equilibrium constants, K_{eq}, are sometimes identical to K_c and K_p values, as in parts (a) and (b) of Example 20-7. In other instances, such as part (c) of Example 20-7, this is not the case. In working through Example 20-7, keep in mind that in this text our sole reason for writing thermodynamic equilibrium constants is to get the proper value to use in Equation (20.9).

EXAMPLE 20-7

Writing Thermodynamic Equilibrium Constant Expressions. For the following reversible reactions, write thermodynamic equilibrium constant expressions, making appropriate substitutions for activities. Then equate K_{eq} to K_c or K_p, where this can be done.

a. The water gas reaction

$$C(s) + H_2O(g) \rightleftharpoons CO(g) + H_2(g)$$

b. The formation of a saturated aqueous solution of lead iodide, a very slightly soluble solute.

$$PbI_2(s) \rightleftharpoons Pb^{2+}(aq) + 2\,I^-(aq)$$

c. The oxidation of sulfide ion by oxygen gas (used in removing sulfides from waste-water, as in pulp and paper mills).

$$O_2(g) + 2\ S^{2-}(aq) + 2\ H_2O(l) \rightleftharpoons 4\ OH^-(aq) + 2\ S(s)$$

Solution

In each case, once we have made the appropriate substitutions for activities, if all terms are molarities, the thermodynamic equilibrium constant is the same as K_c. If all terms are partial pressures, $K_{eq} = K_p$. If both molarities *and* partial pressures appear in the expression, however, the equilibrium constant expression can only be designated as K_{eq}.

a. The activity of solid carbon is 1. Partial pressures are substituted for the activities of the gases.

$$K_{eq} = \frac{(P_{CO})(P_{H_2})}{(1)(P_{H_2O})} = \frac{(P_{CO})(P_{H_2})}{(P_{H_2O})} = K_p$$

b. The activity of solid lead(II) iodide is 1. Molarities are substituted for activities of the ions in aqueous solution.

$$K_{eq} = \frac{[Pb^{2+}][I^-]^2}{1} = [Pb^{2+}][I^-]^2 = K_c = K_{sp}$$

c. The activities of the solid sulfur and of the liquid water are each 1. Molarities are substituted for the activities of $OH^-(aq)$ and $S^{2-}(aq)$. The partial pressure of $O_2(g)$ is substituted for its activity. Thus, the resulting K_{eq} is neither a K_c nor a K_p.

$$K_{eq} = \frac{(1)^2 \cdot [OH^-]^4}{P_{O_2} \cdot [S^{2-}]^2 \cdot (1)^2} = \frac{[OH^-]^4}{P_{O_2} \cdot [S^{2-}]^2}$$

Practice Example A: Write thermodynamic equilibrium constant expressions for the following reactions. Relate these to K_c or K_p where appropriate. **(a)** Si(s) + 2 Cl$_2$(g) $\rightleftharpoons$ SiCl$_4$(g); **(b)** Cl$_2$(g) + H$_2$O(l) $\rightleftharpoons$ HOCl(aq) + H$^+$(aq) + Cl$^-$(aq).

Practice Example B: Write a thermodynamic equilibrium constant expression to represent the reaction of solid lead(II) sulfide with aqueous nitric acid to produce solid sulfur, a solution of lead(II) nitrate, and nitrogen monoxide gas. (*Hint:* Base the expression on the balanced net ionic equation for this redox reaction.)

We have now acquired all the tools with which to perform one of the most practical calculations of chemical thermodynamics—*determining the equilibrium constant for a reaction from tabulated data*. In Example 20-8, where we illustrate this application, we use thermodynamic properties of ions in aqueous solution as well as of compounds. An important idea to note about the thermodynamic properties of ions is that they are relative to H$^+$(aq) which, by convention, is assigned values of *zero* for ΔH_f°, ΔG_f°, *and* S°. This means that entropies listed for ions are not absolute entropies, as they are for compounds. Negative values of S° simply denote an entropy less than that of H$^+$(aq).

EXAMPLE 20-8

Calculating the Equilibrium Constant of a Reaction from the Standard Free Energy Change. Determine the equilibrium constant at 298.15 K for the dissolving of magnesium hydroxide in an acidic solution.

$$Mg(OH)_2(s) + 2\ H^+(aq) \rightleftharpoons Mg^{2+}(aq) + 2\ H_2O(l)$$

Solution

The key to solving this problem is to find a value of $\Delta G°$, and then to use the expression $\Delta G° = -RT \ln K_{eq}$. We can obtain $\Delta G°$ from standard free energies of formation listed in Appendix D. Note that because its value is zero the term $\Delta G_f°[H^+(aq)]$ is not included.

$$\Delta G° = 2 \Delta G_f°[H_2O(l)] + \Delta G_f°[Mg^{2+}(aq)] - \Delta G_f°[Mg(OH)_2(s)]$$

$$= 2(-237.1 \text{ kJ mol}^{-1}) + (-454.8 \text{ kJ mol}^{-1}) - (-833.5 \text{ kJ mol}^{-1})$$

$$= -95.5 \text{ kJ mol}^{-1}$$

Now, solve for $\ln K_{eq}$ and K_{eq}.

$$\Delta G° = -RT \ln K_{eq} = -95.5 \text{ kJ mol}^{-1} = -95.5 \times 10^3 \text{ J mol}^{-1}$$

$$\ln K_{eq} = \frac{-\Delta G°}{RT} = \frac{-(-95.5 \times 10^3 \text{ J mol}^{-1})}{8.3145 \text{ J mol}^{-1} \text{ K}^{-1} \times 298.15 \text{ K}} = 38.5$$

$$K_{eq} = e^{38.5} = 5 \times 10^{16}$$

The value of K_{eq} obtained here is the *thermodynamic equilibrium constant*. According to the conventions we have established, the activities of $Mg(OH)_2(s)$ and of $H_2O(l)$ are 1, and molarities can be substituted for the activities of the ions.

$$K_{eq} = \frac{[Mg^{2+}] \cdot (1)^2}{(1) \cdot [H^+]^2} = \frac{[Mg^{2+}]}{[H^+]^2} = K_c = 5 \times 10^{16}$$

Practice Example A: Determine the equilibrium constant at 298.15 K for AgI(s) $\rightleftharpoons$ Ag$^+$(aq) + I$^-$(aq). Compare your answer to the K_{sp} for AgI in Appendix D.

Practice Example B: At 298.15 K, would you expect manganese dioxide to react to an appreciable extent with 1 M HCl(aq) to produce manganese(II) ion in solution and chlorine gas? (*Hint:* Write a balanced net ionic equation for the redox reaction and then use free energy of formation data from Appendix D.)

When a question requires the use of thermodynamic properties, as in Example 20-8, it is a good idea to think *qualitatively* about the problem before diving into calculations. The dissolving of $Mg(OH)_2(s)$ in acidic solution considered in Example 20-8 is an acid–base reaction. It is an example we used in Chapter 19 to illustrate the effect of pH on solubility. We also mentioned it in Chapter 5 as the basis for using "milk of magnesia" as an antacid. We should certainly expect the reaction to be spontaneous. This means that the value of K_{eq} should be large, which is what we found to be the case. If we had made an error in sign in our calculation (an easy error to make when using the expression $\Delta G° = -RT \ln K_{eq}$) we would have obtained $K_{eq} = 2 \times 10^{-17}$. But we would have seen immediately that this is the wrong answer. This value suggests a reaction in which the concentration of products is extremely low at equilibrium.

The data listed in Appendix D are for 25 °C, and values of $\Delta G°$ and K_{eq} obtained with these data are also at 25 °C. Many chemical reactions are carried out at temperatures other than 25 °C, however, and in the next section we will learn how to calculate values of equilibrium constants at various temperatures.

Are You Wondering . . .

Just what is the difference between ΔG and $\Delta G°$?

We've been dealing with both quantities, and this is a good time to summarize some ideas about them.

- $\Delta G < 0$ signifies that a reaction or process is spontaneous in the forward direction (to the right) for the stated conditions. $\Delta G° < 0$ signifies that the forward reaction is spontaneous when reactants and products are in their standard states. It further signifies that $K_{eq} > 1$, whatever the initial concentrations or pressures of reactants and products.

- $\Delta G = 0$ signifies that the reaction is at equilibrium under the stated conditions. $\Delta G° = 0$ signifies that the reaction is at equilibrium when reactants and products are in their standard states. It further signifies that $K_{eq} = 1$, and this can occur only at a particular temperature.

- $\Delta G > 0$ signifies that the reaction or process is nonspontaneous in the forward direction under the stated conditions. $\Delta G° > 0$ signifies that the forward reaction is nonspontaneous when reactants and products are in their standard states. It further signifies that $K_{eq} < 1$, whatever the initial concentrations or pressures of reactants and products.

- $\Delta G = \Delta G°$ only when all reactants and products are in their standard states. Otherwise, $\Delta G = \Delta G° + RT \ln Q$.

20-7 $\Delta G°$ and K_{eq} as Functions of Temperature

In Chapter 16 we used Le Châtelier's principle to make *qualitative* predictions of the effect of temperature on an equilibrium condition. We can now describe a *quantitative* relationship between the equilibrium constant and temperature. In the method illustrated in Example 20-9, we assume that $\Delta H°$ is practically independent of temperature. Although absolute entropies depend on temperature, we assume that the entropy *change* $\Delta S°$ for a reaction is also independent of temperature. On the other hand, the term "$T \Delta S°$" is strongly temperature-dependent because of the temperature factor T. As a result, $\Delta G°$, which is equal to $\Delta H° - T \Delta S°$, is also dependent on temperature.

EXAMPLE 20-9

Determining the Relationship Between an Equilibrium Constant and Temperature Using Equations for Free Energy Change. At what temperature will the equilibrium constant for the formation of NOCl(g) be $K_{eq} = K_p = 1.00 \times 10^3$? Data for this reaction at 25 °C are

$$2 \text{ NO(g)} + \text{Cl}_2\text{(g)} \rightleftharpoons 2 \text{ NOCl(g)}$$

$$\Delta G° = -41.0 \text{ kJ mol}^{-1} \qquad \Delta H° = -77.1 \text{ kJ mol}^{-1}$$

$$\Delta S° = -121.2 \text{ J mol}^{-1} \text{ K}^{-1}$$

Solution

Because we have a *known* equilibrium constant and an *unknown* temperature, we need an equation in which both of these terms appear. The only equation that we have for this purpose is $\Delta G^\circ = -RT \ln K_{eq}$. If we knew the value of ΔG° at the unknown temperature, we could simply solve this equation for T. We know the value of ΔG° at 25 °C (-41.0 kJ/mol), but we know that this value will be different at other temperatures. We can assume, however, that the values of ΔH° and ΔS° will not change much with temperature. This means that we can obtain a value of ΔG° through the equation $\Delta G^\circ = \Delta H^\circ - T \Delta S^\circ$, where T is the *unknown* temperature and the values of ΔH° and ΔS° are those at 25 °C. Now we have two equations that we can set equal to each other. That is,

$$\Delta G^\circ = \Delta H^\circ - T \Delta S^\circ = -RT \ln K_{eq}$$

We can gather the terms with T on the right

$$\Delta H^\circ = T \Delta S^\circ - RT \ln K_{eq} = T(\Delta S^\circ - R \ln K_{eq})$$

and solve for T.

$$T = \frac{\Delta H^\circ}{\Delta S^\circ - R \ln K_{eq}}$$

Now let's substitute values for ΔH°, ΔS°, R, and $\ln K_{eq}$.

$$T = \frac{-77.1 \times 10^3 \text{ J mol}^{-1}}{-121.2 \text{ J mol}^{-1} \text{ K}^{-1} - [8.3145 \text{ J mol}^{-1} \text{ K}^{-1} \times \ln (1.00 \times 10^3)]}$$

$$= \frac{-77.1 \times 10^3 \text{ J mol}^{-1}}{-121.2 \text{ J mol}^{-1} \text{ K}^{-1} - (8.3145 \times 6.908) \text{ J mol}^{-1} \text{ K}^{-1}}$$

$$= \frac{-77.1 \times 10^3 \text{ J mol}^{-1}}{-178.6 \text{ J mol}^{-1} \text{ K}^{-1}} = 432 \text{ K}$$

Although we have shown three significant figures in our answer, we should probably round the final result to just *two* significant figures. The assumption we made about the constancy of ΔH° and ΔS° is probably no more valid than this.

Practice Example A: At what temperature will the formation of $NO_2(g)$ from $NO(g)$ and $O_2(g)$ have $K_p = 1.50 \times 10^2$? For the reaction $2 NO(g) + O_2(g) \rightleftharpoons 2 NO_2(g)$ at 25 °C, $\Delta H^\circ = -114.1$ kJ/mol and $\Delta S^\circ = -146.4$ J mol^{-1} K^{-1}.

Practice Example B: For the reaction $2 NO(g) + Cl_2(g) \rightleftharpoons 2 NOCl(g)$, what is the value of K_{eq} at **(a)** 25 °C; **(b)** 75 °C? Use data from Example 20-9. [*Hint:* Here you need to solve for values of K_{eq} rather than T. Should these values be greater or less than $K_{eq} = 1.0 \times 10^3$? The solution to (a) can be done somewhat more simply than that for (b).]

An alternative to the method outlined in Example 20-9 is to relate the equilibrium constant and temperature directly, without specific reference to free energy change. To do this, we start with the same two expressions as in Example 20-9,

$$-RT \ln K_{eq} = \Delta G^\circ = \Delta H^\circ - T \Delta S^\circ$$

and divide by $-RT$.

$$\ln K_{eq} = \frac{-\Delta H^\circ}{RT} + \frac{\Delta S^\circ}{R} \qquad (20.10)$$

T, K	1/T, K⁻¹	Kₚ	ln Kₚ

TABLE 20.3 Equilibrium Constants, K_p, for the Reaction
$$2\,SO_2(g) + O_2(g) \rightleftharpoons 2\,SO_3(g)$$
at Several Temperatures

T, K	$1/T$, K⁻¹	K_p	ln K_p
800	12.5×10^{-4}	9.1×10^{2}	6.81
850	11.8×10^{-4}	1.7×10^{2}	5.14
900	11.1×10^{-4}	4.2×10^{1}	3.74
950	10.5×10^{-4}	1.0×10^{1}	2.30
1000	10.0×10^{-4}	3.2×10^{0}	1.16
1050	9.52×10^{-4}	1.0×10^{0}	0.00
1100	9.09×10^{-4}	3.9×10^{-1}	−0.94
1170	8.55×10^{-4}	1.2×10^{-1}	−2.12

If we assume that $\Delta H°$ and $\Delta S°$ are constant, Equation (20.10) is that of a straight line with a slope of $-\Delta H°/R$ and an intercept of $\Delta S°/R$. Table 20.3 lists equilibrium constants for the reaction of $SO_2(g)$ and $O_2(g)$ to form $SO_3(g)$ as a function of the reciprocal of Kelvin temperature. The ln K_p and $1/T$ data from Table 20.3 are plotted in Figure 20-11 and yield the expected straight line.

Now we can follow the procedure used in Appendix A-4 to derive the Clausius–Clapeyron equation shown on page 428: We can write Equation (20.10) twice, for two different temperatures and with the corresponding equilibrium constants. Then, if we subtract one equation from the other, we obtain the result shown here.

$$\ln\frac{K_2}{K_1} = \frac{\Delta H°}{R}\left(\frac{1}{T_1} - \frac{1}{T_2}\right) \qquad (20.11)$$

FIGURE 20-11
Temperature dependence of the equilibrium constant K_p for the reaction

$$2\,SO_2(g) + O_2(g) \rightleftharpoons 2\,SO_3(g)$$

This graph can be used to establish the enthalpy of reaction, $\Delta H°$ (see Equation 20.10).

$$\text{slope} = -\Delta H°/R = 2.2 \times 10^4 \text{ K}$$

$$\Delta H° = -8.3145 \text{ J mol}^{-1}\text{ K}^{-1} \times 2.2 \times 10^4 \text{ K}$$

$$= -1.8 \times 10^5 \text{ J mol}^{-1}$$

$$= -1.8 \times 10^2 \text{ kJ mol}^{-1}$$

$$\text{slope} = \frac{7.6}{3.4 \times 10^{-4}\text{ K}^{-1}} = 2.2 \times 10^4 \text{ K}$$

In Equation (20.11), T_2 and T_1 are two Kelvin temperatures; K_2 and K_1 are the equilibrium constants at these temperatures; $\Delta H°$ is the enthalpy of reaction, expressed in J mol^{-1}; R is the gas constant, expressed as 8.3145 J mol^{-1} K^{-1}. Jacobus van't Hoff (1852–1911) derived Equation (20.11), which is often referred to as the *van't Hoff equation.*

▶ Note that the Clausius–Clapeyron equation (13.2) is just a special case of Equation (20.11) in which the equilibrium constants are equilibrium vapor pressures and $\Delta H°$ is ΔH_{vap}.

EXAMPLE 20-10

Relating Equilibrium Constants and Temperature Through the van't Hoff Equation. Use data from Table 20.3 and Figure 20-11 to estimate the temperature at which $K_p = 1.0 \times 10^6$ for the reaction

$$2\,SO_2(g) + O_2(g) \rightleftharpoons 2\,SO_3(g)$$

Solution

Select one known temperature and equilibrium constant from Table 20.3 and the enthalpy change of the reaction, $\Delta H°$, from Figure 20-11. The data we substitute in Equation (20.11) are $T_1 = ?$, $K_1 = 1.0 \times 10^6$; $T_2 = 800$ K, $K_2 = 9.1 \times 10^2$; $\Delta H° = -1.8 \times 10^5$ J mol^{-1}.

$$\ln \frac{K_2}{K_1} = \frac{\Delta H°}{R}\left(\frac{1}{T_1} - \frac{1}{T_2}\right)$$

$$\ln\frac{9.1 \times 10^2}{1.0 \times 10^6} = \frac{-1.8 \times 10^5}{8.3145}\left(\frac{1}{T_1} - \frac{1}{800}\right)$$

$$-7.00 = -2.2 \times 10^4\left(\frac{1}{T_1} - \frac{1}{800}\right)$$

$$\frac{-7.00}{-2.2 \times 10^4} + \frac{1}{800} = \frac{1}{T_1}$$

$$\frac{1}{T_1} = (3.2 \times 10^{-4}) + (1.25 \times 10^{-3}) = 1.57 \times 10^{-3}$$

$$T_1 = \frac{1}{1.57 \times 10^{-3}} = 6.37 \times 10^2 \text{ K}$$

▶ For simplicity we have dropped the units in this set-up, but you should be able to show that units cancel properly.

Practice Example A: Estimate the temperature at which $K_p = 5.8 \times 10^{-2}$ for the reaction in Example 20-10. Use data from Table 20.3 and Figure 20-11.

Practice Example B: What is the value of K_p for the reaction $2\,SO_2(g) + O_2(g) \rightleftharpoons 2\,SO_3(g)$ at 235 °C? Use data from Table 20.3 and Figure 20-11 and the van't Hoff Equation (20.11).

20-8 Coupled Reactions

When we are faced with a situation in which a reaction is nonspontaneous, we have learned two ways by which we may be able to obtain a desired product: (1) Change the reaction conditions to those that make the reaction spontaneous, mostly by changing the temperature, and (2) carry out the reaction by electrolysis. But there is a third way also. Combine a pair of reactions, called **coupled reactions,** that yield the desired product in a *spontaneous net* reaction.

For example, when copper(I) oxide is heated at 673 K, no copper metal is obtained. The decomposition of Cu_2O to form products in their standard states (for example, $P_{O_2} = 1.00$ atm) is nonspontaneous at 673 K.

Focus On

Heat Engines and Thermal Pollution

Water is heated in a boiler and converted to superheated steam. The steam is allowed to expand in the turbine, where it does work by turning the rotor blades. The turbine shaft drives the electric generator, which produces electric power. Condensation of the steam occurs in the condenser. The water that is used for cooling carries away up to 50% of the heat energy that was released by burning the fuel. It is this waste heat that produces thermal pollution.

Sadi Carnot, a French military engineer, deduced the second law of thermodynamics in a study of the efficiencies of heat engines in 1824. William Thomson (Lord Kelvin) recognized the significance of Carnot's work and developed the absolute temperature scale (the Kelvin scale) necessitated by the second law.

In a heat engine, heat (q_h) is absorbed by the working substance (for example, water) at a high temperature (T_h). This heat is partly converted to work (w), and the remainder (q_l) is released to the surroundings at a lower temperature (T_l). An alternative statement of the *second law of thermodynamics* is that

No heat engine can operate without rejecting waste heat to lower temperature surroundings.

That is, a heat engine cannot operate at a constant temperature.

The efficiency of a heat engine is governed by the ratio w/q_h. If all the heat absorbed could be converted to work, the engine would be 100% efficient. The second law of thermodynamics places an absolute limit on the efficiency of a heat engine, *and it is never 100%*. The expression obtained is

---

$$Cu_2O(s) \xrightarrow{\Delta} 2\,Cu(s) + \tfrac{1}{2}\,O_2(g) \qquad \Delta G^{\circ}_{673\,K} = +125 \text{ kJ} \qquad (20.12)$$

Now, suppose that this nonspontaneous decomposition reaction is coupled with the partial oxidation of carbon to carbon monoxide—a spontaneous reaction. The net reaction (20.13), because of its *negative* value of ΔG°, is spontaneous when reactants and products are in their standard states.

$$Cu_2O(s) \longrightarrow 2\,Cu(s) + \tfrac{1}{2}\,O_2(g) \qquad \Delta G^{\circ}_{673\,K} = +125 \text{ kJ}$$

$$\underline{C(s) + \tfrac{1}{2}\,O_2(g) \longrightarrow CO(g) \qquad\qquad\qquad \Delta G^{\circ}_{673\,K} = -175 \text{ kJ}}$$

$$Cu_2O(s) + C(s) \longrightarrow 2\,Cu(s) + CO(g) \qquad \Delta G^{\circ}_{673\,K} = -50 \text{ kJ} \qquad (20.13)$$

Note that reactions (20.12) and (20.13) are not the same, even though each has Cu(s) as a product. The purpose of coupled reactions, then, is to produce a spontaneous net reaction by combining two other processes: one nonspontaneous and one spontaneous. Many metallurgical processes are viewed as coupled reactions, especially those that use carbon or hydrogen as reducing agents.

▶A graphical representation of coupled reactions in metallurgy is explored in Feature Problem B on page 722.

To sustain life, organisms must synthesize complex molecules from simpler ones. If carried out as single-step reactions, these syntheses would generally be accompanied by increases in enthalpy, decreases in entropy, and increases in free energy—in short, they would be nonspontaneous and would not occur. In

$$\text{efficiency} = \frac{w}{q_h} = \frac{T_h - T_l}{T_h} \qquad (20.14)$$

where temperature is in kelvins.

Thus, if a steam engine operates between 100 °C (the boiler temperature) and 25 °C (the condenser temperature), it can have an efficiency of only 0.20, that is, 20%. Of the heat supplied to the boiler, 80% is given off to the surroundings.

$$\text{efficiency} = \frac{373 - 298}{373} = \frac{75}{373}$$

$$= 0.20$$

Conventional electric power plants are based on the burning of a fossil fuel, which provides the heat energy to convert water to steam. The steam powers a turbine that drives an electric generator. The process is not highly efficient. The efficiency can be improved, however, by using a higher working temperature for the turbine (T_h). This can be accomplished by operating the system under high pres-

sure, at temperatures well in excess of 100 °C, using superheated steam. Still, though, most electric power plants operate at thermal efficiencies of less than 40%. Thus, more than half the heat required to produce electric power is waste heat. The implications of this fact are twofold: wasted heat means wasted fuel, and as the waste heat enters the environment it leads to thermal pollution.

The origin of thermal pollution in a power plant is suggested by the diagram. The best known effects of thermal pollution are those produced on fish and other aquatic life. The differential growth rate of algae with changing temperatures can cause one type of algae, which is ideal food for fish, to be displaced by another, which is a poor food or even toxic to fish. Also, because their metabolic rate goes up with temperature, fish need more food and oxygen as the temperature of their water rises; but the solubility of air in water *decreases*

with increasing temperature. Temperature changes affect other physiological processes too. Some fish are killed by the thermal shock of even relatively small temperature changes.

Even if all other pollution sources could be controlled, release of heat energy to the surroundings by heat engines would still be unavoidable. Means must be found to minimize the quantity of waste heat or to put the waste heat to useful purposes. The first objective can be met through more efficient heat engines and through the direct conversion of other energy forms to electricity (such as with fuel cells and solar energy devices). The second objective can be met by using waste heat from power plants to supply hot water for industrial, commercial, or residential purposes, or using this water for agricultural purposes.

living organisms, changes in temperature and electrolysis are not viable options for dealing with nonspontaneous processes. Here, coupled reactions are crucial.

A simple example of a coupled reaction in living organisms is the first step in the metabolism of glucose ($C_6H_{12}O_6$). In this step glucose is converted to glucose-6-phosphate through the combination of the two reactions shown here. The first of the two reactions has a *positive* $\Delta G°$, whereas the second has a *negative* $\Delta G°$. The sum of the two $\Delta G°$ values is *negative*, and this means that the overall, or net, reaction is spontaneous when reactants and products are in their standard states.

$$HPO_4^{2-} + \text{glucose} \longrightarrow (\text{glucose-6-phosphate})^{2-} + H_2O \qquad \Delta G° = +13.8 \text{ kJ}$$

$$\underline{ATP^{4-} + H_2O \longrightarrow ADP^{3-} + HPO_4^{2-} + H^+ \qquad \Delta G° = -30.5 \text{ kJ}}$$

$$ATP^{4-} + \text{glucose} \longrightarrow (\text{glucose-6-phosphate})^{2-} + ADP^{3-} + H^+$$

$$\Delta G° = -16.7 \text{ kJ}$$

In the net reaction one ATP (adenosine triphosphate) is converted to ADP (adenosine diphosphate). ATP is likened to a "free energy bank" from which "withdrawals" are made by reactions that, if not coupled, would require a free energy increase. Each withdrawal requires coupling to the hydrolysis of an ATP.

Summary

Every spontaneous change produces an increase in the entropy of the universe. The entropy change of a system, ΔS, can be determined mathematically if the path of the reaction is *reversible*. Then, $\Delta S = q_{rev}/T$—that is, the quantity of heat exchanged reversibly with the surroundings, divided by the Kelvin temperature. For a phase transition at equilibrium this becomes $\Delta S_{tr} = \Delta H_{tr}/T_{tr}$. Absolute values can be assigned to entropies of substances. This is made possible by the third law of thermodynamics, which takes as the zero of entropy that of a pure perfect crystal at 0 K.

A criterion for spontaneous change based *just on the system itself* is that the Gibbs free energy, G, decreases, that is, $\Delta G < 0$. At equilibrium, $\Delta G = 0$. Free energy change is related to enthalpy and entropy changes through the expression $\Delta G = \Delta H - T\,\Delta S$.

The standard free energy change, $\Delta G°$, is based on the conversion of reactants in their standard states to products in their standard states. Tabulated free energy data are usually standard molar free energies of formation, $\Delta G_f°$, and usually at 298.15 K (25 °C). The relationship between the standard free energy change and the equilibrium constant for a reaction is $\Delta G° = -RT \ln K_{eq}$. For example, if a value of $\Delta G°$ can be obtained from tabulated data, the corresponding equilibrium constant, K_{eq}, can be calculated. The constant, K_{eq}, is called a thermodynamic equilibrium constant. It is based on the activities of reactants and products, but these activities can be related to solution molarities and gas partial pressures through a few simple conventions.

By starting with the relationship between standard free energy change and the equilibrium constant, the van't Hoff equation relating the equilibrium constant and temperature can be written. With this equation it is possible to use tabulated data at 25 °C to determine equilibrium constants not just at 25 °C but at other temperatures as well.

Summarizing Example

The synthesis of methanol is of great importance because methanol can be used directly as a motor fuel, mixed with gasoline for fuel use, or converted to other organic compounds. The synthesis reaction, carried out at about 500 K, is

$$CO(g) + 2\,H_2(g) \rightleftharpoons CH_3OH(g)$$

What is the value of K_p at 500 K?

1. *Determine $\Delta G°$ at 298 K.* From $\Delta G°$ at 298 K we can obtain K_p at 298 K. To evaluate $\Delta G°$, use free energy of formation data from Appendix D. *Result:* $\Delta G° = -24.8$ kJ mol^{-1}.

2. *Calculate K_p at 298 K.* Use the value of $\Delta G°$ from part 1 in the expression $\Delta G° = -RT \ln K_{eq}$. For this reaction $K_{eq} = K_p$. *Result:* $K_p = 2 \times 10^4$.
3. *Find the value of $\Delta H°$ at 298 K.* To use the van't Hoff equation (20.11) in the next step, a value of $\Delta H°$ is required. This value can be obtained at 298 K from enthalpy of formation data (Appendix D). *Result:* $\Delta H° = -90.2$ kJ mol^{-1}.
4. *Calculate K_p at 500 K.* Assume $\Delta H°$ is constant and has the value determined in step 3. Use this, together with the value of K_p at 298 K (step 2), in the van't Hoff equation (20.11). Solve for K_p at 500 K. *Answer:* $K_p = 8 \times 10^{-3}$.

Key Terms

coupled reactions (20-8)
entropy, S (20-2)
entropy change, ΔS (20-2)
free energy, G (20-3)
free energy change, ΔG (20-3)
nonspontaneous process (20-1)

reversible process (20-2)
second law of thermodynamics (20-3)
spontaneous process (20-1)
standard free energy change, $\Delta G°$ (20-5)

standard free energy of formation, $\Delta G_f°$ (20-5)
thermodynamic equilibrium constant, K_{eq} (20-6)
third law of thermodynamics (20-4)
Trouton's rule (20-4)

Review Questions

1. In your own words define the following symbols: **(a)** ΔS_{univ}; **(b)** $\Delta G_f°$; **(c)** K_{eq}.
2. Briefly describe each of the following ideas, methods, or phenomena: **(a)** *absolute* molar entropy; **(b)** a reversible process; **(c)** Trouton's rule; **(d)** evaluating an equilibrium constant from tabulated thermodynamic data.
3. Explain the important distinctions between each pair of terms: **(a)** spontaneous and nonspontaneous process; **(b)** the second and third laws of thermodynamics; **(c)** ΔG and $\Delta G°$.
4. Indicate whether you would expect the entropy of the system to increase or decrease in each of the following reactions. If you cannot be certain simply by inspecting the equation, explain why.
(a) $CCl_4(l) \rightarrow CCl_4(g)$
(b) $CuSO_4 \cdot 3H_2O(s) + 2\,H_2O(g) \rightarrow CuSO_4 \cdot 5H_2O(s)$
(c) $SO_3(g) + H_2(g) \rightarrow SO_2(g) + H_2O(g)$
(d) $H_2S(g) + O_2(g) \rightarrow H_2O(g) + SO_2(g)$ [not balanced]
5. Which substance in each of the following pairs would you expect to have the greater entropy? Explain.
(a) At 75 °C and 1 atm: 1 mol $H_2O(l)$ *or* 1 mol $H_2O(g)$;

(b) At 5 °C and 1 atm: 50.0 g Fe(s) *or* 0.80 mol Fe(s);
(c) 1 mol Br_2(l, 1 atm, 8 °C) *or* 1 mol Br_2(s, 1 atm, −8 °C);
(d) 0.312 mol SO_2(g, 0.110 atm, 32.5 °C) *or* 0.284 mol O_2(g, 15.0 atm, 22.3 °C).

6. From the data given, indicate which of the four cases in Table 20.1 applies for each of the following reactions.
(a) $H_2(g) \rightarrow 2 H(g)$ $\Delta H° = +436.0$ kJ
(b) $2 SO_2(g) + O_2(g) \rightarrow 2 SO_3(g)$ $\Delta H° = -197.8$ kJ
(c) $N_2H_4(g) \rightarrow N_2(g) + 2 H_2(g)$ $\Delta H° = -95.40$ kJ
(d) $N_2(g) + 3 Cl_2(g) \rightarrow 2 NCl_3(l)$ $\Delta H° = +230$ kJ

7. Which of the following changes in thermodynamic property would you expect to find for the reaction $Br_2(g) \rightarrow 2 Br(g)$ *at all temperatures?* Explain. **(a)** $\Delta H < 0$; **(b)** $\Delta S > 0$; **(c)** $\Delta G < 0$; **(d)** $\Delta S < 0$.

8. If a reaction can only be carried out by electrolysis, which of the following changes in thermodynamic property *must* apply? Explain. **(a)** $\Delta H > 0$; **(b)** $\Delta S > 0$; **(c)** $\Delta G = \Delta H$; **(d)** $\Delta G > 0$.

9. If $\Delta G° = 0$ for a reaction, which of the following statements must also be true? Explain. **(a)** $\Delta H° = 0$; **(b)** $\Delta S° = 0$; **(c)** $K_{eq} = 0$; **(d)** $K_{eq} = 1$.

10. From the data given in the following table, determine $\Delta S°$ for the reaction $NH_3(g) + HCl(g) \rightarrow NH_4Cl(s)$. All data are at 298 K.

	$\Delta H_f°$	$\Delta G_f°$
$NH_3(g)$	− 46.11 kJ mol^{-1}	− 16.48 kJ mol^{-1}
$HCl(g)$	− 92.31	− 95.30
$NH_4Cl(s)$	−314.4	−202.9

11. Use data from Appendix D to determine values of $\Delta G°$ for the following reactions at 25 °C.
(a) $C_2H_2(g) + 2 H_2(g) \rightarrow C_2H_6(g)$
(b) $2 SO_3(g) \rightarrow 2 SO_2(g) + O_2(g)$
(c) $Fe_3O_4(s) + 4 H_2(g) \rightarrow 3 Fe(s) + 4 H_2O(g)$
(d) $2 Al(s) + 6 H^+(aq) \rightarrow 2 Al^{3+}(aq) + 3 H_2(g)$

12. If a graph similar to Figure 20-9 were drawn for the process $I_2(s) \rightarrow I_2(l)$ at 1 atm,
(a) At what temperature would the two lines intersect? (*Hint:* Refer also to Figure 13-17.)
(b) What would be the value of $\Delta G°$ at this temperature? Explain.

13. From the following data determine $\Delta S°$ (in J mol^{-1} K^{-1}) for each transition.
(a) The boiling of HCl(l) at −85.05 °C and 1 atm; $\Delta H_{vap}° = 3.86$ kcal mol^{-1}.
(b) The melting of Na(s) at 97.82 °C; $\Delta H_{fus}° = 27.05$ cal/g.

14. Write thermodynamic equilibrium constant expressions for the following reactions. Do any of these expressions correspond to K_c or K_p?
(a) $2 NO(g) + O_2(g) \rightleftharpoons 2 NO_2(g)$
(b) $MgSO_3(s) \rightleftharpoons MgO(s) + SO_2(g)$
(c) $HC_2H_3O_2(aq) + H_2O \rightleftharpoons H_3O^+(aq) + C_2H_3O_2^-(aq)$
(d) $2 NaHCO_3(s) \rightleftharpoons Na_2CO_3(s) + H_2O(g) + CO_2(g)$

(e) $MnO_2(s) + 4 H^+(aq) + 2 Cl^-(aq) \rightleftharpoons$
$Mn^{2+}(aq) + 2 H_2O(l) + Cl_2(g)$

15. For the reaction $Cl_2(g) \rightleftharpoons 2 Cl(g)$, $K_p = 2.45 \times 10^{-7}$ at 1000 K. What is $\Delta G°$ for this reaction at 1000 K?

16. Use data from Appendix D to determine K_p at 298 K for the reaction $N_2O(g) + \frac{1}{2} O_2(g) \rightleftharpoons 2 NO(g)$.

17. Use data from Appendix D to determine values at 298 K of $\Delta G°$ and K_{eq} for the following reactions. (*Note:* The equations are not balanced.)
(a) $HCl(g) + O_2(g) \rightleftharpoons H_2O(g) + Cl_2(g)$
(b) $Fe_2O_3(g) + H_2(g) \rightleftharpoons Fe_3O_4(s) + H_2O(g)$
(c) $Ag^+(aq) + SO_4^{2-}(aq) \rightleftharpoons Ag_2SO_4(s)$

18. In Example 20-2 we were unable to conclude, by inspection, whether $\Delta S°$ for the reaction $CO(g) + H_2O(g) \rightarrow CO_2(g) + H_2(g)$ should be positive or negative. Use data from Appendix D to obtain $\Delta S°$ at 298 K.

19. *Without performing calculations,* indicate whether you would expect any of the following reactions to occur to a significant extent at 298 K. (*Hint:* What is $\Delta G°$ for each reaction?)
(a) Conversion of dioxygen to ozone:
$$3 O_2(g) \longrightarrow 2 O_3(g)$$
(b) Dissociation of N_2O_4 to NO_2:
$$N_2O_4(g) \longrightarrow 2 NO_2(g)$$
(c) Formation of BrCl:
$$Br_2(l) + Cl_2(g) \longrightarrow 2 BrCl(g)$$

20. What must be the temperature if the following reaction has $\Delta G° = -45.5$ kJ, $\Delta H° = -24.8$ kJ, and $\Delta S° = 15.2$ J K^{-1}?
$$Fe_2O_3(s) + 3 CO(g) \longrightarrow 2 Fe(s) + 3 CO_2(g)$$

21. Use data from Appendix D to establish for the reaction $2 N_2O_4(g) + O_2(g) \rightarrow 2 N_2O_5(g)$
(a) $\Delta G°$ at 298 K for the reaction as written.
(b) The value of K_p at 298 K.

22. Use data from Appendix D to establish at 298 K for the reaction
$$2 NaHCO_3(s) \longrightarrow Na_2CO_3(s) + H_2O(l) + CO_2(g)$$
(a) $\Delta S°$; **(b)** $\Delta H°$; **(c)** $\Delta G°$; **(d)** K_{eq}.

23. A possible reaction for converting methanol to ethanol is
$$CO(g) + 2 H_2(g) + CH_3OH(g) \longrightarrow C_2H_5OH(g) + H_2O(g)$$
(a) Use data from Appendix D to calculate $\Delta H°$, $\Delta S°$, and $\Delta G°$ for this reaction at 25 °C.
(b) Is this reaction thermodynamically favored at high or low temperatures? High or low pressures? Explain.
(c) Estimate a value of K_p for the reaction at 750 K.

24. Estimate the value of K_p at 100 °C for the reaction $2 SO_2(g) + O_2(g) \rightleftharpoons 2 SO_3(g)$. Use data from Table 20.3 and Figure 20-11.

25. For the reaction $2 NO(g) + O_2(g) \rightarrow 2 NO_2(g)$ all but one of the following equations is correct. Which is *incorrect* and why? **(a)** $K_{eq} = K_p$; **(b)** $\Delta S° = (\Delta G° - \Delta H°)/T$; **(c)** $K_p = e^{-\Delta G°/RT}$; **(d)** $\Delta G = \Delta G° + RT \ln Q$.

26. Why is $\Delta G°$ such an important property of a chemical reaction, even though we generally carry out the reaction under *nonstandard* conditions?

Exercises

Spontaneous Change, Entropy, and Disorder

27. Indicate whether each of the following changes represents an increase or decrease in entropy in a system, and explain your reasoning: **(a)** the freezing of ethanol; **(b)** the sublimation of dry ice; **(c)** the burning of a rocket fuel.

28. Arrange the entropy changes of the following processes, all at 25 °C, in the expected order of increasing ΔS, and explain your reasoning. **(a)** $H_2O(l, 1 \text{ atm}) \rightarrow H_2O(g, 1 \text{ atm})$; **(b)** $CO_2(s, 1 \text{ atm}) \rightarrow CO_2(g, 10 \text{ mmHg})$; **(c)** $H_2O(l, 1 \text{ atm}) \rightarrow H_2O(g, 10 \text{ mmHg})$.

29. Use ideas from this chapter to explain this famous remark attributed to Rudolf Clausius (1865). "Die Energie der Welt ist konstant; die Entropie der Welt strebt einem Maximum zu. (The energy of the world is constant; the entropy of the world increases toward a maximum.)"

30. Comment on the difficulties in solving environmental pollution problems from the standpoint of entropy changes associated with the formation of pollutants and with their removal from the environment.

31. For each of the following reactions, indicate whether ΔS for the reaction should be positive or negative. If it is not possible to determine the sign of ΔS from the information given, indicate why.
(a) $CaO(s) + H_2O(l) \rightarrow Ca(OH)_2(s)$
(b) $2 HgO(s) \rightarrow 2 Hg(l) + O_2(g)$
(c) $2 NaCl(l) \rightarrow 2 Na(l) + Cl_2(g)$
(d) $Fe_2O_3(s) + 3 CO(g) \rightarrow 2 Fe(s) + 3 CO_2(g)$
(e) $Si(s) + 2 Cl_2(g) \rightarrow SiCl_4(g)$

32. By analogy to ΔH_f° and ΔG_f°, how would you define an entropy of formation? Which do you think has the largest entropy of formation $CH_4(g)$, $CH_3CH_2OH(l)$, or $CS_2(l)$. First make a qualitative prediction. Then test your prediction with data from Appendix D. (*Hint:* What are the reactions by which these compounds are formed from their elements?)

Phase Transitions

33. In Example 20-4 we dealt with ΔH_{vap}° and ΔS_{vap}° for water at 100 °C.
(a) Use data from Appendix D to determine values for these two quantities at 25 °C.
(b) From what you know of the structure of liquid water, explain the differences in ΔH_{vap}° and in ΔS_{vap}° values between 25 °C and 100 °C.

34. Pentane is one of the most volatile of the hydrocarbons in gasoline. At 298.15 K, the following enthalpies of formation are given for pentane: $\Delta H_f^\circ[C_5H_{12}(l)] = -173.5 \text{ kJ/mol}$; $\Delta H_f^\circ[C_5H_{12}(g)] = -146.9 \text{ kJ/mol}$.
(a) Estimate the normal boiling point of pentane.
(b) Estimate ΔG° for the vaporization of pentane at 298 K.
(c) Comment on the significance of the sign of ΔG° at 298 K.

35. Which of the following substances would you expect to obey Trouton's rule (20.6) most closely: HF, $C_6H_5CH_3$ (toluene), or CH_3OH (methanol)? Explain your reasoning.

36. Estimate the normal boiling point of bromine, Br_2, in the following way: Determine ΔH_{vap}° for Br_2 with data from Appendix D. Assume that ΔH_{vap}° remains constant and that Trouton's rule is obeyed.

Free Energy and Spontaneous Change

37. Indicate which of the four cases in Table 20.1 applies to each of the following reactions. If you are unable to decide just from the information given, state why.
(a) $PCl_3(g) + Cl_2(g) \rightarrow PCl_5(g)$ $\qquad \Delta H^\circ = -87.9 \text{ kJ}$
(b) $CO_2(g) + H_2(g) \rightarrow CO(g) + H_2O(g)$
$\qquad\qquad\qquad\qquad\qquad \Delta H^\circ = +41.2 \text{ kJ}$
(c) $NH_4CO_2NH_2(s) \rightarrow 2 NH_3(g) + CO_2(g)$
$\qquad\qquad\qquad\qquad\qquad \Delta H^\circ = +159.2 \text{ kJ}$

38. Indicate which of the four cases in Table 20.1 applies to each of the following reactions. If you are unable to decide just from the information given, state why.
(a) $H_2O(g) + \frac{1}{2} O_2(g) \rightarrow H_2O_2(g)$ $\qquad \Delta H^\circ = +105.5 \text{ kJ}$
(b) $C_6H_6(l) + \frac{15}{2} O_2(g) \rightarrow 6 CO_2(g) + 3 H_2O(g)$
$\qquad\qquad\qquad\qquad\qquad \Delta H^\circ = -3135 \text{ kJ}$
(c) $NO(g) + \frac{1}{2} Cl_2(g) \rightarrow NOCl(g)$ $\qquad \Delta H^\circ = -38.54 \text{ kJ}$

39. For the mixing of ideal gases (see Figure 20-3), explain whether a positive, negative, or zero value is expected for ΔH, ΔS, and ΔG.

40. What values of ΔH, ΔS, and ΔG would you expect for the formation of an ideal solution of liquid components (that is, is each value positive, negative, or zero)?

41. Explain why
(a) Some exothermic reactions do not occur spontaneously.
(b) Some reactions in which the entropy of the system increases also do not occur spontaneously.

42. Explain why you would expect a reaction of the type $AB(g) \rightarrow A(g) + B(g)$ always to be spontaneous at *high* rather than low temperatures.

Standard Free Energy Change

43. At 298 K, for the reaction $2 PCl_3(g) + O_2(g) \rightarrow 2 POCl_3(l)$, $\Delta H° = -620.2$ kJ and the molar entropies are $PCl_3(g)$, 311.7 J K^{-1}; $O_2(g)$, 205.0 J K^{-1}; and $POCl_3(l)$, 222.4 J K^{-1}. Determine **(a)** $\Delta G°$ at 298 K and **(b)** whether the reaction proceeds spontaneously in the forward or in the reverse direction when reactants and products are in their standard states.

45. The following standard free energy changes are given for 25 °C.
(1) $N_2(g) + 3 H_2(g) \rightarrow 2 NH_3(g)$ $\qquad \Delta G° = -33.0$ kJ
(2) $4 NH_3(g) + 5 O_2(g) \rightarrow 4 NO(g) + 6 H_2O(l)$
$\qquad\qquad\qquad\qquad\qquad\qquad \Delta G° = -1010.5$ kJ
(3) $N_2(g) + O_2(g) \rightarrow 2 NO(g)$ $\qquad \Delta G° = +173.1$ kJ
(4) $N_2(g) + 2 O_2(g) \rightarrow 2 NO_2(g)$ $\qquad \Delta G° = +102.6$ kJ
(5) $2 N_2(g) + O_2(g) \rightarrow 2 N_2O(g)$ $\qquad \Delta G° = +208.4$ kJ
Combine the preceding equations, as necessary, to obtain $\Delta G°$ values for the following reactions.
(a) $N_2O(g) + \frac{3}{2} O_2(g) \rightarrow 2 NO_2(g)$ $\qquad \Delta G° = ?$
(b) $2 H_2(g) + O_2(g) \rightarrow 2 H_2O(l)$ $\qquad \Delta G° = ?$
(c) $2 NH_3(g) + 2 O_2(g) \rightarrow N_2O(g) + 3 H_2O(l)$
$\qquad\qquad\qquad\qquad\qquad\qquad \Delta G° = ?$
Which of the reactions **(a)**, **(b)**, and **(c)** would tend to go to completion at 25 °C, and which would reach an equilibrium condition with significant amounts of all reactants and products present?

47. Write an equation for the combustion of one mole of benzene, $C_6H_6(l)$, and use data from Appendix D to determine $\Delta G°$ at 298 K, if the products of the combustion are **(a)** $CO_2(g)$ and $H_2O(l)$, and **(b)** $CO_2(g)$ and $H_2O(g)$. Describe how you might determine the *difference* between the values obtained in (a) and (b) without having either to write the combustion equation or to determine $\Delta G°$ values for the combustion reactions.

Free Energy Change and Equilibrium

49. Refer to Figures 13-17 and 20-9. Does solid or liquid iodine have the lower free energy at 110 °C and 1 atm pressure? Explain.

The Thermodynamic Equilibrium Constant

51. For one of the following reactions, $K_c = K_p = K_{eq}$. Identify that one. For the other two, what is the relationship between K_c, K_p, and K_{eq}? Explain.
(a) $2 SO_2(g) + O_2(g) \rightleftharpoons 2 SO_3(g)$
(b) $HI(g) \rightleftharpoons \frac{1}{2} H_2(g) + \frac{1}{2} I_2(g)$
(c) $NH_4HCO_3(s) \rightleftharpoons NH_3(g) + CO_2(g) + H_2O(l)$

44. At 298 K, for the reaction $2 H^+(aq) + 2 Br^-(aq) + 2 NO_2(g) \rightarrow Br_2(l) + 2 HNO_2(aq)$, $\Delta H° = -61.6$ kJ and the molar entropies are $H^+(aq)$, 0 J K^{-1}; $Br^-(aq)$, 82.4 J K^{-1}; $NO_2(g)$, 240.0 J K^{-1}; $Br_2(l)$, 152.2 J K^{-1}; $HNO_2(aq)$, 135.6 J K^{-1}. Determine **(a)** $\Delta G°$ at 298 K and **(b)** whether the reaction proceeds spontaneously in the forward or in the reverse direction when reactants and products are in their standard states.

46. The following standard free energy changes are given for 25 °C.
(1) $SO_2(g) + 3 CO(g) \rightarrow COS(g) + 2 CO_2(g)$
$\qquad\qquad\qquad\qquad\qquad\qquad \Delta G° = -246.4$ kJ
(2) $CS_2(g) + H_2O(g) \rightarrow COS(g) + H_2S(g)$
$\qquad\qquad\qquad\qquad\qquad\qquad \Delta G° = -41.5$ kJ
(3) $CO(g) + H_2S(g) \rightarrow COS(g) + H_2(g)$
$\qquad\qquad\qquad\qquad\qquad\qquad \Delta G° = +1.4$ kJ
(4) $CO(g) + H_2O(g) \rightarrow CO_2(g) + H_2(g)$
$\qquad\qquad\qquad\qquad\qquad\qquad \Delta G° = -28.6$ kJ
Combine the preceding equations, as necessary, to obtain $\Delta G°$ values for the following reactions.
(a) $COS(g) + 2 H_2O(g) \rightarrow SO_2(g) + CO(g) + 2 H_2(g)$
$\qquad\qquad\qquad\qquad\qquad\qquad \Delta G° = ?$
(b) $COS(g) + 3 H_2O(g) \rightarrow SO_2(g) + CO_2(g) + 3 H_2(g)$
$\qquad\qquad\qquad\qquad\qquad\qquad \Delta G° = ?$
(c) $COS(g) + H_2O(g) \rightarrow CO_2(g) + H_2S(g)$ $\qquad \Delta G° = ?$
Which of the reactions **(a)**, **(b)**, and **(c)** is spontaneous in the forward direction when reactants and products are present in their standard states?

48. Use molar entropies from Appendix D, together with the following data, to obtain an estimate of the bond dissociation energy of the F_2 molecule.
$$F_2(g) \longrightarrow 2 F(g) \qquad \Delta G° = 123.9 \text{ kJ}$$
Compare your result with the value listed in Table 11.3.

50. Refer to Figures 13-18 and 20-9. Which has the lowest free energy at 1 atm and -60 °C: solid, liquid, or gaseous carbon dioxide? Explain.

52. $H_2(g)$ can be prepared by passing steam over hot iron: $3 Fe(s) + 4 H_2O(g) \rightleftharpoons Fe_3O_4(s) + 4 H_2(g)$.
(a) Write an expression for the thermodynamic equilibrium constant for this reaction.
(b) Explain why the partial pressure of $H_2(g)$ is independent of the amounts of $Fe(s)$ and $Fe_3O_4(s)$ present.
(c) Can we conclude that the production of $H_2(g)$ from $H_2O(g)$ could be accomplished regardless of what proportions of $Fe(s)$ and $Fe_3O_4(s)$ are present? Explain.

Relationships Involving ΔG, $\Delta G°$, Q, and K_{eq}

53. At 1000 K an equilibrium mixture in the reaction $CO_2(g)$ + $H_2(g) \rightleftharpoons CO(g)$ + $H_2O(g)$ contains 0.276 mol H_2, 0.276 mol CO_2, 0.224 mol CO, and 0.224 mol H_2O.
(a) What is the value of K_p at 1000 K?
(b) Calculate $\Delta G°$ at 1000 K.
(c) In what direction will a spontaneous net reaction occur if one brings together at 1000 K: 0.0750 mol CO_2, 0.095 mol H_2, 0.0340 mol CO, and 0.0650 mol H_2O? [*Hint:* Either compare Q with K_p or determine ΔG by using Equation (20.8).]

55. For the following equilibrium reactions calculate the value of $\Delta G°$ at the indicated temperature. (*Hint:* How is each equilibrium constant related to a thermodynamic equilibrium constant, K_{eq}?)
(a) $H_2(g) + I_2(g) \rightleftharpoons 2\ HI(g)$ $K_c = 50.2$ at 445 °C
(b) $N_2O(g) + \frac{1}{2}O_2(g) \rightleftharpoons 2\ NO(g)$
              $K_c = 1.7 \times 10^{-13}$ at 25 °C
(c) $N_2O_4(g) \rightleftharpoons 2\ NO_2(g)$ $K_c = 4.61 \times 10^{-3}$ at 25 °C
(d) $2\ Fe^{3+}(aq) + Hg_2^{2+}(aq) \rightleftharpoons 2\ Fe^{2+}(aq) + 2\ Hg^{2+}(aq)$
              $K_c = 9.14 \times 10^{-6}$ at 25 °C

57. At 298 K, $\Delta G_f°[CO(g)] = -137.2$ kJ/mol and $K_p = 6.5 \times 10^{11}$ for the reaction $CO(g) + Cl_2(g) \rightleftharpoons COCl_2(g)$. Use these data to determine $\Delta G_f°[COCl_2(g)]$ and compare your result with the value in Appendix D.

59. To establish the law of conservation of mass, Lavoisier carefully studied the decomposition of mercury(II) oxide: $HgO(s) \rightarrow Hg(l) + \frac{1}{2}O_2(g)$. At 25 °C, $\Delta H° = +90.83$ kJ and $\Delta G° = +58.54$ kJ.
(a) Show that the partial pressure of $O_2(g)$ in equilibrium with $HgO(s)$ and $Hg(l)$ at 25 °C is extremely low.
(b) What conditions do you suppose Lavoisier used to obtain significant quantities of oxygen?

54. For the reaction $2\ SO_2(g) + O_2(g) \rightleftharpoons 2\ SO_3(g)$, $K_c = 2.8 \times 10^2$ at 1000 K.
(a) What is the value of $\Delta G°$ at 1000 K? (*Hint:* What is K_p?)
(b) If 0.40 mol SO_2, 0.18 mol O_2, and 0.72 mol SO_3 are mixed in a 2.50-L flask at 1000 K, in what direction will a net reaction occur? [*Hint:* Either compare Q with K_c or K_p or determine ΔG with Equation (20.8).]

56. Two different equations can be written for the dissolving of $Mg(OH)_2(s)$ in acidic solution.
$$Mg(OH)_2(s) + 2\ H^+(aq) \rightleftharpoons Mg^{2+}(aq) + 2\ H_2O(l)$$
$$\Delta G° = -95.5\ \text{kJ mol}^{-1}$$
$$\tfrac{1}{2}Mg(OH)_2(s) + H^+(aq) \rightleftharpoons \tfrac{1}{2}Mg^{2+}(aq) + H_2O(l)$$
$$\Delta G° = -47.8\ \text{kJ mol}^{-1}$$
(a) Explain why these two equations have different values for $\Delta G°$.
(b) Will the values of K_{eq} determined for these two equations be the same or different? Explain.
(c) Will the solubilities of $Mg(OH)_2(s)$ in a buffer solution at pH = 8.5 depend on which of the two equations is used as the basis of the calculation? Explain.

58. Use data from Appendix D to determine values of K_{sp} for the following sparingly soluble solutes: **(a)** AgBr; **(b)** $CaSO_4$; **(c)** $Fe(OH)_3$. (*Hint:* Begin by writing solubility equilibrium expressions.)

60. Currently, CO_2 is being studied as a source of carbon atoms for synthesizing organic compounds. One possible reaction involves the conversion of CO_2 to methanol, CH_3OH.
$$CO_2(g) + 3\ H_2(g) \longrightarrow CH_3OH(g) + H_2O(g)$$
With the aid of data from Appendix D, determine
(a) if this reaction proceeds to any significant extent at 25 °C;
(b) if the production of $CH_3OH(g)$ is favored by raising or lowering the temperature from 25 °C;
(c) the value of K_p for this reaction at 500 K;
(d) the partial pressure of $CH_3OH(g)$ at equilibrium if $CO_2(g)$ and $H_2(g)$, each initially at a partial pressure of 1 atm, react at 500 K.

$\Delta G°$ as a Function of Temperature

61. The synthesis of ammonia by the Haber process occurs by the reaction at 400 °C, $N_2(g) + 3\ H_2 \rightleftharpoons 2\ NH_3(g)$. Use data from Appendix D, assume that $\Delta H°$ and $\Delta S°$ are essentially unchanged in the temperature interval from 25 to 400 °C, and estimate the value of K_p at 400 °C.

63. In Example 20-10 we used the van't Hoff equation to determine the temperature at which $K_p = 1.0 \times 10^6$ for the reaction $2\ SO_2(g) + O_2(g) \rightleftharpoons 2\ SO_3(g)$. Obtain another estimate of this temperature with data from Appendix D and Equations (20.5) and (20.9). Compare your result with that obtained in Example 20-10.

62. Use data from Appendix D to determine **(a)** $\Delta H°$, $\Delta S°$, and $\Delta G°$ at 298 K and **(b)** K_p at 875 K for the water-gas shift reaction, used commercially to produce $H_2(g)$: $CO(g) + H_2O(g) \rightleftharpoons CO_2(g) + H_2(g)$. (*Hint:* Assume that $\Delta H°$ and $\Delta S°$ are essentially unchanged in this temperature interval.)

64. The following equilibrium constants have been determined for the reaction $H_2(g) + I_2(g) \rightleftharpoons 2\ HI(g)$: $K_p = 50.0$ at 448 °C and 66.9 at 350 °C. Use these data to estimate $\Delta H°$ for the reaction.

65. For the reaction $N_2O_4(g) \rightleftharpoons 2 NO_2(g)$, $\Delta H° = +57.2$ kJ mol^{-1} and $K_p = 0.113$ at 298 K.
(a) What is the value of K_p at 0 °C?
(b) At what temperature will $K_p = 1.00$?

67. For the reaction
$$CO(g) + 3 H_2(g) \rightleftharpoons CH_4(g) + H_2O(g)$$
$$K_p = 2.15 \times 10^{11} \text{ at } 200 °C$$
$$K_p = 4.56 \times 10^8 \text{ at } 260 °C$$
determine $\Delta H°$ by using the van't Hoff equation (20.11) and by using tabulated data in Appendix D. Compare the two results, and comment on how good the assumption is that $\Delta H°$ is essentially independent of temperature in this case.

Coupled Reactions

69. Titanium is obtained by the reduction of $TiCl_4(l)$, which in turn is produced from the mineral rutile (TiO_2).
(a) With data from Appendix D determine $\Delta G°$ at 298 K for the reaction
$$TiO_2(s) + 2 Cl_2(g) \longrightarrow TiCl_4(l) + O_2(g)$$
(b) Show that the conversion of $TiO_2(s)$ to $TiCl_4(l)$, with reactants and products in their standard states, is spontaneous at 298 K if the reaction in (a) is coupled with the reaction
$$2 CO(g) + O_2(g) \longrightarrow 2 CO_2(g)$$
(*Hint:* Recall the description of coupled reactions on pages 713–715.)

66. Use data from Appendix D and the van't Hoff equation (20.11) to estimate a value of K_p at 100 °C for the reaction $2 NO(g) + O_2(g) \rightleftharpoons 2 NO_2(g)$. (*Hint:* First determine K_p at 25 °C. What is $\Delta H°$ for the reaction?)

68. Sodium carbonate, an important chemical used in the production of glass, is made from sodium hydrogen carbonate by the reaction
$$2 NaHCO_3(s) \rightleftharpoons Na_2CO_3(s) + CO_2(g) + H_2O(g)$$
Data for the temperature variation of K_p for this reaction are $K_p = 1.66 \times 10^{-5}$ at 30 °C; 3.90×10^{-4} at 50 °C; 6.27×10^{-3} at 70 °C; and 2.31×10^{-1} at 100 °C.
(a) Plot a graph similar to Figure 20-10 and determine $\Delta H°$ for the reaction.
(b) Calculate the temperature at which the total gas pressure above a mixture of $NaHCO_3(s)$ and $Na_2CO_3(s)$ is 2.00 atm.

70. Following are some standard free energies of formation, $\Delta G_f°$, per mole of metal oxide at 1000 K: NiO, -115 kJ; MnO, -280 kJ; TiO_2, -630 kJ. The standard free energy of formation of CO at 1000 K is -250 kJ per mol CO. Use the method of coupled reactions (page 714) to determine which of these metal oxides can be reduced to the metal by a spontaneous reaction with carbon at 1000 K, and with all reactants and products in their standard states.

Advanced Exercises

71. Use data from Appendix D to estimate **(a)** the normal boiling point of mercury and **(b)** the vapor pressure of mercury at 25 °C.

72. Dinitrogen pentoxide, N_2O_5, is a solid with a high vapor pressure. Its vapor pressure at 7.5 °C is 100 mmHg, and the solid sublimes at a pressure of 1.00 atm at 32.4 °C. What is the standard free energy change for the process $N_2O_5(s) \rightarrow N_2O_5(g)$ at 25 °C?

73. Consider the vaporization of water: $H_2O(l) \rightarrow H_2O(g)$ at 100 °C, with $H_2O(l)$ in its standard state, but with the partial pressure of $H_2O(g)$ at 2.0 atm. Which of the following statements are true concerning this vaporization at 100 °C: (a) $\Delta G° = 0$, (b) $\Delta G = 0$, (c) $\Delta G° > 0$, (d) $\Delta G > 0$? Explain.

74. 1.00 mol BrCl(g) is introduced into a 10.0-L vessel at 298 K and equilibrium is established in the reaction $BrCl(g) \rightleftharpoons \frac{1}{2} Br_2(g) + \frac{1}{2} Cl_2(g)$. Calculate the amounts of each of the three gases present when equilibrium is established. (*Hint:* You must establish your own value of K_p or K_c using data from Appendix D.)

75. Use data from Appendix D and other information from this chapter to estimate the temperature at which the dissociation of $I_2(g)$ becomes appreciable [for example, with the $I_2(g)$ 50% dissociated into I(g) at 1 atm total pressure].

76. The following table shows the enthalpies and free energies of formation of three different metal oxides at 25 °C.

	$\Delta H_f°$, kJ/mol	$\Delta G_f°$, kJ/mol
PbO(red)	-219.0	-188.9
Ag_2O	-31.05	-11.20
ZnO	-348.3	-318.3

(a) Which of these oxides can be most readily decomposed to the free metal and $O_2(g)$?
(b) For the oxide that is most easily decomposed, to what temperature must it be heated to produce $O_2(g)$ at 1.00 atm pressure?

77. The following data are given for the two solid forms of HgI_2 at 298 K.

	$\Delta H_f°$, kJ/mol	$\Delta G_f°$, kJ/mol	$S°$, J mol^{-1} K^{-1}
HgI_2 (red)	-105.4	-101.7	180
HgI_2 (yellow)	-102.9	(?)	(?)

Estimate values for the two missing entries. To do this, assume that for the transition HgI_2 (red) $\rightarrow$ HgI_2 (yellow) the values of $\Delta H°$ and $\Delta S°$ at 25 °C have the same values that they do at the equilibrium temperature of 127 °C.

78. Oxides of nitrogen are produced in high-temperature combustion processes. The essential reaction is $N_2(g) + O_2(g) \rightleftharpoons 2\ NO(g)$. At what approximate temperature will an *equimolar* mixture of $N_2(g)$ and $O_2(g)$ be 1.0% converted to $NO(g)$? (*Hint:* Use data from Appendix D.)

79. For the dissociation of $CaCO_3(s)$ at 25 °C,

$$CaCO_3(s) \rightleftharpoons CaO(s) + CO_2(g)$$
$$\Delta G° = +131 \text{ kJ/mol}$$

A sample of pure $CaCO_3(s)$ is placed in a flask and connected to an ultrahigh vacuum system capable of reducing the pressure to 10^{-9} mmHg.

(a) Would $CO_2(g)$ produced by the decomposition of $CaCO_3(s)$ at 25 °C be detectable in the vacuum system at 25 °C?

(b) What additional information do you need to determine P_{CO_2} as a function of temperature?

(c) With necessary data from Appendix D, determine the minimum temperature to which $CaCO_3(s)$ would have to be heated for $CO_2(g)$ to become detectable in the vacuum system.

80. 0.100 mol of $PCl_5(g)$ is introduced into a 1.50-L flask and

the flask is held at a temperature of 227 °C until equilibrium is established. What is the total pressure of the gases in the flask at this point?

$$PCl_5(g) \rightleftharpoons PCl_3(g) + Cl_2(g)$$

(*Hint:* Use data from Appendix D and appropriate relationships from this chapter to obtain a value of K_p at 227 °C.)

81. From the data given in Exercise 68, estimate a value of $\Delta S°$ at 298 K for the reaction

$$2\ NaHCO_3(s) \longrightarrow Na_2CO_3(s) + H_2O(g) + CO_2(g)$$

82. The normal boiling point of cyclohexane, C_6H_{12}, is 80.7 °C. Estimate the temperature at which the vapor pressure of cyclohexane is 100.0 mmHg.

83. The decomposition of the poisonous gas phosgene is represented by the equation $COCl_2(g) \rightleftharpoons CO(g) + Cl_2(g)$. Values of K_p for this reaction are $K_p = 6.7 \times 10^{-9}$ at 99.8 °C and $K_p = 4.44 \times 10^{-2}$ at 395 °C. At what temperature is $COCl_2$ 15% dissociated when the total gas pressure is maintained at 1.00 atm?

84. Use data from Appendix D to estimate the aqueous solubility, in milligrams per liter, of $AgBr(s)$ at 100 °C.

Feature Problems

A. A tabulation of thermodynamic data lists the following values for $H_2O(l)$ and $H_2O(g)$ at 298.15 K, but at the new SI standard state pressure of 1 bar rather than 1 atm. A pressure of 1 bar is defined as exactly 1×10^5 Pa.

	$\Delta H_f°$, kJ mol^{-1}	$\Delta G_f°$, kJ mol^{-1}	$S°$, J mol^{-1} K^{-1}
$H_2O(l)$	-285.830	-237.129	69.91
$H_2O(g)$	-241.818	-228.572	188.825

1. Use these data to determine, in *two* different ways, $\Delta G°$ at 298.15 K for the vaporization: $H_2O(l, 1 \text{ bar}) \rightleftharpoons H_2O(g, 1 \text{ bar})$. The value you obtain will differ slightly from that on page 705 because here the standard state pressure is 1 bar and there it is 1 atm.

2. Use the result of part 1 to obtain the value of K_{eq} for this vaporization and, hence, the vapor pressure of water at 298.15 K.

3. The vapor pressure in part 2 is in the unit, bar. Convert the pressure to millimeters of mercury.

4. Start with the value $\Delta G° = 8.590$ kJ, given on page 705, and calculate the vapor pressure of water at 298.15 K in a fashion similar to that in parts 2 and 3. How do the answers in parts 3 and 4 compare? Can you now see that the results obtained in a thermodynamic calculation do not depend on the convention we chose for the standard state pressure, as long as we use data consistent with that choice?

B. The graphs show how $\Delta G°$ varies with temperature for three different oxidation reactions: the oxidations of C(graphite), Zn, and Mg to CO, ZnO, and MgO, respectively. Graphs such as these can be used to show the temperatures at which carbon is an effective reducing agent to reduce metal oxides to the free metals. As a result, such graphs are important to metallurgists. Use these graphs to answer the following questions.

$\Delta G°$ for three reactions as a function of temperature. The reactions are indicated by the equations written above the graphs. The points noted by arrows are the melting points and boiling points of zinc and magnesium.

1. Why is it that Mg can be used to reduce ZnO to Zn at all temperatures, but Zn cannot be used to reduce MgO to Mg at any temperature?

2. Why can C be used to reduce ZnO to Zn at some temperatures but not others? What are these temperatures?

3. Do you think it is possible to produce Zn from ZnO by its direct decomposition, without requiring a coupled reaction? If so, at what approximate temperatures might this occur?

4. Do you think it is possible to decompose CO to C and O_2 in a spontaneous reaction? Explain.

Electrochemistry

21

A conventional gasoline-powered automobile is about 25% efficient in converting chemical energy into kinetic energy (energy of motion). An electric-powered auto is about three times as efficient. Unfortunately, when automotive technology was first being developed, devices for converting chemical energy to electricity did not perform at their intrinsic efficiencies. This fact, together with the availability of high-quality gasoline at a low cost, resulted in the preeminence of the internal combustion automobile. Now, with concern about long-term energy supplies and environmental pollution, there is a renewed interest in electric-powered automobiles.

In this chapter we will see how chemical reactions can be used to produce electricity and how electricity can be used to cause chemical reactions. The practical applications of electrochemistry are countless, ranging from batteries and fuel cells as electric power sources, to the manufacture of key chemicals, to the refining of metals, and to methods of controlling corrosion. Also important, however, are the theoretical implications. Because electricity involves a flow of electric charge, a study of the relationship between chemistry and electricity gives us additional insight into reactions in which electrons are transferred—*oxidation–reduction reactions.*

Steel being coated with zinc to produce galvanized steel. Galvanizing protects the steel from corrosion (see page 748).

21-1 Electrode Potentials and Their Measurement

The criteria for spontaneous change developed in the preceding chapter apply to reactions of all types—precipitation, acid–base, and oxidation–reduction (redox). We can devise an additional useful criterion for redox reactions, however.

Figure 21-1 shows that a redox reaction occurs between Cu(s) and Ag^+(aq), but not between Cu(s) and Zn^{2+}(aq). Specifically, we see that silver ions are reduced to silver atoms on a copper surface, whereas zinc ions are *not* reduced to zinc atoms on a copper surface. We can say that Ag^+ is more readily reduced than is Zn^{2+}. In this section we will introduce a property related to these reduction tendencies; it is called the *electrode potential.*

▶ Sometimes the term *electrode* is used for the entire half-cell assembly.

When used in electrochemical studies, a strip of metal, M, is called an **electrode.** An electrode immersed in a solution containing ions of the same metal, M^{n+}, is called a **half-cell.** Two kinds of interactions are possible between metal atoms on the electrode and metal ions in solution (Figure 21-2).

1. A metal ion M^{n+} from solution may collide with the electrode, gain *n* electrons from it, and be converted to a metal atom M. *The ion is reduced.*

2. A metal atom M on the surface may lose *n* electrons to the electrode and enter the solution as the ion M^{n+}. *The metal atom is oxidized.*

▶ Do not misinterpret expression (21.1). M^{n+}(aq) and ne^- appear together on the right because both are "produced" in the oxidation half-reaction. Only the ion M^{n+} enters the solution, however. The electrons remain on the electrode, M(s). Free electrons are never found in an aqueous solution.

An equilibrium is quickly established between the metal and the solution, which we can represent as

$$M(s) \underset{\text{reduction}}{\overset{\text{oxidation}}{\rightleftharpoons}} M^{n+}(aq) + ne^- \qquad (21.1)$$

If the *oxidation* tendency of the metal is strong, we might expect a negative charge to accumulate on the electrode (from the electrons left behind). In turn, we might expect the solution to acquire a positive charge because of a buildup in concentration of the M^{n+} ions. On the other hand, if the *reduction* tendency is strong, we might expect the reverse situation—a positive charge buildup on

FIGURE 21-1
Behavior of Ag^+(aq) and Zn^{2+}(aq) in the presence of copper.
(a) Silver ions are displaced from *colorless* $AgNO_3$(aq) as a deposit of silver metal. Copper enters the solution as *blue* Cu^{2+}(aq).
$Cu(s) + 2 Ag^+(aq) \longrightarrow$
$\qquad\qquad Cu^{2+}(aq) + 2 Ag(s)$
(b) Cu(s) *does not* displace *colorless* Zn^{2+} from $Zn(NO_3)_2$(aq).
$Cu(s) + Zn^{2+}(aq) \longrightarrow$
$\qquad\qquad$ no reaction

(a)   (b)

FIGURE 21-2
An electrochemical half-cell.
The half-cell consists of a metal electrode, M, partially immersed in an aqueous solution of its ions, M^{n+}. (The anions required to maintain electrical neutrality in the solution are not shown.) The situation illustrated here is limited to metals that do not react with water.

the electrode and a negative charge buildup in the solution. What actually happens is rather different from our expectations: Because an M^{n+} ion cannot readily draw away from an electrode with a net *negative* charge, it is attracted back to the electrode. Also, an M^{n+} ion would be repelled back to the electrode if the solution had a net *positive* charge. As a consequence, any charge buildup on an electrode in a half-cell is very, very slight—too small to be measured.

Although we cannot make a direct measurement on a single electrode, we can connect *two* different electrodes to each other and measure the flow of electrons that occurs between them. Electrons flow from the electrode with the higher buildup of electron charge density to the one with the lower electron charge density. **Electrode potential** is a measure of the electric charge density on an electrode at which an oxidation–reduction equilibrium has been established in half-reactions such as (21.1). Even the slightest *difference* in electrode potential between two electrodes sets up a flow of electrons—an electric current—much like the spontaneous flow of water from a higher to a lower level.

To measure a *difference* in potential, we need to connect *two* half-cells in a special way. *Both* the metal electrodes *and* the solutions have to be connected, so that a continuous circuit is provided for the flow of charged particles. The electrodes are joined by a metal wire to permit the flow of *electrons*. The flow of electric current between the solutions is in the form of a migration of *ions*. Contact between solutions can be through a porous plug or membrane that separates the two solutions, or it can be established through a third solution, usually in a U-tube, that "bridges" the two half-cells. This connection is called a **salt bridge.** An **electrochemical cell** is a properly connected combination of two half-cells.

Consider an electrochemical cell with one half-cell having a copper electrode in 1.00 M $Cu^{2+}(aq)$ and the other half-cell having a silver electrode in 1.00 M $Ag^+(aq)$. As shown in Figure 21-3, the solutions are joined by a salt bridge and the electrodes by a metal wire.

These changes occur: Cu atoms lose electrons at the Cu(s) electrode (the anode) and enter the solution as Cu^{2+} ions. The electrons lost by the Cu atoms pass through the wire and the electrical measuring device (a voltmeter) to the Ag(s) electrode (the cathode). Here Ag^+ ions from solution gain electrons and deposit as silver metal. Anions (NO_3^-) from the salt bridge migrate into the

▶ Anions migrate toward the anode and cations toward the cathode.

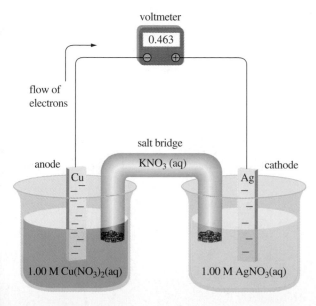

FIGURE 21-3
Measurement of the electromotive force of an electrochemical cell.
An electrochemical cell consists of two half-cells with electrodes joined by a wire and solutions by a salt bridge. (The ends of the salt bridge are plugged with a porous material that allows ions to migrate but prevents the bulk flow of liquid.) The drawing shows a greater density of negative charge (−) on the Cu electrode than on Ag. For precise measurements the amount of electric current drawn from the cell must be kept very small, by using either a specially designed voltmeter or a device called a potentiometer.

copper half-cell to neutralize the positive charge of the excess Cu^{2+} ions. Cations (K^+) migrate into the silver half-cell to neutralize the negative charge of the excess NO_3^- ions. The net reaction that occurs as the electrochemical cell produces electric current is

$$\begin{aligned}
\textit{Oxidation:} \quad & Cu(s) \longrightarrow Cu^{2+}(aq) + 2\,e^- \\
\textit{Reduction:} \quad & \underline{2\,\{Ag^+(aq) + e^- \longrightarrow Ag(s)\}} \\
\textit{Net:} \quad & Cu(s) + 2\,Ag^+(aq) \longrightarrow Cu^{2+}(aq) + 2\,Ag(s) \qquad (21.2)
\end{aligned}$$

▶ Note that the net reaction occurring in the electrochemical cell is identical to that occurring in the direct addition of Cu(s) to $Ag^+(aq)$ pictured in Figure 21-1a. The main difference between the two situations is that reaction (21.2) is divided into two distinct half-reactions.

The spontaneous reaction consists of the oxidation of copper atoms and the reduction of silver ions. The reading on the voltmeter (0.463 V) is significant. It is the *potential difference* between the two half-cells. Because this potential difference is the "driving force" for electrons, it is often called the **electromotive force (emf)** of the cell, or the **cell potential.** The unit for measuring electrode potential and cell potential, also called *cell voltage*, is the *volt* (see Appendix B).

▶ Formulations such as Zn/Zn^{2+} and Cu^{2+}/Cu are called *couples* and are often used as abbreviations for half-cells.

Now let's return to the question raised by Figure 21-1: Why does copper *not* displace Zn^{2+} from solution? If we set up an electrochemical cell consisting of a $Zn(s)/Zn^{2+}(aq)$ half-cell and a $Cu^{2+}(aq)/Cu(s)$ half-cell, we find that electrons flow *from the Zn to the Cu.* The spontaneous reaction in the electrochemical cell in Figure 21-4 is

$$\begin{aligned}
\textit{Oxidation:} \quad & Zn(s) \longrightarrow Zn^{2+}(aq) + 2\,e^- \\
\textit{Reduction:} \quad & \underline{Cu^{2+}(aq) + 2\,e^- \longrightarrow Cu(s)} \\
\textit{Net:} \quad & Zn(s) + Cu^{2+}(aq) \longrightarrow Zn^{2+}(aq) + Cu(s) \qquad (21.3)
\end{aligned}$$

Because (21.3) is a spontaneous reaction, the displacement of $Zn^{2+}(aq)$ by Cu(s)—the *reverse* of (21.3)—does *not* occur spontaneously. This is the observation we made in Figure 21-1. On page 735 we will discuss how to predict the direction of spontaneous change for oxidation–reduction reactions.

Cell Diagrams and Terminology

Drawing sketches of electrochemical cells, as in Figures 21-3 and 21-4, is helpful, but more often a simpler representation is used. A **cell diagram** shows the

FIGURE 21-4
The reaction
$Zn(s) + Cu^{2+}(aq) \longrightarrow Zn^{2+}(aq) + Cu(s)$
occurring in an electrochemical cell.

components of an electrochemical cell in a symbolic way. We will follow the conventions that electrochemists generally use in writing cell diagrams.

- The **anode,** the electrode at which *oxidation* occurs, is placed at the *left* side of the diagram.
- The **cathode,** the electrode at which *reduction* occurs, is placed at the *right* side of the diagram.
- A boundary between different phases (for example, an electrode and a solution) is represented by a *single* vertical line ($|$).
- The boundary between half-cell compartments, often a salt bridge, is represented by a *double* vertical line ($\|$). Species in aqueous solution are placed on either side of the double vertical line, and different species in the same solution are separated from each other by a comma.

The cell diagram corresponding to Figure 21-4 and reaction (21.3) is customarily written as

$$\text{anode} \rightarrow \text{Zn(s)}|\text{Zn}^{2+}(\text{aq}) \quad \| \quad \text{Cu}^{2+}(\text{aq})|\text{Cu(s)} \leftarrow \text{cathode}$$

half-cell   salt   half-cell
(oxidation) bridge (reduction)

The electrochemical cells of Figures 21-3 and 21-4 *produce* electricity as a result of spontaneous chemical reactions; they are called **voltaic** or **galvanic cells.** In Section 21-7 we will consider **electrolytic cells**—electrochemical cells in which electricity is used to accomplish a *nonspontaneous* chemical change—that is, the *reverse* of a spontaneous reaction.

▶ Several memory devices have been proposed for the oxidation/anode and reduction/cathode relationships. Perhaps the simplest is that in the oxidation/anode relationship, both terms begin with a vowel: *o/a*; in the reduction/cathode relationship, both begin with a consonant: *r/c*.

EXAMPLE 21-1

Representing a Redox Reaction Through a Cell Diagram. Aluminum metal displaces zinc(II) ion from aqueous solution.
a. Write oxidation and reduction half-equations and a net equation for this redox reaction.
b. Write a cell diagram for a voltaic cell in which this reaction occurs.

Solution

a. The term *displaces* means that aluminum goes into solution as Al^{3+}(aq), forcing Zn^{2+}(aq) out of solution as zinc metal. Al is oxidized to Al^{3+}, and Zn^{2+} is reduced to Zn. In writing the net equation we must adjust coefficients so that equal numbers of electrons are involved in oxidation and in reduction. (This is the half-reaction method of balancing redox equations that we studied in Section 5-5.)

Oxidation: $2\,\{Al(s) \longrightarrow Al^{3+}(aq) + 3\,e^-\}$
Reduction: $3\,\{Zn^{2+}(aq) + 2\,e^- \longrightarrow Zn(s)\}$

Net: $2\,Al(s) + 3\,Zn^{2+}(aq) \longrightarrow 2\,Al^{3+}(aq) + 3\,Zn(s)$ (21.4)

b. Al(s) is oxidized to Al^{3+}(aq) in the anode half-cell (left). Zn^{2+}(aq) is reduced to Zn(s) in the cathode half-cell (right).

$$Al(s)|Al^{3+}(aq)\|Zn^{2+}(aq)|Zn(s)$$

Practice Example A: Write the net equation for the redox reaction that occurs in the voltaic cell: $Sc(s)|Sc^{3+}(aq)\|Ag^+(aq)|Ag(s)$. (*Hint:* Assume that the customary conventions were used in writing the cell diagram.)

Practice Example B: Draw a voltaic cell in which the displacement of silver ion from solution by aluminum metal occurs. Label the cathode, the anode, and other features of the cell. Show the direction of flow of electrons. Also indicate the direction of flow of cations and anions from a $KNO_3(aq)$ salt bridge. Write an equation for the half-reaction occurring at each electrode, write a balanced equation for the overall reaction, and write a cell diagram.

21-2 Standard Electrode Potentials

> ▶ This is comparable to establishing standard enthalpies or free energies of formation based on an arbitrary zero.

Consider the advantages of measuring a particular half-cell potential just once and then using that value in every electrochemical cell in which the half-cell appears. To do this we arbitrarily choose a particular half-cell to which we assign an electrode potential of *zero*. We then compare other half-cells to this reference. The commonly accepted reference, pictured in Figure 21-5, is the standard hydrogen electrode.

The **standard hydrogen electrode (SHE)** involves an equilibrium between H_3O^+ ions from a solution at unit activity ($a = 1$) and H_2 molecules from the gaseous state at 1 atm pressure, established on the surface of an inert metal such as platinum. The equilibrium reaction produces a particular potential on the metal surface. For simplicity, we usually write H^+ for H_3O^+ and assume that unit activity ($a = 1$) exists at roughly $[H^+] = 1$ M. The SHE is assigned a standard electrode potential of *zero*.

$$2\ H^+(a = 1) + 2\ e^- \xrightleftharpoons{\text{on Pt}} H_2(g, 1\ \text{atm}) \qquad E° = 0\ \text{volt (V)} \qquad (21.5)$$

The diagram for this half-cell is

$$Pt\,|\,H_2(g, 1\ \text{atm})\,|\,H^+(1\ M)$$

FIGURE 21-5
The standard hydrogen electrode (SHE).

$H_2(g)$
1 atm

Pt

H^+ (aq)
($a = 1$)

> ▶ Later in the chapter we also use the symbol E_{cell}. The absence of a superscript degree sign means that reactants and products are not in their standard states.

The two vertical lines signify that three phases are present: solid platinum, gaseous hydrogen, and aqueous hydrogen ion.

By international agreement, a **standard electrode potential, $E°$,** measures the tendency for a *reduction* process to occur at an electrode. In all cases the ionic species are present in aqueous solution at unit activity (approximately 1 M), and gases are at 1 atm pressure. Where no metallic substance is indicated, the potential is established on an inert metallic electrode, such as platinum.

To emphasize that $E°$ refers to a reduction, we will write a reduction couple as a subscript to $E°$, as shown in half-reaction (21.6). The substance being reduced is written on the left of the / sign and the chief reduction product on the right.

$$Cu^{2+}(1\ M) + 2\ e^- \longrightarrow Cu(s) \qquad E°_{Cu^{2+}/Cu} = ? \qquad (21.6)$$

To determine the value of $E°$ for a standard electrode such as (21.6), we compare it with a standard hydrogen electrode (SHE). In the voltaic cell indicated below the measured potential difference is 0.337 V, with electrons flowing from the H_2 to the Cu electrode.

$$\underset{\text{anode}}{Pt\,|\,H_2(g, 1\ \text{atm})\,|\,H^+(1\ M)}\,\|\,\underset{\text{cathode}}{Cu^{2+}(1\ M)\,|\,Cu(s)} \qquad E°_{cell} = 0.337\ V \qquad (21.7)$$

A **standard cell potential, $E°_{cell}$,** is the potential difference, or voltage, of a cell formed from two *standard* electrodes. The net reaction that occurs in the voltaic cell of (21.7) is

$$H_2(g, 1 \text{ atm}) + Cu^{2+}(1 \text{ M}) \longrightarrow 2 \text{ H}^+(1 \text{ M}) + Cu(s) \qquad E^\circ_{cell} = 0.337 \text{ V} \qquad (21.8)$$

The cell reaction (21.8) indicates that $Cu^{2+}(1 \text{ M})$ is *more easily* reduced than is $H^+(1 \text{ M})$. The standard electrode potential representing the reduction of $Cu^{2+}(aq)$ to $Cu(s)$ is assigned a *positive* value, that is, $+0.337$ V.

$$Cu^{2+}(1 \text{ M}) + 2 \text{ e}^- \longrightarrow Cu(s) \qquad E^\circ_{Cu^{2+}/Cu} = +0.337 \text{ V} \qquad (21.9)$$

Suppose we replace the standard copper electrode in the cell diagram of (21.7) by a standard zinc electrode and measure the potential difference between the standard hydrogen and zinc electrodes by using the same voltmeter connections as in (21.7). In this case we will find the voltage to be -0.736 V. The *negative* sign indicates that electrons flow in the *opposite* direction from (21.7), that is, *from* the zinc *to* the hydrogen electrode. $H^+(1 \text{ M})$ is *more easily* reduced than is $Zn^{2+}(1 \text{ M})$. These findings are represented in the following cell diagram in which the zinc anode appears on the *right*.

$$\text{Pt} | H_2(g, 1 \text{ atm}) | H^+(1 \text{ M}) \| Zn^{2+}(1 \text{ M}) | Zn(s) \qquad E^\circ_{cell} = -0.763 \text{ V} \qquad (21.10)$$

cathode   anode

The reduction half-cell reaction and its standard electrode potential are

$$Zn^{2+}(1 \text{ M}) + 2 \text{ e}^- \longrightarrow Zn(s) \qquad E^\circ_{Zn^{2+}/Zn} = -0.763 \text{ V} \qquad (21.11)$$

In summary, the potential of the standard hydrogen electrode is set at *exactly* 0 V. Any electrode at which a reduction half-reaction shows a *greater* tendency to occur than does the reduction of $H^+(1 \text{ M})$ to $H_2(g, 1 \text{ atm})$ has a *positive* value for its standard reduction potential, E°. Any electrode at which a reduction half-reaction shows a *lesser* tendency to occur than does the reduction of $H^+(1 \text{ M})$ to $H_2(g, 1 \text{ atm})$ has a *negative* value for its standard reduction potential, E°. Table 21.1 lists some common reduction half-reactions and their standard reduction potentials at 25 °C.

We will use standard reduction potentials throughout this chapter for many different purposes. Our first objective will be to calculate cell potentials for redox reactions—E°_{cell} values—from standard reduction potentials for half-cell reactions—E° values—by the following procedure.

1. Write the reduction half-equation and its standard reduction potential. This will be an E° value taken *directly* from Table 21.1.

2. Write the *oxidation* half-equation. Its potential will be the *negative* of the standard reduction potential listed in Table 21.1—that is, $-E^\circ$.

3. Combine the half-equations into a net redox equation in the usual manner (recall page 144). *Add* the potentials for the half-cells to obtain E°_{cell}.

There are two important ideas to keep in mind in this three-step procedure: (1) The procedure applies *only* to combining an oxidation and a reduction half-equation into the net oxidation–reduction equation. In particular, the procedure does *not* apply to the occasional need to combine two reduction half-equations and their E° values to obtain a third half-equation and its E° value. (2) E° values are *intensive* properties; they do *not* depend on the quantities of substances involved. Thus, E° values are *unaffected* when the coefficients of half-equations are multiplied by a constant factor, as is often necessary before combining them into a net redox equation in step 3.

In Example 21-2 we predict E°_{cell} for a new battery system. In Example 21-3 we use one known reduction potential and a measured E°_{cell} value to determine an unknown E°.

▶ The proper connection of half cell electrodes to a voltmeter to obtain a positive voltage is: anode to $(-)$ terminal and cathode to $(+)$ terminal. If the connections are reversed (anode to $+$ and cathode to $-$) the voltmeter reading will be negative.

▶ To change the sign of one E° value and add it to another is equivalent to taking the difference between them, hence the term potential *difference* for the voltage of a voltaic cell.

▶ We describe how to use two known E° values to obtain a third, unknown E° value in Feature Problem A on page 762.

> ► A more extensive listing of reduction half-reactions and their potentials is given in Appendix D.

> ► The placement of *oxidizing agents* is: *strongest* oxidizing agents (F_2, O_3, ...), *left* sides, *top* of the list; *poorest* oxidizing agents (Li^+, K^+, ...), *left* sides, *bottom* of list. The placement of *reducing agents* is: *strongest* reducing agents (Li, K, ...), *right* sides, *bottom* of list; *poorest* reducing agents (F^-, O_2, ...), *right* sides, *top* of list.

TABLE 21.1   Some Selected Standard Electrode (Reduction) Potentials at 25 °C

Reduction Half-Reaction	$E°$, V
Acidic solution	
$F_2(g) + 2\,e^- \longrightarrow 2\,F^-(aq)$	+2.866
$O_3(g) + 2\,H^+(aq) + 2\,e^- \longrightarrow O_2(g) + H_2O$	+2.075
$S_2O_8^{2-}(aq) + 2\,e^- \longrightarrow 2\,SO_4^{2-}(aq)$	+2.01
$H_2O_2(aq) + 2\,H^+(aq) + 2\,e^- \longrightarrow 2\,H_2O$	+1.763
$MnO_4^-(aq) + 8\,H^+(aq) + 5\,e^- \longrightarrow Mn^{2+}(aq) + 4\,H_2O$	+1.51
$PbO_2(s) + 4\,H^+(aq) + 2\,e^- \longrightarrow Pb^{2+}(aq) + 2\,H_2O$	+1.455
$Cl_2(g) + 2\,e^- \longrightarrow 2\,Cl^-(aq)$	+1.358
$Cr_2O_7^{2-}(aq) + 14\,H^+(aq) + 6\,e^- \longrightarrow 2\,Cr^{3+}(aq) + 7\,H_2O$	+1.33
$MnO_2(s) + 4\,H^+(aq) + 2\,e^- \longrightarrow Mn^{2+}(aq) + 2\,H_2O$	+1.23
$O_2(g) + 4\,H^+(aq) + 4\,e^- \longrightarrow 2\,H_2O$	+1.229
$2\,IO_3^-(aq) + 12\,H^+(aq) + 10\,e^- \longrightarrow I_2(s) + 6\,H_2O$	+1.20
$Br_2(l) + 2\,e^- \longrightarrow 2\,Br^-(aq)$	+1.065
$NO_3^-(aq) + 4\,H^+(aq) + 3\,e^- \longrightarrow NO(g) + 2\,H_2O$	+0.956
$Ag^+(aq) + e^- \longrightarrow Ag(s)$	+0.800
$Fe^{3+}(aq) + e^- \longrightarrow Fe^{2+}(aq)$	+0.771
$O_2(g) + 2\,H^+(aq) + 2\,e^- \longrightarrow H_2O_2(aq)$	+0.695
$I_2(s) + 2\,e^- \longrightarrow 2\,I^-(aq)$	+0.535
$Cu^{2+}(aq) + 2\,e^- \longrightarrow Cu(s)$	+0.337
$SO_4^{2-}(aq) + 4\,H^+(aq) + 2\,e^- \longrightarrow 2\,H_2O + SO_2(g)$	+0.17
$Sn^{4+}(aq) + 2\,e^- \longrightarrow Sn^{2+}(aq)$	+0.154
$S(s) + 2\,H^+(aq) + 2\,e^- \longrightarrow H_2S(g)$	+0.14
$2\,H^+(aq) + 2\,e^- \longrightarrow H_2(g)$	0
$Pb^{2+}(aq) + 2\,e^- \longrightarrow Pb(s)$	−0.125
$Sn^{2+}(aq) + 2\,e^- \longrightarrow Sn(s)$	−0.137
$Fe^{2+}(aq) + 2\,e^- \longrightarrow Fe(s)$	−0.440
$Zn^{2+}(aq) + 2\,e^- \longrightarrow Zn(s)$	−0.763
$Al^{3+}(aq) + 3\,e^- \longrightarrow Al(s)$	−1.676
$Mg^{2+}(aq) + 2\,e^- \longrightarrow Mg(s)$	−2.356
$Na^+(aq) + e^- \longrightarrow Na(s)$	−2.713
$Ca^{2+}(aq) + 2\,e^- \longrightarrow Ca(s)$	−2.84
$K^+(aq) + e^- \longrightarrow K(s)$	−2.924
$Li^+(aq) + e^- \longrightarrow Li(s)$	−3.040
Basic solution	
$O_3(g) + H_2O + 2\,e^- \longrightarrow O_2(g) + 2\,OH^-(aq)$	+1.246
$OCl^-(aq) + H_2O + 2\,e^- \longrightarrow Cl^-(aq) + 2\,OH^-(aq)$	+0.890
$O_2(g) + 2\,H_2O + 4\,e^- \longrightarrow 4\,OH^-(aq)$	+0.401
$2\,H_2O + 2\,e^- \longrightarrow H_2(g) + 2\,OH^-(aq)$	−0.828

EXAMPLE 21-2

Combining $E°$ Values into $E°_{cell}$ for a Reaction. A new battery system currently under study for possible use in electric vehicles is the zinc–chlorine battery. The net reaction producing electricity in this cell is $Zn(s) + Cl_2(g) \rightarrow ZnCl_2(aq)$. What is $E°_{cell}$ of this voltaic cell?

Solution

From the net reaction we see that $Zn(s)$ is oxidized and $Cl_2(g)$ is reduced. We can write equations for each half-reaction and assign them the appropriate electrode potentials from Table 21.1. $E°_{cell}$ is the sum of these potentials.

Oxidation:	$Zn(s) \longrightarrow Zn^{2+}(aq) + 2\ e^-$	$-E^{\circ}_{Zn^{2+}/Zn} = -(-0.763\ V) = +0.763\ V$
Reduction:	$Cl_2(g) + 2\ e^- \longrightarrow 2\ Cl^-(aq)$	$E^{\circ}_{Cl_2/Cl^-} = +1.358\ V$
Net:	$Zn(s) + Cl_2(g) \longrightarrow ZnCl_2(aq)$	$E^{\circ}_{cell} = 2.121\ V$

$$(21.12)$$

Practice Example A: What is E°_{cell} for the reaction in which $Cl_2(g)$ oxidizes $Fe^{2+}(aq)$ to $Fe^{3+}(aq)$?

$$2\ Fe^{2+}(aq) + Cl_2(g) \longrightarrow 2\ Fe^{3+}(aq) + 2\ Cl^-(aq) \qquad E^{\circ}_{cell} = ?$$

Practice Example B: Use data from Table 21.1 to determine E°_{cell} for the redox reaction in which $Fe^{2+}(aq)$ is oxidized to $Fe^{3+}(aq)$ by $MnO_4^-(aq)$ in acidic solution. (*Hint:* Even though the net redox equation is not given, you can write oxidation and reduction half-equations by the method illustrated in Section 5-4.)

EXAMPLE 21-3

Determining an Unknown E° from an E°_{cell} Measurement. Cadmium is found in small quantities wherever zinc is found. Unlike zinc, which in trace amounts is an essential element, cadmium is an environmental poison. To determine cadmium ion concentrations by electrical measurements, the standard reduction potential for the Cd^{2+}/Cd electrode is needed. The voltage of the following voltaic cell is measured.

$$Cd(s)\,|\,Cd^{2+}(1\ M)\,\|\,Cu^{2+}(1\ M)\,|\,Cu(s) \qquad E^{\circ}_{cell} = 0.740\ V$$

What is the standard reduction potential for the Cd^{2+}/Cd electrode?

Solution

We know one half-cell potential and E°_{cell} for the net redox reaction. We can solve for the unknown standard reduction potential, which we denote as E°. Note that we enter this unknown potential as its *negative* $(-E^{\circ})$ in the set-up below because we are applying it to an *oxidation* half-reaction.

Oxidation:	$Cd(s) \longrightarrow Cd^{2+}(1\ M) + 2\ e^-$	$-E^{\circ}_{Cd^{2+}/Cd}$
Reduction:	$Cu^{2+}(1\ M) + 2\ e^- \longrightarrow Cu(s)$	$E^{\circ}_{Cu^{2+}/Cu} = +0.337\ V$
Net:	$Cd(s) + Cu^{2+}(1\ M) \longrightarrow Cd^{2+}(1\ M) + Cu(s)$	$E^{\circ}_{cell} = -E^{\circ}_{Cd^{2+}/Cd} + 0.337\ V = 0.740\ V$

$$E^{\circ}_{Cd^{2+}/Cd} = 0.337\ V - 0.740\ V = -0.403\ V$$

Practice Example A: In acidic solution, dichromate ion oxidizes oxalic acid, $H_2C_2O_4(aq)$, to $CO_2(g)$ in a reaction with $E^{\circ}_{cell} = 1.81\ V$.

$$Cr_2O_7{}^{2-}(aq) + 3\ H_2C_2O_4(aq) + 8\ H^+(aq) \longrightarrow 2\ Cr^{3+}(aq) + 7\ H_2O + 6\ CO_2(g)$$

Use the value of E°_{cell} for this reaction, together with appropriate data from Table 21.1, to determine the standard electrode potential for the $CO_2(g)/H_2C_2O_4(aq)$ electrode.

Practice Example B: In an acidic solution, $O_2(g)$ oxidizes $Cr^{2+}(aq)$ to $Cr^{3+}(aq)$. The $O_2(g)$ is reduced to the oxidation state -2 in H_2O. The value of E°_{cell} for the reaction is found to be 1.653 V. What is the standard electrode potential for the couple Cr^{3+}/Cr^{2+}? (*Hint:* What is the reduction half-reaction in the cell? Find its standard reduction potential in Table 21.1.)

21-3 E_{cell} and Spontaneous Change

When a reaction occurs in a voltaic cell, the cell does work—electrical work. Think of this as the work of moving electric charges. The total work done is the product of three terms: (a) the emf (voltage) of the cell, that is, E_{cell}; (b) the number of moles of electrons transferred between the electrodes, n; and (c) the electric charge per mole of electrons, called the **Faraday constant** ($\mathscr{F}$) and equal to 96,485 coulombs per mole of electrons (96,485 C/mol e⁻). Because the product volt × coulomb = joule, the unit of w_{elec} is joules (J).

▶ To evaluate n you will generally need to write half-equations and combine them into a net equation. For example, in reaction (21.4) on page 729 the number of moles of electrons appearing in the oxidation half-equation is 3 and in the reduction half-equation, 2; but the number for the net equation is $n = 6$.

$$w_{elec} = n\mathscr{F}E_{cell} \qquad (21.13)$$

Expression (21.13) applies only if the cell operates reversibly.* In the Are You Wondering feature on page 699 we described the amount of available energy (work) that can be derived from a process as equal to $-\Delta G$. Thus,

$$\Delta G = -n\mathscr{F}E_{cell} \qquad (21.14)$$

In the special case in which the reactants and products are in their standard states,

$$\Delta G° = -n\mathscr{F}E°_{cell} \qquad (21.15)$$

Our primary interest is not in calculating quantities of work, but in using expression (21.15) as a means of evaluating free energy changes from measured cell potentials, as illustrated in Example 21-4.

EXAMPLE 21-4

Determining a Free Energy Change from a Cell Potential. Use electrode potential data to determine $\Delta G°$ for the reaction

$$Zn(s) + Cl_2(g, 1 \text{ atm}) \longrightarrow ZnCl_2(aq, 1 \text{ M})$$

Solution

This is the net reaction (21.12) occurring in the voltaic cell described in Example 21-2. In this type of problem, we generally need to separate the net equation into two half-equations. This allows us to determine the value of $E°_{cell}$ and the number of moles of electrons (n) involved in the net cell reaction. If you refer to Example 21-2 you will see that $E°_{cell} = +2.121$ V and $n = 2$ mol e⁻. Now we can use Equation (21.15).

$$\Delta G° = -n\mathscr{F}E°_{cell} = -\left(2 \text{ mol e}^- \times \frac{96,485 \text{ C}}{1 \text{ mol e}^-} \times 2.121 \text{ V}\right)$$

$$= -4.093 \times 10^5 \text{ J} = -409.3 \text{ kJ}$$

Practice Example A: Use electrode potential data to determine $\Delta G°$ for the reaction

$$2 \text{ Al}(s) + 3 \text{ Br}_2(l) \longrightarrow 2 \text{ Al}^{3+}(aq, 1 \text{ M}) + 6 \text{ Br}^-(aq, 1 \text{ M}) \qquad \Delta G° = ?$$

Practice Example B: The hydrogen–oxygen fuel cell is a voltaic cell with a net reaction of $2 \text{ H}_2(g) + O_2(g) \rightarrow 2 \text{ H}_2O(l)$. Calculate $E°_{cell}$ for this reaction. [*Hint:* Use thermodynamic data from Appendix D to calculate $\Delta G°$. To obtain the correct value for n in Equation (21.15), write the two half-equations that yield the net equation.]

---

*The meaning of a *reversible* process was illustrated through Figure 20-5 on page 695. The reversible operation of a voltaic cell requires that electric current be drawn from the cell only very, very slowly.

Spontaneous Change in Oxidation–Reduction Reactions

Our main criterion for spontaneous change is that $\Delta G < 0$. According to Equation (21.14), however, redox equations have the property that if $\Delta G < 0$ then $E_{cell} > 0$. That is, E_{cell} must be *positive* if ΔG is to be negative. Predicting the direction of spontaneous change in a redox reaction is a relatively simple matter when we use the following ideas.

- If E_{cell} is *positive,* a reaction occurs spontaneously in the *forward* direction for the stated conditions. If E_{cell} is *negative,* the reaction occurs spontaneously in the *reverse* direction for the stated conditions. If $E_{cell} = 0$, a reaction is at equilibrium for the stated conditions.
- If a cell reaction is *reversed,* E_{cell} changes sign.

In the special case when reactants and products are in their standard states, we work with $\Delta G°$ and $E_{cell}°$ values, as illustrated in Examples 21-5 and 21-6.

► Recall the relationship between ΔG and $\Delta G°$ given by Equation 20.8: $\Delta G = \Delta G° + RT \ln Q$. *Qualitative* predictions that we make based on $E°$ and $\Delta G°$ values are often applicable over a wide range of nonstandard initial conditions.

EXAMPLE 21-5

Applying the Criterion for Spontaneous Change in a Redox Reaction. Will aluminum metal displace Cu^{2+} ion from aqueous solution? That is, does a spontaneous reaction occur in the forward direction for the following reaction?

$$2 \text{ Al(s)} + 3 \text{ Cu}^{2+}(1 \text{ M}) \longrightarrow 3 \text{ Cu(s)} + 2 \text{ Al}^{3+}(1 \text{ M})$$

Solution

Because the balanced net equation is given, all we need do is separate this equation into two half-equations, and then recombine them, together with their electrode potentials.

Oxidation:	$2 \{\text{Al(s)} \longrightarrow \text{Al}^{3+}(1 \text{ M}) + 3 \text{ e}^-\}$	$-E°_{\text{Al}^{3+}/\text{Al}} = -(-1.676) = +1.676$ V
Reduction:	$3 \{\text{Cu}^{2+}(1 \text{ M}) + 2 \text{ e}^- \longrightarrow \text{Cu(s)}\}$	$E°_{\text{Cu}^{2+}/\text{Cu}} = +0.337$ V
Net:	$2 \text{ Al(s)} + 3 \text{ Cu}^{2+}(1 \text{ M}) \longrightarrow 3 \text{ Cu(s)} + 2 \text{ Al}^{3+}(1 \text{ M})$	$E°_{cell} = +2.013$ V

Because $E°_{cell}$ is positive, the direction of spontaneous change is that of the forward reaction. Al(s) will displace Cu^{2+} from aqueous solution under standard conditions. With this same calculation we can say that Cu(s) will *not* displace Al^{3+} from aqueous solution under standard conditions. This is the reverse reaction and has a *negative* $E°_{cell}$.

Practice Example A: We have just seen that Cu(s) will not displace Al^{3+} from aqueous solution under standard conditions. Name one metal ion that Cu(s) will displace and determine $E°_{cell}$ for the reaction.

Practice Example B: According to $E°$ values, it would seem possible to use sodium metal to displace magnesium metal from seawater that has been concentrated to the point that $[Mg^{2+}] = 1$ M. Use electrode potential data to show this to be the case. Yet, the method is not feasible, even disregarding the prohibitive costs involved. Explain why this is so. (*Hint:* What reaction does occur and why?)

Even though we used electrode potentials and a cell voltage to predict a spontaneous reaction in Example 21-5, we do not have to carry out the reaction in a voltaic cell. This is an important point to keep in mind. Thus, Cu^{2+} is displaced from aqueous solution simply by adding aluminum metal, as shown in Figure 21-6. Another point, illustrated by Example 21-6, is that we can often

FIGURE 21-6
Reaction of Al(s) and Cu²⁺(aq).
Notice the holes in the foil where
Al(s) has dissolved. Notice also
the dark deposit of Cu(s) at the
bottom of the beaker.

give qualitative answers to questions concerning redox reactions without going through a complete calculation of E°_{cell}.

EXAMPLE 21-6

Making Qualitative Predictions with Electrode Potential Data. Peroxodisulfate salts such as $Na_2S_2O_8$ are oxidizing agents used in bleaching. Dichromates such as $K_2Cr_2O_7$ have been used as laboratory oxidizing agents. Which is the better oxidizing agent in acidic solution under standard conditions, $S_2O_8^{2-}$ or $Cr_2O_7^{2-}$?

Solution

In a redox reaction the oxidizing agent is reduced. The greater the tendency for this reduction to occur, the better the oxidizing agent. The reduction tendency, in turn, is measured by the E° value. Because the E° value for the reduction of $S_2O_8^{2-}$(aq) to SO_4^{2-}(aq) (+2.01 V) is larger than that for the reduction of $Cr_2O_7^{2-}$(aq) to Cr^{3+}(aq) (+1.33 V), $S_2O_8^{2-}$(aq) should be the better oxidizing agent.

Practice Example A: A cheap way to produce peroxodisulfates (such as $Na_2S_2O_8$) would be to pass O_2(g) through an acidic solution containing sulfate ion. Is this method feasible under standard conditions? (*Hint:* What would be the reduction half-reaction?)

Practice Example B: Consider the following observations: (1) Aqueous solutions of Sn^{2+} are difficult to maintain because atmospheric oxygen easily oxidizes Sn^{2+} to Sn^{4+}. (2) One way to preserve the Sn^{2+}(aq) solutions is to add some metallic tin. *Without doing detailed calculations* explain these two statements by using E° data. [*Hint:* What are the relevant half-reactions in (1) and in (2)?]

The Behavior of Metals Toward Acids

In discussing redox reactions in Chapter 5 we noted that most metals react with an acid such as HCl, but that a few do not. We are now in a position to explain this observation. When a metal, M, reacts with an acid such as HCl, the metal is oxidized to the metal ion, such as M^{2+}. The reduction involves H^+ being reduced to H_2(g). We can express these ideas as

Oxidation:	$M(s) \longrightarrow M^{2+}(aq) + 2\ e^-$	$-E^\circ_{M^{2+}/M}$
Reduction:	$2\ H^+(aq) + 2\ e^- \longrightarrow H_2(g)$	$E^\circ_{H^+/H_2} = 0\ V$
Net:	$M(s) + 2\ H^+(aq) \longrightarrow M^{2+}(aq) + H_2(g)$	$E^\circ_{cell} = -E^\circ_{M^{2+}/M}$

According to our new criterion for spontaneous change, any metal with a *negative* reduction potential should react with an acid in which $[H^+] = 1$ M. This will make $E^\circ_{cell} > 0$ (positive). Thus, all the metals listed *below* hydrogen in Table 21.1 (Pb through Li) should react with acids.

In acids such as HCl, HBr, and HI, the oxidizing agent is H^+ (that is, H_3O^+). Certain metals that will not react with HCl will react with an acid if there is present an *anion* that is a better oxidizing agent than H^+. Nitrate ion is a good oxidizing agent in acidic solution, and silver metal, which does not react with HCl(aq), readily reacts with nitric acid, HNO_3(aq).

$$3\ Ag(s) + NO_3^-(aq) + 4\ H^+(aq) \longrightarrow 3\ Ag^+(aq) + NO(g) + 2\ H_2O$$

$$E^\circ_{cell} = 0.156\ V$$

Relationship Between E°_{cell} and K_{eq}

We related ΔG° and E°_{cell} through Equation (21.15). In Chapter 20 we related ΔG° and K_{eq} (through Equation 20.9). The three quantities are thus related in this way:

$$\Delta G^\circ = -RT \ln K_{eq} = -n\mathscr{F}E^\circ_{cell}$$

and therefore,

$$E^\circ_{cell} = \frac{RT}{n\mathscr{F}} \ln K_{eq} \qquad (21.16)$$

In Equation (21.16) R has a value of 8.3145 J mol^{-1} K^{-1}, and n represents the number of moles of electrons involved in the reaction. Then, if we specify a temperature of 25 °C = 298.15 K (the temperature at which electrode potentials are generally determined) we can replace the combined terms "$RT/\mathscr{F}$" by the single constant "0.0257." The term $0.0257/n$ has the unit volt.

$$E^\circ_{cell} = \frac{0.0257}{n} \ln K_{eq} \qquad (21.17)$$

▶ If you prefer to work with common logarithms, the equivalent to Equation (21.17) is $E^\circ_{cell} = (0.0592/n) \log K_{eq}$.

As we have noted before, the value of n is not always apparent from the net redox equation. It is usually necessary to write half-equations to determine n, as illustrated in Example 21-7, where the number of moles of electrons is printed in blue.

EXAMPLE 21-7

Relating K_{eq} to E°_{cell} for a Redox Reaction. What is the value of the equilibrium constant K_{eq} for the reaction between copper metal and iron(III) ions in aqueous solution at 25 °C?

$$\text{Cu(s)} + 2\,\text{Fe}^{3+}(\text{aq}) \longrightarrow \text{Cu}^{2+}(\text{aq}) + 2\,\text{Fe}^{2+}(\text{aq}) \qquad K_{eq} = ?$$

Solution

First, use data from Table 21.1 to determine E°_{cell}.

Oxidation:	$\text{Cu(s)} \longrightarrow \text{Cu}^{2+}(\text{aq}) + 2\,\text{e}^-$	$-E^\circ_{\text{Cu}^{2+}/\text{Cu}} = -0.337$ V
Reduction:	$2\,\{\text{Fe}^{3+}(\text{aq}) + \text{e}^- \longrightarrow \text{Fe}^{2+}(\text{aq})\}$	$E^\circ_{\text{Fe}^{3+}/\text{Fe}^{2+}} = +0.771$ V
Net:	$\text{Cu(s)} + 2\,\text{Fe}^{3+}(\text{aq}) \longrightarrow \text{Cu}^{2+}(\text{aq}) + 2\,\text{Fe}^{2+}(\text{aq})$	$E^\circ_{cell} = +0.434$ V

The number of moles of electrons for the cell reaction is 2.

$$E^\circ_{cell} = 0.434 = \frac{0.0257}{2} \ln K_{eq} \qquad \ln K_{eq} = \frac{2 \times 0.434}{0.0257} = 33.8$$

$$K_{eq} = e^{33.8} = 5 \times 10^{14}$$

Practice Example A: Would you expect the displacement of Cu^{2+} from aqueous solution by Al(s) to "go to completion"? (*Hint:* Base your assessment on the value of K_{eq} for the displacement reaction. We determined E°_{cell} for this reaction in Example 21-5.)

Practice Example B: Would you expect the reaction of Sn(s) and $\text{Pb}^{2+}(\text{aq})$ to go to completion? Explain.

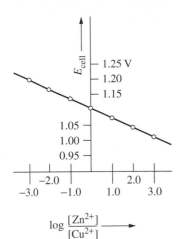

FIGURE 21-7
Variation of E_{cell} with ion concentrations.
The cell reaction is Zn(s) + Cu^{2+}(aq) → Zn^{2+}(aq) + Cu(s) and has $E°_{cell}$ = 1.100 V.

21-4 E_{cell} as a Function of Concentrations

When we combine *standard* electrode potentials, we obtain a *standard* emf for a voltaic cell, such as $E°_{cell}$ = 1.100 V for the voltaic cell of Figure 21-4. For the following cell reaction at *nonstandard* conditions, however, the measured E_{cell} is not 1.100 V.

$$Zn(s) + Cu^{2+}(2.0 \text{ M}) \longrightarrow Zn^{2+}(0.10 \text{ M}) + Cu(s) \qquad E_{cell} = 1.139 \text{ V}$$

Experimental measurements of cell potentials are often made for nonstandard conditions, and these measurements have great significance, especially for performing chemical analyses.

From Le Châtelier's principle we might predict that *increasing* the concentration of a reactant (Cu^{2+}) while *decreasing* the concentration of a product (Zn^{2+}) should favor the forward reaction. Zn(s) should displace Cu^{2+}(aq) even more readily than for standard state conditions and E_{cell} > 1.100 V. E_{cell} is found to vary linearly with log ($[Zn^{2+}]/[Cu^{2+}]$), as illustrated in Figure 21-7.

It is not difficult to establish the relationship between the cell potential, E_{cell}, and the concentrations of reactants and products. From Chapter 20, we can write Equation (20.8)

$$\Delta G = \Delta G° + RT \ln Q$$

For ΔG and $\Delta G°$ we can substitute $-n\mathscr{F}E_{cell}$ and $-n\mathscr{F}E°_{cell}$, respectively.

$$-n\mathscr{F}E_{cell} = -n\mathscr{F}E°_{cell} + RT \ln Q$$

Now, dividing through by $-n\mathscr{F}$, we obtain

$$E_{cell} = E°_{cell} - \frac{RT}{n\mathscr{F}} \ln Q$$

▶ The equivalent expression to Equation (21.18) in natural logarithms is $E_{cell} = E°_{cell} - (0.0257/n) \ln Q$.

This equation was first proposed by Walther Nernst in 1889, and is commonly used by analytical chemists in the form in which Nernst expressed it. We can obtain the **Nernst equation** by switching from natural to common logarithms (log = 2.303 × ln), by specifying a temperature of 298.15 K, and by replacing the term "2.303$RT/n\mathscr{F}$" by "0.0592/n" (which, again, has the unit: volt).

$$E_{cell} = E°_{cell} - \frac{0.0592}{n} \log Q \qquad (21.18)$$

Walther Nernst (1864–1941). Nernst was only 25 years old when he formulated his equation relating cell voltages and concentrations. He is also credited with proposing the solubility product concept in the same year. In 1906 he announced his Heat Theorem, which we now know as the third law of thermodynamics.

In the Nernst equation we make the usual substitutions into Q: a = 1 for the activities of pure solids and liquids; partial pressures (atm) for the activities of gases; and molarities for the activities of solution components.

EXAMPLE 21-8

Applying the Nernst Equation for Determining E_{cell}. What is the value of E_{cell} for the voltaic cell pictured in Figure 21-8 and diagrammed as follows?

$$Pt|Fe^{2+}(0.10 \text{ M}), Fe^{3+}(0.20 \text{ M})\|Ag^{+}(1.0 \text{ M})|Ag(s) \qquad E_{cell} = ?$$

Solution

Two steps are required when using the Nernst equation. First, to determine $E°_{cell}$, use data from Table 21.1 to write

Oxidation:	$Fe^{2+}(1\ M) \longrightarrow Fe^{3+}(1\ M) + e^-$	$-E^\circ_{Fe^{3+}/Fe^{2+}} = -(+0.771)\ V = -0.771\ V$	
Reduction:	$Ag^+(1\ M) + e^- \longrightarrow Ag(s)$	$E^\circ_{Ag^+/Ag} = +0.800\ V$	
Net:	$Fe^{2+}(1\ M) + Ag^+(1\ M) \longrightarrow Fe^{3+}(1\ M) + Ag(s)$	$E^\circ_{cell} = +0.029\ V$	(21.19)

Now, to determine E_{cell} for the reaction

$$Fe^{2+}(0.10\ M) + Ag^+(1.0\ M) \longrightarrow Fe^{3+}(0.20\ M) + Ag(s) \qquad E_{cell} = ? \qquad (21.20)$$

substitute appropriate values into the Nernst equation (21.18), starting with $E^\circ_{cell} = 0.029\ V$ and $n = 1$,

$$E_{cell} = 0.029 - \frac{0.0592}{1} \log \frac{[Fe^{3+}]}{[Fe^{2+}][Ag^+]}$$

and for concentrations: $[Fe^{2+}] = 0.10\ M$; $[Fe^{3+}] = 0.20\ M$; $[Ag^+] = 1.0\ M$.

$$E_{cell} = 0.029 - 0.0592 \log \frac{0.20}{0.10 \times 1.0}$$

$$= 0.029 - 0.0592 \log 2 = 0.029 - 0.018$$

$$= 0.011\ V$$

Practice Example A: Calculate E_{cell} for the following voltaic cell.

$$Al(s)\,|\,Al^{3+}(0.36\ M)\,\|\,Sn^{4+}(0.086\ M),\ Sn^{2+}(0.54\ M)\,|\,Pt$$

Practice Example B: Calculate E_{cell} for the following voltaic cell:

$$Pt(s)\,|\,Cl_2(1\ atm)\,|\,Cl^-(1.0\ M)\,\|\,Pb^{2+}(0.050\ M),\ H^+(0.10\ M)\,|\,PbO_2(s)$$

▶ The Nernst equation can also be used to determine ion concentrations based on measured cell voltages. In Feature Problem B on page 762 calculations of this type are applied to titration reactions.

In the preceding section we developed a criterion for spontaneous change ($E_{cell} > 0$), but we used the criterion only with E° data from Table 21.1. We often find that qualitative conclusions reached with E°_{cell} values hold over a broad range of nonstandard conditions as well. When E°_{cell} is within a few tenths of a volt of zero, however, it is sometimes necessary to determine E_{cell} for nonstandard conditions in order to apply the criterion for spontaneity of redox reactions, as illustrated in Example 21-9.

FIGURE 21-8
A voltaic cell with nonstandard conditions—Example 21-8 illustrated.

EXAMPLE 21-9

Predicting Spontaneous Reactions for Nonstandard Conditions. Will the cell reaction proceed spontaneously as written for the following cell?

$$Sn(s)\,|\,Sn^{2+}(0.50\ M)\,\|\,Pb^{2+}(0.0010\ M)\,|\,Pb(s)$$

Solution

To determine E_{cell}° from E° data we first write

$$
\begin{array}{ll}
Sn(s) \longrightarrow Sn^{2+}(1\ M) + 2\ e^{-} & -E_{Sn^{2+}/Sn}^{\circ} = 0.137\ V \\
\underline{Pb^{2+}(1\ M) + 2\ e^{-} \longrightarrow Pb(s)} & \underline{E_{Pb^{2+}/Pb}^{\circ} = -0.125\ V} \\
Sn(s) + Pb^{2+}(1\ M) \longrightarrow Sn^{2+}(1\ M) + Pb(s) & E_{cell}^{\circ} = 0.012\ V
\end{array}
$$

Then, to determine E_{cell} for the reaction

$$Sn(s) + Pb^{2+}(0.0010\ M) \longrightarrow Pb(s) + Sn^{2+}(0.50\ M) \qquad E_{cell} = ?$$

we substitute the following data into the Nernst equation: $E_{cell}^{\circ} = 0.012$ V, $n = 2$, $[Sn^{2+}] = 0.50$ M and $[Pb^{2+}] = 0.0010$ M.

$$E_{cell} = 0.012 - \frac{0.0592}{2}\log\frac{[Sn^{2+}]}{[Pb^{2+}]}$$

$$= 0.012 - 0.0296\log\frac{0.50}{0.0010} = 0.012 - 0.0296\log 500$$

$$= 0.012 - 0.080 = -0.068\ V$$

Because $E_{cell} < 0$ we conclude that the reaction is nonspontaneous as written.

Practice Example A: Will the cell reaction proceed spontaneously as written for the following cell?

$$Cu(s)\,|\,Cu^{2+}(0.15\ M)\,\|\,Fe^{3+}(0.35\ M),\ Fe^{2+}(0.25\ M)\,|\,Pt(s)$$

Practice Example B: For what ratio of $[Sn^{2+}]/[Pb^{2+}]$ will the cell reaction in Example 21-9 not be spontaneous in either direction?

Concentration Cells

The voltaic cell in Figure 21-9 consists of two hydrogen electrodes. One is a *standard hydrogen electrode* (SHE) and the other is a hydrogen electrode

FIGURE 21-9
A concentration cell.
The cell consists of two hydrogen electrodes. The electrode on the right is a SHE. Oxidation occurs at the anode on the left where $[H^+]$ is less than 1 M. The reading on the voltmeter is directly proportional to the pH of the solution in the anode compartment.

immersed in a solution of unknown $[H^+]$, less than 1 M. The cell diagram is

$$Pt\,|\,H_2(g,\ 1\ atm)\,|\,H^+(x\ M)\,\|\,H^+(1\ M)\,|\,H_2(g,\ 1\ atm)\,|\,Pt$$

The reaction occurring in this cell is

Oxidation:	$\cancel{H_2(g,\ 1\ atm)} \longrightarrow 2\,H^+(x\ M) + \cancel{2\,e^-}$	$-E^\circ_{H^+/H_2} = 0\ V$
Reduction:	$2\,H^+(1\ M) + \cancel{2\,e^-} \longrightarrow \cancel{H_2(g,\ 1\ atm)}$	$E^\circ_{H^+/H_2} = 0\ V$
Net:	$2\,H^+(1\ M) \longrightarrow 2\,H^+(x\ M)$	$E^\circ_{cell} = 0\ V$ (21.21)

Any voltaic cell in which the net cell reaction involves only a change in the concentration of some species (here $[H^+]$) is called a concentration cell. A **concentration cell** consists of two half-cells with *identical electrodes* but different ion concentrations. Because the electrodes are identical, the standard electrode potentials are numerically equal and opposite in sign. This makes $E^\circ_{cell} = 0$. Because the ion concentrations do differ, however, there is a potential difference between the two half-cells. Spontaneous change in a concentration cell always occurs in the direction in which the concentrated solution becomes more dilute, and the dilute solution becomes more concentrated. The final result is the same as if the solutions were simply mixed. In a concentration cell, however, we use the natural tendency for entropy to increase in a mixing process as a means of producing electricity.

The Nernst equation for reaction (21.21) takes the form

$$E_{cell} = E^\circ_{cell} - \frac{0.0592}{2} \log \frac{x^2}{1^2}$$

which simplifies to

$$E_{cell} = 0 - \frac{0.0592}{2} \times 2 \log \frac{x}{1} = -0.0592 \log x$$

Because x is $[H^+]$ in the unknown solution, and $-\log x = -\log [H^+] = pH$, the final result is

$$E_{cell} = 0.0592\ pH \qquad\qquad (21.22)$$

where the pH is that of the unknown solution. If an unknown solution has a pH of 3.50, for example, the measured cell voltage in Figure 21-9 will be $E_{cell} = 0.0592 \times 3.50 = 0.207\ V$.

Constructing and using a hydrogen electrode is difficult. The Pt metal surface must be specially prepared and maintained, gas pressure must be controlled, and the electrode cannot be used in the presence of strong oxidizing or reducing agents. A better approach is to replace the SHE in Figure 21-9 by a different reference electrode and the other hydrogen electrode by a *glass electrode*. A glass electrode is a thin glass membrane enclosing $HCl(aq)$ and a silver wire coated with $AgCl(s)$. When the glass electrode is dipped into a solution, H^+ ions are exchanged across the membrane. The potential established on the silver wire depends on $[H^+]$ in the solution being tested. Glass electrodes are most commonly encountered in laboratory pH meters.

A glass electrode and pH meter. The potential of the glass electrode depends on the hydrogen ion concentration of the solution being tested. The difference in potential between the glass electrode and a reference electrode is converted to a pH reading by the meter.

Measurement of K_{sp}

Recall that in a concentration cell the standard cell voltage, $E^\circ_{cell} = 0.000\ V$. A cell voltage, E_{cell}, exists only if there is a difference in concentration of ions in the two half-cells, and this difference in concentration can provide the basis

▶ Many ΔG°_f and K_{eq} data have been obtained by electrochemical measurements in conjunction with Equations (21.15) and (21.16). K_{sp} values are just special cases of K_{eq}.

FIGURE 21-10

A concentration cell for determining K_{sp} of AgI.
The silver electrode in the anode compartment is in contact with a saturated solution of AgI. In the cathode compartment $[Ag^+] = 0.100$ M.

for determining K_{sp} values for sparingly soluble ionic compounds. Consider this concentration cell.

$$Ag(s)\,|\,Ag^+(satd\ AgI)\,\|\,Ag^+(0.100\ M)\,|\,Ag \qquad E_{cell} = +0.417\ V$$

At the anode a silver electrode is placed in a saturated aqueous solution of silver iodide. At the cathode a second silver electrode is placed in a solution with $[Ag^+] = 0.100$ M. The two half-cells are connected by a salt bridge, and the measured cell voltage is 0.417 V (Figure 21-10). The cell reaction occurring in this *concentration* cell is

Oxidation:	$Ag(s) \longrightarrow Ag^+(satd\ AgI) + e^-$	$-E^{\circ}_{Ag^+/Ag} = -0.800\ V$
Reduction:	$Ag^+(0.100\ M) + e^- \longrightarrow Ag(s)$	$E^{\circ}_{Ag^+/Ag} = +0.800\ V$
Net:	$Ag^+(0.100\ M) \longrightarrow Ag^+(satd\ AgI)$	$E^{\circ}_{cell} = 0.000\ V$ (21.23)

We complete the calculation of K_{sp} of silver iodide in Example 21-10.

EXAMPLE 21-10

Using a Voltaic Cell to Determine K_{sp} of a Slightly Soluble Solute. With the data given for reaction (21.23), calculate K_{sp} for AgI.

$$AgI(s) \rightleftharpoons Ag^+(aq) + I^-(aq) \qquad K_{sp} = ?$$

Solution

Let's represent $[Ag^+]$ in saturated silver iodide solution as x. Then we can apply the Nernst equation to the net reaction (21.23). That is,

$$E_{cell} = E^{\circ}_{cell} - \frac{0.0592}{n} \log \frac{[Ag^+]_{sat'd\ AgI}}{[Ag^+]_{0.100\ M\ soln}}$$

$$E_{cell} = E^{\circ}_{cell} - \frac{0.0592}{1} \log \frac{x}{0.100}$$

$$0.417 = 0 - 0.0592(\log x - \log 0.100)$$

Divide both sides of the equation by 0.0592.

$$\frac{0.417}{0.0592} = -\log x + \log 0.100$$

$$\log x = \log 0.100 - \frac{0.417}{0.0592} = -1.00 - 7.04 = -8.04$$

$$[Ag^+] = 10^{-8.04} = 9.1 \times 10^{-9} \text{ M}$$

Because in saturated AgI the concentrations of Ag^+ and I^- are equal,

$$K_{sp} = [Ag^+][I^-] = (9.1 \times 10^{-9})(9.1 \times 10^{-9}) = 8.3 \times 10^{-17}$$

Practice Example A: K_{sp} for AgCl $= 1.8 \times 10^{-10}$. What would be the measured E_{cell} for the voltaic cell in Example 21-10 if the saturated AgI and AgI(s) in the anode half-cell were replaced by saturated AgCl and AgCl(s)?

Practice Example B: Calculate the K_{sp} for PbI_2 given the following concentration cell information:

$$Pb(s)\,|\,Pb^{2+}(\text{satd } PbI_2)\,\|\,Pb^{2+}(0.100 \text{ M})\,|\,Pb(s) \qquad E_{cell} = +0.0567 \text{ V}$$

21-5   Batteries: Producing Electricity Through Chemical Reactions

A **battery** is a device that stores chemical energy for later release as electricity. Some batteries consist of a single voltaic cell with two electrodes and the appropriate electrolyte(s); an example is a flashlight cell. Other batteries consist of two or more voltaic cells joined in "series" fashion—plus to minus—to increase the total voltage; an example is an automobile battery. In this section we will consider three types of cells and batteries.

▶ Batteries are vitally important to modern society. Annual production in western nations has been estimated at about 10 batteries per person per year.

- **Primary batteries** (or primary cells). The cell reaction is not reversible. When the reactants have been mostly converted to products, no more electricity is produced and the battery is "dead."
- **Secondary batteries** (or secondary cells). The cell reaction can be reversed by passing electricity through the battery ("charging"). This means that a battery may be used through several hundred or more cycles of discharging followed by charging.
- **Flow batteries** and **fuel cells.** Materials (reactants, products, electrolytes) pass through the battery, which is simply a converter of chemical to electric energy.

Leclanché (Dry) Cell

Perhaps the most familiar of all voltaic cells is the "flashlight" cell diagrammed in Figure 21-11. In this cell, oxidation occurs at a zinc anode and reduction at an inert carbon (graphite) cathode. The electrolyte is a moist paste of MnO_2, $ZnCl_2$, NH_4Cl, and carbon black (soot). The maximum cell voltage is 1.55 V. The cell is called a "dry" cell because there is no free liquid present. The anode (oxidation) half-reaction is simple.

Oxidation: $Zn(s) \longrightarrow Zn^{2+}(aq) + 2\,e^-$

FIGURE 21-11
The Leclanché (dry) cell.
The chief components of the cell
are C, carbon rod serving as the
cathode: reduction occurs. Z, zinc
container serving as the anode:
oxidation occurs. E, electrolyte: a
moist paste of MnO_2, $ZnCl_2$,
NH_4Cl, and carbon black.

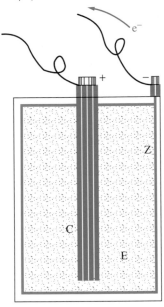

The reduction is more complex. Essentially, it involves the reduction of MnO_2 to compounds having Mn in a $+3$ oxidation state, for example,

Reduction: $\quad 2\,MnO_2(s) + H_2O + 2\,e^- \longrightarrow Mn_2O_3(s) + 2\,OH^-(aq)$

An acid–base reaction occurs between NH_4^+ (from NH_4Cl) and OH^-.

$$NH_4^+(aq) + OH^-(aq) \longrightarrow NH_3(g) + H_2O(l)$$

A buildup of $NH_3(g)$ cannot be permitted to occur around the cathode, because it would disrupt the current by adhering to the cathode. This is prevented by a reaction between Zn^{2+} and $NH_3(g)$ to form the complex ion $[Zn(NH_3)_2]^{2+}$, which crystallizes as the chloride salt.

$$Zn^{2+}(aq) + 2\,NH_3(g) + 2\,Cl^-(aq) \longrightarrow [Zn(NH_3)_2]Cl_2(s)$$

The Leclanché cell is a *primary* cell. The cell reactions cannot be reversed by passing electricity back through the cell. The Leclanché cell is cheap to make, but it has some drawbacks. When current is drawn rapidly from the cell, products such as NH_3 build up on the electrodes, which causes the voltage to drop. Also, because the electrolyte medium is *acidic,* zinc metal slowly dissolves.

A superior form of the Leclanché cell is the *alkaline* cell, which uses NaOH or KOH in place of NH_4Cl as the electrolyte. The oxidation half-reaction involves the formation of $Zn(OH)_2(s)$, which we can think of as occurring in a two-step process.

$$
\begin{array}{r}
Zn(s) \longrightarrow Zn^{2+}(aq) + 2\,e^- \\
Zn^{2+}(aq) + 2\,OH^-(aq) \longrightarrow Zn(OH)_2(s) \\
\hline
Zn(s) + 2\,OH^-(aq) \longrightarrow Zn(OH)_2(s) + 2\,e^-
\end{array}
$$

The advantages of the alkaline battery are that zinc does not dissolve as readily in a *basic* (alkaline) as in an acidic medium, and the battery does a better job of maintaining its voltage as current is drawn from it.

Lead–Acid (Storage) Battery

The most common *secondary* battery, in use since about 1915, is the automobile storage battery pictured in Figure 21-12. The reactants are lead metal at the

FIGURE 21-12
A lead–acid (storage) cell.
The composition of the electrodes is described in the
text. The cell reaction that occurs as the cell is discharged is given in Equation (21.24). The standard voltage of the cell is 2.05 V. In this figure two anode plates
are connected in "parallel" fashion, as are two cathode
plates.

anode, red-brown lead(IV) oxide at the *cathode,* and dilute sulfuric acid. Because ionization of the H_2SO_4 does not go to completion, both $HSO_4^-(aq)$ and $SO_4^{2-}(aq)$ are present, but typically the half-equations and net equation are expressed in terms of $SO_4^{2-}(aq)$.

Oxidation:	$Pb(s) + SO_4^{2-}(aq) \longrightarrow PbSO_4(s) + 2\ e^-$
Reduction:	$PbO_2(s) + 4\ H^+(aq) + SO_4^{2-}(aq) + 2\ e^- \longrightarrow PbSO_4(s) + 2\ H_2O$

$$\text{Net:} \quad Pb(s) + PbO_2(s) + 4\ H^+(aq) + 2\ SO_4^{2-}(aq) \longrightarrow 2\ PbSO_4(s) + 2\ H_2O \qquad E_{cell} = +2.05\ V \qquad (21.24)$$

When an automobile engine is started, the battery is at first discharged. Once the car is in motion, the battery is constantly recharged by an alternator powered by the engine. At times, the plates of the battery become coated with $PbSO_4(s)$ and the electrolyte becomes sufficiently diluted with water that the battery must be recharged by connecting it to an external electric source. This forces the reverse of reaction (21.24), a nonspontaneous reaction.

$$2\ PbSO_4(s) + 2\ H_2O \longrightarrow Pb(s) + PbO_2(s) + 4\ H^+(aq) + 2\ SO_4^-(aq)$$

$$E_{cell} = -2.05\ V$$

▶ The half-reactions in the lead storage cell can be considered in two steps: (1) oxidation of $Pb(s)$ to $Pb^{2+}(aq)$ at the anode and reduction of $PbO_2(s)$ to $Pb^{2+}(aq)$ at the cathode, followed by (2) precipitation of $PbSO_4(s)$ at each electrode.

To prevent short circuiting, alternating anode and cathode plates are separated by sheets of insulating material. A group of anodes is connected together electrically, as is a group of cathodes. This "parallel" connection increases the electrode area in contact with the electrolyte solution and increases the current-delivering capacity of the cell. Cells are then joined in a "series" fashion, + to −, to produce a battery. The typical 12-V battery consists of six cells, each cell with a potential of about 2 V.

The Silver–Zinc Cell: A Button Battery

The cell diagram of a silver–zinc cell (Figure 21-13) is

$$Zn(s),\ ZnO(s)\ |\ KOH(satd)\ |\ Ag_2O(s),\ Ag(s)$$

The half-reactions on discharging are

Oxidation:	$Zn(s) + 2\ OH^-(aq) \longrightarrow ZnO(s) + H_2O + 2\ e^-$
Reduction:	$Ag_2O(s) + H_2O + 2\ e^- \longrightarrow 2\ Ag(s) + 2\ OH^-(aq)$

$$\text{Net:} \quad Zn(s) + Ag_2O(s) \longrightarrow ZnO(s) + 2\ Ag(s) \qquad (21.25)$$

Because no solution species is involved in the net reaction, the quantity of electrolyte is very small, and the electrodes can be maintained very close together. The cell voltage is 1.8 V, and its storage capacity is six times greater than that of a lead–acid battery of the same size. These characteristics make batteries such as the silver–zinc cell useful in "button" batteries. These miniature batteries are used in watches, hearing aids, and cameras.

Lithium ion batteries are a new source of lightweight power for electronic devices. The technology is complex, but these batteries use a $LiCoO_2$ cathode and an anode that contains lithium ions embedded within a graphite electrode. These batteries deliver a potential of about 3.6 V.

FIGURE 21-13
A silver–zinc button (miniature) cell.

Zn anode

metal case

Ag_2O paste

absorbent disc

separator

Fuel Cells

For more than 150 years, scientists have explored the possibility of converting the chemical energy of fuels directly to electricity. The essential process in a fuel cell is *fuel + oxygen → oxidation products.* Applied to natural gas (methane), the half-reactions and cell reaction are

Oxidation: $\quad CH_4(g) + 2\ H_2O(l) \longrightarrow CO_2(g) + 8\ H^+ + 8\ e^-$
Reduction: $\quad 2\ \{O_2(g) + 4\ H^+ + 4\ e^- \longrightarrow 2\ H_2O(l)\}$

Net: $\qquad\qquad CH_4(g) + 2\ O_2(g) \longrightarrow CO_2(g) + 2\ H_2O(l)$ (21.26)

$$\Delta H° = -890\ kJ \qquad \Delta G° = -818\ kJ$$

▶ Recall the meaning of free energy discussed in the Are You Wondering feature on page 699. The ratio $\Delta G°/\Delta H°$ represents the fraction of the liberated energy that is *free* or available to do work.

A fuel cell reaction is often rated in terms of the *efficiency value,* $\epsilon = \Delta G°/\Delta H°$. For the methane fuel cell, $\epsilon = -818\ kJ/-890\ kJ = 0.92$.

The methane fuel cell and others based on the oxidation of a hydrocarbon have been used on a research basis, but they have not yet been commercialized because of technical difficulties. A present alternative is to produce $H_2(g)$ by the reforming of a hydrocarbon (page 262) and to use the $H_2(g)$ in an H_2–O_2 fuel cell (Figure 21-14). In this cell the reaction of $H_2(g)$ and $O_2(g)$ in an alkaline medium (for example, 25% KOH) produces $H_2O(l)$.

Oxidation: $\quad 2\ \{H_2(g) + 2\ OH^-(aq) \longrightarrow 2\ H_2O(l) + 2\ e^-\}$
Reduction: $\quad O_2(g) + 2\ H_2O(l) + 4\ e^- \longrightarrow 4\ OH^-(aq)$

Net: $\qquad\qquad 2\ H_2(g) + O_2(g) \longrightarrow 2\ H_2O(l)$ (21.27)

We should refer to a fuel cell as an *energy converter* rather than a battery. As long as fuel and $O_2(g)$ are available, the cell will produce electricity. It does not have the limited capacity of a primary battery, but neither does it have the storage capacity of a secondary battery. Fuel cells based on reaction (21.27) have had their most notable successes as energy sources in space vehicles. (Water produced in this reaction is also a valuable product of the fuel cell.)

Air Batteries

In a fuel cell $O_2(g)$ is the oxidizing agent that oxidizes a fuel such as $H_2(g)$ or $CH_4(g)$. Another kind of flow battery, because it uses $O_2(g)$ from air, is known as an air battery. The substance that is oxidized is typically a metal.

One heavily studied battery system is the aluminum–air battery. In this battery, oxidation occurs at an aluminum anode and reduction at a carbon–air cathode. The electrolyte circulated through the battery is $NaOH(aq)$. Because it is in the presence of a high concentration of OH^-, Al^{3+} produced at the anode forms the complex ion $[Al(OH)_4]^-$. The operation of the battery is suggested by Figure 21-15. The half-reactions and the net cell reaction are

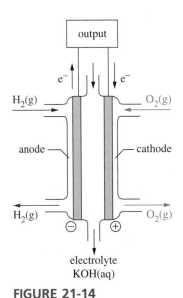

FIGURE 21-14
Schematic representation of a hydrogen–oxygen fuel cell.
A key requirement in fuel cells is porous electrodes that allow for easy access of the gaseous reactants to the electrolyte. The electrodes chosen should also catalyze the electrode reactions.

FIGURE 21-15
Simplified representation of an aluminum–air battery.

Oxidation: $\quad\quad\quad\quad$ 4 {Al(s) + 4 OH$^-$(aq) $\longrightarrow$ [Al(OH)$_4$]$^-$(aq) + 3 e$^-$}
Reduction: $\quad\quad\quad\quad\quad$ 3 {O$_2$(g) + 2 H$_2$O + 4 e$^-$ $\longrightarrow$ 4 OH$^-$(aq)}

Net: $\quad\quad$ 4 Al(s) + 3 O$_2$(g) + 6 H$_2$O + 4 OH$^-$(aq) $\longrightarrow$ 4 [Al(OH)$_4$]$^-$(aq) $\quad\quad\quad$ (21.28)

The battery is kept charged by feeding chunks of Al and water into it. A typical battery can power an automobile several hundred miles before refueling is necessary. The electrolyte is circulated outside the battery, where Al(OH)$_3$(s) is precipitated from the [Al(OH)$_4$]$^-$(aq). This Al(OH)$_3$(s) is collected and can then be converted back to aluminum metal at an aluminum manufacturing facility.

21-6   Corrosion: Unwanted Voltaic Cells

The reactions occurring in voltaic cells (batteries) are important sources of electricity. Similar reactions underlie corrosion processes, but here they are unwanted. First, we consider the electrochemical basis of corrosion, and then we see how electrochemical principles can also be applied to control corrosion.

Figure 21-16(a) demonstrates the basic processes in the corrosion of an iron nail. The nail is embedded in a gel of agar in water. Incorporated in the gel are the acid–base indicator phenolphthalein and the substance K$_3$[Fe(CN)$_6$] (potassium ferricyanide). Here is what we see within hours of starting the experiment: At the head and tip of the nail a deep blue precipitate forms. Along the body of the nail the agar gel turns pink. The blue precipitate, Turnbull's blue, establishes the presence of iron(II). The pink color is that of phenolphthalein in *basic* solution. From these observations we write two simple half-equations.

Oxidation: $\quad\quad\quad\quad\quad$ 2 Fe(s) $\longrightarrow$ 2 Fe^{2+}(aq) + 4 e$^-$
Reduction: $\quad\quad$ O$_2$(g) + 2 H$_2$O(l) + 4 e$^-$ $\longrightarrow$ 4 OH$^-$(aq)

The sum of the electrode potentials is $E_{cell} = (0.440 + 0.401)$ V = 0.841 V, indicating that the corrosion process should be spontaneous when reactants and products are in their standard states. Typically, the corrosion medium has [OH$^-$] $\ll$ 1 M, the reduction half-reaction is even more favorable, and E_{cell} is even greater than 0.841 V. Corrosion is especially significant in acidic solutions, where the reduction half-reaction is

$$O_2(g) + 4\ H^+(aq) + 4\ e^- \longrightarrow 2\ H_2O(l) \quad\quad E_{O_2/H_2O} = 1.229\ \text{V}$$

▶ We run into voltaic cells in a number of unlikely places, as described in Feature Problems C, D, and E on page 763.

(a)

(b)

(c)

(d)

FIGURE 21-16
Demonstration of corrosion and methods of corrosion protection.
The pink and dark blue colors are described in the text. Corrosion (oxidation) of the nail occurs at strained regions: the head and tip (a) and a bend (b) in the nail. Contact with zinc (c) protects the nail from corrosion. Zinc is oxidized instead of the iron (forming the faint white precipitate of zinc ferricyanide). Copper (d) does not protect the nail from corrosion. Electrons lost in the oxidation half-reaction distribute themselves along the copper wire, as seen in the pink color that extends the full length of the wire.

In the corroding nail of Figure 21-16(a), oxidation occurs at the head and tip of the nail. Electrons given up in the oxidation pass along the nail and are used to reduce dissolved O_2. The reduction product, OH^-, is detected by the phenolphthalein. With a bent nail [see Figure 21-16(b)], oxidation occurs at *three* points: the head and tip of the nail and the bend. The nail is preferentially oxidized at these points because the strained metal is more active (more anodic) than the unstrained metal. This is similar to the preferential rusting of a dented automobile fender.

Some metals, such as aluminum, form corrosion products that adhere tightly to the underlying metal and protect it from further corrosion. Iron oxide (rust), on the other hand, flakes off and constantly exposes fresh surface. This difference in corrosion behavior explains why cans made of iron deteriorate rapidly in the environment, whereas aluminum cans have an almost unlimited lifetime. The simplest method of protecting a metal from corrosion is to cover it with paint or some other protective coating impervious to water, an important reactant and solvent in corrosion processes.

Another method of protecting an iron surface is to plate it with a thin layer of a second metal. Iron can be plated with copper by electroplating or with tin by dipping the iron into the molten metal. In either case the underlying metal is protected only as long as the coating remains intact. If the coating is cracked, as when a "tin" can is dented, the underlying iron is exposed and corrodes. Iron, being more active than copper and tin, undergoes oxidation; the reduction half-reaction occurs on the plating (see Figures 21-16d and 21-17b).

When iron is coated with zinc (galvanized iron) the situation is different. Zinc is more active than iron. If a break occurs in the zinc plating, the iron is still protected. Zinc is oxidized instead of the iron, and corrosion products protect the zinc from further corrosion (see Figures 21-16c and 21-17a).

Still another method is used to protect large iron and steel objects in contact with water or moist soils—ships, storage tanks, pipelines, plumbing systems. This involves connecting a chunk of magnesium, aluminum, zinc, or some other active metal to the object, either directly or through a wire. Oxidation occurs at the active metal and it slowly dissolves. The iron surface acquires electrons from the oxidation of the active metal; the iron acts as a cathode and supports a *reduction* half-reaction. As long as some of the active metal remains, the iron is protected. This type of protection is called **cathodic protection,** and the active metal is called, appropriately, a *sacrificial anode.* About 12 million pounds of magnesium is used annually in the United States in sacrificial anodes.

FIGURE 21-17
Protection of iron against electrolytic corrosion.
In the *anodic* reaction, the metal that is more easily oxidized loses electrons to produce metal ions. In case (a) this is zinc; in case (b) it is iron. In the *cathodic* reaction oxygen gas, which is dissolved in a thin film of water on the metal, is reduced to OH^-. Rusting of iron does not occur in (a), but it does in (b). Fe^{2+} and OH^- ions from the half-reactions initiate these further reactions.

$$Fe^{2+} + 2\,OH^- \longrightarrow Fe(OH)_2(s)$$
$$4\,Fe(OH)_2(s) + O_2 + 2\,H_2O \longrightarrow 4\,Fe(OH)_3(s)$$
$$2\,Fe(OH)_3(s) \longrightarrow Fe_2O_3{\cdot}H_2O + 2\,H_2O$$
rust

(a) Galvanized iron

(b) Copper-plated iron

Magnesium sacrificial anode. A large sheet of magnesium attached to the steel hull of a ship provides cathodic protection.

21-7   Electrolysis: Causing Nonspontaneous Reactions to Occur

Until now we have emphasized the type of electrochemical cell, the voltaic cell, in which chemical change is used to produce electricity. Another type of electrochemical cell, the *electrolytic* cell, uses electricity to cause a *nonspontaneous* reaction to take place.

Let's explore the relationship between the two types of electrochemical cells—voltaic and electrolytic—by returning briefly to the cell shown in Figure 21-4. When the cell functions *spontaneously*, electrons flow from the zinc to the copper and the net chemical change in the *voltaic* cell is

$$Zn(s) + Cu^{2+}(aq) \longrightarrow Zn^{2+}(aq) + Cu(s) \qquad E^{\circ}_{cell} = +1.100 \text{ V}$$

Now suppose we connect the same cell to an external electric source of voltage greater than 1.100 V as shown in Figure 21-18. That is, the connection is made so that electrons are forced into the zinc electrode (the cathode) and removed from the copper electrode (the anode). The net reaction in this case is the *reverse* of the voltaic cell reaction, and E°_{cell} is *negative*.

FIGURE 21-18
An electrolytic cell.
The direction of electron flow is the reverse of that in the voltaic cell of Figure 21-4, and so is the cell reaction. The zinc electrode is the *cathode* and the copper electrode, the *anode*. The battery must have a voltage in excess of 1.100 V in order to force electrons to flow in the reverse (nonspontaneous) direction.

Oxidation: $\qquad\qquad$ Cu(s) $\longrightarrow$ Cu^{2+}(aq) + 2 e$^-$ $\qquad$ $-E^\circ_{Cu^{2+}/Cu} = -0.337$ V

Reduction: $\quad$ Zn^{2+}(aq) + 2 e$^-$ $\longrightarrow$ Zn(s) $\qquad\qquad\qquad$ $E^\circ_{Zn^{2+}/Zn} = -0.763$ V

Net: $\qquad\qquad$ Cu(s) + Zn^{2+}(aq) $\longrightarrow$ Cu^{2+}(aq) + Zn(s) $\qquad\quad$ $E^\circ_{cell} = -1.100$ V

We conclude that when the cell reaction in a voltaic cell is reversed by reversing the direction of the electron flow, the *voltaic* cell is changed into an *electrolytic* cell.

Are You Wondering . . .

Why the anode in an electrolytic cell is (+) whereas it is (−) in a voltaic cell?

Assigning the terms anode and cathode is *not* based on the electrode charges; it is based on the half-reactions at the electrode surfaces. Specifically,

- *Oxidation* always occurs at the *anode* of an electrochemical cell. Because of the buildup of electrons *freed* in the oxidation half-reaction, the anode of a *voltaic* cell is (−). Because electrons are *withdrawn* from it, the anode in an *electrolytic* cell is (+).
- *Reduction* always occurs at the *cathode* of an electrochemical cell. Because of the *removal* of electrons by the reduction half-reaction, the cathode of a *voltaic* cell is (+). Because of the electrons *forced* onto it, the cathode of an *electrolytic* cell is (−).

Predicting Electrolysis Reactions

For the cell in Figure 21-18 to function as an electrolytic cell with reactants and products in their standard states, we predicted that the external voltage had to exceed 1.100 V. We can make similar calculations for other electrolyses. What actually happens, however, does not always correspond to these calculations. We need to consider three complicating factors.

First, a voltage significantly in excess of the calculated value, an **overpotential,** may be necessary to cause a particular electrode reaction to occur. Overpotentials are needed to overcome interactions at the electrode surface and are particularly common when gases are involved. For example, the overpotential for the discharge of H$_2$(g) at a mercury cathode is approximately 1.5 V. On the other hand, on a platinum cathode the overpotential is practically zero.

A second factor is that competing electrode reactions may occur. In the electrolysis of *molten* sodium chloride using inert electrodes, only one oxidation and one reduction are possible.

Oxidation: $\qquad\qquad$ 2 Cl$^-$ $\longrightarrow$ Cl$_2$(g) + 2 e$^-$

Reduction: $\quad$ 2 Na$^+$ + 2 e$^-$ $\longrightarrow$ 2 Na(l)

In the electrolysis of *aqueous* sodium chloride using inert electrodes, we must consider *two* oxidation and *two* reduction half-reactions.

Oxidation:	$2 Cl^-(aq) \longrightarrow Cl_2(g) + 2 e^-$	$-E^\circ_{Cl_2/Cl^-} = -1.36$ V	(21.29)
	$2 H_2O \longrightarrow O_2(g) + 4 H^+(aq) + 4 e^-$	$-E^\circ_{O_2/H_2O} = -1.23$ V	(21.30)
Reduction:	$2 Na^+(aq) + 2 e^- \longrightarrow 2 Na(s)$	$E^\circ_{Na^+/Na} = -2.71$ V	(21.31)
	$2 H_2O + 2 e^- \longrightarrow H_2(g) + 2 OH^-(aq)$	$E^\circ_{H_2O/H_2} = -0.83$ V	(21.32)

Because the potential for oxidation half-reaction (21.30) is somewhat less negative than that of (21.29), we might expect $O_2(g)$ as the product at the anode. Actually $Cl_2(g)$ is the predominant product because of the high overpotential of $O_2(g)$ compared with that of $Cl_2(g)$. In the industrial electrolysis of concentrated aqueous NaCl (about 5.5 M NaCl), the $Cl_2(g)$ formed usually contains less than 1% $O_2(g)$. On the other hand, in the electrolysis of very dilute NaCl(aq), $O_2(g)$ is a more important product.

As expected, the reduction half-reaction in the electrolysis of NaCl(aq) is the reduction of H_2O—half-reaction (21.32). The chief exception is if liquid mercury is used as the cathode. Here, because of the high overpotential of $H_2(g)$ on mercury and the fact that sodium is soluble in liquid mercury, half-reaction (21.31) does occur.

The third factor that we must consider when predicting electrolysis reactions concerns the nature of the electrodes. An *inert* electrode, such as platinum, provides a surface on which an electrolysis half-reaction occurs, but the reactants themselves must come from the electrolyte solution. An *active* electrode is one that can itself participate in the oxidation or reduction half-reaction. The distinction between inert and active electrodes is explored in Figure 21-19 and Example 21-11.

▶ In aqueous solutions, when no other electrode reactions are feasible, the combination of half-reactions (21.30) and (21.32) results in the electrolysis of water: $2 H_2O(l) \rightarrow 2 H_2(g) + O_2(g)$.

▶ The difficulty in determining E° values for the alkali metals is the subject of Feature Problem F. The use of a sodium/mercury amalgam electrode in establishing $E^\circ_{Na^+/Na}$ is the subject of Feature Problem G (see page 763).

EXAMPLE 21-11

Predicting Electrode Half-reactions and Net Electrolysis Reactions. Refer to Figure 21-19. Predict the electrode reactions and the net reaction when the anode is made of (a) copper and (b) platinum.

Solution

In both cases the reduction of $Cu^{2+}(aq)$ seems the likely reduction process.

$$Cu^{2+}(aq) + 2 e^- \longrightarrow Cu(s) \qquad E^\circ_{Cu^{2+}/Cu} = +0.337 \text{ V}$$

a. At the anode Cu(s) can be oxidized to $Cu^{2+}(aq)$, as represented by

Oxidation: $\quad Cu(s) \longrightarrow Cu^{2+}(aq) + 2 e^- \qquad -E^\circ_{Cu^{2+}/Cu} = -0.337$ V

If we add the oxidation and reduction half-equations we see that $Cu^{2+}(aq)$ cancels out. The net electrolysis reaction is simply

$$Cu(s)[anode] \longrightarrow Cu(s)[cathode] \qquad E^\circ_{cell} = 0 \qquad (21.33)$$

Only a very slight voltage is required for this electrolysis. For every Cu atom that enters the solution at the anode, an *active* electrode, a Cu^{2+} ion deposits as a Cu atom at the cathode. Copper is transferred from the anode to the cathode through the solution as Cu^{2+}, and the concentration of $CuSO_4(aq)$ is *unchanged*.

b. A platinum electrode is *inert*. It is not easily oxidized. The oxidation of SO_4^{2-} to $S_2O_8^{2-}$ also is not feasible ($-E^\circ_{S_2O_8^{2-}/SO_4^{2-}} = -2.01$ V). The oxidation that occurs most readily is that of H_2O, shown in (21.30).

Oxidation: $\quad 2 H_2O \longrightarrow O_2(g) + 4 H^+(aq) + 4 e^-$

$$-E^\circ_{O_2/H_2O} = -1.23 \text{ V}$$

battery

anode (+)   cathode (−)

FIGURE 21-19
Predicting electrode reactions in electrolysis—Example 21-11 illustrated.

Electrons are forced onto the copper cathode by the external source (battery). Cu^{2+} ions are attracted to the cathode and are reduced to Cu(s). The oxidation half-reaction depends on the metal used for the anode.

Michael Faraday (1791–1867). Faraday, an assistant to Humphry Davy and often called "Davy's greatest discovery," made many contributions to both physics and chemistry, including systematic studies of electrolysis.

The net electrolysis reaction and its E°_{cell} are

$$2\,Cu^{2+}(aq) + 2\,H_2O \longrightarrow 2\,Cu(s) + 4\,H^+(aq) + O_2(g)$$

$$E^\circ_{cell} = -0.89\ V \qquad (21.34)$$

Practice Example A: Use data from Table 21.1 to predict the probable products when Pt electrodes are used in the electrolysis of KI(aq).

Practice Example B: In the electrolysis of $AgNO_3$(aq), what are the expected electrolysis products if the anode is silver metal and the cathode is platinum?

Quantitative Aspects of Electrolysis

We have seen how to calculate the theoretical voltage required to produce electrolysis. Equally important are calculations of the quantities of reactants consumed and products formed in an electrolysis. For these calculations we will continue to use stoichiometric factors from the chemical equation, but another factor enters in as well—the quantity of electric charge associated with one mole of electrons. This factor is provided by the Faraday constant, which we can write as

$$1\ mol\ e^- = 96{,}485\ C$$

Generally, we do not measure electric charge directly; we measure electric current. One ampere (A) of electric current represents the passage of 1 coulomb of charge per second (C/s). The product of current and time yields the total quantity of charge transferred.

$$charge\ (C) = current\ (C/s) \times time\ (s)$$

To determine the number of moles of electrons involved in an electrolysis reaction, we can write

$$number\ of\ mol\ e^- = current\left(\frac{C}{s}\right) \times time\ (s) \times \frac{1\ mol\ e^-}{96{,}485\ C}$$

As illustrated in Example 21-12, to determine the mass of a product in an electrolysis reaction, we can follow this conversion pathway.

$$C/s \longrightarrow C \longrightarrow mol\ e^- \longrightarrow mol\ product \longrightarrow g\ product$$

EXAMPLE 21-12

Calculating Quantities Associated with Electrolysis Reactions. The electrodeposition of copper can be used to determine the copper content of a sample. The sample is dissolved to produce Cu^{2+}(aq), which is electrolyzed. At the cathode the reduction half-reaction is $Cu^{2+}(aq) + 2\,e^- \rightarrow Cu(s)$. What mass of copper can be deposited in 1.00 hour by a current of 1.62 A?

Solution

We can think of this as a two-part calculation. First we can determine the number of moles of electrons involved in the electrolysis in the manner outlined earlier.

$$1.00\ h \times \frac{60\ min}{1\ h} \times \frac{60\ s}{1\ min} \times \frac{1.62\ C}{1\ s} \times \frac{1\ mol\ e^-}{96{,}485\ C} = 0.0604\ mol\ e^-$$

Then we can calculate the mass of Cu(s) produced at the cathode by this number of moles of electrons. The key factor is printed in blue.

$$\text{mass of Cu} = 0.0604 \text{ mol e}^- \times \frac{1 \text{ mol Cu}}{2 \text{ mol e}^-} \times \frac{63.55 \text{ g Cu}}{1 \text{ mol Cu}} = 1.92 \text{ g Cu}$$

Practice Example A: If 12.3 g of Cu is deposited at the cathode of an electrolysis cell after 5.50 h, what was the current used?

Practice Example B: For how long would the electrolysis in Example 21-12 have to be carried out, using Pt electrodes and a current of 2.13 A, to produce 2.62 L O_2(g) at 26.2 °C and 738 mmHg pressure at the anode?

The Chlor-Alkali Process

On page 751 we described the electrolysis of NaCl(aq) through the oxidation half-reaction (21.29) and the reduction half-reaction (21.32).

$$2 \text{ Cl}^-(\text{aq}) + 2 \text{ H}_2\text{O} \longrightarrow 2 \text{ OH}^-(\text{aq}) + \text{H}_2(\text{g}) + \text{Cl}_2(\text{g}) \qquad E^\circ_{\text{cell}} = -2.19 \text{ V}$$

When conducted on an industrial scale, this electrolysis is called the *chlor-alkali* process, named after the two principal products: *chlor*ine and the *alkali* NaOH(aq).

In the *diaphragm cell* pictured in Figure 21-20, Cl_2(g) is produced in the anode compartment and H_2(g) and NaOH(aq) in the cathode compartment. If Cl_2(g) comes into contact with NaOH(aq), the Cl_2 disproportionates into ClO^-(aq), ClO_3^-(aq), and Cl^-(aq). The purpose of the diaphragm is to prevent this contact. The NaCl(aq) in the anode compartment is kept at a slightly higher level than in the cathode compartment. This creates a gradual flow of NaCl(aq) between the compartments and reduces the back flow of NaOH(aq) into the anode compartment. The solution in the cathode compartment, about 10–12% NaOH(aq) and 14–16% NaCl(aq), is concentrated and purified by evaporating some water and crystallizing the NaCl(s). The final product is 50% NaOH(aq) with up to 1% NaCl(aq).

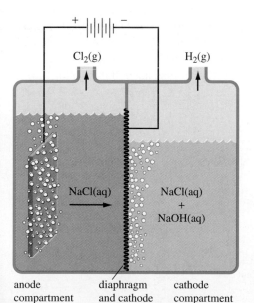

anode   diaphragm   cathode
compartment and cathode compartment

FIGURE 21-20
A diaphragm chlor-alkali cell.
The anode may be made of graphite or, in more modern technology, specially treated titanium metal. The diaphragm and cathode are generally fabricated as a composite unit consisting of asbestos or an asbestos–polymer mixture deposited on a steel wire mesh.

Focus On

Industrial Electrolysis Processes

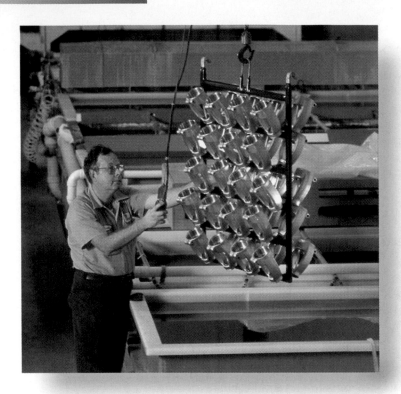

A rack of metal parts being lifted from the electrolyte solution following electroplating.

We are not overstating the case to say that modern industry (and hence modern society) could not function without electrolysis reactions. A number of the elements are produced almost exclusively by electrolysis—for example, aluminum, magnesium, and fluorine. Among the chemical compounds produced by electrolysis are NaOH, $K_2Cr_2O_7$, $KMnO_4$, and $Na_2S_2O_8$. The most important of all commercial electrolysis processes, as we have seen, produces three useful products: NaOH, H_2, and Cl_2. The *chlor-alkali* process consumes about 0.5% of all the electric power generated in the United States.

Electrorefining of metals involves the deposition of pure metal at a cathode, from a solution containing the metal ion. Electrolysis reaction (21.33) on page 751 is the one used commercially to obtain

The theoretical voltage required for the electrolysis is 2.19 V. Actually, because of the internal resistance of the cell and overpotentials at the electrodes, a voltage of about 3.5 V is used. The current is kept very high, typically about 1×10^5 A.

The NaOH(aq) produced in a diaphragm cell is not pure enough for certain uses, such as in rayon manufacture. A higher purity is achieved if electrolysis is carried out in a mercury cell, illustrated in Figure 21-21. This cell takes advantage of the high overpotential for the reduction of H_2O to $H_2(g)$ and $OH^-(aq)$ at a mercury cathode. The reduction that occurs instead is that of $Na^+(aq)$ to

FIGURE 21-21
The mercury-cell chlor-alkali process.
The cathode is a layer of Hg(l) that flows along the bottom of the tank. Anodes, at which $Cl_2(g)$ forms, are situated in the NaCl(aq) just above the Hg(l). Sodium formed at the cathode dissolves in the Hg(l), and the sodium amalgam is decomposed with water to produce NaOH(aq) and $H_2(g)$. The regenerated Hg(l) is recycled.

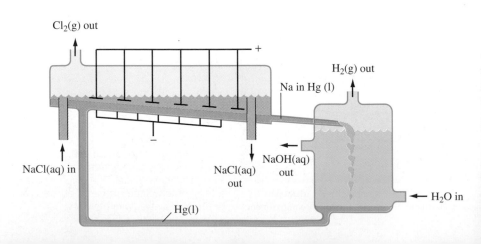

high-purity copper. A chunk of impure copper is taken as the *anode* and a thin sheet of pure copper metal as the cathode. During the electrolysis, copper, as Cu^{2+}, is transported continuously from the anode to the cathode. The pure copper cathode increases in size as the impure chunk of copper is consumed. Gold and silver are commonly found as impurities in copper. These metals are much less active than copper (they are much more difficult to oxidize) and deposit at the bottom of the electrolysis tank as a sludge called *anode mud.* Recovery of Ag and Au from the anode mud offsets the cost of the electrolysis.

In a process similar to electrorefining, *electroplating,* one metal is plated onto another, often less expensive, metal for decorative purposes or to protect the underlying metal from corrosion. Silver-plated flatware, for example, consists of a thin coating of metallic silver on an underlying base of iron. In electroplating, the item to be plated is made the cathode in an electrolytic cell. The electrolyte contains ions of the metal to be plated, which are attracted to the cathode where they are reduced to metal atoms. In the case of copper plating, the electrolyte is usually copper sulfate. In silver plating it is often $K[Ag(CN)_2](aq)$. The concentration of free silver ion in a solution of the complex ion $[Ag(CN)_2]^-(aq)$ is very low, and electroplating under these conditions promotes a strongly adherent microcrystalline deposit of the metal.

Chrome plating is useful for its resistance to corrosion as well as its decorative beauty. Steel can be chrome-plated from an aqueous solution of CrO_3 and H_2SO_4. The plating obtained, however, is thin and porous and tends to develop cracks. In practice, the steel is first plated with a thin coat of copper or nickel, and then the chrome plating is applied. Machine parts can be weatherproofed by chrome or cadmium plating. Plating can even be applied to some plastics. The plastic must first be made electrically conducting, for example, by coating it with graphite powder. The process of electroplating copper to plastic has been important in improving the quality of some microelectronic circuit boards. Electroplating is even used, quite literally, to make money. The current U.S. penny is no longer copper throughout. It is made by electroplating a thin coat of copper onto a zinc plug. The copper-plated plug is then stamped to create a penny.

Na, which dissolves in Hg(l) to form an amalgam (Na–Hg alloy) with about 0.5% Na by mass.

$$2 Na^+(aq) + 2 Cl^-(aq) \longrightarrow 2 Na(in\ Hg) + Cl_2(g) \qquad E^\circ_{cell} = -3.20\ V$$

When the Na amalgam is removed from the cell and treated with water, NaOH(aq) is formed,

$$2 Na(in\ Hg) + 2 H_2O \longrightarrow 2 Na^+(aq) + 2 OH^-(aq) + H_2(g) + Hg(l)$$

and the liquid mercury is recycled back to the electrolysis cell.

Although the mercury cell has the advantage of producing concentrated high-purity NaOH(aq), it has some disadvantages. The mercury cell requires a higher voltage (about 4.5 V) than the diaphragm cell (3.5 V) and consumes more electric energy, about 3400 kWh/ton Cl_2 in a mercury cell compared with 2500 in a diaphragm cell. Another serious drawback is the need to control mercury effluents to the environment. Mercury losses, which at one time were as high as 200 g Hg per ton Cl_2, have been reduced to about 0.25 g Hg per ton of Cl_2 in existing plants and half this amount in new plants.

The ideal chlor-alkali process is one that is energy-efficient and that does not use mercury. A type of cell offering these advantages is the *membrane cell.* Here the porous diaphragm of Figure 21-20 is replaced by a cation-exchange membrane, usually made of a fluorocarbon polymer. The membrane permits hydrated cations (Na^+ and H_3O^+) to pass between the anode and cathode compartments but severely restricts the back flow of Cl^- and OH^- ions.

Summary

An electrochemical cell consists of two half-cells in which electrodes are joined by a wire and the solutions are in contact, as through a salt bridge. The cell reaction involves oxidation at one electrode (anode) and reduction at the other (cathode). A voltaic cell produces electricity from a spontaneous oxidation–reduction reaction. The difference in electric potential between the two electrodes is the voltage of the cell.

The reduction occurring at a standard hydrogen electrode (SHE): $2 H^+(1 M) + 2 e^- \rightarrow H_2(g, 1 atm)$ is arbitrarily assigned a potential of *zero*. A half-cell reaction with a *positive* reduction potential ($E°$) occurs more readily than does reduction of H^+ ions at the SHE. A *negative* standard reduction potential signifies a lesser tendency to undergo reduction. If a half-cell reaction is written as an oxidation, the negative of the standard reduction potential ($-E°$) is used. The voltage of a voltaic cell is the sum of $-E$ for the oxidation and $+E$ for the reduction. If E_{cell} is > 0 the cell reaction is spontaneous in the forward direction for the stated conditions; if $E_{cell} < 0$ the reaction is *nonspontaneous*.

Cell voltages based on *standard* electrode potentials are $E°_{cell}$ values. Important relationships exist between $E°_{cell}$ and $\Delta G°$ and between $E°_{cell}$ and K_{eq}. In the Nernst equation, E_{cell} for *nonstandard* conditions is related to $E°_{cell}$ and the reaction quotient Q. Important applications of voltaic cells are found in various battery systems and in the phenomenon of corrosion and its control.

In an *electrolytic* cell an external source of electricity forces electrons to flow in a direction opposite to that in which they would flow spontaneously. As a result, electrolysis causes a normally *nonspontaneous* reaction to occur. $E°$ values are used to establish the voltage requirements, and the Faraday constant is used in calculating the amounts of reactants and products involved in an electrolysis. Electrolysis is used in the commercial production of many substances.

Summarizing Example

Current fuel cells use the reaction of $H_2(g)$ and $O_2(g)$ to form $H_2O(l)$ (recall page 746). Often, the $H_2(g)$ is obtained by the steam reforming of a hydrocarbon, such as $C_3H_8(g) + 3 H_2O(g) \rightarrow 3 CO(g) + 7 H_2(g)$. A future possibility is a fuel cell that converts a hydrocarbon such as propane directly to $CO_2(g)$ and $H_2O(l)$.

$$C_3H_8(g) + 5 O_2(g) \longrightarrow 3 CO_2(g) + 4 H_2O(l) \qquad (21.35)$$

Based on reaction (21.35), use data from Table 21.1 and Appendix D to determine $E°$ for the reduction of $CO_2(g)$ to $C_3H_8(g)$ in acidic solution.

1. *Express reaction (21.35) as two half-reactions.* Start with the skeleton half-equations $C_3H_8 \rightarrow 3 CO_2$ and $O_2 \rightarrow 2 H_2O$. Complete and balance them by the half-reaction method of Chapter 5. *Result:*

Oxidation: $C_3H_8 + 6 H_2O \longrightarrow 3 CO_2 + 20 H^+ + 20 e^-$
Reduction: $O_2 + 4 H^+ + 4 e^- \longrightarrow 2 H_2O$

2. *Determine $\Delta G°$ for reaction (21.35).* Use the expression $\Delta G° = \Sigma v_p \Delta G_f°(products) - \Sigma v_r \Delta G_f°(reactants)$. *Result:* $\Delta G° = -2108 kJ$.

3. *Evaluate $E°_{cell}$ for reaction (21.35).* Use the expression $\Delta G° = -n\mathscr{F}E°_{cell}$ with the value of $\Delta G°$ from part **2** and a value of n from part **1**. *Result:* $E°_{cell} = 1.092 V$.

4. *Obtain $E°$ for the reduction of CO_2 to C_3H_8.* For the oxidation in part **1** the standard potential is $-E°$. Look up the potential for the reduction half-reaction in Table 21.1. The sum of the two potentials is $E°_{cell}$ from part **3**. Solve for $E°$. *Answer:* $E° = 0.137 V$.

Key Terms

anode (21-1)	**electrode potential** (21-1)	**primary battery** (21-5)
battery (21-5)	**electrolytic cell** (21-1)	**salt bridge** (21-1)
cathode (21-1)	**electromotive force (emf)** (21-1)	**secondary battery** (21-5)
cathodic protection (21-6)	**Faraday constant, $\mathscr{F}$** (21-3)	**standard cell potential, $E°_{cell}$** (21-2)
cell diagram (21-1)	**flow battery** (21-5)	**standard electrode potential, $E°$**
cell potential, E_{cell} (21-1)	**fuel cell** (21-5)	(21-2)
concentration cell (21-4)	**half-cell** (21-1)	**standard hydrogen electrode (SHE)**
electrochemical cell (21-1)	**Nernst equation** (21-4)	(21-2)
electrode (21-1)	**overpotential** (21-7)	**voltaic (galvanic) cell** (21-1)

Review Questions

1. In your own words define the following symbols or terms: **(a)** $E°$; **(b)** $\mathscr{F}$; **(c)** anode; **(d)** cathode.
2. Briefly describe each of the following ideas, methods, or devices: **(a)** salt bridge; **(b)** standard hydrogen electrode

(SHE); **(c)** cathodic protection; **(d)** fuel cell.
3. Explain the important distinctions between each pair of terms: **(a)** half-reaction and net reaction; **(b)** voltaic and electrolytic cell; **(c)** primary and secondary battery; **(d)** E_{cell} and $E°_{cell}$.

4. Only one of the following statements is correct concerning a *voltaic* cell. Identify the correct statement and tell what is wrong with the other statements. **(a)** Electrons move from the cathode to the anode; **(b)** electrons move through the salt bridge; **(c)** electrons leave the cell from either the cathode or the anode, depending on what electrodes are used; **(d)** reduction occurs at the cathode.

5. Write equations for the oxidation and reduction half-reactions and the net reaction describing
(a) the displacement of $Cu^{2+}(aq)$ by $Fe(s)$;
(b) the oxidation of Br^- to $Br_2(aq)$ by $Cl_2(aq)$;
(c) the reduction of $Fe^{3+}(aq)$ to $Fe^{2+}(aq)$ by $Al(s)$;
(d) the oxidation of $Cl^-(aq)$ to $ClO_3^-(aq)$ by MnO_4^- in acidic solution (MnO_4^- is reduced to Mn^{2+});
(e) the oxidation of $S^{2-}(aq)$ to $SO_4^{2-}(aq)$ by $O_2(g)$ in basic solution.

6. The sketch is of a voltaic cell consisting of a standard zinc electrode and a standard electrode of a second metal, M. For the reduction half-reaction $Zn^{2+}(1\ M) + 2\ e^- \rightarrow Zn(s)$, $E° = -0.763\ V$. For the given metals M and measured values of $E°_{cell}$, determine $E°$ for the reduction half-reaction
$$M^{2+}(1\ M) + 2\ e^- \longrightarrow M(s) \qquad E° = ?$$
(a) Mn, $E°_{cell} = +0.417\ V$
(b) Po, $E°_{cell} = -1.13\ V$
(c) Ti, $E°_{cell} = +0.87\ V$
(d) V, $E°_{cell} = +0.37\ V$

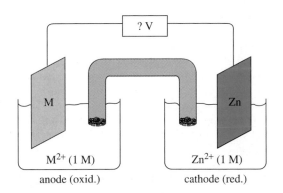

? V

M

Zn

M^{2+} (1 M)

Zn^{2+} (1 M)

anode (oxid.)   cathode (red.)

7. Write cell reactions for the electrochemical cells diagrammed here, and use data from Table 21.1 to calculate $E°_{cell}$ for each reaction.
(a) $Al(s)\,|\,Al^{3+}(aq)\,\|\,Sn^{2+}(aq)\,|\,Sn(s)$
(b) $Pt(s)\,|\,Fe^{2+}(aq),\ Fe^{3+}(aq)\,\|\,Ag^+(aq)\,|\,Ag(s)$

8. From the data listed, together with data from Table 21.1, determine the quantity indicated.
(a) $E°_{cell}$ for the cell
$$Pt(s)\,|\,Cl_2(g)\,|\,Cl^-(aq)\,\|\,Pb^{2+}(aq),\ H^+(aq)\,|\,PbO_2(s)$$
(b) $E°$ for the couple, Sc^{3+}/Sc, given that
$$Mg(s)\,|\,Mg^{2+}(aq)\,\|\,Sc^{3+}(aq)\,|\,Sc(s)$$
$$E°_{cell} = +0.33\ V$$
(c) $E°$ for the couple, Cu^{2+}/Cu^+, given that
$$Pt(s)\,|\,Cu^+(aq),\ Cu^{2+}(aq)\,\|\,Ag^+(aq)\,|\,Ag(s)$$
$$E°_{cell} = +0.641\ V$$
9. Assume that all reactants and products are in their standard states and use data from Table 21.1 to predict whether a spontaneous reaction will occur in the forward direction in each case.

(a) $Sn(s) + Pb^{2+}(aq) \rightarrow Sn^{2+}(aq) + Pb(s)$
(b) $Cu^{2+}(aq) + 2\ I^-(aq) \rightarrow Cu(s) + I_2(s)$
(c) $4\ NO_3^-(aq) + 4\ H^+(aq) \rightarrow 3\ O_2(g) + 4\ NO(g) + 2\ H_2O$
(d) $O_3(g) + Cl^-(aq) \rightarrow OCl^-(aq) + O_2(g)$ (basic solution)
10. For the reduction half-reaction $Hg_2^{2+}(aq) + 2\ e^- \rightarrow 2\ Hg(l)$, $E° = 0.797\ V$. Will $Hg(l)$ react with and dissolve in $HCl(aq)$? in $HNO_3(aq)$? Explain.
11. Use data from Table 21.1 to predict whether to any significant extent
(a) $Mg(s)$ will displace Pb^{2+} from aqueous solution;
(b) tin metal will react with and dissolve in 1 M HCl;
(c) SO_4^{2-} will oxidize Sn^{2+} to Sn^{4+} in acidic solution;
(d) $MnO_4^-(aq)$ will oxidize $H_2O_2(aq)$ to $O_2(g)$ in acidic solution;
(e) I_2 will displace $Br^-(aq)$ to produce $Br_2(l)$.
12. For the reaction $Co(s) + Ni^{2+}(aq) \rightarrow Co^{2+}(aq) + Ni(s)$, $E°_{cell} = +0.02\ V$, if $Co(s)$ is added to a solution with $[Ni^{2+}] = 1\ M$, would you expect the reaction to go to completion? Explain.
13. Dichromate ion ($Cr_2O_7^{2-}$) in acidic solution is a good oxidizing agent. Which of the following oxidations can be accomplished with dichromate ion in acidic solution? Explain.
(a) $Sn^{2+}(aq)$ to $Sn^{4+}(aq)$
(b) $I_2(s)$ to $IO_3^-(aq)$
(c) $Mn^{2+}(aq)$ to $MnO_4^-(aq)$
14. Determine the values of $\Delta G°$ for the following reactions carried out in voltaic cells.
(a) $2\ Al(s) + 3\ Cu^{2+}(aq) \rightarrow 2\ Al^{3+}(aq) + 3\ Cu(s)$
(b) $O_2(g) + 4\ I^-(aq) + 4\ H^+(aq) \rightarrow 2\ H_2O + 2\ I_2(s)$
(c) $Cr_2O_7^{2-}(aq) + 14\ H^+(aq) + 6\ Ag(s) \rightarrow$
$$2\ Cr^{3+}(aq) + 6\ Ag^+(aq) + 7\ H_2O$$
15. Write the equilibrium constant expression for each of the following reactions and determine the numerical value of K_{eq} at 25 °C. Use data from Table 21.1.
(a) $Sn^{4+}(aq) + 2\ Ag(s) \rightarrow Sn^{2+}(aq) + 2\ Ag^+(aq)$
(b) $MnO_2(s) + 4\ H^+(aq) + 2\ Cl^-(aq) \rightarrow$
$$Mn^{2+}(aq) + 2\ H_2O + Cl_2(g)$$
(c) $2\ OCl^-(aq) \rightarrow 2\ Cl^-(aq) + O_2(g)$ (basic solution)
16. For each of the following combinations of electrodes (A and B) and solutions, indicate

- the net cell reaction;

- the direction in which electrons flow spontaneously (from A to B or from B to A);

- the magnitude of the voltage read on the voltmeter, V.

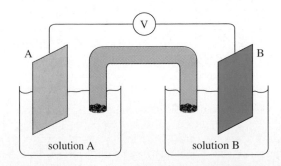

V

A

B

solution A   solution B

A	Solution A	B	Solution B
(a) Cu	1.0 M Cu^{2+}	Fe	1.0 M Fe^{2+}
(b) Pt	1.0 M Sn^{2+}/1.0 M Sn^{4+}	Ag	1.0 M Ag^+
(c) Zn	0.10 M Zn^{2+}	Fe	1.0×10^{-3} M Fe^{2+}

17. A voltaic cell represented by the following cell diagram has $E_{cell} = 1.250$ V. What must be $[Ag^+]$ in the cell?

$$Zn(s)\,|\,Zn^{2+}(1.00\ M)\,\|\,Ag^+(x\ M)\,|\,Ag(s)$$

18. For the cell pictured in Figure 21-9, what is E_{cell} if the unknown solution in the half-cell on the left **(a)** has pH = 5.25; **(b)** is 0.0103 M HCl; **(c)** is 0.158 M $HC_2H_3O_2$ ($K_a = 1.8 \times 10^{-5}$).

19. Use data from Table 21.1, as necessary, to predict the probable products when Pt electrodes are used in the electrolysis of **(a)** $CuCl_2(aq)$; **(b)** $Na_2SO_4(aq)$; **(c)** $BaCl_2(l)$; **(d)** KOH(aq).

20. The commercial production of magnesium involves the electrolysis of molten $MgCl_2$. Why not use the electrolysis of $MgCl_2(aq)$ instead?

21. How many grams of metal are deposited by the passage of 2.15 A of current for 75 min in the electrolysis of an aqueous solution containing **(a)** Zn^{2+}; **(b)** Al^{3+}; **(c)** Ag^+; **(d)** Ni^{2+}.

22. A quantity of electric charge brings about the deposition of 3.28 g Cu at a cathode during the electrolysis of a solution containing $Cu^{2+}(aq)$. What volume of $H_2(g)$, measured at 28.2 °C and 763 mmHg, would be produced by this same quantity of electric charge in the reduction of $H^+(aq)$ at a cathode?

Exercises

(Use data from Table 21.1 and Appendix D, as necessary.)

Standard Electrode Potentials

23. From the observations listed, estimate the value of $E°$ for the half-reaction $M^{2+}(aq) + 2\ e^- \rightarrow M(s)$
(a) The metal M reacts with $HNO_3(aq)$ but not with HCl(aq); it displaces $Ag^+(aq)$ but not $Cu^{2+}(aq)$.
(b) The metal M reacts with HCl(aq) producing $H_2(g)$, but displaces neither $Zn^{2+}(aq)$ nor $Fe^{2+}(aq)$.

25. $E_{cell}° = 0.201$ V for the reaction
$$3\ Pt + 12\ Cl^- + 2\ NO_3^- + 8\ H^+ \longrightarrow$$
$$3\ [PtCl_4]^{2-} + 2\ NO(g) + 4\ H_2O$$
What is $E°$ for the reduction of $[PtCl_4]^{2-}$ to Pt in acidic solution?

27. Given that $E_{cell}°$ for the aluminum–air battery is 2.71 V, what is $E°$ for the reduction half-reaction $[Al(OH)_4]^- + 3\ e^- \rightarrow Al(s) + 4\ OH^-(aq)$? [*Hint:* Refer to the net cell reaction (21.28).]

24. You are given the task of estimating $E°$ for the half-reaction $In^{3+}(aq) + 3\ e^- \rightarrow In(s)$. You do not have any electrical equipment, but you do have all of the metals listed in Table 21.1 and aqueous solutions of their ions, as well as In(s) and $In^{3+}(aq)$. Describe the experiments you would perform and the accuracy you would expect in your result.

26. Given that for the reaction
$$2\ Na\ (in\ Hg) + Cl_2(g) \longrightarrow 2\ Na^+(aq) + 2\ Cl^-(aq)$$
$E_{cell}° = 3.20$ V, what is $E°$ for the reduction $2\ Na^+(aq) + 2\ e^- \rightarrow 2\ Na(in\ Hg)$?

28. The theoretical $E_{cell}°$ for the methane–oxygen fuel cell is 1.06 V. What is $E°$ for the reduction half-reaction $CO_2(g) + 8\ H^+(aq) + 8\ e^- \rightarrow CH_4(g) + 2\ H_2O$? [*Hint:* Refer to the net cell reaction (21.26).]

Predicting Oxidation–Reduction Reactions

29. According to standard reduction potentials, both sodium and aluminum should displace zinc from an aqueous solution of Zn^{2+}. What is observed, however, is that aluminum does displace Zn^{2+} and sodium does not. Explain this difference in behavior. (*Hint:* What are the reactions that occur?)

30. In the photograph, copper and zinc strips are joined and immersed in 1 M HCl. What is the reaction that occurs? Why are gas bubbles evident on both metal strips?

31. Predict whether the following metals will dissolve in the acid indicated. If a reaction does occur, write the net ionic equation for the reaction. Assume that reactants and products are in their standard states. **(a)** Ag in $HNO_3(aq)$; **(b)** Zn in $HI(aq)$; **(c)** Au in HNO_3 (for the couple Au^{3+}/Au, $E° = +1.52$ V).

32. Predict whether, to any significant extent,
(a) $Fe(s)$ will displace $Zn^{2+}(aq)$;
(b) MnO_4^- will oxidize $Cl^-(aq)$ to $Cl_2(g)$ in acidic solution;
(c) $Ag(s)$ will react with 1 M $HCl(aq)$;
(d) $O_2(g)$ will oxidize $Cl^-(aq)$ to $Cl_2(g)$ in acidic solution.

Voltaic Cells

33. In each of the following examples, sketch a voltaic cell that uses the given reaction. Label the anode and cathode; indicate the direction of electron flow; write a balanced equation for the cell reaction; and calculate $E°_{cell}$.
(a) $Cu(s) + Fe^{3+}(aq) \rightarrow Cu^{2+}(aq) + Fe^{2+}(aq)$
(b) $Pb^{2+}(aq)$ is displaced from solution by $Al(s)$.
(c) $Cl_2(g) + H_2O \rightarrow Cl^-(aq) + O_2(g) + H^+(aq)$
(d) $Zn(s) + H^+ + NO_3^- \rightarrow Zn^{2+} + H_2O + NO(g)$

34. Diagram the cell and calculate the value of $E°_{cell}$ for a voltaic cell in which
(a) $Cl_2(g)$ is reduced to $Cl^-(aq)$ and $Fe(s)$ is oxidized to $Fe^{2+}(aq)$;
(b) $Ag^+(aq)$ is displaced from solution by $Zn(s)$;
(c) The net cell reaction is $2\ Cu^+(aq) \rightarrow Cu^{2+}(aq) + Cu(s)$;
(d) $MgBr_2(aq)$ is produced from $Mg(s)$ and $Br_2(l)$.

$\Delta G°$, $\Delta E°_{cell}$, and K_{eq}

35. For the reaction
$$5\ H_2O_2(aq) + 2\ Mn^{2+}(aq) \longrightarrow$$
$$2\ MnO_4^-(aq) + 6\ H^+(aq) + 2\ H_2O$$
use data from Table 21.1 and equations from the chapter to determine **(a)** $E°_{cell}$; **(b)** $\Delta G°$; **(c)** K_{eq}; **(d)** whether the reaction goes substantially to completion when the reactants and products are initially in their standard states.

36. For the reaction that occurs in the voltaic cell
$$Fe(s) | Fe^{2+}(aq) \| Cr^{3+}(aq), Cr^{2+}(aq) | Pt(s)$$
use data from Appendix D and equations from the chapter to determine **(a)** the equation for the net cell reaction; **(b)** $E°_{cell}$; **(c)** $\Delta G°$; **(d)** K_{eq}; **(e)** whether the reaction goes substantially to completion when the reactants and products are initially in their standard states.

37. For the reaction $2\ Cu^+ + Sn^{4+} \rightarrow 2\ Cu^{2+} + Sn^{2+}$, $E°_{cell} = -0.0050$ V.
(a) Can a solution be prepared at 298 K that is 0.500 M in each of the four ions?
(b) If not, in what direction will a net reaction occur?

38. For the reaction $2\ V^{3+} + Ni(s) \rightarrow 2\ V^{2+} + Ni^{2+}$, $E°_{cell} = 0.0020$ V.
(a) Can a solution be prepared at 298 K that has $[V^{2+}] = 0.600$ M and $[V^{3+}] = [Ni^{2+}] = 0.675$ M? **(b)** If not, in what direction will a net reaction occur?

39. Use thermodynamic data from Appendix D to calculate a theoretical voltage of the silver–zinc button cell described on page 745.

40. The theoretical voltage of the aluminum–air battery is $E°_{cell} = 2.71$ V. Use data from Appendix D and Equation (21.28) to determine $\Delta G°_f[Al(OH)_4]^-$.

Concentration Dependence of E_{cell}—The Nernst Equation

41. Use the Nernst equation and Table 21.1 to calculate E_{cell} for each of the following cells.
(a) $Al(s) | Al^{3+}(0.18\ M) \| Fe^{2+}(0.85\ M) | Fe(s)$
(b) $Ag(s) | Ag^+(0.34\ M) \| Cl^-(0.098\ M) | Cl_2(g, 0.55\ atm) | Pt(s)$

42. Use the Nernst equation and data from Appendix D to calculate E_{cell} for each of the following cells.
(a) $Mn(s) | Mn^{2+}(0.40\ M) \| Cr^{3+}(0.35\ M), Cr^{2+}(0.25\ M) | Pt(s)$
(b) $Mg(s) | Mg^{2+}(0.016\ M) \| [Al(OH)_4]^-(0.25\ M), OH^-(0.042\ M) | Al(s)$

43. Consider the reduction half-reactions listed in Appendix D, and give plausible explanations for the following observations:
(a) For some half-reactions E depends on pH and for others it does not.
(b) Wherever H^+ appears in a half-equation, it is on the *left* side.
(c) Wherever OH^- appears in a half-equation, it is on the *right* side.

44. Write an equation to represent the oxidation of $Cl^-(aq)$ to $Cl_2(g)$ by $PbO_2(s)$ in an acidic solution. Will this reaction occur spontaneously in the forward direction if all other reactants and products are in their standard states and **(a)** $[H^+] = 6.0$ M; **(b)** $[H^+] = 1.2$ M; **(c)** pH = 4.25. Explain.

45. If $[Zn^{2+}]$ is maintained at 1.0 M,
(a) What is the minimum $[Cu^{2+}]$ for which reaction (21.3) is spontaneous in the forward direction?
(b) Would you say that the displacement of $Cu^{2+}(aq)$ by $Zn(s)$ goes to completion? Explain.

46. Can the displacement of $Pb(s)$ from 1.0 M $Pb(NO_3)_2$ be carried to completion by tin metal? Explain. (*Hint:* What would be the equilibrium concentration of Sn^{2+}? Also, refer to Exercise 45.)

47. A concentration cell is constructed of two hydrogen electrodes: one immersed in a solution with $[H^+] = 1.0$ M and the other in 0.65 M KOH.
(a) Determine E_{cell} for the reaction that occurs.
(b) Compare this value of E_{cell} with $E°$ for the reduction of H_2O to $H_2(g)$ in basic solution and explain the relationship between them.

49. A voltaic cell is constructed as follows.
$$Ag(s)\,|\,Ag^+(satd\ Ag_2CrO_4)\,\|\,Ag^+(0.125\ M)\,|\,Ag(s)$$
What is the value of E_{cell}? K_{sp} of $Ag_2CrO_4 = 1.1 \times 10^{-12}$.

51. For the voltaic cell
$$Sn(s)\,|\,Sn^{2+}(0.075\ M)\,\|\,Pb^{2+}(0.600\ M)\,|\,Pb(s)$$
(a) What is E_{cell} initially?
(b) If the cell is allowed to operate spontaneously, will E_{cell} increase, decrease, or remain constant with time? Explain.
(c) What will be E_{cell} when $[Pb^{2+}]$ has fallen to 0.500 M?
(d) What will be $[Sn^{2+}]$ at the point where $E_{cell} = 0.020$ V?
(e) What are the ion concentrations when $E_{cell} = 0$?

48. If the 0.65 M KOH of Exercise 47 is replaced by 0.65 M NH_3,
(a) Will E_{cell} be higher or lower than in the cell with 0.65 M KOH?
(b) What will be the value of E_{cell}?

50. A voltaic cell is constructed as follows.
$$Ag(s)\,|\,Ag^+(satd\ Ag_3PO_4)\,\|\,Ag^+(0.140\ M)\,|\,Ag(s)$$
$E_{cell} = 0.180$ V, What is the K_{sp} of Ag_3PO_4?

52. For the voltaic cell
$$Ag(s)\,|\,Ag^+(0.015\ M)\,\|\,Fe^{3+}(0.055\ M),\ Fe^{2+}(0.045\ M)\,|\,Pt(s)$$
(a) What is E_{cell} initially?
(b) As the cell operates, will E_{cell} increase, decrease, or remain constant with time? Explain.
(c) What will be E_{cell} when $[Ag^+]$ has increased to 0.020 M?
(d) What will be $[Ag^+]$ when $E_{cell} = 0.010$ V?
(e) What are the ion concentrations when $E_{cell} = 0$?

Batteries and Fuel Cells

53. The iron–chromium redox battery makes use of the reaction
$$Cr^{2+}(aq) + Fe^{3+}(aq) \longrightarrow Cr^{3+}(aq) + Fe^{2+}(aq)$$
occurring at a chromium anode and an iron cathode.
(a) Write a cell diagram for this battery.
(b) Calculate the theoretical voltage of the battery.

55. What is the theoretical standard cell voltage, $E°_{cell}$, of each of the following voltaic cells: **(a)** the hydrogen–oxygen fuel cell described by Equation (21.27); **(b)** the zinc–air battery; **(c)** a magnesium–iodine battery.

54. Refer to the discussion of the Leclanché cell (page 743).
(a) Combine the several equations written for the operation of the Leclanché cell into a single net equation.
(b) Given that the theoretical voltage of the Leclanché cell is 1.55 V, determine $E°$ for the reduction half-reaction.

56. Describe how you might construct batteries with the following voltages: **(a)** 0.10 V; **(b)** 2.5 V; **(c)** 10.0 V. Be as specific as you can about the electrodes and solution concentrations you would use, and whether the battery would consist of a single cell or two or more cells connected in series.

Electrochemical Mechanism of Corrosion

57. Refer to Figure 21-16 and describe, in words or with a sketch, what you would expect to happen in each of the following cases.
(a) Several turns of copper wire are wrapped around the head and tip of an iron nail.
(b) A deep scratch is filed at the center of an iron nail.
(c) A galvanized nail is substituted for the iron nail.

58. When an iron pipe is partly submerged in water, dissolving of the iron occurs more readily below the water line than it does at the water line. Explain this observation by relating it to the description of corrosion given in Figure 21-17.

Electrolysis Reactions

59. Which of the following reactions occurs spontaneously and which can only be brought about through electrolysis, assuming that all reactants and products are in their standard states? For those requiring electrolysis, what is the *minimum* voltage required?
(a) $2\ H_2O(l) \rightarrow 2\ H_2(g) + O_2(g)$ [in 1 M $H^+(aq)$]
(b) $Zn(s) + Fe^{2+}(aq) \rightarrow Zn^{2+}(aq) + Fe(s)$
(c) $2\ Fe^{2+}(aq) + I_2(s) \rightarrow 2\ Fe^{3+}(aq) + 2\ I^-(aq)$
(d) $Cu(s) + Sn^{4+}(aq) \rightarrow Cu^{2+}(aq) + Sn^{2+}(aq)$

61. If a lead storage battery is charged at too high a voltage, gases are produced at each electrode. (It is possible to recharge a lead storage battery only because of the high overpotential for gas formation on the electrodes.)
(a) What are these gases?
(b) Write a cell reaction to describe their formation.

60. An aqueous solution of K_2SO_4 is electrolyzed using Pt electrodes.
(a) Which of the following gases would you expect to be formed at the *anode*: O_2, H_2, SO_2, SO_3? Explain.
(b) What product would you expect at the *cathode*? Explain.
(c) What is the *minimum* voltage required? Why is the voltage actually needed likely to be higher than this?

62. A dilute aqueous solution of Na_2SO_4 is electrolyzed between Pt electrodes for 3.75 h with a current of 2.83 A. What volume of gas, saturated with water vapor at 25 °C and at a total pressure of 742 mmHg, would be collected at the *anode*? (*Hint:* What is the gas that is collected? Also, use data from Table 13.2.)

63. Calculate the quantity indicated for each of the following electrolyses.
(a) The mass of Zn deposited at the cathode in 42.5 min when 1.87 A of current is passed through an aqueous solution of Zn^{2+}.
(b) The time required to produce 2.79 g I_2 at the anode if a current of 1.75 A is passed through KI(aq).

65. In a silver coulometer, Ag^+(aq) is reduced to Ag(s) at a Pt cathode. If 1.206 g Ag is deposited in 1412 s by a certain quantity of electricity, **(a)** how much electric charge (in C) must have passed, and **(b)** what was the magnitude (in A) of the electric current?

64. Calculate the quantity indicated for each of the following electrolyses.
(a) $[Cu^{2+}]$ *remaining* in 425 mL of a solution that was originally 0.366 M $CuSO_4$, after passage of 2.68 A for 282 s and the deposition of Cu at the cathode.
(b) The time required to reduce $[Ag^+]$ in 255 mL of $AgNO_3$(aq) from 0.196 to 0.175 M by electrolyzing the solution between Pt electrodes with a current of 1.84 A.

66. Electrolysis is carried out for 2.00 h in the cell pictured below. The platinum *cathode,* which has a mass of 25.0782 g, weighs 25.8639 g after the electrolysis. The platinum *anode* weighs the same before and after the electrolysis.
(a) Write plausible equations for the half-reactions occurring at the two electrodes.
(b) What must have been the magnitude of the current used in the electrolysis (assuming a constant current throughout).
(c) A gas is collected at the *anode.* What is this gas, and what volume should it occupy if (when dry) it is measured at 23 °C and 755 mmHg pressure?

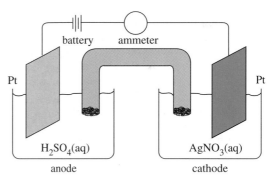

Advanced Exercises

67. Two voltaic cells are assembled in which these reactions occur.
$$V^{2+} + VO^{2+} + 2\,H^+ \longrightarrow 2\,V^{3+} + H_2O$$
$$E^{\circ}_{cell} = 0.616\ \text{V}$$
$$V^{3+} + Ag^+ + H_2O \longrightarrow VO^{2+} + 2\,H^+ + Ag(s)$$
$$E^{\circ}_{cell} = 0.439\ \text{V}$$
Use these data and other values from Table 21.1 to calculate E° for the half-reaction $V^{3+} + e^- \rightarrow V^{2+}$.
68. Show that the oxidation of Cl^-(aq) to Cl_2(g) by $Cr_2O_7{}^{2-}$(aq) in acidic solution with reactants and products in their standard states does not occur spontaneously. Explain why it is still possible to use this method to produce Cl_2(g) in the laboratory. (*Hint:* What experimental conditions would you use?)
69. The concept behind the aluminum–air cell can be extended to other metals, for example zinc–air or iron–air. One of the advantages of the aluminum–air battery is the large quantity of charge transferred per unit mass of Al consumed. Assume that the cell reactions for the zinc–air and iron–air batteries are similar to that for the aluminum–air battery (Equation 21.28). Show that, of the three, the aluminum–air battery produces the greatest amount of charge per unit mass of metal.

70. Suppose that a fully charged lead–acid battery contains 1.50 L of 5.00 M H_2SO_4. What will be the concentration of H_2SO_4 in the battery after 2.50 A of current is drawn from the battery for 6.0 h? [*Hint:* Refer to Equation (21.24).]
71. The energy consumption in electrolysis depends on the *product* of the charge and the voltage [volt × coulomb = $V \cdot C$ = J (joules)]. Determine the theoretical energy consumption per *1000 kg Cl_2* produced in a diaphragm chlor-alkali cell that operates at 3.45 V. Express this energy in **(a)** kJ; **(b)** kilowatt-hours, kWh. [*Hint:* The cell reaction is given on page 753. To relate kilowatt-hours and joules, note that a watt (W) is $1\ \text{J s}^{-1}$.]
72. Given that $E^{\circ} = -0.407$ V for the half-reaction $Cr^{3+} + e^- \rightarrow Cr^{2+}$. If excess Fe(s) is added to a solution in which $[Cr^{3+}] = 1.00$ M, what will be $[Fe^{2+}]$ when equilibrium is reached at 298 K?
$$Fe(s) + 2\,Cr^{3+} \rightleftharpoons Fe^{2+} + 2\,Cr^{2+}$$
73. A voltaic cell is constructed based on the following reaction and initial concentrations.
$$Fe^{2+}(0.0050\ \text{M}) + Ag^+(2.0\ \text{M}) \longrightarrow$$
$$Fe^{3+}(0.0050\ \text{M}) + Ag(s)$$
Calculate $[Fe^{2+}]$ when the cell reaction reaches equilibrium.

74. It is desired to construct a voltaic cell with $E_{cell} = 0.0860$ V. What $[Cl^-]$ must be present in the cathode half-cell to achieve this result? (*Hint:* Obtain K_{sp} values of AgCl and AgI from Table 19.1.)

$$Ag(s)|Ag^+(\text{satd AgI})||Ag^+(\text{satd AgCl}, x \text{ M Cl}^-)|Ag(s)$$

75. Describe a laboratory experiment that you could do to evaluate the Faraday constant, and then show how you could use this value to determine the Avogadro constant, N_A.

76. The hydrazine fuel cell is based on the reaction

$$N_2H_4(aq) + O_2(g) \longrightarrow N_2(g) + 2\ H_2O(l)$$

The theoretical E_{cell}° of this fuel cell is 1.559 V. Use this information and data from Appendix D to calculate a value of $\Delta G_f^\circ[N_2H_4(aq)]$.

77. It is sometimes possible to separate two metal ions through electrolysis. One ion is reduced to the free metal at the cathode and the other remains in solution. In which of these cases would you expect complete or nearly complete separation? Explain. **(a)** Cu^{2+} and K^+; **(b)** Cu^{2+} and Ag^+; **(c)** Pb^{2+} and Sn^{2+}.

78. Show that for some fuel cells the efficiency value, $\epsilon = \Delta G^\circ/\Delta H^\circ$, can have a value greater than 1.00. Can you identify one such reaction with data from Appendix D?

79. In one type of "breathalyzer" (alcohol meter), the quantity of ethanol in a sample is related to the amount of electric current produced by an ethanol–oxygen fuel cell, in which this reaction occurs.

$$CH_3CH_2OH(g) + 3\ O_2(g) \longrightarrow 2\ CO_2(g) + 3\ H_2O(l)$$

Use data from Table 21.1 and Appendix D to determine **(a)** E_{cell}° and **(b)** a value of E° for the reduction of $CO_2(g)$ to $CH_3CH_2OH(g)$.

80. 1.00 L of a *buffer* solution is prepared that is 1.00 M NaH_2PO_4 and 1.00 M Na_2HPO_4. The solution is divided in half between the two compartments of an electrolysis cell. The electrodes used are Pt. Assume that the only electrolysis is that of water. If 1.25 A of current is passed for 212 min, what will be the pH in each cell compartment at the end of the electrolysis?

81. Assume that the volume of each solution in Figure 21-18 is 100.0 mL. The cell is operated as an *electrolytic* cell using a current of 0.500 A. Electrolysis is stopped after 10.00 h, and the cell is allowed to function as a *voltaic* cell. What is E_{cell} at this point?

82. A common reference electrode consists of a silver wire coated with AgCl(s) and immersed in 1 M KCl.

$$AgCl(s) + e^- \longrightarrow Ag(s) + Cl^- \text{ (1 M)}$$
$$E^\circ = 0.2223 \text{ V}$$

(a) What is E_{cell}° when this electrode is a *cathode* in combination with a zinc electrode as an *anode*?
(b) Cite several reasons why this electrode should be easier to use than a standard hydrogen electrode.
(c) By comparing the potential of this silver–silver chloride electrode with that of the silver–silver ion electrode, determine K_{sp} for AgCl.

83. An important source of Ag is recovery as a by-product in the metallurgy of lead. The percentage of Ag in lead was determined as follows. A 1.050-g sample was dissolved in nitric acid to produce $Pb^{2+}(aq)$ and $Ag^+(aq)$. The solution was diluted to 500.0 mL with water, an Ag electrode was immersed in the solution, and the potential difference between this electrode and a SHE was found to be 0.503 V. What was the percent Ag by mass in the lead metal?

Feature Problems

A. Suppose we need to find the value of E° for the reduction half-reaction:

$$Fe^{3+}(aq) + 3\ e^- \longrightarrow Fe(s).$$

Both in Table 21.1 and in Appendix D, the only entries we find are

$Fe^{2+}(aq) + 2\ e^- \longrightarrow Fe(s)$	$E^\circ = -0.440$ V
$Fe^{3+}(aq) + e^- \longrightarrow Fe^{2+}(aq)$	$E^\circ = +0.771$ V

Although we get the desired half-equation by adding these two half-equations together, the E° we seek is *not* the sum of -0.440 V and 0.770 V. The property that *is* additive is ΔG°, and the relationship between ΔG° and E° is $\Delta G^\circ = -nFE^\circ$. That is,

$$Fe^{2+}(aq) + 2\ e^- \longrightarrow Fe(s)$$
$$\Delta G^\circ = -2 \times \mathscr{F} \times (-0.440 \text{ V})$$
$$Fe^{3+}(aq) + e^- \longrightarrow Fe^{2+}(aq)$$
$$\Delta G^\circ = -1 \times \mathscr{F} \times (0.771 \text{ V})$$

$$Fe^{3+}(aq) + 3\ e^- \longrightarrow Fe(s)$$
$$\Delta G^\circ = 0.880\mathscr{F} \text{ V} - 0.771\mathscr{F} \text{ V} = 0.109\mathscr{F} \text{ V}$$

Now, to get $E_{Fe^{3+}/Fe}^\circ$, we can again use the expression: $\Delta G^\circ = -nFE^\circ$, and solve for E°.

$$E_{Fe^{3+}/Fe}^\circ = -\Delta G^\circ/n\mathscr{F} = -0.109\mathscr{F} \text{ V}/3\ \mathscr{F} = -0.0363 \text{ V}$$

1. Use data from Appendix D and the technique just outlined to obtain E° values for the following reduction half-cells.
(a) $Cr^{3+}(aq) + 3\ e^- \rightarrow Cr(s)$ $E^\circ = ?$
(b) $MnO_2(s) + 4\ H^+(aq) + 4\ e^- \longrightarrow Mn(s) + 2\ H_2O$
$$E^\circ = ?$$

2. Use data from Table 21.1 and the technique outlined above (that is, working through ΔG° values) to obtain E_{cell}° for the reaction: $2\ Al(s) + 3\ Cu^{2+}(aq) \rightarrow 2\ Al^{3+}(aq) + 3\ Cu(s)$. Show that the result is the same as that obtained in Example 21-5 by simply adding $E_{Cu^{2+}/Cu}^\circ$ and $-E_{Al^{3+}/Al}^\circ$, and explain why this should be so.

B. Consider the following electrochemical cell:
$$Pt(s)|H_2(g, 1 \text{ atm})|H^+(1 \text{ M})||Ag^+(x \text{ M})|Ag(s)$$
1. What is E_{cell}°, that is, the cell potential when $[Ag^+] = 1$ M?
2. Use the Nernst equation to write an equation for E_{cell} when $[Ag^+] = x$.
3. Now, imagine titrating 50.0 mL of 0.0100 M $AgNO_3$ in the *cathode* half-cell compartment with 0.0100 M KI. The titration reaction is

$$Ag^+(aq) + I^-(aq) \longrightarrow AgI(s)$$

Calculate [Ag$^+$] and then E_{cell} after addition of the following volumes of 0.0100 M KI: **(a)** 0.0 mL; **(b)** 20.0 mL; **(c)** 49.0 mL; **(d)** 50.0 mL; **(e)** 51.0 mL; **(f)** 60.0 mL. [*Hint:* Use a format like that in Example 18-7 (page 637). At the equivalence point neither Ag$^+$ nor I$^-$ is present as a common ion.]
4. Use the results of part 3 to sketch the titration curve: E_{cell} versus volume of titrant.

C. In the original construction of the Statue of Liberty, a framework of iron ribs was covered with thin sheets of copper less than 2.5 mm thick. The copper "skin" and iron framework were separated by a layer of asbestos. Over time, the asbestos wore away and the iron ribs corroded. Some of the ribs lost more than half their mass in the 100 years before the statue was restored. At the same time, the copper skin lost only about 4% of its thickness. Use electrochemical principles to explain these observations.

D. Natural gas transmission pipes are sometimes protected against corrosion by maintaining a small potential difference between the pipe and an inert electrode buried in the ground. Describe how the method works. (*Hint:* Is the inert electrode maintained positively or negatively charged with respect to the pipe?)

E. Silver tarnish is mainly Ag$_2$S, which has the reduction potential
$$Ag_2S(s) + 2\,e^- \longrightarrow 2\,Ag(s) + S^{2-}(aq)$$
$$E° = -0.691 \text{ V}$$
The sketch shows a tarnished silver spoon in contact with a commercially available metallic product in a glass baking dish. Boiling water, to which has been added some NaHCO$_3$, is poured into the dish and the product and spoon are completely covered. Within a short time, the removal of tarnish from the spoon begins.

1. What metal or metals do you think are in the product?
2. What is the probable reaction that occurs?
3. What do you suppose is the function of the NaHCO$_3$?
4. An advertisement for the product appears to make two claims: (a) that there are no chemicals involved, and (b) that the product will never need to be replaced. How valid are these claims? Explain.

F. The fact that the alkali metals react with water, displacing H$_2$(g), greatly complicates the matter of evaluating their standard reduction potentials. One possibility is to use $\Delta G_f°$ values from Appendix D, together with Equation (21.15), to evaluate $E°$ indirectly. In this way, calculate $E°$ for the reduction half-reaction
$$Li^+(aq) + e^- \longrightarrow Li(s) \qquad E° = ?$$
Compare your result with the $E°$ value listed in Appendix D.

G. Ultimately $\Delta G_f°$ values must be based on experimental results, and often these experimental results are themselves obtained from $E°$ values. Early in the twentieth century, G. N. Lewis conceived of an experimental approach for obtaining reduction potentials of the alkali metals. This involved using a solvent with which the alkali metals do not react. Ethylamine was the solvent chosen. In the following cell diagram, Na(amalg, 0.206%) represents a solution of 0.206% Na in liquid mercury.
(1) Na(s) | Na$^+$(in ethylamine) | Na(amalg, 0.206%)
$$E_{cell} = 0.8453 \text{ V}$$
Although Na(s) reacts violently with water to produce H$_2$(g), at least for a short time, a sodium amalgam electrode does not react with water. This makes it possible to determine E_{cell} for the voltaic cell,
(2) Na(amalg, 0.206%) | Na$^+$(1 M) ‖ H$^+$(1 M) | H$_2$(g, 1 atm)
$$E_{cell} = 1.8673 \text{ V}$$
1. Write equations for the cell reactions that occur in the voltaic cells (1) and (2).
2. Use Equation (21.15) to establish ΔG for the cell reactions written in part 1.
3. Write the net equation obtained by combining the equations of part 1, and establish $\Delta G°$ for this net reaction.
4. Use the $\Delta G°$ value from part 3 to obtain $E_{cell}°$ for the net reaction, and from this, obtain $E_{Na^+/Na}°$. Compare your result with the value listed in Appendix D.

Main Group Elements I: Metals

*S*ome of the dramatic colors seen in fireworks displays are the flame colors of the Group 1A and 2A metals. These colors are related to the electronic structures of the metal atoms.

Aluminum production requires prodigious quantities of electricity in a process that is only about a hundred years old. Lead, on the other hand, can be obtained by chemical reduction, a method known for centuries. Concepts of oxidation–reduction and electrochemistry help us to understand which elements can be obtained from their compounds by chemical reactions and which require electrolysis. Principles of acid–base chemistry and solution equilibria help to explain the natural phenomena associated with hardness in water, as well as with methods of water softening.

This and the remaining chapters offer many opportunities to relate new information to principles presented earlier in the text.

Quicklime (CaO) or chalk (a form of $CaCO_3$) is used to outline baseball fields. The relationship between $CaCO_3$, CaO, and $Ca(OH)_2$ (slaked lime) is one of the topics discussed in this chapter.

22-1   Group 1A: The Alkali Metals

By whatever measure we choose—low ionization energy, low electronegativity, large negative electrode potential—the Group 1A metals, the alkali metals, are the most active metals. Table 22.1 lists several of their properties. With only one valence electron available per atom, bonding in the alkali metals is rather weak. The alkali metals have low melting points and are quite soft. For example, a bar of sodium has the consistency of a frozen stick of butter and is easily cut with a knife.

A distinctive feature of the Group 1A metals is their ability to produce *flame colors*. For example, when a sodium compound such as NaCl

▶As described on page 323, hydrogen is often placed in Group 1A of the periodic table, but it is not an alkali metal. Francium is an alkali metal, but this highly radioactive metal is so rare that few of its properties have been measured.

▶The Group 1A metals react with water to displace $H_2(g)$. The special requirements for determining their standard reduction potentials, $E°$, was the subject of Feature Problems F and G on page 763.

TABLE 22.1   Some Properties of the Alkali (Group 1A) Metals

	Li	Na	K	Rb	Cs
atomic number	3	11	19	37	55
valence-shell electron configuration	$2s^1$	$3s^1$	$4s^1$	$5s^1$	$6s^1$
atomic (metallic) radius, pm	152	186	227	248	265
ionic (M^+) radius, pm	59	99	138	149	170
electronegativity	1.0	0.9	0.8	0.8	0.8
first ionization energy, kJ/mol	520.2	495.8	418.8	403.0	375.7
electrode potential $E°$, V[a]	−3.040	−2.713	−2.924	−2.924	−2.923
melting point, °C	180.54	97.81	63.65	39.05	28.4
boiling point, °C	1347	883.0	773.9	687.9	678.5
density, g/cm³ at 20 °C	0.534	0.971	0.862	1.532	1.873
hardness[b]	0.6	0.4	0.5	0.3	0.2
electrical conductivity[c]	17.1	33.2	22.0	12.4	7.76
flame color	carmine	yellow	violet	bluish red	blue
principal visible emission lines, nm	610,671	589	405,767	780,795	456,459

[a]For the reduction $M^+(aq) + e^- \rightarrow M(s)$.

[b]Hardness measures the ability of substances to scratch, abrade, or indent one another. On the Mohs scale, hardnesses of ten minerals range from that of talc (0) to diamond (10). Other values: wax (0 °C), 0.2; asphalt, 1–2; fingernail, 2.5; copper, 2.5–3; iron, 4–5; chromium, 9. Each substance is capable of scratching only others with hardness values lower than its own.

[c]On a scale relative to silver as 100.

is vaporized in a flame, ion pairs are converted to gaseous atoms. Na(g) atoms are excited to high energies, and 589-nm (yellow) light is emitted as the excited atoms (Na*) revert to their ground state electron configurations.

$$Na^+Cl^-(g) \longrightarrow Na(g) + Cl(g)$$

$$\underset{[Ne]3s^1}{Na(g)} \longrightarrow \underset{[Ne]3p^1}{Na*(g)}$$

$$Na*(g) \longrightarrow Na(g) + h\nu \quad \text{(589 nm; yellow)}$$

The color of the sodium flame, and those of lithium and potassium, are shown in Figure 22-1. The sodium and potassium flame colors are both used in pyrotechnic displays—fireworks.

Production and Uses of the Alkali Metals

Lithium and sodium are produced from their molten chlorides by electrolysis. The electrolysis of NaCl(l), for example, is carried out at about 600 °C.

$$2 \text{ NaCl(l)} \xrightarrow{\text{electrolysis}} 2 \text{ Na(l)} + Cl_{2(g)} \qquad (22.1)$$

The melting point of NaCl is 801 °C, too high a temperature to carry out the electrolysis economically. The melting point is reduced by adding $CaCl_2$ to the mixture. [Calcium metal, also produced in the electrolysis, precipitates out from the Na(l) as the liquid metal is cooled. The final product is 99.95% Na.]

Cutting metallic sodium. Note that the sodium, an active metal, is covered with a thick oxide coating.

(a)   **(b)**   **(c)**

FIGURE 22-1
Flame colors of (a) lithium, (b) sodium, and (c) potassium.

Potassium metal is produced by the reduction of molten KCl by liquid sodium.

$$KCl(l) + Na(l) \xrightarrow{850\,°C} NaCl(l) + K(g) \qquad (22.2)$$

Reaction (22.2) is reversible; at low temperatures most of the KCl(l) is unreacted. At 850 °C, however, the equilibrium is displaced far to the right as K(g) escapes from the molten mixture (an application of Le Châtelier's principle).* Rb and Cs can be produced in much the same way, with Ca metal as the reducing agent.

Because sodium is so reactive, its most important use is as a reducing agent—for example, in obtaining metals such as titanium, zirconium, and hafnium.

$$TiCl_4 + 4\,Na \xrightarrow{\Delta} Ti + 4\,NaCl$$

Another important, though diminishing, use of sodium metal is as a heat transfer medium in liquid-metal-cooled fast breeder nuclear reactors (see Section 26-8). Liquid sodium is especially good for this purpose because it has: (1) a low melting point (98 °C), a high boiling point (883 °C), and a low vapor pressure even at high operating temperatures (550 °C); (2) better thermal conductivity and a higher specific heat than most liquid metals; and (3) a low density and low viscosity (making it easy to pump). Sodium is also used in sodium vapor lamps, which are very popular for outdoor lighting (recall Figure 9-6). Because each lamp uses only a few milligrams of Na, however, the total quantity consumed in this application is rather small.

Lithium metal is used as an alloying agent to make high-strength, low-density alloys with aluminum and with magnesium. These alloys are very important in the aerospace and aircraft industries. Lithium is also finding increased use as an anode in batteries, in part because of its ease of oxidation

$$Li(s) \longrightarrow Li^+(aq) + e^- \qquad -E° = +3.040\ V$$

and in part because a small mass of lithium produces a large number of electrons. Only 6.94 g Li (1 mol) needs to be consumed to produce 1 mole of electrons. Lithium batteries are particularly useful where the installed battery must have a high reliability and a long lifetime, as in cardiac pacemakers. Their re-

▶ We introduced the symbol $\xrightarrow{\Delta}$ in Chapter 4 (page 101) to indicate that a reaction is carried out at elevated temperatures.

---

*The vapor obtained is actually a mixture of K(g) and Na(g). The two metals can be separated by condensation of the vapor, followed by fractional distillation.

FIGURE 22-2
Preparation of sodium compounds.
This is a simple and common method of summarizing the reactions of a compound. Most of these reactions are described in this section. A number of these compounds may be prepared by alternative methods. The conversion of Na_2CO_3 to NaOH (broken line arrow) is no longer of commercial importance.

liability and low weight also make lithium batteries useful in computer memory systems, especially for laptop computers. Several lithium compounds (the acetate, carbonate, and chloride, for example) are used in the treatment of manic-depressive psychoses.

Group 1A Compounds

Figure 22-2 introduces a new format for summarizing reaction chemistry. Although it deals with sodium compounds, many of which are discussed in this section, we can construct similar diagrams for other elements. In such a diagram we note a compound of central importance (NaCl in Figure 22-2) and show how several other compounds can be obtained from it. Some of these conversions occur in one step, such as the reaction of NaCl with H_2SO_4 to form Na_2SO_4 (discussed on page 771). Others involve two or more consecutive reactions, as does the preparation of sodium silicate, Na_2SiO_3, for example. The principal reactants required for the conversions are written with the arrow ($\rightarrow$). If it is necessary to heat a reaction mixture, this is noted with a Δ symbol. The reactions may also produce by-products that are not noted in the diagram.

EXAMPLE 22-1

Writing Chemical Equations from a Summary Diagram of Reaction Chemistry. Propose a method for synthesizing sodium carbonate from sodium chloride.

Solution

Let's consult Figure 22-2 to find a route from NaCl to Na_2CO_3. One route involves the conversions $NaCl \rightarrow Na_2SO_4 \rightarrow Na_2S \rightarrow Na_2CO_3$. Other necessary reactants

(H$_2$SO$_4$, C, and CaCO$_3$) are noted. In the following set of equations we must, of course, include other reaction products (such as HCl, CO, and CaS).

First, Na$_2$SO$_4$(s) is produced from NaCl(s) and concentrated sulfuric acid.

$$2\,NaCl(s) + H_2SO_4(concd\,aq) \xrightarrow{\Delta} Na_2SO_4(s) + 2\,HCl(g)$$

Next, the Na$_2$SO$_4$ is reduced to Na$_2$S with carbon.

$$Na_2SO_4(s) + 4\,C(s) \xrightarrow{\Delta} Na_2S(s) + 4\,CO(g)$$

The final step is a reaction between Na$_2$S and CaCO$_3$.

$$Na_2S(s) + CaCO_3(s) \xrightarrow{\Delta} CaS(s) + Na_2CO_3(s)$$

Practice Example A: Propose a method for synthesizing sodium nitrate from sodium chloride.

Practice Example B: Propose a method for synthesizing sodium thiosulfate from sodium chloride.

Harvesting sodium chloride (salt) by evaporation.

Halides. All alkali metals react vigorously, sometimes explosively, with halogens to produce ionic halide salts, the most important of which are NaCl and KCl. Sodium chloride is the primary sodium compound. In fact, it is the most used of all minerals for the production of chemicals. It is not listed among the top chemicals because it is considered a raw material, not a manufactured chemical. Annual use of sodium chloride in the United States amounts to about 50 million tons. Salt is used to preserve meat and fish, control ice on roads, and regenerate water softeners. In the chemical industry, NaCl is a source of many chemicals, including sodium metal, chlorine gas, and sodium hydroxide.

Potassium chloride is obtained from naturally occurring brines (concentrated solutions of salts) and finds its most extensive use in plant fertilizers, because potassium is a major essential element for plant growth. KCl is also used as a raw material in the manufacture of KOH, KNO$_3$, and other industrially important potassium compounds.

Carbonates and Sulfates. Except for Li$_2$CO$_3$, the alkali metal carbonates are all thermally stable compounds. In this and several other ways, lithium and its compounds differ from other members of Groups 1A. The ability of lithium to form a nitride (Li$_3$N), the low aqueous solubility of its carbonate, and the instability of the carbonate with respect to the oxide at high temperature are all properties similar to those of magnesium and its compounds. This similarity is called a **diagonal relationship.** Such relationships also exist between Be and Al and between B and Si (see Figure 22-3). The Li–Mg similarity probably results from the roughly equal sizes of the Li and Mg atoms and of the Li$^+$ and Mg^{2+} ions.

Sodium carbonate (soda ash) is used primarily in the manufacture of glass. The Na$_2$CO$_3$ produced in the United States now comes mostly from *natural* sources, such as the mineral trona, Na$_2$CO$_3 \cdot$NaHCO$_3 \cdot n$H$_2$O, found in dry lakes in California and in immense deposits in western Wyoming. In the past, sodium carbonate was mostly manufactured from NaCl, CaCO$_3$, and NH$_3$, using a process introduced by the Belgian chemist Ernest Solvay in 1863.

The great success of the Solvay process over the synthetic method outlined in Example 22-1 lies in the efficient use of certain raw materials through *recy-*

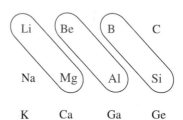

FIGURE 22-3
Diagonal relationships.
The two elements in each encircled pair exhibit many similar properties.

▶ The term "ash" signifies a product obtained by heating or burning. The earliest alkaline materials (notably potassium carbonate, *potash*) were extracted from the ashes of burned plants. The first commercial source of *soda ash* was the Leblanc process featured in Example 22-1.

cling. Recycling makes good economic sense: A process that recycles materials minimizes the use of raw materials (an expense to purchase) and cuts down on the production of by-products (an expense to dispose of).

An outline of the process is shown in Figure 22-4. The key step involves the reaction of $NH_3(g)$ and $CO_2(g)$ in saturated $NaCl(aq)$. Of the possible ionic compounds that could precipitate from such a mixture ($NaCl$, NH_4Cl, $NaHCO_3$, and NH_4HCO_3), the least soluble one is $NaHCO_3$, *sodium hydrogen carbonate* (sodium bicarbonate).

▶ $NH_3(aq)$ and $CO_2(g)$ react to form NH_4HCO_3, known as baker's ammonia. This compound is used in place of $NaHCO_3$ (baking soda) in some recipes.

$$Na^+ + Cl^- + NH_3 + CO_2 + H_2O \longrightarrow NaHCO_3(s) + NH_4^+ + Cl^- \qquad (22.3)$$

The sodium bicarbonate can be isolated and sold or converted to sodium carbonate by heating.

$$2\,NaHCO_3(s) \xrightarrow{\Delta} Na_2CO_3(s) + H_2O(g) + CO_2(g) \qquad (22.4)$$

One noteworthy feature of the Solvay process is that it involves only simple precipitation and acid–base reactions. Another feature is the efficient way materials produced in one step are recycled or used in a subsequent step. Thus, when limestone ($CaCO_3$) is heated to produce the reactant CO_2, the other reaction product, CaO, is also used. It is converted to $Ca(OH)_2$, which is then used to convert NH_4Cl (another reaction by-product) to $NH_3(g)$. The $NH_3(g)$ is recycled into the production of ammoniated brine.

The Solvay process has only one ultimate by-product—$CaCl_2$—for which the demand is very limited. In the past some $CaCl_2$ was used for deicing roads in the winter (page 491) and for dust control on dirt roads in the summer [through the deliquescence of $CaCl_2\cdot6H_2O$, (page 485)]. The bulk of the $CaCl_2$, however, was dumped into local lakes and streams (notably Onondaga Lake, near Solvay, NY), resulting in an environmental tragedy for the local area. Environmental regulations no longer permit such dumping, and partly due to these regulations, but mainly for economic reasons, natural sources of sodium carbonate have sup-

The mineral *trona,* from Green River, Wyoming, is currently the principal source of Na_2CO_3 in the United States.

FIGURE 22-4
The Solvay process for the manufacture of $NaHCO_3$.
The main reaction sequence is traced by solid arrows. Recycling reactions are shown by broken arrows.

planted the Solvay process in the United States. The process is still widely used elsewhere in the world, however.

Sodium sulfate is obtained partly from natural sources, partly from neutralization reactions, and partly through a process discovered by Johann Rudolf Glauber in 1625.

$$H_2SO_4(\text{concd aq}) + NaCl(s) \xrightarrow{\Delta} NaHSO_4(s) + HCl(g)$$
$$NaHSO_4(s) + NaCl(s) \xrightarrow{\Delta} Na_2SO_4(s) + HCl(g) \qquad (22.5)$$

▶ Often, Equations (22.5) are replaced by the equivalent *net* equation: $H_2SO_4(\text{concd aq}) + 2\,NaCl(s) \rightarrow Na_2SO_4(s) + 2\,HCl(g)$.

The strategy behind reactions (22.5) is the production of a *volatile* acid (HCl) by heating one of its salts (NaCl) with a *nonvolatile* acid (H_2SO_4). Several other acids can be produced by similar reactions. The major use of Na_2SO_4 is in the paper industry. In the kraft process for papermaking, undesirable lignin is removed from wood by digesting the wood in an alkaline solution of Na_2S. Na_2S is produced by the reduction of Na_2SO_4 with carbon.

$$Na_2SO_4(s) + 4\,C(s) \xrightarrow{\Delta} Na_2S(s) + 4\,CO(g) \qquad (22.6)$$

About 100 lb of Na_2SO_4 is required for every ton of paper produced.

Oxides and Hydroxides.   The alkali metals react rapidly with oxygen to produce several different ionic oxides. Under appropriate conditions—generally by limiting the supply of oxygen—the oxide M_2O can be prepared for each of the alkali metals. Only for lithium, however, is this the major product of the reaction of the metal with excess oxygen. Li reacts with excess oxygen to give Li_2O and also a small amount of lithium peroxide, Li_2O_2. Na reacts with excess oxygen to give mostly the *peroxide,* Na_2O_2, and a small amount of Na_2O. K, Rb, and Cs react to form the *superoxides,* MO_2.

Oxide ion   Peroxide ion   Superoxide ion

The peroxides contain the O_2^{2-} ion and are fairly stable. Sodium peroxide is used as a bleaching agent and a powerful oxidant. Li_2O_2 and Na_2O_2 are used in emergency breathing devices in submarines and spacecraft because they react with carbon dioxide to produce oxygen.

$$2\,M_2O_2(s) + 2\,CO_2(g) \longrightarrow 2\,M_2CO_3(s) + O_2(g) \qquad (M = Li, Na)$$

KO_2, potassium superoxide, can also be used for this purpose. The oxides, peroxides, and superoxides of the alkali metals react with water to form basic solutions. The reaction of the O^{2-} ion with water is an acid–base reaction that produces hydroxide ions. The peroxide and superoxide ions react with water in oxidation–reduction reactions that produce hydroxide ions and $O_2(g)$.

The hydroxides of the Group 1A metals are strong bases because they dissociate to release hydroxide ions in aqueous solution. Sodium hydroxide is produced commercially by the electrolysis of NaCl(aq). $Na^+(aq)$ goes through the electrolysis unchanged, $Cl^-(aq)$ is oxidized to $Cl_2(g)$, and H_2O is reduced to $H_2(g)$.

$$2\,Na^+ + 2\,Cl^- + 2\,H_2O \xrightarrow{\text{electrolysis}} 2\,Na^+ + 2\,OH^- + H_2 + Cl_2 \qquad (22.7)$$

This is the chlor-alkali process described in Chapter 21 (pages 753–755).

KOH and LiOH are made in a similar fashion. Alkali metal hydroxides can also be prepared by the reaction of the Group 1A metals with water (page 337). An important use of alkali hydroxides is in the manufacture of soaps and detergents, described next.

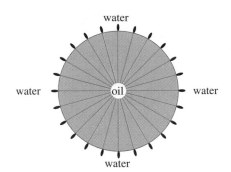

FIGURE 22-5
Structure of a detergent molecule and its cleaning action.
The molecule shown, sodium lauryl sulfate, has a long nonpolar portion and a polar head. Cleaning is accomplished by the nonpolar portion dissolving in the oil while the polar portion dissolves in the water. In this way the oil droplet is emulsified, or solubilized.

Alkali Metal Detergents and Soaps.   A **detergent** is a cleansing agent used primarily because of its ability to emulsify oils. Although the term *detergent* includes common soaps, it is used primarily to describe certain *synthetic* products, such as sodium lauryl sulfate, whose manufacture involves the conversions

$$CH_3(CH_2)_{10}CH_2OH \longrightarrow CH_3(CH_2)_{10}CH_2OSO_3H \longrightarrow CH_3(CH_2)_{10}CH_2OSO_3^-Na^+$$

lauryl alcohol   lauryl hydrogen sulfate   sodium lauryl sulfate

The structure of sodium lauryl sulfate and its detergent action are illustrated in Figure 22-5.

A **soap** is a specific kind of detergent that is the salt of a metal hydroxide and a fatty acid. An example is the sodium soap of palmitic acid, which we can represent as the product of the reaction of palmitic acid and NaOH.

$$CH_3(CH_2)_{14}\overset{O}{\underset{\|}{-C}}-O-H + Na^+ + OH^- \longrightarrow CH_3(CH_2)_{14}\overset{O}{\underset{\|}{-C}}-O^- Na^+ + H_2O$$

palmitic acid   sodium palmitate
                                                (a soap)   (22.8)

Sodium soaps are the familiar hard (bar) soaps. Potassium soaps have low melting points and are soft soaps. High-melting lithium soaps are used in high-temperature lubricating oils and greases. They help hold the oil in contact with moving metal parts under conditions in which the oil by itself would run off.

22-2   Group 2A: The Alkaline Earth Metals

From a chemical standpoint (for example, in their abilities to react with water and acids and to form ionic compounds) the heavier Group 2A metals—Ca, Sr, Ba, and Ra—are nearly as active as the Group 1A metals. In terms of certain physical properties (for example, density, hardness, and melting point) all the Group 2A elements are more typically metallic than the 1A elements, as we can see by comparing the data in Tables 22.1 and 22.2.

Production and Uses of the Alkaline Earth Metals
The preferred method of producing the Group 2A metals (except Mg) is by reducing their salts with other active metals. The mineral *beryl*, $3BeO \cdot Al_2O_3 \cdot 6SiO_2$

TABLE 22.2   Some Properties of the Group 2A Metals

	Be	Mg	Ca	Sr	Ba
atomic number	4	12	20	38	56
atomic (metallic) radius, pm	111	160	197	215	222
ionic (M^{2+}) radius, pm	27	72	100	113	136
electronegativity	1.5	1.2	1.0	1.0	0.9
first ionization energy, kJ/mol	899.4	737.7	589.7	549.5	502.8
electrode potential $E°$, V[a]	-1.85	-2.356	-2.84	-2.89	-2.92
melting point, °C	1278	648.8	839	769	729
boiling point, °C	2970[b]	1090	1483.6	1383.9	1637
density, g/cm³ at 20 °C	1.85	1.74	1.55	2.54	3.60
hardness[c]	~5	2.0	1.5	1.8	~2
electrical conductivity[c]	39.7	35.6	40.6	6.90	3.20
flame color	none	none	orange-red	scarlet	green

[a]For the reduction $M^{2+}(aq) + 2\,e^- \rightarrow M(s)$

[b]Boiling point at 5 mmHg pressure.

[c]See footnotes to Table 22.1.

is the natural source of beryllium compounds. The mineral is processed to produce BeF_2, which is then reduced with Mg. Beryllium metal finds some use as an alloying agent when light weight is a primary requirement. Because it is able to withstand metal fatigue, an alloy of copper with about 2% Be is used in springs, clips, and electrical contacts. The Be atom does not readily absorb X-rays or neutrons, and so beryllium is used to make "windows" for X-ray tubes and for various components in nuclear reactors. Beryllium and its compounds are limited in their use because of their toxicity. In addition, they are suspected of being carcinogens even at a level as low as 0.002 ppm in air.

Calcium, strontium, and barium are obtained by the reduction of their oxides with aluminum; Ca and Sr also can be obtained by electrolysis of their molten chlorides. Calcium metal is primarily used as a reducing agent to prepare, from their oxides or fluorides, other metals such as U, Pu, and most of the lanthanides. Strontium and barium have limited use in alloys, but some of their compounds (discussed later) are quite important. Some salts of Sr and Ba provide vivid colors for pyrotechnic displays.

Magnesium metal is obtained by electrolysis of the molten chloride in the Dow process. The Dow process is outlined in Figure 22-6, and the electrolysis of $MgCl_2(l)$ is pictured in Figure 22-7. Like the Solvay process for making $NaHCO_3$, the Dow process takes advantage of simple chemistry and recycling.

The source of magnesium is seawater or natural brines. The abundance of Mg^{2+} in seawater is about 1350 mg/L. The first step in the Dow process is the precipitation of $Mg(OH)_2(s)$ with slaked lime [$Ca(OH)_2$] as the source of OH^-. Slaked lime is formed by the reaction of quicklime (CaO) with water. The precipitated $Mg(OH)_2(s)$ is washed, filtered, and dissolved in HCl(aq). The resulting concentrated $MgCl_2(aq)$ is dried by evaporation, melted, and electrolyzed, yielding pure Mg metal and $Cl_2(g)$. The $Cl_2(g)$ is converted to HCl, which is recycled.

▶ In Example 19-6 we demonstrated that this precipitation goes to completion.

Magnesium has a lower density than any other metal used for structural purposes and is valued for its light weight. Objects such as aircraft parts are manufactured from magnesium alloyed with aluminum and other metals. Magnesium is a good reducing agent and is used in a number of metallurgical processes,

FIGURE 22-6
The Dow process for the production of Mg.
The main reaction sequence is traced by solid arrows. Recycling of $Cl_2(g)$ is shown by broken arrows.

such as the production of beryllium mentioned earlier. The ease with which magnesium is oxidized also underlies its use in sacrificial anodes for corrosion protection (page 748). Magnesium's most spectacular use, however, may be in fireworks.

Group 2A Compounds

Ionic compounds of the Group 2A metals have properties that differ from those of the Group 1A metals. In some cases this difference is attributable to the smaller ionic size and the larger ionic charge of Group 2A cations. For example, the lattice energy of $Mg(OH)_2$ is about -3000 kJ/mol compared with about

FIGURE 22-7
The electrolysis of molten $MgCl_2$.
The electrolyte is a mixture of molten NaCl, $CaCl_2$, and $MgCl_2$. This mixture has a lower melting point and higher electrical conductivity than $MgCl_2$ alone, but if the voltage is carefully controlled, only Mg^{2+} is reduced in the electrolysis.

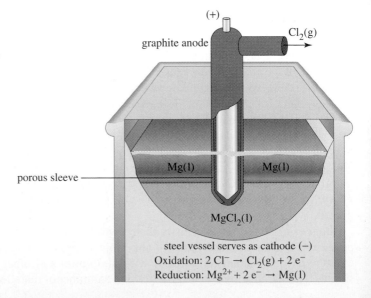

−900 kJ/mol for NaOH. This difference in lattice energy helps to account for the ready solubility of NaOH in water, easily producing solutions up to about 20 M NaOH, and the sparing solubility of $Mg(OH)_2$, which has a K_{sp} of 1.8×10^{-11}. The heavier Group 2A hydroxides are somewhat more soluble than $Mg(OH)_2$. Other alkaline earth compounds that are only slightly soluble include the carbonates, fluorides, and oxides. Another common characteristic of alkali metal compounds is the formation of *hydrates*. Typical hydrates are $MX_2 \cdot 6H_2O$ (where M = Mg, Ca, or Sr and X = Cl or Br).

Beryllium compounds are rather different from the other Group 2A compounds. Because of its small size, the Be atom has a high ionization energy and only a limited tendency to form the ion Be^{2+}. Thus, molten BeF_2 and $BeCl_2$ are poor conductors of electricity, suggesting that the bonding in these substances is essentially covalent.

Magnesium skyrockets

Halides. The Group 2A metals react directly with the halogens to form halides. Except for those of beryllium, these halides are essentially ionic.

$$M + X_2 \longrightarrow MX_2 \qquad (22.9)$$

(M = a Group 2A metal and X = F, Cl, Br, or I)

The halides have varied uses. As an example, $MgCl_2$, in addition to its use in the preparation of magnesium metal, is used in fireproofing wood, in special cements, in ceramics, in treating fabrics, and as a refrigeration brine.

EXAMPLE 22-2

Comparing the Freezing Points of $CaCl_2(aq)$ and $NaCl(aq)$. Two solutions are prepared. One has 5.00 g $CaCl_2$ in 1.000 kg of water, and the other has 5.00 g NaCl in 1.000 kg of water. Which solution would you expect to have the lower freezing point?

Solution

At a guess we might expect the answer to be $CaCl_2$ because three moles of ions are produced per mole of salt dissolved. To check this idea, we need to convert 5.00 g of each salt to moles of ions produced when the salt is dissolved in water.

$$5.00 \text{ g } CaCl_2 \times \frac{1 \text{ mol } CaCl_2}{111.0 \text{ g } CaCl_2} \times \frac{3 \text{ mol ions}}{1 \text{ mol } CaCl_2} = 0.135 \text{ mol ions}$$

$$5.00 \text{ g NaCl} \times \frac{1 \text{ mol NaCl}}{58.44 \text{ g NaCl}} \times \frac{2 \text{ mol ions}}{1 \text{ mol NaCl}} = 0.171 \text{ mol ions}$$

It is clear that there are more ions in the NaCl solution than in the $CaCl_2$, due to the lower molar mass of NaCl. The NaCl(aq) should have the lower freezing point.

Practice Example A: For a solution containing 1.00 g solute per 1.00 kg H_2O, which of the following solutes would produce a solution of the lowest freezing point: NaCl, KF, $MgCl_2$, $CaCl_2$? Explain.

Practice Example B: What mass of $CaCl_2$ is required per kilogram of water to have the same freezing point as a solution of 5.00 g NaCl in 1.000 kg of water? Calculate the freezing point of the solution.

Carbonates and Sulfates.   Carbonates and sulfates are some of the most important compounds of the alkaline earth metals. The Group 2A carbonates are insoluble in water, as are the sulfates of Ca, Sr, and Ba. Because of this in-solubility, these compounds are the most important minerals of the Group 2A metals. The most familiar is probably $CaCO_3$, the principal component of the rock *limestone*. If a limestone contains more than 5% $MgCO_3$ it is usually called dolomitic limestone or *dolomite*. Some clay, sand, or quartz may also be present in limestone. The primary use of limestone (about 70%) is as a building stone. Other uses of limestone include the manufacture of cement, as a flux in metallurgical processes, as a source of quicklime and slaked lime, and as an ingredient in glass.

▶ A metallurgical flux is a material that removes impurities during production of a metal by forming a free flowing liquid called *slag*.

The decomposition (calcination) of limestone is carried out in a long rotary kiln, whether for the production of quicklime, CaO, or the manufacture of Portland cement. Portland cement is a complex mixture of calcium silicates and aluminates formed by heating limestone to about 1500 °C with materials rich in silica (SiO_2) and alumina (Al_2O_3). Because it will set even when completely submerged in water, Portland cement can be used in the construction of bridge piers.

Pure, white $CaCO_3$ is used in a wide variety of products. For example, it is used in papermaking to impart brightness, opacity, smoothness, and good ink-absorbing qualities to paper. It is also used as a filler in plastics, rubber, floor tile, putties, and adhesives, as well as in food, cosmetics, and pharmaceuticals. An important minor use of $CaCO_3$ is as a dietary supplement for the prevention of osteoporosis, a condition in which the bones become porous, brittle, and easily broken.

Three steps are required to obtain pure $CaCO_3$ from limestone: (1) thermal decomposition of limestone (called **calcination**), (2) reaction of CaO with water (slaking), and (3) conversion of an aqueous suspension of $Ca(OH)_2(s)$ to precipitated $CaCO_3$ (carbonation).

$$\text{Calcination:} \quad CaCO_3(s) \xrightarrow{\Delta} CaO(s) + CO_2(g) \tag{22.10}$$

$$\text{Slaking:} \quad CaO(s) + H_2O(l) \longrightarrow Ca(OH)_2(s) \tag{22.11}$$

$$\text{Carbonation:} \quad Ca(OH)_2(s) + CO_2(g) \xrightarrow{aq} CaCO_3(s) + H_2O(l) \tag{22.12}$$

Limestone ($CaCO_3$) is also responsible for the beautiful natural formations found in limestone caves. Natural rainwater is slightly acidic due to dissolved $CO_2(g)$ and is essentially a solution of carbonic acid, H_2CO_3.

$$CO_2 + 2 H_2O \rightleftharpoons H_3O^+ + HCO_3^- \qquad K_{a_1} = 4.4 \times 10^{-7} \tag{22.13}$$

$$HCO_3^- + H_2O \rightleftharpoons H_3O^+ + CO_3^{2-} \qquad K_{a_2} = 4.7 \times 10^{-11} \tag{22.14}$$

Although the carbonates are not very soluble in water, they are bases; thus, they dissolve readily in acidic solutions. As mildly acidic groundwater seeps through limestone beds, insoluble $CaCO_3$ is converted to soluble $Ca(HCO_3)_2$.

▶ K_{eq} for reaction (22.15) can be obtained by appropriately combining values of K_{a_1} and K_{a_2} from reactions (22.13) and (22.14) and K_{sp} of $CaCO_3$ (see Exercise 64).

$$CaCO_3(s) + H_2O + CO_2 \rightleftharpoons Ca(HCO_3)_2(aq) \qquad K_{eq} = 2.6 \times 10^{-5} \tag{22.15}$$

Over time this dissolving action can produce a large cavity in the limestone bed—a limestone cave. Reaction (22.15) is reversible, however, and evaporation of the solution causes a loss of both water and CO_2 and conversion of $Ca(HCO_3)_2(aq)$ back to $CaCO_3(s)$. This process occurs very slowly, but over a period of many years, as $Ca(HCO_3)_2(aq)$ drips from the ceiling of a cave, $CaCO_3(s)$ remains as icicle-like deposits called **stalactites.** Some of the dripping solution may hit the floor of the cave before decomposition occurs and limestone deposits build up from the floor in formations called **stalagmites.** Eventually, some stalactites and stalagmites grow together into limestone columns (Figure 22-8).

EXAMPLE 22-3

Describing Simultaneous Equilibria in Qualitative Terms. Without performing detailed calculations, demonstrate that reaction (22.15) correctly describes the dissolving action of rainwater on limestone. (K_{sp} for $CaCO_3 = 2.8 \times 10^{-9}$.)

Solution

We need to use one idea from Section 17-6, another from Section 19-2, and data from Equations (22.14) and (22.15).

1. $H_2CO_3(aq)$ is a *diprotic* acid. Since $K_{a_1} \gg K_{a_2}$, $[H_3O^+] = [HCO_3^-]$ and $[CO_3^{2-}] = K_{a_2} = 4.7 \times 10^{-11}$ M (see page 600).
2. When pure water is saturated with $CaCO_3$, $[Ca^{2+}] = [CO_3^{2-}]$; $K_{sp} = [Ca^{2+}][CO_3^{2-}] = [CO_3^{2-}]^2 = 2.8 \times 10^{-9}$. In this solution $[CO_3^{2-}] = (2.8 \times 10^{-9})^{1/2} = 5.3 \times 10^{-5}$ M.

   $[CO_3^{2-}]$ in saturated $CaCO_3(aq)$, 5.3×10^{-5} M, is much higher than $[CO_3^{2-}]$ normally produced by the ionization of H_2CO_3, 4.7×10^{-11} M. Carbonate ion derived from $CaCO_3$ acts as a common ion and displaces equilibrium (22.14) *to the left*, converting CO_3^{2-} to HCO_3^- and consuming H_3O^+. Removal of H_3O^+ in this way stimulates reaction (22.13) to shift *to the right* to produce more H_3O^+. The net effect is that CO_2, H_2O, and CO_3^{2-} (from $CaCO_3$) are consumed and HCO_3^- [as $Ca(HCO_3)_2$] is produced.

Practice Example A: As we will learn in Section 22-3, water with significant concentrations of HCO_3^- is called *hard*. Show that the addition of CaO to a solution of $Ca(HCO_3)_2$ should result in the removal of HCO_3^- and the precipitation of $CaCO_3(s)$. [*Hint:* Recall reaction (22.11).]

Practice Example B: Concerning the water with bicarbonate hardness described in Practice Example A, show that the addition of $NH_3(aq)$ to $Ca(HCO_3)_2(aq)$ should also lead to the precipitation of $CaCO_3(s)$.

FIGURE 22-8
Stalactites and stalagmites in Carlsbad Caverns, New Mexico.

Another feature of the Group 2A metal carbonates (MCO_3) that has practical use is their decomposition at high temperatures.

$$MCO_3(s) \xrightarrow{\Delta} MO(s) + CO_2(g) \tag{22.16}$$

This is the chief method of preparing the Group 2A metal oxides. The alkali metal carbonates, except Li_2CO_3, do not share this thermal instability. Thus, in the Solvay process (page 770) heating $NaHCO_3$ produces sodium carbonate, Na_2CO_3, but no further decomposition occurs.

Another important calcium-containing mineral is *gypsum*, $CaSO_4 \cdot 2H_2O$. In the United States, 50 million tons of gypsum are consumed annually. About half of this is converted to the *hemihydrate* ($\frac{1}{2}$-hydrate), **plaster of Paris.**

$$CaSO_4 \cdot 2H_2O(s) \xrightarrow{\Delta} CaSO_4 \cdot \tfrac{1}{2}H_2O(s) + \tfrac{3}{2}H_2O(g) \tag{22.17}$$

When mixed with water, plaster of Paris reverts to gypsum. Because it expands as it sets, a mixture of plaster of Paris and water is useful in making castings where sharp details of an object must be retained. Plaster of Paris is extensively used in jewelry making and in dental work. The most important application, though, is in producing gypsum wallboard (drywall), which has all but supplanted other interior wall coverings in the construction industry.

Barium sulfate has had important applications in medical imaging because barium is opaque to X-rays. Although barium ion is toxic, the very insoluble compound $BaSO_4$ is safe to use to coat the stomach or upper gastrointestinal tract with a "barium milkshake" and the lower tract with a "barium enema."

Oxides and Hydroxides.
The oxides and hydroxides of the Group 2A metals, except those of beryllium, are all bases. Although it is not very soluble in water, calcium hydroxide is the cheapest commercial strong base and is used in a variety of applications, such as in the Solvay and Dow processes.

▶Solids with high melting points, such as CaO (mp 2614 °C), emit light when heated to high temperatures. Before electric lights, lime was heated with an oxyhydrogen flame and used for theatrical lighting. This gave rise to the expression of being "in the limelight."

The term *lime* is familiar, but you may not be aware that it can refer to several different compounds. CaO is called **quicklime** and is produced by the calcination of limestone (reaction 22.10). This reaction is reversible and at room temperature the *reverse* reaction occurs almost exclusively. So, in the calcination of $CaCO_3$ a high temperature must be used and the $CO_2(g)$ produced must be continuously exhausted from the furnace (kiln) in which the reaction occurs. Among the many uses of quicklime, CaO, are water treatment and the removal of $SO_2(g)$ from the smokestack gases in electric power plants (page 611).

$Ca(OH)_2$, called **slaked lime,** is formed by the action of water on CaO (reaction 22.11). A mixture of slaked lime, sand, and water is the familiar mortar used in brick laying. First, excess water is absorbed by the bricks and then lost by evaporation. In the final setting of the mortar, $CO_2(g)$ from the air reacts with $Ca(OH)_2(s)$ and converts it back to $CaCO_3(s)$. The final form of the mortar is a complex mixture of hydrated calcium carbonate and silicate (from the sand).

$$Ca(OH)_2(s) + CO_2(g) \longrightarrow CaCO_3(s) + H_2O(g) \qquad (22.18)$$

This reaction is general for all Group 2A hydroxides. It causes an interesting problem with the Y—Ba—Cu—O superconductors discussed in Chapter 24. The superconductor begins to decompose when water vapor and $CO_2(g)$ in the air convert BaO to $Ba(OH)_2$ and then to $BaCO_3$. On the other hand, art conservationists have used this same reaction to preserve art objects. For example, an aqueous solution of $Ba(NO_3)_2$ is sprayed onto cracking frescos (paintings embedded in plaster). Once the solution has had time to fill small cracks and spaces, an aqueous ammonia solution is applied to the surface of the fresco. The ammonia raises the pH of the solution, resulting in formation of $Ba(OH)_2$. As excess water evaporates, carbon dioxide from the air reacts with the barium hydroxide to produce insoluble barium carbonate, which binds the cracking fresco together and strengthens it without affecting the delicate colors.

22-3 Ions in Natural Waters: Hard Water

Rainwater is not chemically pure. It contains dissolved atmospheric gases and, once rainwater reaches the ground, it dissolves some of the components of soil and rocks. It may pick up anywhere from a few to perhaps 1000 ppm of dissolved substances. If the water contains ions capable of yielding significant quantities of a precipitate, we say that the water is *hard*.

There are two types of hard water: temporary hard water and permanent hard water. **Temporary hard water** contains bicarbonate ion, HCO_3^-. When water containing $HCO_3^-(aq)$ is heated, the bicarbonate ion rapidly decomposes to give CO_3^{2-}, CO_2, and water. The CO_3^{2-} reacts with multivalent cations in the water to form a mixed precipitate of $CaCO_3$, $MgCO_3$, and rust called *boiler scale*. The principal reaction that occurs is the reverse of (22.15).

The formation of boiler scale can be a very serious problem in steam-driven electric power plants and in boilers producing steam for manufacturing processes. Formation of boiler scale lowers the efficiency of water heaters and can eventually cause a boiler to overheat, perhaps even to explode. Closer to home, we are familiar with this scale as a buildup observed on the inside of "hot pots" and other pots used to boil water. Boiler scale can be removed by adding vinegar (acetic acid) to the pot and heating.

▶For pots with an aluminum bottom, adding vinegar might not be such a good idea. Some dissolving of the Al could occur.

Water with temporary hardness can be "softened" at a water treatment plant by adding the base slaked lime [$Ca(OH)_2$] and filtering off the precipitated metal carbonate. The OH^- reacts with bicarbonate ion to produce water and carbonate ion. The carbonate ion reacts with M^{2+} ion, such as Ca^{2+}, to precipitate a metal carbonate.

$$HCO_3^- + OH^- \longrightarrow H_2O + CO_3^{2-}$$

$$CO_3^{2-} + M^{2+} \longrightarrow MCO_3(s)$$

Permanent hard water contains significant concentrations of anions other than HCO_3^-, such as SO_4^{2-}. Adding Na_2CO_3 (washing soda) softens permanent hard water by precipitating cations such as Ca^{2+} and Mg^{2+} as carbonates, leaving an aqueous solution with $Na^+(aq)$. One annoying effect of hard water is encountered in the bath or shower. Water containing Ca^{2+} or Mg^{2+} ions forms a precipitate with soap. This familiar "bathtub ring" is actually a mixture of insoluble calcium and magnesium soaps, such as magnesium palmitate [see Equation (22.8) for the structure of sodium palmitate, a soluble soap]. Formation of these precipitates also makes it difficult for other soaps or shampoos to foam up.

Ion Exchange. One of the best ways to soften water is through **ion exchange,** a process in which the undesirable ions in hard water, typically Ca^{2+}, Mg^{2+}, and Fe^{3+}, are exchanged for ions that are not objectionable, such as Na^+. Ion exchange occurs when the hard water is passed through a column (or bed) containing an ion-exchange material. This material can be either a natural porous sodium aluminosilicate polymer called a *zeolite* or a synthetic resinous material. These polymeric materials ionize to produce two types of ions: *fixed* ions that remain attached to the polymer surface and free, or mobile, *counterions.* The counterions are the ones that exchange places with the undesirable ions when a sample of hard water is passed through the resin or zeolite.

Figure 22-9 depicts a resin with negatively charged fixed ions R^- and positively charged counterions. Initially, the counterions in the resin bed are Na^+. When hard water is passed through the bed, the more highly charged Ca^{2+}, Mg^{2+}, and Fe^{3+} displace Na^+ as counterions. You may have seen large bags of

A sodium soap is added to soft (distilled) water (*left*) and hard water (*right*). The cloudiness in the beaker on the right is caused by the formation of an insoluble calcium soap by the Ca^{2+} ions in the hard water.

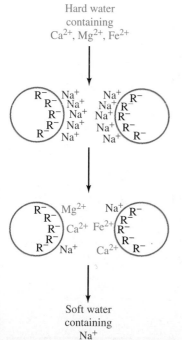

FIGURE 22-9
Ion-exchange process.
The resin pictured here is a cation-exchange resin: multivalent cations are exchanged for Na^+. The exchange can be represented as $2 \, NaR + M^{2+} \rightleftharpoons MR_2 + 2 \, Na^+$. The reaction occurs in the forward direction during water softening. As expected from Le Châtelier's principle, in the presence of concentrated $NaCl(aq)$, the reverse reaction is favored and the resin is recharged.

salt for use in water softeners for sale in stores. The salt is intended for preparing concentrated NaCl(aq) to regenerate the ion-exchange medium. When present in high concentration, Na^+ is able to displace the multivalent cations from the bed and restore the ion-exchange medium to its original condition. The ion-exchange material has an indefinite lifetime.

The only material consumed in water softening by ion exchange is the NaCl required for regenerating the ion-exchange medium. This method of water softening does have the disadvantage that the treated water has a high concentration of Na^+ and would not be appropriate for drinking by a person on a low-sodium diet.

Ion exchange can be used to make **deionized water,** the familiar "DI" water found in chemistry laboratories. Deionized water is used in laboratories because ions present in ordinary tap water may interfere with chemical reactions (by forming precipitates or catalyzing reactions, for example). Deionized water is made by first replacing the cations in a water sample with H^+. Then the water is passed through a second ion-exchange column that exchanges OH^- for all the *anions* present. The H^+ and OH^- combine to form H_2O, resulting in water that is essentially free of all ions.

22-4   Group 3A Metals: Aluminum, Gallium, Indium, and Thallium

Aluminum, gallium, indium, and thallium, in their appearance and physical properties and in much of their chemical behavior, are metallic. On the other hand, boron, the first element in Group 3A, is a nonmetal and will be discussed in Chapter 23. Some properties of the Group 3A metals are listed in Table 22.3.

The most important metal of the group is clearly aluminum. Its main use is in lightweight alloys, and over 5 million tons of aluminum are produced per year, on average, in the United States. Aluminum, like most of the other main group metals, is an active metal. Because it is easily oxidized to the 3+ ion, aluminum is an excellent reducing agent, for example, reacting with acids to reduce $H^+(aq)$ to $H_2(g)$.

$$2\ Al(s) + 6\ H^+(aq) \longrightarrow 2\ Al^{3+}(aq) + 3\ H_2(g) \qquad (22.19)$$

TABLE 22.3   Some Properties of the Group 3A Metals

	Al	Ga	In	Tl
atomic number	13	31	49	81
atomic (metallic) radius, pm	143	122	163	170
ionic (M^{3+}) radius, pm	53	62	79	88
electronegativity	1.5	1.6	1.7	1.8
first ionization energy, kJ/mol	577.6	578.8	558.3	589.3
electrode potential $E°$, V[a]	−1.676	−0.56	−0.34	+0.72
melting point, °C	660.37	29.78	156.17	303.55
boiling point, °C	2467	2403	2080	1457
density, g/cm³ at 20 °C	2.698	5.907	7.310	11.85
hardness[b]	2.75	1.5	1.2	1.25
electrical conductivity[b]	59.7	9.1	19.0	8.82

[a]For the reduction $M^{3+}(aq) + 3\ e^- \rightarrow M(s)$

[b]See footnotes to Table 22.1.

Aluminum is unusual in that it also reacts with basic solutions. This behavior is due to the acid properties of $Al(OH)_3$, a topic discussed further on page 785.

$$2\,Al(s) + 2\,OH^-(aq) + 6\,H_2O \longrightarrow 2[Al(OH)_4]^-(aq) + 3\,H_2(g) \qquad (22.20)$$

You may have had direct experience with this reaction if you have ever attempted to open up a clogged drain. Certain drain cleaners are a mixture of solid sodium hydroxide and granular aluminum. When the mixture is added to water, the reaction gives off heat that helps to melt fat and grease, while the NaOH(aq) reacts to dissolve them. The NaOH(aq) also reacts with the Al to give off gaseous hydrogen, which helps to dislodge obstructions and unplug the stopped-up drain.

Powdered aluminum is easily oxidized by air or other oxidants in highly exothermic reactions that are used in some rocket fuels and explosives.

$$2\,Al(s) + \tfrac{3}{2}\,O_2(g) \longrightarrow Al_2O_3(s) \qquad \Delta H = -1676\;kJ \qquad (22.21)$$

Aluminum is such a good reducing agent that it will extract oxygen from metal oxides to produce aluminum oxide while liberating the other metal in its free state. This reaction is known as the **thermite reaction** and is used in the on-site welding of large metal objects.

$$Fe_2O_3(s) + 2\,Al(s) \longrightarrow Al_2O_3(s) + 2\,Fe(l) \qquad (22.22)$$

Are You Wondering . . .

Why, if aluminum reacts with and dissolves in both acidic and basic solutions, it does not also dissolve in pH-neutral water?

Metallic aluminum reacts rapidly with the oxygen in air to give a thin, tough, water-insoluble coating of Al_2O_3. This oxide layer protects the metal beneath it from further reaction. In either acidic or basic solution, but not in pH-neutral water, the Al_2O_3 layer reacts and dissolves.

$$Al_2O_3(s) + 6\,H^+(aq) \longrightarrow 2\,Al^{3+}(aq) + 3\,H_2O$$

$$Al_2O_3(s) + 2\,OH^-(aq) + 3\,H_2O \longrightarrow 2\,[Al(OH)_4]^-(aq)$$

Once the Al_2O_3 layer has been removed, the underlying metal displays its true reactivity: Aluminum is a sufficiently active metal to displace $H_2(g)$ from pure water. We could never use aluminum metal in aircraft and building construction were it not for the protection of the Al_2O_3 surface coating.

Gallium metal is becoming increasingly important in the electronics industry. It is used to make gallium arsenide (GaAs), a compound that can convert light directly into electricity (photoconduction). This semiconducting material is also used in light-emitting diodes (LEDs; see Focus On, page 788) and in solid-state devices such as transistors.

Indium is a soft, silvery metal used to make low-melting alloys. Like GaAs, InAs also finds use in low-temperature transistors and as a photoconductor in optical devices.

▶A superconducting material loses its electrical resistance below a certain temperature. Metals typically become superconducting only a few degrees above 0 K.

▶Charles Martin Hall was a student at Oberlin College at the time he invented the electrolytic process for the production of aluminum.

Thallium and its compounds are extremely toxic and, because of this, they have few uses in industry. An exciting new possible use, however, is in high-temperature superconductors. Currently, the record maximum temperature for superconducting activity is 125 K for a thallium-based ceramic with the approximate formula $Tl_2Ba_2Ca_2Cu_3O_{8+x}$.

Production of Aluminum

When the aluminum cap was placed atop the Washington Monument in 1884, aluminum was still a semiprecious metal. It cost several dollars a pound to produce and was used mostly in jewelry and artwork. But just two years later all this changed. In 1886, Charles Martin Hall in the United States and Paul Héroult in France independently discovered an economically feasible method of producing aluminum from Al_2O_3 by electrolysis.

The manufacture of Al involves several interesting principles. The chief ore, *bauxite*, contains Fe_2O_3 as an impurity that must be removed. The principle used in the separation is that Al_2O_3 is an *amphoteric* oxide and dissolves in $NaOH(aq)$, whereas the iron oxide is a *basic* oxide and does not (Figure 22-10).

$$Al_2O_3(s) + 2\ OH^-(aq) + 3\ H_2O \longrightarrow 2\ [Al(OH)_4]^-(aq)$$

When the solution containing $[Al(OH)_4]^-$ is slightly acidified, $Al(OH)_3(s)$ precipitates. Pure Al_2O_3 is obtained by heating the $Al(OH)_3$.

$$[Al(OH)_4]^-(aq) + H_3O^+(aq) \longrightarrow Al(OH)_3(s) + 2\ H_2O$$

$$2\ Al(OH)_3(s) \xrightarrow{\Delta} Al_2O_3(s) + 3\ H_2O(g)$$

Al_2O_3 has a very high melting point (2020 °C) and produces a liquid that is a poor electrical conductor. Thus, its electrolysis is not feasible without a better conducting solvent. This was the crux of the discovery by Hall and Héroult. They found, independently, that up to 15% Al_2O_3, by mass, can be dissolved in the molten mineral *cryolite*, Na_3AlF_6, at about 1000 °C. The liquid is a good electrical conductor and so the electrolysis of Al_2O_3 is accomplished in molten cryolite. A typical electrolysis cell is pictured in Figure 22-11 and produces aluminum of 99.6–99.8% purity.

The electrode reactions are not known with certainty, but the net electrolysis reaction is

FIGURE 22-10
Purifying bauxite.
(a) When an excess of $OH^-(aq)$ is added to a solution containing $Al^{3+}(aq)$ and $Fe^{3+}(aq)$, the Fe^{3+} precipitates as $Fe(OH)_3(s)$ and the $Al(OH)_3(s)$ first formed redissolves to produce $[Al(OH)_4]^-(aq)$. (b) The $Fe(OH)_3(s)$ is filtered off, and the $[Al(OH)_4]^-(aq)$ is made slightly acidic through the action of CO_2, here added as dry ice. (c) The precipitated $Al(OH)_3(s)$ collects at the bottom of a clear, colorless solution.

(a)   (b)   (c)

FIGURE 22-11
Electrolysis cell for aluminum production.
The cathode is a carbon lining in a steel tank. The anodes are also made of carbon. Liquid aluminum is denser than the electrolyte medium and collects at the bottom of the tank.

Oxidation:	$3 \{C(s) + 2 \ O^{2-} \longrightarrow CO_2(g) + 4 \ e^-\}$
Reduction:	$4 \{Al^{3+} + 3 \ e^- \longrightarrow Al(l)\}$
Net:	$3 \ C(s) + 4 \ Al^{3+} + 6 \ O^{2-} \longrightarrow 4 \ Al(l) + 3 \ CO_2(g)$ (22.23)

The energy consumed to produce aluminum by electrolysis is very high, about 15,000 kWh per ton Al [compared, for example, with 3000 kWh per ton of Cl_2 in the electrolysis of NaCl(aq)]. This means that aluminum production facilities are generally found near low-cost power sources, typically hydroelectric power. The energy required to recycle Al is only about 5% of that to produce the metal from bauxite, and currently about 45% of the Al produced in the United States is by the recycling of scrap aluminum.

Are You Wondering . . .

Why so much energy is consumed in the electrolytic production of aluminum?

Any process that must be carried out at a high temperature requires large amounts of energy for heating. In the electrolytic production of Al, the electrolysis bath must be kept at about 1000 °C, which is done through electric heating. Two other factors, however, are also involved in the large energy consumption. First, in order to produce one mole of Al, *three* moles of electrons must be transferred: $Al^{3+} + 3 \ e^- \rightarrow Al(l)$. Additionally, the molar mass of Al is relatively low, 27 g/mol. The electric current equivalent to the passage of one mole of electrons produces only 9 g Al. By contrast, one mole of electrons produces 12 g Mg, 20 g Ca, or 108 g Ag. On the other hand, the same factors that make Al production a high energy consumer make Al a high energy producer when it is used in a battery. (Recall the aluminum–air battery described on page 746.)

FIGURE 22-12

Bonding in Al_2Cl_6.
Two Cl atoms bridge the $AlCl_3$ units and produce the dimer Al_2Cl_6. Electrons donated by these Cl atoms to Al atoms are denoted by arrows.

Aluminum Halides

Aluminum fluoride, AlF_3, has considerable ionic character. It has a high melting point (1040 °C) and is a conductor of electric current when molten. By contrast, the other halides exist as *molecular* species with the formula Al_2X_6 (for which X = Cl, Br, or I). We can think of this molecule as comprising two AlX_3 units, thus a **dimer.** The structure consists of two Cl atoms bonded exclusively to each Al atom and two Cl atoms bridging the two metal atoms (Figure 22-12). Bonding in this molecule can be described by sp^3 hybridization of the two Al atoms. Each bridging Cl atom appears to bond to two Al atoms in two ways: through the usual covalent bond to one Al atom, where each atom contributes one electron to the bond, and through a coordinate covalent bond to the other Al atom, where the chlorine atom provides the pair of electrons for the bond, noted by arrows in Figure 22-12.

The aluminum halides are very reactive Lewis acids and easily accept a pair of electrons to form an acid–base compound called an **adduct.** Adduct formation results in the formation of a covalent bond between the Lewis acid and Lewis base. In the following reaction, $AlCl_3$ is the Lewis acid and diethyl ether, $(C_2H_5)_2O$, is the Lewis base.

Aluminum halides are used in a number of organic reactions in which, through adduct formation, they act as catalysts.

The hydride ion, H^-, can be thought of as similar to a halide ion, and the AlH_4^- ion can be thought of as an adduct of AlH_3 and H^-. Lithium aluminum hydride, $LiAlH_4$, is an important reducing agent in organic chemistry.

Another important halide *complex* of aluminum is cryolite, Na_3AlF_6. Cryolite exists as $3 Na^+ + AlF_6^{3-}$ and in the molten state is the solvent and electrolyte in the Hall–Héroult process for producing aluminum metal. For aluminum manufacturing, natural cryolite has largely been displaced by cryolite synthesized in lead-clad vessels by the reaction

$$6 HF + Al(OH)_3 + 3 NaOH \longrightarrow Na_3AlF_6 + 6 H_2O \qquad (22.24)$$

Group 3A Oxides and Hydroxides

Aluminum oxide has several common names. It is often referred to as **alumina** and, when in crystalline form, is also called *corundum.* Corundum, when pure, is also known as the gemstone *white sapphire.* Certain other gemstones consist of corundum with small amounts of transition metal ions: Cr^{3+} in ruby and Fe^{3+} and Ti^{4+} in blue sapphire, for example. The reason for the beautiful colors in these and other gemstones is discussed on page 890.

The bonding and crystal structure of alumina account for its physical properties. The small Al^{3+} ion and small O^{2-} ion make a strong ionic bond. The crystal has a cubic closest packed structure of O^{2-} ions, with Al^{3+} ions filling the octahedral holes. Because of this structure, alumina is a very hard material and is often used as an abrasive. It is also resistant to heat (mp 2020 °C) and is used in linings for high-temperature furnaces and as catalyst supports in industrial chemical processes. Aluminum oxide is relatively unreactive except at very high temperatures. Its stability at high temperatures classifies it as a *refractory* material.

Aluminum hydroxide is *amphoteric*. It reacts with acids in a manner typical of metal hydroxides.

$$Al(OH)_3(s) + 3\ H_3O^+(aq) \longrightarrow [Al(H_2O)_6]^{3+}(aq) \qquad (22.25)$$

Also, it reacts with bases in a reaction best represented as the formation of a hydroxo complex ion.

$$Al(OH)_3(s) + OH^-(aq) \longrightarrow [Al(OH)_4]^-(aq) \qquad (22.26)$$

Thallium, unlike the other Group 3A metals, most often is in the +1 rather than +3 oxidation state in its compounds. Thus, thallium forms the oxide, Tl_2O and the hydroxide, TlOH. These compounds are ionic, and consequently TlOH is a strong base.

Aluminum, at the top of the group of four metals in Group 3A, is commonly in the +3 oxidation state in its compounds. Gallium tends to have the +3 oxidation state as well. Indium compounds can be found with +3 and +1 oxidation states, though the +3 is more common. In thallium, this preference is reversed. The higher stability of the +1 over the +3 oxidation state of thallium is often described as the **inert pair effect.** Thallium has the electron configuration $[Xe]4f^{14}5d^{10}6s^26p^1$. In forming the Tl^+ ion, a Tl atom loses the $6p$ electron and retains two electrons in its $6s$ subshell. It is this pair of electrons—$6s^2$—that is called the inert pair. The electron configuration $(n-1)s^2(n-1)p^6(n-1)d^{10}ns^2$ is commonly encountered in ions of the posttransition elements. One explanation of the inert pair effect is that the small bond energies and lattice energies associated with the large atoms and ions at the bottom of a group are not sufficiently great to offset the ionization energies of the ns^2 electrons.

22-5   Group 4A Metals: Tin and Lead

The properties of the Group 4A elements vary dramatically within the group. Tin and lead at the bottom of the group have mainly metallic properties. Germanium, sometimes referred to as a metalloid, exhibits semiconductor behavior. Silicon, although also exhibiting semiconductor properties, is mostly nonmetallic in its chemical behavior. Carbon, the first member of Group 4A, is a nonmetal.

The data in Table 22.4 suggest that tin and lead are rather similar to each other. Both are soft and malleable and melt at low temperatures. The ionization energies and standard electrode potentials of the two metals are also about the same. This means that their tendencies to be oxidized to the +2 oxidation state are comparable.

The fact that both tin and lead can exist in two oxidation states, +2 and +4, is an example of the inert pair effect mentioned in the previous section. In the +2 oxidation state the inert pair ns^2 is not involved in bond formation, whereas in the +4 oxidation state the pair does participate. Tin displays a stronger tendency to exist in the +4 oxidation state than does lead, an observation consistent with the trend observed in Group 3A, in which the common oxidation state is lower further down a group.

Another difference between tin and lead is that tin exists in two common crystalline forms (α and β), whereas lead has but a single solid form. The α (gray) or "nonmetallic" form of tin is stable below 13 °C, and the β (white) or "metallic" form of tin is stable above 13 °C. Ordinarily, when a sample of β (white) tin is cooled, it must be kept below 13 °C for a long time before the transition to α (gray) tin occurs. Once it does begin, however, the transforma-

TABLE 22.4 Some Properties of Tin and Lead

	Sn	Pb
atomic number	50	82
atomic (metallic) radius, pm	141	175
ionic (M^{2+}) radius, pm	93	118
first ionization energy, kJ/mol	709	716
electrode potential $E°$, V		
$\quad [M^{2+}(aq) + 2\,e^- \rightarrow M(s)]$	-0.137	-0.125
$\quad [M^{4+}(aq) + 2\,e^- \rightarrow M^{2+}(aq)]$	$+0.154$	$+1.5$
melting point, °C	232	327
boiling point, °C	2623	1751
density, g/cm³ at 20 °C	5.77 (α, gray)	11.34
	7.29 (β, white)	
hardness[a]	1.6	1.5
electrical conductivity[a]	14.4	7.68

[a]See footnotes to Table 22.1.

tion takes place rather rapidly and with dramatic results. Because α (gray) tin is less dense than the β (white), the tin expands and crumbles to a powder. This transformation has led to the disintegration of objects made of tin. It has been a particular problem in churches in colder climates because some organ pipes are made of tin or tin alloys. The transformation is known in northern Europe as the *tin disease, tin pest,* or *tin plague.* It added to the troubles of Napoleon's army in the siege of Moscow because the soldiers' buttons were made of tin. As the cold winter wore on, the buttons disintegrated.

The chief tin ore is tin(IV) oxide, SnO_2, known as *cassiterite*. After initial purification, the tin(IV) oxide is reduced with carbon (coke) to produce tin metal.

$$SnO_2(s) + C(s) \xrightarrow{\Delta} Sn(l) + CO_2(g) \tag{22.27}$$

Nearly 50% of the tin metal produced is used in tin plate, especially in plating iron for use in cans for storing foods. The next most important use (about 25%) is in the manufacture of **solders**—low-melting alloys used to join wires or pieces of metal. Other important alloys of tin are *bronze* (90% Cu, 10% Sn) and *pewter* (85% Sn, 7% Cu, 6% Bi, 2% Sb). Alloys of Sn and Pb are used to make organ pipes.

Lead is found chiefly as lead(II) sulfide, PbS, an ore known as *galena*. The lead(II) sulfide is first converted to lead(II) oxide by strongly heating it in air, a process called **roasting.** The oxide is then reduced with coke to produce the metal.

$$2\,PbS(s) + 3\,O_2(g) \xrightarrow{\Delta} 2\,PbO(s) + 2\,SO_2(g) \tag{22.28}$$

$$2\,PbO(s) + C(s) \xrightarrow{\Delta} 2\,Pb(l) + CO_2(g) \tag{22.29}$$

Over half the lead produced is used in lead–acid (storage) batteries. Other uses include the manufacture of solder and other alloys, ammunition, and radiation shields (to protect against X-rays).

Oxides

Tin forms two primary oxides, SnO and SnO_2. By heating it in air, SnO can be converted to SnO_2. One use of SnO_2 is as a jewelry abrasive.

Lead forms a number of oxides and the chemistry of some of these oxides is not completely understood. The best known oxides of lead are yellow PbO, *litharge,* red-brown lead dioxide, PbO_2, and a mixed-valency oxide known as *red lead,* Pb_3O_4. Lead oxides are used in the manufacture of lead–acid (storage) batteries, glass, ceramic glazes, cements (PbO), metal-protecting paints (Pb_3O_4), and matches (PbO_2). Other lead compounds are generally made from the oxides.

Because lead prefers to be in the +2 oxidation state, Pb(IV) compounds tend to undergo reduction to compounds of Pb(II) and are therefore good oxidizing agents. A case in point is PbO_2. In Chapter 21, we noted its use at the cathode in lead-acid storage cells. The reduction of $PbO_2(s)$ can be represented by the half-equation

$$PbO_2(s) + 4\ H^+(aq) + 2\ e^- \longrightarrow Pb^{2+}(aq) + 2\ H_2O \qquad E° = +1.455\ V$$

$PbO_2(s)$ is a better oxidizing agent than $Cl_2(g)$ and nearly as good as $MnO_4^-(aq)$. For example, $PbO_2(s)$ can oxidize HCl(aq) to $Cl_2(g)$.

$$PbO_2(s) + 4\ HCl(aq) \longrightarrow PbCl_2(aq) + 2\ H_2O + Cl_2(g) \qquad E°_{cell} = 0.097\ V$$

▶ Another mixed-valency oxide that we have encountered on several occasions in the text is Fe_3O_4 (see, for example, page 80).

Halides

The chlorides of tin—$SnCl_2$ and $SnCl_4$—both have important uses. Tin(II) chloride, $SnCl_2$, is a good reducing agent and is used in the quantitative analysis of iron ores to reduce Fe(III) to Fe(II) in aqueous solution. Tin(IV) chloride, $SnCl_4$, is formed by the direct reaction of tin and $Cl_2(g)$; it is the form in which tin is recovered from scrap tin plate. Tin(II) fluoride, SnF_2 (stannous fluoride), has an important use as an anticavity additive to toothpaste.

Other Compounds

One of the few soluble lead compounds is lead(II) nitrate, $Pb(NO_3)_2$. It is formed in the reaction of PbO_2 with nitric acid.

$$2\ PbO_2(s) + 4\ HNO_3(aq) \longrightarrow 2\ Pb(NO_3)_2(aq) + 2\ H_2O + O_2(g)$$

Addition of a soluble chromate salt to $Pb(NO_3)_2(aq)$ produces the pigment lead(II) chromate (*chrome yellow*), $PbCrO_4$. Another lead-based pigment used in ceramic glazes and once extensively used in the manufacture of paint is *white lead,* a basic lead carbonate, $2PbCO_3 \cdot Pb(OH)_2$.

Lead Poisoning

Beginning with the ancient Romans and continuing to fairly recent times, lead has been used in plumbing systems, including those designed to transport water. Exposure to lead has also occurred through cooking and eating utensils and pottery glazes. In colonial times, lead poisoning was clearly diagnosed as the cause of "dry bellyache" suffered by North Carolinians who consumed rum made in New England. The distilling equipment used in the manufacture of the rum was found to have components made from lead.

Mild forms of lead poisoning produce nervousness and mental depression. More severe cases can lead to permanent nerve, brain, and kidney damage. Lead interferes with the biochemical reactions that produce the iron-containing heme group in hemoglobin. As little as 10–15 μg Pb/100 mL blood seems to produce physiological effects, especially in small children. The phasing out of leaded gasoline has resulted in a dramatic drop in average lead blood levels. The principal sources of lead contamination now seem to be lead-based painted surfaces found in old buildings and soldered joints in plumbing systems. Lead has been eliminated from modern plumbing solder, which is now a mixture of 95% Sn–5% Sb. Because of lead's toxicity, its disposal is closely monitored, and recycling provides about three-quarters of current lead metal production.

Focus On   Gallium Arsenide

Solar cells employing gallium arsenide (gold and red) placed on a silicon-based solar cell. The cells containing gallium arsenide are more efficient and can be made smaller than a comparable silicon cell.

Until about 15 years ago, the principal application of gallium metal was in high-temperature thermometers. Now an unlikely compound—gallium arsenide, (GaAs)—has become one of the most versatile high-technology materials of our time. Among the properties of gallium arsenide is the ability to convert electric energy into light. A crystal of GaAs can serve as the light-generating component of a light-emitting diode (LED), a type of device found in indicator lights on stereo equipment and calculator displays. GaAs can also be fabricated into diode lasers, very small lasers used in compact disc systems and for transmitting infrared light through fiber-optic cables.

Recall the discussion of the band theory and semiconductors on pages 410–413. Gallium arsenide is an intrinsic semiconductor. This compound with eight valence electrons per formula unit (three from Ga and five from As) has its valence band filled; but if sufficient electric energy is supplied, electrons are

Summary

The alkali (Group 1A) metals are the most active of the metals, as indicated by their low ionization energies and large negative electrode potentials. Most of the metals are prepared by the electrolysis of their molten salts. Electrolysis of NaCl(aq) produces NaOH(aq), from which many other sodium compounds can then be prepared. Na_2CO_3 can be produced from NaCl, NH_3, and $CaCO_3$ by the Solvay process.

The alkaline earth (Group 2A) metals are also very active. Some are prepared by the electrolysis of a molten salt and some by chemical reduction. Among the most important of the alkaline earth compounds are the carbonates, especially $CaCO_3$. Reversible reactions involving CO_3^{2-}, HCO_3^-, $CO_2(g)$, and H_2O account for the formation of limestone caves and temporary hard water. Water may also have noncarbonate, or permanent, hardness. Water can be softened either through chemical reactions or by ion exchange.

The principal metal of Group 3A is aluminum, whose large-scale use is made possible by an effective method of production. The amphoterism of Al_2O_3 is the basis for separating Al_2O_3 from impurities, mostly Fe_2O_3. Electrolysis is carried out in molten Na_3AlF_6 with Al_2O_3 as a solute. Gallium has gained importance in the electronics industry because of the desirable semiconductor properties of gallium arsenide (GaAs). The use of thallium in fabricating high-temperature superconductors is also a possibility.

Tin and lead in Group 4A have some similarities—they are soft metals with low melting points—and some differences. Among the differences is that tin acquires the oxidation state +4 rather easily, whereas the +2 oxidation state is favored by lead.

promoted to the conduction band. When the excited electrons return to the ground state (valence band), they emit photons of light with a wavelength proportional to the energy associated with the band gap (ΔE in Figure 12-28).

What is special about GaAs semiconductors is that the band gap can be "tuned" by adding controlled amounts of another semiconductor—gallium phosphide. GaAs and GaP form solid solutions whose compositions can be varied in all proportions. GaP has a band gap energy corresponding to that of green light (540 nm), whereas the band gap energy in GaAs corresponds to infrared light (890 nm).* A series of solid solutions of GaAs and GaP can be represented by the general formula GaP_xAs_{1-x}. The band gap increases with increased amounts of phosphorus, and so the emitted light increases in energy. Ga–P–As semiconductors, whose compositions fall between those of GaP and GaAs, emit light of wavelength be-

tween 540 and 890 nm. For example, the most common LED, with the formula $GaP_{0.40}As_{0.60}$, emits 660-nm red light.

The effect of the composition of Ga–P–As semiconductors on the color of the emitted light is due in part to electronegativity differences (ΔEN) and in part to differences in atomic sizes. In GaP, ΔEN is slightly greater, and the length of the unit cell in the crystal lattice is smaller than in GaAs. These factors lead to more ionic character in the bonds and a larger band gap energy in GaP than in GaAs. Thus the emitted light from GaP should be of a shorter wavelength than that from GaAs.

An interesting effect is noted when the temperature of an LED is changed. When an LED with the composition $GaP_{0.40}As_{0.60}$ is cooled to the boiling point of liquid nitrogen (77 K), the color shifts from that of the 660-nm red light observed at room temperature to a 639-nm *orange* light. This effect is attributed to a slight contraction of

the crystal lattice of the semiconductor, which brings the atoms into closer contact and increases the band gap energy.

These new semiconductor materials require new methods for their synthesis. Instead of precipitation from solution, the technique used is chemical vapor deposition (CVD). A chemical reaction is carried out between trimethylgallium, $Ga(CH_3)_3(g)$, and highly toxic arsine, $AsH_3(g)$; GaAs is deposited from the vapor phase as a thin film. If an appropriate amount of phosphine, $PH_3(g)$, is included in the reaction mixture, the film produced is GaP_xAs_{1-x}.

The application of gallium arsenide described here and its use in high-speed computers have changed the demand for GaAs from practically none in the early 1980s to a current worldwide market now measured in the hundreds of millions of dollars.

*G. Lisensky, R. Penn, M. Geselbracht, and A. Ellis, *J. Chem. Educ.* **69,** 151 (1992).

Summarizing Example

Not only is the ion-exchange method of Figure 22-9 useful for softening water, it can also be used to determine the hardness of water. A sample of the water is passed through a column filled with the resin, HR. If we assume that the only cation present in the water is Ca^{2+}, the ion-exchange reaction is $Ca^{2+} + 2\ HR \rightarrow CaR_2 + 2\ H^+$.

A 25.00-mL sample of hard water is passed through the ion-exchange column, HR. The water coming off the column requires 7.59 mL of 0.0133 M NaOH for its titration. What is the hardness of the water, expressed as parts per million of Ca^{2+}?

1. *Determine the number of millimoles of H^+ present in the sample.* The titration reaction is simply: $H^+ + OH^- \rightarrow H_2O$.

The number of millimoles of OH^- is the product of the volume (mL) and concentration of the NaOH(aq). *Result:* 0.101 mmol H^+.

2. *Determine the mass of Ca^{2+} in the 25.00-mL water sample.* Convert from millimoles of H^+ to millimoles of Ca^{2+} with a factor from the ion-exchange equation. Convert from millimoles of Ca^{2+} to grams of Ca^{2+}. *Result:* 2.02×10^{-3} g Ca^{2+}.

3. *Express the quantity of Ca^{2+} in the water as parts per million.* The result in part 2 is the mass of Ca^{2+} per 25.00 mL water. Convert this result to the mass of Ca^{2+} per 10^6 mL of water. Assume $d = 1.00$ g/mL for the water. *Answer:* 80.8 ppm Ca^{2+}.

Key Terms

adduct (22-4)
alumina (22-4)
calcination (22-2)
deionized water (22-3)
detergent (22-1)
diagonal relationship (22-1)
dimer (22-4)

inert pair effect (22-4)
ion exchange (22-3)
permanent hard water (22-3)
plaster of Paris (22-2)
quicklime (22-2)
roasting (22-5)
slaked lime (22-2)

soap (22-1)
solders (22-5)
stalactites (22-2)
stalagmites (22-2)
temporary hard water (22-3)
thermite reaction (22-4)

Review Questions

1. In your own words define the following terms: **(a)** dimer; **(b)** adduct; **(c)** calcination; **(d)** amphoteric oxide.

2. Briefly describe each of the following ideas, methods, or phenomena: **(a)** diagonal relationship; **(b)** preparation of deionized water by ion exchange; **(c)** thermite reaction; **(d)** inert pair effect.

3. Explain the important distinction between each pair of terms: **(a)** peroxide and superoxide; **(b)** quicklime and slaked lime; **(c)** temporary and permanent hard water; **(d)** soap and detergent.

4. Provide an acceptable name or formula for each of the following: **(a)** PbO; **(b)** SnF_2; **(c)** $CaSO_4 \cdot \frac{1}{2}H_2O$; **(d)** lithium nitride; **(e)** $Ca(OH)_2$; **(f)** potassium superoxide; **(g)** magnesium hydrogen carbonate.

5. Complete and balance the following. Write the simplest equation possible. If no reaction occurs, so state.

(a) $Li_2CO_3(s) \xrightarrow{\Delta}$
(b) $CaCO_3(s) + HCl(aq) \longrightarrow$
(c) $Al(s) + NaOH(aq) \longrightarrow$
(d) $BaO(s) + H_2O(l) \longrightarrow$
(e) $Na_2O_2(s) + CO_2(g) \longrightarrow$

6. Assuming the availability of water, common reagents (acids, bases, salts), and simple laboratory equipment, give a practical method to prepare **(a)** $MgCl_2$ from $MgCO_3(s)$; **(b)** $NaAl(OH)_4$ from $Na(s)$ and $Al(s)$; **(c)** Na_2SO_4 from $NaCl(s)$.

7. Write the simplest chemical equation to represent the reaction of **(a)** $K_2CO_3(aq)$ and $Ba(OH)_2(aq)$; **(b)** $Mg(HCO_3)_2(aq)$ upon heating; **(c)** tin(II) oxide when heated with carbon; **(d)** $CaF_2(s)$ and $H_2SO_4(concd\ aq)$; **(e)** $NaHCO_3(s)$ and $HCl(aq)$; **(f)** $PbO_2(s)$ and $HBr(aq)$.

8. Write an equation to represent the reaction of gypsum,

$CaSO_4 \cdot 2H_2O$, with ammonium carbonate to produce ammonium sulfate (a fertilizer), calcium carbonate, and water.

9. All but one of the following can be used to soften temporary hard water: NH_3, Na_2CO_3, NH_4Cl, $Ca(OH)_2$. Which one will not, and why?

10. Write chemical equations to represent the most probable outcome in each of the following. If no reaction is likely to occur, so state.

(a) $SrCO_3(s) \xrightarrow{\Delta}$ (b) $Al_2O_3(s) \xrightarrow{\Delta}$
(c) $Li_2CO_3(s) \xrightarrow{\Delta}$

11. A chemical dictionary gives the following descriptions of the production of some compounds. Write plausible chemical equations based on these descriptions.

(a) Lead carbonate: adding a solution of sodium bicarbonate to a solution of lead nitrate.

(b) Lithium carbonate: reaction of lithium oxide with ammonium carbonate solution.

(c) Hydrogen peroxide: by the action of dilute sulfuric acid on barium peroxide.

(d) Lead(IV) oxide: action of an alkaline solution of calcium hypochlorite on lead(II) oxide.

12. One of the following metals does not react with cold water: K, Sr, Al, Na. Indicate which one does not react and explain why.

13. All of the following substances react with water. Which pair yields the same gaseous products: **(a)** Ca and CaH_2; **(b)** Na and Na_2O_2; **(c)** K and KO_2?

14. Name the chemical compound(s) you would expect as the *primary* constituent(s) of **(a)** stalactites; **(b)** gypsum; **(c)** "bathtub ring"; **(d)** "barium milkshake"; **(e)** blue sapphires.

Exercises

Alkali (Group 1A) Metals

15. Use information from the chapter to write chemical equations to represent the following.
(a) Reaction of cesium metal with chlorine gas.
(b) Formation of sodium peroxide (Na_2O_2).
(c) Thermal decomposition of lithium carbonate.
(d) Reduction of sodium sulfate to sodium sulfide.
(e) Combustion of potassium to form potassium superoxide.

16. Use information from the chapter to write chemical equations to represent the following.
(a) Reaction of rubidium metal with water.
(b) Thermal decomposition of aqueous $KHCO_3$.
(c) Combustion of lithium metal in oxygen gas.
(d) Action of concentrated aqueous H_2SO_4 on KCl(s).
(e) Reaction of lithium hydride with water.

17. A pure white solid is either LiCl or KCl. Describe a simple test to determine which it is.

19. Write net ionic equations for the reactions of the normal oxide, peroxide, and superoxide ions with water that are referred to on page 771.

21. A 0.872-L NaCl(aq) solution is electrolyzed for 2.50 min with a current of 0.810 A.
(a) Calculate the pH of the solution after electrolysis.
(b) Does the result depend on the initial concentration of the NaCl(aq)? Explain.

23. An analysis of a Solvay plant shows that for every 1.00 ton of NaCl consumed 1.03 tons of $NaHCO_3$ are obtained. The quantity of NH_3 consumed in the overall process is 1.5 lb.
(a) What is the percent efficiency of this process for converting NaCl to $NaHCO_3$?
(b) Why is so little NH_3 required?

18. An unknown compound is either Li_2CO_3 or K_2CO_3. Describe two methods to determine the identity of the unknown.
20. The first electrolytic process to produce sodium metal used molten NaOH as the electrolyte. Write probable half-equations and a net equation for this electrolysis.
22. A lithium battery used in a cardiac pacemaker has a voltage of 3.0 V and a capacity of 0.50 A h (ampere hour). Assume that to regulate the heartbeat requires 5.0 μW of power. (*Hint:* 1 μW = 1 × 10^{-6} watt, and 1 W = 1 J/s.)
(a) How long will the implanted battery last?
(b) How many grams of lithium must be present in the battery for the lifetime calculated in **(a)**?
24. Consider the reaction
$$Ca(OH)_2(s) + Na_2SO_4(aq) \rightleftharpoons CaSO_4(s) + 2\ NaOH(aq)$$
(a) Write a net ionic equation for the reaction.
(b) Will the reaction go essentially to completion? [*Hint:* What is K_{eq} for the equation in **(a)**?]
(c) What will be $[SO_4^{2-}]$ and $[OH^-]$, *at equilibrium,* if a slurry of $Ca(OH)_2(s)$ is mixed with 1.00 M $Na_2SO_4(aq)$?

Alkaline Earth (Group 2A) Metals

25. In the manner used to construct Figure 22-2, complete the diagram outlined. Specifically, indicate the reactants (and conditions) you would use to produce the indicated substances from $Ca(OH)_2$.

26. Replace the calcium-containing substances shown in the diagram accompanying Exercise 27 by their magnesium-containing equivalents. Then describe the reactants (and conditions) you would use to produce the indicated substances from $MgSO_4$.

27. Write chemical equations to represent the following:
(a) Reduction of BeF_2 to Be metal with Mg as a reducing agent.
(b) Reaction of barium metal with $Br_2(l)$.
(c) Reduction of uranium(IV) oxide to uranium metal with calcium as the reducing agent.
(d) Calcination of dolomite, a mixed calcium magnesium carbonate ($MgCO_3 \cdot CaCO_3$).
(e) Complete neutralization of phosphoric acid with quicklime.

29. *Without performing detailed calculations,* indicate if equilibrium is displaced either far to the left or far to the right for each of the following. Use data from Appendix D, as necessary.
(a) $BaSO_4(s) + CO_3^{2-}(aq) \rightleftharpoons BaCO_3(s) + SO_4^{2-}(aq)$
(b) $Mg_3(PO_4)_2(s) + 3\ CO_3^{2-}(aq) \rightleftharpoons$
$$3\ MgCO_3(s) + 2\ PO_4^{3-}(aq)$$
(c) $Ca(OH)_2(s) + 2\ F^-(aq) \rightleftharpoons CaF_2(s) + 2\ OH^-(aq)$

28. Write chemical equations for the reactions you would expect to occur when
(a) $Mg(HCO_3)_2(s)$ is heated to a high temperature.
(b) $BaCl_2(l)$ is electrolyzed.
(c) Sr(s) is added to cold dilute HBr(aq).
(d) $Ca(OH)_2(aq)$ is added to $H_2SO_4(aq)$.
(e) $CaSO_4 \cdot 2H_2O(s)$ is heated.

30. *Without performing detailed calculations,* indicate why you would expect each of the following reactions to occur to a significant extent as written. Use data from Appendix D, as necessary.
(a) $BaCO_3(s) + 2\ HC_2H_3O_2(aq) \rightarrow$
$$Ba^{2+}(aq) + 2\ C_2H_3O_2^-(aq) + H_2O + CO_2(g)$$
(b) $Ca(OH)_2(s) + 2\ NH_4^+(aq) \rightarrow$
$$Ca^{2+}(aq) + 2\ NH_3(aq) + 2\ H_2O$$
(c) $BaF_2(s) + 2\ H_3O^+(aq) \rightarrow$
$$Ba^{2+}(aq) + 2\ HF(aq) + 2\ H_2O$$

Hard Water

31. A certain hard water contains 185.0 ppm HCO_3^-, and all the cations associated with the HCO_3^- are Ca^{2+}.
(a) What is the concentration of Ca^{2+}, expressed in parts per million?
(b) What is the concentration of Ca^{2+}, expressed in milligrams per liter?
(c) How many grams of $Ca(OH)_2$ are required to soften 1.00×10^6 g of this water?

33. A sample of water has a hardness expressed as 86.4 ppm Ca^{2+}. This sample is passed through an ion-exchange column and the Ca^{2+} is replaced by H^+. What is the pH of the water after it has been so treated? [*Hint:* Base your calculation on 1000 L (1.00×10^6 g) of water.]

35. What mass of "bathtub ring" will form if 32.1 L of water having 82.6 ppm Ca^{2+} is treated with an excess of the soap potassium stearate, $CH_3(CH_2)_{16}COO^- K^+$?

32. Concerning the particular hard water described in Exercise 31,
(a) How many grams of $CaCO_3$ would be precipitated in softening 1.00×10^6 g of this water with $Ca(OH)_2$?
(b) Show that in the $CaCO_3$, half the Ca^{2+} is derived from the $Ca(OH)_2$ used in the water softening and half from the water itself.

34. A sample of water has its hardness due only to $CaSO_4$. When this water is passed through an *anion*-exchange resin, SO_4^{2-} ions are replaced by OH^-. A 25.00-mL sample of the water so treated requires 22.42 mL of 1.00×10^{-3} M H_2SO_4 for its titration. What is the hardness of the water, expressed in parts per million of $CaSO_4$? Assume the density of the water is 1.00 g/mL.

36. One of the components of "bathtub ring" is magnesium palmitate, $Mg(C_{16}H_{31}O_2)_2$. A saturated solution of magnesium palmitate is found to contain 80 ppm $Mg^{2+}(aq)$. What is the K_{sp} of magnesium palmitate?

Aluminum

37. Write chemical equations to represent the
(a) reaction of Al(s) with HCl(aq).
(b) reaction of Al(s) with $Br_2(l)$.
(c) reaction of solid aluminum with NaOH(aq).
(d) oxidation of Al(s) to $Al^{3+}(aq)$ by an aqueous solution of sulfuric acid. The reduction product is $SO_2(g)$.

39. In some foam-type fire extinguishers the reactants are $Al_2(SO_4)_3(aq)$ and $NaHCO_3(aq)$. When the extinguisher is activated, these reactants mix, producing $Al(OH)_3(s)$ and $CO_2(g)$. The $Al(OH)_3$–CO_2 foam extinguishes the fire. Write a net ionic equation to represent this reaction.

41. The maximum resistance to corrosion of aluminum is between pH 4.5 and 8.5. Explain how this observation is consistent with other facts about the behavior of aluminum presented in the text.

43. In the purification of bauxite ore, a preliminary step in the production of aluminum, $[Al(OH)_4]^-(aq)$ can be converted to $Al(OH)_3(s)$ by passing $CO_2(g)$ through the solution. Write an equation for the reaction that occurs.

45. Concerning the thermite reaction,
(a) Use data from Appendix D to calculate $\Delta H°$ at 298 K for the reaction

$$2\,Al(s) + Fe_2O_3(s) \longrightarrow 2\,Fe(s) + Al_2O_3(s)$$

(b) Write an equation for the reaction when $MnO_2(s)$ is substituted for $Fe_2O_3(s)$, and calculate $\Delta H°$ for this reaction.
(c) Show that if MgO were substituted for Fe_2O_3 the reaction would be *endothermic*.

38. Write plausible equations for the
(a) production of aluminum chloride by passing $Cl_2(g)$ over a heated mixture of aluminum oxide and carbon, first accomplished in 1825.
(b) production of Al by heating aluminum chloride with potassium, first accomplished in 1827.
(c) production of Cr from $Cr_2O_3(s)$ by the thermite reaction, with Al as the reducing agent.
(d) separation of Fe_2O_3 impurity from bauxite ore.

40. Some baking powders contain the solids $NaHCO_3$ and $NaAl(SO_4)_2$. When water is added to this mixture of compounds, $CO_2(g)$ and $Al(OH)_3(s)$ are two of the products. Write plausible net ionic equations for the formation of these two products.

42. Describe a series of *simple* chemical reactions that you could use to determine whether a particular metal sample is Aluminum 2S (99.2% Al) or magnalium (70% Al, 30% Mg). You are permitted to destroy the metal sample in the testing.

44. Exercise 43 describes how $Al(OH)_3(s)$ can be obtained by treating $[Al(OH)_4]^-(aq)$ with $CO_2(aq)$. Could HCl(aq) be used instead of $CO_2(g)$? Explain.

46. Use the relationship $\Delta G° = -n\mathcal{F}E°_{cell}$ to estimate the minimum voltage required to electrolyze Al_2O_3 in the Hall-Héroult process (22.23). For this purpose use $\Delta G_f°[Al_2O_3(l)] = -1520$ kJ/mol and $\Delta G_f°[CO_2(g)] = -394$ kJ/mol. Show that the oxidation of the graphite anode to $CO_2(g)$ permits the electrolysis to occur at a lower voltage than if the electrolysis reaction were $Al_2O_3(l) \rightarrow 2\,Al(l) + \frac{3}{2}\,O_2(g)$.

Tin and Lead

47. Write plausible chemical equations for the **(a)** dissolving of lead(II) oxide in nitric acid; **(b)** heating of $SnCO_3(s)$; **(c)** reduction of lead(II) oxide by carbon; **(d)** reduction of $Fe^{3+}(aq)$ to $Fe^{2+}(aq)$ by $Sn^{2+}(aq)$; **(e)** formation of lead(II) sulfate during high-temperature roasting of lead(II) sulfide.

49. Lead dioxide, PbO_2, is a good oxidizing agent. Use appropriate data from Appendix D to determine whether $PbO_2(s)$ in a solution with $[H_3O^+] = 1$ M is a sufficiently good oxidizing agent to carry the following oxidations to the point at which the concentration of the species being oxidized decreases to one-thousandth of its initial value.
(a) $Fe^{2+}(1$ M$)$ to Fe^{3+} **(b)** $SO_4^{2-}(1$ M$)$ to $S_2O_8^{2-}$
(c) $Mn^{2+}(1 \times 10^{-4}$ M$)$ to MnO_4^-

48. Write plausible chemical equations for preparing each compound from the indicated starting material. Discuss what reagents (acids, bases, salts) and equipment commonly available in the laboratory are needed for each reaction. **(a)** $SnCl_2$ from SnO; **(b)** $SnCl_4$ from Sn; **(c)** $PbCrO_4$ from PbO_2.

50. Tin(II) ion, Sn^{2+}, is a good reducing agent. Use data from Appendix D to determine whether Sn^{2+} is a sufficiently good reducing agent to reduce **(a)** I_2 to I^-; **(b)** Fe^{2+} to $Fe(s)$; **(c)** Cu^{2+} to $Cu(s)$; **(d)** $Fe^{3+}(aq)$ to $Fe^{2+}(aq)$. Assume all reactants and products are in their standard states.

Advanced Exercises

51. The melting point of $NaCl(s)$ is 801 °C, much higher than that of $NaOH$ (322 °C). More energy is consumed to melt and maintain molten $NaCl$ than $NaOH$. Yet the preferred commercial process for the production of sodium is electrolysis of $NaCl(l)$ rather than $NaOH(l)$. Give a reason(s) for this.

52. When a 0.200-g sample of Mg is heated in air, 0.315 g of product is obtained. Assume that all the Mg appears in the product.
(a) If the product were pure MgO, what mass should have been obtained?
(b) Show that the 0.315-g product could be a mixture of MgO and Mg_3N_2.
(c) What is the mass percent of MgO in the MgO–Mg_3N_2 mixed product?

53. Comment on the feasibility of using a reaction similar to (22.2) to produce **(a)** lithium metal from $LiCl$ and **(b)** cesium metal from $CsCl$, with $Na(l)$ as the reducing agent in each case. (*Hint:* Consider both physical properties and atomic properties listed in Table 22.1.)

54. The electrolysis of 0.250 L of 0.220 M $MgCl_2$ is conducted until 104 mL of gas (a mixture of H_2 and water vapor) is collected at 23 °C and 748 mmHg. Will $Mg(OH)_2(s)$ precipitate if electrolysis is carried to this point? (Use 21 mmHg as the vapor pressure of the solution.)

55. A particular water sample contains 56.9 ppm SO_4^{2-} and 176 ppm HCO_3^-, with Ca^{2+} as the only cation.
(a) How many parts per million of Ca^{2+} does the water contain?
(b) How many grams of CaO are consumed in removing HCO_3^- from 602 kg of the water?
(c) Show that the Ca^{2+} remaining in the water after the treatment described in **(b)** can be removed by adding Na_2CO_3.
(d) How many grams of Na_2CO_3 are required for the precipitation referred to in **(c)**?

56. An Al production cell of the type pictured in Figure 22-11 operates at a current of 1.00×10^5 A and a voltage of 4.5 V. The cell is 38% efficient in using electric energy to produce chemical change. (The rest of the electric energy is dissipated as thermal energy in the cell.)

(a) What mass of Al can be produced by this cell in 8.00 h?
(b) If the electric energy required to power this cell is produced by burning coal (85% C; heat of combustion of C = 32.8 kJ/g) in a power plant with 35% efficiency, what mass of coal must be burned to produce the mass of Al determined in part **(a)**?

57. At 20 °C a saturated aqueous solution of $Pb(NO_3)_2$ maintains a relative humidity of 97%. What must be the composition of this solution, expressed as g $Pb(NO_3)_2/100.0$ g H_2O? (*Hint:* Review the definition of relative humidity in Section 8-1. Use vapor pressure data from Table 13.2. What is the relationship between the concentration and vapor pressure of a solution?)

58. Use information from this chapter and elsewhere in the text to explain why neither the compound $PbBr_4$ nor the compound PbI_4 exists.

59. To prevent the air oxidation of aqueous solutions of Sn^{2+} to Sn^{4+}, sometimes metallic tin is kept in contact with the $Sn^{2+}(aq)$. Suggest how this helps to prevent the oxidation.

60. Verify the value of K_{eq} for reaction (22.15), that is,
$$CaCO_3(s) + H_2O + CO_2 \rightleftharpoons Ca(HCO_3)_2(aq)$$
$$K_{eq} = 2.6 \times 10^{-5}$$
by combining values of K_{a_1} and K_{a_2} from reactions (22.13) and (22.14) and K_{sp} of $CaCO_3$.

61. The dissolution of $MgCO_3(s)$ in $NH_4^+(aq)$ can be represented as
$$MgCO_3(s) + NH_4^+(aq) \rightleftharpoons$$
$$Mg^{2+}(aq) + HCO_3^-(aq) + NH_3(aq)$$
Calculate the molar solubility of $MgCO_3$ in each of the following solutions: **(a)** 1.00 M $NH_4Cl(aq)$; **(b)** a buffer that is 1.00 M NH_3 and 1.00 M NH_4Cl; **(c)** a buffer that is 0.100 M NH_3 and 1.00 M NH_4Cl.

62. Show that, in principle, $Na_2CO_3(aq)$ can be converted almost completely to $NaOH(aq)$ by the reaction
$$Ca(OH)_2(s) + Na_2CO_3(aq) \longrightarrow CaCO_3(s) + 2\ NaOH(aq)$$
(*Hint:* Use solubility product data from Appendix D.)

63. Assume that the packing of spherical atoms in the crystalline metals is the same for Li, Na, and K, and explain why Na has a higher density than *both* Li and K. (*Hint:* Use data from Table 22.1.)

Main Group Elements II: Nonmetals

23

We have already discussed a number of aspects of the chemistry of the nonmetals. In Chapter 8 we learned that the principal gases of the atmosphere (N_2, O_2, Ar) are nonmetals, and that some of the significant trace components in the atmosphere (for example, the oxides of carbon, nitrogen, and sulfur) are compounds of the nonmetals. Also, in several instances we have encountered chemical reactions of the nonmetals that are of natural or commercial importance, as in the natural and artificial fixation of nitrogen, the depletion of stratospheric ozone, and the formation of acid rain. In this chapter our focus will be entirely on the nonmetals, and we will describe their behavior in a more systematic way.

After a brief survey of a special group of nonmetals—the noble gases—we will first study the most active group of nonmetals, the halogens of Group 7A. Then we will proceed through the periodic table from right to left, moving toward the elements with increasing metallic character. This approach will be a counterpoint to that of Chapter 22, where we progressed from the most metallic elements (Group 1A) toward the elements having more nonmetallic character. Among the fundamental concepts we will emphasize in the discussion are atomic, physical, and thermodynamic properties; bonding and structure; acid–base chemistry; and oxidation states, redox reactions, and electrochemistry.

Prior to the 1800s, paper had a high rag content. In the 1800s wood pulp became the primary source of paper. This paper, however, is easily decomposed by acids, as seen above. One source of acid is SO_2, the acid anhydride of H_2SO_3(aq) and a common air pollutant. Another source is the reaction between O_2 and lignin in the paper to produce various acids. Modern papermaking avoids these problems by reducing the lignin content (see page 808) and incorporating basic materials such as $CaCO_3$ in the paper.

23-1   Group 8A: The Noble Gases

The elements of Group 8A were isolated at the end of the nineteenth century and are now referred to as the *noble gases*. Their production and uses were discussed in Section 8-4. Initially, the noble gases were found to be chemically inert, and this apparent inertness helped to provide a theoretical framework for the Lewis theory of bonding.

In the 1930s, Linus Pauling did theoretical calculations suggesting that xenon should form oxide and fluoride compounds, but attempts to make them failed at that time. In 1962, N. Bartlett and D. H. Lohmann discovered that O_2 and PtF_6 would join in a 1:1 mole ratio to form the compound O_2PtF_6. Properties of this compound suggest it to be ionic: $[O_2]^+[PtF_6]^-$. The energy required to extract an electron from O_2 is 1177 kJ/mol, almost identical to the first ionization energy of Xe, 1170 kJ/mol. The size of the Xe atom is also roughly the same as that of the dioxygen molecule. Thus, it was reasoned that the compound $XePtF_6$ might exist. Bartlett and Lohmann were able to prepare a yellow crystalline solid with this composition.* Soon thereafter, several noble gas compounds were synthesized by chemists around the world. In general, the conditions necessary to form noble gas compounds are, as Pauling predicted,

- a readily ionizable (therefore, high atomic number) noble gas atom, and
- highly electronegative atoms (such as F or O) to bond to it.

Xenon compounds have been synthesized that have Xe in four possible oxidation states.

	+2	**+4**	**+6**	**+8**
Examples:	XeF_2	XeF_4, $XeOF_2$	XeF_6, XeO_3	XeO_4, H_4XeO_6

Because it is *difficult* to oxidize Xe to these positive oxidation states, we should expect Xe compounds to be *easily* reduced, making them very strong oxidizing agents. For example, in aqueous acidic solution

$$XeF_2(aq) + 2\ H^+(aq) + 2\ e^- \longrightarrow Xe(g) + 2\ HF(aq) \qquad E° = +2.64\ V$$

The significance of this large $E°$ value is that XeF_2 is not very stable in aqueous solution; XeF_2 oxidizes the water, producing $O_2(g)$.

Reduction:	$2\ \{XeF_2(aq) + 2\ H^+(aq) + 2\ e^- \longrightarrow Xe(g) + 2\ HF(aq)\}$	$E° = +2.64\ V$
Oxidation:	$2\ H_2O \longrightarrow 4\ H^+(aq) + O_2(g) + 4\ e^-$	$-E° = -1.229\ V$
Net:	$2\ XeF_2(aq) + 2\ H_2O \longrightarrow 2\ Xe(g) + 4\ HF(aq) + O_2(g)$	$E°_{cell} = +1.41\ V$

The fluorides XeF_2, XeF_4, and XeF_6 are colorless crystalline solids and are stable if kept from contact with water. XeO_3 is an explosive white solid and XeO_4 is an explosive colorless gas.

---

*It has since been established that this solid is more complicated than first thought. It has the formula $Xe(PtF_6)_n$, where n is between 1 and 2.

23-2 Group 7A: The Halogens

The halogens exist as diatomic molecules, symbolized by X_2, where X is a generic symbol for a halogen atom. That these elements are found as nonpolar, diatomic molecules accounts for their relatively low melting and boiling points (Table 23.1). As expected, melting and boiling points increase from the smallest and lightest member of the group, fluorine, to the largest and heaviest, iodine.

Chemical reactivity toward other elements and compounds, on the other hand, progresses in the *opposite* order, with fluorine being the most reactive and iodine the least reactive. Fluorine has the highest electronegativity of all the elements and a small atomic radius. Ionic bonds between fluoride and metal ions are strong, as are most covalent bonds between fluorine and other nonmetal atoms.* This tendency to form strong bonds with other atoms makes F_2 a reactive substance. On the other hand, iodine atoms, because of their larger size and lower electronegativity, form weaker ionic and covalent bonds than do the other halogens. I_2 is less reactive than the other halogens.

Much of the reaction chemistry of the halogens involves oxidation–reduction reactions in aqueous solutions. For these reactions, standard electrode potentials are the best guides to the reactivity of the halogens. Among the properties of the halogens listed in Table 23.1 are potentials for the half-reaction

$$X_2 + 2\,e^- \longrightarrow 2\,X^-(aq)$$

By this measure, fluorine is clearly the most reactive element of the group ($E° = +2.866$ V). Of all the elements, it shows the greatest tendency to gain electrons and is, therefore, the most easily reduced. Given this fact, it is not surprising that fluorine occurs naturally only in combination with other elements, and only as the fluoride ion, F^-. Although chlorine and bromine can both exist in a variety of positive oxidation states, they are found in their naturally occur-

▶ We expect astatine ($Z = 85$) also to be a halogen element. Because it is a radioactive element with only short-lived isotopes, only about 0.05 μg has ever been prepared. Nevertheless, its chemical behavior appears to be similar to that of iodine.

▶ Because $F_2(g)$ oxidizes $H_2O(l)$, producing $O_2(g)$ and $F^-(aq)$, the standard reduction potential, $E°_{F_2/F^-}$, is obtained by calculation from thermodynamic properties based on thermal measurements (see Exercise 53).

TABLE 23.1 Group 7A Elements: The Halogens

	Fluorine (F)	Chlorine (Cl)	Bromine (Br)	Iodine (I)
physical form at room temperature	pale yellow gas	yellow-green gas	dark red liquid	violet-black solid
melting point, °C	-220	-101	-7.2	114
boiling point, °C	-188	-35	58.8	184
electron configuration	$[He]2s^22p^5$	$[Ne]3s^23p^5$	$[Ar]3d^{10}4s^24p^5$	$[Kr]4d^{10}5s^25p^5$
covalent radius, pm	71	99	114	133
ionic (X^-) radius, pm	133	181	196	220
first ionization energy, kJ/mol	1681	1251	1140	1008
electron affinity, kJ/mol	-328.0	-349.0	-324.6	-295.2
electronegativity	4.0	3.0	2.8	2.5
standard electrode potential, V ($X_2 + 2\,e^- \rightarrow 2\,X^-$)	$+2.866$	$+1.358$	$+1.065$	$+0.535$

*The weaker-than-expected fluorine-to-fluorine bond in F_2 (159 kJ/mol) is probably the result of repulsions between the fluorine nuclei and between lone pair $2p$ electrons. Both repulsions are rather strong because of the small atomic radius of fluorine. The weakness of the fluorine-to-fluorine bond contributes to the reactivity of F_2.

Salt formations in the Dead Sea. The high concentrations of salts in the Dead Sea make it a good source of bromine and a number of other chemicals.

▶ The hydrogen difluoride ion in KHF_2 features a strong hydrogen bond, with an H^+ midway between two F^- ions, $[F—H—F]^-$.

ring compounds only as chloride and bromide ions. On the other hand, we can find naturally occurring compounds in which iodine is in a *positive* oxidation state (such as the iodate ion, IO_3^-, in $NaIO_3$). In the case of iodine, the tendency for I_2 to be reduced to I^- is not particularly great ($E° = +0.535$ V).

Production and Uses of the Halogens

Although the existence of *fluorine* had been known since early in the nineteenth century, no one was able to devise a chemical reaction to extract the free element from its compounds. Finally, in 1886, H. Moissan was successful in preparing $F_2(g)$ with an *electrolysis* reaction. Moissan's method, which is still the only important commercial method, involves the electrolysis of HF dissolved in molten KHF_2,

$$2 \text{ HF} \xrightarrow[\text{KF} \cdot 2\text{HF(l)}]{\text{electrolysis}} H_2(g) + F_2(g) \qquad (23.1)$$

consisting of the electrode reactions:

$$\textit{anode:} \qquad 2 F^- \longrightarrow F_2(g) + 2 e^-$$

$$\textit{cathode:} \quad 2 H^+ + 2 e^- \longrightarrow H_2(g)$$

Moissan also developed the electric furnace and was honored for both these achievements with the Nobel Prize in chemistry in 1906. Nevertheless, the challenge of producing fluorine through a chemical reaction remained. One century after Moissan isolated fluorine, in 1986, the chemical synthesis of fluorine was announced (see Exercise 18).

Although *chlorine* can be prepared by chemical reactions, electrolysis of NaCl(aq) is the usual industrial method, as we have mentioned previously (see page 753). The net ionic equation is

$$2 \text{ Cl}^-(aq) + 2 H_2O \xrightarrow{\text{electrolysis}} 2 \text{ OH}^-(aq) + H_2(g) + Cl_2(g) \qquad (23.2)$$

Bromine can be extracted from seawater, where it occurs in concentrations of about 70 ppm as Br^-, or from inland brine solutions where it is also found. The seawater or brine solution is adjusted to pH 3.5 and treated with $Cl_2(g)$, which oxidizes Br^- to Br_2 in this displacement reaction

$$Cl_2(g) + 2 Br^-(aq) \longrightarrow Br_2(l) + 2 Cl^-(aq) \qquad E°_{cell} = 0.293 \text{ V} \qquad (23.3)$$

▶ You may have encountered reaction (23.3) in the laboratory. It is a rapid test for the presence of Br^- in aqueous solution.

The liberated Br_2 is swept from seawater with a current of air or from brine with steam. This provides a dilute bromine vapor that can be concentrated by various methods.

Certain marine plants, such as seaweed, absorb and concentrate I^- selectively in the presence of Cl^- and Br^-. *Iodine* is obtained in small quantities from such plants. In the United States, I_2 is obtained from inland brines by a process similar to that for the production of Br_2. Another abundant natural source of iodine is $NaIO_3$, found in large deposits in Chile. Because the oxidation state of iodine must be reduced from +5 in IO_3^- to 0 in I_2, the conversion of IO_3^- to I_2 requires the use of a reducing agent. Sodium hydrogen sulfite (bisulfite) is used as the reducing agent in the first part of a two-step procedure, followed by the reaction of I^- with additional IO_3^- to produce I_2.

$$IO_3^-(aq) + 3 HSO_3^-(aq) \longrightarrow I^-(aq) + 3 SO_4^{2-}(aq) + 3 H^+(aq) \qquad (23.4)$$

$$5 I^-(aq) + IO_3^-(aq) + 6 H^+(aq) \longrightarrow 3 I_2(s) + 3 H_2O \qquad (23.5)$$

The halogen elements form a variety of useful compounds, and the elements themselves are largely used to produce these compounds. All of the halogens are used to make halogenated organic compounds. For example, elemental fluorine is used to produce compounds such as polytetrafluoroethylene, the plastic Teflon. Fluorine has been used to make chlorofluorocarbons (CFCs) as refrigerants, but international treaties have banned the production of CFCs in most countries in order to minimize further damage to the ozone layer from these compounds (see page 254). Now fluorine is used to make HCFCs, which are more environmentally benign alternatives to CFCs. Fluorinated organic compounds tend to be chemically inert, and it is this inertness that makes them useful as components in harsh chemical environments.

Fluorine is a key element in a variety of useful inorganic compounds. Some of these compounds and their uses are listed in Table 23.2.

With an annual production of about 12 million tons, elemental chlorine ranks about tenth in quantity among manufactured chemicals in the United States. It has three main commercial uses: (1) the production of chlorinated organic compounds, including polyvinyl chloride, or PVC (about 70%); (2) as a bleach in the paper and textile industries and for the treatment of swimming pools, municipal water, and sewage (about 20%); (3) the production of chlorine-containing inorganic chemicals (about 10%).

Bromine is used to make brominated organic compounds. Some of these are used as fire retardants and pesticides. Others are used extensively as dyes and pharmaceuticals. An important inorganic bromine compound is AgBr, the primary light-sensitive agent used in photographic film.

Iodine is of much less commercial importance than chlorine, although iodine and its compounds do have applications as catalysts, antiseptics, and germicides, and in the preparation of pharmaceuticals and photographic emulsions (as AgI).

Hydrogen Halides

We have encountered the hydrogen halides from time to time throughout the text. In aqueous solution they are called the hydrohalic acids. Except for HF, the hydrohalic acids are all strong acids in water. The weak acid nature of HF(aq) is difficult to explain and is discussed on page 607.

One well-known property of HF is its ability to etch (and ultimately to dissolve) glass. Glass, for this reaction, may be considered to be related to silicon dioxide, SiO_2.

$$SiO_2(s) + 4 \, HF(aq) \longrightarrow 2 \, H_2O(l) + SiF_4(g) \qquad (23.6)$$

Because of reaction (23.6), HF must be stored in special containers coated with a lining of Teflon or polyethylene.

TABLE 23.2 Some Important Inorganic Compounds of Fluorine

Compound	Uses
Na_3AlF_6	mfr. of aluminum
BF_3	catalyst
CaF_2	optical components, mfr. of HF, metallurgical flux
ClF_3	fluorinating agent, reprocessing nuclear fuels
HF	mfr. of F_2, AlF_3, Na_3AlF_6, and fluorocarbons
LiF	ceramics mfr., welding and soldering
NaF	fluoridating water, dental prophylaxis, insecticide
SnF_2	mfr. of toothpaste
UF_6	mfr. of uranium fuel for nuclear reactors

Hydrogen fluoride is commonly produced by a method discussed in Section 22-1. When a halide salt (such as fluorite, CaF_2) is heated with a *nonvolatile* acid, such as concentrated $H_2SO_4(aq)$—a nonvolatile acid—a sulfate salt and the volatile hydrogen halide are produced.

$$CaF_2(s) + H_2SO_4(\text{concd aq}) \xrightarrow{\Delta} CaSO_4(s) + 2\ HF(g) \qquad (23.7)$$

This method also works for preparing $HCl(g)$, but not for $HBr(g)$ and $HI(g)$. This is because *concentrated* $H_2SO_4(aq)$ is a sufficiently good oxidizing agent to oxidize Br^- to Br_2 and I^- to I_2.

$$2\ NaBr(s) + 2\ H_2SO_4(\text{concd aq}) \xrightarrow{\Delta} Na_2SO_4(s) + 2\ H_2O(l) + Br_2(g) + SO_2(g)$$
$$(23.8)$$

We can get around this difficulty by using a *nonoxidizing* nonvolatile acid, such as phosphoric acid. Also, all the hydrogen halides can be formed by the direct combination of the elements.

$$H_2(g) + X_2(g) \longrightarrow 2\ HX(g) \qquad (23.9)$$

The reaction of $H_2(g)$ and $F_2(g)$ is very fast, however, occurring with explosive violence under some conditions. With $H_2(g)$ and $Cl_2(g)$ the reaction also proceeds rapidly (explosively) in the presence of light (photochemically initiated), although some HCl is made this way commercially. With Br_2 and I_2 the reaction occurs more slowly and a catalyst is required.

From the data in Table 23.3 we see that the standard free energies of formation of $HF(g)$, $HCl(g)$, and $HBr(g)$ are large and negative, suggesting that for them reaction (23.9) goes to completion. For $HI(g)$, on the other hand, ΔG_f° is small and positive. This suggests that even at room temperature $HI(g)$ should dissociate to some extent into its elements. Because of a high *activation energy,* however, the dissociation reaction occurs only very slowly in the absence of a catalyst. As a result, $HI(g)$ is stable at room temperature.

TABLE 23.3 Free Energy of Formation of Hydrogen Halides at 298 K

	ΔG_f°, kJ/mol
HF(g)	−273.2
HCl(g)	−95.30
HBr(g)	−53.43
HI(g)	+1.72

EXAMPLE 23-1

Determining K_p for a Dissociation Reaction from Free Energies of Formation. What is the value of K_p for the dissociation of $HI(g)$ into its elements at 298 K?

Solution

Dissociation of $HI(g)$ into its elements is the reverse of the formation reaction, and ΔG° for the dissociation is the *negative* of ΔG_f° for $HI(g)$ listed in Table 23.3.

$$HI(g) \longrightarrow \tfrac{1}{2} H_2(g) + \tfrac{1}{2} I_2(s) \qquad \Delta G^\circ = -1.72\ \text{kJ} \qquad (23.10)$$

Now we can use the relationship $\Delta G^\circ = -RT \ln K_{eq}$.

$$\ln K_p = \frac{-\Delta G^\circ}{RT} = \frac{-(-1.72 \times 10^3\ \text{J mol}^{-1})}{8.3145\ \text{J mol}^{-1}\ \text{K}^{-1} \times 298\ \text{K}} = 0.694$$

$$K_p = e^{0.694} = 2.00$$

Practice Example A: What is the value of K_p for the dissociation of $HF(g)$ into its elements at 298 K?

Practice Example B: Use data from Table 23.3 to determine K_p and the percent dissociation of $HCl(g)$ into its elements at 298 K. (*Hint:* After you obtain a value of K_p, you need to do a calculation similar to Example 16-12.)

TABLE 23.4 **Oxoacids of the Halogens[a]**			
Oxidation State of Halogen	**Chlorine**	**Bromine**	**Iodine**
+1	HOCl	HOBr	HOI
+3	$HClO_2$	—	—
+5	$HClO_3$	$HBrO_3$	HIO_3
+7	$HClO_4$	$HBrO_4$	HIO_4; H_5IO_6

[a]In all these acids H atoms are bonded to O atoms, as shown for HOCl, HOBr, and HOI, not to the central halogen atom. More accurate representations of the other acids would be HOClO (instead of $HClO_2$), $HOClO_2$ (instead of $HClO_3$), and so on.

Oxoacids and Oxoanions of the Halogens

Fluorine, the most electronegative element, adopts the -1 oxidation state in its compounds. The other halogens, when bonded to a more electronegative element such as oxygen, can have any one of several positive oxidation states: $+1$, $+3$, $+5$, and $+7$. This variability of oxidation states is illustrated by the oxoacids listed in Table 23.4. Chlorine forms a complete set of oxoacids in all these oxidation states, but bromine and iodine do not.

Although the disproportionation of Cl_2 in water is nonspontaneous when all reactants and products are in their standard states, the reaction does occur to a limited extent in solutions that are not strongly acidic.

▶ Feature Problem A on page 828 presents a useful diagrammatic method of displaying the varying oxidation states of an element in its compounds.

Oxidation: $Cl_2(g) + 2\ H_2O \longrightarrow 2\ HOCl(aq) + 2\ H^+(aq) + 2\ e^-$ $\quad -E° = -1.62\ V$

Reduction: $\underline{Cl_2(g) + 2\ e^- \longrightarrow 2\ Cl^-(aq)} \quad\quad\quad\quad\quad\quad\quad\quad E° = 1.358\ V$

Net: $\quad Cl_2(g) + H_2O(l) \longrightarrow HOCl(aq) + H^+(aq) + Cl^-(aq) \quad E°_{cell} = -0.26\ V \quad$ (23.11)

By contrast, the disproportionation is spontaneous for standard state conditions in basic solution.

Oxidation: $Cl_2(g) + 4\ OH^-(aq) \longrightarrow 2\ OCl^-(aq) + 2\ H_2O + 2\ e^-$ $\quad -E° = -0.40\ V$

Reduction: $\underline{\quad Cl_2(g) + 2\ e^- \longrightarrow 2\ Cl^-(aq)} \quad\quad\quad\quad\quad\quad\quad\quad E° = 1.358\ V$

Net: $\quad Cl_2(g) + 2\ OH^-(aq) \longrightarrow OCl^-(aq) + Cl^-(aq) + H_2O \quad E°_{cell} = 0.96\ V \quad$ (23.12)

HOCl(aq) is an effective germicide used in water purification, and aqueous solutions of *hypochlorite* salts, notably NaOCl, are used as common household bleaches.

Chlorine dioxide is an important bleach for paper and fibers. Its reduction with peroxide ion in aqueous solution produces *chlorite* salts.

▶ Notice that E_{Cl_2/Cl^-} is independent of pH, but that both $-E_{HOCl/Cl_2}$ and $-E_{OCl^-/Cl_2}$ become less negative as the pH increases. We expect this to be the case, based on Le Châtelier's principle.

$$2\ ClO_2(g) + O_2^{2-}(aq) \longrightarrow 2\ ClO_2^-(aq) + O_2(g) \quad\quad (23.13)$$

Sodium chlorite is used as a bleaching agent for textiles.

Chlorate salts are formed when $Cl_2(g)$ disproportionates in *hot* alkaline solutions. [Hypochlorites form in cold alkaline solutions (recall reaction 23.12).]

$$3\ Cl_2(g) + 6\ OH^-(aq) \longrightarrow 5\ Cl^-(aq) + ClO_3^-(aq) + 3\ H_2O \quad (23.14)$$

Chlorates are good oxidizing agents. Also, solid chlorates decompose to produce oxygen gas, which makes them useful in matches and fireworks. A simple laboratory method of producing $O_2(g)$ involves heating $KClO_3(s)$ in the presence of $MnO_2(s)$, a catalyst.

$$2\ KClO_3(s) \xrightarrow[MnO_2]{\Delta} 2\ KCl(s) + 3\ O_2(g) \quad\quad (23.15)$$

Bleaching with hypochlorite ion. Both strips of cloth are heavily stained with tomato sauce. Pure water (left) has little ability to remove the stain. NaOCl (right) rapidly bleaches (oxidizes) the colored components of the sauce to coloress products.

Perchlorate salts are mainly prepared by electrolyzing chlorate solutions. Oxidation of ClO_3^- occurs at a Pt anode through the half-reaction

$$ClO_3^-(aq) + H_2O \longrightarrow ClO_4^-(aq) + 2 H^+(aq) + 2 e^- \qquad -E° = -1.19 \text{ V}$$

(23.16)

An interesting laboratory use of perchlorate salts is in aqueous solution studies where complex-ion formation is to be avoided. ClO_4^- has one of the lowest tendencies of any anion to act as a ligand in complex ion formation in water. Compared with the other oxoacid salts, perchlorates are relatively stable. For example, they do not disproportionate, because no oxidation state higher than +7 is available to chlorine. At elevated temperatures or in the presence of a readily oxidizable compound, however, perchlorate salts may react explosively, so caution is advised when using them. Mixtures of ammonium perchlorate and powdered aluminum are used as the propellant in some solid rockets, such as those used on the Space Shuttle. Ammonium perchlorate is especially dangerous to handle because ClO_4^-, the oxidizing agent, can act on NH_4^+, the reducing agent, and an explosive reaction may occur.

Are You Wondering . . .

If there are oxoacids of fluorine?

The hypothetical oxoacid of fluorine, HFO_2 (HOFO), would have a *positive* formal charge on the central F atom.

This is not something that we would expect of fluorine, the most electronegative of all the elements. Hypothetical oxoacids with more O atoms would have the F atom with even higher positive formal charges. Fluorine does form HOF, an oxoacid with no formal charges, H : O—F :, but HOF exists only in the solid and liquid states. In water, HOF decomposes to HF, H_2O_2, and $O_2(g)$.

Polyhalide Ions

▶Iodine is also more soluble in a nonpolar solvent such as $CCl_4(l)$ than it is in water. The differing solubility of iodine in water and in $CCl_4(l)$ is the subject of Feature Problem B on page 829.

Iodine is only slightly soluble in pure water (0.3 g/L) but highly soluble in KI(aq). This increased solubility results from the formation of the triiodide ion, I_3^-.

$$I_2(s) + I^-(aq) \rightleftharpoons I_3^-(aq) \qquad K_c = 710 \text{ (at 298 K)}$$ (23.17)

colorless   yellow-brown

The triiodide ion is one of a group of species, called **polyhalide ions,** that are produced by the reaction of a halide ion with a halogen molecule. In this reaction the halide ion acts as a *Lewis base* (electron pair donor) and the halogen molecule as a *Lewis acid* (electron pair acceptor).

$$: \ddot{I} - \ddot{I} : + : \ddot{I} : ^- \longrightarrow \left[: \ddot{I} - \overset{..}{\underset{..}{I}} - \ddot{I} : \right]^-$$

The structure of the I_3^- ion is shown in Figure 23-1. The familiar iodine solutions used as antiseptics often contain triiodide ion.

23-3   Group 6A: The Oxygen Family

Oxygen and sulfur are clearly nonmetallic in their behavior, but some metallic properties begin to appear toward the bottom of Group 6A with tellurium and polonium, as suggested in the periodic table on the inside front cover.

On the basis of electron configurations alone, we expect oxygen and sulfur to be similar. Both elements form ionic compounds with active metals, and both form similar covalent compounds, such as H_2S and H_2O, CS_2 and CO_2, SCl_2 and Cl_2O.

Even so, there are important differences between oxygen and sulfur compounds. For example, H_2O has a very high boiling point (100 °C) for a compound of such low molecular mass (18 u), whereas the boiling point of H_2S (molecular mass, 34 u) is more normal (−61 °C). This difference in behavior can be explained by the fact that extensive hydrogen bonding occurs in H_2O but not in H_2S (see page 440). A number of properties of sulfur and oxygen are compared in Table 23.5, and in general the differences can be attributed to these characteristics of the oxygen atom: (1) small size, (2) high electronegativity, and (3) the inability to form an expanded octet in Lewis structures.

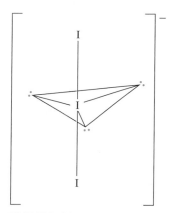

FIGURE 23-1
Structure of the I_3^- ion.
The bonding pairs of electrons are shown as red lines. The lone pair electrons of the central I atom are represented by blue dots.

TABLE 23.5   Some Comparisons of Oxygen and Sulfur

Oxygen	Sulfur
• $O_2(g)$ at 298 K and 1 atm	• $S_8(s)$ at 298 K and 1 atm
• Two allotropes, $O_2(g)$ and $O_3(g)$	• Two solid crystalline forms and many different molecular species in liquid and gaseous states
• Possible oxidation states: $-2, -1, 0$ ($-\frac{1}{2}$ in O_2^-)	• Possible oxidation states: all values from -2 to $+6$
• $O_2(g)$ and $O_3(g)$ are very good oxidizing agents	• $S_8(s)$ is a poor oxidizing agent
• Forms, with metals, oxides that are mostly ionic in character	• Forms ionic sulfides with the most active metals but many metal sulfides have partial covalent character
• O^{2-} completely hydrolyzes in water, producing OH^-	• S^{2-} strongly hydrolyzes in water to HS^- (and OH^-)
• O not often the central atom in a structure, and can never have more than four atoms bonded to it. More commonly has two (as in H_2O) or three (as in H_3O^+)	• S the central atom in many structures. Can easily accommodate up to six atoms around itself (e.g., SO_3, SO_4^{2-}, SF_6)
• Can form only two- and three-atom chains, as in H_2O_2 and O_3. Compounds with O—O bonds decompose readily	• Can form molecules with up to six S atoms per chain in compounds such as H_2S_n, Na_2S_n, $H_2S_nO_6$
• Forms the oxide CO_2, which reacts with $NaOH(aq)$ to produce $Na_2CO_3(aq)$	• Forms the sulfide CS_2, which reacts with $NaOH(aq)$ to produce $Na_2CS_3(aq)$ and $Na_2CO_3(aq)$
• Forms, with hydrogen, the compound H_2O, which is a liquid at 298 K and 1 atm is extensively hydrogen-bonded has a large dipole moment is an excellent solvent for ionic solids forms hydrates and aqua complexes is oxidized with difficulty	• Forms, with hydrogen, the compound H_2S, which is a (poisonous) gas at 298 K and 1 atm is not hydrogen-bonded has a small dipole moment is a poor solvent forms no complexes is easily oxidized

FIGURE 23-2
Some different molecular forms of sulfur.

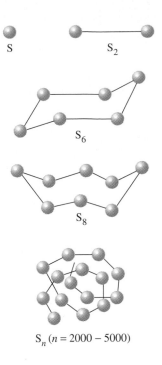

S

S_2

S_6

S_8

S_n $(n = 2000 - 5000)$

Liquid sulfur is poured into water to produce plastic sulfur.

As indicated in Table 23.5, the principal oxidation states of oxygen are -2, -1, and 0. Sulfur, on the other hand, can exhibit all oxidation states from -2 to $+6$, including several "mixed" oxidation states, such as $+2.5$ in the tetrathionate ion, $S_4O_6{}^{2-}$.

Sulfur has more allotropes than any other element (Figure 23-2). The most familiar unit of sulfur is the S_8 ring. This is the predominant form in sulfur vapor and is the constituent of both common solid phases: rhombic and monoclinic sulfur. Rhombic sulfur, the stable form at low temperatures, converts to monoclinic sulfur at 95.5 °C, and monoclinic sulfur melts at 119 °C.

Below 160 °C, liquid sulfur is composed of S_8 molecules and is a yellow, transparent, and mobile liquid. At 160 °C a remarkable transformation occurs. The S_8 rings open up and join together into long spiral-chain molecules. These polymers of sulfur atoms produce a liquid that is dark in color and very thick and viscous. If this dark liquid is poured into cold water, sulfur atoms are temporarily trapped in their polymer chains. This "plastic" sulfur has rubber-like properties when first formed. On standing, however, it becomes brittle and eventually converts to the stable rhombic form.

Production and Uses of the Group 6A Elements

Oxygen is obtained by the fractional distillation of liquid air (Section 8-1), and its main uses were outlined in Table 8.4. Sulfur occurs abundantly in Earth's crust—as elemental sulfur, as mineral sulfides and sulfates, as $H_2S(g)$ in natural gas, and as organosulfur compounds in oil and coal. Extensive deposits of elemental sulfur are found in Texas and Louisiana, some of them in offshore sites. This sulfur is mined in an unusual way, known as the **Frasch process** (Figure 23-3). Superheated water (at about 160 °C and 16 atm) is forced down the outermost of three concentric pipes into an underground bed of sulfur-containing rock. The sulfur melts and forms a liquid pool. Compressed air (at 20–25 atm) is pumped down the innermost pipe and forces the liquid sulfur–water mixture up the remaining pipe.

compressed air

liquid sulfur

superheated water

sulfur-bearing rock

liquid sulfur

FIGURE 23-3
The Frasch process.
Sulfur is melted using superheated water and liquid sulfur is forced to the surface.

Although the Frasch process was once the principal source of elemental sulfur, this is no longer the case. This change has been brought about by the need to control sulfur emissions from industrial operations. H_2S is a common impurity in oil and natural gas. After being removed from the fuel, H_2S is reduced to elemental sulfur in a two-step process. A stream of H_2S gas is split into two parts. One part (about one-third of the stream) is burned to convert H_2S to SO_2. The streams are rejoined in a catalytic converter at 200–300 °C, where the following reaction occurs.

These vast formations of solid sulfur were formed by the solidification of liquid sulfur obtained by the Frasch process.

$$2\ H_2S(g) + SO_2(g) \longrightarrow 3\ S(g) + 2\ H_2O(g) \qquad (23.18)$$

About 90% of all the sulfur produced is burned to form $SO_2(g)$ (Figure 23-4), and in turn, most $SO_2(g)$ is converted to sulfuric acid, H_2SO_4. Elemental sulfur does have a few uses of its own, however. One of these is in vulcanizing rubber (page 91) and another is as a fungicide, for example, in dusting grapevines.

Selenium and *tellurium* have properties similar to those of sulfur, but they are more metallic. For example, sulfur is an electrical insulator, whereas selenium and tellurium are semiconductors. Selenium and tellurium are obtained mostly as by-products of metallurgical processes, such as in the anode mud deposited in the electrolytic refining of copper (page 755). Although there is not much use for tellurium compounds, selenium is used in the manufacture of rectifiers (de-

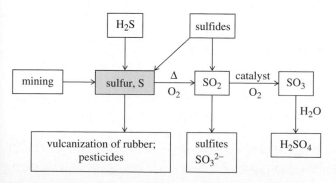

FIGURE 23-4
Diagram showing sources and uses of sulfur and its oxides.

vices used to convert alternating to direct electric current). Both Se and Te find some use in the preparation of alloys, and their compounds are used as additives to control the color of glass.

Selenium also displays the property of *photoconductivity*: The electrical conductivity of selenium increases in the presence of light. This property is used, for example, in photocells in cameras. In some modern photocopying machines, the light-sensitive element is a thin film of Se deposited on aluminum. The light and dark areas of the image being copied are converted into a distribution of charge on the light-sensitive element. A dry black powder (toner) coats the charged portions of the light-sensitive element and this image is transferred to a sheet of paper. Next, the dry powder is fused to the paper. In a final step, the electrostatic charge on the light-sensitive element is neutralized to prepare it for the next cycle.

Polonium is a very rare, radioactive metal, and because of its extremely low abundance, it has not found much practical use. Its main claim to fame is that polonium was the first new radioactive element isolated from uranium ore by Marie Curie and her husband, Pierre, in 1898. Madame Curie named it after her native Poland.

Oxygen Compounds

Oxygen is so central to the study of chemistry that we constantly refer to its physical and chemical properties in developing a framework of chemical principles. For instance, our discussion of stoichiometry began with reactions of substances with $O_2(g)$ to form products such as $CO_2(g)$, $H_2O(l)$, and $SO_2(g)$—combustion reactions. Combustion reactions also figured prominently in the presentation of thermochemistry. Many of the molecules and polyatomic anions described in the chapters on chemical bonding were oxygen-containing species. Water was a primary subject in the discussion of liquids, solids, and intermolecular forces, as well as in the study of acid–base and other solution equilibria. The preparation and uses of oxygen and ozone were presented in Chapter 8, the dual roles of hydrogen peroxide as an oxidizing and a reducing agent were described in Section 5-6, and the kinetics of the decomposition of H_2O_2 was examined in detail in Chapter 15. The acidic, basic, and amphoteric properties of element oxides were outlined in Chapter 17.

A systematic study of oxygen compounds is generally done in conjunction with the study of the other elements. Thus, the oxides of nitrogen were considered in the discussion of nitrogen chemistry in Chapter 8; the oxides of carbon were also considered at that time. The survey of the chemistry of the active metals in the preceding chapter provided an opportunity to describe normal oxides, peroxides, and superoxides, and the important oxides of sulfur, phosphorus, silicon, and boron are discussed in this chapter.

FIGURE 23-5
Structures of some sulfur oxides.
To conform to the observed structures of these molecules, the hybridization scheme proposed for the central S atom is sp^2. S_2O has a similar structure to SO_2 but with an S atom substituted for one O atom.

Oxides and Oxoacids of Sulfur

More than a dozen oxides of sulfur have been reported, but only *sulfur dioxide*, SO_2, and *sulfur trioxide*, SO_3, are commonly encountered. Their structures are shown in Figure 23-5. The main commercial methods of producing $SO_2(g)$ are the direct combustion of sulfur and the "roasting" of metal sulfides.

$$S(s) + O_2(g) \xrightarrow{\Delta} SO_2(g) \tag{23.19}$$

$$2 \, ZnS(s) + 3 \, O_2(g) \xrightarrow{\Delta} 2 \, ZnO(s) + 2 \, SO_2(g) \tag{23.20}$$

The main use of SO_2 is in the synthesis of SO_3 to make sulfuric acid, H_2SO_4. In the modern process, the **contact process,** first $SO_2(g)$ is formed by reaction (23.19) or (23.20). Then sulfur trioxide is produced by oxidizing $SO_2(g)$ in an exothermic, reversible reaction.

$$2\ SO_2(g) + O_2(g) \rightleftharpoons 2\ SO_3(g) \tag{23.21}$$

Reaction (23.21) is the key step in the process, but it occurs very slowly unless catalyzed. The principal catalyst is V_2O_5 mixed with alkali metal sulfates. The catalysis involves adsorption of the $SO_2(g)$ and $O_2(g)$ on the catalyst, followed by reaction at active sites and desorption of SO_3 (recall Figure 15-16).

SO_3 reacts with water to form H_2SO_4, but the direct reaction of $SO_3(g)$ and water produces a fine mist of $H_2SO_4(aq)$ droplets with unreacted $SO_3(g)$ trapped inside the droplets. This would result in a great loss of product and a tremendous pollution problem. To avoid these outcomes, $SO_3(g)$ is instead bubbled through 98% H_2SO_4 in towers packed with a ceramic material. The $SO_3(g)$ readily dissolves in the sulfuric acid and reacts with the small amount of water present to increase the concentration of the sulfuric acid. The result is a form of sulfuric acid sometimes called *oleum* but more commonly called *fuming sulfuric acid.* In a sense the product is "greater than 100% H_2SO_4." Sufficient water is added to the circulating acid in the tower to maintain the required concentration. Later, dilution with water produces sulfuric acid of the strength desired. If we use the formula $H_2S_2O_7$ (disulfuric acid) as an example of a particular oleum, the reactions are

$$SO_3(g) + H_2SO_4(l) \longrightarrow H_2S_2O_7(l) \tag{23.22}$$

$$H_2S_2O_7(l) + H_2O(l) \longrightarrow 2\ H_2SO_4(l) \tag{23.23}$$

$$H_2SO_4(l) + \text{water} \longrightarrow H_2SO_4(aq) \tag{23.24}$$

Dilute sulfuric acid, $H_2SO_4(aq)$, enters into all the common reactions of a strong acid, such as neutralizing bases, reacting with metals to produce $H_2(g)$, and dissolving carbonates to liberate $CO_2(g)$.

Concentrated sulfuric acid has some distinctive properties. It has a very strong affinity for water, strong enough that it will even remove H and O atoms (in the proportion H_2O) from some compounds. In the reaction of concentrated sulfuric acid with a carbohydrate like sucrose, all the H and O atoms are removed and a residue of pure carbon is left.

$$C_{12}H_{22}O_{11}(s) \xrightarrow{\ H_2SO_4\ (concd)\ } 12\ C(s) + 11\ H_2O(l) \tag{23.25}$$

The concentrated acid is a moderately good oxidizing agent and is able, for example, to react with copper.

$$Cu(s) + 2\ H_2SO_4(concd) \longrightarrow Cu^{2+}(aq) + SO_4{}^{2-}(aq) + 2\ H_2O + SO_2(g) \tag{23.26}$$

> ► All acids produce a stinging sensation, but concentrated H_2SO_4 must be handled with extreme care because it produces severe burns as a result of reactions like (23.25).

In keeping with its perennial number 1 ranking among manufactured chemicals, sulfuric acid has many uses. The bulk of H_2SO_4 production (70%) is used in the manufacture of fertilizers. It is also used in various metallurgical processes, the refining of oil, the manufacture of the white pigment titanium dioxide, and in hundreds of other ways, as in storage batteries for automobiles and emergency power supplies.

When $SO_2(g)$ reacts with water, it produces $H_2SO_3(aq)$, but this acid, *sulfurous acid,* has never been isolated in pure form. Salts of sulfurous acid, *sulfites,* are good reducing agents and are easily oxidized by $O_2(g)$, for example,

$$O_2(g) + 2\ SO_3{}^{2-}(aq) \longrightarrow 2\ SO_4{}^{2-}(aq) \tag{23.27}$$

but they can also act as oxidizing agents, as in this reaction with H_2S.

(a)

trigonal pyramidal

(b)

tetrahedral

(c)

tetrahedral

FIGURE 23-6
Structures of the (a) sulfite (SO_3^{2-}), (b) sulfate (SO_4^{2-}), and (c) thiosulfate ($S_2O_3^{2-}$) ions.

$$2 H_2S(g) + 2 H^+(aq) + SO_3^{2-}(aq) \longrightarrow 3 H_2O + 3 S(s) \qquad (23.28)$$

Both H_2SO_3 and H_2SO_4 are *diprotic* acids. They ionize in two steps and produce two types of salts, one in each ionization step. The term **acid salt** is sometimes used for salts such as $NaHSO_3$ and $NaHSO_4$, because their anions undergo a further *acid* ionization. H_2SO_3 is a weak acid in both ionization steps, whereas H_2SO_4 is strong in the first step and somewhat weak in the second. If a solution of H_2SO_4 is sufficiently dilute, however (less than about 0.001 M), we can treat the acid as if both ionization steps go to completion.

Sulfate and *sulfite* salts have a number of important uses. Calcium sulfate dihydrate (gypsum) is used to make the hemihydrate (plaster of Paris) for the building industry (page 777). Aluminum sulfate is used in water treatment and in wallpaper sizing. Copper(II) sulfate is used as a fungicide and algicide and in electroplating. The chief use of sulfites is in the pulp and paper industry. Sulfites solubilize *lignin,* a polymeric substance that coats the cellulose fibers in wood. This treatment frees the fibers for processing into wood pulp and then paper. Sulfites are also used as reducing agents, such as in photography and as scavengers of $O_2(aq)$ in treating boiler water (reaction 23.27).

In addition to sulfite and sulfate ions, another important sulfur–oxygen ion is *thiosulfate ion,* $S_2O_3^{2-}$. The prefix "thio" signifies that an S atom replaces an O atom in a compound. Thus, the thiosulfate ion can be viewed as a sulfate ion, SO_4^{2-}, in which an S atom replaces one of the O atoms. The formal oxidation state of S in $S_2O_3^{2-}$ is +2, but as we can see from Figure 23-6, the two S atoms are not equivalent: The central S atom is in the oxidation state +6 and the terminal S atom, −2.

Thiosulfates can be prepared by boiling elemental sulfur in an alkaline solution of sodium sulfite. The sulfur is oxidized and the sulfite ion is reduced, both to thiosulfate ion.

$$SO_3^{2-}(aq) + S(s) \longrightarrow S_2O_3^{2-}(aq) \qquad E^\circ_{cell} = +0.17 \text{ V} \qquad (23.29)$$

Thiosulfate solutions are important in photographic processing (see page 887). They are also common analytical reagents, often used in conjunction with iodine. For example, in one method of analysis for copper, an excess of iodide ion is added to $Cu^{2+}(aq)$, producing $CuI(s)$ and triiodide ion, I_3^- (recall Figure 23-1).

$$2 Cu^{2+}(aq) + 5 I^-(aq) \longrightarrow 2 CuI(s) + I_3^-(aq) \qquad (23.30)$$

The excess triiodide ion is then titrated with a standard solution of $Na_2S_2O_3(aq)$, forming I^- and $S_4O_6^{2-}$, the *tetrathionate* ion.

$$I_3^-(aq) + 2 S_2O_3^{2-}(aq) \longrightarrow 3 I^-(aq) + S_4O_6^{2-}(aq) \qquad (23.31)$$

An Environmental Issue: SO_2 Emissions

Industrial smog consists primarily of particles (ash and smoke), $SO_2(g)$, and H_2SO_4 mist. A variety of industrial operations produce significant quantities of $SO_2(g)$. The main contributors to atmospheric releases of $SO_2(g)$, however, are power plants burning coal or high-sulfur fuel oils. SO_2 can undergo oxidation to SO_3, especially when the reaction is catalyzed on the surfaces of airborne particles or through reaction with NO_2.

$$SO_2(g) + NO_2(g) \longrightarrow SO_3(g) + NO(g) \qquad (23.32)$$

In turn, SO_3 can react with water vapor in the atmosphere to produce H_2SO_4 mist, a component of acid rain. Also, the reaction of H_2SO_4 with airborne NH_3

► One hundred years ago and more, reaction (23.32) was the principal commercial method for synthesizing SO_3 for use in the manufacture of sulfuric acid. To some extent the reaction has been reintroduced as a means of controlling SO_2 emissions in flue gases.

produces particles of $(NH_4)_2SO_4$. The details of the effect of low concentrations of SO_2 and H_2SO_4 on the body are not well understood, but it is clear that these substances are respiratory irritants. Levels above 0.10 ppm are considered potentially harmful.

The control of industrial smog and acid rain hinges on removing sulfur from fuels and controlling $SO_2(g)$ emissions. Dozens of processes have been proposed for removing SO_2 from smokestack gases, one of which was described on page 612.

23-4 Group 5A: The Nitrogen Family

Perhaps the most interesting feature of the Group 5A elements is that properties of both nonmetals and metals are displayed within the group, even more so than in the Group 6A elements just considered. The electron configurations of the elements provide only a limited clue to their metallic–nonmetallic behavior. The outer shell electron configurations are ns^2np^3. This electron configuration lends itself to a variety of possibilities for forming stable compounds.

One possibility is the gain or sharing of three electrons in the outer shell to provide the Group 5A atom with a complete octet. This is especially likely for the smaller atoms N and P. For the larger atoms—As, Sb, and Bi—the p^3 set of electrons may be involved in compound formation, giving the $+3$ oxidation state. In some cases, all five outer shell electrons may be involved, leading to the oxidation state of $+5$.

Assessment of Metallic–Nonmetallic Character in Group 5A

To aid in this discussion, we have listed several properties of the Group 5A elements in Table 23.6. We note the usual decrease of ionization energy with increasing atomic number. This establishes the order of metallic character within the group. Nitrogen and phosphorus are nonmetallic, arsenic and antimony are metalloids, and bismuth is metallic. As expected, all these ionization energies

TABLE 23.6 Selected Properties of Group 5A Elements

Element	Covalent Radius, pm	Electro-negativity	First Ionization Energy, kJ/mol	Common Physical Form(s)	Density of Solid, g/cm³	Comparative Electrical Conductivity[a]
N	75	3.0	1402	gas	1.03 (−252 °C)	—
P	110	2.1	1012	waxlike white solid;	1.82	—
				red solid	2.20	10^{-17}
As	121	2.0	947	yellow solid;	2.03	—
				gray solid with metallic luster	5.78	6.1
Sb	140	1.9	834	yellow solid;	5.3	—
				silvery white metallic solid	6.69	4.0
Bi	155	1.9	703	pinkish white metallic solid	9.75	1.5

[a]These values are relative to an assigned value of 100 for silver.

The glow of white phosphorus gave the element its name—*phos*, light and *phorus*, bringing. The solid has a relatively high vapor pressure, and the glow results from the slow reaction between phosphorus vapor and oxygen in air.

are high compared with the Group 1A and 2A elements. The electronegativities indicate a high degree of nonmetallic character for nitrogen and less so for the remaining members of the group.

Three of the elements—phosphorus, arsenic, and antimony—exhibit allotropy. The common forms of phosphorus at room temperature, both nonmetallic, are white and red phosphorus. For arsenic and antimony the more stable allotropic forms are the "metallic" ones. These have high densities, fair thermal conductivities, and limited abilities to conduct electricity. Bismuth is a metal in spite of its low electrical conductivity, which is lower even than that of nichrome wire (used as a heating element because of its poor conductivity).

The difference between the nonmetals and metals in Group 5A can also be seen in their oxides. The oxides of nitrogen and phosphorus (for example, N_2O_3 and P_4O_6) are acidic when they react with water. As with the nonmetals in Groups 6A and 7A, this is typical for nonmetal oxides. Arsenic(III) oxide and antimony(III) oxide are amphoteric, whereas bismuth(III) oxide acts only as a base, a property typical of metal oxides.

Production and Uses of the Group 5A Elements

Nitrogen is obtained by the fractional distillation of liquefied air (Section 8-1). Its chief uses were described in Table 8.2.

Phosphorus was originally isolated from putrefied urine—an effective if not particularly pleasant source. Today the principal source of phosphorus compounds is phosphate rock, a class of minerals known as *apatites*, such as fluorapatite [$3Ca_3(PO_4)_2·CaF_2$]. Elemental phosphorus is prepared by heating phosphate rock, silica (SiO_2), and coke (C) in an electric furnace. The net change that occurs is

$$2\ Ca_3(PO_4)_2(s) + 10\ C(s) + 6\ SiO_2(s) \xrightarrow{\Delta} 6\ CaSiO_3(l) + 10\ CO(g) + P_4(g)$$

(23.33)

The $P_4(g)$ is condensed, collected, and stored under water as a white, waxy, phosphorescent solid. (A phosphorescent material glows in the dark.) This solid, which can be cut with a knife, is called *white phosphorus*. White phosphorus is a nonconductor of electricity, can ignite spontaneously in air (hence the reason for storing it under water), and is insoluble in water but soluble in some nonpolar solvents, such as CS_2.

The basic structural units of white phosphorus are P_4 molecules. These molecules, pictured in Figure 23-7a, are *tetrahedral* with a P atom at each corner. The phosphorus-to-phosphorus bonds in P_4 appear to involve the overlap of $3p$ orbitals almost exclusively. Such overlap normally produces 90° bond angles, but in P_4 the P—P—P bond angles are 60°. The bonds are said to be *strained*, and as might be expected, species with strained bonds are reactive.

When white P is heated to about 300 °C, out of contact with air, it transforms to *red phosphorus*. What appears to happen is that one P—P bond per P_4 mol-

FIGURE 23-7
(a) Structure of white phosphorus: the P_4 molecule. (b) The structure of red phosphorus.

(a)

(b)

ecule breaks and the resulting fragments join together into long chains (see Figure 23-7b). Red P is less reactive than white P. Because they have a different atomic arrangement in their basic structural units, red and white phosphorus are allotropic forms of phosphorus, rather than just different solid phases.

Despite the fact that white P is the form obtained by condensing $P_4(g)$ and that the conversion of white P to red P is a very slow process at ambient temperatures, red P is actually the more thermodynamically stable of the two forms at 298.15 K. White P is assigned values of 0 for ΔH_f° and ΔG_f°, and for red P these values are negative.

Although compounds of phosphorus are vitally important to living organisms (DNA and phosphates in bones and teeth, for example), the element itself does not find much use. Almost all the elemental phosphorus produced is reoxidized to give P_4O_{10} for use in the manufacture of high-purity phosphoric acid. The rest is used to make organophosphorus compounds and phosphorus sulfides (P_4S_3) for use in match heads.

Arsenic is obtained by heating arsenic-containing metal sulfides. For example, FeAsS yields FeS and As(g). The As(g) deposits as As(s), which can be used to make other compounds. Some arsenic is also obtained by the reduction of arsenic(III) oxide with CO(g). Antimony is obtained mainly from its sulfide ores. Bismuth is obtained as a by-product of the refining of other metals.

As and Sb are used in making alloys of other metals. For example, the addition of As and Sb to Pb produces an alloy that has desirable properties for use as electrodes in lead–acid batteries. Arsenic and antimony are used to produce semiconductor materials, such as GaAs, GaSb, and InSb, for use in electronic devices. (See page 788 for a discussion of GaAs photocells.)

Nitrogen Compounds

In Chapter 8 we described the synthesis and chemical reactions of ammonia, NH_3. The manufacture of nitric acid from NH_3 was discussed, as were the uses of HNO_3 and several nitrate salts. Our treatment of the oxides of nitrogen emphasized their role in air pollution. We will not repeat that discussion here, but instead, we will discuss some other common compounds of nitrogen.

A number of metals combine directly with nitrogen to form *nitrides,* compounds in which the nitrogen is generally represented as the ion N^{3-}. Thus, when magnesium is burned in air (recall Figure 2-1), a small quantity of magnesium nitride is formed together with the principal product, magnesium oxide.

$$3 \text{ Mg(s)} + N_2(g) \xrightarrow{\Delta} Mg_3N_2(s) \qquad (23.34)$$

The nitride ion is a very strong base, and in aqueous solution accepts protons from water molecules to form ammonia molecules and hydroxide ions.

$$N^{3-}(aq) + 3 \text{ H}_2O(l) \longrightarrow NH_3(aq) + 3 \text{ OH}^-(aq)$$

In the reaction of magnesium nitride with water, magnesium and hydroxide ions combine to form insoluble $Mg(OH)_2$ and the ammonia is released as a gas, easily detectable by its odor.

$$Mg_3N_2(s) + 6 \text{ H}_2O(l) \longrightarrow 3 \text{ Mg(OH)}_2(s) + 2 \text{ NH}_3(g) \qquad (23.35)$$

If an H atom in NH_3 is replaced by the group —NH_2, the resulting molecule is H_2N—NH_2 or N_2H_4, *hydrazine* ($pK_{b_1} = 6.07$; $pK_{b_2} = 15.05$). Replacement of an H atom in NH_3 by —OH produces NH_2OH, *hydroxylamine* ($pK_b = 8.04$). Hydrazine and hydroxylamine are weak bases, and because of its two N atoms,

The upfiring thrusters of the Space Shuttle, *Columbia,* shown here in an in-flight test, use methylhydrazine, CH_3NHNH_2, as a fuel.

N_2H_4 ionizes in two steps. Hydrazine and hydroxylamine form salts analogous to ammonium salts, such as $N_2H_5^+NO_3^-$, $N_2H_6^{2+}SO_4^{2-}$, and $NH_3OH^+Cl^-$. These salts all hydrolyze in water to yield acidic solutions.

Hydrazine and some of its derivatives burn in air with the evolution of large quantities of heat; they are used as rocket fuels. For the combustion of hydrazine,

$$N_2H_4(l) + O_2(g) \longrightarrow N_2(g) + 2\ H_2O(l) \qquad \Delta H° = -622.2\ kJ \qquad (23.36)$$

Reaction (23.36) can also be used to remove dissolved $O_2(g)$ from boiler water. Hydrazine is particularly valued for this purpose because no salts (ionic compounds) are formed that would be objectionable in the water. The industrial preparation of hydrazine was described in Chapter 4 (page 120).

Both hydrazine and hydroxylamine can act either as oxidizing or reducing agents (usually the latter), depending on the pH and the substances with which they react. The oxidation of hydrazine in acidic solution by nitrite ion produces *hydrazoic acid,* $HN_3(aq)$.

$$N_2H_5^+(aq) + NO_2^-(aq) \longrightarrow HN_3(aq) + 2\ H_2O \qquad (23.37)$$

Pure HN_3 is a colorless liquid that boils at 37 °C. It is very unstable and will detonate when subjected to shock. In aqueous solution HN_3 is a weak acid; its salts are called *azides.* Azides resemble chlorides in some properties (for instance, AgN_3 is insoluble in water), but they are unstable. Some azides (such as lead azide) are used to make detonators. The release of $N_2(g)$ by the decomposition of sodium azide, NaN_3, is the basis of the air-bag safety system in automobiles (page 194).

Nitrogen is a key element in many organic and biologically important molecules, such as amines, amino acids, proteins, and DNA, as we will see in Chapters 27 and 28.

Oxides and Oxoacids of Phosphorus

Among the most important compounds of phosphorus are the oxides and oxoacids having P in the oxidation states +3 and +5. The simplest formulas we can write for the oxides are P_2O_3 and P_2O_5, respectively. The corresponding names are "phosphorus trioxide" and "phosphorus pentoxide." P_2O_3 and P_2O_5 are only empirical formulas, however. The true molecular formulas of the oxides are "double" those just written, that is, P_4O_6 and P_4O_{10}.

The structure of each oxide molecule is based on the P_4 tetrahedron and so must have *four* P atoms, not two. As shown in Figure 23-8, in P_4O_6 one O atom bridges each pair of P atoms in the P_4 tetrahedron, which means that there are

FIGURE 23-8
Molecular structures of P_4O_6 and P_4O_{10}.

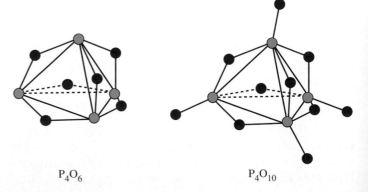

○ phosphorus

● oxygen

P_4O_6 P_4O_{10}

six O atoms per P_4 tetrahedron. In P_4O_{10}, in addition to the six bridging O atoms, one O atom is bonded to each corner P atom. This means that there are a total of *ten* O atoms per P_4 tetrahedron.

The reaction of P_4 with a limited quantity of $O_2(g)$ produces P_4O_6, and if an excess of $O_2(g)$ is used, P_4O_{10} is obtained. Both oxides react with water to form oxoacids—both are *acid anhydrides*.

$$P_4O_6(l) + 6\ H_2O(l) \longrightarrow 4\ H_3PO_3(aq)$$
phosphorous acid

$$P_4O_{10}(s) + 6\ H_2O(l) \longrightarrow 4\ H_3PO_4(l)$$
phosphoric acid   (23.38)

Phosphoric acid ranks about seventh among the chemicals manufactured in the United States, with an annual production of about 13 million tons. It is used mainly to make fertilizers, but also it is used to treat metals to make them more corrosion resistant. Phosphoric acid has many uses in the food industry: It is used to make baking powders and instant cereals, in cheese making, in curing hams, and in soft drinks to impart tartness.

If P_4O_{10} and H_2O are combined in a 1:6 mole ratio in reaction (23.38), the liquid product should be pure H_3PO_4, that is, 100% H_3PO_4, a compound called *orthophosphoric* acid. An analysis of the liquid, however, shows it to be only about 87.3% H_3PO_4. The "missing" phosphorus is still present in the liquid but as $H_4P_2O_7$, a compound called either *diphosphoric* acid or *pyrophosphoric* acid. A molecule of diphosphoric acid is formed when a molecule of H_2O is eliminated from between two molecules of orthophosphoric acid, as pictured in Figure 23-9. If a third molecule of orthophosphoric acid joins in through the elimination of another H_2O molecule, the product is $H_5P_3O_{10}$, *triphosphoric* acid, and so on. As a class, the chain-like phosphoric acid structures are called *polyphosphoric* acids and their salts are called *polyphosphates*. Two especially important derivatives of the polyphosphoric acids present in living organisms are the substances known as ADP and ATP. The "A" portion of the acronym stands for adenosine, a combination of an organic base called adenine and a

FIGURE 23-9
Formation of polyphosphoric acids.
Removal of H_2O molecules results in P—O—P bridges.

five-carbon sugar called ribose. If this adenosine combination is linked to a diphosphate ion, the product is ADP—adenosine diphosphate. Addition of a phosphate ion to ADP yields ATP—adenosine triphosphate.

Most phosphoric acid is made by the action of sulfuric acid on phosphate rock.

$$3Ca_3(PO_4)_2 \cdot CaF_2(s) + 10\ H_2SO_4(concd\ aq) + 20\ H_2O(l) \longrightarrow$$

fluorapatite

$$6\ H_3PO_4(aq) + 10\ CaSO_4 \cdot 2H_2O(s) + 2\ HF(aq)$$

gypsum   (23.39)

The HF is converted to insoluble Na_2SiF_6 and the gypsum formed is filtered off along with other insoluble impurities. The phosphoric acid is concentrated by evaporation. Phosphoric acid obtained by this "wet process" contains a variety of metal ions as impurities and is a dark green or brown color. Nevertheless, it is satisfactory for the manufacture of fertilizers and for metallurgical operations.

If H_3PO_4 from reaction (23.39) is used in place of H_2SO_4 to treat phosphate rock, the principal product is calcium dihydrogen phosphate, a fertilizer containing 20 to 21% P and marketed under the name *triple superphosphate.*

$$3Ca_3(PO_4)_2 \cdot CaF_2(s) + 14\ H_3PO_4(concd\ aq) + 10\ H_2O(l) \longrightarrow$$

$$10\ Ca(H_2PO_4)_2 \cdot H_2O(s) + 2\ HF(aq)   (23.40)$$

triple superphosphate

Triple superphosphate was the first high-phosphorus-content commercial fertilizer, but in recent years its share of the fertilizer market has been in constant decline. The leading phosphorus-containing fertilizers now in use are the ammonium phosphates: mixtures of monoammonium phosphate (MAP), $NH_4H_2PO_4$, and diammonium phosphate (DAP), $(NH_4)_2HPO_4$. They have the advantage of providing two plant nutrients, N and P. They are made by the controlled neutralization of $H_3PO_4(aq)$ with $NH_3(l)$.

An Environmental Issue Concerning Phosphorus

Phosphates are heavily used as fertilizers because phosphorus is an essential nutrient for plant growth. Heavy fertilizer use may lead to phosphate pollution of lakes, ponds, and streams, causing an explosion of plant growth, particularly algae. The algae deplete the oxygen content of the water, eventually killing fish. This type of change, occurring in freshwater bodies as a result of their enrichment by nutrients, is called **eutrophication.** It is a natural process that occurs over geological time periods, but it can be greatly accelerated by human activities.

Natural sources of plant nutrients include animal wastes, decomposition of dead organic matter, and natural nitrogen fixation. Human sources include industrial wastes and municipal sewage plant effluents, in addition to fertilizer runoff. One way to reduce phosphate discharges into the environment is through their removal in sewage treatment plants. In the processing of sewage, polyphosphates are degraded to orthophosphates by bacterial action. The orthophosphates may then be precipitated, either as iron(III) phosphates, aluminum phosphates, or as calcium phosphate or hydroxyapatite $[Ca_5(OH)(PO_4)_3]$. The precipitating agents are generally aluminum sulfate, iron(III) chloride, or calcium hydroxide (slaked lime). In a fully equipped modern sewage treatment plant, up to 98% of the phosphates in sewage can be removed.

The natural eutrophication of a lake is greatly accelerated by phosphates in wastewater and the agricultural runoff of fertilizers.

Other Phosphorus Compounds

The most important compound of phosphorus and hydrogen is *phosphine,* PH_3. This compound is analogous to ammonia, for example, by acting as a base and forming phosphonium (PH_4^+) compounds. But, unlike ammonia, PH_3 is thermally unstable. It is also extremely poisonous and has been used as a fumigant against rodents and insects. Phosphine is produced by the disproportionation of P_4 in aqueous base.

$$P_4(s) + 3\ OH^-(aq) + 3\ H_2O \longrightarrow 3\ H_2PO_2^-(aq) + PH_3(g) \qquad (23.41)$$

A similar phosphorus(III) molecule is *phosphorus trichloride,* PCl_3. One typical reaction of PCl_3 was shown in the introductory photograph of Chapter 5 (page 128)—the hydrolysis of PCl_3 to produce hydrochloric and phosphorous acids.

$$PCl_3(l) + 6\ H_2O(l) \longrightarrow H_3PO_3(aq) + 3\ H_3O^+(aq) + 3\ Cl^-(aq)$$

PCl_3 is the most important phosphorus halide and a variety of phosphorus(III) compounds are produced from it. PCl_3 is produced by the direct action of $Cl_2(g)$ on elemental phosphorus. Although you may never see PCl_3, chemicals produced from it are found everywhere—soaps and detergents, plastics and synthetic rubber, nylon, motor oils, and insecticides and herbicides. A variety of organic groups can replace one or more chlorine atoms of PCl_3 to give a family of phosphine-like compounds. These are good Lewis bases and can act as ligands in complex ion formation.

23-5 Group 4A Nonmetals: Carbon and Silicon

The differences between carbon and silicon, as outlined in Table 23.7, are perhaps the most striking to be found between any second- and third-period elements of a group in the periodic table. As suggested by the approximate bond

TABLE 23.7 Some Comparisons of Carbon and Silicon

Carbon	Silicon
• Two principal allotropes: graphite and diamond	• One stable, diamond-type crystalline modification
• Forms two stable *gaseous* oxides, CO and CO_2, and several less stable ones, such as C_3O_2	• Forms only one *solid* oxide (SiO_2) that is stable at room temperature; a second oxide (SiO) is stable only in the temperature range 1180–2480 °C
• Insoluble in alkaline medium	• Reacts in alkaline medium to form $H_2(g)$ and $SiO_4^{4-}(aq)$
• Principal oxoanion is CO_3^{2-}, which has a trigonal-planar shape	• Principal oxoanion is SiO_4^{2-}, which has a tetrahedral shape
• Strong tendency for catenation, with straight and branched chains and rings containing up to hundreds of C atoms[a]	• Less tendency for catenation, with silicon atom chains limited to about six Si atoms[a]
• Readily forms multiple bonds through use of the orbital sets $sp^2 + p$ and $sp + p^2$	• Multiple bond formation much less common than with carbon
• Approximate single bond energies, kJ/mol C—C, 347 C—H, 414 C—O, 360	• Approximate single bond energies, kJ/mol Si—Si, 226 Si—H, 318 Si—O, 464

[a]Catenation refers to the joining together of like atoms into chains.

energies, strong C—C and C—H bonds account for the central role of carbon-atom chains and rings in establishing the chemical behavior of carbon. A study of these chains and rings and their attached atoms is the focus of organic chemistry (Chapter 27) and biochemistry (Chapter 28). The relatively weaker Si—Si and Si—H bonds imply a less important "organic chemistry" of silicon, and the strength of the Si—O bond accounts for the predominance of the silicates and related compounds among silicon compounds.

In this section we will consider a few carbon compounds that, in addition to the oxides and carbonates discussed in Section 8-5, illustrate the inorganic chemistry of carbon. We will focus most of this section on the inorganic chemistry of silicon.

Production and Uses of Carbon

The most familiar allotropes of carbon are graphite and diamond. The nature of chemical bonding in these allotropes and the crystal structures associated with them were discussed in Section 13-7. Both diamond and graphite occur naturally, and both can be prepared synthetically.

Graphite can be synthesized from any number of carbon-containing materials, for example, anthracite coal. The key requirement is to heat the high-carbon-content material to a temperature of about 3000 °C in an electric furnace. In the United States about 70% of the graphite used is prepared synthetically and the remainder comes from natural sources.

Graphite has excellent lubricating properties, even when dry. This is because the planes of carbon atoms are held together by relatively weak forces and can easily slip past one another (recall Figure 13-32). One handy use of this property is in pencil "lead," actually a thin rod made from a mixture of graphite and clay that glides easily on paper. Because of its ability to conduct electric current, graphite is used for electrodes in batteries and industrial electrolysis. Graphite's use in foundry molds, furnaces, and other high-temperature environments is based on its ability to withstand high temperatures.

A newly developed use of graphite is in the manufacture of high-strength lightweight composites consisting of graphite fibers and various plastics. These composites are used in products ranging from tennis rackets to lightweight aircraft. The graphite fibers can be made by heating a carbon-based fiber such as rayon to a high temperature.

As indicated in the phase diagram for carbon in Figure 23-10, graphite is the more stable form of carbon, not only at room temperature and pressure, but at temperatures ranging up to 3000 °C and pressures of 10^4 atm and higher. Diamond is the more stable form of carbon at very high pressures. Diamonds can be synthesized from graphite by heating the graphite to temperatures of 1000–2000 °C and subjecting it to pressures of 10^5 atm or more. Usually the graphite is mixed with a metal such as iron. The metal melts and the graphite is converted to diamond within the liquid metal. Diamonds can then be picked out of the solidified metal.

According to Figure 23-10, we might expect diamond to revert to graphite at room temperature and pressure. Fortunately for the jewelry industry and those who treasure diamonds as gems, phase changes that require a rearrangement in bond type and crystal structure often occur extremely slowly. This is very much the case with the diamond–graphite transition.

Diamonds for use as gemstones are usually natural diamonds. For industrial purposes, inferior specimens of natural diamonds or, increasingly, synthetic diamonds are used. The industrial use depends on two key properties. Diamonds

Fabric for use in composite materials woven from carbon fibers.

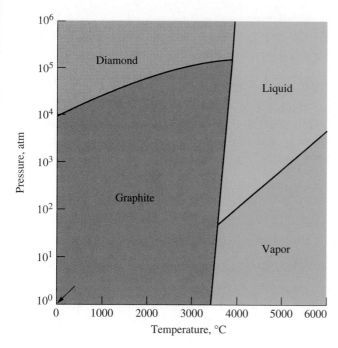

FIGURE 23-10
Phase diagram for carbon.
This is a simplified phase diagram for carbon. To include the vast range of pressures required (from 1 to 1,000,000 atm), pressure is plotted on a logarithmic scale. The arrow marks the point 1 atm, 25 °C.

are used as abrasives because of their extreme hardness (10 on the Mohs hardness scale). No harder substance is known. Diamonds also have a high thermal conductivity (they dissipate heat quickly) and are used in drilling bits for cutting steel and other hard materials. The rapid dissipation of heat makes the drilling process faster and increases the lifetime of the bit. A recent development has been the creation of diamond films, which can be deposited directly onto metals.

▶ Because of their expected future importance, in 1990 the journal *Science* named diamond films the "Molecule of the Year," and in 1991, the fullerenes.

Carbon can be obtained in several other forms of mixed or amorphous structure. Incomplete combustion of natural gas, such as in an improperly adjusted Bunsen burner in the laboratory, produces a smoky flame. This smoke can be deposited as a finely divided soot called **carbon black.** Carbon black is used as a filler in rubber tires (several kilograms per tire), as a pigment in printing inks, and as the transfer material in carbon paper, typewriter ribbons, laser printers, and photocopying machines. Recently, new allotropes of carbon have been isolated from the decomposition of graphite. These allotropes, known as fullerenes, were presented on pages 380 and 381. To summarize that earlier discussion, the molecule C_{60} is remarkably stable and has a shape resembling a soccer ball (12 pentagonal faces, 20 hexagonal faces, and 60 vertices). There are a number of other fullerenes in addition to C_{60}. These include C_{70}, C_{74}, and C_{82}. Fullerenes are typically produced by laser decomposition of graphite under a helium atmosphere. Because nitrogen and oxygen interfere with the process of forming fullerenes, soot (by the combustion of hydrocarbons in air) does not contain fullerenes.

Diamonds are an allotrope of carbon. Their brilliance when cut makes them prized as gems. Their hardness and high thermal conductivity make them useful in industry. Recent advances in coating metal parts with diamond films may increase the efficiency of machines run at high temperatures.

Finally, two other noteworthy carbon-containing materials are coke and charcoal. When coal is heated in the absence of air, volatile substances are driven off to leave a high-carbon residue called *coke.* This same type of destructive distillation of wood produces *charcoal.* Currently, coke is the principal metallurgical reducing agent. It is used in blast furnaces, for instance, to reduce iron oxide to iron metal.

Production and Uses of Silicon

Elemental silicon is produced when quartz or sand (SiO_2) is reduced by reaction with coke in an electric arc furnace.

$$SiO_2 + 2\ C \xrightarrow{\Delta} Si + 2\ CO(g) \tag{23.42}$$

Very high purity Si for solar cells can be made by reducing Na_2SiF_6 with metallic Na. Recall that Na_2SiF_6 is a by-product of the formation of phosphate fertilizers (page 814). High-purity silicon is required in the manufacture of transistors and other semiconductor devices.

Oxides of Silicon

The chemistry of the oxides of carbon (CO and CO_2) was discussed in Chapter 8. Silicon forms only one stable oxide, having the empirical formula SiO_2 (silica). In **silica** each Si atom is bonded to *four* O atoms and each O atom to *two* Si atoms. The structure is that of a network covalent solid, as suggested by Figure 23-11a. This structure is reminiscent of the diamond structure, and silica does have certain properties that resemble those of diamond. For example, quartz, a form of silica, has a Mohs hardness of 7 (compared with 10 for diamond), a high melting point (about 1700 °C), and is a nonconductor of electricity. Silica is the basic raw material of the glass and ceramics industries.

The central feature of all *silicates* is the SiO_4^{4-} tetrahedron depicted in Figure 23-11b. These SiO_4^{4-} tetrahedra may be arranged in a wide variety of forms. The tetrahedra can be strung together in long filaments as in the mineral *asbestos*. The tetrahedra can also form two-dimensional sheets (Figure 23-11c), as in the mineral *mica*.

SiO_2 is a weakly acidic oxide and slowly dissolves in strong bases. It forms a series of silicates, such as Na_2SiO_4 (sodium orthosilicate) and Na_2SiO_3 (sodium metasilicate). These compounds are somewhat soluble in water and because of this are sometimes referred to as "water glass." They are used as adhesives for corrugated cardboard and builders in laundry detergents.

▶ H_4SiO_4 is called *ortho*silicic acid. When a combination of one H atom and one OH group (equivalent to a molecule of H_2O) is eliminated from the *ortho* acid, the resulting acid, H_2SiO_3, is called *meta*silicic acid.

FIGURE 23-11
Structures of silica and silicates.
(a) A three-dimensional network of bonds in silica—SiO_2. (b) The silicate anion, SiO_4^{4-}, commonly found in silicate materials. The Si atom is in the center of the tetrahedron and is surrounded by four O atoms. (c) A depiction of the structure of mica using tetrahedra to represent the SiO_4^{4-} units. The cations K^+ and Al^{3+} are also present but are not shown here. This is the usual way that silicates and similar materials are represented by materials scientists.

● = silicon   ● = oxygen

(a)

(b)

(c)

Are You Wondering . . .

Why silica (SiO_2) does not exist as simple molecules as CO_2 does?

Because they are both in Group 4A of the periodic table and have four valence electrons, we might expect carbon and silicon to form oxides with similar properties. In CO_2 the sidewise overlap of $2p$ orbitals of the C and O atoms is extensive (see page 398). A carbon-to-oxygen double bond proves to be stronger (736 kJ/mol) than two single bonds (2 × 360 kJ/mol). This results in the familiar Lewis structure of CO_2:

$$\ddot{\text{O}}=\text{C}=\ddot{\text{O}}$$

Silicon, being in the third period, would have to use $3p$ orbitals to form double bonds with oxygen. The sidewise overlap of these orbitals with the $2p$ orbitals of oxygen is quite limited. From an energy standpoint, a stronger bonding arrangement results if the Si atoms form *four* single bonds with O atoms (bond energy: 4 × 464 kJ/mol = 1856 kJ total) rather than *two* double bonds (bond energy: 2 × 640 kJ/mol = 1280 kJ total). Because each O atom must be simultaneously bonded to two Si atoms, the result is a network of —Si—O—Si— bonds (see Figure 23-11).

On page 803 we contrasted the second- and third-period members of Group 6A: O and S. Here is another example of how the second-period member of a group (C) differs from the higher period members.

Silicate anions are bases, and when acidified they produce silicic acids, which are unstable and decompose to silica. The silica obtained, however, is not a crystalline solid or powder. Depending on the acidity of the solution, the silica is obtained as a colloidal dispersion, a gelatinous precipitate, or a solid-like gel in which all of the water is entrapped. These hydrated silicates are polymers of silica formed by the elimination of water molecules between neighboring molecules of silicic acid. The process begins with the reaction

$$SiO_4{}^{4-}(aq) + 4\,H^+(aq) \longrightarrow Si(OH)_4$$

and is followed by

where —H and —OH combine to form water (HOH) and Si—O—Si bridges are produced.

Ceramics and Glass.   Hydrated silicate polymers are important in the ceramics industry. A colloidal dispersion of particles in a liquid is called a *sol*. The sol can be poured into a mold and, following removal of some of the liquid, is converted to a *gel*. The gel is then processed into the final ceramic product. Some exceptionally lightweight ceramic materials can be produced by this *sol–gel* process.

A modern ceramic component of a turbine engine.

Uses of these advanced ceramics fall into two general categories: electrical, magnetic, or optical applications (as in the manufacture of integrated circuit components) and applications that take advantage of the ceramic's mechanical and structural properties at high temperatures. These latter properties have been explored in developing ceramic components for gas turbines and automotive engines. Quite possibly the automobile engine of the twenty-first century will be a ceramic engine—lightweight and more fuel-efficient because of higher operating temperatures. Some engines already use several ceramic components. There is some truth to the vision of this ceramic future as the "new stone age."

If sodium and calcium carbonates are mixed with sand and fused at about 1500 °C, the result is a liquid mixture of sodium and calcium silicates. When cooled, the liquid becomes more viscous and eventually becomes a solid that is transparent to light; this solid is called a **glass.** Crystalline solids have a long-range order, whereas glasses are *amorphous* solids in which order is found only over relatively short distances. Think of the structural units in glass (silicate anions) as being in a jumbled rather than a regular arrangement. A glass and a crystalline solid also differ in their melting behavior. A glass softens and melts over a broad temperature range, whereas a crystalline solid has a definite, sharp melting point. Different types of glass and methods of making them are described in the Focus On Glassmaking at the end of the chapter.

Organosilicon Compounds

Several silicon–hydrogen compounds are known, but because Si—Si single bonds are not particularly strong, the chain length in these compounds, called *silanes,* is limited to six.

$$
\begin{array}{llll}
\text{H} & \text{H } \text{ H} & \text{H } \text{ H } \text{ H} & \\
| & | \;\; | & | \;\; | \;\; | & \\
\text{H—Si—H} & \text{H—Si—Si—H} & \text{H—Si—Si—Si—H} \ldots & S_6H_{14} \\
| & | \;\; | & | \;\; | \;\; | & \\
\text{H} & \text{H } \text{ H} & \text{H } \text{ H } \text{ H} & \\
\text{monosilane} & \text{disilane} & \text{trisilane} & \text{hexasilane}
\end{array}
$$

Other atoms or groups of atoms can be substituted for H atoms in silanes to produce compounds called *organosilanes.* Typical is the direct reaction of Si and methyl chloride, CH_3Cl.

$$2\ CH_3Cl + Si \longrightarrow (CH_3)_2SiCl_2$$

The reaction of $(CH_3)_2SiCl_2$, dichlorodimethylsilane, with water produces an interesting compound, *dimethylsilanol,* $(CH_3)_2Si(OH)_2$. Dimethylsilanol undergoes a polymerization reaction in which H_2O molecules are eliminated from among large numbers of silanol molecules. The result of this polymerization is a material consisting of molecules with long silicon–oxygen chains: **silicones.**

$$(CH_3)_2SiCl_2 + 2\ H_2O \longrightarrow (CH_3)_2Si(OH)_2 + 2\ HCl$$

$$
\begin{array}{ccc}
\quad\;\text{CH}_3 & \quad\;\text{CH}_3 & \quad\;\text{CH}_3 \quad\; \left[\text{CH}_3\right] \quad\; \text{CH}_3 \\
\quad\;| & \quad\;| & \quad\;| \qquad\quad | \qquad\quad | \\
\text{HO—Si—O—H} + \text{HO—Si—OH} & \xrightarrow{-\text{H}_2\text{O}} & \text{HO—Si—O} \!\!-\!\!\Big[\text{Si—O}\Big]\!\!-\!\! \text{Si—OH} \\
\quad\;| & \quad\;| & \quad\;| \qquad\quad |_n \qquad\quad | \\
\quad\;\text{CH}_3 & \quad\;\text{CH}_3 & \quad\;\text{CH}_3 \qquad \text{CH}_3 \qquad\; \text{CH}_3
\end{array}
$$

a silicone

Silicones are important polymers because of their versatility. They can be obtained either as oils or as rubber-like materials. Silicone oils are not volatile and may be heated without decomposition. Also, they can be cooled to low temperatures without solidifying or becoming viscous. Silicone oils are excellent high-temperature lubricants. In contrast, hydrocarbon oils break down at high temperatures and become very viscous and then solidify at low temperatures. Silicone rubbers retain their elasticity at low temperatures and are chemically resistant and thermally stable. This makes them useful in caulking around windows when winterizing a home, for instance.

23-6   The Group 3A Nonmetal: Boron

The only Group 3A element that is almost exclusively nonmetallic in its physical and chemical properties is boron. The remaining members of Group 3A—Al, Ga, In, Tl—are metals and were discussed in Chapter 22. Many boron compounds lack an octet about the central boron atom, which makes the compounds *electron-deficient*. This also makes them strong Lewis acids. The electron deficiency of some boron compounds leads to bonding of a type that we have not previously encountered. This type of bonding occurs in the boron hydrides.

Boron Hydrides

The molecule BH_3 (borane) may exist as a reaction intermediate, but it has not been isolated as a stable compound. Notice that in BH_3 the boron lacks a complete octet by having only six electrons surrounding the central B atom. The simplest boron hydride that has been isolated is **diborane,** B_2H_6. To explain the structure and bonding in B_2H_6 requires new ideas in bonding theory, particularly molecular orbital theory. The problem is this: In the B_2H_6 molecule there are only *12* valence electrons (3 each from the two B atoms and 1 each from the six H atoms). The minimum number of valence-shell atomic orbitals required to make a Lewis structure for B_2H_6 resembling that of C_2H_6, however, is *14* (4 each from the two B atoms and 1 each from the six H atoms).

The currently accepted structure of diborane is pictured in Figure 23-12. The two B atoms and four of the H atoms lie in the same plane (the plane of the page). The orbitals used by the B atoms to bond these particular four H atoms can be viewed as sp^2. Eight electrons are involved in these four bonds. This leaves four electrons with which to bond the two remaining H atoms to the two B atoms and also to bond the B atoms together. This is accomplished if each of the two H atoms is simultaneously bonded to *both* B atoms.

Atom "bridges" are actually fairly common, although we have not had much occasion to deal with them before (see the discussion of Al_2Cl_6, page 784). The B–H–B bridges are unusual, however, in having only *two* electrons shared among *three* atoms. For this reason, these bonds are referred to as three-center two-electron bonds.

We can rationalize the bonding in these three-center bonds with molecular orbital theory. *Six* atomic orbitals—an sp^2 and p orbital from each B atom and an s orbital from each H atom—are combined into *six* molecular orbitals in these two "bridge" bonds. Of these six molecular orbitals, *two* are bonding orbitals, and these are the orbitals into which the four electrons are placed. The concept of "bridge" bonds can be extended to B–B–B bonds to describe the structure of higher boranes, such as B_5H_9 shown in Figure 23-13.

Boron hydrides are widely used in reactions for synthesizing organic compounds, and they continue to provide new and exciting developments in chemistry.

Other Boron Compounds of Interest

Boron compounds are widely distributed in Earth's crust, but concentrated ores are found in only a few locations—in Italy, Russia, Tibet, Turkey, and the desert

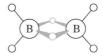

FIGURE 23-12
Structure of diborane, B_2H_6.

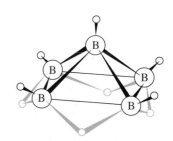

FIGURE 23-13
Structure of pentaborane, B_5H_9.
The boron atoms are joined through multicenter B—B—B bonds. Five of the H atoms are bonded to one B atom each. The other four H atoms bridge pairs of B atoms.

FIGURE 23-14

Preparation of some boron compounds.

$Na_2B_4O_7 \cdot 10H_2O$ (borax) is converted to $B(OH)_3$ by reaction with H_2SO_4. When heated strongly, $B(OH)_3$ is converted to B_2O_3. A variety of boron-containing compounds and boron itself can be prepared from B_2O_3.

regions of California. Typical of these ores is the hydrated borate *borax,* $Na_2B_4O_7 \cdot 10H_2O$. Figure 23-14 suggests how borax can be converted to a variety of boron compounds.

One useful compound that can be obtained by crystallization from a solution of borax and hydrogen peroxide is *sodium perborate,* $NaBO_3 \cdot 4H_2O$. This formula is deceptively simple; a more precise formula is $Na_2[B_2(O_2)_2(OH)_4] \cdot 6H_2O$. This compound is the "bleach alternative" used in many color-safe bleaches. The key to the bleaching action is the presence of the two peroxo groups (—O—O—) bridging the boron atoms in the $[B_2(O_2)_2(OH)_4]^{2-}$ ion.

$$
\left[
\begin{array}{c}
\text{H—O} \qquad \text{O—O} \qquad \text{O—H} \\
\diagdown \diagup \qquad \diagdown \diagup \\
\text{B} \qquad\qquad \text{B} \\
\diagup \diagdown \qquad \diagup \diagdown \\
\text{H—O} \qquad \text{O—O} \qquad \text{O—H}
\end{array}
\right]^{2-}
$$

One of the key compounds from which other boron compounds can be synthesized is *boric acid,* $B(OH)_3$. The weakly acidic nature of boric acid comes about in a rather unusual way: The electron-deficient $B(OH)_3$ accepts an OH^- ion from the self-ionization of water to form the complex ion $[B(OH)_4]^-$. Thus, the source of the H_3O^+ in $B(OH)_3(aq)$ is the water itself. This ionization scheme, together with the fact that $B(OH)_3$ is a *mono*protic acid, not *tri*protic, suggests that the best formula of boric acid is $B(OH)_3$ and not H_3BO_3.

$$B(OH)_3(aq) + 2\,H_2O \rightleftharpoons H_3O^+ + B(OH)_4^- \qquad K_a = 5.6 \times 10^{-10}$$

Borate salts, as expected of the salts of a weak acid, produce basic solutions by hydrolysis, which accounts for their use in cleaning agents. Boric acid also finds use as an insecticide, particularly to kill roaches, and is familiar to some as an antiseptic in eyewash solutions. Boron compounds are used in products as varied as adhesives, cement, disinfectants, fertilizers, fire retardants, glass, herbicides, metallurgical fluxes, and textile bleaches and dyes.

EXAMPLE 23-2

Writing Chemical Equations from a Summary Diagram of Reaction Chemistry. Based on Figure 23-14, write chemical equations for the successive conversions of borax to **(a)** boric acid, **(b)** B_2O_3, and **(c)** impure boron metal.

Solution

Figure 23-14 lists the key substances involved in each reaction. Our task is to identify other plausible reactants and products.

a. Conversion of a salt to the corresponding acid is an acid–base reaction and does not involve changes in oxidation states. Balancing should be possible by inspection. The additional substances required in the equation are Na_2SO_4 and H_2O.

$$Na_2B_4O_7 \cdot 10H_2O + H_2SO_4 \longrightarrow 4\,B(OH)_3 + Na_2SO_4 + 5\,H_2O$$

b. Conversion of a hydroxo compound to an oxide requires that H_2O be driven off.

$$2\,B(OH)_3 \xrightarrow{\Delta} B_2O_3 + 3\,H_2O$$

c. The direct reduction of B_2O_3 with Mg produces impure boron.

$$3\,Mg + B_2O_3 \xrightarrow{\Delta} 2\,B + 3\,MgO$$

Practice Example A: Based on Figure 23-14, write chemical equations for the sequence of reactions to convert borax to diborane.

Practice Example B: Based on Figure 23-4 and reactions in the text, write chemical equations for the sequence of reactions to obtain sulfuric acid from ZnS.

Summary

Principles developed in the first 21 chapters of this text are applied in this chapter to several aspects of the descriptive inorganic chemistry of the nonmetals. Several differences between the first and higher members of a group of the periodic table are encountered with nonmetals. Examples cited are the failure of fluorine to form stable oxoacids and the numerous differences between O and S and between C and Si. Also, by considering several criteria, the progression of properties from nonmetallic to metallic within Group 5A is established. Additional insights into chemical bonding are provided through a study of the boron hydrides.

Oxoacids and oxoanions are studied in terms of the methods required to prepare them, their acid–base properties, their strengths as oxidizing or reducing agents, and their structures. With phosphorus, the elimination of H_2O from simpler acid molecules produces diphosphoric, triphosphoric, and other polyphosphoric acids and their salts.

Throughout the chapter practical uses of the nonmetals and their compounds are described. Quite often it is found that a particular use for a substance is a consequence of some special property that it possesses.

Summarizing Example

For use in analytical chemistry, sodium thiosulfate solutions must be carefully prepared. In particular, the solutions must be kept from becoming acidic. In strongly acidic solutions, thiosulfate ion *disproportionates* into $SO_2(g)$ and $S(s)$.

Show that the disproportionation of $S_2O_3^{2-}(aq)$ is spontaneous for standard state conditions in acidic solution but not in basic solution.

1. *Write an equation for the disproportionation reaction.* Base this on the description of the reaction given above. *Result:* $S_2O_3^{2-}(aq) + 2\,H^+(aq) \rightarrow H_2O + SO_2(g) + S(s)$.
2. *Separate the equation into two half-equations. Result:* oxidation: $S_2O_3^{2-}(aq) + H_2O \rightarrow 2\,SO_2(g) + 2\,H^+(aq) + 4\,e^-$; reduction: $S_2O_3^{2-}(aq) + 6\,H^+(aq) + 4\,e^- \rightarrow 2\,S(s) + 3\,H_2O$.
3. *Obtain $E°$ values for the half-reactions from Appendix D. Result:* oxidation: $-E°_{SO_2/S_2O_3^{2-}} = -0.40\,V$; reduction: $E°_{S_2O_3^{2-}/S} = 0.60\,V$.

4. *Determine $E°_{cell}$ and interpret its meaning. Answer:* $E°_{cell} = +0.20\,V$. The positive value of the cell voltage signifies that, for reactants and products in their standard states in an acidic solution, the forward reaction—the disproportionation of $S_2O_3^{2-}(aq)$—is spontaneous.
5. *Assess the situation in basic solution.* Although we could do a quantitative calculation using the Nernst equation, the simplest approach is to note that $E°_{cell}$ for acidic solution, though positive, is rather small in magnitude $(+0.20\,V)$. Increasing $[OH^-]$, as would be the case in making the solution basic, means decreasing $[H^+]$. In fact, $[OH^-] = 1\,M$ corresponds to $[H^+] = 1 \times 10^{-14}$ M. Because the net equation in part 1 has $H^+(aq)$ on the *left* side of the equation, a decrease in $[H^+]$ favors the *reverse* reaction (Le Châtelier's principle). At some point before the solution becomes basic, the forward reaction is no longer spontaneous. *Answer:* The disproportionation of $S_2O_3^{2-}(aq)$, with reactants and products in their standard states, is nonspontaneous in basic solution.

Focus On   Glassmaking

Stained-glass windows of the Spiral Church in Dallas, Texas.

played in many museums. Today, glass is indispensable in almost every facet of life.

Soda–lime glass is the oldest form of glass. The starting material in its manufacture is a mixture of sodium carbonate (*soda* ash), calcium carbonate (which decomposes to form quick*lime* when heated), and silicon dioxide. The mixture can be fused at a relatively low temperature (1300 °C) compared with the melting point of pure silica (1710 °C), and it is easy to form into the shapes needed. The effect of the sodium ions is to break up the crystalline lattice of the SiO_2, and the calcium ions render the glass insoluble in water, so that it can be used for items such as drinking glasses and windows. At the high temperatures employed, chemical reactions occur that produce a mixture of sodium and cal-

Glass and the art of glassmaking have been known for centuries. Beautiful stained-glass windows can be seen in medieval and modern churches; ancient glass containers for perfume and oil are dis-

Key Terms

acid salt (23-3)
contact process (23-3)
diborane (23-6)

eutrophication (23-4)
Frasch process (23-3)
glass (23-5)

polyhalide ions (23-2)
silica (23-5)
silicones (23-5)

Review Questions

1. In your own words define the following terms: **(a)** polyhalide ion; **(b)** a polyphosphate; **(c)** allotropy; **(d)** disproportionation.

2. Briefly describe each of the following ideas or methods: **(a)** Frasch process; **(b)** contact process; **(c)** eutrophication; **(d)** three-center bond.

3. Explain the important distinctions between each pair of terms: **(a)** acid salt and acid anhydride; **(b)** azide and nitride; **(c)** silane and silicone; **(d)** sol and gel.

4. Provide an acceptable name or formula for each of the following.
(a) $KBrO_3$
(b) I_3^-
(c) sodium hypoiodite
(d) NaH_2PO_4
(e) LiN_3
(f) sodium thiosulfate

5. Complete and balance equations for the following reactions. If no reaction occurs, so state.

(a) $CaCl_2(s) + H_2SO_4(\text{concd aq}) \xrightarrow{\Delta}$
(b) $I_2(s) + Cl^-(aq) \rightarrow$
(c) $NH_3(aq) + HClO_4(aq) \rightarrow$

6. Write a chemical equation to represent the reaction of **(a)** $Cl_2(g)$ with cold NaOH(aq); **(b)** NaI(s) with hot H_2SO_4(concd aq); **(c)** $Cl_2(g)$ with $Br^-(aq)$.

cium silicates as the ultimate glass product.

Glass containing even small amounts of Fe_2O_3 has a distinctive green color (bottle glass). Glass can be made colorless by incorporating MnO_2 into the glassmaking process. The MnO_2 oxidizes green $FeSiO_3$ to yellow $Fe_2(SiO_3)_3$ and is itself reduced to Mn_2O_3, which imparts a violet color. The yellow and violet are complementary colors and the glass appears colorless. Where desired, color can be imparted through appropriate additives, such as CoO for cobalt blue glass. To produce an opaque glass, additives such as calcium phosphate are used. In Bohemian crystal most of the Na^+ is replaced by K^+, and a glass with exceptional transparency and brightness can be made by incorporating lead oxides (as in Waterford crystal).

A problem with soda–lime glass is its high coefficient of thermal expansion—its dimensions change significantly with temperature. The glass cannot resist thermal shock. This posed a particular problem for the lanterns used in the early days of railroads. In the rain, the hot glass in these lanterns would often shatter. The problem was solved by adding B_2O_3 to the glass. A *borosilicate* glass has a low coefficient of thermal expansion and is thus resistant to thermal shock. This is the glass commonly known by its trade name: Pyrex. It is widely used in chemical laboratories and for cookware in the home.

Most glass has small bubbles or impurities in it that decrease its ability to transmit light without scattering, a phenomenon observed in the distorted images produced by the thick bottoms of drinking glasses. In modern fiber-optic cables, light must be transmitted over long distances without distortion or loss of signal. For this purpose a special glass made of pure silica is required. The key to making this glass is in purifying silica, which can be done by a series of chemical reactions. First, impure quartz or sand is reduced to silicon using coke as a reducing agent. The silicon is then allowed to react with $Cl_2(g)$ to form $SiCl_4(g)$.

$$SiO_2(s) + 2\ C(s) \longrightarrow Si(s) + 2\ CO(g)$$

$$Si(s) + 2\ Cl_2(g) \longrightarrow SiCl_4(g)$$

Finally, the $SiCl_4$ is burned in a methane–oxygen flame. SiO_2 deposits as a fine "ash," and chlorocarbon compounds escape as gaseous products. The SiO_2, with impurity levels reduced to parts per billion, can then be melted and drawn into the fine filaments required in fiber-optic cable.

---

7. Write a plausible equation to represent **(a)** the reaction of KI(s) with the nonoxidizing acid, $H_3PO_4(l)$; **(b)** the reduction of Na_2SiF_6 to pure Si, using Na as the reducing agent.

8. Write a plausible equation(s) to represent **(a)** the reduction of As_4O_6 to As, using CO(g) as the reducing agent; **(b)** the production of the fertilizer components ammonium monophosphate (MAP) and diammonium phosphate (DAP) by the reaction of H_3PO_4 and NH_3.

9. A reference work describing the production of phosphorus states that for every 8.00 tons of phosphate rock used 1.00 ton of elemental phosphorus is obtained. The phosphorus content of the phosphate rock is given as 31% P_4O_{10}. What is the percent yield of phosphorus in this reaction? [*Hint:* Is it necessary to use Equation (23.33)?]

10. If Br^- and I^- occur together in an aqueous solution, I^- can be oxidized to IO_3^- with an excess of $Cl_2(aq)$. Simultaneously, Br^- is oxidized to Br_2, which is extracted with $CS_2(l)$. Write chemical equations for the reactions that occur.

11. Imagine that the sulfur present in seawater as SO_4^{2-} (2650 mg/L) could be recovered as elemental sulfur. If this sulfur were then converted to H_2SO_4, how many cubic kilometers of seawater would have to be processed to yield the average U.S. annual consumption of about 38 million tons of H_2SO_4?

12. Use information from this chapter and elsewhere in the text to predict which of the following outcomes is the more likely. Explain your reasoning in each case.
(a) When $Cl_2(g)$ is added to an aqueous solution containing I^-, which is more likely to be produced, $O_2(g)$ or I_2?
(b) When added to an acidic solution containing NH_4^+, is H_2O_2 more likely to be oxidized to $O_2(g)$ or reduced to H_2O?

13. Given the bond energies at 298 K: O_2, 498; N_2, 946; F_2, 159; Cl_2, 243; ClF, 251; OF (in OF_2), 213; ClO (in Cl_2O), 205; and NF (in NF_3), 280 kJ/mol, respectively, calculate ΔH_f° at 298 K for 1 mol of **(a)** ClF(g); **(b)** $OF_2(g)$; **(c)** $Cl_2O(g)$; **(d)** $NF_3(g)$.

14. Although relatively rare, all of the following compounds exist. Based on what you know about related compounds (such as from the periodic table, for example), propose a plausible name or formula for each compound.
(a) silver astatide
(b) CSe_2
(c) magnesium polonide
(d) H_2TeO_3
(e) potassium thioselenate
(f) $KAtO_4$

Exercises

Noble Gas Compounds

15. Use VSEPR theory to predict the probable geometric structures of **(a)** XeO_3; **(b)** XeO_4; **(c)** XeF_5^+.

16. Use VSEPR theory to predict the probable geometric structures of the molecules **(a)** O_2XeF_2; **(b)** O_3XeF_2; **(c)** $OXeF_4$.

The Halogens

17. Freshly prepared solutions containing iodide ion are colorless, but over time they usually develop a yellow color. Describe a plausible chemical reaction (or reactions) to account for this observation?

18. Fluorine can be prepared by the reaction of hexafluoromanganate(IV) ion, MnF_6^{2-}, with antimony pentafluoride to produce manganese(IV) fluoride and SbF_6^-, followed by the disproportionation of manganese(IV) fluoride to manganese(III) fluoride and $F_2(g)$. Write chemical equations for these two reactions.

19. Based on information in this chapter, make a general prediction about which of the elements Cl_2, Br_2, or I_2 displaces other halogens from a solution of halide ions. In other words, will the reaction $Br_2 + 2 I^- \rightarrow 2 Br^- + I_2$ occur? Will the reaction $Br_2 + 2 Cl^- \rightarrow 2 Br^- + Cl_2$ occur?

20. The following properties of astatine have been measured or estimated: **(a)** covalent radius; **(b)** ionic radius (At^-); **(c)** first ionization energy; **(d)** electron affinity; **(e)** electronegativity; **(f)** standard reduction potential. Based on periodic relationships and data in Table 23.1, what values would you expect for these properties?

21. The abundance of F^- in seawater is 1 g F^- per ton of seawater. Suppose that a commercially feasible method could be found to extract fluorine from seawater.
(a) What mass of F_2 could be obtained from 1 km^3 of seawater ($d = 1.03$ g/cm^3)?
(b) Do you think the process would resemble that for extracting bromine from seawater? Explain.

22. Fluorine is chiefly produced from fluorite, CaF_2. Fluorine can also be obtained as a by-product of the production of phosphate fertilizers. These fertilizers are derived from phosphate rock [$3Ca_3(PO_4)_2 \cdot CaF_2$]. What is the maximum mass of fluorine that could be extracted as by-product from 1.00×10^3 kg of phosphate rock?

Oxygen

23. Hydrogen peroxide is a somewhat stronger acid than water. For the ionization
$$H_2O_2(aq) + H_2O \longrightarrow H_3O^+(aq) + HO_2^-(aq)$$
$pK_a = 11.75$. Calculate the expected pH of a 1.0 M $H_2O_2(aq)$ solution.

24. O^{2-} is a strong base in water. If 0.050 g of Li_2O is dissolved in 750.0 mL of aqueous solution, what would be the pH of the solution?

25. For the conversion of $O_2(g)$ to $O_3(g)$, which can be accomplished in an electric discharge, $3\ O_2(g) \rightarrow 2\ O_3(g)$, $\Delta H° = +285$ kJ/mol. The bond energy in O_2, which is essentially the $O{=}O$ double bond energy, is 498 kJ/mol. The $O{-}O$ single bond energy is 142 kJ/mol. Calculate the average $O{-}O$ bond energy in $O_3(g)$.

26. Estimate the average bond energy in $O_3(g)$ from the structure on page 361. Compare this result with that of Exercise 25.

27. Use Lewis structures and other information to explain the observation that
(a) H_2S is a gas at room temperature, whereas H_2O is a liquid;
(b) O_3 is diamagnetic.

28. Use Lewis structures and other information to explain the observation that
(a) The oxygen-to-oxygen bond lengths in O_2, O_3, and H_2O_2 are 121, 128, and 148 pm, respectively.
(b) The oxygen-to-oxygen bond length of O_2 is 121 pm and for O_2^+ is 112 pm. Why is the bond length for O_2^+ so much shorter than for O_2?

Sulfur

29. Give an appropriate name to each of the following: **(a)** ZnS; **(b)** $KHSO_3$; **(c)** $K_2S_2O_3$; **(d)** SF_4.

30. Give an appropriate formula for each of the following: **(a)** calcium sulfate dihydrate; **(b)** hydrosulfuric acid; **(c)** sodium hydrogen sulfate; **(d)** disulfuric acid.

31. Through a chemical equation give a specific example that illustrates the
(a) reaction of a metal sulfide with HCl(aq);
(b) action of a *nonoxidizing* acid on a metal sulfite;
(c) oxidation of SO_2(aq) to SO_4^{2-}(aq) by MnO_2(s) in acidic solution;
(d) disproportionation of $S_2O_3^{2-}$ in acidic solution.

33. Describe a chemical test you could use to determine whether a white solid is Na_2SO_4 or $Na_2S_2O_3$. Explain the basis of this test by writing a chemical equation or equations.

35. Salts like $NaHSO_4$ are called *acid* salts because their anions undergo further ionization. What should be the pH of 250 mL of water solution containing 12.5 g $NaHSO_4$? (*Hint:* Use data from Chapter 17, as necessary.)

32. Show how you would use elemental sulfur, chlorine gas, metallic sodium, water, and air to produce aqueous solutions containing **(a)** Na_2SO_3; **(b)** Na_2SO_4; **(c)** $Na_2S_2O_3$. (*Hint:* You will have to use information from other chapters as well as this one.)

34. Sulfur can occur naturally as sulfates, but not as sulfites. Explain why this is so.

36. What mass of Na_2SO_3 must have been present in a sample that required 26.50 mL of 0.0510 M $KMnO_4$ for its oxidation to Na_2SO_4 in an acidic solution? MnO_4^- is reduced to Mn^{2+}.

Nitrogen Family

37. Use information from this and previous chapters to write chemical equations to represent the following.
(a) Equilibrium between nitrogen dioxide and dinitrogen tetroxide in the gaseous state.
(b) The decomposition of NH_4NO_3(s) by heating.
(c) The neutralization of NH_3(aq) by H_2SO_4(aq).

39. Draw plausible Lewis structures for
(a) hydrazine, N_2H_4
(b) hydrogen azide, HN_3
(c) dinitrogen tetroxide, N_2O_4
41. Supply an appropriate name for each of the following:
(a) HPO_4^{2-}; **(b)** $Ca_2P_2O_7$; **(c)** $H_6P_4O_{13}$.

38. Use information from this and previous chapters to write chemical equations to represent the following.
(a) The reaction of silver metal with HNO_3(aq).
(b) The complete combustion of the rocket fuel, dimethylhydrazine, $(CH_3)_2NNH_2$.
(c) The reduction of nitrogen monoxide with carbon monoxide using a rhodium catalyst.
40. Nitramide and hyponitrous acid both have the formula $H_2N_2O_2$. Hyponitrous acid is a weak diprotic acid; nitramide contains the amide group ($-NH_2$). Based on this information write plausible Lewis structures for these two substances.
42. Write an appropriate formula for each of the following:
(a) hydroxylamine; **(b)** calcium hydrogen phosphate; **(c)** lithium nitride.

Carbon and Silicon

43. Comment on the accuracy of a jeweler's advertising that "diamonds last forever." In what sense is the statement true, and in what ways is it false?

45. Describe what is meant by the terms *silane* and *silanol*. What is their role in the preparation of silicones?

47. Methane and sulfur vapor react to form carbon disulfide and hydrogen sulfide. Carbon disulfide reacts with Cl_2(g) to form carbon tetrachloride and S_2Cl_2. Further reaction of carbon disulfide and S_2Cl_2 produces additional carbon tetrachloride and sulfur. Write a series of equations for the reactions described here.

44. A temporary remedy for a "sticky" lock is to scrape a pencil point across the notches on the key and to work the key in and out of the lock a few times. What is the basis of this remedy?
46. Describe and explain the similarities and differences between the reaction of a silicate with an acid and that of a carbonate with an acid.
48. In a manner similar to that outlined on page 819,
(a) write equations to represent the reaction of $(CH_3)_3SiCl$ with water, followed by the elimination of H_2O from the resulting silanol molecules.
(b) Does a silicone polymer form?
(c) What would be the corresponding product obtained from CH_3SiCl_3?

Boron

49. The molecule tetraborane has the formula B_4H_{10}.
(a) Show that this is an electron-deficient molecule.
(b) How many bridge bonds must occur in the molecule?
(c) Show that the carbon analog, butane, C_4H_{10}, is not electron-deficient.

50. Write Lewis structures for the following species, both of which involve coordinate covalent bonding.
(a) tetrafluoroborate ion, BF_4^-, used in metal cleaning and in electroplating baths;
(b) boron trifluoride ethylamine, used in curing epoxy resins (ethylamine is $C_2H_5NH_2$).

Advanced Exercises

51. Despite the fact that it has the higher molecular mass, XeO_4 exists as a *gas* at 298 K whereas XeO_3 is a *solid*. Give a plausible explanation of this observation.

52. The text mentions on page 802 that ammonium perchlorate is an explosive hazard. Assuming that in the explosion NH_4ClO_4 is the sole reactant, write a plausible equation(s) to represent the reaction that occurs.

53. A marginal note on page 797 states why the standard electrode potential of fluorine cannot be measured directly. Obtain a value of E_{F_2/F^-} using thermodynamic data from Appendix D. (*Hint:* What data do you need? See Chapter 21, Feature Problem F, page 763.)

54. Polonium is the only element known to crystallize in the simple cubic form. In this structure, the interatomic distance between a Po atom and each of its six nearest neighbors is 335 pm. Use this description of the crystal structure to estimate the density of polonium.

55. Refer to Figure 12-20 and arrange the following species in the expected order of increasing (**a**) bond length and (**b**) bond strength (energy): O_2, O_2^+, O_2^-, O_2^{2-}. State the basis of your prediction.

56. One reaction of a chlorofluorocarbon implicated in the destruction of stratospheric ozone is $CFCl_3 + h\nu \rightarrow CFCl_2 + Cl$.
(**a**) What is the energy of the photons ($h\nu$) required to bring about this reaction, expressed in kilojoules per mole?
(**b**) What is the frequency and wavelength of the light necessary to produce the reaction? In what portion of the electromagnetic spectrum will this light be found?

57. The composition of a phosphate mineral can be expressed as % P, % P_4O_{10}, or % BPL [bone phosphate of lime, $Ca_3(PO_4)_2$].
(**a**) Show that % P = $0.436 \times$ (% P_4O_{10}) and % BPL = $2.185 \times$ (% P_4O_{10}).

(**b**) What is the significance of a % BPL greater than 100?
(**c**) What is the % BPL of a typical phosphate rock?

58. Estimate the % dissociation of $Cl_2(g)$ into $Cl(g)$ at 1 atm total pressure and 1000 K. Use data from Appendix D and equations established elsewhere in the text, as necessary.

59. *Peroxonitrous acid* is an unstable intermediate formed in the oxidation of HNO_2 by H_2O_2. It has the same formula as *nitric acid*, HNO_3. Show how you would expect peroxonitrous and nitric acids to differ in structure.

60. The structure of $N(SiH_3)_3$ involves a planar arrangement of N, and Si atoms, whereas that of the related compound $N(CH_3)_3$ has a pyramidal arrangement of N and C atoms. Propose bonding schemes for these molecules that are consistent with this observation.

61. In the extraction of bromine from seawater (reaction 23.3), seawater is first brought to a pH of 3.5 and then treated with $Cl_2(g)$. In actual practice, the pH of the seawater is adjusted with H_2SO_4 and the mass of chlorine used is 15% in excess of the theoretical. Assuming a seawater sample with an initial pH of 7.0, a density of 1.03 g/cm^3, and a bromine content of 70 ppm by mass, what masses of H_2SO_4 and Cl_2 would be used in the extraction of bromine from 1.00×10^3 L of seawater? (*Hint:* How would you describe the ionization of H_2SO_4 in a solution with pH = 3.5? That is, does it go to completion?)

62. The bond energies of Cl_2 and F_2 are 243 and 155 kJ/mol, respectively. Use these data to explain why XeF_2 is a much more stable compound than $XeCl_2$. (*Hint:* Recall that Xe exists as a monatomic gas.)

63. Write plausible half-equations and a balanced oxidation–reduction equation for the disproportionation of XeF_4 to Xe and XeO_3 in aqueous acidic solution. Xe and XeO_3 are produced in a 2:1 mol ratio and $O_2(g)$ is also produced.

Feature Problems

A. The diagrams presented here, called electrode potential diagrams, allow us to summarize electrode potential data more efficiently than in listings such as in Appendix D. In short, in the electrode potential diagram for chlorine in acidic solution, the symbolism $HClO_2 \xrightarrow{1.701\ V} HClO$ signifies that

$$HClO_2(aq) + 2\ H^+ + 2\ e^- \longrightarrow HClO(aq) + H_2O$$
$$E° = 1.701\ V$$

and ClO_3^- | $\underset{1.468\ V}{\rule{3cm}{0.4pt}}$ | Cl_2 signifies that

$$ClO_3^-(aq) + 6\ H^+(aq) + 5\ e^- \longrightarrow \tfrac{1}{2}\ Cl_2(g) + 3\ H_2O$$
$$E° = 1.468\ V$$

Electrode potential diagrams for chlorine
Acidic solution ([H⁺] = 1 M):

Basic solution ([OH⁻] = 1 M):

1. For the electrode potential diagrams for chlorine, replace the entries marked (?) by obtaining $E°$ values for the appropriate half-reactions from Appendix D.
2. For the electrode potential diagrams for chlorine, supply $E°$ values for the entries marked (??), which are not listed in Appendix D, by using the method outlined in Chapter 21, Feature Problem A on page 762.

Electrode potential diagram for sulfur
Basic solution ([OH⁻] = 1 M):

3. For the electrode potential diagram for sulfur in basic solution, supply the missing entries as described in parts 1 and 2.
4. Use the completed potential diagram of part 3 to confirm the conclusion reached in the Summarizing Example on page 823: The disproportionation of $S_2O_3{}^{2-}(aq)$ to $S(s)$ and $SO_3{}^{2-}(aq)$ is *not* spontaneous when reactants and products are in their standard states in basic solution.

B. The photograph in Figure 16-1 (page 546) shows that I_2 is considerably more soluble in $CCl_4(l)$ than it is in $H_2O(l)$. The concentration of I_2 in its saturated aqueous solution is 1.33×10^{-3} M, and the equilibrium achieved when I_2 distributes itself between H_2O and CCl_4 is

$$I_2(aq) \rightleftharpoons I_2(CCl_4) \qquad K_c = \frac{[I_2]_{CCl_4}}{[I_2]_{aq}} = 85.5$$

1. A 10.0-mL sample of saturated $I_2(aq)$ is shaken with 10.0 mL CCl_4. After equilibrium is established, the two liquid layers are separated. How many milligrams of I_2 will be found in the aqueous layer?

2. If the 10.0 mL of aqueous layer from part 1 is extracted with a second 10.0-mL portion of CCl_4, how many milligrams of I_2 will remain in the aqueous layer when equilibrium is re-established?
3. If the 10.0-mL sample of saturated $I_2(aq)$ in part 1 had originally been extracted with 20.0 mL CCl_4, would the mass of I_2 remaining in the aqueous layer have been less than, equal to, or greater than that found in part 2? Explain.

The Transition
Elements

24

*T*he transition elements have a multitude of important functions in modern society. Among the transition metals are the main structural metal, iron (Fe), as well as important alloying metals in the manufacture of steel (V, Cr, Mn, Co, Ni, Mo, W). The best electrical conductors (Ag, Cu) are transition metals. The compounds of several transition metals (Ti, Fe, Cr) are the primary constituents of paint pigments. Compounds of silver (Ag) provide the essential material for photography. Specialized materials for modern applications such as color television screens use compounds of the *f*-block elements (lanthanide oxides). Nine of the transition metals are essential elements for living organisms.

The chemistry of the *d*-block and *f*-block elements has both theoretical and practical significance. These elements and their compounds provide insight into fundamental aspects of bonding, magnetism, and reaction chemistry.

Whiskers of rutile (TiO_2) in quartz, and titanium ore, a source of rutile. Titanium metal, obtained from rutile, is used in industry because of its low density and high strength. Pure TiO_2 is a bright white pigment used in paints and specialty papers.

24-1 General Properties

The transition elements are the *d*-block and *f*-block elements of the periodic table. All of these elements are metals. Their high melting points, good electrical conductivity, and moderate to extreme hardness result from the ready availability of electrons and orbitals for metallic bonding (see Section 12-7). The transition elements have a number of similarities, but each element also has some unique properties that make it and its compounds useful in particular ways. Table 24.1 lists properties of the fourth-period transition elements—the first transition series.

Atomic (Metallic) Radii

In Table 24.1, with the exception of Sc and Ti, we find little variation among the atomic radii across the first transition series. The chief difference in atomic structure between successive elements involves one unit of positive charge on the nucleus and one electron in an orbital of an *inner* electron shell. This is not a major difference and does not cause much of a change in atomic radius, especially in the middle of the series.

When an element in the first transition series is compared with those of the second and third series within the same group, important differences appear. Table 24.2 lists representative data for members of Group 6B—Cr, Mo, and W. As we might expect, the atomic radius of Mo is larger than that of Cr; but contrary to our expectation, the atomic radius of W is the same as that of Mo, not larger. In the Aufbau process, 18 electrons are added in progressing from Cr to Mo, and all of them enter *s*, *p*, and *d* subshells. Between Mo and W, however, 32 electrons must be added, and 14 of them enter the 4*f* subshell. Electrons in an *f* subshell are not very effective in screening outer shell electrons from the nucleus. As a result, the outer shell electrons are held more tightly by the nucleus than we would otherwise expect. Atomic radii do not increase. In fact, in the series of elements in which the 4*f* subshell is filled, atomic radii actually

TABLE 24.1 Selected Properties of Elements of the First Transition Series

	Sc	Ti	V	Cr	Mn	Fe	Co	Ni	Cu	Zn
atomic number	21	22	23	24	25	26	27	28	29	30
electron config.[a]	$3d^14s^2$	$3d^24s^2$	$3d^34s^2$	$3d^54s^1$	$3d^54s^2$	$3d^64s^2$	$3d^74s^2$	$3d^84s^2$	$3d^{10}4s^1$	$3d^{10}4s^2$
metallic radius, pm	161	145	132	125	124	124	125	125	128	133
ioniz. energy, kJ/mol										
first	631	658	650	653	717	759	758	737	745	906
second	1235	1310	1414	1592	1509	1561	1646	1753	1958	1733
third	2389	2653	2828	2987	3248	2957	3232	3393	3554	3833
$E°$, V[b]	−2.03	−1.63	−1.13	−0.90	−1.18	−0.440	−0.277	−0.257	+0.337	−0.763
common positive oxidation states	3	2, 3, 4	2, 3, 4, 5	2, 3, 6	2, 3, 4, 7	2, 3	2, 3	2	1, 2	2
mp, °C	1397	1672	1710	1900	1244	1530	1495	1455	1083	420
density, g/cm³	3.00	4.50	6.11	7.14	7.43	7.87	8.90	8.91	8.95	7.14
hardness[c]	—	—	—	9.0	5.0	4.5	—	—	2.8	2.5
electrical cond'ty[d]	3	4	6	12	1	16	25	23	93	27

[a]Each atom has an argon inner core configuration.

[b]For the reduction process $M^{2+}(aq) + 2 e^- \rightarrow M(s)$ [except for scandium, where the ion is $Sc^{3+}(aq)$].

[c]Hardness values are on the Mohs scale (see Table 22.1).

[d]Electrical conductivity compared with an arbitrarily assigned value of 100 for silver.

TABLE 24.2 The Group 6B Elements: Chromium, Molybdenum, and Tungsten

Transition Series	Element	Atomic Number	Electron Configuration	Atomic Radius, pm	Standard Electrode Potential,[a] V	Positive Oxidation States[b]
first	Cr	24	$[Ar]3d^54s^1$	125	-0.744	2, 3, 6
second	Mo	42	$[Kr]4d^55s^1$	139	-0.20	2, 3, 4, 5, 6
third	W	74	$[Xe]4f^{14}5d^46s^2$	139	-0.11	2, 3, 4, 5, 6

[a]For the reduction process $M^{3+} + 3\,e^- \rightarrow M(s)$.
[b]The most common oxidation states are shown in red.

decrease somewhat. This phenomenon occurs in the lanthanide series ($Z = 58$ to 71) and is called the **lanthanide contraction.**

Electron Configurations and Oxidation States

The elements of the first transition series have electron configurations with the following characteristics:

- an inner core of electrons in the argon configuration;
- two electrons in the $4s$ orbital for eight members and one $4s$ electron for the remaining two (Cr and Cu);
- a number of $3d$ electrons ranging from one in Sc to ten in Cu and Zn.

As we have seen for some of the representative elements in Chapters 22 and 23, it is possible for an element to display several different oxidation states. Often, however, one particular oxidation state is the most common for an element. Ti atoms, with the electron configuration $[Ar]3d^24s^2$, tend to use all four electrons beyond the argon core in compound formation and display the oxidation state $+4$. It is also possible, however, for Ti atoms to use fewer electrons, as through the loss of the $4s^2$ electrons to form the ion Ti^{2+}. With Ti, then, we note two features: (1) several possible oxidation states, as shown in Figure 24-1, and

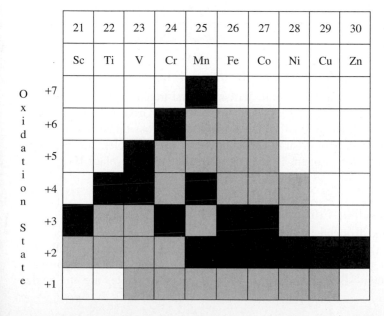

FIGURE 24-1
Positive oxidation states of the elements of the first transition series.
Common oxidation states are shown in red. Some oxidation states are rather rare, and a zero or negative oxidation state is occasionally found, chiefly in transition metal complexes. For example, the oxidation state of Cr is -2 in $Na_2[Cr(CO)_5]$, -1 in $Na_2[Cr_2(CO)_{10}]$, and 0 in $Cr(CO)_6$.

(2) a maximum oxidation state corresponding to the group number, 4B. These two features continue with V, Cr, and Mn, for which the maximum oxidation states are $+5$, $+6$, and $+7$, respectively. A shift in behavior occurs in Group 8B, however. Although Fe, Co, and Ni can all exist in more than one oxidation state, they do not display the wide variety found in the earlier members of the first transition series. Neither do they exhibit a maximum oxidation state corresponding to their group number. As we cross the first transition series, the nuclear charge, number of d electrons, and energy requirement for the successive ionization of d electrons all increase. The loss of a large number of d electrons becomes increasingly unfavorable energetically and only the lower oxidation states are commonly encountered for these later elements of the first transition series.

Although the transition elements display variety in their oxidation states, the possible oxidation states differ in the ease with which they can be attained and in their stabilities. The stability of an oxidation state for a given transition metal depends on a number of factors—other atoms to which the transition metal atom is bonded; whether the compound is in solid form or in solution; and the pH of the solution. For example, $TiCl_2$ is a well-characterized compound as a solid, but the Ti^{2+} ion is readily oxidized to Ti^{3+} by dissolved oxygen in aqueous solutions. On the other hand, $Co^{3+}(aq)$ readily oxidizes water to $O_2(g)$ and is itself reduced to $Co^{2+}(aq)$. $Co^{3+}(aq)$ can be stabilized, however, in complex ions having the appropriate ligands. Generally speaking, higher oxidation states in transition metals are stabilized when oxide or fluoride ions are bound to the metal. In Table 24.1 and elsewhere, the term *common* oxidation state refers to an oxidation state often found in metal compounds in aqueous solution.

Another feature of the transition metals is the progressively increasing stability of higher oxidation states as we descend a group of the periodic table, the reverse of the trend often seen for main group metals. Thus, Table 24.2 lists the most common oxidation states as $+3$ for Cr, $+5$ and $+6$ for Mo, and $+6$ for W. In addition, the electrode potential data suggest that the lower oxidation state, $+3$, is favored at the top of the group. Cr is oxidized to Cr^{3+} ($-E° = 0.744$ V) more readily than Mo to Mo^{3+} ($-E° = 0.20$ V) or W to W^{3+} ($-E° = 0.11$ V). This trend favoring lower oxidation states for the first group member and higher oxidation states for the lower members is also found in other groups of transition metals. For instance, although Fe does not exhibit an oxidation state corresponding to the group number, Os does form the stable oxide OsO_4 with Os in the $+8$ oxidation state.

Ionization Energies and Electrode Potentials

Ionization energies are fairly constant across the first transition series. Values of the first ionization energies are about the same as for the Group 2A metals. Standard electrode potentials gradually increase in value across the series. With the exception of the oxidation of Cu to Cu^{2+}, however, all these elements are more readily oxidized than hydrogen. This means that the metals displace $H_2(g)$ from $H^+(aq)$.

Ionic and Covalent Compounds

We tend to think of metals as forming ionic compounds with nonmetals. This is certainly the case with Group 1A and most Group 2A metal compounds. On the other hand, we have seen that some metal compounds have significant covalent character—$BeCl_2$, and $AlCl_3$ (Al_2Cl_6), for example, are molecular compounds. Transition metal compounds display both ionic and covalent character. In general, compounds with the transition metal in lower oxidation states are essentially ionic, and those in higher oxidation states have covalent character. As an example, MnO is a green ionic solid with a melting point of 1785 °C, whereas Mn_2O_7 is a dark red, oily, molecular liquid that boils at room temperature and is highly explosive.

Catalytic Activity

An unusual ability to adsorb gaseous species makes some transition metals, such as Ni and Pt, good heterogeneous catalysts. The possibility of multiple oxidation states seems to account for the ability of some transition metal ions to serve as catalysts in certain oxidation–reduction reactions. In still other types of catalysis, complex ion formation may play an important role. As we have seen in a limited way in Chapter 19 and will explore more fully in Chapter 25, complex ion formation is a particularly distinctive feature of transition metal chemistry.

Catalysis is an essential aspect of about 90% of all chemical manufacturing processes, and the transition metals are often the key elements in the catalysts used. To name a few: Ni is used in the hydrogenation of oils (page 263); Pt, Pd, and Rh are used in catalytic converters in automobiles (pages 251 and 532); Fe_3O_4 is the main component of the catalyst used in the synthesis of ammonia (page 246); and V_2O_5 is used in the conversion of $SO_2(g)$ to $SO_3(g)$ in the manufacture of sulfuric acid (page 845).

Color and Magnetism

As we will explain more fully in Section 25-6, electronic transitions that occur within partially filled d subshells impart color to solid transition metal compounds and their solutions. Absence of these transitions, in turn, accounts for the fact that so many main group metal compounds are colorless.

Because most transition elements have partially filled d subshells, many transition metals and their compounds are *paramagnetic*—that is, they have unpaired electrons. This description certainly fits Fe, Co, and Ni, but these three metals are unique among the elements in displaying a special magnetic property: the ability to be made into *permanent* magnets, a property known as **ferromagnetism.** A key feature of ferromagnetism is that in the solid state, the metal atoms are thought to be grouped together into small regions—called *domains*—containing rather large numbers of atoms. Instead of the individual magnetic moments of the atoms within a domain being randomly oriented, they are all directed in the same way (Figure 24-2). In an unmagnetized piece of iron the domains are oriented in several directions and their magnetic effects cancel. When the metal is placed in a magnetic field, however, the domains line up and a strong resultant magnetic effect is produced. This alignment of domains may actually involve the growth of domains with favorable orientations at the expense of those with unfavorable orientations (rather like a *recrystallization* of the material). The ordering of domains can persist when the object is removed from the magnetic field, and thus permanent magnetism results.

The key to ferromagnetism involves two basic factors: that the atoms involved have unpaired electrons (a property possessed by many atoms) and that interatomic distances be of just the right magnitude to make possible the ordering of atoms into domains. If atoms are too large, interactions among them are too weak to produce this ordering. With small atoms, the tendency is for atoms to pair and their magnetic moments to cancel. This critical factor of atomic size is

Color-enhanced image of magnetic domains in a ferromagnetic garnet film.

FIGURE 24-2
The phenomenon of ferromagnetism.
(a) In ordinary paramagnetism, the magnetic moments of the atoms or ions are randomly distributed. (b) In a ferromagnetic material, the magnetic moments are aligned into domains. (c) In an unmagnetized piece of the material, the domains are randomly oriented. (d) In a magnetic field the domains are oriented in the direction of the field and the material becomes magnetized.

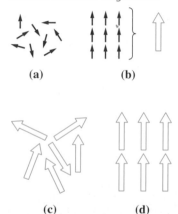

just met in Fe, Co, and Ni. It is possible, however, to prepare *alloys* of other metals in which this condition is also met. Some examples are Al–Cu–Mn, Ag–Al–Mn, and Bi–Mn.

Comparison of Transition and Main Group Elements

With the representative elements, the *s* and *p* orbitals of the outermost electronic shell are the most important in determining the nature of the chemical bonding that occurs. Participation in bonding by *d* orbitals is essentially nonexistent for second-period elements and Group 1A and 2A metals. With the transition elements, *d* orbitals are as important as *s* and *p* orbitals. Most of the observed behavioral differences between the transition and main group elements—multiple versus single oxidation states, complex ion formation, color, magnetic properties, and catalytic activity—can be traced to the orbitals that are most involved in bond formation.

24-2   Principles of Extractive Metallurgy

Many of the transition elements have important uses related to their metallic properties—iron for its structural strength and copper for its excellent electrical conductivity, for example. Unlike the more chemically reactive metals of Groups 1A and 2A and aluminum in Group 3A, which are produced mainly by modern methods of electrolysis, the transition metals are obtained by procedures developed over many centuries.

We use the term *metallurgy* for the general study of metals and **extractive metallurgy** for the winning of metals from their ores. There is no single method of extractive metallurgy, but a few basic operations generally apply. Let us illustrate them with the extractive metallurgy of zinc.

Concentration.   In mining operations, the desired mineral from which a metal is to be extracted often constitutes only a few percent (or occasionally just a fraction of a percent) of the material mined. It is necessary to separate the desired ore from waste rock before proceeding with other metallurgical operations. One useful method, *flotation*, is described in Figure 24-3.

FIGURE 24-3

Concentration of an ore by flotation.
Powdered ore is suspended in water in a large vat, together with suitable additives, and the mixture is agitated with air. Particles of ore become attached to air bubbles, rise to the top of the vat, and are collected in the overflow froth. Particles of undesired waste rock (gangue) fall to the bottom.

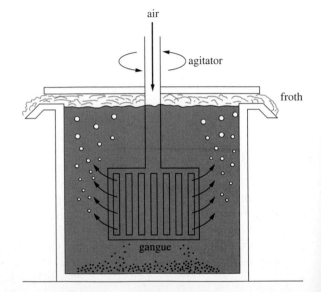

Roasting. An ore is roasted (heated to a high temperature) to convert a metal compound to its oxide, which can then be reduced. For zinc, the commercially important ores are $ZnCO_3$ (smithsonite) and ZnS (sphalerite). $ZnCO_3(s)$, like the carbonates of the Group 2A metals, decomposes to $ZnO(s)$ and $CO_2(g)$ when it is strongly heated. When strongly heated in air, $ZnS(s)$ reacts with $O_2(g)$, producing $ZnO(s)$ and $SO_2(g)$. In modern smelting operations $SO_2(g)$ is converted to sulfuric acid rather than being vented to the atmosphere.

$$ZnCO_3(s) \xrightarrow{\Delta} ZnO(s) + CO_2(g) \tag{24.1}$$

$$2\,ZnS(s) + 3\,O_2(g) \xrightarrow{\Delta} 2\,ZnO(s) + 2\,SO_2(g) \tag{24.2}$$

Reduction. If possible, carbon, in the form of coke or powdered coal, is used as the reducing agent because it is inexpensive and easy to handle. Several reactions occur simultaneously in which both $C(s)$ and $CO(g)$ act as reducing agents. The reduction of ZnO is carried out at about 1100 °C, a temperature above the boiling point of zinc. The zinc is obtained as a vapor and condensed to the liquid.

▶ Feature Problem A on page 862 describes $C(s)$ and $CO(g)$ as reducing agents from a thermodynamic standpoint.

$$ZnO(s) + C(s) \xrightarrow{\Delta} Zn(g) + CO(g) \tag{24.3}$$

$$ZnO(s) + CO(g) \xrightarrow{\Delta} Zn(g) + CO_2(g) \tag{24.4}$$

Refining. The metal produced by chemical reduction is usually not pure enough for its intended uses. Impurities must be removed; that is, the metal must be refined. The refining process chosen depends on the nature of the impurities. The impurities in zinc are mostly Cd and Pb, which can be removed by the fractional distillation of liquid zinc.

Most of the zinc produced worldwide, however, is refined electrolytically, usually in a process that combines reduction and refining. ZnO from the roasting step is dissolved in $H_2SO_4(aq)$. This is represented by the ionic equation

$$ZnO(s) + 2\,H^+(aq) + SO_4{}^{2-}(aq) \longrightarrow Zn^{2+}(aq) + SO_4{}^{2-}(aq) + H_2O(l) \tag{24.5}$$

Powdered Zn is added to the solution to displace less active metals, such as Cd. Then the solution is electrolyzed. The electrode reactions are

Cathode: $\qquad\qquad Zn^{2+}(aq) + 2\,e^- \longrightarrow Zn(s)$

Anode: $\qquad\qquad\qquad H_2O \longrightarrow \tfrac{1}{2}\,O_2(g) + 2\,H^+(aq) + 2\,e^-$

Unchanged: $\qquad\qquad\underline{SO_4{}^{2-}(aq) \longrightarrow SO_4{}^{2-}(aq)}$

Net: $\quad Zn^{2+}(aq) + SO_4{}^{2-}(aq) + H_2O \longrightarrow Zn(s) + 2\,H^+(aq) + SO_4{}^{2-}(aq) + \tfrac{1}{2}\,O_2(g) \tag{24.6}$

Note that in the net electrolysis reaction Zn^{2+} is reduced to pure metallic zinc and sulfuric acid is regenerated. The acid is recycled in reaction (24.5).

EXAMPLE 24-1

Writing Chemical Equations for Metallurgical Processes. Write chemical equations to represent the **(a)** roasting of galena, PbS; **(b)** reduction of $Cu_2O(s)$ with charcoal as a reducing agent; **(c)** deposition of pure silver from an aqueous solution of Ag^+.

Solution

a. We expect this process to be essentially the same as reaction (24.2).

$$2\,PbS\,(s) + 3\,O_2(g) \xrightarrow{\Delta} 2\,PbO\,(s) + 2\,SO_2(g)$$

b. The simplest possible equation we can write is

$$Cu_2O(s) + C(s) \xrightarrow{\Delta} 2\,Cu(l) + CO(g)$$

c. This process involves a reduction half-reaction. The accompanying oxidation half-reaction is not specified. Neither is it specified whether this is an electrolysis process or whether silver is displaced by a more active metal. In either case, the reduction half-reaction is

$$Ag^+(aq) + e^- \longrightarrow Ag(s)$$

Practice Example A: Write plausible chemical equations to represent the **(a)** roasting of Cu_2S; **(b)** reduction of WO_3 with $H_2(g)$; **(c)** thermal decomposition of HgO to its elements.

Practice Example B: Write chemical equations to represent the **(a)** reduction of Cr_2O_3 to chromium with silicon as the reducing agent; **(b)** conversion of $Co(OH)_3(s)$ to $Co_2O_3(s)$ by roasting; **(c)** production of pure $MnO_2(s)$ from $MnSO_4(aq)$ at the anode in an electrolysis cell. [*Hint:* A few simple products are not specifically mentioned and you should propose plausible ones. For reaction **(c)** only a half-reaction is required.]

Zone Refining. In discussing freezing-point depression (Section 14-8), we assumed that a solute is *soluble* in a liquid solvent and *insoluble* in the solid solvent that freezes from solution. This behavior suggests a particularly simple way to purify a solid: Melt the solid and then refreeze a portion of it. Impurities remain in the unfrozen portion and the solid that freezes is pure. In actual practice the method is not quite so simple, because the solid that freezes is wet with unfrozen liquid and thereby retains some impurities. Also, one or more solutes (impurities) might be slightly soluble in the solid solvent. In any case, the impurities do distribute themselves between the solid and liquid, concentrating in the liquid phase. If the solid that freezes from a liquid is remelted and the molten material refrozen, the solid obtained in the second freezing is purer than that in the first. Repeating the melting and refreezing procedure hundreds of times produces a very pure solid product.

The purification procedure we have just described implies that the melting and refreezing is done in batches, but in practice it is done *continuously*. In the method known as **zone refining,** a cylindrical rod of material is alternately melted and refrozen as a series of heating coils passes along the rod (Figure 24-4). Impurities concentrate in the molten zones, and the portions of the rod behind these zones are somewhat purer than the portions in front of the zones. Eventually, impurities are swept to the end of the rod, which is cut off.

Zone refining is capable of producing materials in which the impurity levels are as low as 10 parts per billion (ppb), a common requirement of substances used in semiconductors.

Alternative Methods in Extractive Metallurgy. Some common variations of the methods previously discussed are worth mentioning. First, many ores contain several metals, and it is not always necessary to separate them. For example, a major use of vanadium, chromium, and manganese is in making alloys with iron. Obtaining each metal by itself is not commercially important. Thus, the principal chromium ore *chromite*, $Fe(CrO_2)_2$, can be reduced to give an alloy of Fe and Cr called *ferrochrome*. Ferrochrome may be added directly to iron, together with other metals, to produce one type of steel. Vanadium and manganese can be isolated as the oxides, V_2O_5 and MnO_2, respectively. When

FIGURE 24-4
Zone refining.
As a heating coil moves up the rod of material, melting occurs. Impurities concentrate in the molten zone. The portion of the rod below the molten zone is purer than that in or above the zone. With each successive passage of a heating coil the rod becomes purer.

iron-containing compounds are added to these oxides and the mixtures reduced, ferrovanadium and ferromanganese alloys are formed.

Extensive production of titanium is a rather recent development, spurred first by the needs of the military, and then by the aircraft industry. Steel is unsuitable as the structural metal for aircraft because of its high density ($d = 7.8$ g/cm^3). Aluminum has the advantage of a low density ($d = 2.70$ g/cm^3), but it loses strength at high temperatures. Titanium is a good alternative to aluminum and steel for certain aircraft components because its density is moderately low ($d = 4.50$ g/cm^3) and it does not lose strength at high temperatures.

Titanium metal cannot be produced by reduction of TiO$_2$ with carbon, however, because the metal and carbon react to form titanium carbides. Also, at high temperatures the metal reacts with air to form TiO$_2$ and TiN. The metallurgy of titanium, then, must be conducted out of contact with air and with an active metal rather than carbon as a reducing agent.

The first step in the production of Ti is the conversion of *rutile* ore (TiO$_2$) to TiCl$_4$ by reaction with carbon and Cl$_2$(g).

$$TiO_2(s) + 2\ C(s) + 2\ Cl_2(g) \xrightarrow{800\ °C} TiCl_4(g) + 2\ CO(g) \qquad (24.7)$$

The purified TiCl$_4$ is next reduced to Ti with a good reducing agent. The *Kroll process* uses Mg.

$$TiCl_4(g) + 2\ Mg(l) \xrightarrow{1000\ °C} Ti(s) + 2\ MgCl_2(l) \qquad (24.8)$$

The MgCl$_2$(l) is removed and electrolyzed to produce Cl$_2$ and Mg, which are recycled in reactions (24.7) and (24.8), respectively. The Ti is obtained as a sintered (fused) mass called titanium sponge. The sponge must be subjected to further treatment and alloying with other metals before it can be used.

In 1947, the United States production of Ti was only 2 tons. Today, it exceeds 50,000 tons, and there is concern that continued widespread use of the metal will deplete the world's known reserves of rutile. Fortunately, titanium and its compounds can also be produced from *ilmenite* (FeTiO$_3$), although not as cheaply as from rutile. With titanium and its oxide we see an example of how

Vacuum-distilled metallic titanium sponge produced by the Kroll process.

materials practically unknown in one decade may become major production items in the next.

EXAMPLE 24-2

Writing a Chemical Equation from a Description of a Reaction. The conversion of ilmenite ore, $FeTiO_3$, to $TiCl_4$ is carried out in a similar manner to reaction (24.7). That is, the ore is heated with carbon and chlorine. Iron(III) chloride and carbon monoxide are also produced. Write a balanced equation for this reaction.

Solution

As in several other cases in the preceding two chapters, this type of problem requires that we identify all the reactants and products, substitute correct formulas for names, and balance the equation. The reactants are $FeTiO_3$, C, and Cl_2. The products are $TiCl_4$, $FeCl_3$, and CO.

$$FeTiO_3 + C + Cl_2 \longrightarrow TiCl_4 + FeCl_3 + CO \quad \text{(unbalanced)}$$

Although this is a redox equation, we can balance it by inspection without great difficulty: (1) Fe and Ti are already balanced; C and Cl_2 appear as the free elements on the left; and so we begin by balancing O atoms. (2) We need "3 CO" on the right side of the equation to balance the three O atoms on the left. (3) We need "3 C" on the left to balance the three C atoms on the right. (4) We need "7/2 Cl_2" on the left to balance the seven Cl atoms on the right. (5) Finally, we multiply all coefficients by 2 to get a balanced equation with integral coefficients.

$$2\ FeTiO_3 + 6\ C + 7\ Cl_2 \longrightarrow 2\ TiCl_4 + 2\ FeCl_3 + 6\ CO \quad (24.9)$$

Practice Example A: Vanadium is produced by reducing V_2O_5 with silicon. In a second step, the SiO_2 produced combines with added CaO to form a liquid slag of calcium silicate, $CaSiO_3$. Write equations for these two reactions.

Practice Example B: Tungsten metal can be prepared in the following three-step process: (1) Iron(II) tungstate, $FeWO_4$, is heated with sodium carbonate in the presence of $O_2(g)$ to produce sodium tungstate, iron(III) oxide, and $CO_2(g)$. (2) The sodium tungstate is allowed to react with sulfuric acid to form tungsten(VI) oxide. (3) The tungsten(VI) oxide is reduced to tungsten metal with $H_2(g)$. Write equations for these three reactions.

Metallurgy of Copper. The extraction of copper from its ores (generally sulfides) is rather complicated. The chief reason for this complexity is that copper ores usually contain iron sulfides. The general scheme of extractive metallurgy just discussed produces copper contaminated with iron. For some metals, such as V, Cr, and Mn, contamination with iron is not a problem because the metals are mostly used in the manufacture of steel. Copper, however, is prized commercially for the properties of the pure metal. To avoid contamination with iron, several changes to the usual metallurgical methods are necessary.

Concentration is done by flotation, and roasting converts iron sulfides to iron oxides. The copper remains as the sulfide if the temperature is kept below 800 °C. Smelting of the roasted ore in a furnace at 1400 °C causes the material to melt and separate into two layers. The bottom layer is copper matte, consisting chiefly of the molten sulfides of copper and iron. The top layer is a silicate slag formed by the reaction of oxides of Fe, Ca, and Al with SiO_2 (which typically is present in the ore or can be added).

$$FeO(s) + SiO_2(s) \xrightarrow{\Delta} FeSiO_3(l)$$

A process called *conversion* is carried out in another furnace, where air is blown through the molten copper matte. First, the remaining iron sulfide is converted to the oxide, followed by formation of slag [$FeSiO_3(l)$]. The slag is poured off and air is again blown through the furnace. Now the following reactions occur and yield a product that is about 98–99% Cu.

$$2\,Cu_2S + 3\,O_2(g) \xrightarrow{\Delta} 2\,Cu_2O + 2\,SO_2(g) \qquad (24.10)$$

$$2\,Cu_2O + Cu_2S \xrightarrow{\Delta} 6\,Cu(l) + SO_2(g) \qquad (24.11)$$

The product of reaction (24.11) is called *blister copper* because of the presence of frozen bubbles of $SO_2(g)$. It can be used where high purity is not required (as in plumbing).

Refining to obtain high-purity copper is done electrolytically by the method outlined on page 754. High-purity copper is essential in electrical applications.

The metallurgical method based on roasting an ore followed by reduction of the oxide to the metal is called **pyrometallurgy,** the prefix "pyro" suggesting that high temperatures are involved. Some of the characteristics of pyrometallurgy are

Slag formed during the smelting of copper ore.

- large quantities of waste materials produced in concentrating low-grade ores;
- high energy consumption to maintain high temperatures necessary for roasting and reduction of ores;
- gaseous emissions that must be controlled, such as $SO_2(g)$ in roasting.

Hydrometallurgical Processes. In **hydrometallurgy,** the materials handled are water and aqueous solutions at moderate temperatures rather than dry solids at high temperatures. Generally, three steps are involved in hydrometallurgy.

- *Leaching:* Metal ions are extracted (leached) from the ore by a liquid. Leaching agents include water, acids, bases, and salt solutions. Oxidation–reduction reactions may also be involved.
- *Purification and concentration:* Impurities are separated and the solution produced by leaching may be made more concentrated. Methods include the adsorption of impurities on the surface of activated charcoal, ion exchange, and the evaporation of water.
- *Precipitation:* The desired metal ions are precipitated in an ionic solid or they are reduced to the free metal, often electrolytically.

Hydrometallurgy has long been used in obtaining silver and gold from natural sources. A typical gold ore currently being processed in the United States has only about 10 g Au per ton of ore. The leaching step in gold processing is known as the *cyanidation* process: $O_2(g)$ oxidizes the free metal to Au^+, which complexes with CN^-.

$$4\,Au(s) + 8\,CN^-(aq) + O_2(g) + 2\,H_2O \longrightarrow 4\,[Au(CN)_2]^-(aq) + 4\,OH^-(aq)$$

$$(24.12)$$

The $[Au(CN)_2]^-(aq)$ is then filtered and concentrated. This is followed by displacement of Au(s) from solution by an active metal such as zinc.

$$2\,[Au(CN)_2]^-(aq) + Zn(s) \longrightarrow 2\,Au(s) + [Zn(CN)_4]^{2-}(aq) \qquad (24.13)$$

Waste solution from the leaching operation at a gold-mining facility in the Mojave Desert of California.

24-3 Metallurgy of Iron and Steel

Iron is the most widely used metal from Earth's crust, and for this reason we use this section to explore the metallurgy of iron and its principal alloy—steel—somewhat more fully. A type of steel called wootz steel was first produced in India about 3000 years ago, and this same steel became famous in ancient times as Damascus steel, prized for making swords because of its suppleness and ability to hold a cutting edge. Many technological advances have been made since ancient times. These include introduction of the blast furnace around AD 1300, the Bessemer converter in 1856, the open-hearth furnace in the 1860s, and in the 1950s, the basic oxygen furnace. A true understanding of the iron- and steel-making process has developed only within the past few decades, however. This understanding is based on concepts of thermodynamics, equilibrium, and kinetics.

Pig Iron

The reactions that occur in a blast furnace are complex. A highly simplified representation of the reduction of iron ore to impure iron is

$$Fe_2O_3(s) + 3\ CO(g) \longrightarrow 2\ Fe(l) + 3\ CO_2(g) \tag{24.14}$$

A more complete description of the blast furnace reactions, including the removal of impurities as slag, is given in Table 24.3. Approximate temperatures are given for these reactions so that you may key them to regions of the blast furnace pictured in Figure 24-5.

▶ In the United States slightly over one-half of iron and steel production comes from recycled iron and steel.

The blast furnace charge—that is, the solid reactants—consists of iron ore, coke, a slag-forming flux, and perhaps some scrap iron. The exact proportions depend on the composition of the iron ore and its impurities. The common ores of iron are the oxides and carbonate: hematite (Fe_2O_3), magnetite (Fe_3O_4), limonite ($2Fe_2O_3\cdot3H_2O$), and siderite ($FeCO_3$). The purpose of the flux is to maintain the proper ratio of acidic oxides (SiO_2, Al_2O_3, and P_4O_{10}) to basic oxides (CaO, MgO, and MnO) to obtain an easily liquefied silicate, aluminate, or phosphate slag. Because acidic oxides predominate in most ores, the flux generally employed is limestone, $CaCO_3$, or dolomite, $CaCO_3\cdot MgCO_3$.

The iron obtained from a blast furnace is called **pig iron.** It contains about 95% Fe, 3–4% C, and varying quantities of other impurities. *Cast iron* can be

TABLE 24.3 Some Blast Furnace Reactions

Formation of gaseous reducing agents $CO(g)$ and $H_2(g)$:

$C + H_2O \rightarrow CO + H_2$ (>600 °C)

$C + CO_2 \rightarrow 2\,CO$ (1700 °C)

$2\,C + O_2 \rightarrow 2\,CO$ (1700 °C)

Reduction of iron oxide:

$3\,CO + Fe_2O_3 \rightarrow 2\,Fe + 3\,CO_2$ (900 °C)

$3\,H_2 + Fe_2O_3 \rightarrow 2\,Fe + 3\,H_2O$ (900 °C)

Slag formation to remove impurities from ore:

$CaCO_3 \rightarrow CaO + CO_2$ (800–900 °C)

$CaO + SiO_2 \rightarrow CaSiO_3(l)$ (1200 °C)

$6\,CaO + P_4O_{10} \rightarrow 2\,Ca_3(PO_4)_2(l)$ (1200 °C)

Impurity formation in the iron:

$MnO + C \rightarrow Mn + CO$ (1400 °C)

$SiO_2 + 2\,C \rightarrow Si + 2\,CO$ (1400 °C)

$P_4O_{10} + 10\,C \rightarrow 4\,P + 10\,CO$ (1400 °C)

▶ Feature Problem A on page 862 deals with some thermodynamic aspects of these reactions.

obtained by pouring pig iron directly into molds of the desired shape. Cast iron is very hard and brittle and is used only where it is not subjected to mechanical or thermal shock, such as in engine blocks, brake drums, and transmission housings in automobiles.

Steel

The fundamental changes that must be made to convert pig iron to **steel** are to

1. Reduce the carbon content from 3–4% in pig iron to 0–1.5% in steel.

2. Remove, through slag formation, Si, Mn, and P (each present in pig iron to the extent of 1% or so), together with other minor impurities.

3. Add alloying elements (such as Cr, Ni, Mn, V, Mo, and W) to give the steel its desired end properties.

The most important method of steel-making today is the **basic oxygen process.** Oxygen gas at about 10 atm pressure and a stream of powdered lime-

▶ Thermal shock occurs when an object undergoes a rapid change of temperature. Engine blocks get quite hot, but because they cool slowly, they are not usually subject to thermal shock.

▶ Steel made with 18% Cr and 8% Ni resists corrosion and is commonly known as *stainless steel.*

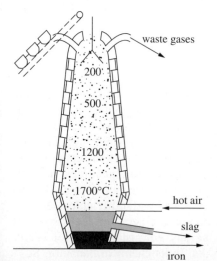

FIGURE 24-5
Typical blast furnace.
Iron ore, coke, and limestone are added at the top of the furnace, and hot air is introduced through the bottom. Maximum temperatures are attained near the bottom of the furnace where molten iron and slag are drained off. The principal reactions occurring in the blast furnace are outlined in Table 24.3.

cooling oxygen
water

exhaust
gases

molten iron

FIGURE 24-6
A basic oxygen furnace.

TABLE 24.4 Some Reactions Occurring in Steel-Making Processes

$$2\,C + O_2 \longrightarrow 2\,CO$$
$$2\,FeO + Si \longrightarrow 2\,Fe + SiO_2$$
$$FeO + Mn \longrightarrow Fe + MnO$$
$$FeO + SiO_2 \longrightarrow \underset{\text{slag}}{FeSiO_3}$$
$$MnO + SiO_2 \longrightarrow \underset{\text{slag}}{MnSiO_3}$$
$$4\,P + 5\,O_2 \longrightarrow P_4O_{10}$$
$$6\,CaO + P_4O_{10} \longrightarrow \underset{\text{slag}}{Ca_3(PO_4)_2}$$

Pouring molten steel.

▶Feature Problem B on page 862 describes one aspect of the qualitative analysis scheme for transition metal ions.

stone are fed through a water-cooled tube (called a *lance*) and discharged above the molten pig iron (Figure 24-6). The reactions that occur (Table 24.4) accomplish the first two objectives on page 843. A typical reaction time is 22 minutes. The reaction vessel is tilted to pour off the liquid slag floating on top of the iron, and then the desired alloying elements are added.

Steel making has been undergoing rapid technological changes. There is the prospect of making iron and steel directly from iron ore in a single-step (perhaps continuous) process at temperatures below the melting point of any of the materials used in the process. In the direct reduction of iron (DRI), CO(g) and H_2(g), obtained in the reaction of steam with natural gas, are used as reducing agents. The economic viability of the DRI process depends on an abundant supply of natural gas. Currently only a small percentage of the world's iron production is by direct reduction, but this is a fast-growing component of the iron and steel industry, particularly in the Middle East and South America.

24-4 First-Row Transition Metal Elements: Scandium to Nickel

The properties and uses of the first-row transition metals span a wide range, strikingly illustrating periodic behavior despite the small variation in some of the atomic properties listed in Table 24.1. The preparation, uses, and reactions of the compounds of these metals illustrate concepts we have previously discussed, including the variability of oxidation states.

Scandium

Scandium is a rather obscure metal, though not especially rare. Scandium constitutes about 25 ppm of Earth's crust, which makes it more abundant than many better known metals, including lead, uranium, molybdenum, tungsten, antimony, silver, mercury, and gold. Its principal mineral form is *thortveitite,* $Sc_2Si_2O_7$. Most scandium is obtained from uranium ores, however, in which it occurs only to the extent of about 0.01% Sc by mass. The commercial uses of scandium are limited, and its production is measured in gram or kilogram quantities, not tonnages. One application is in high-intensity lamps. The pure metal is usually prepared by the electrolysis of a fused mixture of $ScCl_3$ with other chlorides.

Because of its noble gas electron configuration, the Sc^{3+} ion lacks some of the characteristic properties of transition metal ions. For instance, the ion is colorless and diamagnetic, as are most of its salts. In its chemical behavior, Sc^{3+}

most closely resembles Al^{3+}, as in the hydrolysis of $[Sc(H_2O)_6]^{3+}(aq)$ to yield acidic solutions and in the formation of an amphoteric gelatinous hydroxide, $Sc(OH)_3$.

Titanium

Titanium is the ninth most abundant element, constituting 0.6% of Earth's solid crust. The metal is greatly valued for its low density, high structural strength, and corrosion resistance. The first two properties account for its extensive use in the aircraft industry and the third for its uses in the chemical industry: in pipes, component parts of pumps, and reaction vessels. Titanium is also used in dental and other bone implants. The metal provides a strong support, and bone bonds directly to a titanium implant, making it a part of the body.

Several compounds of titanium are of particular commercial importance. *Titanium tetrachloride,* $TiCl_4$, is the starting material for preparing other Ti compounds and plays a central role in the metallurgy of titanium. $TiCl_4$ is also used to formulate catalysts for the production of plastics. The usual method of preparing $TiCl_4$ involves the reaction of naturally occurring rutile (TiO_2) with carbon and $Cl_2(g)$ [reaction (24.7)].

$TiCl_4$ is a colorless liquid (mp -24 °C; bp 136 °C). In the +4 oxidation state all the valence shell electrons of Ti atoms are employed in bond formation. In this oxidation state Ti bears a strong resemblance to the Group 4A elements, with some properties and a molecular shape (tetrahedral) similar to those of CCl_4 and $SiCl_4$. The hydrolysis of $TiCl_4$, when carried out in moist air, is the basis for a type of smoke grenade in which $TiO_2(s)$ is the smoke.

$$TiCl_4(l) + 2\ H_2O(l) \longrightarrow TiO_2(s) + 4\ HCl(g)$$

$SiCl_4$ also fumes in moist air in a similar reaction.

TiO_2, *titanium dioxide,* is bright white, opaque, inert, and nontoxic. Because of these properties, and its relative cheapness, it is now the most widely used white pigment for paints. In this application TiO_2 has displaced toxic basic lead carbonate—so-called white lead. TiO_2 is also used as a paper whitener and in glass, ceramics, floor coverings, and cosmetics.

To produce pure TiO_2 for these and other uses, a gaseous mixture of $TiCl_4$ and O_2 is passed through a silica tube at about 700 °C.

$$TiCl_4(g) + O_2(g) \xrightarrow{\Delta} TiO_2(s) + 2\ Cl_2(g)$$

Vanadium

Vanadium is a fairly abundant element (0.02% of Earth's crust) and is found in several dozen ores. Its principal ores are rather complex, such as *vanadinite,* $3Pb_3(VO_4)_2 \cdot PbCl_2$. The metallurgy of vanadium is not simple, but vanadium of high purity (99.99%) is obtainable. For most of its applications, though, V is prepared as an iron–vanadium alloy, ferrovanadium, containing from 35 to 95% V. About 80% of the vanadium produced is for the manufacture of steel. Vanadium-containing steels are used in applications requiring strength and toughness, such as in springs and high-speed machine tools.

The most important compound of vanadium is the pentoxide, V_2O_5, used mainly as a catalyst, as in the conversion of $SO_2(g)$ to $SO_3(g)$ in the contact method for the manufacture of sulfuric acid. The activity of V_2O_5 as an oxidation catalyst may be linked to its reversible loss of oxygen occurring from 700 to 1100 °C.

In its compounds vanadium can exist in a variety of oxidation states. In each of these oxidation states, vanadium forms an oxide or ion, and the ions display

FIGURE 24-7
Some vanadium species in solution.
The yellow solution has vanadium in the +5 oxidation state, as VO_2^+. In the blue solution the oxidation state is +4, in VO^{2+}. The green solution contains V^{3+}, and the violet solution contains V^{2+}.

distinctive colors in aqueous solution (Figure 24-7). The acid–base properties of vanadium oxides are in accord with the factors discussed in Section 17-8: If the central metal atom is in a low oxidation state, the oxide acts as a base; in higher oxidation states for the central atom, acidic properties become important. Vanadium oxides with V in the +2 and +3 oxidation states are basic, whereas those in the +4 and +5 oxidation states are amphoteric.

Most compounds with vanadium in its highest oxidation state (+5) are good oxidizing agents. In its +2 oxidation state, vanadium (as V^{2+}) is a good reducing agent. The oxidation–reduction relationships between the ionic species pictured in Figure 24-7 are summarized in Table 24.5.

EXAMPLE 24-3

Using Electrode Potential Data to Predict an Oxidation–Reduction Reaction. Can MnO_4^-(aq) be used to oxidize VO^{2+}(aq) to VO_2^+(aq) for standard-state conditions in an acidic solution? If so, write a balanced equation for the redox reaction.

Solution

We start by writing two half-equations, one for the oxidation of VO^{2+} to VO_2^+ and the other for the reduction of MnO_4^- to Mn^{2+}, both in acidic solution. We find one $E°$ value in Table 24.5 and the other in Table 21.1. We then combine the $E°$ values into $E°_{cell}$.

Oxidation:	$5(VO^{2+} + H_2O \longrightarrow VO_2^+ + 2 H^+ + e^-)$	$-E°_{VO_2^+/VO^{2+}} = -1.00$ V
Reduction:	$MnO_4^- + 8 H^+ + 5 e^- \longrightarrow Mn^{2+} + 4 H_2O$	$E°_{MnO_4^-/Mn^{2+}} = +1.51$ V
Net:	$5VO^{2+} + MnO_4^- + H_2O(l) \longrightarrow 5\ VO_2^+ + Mn^{2+} + 2 H^+$	$E°_{cell} = +0.51$ V

Because $E°_{cell}$ is positive, we predict that MnO_4^- should oxidize VO^{2+} to VO_2^+ for standard-state conditions in acidic solution.

Practice Example A: Use data from Tables 21.1 and 24.5 to determine whether nitric acid can be used to oxidize V^{3+}(aq) to VO^{2+}(aq) for standard-state conditions. If so, write a balanced equation for the reaction.

Practice Example B: Select a *reducing* agent from Table 21.1 that can be used to reduce VO^{2+} to V^{2+} for standard-state conditions in acidic solution. [*Hint:* Consider that the reduction occurs in two stages: $VO^{2+} \rightarrow V^{3+} \rightarrow V^{2+}$. Use data from Table 24.5 to determine which of these reductions is more difficult to achieve. Note also that the V^{2+} must not be reduced to V(s).]

▶The relationship among these species is explored in a diagrammatic way in Feature Problem C on page 862.

TABLE 24.5 Oxidation States of Vanadium Species in Acidic Solution

O.S. Change	Reduction Half-Reaction	$E°$
+5 → +4:	VO_2^+(aq) + 2 H$^+$(aq) + e$^-$ → VO^{2+}(aq) + H$_2$O (yellow) (blue)	1.000 V
+4 → +3:	VO^{2+}(aq) + 2 H$^+$(aq) + e$^-$ → V^{3+}(aq) + H$_2$O (green)	0.337 V
+3 → +2:	V^{3+}(aq) + e$^-$ → V^{2+}(aq) (violet)	−0.255 V
+2 → 0:	V^{2+}(aq) + 2 e$^-$ → V(s)	−1.13 V

Chromium

Although it is found only to the extent of 122 parts per million (0.0122%) in Earth's crust, chromium is one of the most important industrial metals. The production of ferrochrome from *chromite*, $Fe(CrO_2)_2$, was discussed in Section 24-2. Chromium metal itself is hard and maintains a bright surface through the protective action of an invisible oxide coating. Because of its corrosion resistance, chromium is extensively used in plating other metals.

Steel is chrome-plated from an aqueous solution containing CrO_3 and H_2SO_4. The plating obtained is thin and porous and tends to develop cracks unless the steel is first plated with copper or nickel, which provides the true protective coating. Then chromium is plated over this for decorative purposes. The efficiency of chrome plating is limited by the fact that reduction of Cr(VI) to Cr(0) produces only $\frac{1}{6}$ mol Cr per mole of electrons. In other words, large quantities of electric energy are required for chrome plating relative to other types of metal plating.

Chromium, like vanadium, has a variety of oxidation states in aqueous solution, each having a different color.

OS +2: $[Cr(H_2O)_6]^{2+}$, blue

OS +3: (acidic) $[Cr(H_2O)_6]^{3+}$, violet   (basic) $[Cr(OH)_4]^-$, green

OS +6: (acidic) $Cr_2O_7{}^{2-}$, orange   (basic) $CrO_4{}^{2-}$, yellow

The oxides and hydroxides of chromium conform to the general principles of acid–base behavior discussed in Section 17-8. CrO is basic, Cr_2O_3 is amphoteric, and CrO_3 is acidic.

Pure chromium reacts with dilute HCl(aq) or H_2SO_4(aq) to produce Cr^{2+}(aq). Nitric acid and other oxidizing agents alter the surface of the metal (perhaps by formation of an oxide coating) and render the metal resistant to further attack— the metal becomes *passive*. A better source of chromium compounds than the pure metal is the alkali metal chromates, which contain Cr(VI) and can be obtained directly from chromite ore by reactions such as

$$4\,Fe(CrO_2)_2 + 8\,Na_2CO_3 + 7\,O_2(g) \xrightarrow{\Delta} 2\,Fe_2O_3 + 8\,Na_2CrO_4 + 8\,CO_2(g) \quad (24.15)$$

The *sodium chromate,* Na_2CrO_4, produced is the basis of many industrially important chromium compounds.

The Cr(VI) oxidation state is also observed in the red oxide, CrO_3. As expected, this oxide dissolves in water to produce a strongly acidic solution. The product of the reaction is not the expected chromic acid, H_2CrO_4, however, which has never been isolated in the pure state. Instead, the observed reaction is

$$2\,CrO_3(s) + H_2O \longrightarrow 2\,H^+(aq) + Cr_2O_7{}^{2-}(aq) \quad (24.16)$$

It is possible to crystallize a *dichromate* salt from a water solution of CrO_3. If the solution is made basic, the color turns from orange to yellow. From basic solutions, only *chromate* salts can be crystallized. Thus, whether a solution contains Cr(VI) as $Cr_2O_7{}^{2-}$ or $CrO_4{}^{2-}$ or a mixture of the two depends on the pH. The relevant equations follow.

$$2\,CrO_4{}^{2-}(aq) + 2\,H^+(aq) \rightleftharpoons Cr_2O_7{}^{2-}(aq) + H_2O \quad (24.17)$$

$$K_c = \frac{[Cr_2O_7{}^{2-}]}{[CrO_4{}^{2-}]^2[H^+]^2} = 3.2 \times 10^{14} \quad (24.18)$$

▶ The word *chromium* is derived from the Greek *chroma* meaning color, an apt name given the range of colors found in chromium compounds.

▶ The colors may also depend on other species present in solution. For example, if $[Cl^-]$ is high, $[Cr(H_2O)_6]^{3+}$ is converted to $[CrCl_2(H_2O)_4]^+$, and the violet color changes to green.

Le Châtelier's principle predicts that the forward reaction of (24.17) is favored in acidic solutions and the predominant Cr(VI) species is $Cr_2O_7^{2-}$. In basic solution, H^+ ions are removed and the reverse reaction is favored, forming CrO_4^{2-} as the principal species. Careful control of the pH is necessary when $Cr_2O_7^{2-}$ is used as an oxidizing agent or CrO_4^{2-} as a precipitating agent. In addition, Equation (24.18) can be used to calculate the relative amounts of the two ions as a function of $[H^+]$.

Chromate ion in basic solution can be used to precipitate metal chromates such as $BaCrO_4(s)$ and $PbCrO_4(s)$, but it is not a good oxidizing agent; it is not readily reduced.

$$CrO_4^{2-}(aq) + 4\ H_2O + 3\ e^- \longrightarrow [Cr(OH)_4]^-(aq) + 4\ OH^-(aq) \qquad E° = -0.13\ V$$

Dichromates are poor precipitating agents but excellent oxidizing agents that are used in a variety of industrial processes. In the chrome tanning process, for example, hides are immersed in $Na_2Cr_2O_7(aq)$, which is then reduced by $SO_2(g)$ to soluble basic chromic sulfate, $Cr(OH)SO_4$. Collagen, a protein in hides, reacts to form an insoluble chromium complex. The hides become *leather*, a tough, pliable material resistant to biological attack.

Dichromates are easily reduced to Cr_2O_3. In the case of ammonium dichromate, simply heating the compound produces Cr_2O_3 in a dramatic reaction (Figure 24-8).

Chromium(II) compounds can be prepared by the reduction of Cr(III) compounds with zinc in acidic solution or electrolytically at a lead cathode. The most distinctive feature of Cr(II) compounds is their reducing power.

$$Cr^{3+}(aq) + e^- \longrightarrow Cr^{2+}(aq) \qquad E° = -0.424\ V$$

That is, the oxidation of $Cr^{2+}(aq)$ occurs readily: $-E° = +0.424\ V$. In fact, Cr(II) solutions can be used to purge gases of trace amounts of $O_2(g)$, through the reaction illustrated in Figure 24-9 and shown next.

FIGURE 24-8
Decomposition of $(NH_4)_2Cr_2O_7$.
Ammonium dichromate (left) contains both an oxidizing agent, $Cr_2O_7^{2-}$, and a reducing agent, NH_4^+. The products of the reaction between these two ions are $Cr_2O_3(s)$, $N_2(g)$, and $H_2O(g)$. Considerable heat and light are also evolved (center). The product is $Cr_2O_3(s)$ (right).

FIGURE 24-9
Relationship between Cr^{2+} and Cr^{3+}.
The solution on the left, containing blue $Cr^{2+}(aq)$, is
prepared by dissolving chromium metal in $HCl(aq)$.
Within minutes, the $Cr^{2+}(aq)$ is oxidized to green
$Cr^{3+}(aq)$ by atmospheric oxygen (right). The green
color is that of the complex ion $[CrCl_2(H_2O)_4]^+(aq)$.

$$4\ Cr^{2+}(aq) + O_2(g) + 4\ H^+(aq) \longrightarrow 4\ Cr^{3+}(aq) + 2\ H_2O \qquad E^{\circ}_{cell} = +1.653\ V$$

Pure Cr can be obtained in small amounts by reducing Cr_2O_3 with Al in a
reaction similar to the thermite reaction.

$$Cr_2O_3(s) + 2\ Al(s) \longrightarrow Al_2O_3(s) + 2\ Cr(l) \qquad (24.19)$$

Manganese

Like V and Cr, Mn finds its most important use in steel production, generally
as the iron–manganese alloy, *ferromanganese*. Mn participates in the purification of iron by reacting with sulfur and oxygen and removing them through slag
formation. In addition, Mn increases the hardness of steel. Steel containing high
proportions of Mn is extremely tough and wear-resistant in such applications as
railroad rails, bulldozers, and road scrapers.

The electron configuration of Mn is $[Ar]3d^5 4s^2$. By employing first the two
$4s$ electrons and then, consecutively, up to all five of its unpaired $3d$ electrons,
manganese exhibits all oxidation states from $+2$ to $+7$. The most important reactions of manganese compounds are oxidation–reduction reactions.

▶ Some of these reactions are
explored through electrode potential data in Feature Problem
C on page 863.

The principal source of manganese compounds is MnO_2. Manganese dioxide itself is used in dry cell batteries, in glass and ceramic glazes, and as a catalyst. When MnO_2 is heated in the presence of an alkali and an oxidizing agent,
a manganate salt is produced.

$$3\ MnO_2 + 6\ KOH + KClO_3 \xrightarrow{\Delta} 3\ K_2MnO_4 + KCl + 3\ H_2O(g)$$

K_2MnO_4 is extracted from the fused mass with water and can then be oxidized
to $KMnO_4$, *potassium permanganate* (with Cl_2 as an oxidizing agent, for instance).

Potassium permanganate, $KMnO_4$, is an important laboratory oxidizing agent.
For chemical analyses it is generally used in acidic solutions, in which it is
reduced to $Mn^{2+}(aq)$. In the analysis of iron by MnO_4^-, a sample of Fe^{2+} is

prepared by dissolving iron in an acid and reducing any Fe^{3+} back to Fe^{2+}. Then the sample is titrated with $MnO_4^-(aq)$.

$$5\ Fe^{2+}(aq) + MnO_4^-(aq) + 8\ H^+(aq) \longrightarrow$$
$$5\ Fe^{3+}(aq) + Mn^{2+}(aq) + 4\ H_2O \qquad (24.20)$$

$Mn^{2+}(aq)$ has a barely discernible pale pink color. $MnO_4^-(aq)$ is an intense purple color. At the endpoint of titration reaction (24.20) the solution acquires a lasting light purple color with just one drop of excess $MnO_4^-(aq)$ (recall Figure 5-10). $MnO_4^-(aq)$ is less satisfactory for titrations in alkaline solutions because the insoluble reduction product, brown $MnO_2(s)$, obscures the endpoint.

Iron

Iron, with an annual worldwide production of over 500 million tons, is, of course, the most important metal in modern civilization. It is found widely distributed in Earth's crust with an abundance of 4.7%. The major commercial use of iron is to make steel (see Section 24-3).

One interesting aspect of iron and many of its compounds is their magnetic behavior, described in Section 24-1. Various iron oxides, because of their ability to have their magnetic polarity altered precisely and "read" back, are used in electronic devices, computer memory systems, and recording tape.

Iron, and the transition elements following it, differ from the early transition metals in the somewhat limited number of common oxidation states they display. For iron, although the +2 oxidation state is commonly encountered, the +3 oxidation state is the more stable.

$$
\begin{array}{ccc}
 & & 3d \\
Fe^{2+}: & [\text{Ar}] & \boxed{\uparrow\downarrow\ \uparrow\ \uparrow\ \uparrow\ \uparrow}
\end{array}
$$

$$
\begin{array}{ccc}
 & & 3d \\
Fe^{3+}: & [\text{Ar}] & \boxed{\uparrow\ \uparrow\ \uparrow\ \uparrow\ \uparrow}
\end{array}
$$

Because of the special stability attributed to the electron configuration of Fe^{3+}, a half-filled *d* subshell with all electrons unpaired, iron does not commonly occur in higher oxidation states.

Iron forms a variety of compounds, particularly those containing complex ions of iron. The iron complexes are often highly colored and some have been used in inks and as dyes. Biologically important complexes of iron are discussed in Chapters 25 and 28.

Cobalt–samarium magnets are used in high-efficiency motors. (The magnet is the grooved cylindrical sleeve in the center of the picture.)

Cobalt

Cobalt is among the rarer metals. It constitutes only 20 ppm of Earth's crust, but it occurs in sufficiently concentrated deposits (ores) that its annual production runs into the millions of pounds. Like iron, cobalt is ferromagnetic. One alloy of cobalt, Co_5Sm, makes a particularly strong and lightweight permanent magnet. Because of the strength of its magnetic field, magnets of this alloy are used in the manufacture of miniature electronic devices.

Unlike iron, the more stable oxidation state for cobalt is +2, although the +3 oxidation state is commonly encountered, particularly in complex ions. Like many transition metal compounds, cobalt salts crystallize from solution as hydrates. The use of $CoCl_2 \cdot 6H_2O$ as an indicator of relative humidity was described on page 89.

Nickel

Nickel ranks twenty-fourth in abundance among the elements in Earth's crust. Its ores are mainly the sulfides, oxides, silicates, and arsenides. Particularly large deposits are found in Canada. Of the 300 million pounds of nickel consumed annually in the United States, about 80% is for the production of alloys. Another 15% is used in electroplating, and the remainder for miscellaneous purposes.

The most stable oxidation state of Ni is +2; the +3 oxidation state is difficult to achieve and is not commonly encountered. The loss of an additional electron by Ni(II) does not lead to an electron configuration with a half-filled d subshell, as it does for Fe(II). As a result the conversion of Ni(II) to Ni(III) is not accomplished as easily as in the case of iron, although it is possible to make nickel compounds in higher oxidation states. These compounds find use as electrode materials in the nickel–cadmium (NiCad) cell, for example.

24-5 Coinage Metals: Copper, Silver, and Gold

Throughout the ages, Cu, Ag, and Au have been the preferred metals for coins because they are so durable and resistant to corrosion. The data in Table 24.6 help us to understand why this is so. The metal ions are easy to reduce to the free metals, which means that the metals are difficult to oxidize.

In Mendeleev's periodic table, the alkali metals (1A) and the coinage metals (1B) appear together as Group I. The only similarity between the two subgroups, however, is that of having a single s electron in the valence shells of their atoms. More significant are the differences between the Group 1A and 1B metals. For example, the first ionization energies for the 1B metals are much larger than for the 1A metals, and the standard reduction potentials, $E°$, are positive for the 1B metals and negative for the 1A metals.

Like the other transition elements that precede them in the periodic table, the 1B metals are able to use d electrons in chemical bonding. Thus the 1B elements can exist in different oxidation states, exhibit paramagnetism and color in some of their compounds, and form complex ions. They also possess to a high degree some of the distinctive physical properties of metals—malleability, ductility, and excellent electrical and thermal conductivity.

The coinage metals are used in jewelry making and the decorative arts. Gold, for instance, is extraordinarily malleable and can be pounded into thin translu-

TABLE 24.6 Some Properties of Copper, Silver, and Gold

	Cu	Ag	Au
electron configuration	[Ar]$3d^{10}4s^1$	[Kr]$4d^{10}5s^1$	[Xe]$4f^{14}5d^{10}6s^1$
metallic radius, pm	128	144	144
first ioniz. energy, kJ/mol	745	731	890
electrode potential, V			
$\quad$ M$^+$(aq) + e$^-$ → M(s)	+0.520	+0.800	+1.83
$\quad$ M^{2+}(aq) + 2 e$^-$ → M(s)	+0.337	+1.39	—
$\quad$ M^{3+}(aq) + 3 e$^-$ → M(s)	—	—	+1.52
oxidation states[a]	+1, +2	+1, +2	+1, +3

[a]The most common oxidation states are shown in red.

Copper wire and gold leaf.

cent sheets known as gold leaf. The coinage metals are valued by the electronics industry for their ability to conduct electricity. Silver has the highest electrical conductivity of any pure element, but both copper and gold are more often used as electrical conductors because copper is inexpensive and gold does not readily corrode. The most important "use" of gold is as the monetary reserve of nations throughout the world.

Generally, the coinage metals are resistant to air oxidation, although silver will tarnish through reactions with sulfur compounds in air to produce black Ag_2S. In moist air copper corrodes to produce green basic copper carbonate. This is the green color associated with copper roofing and gutters and bronze statues. (Bronze is an alloy of Cu and Sn.) Fortunately, this corrosion product forms a tough adherent coating that protects the underlying metal. The corrosion reaction is complex but may be summarized as

$$2\ Cu + H_2O + CO_2 + O_2 \longrightarrow Cu_2(OH)_2CO_3 \tag{24.21}$$
$$\text{basic copper carbonate}$$

The Group 1B metals do not react with HCl(aq), but Cu and Ag both react with concentrated H_2SO_4(aq) or HNO_3(aq). The metals are oxidized to Cu^{2+} and Ag^+, respectively, and the reduction products are SO_2(g) in H_2SO_4 and either NO(g) or NO_2(g) in HNO_3(aq).

Au does not react with either acid, but it will react with "royal water"—*aqua regia* (1 part HNO_3 and 3 parts HCl). The HNO_3(aq) oxidizes the metal and Cl^- from the HCl(aq) promotes the formation of the stable complex ion $[AuCl_4]^-$.

$$Au(s) + 4\ H^+ + NO_3^- + 4\ Cl^- \longrightarrow [AuCl_4]^- + 2\ H_2O + NO(g) \tag{24.22}$$

Trace amounts of Cu are essential to life, but larger quantities are toxic, especially to bacteria, algae, and fungi. Among the many copper compounds used as pesticides are the basic acetate, carbonate, chloride, hydroxide, and sulfate. Commercially, the most important copper compound is $CuSO_4 \cdot 5H_2O$. In addition to its agricultural applications, $CuSO_4$ is employed in batteries and electroplating, in preparing other copper salts, and in a variety of industrial processes.

Silver nitrate is the principal silver compound of commerce and is also an important laboratory reagent for the precipitation of anions, most of which form insoluble silver salts. These precipitation reactions can be used for the quantitative determination of anions, either gravimetrically (by weighing precipitates) or volumetrically (by titration). $AgNO_3$ is the source from which most other Ag compounds are derived. Ag compounds are used in electroplating, in the manufacture of batteries, in medicinal chemistry, as catalysts, and in cloud seeding (AgI). Their most important use by far (mostly as silver halides) is in photography (see Section 25-10).

Gold compounds are used in electroplating, photography, medicinal chemistry (as antiinflammatory agents for severe rheumatoid arthritis, for example), and the manufacture of special glasses and ceramics (such as ruby glass).

24-6 Group 2B: Zinc, Cadmium, and Mercury

The properties of the Group 2B elements are consistent with elements having full subshells, $(n-1)d^{10}ns^2$; some properties are summarized in Table 24.7. The low melting and boiling points of the Group 2B metals can probably be attributed to the fact that, with only the ns^2 electrons participating, metallic bonding is weak. Mercury is the only metal that exists as a liquid at room temperature and below (although liquid gallium can easily be supercooled to room temper-

TABLE 24.7 Some Properties of the Group 2B Metals			
	Zn	**Cd**	**Hg**
density, g/cm^3	7.14	8.64	13.59 (liquid)
melting point, °C	419.6	320.9	−38.87
boiling point, °C	907	765	357
electron configuration	[Ar]$3d^{10}4s^2$	[Kr]$4d^{10}5s^2$	[Xe]$4f^{14}5d^{10}6s^2$
atomic radius, pm	133	149	160
ionization energy, kJ/mol			
first	906	867	1006
second	1733	1631	1809
principal oxidation state(s)	+2	+2	+1, +2
electrode potential $E°$, V			
[$M^{2+}(aq) + 2\,e^- \rightarrow M$]	−0.763	−0.403	+0.854
[$M_2^{2+}(aq) + 2\,e^- \rightarrow 2\,M$]	—	—	+0.796

ature). Mercury differs from Zn and Cd in a number of ways in addition to its physical appearance.

- Mercury has little tendency to combine with oxygen. The oxide, HgO, is thermally unstable.
- Very few mercury compounds are water-soluble and most are not hydrated.
- Many mercury compounds are covalent. Except for HgF_2, mercury halides are only slightly ionized in aqueous solution.
- Mercury(I) forms a common diatomic ion with a metal–metal covalent bond, Hg_2^{2+}.
- Mercury will not displace $H_2(g)$ from $H^+(aq)$.

Some of these differences exhibited by mercury can probably be attributed to the *relativistic effect* discussed in Chapter 10 (page 340). Because their speed reaches a significant fraction of the speed of light, as the $6s$ electrons approach the high positive charge of the mercury nucleus, their masses increase (as predicted by Einstein's theory of relativity), and the $6s$ orbital shrinks in size. The closer approach of the $6s$ electrons to the nucleus subjects them to a greater attractive force than that experienced by the ns^2 electrons in Zn and Cd. As a result, for example, the first ionization energy of Hg is greater than those of Zn or Cd.

Uses of the Group 2B Metals

About one-third of the zinc produced is used in coating iron to give it corrosion protection (Section 21-6). The product is called *galvanized iron.* Large quantities of Zn are consumed in the manufacture of alloys. For example, about 20% of the production of Zn is used in *brass,* a copper alloy having 20–45% Zn and small quantities of Sn, Pb, and Fe. Brass is a good electrical conductor and it is corrosion-resistant. Zinc is also employed in the manufacture of dry cell batteries, in printing (lithography), in the construction industry (roofing materials), and as sacrificial anodes in corrosion protection (Section 21-6).

Although poisonous, cadmium is substituted for zinc as a shiny and protective plating of iron in special applications. It is used in bearing alloys, in low-melting solders, in aluminum solders, and as an additive to impart strength to copper. Another application, based on its neutron-absorbing capacity, is in control rods and shielding for nuclear reactors (see Section 26-8).

The principal uses of mercury take advantage of its metallic and liquid properties and its high density. It is used in thermometers, barometers, gas pressure regulators, and electrical relays and switches, and for electrodes, such as in the chlor-alkali process (Section 21-7). Mercury vapor is used in fluorescent tubes and street lamps. Mercury alloys, called **amalgams,** can be made with most metals, and some of these amalgams are of commercial importance. A typical silver dental filling is an amalgam of mercury with an alloy containing about 70% Ag, 26% Sn, 3% Cu, and 1% Zn.

▶ Iron is one of the few metals that does not form an amalgam. Mercury is generally stored and shipped in iron containers.

Compounds of the Group 2B Elements

Table 24.8 lists a few important compounds of the Group 2B metals and some of their uses. Some of the most interesting compounds are the semiconductors ZnO, CdS, and HgS, which are also the artist's pigments zinc white, cadmium yellow, and vermilion (red), respectively. Like all semiconductor materials, these compounds have an electronic structure consisting of a valence band and a conduction band (see Section 12-7). When light interacts with these compounds, electrons from the valence band may absorb photons and be excited into the conduction band. The energy of the light absorbed must equal or exceed the energy difference between the bands, called the *band gap*. The characteristic colors of these materials depend on the widths of the band gaps, as described on page 412.

Mercury and Cadmium Poisoning

Accumulations of mercury in the body affect the nervous system and cause brain damage. One form of chronic mercury poisoning, "hatter's disease," afflicted the Mad Hatter in *Alice's Adventures in Wonderland.* The disease was fairly common in the nineteenth century. Mercury compounds were used to convert fur to felt for making hats. Many hat makers of the time worked in hot, cramped spaces and used these compounds without any special precautions. Compounds of mercury are *much* more toxic than the pure element, and the hatters inadvertently ingested or inhaled the mercury compounds while they worked.

One proposed mechanism of mercury poisoning involves interference with the functioning of sulfur-containing enzymes; it is based on the fact that Hg has

TABLE 24.8	Some Important Compounds of the Group 2B Metals
Compound	**Uses**
ZnO	reinforcing agent in rubber; pigment; cosmetics; dietary supplement; photoconductors in copying machines.
ZnS	phosphors in X-ray and television screens; pigment; luminous paints.
$ZnSO_4$	rayon manufacture; animal feeds; wood preservative.
CdO	electroplating; batteries; catalyst.
CdS	solar cells; photoconductor in xerography; phosphors; pigment.
$CdSO_4$	electroplating; standard voltaic cells (Weston cell).
HgO	polishing compounds; dry cells; antifouling paints; fungicide; pigment.
$HgCl_2$	manufacture of Hg compounds; disinfectant; fungicide; insecticide; wood preservative.
Hg_2Cl_2	electrodes; pharmaceuticals; fungicide.

a high affinity for sulfur. Organic mercury compounds are generally more poisonous than inorganic ones or the element itself. An insidious aspect of mercury poisoning is that certain microorganisms have the ability to convert mercury(II) compounds to methylmercury (CH_3Hg^+) compounds, which then concentrate in the food chains of fish and other aquatic life. An early discovery of the environmental hazard of mercury was in Japan in the 1950s. Dozens of cases of mercury poisoning, including over 40 deaths, occurred among residents of the shores of Minamata Bay. Local seafood with up to 20 ppm of mercury was a major component of the victims' diet. The source of contamination was traced to a chemical plant discharging mercury waste into the bay.

In the free state, mercury is most poisonous as a vapor. Levels of mercury that exceed 0.05 mg Hg/m^3 air are considered unsafe. Even though we think of mercury as having a low vapor pressure, the concentration of Hg in its saturated vapor far exceeds this limit, and mercury vapor levels sometimes exceed safe limits where mercury is used—chlor-alkali plants, thermometer factories, and smelters.

Although zinc is an essential element in trace amounts, cadmium, which so resembles zinc, is a poison. One effect of cadmium poisoning is an extremely painful skeletal disorder known as "itai-itai kyo" (Japanese for "ouch-ouch" disease). This disorder was discovered in an area of Japan where effluents from a zinc mine became mixed with irrigation water used in rice fields. Cadmium poisoning was discovered in people who ate the rice. Cadmium poisoning can also cause liver damage, kidney failure, and pulmonary disease. The mechanism of cadmium poisoning may involve substitution in certain enzymes of Cd (a poison) for Zn (an essential element). Concern over cadmium poisoning has increased with an awareness that some cadmium is almost always found in zinc and zinc compounds, materials that have many commercial applications.

24-7   Lanthanides

The elements from cerium ($Z = 58$) through lutetium ($Z = 71$) are *inner transition* elements—their electron configurations feature the filling of $4f$ orbitals. These elements, together with lanthanum ($Z = 57$), which closely resembles them, are variously called the lanthanide, lanthanoid, or rare earth elements. The rare earth elements are "rare" only relative to the alkaline earth metals (Group 2A). Otherwise, they are not particularly rare. Ce, Nd, and La, for example, are more abundant than lead, and Tm is about as abundant as iodine. The lanthanides occur primarily as oxides, and mineral deposits containing them are found in various locations. Large deposits near the California–Nevada border are being developed to provide oxides of the lanthanides for use as phosphors in color monitors and television sets.

Because their differences in electron configuration are mainly in $4f$ orbitals, and because $4f$ electrons play a minor role in chemical bonding, strong similarities are found among the lanthanides. For example, $E°$ values for the reduction process $M^{3+}(aq) + 3 e^- \rightarrow M(s)$ do not show much variation. All fall between -2.38 V (La) and -1.99 V (Eu). The differences in properties that do exist among the lanthanides mostly arise from the lanthanide contraction discussed in Section 24-1. This contraction is best illustrated in the radii of the ions M^{3+}. These decrease regularly by about 1 to 2 pm for each unit increase in atomic number, from a radius of 106 pm for La^{3+} to 85 pm for Lu^{3+}.

The lanthanides (Ln) are reactive metals that liberate $H_2(g)$ from hot water and from dilute acids by undergoing oxidation to $Ln^{3+}(aq)$. The lanthanides

▶One use of the lanthanides is in magnets. A cobalt–samarium (Co_5Sm) magnet used in a high-efficiency motor is pictured on page 850.

Focus On High-Temperature Superconductors

The small magnet induces an electric current in the superconductor. Associated with this current is another magnetic field that opposes the field of the small magnet, causing it to be repelled. The magnet remains suspended above the superconductor as long as the superconducting current is present, and the current persists as long as the temperature of the superconductor is maintained at the boiling point of liquid nitrogen (77 K).

Magnetically levitated trains, magnetic resonance imaging (MRI) for medical diagnoses, and particle accelerators used in high-energy physics all require high magnetic fields generated by superconducting electromagnets. Superconductors offer no resistance to an electric current. Electricity is conducted with no loss of energy.

If cooled to near absolute zero, all metals become superconducting. Several metals and alloys superconduct even at marginally higher temperatures of 10–15 K. To maintain a superconductor at these extremely low temperatures requires liquid helium (bp, 4 K) as a coolant.

In the mid-1980s, materials made of lanthanum, strontium, copper, and oxygen were found to become superconducting when the temperature was lowered to 30 K. This was a much higher temperature for superconductivity than had been previously achieved. More surprising, the new material was not a metal at all but a *ceramic!* In short order other types of ceramic superconductors were discovered.

One of these new types was particularly easy to make. When a stoichiometric mixture of yttrium oxide (Y_2O_3), barium carbonate ($BaCO_3$), and copper(II) oxide (CuO) is heated in a stream of $O_2(g)$, a ceramic is produced with the approximate formula $YBa_2Cu_3O_x$ (where x is slightly less than 7). This so-called YBCO ceramic becomes superconducting at the remarkably high temperature of 92 K. Although a temperature of 92 K is still quite low, it is far above the boiling point of helium. In fact, it is above the

combine with $O_2(g)$, sulfur, the halogens, $N_2(g)$, $H_2(g)$, and carbon, in much the same way as expected for metals about as active as the alkaline earths. Preparation of the pure metals can be achieved by electrolytic reduction of Ln^{3+} in a molten salt.

The most common oxidation state for the lanthanides is +3. About half the lanthanides can also be obtained in the oxidation state +2, and about half, +4. The special stability associated with an electron configuration involving half-filled f orbitals (f^7) may account for some of the observed oxidation states, but the reason for the predominance of the +3 oxidation state is less clear. Most of the lanthanide ions are paramagnetic and colored in aqueous solution.

The lanthanide elements are extremely difficult to extract from their natural sources and to separate from one another. All the methods for doing so are based on this principle: Species that are *strongly dissimilar* can often be completely separated in a one-step process, such as separating $Ag^+(aq)$ and $Cu^{2+}(aq)$ by adding $Cl^-(aq)$—AgCl is insoluble. Species that are very *similar* can at best be *fractionated* in a one-step process. That is, the ratio of the concentration of one boiling point of nitrogen (77 K). Thus, inexpensive liquid nitrogen can be used as the coolant.

Many variations of the basic YBCO formula are possible. Almost any lanthanide element can be substituted for yttrium, and combinations of Group 2A elements can be substituted for barium. These variations all yield materials that are superconducting at relatively high temperatures, but of the group, the yttrium compound is superconducting at the highest temperature.

The record high temperature for superconductivity set by the YBCO ceramics was soon eclipsed by another group of ceramics containing bismuth and copper, such as $Bi_2Sr_2CaCu_2O_8$. One of these is superconducting at 110 K. But this record was also short-lived. A ceramic containing thallium and copper, with the approximate formula $TlBa_2Ca_3Cu_4O_y$ (where y is slightly larger than 10), was found to become superconducting at 125 K. Although this material holds the current record, the search continues for materials that might become su-

perconducting at room temperature (about 293 K).

Structurally, ceramic superconductors have a feature in common. Copper and oxygen atoms are bonded together in planar sheets. In the YBCO superconductors the Cu–O planes are widely separated. In the bismuth superconductors, the Cu–O planes occur in "sandwiches" consisting of two closely spaced sheets separated by a layer of Group 2A ions. These "sandwiches" are separated from one another by several layers of bismuth oxide. In the record-holding thallium superconductors the Cu–O planes are stacked in groups of three—"triple-decker sandwiches."

The current theory of superconductivity, developed in the 1950s, explains the superconducting behavior of metals at very low temperatures but not the higher temperature superconductivity of ceramics. It seems that the electrons in superconductors, whether low-temperature or the new high-temperature superconductors, move through the material in pairs—a sort of buddy

system that allows the electrons to move without resistance. The mechanism by which electron pairs form in high-temperature superconductors, however, is clearly different from that in low-temperature superconductors. Lack of a suitable theory complicates the search for higher-temperature superconductors. Once the mechanism for high-temperature superconductors is better understood, it may be that new breakthroughs will be easier to accomplish. Perhaps a room-temperature superconducting material will be possible.

Despite this less-than-complete understanding of high-temperature superconductors, engineers are already building devices that use the new materials. Wires have been made that are superconducting at liquid nitrogen temperatures, and new devices for precise magnetic field measurements using ceramic superconductors are now being produced. Ultimately, ceramic superconductors may find application in low-cost, energy-efficient electric power transmission.

species to that of another can be altered slightly. To achieve a complete separation may mean repeating the same basic step hundreds, or even thousands of times. To achieve the separation of the lanthanides, the methods of fractional crystallization, fractional precipitation, solvent extraction, and ion exchange were brought to their highest level of performance.

Summary

More than half the elements are transition elements. The transition elements are metals, and most are more reactive than hydrogen. Transition metals tend to exist in several different oxidation states in their compounds, and they readily form complex ions (discussed in Chapter 25). Many of the transition metals and their compounds are paramagnetic, and certain of the metals (Fe, Co, and Ni) and their alloys are also ferromagnetic.

Within a group of d-block elements, the members of the second and third transition series resemble one another more than they do the group member in the first transition series. This is a consequence of the phenomenon known as the lanthanide contraction occurring in the sixth period.

The possibility of a variety of oxidation states means that oxidation–reduction reactions are commonly encountered with transition metal compounds. Two common types of oxidizing agents are the dichromates and permanganates. In aqueous solution, dichromate ion is in equilibrium with chromate ion. Although the chromate ion is not a particularly good oxidizing agent, it is a good precipitating agent for a number of metal ions. Most of the oxides and hydroxides of the transition metals are basic if the metal is in one of its lower oxidation states. In higher oxidation states, some of the transition metal oxides and hydroxy compounds are amphoteric, and in the highest oxidation states a few are acidic (as is CrO_3, for example).

The transition metals are among the most widely used metals. Some, such as titanium and iron, display good structural strength. Others, such as copper and silver, are excellent conductors and some metals, such as gold, are highly malleable. General and specialized methods of extractive metallurgy are presented in the chapter, with a closer look at the metallurgy of iron and steel, zinc, copper, and titanium.

Summarizing Example

Although a number of slightly soluble copper(I) compounds (e.g., CuCN) can exist in contact with water, it is not possible to prepare a solution with a high concentration of Cu^+ ion.

Explain why a high $[Cu^+]$ cannot be maintained in aqueous solution. That is, show that Cu^+ disproportionates to $Cu^{2+}(aq)$ and $Cu(s)$.

1. *Describe the disproportionation reaction through half-equations and a net equation. Result:* oxidation: $Cu^+(aq) \rightarrow Cu^{2+}(aq) + e^-$; reduction: $Cu^+(aq) + e^- \rightarrow Cu(s)$; net: $2\,Cu^+(aq) \rightarrow Cu^{2+}(aq) + Cu(s)$.

2. *Obtain $E°$ values for the couples $Cu^+/Cu(s)$ and Cu^{2+}/Cu^+.* Both of these values can be found in Appendix D. *Result:* $E°_{Cu^+/Cu(s)} = 0.520$ V; $E°_{Cu^{2+}/Cu^+} = 0.159$ V.
3. *Obtain $E°_{cell}$ for the disproportionation reaction. Combine $E°$ values. Result:* $E°_{cell} = +0.361$ V.
4. *Interpret the significance of the $E°_{cell}$ value. Answer:* The positive value of $E°_{cell}$ means that the disproportionation reaction will proceed in the forward direction in a solution in which $[Cu^+] = [Cu^{2+}] = 1$ M. If $[Cu^{2+}] < 1$ M or if $[Cu^+] > 1$ M, the reaction is even more spontaneous. A solution cannot be prepared with a high $[Cu^+]$.

Key Terms

amalgam (24-6)	**hydrometallurgy** (24-2)	**pyrometallurgy** (24-2)
basic oxygen process (24-3)	**lanthanide contraction** (24-1)	**steel** (24-3)
extractive metallurgy (24-2)	**pig iron** (24-3)	**zone refining** (24-2)
ferromagnetism (24-1)		

Review Questions

1. In your own words define the following terms: **(a)** domain; **(b)** flotation; **(c)** leaching; **(d)** amalgam.
2. Briefly describe each of the following ideas, phenomena, or methods: **(a)** lanthanide contraction; **(b)** zone refining; **(c)** basic oxygen process; **(d)** slag formation.
3. Explain the important distinctions between each pair of terms: **(a)** ferromagnetism and paramagnetism; **(b)** roasting and reduction; **(c)** hydrometallurgy and pyrometallurgy; **(d)** chromate and dichromate.
4. Provide an acceptable name for each of the following substances.
(a) $Sc(OH)_3$
(b) Cu_2O
(c) $TiCl_4$
(d) V_2O_5
(e) K_2CrO_4
(f) K_2MnO_4
5. Provide an acceptable formula for each of the following substances.
(a) chromium(VI) oxide
(b) iron(II) silicate
(c) barium dichromate

(d) copper(I) cyanide
(e) cobalt(II) chloride hexahydrate
6. Describe the chemical composition of the material called **(a)** pig iron; **(b)** ferromanganese; **(c)** chromite ore; **(d)** brass; **(e)** aqua regia; **(f)** blister copper; **(g)** stainless steel.
7. Complete and balance the following equations. If no reaction occurs, so state.
(a) $TiCl_4(g) + Na(l) \rightarrow$
(b) $Cr_2O_3(s) + Al(s) \rightarrow$
(c) $Ag(s) + HCl(aq) \rightarrow$
(d) $K_2Cr_2O_7(aq) + KOH(aq) \rightarrow$
(e) $MnO_2(s) + C(s) \rightarrow$
8. Balance the following oxidation–reduction equations.
(a) $Fe_2S_3(s) + H_2O + O_2(g) \rightarrow Fe(OH)_3(s) + S(s)$
(b) $Mn^{2+}(aq) + S_2O_8{}^{2-}(aq) + H_2O \rightarrow$
$MnO_4{}^-(aq) + SO_4{}^{2-}(aq) + H^+(aq)$
(c) $Ag(s) + CN^-(aq) + O_2(g) + H_2O \rightarrow$
$[Ag(CN)_2]^-(aq) + OH^-(aq)$
9. Through a chemical equation, give an example to represent the reaction of **(a)** a transition metal with a nonoxidizing acid; **(b)** a transition metal oxide with NaOH(aq); **(c)** an inner transition metal with HCl(aq).

10. Through orbital diagrams, write electron configurations for the following transition element atom and ions.
(a) Ti
(b) V^{3+}
(c) Cr^{2+}
(d) Mn^{4+}
(e) Mn^{2+}
(f) Fe^{3+}

11. Arrange the following species according to the number of unpaired electrons they contain, starting with the one that has the greatest number: Fe, Sc^{3+}, Ti^{2+}, Mn^{4+}, Cr, Cu^{2+}.

12. Transition elements generally are expected to have which of these properties? Explain.
(1) low melting points

(2) high ionization energies
(3) colored ions in solution
(4) positive standard electrode (reduction) potentials
(5) paramagnetic compounds

13. Which of the following ions is diamagnetic: Cr^{2+}, Fe^{3+}, Cu^{2+}, Sc^{3+}?

14. Of the following elements, which one is *not* expected to display an oxidation state of +6 in any of its compounds: Ti, Cr, Mn, or W? Explain.

15. Which of the following ions in aqueous solution is the best oxidizing agent: Na^+, Zn^{2+}, Ag^+, Cu^{2+}? Explain.

16. Why is +3 the most stable oxidation state for Fe, whereas it is +2 for Co and Ni?

Exercises

Properties of the Transition Elements

17. Describe how the transition elements compare with main group metals (e.g., Group 2A) with respect to oxidation states, formation of complexes, colors of compounds, and magnetic properties.

19. Why do the atomic radii vary so much more for two main group elements that differ by one unit in atomic number than they do for two transitions elements that differ by one unit?

21. Which of the first transition series elements exhibits the greatest number of different oxidation states in its compounds? Explain.

23. As a group, the lanthanides are more reactive metals than are those in the first transition series. How do you account for this difference?

18. With but minor irregularities, the melting points of the first series of transition metals rise from that of Sc to that of Cr and then fall to that of Zn. Give a plausible explanation based on atomic structure.

20. The metallic radii of Ni, Pd, and Pt are 125, 138, and 139 pm, respectively. Why is the difference in radius between Pt and Pd so much less than between Pd and Ni?

22. Why is the number of common oxidation states for the elements at the beginning and those at the end of the first transition series less than for elements in the middle of the series?

24. The maximum difference in standard reduction potential, $E^{\circ}_{M^{2+}/M(s)}$, among members of the first transition series is about 2.4 V. For the lanthanides the maximum difference in $E^{\circ}_{M^{3+}/M(s)}$ is only about 0.4 V. How do you account for this fact?

Reactions of Transition Metals and Their Compounds

25. Write balanced chemical equations for the following reactions described in the chapter.
(a) The reaction of $Sc(OH)_3(s)$ with HCl(aq).
(b) Oxidation of $Fe^{2+}(aq)$ by $MnO_4^-(aq)$ in basic solution to give $Fe^{3+}(aq)$ and $MnO_2(s)$.
(c) The reaction of $TiO_2(s)$ with molten KOH to form K_2TiO_3.
(d) Oxidation of Cu(s) to $Cu^{2+}(aq)$ with H_2SO_4(concd aq) to form $SO_2(g)$.

27. Suggest a series of reactions, using common chemicals, by which each of the following syntheses can be performed.
(a) $Fe(OH)_3(s)$ from FeS(s).
(b) $BaCrO_4(s)$ from $BaCO_3(s)$ and $K_2Cr_2O_7(aq)$

26. Write balanced equations for the following reactions described in the chapter.
(a) Sc(l) is produced by the electrolysis of Sc_2O_3 dissolved in $Na_3ScF_6(l)$.
(b) Cr(s) reacts with HCl(aq) to produce a blue solution containing $Cr^{2+}(aq)$.
(c) $Cr^{2+}(aq)$ is readily oxidized by $O_2(g)$ to $Cr^{3+}(aq)$.
(d) Ag(s) reacts with concentrated $HNO_3(aq)$ and $NO_2(g)$ is evolved.

28. Suggest a series of reactions, using common chemicals, by which each of the following syntheses can be performed.
(a) $Cu(OH)_2(s)$ from CuO(s)
(b) $CrCl_3(aq)$ from $(NH_4)_2Cr_2O_7(s)$

Oxidation–Reduction

29. Write plausible half-equations to represent each of the following in acidic solution:
(a) $VO^{2+}(aq)$ as an oxidizing agent;
(b) $Cr^{2+}(aq)$ as a reducing agent.

30. Write plausible half-equations to represent each of the following in basic solution:
(a) oxidation of $Fe(OH)_3(s)$ to FeO_4^{2-};
(b) reduction of $[Ag(CN)_2]^-$ to silver metal.

31. Use electrode potential data from this chapter or Appendix D to predict whether each of the following reactions will occur to any significant extent under standard-state conditions.

(a) $2 VO_2^+ + 6 Br^- + 8 H^+ \rightarrow 2 V^{2+} + 3 Br_2(l) + 4 H_2O$

(b) $VO_2^+ + Fe^{2+} + 2 H^+ \rightarrow VO^{2+} + Fe^{3+} + H_2O$

(c) $MnO_2(s) + H_2O_2 + 2 H^+ \rightarrow Mn^{2+} + 2 H_2O + O_2(g)$

33. Show that $Fe^{2+}(aq)$ should be spontaneously oxidized to $Fe^{3+}(aq)$ by $O_2(g)$ in acidic solution. (*Hint:* What is the reduction half-reaction?)

32. You are given these three reducing agents: $Zn(s)$, $Sn^{2+}(aq)$, and $I^-(aq)$. Use data from Appendix D to determine which of them are able, under standard-state conditions in acidic solution, to reduce

(a) $Cr_2O_7^{2-}(aq)$ to $Cr^{3+}(aq)$;

(b) $Cr^{3+}(aq)$ to $Cr^{2+}(aq)$;

(c) $SO_4^{2-}(aq)$ to $SO_2(g)$.

34. Refer to Example 24-3. Select a reducing agent (from Table 21.1 or Appendix D) that will reduce VO^{2+} to V^{3+} in acidic solution. [*Hint:* The reduction process must stop at V^{3+} and not continue on to V^{2+} or $V(s)$.]

Chromium and Chromium Compounds

35. When a soluble lead compound is added to a solution containing primarily *orange* dichromate ion, *yellow* lead chromate precipitates. Describe the equilibria involved.

37. When $Zn(s)$ is added to $K_2Cr_2O_7$ dissolved in $HCl(aq)$, the color of the solution changes from orange to green, then to blue, and, over a period of time, back to green. Write equations for this series of reactions.

39. Use Equation (24.18) to determine $[Cr_2O_7^{2-}]$ in a solution that has $[CrO_4^{2-}] = 0.20$ M and a pH of (a) 6.62 and (b) 8.85.

41. How many grams of chromium would be deposited on an object in a chrome-plating bath (see page 847) after 1.00 h at a current of 3.4 A?

43. Explain why it is reasonable to expect the chemistry of dichromate ion to involve mainly oxidation–reduction reactions and that of chromate ion, precipitation reactions.

36. When *yellow* $BaCrO_4$ is dissolved in $HCl(aq)$, a *green* solution is obtained. Write a chemical equation to account for the color change.

38. If $CO_2(g)$ under pressure is passed into $Na_2CrO_4(aq)$, $Na_2Cr_2O_7(aq)$ is formed. What is the function of the $CO_2(g)$? Write a plausible equation for the net reaction.

40. If a solution is prepared by dissolving 1.505 g Na_2CrO_4 in 345 mL of a buffer solution with pH = 7.55, what will be $[CrO_4^{2-}]$ and $[Cr_2O_7^{2-}]$?

42. How long would an electric current of 3.5 A have to pass through a chrome-plating bath (see page 847) to produce a deposit of 0.0010 mm thick on an object with a surface area of 0.375 m^2? (The density of Cr is 7.14 g/cm^3.)

44. What are the products obtained when $Mg^{2+}(aq)$ and $Cr^{3+}(aq)$ are each treated with a limited amount of $NaOH(aq)$? With an excess of $NaOH(aq)$? Why are the results different in these two cases?

The Coinage Metals

45. Write plausible equations for these reactions occurring in the hydrometallurgy of the coinage metals.

(a) Copper is precipitated from a solution of copper(II) sulfate by treatment with $H_2(g)$.

(b) Gold is precipitated from a solution of Au^+ by adding iron(II) sulfate.

(c) Copper(II) chloride solution is reduced to copper(I) chloride when treated with $SO_2(g)$ in acidic solution.

47. Use the result of the Summarizing Example to determine the value of K_c for the reaction $2 Cu^+(aq) \rightarrow Cu^{2+}(aq) + Cu(s)$, $K_c = $? (*Hint:* Review Section 21-3.)

46. In the metallurgical extraction of silver and gold, often an alloy of the two metals is obtained. The alloy can be separated into Ag and Au either with concentrated HNO_3 or boiling concentrated H_2SO_4, in a process called *parting*. Write chemical equations to show how these separations work.

48. Refer to Exercise 47, and determine whether a solution can be prepared with $[Cu^+]$ equal to (a) 0.20 M; (b) 1.0×10^{-10} M.

Group 2B Metals

49. In ZnO, the band gap between the valence and conduction bands is 290 kJ/mol, and in CdS it is 250 kJ/mol. Show that CdS absorbs some visible light but that ZnO does not. Explain the observed colors: ZnO is white and CdS is yellow.

50. CdS is yellow, HgS is red, and CdSe is black. Which of these materials has the largest band gap? the smallest? How does the band gap relate to the observed color?

Advanced Exercises

51. A 1.502-g sample of steel is dissolved in sulfuric acid. The acid is neutralized and dimethylglyoxime is added. Nickel can be determined by the precipitation of red nickel dimethylglyoximate, $NiC_8H_{14}N_4O_4$. The yield of nickel dimethylglyoximate is 0.259 g. What is the mass percent Ni in the steel?

52. Although Au reacts with and dissolves in aqua regia (3 parts HCl + 1 part HNO_3), Ag does not dissolve. What is the likely reason(s) for this difference?

53. The text mentions that scandium metal is obtained from its molten chloride by electrolysis and that titanium is obtained from its chloride by reduction with magnesium. What are the reasons why these metals are not obtained by the reduction of their oxides with carbon (coke), as are metals such as zinc and iron?

54. Show that the corrosion reaction in which Cu is converted to its basic carbonate (reaction 24.21) can be thought of in terms of a combination of oxidation–reduction, acid–base, and precipitation reactions.

55. In an atmosphere polluted with industrial smog, Cu corrodes to a basic sulfate, $Cu_2(OH)_2SO_4$. Propose a series of chemical reactions to describe this corrosion.

56. If one attempts to make CuI_2 by the reaction of $Cu^{2+}(aq)$ and $I^-(aq)$, $CuI(s)$ and $I_3^-(aq)$ are obtained instead. Without performing detailed calculations, show why this reaction should occur.

$$2\,Cu^{2+}(aq) + 5\,I^-(aq) \longrightarrow 2\,CuI(s) + I_3^-(aq)$$

57. Without performing detailed calculations, show that significant disproportionation of AuCl occurs if one attempts to make a saturated aqueous solution. Use data from Table 24.6 and K_{sp} (AuCl) = 2.0×10^{-13}.

58. In acidic solution silver(II) oxide first dissolves to produce $Ag^{2+}(aq)$. This is followed by the oxidation of H_2O to $O_2(g)$ and the reduction of Ag^{2+} to Ag^+.

(a) Write equations for the dissolution and oxidation–reduction reactions.

(b) Show that the oxidation–reduction reaction is indeed spontaneous. [*Hint:* You will have to use the method of Feature Problem A of Chapter 21 (page 762) for part of the solution.]

59. Equation (24.17), which represents the chromate–dichromate equilibrium, is actually the sum of two equilibrium expressions. The first is an acid–base reaction, $H^+ + CrO_4^{2-} \rightleftharpoons HCrO_4^-$. The second reaction involves elimination of a water molecule between two $HCrO_4^-$ ions (a dehydration reaction), $2\,HCrO_4^- \rightleftharpoons Cr_2O_7^{2-} + H_2O$. If the ionization constant, K_a, for $HCrO_4^-$ is 3.2×10^{-7}, what is the value of K for the dehydration reaction?

60. Show that under the following conditions $Ba^{2+}(aq)$ can be separated from $Sr^{2+}(aq)$ and $Ca^{2+}(aq)$ by precipitating $BaCrO_4(s)$, with the other ions remaining in solution: $[Ba^{2+}]$ = $[Sr^{2+}] = [Ca^{2+}] = 0.10\,M$; $[HC_2H_3O_2] = [C_2H_3O_2^-] = 1.0\,M$; $[Cr_2O_7^{2-}] = 0.0010$ M. K_{sp} (BaCrO$_4$) = 1.2×10^{-10} and K_{sp} (SrCrO$_4$) = 2.2×10^{-5}. Also, use data from this and previous chapters, as necessary.

61. A 0.589-g sample of pyrolusite ore (impure MnO_2) is treated with 1.651 g of oxalic acid ($H_2C_2O_4\cdot 2H_2O$) in an acidic medium (reaction 1). Following this, the excess oxalic acid is titrated with 30.06 mL of 0.1000 M $KMnO_4$ (reaction 2). What is the mass percent of MnO_2 in the pyrolusite? The following equations are neither complete nor balanced.

(1) $H_2C_2O_4(aq) + MnO_2(s) \rightarrow Mn^{2+}(aq) + CO_2(g)$

(2) $H_2C_2O_4(aq) + MnO_4^-(aq) \rightarrow Mn^{2+}(aq) + CO_2(g)$

62. Both $Cr_2O_7^{2-}(aq)$ and $MnO_4^-(aq)$ can be used to titrate $Fe^{2+}(aq)$ to $Fe^{3+}(aq)$. Suppose you have available as titrants two solutions: 0.1000 M $Cr_2O_7^{2-}(aq)$ and 0.1000 M $MnO_4^-(aq)$,

(a) of which solution would the greater volume of titrant be required for the titration of a particular sample of $Fe^{2+}(aq)$? Explain.

(b) how many mL of 0.1000 M $MnO_4^-(aq)$ would be required for a titration, if the same titration requires 24.50 mL of 0.1000 M $Cr_2O_7^{2-}(aq)$?

63. The only important compounds of Ag(II) are AgF_2 and AgO. Explain why you would expect these two compounds to be stable, but not other silver(II) compounds such as $AgCl_2$, $AgBr_2$, and AgS.

64. A certain steel is to be analyzed for Cr and Mn. By suitable treatment the Cr in the steel is oxidized to $Cr_2O_7^{2-}(aq)$ and the Mn to $MnO_4^-(aq)$. A 10.000-g sample of steel is used to produce 250.0 mL of a solution containing $Cr_2O_7^{2-}(aq)$ and $MnO_4^-(aq)$. A 10.00-mL portion of this solution is added to $BaCl_2(aq)$, and by proper adjustment of the pH the chromium is completely precipitated as $BaCrO_4(s)$; 0.549 g is obtained. A second 10.00-mL portion of the solution requires exactly 15.95 mL of 0.0750 M $Fe^{2+}(aq)$ for its titration in acidic solution. Calculate the % Cr and % Mn in the steel sample. [*Hint:* In the titration $MnO_4^-(aq)$ is reduced to $Mn^{2+}(aq)$ and $Cr_2O_7^{2-}(aq)$ is reduced to $Cr^{3+}(aq)$; the $Fe^{2+}(aq)$ is oxidized to $Fe^{3+}(aq)$.]

65. Nickel can be determined as nickel dimethylglyoximate, a brilliant scarlet precipitate that has the composition 20.31% Ni, 33.26% C, 4.88% H, 22.15% O, and 19.39% N. A 15.020-g steel sample is dissolved in concentrated HCl(aq). The solution obtained is suitably treated to remove interfering ions, to establish the proper pH, and to obtain a final solution volume of 250.0 mL. A 10.00-mL sample of this solution is then treated with dimethylglyoxime. The mass of purified, dry nickel dimethylglyoximate obtained is 0.104 g.

(a) What is the empirical formula of nickel dimethylglyoximate?

(b) What is the mass percent nickel in the steel sample?

66. A solution is believed to contain one or more of the following ions: Cr^{3+}, Zn^{2+}, Fe^{3+}, Ni^{2+}. When the solution is treated with excess NaOH(aq), a precipitate forms. The solution in contact with the precipitate is colorless. The precipitate is dissolved in HCl(aq), and the resulting solution is treated with $NH_3(aq)$. No precipitation occurs. Based solely on these observations, what conclusions can you draw about the ions present in the original solution? That is, which ion or ions are likely present, which are most likely not present, and about which can we not be certain? (*Hint:* Refer to Appendix D for solubility product and complex ion formation data.)

Feature Problems

A. As a continuation of Feature Problem B of Chapter 20 on page 722, the figure represents three graphs of $\Delta G°$ as a function of temperature.

1. Explain the shapes of the three graphs. Specifically, why is one line essentially parallel to the temperature axis, why does one have a positive slope, and why does one have a negative slope? (*Hint:* Recall the expression: $\Delta G° = \Delta H° - T\Delta S°$ and the assumptions that we made in Chapter 20 about ΔH and ΔS as functions of temperature.)

2. One reaction that produces the reducing agents $H_2(g)$ and $CO(g)$ in a blast furnace is the water gas reaction (recall Table 24.3).

$$C(s) + H_2O(g) \rightleftharpoons CO(g) + H_2(g)$$

The following approximate data are given for $\Delta G°$ as a function of temperature for the reaction $2 H_2(g) + O_2(g) \rightarrow 2 H_2O(g)$: $\Delta G° = -465$ kJ at 100 °C; -420 kJ at 500 °C; -360 kJ at 1000 °C; and -240 kJ at 2000 °C. Graph these data on the same figure as the other graphs and estimate the temperature at which $K_{eq} = 1$ for the water gas reaction. (*Hint:* Which two reactions, when coupled, yield the water gas reaction? What must be $\Delta G°$ for the coupled reactions at the temperature in question?)

3. We noted on page 839 that carbon is not used to reduce $TiO_2(s)$ to $Ti(s)$ because of the formation of titanium carbides. Otherwise carbon would be a good enough reducing agent to accomplish this reduction. Use thermodynamic data from Appendix D to sketch a graph of $\Delta G°$ versus temperature for the reaction $Ti(s) + O_2(g) \rightarrow TiO_2(s)$. Then couple this reaction with one of the three shown in the figure to determine the approximate temperature at which the reduction of $TiO_2(s)$ to $Ti(s)$ might be accomplished. (*Hint:* Which of the three reactions graphed in the figure should you use? What must be the relationship between that graph and the one that you sketched?)

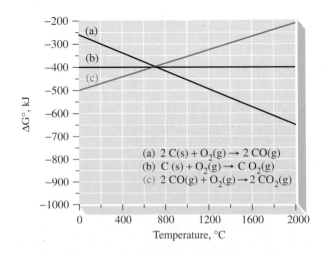

(a) $2 C(s) + O_2(g) \rightarrow 2 CO(g)$
(b) $C(s) + O_2(g) \rightarrow CO_2(g)$
(c) $2 CO(g) + O_2(g) \rightarrow 2 CO_2(g)$

B. Several transition metal ions are found in cation Group 3 of the qualitative analysis scheme outlined in Figure 19-7. At one point in the separation and testing of this group, a solution containing Fe^{3+}, Mn^{2+}, Co^{2+}, Ni^{2+}, Al^{3+}, Cr^{3+}, and Zn^{2+} is treated with an excess of NaOH(aq), together with $H_2O_2(aq)$. (1) The excess NaOH(aq) causes *three* of the cations to precipitate as hydroxides and *three* to form hydroxo complex ions. (2) In the presence of $H_2O_2(aq)$, the cation in one of the insoluble hydroxides is oxidized from the +2 to the +3 oxidation state, and one of the hydroxo complex ions is also oxidized. (3) $MnO_2(s)$ and the three insoluble hydroxides are found as a dark precipitate. (4) The solution above the precipitate has a yellow color. (5) The dark precipitate from (3) reacts with HCl(aq), and all the cations return to solution. (6) The solution from (5) is treated with $NH_3(aq)$, and a precipitate containing one of the cations forms.

1. Write equations for the reactions referred to in item (1).

2. Write an equation for the most likely reaction in which a hydroxide precipitate is oxidized in item (2).

3. What is the ion responsible for the yellow color of the solution in item (4)? Write an equation for its formation.

4. Write equations for the dissolution of the precipitate in item (5).

5. Write an equation for the precipitate formation in item (6). (*Hint:* You may need solubility product and complex ion formation data from Appendix D, together with descriptive information from this chapter and from elsewhere in the text.)

C. To answer the following questions, first refer to Feature Problem A of Chapter 23 on page 828 for a discussion of reduction potential diagrams.

1. Complete the reduction potential diagram for vanadium. As in Feature Problem A of Chapter 23, the entries marked (?) can be obtained directly from tabulated data (Table 24.5), and those marked (??) require the method introduced in Feature Problem A of Chapter 21 on page 762.

Electrode potential diagram for vanadium
Acidic solution ([H⁺] = 1 M):

O.S.   +5   +4   +3   +2   0

VO_2^+ —(?)— VO^{2+} —(?)— V^{3+} —(?)— V^{2+} —(?)— V(s)

(??)

(??)

(??)

2. Use the reduction potential diagrams for manganese to answer the following questions concerning the oxidation–reduction chemistry of manganese.

(a) Explain the observation that in acidic solution $MnO_4^{2-}(aq)$ readily disproportionates to $MnO_4^-(aq)$ and $MnO_2(s)$, but that in strongly basic solution $MnO_4^{2-}(aq)$ is reasonably stable.

(b) Explain why the oxidizing agent $MnO_4^-(aq)$ is commonly reduced to $Mn^{2+}(aq)$ in acidic solutions but to $MnO_2(s)$ in basic solutions. (*Hint:* What are the values of $E^\circ_{MnO_4^-/Mn^{2+}}$ and $E^\circ_{MnO_4^-/Mn(OH)_2}$?)

(c) Estimate a value of K_{sp} for $Mn(OH)_2$. [*Hint:* Consider an electrochemical cell having the net reaction: $Mn(OH)_2(s) \rightleftharpoons Mn^{2+}(aq) + 2\ OH^-(aq)$.]

Electrode potential diagrams for manganese

Acidic solution ($[H^+] = 1\ M$):

Basic solution ($[OH^-] = 1\ M$):

Complex Ions and Coordination Compounds

Turquoise is a mineral of copper, $CuAl_6(PO_4)_4(OH)_8 \cdot 4H_2O$. The distinctive color of this and many other gemstones is a consequence of the nature of metal-ligand bonding in complex ions, a central topic of this chapter.

*I*n the preceding chapter we discussed several situations involving a succession of color changes, and we attributed these to changes in oxidation state. The color changes discussed in this chapter, however, are not caused by oxidation–reduction reactions. Rather, changes in color are observed with changes in the groups (ligands) bound to a metal center, even though the oxidation state of the metal remains unchanged.

To explain this observation, we need to explore more fully the subject of complex ions and coordination compounds, a topic that we briefly introduced in Chapter 19. One topic we will consider in this chapter is the geometric structures of complex ions. In doing so, we will discover possibilities for isomerism: the existence of compounds having identical compositions but different structures and properties. We will also examine the nature of the bonding between ligands and the metal centers to which they are attached. It is through an understanding of bonding in complex ions that we can gain some insight into the origin of their colors.

25-1 Werner's Theory of Coordination Compounds: An Overview

In Chapter 19 we saw how the Swiss chemist Alfred Werner conceived of the series of coordination compounds $CoCl_3 \cdot 6NH_3$, $CoCl_3 \cdot 5NH_3$, and $CoCl_3 \cdot 4NH_3$ as actually being

$$[Co(NH_3)_6]Cl_3 \qquad [CoCl(NH_3)_5]Cl_2 \qquad [CoCl_2(NH_3)_4]Cl$$
$$\text{(a)} \qquad\qquad \text{(b)} \qquad\qquad\qquad \text{(c)}$$

Thus the ionization of coordination compound (a) can be represented as

$$[Co(NH_3)_6]Cl_3(s) \xrightarrow{H_2O} [Co(NH_3)_6]^{3+}(aq) + 3\,Cl^-(aq)$$

Alfred Werner (1866–1919). Werner's success in explaining coordination compounds came in large part through his application of new ideas: the theory of electrolytic dissociation and principles of structural chemistry.

and this ionization scheme helps to explain why aqueous solutions of compound (a) yield *three* moles of AgCl(s) per mole of compound when treated with $AgNO_3(aq)$. Compound (b), on the other hand, yields only *two* moles of AgCl(s) and compound (c), *one* mole of AgCl(s).

Werner's theory also accounts for (1) the coordination compound $[CoCl_3(NH_3)_3]$, which is a *nonelectrolyte* and yields no precipitate with $AgNO_3(aq)$, and (2) the compound $Na[CoCl_4(NH_3)_2]$, in which the complex ion has a net *negative* charge, that is, $[CoCl_4(NH_3)_2]^-$.

The species $[Co(NH_3)_6]^{3+}$, $[CoCl(NH_3)_5]^{2+}$, and $[CoCl_2(NH_3)_4]^+$ are all complex ions, and because they carry a positive charge, they are *cations*. $[CoCl_4(NH_3)_2]^-$ is also a complex ion, but it is an *anion*. $[CoCl_3(NH_3)_3]$ is a *neutral molecule*. The term **complex** describes any species involving coordination of ligands to a metal center; the metal center can be an atom or an ion, and the species can be a cation, an anion, or a neutral molecule. In a chemical formula, a complex—a metal center and attached ligands—is set off by square brackets ([ ]). Compounds that are complexes or contain complex ions are often referred to as **coordination compounds.** In this text our primary interest is in complex cations and anions having a central metal ion.

The **coordination number** is the number of points around the metal center at which bonds to ligands can form. Coordination numbers ranging from 2 to 12 have been observed in complexes, although the number 6 is by far the most common, followed by 4. Coordination number 2 is mostly limited to complexes of Cu(I), Ag(I), and Au(I). Coordination numbers greater than 6 are not often found in members of the first transition series but are more common in those of the second and third series. Stable complexes with coordination numbers 3 and 5 are rarely encountered. The coordination number observed in a complex depends on a number of factors, such as the ratio of the radius of the central metal atom or ion to those of the attached ligands.

Coordination numbers of some common ions are listed in Table 25.1, and the four most commonly observed geometric shapes of complex ions are shown in Figure 25-1. One practical use of the coordination number is to assist in writing and interpreting formulas of complexes, as illustrated in Example 25-1.

TABLE 25.1 Some Common Coordination Numbers of Metal Ions

Cu^+	2, 4		
Ag^+	2		
Au^+	2, 4	Al^{3+}	4, 6
		Sc^{3+}	6
		Cr^{3+}	6
Fe^{2+}	6	Fe^{3+}	6
Co^{2+}	4, 6	Co^{3+}	6
Ni^{2+}	4, 6	Au^{3+}	4
Cu^{2+}	4, 6		
Zn^{2+}	4		

EXAMPLE 25-1

Relating the Formula of a Complex to the Coordination Number and Oxidation State of the Central Metal. What are the coordination number and oxidation state of Co in the complex ion $[CoCl(NO_2)(NH_3)_4]^+$?

linear

square planar

tetrahedral

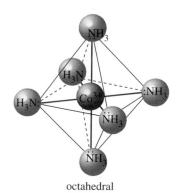

octahedral

FIGURE 25-1
Structures of some complex ions.
Attachment of the NH_3 molecules occurs through the lone pair electrons on the N atoms.

Solution

The complex ion has as ligands *one* Cl^- ion, *one* NO_2^- ion, and *four* NH_3 molecules. The coordination number is 6. Of these six ligands, *two* carry a charge of $1-$ each (the Cl^- and NO_2^- ions) and four are *neutral* (the NH_3 molecules). The total contribution of the anions to the net charge on the complex ion is $2-$. Because the net charge on the complex ion is $1+$, the oxidation state of the central cobalt ion is $+3$. Diagrammatically, we can write

$$
\begin{array}{c}
\text{oxidation state} = x \qquad \text{charge of } 1- \text{ on } Cl^- \\
\text{charge of } 1- \text{ on } NO_2^- \\
[CoCl(NO_2)(NH_3)_4]^+ \\
\text{coordination number} = 6
\end{array}
$$

total negative
charge: $2-$

net charge on complex ion

$x - 2 = +1$
$x = +3$

Practice Example A: What are the coordination number and oxidation state of nickel in the ion $[Ni(CN)_4I]^{3-}$?

Practice Example B: Write the formula of a complex with cyanide ion ligands, an iron ion with an oxidation state of $+3$, and a coordination number of 6.

25-2   Ligands

A common feature shared by the ligands in complexes is the ability to donate electron pairs to central metal atoms or ions. Ligands are *Lewis bases*. In accepting electron pairs, central metal atoms or ions act as *Lewis acids*. A ligand that uses one pair of electrons to form one point of attachment to the central metal atom or ion is called a *monodentate* ligand. Some examples of monodentate ligands on page 868 are monatomic anions like the halide ions, polyatomic anions like nitrite ion, simple molecules such as ammonia (called ammine when it is a ligand), and more complex molecules like methylamine, CH_3NH_2.

TABLE 25.2 Some Common Monodentate Ligands

Formula	Name as Ligand	Formula	Name as Ligand	Formula	Name as Ligand
Neutral molecules		**Anions**		**Anions**	
H_2O	aqua	F^-	fluoro	SO_4^{2-}	sulfato
NH_3	ammine	Cl^-	chloro	$S_2O_3^{2-}$	thiosulfato
CO	carbonyl	Br^-	bromo	NO_2^-	nitro[a]
NO	nitrosyl	I^-	iodo	ONO^-	nitrito[a]
CH_3NH_2	methylamine	O^{2-}	oxo	SCN^-	thiocyanato[b]
C_5H_5N	pyridine	OH^-	hydroxo	NCS^-	isothiocyanato[b]
		CN^-	cyano		

[a]If the nitrite ion is attached through the N atom ($-NO_2$), the name *nitro* is used; if attached through an O atom ($-ONO$), *nitrito*.

[b]If the thiocyanate ion is attached through the S atom ($-SCN$), the name *thiocyanato* is used; if attachment is through the N atom ($-NCS$), *isothiocyanato*.

ligand name: chloro   nitro   ammine   methylamine

Table 25.2 lists some common monodentate ligands.

Some ligands are capable of donating more than a single electron pair from *different* atoms in the ligand and to *different* sites in the geometric structure of a complex. These are called **polydentate** ligands. The molecule *ethylenediamine (en)* can donate *two* electron pairs, one from each N atom. Since en attaches to the metal center at two points, it is called a **bidentate** ligand.

Three common polydentate ligands are shown in Table 25.3.

TABLE 25.3 Some Common Polydentate Ligands (Chelating Agents)

Abbreviation	Name	Formula
en	ethylenediamine	
ox	oxalato	
EDTA	ethylenediaminetetraacetato	

FIGURE 25-2
Two representations of the chelate $[Pt(en)_2]^{2+}$.
The ligands attach at adjacent corners along an edge of the square. They do *not* bridge the square by attaching to opposite corners. Bonds are shown in red, and the square planar shape is indicated by the black parallelogram.

Figure 25-2 represents the attachment of two ethylenediamine (en) ligands to a Pt^{2+} ion. Here is how we can establish that each ligand is attached to two positions in the coordination sphere around the Pt^{2+} ion.

- Because Pt^{2+} ion exhibits a coordination number of 4 with monodentate ligands, and because $[Pt(en)_2]^{2+}$ is unable to attach additional ligands such as NH_3, H_2O, or Cl^-, we conclude that each en group must be attached at two points.

- The en ligands in the complex ion exhibit no further basic properties. They cannot accept protons from water to produce OH^-, as they would if they had an available lone pair of electrons. Both $—NH_2$ groups of each en molecule must be tied up in the complex ion.

Note the two five-member rings (pentagons) outlined in Figure 25-2. They consist of Pt, N, and C atoms. When the bonding of a polydentate ligand to a metal ion produces a ring (usually five- or six-member), we refer to the complex as a **chelate** (pronounced KEY-late). The polydentate ligand is called a **chelating agent,** and the process of chelate formation is called *chelation*.

▶ A Cl^- ion has *four* lone pairs of electrons, but it can only function as a *mono*dentate ligand. The lone pairs of electrons of a polydentate ligand must be far enough apart to attach to the metal center at two or more different points; the donated pairs of electrons must be on different atoms.

Are You Wondering . . .

How terms such as ligand and chelate originated?

Ligand comes from the Latin word *ligare,* which means to bind. Dentate is also derived from a Latin word, *dens,* meaning tooth. Figuratively speaking, a monodentate ligand has one tooth; a bidentate ligand has two teeth; and so on. A ligand attaches itself to the metal center in accordance with the number of "teeth" it possesses. Chelate is derived from the Greek word *chela,* which means a crab's claw. The way in which a chelating agent attaches itself to a metal ion resembles a crab's claw.

25-3   Nomenclature

The system of naming complexes originated with Werner, but it has been modified several times over the years. Even today, usage varies somewhat. Our approach will be to consider a few rules that permit us to relate names and formulas of simple complexes. We will not consider any of the complicated cases for which writing names and formulas is more challenging.

1. *In names and formulas of coordination compounds, cations come first, followed by anions.* This is the same order as used in simple ionic compounds such as sodium chloride, NaCl.

2. *Anions as ligands are named by using the ending "o."* Normally, "ide" endings change to "o," "ite" to "ito," and "ate" to "ato" (see Table 25.2).

3. *Neutral molecules as ligands generally carry the unmodified name.* For example, the name ethylenediamine is used both for the free molecule and for the molecule as a ligand. Aqua, ammine, carbonyl, and nitrosyl are important exceptions (see Table 25.2).

4. *The number of ligands of a given type is denoted by a prefix.* The usual prefixes are mono = 1, di = 2, tri = 3, tetra = 4, penta = 5, and hexa = 6, although as in many other cases the prefix mono is often not used. If the ligand name is a composite name that itself contains a prefix, such as ethylene*di*amine, place parentheses around the name and precede it with *bis* = 2, *tris* = 3, *tetrakis* = 4, and so on. Thus, *dichloro* signifies *two* Cl^- ions as ligands, and *pentaaqua* signifies *five* H_2O molecules. To indicate the presence of *two* ethylenediamine (en) ligands, we write *bis*(ethylenediamine).

5. *In naming a complex, ligands are named first, in alphabetical order, followed by the name of the metal center. The oxidation state of the metal center is denoted by a Roman numeral. If the complex is an anion, the ending "ate" is attached to the name of the metal.* Prefixes (di, tri, bis, tris, . . .) are ignored in establishing the alphabetical order. Thus, $[CrCl_2(H_2O)_4]^+$ is called tetraaquadichlorochromium(III) ion; $[CoCl_2(en)_2]^+$ is dichlorobis-(ethylenediamine)cobalt(III) ion; and $[Cr(OH)_4]^-$ is tetrahydroxochromate(III) ion. For complex anions of a few of the metals, the English name is replaced by the Latin name given in Table 25.4. Thus, $[CuCl_4]^{2-}$ is the tetrachlorocuprate(II) ion.

6. *In writing the formula of a complex, the chemical symbol of the metal center is written first, followed by the formulas of anions and then neutral molecules.* If there are two or more different anions or neutral molecules as ligands, they are written in alphabetical order according to the first chemical symbols of their formulas. Thus, in the formula of the tetraaminechloronitrocobalt(III) ion, Cl^- precedes NO_2^-, and both are placed ahead of the neutral NH_3 molecules: $[CoCl(NO_2)(NH_3)_4]^+$.

▶ Occasionally, the metal center will be in the oxidation state zero, as in $[W(CO)_6]$—hexacarbonyltungsten(0).

TABLE 25.4 Names for Some Metals in Complex Anions

iron ⟶	**ferrate**
copper ⟶	**cuprate**
tin ⟶	**stannate**
silver ⟶	**argentate**
lead ⟶	**plumbate**
gold ⟶	**aurate**

EXAMPLE 25-2

Relating Names and Formulas of Complexes. **(a)** What is the name of the complex $[CoCl_3(NH_3)_3]$? **(b)** What is the formula of the compound pentaaquachlorochromium(III) chloride? **(c)** What is the name of the compound $K_3[Fe(CN)_6]$?

Solution

(a) $[CoCl_3(NH_3)_3]$ consists of *three* ammonia molecules and *three* chloride ions attached to a central Co^{3+} ion; it is electrically neutral. The name of this neutral complex is triamminetrichlorocobalt(III).

(b) The central metal ion is Cr^{3+}. There are five H_2O molecules and one Cl^- ion as ligands. The complex ion carries a net charge of 2+. Two Cl^- ions are required to neutralize the charge on this complex cation. The formula of the coordination compound is $[CrCl(H_2O)_5]Cl_2$.

(c) This compound consists of K^+ cations and complex anions having the formula $[Fe(CN)_6]^{3-}$. Each cyanide ion carries a charge of 1−, so the oxidation state of the iron must be +3. The Latin-based name "ferrate" is used because the complex ion is an anion. The name of the anion is hexacyanoferrate(III) ion. The coordination compound is potassium hexacyanoferrate(III).

> **Practice Example A:** What is the formula of the compound potassium hexachloroplatinate(IV)?
>
> **Practice Example B:** What is the name of the compound $[Co(SCN)(NH_3)_5]Cl_2$?

Although most complexes are named in the manner just outlined, some common or "trivial" names are still in use. Two such trivial names are ferrocyanide for $[Fe(CN)_6]^{4-}$ and ferricyanide for $[Fe(CN)_6]^{3-}$. These common names suggest the oxidation state of the central metal ions through the "o" and "i" designations ("o" for the ferr<u>ous</u> ion, Fe^{2+}, in $[Fe(CN)_6]^{4-}$ and "i" for the fer<u>ric</u> ion, Fe^{3+}, in $[Fe(CN)_6]^{3-}$). These trivial names do not indicate that the metal ions have a coordination number of 6, however. The systematic names—hexacyanoferrate(II) and hexacyanoferrate(III)—are more informative.

25-4   Isomerism

Isomers are substances that have the same formulas but differ in their structures and in their properties. For example, one isomer of C_2H_6O is ethanol, a liquid at room temperature. Common alcoholic beverages are dilute aqueous solutions of ethanol (in the presence of various other components). Another isomer of C_2H_6O is dimethyl ether, a gaseous compound at room temperature. Dimethyl ether is quite toxic and not usually encountered outside the laboratory.

▶ The "ether" that was once commonly used as an anesthetic in medicine is di*ethyl* ether ($C_4H_{10}O$), a less toxic compound than dimethyl ether.

Several kinds of isomerism are found among complex ions and coordination compounds. These can be lumped into two broad categories: **Structural isomers** differ in basic structure or bond type—what ligands are bonded to the metal center and through which atoms. **Stereoisomers** have the same number and types of ligands and the same mode of attachment, but they differ in the way in which the ligands occupy the space around the metal center. Of the following five examples, the first three are types of structural isomerism, and the remaining two are types of stereoisomerism.

Ionization Isomerism

The two coordination compounds whose formulas are shown here have the same central ion (Cr^{3+}), and five of the six ligands (NH_3 molecules) are the same. The way the compounds differ is that one has SO_4^{2-} ion as the sixth ligand, with a Cl^- ion to neutralize the charge of the complex ion; the other has Cl^- as the sixth ligand and SO_4^{2-} to neutralize the charge of the complex ion.

$$[CrSO_4(NH_3)_5]Cl$$
pentaamminesulfatochromium(III) chloride
(a)

$$[CrCl(NH_3)_5]SO_4$$
pentaamminechlorochromium(III) sulfate
(b)

Coordination Isomerism

A somewhat similar situation to that just described can arise when a coordination compound is composed of both complex cations and complex anions. The ligands can be distributed differently between the two complex ions, as are NH_3 and CN^- in these two compounds.

$$[Co(NH_3)_6][Cr(CN)_6]$$
hexaamminecobalt(III) hexacyanochromate(III)
(a)

$$[Cr(NH_3)_6][Co(CN)_6]$$
hexaaminechromium(III) hexacyanocobaltate(III)
(b)

Linkage Isomerism

Some ligands may attach to the central metal ion of a complex ion in different ways. For example, the nitrite ion, a monodentate ligand, has electron pairs available for coordination both on the N and O atoms.

$$\left[\ddot{N} \atop :\ddot{O}: \quad \ddot{O}: \right]^{-}$$

Whether attachment of this ligand is through the N or an O atom, the formula of the complex ion is unaffected. The properties of the complex ion, however, may be affected. When attachment occurs through the N atom, the ligand is referred to as "nitro." Coordination through an O atom produces a "nitrito" complex.

$[CoCl(NO_2)(NH_3)_4]^+$	$[CoCl(ONO)(NH_3)_4]^+$
tetraamminechloronitrocobalt(III) ion	tetraamminechloronitritocobalt(III) ion
(a)	(b)

Geometric Isomerism

If we substitute a single Cl^- ion for an NH_3 molecule in the square-planar complex ion $[Pt(NH_3)_4]^{2+}$ in Figure 23-1, it does not matter at which corner of the square we make this substitution. As shown in Figure 25-3a, all four possibilities are alike. If we substitute a *second* Cl^-, we now have two distinct possibilities (see Figure 25-3b). The two Cl^- ions can either be along the same edge of the square (**cis**) or on opposite corners, across from each other (**trans**). To distinguish clearly between these two possibilities, we must either draw a structure or refer to the appropriate name. The formula alone will not distinguish between them. (Note that this complex is a neutral species, not an ion.)

▶ To help keep the terms *cis* and *trans* straight in your mind, remember that a *trans*continental flight is *across* the continent, and a *trans*Atlantic flight is *across* the Atlantic Ocean.

$$[PtCl_2(NH_3)_2]$$
cis-diamminedichloroplatinum(II)
or
trans-diamminedichloroplatinum(II)

FIGURE 25-3
Geometric isomerism illustrated.
For the square-planar complexes shown here, isomerism exists only when two Cl^- ions have replaced NH_3 molecules.

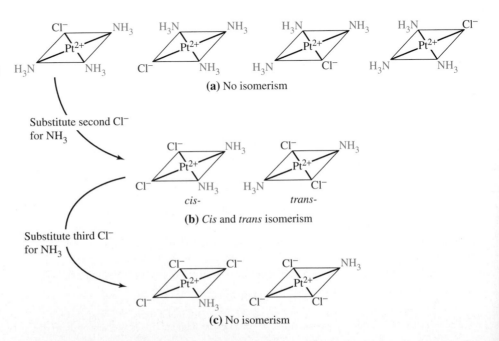

Interestingly, when we substitute a *third* Cl⁻ion, isomerism disappears (Figure 25-3c). There is only one complex ion with the formula $[PtCl_3(NH_3)]^-$.

With an octahedral complex the situation is a bit more complicated. Take the complex ion $[Co(NH_3)_6]^{3+}$ of Figure 25-1 as an example. If we substitute one Cl⁻ for an NH₃, we get a single structure. With *two* Cl⁻ ions substituted for NH₃ molecules we obtain cis and trans isomers. The cis isomer has two Cl⁻ ions along the same edge of the octahedron. The trans isomer has two Cl⁻ ions on opposite corners, that is, at opposite ends of a line drawn through the central metal ion. These two isomers are shown in Figure 25-4. One difference between the two is that the cis isomer has a blue-violet color and the trans has a bright green color.

Let us refer to Figure 25-4a to see what happens when we substitute a *third* Cl⁻ for an NH₃. If we make this third substitution at either position 1 or 6, the result is that three Cl⁻ ions appear on the same face of the octahedron. We call this a *fac* (facial) isomer. If we make the third substitution either at position 4 or 5, the result is three Cl⁻ ions around a perimeter or meridian of the octahedron. We call this a *mer* (meridional) isomer. If we substitute a *fourth* Cl⁻, we again get two isomers, but in this case the isomerism is based on whether the remaining two NH₃ molecules are cis or trans.

(a) *cis*-$[CoCl_2(NH_3)_4]^+$

(b) *trans*-$[CoCl_2(NH_3)_4]^+$

FIGURE 25-4
***Cis* and *trans* isomers of an octahedral complex.**
The Co^{3+} ion is at the center of the octahedron, and one NH₃ ligand is on the far corner (corner 5), out of view.

EXAMPLE 25-3

Identifying Geometric Isomers. Sketch structures of all the possible isomers of $[CoCl(ox)(NH_3)_3]$.

Solution

The Co^{3+} ion exhibits a coordination number of 6. The structure is octahedral. Recall that ox (oxalate ion) is a bidentate ligand carrying a double-negative charge (see Table 25.3). Recall also that such a ligand must be attached in cis positions, not trans (see Figure 25-3). Once the ox ligand is placed, we see that there are two possibilities. The three NH₃ molecules can be situated (1) on the same face of the octahedron (fac isomer) or (2) around a perimeter of the octahedron (mer isomer).

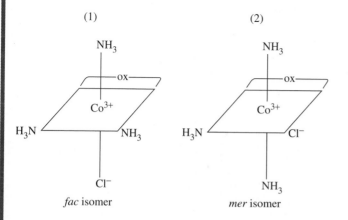

Practice Example A: Sketch the geometric isomers of $[CoCl_2(ox)(NH_3)_2]^-$.

Practice Example B: Sketch the geometric isomers of $[MoCl_2(C_5H_5N)_2(CO)_2]^+$.

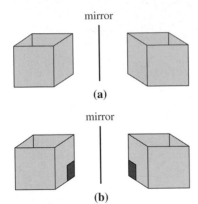

FIGURE 25-5
Superimposable and nonsuperimposable objects—a cubical, open-top box.
(a) You can place the box into its mirror image (hypothetically) in several different ways. (b) In any way that you place the box into its mirror image, the stickers will not appear in the same position. The box and its mirror image are nonsuperimposable.

▶ Here is a process based on chirality: choosing a matched pair of gloves from a bin of gloves of identical sizes, 90 made for the right hand and 10 for the left hand.

▶ The prefixes dextro and levo are derived from the Latin words *dexter* = right and *levo* = left.

Optical Isomerism

To understand optical isomerism we need to understand the relationship between an object and its mirror image. Features on the right side of the object appear on the left side of its image in a mirror, and vice versa. Certain objects can be rotated in such a way as to be *superimposable* on their mirror images, but other objects are *nonsuperimposable* on their mirror images. An unmarked tennis ball is superimposable on its mirror image, but a left hand is nonsuperimposable on its mirror image (a right hand).

As a further example, consider the open-top cubical cardboard box pictured in Figure 25-5(a). There are a number of ways (hypothetical, of course) in which the box can be superimposed on its mirror image. Now, imagine that a distinctive sticker is placed at a corner of one side of the box [Figure 25.5(b)]. In this case there is no way that the box and its mirror image can be superimposed; they are clearly different. This is equivalent to saying that there is no way that a form-fitting left glove can be worn on a right hand (turning it inside out is not allowed).

The two structures of $[Co(en)_3]^{3+}$ depicted in Figure 25-6 are related to each other as are an object and its image in a mirror. Furthermore, the two structures are *nonsuperimposable*, like a left and a right hand. The two structures represent two different complex ions; they are isomers.

Structures that are nonsuperimposable mirror images of each other are called **enantiomers** and are said to be **chiral** (pronounced KYE-rull). (Structures that are superimposable are *achiral.*) Whereas other types of isomers often differ significantly in their physical and chemical properties, enantiomers have identical properties, except in a few specialized situations. These exceptions involve phenomena that are directly linked to chirality, or "handedness," at the molecular level; an example is *optical activity,* pictured in Figure 25-7.

Interactions between a beam of polarized light and the electrons in an enantiomer cause the plane of the polarized light to rotate. One enantiomer rotates the plane of polarized light to the right (clockwise) and is said to be *dextrorotatory* (designated + or *d*). The other enantiomer rotates the plane of polarized light to the same extent, but to the left (counterclockwise). It is said to be *levorotatory* (− or *l*). Because of their ability to rotate the plane of polarized light, the enantiomers are said to be *optically active* and are referred to as **optical isomers.**

When an optically active complex is synthesized, a mixture of the two optical isomers (enantiomers) is obtained, such as the two $[Co(en)_3]^{3+}$ isomers in Figure 25-6. The optical rotation of one isomer just cancels that of the other. The mixture, called a *racemic mixture,* produces no net rotation of the plane of polarized light. Separating the *d* and *l* isomers of a racemic mixture is called

FIGURE 25-6
Optical isomers.
The two structures are nonsuperimposable mirror images. Like a right hand and a left hand, one cannot be superimposed onto the other.

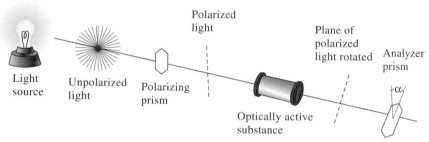

FIGURE 25-7
Optical activity.
Light from an ordinary source consists of electromagnetic waves vibrating in all planes; it is unpolarized. Some substances are able to screen out all waves except those vibrating in a particular plane. Light having all vibrations in the same plane is *polarized*. Optically active substances are able to rotate the plane of polarized light. One optical isomer rotates the plane to the right (here by the angle α); the other optical isomer rotates it by the same angle, but to the left.

resolution. This separation can sometimes be achieved through chemical reactions controlled by chirality, with the two enantiomers behaving differently. Many phenomena of the living state, such as the effectiveness of a drug, the activity of an enzyme, and the ability of a microorganism to promote a reaction, involve chirality. We will return to a discussion of chirality in Chapter 28.

25-5   Bonding in Complex Ions: Crystal Field Theory

The theories of chemical bonding that we found so useful in earlier chapters do not help us much in explaining the characteristic colors and magnetic properties of complex ions. In transition metal ions, we need to focus our attention on how the electrons in the d orbitals of a metal ion are affected when they are in a complex. A theory that provides that focus and an explanation of these properties is crystal field theory.

In the **crystal field theory,** we consider bonding in a complex ion to be an electrostatic attraction between the positively charged nucleus of the central metal ion and electrons in the ligands. Repulsions occur between the ligand electrons and electrons in the central ion. In particular, the crystal field theory focuses on the repulsions between ligand electrons and d electrons of the central ion.

First, a reminder about the d orbitals first presented in Figure 9-23: All five of the orbitals are alike in energy when in an isolated atom or ion, but they are *unlike* in their spatial orientations. One of them, d_{z^2}, is directed along the z axis, and another, $d_{x^2-y^2}$, has lobes along the x and y axes. The remaining three have lobes extending into regions between the perpendicular x, y, and z axes. In the presence of ligands, because of repulsions between ligand electrons and d electrons, the d-orbital energy levels of the central metal ion are raised. As we will soon see, however, they are not all raised to the same extent.

Figure 25-8 pictures six anions (ligands) approaching a central metal ion along the x, y, and z axes. This direction of approach leads to an octahedral complex. Repulsions between ligand electrons and d-orbital electrons are strengthened in the direct, "head-to-head" approach of ligands to the $d_{x^2-y^2}$ and d_{z^2} orbitals. These two orbitals have their energy raised with respect to an average d orbital energy for a central metal ion in the field of the ligands. For the other

▶ Modifications of the simple crystal field theory that take into account such factors as the partial covalency of the metal–ligand bond are called ligand field theory. Often the single term *ligand field theory* is used to signify both the purely electrostatic crystal field theory and its modifications.

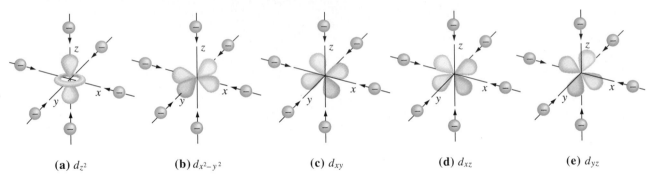

(a) d_{z^2} **(b)** $d_{x^2-y^2}$ **(c)** d_{xy} **(d)** d_{xz} **(e)** d_{yz}

FIGURE 25-8
Approach of six anions to a metal ion to form a complex ion with octahedral structure.
The ligands (anions in this case) approach the central metal ion along the *x*, *y*, and *z* axes. Maximum repulsion occurs with the d_{z^2} and $d_{x^2-y^2}$ orbitals, and their energies are raised. Repulsions with the other *d* orbitals are not as great. A difference in energy results between the two sets of *d* orbitals.

three orbitals—d_{xy}, d_{xz}, and d_{yz}—ligands approach between the lobes of the orbitals and there is a gain in stability over the "head-to-head" approach; these orbital energies are lowered with respect to the average *d* orbital energy. The difference in energy between the two groups of *d* orbitals is represented by the symbol Δ, as shown in Figure 25-9.

In the formation of complex ions of other geometric structures, ligands approach from different directions and produce different patterns of splitting of the *d* energy level. Figure 25-10 shows the pattern for tetrahedral complexes, and Figure 25-11 shows the pattern for square-planar complexes. Suppose we compare hypothetical complexes of different structures but with the same combinations of ligands, metal ions, and metal–ligand distances. We find the greatest energy separation of the *d* levels for the square-planar complex and the smallest energy separation for the tetrahedral complex.

▶ If we call the energy separation for the octahedral complex Δ_o, the comparative value for the tetrahedral complex is 0.44 Δ_o and for the square-planar complex, 1.74 Δ_o.

Ligands differ in their abilities to produce a splitting of the *d* energy levels. Strong Lewis bases like CN^- and NH_3 produce a "strong" field. They repel *d* electrons very strongly and cause a greater separation of the *d* energy levels than do weaker bases like H_2O and F^-. What we normally expect for a distribution of *d* electrons in a metal ion is that to the extent possible, *d* orbitals will be singly occupied. This is the case in the complex ion $[CoF_6]^{3-}$. Sometimes, however, the energy separation between the higher and lower energy *d* levels (Δ) is larger than the energy benefit derived by unpairing electrons. In this case the energetically favored arrangement may be for electrons to remain paired in the lower energy *d* orbitals, which is the case for $[Co(NH_3)_6]^{3+}$.

FIGURE 25-9
Splitting of *d* energy levels in the formation of an octahedral complex ion.
The *d*-orbital energy levels of the free central ion are raised in the presence of ligands to the average level shown, but the five levels are split into two groups.

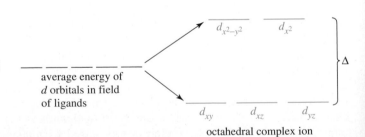

average energy of
d orbitals in field
of ligands

$d_{x^2-y^2}$ d_{z^2}

Δ

d_{xy} d_{xz} d_{yz}

octahedral complex ion

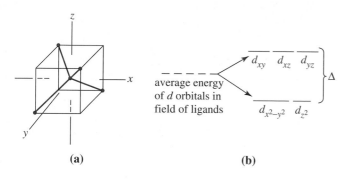

(a)   **(b)**

FIGURE 25-10
Crystal field splitting in a tetrahedral complex ion.
(a) The positions of attachment of ligands to a metal ion leading to the formation of a tetrahedral complex ion. (b) Interference with the d orbitals directed along the x, y, and z axes is not as great as with those that lie between the axes (see Figure 25-8). As a result, the pattern of crystal field splitting is reversed from that of an octahedral complex.

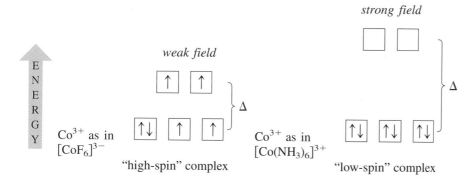

Different ligands can be arranged in order of their abilities to produce a splitting of the d energy levels. This arrangement is known as the **spectrochemical series** and is illustrated in Examples 25-4 and 25-5.

FIGURE 25-11
Comparison of crystal field splitting in a square-planar and an octahedral complex.
(a) The positions of attachment of ligands to a metal ion leading to the formation of a square-planar complex. (b) Splitting of the d energy level in a square-planar complex can be related to that of the octahedral complex. Because there are no ligands along the z axis in a square-planar complex, we should expect the repulsion between ligands and d_{z^2} electrons to be much less than in an octahedral complex. The d_{z^2} energy level is lowered considerably from that in an octahedral complex. Similarly, the energy levels of the d_{xz} and d_{yz} orbitals are lowered slightly because the electrons in these orbitals are concentrated in planes perpendicular to that of the square-planar complex. The energy of the $d_{x^2-y^2}$ orbital is raised, because the x and y axes represent the direction of approach of four ligands to the central ion. The energy of the d_{xy} orbital is also raised because this orbital lies in the plane of the ligands in the square-planar complex.

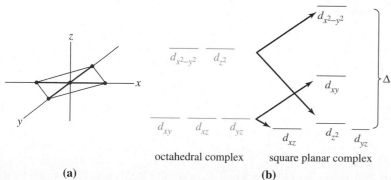

(a)   **(b)**

strong field

(large Δ)

▶ The Spectrochemical Series. $CN^- > NO_2^- > en > py \approx NH_3 > SCN^- > H_2O > OH^- > F^- > Cl^- > Br^- > I^-$

(small Δ)

weak field

(When NO_2^- is attached through an O atom or SCN^- through the S atom, a different placement in the spectrochemical series is found.)

EXAMPLE 25-4

Using the Spectrochemical Series to Predict Magnetic Properties. How many unpaired electrons would you expect to find in the octahedral complex $[Fe(CN)_6]^{3-}$?

Solution

The Fe atom has the electron configuration $[Ar]3d^64s^2$. The Fe^{3+} ion has the configuration $[Ar]3d^5$. CN^- is a strong-field ligand. Because of the large energy separation in the d levels of the metal ion produced by this ligand, we expect all the electrons to be in the lowest energy level. There should be only **one** unpaired electron.

Practice Example A: How many unpaired electrons would you expect to find in the octahedral complex $[MnF_6]^{2-}$?

Practice Example B: How many unpaired electrons would you expect to find in the tetrahedral complex $[CoCl_4]^{2-}$? Would you expect more, fewer, or the same number of unpaired electrons as in the octahedral complex $[Co(H_2O)_6]^{2+}$?

EXAMPLE 25-5

Using the Crystal Field Theory to Predict the Structure of a Complex from Its Magnetic Properties. The complex ion $[Ni(CN)_4]^{2-}$ is found to be diamagnetic. Use ideas from the crystal field theory to speculate on its probable structure.

Solution

We can eliminate an octahedral structure because the coordination number is 4 (not 6). Our choice is between tetrahedral and square-planar.

The electron configuration of Ni is $[Ar]3d^84s^2$ and of Ni(II), $[Ar]3d^8$. Because the complex ion is found to be diamagnetic, all $3d$ electrons must be paired. Let us see how we would distribute these $3d$ electrons if the structure were tetrahedral (recall Figure 25-10). We would place four electrons (all paired) into the two lowest d levels. We would then distribute the remaining four electrons among the three higher level d orbitals. Two of the electrons would be unpaired, and the complex ion would be paramagnetic.

Because the tetrahedral structure would be paramagnetic, we can conclude that the structure of the diamagnetic $[Ni(CN)_4]^{2-}$ ion must be square planar. Let's also demonstrate that this is a reasonable conclusion based on the d-orbital energy level diagram for a square-planar complex (recall Figure 25-11). First we fill the three lowest energy orbitals with electrons (six), and then assume that the energy separation between the d_{xy} and $d_{x^2-y^2}$ orbitals is large enough that the final two electrons remain paired in the d_{xy} orbital. This corresponds to a diamagnetic complex ion.

Practice Example A: The complex ion $[Co(CN)_4]^{2-}$ is found to be paramagnetic with three unpaired electrons. Use ideas from the crystal field theory to speculate on its probable structure.

Practice Example B: Would you expect $[Cu(NH_3)_4]^{2-}$ to be diamagnetic or paramagnetic? Can you use this information about the magnetic properties of $[Cu(NH_3)_4]^{2+}$ to help you determine whether the structure of $[Cu(NH_3)_4]^{2+}$ is tetrahedral or square planar? Explain.

25-6 Color and the Colors of Complexes

To help us understand the nature of color, let's consider the two situations for mixing colors shown in Figure 25-12. Figure 25-12a represents *additive* mixing. It is the type of color mixing that occurs when colored spotlights are superimposed. Figure 25-12b, on the other hand, represents *subtractive* mixing. It is the type of color mixing that occurs when paint pigments or colored solutions are mixed.

For the additive mixing of light beams in Figure 25-12a, we define the **primary colors** as any *three* colors that when combined yield white light (W). Our choice in the figure is *red* (R), *green* (G), and *blue* (B), and we can represent their sum as R + G + B = W. The **secondary colors** are those that are produced by combining *two* primary colors. From Figure 25-12a, we see that the secondary colors are *yellow* (Y = R + G), *cyan* (C = G + B), and *magenta* (M = B + R).

Each secondary color is a **complementary color** of one of the primary colors. Again, from Figure 25-12a, cyan (C) is the complementary color of red (R); magenta (M), of green (G); and yellow (Y), of blue (B). Figure 25-12a demon-

▶Additive color mixing is used in color monitors and color television screens. Subtractive color mixing is used for all the photographs and art in this text, for example.

(a) Additive color mixing

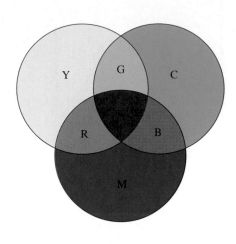

(b) Subtractive color mixing

FIGURE 25-12
The mixing of colors.
(a) The additive mixing of three beams of colored light: red (R), green (G), and blue (B). The secondary colors—yellow (Y), cyan (C), and magenta (M)—are produced in the region where two of the beams overlap. The overlap of all three beams produces white light (W). (b) The subtractive mixing of three pigments with the primary colors magenta (M), yellow (Y), and cyan (C). Here the secondary colors—red (R), green (G), and blue (B)—are formed when two of the primary colors are mixed. A mixture of all three primary colors produces a very dark brown to black color.

In four-color printing, such as in this book and in color printers in computer systems, the basic inks used are magenta (M), yellow (Y), cyan (C), and black. All of the colors you see in photographs and drawings in this book are produced by printing very small dots of these four basic colors, singly and in various combinations. (You can see individual dots of some of these colors with a strong magnifying glass.)

strates that when a primary color and its complementary color are mixed, the result is white light. That this must be the case, we can see as follows: Because cyan, for example, is itself the combination of two of the primary colors—green and blue (C = G + B)—the combination of cyan and red (C + R) is the same as the combination of the three primary colors: G + B + R = W.

In subtractive color mixing, some of the wavelength components of white light are removed by absorption, and the reflected light (for example, from a painted surface or colored fabric) or transmitted light (as seen through glass or a solution) is deficient in some wavelength components. The reflected or transmitted light is colored. In subtractive color mixing (Figure 25-12b) we can again define primary, secondary, and complementary colors. In this case let's choose *magenta* (M), *yellow* (Y), and *cyan* (C) as the primary colors; and *red* (R), *green* (G), and *blue* (B) as the secondary colors. If a material absorbs all three primary colors, there is essentially no light left to be reflected or transmitted and the material appears black, or nearly so. If a material absorbs one color, primary or secondary, the reflected or transmitted light is the complementary color. Thus, a magenta sweater has that color because the dye it contains strongly absorbs green light and reflects magenta, the complement of green. A solution of red food dye has that color because the dye absorbs cyan light and transmits red.

Effect of ligands on the colors of coordination compounds. These compounds all consist of a six-coordinate cobalt complex ion in combination with nitrate ions. In each case the complex ion has five NH_3 molecules and one other group as ligands. From left to right the compounds are $[CoCl(NH_3)_5](NO_3)_2$, $[CoBr(NH_3)_5](NO_3)_2$, $[CoI(NH_3)_5](NO_3)_2$, $[Co(NO_2)(NH_3)_5](NO_3)_2$, $[Co(SO_4)(NH_3)_5]NO_3$, and $[Co(CO_3)(NH_3)_5]NO_3$.

Now let's direct our attention specifically to colored solutions. Colored solutions contain species that can absorb photons of visible light and use the energy of these photons to promote electrons in the species to higher energy levels. The energies of the photons must just match the energy differences through which the electrons are to be promoted. Because the energies of photons are related to the frequencies (and wavelengths) of light (recall Planck's equation: $E = h\nu$), only certain wavelength components are absorbed as white light passes through the solution. The emerging light, because it is lacking some wavelength components, is no longer white; it is colored.

Ions having (1) a noble gas electron configuration, (2) an outer shell of 18 electrons, or (3) the "18 + 2" configuration (18 electrons in the $n - 1$ shell and two in the n, or outermost, shell) do not have electron transitions in the energy range corresponding to visible light. White light passes through these solutions without being absorbed; these ions are colorless in solution. Examples are the alkali and alkaline earth metal ions, the halide ions, Zn^{2+}, Al^{3+}, and Bi^{3+}.

Crystal field splitting of the d energy levels produces the energy difference, Δ, that accounts for the colors of complex ions. Promotion of an electron from a lower to a higher d level results from the absorption of the appropriate components of white light; the transmitted light is colored. The result of subtracting one color from white light is to obtain the complementary color.

A solution containing $[Cu(H_2O)_4]^{2+}$ absorbs most strongly in the yellow region of the spectrum (about 580 nm). The wavelength components of the light transmitted combine to produce the color *blue.* Thus, aqueous solutions of copper(II) compounds usually have a characteristic blue color. In the presence of high concentrations of Cl^-, copper(II) forms the complex ion $[CuCl_4]^{2-}$. This species absorbs strongly in the blue region of the spectrum. The transmitted light, and hence the color of the solution, is *yellow.* Light absorption by these solutions is suggested by Figure 25-13. The colors of some complex ions of chromium are given in Table 25.5.

FIGURE 25-13
Light absorption and transmission.
$[Cu(H_2O)_4]^{2+}$ absorbs in the yellow region of the spectrum and transmits blue light. $[CuCl_4]^{2-}$ absorbs in the blue region of the spectrum and transmits yellow light.

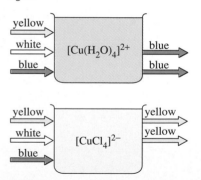

TABLE 25.5 Some Coordination Compounds of Cr^{3+} and Their Colors	
Isomer	**Color**
$[Cr(H_2O)_6]Cl_3$	violet
$[CrCl(H_2O)_5]Cl_2$	blue-green
$[Cr(NH_3)_6]Cl_3$	yellow
$[CrCl(NH_3)_5]Cl_2$	purple

Color differences in structural isomers. The samples all have the formula $CrCl_3 \cdot 6H_2O$. The violet sample contains the $[Cr(H_2O)_6]^{3+}$ ion; the blue-green sample contains the $[CrCl(H_2O)_5]^{2+}$ ion; and the green sample contains the $[CrCl_2(H_2O)_4]^+$ ion.

EXAMPLE 25-6

Relating the Colors of Complexes to the Spectrochemical Series. Table 25.5 lists the color of $[Cr(H_2O)_6]Cl_3$ as violet, whereas that of $[Cr(NH_3)_6]Cl_3$ is yellow. Explain this difference in color.

Solution

Here are the basic facts that we must use: Both chromium(III) complexes in Table 25.5 are octahedral, and the electron configuration of Cr^{3+} is $[Ar]3d^3$. From these facts we can construct the energy level diagram shown here. The three unpaired electrons go into the three lower energy d orbitals. When a photon of light is absorbed, an electron is promoted from the lower to an upper energy level. The quantity of energy required for this promotion depends on the energy level separation, Δ.

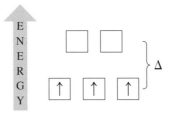

According to the spectrochemical series, NH_3 produces a greater splitting of the d energy level than does H_2O. We should expect $[Cr(NH_3)_6]^{3+}$ to absorb light of a *shorter* wavelength (higher energy) than does $[Cr(H_2O)_6]^{3+}$. If $[Cr(NH_3)_6]^{3+}$ absorbs in the violet region of the spectrum then the *transmitted* light is yellow. If $[Cr(H_2O)_6]^{3+}$ absorbs in the yellow region of the spectrum, then the *transmitted* light is violet.

Practice Example A: The color of $[Co(H_2O)_6]^{2+}$ is pink, whereas that of tetrahedral $[CoCl_4]^{2-}$ is blue. Explain this difference in color.

Practice Example B: One of the following solids has a yellow color and the other, green: $Fe(NO_3)_2 \cdot 6H_2O$; $K_4[Fe(CN)_6] \cdot 3H_2O$. Indicate which is the yellow solid and which, the green. Explain your reasoning.

25-7 Aspects of Complex Ion Equilibria

In Chapter 19 we learned that complex ion formation can have a great effect on the solubilities of substances, as in the ability of $NH_3(aq)$ to dissolve rather large quantities of $AgCl(s)$. To calculate the solubility of $AgCl(s)$ in $NH_3(aq)$, we had to use the formation constant, K_f, of $[Ag(NH_3)_2]^+$. Equations 25.1 and 25.2 illustrate how we dealt with formation constants in Chapter 19, with $[Zn(NH_3)_4]^{2+}$ as an example.

$$Zn^{2+}(aq) + 4 NH_3(aq) \rightleftharpoons [Zn(NH_3)_4]^{2+}(aq) \tag{25.1}$$

$$K_f = \frac{[[Zn(NH_3)_4]^{2+}]}{[Zn^{2+}][NH_3]^4} = 4.1 \times 10^8 \tag{25.2}$$

In fact, though, cations in aqueous solution exist mostly in *hydrated* form. That is, $Zn^{2+}(aq)$ is actually $[Zn(H_2O)_4]^{2+}$. As a result, when NH_3 molecules bond to Zn^{2+} to form an ammine complex ion, they do not enter an empty coordination sphere. They must displace H_2O molecules, and this occurs in a stepwise fashion. The reaction

$$[Zn(H_2O)_4]^{2+} + NH_3 \rightleftharpoons [Zn(H_2O)_3NH_3]^{2+} + H_2O \qquad (25.3)$$

for which

$$K_1 = \frac{[[Zn(H_2O)_3NH_3]^{2+}]}{[[Zn(H_2O)_4]^{2+}][NH_3]} = 3.9 \times 10^2 \qquad (25.4)$$

is followed by

$$[Zn(H_2O)_3NH_3]^{2+} + NH_3 \rightleftharpoons [Zn(H_2O)_2(NH_3)_2]^{2+} + H_2O \qquad (25.5)$$

for which

$$K_2 = \frac{[[Zn(H_2O)_2(NH_3)_2]^{2+}]}{[[Zn(H_2O)_3NH_3]^{2+}][NH_3]} = 2.1 \times 10^2 \qquad (25.6)$$

and so on.

The value of K_1 in Equation (25.4) is often designated as β_1 and called the formation constant for the complex ion $[Zn(H_2O)_3NH_3]^{2+}$. The formation of $[Zn(H_2O)_2(NH_3)_2]^{2+}$ is represented by the *sum* of Equations (25.3) and (25.5),

$$[Zn(H_2O)_4]^{2+} + 2\,NH_3 \rightleftharpoons [Zn(H_2O)_2(NH_3)_2]^{2+} + 2\,H_2O \qquad (25.7)$$

and the formation constant β_2, in turn, is given by the *product* of Equations (25.4) and (25.6).

$$\beta_2 = \frac{[[Zn(H_2O)_2(NH_3)_2]^{2+}]}{[[Zn(H_2O)_4]^{2+}][NH_3]^2} = K_1 \times K_2 = 8.2 \times 10^4 \qquad (25.8)$$

For the next ion in the series, $[Zn(H_2O)(NH_3)_3]^{2+}$, $\beta_3 = K_1 \times K_2 \times K_3$. For the final member, $[Zn(NH_3)_4]^{2+}$, $\beta_4 = K_1 \times K_2 \times K_3 \times K_4$, and it is this product of terms that we called the formation constant in Section 19-9 and listed as K_f in Table 19.2. Additional stepwise formation constants are presented in Table 25.6.

The large numerical value of K_1 for reaction (25.3) indicates that Zn^{2+} has a greater affinity for NH_3 (a stronger Lewis base) than it does for H_2O. Displacement of ligand H_2O molecules by NH_3 occurs even if the number of NH_3 molecules present in aqueous solution is much smaller than the number of H_2O molecules, as in dilute $NH_3(aq)$. The fact that the successive K values decrease regularly in the displacement process, at least for displacements involving neutral molecules as ligands, is due in part to statistical factors: An NH_3 molecule has a better chance of replacing an H_2O molecule in $[Zn(H_2O)_4]^{2+}$, in which

---

TABLE 25.6 Stepwise and Overall Formation (Stability) Constants for Several Complex Ions

Metal Ion	Ligand	K_1	K_2	K_3	K_4	K_5	K_6	β_n (or K_f)[a]
Ag^+	NH_3	2.0×10^3	7.9×10^3					1.6×10^7
Zn^{2+}	NH_3	3.9×10^2	2.1×10^2	1.0×10^2	5.0×10^1			4.1×10^8
Cu^{2+}	NH_3	1.9×10^4	3.9×10^3	1.0×10^3	1.5×10^2			1.1×10^{13}
Ni^{2+}	NH_3	6.3×10^2	1.7×10^2	5.4×10^1	1.5×10^1	5.6	1.1	5.3×10^8
Cu^{2+}	en	5.2×10^{10}	2.0×10^9					1.0×10^{20}
Ni^{2+}	en	3.3×10^7	1.9×10^6	1.8×10^4				1.1×10^{18}
Ni^{2+}	EDTA	4.2×10^{18}						4.2×10^{18}

In many tabulations in the chemical literature, formation constant data are presented as logarithms, i.e., $\log K_1$, $\log K_2$, . . . , and $\log \beta_n$.
[a]The β_n listed is for the number of steps shown, i.e., for $[Ag(NH_3)_2]^+$, $\beta_2 = K_f = K_1 \times K_2$; for $[Ni(en)_3]^{2+}$, $\beta_3 = K_f = K_1 \times K_2 \times K_3$; and for $[Ni(EDTA)]^{2-}$, $\beta_1 = K_f = K_1$.

each coordination position is occupied by H_2O, than in $[Zn(H_2O)_3NH_3]^{2+}$, in which one of the positions is already occupied by NH_3. Also, once the degree of substitution of NH_3 for H_2O has become large, the chances of H_2O molecules replacing NH_3 molecules in a reverse reaction improve; again this tends to reduce the value of K. If irregularities arise in the succession of K values, it is often because of a change in structure of the complex ion at some point in the series of displacement reactions.

If the ligand in a substitution process is *polydentate,* it displaces as many H_2O molecules as there are points of attachment. Thus, ethylenediamine (en) displaces H_2O molecules in $[Ni(H_2O)_6]^{2+}$ two at a time, in three steps. The first step is

$$[Ni(H_2O)_6]^{2+} + en \rightleftharpoons [Ni(en)(H_2O)_4]^{2+} + 2\,H_2O \qquad K_1(\beta_1) = 3.3 \times 10^7$$
$$(25.9)$$

You will note from Table 25.6 that the complex ions with polydentate ligands have much larger formation constants than do those with monodentate ligands. For example, K_f (that is, β_3) for $[Ni(en)_3]^{2+}$ is 1.1×10^{18}, whereas K_f (that is, β_6) for $[Ni(NH_3)_6]^{2+}$ is 5.3×10^8. The additional stability of chelates over complexes with monodentate ligands is known as the **chelation effect.** We can attribute this effect partly to the increase in entropy associated with chelation. In the displacement of H_2O by NH_3, the entropy change is small [two particles on each side of an equation such as (25.3)]. An ethylenediamine molecule, on the other hand, displaces *two* H_2O molecules [two particles on the left and three on the right of Equation (25.9)]. The larger, positive value of $\Delta S°$ for the displacement by ethylenediamine means a more negative $\Delta G°$ and a larger K.

25-8 Acid–Base Reactions of Complex Ions

We have described complex ion formation in terms of Lewis acids and bases. Complex ions may also exhibit acid–base properties in the Brønsted–Lowry sense, that is, they may act as proton donors or acceptors. Figure 25-14 represents the ionization of $[Fe(H_2O)_6]^{3+}$ as an acid. A proton from a *ligand* water molecule in hexaaquairon(III) ion is transferred to a *solvent* water molecule. The H_2O ligand is converted to OH^-.

$$[Fe(H_2O)_6]^{3+} + H_2O \rightleftharpoons [FeOH(H_2O)_5]^{2+} + H_3O^+ \qquad K_{a_1} = 9 \times 10^{-4}$$
$$(25.10)$$

FIGURE 25-14
Ionization of $[Fe(H_2O)_6]^{3+}$.

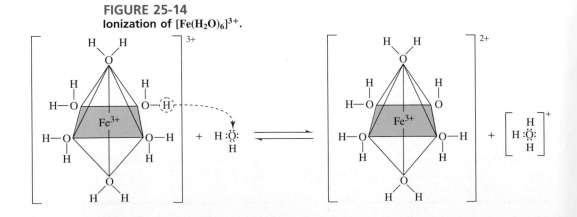

The second ionization step is

$$[FeOH(H_2O)_5]^{2+} + H_2O \rightleftharpoons [Fe(OH)_2(H_2O)_4]^+ + H_3O^+ \qquad K_{a_2} = 5 \times 10^{-4}$$
$$(25.11)$$

From these K_a values we see that $Fe^{3+}(aq)$ is fairly acidic (compared, for example, with acetic acid, with $K_a = 1.8 \times 10^{-5}$). To suppress ionization (hydrolysis) of $[Fe(H_2O)_6]^{3+}$ we need to maintain a low pH by the addition of acids such as HNO_3 or $HClO_4$. The ion $[Fe(H_2O)_6]^{3+}$ is violet in color, but aqueous solutions of $Fe^{3+}(aq)$ are generally yellow owing to the presence of hydroxo complex ions.

The ions Cr^{3+} and Al^{3+} behave in a similar manner to Fe^{3+}, except that with them hydroxo complex ion formation can continue until complex anions are produced. $Cr(OH)_3$ and $Al(OH)_3$, as we have previously noted, are soluble in alkaline as well as acidic solutions; they are amphoteric.

Regarding the acid strengths of aqua complex ions, a critical factor is the charge-to-radius ratio of the central metal ion. Thus, the small, highly charged Fe^{3+} attracts electrons away from an O—H bond in a ligand water molecule more strongly than does Fe^{2+}. Hence $[Fe(H_2O)_6]^{3+}$ is a stronger acid ($K_{a_1} = 9 \times 10^{-4}$) than is $[Fe(H_2O)_6]^{2+}$ ($K_{a_1} = 1 \times 10^{-7}$).

25-9   Some Kinetic Considerations

When we add $NH_3(aq)$ to an aqueous solution containing Cu^{2+}, we see a change in color from pale blue to very deep blue. The reaction involves NH_3 molecules displacing H_2O molecules as ligands.

$$\underset{\text{pale blue}}{[Cu(H_2O)_4]^{2+}} + 4\ NH_3 \longrightarrow \underset{\text{very deep blue}}{[Cu(NH_3)_4]^{2+}} + 4\ H_2O$$

This reaction occurs very rapidly—as rapidly as the two reactants can be brought together. The addition of $HCl(aq)$ to an aqueous solution of Cu^{2+} produces an immediate color change from pale blue to green, or even yellow if the $HCl(aq)$ is sufficiently concentrated.

$$\underset{\text{pale blue}}{[Cu(H_2O)_4]^{2+}} + 4\ Cl^- \longrightarrow \underset{\text{yellow}}{[CuCl_4]^{2-}} + 4\ H_2O$$

Complex ions in which ligands can be interchanged rapidly are said to be **labile**. $[Cu(H_2O)_4]^{2+}$, $[Cu(NH_3)_4]^{2+}$, and $[CuCl_4]^{2-}$ are all labile (Figure 25-15).

In freshly prepared $CrCl_3(aq)$ the ion trans-$[CrCl_2(H_2O)_4]^+$ produces a green color, but the color gradually turns to violet (Figure 25-16). This color change results from the very slow exchange of H_2O for Cl^- ligands. A complex ion that

▶ The terms *labile* and *inert* are not related to the thermodynamic stabilities of complex ions or to the equilibrium constants for ligand-substitution reactions. The terms are *kinetic* terms, referring to the *rates* at which ligands are exchanged.

FIGURE 25-15
Labile complex ions.
The exchange of ligands in the coordination sphere of Cu^{2+} occurs very rapidly. The solution at the extreme left is produced by dissolving $CuSO_4$ in concentrated $HCl(aq)$. Its yellow color is caused by $[CuCl_4]^{2-}$. When a small amount of water is added, the mixture of $[Cu(H_2O)_4]^{2+}$ and $[CuCl_4]^{2-}$ ions produces a yellow-green color. When $CuSO_4$ is dissolved in water, a light blue solution of $[Cu(H_2O)_4]^{2+}$ is formed. NH_3 molecules readily displace H_2O molecules as ligands and produce deep blue $[Cu(NH_3)_4]^{2+}$ (extreme right).

FIGURE 25-16
Inert complex ions.
The solution on the left is obtained by dissolving $CrCl_3 \cdot 6H_2O$ in water. The green color is due to *trans*-$[CrCl_2(H_2O)_4]^+$. A slow exchange of H_2O for Cl^- ligands leads to a violet solution of $[Cr(H_2O)_6]^{3+}$ in one or two days (right).

exchanges ligands slowly is said to be nonlabile, or **inert.** In general, complex ions of the first transition series, except for those of Cr(III) and Co(III), are kinetically labile. Those of the second and third transition series are generally kinetically inert. Whether a complex ion is labile or inert affects the ease with which it can be studied. The inert ones are easiest to obtain and characterize, and this may explain why so many of the early studies of complex ions were based on Cr(III) and Co(III).

25-10 Applications of Coordination Chemistry

The applications of coordination chemistry are numerous and varied. They range from analytical chemistry to biochemistry. The several brief examples in this section give some idea of this diversity.

Hydrates

When a compound is crystallized from an aqueous solution of its ions, the crystals obtained are often hydrated. A hydrate is a substance that has a fixed number of water molecules associated with each formula unit. In some cases the water molecules are ligands bonded directly to a metal ion. The coordination compound $[Co(H_2O)_6](ClO_4)_2$ may be represented as the hexahydrate, $Co(ClO_4)_2 \cdot 6H_2O$. In the hydrate $CuSO_4 \cdot 5H_2O$, four H_2O molecules are associated with copper in the complex ion $[Cu(H_2O)_4]^{2+}$, and the fifth with the SO_4^{2-} anion through hydrogen bonding. Another possibility for hydrate formation is that the water molecules may be incorporated into definite positions in the solid crystal but not associated with any particular cations or anions, as in $BaCl_2 \cdot 2H_2O$. This is called lattice water. Finally, part of the water may be coordinated to an ion and part of it may be lattice water, as appears to be the case with alums such as $KAl(SO_4)_2 \cdot 12H_2O$.

Stabilization of Oxidation States

The standard electrode potential for the reduction of Co(III) to Co(II) is

$$Co^{3+}(aq) + e^- \longrightarrow Co^{2+}(aq) \qquad E^\circ = +1.82 \text{ V}$$

This large positive value suggests that $Co^{3+}(aq)$ is a strong oxidizing agent, strong enough in fact to oxidize water to $O_2(g)$.

$$4\,Co^{3+}(aq) + 2\,H_2O \longrightarrow 4\,Co^{2+}(aq) + 4\,H^+ + O_2(g) \qquad E^\circ_{cell} = +0.59 \text{ V}$$
$$(25.12)$$

Yet one of the complex ions featured in this chapter has been $[Co(NH_3)_6]^{3+}$. This ion is stable in water solution, even though it contains cobalt in the +3 oxidation state. Reaction (25.12) will not occur if the concentration of Co^{3+} is sufficiently low, and $[Co^{3+}]$ is kept very low because of the great stability of the complex ion.

$$Co^{3+}(aq) + 6\,NH_3(aq) \rightleftharpoons [Co(NH_3)_6]^{3+}(aq) \qquad K_f = 4.5 \times 10^{33}$$

In fact the concentration of free Co^{3+} is so low that for the half-reaction

$$[Co(NH_3)_6]^{3+} + e^- \longrightarrow [Co(NH_3)_6]^{2+}$$

E° is only +0.10 V. As a consequence, not only is $[Co(NH_3)_6]^{3+}$ stable but $[Co(NH_3)_6]^{2+}$ is rather easily oxidized to the Co(III) complex.

The ability of strong electron pair donors (strong Lewis bases) to stabilize high oxidation states, such as NH_3 does in Co(III) complexes and O^{2-} in

Mn(VII) complexes (e.g., MnO_4^-), affords a means of attaining certain oxidation states that might otherwise be difficult or impossible to attain.

Photography: Fixing a Photographic Film

A black-and-white photographic film is an emulsion of a finely divided silver halide (typically AgBr) coated on a strip of polymer, such as a modified cellulose. In the *exposure* step, the film is exposed to light, and some of the tiny granules of AgBr(s) absorb photons. The photons promote the oxidation of Br^- to Br and the reduction of Ag^+ to Ag. The Ag and Br atoms remain in the crystalline lattice of AgBr(s) as "defects," in numbers that depend on the intensity of the light absorbed—the brighter the light the more Ag atoms. Because the actual number of Ag atoms produced in the exposure is not large, the silver is invisible to the eye. The pattern of distribution of the Ag atoms, however, creates a *latent* image of the object photographed. To obtain a visible image, the film is developed.

In the *developing* step, the exposed film is placed in a solution of a mild reducing agent such as hydroquinone, $C_6H_4(OH)_2$. An oxidation–reduction reaction occurs in which Ag^+ ions are reduced to Ag and the hydroquinone is oxidized. The action of the developer is such that reduction of Ag^+ to Ag occurs just in those granules of AgBr(s) containing Ag atoms of the latent image. As a result, the number of Ag atoms in the film is greatly increased, and the latent image becomes visible. Bright regions of the photographed object appear as dark regions in the photographic image. At this stage, the film is a photographic *negative*. This negative cannot be exposed to light, however, because reduction of Ag^+ to Ag could still occur in the previously unexposed granules of AgBr(s). The negative must be fixed.

The *fixing* step requires that the black metallic silver of the negative remain on the film and the unexposed AgBr(s) be removed. A common "fixer" is an aqueous solution of sodium thiosulfate (also known as sodium hyposulfite or "hypo"). Because the complex ion $[Ag(S_2O_3)_2]^{3-}$ has a large formation constant, the following reaction is driven to completion: the AgBr(s) dissolves.

$$AgBr(s) + 2\ S_2O_3{}^{2-}(aq) \longrightarrow [Ag(S_2O_3)_2]^{3-}(aq) + Br^-(aq) \qquad (25.13)$$

Once the negative has been fixed, it is used to produce a *positive* image, the final photograph. This is done by projecting light through the negative onto a piece of photographic paper. Regions of the negative that are dark transmit little light to the photographic paper and will appear light when the photographic paper is subsequently developed and fixed. Conversely, light areas of the negative will appear dark in the final print. In this way the areas of light and dark in the final print are the same as in the photographed object.

Qualitative Analysis

In discussing qualitative cation analysis in Section 19-10 (page 675), we showed how the Group 1 precipitate—AgCl(s), $PbCl_2$(s), Hg_2Cl_2(s)—is separated by taking advantage of the stable complex ion formed by Ag^+(aq) and NH_3(aq).

$$AgCl(s) + 2\ NH_3(aq) \longrightarrow [Ag(NH_3)_2]^+](aq) + Cl^-(aq)$$

The qualitative analysis scheme abounds in other examples of complex ion formation. For example, at a point in the procedure for cation Group 3, a test is needed for Co^{2+}. In the presence of SCN^- ion, Co^{2+} forms a blue thiocyanato complex ion, $[Co(SCN)_4]^{2-}$ (Figure 25-17a). A problem develops, however, if there is even a trace amount of Fe^{3+} present in the solution. Fe^{3+} reacts with

FIGURE 25-17
Qualitative tests for Co^{2+} and Fe^{3+}.
(a) $[Co(SCN)_4]^{2-}$ complex ion.
(b) $[FeSCN(H_2O)_5]^{2+}$ complex ion.
(c) Mixture of $[FeF_6]^{3-}$ and
$[Co(SCN)_4]^{2-}$.

(a)   (b)   (c)

SCN^- to produce $[FeSCN(H_2O)_5]^{2+}$, a strongly colored, blood-red complex ion (Figure 25-17b). Fortunately, this complication can be resolved by treating a solution containing both Co^{2+} and Fe^{3+} with an excess of F^-. The Fe^{3+} is converted to the extremely stable, pale yellow $[FeF_6]^{3-}$; the complex ion $[CoF_4]^{2-}$, being much less stable than $[Co(SCN)_4]^{2-}$, does not form; and the $[Co(SCN)_4]^{2-}(aq)$ is detected through the blue-green solution color (Figure 25-17c).

Sequestering Metal Ions

Metal ions can act as catalysts in promoting undesirable chemical reactions in a manufacturing process, or they may alter the properties of the material being manufactured. Thus, for many industrial purposes it is imperative to remove mineral impurities from water. Often these impurities, such as Cu^{2+}, are present only in trace amounts, and precipitation of metal ions is feasible only if K_{sp} for the precipitate is very small. An alternative is to treat the water with a chelating agent. This reduces the free cation concentrations to the point that the cations are no longer able to enter into objectionable reactions. The cations are said to be *sequestered*. Among the chelating agents widely employed are the salts of *ethylenediaminetetraacetic acid* (EDTA), usually as the sodium salt.

$$4\ Na^+ \left[\begin{array}{c} {}^-OOCCH_2 \\ {}^-OOCCH_2 \end{array} \hspace{-0.5em} \begin{array}{c} \\ NCH_2CH_2N \\ \end{array} \hspace{-0.5em} \begin{array}{c} CH_2COO^- \\ CH_2COO^- \end{array} \right]$$

A representative complex ion formed by a metal ion, M^{n+}, with the hexadentate $EDTA^{4-}$ anion is pictured in Figure 25-18. The high stability of such complexes can be attributed to the presence of five, five-member chelate rings.

In the presence of $EDTA^{4-}(aq)$, Ca^{2+}, Mg^{2+}, and Fe^{3+} in hard water are unable to form boiler scale or to precipitate as insoluble soaps. The cations are sequestered in the complex ions: $[Ca(EDTA)]^{2-}$, $[Mg(EDTA)]^{2-}$, and $[Fe(EDTA)]^-$, having K_f values of 4×10^{10}, 4.0×10^8, and 1.7×10^{24}, respectively.

Chelation with EDTA can be used in treating some cases of metal poisoning. If a person suffering from lead poisoning is fed $[Ca(EDTA)]^{2-}$, the following exchange occurs, because $[Pb(EDTA)]^{2-}$ ($K_f = 2 \times 10^{18}$) is even more stable than $[Ca(EDTA)]^{2-}$ ($K_f = 4 \times 10^{10}$).

$$Pb^{2+} + [Ca(EDTA)]^{2-} \longrightarrow [Pb(EDTA)]^{2-} + Ca^{2+}$$

The lead complex is excreted by the body and the Ca^{2+} remains as a body nutrient. A similar method can be used to rid the body of radioactive isotopes, as in the treatment of plutonium poisoning.

FIGURE 25-18

Structure of a metal-EDTA complex.

The central metal ion M^{n+} can be Ca^{2+}, Mg^{2+}, Fe^{2+}, Fe^{3+}, and so on. The ligand is $EDTA^{4-}$, and the net charge on the complex is $+n - 4$.

Biological Applications: Porphyrins

The structure in Figure 25-19 is commonly found in both plant and animal matter. If the eight R groups are all H atoms, the molecule is called porphin. The central N atoms can give up their H atoms and a metal atom can coordinate simultaneously with all four N atoms. The porphin is a tetradentate ligand for the central metal, and the metal-porphin complex is called a *porphyrin*. Specific porphyrins differ in their central metals and in the R groups on the porphin rings.

In the process of *photosynthesis,* carbon dioxide and water, in the presence of inorganic salts, a catalytic agent called *chlorophyll,* and sunlight, combine to form carbohydrates. Carbohydrates are the main structural materials of plants.

FIGURE 25-19
The porphyrin structure.

$$n\,CO_2 + n\,H_2O \xrightarrow[\text{chlorophyll}]{\text{sunlight}} \underset{\text{carbohydrates}}{(CH_2O)_n} + n\,O_2 \qquad (25.14)$$

Chlorophyll is a green pigment that absorbs sunlight and directs the storage of this energy into the chemical bonds of the carbohydrates. The structure of one type of chlorophyll is shown in Figure 25-20, and we see that it is a *porphyrin.* The central metal ion is Mg^{2+}.

Green is the complementary color of magenta—a purplish red—and we should expect chlorophyll to absorb light in the red region of the spectrum (about 670–680 nm). This suggests that plants should grow more readily in red light than in light of other colors, and some experimental evidence indicates that this is the case. For example, the maximum rate of formation of $O_2(g)$ by reaction (25.14) occurs with red light.

In Chapter 28 we will consider another porphyrin structure that is essential to life—hemoglobin.

FIGURE 25-20
Structure of chlorophyll a.

Summary

Many metal atoms or ions, particularly those of the transition elements, have the ability to form bonds with electron pair donors (ligands). Polydentate ligands are able to attach simultaneously to two or more positions at the metal center. This multiple attachment produces complexes with five- or six-member rings of atoms—chelates.

To name complexes requires a set of rules. These rules deal with such matters as denoting the number and kinds of ligands; the oxidation state of the metal center; and whether the complex is neutral, a cation, or an anion. The positions at which ligands are attached to the metal center are not always equivalent. In geometric isomerism different structures with different properties result, depending on where this attachment occurs. Optical isomers differ by being nonsuperimposable mirror images of each other.

A bonding theory for complex ions that is useful in explaining the magnetic properties and characteristic colors of complex ions is the crystal field theory. This theory emphasizes the splitting of the *d* energy level of the central metal ion as a result of repulsions between *d*-orbital electrons of the

Focus On

Colors in Gemstones

Emeralds in matrix.

though the radius of the Cr^{3+} ion is somewhat larger (62 pm) than that of Al^{3+} (53 pm), these values match well enough that Cr^{3+} is able to replace Al^{3+} in the crystal lattice. In ruby, about 1% of the Al^{3+} ions are replaced by Cr^{3+} ions.

Because Al^{3+} has no valence electrons and no readily available electronic transitions for its core electrons, a pure crystal of corundum is colorless. The color of ruby is the result of electron transitions between d orbitals of Cr^{3+} ions. Cr^{3+} has the ground state electron configuration: $[Ar]3d^3$. The six O^{2-} ions create an octahedral crystal field about the Cr^{3+} ion and split the energy levels of the d orbitals into two groups (recall Figure 25-9). In the ground state, the three valence electrons occupy the three lower energy d orbitals and the two higher energy d orbitals are vacant. Two excited states of the Cr^{3+} ion can be reached by absorption of two different frequencies (colors) of visible light. One transition, a "low-energy" transition, requires absorption of a yellow-green light and the other, a "high-energy" transition,

Many of the beautifully colored stones that we consider precious or semiprecious are chemically impure. And it is the impurities—transition metal ions—that are responsible for the brilliant colors. Consider the cases of ruby and emerald. Both of these gems contain small amounts of Cr^{3+} ion as the color-causing impurity. Rubies, however, have a deep red color, whereas emeralds have a distinctive blue-green color. Let us see why.

In a ruby crystal the principal constituent is aluminum oxide (Al_2O_3), or *corundum,* with a structural pattern in which each Al^{3+} ion is surrounded by six O^{2-} ions. Al-

central ion and electrons of the ligands. A prediction of the magnitude of d-level splitting produced by a ligand can be made through the spectrochemical series.

The formation of a complex ion can be viewed as a stepwise equilibrium process in which other ligands displace H_2O molecules from aqua complex ions. The stepwise constants can be combined into an overall formation constant for the complex ion, K_f. The ability of ligand water molecules to ionize causes some aqua complexes to exhibit acidic properties and helps to explain amphoterism. Also important in deter-

mining properties of a complex ion is the rate at which the ion exchanges ligands between its coordination sphere and the solution. Exchange is rapid in a labile complex and slow in an inert complex.

Complex ion formation can be used to stabilize certain oxidation states, such as Co(III). Other applications include dissolving precipitates, such as AgCl by $NH_3(aq)$ in the qualitative analysis scheme and AgBr by $Na_2S_2O_3(aq)$ in the photographic process, and sequestering ions by chelation, as with EDTA.

Summarizing Example

Absorbance is a measure of the proportion of monochromatic (single color) light that is absorbed as the light passes through a solution. An *absorption spectrum* is a graph of absorbance as a function of wavelength. High absorbances correspond to large the absorption of violet light. As a consequence of these absorptions, the light that is transmitted by a ruby is *red,* with a light purple tint.

The basic structure of an emerald is that of a crystal of *beryl,* $3BeO \cdot Al_2O_3 \cdot 6SiO_2$. Here, Cr^{3+} replaces Al^{3+} in the same type of octahedral field as in ruby but with dramatically different results. The presence of BeO and SiO_2 in the beryl crystal weakens the crystal field. The effect of the weaker field is a smaller energy separation (Δ) between the two groups of *d* orbitals in Cr^{3+} than is found in ruby. The energy per photon required to produce electronic transitions between the *d* orbitals is less for Cr^{3+} in emerald than in ruby. The low-energy transition shifts from yellow-green to yellow-red, and the higher energy transition is also lowered slightly in energy. As a result, when light passes through an emerald most of the red and violet colors are absorbed and the blue and green are transmitted; the emerald takes on a distinctive blue-green color.

For rubies and emeralds to exhibit their colors continuously, there must be a mechanism by which electrons in Cr^{3+} can revert from their excited states back to the ground state. This occurs in two stages. First, electrons drop from the higher energy states to a lower, intermediate state somewhat above the ground state. The energies of these transitions are in the infrared region of the spectrum; the emissions are invisible and appear as heat, which stimulates the vibrations of the ions in the crystal. The final transition to the ground state produces a red light. This type of light emission, in which a lower energy light is emitted following the absorption of higher energy light, is called *fluorescence.* Interestingly, both ruby and emerald display a red fluorescence emission. In ruby this red emission adds to the red color of the transmitted light, and in emerald it is of just the right wavelength to enhance the character of the green color. (When rubies and emeralds are observed in ultraviolet, or "black," light, *both* give off a red light.) The transitions referred to in this discussion are suggested in the diagram.

Light absorption and emission in a ruby. The absorption of violet and yellow-green light from an incident white light beam causes the transmitted light to be red in color. In addition, the ruby emits a red light by fluorescence.

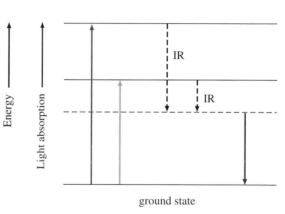

proportions of the light entering a solution being absorbed. Low absorbances signify that large proportions of the light are transmitted. The absorption spectrum of $[Ti(H_2O)_6]^{3+}(aq)$ is plotted in Figure 25-21.

Describe the color of light that $[Ti(H_2O)_6]^{3+}(aq)$ absorbs most strongly and the color of the solution.

1. *Determine the color of the light absorbed most strongly.* Determine the wavelength at the maximum of the absorption spectrum in Figure 25-21. Determine the color corresponding to this wavelength from Figure 9-3. *Answer:* green.
2. *Describe the color of transmitted light.* Colors transmitted are red, most strongly, and some blue. *Answer:* This mixture of red and blue yields a magenta color.

FIGURE 25-21
The absorption spectrum of $[Ti(H_2O)_6]^{3+}(aq)$.

Key Terms

bidentate (25-2)
chelate (25-2)
chelating agent (25-2)
chelation effect (25-7)
chiral (25-4)
cis (25-4)
complex (25-1)
coordination compound (25-1)

coordination isomerism (25-4)
coordination number (25-1)
crystal field theory (25-5)
enantiomer (25-4)
inert complex (25-9)
isomers (25-4)
labile complex (25-9)
monodentate (25-2)

optical isomer (25-4)
polydentate (25-2)
primary color (25-6)
secondary color (25-6)
spectrochemical series (25-5)
stereoisomer (25-4)
structural isomer (25-4)
trans (25-4)

Review Questions

1. In your own words describe the following terms or symbols: **(a)** coordination number; **(b)** Δ_o; **(c)** ammine complex; **(d)** enantiomer.

2. Briefly describe each of the following ideas, phenomena, or methods: **(a)** spectrochemical series; **(b)** crystal field theory; **(c)** optical isomer; **(d)** structural isomerism.

3. Explain the important distinction between each pair: **(a)** coordination number and oxidation number; **(b)** monodentate and polydentate; **(c)** cis and trans isomers; **(d)** dextrorotatory and levorotatory; **(e)** low- and high-spin complexes.

4. Write the formula and name of
(a) a complex ion having Cr^{3+} as the central ion and two NH_3 molecules and four Cl^- ions as ligands.
(b) a complex ion of iron(III) having a coordination number of 6 and CN^- as ligands.
(c) a coordination compound comprising two types of complex ions: One is a complex of Cr(III) with ethylenediamine (en), having a coordination number of 6; the other, a complex of Ni(II) with CN^-, having a coordination number of 4.

5. What is the coordination number and the oxidation state of the central metal ion in each of the following complexes? Name each complex.
(a) $[Co(NH_3)_6]^{2+}$ **(b)** $[AlF_6]^{3-}$
(c) $[Cu(CN)_4]^{2-}$ **(d)** $[CrBr_2(NH_3)_4]^+$
(e) $[Co(ox)_3]^{4-}$ **(f)** $[Ag(S_2O_3)_2]^{3-}$

6. Name the following complex ions. (Do not attempt to distinguish among isomers in this problem.)
(a) $[AgI_2]^-$ **(b)** $[Al(OH)(H_2O)_5]^{2+}$
(c) $[Zn(CN)_4]^{2-}$ **(d)** $[Pt(en)_2]^{2+}$
(e) $[CoCl(NO_2)(NH_3)_4]^+$

7. Name the following coordination compounds. (Do not attempt to distinguish among isomers in this problem.)

(a) $[CoBr(NH_3)_5]SO_4$ **(b)** $[CoSO_4(NH_3)_5]Br$
(c) $[Cr(NH_3)_6][Co(CN)_6]$ **(d)** $Na_3[Co(NO_2)_6]$
(e) $[Co(en)_3]Cl_3$

8. Write appropriate formulas for the following species. (Do not attempt to distinguish among isomers in this problem.)
(a) dicyanoargentate(I) ion
(b) triamminenitroplatinum(II) ion
(c) aquachlorobis(ethylenediamine)cobalt(III) ion
(d) potassium hexacyanochromate(II)

9. Draw Lewis structures for the following monodentate ligands. (*Hint:* Refer to Table 25.2.) **(a)** H_2O; **(b)** CH_3NH_2; **(c)** ONO^- (nitrito); **(d)** SCN^- (thiocyanato).

10. Write formulas for the following hydrates.
(a) manganese(II) sulfate hexahydrate
(b) potassium hexacyanochromate(II) trihydrate

11. Draw structures to represent these four complex ions. **(a)** $[PtCl_4]^{2-}$ **(b)** $[FeCl_4(en)]^-$ **(c)** *cis*-$[FeCl_2(ox)(en)]^-$ **(d)** *trans*-$[CrCl(OH)(NH_3)_4]^+$

12. How many different structures are possible for each of the following complex ions? Draw each structure.
(a) $[Co(H_2O)(NH_3)_5]^{3+}$ **(b)** $[Co(H_2O)_2(NH_3)_4]^{3+}$
(c) $[Co(H_2O)_3(NH_3)_3]^{3+}$ **(d)** $[Co(H_2O)_4(NH_3)_2]^{3+}$

13. Indicate what type of isomerism may be found in each of the following cases. If no isomerism is possible, so indicate.
(a) $[Zn(NH_3)_4][CuCl_4]$ **(b)** $[Fe(CN)_5SCN]^{4-}$
(c) $[NiCl(NH_3)_5]^+$ **(d)** $[PtBrCl_2(py)]^-$
(e) $[Cr(OH)_3(NH_3)_3]^-$

14. Of the complex ions $[Co(H_2O)_6]^{3+}$ and $[Co(en)_3]^{3+}$, one has a yellow color in aqueous solution and the other, blue. Match each ion with its expected color, and state your reason for doing so.

Exercises

Nomenclature

15. Supply acceptable names for the following. (Do not attempt to distinguish among isomers in this exercise.)
(a) $[Co(OH)(H_2O)_4(NH_3)]^{2+}$
(b) $[Co(ONO)_3(NH_3)_3]$
(c) $[Pt(H_2O)_4][PtCl_6]$
(d) $[Fe(ox)_2(H_2O)_2]^-$
(e) $Ag_2[HgI_4]$

16. Write appropriate formulas for the following. (Do not attempt to distinguish among isomers in this exercise.)
(a) potassium hexacyanoferrate(III)
(b) bis(ethylenediamine)copper(II) ion
(c) pentaaquahydroxoaluminum(III) chloride
(d) amminechlorobis(ethylenediamine)chromium(III) sulfate
(e) tris(ethylenediamine)iron(III) hexacyanoferrate(II)

Bonding and Structure in Complex Ions

17. Draw a plausible structure to represent
(a) $[PtCl_4]^{2-}$
(b) *fac*-$[Co(H_2O)_3(NH_3)_3]^{2+}$
(c) $[CrCl(H_2O)_5]^{2+}$

18. Draw plausible structures of the following chelate complexes.
(a) $[Pt(ox)_2]^{2-}$
(b) $[Cr(ox)_3]^{3-}$
(c) $[Fe(EDTA)]^{2-}$

19. Draw plausible structures corresponding to each of the following names.
(a) pentamminesulfatochromium(III) ion
(b) trioxalatocobaltate(III) ion
(c) triamminedichloronitritocobalt(III)

20. Draw plausible structures corresponding to each of the following names.
(a) pentamminenitrocobalt(III) ion
(b) ethylenediaminedithiocyanatocopper(II)
(c) hexaaquanickel(II) ion

Isomerism

21. Which of these general structures for a complex ion would you expect to exhibit *cis* and *trans* isomerism? Explain.
(a) tetrahedral
(b) square planar
(c) linear

22. Which of these octahedral complexes would you expect to exhibit *geometric* isomerism? Explain.
(a) $[CrOH(NH_3)_5]^{2+}$
(b) $[CrCl_2(H_2O)(NH_3)_3]^+$
(c) $[CrCl_2(en)_2]^+$
(d) $[CrCl_4(en)]^-$
(e) $[Cr(en)_3]^{3+}$

23. If A, B, C, and D are four different ligands,
(a) how many geometric isomers will be found for square-planar $[PtABCD]^{2+}$?
(b) will tetrahedral $[ZnABCD]^{2+}$ display optical isomerism?

24. A structure that Werner examined as a possible alternative to the octahedron is the trigonal prism.

(a) Show that for the complex ion $[CoCl_2(NH_3)_4]^+$ this structure predicts three geometric isomers.
(b) Show that this structure does not account for optical isomerism in $[Co(en)_3]^{3+}$.

25. Write the names and formulas of three coordination isomers of $[Co(en)_3][Cr(ox)_3]$.

26. Draw a structure for *cis*-dichlorobis(ethylenediamine)-cobalt(III) ion. Is this ion chiral? Is the trans isomer chiral? Explain.

Crystal Field Theory

27. Describe how the crystal field theory makes possible an explanation of the fact that so many transition metal compounds are colored.
29. If the ion Cr^{2+} is linked with strong-field ligands to produce an octahedral complex, the complex has *two* unpaired electrons. If Cr^{2+} is linked with weak-field ligands, the complex has *four* unpaired electrons. How do you account for this difference?
31. Predict
(a) which of the complex ions, $[MoCl_6]^{3-}$ and $[Co(en)_3]^{3+}$, is diamagnetic and which is paramagnetic.
(b) the number of unpaired electrons expected for the tetrahedral complex ion $[CoCl_4]^{2-}$.
33. In Example 25-5 we chose between a tetrahedral and a square-planar structure for $[Ni(CN)_4]^{2-}$ based on magnetic properties. Could we similarly use magnetic properties to establish whether the ammine complex of Ni(II) is octahedral $[Ni(NH_3)_6]^{2+}$ or tetrahedral $[Ni(NH_3)_4]^{2+}$? Explain.

28. Cyano complexes of transition metal ions (such as Fe^{2+} and Cu^{2+}) are often yellow in color, whereas aqua complexes are often green or blue. Why is there this difference?
30. In contrast to the case of Cr^{2+} considered in Exercise 29, no matter what ligand is linked to Cr^{3+} to form an octahedral complex, the complex always has *three* unpaired electrons. Explain this fact.
32. Predict
(a) whether the square-planar complex ion $[Cu(py)_4]^{2+}$ is diamagnetic or paramagnetic;
(b) whether octahedral $[Mn(CN)_6]^{3-}$ or tetrahedral $[FeCl_4]^-$ has the greater number of unpaired electrons.
34. In both $[Fe(H_2O)_6]^{2+}$ and $[Fe(CN)_6]^{4-}$ ions the iron is present as Fe(II); yet $[Fe(H_2O)_6]^{2+}$ is paramagnetic, whereas $[Fe(CN)_6]^{4-}$ is diamagnetic. Explain this difference.

Complex Ion Equilibria

35. Write equations to represent the following observations. **(a)** A mixture of $Mg(OH)_2(s)$ and $Zn(OH)_2(s)$ is treated with $NH_3(aq)$. The $Zn(OH)_2$ dissolves but the $Mg(OH)_2(s)$ is left behind.
(b) When $NaOH(aq)$ is added to $CuSO_4(aq)$, a pale blue precipitate forms. If $NH_3(aq)$ is added, the precipitate redissolves, producing a solution with an intense deep blue color. If this deep blue solution is made acidic with $HNO_3(aq)$, the color is converted back to pale blue.
37. Which of the following complex ions would you expect to have the largest overall K_f and why? $[Co(NH_3)_6]^{3+}$, $[Co(en)_3]^{3+}$, $[Co(H_2O)_6]^{3+}$, $[Co(H_2O)_4(en)]^{3+}$.
39. Write a series of equations to show the stepwise displacement of H_2O ligands in $[Fe(H_2O)_6]^{3+}$ by ethylenediamine, for which $\log K_1 = 4.34$; $\log K_2 = 3.31$; and $\log K_3 = 2.05$. What is the overall formation constant, $\beta_3 = K_f$, for $[Fe(en)_3]^{3+}$?
41. Explain the following observations in terms of complex ion formation.
(a) $Al(OH)_3(s)$ is soluble in $NaOH(aq)$ but insoluble in $NH_3(aq)$.
(b) $ZnCO_3(s)$ is soluble in $NH_3(aq)$ but $ZnS(s)$ is not.
(c) The molar solubility of $AgCl$ in pure water is about 1×10^{-5} M; in 0.04 M $NaCl(aq)$, it is about 2×10^{-6} M; but in 1 M $NaCl(aq)$, it is about 8×10^{-5} M.

36. Write equations to represent the following observations. **(a)** A quantity of $CuCl_2(s)$ is dissolved in concentrated $HCl(aq)$ and produces a yellow solution. The solution is diluted to twice its volume with water and assumes a green color. On dilution to ten times its original volume, the solution becomes pale blue in color.
(b) When chromium metal is dissolved in $HCl(aq)$ a blue solution is produced that quickly turns a green color. Later the green solution becomes blue-green and then violet in color.
38. Use data from Table 25.6 to determine values of **(a)** β_4 for the formation of $[Zn(NH_3)_4]^{2+}$; **(b)** β_4 for the formation of $[Ni(H_2O)_2(NH_3)_4]^{2+}$.
40. A tabulation of formation constant data lists the following $\log K$ values for the formation of $[CuCl_4]^{2-}$: $\log K_1 = 2.80$; $\log K_2 = 1.60$; $\log K_3 = 0.49$; $\log K_4 = 0.73$. What is the overall formation constant $\beta_4 = K_f$ for $[CuCl_4]^{2-}$?
42. Explain the following observations in terms of complex ion formation.
(a) $CoCl_3$ is unstable in water solution, being reduced to $CoCl_2$ and liberating $O_2(g)$. On the other hand, $[Co(NH_3)_6]Cl_3$ can be easily maintained in aqueous solution.
(b) AgI is insoluble in water and in dilute $NH_3(aq)$, but AgI will dissolve in an aqueous solution of sodium thiosulfate.

Acid–Base Properties

43. Which of the following would you expect to react as a Brønsted–Lowry acid and why? $[Cu(NH_3)_4]^{2+}$, $[FeCl_4]^-$, $[Al(H_2O)_6]^{3+}$, or $[Zn(OH)_4]^{2-}$.
45. For a solution that is 0.100 M in $[Fe(H_2O)_6]^{3+}$,
(a) Calculate the pH of the solution assuming that ionization of the aqua complex ion proceeds only through the first step, Equation (25.10).
(b) Calculate $[[FeOH(H_2O)_5]^{2+}]$ if the solution is also 0.100 M $HClO_4$. (ClO_4^- does not complex with Fe^{3+}.)
(c) Can the pH of the solution be maintained so that $[[FeOH(H_2O)_5]^{2+}]$ does not exceed 1×10^{-6} M? Explain.

44. Write simple chemical equations to show how the complex ion $[CrOH(H_2O)_5]^{2+}$ acts as **(a)** an acid; **(b)** a base.

46. A solution that is 0.010 M in Pb^{2+} is also made to be 0.20 M in a salt of EDTA (that is, having a concentration of the $EDTA^{4-}$ ion of 0.20 M). If this solution is now made 0.10 M in H_2S and 0.10 M in H_3O^+, will $PbS(s)$ precipitate? [*Hint:* You will also have to use data from Chapter 19.]

Applications

47. From data in Chapter 19
(a) Derive an equilibrium constant for reaction (25.13) and explain why this reaction (the fixing of photographic film) is expected to go essentially to completion.
(b) Explain why $NH_3(aq)$ cannot be used in the fixing of photographic film.

48. Show that the oxidation of $[Co(NH_3)_6]^{2+}$ to $[Co(NH_3)_6]^{3+}$ referred to on page 886 should occur spontaneously in alkaline solution with H_2O_2 as an oxidizing agent. (*Hint:* Refer also to Appendix D.)

Advanced Exercises

49. From each of the following names you should be able to deduce the formula of the complex ion or coordination compound intended. Yet, these are not the best systematic names that can be written. Replace each name with one that is more acceptable. (*Hint:* What is wrong with each name as it stands?)
(a) cupric tetraammine ion; **(b)** dichlorotetraammine cobaltic chloride; **(c)** platinic(IV) hexachloride ion; **(d)** disodium copper tetrachloride; **(e)** dipotassium antimony(III) pentachloride.

50. The following compounds were once represented by the empirical formulas given. Write the probable formulas of these coordination compounds according to Werner's theory, and assign them systematic names.
(a) Zeise's salt: $PtCl_2 \cdot KCl \cdot C_2H_4$ (This compound consists of a simple unipositive cation and a uninegative complex anion.)
(b) Magnus's green salt: $PtCl_2 \cdot 2NH_3$ (This compound consists of dipositive and dinegative complex ions.)

51. A theory of coordination compounds that predated Werner's was called the chain theory. According to this theory, in a coordination compound like $CoCl_3 \cdot 6NH_3$, the Co atom was thought to form *three* bonds to other groups—either Cl atoms or NH_3 molecules—and NH_3 molecules were believed to join together into chains, such as

$$Co \underset{NH_3Cl}{\overset{NH_3Cl}{-}} NH_3 - NH_3 - NH_3 - NH_3Cl$$

Cl atoms were thought to be ionizable if bonded to NH_3, but not if bonded to Co. Show how this chain theory can explain the behavior of the compounds (a), (b), and (c) described on page 866. Specifically, show how the theory explains their electrical conductivity and behavior toward $AgNO_3$.

52. Demonstrate why the "chain" theory of Exercise 51 could not explain the existence of compounds such as $[CoCl_3(NH_3)_3]$ and $Na[CoCl_4(NH_3)_2]$.

53. Explain the following series of observations. The green solid $CrCl_3 \cdot 6H_2O$ dissolves in water to form a green solution. After a day or two the solution color is violet. When the violet solution evaporates to dryness, a green solid remains.

54. The cis and trans isomers of $[CoCl_2(en)_2]^+$ can be distinguished through a displacement reaction with oxalate ion. What difference in reactivity toward oxalate ion would you expect between the cis and trans isomers? Explain.

55. We learned in Chapter 17 that for polyprotic acids ionization constants for successive ionization steps decrease rapidly. That is, $K_{a_1} \gg K_{a_2} \gg K_{a_3}$. The ionization constants for the first two steps in the ionization of $[Fe(H_2O)_6]^{3+}$ (reactions 25.10 and 25.11) are more nearly equal in magnitude. Can you think of a reason why this multistep ionization does not seem to follow the pattern for polyprotic acids?

56. Without performing detailed calculations, show why you would expect the concentrations of the various ammine–aqua complex ions to be negligible compared with that of $[Cu(NH_3)_4]^{2+}$ in a solution having a total Cu(II) concentration of 0.10 M and a total concentration of NH_3 of 1.0 M. Under what conditions would the concentrations of these ammine–aqua complex ions (such as $[Cu(H_2O)_3NH_3]^{2+}$) become more significant relative to the concentration of $[Cu(NH_3)_4]^{2+}$? Explain.

57. Verify the statement on page 888 that neither Ca^{2+} nor Mg^{2+} found in natural waters is likely to precipitate from the water on the addition of other reagents, if the ions are complexed with EDTA. [*Hint:* Use data from this chapter and Chapter 19. Assume reasonable values for the total metal ion concentration and that of *free* EDTA, such as 0.10 M each.]

58. Estimate the total $[Cl^-]$ required in a solution that is initially 0.10 M $CuSO_4$ to produce a visible yellow color.

$$[Cu(H_2O)_4]^{2+} + 4 Cl^- \rightleftharpoons [CuCl_4]^- + 4 H_2O$$
$$\text{(blue)} \qquad\qquad \text{(yellow)}$$
$$K_f = 4.2 \times 10^5$$

Assume that 99% conversion of $[Cu(H_2O_4)]^{2+}$ to $[CuCl_4]^{2-}$ is sufficient for this to happen, and ignore the presence of any mixed aqua-chloro complex ions.

59. The structures of four complex ions are given. Each has Co^{3+} as the central ion. The ligands are H_2O, NH_3, and oxalate ion, $C_2O_4^{2-}$. Determine which, *if any,* of these complex ions are isomers (geometric or optical); which, *if any,* are identical (that is, have identical structures); and which, *if any,* are distinctly different.

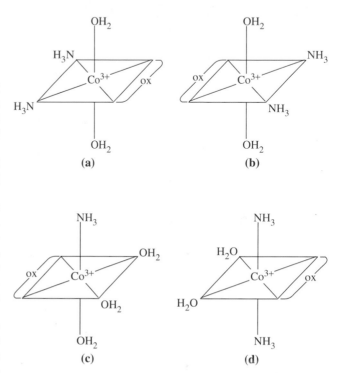

(a)   (b)

(c)   (d)

60. Refer to the Summarizing Example on pages 890–891. If the electronic transition responsible for the maximum at 520 nm in the absorption spectrum for $[Ti(H_2O)_6]^{3+}$ is from one of the lower to one of the upper level d orbitals, what must be the energy separation, Δ, in kilojoules per mole?

61. Refer to the stability of $[Co(NH_3)_6]^{3+}(aq)$ on page 886, and
(a) verify that E°_{cell} for reaction (25.12) is +0.59 V;
(b) calculate $[Co^{3+}]$ in a solution that has a total concentration of cobalt of 1.0 M and $[NH_3] = 0.10$ M;
(c) show that for the value of $[Co^{3+}]$ calculated in part (b), reaction (25.12) will not occur. [*Hint:* What is $[H_3O^+]$ in 0.10 M NH_3? Assume a low, but reasonable, concentration of Co^{2+}, say 1×10^{-4} M, and a partial pressure of $O_2(g)$ of 0.2 atm.]

62. A Cu electrode is immersed in a solution that is 1.00 M NH_3 and 1.00 M in $[Cu(NH_3)_4]^{2+}$. If a standard hydrogen electrode is the cathode, E_{cell} is found to be +0.08 V. What is the value obtained by this method for the formation constant, K_f, of $[Cu(NH_3)_4]^{2+}$?

63. The following concentration cell is constructed.
$$Ag \,|\, Ag^+(0.10 \text{ M } [Ag(CN)_2]^-, 0.10 \text{ M } CN^-) \,\|$$
$$Ag^+(0.10 \text{ M}) \,|\, Ag$$
If K_f for $[Ag(CN)_2]^-$ is 5.6×10^{18}, what value would you expect for E_{cell}? [*Hint:* Recall that the anode is on the left.]

Nuclear Chemistry

26

The sun showing a solar flare. Nuclear reactions in the stars are the source of the various chemical elements. And, energy released in nuclear reactions on the sun—solar energy—makes life on Earth possible.

*T*he origin of all the elements is the stars, including our sun. Nuclear fusion in stars creates heavier elements from lighter ones. The heaviest elements, those with atomic numbers greater than 83, all have unstable nuclei—that is, they are *radioactive.*

Certain isotopes of the lighter elements also have this property of radioactivity. The isotope carbon-14, for example, has chemical and physical properties that are essentially identical to those of the much more abundant isotopes carbon-12 and carbon-13. Carbon-14, however, is radioactive, and it is this property that is used in the technique known as radiocarbon dating.

In this chapter we will consider a variety of phenomena that originate within the nuclei of atoms. Collectively, we refer to these phenomena as nuclear chemistry. While stars are the natural source of all the elements, we will discuss how new heavy elements and radioactive isotopes of existing lighter elements can be made artificially. Another aspect of nuclear chemistry that we will discuss is the effects of ionizing radiation on matter. These effects can have both positive and negative outcomes and are a subject of society's continuing "nuclear debate."

Marie Sklodowska Curie (1867–1934). Marie Curie shared in the 1903 Nobel Prize in physics for studies on radiation phenomena. In 1911 she won the Nobel Prize in chemistry for her discovery of polonium and radium.

26-1   The Phenomenon of Radioactivity

The term *radioactivity* was proposed by Marie Curie to describe the emission of ionizing radiation by some of the heavier elements. Ionizing radiation, as the name implies, interacts with matter to produce ions. This means that the radiation is sufficiently energetic to break chemical bonds. Some ionizing radiation is particulate (consisting of particles), and some is nonparticulate. We introduced α, β, and γ radiation in Section 2-2. Let's describe them again in more detail, together with two other nuclear processes.

Alpha Particles

Alpha (α) particles are the nuclei of helium-4 atoms, $^4_2\text{He}^{2+}$, that have been accelerated artificially or created by ejection from the nucleus of a larger atom. We can think of alpha-particle emission as a process in which a bundle of two protons and two neutrons is emitted by a radioactive nucleus, resulting in a lighter nucleus. Alpha particles produce large numbers of ions through their collisions and near collisions with atoms as they travel through matter, but their penetrating power is low. (Generally they can be stopped by a sheet of paper.) Because of their positive charge, α particles are deflected by electric and magnetic fields.

We can represent the production of alpha particles through a nuclear equation. A **nuclear equation** is written to conform to two rules.

- The sum of mass numbers must be the same on both sides.
- The sum of atomic numbers must be the same on both sides.

In Equation (26.1) the alpha particle is represented as ^4_2He. Mass numbers total 238, and atomic numbers total 92. The loss of an alpha particle produces a *decrease* of 2 in the atomic number and 4 in the mass number of the nucleus.

$$^{238}_{92}\text{U} \longrightarrow {}^{234}_{90}\text{Th} + {}^4_2\text{He} \qquad (26.1)$$

Notice also that, although the alpha particle has a 2+ charge (and thorium has a 2− charge), no charge is shown in the nuclear equation. The reason for this is to focus attention on the nuclei. Unlike other areas of chemistry, nuclear chemistry places little emphasis on the outer electrons. Rather, it is the nucleus of the atom that concerns us.

Beta Particles

Beta (β^-) particles are deflected by electric and magnetic fields in the *opposite* direction from α particles and, because of their smaller mass, they are deflected more strongly than α particles. β^- particles *are* electrons, but electrons that originate from the nuclei of atoms in nuclear decay processes. Electrons that surround the nucleus are given the familiar symbol, e^-.

The simplest decay process producing a β^- particle is the decay of a free neutron, which is unstable outside the nucleus of an atom.

$$^1_0\text{n} \longrightarrow {}^1_1\text{p} + {}^{\,0}_{-1}\beta + \nu \qquad (26.2)$$

A β^- particle does not have an atomic number, but its 1− charge is equivalent to an atomic number of −1. In nuclear equations the β^- particle is represented as $^{\,0}_{-1}\beta$. Also, a β^- particle is small enough, compared to protons and neutrons, that its mass can be ignored in most calculations. Equation 26.2 introduces the symbol ν to represent an entity called a *neutrino*. This particle was first postu-

Alpha particles leave a trail of liquid droplets, artificially colored green in this photograph, as they pass through a supersaturated vapor in a detector known as a cloud chamber. The chamber also contains He(g), and the trail of one α particle, colored yellow, is shown striking the nucleus of an He atom. Following the collision, the α particle and the He atom (red) move apart along lines at about a 90° angle.

lated in the 1930s as necessary for the conservation of certain properties during the beta decay process. Because they interact so weakly with matter, neutrinos were not detected until the 1950s, and still little is known of their properties, including even whether they have a rest mass.

For a typical decay process as represented by Equation (26.3), we can think of a neutron *within* the nucleus of an atom as spontaneously converting to a proton and an electron. The proton remains in the nucleus, whereas the electron is emitted as a β^- particle. The atomic number *increases* by one unit, and the mass number is unchanged. The elusive neutrino is generally not included in the nuclear equation.

▶ Particles, such as electrons, increase in mass as their speed approaches the speed of light. *Rest mass* indicates the mass of a particle when it has no translational motion—that is, when it is *at rest.*

$$\ce{^{234}_{90}Th} \longrightarrow \ce{^{234}_{91}Pa} + \ce{^{0}_{-1}\beta} \qquad (26.3)$$

In a manner similar to that outlined in Equation (26.3), in some decay processes a proton within the nucleus is converted to a neutron, and a β^+ particle and a neutrino* are emitted.

$$\ce{^{1}_{1}p} \longrightarrow \ce{^{1}_{0}n} + \ce{^{0}_{+1}\beta} + \nu \qquad (26.4)$$

The **β^+** particle, also called a **positron,** has properties similar to the β^- particle except that it carries a *positive* charge. This particle is also known as a *positive* electron and is designated $\ce{^{0}_{+1}\beta}$ in nuclear equations. Positron emission is commonly encountered with artificially produced radioactive nuclei of the lighter elements, for example,

$$\ce{^{30}_{15}P} \longrightarrow \ce{^{30}_{14}Si} + \ce{^{0}_{+1}\beta} \qquad (26.5)$$

▶ $\ce{^{30}_{15}P}$ was the first nucleus obtained by artificial means. Its nuclear synthesis is discussed on page 902.

Electron Capture

Another process that achieves the same effect as positron emission is **electron capture (EC).** Here, an electron from an inner electron shell (usually the shell $n = 1$) is absorbed by the nucleus. Inside the nucleus the electron is used to convert a proton to a neutron. When an electron from a higher quantum level drops to the energy level vacated by the captured electron, X-radiation is emitted. For example,

$$\ce{^{202}_{81}Ti} + \ce{^{0}_{-1}e} \longrightarrow \ce{^{202}_{80}Hg} \quad \text{(followed by X-radiation)} \qquad (26.6)$$

Are You Wondering . . .

Just how significant neutrinos are?

To the chemist, the most significant aspects of radioactive decay are the new atomic nuclei produced and the ionizations caused by the ionizing radiation. Neutrinos do not play a role in either of these aspects, and that's why we don't normally represent them in nuclear equations. On the other hand, physicists and astrophysicists are keenly interested in neutrinos and pursue their study vigorously. Some theories of the universe, for example, postulate that up to 90% of the mass of the universe is made up of neutrinos.

---

*There appear to be two related entities: the neutrino and antineutrino. Neutrinos accompany positron emission and electron capture; antineutrinos are associated with β^- emission.

Gamma Rays

Some radioactive decay processes yielding α or β^- particles leave a nucleus in an excited state. The nucleus then loses energy in the form of electromagnetic radiation—a **gamma (γ) ray.** Gamma rays are a highly penetrating form of radiation (see Figure 9-3 on page 274). They are *undeflected* by electric and magnetic fields. In the radioactive decay of $^{234}_{92}\text{U}$, 77% of the nuclei emit α particles having an energy of 4.18 MeV. The remaining 23% of the $^{234}_{92}\text{U}$ nuclei produce α particles with energies of 4.13 MeV. In the latter case the $^{230}_{90}\text{Th}$ nuclei are left with an excess energy of 0.05 MeV. This energy is released as γ rays. If we denote the unstable, excited Th nucleus as $^{230}_{90}\text{Th}^{\ddagger}$, we can write

$$^{234}_{92}\text{U} \longrightarrow {}^{230}_{90}\text{Th}^{\ddagger} + {}^{4}_{2}\text{He} \qquad (26.7)$$

$$^{230}_{90}\text{Th}^{\ddagger} \longrightarrow {}^{230}_{90}\text{Th} + \gamma \qquad (26.8)$$

This γ-emission process is represented diagrammatically in Figure 26-1.

▶ One electronvolt (eV) is the energy acquired by an electron when it falls through an electric potential difference of 1 volt: $1 \text{ eV} = 1.6022 \times 10^{-19}$ J. It is a useful energy unit to describe processes involving individual atoms. The unit MeV represents one million electronvolts, or one megaelectronvolt. $1 \text{ MeV} = 1 \times 10^6$ eV.

FIGURE 26-1
Production of gamma rays.
The transition of a $^{230}_{90}\text{Th}$ nucleus between the two energy states shown results in the emission of 0.05 MeV of energy in the form of γ rays.

EXAMPLE 26-1

Writing Nuclear Equations for Radioactive Decay Processes. Write nuclear equations to represent **(a)** α-particle emission by ^{222}Rn and **(b)** radioactive decay of bismuth-215 to polonium-215.

Solution

a. We can identify two of the species involved in this process simply from the information given. The $^{222}_{86}\text{Rn}$ nucleus ejects an α particle, $^{4}_{2}\text{He}$, as shown in the following incomplete nuclear equation.

$$^{222}_{86}\text{Rn} \longrightarrow {?} + {}^{4}_{2}\text{He}$$

Because the ejected α-particle contains two protons, the unknown product must contain two fewer protons than $^{222}_{86}\text{Rn}$: $Z_{\text{unk}} = 86 - 2 = 84$. This atomic number identifies the element as polonium, $_{84}\text{Po}$. The mass number (A) of the product can be obtained by subtracting the mass number of the α-particle from that of the radon isotope: $A = 222 - 4 = 218$. The completed nuclear equation is

$$^{222}_{86}\text{Rn} \longrightarrow {}^{218}_{84}\text{Po} + {}^{4}_{2}\text{He}$$

b. Bismuth has the atomic number 83, and polonium, 84. We can approach this problem in the same way we approached part (a).

$$^{215}_{83}\text{Bi} \longrightarrow {}^{215}_{84}\text{Po} + {?}$$

There is no change in mass number, so the particle must have zero mass. The atomic number must be $Z = 83 - 84 = -1$. Only a $_{-1}^{0}\beta$ particle fits these parameters: β decay is the only type of emission that leads to an increase of one unit in atomic number without a change in the mass number.

$$^{215}_{83}\text{Bi} \longrightarrow {}^{215}_{84}\text{Po} + {}^{0}_{-1}\beta$$

Practice Example A: Write a nuclear equation to represent β^- particle emission by $^{241}_{94}\text{Pu}$.

Practice Example B: Write a nuclear equation to represent the decay of a radioactive nucleus to produce ^{58}Ni and a positron.

26-2   Naturally Occurring Radioactive Isotopes

Of the stable nuclides, $^{209}_{83}Bi$ has the highest atomic number and mass number. All known nuclides beyond it in atomic and mass numbers are radioactive. Naturally occurring $^{238}_{92}U$ is radioactive and disintegrates by the loss of α particles.

$$^{238}_{92}U \longrightarrow ^{234}_{90}Th + ^{4}_{2}He$$

$^{234}_{90}Th$ is also radioactive; its decay is by β^- emission.

$$^{234}_{90}Th \longrightarrow ^{234}_{91}Pa + ^{0}_{-1}\beta$$

$^{234}_{91}Pa$ also decays by β^- emission to produce $^{234}_{92}U$, which is also radioactive.

$$^{234}_{91}Pa \longrightarrow ^{234}_{92}U + ^{0}_{-1}\beta$$

The chain of radioactive decay that begins with $^{238}_{92}U$ continues through a number of steps of α and β^- emission until it eventually terminates with a stable isotope of lead—$^{206}_{82}Pb$. The entire scheme is outlined in Figure 26-2. All naturally occurring radioactive nuclides of high atomic number belong to one of three **radioactive decay series:** the *uranium* series just described, the *thorium* series, or the *actinium* series.

These radioactive decay schemes can be used to determine the ages of rocks and thereby the age of the Earth (see Section 26-5). The appearance of certain radioactive substances in the environment can also be explained through the radioactive decay series. The environmental hazards of ^{222}Rn were described in Section 8-4. Also, ^{210}Po and ^{210}Pb have been detected in cigarette smoke. These radioactive isotopes are derived from ^{238}U, found in trace amounts in the phosphate fertilizers used in tobacco fields. These α-emitting isotopes have been implicated in the link between cigarette smoking and cancer and heart disease.

Radioactivity, which is so common among isotopes of high atomic number, is a relatively rare phenomenon among the *naturally occurring* lighter isotopes. Even so, ^{40}K is a radioactive isotope, as are ^{50}V and ^{138}La. ^{40}K decays by β^- emission and by electron capture, as shown next.

▶ Recall from page 43 that *nuclide* is the general term for an atom with a particular atomic number and mass number. Different nuclides of an element are referred to as *isotopes*.

▶ The term *daughter* is commonly used to describe the new nuclide produced in a radioactive decay process. Thus ^{234}Th is a daughter of ^{238}U, and ^{234}Pa is the daughter of ^{234}Th.

▶ The actinium series actually begins with uranium-235. This isotope of uranium was once known as actino-uranium, which explains the name of the series.

FIGURE 26-2
The natural radioactive decay series for $^{238}_{92}U$ (uranium series).
The long arrows pointing down and to the left correspond to α-particle emissions. The short horizontal arrows represent β^- emissions. Other natural decay series originate with $^{232}_{90}Th$ (thorium series) and $^{235}_{92}U$ (actinium series).

$$^{40}_{19}K \longrightarrow ^{40}_{20}Ca + ^{0}_{-1}\beta \quad \text{and} \quad ^{40}_{19}K + ^{0}_{-1}e \longrightarrow ^{40}_{18}Ar$$

At the time that Earth was formed, ^{40}K was much more abundant than it is now. It is believed that the high argon content of the atmosphere (0.934%, by volume, and almost all of it as ^{40}Ar) has been derived from the radioactive decay of ^{40}K. Aside from ^{40}K and ^{14}C (produced by cosmic radiation), the most important radioactive isotopes of the lighter elements are those that are *artificially* produced.

26-3   Nuclear Reactions and Artificially Induced Radioactivity

Ernest Rutherford discovered that atoms of one element can be transformed into atoms of another element. He did this by bombarding $^{14}_{7}N$ nuclei with α particles, producing $^{17}_{8}O$ and protons. This process can be represented through a nuclear equation.

$$^{14}_{7}N + ^{4}_{2}He \longrightarrow ^{17}_{8}O + ^{1}_{1}H \tag{26.9}$$

In reaction (26.9), instead of a nucleus disintegrating spontaneously, it must be struck by another small particle to induce a nuclear reaction. $^{17}_{8}O$ is a naturally occurring *nonradioactive* isotope of oxygen (0.037% natural abundance). The situation with $^{30}_{15}P$, which can also be produced by a nuclear reaction, is somewhat different.

In 1934, when bombarding aluminum with α particles, Irène Joliot-Curie and her husband Frédéric observed the emission of two types of particles: neutrons and positrons. The Joliots observed that when bombardment by α particles was stopped, the emission of neutrons also stopped; the emission of positrons continued, however. Their conclusion was that the nuclear bombardment produces $^{30}_{15}P$, which undergoes radioactive decay by the emission of positrons.

$$^{27}_{13}Al + ^{4}_{2}He \longrightarrow ^{30}_{15}P + ^{1}_{0}n$$

$$^{30}_{15}P \longrightarrow ^{30}_{14}Si + ^{0}_{+1}\beta$$

$^{30}_{15}P$ was the first radioactive nuclide obtained by artificial means. Now, over 1000 other radioactive nuclides have been produced, and their number considerably exceeds the number of nonradioactive ones (about 280).

Iréne Joliot-Curie (1897–1956). Iréne Joliot-Curie and her husband, Frédéric, shared the 1935 Nobel Prize in Chemistry for the artificial production of radioactive nuclides.

EXAMPLE 26-2

Writing Equations for Nuclear Bombardment Reactions. Write a nuclear equation for the production of ^{56}Mn by bombardment of ^{59}Co with neutrons.

Solution

First we must realize that a particle is produced along with ^{56}Mn. To find the mass number (A) of the "unknown" particle, we must subtract the mass number of the Mn atom from that of the Co atom plus the neutron that initiates the reaction. Thus, for the unknown particle, $A = 59 + 1 - 56 = 4$. Subtracting the atomic number of Mn from that of Co provides the atomic number of the unknown particle: $Z = 27 - 25 = 2$. The unknown particle must have $A = 4$ and $Z = 2$; it is an α particle.

$$^{59}_{27}Co + ^{1}_{0}n \longrightarrow ^{56}_{25}Mn + ^{4}_{2}He$$

Practice Example A: Write a nuclear equation for the production of ^{147}Eu by bombardment of ^{139}La with ^{12}C.

Practice Example B: Write a nuclear equation for the production of ^{124}I by bombardment of ^{121}Sb with α-particles. Also, write an equation for the subsequent decay of ^{124}I by positron emission.

26-4 Transuranium Elements

Until 1940 the only known elements were those that occur naturally. In 1940, the first synthetic element was produced by bombarding $^{238}_{92}U$ atoms with neutrons. First the unstable nucleus $^{239}_{92}U$ is formed. This nucleus then undergoes β^- decay, yielding the element neptunium, with $Z = 93$.

$$^{238}_{92}U + {}^{1}_{0}n \longrightarrow {}^{239}_{92}U + \gamma$$

$$^{239}_{92}U \longrightarrow {}^{239}_{93}Np + {}^{0}_{-1}\beta$$

Bombardment by neutrons is an effective way to produce nuclear reactions because these heavy, uncharged particles are not repelled as they approach a nucleus.

Since 1940 all the elements from $Z = 93$ to 112 have been synthesized. Many of the new elements of high atomic number have been formed using heavy-atom bombardment, in which a transuranium element is bombarded with the nuclei of lighter elements. For example, in 1970, an isotope of the element $Z = 105$ was produced by bombarding atoms of $^{249}_{98}Cf$ with $^{15}_{7}N$ nuclei.

$$^{249}_{98}Cf + {}^{15}_{7}N \longrightarrow {}^{260}_{105}Ha + 4\,{}^{1}_{0}n \qquad (26.10)$$

Elements 104 and 105, rutherfordium (Rf) and hahnium (Ha) are the first in a series of elements following the actinide series and called *transactinide* elements. The most recently created transactinide element is element 112. Element 112 was formed in March 1996 by bombarding lead with zinc nuclei. The result was an atom with an isotopic mass of 277 u. From our knowledge of the periodic table, we expect element 112 to resemble mercury. It has long been proposed that there is an "island of stability" around $Z = 114$ (as we will see in Section 26-7). Element 112 is a step on the way to these potentially more stable elements. Researchers are now working to produce element 114, which should be similar to lead.

To bring about nuclear reactions such as (26.10) requires bombarding atomic nuclei with energetic particles. Such energetic particles can be obtained in an accelerator. A type of accelerator known as a cyclotron is described in Figure 26-3.

dee — dee

direction of magnetic field

to electric oscillator

FIGURE 26-3

A charged-particle accelerator—the cyclotron.
The accelerator consists of two hollow, flat, semicircular boxes, called dees, that are kept electrically charged. The entire assembly is maintained within a magnetic field. The particles to be accelerated, in the form of positive ions, are produced at the center of the opening between the dees. They are then attracted into the negatively charged dee and forced into a circular path by the magnetic field. When the particles leave the dee and enter the gap, the electric charges on the dees are reversed, so that the particles are attracted into the opposite dee. The particles are accelerated as they pass the gap and travel a wider circular path in the new dee. This process is repeated many times until the particles are brought to the required energy.

A *charged-particle accelerator,* as the name implies, can only produce beams of charged particles as projectiles (such as, $_1^1H^+$). In many cases neutrons are more effective as projectiles for nuclear bombardment. The neutrons required can themselves be generated through a nuclear reaction produced by a charged-particle beam. In the following reaction $_1^2H$ represents a beam of deuterons (actually $_1^2H^+$) from an accelerator.

$$_4^9Be + {}_1^2H \longrightarrow {}_5^{10}B + {}_0^1n$$

Another important source of neutrons for nuclear reactions is a nuclear reactor (see Section 26-8).

26-5   Rate of Radioactive Decay

In time, we can expect every atomic nucleus of a radioactive nuclide to disintegrate, but it is impossible to predict when any given nucleus will do so. Although we cannot make predictions for a particular atom, we can develop a statistical method to make predictions for a collection of atoms. Based on experimental observations, we can establish the **radioactive decay law.**

The rate of disintegration of a radioactive material—called the activity, A, or the decay rate—is directly proportional to the number of atoms present.

In mathematical terms,

$$\text{rate of decay} \propto N \qquad \text{and} \qquad \text{rate of decay} = A = \lambda N \qquad (26.11)$$

The activity is expressed in atoms per unit time, such as atoms per second. N is the number of atoms in the sample being observed. λ is the **decay constant;** its unit is time^{-1}. Consider the case of a 1,000,000-atom sample disintegrating at the rate of 100 atoms per second; $N = 1.0 \times 10^6$ and

$$\lambda = \frac{A}{N} = \frac{100 \text{ atom/s}}{1.0 \times 10^6 \text{ atoms}} = 1.0 \times 10^{-4} \text{ s}^{-1}$$

Radioactive decay is a *first-order* process. To relate it to the first-order kinetics that we studied in Chapter 15, think of the activity as corresponding to a rate of reaction; the number of atoms, to the concentration of a reactant; and the decay constant, λ, to a rate constant, k. We can carry this correspondence further by writing an integrated radioactive decay law and a relationship between the decay constant and the half-life.

$$\ln \frac{N_t}{N_0} = -\lambda t \qquad (26.12)$$

$$t_{1/2} = \frac{0.693}{\lambda} \qquad (26.13)$$

In these equations, N_0 represents the number of atoms at some initial time ($t = 0$) and N_t the number of atoms at some later time, t; λ is the decay constant; $t_{1/2}$ is the **half-life.**

Recall that the half-life of a process is the length of time for half of a substance to disappear, and that the half-life of a first-order process is a *constant.*

TABLE 26.1	Some Representative Half-lives				
Nuclide	**Half-life**[a]	**Nuclide**	**Half-life**[a]	**Nuclide**	**Half-life**[a]
$^{3}_{1}H$	12.26 y	$^{40}_{19}K$	1.25×10^9 y	$^{214}_{84}Po$	1.64×10^{-4} s
$^{14}_{6}C$	5730 y	$^{80}_{35}Br$	17.6 min	$^{222}_{86}Rn$	3.823 d
$^{13}_{8}O$	8.7×10^{-3} s	$^{90}_{38}Sr$	27.7 y	$^{226}_{88}Ra$	1.60×10^3 y
$^{28}_{12}Mg$	21 h	$^{131}_{53}I$	8.040 d	$^{234}_{90}Th$	24.1 d
$^{32}_{15}P$	14.3 d	$^{137}_{55}Cs$	30.23 y	$^{238}_{92}U$	4.51×10^9 y
$^{35}_{16}S$	88 d				

[a] s, second; min, minute; h, hour; d, day; y, year.

Thus, if half of the atoms of a radioactive sample disintegrate in 2.5 min, the number of atoms remaining will be reduced to one-fourth of the original number in 5.0 min, one-eighth in 7.5 min, and so on. The shorter the half-life, the larger the value of λ and the faster the decay process. Half-lives of radioactive nuclides range over periods of time from extremely short to very long, as suggested by the representative data in Table 26.1.

EXAMPLE 26-3

Using the Half-life Concept and the Radioactive Decay Law to Describe the Rate of Radioactive Decay. The phosphorus isotope listed in Table 26.1, ^{32}P, is used in biochemical studies to determine the pathways followed by phosphorus atoms in living organisms. Its presence is detected through its emission of β^- particles. **(a)** What is the decay constant for ^{32}P, expressed in the unit s^{-1}? **(b)** What is the activity of a 1.00-mg sample of ^{32}P (that is, how many ^{32}P atoms disintegrate per second)? **(c)** Approximately what mass of the original 1.00-mg sample of ^{32}P will remain after 57 days? **(d)** What will be the rate of radioactive decay after 57 days?

Solution

a. We can determine λ from $t_{1/2}$ with Equation (26.13). The first result we get has the unit d^{-1}. We must successively convert this to h^{-1}, min^{-1}, and s^{-1}.

$$\lambda = \frac{0.693}{14.3 \text{ d}} \times \frac{1 \text{ d}}{24 \text{ h}} \times \frac{1 \text{ h}}{60 \text{ min}} \times \frac{1 \text{ min}}{60 \text{ s}} = 5.61 \times 10^{-7} \text{ s}^{-1}$$

b. First let us find the number of atoms in 1.00 mg of ^{32}P. Then we can multiply this number by the decay constant to get the activity or decay rate.

$$\text{no. } ^{32}P \text{ atoms} = 0.00100 \text{ g} \times \frac{1 \text{ mol } ^{32}P}{32.0 \text{ g}} \times \frac{6.022 \times 10^{23} \, ^{32}P \text{ atoms}}{1 \text{ mol } ^{32}P}$$

$$= 1.88 \times 10^{19} \, ^{32}P \text{ atoms}$$

$$\text{activity} = \lambda N = 5.61 \times 10^{-7} \text{ s}^{-1} \times 1.88 \times 10^{19} \text{ atoms}$$

$$= 1.05 \times 10^{13} \text{ atoms/s}$$

c. A period of 57 days is $57/14.3 = 4.0$ half-lives. As shown in Figure 26-4, the quantity of radioactive material decreases by one half for every half-life. The quantity remaining is $(\frac{1}{2})^4$ of the original quantity.

$$? \text{ mg } ^{32}P = 1.00 \text{ mg} \times (\tfrac{1}{2})^4 = 1.00 \text{ mg} \times \tfrac{1}{16} = 0.063 \text{ mg } ^{32}P$$

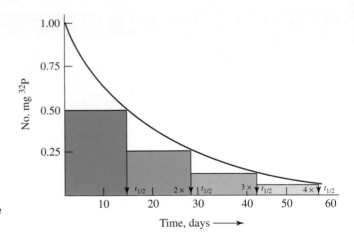

FIGURE 26-4
Radioactive decay of a hypothetical ^{32}P sample—Example
26-3 illustrated.

d. The activity is directly proportional to the number of radioactive atoms remaining
(activity = λN), and the number of atoms is directly proportional to the mass of
^{32}P. When the mass of ^{32}P has dropped to one-sixteenth of its original mass, the
number of ^{32}P atoms also falls to one-sixteenth of the original number, and the
rate of decay is one-sixteenth of the original activity.

$$\text{rate of decay} = \tfrac{1}{16} \times 1.05 \times 10^{13} \text{ atoms/s} = 6.56 \times 10^{11} \text{ atom/s}$$

Practice Example A: ^{131}I is a β^- emitter used as a tracer for radioimmunoas-
says in biological systems. Use information in Table 26.1 to determine **(a)** the decay
constant in s^{-1}; **(b)** The activity of a 2.05-mg sample of ^{131}I; **(c)** The percentage of
^{131}I remaining after 16 days; and **(d)** the rate of β^- emission after 16 days.

Practice Example B: ^{223}Ra has a half-life of 11.4 d. How long would it take
for the activity associated with a sample of ^{223}Ra to decrease to 1.0% of its current
value? [*Hint:* You must use Equation (26.12).]

Radiocarbon Dating

In the upper atmosphere $^{14}_{6}$C is formed at a constant rate by the bombardment
of $^{14}_{7}$N with neutrons.

$$^{14}_{7}\text{N} + ^{1}_{0}\text{n} \longrightarrow ^{14}_{6}\text{C} + ^{1}_{1}\text{H}$$

The neutrons are produced by cosmic rays. $^{14}_{6}$C disintegrates by β^- emission.
    Carbon-containing compounds in living organisms maintain an equilibrium
with ^{14}C in the atmosphere. That is, these organisms replace ^{14}C atoms that have
undergone radioactive decay with "fresh" ^{14}C atoms through interactions with
their environment. The ^{14}C isotope is radioactive and has a half-life of 5730 y.
The activity associated with ^{14}C that is in equilibrium with its environment is
about 15 disintegrations per minute (dis/min) per gram of carbon. When an organism dies (for instance, when a tree is cut down), this equilibrium is destroyed
and the disintegration rate falls off; that is, the dead organism no longer interacts with the atmosphere, and no longer absorbs new ^{14}C. From the measured
disintegration rate at some later time, we can estimate the age (that is, the elapsed
time since the ^{14}C equilibrium was disrupted).

EXAMPLE 26-4

Applying the Integrated Rate Law for Radioactive Decay: Radiocarbon Dating. A wooden object found in an Indian burial mound is subjected to radiocarbon dating. The activity associated with its ^{14}C content is 10 dis min^{-1} g^{-1}. What is the age of the object (in other words, the time elapsed since the tree was cut down)?

Solution

In this example three equations, (26.11), (26.12), and (26.13), are required. Equation (26.13) is used to determine the decay constant.

$$\lambda = \frac{0.693}{5730 \text{ y}} = 1.21 \times 10^{-4} \text{ y}^{-1}$$

Next we use Equation (26.11) to represent the actual number of atoms: N at $t = 0$ (the time when the ^{14}C equilibrium was destroyed) and N_t at time t (the present time). The activity just before the ^{14}C equilibrium was destroyed was 15 dis min^{-1} g^{-1}, and at the time of the measurement it is 10 dis min^{-1} g^{-1}. The corresponding numbers of atoms are equal to these activities divided by λ.

$$N_0 = \frac{A_0}{\lambda} = \frac{15}{\lambda} \quad \text{and} \quad N_t = \frac{A_t}{\lambda} = \frac{10}{\lambda}$$

Finally, we substitute into Equation (26.12).

$$\ln \frac{N_t}{N_0} = \ln \frac{10/\lambda}{15/\lambda} = \ln \frac{10}{15} = -(1.21 \times 10^{-4} \text{ y}^{-1})t$$

$$-0.41 = -(1.21 \times 10^{-4} \text{ y}^{-1})t$$

$$t = \frac{0.41}{1.21 \times 10^{-4} \text{ y}^{-1}} = 3.4 \times 10^3 \text{ y}$$

Practice Example A: What is the age of a mummy, given a ^{14}C activity of 8.5 dis min^{-1} g^{-1}?

Practice Example B: What should be the current activity, in dis min^{-1} g^{-1}, of a wooden object believed to be 1100 years old?

The Age of Earth

The natural radioactive decay scheme of Figure 26-2 suggests the eventual fate awaiting all the $^{238}_{92}U$ found in nature—conversion to lead. Naturally occurring uranium minerals always have associated with them some nonradioactive lead formed by radioactive decay. From the mass ratio of $^{206}_{82}Pb$ to $^{238}_{92}U$ in such a mineral, it is possible to estimate the age of the rock containing the mineral. By the age of the rock we mean the time elapsed since molten magma froze to become a rock. One assumption of this method is that the initial radioactive nuclide, the final stable nuclides, and all the products of a decay series remain in the rock. Another assumption is that any lead present in the rock initially consisted of the several isotopes of lead in their present naturally occurring abundances.

The half-life of $^{238}_{92}U$ is 4.5×10^9 y. According to the natural decay scheme of Figure 26-2, the basic changes that occur as atoms of $^{238}_{92}U$ and its daughters pass through the entire sequence of steps is

$$^{238}_{92}U \longrightarrow ^{206}_{82}Pb + 8 \, ^4_2He + 6 \, ^0_{-1}\beta$$

The decay sequence for $^{238}U \rightarrow {}^{206}Pb$ has 14 steps. The first step, however, has a much longer half-life than any of the other steps in the series. This first step, then, can be thought of as the rate-determining step, with the subsequent steps being "fast." As discussed in Chapter 15, we can ignore the effect on the overall rate of fast steps that occur after the slow, rate-determining step. Thus, the half-life for ^{238}U is essentially equal to the amount of time it takes to convert half the initial ^{238}U to the ^{206}Pb isotope. Discounting the mass associated with the β^- particles, we can see that for every 238 g of uranium that undergoes complete decay, 206 g of lead and 32 g of helium are produced.

Suppose that in a rock containing no lead initially, 1.000 g $^{238}_{92}U$ had disintegrated through one half-life, 4.51×10^9 y. At the end of that time there would be present 0.500 g of undisintegrated $^{238}_{92}U$. The quantity of $^{206}_{82}Pb$ would be

$$0.500 \times \frac{206}{238} = 0.433 \text{ g } {}^{206}_{82}Pb$$

The lead-to-uranium ratio is

$$\frac{^{206}_{82}Pb}{^{238}_{92}U} = \frac{0.433}{0.500} = 0.866$$

If the $^{206}_{82}Pb/^{238}_{92}U$ mass ratio is smaller than 0.866, the age of the rock is less than one half-life of $^{238}_{92}U$. A higher ratio indicates a greater age for the rock. The best estimates of the age of the oldest rocks, and presumably of Earth itself, are in fact about 4.5×10^9 y. These estimates are based on the $^{206}_{82}Pb/^{238}_{92}U$ ratio and on ratios for other pairs of isotopes from natural radioactive decay series.

26-6   Energetics of Nuclear Reactions

To describe the energy change accompanying a nuclear reaction we must use the mass–energy equivalence derived by Albert Einstein.

$$E = mc^2 \tag{26.14}$$

An energy change in a process is always accompanied by a mass change, and the constant that relates them is the square of the speed of light. In chemical reactions energy changes are so small that the equivalent mass changes are undetectable (though real nevertheless). In fact, we base the balancing of equations and stoichiometric calculations on the principle that mass is conserved (unchanged) in a chemical reaction. In nuclear reactions energies are orders of magnitude greater than in chemical reactions. Perceptible changes in mass do occur.

If we know the exact masses of atoms, we can calculate the energy of a nuclear reaction with Equation (26.14). The term m is the net change in mass, in kilograms, and c is expressed in meters per second. The resulting energy is in joules. Another common unit for expressing nuclear energy is the megaelectronvolt (MeV). (Recall from page 900 that 1 eV = 1.6022×10^{-19} J.)

$$1 \text{ MeV} = 1.6022 \times 10^{-13} \text{ J} \tag{26.15}$$

Equation (26.15) is a conversion factor between megaelectronvolts and joules. A conversion factor between atomic mass units (u) and joules (J) is also helpful. We can establish this relationship by determining the energy associated with a mass of 1 u. Let's base our calculation on carbon-12 and note that 1 u is exactly $\frac{1}{12}$ of the mass of a carbon-12 atom. We can calculate the mass, in grams, corresponding to 1 u as follows:

$$1 \text{ u} \times \frac{1 \ ^{12}\text{C atom}}{12 \text{ u}} \times \frac{1 \text{ mol } ^{12}\text{C}}{6.0221 \times 10^{23} \text{ atoms } ^{12}\text{C}} \times \frac{12 \text{ g}}{1 \text{ mol } ^{12}\text{C}} = 1.6606 \times 10^{-24} \text{ g}$$

Converting this value of m to kilograms and using it in Equation (26.14) gives us

$$E = 1 \text{ u} \times \frac{1.6606 \times 10^{-24} \text{ g}}{\text{u}} \times \frac{1 \text{ kg}}{1000 \text{ g}} \times (2.9979 \times 10^8)^2 \frac{\text{m}^2}{\text{s}^2}$$

$$= 1.4924 \times 10^{-10} \text{ J}$$

Thus, the energy equivalent of 1 u is

$$1 \text{ atomic mass unit (u)} = 1.4924 \times 10^{-10} \text{ J} \qquad (26.16)$$

Finally, to express this energy in MeV,

$$1 \text{ atomic mass unit (u)} = 1.4924 \times 10^{-10} \text{ J} \times \frac{1 \text{ MeV}}{1.6022 \times 10^{-13} \text{ J}} = 931.5 \text{ MeV}$$

$$(26.17)$$

These conversion factors will be helpful in solving Example 26-5.

EXAMPLE 26-5

Calculating the Energy of a Nuclear Reaction with the Mass–Energy Relationship. What is the energy, in joules and in megaelectronvolts, associated with the α decay of ^{238}U?

$$^{238}_{92}\text{U} \longrightarrow {}^{234}_{90}\text{Th} + {}^4_2\text{He}$$

The atomic masses in atomic mass units (u) are

$$^{238}_{92}\text{U} = 238.0508 \text{ u} \qquad ^{234}_{90}\text{Th} = 234.0437 \text{ u} \qquad ^4_2\text{He} = 4.0026 \text{ u}$$

Solution

The net change in mass that accompanies the decay of a single nucleus of ^{238}U is $234.0437 + 4.0026 - 238.0508 = -0.0045$ u. This loss of mass appears as kinetic energy carried away by the α-particle. We can use this loss of mass and conversion factors (26.16) and (26.17) to write

$$E = 0.0045 \text{ u} \times \frac{1.49 \times 10^{-10} \text{ J}}{\text{u}} = 6.7 \times 10^{-13} \text{ J}$$

or

$$E = 0.0045 \text{ u} \times \frac{931.5 \text{ MeV}}{\text{u}} = 4.2 \text{ MeV}$$

▶ In this calculation we are using atomic (nuclidic) masses rather than just the masses of the nuclei involved in the nuclear reaction. We can do this because the total number of electrons does not change in the reaction.

Practice Example A: What is the energy associated with the α decay of ^{146}Sm (145.913053 u) to ^{142}Nd (141.907719 u)? Use 4.002603 u as the atomic mass of ^{4}He.

Practice Example B: The decay of ^{222}Rn by α-particle emission is accompanied by a loss of 5.590 MeV of energy. What quantity of mass, in atomic mass units (u), is converted to energy in this process?

Figure 26-5 suggests formation of the nucleus of a $^4_2$He atom from two protons and two neutrons. In this process there is a *mass defect* of 0.0305 u. That is, the experimentally determined mass of a $^4_2$He nucleus is 0.0305 u *less* than

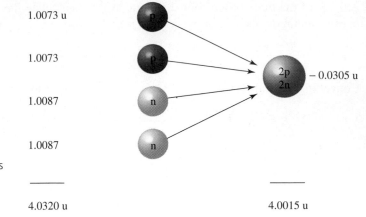

1.0073 u

1.0073

1.0087

1.0087

4.0320 u

2p
2n

− 0.0305 u

4.0015 u

FIGURE 26-5
Nuclear binding energy in $^4_2$He.
The mass of a helium nucleus ($^4_2$He) is 0.0305 u (atomic mass unit) less than the combined masses of two protons and two neutrons. The energy equivalent to this loss of mass (called the mass defect) is the nuclear energy that binds the nuclear particles together.

the combined mass of two protons and two neutrons. This "lost" mass is liberated as energy. With expression (26.17) we can show that 0.0305 u of mass is equivalent to an energy of 28.4 MeV. Because this is the energy released in forming an $^4_2$He nucleus, we can call it the **nuclear binding energy.** Viewed another way, an $^4_2$He nucleus would have to absorb 28.4 MeV to cause its protons and neutrons to become separated. If we consider the binding energy to be apportioned equally among the two protons and two neutrons in $^4_2$He, we obtain a binding energy of 7.10 MeV per nucleon. Similar calculations for other nuclei yield the data needed to plot the graph shown in Figure 26-6.

Figure 26-6 indicates that the maximum binding energy per nucleon is found in a nucleus with a mass number of approximately 60. This leads to two interesting conclusions: (1) If small nuclei are combined into a heavier one (up to about $A = 60$), the binding energy per nucleon increases, and a certain quantity of mass must be converted to energy. The nuclear reaction is highly exothermic.

▶ *Nucleons* are nuclear particles: protons and neutrons.

FIGURE 26-6
Average binding energy per nucleon as a function of atomic number.

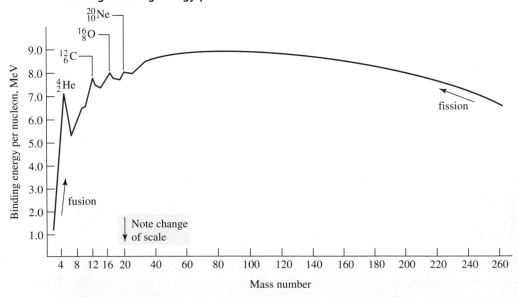

This fusion process serves as the basis of the hydrogen bomb. (2) For nuclei having mass numbers above 60, the addition of extra nucleons to the nucleus would require the expenditure of energy (since the binding energy per nucleon decreases). On the other hand, the *disintegration* of heavier nuclei into lighter ones is accompanied by the release of energy. This nuclear fission process serves as the basis of the atomic bomb and conventional nuclear power reactors. Before considering nuclear fission and fusion, however, let's see what insights Figure 26-6 provides into the question of nuclear stability.

26-7 Nuclear Stability

A number of basic questions have probably occurred to you as we have been describing nuclear decay processes: Why do some radioactive nuclei decay by α emission, some by β^- emission, and so on? Why do the lighter elements have so few naturally occurring radioactive nuclides, whereas those of the heavier elements all seem to be radioactive?

Our first clue to answers for such questions comes from Figure 26-6, in which several nuclides are specifically noted. These nuclides have higher binding energies per nucleon than those of their neighbors. Their nuclei are especially stable. This observation is consistent with a theory of nuclear structure known as the *shell theory*. In the formation of a nucleus, protons and neutrons are believed to occupy a series of nuclear shells. This process is analogous to building up the electronic structure of an atom by the successive addition of electrons to electronic shells. Just as the Aufbau process produces, periodically, electron configurations of exceptional stability, so do certain nuclei acquire a special stability as nuclear shells are closed. This condition of special stability of an atomic nucleus occurs for certain numbers of protons or neutrons known as **magic numbers** (Table 26.2).

Another observation concerning nuclei is that among stable nuclei the most common situation is for the number of protons and the number of neutrons to be *even*. There are fewer stable nuclei with odd numbers of protons and of neutrons. The relationship between numbers of protons (Z), numbers of neutrons (N), and the stability of isotopes is summarized in Table 26.3. Note particularly that stable atoms with the combination Z odd–N odd are very rare. This combination is found only in the nuclides ^2_1H, ^6_3Li, $^{10}_5B$, and $^{14}_7N$. Still another observation is that elements of *odd* atomic number generally have only one or two stable isotopes, whereas those of *even* atomic number have several. Thus, F ($Z = 9$) and I ($Z = 53$) each have only one stable nuclide and Cl ($Z = 17$) and Cu ($Z = 29$) each have two. On the other hand, O ($Z = 8$) has three, Ca ($Z = 20$) has six, and Sn ($Z = 50$) has ten.

**TABLE 26.2
Magic Numbers
for Nuclear Stability**

Number of Protons	Number of Neutrons
2	2
8	8
20	20
28	28
50	50
82	82
114	126
	184

TABLE 26.3 Distribution of Naturally Occurring Stable Nuclides

Combination	Number of Nuclides
Z even–N even	163
Z even–N odd	55
Z odd–N even	50
Z odd–N odd	4

Neutrons are thought to provide a nuclear force to bind protons and neutrons together into a stable unit. Without the neutrons the electrostatic forces of repulsion between positively charged protons would cause the nucleus to fly apart. For the elements of lower atomic numbers (up to about $Z = 20$), the required number of neutrons for a stable nucleus is about equal to the number of protons, for example, ^4_2He, $^{12}_6C$, $^{16}_8O$, $^{28}_{14}Si$, $^{40}_{20}Ca$. For higher atomic numbers, because of increasing repulsive forces between protons, larger numbers of neutrons are required, and the neutron–proton (n/p) ratio increases. For bismuth the ratio is about 1.5:1. Above atomic number 83, no matter how many neutrons are present, the nucleus is unstable. Thus, all isotopes of the known elements with $Z > 83$ are radioactive. Figure 26-7 indicates roughly the range of n/p ratios as a function of atomic number for stable atoms.

Using the ideas outlined here, nuclear scientists have predicted the possible existence of atoms of high atomic number that should have very long half-lives. Currently, a search is on to find such atoms, either naturally or by creating them in a particle accelerator. Figure 26-7 suggests the general range of proton and neutron numbers for these "superheavy" atoms.

FIGURE 26-7
Neutron-to-proton ratio and the stability of nuclides.
(a) The belt of naturally occurring stable nuclides, ranging from ^1_1H to $^{209}_{83}Bi$. (b) Naturally occurring and artificially made radioactive nuclides of the heavier elements. (c) Possible heavy nuclides of high stability (long radioactive half-lives).

EXAMPLE 26-6

Predicting Which Nuclei Are Radioactive. Which of the following nuclides would you expect to be stable and which radioactive? (a) ^{80}As; (b) ^{120}Sn; (c) ^{214}Po.

Solution

a. As has $Z = 33$ and $N = 47$. This is an odd–odd combination that is found only in four of the lighter elements. ^{80}As is radioactive. (Note also that this nuclide is outside the belt of stability in Figure 26-7.)

b. Sn has an atomic number of 50—a magic number. The neutron number is 70 in the nuclide ^{120}Sn. This is an even–even combination, and we should expect the nucleus to be stable. Moreover, Figure 26-7 shows that this nuclide is within the belt of stability. ^{120}Sn is a stable nuclide.

c. ^{214}Po has an atomic number of 84. All known atoms with $Z > 83$ are radioactive. ^{214}Po is radioactive.

Practice Example A: Which of the following nuclides would you expect to be stable and which, radioactive? (a) ^{88}Sr; (b) ^{118}Cs; (c) ^{30}S.

Practice Example B: Of these two fluorine isotopes, ^{17}F and ^{22}F, one decays by β^- emission and the other by β^+ emission. Which does which? (*Hint:* How must the n/p ratios change to make the nuclei more stable?)

26-8   Nuclear Fission

In 1934, Enrico Fermi proposed that transuranium elements might be produced by bombarding uranium with neutrons. He reasoned that the successive loss of β^- particles would cause the atomic number to increase, perhaps to as high as 96. When such experiments were carried out, it was found that in fact the product did emit β^- particles. But in 1938, Otto Hahn, Lise Meitner, and Fritz Strassman found by chemical analysis that the products did not correspond to elements with $Z > 92$. Neither were they the neighboring elements of uranium—Ra, Ac, Th, and Pa. Instead, the products were radioisotopes of much lighter elements, such as Sr and Ba. Neutron bombardment of uranium nuclei causes certain of them to undergo **fission** into smaller fragments, as suggested by Figure 26-8.

The energy equivalent of the mass destroyed in a fission event is somewhat variable, but the average energy is approximately 3.20×10^{-11} J (200 MeV).

$$^{235}_{92}\text{U} + \text{n} \longrightarrow {}^{236}_{92}\text{U} \longrightarrow \text{fission fragments} + \text{neutrons} + 3.20 \times 10^{-11} \text{ J}$$

An energy of 3.20×10^{-11} J may seem small, but this energy is for the fission of a *single* $^{235}_{92}\text{U}$ nucleus. What if 1.00 g $^{235}_{92}\text{U}$ were to undergo fission?

$$? \text{ kJ} = 1.00 \text{ g } {}^{235}\text{U} \times \frac{1 \text{ mol } {}^{235}\text{U}}{235 \text{ g } {}^{235}\text{U}} \times \frac{6.022 \times 10^{23} \text{ atoms } {}^{235}\text{U}}{1 \text{ mol } {}^{235}\text{U}} \times \frac{3.20 \times 10^{-11} \text{ J}}{1 \text{ atom } {}^{235}\text{U}}$$

$$= 8.20 \times 10^{10} \text{ J} = 8.20 \times 10^7 \text{ kJ}$$

This is an enormous quantity of energy! To release this same quantity of energy would require the complete combustion of nearly 3 tons of coal.

▶ Ida Tacke Nodack, co-discoverer of the element rhenium, was the first person to suggest that Fermi's experiments had produced fission. Her explanation was not generally accepted, however, until several years later.

Nuclear Reactors

In the fission of $^{235}_{92}\text{U}$, on average, 2.5 neutrons are released per fission event. These neutrons, on average, produce two or more fission events. The neutrons produced by the second round of fission produce another four or five events,

FIGURE 26-8
Nuclear fission of $^{235}_{92}U$ with thermal neutrons.
A $^{235}_{92}U$ nucleus is struck by a neutron possessing ordinary thermal energy. First the unstable nucleus $^{236}_{92}U$ is produced; this then breaks up into a light and a heavy fragment and several neutrons. A variety of nuclear fragments is possible, but the most probable mass number for the light fragment is 97 and for the heavy one, 137.

^{235}U

^{236}U

and so on. The result is a *chain reaction.* If the reaction is uncontrolled, the released energy causes an explosion; this is the basis of the atomic bomb. Fission leading to an uncontrolled explosion occurs only if the quantity of ^{235}U exceeds the critical mass. The *critical* mass is the quantity of ^{235}U sufficiently large to retain enough neutrons to sustain a chain reaction. Quantities smaller than this are *subcritical;* neutrons escape at too great a rate to produce a chain reaction.

In a nuclear reactor, the release of fission energy is controlled. One common design, called the pressurized water reactor (PWR), is pictured in Figure 26-9. In the core of the reactor, rods of uranium-rich fuel are suspended in water maintained under a pressure of from 70 to 150 atm. The water serves a dual purpose. First, it slows down the neutrons from the fission process so that they possess only normal thermal energy. These thermal neutrons are better able to induce fission than highly energetic ones. In this capacity the water acts as a **moderator.**

FIGURE 26-9
Pressurized water nuclear reactor.

generator

turbine

steam

heat exchanger

water

pump

high-temperature,
high-pressure water

uranium fuel

control rods

concrete shielding

Water also functions as a heat transfer medium. Fission energy maintains the water at a high temperature (about 300 °C). The high-temperature water is brought in contact with colder water in a heat exchanger. The colder water is converted to steam, which drives a turbine, which in turn drives an electric generator. A final component of the nuclear reactor is a set of **control rods,** usually cadmium metal, whose function is to absorb neutrons. When the rods are lowered into the reactor, the fission process is slowed down. When the rods are raised, the density of neutrons and the rate of fission increase.

Breeder Reactors

All that is required to initiate the fission of $^{235}_{92}U$ are neutrons of ordinary thermal energies. On the other hand, nuclei of $^{238}_{92}U$, the abundant nuclide of uranium (99.28%), undergo the following reactions only when struck by energetic neutrons.

$$^{238}_{92}U + ^1_0n \longrightarrow ^{239}_{92}U$$

$$^{239}_{92}U \longrightarrow ^{239}_{93}Np + ^{\ 0}_{-1}\beta$$

$$^{239}_{93}Np \longrightarrow ^{239}_{94}Pu + ^{\ 0}_{-1}\beta$$

A fissionable nuclide such as $^{235}_{92}U$ is called *fissile;* $^{239}_{94}Pu$ is also fissile. A nuclide such as $^{238}_{92}U$, which can be converted into a fissile nuclide, is said to be *fertile.* In a breeder nuclear reactor a small quantity of fissile nuclide provides the neutrons that convert a large quantity of a fertile nuclide into a fissile one. (The newly formed fissile nuclide then participates in a self-sustaining chain reaction.)

An obvious advantage of the breeder reactor is that the amount of uranium "fuel" available immediately jumps by a factor of about 100. This is the ratio of naturally occurring $^{238}_{92}U$ to $^{235}_{92}U$. But the potential advantage is even greater than this. Breeder reactors might use as nuclear fuels materials that have even very low uranium contents, such as shale deposits with about 0.006% U by mass.

There are, however, important disadvantages to breeder reactors. This is especially true of the type known as the liquid-metal-cooled fast breeder reactor (LMFBR). Systems must be designed to handle a liquid metal, such as sodium,

The characteristic blue glow in the water surrounding the core of a nuclear reactor is called Cerenkov radiation. It results when charged particles pass through a transparent medium faster than does light in the same medium. The charged particles are produced by nuclear fission. The radiation is analogous to the shock wave produced in a sonic boom.

which becomes highly radioactive in the reactor. Also, the rates of heat and neutron production are both greater in the LMFBR than in the PWR, so materials deteriorate more rapidly. Perhaps the greatest unsolved problems are those of handling radioactive wastes and reprocessing plutonium fuel. Plutonium is one of the most toxic substances known. It can cause lung cancer when inhaled even in microgram (10^{-6} g) amounts. Furthermore, because of its long half-life (24,000 y), any accident involving plutonium could leave an affected area almost permanently contaminated.

26-9 Nuclear Fusion

The **fusion** of atomic nuclei is the process that produces energy in the sun. An uncontrolled fusion reaction is the basis of the hydrogen bomb. If a fusion reaction can be controlled, this will provide an almost unlimited source of energy. The nuclear reaction that holds the most immediate promise is the deuterium–tritium reaction.

$$\,^2_1H + \,^3_1H \longrightarrow \,^4_2He + \,^1_0n$$

The difficulties in developing a fusion energy source are probably without parallel in the history of technology. In fact, the feasibility of a controlled fusion reaction has yet to be fully demonstrated. There are a number of problems. In order to permit their fusion, the nuclei of deuterium and tritium must be forced into close proximity. Because atomic nuclei repel one another, this close approach requires the nuclei to have very high thermal energies. At the temperatures necessary to initiate a fusion reaction, gases are completely ionized into a mixture of atomic nuclei and electrons known as a *plasma*. Still higher plasma temperatures—over 40,000,000 K—are required to initiate a *self-sustaining* reaction (one that releases more energy than is required to get it started). A method must be devised to confine the plasma out of contact with other materials. The plasma loses thermal energy to any material it strikes. Also, a plasma must be at a sufficiently high density for a sufficient time to permit the fusion reaction to occur. The two methods receiving greatest attention are confinement in a magnetic field and heating of a frozen deuterium–tritium pellet with laser beams.

▶In the hydrogen bomb these high temperatures are attained by exploding an atomic (fission) bomb, which triggers the fusion reaction.

The plasma chamber of a fusion reactor of the magnetic confinement type (called a tokamak). The chamber walls are lined with carbon-fiber composite tiles to protect against the high-temperature plasma.

Another series of technical problems involves the handling of liquid lithium, which is the anticipated heat transfer medium and tritium ($_1^3H$) source.

$$_3^7Li + {}_0^1n \longrightarrow {}_2^4He + {}_1^3H + {}_0^1n$$
$$\text{(fast)} \qquad\qquad\qquad \text{(slow)}$$

Finally, for the magnetic containment method the magnetic field must be produced by superconducting magnets, which currently are very expensive to operate.

The advantages of fusion over fission should be enormous. Since deuterium constitutes about one in every 6500 H atoms, the oceans of the world can supply an almost limitless amount of nuclear fuel. It is estimated that there is sufficient lithium on Earth to provide a source of tritium for about 1 million years.

26-10   Effect of Radiation on Matter

Although there are substantial differences in the way in which α particles, β particles, and γ rays interact with matter, they share an important feature: dislodging electrons from atoms and molecules to produce ions. The ionizing power of radiation may be described in terms of the number of ion pairs formed per centimeter of path through a material. An *ion pair* consists of an ionized electron and the resulting positive ion. Alpha particles have the greatest ionizing power, followed by β particles and then γ rays. The ionized electrons produced directly by the collisions of particles of radiation with atoms are called *primary* electrons. These electrons may themselves possess sufficient energies to cause *secondary* ionizations.

Not all interactions between radiation and matter produce ion pairs. In some cases electrons may simply be raised to higher atomic or molecular energy levels. The return of these electrons to their normal states is then accompanied by radiation—X-rays, ultraviolet light, or visible light, depending on the energies involved.

Some of the possibilities described here are pictured in Figure 26-10.

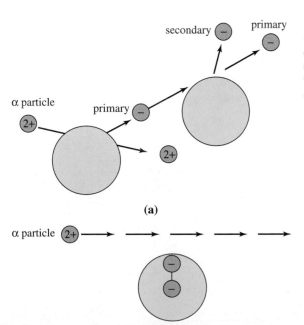

FIGURE 26-10
Some interactions of radiation with matter.
(a) The production of primary and secondary electrons by collisions.
(b) The excitation of an atom by the passage of an α particle. An electron is raised to a higher energy level within the atom. The excited atom reverts to its normal state by emitting radiation.

Radiation Detectors

The interactions of radiation with matter can serve as bases for the detection of radiation and the measurement of its intensity. One of the simplest methods is that used by Henri Becquerel in his discovery of radioactivity—the exposure of a photographic film, as in film badge detectors. The effect of α, β, and γ rays on a photographic emulsion is similar to that of X-rays.

One type of detector used to study high-energy radiation such as γ rays is the **bubble chamber.** In this device a liquid, usually hydrogen, is kept just at its boiling point. As ion pairs are produced by the transit of an ionizing ray, bubbles of vapor form around the ions. The tracks of bubbles can be photographed and analyzed, with different types of radiation producing different tracks. Charged particles, for example, can be detected by their deflection in a magnetic field.

The most common device for detecting and measuring ionizing radiation is the **Geiger-Müller counter** pictured in Figure 26-11. The G-M counter consists of a cylindrical cathode with a wire anode running along its axis. The anode and cathode are sealed in a gas-filled glass tube. The tube is operated in such a way that ions produced by radiation passing through the tube trigger pulses of electric current. It is these pulses that are counted.

Effect of Ionizing Radiation on Living Matter

All life exists against a background of naturally occurring ionizing radiation—cosmic rays, ultraviolet light, and emanations from radioactive elements such as uranium in rocks. The level of this radiation varies from point to point on Earth, being greater, for instance, at higher elevations. Only in recent times have human beings been able to create situations in which living organisms might be exposed to radiation at levels significantly higher than natural background radiation.

The interactions of radiation with living matter are the same as with other forms of matter—ionization, excitation, and dissociation of molecules. There is no question of the effect of large dosages of ionizing radiation on living organisms—the organisms are killed. But even slight exposures to ionizing radiation can cause changes in cell chromosomes. Thus it is believed that even at low dosage rates ionizing radiation can result in birth defects, leukemia, bone cancer, and other forms of cancers. The nagging question that has eluded any definitive answers is how great an increase in the incidence of birth defects and cancers might be caused by certain levels of radiation.

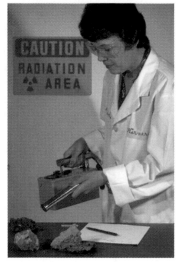

A geologist using a Geiger-Müller counter to check rocks for their radioactivity.

FIGURE 26-11
Geiger-Müller counter.
Radiation enters the G-M tube through the mica window. The radiation ionizes some of the gas (usually argon) in the tube. A pulse of electric current passes through the electric circuit and is counted.

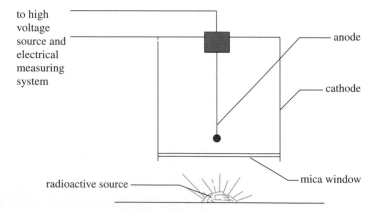

TABLE 26.4	Units of Radiation Dosage
Unit	**Definition**
curie	An amount of radioactive material decaying at the same rate as 1 g of radium (3.7×10^{10} dis/s).
rad	A dosage of radiation corresponding to the absorption of 1×10^{-2} J of energy per kilogram of matter.
rem	A unit related to the rad, but taking into account the varying effects of different types of radiation of the same energy on biological matter. This relationship is through a "quality factor," which may be taken as equal to 1 for X-rays, γ rays, and β^- particles. For protons and slow neutrons the factor has a value of 5; and for α particles, 10. Thus, an exposure to 1 rad of X-rays is about equal to 1 rem, but 1 rad of α particles is about equal to 10 rem.

Sources of α radiation are relatively harmless when external to the body and extremely hazardous when taken internally, as in the lungs or stomach. Other forms of radiation (X-rays, γ rays), because they are highly penetrating, are hazardous even when external to the body.

Radiation Dosage

Several different units are used to describe radiation dosage, that is, the amount of radiation to which matter is exposed. A summary is provided in Table 26.4.

It is thought that exposure in a single short time interval to 1000 rem of radiation would kill 100% of the population. Exposure to 450 rem would probably result in death within 30 days in about 50% of the population. A single dosage of 1 rem delivered to 1 million people would probably produce about 100 cases of cancer within 20–30 years. The total body radiation received by most of the world's population from normal background sources is about 0.13 rem [130 millirem (mrem)] per year. The exposure in a chest X-ray examination is about 20 mrem; and the estimated maximum exposure to the general population associated with the production of nuclear power is about 5 mrem/y.

Some of the foregoing statements about radiation exposures and their anticipated effects are based on (1) medical histories of the survivors of the Hiroshima and Nagasaki atomic blasts, (2) the incidence of leukemia and other cancers in children whose mothers received diagnostic radiation during pregnancy, and (3) the occurrence of lung cancers among uranium miners in the United States. What does all of this tell us about a "safe" level of radiation exposure? One approach has been to extrapolate from these high dosage rates to the lower dosages affecting the general population. This has led the U.S. National Council on Radiation Protection and Measurements to recommend that the dosage rate for the general population be limited to 0.17 rem (170 mrem) per year from all sources above background level. Experts disagree, however, on how the data observed for high-dosage exposures should be extrapolated to low dosages. According to some, the 0.17 rem/y figure is too high, by perhaps a factor of 10. If so, exposure to an additional 0.17 rem/y above normal background levels might cause statistically significant increases in the incidence of birth defects and cancers.

26-11   Applications of Radioisotopes

We have described both the destructive capacity of nuclear reactions and the potential of these reactions to provide new sources of energy. Less heralded but equally important are a variety of practical applications of radioactivity. We close this chapter with a brief survey of some uses of radioisotopes.

Cancer Therapy

Ionizing radiation in low dosages can induce cancers, but this same radiation, particularly γ rays, can also be used in the treatment of cancer. Although ionizing radiation tends to destroy all cells, cancerous cells are more easily destroyed than normal ones. Thus, a carefully directed beam of γ rays or high-energy X rays of the appropriate dosage may be used to arrest the growth of cancerous cells. Also coming into use for some forms of cancer is radiation therapy using beams of protons or neutrons.

Radioactive Tracers

▶ Properties affected by mass, such as rate of diffusion can exhibit differences when one isotope is substituted for another. This is noticed most when deuterium is substituted for hydrogen. This substitution may result in an *isotope effect* on the rates of some reactions.

▶ Solutions of iodide ion were distributed to the general population in some areas of Europe following the Chernobyl accident in 1986.

Radioactive isotopes are different from nonradioactive ones only in the instability of their nuclei, not in physical or chemical properties. Thus, in physical or chemical processes radioactive and nonradioactive isotopes behave in the same way. This is the principle behind the use of radioactive tracers, or "tagged" atoms. For example, if a small quantity of radioactive ^{32}P (as a phosphate) is added to a nutrient solution that is fed to plants, the uptake of the phosphorus can be followed by charting the regions of the plant that become radioactive. Similarly, the fate of iodine (as iodide ion) in the body can be determined by having a person drink a solution of dissolved iodides containing a small quantity of a radioactive iodide as a tracer. Abnormalities in the thyroid gland can be detected in this way. Because I^- concentrates in the thyroid, people can protect themselves by ingesting nonradioactive iodides before being exposed to a radioactive iodide. The thyroid becomes saturated with the nonradioactive iodide and rejects the radioactive iodide.

Industrial applications of tracers are also numerous. The fate of a catalyst in a chemical plant can be followed by incorporating a radioactive tracer in the catalyst, for example, ^{192}Ir in a Pt–Ir catalyst. By monitoring the activity of the ^{192}Ir one can determine the rate at which the catalyst is being carried away and to which parts of the plant.

Structures and Mechanisms

Often the mechanism of a chemical reaction or the structure of a species can be inferred from experiments using radioisotopes as tracers. Consider the following experimental proof that the two S atoms in the thiosulfate ion, $S_2O_3{}^{2-}$, are not equivalent.

$S_2O_3{}^{2-}$ is prepared from radioactive sulfur (^{35}S) and sulfite ion containing the nonradioactive isotope ^{32}S.

$$^{35}S + {}^{32}SO_3{}^{2-} \longrightarrow {}^{35}S{}^{32}SO_3{}^{2-} \tag{26.18}$$

FIGURE 26-12
Structure of thiosulfate ion, $S_2O_3{}^{2-}$.
The central S atom (red) is in the oxidation state +6. The terminal S atom (blue) is in the oxidation state −2.

When the thiosulfate ion is decomposed by acidification, all the radioactivity appears in the precipitated sulfur and none in the $SO_2(g)$. The ^{35}S atoms must be bonded in a different way than the ^{32}S atoms (Figure 26-12).

$$^{35}S{}^{32}SO_3{}^{2-} + 2\ H^+ \longrightarrow H_2O + {}^{32}SO_2(g) + {}^{35}S(s) \tag{26.19}$$

In reaction (26.20) nonradioactive KIO_4 is added to a solution containing iodide ion labeled with the radioisotope ^{128}I. All the radioactivity appears in the I_2 and none in the $IO_3{}^-$. This proves that all the $IO_3{}^-$ is produced by reduction of $IO_4{}^-$ and none by oxidation of I^-.

$$IO_4{}^- + 2\ {}^{128}I^- + H_2O \longrightarrow {}^{128}I_2 + IO_3{}^- + 2\ OH^- \tag{26.20}$$

Analytical Chemistry

The usual procedure of analyzing a substance by precipitation involves filtering, washing, drying, and weighing a pure precipitate. An alternative is to incorporate a radioactive isotope in the precipitating reagent. By measuring the activity of the precipitate and comparing it with that of the original solution, it is possible to calculate the amount of precipitate without having to purify, dry, and weigh it.

Another method of importance in analytical chemistry is *neutron activation analysis*. In this procedure the sample to be analyzed, normally nonradioactive, is bombarded with neutrons; the element of interest is converted to a radioisotope. The activity of this radioisotope is measured. This measurement, together with such factors as the rate of neutron bombardment, the half-life of the radioisotope, and the efficiency of the radiation detector, can be used to calculate the quantity of the element in the sample. The method is especially attractive because (1) trace quantities of elements can be determined (sometimes in parts per billion or less); (2) a sample can be tested without destroying it; and (3) the sample can be in any state of matter, including biological materials. Neutron activation analysis has been used, for instance, to determine the authenticity of old paintings. (Old masters formulated their own paints. Differences between formulations are easily detected through the trace elements they contain.)

Radiation Processing

Radiation processing describes industrial applications of ionizing radiation—γ rays from ^{60}Co or electron beams from electron accelerators. The ionizing radiation is used in the production of certain materials or to modify their properties. Its most extensive current use is in breaking, reforming, and crosslinking polymer chains to affect the physical and mechanical properties of plastics used in foamed products, electrical insulation, and packaging materials. Exposure to ionizing radiation is used to sterilize medical supplies such as sutures, syringes, and hospital garb. In sewage treatment plants radiation processing has been used to decrease the settling time of sewage sludge and to kill pathogens. A currently developing use is in the preservation of foods, as an alternative to canning, freeze-drying, or refrigeration. In radiation processing the irradiated material is *not* rendered radioactive, although the ionizing radiation may produce some chemical changes.

(a)

(b)

Thallium-201 scan of a heart. Nuclear medicine provides methods for diagnosing life-threatening conditions. Here, γ rays released in the radioactive decay of ^{201}Tl are detected and used to provide an image of the blood flow to the heart wall (heart muscle). (a) Normal heart wall; and (b) heart with deficient blood flow.

The mushrooms on the right have been preserved by irradiation.

Focus On   Radioactive Waste Disposal

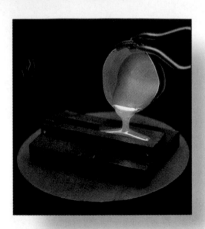

Pouring molten nuclear waste glass to demonstrate methods of immobilizing nuclear waste.

A distinct advantage of nuclear power generation over the use of fossil fuels is that it does not produce oxides of sulfur and nitrogen as air pollutants. Nor does nuclear power generation produce $CO_2(g)$, so it does not contribute to the potential problem of global warming (see Section 7-9). Nuclear power generation has its own waste problem, however. The disposal of nuclear wastes is one of the most difficult environmental problems.

Radioactive waste can be divided into two broad categories. Low-level waste consists of items used in handling radioactive materials (such as gloves and other protective clothing) and waste solutions from medical and laboratory use. The radioactive elements in these items have relatively short half-lives (30 years maximum) and low levels of activity. They can be stored in cement casks and buried. After about 300 years these materials will no longer be radioactive.

The greater challenge is in handling spent fuel rods from nuclear reactors and other high-level radioactive wastes. In addition to significant quantities of uranium and plutonium, spent fuel rods contain a number of daughter isotopes with half-lives of hundreds of years. Plutonium, with a half-life of 24,000 years, is so toxic that any accidental release would contaminate an area for a very long time.

In the United States, high-level wastes are currently stored under water at temporary sites at power plants and elsewhere while awaiting final storage at a permanent site. An appropriate site would have to be geologically stable for tens of thousands of years. One or two such sites are now under careful study, but selection of a final site is still a strongly debated issue.

France, which has an extensive nuclear power program including one breeder reactor, has chosen a

Summary

Radioactivity refers to the ejection of particles (α, β^-, β^+), the capture of electrons from an inner shell, or the emission of electromagnetic radiation (γ) by unstable nuclei. With the exception of γ-ray emission, radioactive decay leads to the transformation (transmutation) of one element into another.

All nuclides with $Z > 83$ are radioactive. Although a few occur naturally, most radioactive nuclides of lower atomic number are produced artificially, by bombarding appropriate target nuclei with energetic particles. The basic rule in writing equations for nuclear reactions is that the sum of the atomic numbers and the sum of the mass numbers must be the same on both sides of the equation.

The rate of radioactive decay—the activity of a sample—is directly proportional to the number of atoms. Calculations of decay rates can be based on equations similar to those for first-order chemical kinetics. Measurements of decay rates of radioactive nuclides have a number of practical applications, ranging from determining the ages of rocks to the dating of certain carbon-containing objects (radiocarbon dating).

The quantity of energy released in the formation of a nucleus from protons and neutrons can be plotted as a function of mass number, yielding a distinctive graph. From this graph one can establish that fission of heavy nuclei and fusion of lighter nuclei yield large quantities of energy. Fission is the basis of nuclear reactors, and fusion is the energy-producing process of the stars.

The stability of a nucleus depends on several factors, including the neutron-to-proton ratio in the nucleus and whether this ratio lies within the belt of stable nuclides (Figure 26-7). Nuclides outside the belt of stability are radioactive.

One of the principal effects of the interaction of radiation with matter is the production of ions. This phenomenon can be used to detect radiation, and it is also the basis of radiation damage to living matter. Several methods have been developed to measure radiation dosages and to predict the biological effects of these dosages, but much uncertainty remains. Despite the hazards associated with radioactivity, radioactive nuclides have beneficial uses in cancer therapy, in basic studies of chemical structures and mechanisms, in analytical chemistry, and in the chemical industry.

different route for handling nuclear reactor waste: *reprocessing.* The first step in nuclear fuel reprocessing is the removal of uranium and plutonium from spent fuel rods. These elements are formed into fuel pellets for loading new fuel rods. Low-level waste is dealt with in the manner already described. What remains at this point are isotopes with long half-lives (up to 100 years). These are stabilized by incorporating them into a borosilicate (Pyrex-like) glass. The boron in the glass is a good neutron absorber. The radioactive glass is placed in sealed containers, and the containers are stored in specially designed silos. The silos are intended to contain the waste for at least 1000 years, and consideration is also being given to storage in caves known to have been geologically stable since the last ice age.

Reprocessing uranium and plutonium is a hazardous operation that must be done with extreme care, mostly through the use of remote-control equipment. An attendant problem with reprocessing is that the plutonium recovered from spent fuel rods is of weapons grade. Strict safeguards are required to ensure that none of this material is diverted for weapons production.

Thus, "clean" nuclear energy, while it does not contribute to problems such as acid rain and global warming, presents its own set of problems. In planning a future energy strategy, the issues are as much social and political as they are scientific. How much risk are we willing to accept in return for the perceived benefits of a given course of action? Should our efforts be directed to developing a new generation of safe nu-

clear reactors to reduce the possible problems of global warming associated with fossil fuels? Should greater emphasis be given to energy conservation and the more efficient use of energy than to increasing our reliance on nuclear energy? Should we largely forego nuclear fission as an energy source while awaiting practical development of less risky fusion power, a technology that has not yet been demonstrated? Is an expanded development of solar energy and other unconventional energy sources a way around the dilemmas posed by fossil fuels and nuclear power? These are all questions that must be seriously debated as we make decisions affecting the energy needs and resources of our society.

Summarizing Example

On April 26, 1986, an explosion at the nuclear power plant at Chernobyl, Ukraine, released the greatest quantity of radioactive material ever associated with an industrial accident. One of the radioisotopes in this emission was ^{131}I, a β^- emitter with a half-life of 8.04 days.

Assume that the total quantity of ^{131}I released was 250 g, and estimate the number of *curies* associated with this ^{131}I one month (30 days) after the accident.

1. *Calculate the mass of ^{131}I remaining after one month.* In Equation (26.12), use $\lambda = 0.693/8.04$ d and $t = 30$ d. Because the mass of ^{131}I is directly proportional to the number of ^{131}I atoms, if N_0 is taken to be 250, the numerical value of N_t will be the same as the mass that is sought. *Result:* 19 g ^{131}I. [Note

that 30 d is slightly less than $4t_{1/2}$; the expected result is slightly larger than $(\frac{1}{2})^4 \times 250$ g $= 16$ g.]

2. *Determine the number of atoms in 19 g ^{131}I.* Use the molar mass, 131 g ^{131}I/mol ^{131}I, and the Avogadro constant. *Result:* 8.7×10^{22} ^{131}I atoms.

3. *Determine the decay constant in s^{-1}.* Use the method outlined in Example 26-3a to obtain λ from $t_{1/2}$. *Result:* 9.98×10^{-7} s^{-1}.

4. *Determine the decay rate (activity) of the ^{131}I after 1 month.* Use the results of parts **2** and **3** in Equation (26.12). *Result:* $A = 8.7 \times 10^{16}$ dis s^{-1}.

5. *Express the activity of the remaining ^{131}I in curies.* Use the definition of a curie in Table 26.4 to convert from dis s^{-1} to curies. *Answer:* 2.4×10^6 curies.

Key Terms

alpha (α) particle (26-1)
beta (β^-) particle (26-1)

control rods (26-8)
curie (26-10)

decay constant (26-5)
electron capture (EC) (26-1)

fission (26-8)	**magic numbers** (26-7)	**rad** (26-10)
fusion (26-9)	**moderator** (26-8)	**radioactive decay law** (26-5)
gamma (γ) ray (26-1)	**nuclear binding energy** (26-6)	**radioactive decay series** (26-2)
Geiger-Müller (G-M) counter (26-10)	**nuclear equation** (26-1)	**rem** (26-10)
half-life (26-5)	**positron (β^+)** (26-1)	

Review Questions

1. In your own words define the following symbols: **(a)** α; **(b)** β^-; **(c)** β^+; **(d)** γ; **(e)** $t_{1/2}$.

2. Briefly describe each of the following ideas, phenomena, or methods: **(a)** radioactive decay series; **(b)** charged-particle accelerator; **(c)** neutron-to-proton ratio; **(d)** mass–energy relationship; **(e)** background radiation.

3. Explain the important distinctions between each pair of terms: **(a)** electron and positron; **(b)** half-life and decay constant; **(c)** mass defect and nuclear binding energy; **(d)** nuclear fission and nuclear fusion; **(e)** primary and secondary ionization.

4. Which of the following—α, β, or γ—generally has the greatest **(a)** penetrating power through matter; **(b)** ionizing power in matter; **(c)** deflection in a magnetic field?

5. Supply the missing information in each of the following nuclear equations representing a radioactive decay process.

(a) $^{160}_{?}W \rightarrow\ ^{?}_{?}Hf + ?$ **(b)** $^{38}_{?}Cl \rightarrow\ ^{?}_{?}Ar + ?$

(c) $^{214}_{?}? \rightarrow\ ^{?}_{?}Po +\ ^{0}_{-1}\beta$ **(d)** $^{32}_{17}Cl \rightarrow\ ^{?}_{16}? + ?$

6. Complete the following nuclear equations.

(a) $^{23}_{11}Na + ? \rightarrow\ ^{24}_{11}Na +\ ^{1}_{1}H$

(b) $^{59}_{27}Co +\ ^{1}_{0}n \rightarrow\ ^{56}_{25}Mn + ?$

(c) $? +\ ^{2}_{1}H \rightarrow\ ^{240}_{94}Pu +\ ^{0}_{-1}\beta$

(d) $^{246}_{96}Cm + ? \rightarrow\ ^{254}_{102}No + 5\ ^{1}_{0}n$

(e) $^{238}_{92}U + ? \rightarrow\ ^{246}_{99}Es + 6\ ^{1}_{0}n$

7. Write nuclear equations to represent
(a) the decay of ^{214}Ra by α-particle emission;
(b) the decay of ^{205}At by positron emission;
(c) the decay of ^{212}Fr by electron capture;
(d) the reaction of two deuterium nuclei (deuterons) to produce a nucleus of ^{3}He;
(e) the production of $^{243}_{97}Bk$ by the α-particle bombardment of $^{241}_{95}Am$.

8. For the radioactive nuclides in Table 26.1,
(a) which one has the largest value of the decay constant, λ?
(b) which one displays a 75% reduction in radioactivity from its current value in approximately one month?
(c) which ones lose more than 99% of their radioactivity in one month?

9. Two radioisotopes are to be compared. Isotope A requires 18.0 h for its decay rate to fall to $\frac{1}{16}$ of its initial value. Isotope B has a half-life that is 2.5 times that of A. How long does it take for the decay rate of isotope B to decrease to $\frac{1}{32}$ of its initial value?

10. A sample of radioactive $^{35}_{16}S$ is found to disintegrate at a rate of 1.00×10^3 atoms/min. The half-life of $^{35}_{16}S$ is 87.9 d. How long will it take for the activity of this sample to decrease to the point of producing **(a)** 253; **(b)** 104; and **(c)** 52 dis/min?

11. With appropriate equations in the text, determine
(a) The energy in joules corresponding to the destruction of 6.02×10^{-23} g of matter.
(b) The energy in megaelectronvolts that would be released if one α particle was completely destroyed.
(c) The number of neutrons that could conceivably be created from 6.75×10^6 MeV of energy. (*Hint:* Refer to Table 2.1.)

12. The measured mass of the nucleus of an atom of silver-107 is 106.879289 u. For this atom, determine the binding energy per nucleon in megaelectronvolts. (*Hint:* Recall the discussion on pages 909 and 910.)

13. Two of the following nuclides do not occur naturally. Which do you think they are? **(a)** ^{2}H; **(b)** ^{32}S; **(c)** ^{80}Br; **(d)** ^{132}Cs; **(e)** ^{184}W.

14. Explain why
(a) Radioactive nuclides with intermediate half-lives are generally more hazardous than those with extremely short or extremely long half-lives.
(b) Some radioactive substances are hazardous from a distance, whereas others must be taken internally to constitute a hazard.
(c) Argon is the most abundant of the noble gases in the atmosphere.
(d) Francium is such a rare element (less than about 30 g is present in Earth's crust at any one time), and why francium cannot be extracted from minerals containing other metals of Group 1A.
(e) Such extremely high temperatures will be required to develop a self-sustaining thermonuclear (fusion) process as an energy source.

Exercises

Radioactive Processes

15. What is the nucleus obtained in each process?
(a) $^{234}_{94}Pu$ decays by α emission.
(b) $^{248}_{97}Bk$ decays by β^- emission.
(c) $^{196}_{82}Pb$ goes through two successive EC processes.

16. What is the nucleus obtained in each process?
(a) $^{214}_{82}Pb$ decays through two successive β^- emissions.
(b) $^{226}_{88}Ra$ decays through three successive α emissions.
(c) $^{69}_{33}As$ decays by β^+ emission.

Radioactive Decay Series

17. The natural decay series starting with the radionuclide $^{232}_{90}\text{Th}$ follows the sequence represented here. Construct a graph of this series, similar to Figure 26-2.

$$^{232}_{90}\text{Th}-\alpha-\beta-\beta-\alpha-\alpha-\alpha\underset{\beta-\alpha}{\overset{\alpha-\beta}{\times}}\overset{\alpha-\beta}{\underset{\beta-\alpha}{\times}}\searrow\,^{208}_{82}\text{Pb}$$

19. The uranium series described in Figure 26-2 is also known as the "$4n + 2$" series because the mass number of each nuclide in the series can be expressed by the equation $A = 4n + 2$, where n is an integer. Show that this equation is indeed applicable to the uranium series.

21. Based on a favorable n:p ratio for the product nucleus, write the most plausible equation for the decay of $^{14}_{6}\text{C}$.

Nuclear Reactions

23. Write equations for the following nuclear reactions.
(a) bombardment of ^{7}Li with protons to produce ^{8}Be and γ rays.
(b) bombardment of ^{9}Be with $^{2}_{1}\text{H}$ to produce ^{10}B.
(c) bombardment of ^{14}N with neutrons to produce ^{14}C.

Rate of Radioactive Decay

25. The disintegration rate for a sample containing $^{60}_{27}\text{Co}$ as the only radioactive nuclide is found to be 6740 dis/h. The half-life of $^{60}_{27}\text{Co}$ is 5.2 y. Estimate the number of atoms of $^{60}_{27}\text{Co}$ in the sample.

27. A sample containing $^{234}_{88}\text{Ra}$, which decays by α-particle emission, is observed to disintegrate at the following rate, expressed as disintegrations per minute or counts per minute (cpm). What is the half-life of this nuclide? $t = 0$, 1000 cpm; $t = 1$ h, 992 cpm; $t = 10$ h, 924 cpm; $t = 100$ h, 452 cpm; $t = 250$ h, 138 cpm.

29. What should be the mass ratio $^{208}\text{Pb}/^{232}\text{Th}$ in a meteorite that is approximately 2.7×10^9 y old? The half-life of ^{232}Th is 1.39×10^{10} y. (*Hint:* One ^{208}Pb atom is the final decay product from one ^{232}Th atom.)

Radiocarbon Dating

31. A wooden object is claimed to have been found in an Egyptian pyramid and is offered for sale to an art museum. Radiocarbon dating of the object reveals a disintegration rate of 10.0 dis min^{-1} g^{-1}. Do you think the object is authentic? Explain.

Energetics of Nuclear Reactions

33. Use the electron mass from Table 2.1 and the measured mass of the nuclide $^{19}_{9}\text{F}$, 18.998403 u, to determine the binding energy per nucleon (in megaelectronvolts) in this atom. (*Hint:* How is the nuclear mass related to the atomic mass of the nuclide?)

35. Calculate the energy, in megaelectronvolts, released in the nuclear reaction

$$^{10}_{5}\text{B} + ^{4}_{2}\text{He} \longrightarrow ^{13}_{6}\text{C} + ^{1}_{1}\text{H}$$

The nuclidic masses are $^{10}_{5}\text{B} = 10.01294$ u; $^{4}_{2}\text{He} = 4.00260$ u; $^{13}_{6}\text{C} = 13.00335$ u; $^{1}_{1}\text{H} = 1.00783$ u.

18. The natural decay series starting with the radionuclide $^{235}_{92}\text{U}$ follows the sequence represented here. Construct a graph of this series, similar to Figure 26-2.

$$^{235}_{92}\text{U}-\alpha-\beta-\alpha\underset{\beta-\alpha}{\overset{\alpha-\beta}{\times}}-\alpha-\alpha-\alpha-\beta\underset{\beta-\alpha}{\overset{\alpha-\beta}{\times}}\nearrow\,^{207}_{82}\text{Pb}$$

20. By the description in Exercise 19, the thorium series can be called the "$4n$" series and the actinium series the "$4n + 3$" series. A "$4n + 1$" series has also been established with $^{241}_{94}\text{Pu}$ as the parent nuclide. To which series does each of the following belong? **(a)** $^{214}_{83}\text{Bi}$; **(b)** $^{216}_{84}\text{Po}$; **(c)** $^{215}_{85}\text{At}$; **(d)** $^{235}_{92}\text{U}$.

22. Tritium, $^{3}_{1}\text{H}$, is the radioactive isotope of hydrogen. Write a plausible equation for its decay.

24. Write equations for the following nuclear reactions.
(a) bombardment of ^{238}U with α-particles to produce ^{239}Pu.
(b) bombardment of tritium with $^{2}_{1}\text{H}$ to produce ^{4}He.
(c) bombardment of ^{33}S with neutrons to produce ^{33}P.

26. How many years must the radioactive sample of Exercise 25 be maintained before the disintegration rate falls to 101 dis/min?

28. Iodine-129 is a product of nuclear fission, whether from an atomic bomb or a nuclear power plant. It is a β^- emitter with a half-life of 1.7×10^7 y. How many disintegrations per second would occur in a sample containing 1.00 mg ^{129}I?

30. Concerning the decay of ^{232}Th described in Exercise 29, a certain rock is found to have a $^{208}\text{Pb}/^{232}\text{Th}$ mass ratio of 0.25:1.00. Estimate the age of the rock.

32. The lowest level of ^{14}C activity that seems possible for experimental detection is 0.03 dis min^{-1} g^{-1}. What is the maximum age of an object that can be determined by the carbon-14 method?

34. Use the electron mass from Table 2.1 and the measured mass of the nuclide $^{56}_{26}\text{Fe}$, 55.934939 u, to determine the binding energy per nucleon (in megaelectronvolts) in this atom. (*Hint:* How is the nuclear mass related to the atomic mass of the nuclide?)

36. You are given the following nuclidic masses; $^{6}_{3}\text{Li} = 6.01513$ u; $^{4}_{2}\text{He} = 4.00260$ u; $^{3}_{1}\text{H} = 3.01604$ u; $^{1}_{0}\text{n} = 1.008665$ u. How much energy, in megaelectronvolts, is released in the nuclear reaction

$$^{6}_{3}\text{Li} + ^{1}_{0}\text{n} \longrightarrow ^{4}_{2}\text{He} + ^{3}_{1}\text{H}$$

Nuclear Stability

37. Which member of the following pairs of nuclides would you expect to be most abundant in natural sources? Explain your reasoning. **(a)** $^{20}_{10}Ne$ or $^{22}_{10}Ne$; **(b)** $^{17}_8O$ or $^{18}_8O$; **(c)** ^6_3Li or ^7_3Li.

39. One member each of the following pairs of radioisotopes decays by β^- emission and the other by positron (β^+) emission. Which is which? Explain your reasoning. **(a)** $^{29}_{15}P$ and $^{33}_{15}P$; **(b)** $^{120}_{53}I$ and $^{134}_{53}I$.

41. Some nuclides are said to be "doubly magic." What do you suppose this term means? Postulate some nuclides that might be doubly magic and locate them in Figure 26-7.

38. Which member of the following pairs of nuclides would you expect to be most abundant in natural sources? Explain your reasoning. **(a)** $^{40}_{20}Ca$ or $^{42}_{20}Ca$; **(b)** $^{31}_{15}P$ or $^{32}_{15}P$; **(c)** $^{63}_{30}Zn$ or $^{64}_{30}Zn$.

40. Each of the following isotopes is radioactive. Which would you expect to decay by β^+ emission? **(a)** $^{28}_{15}P$; **(b)** $^{45}_{19}K$; **(c)** $^{73}_{30}Zn$.

42. Both β^- and β^+ emissions are observed for artificially produced radioisotopes of low atomic numbers, but only β^- emission is observed with naturally occurring radioisotopes of high atomic number. What is the reason for this observation?

Fission and Fusion

43. Refer to the Summarizing Example. In contrast to the Chernobyl accident, the 1979 nuclear accident at Three Mile Island released only 170 curies of ^{131}I. How many milligrams of ^{131}I does this represent?

44. Use data from the text to determine how many metric tons (1 metric ton = 1000 kg) of bituminous coal (85% C) would have to be burned to release as much energy as is produced by the fission of 1.00 kg $^{235}_{92}U$. (*Hint:* What is the heat of combustion of carbon?)

Effect of Radiation on Matter

45. Explain why the rem is more satisfactory than the rad as a unit for measuring radiation dosage.

47. ^{90}Sr is both a product of radioactive fallout and a radioactive waste in a nuclear reactor. This radioisotope is a β^- emitter with a half-life of 27.7 y. Suggest reasons why ^{90}Sr is such a potentially hazardous substance.

46. Discuss briefly the basic difficulties in establishing the physiological effects of low-level radiation.

48. ^{222}Rn is an α-particle emitter with a half-life of 3.82 days. Is it hazardous to be near a flask containing this isotope? Under what conditions might ^{222}Rn be potentially hazardous?

Application of Radioisotopes

49. Describe how you might go about finding a leak in the $H_2(g)$ supply line in an ammonia synthesis plant by using radioactive materials.

51. A small quantity of NaCl containing radioactive $^{24}_{11}Na$ is added to an aqueous solution of $NaNO_3$. The solution is cooled and $NaNO_3$ is crystallized from the solution. Would you expect the $NaNO_3(s)$ to be radioactive? Explain.

50. Explain why neutron activation analysis is so useful in determining trace elements in a sample, in contrast to ordinary methods of quantitative analysis, such as precipitation or titration.

52. The following reactions are carried out with HCl(aq) containing some tritium (^3_1H) as a tracer. Would you expect any of the tritium radioactivity to appear in the $NH_3(g)$? In the H_2O? Explain.

$$NH_3(aq) + HCl(aq) \longrightarrow NH_4Cl(aq)$$
$$NH_4Cl(aq) + NaOH(aq) \longrightarrow NaCl(aq) + H_2O + NH_3(g)$$

Advanced Exercises

53. Sometimes the most abundant isotope of an element can be established by rounding off the atomic mass to the nearest whole number, for example, ^{39}K, ^{85}Rb, and ^{88}Sr. But at other times the isotope corresponding to the "rounded-off" atomic mass does not even occur naturally, for example, ^{64}Cu. Explain the basis of this observation.

54. The net change in the radioactive decay of $^{238}_{92}U$ to $^{206}_{82}Pb$ is the emission of eight α particles. Show that if this loss of eight α particles were not also accompanied by six β^- emissions the product nucleus would still be radioactive. (*Hint:* Recall Figure 26-7.)

55. Based on Figure 26-6, can you explain why more energy is released in a fusion than in a fission process?

56. One method of dating rocks is based on their $^{87}Sr/^{87}Rb$ ratio. The ^{87}Rb is a β^- emitter with a half-life of 5×10^{11} y. A certain rock is found to have a mass ratio $^{87}Sr/^{87}Rb$ of 0.004:1.00. What is the age of the rock?

57. How many millicuries of radioactivity are associated with a sample containing 5.10 mg ^{229}Th, which has a half-life of 7340 y?

58. What mass of ^{90}Sr, with a half-life of 27.7 y, is required to produce 1.00 millicurie of radioactivity?

59. Refer to the Summarizing Example. Another radioisotope produced in the Chernobyl accident was ^{137}Cs. If a 1.00-mg sample of ^{137}Cs is equivalent to 89.8 millicuries, what must be the half-life (in years) of ^{137}Cs?

60. The percent natural abundance of ^{40}K is 0.0117%. The radioactive decay of ^{40}K atoms occurs 89% by β^- emission (the rest is by electron capture and β^+ emission). The half-life of ^{40}K is 1.25×10^9 y. Calculate the number of β^- particles produced per second by the ^{40}K present in a 1.00-g sample of the mineral *microcline*, $KAlSi_3O_8$.

61. The carbon-14 dating method is based on the assumption that the rate of production of ^{14}C by cosmic ray bombardment has remained constant for thousands of years and that the ratio of ^{14}C to ^{12}C has also remained constant. Can you think of any effects of human activities that could invalidate this assumption in the future?

62. Calculate the minimum kinetic energy (in megaelectron-volts) that α particles must possess to produce the nuclear reaction

$$^{4}_{2}He + {}^{14}_{7}N \longrightarrow {}^{17}_{8}O + {}^{1}_{1}H$$

The nuclidic masses are $^{4}_{2}He = 4.00260$ u; $^{14}_{7}N = 14.00307$ u; $^{1}_{1}H = 1.00783$ u; $^{17}_{8}O = 16.99913$ u. [*Hint:* What is the increase in mass in this process?]

63. The packing fraction of a nuclide is related to the fraction of the total mass of a nuclide that is converted to nuclear binding energy. It is defined as the fraction $(M - A)/A$, where M is the actual nuclidic mass and A is the mass number. Use data from a handbook (such as *The Handbook of Chemistry and Physics*, published by the CRC Press) to determine the packing fractions of some representative nuclides. Plot a graph of packing fraction versus mass number and compare it with Figure 26-6. Explain the relationship between the two.

64. ^{40}K undergoes radioactive decay by electron capture to ^{40}Ar and by β^- emission to ^{40}Ca. The fraction of the decay that occurs by electron capture is 0.110. The half-life of ^{40}K is 1.25×10^9 y. Assuming that a rock in which ^{40}K has undergone decay retains all of the ^{40}Ar produced, what would be the $^{40}Ar/^{40}K$ mass ratio in a rock that is 1.5×10^9 y old?

65. Reference is made in the text to using a certain shale deposit containing 0.006% U by mass as a potential fuel in a breeder reactor. Assuming a density of 2.5 g/cm³, how much energy could be released from 1.00×10^3 cm³ of this material? Assume a fission energy of 3.20×10^{-11} J per fission event (that is, per U atom).

Organic Chemistry

27

*T*o early nineteenth century chemists, organic chemistry meant the study of compounds obtainable only from living matter, which was thought to have the "vital force" needed to make these compounds. In 1828, Friedrich Wöhler set out to synthesize ammonium cyanate, NH_4OCN, as in the reaction

$$AgOCN(s) + NH_4Cl(aq) \xrightarrow{\Delta} AgCl(s) + NH_4OCN(aq)$$

The white crystalline solid he obtained from the solution had none of the properties of ammonium cyanate, even though it had the same composition. The compound was not NH_4OCN but $(NH_2)_2CO$—*urea,* an organic compound. As Wöhler excitedly reported to J. J. Berzelius: "I must tell you that I can make urea without the use of kidneys, either man or dog. Ammonium cyanate is urea."

Since that time chemists have synthesized several million organic compounds, and today organic compounds number about 98% of all known chemical substances. In this chapter we explore some of the principal types of organic compounds, and in the next chapter we will study the connection between organic compounds and living matter that once seemed so mysterious.

Vanillin, an aldehyde, gives vanilla beans their characteristic odor. Vanillin can be extracted from vanilla beans or synthesized from waste lignin obtained from the wood pulp industry. Synthetic vanillin is the basis of artificial vanilla flavoring. Vanillin has the structure

929

27-1   Organic Compounds and Structures: An Overview

Organic compounds contain carbon and hydrogen atoms or carbon and hydrogen in combination with a few other types of atoms, such as oxygen, nitrogen, and sulfur. Carbon is singled out for special study because of the ability of carbon atoms to form strong covalent bonds with one another. Carbon atoms can join together into straight chains, branched chains, and rings. The nearly infinite number of possible bonding arrangements of carbon atoms accounts for the vast number and variety of organic compounds.

The simplest organic compounds are those of carbon and hydrogen—**hydrocarbons.** The simplest hydrocarbon is methane, CH_4, the chief constituent of natural gas. Three common representations of methane are

$$H\!:\!\overset{\displaystyle H}{\underset{\displaystyle H}{\overset{..}{\underset{..}{C}}}}\!:\!H \qquad\qquad H\!-\!\overset{\displaystyle H}{\underset{\displaystyle H}{\overset{|}{\underset{|}{C}}}}\!-\!H \qquad\qquad CH_4$$

Lewis structure structural formula condensed structural formula

Three-dimensional representations of the CH_4 molecule are shown in Figure 27-1.

From VSEPR theory we expect the electron group geometry around the central carbon atom in CH_4 to be tetrahedral. The four H atoms are equivalent: They are equidistant from the C atom and attached to it by covalent bonds of equal strength. The angle between any two C—H bonds is $109°28'$.

Imagine removing one H atom from a CH_4 molecule. This leaves the group —CH_3. Then imagine forming a covalent bond between two such —CH_3 groups. The resulting molecule is *ethane,* C_2H_6. By increasing the number of C atoms in the chain, we can obtain still more hydrocarbons. The three-carbon molecule propane, C_3H_8, is pictured in Figure 27-2.

Structural Isomerism

There are *two* ways of assembling a hydrocarbon molecule with four carbon and 10 hydrogen atoms. There are *two* different compounds with the formula C_4H_{10}. One is called butane and the other, isobutane.

FIGURE 27-1
Structural representation of the methane molecule.
(a) Tetrahedral structure showing bond angle. (b) Convention used to suggest a three-dimensional structure through a structural formula. The solid line represents a bond in the plane of the page. The dashed lines project *away* from the viewer, and the heavy wedge projects *toward* the viewer. (c) Ball-and-stick model. (d) Space-filling model.

(a)   **(b)**   **(c)**   **(d)**

FIGURE 27-2
The propane molecule, C_3H_8.
(a) Structural formula. (b) Condensed structural formula. (c) Ball-and-stick model.
(d) Space-filling model.

As we have previously learned, compounds having the same molecular formula but different structural formulas are called *isomers*. In the case of butane, the isomers differ in their carbon chains—one is a straight chain and the other, a branched chain. This type of isomerism is called **structural isomerism.**

EXAMPLE 27-1

Identifying Isomers. Write structural formulas for all the possible isomers with the molecular formula C_5H_{12}.

Solution

A general strategy is first to write the longest chain of carbon atoms from left to right. This gives the straight-chain structure labeled (1). To finish this structure, we add an appropriate number of hydrogen atoms (twelve, in this case) to give each carbon atom four bonds.

Now, let's look for isomers with four carbon atoms in the longest chain and one C atom as a branch (five carbon atoms in all). There is only *one* possibility. Notice that if structure (2) is "flopped" from left to right, the identical structure (2′) is obtained.

Again, we complete the structure of this isomer by adding hydrogen atoms.

$$
\begin{array}{c}
\qquad\qquad \text{H} \\
\qquad\qquad | \\
\text{H}\!-\!\text{C}\!-\!\text{H} \\
\quad \text{H} \quad\;\; | \quad\; \text{H} \;\; \text{H} \\
\;\; | \qquad | \qquad | \quad\; | \\
\text{H}\!-\!\text{C}\!-\!\text{C}\!-\!\text{C}\!-\!\text{C}\!-\!\text{H} \\
\;\; | \qquad | \qquad | \quad\; | \\
\quad \text{H} \quad\; \text{H} \quad\; \text{H} \;\; \text{H}
\end{array}
$$

Finally, let's consider a three-carbon chain with two one-carbon branches. Again, there is only *one* possibility.

$$
\begin{array}{c}
\text{C} \\
| \\
\text{C}\!-\!\text{C}\!-\!\text{C} \\
| \\
\text{C}
\end{array}
\qquad \text{that is,} \qquad
\begin{array}{c}
\text{CH}_3 \\
| \\
\text{CH}_3\!-\!\text{C}\!-\!\text{CH}_3 \\
| \\
\text{CH}_3
\end{array}
$$

The number of isomers with the formula C_5H_{12} is three.

Practice Example A: Write condensed structural formulas for the five possible isomers with the formula C_6H_{14}.

Practice Example B: Write condensed structural formulas for the nine possible isomers with the formula C_7H_{16}.

▶ Do not try to memorize these rules at the outset. Refer to them as you proceed through the examples, and they will become part of your vocabulary of organic chemistry.

Nomenclature

Early organic chemists often assigned names related to the origin or certain properties of new compounds. Some of these names are still in common use. Citric acid is found in citrus fruit; uric acid is present in urine; formic acid is found in ants (from the Latin word for ant, *formica*); and morphine induces sleep (from *Morpheus,* the ancient Greek god of sleep). As thousands upon thousands of new compounds were synthesized, it became apparent that a system of common names was unworkable. Following several interim systems, one recommended by the International Union of Pure and Applied Chemistry (IUPAC) was adopted.

In this introduction to nomenclature we will consider only hydrocarbons with all carbon-to-carbon bonds as single bonds. These are known as **saturated hydrocarbons,** or alkanes. We have no trouble naming the first few: CH_4, methane; C_2H_6, ethane; C_3H_8, propane. We encounter our first difficulty with C_4H_{10}, which has two isomers, but we can resolve this difficulty by assigning the name *butane* to the straight-chain isomer, $CH_3CH_2CH_2CH_3$, and *isobutane* to the branched-chain isomer, $CH_3CH(CH_3)_2$. This method is inadequate for the C_5H_{12} alkanes, for which there are three structural isomers (Example 27-1), and it is even less satisfactory for alkanes with greater numbers of carbon atoms. We avoid unambiguous names, however, with the following rules.

1. Select the *longest* continuous carbon chain in the molecule and use the hydrocarbon name of this chain as the base name. Except for the common names methane, ethane, propane, and butane, standard Greek prefixes relate the name to the number of C atoms in the chain, as in *pent*ane (C_5), *hex*ane (C_6), *hept*ane (C_7), *oct*ane (C_8), ...

2. Consider every branch of the main chain to be a substituent derived from another hydrocarbon. For each of these substituents change the ending of its name from "ane" to "yl." That is, the alkane substituent becomes an **alkyl** group (Table 27.1).

TABLE 27.1 Some Common Alkyl Groups	
Name	**Structural Formula**
methyl	$-CH_3$
ethyl	$-CH_2CH_3$
propyl[a]	$-CH_2CH_2CH_3$
isopropyl	CH_3CHCH_3 $\vert$
butyl[a]	$-CH_2CH_2CH_2CH_3$
isobutyl	$\qquad\quad CH_3$ $\qquad\quad \vert$ $-CH_2CHCH_3$
s-butyl[b]	$CH_3CHCH_2CH_3$ $\qquad\vert$
t-butyl[c]	$\qquad\quad CH_3$ $\qquad\quad \vert$ CH_3CCH_3 $\qquad\quad \vert$

[a]In the past, the prefix *normal* or *n-* was used for a straight-chain alkyl group, such as *n*-propyl or *n*-butyl.

[b]*s* = secondary

[c]*t* = tertiary

3. Number the carbon atoms of the continuous base chain so that the substituents appear *at the lowest numbers possible*.

4. Give each substituent a name and number. For identical substituents use di, tri, tetra, and so on, and *repeat the numbers*.

5. Separate numbers from other numbers by commas and from letters by dashes.

6. List the substituents *alphabetically* by name.

EXAMPLE 27-2

Naming an Alkane Hydrocarbon. Give an appropriate IUPAC name for the following compound, an important constituent of gasoline.

$$\begin{array}{ccccc} & CH_3 & & CH_3 & \\ & \vert & & \vert & \\ CH_3 - & C - CH_2 - & CH - CH_3 \\ {}_1 & {}_{\vert}{}^2 \quad {}_3 & & {}_4 \quad {}_5 \\ & CH_3 & & & \end{array}$$

Solution

The carbon atoms are numbered in red, and the side-chain substituents to be named are shown in blue. Each substituent is a methyl group, $-CH_3$. Two methyl groups are on the second carbon atom, and one methyl group is on the fourth carbon atom. The longest chain of carbon atoms is five. The correct name is

2,2,4-trimethylpentane

If we had numbered the carbon atoms from right to left, we would have obtained the name 2,4,4-trimethylpentane. This is *not* an acceptable name, however, because it does not use the *smallest* numbers possible.

Practice Example A: Give an appropriate IUPAC name for the hydrocarbon

$$
\begin{array}{ccc}
& CH_3 & CH_3 \\
& | & | \\
CH_3CH_2CHCH_2CH_2CCH_2CH_2CH_3 \\
& & | \\
& & CH_3
\end{array}
$$

Practice Example B: Give an appropriate IUPAC name for the hydrocarbon $CH_3CH_2CH(CH_3)CH_2CH_2CH(CH_3)CH_2CH_3$. (*Hint:* Write the complete structural formula from the condensed formula. How many C atoms are present in the longest chain?)

EXAMPLE 27-3

▶ In an alphabetical order, we do not consider the prefixes di, tri, . . . or symbols such as *s* and *t*. Thus, butyl, even though *t*-butyl, precedes methyl in the name 4-*t*-butyl-2-methylheptane.

Writing the Formula to Correspond to the Name of an Alkane Hydrocarbon. Write a condensed structural formula for 4-*t*-butyl-2-methylheptane.

Solution

Because the compound is a heptane, the longest chain of carbon atoms is seven.

$$C—C—C—C—C—C—C$$

Starting on the left, we attach a methyl group to the second C atom.

$$
\begin{array}{c}
CH_3 \\
| \\
C—C—C—C—C—C—C
\end{array}
$$

Next, we attach a *t*-butyl group to the fourth C atom.

$$
\begin{array}{c}
CH_3 \\
| \\
CH_3—C—CH_3 \\
CH_3 \quad\quad | \\
| \\
C—C—C—C—C—C—C
\end{array}
$$

Finally, we add the remaining hydrogen atoms to give each carbon atom four bonds.

$$
\begin{array}{c}
CH_3 \\
| \\
CH_3 \quad CH_3—C—CH_3 \\
| \quad\quad\quad | \\
CH_3—CH—CH_2—CH—CH_2—CH_2—CH_3
\end{array}
$$

Practice Example A: Write a condensed structural formula for 3-ethyl-2,6-dimethylheptane.

Practice Example B: Write a condensed structural formula for 3-isopropyl-2-methylpropane.

Positional Isomerism

A variety of atoms or groups of atoms can be substituents on carbon chains, for example, Br. The three monobromopentanes possess the same carbon skeleton. Because they differ in the position of the bromine atom on the carbon chain, these structural isomers are also called **positional isomers.**

$$CH_3CH_2CH_2CH_2CH_2Br \qquad CH_3CH_2CH_2CHCH_3 \qquad CH_3CH_2CHCH_2CH_3$$
$$\qquad\qquad\qquad\qquad\qquad\qquad | \qquad\qquad\qquad\qquad | $$
$$\qquad\qquad\qquad\qquad\qquad\qquad Br \qquad\qquad\qquad\qquad Br$$

   1-bromopentane   2-bromopentane   3-bromopentane

Functional Groups

Typically, elements in addition to carbon and hydrogen are found in organic compounds. These elements occur as distinctive groupings of one or several atoms. In some cases these groupings of atoms are substituted for H atoms in hydrocarbon chains or rings. In other cases they may substitute for C atoms themselves. These distinctive groupings of atoms are called **functional groups,** and the remainder of the molecule is sometimes referred to by the symbol R. Usually, the R refers to an alkyl group. The physical and chemical properties of organic molecules generally depend on the particular functional groups present. The remainder of the molecule (R) often has little effect on these properties.

 A convenient way to study organic chemistry, then, is to consider the properties associated with specific functional groups. This is what we will do in much of the remainder of the chapter. Table 27.2 lists the major types of organic compounds, and their distinctive functional groups are shown in blue. By combining information from Tables 27.1 and 27.2, for example, you should be able to identify $(CH_3)_2CHOCH_2CH_2CH_3$ as isopropyl propyl ether.

TABLE 27.2   Some Classes of Organic Compounds and Their Functional Groups

Type of Compound	General Structural Formula	Example	Name
alkane	R—H	$CH_3CH_2CH_2CH_2CH_3$	pentane
alkene	$\underset{R}{\overset{R}{\diagdown}}C=C\underset{R}{\overset{R}{\diagup}}$	$CH_3CH_2CH_2CH=CH_2$	1-pentene
alkyne	R—C≡C—R	$CH_3CH_2C≡CCH_3$	2-pentyne
alcohol	R—OH	$CH_3CH_2CH_2CH_2CH_2OH$	1-pentanol

halide	R—X	$CH_3CH_2CHCH_2CH_3$
		    │
		    Br
			3-bromopentane

nitro	$R—NO_2$	$CH_3CH_2CH_2CH_2CH_2NO_2$	1-nitropentane
ether	R—O—R	$CH_3CH_2OCH_2CH_2CH_3$	ethyl propyl ether
aldehyde	$R-\overset{\overset{\textstyle O}{\|\|}}{C}-H$	$CH_3CH_2CH_2CH_2\overset{\overset{\textstyle O}{\|\|}}{C}H$	pentanal
ketone	$R-\overset{\overset{\textstyle O}{\|\|}}{C}-R$	$CH_3CH_2\overset{\overset{\textstyle O}{\|\|}}{C}CH_2CH_3$	3-pentanone
carboxylic acid	$R-\overset{\overset{\textstyle O}{\|\|}}{C}-OH$	$CH_3CH_2CH_2CH_2\overset{\overset{\textstyle O}{\|\|}}{C}OH$	pentanoic acid
ester	$R-\overset{\overset{\textstyle O}{\|\|}}{C}-O-R$	$CH_3CH_2CH_2CH_2\overset{\overset{\textstyle O}{\|\|}}{C}OCH_3$	methyl pentanoate
amine	$R—NH_2$	$CH_3CH_2CH_2CH_2CH_2NH_2$	pentylamine

Some of the functional groups appearing here and discussed later in the chapter have distinctive names:

—OH, hydroxyl; $\diagup\!\!\diagup C=O$, carbonyl; $-\overset{\overset{\textstyle O}{\|\|}}{C}-OH$, carboxyl; $—NH_2$, amino.

27-2 Alkanes

In this section we will explore some properties of the alkanes. The essential characteristic of **alkane** hydrocarbon molecules is that they have only single co-valent bonds. In these compounds the bonds are said to be *saturated.*

The alkanes range in complexity from methane, CH_4 (the primary component of natural gas), to molecules containing 50 carbon atoms or more (found in petro-leum). Most have the formula C_nH_{2n+2}, and each alkane differs from the preced-ing one by a —CH_2— or *methylene* group. Substances whose molecules differ only by a constant unit such as —CH_2— are said to form a **homologous series.** Mem-bers of such a series usually have closely related chemical and physical properties. For example, in Table 27.3 we note that boiling points are related to molecular masses and shapes in the ways discussed in Chapter 13 (recall Figure 13-24).

Ring Structures

Alkanes in chain structures have the formula C_nH_{2n+2} and are called **aliphatic.** Alkanes can also exist in ring, or cyclic, structures called **alicyclic.** Think of these rings as forming by joining together the two ends of an aliphatic chain by eliminating a hydrogen atom from each end. Simple alicyclic compounds have the formula C_nH_{2n}.

► By the nomenclature rules on page 932 we would name

1,1,3-trimethylcyclopentane

cyclopropane
C_3H_6

cyclobutane
C_4H_8

cyclopentane
C_5H_{10}

cyclohexane
C_6H_{12}

Often, when ring structures are drawn, neither C nor H atoms bonded to them are written. Thus the alicyclic compounds shown here can also be represented by an equilateral triangle, a square, a pentagon, and a hexagon.

The bond angles in cyclopropane are $60°$ compared with the normal $109.5°$; the bonds are highly strained. As a result, numerous reactions occur in which the ring breaks open to yield a chain molecule—propane or a propane deriva-tive. Cyclopropane is more reactive than the other alkanes, and its reactivity re-sembles that of alkenes (discussed in Section 27-3) in several ways.

TABLE 27.3	Boiling Points of Some Isomeric Alkanes					
Formula	Isomer	Boiling Point, °C		Formula	Isomer	Boiling Point, °C
C_4H_{10}	butane	−0.5		C_6H_{14}	hexane	68.7
	2-methylpropane	−11.7			3-methylpentane	63.3
C_5H_{12}	pentane	36.1			2-methylpentane	60.3
	2-methylbutane	27.9			2,3-dimethylbutane	58.0
	2,2-dimethylpropane	9.5			2,2-dimethylbutane	49.7

(a)

(b)

FIGURE 27-3
Two important conformations of cyclohexane.
(a) Boat form. (b) Chair form. The H atoms extending laterally from the ring—equatorial H atoms—are shown in red. The H atoms projecting above and below the ring—axial H atoms—are in blue.

The cyclobutane molecule buckles slightly, so that the four C atoms are not all in the same plane. In the cyclopentane molecule the most stable arrangement is for one of the carbon atoms to be buckled out of the plane of the other four.

For cyclohexane there are a number of arrangements or *conformations* of the molecule. Two of these are shown as ball-and-stick models in Figure 27-3. These are the "boat" and the "chair" conformations. The cyclohexane molecule flips between these and other conformations fairly easily at room temperature.

Reactions of the Alkanes

Saturated hydrocarbons have little affinity for most chemical reactants. They are insoluble in water and do not react with aqueous solutions of acids, bases, or oxidizing agents. Halogens react only slowly with alkanes at room temperature, but at higher temperatures, particularly in the presence of light, halogenation occurs. In this reaction, called a **substitution reaction,** a halogen atom *substitutes* for a hydrogen atom. The mechanism of this substitution reaction appears to be by a *chain reaction,* written as follows for the chlorination of methane. (Only the electrons involved in bond breakage or formation are shown.)

$$\text{Initiation:} \qquad \text{Cl:Cl} \xrightarrow[\text{light}]{\text{heat or}} 2\,\text{Cl}\cdot$$

$$\text{Propagation:} \quad \text{H}_3\text{C:H} + \text{Cl}\cdot \longrightarrow \text{H}_3\text{C}\cdot + \text{H:Cl}$$

$$\text{H}_3\text{C}\cdot + \text{Cl:Cl} \longrightarrow \text{H}_3\text{C:Cl} + \text{Cl}\cdot$$

$$\text{Termination:} \qquad \text{Cl}\cdot + \text{Cl}\cdot \longrightarrow \text{Cl:Cl}$$

$$\text{H}_3\text{C}\cdot + \text{Cl}\cdot \longrightarrow \text{H}_3\text{C:Cl}$$

$$\text{H}_3\text{C}\cdot + \text{H}_3\text{C}\cdot \longrightarrow \text{H}_3\text{C:CH}_3$$

▶ The reaction of H_2 with Cl_2 to form HCl, discussed on page 800, proceeds by a similar mechanism.

The reaction is initiated when some Cl_2 molecules absorb sufficient energy to dissociate into Cl atoms (represented above as Cl·). Cl atoms collide with CH_4 molecules to produce methyl *free radicals* ($H_3C·$), which combine with Cl_2 molecules to form the product, CH_3Cl. When any or all of the last three reactions proceed to the extent of consuming the free radicals present, the reaction stops. The net equation for the formation of chloromethane is

$$CH_4 + Cl_2 \xrightarrow[\text{light}]{\text{heat or}} CH_3Cl + HCl \tag{27.1}$$

This free radical chain reaction usually yields a mixture of products, not just chloromethane. Polyhalogenation can also occur to form CH_2Cl_2, dichloromethane (methylene chloride, a solvent and paint remover); $CHCl_3$, trichloromethane (chloroform, a solvent and fumigant); and CCl_4, tetrachloromethane (carbon tetrachloride, a solvent).

The most commonly encountered reaction of alkanes is with oxygen. Alkanes burn; the *oxidation* of hydrocarbons underlies their important use as fuels. For example, octane reacts with oxygen as follows.

$$C_8H_{18}(l) + \tfrac{25}{2} O_2(g) \longrightarrow 8\ CO_2(g) + 9\ H_2O(l) \qquad \Delta H° = -5.48 \times 10^3 \text{ kJ} \tag{27.2}$$

Alkanes from Petroleum

The lower molecular mass alkanes, CH_4 and C_2H_6, are found principally in natural gas. Propane and butane are found dissolved in petroleum, from which they can be extracted as gases and sold as liquefied petroleum gas (LPG). Higher alkanes are obtained by the fractional distillation of petroleum, a complex mixture of at least 500 compounds. The main petroleum fractions are listed in Table 27.4, and the liquid–vapor equilibrium principles that underlie fractional distillation were discussed on page 484.

Not all the gasoline components referred to in Table 27.4 are equally desirable in fuels. Some of them burn more smoothly than others. (Explosive burning results in engine "knocking.") The octane hydrocarbon, *2,2,4-trimethylpentane,* has excellent engine performance; it is given an octane rating of 100. *Heptane* has poor engine performance; its octane rating is set at 0. These two hydrocarbons serve as a basis for establishing the quality of automotive fuels. A gasoline that gives the same performance as a mixture of 87%, 2,2,4-trimethylpentane and 13% heptane is assigned an octane number of 87, for example. In general, branched-chain hydrocarbons have higher octane numbers (burn more smoothly) than their straight-chain counterparts.

TABLE 27.4	**Principal Petroleum Fractions**		
Boiling Range, °C	**Composition**	**Fraction**	**Uses**
below 0	C_1–C_4	gas	gaseous fuel
0–50	C_5–C_7	petroleum ether	solvents
50–100	C_6–C_8	ligroin	solvents
70–150	C_6–C_9	gasoline	motor fuel
150–300	C_{10}–C_{16}	kerosene	jet fuel, diesel oil
over 300	C_{16}–C_{18}	gas–oil	diesel oil, cracking stock
—	C_{18}–C_{20}	wax–oil	lubricating oil, mineral oil, cracking stock
—	C_{21}–C_{40}	paraffin wax	candles, wax paper
—	above C_{40}	residuum	roofing tar, road materials, waterproofing

$$CH_3-\underset{\underset{CH_3}{|}}{\overset{\overset{CH_3}{|}}{C}}-CH_2-\underset{\underset{H}{|}}{\overset{\overset{CH_3}{|}}{C}}-CH_3 \qquad CH_3-CH_2-CH_2-CH_2-CH_2-CH_2-CH_3$$

2,2,4-trimethylpentane
octane rating: 100

n-heptane
octane rating: 0

Gasoline obtained by the fractional distillation of petroleum has an octane number of 50–55 and is not acceptable for use in automobiles, which require fuels with octane number near 90. Extensive modifications of the gasoline fraction are required. In *thermal cracking* large hydrocarbon molecules are broken down into molecules in the gasoline range, and the presence of special catalysts promotes the production of branched-chain hydrocarbons. For example, the molecule $C_{15}H_{32}$ might be broken down into C_8H_{18} and C_7H_{14}. The process of *reforming*, or isomerization, converts straight-chain to branched-chain hydrocarbons and other types of hydrocarbons having higher octane numbers. In thermal and catalytic cracking, some low molecular mass hydrocarbons are rejoined into higher molecular mass hydrocarbons by a process known as *alkylation*.

The octane rating of gasoline can be further improved by adding "antiknock" compounds to prevent premature combustion. At one time the preferred additive was tetraethyllead, $(C_2H_5)_4Pb$. Lead additives have now been phased out of gasoline in the United States, and substitutes like the oxygenated hydrocarbons methanol and ethanol are used instead.

A catalytic cracking unit ("cat cracker") at a petroleum refinery.

▶ Not only do the processes of cracking, reforming, and alkylation produce a higher grade gasoline, but they also increase the yield of gasoline obtained from crude oil.

27-3 Alkenes and Alkynes

A straight or branched-chain alkane has the maximum number of H atoms possible for the number of C atoms it contains. In other classes of hydrocarbons with the same number of C atoms but fewer than this maximum possible number of H atoms, the C atoms must either join into rings, form carbon-to-carbon multiple bonds, or both. This is essential so that the C atoms can acquire valence shell octets. Hydrocarbons whose molecules contain some multiple bonds between carbon atoms are said to be **unsaturated.** If there is *one double bond* in the molecules, the hydrocarbons are the simple **alkenes,** or **olefins;** they have the general formula C_nH_{2n}. *Simple* **alkynes** have *one triple bond* in their molecules; they have the general formula C_nH_{2n-2}.

In the examples that follow, systematic names are shown in blue. The names given in parentheses are also commonly used.

$$CH_2{=}CH_2 \qquad CH_3CH_2CH{=}CH_2 \qquad CH_3CH_2CH_2\underset{\underset{CH_2CH_3}{|}}{C}{=}CH_2$$

ethene
(ethylene)

1-butene

2-ethyl-1-pentene

$$HC{\equiv}CH \qquad CH_3CH_2C{\equiv}CH \qquad CH_3\underset{\underset{CH_3}{|}}{CH}C{\equiv}CCH_3$$

ethyne
(acetylene)

1-butyne
(ethylacetylene)

4-methyl-2-pentyne
(isopropylmethylacetylene)

The minor modifications of the rules on page 932 required in naming alkenes and alkynes are to (1) select as the base chain the longest chain containing the multiple bond, (2) number the carbon atoms of the chain to place the multiple bond at the lowest possible number, and (3) use the ending "ene" for alkenes and "yne" for alkynes. Thus, in the name 2-ethyl-1-pentene, the longest chain *containing the multiple bond* is a five-carbon chain (pent-); This is *not* a hexane. The double bond makes the molecule an alkene (ene), and its location between the first and second carbon makes it a 1-pentene. The ethyl group is attached to the second carbon atom, and so the alkene is named 2-ethyl-1-pentene.

The alkenes are similar to the alkanes in physical properties. At room temperature, those containing two to four carbon atoms are gases; those with 5 to 18 are liquids; those with more than 18 are solids. In general, alkynes have higher boiling points than their alkane and alkene counterparts.

Geometric Isomerism

The molecules 2-butene, $CH_3CH{=}CHCH_3$, and 1-butene, $CH_2{=}CHCH_2CH_3$, differ in the position of the double bond and are *positional* isomers. But another kind of isomerism is possible in 2-butene, represented by these two structures.

H   H   H CH_3

C=C   C=C

CH_3 CH_3 CH_3 H

(a) *cis*-2-butene   (b) *trans*-2-butene

As we learned in Section 12-4, a double bond between C atoms consists of the overlap of sp^2 hybrid orbitals to form a σ bond *and* the sidewise overlap of p orbitals to form a π bond. Rotation about this double bond is severely restricted. Molecule (a) cannot be converted into molecule (b) simply by twisting one end of the molecule through 180°, so the two molecules are distinctly different (see Figure 27-4). To distinguish between these two molecules by name, we call (a) *cis*-2-butene (*cis,* Latin, on the same side), and we call (b) *trans*-2-butene (*trans,* Latin, across). This type of isomerism is called **geometric isomerism.**

FIGURE 27-4
Geometric isomerism in 2-butene.
(a) Ball-and-stick models. (b) Space-filling models.

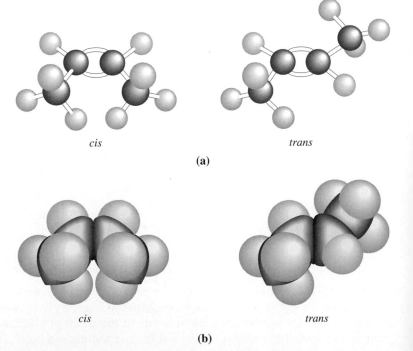

cis   *trans*

(a)

cis   *trans*

(b)

In Chapter 25 we noted that geometric isomerism is only one type of a more general kind of isomerism known as **stereoisomerism** (from the Greek word *stereos,* meaning solid, or three-dimensional, in nature). In stereoisomers the number and types of atoms and bonds are the same, but certain atoms are oriented differently in space. Another type of stereoisomerism that we introduced in Chapter 25 is *optical isomerism.* As we will see in the next chapter, stereoisomerism plays an important role in the unique chemical reactions found in living organisms.

Preparation and Uses

The general laboratory preparation of alkenes uses an **elimination reaction,** a reaction in which atoms are removed from adjacent positions on a carbon chain. A small molecule is produced and an additional bond is formed between the C atoms. For example, H_2O is eliminated in the following reaction.

$$CH_3-\underset{\underset{HO}{|}}{\overset{\overset{H}{|}}{C}}-\underset{\underset{H}{|}}{\overset{\overset{H}{|}}{C}}-H \xrightarrow[H_2SO_4]{\Delta} CH_3CH{=}CH_2 + H_2O \qquad (27.3)$$

The principal alkene of the chemical industry is ethylene (ethene), the highest volume organic chemical and fourth-ranking chemical overall in annual production in the United States. Its chief use is in the manufacture of polymers, although it is also used to manufacture other organic chemicals. Reaction (27.3) is relatively unimportant in the commercial production of ethylene, which is obtained mainly by thermal cracking of other hydrocarbons.

At one time, acetylene, the simplest alkyne, was one of the most important organic raw materials in the chemical industry, but it is no longer among the top 50 industrial chemicals. It can be prepared from coal, water, and limestone in a 3-step process:

$$CaCO_3 \xrightarrow{\Delta} CaO + CO_2$$

$$CaO + 3\,C \xrightarrow[2000\,°C]{\text{electric furnace}} \underset{\substack{\text{calcium acetylide}\\\text{(calcium carbide)}}}{CaC_2} + CO$$

$$CaC_2 + 2\,H_2O \longrightarrow HC{\equiv}CH + Ca(OH)_2$$

Most other alkynes are prepared from acetylene.

The chief use of acetylene is in the manufacture of other chemicals for polymer production, such as vinyl chloride, $H_2C{=}CHCl$, which is polymerized to poly(vinyl chloride) (PVC). Acetylene produces high-temperature flames when burned in excess oxygen, and this is the basis of oxyacetylene torches used for cutting and welding metals.

$$HC{\equiv}CH(g) + \tfrac{5}{2}O_2(g) \longrightarrow 2\,CO_2(g) + H_2O(l) \qquad \Delta H = -1.30 \times 10^3 \text{ kJ}$$

The large negative enthalpy of combustion of acetylene is due to acetylene's large *positive* enthalpy of formation: $\Delta H_f^\circ[C_2H_2(g)] = +226.7$ kJ/mol.

Addition Reactions

Unlike alkanes, which react by *substitution,* alkenes react by *addition.* In an **addition reaction** of an alkene, two new groups become bonded to the C atoms

Cutting steel with an oxyacetylene torch.

at the site of the double bond (and the carbon-to-carbon double bond is converted to a single bond).

$$CH_2{=}CH_2 + Br_2 \longrightarrow \underset{\underset{Br}{|}}{CH_2}{-}\underset{\underset{Br}{|}}{CH_2} \tag{27.4}$$

In this reaction the two Br atoms are identical, as are the two —CH_2 groups. Br_2 and $CH_2{=}CH_2$ are known as *symmetrical* reactants. HBr and $CH_2CH{=}CH_2$ are *unsymmetrical* reactants. When at least one reactant is symmetrical it is a fairly straightforward matter to predict the reaction product. We have a problem, however, in predicting the products when both reactants are unsymmetrical. Let's consider, for example, the reaction of *unsymmetrical* HBr with *unsymmetrical* propene. Which of the two likely products should we expect?

$$CH_3CH{=}CH_2 + H{-}Br \longrightarrow \underset{\underset{Br}{|}\underset{H}{|}}{CH_3CH{-}CH_2} \quad or \quad \underset{\underset{H}{|}\underset{Br}{|}}{CH_3CH{-}CH_2}$$

Because HBr is a polar molecule with its positive end at the H atom and its negative end at the Br atom, 2-bromopropane is the sole product of this reaction. This result is consistent with an empirical rule proposed by Vladimir Markovnikov in 1871 (modified slightly to include reactants other than HBr).

▶ A quick way to remember this rule is "them as has, gets" (referring to H atoms).

> *When an* unsymmetrical *reactant (HX, HOH, HCN, HOSO₃H) is added to an* unsymmetrical *alkene or alkyne, the more positive fragment (usually H) adds to the carbon atom with the greater number of attached H atoms.*

The addition of H_2O to a double bond (as in reaction 27.5) is the reverse of the reaction in which a double bond is formed by the elimination of H_2O (reaction 27.3). In this reversible reaction, the addition reaction is favored in dilute acid, and the elimination reaction is favored in *concentrated* $H_2SO_4(aq)$.

▶ Notice how the H and OH add to the double bond in accordance with Markovnikov's rule. H adds to the C atom with the greater number of attached H atoms—that is, to —CH_2.

$$\underset{CH_3}{\overset{\overset{CH_3}{|}\overset{H}{|}}{CH_3{-}C}}{=}\underset{}{C}{-}H + HOH \xrightarrow{10\% H_2SO_4} \underset{\underset{OH}{|}\underset{H}{|}}{\overset{\overset{CH_3}{|}\overset{H}{|}}{CH_3{-}C{-}\!-\!C{-}H}} \tag{27.5}$$

t-butyl alcohol

27-4 Aromatic Hydrocarbons

August Kekulé (1829–1896), who proposed the hexagonal ring structure for benzene in 1865. His representation of this molecule is still widely used.

Aromatic hydrocarbons have ring structures with unsaturation (multiple bond character) in the carbon-to-carbon bonds in the rings. Most aromatic hydrocarbons are based on the molecule benzene, C_6H_6. In Section 12-6 we discussed bonding in the benzene molecule in some detail, and we showed that there are several ways of representing the molecule, including

Kekulé structures   simplified
                                  molecular orbital
                                  representation

Other examples of aromatic hydrocarbons include

toluene   *o*-xylene   naphthalene   anthracene

▶ As in the case of alicyclic compounds, neither the C atoms at the vertices nor the H atoms bonded to them are shown in these structures.

Toluene and *o*-xylene are *substituted* benzenes, and naphthalene and anthracene feature *fused* benzene rings. When rings are fused together, the resultant structure has two C and four H atoms fewer than the starting structures. Thus, for the formula of naphthalene, $C_6H_6 + C_6H_6 - 2\,C - 4\,H = C_{10}H_8$; for anthracene: $C_{10}H_8 + C_6H_6 - 2\,C - 4\,H = C_{14}H_{10}$. In this text we use the inscribed circle for the simple benzene ring and alternating single and double bonds for fused rings. For the fused ring systems, the bond arrangement represents one of the possible resonance structures for the molecule.

When one of the six equivalent H atoms of a benzene molecule is removed, the species that results is called a **phenyl** group. Two phenyl groups may bond together as in biphenyl, or phenyl groups may be substituents in other molecules, as in phenylhydrazine, used in the detection of sugars.

phenyl group   biphenyl   phenylhydrazine

Other atoms or groups may be substituted for H atoms, and to name these compounds we use a numbering system for the C atoms in the ring. If the name of an aromatic compound is based on a common name other than benzene (such as toluene), the characteristic substituent group (—CH_3) is assigned position "1" on the benzene ring. Otherwise, the substituents are listed alphabetically as in 2-bromochlorobenzene, shown below.

3-bromotoluene   2-bromochlorobenzene   1,4-dichlorobenzene   2-chlorotoluene
(*m*-bromotoluene)   (*o*-bromochlorobenzene)   (*p*-dichlorobenzene)   (*o*-chlorotoluene)

We can also use the terms "ortho," "meta," and "para" (*o*-, *m*-, *p*-) when there are two substituents on the benzene ring. **Ortho** refers to substituents on adjacent carbon atoms, **meta** to substituents with one carbon atom between them, and **para** to substituents opposite one another on the ring.

Benzene and its homologs are similar to other hydrocarbons in being insoluble in water but soluble in organic solvents. The boiling points of the aromatic hydrocarbons are slightly higher than those of the alkanes of similar carbon content. For example, hexane, C_6H_{14}, boils at 69 °C, whereas benzene boils at 80 °C. This can be explained by the planar structure and delocalized electron charge density of benzene, which increases the attractive forces between molecules. The symmetrical structure of benzene permits closer packing of molecules in the crystalline state and results in a higher melting point than for hexane. Benzene melts at 5.5 °C and hexane melts at −95 °C.

Aromatic hydrocarbons are highly flammable and should always be handled with care. Prolonged inhalation of benzene vapor results in a decreased production of both red and white blood corpuscles, which can be fatal. Also, benzene is a carcinogen. Benzene and some other toxic aromatic compounds have been isolated in the tar formed by burning cigarettes, in polluted air, and as a decomposition product of grease in the charcoal grilling of meat.

Current annual production of benzene in the United States is about 16 billion lb. Over 90% of this is produced from petroleum. The process involves dehydrogenation and cyclization of hexane to the aromatic hydrocarbon. The most important use of the petroleum-produced benzene is to manufacture ethylbenzene for the production of styrene plastics. Other applications include the manufacture of phenol, the synthesis of dodecylbenzene (for detergents), and as an octane enhancer in gasoline. The production of aromatic compounds by dehydrogenation (removal of hydrogen) of alkanes yields large amounts of hydrogen gas, an important use for which is in the synthesis of ammonia (page 246).

Aromatic Substitution Reactions

Unlike alkenes and alkynes, aromatic molecules react by *substitution,* not by addition of atoms across a double bond. This difference in reactivity is due to the special stability of the electrons in the aromatic ring. In aromatic rings, certain electrons are *delocalized* into a π electron cloud. Recall the doughnut-shaped regions in Figure 12-24.) In alkene and alkyne molecules, the regions of high electron density associated with multiple bonds are *localized* between specific carbon atoms. These localized π bonds are much more reactive than the delocalized π bonds of the aromatic ring.

The substitution of a single atom or group, X, for an H atom in C_6H_6, can occur at any one of the six positions of the benzene ring. We say that the six positions are equivalent. If a group Y is substituted for an H atom in C_6H_5X, this question arises: To which of the remaining five positions does the Y group go? If all the sites on the benzene ring were equally preferred, the distribution of the products would be a purely statistical one. That is, there are five possibilities for the position where Y can be substituted, and we should get 20% of each one. Since there are two possibilities for substitution that lead to an *ortho* isomer and two that lead to a *meta* isomer, however, we should expect the distribution of products to be 40% ortho, 40% meta, and 20% para.

40% ortho ($\frac{2}{5}$)   40% meta ($\frac{2}{5}$)   20% para ($\frac{1}{5}$)

The following scheme describes the products resulting from nitration followed by chlorination (reaction 27.6) and chlorination followed by nitration (reaction 27.7). It shows that *the substitution is not random.* The —NO$_2$ group directs Cl to a meta position. Almost no ortho or para isomer is formed in reaction (27.6). The Cl group, on the other hand, is an ortho,para director. Essentially no meta isomer is produced in reaction (27.7).

$$(27.6)$$

$$(27.7)$$

Whether a group is an ortho,para or a meta director depends on how the presence of one substituent alters the electron distribution in the benzene ring. As a result, attack by a second group is more likely at one type of position than another. Examination of many reactions leads to the following order.

Ortho,para directors: —NH$_2$, —OR, —OH, —OCOR, —R, —X   (X = halogen)
(from strongest to weakest)

Meta directors: —NO$_2$, —CN, —SO$_3$H, —CHO, —COR, —COOH, —COOR
(from strongest to weakest)

EXAMPLE 27-4

Predicting the Products of an Aromatic Substitution Reaction. Predict the products of the mononitration of

Solution

—OH is an ortho,para director, and we should expect the products

2,6-dinitrophenol   2,4-dinitrophenol

Considering —NO$_2$ as a meta director leads to the same conclusion.

Practice Example A: Predict the major product(s) of the mononitration of benzaldehyde, C$_6$H$_5$CHO.

Practice Example B: Predict the major product(s) of the mononitration of 1,3-dichlorobenzene.

27-5   Alcohols, Phenols, and Ethers

Alcohols and **phenols** feature the **hydroxyl** group, —OH. In alcohols, the hydroxyl group is bound to an aliphatic carbon atom. If the C atom to which the —OH group is attached also has *one* R group (and two H atoms) bonded to it, the alcohol is a *primary* alcohol. If the C atom has *two* R groups (and one H atom), the alcohol is a *secondary* alcohol. Finally, if there are *three* R groups on the C atom (and no H atoms), the alcohol is a *tertiary* alcohol.

$$CH_3CH_2CH_2\!-\!\overset{\displaystyle H}{\underset{\displaystyle H}{\overset{|}{\underset{|}{C}}}}\!-\!OH \qquad CH_3CH_2\!-\!\overset{\displaystyle CH_3}{\underset{\displaystyle H}{\overset{|}{\underset{|}{C}}}}\!-\!OH \qquad CH_3\!-\!\overset{\displaystyle CH_3}{\underset{\displaystyle CH_3}{\overset{|}{\underset{|}{C}}}}\!-\!OH$$

1-butanol   2-butanol   2-methyl-2-propanol
(butyl alcohol)   (*s*-butyl alcohol)   (*t*-butyl alcohol)
(a *primary* alcohol)   (a *secondary* alcohol)   (a *tertiary* alcohol)

In phenols the hydroxyl group is attached to a benzene ring.

phenol   2,4,6-trinitrophenol
(carbolic acid)   (picric acid)

A molecule may have more than one —OH group present. Two common alcohols with more than one —OH are shown in the following formulas. Ethylene glycol is used in automobile antifreeze solutions and glycerol is an important biological molecule used as part of the body's mechanism for fat storage.

$$\overset{\displaystyle CH_2\!-\!CH_2}{\underset{\displaystyle OH \quad\; OH}{\quad|\quad\quad|\quad}} \qquad \overset{\displaystyle CH_2\!-\!CH\!-\!CH_2}{\underset{\displaystyle OH \quad OH \quad OH}{\quad|\quad\quad|\quad\quad|\quad}}$$

1,2-ethanediol   1,2,3-propanetriol
(ethylene glycol)   (glycerol)

Physical properties of aliphatic alcohols are strongly influenced by hydrogen bonding. As the chain length increases, however, the influence of the polar hydroxyl group on the properties of the molecule diminishes. The molecule becomes less like water and more like a hydrocarbon. As a consequence, low molecular mass alcohols tend to be water-soluble; high molecular mass alcohols are not. The boiling points and solubilities of the phenols vary widely, depending on the nature of the other substituents on the benzene ring.

Preparation and Uses of Alcohols

Two methods by which alcohols can be synthesized are by the hydration of alkenes and by the hydrolysis of alkyl halides.

$$CH_3CH\!=\!CH_2 + H_2O \xrightarrow{\;H_2SO_4\;} \overset{\displaystyle OH}{\overset{|}{CH_3CHCH_3}} \qquad (27.8)$$

propene   2-propanol
(propylene)   (isopropyl alcohol)

$$CH_3CH_2CH_2Br + OH^- \longrightarrow CH_3CH_2CH_2OH + Br^- \qquad (27.9)$$

propyl bromide   propyl alcohol

Giant soap bubble.
A soap bubble consists of an air pocket enclosed in a thin film of soap in water. As the water evaporates, the film breaks and the bubble bursts. Glycerol added to the soap–water mixture forms hydrogen bonds to both the soap and water molecules. This slows the rate of evaporation of the water and increases the strength of the film, allowing us to produce very large bubbles.

Methanol (wood alcohol) is a highly toxic substance and can lead to blindness or death if ingested. Most methanol is manufactured from carbon monoxide and hydrogen.

$$CO(g) + 2\,H_2(g) \xrightarrow[\substack{200\ atm \\ ZnO,\ Cr_2O_3}]{350\ ^\circ C} CH_3OH(g)$$

Methanol is the most extensively produced alcohol. It ranks about twenty-first among all industrial chemicals. It is used to synthesize other organic chemicals and as a solvent, but potentially its most important use may be as a motor fuel (recall Section 7-9).

Ethanol, CH_3CH_2OH, is grain alcohol—common "alcohol" to the layperson. It is obtainable by the fermentation of sugar cane juice or from materials containing natural sugars. The industrial method is the hydration of ethylene with sulfuric acid catalyst (similar to reaction 27.8).

Ethylene glycol, CH_2OHCH_2OH, is water-soluble and has a higher boiling point (197 °C) than water. Because of these properties it makes an excellent permanent nonvolatile antifreeze for automobile radiators. It is also used in the manufacture of solvents, paint removers, and plasticizers (softeners).

Glycerol (glycerin), $CH_2OHCHOHCH_2OH$, is obtained commercially as a byproduct in the manufacture of soap. It is a sweet, syrupy liquid that is miscible with water in all proportions. Because it takes up moisture from the air, glycerol can be used to keep skin moist and soft and is found in lotions and cosmetics.

Ethers

An **ether** is a compound with the general formula R—O—R. Structurally, ethers can be pure aliphatic, pure aromatic, or mixed.

dimethyl ether   diphenyl ether   methyl phenyl ether (anisole)

▶ Although they have the same formula (C_2H_6O), dimethyl ether and ethanol have quite different physical and chemical properties. This is because each has a different functional group (see Table 27.2). The two substances are structural isomers.

Symmetrical ethers, such as diethyl ether, can be prepared by the elimination of a water molecule from between two alcohol molecules with a strong dehydrating agent, such as concentrated H_2SO_4.

$$CH_3CH_2OH + HOCH_2CH_3 \xrightarrow[140^\circ C]{\substack{concd \\ H_2SO_4}} CH_3CH_2OCH_2CH_3 + H_2O \qquad (27.10)$$

diethyl ether

Chemically, ethers are comparatively unreactive. The ether linkage is stable to most oxidizing and reducing agents and to action by dilute acids and alkalis.

Diethyl ether has been used extensively as a general anesthetic. It is easy to administer and produces excellent relaxation of the muscles. Also, the pulse rate, rate of respiration, and blood pressure are affected only slightly. However, it is somewhat irritating to the respiratory passages and produces nausea. Methyl propyl ether (neothyl) is also used as an anesthetic and is less irritating to the respiratory passages than diethyl ether. Dimethyl ether, a gas at room temperatures, is used as a propellant for aerosol sprays. Higher molecular mass ethers are used as solvents for varnishes and lacquers. Methyl *t*-butyl ether, an unsymmetric ether, is widely used as an octane enhancer in gasoline and is marketed under the name MTBE.

▶

methyl *t*-butyl ether (MTBE)

This molecule should be named *t*-butyl methyl ether (TBME), but has been referred to by industrial chemists as MTBE for so long that it has become the commonly used name for this compound.

27-6 Aldehydes and Ketones

Aldehydes and ketones contain the **carbonyl** group.

$$\begin{matrix} R \\ {} \\ R \end{matrix}\!\!\!\!C\!\!=\!\!O$$

If one of the R groups is an H atom, the compound is an **aldehyde.** If both R groups are alkyl or aromatic (aryl) groups, the compound is a **ketone.**

methanal
(formaldehyde)

3-chlorobutanal
(β-chlorobutyraldehyde)

benzaldehyde

propanone
(acetone)

3-pentanone
(diethyl ketone)

1-phenylethanone
(acetophenone)

Preparation and Uses

An aldehyde can be produced by the partial oxidation of a *primary* alcohol. Further oxidation yields a carboxylic acid (Section 27-7). Oxidation of a *secondary* alcohol produces a ketone.

$$CH_3CH_2OH \xrightarrow[H^+]{Cr_2O_7^{2-}} CH_3CHO \xrightarrow[H^+]{Cr_2O_7^{2-}} CH_3CO_2H \qquad (27.11)$$

ethanol
(a primary alcohol)

acetaldehyde
(an aldehyde)

acetic acid
(an acid)

$$CH_3CHOHCH_3 \xrightarrow[H^+]{Cr_2O_7^{2-}} CH_3\overset{\displaystyle O}{\overset{\|}{C}}CH_3 \qquad (27.12)$$

2-propanol
(a secondary alcohol)

acetone
(a ketone)

Formaldehyde ($H_2C\!=\!O$), a colorless gas that dissolves readily in water, is the simplest aldehyde. Formaldehyde is used in the manufacture of synthetic resins. A polymer of formaldehyde, called paraformaldehyde, is used as an antiseptic and an insecticide.

Aldehydes and ketones occur widely in nature. Typical natural sources are

False-color scanning electron micrograph of the sticky surface of a 3M Post-It note. The bubbles are 15–40 μm in diameter and consist of a urea–formaldehyde adhesive. Each time the note is pressed to a surface, fresh adhesive is released. The note can be reattached to a surface as long as some of the bubbles remain.

benzaldehyde
(almonds)

cinnamaldehyde
(cinnamon)

camphor
(obtained from
camphor tree)

Acetone is the most important of the ketones. It is a volatile liquid (boiling point, 56 °C) and highly flammable. Acetone is a good solvent for a variety of organic compounds and is widely used in solvents for varnishes, lacquers, and plastics. Unlike many common organic solvents, acetone is miscible with water in all proportions. An industrial method of producing acetone involves the *dehydrogenation* of isopropyl alcohol in the presence of a copper catalyst.

$$CH_3-\underset{\underset{OH}{|}}{CH}-CH_3 \xrightarrow[300\,°C]{Cu} CH_3-\underset{\underset{O}{\|}}{C}-CH_3 + H_2 \qquad (27.13)$$

27-7   Carboxylic Acids and Their Derivatives

Compounds that contain the **carboxyl** group (*car*bonyl and hydr*oxyl*)

$$-\underset{\underset{O}{\|}}{C}-OH$$

are called **carboxylic acids;** they have the general formula R—COOH. Many compounds are known in which R is an aliphatic residue. These are called fatty acids because high molecular weight compounds of this type are readily available from naturally occurring fats and oils. If two carboxyl groups are found on the same molecule the acid is called a dicarboxylic acid. The carboxyl group can also be found attached to the benzene ring.

▶ The carboxyl group is also represented as —COOH and —CO₂H.

acetic acid
(an aliphatic acid)

benzoic acid
(an aromatic acid)

oxalic acid
(an aliphatic dicarboxylic acid)

phthalic acid
(an aromatic dicarboxylic acid)

As noted previously through Equation (27.11), oxidation of a primary alcohol or an aldehyde can be carried to the point of forming a carboxylic acid. For this purpose the oxidizing agent is generally $KMnO_4(aq)$ in an alkaline medium. Because the medium is alkaline the product is the potassium salt, but the free carboxylic acid can be regenerated by making the medium acidic.

Substituted aliphatic acids can be named either by their IUPAC names or by using Greek letters in conjunction with common names. Aromatic acids are named as derivatives of benzoic acid.

3-chlorobutanoic acid
β-chlorobutyric acid

3-methylbenzoic acid
m-methylbenzoic acid
(also *m*-toluic acid)

2-hydroxybenzoic acid
o-hydroxybenzoic acid
(also salicylic acid)

The distinctive aroma and flavor of oranges are due in part to the ester octyl acetate, $CH_3(CH_2)_6CH_2OOCCH_3$.

Because many derivatives of the carboxylic acids involve simple replacement of the hydroxyl groups, special names have been developed for the remaining portion of the molecule, that is,

$$-\overset{\overset{\displaystyle O}{\|}}{C}-R$$

For example, the group

$$-\overset{\overset{\displaystyle O}{\|}}{C}-CH_3$$

is called the **acetyl** group. The acetyl derivative of *o*-hydroxybenzoic (salicylic) acid is ordinary aspirin.

acetylsalicylic acid
(aspirin)

Reactions of the Carboxyl Group

The carboxyl group displays the chemistry of both the carbonyl and the hydroxyl group. Donation of a proton to a base leads to salt formation. The sodium and potassium salts of long-chain fatty acids are known as *soaps*. The structure of the soap sodium palmitate was presented on page 772, and soap formation (saponification) is discussed further in the next chapter.

The product of the reaction of an acid and an alcohol is called an **ester.** It can be viewed as being formed by the elimination of H_2O between the two molecules. The mechanism of this reaction is such that the —OH of the eliminated water comes from the *acid* and the —H from the *alcohol.*

▶ IUPAC names appear below the structural formulas with more common names in parentheses.

$$CH_3CH_2O-H + HO-\overset{\overset{\displaystyle O}{\|}}{C}CH_3 \xrightarrow[\text{heat}]{H^+} H_2O + CH_3CH_2O-\overset{\overset{\displaystyle O}{\|}}{C}CH_3 \quad (27.14)$$

ethanol   ethanoic acid   ethyl ethanoate
(ethyl alcohol)   (acetic acid)   (ethyl acetate)

Unlike the pungent odors of the carboxylic acids from which they are derived, esters have very pleasant aromas. The characteristic fragrances of many flowers and fruits can be traced to the esters they contain. Esters are used in perfumes and in the manufacture of flavoring agents for the confectionery and soft drink industries. Most esters are colorless liquids, insoluble in water. Their melting points and boiling points are generally lower than those of alcohols and acids of comparable carbon content. This is because of the absence of hydrogen bonding in esters.

27-8 Amines

Amines are organic derivatives of ammonia, NH_3, in which one or more organic groups (R) are substituted for H atoms. Their classification is based on the number of R groups bonded to the nitrogen atom—one for primary amines, two for secondary, and three for tertiary.

$$CH_3CH_2NH_2$$

ethylamine
(a primary amine)

diphenylamine
(a secondary amine)

$$CH_3NCH_2CH_2CH_3$$
with CH_2CH_3 group on N

ethylmethylpropylamine
(a tertiary amine)

Amines of low molecular mass are gases that are readily soluble in water, yielding basic solutions. The volatile members have odors similar to ammonia but more "fish-like." Primary and secondary amines form hydrogen bonds, but these bonds are weaker than those in water because nitrogen is less electronegative than oxygen. Like ammonia, amines have trigonal pyramidal structures with a lone pair of electrons on the N atom. Also like ammonia, amines owe their basicity to these lone pair electrons. In aromatic amines, because of unsaturation in the benzene ring, electrons are drawn into the ring, which reduces the electron charge density on the nitrogen atom. As a result, aromatic amines are *weaker* bases than ammonia. Aliphatic amines are somewhat stronger bases than ammonia.

$$NH_3 + H_2O \rightleftharpoons NH_4^+ + OH^- \qquad K_b = 1.8 \times 10^{-5}$$
$$CH_3NH_2 + H_2O \rightleftharpoons CH_3NH_3^+ + OH^- \qquad K_b = 42 \times 10^{-5}$$

methylamine

$$\text{aniline} + H_2O \rightleftharpoons \text{anilinium} + OH^- \qquad K_b = 7.4 \times 10^{-10}$$

Dimethylamine is an accelerator for the removal of hair from hides in the processing of leather. Butyl- and pentylamines are used as antioxidants, corrosion inhibitors, and in the manufacture of oil-soluble soaps. Dimethylamine and trimethylamine are used in the manufacture of ion-exchange resins. Additional applications are found in the manufacture of disinfectants, insecticides, herbicides, drugs, dyes, fungicides, soaps, cosmetics, and photographic developers.

27-9 Heterocyclic Compounds

In the ring structures considered to this point, all the ring atoms have been carbon; these structures are said to be carbocyclic. There are many compounds, however, both natural and synthetic, in which one or more of the atoms in a ring structure is not carbon. These ring structures are said to be **heterocyclic.** The heterocyclic systems most commonly encountered contain N, O, and S atoms, and the rings are of various sizes.

Pyridine is a nitrogen analog of benzene (Figure 27-5), but unlike benzene it is water-soluble and has basic properties (the unshared pair of electrons on the N atom is not part of the π electron cloud of the ring system). Pyridine was once obtained exclusively from coal tar, but it is now used so extensively that several synthetic methods have been developed for its production. It is a liquid with a disagreeable odor used in the production of pharmaceuticals such as sulfa drugs and antihistamines, as a denaturant for ethyl alcohol, as a solvent for organic chemicals, and in the preparation of waterproofing agents for textiles. We consider a number of other examples of heterocyclic compounds in Chapter 28.

FIGURE 27-5
Pyridine.
In the pyridine molecule a nitrogen atom replaces one of the

 C—H units in benzene. Its formula is C_5H_5N. (a) Structural formula. (b) Space-filling model.

(a)

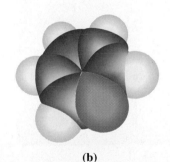

(b)

27-10 Polymerization Reactions

The majority of chemists and chemical engineers work with macromolecular substances, or polymers, and of particular interest to them are the reactions that can be used to make polymers. We will briefly survey the main types of polymerization reactions in this final section of the chapter.

Chain Reaction Polymerization

▶ Molecules of a *poly*mer are formed by the joining together of many simple molecules called *mono*mers.

Monomers with carbon-to-carbon double bonds typically undergo **chain reaction polymerization.** The net result is that the double bonds "open up" and monomer units add to growing chains. As with other chain reactions, the mechanism involves three characteristic steps: initiation, propagation, and termination. Let us illustrate this mechanism for the formation of the polymer polyethylene from the monomer ethylene.

The key to the polymerization reaction is the free radical initiator. In reaction (27.15) an organic peroxide dissociates into two peroxy radicals. The radicals add to the C=C double bonds of ethylene molecules to form radical intermediates. These radical intermediates attack more ethylene molecules and form new intermediates of longer and longer length (27.16). The chains terminate as a result of reactions such as (27.17).

$$\text{\textit{Initiation:}} \qquad \text{R—O:O—R} \longrightarrow 2\ \text{R—O}\cdot \qquad (27.15)$$

an organic peroxide

followed by

$$\text{CH}_2\text{=CH}_2 + \text{RO}\cdot \longrightarrow \text{R—O—CH}_2\text{—CH}_2\cdot$$

$$\text{\textit{Propagation:}} \quad \text{ROCH}_2\text{CH}_2\cdot + \text{CH}_2\text{=CH}_2 \longrightarrow \text{ROCH}_2\text{CH}_2\text{CH}_2\text{CH}_2\cdot \qquad (27.16)$$

$$\text{RO(CH}_2)_3\text{CH}_2\cdot + \text{CH}_2\text{=CH}_2 \longrightarrow \text{RO(CH}_2)_5\text{CH}_2\cdot$$

$$\text{\textit{Termination:}} \qquad \text{RO(CH}_2)_x\text{CH}_2\cdot + \text{RO}\cdot \longrightarrow \text{RO(CH}_2)_x\text{CH}_2\text{OR} \qquad (27.17)$$

or

$$\text{RO(CH}_2)_x\text{CH}_2\cdot + \text{RO(CH}_2)_y\text{CH}_2\cdot \longrightarrow \text{RO(CH}_2)_x\text{CH}_2\text{CH}_2(\text{CH}_2)_y\text{OR}$$

Several polymers formed by chain-reaction polymerization are listed in Table 27.5.

Extruding polyethylene film.

Are You Wondering . . .

What effect the —OR groups have on polymers such as polyethylene?

Essentially, they have no effect on the properties of the polymer. This is because there are hundreds, possibly thousands, of monomer units in the polymer, and the —OR groups are only at the ends of each polymer strand. The long chain of monomer units gives the polymer its particular properties and the —OR group only serves to initiate the reaction and to terminate the polymer chains.

TABLE 27.5 Some Polymers Produced by Chain-Reaction Polymerization

Name	Monomer	Polymer	Uses
polyethylene	$CH_2{=}CH_2$	$-(CH_2-CH_2)_n$	bags, bottles, tubing, packaging film
polypropylene	$CH_2{=}CHCH_3$	$\left(CH_2-\underset{\underset{CH_3}{\mid}}{CH}\right)_n$	laboratory and household ware, artificial turf, surgical casts, toys
poly(vinyl chloride) PVC	$CH_2{=}CHCl$	$\left(CH_2-\underset{\underset{Cl}{\mid}}{CH}\right)_n$	bottles, floor tile, food wrap, piping, hoses
poly(tetrafluoroethylene), Teflon	$CF_2{=}CF_2$	$-(CF_2-CF_2)_n$	bearings, insulation, nonstick surfaces, gaskets, industrial ware
polystyrene	$CH_2{=}CH$ ⬡	$\left(CH_2-CH\right)_n$ ⬡	packaging, refrigerator doors; cups, ice buckets, and coolers (as foam)

Step Reaction Polymerization

In **step reaction polymerization,** also called *condensation* polymerization, the monomers typically have two or more functional groups that react to join the two molecules together. Usually this involves the elimination of a small molecule, such as H_2O. Unlike chain reaction polymerization, in which the reaction of a monomer can only occur at the end of a growing polymer chain, in step reaction polymerization any pair of monomers is free to join into a *dimer*; a dimer may join with a monomer to form a *trimer;* two dimers may join into a *tetramer*; and so on. Step reaction polymerization tends to occur slowly and produces polymers of only moderately high molecular masses (less than 10^5 u). The formation of poly(ethylene glycol terephthalate), Dacron, is illustrated in the following reaction, and several other polymers formed by this method are listed in Table 27.6.

terephthalic acid   ethylene glycol

poly(ethylene glycol terephthalate)
(Dacron)

Stereospecific Polymers

The physical properties of a polymer are determined by a number of factors, such as the average length (average molecular weight) of the polymer chains and the strength of intermolecular forces between chains. Another important factor is whether the polymer chains display any crystallinity—that is, an ordered geometry and spacing of the atoms between polymer chains. In general, amorphous polymers are glass-like or rubbery. A high-strength fiber, on the other

Focus On   Natural and Synthetic Dyes

The synthesis from coal tar chemicals of the purple dye mauveine by William Perkin marked the beginning of the synthetic dye industry.

The production of dyes has been a human activity since the art of weaving was first developed. In the past, materials used in making dyes have been animal, vegetable, and mineral in origin. One of the most famous historical dyes is *Tyrian purple.* This dye, produced from the shells of *Murex brandaris,* was expensive, and in ancient Roman times it was reserved only for the ruling class (hence the term "royal purple").

Tyrian purple

Although it is obtained from a plant rather than a mollusk, *indigo,* the familiar dye providing the blue in "blue jeans," has a molecular structure similar to that of Tyrian purple.

indigo

Tyrian purple and indigo have a feature common to organic dyes: a series of double bonds separated by single bonds, a pattern described as a *conjugated* double bond system. Many organic compounds with conjugated double bonds exhibit color. The orange color of carrots, for example, is due to the presence of β-carotene.

TABLE 27.6   Some Polymers Produced by Step Reaction Polymerization

Name	Monomers	Polymer	Uses
poly(ethylene glycol terephthalate) (Dacron)	$HOCH_2CH_2OH$ and $HOOC\!-\!\bigcirc\!-\!COOH$	$\left[\!-C(O)\!-\!\bigcirc\!-\!C(O)\!-\!O\!-\!CH_2\!-\!CH_2\!-\!O\!-\right]_n$	textile fabrics, twine and rope, fire hoses, plastic containers
poly(hexamethyleneadipamide) nylon 66	$H_2N(CH_2)_6NH_2$ and $HOOC(CH_2)_4COOH$	$\left[\!-C(O)\!-(CH_2)_4\!-\!C(O)\!-\!NH\!-(CH_2)_6\!-\!NH\!-\right]_n$	hosiery, rope, tire cord, parachutes, artificial blood vessels
polyurethane	$HO(CH_2)_4OH$ and $OCN(CH_2)_6NCO$	$\left[\!-O\!-(CH_2)_4O\!-\!C(O)\!-\!N(H)\!-(CH_2)_6N(H)\!-\!C(O)\!-\right]_n$	spandex fibers, bristles for brushes, and cushions and mattresses (as foam)

β-Carotene

In the last half of the nineteenth century, the dye industry underwent an unexpected and fundamental change. Chemical, or *synthetic,* routes were devised for producing dyes. The first synthetic dye was discovered in 1856 by William Perkin, an 18-year-old student at the Royal College of Chemistry in London. Perkin was studying the reactions of substances obtained from coal tar. When he allowed a mixture of substances containing aniline to react with potassium dichromate, he obtained a tarry product from which he ex-

tracted a bright purple, water-soluble substance—the dye *mauveine.* Following this discovery, Perkin went into business as the first synthetic dye manufacturer.

By 1900, the BASF corporation had introduced synthetic indigo. This synthetic version of the dye could be produced more cheaply than the natural product, and the result was a dramatic shift from the natural material to a synthetic one.

Plants and animals on Earth have always been an important source of chemicals. A typical pattern has been that a substance with

desirable medicinal or other commercial uses has first been discovered in a particular organism. Then, chemists have attempted to develop a synthetic route leading to the same substance from cheap raw materials such as petroleum. To date, only a small fraction of Earth's organisms have been studied for the potential beneficial chemicals they contain. To permit an expanded search for useful natural products is an important, if not necessarily the most compelling, reason for preserving biological diversity on Earth.

hand, must possess some crystallinity. Many polymers possess crystalline regions and amorphous regions. The relative amount of each type of region affects the physical properties of the polymer.

The usual designation of a polymer, when applied to polypropylene,

$$\left[\begin{array}{c} H \quad H \\ | \quad | \\ -C-C- \\ | \quad | \\ H \quad CH_3 \end{array} \right]_n$$

is not very revealing about the structure of the polymer. It does not indicate the orientation of the —CH$_3$ groups along the polymer chain.

If propylene (CH$_3$—CH=CH$_2$) is polymerized by the method shown for ethylene on page 952, the orientation of the —CH$_3$ groups is random (Figure 27-6). A polymer of this type is called *atactic.* Because there is no regularity to the structure, atactic polymers are amorphous. In an *isotactic* polymer all —CH$_3$ groups have the same orientation, and in a *syndiotactic* polymer the —CH$_3$ groups have an orientation that alternates back and forth along the chain (see Figure 27-6). Because of their structural regularity, isotactic and syndiotactic polymers possess crystallinity, which makes them stronger and more resistant to chemical attack than an atactic polymer.

FIGURE 27-6
Three representations of a polypropylene chain.
In the *atactic* polymer the —CH₃ groups are randomly distributed on the chain. In the *isotactic* polymer all —CH₃ groups are shown coming out of the plane of the page. In the *syndiotactic* polymer —CH₃ groups alternate along the chain—one in front of the plane of the page, the next one behind the plane, and so on.

atactic

isotactic

syndiotactic

In the 1950s Karl Ziegler and Giulio Natta developed procedures for controlling the spatial orientation of substituent groups on a polymer chain by using special catalysts [such as $(CH_3CH_2)_3Al + TiCl_4$]. This discovery, recognized through the award of a Nobel Prize in 1963, revolutionized polymer chemistry. Through stereospecific polymerization it is literally possible to "tailor-make" large molecules.

Summary

Organic chemistry deals with compounds of carbon. Simplest among these are carbon–hydrogen compounds, hydrocarbons. In hydrocarbons the C atoms are bonded to one another in straight or branched chains or in rings. Some hydrocarbon molecules contain only single bonds (alkanes), others have some double bonds (alkenes), and still others, triple bonds (alkynes). Yet another class of hydrocarbons—aromatic hydrocarbons—is based on the benzene molecule, C_6H_6.

Functional groups are certain atoms or groupings of atoms that can be introduced into hydrocarbon structures through chemical reactions. With alkanes and aromatic hydrocarbons these reactions are based on *substitution*: a functional group replaces an H atom in the hydrocarbon. With alkenes and alkynes chemical reaction occurs by *addition*: functional group atoms are joined to the C atoms at points of unsaturation (double or triple bonds).

Isomerism is frequently found among organic compounds. Some isomers have the same total number of C atoms but different branching of the carbon chain. Other isomers differ in the positions on a hydrocarbon chain or ring at which functional groups may be attached. Still other isomers (such as, cis and trans) arise from different orientations of substituent groups in space.

Compounds of the general formula ROH are alcohols (phenols if R is a phenyl or substituted phenyl group). One method of preparing alcohols is by *hydration* of an alkene (addition of HOH) or by the *hydrolysis* of an alkyl halide. Ethers (R′OR) result from the elimination of HOH from between two alcohol molecules. Aldehydes, RCHO, and ketones, RCOR′, feature the carbonyl group,

$$\diagdown C = O$$

They can be prepared by the controlled oxidation of primary and secondary alcohols, respectively. Carboxylic acids are weak acids having the general formula RCOOH and featuring the carboxyl group,

$$-C\diagup^{O}_{OH}$$

In addition to their typical acid–base reactions, carboxylic acids react with alcohols to form esters. Carboxylic acids can be prepared by the oxidation of a primary alcohol or an aldehyde. Amines are organic derivatives of ammonia, and, like ammonia, they have basic properties.

The substitution of other atoms (such as N, O, or S) for C atoms in ring structures yields heterocyclic compounds. These are widely encountered among molecules of the living state (see Chapter 28).

Summarizing Example

Ethyl acetate, $CH_3CO_2CH_2CH_3$, is an important solvent in paints, lacquers, and plastics. It is also used in organic synthesis and in forming synthetic fruit essences.

Devise a synthesis for ethyl acetate using only inorganic substances as starting materials and reactants.

1. *Synthesize acetylene.* Follow the reaction sequence on page 941.
2. *Convert acetylene to ethylene.* This is an addition reaction similar to (27.4) but with C_2H_2 as the starting material and H_2 as the added reagent.
3. *Convert the ethylene to ethanol.* This is another addition reaction with H_2O as the added reagent.
4. *Convert some of the ethanol to acetic acid.* This is the oxidation–reduction reaction (27.11).
5. *Combine ethanol and acetic acid to form ethyl acetate.* This is the esterification reaction (27.14).

Key Terms

acetyl (27-7)	**carboxyl** (27-7)	**ketone** (27-6)
addition reaction (27-3)	**carboxylic acid** (27-7)	**meta** (*m*) (27-4)
alcohol (27-5)	**chain reaction polymerization** (27-10)	**ortho** (*o*) (27-4)
aldehyde (27-6)	**conformations** (27-2)	**para** (*p*) (27-4)
alicyclic (27-2)	**elimination reaction** (27-3)	**phenol** (27-5)
aliphatic (27-2)	**ester** (27-7)	**phenyl** (27-4)
alkane (27-2)	**ether** (27-5)	**positional isomer** (27-1)
alkene (olefin) (27-3)	**functional group** (27-1)	**saturated hydrocarbon** (27-1)
alkyl (27-1)	**geometric isomer** (27-3)	**step reaction polymerization** (27-10)
alkyne (27-3)	**heterocyclic** (27-9)	**stereoisomerism** (27-3)
amine (27-8)	**homologous series** (27-2)	**structural isomer** (27-1)
aromatic (27-4)	**hydrocarbon** (27-1)	**substitution reaction** (27-2, 27-4)
carbonyl (27-6)	**hydroxyl** (27-5)	**unsaturated hydrocarbon** (27-3)

Review Questions

1. In your own words define the following terms or symbols:

(a) *t*; **(b)** R—; **(c)** ⬡ ; **(d)** carbonyl group; **(e)** primary amine.

2. Briefly describe each of the following ideas, phenomena, or methods: **(a)** substitution reaction; **(b)** octane rating of gasoline; **(c)** stereoisomerism; **(d)** ortho,para director; **(e)** step reaction (condensation) polymerization.

3. Explain the important distinctions between each pair of terms: **(a)** alkane and alkene; **(b)** aliphatic and aromatic compound; **(c)** alcohol and phenol; **(d)** ether and ester; **(e)** amine and ammonia.

4. Describe the characteristics of each of the following types of isomers: **(a)** structural; **(b)** positional; **(c)** cis; **(d)** ortho.

5. Which hydrocarbon has the greater number of isomers, C_4H_8 or C_4H_{10}? Explain your choice.

6. Which of these compounds—C_4H_{10}, $CH_3CH{=}CHCH_3$, $CH_3C{\equiv}CCH_3$, or C_6H_6—has the same carbon-to-hydrogen ratio as cyclobutane? Explain your choice.

7. Identify the functional group in each compound (i.e., whether an alcohol, amine, etc.)

(a) $CH_3CHBrCH_2CH_3$ **(b)** CH_3CH_2COOH
(c) $C_6H_5CH_2CHO$ **(d)** $(CH_3)_2CHCH_2OCH_3$
(e) $CH_3COCH_2CH_3$ **(f)** $CH_3CH(NH_2)CH_2CH_3$
(g) $C_6H_4(OH)_2$ **(h)** CH_3COOCH_3

8. Draw Lewis structures of the following simple organic molecules: **(a)** $CH_3CHClCH_3$; **(b)** $HOCH_2CH_2OH$; **(c)** CH_3CHO.

9. Draw suitable structural formulas to show that there are *four* structural isomers of $C_3H_6Cl_2$.

10. Which of the following pairs of molecules are isomers and which are not? Explain. (Continued on page 958.)
(a) $CH_3CH_2CH_2CH_3$ and $CH_3CH{=}CHCH_3$
(b) $CH_3(CH_2)_5CH(CH_3)_2$ and $CH_3(CH_2)_4CH(CH_3)CH_2CH_3$

(c) $CH_3CHClCH_2CH_3$ and $CH_3CH_2CH_2CH_2CH_2Cl$

(d)
and

(e)
and

(f)
and

11. Give an acceptable name for each of the following.

(a) $CH_3CH_2CH_2\overset{\displaystyle CH_3}{\underset{\displaystyle |}{C}}HCH_2\overset{\displaystyle CH_3}{\underset{\displaystyle |}{C}}HCH_2CH_3$

(b) $CH_3-\overset{\displaystyle CH_3}{\underset{\displaystyle \underset{\textstyle CH_3}{|}}{\overset{|}{C}}}-CH_3$

(c) $CH_3CH_2CHCH_2-\overset{\displaystyle Cl}{\underset{\displaystyle \underset{\textstyle Cl}{|}}{\overset{|}{C}}}-CH_2CH_3$
     $\underset{\textstyle C_2H_5}{}$

12. Give an acceptable name for each of the following.

(a)

(b)

(c) $NH_2-$$-COOH$

13. Draw structural formulas for the following compounds:
(a) 3-isopropyloctane;
(b) 2-chloro-3-methylpentane;
(c) 2-pentene;
(d) dipropyl ether;
(e) *p*-bromophenol.

14. Write a condensed formula for each of the following.
(a) isopropyl alcohol (rubbing alcohol)
(b) 1,1,1-chlorodifluoroethane (a refrigerant)
(c) 2-methyl-1,3-butadiene (used to make elastomers)

15. Supply a structural formula for each of the following:
(a) 1,3,5-trimethylbenzene;
(b) *p*-nitrophenol;
(c) 3-amino-2,5-dichlorobenzoic acid (a plant growth regulator).

16. To prepare methyl ethyl ketone, which of these compounds would you oxidize: 2-propanol, 1-butanol, 2-butanol, or *t*-butyl alcohol? Explain?

17. Indicate the principal product(s) you would expect in
(a) treating $CH_3CH_2CH{=}CH_2$ with *dilute* $H_2SO_4(aq)$;
(b) exposing a mixture of chlorine and propane gases to ultraviolet light;
(c) heating a mixture of isopropyl alcohol and benzoic acid;
(d) oxidizing *s*-butyl alcohol with $Cr_2O_7^{2-}$ in acidic solution.

18. For each of the following pairs, which substance has
(a) the higher boiling point, C_6H_{12} or C_6H_6;
(b) the greater solubility in water, C_3H_7OH or $C_7H_{15}OH$;
(c) the greater acidity in aqueous solution, C_6H_5CHO or C_6H_5COOH?

Exercises

Organic Structures

19. Write structural formulas corresponding to these condensed formulas.
(a) $CH_3CH_2CHBrCHBrCH_3$
(b) $(CH_3)_3CCH_2C(CH_3)_2CH_2CH_2CH_3$
(c) $(C_2H_5)_2CHCH{=}CHCH_2CH_3$

21. With appropriate sketches represent chemical bonding in terms of the overlap of hybridized and unhybridized atomic orbitals in the following molecules.
(a) C_4H_{10} (b) $H_2C{=}CHCl$
(c) $CH_3C{\equiv}CH$

20. Write structural formulas corresponding to these condensed formulas.
(a) $(CH_3)_3CCH_2CH(CH_3)CH_2CH_2CH_3$
(b) $(CH_3)_2CHCH_2C(CH_3)_2CH_2Br$
(c) $Cl_3CCH_2CH(CH_3)CH_2Cl$

22. With appropriate sketches represent chemical bonding in terms of the overlap of hybridized and unhybridized atomic orbitals in the following molecules.

(a) $CH_3\overset{\displaystyle O}{\overset{\displaystyle \|}{C}}CH_3$ (b) $CH_3\overset{\displaystyle O}{\overset{\displaystyle \|}{C}}-OH$ (c) $H_2C{=}C{=}CH_2$

Isomers

23. Indicate the difference in these three types of isomers: structural, positional, and geometric. What term best describes each of the following pairs of isomers?
(a) $CH_3CH_2CH_2Cl$ and $CH_3CHClCH_3$
(b) $CH_3CH(CH_3)CH_2CH_3$ and $CH_3(CH_2)_3CH_3$

(c) and

25. Draw structural formulas for all the isomers of C_7H_{16}.

24. What term—structural, positional, or geometric—best describes each of the following pairs of isomers?
(a) $CHCl{=}CHCl$ and $CH_2{=}CCl_2$
(b) $CH_2{=}CHCH_2CH_3$ and $CH_2{=}C(CH_3)_2$

(c) and

26. Draw and name all the isomers of **(a)** C_6H_{14}; **(b)** C_4H_8; **(c)** C_4H_6. (*Hint:* Do not forget double bonds, rings, and combinations of these.)

Functional Groups

27. The functional groups in each of the following pairs have certain features in common, but what is the essential difference between them?
(a) carbonyl and carboxyl
(b) aldehyde and ketone
(c) acetic acid and acetyl group

28. Give one example of each of the following types of compounds: **(a)** aromatic nitro compound; **(b)** aliphatic amine; **(c)** chlorophenol; **(d)** aliphatic diol; **(e)** unsaturated aliphatic alcohol; **(f)** alicyclic ketone; **(g)** halogenated alkane; **(h)** aromatic dicarboxylic acid.

Nomenclature and Formulas

29. Give an acceptable name for each of the following structures.
(a) $CH_3CH_2C(CH_3)_3$
(b) $(CH_3)_2C{=}CH_2$
(c) $CH_3{-}CH{-}CH{-}CH_3$
$\qquad\qquad CH_2$
(d) $CH_3C{\equiv}CCH(CH_3)_2$
(e) $CH_3CH(C_2H_5)CH(CH_3)CH_2CH_3$
(f) $CH_3CH(CH_3)CH(CH_3)C(C_3H_7){=}CH_2$

31. Does each of the following names convey sufficient information to suggest a specific structure? Explain.
(a) pentene
(b) butanone
(c) butyl alcohol
(d) methylaniline
(e) methylcyclopentane
(f) dibromobenzene

33. Supply condensed structural formulas for the following substances.
(a) 2,4,6-trinitrotoluene (TNT—an explosive)
(b) methyl salicylate (oil of wintergreen) [*Hint:* Salicyclic acid is *o*-hydroxybenzoic acid.]
(c) 2-hydroxy-1,2,3-propanetricarboxylic acid (citric acid, $C_6H_8O_7$)

30. Draw a condensed structure to correspond to each of the following names.
(a) isopentane
(b) cyclohexene
(c) 2-methyl-3-hexyne
(d) 2-butanol
(e) isopropyl ethyl ether
(f) propionaldehyde

32. Indicate why each of these names is incorrect and give a correct name.
(a) 3-pentene
(b) pentadiene
(c) 1-propanone
(d) bromopropane
(e) 2,6-dichlorobenzene
(f) 2-methyl-3-pentyne

34. Supply condensed structural formulas for the following substances.
(a) *o*-*t*-butylphenol (an antioxidant in aviation gasoline)
(b) 1-phenyl-2-aminopropane (benzedrine—an amphetamine, ingredient in "pep pills")
(c) 2-methylheptadecane (a sex pheromone of tiger moths—a chemical used for communication among members of the species) [*Hint:* "Heptadeca" means 17.]

Alkanes

35. Write the structure of each alkane: **(a)** molecular mass 72 u, forms four monochlorination products; **(b)** molecular mass 72 u, forms a single monochlorination product.

36. Write the structure of each alkane.
(a) Molecular mass = 44 u; forms two different monochlorination products.
(b) Molecular mass = 58 u; forms two different monobromination products.

Alkenes

37. Why is it not necessary to refer to ethene and propene as 1-ethene and 1-propene? Can the same be said for butene?

39. Draw the structures of the products of each of the following reactions.
(a) propene + hydrogen (Pt, heat)
(b) 2-butanol + heat (in the presence of sulfuric acid)

38. Alkenes (olefins) and cyclic alkanes (alicyclics) each have the generic formula C_nH_{2n}. In what important ways do these types of compounds differ structurally?

40. Use Markovnikov's rule to predict the product of the reaction of

$$\overset{\text{Cl}}{\underset{|}{}}$$

(a) HCl with $CH_3\overset{|}{C}=CH_2$
(b) HCN with $CH_3C\equiv CH$
(c) HCl with $CH_3CH=C(CH_3)_2$

Aromatic Compounds

41. Supply a name or structural formula for each of the following.
(a) phenylacetylene
(b) *m*-dichlorobenzene

(c)

42. Supply a name or structural formula for each of the following.
(a) *p*-phenylphenol
(b) 2-hydroxy-4-isopropyltoluene (thymol—flavor constituent of the herb thyme)

(c)

Organic Reactions

43. Describe what is meant by each of the following reaction types and illustrate with an example from the text: **(a)** aliphatic substitution reaction; **(b)** aromatic substitution reaction; **(c)** addition reaction; **(d)** elimination reaction.
45. Draw a structure to represent the principal product of each of the following reactions.
(a) 1-pentanol + excess dichromate ion (acid solution)
(b) butyric acid + ethanol (acid solution)
(c) $CH_3CH_2C=CH_2 + H_2O$ (in the presence of H_2SO_4)
$$\underset{\underset{CH_3}{|}}{}$$

44. Starting with acetylene as the only source of carbon, together with any inorganic reagents required, devise a method to synthesize **(a)** 1,1,2,2-tetrabromoethane; **(b)** acetaldehyde.

46. Predict the products of the monobromination of **(a)** *m*-dinitrobenzene; **(b)** aniline; **(c)** *p*-bromoanisole.

Polymerization Reactions

47. In referring to the molecular mass of a polymer, we can speak only of the "average molecular mass." Explain why the molecular mass of a polymer is not a unique quantity, as it is for a substance like benzene.
49. Nylon 66 is produced by the reaction of 1,6-hexanediamine with adipic acid. A different nylon polymer is obtained if sebacyl chloride

$$\overset{\text{O}\quad\quad\text{O}}{\underset{\text{ClC(CH}_2)_8\text{CCl}}{\parallel\quad\quad\parallel}}$$

is substituted for the adipic acid. What is the basic repeating unit of this nylon structure?

48. Explain why Dacron is called a polyester. What is the %O, by mass, in Dacron?

50. Would you expect a polymer to be formed by the reaction of terephthalic acid with ethyl alcohol in place of ethylene glycol? With glycerol in place of ethylene glycol? Explain.

Advanced Exercises

51. In the chlorination of CH_4 some CH_3CH_2Cl is obtained as a product. Explain why this should be so.

52. Supply condensed or structural formulas for the following substances.

(a) 1,5-cyclooctadiene (an intermediate in the manufacture of resins)

(b) 3,7,11-trimethyl-2,6,10-dodecatriene-1-ol (farnesol—odor of lily-of-the-valley) [*Hint:* "Dodeca" means 12.]

(c) 2,6-dimethyl-5-hepten-1-al (used in the manufacture of perfume)

53. Draw structural formulas for all the isomers listed in Table 27.3 and show that, indeed, the substances with more compact structures have lower boiling points.

54. By drawing suitable structural formulas, establish that there are 17 isomers of $C_6H_{13}Cl$. [*Hint:* Refer to Example 27-1].

55. Write the isomers you would expect to obtain in the mononitration of *m*-methoxybenzaldehyde.

What does the fact that no 3-methoxy-5-nitrobenzaldehyde forms imply about the strength of meta and ortho,para directors?

56. The symbol

which is often used to represent benzene, is also the structural formula of cyclohexatriene. Are benzene and cyclohexatriene the same substance? Explain.

57. Use the half-reaction method to balance the following redox equations.

(a) $C_6H_5NO_2 + Fe + H^+ \rightarrow C_6H_5NH_3^+ + Fe^{3+} + H_2O$

(b) $C_6H_5CH_2OH + Cr_2O_7^{2-} + H^+ \rightarrow$
$$C_6H_5CO_2H + Cr^{3+} + H_2O$$

(c) $CH_3CH{=}CH_2 + MnO_4^- + H_2O \rightarrow$
$$CH_3CHOHCH_2OH + MnO_2 + OH^-$$

58. A 10.6-g sample of benzaldehyde was allowed to react with 5.9 g $KMnO_4$ in an excess of $KOH(aq)$. After filtration of the $MnO_2(s)$ and acidification of the solution, 6.1 g of benzoic acid was isolated. What was the percent yield of the re-

action? (*Hint:* Write half-equations for the oxidation and reduction half-reactions.)

59. Combustion of a 0.1908-g sample of a compound gave 0.2895 g CO_2 and 0.1192 g H_2O. Combustion of a second sample weighing 0.1825 g yielded 40.2 mL of $N_2(g)$, collected over 50% $KOH(aq)$ (vapor pressure = 9 mmHg) at 25 °C and 735 mmHg barometric pressure. When 1.082 g of compound was dissolved in 26.00 g benzene (mp 5.50 °C, $K_f = 5.12$ °C m^{-1}), the solution had a freezing point of 3.66 °C. What is the molecular formula of this compound?

60. Explain why a polymer formed by chain reaction polymerization generally has a higher molecular mass than a corresponding polymer formed by step reaction polymerization.

61. The three isomeric tribromobenzenes, I, II, and III, when nitrated, form three, two, and one mononitrotribromobenzenes, respectively. Assign correct structures to I, II, and III.

62. Write the name and structure of each aromatic hydrocarbon.

(a) Formula: C_8H_{10}; forms three ring monochlorination products.

(b) Formula: C_9H_{12}; forms one ring mononitration product.

(c) Formula: C_9H_{12}; forms four ring mononitration products.

63. In the molecule 2-methylbutane, the organic chemist distinguishes the different types of hydrogen and carbon atoms as being primary (1°), secondary (2°), and tertiary (3°). For the monochlorination of hydrocarbons the following ratio of reactivities has been found: 3°/2°/1° = 4.3:3:1. How many different monochloro derivatives of 2-methylbutane are possible and what percent of each would you expect to find?

$$^{1°}H{-}\overset{\overset{\textstyle H^{1°}}{\overset{|}{\underset{}{1°}}}}{\underset{\underset{\textstyle ^{1°}H}{}}{C}}{-}H^{1°}$$

$$^{1°}H{-}\overset{\overset{\textstyle ^{1°}H}{}}{\underset{\underset{\textstyle ^{1°}H}{|}}{C}}{\overset{1°}{}}{-}\overset{3°}{C}{-}\overset{\overset{\textstyle H^{2°}\ H^{1°}}{}}{\underset{\underset{\textstyle H^{2°}\ H^{1°}}{}}{C{-}C}}{\overset{2°\ 1°}{}}{-}H^{1°}$$

64. A particular colorless organic liquid is known to be one of the following compounds: 1-butanol, diethyl ether, methyl propyl ether, butyraldehyde, or propionic acid. Can you identify which it is based on the following tests? If not, what additional test would you perform? (1) A 2.50-g sample dissolved in 100.0 g water has a freezing point of −0.7 °C. (2) An aqueous solution of the liquid does not change the color of blue litmus paper. (3) When alkaline $KMnO_4(aq)$ is added to the liquid and the mixture heated, the purple color of the MnO_4^- disappears.

Chemistry of the Living State

28

*T*he planet Earth teems with life—from the lowly amoeba to the mighty whale. In spite of extraordinary differences in outward appearance, all life forms share similar needs and chemical structures. Raw materials (food) are required for building cells and providing the energy of metabolism and the information of heredity. All forms of life on Earth use the same types of complicated structures to perform specialized functions. In this chapter, we will stress the structures of the macromolecular substances common to all organisms. Our discussion will emphasize how fundamental principles of chemistry acquired throughout the text can contribute to a knowledge of the living state.

Human insulin protein shown as a ribbon molecular model. A wireframe model of the protein is superimposed on the ribbon model. The ribbon model clearly shows the α-helix portions of the protein.

963

28-1   Chemical Structure of Living Matter: An Overview

From one-celled plants and animals to humans, the living state includes some highly complex forms of matter. Of the known elements, about 50 occur in measurable concentrations in living matter. Of these, about 25 have functions that are definitely known. Four elements together—oxygen, carbon, hydrogen, and nitrogen—account for 96% of human body mass. Table 28.1 lists the chief elements found in living matter.

In addition to water, which is the most abundant compound in most living organisms, the next most important constituents are lipids, carbohydrates, proteins, and nucleic acids. These four types of macromolecular substances are the principal topics of this chapter.

The **cell** is the fundamental unit of all life. Cells are made up of components such as a nucleus, mitochondria, and chloroplasts (plant cells). Cells combine to form tissues; tissues may be grouped into organs; organs combine into organisms.

28-2   Lipids

Lipids are best described through their physical properties rather than in precise structural terms. **Lipids** are those constituents of plant and animal tissue that are soluble in solvents of low polarity, such as chloroform, carbon tetrachloride, diethyl ether, and benzene. Many compounds fit this description, but we will discuss only the most common type, those known as triglycerides.

Triglycerides are esters of glycerol (1,2,3-propanetriol) and long-chain monocarboxylic acids (fatty acids). *Glycerol* provides the three-carbon backbone, and the fatty acids provide *acyl* groups (in blue).

glycerol   fatty acid

The systematic name for these esters is triacylglycerols, but the common name that has long been used is triglycerides, the name we use here. If all acyl groups are the same, the triglyceride is a *simple* triglyceride; otherwise the triglyceride

TABLE 28.1 Chemical Elements of the Living State[a]		
Essential to All Organisms	**Essential to Some Organisms**	**Believed to Have Some Biological Function**
H, C, N, O, Mg, P, S, K, Ca, Mn, Fe, Cu, Zn, Mo	B, F, Na, Si, Cl, V, Cr, Co, Ni, As, Se, I, Pb	Br, Cd, Sn, W

[a]To some extent, metallic elements are incorporated into the structures of large coordination complexes, such as Mg in chlorophyll and Fe in hemoglobin, but metals can also exist as free ions, such as Na^+ and K^+.

TABLE 28.2 Some Common Fatty Acids

Common Name	IUPAC Name	Formula
Saturated acids		
lauric acid	dodecanoic acid	$C_{11}H_{23}CO_2H$
myristic acid	tetradecanoic acid	$C_{13}H_{27}CO_2H$
palmitic acid	hexadecanoic acid	$C_{15}H_{31}CO_2H$
stearic acid	octadecanoic acid	$C_{17}H_{35}CO_2H$
Unsaturated acids		
oleic acid	9-octadecenoic acid	$C_{17}H_{33}CO_2H$
linoleic acid	9,12-octadecadienoic acid	$C_{17}H_{31}CO_2H$
linolenic acid	9,12,15-octadecatrienoic acid	$C_{17}H_{29}CO_2H$
eleostearic acid	9,11,13-octadecatrienoic acid	$C_{17}H_{29}CO_2H$

is a *mixed* triglyceride. Some long-chain, or fatty, acids commonly encountered in triglycerides are listed in Table 28.2.

glycerol

glyceryl tripalmitate
tripalmitin
(a simple triglyceride; a fat)

glyceryl trioleate
triolein
(a simple triglyceride; an oil)

glyceryl palmitooleostearate
(a mixed triglyceride)

EXAMPLE 28-1

Drawing Structures of Triglycerides. What is the structural formula for glyceryl butyropalmitooleate?

Solution

First, identify the acids from which the acyl groups in this *mixed* triglyceride are derived and write the structures of these acyl groups. We can base the formulas of the parent carboxylic acids on their names. Recall from Section 27-7 that in forming an ester, a carboxylic acid loses its —OH group. The acyl groups are

butyric

palmitic

oleic

Then, attach the acyl groups to the glycerol backbone with the first-named acyl group at the top and the last-named acyl group of the triglyceride at the bottom.

$$CH_2OC(CH_2)_2CH_3$$

$$CHOC(CH_2)_{14}CH_3$$

$$CH_2OC(CH_2)_7CH{=}CH(CH_2)_7CH_3$$

Practice Example A: What is the name for the following triglyceride?

$$CH_2OC(CH_2)_{10}CH_3$$

$$CHOC(CH_2)_{10}CH_3$$

$$CH_2OC(CH_2)_{10}CH_3$$

Practice Example B: What is the structural formula for glyceryl lauromyristolinoleate? (*Hint:* Refer to the IUPAC names in Table 28.2 for information about the structural features of the fatty acids.)

Triglycerides can be *hydrolyzed* in alkaline solution to produce glycerol and salts of the fatty acids. The hydrolysis process is called **saponification,** and the salts are commonly known as **soaps.** For example, the hydrolysis of tristearin with aqueous KOH gives glycerol and the soap potassium stearate.

Soapmaking was a common frontier activity in colonial times.

$$
\begin{array}{c}
\mathrm{CH_2OC(CH_2)_{16}CH_3} \\[-2pt]
\overset{\textstyle O}{\|} \\[-4pt]
\mathrm{CHOC(CH_2)_{16}CH_3} \;+\; 3\,\mathrm{KOH} \longrightarrow \;\; \mathrm{CHOH} \;+\; 3\,\mathrm{CH_3(CH_2)_{16}COK} \qquad (28.1) \\[-2pt]
\overset{\textstyle O}{\|} \\[-4pt]
\mathrm{CH_2OC(CH_2)_{16}CH_3} \qquad\qquad\qquad \mathrm{CH_2OH}
\end{array}
$$

         tristearin   glycerol   potassium stearate
                                                    (a soap)

The cleansing action of soaps was described in Section 22-1.

Fats and oils are both triglycerides (*glyceryl esters*) and differ from each other by the nature of the acid components in the triglycerides. **Fats** are glyceryl esters in which *saturated* acid components (containing no carbon-to-carbon double bonds) predominate; they are solids at room temperature. **Oils** have a predominance of *unsaturated* fatty acids (which contain carbon-to-carbon double bonds) and are liquids at room temperature. The compositions of fats and oils are variable and depend not only on the particular plant or animal species involved but also on dietary and climatic factors. Some common fats and oils are listed in Table 28.3.

When pure, fats and oils are colorless, odorless, and tasteless. The characteristic colors, odors, and flavors commonly associated with them are imparted by other organic substances present in the impure materials. The yellow color of butter is that of β-carotene (a yellow pigment also found in carrots and marigolds; recall page 955). The taste of butter is attributed to the following two compounds produced in the aging of cream.

$$
\begin{array}{cc}
\overset{\textstyle O}{\|} & \overset{\textstyle O}{\|}\;\;\overset{\textstyle O}{\|} \\
\mathrm{CH_3-C-CHCH_3} & \mathrm{CH_3-C-C-CH_3} \\
\;\;\;\;\;\;\;\;\;\;| & \\
\;\;\;\;\;\;\;\;\;\mathrm{OH} &
\end{array}
$$

    3-hydroxy-2-butanone   diacetyl

Corn oil has the approximate composition listed in Table 28.3. Hydrogenation converts some of the unsaturated to saturated fatty acid components—the primary change required in making margarine from vegetable oils.

TABLE 28.3 Some Common Fats and Oils

| | Component Acids, % by Mass | | | | | |
| | Saturated | | | Unsaturated | | |
Lipid	Myristic	Palmitic	Stearic	Oleic	Linoleic	Linolenic
fats						
butter	7–10	24–26	10–13	28–31	1–3	0.2–0.5
lard	1–2	28–30	12–18	40–50	7–13	0–1
edible oils						
corn	1–2	8–12	2–5	19–49	34–62	—
safflower	—	6–7	2–3	12–14	75–80	0.5–1.5

The formulas of the individual acids are listed in Table 28.2.

Unsaturation in a fat or oil can be removed (that is, the fat or oil can become saturated) by the catalytic addition of hydrogen (hydrogenation). Thus, oils or low-melting fats can be changed to higher melting fats. These higher melting fats, when mixed with skim milk, fortified with vitamin A, and artificially colored, are known as *margarines*. Unsaturation in a fat or oil is also removed when the fat or oil decomposes. Edible fats and oils both hydrolyze and cleave at the double bonds by oxidation on exposure to heat, air, and light. When this happens the fat becomes *rancid*. The low-molecular-mass fatty acids produced by this cleavage have offensive odors, such as butyric acid in rancid butter. Antioxidants, such as 3-*t*-butyl-4-hydroxyanisole (BHA), retard this oxidative rancidity and are commonly added to oils used in the high-temperature cooking of potato chips and other foods.

Medical evidence suggests a relationship between a high intake of saturated fats and the incidence of coronary heart disease. For this reason, many diets call for the substitution of unsaturated for saturated fatty acids in foods. In general, mammal fats are saturated, whereas those derived from vegetables, seafood, and poultry are unsaturated.

3-*t*-butyl-4-hydroxyanisole (BHA)

28-3   Carbohydrates

The literal meaning of "carbohydrate" is hydrate of carbon: $C_x(H_2O)_y$. Thus, sucrose, or cane sugar, $C_{12}H_{22}O_{11}$, is equivalent to $C_{12}(H_2O)_{11}$. A more useful definition is that **carbohydrates** are polyhydroxy aldehydes, polyhydroxy ketones, their derivatives, and substances that yield them on hydrolysis. Carbohydrates that are aldehydes are called aldoses; those that are ketones are called ketoses. A five-carbon carbohydrate is a pentose, a six-carbon one, a hexose, and so on. Two familiar hexoses are glucose and fructose, an aldose and a ketose, respectively.

FIGURE 28-1
Optical isomers of glyceraldehyde.
The structure in (a) is not superimposable on (b), just as a right hand and a left hand are not superimposable. (A right-handed glove does not fit on a left hand.)

mirror

(a)   **(b)**

$$
\begin{array}{ll}
\text{HC}{=}\text{O} & \text{CH}_2\text{OH} \\
| & | \\
\text{CHOH} & \text{C}{=}\text{O} \\
| & | \\
\text{CHOH} & \text{CHOH} \\
| & | \\
\text{CHOH} & \text{CHOH} \\
| & | \\
\text{CHOH} & \text{CHOH} \\
| & | \\
\text{CH}_2\text{OH} & \text{CH}_2\text{OH} \\
\text{glucose} & \text{fructose}
\end{array}
$$

The simplest carbohydrates are the **monosaccharides. Oligosaccharides** contain from two to ten monosaccharide units bonded together. Names can be assigned to reflect the actual number of such units present, such as *di*saccharide and *tri*saccharide. Mono- and oligosaccharides are also called **sugars. Polysaccharides** contain more than ten monosaccharide units. The general term for all carbohydrates is glycoses. In summary,

Glycoses
{
   Monosaccharides
      aldose (aldotriose, aldotetrose, . . .)
      ketoses (ketotriose, ketotetrose, . . .)
   Oligosaccharides (from two to ten monosaccharide units)
      disaccharides (e.g., sucrose)
      trisaccharides (e.g., raffinose)
      and so on.
   Polysaccharides (more than ten monosaccharide units)
      (e.g., starch and cellulose)

The simplest glycose is 2,3-dihydroxypropanal (glyceraldehyde), an *aldotriose*. Figure 28-1 illustrates an interesting form of stereoisomerism in glyceraldehyde—*optical isomerism*. There are *two* nonsuperimposable structures for glyceraldehyde. As we discovered in Section 25-4, such structures are related to each other like a right and a left hand, or like an object and its nonsuperimposable mirror image; they are called *enantiomers*.

As we also learned in Section 25-4, optically active molecules affect plane-polarized light (Figure 28-2). Interactions between a beam of polarized light and the electrons in an enantiomer cause a rotation of the plane of the polarized light. One enantiomer rotates the plane of polarized light to the right (clock-

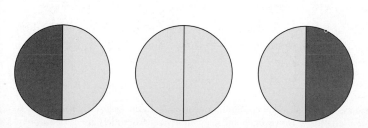

FIGURE 28-2
View through the analyzer prism of a polarimeter (see also Figure 25-7).
The field of view of polarized light from a sodium vapor lamp is split in half. The analyzer prism in the eyepiece has been rotated in the correct direction—clockwise (to the right) or counterclockwise (to the left)—and through the appropriate angle, α, when the two halves transmit light of equal intensity (center). For incorrect angles of rotation, one semicircle is darker than the other.

▶ The prefixes dextro and levo are derived from the Latin words *dexter* = right and *laevus* = left.

Emil Fischer (1852–1919). Fischer was awarded the Nobel Prize in 1902 for his research on the structures of sugars. Later he also elucidated how amino acid molecules join to form proteins.

FIGURE 28-3
**The structure of
D-(−)-erythrose.**
The three-dimensional structure (a) is represented in two dimensions by (b).

(a)

$$\begin{array}{c} CHO \\ | \\ H-C-OH \\ | \\ H-C-OH \\ | \\ CH_2OH \end{array}$$

(b)

wise) and is said to be **dextrorotatory** (designated +). The other enantiomer rotates the plane of polarized light to the same extent, but to the left (counterclockwise). It is said to be **levorotatory** (−). Because of their ability to rotate the plane of polarized light, isomers of these types are said to be *optically active,* and they are called *optical isomers.* Almost all molecules exhibiting optical isomerism possess at least one carbon atom with four different groups attached to it. Such a carbon atom is said to be *asymmetric,* or *chiral.*

The arrangement of groups at an asymmetric carbon atom is called the **absolute configuration.** Carbohydrate chemists use a system of assigning absolute configurations based on X-ray studies of glyceraldehyde. The dextrorotatory (+) isomer has the configuration shown in Figure 28-1a. Its mirror image, the levorotatory (−) isomer, has the configuration in Figure 28-1b. This system, extended to other chiral structures, is called the D,L convention.*

We cannot easily represent a three-dimensional structure in a two-dimensional drawing, but Emil Fischer, a pioneer in carbohydrate chemistry, devised a helpful scheme. The Fischer notation places the structural formula so that the backbone of the molecule is arranged from top to bottom, with the most oxidized portion of the molecule (—CHO) at the top and the least oxidized (—CH$_2$OH) at the bottom. Attached groups (—H and —OH) are written to the sides. The end groups of the backbone are considered to extend *behind* the plane of the page, *away* from the viewer. The glyceraldehyde enantiomers are written as

$$\begin{array}{cc} CHO & CHO \\ | & | \\ H-C-OH & HO-C-H \\ | & | \\ CH_2OH & CH_2OH \end{array}$$

D-(+)-glyceraldehyde   L-(−)-glyceraldehyde

and establish the D,L convention to be used for other sugars: The —H and —OH groups on the next-to-last (penultimate) carbon atom extend in *front* of the page, *toward* the viewer. The penultimate carbon atoms are shown in blue. If the —OH group on this penultimate carbon atom is to the *right,* the configuration is D. If the —OH is to the left, the configuration is L. This convention is applied below to the four-carbon aldoses. All D-sugars have the same configuration at this penultimate carbon. Figure 28-3 may help you to picture the relationship between a three-dimensional structure and its two-dimensional representation. Note that the letters D and L have no relationship to the + and − notation for the compound (see Are You Wondering . . . on page 971).

$$\begin{array}{cccc} CHO & CHO & CHO & CHO \\ | & | & | & | \\ H-C-OH & HO-C-H & HO-C-H & H-C-OH \\ | & | & | & | \\ H-C-OH & HO-C-H & H-C-OH & HO-C-H \\ | & | & | & | \\ CH_2OH & CH_2OH & CH_2OH & CH_2OH \end{array}$$

mirror   mirror

D-(−)-erythrose   L-(+)-erythrose   D-(−)-threose   L-(+)-threose

D-Erythrose and L-erythrose are enantiomers, as are also D-threose and L-threose. If we compare the configurations of D-erythrose and D-threose, we note that

---

*A more generalized system for absolute configurations uses *R* (Latin, *rectus,* right) in place of D and *S* (Latin, *sinister,* left) for L. However, D and L symbols are still widely used for carbohydrates.

these two molecules are *not* mirror images. On the other hand, they are isomers of each other and both are optically active. Optical isomers that are *not* mirror images of each other are called **diastereomers.**

$$
\begin{array}{ccc}
\text{CHO} & & \text{CHO} \\
| & & | \\
\text{H}-\text{C}-\text{OH} & & \text{HO}-\text{C}-\text{H} \\
| & & | \\
\text{H}-\text{C}-\text{OH} & & \text{H}-\text{C}-\text{OH} \\
| & & | \\
\text{CH}_2\text{OH} & & \text{CH}_2\text{OH}
\end{array}
$$

mirror

D-erythose   D-threose

not mirror images

Are You Wondering . . .

What is the relationship between the D,L and (+), (−) in names such as erythrose and threose?

The designations D and L indicate how the H and OH groups are attached to the penultimate carbon atoms, based on the glyceraldehyde molecule as a reference. These designations are unrelated to the direction in which the plane of polarized light is rotated, which is noted as (+) or (−). In the case of glyceraldehyde, it is simply a matter of chance that the isomer with the configuration D does rotate the plane of polarized light to the *right* (*dextro*rotatory), and so it is also (+) in terms of its optical activity.

But the situation we find with erythrose and threose is equally likely. Here the isomer that has the absolute configuration corresponding to D rotates the plane of polarized light to the *left* (*levo*rotatory), and so it is a (−) isomer. Similarly, the one with an L configuration is (+) in its optical activity.

Enantiomers differ only in the *direction,* not the extent to which, they rotate plane-polarized light. Diastereomers do differ in the extent to which they rotate plane-polarized light. They also differ in physical and chemical properties.

A mixture of equal amounts of the D and L configurations of a particular substance, called a **racemic mixture,** does not rotate the plane of polarized light either to the left or to the right. The designation DL-erythrose, for example, signifies a racemic mixture. Often, when molecules with chiral centers are synthesized, the product is a racemic mixture. This is because the creation of these centers is a random process like flipping a coin (an equal probability for "heads" or "tails"). If optically pure isomers are desired, the racemic mixture must be separated into the component enantiomers by a process called *resolution.* Sometimes this is carried out through an enzyme reaction, with a particular enzyme that reacts with one enantiomer but not the other.

Monosaccharides

Of the 16 possible aldohexoses, only three occur widely in nature: D-glucose, D-mannose, and D-galactose.

FIGURE 28-4

Representation of ring closure in glucose molecule.
The straight chain D-(+)-glucose molecule folds back on itself to bring the —OH group on the C-5 atom close to the aldehyde group of the C-1 atom. A proton is transferred, and this is followed by the formation of a six-member ring with an O atom joining the C-1 and C-5 atoms.

D-(+)-glucose

α-D-(+)-glucose

D-glucose   D-mannose   D-galactose

To a very small extent (less than 0.5% for glucose) these three sugar molecules do exist in the straight-chain forms pictured here, but in the main they occur in *cyclic* form. In this ring formation, the —OH group of the fifth carbon atom (C-5) adds to the carbonyl of the C-1 atom and produces a ring composed of five C atoms and one O atom, as illustrated in Figure 28-4. The conformation of the six-member ring is of the "chair" type (recall Figure 27-3).

When the chain form of a sugar is converted to the ring form, a new chiral (asymmetrical) center is produced at the C-1 atom. There are two possible orientations at this center. In the α form the —OH at C-1 is *axial* (directed down); in the β form it is *equatorial* (extends out from the ring). The α and β forms of glucose are pictured in Figure 28-5.

The naming of monosaccharides in the ring form is complicated, but each item in a name conveys precise information. Thus, D-(+)-glucose refers to the straight-chain form of glucose in the D configuration; the (+) indicates this form is dextrorotatory. The name α-D-(+)-glucose denotes the ring form derived from D-glucose with the α configuration at the C-1 atom.

With many sugars, known as **reducing sugars,** a sufficient amount of the straight-chain form is in equilibrium with the cyclic form that the sugar engages in an oxidation–reduction reaction with $Cu^{2+}(aq)$. The $Cu^{2+}(aq)$ is reduced to insoluble red Cu_2O, and the aldehyde portion of the sugar is oxidized (to an

FIGURE 28-5

α and β forms of D-glucose.
In the α form the —OH group at C-1 (in red) is axial, extending below the chair-like ring. In the β form the —OH group is equatorial and extends out from the ring.

α-D-(+)-glucose   β-D-(+)-glucose

acid). The test for a reducing sugar is conducted with alkaline copper ion complexed with tartrate (Fehling's solution) or citrate ion (Benedict's solution).

$$\text{certain cyclic sugars} \rightleftharpoons \underset{\text{straight-chain form}}{\begin{array}{c}\text{CHO}\\|\\\text{CHOH}\\|\end{array}} \xrightarrow{Cu^{2+}} \underset{\text{red ppt.}}{\begin{array}{c}\text{COOH}\\|\\\text{CHOH}+Cu_2O(s)\\|\end{array}} \quad (28.2)$$

Disaccharides

Two monosaccharides can join together by eliminating a H_2O molecule between them—a condensation reaction. This combination is called a disaccharide. We must consider two points in describing a disaccharide: What are the component monosaccharide units, and is the configuration of the linkage between the monosaccharide units α or β? The important naturally occurring disaccharides—maltose, cellobiose, lactose, and sucrose—are presented in Figure 28-6.

In *maltose* an H atom on the C-1 hydroxyl group of one glucose unit reacts with the hydroxyl group on the C-4 atom of a second glucose unit. The two units are linked in the α manner. Equilibrium is possible between the cyclic and straight-chain forms of maltose, so it is a reducing sugar. Maltose is produced by the action of malt enzyme on starch. Maltose undergoes fermentation, in the presence of yeast, first to glucose, and then to ethanol and $CO_2(g)$.

Cellobiose can be obtained by the careful hydrolysis of cellulose. It is a glucose–glucose disaccharide with β linkages. *Lactose,* or milk sugar, is naturally present in milk, where its concentration may range from 0% to 7%. It is a galactose–glucose disaccharide with β linkages. *Sucrose* is ordinary table sugar (cane or beet sugar). It is a glucose–fructose disaccharide linked $1\alpha, 2\beta$. Neither of the two cyclic sugar units can open up into a chain form, and as a result sucrose is *not* a reducing sugar.

FIGURE 28-6
Some common disaccharides.

maltose (α-form)

lactose (β-form)

cellobiose

sucrose

■ Glucose   ■ Galactose   ■ Fructose

Polysaccharides

Polysaccharides are composed of monosaccharide units joined into long chains by oxygen linkages. *Starch,* with a molecular mass between 20,000 and 1,000,000 u, is the reserve carbohydrate of many plants and is the bulk constituent of cereals, rice, corn, and potatoes. Its structural features are brought out in Figure 28-7. *Glycogen* is the reserve carbohydrate of animals, occurring in liver and muscle tissue. It has a higher molecular mass than starch, and the polysaccharide chains are more branched. *Cellulose* is the main structural material of plants. It is the chief component of wood pulp, cotton, and straw. Complete hydrolysis of cellulose produces glucose. Cellulose has a molecular mass between 300,000 and 500,000 u, corresponding to 1800–3000 glucose units. Most animals, including human beings, do not possess the necessary enzymes to hydrolyze β linkages. As a result they cannot digest cellulose. Certain bacteria in ruminants (cows, sheep, horses) and termites can hydrolyze cellulose, allowing them to use it as a food. Termites, as we know, subsist on a diet of wood.

Photosynthesis

As we have noted on previous occasions (Sections 7-9 and 25-10), the process of photosynthesis involves the conversion by plants of carbon dioxide and water into carbohydrates. Key roles are played in this conversion by the catalyst chlorophyll (see Figure 25-20) and by sunlight.

$$n\,CO_2 + n\,H_2O \xrightarrow[\text{chlorophyll}]{\text{sunlight}} C_n(H_2O)_n + n\,O_2 \tag{28.3}$$

Equation (28-3) is greatly oversimplified. The currently accepted mechanism, proposed by Melvin Calvin (Nobel Prize, 1961), involves as many as 100 sequential steps for the conversion of 6 mol of carbon dioxide to 1 mol of glu-

FIGURE 28-7
Two common polysaccharides.

starch

cellulose

cose. The elucidation of this mechanism was greatly aided by the use of carbon-14 as a radioactive tracer. For simplicity, the overall photosynthetic process is divided into two phases: (1) the conversion of solar energy to chemical energy—the light reaction; and (2) the synthesis, promoted by enzymes, of carbohydrates. This latter reaction can occur in the absence of light and is called the dark reaction.

28-4   Proteins

When a protein is hydrolyzed by dilute acids, bases, or hydrolytic enzymes, the result is a mixture of α-amino acids. An *amino acid* is a carboxylic acid that also contains an amine group, $—NH_2$; an **α-amino acid** has the amino group on the α carbon atom—the carbon atom next to the carboxyl group. Thus proteins are high-molecular-mass polymers composed of α-amino acids. Of the known α-amino acids, about 20 have been identified as building blocks of most plant and animal proteins. Some are listed in Table 28.4.

Proteins are the basis of protoplasm and are found in all living organisms. Proteins, as muscle, skin, hair, and other tissue, make up the bulk of the body's nonskeletal structure. As enzymes, proteins catalyze biochemical reactions; as hormones, they regulate metabolic processes; and as antibodies, they counteract the effect of invading organisms.

Other than glycine ($H_2NCH_2CO_2H$), naturally occurring amino acids are optically active, mostly with an L configuration.

▶ The name "protein" is derived from the Greek word *proteios,* meaning "of first importance" (similar to the derivation of "proton").

$$
\underset{\substack{H_2N \quad \ \ \ H \\ R}}{\overset{CO_2H}{C}} \equiv \underset{R}{\overset{CO_2H}{H_2NCH}}
$$

The reference structure for establishing the absolute configurations of amino acids is again glyceraldehyde, with the $—NH_2$ group substituting for $—OH$ and $—CO_2H$, for $—CHO$. The molecule shown above has an L configuration because the $—NH_2$ group appears on the *left*.

Certain amino acids are required for proper health and growth in human beings, yet the body is unable to synthesize them. These must be ingested through foods and are called *essential* amino acids. Eight are known to be essential; the case of three others is less certain (see Table 28.4).

The amino acids are colorless, crystalline, high-melting solids that are moderately soluble in water. In a strongly acidic solution (low pH) the amino acid exists as a cation: a proton from solution attaches itself to the unshared pair of electrons on the nitrogen atom in the group $—NH_2$. In a strongly basic solution (high pH), an anion forms through the loss of protons by the $—CO_2H$ and $—NH_3^+$ groups. At an intermediate point, a proton is lost from the $—CO_2H$ but retained by the $—NH_3^+$ group. The product is a dipolar ion or a "zwitterion."

$$
\underset{\substack{\text{acidic soln}}}{\underset{R}{NH_3^+—CH—CO_2H}} \underset{H^+}{\overset{OH^-}{\rightleftharpoons}} \underset{\substack{\text{isoelectric point}}}{\underset{R}{NH_3^+—CH—CO_2^-}} \underset{H^+}{\overset{OH^-}{\rightleftharpoons}} \underset{\substack{\text{basic soln}}}{\underset{R}{NH_2—CH—CO_2^-}} \qquad (28.4)
$$

Amino acids are amphoteric. In most amino acids, the acidity of the $—NH_3^+$ group is slightly greater than the basicity of the $—CO_2^-$ group. The largest group of amino acids are close to pH neutral. The pH at which the dipolar structure predominates is called the **isoelectric point,** or p*I*. At this pH the molecule

TABLE 28.4 Some Common Amino Acids

Name	Symbol	Formula	pI[a]
Neutral amino acids			
glycine	Gly	$HCH(NH_2)CO_2H$	6.03
alanine	Ala	$CH_3CH(NH_2)CO_2H$	6.10
valine[b]	Val	$(CH_3)_2CHCH(NH_2)CO_2H$	6.04
leucine[b]	Leu	$(CH_3)_2CHCH_2CH(NH_2)CO_2H$	6.04
isoleucine[b]	Ileu or Ile	$CH_3CH_2CH(CH_3)CH(NH_2)CO_2H$	6.04
serine	Ser	$HOCH_2CH(NH_2)CO_2H$	5.70
threonine[b]	Thr	$CH_3CHOHCH(NH_2)CO_2H$	5.6
phenylalanine[b]	Phe	$C_6H_5CH_2CH(NH_2)CO_2H$	5.74
methionine[b]	Met	$CH_3SCH_2CH_2CH(NH_2)CO_2H$	5.71
cysteine	Cys	$HSCH_2CH(NH_2)CO_2H$	5.05
cystine	$(Cys)_2$	$[SCH_2CH(NH_2)CO_2H]_2$	5.1
tyrosine	Tyr	$4\text{-}HOC_6H_4CH_2CH(NH_2)CO_2H$	5.70
tryptophan[b]	Trp		5.89
proline[c]	Pro		6.21
Acidic amino acids			
aspartic acid	Asp	$HO_2CCH_2CH(NH_2)CO_2H$	2.96
glutamic acid	Glu	$HO_2CCH_2CH_2CH(NH_2)CO_2H$	3.22
Basic amino acids			
lysine[b]	Lys	$H_2N(CH_2)_4CH(NH_2)CO_2H$	9.74
arginine	Arg	$H_2NC(=NH)NH(CH_2)_3CH(NH_2)CO_2H$	10.73
histidine	His		7.58

[a]pH of *isoelectric point.*

[b]Essential amino acids. In addition, arginine and glycine are required by the chick, arginine by the rat, and histidine by human infants.

[c]The secondary amino group makes proline an *α-imino* acid. Nevertheless, it is commonly listed with amino acids.

does not migrate in an electric field. At a pH above the p*I*, the molecule migrates to the anode (positive electrode), and below the p*I*, to the cathode (negative electrode). Most basic amino acids have a p*I* well above 7, acidic ones well below 7, and most neutral ones slightly less than 7 (5.7–6.1).

Peptides

Two amino acid molecules can be joined by the elimination of a water molecule between them. The amino acids thus joined form a dipeptide. The bond between the two amino acid units is called a **peptide bond.**

$$NH_2-CH-\overset{\overset{\displaystyle O}{\|}}{C}-OH + HN-\overset{\overset{\displaystyle R'}{|}}{CH}-CO_2H \longrightarrow$$

with H on the nitrogen and R above

$$\text{peptide bond}$$

$$H_2O + NH_2-\underset{\underset{\displaystyle R}{|}}{CH}-\overset{\overset{\displaystyle O}{\|}}{C}-NH-\underset{}{\overset{\overset{\displaystyle R'}{|}}{CH}}-CO_2H \qquad (28.5)$$

a dipeptide

A tripeptide has three amino acid residues and two peptide linkages. A large number of amino acid units may join to form a **polypeptide.**

The amino acid unit present at one end of the polypeptide chain has a free —NH_2 group; this is the "N-terminal" end. The other end of the chain has a free —CO_2H group; it is the "C-terminal" end. The polypeptide structure is written with the N-terminal end to the left and the C-terminal end to the right. The base name of the polypeptide is that of the C-terminal amino acid. The other amino acid units in the chain are named as substituents of this acid. Their names change from an "ine" to the "yl" ending. Abbreviations are also commonly used in writing polypeptide names, as illustrated in Example 28-2.

EXAMPLE 28-2

Naming a Polypeptide. What is the name of the polypeptide whose structure is shown below?

$$H_2N-CH_2-\overset{\overset{\displaystyle O}{\|}}{C}-NH-\underset{\underset{\displaystyle CH_3}{|}}{CH}-\overset{\overset{\displaystyle O}{\|}}{C}-NH-\underset{\underset{\displaystyle CH_2OH}{|}}{CH}-\overset{\overset{\displaystyle O}{\|}}{C}-OH$$

(a)   (b)   (c)

Solution

We can identify the three amino acids in this tripeptide through Table 28.4. (a) = glycine; (b) = alanine; (c) = serine. The C-terminal amino acid is serine. The name is

glycylalanylserine (Gly-Ala-Ser)

Practice Example A: What is the name of the polypeptide shown below?

$$H_2N-\underset{\underset{\displaystyle CH_3CHOH}{|}}{CH}-\overset{\overset{\displaystyle O}{\|}}{C}-NH-\underset{\underset{\displaystyle CH_3CHOH}{|}}{CH}-\overset{\overset{\displaystyle O}{\|}}{C}-NH-\underset{\underset{\displaystyle CH_2CH_2SCH_3}{|}}{CH}-\overset{\overset{\displaystyle O}{\|}}{C}-OH$$

Practice Example B: Write the structural formula of the polypeptide serylglycylvaline.

Sequencing.   Suppose that a tripeptide is known to consist of the three amino acids: A, B, and C. What is the correct structure: ABC?, BAC?, . . . Can you see that there are six possibilities? For longer chains, of course, the number of possibilities is enormous. Determining the sequence of amino acids in a polypep-

2,4-dinitrofluorobenzene
DNFB
(yellow)

a colorless
peptide chain

DNP-amino acid
(yellow)

FIGURE 28-8
Experimental determination of amino acid sequence.
In the reaction between DNFB and a polypeptide the N-terminal amino acid ends up with the yellow "marker" (a dinitrophenyl group, DNP) attached to it. By gentle hydrolysis and repeated use of the marker, a polypeptide chain can be broken down and the sequence of the individual units determined.

tide chain is one of the most significant problems in all of biochemistry. The method employed is outlined in Figure 28-8, and the structure of a typical polypeptide, beef insulin, is shown in Figure 28-9.

EXAMPLE 28-3

Determining the Sequence of Amino Acids in a Polypeptide. A polypeptide, on complete hydrolysis, yielded the amino acids A, B, C, D, and E. Partial hydrolysis and sequence proof gave single amino acids together with the following larger fragments: AD, CD, DCB, BE, and BC. What must be the sequence of amino acids in the polypeptide?

Solution

By arranging the fragments in the following manner,

$$AD$$
$$DC$$
$$DCB$$
$$BE$$
$$CB$$

we see that the sequence ADCBE is consistent with the fragments observed.

▶The sequence EBCDA is also possible, because we do not know whether A or E is the N-terminal end of the polypeptide.

Practice Example A: On complete hydrolysis a pentapeptide yields the amino acids Val, Phe, Gly, Cys, and Tyr. Partial hydrolysis yields the fragments Val-Phe, Gly-Cys, Cys-Val-Phe, and Phe-Tyr. Glycine (Gly) is the N-terminal acid. What is the sequence of amino acids in the polypeptide?

FIGURE 28-9
Amino acid sequence in beef insulin—primary structure of a protein.
There are two polypeptide chains joined by disulfide (—S—S—) linkages. One chain has 21 amino acids, and the other 30. In chain A, the Gly at the left end is N-terminal, and the Asp is C-terminal. In chain B, Phe is N-terminal, and Ala is C-terminal.

A

Gly-Ile-Val-Glu-Glu-Cys-Cys-Ala-Ser-Val-Cys-Ser-Leu-Tyr-Glu-Leu-Glu-Asp-Tyr-Cys-Asp
1   2   3   4   5   6   7   8   9   10 11 12 13 14 15 16 17 18 19 20 21

B

Phe-Val-Asp-Glu-His-Leu-Cys-Gly-Ser-His-Leu-Val-Glu-Ala-Leu-Tyr-Leu-Val-Cys-Gly-Glu-Arg-Gly-Phe-Phe-Tyr-Thr-Pro-Lys-Ala
1   2   3   4   5   6   7   8   9   10 11 12 13 14 15 16 17 18 19 20 21 22 23 24 25 26 27 28 29 30

Practice Example B: On complete hydrolysis a hexapeptide yields the amino acids Ala, Gly, Ser, Trp, and Val. Partial hydrolysis yields the fragments Val-Trp, Gly-Gly-Ala, Ser-Gly-Gly, and Ala-Val-Trp. Serine (Ser) is the N-terminal acid. What is the sequence of amino acids in the polypeptide?

The distinction between large polypeptides and proteins is arbitrary. It is generally accepted that if the molecular mass is over 10,000 u (roughly 50–75 amino acid units) the substance is a protein. Proteins possess characteristic isoelectric points, and their acidity or basicity depends on their amino acid composition. When proteins are heated, treated with salts, or exposed to ultraviolet light, profound and complex changes called **denaturation** occur. Denaturation usually brings about a lowering of solubility and loss of biological activity. The frying or boiling of an egg involves the denaturation (coagulation) of the egg albumin, a protein. The beauty shop "permanent wave" takes advantage of a denaturation process that is reversible. The proteins found in hair (e.g., keratin) contain disulfide linkages (—S—S—). When hair is treated with a reducing agent, these linkages break—a denaturation process. Following this step the hair is set into the desired shape. Next, the hair is treated with a mild oxidizing agent. The disulfide linkages are reestablished, and the hair remains in the style in which it was set.

The Structure of Proteins

The **primary structure** of a protein refers to the exact sequence of amino acids in the polypeptide chains that make up the protein. But what are the shapes of the long polymeric chains themselves? Are they simply limp and entangled like a plate of spaghetti or is there some order within chains and among chains? The structure or shape of an entire protein chain is referred to as **secondary structure.** In 1951, Linus Pauling and R. B. Corey, based on X-ray-diffraction studies on polylysine, a synthetic polypeptide, proposed that the orientation of this polypeptide and thus of protein chains is *helical.* A spiral, helical, or spring-like shape can be either left- or right-handed, but because proteins are composed of L-amino acids, their helical structure is right-handed (Figure 28-10).

FIGURE 28-10
Secondary structure of a protein—an alpha helix.
The helical structure is stabilized by hydrogen bonds between
$$\overset{\text{O}}{\underset{\|}{\text{—C—}}}$$ groups in one turn and $\overset{\text{H}}{\underset{|}{\text{—N—}}}$ groups in the next turn above. The bulky R groups are directed outward from the atoms in the spiral.

(a)   **(b)**

FIGURE 28-11
Pleated-sheet model of β-keratin.
(a) A polypeptide chain showing the direction of interpolypeptide hydrogen bonds (other polypeptide chains lie to the left and to the right of the chain shown). Bulky R groups extend above and below the pleated sheet. (b) The stacking of pleated sheets.

Other types of orientations are also possible. For example, β-keratin and silk fibroin are arranged in pleated sheets. In these proteins the side chains extend above and below the pleated sheets and hydrogen bonding is between *different* molecules (interpeptide bonding) lying next to each other and about 0.47 nm apart in the same sheet. These sheets are stacked on top of one another about 1.0 nm apart, rather like a pile of sheets of corrugated roofing (Figure 28-11). Some proteins, such as gamma globulin, are amorphous; they do not have a definite secondary structure.

Many proteins possess additional structural features: rather than being elongated, the coils may be twisted, knotted, and so forth. The final statement regarding the shape of a protein molecule lies in a description of its **tertiary structure.** Because the internal hydrogen bonding that occurs between atoms in successive turns of a protein helix is weak, these hydrogen bonds ought to be easily broken. In particular, we should expect them to be replaced by hydrogen bonds to water molecules when the protein is placed in water. That is, the α helix should open up and become a randomized structure when placed in water. But experimental evidence indicates that this does not happen. We are led to the conclusion that other forces must be involved in entwining the long α-helical chains into definite geometric shapes. The tertiary structure of myoglobin is shown in Figure 28-12.

The hemoglobin molecule consists of four separate polypeptide chains or subunits. The arrangement of these four subunits constitutes a still higher order of structure referred to as the **quaternary structure.** The levels of protein structure in hemoglobin, from primary to quaternary, are depicted in Figure 28-13.

Even minor changes in the structure of a protein can have profound effects. Hemoglobin contains four polypeptide chains, each with 146 amino acid units. The substitution of valine for glutamic acid at one site in two of these chains gives rise to the sometimes fatal blood disease known as sickle cell anemia. Apparently, the altered hemoglobin has a reduced ability to transport oxygen through the blood, although it does seem to provide some defense against malaria.

FIGURE 28-12
Representation of the tertiary structure of myoglobin.
The primary structure is that of a peptide of 153 units in a single chain. Secondary structure involves coiling of 70% of the chain into an α helix (shown as blue rods).

Enzymes

An **enzyme** is a biological catalyst that contains protein. Enzymes are specific for each biological transformation and catalyze a reaction without requiring a change in temperature or pH. Originally, enzymes were assigned common or

FIGURE 28-13
The four levels of protein structure.
Hemoglobin is the oxygen-carrying protein in red blood cells. The iron-containing heme system is shown in red. The primary structure of the protein is determined by its amino acid sequence. The secondary structure (α helix) is stabilized through formation of hydrogen bonds, as illustrated in Figure 28-10. The tertiary structure is determined by interactions between the R groups and their surroundings, causing the polypeptide chains to fold in a particular manner. Finally, the quaternary structure is an aggregation of two or more folded chains—four chains in the case of hemoglobin. Not all proteins have a quaternary structure.

trivial names, such as pepsin and catalase. Present practice, however, is to name them after the processes they catalyze, usually employing an "ase" ending.

We introduced a model for enzyme action in our discussion of enzyme kinetics in Section 15-11. According to this model, an enzyme can exert its catalytic activity only after combining with the reacting substance, the *substrate,* to form a complex. There appears to be a definite site on the enzyme where the substrate combines; for some enzymes more than one active site exists. Reaction of the substrate (S) with the enzyme (E) to form a complex (ES) permits the reaction to proceed via a path of lower activation energy than the noncatalyzed path. When the complex decomposes, products (P) are formed and the enzyme is regenerated. This general reaction scheme and a specific example follow. The example is the hydrolysis of sucrose. Recall from Figure 28-6 that sucrose is a disaccharide of glucose and fructose.

$$E \; + \; S \; \rightleftharpoons \; ES \; \longrightarrow \; E \; + \; P \; + \; P$$

$$\text{sucrase} + \text{sucrose} \rightleftharpoons \underset{\text{complex}}{\text{sucrase–sucrose}} \xrightarrow{H_2O} \text{sucrase} + \text{glucose} + \text{fructose}$$

The optimum temperature for enzyme activity is about 37 °C (98 °F) for the enzymes present in warm-blooded animals. Above this temperature, enzyme activity declines. This decline in activity results from the fact that enzymes are proteins and proteins are denatured by heat. This denaturation disrupts the secondary and tertiary structure and distorts the active site on the enzyme.

Protein behavior is extremely sensitive to changes in pH, and most enzymes in the body have their maximum activity between pH 6 and 8. Specific inhibition can occur when a molecule other than the substrate competes for an active enzyme site. For example, heavy metal ions (Hg^{2+}, Pb^{2+}, and Ag^+) may combine in a nonreversible way with active site groups, such as —OH, —SH, —CO_2^-, and —NH_3^+, and deactivate the enzyme. It has been suggested that antibiotics function by inhibiting enzyme–coenzyme reactions in microorganisms, and a similar mechanism explains the functioning of some insecticides.

Pentose sugars | Purine bases | Pyrimidine bases

ribose   2-deoxyribose   adenine (A)   guanine (G)   thymine (T)   cytosine (C)   uracil (U)

FIGURE 28-14
Constituents of nucleic acids.
If the sugar is 2-deoxyribose and the bases are A, G, T, and C, the nucleic acid is DNA. If the sugar is ribose and the bases are A, G, U, and C, the nucleic acid is RNA. (The term 2-deoxy means without an oxygen atom on the second carbon atom.)

28-5   Nucleic Acids

Lipids, carbohydrates, and proteins, taken together with water, constitute about 99% of most living organisms. The remaining 1% includes compounds of vital importance to the existence of life. Among these are the **nucleic acids,** which carry the information that directs the metabolic activity of cells.

The nucleus of the cell contains *chromosomes* that cause replication from one generation to the next. The portions of the chromosomes that carry specific traits are known as *genes.* **DNA,** or **deoxyribonucleic acid,** is the actual substance constituting the genes. Figure 28-14 presents the primary constituents of the nucleic acids: heterocyclic amines known as purines and pyrimidines and a five-carbon sugar, either ribose or 2-deoxyribose. Figure 28-15 represents a portion of a nucleic acid chain and shows how the constituents are joined together by phosphate groups. The combination of a phosphate group, a sugar and a heterocyclic amine is called a *nucleotide.*

FIGURE 28-15
A portion of a nucleic acid chain.
The combination of a pentose sugar and a purine or pyrimidine base is called a *nucleoside.* The combination of the nucleoside and a phosphate group is a *nucleotide.* The nucleic acid is a polymer of nucleotides.

FIGURE 28-16
DNA model.

The usual form for DNA is a *double helix,* as shown in Figure 28-16. The postulation of this structure by Francis Crick and James Watson in 1953 was one of the great scientific breakthroughs of modern times. Their work was critically dependent on precise X-ray-diffraction studies on DNA by Maurice Wilkins and Rosalind Franklin and on a set of regularities regarding the purine and pyrimidine bases in nucleic acids discovered by Erwin Chargaff. These regularities, known as the *base-pairing rules,* require the following.

1. The amount of adenine is equal to the amount of thymine (A = T).
2. The amount of guanine is equal to the amount of cytosine (G = C).
3. The total amount of purine bases is equal to the total amount of pyrimidine bases (G + A = C + T).

To maintain the structure of a double helix, hydrogen bonding must occur between the two single strands. The necessary conditions for hydrogen bonding exist only if an A on one strand appears opposite a T on the other, or if a G is opposite a C. C cannot be paired with T because the relatively small molecules (single ring) would not approach each other closely enough. The combination of G and A cannot occur because the molecules are too large (double rings).

The critical step in the replication of a DNA molecule requires the molecule to unwind into single strands. As the unwinding occurs, nucleotides present in the cell nucleus, through the action of enzymes, become attached to the exposed portions of the two single strands, converting each to a new double helix of DNA. As suggested by Figure 28-17, when the original DNA (parent) molecule is completely unwound, two new molecules (daughters) appear in its place!

FIGURE 28-17
Replication of a DNA molecule visualized.
As the unzipping process occurs from right to left, hydrogen bonds between the old DNA strands are broken. The new strands grow in the direction of the arrows by attaching nucleotides, which then form hydrogen bonds to the old DNA strands.

There are two pieces of evidence, each very convincing, that the process just outlined does indeed occur. First, electron micrographs of the DNA molecule have been obtained that capture DNA in the act of replication. Another elegant experiment involves growing bacteria in a medium containing ^{15}N atoms, so that all the N atoms of the bases of the DNA molecules are ^{15}N. The bacteria are then transferred to a nutrient with nucleotides containing normal ^{14}N. Here the bacteria are allowed to divide and reproduce themselves. The DNA of the offspring cells are then analyzed. Those of the first generation, for example, consist of DNA molecules with one strand having ^{15}N and the other ^{14}N atoms. This is exactly the result to be expected if replication occurs by the unzipping process described in Figure 28-17.

Summary

Four categories of substances found in living organisms are considered in this chapter: lipids, carbohydrates, proteins, and nucleic acids.

One familiar group of lipids are the triglycerides, esters of glycerol with long-chain monocarboxylic (fatty) acids. If saturated fatty acids predominate, the triglyceride is a fat. If unsaturation (in the form of double bonds) occurs in some of the fatty acid components, the triglyceride is an oil. Hydrolysis of a triglyceride with a strong base (saponification) yields glycerol and salts of the fatty acids—soaps.

The simplest carbohydrate molecules are five- and six-carbon-chain polyhydroxy aldehydes and ketones. These molecules, however, convert readily into cyclic structures: five- and six-member rings. The subject is further complicated by the existence of chiral carbon atoms in these molecules, leading to optical isomers—molecules of identical composition that differ in their spatial orientations and in their optical activity.

Monosaccharides, or simple sugars, can be bound together to form oligosaccharides. The most common of these are the disaccharides (such as sucrose) and trisaccharides. Polysaccharides contain from a few to a few thousand monosaccharide units. The most common of these are starch and cellulose.

The basic building blocks of proteins are some 20 different α-amino acids. Two amino acid molecules may join by eliminating a H_2O molecule between them. The result is a peptide bond. Long polypeptide chains—protein molecules—are formed as this process is repeated. Additional structural features include the twisting of a polypeptide chain into a helical coil and the twisting and folding of coils.

Nucleic acids, the substances that make up the genes in chromosomes, carry the information that directs the metabolic activity of cells and the replication of cells from one generation to another.

Summarizing Example

Hydrogenation is an important industrial process for converting unsaturated oils to saturated fats, as in the manufacture of solid shortenings used in cooking and baking.

What volume of $H_2(g)$, at 25.5 °C and 756 mmHg pressure, is consumed in converting 15.5 kg of the oil glyceryl trioleate (triolein) to the fat glyceryl tristearate (tristearin)?

1. *Draw the structural formula of triolein.* Use the formula and systematic name of oleic acid from Table 28.2 as a guide in doing this. *Result:* The structure is that shown on page 965.
2. *Determine the molar mass of triolein.* Base this on the structural formula written in part **1.** *Result:* 885.5 g/mol.

3. *Determine the stoichiometric factor relating H_2 and triolein.* Note that tristearin is produced from triolein by adding H_2 across each double bond. *Result:* 3 mol H_2/1 mol triolein.
4. *Calculate the number of moles of H_2 consumed.* The required conversions are kg triolein $\rightarrow$ g triolein $\rightarrow$ mol triolein $\rightarrow$ mol H_2. *Result:* 52.5 mol H_2.

5. *Calculate the volume of $H_2(g)$.* Use the result of part **4,** together with the temperature and pressure data and the ideal gas equation. *Answer:* 1.29×10^3 L $H_2(g)$.

Key Terms

absolute configuration (28-3)	**levorotatory** (28-3)	**protein** (28-4)
α-amino acid (28-4)	**lipid** (28-2)	**quaternary structure** (28-4)
carbohydrate (28-3)	**monosaccharide** (28-3)	**racemic mixture** (28-3)
denaturation (28-4)	**nucleic acid** (28-5)	**reducing sugar** (28-3)
deoxyribonucleic acid (DNA) (28-5)	**oil** (28-2)	**saponification** (28-2)
dextrorotatory (28-3)	**oligosaccharide** (28-3)	**secondary structure** (28-4)
diastereomer (28-3)	**peptide bond** (28-4)	**sugar** (28-3)
enzyme (28-4)	**polypeptide** (28-4)	**tertiary structure** (28-4)
fat (28-2)	**polysaccharide** (28-3)	**triglyceride** (28-2)
isoelectric point (28-4)	**primary structure** (28-4)	

Review Questions

1. In your own words define the following terms or symbols: **(a)** (+); **(b)** L; **(c)** sugar; **(d)** α-amino acid; **(e)** isoelectric point.
2. Briefly describe each of the following ideas, phenomena, or methods: **(a)** saponification; **(b)** chiral carbon atom; **(c)** racemic mixture; **(d)** denaturation of a protein.
3. Explain the important distinctions between each pair of terms: **(a)** fat and oil; **(b)** enantiomer and diastereomer; **(c)** primary and secondary structure of a protein; **(d)** DNA and RNA.
4. Name the following compounds.

(a)
$$H_2CO-\overset{\displaystyle O}{\overset{\|}{C}}-C_{15}H_{31}$$
$$HCO-\overset{\displaystyle O}{\overset{\|}{C}}-C_{17}H_{29}$$
$$H_2CO-\overset{\displaystyle O}{\overset{\|}{C}}-C_{11}H_{23}$$

(b)
$$H_2CO-\overset{\displaystyle O}{\overset{\|}{C}}-C_{17}H_{33}$$
$$HCO-\overset{\displaystyle O}{\overset{\|}{C}}-C_{17}H_{33}$$
$$H_2CO-\overset{\displaystyle O}{\overset{\|}{C}}-C_{17}H_{33}$$

(c) $C_{13}H_{27}CO_2^- \ Na^+$
5. Write structural formulas for the following.
(a) glyceryl palmitolauroeleostearate
(b) tripalmitin
(c) potassium myristate
(d) butyl oleate

6. Which of the following statements best describes the way a sugar sample labeled DL-erythrose rotates the plane of polarized light: (1) to the left; (2) to the right; (3) first to the right and then to the left; (4) neither to the left nor to the right? Explain.
7. From the given structure of L-(+)-arabinose derive the structure of **(a)** D-(−)-arabinose; **(b)** a diastereomer of L-(+)-arabinose.

$$
\begin{array}{c}
H-C=O \\
| \\
H-C-OH \\
| \\
HO-C-H \\
| \\
HO-C-H \\
| \\
CH_2OH
\end{array}
$$

L-(+)-arabinose

8. Write the formulas of the species expected if the amino acid phenylalanine is maintained in **(a)** 1.0 M HCl; **(b)** 1.0 M NaOH; **(c)** a buffer solution with pH 5.7.
9. Write the structures of **(a)** alanylcysteine; **(b)** threonyl-valylglycine.
10. For the polypeptide Met-Val-Thr-Cys, **(a)** write the structural formula; **(b)** name the polypeptide. (*Hint:* Which is the N-terminal and which is the C-terminal amino acid?)
11. Explain why enzyme action is so dependent on factors such as temperature, pH, and the presence of metal ions. Why are enzymes so specific in the reactions they catalyze? (That is, why doesn't an enzyme catalyze a variety of reactions, as does platinum metal, for example.)
12. With reference to Figure 28-15, identify the purine bases, the pyrimidine bases, the pentose sugars, and the phosphate groups. Is this a chain of DNA or RNA? Explain.

Focus On

Protein Synthesis and the Genetic Code

The Genetic Code

Second base

First base		U	C	A	G
U	U	UUU UUC } Phe UUA UUG } Leu	UCU UCC UCA UCG } Ser	UAU UAC } Tyr UAA Stop UAG Stop	UGU UGC } Cys UGA Stop UGG Trp
	C	CUU CUC CUA CUG } Leu	CCU CCC CCA CCG } Pro	CAU CAC } His CAA CAG } Gln	CGU CGC CGA CGG } Arg
	A	AUU AUC AUA } Ile AUG Met	ACU ACC ACA ACG } Thr	AAU AAC } Asn AAA AAG } Lys	AGU AGC } Ser AGA AGG } Arg
	G	GUU GUC GUA GUG } Val	GCU GCC GCA GCG } Ala	GAU GAC } Asp GAA GAG } Glu	GGU GGC GGA GGG } Gly

The Genetic Code.
Trinucleotide sequences (*codons*) are used by ribosomes to identify the amino acids needed in protein synthesis. Note that the codons UAA, UAG, and UGA are signal codons. They signal the ribosome to stop protein synthesis.

The puzzle of protein synthesis is that it occurs outside the nucleus of the cell but is directed by DNA, which is found only inside the nucleus. How is the necessary information transmitted? Here is where different forms of RNA play a key role. The following discussion relates to the process outlined in Figure 28-18.

First, a molecule of *messenger* RNA, mRNA, is synthesized on a portion of a DNA strand in the nucleus. This probably involves an unzipping of the DNA molecule similar to the process of DNA replication described through Figure 28-17. The mRNA migrates out of the nucleus into the cytoplasm. The mRNA has a high affinity for cell components called ribosomes and gathers them up along the chain. The combination of the mRNA and its ribosomes is called a polysome. *Transfer* RNA, tRNA, refers to a variety of rather short RNA chains, each of which can have only a specific amino acid attached to it. The function of tRNA is to bring a specific amino acid to a site on the ribosome where the amino acid can form a polypeptide bond and become part of a growing polypeptide chain. The ribosome moves along the mRNA chain, picking up different tRNA molecules and incorporating their amino acids into the polypeptide chain. When the ribosome reaches the end of the mRNA chain it falls off and releases the protein molecule that has been synthesized. The entire process is like the stringing of beads.

The code (set of directions) that determines the exact sequence of amino acids in the synthesis of a protein is incorporated in the chromosomal DNA. It is found in the particular pattern of base molecules on the double helix. Since there are only four different bases possible in an mRNA molecule—U, A, G, and C—but 20 different amino acids, it is clear that the code cannot correspond to individual base molecules. There are $4^2 = 16$ combinations of base molecules taken two at a time (i.e., UU, UA, UG, UC, etc.). But these could only account for 16 amino acids. When the base molecules are taken three at a time, there are $4^3 = 64$ possible combinations.

Exercises

Structure and Composition of the Cell

Exercises 13 to 16 refer to a typical *Escherichia coli* bacterium. This is a cylindrical cell about 2 μm long and 1 μm in diameter, weighing about 2×10^{-12} g and containing about 80% water by volume.

13. The intracellular pH is 6.4 and $[K^+] = 1.5 \times 10^{-4}$ M. Determine the number of (a) H_3O^+ ions and (b) K^+ ions in a typical cell.

14. Calculate the number of lipid molecules present, assuming their average molecular mass to be 700 u and the lipid content to be 2% by mass.

FIGURE 28-18
A representation of protein synthesis.
(a) Through the action of an enzyme, a tRNA molecule brings a single amino acid to a site on a ribosome. The anticodon of the tRNA (AAA for the example shown) must be complementary to the codon of the mRNA (UUU). (The amino acid carried by this tRNA is phenylalanine.) The amino acid is added to the chain in the manner shown; the chain moves from the tRNA on the left to the one on the right. (b) As the ribosome moves along the mRNA strand, more and more amino acid units are added through the proper matching of tRNA molecules with the code on the mRNA. (c) When the ribosome reaches the end of the mRNA strand, it and the completed protein are released. The ribosome is free to repeat the process.

A group of three base molecules in a DNA strand, called a triplet, causes a complementary set of base molecules to appear in the mRNA formed on it. This triplet, or *codon*, on the mRNA must be matched by a complementary triplet, called an *anticodon,* in a tRNA molecule. The particular tRNA with this anticodon carries a specific amino acid to the site of protein synthesis.

The significant features of the genetic code shown in the table are
- There is more than one triplet code for most amino acids.
- The first two letters of the codon are most significant. There is considerable variation in the third.
- There are three codons that do not correspond to any amino acids. Although they

are sometimes referred to as nonsense codons, these do seem to play a role in stopping protein synthesis (rather like the word "STOP" used to separate phrases in a telegram).
- The various codons direct the same protein synthesis, whether in bacteria, plants, lower animals, or humans.

15. The cell is about 15% protein, by mass, with 90% of this protein in the cytoplasm. Assuming an average molecular mass of 3×10^4 u, how many protein molecules are present in the cytoplasm?

16. A single chromosomal DNA molecule contains about 4.5 million mononucleotide units. If this molecule were extended so that the mononucleotide units were 450 pm apart, what would be the length of the molecule? How does this compare with the length of the cell itself? What does this result suggest about the shape of the DNA molecule?

Lipids

17. Describe briefly what is meant by each of the following terms, using specific examples where appropriate: **(a)** triglyceride; **(b)** simple glyceride; **(c)** mixed glyceride.

18. Describe briefly what is meant by each of the following terms, using specific examples where appropriate: **(a)** lipid; **(b)** fatty acid; **(c)** soap.

19. Oleic acid is a moderately unsaturated fatty acid. Linoleic acid belongs to a group called *polyunsaturated*. What structural feature characterizes polyunsaturated fatty acids? Is stearic acid polyunsaturated? Is eleostearic acid? Why do you suppose safflower oil is so highly recommended in dietary programs? *(Hint:* Refer to Tables 28.2 and 28.3.)

21. Write structural formulas to represent the products of the saponification of tripalmitin with NaOH(aq).

20. Corn oil and safflower oil are popular cooking oils containing mostly unsaturated fatty acids. They can be converted to solid fats by hydrogenation. Which would consume the greater amount of $H_2(g)$ in its hydrogenation to a solid fat, 1 kg of corn oil or 1 kg of safflower oil? Explain. *(Hint:* Refer to Tables 28.2 and 28.3.)

22. Calculate the maximum mass of the sodium soap that can be prepared from 105 g of glyceryl trimyristate.

Carbohydrates

23. Describe what is meant by each of the following terms, using specific examples where appropriate: **(a)** monosaccharide; **(b)** disaccharide; **(c)** oligosaccharide; **(d)** polysaccharide; **(e)** sugar.

25. The following terms are all related to stereoisomers and their optical activity. Explain the meaning of each. **(a)** dextrorotatory; **(b)** levorotatory; **(c)** racemic mixture; **(d)** (+).

27. Explain the meaning of the term *reducing sugar*. What structural feature characterizes a reducing sugar?

29. The pure α and β forms of D-glucose rotate the plane of polarized light to the right by 112° and 18.7°, respectively (denoted as +112 and +18.7). Are these two forms of glucose enantiomers or diastereomers? (Consider also the structures shown in Figure 28-5.)

24. Describe what is meant by each of the following terms, using specific examples where appropriate: **(a)** glycose; **(b)** aldose; **(c)** ketose; **(d)** pentose; **(e)** hexose.

26. The following terms are all related to optical isomers. Explain the meaning of each. **(a)** diastereomers; **(b)** enantiomers; **(c)** (−); **(d)** D configuration.

28. Write the structure for the straight-chain form of L-glucose. Does the structure determine if this isomer is levorotatory? Explain.

30. When a mixture of the pure α and β forms of D-glucose is allowed to reach equilibrium in solution, the rotation changes to +52.7 (a phenomenon known as mutarotation). What are the percentages of the α and β forms in the equilibrium mixture? [*Hint:* Refer to Exercise 29, and also to the discussion of weighted average on page 46.]

Amino Acids, Polypeptides, and Proteins

31. Describe what is meant by each of the following terms, using specific examples where appropriate: **(a)** α-amino acid; **(b)** zwitterion; **(c)** isoelectric point; **(d)** peptide bond; **(e)** tertiary structure.

33. A mixture of the amino acids lysine, proline, and aspartic acid is placed in a gel buffered at pH 6.3. An electric current is applied between an anode and a cathode immersed in the gel. Toward which electrode will each amino acid migrate?

35. Draw condensed structural formulas showing what form you would expect for the essential amino acid threonine **(a)** in strongly acidic solutions, **(b)** at the isoelectric point, and **(c)** in strongly basic solutions.

37. Write the structures of **(a)** the different tripeptides that can be obtained from a combination of alanine, serine, and lysine; **(b)** the tetrapeptides containing two serine and two alanine amino acid units.

39. After undergoing complete hydrolysis a polypeptide yields the following amino acids: Gly, Leu, Ala, Val, Ser, Thr. Partial hydrolysis yields the following fragments: Ser-Gly-Val, Thr-Val, Ala-Ser, Leu-Thr-Val, Gly-Val-Thr. An experiment using a marker establishes that Ala is the N-terminal amino acid.
(a) Establish the amino acid sequence in this polypeptide.
(b) What is the name of the polypeptide?

41. Describe what is meant by the primary, secondary, and tertiary structure of a protein. What is the quaternary structure? Do all proteins have a quaternary structure? Explain.

32. Describe what is meant by each of the following terms, using specific examples where appropriate: **(a)** polypeptide; **(b)** protein; **(c)** N-terminal amino acid; **(d)** α helix; **(e)** denaturation.

34. A mixture of the amino acids histidine, glutamic acid, and phenylalanine is placed in a gel buffered at pH 5.7. An electric current is applied between an anode and a cathode immersed in the gel. Toward which electrode will each amino acid migrate?

36. Draw condensed structural formulas for the following amino acids buffered at pH 6.0: **(a)** aspartic acid; **(b)** lysine; and **(c)** alanine.

38. Write the structures of the different tetrapeptides that can be obtained from a combination of alanine, lysine, serine, and phenylalanine. Give the abbreviated formula of each (such as Ala-Lys-Ser-Phe), starting at the N-terminal end.

40. After undergoing complete hydrolysis a polypeptide yields the following amino acids: Ala, Gly, Lys, Ser, Phe. Partial hydrolysis yields the following fragments: Ala-Lys-Ser, Gly-Phe-Gly, Ser-Gly, Gly-Phe, Lys-Ser-Gly. An experiment using a marker establishes that Ala is the N-terminal amino acid.
(a) Establish the amino acid sequence in this polypeptide.
(b) What is the name of the polypeptide?

42. Sickle cell anemia is sometimes referred to as a "molecular" disease. Comment on the appropriateness of this term.

Nucleic Acids

43. What are the two major types of nucleic acids? List their principal components.

45. If one strand of a DNA molecule has the sequence of bases AGC, what must be the sequence on the opposite strand? Draw a structure of this portion of the double helix, showing all hydrogen bonds.

44. DNA has been called the "thread of life." Comment on the appropriateness of this expression.

46. If one strand of a DNA molecule has the sequence of bases TCT, what must be the sequence on the opposite strand? Draw a structure of this portion of the double helix, showing all hydrogen bonds.

Advanced Exercises

47. The protein molecule hemoglobin contains four iron atoms (recall Figure 28-13). The mass percent iron in hemoglobin is 0.34%. What is the molecular mass of hemoglobin?

48. A 1.00-mL solution containing 1.00 mg of an enzyme was deactivated by the addition of 0.346 μmol $AgNO_3$ (1 μmol = 1×10^{-6} mol). What is the *minimum* molecular mass of the enzyme? Why does this calculation yield only a minimum value?

49. The *saponification value* of a triglyceride is the number of milligrams of KOH required to saponify 1.00 g of the triglyceride. The *iodine number* is the number of grams of I_2 that adds to double bonds in 100.0 g of a triglyceride. What are the saponification value and iodine number of **(a)** glyceryl tristearate; **(b)** glyceryl trioleate?

50. Castor oil is a mixture of triglycerides having about 90% of its fatty acid content in an unsaturated hydroxy aliphatic acid, ricinoleic acid.

$$CH_3(CH_2)_5CHOHCH_2CH{=}CH(CH_2)_7COOH$$

Refer to the definitions given in Exercise 49 and estimate the saponification value and iodine number of castor oil.

51. There are eight aldopentoses. Draw their structures and indicate which are enantiomers.

52. The term **epimer** is used to describe diastereomers that differ in the configuration about a *single* carbon atom. Which pairs of the three naturally occurring aldohexoses shown on page 972 are epimers (i.e., are D-galactose and D-mannose epimers, etc.)?

53. The amino acid ornithine, not normally found in proteins, has the following structure at a pH of 1.0.

$$\overset{+}{H_3N}{-}CH_2{-}CH_2{-}CH_2{-}\underset{\underset{+NH_3}{|}}{CH}{-}\overset{\overset{O}{\|}}{C}{-}OH$$

$pK_{a_1} = 1.94$; $pK_{a_2} = 8.65$; $pK_{a_3} = 10.76$
What is the p*I* value of this amino acid? (*Hint:* If you add

NaOH(aq) to the solution at what point do you obtain the zwitterion?)

54. In the experiment described on page 983, the first-generation offspring of DNA molecules each contained one strand with ^{15}N atoms and one with ^{14}N. If the experiment were carried through a second, third, and fourth generation, what fractions of the DNA molecules would still have strands with ^{15}N atoms?

55. Bradykinin is a nonapeptide that is obtained by the partial hydrolysis of blood serum protein. It causes a lowering of blood pressure and an increase in capillary permeability. Complete hydrolysis of bradykinin yields three proline (Pro), two arginine (Arg), two phenylalanine (Phe), one glycine (Gly), and one serine (Ser) amino acid units. The N-terminal and C-terminal units are both arginine (Arg). In a hypothetical experiment, partial hydrolysis and sequence proof reveals the following fragments: Gly-Phe-Ser-Pro; Pro-Phe-Arg; Ser-Pro-Phe; Pro-Pro-Gly; Pro-Gly-Phe; Arg-Pro-Pro; Phe-Arg. Deduce the sequence of amino acid units in bradykinin.

56. A pentapeptide was isolated from a cell extract and purified. A portion of the compound was treated with 2,4-dinitrofluorobenzene (DNFB) and the resulting material hydrolyzed (recall Figure 28-8). Analysis of the hydrolysis products revealed 1 mol of DNP-methionine, 2 mol of methionine, and 1 mol each of serine and glycine. A second portion of the original compound was partially hydrolyzed and separated into four products. Separately, the four products were hydrolyzed further, giving the following four sets of compounds: (1) 1 mol of DNP-methionine, 1 mol of methionine, and 1 mol of glycine; (2) 1 mol of DNP-methionine and 1 mol of methionine; (3) 1 mol of DNP-serine and 1 mol of methionine; (4) 1 mol of DNP-methionine, 1 mol of methionine, and 1 mol of serine. What is the amino acid sequence of the pentapeptide?

Mathematical Operations

A-1 Exponential Arithmetic

Measured quantities in this text range from very small to very large. For example, the mass of an individual hydrogen atom is 0.00000000000000000000000167 g, and the number of molecules in 18.0153 g of the substance water is 602,214,000,000,000,000,000,000. These numbers are difficult to write in conventional form and are even more cumbersome to handle in numerical calculations. We can greatly simplify them by expressing them in exponential form. The *exponential form* of a number consists of a coefficient (a number with value between 1 and 10) multiplied by a power of 10.

The number 10^n is the *nth power* of 10. If n is a *positive* quantity, 10^n is *greater than 1*. If n is a *negative* quantity, 10^n is *between 0 and 1*. The value of $10^0 = 1$.

Positive powers	*Negative powers*
$10^0 = 1$	$10^0 = 1$
$10^1 = 10$	$10^{-1} = \dfrac{1}{10} = 0.1$
$10^2 = 10 \times 10 = 100$	$10^{-2} = \dfrac{1}{10 \times 10} = \dfrac{1}{10^2} = 0.01$
$10^3 = 10 \times 10 \times 10 = 1000$	$10^{-3} = \dfrac{1}{10 \times 10 \times 10} = \dfrac{1}{10^3} = 0.001$

To express the number 3170 in exponential form, we write

$$3170 = 3.17 \times 1000 = 3.17 \times 10^3$$

For the number 0.00046 we write

$$0.00046 = 4.6 \times 0.0001 = 4.6 \times 10^{-4}$$

A simpler method of converting a number to exponential form that avoids intermediate steps is illustrated below.

$$\underset{3\;2\;1}{3\,1\,7\,0} = 3.17 \times 10^3$$

$$\underset{1\;2\;3\;4}{0.0\,0\,0\,4\,6} = 4.6 \times 10^{-4}$$

That is, to convert a number to exponential form,

- Move the decimal point to obtain a coefficient with value between 1 and 10.
- The exponent (power) of 10 is equal to the number of places the decimal point is moved.
- If the decimal point is moved *to the left*, the exponent of 10 is *positive*.
- If the decimal point is moved *to the right*, the exponent of 10 is *negative*.

To convert a number from exponential form to conventional form, move the decimal point the number of places indicated by the power of 10. That is,

$$6.1 \times 10^6 = 6.\underset{1\,2\,3\,4\,5\,6}{100000} = 6{,}100{,}000$$

$$8.2 \times 10^{-5} = \underset{5\,4\,3\,2\,1}{00000}8.2 = 0.000082$$

Electronic calculators designed for scientific and engineering work easily accommodate exponential numbers. A typical procedure is to key in the number, followed by the key "EXP" or "EE." Thus, the keystrokes required for the number 6.57×10^3 are

$$\boxed{6}\ \boxed{.}\ \boxed{5}\ \boxed{7}\ \boxed{\text{EXP}}\ \boxed{3}$$

and the result displayed is $\boxed{6.57^{03}}$

For the number 6.25×10^{-4}, the keystrokes are

$$\boxed{6}\ \boxed{.}\ \boxed{2}\ \boxed{5}\ \boxed{\text{EXP}}\ \boxed{4}\ \boxed{\pm}$$

and the result displayed is $\boxed{6.25^{-04}}$

▶The instructions given here are for a typical electronic calculator. The keystrokes required with your calculator may be somewhat different. Look for specific instructions in the instruction manual supplied with the calculator.

Some calculators have a mode setting that automatically converts all numbers and calculated results to the exponential form, regardless of the form in which numbers are entered. In this mode setting you can generally also set the number of significant figures to be carried in displayed results.

Addition and Subtraction. To add or subtract numbers written in exponential form, first express each quantity as *the same power of 10*. Then add and/or subtract the coefficients as indicated. That is, treat the power of 10 as you would a unit common to the terms being added and/or subtracted. In the example that follows, convert 3.8×10^{-3} to 0.38×10^{-2} and use 10^{-2} as the common power of 10.

$$(5.60 \times 10^{-2}) + (3.8 \times 10^{-3}) - (1.52 \times 10^{-2}) = (5.60 + 0.38 - 1.52) \times 10^{-2}$$

$$= 4.46 \times 10^{-2}$$

Multiplication. Consider the numbers $a \times 10^y$ and $b \times 10^z$. Their product is $a \times b \times 10^{(y+z)}$. *Coefficients are multiplied, and exponents are added.*

$$0.0220 \times 0.0040 \times 750 = (2.20 \times 10^{-2})(4.0 \times 10^{-3})(7.5 \times 10^2)$$

$$= (2.20 \times 4.0 \times 7.5) \times 10^{(-2-3+2)} = 66 \times 10^{-3}$$

$$= 6.6 \times 10^1 \times 10^{-3} = 6.6 \times 10^{-2}$$

Division. Consider the numbers $a \times 10^y$ and $b \times 10^z$. Their quotient is

$$\frac{a \times 10^y}{b \times 10^z} = (a/b) \times 10^{(y-z)}$$

Coefficients are divided, and the exponent of the denominator is subtracted from the exponent of the numerator.

$$\frac{20.0 \times 636 \times 0.150}{0.0400 \times 1.80} = \frac{(2.00 \times 10^1)(6.36 \times 10^2)(1.50 \times 10^{-1})}{(4.00 \times 10^{-2}) \times 1.80}$$

$$= \frac{2.00 \times 6.36 \times 1.50 \times 10^{(1+2-1)}}{(4.00 \times 1.80) \times 10^{-2}} = \frac{19.1 \times 10^2}{7.20 \times 10^{-2}}$$

$$= 2.65 \times 10^{(2-(-2))} = 2.65 \times 10^4$$

Raising a Number to a Power. To "square" the number $a \times 10^y$ means to determine the value $(a \times 10^y)^2$, or the product $(a \times 10^y)(a \times 10^y)$. According to the rule for multiplication, this product is $(a \times a) \times 10^{(y+y)} = a^2 \times 10^{2y}$. When an exponential number is raised to a power, *the coefficient is raised to that power and the exponent is multiplied by the power.* For example,

$$(0.0034)^3 = (3.4 \times 10^{-3})^3 = (3.4)^3 \times 10^{3 \times (-3)} = 39 \times 10^{-9} = 3.9 \times 10^{-8}$$

Extracting the Root of an Exponential Number. To extract the root of a number is the same as raising the number to a fractional power. This means that the square root of a number is the number to the one-half power; the cube root is the number to the one-third power; and so on. Thus,

$$\sqrt{a \times 10^y} = (a \times 10^y)^{1/2} = a^{1/2} \times 10^{y/2}$$

$$\sqrt{156} = \sqrt{1.56 \times 10^2} = (1.56)^{1/2} \times 10^{2/2} = 1.25 \times 10^1 = 12.5$$

In the following example, where the cube root is sought, the exponent (-5) is not divisible by 3; the number is rewritten with an exponent (-6) that is divisible by 3.

$$(3.52 \times 10^{-5})^{1/3} = (35.2 \times 10^{-6})^{1/3} = (35.2)^{1/3} \times 10^{-6/3} = 3.28 \times 10^{-2}$$

A-2 Logarithms

The *common* logarithm (log) of a number (N) is the exponent (x) to which the base 10 must be raised to yield the number N. That is, $\log N = x$ means that $N = 10^x = 10^{\log N}$. For simple powers of ten, for example,

$$\log 1 = \log 10^0 = 0$$

$$\log 10 = \log 10^1 = 1 \qquad \log 0.10 = \log 10^{-1} = -1$$

$$\log 100 = \log 10^2 = 2 \qquad \log 0.01 = \log 10^{-2} = -2$$

Most of the numbers that result from measurements and appear in calculations are not simple powers of 10, but it is not difficult to obtain logarithms of these numbers with an electronic calculator. To find the logarithm of a number, enter the number, followed by the "LOG" key.

$$\log 734 = 2.866$$

$$\log 0.0150 = -1.824$$

Another common example requires us to find the number having a certain logarithm. This number is often called the *antilogarithm* or the *inverse* logarithm. For example, if $\log N = 4.350$, what is N? N, the antilogarithm, is simply $10^{4.350}$, and to find its value we enter 4.350, followed by the key "10^x." Depending on the calculator used, it is usually necessary to press the key "INV" or "2nd F" before the log key.

$$\log N = 4.350$$

$$N = 10^{4.350}$$

$$N = 2.24 \times 10^4$$

If the task is to find the antilogarithm of -4.350, we again note that $N = 10^{-4.350}$, and $N = 4.47 \times 10^{-5}$. The required keystrokes on a typical electronic calculator are

$$\boxed{4}\ \boxed{.}\ \boxed{3}\ \boxed{5}\ \boxed{0}\ \boxed{\pm}\ \boxed{\text{INV}}\ \boxed{\text{log}}$$

and the display, to three significant figures, is

$$\boxed{4.47^{-05}}$$

Some Useful Relationships. From the definition of a logarithm we can write $M = 10^{\log M}$, $N = 10^{\log N}$, and $M \times N = 10^{\log(M \times N)}$. This means that

$$\log(M \times N) = \log M + \log N$$

Similarly, it is not difficult to show that

$$\log \frac{M}{N} = \log M - \log N$$

Finally, because $N^2 = N \times N$, $10^{\log N^2} = 10^{\log N} \times 10^{\log N}$, and

$$\log N^2 = \log N + \log N = 2 \log N$$

Or, in more general terms,

$$\log N^a = a \log N$$

This expression is especially useful for extracting the roots of numbers. Thus, to determine $(2.5 \times 10^{-8})^{1/5}$, we write

$$\log(2.5 \times 10^{-8})^{1/5} = \tfrac{1}{5}\log(2.5 \times 10^{-8}) = \tfrac{1}{5}(-7.60) = -1.52$$

$$(2.5 \times 10^{-8})^{1/5} = 10^{-1.52} = 0.030$$

Significant Figures in Logarithms. To establish the number of significant figures to use in a logarithm or antilogarithm, use this fundamental rule: All digits to the *right* of the decimal point in a logarithm are significant. Digits to the *left* are used to establish the power of 10. Thus, the logarithm -2.08 is expressed to *two* significant figures. The antilogarithm of -2.08 should also be expressed to *two* significant figures; it is 8.3×10^{-3}. To help settle this point, take the antilogarithms of -2.07, -2.08, and -2.09. You will find these antilogs to be 8.5×10^{-3}, 8.3×10^{-3}, and 8.1×10^{-3}, respectively. Only *two* significant figures are justified.

▶ The "ln" function arises in situations where the rate of change of some quantity is proportional to its present value.

Natural Logarithms. Logarithms can be expressed to a base other than 10. For instance, because $2^3 = 8$, $\log_2 8 = 3$ (read as "the logarithm of 8 to the base 2 is equal to 3"). Similarly, $\log_2 10 = 3.322$. Several equations in this text are derived by the methods of calculus and involve logarithms. These equations require that the logarithm be a "natural" one. A *natural* logarithm has the base $e = 2.71828\ldots$. A logarithm to the base "e" is usually denoted as *ln*.

The relationship between a "natural" and "common" logarithm simply involves the factor $\log_e 10 = 2.303$. That is, for the number N, $\ln N = 2.303 \log N$. The methods and relationships described for logarithms and antilogarithms to the base 10 all apply to the base e as well, except that the relevant keys on an electronic calculator are "ln" and "e^x" rather than "LOG" and "10^x."

A-3   Algebraic Operations

An algebraic equation is solved when one of the quantities, the unknown, is expressed in terms of all the other quantities in the equation. This effect is achieved when the unknown is present, *alone,* on one side of the equation, and the rest of the terms are on the other side. To solve an equation, a rearrangement of terms may be necessary. The basic principle governing these rearrangements is quite simple. *Whatever is done to one side of the equation must be done to the other as well.*

$$3x^2 + 6 = 33 \qquad \textbf{\textit{Solve for x.}}$$

$$3x^2 + \cancel{6} - \cancel{6} = 33 - 6 \qquad \text{(1) Subtract 6 from each side.}$$

$$3x^2 = 27$$

$$\frac{3x^2}{3} = \frac{27}{3} \qquad \text{(2) Divide each side by 3.}$$

$$x^2 = 9$$

$$\sqrt{x^2} = \sqrt{9} \qquad \text{(3) Extract the square root of each side.}$$

$$x = 3 \qquad \text{(4) Simplify. The square root of 9 is 3.}$$

Quadratic Equations.   A quadratic equation has the form $ax^2 + bx + c = 0$, where a, b, and c are constants (a cannot be equal to 0). A number of calculations in the text require us to solve a quadratic equation. At times, quadratic equations are of the form

$$(x + n)^2 = m^2$$

Such equations can be solved by extracting the square root of each side.

$$x + n = \pm m \quad \text{and} \quad x = m - n \quad \text{or} \quad x = -m - n$$

More likely, however, the *quadratic formula* will be needed.

$$x = \frac{-b \pm \sqrt{b^2 - 4ac}}{2a}$$

In Example 16-13 on page 570, the following equation must be solved.

$$\frac{(0.300 - x)}{(0.200 + x)(0.100 + x)} = 2.98$$

This is a quadratic equation, but before the quadratic formula can be applied, the equation must be rearranged to the standard form: $ax^2 + bx + c = 0$. This is accomplished in the steps that follow.

$$(0.300 - x) = 2.98(0.200 + x)(0.100 + x)$$

$$0.300 - x = 2.98(0.0200 + 0.300x + x^2)$$

$$0.300 - x = 0.0596 + 0.894x + 2.98x^2$$

$$2.98x^2 + 1.894x - 0.240 = 0$$

Now we can apply the quadratic formula.

$$x = \frac{-1.894 \pm \sqrt{(1.894)^2 + (4 \times 2.98 \times 0.240)}}{2 \times 2.98}$$

$$= \frac{-1.894 \pm \sqrt{3.587 + 2.86}}{2 \times 2.98}$$

$$= \frac{-1.894 \pm \sqrt{6.45}}{2 \times 2.98} = \frac{-1.894 \pm 2.54}{2 \times 2.98}$$

$$= \frac{-1.894 + 2.54}{2 \times 2.98} = \frac{0.65}{5.96} = 0.11$$

Note that only the $(+)$ value of the $(\pm)$ sign was used in solving for x. If the $(-)$ value had been used, a negative value of x would have resulted. However, for the given situation a negative value of x is meaningless.

A-4   Graphs

Suppose the following sets of numbers are obtained for two quantities x and y by laboratory measurement.

$$x = 0, 1, 2, 3, 4, \ldots$$

$$y = 2, 4, 6, 8, 10, \ldots$$

The relationship between these sets of numbers is not difficult to establish.

$$y = 2x + 2$$

Ideally, the results of experimental measurements are best expressed through a mathematical equation. Sometimes, however, an exact equation cannot be written, or its form is not clear from the experimental data. The graphing of data is very useful in such cases. In Figure A-1 the points listed above are located on a coordinate grid in which x values are placed along the horizontal axis (abscissa) and y values along the vertical axis (ordinate). For each point the x and y values are indicated in parentheses.

The data points are seen to define a straight line. A mathematical equation of a straight line has the form

$$y = mx + b$$

Values of m, the *slope* of the line, and b, the *intercept,* can be obtained from the straight-line graph.

When $x = 0$, $y = b$. The intercept is the point where the straight line intersects the y-axis. The slope can be obtained from two points on the graph.

$$y_2 = mx_2 + b \qquad \text{and} \qquad y_1 = mx_1 + b$$

$$y_2 - y_1 = m(x_2 - x_1) + \cancel{b} - \cancel{b}$$

$$m = \frac{y_2 - y_1}{x_2 - x_1}$$

From the straight line in Figure A-1, can you establish that $m = b = 2$?

The technique used above to eliminate the constant b is applied to logarithmic functions in several places in the text. For example, the expression written below is from page 428. In this expression P is a pressure, T is a Kelvin temperature, and A and B are constants. The equation is that of a straight line.

$$\ln P = -A\left(\frac{1}{T}\right) + B$$

equation of straight line: $\quad y \;=\; m \cdot x \;+\; b$

We can write this equation twice, for the point (P_1, T_1) and the point (P_2, T_2).

$$\ln P_1 = -A\left(\frac{1}{T_1}\right) + B \qquad \text{and} \qquad \ln P_2 = -A\left(\frac{1}{T_2}\right) + B$$

FIGURE A-1
A straight-line graph:
$y = mx + b$.

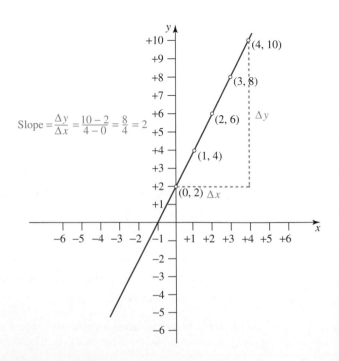

$$\text{Slope} = \frac{\Delta y}{\Delta x} = \frac{10 - 2}{4 - 0} = \frac{8}{4} = 2$$

The difference between these equations is

$$\ln P_2 - \ln P_1 = -A\left(\frac{1}{T_2}\right) + \cancel{B} + A\left(\frac{1}{T_1}\right) - \cancel{B}$$

$$\ln \frac{P_2}{P_1} = A\left(\frac{1}{T_1} - \frac{1}{T_2}\right)$$

A-5   Using Conversion Factors (Dimensional Analysis)

Some calculations in general chemistry require that a quantity measured in one set of units be converted to another set of units. Consider this fact.

$$1 \text{ m} = 100 \text{ cm}$$

Divide each side of the equation by 1 m.

$$\frac{1 \cancel{m}}{1 \cancel{m}} = \frac{100 \text{ cm}}{1 \text{ m}}$$

On the left side of the equation, the numerator and denominator are identical; they cancel.

$$1 = \frac{100 \text{ cm}}{1 \text{ m}} \tag{A.1}$$

On the right side they are not identical, but they are equal because they do represent the *same length*. The ratio 100 cm/1 m, when multiplied by a length in meters, converts that length to centimeters. The ratio is called a *conversion factor*.

Consider the question, "how many centimeters are there in 6.22 m?" The measured quantity is 6.22 m, and multiplying this quantity by 1 does not change its value.

$$6.22 \text{ m} \times 1 = 6.22 \text{ m}$$

Now replace the factor "1" by its equivalent—the conversion factor (A.1). Cancel the unit, m, and carry out the multiplication.

$$6.22 \cancel{m} \times \underbrace{\frac{100 \text{ cm}}{1 \cancel{m}}}_{\substack{\text{this factor} \\ \text{converts} \\ \text{m to cm}}} = 622 \text{ cm}$$

Next consider the question, "how many meters are there in 576 cm?" If we use the same factor (A.1) as before, the result is nonsensical.

$$576 \text{ cm} \times \frac{100 \text{ cm}}{1 \text{ m}} = 5.76 \times 10^4 \text{ cm}^2/\text{m}$$

Factor (A.1) must be rearranged to 1 m/100 cm.

$$576 \cancel{cm} \times \underbrace{\frac{1 \text{ m}}{100 \cancel{cm}}}_{\substack{\text{this factor} \\ \text{converts} \\ \text{cm to m}}} = 5.76 \text{ m}$$

This second example emphasizes two points.
1. There are two ways to write a conversion factor—in one form or its reciprocal (inverse). Because a conversion factor is equal to 1, its value is not changed by the inversion, but

▶Because of the importance of the cancellation of units, this problem-solving method is often called *unit analysis* or *dimensional analysis*.

2. A conversion factor must be used in such a way as to produce the necessary cancellaton of units.

Calculations based on conversion factors are always of the form

$$information\ sought = information\ given \times conversion\ factor(s) \qquad (A.2)$$

Often we must make several conversions in sequence in order to get to the desired result. For example, if we want to know how many yards (yd) there are in 576 cm, we find that there is no direct cm → yd conversion factor available. From the inside back cover of the text, however, we do find a conversion factor for cm → in. Thus, we can develop a *conversion pathway,* that is, a series of conversion factors that will take us from centimeters to yards:

$$? \text{ yards} = 576 \text{ cm} \times \overset{\text{cm} \rightarrow \text{ in.} \rightarrow \text{ ft} \rightarrow \text{ yd}}{\frac{1 \text{ in.}}{2.54 \text{ cm}} \times \frac{1 \text{ ft}}{12 \text{ in.}} \times \frac{1 \text{ yd}}{3 \text{ ft}}}$$

$$= 6.30 \text{ yd}$$

We can use the same idea of a conversion pathway to deal with the somewhat more challenging situation faced when the units are squared (or cubed). Consider the question, "how many square feet (ft^2) correspond to an area of 1.00 square meter (m^2), given that 1 m = 39.37 in. and 12 in. = 1 ft?" Here, it may be helpful to begin by drawing a sketch or outline of the situation. Figure A-2 represents a 1.00-m^2 area. Think of it as a square with sides 1 m long. Figure A-2 also represents the length 1 ft and an area of 1.00 ft^2. Do you see that there is somewhat more than 9 ft^2 in 1m^2?

We can write expression (A.2) as follows:

$$? \text{ ft}^2 = 1.00 \text{ m}^2 \times \underbrace{\left(\frac{39.37 \text{ in.}}{1 \text{ m}}\right)\left(\frac{39.37 \text{ in.}}{1 \text{ m}}\right)}_{\substack{\text{to convert} \\ \text{m}^2 \text{ to in.}^2}} \times \underbrace{\left(\frac{1 \text{ ft}}{12 \text{ in.}}\right)\left(\frac{1 \text{ ft}}{12 \text{ in.}}\right)}_{\substack{\text{to convert} \\ \text{in.}^2 \text{ to ft}^2}}$$

This is the same as writing

$$? \text{ ft}^2 = 1.00 \text{ m}^2 \times \frac{(39.37)^2 \text{ in.}^2}{1 \text{ m}^2} \times \frac{1 \text{ ft}^2}{(12)^2 \text{ in.}^2} = 10.8 \text{ ft}^2$$

Another way to look at the problem is to convert the length 1.00 m to feet,

$$? \text{ ft} = 1.00 \text{ m} \times \frac{39.37 \text{ in.}}{1 \text{ m}} \times \frac{1 \text{ ft}}{12 \text{ in.}} = 3.28 \text{ ft}$$

and square the result

$$? \text{ ft}^2 = 3.28 \text{ ft} \times 3.28 \text{ ft} = 10.8 \text{ ft}^2$$

Our last example incorporates several ideas discussed above. Here we will examine the situation in which the units in both the numerator and denominator must be converted. Consider the question, "how many meters per second (m/s) correspond to a speed of 63 mph, given that 1 mi = 5280 ft?"

We need to convert from miles to meters in the numerator and from hours to seconds in the denominator. We will need to use conversion factors from elsewhere in Appendix A-5 in addition to the given value. Also, we must be careful that our conversion factors produce the correct cancellation of units.

$$? \frac{\text{m}}{\text{s}} = \frac{63 \text{ mi}}{1 \text{ h}} \times \frac{1 \text{ h}}{60 \text{ min}} \times \frac{1 \text{ min}}{60 \text{ s}} \times \frac{5280 \text{ ft}}{1 \text{ mi}} \times \frac{12 \text{ in.}}{1 \text{ ft}} \times \frac{1 \text{ m}}{39.37 \text{ in.}}$$

$$= 28 \frac{\text{m}}{\text{s}}$$

FIGURE A-2
Comparison of one square foot and one square meter. One meter is slightly longer than 3 ft; 1 m^2 is somewhat larger than 9 ft^2.

In an alternative approach we break down the problem into three steps: (1) Convert 63 miles to a distance in meters; (2) convert 1 hour to a time in seconds; and (3) express the speed as a ratio of distance over time.

Step 1

$$\text{distance} = 63\,\text{mi} \times \frac{5280\,\text{ft}}{1\,\text{mi}} \times \frac{12\,\text{in.}}{1\,\text{ft}} \times \frac{1\,\text{m}}{39.37\,\text{in.}} = 1.0 \times 10^5\,\text{m}$$

Step 2

$$\text{time} = 1\,\text{h} \times \frac{60\,\text{min}}{1\,\text{h}} \times \frac{60\,\text{s}}{1\,\text{min}} = 3.6 \times 10^3\,\text{s}$$

Step 3

$$\text{speed} = \frac{\text{distance}}{\text{time}} = \frac{1.0 \times 10^5\,\text{m}}{3.6 \times 10^3\,\text{s}} = 28\,\frac{\text{m}}{\text{s}}$$

In conclusion, we have shown (1) how to make a conversion factor; (2) that a conversion factor may be inverted; (3) that a series of conversion factors may be used to make a conversion pathway; (4) that conversion factors may be raised to powers, if necessary; and (5) that conversions of values with units in both the numerator and the denominator (such as miles per hour or pounds per square inch) can be performed in one step or in several steps.

Some Basic Physical Concepts

B-1   Velocity and Acceleration

Time elapses as an object moves from one point to another. The *velocity* of the object is defined as the distance traveled per unit of time. An automobile that travels a distance of 60.0 km in exactly one hour has a velocity of 60.0 km/h (or 16.7 m/s).

Table B-1 contains data on the velocity of a free-falling body. For this motion velocity is not constant—it increases with time. The falling body "speeds up" continuously. The rate of change of velocity with time is called *acceleration*. Acceleration has the units of distance per unit time per unit time. With the methods of calculus, mathematical equations can be derived for the velocity (u) and distance (d) traveled in a time (t) by an object that has a constant acceleration (a).

$$u = at \tag{B.1}$$

$$d = \tfrac{1}{2}at^2 \tag{B.2}$$

For a free-falling body, the constant acceleration, called the *acceleration due to gravity*, is $a = g = 9.8 \text{ m/s}^2$. Equations (B.1) and (B.2) can be used to calculate the velocity and distance traveled by a free-falling body.

B-2   Force and Work

Newton's *first law* of motion states that an object at rest remains at rest, and that an object in motion remains in uniform motion, unless acted upon by an external force. The tendency for an object to remain at rest or in uniform motion is called *inertia;* a *force* is what is required to overcome inertia. Since the application of a force either gives an object motion or changes its motion, the actual effect of a force is to change the velocity of an object. Change in velocity is an *acceleration,* so force is what provides an object with acceleration.

Newton's *second law* of motion describes the force F required to produce an acceleration a in an object of mass m.

$$F = ma \tag{B.3}$$

The basic unit of force in the SI system is the *newton* (N). It is the force required to provide a one-kilogram mass with an acceleration of one meter per second per second.

$$1 \text{ N} = 1 \text{ kg} \times 1 \text{ m s}^{-2} \tag{B.4}$$

The force of gravity on an object (its weight) is the product of the mass of the object and the acceleration due to gravity, g.

$$F = mg \tag{B.5}$$

Work (w) is performed when a force acts through a distance.

$$\text{work } (w) = \text{force } (F) \times \text{distance } (d) \tag{B.6}$$

TABLE B-1	VELOCITY AND ACCELERATION OF A FREE-FALLING BODY		
Time Elapsed, s	**Total Distance, m**	**Velocity, m/s**	**Acceleration, m/s^2**
0	0		
		4.9	
1	4.9		9.8
		14.7	
2	19.6		9.8
		24.5	
3	44.1		9.8
		34.3	
4	78.4		

The *joule* (J) is the amount of work associated with a force of one newton (N) acting through a distance of one meter.

$$1 \text{ J} = 1 \text{ N} \times 1 \text{ m} \tag{B.7}$$

From the definition of the newton in expression (B.4), we can also write

$$1 \text{ J} = 1 \text{ kg} \times 1 \text{ m s}^{-2} \times 1 \text{ m} = 1 \text{ kg m}^2 \text{ s}^{-2} \tag{B.8}$$

B-3 Energy

Energy is defined as the capacity to do work, but there are other useful descriptions of energy as well. For example, a moving object possesses a kind of energy known as *kinetic energy*. We can obtain a useful equation for kinetic energy by combining some of the other simple equations in this appendix. Thus, because work is the product of a force and distance (Equation B.6), and force is the product of a mass and acceleration (Equation B.3), we can write

$$w \text{ (work)} = m \times a \times d \tag{B.9}$$

Now, if we substitute Equation (B.2) relating acceleration (a), distance (d), and time (t), into Equation (B.9), we obtain

$$w \text{ (work)} = m \times a \times \tfrac{1}{2}at^2 \tag{B.10}$$

Finally, let's substitute expression (B.1) relating acceleration (a) and velocity (u) into (B.10). That is, because $a = u/t$,

$$w \text{ (work)} = \tfrac{1}{2}m \left(\frac{u}{t} \right)^2 t^2 \tag{B.11}$$

Think of the work in (B.11) as the amount of work necessary to produce a velocity of u in an object of mass m. This amount of work is the energy that appears in the object as kinetic energy (e_k).

$$e_k \text{ (kinetic energy)} = \tfrac{1}{2}mu^2 \tag{B.12}$$

An object at rest may also have the capacity to do work by changing its position. The energy it possesses, which can be transformed into actual work, is called *potential energy*. Think of potential energy as energy "stored" within an object. Equations can be written for potential energy, but the exact forms of these equations depend on the manner in which the energy is "stored."

B-4   **Magnetism**

Attractive and repulsive forces associated with a magnet are centered at regions called *poles.* A magnet has a north and a south pole. If two magnets are aligned such that the north pole of one is directed toward the south pole of the second, an attractive force develops. If the alignment brings like poles into proximity, either both north or both south, a repulsive force develops. *Unlike poles attract; like poles repel.*

A *magnetic field* exists in that region surrounding a magnet in which the influence of the magnet can be felt. Internal changes produced within an iron object by a magnetic field, not produced in a field-free region, are responsible for the attractive force that the object experiences.

B-5   **Static Electricity**

Another property with which certain objects may be endowed is *electric charge.* Analogous to the case of magnetism, *unlike charges attract, and like charges repel* (recall Figure 2-2). In Coulomb's law, stated below, a *positive* force between electrically charged objects is *repulsive;* a *negative* force is *attractive.*

$$F = \frac{Q_1 Q_2}{\epsilon r^2} \tag{B.13}$$

where Q_1 is the magnitude of the charge on object 1,

Q_2 is the magnitude of the charge on object 2,

r is the distance between the objects, and

ϵ is a proportionality constant called the *dielectric constant,* whose numerical value reflects the effect that the medium separating two charged objects has on the force existing between them. For vacuum, $\epsilon = 1$; for other media, ϵ is greater than 1 (for example, for water $\epsilon = 78.5$).

An *electric field* exists in that region surrounding an electrically charged object in which the influence of the electric charge is felt. If an uncharged object is brought into the field of a charged object, the uncharged object may undergo internal changes that it would not experience in a field-free region. These changes may lead to the production of electric charges in the formerly uncharged object, a phenomenon called *induction* (illustrated in Figure B-1).

FIGURE B-1

Production of electric charges by induction in a gold leaf electroscope.

The glass rod acquires a positive electric charge by being rubbed with a silk cloth. As the rod is brought near the electroscope, a separation of charge occurs in the electroscope. The leaves become positively charged and repel one another. Negative charge is attracted to the spherical terminal at the end of the metal rod. If the glass rod is removed, the charges on the electroscope redistribute themselves, and the leaves collapse. If the spherical ball is touched by an electric conductor before the glass rod is removed, negative charge is removed from the ball. The electroscope retains a net positive charge, and the leaves remain outstretched.

B-6   Current Electricity

Current electricity is a flow of electrically charged particles. In electric currents in metallic conductors, the charged particles are electrons; in molten salts or in aqueous solutions, the particles are both negatively and positively charged ions.

The unit of electric charge is called a *coulomb* (C). The unit of electric current known as the *ampere* (A) is defined as a flow of one coulomb per second through an electrical conductor. Two variables determine the magnitude of the electric current I flowing through a conductor. These are the potential difference, or voltage drop, E, along the conductor and the electrical resistance of the conductor, R. The units of voltage and resistance are the *volt* (V) and *ohm,* respectively. The relationship of electric current, voltage, and resistance is given by Ohm's law.

$$I = \frac{E}{R} \tag{B.14}$$

One joule of energy is associated with the passage of one coulomb of electric charge through a potential difference (voltage) of one volt. That is, one joule = one volt-coulomb. Electric *power* refers to the rate of production (or consumption) of electric energy. It has the unit *watt* (W).

$$1\ \text{W} = 1\ \text{J s}^{-1} = 1\ \text{V C s}^{-1}$$

Since one coulomb per second is a current of one ampere,

$$1\ \text{W} = 1\ \text{V} \times 1\ \text{A} \tag{B.15}$$

Thus, a 100-watt light bulb operating at 110 V draws a current of 100 W/110 V = 0.91 A.

B-7   Electromagnetism

The relationship between electricity and magnetism is an intimate one. Interactions of electric and magnetic fields result in (1) magnetic fields associated with the flow of electric current (as in electromagnets), (2) forces experienced by current-carrying conductors when placed in a magnetic field (as in electric motors), and (3) electric current being induced when an electric conductor is moved through a magnetic field (as in electric generators). Several observations described in this text can be understood in terms of electromagnetic phenomena.

SI Units

The system of units that will in time be used universally for expressing all measured quantities is Le Système International d'Unités (The International System of Units), adopted in 1960 by the Conference Générale des Poids et Measures (General Conference of Weights and Measures). A summary of some of the provisions of the SI convention is provided here.

C-1 SI Base Units

A single unit has been established for each of the basic quantities involved in measurement. These are as follows:

Physical Quantity	Unit	Abbreviation
length	meter	m
mass	kilogram	kg
time	second	s
electric current	ampere	A
temperature	kelvin	K
luminous intensity	candela	cd
amount of substance	mole	mol
plane angle	radian	rad
solid angle	steradian	sr

C-2 SI Prefixes

Distinctive prefixes are attached to the base unit to express quantities that are *multiples* (greater than) or *submultiples* (less than) of the base unit. The multiples and submultiples are obtained by multiplying the base unit by powers of ten.

Multiple	Prefix	Abbreviation	Submultiple	Prefix	Abbreviation
10^{12}	tera	T	10^{-1}	deci	d
10^{9}	giga	G	10^{-2}	centi	c
10^{6}	mega	M	10^{-3}	milli	m
10^{3}	kilo	k	10^{-6}	micro	μ
10^{2}	hecto	h	10^{-9}	nano	n
10^{1}	deka	da	10^{-12}	pico	p
			10^{-15}	femto	f
			10^{-18}	atto	a

C-3   Derived SI Units

A number of quantities must be derived from measured values of the SI base quantities [for example, volume has the unit (length)^3]. Two sets of derived units are given, those whose names follow directly from the base units and those that are given special names. Notice that the units used in the text differ in some respects from those in the table. For example, the text expresses density as g cm^{-3}, molar mass as g mol^{-1}, molar volume as mL mol^{-1} or L mol^{-1}, and molar concentration (molarity) as mol L^{-1} or M.

▶ Two other SI conventions are illustrated through this table: (a) Units are written in singular form—meter or m, *not* meters or ms; (b) negative exponents are preferred to the shilling bar or solidus (/), that is, m s^{-1} and m s^{-2}, *not* m/s and m/s/s.

Physical Quantity	Unit	Abbreviation
area	square meter	m^2
volume	cubic meter	m^3
velocity	meter per second	m s^{-1}
acceleration	meter per second squared	m s^{-2}
density	kilogram per cubic meter	kg m^{-3}
molar mass	kilogram per mole	kg mol^{-1}
molar volume	cubic meter per mole	m^3 mol^{-1}
molar concentration	mole per cubic meter	mol m^{-3}

Physical Quantity	Unit	Abbreviation	In Terms of SI Units
frequency	hertz	Hz	s^{-1}
force	newton	N	kg m s^{-2}
pressure	pascal	Pa	N m^{-2}
energy	joule	J	kg m^2 s^{-2}
power	watt	W	J s^{-1}
electric charge	coulomb	C	A s
electric potential difference	volt	V	J A^{-1} s^{-1}
electric resistance	ohm	Ω	V A^{-1}

C-4   Units to Be Discouraged or Abandoned

There are several commonly used units whose use is to be discouraged and ultimately abandoned. Their gradual disappearance is to be expected, though some are used in this text. A few such units are listed.

▶ Another SI convention is implied here. No commas are used in expressing large numbers. Instead, spaces are left between groupings of three digits, that is, 101 325 rather than 101,235. Decimal points are written either as periods or commas.

Physical Quantity	Unit	Abbreviation	Definition in SI Units
length	angstrom	Å	1×10^{-10} m
force	dyne	dyn	1×10^{-5} N
energy	erg	erg	1×10^{-7} J
energy	calorie	cal	4.184 J
pressure	atmosphere	atm	101 325 Pa
pressure	millimeter of mercury	mmHg	133.322 Pa
pressure	torr	torr	133.322 Pa

Data Tables

TABLE D-1 Ground-State Electron Configurations

Z	Element	Configuration	Z	Element	Configuration	Z	Element	Configuration
1	H	$1s^1$	37	Rb	[Kr] $5s^1$	72	Hf	[Xe] $4f^{14}5d^26s^2$
2	He	$1s^2$	38	Sr	[Kr] $5s^2$	73	Ta	[Xe] $4f^{14}5d^36s^2$
3	Li	[He] $2s^1$	39	Y	[Kr] $4d^15s^2$	74	W	[Xe] $4f^{14}5d^46s^2$
4	Be	[He] $2s^2$	40	Zr	[Kr] $4d^25s^2$	75	Re	[Xe] $4f^{14}5d^56s^2$
5	B	[He] $2s^22p^1$	41	Nb	[Kr] $4d^45s^1$	76	Os	[Xe] $4f^{14}5d^66s^2$
6	C	[He] $2s^22p^2$	42	Mo	[Kr] $4d^55s^1$	77	Ir	[Xe] $4f^{14}5d^76s^2$
7	N	[He] $2s^22p^3$	43	Tc	[Kr] $4d^55s^2$	78	Pt	[Xe] $4f^{14}5d^96s^1$
8	O	[He] $2s^22p^4$	44	Ru	[Kr] $4d^75s^1$	79	Au	[Xe] $4f^{14}5d^{10}6s^1$
9	F	[He] $2s^22p^5$	45	Rh	[Kr] $4d^85s^1$	80	Hg	[Xe] $4f^{14}5d^{10}6s^2$
10	Ne	[He] $2s^22p^6$	46	Pd	[Kr] $4d^{10}$	81	Tl	[Xe] $4f^{14}5d^{10}6s^26p^1$
11	Na	[Ne] $3s^1$	47	Ag	[Kr] $4d^{10}5s^1$	82	Pb	[Xe] $4f^{14}5d^{10}6s^26p^2$
12	Mg	[Ne] $3s^2$	48	Cd	[Kr] $4d^{10}5s^2$	83	Bi	[Xe] $4f^{14}5d^{10}6s^26p^3$
13	Al	[Ne] $3s^23p^1$	49	In	[Kr] $4d^{10}5s^25p^1$	84	Po	[Xe] $4f^{14}5d^{10}6s^26p^4$
14	Si	[Ne] $3s^23p^2$	50	Sn	[Kr] $4d^{10}5s^25p^2$	85	At	[Xe] $4f^{14}5d^{10}6s^26p^5$
15	P	[Ne] $3s^23p^3$	51	Sb	[Kr] $4d^{10}5s^25p^3$	86	Rn	[Xe] $4f^{14}5d^{10}6s^26p^6$
16	S	[Ne] $3s^23p^4$	52	Te	[Kr] $4d^{10}5s^25p^4$	87	Fr	[Rn] $7s^1$
17	Cl	[Ne] $3s^23p^5$	53	I	[Kr] $4d^{10}5s^25p^5$	88	Ra	[Rn] $7s^2$
18	Ar	[Ne] $3s^23p^6$	54	Xe	[Kr] $4d^{10}5s^25p^6$	89	Ac	[Rn] $6d^17s^2$
19	K	[Ar] $4s^1$	55	Cs	[Xe] $6s^1$	90	Th	[Rn] $6d^27s^2$
20	Ca	[Ar] $4s^2$	56	Ba	[Xe] $6s^2$	91	Pa	[Rn] $5f^26d^17s^2$
21	Sc	[Ar] $3d^14s^2$	57	La	[Xe] $5d^16s^2$	92	U	[Rn] $5f^36d^17s^2$
22	Ti	[Ar] $3d^24s^2$	58	Ce	[Xe] $4f^26s^2$	93	Np	[Rn] $5f^46d^17s^2$
23	V	[Ar] $3d^34s^2$	59	Pr	[Xe] $4f^36s^2$	94	Pu	[Rn] $5f^67s^2$
24	Cr	[Ar] $3d^54s^1$	60	Nd	[Xe] $4f^46s^2$	95	Am	[Rn] $5f^77s^2$
25	Mn	[Ar] $3d^54s^2$	61	Pm	[Xe] $4f^56s^2$	96	Cm	[Rn] $5f^76d^17s^2$
26	Fe	[Ar] $3d^64s^2$	62	Sm	[Xe] $4f^66s^2$	97	Bk	[Rn] $5f^97s^2$
27	Co	[Ar] $3d^74s^2$	63	Eu	[Xe] $4f^76s^2$	98	Cf	[Rn] $5f^{10}7s^2$
28	Ni	[Ar] $3d^84s^2$	64	Gd	[Xe] $4f^75d^16s^2$	99	Es	[Rn] $5f^{11}7s^2$
29	Cu	[Ar] $3d^{10}4s^1$	65	Tb	[Xe] $4f^96s^2$	100	Fm	[Rn] $5f^{12}7s^2$
30	Zn	[Ar] $3d^{10}4s^2$	66	Dy	[Xe] $4f^{10}6s^2$	101	Md	[Rn] $5f^{13}7s^2$
31	Ga	[Ar] $3d^{10}4s^24p^1$	67	Ho	[Xe] $4f^{11}6s^2$	102	No	[Rn] $5f^{14}7s^2$
32	Ge	[Ar] $3d^{10}4s^24p^2$	68	Er	[Xe] $4f^{12}6s^2$	103	Lr	[Rn] $5f^{14}6d^17s^2$
33	As	[Ar] $3d^{10}4s^24p^3$	69	Tm	[Xe] $4f^{13}6s^2$	104	Rf	[Rn] $5f^{14}6d^27s^2$
34	Se	[Ar] $3d^{10}4s^24p^4$	70	Yb	[Xe] $4f^{14}6s^2$	105	Ha	[Rn] $5f^{14}6d^37s^2$
35	Br	[Ar] $3d^{10}4s^24p^5$	71	Lu	[Xe] $4f^{14}5d^16s^2$	106	Sg	[Rn] $5f^{14}6d^47s^2$
36	Kr	[Ar] $3d^{10}4s^24p^6$						

The electron configurations printed in blue are those of the noble gases. Each noble gas configuration serves as the core of the electron configurations of the elements that follow it, until the next noble gas is reached. Thus, [He] represents the core configuration of the 2nd period elements; [Ne], the third period; [Ar], the fourth period; [Kr], the fifth period; [Xe], the sixth period; and [Rn], the seventh period.

TABLE D-2 Thermodynamic Properties of Substances at 298.15 K*
Substances are at 1 atm pressure. For aqueous solutions, solutes are at unit activity (roughly 1 M). Data for ions in aqueous solution are relative to values of *zero* for ΔH_f°, ΔG_f°, and S° for H^+.

Inorganic Substances

	ΔH_f°, kJ mol^{-1}	ΔG_f°, kJ mol^{-1}	S°, J mol^{-1} K^{-1}
Aluminum			
Al(s)	0	0	28.33
Al^{3+}(aq)	−531	−485	−321.7
AlCl$_3$(s)	−704.2	−628.8	110.7
Al$_2$Cl$_6$(g)	−1291	−1220	490
AlF$_3$(s)	−1504	−1425	66.44
Al$_2$O$_3$(α solid)	−1676	−1582	50.92
Al(OH)$_3$(s)	−1276	—	—
Al$_2$(SO$_4$)$_3$(s)	−3441	−3100	239
Barium			
Ba(s)	0	0	62.8
Ba^{2+}(aq)	−537.6	−560.8	9.6
BaCO$_3$(s)	−1216	−1138	112
BaCl$_2$(s)	−858.6	−810.4	123.7
BaF$_2$(s)	−1207	−1157	96.36
BaO(s)	−553.5	−525.1	70.42
Ba(OH)$_2$(s)	−944.7	—	—
Ba(OH)$_2$·8H$_2$O(s)	−3342	−2793	427
BaSO$_4$(s)	−1473	−1362	132.2
Beryllium			
Be(s)	0	0	9.50
BeCl$_2$(α solid)	−490.4	−445.6	82.68
BeF$_2$(α solid)	−1027	−979.4	53.35
BeO(s)	−609.6	−580.3	14.14
Bismuth			
Bi(s)	0	0	56.74
BiCl$_3$(s)	−379.1	−315.0	177.0
Bi$_2$O$_3$(s)	−573.9	−493.7	151.5
Boron			
B(s)	0	0	5.86
BCl$_3$(l)	−427.2	−387.4	206.3
BF$_3$(g)	−1137	−1120	254.0
B$_2$H$_6$(g)	35.6	86.6	232.0
B$_2$O$_3$(s)	−1273	−1194	53.97
Bromine			
Br(g)	111.9	82.42	174.9
Br$^-$(aq)	−121.6	−104.0	82.4
Br$_2$(g)	30.91	3.14	245.4
Br$_2$(l)	0	0	152.2
BrCl(g)	14.64	−0.98	240.0
BrF$_3$(g)	−255.6	−229.4	292.4
BrF$_3$(l)	−300.8	−240.5	178.2

*Data for inorganic substances and for organic compounds with up to two carbon atoms per molecule are adapted from D. D. Wagman, *et. al.*, *The NBS Tables of Chemical Thermodynamic Properties: Selected Values for Inorganic and C_1 and C_2 Organic Substances in SI Units*, *Journal of Physical and Chemical Reference Data*, Volume 11, 1982, Supplement 2. Data for other organic compounds are from J. A. Dean, *Lange's Handbook of Chemistry*, 14/e, McGraw-Hill, Inc., 1992.

Inorganic Substances

	ΔH_f°, kJ mol^{-1}	ΔG_f°, kJ mol^{-1}	S°, J mol^{-1} K^{-1}
Cadmium			
Cd(s)	0	0	51.76
Cd^{2+}(aq)	−75.90	−77.61	−73.2
CdCl$_2$(s)	−391.5	−343.9	115.3
CdO(s)	−258.2	−228.4	54.8
Calcium			
Ca(s)	0	0	41.42
Ca^{2+}(aq)	−542.8	−553.6	−53.1
CaCO$_3$(s)	−1207	−1129	92.9
CaCl$_2$(s)	−795.8	−748.1	104.6
CaF$_2$(s)	−1220	−1167	68.87
CaH$_2$(s)	−186.2	−147.2	42
Ca(NO$_3$)$_2$(s)	−938.4	−743.1	193.3
CaO(s)	−635.1	−604.0	39.75
Ca(OH)$_2$(s)	−986.1	−898.5	83.39
Ca$_3$(PO$_4$)$_2$(s)	−4121	−3885	236.0
CaSO$_4$(s)	−1434	−1322	106.7
Carbon (See also the table of organic substances.)			
C(g)	716.7	671.3	158.0
C(diamond)	1.90	2.90	2.38
C(graphite)	0	0	5.74
CCl$_4$(g)	−102.9	−60.62	309.7
CCl$_4$(l)	−135.4	−65.21	216.4
C$_2$N$_2$(g)	309.0	297.4	241.8
CO(g)	−110.5	−137.2	197.6
CO$_2$(g)	−393.5	−394.4	213.6
CO$_3^{2-}$(aq)	−677.1	−527.8	−56.9
C$_3$O$_2$(g)	−93.72	−109.8	276.4
C$_3$O$_2$(l)	−117.3	−105.0	181.1
COCl$_2$(g)	−218.8	−204.6	283.4
COS(g)	−142.1	−169.3	231.5
CS$_2$(l)	89.70	65.27	151.3
Chlorine			
Cl(g)	121.7	105.7	165.1
Cl$^-$(aq)	−167.2	−131.2	56.5
Cl$_2$(g)	0	0	223.0
ClF$_3$(g)	−163.2	−123.0	281.5
ClO$_2$(g)	102.5	120.5	256.7
Cl$_2$O(g)	80.3	97.9	266.1
Chromium			
Cr(s)	0	0	23.77
[Cr(H$_2$O)$_6$]$^{3+}$(aq)	−1999	—	—
Cr$_2$O$_3$(s)	−1140	−1058	81.2
CrO$_4^{2-}$(aq)	−881.2	−727.8	50.21
Cr$_2$O$_7^{2-}$(aq)	−1490	−1301	261.9
Cobalt			
Co(s)	0	0	30.04
CoO(s)	−237.9	−214.2	52.97
Co(OH)$_2$(pink solid)	−539.7	−454.3	79
Copper			
Cu(s)	0	0	33.15
Cu^{2+}(aq)	64.77	65.49	−99.6
CuCO$_3$·Cu(OH)$_2$(s)	−1051	−893.6	186.2

Inorganic Substances

	ΔH_f°, kJ mol^{-1}	ΔG_f°, kJ mol^{-1}	S°, J mol^{-1} K^{-1}
CuO(s)	−157.3	−129.7	42.63
Cu(OH)$_2$(s)	−449.8	—	—
CuSO$_4$·5H$_2$O(s)	−2280	−1880	300.4
Fluorine			
F(g)	78.99	61.93	158.7
F$^-$(aq)	−332.6	−278.8	−13.8
F$_2$(g)	0	0	202.7
Helium			
He(g)	0	0	126.0
Hydrogen			
H(g)	218.0	203.3	114.6
H$^+$(aq)	0	0	0
H$_2$(g)	0	0	130.6
HBr(g)	−36.40	−53.43	198.6
HCl(g)	−92.31	−95.30	186.8
HCl(aq)	−167.2	−131.2	56.5
HCN(g)	135.1	124.7	201.7
HF(g)	−271.1	−273.2	173.7
HI(g)	26.48	1.72	206.5
HNO$_3$(l)	−174.1	−80.71	155.6
HNO$_3$(aq)	−207.4	−111.3	146.4
H$_2$O(g)	−241.8	−228.6	188.7
H$_2$O(l)	−285.8	−237.2	69.91
H$_2$O$_2$(g)	−136.3	−105.6	232.6
H$_2$O$_2$(l)	−187.8	−120.4	109.6
H$_2$S(g)	−20.63	−33.56	205.7
H$_2$SO$_4$(l)	−814.0	−690.0	156.9
H$_2$SO$_4$(aq)	−909.3	−744.5	20.1
Iodine			
I(g)	106.8	70.28	180.7
I$^-$(aq)	−55.19	−51.57	111.3
I$_2$(g)	62.44	19.36	260.6
I$_2$(s)	0	0	116.1
IBr(g)	40.84	3.72	258.7
ICl(g)	17.78	−5.44	247.4
ICl(l)	−23.89	−13.58	135.1
Iron			
Fe(s)	0	0	27.28
Fe^{2+}(aq)	−89.1	−78.90	−137.7
Fe^{3+}(aq)	−48.5	−4.7	−315.9
FeCO$_3$(s)	−740.6	−666.7	92.9
FeCl$_3$(s)	−399.5	−334.0	142.3
FeO(s)	−272	—	—
Fe$_2$O$_3$(s)	−824.2	−742.2	87.40
Fe$_3$O$_4$(s)	−1118	−1015	146.4
Fe(OH)$_3$(s)	−823.0	−696.5	106.7
Lead			
Pb(s)	0	0	64.81
Pb^{2+}(aq)	−1.7	−24.43	10.5
PbI$_2$(s)	−175.5	−173.6	174.9
PbO$_2$(s)	−277.4	−217.3	68.6
PbSO$_4$(s)	−919.9	−813.1	148.6

Inorganic Substances

	ΔH_f°, kJ mol^{-1}	ΔG_f°, kJ mol^{-1}	S°, J mol^{-1} K^{-1}
Lithium			
Li(s)	0	0	29.12
Li$^+$(aq)	−278.5	−293.3	13.4
LiCl(s)	−408.6	−384.4	59.33
LiOH(s)	−484.9	−439.0	42.80
LiNO$_3$(s)	−483.1	−381.1	90.0
Magnesium			
Mg(s)	0	0	32.68
Mg^{2+}(aq)	−466.9	−454.8	−138.1
MgCl$_2$(s)	−641.3	−591.8	89.62
MgCO$_3$(s)	−1096	−1012	65.7
MgF$_2$(s)	−1123	−1070	57.24
MgO(s)	−601.7	−569.4	26.94
Mg(OH)$_2$(s)	−924.5	−833.5	63.18
MgSO$_4$(s)	−1285	−1171	91.6
Manganese			
Mn(s)	0	0	32.01
Mn^{2+}(aq)	−220.8	−228.1	−73.6
MnO$_2$(s)	−520.0	−465.1	53.05
MnO$_4^-$(aq)	−541.4	−447.2	191.2
Mercury			
Hg(g)	61.32	31.82	175.0
Hg(l)	0	0	76.02
HgO(s)	−90.83	−58.54	70.29
Nitrogen			
N(g)	472.7	455.6	153.2
N$_2$(g)	0	0	191.5
NF$_3$(g)	−124.7	−83.2	260.6
NH$_3$(g)	−46.11	−16.48	192.3
NH$_3$(aq)	−80.29	−26.50	111.3
NH$_4^+$(aq)	−132.5	−79.31	113.4
NH$_4$Br(s)	−270.8	−175.2	113
NH$_4$Cl(s)	−314.4	−202.9	94.6
NH$_4$F(s)	−464.0	−348.7	71.96
NH$_4$HCO$_3$(s)	−849.4	−666.0	120.9
NH$_4$I(s)	−201.4	−112.5	117
NH$_4$NO$_3$(s)	−365.6	−183.9	151.1
NH$_4$NO$_3$(aq)	−339.9	−190.6	259.8
(NH$_4$)$_2$SO$_4$(s)	−1181	−901.7	220.1
N$_2$H$_4$(g)	95.40	159.3	238.4
N$_2$H$_4$(l)	50.63	149.3	121.2
NO(g)	90.25	86.55	210.7
N$_2$O(g)	82.05	104.2	219.7
NO$_2$(g)	33.18	51.29	240.0
N$_2$O$_4$(g)	9.16	97.82	304.2
N$_2$O$_4$(l)	−19.50	97.54	209.2
N$_2$O$_5$(g)	11.3	115.0	355.6
NO$_3^-$(aq)	−205.0	−108.7	146.4
NOBr(g)	82.17	82.42	273.6
NOCl(g)	51.71	66.06	261.6
Oxygen			
O(g)	249.2	231.7	160.9
O$_2$(g)	0	0	205.0

Inorganic Substances

	ΔH_f°, kJ mol^{-1}	ΔG_f°, kJ mol^{-1}	S°, J mol^{-1} K^{-1}
$O_3(g)$	142.7	163.2	238.8
$OH^-(aq)$	−230.0	−157.2	−10.75
$OF_2(g)$	24.7	41.9	247.3
Phosphorus			
$P(\alpha \text{ white})$	0	0	41.09
$P(\text{red})$	−17.6	−12.1	22.80
$P_4(g)$	58.91	24.47	279.9
$PCl_3(g)$	−287.0	−267.8	311.7
$PCl_5(g)$	−374.9	−305.0	364.5
$PH_3(g)$	5.4	13.4	210.1
$P_4O_{10}(s)$	−2984	−2698	228.9
$PO_4^{3-}(aq)$	−1277	−1019	−222
Potassium			
$K(g)$	89.24	60.62	160.2
$K(s)$	0	0	64.18
$K^+(aq)$	−252.4	−283.3	102.5
$KBr(s)$	−393.8	−380.7	95.90
$KCN(s)$	−113.0	−101.9	128.5
$KCl(s)$	−436.7	−409.1	82.59
$KClO_3(s)$	−397.7	−296.3	143.1
$KClO_4(s)$	−432.8	−303.1	151.0
$KF(s)$	−567.3	−537.8	66.57
$KI(s)$	−327.9	−324.9	106.3
$KNO_3(s)$	−494.6	−394.9	133.1
$KOH(s)$	−424.8	−379.1	78.9
$KOH(aq)$	−482.4	−440.5	91.6
$K_2SO_4(s)$	−1438	−1321	175.6
Silicon			
$Si(s)$	0	0	18.83
$SiH_4(g)$	34.3	56.9	204.5
$Si_2H_6(g)$	80.3	127.2	272.6
$SiO_2(\text{quartz})$	−910.9	−856.6	41.84
Silver			
$Ag(s)$	0	0	42.55
$Ag^+(aq)$	105.6	77.11	72.68
$AgBr(s)$	−100.4	−96.90	107.1
$AgCl(s)$	−127.1	−109.8	96.2
$AgI(s)$	−61.84	−66.19	115.5
$AgNO_3(s)$	−124.4	−33.41	140.9
$Ag_2O(s)$	−31.05	−11.20	121.3
$Ag_2SO_4(s)$	−715.9	−618.4	200.4
Sodium			
$Na(g)$	107.3	76.79	153.6
$Na(s)$	0	0	51.21
$Na^+(aq)$	−240.1	−261.9	59.0
$Na_2(g)$	142.1	104.0	230.1
$NaBr(s)$	−361.1	−349.0	86.82
$Na_2CO_3(s)$	−1131	−1044	135.0
$NaHCO_3(s)$	−950.8	−851.0	101.7
$NaCl(s)$	−411.2	−384.1	72.13
$NaCl(aq)$	−407.3	−393.1	115.5
$NaClO_3(s)$	−365.8	−262.3	123.4
$NaClO_4(s)$	−383.3	−254.9	142.3

Inorganic Substances

	ΔH_f°, kJ mol^{-1}	ΔG_f°, kJ mol^{-1}	S°, J mol^{-1} K^{-1}
NaF(s)	−573.6	−543.5	51.46
NaH(s)	−56.28	−33.46	40.02
NaI(s)	−287.8	−286.1	98.53
NaNO$_3$(s)	−467.9	−367.0	116.5
NaNO$_3$(aq)	−447.5	−373.2	205.4
Na$_2$O$_2$(s)	−510.9	−447.7	95.0
NaOH(s)	−425.6	−379.5	64.46
NaOH(aq)	−470.1	−419.2	48.1
NaH$_2$PO$_4$(s)	−1537	−1386	127.5
Na$_2$HPO$_4$(s)	−1748	−1608	150.5
Na$_3$PO$_4$(s)	−1917	−1789	173.8
NaHSO$_4$(s)	−1126	−992.8	113.0
Na$_2$SO$_4$(s)	−1387	−1270	149.6
Na$_2$SO$_4$(aq)	−1390	−1268	138.1
Na$_2$SO$_4$·10H$_2$O(s)	−4327	−3647	592.0
Na$_2$S$_2$O$_3$(s)	−1123	−1028	155
Sulfur			
S(rhombic)	0	0	31.80
S$_8$(g)	102.3	49.66	430.9
S$_2$Cl$_2$(g)	−18.4	−31.8	331.4
SF$_6$(g)	−1209	−1105	291.7
SO$_2$(g)	−296.8	−300.2	248.1
SO$_3$(g)	−395.7	−371.1	256.7
SO$_4^{2-}$(aq)	−909.3	−744.5	20.1
S$_2$O$_3^{2-}$(aq)	−648.5	−522.5	67
SO$_2$Cl$_2$(g)	−364.0	−320.0	311.8
SO$_2$Cl$_2$(l)	−394.1	—	—
Tin			
Sn(white)	0	0	51.55
Sn(gray)	−2.09	0.13	44.14
SnCl$_4$(l)	−511.3	−440.1	259
SnO(s)	−285.8	−256.9	56.5
SnO$_2$(s)	−580.7	−519.6	52.3
Titanium			
Ti(s)	0	0	30.63
TiCl$_4$(g)	−763.2	−726.7	354.9
TiCl$_4$(l)	−804.2	−737.2	252.3
TiO$_2$(s)	−944.7	−889.5	50.33
Uranium			
U(s)	0	0	50.21
UF$_6$(g)	−2147	−2064	377.8
UF$_6$(s)	−2197	−2069	227.6
UO$_2$(s)	−1085	−1032	77.03
Zinc			
Zn(s)	0	0	41.63
Zn^{2+}(aq)	−153.9	−147.1	−112.1
ZnO(s)	−348.3	−318.3	43.64

Organic Substances

	Name	ΔH_f°, kJ mol^{-1}	ΔG_f°, kJ mol^{-1}	S°, J mol^{-1} K^{-1}
$CH_4(g)$	methane(g)	-74.81	-50.75	186.2
$C_2H_2(g)$	acetylene(g)	226.7	209.2	200.8
$C_2H_4(g)$	ethylene(g)	52.26	68.12	219.5
$C_2H_6(g)$	ethane(g)	-84.68	-32.89	229.5
$C_3H_8(g)$	propane(g)	-104.7	-23.6	270.2
$C_4H_{10}(g)$	butane(g)	-125.6	-17.2	310.1
$C_6H_6(g)$	benzene(g)	82.6	129.7	269.2
$C_6H_6(l)$	benzene(l)	49.0	124.4	173.4
$C_6H_{12}(g)$	cyclohexane(g)	-123.4	31.8	298.3
$C_6H_{12}(l)$	cyclohexane(l)	-156.4	26.7	204.4
$C_{10}H_8(g)$	naphthalene(g)	149	223.6	335.6
$C_{10}H_8(s)$	naphthalene(s)	77.9	201.0	166.9
$CH_2O(g)$	formaldehyde(g)	-108.6	-102.5	218.7
$CH_3CHO(g)$	acetaldehyde(g)	-166.1	-133.4	246.4
$CH_3CHO(l)$	acetaldehyde(l)	-191.8	-128.3	160.4
$CH_3OH(g)$	methanol(g)	-200.7	-162.0	239.7
$CH_3OH(l)$	methanol(l)	-238.7	-166.3	126.8
$CH_3CH_2OH(g)$	ethanol(g)	-235.1	-168.6	282.6
$CH_3CH_2OH(l)$	ethanol(l)	-277.7	-174.8	160.7
$C_6H_5OH(s)$	phenol(s)	-165.1	-50.4	144.0
$(CH_3)_2CO(g)$	acetone(g)	-216.6	-153.1	294.9
$(CH_3)_2CO(l)$	acetone(l)	-242.1	-155.8	200.6
$CH_3COOH(g)$	acetic acid(g)	-432.3	-374.1	282.4
$CH_3COOH(l)$	acetic acid(l)	-484.5	-389.9	159.8
$CH_3COOH(aq)$	acetic acid(aq)	-485.8	-396.5	178.7
$C_6H_5COOH(s)$	benzoic acid(s)	-385.2	-245.3	167.6
$CH_3NH_2(g)$	methylamine(g)	-22.97	32.09	243.4
$C_6H_5NH_2(g)$	aniline(g)	86.86	166.7	319.2
$C_6H_5NH_2(l)$	aniline(l)	31.3	149.2	191.4

TABLE D-3 Equilibrium Constants

A. Ionization Constants of Weak Acids at 25 °C

Name of acid	Formula	K_a	Name of acid	Formula	K_a
acetic	$HC_2H_3O_2$	1.8×10^{-5}	hyponitrous	$HON{=}NOH$	8.9×10^{-8}
acrylic	$HC_3H_3O_2$	5.5×10^{-5}		$HON{=}NO^-$	4×10^{-12}
arsenic	H_3AsO_4	6.0×10^{-3}	iodic	HIO_3	1.6×10^{-1}
	$H_2AsO_4^-$	1.0×10^{-7}	iodoacetic	$HC_2H_2IO_2$	6.7×10^{-4}
	$HAsO_4^{2-}$	3.2×10^{-12}	malonic	$H_2C_3H_2O_4$	1.5×10^{-3}
arsenous	H_3AsO_3	6.6×10^{-10}		$HC_3H_2O_4^-$	2.0×10^{-6}
benzoic	$HC_7H_5O_2$	6.3×10^{-5}	nitrous	HNO_2	7.2×10^{-4}
bromoacetic	$HC_2H_2BrO_2$	1.3×10^{-3}	oxalic	$H_2C_2O_4$	5.4×10^{-2}
butyric	$HC_4H_7O_2$	1.5×10^{-5}		$HC_2O_4^-$	5.3×10^{-5}
carbonic	H_2CO_3	4.4×10^{-7}	phenol	HOC_6H_5	1.0×10^{-10}
	HCO_3^-	4.7×10^{-11}	phenylacetic	$HC_8H_7O_2$	4.9×10^{-5}
chloroacetic	$HC_2H_2ClO_2$	1.4×10^{-3}	phosphoric	H_3PO_4	7.1×10^{-3}
chlorous	$HClO_2$	1.1×10^{-2}		$H_2PO_4^-$	6.3×10^{-8}
citric	$H_3C_6H_5O_7$	7.4×10^{-4}		HPO_4^{2-}	4.2×10^{-13}
	$H_2C_6H_5O_7^-$	1.7×10^{-5}	phosphorous	H_3PO_3	3.7×10^{-2}
	$HC_6H_5O_7^{2-}$	4.0×10^{-7}		$H_2PO_3^-$	2.1×10^{-7}
cyanic	$HOCN$	3.5×10^{-4}	propionic	$HC_3H_5O_2$	1.3×10^{-5}
dichloroacetic	$HC_2HCl_2O_2$	5.5×10^{-2}	pyrophosphoric	$H_4P_2O_7$	3.0×10^{-2}
fluoroacetic	$HC_2H_2FO_2$	2.6×10^{-3}		$H_3P_2O_7^-$	4.4×10^{-3}
formic	$HCHO_2$	1.8×10^{-4}		$H_2P_2O_7^{2-}$	2.5×10^{-7}
hydrazoic	HN_3	1.9×10^{-5}		$HP_2O_7^{3-}$	5.6×10^{-10}
hydrocyanic	HCN	6.2×10^{-10}	selenic	H_2SeO_4	strong acid
hydrofluoric	HF	6.6×10^{-4}		$HSeO_4^-$	2.2×10^{-2}
hydrogen peroxide	H_2O_2	2.2×10^{-12}	selenous	H_2SeO_3	2.3×10^{-3}
hydroselenic	H_2Se	1.3×10^{-4}		$HSeO_3^-$	5.4×10^{-9}
	HSe^-	1×10^{-11}	succinic	$H_2C_4H_4O_4$	6.2×10^{-5}
hydrosulfuric	H_2S	1.0×10^{-7}		$HC_4H_4O_4^-$	2.3×10^{-6}
	HS^-	1×10^{-19}	sulfuric	H_2SO_4	strong acid
hydrotelluric	H_2Te	2.3×10^{-3}		HSO_4^-	1.1×10^{-2}
	HTe^-	1.6×10^{-11}	sulfurous	H_2SO_3	1.3×10^{-2}
hypobromous	$HOBr$	2.5×10^{-9}		HSO_3^-	6.2×10^{-8}
hypochlorous	$HOCl$	2.9×10^{-8}	thiophenol	HSC_6H_5	3.2×10^{-7}
hypoiodous	HOI	2.3×10^{-11}	trichloroacetic	$HC_2Cl_3O_2$	3.0×10^{-1}

B. Ionization Constants of Weak Bases at 25 °C

Name of base	Formula	K_b	Name of base	Formula	K_b
ammonia	NH_3	1.8×10^{-5}	isoquinoline	C_9H_7N	2.5×10^{-9}
aniline	$C_6H_5NH_2$	7.4×10^{-10}	methylamine	CH_3NH_2	4.2×10^{-4}
codeine	$C_{18}H_{21}O_3N$	8.9×10^{-7}	morphine	$C_{17}H_{19}O_3N$	7.4×10^{-7}
diethylamine	$(C_2H_5)_2NH$	6.9×10^{-4}	piperdine	$C_5H_{11}N$	1.3×10^{-3}
dimethylamine	$(CH_3)_2NH$	5.9×10^{-4}	pyridine	C_5H_5N	1.5×10^{-9}
ethylamine	$C_2H_5NH_2$	4.3×10^{-4}	quinoline	C_9H_7N	6.3×10^{-10}
hydrazine	NH_2NH_2	8.5×10^{-7}	triethanolamine	$C_6H_{15}O_3N$	5.8×10^{-7}
	$NH_2NH_3^+$	8.9×10^{-16}	triethylamine	$(C_2H_5)_3N$	5.2×10^{-4}
hydroxylamine	NH_2OH	9.1×10^{-9}	trimethylamine	$(CH_3)_3N$	6.3×10^{-5}

C. Solubility Product Constants[a]

Name of solute	Formula	K_{sp}	Name of solute	Formula	K_{sp}
aluminum hydroxide	$Al(OH)_3$	1.3×10^{-33}	lead(II) hydroxide	$Pb(OH)_2$	1.2×10^{-15}
aluminum phosphate	$AlPO_4$	6.3×10^{-19}	lead(II) iodide	PbI_2	7.1×10^{-9}
barium carbonate	$BaCO_3$	5.1×10^{-9}	lead(II) sulfate	$PbSO_4$	1.6×10^{-8}
barium chromate	$BaCrO_4$	1.2×10^{-10}	lead(II) sulfide[b]	PbS	3×10^{-28}
barium fluoride	BaF_2	1.0×10^{-6}	lithium carbonate	Li_2CO_3	2.5×10^{-2}
barium hydroxide	$Ba(OH)_2$	5×10^{-3}	lithium fluoride	LiF	3.8×10^{-3}
barium sulfate	$BaSO_4$	1.1×10^{-10}	lithium phosphate	Li_3PO_4	3.2×10^{-9}
barium sulfite	$BaSO_3$	8×10^{-7}	magnesium ammonium phosphate	$MgNH_4PO_4$	2.5×10^{-13}
barium thiosulfate	BaS_2O_3	1.6×10^{-5}			
bismuthyl chloride	$BiOCl$	1.8×10^{-31}	magnesium carbonate	$MgCO_3$	3.5×10^{-8}
bismuthyl hydroxide	$BiOOH$	4×10^{-10}	magnesium fluoride	MgF_2	3.7×10^{-8}
cadmium carbonate	$CdCO_3$	5.2×10^{-12}	magnesium hydroxide	$Mg(OH)_2$	1.8×10^{-11}
cadmium hydroxide	$Cd(OH)_2$	2.5×10^{-14}	magnesium phosphate	$Mg_3(PO_4)_2$	1×10^{-25}
cadmium sulfide[b]	CdS	8×10^{-28}	manganese(II) carbonate	$MnCO_3$	1.8×10^{-11}
calcium carbonate	$CaCO_3$	2.8×10^{-9}	manganese(II) hydroxide	$Mn(OH)_2$	1.9×10^{-13}
calcium chromate	$CaCrO_4$	7.1×10^{-4}	manganese(II) sulfide[b]	MnS	3×10^{-14}
calcium fluoride	CaF_2	5.3×10^{-9}	mercury(I) bromide	Hg_2Br_2	5.6×10^{-23}
calcium hydroxide	$Ca(OH)_2$	5.5×10^{-6}	mercury(I) chloride	Hg_2Cl_2	1.3×10^{-18}
calcium hydrogen phosphate	$CaHPO_4$	1×10^{-7}	mercury(I) iodide	Hg_2I_2	4.5×10^{-29}
			mercury(II) sulfide[b]	HgS	2×10^{-53}
calcium oxalate	CaC_2O_4	4×10^{-9}	nickel(II) carbonate	$NiCO_3$	6.6×10^{-9}
calcium phosphate	$Ca_3(PO_4)_2$	2.0×10^{-29}	nickel(II) hydroxide	$Ni(OH)_2$	2.0×10^{-15}
calcium sulfate	$CaSO_4$	9.1×10^{-6}	scandium fluoride	ScF_3	4.2×10^{-18}
calcium sulfite	$CaSO_3$	6.8×10^{-8}	scandium hydroxide	$Sc(OH)_3$	8.0×10^{-31}
chromium(II) hydroxide	$Cr(OH)_2$	2×10^{-16}	silver arsenate	Ag_3AsO_4	1.0×10^{-22}
chromium(III) hydroxide	$Cr(OH)_3$	6.3×10^{-31}	silver azide	AgN_3	2.8×10^{-9}
cobalt(II) carbonate	$CoCO_3$	1.4×10^{-13}	silver bromide	$AgBr$	5.0×10^{-13}
cobalt(II) hydroxide	$Co(OH)_2$	1.6×10^{-15}	silver chloride	$AgCl$	1.8×10^{-10}
cobalt(III) hydroxide	$Co(OH)_3$	1.6×10^{-44}	silver chromate	Ag_2CrO_4	1.1×10^{-12}
copper(I) chloride	$CuCl$	1.2×10^{-6}	silver cyanide	$AgCN$	1.2×10^{-16}
copper(I) cyanide	$CuCN$	3.2×10^{-20}	silver iodate	$AgIO_3$	3.0×10^{-8}
copper(I) iodide	CuI	1.1×10^{-12}	silver iodide	AgI	8.5×10^{-17}
copper(II) arsenate	$Cu_3(AsO_4)_2$	7.6×10^{-36}	silver nitrite	$AgNO_2$	6.0×10^{-4}
copper(II) carbonate	$CuCO_3$	1.4×10^{-10}	silver sulfate	Ag_2SO_4	1.4×10^{-5}
copper(II) chromate	$CuCrO_4$	3.6×10^{-6}	silver sulfide[b]	Ag_2S	6×10^{-51}
copper(II) ferrocyanide	$Cu_2[Fe(CN)_6]$	1.3×10^{-16}	silver sulfite	Ag_2SO_3	1.5×10^{-14}
copper(II) hydroxide	$Cu(OH)_2$	2.2×10^{-20}	silver thiocyanate	$AgSCN$	1.0×10^{-12}
copper(II) sulfide[b]	CuS	6×10^{-37}	strontium carbonate	$SrCO_3$	1.1×10^{-10}
iron(II) carbonate	$FeCO_3$	3.2×10^{-11}	strontium chromate	$SrCrO_4$	2.2×10^{-5}
iron(II) hydroxide	$Fe(OH)_2$	8.0×10^{-16}	strontium fluoride	SrF_2	2.5×10^{-9}
iron(II) sulfide[b]	FeS	6×10^{-19}	strontium sulfate	$SrSO_4$	3.2×10^{-7}
iron(III) arsenate	$FeAsO_4$	5.7×10^{-21}	thallium(I) bromide	$TlBr$	3.4×10^{-6}
iron(III) ferrocyanide	$Fe_4[Fe(CN)_6]_3$	3.3×10^{-41}	thallium(I) chloride	$TlCl$	1.7×10^{-4}
iron(III) hydroxide	$Fe(OH)_3$	4×10^{-38}	thallium(I) iodide	TlI	6.5×10^{-8}
iron(III) phosphate	$FePO_4$	1.3×10^{-22}	thallium(III) hydroxide	$Tl(OH)_3$	6.3×10^{-46}
lead(II) arsenate	$Pb_3(AsO_4)_2$	4.0×10^{-36}	tin(II) hydroxide	$Sn(OH)_2$	1.4×10^{-28}
lead(II) azide	$Pb(N_3)_2$	2.5×10^{-9}	tin(II) sulfide[b]	SnS	1×10^{-26}
lead(II) bromide	$PbBr_2$	4.0×10^{-5}	zinc carbonate	$ZnCO_3$	1.4×10^{-11}
lead(II) carbonate	$PbCO_3$	7.4×10^{-14}	zinc hydroxide	$Zn(OH)_2$	1.2×10^{-17}
lead(II) chloride	$PbCl_2$	1.6×10^{-5}	zinc oxalate	ZnC_2O_4	2.7×10^{-8}
lead(II) chromate	$PbCrO_4$	2.8×10^{-13}	zinc phosphate	$Zn_3(PO_4)_2$	9.0×10^{-33}
lead(II) fluoride	PbF_2	2.7×10^{-8}	zinc sulfide[b]	ZnS	2×10^{-25}

[a]Data are at various temperatures around "room" temperature, from 18 to 25 °C.
[b]For a solubility equilibrium of the type $MS(s) + H_2O \rightleftharpoons M^{2+}(aq) + HS^-(aq) + OH^-(aq)$.

D. Complex-Ion Formation Constants[c,d]

Formula	K_f	Formula	K_f	Formula	K_f
$[Ag(CN)_2]^-$	5.6×10^{18}	$[Co(ox)_3]^{3-}$	10^{20}	$[HgI_4]^{2-}$	6.8×10^{29}
$[Ag(EDTA)]^{3-}$	2.1×10^7	$[Cr(EDTA)]^-$	10^{23}	$[Hg(ox)_2]^{2-}$	9.5×10^6
$[Ag(en)_2]^+$	5.0×10^7	$[Cr(OH)_4]^-$	8×10^{29}	$[Ni(CN)_4]^{2-}$	2×10^{31}
$[Ag(NH_3)_2]^+$	1.6×10^7	$[CuCl_3]^{2-}$	5×10^5	$[Ni(EDTA)]^{2-}$	3.6×10^{18}
$[Ag(SCN)_4]^{3-}$	1.2×10^{10}	$[Cu(CN)_4]^{3-}$	2.0×10^{30}	$[Ni(en)_3]^{2+}$	2.1×10^{18}
$[Ag(S_2O_3)_2]^{3-}$	1.7×10^{13}	$[Cu(EDTA)]^{2-}$	5×10^{18}	$[Ni(NH_3)_6]^{2+}$	5.5×10^8
$[Al(EDTA)]^-$	1.3×10^{16}	$[Cu(en)_2]^{2+}$	1×10^{20}	$[Ni(ox)_3]^{4-}$	3×10^8
$[Al(OH)_4]^-$	1.1×10^{33}	$[Cu(NH_3)_4]^{2+}$	1.1×10^{13}	$[PbCl_3]^-$	2.4×10^1
$[Al(ox)_3]^{3-}$	2×10^{16}	$[Cu(ox)_2]^{2-}$	3×10^8	$[Pb(EDTA)]^{2-}$	2×10^{18}
$[CdCl_4]^{2-}$	6.3×10^2	$[Fe(CN)_6]^{4-}$	10^{37}	$[PbI_4]^{2-}$	3.0×10^4
$[Cd(CN)_4]^{2-}$	6.0×10^{18}	$[Fe(EDTA)]^{2-}$	2.1×10^{14}	$[Pb(OH)_3]^-$	3.8×10^{14}
$[Cd(en)_3]^{2+}$	1.2×10^{12}	$[Fe(en)_3]^{2+}$	5.0×10^9	$[Pb(ox)_2]^{2-}$	3.5×10^6
$[Cd(NH_3)_4]^{2+}$	1.3×10^7	$[Fe(ox)_3]^{4-}$	1.7×10^5	$[Pb(S_2O_3)_3]^{4-}$	2.2×10^6
$[Co(EDTA)]^{2-}$	2.0×10^{16}	$[Fe(CN)_6]^{3-}$	10^{42}	$[PtCl_4]^{2-}$	1×10^{16}
$[Co(en)_3]^{2+}$	8.7×10^{13}	$[Fe(EDTA)]^-$	1.7×10^{24}	$[Pt(NH_3)_6]^{2+}$	2×10^{35}
$[Co(NH_3)_6]^{2+}$	1.3×10^5	$[Fe(ox)_3]^{3-}$	2×10^{20}	$[Zn(CN)_4]^{2-}$	1×10^{18}
$[Co(ox)_3]^{4-}$	5×10^9	$[Fe(SCN)]^{2+}$	8.9×10^2	$[Zn(EDTA)]^{2-}$	3×10^{16}
$[Co(SCN)_4]^{2-}$	1.0×10^3	$[HgCl_4]^{2-}$	1.2×10^{15}	$[Zn(en)_3]^{2+}$	1.3×10^{14}
$[Co(EDTA)]^-$	10^{36}	$[Hg(CN)_4]^{2-}$	3×10^{41}	$[Zn(NH_3)_4]^{2+}$	4.1×10^8
$[Co(en)_3]^{3+}$	4.9×10^{48}	$[Hg(EDTA)]^{2-}$	6.3×10^{21}	$[Zn(OH)_4]^{2-}$	4.6×10^{17}
$[Co(NH_3)_6]^{3+}$	4.5×10^{33}	$[Hg(en)_2]^{2+}$	2×10^{23}	$[Zn(ox)_3]^{4-}$	1.4×10^8

[c]The ligands referred to in this table are monodentate: Cl^-, CN^-, I^-, NH_3, OH^-, SCN^-, $S_2O_3^{2-}$; bidentate: ethylenediamine (en), oxalate ion (ox); tetradentate: ethylenediaminetetraacetato ion, $EDTA^{4-}$.

[d]The K_f values are cumulative or overall formation constants (see page 671).

TABLE D-4 Standard Electrode (Reduction) Potentials at 25 °C

Reduction half-reaction	$E°$, V
$F_2(g) + 2\,e^- \longrightarrow 2\,F^-(aq)$	+2.866
$OF_2(g) + 2\,H^+(aq) + 4\,e^- \longrightarrow H_2O + 2\,F^-(aq)$	+2.1
$O_3(g) + 2\,H^+(aq) + 2\,e^- \longrightarrow O_2(g) + H_2O$	+2.075
$S_2O_8^{2-}(aq) + 2\,e^- \longrightarrow 2\,SO_4^{2-}(aq)$	+2.01
$Ag^{2+}(aq) + e^- \longrightarrow Ag^+(aq)$	+1.98
$H_2O_2(aq) + 2\,H^+(aq) + 2\,e^- \longrightarrow 2\,H_2O$	+1.763
$MnO_4^-(aq) + 4\,H^+(aq) + 3\,e^- \longrightarrow MnO_2(s) + 2\,H_2O$	+1.70
$PbO_2(s) + SO_4^{2-}(aq) + 4\,H^+(aq) + 2\,e^- \longrightarrow PbSO_4(s) + 2\,H_2O$	+1.69
$Au^{3+}(aq) + 3\,e^- \longrightarrow Au(s)$	+1.52
$MnO_4^-(aq) + 8\,H^+(aq) + 5\,e^- \longrightarrow Mn^{2+}(aq) + 4\,H_2O$	+1.51
$2\,BrO_3^-(aq) + 12\,H^+(aq) + 10\,e^- \longrightarrow Br_2(l) + 6\,H_2O$	+1.478
$PbO_2(s) + 4\,H^+(aq) + 2\,e^- \longrightarrow Pb^{2+}(aq) + 2\,H_2O$	+1.455
$ClO_3^-(aq) + 6\,H^+(aq) + 6\,e^- \longrightarrow Cl^-(aq) + 3\,H_2O$	+1.450
$Au^{3+}(aq) + 2\,e^- \longrightarrow Au^+(aq)$	+1.36
$Cl_2(g) + 2\,e^- \longrightarrow 2\,Cl^-(aq)$	+1.358
$Cr_2O_7^{2-}(aq) + 14\,H^+(aq) + 6\,e^- \longrightarrow 2\,Cr^{3+}(aq) + 7\,H_2O$	+1.33
$MnO_2(s) + 4\,H^+(aq) + 2\,e^- \longrightarrow Mn^{2+}(aq) + 2\,H_2O$	+1.23
$O_2(g) + 4\,H^+(aq) + 4\,e^- \longrightarrow 2\,H_2O$	+1.229
$2\,IO_3^-(aq) + 12\,H^+(aq) + 10\,e^- \longrightarrow I_2(s) + 6\,H_2O$	+1.20
$ClO_4^-(aq) + 2\,H^+(aq) + 2\,e^- \longrightarrow ClO_3^-(aq) + H_2O$	+1.19
$ClO_3^-(aq) + 2\,H^+(aq) + e^- \longrightarrow ClO_2(g) + H_2O$	+1.175
$NO_2(g) + H^+(aq) + e^- \longrightarrow HNO_2(aq)$	+1.07
$Br_2(l) + 2\,e^- \longrightarrow 2\,Br^-(aq)$	+1.065
$NO_2(g) + 2\,H^+(aq) + 2\,e^- \longrightarrow NO(g) + H_2O$	+1.03
$[AuCl_4]^-(aq) + 3\,e^- \longrightarrow Au(s) + 4\,Cl^-(aq)$	+1.002
$VO_2^+(aq) + 2\,H^+(aq) + e^- \longrightarrow VO^{2+}(aq) + H_2O$	+1.000
$NO_3^-(aq) + 4\,H^+(aq) + 3\,e^- \longrightarrow NO(g) + 2\,H_2O$	+0.956
$Cu^{2+}(aq) + I^-(aq) + e^- \longrightarrow CuI(s)$	+0.86
$Hg^{2+}(aq) + 2\,e^- \longrightarrow Hg(l)$	+0.854
$Ag^+(aq) + e^- \longrightarrow Ag(s)$	+0.800
$Fe^{3+}(aq) + e^- \longrightarrow Fe^{2+}(aq)$	+0.771
$O_2(g) + 2\,H^+(aq) + 2\,e^- \longrightarrow H_2O_2(aq)$	+0.695
$2\,HgCl_2(aq) + 2\,e^- \longrightarrow Hg_2Cl_2(s) + 2\,Cl^-(aq)$	+0.63
$MnO_4^-(aq) + e^- \longrightarrow MnO_4^{2-}(aq)$	+0.56
$I_2(s) + 2\,e^- \longrightarrow 2\,I^-(aq)$	+0.535
$Cu^+(aq) + e^- \longrightarrow Cu(s)$	+0.520
$H_2SO_3(aq) + 4\,H^+(aq) + 4\,e^- \longrightarrow S(s) + 3\,H_2O$	+0.449
$C_2N_2(g) + 2\,H^+(aq) + 2\,e^- \longrightarrow 2\,HCN(aq)$	+0.37
$[Fe(CN)_6]^{3-}(aq) + e^- \longrightarrow [Fe(CN)_6]^{4-}(aq)$	+0.361
$VO^{2+}(aq) + 2\,H^+(aq) + e^- \longrightarrow V^{3+}(aq) + H_2O$	+0.337
$Cu^{2+}(aq) + 2\,e^- \longrightarrow Cu(s)$	+0.337
$PbO_2(s) + 2\,H^+(aq) + 2\,e^- \longrightarrow PbO(s) + H_2O$	+0.28
$Hg_2Cl_2(s) + 2\,e^- \longrightarrow 2\,Hg(l) + 2\,Cl^-(aq)$	+0.2676
$HAsO_2(aq) + 3\,H^+(aq) + 3\,e^- \longrightarrow As(s) + 2\,H_2O$	+0.240
$AgCl(s) + e^- \longrightarrow Ag(s) + Cl^-(aq)$	+0.2223
$SO_4^{2-}(aq) + 4\,H^+(aq) + 2\,e^- \longrightarrow 2\,H_2O + SO_2(g)$	+0.17
$Cu^{2+}(aq) + e^- \longrightarrow Cu^+(aq)$	+0.159
$Sn^{4+}(aq) + 2\,e^- \longrightarrow Sn^{2+}(aq)$	+0.154
$S(s) + 2\,H^+(aq) + 2\,e^- \longrightarrow H_2S(g)$	+0.14
$AgBr(s) + e^- \longrightarrow Ag(s) + Br^-(aq)$	+0.071
$2\,H^+(aq) + 2\,e^- \longrightarrow H_2(g)$	0
$Pb^{2+}(aq) + 2\,e^- \longrightarrow Pb(s)$	−0.125
$Sn^{2+}(aq) + 2\,e^- \longrightarrow Sn(s)$	−0.137
$AgI(s) + e^- \longrightarrow Ag(s) + I^-(aq)$	−0.152

Reduction half-reaction	$E°$, V
$V^{3+}(aq) + e^- \longrightarrow V^{2+}(aq)$	−0.255
$Ni^{2+}(aq) + 2\,e^- \longrightarrow Ni(s)$	−0.257
$H_3PO_4(aq) + 2\,H^+(aq) + 2\,e^- \longrightarrow H_3PO_3(aq) + H_2O$	−0.276
$Co^{2+}(aq) + 2\,e^- \longrightarrow Co(s)$	−0.277
$In^{3+}(aq) + 3\,e^- \longrightarrow In(s)$	−0.338
$PbSO_4(s) + 2\,e^- \longrightarrow Pb(s) + SO_4^{2-}(aq)$	−0.356
$Cd^{2+}(aq) + 2\,e^- \longrightarrow Cd(s)$	−0.403
$Cr^{3+}(aq) + e^- \longrightarrow Cr^{2+}(aq)$	−0.424
$Fe^{2+}(aq) + 2\,e^- \longrightarrow Fe(s)$	−0.440
$2\,CO_2(g) + 2\,H^+(aq) + 2\,e^- \longrightarrow H_2C_2O_4(aq)$	−0.49
$Zn^{2+}(aq) + 2\,e^- \longrightarrow Zn(s)$	−0.763
$Cr^{2+}(aq) + 2\,e^- \longrightarrow Cr(s)$	−0.90
$Mn^{2+}(aq) + 2\,e^- \longrightarrow Mn(s)$	−1.18
$Ti^{2+}(aq) + 2\,e^- \longrightarrow Ti(s)$	−1.63
$U^{3+}(aq) + 3\,e^- \longrightarrow U(s)$	−1.66
$Al^{3+}(aq) + 3\,e^- \longrightarrow Al(s)$	−1.676
$Mg^{2+}(aq) + 2\,e^- \longrightarrow Mg(s)$	−2.356
$La^{3+}(aq) + 3\,e^- \longrightarrow La(s)$	−2.38
$Na^+(aq) + e^- \longrightarrow Na(s)$	−2.713
$Ca^{2+}(aq) + 2\,e^- \longrightarrow Ca(s)$	−2.84
$Sr^{2+}(aq) + 2\,e^- \longrightarrow Sr(s)$	−2.89
$Ba^{2+}(aq) + 2\,e^- \longrightarrow Ba(s)$	−2.92
$Cs^+(aq) + e^- \longrightarrow Cs(s)$	−2.923
$K^+(aq) + e^- \longrightarrow K(s)$	−2.924
$Rb^+(aq) + e^- \longrightarrow Rb(s)$	−2.924
$Li^+(aq) + e^- \longrightarrow Li(s)$	−3.040
Basic Solution	
$O_3(g) + H_2O + 2\,e^- \longrightarrow O_2(g) + 2\,OH^-(aq)$	+1.246
$ClO^-(aq) + H_2O + 2\,e^- \longrightarrow Cl^-(aq) + 2\,OH^-(aq)$	+0.890
$H_2O_2(aq) + 2\,e^- \longrightarrow 2\,OH^-(aq)$	+0.88
$BrO^-(aq) + H_2O + 2\,e^- \longrightarrow Br^-(aq) + 2\,OH^-(aq)$	+0.766
$ClO_3^-(aq) + 3\,H_2O + 6\,e^- \longrightarrow Cl^-(aq) + 6\,OH^-(aq)$	+0.622
$2\,AgO(s) + H_2O + 2\,e^- \longrightarrow Ag_2O(s) + 2\,OH^-(aq)$	+0.604
$MnO_4^-(aq) + 2\,H_2O + 3\,e^- \longrightarrow MnO_2(s) + 4\,OH^-(aq)$	+0.60
$BrO_3^-(aq) + 3\,H_2O + 6\,e^- \longrightarrow Br^-(aq) + 6\,OH^-(aq)$	+0.584
$Ni(OH)_3(s) + e^- \longrightarrow Ni(OH)_2(s) + OH^-(aq)$	+0.48
$2\,BrO^-(aq) + 2\,H_2O + 2\,e^- \longrightarrow Br_2(l) + 4\,OH^-(aq)$	+0.455
$2\,IO^-(aq) + 2\,H_2O + 2\,e^- \longrightarrow I_2(s) + 4\,OH^-(aq)$	+0.42
$O_2(g) + 2\,H_2O + 4\,e^- \longrightarrow 4\,OH^-(aq)$	+0.401
$Ag_2O(s) + H_2O + 2\,e^- \longrightarrow 2\,Ag(s) + 2\,OH^-(aq)$	+0.342
$Co(OH)_3(s) + e^- \longrightarrow Co(OH)_2(s) + OH^-(aq)$	+0.17
$NO_3^-(aq) + H_2O + 2\,e^- \longrightarrow NO_2^-(aq) + 2\,OH^-(aq)$	+0.01
$CrO_4^{2-}(aq) + 4\,H_2O + 3\,e^- \longrightarrow Cr(OH)_3(s) + 5\,OH^-(aq)$	−0.11
$HPbO_2^-(aq) + H_2O + 2\,e^- \longrightarrow Pb(s) + 3\,OH^-(aq)$	−0.54
$HCHO(aq) + 2\,H_2O + 2\,e^- \longrightarrow CH_3OH(aq) + 2\,OH^-(aq)$	−0.59
$SO_3^{2-}(aq) + 3\,H_2O + 4\,e^- \longrightarrow S(s) + 6\,OH^-(aq)$	−0.66
$AsO_4^{3-}(aq) + 2\,H_2O + 2\,e^- \longrightarrow AsO_2^-(aq) + 4\,OH^-(aq)$	−0.67
$AsO_2^-(aq) + 2\,H_2O + 3\,e^- \longrightarrow As(s) + 4\,OH^-(aq)$	−0.68
$2\,H_2O + 2\,e^- \longrightarrow H_2(g) + 2\,OH^-(aq)$	−0.828
$OCN^-(aq) + H_2O + 2\,e^- \longrightarrow CN^-(aq) + 2\,OH^-(aq)$	−0.97
$As(s) + 3\,H_2O + 3\,e^- \longrightarrow AsH_3(g) + 3\,OH^-(aq)$	−1.21
$Zn(OH)_2(s) + 2\,e^- \longrightarrow Zn(s) + 2\,OH^-(aq)$	−1.246
$Sb(s) + 3\,H_2O + 3\,e^- \longrightarrow SbH_3(g) + 3\,OH^-(aq)$	−1.338
$Al(OH)_4^-(aq) + 3\,e^- \longrightarrow Al(s) + 4\,OH^-(aq)$	−2.310
$Mg(OH)_2(s) + 2\,e^- \longrightarrow Mg(s) + 2\,OH^-(aq)$	−2.687

Glossary

Absolute configuration refers to the spatial arrangement of the groups attached to a chiral carbon atom. The two possibilities are D and L.

The **absolute zero of temperature** is the temperature at which molecular motion is hypothesized to cease: $0\ K = -273.15\ °C$.

Accuracy is the "closeness" of a measured value to the true or accepted value of a quantity.

Acetyl (See **acyl.**)

An **acid** is (1) a hydrogen-containing compound that, under appropriate conditions, can produce hydrogen ions, H^+ (Arrhenius theory); (2) a proton donor (Brønsted–Lowry theory); (3) an atom, ion, or molecule that can accept a pair of electrons to form a covalent bond (Lewis theory).

An **acid anhydride** is an oxide that reacts with water to form a ternary acid as the sole product.

An **acid–base indicator** is a substance that can be used to measure the pH of a solution or to signal the equivalence point in an acid–base titration. The indicator takes on one color when in its nonionized weak acid form and a different color in its anion form.

An **acid ionization constant,** K_a, is the equilibrium constant for the ionization reaction of a weak acid.

An **acid salt** contains an anion that can act as an acid (proton donor); examples are $NaHSO_4$ and NaH_2PO_4.

The **actinides** are a series of elements ($Z = 90$–103) characterized by partially filled $5f$ orbitals in their atoms. All actinide elements are radioactive.

An **activated complex** is a species formed as a result of collisions between energetic molecules that is an intermediate between the reactants and products of a reaction. Once formed, the activated complex dissociates either into the products or back to the reactants.

Activation energy is the minimum total kinetic energy that molecules must bring to their collisions for a chemical reaction to occur.

Active sites are the locations at which catalysis occurs, whether on the surface of a heterogeneous catalyst or an enzyme.

The **actual yield** is the *measured* quantity of a product obtained in a chemical reaction.

(See also **theoretical yield** and **percent yield.**)

The **acyl group** is $-\overset{\overset{\displaystyle O}{\|}}{C}-R$. If $R = H$, this is called the **formyl** group; $R = CH_3$, **acetyl;** and $R = C_6H_5$, **benzoyl.**

In an **addition reaction** functional group atoms are joined to the carbon atoms at points of unsaturation in alkene and alkyne hydrocarbon molecules.

An **adduct** is a compound formed by joining together two simpler molecules through a coordinate covalent bond, such as the adduct of $AlCl_3$ and $(C_2H_5)_2O$ pictured on page 784.

Adenosine diphosphate (ADP) and **adenosine triphosphate (ATP)** are agents involved in energy transfers during metabolism. The hydrolysis of ATP produces ADP, the ion HPO_4^{2-}, and a release of energy.

Adhesive forces are intermolecular forces between unlike molecules, such as molecules of a liquid and of a surface with which it is in contact.

Adsorption refers to the attachment of ions or molecules to the surface of a material.

ADP (See **adenosine diphosphate.**)

Alcohols contain the functional group $-OH$ and have the general formula ROH.

Aldehydes have the general formula $R-\overset{\overset{\displaystyle O}{\|}}{C}-H$.

Alicyclic hydrocarbon molecules have their carbon atom skeletons arranged in rings and resemble aliphatic (rather than aromatic) hydrocarbons.

Aliphatic hydrocarbon molecules have their carbon atom skeletons arranged in straight or branched chains.

Alkane hydrocarbon molecules have only single covalent bonds between carbon atoms. In their chain structures alkanes have the general formula C_nH_{2n+2}.

Alkene hydrocarbons have one or more carbon-to-carbon double bonds in their molecules. The simple alkenes have the general formula C_nH_{2n}.

Alkyl groups are alkane hydrocarbon molecules from which one hydrogen atom has been extracted. For example, the group —CH_3 is the **methyl** group; —CH_2CH_3 is the **ethyl** group. (See also Table 27.1.)

Alkyne hydrocarbons have one or more carbon-to-carbon triple bonds in their molecules. The simple alkynes have the general formula C_nH_{2n-2}.

Allotropy refers to the existence of an *element* in two or more different *molecular* forms, such as O_2 and O_3 or red and white phosphorus.

An **alloy** is a mixture of two or more metals. Some alloys are solid solutions, some are heterogeneous mixtures, and some are intermetallic compounds.

An **alpha (α) particle** is a combination of two protons and two neutrons identical to the helium ion, that is, $^4He^{2+}$. Alpha particles are emitted in some radioactive decay processes.

Alumina is a general term used for one of the solid forms of aluminum oxide, Al_2O_3.

Amalgams are metal alloys containing mercury. Depending on their compositions, some are liquid and some are solid.

An **amine** is an organic base having the formula RNH_2 (primary), R_2NH (secondary), or R_3N (tertiary), depending on the number of hydrogen atoms of an NH_3 molecule that are replaced by R groups.

An **α-amino acid** is a carboxylic acid that has an amino group (—NH_2) attached to the carbon atom adjacent to the **carboxyl group** (—COOH).

Amphoterism refers to the ability of certain oxides and hydroxo compounds to act as either acids or bases.

An **anion** is a negatively charged ion. An anion migrates toward the anode in an electrochemical cell.

The **anode** is the electrode in an electrochemical cell at which an oxidation half-reaction occurs.

An **antibonding molecular orbital** describes regions in a molecule in which there is a low electron probability or charge density between two bonded atoms.

Aromatic compounds are organic substances whose carbon atom skeletons are arranged in hexagonal rings, based on benzene, C_6H_6.

In the **Arrhenius acid–base theory** an acid produces H^+ and a base produces OH^- in aqueous solutions.

The **Arrhenius equation** relates the rate constant of a reaction to the temperature and activation energy.

One **atmosphere** (standard atmosphere) is the pressure exerted, under carefully specified conditions, by a column of mercury 760 mm high.

The **atmosphere** is the mixture of gases (nitrogen, oxygen, argon, and traces of others) that lies above the solid crust and oceans of Earth.

The **atom** is the basic building block of matter. The number of different atoms currently known is 112. A chemical element consists of a single type of atom and a chemical compound, of two or more different kinds of atoms.

The **atomic mass (weight)** of an element is the average of the isotopic masses weighted according to the naturally occurring abundances of the isotopes of the element. The atomic mass is expressed relative to the value of exactly 12 u for a carbon-12 atom.

An **atomic mass unit, u,** is used to express the masses of individual atoms. One u is 1/12 the mass of a carbon-12 atom.

The **atomic number, Z,** is the number of protons in the nucleus of an atom. It is also the number of electrons outside the nucleus of an electrically neutral atom.

Atomic (line) spectra are produced by dispersing light emitted by excited gaseous atoms. Only a discrete set of wavelength components (seen as colored lines) is present in a line spectrum.

ATP (See **adenosine triphosphate.**)

The **Aufbau process** is a method of writing the probable electron configurations of the elements. Each element is described as differing from the preceding one in terms of the orbital to which the one additional electron is assigned.

An **average bond energy** is the average of bond-dissociation energies for a number of different species containing a particular covalent bond. (See also **bond dissociation energy.**)

The **Avogadro constant, N_A,** has a value of $6.02214 \times 10^{23} \, mol^{-1}$. It is the number of elementary units in one mole.

Avogadro's law (hypothesis) states that at a fixed temperature and pressure, the volume of a gas is directly proportional to the amount of gas and that equal volumes of different gases, compared under identical conditions of temperature and pressure, contain equal numbers of molecules.

An **azeotrope** is a solution that boils at a constant temperature, producing vapor of the same composition as the liquid. In some cases, the boiling point of the azeotrope is lower than that of either solution component (minimum boiling point), and in other cases it is higher (maximum boiling point).

A **balanced equation** has the same number of atoms of each type on both sides. (See also **chemical equation.**)

Band theory is a form of molecular orbital theory to describe bonding in metals and semiconductors.

A **barometer** is a device used to measure the pressure of the atmosphere.

Barometric pressure is the prevailing pressure of the atmosphere as indicated by a barometer.

A **base** is (1) a compound that produces hydroxide ions, OH^-, in water solution (Arrhenius theory); (2) a proton acceptor (Brønsted–Lowry theory); (3) an atom, ion, or molecule that can donate a pair of electrons to form a covalent bond (Lewis theory).

A **base anhydride** is an oxide that reacts with water to form a base.

A **base ionization constant, K_b,** is the equilibrium constant for the ionization reaction of a weak base.

The **basic oxygen process** is the principal process used to convert impure iron (pig iron) into steel.

A **battery** is a voltaic cell [or a group of voltaic cells connected in series (+ to −)] used to produce electricity from chemical change.

bcc (See **body-centered cubic.**)

A **beta (β^-) particle** is an electron emitted as a result of the conversion of a neutron to a proton in certain atomic nuclei undergoing radioactive decay.

A **bidentate** ligand attaches itself to the central atom of a complex at two points in the coordination sphere.

A **bimolecular process** is an elementary process involving the collision of two molecules.

Binary acids are compounds of hydrogen and another nonmetal that are capable of acting as acids.

Binary compounds are compounds composed of *two* elements.

A **body-centered cubic (bcc)** crystal structure is one in which the unit cell has structural units at each corner and one in the center of the cube.

Boiling is a process in which vaporization occurs throughout a liquid. Pockets of vapor rise through the liquid and escape. Boiling occurs when the vapor pressure of a liquid is equal to barometric pressure.

A **bomb calorimeter** is a device used to measure the heat of a combustion reaction. The quantity measured is the heat of reaction at constant volume, $q_V = \Delta E$.

A **bond angle** is the angle between two covalent bonds. It is the angle between hypothetical lines joining the nuclei of two atoms to the nucleus of a third atom to which each of the two atoms is covalently bonded.

Bond dissociation energy, *D*, is the quantity of energy required to break one mole of covalent bonds in a gaseous species, usually expressed in kJ/mol. (See also **average bond energy.**)

Bond length (bond distance) is the distance between the centers of two atoms joined by a covalent bond.

Bond order is one-half the difference between the numbers of electrons in bonding and in antibonding molecular orbitals in a covalent bond. A single bond has a bond order of 1; a double bond, 2; and a triple bond, 3.

A **bond pair** is a pair of electrons involved in covalent bond formation.

A **bonding molecular orbital** describes regions of high electron probability or charge density in the internuclear region between two bonded atoms.

The **Born–Fajans–Haber cycle** relates lattice energies of crystalline ionic solids to ionization energies, electron affinities, and enthalpies of sublimation, dissociation, and formation.

Boyle's law states that the volume of a fixed amount of gas at a constant temperature is inversely proportional to the gas pressure.

A **breeder reactor** is a nuclear reactor that creates more nuclear fuel than it consumes, for example, by converting ^{238}U to ^{239}Pu.

The **Brønsted–Lowry theory** describes acids as proton donors and bases as proton acceptors. An acid–base reaction involves the transfer of protons from an acid to a base.

Buffer capacity refers to the amount of acid and/or base that a buffer solution can neutralize and still maintain an essentially constant pH.

Buffer range is the range of pH values over which a buffer solution can maintain a fairly constant pH.

A **buffer solution** resists a change in its pH. It contains components capable of reacting with (neutralizing) small added amounts of acids and base.

By-products are substances produced along with the principal product in a chemical process, either through the main reaction or a side reaction.

Calcination refers to the decomposition of a solid by heating at temperatures below its melting point, such as the decomposition of calcium carbonate to calcium oxide and $CO_2(g)$.

The **calorie (cal)** is the quantity of heat required to change the temperature of one gram of water by one degree Celsius.

A **calorimeter** is a device (of which there are numerous types) used to measure a quantity of heat.

A **carbohydrate** is a polyhydroxy aldehyde, a polyhydroxy ketone, a derivative of these, or a substance that yields them upon hydrolysis. Carbohydrates can be viewed as "hydrates" of carbon, in the sense that their general formulas are $C_x(H_2O)_y$.

The **carbonyl group,** $\diagdown C{=}O \diagup$, is found in aldehydes, ketones, and carboxylic acids.

The **carboxyl group** is $-\overset{\displaystyle O}{\overset{\|}{C}}-OH$.

A **carboxylic acid** has a carboxyl group attached to a hydrocarbon chain or ring structure.

A **catalyst** is a substance that provides an alternative mechanism of lower activation energy for a chemical reaction. The reaction is speeded up, and the catalyst is regenerated.

The **cathode** is the electrode of an electrochemical cell where a reduction half-reaction occurs.

Cathode rays are negatively charged particles (electrons) emitted at the negative electrode (cathode) in the passage of electricity through gases at very low pressures.

Cathodic protection is a method of corrosion control in which the metal to be protected is joined to a more active metal that corrodes instead. The protected metal acts as the cathode of a voltaic cell.

A **cation** is a positively charged ion. A cation migrates toward the cathode in an electrochemical cell.

A **cell diagram** is a symbolic representation of an electrochemical cell that indicates the substances entering into the cell reaction, electrode materials, solution concentrations, etc.

The **cell potential, E_{cell},** is the potential difference (voltage) between the two electrodes of an electrochemical cell. If E_{cell} is *positive,* the cell reaction is spontaneous. If E_{cell} is *negative,* the cell reaction is nonspontaneous.

A **central atom** in a structure is an atom that is bonded to two or more other atoms.

The **Celsius** temperature scale is based on a value of 0 °C for the normal melting point of ice and 100 °C for the normal boiling point of water.

In **chain reaction polymerization,** a reaction is initiated by "opening up" a carbon-to-carbon double bond. Monomer units add to free-radical intermediates to produce a long-chain polymer. An example is the polymerization of ethylene to polyethylene.

Charles's law states that the volume of a fixed amount of gas at a constant pressure is directly proportional to the Kelvin (absolute) temperature.

A **chelate** results from the attachment of polydentate ligands to the central atom of a complex ion. Chelates are five- or six-membered rings that include the central atom and atoms of the ligands.

A **chelating agent** is a polydentate ligand. It simultaneously attaches to two or more positions in the coordination sphere of the central atom of a complex ion.

The **chelation effect** refers to an exceptional stability conferred to a complex ion when polydentate ligands are present.

Chemical change (See **chemical reaction.**)

Chemical energy is the energy associated with chemical bonds and intermolecular forces.

A **chemical equation** is a symbolic representation of a chemical reaction. Symbols and formulas are used to represent reactants and products, and stoichiometric coefficients are used to balance the equation. (See also **balanced equation.**)

A **chemical formula** represents the relative numbers of atoms of each kind in a substance through symbols and numerical subscripts.

A **chemical property** is the ability (or inability) of a sample of matter to undergo a particular chemical reaction, for example, to combine with oxygen or react with an acid.

A **chemical reaction** is a process in which one set of substances (reactants) is transformed into a new set of substances (products).

Chemical symbols are abbreviations of the names of the elements consisting of one or two letters (e.g., N = nitrogen and Ne = neon).

Chiral refers to the nonsuperimposability of enantiomers and situations affected by this property. (See also **enantiomers.**)

A **chlor-alkali process** involves the electrolysis of NaCl(aq) to produce NaOH(aq), $Cl_2(g)$, and $H_2(g)$.

The term **cis** is used to describe geometric isomers in which two groups are attached on the same side of a double bond in an organic molecule, or along the same edge of a square in a square-planar, or at two adjacent vertices of an octahedral complex ion. (See also **geometric isomerism.**)

A **closed system** is one that can exchange energy but not matter with its surroundings.

Cohesive forces are intermolecular forces between like molecules, such as within a drop of liquid.

Coke is a relatively pure form of carbon produced by heating coal out of contact with air (destructive distillation).

Colligative properties—vapor pressure lowering, freezing point depression, boiling point elevation, osmotic pressure—have values that depend only on the number of solute particles in a solution and not on the identity of these particles.

Collision frequency is the number of collisions occurring between molecules in a unit of time.

Collision theory describes reactions in terms of molecular collisions—the frequency of collisions, the fraction of activated molecules, and the probability that collisions will be effective.

A **colloidal mixture** contains particles that are intermediate in size to those of a true solution and an ordinary heterogeneous mixture.

Combustion analysis is a method of determining the carbon, hydrogen, and oxygen contents of a compound from its complete combustion in oxygen gas.

The **common ion effect** describes the effect on an equilibrium by a substance that furnishes ions that can participate in the equilibrium. For example, sodium acetate, $NaC_2H_3O_2$, furnishes the common ion, $C_2H_3O_2^-$, to the ionization equilibrium of acetic acid, $HC_2H_3O_2$.

A **complex** is a polyatomic cation, anion, or neutral molecule in which groups (molecules or ions) called ligands are bonded to a central metal atom or ion.

A **complex ion** is a complex having a net electrical charge.

Composition refers to the components and their relative proportions in a sample of matter.

A **compound** is a substance made up of two or more elements. It does not change its identity in physical changes, but it can be broken down into its constituent elements by chemical changes.

Concentration (1) refers to the composition of a solution. (2) (see **extractive metallurgy.**)

In a **concentration cell** identical electrodes are immersed in solutions of different concentrations. The voltage (emf) of the cell is a function simply of the concentrations of the two solutions.

Condensation is the passage of molecules from the gaseous state to the liquid state.

A **condensed formula** is a simplified representation of a structural formula.

Conformations refer to the different spatial arrangements possible in a molecule. Examples are the "boat" and "chair" forms of cyclohexane.

A **conjugate acid** is formed when a Brønsted–Lowry base gains a proton. Every base has a conjugate acid.

A **conjugate base** remains after a Brønsted–Lowry acid has lost a proton. Every acid has a conjugate base.

Consecutive reactions are two or more reactions carried out in sequence. A product of each reaction becomes a reactant in a following reaction, until a final product is formed.

A **continuous spectrum** is one in which all wavelength components of the visible portion of the electromagnetic spectrum are present.

The **contact process** is a process for the manufacture of sulfuric acid that has as its key reaction the oxidation of $SO_2(g)$ to $SO_3(g)$ in contact with a catalyst.

Control rods are neutron-absorbing metal rods (e.g., Cd) that are used to control the neutron flux in a nuclear reactor and thereby control the rate of the fission reaction.

A **conversion factor** is a relationship between quantities expressed in different units. It is written as a *ratio* of two quantities and has a value of 1. For example, the relationship 1 in. = 2.54 cm can be written as 2.54 cm/1 in. when converting from in. to cm, or as 1 in./2.54 cm when converting from cm to in.

In a **coordinate covalent bond** the electrons shared between atoms are contributed by just one of the atoms. As a result, the bonded atoms exhibit formal charges.

Coordination compounds are neutral complexes or compounds containing complex ions.

Coordination isomerism arises in certain coordination compounds having both a complex cation and a complex anion. An interchange of ligands between the two complex ions leaves the composition of the compound unchanged.

Coordination number is the number of positions in the coordination sphere of a central atom to which ligands can be attached in the formation of a complex. Applied to a crystalline solid, coordination number signifies the number of nearest neighboring atoms (or ions of opposite charge) to any given atom (or ion) in a crystal.

The **coordination sphere** is the region around a central atom or ion where linkage to ligands can occur to produce a complex.

Corrosion is the oxidation of a metal through the action of water, air, and/or salt solutions. A corrosion reaction consists of an oxidation and a reduction half-reaction.

Coupled reactions are sets of chemical reactions that occur together. One (or more) of the reactions taken alone is (are) *nonspontaneous* and other(s), *spontaneous.* The *net* reaction is *spontaneous,* with the increase in free energy of the nonspontaneous reaction(s) offset by the decrease in free energy of the spontaneous reaction(s).

A **covalent bond** is formed when electrons are shared between atoms. In valence bond theory, the covalent bond is described as the sharing of a pair of electrons in the region in which atomic orbitals overlap.

Covalent radius is one-half the distance between the centers of two atoms that are bonded together covalently. It is the atomic radius associated with an element in its covalent compounds.

The **critical point** refers to the temperature and pressure at which a liquid and its vapor become identical. It is the highest temperature point on the vapor pressure curve.

Crystal field theory describes bonding in complex ions in terms of electrostatic attractions between ligands and the nucleus of the central metal ion. Particular attention is focused on the splitting of the *d* energy level of the central metal ion that results from electron repulsions.

Cubic closest packed is one of the two ways in which spheres can be packed to minimize the amount of free space or voids among them. (See Figures 13-36 and 13-37.)

A **curie** is a quantity of radioactive material producing 3.70×10^{10} disintegrations per second. (This is the decay rate for 1.00 g Ra.)

Dalton's law of partial pressures states that in a mixture of gases the total pressure is the sum of the partial pressures of the gases present. (See also **partial pressure.**)

The **d-block** refers to that section of the periodic table in which the process of orbital filling (Aufbau process) involves a *d* subshell.

A **decay constant** is a first-order rate constant describing radioactive decay.

Degree of ionization refers to the extent to which molecules of a weak acid or weak base ionize. The degree of ionization increases as the weak electrolyte solution is diluted. (See also **percent ionization.**)

Deionized water is water that has been freed of most of its impurity ions by being passed first through a cation and then an anion exchange resin.

Deliquescence is a process in which atmospheric water vapor condenses on a highly soluble solid and the solid dissolves in the water to produce an aqueous solution.

A **delocalized molecular orbital** describes a region of high electron probability or charge density that extends over three or more atoms.

Denaturation refers to the loss of biological activity of a protein brought about by changes in its secondary and tertiary structures.

Density is a physical property obtained by dividing the mass of a material or object by its volume (i.e., mass per unit volume).

Deposition is the passage of molecules from the gaseous to the solid state.

Deoxyribonucleic acid (DNA) is the substance that makes up the genes of the chromosomes in the nuclei of cells.

Detergents are cleansing agents that act by emulsifying oils. Most common among synthetic detergents are the salts of organic sulfonic acids, $RSO_3^-Na^+$, in which R is a hydrocarbon chain or a combination of a benzene ring and a hydrocarbon chain.

Dextrorotatory means the ability to rotate the plane of polarized light to the *right,* designated (+).

Diagonal relationships refer to similarities that exist between certain pairs of elements in different groups and periods of the periodic table, such as Li and Mg, Be and Al, and B and Si.

Dialysis is a process, similar to osmosis, in which ions or molecules in solution pass through a semipermeable membrane but colloidal particles do not.

A **diamagnetic** substance has all its electrons paired and is slightly repelled by a magnetic field.

Diastereomers are optically active isomers of a compound, but their structures are *not* mirror images (as are enantiomers).

Diborane is the first in a series of electron-deficient boron hydrides. Its formula is B_2H_6.

Diffusion refers to the spreading of a substance (usually a gas or liquid) into a region where it is not originally present as a result of random molecular motion.

Dilution is the process of reducing the concentration of a solution by adding more solvent.

A **dimer** is a molecule comprised of two simpler formula units, such as Al_2Cl_6, which is a dimer of $AlCl_3$.

Dipole moment, μ, is a measure of the extent to which a separation exists between the centers of positive and negative charge within a molecule. The unit used to measure dipole moment is the **debye,** 3.34×10^{-30} C m.

A **disaccharide** is a sugar molecule formed by the joining of two monosaccharide molecules through the elimination of an H_2O molecule between them.

Dispersion (London) forces are intermolecular forces associated with instantaneous and induced dipoles.

In a **disproportionation reaction** the same substance is both oxidized and reduced.

Distillation is a procedure for separating a liquid from a mixture by vaporizing the liquid and condensing the vapor. In **simple** distillation a volatile liquid is separated from nonvolatile solutes dissolved in it. In **fractional** distillation the components of a liquid solution are separated from each other on the basis of their different volatilities.

In a **double covalent bond** *two pairs* of electrons are shared between bonded atoms. The bond is represented by a double-dash sign (=).

Effective nuclear charge, Z_{eff}, is the positive charge acting on a particular electron in an atom. Its value is the charge on the nucleus, reduced to the extent that other electrons screen the particular electron from the nucleus.

Effusion is the escape of a gas through a tiny hole in its container.

An **electrochemical cell** is a device in which the electrons transferred in an oxidation–reduction reaction are made to pass through an electrical circuit. (See also **electrolytic cell** and **voltaic cell.**)

An **electrode** is a metal surface on which an oxidation–reduction equilibrium is established between a substance on the surface and substances in solution.

An **electrode potential** is the electric potential developed on a metal electrode when an oxidation and a reduction half-reaction have reached equilibrium on the surface of the electrode.

Electrolysis is the decomposition of a substance, either in the molten state or in an electrolyte solution, by means of electric current.

An **electrolytic cell** is an electrochemical cell in which a nonspontaneous reaction is carried out by electrolysis.

Electromagnetic radiation is a form of energy propagated as mutually perpendicular electric and magnetic fields. It includes visible light, infrared, ultraviolet, X-ray, and radio waves.

Electromotive force (emf) is the potential difference between two electrodes in a voltaic cell, expressed in volts.

Electron affinity is the energy change associated with the gain of an electron by a neutral gaseous atom.

Electron capture (E.C.) is a form of radioactive decay in which an electron from an inner electronic shell is absorbed by a nucleus. In the nucleus the electron is used to convert a proton to a neutron.

An **electron configuration** is a designation of how electrons are distributed among various orbitals in an atom.

Electronegativity is a measure of the electron-attracting power of a bonded atom; metals have low electronegativities, and nonmetals have high electronegativities.

The **electronegativity difference** between two atoms that are bonded together is used to assess the degree of polarity in the bond.

Electron group geometry refers to the geometrical distribution about a central atom of the electron pairs in its valence shell.

Electron spin is a characteristic of electrons that gives rise to the magnetic properties of atoms. The two possibilities for electron spin are $+\frac{1}{2}$ and $-\frac{1}{2}$. Electrons having opposing spins are said to be paired, and those having like spins are said to have parallel spins.

Electrons are particles carrying the fundamental unit of negative electric charge and found outside the nuclei of all atoms.

An **element** is a substance composed of a single type of atom. It cannot be broken down into simpler substances by chemical reactions.

An **elementary process** is an event that significantly alters a molecule's energy or geometry or produces a new molecule(s). It represents a single step in a reaction mechanism.

An **elimination reaction** is one in which atoms are removed from adjacent positions on a hydrocarbon chain to produce a small molecule (e.g., H_2O) and an additional bond between carbon atoms.

An **empirical formula** is the simplest chemical formula that can be written for a compound, that is, having the smallest integral subscripts possible.

Enantiomers (optical isomers) are molecules whose structures are nonsuperimposable mirror images. The molecules are optically active, that is, able to rotate the plane of polarized light.

An **endothermic** process results in a lowering of the temperature of an isolated system or, in a system that interacts with its surroundings, the absorption of heat.

The **end point** of a titration is the point in the titration where the indicator used changes color. A properly chosen indicator for a titration must have its end point correspond as closely as possible to the equivalence point of the titration reaction.

Energy is the capacity to do work. (See also **work**.)

An **energy-level diagram** is a representation of the allowed energy states for the electrons in atoms. The simplest energy-level diagram—that of the hydrogen atom—is shown in Figure 9-14.

The **English system** is a system of measurement in which the unit of length is the yard, the unit of mass is the pound, and the unit of time is the second.

Enthalpy, H, is a thermodynamic function used to describe constant-pressure processes: $H = E + PV$, and at constant pressure, $\Delta H = \Delta E + P\Delta V$.

Enthalpy change, ΔH, is the difference in enthalpy between two states of a system. If the process by which the change occurs is a chemical reaction carried out at constant temperature and pressure and with work limited to pressure–volume work, the enthalpy change is called the *heat of reaction at constant pressure.*

An **enthalpy diagram** is a diagrammatic representation of the enthalpy changes in a process.

Enthalpy (heat) of formation (See **standard enthalpy of formation.**)

Entropy, S, is a measure of the degree of *disorder* in a system. The greater the disorder found in a system, the greater is its entropy.

Entropy change, ΔS, is the difference in entropy between two states of a system. Its value indicates the extent to which the degree of order changes as the result of some process. A *positive ΔS* means an *increase in disorder.*

An **enzyme** is a high molar mass protein that catalyzes biological reactions.

An **equation of state** is a mathematical expression relating the amount, volume, temperature, and pressure of a substance (usually applied to gases).

Equilibrium refers to a condition where forward and reverse processes proceed at equal rates and no further net change occurs. For example, amounts of reactants and products in a reversible reaction remain constant over time.

An **equilibrium constant expression** describes the relationship among the concentrations (or partial pressures) of the substances present in a system at equilibrium. The numerical value of this expression, called the **equilibrium constant, K,** depends

on the temperature but *not* on how the equilibrium condition was established.

The **equivalence point** of a titration is the condition in which the reactants are in stoichiometric proportions. They consume each other, and neither reactant is in excess.

An **ester** is the product of the elimination of H_2O from between an acid and an alcohol molecule. Esters have the general formula

$$R-\overset{\displaystyle O}{\overset{\|}{C}}-O-R'.$$

An **ether** has the general formula $R-O-R'$.

Eutrophication is the deterioration of a freshwater body caused by nutrients such as nitrates and phosphates that results in growth of algae, oxygen depletion, and fish kills.

In an **excited state** of an atom, one or more electrons are promoted to a higher energy level than in the ground state. (See also **ground state.**)

An **exothermic** process produces an increase in temperature in an isolated system or, for a system that interacts with its surroundings, the evolution of heat.

Expanded octet is a term used to describe Lewis structures in which certain atoms in the third or higher period of the periodic table appear to require 10 or 12 electrons in their valence shells.

An **extensive property** is one, like mass or volume, whose value depends on the quantity of matter observed.

Extractive metallurgy refers to the process of extracting a metal from its ores. Generally this occurs in four steps. **Concentration** separates the ore from waste rock (gangue). **Roasting** converts the ore to the metal oxide. **Reduction** (usually with carbon) converts the oxide to the metal. **Refining** removes impurities from the metal.

A **face-centered cubic (fcc)** crystal structure is one in which the unit cell has structural units at the eight corners and in the center of each face of the unit cell. It is derived from the cubic closest packed arrangement of spheres (see Figure 13-37).

The **Fahrenheit** temperature scale is based on a value of 32 °F as the melting point of ice and 212 °F as the boiling point of water.

A **family** of elements is a numbered group from the periodic table, sometimes carrying a distinctive name. For example, Group 7A is the halogen family. Members of a family of elements have a number of similarities in physical and chemical properties.

The **Faraday constant, $\mathscr{F}$,** is the charge associated with one mole of electrons, 96,485 C/mol e⁻.

Fats are triglycerides in which saturated fatty acid components predominate.

The **f-block** is that portion of the periodic table where the process of filling of electron orbitals (Aufbau process) involves f subshells. These are the lanthanide and actinide elements.

fcc (See **face-centered cubic.**)

Ferromagnetism is a property that permits certain materials (notably Fe, Co, and Ni) to be made into permanent magnets. The magnetic moments of individual atoms are aligned into domains. In the presence of a magnetic field, these domains orient themselves to produce a permanent magnetic moment.

Filtration is a method of separating a solid from a liquid in which solid is retained by the filtering device and liquid passes through.

The **first law of thermodynamics,** expressed as $\Delta E = q + w$, is an alternative statement of the law of conservation of energy. (See also **law of conservation of energy.**)

A **first-order** reaction is one for which the sum of the concentration-term exponents in the rate equation is 1.

Fission (See **nuclear fission.**)

A **flow battery** is a battery in which materials (reactants, products, electrolyte) pass continuously through the battery. The battery is simply a converter of chemical to electrical energy.

Formal charge is the number of outer-shell (valence) electrons in an isolated atom minus the number of electrons assigned to that atom in a Lewis structure.

The **formation constant, K_f,** is the equilibrium constant describing equilibrium among a complex ion, the free metal ion, and ligands. It is obtained by combining equilibrium constants for the stepwise displacement of H_2O molecules from the coordination sphere of a metal ion by other ligands.

Formula mass is the mass of a formula unit of a compound relative to a mass of exactly 12 u for carbon-12.

A **formula unit** is the smallest collection of atoms from which the empirical formula of a compound can be established.

Fractional crystallization (recrystallization) is a method of purifying a substance by crystallizing the pure solid from a saturated solution while impurities remain in solution.

Fractional distillation (See **distillation.**)

Fractional precipitation is a technique in which two or more ions in solution, each capable of being precipitated by the same reagent, are separated by the use of that reagent.

The **Frasch process** is a method of extracting sulfur from underwater deposits. It is based on the use of superheated water to melt the sulfur.

Free energy, G, is a thermodynamic function designed to produce a criterion for spontaneous change. It is defined through the equation $G = H - TS$.

Free energy change, ΔG, is the change in free energy that accompanies a process and can be used to indicate the direction of spontaneous change. For a spontaneous process at constant temperature and pressure, $\Delta G < 0$. (See also **standard free energy change.**)

Free radicals are highly reactive molecular fragments containing unpaired electrons.

Freezing is the conversion of a liquid to a solid that occurs at a fixed temperature known as the **freezing point.**

The **frequency** of a wave motion is the number of wave crests or troughs that pass through a given point in a unit of time. It is expressed by the unit $time^{-1}$ (e.g., s^{-1}, also called a hertz, Hz).

A **fuel cell** is a voltaic cell in which the cell reaction is the equivalent of the combustion of a fuel. Chemical energy of the fuel is converted to electricity.

A **function of state (state function)** is a property that assumes a unique value when the state or present condition of a system is defined. This value is *independent* of how the state is attained.

A **functional group** is an atom or grouping of atoms attached to a hydrocarbon residue, R. The functional group often confers specific properties to an organic molecule.

Fusion (See **nuclear fusion.**)

Galvanic cell (See **voltaic cell.**)

Galvanizing is the name given to any process in which an iron surface is coated with zinc to give it corrosion protection.

Gamma (γ) rays are a form of electromagnetic radiation of high penetrating power emitted by certain radioactive nuclei.

A **gas** is a form or state of matter in which a material assumes the shape of its container and expands to fill the container, thus having neither definite shape nor volume.

The **gas constant, R,** is the numerical constant appearing in the ideal gas equation ($PV = nRT$) and in several other equations as well.

A **Geiger–Müller counter** is a device used to detect ionizing radiation. Ionizing events that occur in the counter produce electric discharges that can be recorded.

The **general gas equation** is an expression based on the ideal gas equation and written in the form $P_1V_1/n_1T_1 = P_2V_2/n_2T_2$.

The **genetic code** relates to the sequences of bases in DNA molecules that determine complementary sequences in mRNA and, ultimately, sequences of amino acids in proteins.

Geometric isomerism in organic compounds refers to the existence of nonequivalent structures (cis and trans) that differ in the positioning of substituent groups relative to a double bond. In complexes the nonequivalent structures are based on the positions at which ligands are attached to the metal center.

Glass is a transparent, amorphous solid consisting of Na^+ and Ca^{2+} ions in a network of SiO_4^{4-} anions. It is made by fusing together a mixture of sodium and calcium carbonates with sand.

Global warming refers to the warming of Earth that results from an accumulation in the atmosphere of gases such as CO_2 that absorb infrared radiation radiated from Earth's surface.

Graham's law states that the rates of effusion or diffusion of two different gases are inversely proportional to the square roots of their molar masses.

The **ground state** is the lowest energy state for the electrons in an atom or molecule.

A **group** is a vertical column of elements in the periodic table. Members of a group have similar properties.

A **half-cell** is a combination of an electrode and a solution. An oxidation–reduction equilibrium is established on the electrode. An electrochemical cell is a combination of two half-cells.

The **half-life** of a reaction is the time required for one-half of a reactant to be consumed. In a nuclear decay process, it is the time required for one-half of the atoms present in a sample to undergo radioactive decay.

A **half-reaction** describes one portion of a net oxidation–reduction reaction, either the oxidation or the reduction.

The **half-reaction method** is a method of balancing an oxidation–reduction equation by dividing it into two half-equations that are balanced separately and then recombined.

Hard water contains dissolved minerals in significant concentrations. If the hardness is primarily due to HCO_3^- and associated cations, the water is said to have **temporary hardness.** If the hardness is due to anions other than HCO_3^- (e.g., SO_4^{2-}), it is referred to as **permanent hardness.**

hcp (See **hexagonal closest packed.**)

Heat is a transfer of thermal energy as a result of a temperature difference.

Heat capacity is the quantity of heat required to change the temperature of an object or substance by one degree, usually expressed as $J\,°C^{-1}$ or $cal\,°C^{-1}$. **Specific heat capacity** is the heat capacity per gram of substance, i.e., $J\,°C^{-1}\,g^{-1}$, and **molar heat capacity** is the heat capacity per mole, i.e., $J\,°C^{-1}\,mol^{-1}$.

A **heat of reaction** is energy converted from chemical to thermal (or vice versa) in a reaction. In an isolated system, this energy conversion causes a temperature change, and in a system that interacts with its surroundings, heat (q) is either evolved to the surroundings (exothermic reaction) or absorbed (endothermic reaction). In either case, products must be returned to the initial temperature.

The **Heisenberg uncertainty principle** states that, when measuring the position and momentum of fundamental particles of matter, uncertainties in measurement are inevitable.

Henry's law relates the solubility of a gas to the gas pressure maintained above a solution of the gaseous solute: $M = k \cdot P_{gas}$. That is, the solubility is directly proportional to the pressure of the gas above the solution.

The **hertz (Hz)** is the SI unit of frequency, equal to s^{-1}.

Hess's law states that the enthalpy change for an overall or net process is the sum of enthalpy changes for individual steps in the process.

Heterocyclic compounds are based on hydrocarbon ring structures in which one or more C atoms is replaced by atoms such as N, O, or S.

In a **heterogeneous mixture** the components separate into physically distinct regions of differing properties.

Hexagonal closest packed is one of the two ways in which spheres can be packed to minimize the amount of free space or voids among them. (See Figures 13-36 and 13-38.) The crystal structure based on this type of packing is referred to as **hcp.**

A **homogeneous mixture (solution)** is a mixture of elements and/or compounds that has a uniform composition and properties within a given sample, but the composition and properties may vary from one sample to another.

A **homologous series** is a group of compounds that differ in composition by some constant unit (—CH_2 in the case of alkanes).

Hund's rule (rule of maximum multiplicity) states that whenever orbitals of equal energy are available, electrons occupy these orbitals singly before any pairing of electrons occurs.

A **hybrid orbital** is one of a set of identical orbitals reformulated from pure atomic orbitals and used to describe certain covalent bonds.

Hybridization refers to the combining of pure atomic orbitals to generate hybrid orbitals in the valence bond approach to covalent bonding.

A **hydrate** is a compound in which a fixed number of water molecules is associated with each formula unit, such as $CuSO_4 \cdot 5H_2O$.

Hydration energy is the energy associated with dissolving gaseous ions in water, usually expressed per mole of ions.

Hydrides are compounds of hydrogen, usually divided into the categories of covalent (e.g., H_2O and HCl), ionic (e.g., LiH and CaH_2), and metallic (mostly nonstoichiometric compounds with the transition metals).

A **hydrocarbon** is a compound containing the two elements carbon and hydrogen. The C atoms are arranged in straight or branched chains or ring structures.

A **hydrogen bond** is an intermolecular force of attraction in which an H atom covalently bonded to one atom is attracted simultaneously to another highly nonmetallic atom of the same or a nearby molecule.

The **hydrogen economy** refers to a possible future in which hydrogen may become the principal fuel and metallurgical reducing agent, supplanting fossil fuels.

A **hydrogenation reaction** is one in which H atoms are added to multiple bonds between carbon atoms, converting carbon-to-carbon double bonds to single bonds and carbon-to-carbon triple bonds to double or single bonds. It is a reaction, for example, that converts an unsaturated to a saturated fatty acid.

Hydrolysis is a special name given to acid–base reactions in which ions act as acids or bases. As a result of hydrolysis, for many salt solutions, $pH \neq 7$.

Hydrometallurgy refers to metallurgical procedures where water and aqueous solutions are used to extract metals from their ores. The first step is a process of **leaching,** in which the metal of interest is obtained in soluble form in aqueous solution. Other steps include purifying the leached solution and then depositing the metal from solution.

Hydronium ion, H_3O^+, is the form in which protons are found in aqueous solution. The terms "hydrogen ion" and "hydronium ion" are often used synonymously.

The **hydroxyl** group is —OH and is usually found attached to a straight or branched hydrocarbon chain (an alcohol) or a ring structure (a phenol).

A **hypothesis** is a tentative explanation of a series of observations or of a natural law.

An **ICE table** refers to a format for organizing the data in an equilibrium calculation. The table is based on the initial concentrations of reactants and products, changes in concentrations to attain equilibrium, and equilibrium concentrations.

An **ideal gas** is one whose behavior can be predicted by the ideal gas equation.

Ideal gas constant (See **gas constant**.)

The **ideal gas equation** relates the pressure, volume, temperature, and number of moles of ideal gas (n) through the expression $PV = nRT$.

An **ideal solution** has $\Delta H_{soln} = 0$ and certain properties (notably vapor pressure) that are predictable from the properties of the solution components.

Incomplete octet is a term used to describe situations in which an atom fails to acquire eight outer-shell electrons when it forms bonds.

An **indicator** is a substance that is added in a small amount to a titration mixture and changes color at the equivalence point.

An **induced dipole** is an atom or molecule in which a separation of charge is produced by a neighboring dipole.

Industrial smog is air pollution in which the chief pollutants are $SO_2(g)$, $SO_3(g)$, H_2SO_4 mist, and smoke.

Inert complex is the term used to describe a complex ion in which the exchange of ligands occurs very slowly.

The **inert pair effect** refers to the effects on the properties of certain post-transition elements that result from a pair of electrons in the s orbital of the valence shells of their atoms, that is, ns^2.

The **initial rate of a reaction** is the rate of a reaction immediately after the reactants are brought together.

Inner-transition elements are those of the f-block series, that is, the lanthanide and actinide elements.

An **instantaneous dipole** is an atom or molecule with a separation of charge produced by a momentary displacement of electrons from their normal distribution.

An **instantaneous rate of reaction** is the exact rate of a reaction at some precise point in the reaction. It is obtained from the slope of a tangent line to a concentration–time graph.

An **integrated rate equation** is derived from a rate equation by the calculus technique of integration and relates the concentration of a reactant (or product) to elapsed time from the start of a reaction. The equation has different forms depending on the order of the reaction.

An **intensive property** is *independent* of the quantity of matter involved in the observation. Density and temperature are examples of intensive properties.

An **interhalogen compound** is a covalent compound between two or more halogen elements, such as ICl and BrF_3.

An **intermediate** is the product of one reaction that is consumed in a following reaction in a process that proceeds through several steps.

An **intermolecular force** is an attraction *between* molecules.

The **internal energy, E,** of a system is the total energy attributed to the particles of matter and their interactions within a system.

An **ion** is a charged species consisting of a single atom or a group of atoms. It is formed when a neutral atom or a covalently bonded group of atoms either gains or loses electrons.

Ion exchange is a process in which ions held to the surface of an ion exchange material are exchanged for other ions in solution. For example, Na^+ may be exchanged for Ca^{2+} and Mg^{2+}, or OH^- may be exchanged for SO_4^{2-}.

An **ion pair** is an association of a cation and an anion in solution. Such combinations, when they occur, can have a significant effect on solution equilibria.

An **ion product, Q_{sp},** is formulated in the same manner as a solubility product constant, K_{sp}, but with the concentration terms in the product equal to actual *nonequilibrium* values. A comparison of Q_{sp} and K_{sp} serves as a criterion for precipitation from solution.

The **ion product of water, K_w,** is the product of $[H_3O^+]$ and $[OH^-]$ in pure water or in an aqueous solution. This product has a unique value that depends only on temperature. At 25 °C, $K_w = 1.0 \times 10^{-14}$.

An **ionic bond** results from the transfer of electrons between metal and nonmetal atoms. Positive and negative ions are formed and held together by electrostatic attractions.

An **ionic compound** is a compound consisting of positive and negative ions that are held together by electrostatic forces of attraction.

Ionic radius is the radius of a spherical ion. It is the atomic radius associated with an element in its ionic compounds.

The first **ionization energy, I_1,** is the energy required to remove the most loosely held electron from a *gaseous* atom. The second ionization energy, I_2, is the energy required to remove an electron from a gaseous unipositive ion, and so on for higher ionization energies.

The **isoelectric point, p***I*, of an amino acid is the pH at which the dipolar structure or "zwitterion" predominates.

Isoelectric species have the same number of electrons (usually in the same configuration). Na^+ and Ne are isoelectronic, as are CO and N_2.

An **isolated system** is one that exchanges neither energy nor matter with its surroundings.

Isomers are two or more compounds having the same formulas but different structures and therefore different properties.

Isotopes of an element are atoms with different numbers of neutrons in their nuclei. That is, isotopes of an element have the same atomic numbers but different mass numbers.

Isotopic mass (See **nuclidic mass**.)

IUPAC (or **IUC**) refers to the International Union of Pure and Applied Chemistry. Among its activities, the IUPAC makes recommendations on chemical terminology and nomenclature.

The **joule**, J, is the basic SI unit of energy. It is the quantity of work done when a force of one newton acts through a distance of one meter.

K_c is a relationship that exists among the *concentrations* of the reactants and products in a reversible reaction at equilibrium. Concentrations are expressed as molarities.

K_p, the partial pressure equilibrium constant, is a relationship that exists among the partial pressures of gaseous reactants and products in a reversible reaction at equilibrium. Partial pressures are expressed in atm.

The **Kelvin** temperature is an *absolute* temperature. That is, the lowest attainable temperature is 0 K = −273.15 °C (the temperature at which molecular motion ceases). Temperatures on the Kelvin scale are related to Celsius temperatures through the expression T (K) = t (°C) + 273.15.

A **ketone** has the general formula
$$R—\overset{\overset{\displaystyle O}{\|}}{C}—R'.$$

Kinetic energy is energy of motion. The kinetic energy of an object with mass m and velocity u is K.E. = $\frac{1}{2}mu^2$.

The **kinetic molecular theory of gases** is a model for describing gas behavior. It is based on a set of assumptions and yields equations from which various properties of gases can be deduced.

The **Kroll process** is an industrial process for the manufacture of titanium metal from TiO_2. Key steps involve conversion of TiO_2 to $TiCl_4$ and reduction of $TiCl_4$ to Ti with Mg as the reducing agent.

Labile complex is the term used to describe a complex ion in which a rapid exchange of ligands occurs.

The **lanthanide contraction** refers to the decrease in atomic size in a series of elements in which an f subshell fills with electrons (an inner transition series). It results from the ineffectiveness of f electrons in shielding outer-shell electrons from the nuclear charge of an atom.

The **lanthanides** are the elements (Z = 58–71) characterized by a partially filled $4f$ subshell in their atoms. Because lanthanum resembles them, La (Z = 57) is generally considered together with them.

Lattice energy is the quantity of energy released in the formation of one mole of a crystalline ionic solid from its separated gaseous ions.

The **law of conservation of energy** states that energy can neither be created nor destroyed in ordinary processes.

The **law of conservation of mass** states that the total mass of the products of a chemical reaction is the same as the total mass of the reactants entering into the reaction.

The **law of constant composition (definite proportions)** states that all samples of a compound have the same composition, for example, the same proportions by mass of the constituent elements.

The **law of multiple proportions** states that if two elements form more than a single compound, the masses of one element combined with a fixed mass of the second are in the ratio of small whole numbers.

Le Châtelier's principle states that an action that tends to change the temperature, pressure, or concentrations of reactants in a system at equilibrium stimulates a response that partially offsets the change while a new equilibrium condition is established.

Levorotatory means the ability to rotate the plane of polarized light to the *left,* designated (−).

Lewis acid (See **acid**.)

Lewis base (See **base**.)

A **Lewis structure** is a combination of Lewis symbols that depicts the transfer or sharing of electrons in a chemical bond.

In the **Lewis symbol** of an element, valence electrons are represented by dots placed around the chemical symbol of the element.

The **Lewis theory** refers to a description of chemical bonding through Lewis symbols and Lewis structures in accordance with a particular set of rules.

Ligands are the groups that are coordinated (bonded) to the central atom in a complex.

The **limiting reactant (reagent)** in a reaction is the reactant that is consumed completely. The quantity of product(s) formed depends on the quantity of the limiting reactant.

Lipids include a variety of naturally occurring substances (e.g., fats and oils) sharing the property of solubility in solvents of low polarity [such as in $CHCl_3$, CCl_4, C_6H_6, and $(C_2H_5)_2O$].

A **liquid** is a form or state of matter in which a material occupies a definite volume but has the ability to flow and assume the shape of its container.

Liquid crystals are a form of matter with some of the properties of a liquid and some, of a crystalline solid.

London forces (See **dispersion forces**.)

A **lone pair** is a pair of electrons found in the valence shell of an atom and *not* involved in bond formation.

Magic numbers is a term used to describe numbers of protons and neutrons that confer a special stability to an atomic nucleus.

The **main group elements** are those in which s or p subshells are being filled in the Aufbau process. They are also referred to as the s-block and p-block elements. They are the A groups in the periodic table.

A **manometer** is a device used to measure the pressure of a gas, usually by comparing the gas pressure with barometric pressure.

Mass describes the quantity of matter in an object.

The **mass number, A,** is the total of the number of protons and neutrons in the nucleus of an atom.

A **mass spectrometer (mass spectrograph)** is a device used to separate and to measure the quantities and masses of different ions in a beam of positively charged gaseous ions.

Matter is anything that occupies space and has the property known as mass.

Melting is the transition of a solid to a liquid and occurs at the **melting point.** The melting point and freezing point of a substance are identical.

A **meta (*m*-) isomer** has two substituents on a benzene ring separated by one C atom.

A **metal** is an element whose atoms have small numbers of electrons in the electronic shell of highest principal quantum number. Removal of an electron(s) from a metal atom occurs without great difficulty, producing a positive ion (cation). Metals generally are malleable and ductile solids with a lustrous appearance and an ability to conduct heat and electricity.

Metallic radius is one-half the distance between the centers of adjacent atoms in a solid metal.

A **metalloid** is an element that may display both metallic and nonmetallic properties under the appropriate conditions (e.g., Si, Ge, As).

The **method of initial rates** can be used to establish the order of a reaction from a series of measurements of the initial rate of the reaction. The initial concentration of one reactant is varied while the initial concentrations of all the other reactants are held constant.

The **metric system** is a system of measurement in which the unit of length is the meter, the unit of mass is the kilogram (1 kg = 1000 g), and the unit of time is the second.

A **millimole** is one-thousandth of a mole (0.001 mol). It is especially useful in titration calculations.

A **mixture** is any sample of matter that is not pure, that is, not an element or compound. The composition of a mixture, unlike that of a substance, can be varied. Mixtures are either *homogeneous* or *heterogeneous*.

A **moderator** slows down energetic neutrons from a fission process so that they are able to induce additional fission.

Molality, *m*, is a solution concentration expressed as the amount of solute, in moles, divided by the mass of solvent, in kg.

Molar mass, *M*, is the mass of one mole of atoms, formula units, or molecules of a substance.

Molar solubility is the molarity of solute (mol L^{-1}) in a saturated solution.

Molarity, M, is the concentration of a solution expressed as the amount of solute, in moles, per liter of solution.

A **mole** is an amount of substance containing 6.02214×10^{23} (the Avogadro constant) atoms, formula units, or molecules.

Mole fraction describes a mixture in terms of the fraction of all the molecules that are of a particular type. It is the amount of one component, in moles, divided by the total amount of all the substances in the mixture.

A **mole percent** is a mole fraction expressed on a percentage basis, that is, mole fraction $\times 100\%$.

A **molecular compound** is a compound consisting of discrete molecules.

A **molecular formula** denotes the numbers of the different atoms present in a molecule. In some cases the molecular formula is the same as the empirical formula; in others it is an integral multiple of that formula.

Molecular geometry refers to the geometrical shape of a molecule or polyatomic ion. In a species in which all electron pairs are bond pairs, the molecular geometry is the same as the electron group geometry. In other cases the two properties are related but not the same.

Molecular mass is the mass of a molecule relative to a mass of exactly 12 u for carbon-12.

Molecular orbital theory describes the covalent bonds in a molecule by considering that the atomic orbitals of the component atoms are replaced by electron orbitals belonging to the molecule as a whole—molecular orbitals. A set of rules is used to assign electrons to these molecular orbitals, thereby yielding the electronic structure of the molecule.

A **molecule** is a group of bonded atoms held together by covalent bonds and existing as a separate entity.

A **monodentate ligand** is a ligand that is able to attach to a metal center in a complex at only one position and using just one lone pair of electrons.

A **monomer** is a simple molecule that is capable of joining with others to form a complex long-chain molecule called a **polymer.**

A **monosaccharide** is a single, simple molecule having the structural features of a carbohydrate. It can also be called a simple sugar.

A **multiple covalent bond** is a bond in which more than two electrons are shared between the bonded atoms.

A **natural law** is a general statement that can be used to summarize observations of natural phenomena.

The **Nernst equation** is used to relate E_{cell}, E°_{cell}, and the activites of the reactants and products in a cell reaction.

A **net equation** is the resulting equation when equations for individual steps in a multistep reaction are combined.

A **net ionic equation** represents a reaction between ions in solution in such a way that all nonparticipant (spectator) ions are eliminated from the equation. The equation must be balanced both atomically and for net electric charge.

A **net reaction** is the overall change that occurs in a process involving two or more steps.

A **network covalent solid** is a substance in which covalent bonds extend throughout the crystal, that is, in which the covalent bond is both an *intra*molecular and an *inter*molecular force.

A **neutralization** reaction is one in which an acid and a base react in stoichiometric proportions, so that there is no excess of either acid or base in the final solution. The products are water and a salt.

Neutrons are electrically neutral fundamental particles of matter found in all atomic nuclei except that of the simple hydrogen atom, protium, ^1H.

The **neutron number** is the number of neutrons in the nucleus of an atom. It is equal to the mass number (A) minus the atomic number (Z).

The **nitrogen cycle** is a series of processes by which atmospheric N_2 is fixed, enters into the food chain of animals, and eventually is returned to the atmosphere by bacteria.

Noble gases are elements whose atoms have the electron configuration ns^2np^6 in the electronic shell of highest principal quantum number. (The noble gas helium has the configuration $1s^2$.)

Nomenclature refers to the writing of chemical names and formulas by some systematic method.

A **nonbonding molecular orbital** is a molecular orbital that neither contributes to nor detracts from bond formation in a molecule. (It does not affect the bond order.)

A **nonelectrolyte** is a substance that is essentially un-ionized, both in the pure state and in solution.

A **nonideal gas** departs from the behavior predicted by the ideal gas equation. Its behavior can only be predicted by other equations of state (e.g., the van der Waals equation).

A **nonmetal** is an element whose atoms tend to gain small numbers of electrons to form negative ions (anions) with the electron configuration of a noble gas. Nonmetal atoms may also alter their electron configurations by sharing electrons. Nonmetals are mostly gases, liquid (bromine), or low melting point solids and are very poor conductors of heat and electricity.

In a **nonpolar molecule** the centers of positive and negative charge coincide. That is, there is no net separation of charge within the molecule.

A **nonspontaneous** process is one that will not occur naturally. A nonspontaneous process can be brought about only by intervention from outside the system, as in the use of electricity to decompose a chemical compound (electrolysis).

The **normal boiling point** is the temperature at which the vapor pressure of a liquid is 1 atm. It is the temperature at which the liquid boils in a container open to the atmosphere at a pressure of 1 atm.

The **normal melting point** is the temperature at which the melting of a solid occurs at 1 atm pressure. This is the same temperature as the **normal freezing point.**

Nuclear binding energy is the energy released when nucleons (protons and neutrons) are fused into an atomic nucleus. This energy replaces an equivalent quantity of matter.

A **nuclear equation** represents the changes that occur during a nuclear process. The target nucleus and bombarding particle are represented on the left side of the equation, and the product nucleus and ejected particle on the right side.

Nuclear fission is a radioactive decay process in which a heavy nucleus breaks up into two lighter nuclei and several neutrons, accompanied by the release of energy.

In **nuclear fusion** small atomic nuclei are fused into larger ones, with some of their mass being converted to energy.

A **nuclear reactor** is a device in which nuclear fission is carried out as a controlled chain reaction. That is, neutrons produced in one fission event trigger the fission of other nuclei, and so on.

Nucleic acids are cell components comprised of purine and pyrimidine bases, pentose sugars, and phosphoric acid.

Nuclide is a term used to designate an atom with a specific atomic number and mass number. It is represented by the symbolism $_{Z}^{A}E$.

Nuclidic mass is the mass, in atomic mass units, u, of an individual atom, relative to an arbitrarily assigned value of exactly 12 u as the nuclidic mass of carbon-12.

An **octet** refers to the presence of *eight* electrons in the outermost (valence) electronic shell of an atom in a Lewis structure.

The **octet rule** states that the number of electrons associated with bond pairs and lone pairs of electrons for each of the Lewis symbols in a Lewis structure will be eight (an **octet**). The principal exception is the H atom, which can accommodate only two electrons in its valence shell, but there are other well-known exceptions as well.

An **odd-electron species** is one in which the total number of valence electrons is an *odd*

number. At least one *unpaired* electron is present in such a species.

Oils are triglycerides in which unsaturated fatty acid components predominate.

Oligosaccharides are carbohydrates consisting of two to ten monosaccharide units. (See also **sugar**.)

An **open system** is one that can exchange both matter and energy with its surroundings.

Optical isomers, also called enantiomers (nonsuperimposable mirror images), are isomers that differ only in the way they rotate the plane of polarized light.

An **orbital** describes a region in an atom where the electron charge density or the probability of finding an electron is high. The several kinds of orbitals ($s, p, d, f, \ldots$) differ from one another in the shapes of the regions of high electron charge density they describe.

An **orbital diagram** is a representation of an electron configuration in which the most probable orbital designation and spin of each electron in the atom are indicated.

The **order of a reaction** relates to the exponents of the concentration terms in the rate law for a chemical reaction. The order can be stated with respect to a particular reactant (first order in A, second order in B, ...) or, more commonly, as the overall order. The overall order is the sum of the concentration-term exponents.

An **ortho (*o-*) isomer** has two substituents attached to adjacent C atoms in a benzene ring.

Osmosis is the net flow of solvent molecules through a semipermeable membrane, from a more dilute solution (or from the pure solvent) into a more concentrated solution.

Osmotic pressure is the pressure that would have to be applied to a solution to stop the passage through a semipermeable membrane of solvent molecules from the pure solvent.

An **overpotential** is the voltage in excess of the theoretical value required to produce a particular electrode reaction in electrolysis.

Oxidation is a process in which electrons are "lost" and the oxidation state of some atom increases. (Oxidation can occur only in combination with reduction.)

An **oxidation–reduction** reaction is one in which certain atoms undergo changes in oxidation state. The substance containing atoms whose oxidation states *increase* is **oxidized.** The substance containing atoms whose oxidation states *decrease* is **reduced.**

Oxidation states relate to the number of electrons an atom loses, gains, or shares in combining with other atoms to form molecules or polyatomic ions. The assignment of

oxidation states requires applying a set of rules; once assigned, oxidation states are useful in naming compounds and balancing chemical equations.

An **oxidizing agent (oxidant)** makes possible an oxidation process by itself being *reduced.*

An **oxoacid** is an acid in which an ionizable hydrogen atom(s) is bonded through an oxygen atom to a central atom, that is, E—O—H. Other groups bonded to the central atom are either additional —OH groups or O atoms (or in a few cases H atoms).

An **oxoanion** is a polyatomic anion containing a nonmetal such as Cl, N, P, or S in combination with some number of oxygen atoms. An oxoanion can be obtained by neutralizing an oxoacid.

The **ozone layer** refers to a region of the stratosphere from about 25 to 35 km above Earth's surface having a concentration of ozone, O_3, considerably greater than that at Earth's surface.

A **para (*p-*) isomer** has two substituents located opposite to one another on a benzene ring.

A **paramagnetic** substance has one or more unpaired electrons in its atoms or molecules. It is attracted into a magnetic field.

A **partial pressure** is the pressure exerted by an individual gas in a mixture, independently of other gases. Each gas in the mixture expands to fill its container and exerts its own partial pressure.

The **Pauli exclusion principle** states that no two electrons may have all four quantum numbers alike. This limits occupancy of an orbital to two electrons with opposing spins.

The **p-block** is that portion of the periodic table in which the filling of electron orbitals (Aufbau process) involves *p* subshells.

A **peptide bond** is formed by the elimination of a water molecule from between two amino acid molecules. The H atom comes from the —NH_2 group of one amino acid and the —OH group, from the —COOH group of the other acid.

Percent is the number of parts of a constituent in 100 parts of the whole.

The **percent ionization** of a weak acid or a weak base is the percent of its molecules that ionize in an aqueous solution.

Percent natural abundances refer to the relative proportions, expressed as percentages, by number, in which the isotopes of an element are found in natural sources.

Percent yield is the percent of the theoretical yield of product that is actually obtained in a chemical reaction. (See also **actual yield** and **theoretical yield**.)

A **period** is a horizontal row of the periodic table. All members of a period have atoms with the same highest principal quantum number.

The **periodic law** refers to the periodic recurrence of certain physical and chemical properties when the elements are considered in terms of increasing atomic number.

The **periodic table** is an arrangement of the elements, by atomic number, in which elements with similar physical and chemical properties are grouped together in vertical columns.

Permanent hard water (See **hard water**.)

The **peroxide** ion has the structure $[:\ddot{O}-\ddot{O}:]^{2-}$

pH is a shorthand designation for $[H_3O^+]$ in a solution. It is defined as $pH = -\log [H_3O^+]$.

A **phase diagram** is a graphical representation of the conditions of temperature and pressure at which solids, liquids, and gases (vapors) exist, either as single phases or states of matter or as two or more phases in equilibrium.

A **phenol** has the functional group —OH as part of an aromatic hydrocarbon structure.

A **phenyl group** is a benzene ring from which one H atom has been removed: $-C_6H_5$.

Photochemical smog is air pollution resulting from reactions involving sunlight, oxides of nitrogen, ozone, and hydrocarbons.

The **photoelectric effect** is the emission of electrons by certain materials when their surfaces are struck by electromagnetic radiation of the appropriate frequency.

A **photon** is a "particle" of light. The energy of a beam of light is concentrated into these photons.

In a **physical change** one or more physical properties of a sample of matter change, but the composition remains unchanged.

A **physical property** is a characteristic that a substance can display without undergoing a change in its composition.

A **pi (π) bond** results from the "sidewise" overlap of p orbitals, producing a high electron charge density above and below the line joining the bonded atoms.

Pig iron is an impure form of iron (about 95% Fe and 3–4% C, together with small quantities of Mn, Si, and P) produced in a blast furnace.

pK is a shorthand designation for an ionization constant: $pK = -\log K$. pK values are useful when comparing the relative strengths of acids or bases.

Planck's constant, h, is the proportionality constant that relates the energy of a photon of light to its frequency. Its value is 6.626×10^{-34} J s.

Plaster of paris, $CaSO_4 \cdot \frac{1}{2}H_2O$, a hemihydrate of calcium sulfate, is obtained by heating gypsum, $CaSO_4 \cdot 2H_2O$. It is a widely used material in the construction industry.

pOH is a shorthand designation for $[OH^-]$ in a solution: $pOH = -\log [OH^-]$.

In a **polar covalent bond** a separation exists between the centers of positive and negative charge in the bond.

In a **polar molecule,** the presence of one or more polar covalent bonds leads to a separation of the positive and negative charge centers for the molecule as a whole. A polar molecule has a resultant dipole moment.

Polarizability describes the ease with which the electron cloud in an atom or molecule can be distorted in an electric field, that is, the ease with which a dipole can be induced.

A **polyatomic ion** is a combination of two or more covalently bonded atoms that exists as an ion.

A **polydentate ligand** is capable of donating more than a single electron pair to the metal center of a complex, from different atoms in the ligand and to different sites in the geometric structure.

In a **polyhalide ion** two or more halogen atoms are covalently bonded into a polyatomic anion, e.g., I_3^-.

A **polymer** is a complex, long-chain molecule made up of many (hundreds, thousands) smaller units called *monomers.*

Polymerization is the process of producing a giant molecule (polymer) from simpler molecular units (monomers).

A **polypeptide** is formed by the joining of a large number of amino acid units through peptide bonds.

A **polyprotic acid** is capable of losing more than a single proton per molecule in acid–base reactions. Protons are lost in a stepwise fashion, with the first proton being the most readily lost.

A **polysaccharide** is a carbohydrate (such as starch or cellulose) consisting of more than ten monosaccharide units.

Positional isomers differ in the position on a hydrocarbon chain or ring where a functional group(s) is attached.

A **positron (β^+)** is a *positive* electron emitted as a result of the conversion of a proton to a neutron in a radioactive nucleus.

Potential energy is energy due to position or arrangement. It is the energy associated with forces of attraction and repulsion between objects.

A **precipitate** is an insoluble solid that deposits from a solution as a result of a chemical reaction.

Precision is the degree of reproducibility of a measured quantity—the closeness of agreement among repeated measurements.

Pressure is a force per unit area. Applied to gases, pressure is most easily understood in terms of the height of a liquid column that can be maintained by the gas.

Pressure–volume work is work associated with the expansion or compression of gases.

A **primary battery** produces electricity from a chemical reaction that cannot be reversed. As a result the battery cannot be recharged.

A **primary color** is one of a set of colors that when added together as light produce white light. Subtractive mixing leads to an absence of color (black). Red, yellow, and blue are a set of primary colors.

Primary structure refers to the sequence of amino acids in the polypeptide chains that make up a protein.

A **principal electronic shell (level)** refers to the collection of all orbitals having the same value of the principal quantum number, n. For example, the $3s$, $3p$, and $3d$ orbitals comprise the third principal shell ($n = 3$).

The **products** are the substances formed in a chemical reaction.

Properties are qualities or attributes that can be used to distinguish one sample of matter from others.

A **protein** is a large polypeptide, that is, having a molecular mass of 10,000 u or more.

A **proton acceptor** is a base in the Brønsted–Lowry acid–base theory.

A **proton donor** is an acid in the Brønsted–Lowry acid–base theory.

Protons are fundamental particles carrying the basic unit of positive electric charge and found in the nuclei of all atoms.

A **pseudo-first-order** reaction is a reaction of higher order (e.g., second-order) that is made to behave like a first-order reaction by using such large initial concentrations of all the reactants save one that only the concentration of that one reactant changes during the reaction. As a result the reaction behaves like a first-order reaction.

Pyrometallurgy is the traditional approach to extractive metallurgy that uses dry solid materials heated to high temperatures. (See also **extractive metallurgy** and **hydrometallurgy**.)

Qualitative cation analysis is a laboratory method, based on a variety of solution equilibrium concepts, for determining the presence or absence of certain cations in a sample.

Quantitative analysis refers to the analysis of substances or mixtures to determine the *quantities* of the various components rather than their mere presence or absence.

Quantum numbers are integral numbers whose values must be specified in order to solve the equations of wave mechanics. Three different quantum numbers are required: the *principal quantum number, n*; the *orbital angular momentum quantum number, ℓ*; and the *magnetic quantum number, m_ℓ*. The permitted values of these numbers are interrelated.

The **quantum theory** is based on the proposition that energy exists in the form of tiny, discrete units called quanta. Whenever an energy transfer occurs, it must involve an entire **quantum.**

Quaternary structure is the highest order structure that is found in some proteins. It describes how separate polypeptide chains may be assembled into a larger, more complex structure.

Quicklime is a common name for calcium oxide, CaO.

A **racemic mixture** is a mixture containing equal amounts of the D and L isomers of an optically active substance.

A **rad** is a quantity of radiation able to deposit 1×10^{-2} J of energy per kilogram of matter.

Radical (See **free radical**.)

The **radioactive decay law** states that the rate of decay of a radioactive material—the activity, *A*—is directly proportional to the number of atoms present.

A **radioactive decay series** is a succession of individual steps whereby an initial radioactive isotope (e.g., ^{238}U) is ultimately converted to a stable isotope (e.g., ^{206}Pb).

Radioactivity is a phenomenon in which small particles of matter (α or β particles) and/or electromagnetic radiation (γ rays) are emitted by unstable atomic nuclei.

Radiocarbon dating is a method of determining the age of a carbon-containing material based on the rate of decay of radioactive carbon-14.

A **random error** is an error made by the experimenter in performing an experimental technique or measurement, such as the error in estimating a temperature reading on a thermometer.

Raoult's law states that the vapor pressure of a solution component is equal to the product of the vapor pressure of the pure liquid and its mole fraction in solution: $P_A = \chi_A P_A^\circ$.

The **rate constant,** *k*, is the proportionality constant in a rate law that permits the rate of a reaction to be related to the concentrations of the reactants.

A **rate-determining step** in a reaction mechanism is an elementary process that is instrumental in establishing the rate of the overall reaction, usually because it is the slowest step in the mechanism.

The **rate law (rate equation)** for a reaction relates the reaction rate to the concentrations of the reactants. It has the form: rate = $k[A]^m[B]^n \cdots$.

The **rate of a chemical reaction** describes how fast reactants are consumed and products are formed, usually expressed as change of concentration per unit time.

Reactants are the substances that enter into a chemical reaction. This term is often applied to *all* the substances involved in a reversible reaction, but it can also be limited to the substances that appear on the *left* side of a chemical equation—the starting substances. (Substances on the *right* side of the equation are often called products.)

A **reaction mechanism** is a set of elementary steps or processes by which a reaction is proposed to occur. The mechanism must be consistent with the stoichiometry of the net equation and with the rate law for the net reaction.

A **reaction profile** is a graphical representation of a chemical reaction in terms of the energies of the reactants, activated complex(es), and products.

The **reaction quotient, *Q*,** is a ratio of concentration terms (or partial pressures) having the same form as an equilibrium constant expression, but usually applied to *nonequilibrium* conditions. It is used to determine the direction in which a net reaction occurs to establish equilibrium.

A **reducing agent (reductant)** makes possible a reduction process by itself becoming *oxidized.*

A **reducing sugar** is one that is able to reduce $Cu^{2+}(aq)$ to red, insoluble Cu_2O. The sugar must have available an aldehyde group, which is oxidized to an acid.

A **reduction** process is one in which electrons are "gained" and the oxidation state of some atom decreases. (Reduction can only occur in combination with oxidation.) (See also **extractive metallurgy**.)

Refining (See **extractive metallurgy**.)

The **relative humidity** is the ratio of the partial pressure of water vapor in air to the vapor pressure of water at the same temperature, expressed on a percent basis.

A **rem** is a unit of radiation related to the rad, but taking into account the varying effects on biological matter of different types of radiation of the same energy.

Representative elements (See **main group elements**.)

Resonance occurs when two or more plausible Lewis structures can be written for a species. The true structure is a composite or *hybrid* of these different contributing structures.

Rest mass is the mass of a particle (e.g., an electron) when it is essentially at rest. As its velocity approaches the speed of light, the mass of a particle increases.

Reverse osmosis is the passage through a semipermeable membrane of solvent molecules *from a solution into a pure solvent*. It can be achieved by applying to the solution a pressure in excess of its osmotic pressure.

A **reversible process** is one that can be made to reverse direction by just an infinitesimal change in a system property.

Ribonucleic acid (RNA), through its **messenger RNA (mRNA)** and **transfer RNA (tRNA)** forms, is involved in the synthesis of proteins.

Roasting (See **extractive metallurgy**.)

The **root-mean-square speed** is the square root of the average of the squares of the speeds of all the gas molecules in a gaseous sample.

A **salt bridge** is a device (a ∪-tube filled with a salt solution) used to join two half-cells in an electrochemical cell. The salt bridge permits the flow of ions between the two half-cells.

The **salt effect** is that of ions *different* from those directly involved in a solution equilibrium. The salt effect is also known as the diverse or "uncommon" ion effect.

Salts are ionic compounds in which hydrogen atoms of acids are replaced by metal ions. Salts are produced by the neutralization of acids with bases.

Saponification is the hydrolysis of a triglyceride by a strong base. The products are glycerol and a soap.

Saturated hydrocarbon molecules contain only single bonds between carbon atoms.

A **saturated solution** is one that contains the maximum quantity of solute that is normally possible at the given temperature.

The **s-block** refers to the portion of the periodic table in which the filling of electron orbitals (Aufbau process) involves the *s* subshell of the electronic shell of highest principal quantum number.

The **scientific method** refers to the general sequence of activities—observation, experimentation, and the formulation of laws and theories—that lead to the advancement of scientific knowledge.

The **second law of thermodynamics** relates to the direction of spontaneous change. One statement of the law is that all spontaneous processes produce an increase in the entropy of the universe.

A **secondary battery** produces electricity from a reversible chemical reaction. When electricity is passed through the battery in the reverse direction the battery is recharged.

A **secondary color** is the complement of a **primary color**. When light of a primary color and its complement (secondary) color are added, the result is white light. When they are subtracted, the result is an absence of color (black).

The **secondary structure** of a protein describes the structure or shape of a polypeptide chain, for example, a coiled helix.

A **second-order** reaction is one for which the sum of the concentration-term exponents in the rate equation is 2.

Self-ionization is an acid–base reaction in which one molecule acts as an acid and donates a proton to another molecule of the same kind acting as a base.

A **semiconductor** is characterized by a small energy gap between a filled valence band and an empty conduction band.

A **semipermeable membrane** permits the passage of some solution species but restricts the flow of others. It is a film of material containing submicroscopic holes.

The **shielding effect** refers to the effect of inner-shell electrons in shielding or screening outer-shell electrons from the full effects of the nuclear charge. In effect the inner electrons partially reduce the nuclear charge. (See also **effective nuclear charge**.)

SI units are the units of expressing measured quantities preferred by various international scientific agencies. (See Appendix C.)

A **side reaction** is a reaction that occurs at the same time as the main reaction in a chemical process.

A **sigma (σ) bond** results from the end-to-end overlap of simple or hybridized atomic orbitals along the straight line joining the nuclei of the bonded atoms.

Significant figures are those digits in an experimentally measured quantity that establish the precision with which the quantity is known.

Silica is a term used to describe various solid forms of silicon dioxide, SiO_2.

A **silicone** is an organosilicon polymer containing O—Si—O bonds.

Simultaneous reactions are two or more reactions that occur at the same time.

A **single covalent** bond results from the sharing of *one pair* of electrons between bonded atoms. It is represented by a single dash sign (—).

A **skeleton structure** is an arrangement of atoms in a Lewis structure to correspond to the actual arrangement found by experiment.

Slaked lime is a common name for calcium hydroxide, $Ca(OH)_2$.

Smog is the general term used to refer to a condition in which polluted air reduces visibility, causes stinging eyes and breathing difficulties, and produces additional minor and major health problems. (See also **industrial smog** and **photochemical smog**.)

Soaps are the salts of fatty acids, e.g., $RCOO^-Na^+$, where the R group is a hydrocarbon chain containing from 3 to 21 C atoms. Sodium and potassium soaps are the common soaps used as cleansing agents.

Solders are low-melting alloys used for joining wires or pieces of metal. They usually contain metals such as Sn, Pb, Bi, and Cd.

A **solid** is a form or state of matter in which a material has a definite shape and occupies a definite volume.

The **solubility** of a substance is the concentration of its saturated solution.

The **solubility product constant, K_{sp},** is the equilibrium constant that describes the formation of a saturated solution of a slightly soluble ionic compound. It is the product of ionic concentration terms, with each term raised to an appropriate power.

A **solute** is a solution component that is dissolved in a solvent. A solution may have several solutes, with the solutes generally present in lesser amounts than is the solvent.

The **solvent** is the solution component in which one or more solutes are dissolved. Usually the solvent is present in greater amount than are the solutes and determines the state of matter in which the solution exists.

An **sp hybrid orbital** is one of the pair of orbitals formed by the hybridization of one *s* and one *p* orbital. The angle between the two orbitals is 180°.

An **sp^2 hybrid orbital** is one of the three orbitals formed by the hybridization of one *s*

and two *p* orbitals. The angle between any two of the orbitals is 120°.

An **sp^3 hybrid orbital** is one of the four orbitals formed by the hybridization of one *s* and three *p* orbitals. The angle between any two of the orbitals is the tetrahedral angle—109.5°.

An **sp^3d hybrid orbital** is one of the five orbitals formed by the hybridization of one *s*, three *p*, and one *d* orbital. The five orbitals are directed to the corners of a trigonal bipyramid.

An **sp^3d^2 hybrid orbital** is one of the six orbitals formed by the hybridization of one *s*, three *p*, and two *d* orbitals. The six orbitals are directed to the corners of a regular octahedron.

spdf notation is a method of describing electron configurations in which the numbers of electrons assigned to each orbital are denoted as superscripts. For example, the electron configuration of Cl is $1s^2 2s^2 2p^6 3s^2 3p^5$.

The **specific heat** of a substance is the quantity of heat required to change the temperature of one gram of the substance by one degree Celsius.

Spectator ions are ionic species that are present in a reaction mixture but do not take part in the reaction. They are usually eliminated from a chemical equation, as in $\text{N\!a}^+ + Cl^- + Ag^+ + \text{N\!O}_3^- \rightarrow AgCl(s) + \text{N\!a}^+ + \text{N\!O}_3^-$

The **spectrochemical series** is a ranking of ligand abilities to produce a splitting of the *d* energy level of a central metal ion in a complex ion.

A **spontaneous (natural) process** is one that is able to take place in a system left to itself. No external action is required to make the process go, although in some cases the process may take a very long time.

Stalactites and **stalagmites** are limestone ($CaCO_3$) formations in limestone caves produced by the slow decomposition of $Ca(HCO_3)_2(aq)$.

The **standard atmosphere** is the pressure exerted by a column of mercury exactly 760 mm high when the density of mercury is 13.5951 g cm^{-3} and the acceleration due to gravity is $g = 9.80665$ m s^{-2}.

A **standard cell potential, $E°_{cell}$,** is the voltage of an electrochemical cell in which all species are in their standard states. (See also **cell potential**.)

Standard conditions of temperature and pressure (STP) refers to a gas maintained at a temperature of exactly 0 °C (273.15 K) and 760 mmHg (1 atm).

A **standard electrode potential, $E°$,** is the electric potential that develops on an electrode when the oxidized and reduced forms of some substance are in their *standard* states. Tabulated data are expressed in terms of the reduction process, that is, standard electrode potentials are standard reduction potentials.

The **standard enthalpy of formation, ΔH_f°,** of a substance is the enthalpy change that occurs in the formation of 1 mol of the substance in its standard state from the reference forms of its elements in their standard states. The reference forms of the elements are their most stable forms at the given temperature and 1 atm pressure.

The **standard enthalpy of reaction, ΔH°,** is the enthalpy change of a reaction in which all reactants and products are in their standard states.

Standard free energy change, ΔG°, is the free energy change of a process when the reactants and products are all in their standard states. The equation relating standard free energy change to the equilibrium constant is $\Delta G^\circ = -RT \ln K_{eq}$.

The **standard free energy of formation, ΔG_f°,** is the standard free energy change associated with the formation of 1 mol of compound from its elements in their most stable forms at 1 atm pressure.

The **standard hydrogen electrode (SHE)** is an electrode at which equilibrium is established between H_3O^+ ($a = 1$) and $H_2(g, 1 \text{ atm})$ on an inert (Pt) surface. The standard hydrogen electrode is *arbitrarily* assigned an electrode potential of exactly 0 V.

The **standard state** of a substance refers to that substance when it is maintained at 1 atm pressure and at the temperature of interest. For a gas it is the (hypothetical) pure gas behaving as an ideal gas at 1 atm pressure and the temperature of interest.

Standardization of a solution refers to establishing the exact concentration of the solution, usually through a titration.

A **standing wave** is a wave motion that reflects back on itself in such a way that the wave contains a certain number of points (nodes) that undergo no motion. A common example is the vibration of a plucked guitar string, and a related example is the description of electrons as matter waves.

Steel is a term used to describe iron alloys containing from 0 to 1.5% C together with other key elements, such as V, Cr, Mn, Ni, W, and Mo.

Step-reaction polymerization is a type of polymerization reaction in which monomers are joined together by the elimination of small molecules between them. For example, an H_2O molecule might be eliminated by the reaction of an H atom from one monomer with an —OH group from another.

In **stereoisomers** the number and types of atoms and bonds in molecules are the same, but certain atoms are oriented differently in space. Cis and trans isomerism is one type of stereoisomerism; optical isomerism is another.

Stoichiometric coefficients are the coefficients used to balance an equation.

A **stoichiometric factor** is a conversion factor relating molar amounts of two species involved in a chemical reaction (i.e., a reactant to a product, one reactant to another, etc.). The numerical values used in formulating the factor are stoichiometric coefficients.

Stoichiometric proportions refer to relative amounts of reactants that are in the same mole ratio as implied by the balanced equation for a chemical reaction. For example, a mixture of 2 mol H_2 and 1 mol O_2 is in stoichiometric proportions, and a mixture of 1 mol H_2 and 1 mol O_2 is not, for the reaction $2 H_2 + O_2 \rightarrow 2 H_2O$.

Stoichiometry refers to quantitative measurements and relationships involving substances and mixtures of chemical interest.

The **stratosphere** is the region of the atmosphere that extends from approximately 12 to 55 km above Earth's surface.

A **strong acid** is an acid that is completely ionized in aqueous solution.

A **strong base** is a base that is completely ionized in aqueous solution.

A **strong electrolyte** is a substance that is completely ionized in solution.

A **structural formula** for a compound indicates which atoms in a molecule are bonded together, and whether by single, double, or triple bonds.

Structural isomers have the same number and kinds of atoms but they differ in their structural formulas.

Sublimation is the passage of molecules from the solid to the gaseous state.

A **subshell** refers to a collection of orbitals of the same type. For example, the three $2p$ orbitals constitute the $2p$ subshell.

A **substance** has a constant composition and properties throughout a given sample and from one sample to another. All substances are either elements or compounds.

Substitution reactions are typical of those involving alkane and aromatic hydrocarbons. In such a reaction a functional group replaces an H atom on a chain or ring.

A **substrate** is the substance that is acted upon by an enzyme in an enzyme-catalyzed reaction. The substrate is converted to products, and the enzyme is regenerated.

A **sugar** is a monosaccharide (simple sugar), a disaccharide, or an oligosaccharide containing up to ten monosaccharide units.

The **superoxide** ion has the structure $[:\overset{..}{O}—\overset{..}{O}:]^-$

Superphosphate is a mixture of $Ca(H_2PO_4)_2$ and $CaSO_4$ produced by the action of H_2SO_4 on phosphate rock.

A **supersaturated solution** contains more solute than normally expected for a saturated solution. The usual manner of preparation is to start with a solution that is saturated at one temperature and to change its temperature to one where supersaturation can occur. Another requirement is the lack of nuclei on which crystallization can occur.

Surface tension is the energy or work required to extend the surface of a liquid.

The **surroundings** represent that portion of the universe with which a system interacts.

Synthesis gas is a mixture of $CO(g)$ and $H_2(g)$, generally made from coal or natural gas, that can be used as a fuel or in the synthesis of organic compounds.

A **system** is the portion of the universe selected for a thermodynamic study. (See also **open, closed,** and **isolated** systems.)

A **systematic error** is one that recurs regularly in a series of measurements because of an inherent error in the measuring system (e.g., through faulty calibration of a measuring device).

Temperature is a measure of the average molecular kinetic energy of a substance (translational kinetic energy for gases and liquids, and vibrational kinetic energy for solids).

Temporary hard water (See **hard water**.)

A **terminal atom** is any atom that is bonded to only one other atom in a molecule or polyatomic ion.

A **termolecular process** is an elementary process in a reaction mechanism in which three atoms or molecules must collide simultaneously.

A **ternary compound** is comprised of *three* elements.

The **tertiary structure** of a protein refers to its three-dimensional structure—for example, the twisting and folding of coils.

The **theoretical yield** is the quantity of product *calculated* to result from a chemical reaction. (See also **actual yield** and **percent yield**.)

A **theory** is a model or conceptual framework with which one is able to explain and make further predictions about natural phenomena.

Thermal energy is energy associated with random molecular motion.

The **thermite reaction** is an oxidation–reduction reaction that uses powdered aluminum metal as a reducing agent to reduce a metal oxide, such as Fe_2O_3, to the free metal.

The **thermodynamic equilibrium constant,** K_{eq}, is an equilibrium constant expression based on activities. In dilute solutions activities can be replaced by molarities and in ideal gases, by partial pressures in atm. The activities of pure solids and liquids are 1.

A **thio** compound is one in which an S atom replaces an O atom. For example, replacement of an O by S converts SO_4^{2-} to $S_2O_3^{2-}$ (thiosulfate ion).

The **third law of thermodynamics** states that the entropy of a pure perfect crystal is *zero* at the absolute zero of temperature, 0 K.

Titration is a procedure for carrying out a chemical reaction between two solutions by the controlled addition (from a buret) of one solution to the other. In a titration a means must be found, as by the use of an indicator, to locate the equivalence point.

A **titration curve** is a graph of solution pH versus volume of titrant. It outlines how pH changes during an acid–base titration, and it can be used to establish such features as the equivalence point of the titration.

The term **trans** is used to describe geometric isomers in which two groups are attached on opposite sides of a double bond in an organic molecule, or at opposite corners of a square in a square-planar, or at positions above and below the central plane of an octahedral complex. (See also **geometric isomerism**.)

Transition elements are those elements whose atoms feature the filling of a *d* or *f* subshell of an inner electronic shell. If the filling of an *f* subshell occurs, the elements are sometimes referred to as *inner* transition elements.

The **transition state** in a chemical reaction is an intermediate state between the reactants and products. (See also **activated complex** and **reaction profile**.)

Transmutation is the process in which one element is converted to another as a result of a change affecting the nuclei of atoms (such as in radioactive decay).

A **transuranium element** is one with an atomic number $Z > 92$.

Triglycerides are esters of glycerol (1,2,3-propanetriol) with long-chain monocarboxylic (fatty) acids.

In a **triple covalent bond** *three pairs* of electrons are shared between the bonded atoms. It is represented by a triple-dash sign ($\equiv$).

A **triple point** is a condition of temperature and pressure at which three phases of a substance (usually solid, liquid, and vapor) coexist at equilibrium.

The **troposphere** is the region of the atmosphere that extends from Earth's surface to a height of about 12 km.

Trouton's rule states that at their normal boiling points the entropies of vaporization of many liquids have about the same value: $87 \text{ J mol}^{-1} \text{ K}^{-1}$.

A **unimolecular process** is an elementary process in a reaction mechanism in which a single molecule, when sufficiently energetic, dissociates.

A **unit cell** is a small collection of atoms, ions, or molecules occupying positions in a parallelepiped. An entire crystal can be generated by straight-line displacements of the unit cell in the three perpendicular directions.

Unsaturated hydrocarbon molecules contain one or more carbon-to-carbon multiple bonds.

An **unsaturated solution** contains less solute than the solvent is capable of dissolving under the given conditions.

The **valence bond method** treats a covalent bond in terms of the overlap of pure or hybridized atomic orbitals. Electron probability (or electron charge density) is concentrated in the region of overlap.

Valence electrons are electrons in the electronic shell of highest principal quantum number, that is, electrons in the outermost shell.

The **valence-shell electron-pair repulsion (VSEPR) theory** is a theory used to predict probable shapes of molecules and polyatomic ions based on the mutual repulsions of electron pairs found in the valence shell of the central atom in the structure.

The **van der Waals equation** is an equation of state for nonideal gases. It includes correction terms to account for intermolecular forces of attraction and for the volume occupied by the gas molecules themselves.

van der Waals forces is a term used to describe, collectively, intermolecular forces of the London type and interactions between permanent dipoles.

Vaporization is the passage of molecules from the liquid to the gaseous state.

Vapor pressure is the pressure exerted by a vapor when it is in dynamic equilibrium with its liquid at a fixed temperature.

A **vapor-pressure curve** is a graph of vapor pressure as a function of temperature.

Viscosity refers to a liquid's resistance to flow. Its magnitude depends on intermolecular forces of attraction and, in some cases, on molecular sizes and shapes.

A **voltaic (galvanic) cell** is an electrochemical cell in which a *spontaneous* chemical reaction produces electricity.

Water gas is a mixture of $CO(g)$ and $H_2(g)$, together with some of the noncombustible gases CO_2 and N_2, produced by passing steam [$H_2O(g)$] over heated coke.

A **wave** is a disturbance that transmits energy through a medium.

The **wavelength** is the distance between successive crests or troughs of a wave motion.

Wave mechanics is a form of quantum theory based on the concepts of wave–particle duality, the Heisenberg uncertainty principle, and the treatment of electrons as matter waves. Mathematical solutions of the equations of wave mechanics are known as **wave functions** (ψ).

A **weak acid** is an acid that is only partially ionized in aqueous solution in a reversible reaction.

A **weak base** is a base that it only partially ionized in aqueous solution in a reversible reaction.

A **weak electrolyte** is a substance that is only partially ionized in solution in a reversible reaction.

Weight refers to the force exerted on an object when it is placed in a gravitational field (the "force of gravity"). The terms *weight* and *mass* are often used interchangeably.

Work is a form of energy transfer between a system and its surroundings that can be expressed as a force acting through a distance.

X-ray diffraction is a method of crystal structure determination based on the interaction of a crystal with X-rays.

A **zero-order** reaction proceeds at a rate that is *independent* of reactant concentrations. The sum of the concentration-term exponent(s) in the rate equation is equal to *zero*.

Zone refining is a purification process in which a rod of material is subjected to successive melting and freezing cycles. Impurities are swept by a moving molten zone to the end of the rod, which is cut off.

Answers to Practice Examples and Selected Exercises

Note: Your answers may differ slightly from those given here, depending on the number of steps used to solve a problem and whether any intermediate results were rounded off.

Chapter 1

Practice Examples. 1A. The New Delhi 41 °C temperature is higher than the predicted Phoenix high of 103 °F = 39.4 °C. **1B.** The antifreeze protects only to −22 °C, not to −15 °F = −26.1 °C. **2A.** 1.46 g/mL. **2B.** 2.76 g/mL. **3A.** 135 g. **3B.** 63.4 L. **4A.** 1.8 kg. **4B.** 0.856 g/mL. **5A.** 21.3. **5B.** 1.1×10^6. **6A.** 26.7. **6B.** 15.6.

Review Questions. 4. (a) 1.55×10^3 g; **(b)** 0.642 kg; **(c)** 289.6 cm; **(d)** 0.86 mm. **5. (a)** 127 mL; **(b)** 0.0158 L; **(c)** 0.981 L; **(d)** 2.65×10^6 cm³. **6. (a)** 173 cm; **(b)** 29 m; **(c)** 644 g; **(d)** 112 kg; **(e)** 7.00 L; **(f)** 3.52×10^3 mL. **7. (a)** 1.00×10^6 m²; **(b)** 1.00×10^4 cm²; **(c)** 2.59×10^6 m². **8.** 102 °C. **9.** 90.0 mL carbon disulfide. **10.** 0.958 g/mL. **11.** 13.6 g/mL **12. (a)** 502 g; **(b)** 20.6 kg; **(c)** 58.6 mL; **(d)** 21.5 L. **13.** 0.629 kg. **14.** 24 g. **15.** 7.92×10^3 g. **16.** 3.69 kg. **17. (a)** 8.950×10^3; **(b)** 1.0700×10^4; **(c)** 2.40×10^{-2}; **(d)** 4.7×10^{-3}; **(e)** 9.383×10^2; **(f)** 2.75482×10^5. **18. (a)** 0.0321; **(b)** 0.000508; **(c)** 0.001219; **(d)** 0.162. **19. (a)** two or three; **(b)** three; **(c)** two; **(d)** five; **(e)** four; **(f)** one to four; **(g)** three; **(h)** two to eight. **20. (a)** 3985; **(b)** 422.0; **(c)** 1.860×10^5; **(d)** 3.390×10^4; **(e)** 6.321×10^4; **(f)** 5.047×10^{-4}. **21. (a)** 9.3×10^{-4}; **(b)** 2.2×10^{-1}; **(c)** 5.4×10^{-1}; **(d)** 3.058×10^1. **22. (a)** 1.0×10^5; **(b)** 1.0; **(c)** 4.46×10^{-1}; **(d)** 5.79×10^{-1}. **23.** 1.04×10^4 g. **24.** 1.58×10^3 g.

Exercises. 25. There is no point at which the law is ever verified with certainty. **27.** Although cause-and-effect relationships may be difficult to establish at times ("God is subtle"), they nevertheless do exist ("He is not malicious"). **29. (a)** physical; **(b)** chemical; **(c)** chemical; **(d)** physical. **31. (a)** heterogeneous; **(b)** homogeneous; **(c)** substance; **(d)** heterogeneous. **33. (a)** physical; **(b)** chemical; **(c)** chemical; **(d)** physical. **35. (a)** 3.4×10^4 cm/s; **(b)** 6.378×10^3 km; **(c)** 4.7×10^5. **37. (a)** exact; **(b)** measured; **(c)** measured. **39. (a)** 2.44×10^4; **(b)** 1.5×10^3; **(c)** 40.0; **(d)** 2.131×10^3. **41. (a)** 115.76 mi./h; **(b)** 2.80 mi./lb. **43.** 2172 μg. **45.** 1.5 m. **47. (a)** 10. s; **(b)** 9.8 m/s; **(c)** 2.5 min. **49.** 2.47 acres. **51.** 2.25×10^3 g/cm², 2.25×10^4 kg/m². **53.** 47.8 °C, −8.3 °C. **55.** No, 240 °F = 116 °C is above the range of the thermometer. **57.** 0.790 g/mL. **59.** iron bar < aluminum foil < water. **61.** 0.04 g/shot. **63.** 12% A, 28% B, 47% C, 11% D, 3% F. **65.** 8.60×10^2 g sucrose.

Advanced Exercises. 68. 0.9 L. **71.** 38.8 m. **74. (a)** 1.54×10^{11} ft³; **(b)** 4.36×10^9 m³; **(c)** 1.15×10^{12} gal. **78.** 11 g/mL.

Feature Problems. A. Pre-1982 pennies weigh more than 3.00 g. The composition changed during 1982, and post-1982 pennies weigh less than 2.60 g. **B.** The crown immersed in water displaces a volume of water equal to its volume. The Au-Ag alloy is less dense than pure Au and displaces a larger volume of water than does pure Au. **C.** Each ball and its contents have a slightly different density. As temperature changes, so does the density of the liquid in the cylinder; different balls float or sink, depending on the temperature of the liquid. **D.** The air inside the canoe ensures that the entire vessel (canoe plus air) is less dense than water.

Chapter 2

Practice Examples. 1A. 0.529 g. **1B.** 6.85 g. **2A.** 0.301 g. **2B.** 1.20 g magnesium, 0.80 g oxygen. **3A.** $^{108}_{74}$ Ag. **3B.** 16

protons, 19 neutrons, 18 electrons. **4A.** 15.0001 u. **4B.** 16.830884. **5A.** boron-11. **5B.** 7.5% lithium-6 and 92.5% lithium-7. **6A.** Li^+, S^{2-}, Ra^{2+}, F^-, I^-, Al^{3+}. **6B.** main group metals: Na, Mg, Al; main group nonmetals: S, Kr (noble gas), B, I, H; main group metalloids: As, Si; transition metals: U, Re. **7A.** 3.05×10^{21} Au atoms. **7B.** 1.20×10^{21} ^{206}Pb atoms. **8A.** 248 g Cu. **8B.** 3.40×10^{24} He atoms. **9A.** 3.46×10^{21}. **9B.** 62.5%.

Review Questions. 4. 0.268 g. **5.** 4.720 g. **6.** The product of combustion is sulfur dioxide gas. **7.** 39.34% Na. **8. (a)** 0.066 g oxygen/0.166 g magnesium oxide; **(b)** 0.66 g oxygen/1 g magnesium; **(c)** 60.2% Mg. **9. (a)** yes, 27.3% C; **(b)** 72.7% O. **10.** No, we need to know the proportions in which the elements react. **11.** minimum information: means of establishing numbers of p, n, and e. **12. (a)** $^{40}_{18}Ar < ^{39}_{19}K < ^{58}_{27}Co < ^{59}_{29}Cu < ^{120}_{48}Cd < ^{112}_{50}Sn < ^{122}_{52}Te$; **(b)** $^{39}_{19}K < ^{40}_{18}Ar < ^{59}_{29}Cu < ^{58}_{27}Co < ^{112}_{50}Sn < ^{122}_{52}Te < ^{120}_{48}Cd$; **(c)** $^{39}_{19}K < ^{40}_{18}Ar < ^{58}_{27}Co < ^{59}_{29}Cu < ^{112}_{50}Sn < ^{120}_{48}Cd < ^{122}_{52}Te$. **13. (a)** $^{60}_{27}Co$; **(b)** $^{32}_{15}P$; **(c)** $^{131}_{53}I$; **(d)** $^{35}_{16}S$. **14.** 59%. **15.** ^{193}Ir. **16.** argon. **17. (a)** 2.914071; **(b)** 2.165216; **(c)** 18.50146. **18.** 80.916 u. **19.** 39.948 u. **20. (a)** In; **(b)** like S: O, Se; unlike S: Na, Ba; **(c)** Cs; **(d)** I; **(e)** Xe; **(f)** 5A; **(g)** 2A. **21.** noble gas = 118, alkali metal = 119. **22. (a)** 7.65×10^{24} Ca atoms; **(b)** 2.17×10^{21} Ne atoms; **(c)** 1.1×10^{12} Pu atoms. **23. (a)** 362 mol; **(b)** 646 g; **(c)** 20.1 mg; **(d)** 9.49×10^{24} Fe atoms. **24.** 25.0 g N. **25.** 8.75×10^{18} atoms ^{204}Pb. **26.** 152 g.

Exercises. 27. No: Iron and oxygen from air combine to form iron oxide, and carbon dioxide escapes from the burning match. **29.** 21.625 g before and after reaction. **31.** Both samples have 39.4% Na. **33.** 0.422 g. **35.** cathode rays have the same properties no matter how they are produced. **37.** All values given are multiples of e. **39. (a)** The ratio is about 1:1800; **(b)** 1.044×10^{-5} g/C for $^1_1H^+$. **41. (a)** 46p, 62n, 46e; **(b)** 8.9919908. **43.** 106.906 u. **45. (a)** $^{24}_{12}Mg^{2+}$; **(b)** $^{47}_{24}Cr$; **(c)** $^{226}_{90}Th$. **47.** Unlikely; masses of p, n, and e are not integral. **49.** 24.31 u. **51.** 108.9 u. **53.** 40.962 u. **55. (b)** 73, two significant-figure isotopic masses. **57. (a)** 1.954 mol; **(b)** 3.916×10^{27} Fe atoms; **(c)** 0.18 ng

Ag. **59.** 2.00×10^{23} Ag atoms. **61. (a)** 1.4×10^{-6} mol Pb/L; **(b)** 8.4×10^{14} Pb atoms/mL.

Advanced Exercises. 65. 3×10^{15} g/cm³. **70.** ^{210}Po.

Feature Problems. A. There was only a 0.0409% mass difference before and after heating. **B.** All measured quantities are multiples of 4.915. **C.** generally higher; 16.000... is larger than 15.9994, the current value. **D.** 3.09. **E.** 2.95×10^5 km³.

Chapter 3

Practice Examples. 1A. 1.5×10^{22} ions. **1B.** 4.0×10^1 g $MgCl_2$. **2A.** 3.69×10^{21} Au atoms. **2B.** the vapor will be detectable. **3A.** 17.1 g C. **3B.** 132.0 mL $C_2HBrClF_3$. **4A.** 40.001% C, 6.714% H, 53.284% O. **4B.** 23.919% C, 2.208% H, 13.95% N, 18.50% P, 41.420% O. **5A.** $C_6H_{10}O_3$ and $C_{12}H_{20}O_6$. **5B.** $C_3H_7O_3$ and $C_6H_{14}O_6$. **6A.** $C_7H_{14}O_2$. **6B.** C_4H_4S. **7A.** 0 for S; +6 for Cr; +1 for Cl; $-\frac{1}{2}$ for O. **7B.** +2 for S; +1 for Hg; +7 for Mn; 0 for C. **8A.** Li_2O; SnF_2; Li_3N. **8B.** Al_2S_3; Mg_3N_2; V_2O_3. **9A.** cesium iodide; calcium fluoride; iron(II) oxide; chromium(III) chloride. **9B.** calcium hydride; copper(I) chloride; silver sulfide; mercury(I) chloride. **10A.** sulfur hexafluoride; nitrous acid; calcium hydrogen carbonate; iron(II) sulfate. **10B.** ammonium nitrate; phosphorus trichloride; hypobromous acid; silver perchlorate; iron(III) sulfate. **11A.** BF_3; $K_2Cr_2O_7$; H_2SO_4; $CaCl_2$. **11B.** $Al(NO_3)_3$; P_4O_{10}; $Cr(OH)_3$; HIO_3.

Review Questions. 4. (a) mass of one (average) atom; **(b)** mass of one (average) molecule; **(c)** mass of one mole of oxygen molecules or atoms. **5. (a)** 20 atoms; **(b)** 4.91×10^{21} atoms; **(c)** 2.195×10^{25} F atoms. **6. (a)** 675 g; **(b)** 168 g; **(c)** 4.64×10^3 g; **(d)** 431 g. **7. (a)** 0.134 mol; **(b)** 1.80 mol; **(c)** 70.7 mol; **(d)** 51.4 mol. **8. (a)** 149.2 u; **(b)** 11 mol; **(c)** 60.055 g; **(d)** 2.73×10^{25} C atoms. **9.** 15.585%. **10.** 36.18%. **11.** 40.53%. **12. (a)** 64.07%; **(b)** 45.50%. **13.** SO_2. **14.** Cr_2O_3. **15.** $C_4H_{10}O_3$. **16.** $NaC_5H_8NO_4$. **17.** $C_8H_6O_4$. **18. (a)** 75.71% C, 8.795% H, 15.50% O; **(b)** $C_{13}H_{18}O_2$. **19.** CrO_3. **20. (a)** Pb^{2+}; **(b)** cobalt(III) ion; **(c)** barium ion; **(d)** Cr^{2+};

(e) IO_4^-; **(f)** chlorite ion; **(g)** Au^{3+}; **(h)** hydrogen sulfite ion; **(i)** HCO_3^-; **(j)** CN^-. **21. (a)** potassium bromide; **(b)** strontium chloride; **(c)** chlorine trifluoride; **(d)** dinitrogen tetroxide; **(e)** phosphorus pentachloride. **22. (a)** potassium cyanide; **(b)** hypochlorous acid; **(c)** ammonium sulfate **(d)** potassium iodate. **23. (a)** Zn = 0; **(b)** S = -2; **(c)** N = +4; **(d)** N = +3; **(e)** V = +4; **(f)** P = +5. **24. (a)** $MgBr_2$; **(b)** BaO; **(c)** $Hg(C_2H_3O_2)_2$; **(d)** $Fe_2(C_2O_4)_3$; **(e)** $Sr(ClO_4)_2$; **(f)** $KHSO_4$; **(g)** NCl_3; **(h)** BrF_5. **25. (a)** chlorous acid; **(b)** sulfurous acid; **(c)** hydroselenic acid; **(d)** nitrous acid. **26. (a)** HI; **(b)** HNO_3; **(c)** H_3PO_4; **(d)** H_2SO_4.

Exercises. 27. SO_2. **29. (a)** 7.74 g; **(b)** 1.90×10^3 g; **(c)** 84.1 g. **31. (a)** 4.58×10^{-6} mol; **(b)** 2.21×10^{19} S atoms. **33.** only (2) is true. **35. (a)** 8 atoms; **(b)** 3 F atom/2 C atom; **(c)** 1.402 g Br/g F; **(d)** Br; **(e)** 3.45 g. **37.** 75.998% C, 12.755% H, 11.248% O. **39. (a)** 49.765%; **(b)** 5.0300%; **(c)** 33.659%; **(d)** 6.6635%. **41.** $CrO_3 < CrO_2 < Cr_2O_3 < CrO$. **43.** $C_{10}H_8$. **45. (a)** $C_{19}H_{16}O_4$; **(b)** $C_4H_8Cl_2S$. **47.** $C_{16}H_{10}N_2O_2$. **49.** 31 g/mol. **51. (a)** 90.51% C, 9.491% H; **(b)** C_4H_5; **(c)** C_8H_{10}. **53.** CH_4N. **55.** $C_{10}H_8$. **57. (a)** C = -4; **(b)** S = $+4$; **(c)** O = -1; **(d)** C = 0; **(e)** Fe = $+6$. **59.** Cr_2O_3; CrO_2; CrO_3. **61. (a)** strontium oxide; **(b)** zinc sulfide; **(c)** potassium chromate; **(d)** cesium sulfate; **(e)** chromium(III) oxide; **(f)** iron(III) sulfate; **(g)** magnesium hydrogen carbonate; **(h)** ammonium hydrogen phosphate; **(i)** calcium hydrogen sulfite; **(j)** copper(II) hydroxide; **(k)** nitric acid; **(l)** potassium perchlorate; **(m)** bromic acid; **(n)** phosphorous acid. **63. (a)** carbon disulfide; **(b)** silicon tetrafluoride; **(c)** chlorine pentafluoride; **(d)** dinitrogen pentoxide; **(e)** sulfur hexafluoride; **(f)** diiodine hexachloride. **65. (a)** $Al_2(SO_4)_3$; **(b)** $(NH_4)_2Cr_2O_7$; **(c)** SiF_4; **(d)** Fe_2O_3; **(e)** C_3S_2; **(f)** $Co(NO_3)_2$; **(g)** $Sr(NO_2)_2$; **(h)** HBr; **(i)** HIO_3; **(j)** PCl_2F_3. **67. (a)** $TiCl_4$; **(b)** $Fe_2(SO_4)_3$; **(c)** Cl_2O_7; **(d)** $S_2O_8^{2-}$. **69.** $MgCl_2 \cdot 6 H_2O$. **71.** 15 g.

Advanced Exercises. 73. 1.24×10^{23}. **75.** 894 g/mol. **79.** 4.802 g CO_2 and 2.247 g H_2O. **83.** CH_4. **86.** 26.9 u. **90.** $ZnSO_4 \cdot 7 H_2O$.

Feature Problems. A. 6.023×10^{23} electrons/mol. **B.** 1:2:4; no oxide with 26.40% N. **C. 1.** 5% N, 4.4% P, 4.2% K; **2. (a)** 60.6%; **(b)** 53.7%.

Chapter 4

Practice Examples. 1A. $C_2H_4 + 3\,O_2 \rightarrow 2\,CO_2 + 2\,H_2O$. **1B.** $C_4H_4S + 6\,O_2 \rightarrow 4\,CO_2 + 2\,H_2O + SO_2$. **2A.** $2\,C_3H_8O_3 + 7\,O_2 \rightarrow 6\,CO_2 + 8\,H_2O$. **2B.** $2\,C_7H_6O_2S + 17\,O_2 \rightarrow 14\,CO_2 + 6\,H_2O + 2\,SO_2$. **3A.** 2.64 mol. **3B.** 8.63 mol. **4A.** 5.29 g. **4B.** 126 g. **5A.** 0.126 g. **5B.** 3.50 g. **6A.** 3.34 cm³. **6B.** 0.78 g. **7A.** 0.4 mg. **7B.** 0.15 g. **8A.** 0.307 M. **8B.** 0.524 M. **9A.** 115 g. **9B.** 50.9 g. **10A.** 0.0675 M. **10B.** 0.122 M. **11A.** 18.1 mL. **11B.** 40.2 mL. **12A.** 936 g. **12B.** 1.86 kg. **13A.** 3.8 g P_4. **13B.** 57 g O_2. **14A. (a)** 30.03 g; **(b)** 25.7 g; **(c)** 85.6%. **14B.** 93.7%. **15A.** 38.0 g. **15B.** 69.0 g. **16A.** 2.47×10^3 g. **16B.** 0.0734 g.

Review Questions. 4. (a) $Na_2SO_4(s) + 2\,C(s) \rightarrow Na_2S(s) + 4\,CO(g)$; **(b)** $4\,HCl(g) + O_2(g) \rightarrow 2\,H_2O(l) + 2\,Cl_2(g)$; **(c)** $PCl_5(l) + 4\,H_2O(l) \rightarrow H_3PO_4(aq) + 5\,HCl(aq)$; **(d)** $3\,PbO(s) + 2\,NH_3(g) \rightarrow 3\,Pb(s) + N_2(g) + 3\,H_2O(l)$; **(e)** $Mg_3N_2(s) + 6\,H_2O(l) \rightarrow 3\,Mg(OH)_2(s) + 2\,NH_3(g)$. **5. (a)** $2\,Mg(s) + O_2(g) \rightarrow 2\,MgO(s)$; **(b)** $2\,NO(g) + O_2(g) \rightarrow 2\,NO_2(g)$; **(c)** $2\,C_2H_6(g) + 7\,O_2(g) \rightarrow 4\,CO_2(g) + 6\,H_2O(l)$; **(d)** $Ag_2SO_4(aq) + BaI_2(aq) \rightarrow BaSO_4(s) + 2\,AgI(s)$. **6. (a)** $C_7H_{16}(l) + 11\,O_2(g) \rightarrow 7\,CO_2(g) + 8\,H_2O(l)$; **(b)** $C_4H_9OH(l) + 6\,O_2(g) \rightarrow 4\,CO_2(g) + 5\,H_2O(l)$; **(c)** $2\,HI(aq) + Na_2CO_3(aq) \rightarrow 2\,NaI(aq) + H_2O(l) + CO_2(g)$; **(d)** $3\,NaOH(aq) + FeCl_3(aq) \rightarrow Fe(OH)_3(s) + 3\,NaCl(aq)$. **7. (a)** and **(b)** O(g) is not produced in reactions. **(c)** KClO is not a product. **8.** (3) and (4) are true. **9.** 4.84 mol. **10.** 17.0 g. **11.** 35.8 g Cl_2, 10.4 g P_4. **12. (a)** 5.32 mol; **(b)** 323.1 g; **(c)** 6.35×10^{18} O_2 molecules. **13. (a)** 0.408 M; **(b)** 0.154 M; **(c)** 1.53 M; **(d)** 0.676 M. **14. (a)** 319 mol; **(b)** 20.0 g; **(c)** 16.4 mg. **15.** 373 g KCl in 5.00 L. **16.** 163 mL. **17.** 0.242 M. **18.** 14.1 g. **19.** 14.9 mL. **20.** 1.52 mol. **21.** Some Cu does not react; HNO_3 is limiting. **22. (a)** 1.80 mol CCl_2F_2; **(b)** 1.55 mol CCl_2F_2; **(c)** 86.1%. **23. (a)** 82.01 g C_6H_{10}; **(b)** 78.0%; **(c)** 156 g. **24.** 68.0% $CaCO_3$.

Exercises. 25. (a) $Cr_2O_3(s) + 2\,Al(s) \xrightarrow{\Delta} Al_2O_3(s) + 2\,Cr(l)$; **(b)** $CaC_2(s) +$ $2\,H_2O(l) \rightarrow Ca(OH)_2(s) + C_2H_2(g)$; **(c)** $3\,H_2(g) + Fe_2O_3(s) \xrightarrow{\Delta} 2\,Fe(l) + 3\,H_2O(g)$; **(d)** $NCl_3(g) + 3\,H_2O(l) \rightarrow NH_3(g) + 3\,HOCl(aq)$. **27. (a)** $2\,C_4H_{10} + 13\,O_2(g) \rightarrow 8\,CO_2(g) + 10\,H_2O(l)$; **(b)** $2\,C_3H_7OH + 9\,O_2(g) \rightarrow 6\,CO_2(g) + 8\,H_2O(l)$; **(c)** $HC_3H_5O_3(s) + 3\,O_2(g) \rightarrow 3\,CO_2(g) + 3\,H_2O(l)$. **29. (a)** $NH_4NO_3(s) \xrightarrow{\Delta} N_2O(g) + 2\,H_2O(g)$; **(b)** $Na_2CO_3(aq) + 2\,HCl(aq) \rightarrow 2\,NaCl(aq) + H_2O(l) + CO_2(g)$; **(c)** $2\,CH_4(g) + 2\,NH_3(g) + 3\,O_2(g) \rightarrow 2\,HCN(g) + 6\,H_2O(g)$. **31. (a)** 0.401 mol; **(b)** 128 g; **(c)** 44.0 g. **33.** 96.0 g. **35.** 79.7%. **37.** 1.03 g. **39.** NH_4NO_3. **41.** 474 g. **43.** 74.4 mol. **45. (a)** 1.753 M; **(b)** 0.320 M; **(c)** 0.206 M. **47. (a)** 4.73 g; **(b)** 44.1 mL. **49.** the 46% solution (1.63 M). **51.** 0.0820 M. **53.** a 100.0-mL flask and a 5.00-mL pipet, for example. **55. (a)** 0.177 g; **(b)** 0.562 g. **57. (a)** 0.0693 mol; **(b)** 2.91 M. **59.** 59.4 mL. **61.** 0.624 g. **63.** 0.2649 M. **65.** 24.0 g. **67.** 143 g. **69.** 10.5 g NH_3, 10.1 g excess $Ca(OH)_2$. **71. (a)** 27.7 g; **(b)** 17.1 g; **(c)** 61.7%. **73.** 76 g. **75.** In a synthesis reaction, yield may be sacrificed to cut costs. A *quantitative* analysis requires that no product be lost (100% yield).

Advanced Exercises. 77. (a) $CaCO_3(s) \xrightarrow{\Delta} CaO(s) + CO_2(g)$; **(b)** $2\,ZnS(s) + 3\,O_2 \xrightarrow{\Delta} 2\,ZnO(s) + 2\,SO_2(g)$; **(c)** $C_3H_8(g) + 3\,H_2O(g) \rightarrow 3\,CO(g) + 7\,H_2(g)$. **80.** $3\,FeS(s) + 5\,O_2(g) \xrightarrow{heat} Fe_3O_4(s) + 3\,SO_2(g)$. **83.** 0.2 cm². **85.** 118 mL. **87.** 18.2 g. **89.** 20.% Al. **90.** 24%. **91.** a mixture. **94.** $C_3H_4O_4 + 2\,O_2(g) \rightarrow 3\,CO_2(g) + 2\,H_2O(l)$.

Feature Problems. A. 360 fish. **B. 1.** The graph peaks at 2.50 g $Pb(NO_3)_2$; **2.** 3.476 g PbI_2 maximum. **C. 1.** 20 r + 20 b + 30 g → 1 necklace; **2.** 163 necklaces.

Chapter 5

Practice Examples. 1A. 1.07 M Na^+, 0.358 M PO_4^{3-}. **1B.** 0.540 M. **2A. (a)** $Al^{3+}(aq) + 3\,OH^-(aq) \rightarrow Al(OH)_3(s)$; **(b)** No reaction; **(c)** $Pb^{2+}(aq) + 2\,I^-(aq) \rightarrow PbI_2(s)$. **2B. (a)** $Al^{3+}(aq) + PO_4^{3-}(aq) \rightarrow AlPO_4(s)$; **(b)** $Ba^{2+}(aq) + SO_4^{2-}(aq) \rightarrow BaSO_4(s)$; **(c)** $Pb^{2+}(aq) + CO_3^{2-}(aq) \rightarrow PbCO_3(s)$. **3A.** $NH_3(aq) + HC_3H_5O_2(aq) \rightarrow NH_4^+(aq) + C_3H_5O_2^-(aq)$ (ammonium propionate). **3B.** $CaCO_3(s) + 2\,HC_2H_3O_2(aq) \rightarrow$ $CO_2(g) + H_2O(l) + Ca^{2+}(aq) + 2\,C_2H_3O_2^-(aq)$. **4A. (a)** no; **(b)** yes. **4B.** VO^{2+} is oxidized; MnO_4^- is reduced. **5A.** ox: $Al \rightarrow Al^{3+} + 3\,e^-$; red.: $2\,H^+ + 2\,e^- \rightarrow H_2$; net: $2\,Al(s) + 6\,H^+(aq) \rightarrow 2\,Al^{3+}(aq) + 3\,H_2(g)$. **5B.** ox.: $2\,Br^- \rightarrow Br_2 + 2\,e^-$; red.: $Cl_2 + 2\,e^- \rightarrow 2\,Cl^-$; net: $2\,Br^-(aq) + Cl_2(g) \rightarrow Br_2(l) + 2\,Cl^-(aq)$. **6A.** $MnO_4^-(aq) + 8\,H^+ + 5\,Fe^{2+}(aq) \rightarrow Mn^{2+}(aq) + 4\,H_2O + 5\,Fe^{3+}(aq)$. **6B.** $3\,UO^{2+}(aq) + Cr_2O_7^{2-}(aq) + 8\,H^+(aq) \rightarrow 3\,UO_2^{2+}(aq) + 2\,Cr^{3+}(aq) + 4\,H_2O$. **7A.** $S(s) + 2\,OH^- + 2\,OCl^-(aq) \rightarrow SO_3^{2-}(aq) + H_2O + 2\,Cl^-(aq)$. **7B.** $2\,MnO_4^-(aq) + 3\,SO_3^{2-}(aq) + H_2O \rightarrow 2\,MnO_2(s) + 3\,SO_4^{2-}(aq) + 2\,OH^-(aq)$. **8A.** $H_2(g)$ is a reducing agent. **8B.** Au(s) is the reducing agent, $O_2(g)$ is the oxidizing agent. **9A.** 0.102 M. **9B.** 0.0895 M. **10A.** 0.03129 M. **10B.** 65.4% Fe.

Review Questions. 4. (a) 0.10 M NaCl; **(b)** 0.10 M C_2H_5OH. **5. (a)** salt; **(b)** strong base; **(c)** salt; **(d)** weak acid; **(e)** strong acid; **(f)** weak acid; **(g)** weak base; **(h)** salt; **(i)** strong base. **6.** 0.080 M $Al_2(SO_4)_3$. **7. (a)** 0.238 M K^+; **(b)** 0.334 M NO_3^-; **(c)** 0.17 M Al^{3+}; **(d)** 0.627 M Na^+. **8.** 200.0 mL of 0.035 M NaCl. **9.** 3.04×10^{-3} M. **10.** 0.126 M K^+; 0.148 M Mg^{2+}; 0.422 M Cl^-. **11.** 4.3×10^2 mg. **12.** $BaSO_4(s)$ and $Al(OH)_3(s)$. **13.** Ca(s) and $KHSO_3(s)$. **14. (a)** $Pb^{2+}(aq) + 2\,Br^-(aq) \rightarrow PbBr_2(s)$; **(b)** No reaction occurs; **(c)** $Fe^{3+}(aq) + 3\,OH^-(aq) \rightarrow Fe(OH)_3(s)$; **(d)** $Ca^{2+}(aq) + CO_3^{2-}(aq) \rightarrow CaCO_3(s)$; **(e)** $Ba^{2+}(aq) + SO_4^{2-}(aq) \rightarrow BaSO_4(s)$; **(f)** no reaction occurs. **15. (a)** $OH^-(aq) + HC_2H_3O_2(aq) \rightarrow H_2O(l) + C_2H_3O_2^-(aq)$; **(b)** No reaction occurs; **(c)** $FeS(s) + 2\,H^+(aq) \rightarrow H_2S(g) + Fe^{2+}(aq)$; **(d)** $HCO_3^-(aq) + H^+(aq) \rightarrow H_2O(l) + CO_2(g)$; **(e)** $Mg(s) + 2\,H^+(aq) \rightarrow Mg^{2+}(aq) + H_2(g)$; **(f)** No reaction occurs. **16.** 13.3 mL. **17.** 0.08683 M. **18.** acidic. **19. (a)** not a redox reaction; **(b)** O.S. of Cl: Cl_2, 0; Cl^-, -1; O.S. of Br: Br^-, -1; Br_2, 0; **(c)** O.S. of Ag: Ag, 0; Ag^+, $+1$; O.S. of N: NO_3^-, $+5$; NO_2, $+4$; **(d)** not a redox reaction. **20. (a)** NO is oxidizing agent, H_2 is reducing agent; **(b)** Cu is reducing agent; NO_3^- is oxidizing agent; **(c)** Cl_2 is both oxidizing and reducing agent. **21.** reductions: **(a)** $2\,SO_3^{2-}(aq) + 6\,H^+(aq)$

$+ 4 e^- \rightarrow S_2O_3^{2-}(aq) + 3 H_2O$; **(b)** 2 $NO_3^-(aq) + 10 H^+(aq) + 8 e^- \rightarrow N_2O(g) + 5 H_2O(l)$; oxidations: **(c)** $I^-(aq) + 3 H_2O(l) \rightarrow IO_3^-(aq) + 6 H^+(aq) + 6 e^-$; **(d)** $Al(s) + 4 OH^-(aq) \rightarrow Al(OH)_4^-(aq) + 3 e^-$. **22. (a)** $3 Zn(s) + 2 NO_3^-(aq) + 8 H^+(aq) \rightarrow 3 Zn^{2+}(aq) + 2 NO(g) + 4 H_2O$; **(b)** 4 $Zn(s) + NO_3^-(aq) + 10 H^+(aq) \rightarrow 4 Zn^{2+}(aq) + NH_4^+(aq) + 3 H_2O(l)$; **(c)** $Cr_2O_7^{2-}(aq) + 14 H^+(aq) + 6 Fe^{2+}(aq) \rightarrow 6 Fe^{3+}(aq) + 2 Cr^{3+}(aq) + 7 H_2O$; **(d)** $2 MnO_4^-(aq) + 6 H^+(aq) + 5 H_2O_2(aq) \rightarrow 2 Mn^{2+}(aq) + 8 H_2O + 5 O_2(g)$. **23. (a)** $2 MnO_2(s) + ClO_3^-(aq) + 2 OH^-(aq) \rightarrow 2 MnO_4^-(aq) + Cl^-(aq) + H_2O$; **(b)** $2 Fe(OH)_3(s) + 3 OCl^-(aq) + 4 OH^-(aq) \rightarrow 2 FeO_4^{2-}(aq) + 3 Cl^-(aq) + 5 H_2O$; **(c)** $6 ClO_2 + 6 OH^-(aq) \rightarrow 5 ClO_3^-(aq) + Cl^-(aq) + 3 H_2O$. **24.** 0.03421 M.

Exercises. 25. (a) weak electrolyte; **(b)** strong electrolyte; **(c)** strong electrolyte; **(d)** nonelectrolyte; **(e)** strong electrolyte. **27. (a)** 8.73×10^{-4} M; **(b)** 6.55×10^{-3} M; **(c)** 2.57×10^{-3} M. **29. (c)** 14.7 mg Na^+/mL is highest. **31.** 0.732 M. **33. (a)** no reaction; **(b)** $Cu^{2+}(aq) + CO_3^{2-}(aq) \rightarrow CuCO_3(s)$; **(c)** $3 Cu^{2+}(aq) + 2 PO_4^{3-}(aq) \rightarrow Cu_3(PO_4)_2(s)$. **35. (a)** $K_2SO_4(aq)$; **(b)** $H_2O(l)$; **(c)** $KCl(aq)$. **37. (a)** $Sr^{2+}(aq) + SO_4^{2-}(aq) \rightarrow SrSO_4(s)$; **(b)** $Mg^{2+}(aq) + 2 OH^-(aq) \rightarrow Mg(OH)_2(s)$; **(c)** $Ba^{2+}(aq) + SO_4^{2-}(aq) \rightarrow BaSO_4(s)$; $KCl(aq)$ left. **39. (a)** $NaHCO_3(s) + H^+(aq) \rightarrow Na^+(aq) + H_2O(l) + CO_2(g)$; **(b)** $CaCO_3(s) + 2 H^+(aq) \rightarrow Ca^{2+}(aq) + H_2O(l) + CO_2(g)$; **(c)** $Mg(OH)_2(s) + 2 H^+(aq) \rightarrow Mg^{2+}(aq) + 2 H_2O(l)$; **(d)** $Mg(OH)_2(s) + 2 H^+(aq) \rightarrow Mg^{2+}(aq) + 2 H_2O(l)$; $Al(OH)_3(s) + 3 H^+(aq) \rightarrow Al^{3+}(aq) + 3 H_2O(l)$; **(e)** $NaAl(OH)_2CO_3(s) + 4 H^+(aq) \rightarrow Al^{3+}(aq) + Na^+(aq) + 3 H_2O(l) + CO_2(g)$. **41.** salt: $NaHSO_4(aq) \rightarrow Na^+(aq) + HSO_4^-(aq)$; acid: $HSO_4^-(aq) + OH^-(aq) \rightarrow H_2O(l) + SO_4^{2-}(aq)$. **43. (a)** There are two reductions and no oxidation; **(b)** There are two oxidations and no reduction. **45. (a)** $(NH_4)_2Cr_2O_7(s) \rightarrow N_2(g) + Cr_2O_3(s) + 4 H_2O$; **(b)** $2 Pb(NO_3)_2(s) \rightarrow 2 PbO(s) + 4 NO_2(g) + O_2(g)$. **47. (a)** $10 I^-(aq) + 2 MnO_4^-(aq) + 16 H^+(aq) \rightarrow 5 I_2(s) + 2 Mn^{2+}(aq) + 8 H_2O(l)$; **(b)** $3 N_2H_4(l) + 2 BrO_3^-(aq) \rightarrow 3 N_2(g) + 2 Br^-(aq) + 6 H_2O(l)$; **(c)** $Fe^{2+}(aq) +$

$VO_4^{3-}(aq) + 6 H^+(aq) \rightarrow Fe^{3+}(aq) + VO^{2+}(aq) + 3 H_2O(l)$; **(d)** $3 UO^{2+}(aq) + 2 NO_3^-(aq) + 2 H^+(aq) \rightarrow 3 UO_2^{2+}(aq) + 2 NO(g) + H_2O(l)$. **49. (a)** $3 CN^-(aq) + 2 MnO_4^-(aq) + H_2O(l) \rightarrow 3 CNO^-(aq) + 2 MnO_2(s) + 2 OH^-(aq)$; **(b)** $4 [Fe(CN)_6]^{3-}(aq) + N_2H_4(l) + 4 OH^-(aq) \rightarrow 4 [Fe(CN)_6]^{4-}(aq) + N_2(g) + 4 H_2O(l)$; **(c)** $4 Fe(OH)_2(s) + O_2(g) + 2 H_2O(l) \rightarrow 4 Fe(OH)_3(s)$; **(d)** $3 C_2H_5OH(aq) + 4 MnO_4^-(aq) \rightarrow 3 C_2H_3O_2^-(aq) + 4 MnO_2(s) + OH^-(aq) + 4 H_2O(l)$. **51. (a)** $S_2O_3^{2-}(aq) + 5 H_2O(l) + 4 Cl_2(g) \rightarrow 2 SO_4^{2-}(aq) + 8 Cl^-(aq) + 10 H^+(aq)$; **(b)** $Cr_2O_7^{2-}(aq) + 14 H^+(aq) + 3 Sn^{2+}(aq) \rightarrow 3 Sn^{4+}(aq) + 2 Cr^{3+}(aq) + 7 H_2O$; **(c)** $S_8(s) + 12 OH^-(aq) \rightarrow 4 S^{2-}(aq) + 2 S_2O_3^{2-}(aq) + 6 H_2O(l)$; **(d)** $As_2S_3(s) + 12 OH^-(aq) + 14 H_2O_2(aq) \rightarrow 2 AsO_4^{3-}(aq) + 3 SO_4^{2-}(aq) + 20 H_2O$. **53. (a)** $SO_3^{2-}(aq)$ is the reducing agent; $MnO_4^-(aq)$ is the oxidizing agent; **(b)** $H_2(g)$ is the reducing agent; $NO_2(g)$ is the oxidizing agent; **(c)** $[Fe(CN)_6]^{4-}(aq)$ is the reducing agent; $H_2O_2(aq)$ is the oxidizing agent. **55.** 0.1230 M. **57.** 35.46 mL. **59.** 0.077 M. **61.** acidic. **63.** 34 mL base. **65.** 0.01968 M. **67.** 53.23%. **69.** 3.70 g.

Advanced Exercises. 72. $3 Ca^{2+}(aq) + 2 HPO_4^{2-}(aq) \rightarrow Ca_3(PO_4)_2(s) + 2 H^+(aq)$. **76.** 0.0874 L. **81.** 32.1%. **84.** 24.63 mL.

Feature Problems. A. (a) $2 Na(s) + 2 H_2O(l) \rightarrow 2 NaOH(aq) + H_2(g)$; **(b)** $Fe^{3+}(aq) + 3 OH^-(aq) \rightarrow Fe(OH)_3(s)$; **(c)** $Fe(OH)_3(s) + 3 H^+(aq) \rightarrow Fe^{3+}(aq) + 3 H_2O(l)$. **B.** $x = 1.07$. **C.** 91.0% MnO_2.

Chapter 6

Practice Examples. 1A. 760 mmHg. **1B.** 0.790 g/cm^3. **2A.** 756.0 mmHg. **2B.** 93 mm. **3A.** 4.40 L. **3B.** 317 mmHg. **4A.** 1.69×10^3 °C. **4B.** 7.45 atm. **5A.** $O_2(g)$. **5B.** (b). **6A.** 59.0 g. **6B.** 65.2 L. **7A.** 24.4 L. **7B.** 464 K. **8A.** 2.11 mol. **8B.** 1.52 mmHg. **9A.** 2.11 mL. **9B.** 0.383 g. **10A.** 86.46 g/mol. **10B.** NO. **11A.** 0.162 g/L. **11B.** 382 K. **12A.** 35.6 g Na_3N. **12B.** 0.619 g $Na(l)$. **13A.** 1.25 L. **13B.** 150. L. **14A.** 13 atm. **14B.** 5.46 atm. **15A.** 0.0348 atm $H_2O(g)$, 2.47 atm $CO_2(g)$. **15B.** 584 mmHg N_2, 157 mmHg O_2,

0.27 mmHg CO_2, 7.0 mmHg Ar. **16A.** 0.00278 mol. **16B.** 0.395 L. **17A.** $NH_3(g)$: 661 m/s. **17B.** 76.75 K. **18A.** 2.1×10^{-4} mol. **18B.** 28.2 s. **19A.** 189.6 g/mol. **19B.** 52.3 s. **20A.** $Cl_2(g)$. **20B.** $Cl_2(g)$.

Review Questions. 4. (a) 0.968 atm; **(b)** 0.766 atm; **(c)** 1.17 atm; **(d)** 3.3 atm. **5. (a)** 748 mmHg; **(b)** 928 mmHg; **(c)** 10.4 ft Hg. **6.** 753 mmHg. **7. (a)** 52.8 L; **(b)** 7.27 L. **8. (a)** 1.10×10^3 mL; **(b)** 7.50×10^2 mL. **9.** 576 °C. **10.** 42.7 L C_2H_2. **11.** 79.03 L Cl_2. **12.** PF_3. **13.** 5.32×10^4 mL. **14.** 2.74 atm. **15.** 103 g/mol. **16.** 1.75 g/L. **17.** 1.25 L $H_2(g)$. **18.** 519 L $CO_2(g)$. **19.** 5.58 L. **20.** 6.1 L. **21. (a)** 737 mmHg; **(b)** 97.5%; **(c)** 0.111 g O_2. **22.** (1). **23.** 18.6 s. **24.** (3).

Exercises. 25. 11.4 m. **27.** 976 mmHg. **29.** 50.7 atm. **31.** 9 °C. **33. (a)** 41.8 mg; **(b)** 7.41×10^{20} molecules. **35.** Air in the bag expands at the lower pressures at higher altitudes. **37.** 2.29 atm. **39.** 1.41×10^3 °C. **41.** 253 g O_2. **43.** C_3H_6. **45.** C_4H_8. **47.** 1.22×10^3 mmHg. **49. (a)** $d = 1.18$ g/L air; **(b)** $d = 1.80$ g/L CO_2. **51.** P_4. **53.** 378 L O_2. **55.** 3.3×10^7 L. **57.** 109 mL. **59.** 2.1×10^3 g. **61. (a)** 23.5 atm; **(b)** 10.4 atm H_2, 13.1 atm He. **63. (a)** less; **(b)** 1.1×10^2 CO_2 (expired air to ordinary air). **65.** 2.37 L. **67.** 751 mmHg. **69.** 326 m/s. **71.** 7.8 u. **73.** 1.51×10^3 K. **75.** 0.00473 mol. **77. (a)** 1.07; **(b)** 1.05; **(c)** 0.978; **(d)** 1.004. **79.** 6.86 atm, van der Waals; 7.34 atm, ideal.

Advanced Exercises. 81. The mass of a liquid column depends on the liquid's density, and the downward force exerted by this mass depends on the acceleration due to gravity, g, which varies slightly from point to point on Earth. **83.** C_4H_6. **85.** Squeezing the bottle increases the pressure in a small air pocket in the diver. The density of the diver increases slightly, and the diver sinks. **87.** 2.24 atm. **90.** 153 mmHg. **91.** 19.9%. **95. (a)** and **(b)** 3.42% by volume and by number; **(c)** 1.95%. **97. (b)** 7.39 L

Feature Problems. A. $P_{gas} \times A = 2.27$ (± 0.02) mmHg·cm, reasonably constant. **B.** Slope depends on T and n. **C.** V (container volume), $\bar{u}$ (average molecular speed), and N (number of molecules) are unchanged, except as follows. **1.** V triples; **2.** V halves $\bar{u}^2$ halves; **3.** V quadruples, u^2 doubles; **4.** N and $\bar{u}^2$ increase by 50%. **D.** 15.999 u.

Chapter 7

Practice Examples. 1A. $+1.15 \times 10^3$ J. **1B.** 3.00×10^2 J. **2A.** 32.7 kJ. **2B.** 4.89 kJ. **3A.** 3.0×10^2 g. **3B.** 37.9 °C. **4A.** -3.83×10^3 kJ/mol. **4B.** 6.26 kJ/°C. **5A.** 65.2 kJ/mol. **5B.** 33.6 °C. **6A.** $+1.70 \times 10^2$ J. **6B.** 179 J of work is done by the system. **7A.** 60.6 g. **7B.** 0.15 kJ of heat evolved. **8A.** -125 kJ. **8B.** -2021 kJ. **9A.** 6 C(graphite) $+ \frac{13}{2} H_2(g)$ $+ O_2(g) + \frac{1}{2} N_2(g) \rightarrow C_6H_{13}O_2N(s)$. **9B.** twice the negative of the value in Table 7.2: $+92.22$ kJ. **10A.** -2.5×10^{-3} kJ. **10B.** $+49$ kJ/mol. **11A.** -112.3 kJ/mol AgI. **11B.** $+2.4$ kJ/mol Mg(OH)$_2$. **Review Questions. 4. (a)** $+68.5$ kcal; **(b)** -177 kJ. **5. (a)** 39.5 °C; **(b)** 57 °C. **6. (a)** 1.7 J g^{-1} °C^{-1}; **(b)** 21.9 °C. **7.** $+49.1$ kJ. **8.** 0.450 J g^{-1} °C^{-1}. **9.** 46 °C. **10. (a)** 0 J; **(b)** -236 J; **(c)** $+133$ J; **(d)** -416 J. **11. (a)** -2.22×10^3 kJ/mol; **(b)** -5.90×10^3 kJ/mol; **(c)** -1.82×10^3 kJ/mol. **12.** 4.98 kJ/°C. **13. (a)** 23.82 °C; **(b)** 29.60 °C. **14.** $C_5H_{10}O_5(g) + 5$ $O_2(g) \rightarrow 5 CO_2(g) + 5 H_2O(l) \Delta H =$ -2.34×10^3 kJ. **15.** -299 kJ/mol. **16. (a)** endothermic; **(b)** $+18$ kJ/mol. **17.** 26.0 °C. **18. (a)** $N_2(g) + \frac{1}{2} O_2(g)$ $\rightarrow N_2O(g)$; **(b)** S(rhombic) $+ O_2(g)$ $+ Cl_2(g) \rightarrow SO_2Cl_2(l)$; **(c)** $CH_3CH_2COOH(l) + \frac{7}{2} O_2(g) \rightarrow 3$ $CO_2(g) + 3 H_2O(l)$. **19. (a)** 65.59 kJ; **(b)** 3.65×10^3 kJ; **(c)** 1.45×10^3 kJ. **20.** $+30.74$ kJ. **21.** -282.97 kJ. **22.** -290 kJ. **23.** $2 \Delta H_1 + 2 \Delta H_2 - 3 \Delta H_3$. **24. (a)** -54.8 kJ; **(b)** -1123.9 kJ. **25.** $+30.6$ kJ. **26.** -206.0 kJ/mol.
Exercises. 27. (a) 3.4 L atm; **(b)** $3.4 \times$ 10^2 J; **(c)** 82 cal. **29. (a)** 0.385 J g^{-1} °C^{-1}; **(b)** 0.13 J g^{-1} °C^{-1}; **(c)** 0.905 J g^{-1} °C^{-1}. **31.** 5.5×10^2 °C. **33.** 24.0 °C. **35.** 2.3×10^2 J mol^{-1} °C^{-1}. **37.** -2.49×10^5 kJ. **39. (a)** 504 kg; **(b)** -6.21×10^5 kJ; **(c)** 2.90×10^3 L. **41.** 1.4×10^3 kJ. **43. (a)** -5×10^1 kJ/mol; **(b)** Use more solute and less water to increase ΔT. **45.** 4.2×10^2 g. **47.** -56 kJ/mol. **49.** 7.72 kJ/°C. **51.** 5.23 °C. **53.** -217.5 kJ. **55.** -747.5 kJ. **57.** -25.5 kJ. **59.** Some compounds are more stable and some, less stable, than their elements. $\Delta H_f^\circ = 0$ is unlikely for a compound. **61.** $+202.4$ kJ. **63.** -1366.7 kJ. **65.** -102.9 kJ. **67.** -55 kJ. **69.** 2.40×10^6 kJ.
Advanced Exercises. 72. 3.5 °C. **73. (a)**

yes; **(b)** yes; **(c)** no change; **(d)** remains constant. **75.** -1.95×10^3 kJ/mol $C_6H_8O_7$. **79.** 1.61×10^4 kJ. **82. (a)** -890.3 kJ; **(b)** -892 kJ/mol CH_4. **84.** 69.1%. **85.** 303 g.
Feature Problems. A. 69 s. **B. 1.** sp ht $= 0.011440 + (23.967/\text{atomic mass})$; **2.** 110 u; **3.** 62 u, likely Cu. **C. 1.** 45.0 mL of 1.00 M NaOH(aq) and 15.0 mL of 1.00 M citric acid; **2.** maximum heat is evolved at the stoichiometric point; **3.** $H_3C_6H_5O_7(aq) + 3 OH^-(aq) \rightarrow 3$ $H_2O(l) + C_6H_5O_7^{3-}(aq)$. **D. 1.** $NH_3(\text{conc. aq}) \rightarrow NH_3(g)$, $+2.54$ kJ $= \Delta H_{2a}$; $NH_3(g) + HCl(aq) \rightarrow$ $NH_4Cl(aq)$, -7.98 kJ $= \Delta H_{2b}$; $NH_3(\text{conc. aq}) + HCl(aq) \rightarrow$ $NH_4Cl(aq)$, -5.42 kJ $= \Delta H_1$. **2.** ΔH_{2a} $+ \Delta H_{2b} = -5.44$ kJ $\approx \Delta H_1$.

Chapter 8

Practice Examples. 1A. 535 L. **1B.** 7.7×10^3 L. **2A.** $2 PbO_2(s) + 4 HNO_3(aq)$ $\rightarrow 2 Pb(NO_3)_2(aq) + O_2(g) + 2 H_2O(l)$. **2B.** $4 Zn(s) + 10 HNO_3(aq) \rightarrow 4$ $Zn(NO_3)_2(aq) + NH_4NO_3(aq) + 3 H_2O$. **3A.** $2 H_2O + 2 e^- \xrightarrow[\text{reduction}]{\text{cathode}} H_2(g) +$ $2 OH^-(aq)$, $4 OH^-(aq) \xrightarrow[\text{oxidation}]{\text{anode}} O_2(g)$ $+ 2 H_2O + 4 e^-$. **3B.** $2 Li_2O_2(s) +$ $2 CO_2(g) \rightarrow 2 Li_2CO_3(s) + O_2(g)$.
Review Questions. 4. (a) ozone; **(b)** N_2O; **(c)** KO_2; **(d)** calcium hydride; **(e)** magnesium nitride; **(f)** K_2CO_3; **(g)** $(NH_4)H_2PO_4$. **5. (a)** NH_3; **(b)** almost pure C; **(c)** $CO(NH_2)_2$; **(d)** principally $CaCO_3$; **(e)** a mixture of CO(g) and $H_2(g)$, produced by steam reforming of a hydrocarbon. **6. (a)** NO_2; **(b)** KO_2; **(c)** N_2O; **(d)** CaH_2; **(e)** CO. **7. (a)** 2 $KClO_3(s) \xrightarrow{\text{heat, MnO2}} 2 KCl(s) + 3$ $O_2(g)$; **(b)** $NH_4NO_3(s) \xrightarrow{200-260°C}$ $N_2O(g) + 2 H_2O(g)$; **(c)** $Zn(s) + 2$ $HCl(aq) \rightarrow ZnCl_2(aq) + H_2(g)$; **(d)** $CaCO_3(s) + 2 HCl(aq) \rightarrow CaCl_2(aq) +$ $H_2O(l) + CO_2(g)$. **8.** 9.54×10^5 L air. **9. (a)** $LiH(s) + H_2O \rightarrow Li^+(aq) +$ $OH^-(aq) + H_2(g)$; **(b)** $C(s) + H_2O(g)$ $\xrightarrow{\Delta} CO(g) + H_2(g)$; **(c)** $3 NO_2(g) +$ $H_2O(l) \rightarrow 2 HNO_3(aq) + NO(g)$. **10. (a)** $Mg(s) + 2 HCl(aq) \rightarrow MgCl_2(aq) +$ $H_2(g)$; **(b)** $NH_3(g) + HNO_3(aq) \rightarrow$ $NH_4NO_3(aq)$; **(c)** $MgCO_3(s) + 2$ $HCl(aq) \rightarrow MgCl_2(aq) + H_2O +$ $CO_2(g)$; **(d)** $NaHCO_3(s) +$ $HC_2H_3O_2(aq) \rightarrow NaC_2H_3O_2(aq) + H_2O$ $+ CO_2(g)$. **11.** $H_2SO_4(aq) + 2 NH_3(aq)$

$\rightarrow (NH_4)_2SO_4(aq)$. **12.** (8.4) $NH_3(g)$ is the reducing agent, $O_2(g)$ is the oxidizing agent; (8.8) $NO_2(g)$ is the reducing agent and also the oxidizing agent; (8.16) $C_8H_{18}(l)$ is the reducing agent, $O_2(g)$ is the oxidizing agent. **13.** $Fe_2O_3(s) + 3 H_2(g) \rightarrow 2 Fe(s) + 3$ $H_2O(g)$. **14.** $3 Cu(s) + 2 NO_3^-(aq) + 8$ $H^+(aq) \rightarrow 3 Cu^{2+}(aq) + 2 NO(g) + 4$ H_2O. **15.** $2 KClO_3(s) \xrightarrow{\text{heat, MnO2}}$ $2 KCl(s) + 3 O_2(g)$; $CaCO_3(s) \xrightarrow{\Delta}$ $CaO(s) + CO_2(g)$; $NH_4NO_3(s)$ $\xrightarrow{200-260°C} 2 H_2O(g) + N_2O(g)$. **16.** $6 \times$ 10^9 kg. **17.** result of alpha particle emission. **18.** natural gas. **19.** 196 g. **20.** -316.6 kJ.
Exercises. 21. Because of the ideal gas law ($V = nRT/P$), volumes measured at the same T and P are proportional to moles. **23. (a)** $N_2(g) + 3 H_2(g) \rightleftharpoons$ $2 NH_3(g)$; **(b)** $4 NH_3(g) + 5 O_2(g)$ $\xrightarrow{450°C, Pt} 4 NO(g) + 6 H_2O(g)$; **(c)** $2 NO(g) + O_2(g) \rightarrow 2 NO_2(g)$ and 3 $NO_2(g) + H_2O(l) \rightarrow 2 HNO_3(aq) +$ $NO(g)$. **25.** $4 HNO_3(l) \rightarrow 2 N_2O_4(g) +$ $2 H_2O(l) + O_2(g)$. **27.** 3.32×10^{10} kg N_2. **29. (a)** $2 HgO(s) \xrightarrow{\Delta} 2 Hg(l) +$ $O_2(g)$; **(b)** $2 KClO_4(s) \xrightarrow{\Delta} 2 KClO_3(s)$ $+ O_2(g)$. **31.** CO_2. **33.** 3×10^{-5} mmHg. **35.** 2 moles of H_2 are produced for each mole of $O_2(g)$. **37.** 7.9×10^5 L. **39.** 0.40 g/L. **41. (a)** $2 C_6H_{14}(l) + 19 O_2(g)$ $\rightarrow 12 CO_2(g) + 14 H_2O(l)$; **(b)** PbO(s) $+ CO(g) \xrightarrow{\text{heat}} Pb(s) + CO_2(g)$; **(c)** 2 $KOH(aq) + CO_2(g) \rightarrow K_2CO_3(aq) +$ $H_2O(l)$; **(d)** $MgCO_3(s) + 2 HCl(aq) \rightarrow$ $MgCl_2(aq) + H_2O(l) + CO_2(g)$. **43.** no CO(g) will form. **45.** 1.27×10^5 kJ of heat evolved. **47. (a)** $2 Al(s) + 6$ $HCl(aq) \rightarrow 2 AlCl_3(aq) + 3 H_2(g)$; **(b)** $C_3H_8(g) + 3 H_2O(g) \rightarrow 3 CO(g) + 7$ $H_2(g)$; **(c)** $MnO_2(s) + 2 H_2(g) \xrightarrow{\Delta} Mn(s)$ $+ 2 H_2O(g)$. **49. (a)** $CaH_2(s)$; **(b)** CaH_2.
Advanced Exercises. 54. approximately 19°C. **56.** CH_4 is the better fuel. **60.** 0.1716 M.
Feature Problems. A. 1. Fuel lean: Excess air leads to high NO(g) levels. Fuel rich: Unburned and partially burned hydrocarbons lead to high RH(g) and CO(g) levels. **B. 1.** N_2 from liquid air contains some Ar. **2.** N_2 from liquid air. **3.** Mg does not react with Ar. **4.** 0.50%. **C.** First, balance each equation. Then, double the first equation, and add the second equation and three times the third equation.

Chapter 9

Practice Examples. 1A. 4.34×10^{14} Hz. **1B.** 3.28 m. **2A.** X-ray machine. **2B.** Hg lamp < "go" < "caution" < "stop." **3A.** 400 to 520 kJ/mol. **3B.** 4.612×10^{14} Hz (orange), 6.662×10^{14} Hz (indigo). We see the colors not absorbed, principally green. **4A.** $E_9 = -2.69 \times 10^{-20}$ J. **4B.** no. **5A.** 486.2 nm. **5B.** 121.6 nm. **6A.** 1.21×10^{-43} m. **6B.** 39.6 km/s. **7A.** yes. **7B.** $\ell = 2$ and $\ell = 1$. **8A.** $3p$. **8B.** $n = 5$, $\ell = 0$, $m_\ell = 0$. **9A.** Ti. **9B.** $1s^2 2s^2 2p^6 3s^2 3p^6 3d^{10} 4s^2 4p^6 4d^{10} 5s^2 5p^5$; ten $3d$ electrons; one unpaired electron.
10A. Fe: [Ar] $3d$ ⇅↑↑↑↑ $4s$ ⇅.
10B. Bi: [Xe] $4f$ ⇅⇅⇅⇅⇅⇅⇅ $5d$ ⇅ ⇅⇅⇅⇅⇅$6s$⇅ $6p$ ↑ ↑ ↑. **11A. (a)** 5; **(b)** 6; **(c)** 0; **(d)** 2. **11B. (a)** 10; **(b)** 2; **(c)** 1. **Review Questions. 4. (a)** 162.5 nm; **(b)** 0.388 μm; **(c)** 7.27×10^6 nm; **(d)** 5.46×10^{-7} m; **(e)** 1.12×10^7 nm; **(f)** 2.6×10^{-4} cm. **5. (a)** 4.4×10^{-5} m, infrared; **(b)** 3.1×10^{-8} m, ultraviolet; **(c)** 11.8 m, radio; **(d)** 2.80 m, radio. **6.** 860 Å. **7.** 4.68 nm. **8. (a)** 6.41×10^{16} Hz; **(b)** 4.25×10^{-17} J. **9. (a)** independent; **(b)** inversely; **(c)** directly. **10. (a)** 6.9050×10^{14} s^{-1}; **(b)** 397.11 nm; **(c)** $n = 10$. **11.** $\nu = 3.2881 \times 10^{15}$ s^{-1} $[(1/3^2) - (1/n^2)]$ where $n = 4, 5, \dots$. **12. (a)** 5.71×10^{-18} J/photon; **(b)** 61.1 kJ/mol. **13. (a)** 6.31×10^{12} Hz; **(b)** 557 nm. **14. (b)** most; **(d)** least. **15. (a)** 8.5 Å; **(b)** no; **(c)** -3.405×10^{-20} J; **(d)** no. **16.** $\Delta E = -1.5495 \times 10^{-19}$ J; $\nu = 2.339 \times 10^{14}$ s^{-1}. **17.** (a). **18.** (a). **19. (a)** $m_\ell = 0, \pm 1$; **(b)** $\ell = 1, 2, 3$; **(c)** $n = 2, 3, \dots$. **20. (a)** $n = 4$, $\ell = 0$; **(b)** $n = 3$, $\ell = 1$; **(c)** $n = 5$, $\ell = 3$; **(d)** $n = 3$, $\ell = 2$. **21. (b)** it has $\ell > n$.; **(c)** it has $m_\ell > \ell$; **(e)** it has $\ell = n$. **22. (a)** one; **(b)** no such subshell; **(c)** three; **(d)** five; **(e)** seven; **(f)** three. **23. (a)** $1s^2 2s^2 2p^6 3s^2 3p^6 3d^{10} 4s^2 4p^5$; **(b)** $1s^2 2s^2 2p^6 3s^2 3p^4$; **(c)** $1s^2 2s^2 2p^6 3s^2 3p^6 3d^{10} 4s^2 4p^6 4d^{10} 5s^2 5p^3$; **(d)** $1s^2 2s^2 2p^6 3s^2 3p^2$. **24. (a)** [Ar] $4s$ ⇅; **(b)** [Ar] $3d$ ↑ ▯▯▯▯ $4s$ ⇅; **(c)** [Ar] $3d$ ⇅⇅⇅⇅⇅ $4s$ ⇅ $4p$ ↑ ▯ ▯; **(d)** [Ar] $3d$ ⇅⇅⇅↑↑ $4s$ ⇅; **(e)** [Xe] $4f$ ⇅⇅⇅⇅⇅⇅⇅ $5d$ ⇅⇅⇅⇅⇅ $6s$ ⇅ $6p$ ↑ ↑ ↑. **25. (a)** 0; **(b)** 1; **(c)** 2; **(d)** 1. **26. (a)** those in Groups 1A and 2A, along with Al, Ga, In, Tl, Sn, Pb, and Bi; **(b)** H, F, Cl, Br, I, O, S, Se, N,

P, C, and B; **(c)** He, Ne, Ar, Kr, Xe, and Rn; **(d)** those in Groups 1B, 2B, 3B, 4B, 5B, 6B, 7B, and 8B; **(e)** those with atomic numbers from $Z = 58$ (Ce) through $Z = 71$ (Lu) and those from $Z = 90$ (Th) through $Z = 103$ (Lr). **27. (a)** [Kr] $4d^{10} 5s^2 5p^1$; **(b)** [Kr] $4d^{10} 5s^2$; **(c)** [Kr] $4d^{10} 5s^2 5p^3$; **(d)** [Xe] $4f^{14} 5d^{10} 6s^1$. **28. (a)** 1; **(b)** 5; **(c)** 10; **(d)** 6; **(e)** 14; **(f)** 8.

Exercises. 29. (a) and **(d)** are true. **31.** 8.3 min. **33.** 656.46 nm, 486.26 nm, 434.16 nm, 410.28 nm. **35.** $n = 8$. **37. (a)** 4.19×10^{-19} J/photon; **(b)** 252 kJ/mol. **39.** 3.03×10^{16} s^{-1}, ultraviolet. **41. (a)** 6.60×10^{-19} J/photon; **(b)** ultraviolet yes, not infrared. **43. (a)** 1.9 nm; **(b)** -6.053×10^{-20} J. **45. (a)** 1.384×10^{14} s^{-1}; **(b)** 2166 nm; **(c)** infrared. **47.** $n = 4$. **49.** electrons. **51.** 9.79×10^{-35} m. **53.** The electron's position and velocity are both exactly known in the Bohr theory. This is not allowed by the Heisenberg uncertainty principle. **55.** 16.8 cm. **57.** Discrete pathways versus regions of high probability. Both theories give the same value of the most probable distance of the electron from the nucleus. **59.** (c). **61. (a)** $5p$; **(b)** $4d$; **(c)** $2s$. **63. (a)** four; **(b)** $3s$, $3p$, $3d$; **(c)** 7; **(d)** 1; **(e)** 16. **65. (a)** $3s$ electrons with same spin; **(b)** correct; **(c)** one electron needed in each $3p$ orbital; **(d)** all $3p$ spins must be the same. **67. (a)** 3; **(b)** 10; **(c)** 2; **(d)** 2; **(e)** 14. **69. (a)** 3; **(b)** 1; **(c)** 0; **(d)** 14; **(e)** 2; **(f)** 5; **(g)** 32. **71. (a)** [Xe] $4f^{14} 5d^{10} 6s^2 6p^2$; **(b)** [Rn] $5f^{14} 6d^{10} 7s^2 7p^2$.

Advanced Exercises. 74. (a) no, wavelengths too long; **(b)** 2.02×10^{-19} J; **(c)** 6.66×10^5 m s^{-1}. **77.** $n = 4$ and $n = 3$. **81.** 1.0×10^{20} photons/sec. **84.** 1.4×10^7 m s^{-1}. **86.** $\lambda = 16$ in. **88.** 4.8×10^{-3} J.

Feature Problems. A. Cr, Fe, and Ni, likely, and possibly other elements not listed. **B.** plot ν on the vertical axis, and $1/n^2$ on the horizontal axis. The slope is -3.2881×10^{15} Hz and the intercept is 8.2203×10^{14} Hz. **C. 1.** The graph sums the probabilities for all points equidistant from the nucleus. **D. 1.** [Mn] i, f, k (or l); **2.** [Kr] j, d; **3.** (I) [Kr] b, c; **4.** (U) [Rn] a, h, l (or k).

Chapter 10

Practice Examples. 1A. S. **1B.** Ca. **2A.**

$V^{3+} < Ti^{2+} < Ca^{2+} < Sr^{2+} < Br^-$. **2B.** As. **3A.** $K < Mg < S < Cl$. **3B.** Sb. **4A.** Cl and Al are paramagnetic. **4B.** Cr^{2+}. **5A.** 280 K. **5B.** 570 K.

Review Questions. 4. Ar—K, Te—I, Co—Ni. **5. (a)** 50; **(b)** 69; **(c)** 10; **(d)** 2; **(e)** 2; **(f)** 4. **6.** Fe^{2+} and Co^{3+}, Sc^{3+} and Ca^{2+}, F^- and Al^{3+}, Zn^{2+} and Cu^+. **7.** Two atoms may not be, but the other combinations may be isoelectronic. **8. (a)** fluorine; **(b)** scandium; **(c)** silicon. **9. (a)** Te; **(b)** K; **(c)** Cs; **(d)** N; **(e)** P; **(f)** Au. **10.** smallest: F; largest: I^-. **11.** Cs $< Sr < As < S < F$. **12. (a)** C; **(b)** Rb; **(c)** At. **13.** 21.6 J. **14. (a)** Ba; **(b)** S; **(c)** Bi. **15.** Rb > Ca > Sc > Fe > Te > Br > O > F. **16.** paramagnetic, only Cr^{3+}, Co^{3+}, and Br. **17. (a)** 6; **(b)** 8; **(c)** 5; **(d)** 1; **(e)** 2; **(f)** 4. **18.** 210 K.

Exercises. 19. 14 to 16 g/cm^3. **21.** plot of density vs. atomic number has similar shape for both periods. **23.** Moseley permitted only integral atomic numbers. **25. (a)** 118; **(b)** 119; **(c)** about 300 u. **27.** Electrons are being added to the same subshell in some instances but to a new subshell in others. **29. (a)** B; **(b)** Te. **31.** yes, like Cu^+ and Zn^{2+}. **33.** $Li^+ < Br < Se < I^-$. **35.** No, the second electron is being removed from a positive ion. **37.** 9952 kJ. **39.** -456.2 kJ, exothermic. **41.** similar answer to Exercise 35. **43.** Fe^{2+}. **45.** All atoms with odd Z are paramagnetic; some with even Z are also paramagnetic, for example, C. **47. (a)** yes: Cs, Rb, and K; no; Zn, Cd, Hg; **(b)** Rn; **(c)** $+600$ kJ/mol; **(d)** 5.7 g/cm^3. **49. (a)** 5.6 g/cm^3; **(b)** 74% Ga, Ga_2O_3. **51. (a)** Ba; **(b)** Sr; **(c)** Cl; **(d)** N. **Advanced Exercises. 55.** Data yield an atomic mass of 75.7 u, suggesting either Gruppe V or VI. **59.** BrCl gas, ICl liquid. **62.** 5249 kJ/mol.

Chapter 11

Practice Examples. 1A. ·Mg· ·Ge· K· :Ne: **1B.** ·Sn· [:Br:]$^-$ [·Tl·]$^+$ [:S:]$^{2-}$ **2A. (a)** 2 [Na]$^+$ [:S:]$^{2-}$; **(b)** 3 [Mg]$^{2+}$ 2[:N:]$^{3-}$. **2B. (a)** [Ca]$^{2+}$ 2 [:I:]$^-$ **(b)** [Ba]$^{2+}$ [:S:]$^{2-}$; **(c)** 2 [Li]$^+$ [:O:]$^{2-}$ **3A.** C—H = 110 pm, C—Br = 191 pm. **3B.** :O=C=O: **4A.** P—Cl or N—H;

:I:
|
4B. P—O **5A. (a)** :I—N—I:

(b) S=C=S **5B.** : F—N=N—F :

H
|
6A. (b). **6B.** H—N—C≡N : **7A.** boron.

7B. 2 [Na]⁺ [: O—O :]²⁻.

8A.

$$\left[\begin{array}{c} H\ :O: \\ | \quad | \\ H-C-C=O \\ | \\ H \end{array}\right]^{-} \longleftrightarrow$$

$$\left[\begin{array}{c} H\ :O: \\ | \quad \| \\ H-C-C-O: \\ | \\ H \end{array}\right]^{-}$$

8B.   [: N—N≡N :]⁻   ↔
[: N≡N—N :]⁻ ↔ [: N=N=N :]⁻
9A. trigonal-pyramidal. **9B.** tetrahedral.
10A. linear. **10B.** linear. **11A.**

H
|
H—C—O—H bond angles are about
|
H

109.5°. **11B.**

H H :O:
| | \|
H—N—C—C—O—H
|
H

all bond angles are about 109.5° except
those around the carbon double-bonded
to O, which has 120° bond angles. **12A.**
H₂O₂. **12B.** PCl₅. **13A.** −486 kJ. **13B.**
−4 × 10¹ kJ/mol NH₃. **14A.** exother-
mic. **14B.** endothermic.
Review Questions. 4. (a) [H :]⁻ **(b)**
: Kr :   **(c)** [· Sn ·]²⁺ **(d)** [K]⁺ **(e)**
[: Br :]⁻ **(f)** · Ge · **(g)** : N · **(h)** · Ca ·
(i) [: Se :]²⁻ **(j)** [Sc]³⁺ **5. (a)** [Ca]²⁺ 2
[: Cl :]⁻ **(b)** [Ba]²⁺ [: S :]²⁻ **(c)** 2
[Li]⁺ [: O :]²⁻ **(d)** [Na]⁺ [: F :]⁻ **(e)**
3 [Mg]²⁺ 2 [: N :]³⁻ **6. (a)** : I—Cl :
(b) : Br—Br : **(c)** : F—O—F : **(d)**
: I—N—I **(e)** H—Se—H **7. (a)**
|
: I :

: S=C=S : **(b)**

H :O: H
| \| |
H—C—C—C—H
| |
H   H

:O:
\|
(c) : Cl—C—Cl : **(d)** : F—N=O : **8.**
(a) 2 bonds to H; **(b)** should have 19,

---

not 20 valence electrons; **(c)** only 6 electrons on C, 2 unpaired; **(d)** should
be ionic. **9. (a)** Ca²⁺ 2 [: O—H]⁻ ;

(b)
$$\left[\begin{array}{c} H \\ | \\ H-N-H \\ | \\ H \end{array}\right]^{+} \quad [: Br :]^{-};$$

(c) Ca²⁺ 2 [: O—Cl :]⁻. **10.** All formal
charges are zero, except: **(a)** ≡C: = −1;
(b) —O = −1; **(c)** central C = +1; **(e)**
—O = −1, =S— = +1; **(f)** —O =
−1, N = +1. **11.** diamagnetic: **(a)**, **(d)**,
(e). **12. (a)** : Cl—O—Cl : **(b)**

: F—P—F : **(c)** $\left[: O-C=O : \right]^{2-}$ **(d)**
| |
: F :   : O :

: F   F :
\ /
: F—Br—F :   **13. (a)** bent; **(b)** tri-
|
: F :

gonal-planar; **(c)** linear; **(d)** octahedral;
(e) tetrahedral. **14. (a)** bent; **(b)** trigonal-
pyramidal; **(c)** tetrahedral. **15. (a)** linear;
(b) tetrahedral; **(c)** trigonal-pyramidal;
(d) T-shaped; **(e)** trigonal-bipyramidal;
(f) bent; **(g)** octahedral. **16.** H—C 110
pm, 414 kJ/mol, C=O 120 pm, 736
kJ/mol; C—C 154 pm, 347 kJ/mol;
C—Cl 178 pm, 339 kJ/mol. **17. (c). 18.**
endothermic. **19. (a)** endothermic; **(b)**
exothermic. **20.** −7.4 × 10² kJ. **21.** Bi.
22. C—H < Br—H < F—H < Na—Cl
< K—F. **23.** polar: **(b)**, **(d)**, **(e)**, **(g)**. **24.**
(a) polyatomic ions must have formal
charges; **(b)** some are linear; **(c)** Bond
moments can cancel, leaving no resultant dipole moment.
Exercises. 25. H₂, ICl₃, BF₃. **27. (a)** Cs⁺
[: Br :]⁻;   **(b)**   H—Sb—H ;   **(c)**
|
H

: Cl—B—Cl : ;   **(d)**   Cs⁺   [: Cl :]⁻;
|
: Cl :

(e) 2 Li⁺ [: O :]²⁻; **(f)** : I—Cl :. **29.**
(c). 31. (a) 2 Li⁺ [: S :]²⁻; **(b)** Na⁺
[: F :]⁻;   **(c)**   2   [: I :]⁻   Ca²⁺;
(d) Sc³⁺ 3 [: Cl :]⁻ **33.** most significant
difference: O.S. of an element in a compound rarely is zero, F.C. often is. **35.**
(a) +1; **(b)** −1; **(c)** +1; **(d)** 0; **(e)** −1.
37. (a) H—N—O—H ;
|
H

(b) H—O—Cl—O : ; **(c)** H—O—N=O ;
|
:O:

:O:
\|
(d) : Cl—S—Cl :
\|
:O:

39. (a) $\left[: O-S-O : \right]^{2-}$;
|
:O:

(b) [: O—N=O]⁻ ↔ [O=N—O :]⁻;

(c) $\left[: O-C=O \right]^{2-} \longleftrightarrow$
|
:O:

$\left[O=C-O : \right]^{2-} \longleftrightarrow \left[: O-C-O : \right]^{2-}$;
| \|
:O: :O:

(d) [H—O—O :]⁻

H H H :O:
| | | \|
41. H—C—C=C—C—H
|
H

43. (a) 233 pm; **(b)** 172 pm; **(c)** 149 pm;
(d) 191 pm. **45.** 144 pm. **47. (a)** 4%; **(b)**
5%; **(c)** 60%; **(d)** 33%. **49.**
[: O—N=O]⁻ ↔ [O=N—O :]⁻. **51.**
most plausible: (1) and (2).

53. (a) H—C—H ;
|
H

(b) : O—Cl—O · ; **(c)** · O—N=O
|
:O:

55. necessary in: ICl₃ and SF₄. **57. (a)**
linear; **(b)** linear; **(c)** tetrahedral; **(d)**
trigonal-planar; **(e)** bent. **59.** CO₃²⁻.
61. (a) O=C=O linear;

(b) O=C—Cl : trigonal-planar; **(c)**
|
: Cl :

: Cl—N=O   trigonal-planar. **63. (a)**
|
:O:

tetrahedral; **(b)** tetrahedral; **(c)** octahedral; **(d)** linear. **65.** tetrahedral. **67. (a)**
bent, polar; **(b)** trigonal-pyramidal, polar; **(c)** bent, polar; **(d)** planar, nonpolar;
(e) octahedral, nonpolar; **(f)** tetrahedral,
polar. **69.** must be bent if bond moments
are not to cancel. **71.** −113 kJ/mol.
73. (a) 3 kJ/mol; **(b)** 99 kJ. **75.**

-9.9×10^2 kJ.
Advanced Exercises. 79.
H—C=C—C=C—H **82.** ICl_2^+ has
  | | | |
  H   H   H   H

tetrahedral electron group geometry and bent molecular shape, ICl_2^- has trigonal-bipyramidal electron group geometry and a linear molecular shape. **84.** Lewis structure: H—N̈=N=N̈, shape has three Ns in a line, with 120° H—N—N bond angle. **87.** −166 kJ. **89.** 142 kJ/mol.

Feature Problems. A. 1. 91 kJ/mol; **2.** 0.97; **3.** 23% ionic. **B. 1.** 1.86 D; **2.** 92°.

Chapter 12

Practice Examples. 1A. trigonal-pyramidal. **1B.** trigonal-pyramidal. VSEPR has 109.5° bond angles, valence bond has 90° angles. **2A.** All central atoms have sp^3 hybridization, approximate 109.5° bond angles. **2B.** The C=O carbon is sp^2-hybridized with 120° bond angles; all other central atoms are sp^3-hybridized with 109.5° bond angles. **3A.** The triple-bonded C is sp-hybridized with 180° bond angles; the other C is sp^3-hybridized with 109.5° bond angles. **3B.** :N≡N—Ö: ↔ :N̈=N=Ö the central N is sp hybridized, with 180° bond angles. **4A.** 53 kJ/mol. **4B.** yes, bond order = 0.5. **5A.** bond orders: (a) 2.5; (b) 0.5; (c) 3.0. **5B.** bond orders: (a) 2.5; (b) 2.0; (c) 1.5; (d) 1.0.

Review Questions. 4. $1s$ on H and $4p$ on Se. **5.** CO_3^{2-}, SO_2, and NO_2^-. **6. (b)** is true. **7.** overlaps are: (a) H($1s$)—I($5p$); (b) Br($4p$)—Cl($3p$); (c) Se($4p$)—H($1s$); (d) O($2p$)—Cl($3p$). **8.** experimental 120° bond angles contradict predicted 90° p-orbital angles. **9. (a)** C(sp^3) bonds H($1s$) or Cl($3p$), tetrahedral molecule; (b) sp hybrid C forms two σ and two π bonds, linear molecule; (c) B(sp^2) bonds with F($2p$), trigonal-planar molecule. **10. (a)** sp^3d^2; (b) sp; (c) sp^3; (d) sp^2; (e) sp^3d. **11. (a)** sp on C; (b) sp^3 on C and O; (c) sp^2 on central C, sp^3 on other C atoms; (d) sp^2 on C, sp^3 on N and O. **12. (a)** planar, sp^2 on each C; (b) linear, sp on each C; (c) sp^3 on F-surrounded C, sp on other C; (d) linear, sp on C. **13.** A single bond is a σ bond, a double bond is a σ and a π bond, a triple bond is a σ and two π bonds. (a) H—C≡N̈: ;

(b) :N≡C—C≡N: ;

      H   :Cl̈:
      | |
(c) H—C—C=C—C—Cl̈: ; (d)
      | | | |
      H   H   H :Cl̈:

H—Ö—N̈=O **14. (a)** Ö=C=Ö ; **(b)** C is sp-hybridized, and to each O is a σ bond C(sp)—O($2p$) and one π bond C($2p$)—O($2p$). **15. (a)** 6 σ; **(b)**

      H
      |
2 π; H—C—N̈=C=Ö . **16. (a)** σ_{1s};
      |
      H

(b) σ_{2s}; **(c)** σ_{1s}^*; **(d)** σ_{2p}. **17.** C_2. **18.** C_2^+
σ_{2s}[↑↓] σ_{2s}^*[↑↓] π_{2p}[↑↓|↑] σ_{2p}[ ]
π_{2p}^*[|] σ_{2p}^*[ ], 1.5 bond order, paramagnetic; O_2^{2-} σ_{2s}[↑↓] σ_{2s}^*[↑↓]
σ_{2p}[↑↓] π_{2p}[↑↓|↑↓] π_{2p}^*[↑↓|↑↓]
σ_{2p}^*[ ], 1.0 bond order, diamagnetic;
F_2^+ σ_{2s}[↑↓] σ_{2s}^*[↑↓] σ_{2p}[↑↓] π_{2p}[↑↓|↑↓]
π_{2p}^*[↑↓|↑] σ_{2p}^*[ ], 1.5 bond order, paramagnetic. **19.** electron sea is the "glue" between atomic cores; no explanation of luster or semiconductors. **20.** insulator.

Exercises. 21. VB method distinguishes between σ and π bonds, predicts molecular shape, explains hindered rotation around double bonds. **23. (a)** no prediction; (b) 90° bond angles; (c) and (d) 109.5° bond angles. **25. (a)** sp on C; (b) sp^2 on N, sp^3 on O; (c) sp^3 on Cl; (d) sp^3 on B. **27.** sp^3d on Cl holds two lone pair and forms three overlaps with $2p$ on F. **29. (a)** four C(sp^3)—Cl($3p$) σ bonds; **(b)** σ: O($2p$)—N(sp^2), σ: N(sp^2)—Cl($3p$), σ: O($2p$)—N($2p$); **(c)** σ: H($1s$)—O_a(sp^3), σ: O_a(sp^3)—N(sp^2), σ: N(sp^2)—O_b($2p$), π: N($2p$)—O_b($2p$); **(d)** 2σ: Cl($3p$)—C(sp^2), σ: O($2p$)—C(sp^2), π: O($2p$)—C($2p$). **31.** all O—H are O(sp^3)—H($1s$), all C—H are C(sp^3)—H($1s$), C=O are σ C(sp^2)—O($2p$) and π C($2p$)—O($2p$), central C—O is C(sp^3)—O(sp^3), side C—O are C(sp^3)—O(sp^3), the two central C's are sp^3-hybridized and σ bond to adjacent C atoms. **33.** σ: C_a(sp)—H_a($1s$), σ: C_a(sp)—C_b(sp), σ: C_b(sp)—C_c(sp^2), σ: C_c(sp^2)—H_b($1s$), σ: C_c(sp^2)—O($2p$), π: C_a($2p$)—C_b($2p$), π: C_a($2p$)—C_b($2p$), π: C_c($2p$)—O($2p$). **35.** Valence bond

theory describes bonds through orbital overlaps between atoms. Molecular orbital theory describes bonds through the electron occupancies of orbitals belonging to a molecule as a whole. **37.** yes, N_2^- bond order = 2.5, N_2^{2-} bond order = 2. **39.** Four consecutive bonding molecular orbitals with increasing energies and no intervening antibonding orbitals would be required, and this does not occur. **41. (a)** σ_{2s}[↑↓] σ_{2s}^*[↑↓] π_{2p}
[↑↓][↑↓] σ_{2p}[↑↓] π_{2p}^*[↑|] σ_{2p}^*[ ]; **(b),** **(c),** and **(e)** σ_{2s}[↑↓] σ_{2s}^*[↑↓] π_{2p}[↑↓|↑↓]
σ_{2p}[↑↓] π_{2p}^*[|] σ_{2p}^*[ ]; **(d)** σ_{2s}[↑↓]
σ_{2s}^*[↑↓] π_{2p}[↑↓|↑] σ_{2p}[↑] π_{2p}^*[|]
σ_{2p}^*[ ]; **(f)** and **(g)** σ_{2s}[↑↓] σ_{2s}^*[↑↓]
π_{2p}[↑↓|↑↓] σ_{2p}[ ] π_{2p}^*[|] σ_{2p}^*[ ].
43. NO^+ bond order = 3, diamagnetic, shorter bond; N_2^+ bond order = 2.5, paramagnetic, longer bond. **45.** to account for the resonance structures required by VB theory. **47. (b)** is the only one with resonance forms and thus delocalized orbitals. **49. (c)** and **(d)** need few valence electrons and empty orbitals for metallic bonding. **51.** 7.02×10^{20} of each.

Advanced Exercises. 54. MO theory: First and second shells filled, followed by σ_{3s}[↑↓]; VB theory: overlap of two $3s$ orbitals; Lewis theory: No, the structure would have incomplete octets. **56. (a)** O_2^- has a bond order of 1.5, O_2^{2-} has a bond order of 1.0; (b) [:Ö—Ö·]⁻, [:Ö—Ö:]²⁻.

              :O:
               ‖
60. :F̈—Ö_a—N—Ö: ↔

              :O:
               |
       :F̈—Ö_a—N=Ö ;

bond angles around N are 120°, N—O—F bond angles are about 109°; overlaps are σ: F($2p$)—O_a(sp^3), σ: O_a(sp^3)—N(sp^2), σ: N(sp^2)—O($2p$), π: N($2p$)—O($2p$) **64.** He_2^* σ_{1s}[↑↓] σ_{1s}^*
[↑], σ_{2s}[↑] σ_{2s}^*[ ] has a bond order of 1.0. **68.** Cl in F_2Cl^- is sp^3d-hybridized; Cl in F_2Cl^+ is sp^3-hybridized.

Chapter 13

Practice Examples. 1A. 0.923 kJ. **1B.**

260 kJ. **2A.** vapor only. **2B.** 0.0435 g H_2O vapor, 0.089 g liquid water. **3A.** 121 mmHg. **3B.** 42.9 mmHg. **4A. (a)** graphite becomes diamond at about 2.5 $\times 10^4$ atm; **(b)** At 3800 K (3527 °C) the carbon starts out as vapor that, with the application of pressure, should condense to C(graphite) at about 10 atm pressure. The C(graphite) converts to C(diamond) at somewhat above 10^5 atm pressure. The liquid area of the diagram is just missed in this transition. **4B.** The first halt is at the melting point of ice; the second, much longer halt is at 100 °C. **5A.** CH_3CN. **5B.** $(CH_3)_3CH <$ $CH_3CH_2CH_2CH_3 < SO_3 < C_8H_{18} <$ C_6H_5CHO. **6A.** He $<$ Ne $< O_2 < O_3$ $< Cl_2 < (CH_3)_2CO$. **6B.** Higher molar masses produce larger exchange forces; polar CH_3NO_3 has additional intermolecular forces. **7A.** lower: RbI or CsI; higher: MgO. **7B.** NaI. **8A.** 524 pm. **8B.** 6.628×10^7 pm^3. **9A.** 0.903 g/cm^3. **9B.** 6.035×10^{23}/mol. **10A.** 402 pm. **10B.** 2.31 g/cm^3. **11A.** -669.2 kJ/mol. **11B.** -796 kJ/mol.

Review Questions. 5. (a) and **(b)** not necessarily, CO_2 has neither; **(c)** yes. **6. (d):** The enthalpy of sublimation is the sum of the enthalpies of fusion and vaporization. **7.** Intermolecular forces (stonger ones produce lower vapor pressure) and temperature (vapor pressure is higher at high temperatures). **8.** Yes, it lowers surface tension. **9. (a)** exothermic condensation; **(b)** 100 °C. **10. (a)** 26.8 g; **(b)** 29.5 kJ/mol; **(c)** 4.84 kJ. **11. (a)** 45 mmHg; **(b)** 110 °C. **12. (a)** 459 K; **(b)** 464 mmHg. **13.** 226 mmHg. **14.** 49.7 kJ/mol. **15.** $T_{nbp} = 339$ K. **16.** 40. kJ. **17. (a)** 776 kJ; **(b)** 2.5×10^4 kJ. **18.** liquid and vapor [assuming that for $CO_2(l)$, $d > 0.16$ g/mL]. **19. (a)** $C_{10}H_{22}$; **(b)** CH_3OCH_3; **(c)** CH_3CH_2OH. **20.** N_2 $< F_2 < Ar < O_3 < Cl_2$, based on increasing molar mass. **21.** Ne $< C_3H_8 <$ $CH_3CH_2OH < CH_2OHCHOHCH_2OH$ $< KI < K_2SO_4 < MgO$. **22.** Yes, it boils at a high temperature compared to HCl, for example. **23. (a)** network covalent solid; **(b)** molecular solid; **(c)** ionic solid; **(d)** metallic solid; **(e)** molecular solid. **24. (c)** is correct. **25.** no. Cl^- ions $= (8 \times 1/8) = 1$ Cl^- ion plus 1 Cs^+ ion. **26. (a)** 362 pm; **(b)** 4.74×10^7 pm^3; **(c)** 4 Cu atoms; **(d)** 4.221×10^{-22} g Cu; **(e)** 8.91 g/cm^3.

Exercises. 27. Nonpolar silicone adheres to cloth or leather and repels polar, hydrogen-bonded water. **29.** Vaporization is endothermic; in the open container ΔH_{vap} comes from the surroundings, and in the insulated container ΔH_{vap} must come from the liquid. **31.** 8.88 L. **33.** 221 L. **35.** 120.7 °C. **37.** 0.907 mmHg. **39. (b)** 6.5 atm. **41.** 33 °C. **43.** CO_2, HCl, NH_3, SO_2, and H_2O. **45. (a)** melts at 0 °C, vaporizes at 93.5 °C; **(b)** liquefies at slightly above 91 mmHg and probably solidifies by 50 atm; **(c)** liquefies at about 20 °C, solidifies at about -56 °C. **47. (a)** 0.0418 atm; **(b)** 0.110 atm; **(c)** 0.117 atm. **49.** $SO_2(g)$ only. **51.** 27 g. **53.** Supercooled liquid solidifies. **55. (a)** weak London forces, polar molecule; **(b)** moderate London forces; **(c)** moderate London forces, polar molecule; **(d)** strong hydrogen bonds; **(e)** All forces are weak. **57.** CH_3OH (hydrogen bonding). **59.** diamond. **61. (b)** sp^2 hybridization for B and N; full $2p$ orbitals on N overlap with empty $2p$ orbitals on B to forms π bond framework. **63.** Yes, smaller ions produce larger interionic forces. **65.** There are two different ways of placing the third layer in the packing-of-spheres arrangement of Figure 13-36. **67. (a)** a diamond-shaped cell consisting of nine symbols, with a colored diamond at each corner and a gray square at the center; other unit cells are possible; **(b)** 1 gray square, 2 open circles, 1 colored diamond. **69.** 18.5 g/cm^3. **71.** $(8 \times 1/8) +$ $(6 \times 1/2) =$ four Ca^{2+} ions and eight F^- ions; $(8 \times 1/8) + 1 =$ two Ti^{4+} ions and $(4 \times 1/2) + 2 =$ four O^{2-} ions. **73. (a)** 6 for each; **(b)** 4; **(c)** 7.62×10^{-23} cm^3; **(d)** 3.51 g/cm^3. **75.** LiCl(s) (most negative) $<$ NaCl(s) $<$ KCl(s) $<$ RbCl(s) $<$ CsCl(s) (least negative). **77.** -645 kJ/mol.

Advanced Exercises. 82. Isooctane will be more volatile. **84.** 62.4 kJ. **85.** ΔH_{vap} decreases as temperature rises. **86.** at 1.00 bar, bp $= 99.65$ °C. **89. (a)** 31 atm; **(b)** -0.25 °C. **90.** 89 °C. **92.** 54.3%, decrease with temperature. **93.** 1.84 kg. **99. (b)** $EA_2 = +881$ kJ.

Feature Problems. A. 21 $g/s^2 = 2.1 \times 10^{-2}$ J/m^2. **B.** water vapor in the can condenses to liquid; external pressure crushes the can. **C. 2.** 42.9 kJ/mol; **3.** 55.6 kJ/mol; **4.** 2.83 mmHg. **D. 1.** Divide the central region with a line from its peak; extend the sublimation curve of S_α and the vapor pressure curve to a new triple point in the area occupied by S_β. **2.** It freezes into monoclinic sulfur. **E. (a)** -105 °C; **(b)** -79 °C; **(c)** -105 °C; **(d)** solid more dense than liquid, and no life as we know it.

Chapter 14

Practice Examples. 1A. 16.2%. **1B. (a)** $\chi[H_2O] = 0.927$; **(b)** 3.73 M; **(c)** 4.34 m. **2A.** 0.03593. **2B. (a)** 0.3038 M; **(b)** 0.3426 m; **(c)** 0.005815. **3A.** oxalic acid. **3B.** CCl_4. **4A. (1)** 31 g; **(2)** 46 g NH_4Cl crystallized if both suggestions (1) and (2) are followed. **4B.** $KClO_4 > KNO_3$ $> K_2SO_4$. **5A.** 0.383 atm. **5B.** 6.33 atm. **6A.** 112 mmHg hexane, 239 mmHg total. **6B.** 13.0 mmHg toluene, 64.5 mmHg total. **7A.** $\chi_{hexane} = 0.469$. **7B.** $\chi_{toluene} = 0.202$. **8A.** 0.857 atm. **8B.** 8.24 mg. **9A.** 8.52×10^4 g/mol. **9B.** 0.0105 atm. **10A. (a)** 0.126 m; **(b)** 365 g/mol; **(c)** $C_{17}H_{20}O_6N_4$. **10B.** lower than 760.0 mmHg. **11A.** 3.9 atm. **11B.** 0.56 mL.

Review Questions. 4. 53.7%. **5. (a)** 17.1% (V/V); **(b)** 13.9% (m/m); **(c)** 13.7% (m/V). **6.** 1.83×10^4 L. **7.** 0.683 g. **8.** 0.410 m. **9.** 1.85 M. **10.** $\chi_7 = 0.187$, 18.7%, $\chi_8 = 0.427$, 42.7%, $\chi_9 = 0.386$, 38.6%. **11. (a)** 7.80 M; **(b)** 24.4 M; **(c)** 34.0 m; **(d)** $\chi_{glycerol} = 0.242$; **(e)** 75.8 mol % H_2O. **12.** unsaturated. **13. (d)**, both are nonpolar. **14.** $NH_2OH(s)$, because of hydrogen bonding. **15. (a)** The hydrocarbon chain promotes the solubility of C_4H_9OH in benzene, and hydrogen bonding promotes its solubility in water. **16.** 4.48×10^{-3} M. **17.** 40.6 mmHg benzene, 56.9 mmHg total. **18.** $\chi_{benzene} = 0.713$. **19.** 0.050 m C_2H_5OH. **20.** $C_2H_5OH > HC_2H_3O_2 > NaCl >$ $MgBr_2 > Al_2(SO_4)_3$. **21. (a)** 0.16 M Na^+; **(b)** 0.32 M; **(c)** 8.1 atm; **(d)** -0.60 °C. **22.** 2.7×10^4 g/mol. **23.** 1.2×10^2 g/mol. **24.** 100.02 °C.

Exercises. 25. salicyl alcohol. **27. (c)** formic acid and **(f)** propylene glycol are soluble in water; **(b)** benzoic acid and **(d)** butyl alcohol are only slightly soluble; **(a)** iodoform and **(e)** chlorobenzene are insoluble. **29.** solubilities: $CaF_2 <$ $MgF_2 <$ NaF $<$ KF. **31.** only true in cases when the other component's density (for example, H_2O) is greater than ethanol's. **33.** 21.6 g. **35.** 135 mL. **37.** 54.8 g. **39.** 4.19 M, 5.25 m. **41. (a)**

$\chi_{ethanol}$ = 0.0979; **(b)** χ_{urea} = 0.0122. **43.** 4.93 g. **45.** 8.66 *m*. **47. (a)** unsaturated; **(b)** about 3 g. **49.** 4×10^2 g. **51.** 1.4×10^{-5} M. **53.** 23.2 mmHg. **55.** $\chi_{styrene}$ = 0.32. **57.** Water is drawn out by osmosis. **59.** 22.4 L; ideal gas law. **61.** 2.8×10^5 g/mol. **63.** 21 atm. **65. (a)** 20. °C/*m*; **(b)** cyclohexane, because of larger K_f. **67.** $C_6H_4N_2O_4$. **69.** C_4H_4S. **71. (a)** −0.19 °C; **(b)** −0.37 °C; **(c)** −0.37 °C; **(d)** −0.56 °C; **(e)** −0.37 °C; **(f)** −0.19 °C; **(g)** < −0.19 °C. **73.** $NH_3(aq)$ + $HC_2H_3O_2(aq)$ → $NH_4^+(aq)$ + $C_2H_3O_2^-(aq)$ (a salt: stong electrolyte).

Advanced Exercises. 76. (a) unsaturated solution; **(b)** saturated solution—its concentration does not change; **(c)** saturated solution. **77.** 1.2×10^2 g NaCl. **80.** 29.3%. **81.** 682 g H_2O. **86.** 4.5 kg H_2O. **88.** about 10%. **94.** 34.62% O_2 by volume. **95.** 1.3×10^2 g.

Feature Problems. A. solutions that are more than 8.6 M. **B. 1.** > 0.50; **2.** < 0.05; **3.** The more volatile component is depleted as the solution boils; **4.** χ_{HCl} = 0.12 and 110 °C; **5.** χ_{HCl} = 0.111. **C. 1.** 90.66%; **2.** yes; **3.** high water solubility. **D. 1.** 0.92 % m/V has t_f = −0.60 °C; **2.** t_f = −0.62 °C.

Chapter 15

Practice Examples. 1A. 8.93×10^{-5} M s⁻¹. **1B.** 1.62×10^{-4} M s⁻¹. **2A. (a)** 3.3×10^{-4} M s⁻¹; **(b)** 0.37 M. **2B.** 2.17 M. **3A.** first-order in N_2O_5. **3B.** 3.8×10^{-7} M min⁻¹. **4A.** 4.4×10^{-2} M⁻² s⁻¹. **4B.** 2.4×10^{-7} M s⁻¹. **5A.** 1.0 M. **5B.** first-order, with $k = 7.3 \times 10^{-4}$ s⁻¹. **6A.** 64.2%. **6B.** 25.2 min. **7A.** 1.9×10^{-7} mmHg. **7B.** 272 mmHg. **8A.** first-order (constant half-life) with $k = 6.93 \times 10^{-3}$ s⁻¹. **8B.** zero-order from a plot of [A] vs. *t*, with $k = 9.28 \times 10^{-3}$ M/min. **9A.** 43 s. **9B.** T_2 = 311 K. **10A.** $2 NO_2 \rightarrow NO + NO_3$; rate = $k_1[NO_2]^2$. **10B.** Rate = $(k_1k_3/k_2)[NO_2F_2]$, agrees with experiment.

Review Questions. 4. (a) 2.3×10^{-3} M min⁻¹; **(b)** 3.8×10^{-5} M sec⁻¹. **5. (a)** 3.1×10^{-4} mol B L⁻¹ s⁻¹; **(b)** 9.3×10^{-4} mol D L⁻¹ s⁻¹. **6. (a)** 8.7×10^{-4} M s⁻¹; **(b)** 4.4×10^{-4} M s⁻¹. **7. (b)** is correct. **8.** 1.39×10^{-4} M/min. **9.** 0.0221 M min⁻¹. **10. (a)** 60.0 min; **(b)** 45 min. **11. (a)** 42.0 min; **(b)** 0.0165 min⁻¹. **12. (a)** 19 min; **(b)** 0.18 g. **13.**

(a) 0.0160 min⁻¹; **(b)** 43.3 min; **(c)** 77.8 min; **(d)** 0.074 M. **14. (a)** first order in A, second order in B; **(b)** third order overall; **(c)** 0.102 M⁻² s⁻¹. **15.** second order: non-constant rate, doubling of half-life for each half-life period. **16.** (d). **17. (a)** 160 kJ/mol; **(b)** *T* = 645 K. **18. (a)** II; **(b)** I; **(c)** III. **19.** 0.0100 M/s. **20.** slightly less than 75 s. **21.** 0.0080 M s⁻¹. **22. (a)** 0.010 M s⁻¹; **(b)** 0.0048 M s⁻¹; **(c)** 0.0034 M s⁻¹. **23. (a)** 0.00 M; **(b)** 0.32 M; **(c)** 0.47 M. **24. (a)** rate = $k_{slow}[A][B]$; **(b)** rate = (k_1k_3/k_2) $[A][B]^2$.

Exercises. 25. 1.0×10^{-3} M s⁻¹. **27. (a)** 0.575 M; **(b)** 5.4 min. **29. (a)** 3.52×10^{-5} M/s; **(b)** 0.3569 M; **(c)** 4.5×10^2 s. **31. (a)** 3000 mmHg; **(b)** 1400 mmHg. **33.** Rate = $(5.8 \times 10^{-3}$ L mol⁻¹ min⁻¹$)[A]^1[B]^1$. **35. (a)** true; **(b)** false, half-life is independent of [A]. **37. (a)** 3.13%; **(b)** 0.00193 M/s. **39.** 20.6 min. **41. (a)** 218 min; **(b)** 2.1 L. **43. (a)** first order, constant rate constant; **(b)** 1.86×10^{-3} s⁻¹; **(c)** 0.148 M. **45. (a)** graph 1/[ArSO₂H] vs. time; **(b)** 0.136 L mol⁻¹ min⁻¹; **(c)** 73.5 min; **(d)** 221 min; **(e)** 137 min. **47. (a)** 0.022 M/min and 0.089 M/min; **(b)** second order. **49.** 6.5×10^{-4} s⁻¹. **51.** Rate = $k[A]^2$ and k = 0.020 L mol⁻¹ min⁻¹. **53. (a)** need fraction of successful collisions also; **(b)** fraction with sufficient energy to react increases dramatically, whereas average kinetic energy increases more gradually; **(c)** catalyst effectively decreases activation energy. **55. (a)** 63 kJ/mol. **57. (b)** 113 kJ/mol; **(c)** 2.2×10^2 s. **59. (a)** 34.8 kJ/mol; **(b)** 61 °C. **61. (a)** 51 kJ/mol; **(b)** no, it does not. **63. (a)** A catalyst reacts but is recovered unchanged. **(b)** A catalyst lowers E_a by *changing* the mechanism of a reaction. **65.** Platinum, unlike an enzyme, is rather nonspecific in the reactions it catalyzes. **67.** Molecularity and reaction order correspond only for elementary processes. **69.** second step: $N_2O_2 + H_2$ $\xrightarrow{k_3} N_2O + H_2O$; (k_1k_3/k_2) $[H_2]$ $[NO]^2$. **Advanced Exercises. 71. (a)** first-order; **(b)** k = 0.29 min⁻¹; **(c)** 0.10 M/min; **(d)** 0.064 M/min; **(e)** 0.29 M/min. **72.** 1.86 M. **75.** M^{1-o} s⁻¹. **80.** 1.9×10^2 min.

Feature Problems. A. 4. 1.1×10^{-3} M min⁻¹; **5.** 5.5×10^{-3} M min⁻¹; **6.** k_{avg} = 0.071 min⁻¹; **7.** about 10.5 min; **8.** about 21 min; **10.** 0.0661 min⁻¹. **B.**

1. first-order in each reactant, second-order overall; **2.** 3.7×10^{-5} M s⁻¹; **3.** 6.2×10^{-3} M⁻¹ s⁻¹; **4.** k_3 = 0.0028 M⁻¹ s⁻¹, k_{13} = 0.0060 M⁻¹ s⁻¹, k_{24} = 0.013 M⁻¹ s⁻¹, k_{33} = 0.025 M⁻¹ s⁻¹; **5.** 51 kJ/mol; **6.** Two anions will repel each other.

Chapter 16

Practice Examples. 1A. 0.263 M. **1B.** 0.25 M. **2A.** 7.1×10^2. **2B.** 1.7×10^{-6}. **3A.** 1.7×10^{-6}. **3B.** 1.1×10^{-4}. **4A.** K_c = $[Ca^{2+}]^5[HPO_4^{2-}]^3/[H^+]^4$. **4B.** K_c = $[H_2]^4/[H_2O]^4$; K_p = $(P_{H_2})^4/(P_{H_2O})^4$; $K_p = K_c$. **5A.** to the right, forming more products. **5B.** At equilibrium there will be more of reactants, and less of products than initially. **6A.** Equilibrium shifts to the right. **6B. (a)** no effect; **(b)** shifts to the left. **7A.** $[N_2O_4]$ will increase and $[NO_2]$ will decrease. **7B.** no effect. **8A.** high. **8B.** $[NH_3(g)]$ will be greater at −100 °C. **9A.** 2.3×10^{-4}. **9B.** 25.2 g. **10A.** K_c = 1.5×10^{-4}, K_p = 3.7×10^{-3}. **10B.** 0.022. **11A.** 0.481 atm. **11B.** 0.716 atm. **12A.** 0.262 mol. **12B. (a)** decrease. **(b)** 0.0118 mol. **13A.** $[Ag^+]$ = $[Fe^{2+}]$ = 0.489 M; $[Fe^{3+}]$ = 0.71 M. **13B.** $[V^{3+}]$ = $[Cr^{2+}]$ = 0.0058 M; $[V^{2+}]$ = $[Cr^{3+}]$ = 0.154 M.

Review Questions. 4. (a) $[NO_2]^2/$ $[NO]^2[O_2]$; **(b)** $[Zn^{2+}]/[Ag^+]^2$; **(c)** $[OH^-]^2/[CO_3^{2-}]$. **5. (a)** $P\{CH_4\}P\{H_2S\}^2/(P\{CS_2\}P\{H_2\}^4)$; **(b)** $P\{O_2\}^{1/2}$; **(c)** $P\{CO_2\}P\{H_2O\}$. **6. (a)** $[N_2O]/([N_2][O_2]^{1/2})$; **(b)** $[HBr]/([H_2]^{1/2}[Br_2]^{1/2})$; **(c)** $[NH_3]/([N_2]^{1/2}[H_2]^{3/2})$; **(d)** $[ClF_3]/([Cl_2]^{1/2}[F_2]^{3/2})$; **(e)** $[NOF_2]/([N_2]^{1/2}[O_2]^{1/2}[F_2])$. **7. (a)** 0.0431; **(b)** 3.1×10^3; **(c)** 66. **8.** 9.7×10^{-16}. **9.** (d): 1 mol I_2 is the limiting reactant. **10.** 1.9. **11.** 2.0×10^{-19}. **12. (a)** 0.0578 mol; **(b)** 0.232 mol. **13. (a)** 0.0431; **(b)** 43; **(c)** 6.1. **14.** 1.2×10^2. **15. (b)** 0.659. **16. (a)** 26.3; **(b)** 0.613. **17. (a)** 0.011; **(b)** 1.3. **18. (a)** not at equilibrium; **(b)** to the right. **19. (a)** shift right; **(b)** no shift; **(c)** shift right. **20. (b)** and **(d)** shift right with raised temperatures. **21. (a)** increase; **(b)** decrease; **(c)** decrease; **(d)** no effect; **(e)** decrease. **22.** 0.234 mol HI, 0.033 mol each of H_2 and I_2. **23.** 0.0165 mol I_2. **24.** 1.154 atm.

Exercises. 25. (a) $2 COF_2(g) \rightleftharpoons CO_2(g) + CF_4(g)$, $K_c = [CO_2][CF_4]/[COF_2]^2$; **(b)** $Cu(s) + 2 Ag^+(aq) \rightleftharpoons Cu^{2+}(aq) + 2 Ag(s)$, $K_c = [Cu^{2+}]/[Ag^+]^2$; **(c)** $S_2O_8^{2-}(aq) + 2 Fe^{2+}(aq) \rightleftharpoons 2 SO_4^{2-}(aq) + 2 Fe^{3+}(aq)$, $K_c = [SO_4^{2-}]^2 [Fe^{3+}]^2/([S_2O_8^{2-}] [Fe^{2+}]^2)$. **27. (a)** 0.0012; **(b)** 5.55×10^5; **(c)** 0.429. **29.** $K_p = 0.0313$, $K_c = 1.28 \times 10^{-3}$. **31.** 1.8×10^6. **33.** 0.106. **35.** 0.516. **37. (a)** $NH_3(g) + \frac{7}{4} O_2(g) \rightleftharpoons NO_2(g) + \frac{3}{2} H_2O(g)$; **(b)** 4.03×10^{19}. **39.** no, a net forward reaction. **41.** 0.116 mol $SbCl_5$, 0.17 mol $SbCl_3$, 0.0440 mol Cl_2. **43.** 0.949 g. **45. (a)** no; **(b)** to the right; **(c)** 0.112 mol COF_2, 0.279 mol CO_2, 0.0905 mol CF_4. **47.** 78.0% HI. **49.** 5.58 atm. **51.** $[Fe^{3+}] = 0.56$ M, $[Ag^+] = [Fe^{2+}] = 0.44$ M. **53.** 0.525 atm O_2, 0.597 atm total. **55.** 30.1 atm $COCl_2$, 32.4 atm total. **57.** causes the reaction to continually shift right. **59. (a)** decrease; **(b)** increase; **(c)** no effect; **(d)** no effect. **61. (a)** increased [NO]; **(b)** more rapid. **63.** to increase greatly the reaction rate.
Advanced Exercises. 68. 2.79 atm. **75.** 0.177. **77.** Substitute expressions such as $\chi\{NH_3\} P_{tot}$ for $P\{NH_3\}$. **79.** 3.36×10^{-4}. **81.** 0.085 mol CH_4, 0.070 mol H_2O, 0.12 mol CO_2, 0.16 mol H_2.
Feature Problems. A. $K_c = 4.0$. **B.** Use a high temperature and a high pressure; cool equilibrium mixture to liquefy and remove NH_3; recycle unreacted N_2 and H_2. **C. 1.** 9.23×10^{-3}. **5.** $P\{CH_3OH\} = 24.6$ atm.

Chapter 17

Practice Examples. 1A. acids: **(a)** HF and H_3O^+; **(b)** HSO_4^- and NH_4^+; **(c)** HCl and $HC_2H_3O_2$; the other species are bases. **1B.** $HNO_2(aq) + H_2O \rightleftharpoons NO_2^-(aq) + H_3O^+(aq)$, $HCO_3^-(aq) + H_2O \rightleftharpoons CO_3^{2-}(aq) + H_3O^+(aq)$, $PO_4^{3-}(aq) + H_2O \rightleftharpoons HPO_4^{2-}(aq) + OH^-(aq)$, $HCO_3^-(aq) + H_2O \rightleftharpoons H_2CO_3(aq) + OH^-(aq)$. **2A.** $[H_3O^+] = 1.4 \times 10^{-3}$ M, $[OH^-] = 7.1 \times 10^{-12}$ M. **2B.** 3.1 L. **3A.** $[I^-] = 0.0025$ M, $[OH^-] = 4.0 \times 10^{-12}$ M, pH = 2.60. **3B.** 1.466. **4A.** 11.52. **4B.** 13.74. **5A.** 2.9×10^{-8}. **5B.** 2.6×10^{-6}. **6A.** 1.80. **6B.** 2.66. **7A.** 2.29. **7B.** 11.28. **8A.** 5.5% in 0.20 M, 17% in 0.020 M. **8B.** 1.4×10^{-4}. **9A.** 3.7×10^{-2}

M = $[H_3O^+]$ = $[HOOCCH_2COO^-]$, $[^-OOCCH_2COO^-] = 2.0 \times 10^{-6}$ M. **9B.** $Ka_2 = 5.3 \times 10^{-5}$, $Ka_1 = 5.3 \times 10^{-2}$. **10A.** 0.011 M = $[SO_4^{2-}]$, $[H_3O^+]$ = 0.21 M, $[HSO_4^-]$ = 0.19 M. **10B.** 0.0060 M = $[SO_4^{2-}]$, $[HSO_4^-]$ = 0.014 M, $[H_3O^+]$ = 0.026 M. **11A. (a)** acidic; **(b)** neutral; **(c)** basic. **11B.** $H_2PO_4^-(aq) + H_2O \rightleftharpoons H_3O^+(aq) + HPO_4^-(aq)$, $H_2PO_4^-(aq) + H_2O \rightleftharpoons OH^-(aq) + H_3PO_4(aq)$; acid ionization occurs to the greater extent. **12A.** codeine hydrochlorine. **12B.** basic. **13A.** 8.08. **13B.** 3.6×10^{-3} M. **14A.** $HClO_4$; CH_2FCOOH. **14B.** H_2SO_3; CCl_2FCH_2COOH. **15A. (a)** BF_3 is a Lewis acid, NH_3 is a Lewis base; **(b)** H_2O is a Lewis base, Cr^{3+} is a Lewis acid. **15B.** Hydroxide ion and chloride ion are the Lewis bases; $Al(OH)_3$ and $SnCl_4$ are the Lewis acids.
Review Questions. 4. (a) acid; **(b)** base; **(c)** base; **(d)** acid; **(e)** acid. **5. (a)** IO_3^-; **(b)** $C_6H_5COO^-$; **(c)** PO_4^{3-}; **(d)** $C_2H_5NH_2$. **6.** $[H_3O^+]$ = **(a)** 0.00165 M; **(b)** 1.1×10^{-12} M; **(c)** 2.3×10^{-12} M; **(d)** 5.8×10^{-4} M; $[OH^-]$ = **(a)** 6.1×10^{-12} M; **(b)** 0.0087 M; **(c)** 0.00426 M; **(d)** 1.7×10^{-11} M. **7. (a)** 2.35; **(b)** 3.212; **(c)** 11.83; **(d)** 11.98. **8.** $[H_3O^+]$ = 0.11 M. **9.** 2.7. **10.** pH < 13. **11.** 1.30×10^{-5}. **12. (a)** 0.0030 M; **(b)** 2.62. **13.** 1.4 g. **14.** 0.028 M. **15. (a)** 1.4×10^{-4} M; **(b)** 1.4×10^{-4} M; **(c)** 4.7×10^{-11} M. **16.** Just products are written: **(a)** $NH_3(aq) + NO_3^-(aq) + H_3O^+(aq)$; **(b)** $Na^+(aq) + HNO_2(aq) + OH^-(aq)$; **(c)** $K^+(aq) + HC_7H_5O_2(aq) + OH^-(aq)$; **(d)** no reaction; **(e)** $C_6H_5NH_2(aq) + Cl^-(aq) + H_3O^+(aq)$. **17.** 4.51. **18. (a)** 6.7×10^{-6}; **(b)** 5.6×10^{-11}; **(c)** 1.0×10^{-4}. **19.** 8.58. **20.** Electronegative Cl atoms withdraw electrons from the O—H bond. **21. (a)** HI; **(b)** HOClO; **(c)** $H_3CCH_2CCl_2COOH$. **22. (a)** base; **(b)** acid; **(c)** base. **23.** Acids are: **(a)** SO_3; **(b)** $Zn(OH)_2$. **24.** decreasing pH: (b) > (c) > (d) > (a).
Exercises. 25. Bases are: **(a)** H_2O, OBr^-; **(b)** H_2O, SO_4^{2-}; **(c)** HS^-, OH^-; **(d)** OH^-, $C_6H_5NH_2$. **27.** NH_3. **29. (a)** forward; **(b)** reverse; **(c)** reverse. **31.** 4.0×10^{-13} M, 12.40. **33.** 1.97×10^{-3} M. **35.** 8.5 mL. **37.** 0.0539 L. **39.** 10.20. **41.** 9.8×10^{-3} M; 2.01. **43.** 2.7×10^{-3}. **45.** 0.073 M = $[H_3O^+]$, $[OH^-]$ = 1.4×10^{-13} M, pH = 1.14, pOH = 12.86. **47.** 11.04. **49.** 0.063 g. **51. (a)** 0.0053; **(b)** 0.53%. **53.** 0.0098 M. **55.** little PO_4^{3-}

is produced in the third ionization. **57.** $[S^{2-}] = K_2 = 1 \times 10^{-19}$ M, **(a)** 8.7×10^{-5} M = $[H_3O^+]$ = $[HS^-]$; **(b)** 2.2×10^{-5} M = $[H_3O^+]$ = $[HS^-]$; **(c)** 9.5×10^{-7} M = $[H_3O^+]$ = $[HS^-]$. **59.** 2.0×10^{-3} M = $[H_3O^+]$ = $[HOOC(CH_2)_4COO^-]$, $[OH^-] = 5.0 \times 10^{-12}$ M, $[HOOC(CH_2)_4COOH] = 0.10$ M, $[^-OOC(CH_2)_4COO^-] = 3.9 \times 10^{-6}$ M. **61. (a)** neutral; **(b)** basic; **(c)** neutral; **(d)** basic; **(e)** acidic. **63.** 10.23. **65.** 9.18. **67. (a)** $HSO_3^- + H_2O \rightleftharpoons H_3O^+ + SO_3^{2-}$, $HSO_3^- + H_2O \rightleftharpoons OH^- + H_2SO_3$, acidic; **(b)** $HS^- + H_2O \rightleftharpoons H_3O^+ + S^{2-}$, $HS^- + H_2O \rightleftharpoons OH^- + H_2S$, basic; **(c)** $HPO_4^{2-} + H_2O \rightleftharpoons H_3O^+ + PO_4^{3-}$; $HPO_4^{2-} + H_2O \rightleftharpoons OH^- + H_2PO_4^-$; basic. **69. (a)** $HClO_3$; **(b)** HNO_2; **(c)** H_3PO_4. **71.** Lewis bases: **(a)** OH^-; **(b)** N_2H_4; **(c)** $(C_2H_5)_2O$.
Advanced Exercises. 75. (a) 13.11; **(b)** 0.13 M. **76.** 8.6. **80.** 6.70. **81.** 5×10^{-5}. **82. (b)** 2.7×10^{-5} mL; **(c)** 0.096 g; **(d)** 2.3×10^{-5} g. **87.** 9.73 g.
Feature Problems. A. 1. CHO and $C_4H_4O_4$; **2.** $H_2C_4H_2O_4$; **3.** $HOOCCH_2CH_2COOH$; **4.** 1.2×10^{-2}; **5.** 1.72. **B.** 1.14. **C. 1.** 2.15; **2.** 11.95; **3.** pH = 9.23.

Chapter 18

Practice Examples. 1A. 0.500 M HF: $[H_3O^+]$ = 0.018 M, $[HF]$ = 0.482 M; 0.500 M HF/0.100 M HCl: $[H_3O^+]$ = 0.103 M; $[HF]$ = 0.497 M. **1B.** 30. drops. **2A.** $[H_3O^+] = 1.2 \times 10^{-4}$ M, $[CHO_2^-]$ = 0.150 M. **2B.** 15 g. **3A.** A reaction occurs producing acetic acid, forming an acetic acid–acetate ion buffer. **3B.** A reaction occurs producing an ammonia–ammonium chloride buffer. **4A.** 3.96. **4B.** 4.51. **5A.** 20.8 g. **5B.** The calculated pH is 5.09. **6A. (a)** 3.54; **(b)** 3.53; **(c)** 3.55. **6B.** 1.3 mL. **7A. (a)** 0.824; **(b)** 1.238; **(c)** 7.00; **(d)** 11.98. **7B. (a)** 12.21; **(b)** 11.79; **(c)** 7.00. **8A. (a)** 2.00; **(b)** 2.70; **(c)** 3.18; **(d)** 8.08. **8B. (a)** 11.14; **(b)** 9.74; **(c)** 9.26; **(d)** 5.20. **9A.** 12.18. **9B.** 10.45.
Review Questions. 4. (a) 0.0892 M; **(b)** 1.1×10^{-13} M; **(c)** 4.0×10^{-5} M; **(d)** 0.0892 M. **5. (a)** 2.9×10^{-5} M; **(b)** 0.102 M; **(c)** 0.102 M; **(d)** 3.4×10^{-10} M. **6. (a)** $CHO_2^-(aq) + H_3O^+(aq) \rightarrow HCHO_2(aq) + H_2O$, $HCHO_2(aq) + OH^-(aq) \rightarrow CHO_2^-(aq) + H_2O$; **(b)** $C_6H_5NH_2(aq) + H_3O^+(aq) \rightarrow$

$C_6H_5NH_3^+$(aq) + H_2O, $C_6H_5NH_3^+$(aq) + OH^-(aq) → $C_6H_5NH_2$(aq) + H_2O; **(c)** HPO_4^{2-}(aq) + H_3O^+(aq) → $H_2PO_4^-$(aq) + H_2O, $H_2PO_4^-$(aq) + OH^-(aq) → HPO_4^{2-}(aq) + H_2O. **7. (a)** 4.64; **(b)** 9.68. **8.** 0.77 M. **9.** 0.53 M. **10.** 0.60 mol $NaC_2H_3O_2$. **11.** 1.4×10^{-4}. **12. (a)** 9.5 g; **(b)** 3.86. **13. (a)** neutral: bromothymol blue, basic: thymolphthalein, remainder in acidic solution; **(b)** 4.7, 6.0. **14. (a)** colorless; **(b)** red; **(c)** blue; **(d)** blue. **15. (a)** 36.3 mL; **(b)** 25.8 mL. **16.** equivalence point **(a)** pH = 7.00, bromothymol blue; **(b)** pH = 4.62, methyl red; **(c)** pH = 9.38, phenolphthalein; **(d)** pH = 10.6, alizarin yellow R; neutralization to PO_4^{3-}(aq) is not possible. **17. (a)** 1.346; **(b)** 2.034. **18. (a)** 11.85; **(b)** 1.426. **19. (a)** 2.87; **(b)** 3.51. **20. (a)** 8.79; **(b)** 1.66. **21. (a)** 3.10; **(b)** 4.20; **(c)** 8.00; **(d)** 11.11. **22.** $NaHSO_4$. **Exercises. 23. (a)** +1.07 units; **(b)** pH unaffected. **25. (a)** 0.035 M; **(b)** 4.0×10^{-4} M; **(c)** 3.9×10^{-5} M. **27.** Buffer solutions are: **(c), (e),** and **(f). 29.** 8.88. **31. (a)** 4.8 g; **(b)** 0.4 g $(NH_4)_2SO_4$. **33.** 9.16. **35.** 100. mL 0.100 M $HCHO_2$, 58 mL 0.100 M $NaCHO_2$. **37. (a)** pH = 3.89 to pH = 5.89; **(b)** 0.100 mol of acid or base per liter. **39. (a)** 3.48; **(b)** 3.52; **(c)** 1.23. **41. (a)** 9.01; **(b)** Yes for dilution to 1.00 L but not to 1000 L, because the buffer components are too dilute; **(c)** 8.99; **(d)** 1.1 mL. **43. (a)** Each indicator only serves to fix the pH over a quite small region; **(b)** An indicator is a weak acid and consumes some base in an acid–base titration. **45. (a)** red; **(b)** yellow; **(c)** yellow; **(d)** yellow; **(e)** red; **(f)** yellow. **47.** 0.1508 M. **49.** 11.63. **51.** In each case, the volume of acid and its molarity are the same. **53. (a)** initial pH = 13.00, equivalence point pH = 7.00, 12.5 mL HI soln, phenophthalien; **(b)** initial pH = 11.62, equivalence point pH = 4.96, 40.0 mL HCl soln, methyl red. **55. (a)** 11.9 mL; **(b)** 17.4 mL; **(c)** 17.5 mL. **57. (a)** initial pH = 1.59, equiv pH = 7.43, bromothymol blue; **(b)** initial pH = 2.26, equiv pH = 8.11, thymol blue; **(c)** initial pH = 4.35, equiv pH = 10.20, alizarin yellow R. **59.** basic. **61. (a)** H_3PO_4(aq) + CO_3^{2-}(aq) → $H_2PO_4^-$(aq) + HCO_3^-(aq), $H_2PO_4^-$(aq) + CO_3^{2-}(aq) → HPO_4^{2-}(aq) + HCO_3^-(aq), HPO_4^{2-}(aq) + OH^-(aq) → PO_4^{3-}(aq)

+ H_2O; **(b)** Na_2CO_3 is not a strong enough base to cause the third ionization of H_3PO_4. **63. (a)** 0.0038 M; **(b)** 0.49 M. **65. (a)** not possible; **(b)** Solution ends up with an appreciable $[HC_2H_3O_2]$; **(c)** yes; **(d)** We end up with $BaCl_2$(aq), NH_3(aq), unreacted NH_4Cl(aq); **(e)** yes; **(f)** not possible, solution would be basic. **Advanced Exercises. 69. (a)** HSO_4^-(aq) + OH^-(aq) → SO_4^{2-}(aq) + H_2O; **(b)** 2.8%; **(c)** bromothymol blue or phenol red. **70.** 4.20. **71. (a)** no sudden change in color; **(b)** 18%. **74. (a)** pH = pK_a + log $[f/(1 - f)]$; **(b)** 9.57. **76.** 3 drops. **78. (a)** rxn(1): 1.0×10^{-14}, rxn(2): 1.8×10^9; **(b)** the large size of the equilibrium constant.

Feature Problems. A. 1. the half equivalence point, each fraction is 0.50, pH = 4.74. **C. 2.** 0.216 M HCl and 0.216 M H_3PO_4; **3.** equivalence points come at 20.00 mL NaOH (pH = 4.68) and 47.50 mL NaOH (pH = 9.79).

Chapter 19

Practice Examples. 1A. $[Mg^{2+}][CO_3^{2-}]$, $[Ag^+]^3[PO_4^{3-}]$. **1B.** $CaHPO_4$(s) ⇌ Ca^{2+}(aq) + HPO_4^{2-}(aq), K_{sp} = $[Ca^{2+}][HPO_4^{2-}]$. **2A.** 3×10^{-7}. **2B.** 1.9×10^{-9}. **3A.** 2×10^{-10} M. **3B.** 0.55 mg. **4A.** 1.3×10^{-4} M. **4B.** 1×10^{-20} M. **5A.** Precipitation should occur. **5B.** 10 drops. **6A.** Precipitation is not complete. **6B.** 0.02 M. **7A.** 0.12%. **7B.** $[CO_3^{2-}]$ = 2.1×10^{-5} M. **8A.** Precipitation will not occur. **8B.** Precipitation should occur. **9A.** 0.074 M. **9B. (a)** 3.71; **(b)** 0.093. **10A. (a)** Cu^{2+}(aq) + 2 OH^-(aq) → $Cu(OH)_2$(s); **(b)** $Cu(OH)_2$(s) + 4 NH_3(aq) ⇌ $[Cu(NH_3)_4]^{2+}$(aq) + 2 OH^-(aq); **(c)** $[Cu(NH_3)_4]^{2+}$(aq) + 4 H_3O^+(aq) ⇌ $[Cu(H_2O)_4]^{2+}$(aq) + 4 NH_4^+(aq). **10B. (a)** Zn^{2+}(aq) + 4 NH_3(aq) ⇌ $[Zn(NH_3)_4]^{2+}$(aq); **(b)** $[Zn(NH_3)_4]^{2+}$(aq) + 4 H_3O^+(aq) ⇌ $[Zn(H_2O)_4]^{2+}$(aq) + 4 NH_4^+(aq); **(c)** $[Zn(H_2O)_4]^{2+}$(aq) + 2 OH^-(aq) ⇌ $Zn(OH)_2$(s) + 4 H_2O(l); **(d)** $Zn(OH)_2$(s) + 2 OH^-(aq) ⇌ $[Zn(OH)_4]^{2-}$(aq). **11A.** Precipitation should occur. **11B.** Precipitation will not occur. **12A.** 0.84 M. **12B.** 0.20 M. **13A.** 4×10^{-6} M. **13B.** Solubility decreases, as does K_{sp}. **14A.**

Compute the values of Q_{spa} and compare with K_{spa} values. **14B.** 2.7. **Review Questions. 4. (a)** $[Ag^+]^2[SO_4^{2-}]$; **(b)** $[Ra^{2+}][IO_3^-]^2$; **(c)** $[Ni^{2+}]^3[PO_4^{3-}]^2$; **(d)** $[PuO_2^+]^2[CO_3^{2-}]$. **5. (a)** $Fe(OH)_3$(s) ⇌ Fe^{3+}(aq) + 3 OH^-(aq); **(b)** $BiOOH$(s) ⇌ BiO^+(aq) + OH^-(aq); **(c)** Hg_2I_2(s) ⇌ Hg_2^{2+}(aq) + 2 I^-(aq); **(d)** $Pb_3(AsO_4)_2$(s) ⇌ 3 Pb^{2+}(aq) + 2 AsO_4^{3-}(aq). **6. (a)** $[Cr^{3+}][F^-]^3$; **(b)** $[Au^{3+}]^2[C_2O_4^{2-}]^3$; **(c)** $[Cd^{2+}]^3[PO_4^{3-}]^2$; **(d)** $[Sr^{2+}][F^-]^2$. **7. (a)** $s = 1.1 \times 10^{-5}$ M; **(b)** $s = 2.2 \times 10^{-2}$ M; **(c)** $s = 7 \times 10^{-5}$ M; **(d)** $s = 4.5 \times 10^{-5}$ M. **8. (a)** 1.4×10^{-5}; **(b)** 8.8×10^{-8}; **(c)** 1.9×10^{-9}. **9. (d)** is correct. **10. (a)** $s = 1.7 \times 10^{-4}$ M; **(b)** $s = 7.3 \times 10^{-6}$ M; **(c)** $s = 1.4 \times 10^{-8}$ M. **11. (a)** lowers solubility; **(b)** enhances solubility; **(c)** enhances solubility. **12. (a)** Precipitation should occur; **(b)** Precipitation should not occur; **(c)** Precipitation should not occur. **13.** 0.012% unprecipitated. **14. (a)** AgI; **(b)** $[I^-]$ = 2.7×10^{-4} M; **(c)** $[Ag^+]$ = 3.1×10^{-13} M; **(d)** Ag^+ and Pb^{2+} can be separated. **15.** $NaHSO_4$(aq). **16. (a)** Ag^+(aq) + NO_3^-(aq) + Na^+(aq) + Br^-(aq) → $AgBr$(s) + Na^+(aq) + NO_3^-(aq); **(b)** no reaction; **(c)** no reaction; **(d)** $Cu(OH)_2$(s) + 4 NH_3(aq) → $[Cu(NH_3)_4]^{2+}$(aq) + 2 OH^-(aq); **(e)** Fe^{3+}(aq) + 3 NH_3(aq) + 3 H_2O → $Fe(OH)_3$(s) + 3 NH_4^+(aq); **(f)** Ag_2SO_4(s) + 4 NH_3(aq) → 2 $[Ag(NH_3)_2]^+$(aq) + SO_4^{2-}(aq); **(g)** $CaSO_3$(s) + 2 H_3O^+(aq) → Ca^{2+}(aq) + 3 H_2O + SO_2(g). **17.** $Cu(OH)_2$(s) + 2 H_3O^+(aq) → Cu^{2+}(aq) + 4 H_2O (with both HCl and HNO_3); $Cu(OH)_2$(s) + 4 NH_3(aq) → $[Cu(NH_3)_4]^{2+}$(aq) + 2 OH^-(aq). **18.** HCl(aq). **19.** NH_3(aq). **20.** Precipitation should occur. **21.** Precipitation should not occur. **22.** 2.8×10^{-3} M.

Exercises. 23. AgI < AgCN < $AgIO_3$ < Ag_2SO_4 < $AgNO_2$. **25.** yes. **27.** 11 mg. **29.** 4.5×10^{-3} M. **31.** 5.30×10^{-5}. **33.** Both KI and $AgNO_3$ reduce the solubility through the common ion effect. **35.** 1.5×10^{-5}. **37.** 0.079 M. **39.** not with CrO_4^{2-}; yes with Ag^+. **41.** 27 ppm. **43.** precipitation should occur. **45.** 8.32. **47. (a)** precipitation of AgCl(s) should occur; **(b)** Precipitation of AgBr(s) should occur; **(c)** Precipitation of $Mg(OH)_2$(s) will not occur. **49.** CaC_2O_4

should precipitate from this solution. **51.** 0.052%. **53.** The concentration of the $AgNO_3$ establishes the volume required but does not otherwise affect the separation. **55. (a)** CrO_4^{2-}; **(b)** 1.6×10^{-6} M; **(c)** The two anions are effectively separated. **57.** $MgCO_3$, FeS, $Ca(OH)_2$. **59.** 9.60. **61. (a)** 3.41; **(b)** 0.96 g. **63.** $[PbCl_3]^-$(aq) forms in HCl(aq). **65.** 2.0×10^{30}. **67.** Precipitation of AgI(s) should occur. **69.** 1.4×10^{-6} g. **71.** With $[H_3O^+]$ a bit higher than 1.8×10^{-5} M, separation is complete. **73. (a)** Q_{spa} $(1.7 \times 10^7) < K_{spa}$. **(b)** raise $[C_2H_3O_2^-]$ to 0.21 M. **75.** Pb^{2+} would be missed and might interfere with the test for Ag^+. **77. (a)** and **(c)** are the valid conclusions. **Advanced Exercises. 79.** 68%. **81.** 2.1%. **86.** 2 g MnS/L. **89.** 1.2×10^{-16}. **92.** 6.5×10^{-4} M. **Feature Problems. A.** 3.0×10^{-4}. **B. 1.** 8.32 mL; **2.** 1.3×10^{-5} M; **3.** 0.0065 M; **4. (a)** falsely low value of $[Cl^-]$; **(b)** falsely high value of $[Cl^-]$; **5.** The disappearance of Ag_2CrO_4(s) is hard to establish as an endpoint. **C. 1.** 1.7×10^{-4} M; **2.** 1.7×10^{-4} M; **3.** 2.8×10^{-5} M; **4.** $[Mg^{2+}] = 0.059$ M; **5.** 1.4×10^{-9} M.

Chapter 20

Practice Examples. 1A. (a) and **(c)** are spontaneous. **1B. (a)** and **(d)** are spontaneous. **2A. (a)** $\Delta S < 0$; **(b)** $\Delta S > 0$. **2B. (a)** uncertain; **(b)** $\Delta S > 0$. **3A. (a)** case 2; **(b)** case 4. **3B. (a)** high temperatures; **(b)** low temperatures. **4A.** 83.0 J mol^{-1} K^{-1}. **4B.** 402 J/mol. **5A.** per mol NH_3(g): -99.4 J K^{-1}. **5B.** 312.2 J mol^{-1} K. **6A.** -1484 kJ. **6B.** -70.52 kJ. **7A. (a)** $K_{eq} = P\{SiCl_4\}/P\{Cl_2\}^2 = K_p$; **(b)** $K_{eq} = [HOCl][H^+][Cl^-]/P\{Cl_2\}$. **7B.** $K = [Pb^{2+}]^3 P\{NO\}^2/\{[NO_3^-]^2[H^+]^8\}$. **8A.** 8.5×10^{-17}. **8B.** $K_{eq} = 4 \times 10^{-5}$, no appreciable reaction. **9A.** 607 K. **9B. (a)** 1.5×10^7; **(b)** 1.8×10^5. **10A.** 315 °C. **10B.** 5×10^9. **Review Questions. 4. (a)** increase; **(b)** decrease; **(c)** uncertain; **(d)** decrease. **5. (a)** H_2O(g); **(b)** 50.0 g Fe; **(c)** Br_2(l); **(d)** SO_2. **6. (a)** 3; **(b)** 2; **(c)** 1; **(d)** 4. **7. (b)** is correct. **8. (d)** is correct. **9. (d)** is correct. **10.** -285 J mol^{-1} K^{-1}. **11. (a)** -242.1 kJ; **(b)** $+141.8$ kJ; **(c)** $+101$ kJ; **(d)** -970 kJ. **12. (a)** 114 °C; **(b)** $\Delta G°$ $= 0$. **13. (a)** 85.9 J mol^{-1} K^{-1}; **(b)** 7.014

J mol^{-1} K^{-1}. **14. (a)** $K_{eq} = P\{NO_2(g)\}^2/[P\{NO(g)\}^2 P\{O_2\}] = K_p$ **(b)** $K_{eq} = P\{SO_2(g)\} = K_p$; **(c)** $K_{eq} = [H_3O^+][C_2H_3O_2^-]/[HC_2H_3O_2] = K_c$; **(d)** $K_{eq} = P\{H_2O(g)\} P\{CO_2(g)\} = K_p$; **(e)** $K_{eq} = [Mn^{2+}(aq)] P\{Cl_2(g)\}/\{[H^+]^4 [Cl^-]^2\}$. **15.** 126.6 kJ. **16.** 8.4×10^{-13}. **17. (a)** -76.0 kJ, 2.2×10^{13}; **(b)** -32 kJ, $K_{eq} = 4 \times 10^5$; **(c)** -28.1 kJ, 8.1×10^4. **18.** -42.1 J mol^{-1} K^{-1}. **19. (a)** does not occur to any measurable extent; **(b)** occurs only to a very small extent; **(c)** occurs to some small extent. **20.** 1.36×10^3 K. **21. (a)** 34.4 kJ; **(b)** 9×10^{-7}. **22. (a)** 215 J K^{-1}; **(b)** $+ 91$ kJ; **(c)** 27 kJ; **(d)** 2×10^{-5}. **23. (a)** $\Delta S° = -227.2$ J K^{-1}, $\Delta H° = -165.7$ kJ, $\Delta G° = -98.0$ kJ; **(b)** low temperatures, high pressures; **(c)** 0.55. **24.** 3×10^{16}. **25. (b)** is incorrect; the negative of this expression is correct. **26.** $\Delta G° = - RT \ln K$. **Exercises. 27. (a)** decrease; **(b)** increase; **(c)** increase. **29.** First law states energy is neither created nor destroyed; second law states that total entropy increases in all spontaneous processes. **31. (a)** negative; **(b)** positive; **(c)** positive; **(d)** uncertain; **(e)** negative. **33. (a)** $\Delta H°_{vap} = +44.0$ kJ/mol, $\Delta S°_{vap} = 118.8$ J mol^{-1} K^{-1}; **(b)** Hydrogen bonding in the liquid weakens as the temperature is raised, causing ΔS_{vap} to increase and ΔH_{vap} to decrease. **35.** $C_6H_5CH_3$. **37. (a)** 2; **(b)** no prediction; **(c)** 3. **39.** $\Delta G < 0$, $\Delta H = 0$, $\Delta S > 0$. **41. (a)** it may be that $\Delta S° < 0$; **(b)** the process is endothermic. **43. (a)** -506 kJ; **(b)** forward. **45. (a)** -1.6 kJ, equilibrium; **(b)** -474.23 kJ, go to completion; **(c)** -574.15 kJ, go to completion. **47. (a)** -3176 kJ; **(b)** -3202 kJ; **(c)** $3 \times\{\Delta G_f°[H_2O(g)] - \Delta G_f°[H_2O(l)]\}$ **49.** solid. **51. (a)** $K_{eq} = K_c(RT)^{-1}$; **(b)** $K_{eq} = K_p = K_c$; **(c)** $K_{eq} = K_p = K_c(RT)^2$. **53. (a)** 0.659; **(b)** 3.47 kJ; **(c)** to the right. **55. (a)** -23.4 kJ; **(b)** $+68.89$ kJ; **(c)** $+5.40$ kJ; **(d)** $+28.75$ kJ. **57.** -204.6 kJ mol^{-1}. **59. (a)** $P_{O_2} = 3.0 \times 10^{-21}$ atm; **(b)** high temperature, oxygen was continuously removed. **61.** 6.0×10^{-4}. **63.** 653.5 K. **65. (a)** 0.014; **(b)** 329 K. **67.** -2.1×10^2 kJ, -206.1 kJ. **69. (a)** $\Delta G° = +152.1$ kJ; **(b)** $\Delta G° = -362.3$ kJ. **Advanced Exercises. 73.** (a) and (d) are true. **74.** 0.208 mol Br_2, 0.208 mol Cl_2, 0.584 mol BrCl. **78.** 2.01×10^3 K. **80.** 3.70 atm. **84.** 4 mg AgBr/L.

Feature Problems. A. 1. 8.557 kJ; **2.** 0.03168 bar; **3.** 23.78 mmHg; **4.** 0.03127 atm. **B. 1.** The graph for the formation of MgO(s) is below that of ZnO(s) at all temperatures. **2.** The graph for CO(g) must be below that for ZnO(s), that is, at temperatures above 1000 °C. **3.** Yes, but only above 1850 °C. **4.** No, $\Delta G° > 0$ at all temperatures for this reaction.

Chapter 21

Practice Examples. 1A. Sc(s) + 3 Ag^+(aq) → Sc^{3+}(aq) + 3 Ag(s). **1B.** Al(s) + 3 Ag^+(aq) → Al^{3+}(aq) + 3 Ag(s), Al|Al^{3+}(aq)||Ag^+(aq)|Ag. **2A.** +0.587 V. **2B.** +0.74 V. **3A.** -0.48 V. **3B.** -0.424 V. **4A.** -1587 kJ. **4B.** +1.229 V. **5A.** Copper will displace Ag^+(aq); $E°_{cell} = +0.463$ V. **5B.** Na(s) reacts with H_2O to produce H_2(g) ($E°_{cell} = +1.885$ V) rather than Mg(s) ($E°_{cell} = +0.357$ V). **6A.** $E°_{cell} = -0.78$ V, not a feasible method. **6B.** (1) oxidation of Sn^{2+} to Sn^{4+} by O_2(g) ($E°_{cell} = +1.075$ V); (2) oxidation of Sn(s) to Sn^{2+} ($E°_{cell} = +1.336$ V), and also reaction of Sn(s) with Sn^{4+} to form Sn^{2+} ($E°_{cell} = +0.357$ V). **7A.** $K_{eq} = 10^{204}$; reaction will go to completion. **7B.** $K_{eq} = 2.5$; reaction does not go to completion. **8A.** +1.815 V. **8B.** +0.017 V. **9A.** + 0.467 V, spontaneous as written. **9B.** 2.6. **10A.** +0.23 V. **10B.** 6.9×10^{-9}. **11A.** I_2(s), KOH(aq), and H_2(g). **11B.** Ag(s, anode) → Ag(s, cathode). **12A.** 1.89 A. **12B.** 5.23 h.

Review Questions. 4. (d) is correct. **5. (a)** Fe(s) + Cu^{2+}(aq) → Fe^{2+}(aq) + Cu(s); **(b)** 2 Br^-(aq) + Cl_2(aq) → Br_2(aq) + 2 Cl^-(aq); **(c)** Al(s) + 3 Fe^{3+}(aq) → Al^{3+}(aq) + 3 Fe^{2+}(aq); **(d)** 5 Cl^-(aq) + 6 MnO_4^-(aq) + 18 H^+(aq) → 5 ClO_3^-(aq) + 6 Mn^{2+}(aq) + 9 H_2O; **(e)** S^{2-}(aq) + 2 O_2(g) → SO_4^{2-}(aq). **6. (a)** -1.180 V; **(b)** +0.37 V; **(c)** -1.63 V; **(d)** -1.13 V. **7. (a)** 2 Al(s) + 3 Sn^{2+}(aq) → 2 Al^{3+}(aq) + 3 Sn(s), +1.539 V; **(b)** Fe^{2+}(aq) + Ag^+(aq) → Fe^{3+}(aq) + Ag(s), +0.029 V. **8. (a)** 0.097 V; **(b)** -2.03 V; **(c)** +0.159 V. **9. (a)** spontaneous; **(b)** nonspontaneous; **(c)** nonspontaneous; **(d)** spontaneous. **10.** 1 M HCl: no; 1 M HNO_3: yes. **11. (a)** significant; **(b)** significant; **(c)** barely; **(d)** significant; **(e)** will not. **12.** Reaction stops well short of completion, $K_{eq} = 7$. **13. (a)** and **(b)** will occur. **14. (a)** -1.165×10^3 kJ; **(b)** -268 kJ; **(c)** -3.1×10^2 kJ. **15. (a)** $1 \times 10^{-22} = [Sn^{2+}][Ag^+]^2/[Sn^{4+}]$; **(b)** $4 \times 10^{-5} = [Mn^{2+}] \, P\{Cl_2(g)\}/\{[Cl^-]^2[H^+]^4\}$; **(c)** $1 \times 10^{33} = [Cl^-]^2 \, P\{O_2(g)\}/[OCl^-]^2$. **16. (a)** Fe(s) + Cu^{2+}(aq) → Fe^{2+}(aq) + Cu(s), +0.777 V, electron flow from B to A; **(b)** Sn^{2+}(aq) + 2 Ag^+(aq) → Sn^{4+}(aq) + 2 Ag(s), +0.646 V, electron flow from A to B; **(c)** Zn(s) + Fe^{2+}(aq) → Zn^{2+}(aq) + Fe(s), +0.264V, electron flow from A to B. **17.** 5.0×10^{-6} M. **18. (a)** 0.311 V; **(b)** 0.118 V; **(c)** 0.164 V. **19. (a)** Cl_2(g), Cu(s); **(b)** O_2(g), H_2(g), OH^-(aq); **(c)** Cl_2(g), Ba(l); **(d)** O_2(g), H_2(g). **20.** H_2(g) is produced at the cathode instead of Mg(s). **21. (a)** 3.3 g; **(b)** 0.90 g; **(c)** 11 g; **(d)** 2.9 g. **22.** 1.27 L.

Exercises. 23. (a) $0.337 \, V < E° < 0.800 \, V$; **(b)** $-0.440 \, V < E° < 0.000 \, V$. **25.** +0.755 V. **27.** -2.31 V. **29.** Sodium reacts with water, producing H_2(g) rather than Zn(s). **31. (a)** 3 Ag(s) + NO_3^-(aq) + 4 H^+(aq) → 3 Ag^+(aq) + NO(g) + 2 H_2O; **(b)** Zn(s) + 2 H^+(aq) → Zn^{2+}(aq) + H_2(g). **(c)** Au(s) does not react with 1.00 M HNO_3(aq). **33. (a)** Cu(s) + 2 Fe^{3+}(aq) → Cu^{2+}(aq) + 2 Fe^{2+}(aq), +0.434 V; **(b)** 2 Al(s) + 3 Pb^{2+}(aq) → 2 Al^{3+}(aq) + 3 Pb(s), +1.551 V; **(c)** 2 H_2O + 2 Cl_2(g) → O_2(g) + 4 H^+(aq) + 4 Cl^-(aq), +0.129 V; **(d)** 3 Zn(s) + 2 NO_3^-(aq) + 8 H^+(aq) → 3 Zn^{2+}(aq) + 2 NO(g) + 4 H_2O, +1.719 V. **35. (a)** +0.25 V; **(b)** -2.4×10^2 kJ; **(c)** 1×10^{42}; **(d)** to completion. **37. (a)** not possible; **(b)** reaction to the left. **39.** 1.591 V. **41. (a)** 1.249 V; **(b)** 0.638 V. **43. (a)** E depends on pH only if the reduction half-reaction involves either H^+(aq) or OH^-(aq). **(b)** Reduction of oxoanions is accompanied by the loss of oxygen atoms, requiring H^+ ions on the left. **(c)** Switch from acidic to basic solution by adding OH^- to both sides. On the left side H^+ and OH^- combine to H_2O; OH^- remains on the right. **45. (a)** 6×10^{-38} M; **(b)** goes to completion. **47. (a)** +0.818 V; **(b)** -0.828 V; difference due to reaction reversal and standard (1 M) rather than nonstandard concentration (0.65 M). **49.** 0.177 V. **51. (a)** 0.039 V; **(b)** decrease; **(c)** 0.025 V; **(d)** 0.237 M; **(e)** $[Sn^{2+}] =$

0.50 M, $[Pb^{2+}] = 0.18$ M. **53. (a)** Cr(s) | Cr^{2+}(aq), Cr^{3+}(aq) ‖ Fe^{2+}(aq), Fe^{3+}(aq) | Fe(s); **(b)** +1.195 V. **55. (a)** 1.229 V; **(b)** +1.992 V; **(c)** +2.891 V. **57. (a)** oxidation of iron metal should be enhanced in the body of the nail; **(b)** blue precipitate in the vicinity of the scratch; **(c)** a faint white and no blue precipitate. **59. (a)** electrolysis, 1.229 V; **(b)** spontaneous; **(c)** electrolysis, 0.236 V; **(d)** electrolysis, 0.183 V. **61. (a)** H_2(g), O_2(g); **(b)** 2 H_2O → 2 H_2(g) + O_2(g), $E°_{cell} = -1.229$ V. **63. (a)** 1.62 g; **(b)** 20.2 min. **65. (a)** 1079 C; **(b)** 0.7642 A. **Advanced Exercises. 67.** -0.255 V. **71. (a)** 9.38×10^6 kJ; **(b)** 2.61×10^3 kWh. **74.** 6.9×10^{-4} M. **81.** 1.143 V.

Feature Problems. A. 1.(a) -0.74 V, **(b)** 0.025 V; **2.** 2.013 V. **B. 1.** 0.800 V; **2.** $E = 0.800 + 0.0592 \log [Ag^+]$; **3. (a)** 0.682 V; **(b)** 0.660 V; **(c)** 0.563 V; **(d)** 0.324 V; **(e)** 0.086 V; **(f)** 0.029 V. **C.** Iron behaved as a sacrificial anode. **D.** The pipe will be the negative electrode, where reduction occurs. **E. 1.** aluminum; **2.** 2 Al(s) + 3 Ag_2S(s) → 6 Ag(s) + 3 S^{2-}(aq) + 2 Al^{3+}(aq); **3.** electrolyte. **4.** Several chemicals are involved, and Al is consumed, but with very little loss of mass. **F.** -3.040 V. **G. 1.** Na(s) → Na(amalg, 0.206%), 2 Na(amalg, 0.206%) + 2 H^+(1 M) → 2 Na^+(1 M) + H_2(g, 1 atm); **2.** (1) -81.56 kJ, (2) -360.33 kJ; **3.** 2 Na(s) + 2 H^+(1 M) → 2 Na^+(1 M) + H_2(g, 1 atm) $\Delta G° = -523.45$ kJ; **4.** -2.713 V.

Chapter 22

Practice Examples. 1A. electrolysis of NaCl(aq) to form NaOH(aq), followed by addition of NO_2(g) to NaOH(aq). **1B.** electrolysis of NaCl(aq) to produce NaOH(aq), the reaction of SO_2(g) with NaOH(aq), and the addition of S to the boiling solution. **2A.** KF. **2B.** 6.33 g, -0.318 °C. **3A.** CaO(s) + H_2O ⇌ Ca^{2+}(aq) + 2 OH^-(aq), followed by HCO_3^-(aq) + OH^-(aq) → H_2O + CO_3^{2-}(aq). **3B.** $Ca(HCO_3)_2$(aq) + NH_3(aq) → $CaCO_3$(s) + HCO_3^-(aq) + NH_4^+(aq).

Review Questions. 4. (a) lead(II) oxide; **(b)** tin(II) fluoride; **(c)** calcium sulfate hemihydrate; **(d)** Li_3N; **(e)** calcium hydroxide; **(f)** KO_2; **(g)** $Mg(HCO_3)_2$. **5. (a)** Li_2CO_3(s) $\xrightarrow{\Delta}$ Li_2O(s) + CO_2(g);

(b) $CaCO_3$(s) + 2 HCl(aq) → $CaCl_2$(aq) + H_2O + CO_2(g); **(c)** 2 Al(s) + 2 Na^+(aq) +2 OH^-(aq) + 6 H_2O → 2 Na^+(aq) + 2 $[Al(OH)_4]^-$(aq) + 3 H_2(g); **(d)** BaO(s) + H_2O(l) → $Ba(OH)_2$(s); **(e)** 2 Na_2O_2(s) + 2 CO_2(g) → 2 Na_2CO_3(s) + O_2(g). **6. (a)** $MgCO_3$(s) + 2 HCl(aq) → $MgCl_2$(aq) + CO_2(g) + H_2O(l); **(b)** 2 Na(s) + 2 H_2O(l) → 2 NaOH(aq) + H_2(g), followed by 2 Al(s) + 2 NaOH(aq) + 6 H_2O → 2 $Na[Al(OH)_4]$(aq) + 3 H_2(g); **(c)** 2 NaCl(s) + H_2SO_4(conc., aq) → 2 HCl(g) + Na_2SO_4(s). **7. (a)** K_2CO_3(aq) + $Ba(OH)_2$(aq) → $BaCO_3$(s) + 2 KOH(aq); **(b)** $Mg(HCO_3)_2$(aq) $\xrightarrow{\Delta}$ $MgCO_3$(s) + CO_2(g) + H_2O; **(c)** SnO(s) + C(s) $\xrightarrow{\Delta}$ Sn(l) + CO(g); **(d)** CaF_2(s) + H_2SO_4(conc., aq) → 2 HF(g) + $CaSO_4$(s); **(e)** $NaHCO_3$(aq) + HCl(aq) → NaCl(aq) + H_2O + CO_2(g); **(f)** PbO_2(s) + 4 HBr(aq) → $PbBr_2$(s) + Br_2(l) + 2 H_2O. **8.** $CaSO_4 \cdot 2H_2O$(s) + $(NH_4)_2CO_3$(aq) → $(NH_4)_2SO_4$(aq) + $CaCO_3$(s) + 2 H_2O. **9.** NH_4Cl (because it lacks basic properties). **10. (a)** $SrCO_3$(s) $\xrightarrow{\Delta}$ SrO(s) + CO_2(g); **(b)** Al_2O_3(s) $\xrightarrow{\Delta}$ no reaction; **(c)** Li_2CO_3(g) $\xrightarrow{\Delta}$ Li_2O(s) + CO_2(g). **11. (a)** $Pb(NO_3)_2$(aq) + 2 $NaHCO_3$(aq) → $PbCO_3$(s) + H_2O + CO_2(g) + 2 $NaNO_3$(aq); **(b)** Li_2O(s) + $(NH_4)_2CO_3$(aq) → Li_2CO_3(s) + 2 NH_3(aq) + H_2O; **(c)** H_2SO_4(aq) + BaO_2(s) → H_2O_2(aq) + $BaSO_4$(s); **(d)** PbO(s) + OCl^-(aq) → PbO_2(s) + Cl^-(aq). **12.** Al, because it is protected by a water-insoluble oxide coating. **13. (a)** both yield H_2(g). **14. (a)** $CaCO_3$(s); **(b)** $CaSO_4 \cdot 2H_2O$; **(c)** $Ca[CH_3(CH_2)_{14}COO]_2$ (s); **(d)** $BaSO_4$(s); **(e)** Al_2O_3 with Fe^{3+} and Ti^{4+}.

Exercises. 15. (a) 2 Cs(s) + Cl_2(g) → 2 CsCl(s); **(b)** 2 Na(s) + O_2(g) → Na_2O_2(s); **(c)** Li_2CO_3(s) $\xrightarrow{\Delta}$ Li_2O(s) + CO_2(g); **(d)** Na_2SO_4(s) + 4 C(s) → Na_2S(s) + 4 CO(g); **(e)** K(s) + O_2(g) → KO_2(s). **17.** flame test: LiCl (red), KCl (violet). **19.** O^{2-} + H_2O → 2 OH^-(aq); 2 O_2^{2-} + 2 H_2O → 4 OH^-(aq) + O_2(g); 4 O_2^- + 2 H_2O → 4 OH^-(aq) + 3 O_2. **21. (a)** 11.16; **(b)** no. **23. (a)** 71.5%; **(b)** NH_3 in intermediate steps is recycled. **25.** $Ca(OH)_2$(s) + 2 HCl(aq) → $CaCl_2$(aq) + 2 H_2O; $CaCl_2$(l) $\xrightarrow{\Delta, \text{ electrolysis}}$ Ca(l) + Cl_2(g); $Ca(OH)_2$(s) + CO_2(g) → $CaCO_3$(s) +

$H_2O(g)$; $CaCO_3(s) \xrightarrow{\Delta} CaO(s) +$ $CO_2(g)$; $Ca(OH)_2(s) + H_2SO_4(aq) \rightarrow$ $CaSO_4(s) + 2 H_2O$; $Ca(OH)_2(s) +$ $H_3PO_4(aq) \rightarrow CaHPO_4(aq) + 2 H_2O$. **27.** **(a)** $BeF_2 + Mg \rightarrow Be + MgF_2$; **(b)** $Ba(s) + Br_2(l) \rightarrow BaBr_2(s)$; **(c)** $UO_2(s) + 2 Ca(s) \rightarrow U(s) + 2 CaO(s)$; **(d)** $MgCO_3 \cdot CaCO_3(s) \xrightarrow{\Delta} MgO(s) + CaO(s) + 2 CO_2(g)$; **(e)** $2 H_3PO_4(aq) + 3 CaO(s) \rightarrow Ca_3(PO_4)_2(s) + 3 H_2O(l)$. **29.** **(a)** slightly to the left; **(b)** to the left; **(c)** far to the right. **31.** **(a)** 60.76 ppm; **(b)** 60.76 mg Ca^{2+}/L; **(c)** 112.3 g. **33.** 2.366. **35.** 40.2 g. **37.** **(a)** $2 Al(s) + 6 HCl(aq) \rightarrow 2 AlCl_3(aq) + 3 H_2(g)$; **(b)** $2 Al(s) + 3 Br_2(l) \rightarrow 2 AlBr_3(s)$; **(c)** $2 NaOH(aq) + 2 Al(s) + 6 H_2O(l) \rightarrow 2 Na^+(aq) + 2 [Al(OH)_4]^-(aq) + 3 H_2(g)$; **(d)** $2 Al(s) + 3 SO_4^{2-}(aq) + 12 H^+(aq) \rightarrow 2 Al^{3+}(aq) + 3 SO_2(aq) + 6 H_2O$. **39.** $Al^{3+}(aq) + 3 HCO_3^-(aq) \rightarrow Al(OH)_3(s) + 3 CO_2(g)$. **41.** Either acid or base can attack aluminum, but in pH neutral solutions the metal is protected by $Al_2O_3(s)$. **43.** $[Al(OH)_4]^-(aq) + CO_2(aq) \rightarrow Al(OH)_3(s) + HCO_3^-(aq)$. **45.** **(a)** -852 kJ; **(b)** -1792 kJ; **(c)** $+129$ kJ. **47.** **(a)** $PbO(s) + 2 HNO_3(aq) \rightarrow Pb(NO_3)_2(s) + H_2O$; **(b)** $SnCO_3(s) \xrightarrow{\Delta} SnO(s) + CO_2(g)$; **(c)** $PbO(s) + 2 C(s) \xrightarrow{\Delta} Pb(l) + CO(g)$; **(d)** $2 Fe^{3+}(aq) + Sn^{2+}(aq) \rightarrow 2 Fe^{2+}(aq) + Sn^{4+}(aq)$; **(e)** $PbS(s) + 2 O_2(g) \rightarrow PbSO_4(s)$ and possibly other schemes. **49.** **(a)** yes; **(b)** no; **(c)** no.
Advanced Exercises. 52. **(a)** 0.332 g; **(c)** 69%. **55.** **(a)** 81.4 ppm Ca^{2+}; **(b)** 48.7 g CaO; **(c)** $CaCO_3(s)$ can be precipitated if $[CO_3^{2-}] > 0.0019$ M; **(d)** 2.1×10^2 g. **57.** 19 g $Pb(NO_3)_2$. **61.** **(a)** 7.4×10^{-3} M; **(b)** 6.4×10^{-4} M; **(c)** 2.0×10^{-3} M.

Chapter 23

Practice Examples. 1A. $K_p = 2 \times 10^{-48}$. **1B.** $K_p = 4 \times 10^{-34}$, $4 \times 10^{-15}\%$. **2A.** $Na_2B_4O_7 \cdot 10H_2O + H_2SO_4 \rightarrow 4 B(OH)_3 + Na_2SO_4 + 5 H_2O$, $2 B(OH)_3 \xrightarrow{\Delta} B_2O_3 + 3 H_2O$, $2 B_2O_3(s) + 3 C(s) + 6 Cl_2(g) \xrightarrow{\Delta} 4 BCl_3(l) + 3 CO_2(g)$, $4 BCl_3(l) + 3 LiAlH_4 \rightarrow 2 B_2H_6(g) + 3 LiCl + 3 AlCl_3$. **2B.** $2 ZnS(s) + 3 O_2(g) \rightarrow 2 ZnO(s) + 2 SO_2(g)$, $2 SO_2(g) + O_2(g) \rightarrow 2 SO_3(g)$, V_2O_5 catalyst; $H_2SO_4(l) +$

$SO_3(g) \rightarrow H_2S_2O_7(l)$, $H_2S_2O_7(l) + H_2O(l) \rightarrow 2 H_2SO_4(l)$.
Review Questions. 4. **(a)** potassium bromate; **(b)** triiodide ion; **(c)** NaIO; **(d)** sodium dihydrogen phosphate; **(e)** lithium azide; **(f)** $Na_2S_2O_3$. **5.** **(a)** $\xrightarrow{\Delta} CaSO_4(s) + 2 HCl(g)$; **(b)** no reaction; **(c)** $\rightarrow NH_4ClO_4(aq)$. **6.(a)** $Cl_2(aq) + 2 NaOH(aq) \rightarrow NaCl(aq) + NaOCl(aq) + H_2O$; **(b)** $2 I^-(aq) + SO_4^{2-}(aq) + 4 H^+(aq) \rightarrow I_2(g) + SO_2(g) + 2 H_2O$; **(c)** $Cl_2(g) + 2 Br^-(aq) \rightarrow 2 Cl^-(aq) + Br_2(l)$. **7.** **(a)** $KI(s) + H_3PO_4(l) \rightarrow HI(g) + KH_2PO_4(s)$; **(b)** $Na_2SiF_6 + 4 Na \rightarrow Si + 6 NaF$. **8.** **(a)** $As_4O_6(s) + 6 CO(g) \rightarrow 4 As(s) + 6 CO_2(g)$; **(b)** $NH_3(g) + H_3PO_4(aq) \rightarrow NH_4H_2PO_4(aq)$, $2 NH_3(g) + H_3PO_4(aq) \rightarrow (NH_4)_2 HPO_4(aq)$. **9.** 91%. **10.** $3 Cl_2(g) + I^-(aq) + 3 H_2O \rightarrow 6 Cl^-(aq) + IO_3^-(aq) + 6 H^+(aq)$, $Cl_2(g) + 2 Br^-(aq) \rightarrow 2 Cl^-(aq) + Br_2(aq)$. **11.** 13 km³. **12.** **(a)** $I_2(s)$; **(b)** H_2O. **13.** **(a)** $-50.$ kJ/mol; **(b)** -18 kJ/mol; **(c)** $+82$ kJ/mol; **(d)** -128 kJ/mol. **14.** **(a)** AgAt; **(b)** carbon diselenide; **(c)** MgPo; **(d)** tellurous acid; **(e)** K_2SeSO_3; **(f)** perastatic acid.
Exercises. 15. **(a)** trigonal-pyramidal; **(b)** tetrahedral; **(c)** square-pyramidal. **17.** $4 I^-(aq) + O_2(g) + 4 H^+(aq) \rightarrow 2 I_2(aq) + 2 H_2O$, possibly followed by $I_2(aq) + I^-(aq) \rightarrow I_3^-(aq)$. **19.** elemental forms of the lighter elements in the halogen family will displace the ions of the heavier elements. **21.** **(a)** 1×10^6 kg F_2; **(b)** No, common oxidizing agents cannot oxidize F^- to F_2. **23.** 5.89. **25.** 302 kJ/mol. **27.** **(a)** H_2S forms very weak hydrogen bonds compared to H_2O; **(b)** All electrons are paired. **29.** **(a)** zinc sulfide; **(b)** potassium hydrogen sulfite; **(c)** potassium thiosulfate; **(d)** sulfur tetrafluoride. **31.** **(a)** $FeS(s) + 2 HCl(aq) \rightarrow FeCl_2(aq) + H_2S(g)$; **(b)** $CaSO_3(s) + 2 HCl(aq) \rightarrow CaCl_2(aq) + H_2O + SO_2(g)$; **(c)** $SO_2(aq) + MnO_2(s) \rightarrow Mn^{2+}(aq) + SO_4^{2-}(aq)$; **(d)** $S_2O_3^{2-}(aq) + 2 H^+(aq) \rightarrow S(s) + SO_2(g) + H_2O$. **33.** $S_2O_3^{2-}(aq) + 2 H^+(aq) \rightarrow S(s) + SO_2(g) + H_2O$; no reaction with SO_4^{2-}. **35.** 1.21. **37.** **(a)** $2 NO_2(g) \rightleftharpoons N_2O_4(g)$; **(b)** $NH_4NO_3(s) \xrightarrow{\Delta} N_2O(g) + 2 H_2O(g)$; **(c)** $2 NH_3(g) + H_2SO_4(aq) \rightarrow (NH_4)_2SO_4(aq)$. **39.** **(a)**

(b)

(c)

41. (a) hydrogen

phosphate ion; **(b)** calcium pyrophosphate; **(c)** tetraphosphoric acid. **43.** Graphite is the most stable form of carbon at room temperature and room pressure, but diamonds convert to graphite imperceptibly slowly under these conditions. **45.** Silane has general formula Si_nH_{2n+2}, silanols substitute —OH for one or more of the H atoms (used to produce silicones). **47.** $2 CH_4(s) + S_8(g) \rightarrow 2 CS_2(g) + 4 H_2S(g)$; $CS_2(g) + 3 Cl_2(g) \rightarrow CCl_4(l) + S_2Cl_2(l)$; $4 CS_2(g) + 8 S_2Cl_2(g) \rightarrow 4 CCl_4(l) + 3 S_8$. **49.** **(a)** Ten B—H bonds require 20 of the 22 valence electrons, leaving just two electrons to bond four B atoms together. **(b)** 4. **(c)** C_4H_{10} has a normal Lewis structure.
Advanced Exercises. 52. $2 NH_4ClO_4(s) \rightarrow N_2(g) + 4 H_2O(g) + Cl_2(g) + 2 O_2(g)$. **56.** **(a)** 339 kJ/mol; **(b)** 353 nm. **57.** **(b)** larger %P than pure $Ca_3(PO_4)_2$; **(c)** 92.4% BPL. **59.**

and **63.** $3 XeF_4(g) + 6 H_2O \rightarrow 2 Xe(g) + XeO_3(g) + 12 HF(aq) + \frac{3}{2} O_2(g)$.
Feature Problems. A. 1.
$E°[ClO_4^-|ClO_3^-] = +1.19$ V,
$E°[ClO_3^-|ClO_2] = +1.175$ V,
$E°[Cl_2|Cl^-] = +1.358$ V,
$E°[ClO_3^-|Cl^-] = +1.450$ V all in acidic solution; $E°[ClO_3^-|Cl^-] = +0.622$ V, $E°[ClO^-|Cl^-] = +0.890$ V, both in basic solution; **2.** $E°[ClO_3^-|HClO_2] = +1.182$ V in acidic solution; $E°[ClO_3^-|ClO_2^-] = +0.295$ V and $E°[ClO_3^-|ClO^-] = +0.488$ V in basic solution; **3.** $E°[SO_3^{2-}|S] = -0.66$ V; $E°[S_2O_3^{2-}|S_4O_6^{2-}] = -0.795$ V. **B. 1.** about 0.04 mg; **2.** about 0.0003 mg; **3.** less.

Chapter 24

Practice Examples. 1A. **(a)** $2 Cu_2S(s) + 3 O_2(g) \rightarrow 2 Cu_2O(s) + 2 SO_2(g)$; **(b)** $WO_3(s) + 3 H_2(g) \rightarrow W(s) + 3 H_2O(g)$; **(c)** $2 HgO(s) \xrightarrow{\Delta} 2 Hg(l) + O_2(g)$. **1B.** **(a)** $3 Si(s) + 2 Cr_2O_3(s) \xrightarrow{\Delta} 3 SiO_2(s) + 4 Cr(l)$; **(b)** 2

$Co(OH)_3(s) \xrightarrow{\Delta} Co_2O_3(s) + 3 H_2O(g)$; **(c)** $Mn^{2+}(aq) + 2 H_2O \rightarrow MnO_2(s) + 4 H^+(aq) + 2 e^-$. **2A.** $2 V_2O_5 + 5 Si \xrightarrow{\Delta} 4 V + 5 SiO_2$; $CaO(s) + SiO_2(s) \xrightarrow{\Delta} CaSiO_3(l)$. **2B.** $4 FeWO_4(s) + 4 Na_2CO_3(s) + O_2(g) \rightarrow 2 Fe_2O_3(s) + 4 Na_2WO_4(aq) + 4 CO_2(g)$; $Na_2WO_4(s) + H_2SO_4(aq) \rightarrow Na_2SO_4(aq) + H_2O(l) + WO_3(s)$; $WO_3(s) + 3 H_2(g) \rightarrow W(s) + 3 H_2O(g)$. **3A.** $NO_3^-(aq) + 3 V^{3+}(aq) + H_2O \rightarrow NO(g) + 3 VO^{2+}(aq) + 2 H^+(aq)$, $E°_{cell} = +0.619$ V. **3B.** possibilities: $-E°[Cr^{2+}(aq)|Cr^{3+}(aq)] = +0.424$ V, $-E°[Fe(s)|Fe^{2+}(aq)] = +0.440$ V, $-E°[Zn(s)|Zn^{2+}(aq)] = +0.763$ V.

Review Questions. 4. (a) scandium hydroxide; **(b)** copper(I) oxide; **(c)** titanium(IV) chloride; **(d)** vanadium(V) oxide; **(e)** potassium chromate; **(f)** potassium manganate. **5. (a)** CrO_3; **(b)** $FeSiO_3$; **(c)** $BaCr_2O_7$; **(d)** $CuCN$; **(e)** $CoCl_2 \cdot 6H_2O$. **6. (a)** about 95% Fe, with 3–4% C and other impurities; **(b)** alloy of iron and manganese; **(c)** $Fe(CrO_2)_2$; **(d)** copper and zinc; **(e)** $HCl(aq)$ and $HNO_3(aq)$; **(f)** impure copper; **(g)** iron, and chromium and nickel. **7. (a)** $TiCl_4(g) + 4 Na(l) \xrightarrow{\Delta} Ti(s) + 4 NaCl(l)$; **(b)** $Cr_2O_3(s) + 2 Al(s) \xrightarrow{\Delta} 2 Cr(l) + Al_2O_3(s)$; **(c)** no reaction; **(d)** $K_2Cr_2O_7(aq) + 2 KOH(aq) \rightarrow 2 K_2CrO_4(aq) + H_2O$; **(e)** $MnO_2(s) + 2 C(s) \xrightarrow{\Delta} Mn(l) + 2 CO(g)$. **8. (a)** $2 Fe_2S_3(s) + 3 O_2(g) + 6 H_2O \rightarrow 4 Fe(OH)_3(s) + 6 S(s)$; **(b)** $2 Mn^{2+}(aq) + 8 H_2O + 5 S_2O_8^{2-}(aq) \rightarrow 2 MnO_4^-(aq) + 16 H^+(aq) + 10 SO_4^{2-}(aq)$; **(c)** $4 Ag(s) + 8 CN^-(aq) + O_2(g) + 2 H_2O \rightarrow 4 [Ag(CN)_2]^-(aq) + 4 OH^-(aq)$. **9. (a)** $Cr(s) + 2 HCl(aq) \rightarrow CrCl_2(aq) + H_2(g)$; **(b)** $Cr_2O_3(s) + 2 OH^-(aq) + 3 H_2O \rightarrow 2 Cr(OH)_4^-(aq)$; **(c)** $2 La(s) + 6 HCl(aq) \rightarrow 2 LaCl_3(aq) + 3 H_2(g)$. **10. (a)** [Ar] $3d$ [↑][↑][ ][ ][ ] $4s$ [↑↓]; **(b)** [Ar] $3d$ [↑][↑][ ][ ][ ] $4s$ [ ]; **(c)** [Ar] $3d$ [↑][↑][↑][↑][ ] $4s$ [ ]; **(d)** [Ar] $3d$ [↑][↑][↑][ ][ ] $4s$ [ ]; **(e)** [Ar] $3d$ [↑][↑][↑][↑][↑] $4s$ [ ]; **(f)** [Ar] $3d$ [↑][↑][↑][↑][↑] $4s$ [ ]. **11.** $Cr > Fe > Mn^{4+} > Ti^{2+} > Cu^{2+} > Sc^{3+}$. **12.** (3) and (5). The metals tend to have high melting points, low ionization energies, and negative $E°$ values. **13.** Sc^{3+}. **14.** Ti. **15.** Ag^+. **16.** Fe^{3+} has a d^5 electron configuration.

Exercises. 17. Unlike the main group metals, transition metals usually display several oxidation states, readily form complexes and colored compounds, are mostly paramagnetic, and some are ferromagnetic. **19.** Main group elements have electrons added to the outermost shell. **21.** Mn. **23.** The lanthanide atoms are larger and more easily ionized. **25. (a)** $Sc(OH)_3(s) + 3 H^+(aq) \rightarrow Sc^{3+}(aq) + 3 H_2O$; **(b)** $3 Fe^{2+}(aq) + MnO_4^-(aq) + 2 H_2O \rightarrow 3 Fe^{3+}(aq) + MnO_2(s) + 4 OH^-(aq)$; **(c)** $2 KOH(l) + TiO_2(s) \xrightarrow{\Delta} K_2TiO_3(s) + H_2O(g)$; **(d)** $Cu(s) + 2 H_2SO_4(concd aq) \rightarrow CuSO_4(aq) + SO_2(g) + 2 H_2O$. **27. (a)** $FeS(s) + 2 HCl(aq) \rightarrow FeCl_2(aq) + H_2S(g)$, $4 Fe^{2+}(aq) + O_2(g) + 4 H^+(aq) \rightarrow 4 Fe^{3+}(aq) + 2 H_2O$, $Fe^{3+}(aq) + 3 OH^-(aq) \rightarrow Fe(OH)_3(s)$; **(b)** $BaCO_3(s) + 2 HCl(aq) \rightarrow BaCl_2(aq) + H_2O + CO_2(g)$, $2 BaCl_2(aq) + K_2Cr_2O_7(aq) + 2 NaOH(aq) \rightarrow 2 BaCrO_4(s) + 2 KCl(aq) + 2 NaCl(aq) + H_2O$. **29. (a)** $VO^{2+}(aq) + 2 H^+(aq) + e^- \rightarrow V^{3+}(aq) + H_2O$; **(b)** $Cr^{2+}(aq) \rightarrow Cr^{3+}(aq) + e^-$. **31. (a)** no; **(b)** yes; **(c)** yes. **33.** $4 Fe^{2+}(aq) + O_2(g) + 4 H^+(aq) \rightarrow 4 Fe^{3+}(aq) + 2 H_2O$, $E°_{cell} = +0.458$ V. **35.** $Cr_2O_7^{2-}(aq) + H_2O \rightleftharpoons 2 CrO_4^{2-}(aq) + 2 H^+(aq)$, $Pb^{2+}(aq) + CrO_4^{2-}(aq) \rightleftharpoons PbCrO_4(s)$. **37.** $3 Zn(s) + Cr_2O_7^{2-}(aq, orange) + 14 H^+(aq) \rightarrow 3 Zn^{2+}(aq) + 2 Cr^{3+}(aq, green) + 7 H_2O$, $Zn(s) + 2 Cr^{3+}(aq) \rightarrow Zn^{2+}(aq) + 2 Cr^{2+}(aq, blue)$; $4 Cr^{2+}(aq) + O_2(g) + 4H^+(aq) \rightarrow 4 Cr^{3+}(aq) + 2 H_2O$. **39. (a)** 0.74 M; **(b)** 2.6×10^{-5} M. **41.** 1.10 g. **43.** $E°[Cr_2O_7^{2-}|Cr^{3+}] = +1.33$ V, making $Cr_2O_7^{2-}$ a good oxidizing agent; $E°[CrO_4^{2-}|Cr(OH)_3] = -0.11$ V, making CrO_4^- a poor oxidizing agent. However, many metal chromates are insoluble and can be precipitated from aqueous solution. **45. (a)** $Cu^{2+}(aq) + H_2(g) \rightarrow Cu(s) + 2 H^+(aq)$; **(b)** $Au^+(aq) + Fe^{2+}(aq) \rightarrow Au(s) + Fe^{3+}(aq)$; **(c)** $2 Cu^{2+}(aq) + SO_2(g) + 2 H_2O \rightarrow 2 Cu^+(aq) + SO_4^{2-}(aq) + 4 H^+(aq)$. **47.** 1.3×10^6. **49.** ZnO, $\lambda = 413$ nm; CdS, $\lambda = 479$ nm.

Advanced Exercises. 51. 3.50%. **54.** Redox and acid–base reactions are followed by the reaction $2 Cu^{2+}(aq) + 2 OH^-(aq) + CO_3^{2-}(aq) \rightarrow Cu_2(OH)_2CO_3(s)$. **58. (a)** net: $2 H_2O + 4 Ag^{2+}(aq) \rightarrow O_2(g) + 4 H^+(aq) + 4$

$Ag^+(aq)$; **(b)** $E°_{cell} = +0.75$ V. **64.** 28.2% Cr, 3.286% Mn.

Feature Problems. A. 1. If $n_{gas} = 0$, $\Delta S° \approx 0$ and $\Delta G°$ is essentially independent of temperature. If $n_{gas} > 0$, $\Delta S° > 0$, and $\Delta G°$ decreases with temperature. If $n_{gas} < 0$, $\Delta S° < 0$, and $\Delta G°$ increases with temperature. **2.** 690 °C; **3.** 1310 °C. **B. 1.** $Fe^{3+}(aq) + 3 OH^-(aq) \rightarrow Fe(OH)_3(s)$, $Al^{3+}(aq) + 4 OH^-(aq) \rightarrow [Al(OH)_4]^-(aq)$, $Co^{2+}(aq) + 2 OH^-(aq) \rightarrow Co(OH)_2(s)$, $Cr^{3+}(aq) + 4 OH^-(aq) \rightarrow [Cr(OH)_4]^-(aq)$, $Ni^{2+}(aq) + 2 OH^-(aq) \rightarrow Ni(OH)_2(aq)$, $Zn^{2+}(aq) + 4 OH^-(aq) \rightarrow [Zn(OH)_4]^{2-}(aq)$; **2.** $2 Co(OH)_2(s) + H_2O_2(aq) \rightarrow 2 Co(OH)_3(s)$; **3.** $2 [Cr(OH)_4]^-(aq) + 3 H_2O_2(aq) + 2 OH^-(aq) \rightarrow 2 CrO_4^{2-}(aq, yellow) + 8 H_2O$; **4.** $Fe(OH)_3(s) + 3 H^+(aq) \rightarrow Fe^{3+}(aq) + 3 H_2O$, $Co(OH)_3(s) + 3 H^+(aq) \rightarrow Co^{3+}(aq) + 3 H_2O$, $Ni(OH)_2(s) + 2 H^+(aq) \rightarrow Ni^{2+}(aq) + 2 H_2O$, $MnO_2(s) + 4 HCl(aq) \rightarrow MnCl_2(aq) + 2 H_2O + Cl_2(g)$; **5.** $Fe^{3+}(aq) + 3 NH_3(aq) + 3 H_2O \rightarrow Fe(OH)_3(s) + 3 NH_4^+(aq)$. **C. 1.** Data marked (?) can be found in Table 24.5; for data marked (??): $+0.669$ V $= E°[VO_2^+|V^{3+}]$, $+0.361$ V $= E°[VO_2^+|V^{2+}]$, -0.24 V $= E°[VO_2^+|V^0]$. **2. (a)** The disproportionation reaction cited is highly spontaneous in acidic solution ($E°_{cell} = +1.71$ V) but only slightly so in basic solution ($E°_{cell} = +0.06$ V). At low concentrations of MnO_4^{2-} and high concentrations of OH^-, the reaction is not spontaneous; **(b)** In acidic solution, MnO_4^- is readily reduced to $MnO_2(s)$, which is also readily reduced to Mn^{2+}. In basic solution, the reduction of $MnO_2(s)$ to $Mn(OH)_2(s)$ does not occur readily; **(c)** 1.4×10^{-13}.

Chapter 25

Practice Examples. 1A. CN = 5, OS = +2. **1B.** $[Fe(CN)_6]^{3-}$. **2A.** $K_2[PtCl_6]$. **2B.** pentaamminethiocyanatocobalt(III) chloride. **3A.** The oxalate ion occupies *cis* positions. One isomer has *trans* NH_3 ligands, one has *cis* NH_3 ligands, and one has an NH_3 and a Cl^- in *cis* positions. **3B.** There are five: three with the two pyridines *cis*, and two with them *trans*. **4A.** 5. **4B.** 3 in each case. **5A.** tetrahedral. **5B.** Paramagnetic, but the structure cannot be determined from the magnetic properties. **6A.** greater d orbital energy splitting for $[Co(H_2O)_6]^{2+}$ than for $[CoCl_4]^{2-}$. **6B.** $[Fe(CN)_6]^{4-}$ absorbs light of *shorter* wavelength than does $[Fe(H_2O)_6]^{2+}$; $K_4[Fe(CN)_6]$ is yellow. **Review Questions. 4. (a)** $[CrCl_4(NH_3)_2]^-$, diamminetetrachlorochromate(III) ion; **(b)** $[Fe(CN)_6]^{3-}$, hexacyanoferrate(III) ion; **(c)** $[Cr(en)_3][Ni(CN)_4]$, tris(ethylenediamine)chromium(III) tetracyanonickelate(II). **5. (a)** C.N. = 6, O.S. = +2; **(b)** C.N. = 6, O.S. = +3; **(c)** C.N. = 4, O.S. = +2; **(d)** C.N. = 6, O.S. = +3; **(e)** C.N. = 6, O.S. = +2; **(f)** C.N. = 2, O.S. = +1. **6. (a)** diiodoargentate(I) ion; **(b)** pentaaquahydroxoaluminum(III) ion; **(c)** tetracyanozincate(II) ion; **(d)** bis(ethylenediamine)platinum(II) ion; **(e)** tetraamminechloronitrocobalt(III) ion. **7. (a)** pentaamminebromocobalt(III) sulfate; **(b)** pentaamminesulfatocobalt(III) bromide; **(c)** hexaamminechromium(III) hexacyanocobaltate(III); **(d)** sodium hexanitrocobaltate(III); **(e)** tris(ethylenediamine)cobalt(III) chloride. **8. (a)** $[Ag(CN)_2]^-$; **(b)** $[Pt(NO_2)(NH_3)_3]^+$; **(c)** $[CoCl(en)_2(H_2O)]^{2+}$; **(d)** $K_4[Cr(CN)_6]$.

9. (a) H—Ö—H ; **(b)** H—C—N—H ; (with H's) **(c)** $[\ddot{O}—N{=}\ddot{O}]^-$; **(d)** $[\ddot{S}{=}C{=}\ddot{N}]^-$ **10. (a)** $[Mn(H_2O)_6]SO_4$; **(b)** $K_4[Cr(CN)_6]\cdot 3H_2O$. **11. (a)** square-planar; **(b)** octahedral; **(c)** octahedral, Cl^- are *trans;* **(d)** octahedral, Cl^- *trans* to OH^-. **12. (a)** one; **(b)** two; **(c)** two; **(d)** two. **13. (a)** coordination; **(b)** linkage; **(c)** none; **(d)** geometric; **(e)** geometric. **14.** $[Co(en)_3]^{3+}$ is yellow. **Exercises. 15. (a)** amminetetraaquahydroxocobalt(III) ion; **(b)** triamminetrinitritocobalt(III); **(c)** tetraaquaplatinum(II) hexachloroplatinate(IV); **(d)** diaquadioxalatoferrate(III) ion; **(e)** silver(I) tetraiodomercurate(II). **17. (a)** square-planar; **(b)** octahedral; *fac* (or *cis*) isomer has three NH_3 on the same face; **(c)** octahedral. **19. (a)** octahedral; **(b)** octahedral, three bidentate ligands; **(c)** octahedral, three isomers. **21. (a)** not possible; all positions in coordination sphere equivalent; **(b)** readily possible;

(c) not possible [same reason as in (a)]. **23. (a)** three isomers; **(b)** yes. **25.** $[Co(en)_3][Cr(ox)_3]$, $[Co(ox)(en)_2][Cr(ox)_2(en)]$, $[Cr(ox)(en)_2][Co(ox)_2(en)]$, $[Cr(en)_3][Co(ox)_3]$. **27.** Energy differences between the d orbitals of the central metal in the field of the ligands correspond to light of different colors. **29.** Because of the larger Δ in the strong-field complex, an additional two d electrons pair, leaving only two unpaired. **31. (a)** $[Co(en)_3]^{3+}$ is diamagnetic; **(b)** three. **33.** no, can only confirm or eliminate square-planar. **35. (a)** $Zn(OH)_2(s) + 4\,NH_3(aq) \rightleftharpoons [Zn(NH_3)_4]^{2+}(aq) + 2\,OH^-(aq)$; **(b)** $Cu^{2+}(aq) + 2\,OH^-(aq) \rightleftharpoons Cu(OH)_2(s)$, $Cu(OH)_2(s) + 4\,NH_3(aq) \rightleftharpoons [Cu(NH_3)_4]^{2+}(aq,$ dark blue) $+ 2\,OH^-(aq)$, $[Cu(NH_3)_4]^{2+}(aq) + 4\,H_3O^+(aq) \rightleftharpoons [Cu(H_2O)_4]^{2+}(aq) + 4\,NH_4^+(aq)$. **37.** $[Co(en)_3]^{3+}$ chelate effect. **39.** 5.0×10^9. **41. (a)** Al^{3+} forms a hydroxo complex but not an ammine complex. **(b)** $ZnCO_3(s)$ produces a high enough $[Zn^{2+}]$ to form $[Zn(NH_3)_4]^{2+}$, but ZnS(s) is too insoluble. **(c)** At low $[Cl^-]$ the solubility of AgCl is suppressed by the common ion effect, but at higher $[Cl^-]$ the solubility increases due to the formation of $[AgCl_2]^-$. **43.** $[Al(H_2O)_6]^{3+}(aq)$. **45. (a)** 2.0; **(b)** 9×10^{-4} M; **(c)** no. **47. (a)** $K = 8.5$; **(b)** K for the dissolution of AgBr(s) in $NH_3(aq)$ is only 8.0×10^{-6}.

Advanced Exercises. 49. (a) tetraamminecopper(II) ion; **(b)** tetraamminedichlorocobalt(III) chloride; **(c)** hexachloroplatinate(IV) ion; **(d)** sodium tetrachlorocuprate(II); **(e)** potassium pentachloroantimonate(III). **50. (a)** $K[PtCl_3(C_2H_4)]$; **(b)** $[Pt(NH_3)_4][PtCl_4]$. **55.** because the complex ion has a positive charge, which repels the ionized H^+. **59. (a)** and **(b)** are identical. **(d)** is a geometric isomer of (a). **(c)** is a different compound. **60.** 230 kJ/mol. **62.** 1×10^{14}.

Chapter 26

Practice Examples. 1A. $^{241}_{94}Pu \rightarrow\ ^{241}_{95}Am + {}^{0}_{-1}\beta$. **1B.** $^{58}_{29}Cu \rightarrow\ ^{58}_{28}Ni + {}^{0}_{+1}\beta$. **2A.** $^{139}_{57}La + {}^{12}_{6}C \rightarrow\ ^{147}_{63}Eu + 4\,{}^{1}_{0}n$. **2B.** $^{121}_{51}Sb + {}^{4}_{2}He \rightarrow\ ^{124}_{53}I + {}^{1}_{0}n,\ ^{124}_{53}I \rightarrow\ ^{0}_{+1}\beta +\ ^{124}_{52}Te$. **3A. (a)** $9.98 \times 10^{-7}\,s^{-1}$; **(b)** 9.40×10^{12} dis s^{-1}; **(c)** 25%; **(d)** 2.36×10^{12}

dis s^{-1}. **3B.** 75.8 d. **4A.** 4.69×10^3 y. **4B.** 13 dis $min^{-1}\,g^{-1}$ C. **5A.** 2.544 MeV. **5B.** 0.006001 u. **6A. (a)** ^{88}Sr stable; **(b)** ^{118}Cs radioactive; **(c)** ^{30}S radioactive. **6B.** positron emission by ^{17}F; β^- emission by ^{22}F.

Review Questions. 4. (a) γ rays; **(b)** α particles **(c)** β particles. **5. (a)** $^{160}_{74}W \rightarrow\ ^{156}_{72}Th + {}^{4}_{2}He$; **(b)** $^{38}_{17}Cl \rightarrow\ ^{38}_{18}Ar + {}^{0}_{-1}\beta$; **(c)** $^{214}_{83}Bi \rightarrow\ ^{214}_{84}Po + {}^{0}_{-1}\beta$; **(d)** $^{32}_{17}Cl \rightarrow\ ^{32}_{16}S + {}^{0}_{+1}\beta$. **6. (a)** $^{23}_{11}Na + {}^{2}_{1}H \rightarrow\ ^{24}_{11}Na + {}^{1}_{1}H$; **(b)** $^{59}_{27}Co + {}^{1}_{0}n \rightarrow\ ^{56}_{25}Mn + {}^{4}_{2}He$; **(c)** $^{238}_{92}U + {}^{2}_{1}H \rightarrow\ ^{240}_{94}Pu + {}^{0}_{-1}\beta$; **(d)** $^{246}_{96}Cm + {}^{13}_{6}C \rightarrow\ ^{254}_{102}No + 5\,{}^{1}_{0}n$; **(e)** $^{238}_{92}U + {}^{14}_{7}N \rightarrow\ ^{246}_{99}Es + 6\,{}^{1}_{0}n$. **7. (a)** $^{214}Ra \rightarrow\ ^{210}Rn + {}^{4}_{2}He$; **(b)** $^{205}At \rightarrow\ ^{205}Po + \beta^+$; **(c)** $^{212}Fr + {}^{0}e^- \rightarrow\ ^{212}Rn$; **(d)** $^{2}H + {}^{2}H \rightarrow\ ^{3}He + {}^{1}n$; **(e)** $^{241}_{95}Am + {}^{4}_{2}He \rightarrow\ ^{243}_{97}Bk + 2\,{}^{1}_{0}n$. **8. (a)** $^{214}_{84}Po$; **(b)** $^{32}_{15}P$; **(c)** ^{13}O, ^{28}Mg, ^{80}Br, ^{214}Po, and ^{222}Rn. **9.** 56.3 h. **10. (a)** 174 d; **(b)** 287 d; **(c)** 375 d. **11. (a)** 5.42×10^{-9} J; **(b)** 3727.4 MeV; **(c)** 7.18×10^3 neutrons. **12.** 8.36 MeV/nucleon. **13. (c)** ^{80}Br; **(d)** ^{132}Cs. **14. (a)** intermediate half-life means continuous moderate activity; **(b)** Those that give off γ or high energy β are hazardous at a distance; α-emitters are dangerous internally. **(c)** Potassium-40 decays by electron capture to argon-40; **(d)** Francium is produced in a radioactive decay series and is found in radioactive minerals rather than in Group 1A minerals. **(e)** Fusion involves joining positively charged nuclei and requires extreme thermal energies to initiate it.

Exercises. 15. (a) $^{230}_{92}U$; **(b)** $^{248}_{98}Cf$; **(c)** $^{196}_{80}Hg$. **17.** $^{232}_{90}Th \xrightarrow{\alpha} ^{228}_{88}Ra \xrightarrow{\beta} ^{228}_{89}Ac \xrightarrow{\beta} ^{228}_{90}Th \xrightarrow{\alpha} ^{224}_{88}Ra \xrightarrow{\alpha} ^{220}_{86}Rn \xrightarrow{\alpha} ^{216}_{84}Po$; first branch, first fork $[^{216}_{84}Po \xrightarrow{\alpha} ^{212}_{82}Pb \xrightarrow{\beta} ^{212}_{83}Bi]$, second fork $[^{216}_{84}Po \xrightarrow{\beta} ^{216}_{85}At \xrightarrow{\alpha} ^{212}_{83}Bi]$; second branch, first fork $[^{212}_{83}Bi \xrightarrow{\alpha} ^{208}_{81}Tl \xrightarrow{\beta} ^{208}_{82}Pb]$, 2nd fork $[^{212}_{83}Bi \xrightarrow{\beta} ^{212}_{84}Po \xrightarrow{\alpha} ^{208}_{82}Pb]$. **19.** "$4n + 2$" is only consistent with the mass numbers: 206, 210, 214, 218, 222, 226, 230, 234, and 238— as seen in Figure 26-2. **21.** $^{14}_{6}C \rightarrow\ ^{14}_{7}N + {}^{0}_{-1}\beta^-$. **23. (a)** $^{7}_{3}Li + {}^{1}_{1}H \rightarrow\ ^{8}_{4}Be + \gamma$; **(b)** $^{9}_{4}Be + {}^{2}_{1}H \rightarrow\ ^{10}_{5}B + {}^{1}_{0}n$; **(c)** $^{14}_{7}N + {}^{1}_{0}n \rightarrow\ ^{14}_{6}C + {}^{1}_{1}H$. **25.** $4.4 \times 10^8\ ^{60}_{27}Co$ atoms. **27.** 87.5 h. **29.** 0.13 g ^{208}Pb/1 g ^{232}Th. **31.** 3.35×10^3 y. **33.** 7.805 MeV/nucleon. **35.** 4.06 MeV. **37. (a)** ^{20}Ne; **(b)** ^{18}O; **(c)** ^{7}Li. **39.** β^- emission: **(a)** ^{33}P; **(b)** ^{134}I. **41.** Atomic number and mass number both are magic numbers: ^{4}He, ^{16}O, ^{40}Ca,

^{56}Ni, ^{208}Pb. **43.** 1.35 mg. **45.** takes into account the quantity of biological damage. **47.** Sr is in the same periodic family as Ca, concentrates in bones. **49.** Mix in a small amount of tritium with the $H_2(g)$. **51.** yes.
Advanced Exercises. 56. 3×10^9 y. **59.** 29 y. **65.** 1.2×10^7 kJ.

Chapter 27

Practice Examples.
1A. $CH_3CH_2CH_2CH_2CH_2CH_3$,
$(CH_3)_2CHCH_2CH_2CH_3$,
$CH_3CH_2CH(CH_3)CH_2CH_3$,
$(CH_3)_2CHCH(CH_3)_2$,
$(CH_3)_3CCH_2CH_3$.
1B. $CH_3CH_2CH_2CH_2CH_2CH_2CH_3$,
$(CH_3)_2CHCH_2CH_2CH_2CH_3$,
$CH_3CH_2CH(CH_3)CH_2CH_2CH_3$,
$(CH_3)_2CHCH_2CH(CH_3)_2$,
$(CH_3)_2CHCH(CH_3)CH_2CH_3$,
$(CH_3)_3CCH_2CH_2CH_3$, $CH(CH_2CH_3)_3$,
$(CH_3)_3CCH(CH_3)_2$,
$CH_3CH_2C(CH_3)_2CH_2CH_3$.
2A. 3,3,6-trimethylnonane. **2B.** 3,6-dimethyloctane.
3A. $CH_3CH(CH_3)CH(CH_2CH_3)CH_2$ $CH_2CH(CH_3)_2$.
3B. $CH_3CH(CH_3)CH[CH(CH_3)_2]CH_2$ CH_3.
4A. 3-nitrobenzaldehyde. **4B.** 1,3-dichloro-2-nitrobenzene, 2,4-dichloro-1-nitrobenzene.

Review Questions. 5. C_4H_8 can be a cyclic alkane or an alkene with different positions for double bonds; C_4H_{10} has only one straight-chain and one branched-chain isomer. **6.** CH_3CH $=CHCH_3$. **7. (a)** alkyl halide; **(b)** carboxylic acid; **(c)** aldehyde; **(d)** ether; **(e)** ketone; **(f)** amine; **(g)** alcohol, phenyl group; **(h)** ester.

8. (a)

(b)

(c)

9.

10. (a) different formulas; **(b)** structural isomers; **(c)** different formulas; **(d)** identical; **(e)** identical; **(f)** ortho-para isomers. **11. (a)** 3,5-dimethyloctane; **(b)** 2,2-dimethylpropane; **(c)** 3,3-dichloro-5-ethylheptane. **12. (a)** *m*-dibromobenzene; **(b)** *m*-nitrotoluene; **(c)** *p*-aminobenzoic acid.

13. (a)

(b)

(c)

(d) C—C—C—O—C—C—C;

(e)

14. (a) $CH_3CH(OH)CH_3$; **(b)** $CClF_2$ CH_3; **(c)** $CH_2=C(CH_3)CH=CH_2$

15. (a)

(b)

(c)

16. 2-butanol.

17. (a) $CH_3CH_2CH(OH)CH_3$; **(b)** $CH_3CH_2CH_2Cl + CH_3CHClCH_3 +$ other chlorinated propanes
(c) $C_6H_5-COOCH(CH_3)_2$;
(d) **18. (a)** C_6H_6;
(b) C_3H_7OH; **(c)** C_6H_5COOH.

Exercises. 19. (a)

(b)

(c)

21. (a) C are sp^3-hybridized; **(b)** C are sp^2-hybridized; **(c)** left C is sp^3, others are sp-hybridized. **23. (a)** positional; **(b)** structural; **(c)** positional.
25. nine possible structures. **27. (a)** and **(b)**: Refer to Table 27.2; **(c)** CH_3COOH and CH_3CO-. **29. (a)** 2,2-dimethylbutane; **(b)** 2-methylpropene; **(c)** 1,2-dimethylcyclopropane; **(d)** 4-methyl-2-pentyne; **(e)** 3,4-dimethylhexane; **(f)** 3,4-dimethyl-2-propyl-1-pentene. **31. (a)** 1-pentene or 2-pentene; **(b)** sufficient; **(c)** 1-butanol, 2-butanol, isobutanol, and *t*-butanol; **(d)** need position on ring; **(e)** sufficient; **(f)** need positions on ring.

33. (a)

(b)

(c) $HOOCCH_2C(OH)$ $(COOH)CH_2COOH$

35. (a) **(b)**
37. There is only one position for the double bond in ethene and propene but two in butene. **39. (a)** $CH_3CH_2CH_3$ **(b)** $CH_3CH = CHCH_3$
41. (a) $C_6H_5-C\equiv CH$;

(b)

; **(c)** 3,5-dihydroxy-

phenol. **43. (a)** H on a carbon chain is displaced by another group; **(b)** H on an aromatic ring is replaced by another group; **(c)** two fragments of a small molecule add across a double bond; **(d)** Two groups from within the same

molecule join to produce a small molecule, such as H and OH to form H_2O.
45. (a) $CH_3CH_2CH_2CH_2COOH$;

(b) $CH_3CH_2CH_2\overset{\overset{\textstyle O}{\|}}{C}OCH_2CH_3$

(c) $CH_3CH_2\overset{\overset{\textstyle OH}{|}}{\underset{\underset{\textstyle CH_3}{|}}{C}}{-}CH_3$

47. polymer chains are of various

lengths. **49.** $\left(\overset{\overset{\textstyle O}{\|}}{C}{-}(CH_2)_8{-}\overset{\overset{\textstyle O}{\|}}{C}{-}\overset{\overset{\textstyle H}{|}}{N}{-}\right.$

$\left.(CH_2)_6{-}\overset{\overset{\textstyle H}{|}}{N}{-}\right)_x$

Advanced Exercises. 57. (a) $C_6H_5NO_2 + 7\,H^+ + 2\,Fe \rightarrow C_6H_5NH_3^+ + 2\,H_2O + 2\,Fe^{3+}$; **(b)** $3\,C_6H_5CH_2OH + 2\,Cr_2O_7^{2-} + 16\,H^+ \rightarrow 3\,C_6H_5COOH + 4\,Cr^{3+} + 11\,H_2O$, **(c)** $3\,CH_3CH{=}CH_2 + 2\,MnO_4^- + 4\,H_2O \rightarrow 3\,CH_3CHOHCH_2OH + 2\,MnO_2 + 2\,OH^-$. **59.** $C_4H_8N_2O_2$; **62. (a)** 1,3-dimethylbenzene or ethylbenzene; **(b)** 1,3,5-trimethylbenzene; **(c)** 2-ethyltoluene.

Chapter 28

Practice Examples. 1A. glyceryl trilaurate or trilaurin.
1B.
$CH_2OOC(CH_2)_{10}CH_3$
$|$
$CHOOC(CH_2)_{12}CH_3$
$|$
$CH_2OOC(CH_2)_7CH{=}CHCH_2CH{=}CH(CH_2)_4CH_3$
2A. dithreonylmethionine.
2B.

$H_2N{-}\overset{\overset{\textstyle }{}}{\underset{\underset{\textstyle CH_2OH}{|}}{CH}}{-}\overset{\overset{\textstyle O}{\|}}{C}{-}NH{-}\overset{\overset{\textstyle }{}}{\underset{\underset{\textstyle H}{|}}{CH}}{-}\overset{\overset{\textstyle O}{\|}}{C}{-}NH{-}\overset{\overset{\textstyle }{}}{\underset{\underset{\textstyle CH(CH_3)_2}{|}}{CH}}{-}\overset{\overset{\textstyle O}{\|}}{C}{-}OH$

3A. Gly-Cys-Val-Phe-Tyr. **3B.** Ser-Gly-Gly-Ala-Val-Trp.
Review Questions. 4. (a) glyceryl palmitolinolenolaurate or glyceryl palmitoeleosterolaurate; **(b)** glyceryl trioleate or triolein; **(c)** sodium myristate.

5. (c) $CH_3(CH_2)_{12}{-}\overset{\overset{\textstyle O}{\|}}{C}{-}O^-K^+$

(d) $CH_3(CH_2)_3{-}O{-}\overset{\overset{\textstyle O}{\|}}{C}{-}(CH_2)_7{-}$

$CH{=}CH{-}(CH_2)_7{-}CH_3$. **6.** neither to the right nor to the left (a racemic mixture). **7. (a)** the mirror image of L-(+)-arabinose; **(b)** one possibility: switch positions of H and OH on the first and second C atoms.

8. (a) $C_6H_5CH_2\overset{\overset{\textstyle NH_3^+}{|}}{CH}COOH$;

(b) $C_6H_5CH_2\overset{\overset{\textstyle NH_2}{|}}{CH}COO^-$;

(c) $C_6H_5CH_2\overset{\overset{\textstyle NH_3^+}{|}}{CH}COO^-$

9. (a)

$H_2N{-}CH{-}\overset{\overset{\textstyle O}{\|}}{C}{-}NH{-}CH{-}\overset{\overset{\textstyle O}{\|}}{C}{-}OH$
$\quad\quad\underset{\textstyle CH_3}{|}\quad\quad\quad\quad\quad\underset{\textstyle CH_2SH}{|}$

(b)

$H_2N{-}CH{-}\overset{\overset{\textstyle O}{\|}}{C}{-}NH{-}CH{-}\overset{\overset{\textstyle O}{\|}}{C}{-}NH{-}CH{-}\overset{\overset{\textstyle O}{\|}}{C}{-}OH$
$\quad\quad\underset{\textstyle CH_3CHOH}{|}\quad\quad\quad\underset{\textstyle CH(CH_3)_2}{|}\quad\quad\underset{\textstyle H}{|}$

10. (a)

$H_2N{-}CH{-}\overset{\overset{\textstyle O}{\|}}{C}{-}NH{-}CH{-}\overset{\overset{\textstyle O}{\|}}{C}{-}NH{-}CH{-}\overset{\overset{\textstyle O}{\|}}{C}NH{-}CH{-}\overset{\overset{\textstyle O}{\|}}{C}{-}OH$;
$\quad\quad\underset{\textstyle CH_2SCH_3}{|}\quad\quad\quad\underset{\textstyle CH(CH_3)_2}{|}\quad\quad\underset{\textstyle CHOH}{|}\quad\quad\underset{\textstyle CH_2SH}{|}$
$\quad\quad\quad\quad\quad\quad\quad\quad\quad\quad\quad\quad\quad\quad\quad\quad\quad\quad\quad\underset{\textstyle CH_3}{|}$

(b) methionylvalylthreonylcysteine. **11.** Increases in temperature can disrupt weak interactions between protein chains; pH changes can alter ionic bonding, as can the presence of metal ions. Also, metal ions are essential in the structures of some enzymes. Only certain molecules fit into the active sites of enzymes ("lock-and-key"). **12.** RNA. **Exercises. 13. (a)** $3 \times 10^2\ H_3O^+$ ions; **(b)** $1.2 \times 10^5\ K^+$ ions. **15.** 5×10^6 protein molecules. **19.** Polyunsaturated fatty acids have two or more carbon-to-carbon double bonds. **21.** $CH_2OHCHOHCH_2OH$ and Na^+ $^-OOC(CH_2)_{14}CH_3$. **27.** A reducing sugar will reduce Cu^{2+}(aq) to insoluble, red Cu_2O(s). **29.** not enantiomers. **33.** proline will not migrate; lysine toward the cathode; aspartic acid toward the anode.

35. (a) $^+H_3NCH(CHOHCH_3)COOH$; **(b)** $^+H_3NCH(CHOHCH_3)COO^-$; **(c)** $H_2NCH(CHOHCH_3)COO^-$. **37. (a)** Lys-Ser-Ala, Lys-Ala-Ser, Ser-Lys-Ala, Ser-Ala-Lys, Ala-Ser-Lys, Ala-Lys-Ser; **(b)** Ala-Ser-Ala-Ser, Ala-Ala-Ser-Ser, Ala-Ser-Ser-Ala, Ser-Ser-Ala-Ala, Ser-Ala-Ser-Ala, Ser-Ala-Ala-Ser. **39. (a)** Ala$-$Ser$-$Gly$-$Val$-$Thr$-$Leu; **(b)** alanylserylglycylvalylthreonylleucine.
43. DNA and RNA: sugars (deoxyribose and ribose), phosphate groups, and purine and pyrimidine bases. **45.** TCG.
Advanced Exercises. 48. 2.9×10^3 g/mol; more than one Ag^+ per protein might be needed. **49. (a)** iodine number of tristearin is 0; saponification value: 189 mg KOH; **(b)** 86.0 g I_2 = iodine number of triolein; saponification value: 190 mg KOH. **53.** 9.71. **55.** Arg-Pro-Pro-Gly-Phe-Ser-Pro-Phe-Arg. **56.** Met-Ser-Met-Met-Gly.

Photo Credits

Van Loon; **p. 591,** Carey Van Loon; **p. 592,** Carey Van Loon; **p. 598,** Carey Van Loon; **p. 601,** Carey Van Loon; **17-6,** Carey Van Loon; **17-7,** Energy Technology Visuals Collection/Department of Energy; **p. 614** (left), NYC Parks Photo Archive/ Fundamental Photographs; **p. 614** (right), Kristen Brochmann/Fundamental Photographs

CHAPTER 18 p. 620 and p. 621, Richard Megna/ Fundamental Photographs; **18-1,** Carey Van Loon; **18-2,** Carey Van Loon; **18-3,** Carey Van Loon; **18-4** (top), Carey Van Loon; **18-4** (center), Carey Van Loon; **18-4** (bottom), Carey Van Loon; **18-5a,** Carey Van Loon; **18-5b,** Carey Van Loon; **18-5c,** Carey Van Loon; **18-5d,** Carey Van Loon; **18-6** (top), Carey Van Loon; **18-6** (center), Carey Van Loon; **18-6** (bottom), Carey Van Loon; **p. 646,** Dr. John Finley/Nabisco, Inc.

CHAPTER 19 p. 654 and p. 655, E. Nagele/FPG International; **19-1,** Carey Van Loon; **19-3a,** Carey Van Loon; **19-3b,** Carey Van Loon; **19-5,** Carey Van Loon; **19-6,** Richard Megna/Fundamental Photographs; **19-8,** Carey Van Loon; **19-9a,** Richard Megna/Fundamental Photographs; **19-9b,** Richard Megna/Fundamental Photographs; **p. 680,** Jane Burton/Bruce Coleman, Inc.

CHAPTER 20 p. 688 and p. 689, Leonard Lessin/ Peter Arnold, Inc.; **p. 697,** Science Photo Library/ Photo Researchers, Inc.

CHAPTER 21 p. 724 and p. 725, Tom Carroll/ Phototake NYC; **21-1,** Carey Van Loon; **21-6,** Diane Schiumo/Fundamental Photographs; **p. 738,** Stamp from the private collection of Professor C. M. Lang, photography by Gary J. Shulfer, University of Wisconsin, Stevens Point. "1980, Sweden (Scott #1344); " Scott Standard Postage Stamp Catalogue, Scott Pub. Co., Sidney, Ohio; **p. 741,** Tom Pantages; **p. 745,** Courtesy of Motorola; **21-16,** Carey Van Loon; **p. 749,** Robert Earlbacher/Missouri Dry Dock and Repair Company, Inc.; **p. 752,** Corbis-Bettmann; **p. 754,** Richard Pasley/Stock Boston; **p. 758,** Carey Van Loon; **p. 763,** Judy & Chas Walker/Peter Arnold, Inc.

CHAPTER 22 p. 764 and p. 765, Don Mason/The Stock Market; **p. 766,** Dr. E. R. Degginger; **22-1a,** Carey Van Loon; **22-1b,** Carey Van Loon; **22-1c,** Carey Van Loon; **p. 769,** Peter M. Fisher/The Stock Market; **p. 770,** Dr. E. R. Degginger; **p. 775,** Dr. E. R. Degginger; **p. 776,** William E. Ferguson; **22-8,** Bjorn Bolstad/Peter Arnold, Inc.; **p. 779,** Carey Van Loon; **22-10a,** Carey Van Loon; **22-10b,** Carey Van Loon; **22-10c,** Carey Van Loon; **p. 788,** Bruce Frisch/Photo Researchers, Inc.

CHAPTER 23 p. 794 and p. 795, Paul Silverman/ Fundamental Photographs; **p. 798,** Richard Dunoff/ The Stock Market; **p. 802,** Carey Van Loon; **p. 804,** Stephen Frisch/Stock Boston; **p. 805,** Bill Pierce/ Rainbow; **p. 810,** Dr. E. R. Degginger; **p. 812,** NASA Headquarters; **p. 814,** Charlie Ott/Photo Researchers, Inc.; **p. 816,** Bruce Frisch/Photo Researchers, Inc.; **p. 817,** The Stock Market; **p. 820,** Hank Morgan/Photo Researchers, Inc.; **p. 824,** Jan Halaska/Photo Researchers, Inc.

CHAPTER 24 p. 830 and p. 831, Michael Dalton/ Fundamental Photographs; **p. 835,** M. Seul, R. Wolfe, L. R. Monar, & L. O'Gorman/AT&T Bell Laboratories; **24-4,** Sol Mednick; **p. 839,** OREMET; **p. 841,** Day Williams/Photo Researchers, Inc.; **p. 842,** Robert Semeniuk/The Stock Market; **p. 844,** Keith Wood/Tony Stone Images; **24-7,** Richard Megna/Fundamental Photographs; **24-8** (left), Carey Van Loon; **24-8** (center), Carey Van Loon; **24-8** (right), Carey Van Loon; **24-9** (left), Carey Van Loon; **24-9** (right), Carey Van Loon; **p. 850,** Pacific Scientific; **p. 852,** Richard Megna/ Fundamental Photographs; **p. 856,** David Parker/

IMI/Univ. of Birmingham High TC Consortium/ Science Photo Library/Photo Researchers, Inc.

CHAPTER 25 p. 864 and p. 865, Karl Hartmann/Sachs/Phototake NYC; **p. 866,** Science Photo Library/Photo Researchers, Inc.; **25-12a,** Fritz Goro/Life Magazine, Time Warner, Inc.; **p. 881,** Carey Van Loon; **p. 882,** Jerry Mason/ Science Photo Library/Photo Researchers, Inc.; **25-15,** Carey Van Loon; **25-16,** Carey Van Loon; **25-17,** Carey Van Loon; **p. 890,** Dr. E. R. Degginger

CHAPTER 26 p. 896 and p. 897, NASA/Grant Heilman Photography, Inc.; **p. 898** (top), Corbis–Bettmann; **p. 898** (bottom), Science Photo Library/Photo Researchers, Inc.; **p. 902,** UPI/ Corbis-Bettmann; **p. 915,** Y. Arthus Bertrand/Photo Researchers, Inc.; **p. 916,** Jerry Mason/Science Photo Library/Photo Researchers, Inc.; **p. 918,** Dr. E. R. Degginger; **p. 921** (top), Perrin/CNRI/ Phototake NYC; **p. 921** (center), Perrin/CNRI/ Phototake NYC; **p. 921** (bottom), Peticolas/Megna/ Fundamental Photographs; **p. 922,** D.O.E./Science Source/Photo Researchers, Inc.

CHAPTER 27 p. 928 and p. 929, Kristen Brochmann/Fundamental Photographs; **p. 932,** Government of Switzerland; **p. 939,** Dr. E. R. Degginger; **p. 941,** Dr. E. R. Degginger; **p. 942,** C. Marvin Lang; **p. 946,** Pete Saloutos/The Stock Market; **p. 948,** Science Photo Library/Photo Researchers, Inc.; **p. 950,** Bill Ross/Westlight; **p. 952,** Christopher Springmann/The Stock Market; **p. 954,** Michael Holford Photographs

CHAPTER 28 p. 962 and p. 963, Leonard Lessin/ Peter Arnold, Inc.; **p. 966,** Corbis-Bettmann; **p. 967,** Richard Megna/Fundamental Photographs; **p. 970,** Corbis-Bettmann; **28-12,** Dr. Arthur Lesk/Science Photo Library/Photo Researchers, Inc.

Index